W9-CLW-032

About the Cover

The cover image illustrates a Seifert surface, named for German mathematician Herbert Seifert. The boundary of a Seifert surface is a simple knot called a trefoil. You could make a trefoil by joining together the two loose ends of a common overhand knot, resulting in a knotted loop.

Meet the Artist

The cover art was generated by Paul Nylander, a mechanical engineer with strong programming and mathematical skills that he uses to design complex engineering systems. Nylander says that he always enjoyed science and art as a hobby. When he was in high school, he had some aptitude for math, but programming was difficult for him. However, he became much more interested in programming when he began studying computer graphics. Most of Nylander's artwork is created in Mathematica, POV-Ray, and C++.

You can find a short bio and description of his work at: http://virtualmathmuseum.org/mathart/ArtGalleryNylander/Nylanderindex.html.

Cover: Seifert Surface. By Paul Nylander (bugman123.com)

GLENCOE

GEOMETRY

McGraw Hill

mheducation.com/prek-12

Send all inquiries to:
McGraw-Hill Education
8787 Orion Place
Columbus, OH 43240

ISBN: 978-0-07-903994-1
MHID: 0-07-903994-4

Printed in the United States of America.

7 8 9 10 LWI 23 22 21 20 19

McGraw-Hill is committed to providing instructional materials in Science, Technology, Engineering, and Mathematics (STEM) that give all students a solid foundation, one that prepares them for college and careers in the 21st century.

Contents in Brief

Authors

Our lead authors ensure that the Macmillan/McGraw-Hill and Glencoe/McGraw-Hill mathematics programs are truly vertically aligned by beginning with the end in mind — success in Geometry and beyond. By "backmapping" the content from the high school programs, all of our mathematics programs are well articulated in their scope and sequence.

LEAD AUTHORS

John A. Carter, Ph.D.

Mathematics Teacher
WINNETKA, ILLINOIS

Areas of Expertise:
Using technology and manipulatives to visualize concepts; mathematics achievement of English-language learners

Gilbert J. Cuevas, Ph.D.

Professor of Mathematics Education, Texas State University—San Marcos
SAN MARCOS, TEXAS

Areas of Expertise:
Applying concepts and skills in mathematically rich contexts; mathematical representations; use of technology in the development of geometric thinking

Roger Day, Ph.D., NBCT

Mathematics Department Chairperson, Pontiac Township High School
PONTIAC, ILLINOIS

Areas of Expertise:
Understanding and applying probability and statistics; mathematics teacher education

In Memoriam

Carol Malloy, Ph.D.

Dr. Carol Malloy was a fervent supporter of mathematics education. She was a Professor at the University of North Carolina, Chapel Hill, NCTM Board of Directors member, President of the Benjamin Banneker Association (BBA), and 2013 BBA Lifetime Achievement Award for Mathematics winner. She joined McGraw-Hill in 1996. Her influence significantly improved our programs' focus on real-world problem solving and equity. We will miss her inspiration and passion for education.

PROGRAM AUTHOR

In Memoriam
Jerry Cummins

Jerry Cummins was a passionate and enthusiastic mathematics educator who taught math and computer science in the Chicago area for 32 years. Mr. Cummins was on the forefront of graphing technology usage in the classroom. His influence greatly improved the integration of technology within our programs. In 1997, Mr. Cummins received the Distinguished Life Achievement Award from the Illinois Council of Teachers of Mathematics. He also served as the President of the National Council of Supervisors of Mathematics from 1999-2001. We will miss his exuberance and devotion to education.

CONTRIBUTING AUTHORS

Dinah Zike FOLDABLES

Educational Consultant
Dinah-Might Activities, Inc.
SAN ANTONIO, TEXAS

Jay McTighe

Educational Author and Consultant
COLUMBIA, MARYLAND

Consultants and Reviewers

These professionals were instrumental in providing valuable input and suggestions for improving the effectiveness of the mathematics instruction.

LEAD CONSULTANT

Viken Hovsepian
Professor of Mathematics
Rio Hondo College
WHITTIER, CALIFORNIA

CONSULTANTS

MATHEMATICAL CONTENT

Lead Consultant
Viken Hovsepian
Professor of Mathematics
Rio Hondo College
WHITTIER, CALIFORNIA

Grant A. Fraser, Ph.D.
Professor of Mathematics
California State University, Los Angeles
LOS ANGELES, CALIFORNIA

Arthur K. Wayman, Ph.D.
Professor of Mathematics Emeritus
California State University, Long Beach
LONG BEACH, CALIFORNIA

GIFTED AND TALENTED

Shelbi K. Cole
Research Assistant
University of Connecticut
STORRS, CONNECTICUT

COLLEGE READINESS

Robert Lee Kimball, Jr.
Department Head, Math and Physics
Wake Technical Community College
RALEIGH, NORTH CAROLINA

DIFFERENTIATION FOR ENGLISH-LANGUAGE LEARNERS

Susana Davidenko
State University of New York
CORTLAND, NEW YORK

Alfredo Gómez
Mathematics/ESL Teacher
George W. Fowler High School
SYRACUSE, NEW YORK

GRAPHING CALCULATOR

Ruth M. Casey
T³ National Instructor
FRANKFORT, KENTUCKY

MATHEMATICAL FLUENCY

Robert M. Capraro
Associate Professor
Texas A&M University
COLLEGE STATION, TEXAS

PRE-AP

Dixie Ross
Lead Teacher for Advanced Placement Mathematics
Pflugerville High School
PFLUGERVILLE, TEXAS

READING AND WRITING

ReLeah Cossett Lent
Author and Educational Consultant
MORGANTON, GEORGIA

Lynn T. Havens
Director of Project CRISS
KALISPELL, MONTANA

REVIEWERS

Corey Andreasen
Mathematics Teacher
North High School
SHEBOYGAN, MICHIGAN

Mark B. Baetz
Mathematics Coordinating Teacher
Salem City Schools
SALEM, VIRGINIA

Kathryn Ballin
Mathematics Supervisor
Newark Public Schools
NEWARK, NEW JERSEY

Kevin C. Barhorst
Mathematics Department Chair
Independence High School
COLUMBUS, OHIO

Brenda S. Berg
Mathematics Teacher
Carbondale Community High School
CARBONDALE, ILLINOIS

Sheryl Pernell Clayton
Mathematics Teacher
Hume Fogg Magnet School
NASHVILLE, TENNESSEE

Bob Coleman
Mathematics Teacher
Cobb Middle School
TALLAHASSEE, FLORIDA

Jane E. Cotts
Mathematics Teacher
O'Fallon Township High School
O'FALLON, ILLINOIS

Michael D. Cuddy
Mathematics Instructor
Zypherhills High School
ZYPHERHILLS, FLORIDA

Melissa M. Dalton, NBCT
Mathematics Instructor
Rural Retreat High School
RURAL RETREAT, VIRGINIA

Trina Louise Davis
Teacher
Fort Mill High School
FORT MILL, SOUTH CAROLINA

Tina S. Dohm
Mathematics Teacher
Naperville Central High School
NAPERVILLE, ILLINOIS

Laurie L.E. Ferrari
Teacher
L'Anse Creuse High School—North
MACOMB, MICHIGAN

Patricia R. Frazier
Mathematics Department Chair/Instructor
Celina High School
CELINA, OHIO

Steve Freshour
Mathematics Teacher
Parkersburg South High School
PARKERSBURG, WEST VIRGINIA

Shirley D. Glover
Mathematics Teacher
TC Roberson High School
ASHEVILLE, NORTH CAROLINA

Caroline W. Greenough
Mathematics Teacher
Cape Fear Academy
WILMINGTON, NORTH CAROLINA

Michelle Hanneman
Mathematics Teacher
Moore High School
MOORE, OKLAHOMA

Theresalynn Haynes
Mathematics Teacher
Glenbard East High School
LOMBARD, ILLINOIS

Sandra Hester
Mathematics Teacher/AIG Specialist
North Henderson High School
HENDERSONVILLE, NORTH CAROLINA

Jacob K. Holloway
Mathematics Teacher
Capitol Heights Junior High School
MONTGOMERY, ALABAMA

Robert Hopp
Mathematics Teacher
Harrison High School
HARRISON, MICHIGAN

Eileen Howanitz
Mathematics Teacher/Department Chairperson
Valley View High School
ARCHBALD, PENNSYLVANIA

Charles R. Howard, NBCT
Mathematics Teacher
Tuscola High School
WAYNESVILLE, NORTH CAROLINA

Sue Hvizdos
Mathematics Department Chairperson
Wheeling Park High School
WHEELING, WEST VIRGINIA

Elaine Keller
Mathematics Teacher
Mathematics Curriculum Director K–12
Northwest Local Schools
CANAL FULTON, OHIO

Sheila A. Kotter
Mathematics Educator
River Ridge High School
NEW PORT RICHEY, FLORIDA

Frank Lear
Mathematics Department Chair
Cleveland High School
CLEVELAND, TENNESSEE

Jennifer Lewis
Mathematics Teacher
Triad High School
TROY, ILLINOIS

Catherine McCarthy
Mathematics Teacher
Glen Ridge High School
GLEN RIDGE, NEW JERSEY

Jacqueline Palmquist
Mathematics Department Chair
Waubonsie Valley High School
AURORA, ILLINOIS

Thom Schacher
Mathematics Teacher
Otsego High School
OTSEGO, MICHIGAN

Laurie Shappee
Teacher/Mathematics Coordinator
Larson Middle School
TROY, MICHIGAN

Jennifer J. Southers
Mathematics Teacher
Hillcrest High School
SIMPSONVILLE, SOUTH CAROLINA

Sue Steinbeck
Mathematics Department Chair
Parkersburg High School
PARKERSBURG, WEST VIRGINIA

Kathleen D. Van Sise
Mathematics Teacher
Mandarin High School
JACKSONVILLE, FLORIDA

Karen Wiedman
Mathematics Teacher
Taylorville High School
TAYLORVILLE, ILLINOIS

Digital Tools to Enhance Your Learning

Students today have an unprecedented access to and appetite for technology and new media. You see technology as your friend and rely on it to study, work, play, relax, and communicate.

You are accustomed to the role that computers play in today's world. You are the first generation whose primary educational tool is a computer or a cell phone. The eStudentEdition gives you access to your math curriculum anytime, anywhere.

Investigate

Sketchpad Discover concepts using The Geometer's Sketchpad®.

Vocabulary tools include fun Vocabulary Review Games.

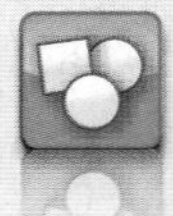

Tools enhance understanding through exploration.

The Geometer's Sketchpad

The Geometer's Sketchpad gives you a tangible, visual way to see the math in action through dynamic model manipulation of lines, shapes, and functions.

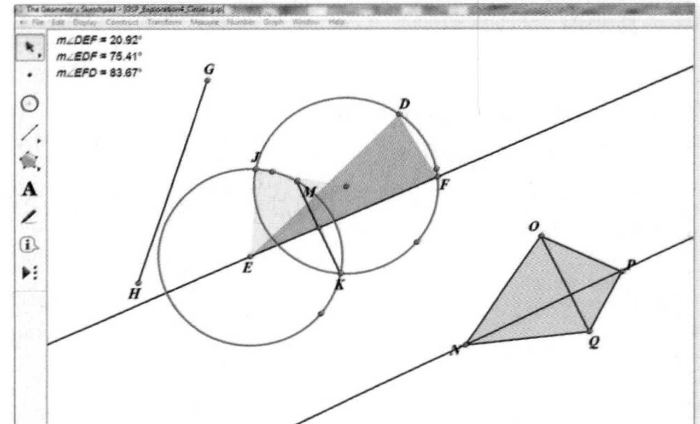

eToolkit

eToolkit helps you to use virtual manipulative's to extend your learning beyond the classroom by modifying concrete models in a real-time, interactive format focused on problem-based learning.

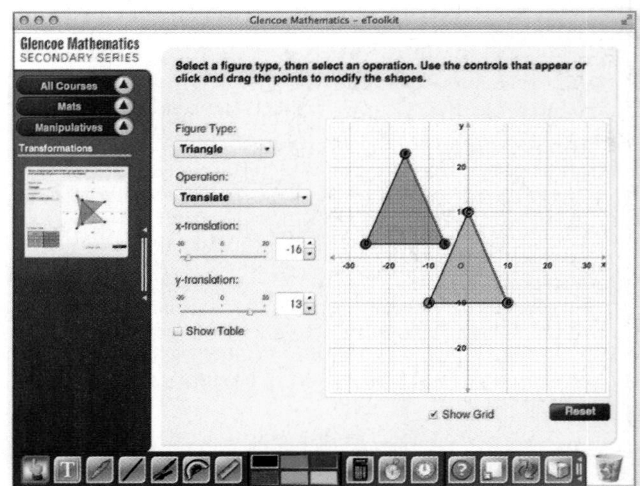

Learn

LearnSmart
Topic based online assessment

Animations
illustrate key concepts through step-by-step tutorials and videos.

Tutors
See and hear a teacher explain how to solve problems.

Calculator Resources
provides other calculator keystrokes for each Graphing Technology Lab.

Practice

Self-Check Practice
allows students to check their understanding and send results to their teacher.

eBook
Interactive learning experience with links directly to assets

eStudentEdition

Use your eStudentEdition to access your print text 24/7. This interactive eBook gives you access to all the resources that help you learn the concepts. You can take notes, highlight, digitally write on the pages, and bookmark where you are.

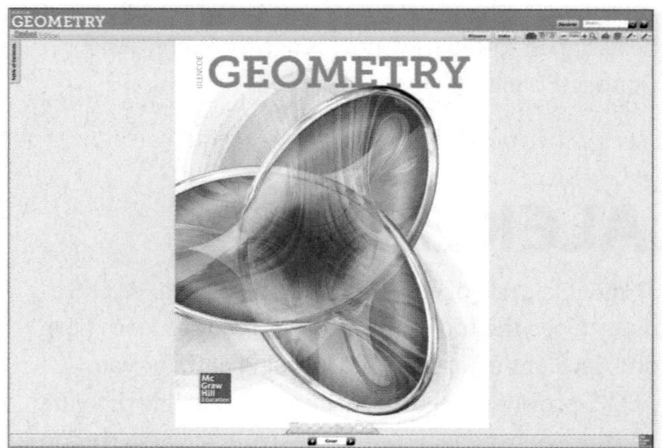

Interactive Student Guide

Interactive Student Guide is a dynamic resource to help you meet the challenges of content standards.

This guide works together with the student edition to ensure that you can reflect on comprehension and application, apply math concepts to the real world, and internalize concepts to develop "second nature" recall.

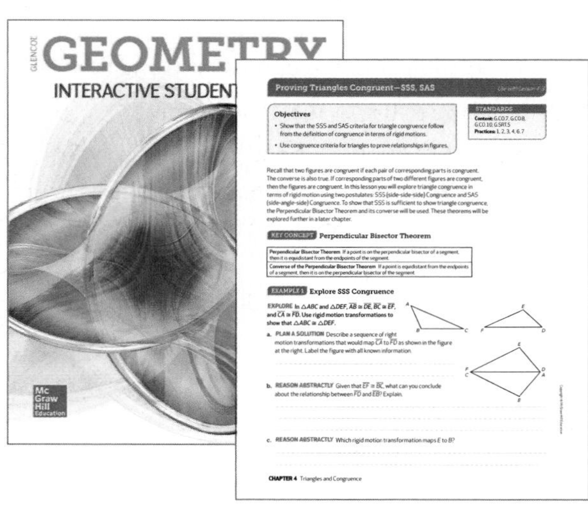

Adaptive Learning

LEARNSMART®

LearnSmart is your online test-prep solution with adaptive capability and resources for end-of-course assessment. Through a series of adaptive questions, LearnSmart identifies areas where you need to focus your learning. It provides you with learning resources such as slides, videos, kaleidoscopes, and label games to encourage you to review material again. This adaptive learning will help increase your likelihood that you will retain your new knowledge.

LearnSmart can be used as a self-study tool or it can be assigned to you by your teacher through your ConnectED platform.

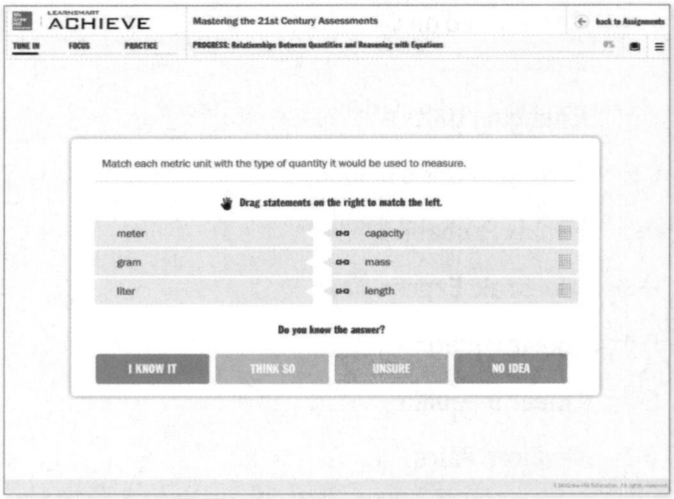

ALEKS®

Through a cycle of assessment and learning, *ALEKS determines the topics that you are ready to learn next and develops a personalized learning path for you. ALEKS provides you with real-time, actionable data that informs you what you need to learn. It will help you review and master the skills needed to be successful in your math class. Use your ALEKS Pie to see a snapshot of your current progress of the course. Click on the Pie slice to see the number of topics mastered and ready to learn next.

Use the timeline to watch your progress toward your learning goals, and if needed you can toggle between English and Spanish translations of the content and interface.

*Ask your teacher if you have access to ALEKS.

CHAPTER 0

Preparing for Geometry

Sigrid Olsson/PhotoAlto sas/Alamy

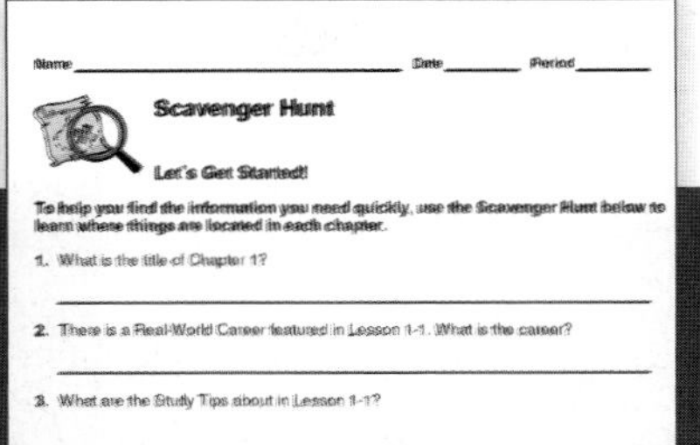

Name ____________ Date ______ Period ______

Scavenger Hunt

Let's Get Started!

To help you find the information you need quickly, use the Scavenger Hunt below to learn where things are located in each chapter.

1. What is the title of Chapter 1?

____

2. There is a Real-World Career featured in Lesson 1-1. What is the career?

____

3. What are the Study Tips about in Lesson 1-1?

Worksheets help to explain key concepts, let you practice your skills, and offer opportunities for extending the lessons. Find them in the Resources in ConnectED.

Go Online!
connectED.mcgraw-hill.com

CHAPTER 1
Tools of Geometry

ASSESSMENT

Geometer's Sketchpad® allows you to interact with geometry in a visual way. Investigate geometric concepts with sketches in ConnectED.

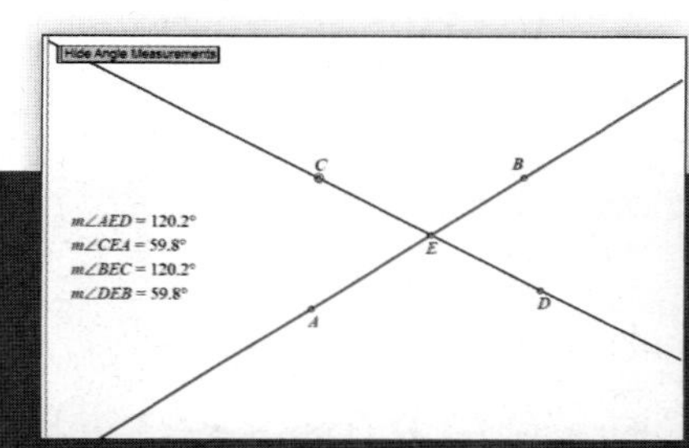

CHAPTER 2
Logical Arguments and Line Relationships

ASSESSMENT

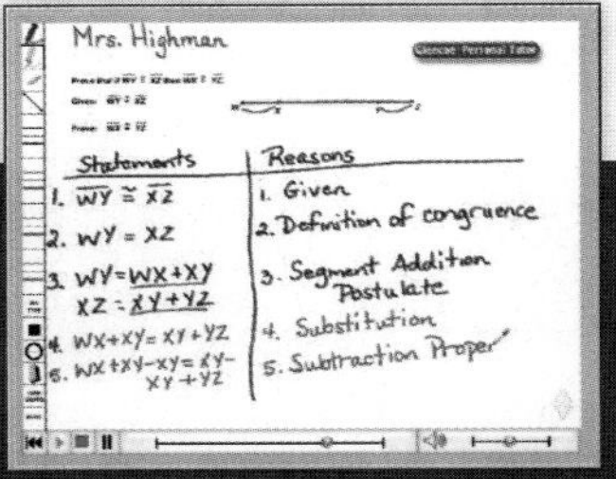

Watch as a real teacher solves a problem, pause to work ahead, or rewind to watch again. A **Personal Tutor** for each example is just a click away in ConnectED.

Go Online!
connectED.mcgraw-hill.com

CHAPTER 3
Rigid Transformations and Symmetry

ASSESSMENT

Michael DeYoung/Blend Images

Go Online!
connectED.mcgraw-hill.com

With the **Geometry Tools** in ConnectED, you can explore the effects of transformations on geometric figures.

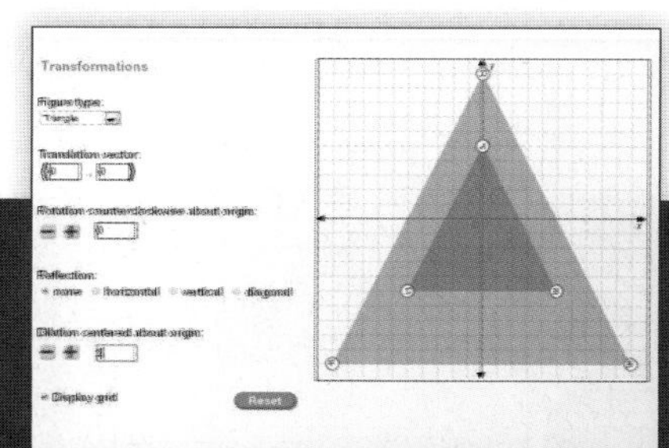

CHAPTER 4

Triangles and Congruence

John Kelly/Media Bakery

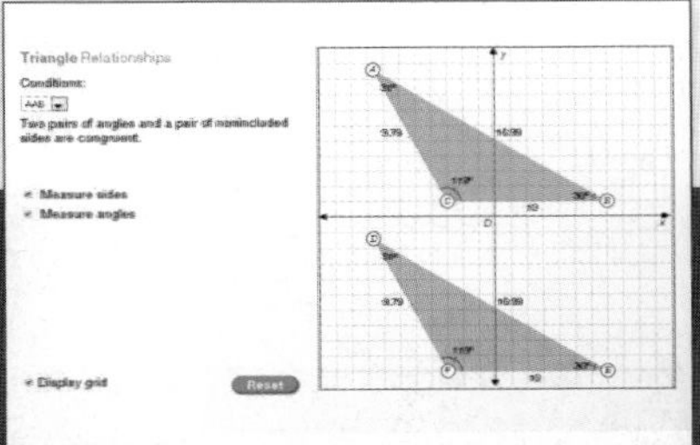

With the **Graphing Tools** in ConnectED, explore when two triangles are congruent and when they are not.

CHAPTER 5
Relationships in Triangles

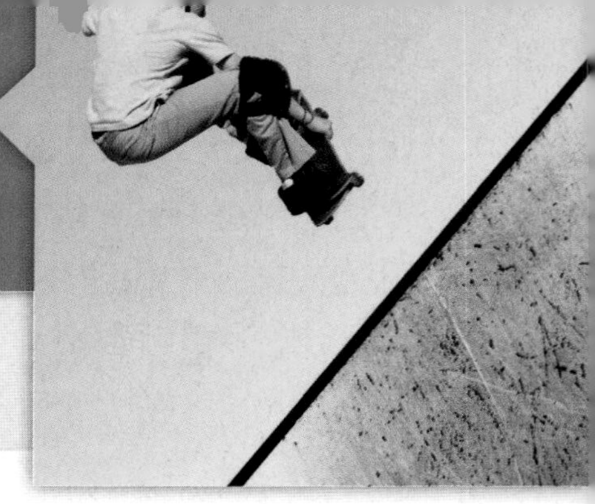

ASSESSMENT

Galina Barskaya/Alamy Stock Photo

Animations demonstrate Key Concepts and topics from the chapter. Click to watch animations in ConnectED.

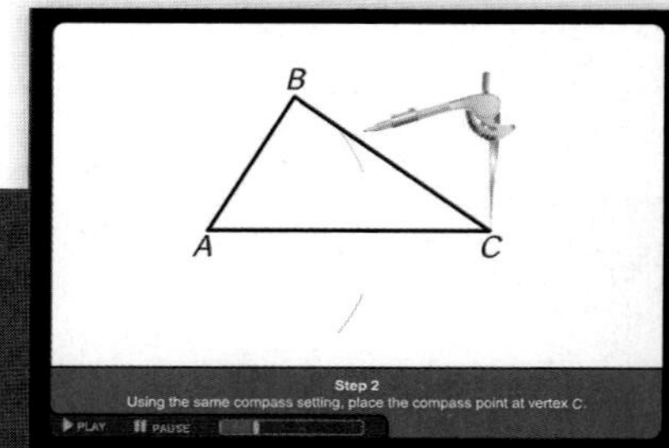

CHAPTER 6
Quadrilaterals

ASSESSMENT

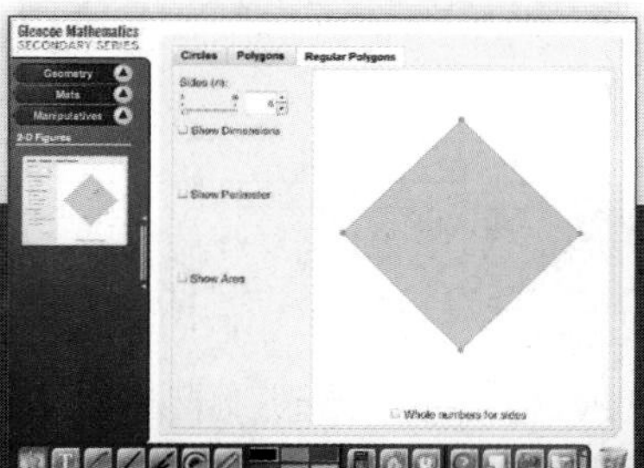

Create, change, and investigate quadrilaterals using the 2-D Figures tool in the eToolkit. Find these **virtual manipulatives** in ConnectED.

Go Online!
connectED.mcgraw-hill.com

CHAPTER 7
Similarity

ASSESSMENT

Worksheets help to explain key concepts, let you practice your skills, and offer opportunities for extending the lessons. Find them in the Resources in ConnectED.

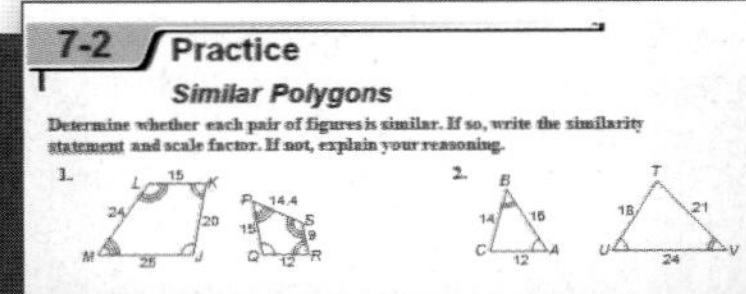

CHAPTER 8

Right Triangles and Trigonometry

ASSESSMENT

RHIMAGE/Shutterstock.com

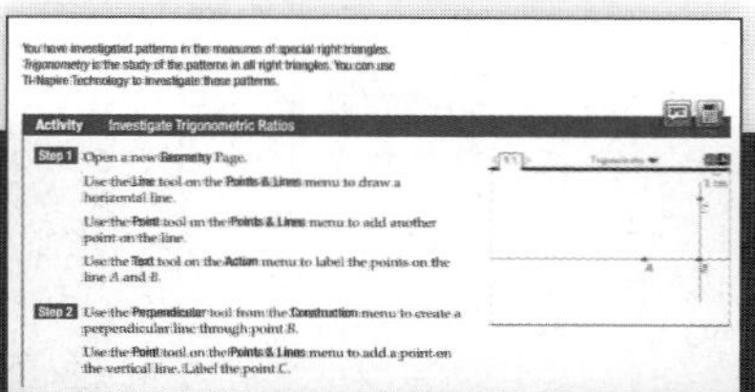

You have investigated patterns in the measures of special right triangles. *Trigonometry* is the study of the patterns in all right triangles. You can use TI-Nspire Technology to investigate these patterns.

Activity Investigate Trigonometric Ratios

Step 1 Open a new **Geometry** Page.

Use the **Line** tool on the **Points & Lines** menu to draw a horizontal line.

Use the **Point** tool on the **Points & Lines** menu to add another point on the line.

Use the **Text** tool on the **Action** menu to label the points on the line *A* and *B*.

Step 2 Use the **Perpendicular** tool from the **Construction** menu to create a perpendicular line through point *B*.

Use the **Point** tool on the **Points & Lines** menu to add a point on the vertical line. Label the point *C*.

Using a graphing calculator allows you to quickly visualize concepts. **Graphing Calculator Keystrokes** help you navigate your calculator. Find the keystrokes for your calculator in the Resources in ConnectED.

Go Online!
connectED.mcgraw-hill.com

CHAPTER 9
Circles

ASSESSMENT

MacMag83/Alamy Stock Photo

Go Online!
connectED.mcgraw-hill.com

Tap into the power of your graphing calculator with Graphing Calculator Easy Files™. Practice vocabulary in English or Spanish with Lesson Vocabulary Review Files. Or review with a 5-Minute Check. Ask your teacher to assign them to you in ConnectED.

CHAPTER 10
Extending Area

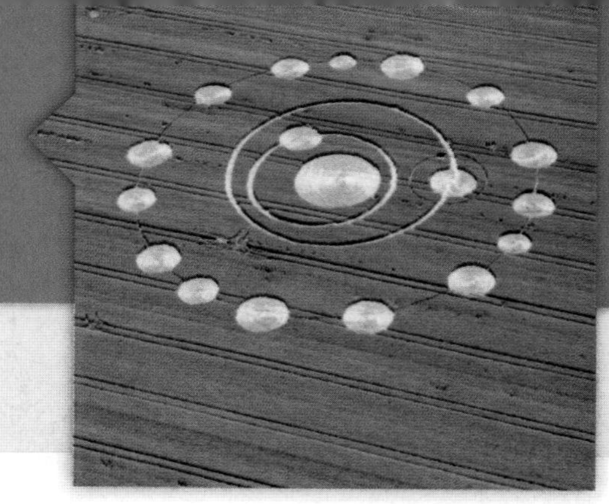

ASSESSMENT

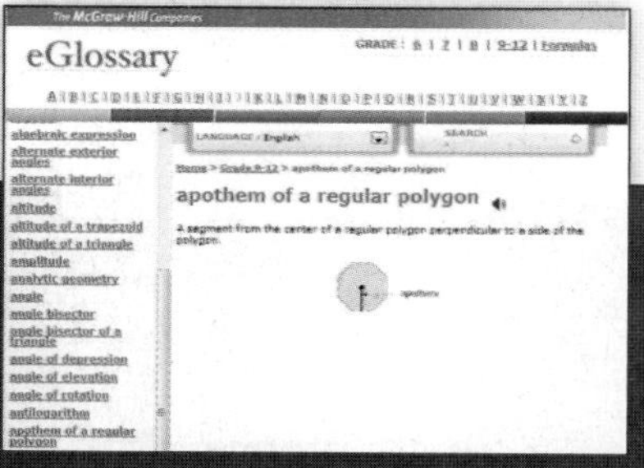

Vocabulary is important to learning the key concepts in this chapter. Find all the terms with animations, English pronunciations, and translations into 13 languages in the eGlossary in ConnectED.

Go Online!
connectED.mcgraw-hill.com

CHAPTER 11
Extending Volume

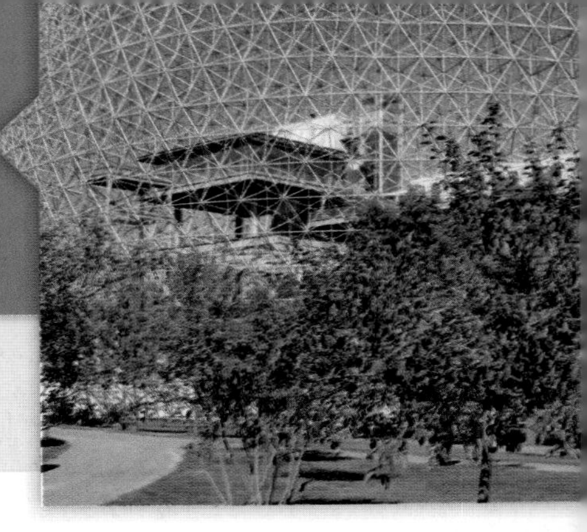

ASSESSMENT

Norman Pogson/Alamy Stock Photo

Go Online!
connectED.mcgraw-hill.com

Visualize and interact with three-dimensional figures as you study this chapter. Find the **Virtual Manipulatives** in the eToolkit in ConnectED.

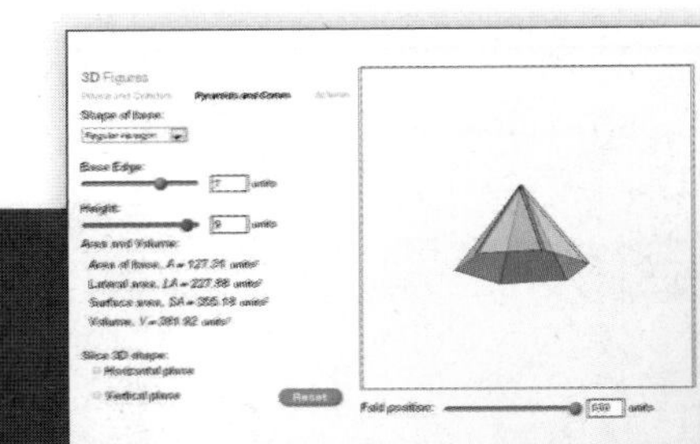

CHAPTER 12
Probability

ASSESSMENT

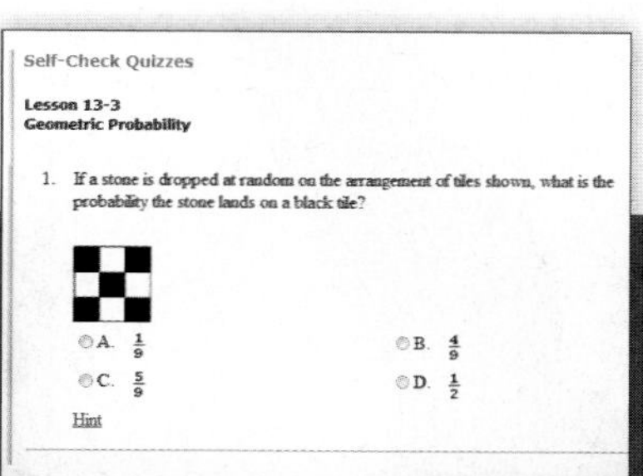

Check your understanding of probability and measurement using **Self-Check Quizzes** in ConnectED.

Student Handbook

Built-In Workbook

Reference

 Standards for Mathematical Practice

Glencoe Geometry exhibits these practices throughout the entire program. All of the Standards for Mathematical Practice will be covered in each chapter. The MP icon notes specific areas of coverage.

Mathematical Practices	What does it mean?
1. Make sense of problems and persevere in solving them.	Solving a mathematical problem takes time. Use a logical process to make sense of problems, understand that there may be more than one way to solve a problem, and alter the process if needed.
2. Reason abstractly and quantitatively.	You can start with a concrete or real-world context and then represent it with abstract numbers or symbols (decontextualize), find a solution, then refer back to the context to check that the solution makes sense (contextualize).
3. Construct viable arguments and critique the reasoning of others.	Sound mathematical arguments require a logical progression of statements and reasons. Mathematically proficient students can clearly communicate their thoughts and defend them.
4. Model with mathematics.	Modeling links classroom mathematics and statistics to everyday life, work, and decision-making. High school students at this level are expected to apply key takeaways from earlier grades to high-school level problems.
5. Use appropriate tools strategically.	Certain tools, including estimation and virtual tools are more appropriate than others. You should understand the benefits and limitations of each tool.
6. Attend to precision.	Precision in mathematics is more than accurate calculations. It is also the ability to communicate with the language of mathematics. In high school mathematics, precise language makes for effective communication and serves as a tool for understanding and solving problems.
7. Look for and make use of structure.	Mathematics is based on a well-defined structure. Mathematically proficient students look for that structure to find easier ways to solve problems.
8. Look for and express regularity in repeated reasoning.	Mathematics has been described as the study of patterns. Recognizing a pattern can lead to results more quickly and efficiently.

by Dinah Zike

Folding Instructions

The following pages offer step-by-step instructions to make the Foldables® study guides.

Layered-Look Book

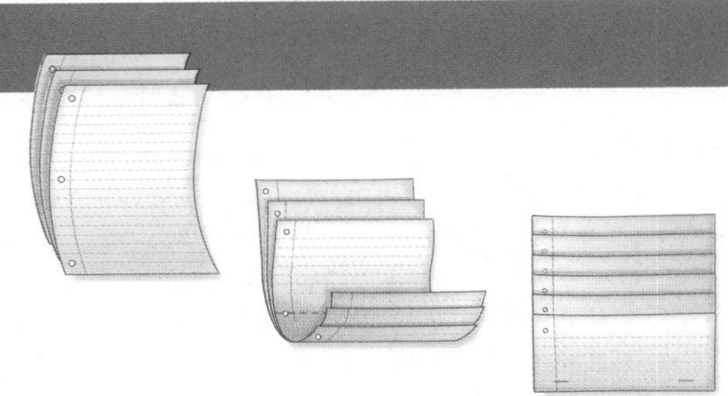

1. Collect three sheets of paper and layer them about 1 cm apart vertically. Keep the edges level.
2. Fold up the bottom edges of the paper to form 6 equal tabs.
3. Fold the papers and crease well to hold the tabs in place. Staple along the fold. Label each tab.

Shutter-Fold and Four-Door Books

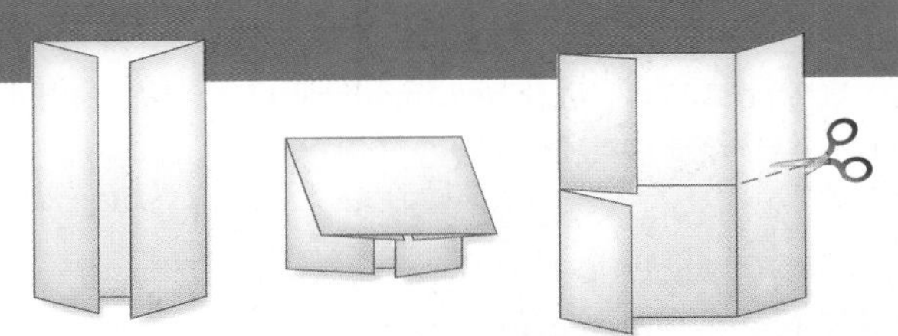

1. Find the middle of a horizontal sheet of paper. Fold both edges to the middle and crease the folds. Stop here if making a shutter-fold book. For a four-door book, complete the steps below.
2. Fold the folded paper in half, from top to bottom.
3. Unfold and cut along the fold lines to make four tabs. Label each tab.

Concept-Map Book

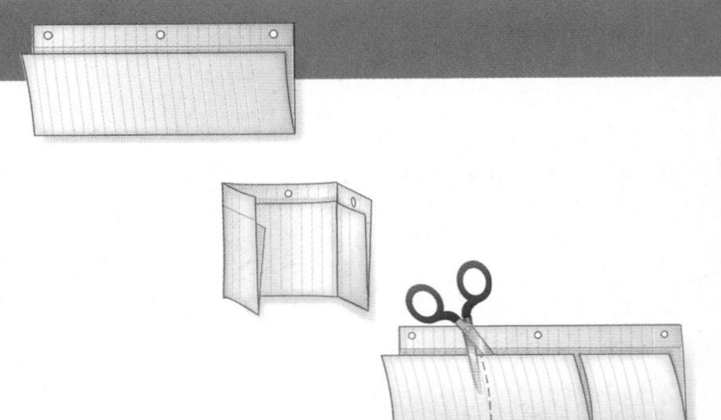

1. Fold a horizontal sheet of paper from top to bottom. Make the top edge about 2 cm shorter than the bottom edge.
2. Fold width-wise into thirds.
3. Unfold and cut only the top layer along both folds to make three tabs. Label the top and each tab.

Vocabulary Book

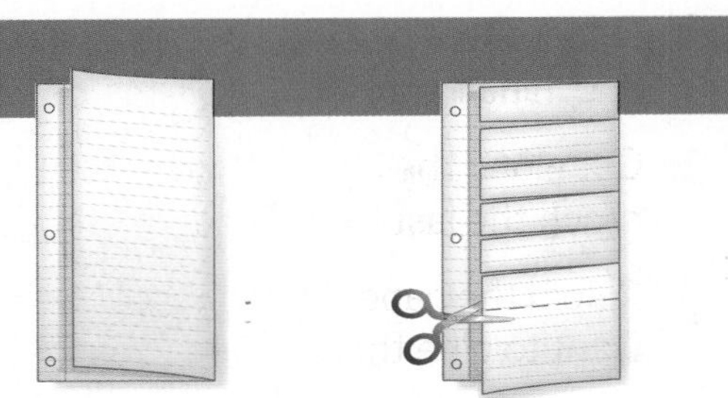

1. Fold a vertical sheet of notebook paper in half.
2. Cut along every third line of only the top layer to form tabs. Label each tab.

Pocket Book

1. Fold the bottom of a horizontal sheet of paper up about 3 cm.
2. If making a two-pocket book, fold in half. If making a three-pocket book, fold in thirds.
3. Unfold once and dot with glue or staple to make pockets. Label each pocket.

Bound Book

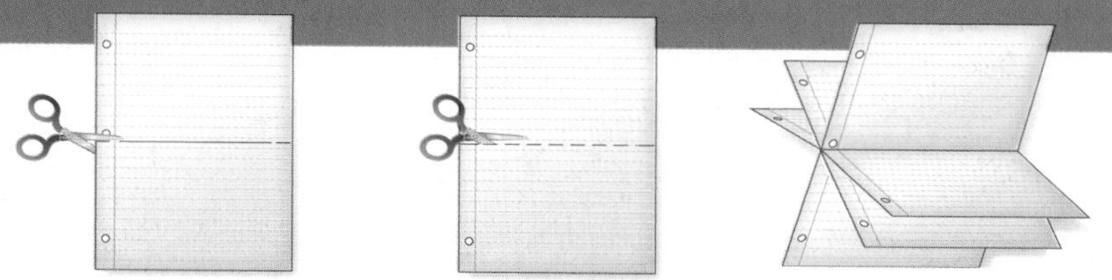

1. Fold several sheets of paper in half to find the middle. Hold all but one sheet together and make a 3-cm cut at the fold line on each side of the paper.
2. On the final page, cut along the fold line to within 3-cm of each edge.
3. Slip the first few sheets through the cut in the final sheet to make a multi-page book.

Top-Tab Book

1. Layer multiple sheets of paper so that about 2–3 cm of each can be seen.
2. Make a 2–3-cm horizontal cut through all pages a short distance (3 cm) from the top edge of the top sheet.

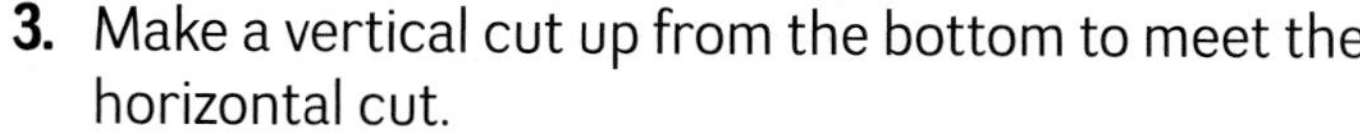

3. Make a vertical cut up from the bottom to meet the horizontal cut.
4. Place the sheets on top of an uncut sheet and align the tops and sides of all sheets. Label each tab.

Accordion Book

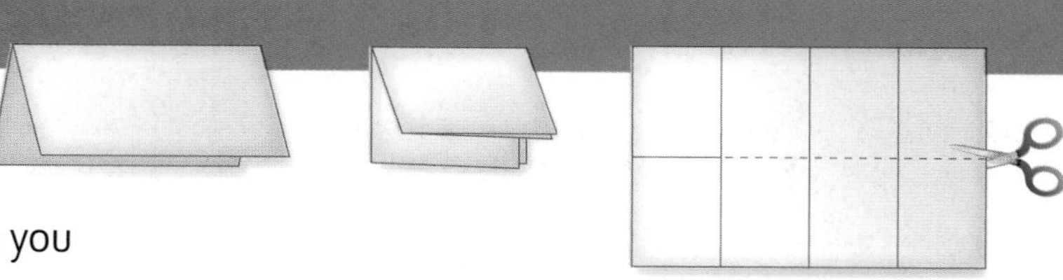

1. Fold a sheet of paper in half. Fold in half and in half again to form eight sections.

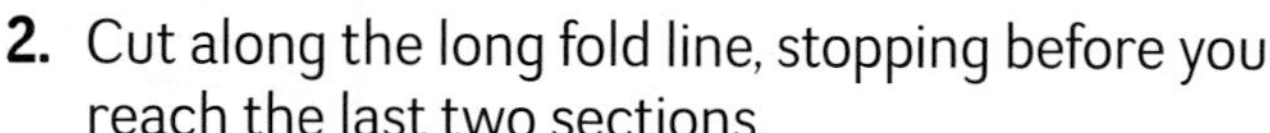

2. Cut along the long fold line, stopping before you reach the last two sections.
3. Refold the paper into an accordion book. You may want to glue the double pages together.

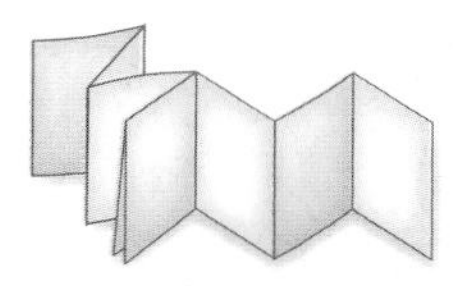

CHAPTER 0

Preparing for Geometry

NOW

Chapter 0 contains lessons on topics from previous courses. You can use this chapter in various ways.

- Begin the school year by taking the Pretest. If you need additional review, complete the lessons in this chapter. To verify that you have successfully reviewed the topics, take the Posttest.
- As you work through the text, you may find that there are topics you need to review. When this happens, complete the individual lessons that you need.
- Use this chapter for reference. When you have questions about any of these topics, flip back to this chapter to review definitions or key concepts.

MP WHY

GARDENING The height of plants can be modeled with mathematics.

Use the Mathematical Practices to complete the activity.

1. Sense-making If you know that one plant grows twice as fast as another and the total height of the plants together, could you determine the plants' heights?

2. Apply Math Assign a variable to the larger plant height and write an equation, picking any height for the total of the two plants together.

3. Use Tools Use a Graph Mat to sketch a graph of the equation as well as of another equation you create that represents plant growth (i.e., the difference in plant heights) and find the solutions.

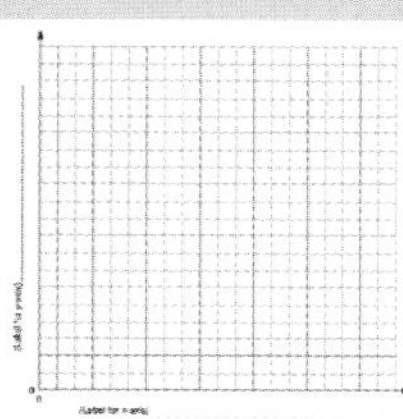

Sigrid Olsson/PhotoAlto sas/Alamy

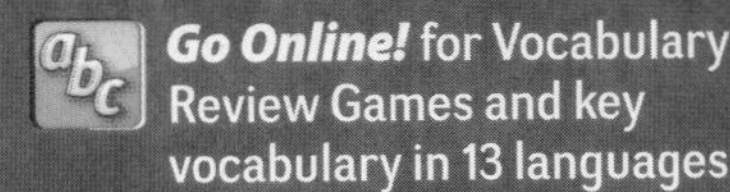

Go Online to Guide Your Learning

Organize

Throughout this text, you will be invited to use Foldables to organize your notes.

FOLDABLES®

Why should you use them?

- They help you organize, display, and arrange information.
- They make great study guides, specifically designed for you.
- They give you a chance to improve your math vocabulary.

How should you use them?

- Write general information—titles, vocabulary terms, concepts, questions, and main ideas—on the front tabs of your Foldable.
- Write specific information—ideas, your thoughts, answers to questions, steps, notes, and definitions—under the tabs.

When should you use them?

- Set up your Foldable as you begin a chapter, or when you start learning a new concept.
- Write in your Foldable every day.
- Use your Foldable to review for homework, quizzes, and tests.

New Vocabulary

English		Español
experiment	p. P8	experimento
trial	p. P8	prueba
outcome	p. P8	resultado
event	p. P8	evento
probability	p. P8	probabilidad
theoretical probability	p. P9	probabilidad teórica
experimental probability	p. P9	probabilidad experimental
ordered pair	p. P15	par ordenado
x-coordinate	p. P15	coordenada *x*
y-coordinate	p. P15	coordenada *y*
quadrant	p. P15	cuadrante
origin	p. P15	origen
system of equations	p. P17	sistema de ecuaciones
substitution	p. P17	sustitución
elimination	p. P18	eliminación
Product Property	p. P19	Propriedad de Producto
Quotient Property	p. P19	Propriedad de Cociente

Explore & Explain

Vocab

Investigate using the **Glossary.** Which geometry terms do you remember?

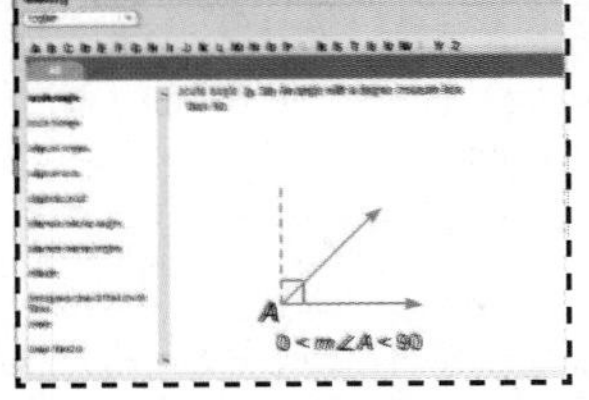

Tools

Investigate using the **KWL chart** tool. How do you organize information to express and solve equations?

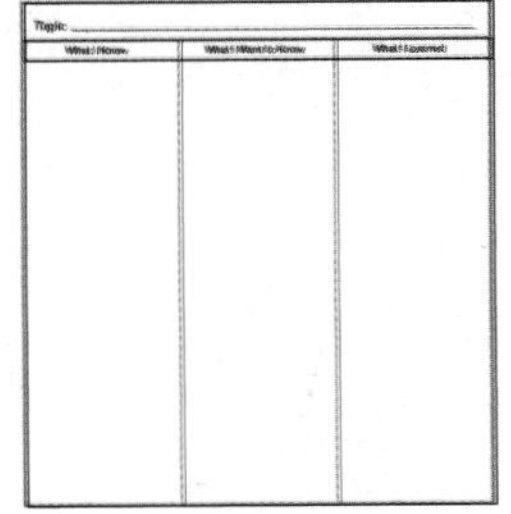

CHAPTER 0
Pretest

State which metric unit you would probably use to measure each item.

1. length of a computer keyboard
2. mass of a large dog

Complete each sentence.

3. 4 ft = __?__ in.
4. 21 ft = __?__ yd
5. 180 g = __?__ kg
6. 3 T = __?__ lb
7. 32 g ≈ __?__ oz
8. 3 mi ≈ __?__ km
9. 35 yd ≈ __?__ m
10. 5.1 L ≈ __?__ qt
11. **TUNA** A can of tuna is 6 ounces. About how many grams is it?
12. **CRACKERS** A box of crackers is 453 grams. About how many pounds is it? Round to the nearest pound.
13. **DISTANCE** A road sign in Texas gives the distance to Odessa as 87 miles. What is this distance to the nearest kilometer?

PROBABILITY A bag contains 3 blue chips, 7 red chips, 4 yellow chips, and 5 green chips. A chip is randomly drawn from the bag. Find each probability.

14. $P(\text{yellow})$
15. $P(\text{green})$
16. $P(\text{red or blue})$
17. $P(\text{not red})$

Evaluate each expression if $r = 3$, $q = 1$, and $w = -2$.

18. $4r + q$
19. $rw - 6$
20. $\frac{r + 3q}{4r}$
21. $\frac{5w}{3r + q}$
22. $|2 - r| + 17$
23. $8 + |q - 5|$

Solve each equation.

24. $k + 3 = 14$
25. $a - 7 = 9$
26. $5c = 20$
27. $n + 2 = -11$
28. $6t - 18 = 30$
29. $4x + 7 = -1$
30. $\frac{r}{4} = -8$
31. $\frac{3}{5}b = -2$
32. $-\frac{w}{2} = -9$
33. $3y - 15 = y + 1$
34. $27 - 6d = 7 + 4d$
35. $2(m - 16) = 44$

Solve each inequality.

36. $y - 13 < 2$
37. $t + 8 \geq 19$
38. $\frac{n}{4} > -6$
39. $9a \leq 45$
40. $x + 12 > -14$
41. $-2w < 24$
42. $-\frac{n}{7} \geq 3$
43. $-\frac{b}{5} \leq -6$

Write the ordered pair for each point shown.

44. F
45. H
46. A
47. D

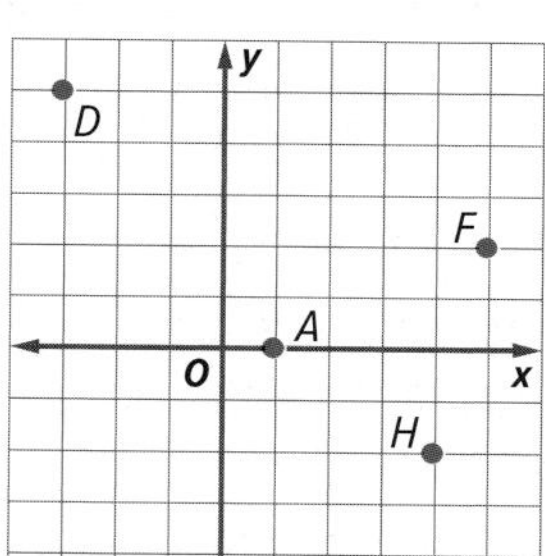

Graph and label each point on the coordinate plane above.

48. $B(4, 1)$
49. $G(0, -3)$
50. $R(-2, -4)$
51. $P(-3, 3)$
52. Graph the triangle with vertices $J(1, -4)$, $K(2, 3)$, and $L(-1, 2)$.
53. Graph four points that satisfy the equation $y = 2x - 1$.

Solve each system of equations.

54. $y = 2x$
    $y = -x + 6$
55. $-3x - y = 4$
    $4x + 2y = -8$
56. $y = 2x + 1$
    $y = 3x$
57. $\frac{1}{2}x - y = -1$
    $x - 2y = 5$
58. $x + y = -6$
    $2x - y = 3$
59. $\frac{1}{3}x - 3y = -4$
    $x - 9y = -12$

Simplify.

60. $\sqrt{18}$
61. $\sqrt{\frac{25}{49}}$
62. $\sqrt{24x^2y^3}$
63. $\frac{3}{4 - \sqrt{5}}$

LESSON 1

Changing Units of Measure Within Systems

Objective

- Convert units of measure within the customary and metric systems.

Example 1 Choose Best Unit of Measure

State which metric unit you would use to measure the length of your pen.

A pen has a small length, but not very small. The *centimeter* is the appropriate unit of measure.

Metric Units of Length
1 kilometer (km) = 1000 meters (m)
1 m = 100 centimeters (cm)
1 cm = 10 millimeters (mm)

Customary Units of Length
1 foot (ft) = 12 inches (in.)
1 yard (yd) = 3 ft
1 mile (mi) = 5280 ft

- To convert from larger units to smaller units, multiply.
- To convert from smaller units to larger units, divide.
- To use dimensional analysis, multiply by the ratio of the units.

Example 2 Convert from Larger Units to Smaller Units of Length

Complete each sentence.

a. 4.2 km = __?__ m

There are 1000 meters in a kilometer.
4.2 km × 1000 = 4200 m

b. 13 yd = __?__ ft

There are 3 feet in a yard.
13 yd × 3 = 39 ft

Example 3 Convert from Smaller Units to Larger Units of Length

Complete each sentence.

a. 17 mm = __?__ m

There are 100 centimeters in a meter. First change *millimeters* to *centimeters.*

17 mm = __?__ cm — smaller unit → larger unit

17 mm ÷ 10 = 1.7 cm — Since 10 mm = 1 cm, divide by 10.

Then change *centimeters* to *meters.*

1.7 cm = __?__ m — smaller unit → larger unit

1.7 cm ÷ 100 = 0.017 m — Since 100 cm = 1 m, divide by 100.

b. 6600 yd = __?__ mi

Use dimensional analysis.

$$6600 \cancel{\text{yd}} \times \frac{3 \cancel{\text{ft}}}{1 \cancel{\text{yd}}} \times \frac{1 \text{ mi}}{5280 \cancel{\text{ft}}} = 3.75 \text{ mi}$$

Metric Units of Capacity
1 liter (L) = 1000 milliliters (mL)

Customary Units of Capacity	
1 cup (c) = 8 fluid ounces (fl oz)	1 quart (qt) = 2 pt
1 pint (pt) = 2 c	1 gallon (gal) = 4 qt

Study Tip

Dimensional Analysis You can use dimensional analysis for any conversion in this lesson.

Example 4 Convert Units of Capacity

Complete each sentence.

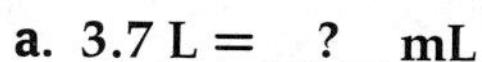

a. **3.7 L = ___?___ mL**

There are 1000 milliliters in a liter.

3.7 L × 1000 = 3700 mL

b. **16 qt = ___?___ gal**

There are 4 quarts in a gallon.

16 qt ÷ 4 = 4 gal

c. **7 pt = ___?___ fl oz**

There are 8 fluid ounces in a cup.

First change *pints* to *cups*.

7 pt = ___?___ c

7 pt × 2 = 14 c

Then change *cups* to *fluid ounces*.

14 c = ___?___ fl oz

14 c × 8 = 112 fl oz

d. **4 gal = ___?___ pt**

There are 4 quarts in a gallon.

First change *gallons* to *quarts*.

4 gal = ___?___ qt

4 gal × 4 = 16 qt

Then change *quarts* to *pints*.

16 qt = ___?___ pt

16 qt × 2 = 32 pt

The mass of an object is the amount of matter that it contains.

Metric Units of Mass
1 kilogram (kg) = 1000 grams (g)
1 g = 1000 milligrams (mg)

Customary Units of Weight
1 pound (lb) = 16 ounces (oz)
1 ton (T) = 2000 lb

Example 5 Convert Units of Mass

Complete each sentence.

a. **5.47 kg = ___?___ mg**

There are 1000 milligrams in a gram.

Change *kilograms* to *grams*.

5.47 kg = ___?___ g

5.47 kg × 1000 = 5470 g

Then change *grams* to *milligrams*.

5470 g = ___?___ mg

5470 g × 1000 = 5,470,000 mg

b. **5 T = ___?___ oz**

There are 16 ounces in a pound.

Change *tons* to *pounds*.

5 T = ___?___ lb

5 T × 2000 = 10,000 lb

Then change *pounds* to *ounces*.

10,000 lb = ___?___ oz

10,000 lb × 16 = 160,000 oz

Exercises

State which metric unit you would probably use to measure each item.

1. radius of a tennis ball

2. length of a notebook

3. mass of a textbook

4. mass of a beach ball

5. liquid in a cup

6. water in a bathtub

Complete each sentence.

7. 120 in. = ___?___ ft

8. 18 ft = ___?___ yd

9. 10 km = ___?___ m

10. 210 mm = ___?___ cm

11. 180 mm = ___?___ m

12. 3100 m = ___?___ km

13. 90 in. = ___?___ yd

14. 5280 yd = ___?___ mi

15. 8 yd = ___?___ ft

16. 0.62 km = ___?___ m

17. 370 mL = ___?___ L

18. 12 L = ___?___ mL

19. 32 fl oz = ___?___ c

20. 5 qt = ___?___ c

21. 10 pt = ___?___ qt

22. 48 c = ___?___ gal

23. 4 gal = ___?___ qt

24. 36 mg = ___?___ g

25. 13 lb = ___?___ oz

26. 130 g = ___?___ kg

27. 9.05 kg = ___?___ g

LESSON 2

Changing Units of Measure Between Systems

Objective

- Convert units of measure between the customary and metric systems.

The table below shows approximate equivalents between customary units of length and metric units of length.

Units of Length	
Customary → Metric	Metric → Customary
1 in. ≈ 2.5 cm	1 cm ≈ 0.4 in.
1 yd ≈ 0.9 m	1 m ≈ 1.1 yd
1 mi ≈ 1.6 km	1 km ≈ 0.6 mi

Example 1 Convert Units of Length Between Systems

Complete each sentence.

a. 30 in. ≈ __?__ cm

There are approximately 2.5 centimeters in an inch.

30 in. × 2.5 = 75 cm

b. 5 km ≈ __?__ mi

There is approximately 0.6 mile in a kilometer.

5 km × 0.6 = 3 mi

Example 2 Convert Units of Length Between Systems

Complete: 2000 yd ≈ __?__ km.

There is approximately 0.9 meter in a yard. First find the number of meters in 2000 yards.

2000 yd × 0.9 = 1800 m

Then change *meters* to *kilometers*. There are 1000 meters in a kilometer.

1800 m ÷ 1000 = 1.8 km

The table below shows approximate equivalents between customary units of capacity and metric units of capacity.

Units of Capacity	
Customary → Metric	Metric → Customary
1 qt ≈ 0.9 L	1 L ≈ 1.1 qt
1 pt ≈ 0.5 L	1 L ≈ 2.1 pt

Example 3 Convert Units of Capacity Between Systems

Complete each sentence.

a. 7 qt ≈ __?__ L

There is approximately 0.9 liter in a quart.

7 qt × 0.9 = 6.3 L

b. 2 L ≈ __?__ pt

There are approximately 2.1 pints in a liter.

2 L × 2.1 = 4.2 pt

Example 4 Convert Units of Capacity Between Systems

Complete 10 L ≈ __?__ gal.

There are approximately 1.1 quarts in a liter. First find the number of quarts in 10 liters.

$10 \text{ L} \times 1.1 = 11 \text{ qt}$

Then change *quarts* to *gallons*. There are 4 quarts in a gallon.

$11 \text{ qt} \div 4 = 2.75 \text{ gal}$

You can also use dimensional analysis.

$$10 \cancel{\text{L}} \times \frac{1.1 \cancel{\text{qt}}}{1 \cancel{\text{L}}} \times \frac{1 \text{ gal}}{4 \cancel{\text{qt}}} = 2.75 \text{ gal}$$

Study Tip

Dimensional Analysis If the unit that you want to eliminate is in the numerator, make sure it is in the denominator of the ratio when you multiply. If it is in the denominator, make sure that it is in the numerator of the ratio.

The table below shows approximate equivalents between customary units of weight and metric units of mass.

Units of Weight/Mass	
Customary ⟶ Metric	**Metric ⟶ Customary**
1 oz ≈ 28.3 g	1 g ≈ 0.04 oz
1 lb ≈ 0.5 kg	1 kg ≈ 2.2 lb

Example 5 Convert Units of Mass Between Systems

Complete each sentence.

a. 58.5 kg ≈ __?__ lb

There are approximately 2.2 pounds in a kilogram.

$58.5 \text{ kg} \times 2.2 = 128.7 \text{ lb}$

b. 14 oz ≈ __?__ g

There are approximately 28.3 grams in an ounce.

$14 \text{ oz} \times 28.3 = 396.2 \text{ g}$

Exercises

Complete each sentence.

1. 8 in. ≈ __?__ cm

2. 15 m ≈ __?__ yd

3. 11 qt ≈ __?__ L

4. 25 oz ≈ __?__ g

5. 10 mi ≈ __?__ km

6. 32 cm ≈ __?__ in.

7. 20 km ≈ __?__ mi

8. 9.5 L ≈ __?__ qt

9. 6 yd ≈ __?__ m

10. 4.3 kg ≈ __?__ lb

11. 10.7 L ≈ __?__ pt

12. 82.5 g ≈ __?__ oz

13. $2\frac{1}{4}$ lb ≈ __?__ kg

14. 10 ft ≈ __?__ m

15. $1\frac{1}{2}$ gal ≈ __?__ L

16. 350 g ≈ __?__ lb

17. 600 in. ≈ __?__ m

18. 2.1 km ≈ __?__ yd

19. CEREAL A box of cereal is 13 ounces. About how many grams is it?

20. FLOUR A bag of flour is 2.26 kilograms. How much does it weigh? Round to the nearest pound.

21. SAUCE A jar of tomato sauce is 1 pound 10 ounces. About how many grams is it?

LESSON 3

Simple Probability

Objective

- Find the probability of simple events.

New Vocabulary

experiment
trial
outcome
event
probability
theoretical probability
experimental probability

A situation involving chance, such as flipping a coin or rolling a die, is an **experiment**. A single performance of an experiment, such as rolling a die one time, is a **trial**. The result of a trial is called an **outcome**. An **event** is one or more outcome(s) of an experiment.

When each outcome is equally likely to happen, the **probability** of an event is the ratio of the number of favorable outcomes to the number of possible outcomes. The probability of an event is always between 0 and 1, inclusive.

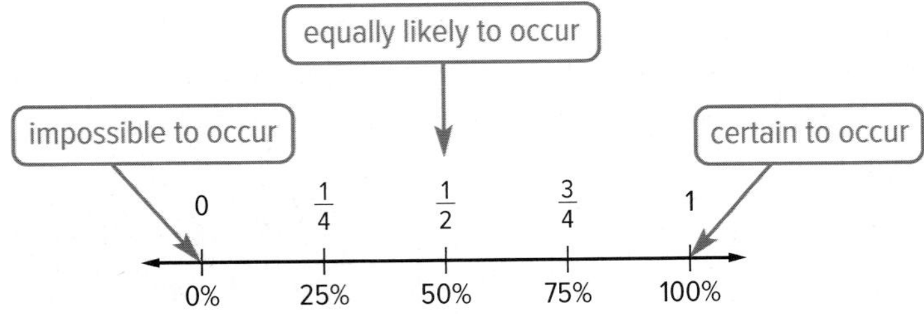

Example 1 Find Probability

Suppose a die is rolled. What is the probability of rolling an odd number?

There are 3 odd numbers on a die: 1, 3, and 5.

There are 6 possible outcomes: 1, 2, 3, 4, 5, and 6.

$$P(\text{odd}) = \frac{\text{number of favorable outcomes}}{\text{number of possible outcomes}}$$

$$= \frac{3}{6} \text{ or } \frac{1}{2}$$

The probability of rolling an odd number is $\frac{1}{2}$ or 50%.

For a given experiment, the sum of the probabilities of all possible outcomes must sum to 1.

Example 2 Find Probability

Suppose a bag contains 4 red, 3 green, 6 blue, and 2 yellow marbles. What is the probability a randomly chosen marble will not be yellow?

Since the sum of the probabilities of all of the colors must sum to 1, subtract the probability that the marble will be yellow from 1.

The probability that the marble will be yellow is $\frac{2}{15}$ because there are 2 yellow marbles and 15 total marbles.

$$P(\text{not yellow}) = 1 - P(\text{yellow})$$

$$= 1 - \frac{2}{15}$$

$$= \frac{13}{15}$$

The probability that the marble will not be yellow is $\frac{13}{15}$ or about 87%.

The probabilities in Examples 1 and 2 are called theoretical probabilities. The **theoretical probability** is what *should* occur. The **experimental probability** is what *actually* occurs when a probability experiment is repeated many times.

Study Tip

Experimental Probability The experimental probability of an experiment is not necessarily the same as the theoretical probability, but when an experiment is repeated many times, the experimental probability should be close to the theoretical probability.

Example 3 Find Experimental Probability

The table shows the results of an experiment in which a number cube was rolled. Find the experimental probability of rolling a 3.

Outcome	Tally	Frequency
1	卌 I	6
2	IIII	4
3	卌 II	7
4	III	3
5	IIII	4
6	I	1

$$P(3) = \frac{\text{number of times 3 occurs}}{\text{total number of outcomes}} \text{ or } \frac{7}{25}$$

The experimental probability for getting a 3 in this case is $\frac{7}{25}$ or 28%.

Exercises

A die is rolled. Find the probability of each outcome.

1. P(less than 3)
2. P(even)
3. P(greater than 2)
4. P(prime)
5. P(4 or 2)
6. P(integer)

A jar contains 65 pennies, 27 nickels, 30 dimes, and 18 quarters. A coin is randomly selected from the jar. Find each probability.

7. P(penny)
8. P(quarter)
9. P(not dime)
10. P(penny or dime)
11. P(value greater than \$0.15)
12. P(not nickel)
13. P(nickel or quarter)
14. P(value less than \$0.20)

PRESENTATIONS **The students in a class are randomly drawing cards numbered 1 through 28 from a hat to determine the order in which they will give their presentations. Find each probability.**

15. P(13)
16. P(1 or 28)
17. P(less than 14)
18. P(not 1)
19. P(not 2 or 17)
20. P(greater than 16)

The table shows the results of an experiment in which three coins were tossed.

Outcome	HHH	HHT	HTH	THH	TTH	THT	HTT	TTT
Tally	卌	卌	卌 I	卌 I	卌 II	卌	卌 III	卌 III
Frequency	5	5	6	6	7	5	8	8

21. What is the experimental probability that all three of the coins will be heads? the theoretical probability?
22. What is the experimental probability that at least two of the coins will be heads? the theoretical probability?
23. **DECISION MAKING** You and two of your friends have pooled your money to buy a new video game. Describe a method that could be used to make a fair decision as to who gets to play the game first.
24. **DECISION MAKING** A new study finds that the incidence of heart attack while taking a certain diabetes drug is less than 5%. Should a person with diabetes take this drug? Should he or she take the drug if the risk is less than 1%? Explain your reasoning.

LESSON 4

Algebraic Expressions

Objective

- Use the order of operations to evaluate algebraic expressions.

An expression is an algebraic expression if it contains sums and/or products of variables and numbers. To evaluate an algebraic expression, replace the variable or variables with known values, and then use the order of operations.

Order of Operations
Step 1 Evaluate expressions inside grouping symbols.
Step 2 Evaluate all powers.
Step 3 Do all multiplications and/or divisions from left to right.
Step 4 Do all additions and/or subtractions from left to right.

Example 1 Addition/Subtraction Algebraic Expressions

Evaluate $x - 5 + y$ if $x = 15$ and $y = -7$.

$x - 5 + y = 15 - 5 + (-7)$ Substitute.

$= 10 + (-7)$ or 3 Subtract.

Example 2 Multiplication/Division Algebraic Expressions

Evaluate each expression if $k = -2$, $n = -4$, and $p = 5$.

a. $\dfrac{2k + n}{p - 3}$

$\dfrac{2k + n}{p - 3} = \dfrac{2(-2) + (-4)}{5 - 3}$ Substitute.

$= \dfrac{-4 - 4}{5 - 3}$ Multiply.

$= \dfrac{-8}{2}$ or -4 Subtract.

b. $-3(k^2 + 2n)$

$-3(k^2 + 2n) = -3[(-2)^2 + 2(-4)]$

$= -3[4 + (-8)]$

$= -3(-4)$ or 12

Example 3 Absolute Value Algebraic Expressions

Evaluate $3|a - b| + 2|c - 5|$ if $a = -2$, $b = -4$, and $c = 3$.

$3|a - b| + 2|c - 5| = 3|-2 - (-4)| + 2|3 - 5|$ Substitute for *a*, *b*, and *c*.

$= 3|2| + 2|-2|$ Simplify.

$= 3(2) + 2(2)$ or 10 Find absolute values.

Exercises

Evaluate each expression if $a = 2$, $b = -3$, $c = -1$, and $d = 4$.

1. $2a + c$

2. $\dfrac{bd}{2c}$

3. $\dfrac{2d - a}{b}$

4. $3d - c$

5. $\dfrac{3b}{5a + c}$

6. $5bc$

7. $2cd + 3ab$

8. $\dfrac{c - 2d}{a}$

Evaluate each expression if $x = 2$, $y = -3$, and $z = 1$.

9. $24 + |x - 4|$

10. $13 + |8 + y|$

11. $|5 - z| + 11$

12. $|2y - 15| + 7$

LESSON 5

Linear Equations

Objective

- Use algebra to solve linear equations.

If the same number is added to or subtracted from each side of an equation, the resulting equation is true.

Example 1 Addition/Subtraction Linear Equations

Solve each equation.

a. $x - 7 = 16$

$x - 7 = 16$	Original equation
$x - 7 + 7 = 16 + 7$	Add 7 to each side.
$x = 23$	Simplify.

b. $m + 12 = -5$

$m + 12 = -5$	Original equation
$m + 12 + (-12) = -5 + (-12)$	Add −12 to each side.
$m = -17$	Simplify.

c. $k + 31 = 10$

$k + 31 = 10$	Original equation
$k + 31 - 31 = 10 - 31$	Subtract 31 from each side.
$k = -21$	Simplify.

If each side of an equation is multiplied or divided by the same number, the resulting equation is true.

Example 2 Multiplication/Division Linear Equations

Solve each equation.

a. $4d = 36$

$4d = 36$	Original equation
$\frac{4d}{4} = \frac{36}{4}$	Divide each side by 4.
$x = 9$	Simplify.

b. $-\frac{t}{8} = -7$

$-\frac{t}{8} = -7$	Original equation
$-8\left(-\frac{t}{8}\right) = -8(-7)$	Multiply each side by −8.
$t = 56$	Simplify.

c. $\frac{3}{5}x = -8$

$\frac{3}{5}x = -8$	Original equation
$\frac{5}{3}\left(\frac{3}{5}\right)x = \frac{5}{3}(-8)$	Multiply each side by $\frac{5}{3}$.
$x = -\frac{40}{3}$	Simplify.

To solve equations with more than one operation, often called *multi-step equations,* undo operations by working backward.

Example 3 Multi-Step Linear Equations

Solve each equation.

a. $8q - 15 = 49$

$8q - 15 = 49$ Original equation

$8q = 64$ Add 15 to each side.

$q = 8$ Divide each side by 8.

b. $12y + 8 = 6y - 5$

$12y + 8 = 6y - 5$ Original equation

$12y = 6y - 13$ Subtract 8 from each side.

$6y = -13$ Subtract $6y$ from each side.

$y = -\frac{13}{6}$ Divide each side by 6.

Watch Out!

Order of Operations Remember that the order of operations applies when you are solving linear equations.

When solving equations that contain grouping symbols, first use the Distributive Property to remove the grouping symbols.

Example 4 Multi-Step Linear Equations

Solve $3(x - 5) = 13$.

$3(x - 5) = 13$ Original equation

$3x - 15 = 13$ Distributive Property

$3x = 28$ Add 15 to each side.

$x = \frac{28}{3}$ Divide each side by 3.

Exercises

Solve each equation.

1. $r + 11 = 3$ **2.** $n + 7 = 13$ **3.** $d - 7 = 8$

4. $\frac{8}{5}a = -6$ **5.** $-\frac{p}{12} = 6$ **6.** $\frac{x}{4} = 8$

7. $\frac{12}{5}f = -18$ **8.** $\frac{y}{7} = -11$ **9.** $\frac{6}{7}y = 3$

10. $c - 14 = -11$ **11.** $t - 14 = -29$ **12.** $p - 21 = 52$

13. $b + 2 = -5$ **14.** $q + 10 = 22$ **15.** $-12q = 84$

16. $5t = 30$ **17.** $5c - 7 = 8c - 4$ **18.** $2\ell + 6 = 6\ell - 10$

19. $\frac{m}{10} + 15 = 21$ **20.** $-\frac{m}{8} + 7 = 5$ **21.** $8t + 1 = 3t - 19$

22. $9n + 4 = 5n + 18$ **23.** $5c - 24 = -4$ **24.** $3n + 7 = 28$

25. $-2y + 17 = -13$ **26.** $-\frac{t}{13} - 2 = 3$ **27.** $\frac{2}{9}x - 4 = \frac{2}{3}$

28. $9 - 4g = -15$ **29.** $-4 - p = -2$ **30.** $21 - b = 11$

31. $-2(n + 7) = 15$ **32.** $5(m - 1) = -25$ **33.** $-8a - 11 = 37$

34. $\frac{7}{4}q - 2 = -5$ **35.** $2(5 - n) = 8$ **36.** $-3(d - 7) = 6$

LESSON 6
Linear Inequalities

Objective

- Use algebra to solve linear inequalities.

Statements with greater than (>), less than (<), greater than or equal to (≥), or less than or equal to (≤) are inequalities. If any number is added or subtracted to each side of an inequality, the resulting inequality is true.

Example 1 **Addition/Subtraction Linear Inequalities**

Solve each inequality.

a. $x - 17 > 12$

$x - 17 > 12$	Original inequality
$x - 17 + \mathbf{17} > 12 + \mathbf{17}$	Add 17 to each side.
$x > 29$	Simplify.

The solution set is $\{x|x > 29\}$.

b. $y + 11 \leq 5$

$y + 11 \leq 5$	Original inequality
$y + 11 - \mathbf{11} \leq 5 - \mathbf{11}$	Subtract 11 from each side.
$y \leq -6$	Simplify.

The solution set is $\{y|y \leq -6\}$.

If each side of an inequality is multiplied or divided by a positive number, the resulting inequality is true.

Example 2 **Multiplication/Division Linear Inequalities**

Solve each inequality.

a. $\frac{t}{6} \geq 11$

$\frac{t}{6} \geq 11$	Original inequality
$(\mathbf{6})\frac{t}{6} \geq (\mathbf{6})11$	Multiply each side by 6.
$t \geq 66$	Simplify.

The solution set is $\{t \mid t \geq 66\}$.

b. $8p < 72$

$8p < 72$	Original inequality
$\frac{8p}{\mathbf{8}} < \frac{72}{\mathbf{8}}$	Divide each side by 8.
$p < 9$	Simplify.

The solution set is $\{p|p < 9\}$.

If each side of an inequality is multiplied or divided by the same negative number, the direction of the inequality symbol must be reversed so that the resulting inequality is true.

Example 3 **Multiplication/Division Linear Inequalities**

Solve each inequality.

a. $-5c > 30$

$-5c > 30$	Original inequality
$\frac{-5c}{\mathbf{-5}} < \frac{30}{\mathbf{-5}}$	Divide each side by −5. Change > to <.
$c < -6$	Simplify.

The solution set is $\{c|c < -6\}$.

(continued on the next page)

b. $-\frac{d}{13} \leq -4$

$-\frac{d}{13} \leq -4$	Original inequality
$(-13)\left(\frac{-d}{13}\right) \geq (-13)(-4)$	Multiply each side by −13. Change ≤ to ≥.
$d \geq 52$	Simplify.

The solution set is $\{d \mid d \geq 52\}$.

Inequalities involving more than one operation can be solved by undoing the operations in the same way you would solve an equation with more than one operation.

Example 4 Multi-Step Linear Inequalities

Solve each inequality.

a. $-6a + 13 < -7$

$-6a + 13 < -7$	Original inequality
$-6a + 13 - 13 < -7 - 13$	Subtract 13 from each side.
$-6a < -20$	Simplify.
$\frac{-6a}{-6} > \frac{-20}{-6}$	Divide each side by −6. Change < to >.
$a > \frac{10}{3}$	Simplify.

The solution set is $\left\{a \middle| a > \frac{10}{3}\right\}$.

Watch Out!

Dividing by a Negative Remember that any time you divide an inequality by a negative number you reverse the direction of the sign.

b. $4z + 7 \geq 8z - 1$

$4z + 7 \geq 8z - 1$	Original inequality
$4z + 7 - 7 \geq 8z - 1 - 7$	Subtract 7 from each side.
$4z \geq 8z - 8$	Simplify.
$4z - 8z \geq 8z - 8 - 8z$	Subtract $8z$ from each side.
$-4z \geq -8$	Simplify.
$\frac{-4z}{-4} \leq \frac{-8}{-4}$	Divide each side by −4. Change ≥ to ≤.
$z \leq 2$	Simplify.

The solution set is $\{z \mid z \leq 2\}$.

Exercises

1. $x - 7 < 6$
2. $a + 7 \geq -5$
3. $4y < 20$
4. $-\frac{a}{8} < 5$
5. $\frac{t}{6} > -7$
6. $\frac{a}{11} \leq 8$
7. $d + 8 \leq 12$
8. $m + 14 > 10$
9. $12k \geq -36$
10. $6t - 10 \geq 4t$
11. $3z + 8 < 2$
12. $4c + 23 \leq -13$
13. $m - 21 < 8$
14. $x - 6 \geq 3$
15. $-3b \leq 48$
16. $-\frac{p}{5} \geq 14$
17. $2z - 9 < 7z + 1$
18. $-4h > 36$
19. $\frac{2}{5}b - 6 \leq -2$
20. $\frac{8}{3}t + 1 > -5$
21. $7q + 3 \geq -4q + 25$
22. $-3n - 8 > 2n + 7$
23. $-3w + 1 \leq 8$
24. $-\frac{4}{5}k - 17 > 11$

LESSON 7

Ordered Pairs

:: Objective

- Name and graph points in the coordinate plane.

New Vocabulary

ordered pair
x-coordinate
y-coordinate
quadrant
origin

Points in the coordinate plane are named by **ordered pairs** of the form (x, y). The first number, or **x-coordinate**, corresponds to a number on the x-axis. The second number, or **y-coordinate**, corresponds to a number on the y-axis.

Example 1 Writing Ordered Pairs

Write the ordered pair for each point.

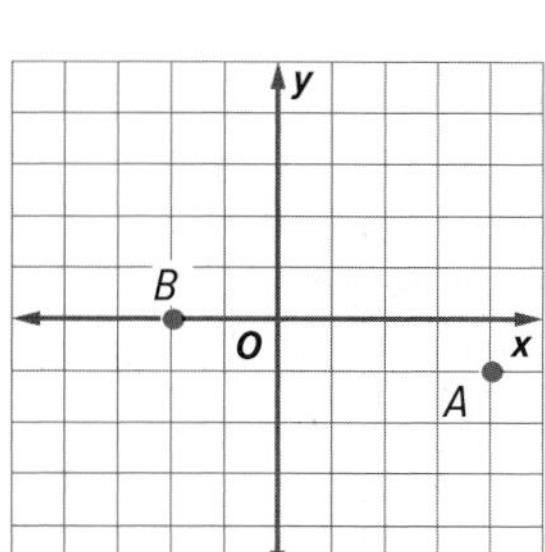

a. A

The x-coordinate is 4.

The y-coordinate is -1.

The ordered pair is $(4, -1)$.

b. B

The x-coordinate is -2.

The point lies on the x-axis, so its y-coordinate is 0.

The ordered pair is $(-2, 0)$.

The x-axis and y-axis separate the coordinate plane into four regions, called **quadrants**. The point at which the axes intersect is called the **origin**. The axes and points on the axes are not located in any of the quadrants.

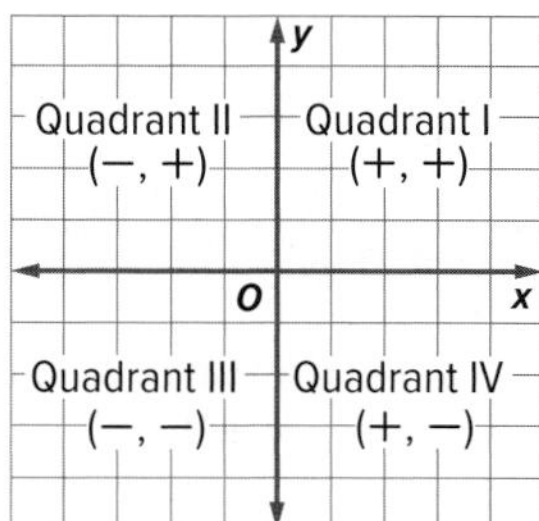

Example 2 Graphing Ordered Pairs

Graph and label each point on a coordinate plane. Name the quadrant in which each point is located.

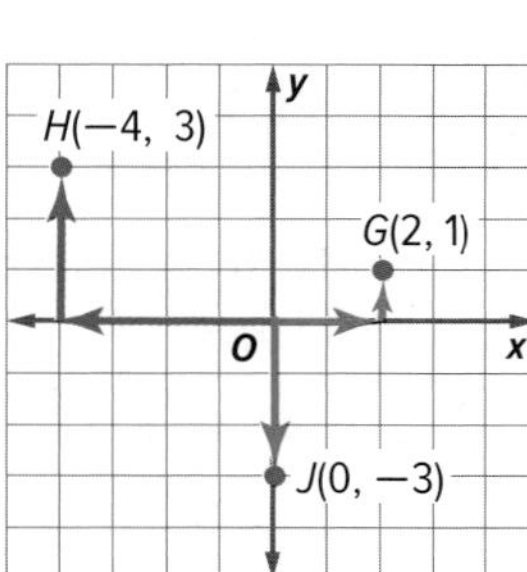

a. $G(2, 1)$

Start at the origin. Move 2 units right, since the x-coordinate is 2. Then move 1 unit up, since the y-coordinate is 1. Draw a dot, and label it G. Point G(2, 1) is in Quadrant I.

b. $H(-4, 3)$

Start at the origin. Move 4 units left, since the x-coordinate is -4. Then move 3 units up, since the y-coordinate is 3. Draw a dot, and label it H. Point $H(-4, 3)$ is in Quadrant II.

c. $J(0, -3)$

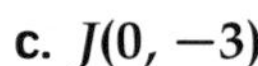

Start at the origin. Since the x-coordinate is 0, the point lies on the y-axis. Move 3 units down, since the y-coordinate is -3. Draw a dot, and label it J. Because it is on one of the axes, point $J(0, -3)$ is not in any quadrant.

Example 3 Graphing Multiple Ordered Pairs

Graph a polygon with vertices $A(-3, 3)$, $B(1, 3)$, $C(0, 1)$, and $D(-4, 1)$.

Graph the ordered pairs on a coordinate plane. Connect each pair of consecutive points. The polygon is a parallelogram.

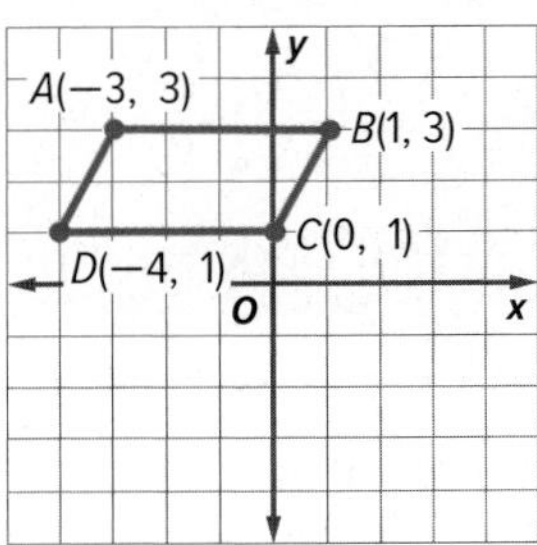

Example 4 Graphing and Solving for Ordered Pairs

Study Tip

Lines There are infinitely many points on a line, so when you are asked to find points on a line, there are many answers.

Graph four points that satisfy the equation $y = 4 - x$.

Make a table.

Choose four values for x.

Evaluate each value of x for $4 - x$.

x	$4 - x$	y	(xy)
0	4 − 0	4	(0, 4)
1	4 − 1	3	(1, 3)
2	4 − 2	2	(2, 2)
3	4 − 3	1	(3, 1)

Plot the points.

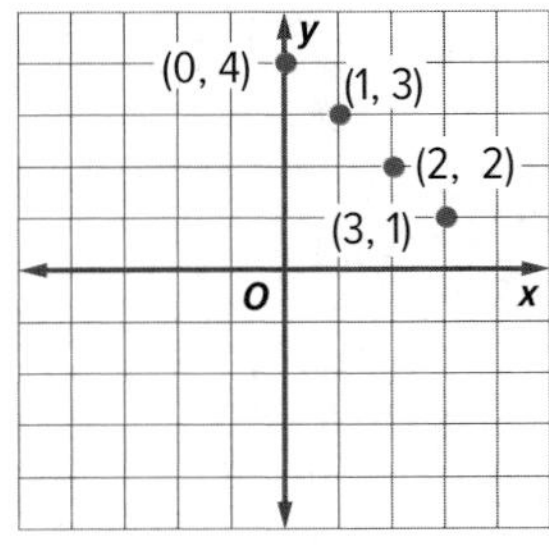

Exercises

Write the ordered pair for each point shown at the right.

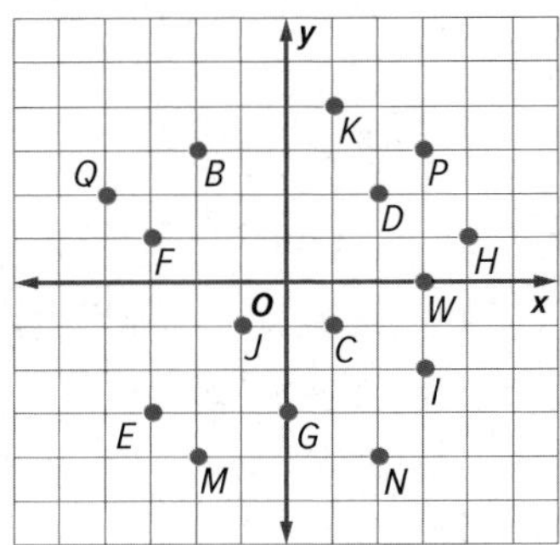

1. B **2.** C **3.** D

4. E **5.** F **6.** G

7. H **8.** I **9.** J

10. K **11.** W **12.** M

13. N **14.** P **15.** Q

Graph and label each point on a coordinate plane. Name the quadrant in which each point is located.

16. $M(-1, 3)$ **17.** $S(2, 0)$ **18.** $R(-3, -2)$ **19.** $P(1, -4)$

20. $B(5, -1)$ **21.** $D(3, 4)$ **22.** $T(2, 5)$ **23.** $L(-4, -3)$

Graph the following geometric figures.

24. a square with vertices $W(-3, 3)$, $X(-3, -1)$, $Z(1, 3)$, and $Y(1, -1)$

25. a polygon with vertices $J(4, 2)$, $K(1, -1)$, $L(-2, 2)$, and $M(1, 5)$

26. a triangle with vertices $F(2, 4)$, $G(-3, 2)$, and $H(-1, -3)$

Graph four points that satisfy each equation.

27. $y = 2x$ **28.** $y = 1 + x$ **29.** $y = 3x - 1$ **30.** $y = 2 - x$

LESSON 8

Systems of Linear Equations

Objective

- Use graphing, substitution, and elimination to solve systems of linear equations.

New Vocabulary

system of equations
substitution
elimination

Two or more equations that have common variables are called a **system of equations**. The solution of a system of equations in two variables is an ordered pair of numbers that satisfies both equations. A system of two linear equations can have zero, one, or an infinite number of solutions. There are three methods by which systems of equations can be solved: graphing, elimination, and substitution.

Example 1 Graphing Linear Equations

Solve each system of equations by graphing. Then determine whether each system has *no* solution, *one* solution, or *infinitely many* solutions.

a. $y = -x + 3$
$y = 2x - 3$

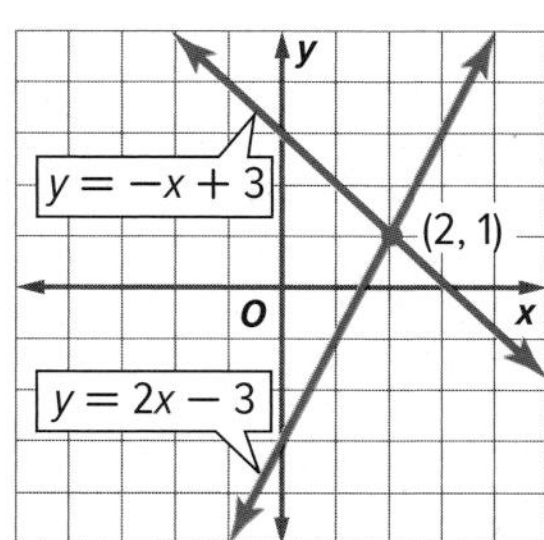

The graphs appear to intersect at (2, 1). Check this estimate by replacing x with 2 and y with 1 in each equation.

CHECK $y = -x + 3$ $\qquad y = 2x - 3$

$1 \stackrel{?}{=} -2 + 3 \qquad 1 \stackrel{?}{=} 2(2) - 3$

$1 = 1\ \checkmark \qquad 1 = 1\ \checkmark$

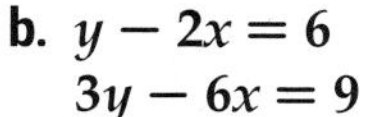

The system has one solution at (2, 1).

b. $y - 2x = 6$
$3y - 6x = 9$

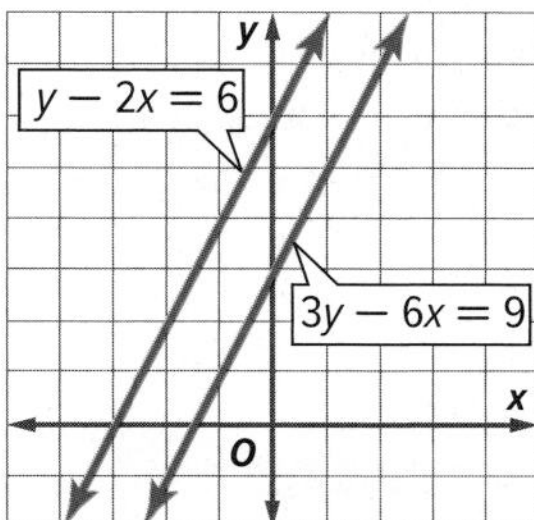

The graphs of the equations are parallel lines. Since they do not intersect, there are no solutions of this system of equations. Notice that the lines have the same slope but different y-intercepts. Equations with the same slope *and* the same y-intercepts have an infinite number of solutions.

It is difficult to determine the solution of a system when the two graphs intersect at noninteger values. There are algebraic methods by which an exact solution can be found. One such method is **substitution**.

Example 2 Substitution

Use substitution to solve the system of equations.

$y = -4x$
$2y + 3x = 8$

Since $y = -4x$, substitute $-4x$ for y in the second equation.

$2y + 3x = 8$	Second equation
$2(-4x) + 3x = 8$	$y = -4x$
$-8x + 3x = 8$	Simplify.
$-5x = 8$	Combine like terms.
$\frac{-5x}{-5} = \frac{8}{-5}$	Divide each side by −5.
$x = -\frac{8}{5}$	Simplify.

Use $y = -4x$ to find the value of y.

$y = -4x$	First equation
$= -4\left(-\frac{8}{5}\right)$	$x = -\frac{8}{5}$
$= \frac{32}{5}$	Simplify.

The solution is $\left(-\frac{8}{5}, \frac{32}{5}\right)$.

Sometimes adding or subtracting two equations together will eliminate one variable. Using this step to solve a system of equations is called **elimination**.

Example 3 Elimination

Use elimination to solve the system of equations.

$3x + 5y = 7$

$4x + 2y = 0$

Either x or y can be eliminated. In this example, we will eliminate x.

$3x + 5y = 7$ → Multiply by 4. → $12x + 20y = 28$

$4x + 2y = 0$ → Multiply by −3. → $+ (-12x) - 6y = 0$

$14y = 28$ Add the equations.

$\frac{14y}{14} = \frac{28}{14}$ Divide each side by 14.

$y = 2$ Simplify.

Now substitute 2 for y in either equation to find the value of x.

$4x + 2y = 0$ Second equation

$4x + 2(2) = 0$ $y = 2$

$4x + 4 = 0$ Simplify.

$4x + 4 - 4 = 0 - 4$ Subtract 4 from each side.

$4x = -4$ Simplify.

$\frac{4x}{4} = \frac{-4}{4}$ Divide each side by 4.

$x = -1$ Simplify.

The solution is $(-1, 2)$.

Study Tip

Checking Solutions You can confirm that your solutions are correct by substituting the values into both of the original equations.

Exercises

Solve by graphing.

1. $y = -x + 2$; $y = -\frac{1}{2}x + 1$
2. $y = 3x - 3$; $y = x + 1$
3. $y - 2x = 1$; $2y - 4x = 1$

Solve by substitution.

4. $-5x + 3y = 12$; $x + 2y = 8$
5. $x - 4y = 22$; $2x + 5y = -21$
6. $y + 5x = -3$; $3y - 2x = 8$

Solve by elimination.

7. $-3x + y = 7$; $3x + 2y = 2$
8. $3x + 4y = -1$; $-9x - 4y = 13$
9. $-4x + 5y = -11$; $2x + 3y = 11$

Name an appropriate method to solve each system of equations. Then solve the system.

10. $4x - y = 11$; $2x - 3y = 3$
11. $4x + 6y = 3$; $-10x - 15y = -4$
12. $3x - 2y = 6$; $5x - 5y = 5$
13. $3y + x = 3$; $-2y + 5x = 15$
14. $4x - 7y = 8$; $-2x + 5y = -1$
15. $x + 3y = 6$; $4x - 2y = -32$

LESSON 9

Square Roots and Simplifying Radicals

Objective

- Evaluate square roots and simplify radical expressions.

New Vocabulary

Product Property
Quotient Property

A radical expression is an expression that contains a square root. The expression is in simplest form when the following three conditions have been met.

- No radicands have perfect square factors other than 1.
- No radicands contain fractions.
- No radicals appear in the denominator of a fraction.

The **Product Property** states that for two numbers a and $b \geq 0$, $\sqrt{ab} = \sqrt{a} \cdot \sqrt{b}$.

Example 1 Product Property

Simplify.

a. $\sqrt{45}$

$\sqrt{45} = \sqrt{3 \cdot 3 \cdot 5}$ Prime factorization of 45

$= \sqrt{3^2} \cdot \sqrt{5}$ Product Property of Square Roots

$= 3\sqrt{5}$ Simplify.

b. $\sqrt{6} \cdot \sqrt{15}$

$\sqrt{6} \cdot \sqrt{15} = \sqrt{6 \cdot 15}$ Product Property

$= \sqrt{3 \cdot 2 \cdot 3 \cdot 5}$ Prime factorization

$= \sqrt{3^2} \cdot \sqrt{10}$ Product Property

$= 3\sqrt{10}$ Simplify.

For radical expressions in which the exponent of the variable inside the radical is *even* and the resulting simplified exponent is *odd,* you must use absolute value to ensure nonnegative results.

Example 2 Product Property

Simplify $\sqrt{20x^3y^5z^6}$.

$\sqrt{20x^3y^5z^6} = \sqrt{2^2 \cdot 5 \cdot x^3 \cdot y^5 \cdot z^6}$ Prime factorization

$= \sqrt{2^2} \cdot \sqrt{5} \cdot \sqrt{x^3} \cdot \sqrt{y^5} \cdot \sqrt{z^6}$ Product Property

$= 2 \cdot \sqrt{5} \cdot x \cdot \sqrt{x} \cdot y^2 \cdot \sqrt{y} \cdot |z^3|$ Simplify.

$= 2xy^2|z^3|\sqrt{5xy}$ Simplify.

The **Quotient Property** states that for any numbers a and b, where $a \geq 0$ and $b \geq 0$, $\sqrt{\frac{a}{b}} = \frac{\sqrt{a}}{\sqrt{b}}$.

Example 3 Quotient Property

Simplify $\sqrt{\frac{25}{16}}$.

$\sqrt{\frac{25}{16}} = \frac{\sqrt{25}}{\sqrt{16}}$ Quotient Property

$= \frac{5}{4}$ Simplify.

Rationalizing the denominator of a radical expression is a method used to eliminate radicals from the denominator of a fraction. To rationalize the denominator, multiply the expression by a fraction equivalent to 1 such that the resulting denominator is a perfect square.

Example 4 Rationalize the Denominator

Simplify.

a. $\frac{2}{\sqrt{3}}$

$\frac{2}{\sqrt{3}} = \frac{2}{\sqrt{3}} \cdot \frac{\sqrt{3}}{\sqrt{3}}$ Multiply by $\frac{\sqrt{3}}{\sqrt{3}}$.

$= \frac{2\sqrt{3}}{3}$ Simplify.

b. $\frac{\sqrt{13y}}{\sqrt{18}}$

$\frac{\sqrt{13y}}{\sqrt{18}} = \frac{\sqrt{13y}}{\sqrt{2 \cdot 3 \cdot 3}}$ Prime factorization

$= \frac{\sqrt{13y}}{3\sqrt{2}}$ Product Property

$= \frac{\sqrt{13y}}{3\sqrt{2}} \cdot \frac{\sqrt{2}}{\sqrt{2}}$ Multiply by $\frac{\sqrt{2}}{\sqrt{2}}$.

$= \frac{\sqrt{26y}}{6}$ Product Property

Watch Out

Rationalizing the Denominator Don't forget to multiply both the numerator and denominator by the radical when you rationalize the denominator.

Sometimes, conjugates are used to simplify radical expressions. Conjugates are binomials of the form $p\sqrt{q} + r\sqrt{t}$ and $p\sqrt{q} - r\sqrt{t}$.

Example 5 Conjugates

Simplify $\frac{3}{5 - \sqrt{2}}$.

$\frac{3}{5 - \sqrt{2}} = \frac{3}{5 - \sqrt{2}} \cdot \frac{5 + \sqrt{2}}{5 + \sqrt{2}}$ $\frac{5 + \sqrt{2}}{5 + \sqrt{2}} = 1$

$= \frac{3(5 + \sqrt{2})}{5^2 - (\sqrt{2})^2}$ $(a - b)(a + b) = a^2 - b^2$

$= \frac{15 + 3\sqrt{2}}{25 - 2}$ Multiply. $(\sqrt{2})^2 = 2$

$= \frac{15 + 3\sqrt{2}}{23}$ Simplify.

Exercises

Simplify.

1. $\sqrt{32}$
2. $\sqrt{75}$
3. $\sqrt{50} \cdot \sqrt{10}$
4. $\sqrt{12} \cdot \sqrt{20}$
5. $\sqrt{6} \cdot \sqrt{6}$
6. $\sqrt{16} \cdot \sqrt{25}$
7. $\sqrt{98x^3y^6}$
8. $\sqrt{56a^2b^4c^5}$
9. $\sqrt{\frac{81}{49}}$
10. $\sqrt{\frac{121}{16}}$
11. $\sqrt{\frac{63}{8}}$
12. $\sqrt{\frac{288}{147}}$
13. $\frac{\sqrt{10p^3}}{\sqrt{27}}$
14. $\frac{\sqrt{108}}{\sqrt{2q^6}}$
15. $\frac{4}{5 - 2\sqrt{3}}$
16. $\frac{7\sqrt{3}}{5 - 2\sqrt{6}}$
17. $\frac{3}{\sqrt{48}}$
18. $\frac{\sqrt{24}}{\sqrt{125}}$
19. $\frac{3\sqrt{5}}{2 - \sqrt{2}}$
20. $\frac{3}{-2 + \sqrt{13}}$

CHAPTER 0

Posttest

Go Online! for another Chapter Test

State which metric unit you would probably use to measure each item.

1. mass of a book
2. length of a highway

Complete each sentence.

3. 8 in. = __?__ ft
4. 6 yd = __?__ ft
5. 24 fl oz = __?__ pt
6. 3.7 kg = __?__ lb
7. 4.2 km = __?__ m
8. 285 g = __?__ kg
9. 0.75 kg = __?__ mg
10. 1.9 L = __?__ qt

11. **PROBABILITY** The table shows the results of an experiment in which a number cube was rolled. Find the experimental probability of rolling a 4.

Outcome	Tally	Frequency
1	||||	4
2	~~||||~~ |	6
3	~~||||~~	5
4	|||	3
5	~~||||~~ ||	7
6		0

CANDY A bag of candy contains 3 lollipops, 8 peanut butter cups, and 4 chocolate bars. A piece of candy is randomly drawn from the bag. Find each probability.

12. P(peanut butter cup)
13. P(lollipop or peanut butter cup)
14. P(not chocolate bar)
15. P(chocolate bar or lollipop)

Evaluate each expression if $x = 2$, $y = -3$, and $z = 4$.

16. $6x - z$
17. $6y + xz$
18. $3yz$
19. $\frac{6z}{xy}$
20. $\frac{y + 2x}{10z}$
21. $7 + |y - 11|$

Solve each equation.

22. $9 + s = 21$
23. $h - 8 = 12$
24. $\frac{4m}{14} = 18$
25. $\frac{2}{9}d = 10$
26. $3(20 - b) = 36$
27. $37 + w = 5w - 27$
28. $\frac{x}{6} = 7$
29. $\frac{1}{4}(n + 5) = 16$

Solve each inequality.

30. $4y - 9 > 1$
31. $-2z + 15 \geq 4$
32. $3r + 7 < r - 8$
33. $-\frac{2}{5}k - 20 \leq 10$
34. $-3(b - 4) > 33$
35. $2 - m \leq 6m - 12$
36. $8 \leq r - 14$
37. $\frac{2}{3}n < \frac{3}{9}n - 5$

Write the ordered pair for each point shown.

38. M
39. N
40. P
41. Q

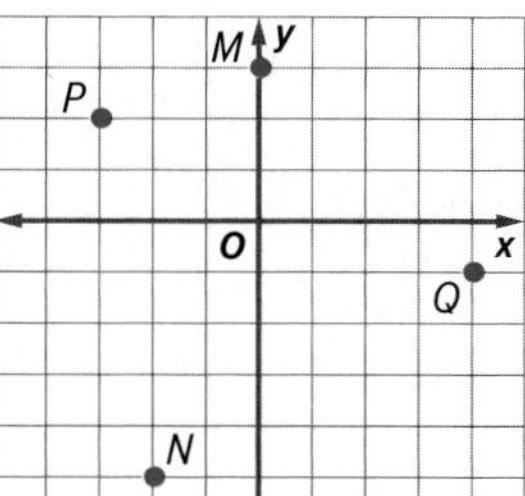

Graph and label each point on the coordinate plane above.

42. $A(-2, 0)$
43. $C(1, 3)$
44. $D(-4, -4)$
45. $F(3, -5)$

46. Graph the quadrilateral with vertices $R(2, 0)$, $S(4, -2)$, $T(4, 3)$, and $W(2, 5)$.

47. Graph three points that satisfy the equation $y = \frac{1}{2}x - 5$.

Solve each system of equations.

48. $2r + m = 11$
$6r - 2m = -2$
49. $2x + 4y = 6$
$7x = 4 + 3y$
50. $2c + 6d = 14$
$-\frac{7}{3} + \frac{1}{3}c = -d$
51. $5a - b = 17$
$3a + 2b = 5$
52. $6d + 3f = 12$
$2d = 8 - f$
53. $4x - 5y = 17$
$3x + 4y = 5$

Simplify.

54. $\sqrt{80}$
55. $\sqrt{\frac{128}{5}}$
56. $\sqrt{36} \cdot \sqrt{81}$
57. $\sqrt{\frac{7x^3}{3}}$
58. $\sqrt{\frac{5}{81}}$
59. $\sqrt{12x^5y^2}$

CHAPTER 1

Tools of Geometry

THEN

You graphed points on the coordinate plane and evaluated mathematical expressions.

NOW

In this chapter, you will:

- Find distances between points and midpoints of line segments.
- Identify angle relationships.
- Find perimeters, areas, surface areas, and volumes.

MP WHY

MAPS Geometric figures and terms can be used to represent and describe real-world situations. On a map, locations of cities can be represented by points and highways or streets by lines. Specific places can be represented by polygons that have both perimeter and area. The map itself is representative of a plane.

Use the Mathematical Practices to complete the activity.

1. **Using Tools** Use the Internet and a mapping application to locate the map of a national park you have visited or would like to visit.
2. **Sense-Making** What basic geometric shapes do you already know? Do you see any of these shapes in the map?
3. **Reasoning** How are line segments useful in determining distances? How can the edges of a shape be used to determine the area of a shape?
4. **Model with Mathematics** Use the **2D Tool** to approximate the shape and size of the national park you selected. How can you use this to approximate the area of the park?

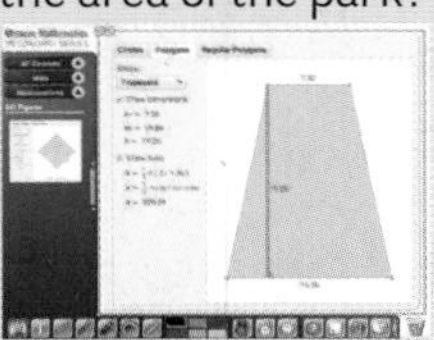

Go Online to Guide Your Learning

Explore & Explain

The Geometer's Sketchpad

Use **The Geometer's Sketchpad** to illustrate how to copy a line segment and create, drag, measure, and animate angles, to explore the relationships between pairs of angles formed by intersecting lines, and to explore the coordinates of points that have been reflected in the *x*-axis or the *y*-axis.

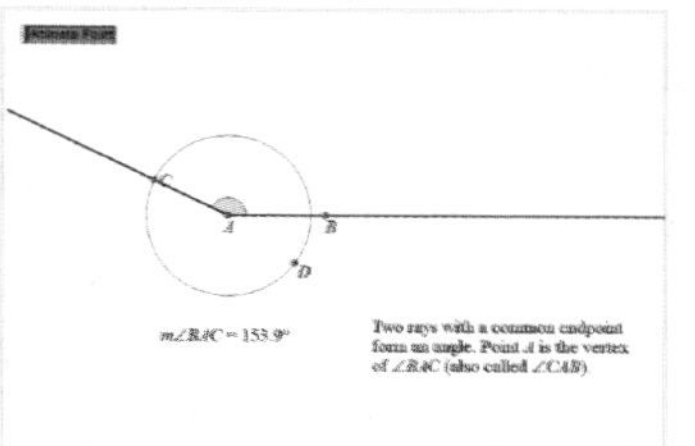

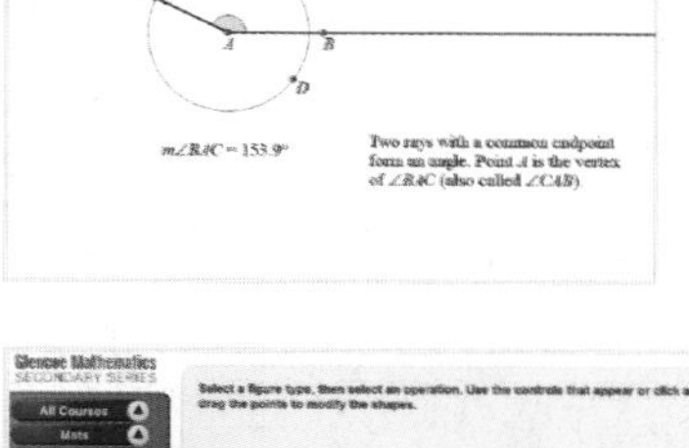

Transformations Tool

Use the **Transformations** tool in eToolkit to practice reflections in the coordinate plane in Lesson 1-7.

eBook

Interactive Student Guide

Before starting the chapter, answer the **Chapter Focus** preview questions. Check your answers as you complete each lesson. At the end of the chapter, try the **Performance Task**.

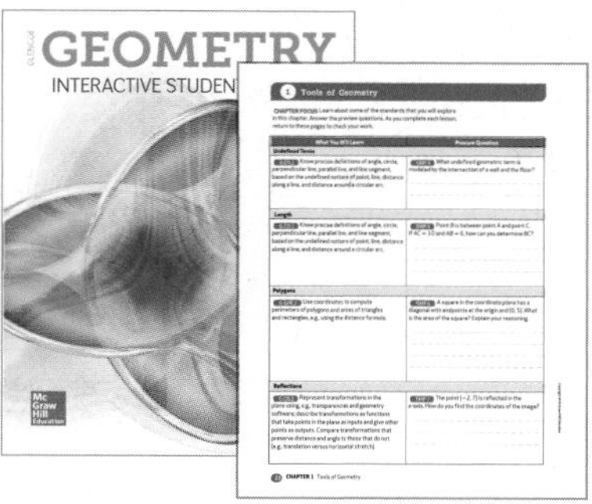

Organize

Foldables

Make this Foldable to organize your notes. Begin with a sheet of 11″× 17″ paper.

Fold the short sides to meet in the middle. Cut the front sides in fifths and label the tabs with lesson titles.

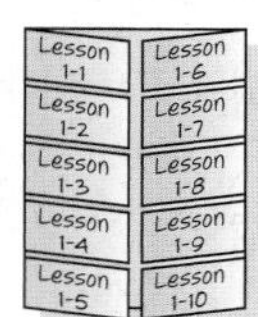

Collaborate

Chapter Project

In the **Urban Planning** project, you will use what you have learned about angles and polygons to complete a project that addresses civic literacy.

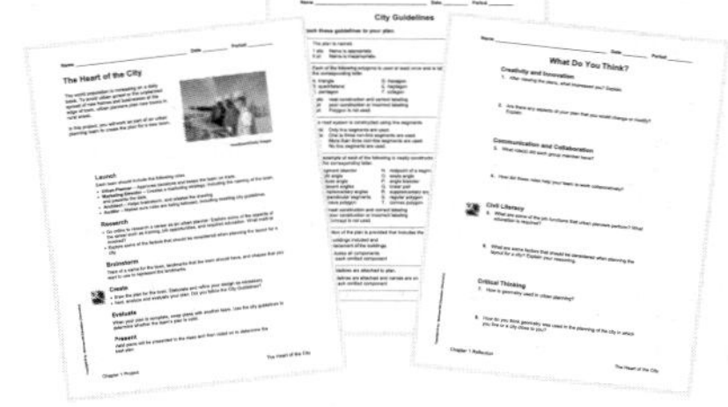

Focus

LEARNSMART®

Need help studying? Complete the **Congruence, Proof, and Constructions** topic in LearnSmart to review for the chapter test.

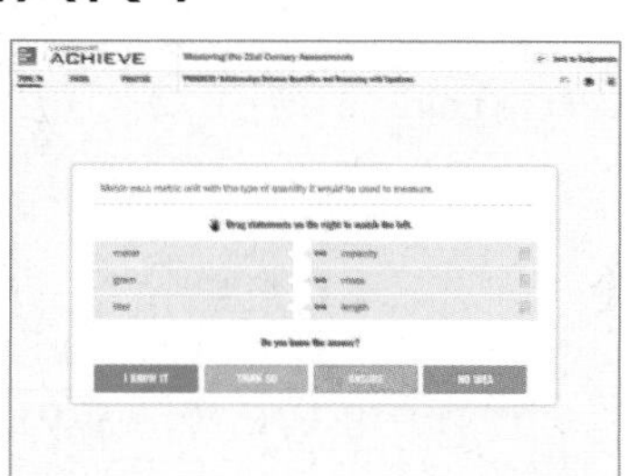

ALEKS®

You can use the **Coordinate Geometry** and **Lines and Angles** topics in ALEKS to explore what you know about coordinate planes, lines, and angles and what you are ready to learn.*

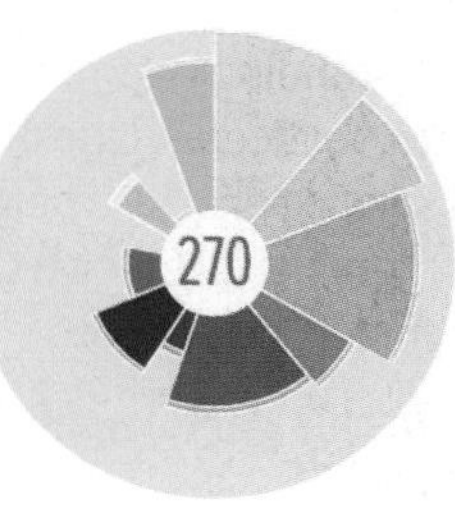

* Ask your teacher if this is part of your program.

Go Online! for Vocabulary Review Games and key vocabulary in 13 languages.

Get Ready for the Chapter

Connecting Concepts

Concept Check

Review the concepts used in this chapter by answering the questions below.

1. If you were told to graph and label a point in a coordinate plane, where would you start?
2. If you are graphing a point with a negative *x*-coordinate, what direction would you move in a coordinate plane?
3. If the location of the knight shown on the chessboard is considered the origin of the coordinate grid and you are told to move it to (−1, 2), where would the knight end up?

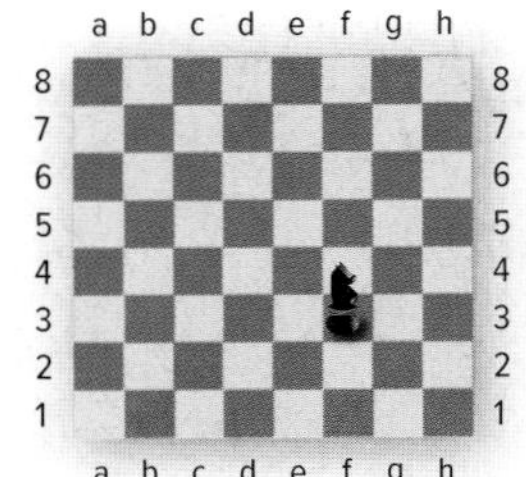

4. When you are given two fractions that do not have a common denominator to add or subtract, what is the first step you need to take to solve the problem?
5. If the denominator of one fraction is 6 and of another fraction is 4, what is the least common denominator for these two fractions?
6. When presented with an expression such as $[-2 - (-7)]^2 + (1 - 8)^2$, what rules are followed to evaluate the expression?

Performance Task Preview

You can use the concepts and skills in the chapter to solve problems in a real-world setting. Understanding coordinate planes, vertices, shapes, volume, and surface area will help you finish the Performance Task at the end of the chapter.

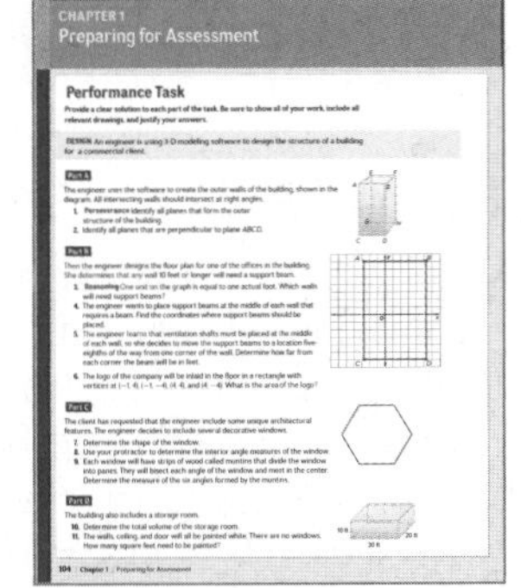

MP In this Performance Task you will:

- make sense of problems and persevere in solving them
- reason abstractly and quantitatively
- model with mathematics
- look for and express regularity in repeated reasoning

New Vocabulary

English		Español
collinear	p. 5	colineal
coplanar	p. 5	coplanar
congruent	p. 15	congruente
irrational numbers	p. 17	números irracionales
midpoint	p. 26	punto medio
segment bisector	p. 27	bisectriz de segmento
angle	p. 36	angulo
vertex	p. 36	vertice
angle bisector	p. 39	bisectriz de un angulo
perpendicular	p. 48	perpendicular
polygon	p. 56	poligono
perimeter	p. 58	perimetro
surface area	p. 78	área de superficie
volume	p. 78	volumen
accuracy	p. 92	exactitud

Review Vocabulary

ordered pair par ordenado a set of numbers or coordinates used to locate any point on a coordinate plane, written in the form (*x*, *y*)

origin origen the point where the two axes intersect at their zero points

quadrants cuadrantes the four regions into which the *x*-axis and *y*-axis separate the coordinate plane

x-coordinate coordenada *x* the first number in an ordered pair

y-coordinate coordenada *y* the second number in an ordered pair

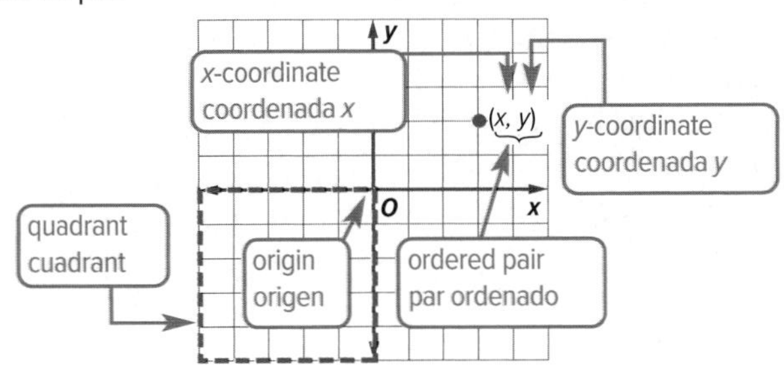

LESSON 1

Points, Lines, and Planes

Then

- You used basic geometric concepts and properties to solve problems.

Now

1. Identify and model points, lines, and planes.
2. Identify intersecting lines and planes.

Why?

- On a subway map, the locations of stops are represented by *points*. The route the train can take is modeled by a series of connected paths that look like *lines*. The flat surface of the map on which these points and lines lie is representative of a *plane*.

New Vocabulary

undefined term
point
line
plane
collinear
coplanar
intersection
definition
defined term
space

Mathematical Practices

3 Construct viable arguments and critique the reasoning of others.
4 Model with mathematics.
6 Attend to precision.

1 Points, Lines, and Planes

Unlike the real-world objects that they model, points, lines, and planes do not have any actual size. In geometry, *point*, *line*, and *plane* are considered **undefined terms** because they are only explained using examples and descriptions.

You are already familiar with the terms point, line, and plane from algebra. You graphed on a coordinate *plane* and found ordered pairs that represented *points* on *lines*. In geometry, these terms have a similar meaning.

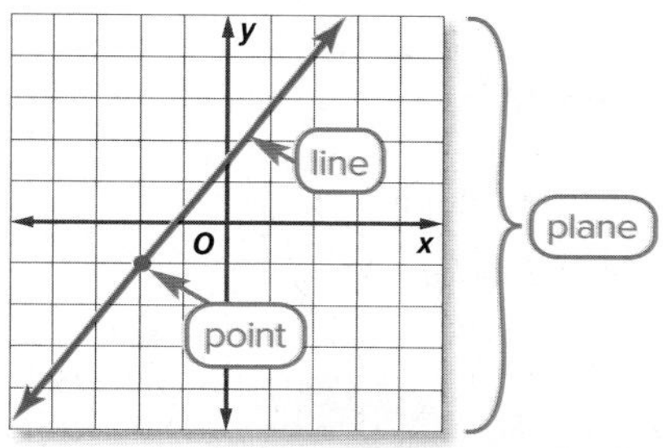

The phrase *exactly one* in a statement such as, "There is exactly one line through any two points," means that there is *one and only one*.

Key Concept Undefined Terms

A **point** is a location. It has neither shape nor size.

Named by	a capital letter
Example	point *A*

A **line** is made up of points and has no thickness or width. There is exactly one line through any two points.

Named by	the letters representing two points on the line or a lowercase script letter
Example	line *m*, line *PQ* or $\overleftrightarrow{PQ}$, line *QP* or $\overleftrightarrow{QP}$

A **plane** is a flat surface made up of points that extends infinitely in all directions. There is exactly one plane through any three points not on the same line.

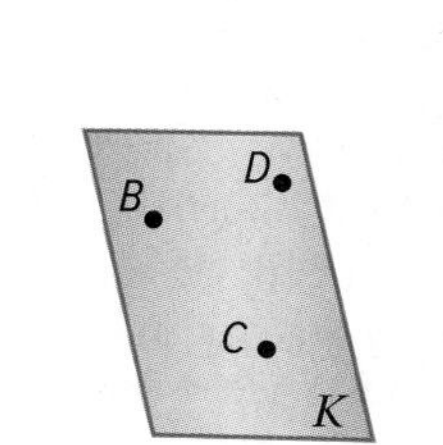

Named by	a capital script letter or by the letters naming three points that are not all on the same line
Example	plane *K*, plane *BCD*, plane *CDB*, plane *DCB*, plane *DBC*, plane *CBD*, plane *BDC*

Collinear points are points that lie on the same line. *Noncollinear* points do not lie on the same line. **Coplanar** points are points that lie in the same plane. *Noncoplanar* points do not lie in the same plane.

Example 1 Name Lines and Planes

Use the figure to name each of the following.

a. a line containing point *W*

The line can be named as line *n*, or any two of the four points on the line can be used to name the line.

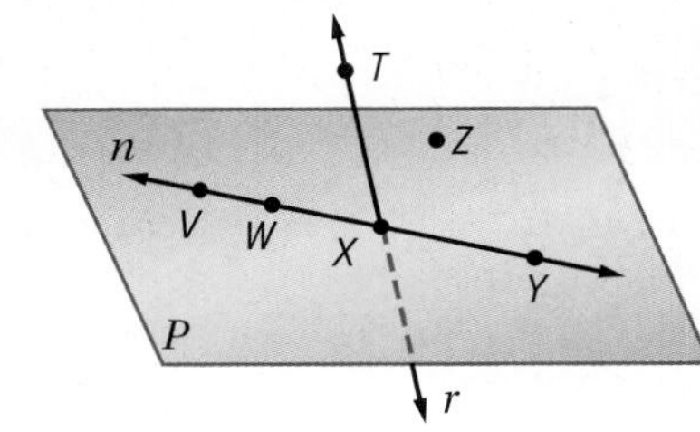

$\overleftrightarrow{VW}$ $\overleftrightarrow{WV}$ $\overleftrightarrow{VX}$ $\overleftrightarrow{XV}$ $\overleftrightarrow{VY}$ $\overleftrightarrow{YV}$

$\overleftrightarrow{WX}$ $\overleftrightarrow{XW}$ $\overleftrightarrow{WY}$ $\overleftrightarrow{YW}$ $\overleftrightarrow{XY}$ $\overleftrightarrow{YX}$

Study Tip

Additional Planes Although not drawn in Example 1b, there is another plane that contains point *X*. Since points *W*, *T*, and *X* are noncollinear, point *X* is also in plane *WTX*.

b. a plane containing point *X*

One plane that can be named is plane *P*. You can also use the letters of any three *noncollinear* points to name this plane.

plane *XZY*	plane *VZW*	plane *VZX*
plane *VZY*	plane *WZX*	plane *WZY*

The letters of each of these names can be reordered to create other acceptable names for this plane. For example, *XZY* can also be written as *XYZ*, *ZXY*, *ZYX*, *YXZ*, and *YZX*. In all, there are 36 different three-letter names for this plane.

Guided Practice

1A. a plane containing points *T* and *Z* **1B.** a line containing point *T*

Real-World Example 2 Model Points, Lines, and Planes

Real-World Career

Drafter Drafters use perspective to create drawings to build everything from toys to school buildings. Drafters need skills in math and computers. They get their education at trade schools, community colleges, and some 4-year colleges. Refer to Exercises 50 and 51.

MESSAGE BOARD Name the geometric terms modeled by the objects in the picture.

The push pin models point *G*.

The maroon border on the card models line *GH*.

The edge of the card models line *HJ*.

The card itself models plane *FGJ*.

Guided Practice

Name the geometric term modeled by each object.

2A. stripes on a sweater **2B.** the corner of a box

2 Intersections of Lines and Planes

The **intersection** of two or more geometric figures is the set of points they have in common. Two lines intersect in a point. Lines can intersect planes, and planes can intersect each other.

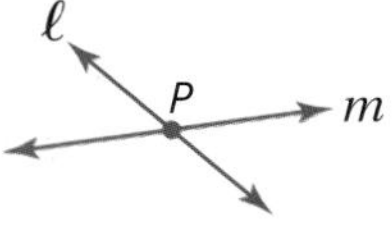

P represents the intersection of lines *ℓ* and *m*.

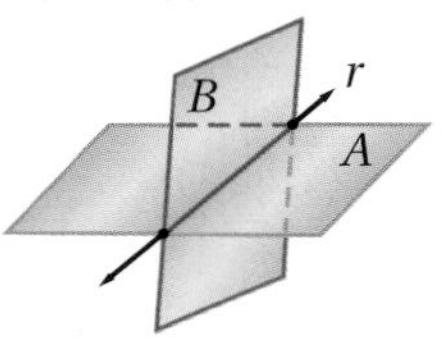

Line *r* represents the intersection of planes *A* and *B*.

Example 3 Draw Geometric Figures

Draw and label a figure for each relationship.

a. ALGEBRA Lines AB and CD intersect at E for $A(-2, 4)$, $B(0, -2)$, $C(-3, 0)$, and $D(3, 3)$ on a coordinate plane. Point F is coplanar with these points, but not collinear with $\overleftrightarrow{AB}$ or $\overleftrightarrow{CD}$.

Graph each point and draw $\overleftrightarrow{AB}$ and $\overleftrightarrow{CD}$.

Label the intersection point as E.

An infinite number of points are coplanar with A, B, C, D, and E but not collinear with $\overleftrightarrow{AB}$ and $\overleftrightarrow{CD}$. In the graph, one such point is $F(2, -3)$.

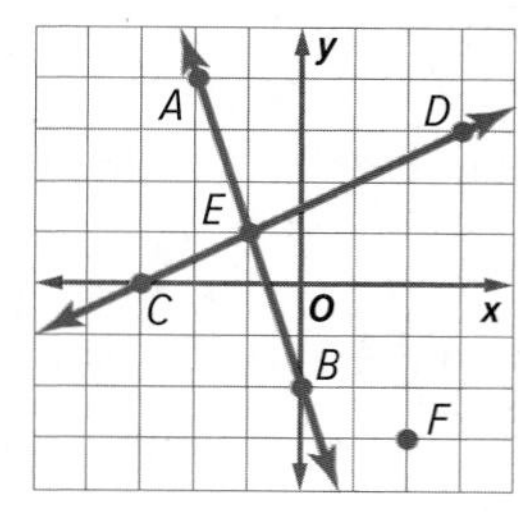

b. QR intersects plane T at point S.

Draw a surface to represent plane T and label it.

Draw a dot for point S anywhere on the plane and a dot that is not on plane T for point Q.

Draw a line through points Q and S. Dash the line to indicate the portion hidden by the plane. Then draw another dot on the line and label it R.

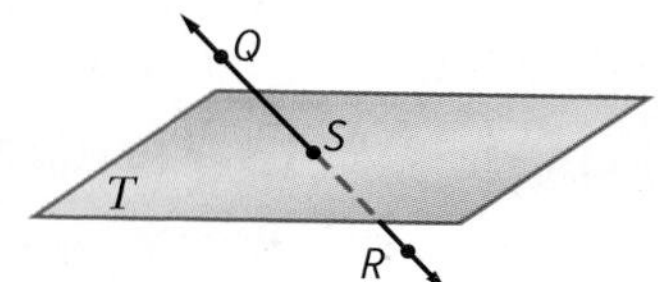

Study Tip

Three-Dimensional Drawings Because it is impossible to show an entire plane in a figure, edged shapes with different shades of color are used to represent planes.

Guided Practice

3A. Points $J(-4, 2)$, $K(3, 2)$, and L are collinear.

3B. Line p lies in plane N and contains point L.

Study Tip

Using Your Text Notice that new terms are listed at the beginning of the lesson and also highlighted in context.

Definitions or **defined terms** are explained using undefined terms and/or other defined terms. **Space** is defined as a boundless, three-dimensional set of all points. Space can contain lines and planes.

Example 4 Interpret Drawings

a. How many planes appear in this figure?

six: plane X, plane JDH, plane JDE, plane EDF, plane FDG, and plane HDG

b. Name three points that are collinear.

Points J, K, and D are collinear.

c. Name the intersection of plane HDG with plane X.

Plane HDG intersects plane X in $\overleftrightarrow{HG}$.

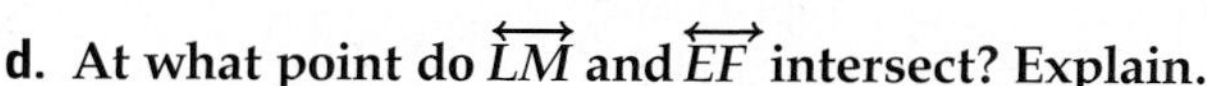

d. At what point do $\overleftrightarrow{LM}$ and $\overleftrightarrow{EF}$ intersect? Explain.

It does not appear that these lines intersect. $\overleftrightarrow{EF}$ lies in plane X, but only point L of $\overleftrightarrow{LM}$ lies in X.

Study Tip

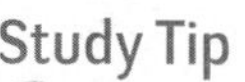

Precision A point has no dimension. A line exists in one dimension. However, a circle is two-dimensional, and a pyramid is three-dimensional.

Guided Practice

Explain your reasoning.

4A. Are points E, D, F, and G coplanar?

4B. At what point or in what line do planes JDH, JDE, and EDF intersect?

Check Your Understanding

◯ = Step-by-Step Check begin on page R13.

Example 1 **Use the figure to name each of the following.**

1. a line containing point *X*
2. a line containing point *Z*
3. a plane containing points *W* and *R*

Example 2 **Name the geometric term modeled by each object.**

4. a tightrope
5. a floor

Example 3 **Draw and label a figure for each relationship.**

6. A line in a coordinate plane contains $A(0, -5)$ and $B(3, 1)$ and a point *C* that is not collinear with $\overleftrightarrow{AB}$.
7. Plane *Z* contains lines *x*, *y*, and *w*. Lines *x* and *y* intersect at point *V* and lines *x* and *w* intersect at point *P*.

Example 4 **Refer to the figure.**

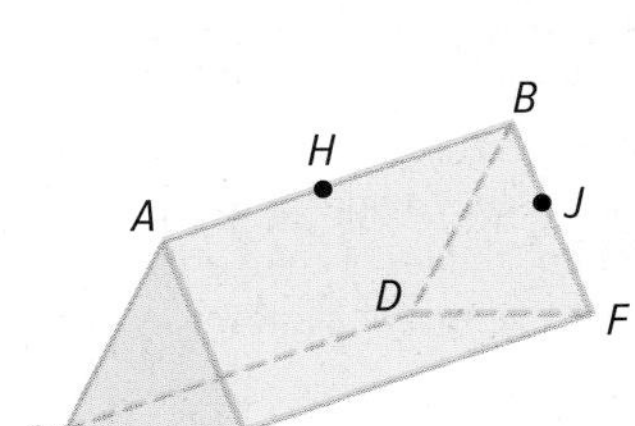

8. How many planes are shown in the figure?
9. Name three points that are collinear.
10. Are points *A*, *H*, *J*, and *D* coplanar? Explain.
11. Are points *B*, *D*, and *F* coplanar? Explain.
12. **ASTRONOMY** Ursa Minor, or the Little Dipper, is a constellation made up of seven stars in the northern sky including the star Polaris. The diagram at the right shows the arrangement of the stars in the constellation.
    a. What geometric figures are modeled by the stars?
    b. Are Star 1, Star 2, and Star 3 collinear on the constellation map? Explain.
    c. Are Polaris, Star 2, and Star 6 coplanar on the map?

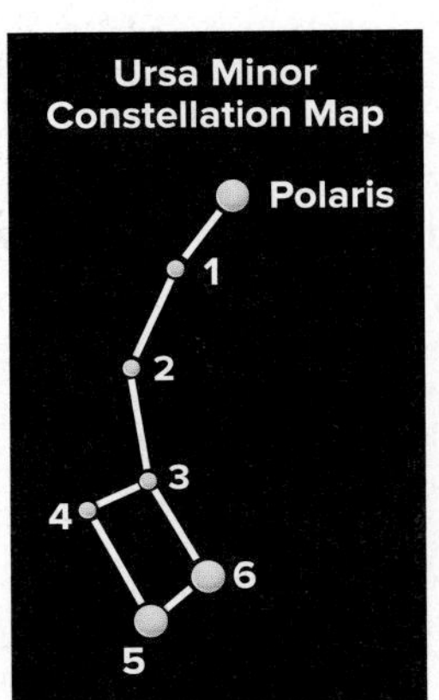

Practice and Problem Solving

Extra Practice is on page R1.

Example 1 **Refer to the figure.**

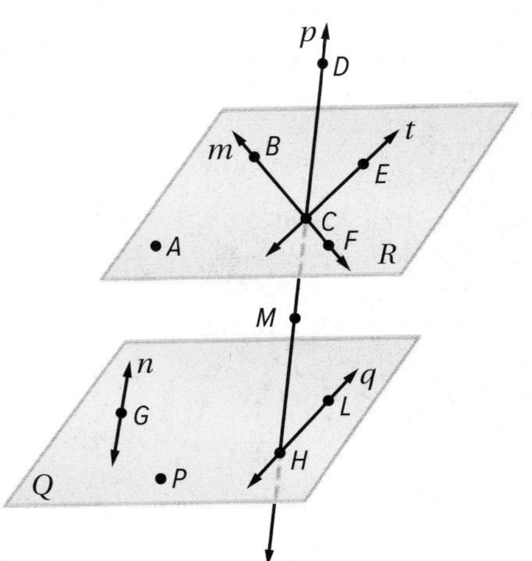

13. Name the lines that are only in plane *Q*.
14. How many planes are labeled in the figure?
15. Name the plane containing the lines *m* and *t*.
16. Name the intersection of lines *m* and *t*.
17. Name a point that is not coplanar with points *A*, *B*, and *C*.
18. Are points *F*, *M*, *G*, and *P* coplanar? Explain.
19. Name the points not contained in a line shown.
20. What is another name for line *t*?
21. Does line *n* intersect line *q*? Explain.

Example 2 **Name the geometric term(s) modeled by each object.**

22.

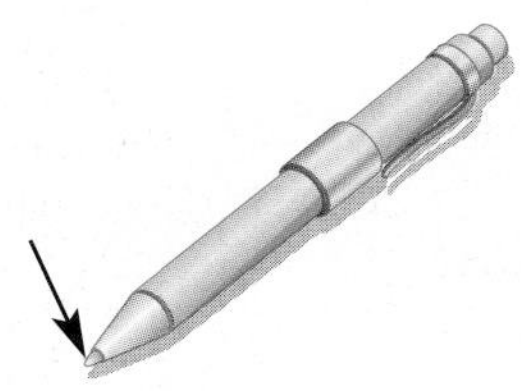

23.

24.

25.

26. a blanket

27. a knot in a rope

28. a telephone pole

29. the edge of a desk

30. two connected walls

31. a partially opened folder

Example 3 **Draw and label a figure for each relationship.**

32. Line m intersects plane R at a single point.

33. Two planes do not intersect.

34. Points X and Y lie on $\overleftrightarrow{CD}$.

35. Three lines intersect at point J but do not all lie in the same plane.

36. Points $A(2, 3)$, $B(2, -3)$, C, and D are collinear, but A, B, C, D, and F are not.

37. Lines $\overleftrightarrow{LM}$ and $\overleftrightarrow{NP}$ are coplanar but do not intersect.

38. $\overleftrightarrow{FG}$ and $\overleftrightarrow{JK}$ intersect at $P(4, 3)$, where point F is at $(-2, 5)$ and point J is at $(7, 9)$.

39. Lines s and t intersect, and line v does not intersect either one.

Example 4 **MP MODELING** **When packing breakable objects such as glasses, movers frequently use boxes with inserted dividers like the one shown.**

40. How many planes are modeled in the picture?

41. What parts of the box model lines?

42. What parts of the box model points?

Refer to the figure at the right.

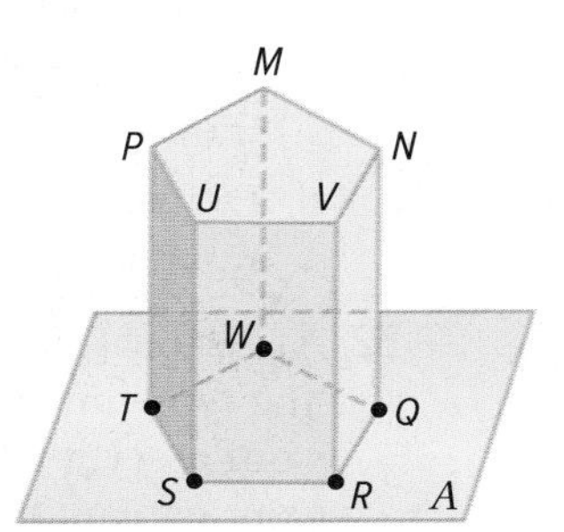

43. Name two collinear points.

44. How many planes appear in the figure?

45 Do plane A and plane MNP intersect? Explain.

46. In what line do planes A and QRV intersect?

47. Are points T, S, R, Q, and V coplanar? Explain.

48. Are points T, S, R, Q, and W coplanar? Explain.

49 **FINITE PLANES** A *finite plane* is a plane that has boundaries, or does not extend indefinitely. The street signs shown are finite planes.

a. If the pole models a line, name the geometric term that describes the intersection between the signs and the pole.

b. What geometric term(s) describes the intersection between the two finite planes? Explain your answer with a diagram if necessary.

50. **ONE-POINT PERSPECTIVE** One-point perspective drawings use lines to convey depth. Lines representing horizontal lines in the real object can be extended to meet at a single point called the *vanishing point*. Suppose you want to draw a tiled ceiling in the room below with nine tiles across.

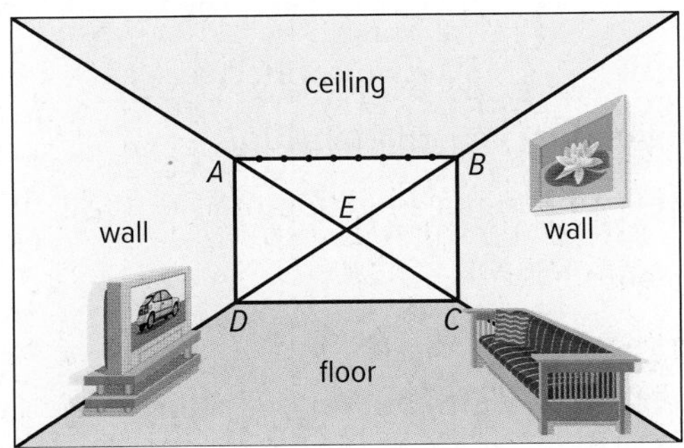

a. What point represents the vanishing point in the drawing?

b. Trace the figure. Then draw lines from the vanishing point through each of the eight points between *A* and *B*. Extend these lines to the top edge of the drawing.

c. How could you change the drawing to make the back wall of the room appear farther away?

51. **TWO-POINT PERSPECTIVE** Two-point perspective drawings use two vanishing points to convey depth.

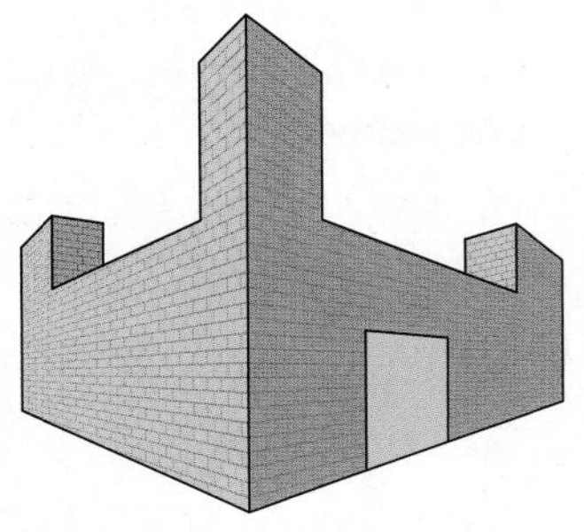

a. Trace the drawing of the castle shown. Draw five of the vertical lines used to create the drawing.

b. Draw and extend the horizontal lines to locate the vanishing points. Be sure to label each of the vanishing points.

c. What do you notice about the vertical lines as they get closer to the vanishing point?

d. Draw a two-point perspective of a home or room in a home.

a

52. **MP** **CONSTRUCT ARGUMENTS** Name two points on the same line in the figure. How can you support your assertion? Use specific geometry terms in your explanation.

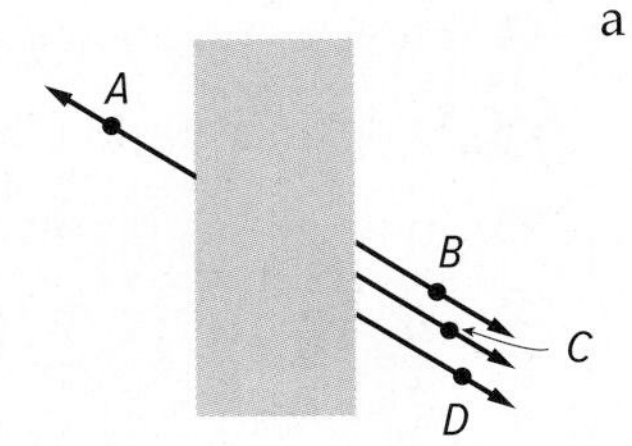

53. **TRANSPORTATION** When two cars enter an intersection at the same time on opposing paths, one of the cars must adjust its speed or direction to avoid a collision. Two airplanes, however, can cross paths while traveling in different directions without colliding. Explain how this is possible.

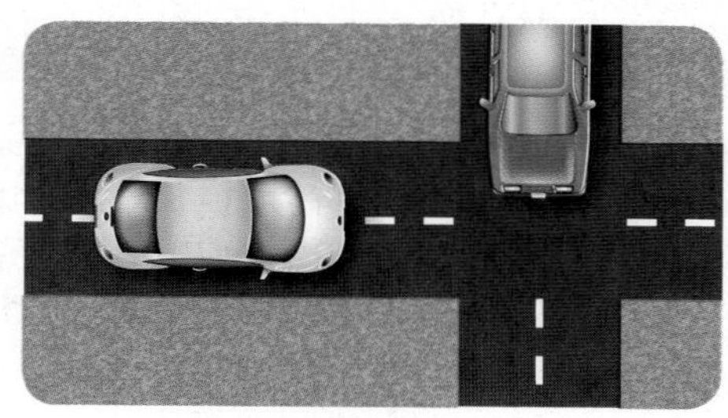

54. **MULTIPLE REPRESENTATIONS** Another way to describe a group of points is called a locus. A **locus** is a set of points that satisfy a particular condition. In this problem, you will explore the locus of points that satisfy an equation.

    a. **Tabular** Represent the locus of points satisfying the equation $2 + x = y$ using a table of at least five values.

    b. **Graphical** Represent this same locus of points using a graph.

    c. **Verbal** Describe the geometric figure that the points suggest.

55. **PROBABILITY** Three of the labeled points are chosen at random.

    a. What is the probability that the points chosen are collinear?

    b. What is the probability that the points chosen are coplanar?

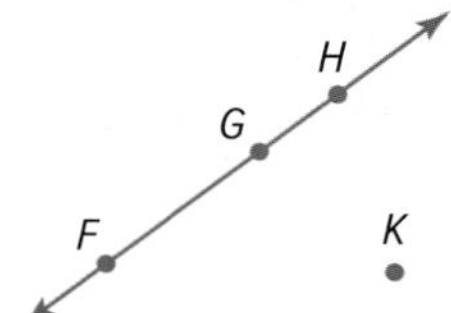

56. **MULTIPLE REPRESENTATIONS** In this problem, you will explore the locus of points that satisfy an inequality.

    a. **Tabular** Represent the locus of points satisfying the inequality $y < -3x - 1$ using a table of at least 10 values.

    b. **Graphical** Represent this same locus of points using a graph.

    c. **Verbal** Describe the geometric figure that the points suggest.

H.O.T. Problems Use Higher-Order Thinking Skills

57. **OPEN-ENDED** Sketch three planes that intersect in a line.

58. **ERROR ANALYSIS** Camille and Hiroshi are trying to determine the most number of lines that can be drawn using any two of four random points. Is either correct? Explain.

Camille	Hiroshi
Because there are four points, $4 \cdot 3$ or 12 lines can be drawn between the points.	You can draw $3 \cdot 2 \cdot 1$ or 6 lines between the points.

59. MP **CONSTRUCT ARGUMENTS** What is the greatest number of planes determined using any three of the points *A*, *B*, *C*, and *D* if no three points are collinear?

60. MP **REASONING** Is it possible for two points on the surface of a prism to be neither collinear nor coplanar? Justify your answer.

61. **WRITING IN MATH** Refer to Exercise 49. Give a real-life example of a finite plane. Is it possible to have a real-life object that is an infinite plane? Explain your reasoning.

62. The figure illustrates the intersection of plane P and plane T. The planes extend infinitely in all directions. What undefined term best describes the intersection? MP 6

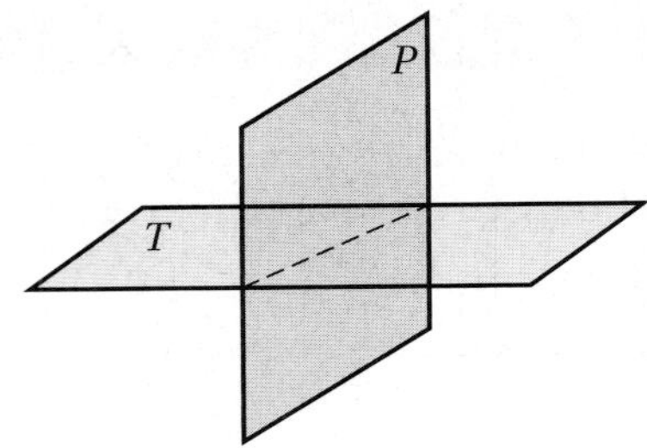

- A line
- B plane
- C point
- D segment

63. Four lines are coplanar. What is the greatest number of intersection points that can exist? MP 4

64. Which of the following terms are undefined? Select all that apply. MP 6

- A distance
- B line
- C point
- D plane
- E space

65. Dwayne is walking on a straight sidewalk. He spots a vertical flagpole 10 feet to his right. His friend Ursula says that the flagpole and sidewalk can be used to model two lines. Which term best describes the sidewalk and the flagpole? MP 4, 6

- A collinear
- B skew
- C parallel
- D intersecting

66. Samir is using a compass to draw a circle on a piece of paper. He places the metal tip of the compass at one location on the paper, and then moves the pencil around the tip to draw the figure. Which of the following models a term that can be defined? MP 4, 6

- A the location of the metal tip on the paper
- B the plane that contains the piece of paper
- C the tip of the pencil
- D the circle drawn by the pencil

67. Which undefined term is best modeled by a laptop screen? MP 6

- A line
- B parallel
- C plane
- D rectangle

68. **MULTI-STEP** Name the geometric term(s) modeled by each object. MP 4, 6

a. two sides of a roof

b. a driveway

c. two connected desks

d. end of a pencil

e. cell phone screen

f. fence post

Geometry Lab

Describing What You See

When you are learning geometric concepts, it is critical to have accurate drawings to represent the information. It is helpful to know what words and phrases can be used to describe figures. Likewise, it is important to know how to read a geometric description and be able to draw the figure it describes.

Mathematical Practices
MP 6 Attend to precision.

The figures and descriptions below help you visualize and write about points, lines, and planes.

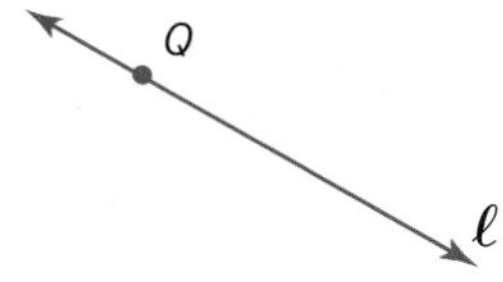

Point Q is on ℓ.

Line ℓ contains Q.

Line ℓ passes through Q.

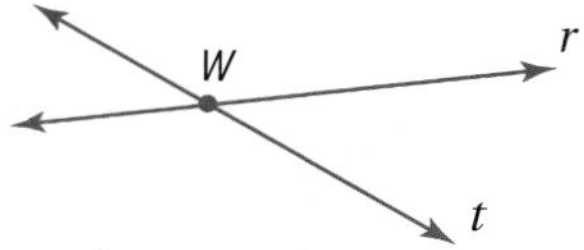

Lines r and t intersect at W.

Point W is the intersection of r and t.

Point W is on r. Point W is on t.

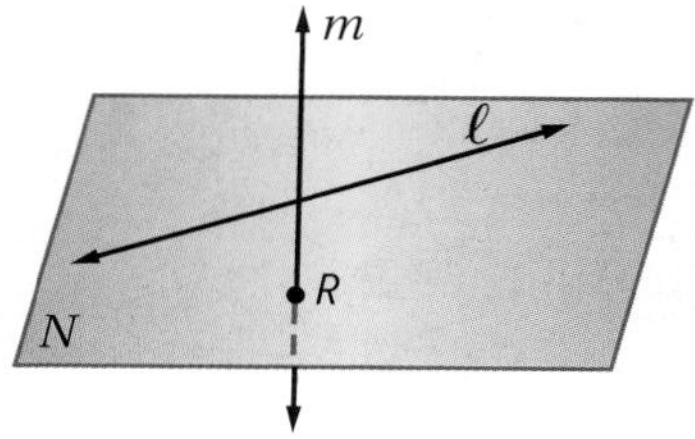

Line ℓ and point R are in N.

Point R lies in N.

Plane N contains R and ℓ.

Line m intersects N at R.

Point R is the intersection of m with N.

Lines ℓ and m do not intersect.

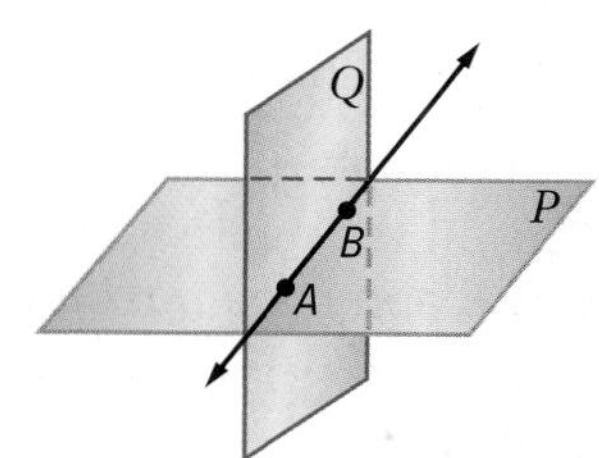

$\overleftrightarrow{AB}$ is in P and Q.

Points A and B lie in both P and Q.

Planes P and Q both contain $\overleftrightarrow{AB}$.

Planes P and Q intersect in $\overleftrightarrow{AB}$.

$\overleftrightarrow{AB}$ is the intersection of P and Q.

Exercises

Work cooperatively. Write a description for each figure.

1.

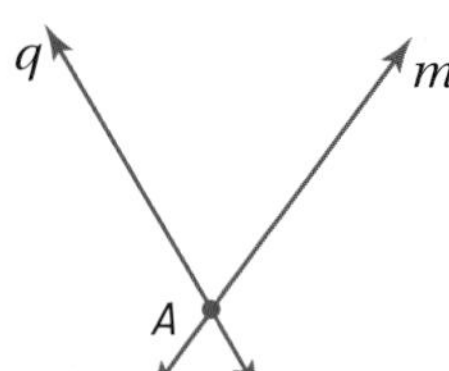

2.

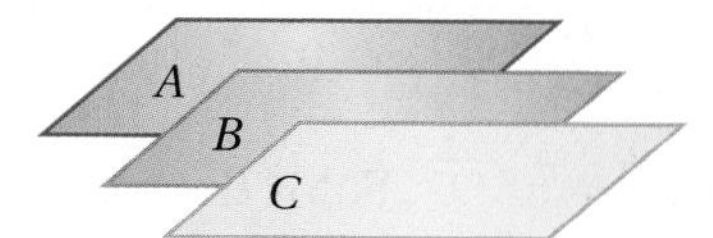

3.

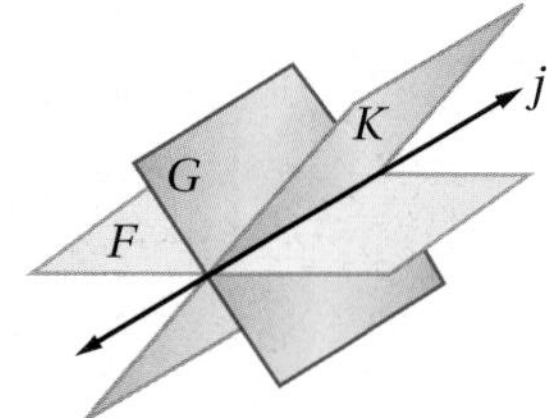

4. Draw and label a figure for the statement *planes N and P contain line a*.

LESSON 2

Line Segments and Distance

Then

- You identified points, lines, and planes. (Lesson 1-1)

Now

1. Calculate with measures.
2. Find the distance between two points.

Why?

- Indianapolis, Indiana is between St. Louis, Missouri and Columbus, Ohio. If you drive on Interstate 70 from St. Louis to Columbus, you will pass through Indianapolis.

New Vocabulary

line segment
betweenness of points
between
congruent
rigid transformation
congruent segments
construction
distance
irrational number

Mathematical Practices

6 Attend to precision.

1 Calculate Measures

Unlike a line, a **line segment** can be measured because it has two endpoints. These endpoints are used to name the segment.

Meaning	Notation
a segment with endpoints *A* and *B*	$\overline{AB}$ or $\overline{BA}$
a line that contains points *A* and *B*	$\overleftrightarrow{AB}$ or $\overleftrightarrow{BA}$

You know that for any two real numbers a and b, there is a real number n between a and b such that $a < n < b$. This relationship also applies to points on a line and is called **betweenness of points**. In the figure, point N is between points A and B, but points R and P are not between A and B.

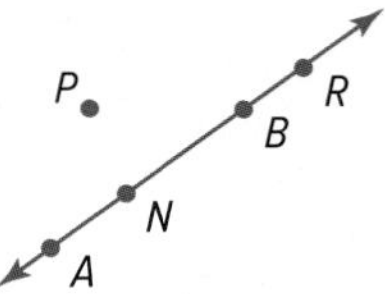

Key Concept Betweenness of Points

Words

Point M is **between** points P and Q if and only if P, Q, and M are collinear and $PM + MQ = PQ$.

Model

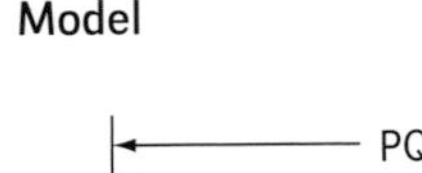
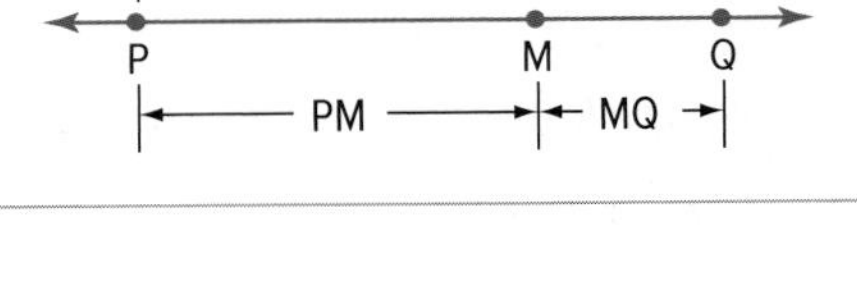

The measure of $\overline{AB}$ is written as AB. Measures are real numbers, so all arithmetic operations can be used with them. You know that the whole equals the sum of its parts. This is also true of line segments in geometry.

Example 1 Finding Measurements by Adding or Subtracting

Find each measure. Assume that the figures are not drawn to scale.

a. EG

EG is the measure of $\overline{EG}$. Point F is between E and G. Find EG by adding EF and FG.

$EF + FG = EG$ Betweenness of points

$2\frac{3}{4} + 2\frac{3}{4} = EG$ Substitution

$5\frac{1}{2}\text{ in.} = EG$ Add.

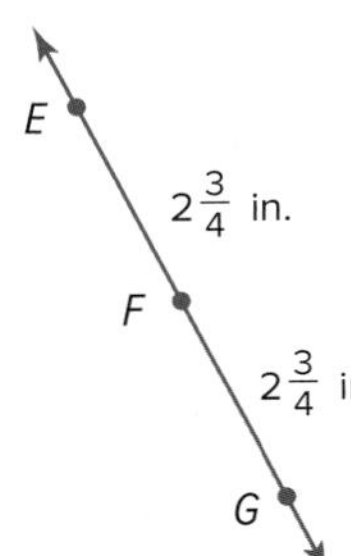

b. AB

A B 5.8 m C
13.2 m

Point B is between points A and C.

$AB + BC = AC$	Betweenness of points
$AB + 5.8 = 13.2$	Substitution
$AB + 5.8 - 5.8 = 13.2 - 5.8$	Subtract 5.8 from each side.
$AB = 7.4$ m	Simplify.

Guided Practice

1A. JL

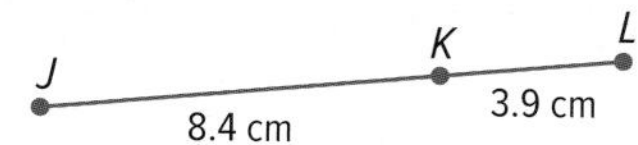

1B. QR

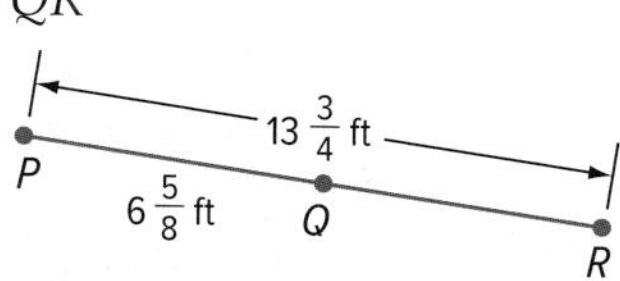

Example 2 Write and Solve Equations to Find Measurements

ALGEBRA **Find the value of a and XY if Y is between X and Z, $XY = 3a$, $XZ = 5a - 4$, and $YZ = 14$.**

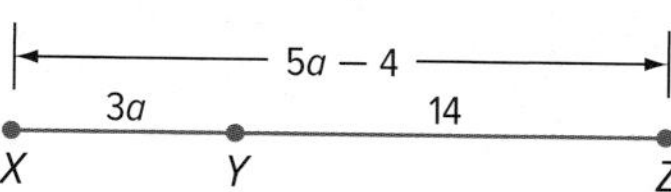

Draw a figure to represent the information.

$XZ = XY + YZ$	Betweenness of points
$5a - 4 = 3a + 14$	Substitution
$5a - 4 - 3a = 3a + 14 - 3a$	Subtract $3a$ from each side.
$2a - 4 = 14$	Simplify.
$2a - 4 + 4 = 14 + 4$	Add 4 to each side.
$2a = 18$	Simplify.
$\frac{2a}{2} = \frac{18}{2}$	Divide each side by 2.
$a = 9$	Simplify.

CHECK

$5a - 4 = 3a + 14$	Original Equation
$5(9) - 4 \stackrel{?}{=} 3(9) + 14$	Substitution
$45 - 4 \stackrel{?}{=} 27 + 14$	Multiply.
$41 = 41$ ✓	Simplify.

Now find XY.

$XY = 3a$	Given
$= 3(9)$ or 27	$a = 9$

Guided Practice

2. Find x and BC if B is between A and C, $AC = 4x - 12$, $AB = x$, and $BC = 2x + 3$.

Study Tip

Equal vs. Congruent Lengths Lengths are *equal* and segments are *congruent*. It is correct to say that $AB = CD$ and $\overline{AB} \cong \overline{CD}$. However, it is not correct to say that $\overline{AB} = \overline{CD}$ or that $AB \cong CD$.

If two geometric figures have exactly the same shape and size, they are **congruent**. Recall that a *transformation* is an operation that maps a geometric figure, the *preimage*, onto a new figure called the *image*. A **rigid transformation** is a transformation in which the position of the image may differ from that of the preimage, but the two figures remain congruent. If one segment can be mapped onto another segment using rigid transformations, then the two segments are called **congruent segments**. If two segments are congruent, then they have the same measure.

Key Concept Congruent Segments

Words Congruent segments have the same measure.

Symbols ≅ is read *is congruent to*. Red slashes on the figure also indicate congruence.

Example $\overline{AB} \cong \overline{CD}$

Meaning Segment *AB* is congruent to segment *CD*.

A, B, C, D, 1.7 cm, 1.7 cm

Tracing paper or a transparency sheet can be used to verify congruence by rigid transformations.

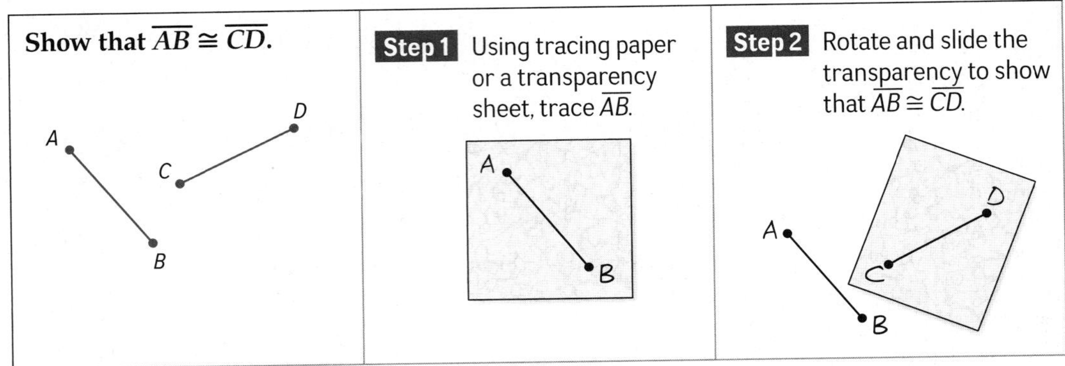

Since $\overline{AB}$ maps to $\overline{CD}$ exactly, $\overline{AB} \cong \overline{CD}$.

Drawings of geometric figures are created using measurement tools such as a ruler and a protractor. **Constructions** are methods of creating these figures without the benefit of measuring tools. Generally, only a pencil, straightedge, and compass are used in constructions. *Sketches* are created using pencil only.

> **Study Tip**
> **Lines vs. Segments** Remember that a segment has two endpoints, and that a line extends indefinitely in both directions.

You can construct a segment by first using a compass to establish the length of the given segment.

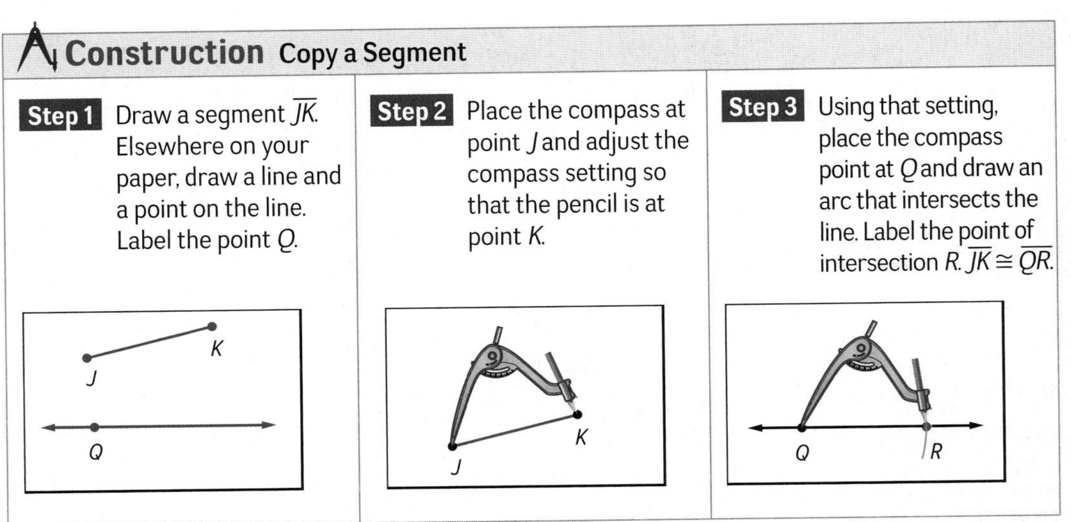

2 Distance Between Two Points

The **distance** between two points is the length of the segment with those points as its endpoints. The coordinates of the points can be used to find this length.

Study Tip

Distance Formula The Distance Formula uses absolute values because distances cannot be negative.

Key Concept Distance Formula (on Number Line)

Words The distance between two points is the absolute value of the difference between their coordinates.

Symbols If P has coordinate x_1 and Q has coordinate x_2, $PQ = |x_2 - x_1|$ or $|x_1 - x_2|$.

Because $\overline{PQ}$ is the same as $\overline{QP}$, the order in which you name the endpoints is not important when calculating distance.

Example 3 Find Distance on a Number Line

Use the number line.

A B C D E F

−7 −6 −5 −4 −3 −2 −1 0 1 2 3 4 5 6 7

a. Find BE.

The coordinates of B and E are −6 and 2.

$BE = |x_2 - x_1|$ Distance Formula

$= |2 - (-6)|$ $x_1 = -6$ and $x_2 = 2$

$= |8|$ or 8 Simplify.

b. Determine whether $\overline{CA}$ and $\overline{ED}$ are congruent.

The coordinates of C and A are −4 and −7. Those of E and D are 2 and −2.

$CA = |x_2 - x_1|$ Distance Formula

$= |-7 - (-4)|$ Substitute.

$= |-3|$ Subtract.

$= 3$ Simplify.

$ED = |x_2 - x_1|$

$= |-2 - 2|$

$= |-4|$

$= 4$

No, the segments are not congruent.

Guided Practice

3A. Find AD.

3B. Determine whether $\overline{CF}$ and $\overline{AE}$ are congruent.

Study Tip

MP Structure The Pythagorean Theorem is often expressed as $a^2 + b^2 = c^2$, where a and b are the measures of the shorter sides (legs) of a right triangle, and c is the measure of the longest side (hypotenuse). You will prove and learn about other applications of the Pythagorean Theorem in Lesson 8-2.

To find the distance between two points A and B in the coordinate plane, you can form a right triangle with $\overline{AB}$ as its hypotenuse and point C as its vertex as shown. Then use the Pythagorean Theorem to find AB.

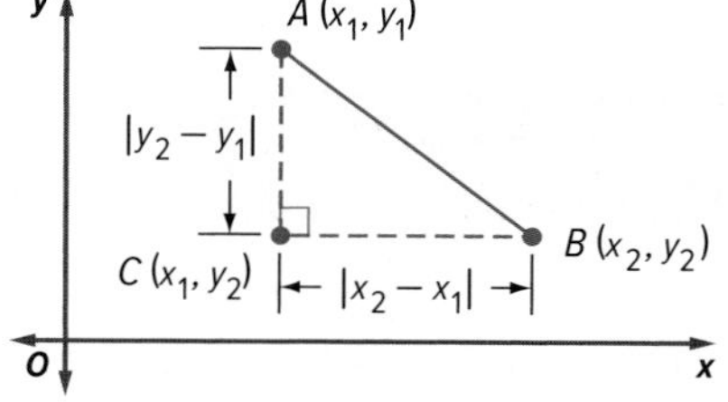

$(CB)^2 + (AC)^2 = (AB)^2$ Pythagorean Theorem

$(|x_2 - x_1|)^2 + (|y_2 - y_1|)^2 = (AB)^2$ $CB = |x_2 - x_1|, AC = |y_2 - y_1|$

$(x_2 - x_1)^2 + (y_2 - y_1)^2 = (AB)^2$ The square of a number is always positive.

$\sqrt{(x_2 - x_1)^2 + (y_2 - y_1)^2} = AB$ Take the positive square root of each side.

This gives us the Distance Formula for points in the coordinate plane. Because this formula involves taking the square root of a number, distances can be irrational. Recall that an **irrational number** is a number that cannot be expressed as a terminating or repeating decimal.

Key Concept Distance Formula (in Coordinate Plane)

If P has coordinates (x_1, y_1) and Q has coordinates (x_2, y_2), then

$$PQ = \sqrt{(x_2 - x_1)^2 + (y_2 - y_1)^2}.$$

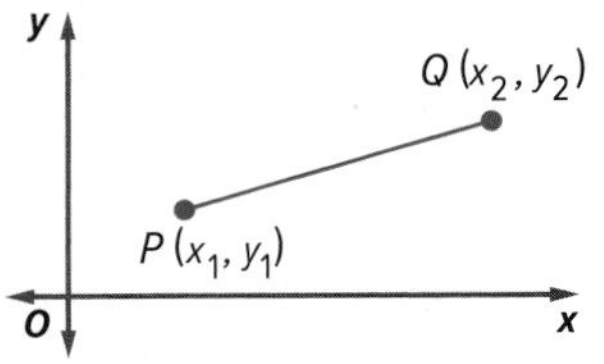

The order of the x- and y-coordinates in each set of parentheses is not important.

Example 4 Find Distance on a Coordinate Plane

Find the distance between each pair of points.

a. $C(-4, -6)$ **and** $D(5, -1)$

$CD = \sqrt{(x_2 - x_1)^2 + (y_2 - y_1)^2}$ Distance Formula

$= \sqrt{[5 - (-4)]^2 + [-1 - (-6)]^2}$ $(x_1, y_1) = (-4, -6)$ and $(x_2, y_2) = (5, -1)$

$= \sqrt{9^2 + 5^2}$ or $\sqrt{106}$ Subtract.

The distance between C and D is $\sqrt{106}$ or approximately 10.3 units.

CHECK Graph the ordered pairs and check by using the Pythagorean Theorem.

$(CD)^2 = (EC)^2 + (ED)^2$

$(CD)^2 = 5^2 + 9^2$

$(CD)^2 = 106$

$CD = \sqrt{106}$ ✓

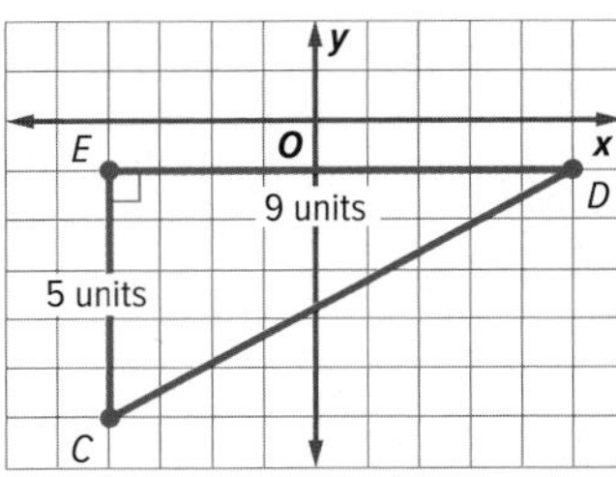

Study Tip

Distance on a Coordinate Plane To find the distance between E and D in Example 4 count the squares on the grid from E to D or, find the absolute value of the difference between the x-coordinates, $|-4 - 5| = 9$.

b. $M(2, 4)$ **and** $N(-3, -2)$

$MN = \sqrt{(x_2 - x_1)^2 + (y_2 - y_1)^2}$ Distance Formula

$= \sqrt{[(-3) - 2]^2 + [(-2) - 4]^2}$ $(x_1, y_1) = (2, 4)$ and $(x_2, y_2) = (-3, -2)$

$= \sqrt{(-5)^2 + (-6)^2}$ or $\sqrt{61}$ Subtract.

The distance between M and N is $\sqrt{61}$ or approximately 7.81 units.

CHECK Graph the ordered pairs and check by using the Pythagorean Theorem.

$(MN)^2 = (LM)^2 + (LN)^2$

$(MN)^2 = 6^2 + 5^2$

$(MN)^2 = 61$

$MN = \sqrt{61}$ ✓

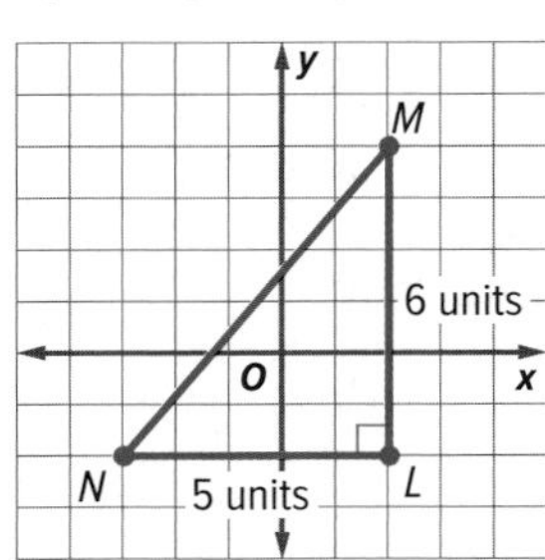

Go Online!

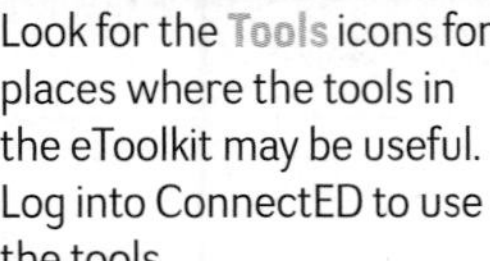

Look for the Tools icons for places where the tools in the eToolkit may be useful. Log into ConnectED to use the tools.

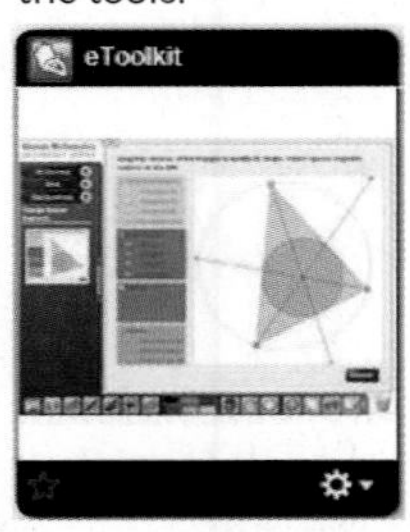

Guided Practice

Find the distance between each pair of points.

4A. $E(-5, 6)$ and $F(8, -4)$

4B. $J(4, 3)$ and $K(-3, -7)$

Real-World Example 5 Find Distance

FOOTBALL Luke is standing on his team's 20-yard line, 5 yards from the sideline, when he throws the football. Xavier catches it on the other team's 40-yard line, 20 yards from the same sideline. How far did Luke throw the football?

Understand First draw a diagram to represent the situation.

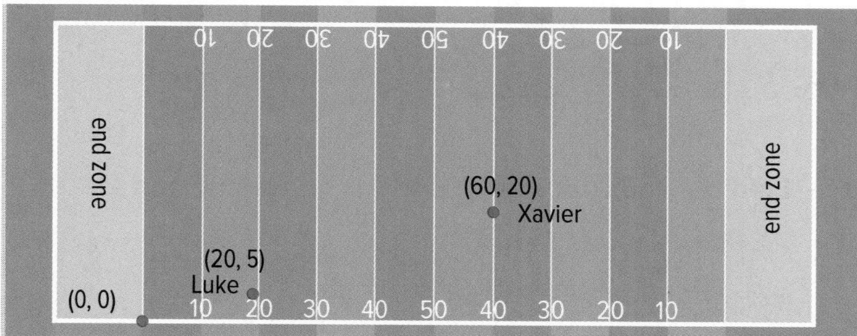

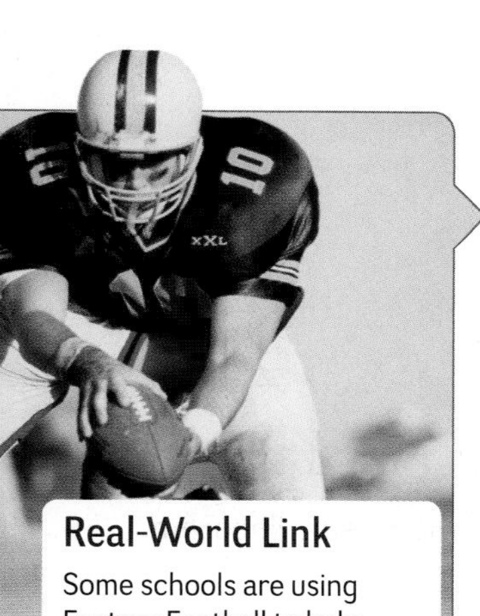

Real-World Link

Some schools are using Fantasy Football to help teach mathematics. Students calculate statistics from that week's games in order to determine their team's score.

Source: ESPN

Place a point where one of the end zones intersects a sideline. Label this point (0, 0). Place a second point on the 20-yard line that is on the same side of the field as the origin and is about 5 yards away from the sideline. Label the point (20, 5). Next, place a third point on the opposite 40 yard line about 20 yards away from the same sideline. Label the point (60, 20) because it is 60 yards away from the origin.

Plan Use the Distance Formula knowing $(x_1, y_1) = (20, 5)$ and $(x_2, y_2) = (60, 20)$.

Solve

$$
\begin{aligned}
D &= \sqrt{(x_2 - x_1)^2 + (y_2 - y_1)^2} && \text{Distance Formula} \\
&= \sqrt{(60 - 20)^2 + (20 - 5)^2} && (x_1, y_1) = (20, 5) \text{ and } (x_2, y_2) = (60, 20) \\
&= \sqrt{40^2 + 15^2} && \text{Subtract.} \\
&= \sqrt{1600 + 225} && \text{Square each term.} \\
&= \sqrt{1825} && \text{Add.} \\
&\approx 42.7 && \text{Take the positive square root.}
\end{aligned}
$$

Luke threw the football approximately 42.7 yards.

Check Luke is 40 yards from the other team's 40-yard line where Xavier is located, and Xavier is 15 yards farther from the sideline than Luke. Use the Pythagorean Theorem to justify the answer.

$c^2 = a^2 + b^2$

$c^2 = 40^2 + 15^2$

$c^2 = 1825$

$c \approx 42.7$ yards

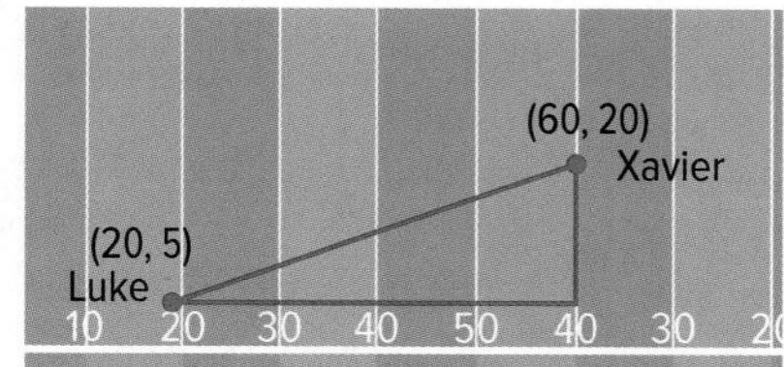

We were able to determine Luke's and Xavier's locations in terms of points on a coordinate plane and we know how to use the Distance Formula to calculate the distance between two points on a coordinate plane. The distance we calculated seems reasonable.

Guided Practice

5. Using the diagram above, find the distance that Luke threw the football if he was standing on the 30-yard line about 10 yards from the sideline, and Xavier was standing at the 50-yard line about 40 yards from the same sideline.

Check Your Understanding

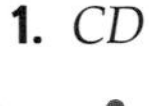 = Step-by-Step Solutions begin on page R13.

Example 1 **Find each measure. Assume that the figures are not drawn to scale.**

1. CD

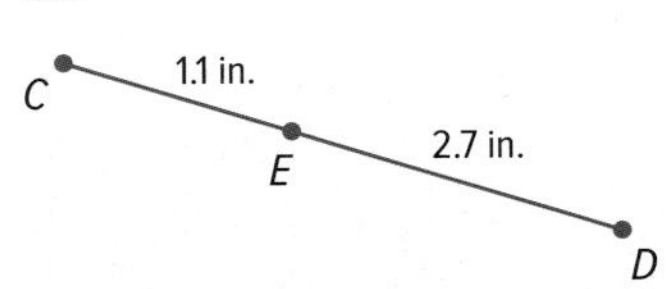

2. RS

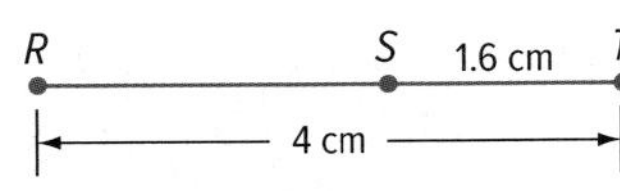

Example 2 **ALGEBRA Find the value of x and BC if B is between C and D.**

3. $CB = 2x$, $BD = 4x$, and $BD = 12$

4. $CB = 4x - 9$, $BD = 3x + 5$, and $CD = 17$

Example 3 **Use the number line to find each measure.**

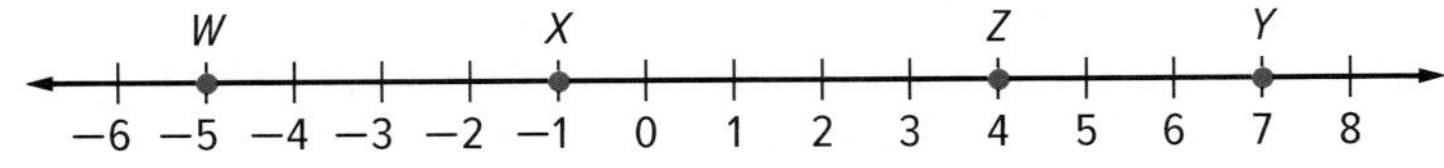

5. XY

6. WZ

Example 4 **TIME CAPSULE** Graduating classes have buried time capsules on the campus of East Side High School for over 20 years. The points on the diagram show the position of three time capsules. Find the distance between each pair of time capsules.

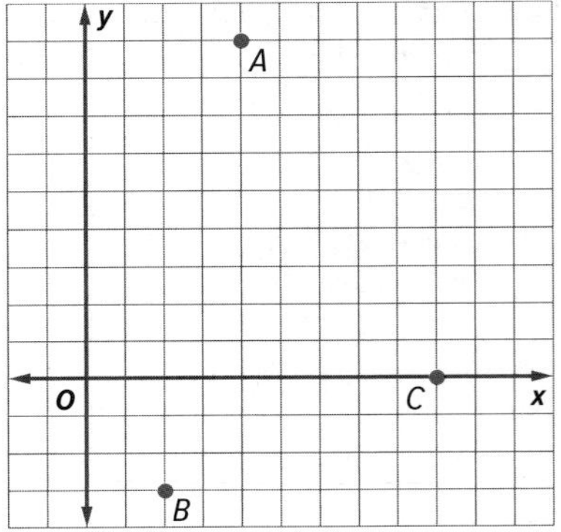

7. $A(4, 9)$, $B(2, -3)$

8. $A(4, 9)$, $C(9, 0)$

9. $B(2, -3)$, $C(9, 0)$

Example 5 10. 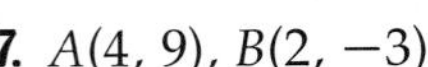 **STRUCTURE** Which two time capsules are the closest to each other? Which are farthest apart?

Practice and Problem Solving

Extra Practice is found on page R1.

Example 1 **Find each measure. Assume that the figures are not drawn to scale.**

11. JL

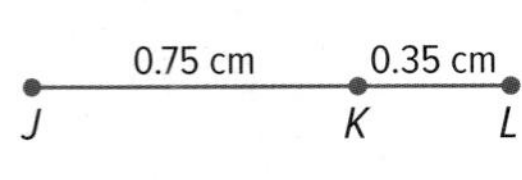

12. EF

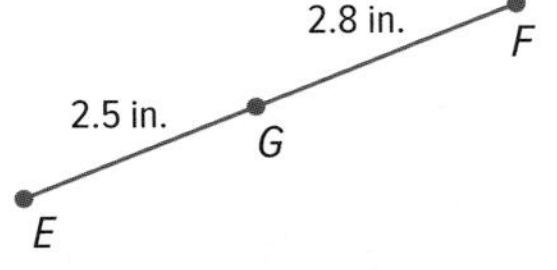

13. SV

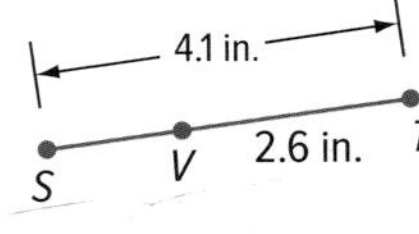

14. PR

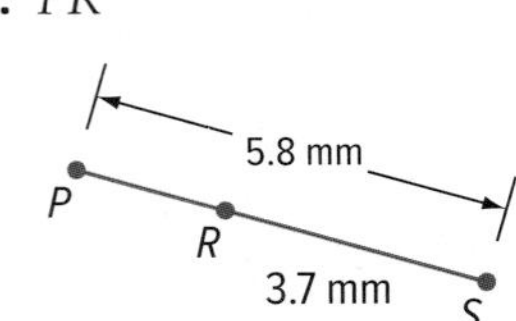

15. FG

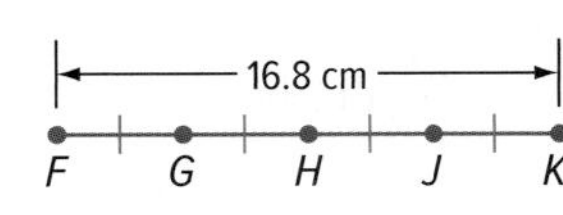

16. WY

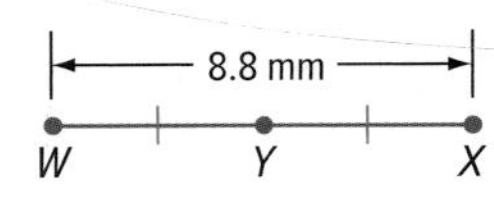

17. **SENSE-MAKING** The stacked bar graph shows the number of canned food items donated by the students in a homeroom class over three years. Use the concept of betweenness of points to find the number of cans donated by the boys for each year. Explain your method.

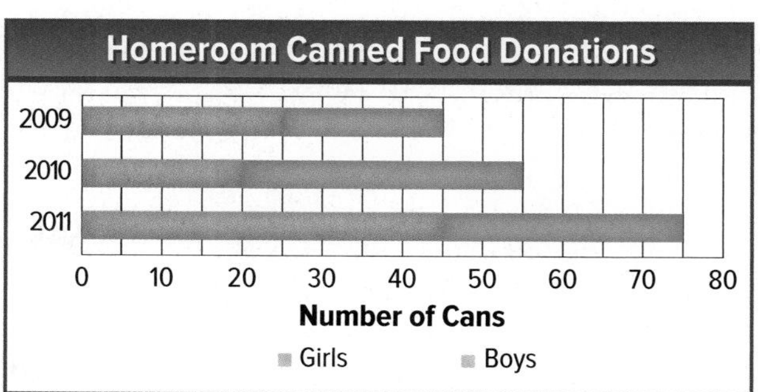

Example 2 **ALGEBRA Find the value of the variable and YZ if Y is between X and Z.**

18. $XY = 11$, $YZ = 4c$, $XZ = 83$

19. $XY = 6b$, $YZ = 8b$, $XZ = 175$

20. $XY = 7a$, $YZ = 5a$, $XZ = 6a + 24$

21. $XY = 11d$, $YZ = 9d - 2$, $XZ = 5d + 28$

22. $XY = 4n + 3$, $YZ = 2n - 7$, $XZ = 22$

23. $XY = 3a - 4$, $YZ = 6a + 2$, $XZ = 5a + 22$

Example 3 **Use the number line to find each measure.**

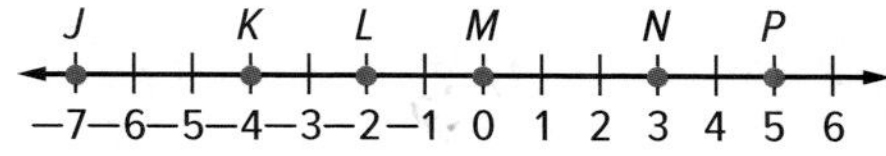

24. JL

25. JK

26. KP

27. NP

28. JP

29. LN

Example 4 **Find the distance between each pair of points.**

30.

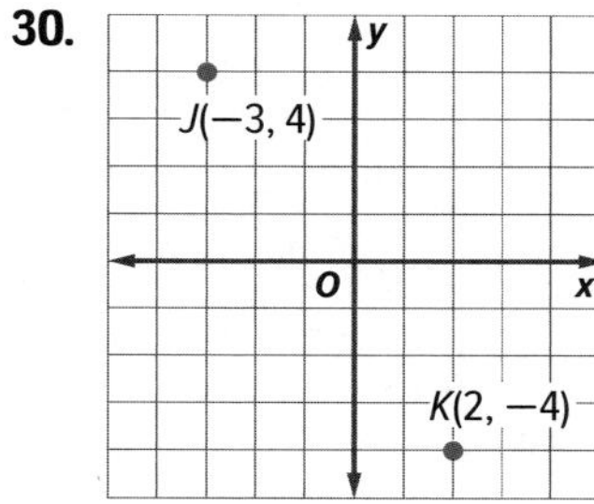

31.

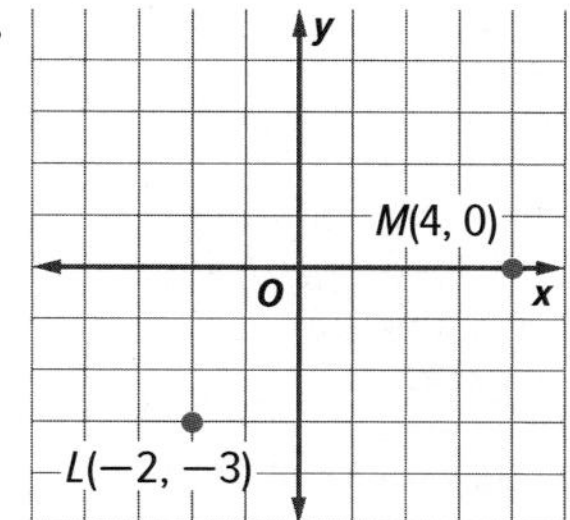

32.

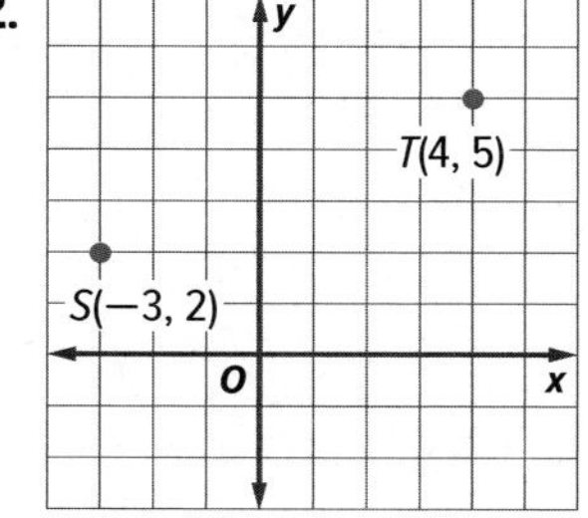

33.
y
V (5, 7)
U (2, 3)
x
O

34.

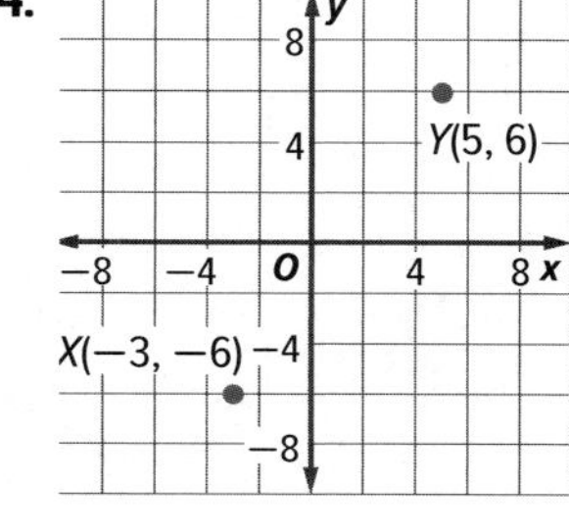

35.

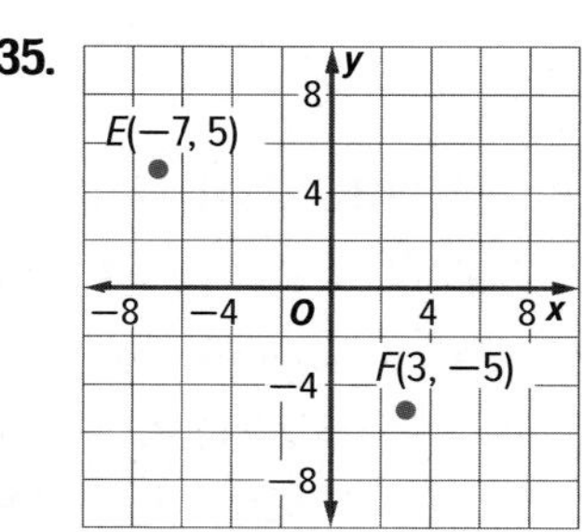

Find the distance between each pair of points.

36. $X(1, 4), Y(6, 9)$

37. $P(3, 4), Q(7, 2)$

38. $M(-3, 8), N(-5, 1)$

39. $Y(-4, 9), Z(-5, 3)$

40. $A(2, 4), B(5, 7)$

41. $C(5, 1), D(3, 6)$

Examples 5

42. MP **REASONING** Vivian is planning to hike to the top of Guadalupe Peak on her family vacation. The coordinates of the peak of the mountain and of the base of the trail are shown in feet. If the trail can be approximated by a straight line, estimate the length of the trail to the nearest tenth of a mile. (*Hint:* 1 mi = 5280 ft)

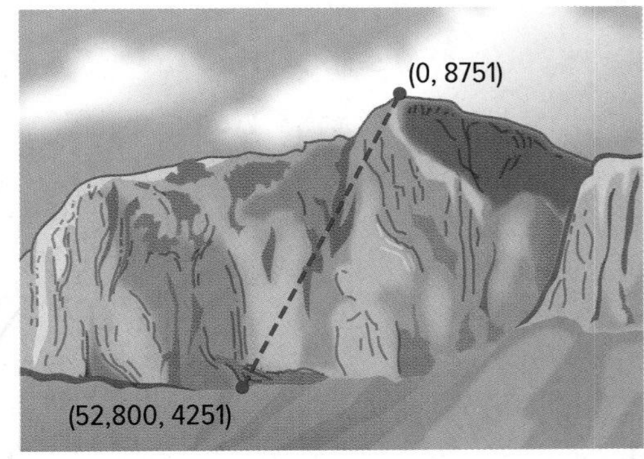

Determine whether each pair of segments is congruent.

43. $\overline{KJ}, \overline{HL}$

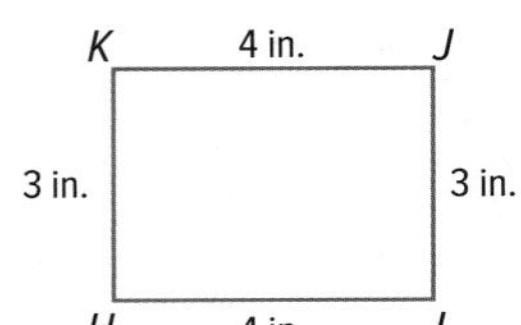

44. $\overline{AC}, \overline{BD}$

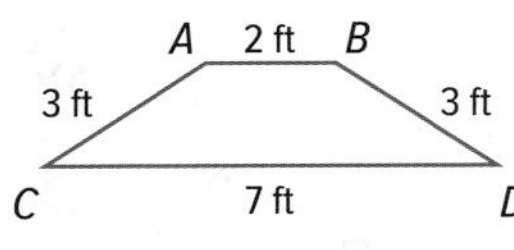

45. $\overline{EH}, \overline{FG}$

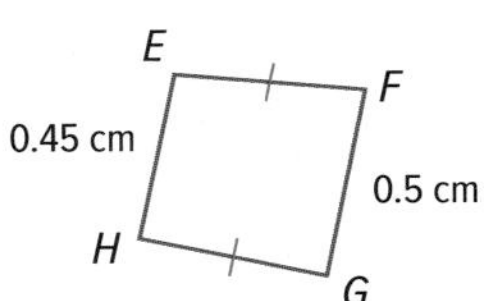

46. $\overline{VW}, \overline{UZ}$

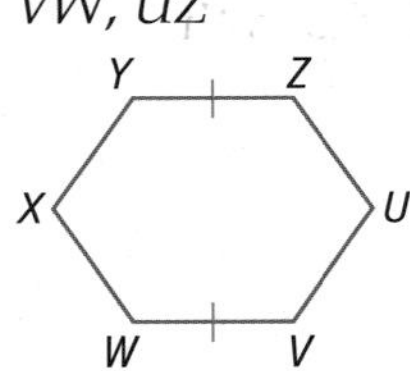

47. $\overline{MN}, \overline{RQ}$

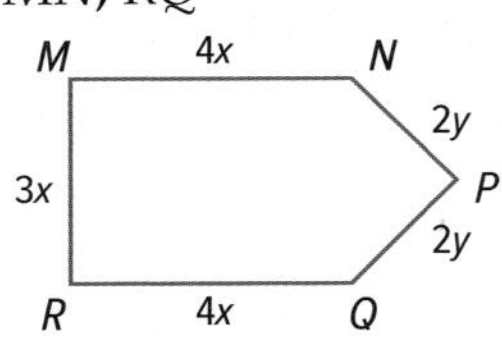

48. $\overline{SU}, \overline{VT}$

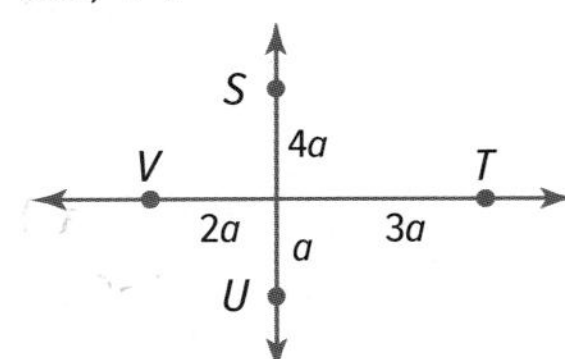

49. **TRUSSES** A truss is a structure used to support a load over a span, such as a bridge or the roof of a house. List all of the congruent segments in the figure.

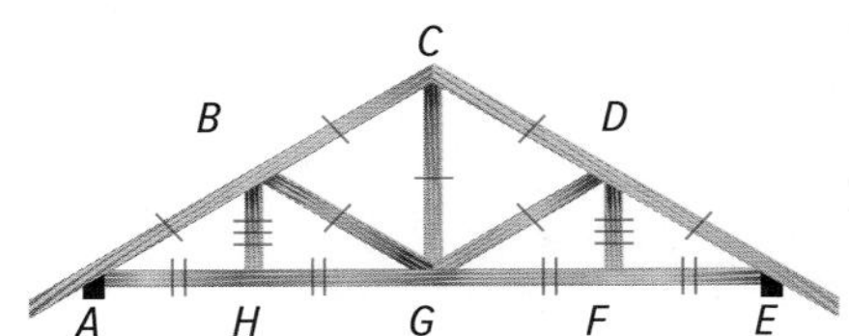

50. **CONSTRUCTION** For each expression:

- construct a segment with the given measure,
- explain the process you used to construct the segment, and
- verify that the segment you constructed has the given measure.

a. WZ
b. $2(XY)$
c. $6(WZ) - XY$

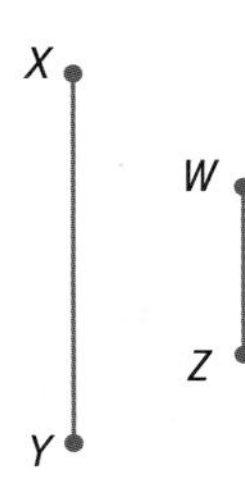

51 **BLUEPRINTS** Use a ruler to determine at least five pairs of congruent segments with labeled endpoints in the blueprint shown.

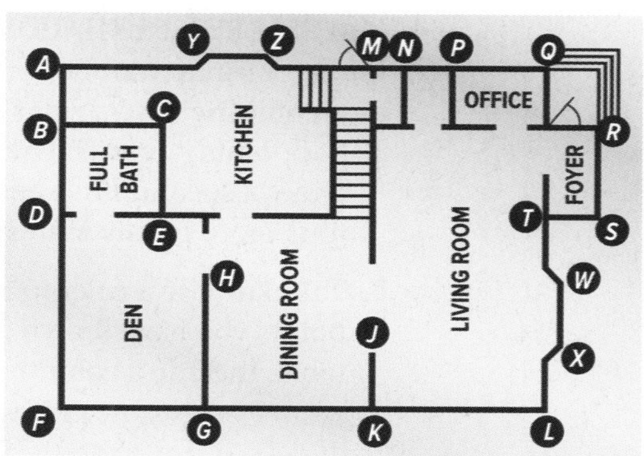

52. **MODELING** Penny and Akiko live in the locations shown on the map below.

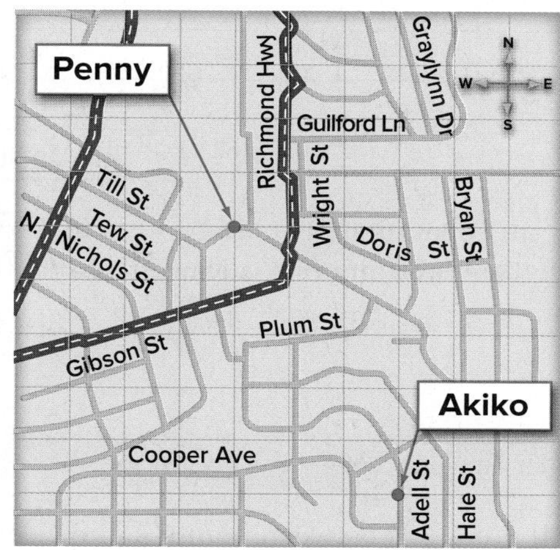

a. If each square on the grid represents one block and the bottom left corner of the grid is the location of the origin, what is the distance from Penny's house to Akiko's?

b. If Penny moves 3 blocks to the north and Akiko moves 5 blocks to the west, how far apart will they be?

53. **MULTI-STEP** Coach Willis designs a play that requires the ball to be passed from point A to point E as shown below. The arrows represent quick passes to different members of his team. Randi can throw the ball from under the basket to midcourt, Jen and Mandy can throw the ball half the width of the court, Makayla can throw the ball to the free throw line from under the basket, and Kim can throw the ball farther than Jen.

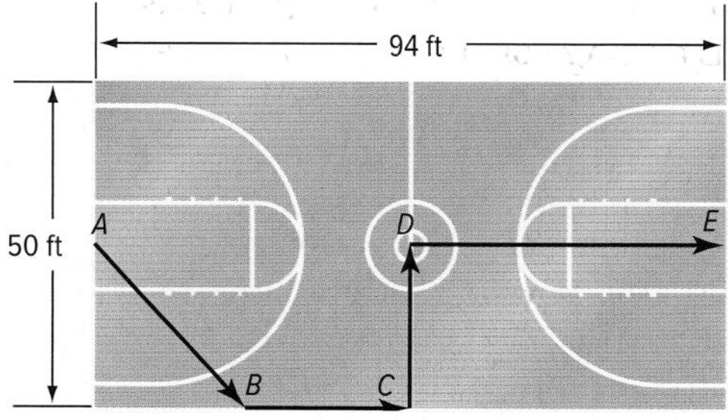

a. In which position should each girl be?

b. Describe your solution process.

c. What assumptions did you make?

54. **MULTIPLE REPRESENTATIONS** Betweenness of points ensures that a line segment may be divided into an infinite number of line segments.

a. **Geometric** Use a ruler to draw a line segment 3 centimeters long. Label the endpoints A and D. Draw two more points along the segment and label them B and C. Draw a second line segment 6 centimeters long. Label the endpoints K and P. Add four more points along the line and label them L, M, N, and O.

b. **Tabular** Use a ruler to measure the length of the line segment between each of the points you have drawn. Organize the lengths of the segments in $\overline{AD}$ and $\overline{KP}$ into a table. Include a column in your table to record the sum of these measures.

$\overline{AD}$		$\overline{KP}$	
Segment	Length (cm)	Segment	Length (cm)
$\overline{AB}$		$\overline{KL}$	
$\overline{BC}$		$\overline{LM}$	
$\overline{CD}$		$\overline{MN}$	
Total		$\overline{NO}$	
		$\overline{OP}$	
		Total	

c. **Algebraic** Write an equation that could be used to find the lengths of $\overline{AD}$ and $\overline{KP}$. Compare the lengths determined by your equation to the actual lengths.

H.O.T. Problems Use Higher-Order Thinking Skills

55. **WRITING IN MATH** If point B is between points A and C, explain how you can find AC if you know AB and BC. Explain how you can find BC if you know AB and AC.

56. **MP TOOLS** Draw a segment $\overline{AB}$ that measures between 2 and 3 inches long. Then sketch a segment $\overline{CD}$ congruent to $\overline{AB}$, draw a segment $\overline{EF}$ congruent to $\overline{AB}$, and construct a segment $\overline{GH}$ congruent to $\overline{AB}$. Compare your methods.

57. **MP REASONING** Determine whether the statement *If point M is between points C and D, then CD is greater than either CM or MD* is *sometimes, never,* or *always* true. Explain.

58. **CHALLENGE** Point P is located on the segment between point $A(1, 4)$ and point $D(7, 13)$. The distance from A to P is twice the distance from P to D. What are the coordinates of point P?

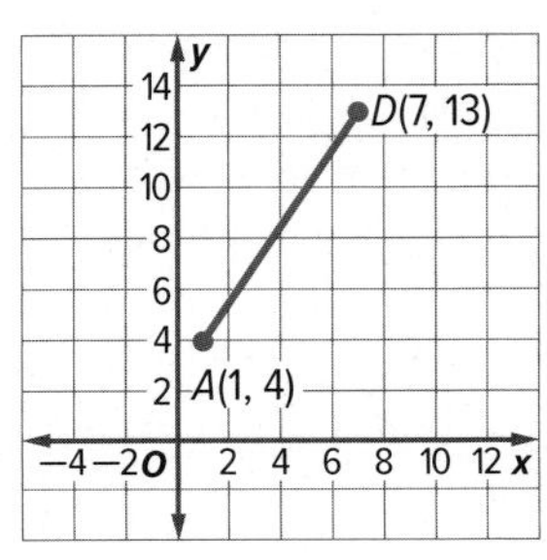

59. **WRITING IN MATH** Explain how the Pythagorean Theorem is used to derive the Distance Formula.

60. Jonah draws $\overline{PQ}$. Then, he draws a line and labels point X on the line. Next, he places the compass at point P and opens the compass so the pencil is at point Q. Using that setting, he places the compass at point X and draws an arc that intersects the line at point Y.

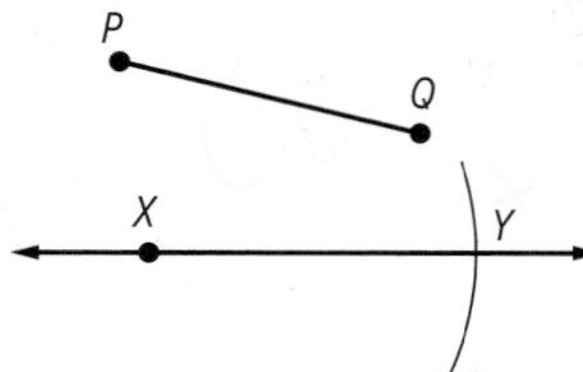

Which of the following must be true? Select all that apply. MP 6

- ☐ **A** $PQ = XY$
- ☐ **B** $PX = QY$
- ☐ **C** $\overline{PQ} \cong \overline{XY}$
- ☐ **D** $\overline{PX} \cong \overline{QY}$

61. Shauntay has a thumbtack that is $\frac{1}{3}$ as long as the paper clip shown.

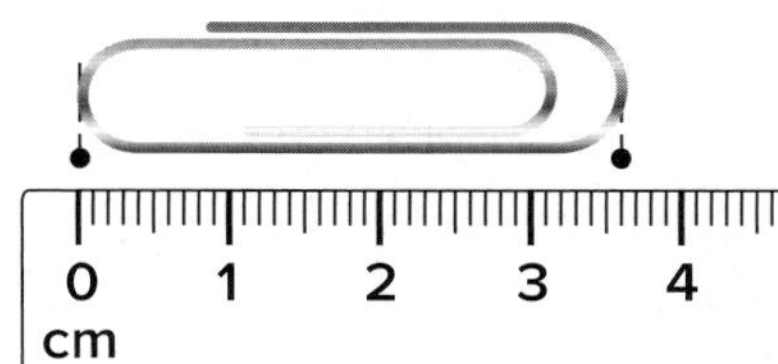

Which of the following represents the length of the thumbtack? MP 4

- ○ **A** 0.33 cm
- ○ **B** 1.2 cm
- ○ **C** 3.6 cm
- ○ **D** 10.8 cm

62. What is the length of $\overline{RU}$? MP 6

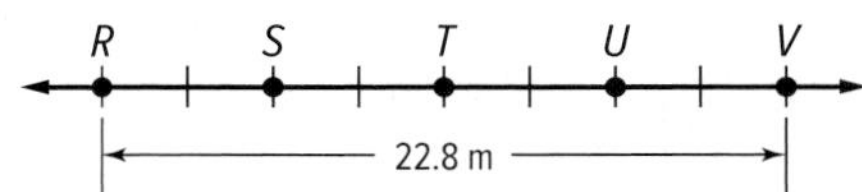

63. $\overline{AB}$ is 7.8 m long. If B lies between A and C and AB is three times the length of $\overline{BC}$, what is AC? MP 1, 6

- ○ **A** 2.6 m
- ○ **B** 5.2 m
- ○ **C** 10.4 m
- ○ **D** 23.4 m

64. Xavier draws $\overline{XZ}$ and point Y, which lies between X and Z. If $\overline{XY}$ is $\frac{3}{2}$ as long as $\overline{YZ}$, which statement correctly describes $\overline{YZ}$ in terms of $\overline{XZ}$? MP 6, 7

- ○ **A** $\overline{YZ}$ is $\frac{2}{5}$ as long as $\overline{XZ}$.
- ○ **B** $\overline{YZ}$ is $\frac{5}{2}$ as long as $\overline{XZ}$.
- ○ **C** $\overline{YZ}$ is $\frac{2}{3}$ as long as $\overline{XZ}$.
- ○ **D** $\overline{YZ}$ is $\frac{3}{2}$ as long as $\overline{XZ}$.

65. Isaac draws $\overline{MN}$ so that $MN = 5.2$ centimeters. Then, he draws a point P between M and N so that P is $\frac{1}{4}$ of the distance from M and N. What is the length of $\overline{PN}$? MP 6, 7

- ○ **A** 1.3 cm
- ○ **B** 3.9 cm
- ○ **C** 15.6 cm
- ○ **D** 20.8 cm

66. **MULTI-STEP** Perform the geometric constructions of the following figures using a compass and a straightedge. Then, provide an explanation on how each construction was formed MP 6, 7

a. Construct a segment.

b. Construct a congruent segment.

c. Construct a segment that is one half of the original segment.

d. Construct a segment that is three times the length of the original segment.

LESSON 3

Locating Points and Midpoints

Then

- You found the distance between two points on a line segment.

Now

1. Find the midpoint of a segment.
2. Locate a point on a segment a given fractional distance from one endpoint.

Why?

- Lexington, Nebraska is located on Interstate Route 80. It is the midpoint of Interstate 80, 1450 miles from its endpoints, San Francisco, California and Teaneck, New Jersey.

New Vocabulary
midpoint
segment bisector

Mathematical Practices
6 Attend to precision.
7 Look for and make use of structure.

1 Midpoint of a Segment

The **midpoint** of a segment is the point halfway between the endpoints of the segment. If M is the midpoint of $\overline{AB}$, then $AM = MB$ and $\overline{AM} \cong \overline{MB}$. To find the midpoint of a segment on a number line, find the *mean*, or average, of the coordinates of its endpoints.

Key Concept Midpoint Formula (on Number Line)

If $\overline{AB}$ has endpoints at x_1 and x_2 on a number line, then the midpoint M of $\overline{AB}$ has coordinate

$$\frac{x_1 + x_2}{2}.$$

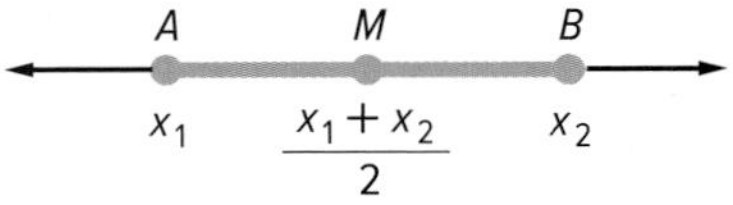

(Number line: A at x_1, M at $\frac{x_1 + x_2}{2}$, B at x_2)

Real-World Example 1 Find the Midpoint on a Number Line

DECORATING Jacinta is hanging a picture 15 inches from the left side of a wall. How far from the edge of the wall should she mark the location for the nail the picture will hang on if the right edge is 37.5 inches from the wall's left side?

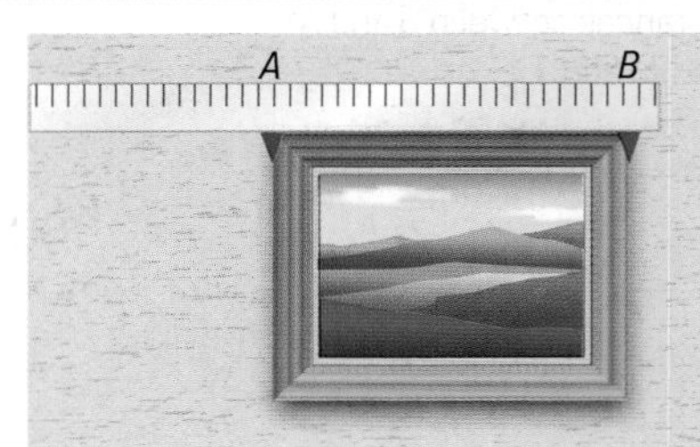

The coordinates of the endpoints of the top of the picture frame are 15 inches and 37.5 inches. Let M be the midpoint of $\overline{AB}$.

$M = \frac{x_1 + x_2}{2}$ Midpoint Formula

$= \frac{15 + 37.5}{2}$ $x_1 = 15, x_2 = 37.5$

$= \frac{52.5}{2}$ or 26.25 Simplify.

The midpoint is located 26.25 inches from the left edge of the wall.

Guided Practice

1. **TEMPERATURE** The temperature on a thermometer dropped from a reading of 25° to −8°. Find the midpoint of these temperatures.

The x-coordinate of the midpoint of a segment on the coordinate plane is the average of the x-coordinates of the endpoints of the segment. Similarly, the y-coordinate of the midpoint of a segment is the average of the y-coordinates of the endpoints.

Study Tip

Using Your Text Look for Key Concepts to learn important properties, definitions, and concepts.

Key Concept Midpoint Formula (in Coordinate Plane)

If $\overline{PQ}$ has endpoints at $P(x_1, y_1)$ and $Q(x_2, y_2)$ in the coordinate plane, then the midpoint M of $\overline{PQ}$ has coordinates

$$M\left(\frac{x_1 + x_2}{2}, \frac{y_1 + y_2}{2}\right).$$

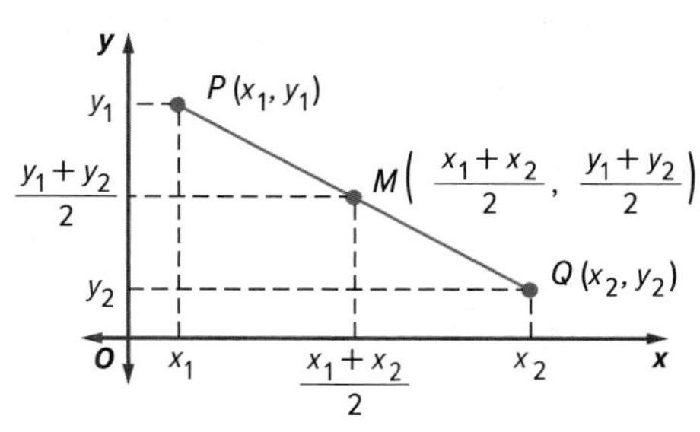

When finding the midpoint of a segment, the order of the endpoints is not important.

Example 2 Find the Midpoint in the Coordinate Plane

Find the coordinates of M, the midpoint of $\overline{ST}$, for $S(-6, 3)$ and $T(1, 0)$.

$M = \left(\frac{x_1 + x_2}{2}, \frac{y_1 + y_2}{2}\right)$ Midpoint Formula

$= \left(\frac{-6 + 1}{2}, \frac{3 + 0}{2}\right)$ $(x_1, y_1) = S(-6, 3)$, $(x_2, y_2) = T(1, 0)$

$= \left(\frac{-5}{2}, \frac{3}{2}\right)$ or $\left(-2\frac{1}{2}, 1\frac{1}{2}\right)$ Simplify.

Study Tip

Comparing Measures Another way to check your answer is to use the Distance Formula to compare the distances between S and M and M and T.

CHECK Graph S, T, and M. The distance from S to M does appear to be the same as the distance from M to T, so our answer is reasonable.

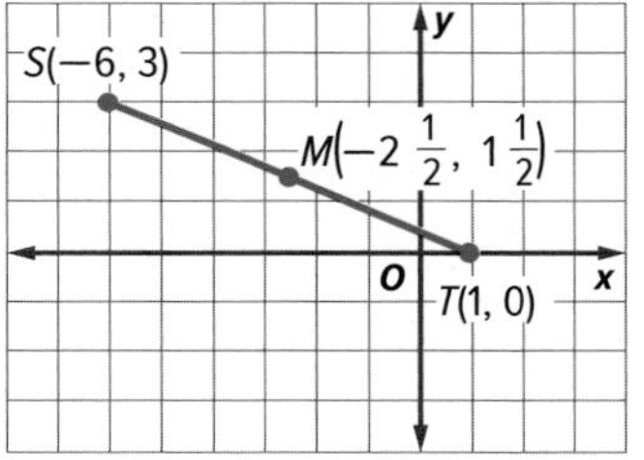

Guided Practice

Find the coordinates of the midpoint of a segment with the given coordinates.

2A. $A(5, 12)$, $B(-4, 8)$

2B. $C(-8, -2)$, $D(5, 1)$

Study Tip

Segment Bisectors Each segment has an infinite number of bisectors. Each bisector must contain the midpoint of the segment.

Any segment, line, or plane that intersects a segment at its midpoint is called a **segment bisector**. In the figure, M is the midpoint of $\overline{PQ}$. Plane A, $\overline{MJ}$, $\overleftrightarrow{KM}$, and point M are all bisectors of $\overline{PQ}$. We say that they each bisect $\overline{PQ}$.

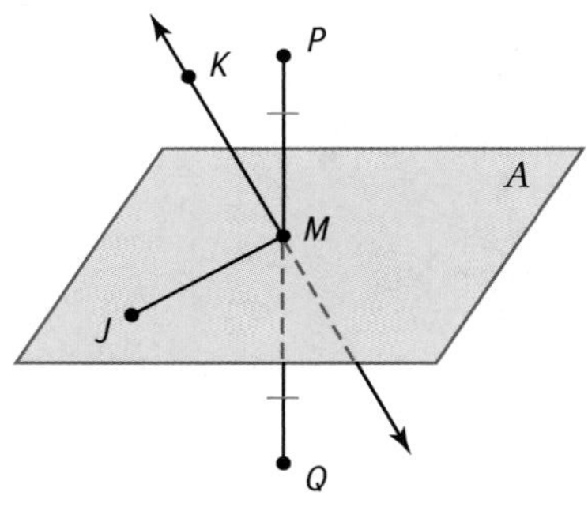

The construction shows how to construct a line that bisects a segment.

Construction Bisect a Segment

Step 1 Draw a segment and name it $\overline{AB}$. Place the compass at point A. Adjust the compass so that its width is greater than $\frac{1}{2}AB$. Draw arcs above and below $\overline{AB}$.

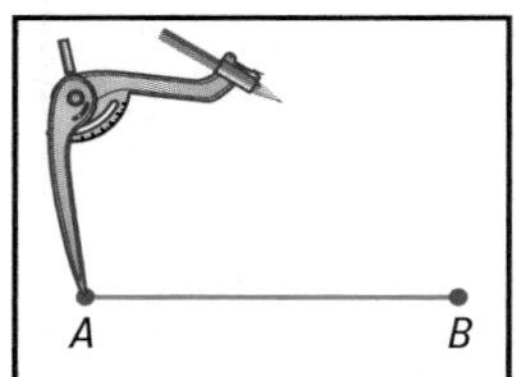

Step 2 Using the same compass setting, place the compass at point B and draw arcs above and below $\overline{AB}$ so that they intersect the two arcs previously drawn. Label the points of the intersection of the arcs as C and D.

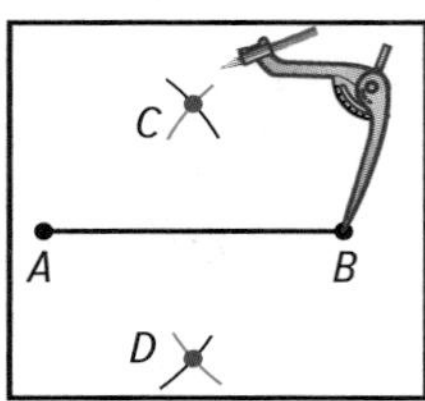

Step 3 Use a straightedge to draw $\overline{CD}$. Label the point where it intersects $\overline{AB}$ as M. Point M is the midpoint of $\overline{AB}$, and $\overline{CD}$ is a bisector of $\overline{AB}$.

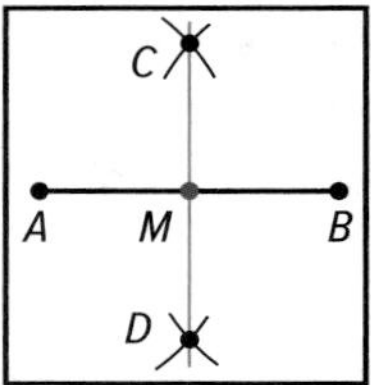

If you know that a given point is the midpoint of a segment, then you can use algebra and other key information to find missing measures or coordinates.

Example 3 Find Missing Coordinates

Find the coordinates of J if $K(-1, 2)$ is the midpoint of $\overline{JL}$ and L has the coordinates $(3, -5)$.

Step 1 Substitute the information you are given into the Midpoint Formula. Let J be (x_1, y_1) and L be (x_2, y_2).

$$\left(\frac{x_1 + 3}{2}, \frac{y_1 + (-5)}{2}\right) = (-1, 2) \qquad (x_2, y_2) = (3, -5)$$

Step 2 Write two equations to find the coordinates of J.

$$\frac{x_1 + 3}{2} = -1 \qquad \text{Midpoint Formula}$$

$$x_1 + 3 = -2 \qquad \text{Multiply each side by 2.}$$

$$x_1 = -5 \qquad \text{Subtract 3 from each side.}$$

$$\frac{y_1 + (-5)}{2} = 2 \qquad \text{Midpoint Formula}$$

$$y_1 - 5 = 4 \qquad \text{Multiply each side by 2.}$$

$$y_1 = 9 \qquad \text{Add 5 to each side.}$$

The coordinates of J are $(-5, 9)$.

CHECK Graph J, K, and L. The distance from J to K does appear to be the same as the distance from K to L, so our answer is reasonable.

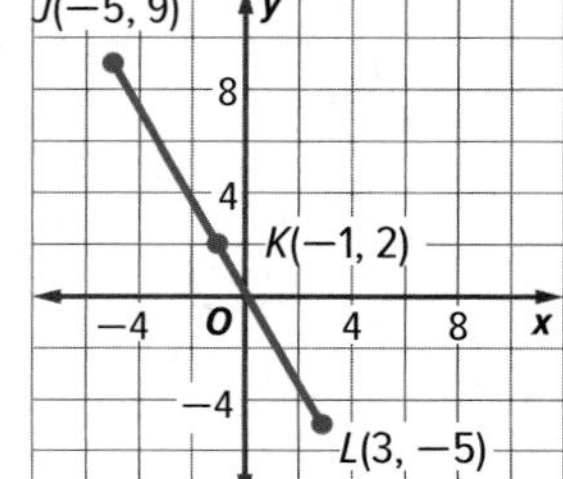

Study Tip

MP Sense-Making Always graph the given information and the calculated coordinates of the third point to check the reasonableness of your answer.

Guided Practice

3. Find the coordinates of G if P is the midpoint of $\overline{EG}$ for $E(-8, 6)$ and $P(-5, 10)$.

Example 4 Find Missing Measures

Find the measure of $\overline{PQ}$ if Q is the midpoint of $\overline{PR}$.

Because Q is the midpoint, you know that $PQ = QR$. Use this equation to find the value of y.

$PQ = QR$	Definition of midpoint
$9y - 2 = 14 + 5y$	$PQ = 9y - 2$, $QR = 14 + 5y$
$4y - 2 = 14$	Subtract $5y$ from each side.
$4y = 16$	Add 2 to each side.
$y = 4$	Divide each side by 4.

Now substitute 4 for y in the expression for PQ.

$PQ = 9y - 2$	Given
$= 9(4) - 2$	$y = 4$
$= 36 - 2$ or 34	Simplify.

The measure of $\overline{PQ}$ is 34.

Guided Practice

4. Find the value of x if C is the midpoint of $\overline{AB}$, $AC = 4x + 5$, and $AB = 78$.

2 Locate Points

You can think of the midpoint of a segment as being half the distance from one endpoint to the other. You can also find points that are other fractional distances from an endpoint.

Example 5 Locating a Point at Fractional Distances

Find X on $\overline{AF}$ that is $\frac{1}{6}$ of the distance from A to F.

Find the distance between A and F.

$AF = \lvert x_2 - x_1 \rvert$	Distance Formula
$= \lvert 5 - (-7) \rvert$	$x_1 = -7$ and $x_2 = 5$
$= \lvert 12 \rvert$ or 12	Subtract.

The distance between A and F is 12 units.

To find the point $\frac{1}{6}$ of the distance from A to F find $\frac{1}{6}AF$.

$$12\left(\frac{1}{6}\right) = 2$$

X is 2 units from point A on $\overline{AF}$ so, point X is located at $-7 + 2 = -5$ on the number line.

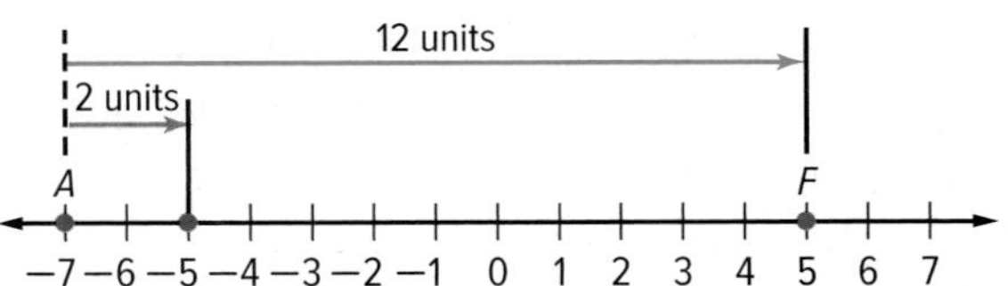

Study Tip

Direction In Example 5, the phrase "from A to F" indicates the direction of the line segment. A is the starting point.

Guided Practice

5. Use the number line above to find the point on $\overline{CE}$ that is $\frac{1}{8}$ of the distance from C to E.

You can also find a point a fractional distance from an endpoint on the coordinate plane.

Example 6 Fractional Distances on a Coordinate Plane

Find R on $\overline{NM}$ that is $\frac{1}{4}$ the distance from N to M.

Find the distance between the x-coordinates of N and M.

$|x_2 - x_1| = |2 - (-3)|$ $\quad x_1 = -3, x_2 = 2$

$= |5|$ or 5 $\quad$ Subtract.

Multiply this distance by the fractional distance.

$5\left(\frac{1}{4}\right) = 1.25$ Add this to the x-coordinate of N to determine the x-coordinate of R.

$-3 + 1.25 = -1.75$ $\quad$ The x-coordinate of R is -1.75.

Then find the distance between the y-coordinates of N and M.

$|y_2 - y_1| = |3 - (-3)|$ $\quad y_1 = -3, y_2 = 3$

$= |6|$ or 6 $\quad$ Subtract

Next multiply by the fractional distance $\frac{1}{4}$ to get $6\left(\frac{1}{4}\right) = 1.5$. Add this to the y-coordinate of N to find the y-coordinate of R.

$-3 + 1.5 = -1.5$ $\quad$ The y-coordinate of R is -1.5.

The point R is located at $(-1.75, -1.5)$.

Guided Practice

6. Find P on $\overline{NM}$ that is $\frac{1}{5}$ the distance from N to M.

A line segment can also be separated into two or more segments with lengths in a given ratio.

Go Online!

Personal Tutors for each example let you follow along as a teacher solves a problem. Pause and rewind as you need.

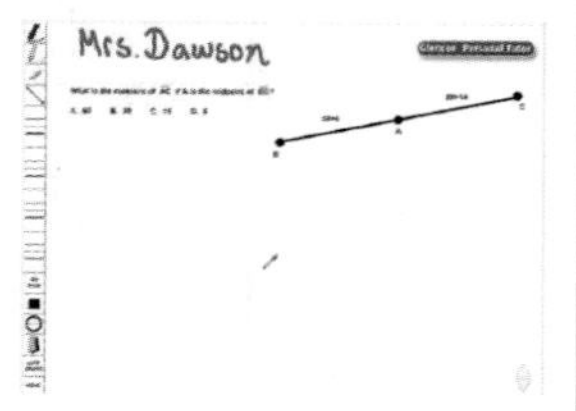

Example 7 Locating a Point Given a Ratio

Find C on $\overline{AB}$ such that the ratio of AC to CB is 1:4.

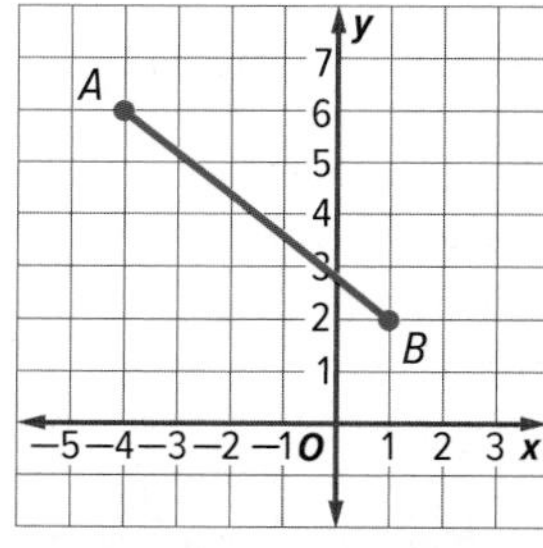

Since the ratio of the measures is 1:4, $4AC = CB$. So, $AB = AC + CB = AC + 4AC$ or $5AC$. Thus, AC is $\frac{1}{5}$ of AB.

Knowing the fractional distance, we can solve as we did in Example 6.

$|x_2 - x_1| = |1 - (-4)| = |5|$ or 5 $\quad x_1 = -4, x_2 = 1$

$|y_2 - y_1| = |2 - 6| = |-4|$ or 4 $\quad y_1 = 6, y_2 = 2$

The distance from the x-coordinate is $5\left(\frac{1}{5}\right) = 1$ and the distance from the y-coordinate is $4\left(\frac{1}{5}\right) = 0.8$. So, C is located at $(-4 + 1, 6 - 0.8)$ or $(-3, 5.2)$.

Guided Practice

7. Find D on $\overline{AB}$ such that the ratio of AD to DB is 1:3.

Check Your Understanding

◯ = Step-by-Step Solutions begin on page R13.

Example 1

1. **RAINFALL** Mrs. Smith's class measured the amount of rainfall each month during the school year. If the minimum rainfall was 1.7 centimeters and the maximum rainfall was 6.9 centimeters, find the midpoint.

Example 2

Use the number line to find the coordinate of the midpoint of each segment.

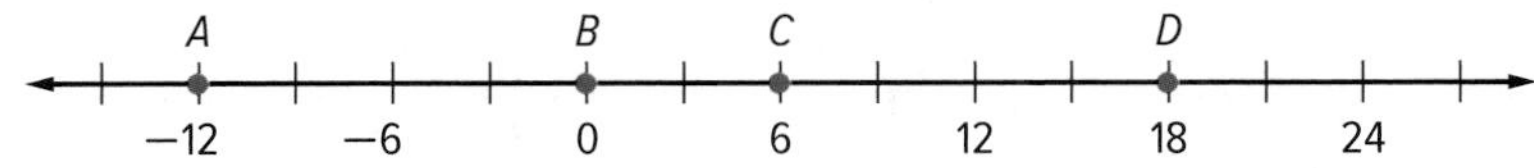

2. $\overline{BD}$
3. $\overline{AC}$

Find the coordinates of the midpoint of a segment with the given endpoints.

4. $M(7, 1)$, $N(4, -1)$
5. $J(5, -3)$, $K(3, -8)$

Examples 3–4

6. Find the measure of $\overline{LM}$ if M is the midpoint of $\overline{LN}$ and $LM = 3x - 2$ and $MN = 2x + 1$.

7. Find the coordinates of G if $F(1, 3.5)$ is the midpoint of $\overline{GJ}$ and J has coordinates $(6, -2)$.

Example 5

Refer to the number line.

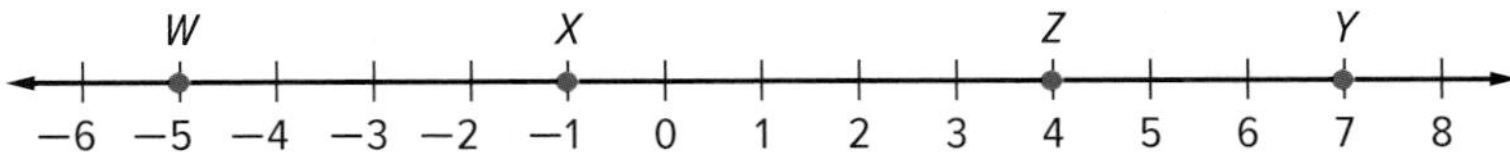

8. Find the point N on $\overline{WZ}$ that is $\frac{1}{3}$ of the distance from W to Z.

9. Find the point M on $\overline{XY}$ that is $\frac{1}{4}$ of the distance from X to Y.

For Exercises 10 and 11, refer to the coordinate grid shown.

Examples 6–7

10. Find M on $\overline{XY}$ that is $\frac{1}{3}$ the distance from X to Y.

11. Find N on $\overline{XY}$ such that the ratio of XN to NY is 1:3.

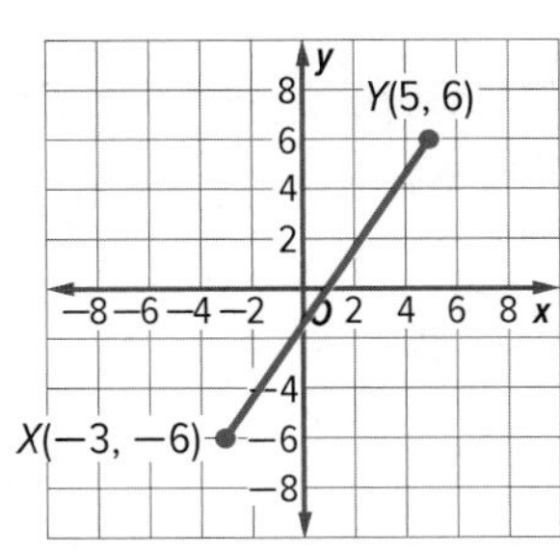

Practice and Problem Solving

Extra Practice is on page R1.

Example 1

12. **RACE** A city hosts a race annually. They decide to place a camera halfway between the start and finish lines. If A represents the starting line and Z represents the finish line, find the coordinate of the midpoint.

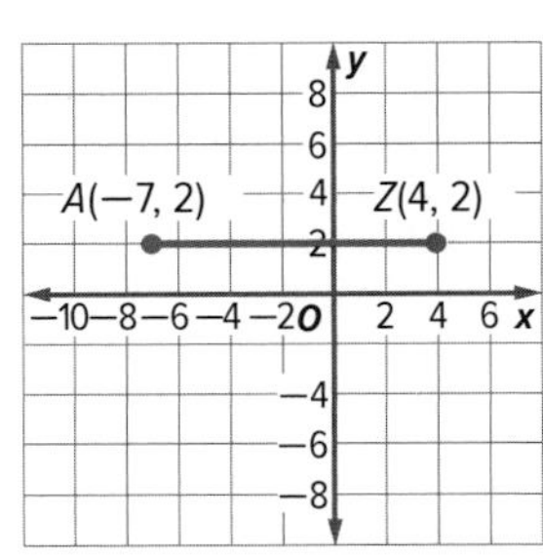

13. **MP REASONING** A business is trying to decide where to build an office. The business wants to place the office halfway between city B and city C. If city B is at (3, 9) and city C is at (3, −5), find the coordinates of the midpoint.

Example 2 **Use the number line to find the coordinate of the midpoint of each segment.**

E F G H J K L

−6 −4 −2 0 2 4 6 8 10

14. $\overline{JL}$ **15.** $\overline{HK}$ **16.** $\overline{FG}$

17. $\overline{EF}$ **18.** $\overline{EL}$ **19.** $\overline{FK}$

Find the coordinates of the midpoint of a segment with the given endpoints.

20. $W(12, 2)$, $X(7, 9)$

21. $C(22, 4)$, $B(15, 7)$

22. $V(-2, 5)$, $Z(3, -17)$

23. $D(-15, 4)$, $E(2, -10)$

24. $J(-11.2, -3.4)$, $K(-5.6, -7.8)$

25. $X(-2.4, -14)$, $Y(-6, -6.8)$

26. 

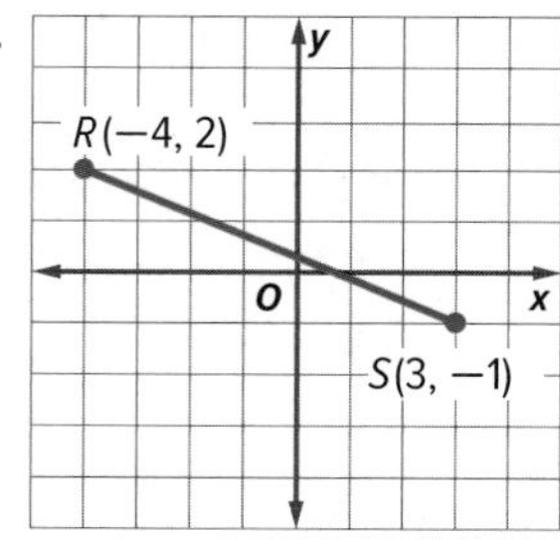

27.

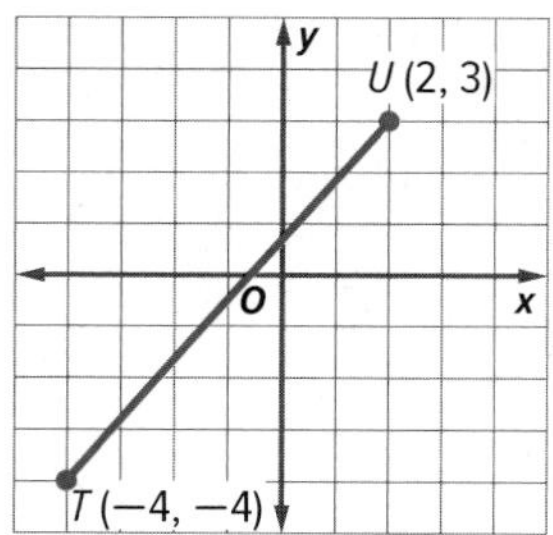

Example 3 **Find the coordinates of the missing endpoint if B is the midpoint of $\overline{AC}$.**

28. $A(1, 7)$, $B(-3, 1)$ **29.** $C(-5, 4)$, $B(-2, 5)$ **30.** $C(-6, -2)$, $B(-3, -5)$

31. $A(-4, 2)$, $B(6, -1)$ **32.** $C\left(\frac{5}{3}, -6\right)$, $B\left(\frac{8}{3}, 4\right)$ **33.** $A(4, -0.25)$, $B(-4, 6.5)$

Example 4 **Suppose M is the midpoint of $\overline{FG}$. Find each missing measure.**

34. $FM = 5y + 13$, $MG = 5 - 3y$, $FG = ?$

35. $FM = 3x - 4$, $MG = 5x - 26$, $FG = ?$

36. $FM = 8a + 1$, $FG = 42$, $a = ?$

37. $MG = 7x - 15$, $FG = 33$, $x = ?$

Example 5 MP **PERSEVERENCE** Refer to the number line.

A B C D E F

−7 −6 −5 −4 −3 −2 −1 0 1 2 3 4 5 6 7

38. Find the point X on $\overline{CF}$ that is $\frac{1}{5}$ of the distance from C to F.

39. Find the point X on $\overline{BD}$ that is $\frac{2}{3}$ of the distance from B to D.

40. Find the point X on $\overline{AE}$ that is $\frac{1}{6}$ of the distance from A to E.

41. Find the point X on $\overline{AF}$ that is $\frac{4}{5}$ of the distance from A to F.

Examples 6–7 **42.** Find X on $\overline{AB}$ that is $\frac{1}{5}$ the distance from A to B.

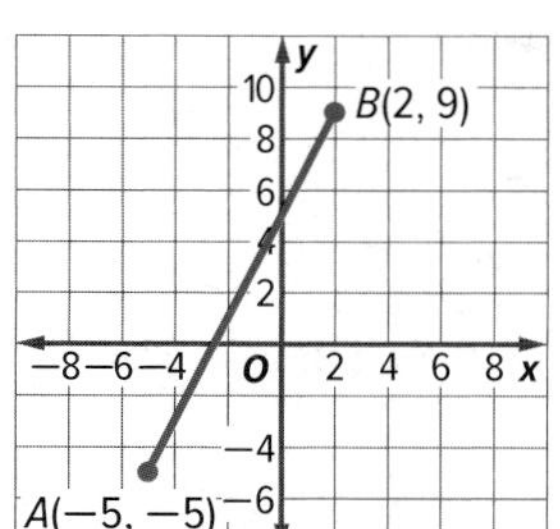

43. Find X on $\overline{RS}$ that is $\frac{1}{6}$ the distance from R to S.

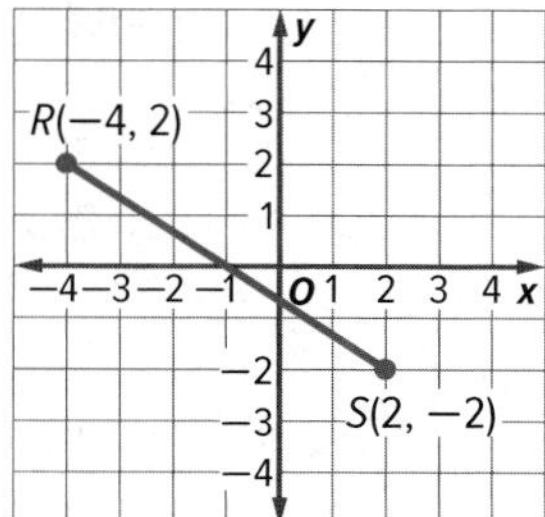

44. Find X on $\overline{JK}$ such that the ratio of JX to XK is 1:2.

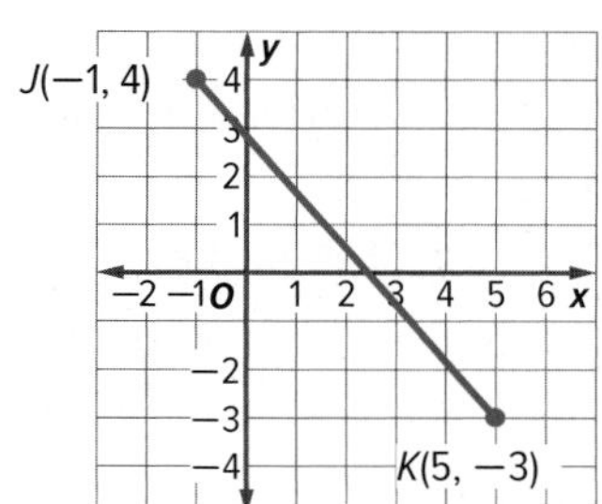

45. Find X on $\overline{MN}$ such that the ratio of MX to XN is 2:1.

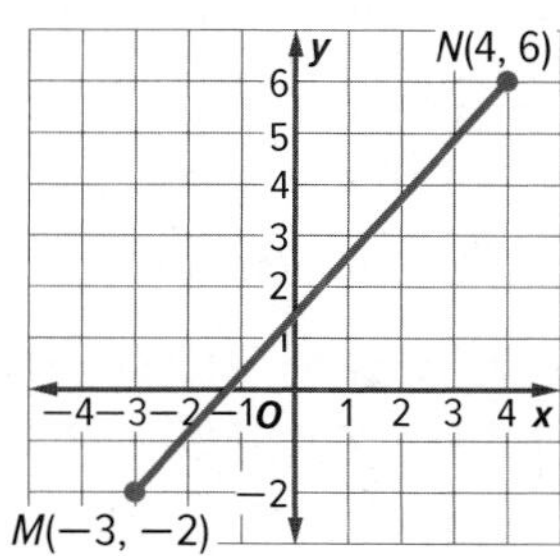

Determine the coordinates of the points that satisfy each condition.

46. Two points on the y-axis are 25 units from (−24, 3).

47 Two points on the x-axis are 10 units from (1, 8).

48. MP **MODELING** Points A and B represent two cities. Where should the state place a rest area so it is halfway between cities A and B?

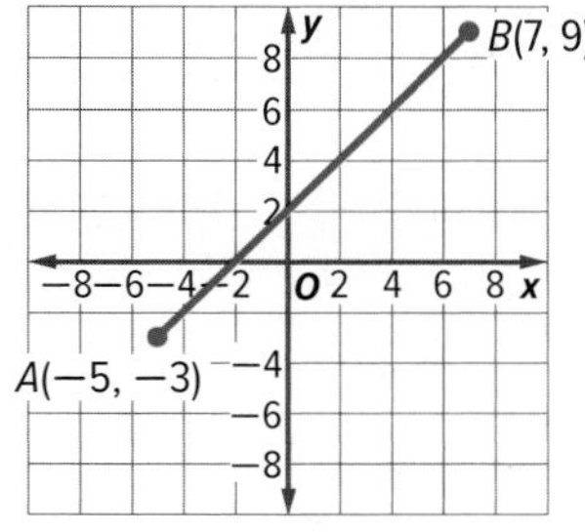

49. **COORDINATE GEOMETRY** Find the coordinates of B if B is halfway between $\overline{AC}$ and C is halfway between $\overline{AD}$.

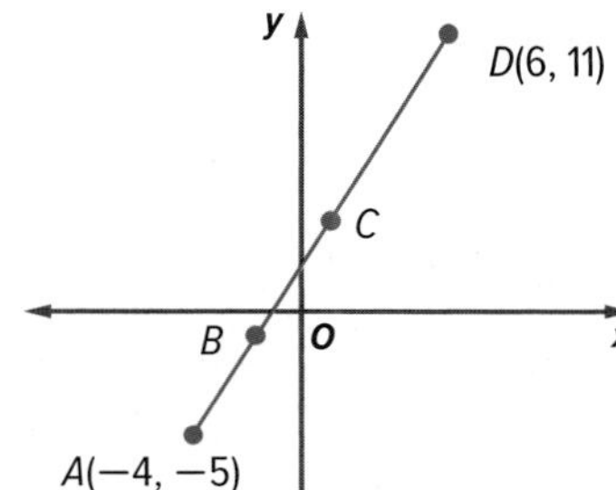

50. **GEOMETRY** One endpoint of $\overline{AB}$ has coordinates (−3, 5). If the coordinates of the midpoint of $\overline{AB}$ are (2, −6), what is the length of $\overline{AB}$?

51. **CONSTRUCTION** Copy the figure. Use a compass and straightedge to determine whether B is the midpoint of $\overline{AD}$. Explain.

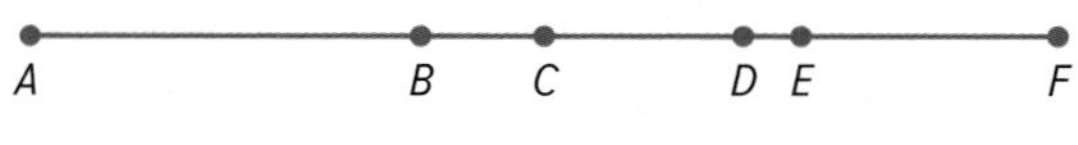

52. MP **PERSEVERANCE** Wilmington, North Carolina, is located at (34.3°, 77.9°), which represents north latitude and west longitude. Winston-Salem is in the northern part of the state at (36.1°, 80.2°).

a. Find the latitude and longitude of the midpoint of the segment between Wilmington and Winston-Salem.

b. Use an atlas or the Internet to find a city near the location of the midpoint.

c. If Winston-Salem is the midpoint of the segment with one endpoint at Wilmington, find the latitude and longitude of the other endpoint.

d. Use an atlas or the Internet to find a city near the location of the other endpoint.

53. **MULTIPLE REPRESENTATIONS** In this problem, you will explore the relationship between the midpoint of a segment and the midpoint of the segment between an endpoint and the midpoint.

a. Geometric Use a straightedge to draw three different line segments. Label each pair of endpoints A and B.

b. Geometric For each line segment, use a compass and straightedge to find the midpoint of $\overline{AB}$ and label it C. Then find the midpoint of $\overline{AC}$ and label it D.

c. Tabular Measure and record AB, AC, and AD for each line segment. Organize your results into a table.

d. Algebraic If $AB = x$, write an expression for the measures AC and AD.

e. Verbal Make a conjecture about the relationship between AB and each segment if you were to continue to find the midpoints of a segment and a midpoint you previously found.

54. **MULTI-STEP** John wants to center a canvas, which is 8 feet wide, on his living room wall, which is 17 feet wide. Where on the wall should John mark the location of the nails, if the canvas requires nails every $\frac{1}{5}$ of its length, excluding the edges? Explain your solution process.

H.O.T. Problems Use Higher-Order Thinking Skills

55. MP **REASONING** Is the point one third of the way from (x_1, y_1) to (x_2, y_2) *sometimes, always,* or *never* the point $\left(\frac{x_1 + x_2}{3}, \frac{y_1 + y_2}{3}\right)$? Explain.

56. **CHALLENGE** Point P is located on the segment between point $A(1, 4)$ and point $D(7, 13)$. The distance from A to P is twice the distance from P to D. What are the coordinates of point P?

57. MP **PRECISION** Draw a segment and label it $\overline{AB}$. Using only a compass and a straightedge, construct a segment $\overline{CD}$ such that $CD = 5\frac{1}{4}AB$. Explain and then justify your construction.

58. **WRITING IN MATH** Describe a method of finding the midpoint of a segment that has one endpoint at (0, 0). Derive the midpoint formula, give an example using your method, and explain why your method works.

59. Jamar plots two points P and Q on a coordinate plane. The midpoint of $\overline{PQ}$ is $M(-3, 4)$. Which of the following could be the points that Jamar plots? MP 6

- ○ A $P(-5, 10)$ and $Q(1, 2)$
- ○ B $P(-2, 6)$ and $Q(-4, 2)$
- ○ C $P(-7, 1)$ and $Q(4, 3)$
- ○ D $P(-1, 7)$ and $Q(2, 3)$

60. Points A, B, C, and D are on a number line as shown.

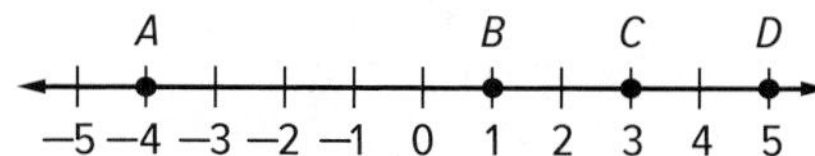

Which of the following is the distance from the midpoint of $\overline{AB}$ to the midpoint of $\overline{CD}$? MP 6

- ○ A $1\frac{1}{4}$
- ○ B $2\frac{1}{2}$
- ○ C $5\frac{1}{2}$
- ○ D $6\frac{1}{2}$

61. Isaac draws $\overline{MN}$ so that $MN = 5.2$ cm. Then, he draws a point P between M and N so that P is $\frac{1}{4}$ of the distance from M to N. What is the length of $\overline{PN}$? MP 6

- ○ A 3.9 cm
- ○ B 3.5 cm
- ○ C 2.6 cm
- ○ D 1.3 cm

62. $\overline{PQ}$ has endpoints $P(2, 1)$ and $Q(5, 5)$. If $\overline{PQ}$ and $\overline{PR}$ have the same length and P, Q, and R are not collinear, which could be the coordinates of R? Select all that apply. MP 1, 7

- ☐ A $(-1, -3)$
- ☐ B $(-1, 5)$
- ☐ C $(6, -3)$
- ☐ D $(6, 4)$
- ☐ E $(8, 9)$

63. $\overline{XY}$ has one endpoint at $X(-7, -2)$ and its midpoint is $M(-3, 7)$. What are the coordinates of Y? MP 6

64. Each unit of the coordinate plane represents 1 mile. Kate drives along a straight road from city hall C to the library L. Which statement best describes the distance Kate drives? MP 4, 7

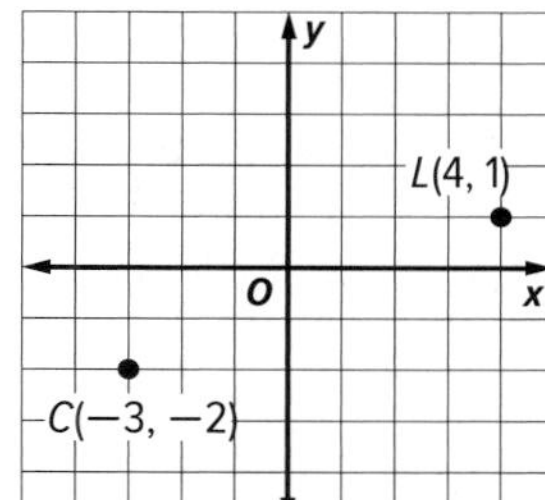

- ○ A Kate drives exactly 8 miles.
- ○ B Kate drives less than 8 miles.
- ○ C Kate drives exactly 58 miles.
- ○ D Kate drives less than 58 miles.

65. **MULTI-STEP** A video game designer places towers at J, K, and L, as shown. She plans to place a straight bridge between each pair of towers. MP 4, 7

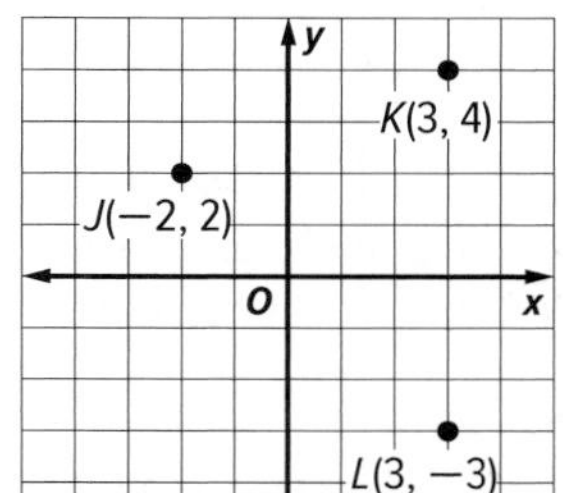

a. Which bridge is the shortest? Find its length and to the nearest tenth.

b. Which bridge is the longest? Find its length and to the nearest tenth.

c. What is the total length of the bridges?

LESSON 4

Angle Measure

∴Then	∴Now	∴Why?
• You measured line segments.	• **1** Measure and classify angles. **2** Identify and use congruent angles and the bisector of an angle.	• One of the skills Dale must learn in carpentry class is how to cut a *miter* joint. This joint is created when two boards are cut at an angle to each other. He has learned that one miscalculation in angle measure can result in mitered edges that do not fit together.

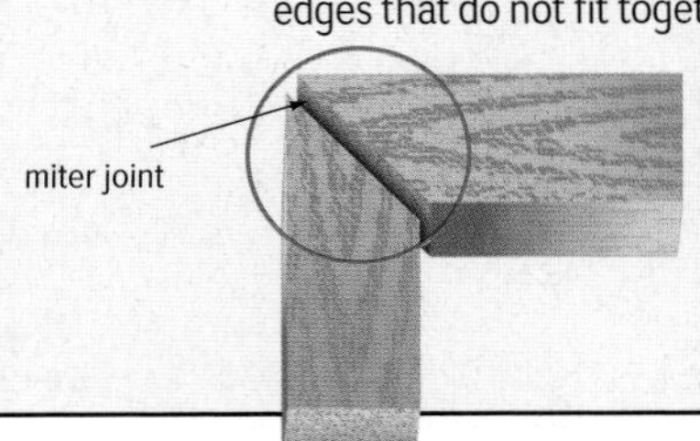

New Vocabulary

ray
opposite rays
angle
side
vertex
interior
exterior
degree
right angle
acute angle
obtuse angle
angle bisector

Mathematical Practices

5 Use appropriate tools strategically.
6 Attend to precision.

1 Measure and Classify Angles

A **ray** is a part of a line. It has one endpoint and extends indefinitely in one direction. Rays are named by stating the endpoint first and then any other point on the ray. The ray shown cannot be named as $\overrightarrow{OM}$ because O is not the endpoint of the ray.

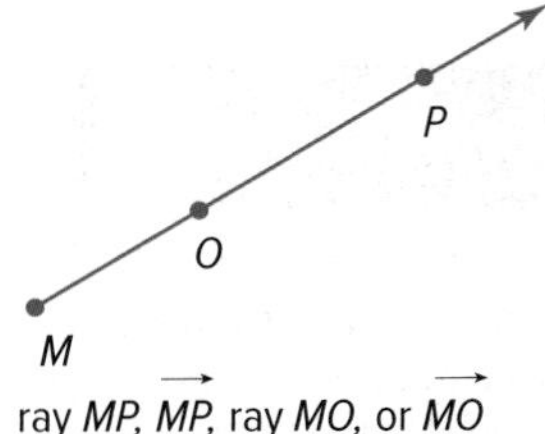

ray *MP*, $\overrightarrow{MP}$, ray *MO*, or $\overrightarrow{MO}$

If you choose a point on a line, that point determines exactly two rays called **opposite rays**. Since both rays share a common endpoint, opposite rays are collinear

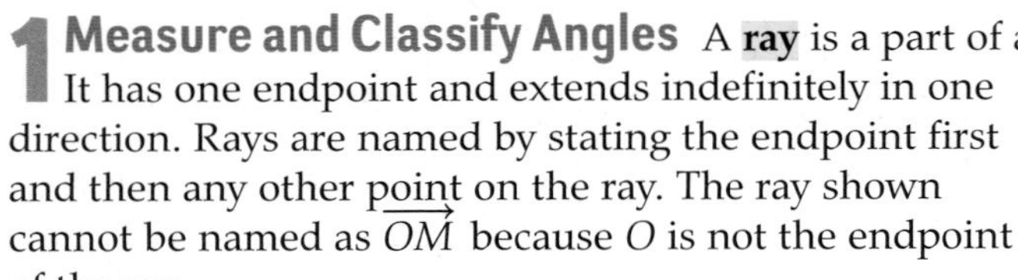

$\overrightarrow{JH}$ and $\overrightarrow{JK}$ are opposite rays.

An **angle** is formed by two *noncollinear* rays that have a common endpoint. The rays are called **sides** of the angle. The common endpoint is the **vertex**.

When naming angles using three letters, the vertex must be the second of the three letters. You can name an angle using a single letter only when there is exactly one angle located at that vertex. The angle shown can be named as $\angle X$, $\angle YXZ$, $\angle ZXY$, or $\angle 3$.

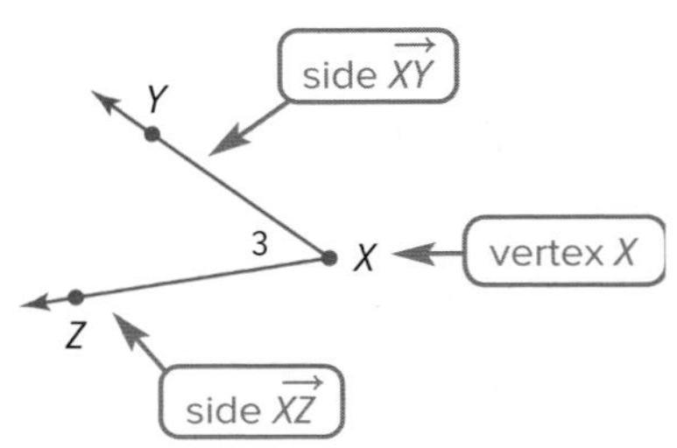

An angle divides a plane into three distinct parts.

- Points Q, M, and N lie on the angle.
- Points S and R lie in the **interior** of the angle.
- Points P and O lie in the **exterior** of the angle.

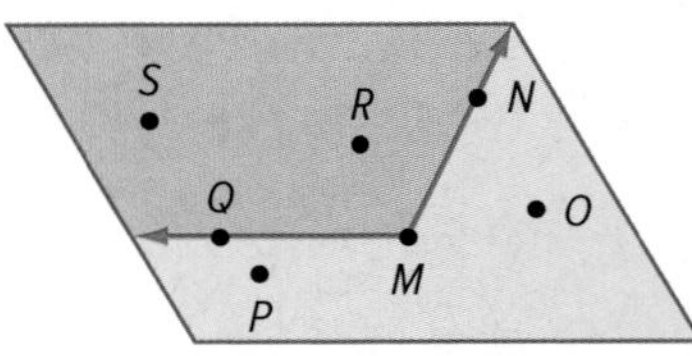

Real-World Example 1 Angles and Their Parts

MAPS Use the map of a high school shown.

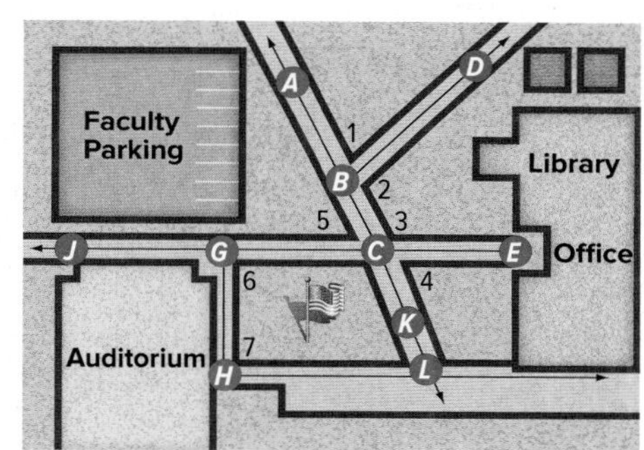

a. Name all angles that have *B* as a vertex.

$\angle 1$ or $\angle ABD$, and $\angle 2$ or $\angle DBC$

b. Name the sides of $\angle 3$.

$\overrightarrow{CA}$ and $\overline{CE}$, or $\overrightarrow{CB}$ and $\overline{CE}$

c. What is another name for $\angle GHL$?

$\angle 7$, $\angle H$, or $\angle LHG$

d. Name a point in the interior of $\angle DBK$.

point E

Study Tip

Segments as Sides Because a ray can contain a line segment, the side of an angle can be a segment.

Go Online!

Look for the **Animations** icons for concepts that have video animations. Log into ConnectED to see them.

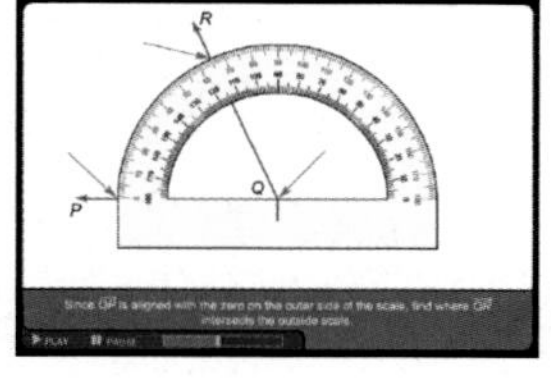

Guided Practice

1A. What is the vertex of $\angle 5$?

1B. Name the sides of $\angle 5$.

1C. Write another name for $\angle ECL$.

1D. Name a point in the exterior of $\angle CLH$.

Angles are measured in units called degrees. The **degree** results from dividing the distance around a circle into 360 parts.

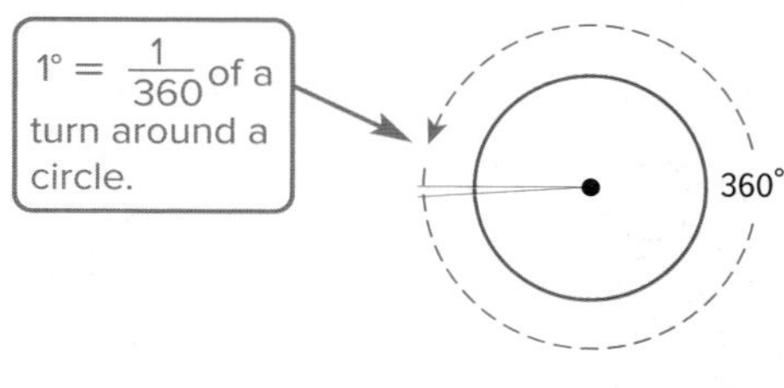

To measure an angle, you can use a *protractor*. Angle DEF below is a 50 degree (50°) angle. We say that the *degree measure* of $\angle DEF$ is 50, or $m\angle DEF = 50$.

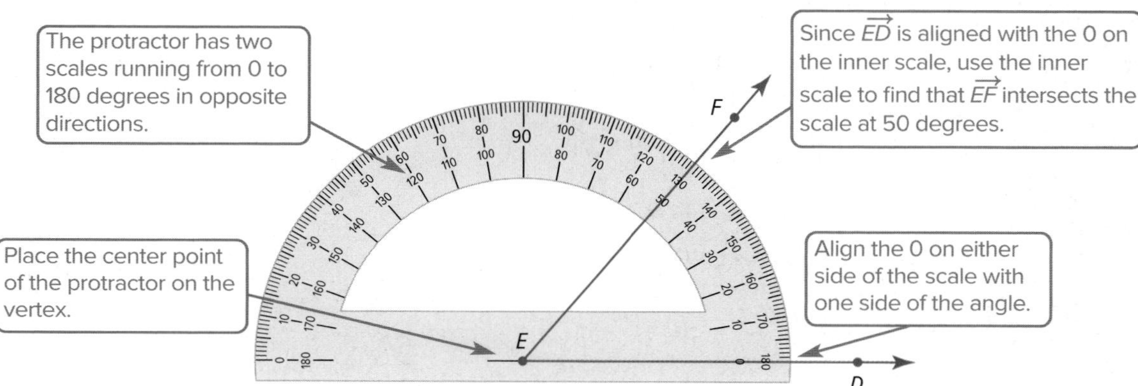

Reading Math

Straight Angle Opposite rays with the same vertex form a *straight angle*. Its measure is 180. Unless otherwise specified in this book, however, the term *angle* means a nonstraight angle.

Angles can be classified by their measures as shown below.

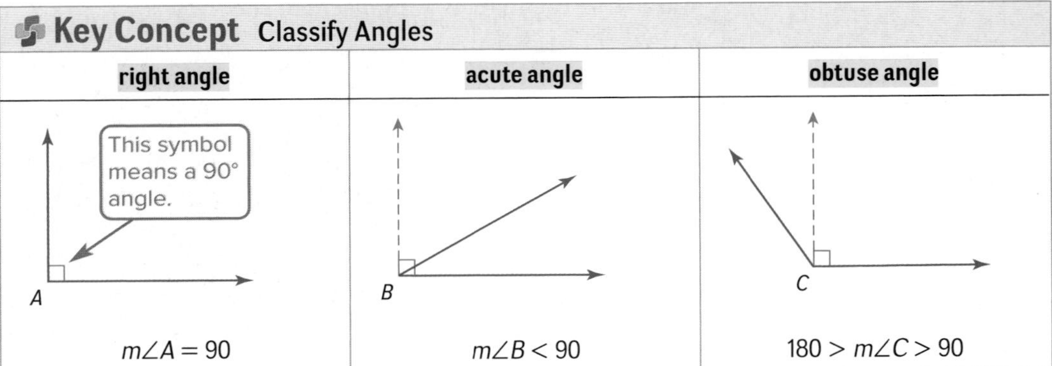

Example 2 Measure and Classify Angles

Copy the diagram below, and extend each ray. Classify each angle as *right*, *acute*, or *obtuse*. Then use a protractor to measure the angle to the nearest degree.

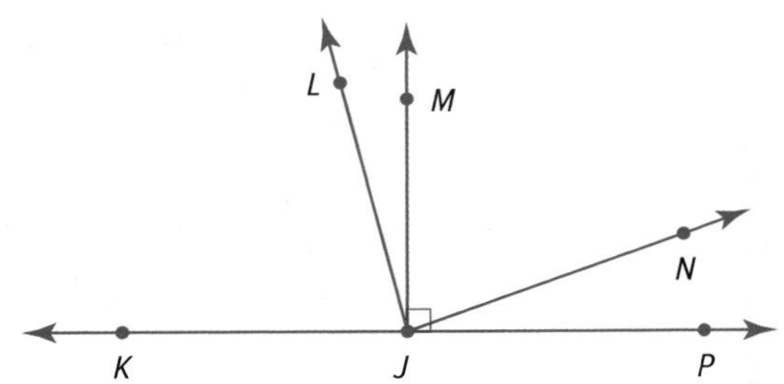

a. $\angle MJP$

$\angle MJP$ is marked as a right angle, so $m\angle MJP = 90$.

b. $\angle LJP$

Point L on angle $\angle LJP$ lies on the exterior of right angle $\angle MJP$, so $\angle LJP$ is an obtuse angle. Use a protractor to find that $m\angle LJP = 105$

CHECK Since $105 > 90$, $\angle LJP$ is an obtuse angle. ✓

c. $\angle NJP$

Point N on angle $\angle NJP$ lies on the interior of right angle $\angle MJP$, so $\angle NJP$ is an acute angle. Use a protractor to find that $m\ \angle NJP = 20$.

CHECK Since $20 < 90$, $\angle NJP$ is an acute angle. ✓

Watch Out!

Structure Classifying an angle before measuring it can prevent you from choosing the wrong scale on your protractor. In Example 2b, you must decide whether $\angle LJP$ measures 75 or 105. Since $\angle LJP$ is an obtuse angle, you can reason that the correct measure must be 105.

Guided Practice

2A. $\angle AFB$

2B. $\angle CFA$

2C. $\angle AFD$

2D. $\angle CFD$

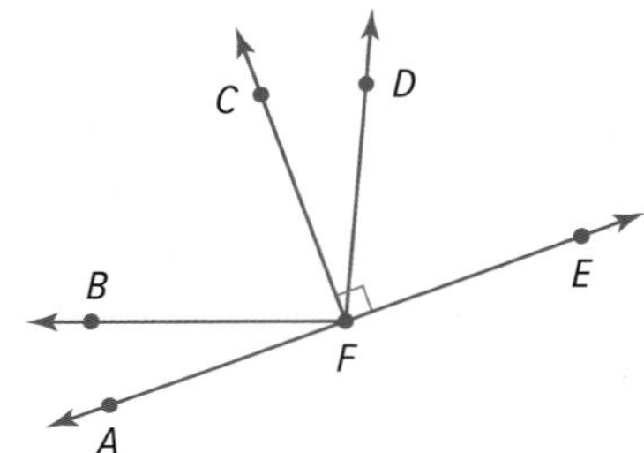

2 Congruent Angles

Congruent Angles Just as segments that have the same measure are congruent segments, angles that have the same measure are *congruent angles*.

In the figure, since $m\angle ABC = m\angle FED$, then $\angle ABC \cong \angle FED$. Matching numbers of arcs on a figure also indicate congruent angles, so $\angle CBE \cong \angle DEB$.

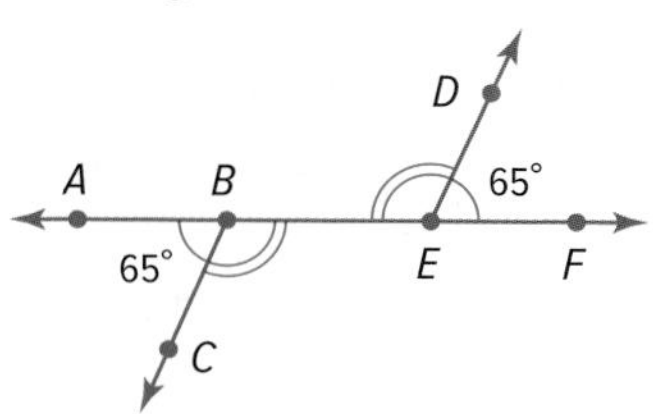

You can produce an angle congruent to a given angle using a construction.

Construction Copy an Angle

Step 1 Draw an angle like $\angle B$ on your paper. Use a straightedge to draw a ray on your paper. Label its endpoint G.

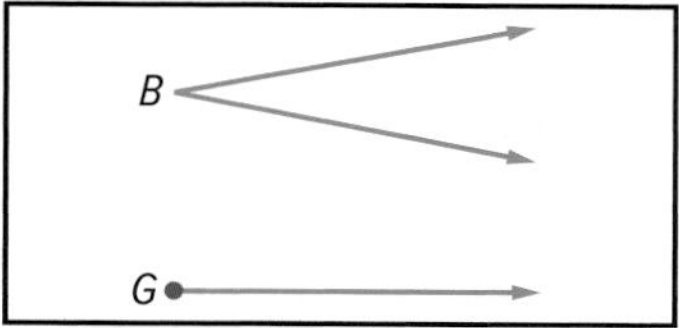

Step 2 Place the tip of the compass at point B and draw a large arc that intersects both sides of $\angle B$. Label the points of intersection A and C.

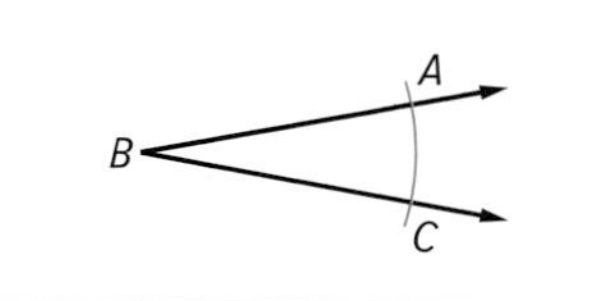

Step 3 Using the same compass setting, put the compass at point G and draw a large arc that starts above the ray and intersects the ray. Label the point of intersection H.

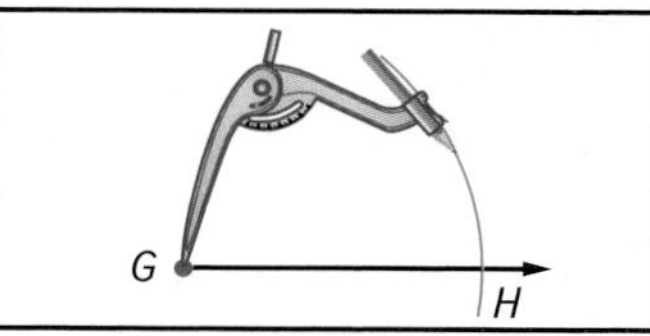

Step 4 Place the point of your compass on C and adjust so that the pencil tip is on A.

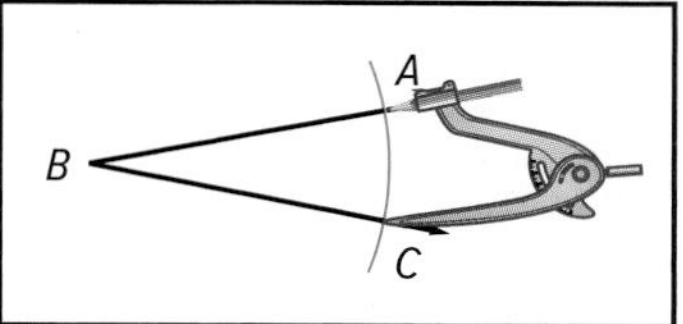

Step 5 Without changing the setting, place the compass at point H and draw an arc to intersect the larger arc you drew in Step 3. Label the point of intersection F.

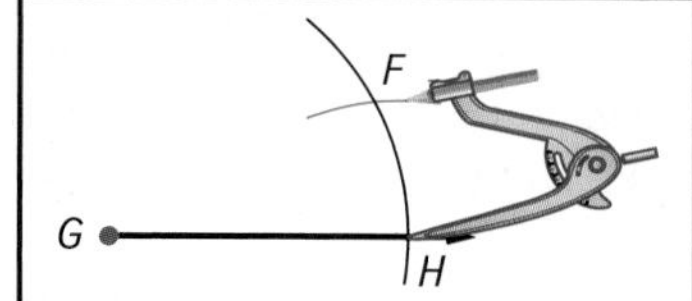

Step 6 Use a straightedge to draw $\overrightarrow{GF}$. $\angle ABC \cong \angle FGH$

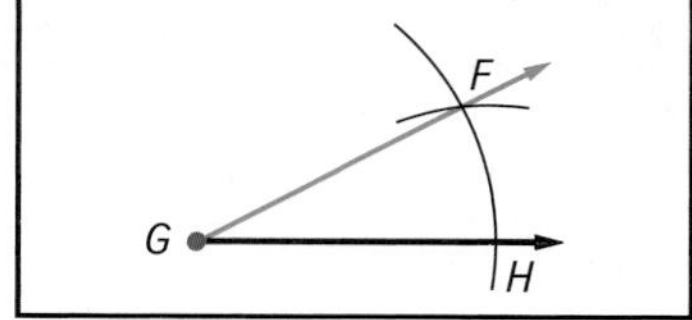

Study Tip

Segments A line segment can also bisect an angle.

A ray that divides an angle into two congruent angles is called an **angle bisector**. If $\overrightarrow{YW}$ is the angle bisector of $\angle XYZ$, then point W lies in the interior of $\angle XYZ$ and $\angle XYW \cong \angle WYZ$.

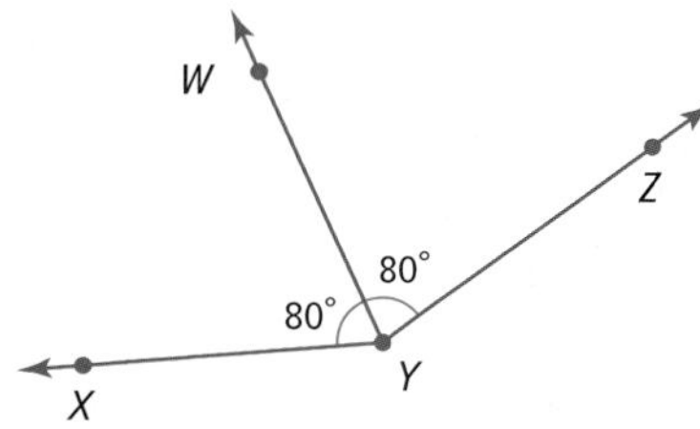

Just as with segments, when a line, segment, or ray divides an angle into smaller angles, the sum of the measures of the smaller angles equals the measure of the largest angle. So in the figure, $m\angle XYW + m\angle WYZ = m\angle XYZ$.

Example 3 Find the Angle Measure

ALGEBRA In the figure, $\overrightarrow{KJ}$ and $\overrightarrow{KM}$ are opposite rays, and $\overrightarrow{KN}$ bisects $\angle JKL$. If $m\angle JKN = 8x - 13$ and $m\angle NKL = 6x + 11$, find $m\angle JKN$.

Step 1 Solve for x.

Since $\overrightarrow{KN}$ bisects $\angle JKL$, $\angle JKN \cong \angle NKL$.

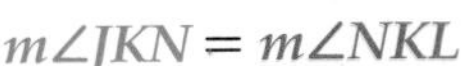

$m\angle JKN = m\angle NKL$	Definition of congruent angles
$8x - 13 = 6x + 11$	Substitution
$8x = 6x + 24$	Add 13 to each side.
$2x = 24$	Subtract $6x$ from each side.
$x = 12$	Divide each side by 2.

Step 2 Use the value of x to find $m\angle JKN$.

$m\angle JKN = 8x - 13$	Given
$= 8(12) - 13$	$x = 12$
$= 96 - 13$ or 83	Simplify.

Study Tip

Checking Solutions Check that you have computed the value of x correctly by substituting the value into the expression for $\angle NKL$. If you don't get the same measure as $\angle JKN$, you have made an error.

Guided Practice

3. Suppose $m\angle JKL = 9y + 15$ and $m\angle JKN = 5y + 2$. Find $m\angle JKL$.

You can produce the angle bisector of any angle without knowing the measure of the angle.

Construction Bisect an Angle

Step 1 Draw an angle. Put your compass at P and draw a large arc that intersects both sides of $\angle P$. Label the points of intersection Q and R.

Step 2 With the compass at point Q, draw an arc in the interior of the angle.

Step 3 Keeping the same compass setting, place the compass at point R and draw an arc that intersects the arc drawn in Step 2. Label the point of intersection T.

Step 4 Draw $\overrightarrow{PT}$. $\overrightarrow{PT}$ is the bisector of $\angle P$.

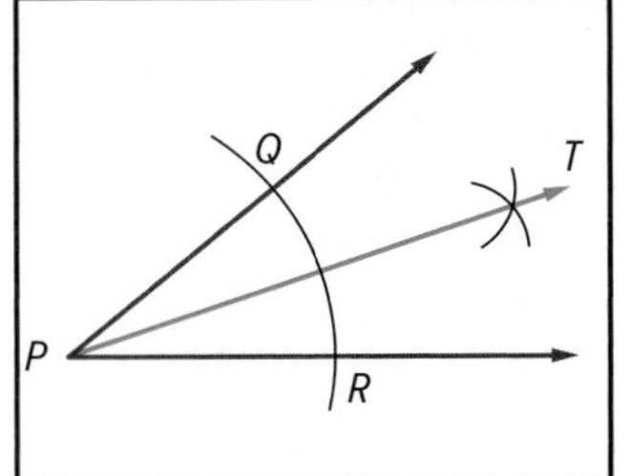

MAKE A CONJECTURE about the angles that result when you bisect an obtuse angle, a right angle, and an acute angle.

Check Your Understanding

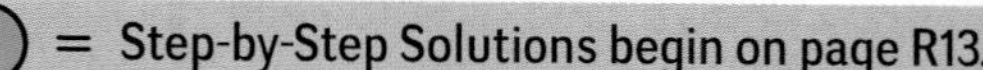

Example 1 **Use the figure at the right.**

1. Name the vertex of $\angle 4$.
2. Name the sides of $\angle 3$.
3. What is another name for $\angle 2$?
4. What is another name for $\angle UXY$?

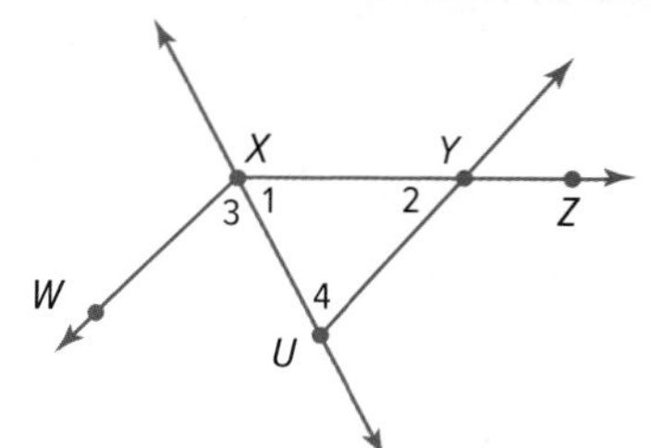

Example 2 **Copy the diagram shown, and extend each ray. Classify each angle as *right*, *acute*, or *obtuse*. Then use a protractor to measure the angle to the nearest degree.**

5. $\angle CFD$
6. $\angle AFD$
7. $\angle BFC$
8. $\angle AFB$

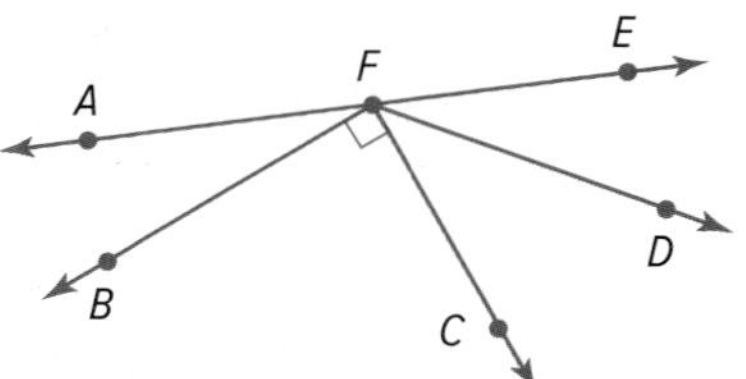

Example 3 **ALGEBRA In the figure, $\overrightarrow{KJ}$ and $\overrightarrow{KL}$ are opposite rays. $\overrightarrow{KN}$ bisects $\angle LKM$.**

9. If $m\angle LKM = 7x - 5$ and $m\angle NKM = 3x + 9$, find $m\angle LKM$.
10. If $m\angle NKL = 7x - 9$ and $m\angle JKM = x + 3$, find $m\angle JKN$.

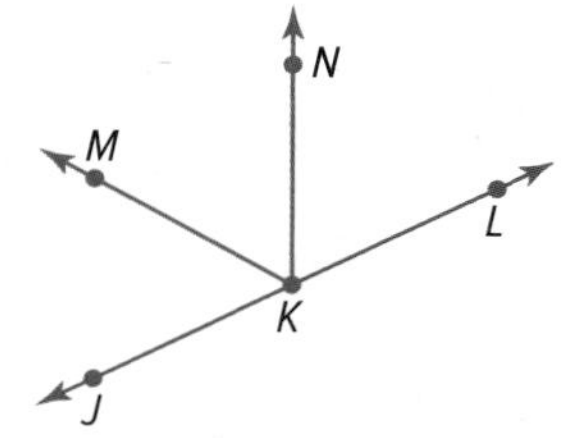

11. **MP PRECISION** A miter cut is used to build picture frames with corners that meet at right angles.
    a. José miters the ends of some wood for a picture frame at congruent angles. What is the degree measure of his cut? Explain and classify the angle.
    b. What does the joint represent in relation to the angle formed by the two pieces?

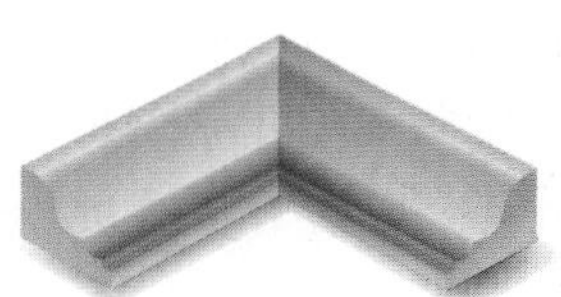

Practice and Problem Solving

Extra Practice is on page R1.

Example 1 **For Exercises 12–29, use the figure at the right.**

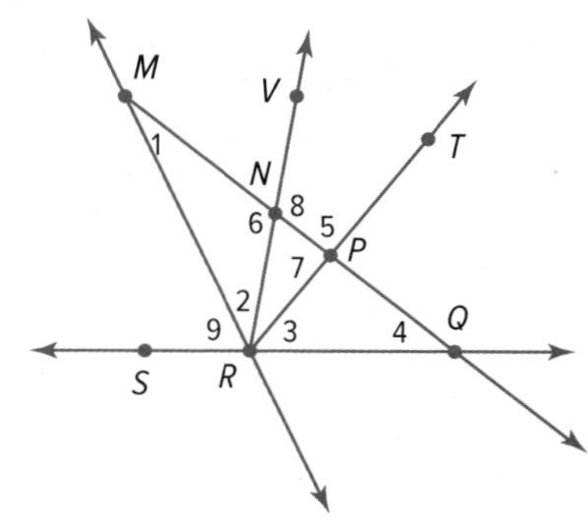

Name the vertex of each angle.

12. $\angle 4$
13. $\angle 7$
14. $\angle 2$
15. $\angle 1$

Name the sides of each angle.

16. $\angle TPQ$
17. $\angle VNM$
18. $\angle 6$
19. $\angle 3$

Write another name for each angle.

20. $\angle 9$
21. $\angle QPT$
22. $\angle MQS$
23. $\angle 5$

24. Name an angle with vertex N that appears obtuse.
25. Name an angle with vertex Q that appears acute.
26. Name a point in the interior of $\angle VRQ$.
27. Name a point in the exterior of $\angle MRT$.
28. Name a pair of angles that share exactly one point.
29. Name a pair of angles that share more than one point.

Example 2

Copy the diagram shown, and extend each ray. Classify each angle as *right*, *acute*, or *obtuse*. Then use a protractor to measure the angle to the nearest degree.

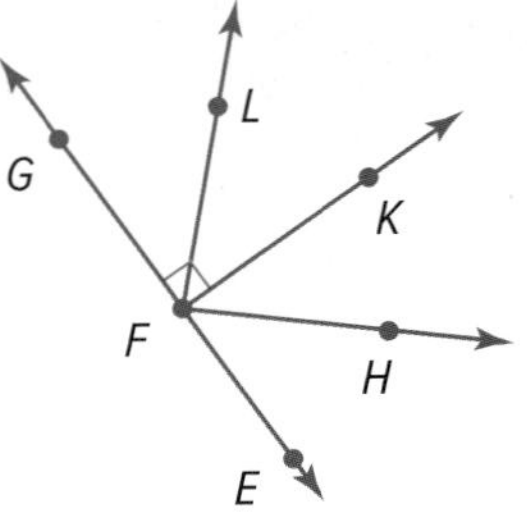

30. $\angle GFK$

31. $\angle EFK$

32. $\angle LFK$

33. $\angle EFH$

34. $\angle GFH$

35. $\angle EFL$

36. CLOCKS Determine at least three different times during the day when the hands on a clock form each of the following angles. Explain.

a. right angle

b. obtuse angle

c. congruent acute angles

Example 3

ALGEBRA In the figure, $\overrightarrow{BA}$ and $\overrightarrow{BC}$ are opposite rays. $\overrightarrow{BH}$ bisects $\angle EBC$.

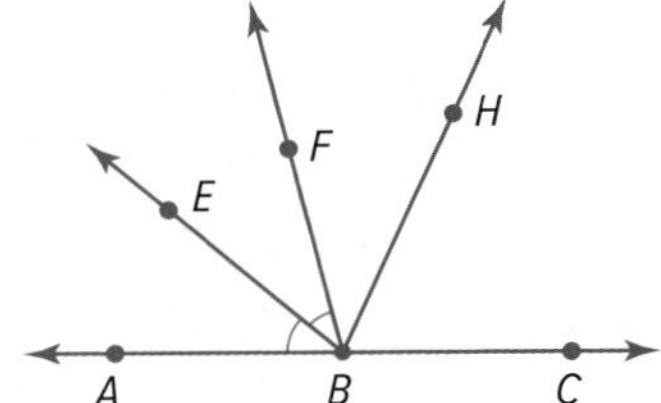

37 If $m\angle ABE = 2n + 7$ and $m\angle EBF = 4n - 13$, find $m\angle ABE$.

38. If $m\angle EBH = 6x + 12$ and $m\angle HBC = 8x - 10$, find $m\angle EBH$.

39. If $m\angle ABF = 7b - 24$ and $m\angle ABE = 2b$, find $m\angle EBF$.

40. If $m\angle EBC = 31a - 2$ and $m\angle EBH = 4a + 45$, find $m\angle HBC$.

41. If $m\angle ABF = 8s - 6$ and $m\angle ABE = 2(s + 11)$, find $m\angle EBF$.

42. If $m\angle EBC = 3r + 10$ and $m\angle ABE = 2r - 20$, find $m\angle EBF$.

43. MAPS Estimate the measure of the angle formed by each city or location listed, the North Pole, and the Prime Meridian.

a. Nuuk, Greenland

b. Fairbanks, Alaska

c. Reykjavik, Iceland

d. Prime Meridian

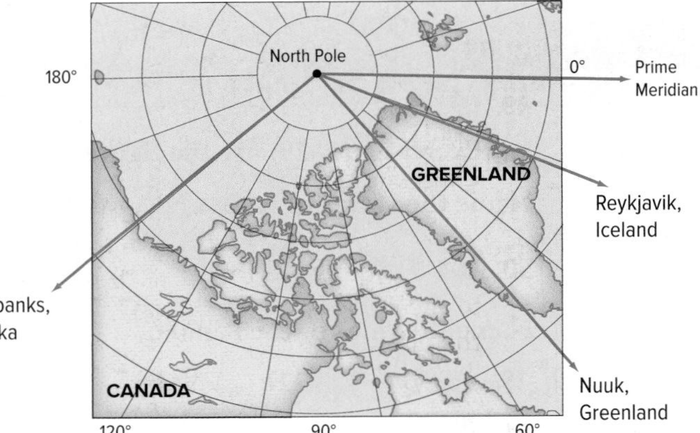

44. MP TOOLS A compass rose is a design on a map that shows directions. In addition to the directions of north, south, east, and west, a compass rose can have as many as 32 markings.

a. With the center of the compass as its vertex, what is the measure of the angle between due west and due north?

b. What is the measure of the angle between due north and northwest?

c. How does the northwest ray relate to the angle in part **a**?

Plot the points in a coordinate plane and sketch ∠XYZ. Then classify it as *right, acute,* or *obtuse.*

45. $X(5, -3)$, $Y(4, -1)$, $Z(6, -2)$

46. $X(6, 7)$, $Y(2, 3)$, $Z(4, 1)$

47 **PHYSICS** When you look at a pencil in water, it looks bent. This illusion is due to *refraction,* or the bending of light when it moves from one substance to the next.

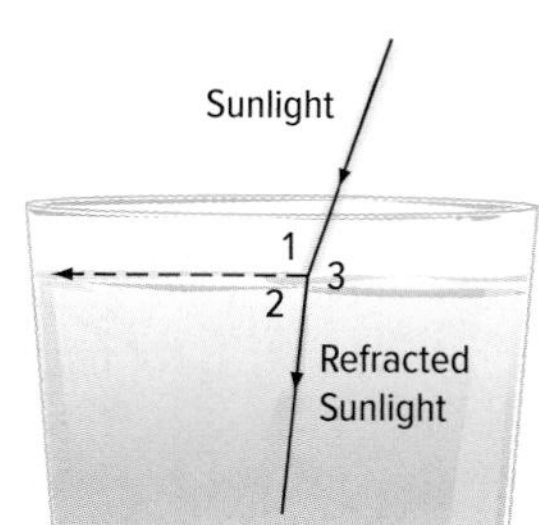

a. What is $m\angle 1$? Classify this angle as *acute, right,* or *obtuse.*

b. What is $m\angle 2$? Classify this angle as *acute, right,* or *obtuse.*

c. Without measuring, determine how many degrees the path of the light changes after it enters the water. Explain your reasoning.

48. MULTIPLE REPRESENTATIONS In this problem, you will explore the relationship of angles that compose opposite rays.

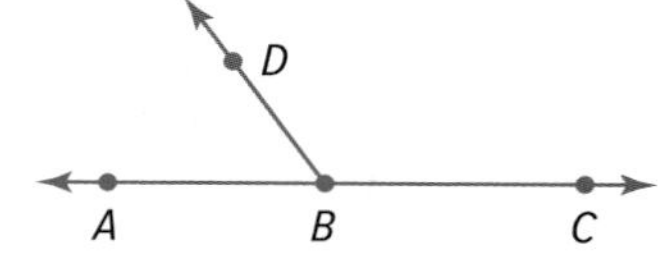

a. Geometric Copy the figure shown. Draw three additional figures, varying the placement of point *D*. Use a protractor to measure ∠*ABD* and ∠*DBC* for each figure.

b. Tabular Organize the measures for each figure into a table. Include a column in your table to record the sum of these measures.

c. Verbal Make a conjecture about the sum of the measures of the two angles. Explain your reasoning.

d. Algebraic If *x* is the measure of ∠*ABD* and *y* is the measure of ∠*DBC*, write an equation that relates the two angle measures.

H.O.T. Problems Use Higher-Order Thinking Skills

49. OPEN-ENDED Draw an obtuse angle named *ABC*. Measure ∠*ABC*. Construct an angle bisector $\overrightarrow{BD}$ of ∠*ABC*. Explain the steps in your construction and justify each step. Classify the two angles formed by the angle bisector.

50. MP **TOOLS** Use a compass and a straightedge to construct an angle congruent to the angle shown. How would you use a protractor to find the measure of the angle?

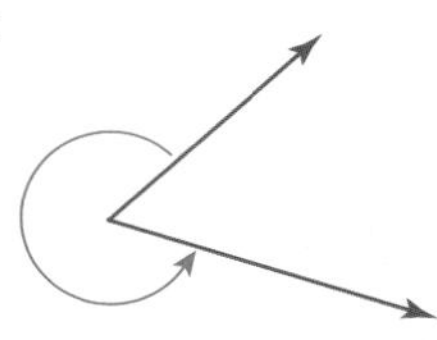

51. MP **ARGUMENTS** Is the sum of two acute angles *sometimes, always,* or *never* an obtuse angle? Explain.

52. CHALLENGE $\overrightarrow{MP}$ bisects ∠*LMN*, $\overrightarrow{MQ}$ bisects ∠*LMP*, and $\overrightarrow{MR}$ bisects ∠*QMP*. If $m\angle RMP = 21$, find $m\angle LMN$. Explain your reasoning.

53. WRITING IN MATH Rashid says that he can estimate the measure of an acute angle using a piece of paper to within six degrees of accuracy. Explain how this would be possible. Then use this method to estimate the measure of the angle shown.

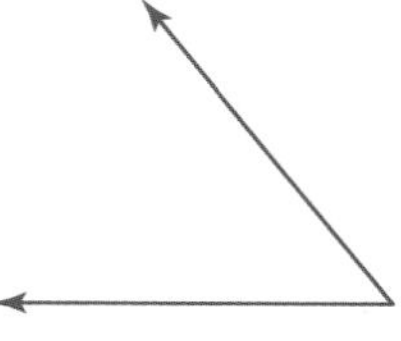

Preparing for Assessment

54. **MULTI-STEP** Mei used a compass and straightedge to make the construction shown here. MP 7

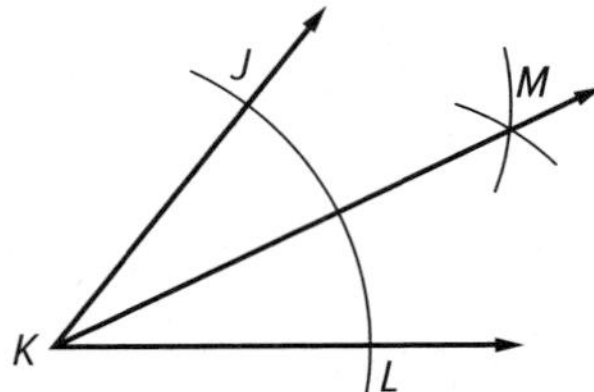

a. Which of the following statements must be true?

- A $m\angle JKM = \frac{1}{2}m\angle JKL$
- B $m\angle JKL = \frac{1}{2}m\angle MKL$
- C $m\angle JKM = m\angle JKL$
- D $m\angle MKL = 2m\angle JKL$

b. How could you construct an angle with measure that is one-fourth of the measure of $\angle JKL$?

c. How could you construct a 15° angle from a given 60° angle

55. Tyrell is constructing an angle congruent to $\angle Q$. The work he has done so far is shown below.

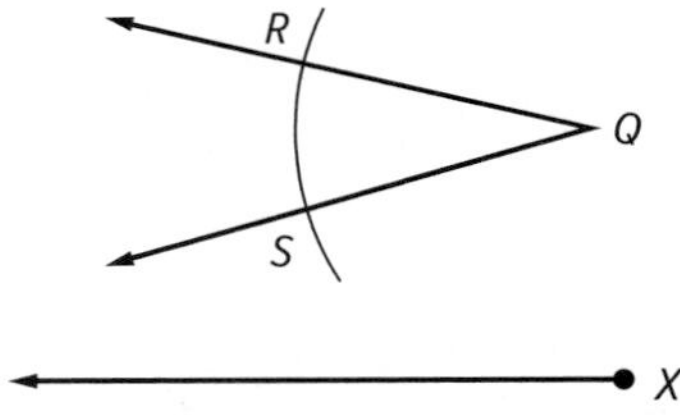

Which of the following best describes what Tyrell should do next? MP 7

- A Adjust the compass setting to measure the distance from point R to point S.
- B Place the point of the compass on point R.
- C Place the point of the compass on point S.
- D Place the point of the compass on point X.

56. Which construction requires you to draw only one arc? MP 7

- A copy a segment
- B bisect a segment
- C copy an angle
- D bisect an angle

57. A student was asked to construct the angle bisector of $\angle X$. The figure shows the work the student has done so far.

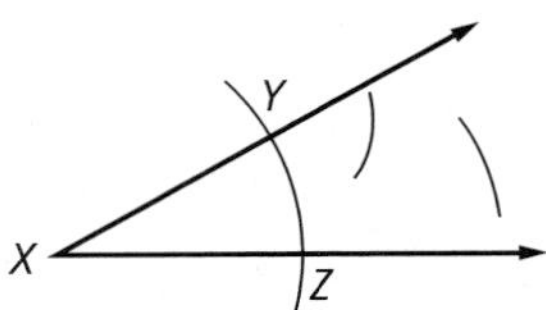

Which of the following best describes the error the student made? MP 5

- A The arcs are not long enough.
- B The student used the steps for constructing an angle that is congruent to $\angle X$.
- C The student changed the compass setting when drawing the arcs.
- D The student should not have drawn the arc that passes through points Y and Z.
- E The student should have used a protractor.

58. Naomi constructed a ray $\overrightarrow{BD}$ that is the bisector of $\angle ABC$. Given that $m\angle ABD = 65.2$, what is $m\angle ABC$? MP 5

59. A right angle is bisected by a ray. The angle between one of the sides and the bisector measures $(2x + 3)°$. Find the value of x. MP 7

60. $\angle ABC$ is bisected by ray BD. $\angle ABC$ is bisected by ray BE. If $m\angle ABE = 30$, what is $m\angle ABC$? MP 7

Mid-Chapter Quiz

Lessons 1-1 through 1-4

Use the figure to complete each of the following. (Lesson 1-1)

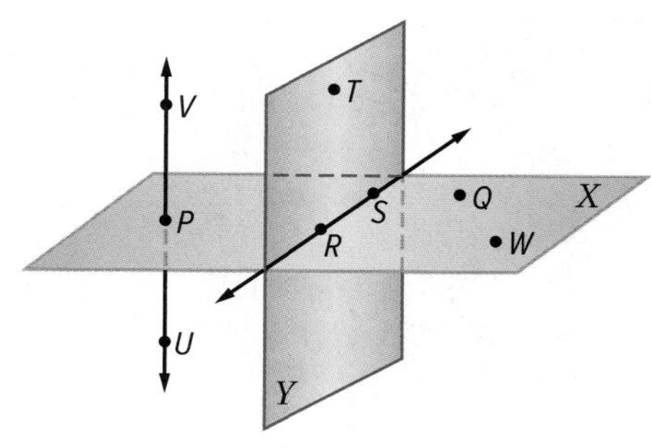

1. Name another point that is collinear with points U and V.
2. What is another name for plane Y?
3. Name a line that is coplanar with points P, Q, and W.

Find the value of x and AC if B is between points A and C. (Lesson 1-2)

4. $AB = 12$, $BC = 8x - 2$, $AC = 10x$
5. $AB = 5x$, $BC = 9x - 2$, $AC = 11x + 7.6$
6. Find CD and the coordinate of the midpoint of $\overline{CD}$. (Lesson 1-3)

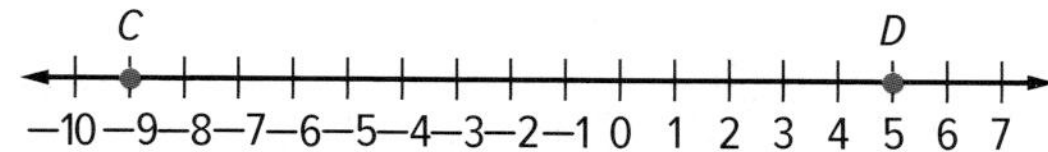

Find the coordinates of the midpoint of each segment. Then find the length of each segment. (Lesson 1-3)

7.

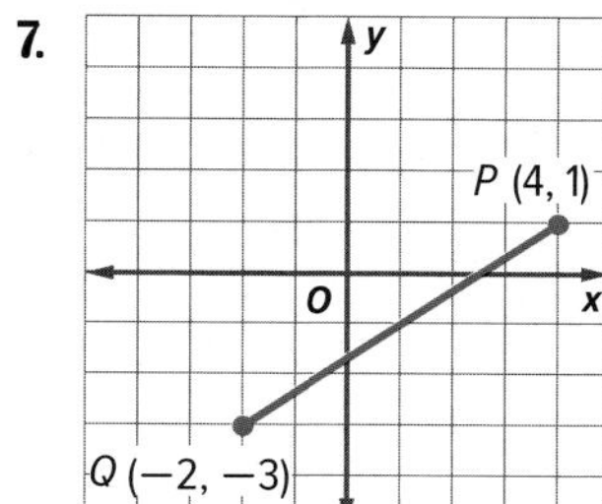

8.

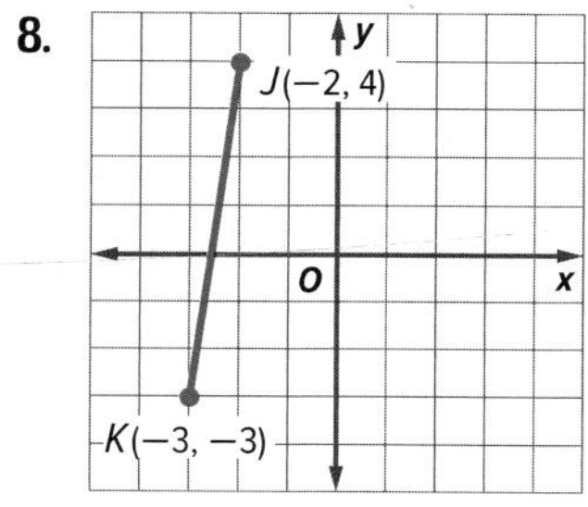

Find the coordinates of the midpoint of a segment with the given endpoints. Then find the distance between each pair of points. (Lesson 1-3)

9. $P(26, 12)$ and $Q(8, 42)$
10. $M(6, -41)$ and $N(-18, -27)$
11. **MAPS** A map of a town is drawn on a coordinate grid. The high school is found at point (3, 1) and town hall is found at (−5, 7). (Lesson 1-3)
    a. If the high school is at the midpoint between the town hall and the town library, at which ordered pair should you find the library?
    b. If one unit on the grid is equivalent to 50 meters, how far is the high school from town hall?
    c. MP What mathematical practice did you use to solve this problem?
12. **MULTIPLE CHOICE** The vertex of $\angle ABC$ is located at the origin. Point A is located at (5, 0) and Point C is located at (0, 2). Classify $\angle ABC$. (Lesson 1-4)

    A acute　　**C** right

    B obtuse　　**D** scalene

In the figure, $\overrightarrow{XA}$ and $\overrightarrow{XE}$ are opposite rays, and $\angle AXC$ is bisected by $\overrightarrow{XB}$. (Lesson 1-4)

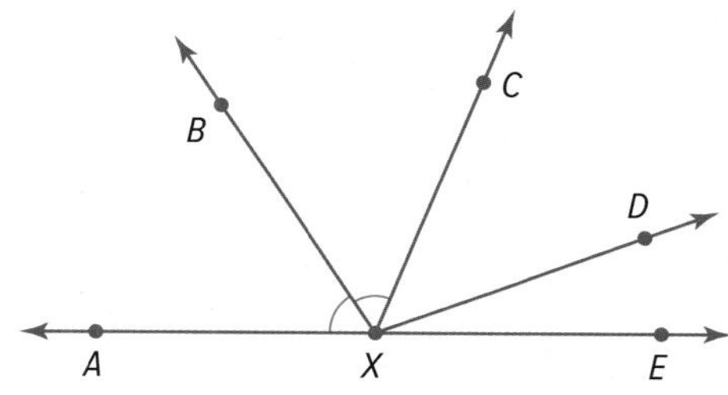

13. If $m\angle AXC = 8x - 7$ and $m\angle AXB = 3x + 10$, find $m\angle AXC$.
14. If $m\angle CXD = 4x + 6$, $m\angle DXE = 3x + 1$, and $m\angle CXE = 8x - 2$, find $m\angle DXE$.

Classify each angle as *acute*, *right*, or *obtuse*. (Lesson 1-4)

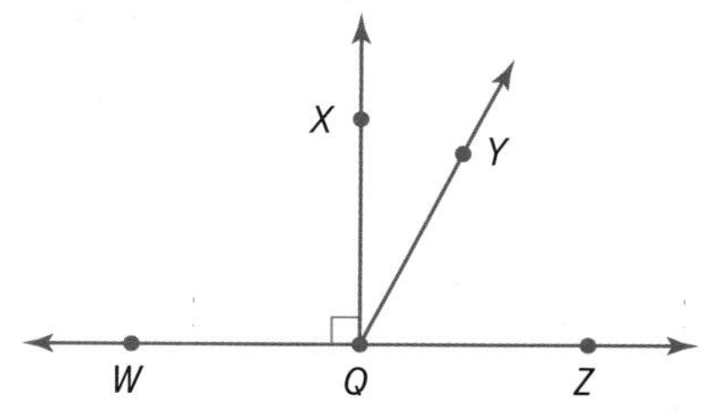

15. $\angle WQY$　　16. $\angle YQZ$

LESSON 5

Angle Relationships

Then

- You measured and classified angles.

Now

1. Identify and use special pairs of angles.
2. Identify perpendicular lines.

Why?

- Cheerleaders position their arms and legs at specific angles to create various formations when performing at games and competitions. Certain pairs of angles have special names and share specific relationships.

New Vocabulary

adjacent angles
linear pair
vertical angles
complementary angles
supplementary angles
perpendicular

Mathematical Practices

5 Use appropriate tools strategically.
6 Attend to precision.

1 Pairs of Angles

Some pairs of angles are special because of how they are positioned in relationship to each other. Three of these angle pairs are described below.

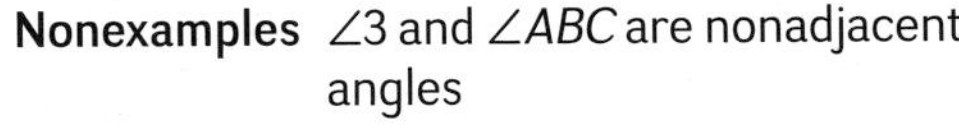

Key Concept Special Angle Pairs

Adjacent angles are two angles that lie in the same plane and have a common vertex and a common side, but no common interior points.

Examples ∠1 and ∠2 are adjacent angles.

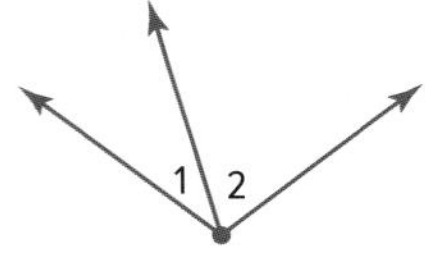

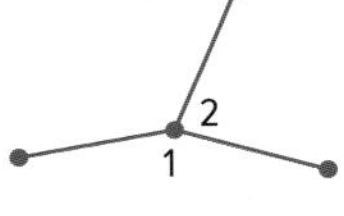

Nonexamples ∠3 and ∠*ABC* are nonadjacent angles

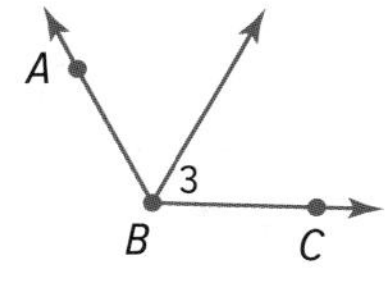

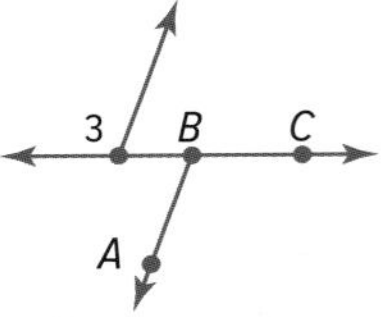

A **linear pair** is a pair of adjacent angles with noncommon sides that are opposite rays.

Example ∠1 and ∠2

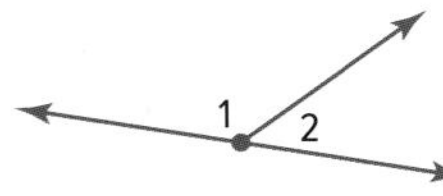

Nonexample ∠*ADB* and ∠*ADC*

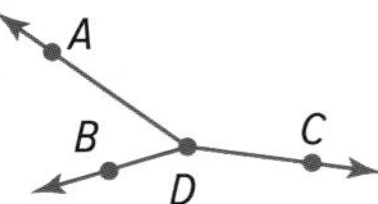

Vertical angles are two nonadjacent angles formed by two intersecting lines.

Examples ∠1 and ∠2; ∠3 and ∠4

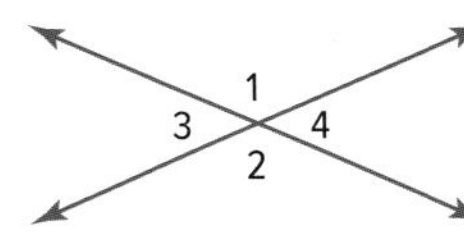

Nonexample ∠*AEB* and ∠*DEC*

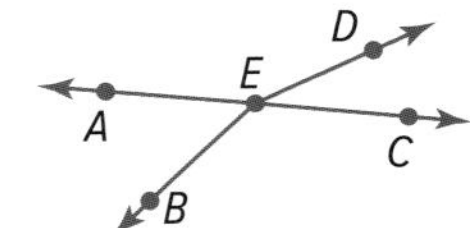

U.S. Air Force photo by Danny Meyer

Real-World Example 1 Identify Angle Pairs

CHEERLEADING Name an angle pair that satisfies each condition.

a. two acute adjacent angles

∠*HJK*, ∠*LJM*, ∠*MJN*, and ∠*NJO* are acute angles.

∠*LJM* and ∠*MJN* are acute adjacent angles, and ∠*MJN* and ∠*NJO* are acute adjacent angles.

b. two obtuse vertical angles

∠*HJN* and ∠*KJM* are obtuse vertical angles.

Guided Practice

1A. a linear pair

1B. two acute vertical angles

Go Online! 

When two lines intersect, they form four angles. Investigate relationships between pairs of these angles with a Geometer's Sketchpad® sketch in ConnectED.

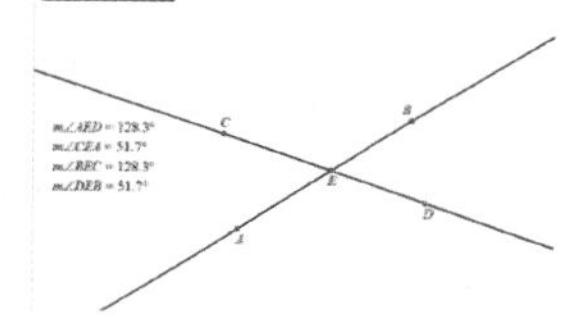

Some pairs of angles are special because of the relationship between their angle measures.

Key Concept Angle Pair Relationships

Vertical angles are congruent.

Examples ∠*ABC* ≅ ∠*DBE* and ∠*ABD* ≅ ∠*CBE*

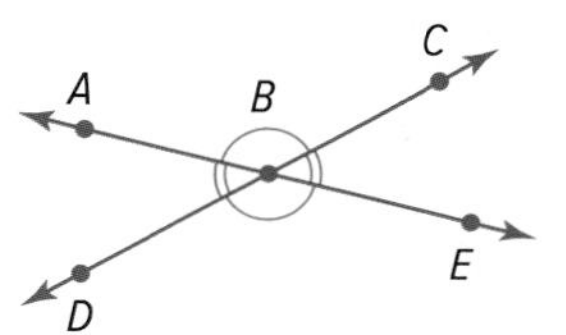

Complementary angles are two angles with measures that have a sum of 90.

Examples ∠1 and ∠2 are complementary.
∠*A* is complementary to ∠*B*.

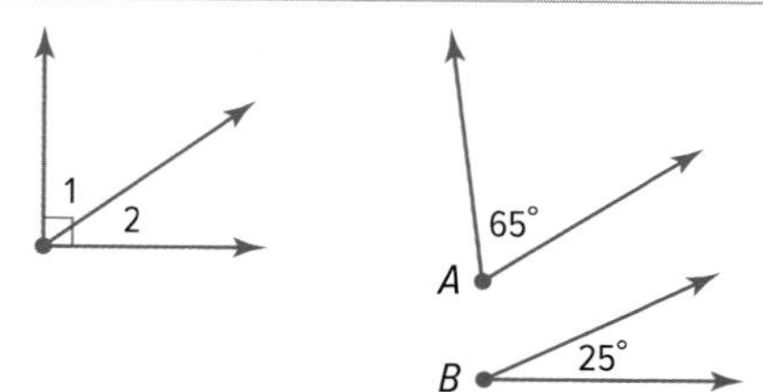

Supplementary angles are two angles with measures that have a sum of 180.

Examples ∠3 and ∠4 are supplementary.
∠*P* and ∠*Q* are supplementary.

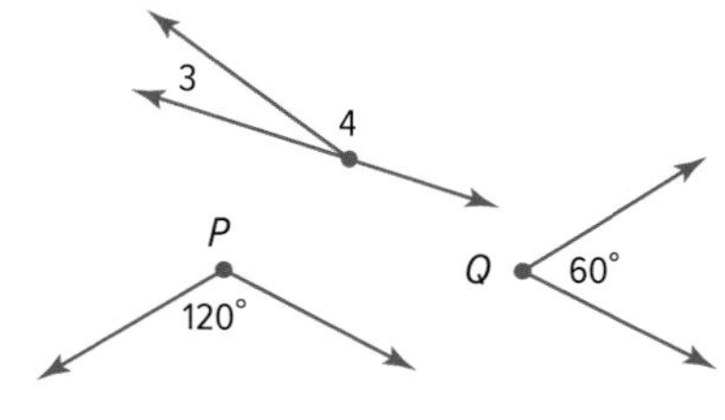

The angles in a linear pair are supplementary.

Example $m\angle 1 + m\angle 2 = 180$

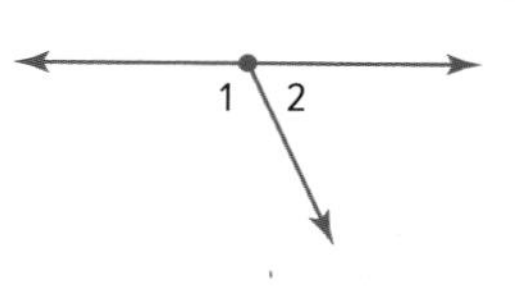

Study Tip

Linear Pair vs. Supplementary Angles While the angles in a linear pair are always supplementary, some supplementary angles do not form a linear pair.

Remember that angle measures are real numbers. So the operations for real numbers and algebra can be used with angle measures.

Example 2 Angle Measure

ALGEBRA Find the measures of two supplementary angles if the difference in the measures of the two angles is 18.

Understand You know that the sum of the measures of supplementary angles is 180. You need to find the measure of each angle.

Plan Draw two figures to represent the angles. Let the measure of one angle be x. If $m\angle A = x$, then because $\angle A$ and $\angle B$ are supplementary, $m\angle B + x = 180$ or $m\angle B = 180 - x$.

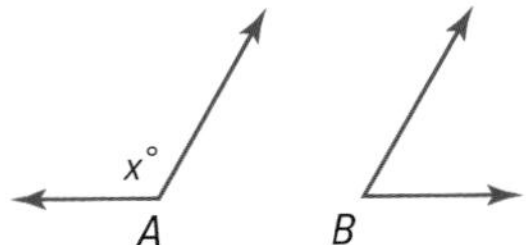

The problem states that the difference of the two angle measures is 18, or $m\angle B - m\angle A = 18$.

Solve

$m\angle B - m\angle A = 18$	Given
$(180 - x) - x = 18$	$m\angle A = x$, $m\angle B = 180 - x$
$180 - 2x = 18$	Simplify.
$-2x = -162$	Subtract 180 from each side.
$x = 81$	Divide each side by -2.

Use the value of x to find each angle measure.

$m\angle A = x$ $\qquad$ $m\angle B = 180 - x$

$= 81$ $\qquad$ $= 180 - 81$ or 99

Check Add the angle measures to verify that the angles are supplementary.

$m\angle A + m\angle B \stackrel{?}{=} 180$

$81 + 99 = 180$ ✓

The angles add up to 180 and their difference is 18.

Guided Practice

2. Find the measures of two complementary angles if the measure of the larger angle is 12 more than twice the measure of the smaller angle.

> **Problem Solving**
>
> **MP Structure** While you could use the guess-and-check strategy to find two measures with a sum of 180 and a difference of 18, writing an equation is a more efficient approach to this problem.

2 Perpendicular Lines

Lines, segments, or rays that form right angles are **perpendicular.**

Key Concept Perpendicular Lines

- Perpendicular lines intersect to form four right angles.
- Perpendicular lines intersect to form congruent adjacent angles.
- Segments and rays can be perpendicular to lines or other line segments and rays.
- The right angle symbol in the figure indicates that the lines are perpendicular.

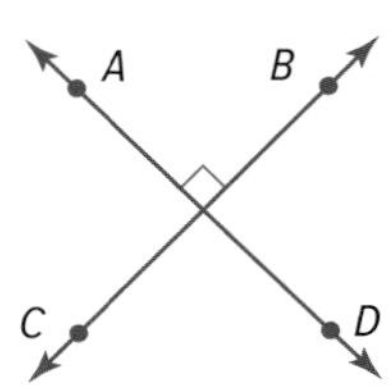

Symbol $\perp$ is read *is perpendicular to.* **Example** $\overleftrightarrow{AD} \perp \overleftrightarrow{CB}$

Example 3 Perpendicular Lines

ALGEBRA Find x and y so that $\overleftrightarrow{PR}$ and $\overleftrightarrow{SQ}$ are perpendicular.

If $\overleftrightarrow{PR} \perp \overleftrightarrow{SQ}$, then $m\angle STR = 90$ and $m\angle PTQ = 90$.

To find x, use $\angle STW$ and $\angle WTR$.

$m\angle STR = m\angle STW + m\angle WTR$	Sum of parts = whole
$90 = 2x + (5x + 6)$	Substitution
$90 = 7x + 6$	Combine like terms.
$84 = 7x$	Subtract 6 from each side.
$12 = x$	Divide each side by 7.

To find y, use $m\angle PTQ$.

$m\angle PTQ = 4y - 2$	Given
$90 = 4y - 2$	Substitution
$92 = 4y$	Add 2 to each side.
$23 = y$	Divide each side by 4.

Guided Practice

3. Suppose $m\angle D = 3x - 12$. Find x so that $\angle D$ is a right angle.

In the figure at the right, it *appears* that $\overleftrightarrow{FG} \perp \overleftrightarrow{JK}$. However, you cannot assume this is true unless other information, such as $m\angle FHJ = 90$, is given.

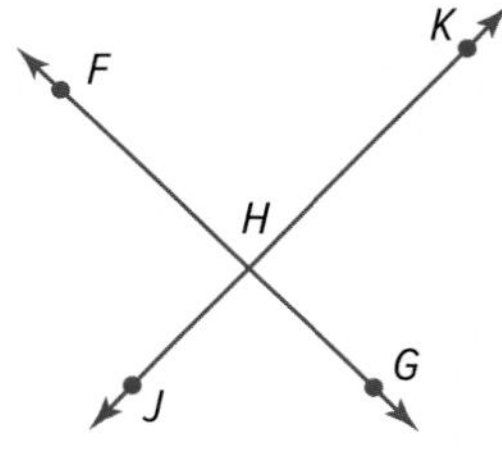

In geometry, figures are sketches used to depict a situation. They are not drawn to reflect total accuracy. There are certain relationships that you can assume to be true, but others you cannot. Study the figure and the lists below.

Key Concept Interpreting Diagrams

CAN be Assumed

- All points shown are coplanar.
- G, H, and J are collinear.
- $\overrightarrow{HM}$, $\overrightarrow{HL}$, $\overrightarrow{HK}$, and $\overleftrightarrow{GJ}$ intersect at H.
- H is between G and J.
- L is in the interior of $\angle MHK$.
- $\angle GHM$ and $\angle MHL$ are adjacent angles.
- $\angle GHL$ and $\angle LHJ$ are a linear pair.
- $\angle JHK$ and $\angle KHG$ are supplementary.

CANNOT be Assumed

- Perpendicular lines: $\overrightarrow{HM} \perp \overrightarrow{HL}$
- Congruent angles: $\angle JHK \cong \angle GHM$
  - $\angle JHK \cong \angle KHL$
  - $\angle KHL \cong \angle LHM$
- Congruent segments: $\overline{GH} \cong \overline{HJ}$
  - $\overline{HJ} \cong \overline{HK}$
  - $\overline{HK} \cong \overline{HL}$
  - $\overline{HL} \cong \overline{HG}$

The list of statements that can be assumed is not a complete list. There are more special pairs of angles than those listed.

Study Tip

Additional Information Additional information for a figure may be given using congruent angle markings, congruent segment markings, or right angle symbols.

Example 4 Interpret Figures

Determine whether each statement can be assumed from the figure. Explain.

a. $\angle DBC$ **and** $\angle ABG$ **are complementary.**

No; they are congruent, but we do not know anything about their exact measures.

b. $\angle ABD$ **and** $\angle CBD$ **are a linear pair.**

Yes; they are adjacent angles whose noncommon sides are opposite rays.

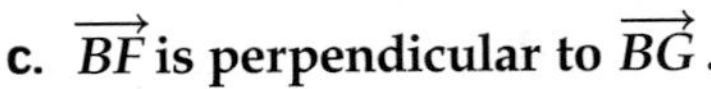
c. $\overrightarrow{BF}$ **is perpendicular to** $\overrightarrow{BG}$.

Yes; the right angle symbol in the figure indicates that $\overrightarrow{BF} \perp \overrightarrow{BG}$.

Guided Practice

4A. $\angle ABF$ and $\angle FBC$ are supplementary.

4B. $\angle ABG$ and $\angle GBD$ are adjacent angles.

Check Your Understanding

= Step-by-Step Solutions begin on page R13.

Example 1 **Name an angle pair that satisfies each condition.**

1. two acute vertical angles

2. two obtuse adjacent angles

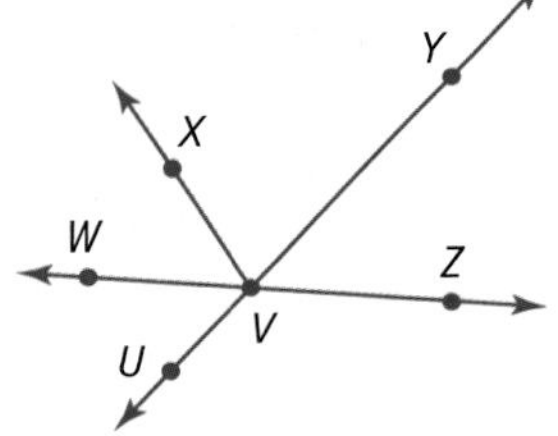

Examples 1–2 **3. CAMERAS** Cameras use lenses and light to capture images.

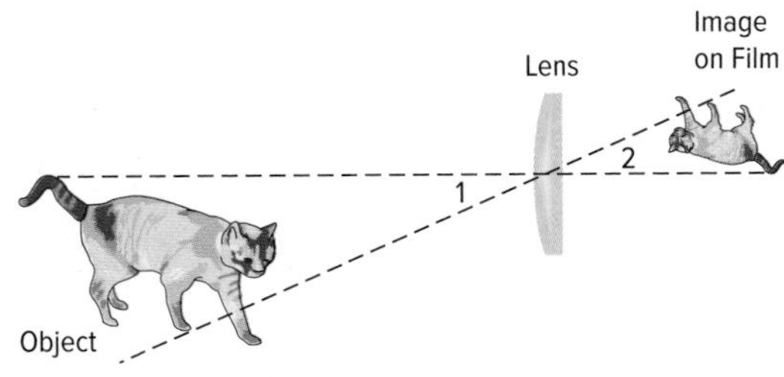

a. What type of angles are formed by the object and its image?

b. If the measure of $\angle 2$ is 15, what is the measure of $\angle 1$?

Examples 2–3 **4. ALGEBRA** The measures of two complementary angles are $7x + 17$ and $3x - 20$. Find the measures of the angles.

5. ALGEBRA Lines x and y intersect to form adjacent angles 2 and 3. If $m\angle 2 = 3a - 27$ and $m\angle 3 = 2b + 14$, find the values of a and b so that x is perpendicular to y.

Example 4 **Determine whether each statement can be assumed from the figure. Explain.**

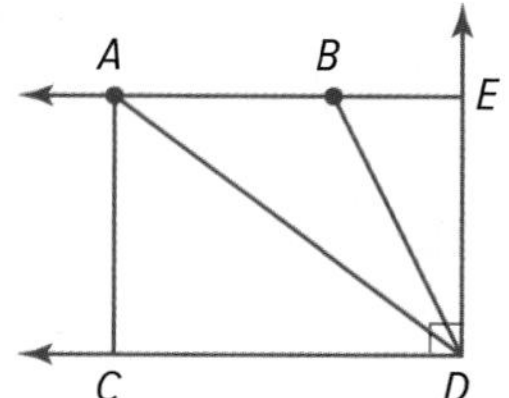

6. $\angle CAD$ and $\angle DAB$ are complementary.
7. $\angle EDB$ and $\angle BDA$ are adjacent, but they are neither complementary nor supplementary.

Practice and Problem Solving

Extra Practice is on page R1.

Examples 1–2 **Name an angle or angle pair that satisfies each condition.**

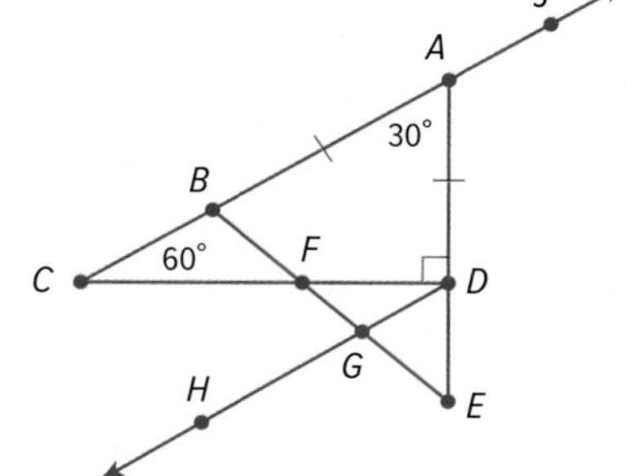

8. two adjacent angles
9. two acute vertical angles
10. two obtuse vertical angles
11. two complementary adjacent angles
12. two complementary nonadjacent angles
13. two supplementary adjacent angles
14. a linear pair whose vertex is F
15. an angle complementary to $\angle FDG$
16. an angle supplementary to $\angle CBF$
17. an angle supplementary to $\angle JAE$

18. **MP REASONING** You are using a compass to drive 23° east of north. Express your direction in another way using an acute angle and two of the four directions: north, south, east, and west. Explain your reasoning.

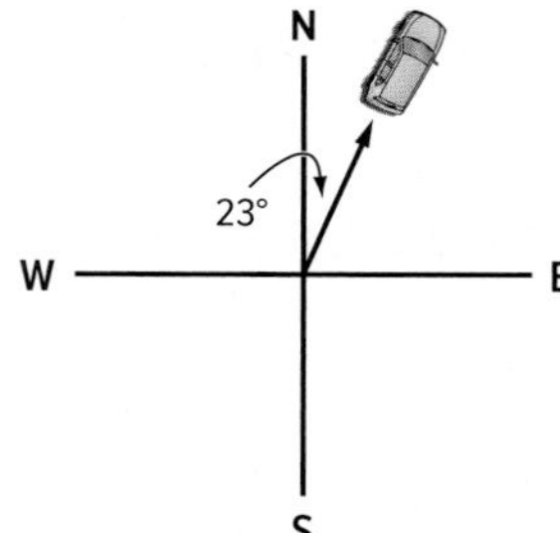

Example 2 **Find the value of each variable.**

19.

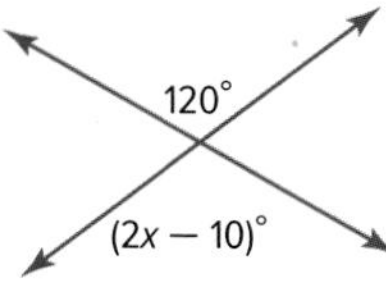

20.

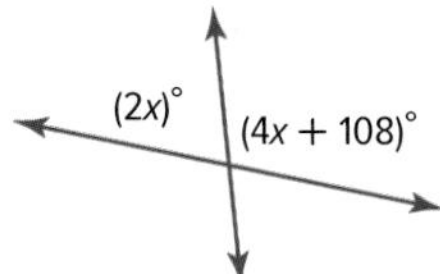

21.

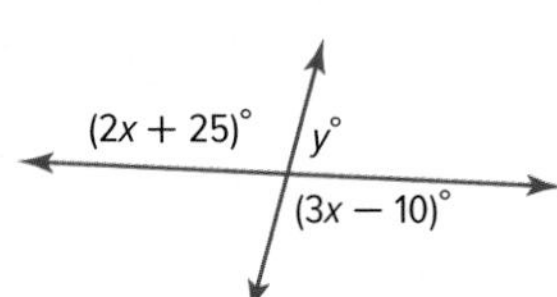

22.

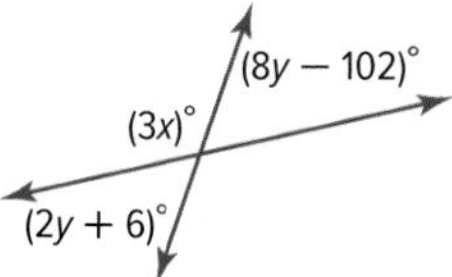

23.

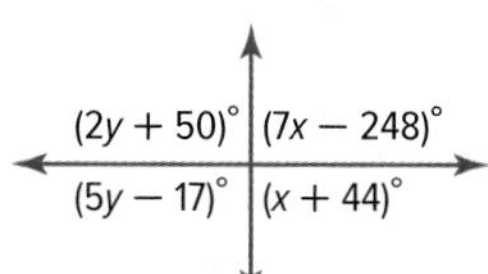

24.

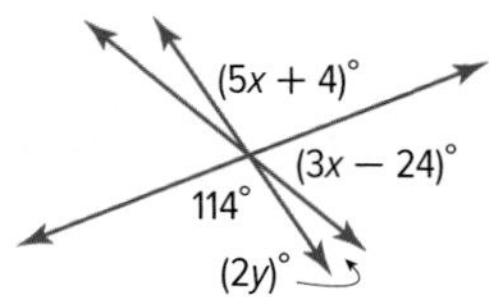

25. **ALGEBRA** $\angle E$ and $\angle F$ are supplementary. The measure of $\angle E$ is 54 more than the measure of $\angle F$. Find the measures of each angle.

26. **ALGEBRA** The measure of an angle's supplement is 76 less than the measure of the angle. Find the measure of the angle and its supplement.

27. **ALGEBRA** The measure of the supplement of an angle is 40 more than two times the measure of the complement of the angle. Find the measure of the angle.

28. **ALGEBRA** $\angle 3$ and $\angle 4$ form a linear pair. The measure of $\angle 3$ is four more than three times the measure of $\angle 4$. Find the measure of each angle.

Example 3 **ALGEBRA** Use the figure at the right.

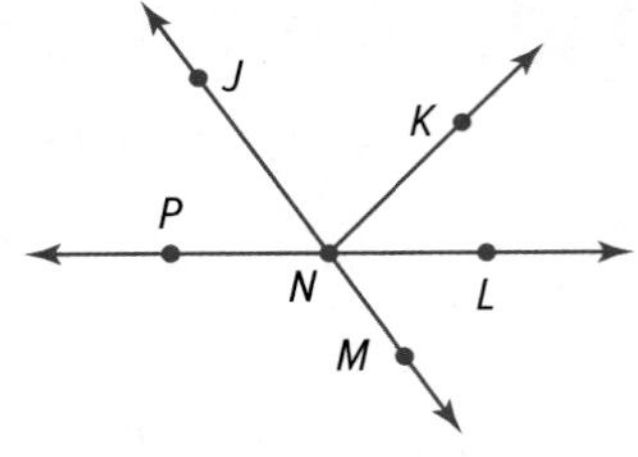

29. If $m\angle KNL = 6x - 4$ and $m\angle LNM = 4x + 24$, find the value of x so that $\angle KNM$ is a right angle.

30. If $m\angle JNP = 3x - 15$ and $m\angle JNL = 5x + 59$, find the value of x so that $\angle JNP$ and $\angle JNL$ are supplements of each other.

31. If $m\angle LNM = 8x + 12$ and $m\angle JNL = 12x - 32$, find $m\angle JNP$.

32. If $m\angle JNP = 2x + 3$, $m\angle KNL = 3x - 17$, and $m\angle KNJ = 3x + 34$, find the measure of each angle.

33. **PHYSICS** As a ray of light meets a mirror, the light is reflected. The angle at which the light strikes the mirror is the *angle of incidence*. The angle at which the light is reflected is the *angle of reflection*. The angle of incidence and the angle of reflection are congruent. In the diagram at the right, if $m\angle RMI = 106$, find the angle of reflection and $m\angle RMJ$.

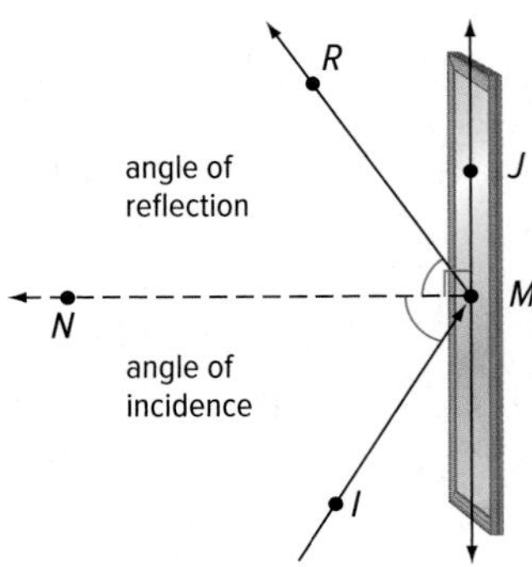

34. **ALGEBRA** Rays AB and BC are perpendicular. Point D lies in the interior of $\angle ABC$. If $m\angle ABD = 3r + 5$ and $m\angle DBC = 5r - 27$, find $m\angle ABD$ and $m\angle DBC$.

35. **ALGEBRA** $\overleftrightarrow{WX}$ and $\overleftrightarrow{YZ}$ intersect at point V. If $m\angle WVY = 4a + 58$ and $m\angle XVY = 2b - 18$, find the values of a and b so that $\overleftrightarrow{WX}$ is perpendicular to $\overleftrightarrow{YZ}$.

Example 4 **Determine whether each statement can be assumed from the figure. Explain.**

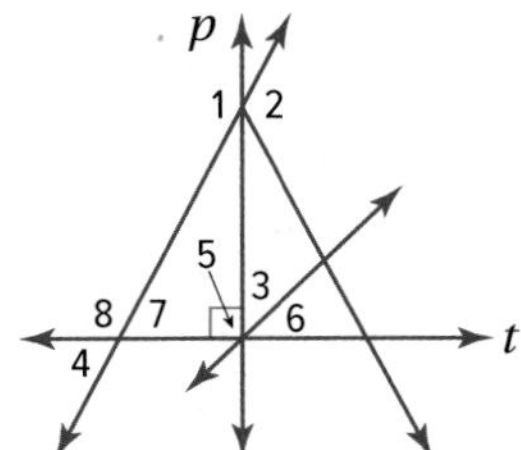

36. $\angle 4$ and $\angle 7$ are vertical angles.

37. $\angle 4$ and $\angle 8$ are supplementary.

38. $p \perp t$

39. $\angle 3 \cong \angle 6$

40. $\angle 5 \cong \angle 3 + \angle 6$

41. $\angle 5$ and $\angle 7$ form a linear pair.

42. MP **CRITIQUE ARGUMENTS** In the diagram of the pruning shears shown, $m\angle 1 = m\angle 3$. What conclusion can you reach about the relationship between $\angle 4$ and $\angle 2$? Explain.

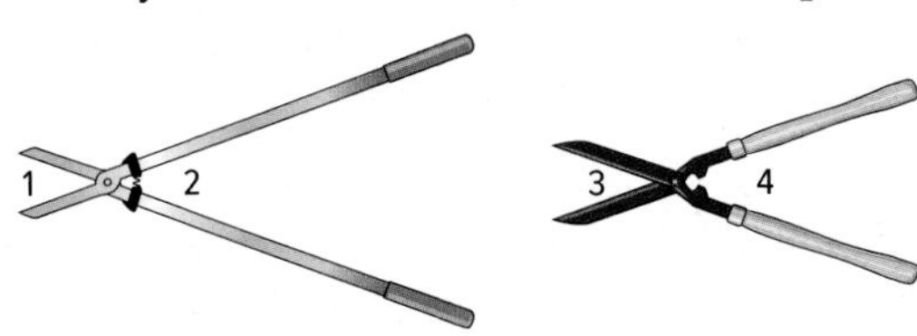

FLIGHT The wing of the aircraft shown can pivot up to 60° in either direction from the perpendicular position.

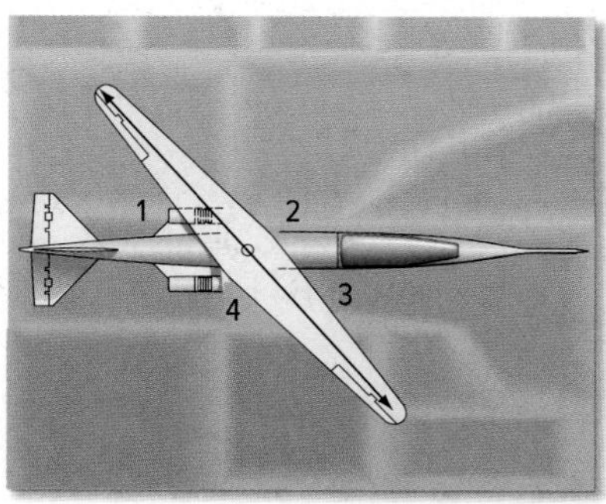

43. Identify a pair of vertical angles.

44. Identify two pairs of supplementary angles.

45. If $m\angle 1 = 110$, what is $m\angle 3$? $m\angle 4$?

46. What is the minimum possible value for $m\angle 2$? the maximum?

47. Is there a wing position in which none of the angles are obtuse? Explain.

48. **MULTIPLE REPRESENTATIONS** In this problem, you will explore the relationship between the sum of the interior angles of a triangle and the angles vertical to them.

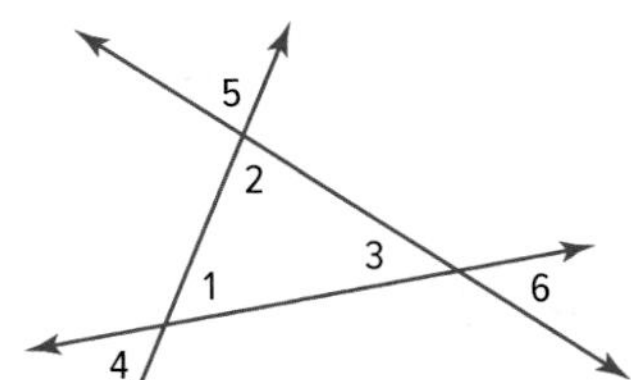

   a. **Geometric** Draw three sets of three intersecting lines and label each as shown.

   b. **Tabular** For each set of lines, measure and record $m\angle 1$, $m\angle 2$, and $m\angle 3$ in a table. Record $m\angle 1 + m\angle 2 + m\angle 3$ in a separate column.

   c. **Verbal** Explain how you can find $m\angle 4$, $m\angle 5$, and $m\angle 6$ when you know $m\angle 1$, $m\angle 2$, and $m\angle 3$.

   d. **Algebraic** Write an equation that relates $m\angle 1 + m\angle 2 + m\angle 3$ to $m\angle 4 + m\angle 5 + m\angle 6$. Then use substitution to write an equation that relates $m\angle 4 + m\angle 5 + m\angle 6$ to an integer.

H.O.T. Problems Use Higher-Order Thinking Skills

49. MP **REASONING** Are there angles that do not have a complement? Explain.

50. **OPEN ENDED** Draw a pair of intersecting lines that forms a pair of complementary angles. Explain your reasoning.

51. **CHALLENGE** If a line, line segment, or ray is perpendicular to a plane, it is perpendicular to every line, line segment, or ray in the plane that intersects it.

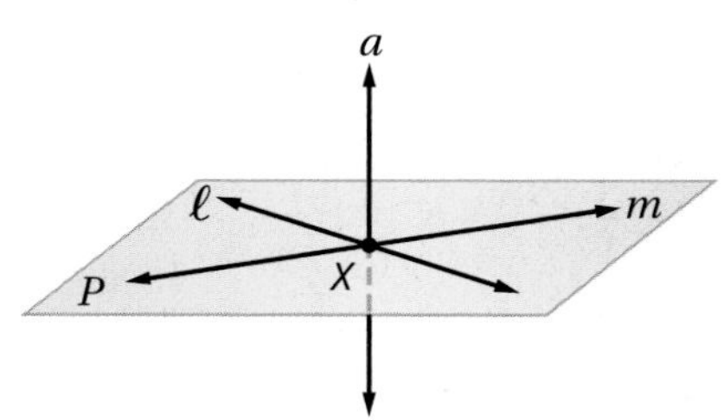

   a. If a line is perpendicular to each of two intersecting lines at their point of intersection, then the line is perpendicular to the plane determined by them. If line a is perpendicular to line ℓ and line a at point X, what must also be true?

   b. If a line is perpendicular to a plane, then any line perpendicular to the given line at the point of intersection with the given plane is in the given plane. If line a is perpendicular to plane P and line m at point X, what must also be true?

   c. If a line is perpendicular to a plane, then every plane containing the line is perpendicular to the given plane. If line a is perpendicular to plane P, what must also be true?

52. **WRITING IN MATH** Describe three different ways you can determine that an angle is a right angle.

53. **MULTI-STEP** Tyler was given a figure consisting of the perpendicular lines $\overleftrightarrow{LM}$ and $\overleftrightarrow{NP}$. Then he used a compass and straightedge to make the construction shown here. MP 6

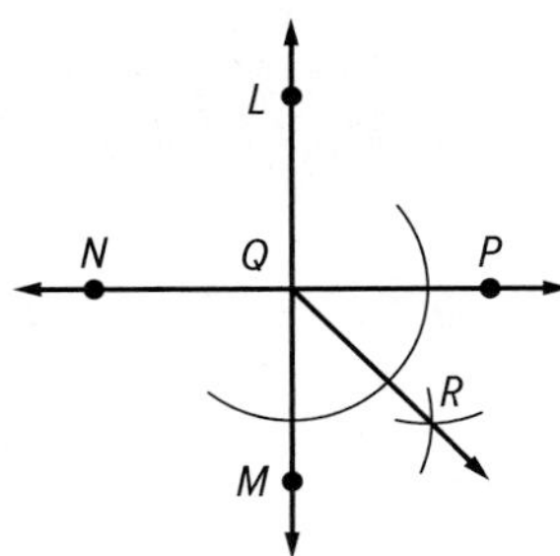

a. Which of the following is $m\angle LQR$?

- ○ A 45
- ○ B 90
- ○ C 135
- ○ D 145

b. How could you use the figure to construct an angle with measure 67.5°?

c. How could you use the figure to construct an angle with measure 22.5°?

54. In the figure, $\angle JKL$ and $\angle LKM$ are complementary. Jaycee used the figure to construct $\angle G$, as shown below. MP 5

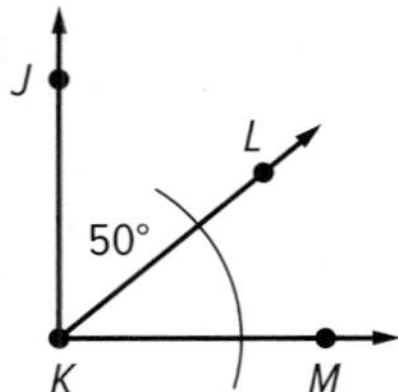

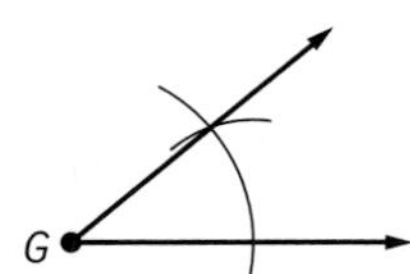

What is $m\angle G$?

- ○ A 20
- ○ B 40
- ○ C 50
- ○ D 130

55. A student drew two lines that intersect at an angle of 78°, as shown. Then the student used a compass and straightedge as shown to create $\overrightarrow{EY}$. MP 7

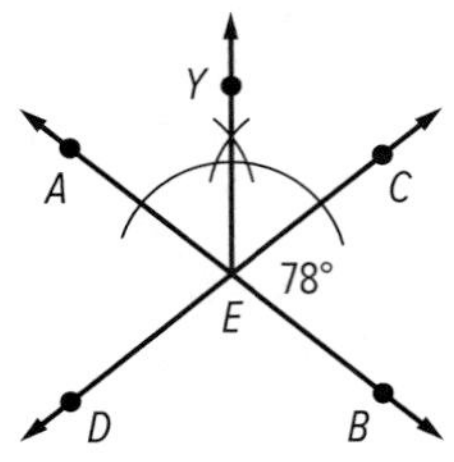

What is $m\angle AEY$?

- ○ A 39
- ○ B 51
- ○ C 78
- ○ D 102
- ○ E cannot be determined

56. The measures of two complementary angles are in a 5 : 7 ratio. Find the measure of the smaller angle. MP 1

57. There is half of an apple pie left. You want to eat twice as much as your little sister, but you also want to save a piece for your mom. You cut your mom a 30° piece. What is the measure of your piece of pie in degrees? MP 4

58. The measures of two supplementary angles are in the ratio of 11 : 4. What is the measure of the larger angle? MP 1

- ○ A 12
- ○ B 48
- ○ C 132
- ○ D 168

EXTEND 1-5

Geometry Lab

Constructing Perpendiculars

You can use a compass and a straightedge to construct a line perpendicular to a given line through a point on the line, or through a point *not* on the line.

Mathematical Practices

 **5** Use appropriate tools strategically.

Activity Constructing Perpendiculars

a. Construct a line perpendicular to line ℓ and passing through point *P* on ℓ.

Step 1

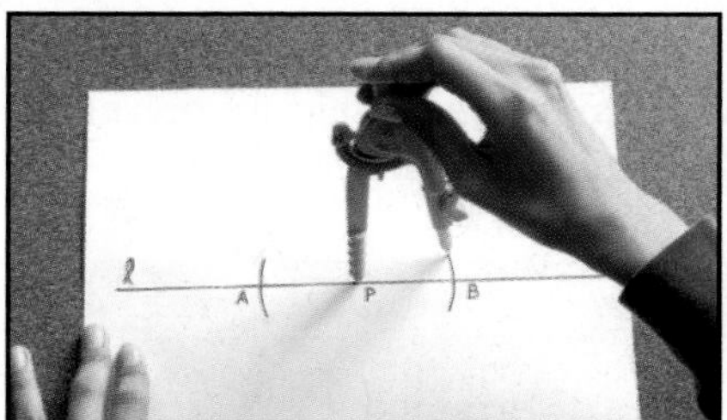

Place the compass at *P*. Draw arcs to the right and left of *P* that intersect line ℓ using the same compass setting. Label the points of intersection *A* and *B*.

Step 2

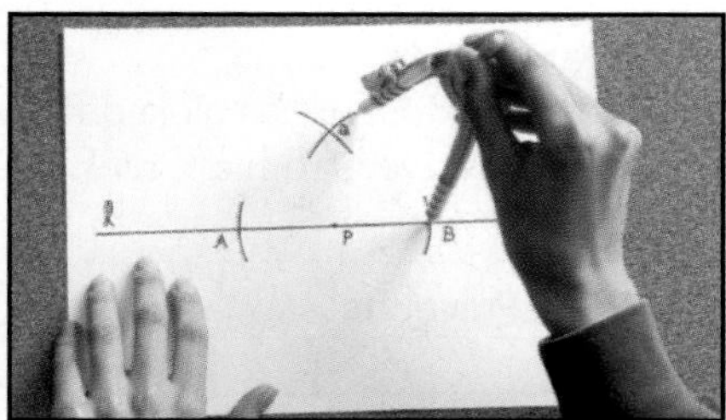

With the compass at *A*, draw an arc above line ℓ using a setting greater than *AP*. Using the same compass setting, draw an arc from *B* that intersects the previous arc at a point *Q*.

Step 3

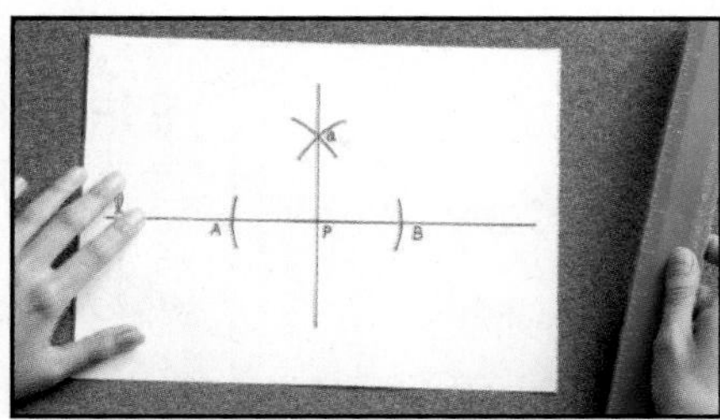

Use a straightedge to draw $\overleftrightarrow{QP}$.

b. Construct a line perpendicular to line *k* and passing through point *P not* on *k*.

Step 1

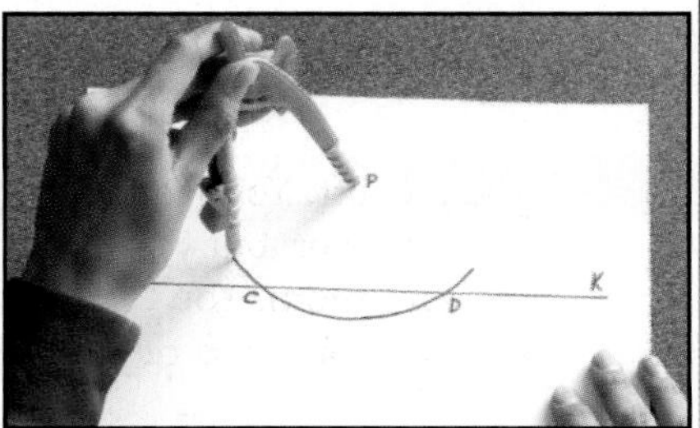

Place the compass at *P*. Draw an arc that intersects line *k* in two different places. Label the points of intersection *C* and *D*.

Step 2

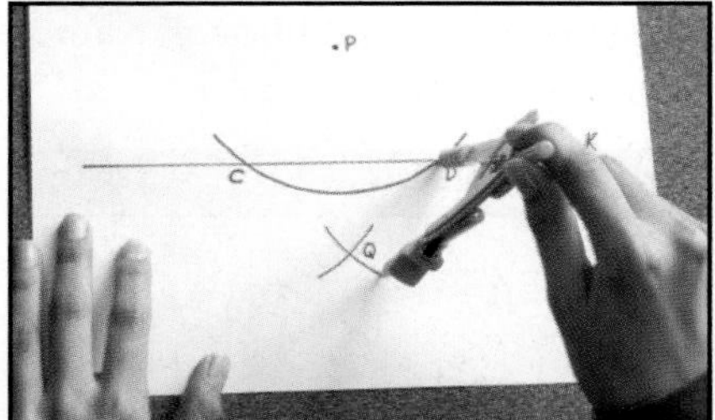

With the compass at *C*, draw an arc below line *k* using a setting greater than $\frac{1}{2}CD$. Using the same compass setting, draw an arc from *D* that intersects the previous arc at a point *Q*.

Step 3

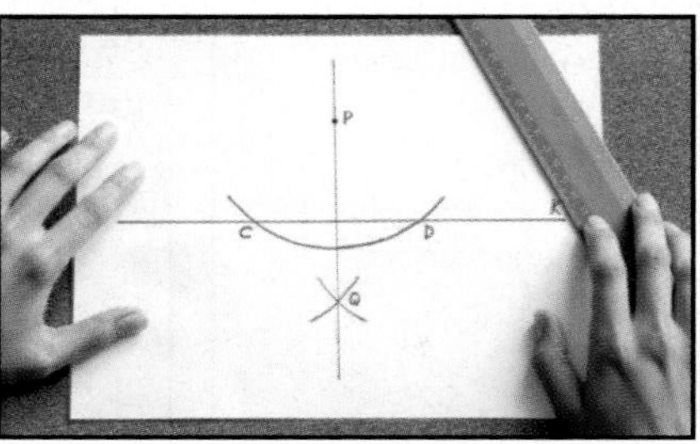

Use a straightedge to draw $\overleftrightarrow{PQ}$.

Model and Analyze the Results

Work cooperatively.

1. Draw a line and construct a line perpendicular to it through a point on the line.
2. Draw a line and construct a line perpendicular to it through a point not on the line.
3. How is the second construction similar to the first one?
4. Compare this construction to the construction of a segment bisector from Lesson 1-3. What can you conclude about the segment bisector you constructed? Explain.

LESSON 6

Two-Dimensional Figures

∴ Then	∴ Now	∴ Why?
• You measured one-dimensional figures.	• **1** Identify and name polygons. **2** Find perimeter, circumference, and area of two-dimensional figures.	• Mosaics are patterns or pictures created using small bits of colored glass or stone. They are usually set into a wall or floor and often make use of polygons.

New Vocabulary

polygon
vertex of a polygon
concave
convex
n-gon
equilateral polygon
equiangular polygon
regular polygon
perimeter
circumference
area

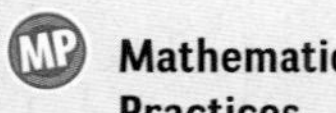

Mathematical Practices

2 Reason abstractly and quantitatively.
6 Attend to precision.
8 Look for and express regularity in repeated reasoning.

1 Identify Polygons

Most of the closed figures shown in the mosaic are polygons. The term *polygon* is derived from a Greek word meaning *many angles*.

Key Concept Polygons

A **polygon** is a closed figure formed by a finite number of coplanar segments called ***sides*** such that

- the sides that have a common endpoint are noncollinear, and
- each side intersects exactly two other sides, but only at their endpoints.

The vertex of each angle is a **vertex of the polygon**. A polygon is named by the letters of its vertices, written in order of consecutive vertices.

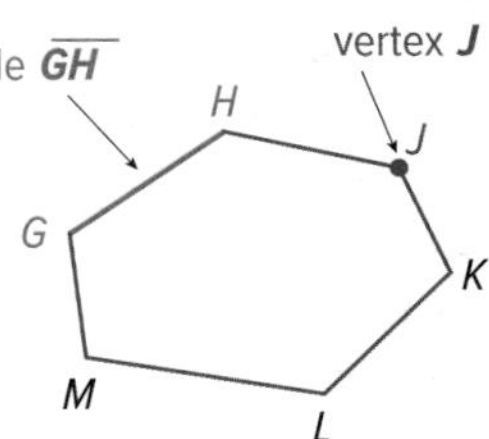

polygon ***GHJKLM***

The table below shows some additional examples of polygons and some examples of figures that are not polygons.

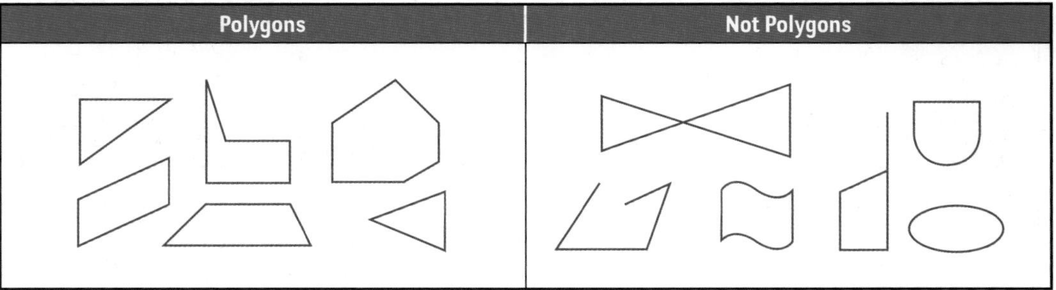

Polygons can be **concave** or **convex**. Suppose the line containing each side is drawn. If any of the lines contain any point in the interior of the polygon, then it is concave. Otherwise it is convex.

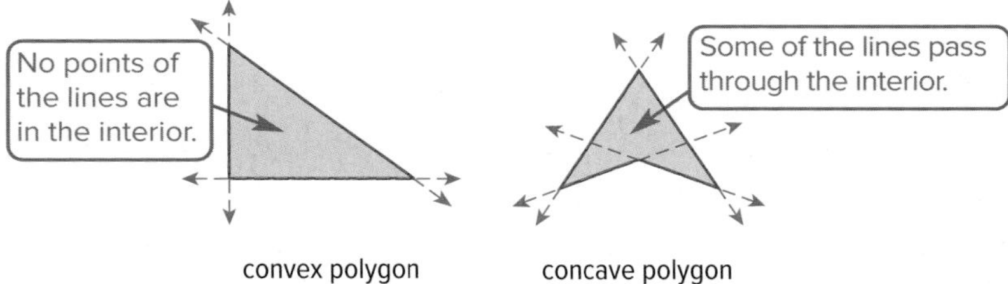

convex polygon   concave polygon

Study Tip

Naming Polygons The Greek prefixes used to name polygons are also used to denote number. For example, a bicycle has two wheels, and a tripod has three legs.

In general, a polygon is classified by its number of sides. The table lists some common names for various categories of polygon. A polygon with n sides is an ***n*-gon**. For example, a polygon with 15 sides is a 15-gon.

Number of Sides	Polygon
3	triangle
4	quadrilateral
5	pentagon
6	hexagon
7	heptagon
8	octagon
9	nonagon
10	decagon
11	hendecagon
12	dodecagon
n	*n*-gon

An **equilateral polygon** is a polygon in which all sides are congruent. An **equiangular polygon** is a polygon in which all angles are congruent.

A convex polygon that is both equilateral and equiangular is called a **regular polygon**. An *irregular polygon* is a polygon that is *not* regular.

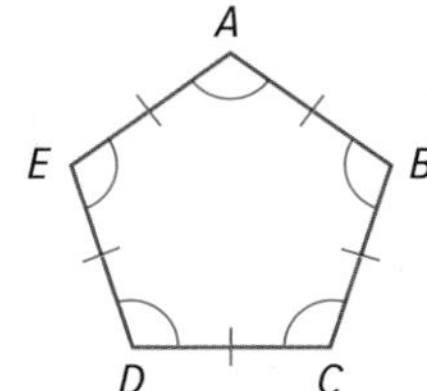

regular pentagon *ABCDE*

Example 1 Name and Classify Polygons

Name each polygon by its number of sides. Then classify it as *convex* or *concave* and *regular* or *irregular*.

Reading Math

Simple Closed Curves Polygons and circles are examples of ***simple closed curves***. Such a curve begins and ends at the same point without crossing itself. The figures below are not simple closed curves.

a.

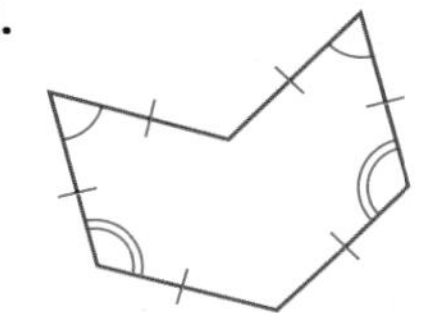

The polygon has six sides, so it is a hexagon.

Two of the lines containing the sides of the polygon will pass through the interior of the hexagon, so it is concave.

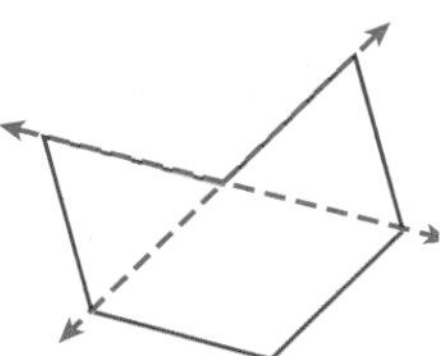

Only convex polygons can be regular, so this is an irregular hexagon.

b.

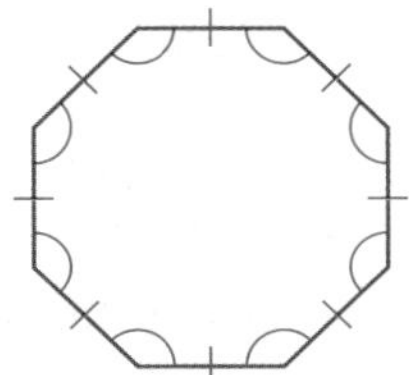

There are eight sides, so this is an octagon.

No line containing any of the sides will pass through the interior of the octagon, so it is convex.

All of the sides are congruent, so it is equilateral. All of the angles are congruent, so it is equiangular.

Because the polygon is convex, equilateral, and equiangular, it is regular. So this is a regular octagon.

Guided Practice

1A.

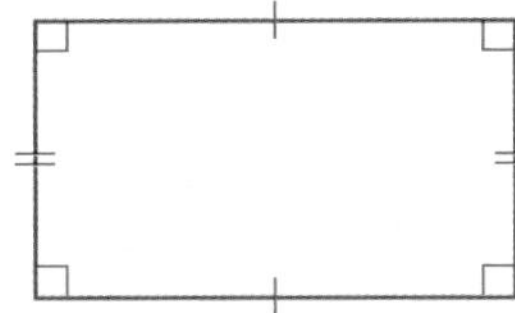

1B.

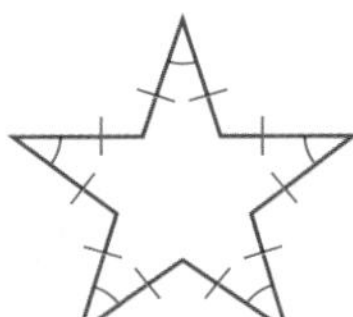

1C.

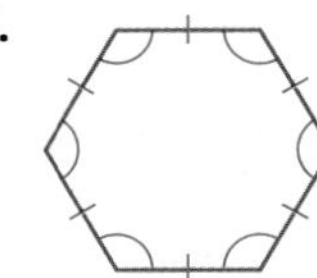

2 Perimeter, Circumference, and Area

The **perimeter** of a polygon is the sum of the lengths of the sides of the polygon. Some shapes have special formulas for perimeter, but all are derived from the basic definition of perimeter. You will derive these formulas in Chapter 11. The **circumference** of a circle is the distance around the circle.

The **area** of a figure is the number of square units needed to cover a surface. Review the formulas for the perimeter and area of three common polygons and circle given below.

Key Concept Perimeter, Circumference, and Area

Triangle	Square	Rectangle	Circle
$P = b + c + d$	$P = s + s + s + s$ $= 4s$	$P = \ell + w + \ell + w$ $= 2\ell + 2w$	$C = 2\pi r$ or $C = \pi d$
$A = \frac{1}{2}bh$	$A = s^2$	$A = \ell w$	$A = \pi r^2$

P = perimeter of polygon; A = area of figure; C = circumference

b = base, h = height; ℓ = length, w = width; r = radius, d = diameter

Reading Math

Pi The symbol π is read *pi*. This is not a variable but an irrational number. The most accurate way to perform a calculation with π is to use a calculator. If no calculator is available, 3.14 is a good estimate for π.

Example 2 Find Perimeter and Area

Find the perimeter or circumference and area of each figure.

a.

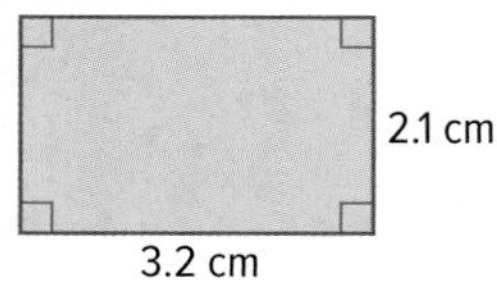

$P = 2\ell + 2w$ Perimeter of rectangle

$= 2(3.2) + 2(2.1)$ $\ell = 3.2, w = 2.1$

$= 10.6$ Simplify.

The perimeter is 10.6 centimeters.

$A = \ell w$ Area of rectangle

$= (3.2)(2.1)$ $\ell = 3.2, w = 2.1$

$= 6.72$ Simplify.

The area is about 6.7 square centimeters.

b.

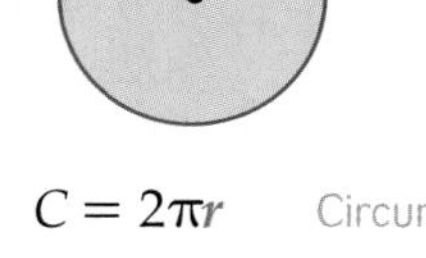

$C = 2\pi r$ Circumference

$= 2\pi(3)$ $r = 3$

≈ 18.85 Use a calculator.

The circumference is about 18.9 inches.

$A = \pi r^2$ Area of circle

$= \pi(3)^2$ $r = 3$

≈ 28.3 Use a calculator.

The area is about 28.3 square inches.

Study Tip

Perimeter vs. Area Since calculating the area of a figure involves multiplying two dimensions (unit × unit), *square units* are used. There is only one dimension used when finding the perimeter (the distance around), thus, it is given simply in *units*.

Guided Practice

2A.

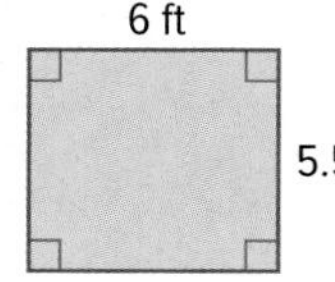

2B.

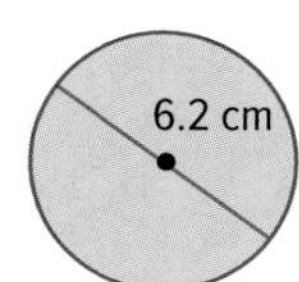

2C.

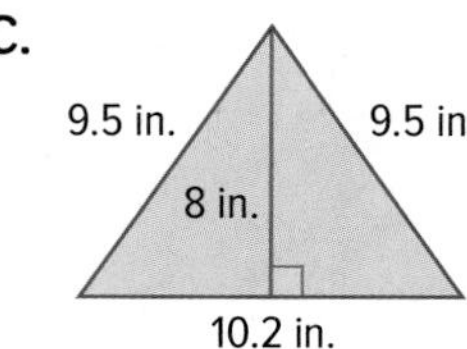

Example 3 Largest Area

Yolanda has 26 centimeters of cording to frame a photograph. Which of these shapes would use most or all of the cording and enclose the largest area?

A right triangle with each leg about 7 centimeters long

B circle with a radius of about 4 centimeters

C rectangle with a length of 8 centimeters and a width of 4.5 centimeters

D square with a side length of 6 centimeters

> **Study Tip**
> **Mental Math** When you are asked to compare measures for varying figures, it can be helpful to use mental math. Estimate the perimeter or area of each figure, and then check your calculations.

Read the Item

You are asked to compare the area and perimeter of four different shapes.

Solve the Item

Find the perimeter and area of each shape.

Right Triangle

Use the Pythagorean Theorem to find the length of the hypotenuse.

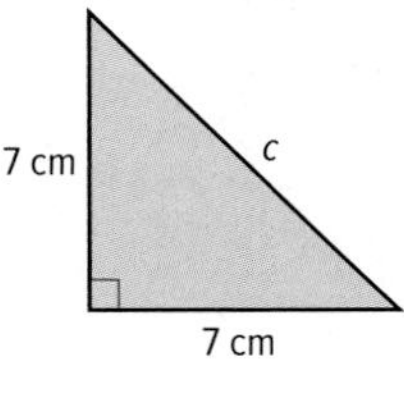

$c^2 = a^2 + b^2$ Pythagorean Theorem

$c^2 = 7^2 + 7^2$ or 98 $a = 7, b = 7$

$c = \sqrt{98}$ or about 9.9 Simplify.

$P = a + b + c$ Perimeter of a triangle

$\approx 7 + 7 + 9.9$ or about 23.9 cm Substitution

$A = \frac{1}{2}bh$ Area of a triangle

$= \frac{1}{2}(7)(7)$ or 24.5 cm^2 Substitution

> **Study Tip**
> **Irrational Measures** Notice that the triangle perimeter given in Example 3 is only an approximation. Because the length of the hypotenuse is an irrational number, the actual perimeter of the triangle is the irrational measure $(14 + \sqrt{98})$ centimeters.

Circle	Rectangle	Square
$C = 2\pi r$	$P = 2\ell + 2w$	$P = 4s$
$= 2\pi(4)$	$= 2(8) + 2(4.5)$	$= 4(6)$
≈ 25.1 cm	$= 25$ cm	$= 24$ cm
$A = \pi r^2$	$A = \ell w$	$A = s^2$
$= \pi(4)^2$	$= (8)(4.5)$	$= 6^2$
≈ 50.3 cm^2	$= 36$ cm^2	$= 36$ cm^2

The shape that uses the most cording and encloses the largest area is the circle. The answer is B.

Guided Practice

3. Dasan has 32 feet of fencing to fence in a play area for his dog. Which shape of play area uses most or all of the fencing and encloses the largest area?

F circle with radius of about 5 feet

G rectangle with length 5 feet and width 10 feet

H right triangle with legs of length 10 feet each

J square with side length 8 feet

Go Online! 

Work with a partner to complete the **Self-Check Quiz** to help your understanding. Take turns describing how to solve each problem about two-dimensional figures. Ask for clarification as you need it.

You can use the Distance Formula to find the perimeter of a polygon graphed on a coordinate plane.

Example 4 Perimeter and Area on the Coordinate Plane

COORDINATE GEOMETRY Find the perimeter and area of ▱*ABCD* with vertices *A*(−1, −1), *B*(2, 2), *C*(5, −1), and *D*(2, −4).

y
B(2, 2)
O
x
A(−1, −1)
C(5, −1)
D(2, −4)

Step 1 Find the perimeter of *ABCD*.

Graph ▱*ABCD*.

To find the perimeter of ▱*ABCD*, use the Distance Formula to find the lengths of each side.

$\overline{AB}$ has endpoints at *A*(−1, −1) and *B*(2, 2).

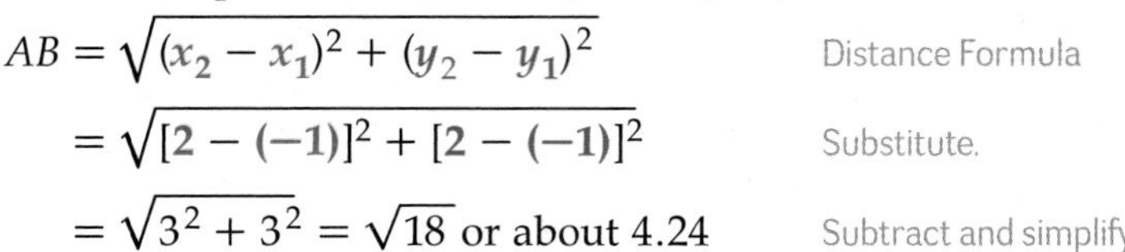

$AB = \sqrt{(x_2 - x_1)^2 + (y_2 - y_1)^2}$ Distance Formula

$= \sqrt{[2 - (-1)]^2 + [2 - (-1)]^2}$ Substitute.

$= \sqrt{3^2 + 3^2} = \sqrt{18}$ or about 4.24 Subtract and simplify.

$\overline{AD}$ has endpoints at *A*(−1, −1) and *D*(2, −4).

$AD = \sqrt{(x_2 - x_1)^2 + (y_2 - y_1)^2}$ Distance Formula

$= \sqrt{[2 - (-1)]^2 + [-4 - (-1)]^2}$ Substitute.

$= \sqrt{3^2 + (-3)^2} = \sqrt{18}$ or about 4.24 Subtract and simplify.

$\overline{BC}$ has endpoints at *B*(2, 2) and *C*(5, −1). $\overline{CD}$ has endpoints at *C*(5, −1) and *D*(2, −4).

$BC = \sqrt{(x_2 - x_1)^2 + (y_2 - y_1)^2}$ Distance Formula

$= \sqrt{(5 - 2)^2 + (-1 - 2)^2}$ Substitute.

$= \sqrt{3^2 + (-3)^2}$ Subtract.

$= \sqrt{18}$ or about 4.24 Simplify.

$CD = \sqrt{(x_2 - x_1)^2 + (y_2 - y_1)^2}$

$= \sqrt{(2 - 5)2 + [-4 - (-1)]^2}$

$= \sqrt{(-3)^2 + (-3)^2}$

$= \sqrt{18}$ or about 4.24

The perimeter of ▱*ABCD* is $4\sqrt{18}$ or about 17 units.

Step 2 Find the area of ▱*ABCD*.

Since all four angles are right angles and all four sides are congruent, ▱*ABCD* is a square. The area of a square is given by $A = s^2$ where *s* is the side length.

$A = s^2$ Area of a square

$= (\sqrt{18})^2$ or 18 Substitute and simplify.

The area of ▱*ABCD* is 18 square units.

Study Tip

MP Precision Remember to use linear units with perimeter and square units with area.

Guided Practice

4. Find the perimeter and area of △*ABC* with vertices *A*(−1, 4), *B*(−1, −1), and *C*(6, −1).

Check Your Understanding

(circled number) = Step-by-Step Solutions begin on page R13.

Example 1 **Name each polygon by its number of sides. Then classify it as *convex* or *concave* and *regular* or *irregular*.**

1.

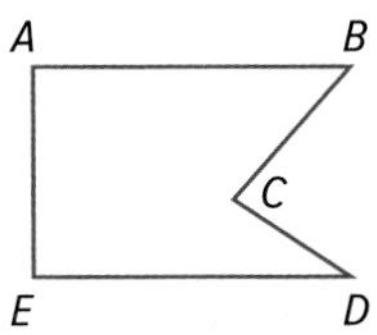

2.

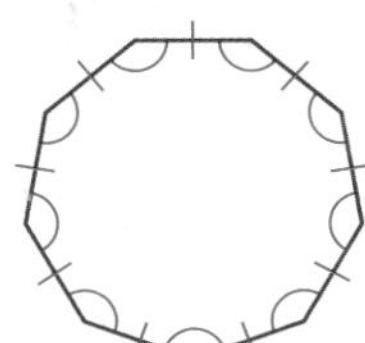

SIGNS Identify the shape of each traffic sign and classify it as *regular* or *irregular*.

3. stop

4. caution or warning

5. slow-moving vehicle

Example 2 **Find the perimeter or circumference and area of each figure. Round to the nearest tenth.**

6.

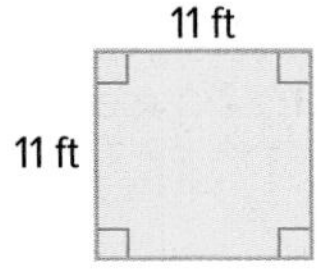

7.

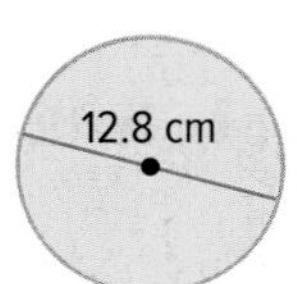

8.

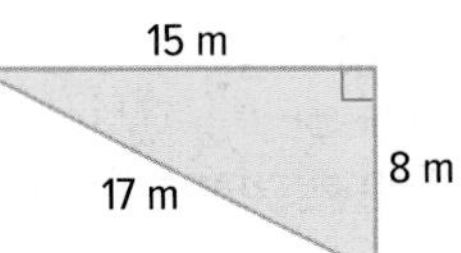

Example 3 **9. MULTIPLE CHOICE** Vanesa is making a banner for the game. She has 20 square feet of fabric. What shape will use most or all of the fabric?

A a square with a side length of 4 feet

B a rectangle with a length of 4 feet and a width of 3.5 feet

C a circle with a radius of about 2.5 feet

D a right triangle with legs of about 5 feet each

Example 4 **10.** MP **REASONING** Find the perimeter and area of $\triangle ABC$ with vertices $A(-1, 2)$, $B(3, 6)$, and $C(3, -2)$.

Practice and Problem Solving

Extra Practice is on page R1.

Example 1 **Name each polygon by its number of sides. Then classify it as convex or concave and regular or irregular.**

11.

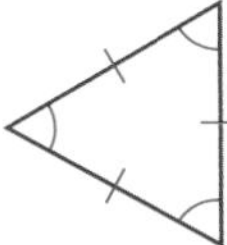

12.

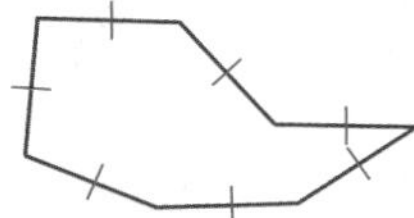

13

14.

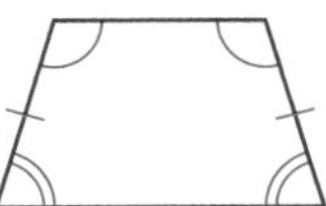

15.

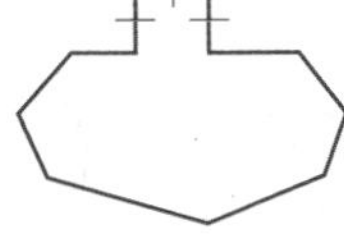

16.

Example 2–3 **Find the perimeter or circumference and area of each figure. Round to the nearest tenth.**

17.

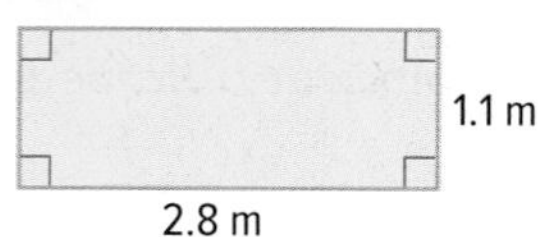

18.

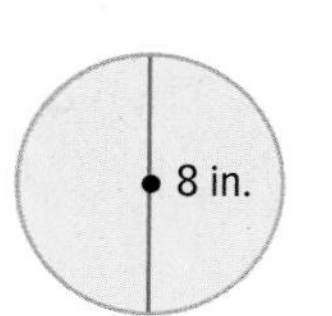

19.

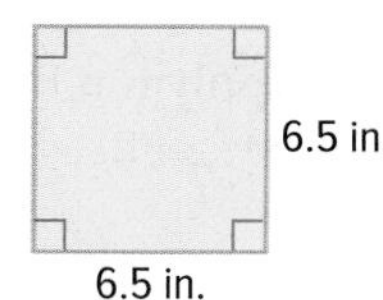

20.

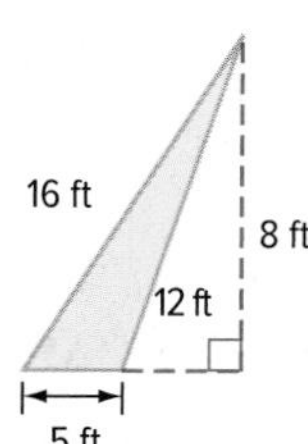

21.

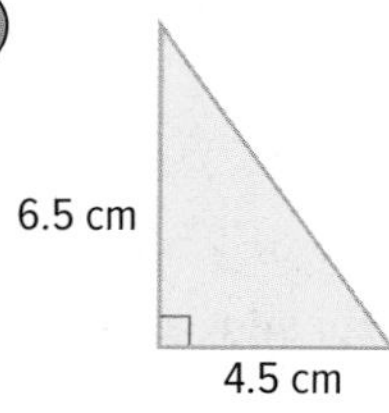

22.

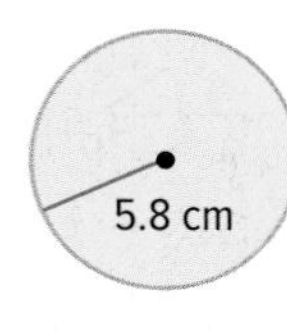

23. SEWING Elaine purchased 4.5 yards of lace trim from her local fabric shop. She has a circular tablecloth that is 54 inches in diameter and a square tablecloth that is 40 inches on each side. Can she trim either tablecloth?

24. DECORATING Ms. Bell decorates her gazebo with colored lights. The base of the gazebo is a circle with a diameter of 12 feet. Using the same length of lights, she is going to decorate the sides of her garden shed which has a square base. What is the maximum side length of the shed?

Example 4 MP **Reasoning** **Graph each figure with the given vertices and identify the figure. Then find the perimeter and area of the figure.**

25. $D(-2, -2)$, $E(-2, 3)$, $F(2, -1)$

26. $J(-3, -3)$, $K(3, 2)$, $L(3, -3)$

27. $P(-1, 1)$, $Q(3, 4)$, $R(6, 0)$, $S(2, -3)$

28. $T(-2, 3)$, $U(1, 6)$, $V(5, 2)$, $W(2, -1)$

29. CHANGING DIMENSIONS Use the rectangle at the right.

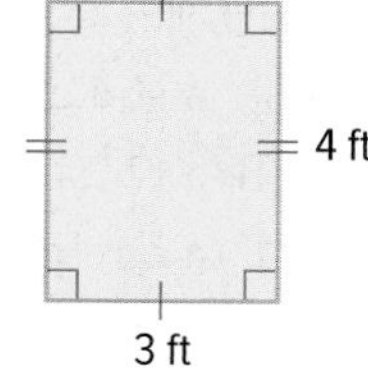

a. Find the perimeter of the rectangle.

b. Find the area of the rectangle.

c. Suppose the 4-foot length of the rectangle is doubled. What effect would this have on the perimeter? the area? Justify your answer.

d. Suppose the length and width of the rectangle are doubled. What effect does this have on the perimeter? the area? Justify your answer.

30. CHANGING DIMENSIONS Use the triangle at the right.

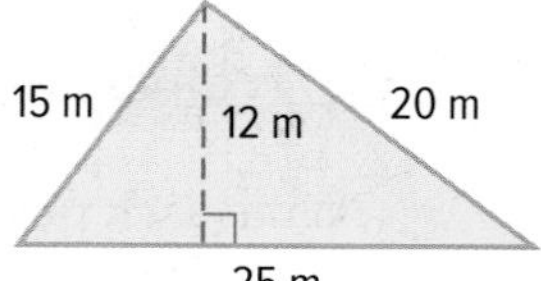

a. Find the perimeter of the triangle.

b. Find the area of the triangle.

c. Suppose the side lengths and height of the triangle were doubled. What effect would this have on the perimeter? the area? Justify your answer.

d. Suppose the side lengths and height of the triangle were divided by three. What effect would this have on the perimeter? the area? Justify your answer.

31. ALGEBRA A rectangle of area 360 square yards is 10 times as long as it is wide. Find its length and width.

32. ALGEBRA A rectangle of area 350 square feet is 14 times as wide as it is long. Find its length and width.

33. DISC GOLF The diameter of the most popular brand of flying disc used in disc golf measures between 8 and 10 inches. Find the range of possible circumferences and areas for these flying discs to the nearest tenth.

ALGEBRA Find the perimeter or circumference for each figure described.

34. The area of a square is 36 square units.

35. The length of a rectangle is half the width. The area is 25 square meters.

36. The area of a circle is 25π square units.

37. The area of a circle is 32π square units.

38. A rectangle's length is 3 times its width. The area is 27 square inches.

39. A rectangle's length is twice its width. The area is 48 square inches.

MP PRECISION Find the perimeter in inches and area in square inches of each figure. Round to the nearest hundredth, if necessary.

40.

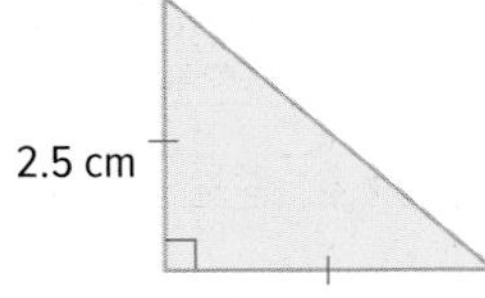

2.5 cm

41.

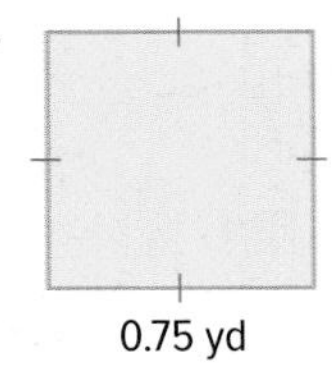

0.75 yd

42.

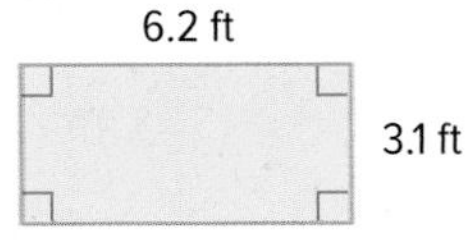

6.2 ft

3.1 ft

43. MULTIPLE REPRESENTATIONS Collect and measure the diameter and circumference of ten round objects using a millimeter measuring tape.

a. Tabular Record the measures in a table as shown.

b. Algebraic Compute the value of $\frac{C}{d}$ to the nearest hundredth for each object and record the result.

c. Graphical Make a scatter plot of the data with d values on the horizontal axis and C values on the vertical axis.

d. Verbal Find an equation for a line of best fit for the data. What does this equation represent? What does the slope of the line represent?

Object	d	C	$\frac{C}{d}$
1			
2			
3			
⋮			
10			

H.O.T. Problems Use Higher-Order Thinking Skills

44. CHALLENGE A triangle has a base of 4 inches and a height of 6 inches. Describe the change in area for each of the following. **a.** The base is doubled. **b.** The height is doubled. **c.** The base is doubled and the height is tripled.

45. CHALLENGE The vertices of a rectangle with side lengths of 10 and 24 units are on a circle of radius 13 units. Find the area between the figures.

46. MP REASONING Name a polygon that is always regular and a polygon that is sometimes regular. Explain your reasoning.

47. OPEN-ENDED Draw a pentagon. Is your pentagon *convex* or *concave*? Is your pentagon *regular* or *irregular*? Justify your answers.

48. CHALLENGE A rectangular room measures 20 feet by 12.5 feet. How many 5-inch square tiles will it take to cover the floor of this room? Explain.

49. WRITING IN MATH Describe two possible ways that a polygon can be equiangular but not a regular polygon.

50. Use the right triangle graphed below.

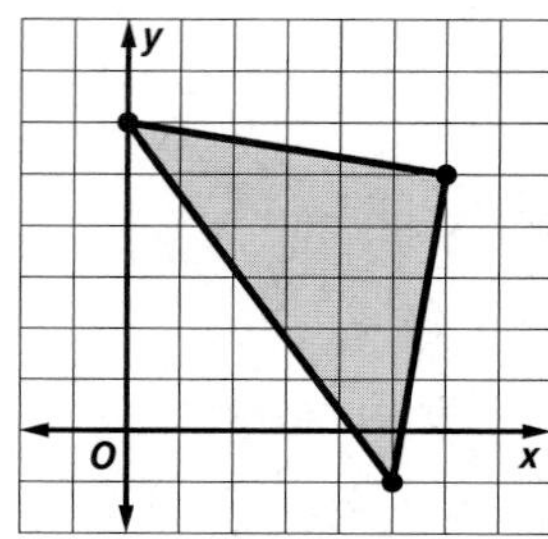

To the nearest tenth, what is the perimeter of the right triangle? MP 6

A 18.5 units

B 20.8 units

C 37.0 units

D 41.6 units

51. MULTI-STEP Cassie drew a plan for her landscape architecture class that includes a square concrete patio. The vertices of the patio are at (18, 5), (28, 18), (5, 15) and (15, 28). Each unit on the plan represents 1 foot. MP 1

a. To the nearest tenth, what are the perimeter and area of the patio?

b. The edges of the patio will be enclosed with metal trim. If the trim is sold in sections that are 6 feet long, how many sections of trim will be needed to edge the patio?

c. The surface of the patio will be stained with two coats of stain and then sealed with concrete sealer. A gallon of stain covers 100 square feet and a gallon of concrete sealer covers 225 square feet. How many gallons of stain and sealer will be needed for the patio?

52. The center of circle C is at $(9, -3)$. Point P on circle C is at $(4, 7)$. MP 6

a. To the nearest tenth, what is the circumference of $\odot C$?

b. To the nearest tenth, area of $\odot C$?

53. Points $A(-2, 5)$ and $B(6, 11)$ are the endpoints of a diameter of circle N. MP 6

a. To the nearest tenth, what is the circumference of $\odot N$?

b. To the nearest tenth, area of $\odot N$?

54. Use the rectangle graphed below.

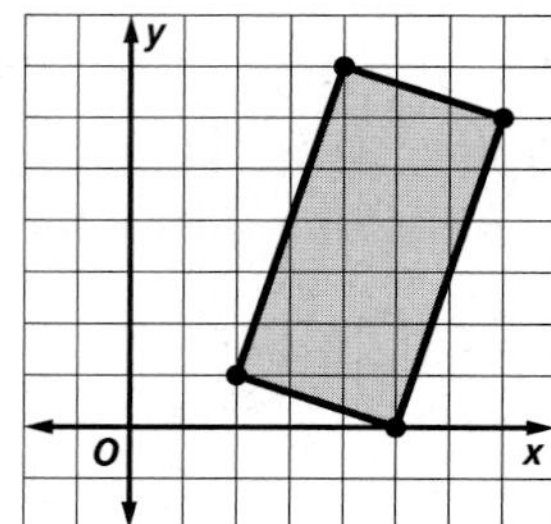

To the nearest tenth, what is the area of the rectangle? MP 6

A 12.6 units^2

B 19.0 units^2

C 20 units^2

D 25.3 units^2

55. ARCHAEOLOGY An archeologist at a dig has laid out a grid to track where artifacts are found. The corners of an old structure have been found at (16, 22), (4, 24), (0, 4), and (12, 2). Each unit on the grid is 1 foot. MP 4

a. To the nearest tenth, what was the perimeter of the structure?

b. To the nearest tenth, what was the area of the structure?

c. Suppose the archaeological team spends an average of 4 hours excavating 1 square foot of ground. If they work 8 hours each day, about how many days will it take them to excavate the interior of the old structure?

EXTEND 1-6

Geometry Software Lab

Two-Dimensional Figures

You can use The Geometer's Sketchpad® to draw and investigate polygons.

Mathematical Practices

5 Use appropriate tools strategically.

Activity 1 Measure Sides

Work cooperatively. Find XY, YZ, and ZX.

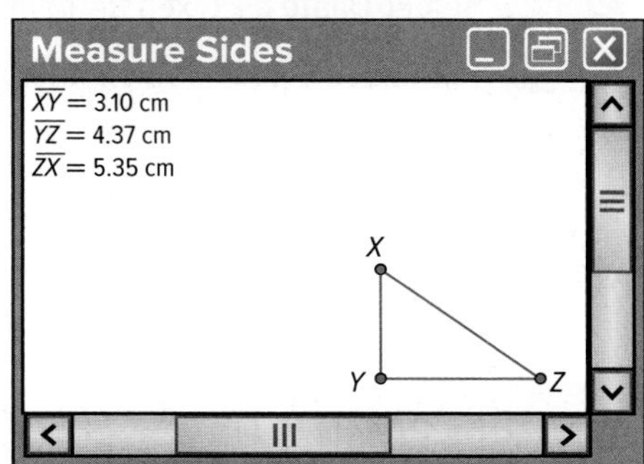

Step 1 Draw right triangle XYZ.

Step 2 Use the pointer tool to select $\overline{XY}$, $\overline{YZ}$, and $\overline{ZX}$.

Step 3 Select the Length command under the **Measure** menu to display the lengths of $\overline{XY}$, $\overline{YZ}$, and $\overline{ZX}$.

$XY = 3.10$ cm

$YZ = 4.37$ cm

$ZX = 5.35$ cm

Activity 2 Find Perimeter and Area

Work cooperatively. Find the perimeter of $\triangle XYZ$.

Perimeter and Area

$\overline{XY} = 3.10$ cm
$\overline{YZ} = 4.37$ cm
$\overline{ZX} = 5.35$ cm
Perimeter $\triangle XYZ = 12.81$ cm
Area $\triangle XYZ = 6.76$ cm^2

Step 1 Use the pointer tool to select points X, Y, and Z.

Step 2 Under the **Construct** menu, select **Triangle Interior**. The triangle will now be shaded.

Step 3 Select the triangle interior using the pointer.

Step 4 Choose the **Perimeter** command under the **Measure** menu to find the perimeter of $\triangle XYZ$.

Step 5 Choose the **Area** command under the **Measure** menu to find the area of $\triangle XYZ$.

The perimeter of $\triangle XYZ$ is 12.81 centimeters and the area of $\triangle XYZ$ is 6.76 square centimeters.

Activity 3 Measure Angles

Find $m\angle X$, $m\angle Y$, and $m\angle Z$.

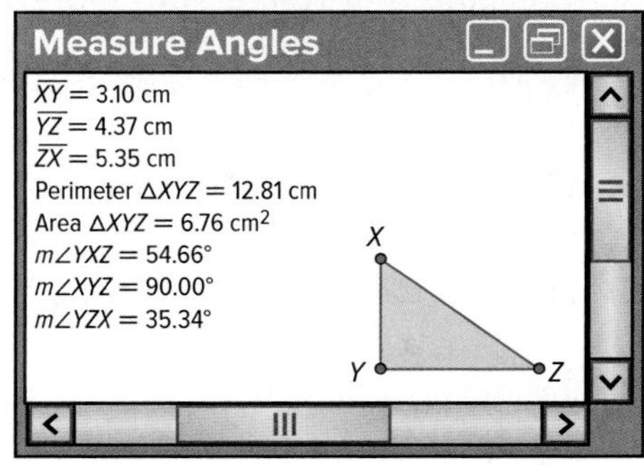

Step 1 Recall that $\angle X$ can also be named $\angle YXZ$ or $\angle ZXY$. Use the pointer to select points Y, X, and Z in order.

Step 2 Select the **Angle** command from the **Measure** menu to find $m\angle X$.

Step 3 Select points X, Y, and Z. Find $m\angle Y$.

Step 4 Select points X, Z, and Y. Find $m\angle Z$.

$m\angle X = 54.66$, $m\angle Y = 90.00$, and $m\angle Z = 35.34$.

(continued on the next page)

Geometry Software Lab

Two-Dimensional Figures *Continued*

Activity 4 Enlarge a Polygon

Enlarge $\triangle XYZ$.

Step 1 Highlight $\triangle XYZ$.

Step 2 Select the **Dilate** command from the **Transform menu**.

Step 3 Select **Dilate by: Fixed Ratio** and input $\frac{2.0}{1.0}$.

Step 4 Label the vertices of the new triangle X', Y', and Z.

Dilations
$\overline{XY} = 3.10$ cm
$\overline{YZ} = 4.37$ cm
$\overline{ZX} = 5.35$ cm
Perimeter $P_2 = 12.81$ cm
Area $P_2 = 6.76$ cm^2
$m\angle YXZ = 54.66°$
$m\angle XYZ = 90.00°$
$m\angle YZX = 35.34°$
X'
X
Y'
Y
Z

Analyze the Results

Work cooperatively.

1. Find the measures for $\triangle X'Y'Z$ that were found for $\triangle XYZ$ in Activities 1–3.
   - **a.** How do the side lengths of $\triangle X'Y'Z$ compare to the side lengths of $\triangle XYZ$?
   - **b.** How does the perimeter of $\triangle X'Y'Z$ compare to the perimeter of $\triangle XYZ$?
   - **c.** How does the area of $\triangle X'Y'Z$ compare to the area of $\triangle XYZ$?
   - **d.** How do the angle measures of $\triangle X'Y'Z$ compare to the angle measures of $\triangle XYZ$?
2. Add the side measures from Activity 1. How does this compare to the result in Activity 2?
3. What is the sum of the angle measures of $\triangle XYZ$?
4. Repeat the activities for each figure.
   a. irregular quadrilateral **b.** square **c.** pentagon **d.** hexagon
5. Draw another right triangle and find its perimeter and area. Select one vertex and double the side length. How does changing a side affect the perimeter and area?
6. Compare your results with those of your classmates.
7. Make a conjecture about the sum of the measures of the angles in any triangle.
8. What is the sum of the measures of the angles of a quadrilateral? pentagon? hexagon?
9. How are the sums of the angles of polygons related to the number of sides?
10. Test your conjecture on other polygons. Does your conjecture hold? Explain.
11. When the sides of a polygon are changed by a common factor, how does the perimeter of the polygon change? How does the area change?

LESSON 7

Transformations in the Plane

Then

- You identified and named polygons.

Now

1. Identify reflections, translations, and rotations.
2. Calculate the coordinates of the vertices of images after reflection, translation, and rotation given the coordinates of the preimage.

Why?

- The fashion industry often uses prints that display patterns. Many of these patterns are created by taking one figure and sliding it to create another figure in a different location, flipping the figure to create a mirror image of the original, or turning the original figure to create a new one.

New Vocabulary

transformation
preimage
image
isometry
reflection
translation
translation vector
rotation

Mathematical Practices

1 Make sense of problems and persevere in solving them.

2 Reason abstractly and quantitatively.

1 Identify Rigid Transformations

A **transformation** is a function that maps a figure, the **preimage**, onto a new figure called the **image**. A transformation can change the position, size, or shape of a figure.

A transformation can be noted using an arrow. The transformation statement $\triangle ABC \rightarrow \triangle XYZ$ tells you that A is mapped to X, B is mapped to Y, and C is mapped to Z.

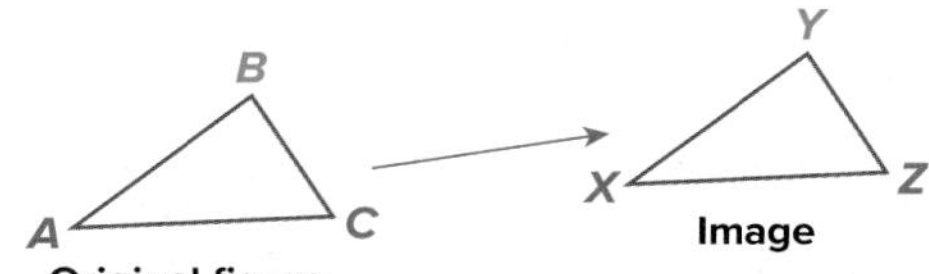

In a rigid transformation, also called a *congruence transformation* or an **isometry**, the position of the image may differ from that of the preimage, but the two figures remain congruent. The three main types of rigid transformations are shown below.

Key Concept Reflections, Translations, and Rotations

A **reflection** or *flip* is a transformation over a line called the *line of reflection.* Each point of the preimage and its image are the same distance from the line of reflection.	A **translation** or *slide* is a transformation that moves all points of the original figure the same distance in the same direction.	A **rotation** or *turn* is a transformation around a fixed point called the *center of rotation*, through a specific angle, and in a specific direction. Each point of the original figure and its image are the same distance from the center.
Example 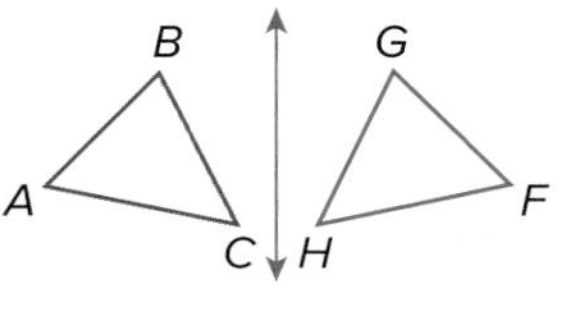 $\triangle ABC \longrightarrow \triangle FGH$	Example 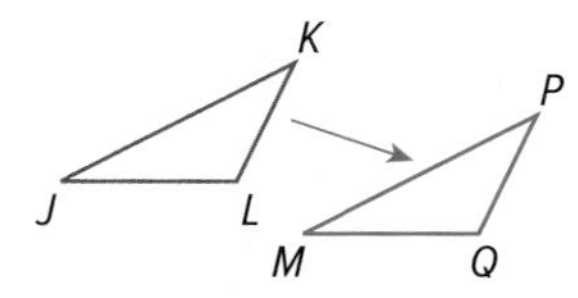 $\triangle JKL \longrightarrow \triangle MPQ$	Example 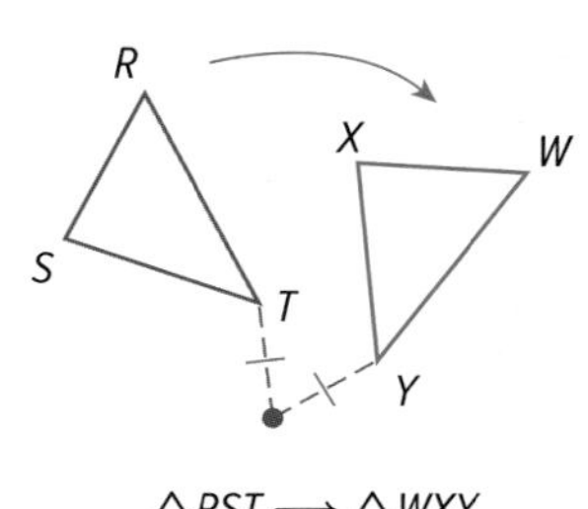 $\triangle RST \longrightarrow \triangle WXY$

Example 1 Identify Rigid Transformations

Identify the type of rigid transformation shown as a *reflection, translation,* or *rotation.*

a.

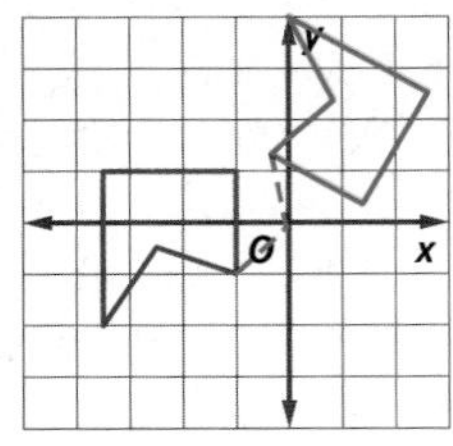

Each vertex and its image are the same distance from the origin. The angles formed by each pair of corresponding points and the origin are congruent. This is a rotation.

b.

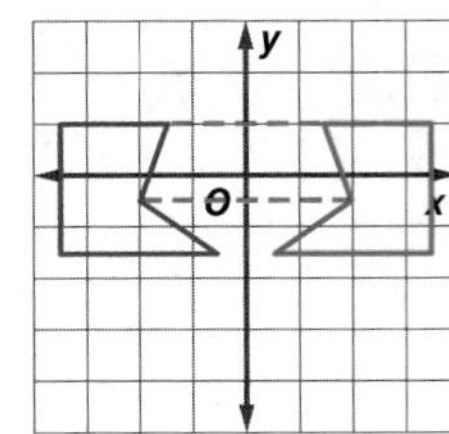

Each vertex and its image are the same distance from the y-axis. This is a reflection.

c.

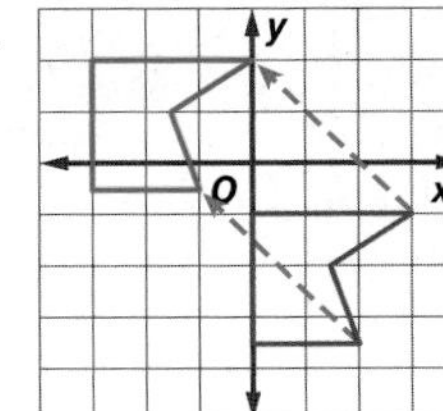

Each vertex and its image are in the same position, just 3 units left and 3 units up. This is a translation.

Study Tip

MP Precision Not all transformations preserve congruence. Only transformations that do not change the size or shape of the figure are rigid transformations. You will learn more about transformations in Chapter 3.

Guided Practice

1A.

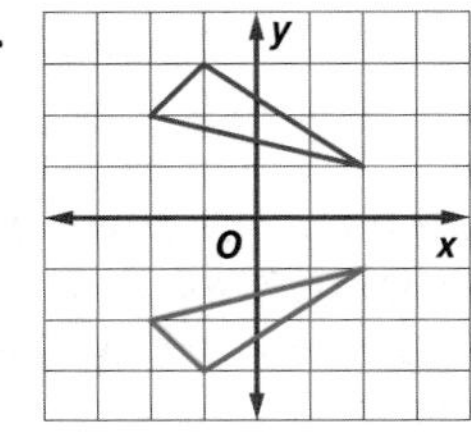

1B.

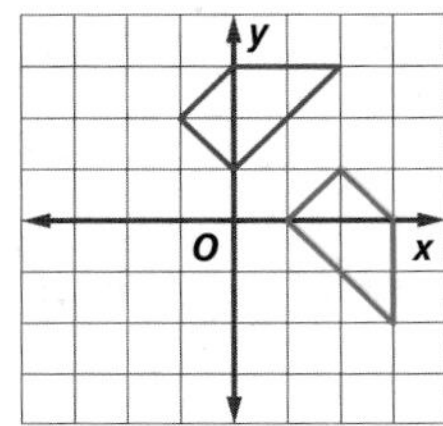

1C.

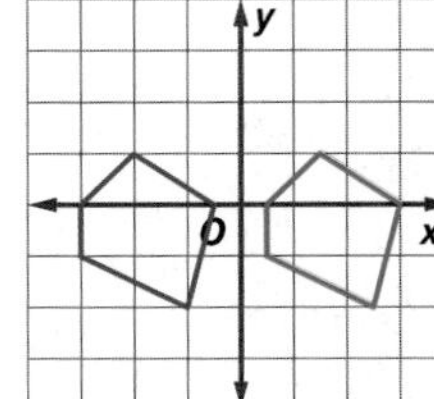

Some real-world motions or objects can be represented by transformations.

Real-World Example 2 Identify a Real-World Transformation

GAMES Refer to the information at the left. Identify the type of rigid transformation shown in the diagram as a ***reflection, translation,* or *rotation.***

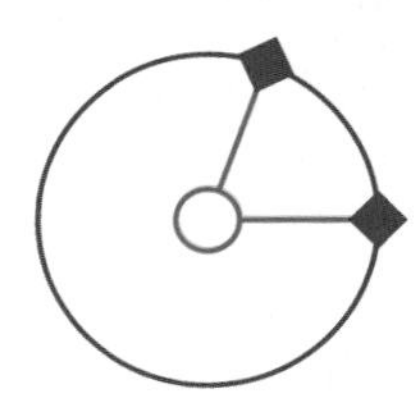

The position of the weight at different times is an example of a rotation. The center of rotation is the person's ankle.

Real-World Link

The game shown above involves a weight attached to a ring that you can place around your ankle. As the rope passes in front of your other foot, you skip over it.

Guided Practice

Identify the type of rigid transformation shown as a *reflection, translation,* or *rotation.*

2A.

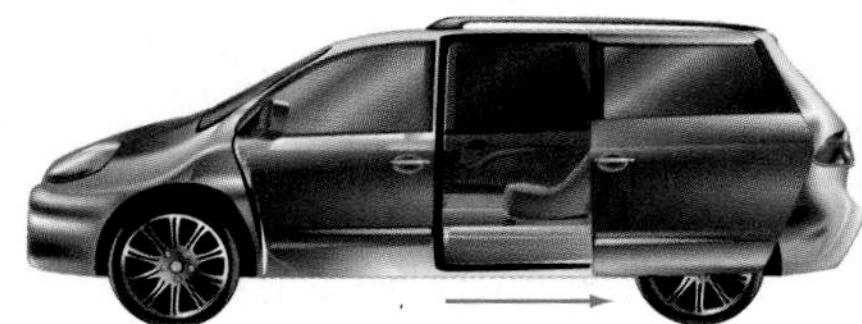

2B.

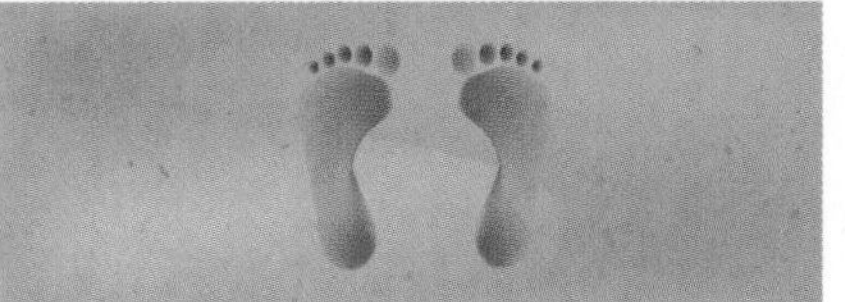

2 Transformations on the Coordinate Plane

When figures are transformed on the coordinate plane, the exact method in which one figure is transformed onto another can be described functionally.

Reflection in the x-axis

$(x, y) \rightarrow (x, -y)$

Reflection in the y-axis

$(x, y) \rightarrow (-x, y)$

Example 3 Reflections on the Coordinate Plane

Consider $\triangle ABC$ that has coordinates $A(3, 2)$, $B(2, -2)$, and $C(4, -5)$. Find the coordinates of the image after each transformation.

a. a reflection in the x-axis

$A(3, 2) \rightarrow A'(3, -2)$

$B(2, -2) \rightarrow B'(2, 2)$

$C(4, -5) \rightarrow C'(4, 5)$

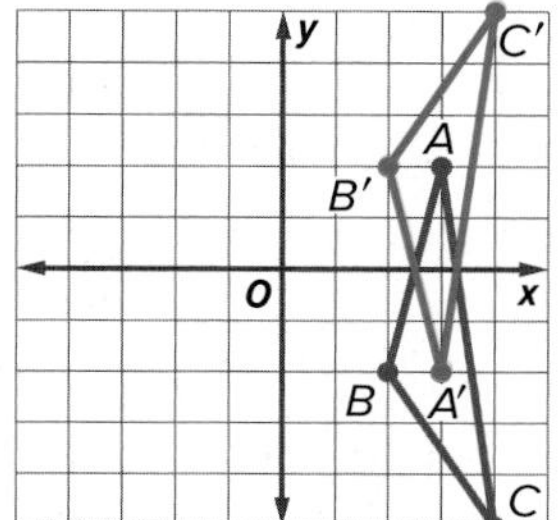

b. a reflection in the y-axis

$A(3, 2) \rightarrow A'(-3, 2)$

$B(2, -2) \rightarrow B'(-2, -2)$

$C(4, -5) \rightarrow C'(-4, -5)$

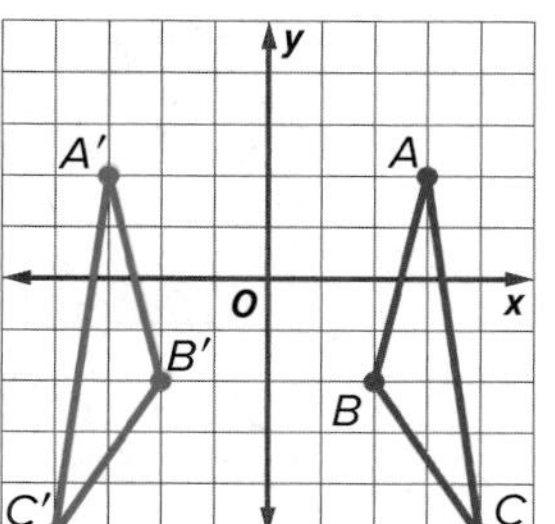

Study Tip

MP Precision To find the image of A, flip it across the x-axis so that the x-coordinate stays the same. The y-coordinate of the image is the opposite of the y-coordinate of A.

Guided Practice

3. Consider rectangle $ABCD$ that has coordinates $A(1, 1)$, $B(1, 5)$, $C(3, 5)$, and $D(3, 1)$. Determine the coordinates of the vertices of the image after a reflection in the x-axis.

A preimage is translated along a **translation vector**. A vector in **component form** is written as $\langle x, y \rangle$, which describes the vector in terms of its horizontal component x and vertical component y.

Translation along $\langle a, b \rangle$

$(x, y) \rightarrow (x + a, y + b)$

Example 4 Translations on the Coordinate Plane

For quadrilateral $QRST$ with vertices $Q(-8, -2)$, $R(-9, -5)$, $S(-4, -7)$, and $T(-4, -2)$, find the coordinates of the vertices of the image after a translation along the vector $\langle 7, 1 \rangle$.

Translating along $\langle 7, 1 \rangle$ moves the figure 7 units to the right and 1 unit up.

$Q(-8, -2) \rightarrow Q'(-1, -1)$

$R(-9, -5) \rightarrow R'(-2, -4)$

$S(-4, -7) \rightarrow S'(3, -6)$

$T(-4, -2) \rightarrow T'(3, -1)$

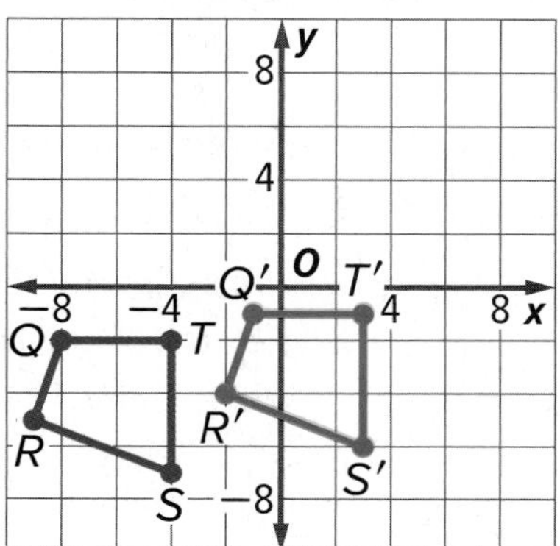

Guided Practice

4. For $\triangle MNP$ with vertices $M(2, 3)$, $N(4, 6)$, and $P(1, 8)$, find the coordinates of the vertices of the image after a translation along the vector $\langle 2, -3 \rangle$.

A rotation is a function that moves every point of a preimage through a specified angle and direction about a fixed point called the center of rotation. In this lesson, use the origin as the center when you perform a rotation.

<u>90° Rotation Counterclockwise</u>	<u>180° Rotation</u>	<u>270° Rotation Counterclockwise</u>
$(x, y) \rightarrow (-y, x)$	$(x, y) \rightarrow (-x, -y)$	$(x, y) \rightarrow (y, -x)$

Example 5 Rotations on the Coordinate Plane

> **Watch Out!**
>
> Verify Vertices Some transformations may look equivalent but are not. The transformation on the plane of *FGHJ* may look like a translation, but the vertices are in different positions. Be sure to verify that corresponding vertices are labeled correctly.

Parallelogram *FGHJ* has vertices *F*(2, 1), *G*(7, 1), *H*(6, −3), and *J*(1, −3). What are the coordinates of the vertices of its image after a rotation 180° about the origin?

The figure is in Quadrants I and IV. Its image will be oriented differently and it will be in Quadrants II and III.

$F(2, 1) \rightarrow F'(-2, -1)$

$G(7, 1) \rightarrow G'(-7, -1)$

$H(6, -3) \rightarrow H'(-6, 3)$

$J(1, -3) \rightarrow J'(-1, 3)$

Guided Practice

5. Consider $\triangle XYZ$ with vertices $X(-4, 2)$, $Y(-1, 1)$, and $Z(2, 3)$. What are the coordinates of the vertices of its image after a rotation 90° counterclockwise about the origin?

Check Your Understanding

 = Step-by-Step Solutions begin on page R13.

Examples 1, 2 **Identify the type of rigid transformation shown as a *reflection*, *translation*, or *rotation*.**

1.

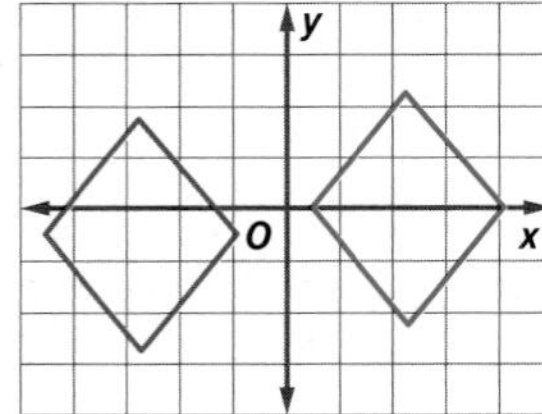

2.

Find the coordinates of the image after each rigid transformation.

Examples 3-5 **3.** $\triangle XYZ$ with vertices $X(-9, 4)$, $Y(-9, -3)$, $Z(-1, -3)$

a. reflection in the x-axis

b. translation along $\langle 4, 6 \rangle$

c. counterclockwise rotation of 270° about the origin

4. parallelogram *PQRS* with vertices $P(-2, 5)$, $Q(-9, 5)$, $R(-9, -1)$, $S(-2, -1)$

a. reflection in the y-axis

b. translation along $-3, 1$

c. rotation of 180° about the origin

Practice and Problem Solving

Extra Practice is on page R1.

Example 1 **Identify the type of rigid transformation shown as a *reflection, translation,* or *rotation*.**

5. 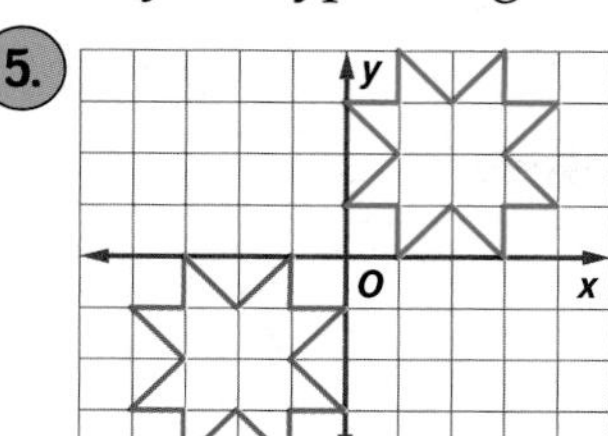

6. 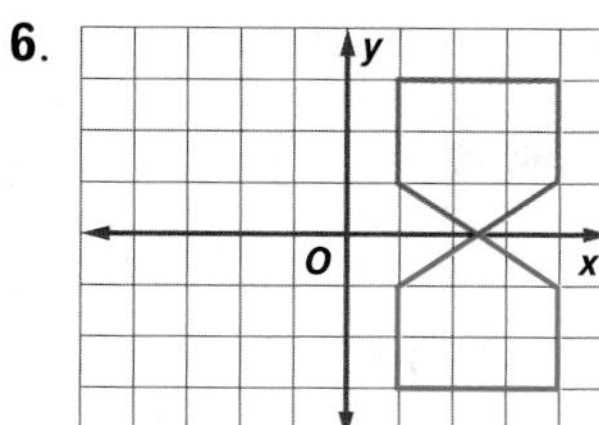

7. 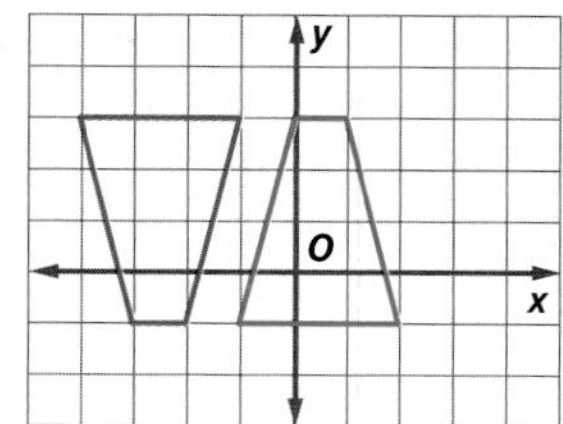

Example 2 **Identify the type of rigid transformation shown in each picture as a *reflection, translation,* or *rotation*.**

8.

9. 

Examples 3-5 **COORDINATE GEOMETRY** Find the coordinates of the figure with the given coordinates after the transformation on the plane. Then graph the preimage and image.

10. $A(3, 9)$, $B(3, 7)$, $C(7, 7)$; translation along the vector $\langle 0, -4\rangle$

11. $M(-7, -1)$, $P(-7, -7)$, $R(-1, -4)$; reflection in the y-axis

12. $X(-5, -4)$, $Y(-2, 0)$, $Z(-2, -4)$; rotation by 270° counterclockwise

13. $A(2, 2)$, $B(4, 7)$, $C(6, 2)$; reflection in the x-axis

CONSTRUCTION **Identify the type of rigid transformation performed on each given triangle to generate the other triangle in the truss with matching left and right sides shown below.**

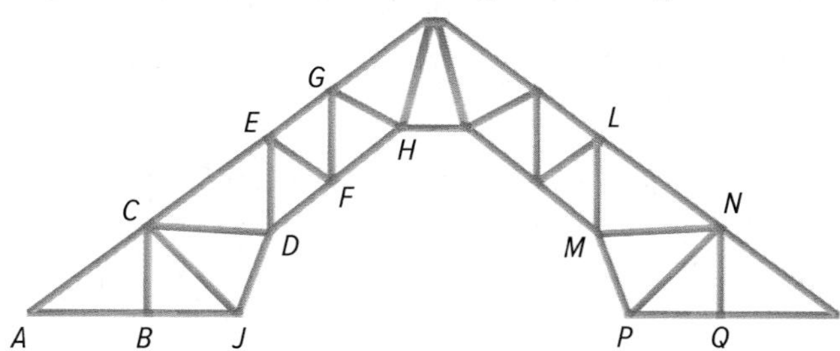

14. $\triangle NMP$ to $\triangle CJD$

15. $\triangle EFD$ to $\triangle GHF$

16. $\triangle CBJ$ to $\triangle NQP$

17. **SCHOOL** Identify the transformations that are used to open a combination lock on a locker. If appropriate, identify the line of symmetry or center of rotation.

18. Determine which capital letters of the alphabet have the following:

 a. only vertical lines of reflection.
 b. only horizontal lines of reflection
 c. both vertical and horizontal lines of reflection
 d. no lines of reflection

19 Consider $\triangle ABC$ with vertices $A(-4, 4)$, $B(-2, 8)$, and $C(2, 6)$.

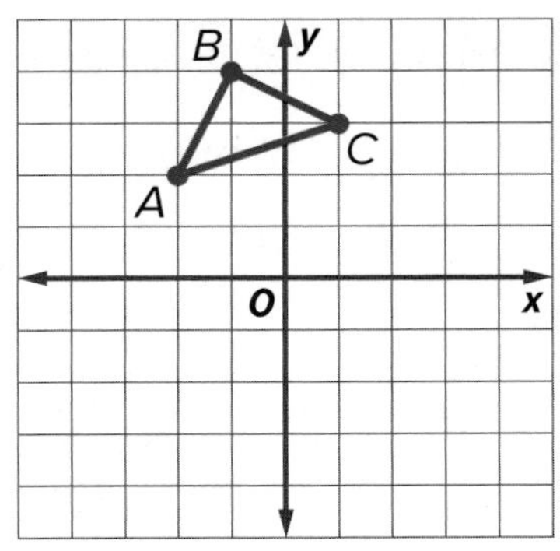

a. Find the coordinates of the image $\triangle A'B'C'$ after a reflection of $\triangle ABC$ in the y-axis.

b. Use the graph to determine the distance from each vertex of the preimage to the line of reflection.

c. How does the distance of the vertices of the image change after the reflection?

20. MULTIPLE REPRESENTATIONS In this problem, you will investigate the relationship between the ordered pairs of a figure and its translated image.

a. **Geometric** Draw congruent rectangles $ABCD$ and $WXYZ$ on a coordinate plane.

b. **Verbal** How do you get from a vertex on $ABCD$ to the corresponding vertex on $WXYZ$ using only horizontal and vertical movement?

c. **Tabular** Copy the table shown. Use your rectangles to fill in the x-coordinates, the y-coordinates, and the unknown value in the transformation column.

Rectangle *ABCD*	Transformation	Rectangle *WXYZ*
A(?, ?)	$(x_1 + ?, y_1 + ?)$	W(?, ?)
B(?, ?)	$(x_1 + ?, y_1 + ?)$	X(?, ?)
C(?, ?)	$(x_1 + ?, y_1 + ?)$	Y(?, ?)
D(?, ?)	$(x_1 + ?, y_1 + ?)$	Z(?, ?)

d. **Algebraic** Function notation $(x, y) \rightarrow (x + a, y + b)$, where a and b are real numbers, represents a mapping from one set of coordinates onto another. Complete the following notation that represents the rule for the translation $ABCD \rightarrow WXYZ$: $(x, y) \rightarrow (x + a, y + b)$.

H.O.T. Problems Use Higher-Order Thinking Skills

21. **CHALLENGE** Use the diagram at the right.

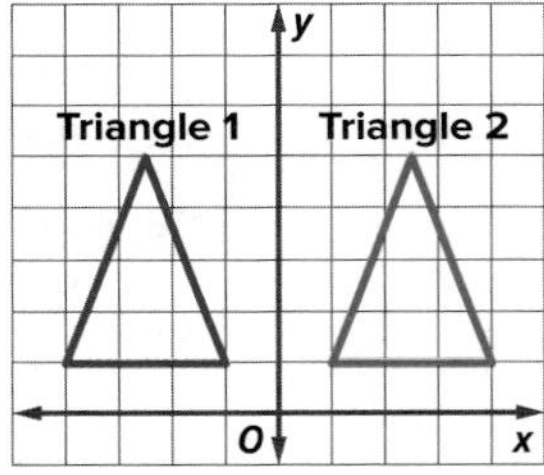

a. Identify two transformations on the plane of Triangle 1 that can result in Triangle 2.

b. What must be true of the triangles so that more than one transformation on a preimage results in the same image? Explain your reasoning.

22. **REASONING** A *dilation* is another type of transformation. In the diagram, a small paper clip has been dilated to produce a larger paper clip. Explain why dilations are not rigid transformations.

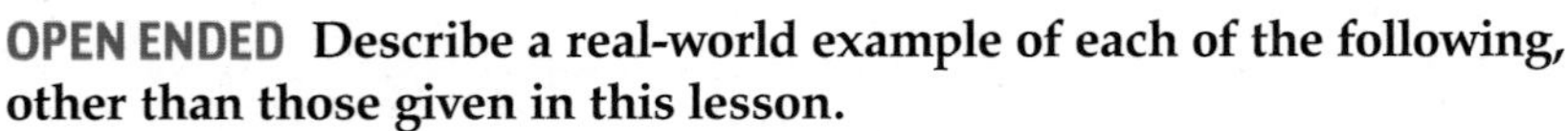

OPEN ENDED Describe a real-world example of each of the following, other than those given in this lesson.

23. reflection **24.** translation **25.** rotation

26. WRITING IN MATH Use the diagram at the right.

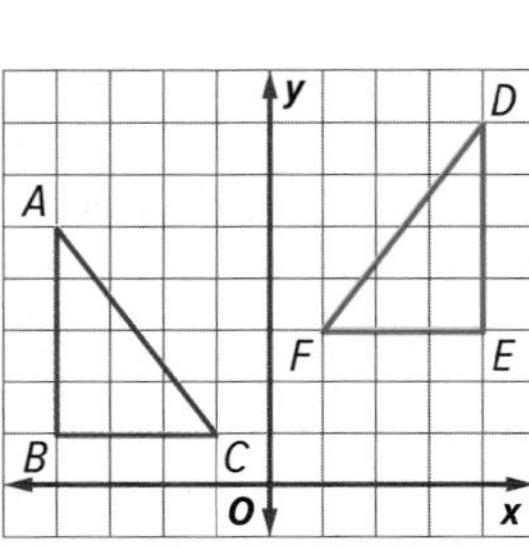

a. $\triangle DEF$ is called a *glide reflection* of $\triangle ABC$.
Based on the diagram, define a glide reflection.
Is a glide reflection a rigid transformation?
Include a definition of rigid transformation in your response.
Explain your reasoning.

b. Does it matter whether you reflect or glide first in a *glide reflection*? Explain.

27. Which of the following shows a rotation of $\triangle ABC$? MP 2

○ **A**

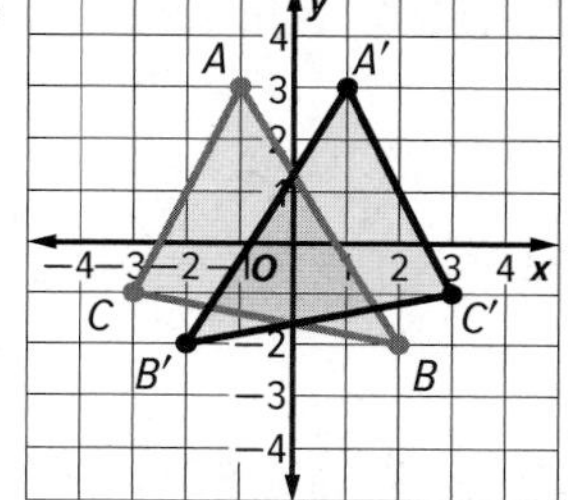

○ **B**

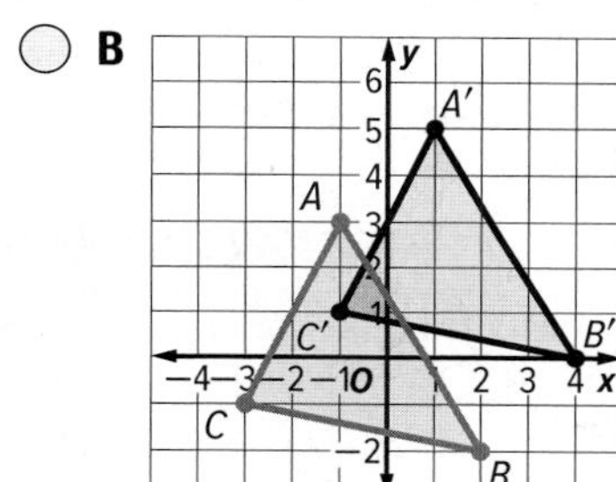

○ **C**

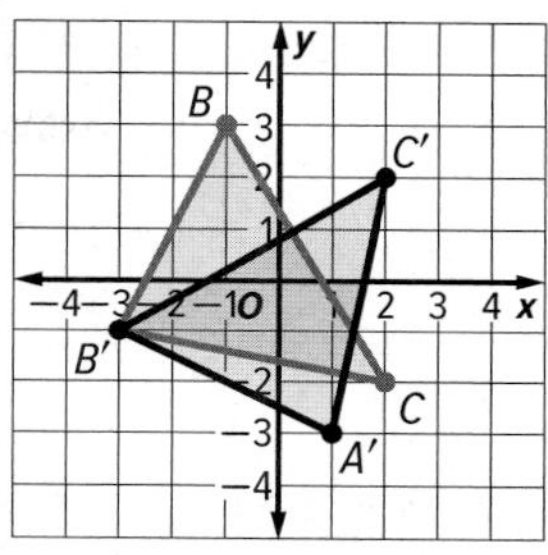

○ **D**

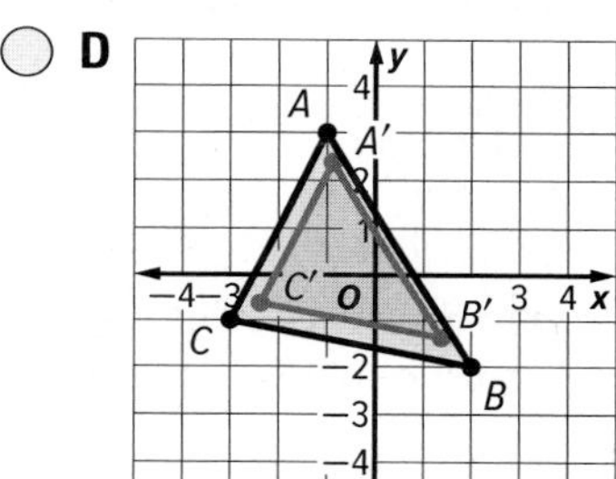

28. $\triangle MNP$ has vertices at $M(-2, 1)$, $N(-1, 4)$, and $P(1, 2)$. What are the coordinates of the image after a translation along vector $\langle 4, -9 \rangle$? MP 2

○ **A** $M'(-6, 10)$, $N'(-5, 13)$, $P'(-3, 11)$

○ **B** $M'(-11, 5)$, $N'(-10, 8)$, $P'(-8, 6)$

○ **C** $M'(2, -8)$, $N'(3, -5)$, $P'(5, -7)$

○ **D** $M'(7, 5)$, $N'(8, 8)$, $P'(10, 6)$

29. Which transformation on the plane is shown on the graph below? MP 2

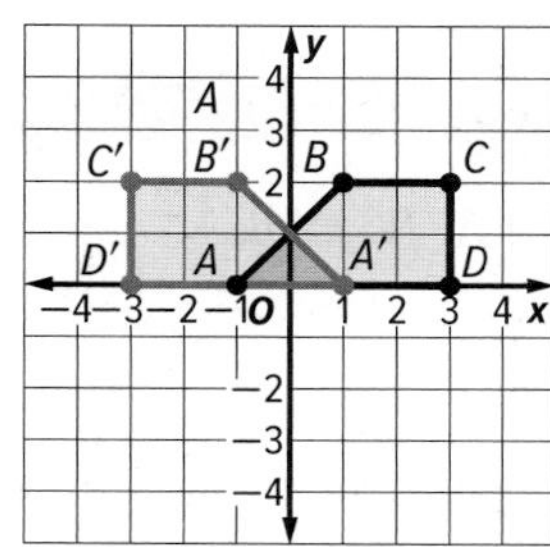

○ **A** reflection in the x-axis

○ **B** reflection in the y-axis

○ **C** rotation by 180°

○ **D** rotation by 90°

○ **E** translation along $\langle 2, 0 \rangle$

30. **MULTI-STEP** $\triangle PQR$ has coordinates $P(1, 1)$, $Q(4, 5)$, and $R(5, 1)$.

a. Determine the coordinates of the vertices of the image after a 180° rotation about the origin.

○ **A** $P'(1, 1)$, $Q'(5, 4)$, $R'(1, 5)$

○ **B** $P'(-1, 1)$, $Q'(-5, 4)$, $R'(-1, 5)$

○ **C** $P'(-1, -1)$, $Q'(-4, -5)$, $R'(-5, -1)$

○ **D** $P'(1, -1)$, $Q'(5, -4)$, $R'(1, -5)$

b. What are two consecutive transformations on the plane that would result in the same image of $\triangle PQR$? MP 2

31. Which of the following best describes the transformation of the quadrilateral from 1 to 2 to 3? MP 7

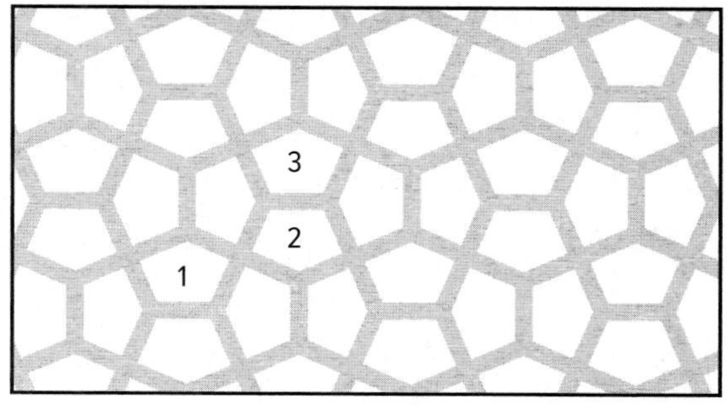

○ **A** rotation, translation

○ **B** rotation, reflection

○ **C** reflection, reflection

○ **D** translation, reflection

EXTEND 1-7

Geometry Lab

Transformations

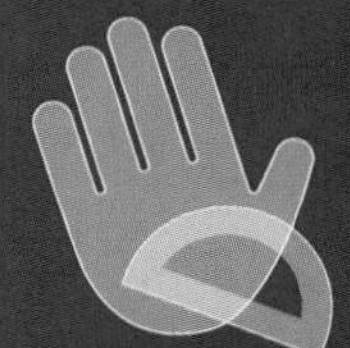

You can use transparencies to find the images of transformations.

Mathematical Practices
MP **5** Use appropriate tools strategically.

Activity 1 Reflections on the Coordinate Plane

Given $\triangle ABC$ with vertices $A(-2, 2)$, $B(-1, 4)$, and $C(1, 3)$, find the image $\triangle A'B'C'$ of a reflection of $\triangle ABC$ in the y-axis.

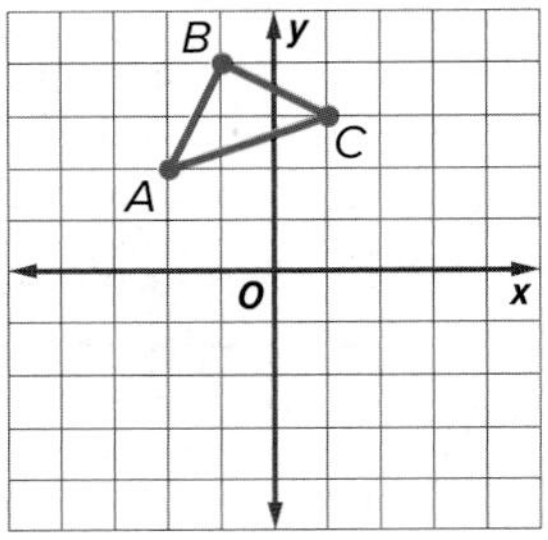

Step 1 Trace $\triangle ABC$ and the origin onto the transparency.

Step 2 Flip the transparency over vertically. Align the marked y-axis with the given y-axis and align the marked positive x-axis with the given negative x-axis.

1. Use the given axes to find the coordinates of the vertices of the image. Record your findings in a table like the one shown.

Coordinates of Vertices of Image	
A′	(______, ______)
B′	(______, ______)
C′	(______, ______)

2. Use the graph to determine the distance from each vertex to the line of reflection. Record your findings in a table like the one shown.

Triangle	Distance from Vertex to Line of Reflection		
△ABC	A ______ unit(s)	B ______ unit(s)	C ______ unit(s)
△A′B′C′	A′ ______ unit(s)	B′ ______ unit(s)	C′ ______ unit(s)

3. What do you observe about the distances from each vertex to the line of reflection?

Activity 2 Translations on the Coordinate Plane

Given $\triangle ABC$ with vertices $A(-2, 2)$, $B(-1, 4)$, and $C(1, 3)$, find the image $\triangle A''B''C''$ of a translation along vector $\langle 3, 4 \rangle$.

Step 1 Trace $\triangle ABC$ and the origin onto the transparency.

Step 2 Slide the transparency along vector $\langle 3, 4 \rangle$ by sliding the origin of your drawing to $(3, 4)$ on the given graph.

1. Use the given axes to find the coordinates of the vertices of the translated image. Record your findings in a table like the one shown.

Coordinates of Vertices of Image	
A′	(______, ______)
B′	(______, ______)
C′	(______, ______)

(continued on the next page)

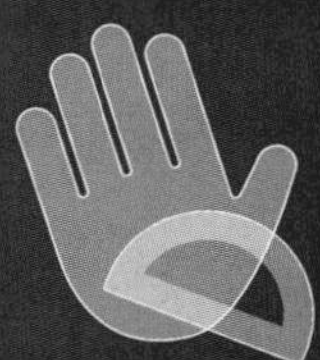

2. Use the graph to determine the slope of each segment between corresponding vertices. Record your findings in a table like the one shown.

3. What do you observe about the slopes of each segment?

Segment	Slope = $\frac{\text{change in } y}{\text{change in } x}$
$\overline{AA''}$	
$\overline{BB''}$	
$\overline{CC''}$	

Activity 3 Rotations on the Coordinate Plane

Given $\triangle ABC$ with vertices $A(-2, 2)$, $B(-1, 4)$, and $C(-1, 3)$, find the image $\triangle A'''B'''C'''$ of a rotation 90° counterclockwise about the origin.

Step 1 Trace $\triangle ABC$ and the origin onto the transparency.

Step 2 Keep the marked origin aligned with the given origin. Rotate the transparency 90° counterclockwise.

1. Use the given axes to find the coordinates of the vertices of the rotated image. Record your findings in the table.

Coordinates of Vertices of Image	
A'''	(______, ______)
B'''	(______, ______)
C'''	(______, ______)

2. Draw segments to connect each vertex to the center of rotation, P. Use the Distance Formula to find the length of each segment.

Preimage	Image
$AP =$ ______ units	$A'''P =$ ______ units
$BP =$ ______ units	$B'''P =$ ______ units
$CP =$ ______ units	$C'''P =$ ______ units

3. What do you observe about the lengths of each segment?

Exercises

1. $\triangle ABC$ has vertices at $A(3, 5)$, $B(-2, 1)$, and $C(7, 2)$. Find the coordinates of the following images.

   a. $\triangle A'\,B'\,C'$ as a reflection in the x-axis

   b. $\triangle A''\,B''\,C''$ as a translation along vector $\langle -4, 7 \rangle$

   c. $\triangle A'''\,B'''\,C'''$ as a rotation 270° counterclockwise about the origin

2. $\triangle ABC$ has vertices at $A(x_1, y_1)$, $B(x_2, y_2)$, and $C(x_3, y_3)$. Find the coordinates of the following images.

   a. $\triangle A'B'C'$ as a reflection in the x-axis

   b. $\triangle A''B''C''$ as a translation along vector $\langle a, b \rangle$

   c. $\triangle A'''B'''C'''$ as a rotation 270° counterclockwise about the origin

LESSON 8

Three-Dimensional Figures

Then

- You identified and named two-dimensional figures.

Now

1. Identify and name three-dimensional figures.
2. Find surface area and volume.

Why?

- Architects often provide three-dimensional models of their ideas to clients. These models give their clients a better idea of what the completed structure will look like than a two-dimensional drawing. Three-dimensional figures, or *solids*, are made up of flat or curved surfaces.

New Vocabulary

polyhedron
face
edge
vertex
prism
base
pyramid
cylinder
cone
sphere
regular polyhedron
Platonic solid
surface area
volume

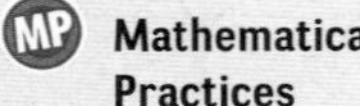

Mathematical Practices

2 Reason abstractly and quantitatively.
6 Attend to precision.

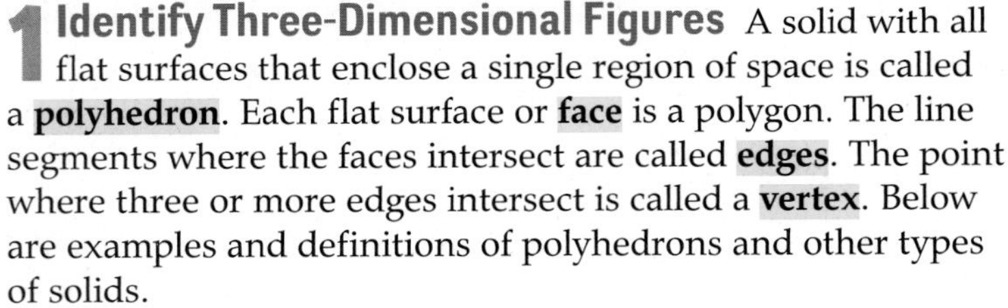

1 Identify Three-Dimensional Figures A solid with all flat surfaces that enclose a single region of space is called a **polyhedron**. Each flat surface or **face** is a polygon. The line segments where the faces intersect are called **edges**. The point where three or more edges intersect is called a **vertex**. Below are examples and definitions of polyhedrons and other types of solids.

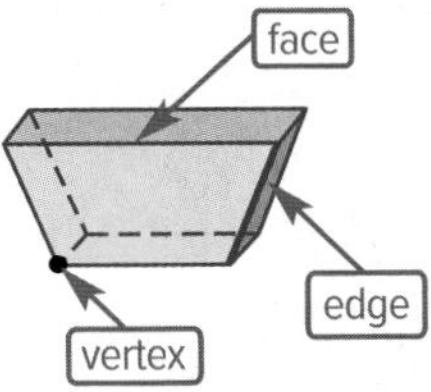

Key Concept Types of Solids

Polyhedrons

A **prism** is a polyhedron with two parallel congruent faces called **bases** connected by parallelogram faces.

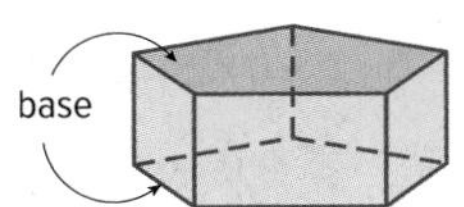

A **pyramid** is a polyhedron that has a polygonal base and three or more triangular faces that meet at a common vertex.

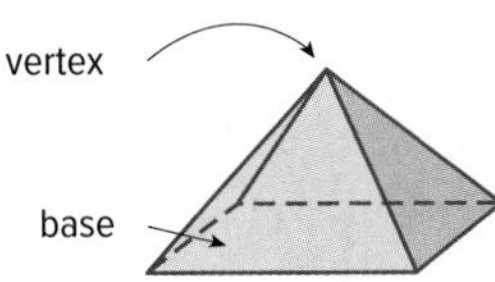

Not Polyhedrons

A **cylinder** is a solid with congruent parallel circular bases connected by a curved surface.

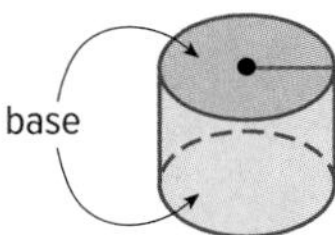

A **cone** is a solid with a circular base connected by a curved surface to a single vertex.

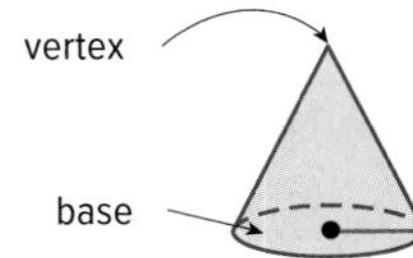

A **sphere** is a set of points in space that are the same distance from a given point. A sphere has no faces, edges, or vertices.

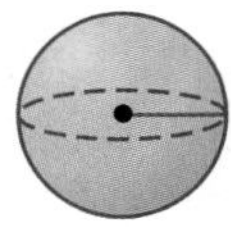

Polyhedrons or *polyhedra* are named by the shape of their bases.

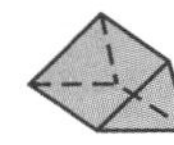
triangular prism

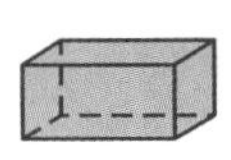
rectangular prism

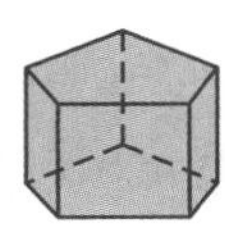
pentagonal prism

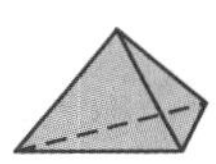
triangular pyramid

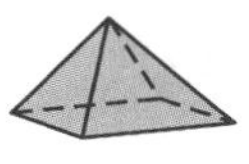
rectangular pyramid

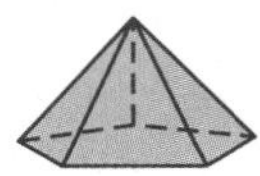
pentagonal pyramid

Example 1 Identify Solids

Determine whether each solid is a polyhedron. Then identify the solid. If it is a polyhedron, name the bases, faces, edges, and vertices.

a.

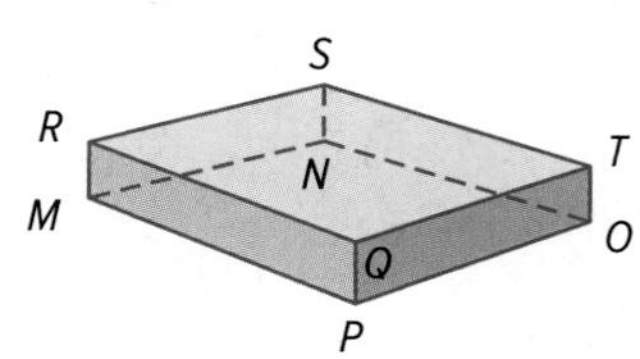

The solid is formed by polygonal faces, so it is a polyhedron. There are two parallel congruent rectangular bases, so it is a rectangular prism.

bases: ▭*MNOP*, ▭*RSTQ*

faces: ▭*RQPM*, ▭*RSNM*, ▭*STON*, ▭*QTOP*, ▭*RSTQ*, ▭*MNOP*

edges: $\overline{MN}$, $\overline{NO}$, $\overline{OP}$, $\overline{PM}$, $\overline{RS}$, $\overline{ST}$, $\overline{TQ}$, $\overline{QR}$, $\overline{RM}$, $\overline{SN}$, $\overline{TO}$, $\overline{QP}$

vertices: *M*, *N*, *O*, *P*, *Q*, *R*, *S*, *T*

Reading Math

Symbols Symbols can be used in naming the faces of solids. The symbol ▭ means rectangle. The symbol △ means triangle. The symbol ⊙ means circle.

b.

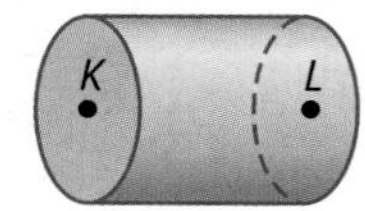

The solid is a curved surface, so it is not a polyhedron. It has two congruent circular bases, so it is a cylinder.

c.

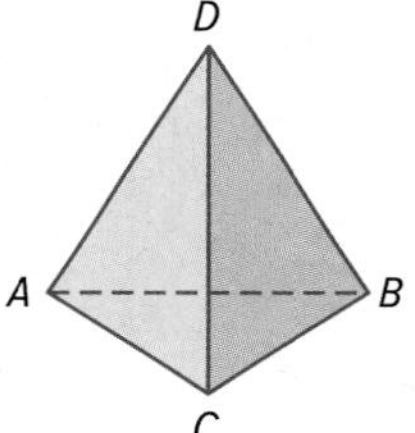

The solid has formed by polygonal faces, so it is a polyhedron. The base is a triangle, and the three faces meet in a vertex, so it is a triangular pyramid.

bases: △*ABC*

faces: △*ABC*, △*ADC*, △*CDB*, △*BDA*

edges: $\overline{AB}$, $\overline{BC}$, $\overline{CA}$, $\overline{DA}$, $\overline{DB}$, $\overline{DC}$

vertices: *A*, *B*, *C*, *D*

Guided Practice

1A.

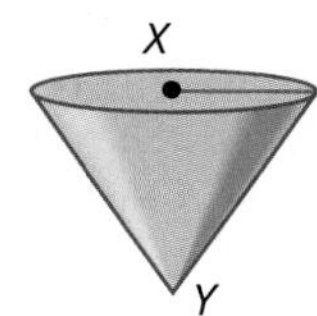

1B.

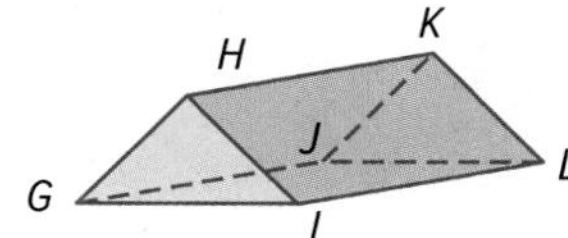

Math History Link

Plato (427–347 B.C.) Plato, a philosopher, mathematician, and scientist, lived in Athens, Greece. He is best known for founding a school known as "The Academy." In mathematics, he was concerned with the idea of proofs, and he insisted that definitions must be accurate and hypotheses must be clear.

Rijksmuseum, Amsterdam

A polyhedron is a **regular polyhedron** if all of its faces are regular congruent polygons and all of the edges are congruent. There are exactly five types of regular polyhedrons, called **Platonic solids** because Plato used them extensively.

Key Concept Platonic Solids

Tetrahedron	Hexahedron or Cube	Octahedron	Dodecahedron	Icosahedron
4 equilateral triangle faces	6 square faces	8 equilateral triangular faces	12 regular pentagonal faces	20 equilateral triangular faces

> **Study Tip**
> **Euclidean Solids** The *Euclidean solids* include the cube, the pyramid, the cylinder, the cone, and the sphere.

2 Surface Area and Volume

Surface area is a two-dimensional measurement of the surface of a solid figure. The surface area of a polyhedron is the sum of the areas of each face. **Volume** is the measure of the amount of space enclosed by a solid figure.

Review the formulas for the surface area and volume of five common solids given below. You will derive these formulas in Chapter 11.

Key Concept Surface Area and Volume

Prism	Regular Pyramid	Cylinder	Cone	Sphere
P, h	ℓ	r, h	r, h, ℓ	r
$S = Ph + 2B$	$S = \frac{1}{2}P\ell + B$	$S = 2\pi rh + 2\pi r^2$	$S = \pi r\ell + \pi r^2$	$S = 4\pi r^2$
$V = Bh$	$V = \frac{1}{3}Bh$	$V = \pi r^2 h$	$V = \frac{1}{3}\pi r^2 h$	$V = \frac{4}{3}\pi r^3$

S = total surface area, V = volume, h = height of a solid

P = perimeter of the base, B = area of base, ℓ = slant height, r = radius

> **WatchOut!**
> **Height vs. Slant Height** The *height* of a pyramid or cone is not the same as its *slant height.*
>
> 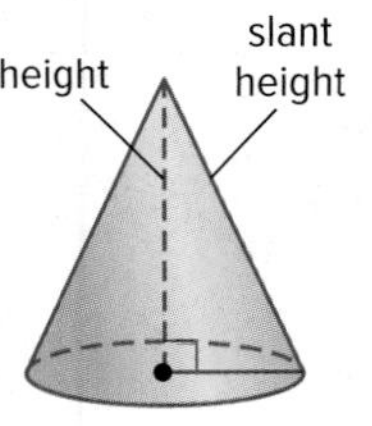
>

Example 2 Find Surface Area and Volume

Find the surface area and volume of the square pyramid.

5 cm, 4 cm, 6 cm

Surface Area

Since the base of the pyramid is a square, the perimeter P of the base is $4 \cdot 6$ or 24 centimeters. The area of the base B is $6 \cdot 6$ or 36 square centimeters. The slant height is 5 centimeters.

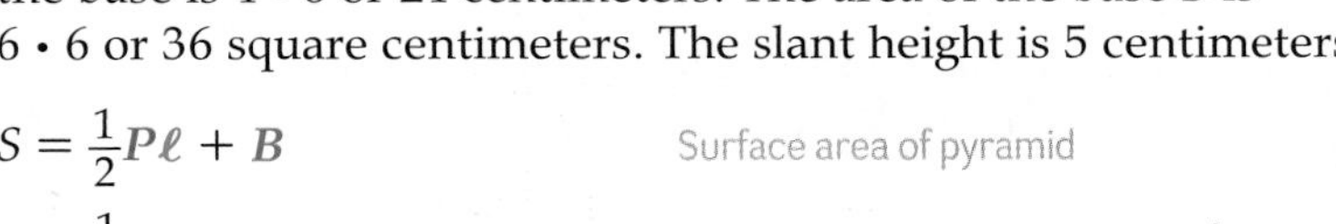

$S = \frac{1}{2}P\ell + B$ — Surface area of pyramid

$= \frac{1}{2}(24)(5) + 36$ or 96 — $P = 24$ cm, $\ell = 5$ cm, $B = 36$ cm^2

The surface area of the square pyramid is 96 square centimeters.

Volume

The height of the pyramid is 4 centimeters.

$V = \frac{1}{3}Bh$ — Volume of pyramid

$= \frac{1}{3}(36)(4)$ or 48 — $B = 36$ cm^2, $h = 4$ cm

The volume is 48 cubic centimeters.

Guided Practice

Find the surface area and volume of each solid to the nearest tenth.

2A. **2B.** **2C.**

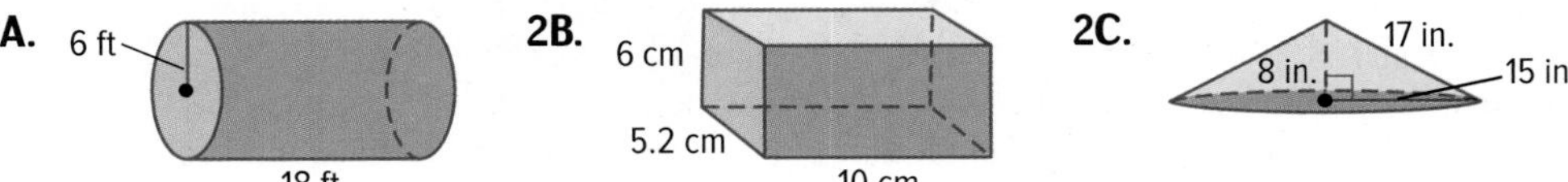

Study Tip

MP Precision Be sure that you have converted all units of measure to be consistent before you begin volume or surface area calculations.

Real-World Example 3 Surface Area and Volume

PEP RALLY Joanne is making noisemakers out of rice and empty paper towel tubes capped at both ends for an upcoming pep rally. The paper towel tubes are 12 inches long and 1.5 inches in diameter. Find the surface area of the noisemaker and the volume of rice needed to fill the noisemaker $\frac{1}{4}$ of the depth. Round to the nearest tenth.

The noisemaker is a cylinder.

$S = 2\pi rh + 2\pi r^2$ Surface area of cylinder

$= 2\pi(0.75)(12) + 2\pi(0.75)^2$ $r = 0.75$ in., $h = 12$ in.

≈ 60.1 Use a calculator.

The surface area of the noisemaker is about 60.1 square inches.

Since the cylinder will only be filled $\frac{1}{4}$ full, let $h = \frac{1}{4}$ (12) or 3 inches.

$V = \pi r^2 h$ Volume of a cylinder

$= \pi(0.75)^2(3)$ $r = \frac{1}{2}d$ or 0.75 in., $h = \frac{1}{4}$ (12 in.) or 3 in.

≈ 5.3 Use a calculator.

The volume of rice needed is approximately 5.3 cubic inches.

Guided Pracatice

3. **ORNAMENTS** Leslie is decorating spherical ornaments that are 18 centimeters in diameter. Find the volume of glitter needed to fill the ornament and the surface area of the finished ornament. Round each measure to the nearest tenth.

Check Your Understanding

 = Step-by-Step Solutions begin on page R13.

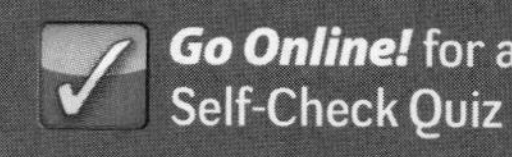

Go Online! for a Self-Check Quiz

Example 1 **Determine whether the solid is a polyhedron. Then identify the solid. If it is a polyhedron, name the bases, faces, edges, and vertices.**

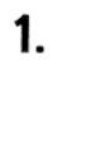

1. 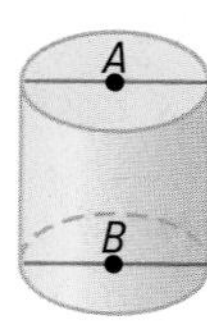

2. 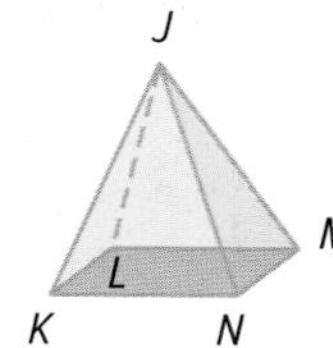

Example 2 **Find the surface area and volume of each solid to the nearest tenth.**

3. 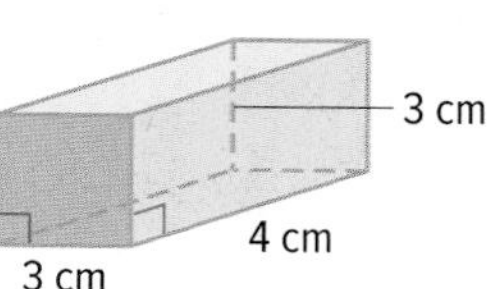

4. 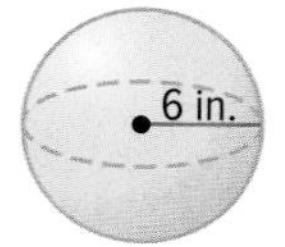

Example 3 5. **CUPCAKES** LaMea is icing cupcakes with a cone-shaped icing bag 3.5 inches in diameter, 5 inches tall, with a slant height of 5.3 inches. The icing bag has no top. Find each measure to the nearest tenth.

a. the volume of icing that will fill the bag

b. the area of plastic used to make the icing bag

Practice and Problem Solving

Extra Practice is on page R1.

Example 1 **Identify the solid modeled by each object. State whether the solid modeled is a polyhedron.**

6. 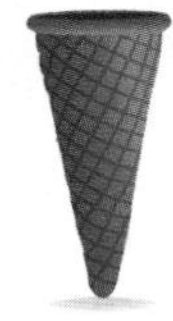

7. 

8. 

9. 

10. 

11. 

STRUCTURE Determine whether the solid is a polyhedron. Then identify the solid. If it is a polyhedron, name the bases, faces, edges, and vertices.

12. 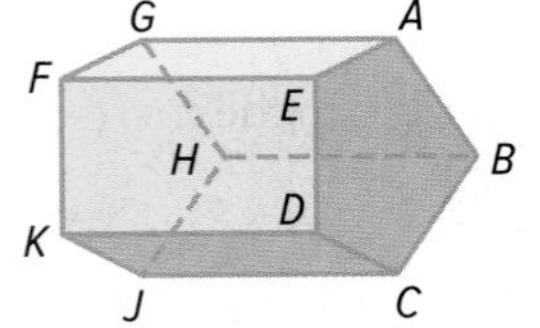

13. 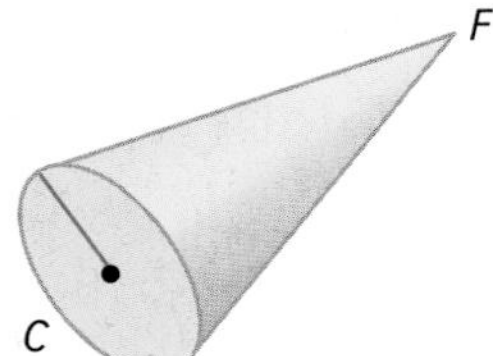

14. 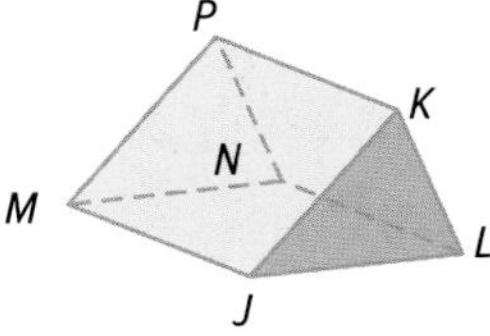

15. 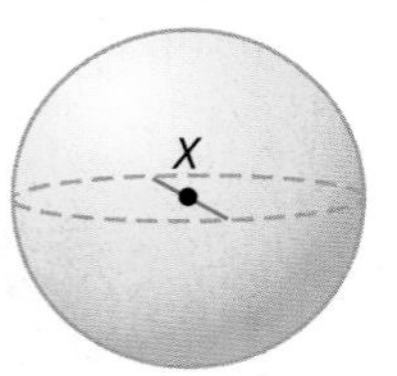

16. 

17. 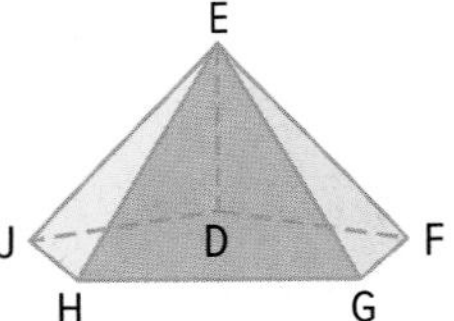

Example 2 **Find the surface area and volume of each solid to the nearest tenth.**

18. 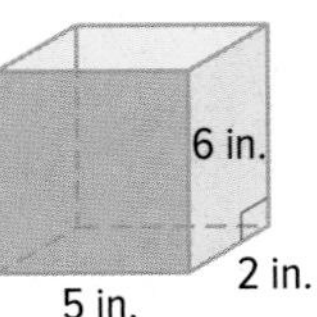

19. 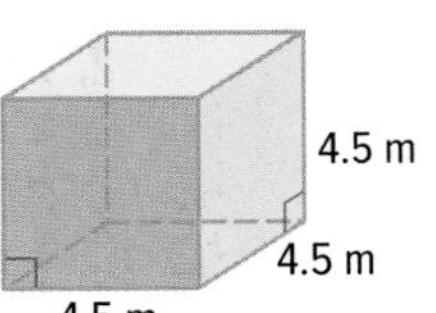

20. 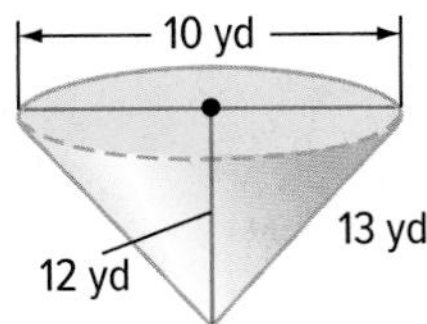

21. 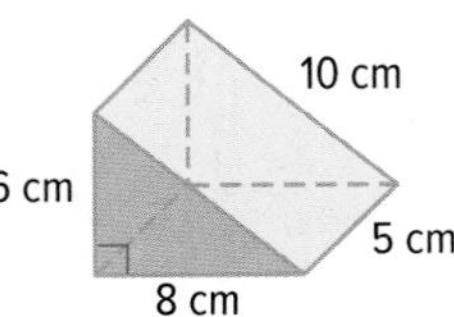

22. 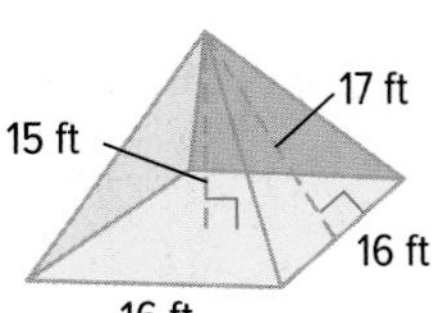

23. 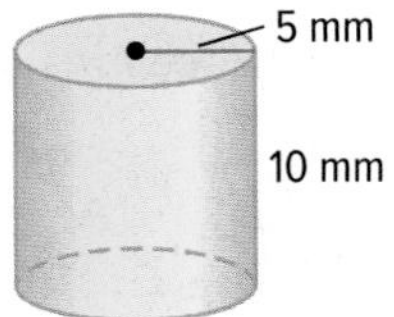

Example 3 24. **CARDS** A rectangular card box is 2.5 inches by 3.5 inches. The depth of the box is 0.75 inch but the depth of cards in the box is only $\frac{1}{2}$ the depth of the box. Find each measure to the nearest tenth.

a. the surface area of the card box

b. the volume of cards in the card box

25. **DRUMS** Drum shell size is important to the tone of the drum. The bigger the diameter is, the deeper the sound will be. Shaun's drum has a diameter of 14 inches. Suppose the height of the drum is 8.5 inches. Find each measure to the nearest tenth.

a. the volume of air within the drum

b. the surface area of the drum

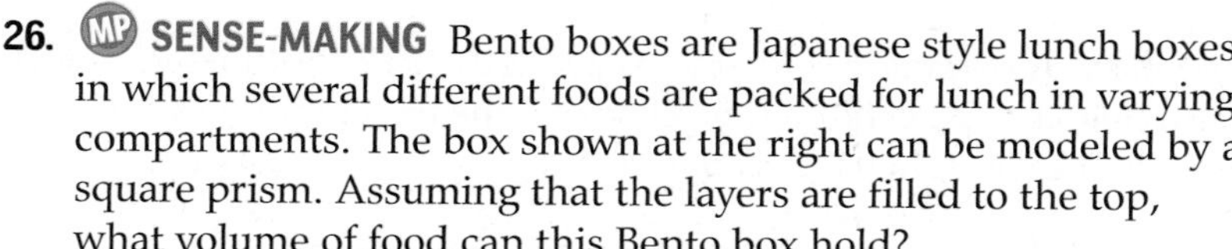

26. **MP SENSE-MAKING** Bento boxes are Japanese style lunch boxes in which several different foods are packed for lunch in varying compartments. The box shown at the right can be modeled by a square prism. Assuming that the layers are filled to the top, what volume of food can this Bento box hold?

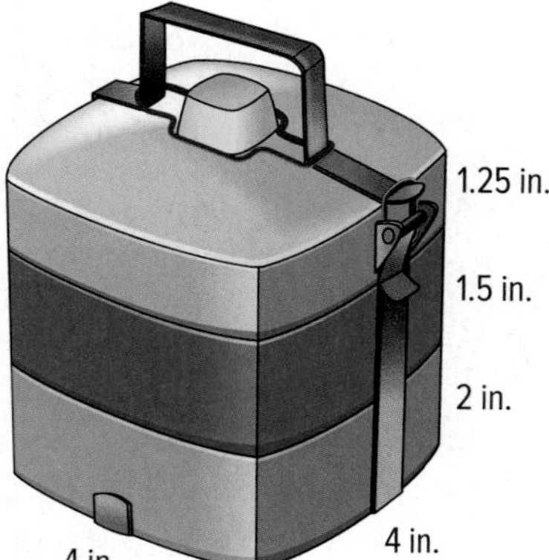

27. **ALGEBRA** The surface area of a cube is 54 square inches. Find the length of each edge.

28. **ALGEBRA** The volume of a cube is 729 cubic centimeters. Find the length of each edge.

29. **PAINTING** Tara is painting her family's fence. Each post is composed of a square prism and a square pyramid. The slant height of the pyramid is 4 inches. Determine the surface area and volume of each post.

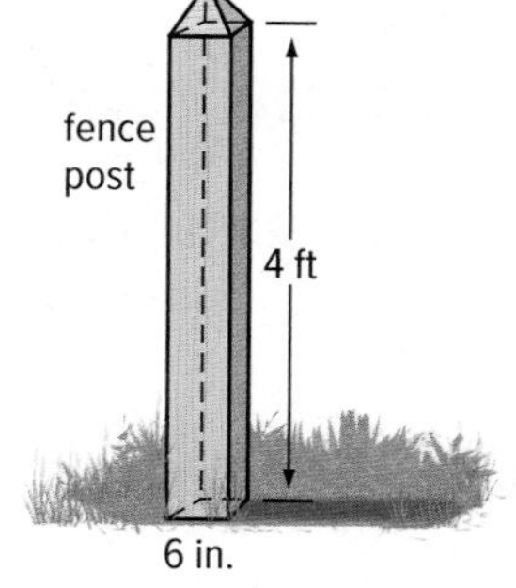

30. **COLLECT DATA** Use a ruler or tape measure and what you have learned in this lesson to find the surface area and volume of a soup can.

31. **CAKES** Cakes come in many shapes and sizes. Often they are stacked in two or more layers, like those in the diagrams shown below.

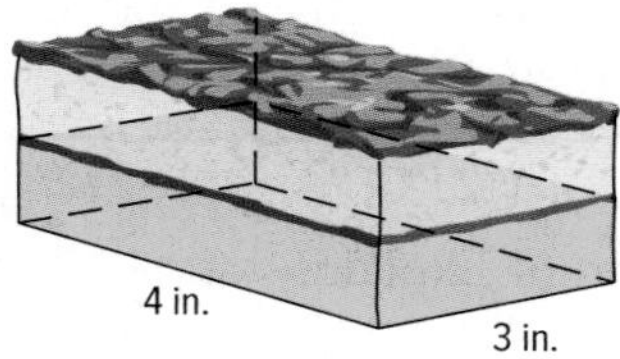

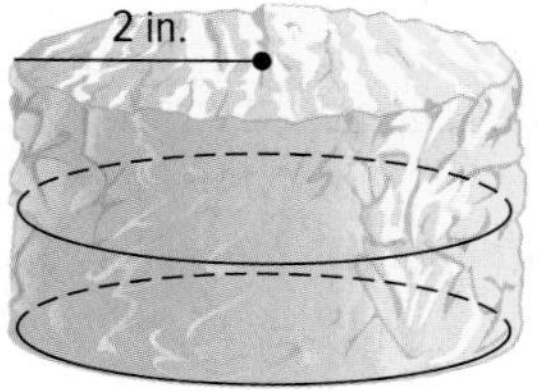

a. If each layer of the rectangular prism cake is 3 inches high, calculate the area of the cake that will be frosted assuming there is no frosting between layers.

b. Calculate the area of the cylindrical cake that will be frosted, if each layer is 4 inches in height.

c. If one can of frosting will cover 50 square inches of cake, how many cans of frosting will be needed for each cake?

d. If the height of each layer of cake is 5 inches, what does the radius of the cylindrical cake need to be, so the same amount of frosting is used for both cakes? Explain your reasoning.

32. **CHANGING UNITS** A gift box has a surface area of 6.25 square feet. What is the surface area of the box in square inches?

33. **CHANGING UNITS** A square pyramid has a volume of 4320 cubic inches. What is the volume of this pyramid in cubic feet?

34. **EULER'S FORMULA** The number of faces F, vertices V, and edges E of a polyhedron are related by Euler's (OY luhrz) Formula: $F + V = E + 2$. Determine whether Euler's Formula is true for each of the figures in Exercises 18–23.

35. **CHANGING DIMENSIONS** A rectangular prism has a length of 12 centimeters, width of 18 centimeters, and height of 22 centimeters. Describe the effect on the volume of a rectangular prism when each dimension is doubled.

36. **MULTIPLE REPRESENTATIONS** In this problem, you will investigate how changing the length of the radius of a cone affects the cone's volume.

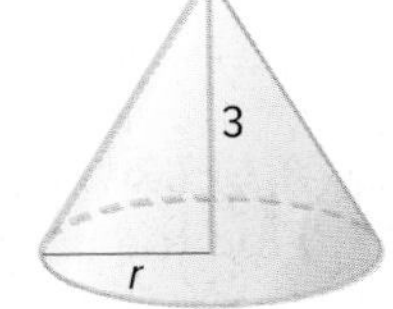

   a. **Tabular** Create a table showing the volume of a cone when doubling the radius. Use radius values between 1 and 8.

   b. **Graphical** Use the values from your table to create a graph of radius versus volume.

   c. **Verbal** Make a conjecture about the effect of doubling the radius of a cone on the volume. Explain your reasoning.

   d. **Algebraic** If r is the radius of a cone, write an expression showing the effect doubling the radius has on the cone's volume.

H.O.T. Problems Use Higher-Order Thinking Skills

37. **CRITIQUE** Alex and Emily are calculating the surface area of the rectangular prism shown. Is either of them correct? Explain your reasoning.

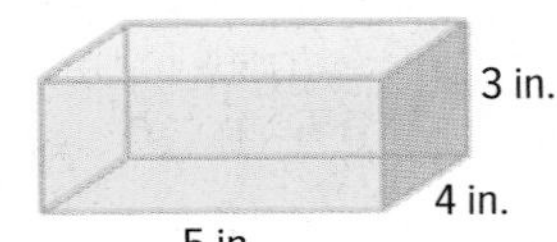

Alex

$(5 \cdot 3) \cdot 6$ faces

$= 90 \text{ in}^2$

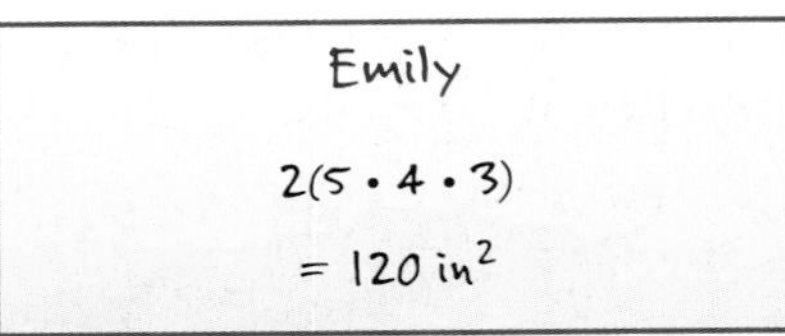

38. **REASONING** Is a cube a regular polyhedron? Explain.

39. **CHALLENGE** Describe the solid that results if the number of sides of each base increases infinitely. The bases of each solid are regular polygons inscribed in a circle.

   a. pyramid

   b. prism

40. **OPEN-ENDED** Draw an irregular 14-sided polyhedron which has two congruent bases.

41. **CHALLENGE** Find the volume of a cube that has a total surface area of 54 square millimeters.

42. **WRITING IN MATH** A reference sheet listed the formula for the surface area of a prism as $SA = Bh + 2B$. Use units of measure to explain why there must be a typographical error in this formula.

43. **MULTI-STEP** The radius of a cylindrical vase is 5 centimeters. The height of the vase is 21 centimeters. Jorge fills the vase to a height of 15 centimeters, as shown.

a. Which of the following is the best estimate of the volume of water Jorge must add to fill the vase completely? MP 5

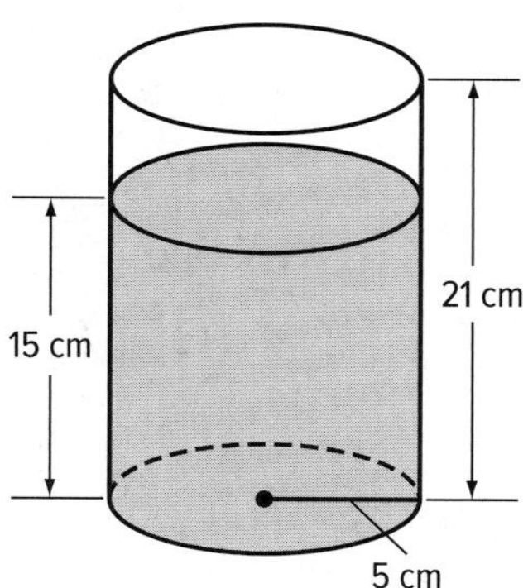

○ **A** 188 cm^3

○ **B** 471 cm^3

○ **C** 1178 cm^3

○ **D** 565 cm^3

b. Instead of filling the vase to the brim, Jorge decides to use the water in the vase to fill a cubic vase with side length 5 centimeters. How much water is left over?

c. Is the water left over enough to fill a spherical bowl of radius 6 centimeters? Explain.

44. An aquarium is a rectangular prism with an open top. The height and width of the aquarium are both 10 inches, and its length is 20 inches. What is the surface area of the aquarium in square inches? MP 5

45. A stand at the state fair sells peanuts in containers shaped like a square pyramid. The dimensions of the containers are shown.

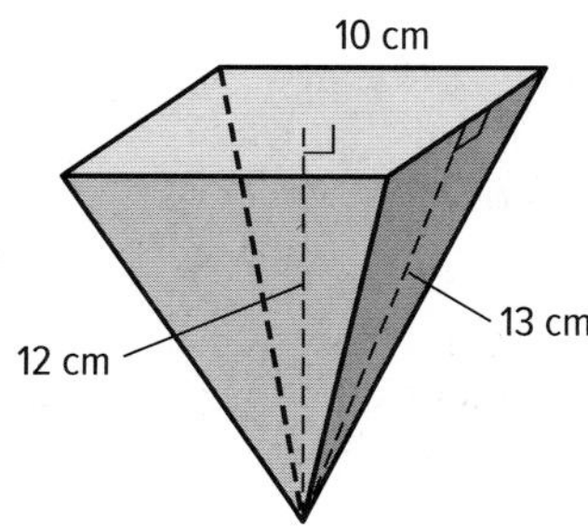

Which of the following shows the amount of peanuts that can fit in each container? MP 5

☐ **A** 260 cm^3

☐ **B** 360 cm^3

☐ **C** 400 cm^3

☐ **D** 600 cm^3

☐ **E** 1200 cm^3

46. A toy store sells beach balls with the dimensions shown in the figure.

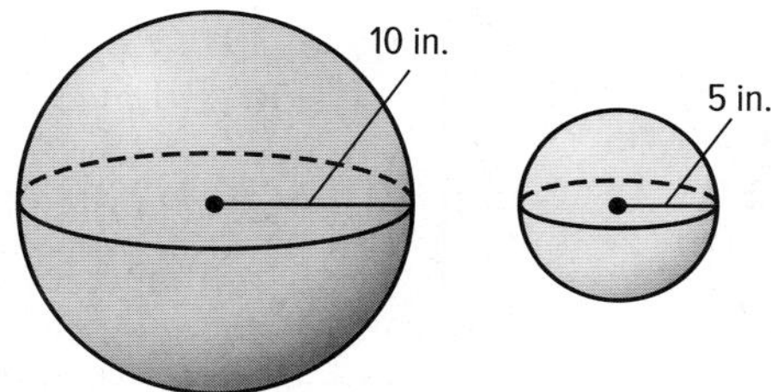

Based on this information, which of the following statements is true? MP 5

○ **A** The surface area of the larger beach ball is two times the surface area of the smaller beach ball.

○ **B** The surface area of the larger beach ball is four times the surface area of the smaller beach ball.

○ **C** The surface area of the larger beach ball is eight times the surface area of the smaller beach ball.

○ **D** The surface area of the larger beach ball is 5 square inches more than the surface area of the smaller beach ball.

47. Each dimension of a rectangular prism is doubled. By what factor does the surface area and volume increase, respectively? MP 2

48. The radius of a cone is doubled and the volume stays the same. What can you say about the height of the cone? MP 2

LESSON 9

Two-Dimensional Representations of Three-Dimensional Figures

Then

- You named and drew different three-dimensional figures.

Now

- You will represent the three-dimensional figures as two-dimensional figures with orthographic drawings and nets.

Why?

- You can use orthographic drawings similar to the ones shown below to help determine the shape of the base of an ancient Egyptian pyramid.

New Vocabulary

orthographic drawing
net

Mathematical Practices

4 Model with mathematics.

1 Orthographic Drawings and Nets

If you see a three-dimensional object from only one viewpoint, you may not know its true shape. Here are four views of a square pyramid.

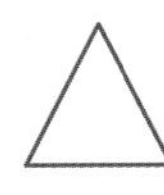

top view | left view | front view | right view

The two-dimensional views of the top, left, front, and right sides of an object are called an **orthographic drawing.**

Example 1 Use Orthographic Drawings

Make a model of a figure for the orthographic drawing shown.

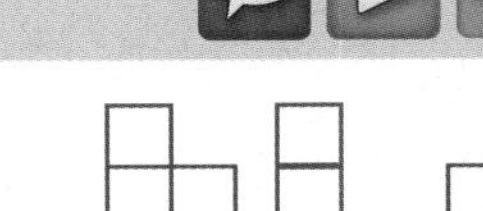

top view | left view | front view | right view

Step 1 Start with a base that matches the top view.

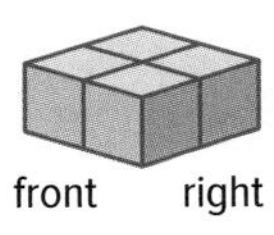

Step 2 The front view indicates that the front left side is 5 blocks high and that the right side is 3 blocks high. However, the dark segments indicate breaks in the surface.

Step 3 The break on the left side of the front view indicates that the back left column is 5 blocks high, but that the front left column is only 4 blocks high, so remove 1 block from the front left column.

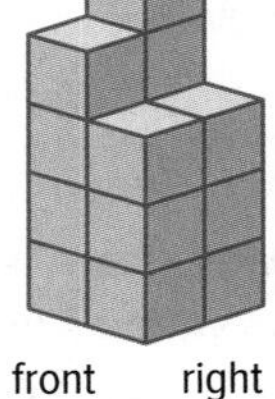

Step 4 The break on the right side of the front view indicates that the back right column is 3 blocks high, but that the front right column is only 1 block high, so remove 2 blocks from the front right column.

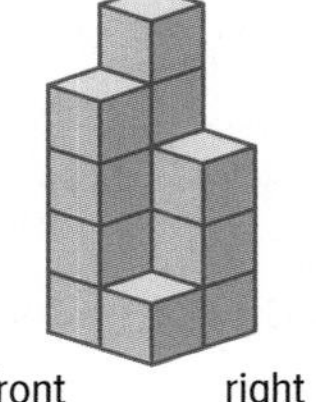

Step 5 Use the left and right views and the breaks in those views to confirm that you have made the correct figure.

Guided Practice

1A. Make a model of a figure for the orthographic drawing shown.

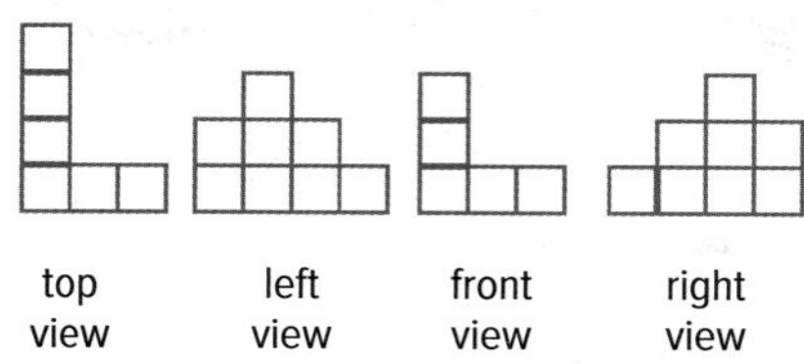

1B. Make an orthographic drawing of the figure shown.

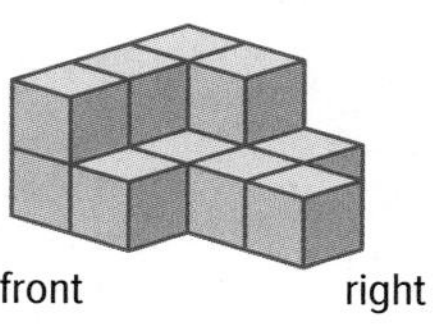

If you cut a cardboard box at the edges and lay it flat, you will have a two-dimensional diagram called a **net** that you can fold to form a three-dimensional solid.

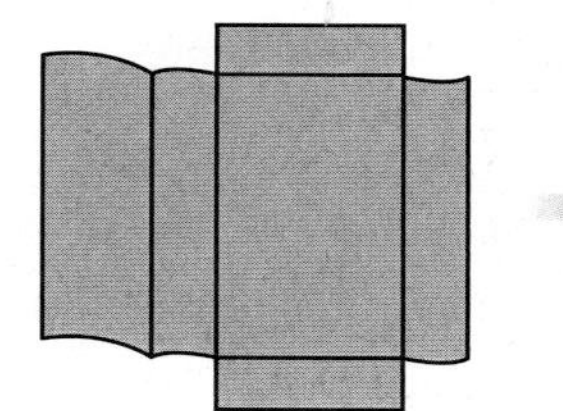

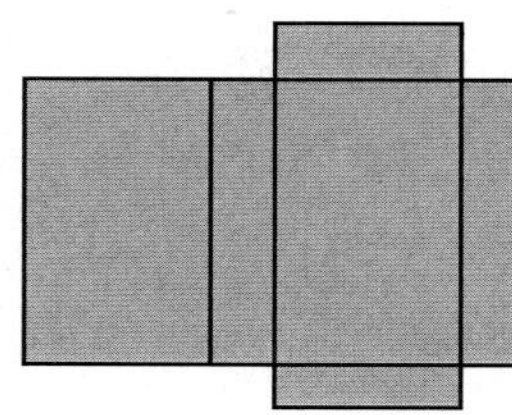

Study Tip

Precision Note that both triangles in the net are identical. It's important to check your measurements before cutting.

Example 2 Use Nets to Identify Solids

Make a model of a figure for the given net. Then identify the solid formed.

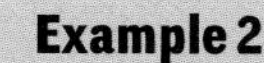

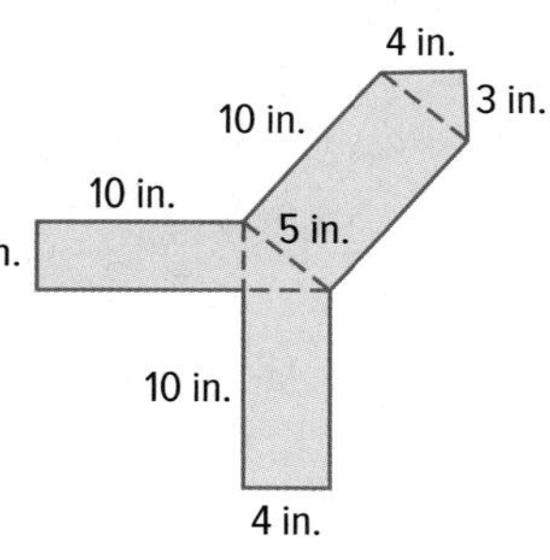

Use a large sheet of paper, a ruler, scissors, and tape. Draw the net on the paper. Cut along the solid lines. Fold the pattern on the dashed lines and secure the edges with tape. This is the net of a triangular prism.

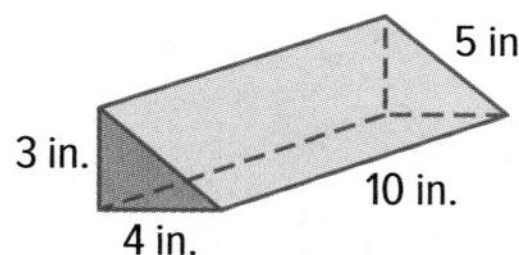

Guided Practice

Make a model of a figure for each net. Then identify the solid formed. If the solid has more than one name, list both.

2A.

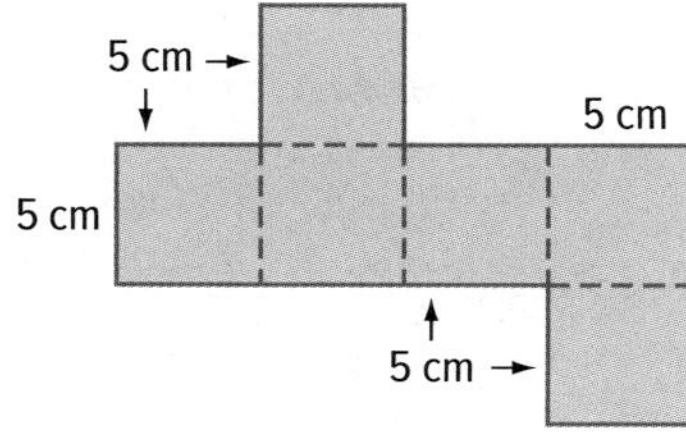

2B.

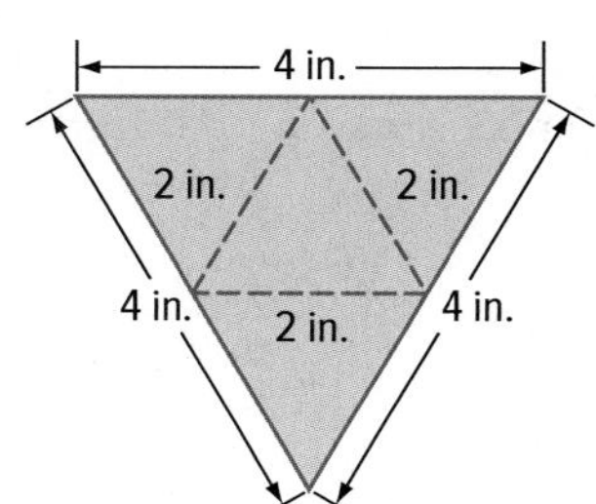

To draw the net of a three-dimensional solid, visualize cutting the solid along one or more of its edges, opening up the solid, and flattening it completely.

Study Tip

MP Model The congruence marks mean that each line segment is the same length as the other ones. You can't always assume segments that look the same are the same length unless these marks are there.

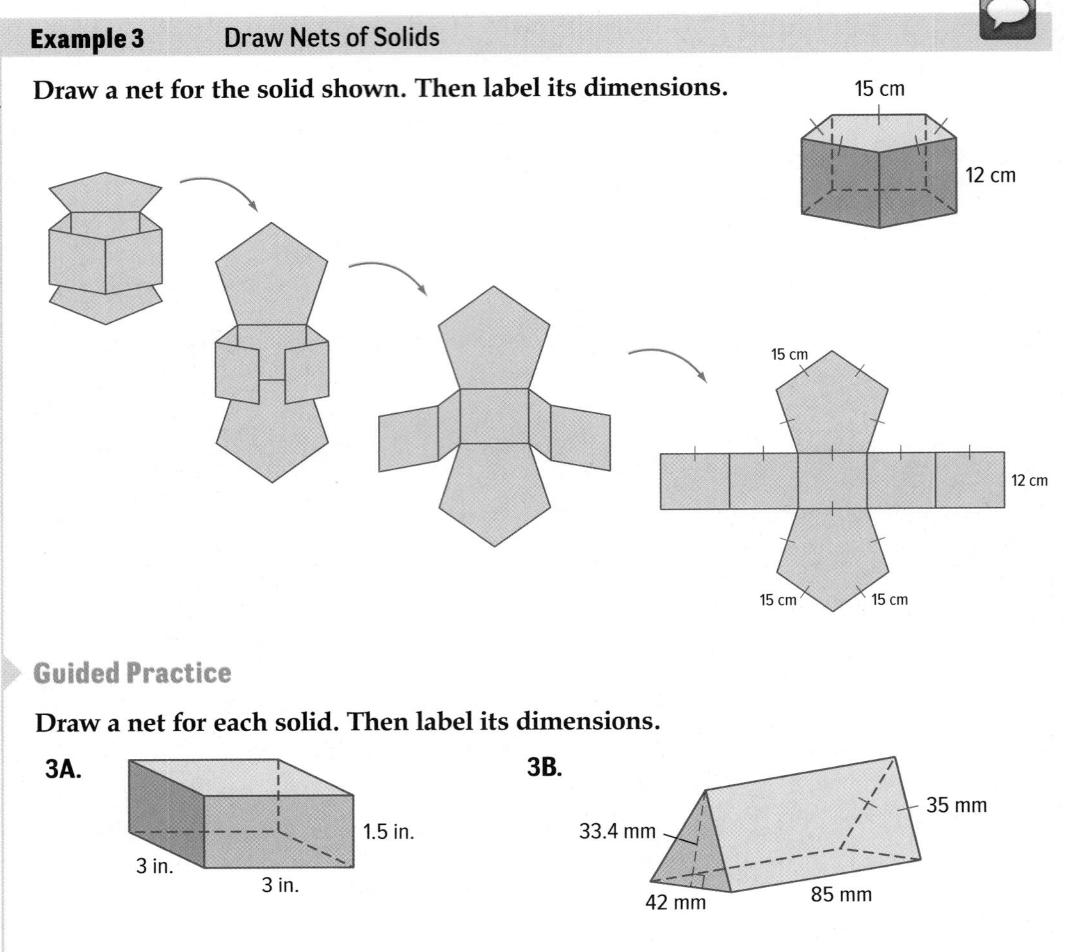

Example 3 Draw Nets of Solids

Draw a net for the solid shown. Then label its dimensions.

Guided Practice

Draw a net for each solid. Then label its dimensions.

3A.

3B.

Check Your Understanding

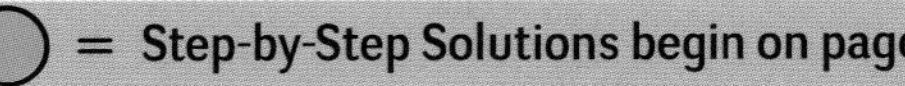

= Step-by-Step Solutions begin on page R13.

Go Online! for a Self-Check Quiz

Example 1

1. Make a model of a figure for the orthographic drawing shown.

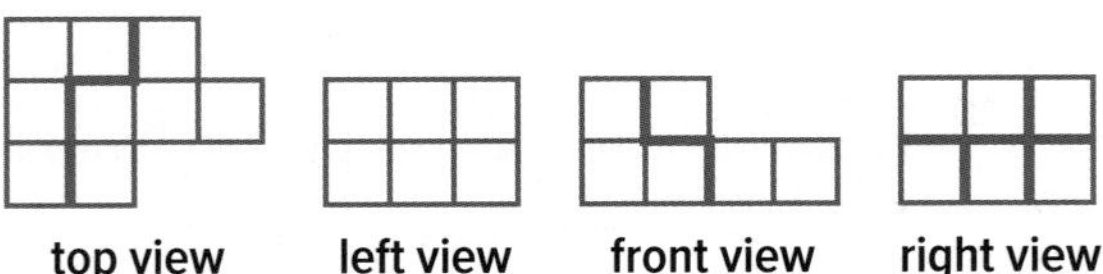

Example 2

Make a model of a figure for the given net. Then identify the solid formed.

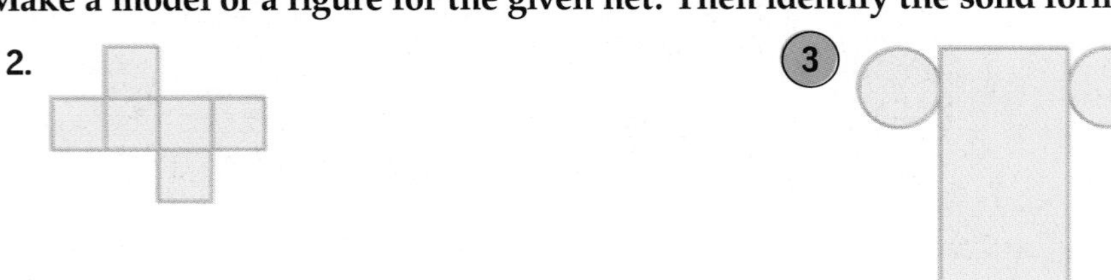

Example 3

4. MP **TOOLS** Draw a net for the figure.

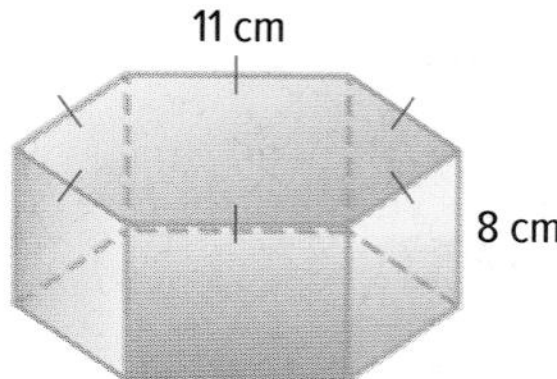

Practice and Problem Solving

Extra Practice is on page R1.

Example 1

Draw the requested view of each model. Be sure to include breaks where necessary.

5. left view

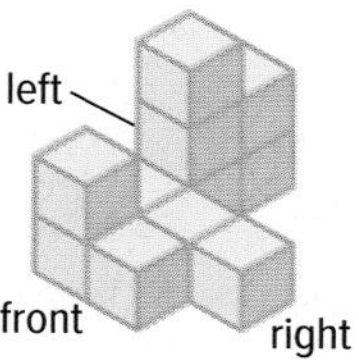

6. front view

Example 2

Draw a figure for each net.

7.

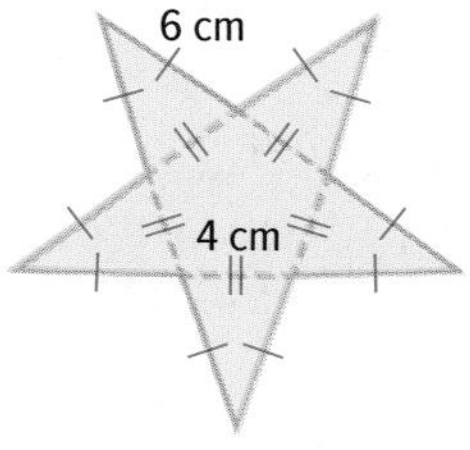

8.

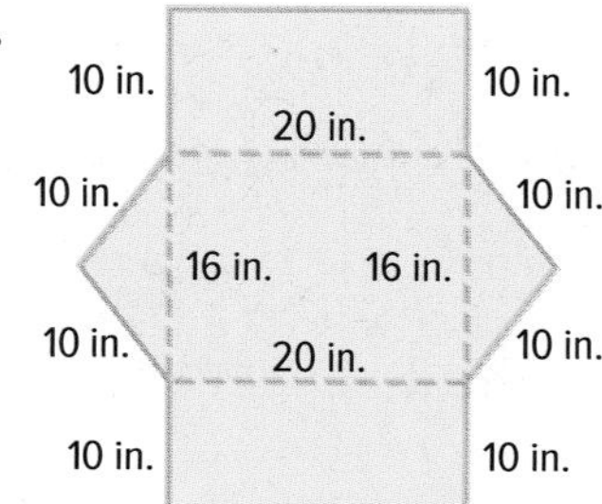

Example 3

Draw a net for each figure.

9.

10.

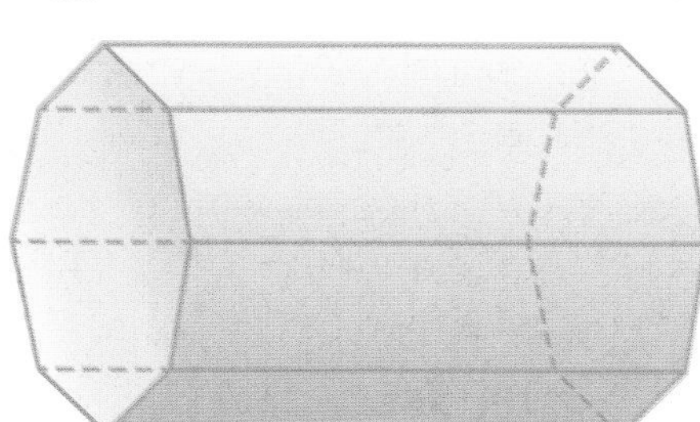

MP **MODELING** **Draw a net for the gift box and the camping tent.**

11.

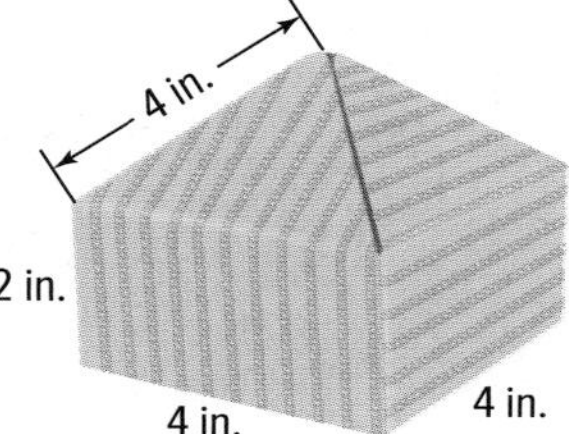

12.

5 ft

4 ft

5 ft

7 ft

13. **MANUFACTURING DRAWINGS** The project manager for a new statue at city hall sent orthographic drawings of the platform for the statue to a manufacturer. Draw the platform from the drawings.

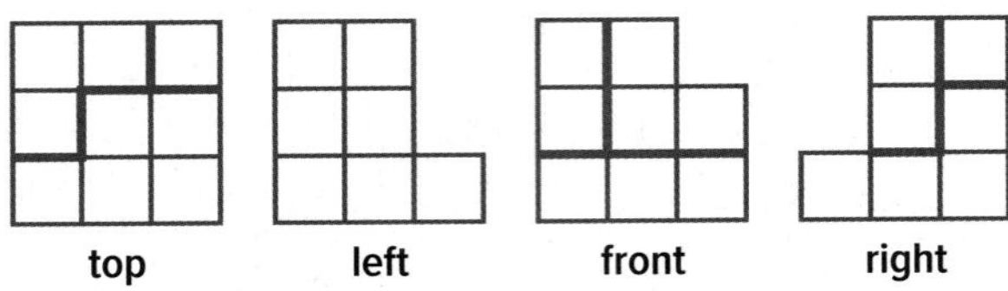

14. Explain why each net will or will not make the fish tank shown.

a.

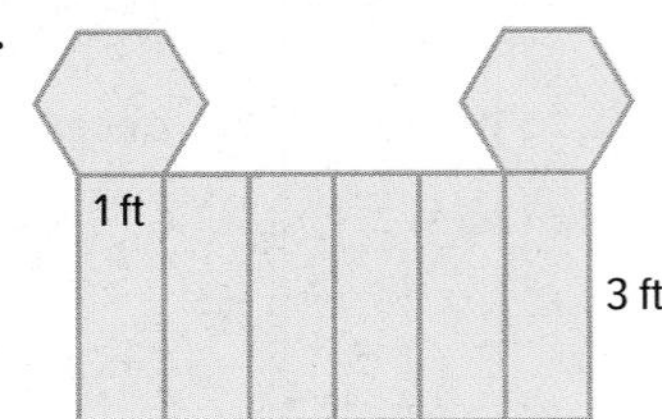

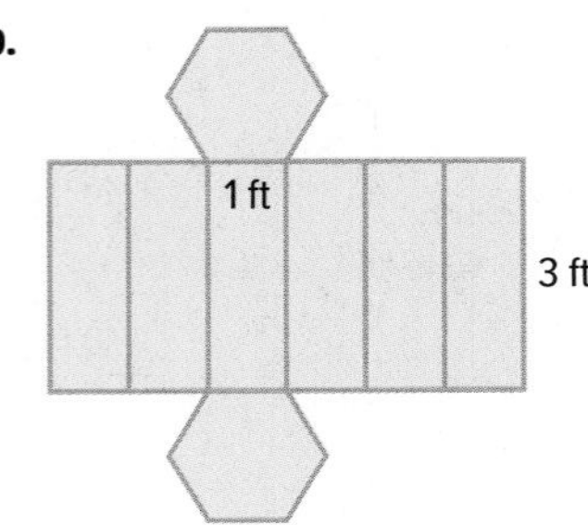

b.

c.

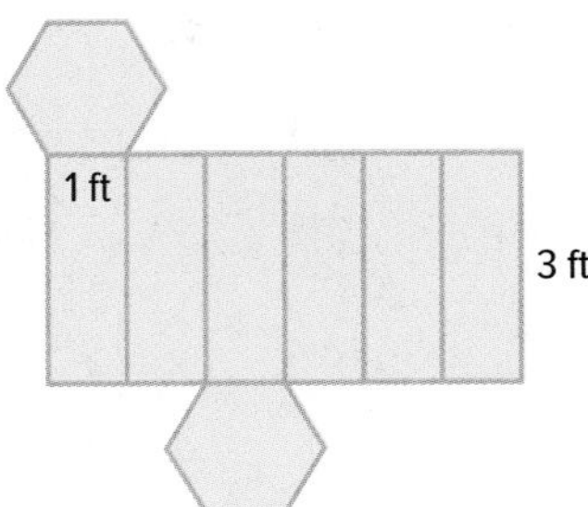

15. Which figure could be made with the net shown?

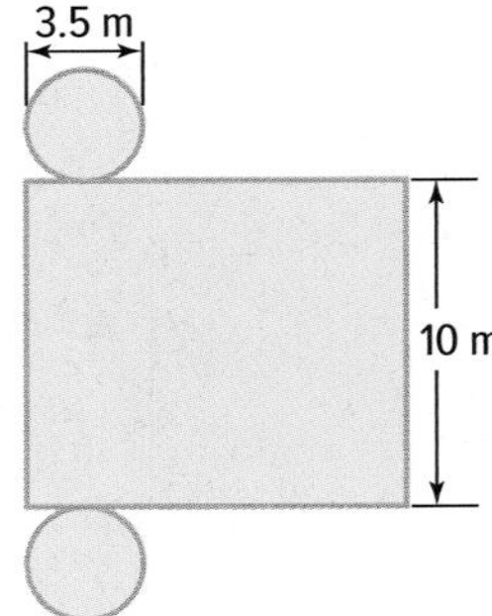

A.

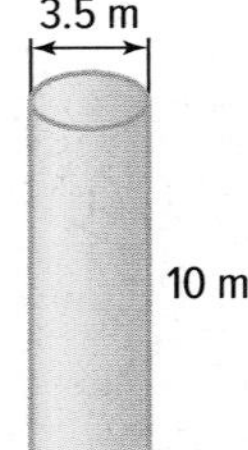

B.

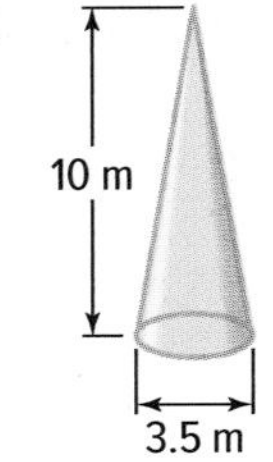

16. **MP PERSEVERANCE** Draw five different nets that would form the cube. How many different nets are there?

17. **MP MODELING** A can of pineapple is shown.

   a. What shape is the top and bottom of the can?

   b. If you remove the top and bottom and then make a vertical cut down the side of the can, what shape will you get when you uncurl the remaining body of the can and flatten it?

   c. If the diameter of the can is 3 inches and its height is 2 inches, draw a net of the can and label its dimensions. Explain your reasoning.

H.O.T. Problems Use Higher-Order Thinking Skills

18. **WRITING IN MATH** Explain when orthographic drawings might be useful and when nets might be useful in real-life.

19. **MP REASONING** Identify the Platonic solid that can be formed by the given net.

   a.

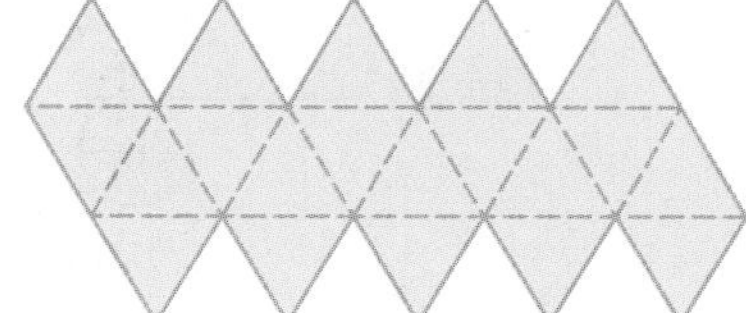

   b.

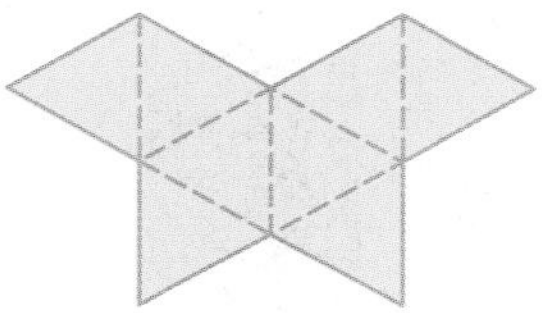

   c.

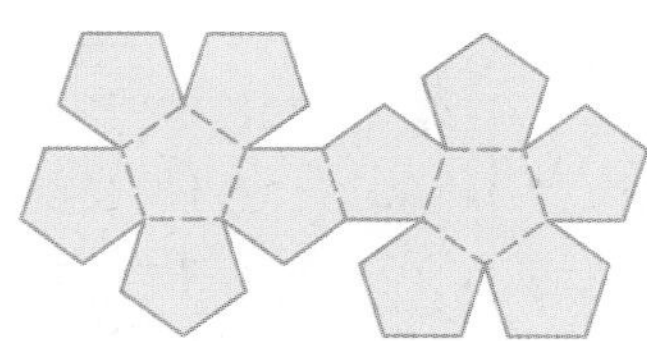

20. **CHALLENGE** The surface area of a figure is the sum of the area of each face. The center face of the net shown is a square.

   a. How many faces are there in this net?

   b. What type of figure does the net form?

   c. What is the area of the square?

   d. What is the area of one triangle? four triangles?

   e. What is the sum of the areas of the surface?

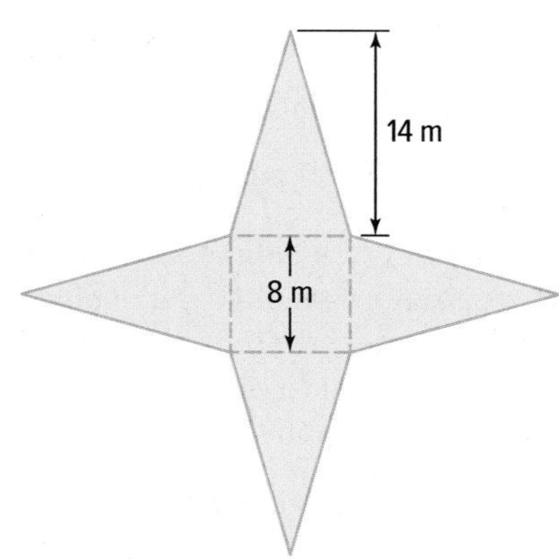

21. **MULTI-STEP** Which model corresponds to the orthographic drawing? MP 4

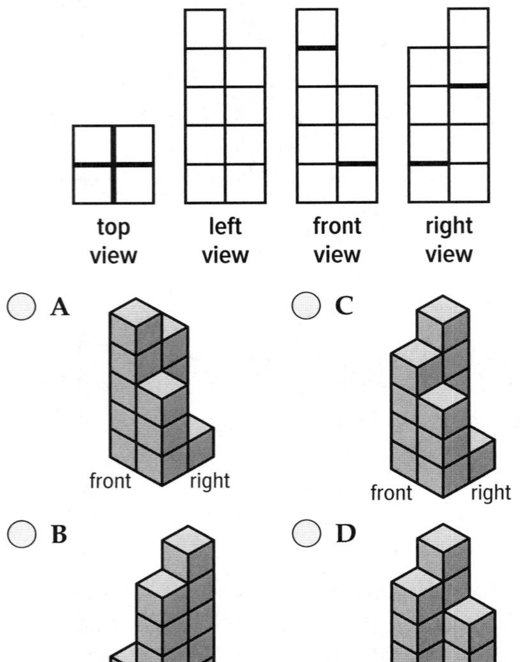

22. The design of a building was sent from the architect to the investors. The investors asked for orthographic drawings as well. Create an orthographic front view of the building. Include breaks where necessary. MP 4

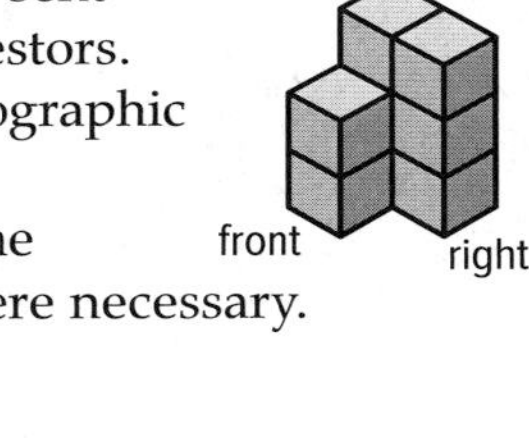

23. Which of the following is the most precise description for the net of a square pyramid? MP 6

- A 1 square and 3 triangles
- B 1 square and 4 triangles
- C 4 triangles
- D 3 triangles

24. The net of a solid consists of six congruent rectangles stacked vertically. On one of the rectangles, there is a regular polygon on the left and right side of the rectangle. What is the resulting solid? MP 1, 6

- A right triangular prism
- B right pentagonal prism
- C right hexagonal prism
- D right octagonal prism

25. Which net or nets correspond to the cylinder? Choose all that apply. MP 4

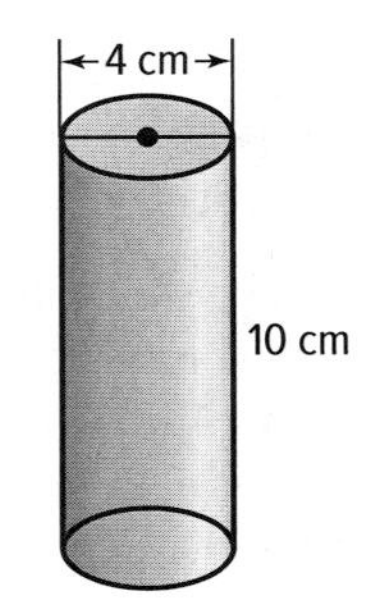

A
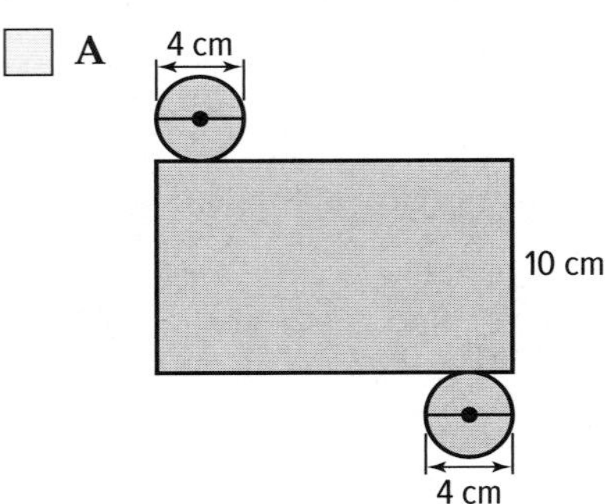

B
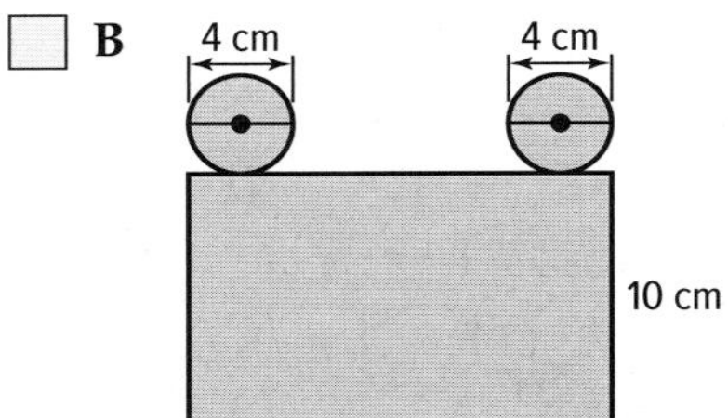

C
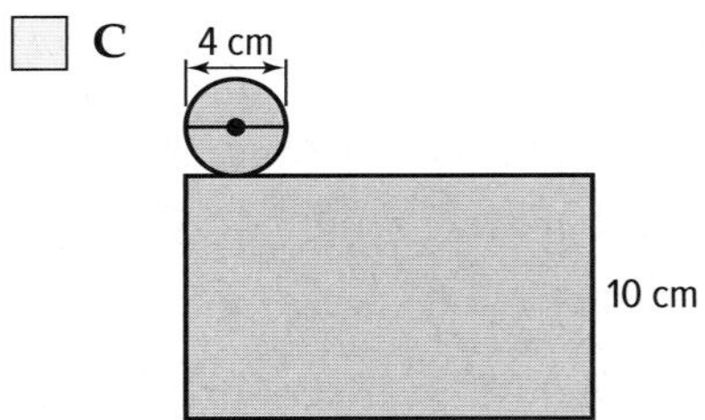

D
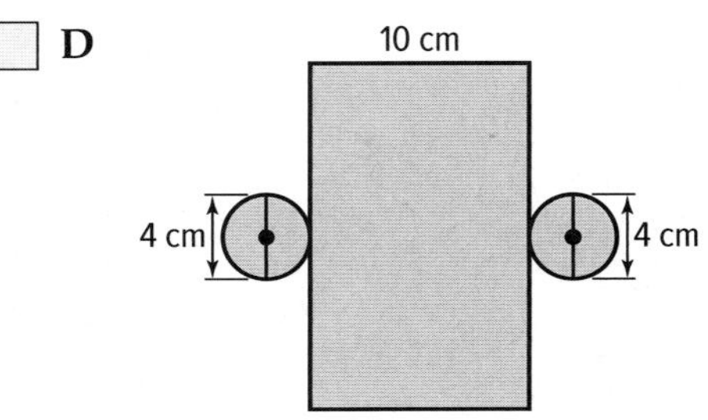

LESSON 10

Precision and Accuracy

Then

- You represented three-dimensional figures.

Now

1. Determine precision of measurements.
2. Determine accuracy of measurements.

Why?

- Precision and accuracy play an important role in archery. In archery competitions, each arrow's distance from the center of the target is measured to within 0.2 millimeter.

 **New Vocabulary**

precision
absolute error
significant digits
accuracy
relative error

 **Mathematical Practices**

5 Use appropriate tools strategically.
6 Attend to precision.

1 Precision

Precision refers to the clustering of a group of measurements. It depends only on the smallest unit of measure available on a measuring tool. Suppose you are told that a segment measures 8 centimeters. The length, to the nearest centimeter, of each segment shown below is 8 centimeters.

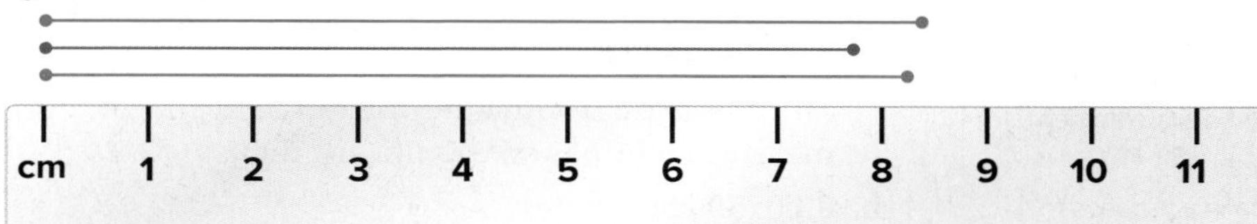

Notice that the exact length of each segment above is between 7.5 and 8.5 centimeters, or within 0.5 centimeter of 8 centimeters. The **absolute error** of a measurement is equal to one half the unit of measure. A smaller unit of measure provides a more precise measurement.

Example 1 Find Absolute Error

Find the absolute error of each measurement. Then explain its meaning.

a. 6.4 centimeters

The measure is given to the nearest 0.1 centimeter, so the absolute error of this measurement is $\frac{1}{2}(0.1)$ or 0.05 centimeter. Therefore, the exact measurement could be between 6.35 and 6.45 centimeters. The two segments below measure 6.4 ± 0.05 centimeters.

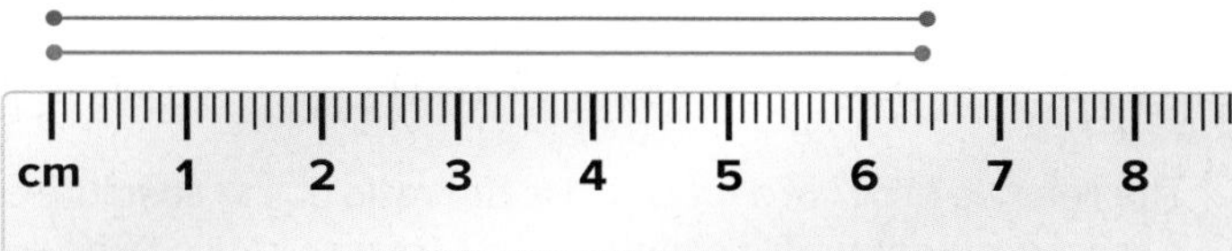

b. $2\frac{1}{4}$ inches

The measure is given to the nearest $\frac{1}{4}$ inch, so the absolute error of this measurement is $\frac{1}{2}\left(\frac{1}{4}\right)$ or $\frac{1}{8}$ inch. Therefore, the exact measurement could be between $2\frac{1}{8}$ and $2\frac{3}{8}$ inches. The two segments below measure $2\frac{1}{4} \pm \frac{1}{8}$ inches.

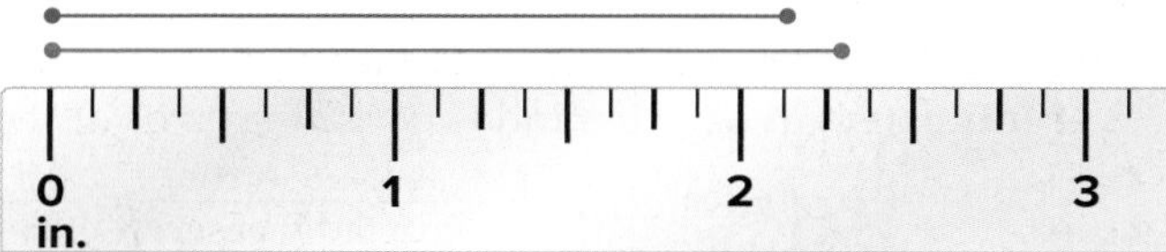

Guided Practice

1A. $1\frac{1}{2}$ inches

1B. 4 centimeters

Precision in a measurement is usually expressed by the number of **significant digits** reported. Reporting that the measure of $\overline{AB}$ is 4 centimeters is *less precise* than reporting that the measure of $\overline{AB}$ is 4.1 centimeters.

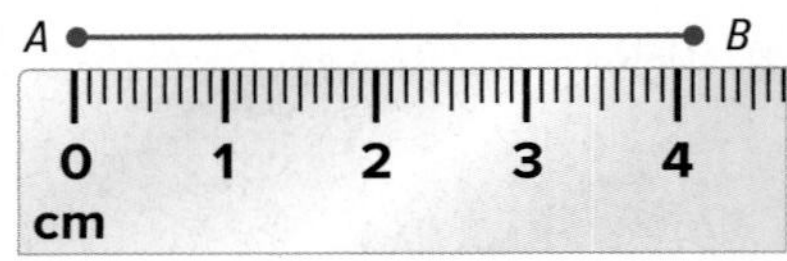

Real-World Link

Precision in measurement in the real world usually comes at a price.

- Precision in a process to 3 significant digits, commercial quality, can cost $100.
- Precision in a process to 4 significant digits, industrial quality, can cost $500.
- Precision in a process to 5 significant digits, scientific quality, can cost $2500.

Source: Southwest Texas Junior College

Key Concept Significant Digits

To determine whether digits are considered significant, use the following rules.

- Nonzero digits are always significant.
- In whole numbers, zeros are significant if they fall between nonzero digits.
- In decimal numbers greater than or equal to 1, every digit is significant.
- In decimal numbers less than 1, the first nonzero digit and every digit to the right are significant.

Example 2 Determine the Number of Significant Digits

Determine the number of significant digits in each measurement.

a. 430.008 meters

Since this is a decimal number greater than 1, every digit is significant. So, this measurement has six significant digits.

b. 0.00750 centimeter

This is a decimal number less than 1. The first nonzero digit is 7, and there are two digits to the right of 7, 5, and 0. So, this measurement has three significant digits.

Guided Practice

2A. 779,000 mi **2B.** 50,008 ft **2C.** 230.004500 m

2 Accuracy

Accuracy refers to how close a measured value comes to the actual or desired value. Consider the target practice results shown below.

accurate and precise

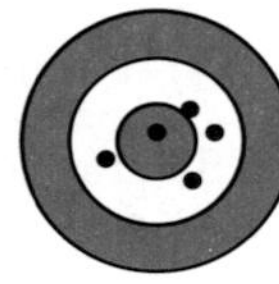
accurate but not precise

precise but not accurate

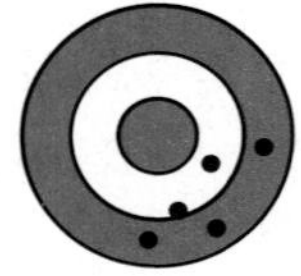
not accurate and not precise

The **relative error** of a measure is the ratio of the absolute error to the expected measure. A measurement with a smaller relative error is said to be more accurate.

Key Concept Relative Error

$$\text{relative error} = \frac{\text{absolute error}}{\text{expected measurement}}$$

Real-World Example 3 Find Relative Error

MANUFACTURING A manufacturer measures each part for a piece of equipment to be 23 centimeters in length. Find the relative error of this measurement.

$$\text{relative error} = \frac{\text{absolute error}}{\text{expected measure}} = \frac{0.5 \text{ cm}}{23 \text{ cm}} \approx 0.022 \text{ or } 2.2\%$$

Guided Practice

3. CITY PLANNING A city planner makes a proposal for two new parks. The distance between the parks is 3.2 miles. Find the relative error of this measurement.

Study Tip

Accuracy The accuracy or relative error of a measurement depends on both the absolute error and the size of the object being measured.

Check Your Understanding

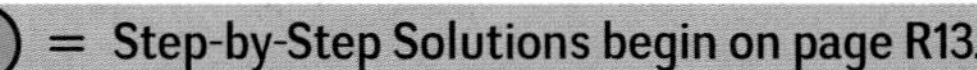

Example 1 **Find the absolute error of each measurement. Then explain its meaning.**

1. $2\frac{1}{8}$ ft **2.** 4.81 mm **3.** 9 mi

Example 2 **Determine the number of significant digits in each measurement.**

4. 0.00503 m **5.** 95,001 yd **6.** 7.0050 cm

Example 3 **7.** **PHOTOGRAPHY** For an art show, a photographer makes prints that are each 1 foot tall. Find the relative error of this measurement.

Practice and Problem Solving

Extra Practice is found on Page R1.

Example 1 **Find the absolute error of each measurement. Then explain its meaning.**

8. 12 yd **9.** $50\frac{4}{16}$ in. **10.** 3.28 ft

11. 2.759 cm **12.** $14\frac{7}{8}$ mi **13.** 8.001 m

Example 2 **Determine the number of significant digits in each measurement.**

14. 4.05 in. **15.** 53,000 mi **16.** 0.0005 mm

17. 750,001 ft **18.** 470 yd **19.** 0.04005 cm

Example 3 **20.** **ARCHITECTURE** An architect designs a lobby for an apartment building. The lobby is 26 feet wide. Find the relative error of this measurement.

21. **GARDENS** A rectangular flower bed at a botanical garden is 4.5 meters long and 3.2 meters wide. Find the relative error of these measurements.

Find the relative error of each measurement.

22. 48 in. **23.** 2.0 mi **24.** 11.14 cm **25.** 0.6 m

Determine which measurement is more precise and which is more accurate. Explain your reasoning.

26. 22.4 ft; 5.82 ft **27.** 25 mi; 8 mi **28.** 9.2 cm; 42 mm **29.** $18\frac{1}{4}$ in.; 125 yd

For each situation, determine the level of accuracy needed. Explain.

30. You are estimating the height of a person. Which unit of measure should you use: 1 foot, 1 inch, or $\frac{1}{16}$ inch?

31. You are estimating the height of a mountain. Which unit of measure should you use: 1 foot, 1 inch, or $\frac{1}{16}$ inch?

32. **VOLUME** When multiplying or dividing measures, the product or quotient should have only as many significant digits as the multiplied or divided measurement showing the least number of significant digits. To how many significant digits should the volume of the rectangle prism shown be reported? Report the volume to this number of significant digits.

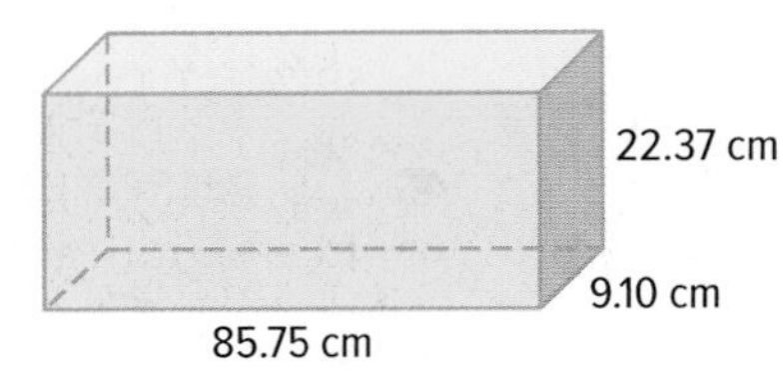

TOOLS Use a ruler to measure the given line segment to the indicated degree of precision.

33. nearest $\frac{1}{2}$ inch

34. nearest $\frac{1}{4}$ inch

35. nearest centimeter

36. nearest 0.5 centimeter

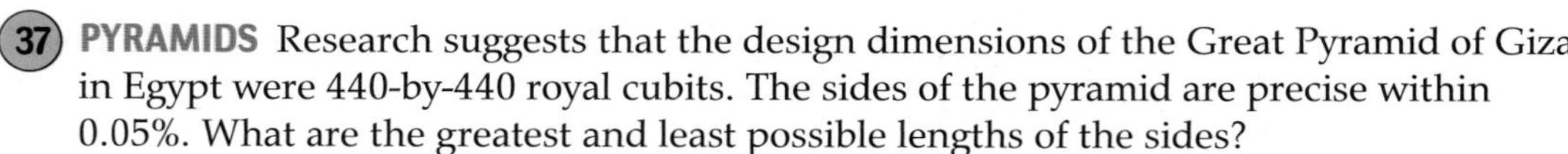

37. **PYRAMIDS** Research suggests that the design dimensions of the Great Pyramid of Giza in Egypt were 440-by-440 royal cubits. The sides of the pyramid are precise within 0.05%. What are the greatest and least possible lengths of the sides?

38. **PERIMETER** The *perimeter* of a geometric figure is the sum of the lengths of its sides. Jermaine uses a ruler divided into quarter-inches and measures the sides of a rectangle to be $2\frac{1}{4}$ inches and $4\frac{3}{4}$ inches. What are the least and greatest possible perimeters of the rectangle? Explain.

39. **TOOLS** Eduardo is planning to measure the width of a window in his bedroom so he can buy new curtains. He wants the absolute error of the measurement to be at most $\frac{1}{32}$ inch.
   a. What type of tool should Eduardo use to make the measurement? Explain.
   b. Eduardo finds that the width of the window is 30 inches. Assuming he uses the tool you described in part **a**, what are the minimum and maximum possible widths of the window?

When you add or subtract measures, you should round the sum or difference to the same place as the last significant digit of the least precise measurement. Add or subtract and use this rule to write each answer with the correct number of significant digits.

40. 23.48 m + 12.2 m

41. 16.11 km − 9.502 km

42. 7.1 cm + 8.5 cm + 12.71 cm

43. **PERSEVERANCE** Aisha measures the width of a table to the nearest centimeter. She finds that the relative error for the measurement is 1.25%. What is the width of the table according to Aisha's measurement?

H.O.T. Problems Use Higher-Order Thinking Skills

44. **OPEN-ENDED** Write a measurement that is less than 1 centimeter and that has three significant digits. Then write a measurement greater than 1 centimeters that has three significant digits.

45. **ERROR ANALYSIS** In biology class, Manuel and Jocelyn measure a beetle as shown. Manuel says that the beetle measures between $1\frac{5}{8}$ and $1\frac{3}{4}$ inches. Jocelyn says that it measures between $1\frac{9}{16}$ and $1\frac{5}{8}$ inches. Is either of their statements about the beetle's measure correct? Explain your reasoning.

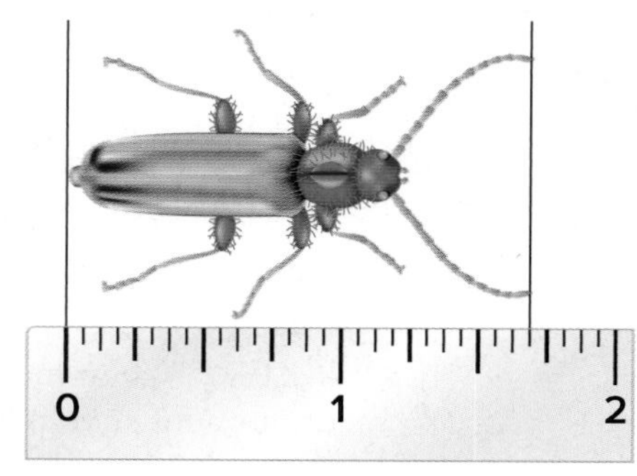

46. **REASONING** What is the least whole number with four significant digits?

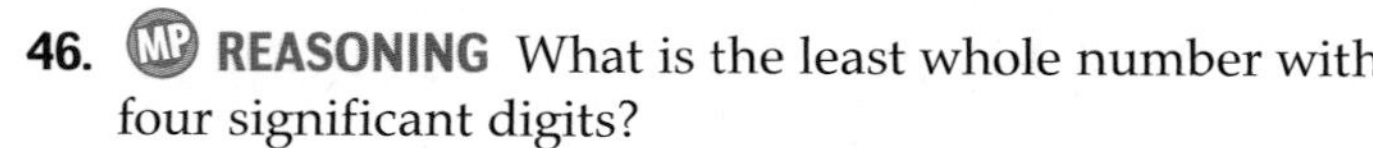

47. **WRITING IN MATH** How precise is precise enough?

48. David measures a coin and finds that its diameter is 2.5 centimeters. What is the absolute error of the measurement? MP 6

- A 0.02 cm
- B 0.05 cm
- C 0.1 cm
- D 1.25 cm

49. Which of these measurements have three significant digits? Select all that apply. MP 6

- A 0.00282 km
- B 800 yd
- C 1.04 cm
- D 5280 ft
- E 2.4500 mm
- F 905 mi

50. A geologist finds that the mass of a rock is 1.8 kilograms. What is the relative error for the measurement to the nearest tenth of a percent? MP 6

relative error = ☐

51. Four students were asked to choose an appropriate tool and measure the length of a wall in their classroom. The table shows their results. Which student's measurement is the most precise? MP 5, 6

Student	Measurement
Alyssa	4 m
Benjamin	422 cm
Carmina	400.1 cm
DeShawn	4.2 m

- A Alyssa
- B Benjamin
- C Carmina
- D DeShawn

52. **MULTI-STEP** Tamiko measures a picture frame and finds that its length is 20 centimeters. MP 6

a. Determine the number of significant digits in Tamiko's measurement.

b. Find the relative error of this measurement.

c. Suppose Tamiko reports the length of the frame in millimeters. Does this increase or decrease the relative error of the measurement? Explain.

53. A brochure lists the height of a new office building as 202.0 meters. What is the number of significant digits in this measurement? MP 6

- A 1
- B 2
- C 3
- D 4

54. Which expression can be used to find the relative error for the measurement 4.2 kilograms? MP 6

- A $\frac{0.05}{4.2}$
- B $\frac{0.5}{4.2}$
- C $\frac{4.2}{0.05}$
- D $\frac{4.2}{0.5}$

55. Which of the following tools allow you to make more precise measurements than a ruler that shows only $\frac{1}{4}$-inch marks? Select all that apply. MP 5

- A a ruler that shows $\frac{1}{8}$-inch marks
- B a ruler that shows $\frac{1}{2}$-inch marks
- C a ruler that shows millimeters
- D a ruler that shows tenths of a centimeter
- E a ruler that shows tenths of a meter

Study Guide and Review

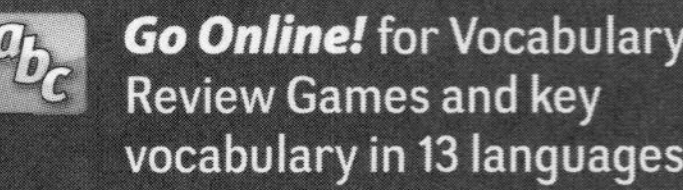

Study Guide

Key Concepts

Points, Lines, and Planes (Lesson 1-1)

- There is exactly one line through any two points.
- There is exactly one plane through any three noncollinear points.

Distance and Midpoints (Lessons 1-2 and 1-3)

- On a number line, the measure of a segment with endpoint coordinates a and b is $|a - b|$.
- In the coordinate plane, the distance between two points (x_1, y_1) and (x_2, y_2) is given by $d = \sqrt{(x_2 - x_1)^2 + (y_2 - y_1)^2}$.
- On a number line, the coordinate of the midpoint of a segment with endpoints a and b is $\frac{a + b}{2}$.
- In the coordinate plane, the coordinates of the midpoint of a segment with endpoints that are (x_1, y_1) and (x_2, y_2) are $\left(\frac{x_1 + x_2}{2}, \frac{y_1 + y_2}{2}\right)$.

Angles (Lessons 1-4 and 1-5)

- An angle is formed by two noncollinear rays that have a common endpoint, called its vertex. Angles can be classified by their measures.
- Adjacent angles are two coplanar angles that lie in the same plane and have a common vertex and a common side but no common interior points.
- Vertical angles are two nonadjacent angles formed by two intersecting lines.
- A linear pair is a pair of adjacent angles with noncommon sides that are opposite rays.
- Complementary angles are two angles with measures that have a sum of 90.
- Supplementary angles are two angles with measures that have a sum of 180.

Transformations (Lesson 1-7)

- In a congruence transformation, the position of the image may differ from the preimage, but the two figures remain congruent.

FOLDABLES® Study Organizer

Use your Foldable to review the chapter. Working with a partner can be helpful. Ask for clarification of concepts as needed.

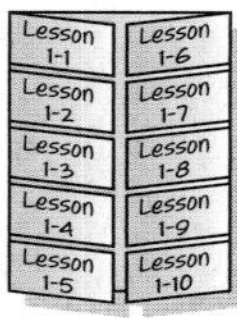

Key Vocabulary

acute angle (p. 38)
adjacent angles (p. 46)
angle (p. 36)
angle bisector (p. 39)
area (p. 58)
base (p. 76)
circumference (p. 58)
collinear (p. 5)
complementary angles (p. 47)
concave (p. 56)
cone (p. 76)
congruent (p. 15)
convex (p. 56)
coplanar (p. 5)
cylinder (p. 76)
degree (p. 37)
distance (p. 16)
edge (p. 76)
equiangular polygon (p. 57)
equilateral polygon (p. 57)
exterior (p. 36)
face (p. 76)
interior (p. 36)
intersection (p. 6)
line (p. 5)
line segment (p. 14)
linear pair (p. 46)
midpoint (p. 26)
***n*-gon** (p. 57)
obtuse angle (p. 38)
opposite rays (p. 36)
perimeter (p. 58)
perpendicular (p. 48)
plane (p. 5)
Platonic solid (p. 77)
point (p. 5)
polygon (p. 56)
polyhedron (p. 76)
prism (p. 76)
pyramid (p. 76)
ray (p. 36)
regular polygon (p. 57)
regular polyhedron (p. 77)
right angle (p. 38)
segment bisector (p. 27)
side (p. 36)
sphere (p. 76)
supplementary angles (p. 47)
surface area (p. 78)
undefined term (p. 5)
vertex (pp. 36, 76)
vertex of a polygon (p. 56)
vertical angles (p. 46)
volume (p. 78)

Vocabulary Check

Fill in the blank in each sentence with the vocabulary term that best completes the sentence.

1. A _____ is a flat surface made up of points that extends infinitely in all directions.
2. A set of points that all lie on the same line are said to be _____.

Concept Check

3. If two lines are perpendicular, then how many right angles are formed at the intersection?
4. What is the difference between complementary and supplementary angles?

Lesson-by-Lesson Review 

1-1 Points, Lines, and Planes

Use the figure to complete each of the following.

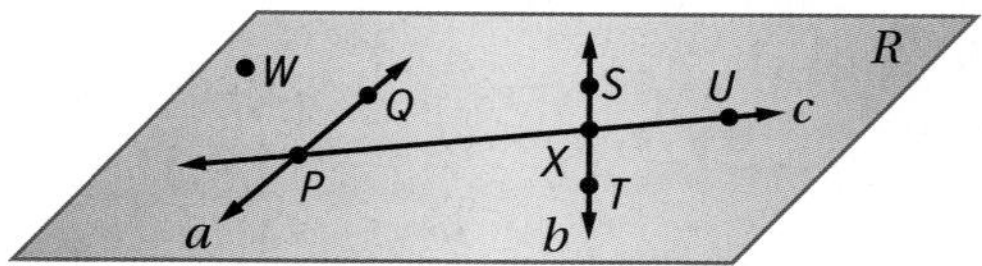

5. Name the intersection of lines a and c.
6. Give another name for line b.
7. Name a point that is not contained in any of the three lines a, b, or c.
8. Give another name for plane WPX.

Name the geometric term that is best modeled by each item.

9. 

10. 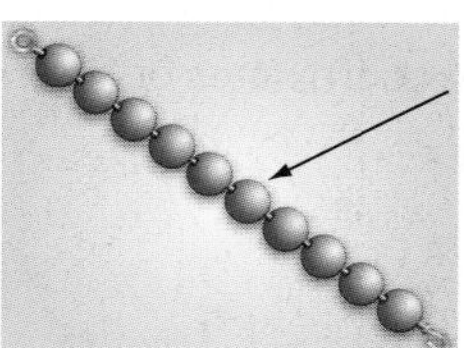

Example 1

Draw and label a figure for the relationship below.

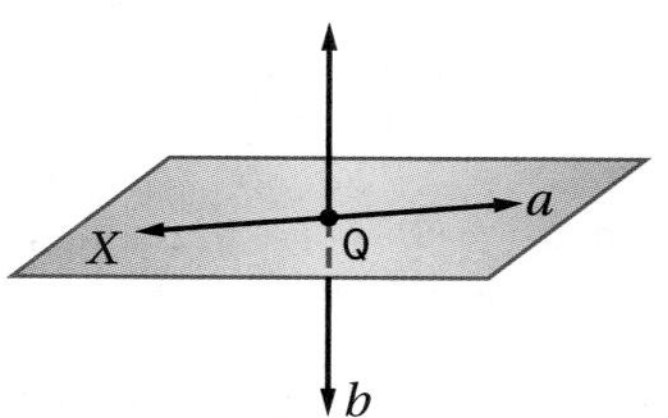

Plane X contains line a, line b intersects line a at point Q, but line b is not in plane X.

Draw a surface to represent plane X and label it.

Draw a line in plane X and label it line a.

Draw a line b intersecting both the plane and line a and label the point of intersection Q.

1-2 Line Segments and Distance

Find the value of the variable and XP, if X is between P and Q.

11. $XQ = 13$, $XP = 5x - 3$, $PQ = 40$
12. $XQ = 3k$, $XP = 7k - 2$, $PQ = 6k + 16$

Find the distance between each pair of points.

13. $A(-3, 1)$ and $B(7, 13)$
14. $P(2, -1)$ and $Q(10, -7)$
15. $M(9, -2)$ and $N(-1, 4)$
16. $J(3, 2)$ and $K(6, -5)$
17. **DISTANCE** The distance from Salvador's job to his house is 3 times greater than the distance from his house to school. If his house is between his job and school and the distance from his job to school is 6 miles, how far is it from Salvador's house to school?

Example 2

Use the figure to find the value of the variable and the length of $\overline{YZ}$.

X ——10—— Y ——3x + 7—— Z
(XZ = 29)

$XZ = XY + YZ$	Betweenness of points
$29 = 10 + 3x + 7$	Substitution
$29 = 3x + 17$	Simplify.
$12 = 3x$	Subtract 17 from each side.
$4 = x$	Divide each side by 3.
$YZ = 3x + 7$	Given
$= 3(4) + 7$ or 19	Substitution

So, $x = 4$ and $YZ = 19$.

Study Guide and Review *Continued*

1-3 Locating Points and Midpoints

Find the coordinates of the midpoint of a segment with the given endpoints.

18. $L(-3, 16)$, $M(17, 4)$
19. $C(32, -1)$, $D(0, -12)$

Find the coordinates of the missing endpoint if M is the midpoint of $\overline{XY}$.

20. $X(-11, -6)$, $M(15, 4)$
21. $M(-4, 8)$, $Y(19, 0)$
22. **HIKING** Carol and Marita are hiking in a state park and decide to take separate trails. The map of the park is set up on a coordinate grid. Carol's location is at the point (7, 13) and Marita is at (3, 5).
    a. Find the coordinates of the point midway between their locations.
    b. Find the coordinates of the point $\frac{1}{3}$ the distance from Carol to Marita.
23. Find C that is $\frac{1}{5}$ the distance from $A(4, 2)$ to $B(8, 7)$.
24. Find K between $M(1, -3)$ and $N(4, 6)$ such that the ratio of MK to KN is 1 : 2.

Example 3

Find the coordinates of the midpoint between $P(-4, 13)$ and $Q(6, 5)$.

Let $(x_1, y_1) = (-4, 13)$ and $(x_2, y_2) = (6, 5)$.

$$M\left(\frac{x_1 + x_2}{2}, \frac{y_1 + y_2}{2}\right) = M\left(\frac{-4 + 6}{2}, \frac{13 + 5}{2}\right)$$
$$= M(1, 9)$$

The coordinates of the midpoint are (1, 9).

Example 4

Find F on $\overline{AB}$ that is $\frac{1}{4}$ the distance from A to B.

$|x_2 - x_1| = |0 - (-3)| = 3$

$|y_2 - y_1| = |1 - 5| = |-4| = 4$

The coordinates of F are

$\left(-3 + \frac{3}{4}, 5 - \frac{4}{4}\right) = (-2.25, 4)$.

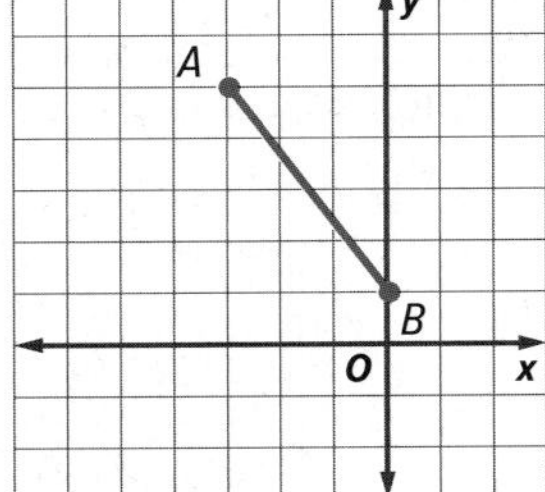

1-4 Angle Measure

For Exercises 25–28, refer to the figure below.

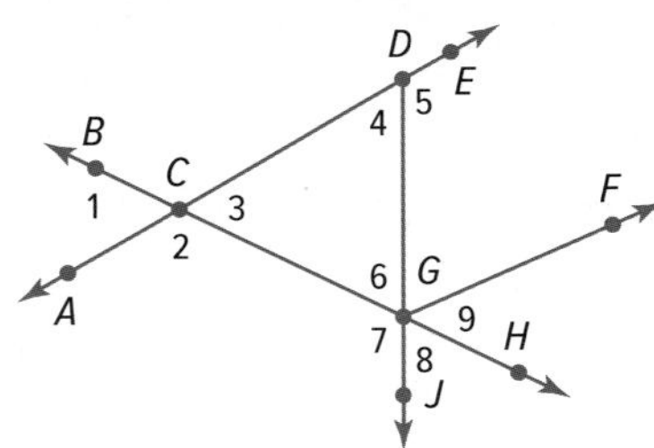

25. Name the vertex of $\angle 7$.
26. Write another name for $\angle 4$.
27. Name the sides of $\angle 2$.
28. Name a pair of opposite rays.
29. **SIGNS** A sign at Mesquite High School has the shape shown. Measure each of the angles and classify them as *right*, *acute*, or *obtuse*.

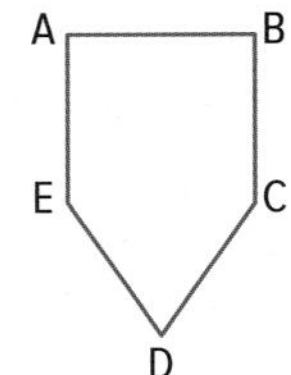

Example 5

Refer to the figure below. Name all angles that have Q as a vertex.

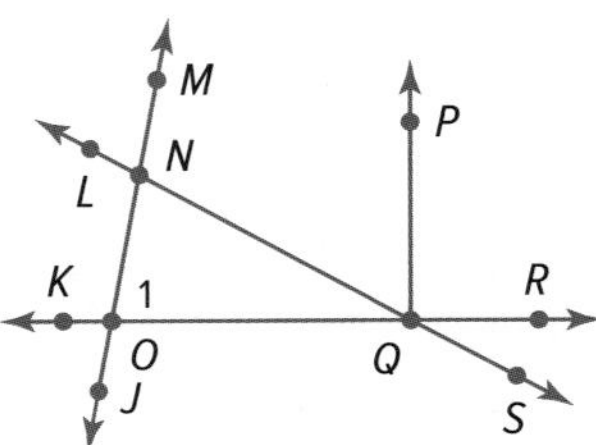

$\angle OQN$, $\angle NQP$, $\angle PQR$, $\angle RQS$, $\angle SQO$, $\angle OQP$, $\angle NQR$, $\angle PQS$, $\angle OQR$

Example 6

In the figure above, list all other names for $\angle 1$.

$\angle NOQ$, $\angle QON$, $\angle MOQ$, $\angle QOM$, $\angle MOR$, $\angle ROM$, $\angle NOR$, $\angle RON$

Study Guide and Review *Continued*

1-5 Angle Relationships

For Exercises 30–32, refer to the figure below.

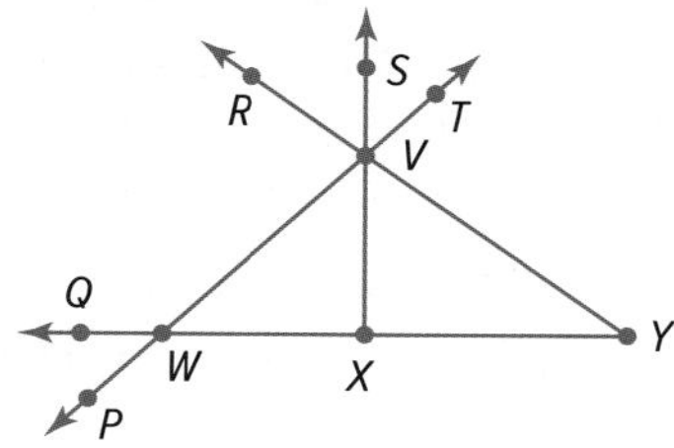

30. Name an angle supplementary to $\angle TVY$.
31. Name a pair of vertical angles with vertex W.
32. If $m\angle SXW = 5x - 16$, find the value of x so that $\overline{SX} \perp \overline{WY}$.
33. **PARKING** The parking arm shown below rests in a horizontal position and opens to a vertical position. After the arm has moved 24°, how many more degrees does it have to move so that it is vertical?

Example 7

Name a pair of supplementary angles and a pair of complementary angles in the figure below.

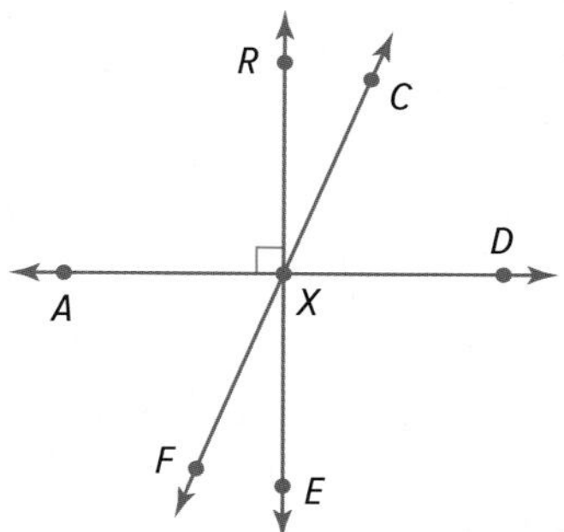

Sample answers:

Supplementary angles: $\angle RXA$ and $\angle RXD$

Complementary angles: $\angle RXC$ and $\angle CXD$

1-6 Two-Dimensional Figures

Name each polygon by its number of sides. Then classify it as *convex* or *concave* and *regular* or *irregular*.

34.

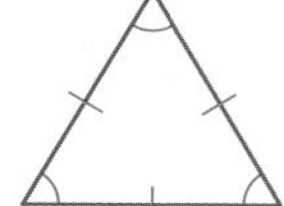

35.

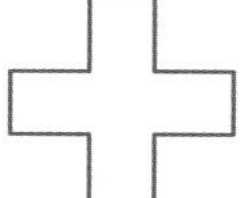

36. Find the perimeter of quadrilateral $ABCD$ with vertices $A(-3, 5)$, $B(0, 5)$, $C(2, 0)$, and $D(-5, 0)$.
37. **PARKS** Westside Park received 440 feet of chain-link fencing as a donation to build an enclosed play area for dogs. The park administrators need to decide what shape the area should have. They have three options: (1) a rectangle with length of 100 feet and width of 120 feet, (2) a square with sides of length 110 feet, or (3) a circle with radius of approximately 70 feet. Find the areas of all three enclosures and determine which would provide the largest area for the dogs.

Example 8

Name the polygon by its number of sides. Then classify it as *convex* or *concave* and *regular* or *irregular*.

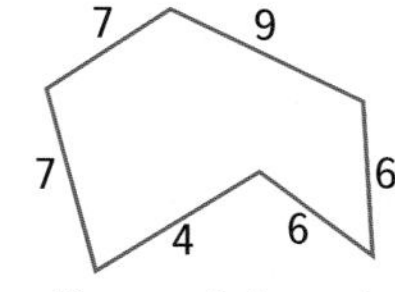

There are six sides, so this is a hexagon. If two of the sides are extended to make lines, they will pass through the interior of the hexagon, so it is concave. Since it is concave, it cannot be regular.

Example 9

Find the perimeter of the polygon in the figure above.

$P = s_1 + s_2 + s_3 + s_4 + s_5 + s_6$ Definition of perimeter

$= 7 + 7 + 9 + 6 + 6 + 4$ Substitution

$= 39$ Simplify.

The perimeter of the polygon is 39 units.

Study Guide and Review *Continued*

1-7 Transformations

Identify the type of congruence transformation shown as a *reflection*, *translation*, or *rotation*.

38.

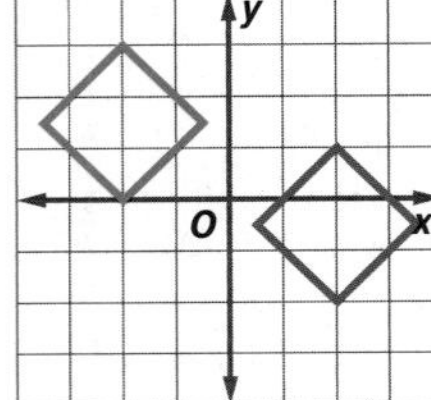

39.

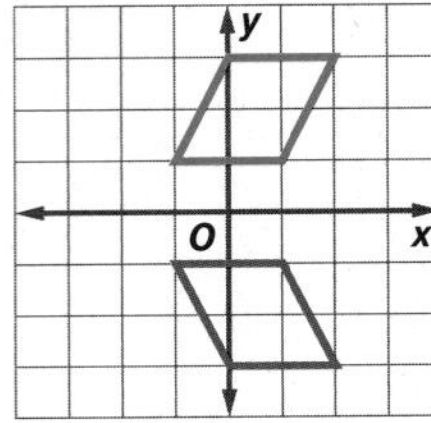

40.

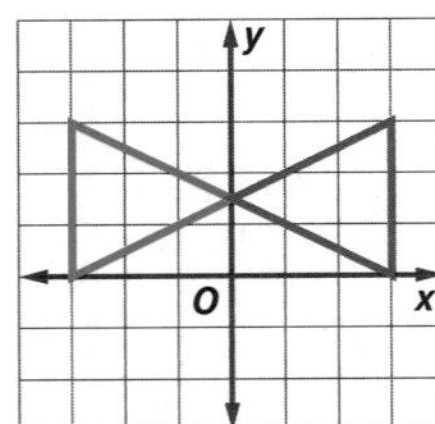

41.

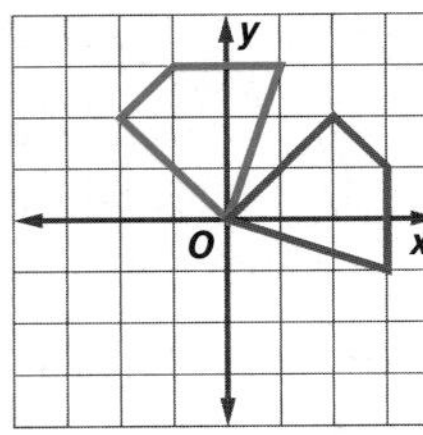

42. Triangle *ABC* with vertices $A(1, 1)$, $B(2, 3)$, and $C(3, -1)$ is a transformation of $\triangle MNO$ with vertices $M(-1, 1)$, $N(-2, 3)$, and $O(-3, -1)$. Graph the original figure and its image. Identify the transformation and verify that it is a congruence transformation.

Example 10

Triangle *RST* with vertices $R(4, 1)$, $S(2, 5)$, and $T(-1, 0)$ is a transformation of $\triangle CDF$ with vertices $C(1, -3)$, $D(-1, 1)$, and $F(-4, -4)$. Identify the transformation and verify that it is a congruence transformation.

Graph each figure. The transformation appears to be a translation. Find the lengths of the sides of each triangle.

$RS = \sqrt{(4-2)^2 + (1-5)^2}$ or $\sqrt{20}$

$TS = \sqrt{(-1-2)^2 + (0-5)^2}$ or $\sqrt{34}$

$RT = \sqrt{(-1-4)^2 + (0-1)^2}$ or $\sqrt{26}$

$CD = \sqrt{(-1-1)^2 + [1-(-3)]^2}$ or $\sqrt{20}$

$DF = \sqrt{[-4-(-1)]^2 + (-4-1)^2}$ or $\sqrt{34}$

$CF = \sqrt{(-4-1)^2 + [-4-(-3)]^2}$ or $\sqrt{26}$

Since each vertex of $\triangle CDF$ has undergone a transformation 3 units to the right and 4 units up, this is a translation.

Since $RS = CD$, $TS = DF$, and $RT = CF$, $\overline{RS} \cong \overline{CD}$, $\overline{TS} \cong \overline{DF}$, and $\overline{RT} \cong \overline{CF}$. By SSS, $\triangle RST \cong \triangle CDF$.

CHAPTER 1

Study Guide and Review *Continued*

1-8 Three-Dimensional Figures

Identify each solid. Name the bases, faces, edges, and vertices.

43.

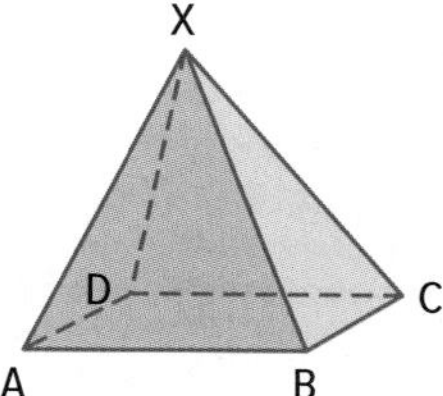

44.

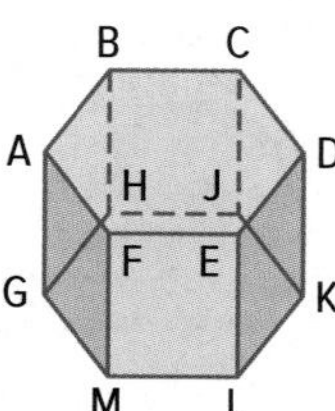

Find the surface area and volume of each solid.

45.

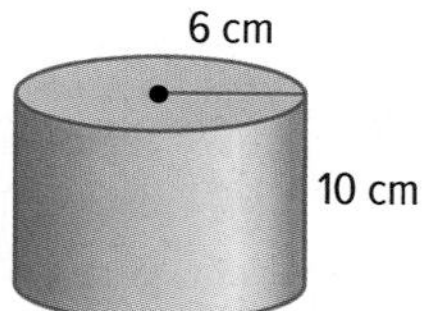

46.

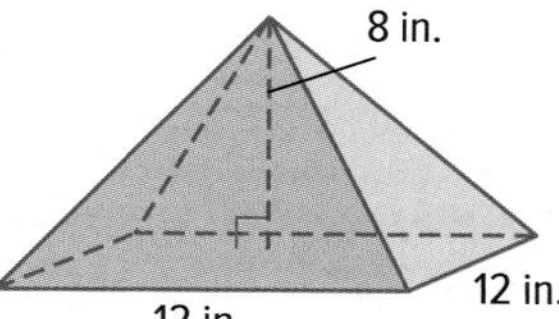

47.

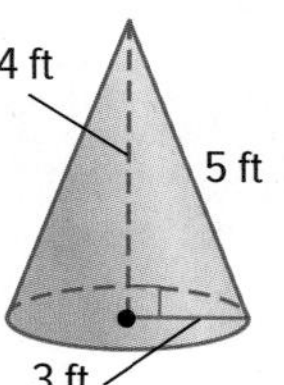

48.

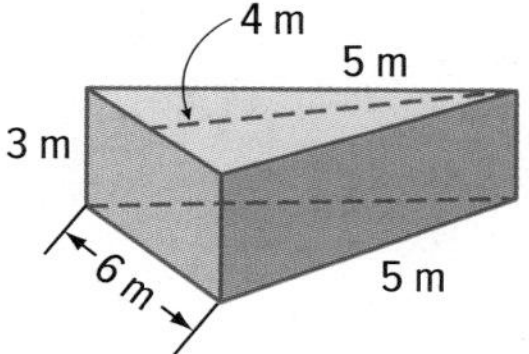

49. **BUILDING** Chris is building a trunk like the one shown below. His design is a square prism. What is the volume of the trunk?

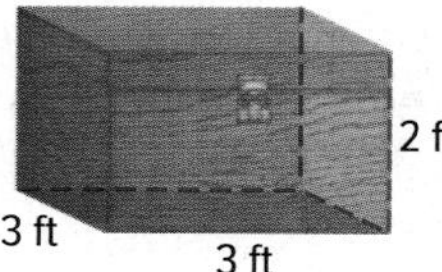

50. **HOCKEY** A regulation hockey puck is a cylinder made of vulcanized rubber 1 inch thick and 3 inches in diameter. Find the surface area and volume of a hockey puck.

Example 11

Identify the solid below. Name the bases, faces, edges, and vertices.

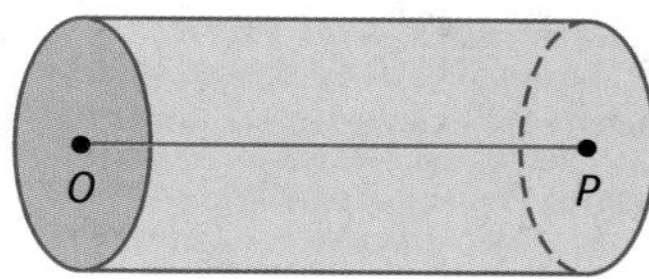

This solid has congruent circular bases in a pair of parallel planes. So, it is a cylinder.

Bases: circle O and circle P

A cylinder has no faces, edges, or vertices.

Example 12

Find the surface area and volume of the rectangular prism below.

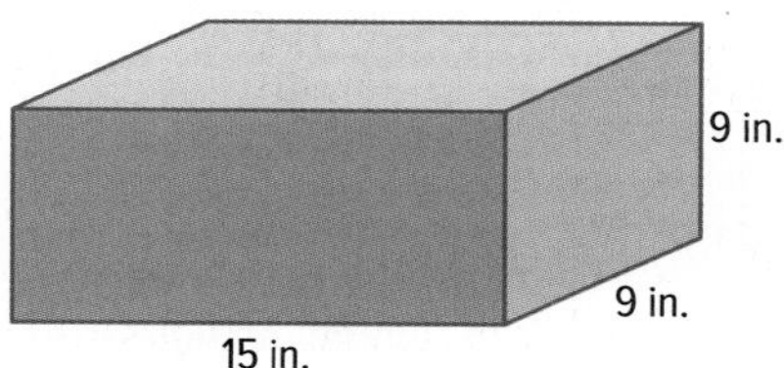

$S = Ph + 2B$	Surface area of a prism
$= (48)(9) + 2(135)$	Substitution
$= 702$	Simplify.

The surface area is 702 square inches.

$V = Bh$	Volume of a prism
$= (135)(9)$	Substitution
$= 1215$	Simplify.

The volume is 1215 cubic inches.

Study Guide and Review *Continued*

1-9 Two-Dimensional Representations of Three-Dimensional Figures

What solids would these nets make?

51.

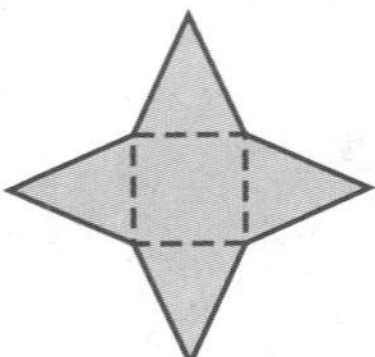

52.

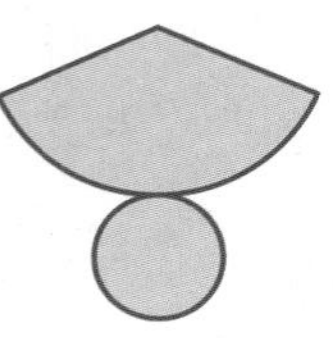

53. Make an orthographic drawing for the figure below.

Example 13

What solid would this net make?

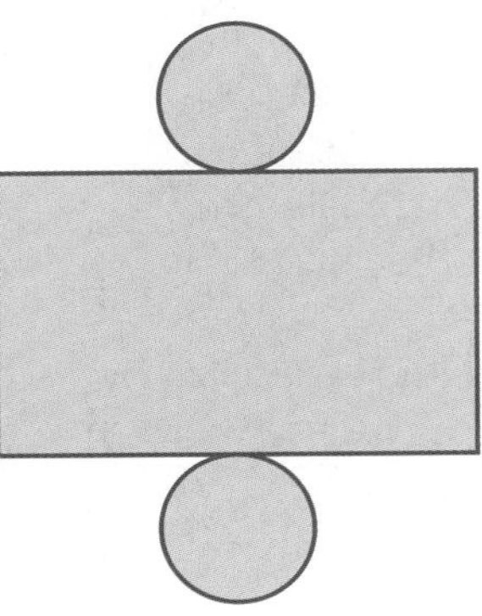

The rectangle can be curved around the circles to make a cylinder.

1-10 Precision and Accuracy

Determine the number of significant digits in each measurement.

54. 4.02 in.

55. 53,200 mi

56. 0.004 m

57. 7308 ft

Find the relative error in each measurement.

58. 7 ft

59. 2.50 g

60. 421 ml

61. 6.3

Example 14

Find the relative error in each measurement.

$$\text{relative error} = \frac{\text{absolute error}}{\text{expected measure}}$$

a. 8.5 m Solution: $\frac{0.05}{8.5} \cdot 100\% = 0.59\%$

b. 7 ft Solution: $\frac{0.5}{7} \cdot 100 = 7.1\%$

Example 15

Determine the number of significant digits in each measurement.

a. 0.0056 cm
There are two nonzero digits after a number of zeros, thus, the number of significant digits is 2.

b. 0.1056 cm
There are four digits the first zero, thus, the number of significant digits is 4.

CHAPTER 1

Practice Test

Use the figure to name each of the following.

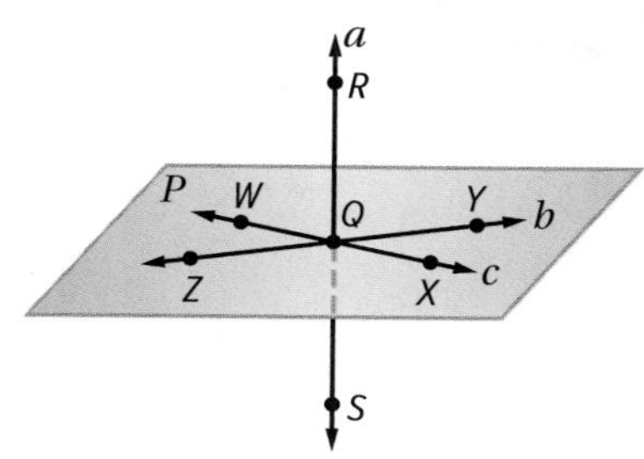

1. the line that contains points Q and Z
2. two points that are coplanar with points W, X, and Y
3. the intersection of lines a and b

Find the value of the variable if P is between J and K.

4. $JP = 2x$, $PK = 7x$, $JK = 27$
5. $JP = 3y + 1$, $PK = 12y - 4$, $JK = 75$
6. $JP = 8z - 17$, $PK = 5z + 37$, $JK = 17z - 4$

Find the coordinates of the midpoint of a segment with the given endpoints.

7. (16, 5) and (28, −13)
8. (−11, 34) and (47, 0)
9. (. −4, −14) and (−22, 9)

Find the distance between each pair of points.

10. (43, −15) and (29, −3)
11. (21, 5) and (28, −1)
12. (0, −5) and (18, −10)

13. **ALGEBRA** The measure of $\angle X$ is 18 more than three times the measure of its complement. Find the measure of $\angle X$.

14. Find the value of x that will make lines a and b perpendicular in the figure below.

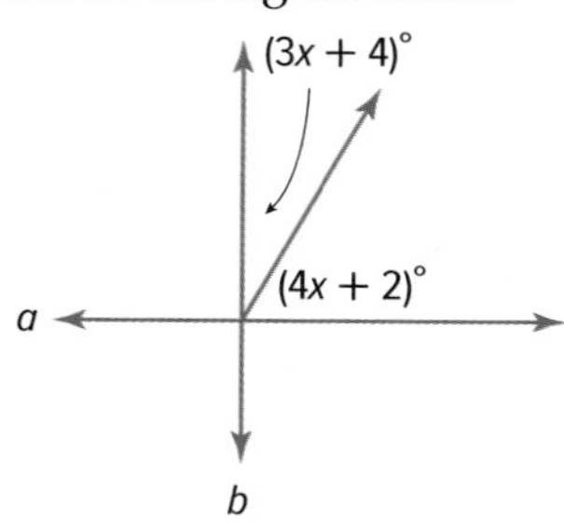

For Exercises 15–18, use the figure below.

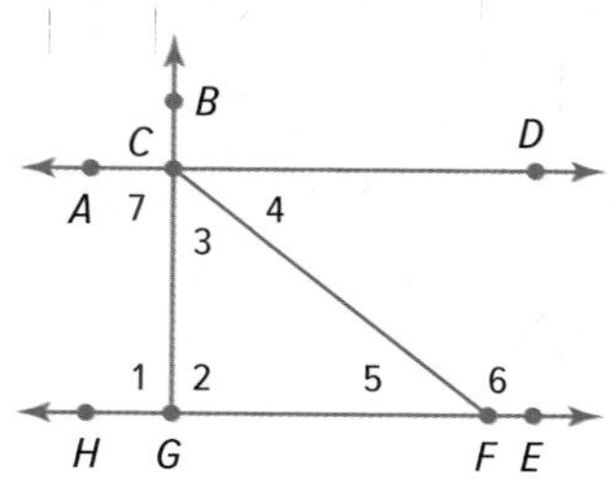

15. Name the vertex of $\angle 3$.
16. Name the sides of $\angle 1$.
17. Write another name for $\angle 6$.
18. Name a pair of angles that share exactly one point.

19. **MULTIPLE CHOICE** If $m\angle 1 = m\angle 2$, which of the following statements is true?

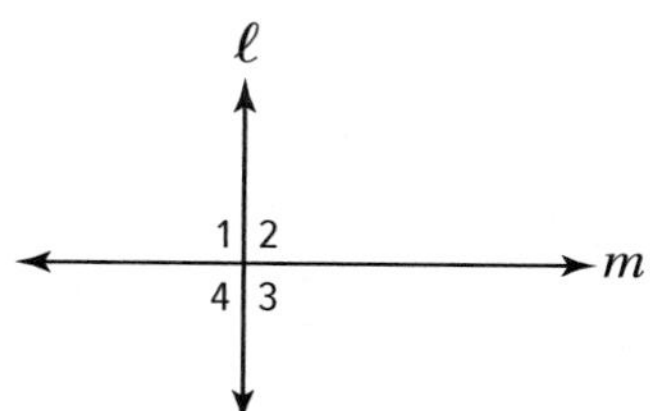

A $\angle 2 \cong \angle 4$ **C** $\ell \perp m$

B $\angle 2$ is a right angle. **D** all of the above

Find the perimeter of each polygon.

20. triangle XYZ with vertices $X(3, 7)$, $Y(-1, -5)$, and $Z(6, -4)$
21. rectangle $PQRS$ with vertices $P(0, 0)$, $Q(0, 7)$, $R(12, 7)$, and $S(12, 0)$

Refer to the figure at the right.

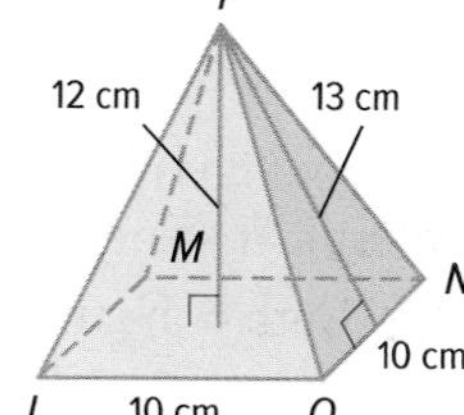

22. Name the base.
23. Find the surface area.
24. Find the volume.

25. **LANDSCAPING** Angie has laid out a design for a garden consisting of two triangular areas as shown below. The points are $A(0, 0)$, $B(0, 5)$, $C(3, 5)$, $D(6, 5)$, and $E(6, 0)$. Name the type of congruence transformation for the preimage $\triangle ABC$ to $\triangle EDC$.

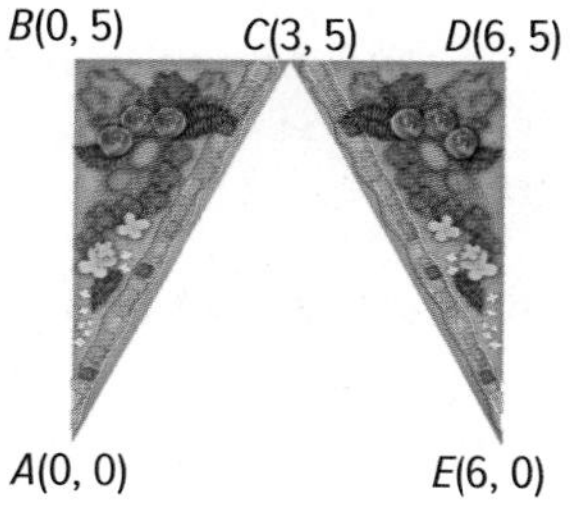

B(0, 5) C(3, 5) D(6, 5)

A(0, 0) E(6, 0)

Preparing for Assessment

Performance Task

Provide a clear solution to each part of the task. Be sure to show all of your work, include all relevant drawings, and justify your answers.

DESIGN An engineer is using 3-D modeling software to design the structure of a building for a commercial client.

Part A

The engineer uses the software to create the outer walls of the building, shown in the diagram. All intersecting walls should intersect at right angles.

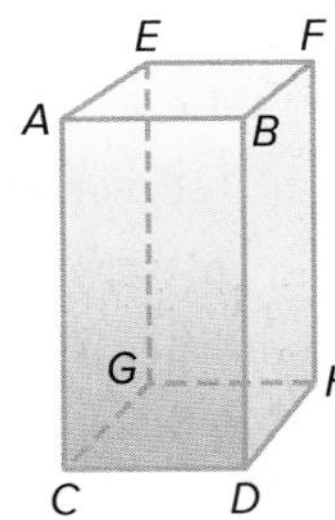

1. **Perseverance** Identify all planes that form the outer structure of the building.
2. Identify all planes that are perpendicular to plane *ABCD*.

Part B

Then the engineer designs the floor plan for one of the offices in the building. She determines that any wall 10 feet or longer will need a support beam.

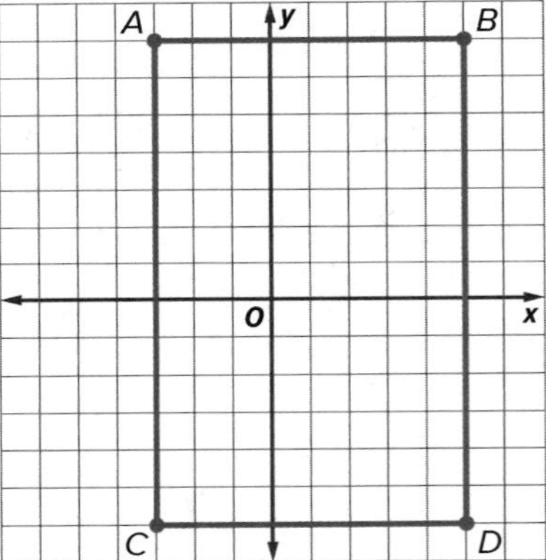

3. **Reasoning** One unit on the graph is equal to one actual foot. Which walls will need support beams?
4. The engineer wants to place support beams at the middle of each wall that requires a beam. Find the coordinates where support beams should be placed.
5. The engineer learns that ventilation shafts must be placed at the middle of each wall, so she decides to move the support beams to a location five-eighths of the way from one corner of the wall. Determine how far from each corner the beam will be in feet.
6. The logo of the company will be inlaid in the floor in a rectangle with vertices at (−1, 4), (−1, −4), (4, 4), and (4, −4). What is the area of the logo?

Part C

The client has requested that the engineer include some unique architectural features. The engineer decides to include several decorative windows.

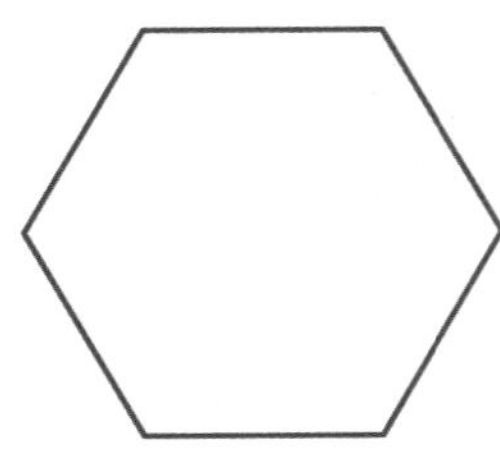

7. Determine the shape of the window.
8. Use your protractor to determine the interior angle measures of the window.
9. Each window will have strips of wood called muntins that divide the window into panes. They will bisect each angle of the window and meet in the center. Determine the measure of the six angles formed by the muntins.

Part D

The building also includes a storage room.

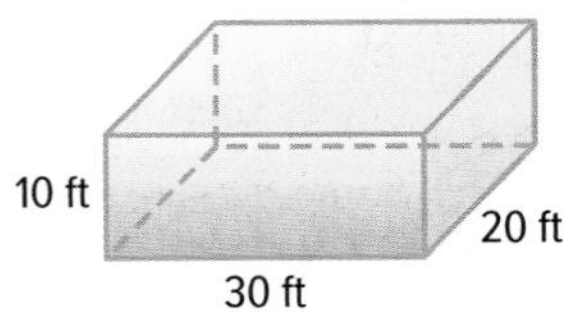

10. Determine the total volume of the storage room.
11. The walls, ceiling, and door will all be painted white. There are no windows. How many square feet need to be painted?

Test-Taking Strategy

Example

Read the problem. Identify what you need to know. Then use the information in the problem to solve.

Carmen is using a coordinate grid to make a map of her backyard. She plots the swing set at point $S(2, 5)$ and the big oak tree at point $O(-3, -6)$. If each unit on the grid represents 5 feet, what is the distance between the swing set and the oak tree? Round your answer to the nearest whole foot.

A 12 ft B 25 ft C 60 ft D 74 ft

Step 1 **What is the problem asking you to solve? What information is given that will be useful in solving the problem?**
I need to find the distance between two points. The problem gives the two points.

Step 2 **Are there any key words that give you clues? Would drawing a diagram be useful? Is there a formula you can use to help you?**
The word *distance* and the presence of coordinate points indicate that I need to use the distance formula.

Step 3 **What is your plan for solving this problem?**
I'm going to substitute the two points into the distance formula and simplify.

Step 4 **How can you be sure your answer is correct?**
I can reread the problem to make sure I answered the right question and that my answer contains the right units, if needed.

Step 5 **What is the correct answer?**
The correct answer is C.

Test-Taking Tip

Steps for Solving Math Problems

The first step to solving any math problem is to read the problem. When reading a math problem to get the information you need to know, it is helpful to use special reading strategies, such as underlining or drawing a diagram.

Apply the Strategy

Read the problem. Identify what you need to know. Then use the information in the problem to solve.

A farmer has laid out his farm in a grid pattern and located all of the structures relative to the farmhouse. The chicken coop is 70 feet north and 30 feet west of the farmhouse, and the grain silo is 90 feet south and 40 feet east of the farmhouse. To the nearest foot, what is the distance between the chicken coop and the grain silo?

A 17

B 22

C 160

D 175

Answer the questions below.

a. What problem are you asked to solve? What useful information is given?

b. Are there any key words? Would a diagram help? Is there an applicable formula?

c. What is your plan for solving the problem?

d. What is the correct answer?

CHAPTER 1

Preparing for Assessment

Cumulative Review

Read each question. Then fill in the correct answer on the answer document provided by your teacher or on a sheet of paper.

1. The number line shows the locations of points A and C.

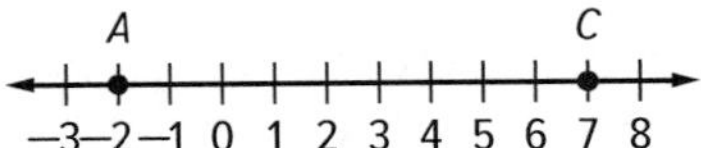

If point B is located somewhere between $\frac{1}{4}$ and $\frac{2}{3}$ of the way from point A to point C, which of the following coordinates could be the location of B? Select all that apply.

☐ **A** 0

☐ **B** 2

☐ **C** 4

☐ **D** 5

2. The midpoint of the segment with endpoints $A(6, -4)$ and $B(2, y)$ is $(4, 1)$. What is the value of y?

3. Ellis claims that the triangle shown is equilateral. Which of the following best refutes his claim?

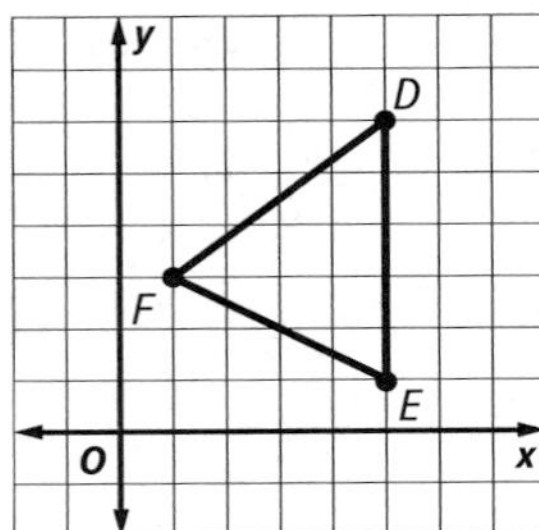

○ **A** $ED = DF = 5$

○ **B** $ED = 5$, $FE = 2\sqrt{5}$

○ **C** $m\angle E = m\angle F$

○ **D** $ED = 5$, $FE = 6$

4. Melanie wants to construct $\overline{PT}$ congruent to $\overline{MN}$. She begins by placing the compass on $\overline{MN}$ as shown.

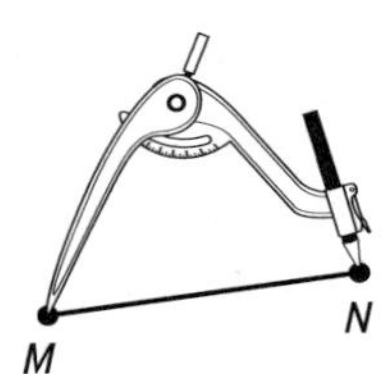

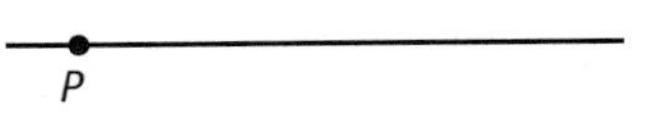

Which of the following best describes what Melanie should do next?

○ **A** Put the compass tip on point P.

○ **B** Put the compass tip on the line through P, opposite point P.

○ **C** Draw an arc intersecting the line through P.

○ **D** Measure with a ruler.

5. A large pile of sand is approximately in the shape of a cone.

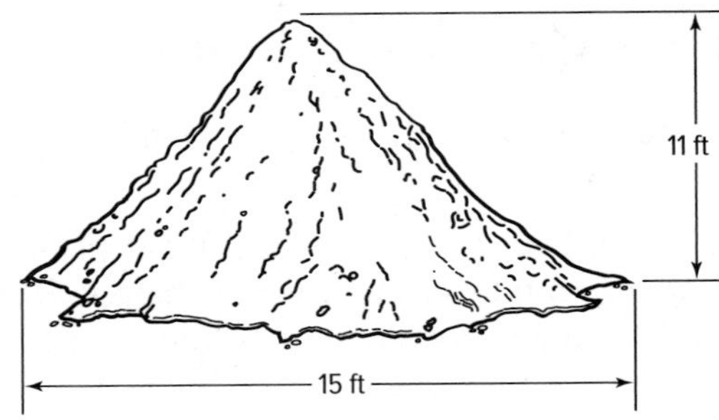

To the nearest cubic foot, what is the volume of sand in the pile, in cubic feet?

Test-Taking Tip

Question 5 Remember that the radius of the base of the cone is half the diameter.

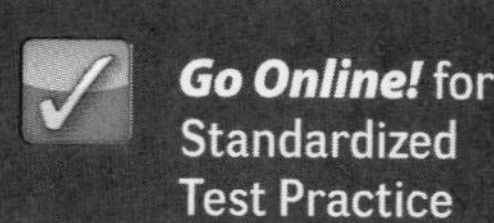

6. Emilio is constructing $\angle V$ congruent to $\angle W$. He uses a compass to draw an arc through $\angle W$ centered at point W.

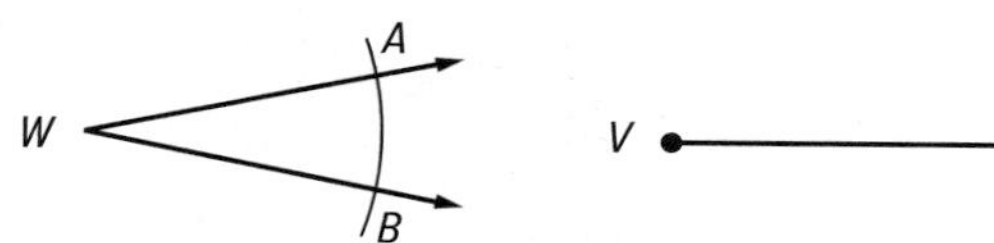

Which best describes the next step that Emilio should perform?

- A Set the compass to length AB; put the compass point on V and draw an arc on the line through V.
- B Set the compass to length AB; put the compass point on the line through V and draw an arc through V.
- C Set the compass to length WB; put the compass point on V and draw an arc on the line through V.
- D Set the compass to length WB; put the compass point on the line through V and draw an arc through V.

7. The diagram shows a design that a landscaper wants to use for a section of a patio.

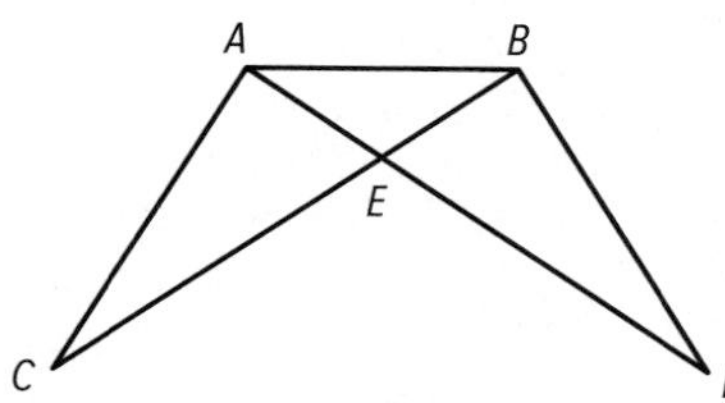

List all pairs of angles that can be determined to be congruent using the Vertical Angles Theorem.

8. The figure below consists of two intersecting lines m and n.

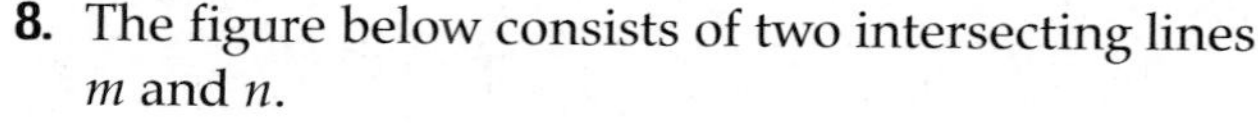
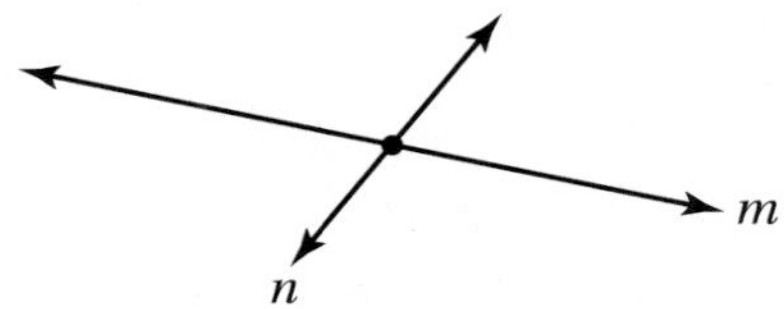

Which of the following is an undefined term that would include the entire figure?

- A line
- B plane
- C point
- D not here

9. The graph shows one side of a regular hexagon on the coordinate plane. What is the perimeter of the hexagon in units? Round to the nearest tenth.

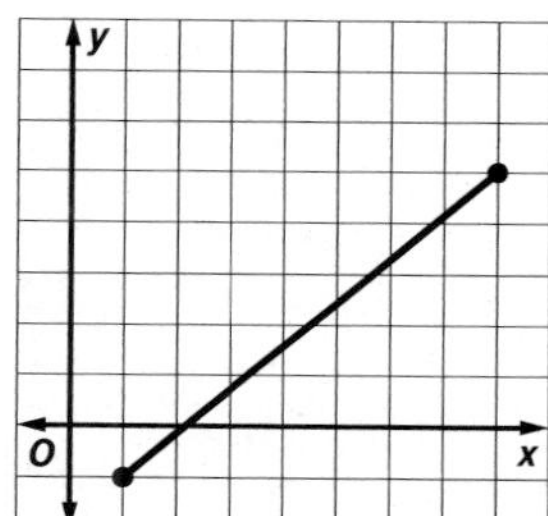

10. A spherical porcelain ornament has a radius of 6 inches. It is shipped in a cube-shaped box that has an edge length of 15 inches.

If all of the space inside the box surrounding the ornament is filled with packing material, how much packing material is there? Round to the nearest cubic inch.

- A 905 in^3
- B 2470 in^3
- C 3149 in^3
- D 3224 in^3

Need Extra Help?

If you missed Question...	1	2	3	4	5	6	7	8	9	10
Go to Lesson...	1-2	1-3	1-6	1-2	1-7	1-4	1-5	1-1	1-6	1-7

CHAPTER 2

Logical Arguments and Line Relationships

THEN

You used segment and angle relationships.

NOW

In this chapter, you will:

- Make conjectures and find counterexamples for statements.
- Use deductive reasoning to reach valid conclusions.
- Write proofs involving segment and angle theorems.

MP WHY

ARCHITECTURE Architects use geometry every day—in both simple and complex ways. You can see evidence of this in the structures all around us.

Use the Mathematical Practices to complete the activity.

1. Sense-Making What do you already know about lines? Can you identify parallel or perpendicular lines in the buildings you see?

2. Reasoning In what other ways might architects use geometry? How might logic be used in their work?

3. Discuss Discuss the geometry of ancient architecture as it compares to the geometry of modern architecture.

4. Apply Math Use the KWL Chart tool in ConnectED to organize what you know and what you want to know about using geometry and reasoning in architecture. Add to this chart as you learn more in this chapter.

Topic Geometry in Architecture

What I Know	What I Want to Know	What I Learned

Go Online to Guide Your Learning

Explore & Explain

The Geometer's Sketchpad

Use **The Geometer's Sketchpad** to explore angle relationships formed when a transversal cuts parallel lines, to illustrate the properties of slopes of lines on the coordinate plane and how solving systems by graphing does not provide an exact answer, and to explore the distance from a point to a line.

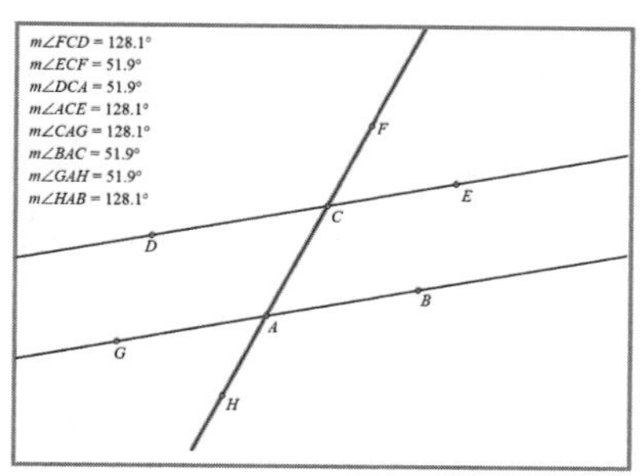

Protractor Tool

Use the **Protractor** tool to demonstrate postulates in Lesson 2-6.

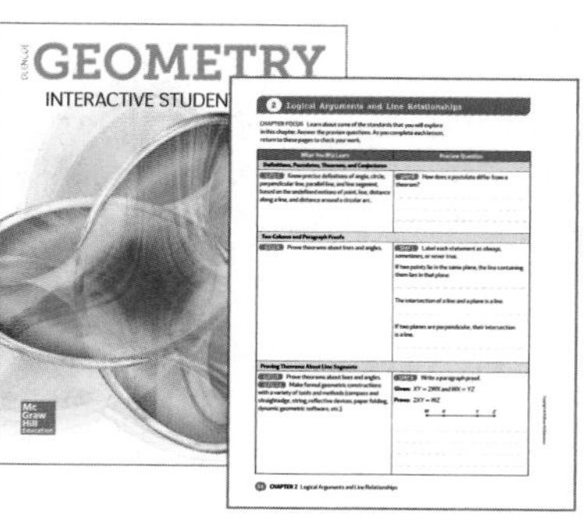

eBook

Interactive Student Guide

Before starting the chapter, answer the **Chapter Focus** preview questions. Check your answers as you complete each lesson. At the end of the chapter, try the **Performance Task**.

GEOMETRY
INTERACTIVE STUDEN

Organize

Foldables

Reasoning and Proof Make this Foldable to help you organize your notes about logic, reasoning, and proofs. Begin with one sheet of notebook paper, folded lengthwise. Cut five tabs in the top sheet and label the tabs as shown.

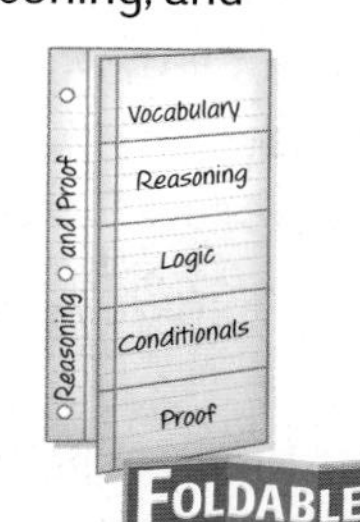

Collaborate

Chapter Project

In the **For the Birds** project, you will use what you have learned about conditional statements to complete a project that addresses business literacy.

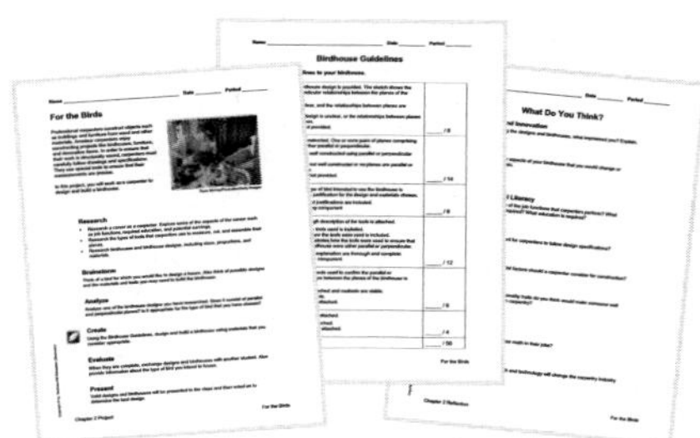

Focus

LEARNSMART®

Need help studying? Complete the **Congruence, Proof, and Constructions** topic in LearnSmart to review for the chapter test.

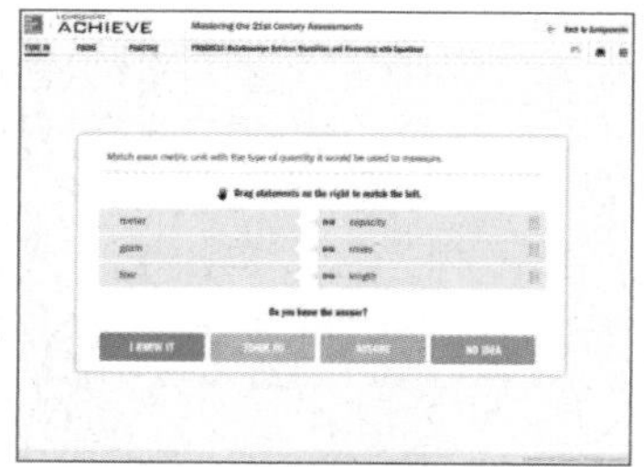

ALEKS®

You can use the **Algebra and Deductive Reasoning** topic in ALEKS to explore what you know about deductive reasoning and what you are ready to learn.*

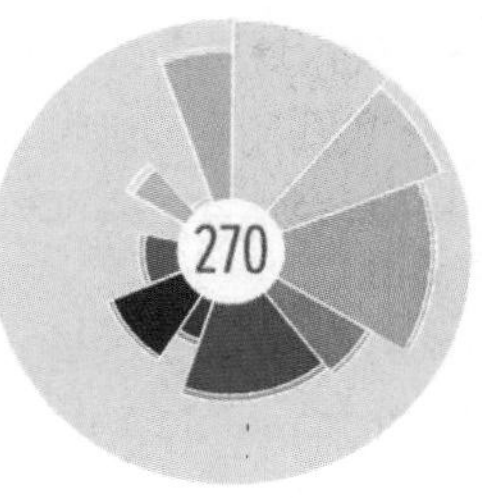

* Ask your teacher if this is part of your program.

Go Online! for Vocabulary Review Games and key vocabulary in 13 languages.

Get Ready for the Chapter

Connecting Concepts

Concept Check

Review the concepts used in this chapter by answering the questions below.

1. What would be the first step to evaluate $4x + 7$ when $x = 6$?
2. Given the equation $8x - 10 = 6x$, how would you begin to solve for x?
3. What would be your first step to solve $3(11x - 7) = 13x + 25$?
4. Nancy bought 4 shirts at the mall for $52. What equation could you write to find the average cost of one shirt?
5. Based on the figure, what can you conclude about $\angle BXD$ and $\angle AXE$?
6. Based on the figure, what can you conclude about $\angle CXD$ and $\angle DXE$?
7. Based on the figure, what can you conclude about $\angle DXB$ and $\angle DXE$?
8. Based on the figure, what equation can be written to show the relationship between $\angle BXA$ and $\angle DXE$?

Performance Task Review

You can use the concepts and skills in the chapter to solve problems in a real-world setting. Understanding proofs and identifying different types of angles will help you finish the Performance Task at the end of the chapter.

In this Performance Task you will:

- make sense of problems and persevere in solving them
- construct viable arguments and critique the reasoning of others

New Vocabulary

English		Español
inductive reasoning	p. 111	razonamiento inductivo
conjecture	p. 111	conjetura
counterexample	p. 114	contraejemplo
negation	p. 119	negación
deductive reasoning	p. 131	razonamiento deductivo
postulate	p. 141	postulado
proof	p. 143	demostración
parallel lines	p. 170	rectas paralelas
slope	p. 178	pendiente
point-slope form	p. 179	forma punto-pendiente

Review Vocabulary

complementary angles ángulos complementarios two angles with measures that have a sum of 90

supplementary angles ángulos suplementarios two angles with measures that have a sum of 180

vertical angles ángulos opuestos por el vértice two nonadjacent angles formed by intersecting lines

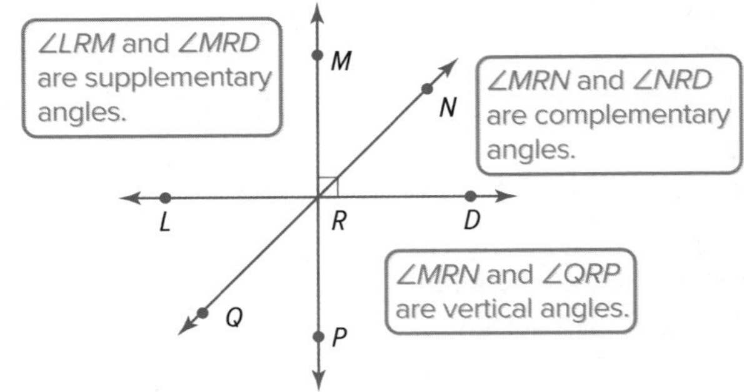

LESSON 1

Conjectures and Counterexamples

Then

- You used data to find patterns and make predictions.

Now

1. Write and analyze conjectures by using inductive reasoning.
2. Disprove conjectures by using counterexamples.

Why?

- Ana claims that it never snows in Florida, because she has never seen it snow. In January of 2016, snow flurries hit parts of North Florida. This is a counterexample. Ana came to her conclusion based on specific examples.

New Vocabulary
inductive reasoning
conjecture
counterexample

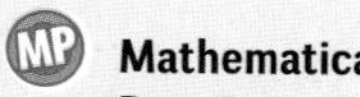

Mathematical Practices

3 Construct viable arguments and critique the reasoning of others.

1 Make Conjectures

Inductive reasoning is reasoning that uses a number of specific examples to arrive at a conclusion. When you assume that an observed pattern will continue, you are applying inductive reasoning. A concluding statement reached using inductive reasoning is called a **conjecture**.

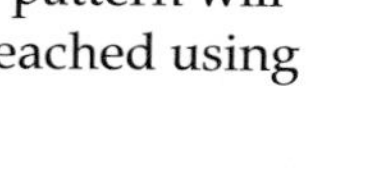

Example 1 Patterns and Conjecture

Write a conjecture that describes the pattern in each sequence. Then use your conjecture to find the next item in the sequence.

a. Movie show times: 8:30 A.M., 9:45 A.M., 11:00 A.M., 12:15 P.M., . . .

Step 1 Look for a pattern.

8:30 A.M., 9:45 A.M., 11:00 A.M., 12:15 P.M., . . .

+1 hr 15 min +1 hr 15 min +1 hr 15 min

Step 2 Make a conjecture.

The show time is 1 hour and fifteen minutes greater than the previous show time. The next show time will be 12:15 P.M. + 1:15 or 1:30 P.M.

b.

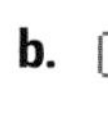

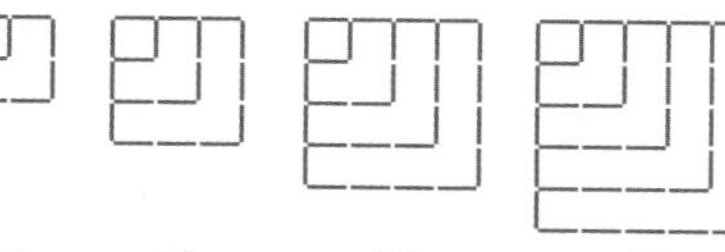

4 10 18 28 40 . . .

Step 1 4, 10, 18, 28, 40

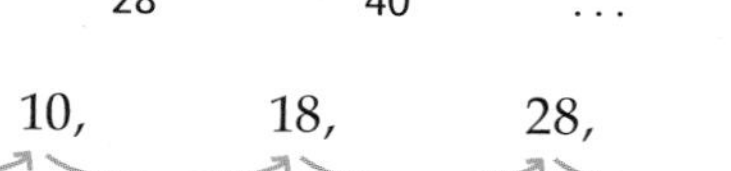

+6 +8 +10 +12

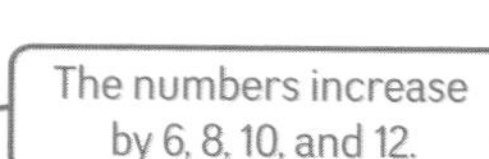

Step 2 The next figure will increase by 12 + 2 or 14 segments. So, the next figure will have 40 + 14 or 54 segments.

CHECK Draw the next figure to check your conjecture ✔

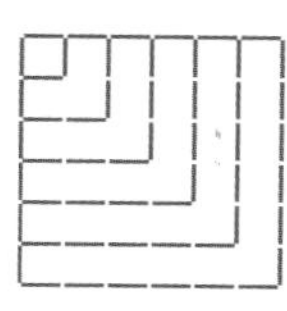

54

NPS Photo - Guadalupe Mountains National Park

Guided Practice

Write a conjecture that describes the pattern in each sequence. Then use your conjecture to find the next item in the sequence.

1A. Follow-up visits: Dec., May, Oct., Mar., . . .

1B. 10, 4, −2, −8, . . .

1C.

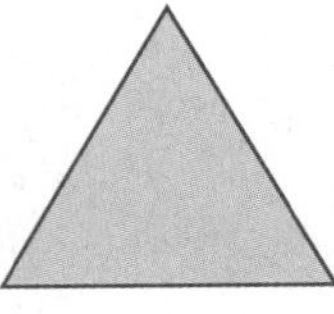
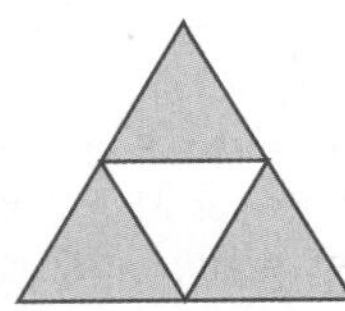

Go Online!

In Personal Tutor videos for this lesson, teachers describe how to make conjectures. Working with a partner, take turns describing how to make a conjecture. Ask each other questions to help your understanding.

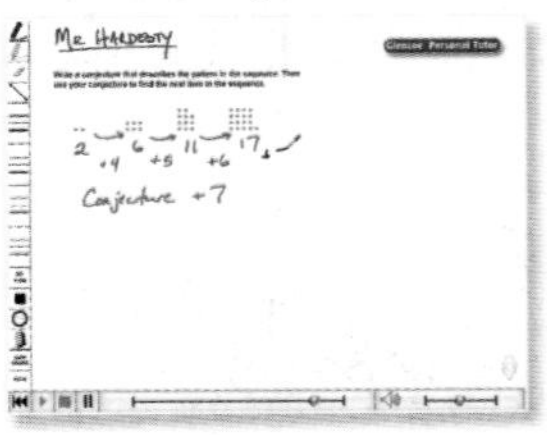

To make some algebraic and geometric conjectures, you will need to provide examples.

Example 2 Algebraic and Geometric Conjectures

Make a conjecture about each value or geometric relationship. List or draw some examples that support your conjecture.

a. the sum of two odd numbers

Step 1 List examples.

$1 + 3 = 4$ $\quad 1 + 5 = 6$ $\quad 3 + 5 = 8$ $\quad 7 + 9 = 16$

Step 2 Look for a pattern.

Notice that the sums 4, 6, 8, and 16 are all even numbers.

Step 3 Make a conjecture.

The sum of two odd numbers is an even number.

b. segments joining opposite vertices of a rectangle

Step 1

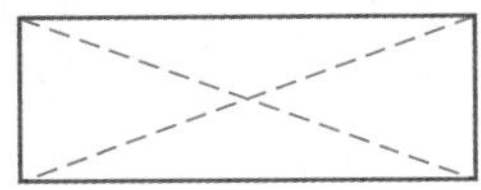
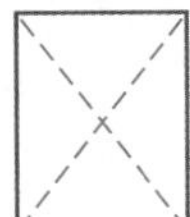
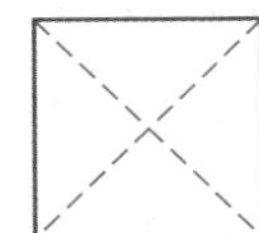
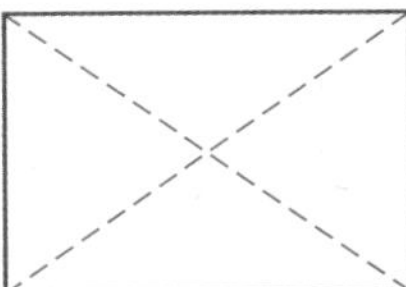

Step 2 Notice that the segments joining opposite vertices of each rectangle appear to have the same measure. Use a ruler or compass to confirm this.

Step 3 Conjecture: the segments joining opposite vertices of a rectangle are congruent.

Study Tip

MP Construct Arguments Examples that support a conjecture are not enough to show that a conjecture is true. To show that an algebraic or geometric conjecture is true, you must offer a logical argument called a proof. You will learn more about proofs in Lesson 2-4.

Guided Practice

2A. the sum of two even numbers

2B. the relationship between AB and EF, if $AB = CD$ and $CD = EF$

2C. the sum of the squares of two consecutive natural numbers

Real-world conjectures are often made based on data gathered about a specific topic of interest.

Real-Real-World Career

Chef Chefs direct the kitchen staff and oversee food preparation at restaurants. The work can involve long hours and be very demanding and high-stress. The job outlook for chefs is positive; according to the Bureau of Labor Statistics, the percent change in employment for chefs from 2014 to 2024 is expected to be 9%, as compared to the average growth rate for all occupations of 7%.

Real-World Example 3 Make Conjectures from Data

BUSINESS The owner of a restaurant collected data on the number of customers her restaurant had each Friday, Saturday, and Sunday for 6 months to decide whether she should increase the number of servers working each weekend. The data she collected are shown below.

Number of Customers on the Weekend						
Day	Month 1	Month 2	Month 3	Month 4	Month 5	Month 6
Friday	225	255	321	406	540	450
Saturday	603	658	652	712	746	832
Sunday	552	635	642	692	685	705
Total	1380	1548	1615	1810	1971	1987

a. Make a statistical graph that best displays the data.

Because you want to look for a pattern over time, use a scatter plot to display the data. Label the horizontal axis with the months and the vertical axis with the number of customers. Plot each set of data using a different color and include a legend.

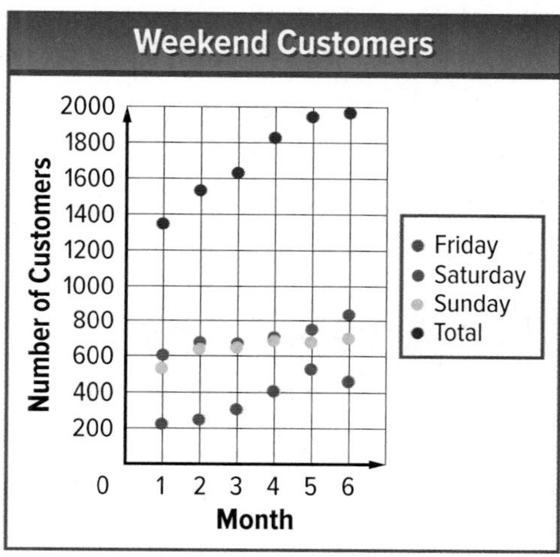

b. Make a conjecture based on the data and explain how this conjecture is supported by your graph.

Look for patterns in the data. The number of customers on each day usually increases each month, and the total number of customers increases every single month.

Survey data supports a conjecture that the amount of business on the weekends has increased, so the owner should schedule more servers to work on those days.

Guided Practice

3. **POSTAGE** The table at the right shows the price of postage for the years 1987 through 2016.

   A. Make a statistical graph that best displays the data.

   B. Predict the postage rate in 2020 based on the graph.

   C. Does it make sense that the pattern of the data will continue over time? If not, how will it change? Explain your reasoning.

Year	Rate (cents)
1987	22
1992	29
1997	32
2002	37
2007	41
2009	44
2012	45
2013	46
2014	49
2016	47

2 Find Counterexamples

To show that a conjecture is true for all cases, you must prove it. It takes only one false example, however, to show that a conjecture is not true. This false example is called a **counterexample**, and it can be a number, a drawing, or a statement.

> **Vocabulary Link**
>
> **Counterexample**
>
> **Everyday Use** The prefix *counter-* means *the opposite of.*
>
> **Math Use** A counterexample is the opposite of an example.

Example 4 Find Counterexamples

Find a counterexample to show that each conjecture is false.

a. If n is a real number, then $n^2 > n$.

Use number sense. When n is greater than 1, the statement is true. When n is 1, the conjecture is false, because $1^2 \not> 1$.

b. If $JK = KL$, then K is the midpoint of $\overline{JL}$.

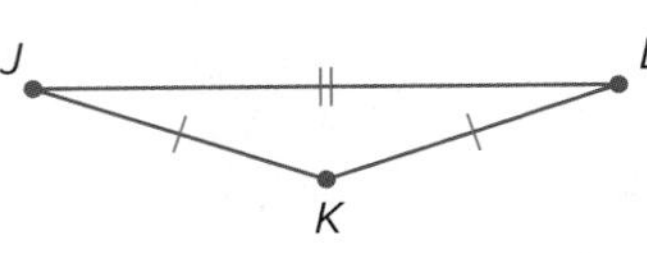

When J, K, and L are noncollinear, the conjecture is false. In the figure, $JK = KL$, but K is not the midpoint of $\overline{JL}$.

Guided Practice

4A. If n is a real number, then $-n$ is a negative number.

4B. If $\angle ABC \cong \angle DBE$, then $\angle ABC$ and $\angle DBE$ are vertical angles.

Go Online! for a Self-Check Quiz

Check Your Understanding

◯ = Step-by-Step Solutions begin on page R13.

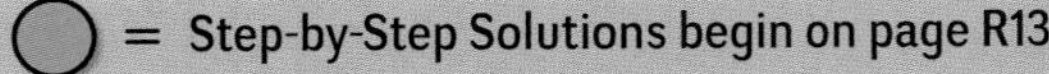

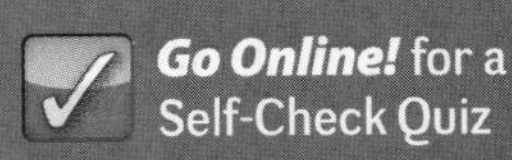

Example 1 **Write a conjecture that describes the pattern in each sequence. Then use your conjecture to find the next item in the sequence.**

1. Costs: \$4.50, \$6.75, \$9.00, . . .

2. Appointment times: 10:15 A.M., 11:00 A.M., 11:45 A.M., . . .

3.

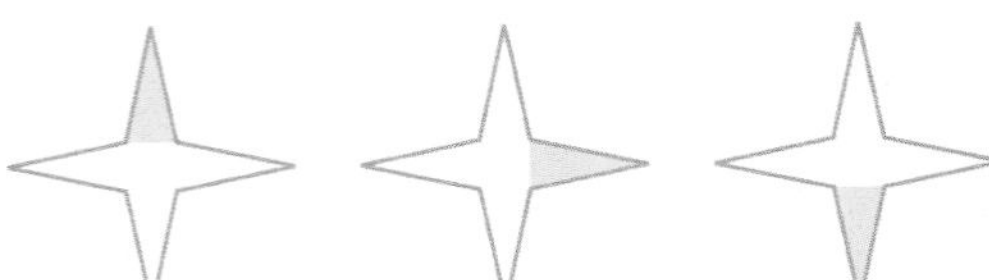

4.

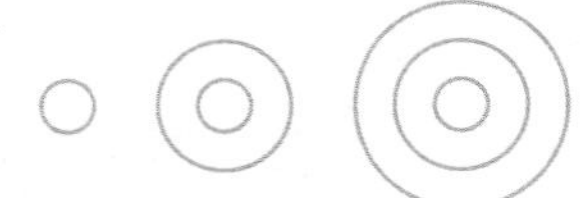

5. 3, 3, 6, 9, 15, . . .

6. 2, 6, 14, 30, 62, . . .

Example 2 **Make a conjecture about each value or geometric relationship.**

7. the product of two even numbers

8. the relationship between a and b if $a + b = 0$

9. the relationship between the set of points in a plane equidistant from point A

10. the relationship between $\overline{AP}$ and $\overline{PB}$ if M is the midpoint of $\overline{AB}$ and P is the midpoint of $\overline{AM}$

Example 3

11. **SMARTPHONES** Refer to the table of the number of smartphone users in the United States by year.

   a. Make a graph that shows U.S. smartphone use from 2010–2015.

   b. Make a conjecture about U.S. smartphone use in 2018.

Year	Users (in millions)
2010	62
2011	93
2012	122
2013	145
2014	171
2015	191

Example 4

MP CRITIQUE ARGUMENTS Find a counterexample to show that each conjecture is false.

12. If $\angle A$ and $\angle B$ are complementary angles, then they share a common side.

13. If a ray intersects a segment at its midpoint, then the ray is perpendicular to the segment.

Practice and Problem Solving

Extra Practice is on page R2.

Example 1

Write a conjecture that describes the pattern in each sequence. Then use your conjecture to find the next item in the sequence.

14. 0, 2, 4, 6, 8

15. 3, 6, 9, 12, 15

16. 4, 8, 12, 16, 20

17. 2, 22, 222, 2222

18. 1, 4, 9, 16

19. $1, \frac{1}{2}, \frac{1}{4}, \frac{1}{8}$

20. Arrival times: 3:00 P.M., 12:30 P.M., 10:00 A.M., . . .

21. Percent humidity: 100%, 93%, 86%, . . .

22. Workout days: Sunday, Tuesday, Thursday, . . .

23. Club meetings: January, March, May, . . .

24. 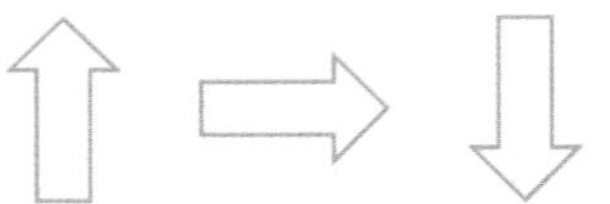

25. 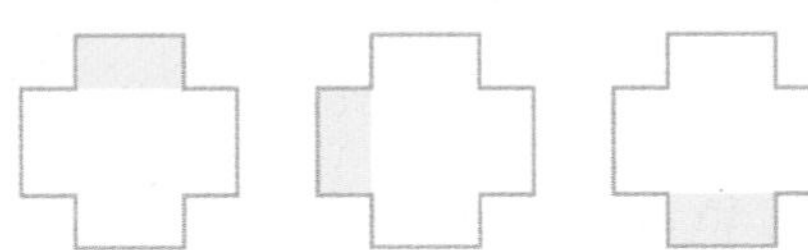

26. 

27. 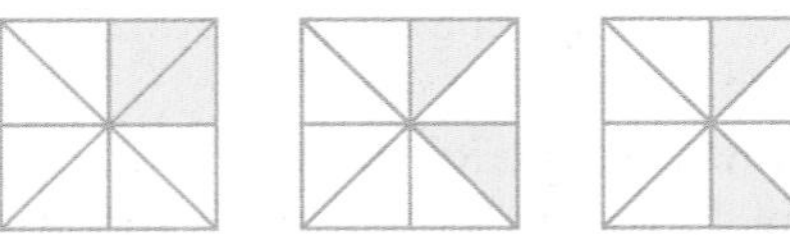

28. **FITNESS** Gabriel started training with the track team five weeks ago. During the first week, he ran 0.5 mile at each practice. The next three weeks he ran 0.75 mile, 1 mile, and 1.25 miles at each practice. If he continues this pattern, how many miles will he be running at each practice during the 7th week?

29. **CONSERVATION** When there is a shortage of water, some municipalities limit the amount of water each household is allowed to consume. Most cities that experience water restrictions are in the western and southern parts of the United States. Make a conjecture about why water restrictions occur in these areas.

30. **VOLUNTEERING** Carrie collected canned food for a homeless shelter in her area each day for one week. On day one, she collected 7 cans of food. On day two, she collected 8 cans. On day three, she collected 10 cans. On day four, she collected 13 cans. If Carrie wanted to give at least 100 cans of food to the shelter and this pattern of can collecting continued, did she meet her goal?

Example 2 **Make a conjecture about each value or geometric relationship.**

31. the product of two odd numbers

32. the product of two and a number, plus one

33. the relationship between a and c if $ab = bc$, $b \neq 0$

34. the relationship between a and b if $ab = 1$

35. the relationship between $\overline{AB}$ and the set of points equidistant from A and B

36. the relationship between the angles of a triangle with all sides congruent

37. the relationship between the areas of a square with side x and a rectangle with sides x and $2x$

38. the relationship between the volume of a prism and a pyramid with the same base and equal heights

Example 3 39. **SPORTS** Nontraditional races, like obstacle races and mud races, have grown rapidly in popularity in recent years. Use the data to make a conjecture about the future of these races.

a. Make a statistical graph that best displays the data.

b. Make a conjecture based on the data and explain how this conjecture is supported by your graph.

Estimated Number of Participants in U.S. Nontraditional Races

Year	Number of Participants (millions)
2009	0.1
2010	0.3
2011	1.0
2012	2.0
2013	4.0
2014	4.2
2015	4.5

Source: Running USA/Athlinks

Example 4 MP **REASONING Determine whether each conjecture is *true* or *false*. Give a counterexample for any false conjecture.**

40. If n is a prime number, then $n + 1$ is not prime.

41. If x is an integer, then $-x$ is positive.

42. If $\angle 2$ and $\angle 3$ are supplementary angles, then $\angle 2$ and $\angle 3$ form a linear pair.

43. If you have three points A, B, and C, then A, B, and C are noncollinear.

44. If in $\triangle ABC$, $(AB)^2 + (BC)^2 = (AC)^2$, then $\triangle ABC$ is a right triangle.

45. If the area of a rectangle is 20 square meters, then the length is 10 meters and the width is 2 meters.

FIGURAL NUMBERS Numbers that can be represented by evenly spaced points arranged to form a geometric shape are called figural numbers. For each figural pattern below,

a. write the first four numbers that are represented,

b. write a conjecture that describes the pattern in the sequence,

c. explain how this numerical pattern is shown in the sequence of figures,

d. find the next two numbers, and draw the next two figures.

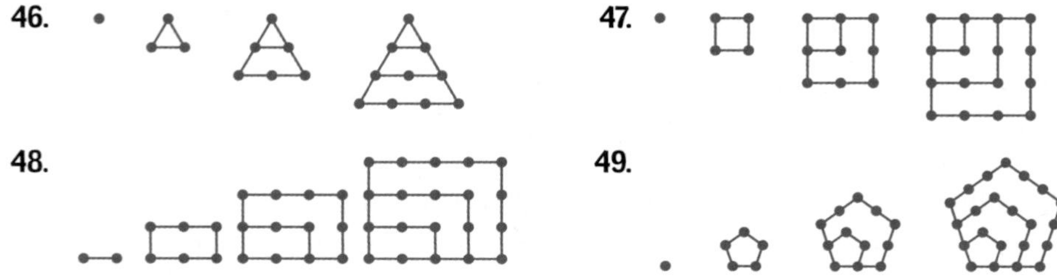

50. The sequence of odd numbers, 1, 3, 5, 7, . . . can also be a sequence of figural numbers. Use a figural pattern to represent this sequence.

51 **GOLDBACH'S CONJECTURE** Goldbach's conjecture states that every even number greater than 2 can be written as the sum of two primes. For example, $4 = 2 + 2$, $6 = 3 + 3$, and $8 = 3 + 5$.

a. Show that the conjecture is true for the even numbers from 10 to 20.

b. Given the conjecture *All odd numbers greater than 2 can be written as the sum of two primes*, is the conjecture *true* or *false*? Give a counterexample if the conjecture is false.

52. **SEGMENTS** Two collinear points form one segment, as shown for $\overline{AB}$. If a collinear point is added to $\overline{AB}$, the three collinear points form three segments.

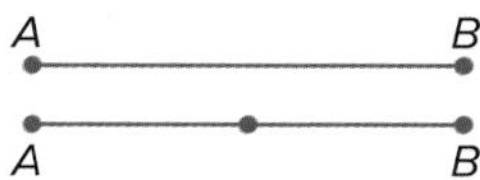

a. How many distinct segments are formed by four collinear points? by five collinear points?

b. Make a conjecture about the number of distinct segments formed by n collinear points.

c. Test your conjecture by finding the number of distinct segments formed by six points.

53. MP **TOOLS** Using dynamic geometry software, Nora calculates the perimeter P and area A of a regular hexagon with a side length of 2 units. The change to the perimeter and area after three doublings of this side length are listed in the table. Make a conjecture as to the effects on the perimeter and area of a regular hexagon when the side length is doubled. Explain.

Side (units)	P (units)	A (units^2)
2	12	$6\sqrt{3}$
4	24	$24\sqrt{3}$
8	48	$96\sqrt{3}$
16	96	$384\sqrt{3}$

H.O.T. Problems Use Higher-Order Thinking Skills

54. **CHALLENGE** If you draw points on a circle and connect every pair of points, the circle is divided into regions. For example, two points form two regions, three points form four regions, and four points form eight regions.

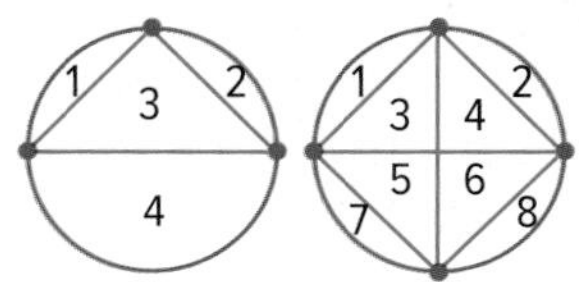

a. Make a conjecture about the relationship between the number of points on a circle and the number of regions formed in the circle.

b. Does your conjecture hold true when there are six points? Support your answer with a diagram.

55. **ERROR ANALYSIS** Juan and Jack are discussing prime numbers. Juan states a conjecture that all prime numbers are odd. Jack disagrees with the conjecture and states that not all prime numbers are odd. Is either of them correct? Explain.

56. **OPEN ENDED** Write a number sequence that can be generated by two different patterns. Explain your patterns.

57. MP **REASONING** Consider the conjecture *If two points are equidistant from a third point, then the three points are collinear.* Is the conjecture *true* or *false*? If false, give a counterexample.

58. **WRITING IN MATH** Suppose you are conducting a survey. Choose a topic and write three questions you would include in your survey. How would you use inductive reasoning with your responses?

59. Terrence claims that if two angles are supplementary, then they are a linear pair. Which of the following figures is a counterexample to his conjecture? MP 3

○ A

○ B

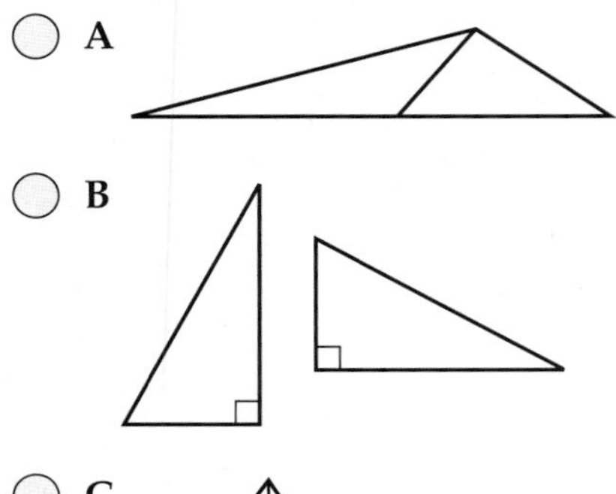

○ C

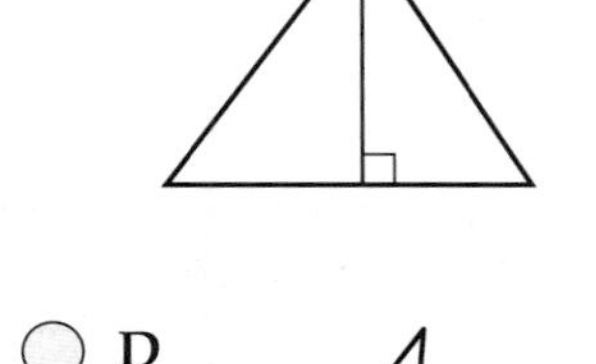

○ D

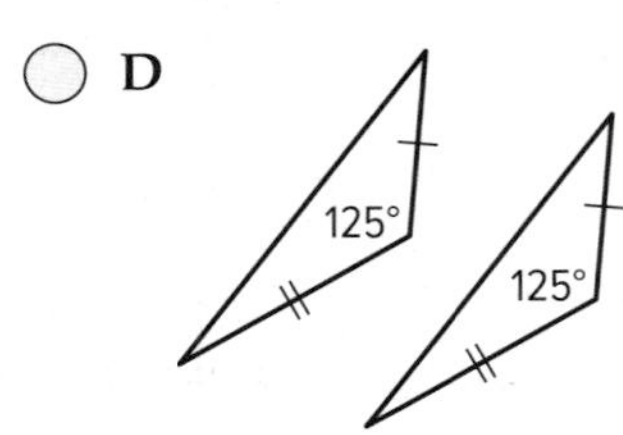

60. Ray made the following conjecture: "If four points lie in a plane, then the points are collinear." Which figure is a counterexample to Ray's conjecture? MP 3

○ A

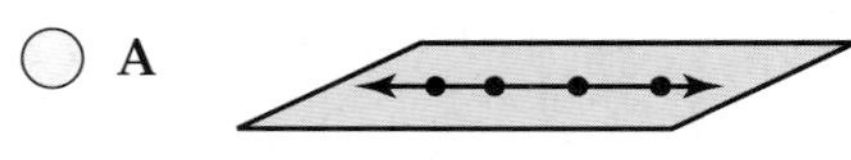

○ B

○ C

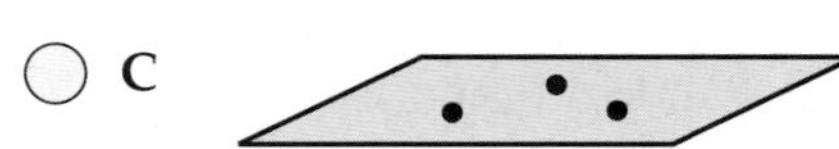

○ D

○ E

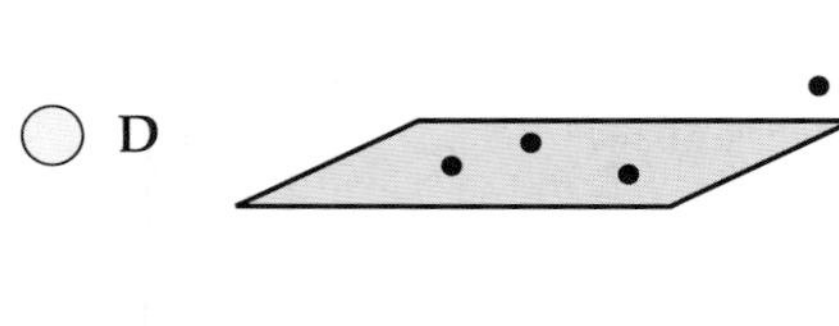

61. Maria drew the sequence of figures shown. MP 2, 3

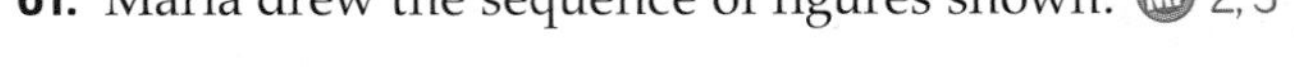

Which of the following is the best conjecture Maria can make based on the figures for $n > 2$?

○ **A** Given n noncollinear points, you can draw $n - 1$ distinct line segments with the points as endpoints.

○ **B** Given n noncollinear points, you can draw n distinct line segments with the points as endpoints.

○ **C** Given n noncollinear points, you can draw $n + 2$ distinct line segments with the points as endpoints.

○ **D** Given n noncollinear points, you can draw $\frac{n(n-1)}{2}$ distinct line segments with the points as endpoints.

62. Study the appointment times of 8:30 A.M., 9:10 A.M., 9:50 A.M., 10:30 A.M., . . . and make a conjecture about when the next appointment time will be. Explain your answer. MP 3

63. **MULTI-STEP** Study the pattern to make conjectures about number relationships. MP 2, 3

a. Complete the table to show the value of x^2 and $(x - 1)(x + 1)$.

x	x^2	$(x - 1)(x + 1)$
1		
2		
3		
4		
5		

b. What pattern do you observe?

c. Predict the product of 79×81.

d. Do you think that this rule will work for all real numbers? If not, provide a counter example.

LESSON 2

Statements, Conditionals, and Biconditionals

Then

- You found counterexamples for false conjectures.

Now

1. Write compound statements and determine truth values of compound statements.
2. Write conditional statements and determine truth values of conditional statements.

Why?

- Many electrical circuits operate by evaluating a series of tests that are either true or false. For example, a single light can be controlled by two different switches connected on a circuit. The positions of both switches, either up or down, determine whether the light is on or off.

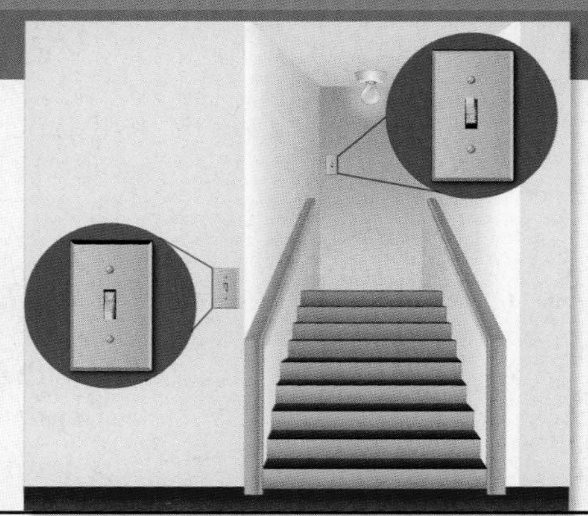

New Vocabulary

statement
truth value
negation
compound statement
conjunction
disjunction
conditional statement
if-then statement
hypothesis
conclusion
related conditionals
converse
inverse
contrapositive
logically equivalent
biconditional statement

Mathematical Practices

3 Construct viable arguments and critique the reasoning of others.

1 Determine Truth Values

A **statement** is a sentence that is either true or false. The **truth value** of a statement is either true (T) or false (F). Statements are often represented using a letter such as p or q.

p: A rectangle is a quadrilateral. Truth value: T

The **negation** of a statement has the opposite meaning, as well as an opposite truth value. For example, the negation of the statement above is *not* p or $\sim p$.

$\sim p$: A rectangle is not a quadrilateral. Truth value: F

Two or more statements joined by the word *and* or *or* form a **compound statement**. A compound statement using the word *and* is called a **conjunction**. A conjunction is true only when both statements that form it are true.

p: A rectangle is a quadrilateral. Truth value: T

q: A rectangle is convex. Truth value: T

p and q: **A rectangle is a quadrilateral,** and **a rectangle is convex.**

Because both p and q are true, the conjunction p and q, also written $p \wedge q$, is true.

Example 1 Truth Values of Conjunctions

Use the following statements to write a compound statement for each conjunction. Then find its truth value. Explain your reasoning.

p: The figure is a triangle.

q: The figure has two congruent sides.

r: The figure has three acute angles.

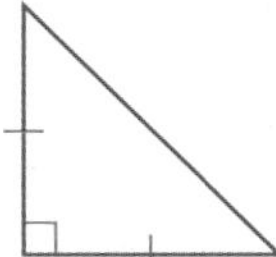

a. p **and** r

p and r: The figure is a triangle, and the figure has three acute angles.

Although p is true, r is false. So, p and r is false.

b. $q \wedge \sim r$

$q \wedge \sim r$: The figure has two congruent sides, and the figure does not have three acute angles.

Both q and $\sim r$ are true, so $q \wedge \sim r$ is true.

Guided Practice

1A. $p \wedge q$

1B. not p and not r

A compound statement that uses the word *or* is called a **disjunction**.

p: **Malik studies geometry.**

q: **Malik studies chemistry.**

p or *q*: **Malik studies geometry,** or **Malik studies chemistry**.

A disjunction is true if at least one of the statements is true. If Malik studies either geometry or chemistry or both subjects, the disjunction p or q, also written as $p \vee q$, is true. If Malik studies neither geometry nor chemistry, p or q is false.

Example 2 Truth Values of Disjunctions

Use the following statements to write a compound statement for each disjunction. Then find its truth value. Explain your reasoning.

***p*: January is a fall month.**

***q*: January has only 30 days.**

***r*: January 1 is the first day of a new year.**

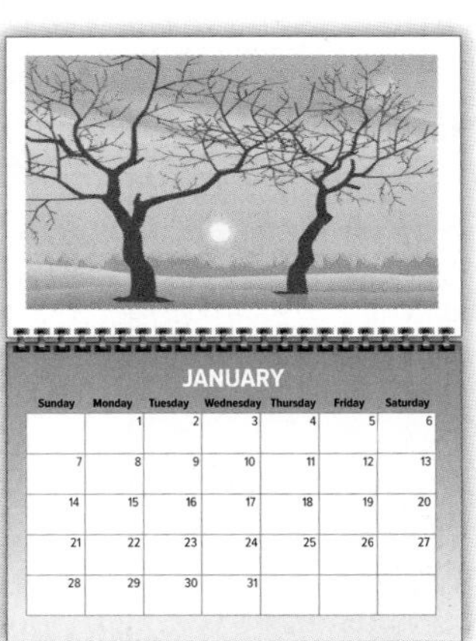

a. q **or** r

q or r: **January has only 30 days, or January 1 is the first day of a new year.**

q or r is true because r is true. It does not matter that q is false.

b. 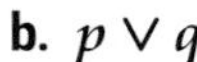 $p \vee q$

$p \vee q$: **January is a fall month, or January has only 30 days.**

Because both p and q are false, $p \vee q$ is false.

c. $\sim p \vee r$

$\sim p \vee r$: January is *not* a fall month, or January 1 is the first day of a new year.

Not p or r is true, because not p is true and r is true.

Watch Out!

Negation Just as the opposite of an integer is not always negative, the negation of a statement is not always false. The negation of a statement has the opposite truth value of the original statement.

Guided Practice

2A. r or p **2B.** $q \vee \sim r$ **2C.** $p \vee \sim q$

Concept Summary Negation, Conjunction, Disjunction

Statement	Words	Symbols
negation	a statement that has the opposite meaning and truth value of an original statement	$\sim p$, read not p
conjunction	a compound statement formed by joining two or more statements using the word *and*	$p \wedge q$, read p and q
disjunction	a compound statement formed by joining two or more statements using the word *or*	$p \vee q$, read p or q

2 Conditional Statements

A **conditional statement** is a statement that can be written in *if-then form*. The sentence given below is an example of a conditional statement.

If the forecast calls for rain, then I will take an umbrella.

Key Concept Conditional Statement

Words	Symbols	Examples
An **if-then statement** is of the form *if p, then q.*	$p \rightarrow q$ read *if p then q,* or *p implies q*	If the forecast calls for rain, then I will take an umbrella.
The **hypothesis** of a conditional statement is the phrase immediately following the word *if.*	p	The forecast calls for rain.
The **conclusion** of a conditional statement is the phrase immediately following the word *then.*	q	I will take an umbrella.

Many conditional statements are written without using the words *if* and *then*. To write these statements in if-then form, identify the hypothesis and conclusion.

> **Reading Math**
> **If and Then** The word *if* is not part of the hypothesis. The word *then* is not part of the conclusion.

Points will be deducted (Conclusion) from any **paper turned in after Wednesday's deadline.** (Hypothesis)

If **a paper is turned in after Wednesday's deadline,** then **points will be deducted.**

Remember, the conclusion depends upon the hypothesis.

Example 3 Write a Conditional in If-Then Form

Write each statement in if-then form.

a. A mammal is a warm-blooded animal.

Hypothesis: An animal is a mammal.

Conclusion: It is warm-blooded.

If an animal is a mammal, then it is warm-blooded.

b. A prism with bases that are regular polygons is a regular prism.

Hypothesis: A prism has bases that are regular polygons.

Conclusion: It is a regular prism.

If a prism has bases that are regular polygons, then it is a regular prism.

Guided Practice

3A. Four quarters can be exchanged for a $1 bill.

3B. The sum of the measures of two supplementary angles is 180.

WatchOut!

Analyzing Conditionals When analyzing a conditional, do not try to determine whether the argument makes sense. Instead, analyze the form of the argument to determine whether the conclusion follows logically from the hypothesis.

To show that a conditional is true, you must show that for each case when the hypothesis is true, the conclusion is also true. To show that a conditional is false, you only need to find one counterexample.

When the hypothesis of a conditional is not met, the truth of a conditional cannot be determined. When the truth of a conditional statement cannot be determined, it is considered true by default. If a statement is not false, logic dictates it must be true.

Example 4 Truth Values of Conditionals

Determine the truth value of each conditional statement. If *true*, explain your reasoning. If *false*, give a counterexample.

a. If you divide an integer by another integer, the result is also an integer.

Counterexample: When you divide 1 by 2, the result is 0.5.

Because 0.5 is not an integer, the conclusion is false.
Because you can find a counterexample, the conditional statement is false.

b. If next month is August, then this month is July.

When the hypothesis is true, the conclusion is also true, because August is the month that follows July. So, the conditional statement is true.

c. If a triangle has four sides, then it is concave.

The hypothesis is false, because a triangle can never have four sides. A conditional with a false hypothesis is always true.

Guided Practice

4A. If $\angle A$ is an acute angle, then $m\angle A$ is 35. **4B.** If $\sqrt{x} = -1$, then $(-1)^2 = -1$.

There are other statements that are based on a given conditional statement. These are known as **related conditionals**.

Go Online!

Log into ConnectED to watch an Animation about conditional statements.

Key Concept Related Conditionals

Words	Symbols	Examples
A conditional statement is a statement that can be written in the form *if p, then q.*	$p \rightarrow q$	If $m\angle A$ is 35, then $\angle A$ is an acute angle.
The **converse** is formed by exchanging the hypothesis and conclusion of the conditional.	$q \rightarrow p$	If $\angle A$ is an acute angle, then $m\angle A$ is 35.
The **inverse** is formed by negating both the hypothesis and conclusion of the conditional.	$\sim p \rightarrow \sim q$	If $m\angle A$ is *not* 35, then $\angle A$ is *not* an acute angle.
The **contrapositive** is formed by negating both the hypothesis and the conclusion of the converse of the conditional.	$\sim q \rightarrow \sim p$	If $\angle A$ is *not* an acute angle, then $m\angle A$ is *not* 35.

A conditional and its contrapositive are either both true or both false. Similarly, the converse and inverse of a conditional are either both true or both false. Statements with the same truth values are said to be **logically equivalent**.

Key Concept Logically Equivalent Statements

- A conditional and its contrapositve are logically equivalent.
- The converse and inverse of a conditional are logically equivalent.

If a conditional is true, the converse may or may not be true.

You can use logical equivalence to check the truth values of statements. Notice that in Example 5, both the conditional and contrapositive are true. Also, both the converse and inverse are false.

Real-World Example 5 Related Conditionals

Real-WorldLink

Cats in the genus *Panthera* include the leopard, jaguar, lion, and tiger. These are the only cats that can roar. They cannot, however, purr.

Source: *Encyclopaedia Britannica*

NATURE Write the converse, inverse, and contrapositive of the following true conditional statement. Then use the information at the left to determine whether each related conditional is ***true*** or ***false***. If a statement is false, find a counterexample.

Lions are cats that can roar.

Conditional: First, rewrite the conditional in if-then form.

If an animal is a lion, then it is a cat that can roar.

Based on the information at the left, this statement is true.

Converse: If an animal is a cat that can roar, then it is a lion.

Counterexample: A tiger is a cat that can roar, but it is not a lion.

Therefore, the converse is false.

Inverse: If an animal is not a lion, then it is not a cat that can roar.

Counterexample: A tiger is not a lion, but it is a cat that can roar.

Therefore, the inverse is false.

Contrapositive: If an animal is not a cat that can roar, then it is not a lion.

Based on the information at the left, this statement is true.

CHECK Check to see that logically equivalent statements have the same truth value.

Both the conditional and contrapositive are true. ✓

Both the converse and inverse are false. ✓

Guided Practice

Write the converse, inverse, and contrapositive of each true conditional statement. Determine whether each related conditional is *true* or *false*. If a statement is false, find a counterexample.

5A. Two angles that have the same measure are congruent.

5B. A hamster is a rodent.

Biconditionals Inverses, converses, and contrapositives are statements related to conditionals. If you make a compound statement from a conditional statement and its converse, then you can write them as a single **biconditional statement**. A biconditional statement often connects a conditional with its converse by using the words "if and only if." A conditional statement and its converse must both be true for a biconditional to be true.

Key Concept Biconditional Statement

Words A biconditional statement is a conjunction of a conditional statement and its converse.

Symbols $(p \rightarrow q) \wedge (q \rightarrow p) \rightarrow (p \rightarrow q)$ (Read "*p if and only if q*.")

Example *p:* Angle *O* has measure between 90° and 180°.

q: Angle *O* is an obtuse angle.

Biconditional: Angle *O* has measure between 90° and 180° if and only if angle *O* is an obtuse angle.

Study Tip

MP Structure The phrase *if and only if* can be abbreviated with *iff.*

Definitions in geometry can always be written as biconditional statements.

Example 6 Write Biconditional Statements

Rewrite each statement as a biconditional statement. Then determine whether the biconditional is *true* or *false*.

a. A number of the form $2k + 1$, where k is some integer, is an odd number.

Step 1 Rewrite the statement as a conditional and its converse. Give the truth value of each.

Conditional: If a number has the form $2k + 1$, where k is some integer, then the number is an odd number. True.

Converse: If a number is odd, then it has the form $2k + 1$, where k is some integer. True.

Step 2 Combine the statements as a biconditional. Tell whether the biconditional is true or false.

A number has the form $2k + 1$, where k is some integer, if and only if the number is an odd number.

The biconditional is true because the conditional and its converse are both true.

b. Lines that do not intersect are parallel.

Step 1 Rewrite the statement as a conditional and its converse. Give the truth value of each.

Conditional: If two lines do not intersect, then the lines are parallel. False. *Counterexample:* Skew lines do not intersect and are not parallel.

Converse: If two lines are parallel, then they do not intersect. True.

Step 2 Combine the statements as a biconditional. Tell whether the biconditional is true or false.

Lines do not intersect if and only if they are parallel.

The biconditional is false because the conditional is false.

Guided Practice

6A. A number less than or equal to its opposite is a negative number.

6B. The mean of a data set is between the lowest and highest values in the data set.

Check Your Understanding

○ = Step-by-Step Solutions begin on page R13.

Go Online! for a Self-Check Quiz

Examples 1–2 **Use the following statements to write a compound statement for each conjunction or disjunction. Then find its truth value. Explain your reasoning.**

p: A week has seven days.

q: There are 20 hours in a day.

r: There are 60 minutes in an hour.

1. p and r
2. $p \wedge q$
3. $q \vee r$
4. $\sim p$ or q
5. $p \vee r$
6. $\sim p \wedge \sim r$

Example 3 **Write each statement in if-then form.**

7. Sixteen-year-olds are eligible to drive.
8. Cheese contains calcium.
9. The measure of an acute angle is between 0 and 90.
10. Equilateral triangles are equiangular.

Example 4 **Determine the truth value of each conditional statement. If *true*, explain your reasoning. If *false*, give a counterexample.**

11. If $x^2 = 16$, then $x = 4$.
12. If you live in Atlanta, then you live in Georgia.
13. If tomorrow is Friday, then today is Thursday.
14. If an animal is spotted, then it is a Dalmatian.
15. If the measure of a right angle is 95, then bees are lizards.
16. If pigs can fly, then $2 + 5 = 7$.
17. If Jacqueline turned 14 years old last year, then she will turn 15 this year.
18. If a number is between 10 and 12, then it is 11.

Example 5 MP **REASONING** **Write the converse, inverse, and contrapositive of each true conditional statement. Determine whether each related conditional is *true* or *false*. If a statement is false, find a counterexample.**

19. If a number is divisible by 4, then it is divisible by 2.
20. All whole numbers are integers.

Example 6 **Rewrite each statement as a biconditional statement. Then determine whether the biconditional is *true* or *false*.**

21. Two angles whose measures add to 90° are complementary.
22. A number x such that $x > -0.5$ is positive.
23. An angle with a measure between 0° and 90° is an acute angle.
24. There is no school on Saturday.
25. An integer is a rational number.
26. An obtuse triangle has one obtuse angle.
27. A dodecagon is a polygon with 12 sides.

Practice and Problem Solving

Extra Practice is on page R2.

Examples 1–2 **Use the following statements and figure to write a compound statement for each conjunction or disjunction. Then find its truth value. Explain your reasoning.**

p: $\overrightarrow{DB}$ is the angle bisector of $\angle ADC$.

q: Points C, D, and B are collinear.

r: $\overline{AD} \cong \overline{DC}$

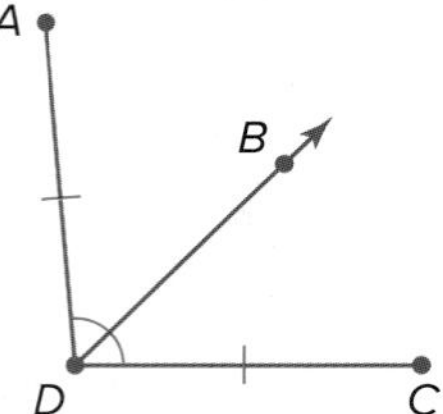

28. p and r **29.** q or p **30.** r or $\sim p$

31. r and q **32.** $\sim p$ or $\sim r$ **33.** $\sim p$ and $\sim r$

MP REASONING **Use the following statements to write a compound statement for each conjunction or disjunction. Then find its truth value. Explain your reasoning.**

p: Springfield is the capital of Illinois.

q: Illinois borders the Atlantic Ocean.

r: Illinois shares a border with Kentucky.

s: Illinois is to the west of Missouri.

34. $p \wedge r$ **35.** $p \wedge q$ **36.** $\sim r \vee s$

37. $r \vee q$ **38.** $\sim p \wedge \sim r$ **39.** $\sim s \vee \sim p$

Example 3 **Write each statement in if-then form.**

40. Get a free milkshake with any combo purchase.

41. Everybody at the party received a gift.

42. The intersection of two planes is a line.

43. The area of a circle is πr^2.

44. Collinear points lie on the same line.

45. A right angle measures 90 degrees.

Example 4 **MP CONSTRUCT ARGUMENTS** **Determine the truth value of each conditional statement. If *true*, explain your reasoning. If *false*, give a counterexample.**

46. If a banana is blue, then an apple is a vegetable.

47. If a number is odd, then it is divisible by 5.

48. If a dog is an amphibian, then the season is summer.

49. If an angle is acute, then it has a measure of 45.

50. If a polygon has six sides, then it is a regular polygon.

51. If an angle's measure is 25, then the measure of the angle's complement is 65.

52. If North Carolina is south of Florida, then the capital of Ohio is Columbus.

53. If red paint and blue paint mixed together make white paint, then $3 - 2 = 0$.

54. If two angles are congruent, then they are vertical angles.

55. If an animal is a bird, then it is an eagle.

56. If two angles are acute, then they are supplementary.

57. If two lines intersect, then they form right angles.

Example 5 **Write the converse, inverse, and contrapositive of each true conditional statement. Determine whether each related conditional is *true* or *false*. If a statement is false, find a counterexample.**

58. A right triangle has an angle that measures 90°.
59. If you live in Chicago, you live in Illinois.
60. If a bird is an ostrich, then it cannot fly.
61. If two angles have the same measure, then the angles are congruent.
62. All squares are rectangles.
63. All congruent segments have the same length.

Example 6 **Rewrite each statement as a biconditional statement. Then determine whether the biconditional is *true* or *false*.**

64. Lines that do not intersect are horizontal.
65. Points that lie in the same plane are coplanar.
66. Right angles measure 90°.
67. The midpoint of a segment bisects the segment.
68. Real numbers are irrational numbers.
69. Perpendicular lines meet at right angles.

MP **CONSTRUCT ARGUMENTS** Write the statement indicated and determine the truth value of each statement. If a statement is false, give a counterexample.

Animals with stripes are zebras.

70. conditional

71. converse

72. contrapositive

inverse

Write the conditional and converse for each statement. Determine the truth values of the conditionals and converses. If false, write a counterexample. Write a biconditional if possible.

74. Regular quadrilaterals are squares.
75. Equilateral triangles have all sides the same length.
76. Right angles have measures greater than the measures of acute angles.
77. Integers are rational numbers.
78. **VEHICLES** Different vehicles are characterized by different structural features. Write each statement in if-then form.
    - A convertible describes any vehicle with a fully retractable top.
    - A coupe is any car with only two full-size passenger doors.
    - A pickup is any vehicle with an open cargo bed in the rear.
79. **ART** Write the following statement in if-then form: At the Andy Warhol Museum in Pittsburgh, Pennsylvania, most of the collection is Andy Warhol's artwork.

80. SCIENCE The water on Earth is constantly changing through a process called the *water cycle*. Write the three conditionals below in if-then form.

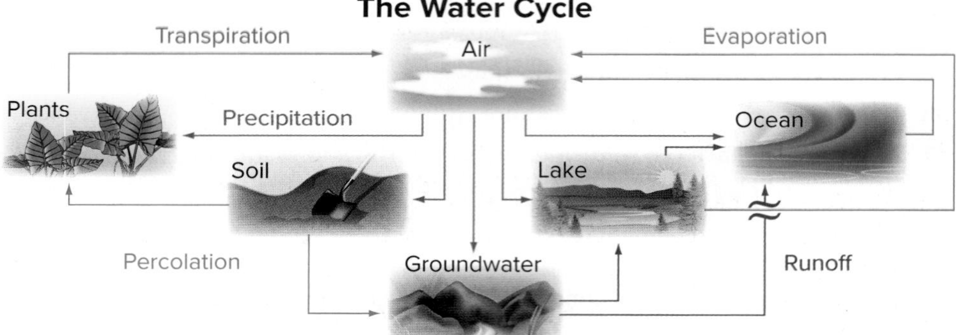

a. As runoff, water flows into bodies of water.

b. Plants return water to the air through transpiration.

c. Water bodies return water to the air through evaporation.

81. SPORTS In football, touchdowns are worth 6 points, extra point conversions are worth 2 points, and safeties are worth 2 points.

a. Write three conditional statements in if-then form for scoring in football.

b. Write the converse of the three true conditional statements. State whether each is *true* or *false*. If a statement is false, find a counterexample.

82. SCIENCE Chemical compounds are grouped and described by the elements that they contain. Acids contain hydrogen (H). Bases contain hydroxide (OH). Hydrocarbons contain only hydrogen (H) and carbon (C).

Compound	Example	Chemical Formula
acid	hydrochloric acid	HCl
base	sodium hydroxide	$NaOH$
hydrocarbon	methane	CH_4

a. Write three conditional statements in if-then form for classifying chemical compounds.

b. Write the contrapositive of the three true conditional statements. State whether each is *true* or *false*. If a statement is false, find a counterexample.

83. Can the conditional, "If $x^2 - 36 = 0$, then $x = 6$," be combined with its converse to form a true biconditional? Explain.

84. If the contrapositive of a conditional is true, can you rewrite the conditional as a true biconditional? Explain.

85. Compare the mathematical meanings of the symbols $\rightarrow$ and $\leftrightarrow$ in $p \rightarrow q$ and $p \leftrightarrow q$.

86. MULTIPLE REPRESENTATIONS In this problem, you will investigate a law of logic by using conditionals.

a. **Logical** Write three true conditional statements, using each consecutive conclusion as the hypothesis for the next statement.

b. **Logical** Write a conditional using the hypothesis of your first conditional and the conclusion of your third conditional. Is the conditional true if the hypothesis is true?

c. **Verbal** Given two conditionals *If a, then b* and *If b, then c,* make a conjecture about the truth value of *c* when *a* is true. Explain your reasoning.

LOGIC To negate a statement containing the words *all* or *for every*, you can use the phrase *at least one* or *there exists*. To negate a statement containing the phrase *there exists*, you can use the phrase *for all* or *for every*.

p:	All polygons are convex.	$\sim p$:	*At least one* polygon is *not* convex.
q:	*There exists* a problem that has no solution.	$\sim q$:	*For every* problem, there is a solution.

Sometimes these phrases may be implied. For example, *The square of a real number is nonnegative* implies the following conditional and its negation.

p: *For every* real number x, $x^2 \geq 0$.

$\sim p$: *There exists* a real number x such that $x^2 < 0$.

Use the information above to write the negation of each statement.

87. There exists a segment that has no midpoint.

88. Every student at Hammond High School has a locker.

89. All squares are rectangles.

90. There exists a real number x such that $x^2 = x$.

91. There exists an even number x such that $2x - 2 = x$.

92. Every real number has a real square root.

H.O.T. Problems Use Higher-Order Thinking Skills

93. CRITIQUE ARGUMENTS Nicole and Kiri are evaluating the conditional *If 15 is prime, then 20 is divisible by 4*. Both think that the conditional is true, but their reasoning differs. Is either of them correct? Explain.

Nicole	Kiri
The conclusion is true, because 20 is divisible by 4, so the conditional is true.	The hypothesis is false, because 15 is not prime, so the conditional is true.

94. CHALLENGE You have learned that statements with the same truth value are logically equivalent. Use logical equivalence to summarize the conditional, converse, inverse, and contrapositive for the statements p and q.

95. MP **REASONING** You are evaluating a conditional statement in which the hypothesis is true, but the conclusion is false. Is the inverse of the statement true or false? Explain your reasoning.

96. OPEN ENDED Write a conditional statement in which the converse, inverse, and contrapositive are all true. Explain your reasoning.

97. CHALLENGE The inverse of conditional A is given below. Write conditional A, its converse, and its contrapositive. Explain your reasoning.

If I received a detention, then I did not arrive at school on time.

98. WRITING IN MATH Describe the relationship between a conditional, its converse, its inverse, and its contrapositive.

99. WRITING IN MATH If a biconditional is true, what do you know about the conditional and the converse? If the biconditional is false, what do you know about the conditional and converse?

100. REASONING Because a conditional statement and its converse must both be true for a biconditional to be true, the order of the hypothesis and conclusion do not matter. How does this affect the truth values of the statements? Explain.

Preparing for Assessment

101. MULTI-STEP Use conditional statements I through IV to answer the following questions. MP 2

I. Two lines are perpendicular, and the lines intersect.

II. All triangles have an acute angle.

III. The sum of the measures of two supplementary angles is 180°.

IV. An angle is a right angle, and it has a measurement of 90°.

a. Rewrite each of the conditional statements in if-then form. Then, write the converse statement.

b. Which conditional statements have a true converse?

- ☐ **A** I
- ☐ **B** II
- ☐ **C** III
- ☐ **D** IV

c. Which statements, paired with their converses, can be written as a biconditional? Select all of the true statements.

- ☐ **A** I
- ☐ **B** II
- ☐ **C** III
- ☐ **D** IV

102. Which of the following can be used to prove that a conditional statement is false? MP 2

- ○ **A** counterexample
- ○ **B** converse
- ○ **C** conclusion
- ○ **D** contrapositive

103. Consider the conditional statements below. MP 4

I. Every eagle is a bird.

II. A butterfly is an insect.

III. If an animal is a tiger, then it lives underwater.

Which of the conditional statements has a true contrapositive?

- ○ **A** I only
- ○ **B** II only
- ○ **C** III only
- ○ **D** I and II only
- ○ **E** I, II, and III

104. Which of the following statements is logically equivalent to the statement below. Select all logically equivalent statements. MP 2

All octagons are polygons.

- ☐ **A** If a figure is not a polygon, then it is not an octagon.
- ☐ **B** All polygons are octagons.
- ☐ **C** If a figure is not an octagon, then it is not a polygon.
- ☐ **D** Every polygon is also an octagon.
- ☐ **E** A figure is an octagon if and only if it is a polygon.

105. Use the following statements to write a compound statement. Then find the truth value. MP 2

p: A triangle has two congruent sides.

q: A triangle has no congruent sides.

r: A triangle is equilateral.

a. $p \lor q \lor r$

b. $p \land q$

LESSON 3

Deductive Reasoning

Then

- You used inductive reasoning to analyze patterns and make conjectures.

Now

1. Use the Law of Detachment.
2. Use the Law of Syllogism.

Why?

- When detectives are trying to solve a case, they use techniques like fingerprinting to analyze evidence. Then they use this evidence to eliminate suspects and eventually identify the person responsible for the crime.

New Vocabulary

deductive reasoning
valid
Law of Detachment
Law of Syllogism

Mathematical Practices

3 Construct viable arguments and critique the reasoning of others.

1 Law of Detachment

The process that detectives use to identify who is most likely responsible for a crime is called deductive reasoning. Unlike inductive reasoning, which uses a pattern of examples or observations to make a conjecture, **deductive reasoning** uses facts, rules, definitions, or properties to reach logical conclusions from given statements.

Real-World Example 1 Inductive and Deductive Reasoning

Determine whether each conclusion is based on *inductive* or *deductive* reasoning.

a. Every time Lauren has a healthy meal before her run, she is able to run farther than her sister who does not like healthy foods. Lauren has eaten a healthy meal, so she concludes that she will run farther than her sister.

Lauren is basing her conclusion on a pattern of observations, so she is using inductive reasoning.

b. If John is late making his car insurance payment, he will be assessed a late fee of $50. John's payment is late this month, so he concludes that he will be assessed a late fee of $50.

John is basing his conclusion on facts provided to him by his insurance company, so he is using deductive reasoning.

Guided Practice

1A. All of the signature items on the restaurant's menu shown are noted with a special symbol. Kevin orders a menu item that has this symbol next to it, so he concludes that the menu item that he has ordered is a signature item.

1B. Every time Raul goes to the restaurant on a Thursday, Cheese Supreme is the special. Raul goes to the restaurant on a Thursday, so he concludes that Cheese Supreme will be the special.

While one counterexample is enough to disprove a conjecture reached using inductive reasoning, it is not a logically correct, or **valid**, method of proving a conjecture. To prove a conjecture requires deductive reasoning. One valid form of deductive reasoning is the **Law of Detachment**.

Key Concept Law of Detachment

Words If $p \rightarrow q$ is a true statement and p is true, then q is true.

Example *Given*: If a car is out of gas, then it will not start.
Sarah's car is out of gas.

Valid Conclusion: Sarah's car will not start.

As long as the facts given are true, the conclusion reached using deductive reasoning will also be true.

Example 2 Law of Detachment

Study Tip

Given Information From this point forward in this text, all given information can be assumed true.

Determine whether each conclusion is valid based on the given information. If not, write *invalid*. Explain your reasoning.

a. Given: If two angles form a linear pair, then their noncommon sides are opposite rays.
$\angle AED$ and $\angle AEB$ form a linear pair.

Conclusion: $\overrightarrow{ED}$ and $\overrightarrow{EB}$ are opposite rays.

Step 1 Identify the hypothesis p and the conclusion q of the true conditional.

p: Two angles form a linear pair.

q: Their noncommon sides are opposite rays.

Step 2 Analyze the conclusion.

The given statement *$\angle AED$ and $\angle AEB$ form a linear pair* satisfies the hypothesis, so p is true. By the Law of Detachment, $\overrightarrow{ED}$ and $\overrightarrow{EB}$ *are opposite rays*, which matches q, is a true or valid conclusion.

b. Given: If Mika goes to the beach, she will wear sunscreen.
Mika is wearing sunscreen.

Conclusion: Mika is at the beach.

Step 1 **p: Mika goes to the beach.**

q: Mika wears sunscreen.

Step 2 The given statement *Mika is wearing sunscreen* satisfies the conclusion q of the true conditional. However, knowing that a conditional statement and its conclusion are true does not make the hypothesis true. Mika could be wearing sunscreen because she is at the pool. The conclusion is invalid.

Guided Practice

2A. Given: If three points are noncollinear, then they determine a plane.
Points A, B, and C lie in plane G.

Conclusion: Points A, B, and C are noncollinear.

2B. Given: If a student turns in a permission slip, then the student can go on the field trip.
Felipe turned in his permission slip.

Conclusion: Felipe can go on the field trip.

Real-World Link

The easiest way to distinguish monkeys from other primates is to look for a tail. Most monkey species have tails, but apes do not.

You can also use a Venn diagram to test the validity of a conclusion.

Example 3 Judge Conclusions Using Venn Diagrams

NATURE Determine whether each conclusion is valid based on the given information. If not, write *invalid*. Explain your reasoning using a Venn diagram.

Given: **If a primate is an ape, then it does not have a tail. Koko is a primate who does not have a tail.**

Conclusion: **Koko is an ape.**

Understand Draw a Venn diagram to represent the conditional.

Plan Because we are only given that Koko does not have a tail, we can only conclude that Koko belongs outside the circle for primates with tails.

Solve This could put her in the area inside or outside of the Apes circle, so the conclusion is invalid.

Check It is possible for Koko to be a primate without a tail and still not be an ape. Therefore, the conclusion *is* invalid. ✓

Venn diagrams are a great way to visualize groups and subgroups. By representing the conditional as a Venn diagram, it becomes clear that there are two possible subgroups, apes and primates without tails, in which Koko could be a member. Thus, we cannot definitively prove that Koko is an ape and thus, our answer is reasonable.

Guided Practice

3. **Given:** If a figure is a square, then it is a polygon. Figure A is a square.

   Conclusion: Figure A is a polygon.

Study Tip

MP Construct Arguments An *argument* consists of reasons, proof, or evidence to support a position. A *logical argument* such as the one shown is supported by the rules of logic. This is different from a *statistical argument*, which is supported by examples or data.

2 Law of Syllogism

The **Law of Syllogism** is another valid form of deductive reasoning. This law allows you to draw conclusions from two true conditional statements when the conclusion of one statement is the hypothesis of the other.

Key Concept Law of Syllogism

Words	If $p \rightarrow q$ and $q \rightarrow r$ are true statements, then $p \rightarrow r$ is a true statement.
Example	*Given:* If you get a job, then you will earn money. If you earn money, then you will buy a car. *Valid Conclusion:* If you get a job, then you will buy a car.

It is important to remember that if the conclusion of the first statement is *not* the hypothesis of the second statement, then no valid conclusion can be drawn.

Example 4 Law of Syllogism

Determine whether each statement is valid based on the information. If not, write *invalid.* Explain your reasoning.

(1) If you are an actor, then you enjoy musicals.
(2) If you enjoy musicals, then you enjoy theater productions.

a. If you enjoy musicals, then you are an actor.

Step 1 Let p, q, and r represent the parts of the given conditional statements.

p: You are an actor.
q: You enjoy musicals.
r: You enjoy theater productions.

Step 2 Analyze the logic of the given conditional statements using symbols.

Statement (1): $p \rightarrow q$ Statement (2): $q \rightarrow r$

Both given statements are considered true. Because the given statements are of the form Statement (1): $p \rightarrow q$; Statement (2): $q \rightarrow r$, they have a valid conclusion by the Law of Syllogism. A valid conclusion would be of the form $p \rightarrow r$.

Step 3 Analyze the logic of the conclusion. The conclusion is of the form $q \rightarrow p$ and is invalid by the Law of Syllogism.

b. If you are an actor, then you enjoy theater productions.
The conclusion is of the form $p \rightarrow r$ and is valid by the Law of Syllogism.

Study Tip

True vs. Valid Conclusions A true conclusion is not the same as a valid conclusion. True conclusions that are reached using invalid deductive reasoning are still invalid.

Guided Practice

Determine whether each statement is valid based on the information. If not, write *invalid.* Explain your reasoning.

(1) If you do not get enough sleep, then you will be tired.

(2) If you are tired, then you will not do well on the test.

4A. If you are tired, then you will not get enough sleep.

4B. If you do not get enough sleep, then you will not do well on the test.

Example 5 Apply Laws of Deductive Reasoning

Draw a valid conclusion from the given statements, if possible. Then state whether your conclusion was drawn using the Law of Detachment or the Law of Syllogism. If no valid conclusion can be drawn, write *no valid conclusion* and explain your reasoning.

Given: If you are 16 years old, then you can apply for a driver's license. Nate is 16 years old.

p: You are 16 years old.

q: You can apply for a driver's license.

Because *Nate is 16 years old* satisfies the hypothesis, p is true. By the Law of Detachment, a valid conclusion is *Nate can apply for a driver's license.*

Go Online!

When working with conditional statements, it may be helpful to mark the hypothesis and conclusion with different colored highlighting. Log into your eStudent Edition to use the highlighting tools with your text.

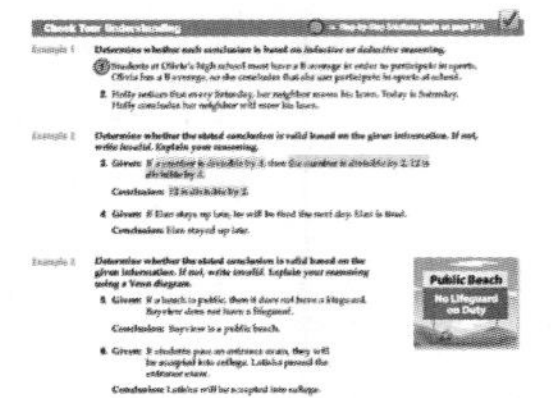

Guided Practice

5. Given: The midpoint divides a segment into two congruent segments. If two segments are congruent, then their measures are equal. M is the midpoint of $\overline{AB}$.

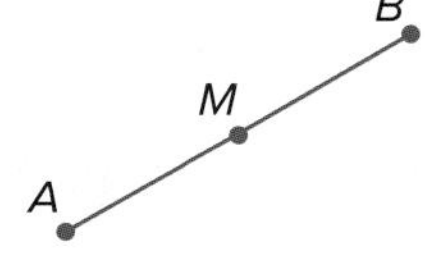

Check Your Understanding

◯ = Step-by-Step Solutions begin on page R13.

Example 1 **Determine whether each conclusion is based on *inductive* or *deductive* reasoning.**

1. Students at Olivia's high school must have a B average in order to participate in sports. Olivia has a B average, so she concludes that she can participate in sports at school.

2. Holly notices that every Saturday, her neighbor mows his lawn. Today is Saturday. Holly concludes her neighbor will mow his lawn.

Example 2 **Determine whether the stated conclusion is valid based on the given information. If not, write *invalid*. Explain your reasoning.**

3. **Given:** If a number is divisible by 4, then the number is divisible by 2. 12 is divisible by 4.

   Conclusion: 12 is divisible by 2.

4. **Given:** If Elan stays up late, he will be tired the next day. Elan is tired.

   Conclusion: Elan stayed up late.

Example 3 **Determine whether the stated conclusion is valid based on the given information. If not, write *invalid*. Explain your reasoning using a Venn diagram.**

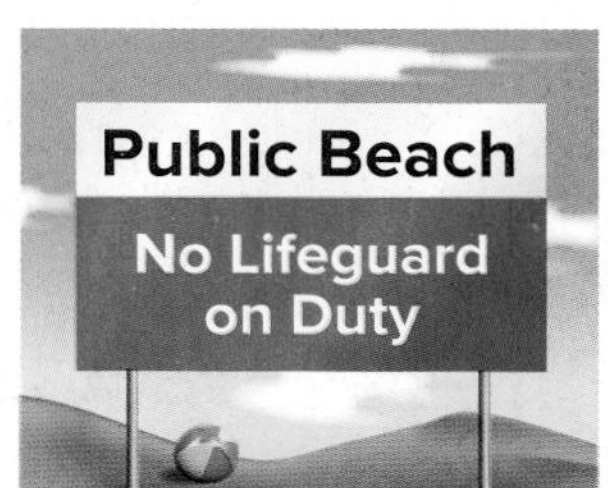

5. **Given:** If a beach is public, then it does not have a lifeguard. Bayview does not have a lifeguard.

   Conclusion: Bayview is a public beach.

6. **Given:** If students pass an entrance exam, they will be accepted into college. Latisha passed the entrance exam.

   Conclusion: Latisha will be accepted into college.

Example 4 **Determine whether each statement is valid based on the information. If not, write *invalid*. Explain your reasoning.**

(1) If a triangle is a right triangle, then it has an angle that measures 90.
(2) If a triangle has an angle that measures 90, then its acute angles are complementary.

7. If the acute angles of a triangle are complementary, then it is a right triangle.

8. If a triangle is a right triangle, then its acute angles are complementary.

9. If a triangle has an angle that measures 90, then its acute angles are complementary.

Example 5 MP **CONSTRUCT ARGUMENTS** **Draw a valid conclusion from the given statements, if possible. Then state whether your conclusion was drawn using the Law of Detachment or the Law of Syllogism. If no valid conclusion can be drawn, write *no valid conclusion* and explain your reasoning.**

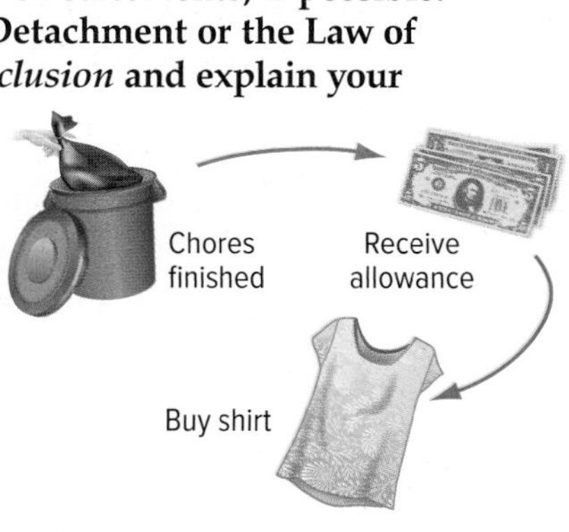

10. **Given:** If Dalila finishes her chores, she will receive her allowance.

    If Dalila receives her allowance, she will buy a new shirt.

11. **Given:** Vertical angles are congruent.

    $\angle 1 \cong \angle 2$

Practice and Problem Solving

Extra Practice is on page R2.

Example 1 **Determine whether each conclusion is based on *inductive* or *deductive* reasoning.**

12. At Fumio's school if you are late five times, you will receive a detention. Fumio has been late to school five times; therefore, he will receive a detention.

13. A dental assistant notices a patient has never been on time for an appointment. She concludes the patient will be late for her next appointment.

14. A person must have a membership to work out at a gym. Jesse is working out at a gym. Jesse has a membership to the gym.

15. If Eduardo decides to go to a concert tonight, he will miss football practice. Tonight, Eduardo went to a concert. Eduardo missed football practice.

16. Every Wednesday Lucy's mother calls. Today is Wednesday, so Lucy concludes her mother will call.

17. Whenever Juanita has attended a tutoring session she notices that her grades have improved. Juanita attends a tutoring session, and she concludes her grades will improve.

Example 2 MP **CRITIQUE ARGUMENTS Determine whether the stated conclusion is valid based on the given information. If not, write *invalid*. Explain your reasoning.**

18. **Given:** Right angles are congruent. $\angle 1$ and $\angle 2$ are right angles.

    Conclusion: $\angle 1 \cong \angle 2$

19. **Given:** If a figure is a square, it has four right angles. Figure *ABCD* has four right angles.

    Conclusion: Figure *ABCD* is a square.

20. **Given:** An angle bisector divides an angle into two congruent angles. $\overrightarrow{KM}$ is an angle bisector of $\angle JKL$.

    Conclusion: $\angle JKM \cong \angle MKL$

    J M K L

21. **Given:** If you leave your lights on while your car is off, your battery will die. Your battery is dead.

    Conclusion: You left your lights on while the car was off.

22. **Given:** If Dante obtains a part-time job, he can afford a car payment. Dante can afford a car payment.

    Conclusion: Dante obtained a part-time job.

23. **Given:** If 75% of the prom tickets are sold, the prom will be held at the country club. 75% of the prom tickets were sold.

    Conclusion: The prom will be held at the country club.

24. **COMPUTER GAMES** Refer to the game ratings at the right. Determine whether the stated conclusion is valid based on the given information. If not, write *invalid*. Explain your reasoning.

    Given: If a title is rated E, then it has content that may be suitable for ages 6 and older. Cesar buys a computer game that he believes is suitable for his younger sister, who is 7.

    Conclusion: The game Cesar purchased has a rating of E.

Game Ratings	
Rating	**Age**
EC	3 and older
E	6 and older
E10+	10 and older
T	13 and older
M	17 and older

Example 3 **Determine whether the stated conclusion is valid based on the given information. If not, write *invalid*. Explain your reasoning using a Venn diagram.**

25. **Given:** If the temperature drops below $-15°F$, then school will be cancelled. The temperature did not drop below $-15°F$ on Monday.

    Conclusion: School was not cancelled on Monday.

26. **Given:** If a person is a Missouri resident, he or she does not live by a beach. Michelle does not live by the beach.

    Conclusion: Michelle is a Missouri resident.

27. **Given:** Some nurses wear blue uniforms. Sabrina is a nurse.

    Conclusion: Sabrina wears a blue uniform.

28. **Given:** All vegans do not eat meat. Theo is a vegan.

    Conclusion: Theo does not eat meat.

29. **TRANSPORTATION** There are many types of vehicles, and they are classified using different sets of criteria. Determine whether the stated conclusion is valid based on the given information. If not, write *invalid*. Explain your reasoning using a Venn diagram.

    Given: If a vehicle is a sport-utility vehicle, then it is a four-wheel-drive car built on a truck chassis. Ms. Rodriguez has just purchased a vehicle that has four-wheel drive.

    Conclusion: Ms. Rodriguez has just purchased a sport-utility vehicle.

Examples 4–5 30. **GOLF** Sergio Garcia won the Masters Tournament in 2017. Use the Law of Syllogism to draw a valid conclusion from each set of statements, if possible. If no valid conclusion can be drawn, write *no valid conclusion* and explain your reasoning.

    (1) If Sergio Garcia's score is lower than the other golfers at the end of the tournament, then he wins the tournament.

    (2) If a golfer wins the Masters Tournament, then he gets a green jacket.

MP **CONSTRUCT ARGUMENTS Use the Law of Syllogism to draw a valid conclusion from each set of statements, if possible. If no valid conclusion can be drawn, write *no valid conclusion* and explain your reasoning.**

31. If you interview for a job, then you wear a suit.

    If you interview for a job, then you will update your resume.

32. If Tina has a grade point average of 3.0 or greater, she will be on the honor roll.

    If Tina is on the honor roll, then she will have her name in the school paper.

33. If two lines are perpendicular, then they intersect to form right angles.

    Lines r and s form right angles.

34. If the measure of an angle is between 90 and 180, then it is obtuse.

    If an angle is obtuse, then it is not acute.

35. If two lines in a plane are not parallel, then they intersect.

    If two lines intersect, then they intersect in a point.

36. If a number ends in 0, then it is divisible by 2.

    If a number ends in 4, then it is divisible by 2.

Draw a valid conclusion from the given statements, if possible. Then state whether your conclusion was drawn using the Law of Detachment or the Law of Syllogism. If no valid conclusion can be drawn, write *no valid conclusion* and explain your reasoning.

37. **Given:** If a figure is a square, then all the sides are congruent.

    Figure $ABCD$ is a square.

38. **Given:** If two angles are complementary, the sum of the measures of the angles is 90.

    $\angle 1$ and $\angle 2$ are complements of each other.

39. **Given:** Ballet dancers like classical music.

    If you like classical music, then you enjoy the opera.

40. **Given:** If you are athletic, then you enjoy sports.

    If you are competitive, then you enjoy sports.

41. **Given:** If a polygon is regular, then all of its sides are congruent.

    All sides of polygon $WXYZ$ are congruent.

42. **Given:** If Bob completes a course with a grade of C, then he will not receive credit.

    If Bob does not receive credit, he will have to take the course again.

43. **DATA ANALYSIS** The table shows the number of at bats and hits for some of the members of the Miami Marlins in a recent season.

    a. Construct a scatter plot to represent the data.
    b. Predict the number of hits a player with 300 at bats would get. Identify and explain your reasoning.
    c. Did the player with 157 at bats or the player with 240 at bats get more hits? What type of reasoning did you use? Explain.

At Bats	Hits
13	6
576	195
240	79
502	139
157	36
64	11

Source: ESPN

H.O.T. Problems Use **H**igher-**O**rder **T**hinking Skills

44. **WRITING IN MATH** Explain why the Law of Syllogism cannot be used to draw a conclusion from these conditionals.

    If you wear winter gloves, then you will have warm hands.

    If you do not have warm hands, then your gloves are too thin.

45. **CHALLENGE** Use the symbols from Lesson 2-2 for *conjunction* and *disjunction* and the symbol for *implies* to represent the Law of Detachment and the Law of Syllogism symbolically. Let p represent the hypothesis and let q represent the conclusion.

46. **OPEN ENDED** Write a pair of statements in which the Law of Syllogism can be used to reach a valid conclusion. Specify the conclusion that can be reached.

47. MP **REASONING** Students in Mr. Kendrick's class are divided into two groups for an activity. Students in group A must always tell the truth. Students in group B must always lie. Jonah and Janeka are in Mr. Kendrick's class. When asked if he and Janeka are in group A or B, Jonah says, "We are both in Group B." To which group does each student belong? Explain your reasoning.

48. **WRITING IN MATH** Compare and contrast inductive and deductive reasoning when making conclusions and proving conjectures.

49. Consider the given statements below.

I. If the coordinates of a point are (5, 3), then its distance from the origin is $\sqrt{34}$ units.

II. The distance of point P from the origin is $\sqrt{34}$ units.

Natalie concluded that the coordinates of point P are (5, 3).

Which of the following best describes Natalie's conclusion? MP 2, 3

- ○ **A** It is false because Natalie used invalid reasoning.
- ○ **B** It is false because statement I is false.
- ○ **C** It is valid by the Law of Detachment.
- ○ **D** It is valid by the Law of Syllogism.

50. Consider the given statements below.

I. If a sphere has a radius of 3 inches, then its surface area is 36π square inches.

II. If a sphere has a surface area of 36π square inches, then its volume is 36π cubic inches.

Javier concluded that if a sphere has a radius of 3 inches, then its volume is 36π cubic inches.

Which of the following best describes Javier's conclusion? MP 2, 3

- ○ **A** It is false because statement I is false.
- ○ **B** It is false, because statement II is false.
- ○ **C** It is false, because Javier used invalid reasoning.
- ○ **D** It is valid by the Law of Detachment.
- ○ **E** It is valid by the Law of Syllogism.

51. **MULTI-STEP** The two statements below are true. Answer each question about the conclusions based on these statements. MP 2, 3

Given Statements

I. If three points lie on the same line, then they are collinear.

II. If points are collinear, then they are coplanar.

Conclusion

If three points lie on the same line, then they are coplanar.

a. Which law was used to write the conclusion?

b. Which of the following best describes the conclusion?

- ○ **A** The conclusion is false, because it follows invalid reasoning.
- ○ **B** The contrapositive of the conclusion is true.
- ○ **C** The converse of the conclusion is true.
- ○ **D** The inverse of the conclusion is true.

c. If three points are not coplanar, then they do not lie on the same line.

Which of the following best describes this statement using the two given statements? Check all that apply.

- ☐ **A** The statement is always true.
- ☐ **B** The statement is sometimes true.
- ☐ **C** The statement is always false.
- ☐ **D** It is the contrapositive of the conclusion statement above.
- ☐ **E** It is the converse of the conclusion statement above.
- ☐ **F** It is the inverse of the conclusion statement above.

d. If three points do not lie on the same line, then they are not coplanar.

Which of the following best describes this statement? Check all that apply.

- ☐ **A** The statement is always true.
- ☐ **B** The statement is sometimes true.
- ☐ **C** The statement is always false.
- ☐ **D** It is the contrapositive of the statement in part **c**.
- ☐ **E** It is the converse of the statement in part **c**.
- ☐ **F** It is the inverse of the statement in part **c**.

EXPLORE 2-4

Geometry Lab
Necessary and Sufficient Conditions

We all know that water is a *necessary* condition for plants to survive. However, it is not a *sufficient* condition. For example, plants also need sunlight to survive.

Mathematical Practices
MP 3 Construct viable arguments and critique the reasoning of others.

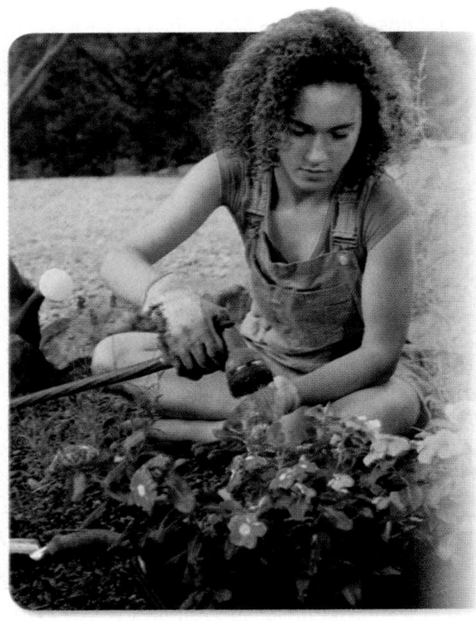

Necessary and sufficient conditions are important in mathematics. Consider the property of having four sides. While *having four sides* is a necessary condition for something being a square, that single condition is not, by itself, a sufficient condition to guarantee that it is a square. Trapezoids are four-sided figures that are not squares.

Condition	Definition	Examples
necessary	A condition *A* is said to be *necessary* for a condition *B*, if and only if the falsity or nonexistence of *A* guarantees the falsity or nonexistence of *B*.	Having opposite sides parallel is a necessary condition for something being a square.
sufficient	A condition *A* is said to be *sufficient* for a condition *B*, if and only if the truth or existence of *A* guarantees the truth or existence of *B*.	Being a square is a sufficient condition for something being a rectangle.

Exercises

Work cooperatively. Determine whether each statement is *true* or *false*. If false, give a counterexample.

1. Being a square is a necessary condition for being a rectangle.
2. Being a rectangle is a necessary condition for being a square.
3. Being greater than 5 is a necessary condition for being less than 10.
4. Being less than 18 is a sufficient condition for being less than 25.
5. Walking on four legs is a sufficient condition for being a dog.
6. Breathing air is a necessary condition for being a human being.
7. Being an equilateral rectangle is both a necessary and sufficient condition for being a square.

Work cooperatively. Determine whether I is a *necessary* condition for II, a *sufficient* condition for II, or *both*. Explain.

8. I. Two points are given.
   II. An equation of a line can be written.
9. I. Two planes are parallel.
   II. Two planes do not intersect.
10. I. Two angles are acute.
    II. Two angles are complementary.

Marc Romanelli/Photodisc/Getty Images

LESSON 4

Writing Proofs

Then

- You used deductive reasoning to prove statements.

Now

1. Analyze figures to identify and use postulates about points, lines, and planes.
2. Analyze and construct viable arguments in several proof formats.

Why?

- If a rock and a feather are dropped from the same height in a vacuum chamber, the two objects fall at the same rate. This demonstrates one of Sir Isaac Newton's laws of gravity and inertia. These laws are accepted as fundamental truths of physics. Some laws in geometry must also be accepted as true.

New Vocabulary

postulate
axiom
proof
flow proof
deductive argument
algebraic proof
theorem
paragraph proof
informal proof

Mathematical Practices

2 Reason abstractly and quantitatively.

3 Construct viable arguments and critique reasoning of others.

1 Postulates

A **postulate**, or **axiom**, is a statement that is accepted as true without proof. Basic ideas about points, lines, and planes can be stated as postulates.

Postulates Points, Lines, and Planes

Postulate	Diagram	Explanation
2.1 Through any two points, there is exactly one line.		Line *n* is the only line through points *P* and *R*.
2.2 Through any three noncollinear points, there is exactly one plane.		Plane *K* is the only plane through noncollinear points *A*, *B*, and *C*.
2.3 A line contains at least two points.		Line *n* contains points *P*, *Q* and *R*.
2.4 A plane contains at least three noncollinear points.		Plane *K* contains noncollinear points *L*, *B*, *C*, and *E*.
2.5 If two points lie in a plane, then the entire line containing those points lies in that plane.		Points *A* and *B* lie in plane *K*, and line *m* contains points *A* and *B*, so line m is in plane *K*.
2.6 If two lines intersect, then their intersection is exactly one point.		Lines *s* and *t* intersect at point *P*.
2.7 If two planes intersect, then their intersection is a line.		Planes *F* and *G* intersect in line *w*.

These basic geometry postulates form a foundation for proofs and reasoning about points, lines, and planes.

Study Tip

Undefined Terms Recall from Lesson 1-1 that points, lines, and planes are *undefined terms.* The postulates you have learned in this lesson describe special relationships between these undefined terms.

Real-World Example 1 Identifying Postulates

ARCHITECTURE Explain how the photo illustrates that each statement is true. Then state the postulate that can be used to show each statement is true.

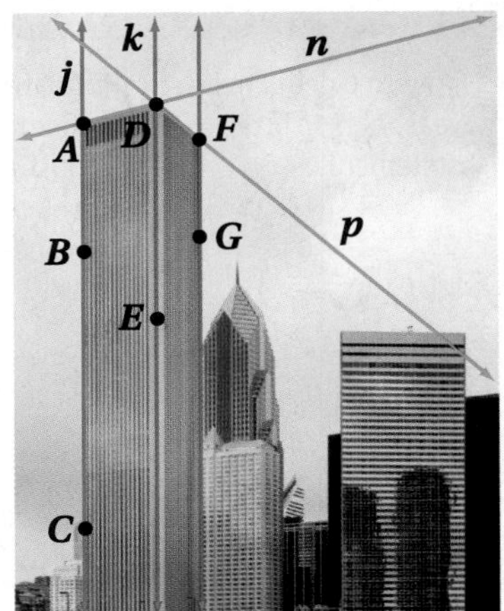

a. Lines n and p intersect at point D.

The top edges of the building are represented by lines n and p. The lines intersect at the corner point, D. Postulate 2.6 states that if two lines intersect, their intersection is exactly one point.

b. Points A, B, and D form a plane.

Points A, B, and D are three noncollinear points on the front face of the building. By Postulate 2.2, through any three noncollinear points, there is exactly one plane.

Guided Practice

Explain how the photo illustrates that each statement is true. Then state the postulate that can be used to show that each statement is true.

1A. Points A, B, and C are points that make up line j.

1B. The plane that contains A, D, and E intersects the plane that contains F, G, and E in line k.

You can use postulates to explain your reasoning when analyzing statements.

Example 2 Analyze Statements Using Postulates

Determine whether each statement is *always*, *sometimes*, or *never* true. Explain your reasoning.

a. If two coplanar lines intersect, then the point of intersection lies in the same plane as the two lines.

Always; Postulate 2.5 states that if two points lie in a plane, then the entire line containing those points lies in that plane. So, because both lines lie in the plane, any points on those lines, including their point of intersection, also lie in the plane.

b. Four points are noncollinear.

Sometimes; Postulate 2.3 states that a line contains at least two points. This means that a line contains two or *more* points. A line can contain four labeled points or three points on the line and one point not on the line.

Guided Practice

Determine whether each statement is *always*, *sometimes*, or *never* true. Explain your reasoning.

2A. Two intersecting lines determine a plane.

2B. Three lines intersect in two points.

2 Proofs

Proofs To prove a conjecture, you use deductive reasoning to move from a hypothesis to the conclusion of the conjecture you are trying to prove. This is done by writing a **proof**, which is a logical argument in which each statement you make is supported by a statement that is accepted as true.

A **flow proof** uses statements written in boxes and arrows to show the logical progression of an argument. The reason justifying each statement is written below the box. This type of proof allows you to see how each statement is deduced from the other statements.

Study Tip

Flow Proof Flow proofs can be written vertically or horizontally.

Key Concept How to Write a Flow Proof

Step 1 Write the given statement in a box and the word *Given* underneath.

Step 2 Create a **deductive argument** by forming a logical chain of statements linking the given to what you are trying to prove. Write these statements in linked boxes and justify each statement with a reason written beneath the statement.

Step 3 State what it is that you have proved.

An **algebraic proof** is a proof that is made up of a series of algebraic statements. The following table summarizes several properties of real numbers that you studied in algebra.

Key Concept Properties of Real Numbers

The following properties are true for any real numbers a, b, and c.	
Addition Property of Equality	If $a = b$, then $a + c = b + c$.
Subtraction Property of Equality	If $a = b$, then $a - c = b - c$.
Multiplication Property of Equality	If $a = b$, then $a \cdot c = b \cdot c$.
Division Property of Equality	If $a = b$ and $c \neq 0$, then, $\frac{a}{c} = \frac{b}{c}$.
Reflexive Property of Equality	$a = a$
Symmetric Property of Equality	If $a = b$, then $b = a$.
Transitive Property of Equality	If $a = b$ and $b = c$, then $a = c$.
Substitution Property of Equality	If $a = b$, then a may be replaced by b in any equation or expression.
Distributive Property of Equality	$a(b + c) = ab + ac$

Writing an algebraic proof is closely related to solving an equation or inequality. The properties of equality provide justification for many statements in algebraic proofs.

Example 3 Write an Algebraic Flow Proof

Prove that if $\frac{y+2}{3} = 3$, then $y = 7$. Write a flow proof.

Begin by stating what is given and what you are to prove.

Given: $\frac{y+2}{3} = 3$

Prove: $y = 7$

Flow Proof:

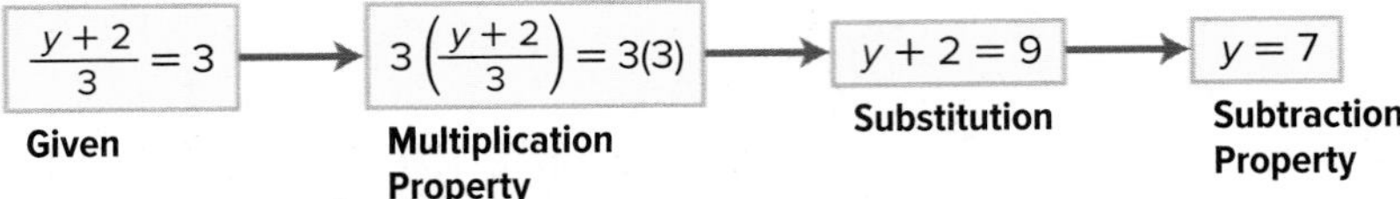

Guided Practice

3. Prove that if $-5(x + 4) = 70$, then $x = -18$. Write a flow proof.

Given: $-5(x + 4) = 70$

Prove: $x = -18$

Flow Proof:

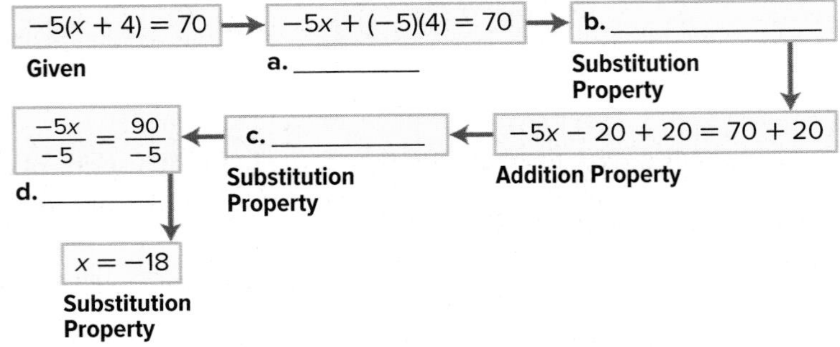

Once a statement or conjecture has been proved, it is called a **theorem**, and it can be used as a reason to justify statements in other proofs. A geometric proof uses undefined terms, definitions, postulates, and theorems to prove geometric conjectures.

Example 4 Write a Geometric Flow Proof

Given that Q is the midpoint of $\overline{PR}$, write a flow proof to show that $\overline{PQ} \cong \overline{QR}$.

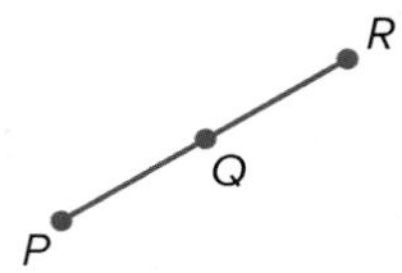

Begin by stating what is given and what you are to prove.

Given: Q is the midpoint of $\overline{PR}$.

Prove: $\overline{PQ} \cong \overline{QR}$

Flow Proof:

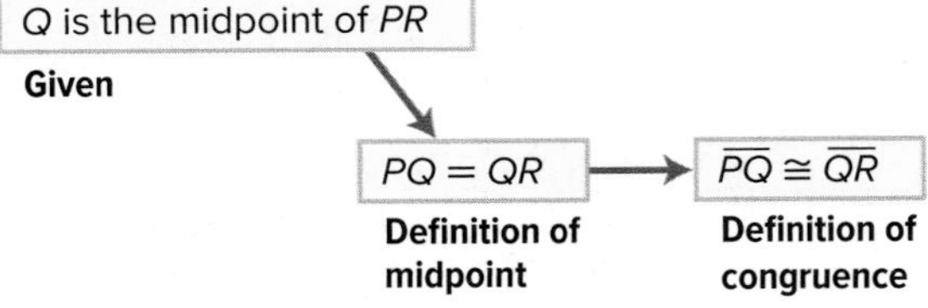

Guided Practice

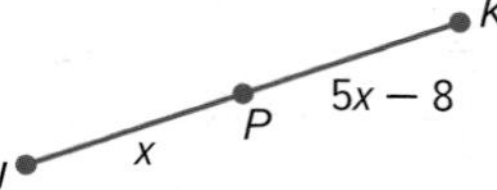

4. Given that P is the midpoint of $\overline{JK}$, write a flow proof to show that $x = 2$.

Given: P is the midpoint of $\overline{JK}$.

Prove: $x = 2$

Flow Proof:

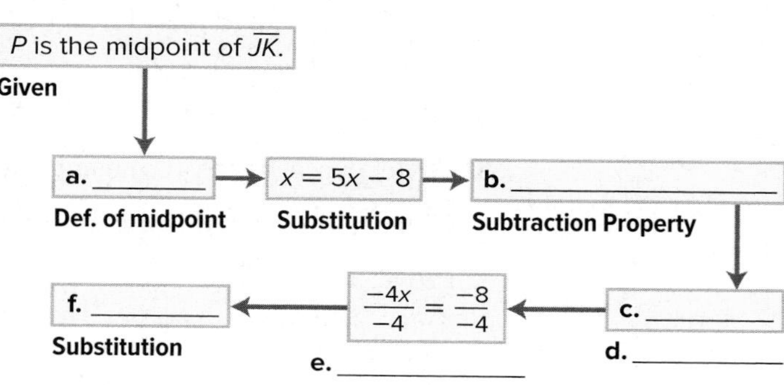

Study Tip

Midpoint Theorem and Definition The definition of midpoint is in terms of equality, while the Midpoint Theorem is in terms of congruence.

The conjecture in Example 4 is known as the Midpoint Theorem.

Theorem 2.1 Midpoint Theorem

If M is the midpoint of $\overline{AB}$, then $\overline{AM} \cong \overline{MB}$.

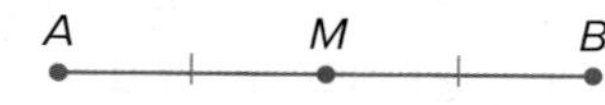

Another way to prove a conjecture is to write a paragraph that explains why the conjecture for a given situation is true. This is called a **paragraph proof**. Paragraph proofs are also called **informal proofs**, although the term *informal* is not meant to imply that this form of proof is any less valid than any other type of proof. Similar to a flow proof, a paragraph proof includes the theorems, definitions, or postulates that support each statement.

Avoid a Common Error

A common error in writing proofs is to skip steps in the logical progression of the argument.

Key Concept How to Write a Paragraph Proof

Step 1 Write the *Given* and *Prove* statements.

Step 2 Draw a diagram and label any given information.

Step 3 Write the proof. Create a deductive argument by forming a logical chain of statements. Justify each statement with a reason that includes definitions, postulates, algebraic properties, postulates, or theorems.

Step 4 State what it is you have proved.

Example 5 Write a Paragraph Proof

Given that C is between A and B and $\overline{AC} \cong \overline{CB}$, write a paragraph proof to show that C is the midpoint of $\overline{AB}$.

Step 1 **Write the *Given* and *Prove* statements.**

Given: C is between A and B and $\overline{AC} \cong \overline{CB}$.

Prove: C is the midpoint of $\overline{AB}$.

Step 2 **Draw a diagram and label any given information.**

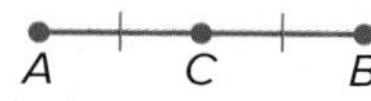

Step 3 **Write the proof.**

Proof: If C is between points A and B, then by the definition of betweenness, A, B, and C are collinear and $AC + CB = AB$. If $\overline{AC} \cong \overline{CB}$, then, by the definition of congruence, the segments have the same measure, which means that $AC = CB$. From the definition of midpoint of a segment, if C is between points A and B and $AC = CB$, then C is the midpoint of $\overline{AB}$.

Guided Practice

5. Write a paragraph proof using the following information.

Given: $\overline{AE} \cong \overline{DB}$; C is the midpoint of $\overline{AE}$ and $\overline{DB}$.

Prove: $CA = CB$

Check Your Understanding

◯ = Step-by-Step Solutions begin on page R13.

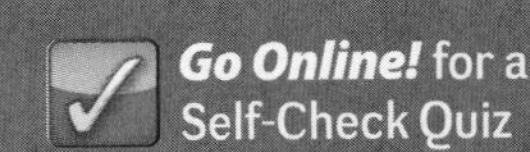

Example 1 **Explain how the figure illustrates that each statement is true. Then state the postulate that can be used to show each statement is true.**

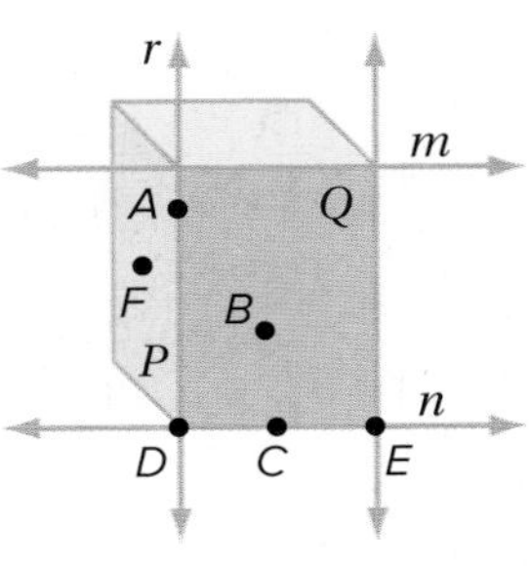

1. Planes P and Q intersect in line r.
2. Lines r and n intersect at point D.
3. Line n contains the points C, D, and E.
4. Plane P contains the points A, F, and D.
5. Line n lies in plane Q.

Example 2 **Determine whether each statement is *always*, *sometimes*, or *never* true. Explain your reasoning.**

6. If the plane contains a line and a line contains a point, then the plane contains the point.
7. A line contains three noncollinear points.
8. Line r contains only a point B.

In the figure, $\overrightarrow{AK}$ is in plane P and M is on $\overleftrightarrow{NE}$. State the postulate that can be used to show each statement as true.

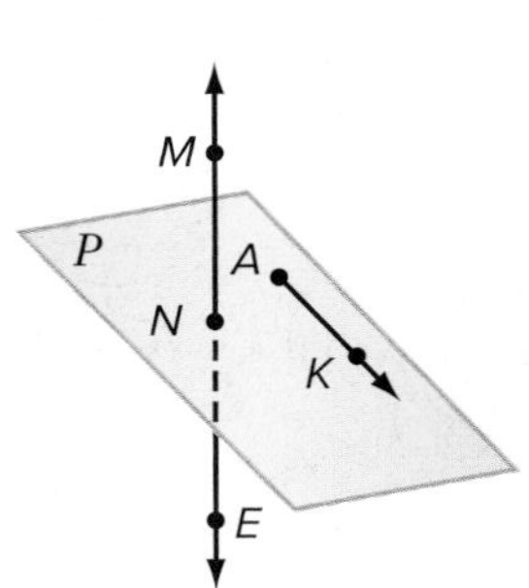

9. M, K, and N are coplanar.
10. $\overleftrightarrow{NE}$ contains points N and M.
11. N and K are collinear.
12. Points N, A, and K are coplanar.

Example 3 **13** Complete the following flow proof.

Given: $5x - 12 = 8$

Prove: $x = 4$

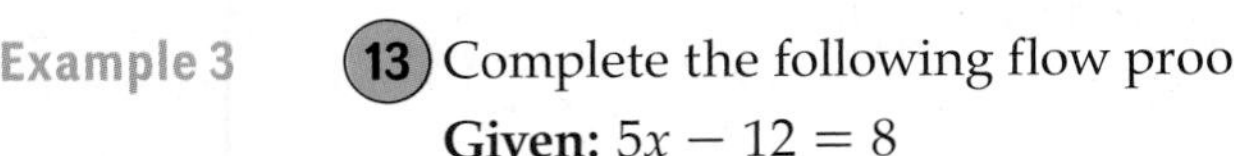

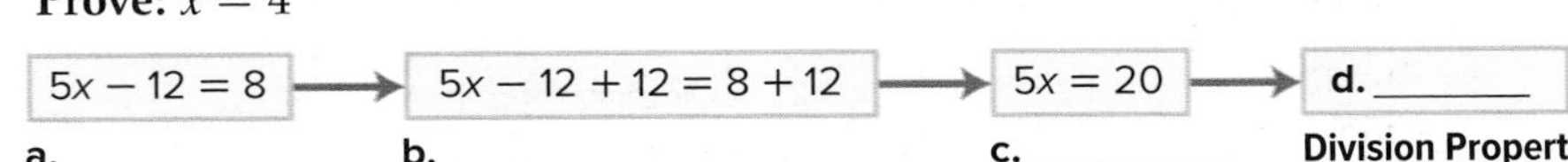

Example 4 **14.** Write a flow proof.

Given: $\overline{JK} \cong \overline{KL}$.

Prove: $y = 8$

Example 5 **15.** Write a paragraph proof.

Given: Y is the midpoint of $\overline{XZ}$.

$\overline{XY} \cong \overline{WY}$.

Prove: $\overline{WY} \cong \overline{YZ}$

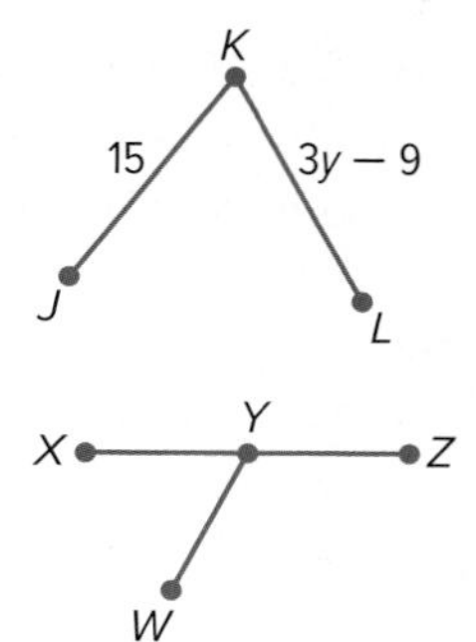

Practice and Problem Solving

Extra Practice is found on page R2.

Example 1 **CAKES Explain how the figure illustrates that each statement is true. Then state the postulate that can be used to show each statement is true.**

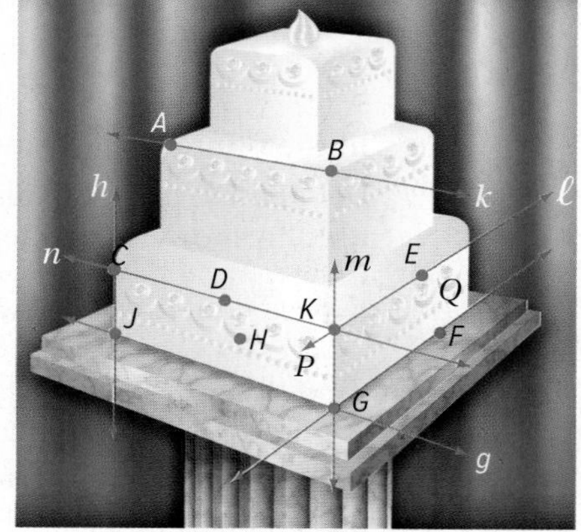

16. Lines n and l intersect at point K.

17. Planes P and Q intersect in line m.

18. Points D, K, and H determine a plane.

19. Point D is also on the line n through points C and K.

20. Points D and H are collinear.

21. Points E, F, and G are coplanar.

22. $\overleftrightarrow{EF}$ lies in plane Q.

23. Lines h and g intersect at point J.

Example 2 **Determine whether each statement is *always*, *sometimes*, or *never* true. Explain your reasoning.**

24. There is exactly one plane that contains noncollinear points A, B, and C.

25. There are at least three lines through points J and K.

26. If points M, N, and P lie in plane X, then they are collinear.

27. Points X and Y are in plane Z. Any point collinear with X and Y is in plane Z.

28. The intersection of two planes can be a point.

29. Points A, B, and C determine a plane.

Example 3 **30.** Write a flow proof.

Given: $5(y - 9) = 25$

Prove: $y = 14$

Example 4 **31.** Write a flow proof.

Given: L is the midpoint of $\overline{JK}$.

$\overline{MK} \cong \overline{JL}$.

Prove: $\overline{LK} \cong \overline{MK}$

J L K M

Example 5 **32.** Write a paragraph proof.

Given: C is the midpoint of $\overline{AB}$.

B is the midpoint of $\overline{CD}$.

Prove: $\overline{AC} \cong \overline{BD}$

A C B D

33. **ARGUMENTS** Emilio and his friends spent Saturday afternoon at the park. There were several people there with bikes and skateboards. There were a total of 11 bikes and skateboards that had a total of 36 wheels. Use a paragraph proof to show how many bikes and how many skateboards were there.

34. **DRIVING** Kiesha is traveling from point A to point B. Two possible routes are shown on the map. Assume that the speed limit on Southside Boulevard is 55 miles per hour and the speed limit of I-295 is 70 miles per hour.

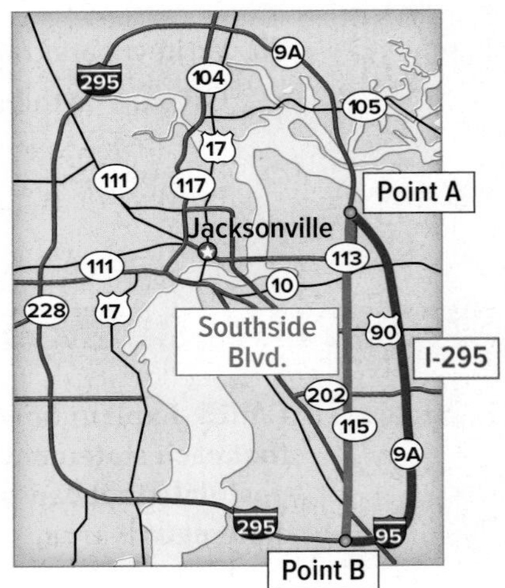

a. Which of the two routes covers the shortest distance? Explain your reasoning.

b. If the distance from point A to point B along Southside Boulevard is 10.5 miles and the distance along I-295 is 11.6 miles, which route is faster, assuming that Kiesha drives the speed limit?

In the figure at the right, $\overleftrightarrow{CD}$ and $\overleftrightarrow{CE}$ lie in the plane P and $\overleftrightarrow{DH}$ and $\overleftrightarrow{DJ}$ lie in the plane Q. State the postulate that can be used to show each statement as true.

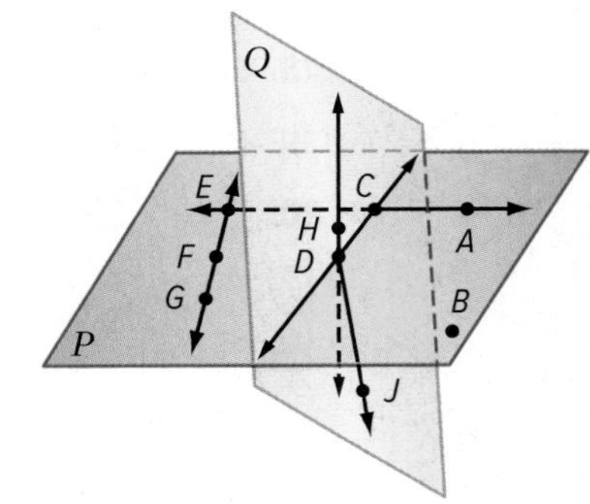

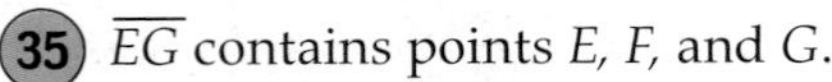

35. $\overline{EG}$ contains points E, F, and G.

36. Points C and B are collinear.

37. $\overleftrightarrow{AD}$ lies in plane P.

38. Points D and F are collinear.

39. Points C, B, and D are coplanar.

40. Plane Q contains the points C, H, D, and J.

41. $\overleftrightarrow{AC}$ and $\overleftrightarrow{FG}$ intersect at point E.

42. Plane P and plane Q intersect at $\overleftrightarrow{CD}$.

43. **ARGUMENTS** Roofs are designed based on the materials used to ensure that water does not leak into the buildings they cover. Some roofs are constructed from waterproof material, and others are constructed for watershed, or gravity removal of water. The pitch of a roof is the rise over the run, which is generally measured in rise per foot of run. Use the statements below to write a paragraph proof showing that the pitch in Den's design is not steep enough.

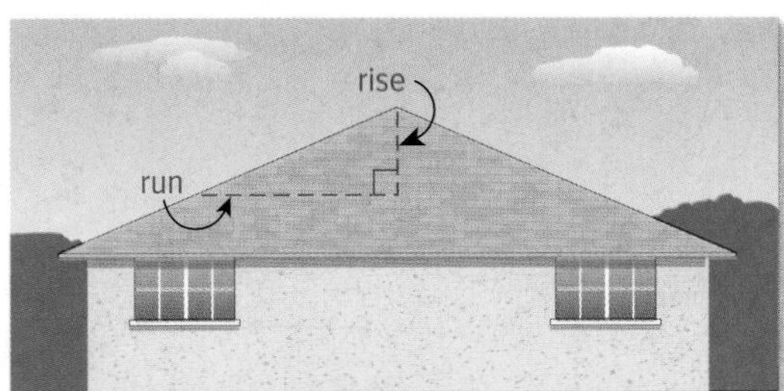

- Waterproof roofs should have a minimum slope of 0.25 inch per foot.
- Watershed roofs should have a minimum slope of 4 inches per foot.
- Den is designing a house with a watershed roof.
- The pitch in Den's design is two inches per foot.

44. **ARGUMENTS** If $\overline{AB}$ is congruent to $\overline{BD}$, and A, B, and D are collinear, then B is the midpoint of $\overline{AD}$. Write a flow proof to prove this statement.

45. **SENSE-MAKING** Study the photo of a rotunda in the capitol building in St. Paul, Minnesota. A rotunda is a round building, usually covered by a dome. Use Postulate 2.1 to answer parts **a-c**.

   a. If you were standing in the middle of the rotunda, which arched exit is the closest to you?

   b. What information did you use to formulate your answer?

   c. What term describes the shortest distance from the center of a circle to a point on the circle?

H.O.T. Problems Use Higher-Order Thinking Skills

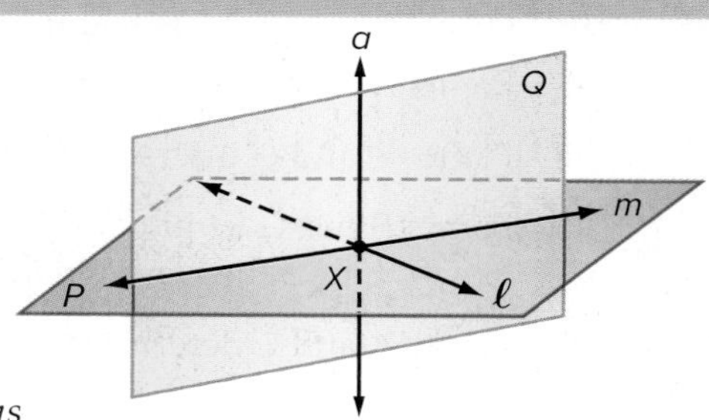

46. **OPEN ENDED** Draw a figure that satisfies five of the seven postulates you have learned. Explain how your figure satisfies the chosen postulates.

47. **CHALLENGE** Use the following true statement and the definitions and postulates you have learned to answer each question.

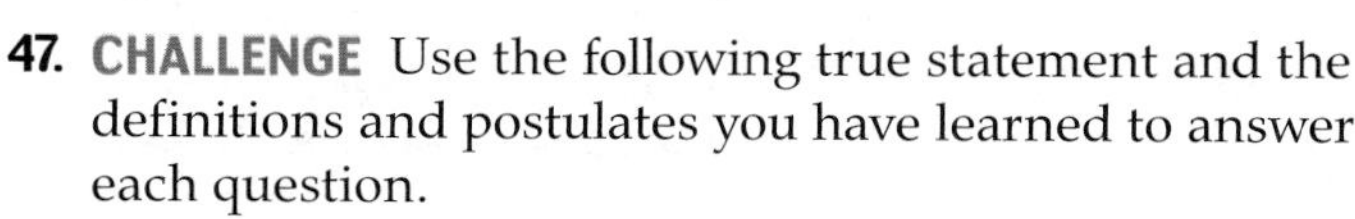

Two planes are perpendicular if and only if one plane contains a line perpendicular to the second plane.

   a. Through a given point, there passes one and only one plane perpendicular to a given line. If plane Q is perpendicular to line l at point X and line l lies in plane P, what must also be true?

   b. Through a given point, there passes one and only one line perpendicular to a given plane. If plane Q is perpendicular to plane P at point X and line a lies in plane Q, what must also be true?

48. **REASONING Determine if each statement is *sometimes*, *always*, or *never* true. Explain your reasoning or provide a counterexample.**

   a. Through any three points, there is exactly one plane.

   b. Three coplanar lines have two points of intersection.

49. **ERROR ANALYSIS** Blake and Katrina were asked to write a flow proof for the following. Is either of them correct? Explain.

Given: $2(x - 5) = -16$

Prove: $x = -3$

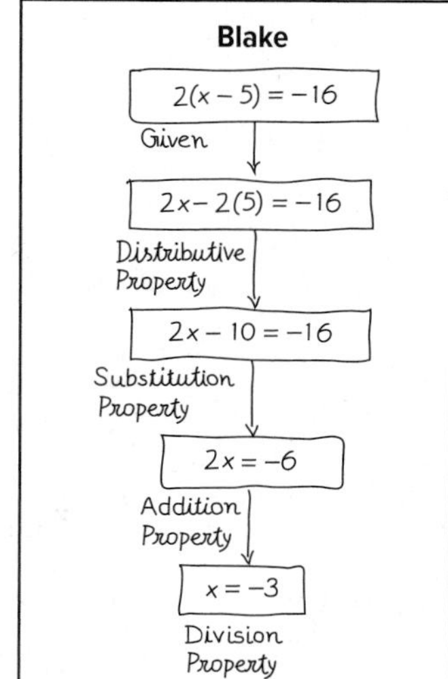

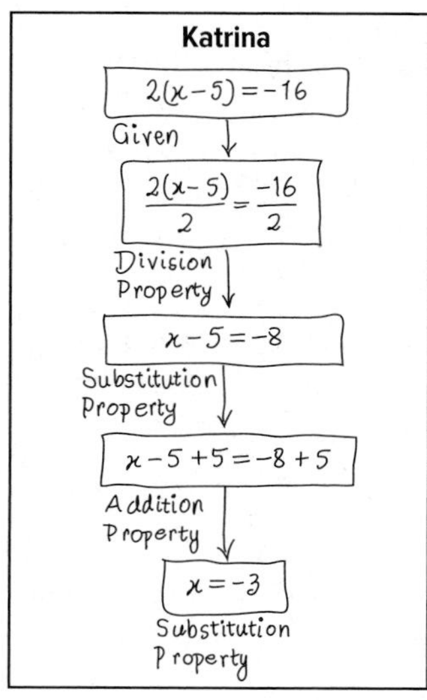

50. **WRITING IN MATH** Describe the differences between undefined terms, definitions, postulates, conjectures, and theorems.

51. Using the point, line, and plane postulates, determine the true statements from the diagram. Check all that apply. MP 2

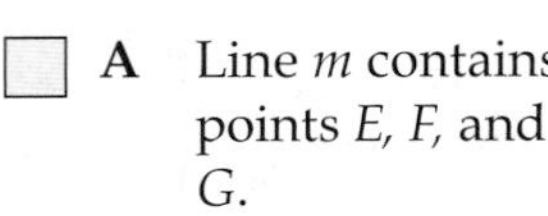

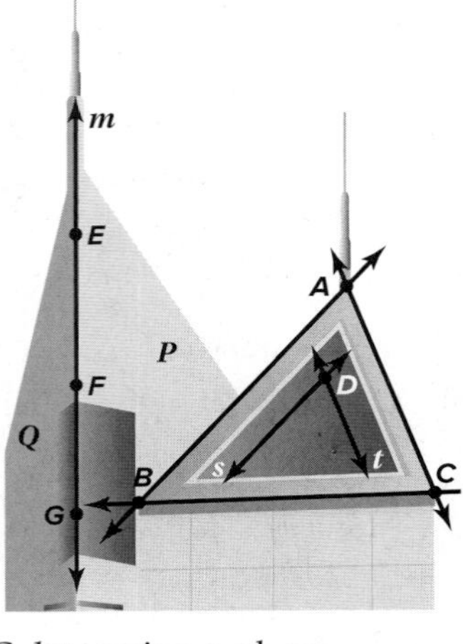

☐ **A** Line m contains points E, F, and G.

☐ **B** Lines s and t intersect at point A.

☐ **C** Points A, B, and C determine a plane.

☐ **D** Planes P and Q intersect at line m.

☐ **E** Points F and G lie in plane Q and on line m. Line m lies entirely in plane P.

☐ **F** Points A and C determine a line.

52. Which statement *cannot* be true? MP 2

○ **A** Three noncollinear points can determine a plane.

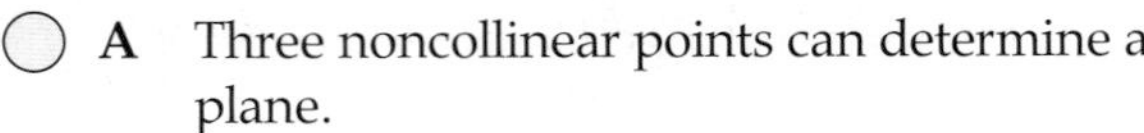

○ **B** If two lines intersect, it is at one and only one point.

○ **C** At least two lines can contain the same two points.

○ **D** A midpoint divides a segment into two congruent segments.

53. The image illustrates the statement $\overleftrightarrow{AB}$ *is the only line through points A and B.* Which postulate proves this statement is true? MP 2

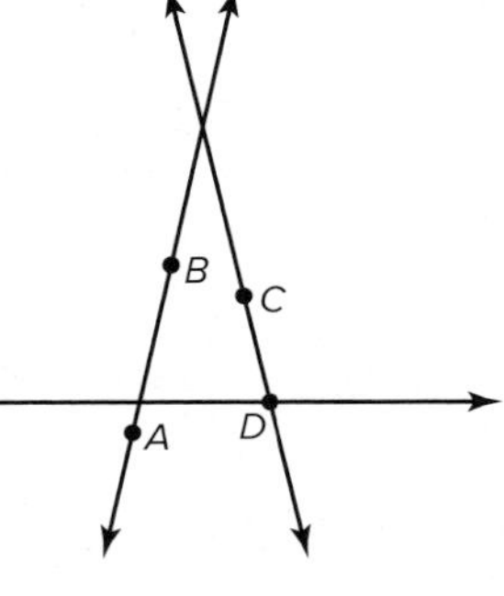

○ **A** A line contains at least two points.

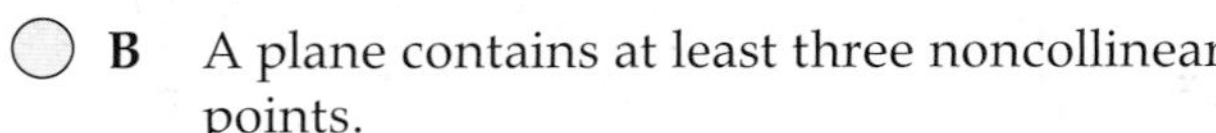

○ **B** A plane contains at least three noncollinear points.

○ **C** Through any two points, there is exactly one line.

○ **D** Through any three noncollinear points, there is exactly one plane.

54. Complete the following statement:
Three lines ________ intersect at two different points. MP 3

○ **A** always

○ **B** sometimes

○ **C** never

○ **D** not enough information

55. Which statement is missing from the following proof? MP 3

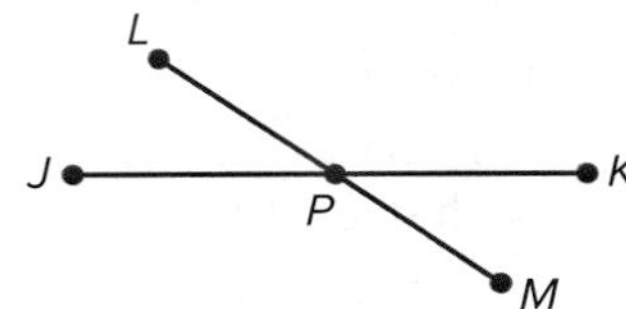

Given: $LP = PM$, $JP = PK$, $LP = JP$

Prove: $\overline{LM} \cong \overline{JK}$

Proof:

Because $LP = PM$ and $JP = PK$, P is the midpoint of $\overline{LM}$ and $\overline{JK}$ by definition of midpoint. By the Midpoint Theorem, $\overline{LP} \cong \overline{PM}$ and $\overline{JP} \cong \overline{PK}$. Therefore, $\overline{LM} \cong \overline{JK}$ by the Transitive Property.

○ **A** $LP = JP$

○ **B** $LP = PK$

○ **C** $PM = JP$

○ **D** $PM = PK$

56. MULTI-STEP MP 3

Given: $-4(x - 3) + 5x = 24$

Prove: $x = 12$

a. Write a proof using a flow proof or a paragraph proof.

b. Explain your reasoning on why you chose the specific form of the proof.

CHAPTER 2

Mid-Chapter Quiz

Lessons 2-1 through 2-4

Write a conjecture that describes the pattern in each sequence. Then use your conjecture to find the next item in the sequence. (Lesson 2-1)

1. 5, 5, 10, 15, 25, . . .
2. 

Find a counterexample to show that each conjecture is false. (Lesson 2-1)

3. If $AB = BC$, then B is the midpoint of $\overline{AC}$.
4. If n is a real number, then $n^3 > n$.

Use the following statements to write a compound statement for each conjunction or disjunction. Then find its truth value. Explain your reasoning. (Lesson 2-2)

p: A dollar is equal to 100 cents.
q: There are 4 quarters in a dollar.
r: February is the month before January.

5. $p \wedge r$
6. p and q
7. $p \wedge {\sim}r$

Rewrite the statement as a biconditional statement. Then determine whether the biconditional is *true* or *false*.

8. You will get a speeding ticket for driving faster than 30 miles per hour. (Lesson 2-2)

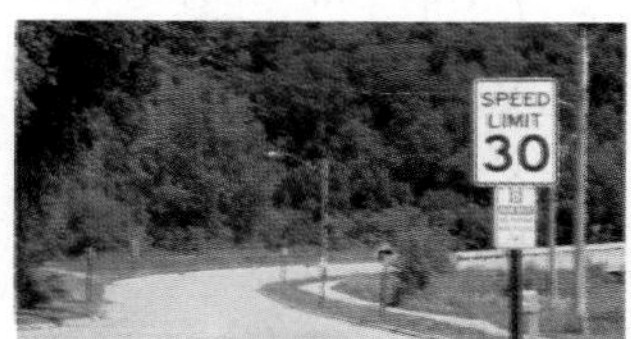

Write each statement in if-then form. (Lesson 2-2)

9. A polygon with five sides is a pentagon.
10. For the equation $4x - 6 = 10$, $x = 4$.
11. An angle with a measure less than 90 is an acute angle.

Determine the truth value of each conditional statement. If *true*, explain your reasoning. If *false*, give a counterexample. (Lesson 2-2)

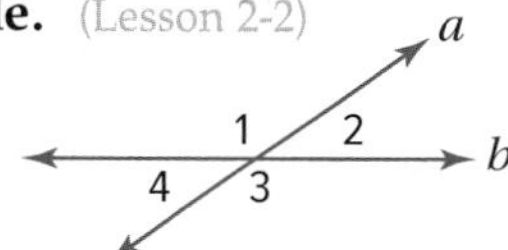

12. If ∠1 and ∠2 form a linear pair, they are supplementary angles.
13. If ∠1 and ∠4 form a linear pair, they are congruent angles.

Use the Venn diagrams below to determine the truth value of each conditional. Explain your reasoning. (Lesson 2-3)

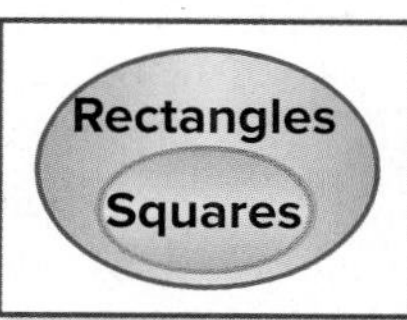

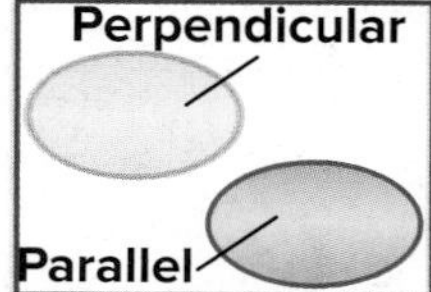

14. If a polygon is a square, then it is a rectangle.
15. If two lines are perpendicular, then they cannot be parallel.
16. **FOOTBALL** The New England Patriots played the Atlanta Falcons in the 2017 Super Bowl. (Lesson 2-3)

    a. Determine whether the stated conclusion is valid based on the given information. If not, write *invalid*. Explain your reasoning.

    Given: The Super Bowl winner has the highest score at the end of the game. At the end of the game, the Patriots had a score of 34 and the Falcons had a score of 28.

    Conclusion: The Patriots won the Super Bowl.

    b. MP What mathematical practice did you use to solve this problem?

17. **MULTIPLE CHOICE** Determine which statement follows logically from the given statements. (Lesson 2-3)

    (1) If you are a junior in high school, then you are at least 16 years old.

    (2) If you are at least 16 years old, then you are old enough to drive.

    A If you are old enough to drive, then you are a junior in high school.

    B If you are not old enough to drive, then you are a sophomore in high school.

    C If you are a junior in high school, then you are old enough to drive.

    D No valid conclusion is possible.

Determine whether each statement is *always*, *sometimes*, or *never* true. Explain your reasoning. (Lesson 2-4)

18. Points J, K, L, and N are noncollinear and lie in the same plane M.
19. There is exactly one line through points R and S.

LESSON 5

Proving Segment Relationships

Then

- You wrote algebraic and two-column proofs.

Now

1. Write proofs involving segment addition.
2. Write proofs involving segment congruence.

Why?

- You can measure segments with a ruler by matching the mark for zero with one endpoint and then finding the number on the ruler that corresponds to the other endpoint. To measure lengths such as 39 inches, which is longer than the yardstick, Kai marks a length of 36 inches. From the end of that mark, she measures an additional length of 3 inches. This ensures that the total length of fabric is 36 + 3 inches or 39 inches.

New Vocabulary

two-column proof
formal proof

Mathematical Practices

2 Reason abstractly and quantitatively.
3 Construct viable arguments and critique the reasoning of others.

1 Segment Addition

In Lesson 1-2, you measured segments with a ruler by matching the mark for zero with one endpoint and then finding the number on the ruler that corresponded to the other endpoint. This illustrates the Ruler Postulate.

Postulate 2.8 Ruler Postulate

Words The points on any line or line segment can be put into one-to-one correspondence with real numbers.

Symbols Given any two points *A* and *B* on a line, if *A* corresponds to zero, then *B* corresponds to a positive real number.

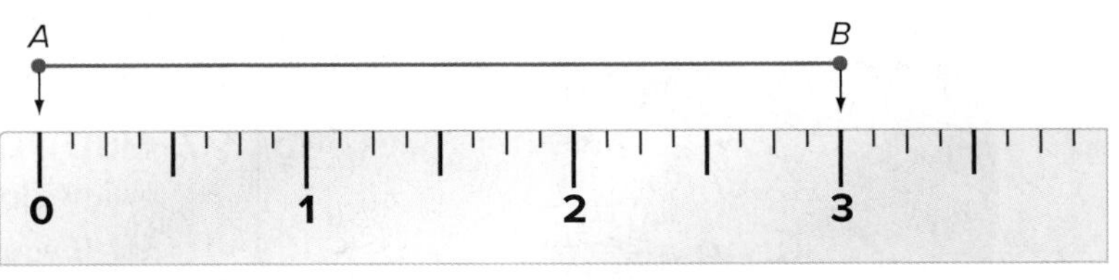

In Lesson 1-2, you learned about what it means for a point to be *between* two other points.

Postulate 2.9 Segment Addition Postulate

Words If *A*, *B*, and *C* are collinear, then point *B* is between *A* and *C* if and only if $AB + BC = AC$.

Symbols

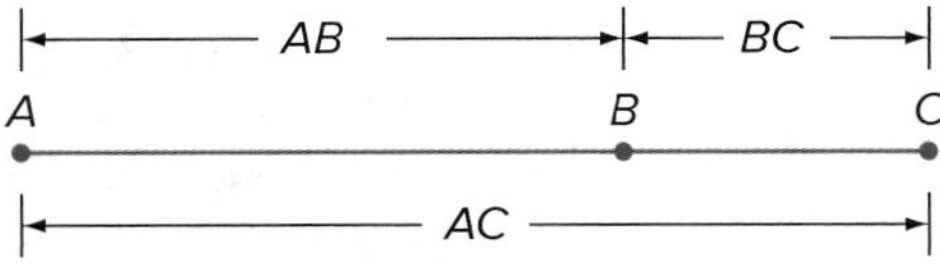

The Segment Addition Postulate is used as a justification in many geometric proofs. A **two-column proof**, or **formal proof**, contains *statements* and *reasons* organized in two columns.

Example 1 Use the Segment Addition Postulate

Prove that if $\overline{CE} \cong \overline{FE}$ and $\overline{ED} \cong \overline{EG}$, then $\overline{CD} \cong \overline{FG}$.

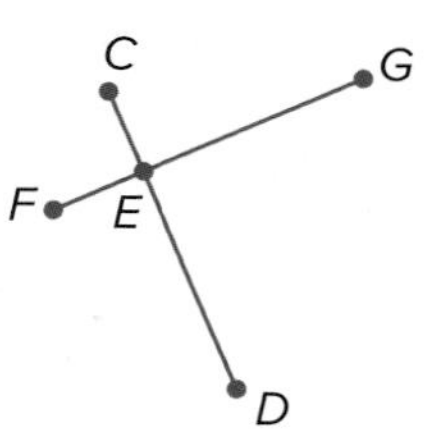

Given: $\overline{CE} \cong \overline{FE}$; $\overline{ED} \cong \overline{EG}$

Prove: $\overline{CD} \cong \overline{FG}$

Proof:

Statements	Reasons
1. $\overline{CE} \cong \overline{FE}$; $\overline{ED} \cong \overline{EG}$	**1.** Given
2. $CE = FE$; $ED = EG$	**2.** Definition of congruence
3. $CE + ED = CD$	**3.** Segment Addition Postulate
4. $FE + EG = CD$	**4.** Substitution (Steps 2 & 3)
5. $FE + EG = FG$	**5.** Segment Addition Postulate
6. $CD = FG$	**6.** Substitution (Steps 4 & 5)
7. $\overline{CD} \cong \overline{FG}$	**7.** Definition of congruence

Reading Math

MP Precision The Substitution Property of Equality is often just written as *Substitution*.

Guided Practice

Copy and complete the proof.

1. Given: $\overline{JL} \cong \overline{KM}$

Prove: $\overline{JK} \cong \overline{LM}$

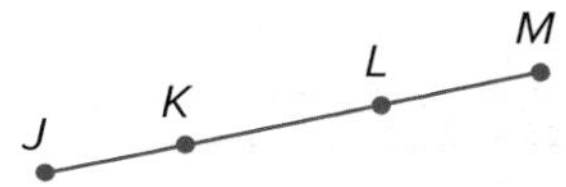

Proof:

Statements	Reasons
a. $\overline{JL} \cong \overline{KM}$	**a.** Given
b. $JL = KM$	**b.** ___?___
c. $JK + KL =$ ___?___ ; $KL + LM =$ ___?___	**c.** Segment Addition Postulate
d. $JK + KL = KL + LM$	**d.** ___?___
e. $JK + KL - KL = KL + LM - KL$	**e.** Subtraction Property of Equality
f. ___?___	**f.** Substitution
g. $\overline{JK} \cong \overline{LM}$	**g.** Definition of congruence

2 Segment Congruence

Segment measures are reflexive, symmetric, and transitive. Because segments with the same measure are congruent, congruence of segments is also reflexive, symmetric, and transitive.

Vocabulary Link

Symmetric

Everyday Use balanced or proportional

Math Use If $a = b$, then $b = a$.

Theorem 2.2 Properties of Segment Congruence

Reflexive Property of Congruence	$\overline{AB} \cong \overline{AB}$
Symmetric Property of Congruence	If $\overline{AB} \cong \overline{CD}$, then $\overline{CD} \cong \overline{AB}$.
Transitive Property of Congruence	If $\overline{AB} \cong \overline{CD}$ and $\overline{CD} \cong \overline{EF}$, then $\overline{AB} \cong \overline{EF}$.

You will prove the Symmetric and Reflexive Properties in Exercises 6 and 7, respectively.

Proof Transitive Property of Congruence

Given: $\overline{AB} \cong \overline{CD}$; $\overline{CD} \cong \overline{EF}$

Prove: $\overline{AB} \cong \overline{EF}$

Paragraph Proof:

Because $\overline{AB} \cong \overline{CD}$ and $\overline{CD} \cong \overline{EF}$, $AB = CD$ and $CD = EF$ by the definition of congruent segments. By the Transitive Property of Equality, $AB = EF$. Thus, $\overline{AB} \cong \overline{EF}$ by the definition of congruence.

Real-World Example 2 Proof Using Segment Congruence

VOLUNTEERING The route for a charity fitness run is shown. Checkpoints X and Z are the midpoints between the starting line and Checkpoint Y and Checkpoint Y and the finish line F, respectively. If Checkpoint Y is the same distance from Checkpoints X and Z, prove that the route from Checkpoint Z to the finish line is congruent to the route from the starting line to Checkpoint X.

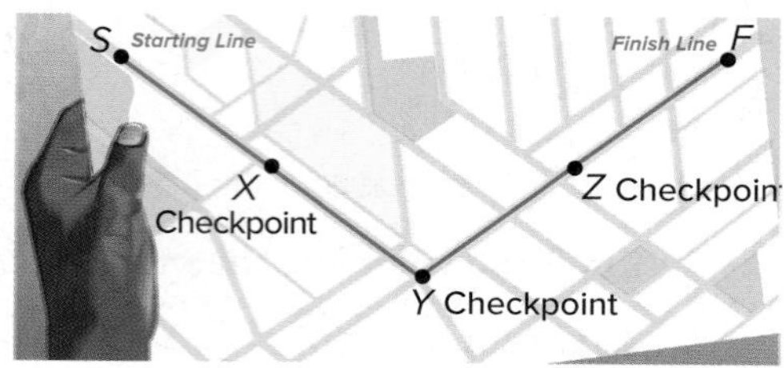

Given: X is the midpoint of $\overline{SY}$. Z is the midpoint of $\overline{YF}$. $XY = YZ$

Prove: $\overline{ZF} \cong \overline{SX}$

Two-Column Proof:

Statements	Reasons
1. X is the midpoint of $\overline{SY}$. Z is the midpoint of $\overline{YF}$. $XY = YZ$	1. Given
2. $\overline{SX} \cong \overline{XY}$; $\overline{YZ} \cong \overline{ZF}$	2. Definition of midpoint
3. $\overline{XY} \cong \overline{YZ}$	3. Definition of congruence
4. $\overline{SX} \cong \overline{YZ}$	4. Transitive Property of Congruence
5. $\overline{SX} \cong \overline{ZF}$	5. Transitive Property of Congruence
6. $\overline{ZF} \cong \overline{SX}$	6. Symmetric Property of Congruence

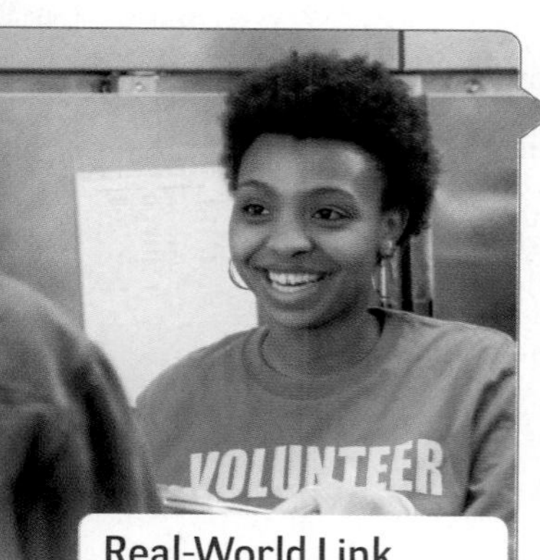

Real-World Link

According to a recent poll, 70% of teens who volunteer began doing so before age 12. Others said they would volunteer if given more opportunities to do so.

Source: Youth Service America

Go Online!

Use the Self-Check Quiz in ConnectED to check your own progress as you complete each lesson.

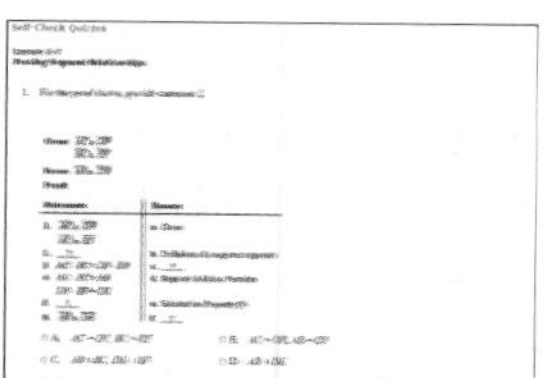

Guided Practice

2. **CARPENTRY** A carpenter cuts a 2″ × 4″ board to a desired length. He then uses this board as a pattern to cut a second board congruent to the first. Similarly, he uses the second board to cut a third board and the third board to cut a fourth board. Prove that the last board cut has the same measure as the first.

Ariel Skelley/Blend Images

Check Your Understanding

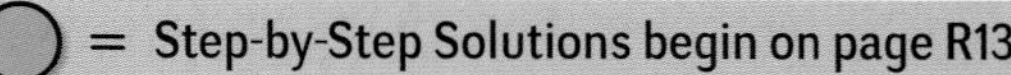

◯ = Step-by-Step Solutions begin on page R13.

Go Online! for a Self-Check Quiz

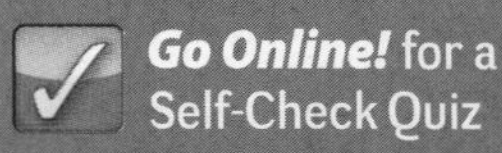

Example 1

1. MP **CONSTRUCT ARGUMENTS** **Copy and complete the proof.**

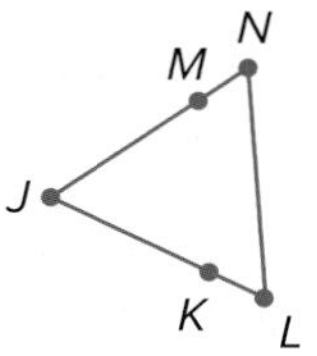

Given: $\overline{LK} \cong \overline{NM}$, $\overline{KJ} \cong \overline{MJ}$

Prove: $\overline{LJ} \cong \overline{NJ}$

Proof:

Statements	Reasons
a. $\overline{LK} \cong \overline{NM}$, $\overline{KJ} \cong \overline{MJ}$	**a.** ?
b. ?	**b.** Def. of congruent segments
c. $LK + KJ = NM + MJ$	**c.** ?
d. ?	**d.** Segment Addition Postulate
e. $LJ = NJ$	**e.** ?
f. $\overline{LJ} \cong \overline{NJ}$	**f.** ?

Example 2

2. **PROOF** Prove the following.

Given: $\overline{WX} \cong \overline{YZ}$

Prove: $\overline{WY} \cong \overline{XZ}$

3 **SCISSORS** Refer to the diagram shown. $\overline{AR}$ is congruent to $\overline{CR}$. $\overline{DR}$ is congruent to $\overline{BR}$. Prove that $AR + DR = CR + BR$.

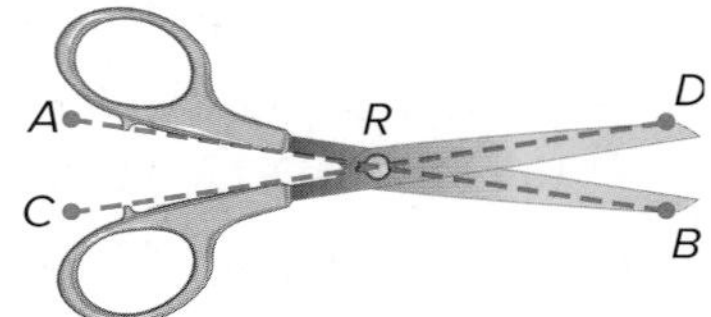

Practice and Problem Solving

Extra Practice is on page R2.

Example 1

4. MP **CONSTRUCT ARGUMENTS** **Copy and complete the proof.**

Given: C is the midpoint of $\overline{AE}$.

C is the midpoint of $\overline{BD}$.

$\overline{AE} \cong \overline{BD}$

Prove: $\overline{AC} \cong \overline{CD}$

Proof:

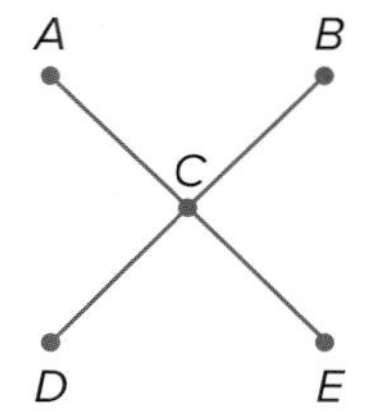

Statements	Reasons
a. ?	**a.** Given
b. $AC = CE$, $BC = CD$	**b.** ?
c. $AE = BD$	**c.** ?
d. ?	**d.** Segment Addition Postulate
e. $AC + CE = BC + CD$	**e.** ?
f. $AC + AC = CD + CD$	**f.** ?
g. ?	**g.** Simplify.
h. ?	**h.** Division Property
i. $\overline{AC} \cong \overline{CD}$	**i.** ?

Example 2

5. **TILING** A tile setter cuts a piece of tile to a desired length. He then uses this tile as a pattern to cut a second tile congruent to the first. He uses the first two tiles to cut a third tile whose length is the sum of the measures of the first two tiles. Prove that the measure of the third tile is twice the measure of the first tile.

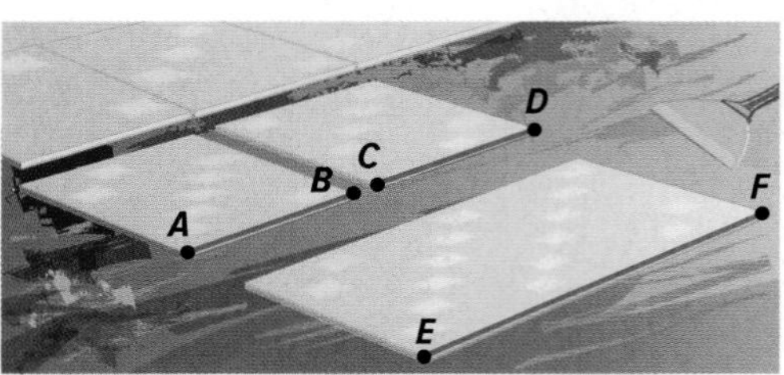

CONSTRUCT ARGUMENTS **Prove each theorem.**

6. Symmetric Property of Congruence (Theorem 2.2)

7. Reflexive Property of Congruence (Theorem 2.2)

8. **TRAVEL** Four cities in New York are connected by Interstate 90: Buffalo, Utica, Albany, and Syracuse. Buffalo is the farthest west.
   - Albany is 126 miles from Syracuse and 263 miles from Buffalo.
   - Buffalo is 137 miles from Syracuse and 184 miles from Utica.

   a. Draw a diagram to represent the locations of the cities in relation to each other and the distances between each city. Assume that Interstate 90 is straight.

   b. Write a paragraph proof to support your conclusion.

PROOF Prove the following.

9. If $\overline{SC} \cong \overline{HR}$ and $\overline{HR} \cong \overline{AB}$, then $\overline{SC} \cong \overline{AB}$.

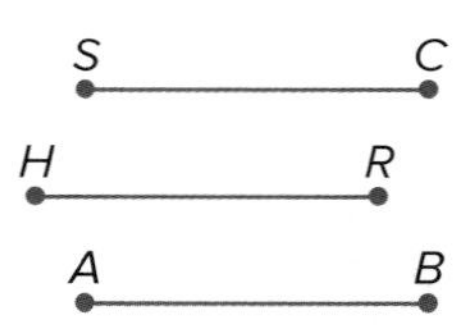

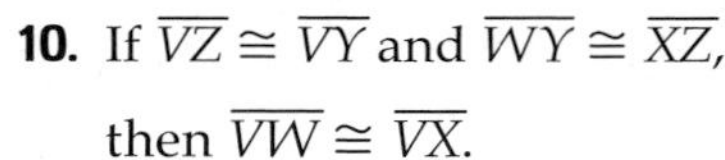

10. If $\overline{VZ} \cong \overline{VY}$ and $\overline{WY} \cong \overline{XZ}$, then $\overline{VW} \cong \overline{VX}$.

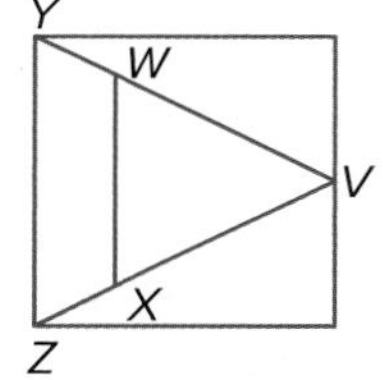

11. If E is the midpoint of $\overline{DF}$ and $\overline{CD} \cong \overline{FG}$, then $\overline{CE} \cong \overline{EG}$.

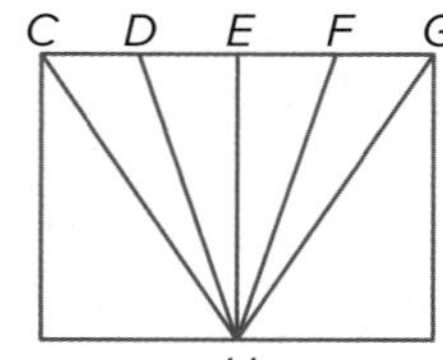

12. If B is the midpoint of $\overline{AC}$, D is the midpoint of $\overline{CE}$, and $\overline{AB} \cong \overline{DE}$, then $AE = 4AB$.

13. **OPTICAL ILLUSION** $\overline{AC} \cong \overline{GI}$, $\overline{FE} \cong \overline{LK}$, and $AC + CF + FE = GI + IL + LK$.

   a. Prove that $\overline{CF} \cong \overline{IL}$.

   b. Justify your proof using measurement. Explain your method.

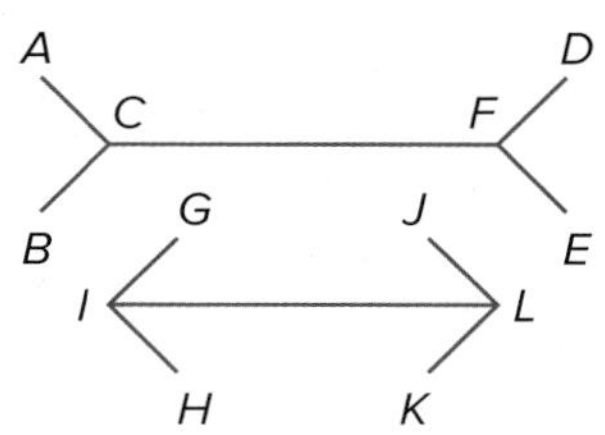

14. **CONSTRUCTION** Construct a segment that is twice as long as $\overline{PQ}$. Explain how the Segment Addition Postulate can be used to justify your construction.

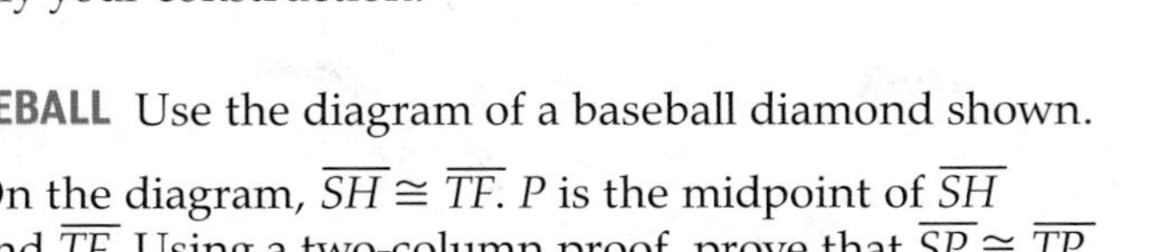

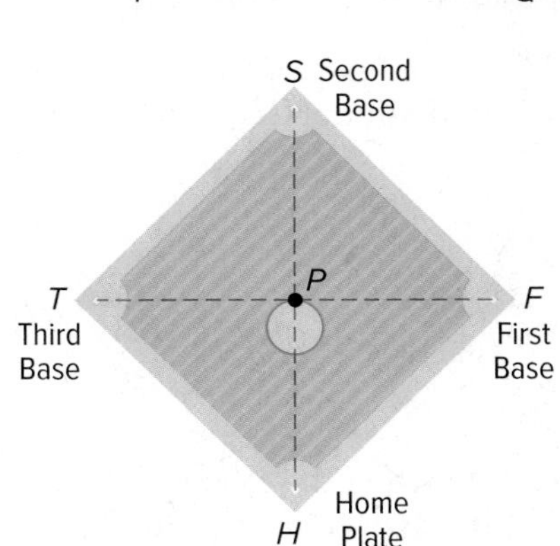

15. **BASEBALL** Use the diagram of a baseball diamond shown.

a. On the diagram, $\overline{SH} \cong \overline{TF}$. P is the midpoint of $\overline{SH}$ and $\overline{TF}$. Using a two-column proof, prove that $\overline{SP} \cong \overline{TP}$.

b. The distance from home plate to second base is 127.3 feet. What is the distance from first base to second base?

16. **MULTIPLE REPRESENTATIONS** A is the midpoint of $\overline{PQ}$, B is the midpoint of $\overline{PA}$, and C is the midpoint of $\overline{PB}$.

a. **Geometric** Make a sketch to represent this situation.

b. **Algebraic** Make a conjecture as to the algebraic relationship between PC and PQ.

c. **Geometric** Copy segment $\overline{PQ}$ from your sketch. Then construct points B and C on $\overline{PQ}$. Explain how you can use your construction to support your conjecture.

d. **Concrete** Use a ruler to draw a segment congruent to $\overline{PQ}$ from your sketch and to draw points B and C on $\overline{PQ}$. Use your drawing to support your conjecture.

e. **Logical** Prove your conjecture.

H.O.T. Problems Use Higher-Order Thinking Skills

17. **MP CRITIQUE ARGUMENTS** In the diagram, $\overline{AB} \cong \overline{CD}$ and $\overline{CD} \cong \overline{BF}$. Examine the conclusions made by Leslie and Shantice. Is either of them correct?

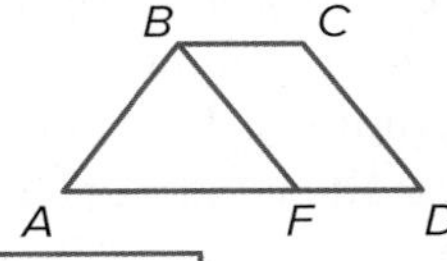

Leslie

Since $\overline{AB} \cong \overline{CD}$ and $\overline{CD} \cong \overline{BF}$, then $\overline{AB} \cong \overline{AF}$ by the Transitive Property of Congruence

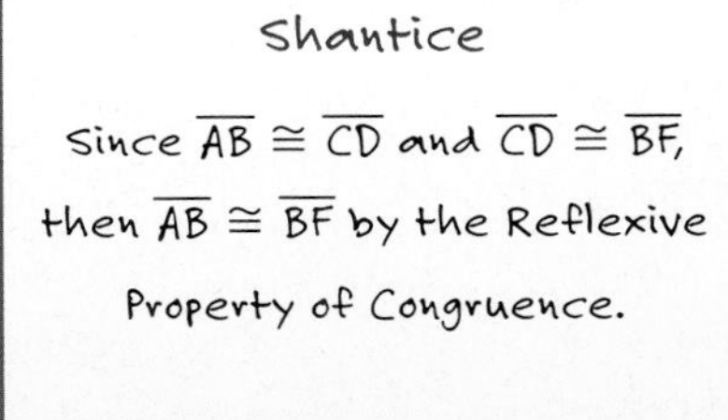
Shantice

Since $\overline{AB} \cong \overline{CD}$ and $\overline{CD} \cong \overline{BF}$, then $\overline{AB} \cong \overline{BF}$ by the Reflexive Property of Congruence.

18. **CHALLENGE** $ABCD$ is a square. Prove that $\overline{AC} \cong \overline{BD}$.

19. **WRITING IN MATH** Does there exist an Addition Property of Congruence? Explain.

20. **MP REASONING** Classify the following statement as *true* or *false*. If false, provide a counterexample.

If A, B, C, D, and E are collinear with B between A and C, C between B and D, and D between C and E, and AC = BD = CE, then AB = BC = DE.

21. **OPEN ENDED** Draw a representation of the Segment Addition Postulate in which the segment is two inches long, contains four collinear points, and contains no congruent segments.

22. **WRITING IN MATH** Compare and contrast paragraph proofs and two-column proofs.

23. In the figure, N is the midpoint of $\overline{EF}$, N is the midpoint of $\overline{GH}$, and $\overline{EN} \cong \overline{GN}$.

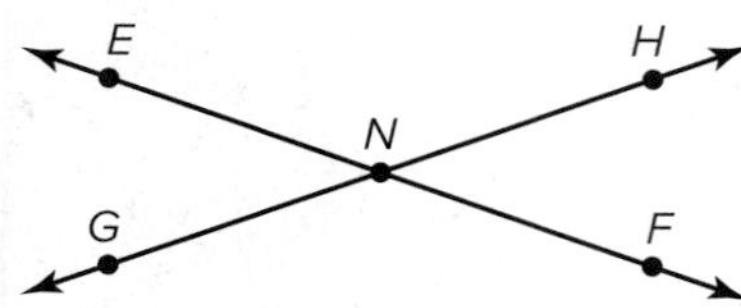

Which of the following will appear as a reason in a two-column proof that $\overline{HN} \cong \overline{FN}$? MP 2, 3

I. Given

II. Transitive Property of Congruence

III. Definition of midpoint

○ **A** III only

○ **B** I and II only

○ **C** II and III only

○ **D** I and III only

○ **E** I, II, and III

24. Suppose line m contains points D, E, and F. If $DE = 12$ millimeters, $EF = 15$ millimeters, and point D is between points E and F, what is the length of $\overline{DF}$ in millimeters? MP 1

25. **Given:** $\overline{AB} \cong \overline{AD}$; $\overline{BC} \cong \overline{DE}$

Prove: $\overline{AC} \cong \overline{AE}$

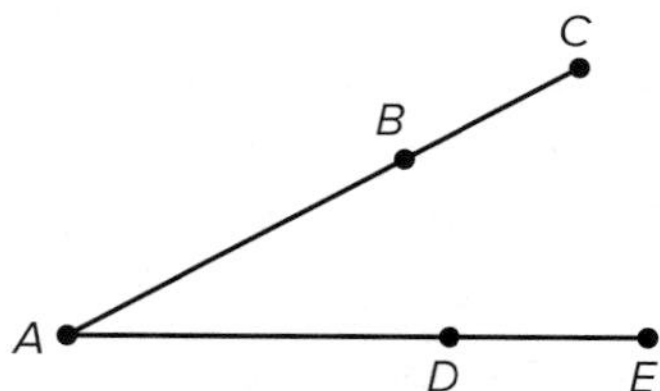

Which of the following statements can be justified by the Segment Addition Postulate? MP 2

○ **A** $AB = AD$

○ **B** $\overline{AB} \cong \overline{AD}$; $\overline{BC} \cong \overline{DE}$

○ **C** $AD + DE = AC$

○ **D** $AB + BC = AC$

26. What is the measure of $\angle CFD$? MP 1

○ **A** 66

○ **B** 72

○ **C** 108

○ **D** 138

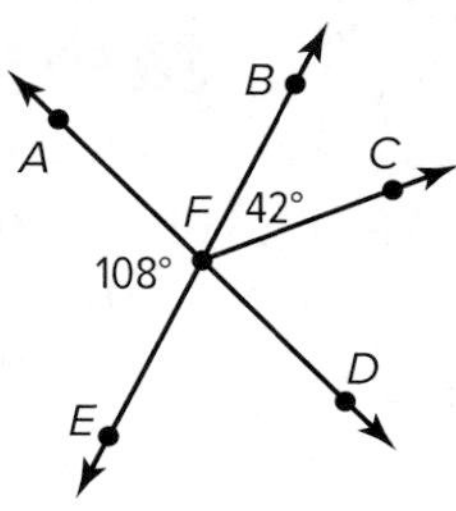

27. **MULTI-STEP** Look at the figure. It is given that $\overline{ST} \cong \overline{MS}$ and M is the midpoint of $\overline{RS}$. MP 2, 3

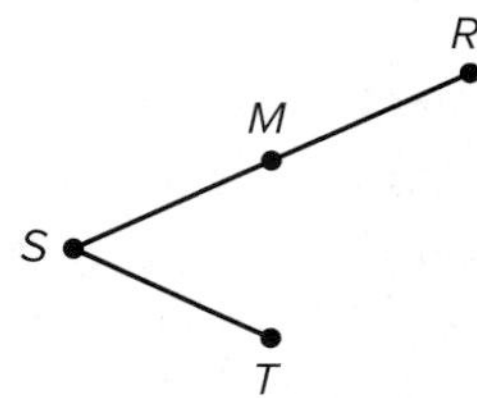

a. Which statements are true? Check all that apply.

☐ **A** $\overline{RS} \cong \overline{TS}$

☐ **B** $\overline{RM} \cong \overline{TM}$

☐ **C** $\overline{RM} \cong \overline{SM}$

☐ **D** $\overline{RM} \cong \overline{TS}$

☐ **E** $\overline{TS} \cong \overline{SM}$

b. Which property could prove that $\overline{ST} \cong \overline{RM}$?

○ **A** Reflective Property of Congruence

○ **B** Transitive Property of Congruence

○ **C** Ruler Postulate

○ **D** Multiplication Property of Equality

c. Which statement is true because of the Midpoint Theorem?

○ **A** $\overline{RS} \cong \overline{TS}$

○ **B** $\overline{RM} \cong \overline{TM}$

○ **C** $\overline{RM} \cong \overline{SM}$

○ **D** $\overline{RM} \cong \overline{TS}$

○ **E** $\overline{TS} \cong \overline{SM}$

LESSON 6

Proving Angle Relationships

Then

- You identified and used special pairs of angles.

Now

1. Write proofs involving supplementary and complementary angles.
2. Write proofs involving congruent and right angles.

Why?

- Jamal's school is building a walkway that will include bricks with the names of graduates from each class. All of the bricks are rectangular, so when the bricks are laid, all of the angles form linear pairs.

Mathematical Practices

3 Construct viable arguments and critique the reasoning of others.

6 Attend to precision.

1 Supplementary and Complementary Angles

The Protractor Postulate illustrates the relationship between angle measures and real numbers.

Postulate 2.10 Protractor Postulate

Words Given any angle, the measure can be put into one-to-one correspondence with real numbers between 0 and 180.

Example If $\overrightarrow{BA}$ is placed along the protractor at 0°, then the measure of $\angle ABC$ corresponds to a positive real number.

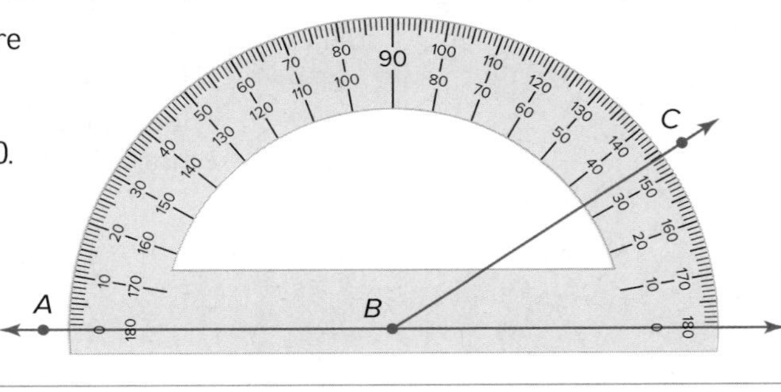

In Lesson 2-5, you learned about the Segment Addition Postulate. A similar relationship exists between the measures of angles.

Postulate 2.11 Angle Addition Postulate

D is in the interior of $\angle ABC$ if and only if $m\angle ABD + m\angle DBC = m\angle ABC$.

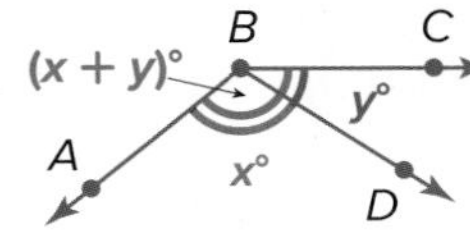

Example 1 Use the Angle Addition Postulate

Find $m\angle 1$ if $m\angle 2 = 56$ and $m\angle JKL = 145$. Justify each step.

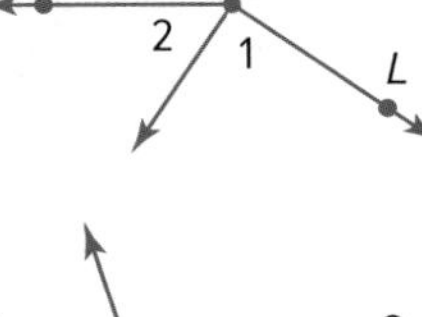

$m\angle 1 + m\angle 2 = m\angle JKL$	Angle Addition Postulate
$m\ 1 + 56 = 145$	$m\angle 2 = 56$ $m\angle JKL = 145$
$m\angle 1 + 56 - 56 = 145 - 56$	Subtraction Property of Equality
$m\angle 1 = 89$	Substitution

Guided Practice

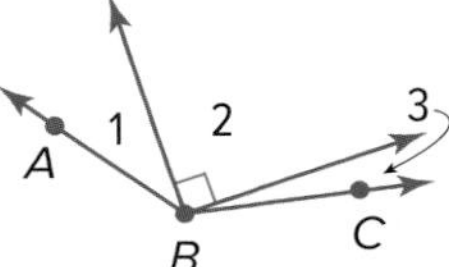

1. Find $m\angle 3$ if $m\angle 1 = 23$ and $m\angle ABC = 131$. Justify each step.

ReviewVocabulary

supplementary angles two angles with measures that add to 180

complementary angles two angles with measures that add to 90

linear pair a pair of adjacent angles with noncommon sides that are opposite rays

The Angle Addition Postulate can be used with other angle relationships to provide additional theorems relating to angles.

Theorems

2.3 Supplement Theorem If two angles form a linear pair, then they are supplementary angles.

Example $m\angle 1 + m\angle 2 = 180$

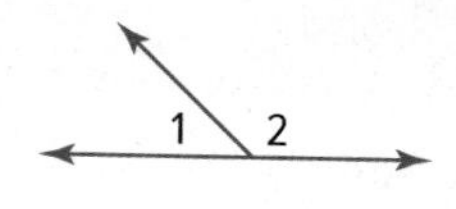

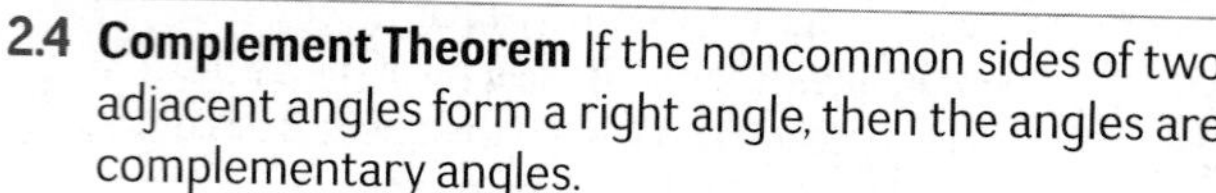

2.4 Complement Theorem If the noncommon sides of two adjacent angles form a right angle, then the angles are complementary angles.

Example $m\angle 1 + m\angle 2 = 90$

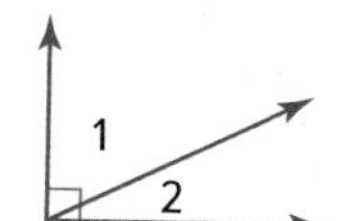

You will prove Theorems 2.3 and 2.4 in Exercises 16 and 17, respectively.

Real-World Example 2 Use the Supplement or Complement Theorem

SURVEYING Using a transit, a surveyor sights the top of a hill and records an angle measure of about 73°. What is the measure of the angle the top of the hill makes with the horizon? Justify each step.

Understand Make a sketch of the situation. The surveyor is measuring the angle of his line of sight. Draw a vertical ray and a horizontal ray from the surveyor's location and label the angles formed. We know that the vertical and horizontal rays form a right angle.

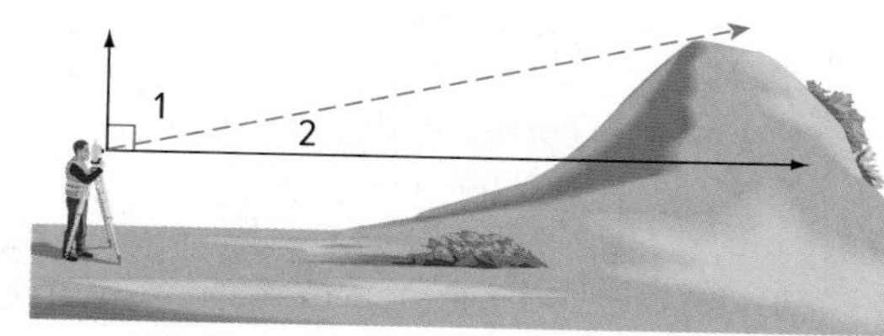

Plan Because $\angle 1$ and $\angle 2$ form a right angle, use the Complement Theorem.

Solve

$m\angle 1 + m\angle 2 = 90$	Complement Theorem
$73 + m\angle 2 = 90$	$m\angle 1 = 73$
$73 + m\angle 2 - 73 = 90 - 73$	Subtraction Property of Equality
$m\angle 2 = 17$	Substitution

The top of the hill makes a 17° angle with the horizon.

Check The sum of 17 and 73 is 90. ✓

The sketch we drew helps us determine an appropriate solution method. We expect the angle to be small, so our answer is reasonable.

Go Online!

You can use the protractor in the **eToolkit** in ConnectED to investigate congruent angles.

Guided Practice

2. $\angle 6$ and $\angle 7$ form a linear pair. If $m\angle 6 = 3x + 32$ and $m\angle 7 = 5x + 12$, find x, $m\angle 6$, and $m\angle 7$. Justify each step.

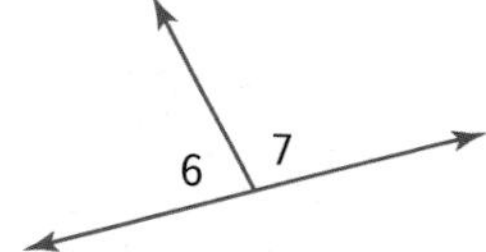

2 Congruent and Right Angles

The properties of algebra that applied to the congruence of segments and the equality of their measures also hold true for the congruence of angles and the equality of their measures.

Theorem 2.5 Properties of Angle Congruence

Reflexive Property of Congruence

$\angle 1 \cong \angle 1$

Symmetric Property of Congruence

If $\angle 1 \cong \angle 2$, then $\angle 2 \cong \angle 1$.

Transitive Property of Congruence

If $\angle 1 \cong \angle 2$ and $\angle 2 \cong \angle 3$, then $\angle 1 \cong \angle 3$.

You will prove the Reflexive and Transitive Properties of Angle Congruence in Exercises 18 and 19, respectively.

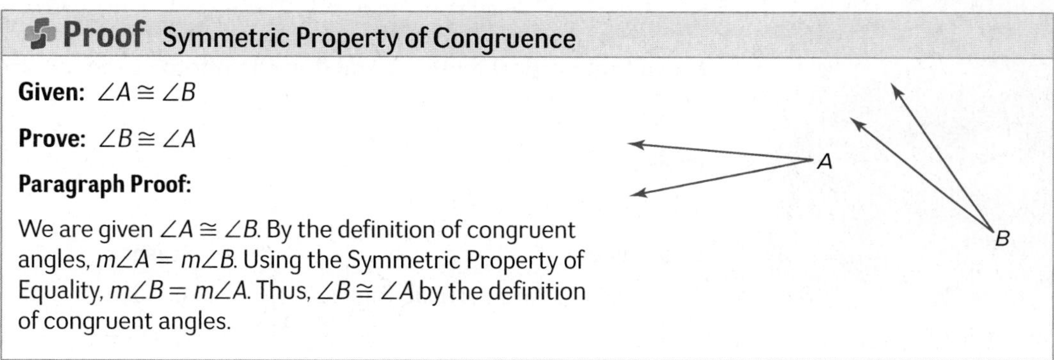

Proof Symmetric Property of Congruence

Given: $\angle A \cong \angle B$

Prove: $\angle B \cong \angle A$

Paragraph Proof:

We are given $\angle A \cong \angle B$. By the definition of congruent angles, $m\angle A = m\angle B$. Using the Symmetric Property of Equality, $m\angle B = m\angle A$. Thus, $\angle B \cong \angle A$ by the definition of congruent angles.

Algebraic properties can be applied to prove theorems for congruence relationships involving supplementary and complementary angles.

Reading Math

Abbreviations and Symbols The notation ∡ means angles.

Theorems

2.6 Congruent Supplements Theorem
Angles supplementary to the same angle or to congruent angles are congruent.

Abbreviation *∡ suppl. to same ∠ or ≅ ∡ are ≅.*

Example If $m\angle 1 + m\angle 2 = 180$ and $m\angle 2 + m\angle 3 = 180$, then $\angle 1 \cong \angle 3$.

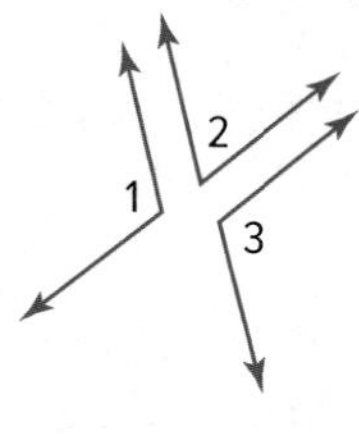

2.7 Congruent Complements Theorem
Angles complementary to the same angle or to congruent angles are congruent.

Abbreviation *∡ compl. to same ∠ or ≅ ∡ are ≅.*

Example If $m\angle 4 + m\angle 5 = 90$ and $m\angle 5 + m\angle 6 = 90$, then $\angle 4 \cong \angle 6$.

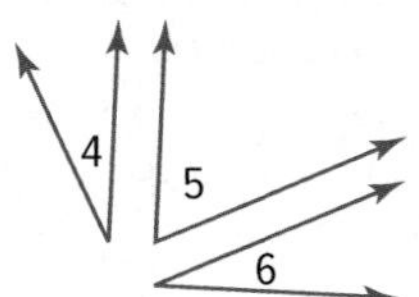

You will prove one case of Theorem 2.7 in Exercise 6.

Proof Congruent Supplements Theorem

Given: $\angle 1$ and $\angle 2$ are supplementary.
$\angle 2$ and $\angle 3$ are supplementary.

Prove: $\angle 1 \cong \angle 3$

Proof:

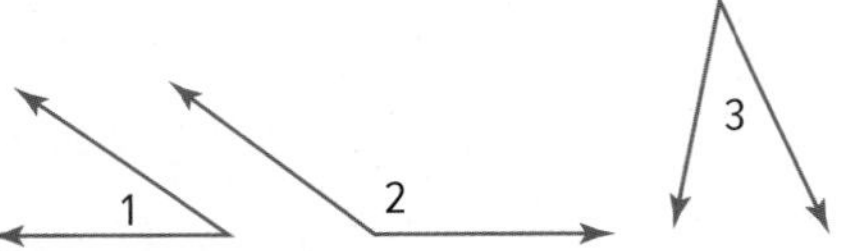

Statements	Reasons
1. $\angle 1$ and $\angle 2$ are supplementary. $\angle 2$ and $\angle 3$ are supplementary.	**1.** Given
2. $m\angle 1 + m\angle 2 = 180$; $m\angle 2 + m\angle 3 = 180$	**2.** Definition of supplementary angles
3. $m\angle 1 + m\angle 2 = m\angle 2 + m\angle 3$	**3.** Substitution
4. $m\angle 2 = m\angle 2$	**4.** Reflexive Property
5. $m\angle 1 = m\angle 3$	**5.** Subtraction Property
6. $\angle 1 \cong \angle 3$	**6.** Definition of congruent angles

Example 3 Proofs Using Congruent Compl. or Suppl. Theorems

Prove that vertical angles 2 and 4 in the photo at the left are congruent.

Given: $\angle 2$ and $\angle 4$ are vertical angles.

Prove: $\angle 2 \cong \angle 4$

Proof:

Statements	Reasons
1. $\angle 2$ and $\angle 4$ are vertical angles.	**1.** Given
2. $\angle 2$ and $\angle 4$ are nonadjacent angles formed by intersecting lines.	**2.** Definition of vertical angles
3. $\angle 2$ and $\angle 3$ form a linear pair. $\angle 3$ and $\angle 4$ form a linear pair.	**3.** Definition of a linear pair
4. $\angle 2$ and $\angle 3$ are supplementary. $\angle 3$ and $\angle 4$ are supplementary.	**4.** Supplement Theorem
5. $\angle 2 \cong \angle 4$	**5.** $\measuredangle$ suppl. to same $\angle$ or $\cong$ $\measuredangle$ are $\cong$.

Real-World Link

The 100-story John Hancock Building uses huge X-braces in its design. These diagonals are connected to the exterior columns, making it possible for strong wind forces to be carried from the braces to the exterior columns and back.

Source: PBS

Guided Practice

3. In the figure, $\angle ABE$ and $\angle DBC$ are right angles. Prove that $\angle ABD \cong \angle EBC$.

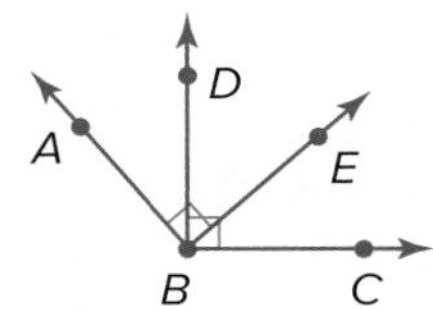

Review Vocabulary

Vertical Angles two nonadjacent angles formed by intersecting lines

Note that in Example 3, $\angle 1$ and $\angle 3$ are vertical angles. The conclusion in the example supports the following Vertical Angles Theorem.

Theorem 2.8 Vertical Angles Theorem

If two angles are vertical angles, then they are congruent.

Abbreviation: *Vert. $\measuredangle$ are $\cong$.*

Example: $\angle 1 \cong \angle 3$ and $\angle 2 \cong \angle 4$

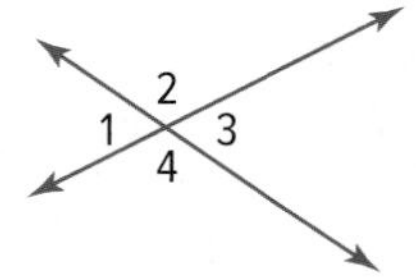

You will prove Theorem 2.8 in Exercise 28.

Example 4 Use Vertical Angles

Prove that if $\overrightarrow{DB}$ bisects $\angle ADC$, then $\angle 2 \cong \angle 3$.

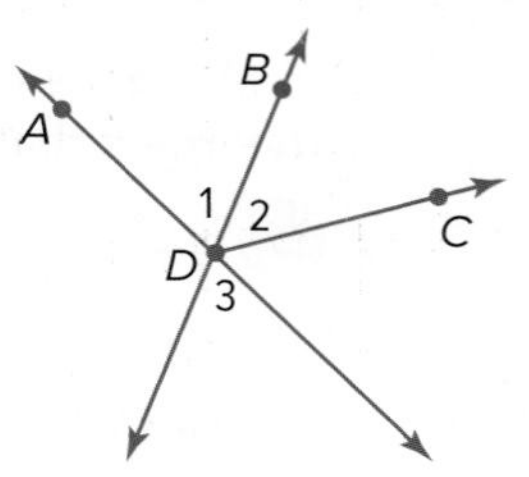

Given: $\overrightarrow{DB}$ bisects $\angle ADC$.

Prove: $\angle 2 \cong \angle 3$

Proof:

Statements	Reasons
1. $\overrightarrow{DB}$ bisects $\angle ADC$.	**1.** Given
2. $\angle 1 \cong \angle 2$	**2.** Definition of angle bisector
3. $\angle 1$ and $\angle 3$ are vertical angles.	**3.** Definition of vertical angles
4. $\angle 3 \cong \angle 1$	**4.** Vert. $\angle$s are $\cong$.
5. $\angle 3 \cong \angle 2$	**5.** Transitive Property of Congruence
6. $\angle 2 \cong \angle 3$	**6.** Symmetric Property of Congruence

Guided Practice

4. If $\angle 3$ and $\angle 4$ are vertical angles, $m\angle 3 = 6x + 2$, and $m\angle 4 = 8x - 14$, find $m\angle 3$ and $m\angle 4$. Justify each step.

The theorems in this lesson can be used to prove the following right angle theorems.

Reading Math

Perpendicular Recall from Lesson 1-5 that the symbol $\perp$ means *is perpendicular to.*

Theorems Right Angle Theorems

Theorem	Example
2.9 Perpendicular lines intersect to form four right angles. **Example** If $\overrightarrow{AC} \perp \overrightarrow{DB}$, then $\angle 1$, $\angle 2$, $\angle 3$, and $\angle 4$ are rt. $\angle$s.	
2.10 All right angles are congruent. **Example** If $\angle 1$, $\angle 2$, $\angle 3$, and $\angle 4$ are rt. $\angle$s, then $\angle 1 \cong \angle 2 \cong \angle 3 \cong \angle 4$.	
2.11 Perpendicular lines form congruent adjacent angles. **Example** If $\overrightarrow{AC} \perp \overrightarrow{DB}$, then $\angle 1 \cong \angle 2$, $\angle 2 \cong \angle 4$, $\angle 3 \cong \angle 4$, and $\angle 1 \cong \angle 3$.	
2.12 If two angles are congruent and supplementary, then each angle is a right angle. **Example** If $\angle 5 \cong \angle 6$ and $\angle 5$ is suppl. to $\angle 6$, then $\angle 5$ and $\angle 6$ are rt. $\angle$s.	
2.13 If two congruent angles form a linear pair, then they are right angles. **Example** If $\angle 7 \cong \angle 8$ and $\angle 7$ and $\angle 8$ form a linear pair, then $\angle 7$ and $\angle 8$ are rt. $\angle$s.	

You will prove Theorems 2.9–2.13 in Exercises 22–26.

Check Your Understanding

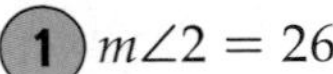 = Step-by-Step Solutions begin on page R13.

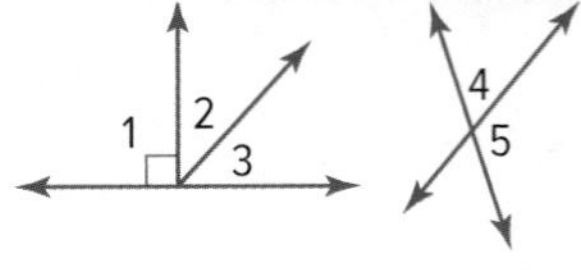

Example 1 **Find the measure of each numbered angle and name the theorems that justify your work.**

1. $m\angle 2 = 26$
2. $m\angle 2 = x$, $m\angle 3 = x - 16$
3. $m\angle 4 = 2x$, $m\angle 5 = x + 9$
4. $m\angle 4 = 3(x - 1)$, $m\angle 5 = x + 7$

Example 2

5. **PARKING** Refer to the diagram of the parking lot at the right. Given that $\angle 2 \cong \angle 6$, prove that $\angle 4 \cong \angle 8$.

Example 3

6. **PROOF** Copy and complete the proof of one case of Theorem 2.7.

Given: $\angle 1$ and $\angle 3$ are complementary.
$\angle 2$ and $\angle 3$ are complementary.

Prove: $\angle 1 \cong \angle 2$

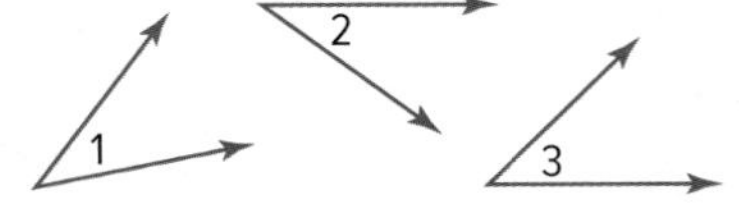

Proof:

Statements	Reasons
a. $\angle 1$ and $\angle 3$ are complementary. $\angle 2$ and $\angle 3$ are complementary.	**a.** ?
b. $m\angle 1 + m\angle 3 = 90$; $m\angle 2 + m\angle 3 = 90$	**b.** ?
c. $m\angle 1 + m\angle 3 = m\angle 2 + m\angle 3$	**c.** ?
d. ?	**d.** Reflexive Property
e. $m\angle 1 = m\angle 2$	**e.** ?
f. $\angle 1 \cong \angle 2$	**f.** ?

Example 4

7. **CONSTRUCT ARGUMENTS** Write a two-column proof.

Given: $\angle 4 \cong \angle 7$

Prove: $\angle 5 \cong \angle 6$

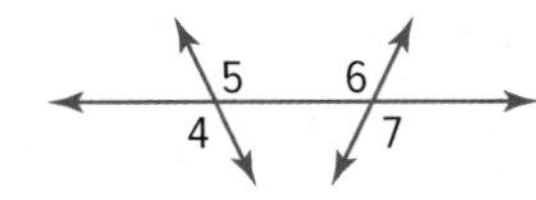

Practice and Problem Solving

Extra Practice is on page R2.

Examples 1–3 **Find the measure of each numbered angle and name the theorems used that justify your work.**

8. $m\angle 5 = m\angle 6$

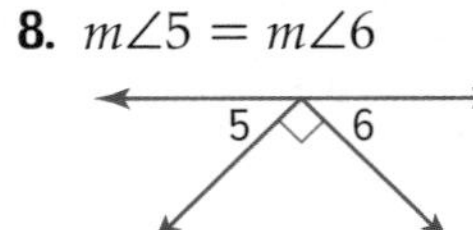

9. $\angle 2$ and $\angle 3$ are complementary. $\angle 1 \cong \angle 4$ and $m\angle 2 = 28$

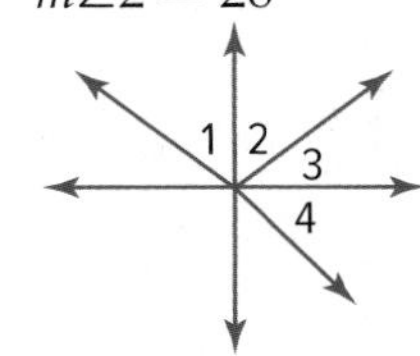

10. $\angle 2$ and $\angle 4$ and $\angle 4$ and $\angle 5$ are supplementary. $m\angle 4 = 105$

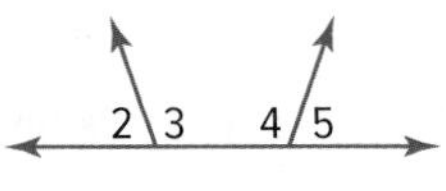

Find the measure of each numbered angle and name the theorems used that justify your work.

11. $m\angle 9 = 3x + 12$
$m\angle 10 = x - 24$

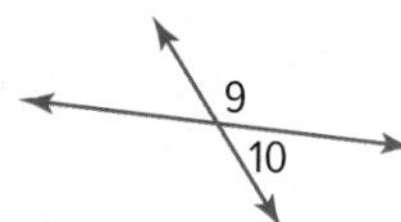

12. $m\angle 3 = 2x + 23$
$m\angle 4 = 5x - 112$

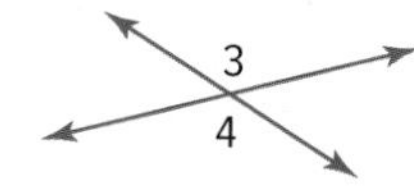

13. $m\angle 6 = 2x - 21$
$m\angle 7 = 3x - 34$

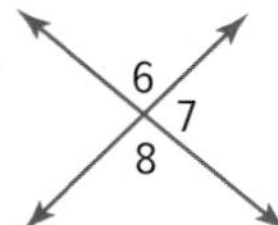

Example 4

PROOF **Write a two-column proof.**

14. Given: $\angle ABC$ is a right angle.
Prove: $\angle ABD$ and $\angle CBD$ are complementary.

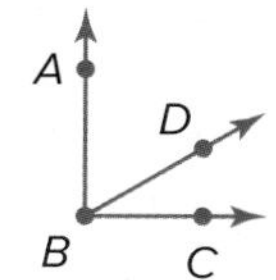

15. Given: $\angle 5 \cong \angle 6$
Prove: $\angle 4$ and $\angle 6$ are supplementary.

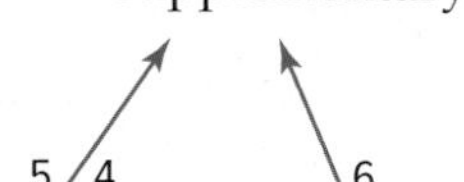

Write a proof for each theorem.

16. Supplement Theorem

17. Complement Theorem

18. Reflexive Property of Angle Congruence

19. Transitive Property of Angle Congruence

20. **FLAGS** Refer to the Florida state flag at the right. Prove that the sum of the four angle measures is 360.

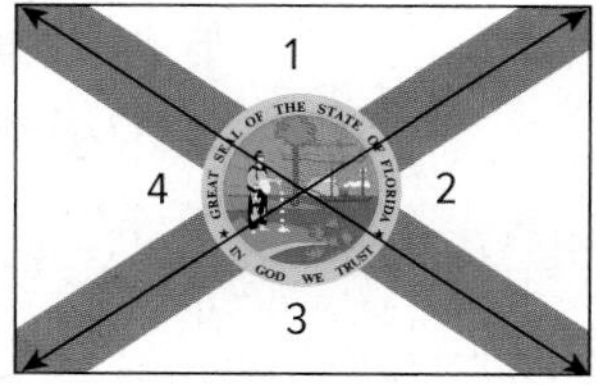

21. **CONSTRUCT ARGUMENTS** The diamondback rattlesnake is a pit viper with a diamond pattern on its back. An enlargement of a skin is shown below. If $\angle 1 \cong \angle 4$, prove that $\angle 2 \cong \angle 3$.

PROOF **Use the figure to write a proof of each theorem.**

22. Theorem 2.9

23. Theorem 2.10

24. Theorem 2.11

25. Theorem 2.12

26. Theorem 2.13

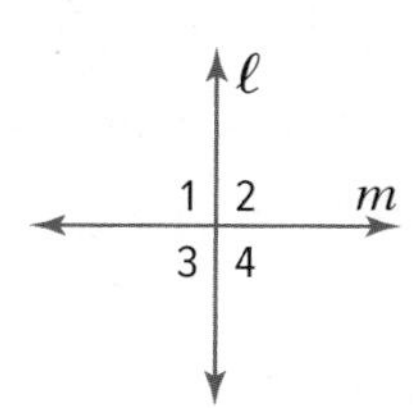

27. **CONSTRUCT ARGUMENTS** To mark a specific tempo, the weight on the pendulum of a metronome is adjusted so that it swings at a specific rate. Suppose $\angle ABC$ in the photo is a right angle. If $m\angle 1 = 45$, write a paragraph proof to show that $\overrightarrow{BR}$ bisects $\angle ABC$.

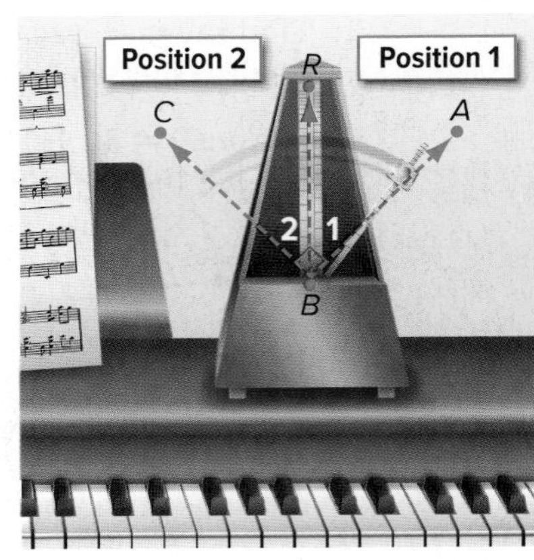

28. **PROOF** Write a proof of Theorem 2.8.

29. **GEOGRAPHY** Utah, Colorado, Arizona, and New Mexico all share a common point on their borders called Four Corners. This is the only place where four states meet in a single point. If $\angle 2$ is a right angle, prove that lines ℓ and m are perpendicular.

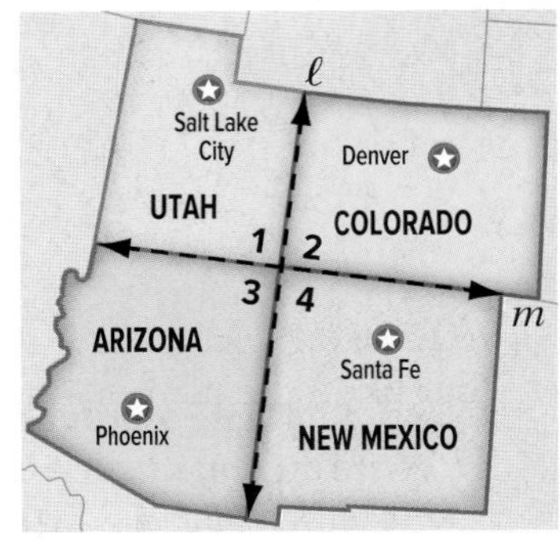

30. **MULTIPLE REPRESENTATIONS** In this problem, you will explore angle relationships.

   a. **Geometric** Draw a right angle ABC. Place point D in the interior of this angle and draw $\overrightarrow{BD}$. Draw $\overrightarrow{KL}$ and construct $\angle JKL$ congruent to $\angle ABD$.

   b. **Verbal** Make a conjecture as to the relationship between $\angle JKL$ and $\angle DBC$.

   c. **Logical** Prove your conjecture.

H.O.T. Problems Use Higher-Order Thinking Skills

31. **OPEN ENDED** Draw an angle WXZ such that $m\angle WXZ = 45$. Construct $\angle YXZ$ congruent to $\angle WXZ$. Make a conjecture as to the measure of $\angle WXY$ and then prove your conjecture.

32. **WRITING IN MATH** Write the steps that you would use to complete the proof below.

   Given: $\overline{BC} \cong \overline{CD}$, $AB = \frac{1}{2}BD$

   Prove: $\overline{AB} \cong \overline{CD}$

   A B C D

33. **CHALLENGE** In this lesson, one case of the Congruent Supplements Theorem was proven. In Exercise 6, you proved the same case for the Congruent Complements Theorem. Explain why there is another case for each of these theorems. Then write a proof of this second case for each theorem.

34. **REASONING** Determine whether the following statement is *sometimes*, *always*, or *never* true. Explain your reasoning.

   If one of the angles formed by two intersecting lines is acute, then the other three angles formed are also acute.

35. **WRITING IN MATH** Explain how you can use your protractor to quickly find the measure of the supplement of an angle.

36. What is the measure of $\angle KLH$? MP 7

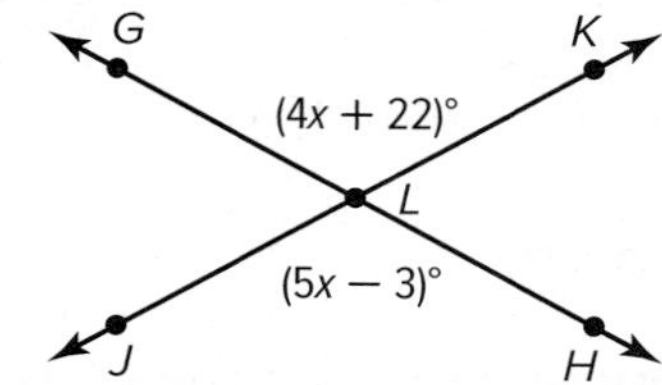

- A 155°
- B 122°
- C 58°
- D 25°

37. MULTI-STEP An angle measures 5 degrees less than four times the measure of its supplement. MP 1

a. Write an equation to represent the sum of the measures of these two angles.

b. Solve the equation to find the measures of the angles.

38. Lyon Street and Baker Avenue intersect as shown in the figure. A city planner knows that the measure of $\angle ABD$ is three times the measure of $\angle CBD$. What is the measure of $\angle ABD$ in degrees? MP 1

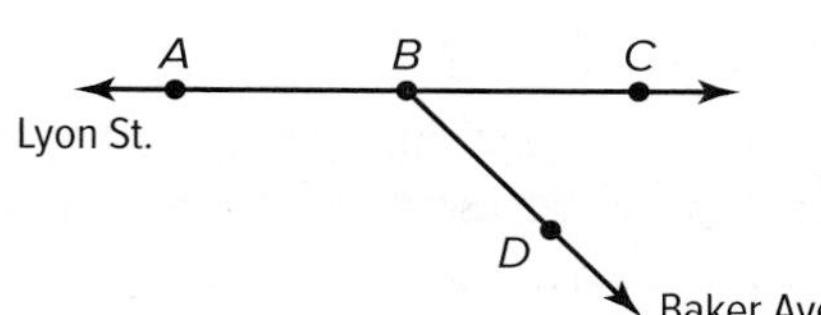

39. $\angle P$ is complementary to $\angle X$, and $\angle Q$ is complementary to $\angle X$. What is $m\angle P$? MP 7

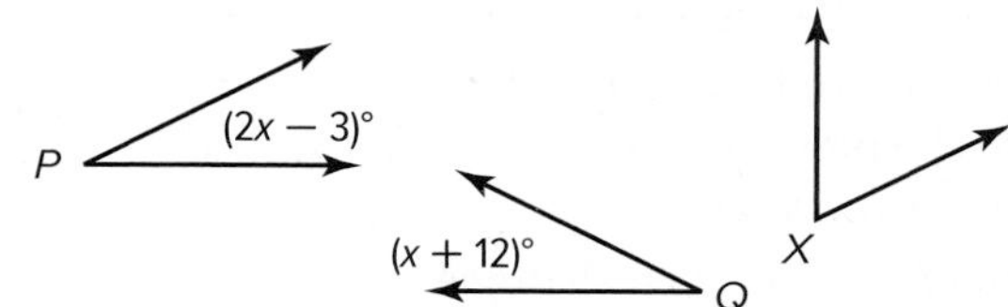

- A 15°
- B 27°
- C 51°
- D 63°

40. Which pairs of angles in the figure must be congruent? MP 2

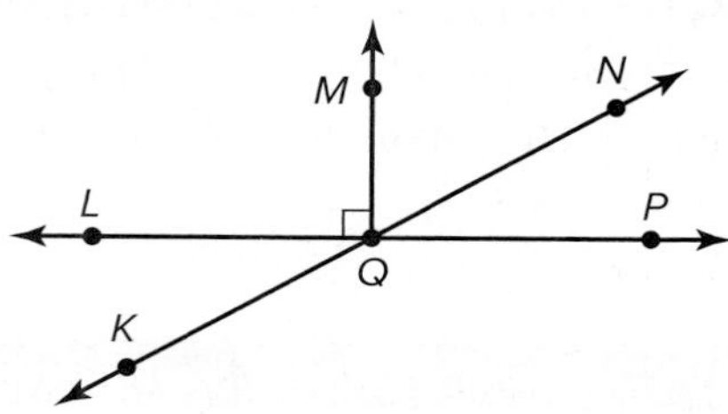

I. $\angle LQK$ and $\angle PQN$

II. $\angle LQM$ and $\angle MQP$

III. $\angle LQN$ and $\angle PQK$

- A I only
- B II only
- C I and II only
- D I and III only
- E I, II, and III

41. Which of the following is accepted as true without proof? MP 2

- A conjecture
- B postulate
- C theorem
- D hypothesis

42. MULTI-STEP Use the following diagram to answer the following questions. MP 1

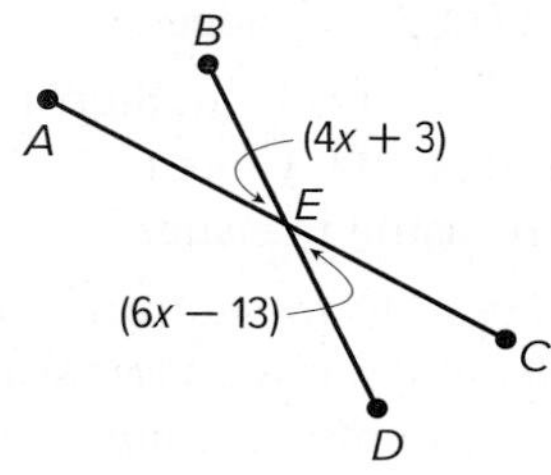

- A What is the measure of $\angle AEB$?
- B What is the measure of $\angle CED$?
- C What is the measure of $\angle BEC$?
- D What is the measure of $\angle AED$?
- E What are two ways to find the measure of $\angle AED$?

Geometry Software Lab

Angles and Parallel Lines

You can use The Geometer's Sketchpad® to explore the angles formed by two parallel lines and a transversal.

Mathematical Practices
MP 5 Use appropriate tools strategically.

Activity Parallel Lines and a Transversal

Work cooperatively.

Step 1 **Draw a line.**

Draw and label points F and G. Then use the line tool to draw $\overleftrightarrow{FG}$.

Step 2 **Draw a parallel line.**

Draw a point that is not on $\overleftrightarrow{FG}$ and label it J. Select $\overleftrightarrow{FG}$ and point J, and then choose **Parallel Line** from the **Construct** menu. Draw and label a point K on this parallel line.

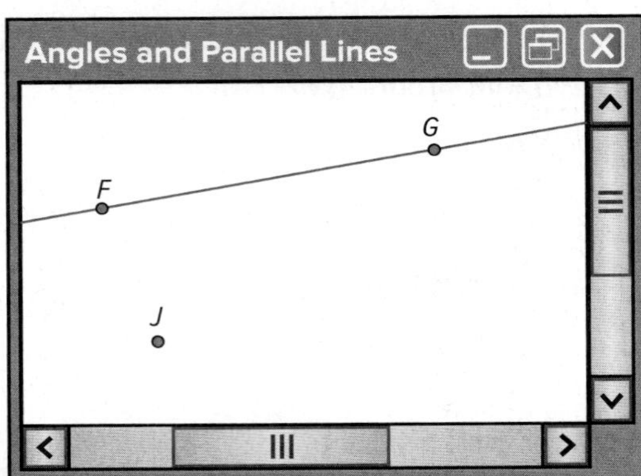

Step 3 **Draw a transversal.**

Draw and label point A on $\overleftrightarrow{FG}$ and point B on $\overleftrightarrow{JK}$. Select A and B and then choose **Line** from the **Construct** menu to draw transversal $\overleftrightarrow{AB}$. Then draw and label points C and D on $\overleftrightarrow{AB}$ as shown.

Step 4 **Measure each angle.**

Measure all eight angles formed by these lines. For example, select points F, A, then C, and choose **Angle** from the **Measure** menu to find $m\angle FAC$.

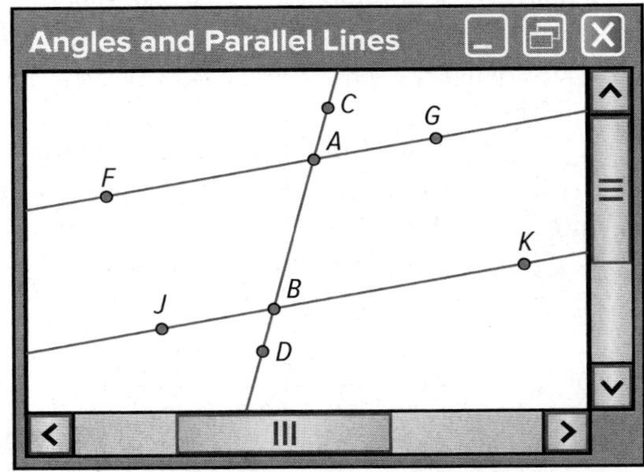

Analyze the Results Work cooperatively.

1. Organize the measures from Step 4 in a table like this one. Which angles have the same measure?

Angle	$\angle FAC$	$\angle CAG$	$\angle GAB$	$\angle FAB$	$\angle JBA$	$\angle ABK$	$\angle KBD$	$\angle JBD$
1st Measure								

2. Drag point C or D to move transversal $\overleftrightarrow{AB}$ so that it intersects the two parallel lines at a different angle. Add a row **2nd Measure** to your table and record the new measures. Repeat these steps until your table has 3rd, 4th, and 5th Measure rows of data.
3. Using the angles listed in the table, identify and describe the relationship between all angle pairs that have the following special names. Then write a conjecture in if-then form about each angle pair when formed by any two parallel lines cut by a transversal.
   a. corresponding **b.** alternate interior **c.** alternate exterior **d.** consecutive interior
4. Drag point C or D so that the measure of any of the angles is 90.
   a. What do you notice about the measures of the other angles?
   b. Make a conjecture about a transversal that is perpendicular to one of two parallel lines.

LESSON 7

Parallel Lines and Transversals

Then

- You used angle and line segment relationships to prove theorems.

Now

1. Name angle pairs formed by lines and transversals.
2. Use theorems to determine the relationships between specific pairs of angles.

Why?

- City planners often use a grid layout. In this arrangement, streets are parallel and are intersected by cross streets. This creates square or rectangular city blocks. The size of a city block is not standardized. Within a city, there can be variances in the size of a block.

New Vocabulary

transversal
interior angles
exterior angles
consecutive interior angles
alternate interior angles
alternate exterior angles
corresponding angles
parallel lines

Mathematical Practices

1 Make sense of problems and persevere in solving them.

3 Construct viable arguments and critique the reasoning of others.

1 Transversal Angle Pair Relationships

A line that intersects two or more coplanar lines at two different points is called a **transversal**. In the diagram below, line t is a transversal of lines q and r. Notice that line t forms a total of eight angles with lines q and r. These angles, and specific pairings of these angles, are given special names.

Key Concept Transversal Angle Pair Relationships

Four **interior angles** lie in the region between lines q and r.	$\angle 3, \angle 4, \angle 5, \angle 6$	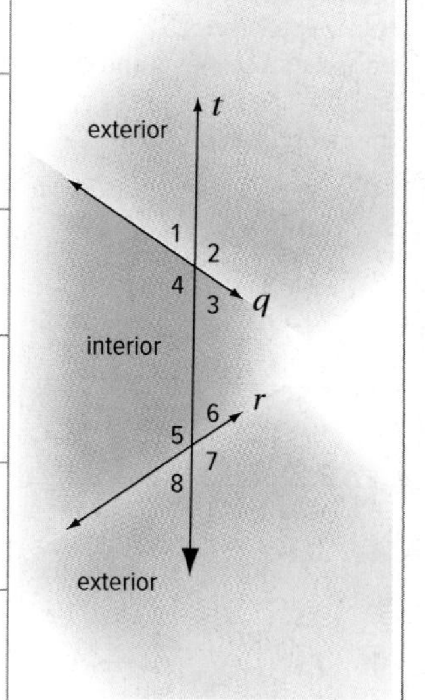
Four **exterior angles** lie in the two regions that are not between lines q an r.	$\angle 1, \angle 2, \angle 7, \angle 8$	
Consecutive interior angles are interior angles that lie on the same side of transversal t.	$\angle 4$ and $\angle 5$, $\angle 3$ and $\angle 6$	
Alternate interior angles are nonadjacent interior angles that lie on opposite sides of transversal t.	$\angle 3$ and $\angle 5$, $\angle 4$ and $\angle 6$	
Alternate exterior angles are nonadjacent exterior angles that lie on opposite sides of transversal t.	$\angle 1$ and $\angle 7$, $\angle 2$ and $\angle 8$	
Corresponding angles lie on the same side of transversal t and on the same side of lines q and r.	$\angle 1$ and $\angle 5$, $\angle 2$ and $\angle 6$ $\angle 3$ and $\angle 7$, $\angle 4$ and $\angle 8$	

Example 1 Classify Angle Pair Relationships

Refer to the figure below. Classify the relationship between each pair of angles as *alternate interior, alternate exterior, corresponding,* or *consecutive interior* angles.

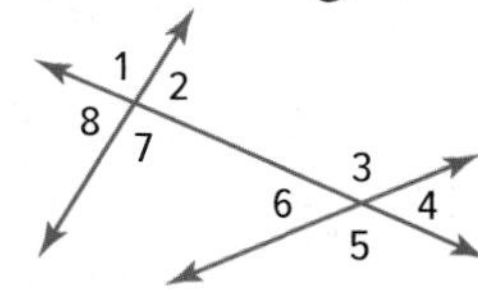

a. $\angle 1$ and $\angle 5$
alternate exterior

b. $\angle 6$ and $\angle 7$
consecutive interior

c. $\angle 2$ and $\angle 4$
corresponding

d. $\angle 2$ and $\angle 6$
alternate interior

Guided Practice

1A. $\angle 3$ and $\angle 7$ **1B.** $\angle 5$ and $\angle 7$ **1C.** $\angle 4$ and $\angle 8$ **1D.** $\angle 2$ and $\angle 3$

Stuart Dee/Getty Images

2 Parallel Lines and Angle Pairs

If two lines do not intersect, then they are either parallel or skew.

Key Concepts Parallel and Skew

Parallel lines are coplanar lines that do not intersect.

Example $\overleftrightarrow{JK} \parallel \overleftrightarrow{LM}$

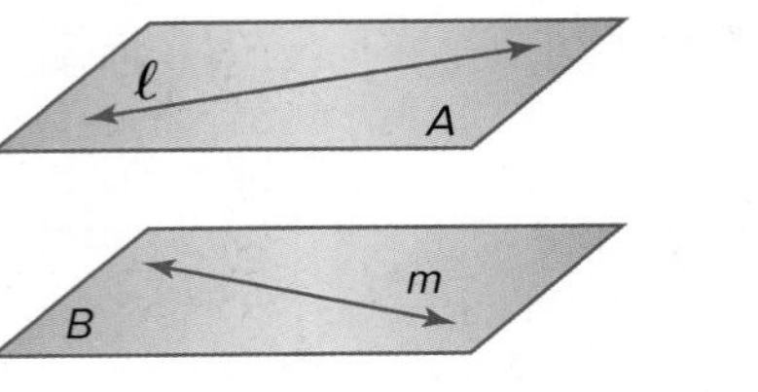

Arrows are used to indicate that lines are parallel.

Skew lines are lines that do not intersect and are not coplanar.

Example Lines ℓ and m are skew.

Parallel planes are planes that do not intersect.

Example Planes $\mathcal{A}$ and $\mathcal{B}$ are parallel.

When two parallel lines are intersected by a transversal, there is a special relationship between corresponding angle pairs.

Postulate 2.12 Corresponding Angles Postulate

If two parallel lines are cut by a transversal, then each pair of corresponding angles is congruent.

Examples $\angle 1 \cong \angle 3$, $\angle 2 \cong \angle 4$, $\angle 5 \cong \angle 7$, $\angle 6 \cong \angle 8$

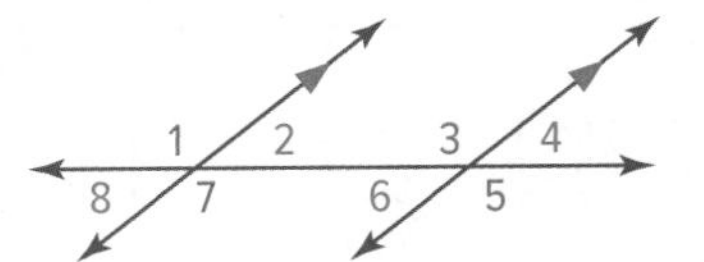

Study Tip

Nonexample In the figure below, line c is *not* a transversal of lines a and b, because line c intersects lines a and b in only one point.

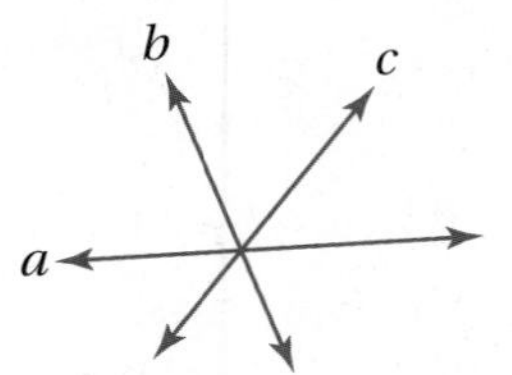

Example 2 Use Corresponding Angles Postulate

In the figure, $m\angle 5 = 72$. Find the measure of each angle. Tell which postulate(s) or theorem(s) you used.

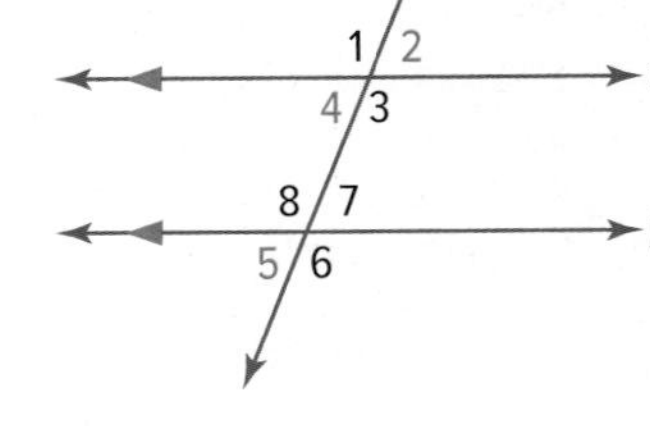

a. $\angle 4$

$\angle 4 \cong \angle 5$	Corresponding Angles Postulate
$m\angle 4 = m\angle 5$	Definition of congruent angles
$m\angle 4 = 72$	Substitution

b. $\angle 2$

$\angle 2 \cong \angle 4$	Vertical Angles Theorem
$\angle 4 \cong \angle 5$	Corresponding Angles Postulate
$\angle 2 \cong \angle 5$	Transitive Property of Congruence
$m\angle 2 = m\angle 5$	Definition of congruent angles
$m\angle 2 = 72$	Substitution

Guided Practice

In the figure, suppose that $m\angle 8 = 105$. Find the measure of each angle. Tell which postulate(s) or theorem(s) you used.

2A. $\angle 1$ **2B.** $\angle 2$ **2C.** $\angle 3$

In Example 2, $\angle 2$ and $\angle 5$ are congruent alternate exterior angles. This and other examples suggest the following theorems about the other angle pairs formed by two parallel lines cut by a transversal.

Go Online!

Discover relationships among the angles formed when parallel lines are intersected by a transversal with the **Geometer's Sketchpad®** activity in ConnectED.

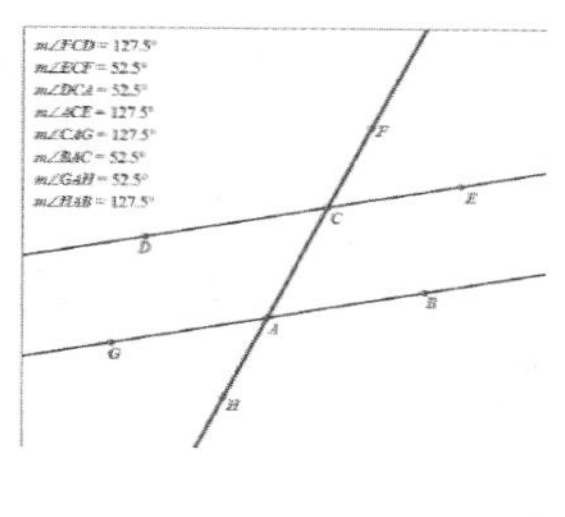

Theorems Parallel Lines and Angle Pairs

2.14 Alternate Interior Angles Theorem If two parallel lines are cut by a transversal, then each pair of alternate interior angles is congruent.

Examples $\angle 1 \cong \angle 3$ and $\angle 2 \cong \angle 4$

2.15 Consecutive Interior Angles Theorem If two parallel lines are cut by a transversal, then each pair of consecutive interior angles is supplementary.

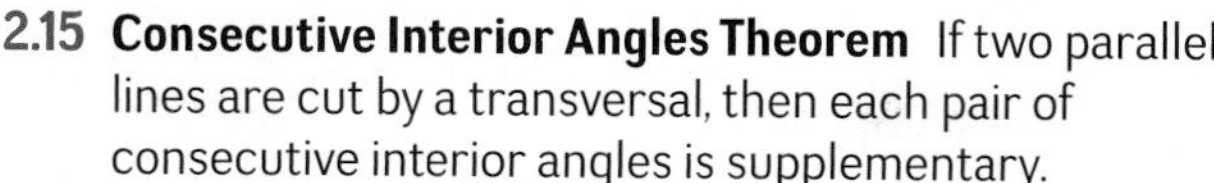

Examples $\angle 1$ and $\angle 2$ are supplementary.
$\angle 3$ and $\angle 4$ are supplementary.

2.16 Alternate Exterior Angles Theorem If two parallel lines are cut by a transversal, then each pair of alternate exterior angles is congruent.

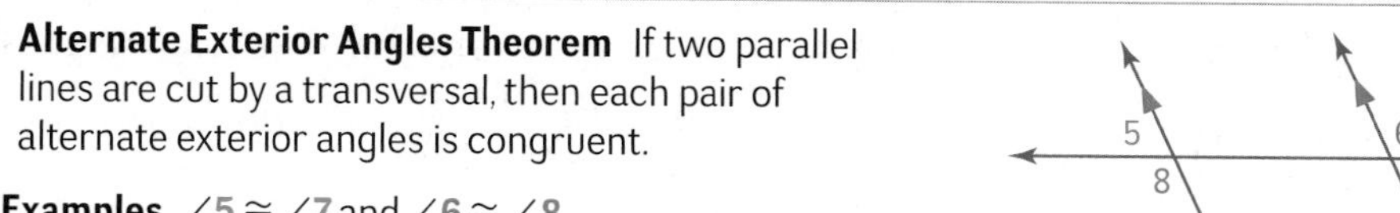

Examples $\angle 5 \cong \angle 7$ and $\angle 6 \cong \angle 8$

You will prove Theorems 2.15 and 2.16 in Exercises 40 and 45, respectively.

Because postulates are accepted without proof, you can use the Corresponding Angles Postulate to prove each of the theorems above.

Proof Alternate Interior Angles Theorem

Given: $a \parallel b$
t is a transversal of a and b.

Prove: $\angle 4 \cong \angle 5$, $\angle 3 \cong \angle 6$

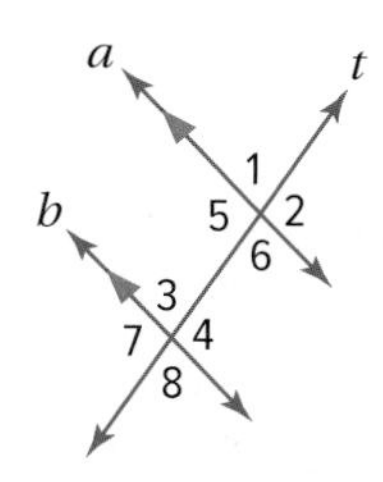

Paragraph Proof: We are given that $a \parallel b$ with a transversal t. By the Corresponding Angles Postulate, corresponding angles are congruent. So, $\angle 2 \cong \angle 4$ and $\angle 6 \cong \angle 8$. Also, $\angle 5 \cong \angle 2$ and $\angle 8 \cong \angle 3$ because vertical angles are congruent. Therefore, $\angle 5 \cong \angle 4$ and $\angle 3 \cong \angle 6$ because congruence of angles is transitive.

Real-World Example 3 Use Theorems about Parallel Lines

COMMUNITY PLANNING Redding Lane and Creek Road are parallel streets that intersect Park Road along the west side of Wendell Park. If $m\angle 1 = 118$, find $m\angle 2$.

$\angle 2 \cong \angle 1$	Alternate Interior Angles Theorem
$m\angle 2 = m\angle 1$	Definition of congruent angles
$m\angle 2 = 118$	Substitution

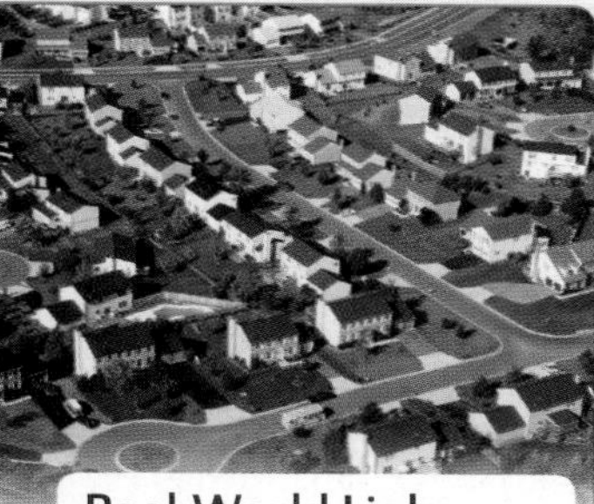

Real-World Link

Some cities require that streets in newly planned subdivisions intersect at no less than a 60° angle.

Glowimages/Getty Images

Guided Practice

COMMUNITY PLANNING Refer to the diagram above to find each angle measure. Tell which postulate(s) or theorem(s) you used.

3A. If $m\angle 1 = 100$, find $m\angle 4$.

3B. If $m\angle 3 = 70$, find $m\angle 4$.

The special relationships between the angles formed by two parallel lines and a transversal can be used to find unknown values.

Example 4 Find Values of Variables

ALGEBRA If $m\angle 4 = 2x - 17$ and $m\angle 1 = 85$, find x. Explain your reasoning.

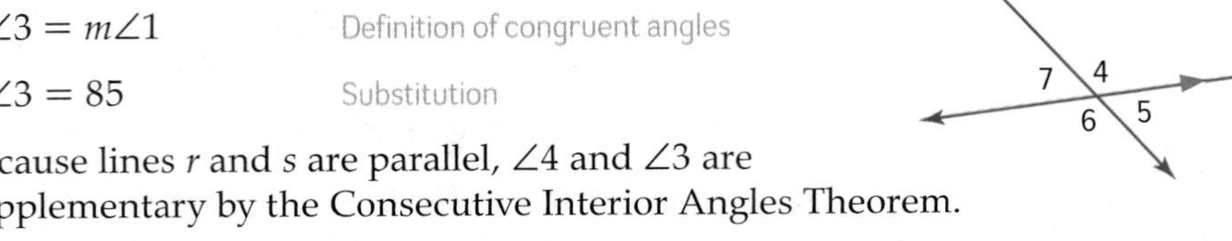

$\angle 3 \cong \angle 1$	Vertical Angles Theorem
$m\angle 3 = m\angle 1$	Definition of congruent angles
$m\angle 3 = 85$	Substitution

Because lines r and s are parallel, $\angle 4$ and $\angle 3$ are supplementary by the Consecutive Interior Angles Theorem.

$m\angle 3 + m\angle 4 = 180$	Definition of supplementary angles
$85 + 2x - 17 = 180$	Substitution
$2x + 68 = 180$	Simplify.
$2x = 112$	Subtract 68 from each side.
$x = 56$	Divide each side by 2.

Study Tip

MP Precision The postulates and theorems you will be studying in this lesson only apply to *parallel* lines cut by a transversal. You should assume that lines are parallel only if the information is given or the lines are marked with parallel arrows.

Guided Practice

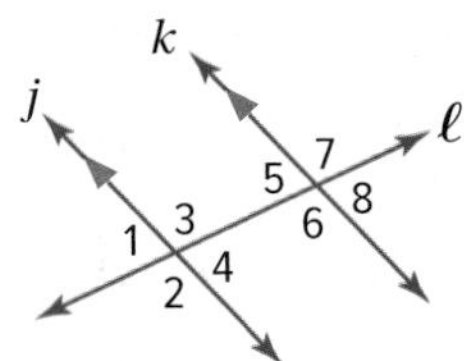

4A. If $m\angle 2 = 4x + 7$ and $m\angle 7 = 5x - 13$, find x. Explain your reasoning.

4B. Find y if $m\angle 5 = 68$ and $m\angle 3 = 3y - 2$. Explain your reasoning.

A special relationship exists when the transversal of two parallel lines is a perpendicular line.

Theorem 2.17 Perpendicular Transversal Theorem

In a plane, if a line is perpendicular to one of two parallel lines, then it is perpendicular to the other.

Examples If line $a \parallel$ line b and line $a \perp$ line t, then line $b \perp$ line t.

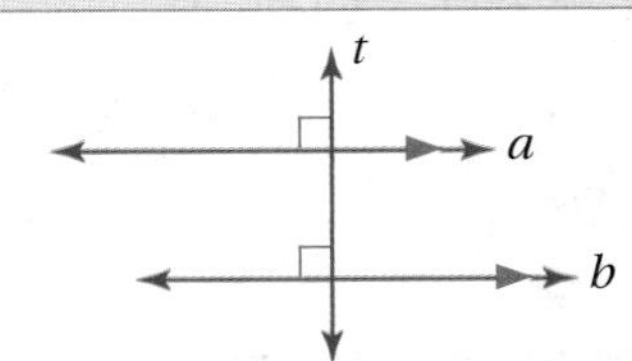

You will prove Theorem 2.17 in Exercise 47.

Reading Math

perpendicular Recall from Lesson 1-5 that line $b \perp$ line t is read as *Line b is perpendicular to line t.*

Check Your Understanding

 = Step-by-Step Solutions begin on page R13.

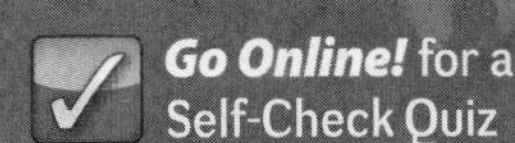

Example 1 **Classify the relationship between each pair of angles as *alternate interior, alternate exterior, corresponding,* or *consecutive interior* angles.**

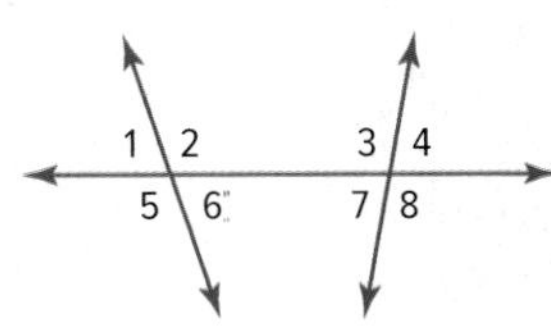

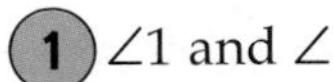 **1.** $\angle 1$ and $\angle 8$

2. $\angle 2$ and $\angle 4$

3. $\angle 3$ and $\angle 6$

4. $\angle 6$ and $\angle 7$

Example 2 **In the figure, $m\angle 1 = 94$. Find the measure of each angle. Tell which postulate(s) or theorem(s) you used.**

5. $\angle 3$ **6.** $\angle 5$ **7.** $\angle 4$

Example 3 **In the figure, $m\angle 4 = 101$. Find the measure of each angle. Tell which postulate(s) or theorem(s) you used.**

8. $\angle 6$ **9.** $\angle 7$ **10.** $\angle 5$

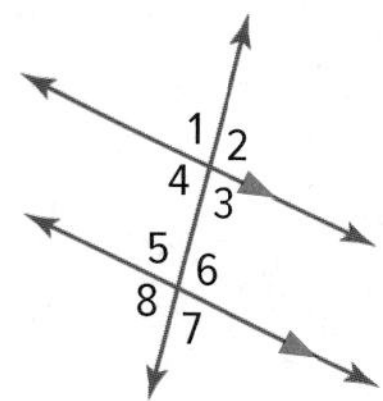

11. ROADS In the diagram, the guard rail is parallel to the surface of the roadway and the vertical supports are parallel to each other. Find the measures of angles 2, 3, and 4.

Example 4 **Find the value of the variable(s) in each figure. Explain your reasoning.**

12.

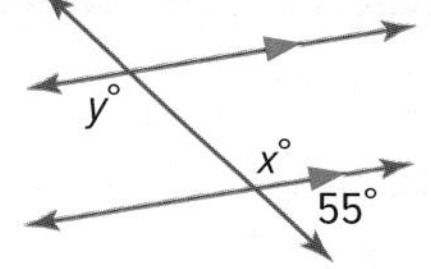

13.

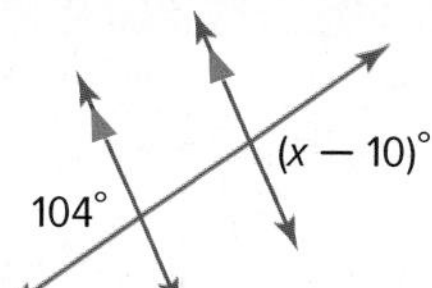

14.

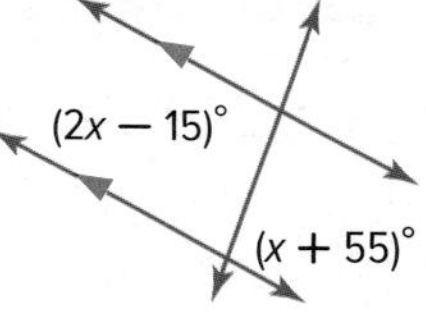

Practice and Problem Solving

Extra Practice is on page R2.

Example 1 **MP PRECISION** **Classify the relationship between each pair of angles as *alternate interior, alternate exterior, corresponding,* or *consecutive interior* angles.**

15. $\angle 4$ and $\angle 9$ **16.** $\angle 5$ and $\angle 7$

17. $\angle 3$ and $\angle 5$ **18.** $\angle 10$ and $\angle 11$

19. $\angle 1$ and $\angle 6$ **20.** $\angle 6$ and $\angle 8$

21. $\angle 2$ and $\angle 3$ **22.** $\angle 9$ and $\angle 10$

23. $\angle 4$ and $\angle 11$ **24.** $\angle 7$ and $\angle 11$

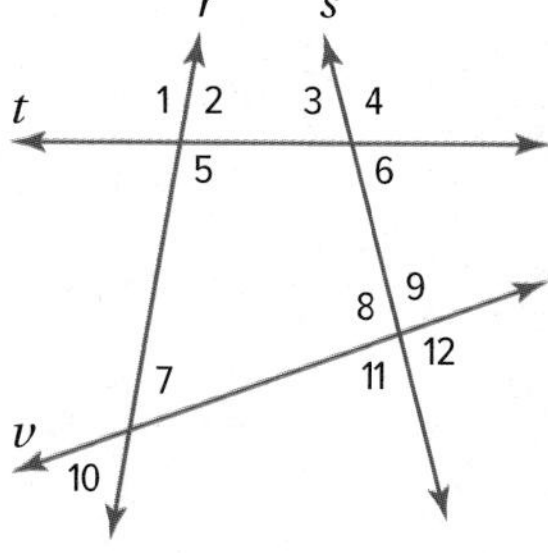

Examples 2–3 **In the figure, $m\angle 11 = 62$ and $m\angle 14 = 38$. Find the measure of each angle. Tell which postulate(s) or theorem(s) you used.**

25. $\angle 4$ **26.** $\angle 3$ **27.** $\angle 12$

28. $\angle 8$ **29.** $\angle 6$ **30.** $\angle 2$

31. $\angle 10$ **32.** $\angle 5$ **33.** $\angle 1$

Example 4 **Find the value of the variable(s) in each figure. Explain your reasoning.**

34.

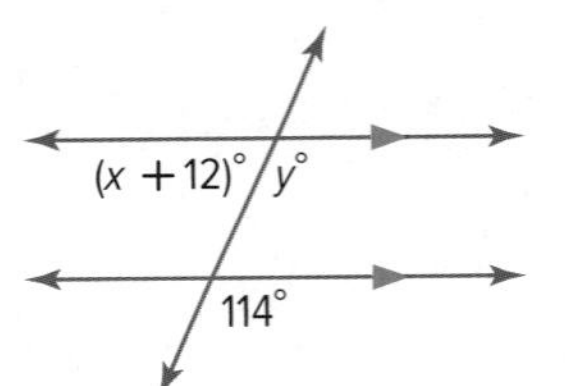

35.

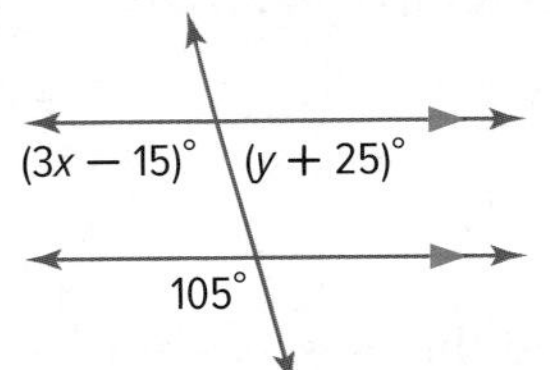

36.

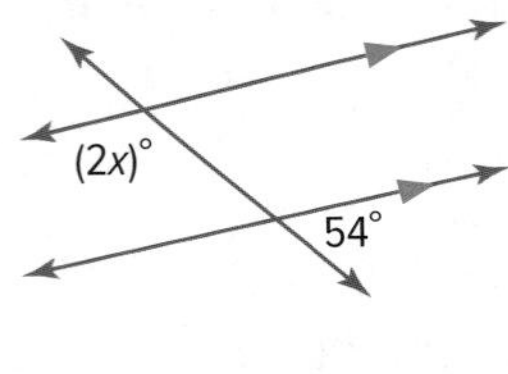

37.

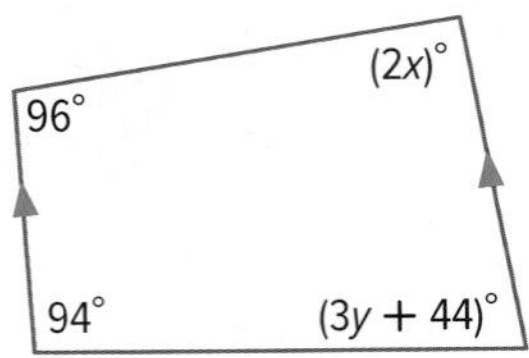

38.

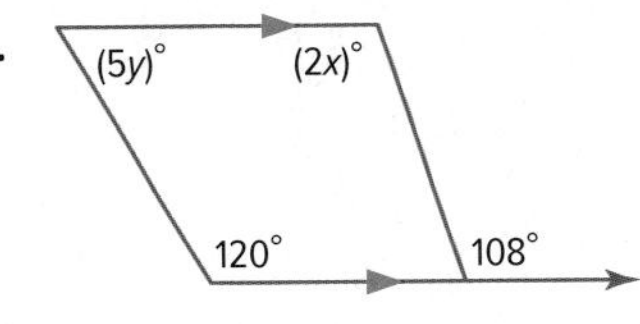

39.

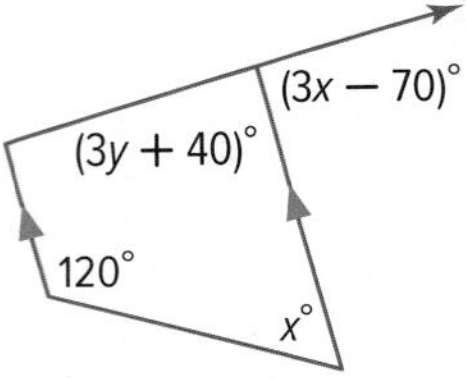

40. PROOF Copy and complete the proof of Theorem 2.15.

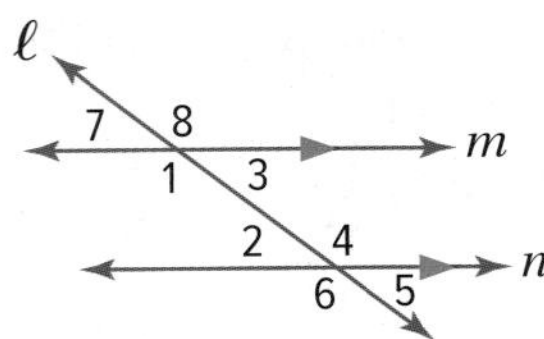

Given: $m \parallel n$; ℓ is a transversal.

Prove: $\angle 1$ and $\angle 2$ are supplementary; $\angle 3$ and $\angle 4$ are supplementary.

Proof:

Statements	Reasons
a. ?	**a.** Given
b. $\angle 1$ and $\angle 3$ form a linear pair; $\angle 2$ and $\angle 4$ form a linear pair.	**b.** ?
c. ?	**c.** If two angles form a linear pair, then they are supplementary.
d. $\angle 1 \cong \angle 4$, $\angle 2 \cong \angle 3$	**d.** ?
e. $m\angle 1 = m\angle 4$, $m\angle 2 = m\angle 3$	**e.** Definition of Congruence
f. ?	**f.** ?

STORAGE When industrial shelving needs to be accessible from either side, additional support is provided on the side by transverse members. Determine the relationship between each pair of angles and explain your reasoning.

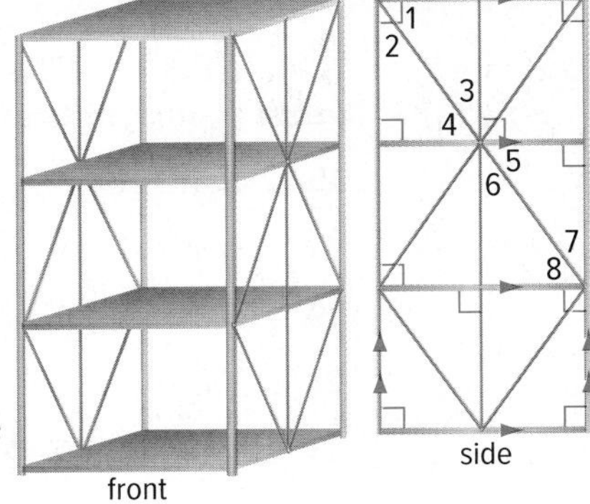

41. $\angle 1$ and $\angle 8$

42. $\angle 1$ and $\angle 5$

43. $\angle 3$ and $\angle 6$

44. $\angle 1$ and $\angle 2$

45. MP **CONSTRUCT ARGUMENTS** Write a two-column proof of the Alternate Exterior Angles Theorem. (Theorem 2.16)

46. BRIDGES Refer to the diagram of the double decker Michigan Avenue Bridge in Chicago, Illinois at the right. The top and bottom beams and its diagonal braces are parallel.

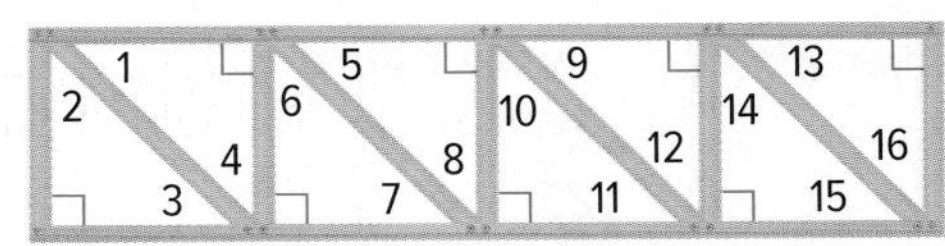

a. How are the measures of the odd-numbered angles related? Explain.

b. How are the measures of the even-numbered angles related? Explain.

c. How are any pair of angles in which one is odd and the other is even related?

d. What geometric term(s) can be used to relate the two roadways contained by the bridge?

47. **PROOF** In a plane, prove that if a line is perpendicular to one of two parallel lines, then it is perpendicular to the other. (Theorem 2.17)

MP TOOLS **Find x.**

48.

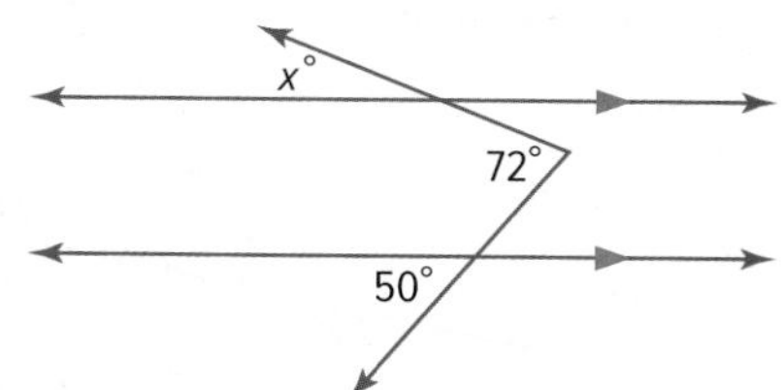

49.

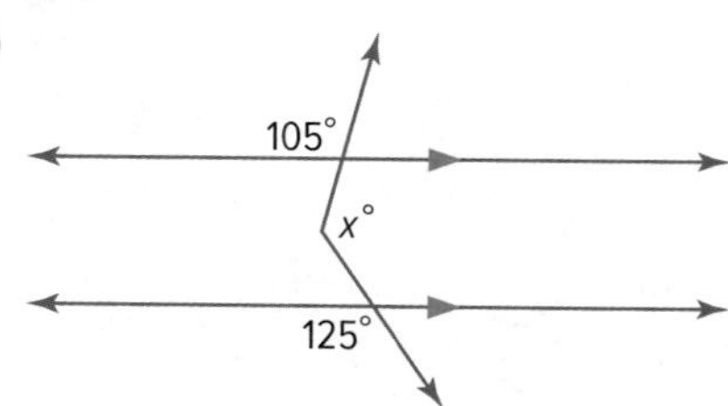

50. **PROBABILITY** Suppose you were to pick any two angles in the figure below.

   a. How many possible angle pairings are there? Explain.

   b. Describe the possible relationships between the measures of the angles in each pair. Explain.

   c. Describe the likelihood of randomly selecting a pair of congruent angles. Explain your reasoning.

51. **MULTIPLE REPRESENTATIONS** In this problem, you will investigate the relationship between same-side exterior angles.

   a. **Geometry** Draw five pairs of parallel lines, m and n, a and b, r and s, j and k, and x and y, cut by a transversal t, and measure the four angles on one side of t.

   b. **Tabular** Record your data in a table.

   c. **Verbal** Make a conjecture about the relationship between the pair of angles formed on the exterior of parallel lines and on the same side of the transversal.

   d. **Logical** What type of reasoning did you use to form your conjecture? Explain.

   e. **Proof** Write a proof of your conjecture.

H.O.T. Problems Use Higher-Order Thinking Skills

52. **WRITING IN MATH** If line a is parallel to line b and $\angle 1 \cong \angle 2$, describe the relationship between lines b and c. Explain your reasoning.

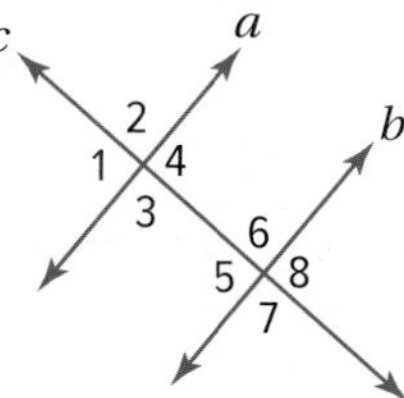

53. **WRITING IN MATH** Compare and contrast the Alternate Interior Angles Theorem and the Consecutive Interior Angles Theorem.

54. **OPEN ENDED** Draw a pair of parallel lines cut by a transversal and measure the two exterior angles on the same side of the transversal. Include the measures on your drawing. Based on the pattern you have seen for naming other pairs of angles, what do you think the name of the pair you measured would be?

55. **CHALLENGE** Find x and y.

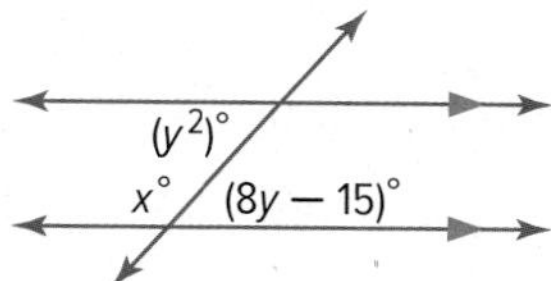

56. **MP REASONING** Determine the minimum number of angle measures you would have to know to find the measures of all the angles formed by two parallel lines cut by a transversal. Explain.

57. In the diagram, $m \parallel n$. If $m\angle 1 = 4x - 6$ and $m\angle 7 = 2x + 40$, what is $m\angle 4$? MP 1

- ○ **A** 23
- ○ **B** 86
- ○ **C** 94
- ○ **D** 157

58. In the diagram, $m\angle 8 = 11y + 7$ and $m\angle 3 = y + 17$. What is $m\angle 2$? MP 1

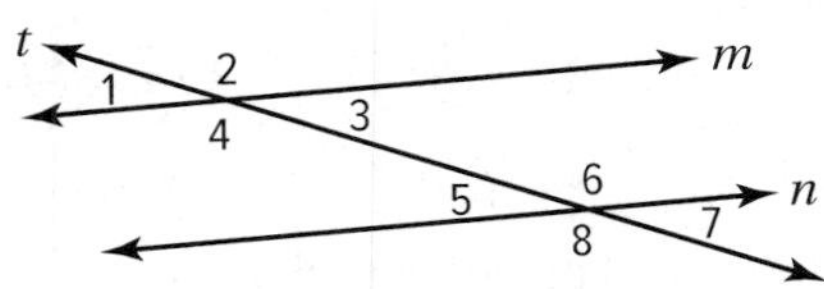

- ○ **A** 13
- ○ **B** 30
- ○ **C** 150
- ○ **D** 167

59. Which angles are consecutive angles? MP 1

- ○ **A** $\angle 1$ and $\angle 5$
- ○ **B** $\angle 3$ and $\angle 4$
- ○ **C** $\angle 4$ and $\angle 6$
- ○ **D** $\angle 4$ and $\angle 5$

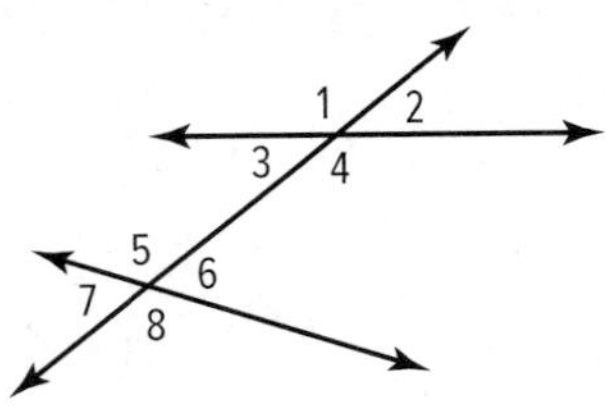

60. Which of the following descriptions for pairs of angles would not necessarily be supplementary? MP 1

- ○ **A** angles that form a straight line
- ○ **B** corresponding angles
- ○ **C** any pair of angles in a rectangle
- ○ **D** consecutive interior angles

61. In the figure, which pairs of angles are corresponding angles? Select all of the pairs in the figure. MP 6

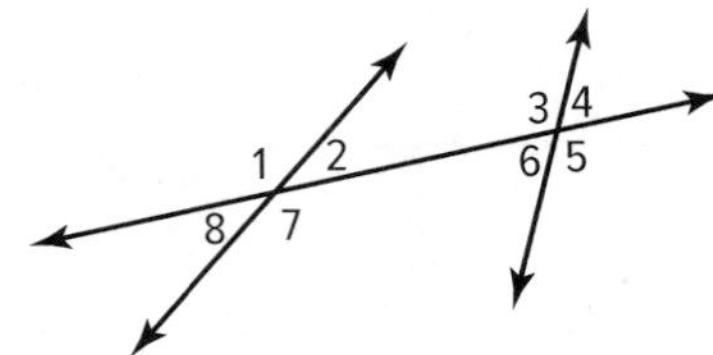

- ☐ **A** $\angle 1$ and $\angle 5$
- ☐ **B** $\angle 2$ and $\angle 8$
- ☐ **C** $\angle 3$ and $\angle 7$
- ☐ **D** $\angle 4$ and $\angle 8$
- ☐ **E** $\angle 5$ and $\angle 7$
- ☐ **F** $\angle 6$ and $\angle 8$

62. **MULTI-STEP** Mitchell is designing the parking lot for a new shopping center. The figure below shows a plan for several of the spaces in the parking lot. In the figure, lines m, n, and p are parallel to each other. MP 4

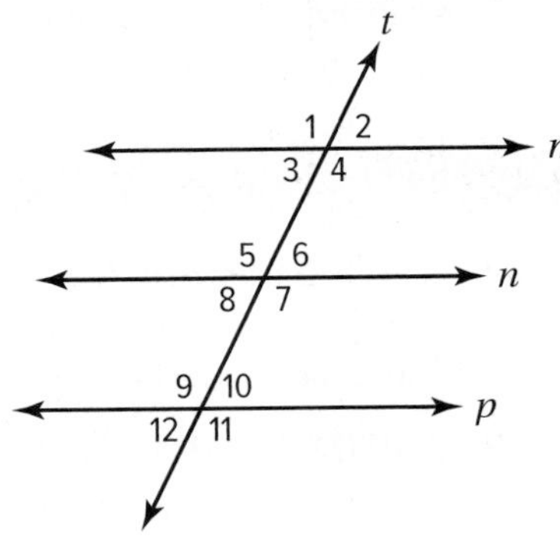

a. Classify the relationship between angle 4 and angle 9.

b. Suppose Mitchell decides that $m\angle 9 = 150$. Find $m\angle 3$.

c. Is it possible for $\angle 4$ and $\angle 10$ to be congruent? If so, what must be true about the angles? If not, why not? Explain.

d. Suppose Mitchell decides that $m\angle 2 = 90$. In this case, what could he conclude about lines t and p? Explain.

EXPLORE 2-8

Graphing Technology Lab

Investigating Slope

The rate of change of the steepness of a line is called the *slope*. Slope can be used to investigate the relationship between real-world quantities.

Mathematical Practices
MP **5** Use appropriate tools strategically.

Set Up the Lab

Work cooperatively.

- Connect a data-collection device to a graphing calculator. Place the device on a desk or table so that it can read the motion of a walker.
- Mark the floor at distances of 1 meter and 6 meters from the device.

Activity

Step 1 Have one group member stand at the 1-meter mark. When another group member presses the button to begin collecting data, the walker should walk away from the device at a slow, steady pace.

Step 2 Stop collecting data when the walker passes the 6-meter mark. Save the data as Trial 1.

Step 3 Repeat the experiment, walking more quickly. Save the data as Trial 2.

Step 4 For Trial 3, repeat the experiment by slowly walking toward the data collection device.

Step 5 Repeat the experiment, walking quickly toward the device. Save the data as Trial 4.

Analyze the Results

Work cooperatively.

1. Compare and contrast the graphs for Trials 1 and 2. How do the graphs for Trials 1 and 3 compare?

2. Use the **TRACE** feature of the calculator to find the coordinates of two points on each graph. Record the coordinates in a table like the one shown. Then use the points to find the slope of the line.

Trial	Point *A* (x_1, y_1)	Point *B* (x_2, y_2)	Slope $= \frac{y_2 - y_1}{x_2 - x_1}$
1			
2			
3			
4			

3. Compare and contrast the slopes for Trials 1 and 2. How do the slopes for Trials 1 and 2 compare to the slopes for Trials 3 and 4?

4. The slope of a line describes the rate of change of the quantities represented by the x- and y value. What is represented by the rate of change in this experiment?

5. **MAKE A CONJECTURE** What would the graph look like if you were to collect data while the walker was standing still? Use the data collection device to test your conjecture.

LESSON 8

Slope and Equations of Lines

:: Then

- You used the properties of parallel lines to determine congruent angles.

:: Now

- **1** Find the slope of a line and use slope to write the equation of a line.
- **2** Use slope to identify parallel and perpendicular lines.

:: Why?

- Ski resorts assign ratings to their ski trails according to their difficulty. A primary factor in determining this rating is a trail's steepness or *slope gradient*. A trail with a 6% or $\frac{6}{100}$ grade falls 6 feet vertically for every 100 feet traveled horizontally.

The easiest trails, labeled ●, have slopes ranging from 6% to 25%, while more difficult trails, labeled ♦ or ♦♦, have slopes of 40% or greater.

New Vocabulary
slope
slope-intercept form
point-slope form

Mathematical Practices

1 Make sense of problems and persevere in solving them.
4 Model with mathematics.
7 Look for and make use of structure.

1 Slope of a Line

The steepness or slope of a hill is described by the ratio of the hill's vertical rise to its horizontal run. In algebra, you learned that the slope of a line in the coordinate plane can be calculated using any two points on the line.

Key Concept Slope of a Line

In a coordinate plane, the **slope** of a line is the ratio of the change along the y-axis to the change along the x-axis between any two points on the line.

The slope m of a line containing two points with coordinates (x_1, y_1) and (x_2, y_2) is given by the formula

$$m = \frac{y_2 - y_1}{x_2 - x_1}, \text{ where } x_1 \neq x_2.$$

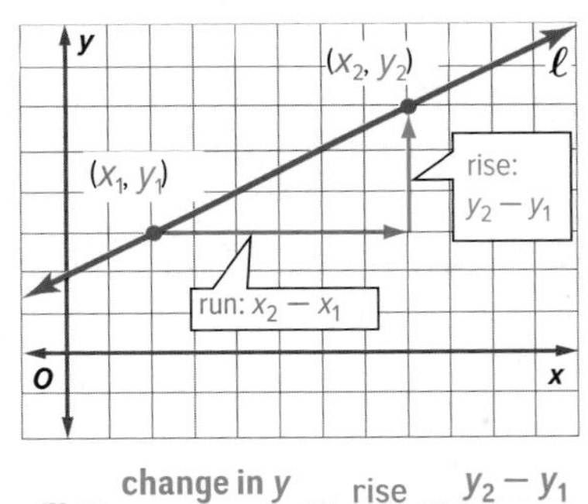

$$m = \frac{\text{change in } y}{\text{change in } x} = \frac{\text{rise}}{\text{run}} = \frac{y_2 - y_1}{x_2 - x_1}$$

Example 1 Find the Slope of a Line

Find the slope of each line.

a.

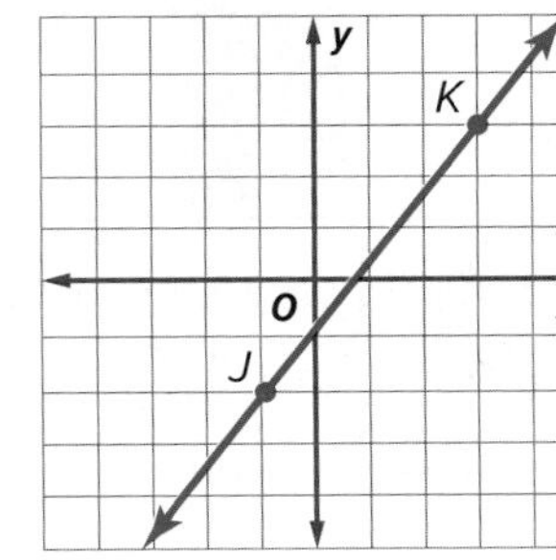

Substitute $(-1, -2)$ for (x_1, y_1) and $(3, 3)$ for (x_2, y_2).

$m = \frac{y_2 - y_1}{x_2 - x_1}$ Slope Formula

$= \frac{3 - (-2)}{3 - (-1)}$ Substitution

$= \frac{5}{4}$ Simplify.

b.

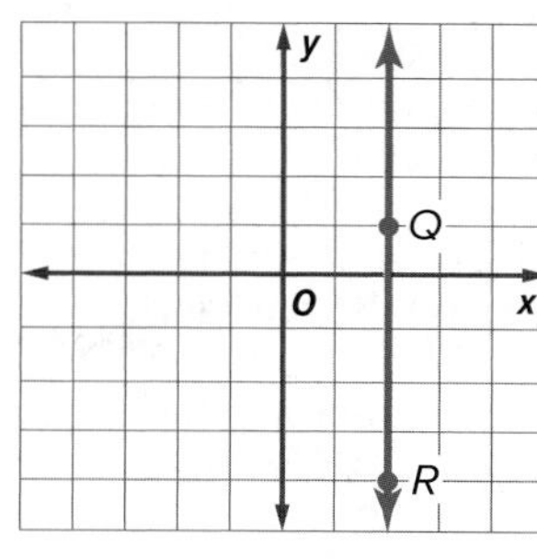

Substitute (2, 1) for (x_1, y_1) and (2, −4) for (x_2, y_2).

$m = \frac{y_2 - y_1}{x_2 - x_1}$ Slope Formula

$= \frac{-4 - 1}{2 - 2}$ Substitution

$= \frac{-5}{0}$ Simplify.

This slope is **undefined**.

Study Tip

Dividing by 0 The slope $\frac{-5}{0}$ is undefined because there is no number that you can multiply by 0 and get −5. Because this is true for any number, all numbers divided by 0 will have an undefined slope. All vertical lines have undefined slopes.

Guided Practice

1A. the line containing (6, −2) and (−3, −5)

1B. the line containing (8, −3) and (−6, −2)

1C. the line containing (4, 2) and (4, −3)

1D. the line containing (−3, 3) and (4, 3)

You may remember from algebra that an equation of a nonvertical line can be written in different but equivalent forms.

Go Online!

Slope is a key concept in mathematics. Search *slope* in ConnectED to find resources to review.

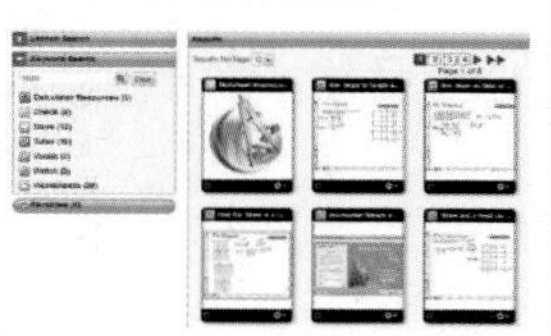

Key Concept Nonvertical Line Equations

The **slope-intercept form** of a linear equation is $y = mx + b$, where m is the slope of the line and b is the y-intercept.

slope
$y = mx + b$ $y = 3x + 8$
y-intercept

The **point-slope form** of a linear equation is $y - y_1 = m(x - x_1)$, where (x_1, y_1) is any point on the line and m is the slope of the line.

point of line (3, 5)
$y - 5 = -2(x - 3)$
slope

When given the slope and either the y-intercept or a point on a line, you can use these forms to write the equation of the line.

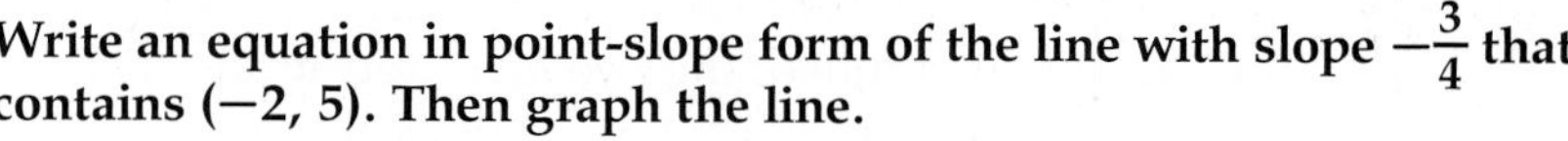

Example 2 Use Slope and a Point on the Line

Write an equation in point-slope form of the line with slope $-\frac{3}{4}$ that contains (−2, 5). Then graph the line.

$y - y_1 = m(x - x_1)$ Point-Slope form

$y - 5 = -\frac{3}{4}[x - (-2)]$ $m = -\frac{3}{4}$, $(x_1, y_1) = (-2, 5)$

$y - 5 = -\frac{3}{4}(x + 2)$ Simplify.

Watch out!

Substituting Negative Coordinates When substituting negative coordinates, use parentheses to avoid making errors with the signs.

Graph the given point (−2, 5). Use the slope $-\frac{3}{4}$ or $\frac{-3}{4}$ to find another point 3 units down and 4 units to the right. Then draw the line through these two points.

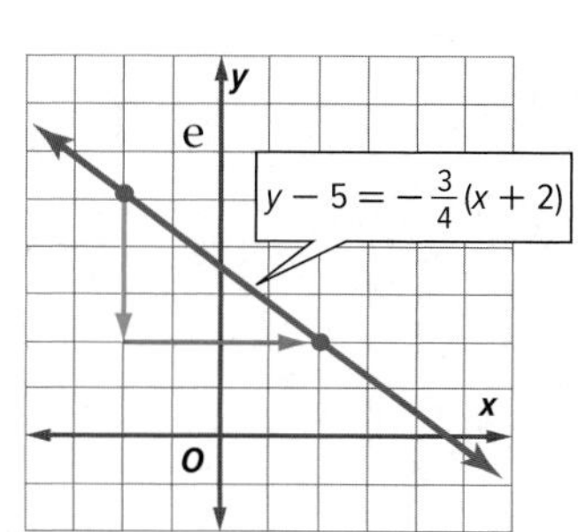

Guided Practice

2. Write an equation in point-slope form of the line with slope 4 that contains (−3, −6). Then graph the line.

2 Parallel and Perpendicular Lines

You can use the slopes of two lines to determine whether the lines are parallel or perpendicular. Lines with the same slope are parallel.

Theorems Parallel and Perpendicular Lines

2.18 Slopes of Parallel Lines Two distinct nonvertical lines have the same slope if and only if they are parallel. All vertical lines are parallel.

Example Parallel lines ℓ and m have the same slope, 4.

2.19 Slopes of Perpendicular Lines Two nonvertical lines are perpendicular if and only if the product of their slopes is −1. Vertical and horizontal lines are perpendicular.

Example line $m \perp$ line p
product of slopes $= 4 \cdot -\frac{1}{4}$ or -1

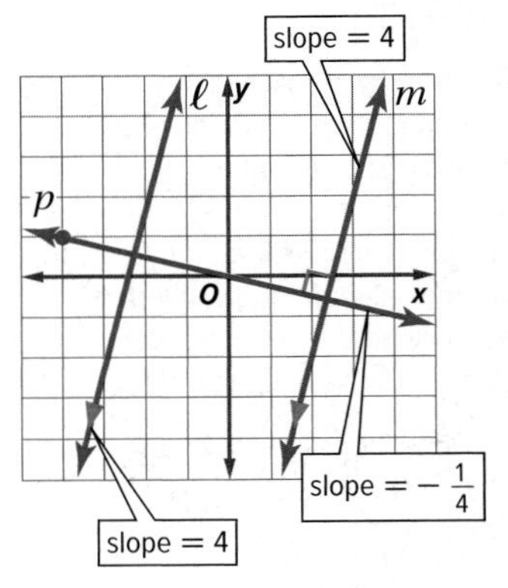

You will prove Theorems 2.18 and 2.19 in Extend 7-4.

Example 3 Determine Line Relationships

Determine whether $\overleftrightarrow{AB}$ and $\overleftrightarrow{CD}$ are *parallel, perpendicular,* or *neither* for $A(1, 1)$, $B(-1, -5)$, $C(3, 2)$, and $D(6, 1)$. Graph each line to verify your answer.

Step 1 Find the slope of each line.

slope of $\overleftrightarrow{AB} = \frac{-5-1}{-1-1} = \frac{-6}{-2}$ or 3 slope of $\overleftrightarrow{CD} = \frac{1-2}{6-3}$ or $\frac{-1}{3}$

Step 2 Determine the relationship, if any, between the lines.

The two lines do not have the same slope, so they are *not* parallel. To determine if the lines are perpendicular, find the product of their slopes.

$3\left(-\frac{1}{3}\right) = -1$ Product of slopes for $\overleftrightarrow{AB}$ and $\overleftrightarrow{CD}$

Because the product of their slopes is −1, $\overleftrightarrow{AB}$ is perpendicular to $\overleftrightarrow{CD}$.

CHECK When graphed, the two lines appear to intersect and form four right angles. ✔

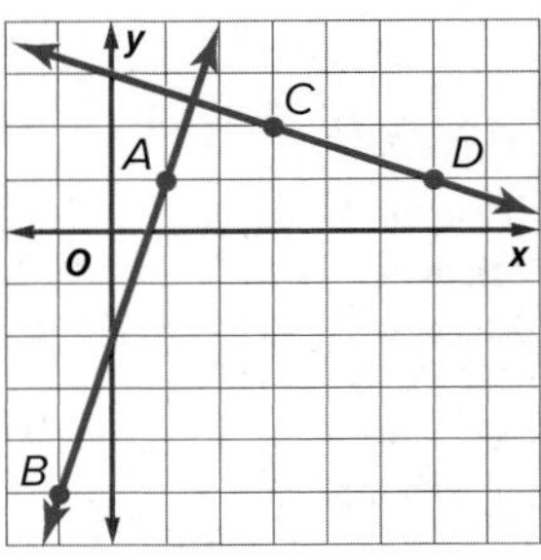

Study Tip

MP Structure If a line ℓ has a slope of $\frac{a}{b}$, then the slope of a line perpendicular to line ℓ is the opposite reciprocal, $-\frac{b}{a}$, because $\frac{a}{b}\left(-\frac{b}{a}\right) = -1$.

Guided Practice

Determine whether $\overleftrightarrow{AB}$ and $\overleftrightarrow{CD}$ are *parallel, perpendicular,* or *neither.* Graph each line to verify your answer.

3A. $A(14, 13)$, $B(-11, 0)$, $C(-3, 7)$, $D(-4, -5)$

3B. $A(3, 6)$, $B(-9, 2)$, $C(5, 4)$, $D(2, 3)$

Real-World Example 4 Write Equations of Parallel or Perpendicular Lines

MAPS On a map, Basin Street is represented by the line $y = -3x + 2$. Waller Street is perpendicular to Basin Street and passes through (4, 0). Write an equation in slope-intercept form for Waller Street.

The slope of $y = -3x + 2$ is -3, so the slope of a line perpendicular to it is $\frac{1}{3}$.

$y = mx + b$	Slope-Intercept form
$0 = \frac{1}{3}(4) + b$	$m = \frac{1}{3}$ and $(x, y) = (4, 0)$
$0 = \frac{4}{3} + b$	Simplify.
$-\frac{4}{3} = b$	Subtract $\frac{4}{3}$ from each side.

So, the equation for Waller Street is $y = \frac{1}{3}x + \left(-\frac{4}{3}\right)$ or $y = \frac{1}{3}x - 1\frac{1}{3}$.

Guided Practice

4. MURALS An artist is using a coordinate plane to plan a mural. A red line on the mural is represented by the equation $y = -\frac{3}{4}x + 3$. A blue line is parallel to the red line and contains $(-3, 6)$. Write an equation in slope-intercept form for the blue line.

Check Your Understanding

 = Step-by-Step Solutions begin on page R13.

Example 1 **Find the slope of each line.**

1. 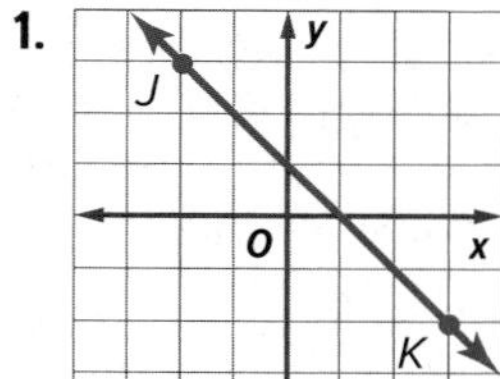

2. 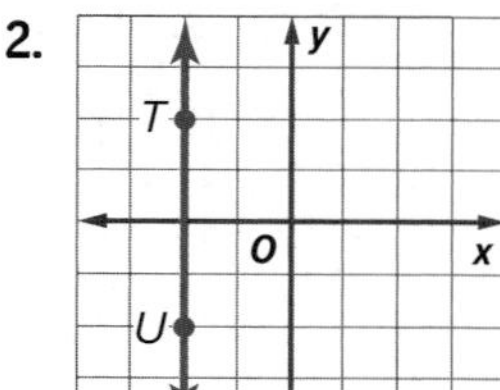

3. 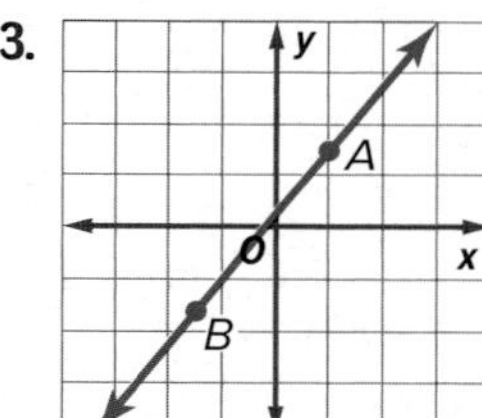

Example 2 **Write an equation in point-slope form of the line having the given slope that contains the given point. Then graph the line.**

4. $m = 5, (3, -2)$ **5.** $m = \frac{1}{4}, (-2, -3)$ **6.** $m = -4.25, (-4, 6)$

Example 3 **Determine whether $\overleftrightarrow{WX}$ and $\overleftrightarrow{YZ}$ are *parallel, perpendicular,* or *neither.* Graph each line to verify your answer.**

7. $W(2, 4), X(4, 5), Y(4, 1), Z(8, -7)$

8. $W(1, 3), X(-2, -5), Y(-6, -2), Z(8, 3)$

9. $W(-7, 6), X(-6, 9), Y(6, 3), Z(3, -6)$

10. $W(1, -3), X(0, 2), Y(-2, 0), Z(8, 2)$

Example 4

11. QUILTS Tamiko uses a coordinate plane to help her cut out patches for a quilt. Her first cut is along the line $y = -2x + 6$. Her second cut will be perpendicular to the first and will pass through (3, 2). Write an equation in slope-intercept form for the second cut.

12. A line segment has endpoints at $A(-2, 4)$ and $B(-4, 2)$. Write an equation in slope-intercept form for a line that bisects this segment that is parallel to the line $y = -2x + 5$.

Practice and Problem Solving

Extra Practice is on page R2.

Example 1 **Find the slope of each line.**

13

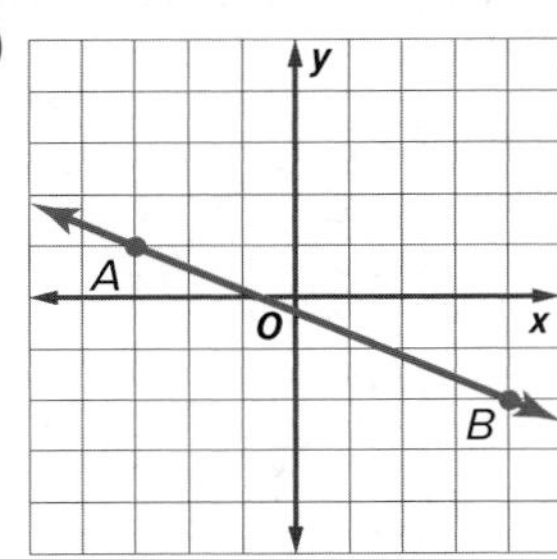

14.

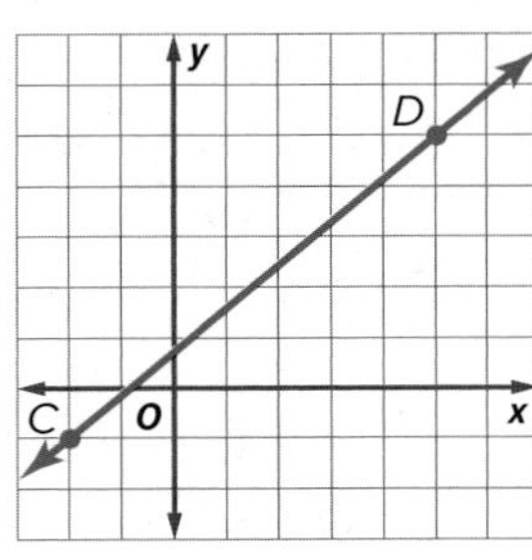

15.

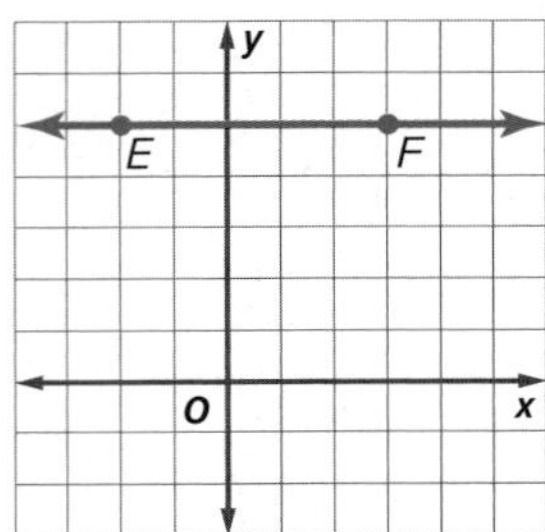

Determine the slope of the line that contains the given points.

16. $C(3, 1), D(-2, 1)$

17. $E(5, -1), F(2, -4)$

18. $G(-4, 3), H(-4, 7)$

19. $J(7, -3), K(-8, -3)$

20. $L(8, -3), M(-4, -12)$

21. $P(-3, -5), Q(-3, -1)$

22. $R(2, -6), S(-6, 5)$

23. $T(-6, -11), V(-12, -10)$

Example 2 **Write an equation in point-slope form of the line having the given slope that contains the given point. Then graph the line.**

24. $m = \frac{5}{7}, (-2, -5)$

25. $m = 4, (-4, 8)$

26. $m = -7, (1, 9)$

27 $m = 2, (3, 11)$

28. $m = -\frac{4}{5}, (-3, -6)$

29. $m = -2.4, (14, -12)$

Example 3 **Determine whether $\overleftrightarrow{AB}$ and $\overleftrightarrow{CD}$ are *parallel, perpendicular,* or *neither.* Graph each line to verify your answer.**

30. $A(1, 5), B(4, 4), C(9, -10), D(-6, -5)$

31. $A(-6, -9), B(8, 19), C(0, -4), D(2, 0)$

32. $A(4, 2), B(-3, 1), C(6, 0), D(-10, 8)$

33. $A(8, -2), B(4, -1), C(3, 11), D(-2, -9)$

34. $A(8, 4), B(4, 3), C(4, -9), D(2, -1)$

35. $A(4, -2), B(-2, -8), C(4, 6), D(8, 5)$

Example 4 **36. MARCHING BANDS** A band director uses a coordinate plane to plan a show for a football game. During the show, the drummers will march along the line $y = -5x - 8$. The trumpet players will march along a perpendicular line that passes through $(-2, 2)$. Write an equation in slope-intercept form for the path of the trumpet players.

Name the line(s) on the graph shown that match each description.

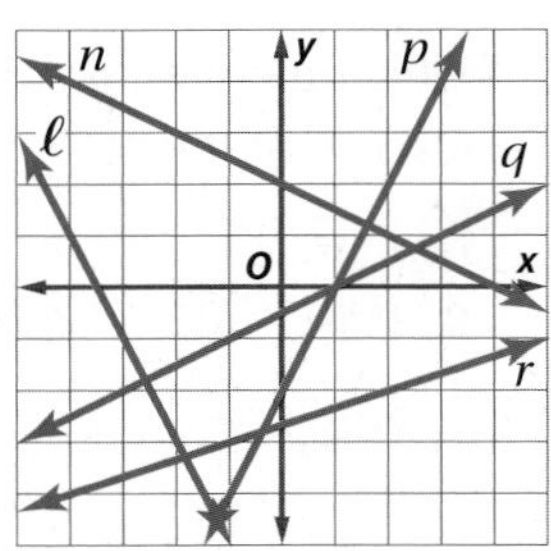

37. parallel to $y = 2x - 3$

38. perpendicular to $y = \frac{1}{2}x + 7$

39. intersecting, but not perpendicular to $y = \frac{1}{2}x - 5$

Determine whether the lines are *parallel, perpendicular,* or *neither.*

40. $y = 2x + 4, y = 2x - 10$

41. $y = -\frac{1}{2}x - 12, y = 2x + 7$

42. $y - 4 = 3(x + 5), y + 3 = -\frac{1}{3}(x + 1)$

43. $y - 3 = 6(x + 2), y + 3 = -\frac{1}{3}(x - 4)$

44. Write an equation in slope-intercept form for a line containing (4, 2) that is parallel to the line $y - 2 = 3(x + 7)$.

45. Write an equation for a line containing (−8, 12) that is perpendicular to the line containing the points (3, 2) and (−7, 2).

46. Write an equation in slope-intercept form for a line containing (5, 3) that is parallel to the line $y + 11 = \frac{1}{2}(4x + 6)$.

47. **SKATING** A skating rink charges $5.50 to rent skates and $4 per hour to use the rink.

a. Write a function $C(t)$ that gives the cost of renting skates and using the rink for t hours.

b. The manager of the rink decides to increase the price of renting skates to $6.25. Write a function $D(t)$ that gives the new cost of renting skates and using the rink for t hours.

c. How are the graphs of $C(t)$ and $D(t)$ related? How do the functions you wrote justify your answer?

48. Line ℓ passes through the origin and through the point (a, b). Write the equation of the line that passes through the origin that is perpendicular to line ℓ.

49. Without graphing, what type of geometric shape is enclosed by the lines $x - 2y = -6$, $2x + y = 13$, $x - 2y = 4$, and $2x + y = -2$? Explain.

50. **MAPS** On a map, Main Street is represented by the line $y = 3x + 2$. Grand Avenue is perpendicular to Main Street and passes through the point (4, 4). At what point do Main Street and Grand Avenue intersect?

H.O.T. Problems Use Higher-Order Thinking Skills

51. **CHALLENGE** Find the value of n so that the line perpendicular to the line with the equation $-2y + 4 = 6x + 8$ passes through the points at $(n, -4)$ and $(2, -8)$.

52. MP **REASONING** Determine whether the points at (−2, 2), (2, 5), and (6, 8) are collinear. Justify your answer.

53. **OPEN ENDED** Write equations for a pair of perpendicular lines that intersect at the point (−3, −7).

54. MP **CRITIQUE ARGUMENTS** Mark and Josefina wrote an equation of a line with slope −5 that passes through the point (−2, 4). Is either of them correct? Explain your reasoning.

Mark	Josefina
$y - 4 = -5(x - (-2))$	$y - 4 = -5(x - (-2))$
$y - 4 = -5(x + 2)$	$y - 4 = -5(x + 2)$
$y - 4 = -5x - 10$	
$y = -5x - 6$	

55. **WRITING IN MATH** When is it easier to use the point-slope form to write an equation of a line, and when is it easier to use the slope-intercept form?

Preparing for Assessment

56. What is an equation, in slope-intercept form, for a line parallel to $y = -x + 11$ through the point $(-3, 5)$? MP 1

○ **A** $y = x + 8$

○ **B** $y = -x - 8$

○ **C** $y = -x + 2$

○ **D** $y = -x + 5$

57. Which equation represents a line through (2, 2) that is perpendicular to the line through (−3, 9) and (4, −5)? MP 1

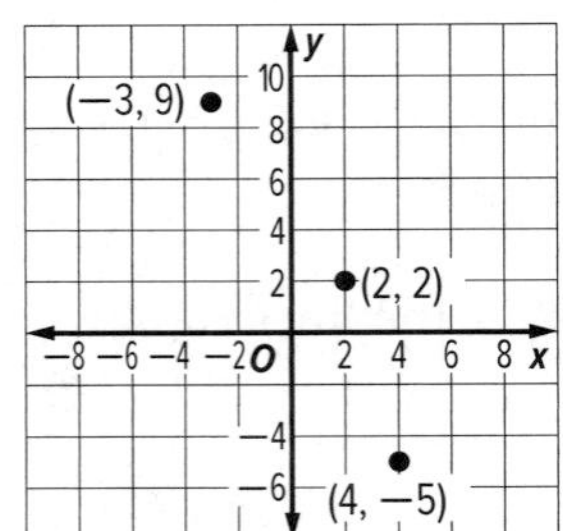

○ **A** $y = -2x + 6$

○ **B** $y = -0.5x + 3$

○ **C** $y = 0.5x + 1$

○ **D** $y = 2x - 2$

58. What is an equation for a line perpendicular to $y = -\frac{1}{3}x + 3$ that contains (6, 16)? MP 1

○ **A** $y = -3x + 34$

○ **B** $y = -\frac{1}{3}x + 18$

○ **C** $y = \frac{1}{3}x + 14$

○ **D** $y = 3x - 2$

59. A line goes through the points (−6, −2) and (2, 4). MP 1

a. What is the slope of a line that is parallel to the given line?

b. What is the slope of a line that is perpendicular to the given line?

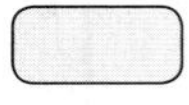

60. Which lines are parallel to line p? Select all lines that apply. MP 2

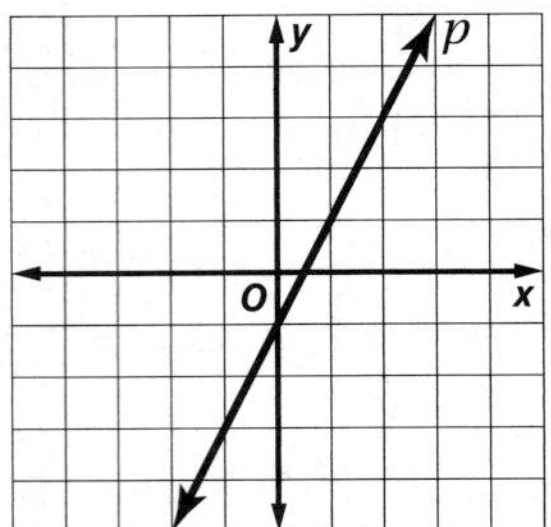

☐ **A** $y = -\frac{1}{2}x + 3$

☐ **B** $y = 2x + 1$

☐ **C** $-2x + y = -4$

☐ **D** $2y = -x + 8$

☐ **E** $y = \frac{1}{2}x - 5$

☐ **F** $2x + y = 6$

61. **MULTI-STEP** Ariella is using a coordinate plane to program a video game. In the game, a mouse moves along a path represented by line m in the graph below. MP 4

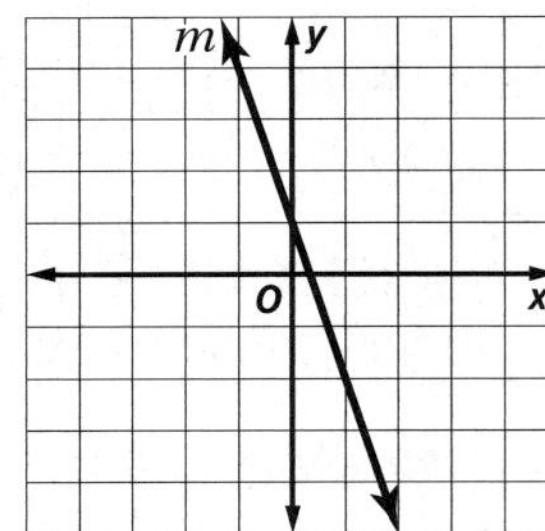

a. Write the equation of line m.

b. In the game, a cat moves along a line through (−6, −1) that is perpendicular to the path of the mouse. What is the equation of the line that represents the path of the cat?

c. If the cat catches the mouse along its path, at what point will this occur? Explain your answer.

d. Ariella decides to shift the path of the mouse so that the new path is parallel to the old path and so that it passes through (−2, 2). What is the new equation of the path?

EXTEND 2-8

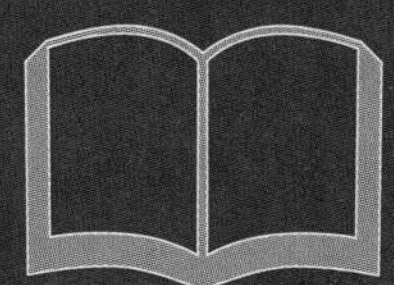

Geometry Lab

Equations of Perpendicular Bisectors

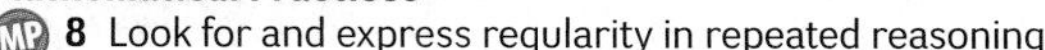

You can apply what you have learned about slope and equations of lines to geometric figures on a plane.

Mathematical Practices

MP **8** Look for and express regularity in repeated reasoning.

Activity

Find the equation of a line that is a perpendicular bisector of a segment *AB* with endpoints *A*(–3, 3) and *B*(4, 0).

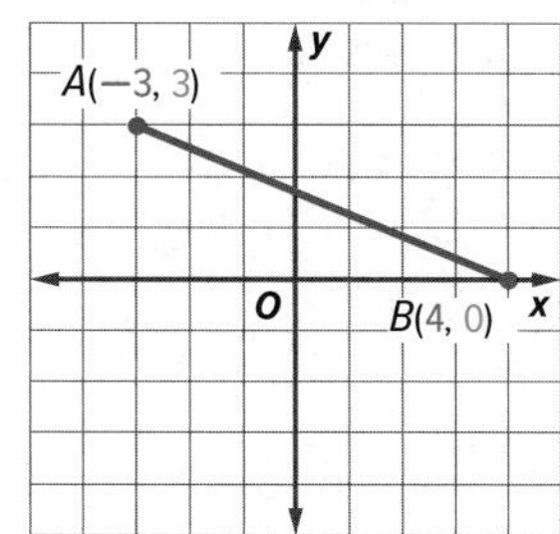

Step 1 A segment bisector contains the midpoint of the segment. Use the Midpoint Formula to find the midpoint *M* of $\overline{AB}$.

$$M\left(\frac{x_1 + x_2}{2}, \frac{y_1 + y_2}{2}\right) = M\left(\frac{-3 + 4}{2}, \frac{3 + 0}{2}\right)$$

$$= M\left(\frac{1}{2}, \frac{3}{2}\right)$$

Step 2 A perpendicular bisector is perpendicular to the segment through the midpoint. In order to find the slope of the bisector, first find the slope of $\overline{AB}$.

$m = \frac{y_2 - y_1}{x_2 - x_1}$ Slope Formula

$= \frac{0 - 3}{4 - (-3)}$ $x_1 = -3, x_2 = 4, y_1 = 3, y_2 = 0$

$= -\frac{3}{7}$ Simplify.

Step 3 Now use the point-slope form to write the equation of the line. The slope of the bisector is $\frac{7}{3}$ because $-\frac{3}{7}\left(\frac{7}{3}\right) = -1$

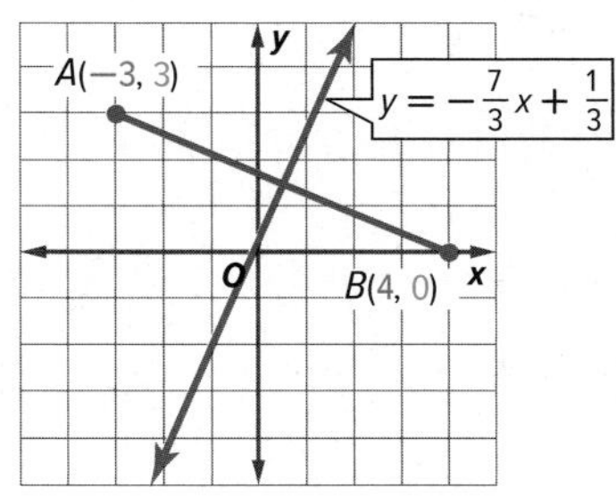

$y - y_1 = m(x - x_1)$ Point-slope form

$y - \frac{3}{2} = \frac{7}{3}\left(x - \frac{1}{2}\right)$ $m = \frac{7}{3}, (x_1, y_1) = \left(\frac{1}{2}, \frac{3}{2}\right)$

$y - \frac{3}{2} = \frac{7}{3}x - \frac{7}{6}$ Distributive Property

$y = \frac{7}{3}x + \frac{1}{3}$ Add $\frac{3}{2}$ to each side.

Exercises

Find the equation of a line that is the perpendicular bisector of $\overline{PQ}$ for the given endpoints.

1. *P*(5, 2), *Q*(7, 4)
2. *P*(–3, 9), *Q*(–1, 5)
3. *P*(–6, –1), *Q*(8, 7)
4. *P*(–2, 1), *Q*(0, –3)
5. *P*(0, 1.6), *Q*(0.5, 2.1)
6. *P*(–7, 3), *Q*(5, 3)
7. **CHALLENGE** Find the equations of the lines that contain the sides of △*XYZ* with vertices *X*(–2, 0), *Y*(1, 3), and *Z*(3, –1).

LESSON 9

Proving Lines Parallel

:: Then

- You used slopes to identify parallel and perpendicular lines.

:: Now

1. Recognize angle pairs that occur with parallel lines.
2. Prove that two lines are parallel.

:: Why?

- When you see a roller coaster track, the two sides of the track are always the same distance apart, even though the track curves and turns. The tracks are carefully constructed to be parallel at all points so that the car is secure on the track.

 **Mathematical Practices**

1 Make sense of problems and persevere in solving them.

3 Construct viable arguments and critique the reasoning of others.

1 Identify Parallel Lines

The two sides of the track of a roller coaster are parallel, and all of the supports along the track are also parallel. Each of the angles formed between the track and the supports are corresponding angles. We have learned that corresponding angles are congruent when lines are parallel. The converse of this relationship is also true.

Postulate 2.13 Converse of Corresponding Angles Postulate

If two lines are cut by a transversal so that corresponding angles are congruent, then the lines are parallel.

Example If $\angle 1 \cong \angle 3$, $\angle 2 \cong \angle 4$, $\angle 5 \cong \angle 7$, $\angle 6 \cong \angle 8$, then $a \parallel b$.

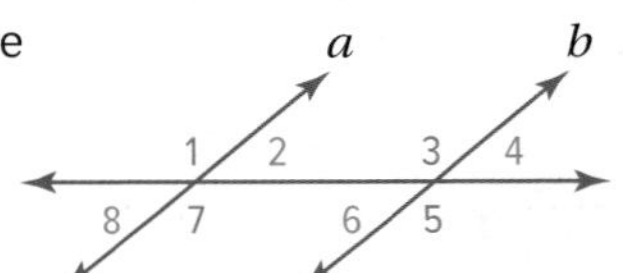

Construction Parallel Line Through a Point Not on the Line

Step 1 Use a straightedge to draw $\overleftrightarrow{AB}$. Draw a point C that is not on $\overleftrightarrow{AB}$. Draw $\overleftrightarrow{CA}$.

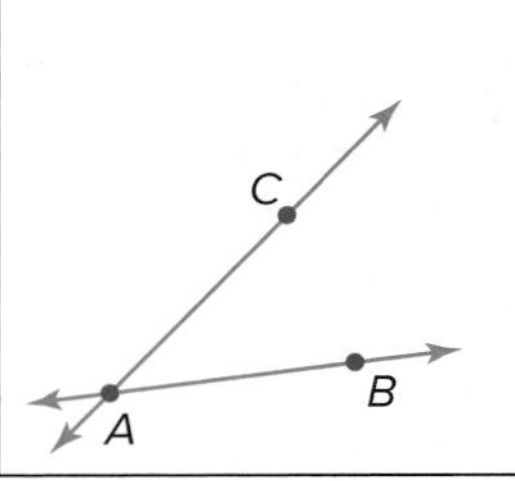

Step 2 Copy $\angle CAB$ so that C is the vertex of the new angle. Label the intersection points D and E.

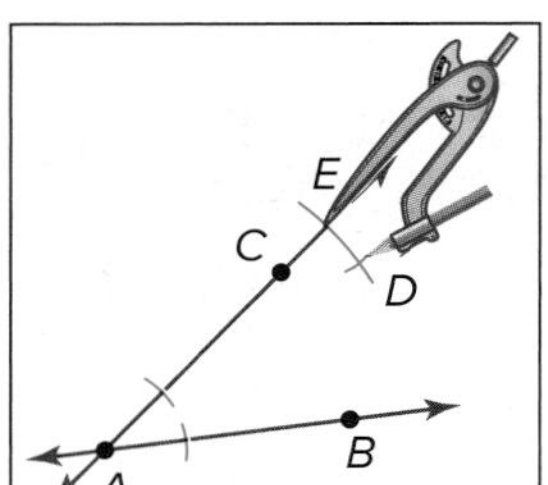

Step 3 Draw CD.

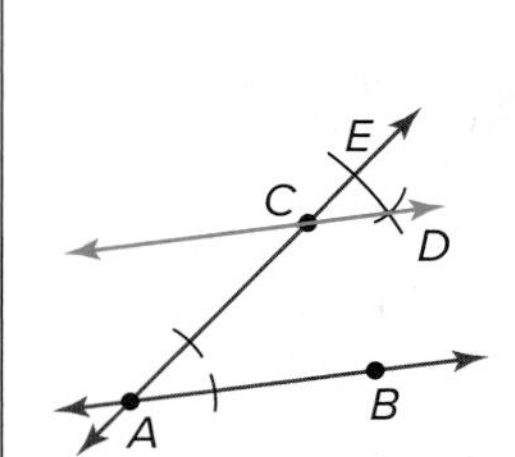

Make a conjecture about the relationship between $\overline{AB}$ and $\overline{CD}$.

The construction establishes that there is *at least* one line through C that is parallel to $\overleftrightarrow{AB}$. The following postulate guarantees that this line is the *only* one.

Study Tip

Euclid's Postulates The father of modern geometry, Euclid (c. 300 B.C.) realized that only a few postulates were needed to prove the theorems in his day. Postulate 2.14 is one of Euclid's five original postulates. Postulate 2.1 and Theorem 2.10 also reflect two of Euclid's postulates.

Postulate 2.14 Parallel Postulate

If given a line and a point not on the line, then there exists exactly one line through the point that is parallel to the given line.

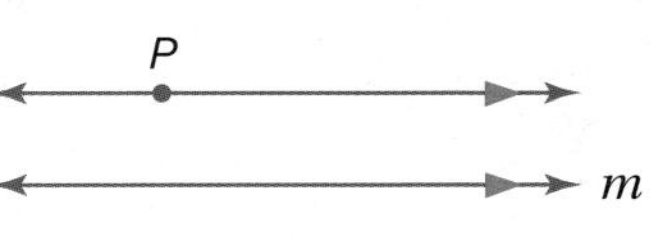

Parallel lines that are cut by a transversal create several pairs of congruent angles. These special angle pairs can also be used to prove that two lines are parallel.

Theorems Proving Lines Parallel

2.20 Alternate Exterior Angles Converse If two lines in a plane are cut by a transversal so that a pair of alternate exterior angles is congruent, then the two lines are parallel.

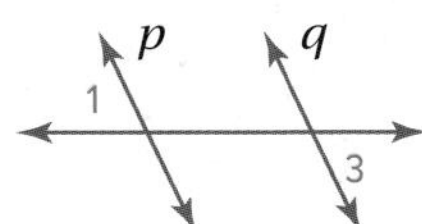

If $\angle 1 \cong \angle 3$, then $p \parallel q$.

2.21 Consecutive Interior Angles Converse If two lines in a plane are cut by a transversal so that a pair of consecutive interior angles is supplementary, then the lines are parallel.

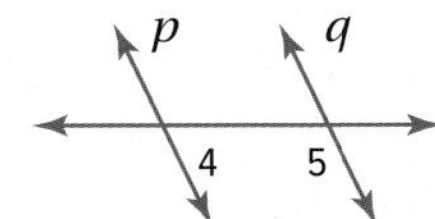

If $m\angle 4 + m\angle 5 = 180$, then $p \parallel q$.

2.22 Alternate Interior Angles Converse If two lines in a plane are cut by a transversal so that a pair of alternate interior angles is congruent, then the lines are parallel.

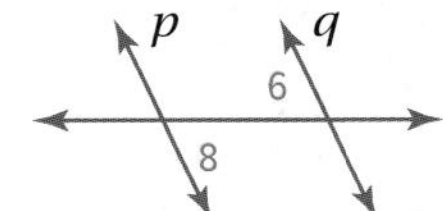

If $\angle 6 \cong \angle 8$, then $p \parallel q$.

2.23 Perpendicular Transversal Converse In a plane, if two lines are perpendicular to the same line, then they are parallel.

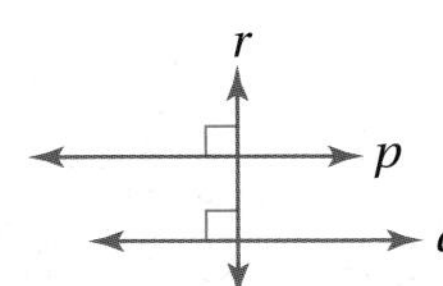

If $p \perp r$ and $q \perp r$, then $p \parallel q$.

You will prove Theorems 2.20, 2.21, 2.22, 2.23 in Exercises 6, 23, 31, and 30, respectively.

Go Online!

When solving problems involving geometric figures, it may be helpful to mark congruent angles with the same color highlighting. Log into your eStudent Edition to use the highlighting tools with your text.

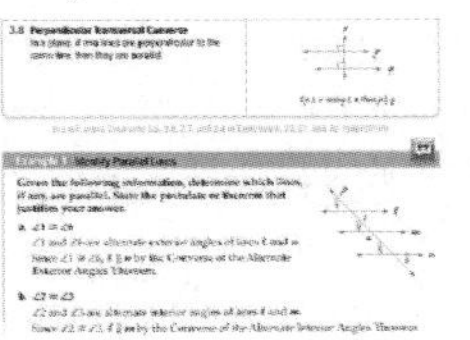

Example 1 Identify Parallel Lines

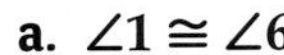

Given the following information, determine which lines, if any, are parallel. State the postulate or theorem that justifies your answer.

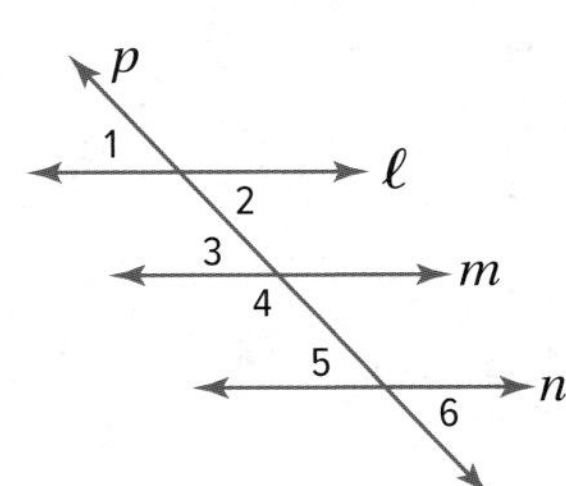

a. $\angle 1 \cong \angle 6$

$\angle 1$ and $\angle 6$ are alternate exterior angles of lines ℓ and n.

Because $\angle 1 \cong \angle 6$, $\ell \parallel n$ by the Converse of the Alternate Exterior Angles Theorem.

b. $\angle 2 \cong \angle 3$

$\angle 2$ and $\angle 3$ are alternate interior angles of lines ℓ and m.

Because $\angle 2 \cong \angle 3$, $\ell \parallel m$ by the Converse of the Alternate Interior Angles Theorem.

Guided Practice

1A. $\angle 2 \cong \angle 8$

1B. $\angle 3 \cong \angle 11$

1C. $\angle 12 \cong \angle 14$

1D. $\angle 1 \cong \angle 15$

1E. $m\angle 8 + m\angle 13 = 180$

1F. $\angle 8 \cong \angle 6$

Angle relationships can be used to solve problems involving unknown values.

Example 2 Use Angle Relationships

Find $m\angle MRQ$ so that $a \parallel b$. Identify the postulate or theorem you used.

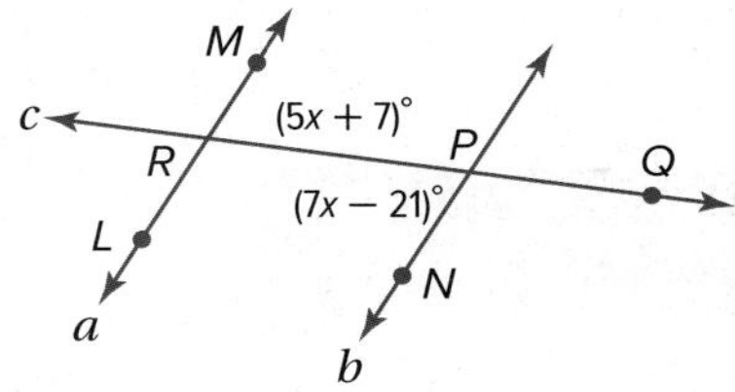

Read the Item

From the figure, you know that $m\angle MRQ = 5x + 7$ and $m\angle RPN = 7x - 21$. You are asked to find the measure of $\angle MRQ$ so that $a \parallel b$.

Solve the Item

$\angle MRQ$ and $\angle RPN$ are alternate interior angles. For lines a and b to be parallel, alternate interior angles must be congruent by the Alternate Interior Angles Converse, so $\angle MRQ \cong \angle RPN$. By the definition of congruence, $m\angle MRQ = m\angle RPN$. Substitute the given angle measures into this equation and solve for x.

$m\angle MRQ = m\angle RPN$	Alternate interior angles
$5x + 7 = 7x - 21$	Substitution
$7 = 2x - 21$	Subtract $5x$ from each side.
$28 = 2x$	Add 21 to each side.
$14 = x$	Divide each side by 2.

Now, use the value of x to find $\angle MRQ$.

$m\angle MRQ = 5x + 7$	Substitution
$= 5(14) + 7$ or 77	Simplify.

CHECK Check your answer by using the value of x to find $m\angle RPN$.

$m\angle RPN = 7x - 21$

$= 7(14) - 21$ or 77 ✓

Because $m\angle MRQ = m\angle RPN$, $\angle MRQ \cong \angle RPN$ and $a \parallel b$. ✓

Guided Practice

2. Find y so that $e \parallel f$. Show your work.

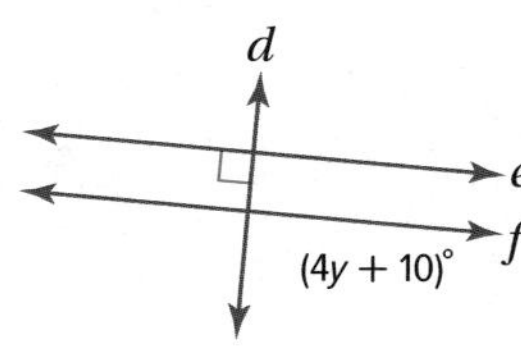

> **Study Tip**
> **Proving Lines Parallel** When two parallel lines are cut by a transversal, the angle pairs formed are either congruent or supplementary. When a pair of lines forms angles that do not meet this criterion, the lines cannot possibly be parallel.

2 Prove Lines Parallel

The angle pair relationships formed by a transversal can be used to prove that two lines are parallel.

Real-World Example 3 Prove Lines Parallel

HOME FURNISHINGS In the ladder shown, each rung is perpendicular to the two rails. Is it possible to prove that the two rails are parallel and that all of the rungs are parallel? If so, explain how. If not, explain why not.

Because both rails are perpendicular to each rung, the rails are parallel by the Perpendicular Transversal Converse. Because any pair of rungs is perpendicular to the rails, they are also parallel.

Guided Practice

3. **ROWING** In order to move in a straight line with maximum efficiency, rower's oars should be parallel. Refer to the photo at the right. Is it possible to prove that any of the oars are parallel? If so, explain how. If not, explain why not.

Check Your Understanding

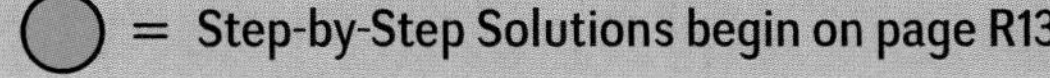

Example 1 Given the following information, determine which lines, if any, are parallel. State the postulate or theorem that justifies your answer.

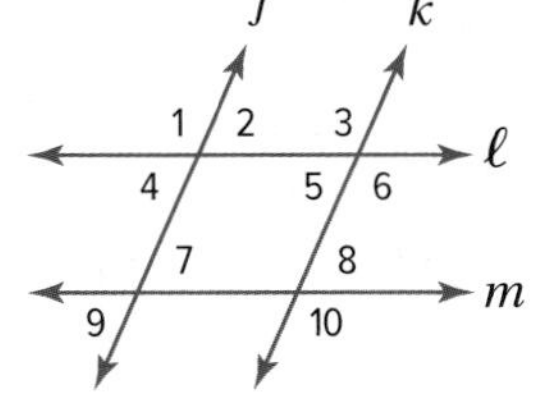

1. $\angle 1 \cong \angle 3$
2. $\angle 2 \cong \angle 5$
3. $\angle 3 \cong \angle 10$
4. $m\angle 6 + m\angle 8 = 180$

Example 2

5. Find x so that $m \parallel n$. Identify the postulate or theorem you used.

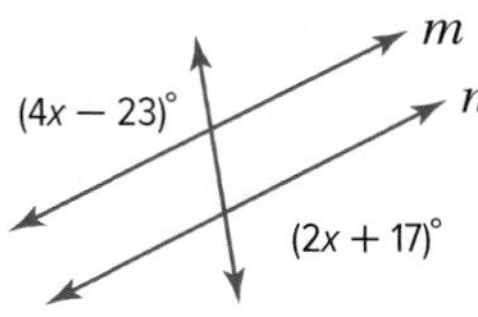

Example 3

6. **PROOF** Copy and complete the proof of Theorem 2.20.

Given: $\angle 1 \cong \angle 2$

Prove: $\ell \parallel m$

Proof:

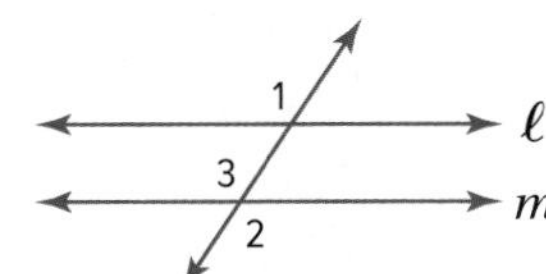

Statements	Reasons
a. $\angle 1 \cong \angle 2$	a. Given
b. $\angle 2 \cong \angle 3$	b. ___?___
c. $\angle 1 \cong \angle 3$	c. Transitive Property
d. ___?___	d. ___?___

7. **RECREATION** Is it possible to prove that the backrest and footrest of the lounging beach chair are parallel? If so, explain how. If not, explain why not.

Practice and Problem Solving

Extra Practice is on page R2.

Example 1

Given the following information, determine which lines, if any, are parallel. State the postulate or theorem that justifies your answer.

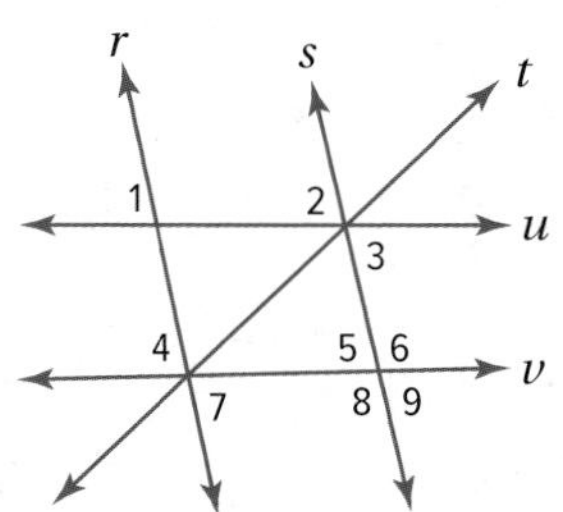

8. $\angle 1 \cong \angle 2$
9. $\angle 2 \cong \angle 9$
10. $\angle 5 \cong \angle 7$
11. $m\angle 7 + m\angle 8 = 180$
12. $m\angle 3 + m\angle 6 = 180$
13. $\angle 3 \cong \angle 5$
14. $\angle 3 \cong \angle 7$
15. $\angle 4 \cong \angle 5$

Example 2

Find x so that $m \parallel n$. Identify the postulate or theorem you used.

16.

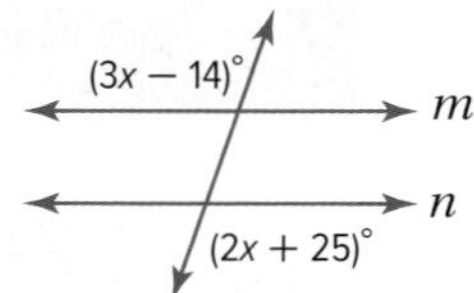

17.

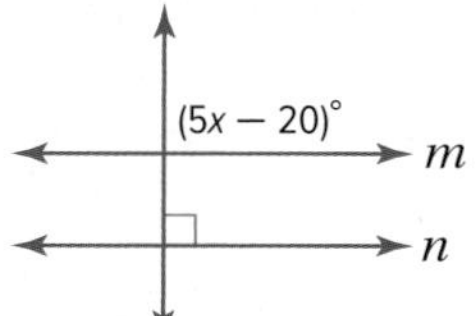

18.

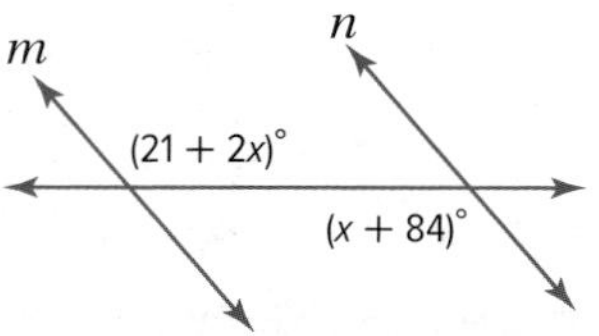

19.

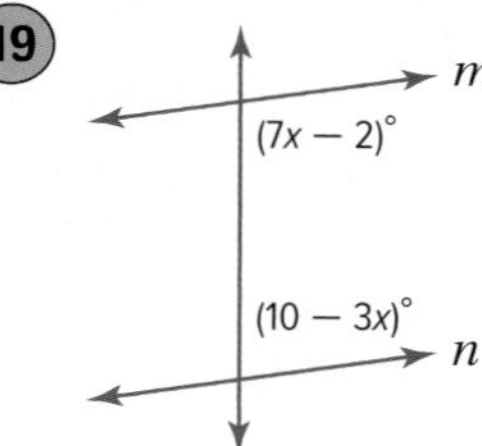

20.

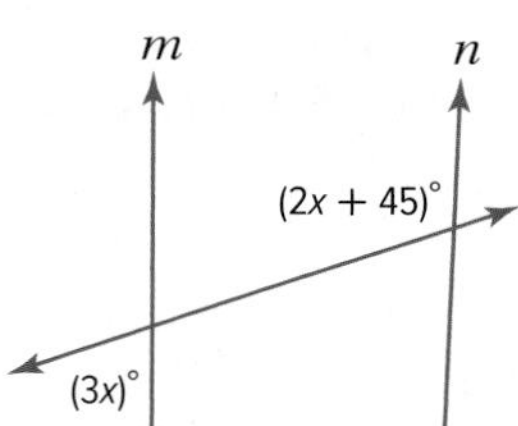

21.

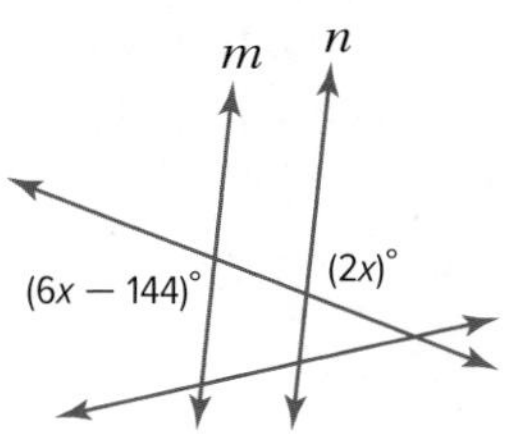

22. **SENSE-MAKING** Wooden picture frames are often constructed using a miter box or miter saw. These tools allow you to cut at an angle of a given size. If each of the four pieces of framing material is cut at a 45° angle, will the sides of the frame be parallel? Explain your reasoning.

Example 3

23. **PROOF** Copy and complete the proof of Theorem 2.21.

Given: $\angle 1$ and $\angle 2$ are supplementary.

Prove: $\ell \parallel m$

Proof:

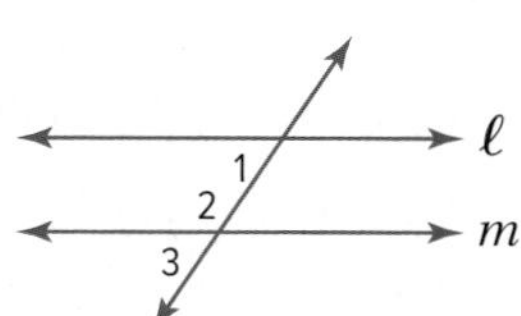

Statements	Reasons
a. ___?___	a. Given
b. $\angle 2$ and $\angle 3$ form a linear pair.	b. ___?___
c. ___?___	c. ___?___
d. $\angle 1 \cong \angle 3$	d. ___?___
e. $\ell \parallel m$	e. ___?___

24. **CRAFTS** Jacqui is making a stained glass piece. She cuts the top and bottom pieces at a 30° angle. If the corners are right angles, explain how Jacqui knows that each pair of opposite sides are parallel.

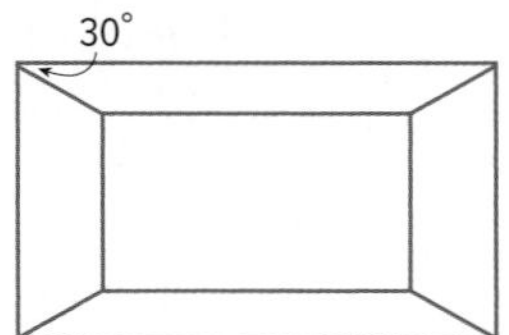

PROOF Write a two-column proof for each of the following.

25. **Given:** $\angle 1 \cong \angle 3$

$\overline{AC} \parallel \overline{BD}$

Prove: $\overline{AB} \parallel \overline{CD}$

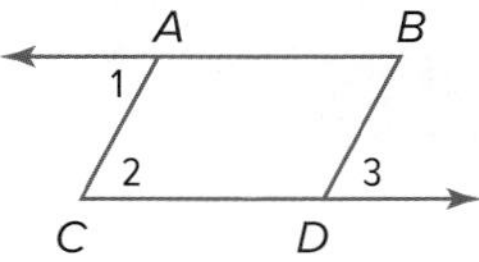

26. **Given:** $\overline{WX} \parallel \overline{YZ}$

$\angle 2 \cong \angle 3$

Prove: $\overline{WY} \parallel \overline{XZ}$

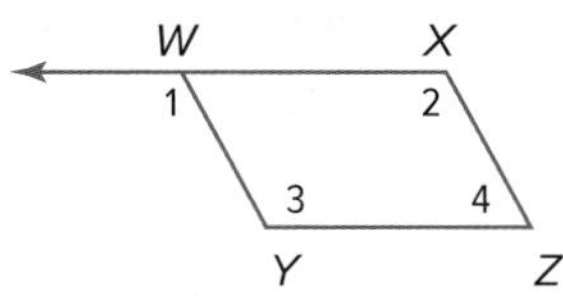

27. **Given:** $\angle ABC \cong \angle ADC$

$m\angle A + m\angle ABC = 180$

Prove: $\overline{AB} \parallel \overline{CD}$

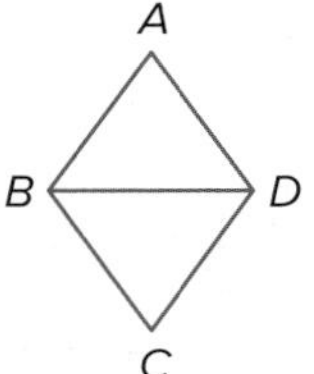

28. **Given:** $\angle 1 \cong \angle 2$

$\overline{LJ} \perp \overline{ML}$

Prove: $\overline{KM} \perp \overline{ML}$

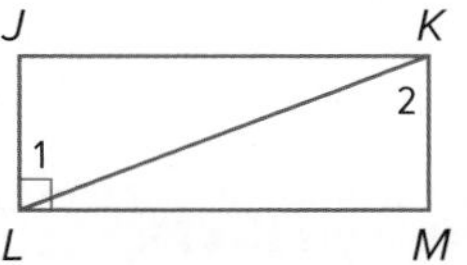

29. **MAILBOXES** Mail slots are used to make the organization and distribution of mail easier. In the mail slots shown, each slot is perpendicular to each of the sides. Explain why you can conclude that the slots are parallel.

30. **PROOF** Write a paragraph proof of Theorem 2.23.

31. **PROOF** Write a two-column proof of Theorem 2.22.

32. **MP REASONING** Based upon the information given in the photo of the staircase at the right, what is the relationship between each step? Explain your answer.

Determine whether lines *r* and *s* are parallel. Justify your answer.

33.

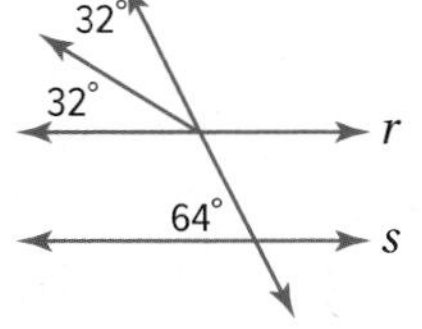

34.

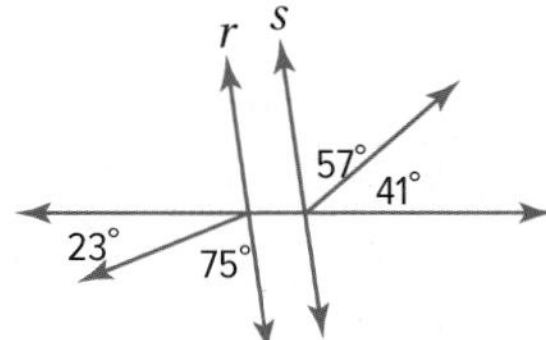

35.

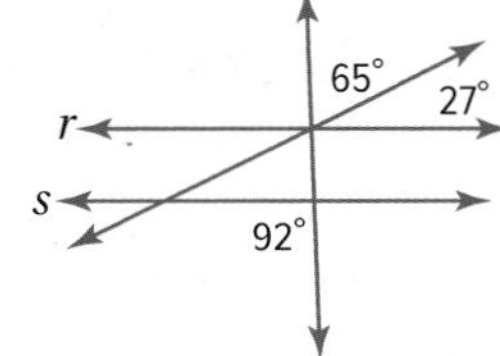

(t to b) Corey Hochachka/Design Pics Inc/Alamy, mtv2020/iStock/Getty Images

36. MULTIPLE REPRESENTATIONS In this problem, you will explore the shortest distance between two parallel lines. Use a straightedge and a protractor or dynamic geometry software.

a. Geometric Draw three sets of parallel lines k and ℓ, s and t, and x and y. For each set, draw the shortest segment $\overline{BC}$ and label points A and D as shown below.

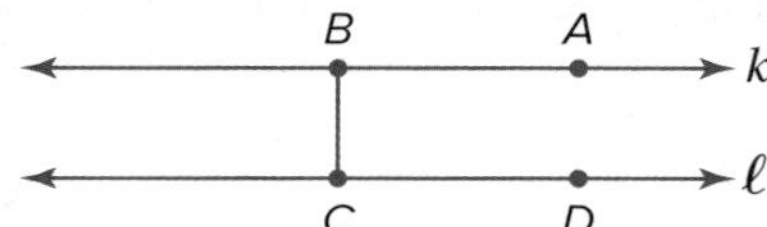

b. Tabular Copy the table below, measure $\angle ABC$ and $\angle BCD$, and complete the table.

Set of Parallel Lines	$m\angle ABC$	$m\angle BCD$
k and ℓ		
s and t		
x and y		

c. Verbal Make a conjecture about the angle the shortest segment forms with both parallel lines.

H.O.T. Problems Use Higher-Order Thinking Skills

37. ERROR ANALYSIS Sumi and Daniela are determining which lines are parallel in the figure at the right. Sumi says that because $\angle 1 \cong \angle 2$, $\overline{WY} \parallel \overline{XZ}$. Daniela disagrees and says that because $\angle 1 \cong \angle 2$, $\overline{WX} \parallel \overline{YZ}$. Is either of them correct? Explain.

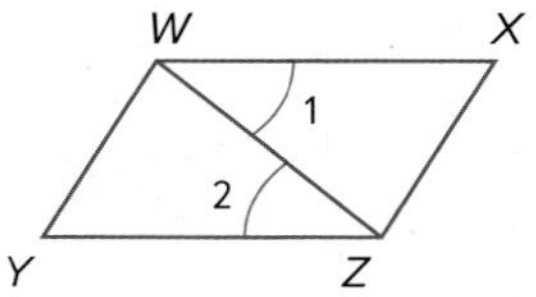

38. MP REASONING Is Theorem 2.23 still true if the two lines are not coplanar? Draw a figure to justify your answer.

39. CHALLENGE Use the figure at the right to prove that two lines parallel to a third line are parallel to each other.

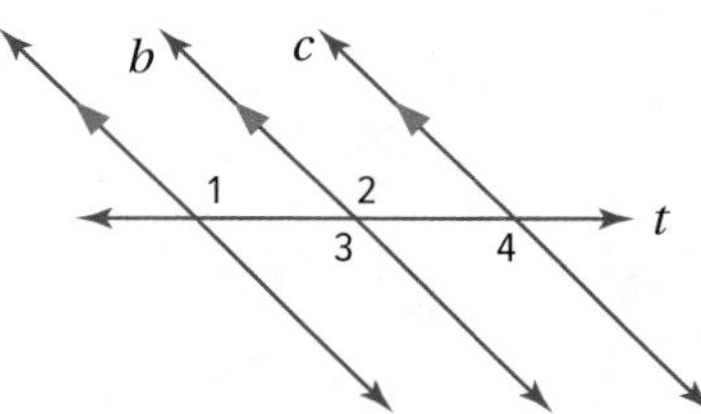

40. OPEN ENDED Draw a triangle ABC.

a. Construct the line parallel to $\overline{BC}$ through point A.

b. Use measurement to justify that the line you constructed is parallel to $\overline{BC}$.

c. Use mathematics to justify this construction.

41. CHALLENGE Refer to the figure at the right.

a. If $m\angle 1 + m\angle 2 = 180$, prove that $a \parallel c$.

b. Given that $a \parallel c$, if $m\angle 1 + m\angle 3 = 180$, prove that $t \perp c$.

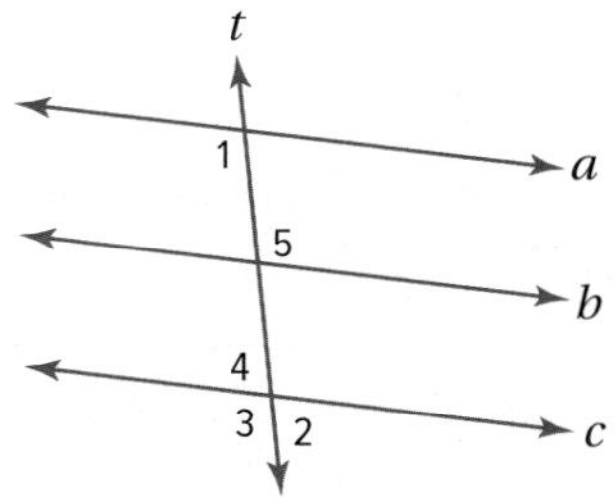

42. WRITING IN MATH Summarize the five methods used in this lesson to prove that two lines are parallel.

43. WRITING IN MATH Can a pair of angles be supplementary and congruent? Explain your reasoning.

44. In the diagram, $\angle 1 \cong \angle 2$. What is the best reason for concluding that line a is parallel to line b? MP 3

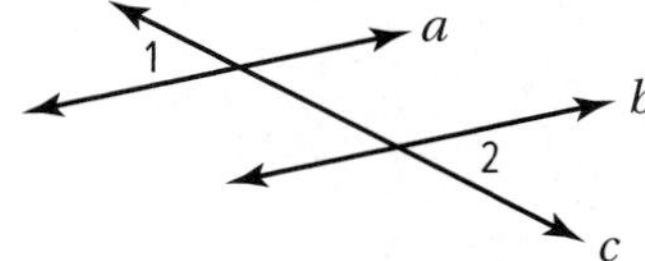

A Alternate Interior Angles Theorem

B Converse of Corresponding Angles Postulate

C Alternate Exterior Angles Theorem

D Converse of Alternate Exterior Angles Theorem

45. **MULTI-STEP** Parallel lines c and d are cut by a transversal. Two of the corresponding angles have measures of $(4x - 6)^\circ$ degrees and $(3x + 18)^\circ$ degrees. MP 7

a. Find x.

b. What is the best classification for the corresponding angles?

A Acute

B Right

C Obtuse

D Not enough information

46. In this diagram, $\angle 3 \cong \angle 4$. What is the appropriate reason for concluding $q \parallel r$? MP 3

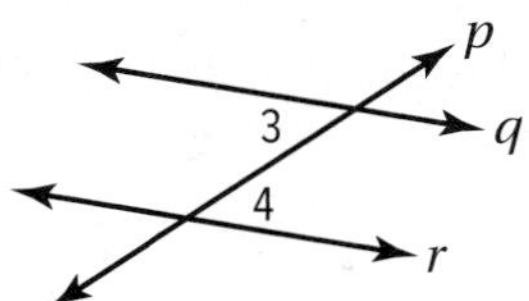

A Alternate Interior Angles Theorem

B Converse of Alternate Interior Angles Theorem

C Converse of Corresponding Angles Postulate

D Converse of Consecutive Interior Angles Theorem

47. Two lines a and b are cut by a transversal line. The measures of two consecutive interior angles are $(4x + 1)^\circ$ and $(5x - 10)^\circ$. What value of x will make line a parallel to line b? MP 7

A 11

B 21

C 85

D 95

E 159

48. Engineers designed streets in a city so that $\angle 1 \cong \angle 2$. Which of the following angle statements is true? MP 2

A $\angle 1 \cong \angle 3$

B $\angle 1 \cong \angle 4$

C $\angle 2 \cong \angle 4$

D $\angle 3 \cong \angle 4$

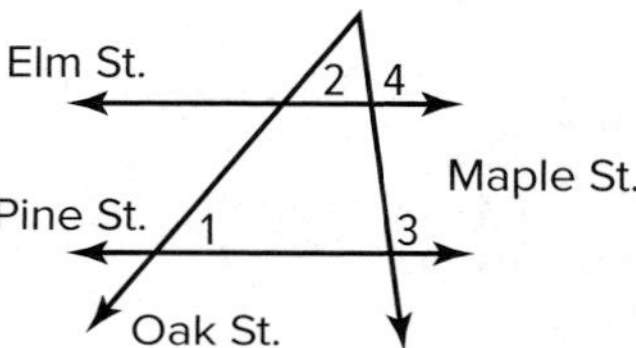

49. A student started the diagram below by drawing $\overleftrightarrow{AB}$ and $\overleftrightarrow{CD}$. To construct $\overleftrightarrow{PQ}$ so that it is parallel to $\overleftrightarrow{AB}$, which construction should the student perform? MP 2

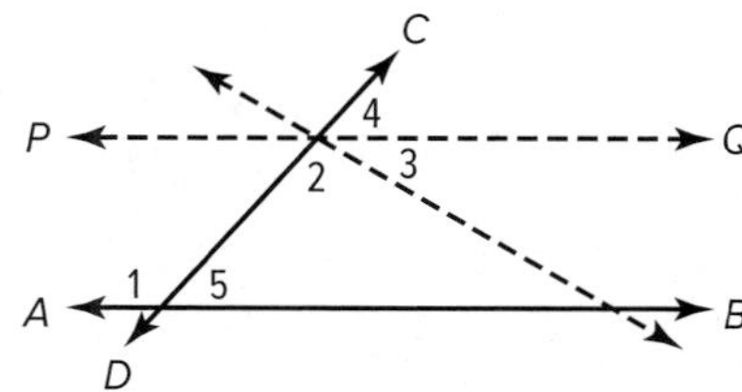

A Construct $\angle 2$ congruent to $\angle 1$.

B Construct $\angle 3$ congruent to $\angle 5$.

C Construct $\angle 3$ congruent to $\angle 1$.

D Construct $\angle 4$ congruent to $\angle 5$.

50. Parallel lines ℓ and m are cut by a transversal. Two alternate interior angles have a measurement of $(10x - 3)^\circ$ and $(8x + 21)^\circ$. What is the measurement for each of these angles? MP 1

LESSON 10

Perpendiculars and Distance

Then

- You proved that two lines are parallel using angle relationships.

Now

1. Find the distance between a point and a line.
2. Find the distance between parallel lines.

Why?

- A *plumb bob* is made of string with a specially designed weight. When the weight is suspended and allowed to swing freely, the point of the bob is precisely below the point to which the string is fixed.

  The plumb bob is useful in establishing what is the true vertical or *plumb* when constructing a wall or when hanging wallpaper.

New Vocabulary

equidistant

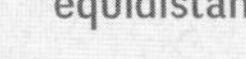

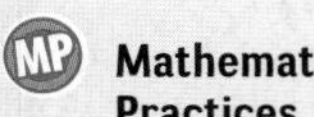

Mathematical Practices

2 Reason abstractly and quantitatively.

4 Model with mathematics.

1 Distance from a Point to a Line

The plumb bob also indicates the shortest distance between the point at which it is attached on the ceiling and a level floor below. This perpendicular distance between a point and a line is the shortest in all cases.

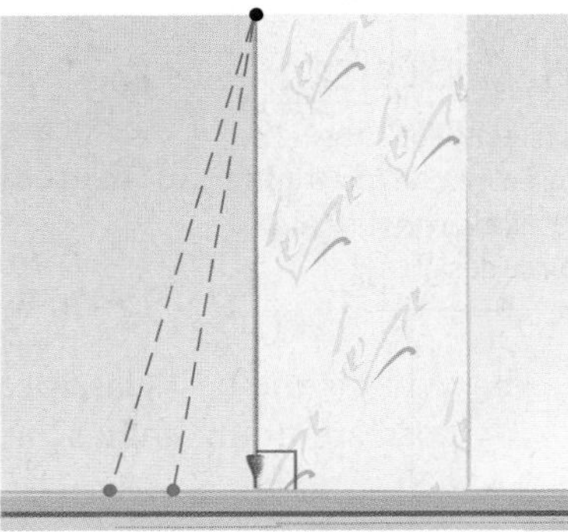

Key Concept Distance Between a Point and a Line

Words The distance between a line and a point not on the line is the length of the segment perpendicular to the line from the point.

Model

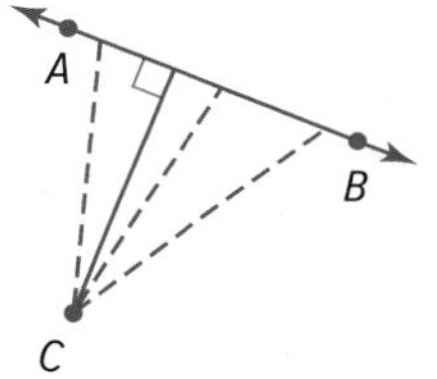

The construction of a line perpendicular to an existing line through a point not on the existing line in Extend Lesson 1-5 establishes that there is *at least one* line through a point P that is perpendicular to a line AB. The following postulate states that this line is the *only* line through P perpendicular to $\overleftrightarrow{AB}$.

Postulate 2.15 Perpendicular Postulate

Words If given a line and a point not on the line, then there exists exactly one line through the point that is perpendicular to the given line.

Model

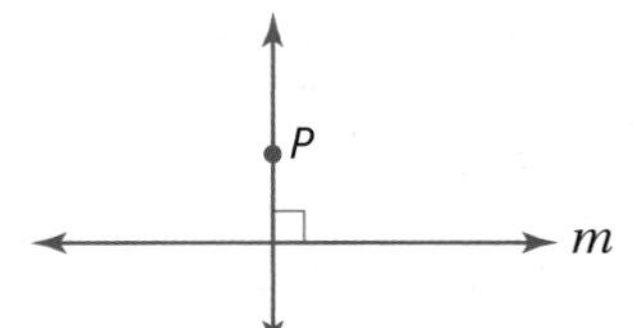

Real-WorldCareer

Landscape Architect
Landscape architects enjoy working with their hands and possess strong analytical skills. Creative vision and artistic talent are also desirable qualities. Typically, a bachelor's degree is required of landscape architects, but a master's degree may be required for specializations such as golf course design.

Real-World Example 1 Construct Distance from a Point to a Line

LANDSCAPING A landscape architect notices that one part of a yard does not drain well. She wants to tap into an existing underground drain represented by line *m*. Construct and name the segment with the length that represents the shortest amount of pipe she will need to lay to connect this drain to point *A*.

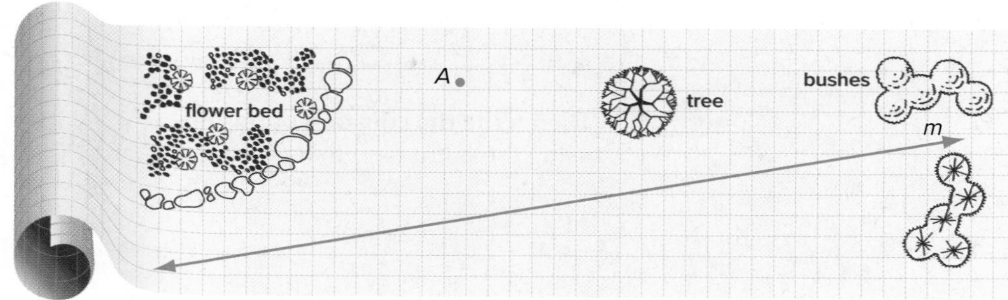

The distance from a line to a point not on the line is the length of the segment perpendicular to the line from the point. Locate points B and C on line m equidistant from point A.

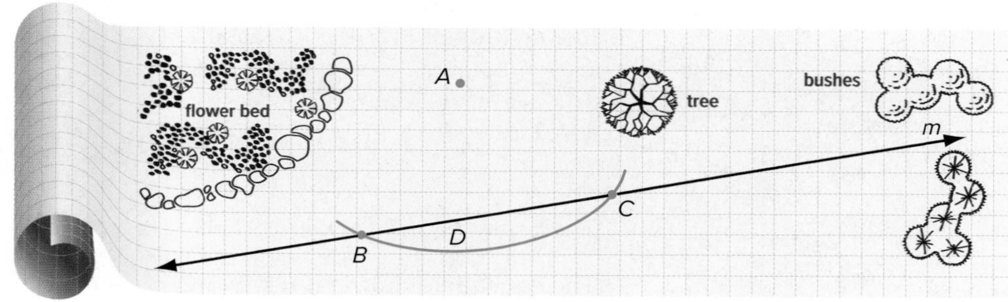

Locate a third point on line m equidistant from B and C. Label this point D. Then draw $\overleftrightarrow{AD}$ so that $\overleftrightarrow{AD} \perp \overleftrightarrow{BC}$.

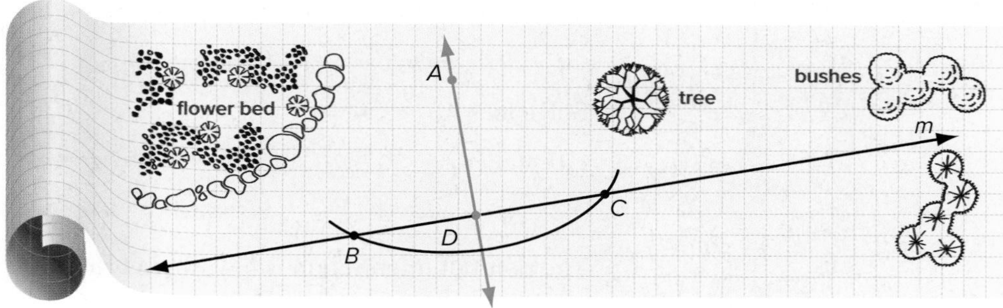

The measure of $\overline{AD}$ represents the shortest amount of pipe the architect will need to lay to connect the drain to point A.

Study Tip

MP Sense-Making
You can use tools like the corner of a piece of paper to help you draw a perpendicular segment from a point to a line, but only a compass and a straightedge can be used to construct this segment.

Guided Practice

1. Copy the figure. Then construct and name the segment that represents the shortest distance from Q to $\overleftrightarrow{PR}$.

P
Q R

> **Study Tip**
>
> **Distance to Axes** Note that the distance from a point to the x-axis can be determined by looking at the y-coordinate, and the distance from a point to the y-axis can be determined by looking at the x-coordinate.

Example 2 Distance from a Point to a Line on a Coordinate Plane

COORDINATE GEOMETRY **Line ℓ contains points at $(-5, 3)$ and $(4, -6)$. Find the distance between line ℓ and point $P(2, 4)$.**

Step 1 Find the equation of the line ℓ.

Begin by finding the slope of the line through points $(-5, 3)$ and $(4, -6)$.

$$m = \frac{y_2 - y_1}{x_2 - x_1} = \frac{-6 - 3}{4 - (-5)} = \frac{-9}{9} \text{ or } -1$$

Then write the equation of this line using the point $(4, -6)$ on the line.

$y = mx + b$	Slope-intercept form
$-6 = -1(4) + b$	$m = -1, (x, y) = (4, -6)$
$-6 = -4 + b$	Simplify.
$-2 = b$	Add 4 to each side.

The equation of line ℓ is $y = -x + (-2)$ or $y = -x - 2$.

Step 2 Write an equation of the line w perpendicular to line ℓ through $P(2, 4)$.

Because the slope of line ℓ is -1, the slope of line w is 1. Write the equation of line w through $P(2, 4)$ with slope 1.

$y = mx + b$	Slope-intercept form
$4 = 1(2) + b$	$m = 1, (x, y) = (2, 4)$
$4 = 2 + b$	Simplify.
$2 = b$	Subtract 2 from each side.

The equation of line w is $y = x + 2$.

> **Study Tip**
>
> **Elimination Method** To review **solving systems of equations** using the elimination method, see Lesson 0-8.

Step 3 Solve the system of equations to determine the point of intersection.

line ℓ: $y = -x - 2$

line w: $(+)\ y = x + 2$

$2y = 0$	Add the two equations.
$y = 0$	Divide each side by 2.

Solve for x.

$0 = x + 2$	Substitute 0 for y in the second equation.
$-2 = x$	Subtract 2 from each side.

The point of intersection is $(-2, 0)$. Let this be point Q.

Step 4 Use the Distance Formula to determine the distance between $P(2, 4)$ and $Q(-2, 0)$.

$d = \sqrt{(x_2 - x_1)^2 + (y_2 - y_1)^2}$	Distance formula
$= \sqrt{(-2 - 2)^2 + (0 - 4)^2}$	$x_2 = -2, x_1 = 2, y_2 = 0, y_1 = 4$
$= \sqrt{32}$	Simplify.

The distance between the point and the line is $\sqrt{32}$ or about 5.66 units.

Guided Practice

2. Line ℓ contains points at (1, 2) and (5, 4). Construct a line perpendicular to ℓ through $P(1, 7)$. Then find the distance from P to ℓ.

Study Tip

Equidistant You will use this concept of *equidistant* to describe special points and lines relating to the sides and angles of triangles in Lesson 5-1.

2 Distance Between Parallel Lines

By definition, parallel lines do not intersect. An alternate definition states that two lines in a plane are parallel if they are everywhere **equidistant**. Equidistant means that the distance between two lines measured along a perpendicular line to the lines is always the same.

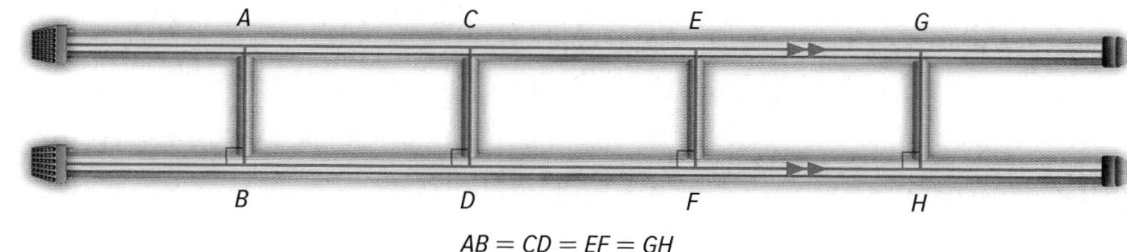

$AB = CD = EF = GH$

This leads to the definition of the distance between two parallel lines.

Key Concept Distance Between Parallel Lines

The distance between two parallel lines is the perpendicular distance between one of the lines and any point on the other line.

Study Tip

Locus of Points Equidistant from Two Parallel Lines Conversely, the locus of points in a plane that are equidistant from two parallel lines is a third line that is parallel to and centered between the two parallel lines.

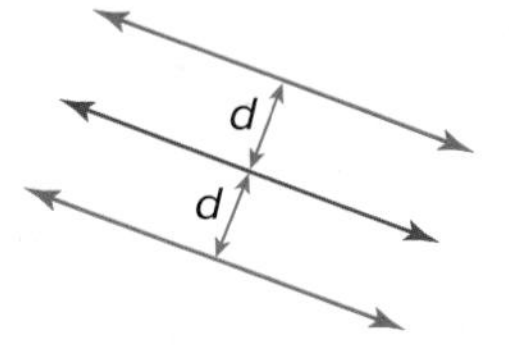

Recall from Lesson 1-1 that a *locus* is the set of all points that satisfy a given condition. Parallel lines can be described as the locus of points in a plane equidistant from a given line.

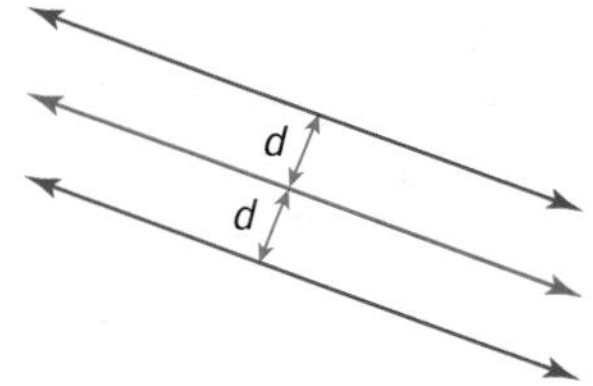

Theorem 2.24 Two Lines Equidistant from a Third

In a plane, if two lines are each equidistant from a third line, then the two lines are parallel to each other.

You will prove Theorem 2.24 in Exercise 30.

Example 3 Distance Between Parallel Lines

Find the distance between the parallel lines ℓ and m with equations $y = 2x + 1$ and $y = 2x - 3$, respectively.

You will need to solve a system of equations to find the endpoints of a segment that is perpendicular to both ℓ and m. From their equations, we know that the slope of line ℓ and line m is 2.

Sketch line p through the y-intercept of line m, (0, −3), perpendicular to lines m and ℓ.

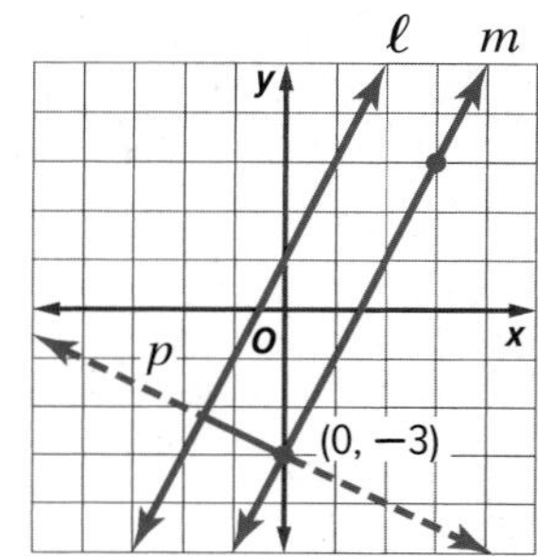

Step 1 Write an equation of line p. The slope of p is the opposite reciprocal of 2, or $-\frac{1}{2}$. Use the y-intercept of line m, $(0, -3)$, as one of the endpoints of the perpendicular segment.

$(y - y_1) = m(x - x_1)$	Point-slope form
$[y - (-3)] = -\frac{1}{2}(x - 0)$	$x_1 = 0$, $y_1 = 3$, and $m = -\frac{1}{2}$
$y + 3 = -\frac{1}{2}x$	Simplify.
$y = -\frac{1}{2}x - 3$	Subtract 3 from each side.

Step 2 Use a system of equations to determine the point of intersection of lines ℓ and p.

ℓ: $y = 2x + 1$

p: $y = -\frac{1}{2}x - 3$

$2x + 1 = -\frac{1}{2}x - 3$	Substitute $2x + 1$ for y in the second equation.
$2x + \frac{1}{2}x = -3 - 1$	Group like terms on each side.
$\frac{5}{2}x = -4$	Simplify on each side.
$x = -\frac{8}{5}$	Multiply each side by $\frac{2}{5}$.
$y = -\frac{1}{2}\left(-\frac{8}{5}\right) - 3$	Substitute $-\frac{8}{5}$ for x in the equation for p.
$= -\frac{11}{5}$	Simplify.

The point of intersection is $\left(-\frac{8}{5}, -\frac{11}{5}\right)$ or $(-1.6, -2.2)$.

Step 3 Use the Distance Formula to determine the distance between $(0, -3)$ and $(-1.6, -2.2)$.

$d = \sqrt{(x_2 - x_1)^2 + (y_2 - y_1)^2}$	Distance Formula
$= \sqrt{(-1.6 - 0)^2 + [-2.2 - (-3)]^2}$	$x_2 = -1.6$, $x_1 = 0$, $y_2 = -2.2$, and $y_1 = -3$
≈ 1.8	Simplify using a calculator.

The distance between the lines is about 1.8 units.

Study Tip

Substitution Method To review solving systems of equations using the substitution method, see p. P17.

Guided Practice

3A. Find the distance between the parallel lines r and s whose equations are $y = -3x - 5$ and $y = -3x + 6$, respectively.

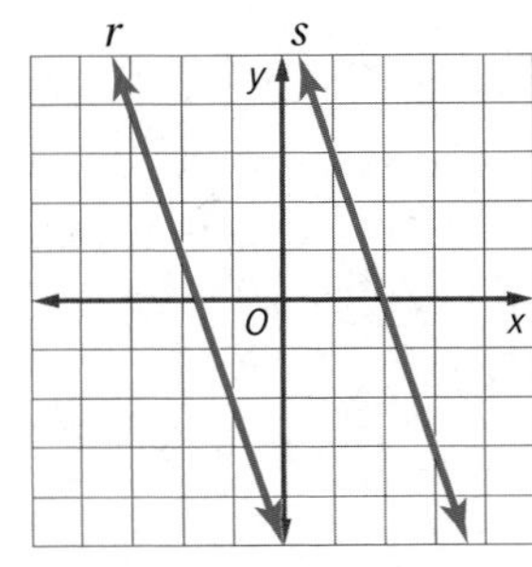

3B. Find the distance between parallel lines a and b with equations $x + 3y = 6$ and $x + 3y = -14$, respectively.

Check Your Understanding

◯ = Step-by-Step Solutions begin on page R13.

Example 1 **Copy each figure. Construct the segment that represents the distance indicated.**

1. Y to $\overleftrightarrow{TS}$

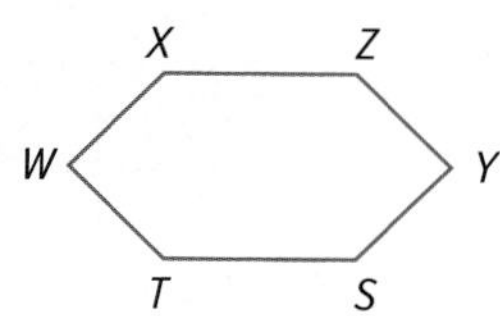

2. C to $\overleftrightarrow{AB}$

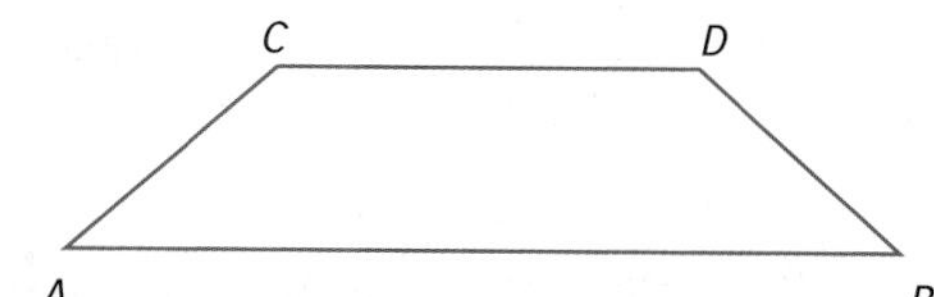

3. **MP STRUCTURE** After forming a line, every even member of a marching band turns to face the home team's end zone and marches 5 paces straight forward. At the same time, every odd member turns in the opposite direction and marches 5 paces straight forward. Assuming that each band member covers the same distance, what formation should result? Justify your answer.

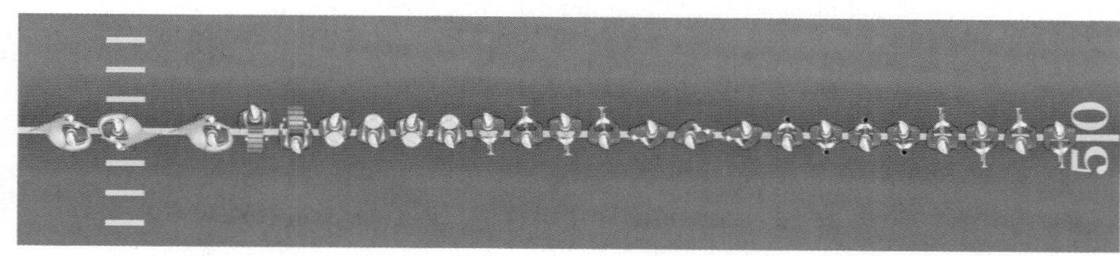

Example 2 **COORDINATE GEOMETRY** Find the distance from P to ℓ.

4. Line ℓ contains points (4, 3) and (−2, 0). Point P has coordinates (3, 10).

5. Line ℓ contains points (−6, 1) and (9, −4). Point P has coordinates (4, 1).

6. Line ℓ contains points (4, 18) and (−2, 9). Point P has coordinates (−9, 5).

Example 3 **Find the distance between each pair of parallel lines with the given equations.**

7. $y = -2x + 4$
   $y = -2x + 14$

8. $y = 7$
   $y = -3$

Practice and Problem Solving

Extra Practice is on page R2.

Example 1 **Copy each figure. Construct the segment that represents the distance indicated.**

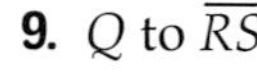

9. Q to $\overline{RS}$

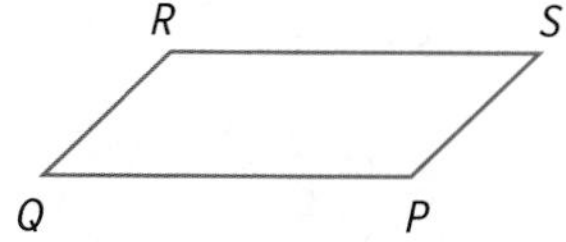

10. A to $\overline{BC}$

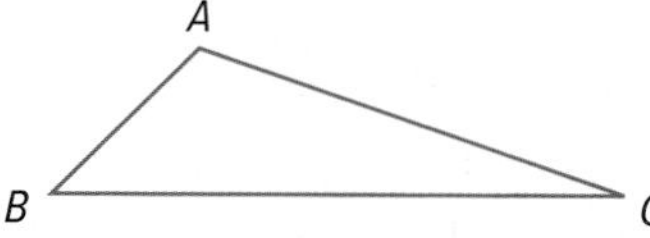

11. H to $\overline{FG}$

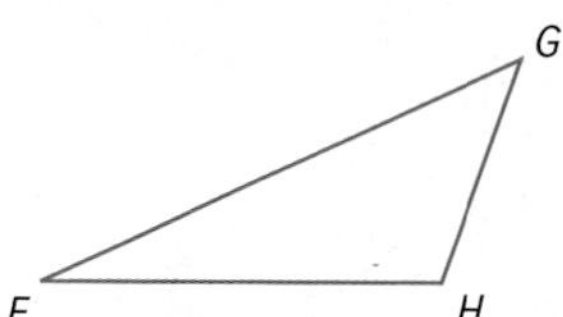

12. K to $\overline{LM}$

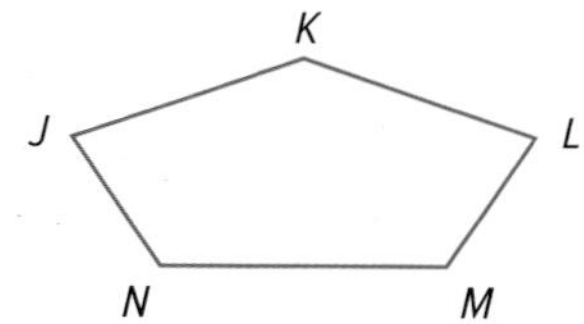

13. **DRIVEWAYS** In the diagram at the right, is the driveway shown the shortest possible one from the house to the road? Explain why or why not.

14. **MP MODELING** Rondell is crossing the courtyard in front of his school. Three possible paths are shown in the diagram at the right. Which of the three paths shown is the shortest? Explain your reasoning.

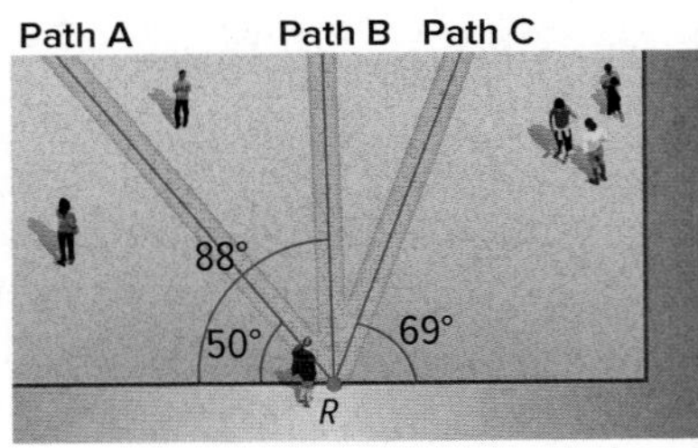

Example 2 **COORDINATE GEOMETRY** Find the distance from P to ℓ.

15. Line ℓ contains points $(0, -3)$ and $(7, 4)$. Point P has coordinates $(4, 3)$.
16. Line ℓ contains points $(11, -1)$ and $(-3, -11)$. Point P has coordinates $(-1, 1)$.
17. Line ℓ contains points $(-2, 1)$ and $(4, 1)$. Point P has coordinates $(5, 7)$.
18. Line ℓ contains points $(4, -1)$ and $(4, 9)$. Point P has coordinates $(1, 6)$.
19. Line ℓ contains points $(1, 5)$ and $(4, -4)$. Point P has coordinates $(-1, 1)$.
20. Line ℓ contains points $(-8, 1)$ and $(3, 1)$. Point P has coordinates $(-2, 4)$.

Example 3 **Find the distance between each pair of parallel lines with the given equations.**

21. $y = -2$, $y = 4$
22. $x = 3$, $x = 7$
23. $y = 5x - 22$, $y = 5x + 4$
24. $y = \frac{1}{3}x - 3$, $y = \frac{1}{3}x + 2$
25. $x = 8.5$, $x = -12.5$
26. $y = 15$, $y = -4$
27. $y = \frac{1}{4}x + 2$, $4y - x = -60$
28. $3x + y = 3$, $y + 17 = -3x$
29. $y = -\frac{5}{4}x + 3.5$, $4y + 10.6 = -5x$

30. **PROOF** Write a two-column proof of Theorem 2.24.

Find the distance from the line to the given point.

31. $y = -3, (5, 2)$
32. $y = \frac{1}{6}x + 6, (-6, 5)$
33. $x = 4, (-2, 5)$

34. **POSTERS** Alma is hanging two posters on the wall in her room as shown. How can Alma use perpendicular distances to confirm that the posters are parallel?

 **SCHOOL SPIRIT** Brock is decorating a hallway bulletin board to display pictures of students demonstrating school spirit. He cuts off one length of border to match the width of the top of the board, and then uses that strip as a template to cut a second strip that is exactly the same length for the bottom.

DON'T MESS WITH THE BEST!

When stapling the bottom border in place, he notices that the strip he cut is about a quarter of an inch too short. Describe what he can conclude about the bulletin board. Explain your reasoning.

CONSTRUCTION Line ℓ contains points at $(-4, 3)$ and $(2, -3)$. Point P at $(-2, 1)$ is on line ℓ. Complete the following construction.

Step 1

Graph line ℓ and point P and put the compass at point P. Using the same compass setting, draw arcs to the left and right of P. Label these points A and B.

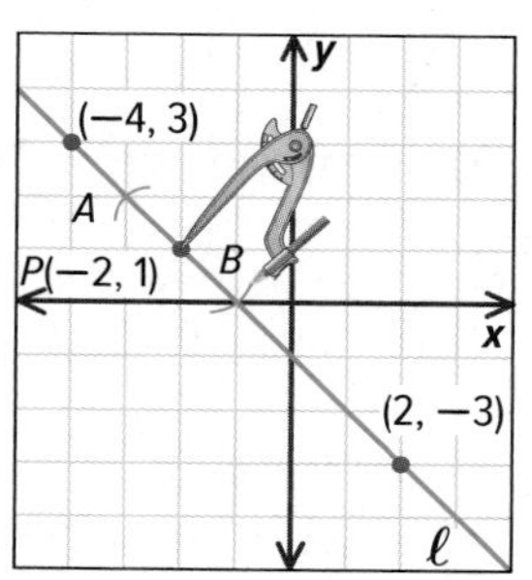

Step 2

Open the compass to a setting greater than AP. Put the compass at point A and draw an arc above line ℓ.

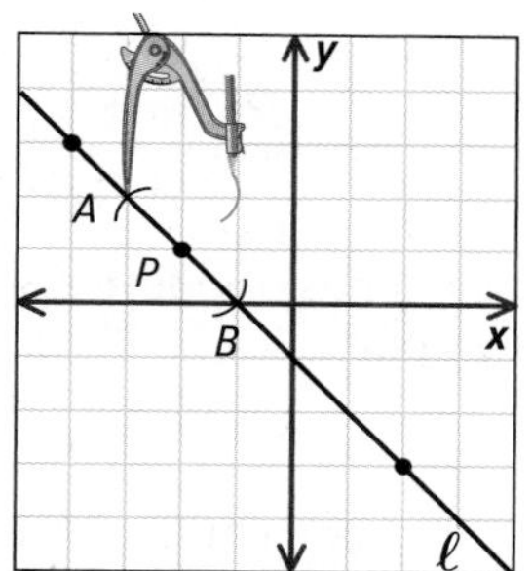

Step 3

Using the same compass setting, put the compass at point B and draw an arc above line ℓ. Label the point of intersection Q. Then draw $\overleftrightarrow{PQ}$.

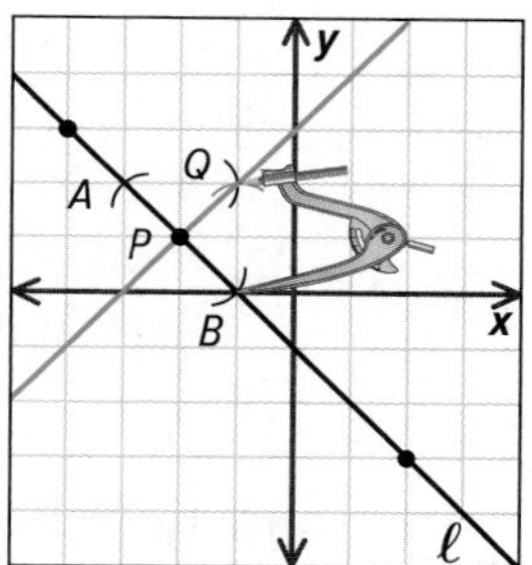

36. What is the relationship between line ℓ and $\overleftrightarrow{PQ}$? Verify your conjecture using the slopes of the two lines.

37. Repeat the construction above using a different line and point on that line.

38. MP **SENSE-MAKING** $\overline{AB}$ has a slope of 2 and midpoint $M(3, 2)$. A segment perpendicular to $\overline{AB}$ has midpoint $P(4, -1)$ and shares endpoint B with $\overline{AB}$.

a. Graph the segments.

b. Find the coordinates of A and B.

39. **MULTIPLE REPRESENTATIONS** In this problem, you will explore the areas of triangles formed by points on parallel lines.

a. Geometric Draw two parallel lines and label them as shown.

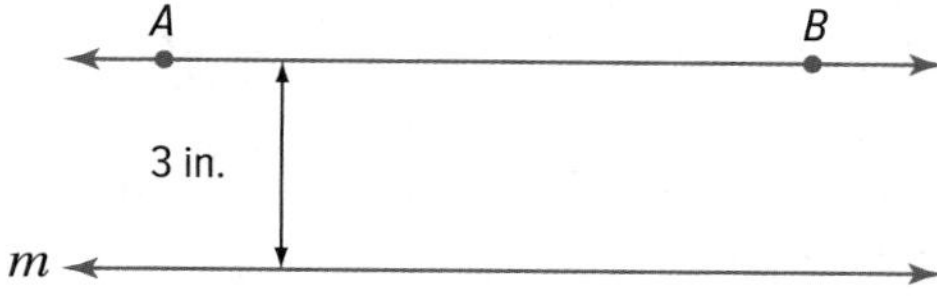

b. Verbal Where would you place point C on line m to ensure that triangle ABC would have the largest area? Explain your reasoning.

c. Analytical If $AB = 11$ inches, what is the maximum area of $\triangle ABC$?

40. **MULTI-STEP** Draw a diagram that represents each statement, marking the diagram with the given information. Then, use your diagram to answer the questions.

- Lines a and b are perpendicular to plane P.
- Plane P is perpendicular to planes R and Q.
- Plane Q is perpendicular to line ℓ.

a. If planes R and Q are parallel and they intersect plane P, what must also be true?

b. If line ℓ is perpendicular to plane Q, what must also be true?

c. If both line a and line b are perpendicular to plane P, what must also be true?

H.O.T. Problems Use Higher-Order Thinking Skills

41. **ERROR ANALYSIS** Han draws the segments $\overline{AB}$ and $\overline{CD}$ shown below using a straightedge. He claims that these two lines, if extended, will never intersect. Shenequa claims that they will. Is either of them correct? Justify your answer.

A ——— B

C ——— D

42. **CHALLENGE** Describe the locus of points that are equidistant from two intersecting lines and sketch an example.

43. **CHALLENGE** Suppose a line perpendicular to a pair of parallel lines intersects the lines at the points $(a, 4)$ and $(0, 6)$. If the distance between the parallel lines is $\sqrt{5}$, find the value of a and the equations of the parallel lines.

44. **MP REASONING** Determine whether the following statement is *sometimes, always,* or *never* true. Explain.

The distance between a line and a plane can be found.

45. **OPEN ENDED** Draw an irregular convex pentagon using a straightedge.

a. Use a compass and straightedge to construct a line between one vertex and a side opposite the vertex.

b. Use measurement to justify that the line constructed is perpendicular to the side chosen.

c. Use mathematics to justify this conclusion.

46. **MP SENSE-MAKING** Rewrite Theorem 2.25 in terms of two planes that are equidistant from a third plane. Sketch an example.

47. **WRITING IN MATH** Describe how you can use the Distance Formula to verify that two lines are parallel.

48. MULTI-STEP The diagram shows line $\overleftrightarrow{XY}$ and point P. MP 1

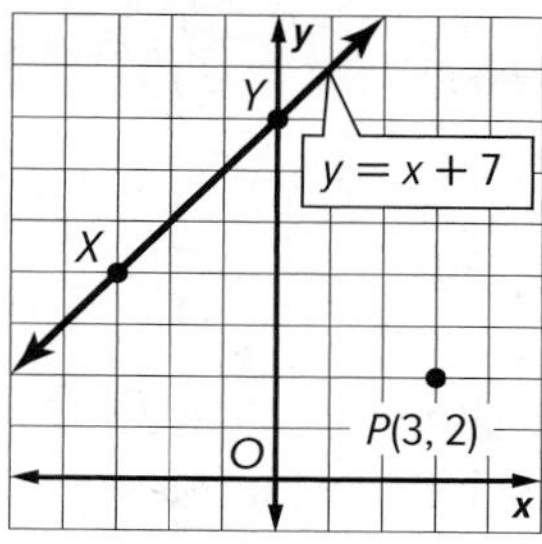

a. Find the coordinates a and b.

b. Find the distance between point P and line $\overleftrightarrow{XY}$.

c. Find the coordinates of the point Q where $\overleftrightarrow{XY}$ is the perpendicular bisector of $\overleftrightarrow{PQ}$.

49. In the diagram, the dashed line through (0, −2) is perpendicular to line q.

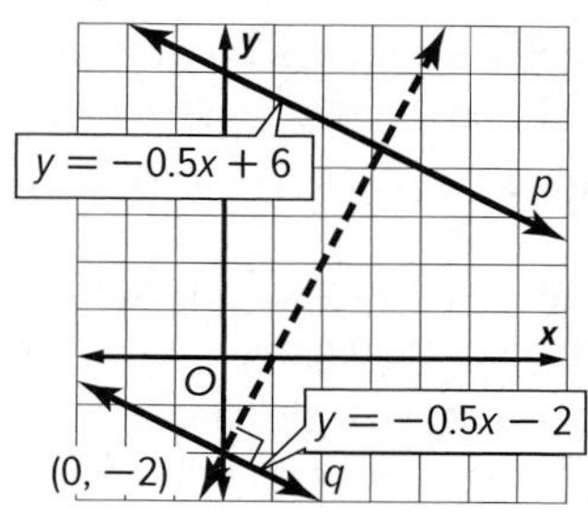

What are the coordinates of the intersection of the dashed line and line p? MP 1

- A (8.0, 2.0)
- B (4.4, 3.2)
- C (3.2, 4.4)
- D (3.2, −3.6)

50. What is the distance from the point (4, 9) to the line $y = -x + 6$ to the nearest tenth? MP 1

51. The manager of a park wants to find the distance from the tree at point T to the path of the line. What are the first two construction steps to find that distance? MP 2

- **A** Measure the distance TP and then measure the distance TH.
- **B** Use an arc to find two points X and Y on line $\overleftrightarrow{PH}$ that are the same distance from T. Then find another point that is equidistant from points X and Y.
- **C** Draw a line segment from T to H and then bisect segment $\overleftrightarrow{TH}$.
- **D** Find the midpoint M of $\overleftrightarrow{PH}$ and then draw $\overleftrightarrow{TM}$.
- **E** Construct the line parallel to $\overleftrightarrow{PH}$ through T and then use a ruler to measure the exact distance.

52. Given $y = 3x + 2$, write the following equations: MP 1

a. an equation parallel to the given line

b. an equation perpendicular to the given line with the same y-intercept

c. an equation perpendicular to the given line with a different y-intercept

53. Graph the lines $y = 2x + 4$ and $y = \frac{1}{2}x - 4$. What is the point of intersection of the two lines? MP 1

- A (−5, −6)
- B (−6, −5)
- C (5, 6)
- D (6, 5)

CHAPTER 2

Study Guide and Review

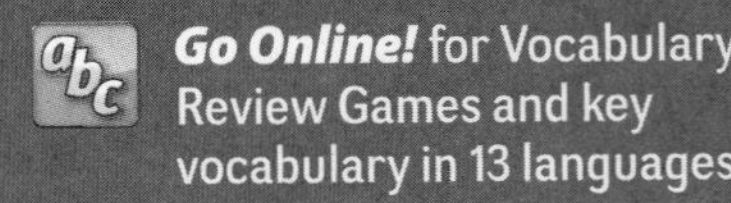

Study Guide

Key Concepts

Logic (Lessons 2-1 through 2-3)

- Negation of statement p: ~p
- An if-then statement is written in the form if p, then q.

statement	$p \rightarrow q$
converse	$q \rightarrow p$
inverse	$\sim p \rightarrow \sim q$
contrapositive	$\sim q \rightarrow \sim p$

- Law of Detachment: If $p \rightarrow q$ is true and p is true, then q is also true.
- Law of Syllogism: If $p \rightarrow q$ and $q \rightarrow r$ are true, then $p \rightarrow r$ is also true.

Proof (Lessons 2-4 through 2-6)

Step 1 List the given information and draw a diagram, if possible.

Step 2 State what is to be proved.

Step 3 Create a deductive argument.

Step 4 Justify each statement with a reason.

Step 5 State what you have proved.

Parallel Lines and Angles (Lessons 2-7 and 2-9)

- If two parallel lines are cut by a transversal, then
  - each pair of corresponding angles is congruent
  - each pair of alternate exterior angles is congruent
  - each pair of alternate interior angles is congruent, and
  - each pair of consecutive interior angles is supplementary.

Slope (Lesson 2-8)

- The slope m of a line containing two points with coordinates (x_1, y_1) and (x_2, y_2) is $m = \frac{y_2 - y_1}{x_2 - x_1}$, where $x_1 \neq x_2$.

Distance (Lesson 2-10)

- The distance from a line to a point not on the line is the length of the perpendicular segment to the line from the point.

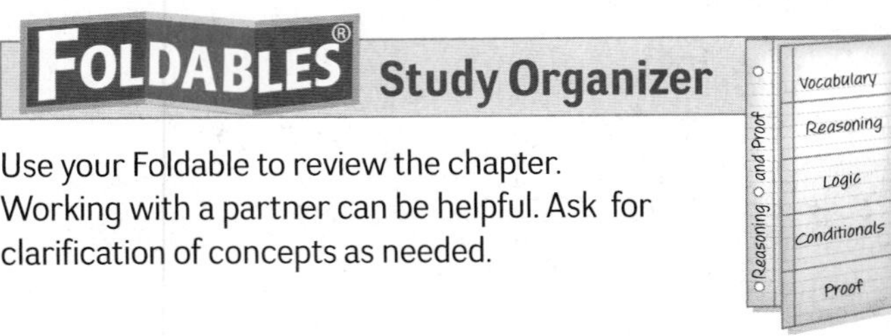

Study Organizer

Use your Foldable to review the chapter. Working with a partner can be helpful. Ask for clarification of concepts as needed.

Key Vocabulary

alternate exterior angles (p. 169)
alternate interior angles (p. 169)
compound statement (p. 119)
conclusion (p. 121)
conditional statement (p. 121)
conjunction (p. 119)
contrapositive (p. 122)
converse (p. 122)
counterexample (p. 114)
deductive reasoning (p. 131)
disjunction (p. 120)
flow proof (p. 143)
hypothesis (p. 121)
if-then statement (p. 121)
inductive reasoning (p. 111)
inverse (p. 122)
logically equivalent (p. 123)
negation (p. 119)
paragraph proof (p. 145)
parallel lines (p. 170)
point-slope form (p. 179)
related conditionals (p. 122)
slope-intercept form (p. 179)
statement (p. 119)
theorem (p. 144)
transversal (p. 169)
truth value (p. 119)
two-column proof (p. 152)

Vocabulary Check

State whether each sentence is *true* or *false*. If *false*, replace the underlined term to make a true sentence.

1. The first part of an if-then statement is the <u>conjecture</u>.
2. The <u>contrapositive</u> is formed by negating the hypothesis and conclusion of a conditional.
3. A <u>conjunction</u> is formed by joining two or more statements with the word *and*.
4. To show that a conjecture is false, you would provide a <u>disjunction</u>.
5. The <u>inverse</u> of a statement p would be written in the form *not p*.
6. In a two-column proof, the properties that justify each step are called <u>reasons</u>.
7. The <u>slope-intercept</u> form of a linear equation is $y - y_1 = m(x - x_1)$.
8. The distance from point X to the line q is the length of the segment <u>perpendicular</u> to line q from X.

Concept Check

9. Explain the difference between inductive and deductive reasoning.
10. Explain how to prove two lines cut by a transversal are parallel.

Lesson-by-Lesson Review 

2-1 Conjectures and Counterexamples

Determine whether each conjecture is *true* or *false*. If false, give a counterexample.

11. If $\angle 1$ and $\angle 2$ are supplementary angles, then $\angle 1$ and $\angle 2$ form a linear pair.

12. If $W(-3, 2)$, $X(-3, 7)$, $Y(6, 7)$, and $Z(6, 2)$, then quadrilateral $WXYZ$ is a rectangle.

13. **PARKS** Jacinto enjoys hiking with his dog in the forest at his local park. While on vacation in Smoky Mountain National Park in Tennessee, he was disappointed that dogs were not allowed on most hiking trails. Make a conjecture about why his local park and the national park have differing rules with regard to pets.

Example 1

Determine whether each conjecture is *true* or *false*. If false, give a counterexample.

a. **$c = d$, $d = c$ is an example of a property of real numbers.**

$c = d$, $d = c$ is an example of the Symmetric Property of real numbers, so the conjecture is true.

b. **If x is a positive integer and y is a negative integer, then $x + y$ is a negative integer.**

This conjecture is false. If $x = 3$ and $y = -1$, then $x + y = 2$, a positive integer.

2-2 Statements, Conditionals, and Biconditionals

Use the following statements to write a compound statement for each conjunction or disjunction. Then find its truth value. Explain.

p: A plane contains at least three noncollinear points.

q: A square yard is equivalent to three square feet.

r: The sum of the measures of two complementary angles is 180.

14. $\sim q \vee r$ **15.** $p \wedge \sim r$ **16.** $\sim p \vee q$

Determine the truth value of each conditional statement. If *true*, explain your reasoning. If *false*, give a counterexample.

17. If you square an integer, then the result is a positive integer.

18. If a hexagon has eight sides, then all of its angles will be obtuse.

19. Write the converse, inverse, and contrapositive of the following true conditional. Then, determine whether each related conditional is *true* or *false*. If a statement is false, find a counterexample.

If two angles are congruent, then they have the same degree measure.

Example 2

Use the following statements to write a compound statement for each conjunction or disjunction. Then find its truth value. Explain.

p: x^2 is a nonnegative number.

q: Adjacent angles lie in the same plane.

r: A negative number is not a real number.

a. $\sim q \wedge r$

$\sim q \wedge r$: Adjacent angles do not lie in the same plane, and a negative number is not a real number.

Because both $\sim q$ and r are false, $\sim q \wedge r$ is false.

b. **p or r**

p or r: x^2 is a nonnegative number, or a negative number is not a real number.

p or r is true because p is true. It does not matter that r is false.

Example 3

Write the *converse*, *inverse*, and *contrapositive* of the following true conditional.

If a figure is a square, then it is a parallelogram.

Converse: If a figure is a parallelogram, then it is a square.

Inverse: If a figure is not a square, then it is not a parallelogram.

Contrapositive: If a figure is not a parallelogram, then it is not a square.

2-3 Deductive Reasoning

Draw a valid conclusion from the given statements, if possible. Then state whether your conclusion was drawn using the Law of Detachment or the Law of Syllogism. If no valid conclusion can be drawn, write *no valid conclusion* and explain your reasoning.

20. **Given:** If a quadrilateral has diagonals that bisect each other, then it is a parallelogram.

The diagonals of quadrilateral *PQRS* bisect each other.

21. **Given:** If Liana struggles in science class, then she will receive tutoring.

If Liana stays after school on Thursday, then she will receive tutoring.

22. **EARTHQUAKES** Determine whether the stated conclusion is valid based on the given information. If not, write *invalid*. Explain.

Given: If an earthquake measures a 7.0 or higher on the Richter scale, then it is considered a major earthquake that could cause serious damage. The 1906 San Francisco earthquake measured 8.0 on the Richter scale.

Conclusion: The 1906 San Francisco earthquake was a major earthquake that caused serious damage.

Example 4

Use the Law of Syllogism to determine whether a valid conclusion can be reached from the following statements.

(1) **If the measure of an angle is greater than 90, then it is an obtuse angle.**

(2) **If an angle is an obtuse angle, then it is not a right angle.**

p: the measure of an angle is greater than 90

q: the angle is an obtuse angle

r: the angle is not a right angle

Statement (1): $p \rightarrow q$

Statement (2): $q \rightarrow r$

Because the given statements are true, use the Law of Syllogism to conclude that $p \rightarrow r$. That is, *If the measure of an angle is greater than 90, then it is not a right angle.*

2-4 Writing Proofs

Determine whether each statement is *always*, *sometimes*, or *never* true. Explain.

23. Two planes intersect at a point.

24. Three points are contained in more than one plane.

25. If line *m* lies in plane *X* and line *m* contains a point *Q*, then point *Q* lies in plane *X*.

26. If two angles are complementary, then they form a right angle.

27. **NETWORKING** Six people are introduced at a business convention. If each person shakes hands with each of the others, how many handshakes will be exchanged? Include a model to support your reasoning.

Example 5

Determine whether each statement is *always*, *sometimes*, or *never* true. Explain.

a. ***If points X, Y, and Z lie in plane R, then they are not collinear.***

Sometimes; the fact that *X*, *Y*, and *Z* are contained in plane *R* has no bearing on whether those points are collinear or not.

b. **For any two points *A* and *B*, there is exactly one line that contains them.**

Always; according to Postulate 2-1, there is exactly one line through any two points.

2-5 Proving Segment Relationships

Write a two-column proof.

28. Given: X is the midpoint of $\overline{WY}$ and $\overline{VZ}$.

Prove: $VW = ZY$

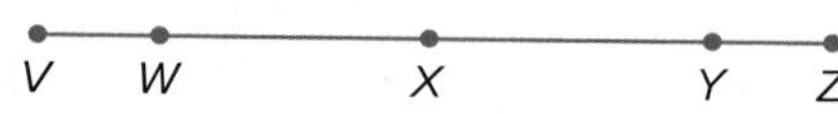

29. Given: $AB = DC$

Prove: $AC = DB$

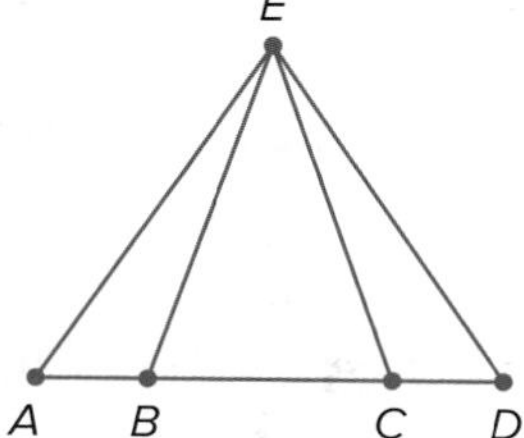

30. GEOGRAPHY Leandro is planning to drive from Kansas City to Minneapolis along Interstate 35. The map he is using gives the distance from Kansas City to Des Moines as 194 miles and from Des Moines to Minneapolis as 243 miles. What allows him to conclude that the distance he will be driving is 437 miles from Kansas City to Minneapolis? Assume that Interstate 35 forms a straight line.

Example 6

Write a two-column proof.

Given: B is the midpoint of $\overline{AC}$.

C is the midpoint of $\overline{BD}$.

Prove: $\overline{AB} \cong \overline{CD}$

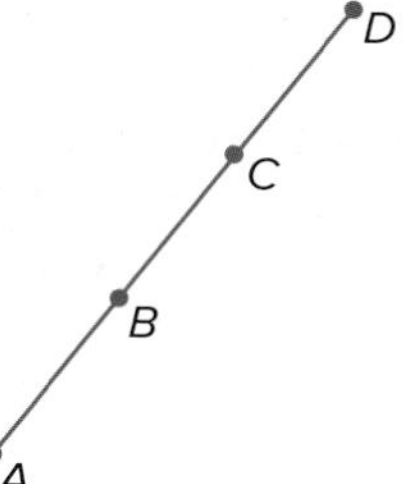

Proof:

Statements	Reasons
1. B is the midpoint of $\overline{AC}$.	1. Given
2. $\overline{AB} \cong \overline{BC}$	2. Definition of midpoint
3. C is the midpoint of $\overline{BD}$.	3. Given
4. $\overline{BC} \cong \overline{CD}$	4. Definition of midpoint
5. $\overline{AB} \cong \overline{CD}$	5. Transitive Property of Equality

2-6 Proving Angle Relationships

Find the measure of each angle.

31. $\angle 5$

32. $\angle 6$

33. $\angle 7$

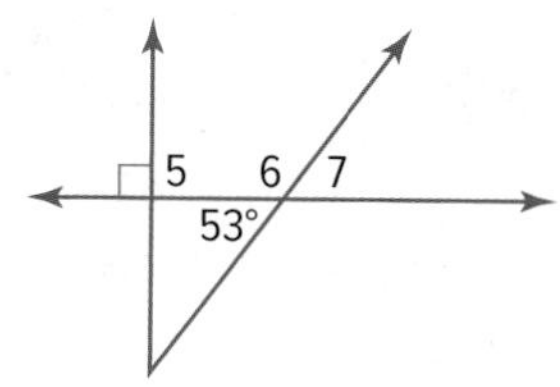

34. PROOF Write a two-column proof.

Given: $\angle 1 \cong \angle 4$, $\angle 2 \cong \angle 3$

Prove: $\angle AFC \cong \angle EFC$

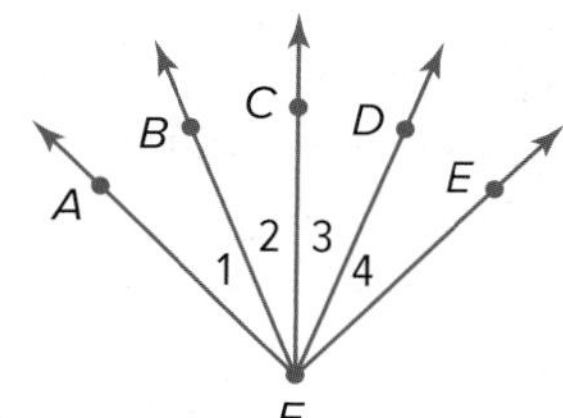

Example 7

Find the measure of each numbered angle if $m\angle 1 = 72$ and $m\angle 3 = 26$.

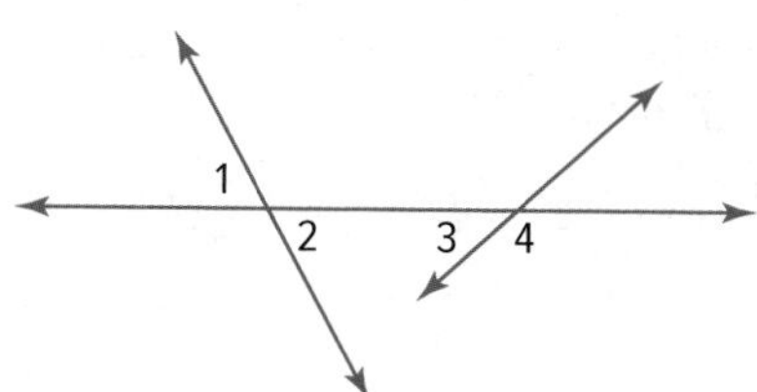

$m\angle 2 = 72$, because $\angle 1$ and $\angle 2$ are vertical angles.

$\angle 3$ and $\angle 4$ form a linear pair and must be supplementary angles.

$26 + m\angle 4 = 180$ Definition of supplementary angles

$m\angle 4 = 154$ Subtract 26 from each side.

Study Guide and Review *Continued*

2-7 Parallel Lines and Transversals

Classify the relationship between each pair of angles as *alternate interior*, *alternate exterior*, *corresponding*, or *consecutive interior* angles.

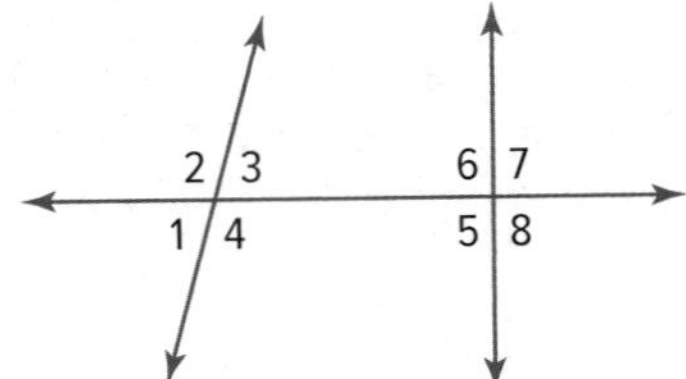

35. ∠1 and ∠5

36. ∠4 and ∠6

37. ∠2 and ∠8

38. ∠4 and ∠5

39. **BRIDGES** The Roebling Suspension Bridge extends over the Ohio River connecting Cincinnati, Ohio, to Covington, Kentucky. Describe the type of lines formed by the bridge and the river.

In the figure, $m\angle 1 = 123$. Find the measure of each angle. Tell which postulate(s) or theorem(s) you used.

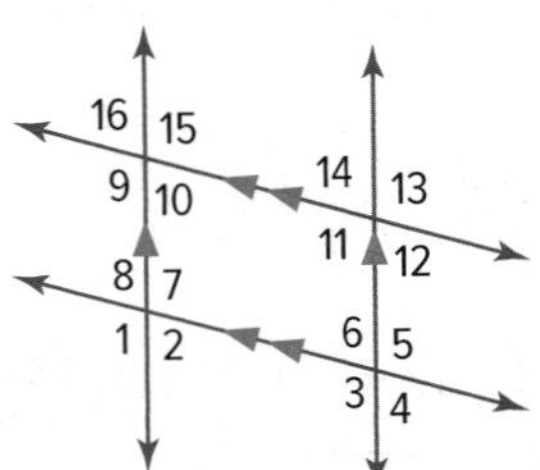

40. ∠5

41. ∠14

42. ∠16

43. ∠11

44. ∠4

45. ∠6

46. **MAPS** The diagram shows the layout of Elm, Plum, and Oak streets. Find the value of *x*.

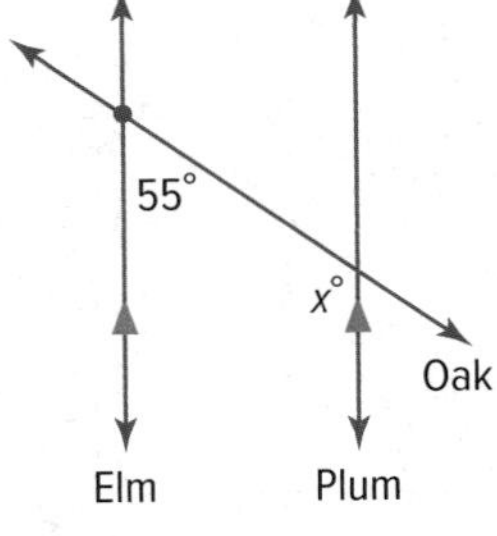

Example 8

Refer to the figure below. Classify the relationship between each pair of angles as *alternate interior*, *alternate exterior*, *corresponding*, or *consecutive interior* angles.

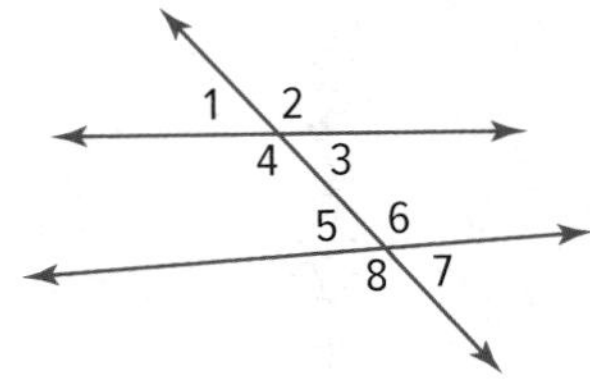

a. ∠3 and ∠6
consecutive interior

b. ∠2 and ∠6
corresponding

c. ∠1 and ∠7
alternate exterior

d. ∠3 and ∠5
alternate interior

Example 9

ALGEBRA If $m\angle 5 = 7x - 5$ and $m\angle 4 = 2x + 23$, find *x*. Explain your reasoning.

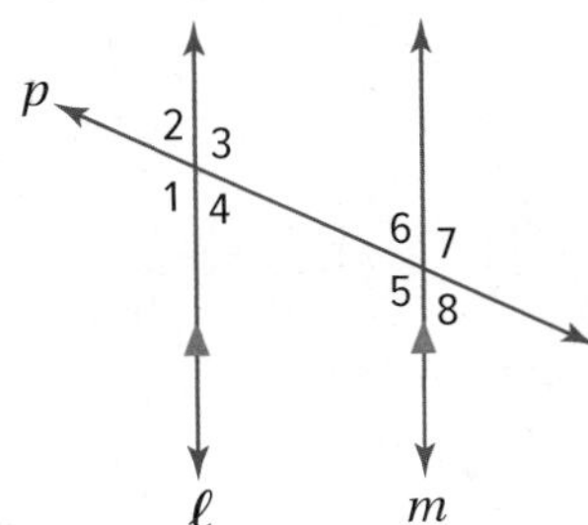

$m\angle 4 + m\angle 5 = 180$	Def. of Supp. ∠s
$(2x + 23) + (7x - 5) = 180$	Substitution
$9x + 18 = 180$	Simplify.
$9x = 162$	Subtract.
$x = 18$	Divide.

Because lines ℓ and *m* are parallel, ∠4 and ∠5 are supplementary by the Consecutive Interior Angles Theorem.

Determine whether $\overleftrightarrow{AB}$ and $\overleftrightarrow{XY}$ are *parallel, perpendicular,* or *neither*. Graph each line to verify your answer.

47. $A(5, 3), B(8, 0), X(-7, 2), Y(1, 10)$

48. $A(-3, 9), B(0, 7), X(4, 13), Y(-5, 7)$

49. $A(8, 1), B(-2, 7), X(-6, 2), Y(-1, -1)$

Graph the line that satisfies each condition.

50. contains $(-3, 4)$ and is parallel to $\overleftrightarrow{AB}$ with $A(2, 5)$ and $B(9, 2)$

51. contains $(1, 3)$ and is perpendicular to $\overleftrightarrow{PQ}$ with $P(4, -6)$ and $Q(6, -1)$

52. **AIRPLANES** Two Oceanic Airlines planes are flying at the same altitude. Using satellite imagery, each plane's position can be mapped onto a coordinate plane. Flight 815 was mapped at (23, 17) and (5, 11) while Flight 44 was mapped at (3, 15) and (9, 17). Determine whether their paths are *parallel, perpendicular,* or *neither.*

Write an equation in point-slope form of the line having the given slope that contains the given point.

53. $m = 2, (4, -9)$

54. $m = -\frac{3}{4}, (8, -1)$

Write an equation in slope-intercept form of the line having the given slope and y-intercept.

55. m: 5, y-intercept: -3

56. m: $\frac{1}{2}$, y-intercept: 4

Write an equation in slope-intercept form for each line.

57. $(-3, 12)$ and $(15, 0)$

58. $(-7, 2)$ and $(5, 8)$

59. **WINDOW CLEANING** Ace Window Cleaning Service charges $50 for the service call and $20 for each hour spent on the job. Write an equation in slope-intercept form that represents the total cost C in terms of the number of hours h.

Example 10

Graph the line that contains $C(0, -4)$ and is perpendicular to $\overleftrightarrow{AB}$ with $A(5, -4)$ and $B(0, -2)$.

The slope of $\overleftrightarrow{AB}$ is $\frac{-2-(-4)}{0-5}$ or $-\frac{2}{5}$.

Because $-\frac{2}{5}\left(\frac{5}{2}\right) = -1$, the slope of the line perpendicular to $\overleftrightarrow{AB}$ through C is $\frac{5}{2}$.

To graph the line, start at C. Move up 5 units and then right 2 units. Label the point D and draw $\overleftrightarrow{CD}$.

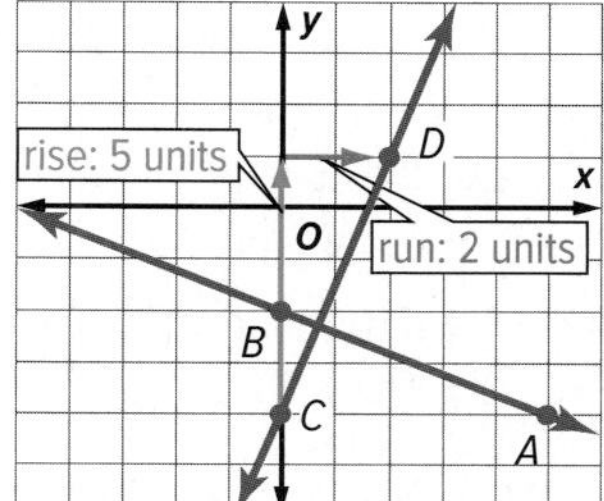

Example 11

Write an equation of the line through (2, 5) and (6, 3) in slope-intercept form.

Step 1 Find the slope of the line through the points.

$m = \frac{y_2 - y_1}{x_2 - x_1}$ Slope Formula

$= \frac{3-5}{6-2}$ $x_1 = 2, y_1 = 5, x_2 = 6$, and $y_2 = 3$

$= \frac{-2}{4}$ or $-\frac{1}{2}$ Simplify.

Step 2 Write an equation of the line.

$y - y_1 = m(x - x_1)$ Point-slope form

$y - 5 = -\frac{1}{2}[x - (2)]$ $m = -\frac{1}{2}, (x_1, y_1) = (2, 5)$

$y - 5 = -\frac{1}{2}x + 1$ Simplify.

$y = -\frac{1}{2}x + 6$ Add 5 to each side.

Study Guide and Review *Continued*

2-9 Proving Lines Parallel

Given the following information, determine which lines, if any, are parallel. State the postulate or theorem that justifies your answer.

60. $m\angle 7 + m\angle 10 = 180$

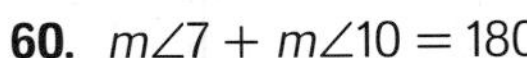

61. $\angle 2 \cong \angle 10$

62. $\angle 1 \cong \angle 3$

63. $\angle 3 \cong \angle 11$

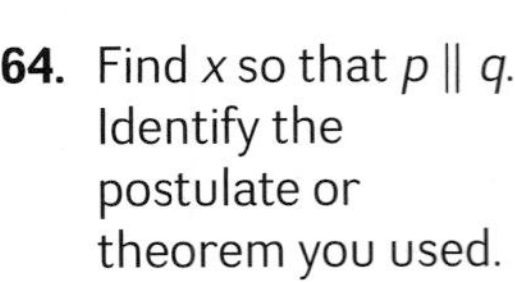

64. Find x so that $p \parallel q$. Identify the postulate or theorem you used.

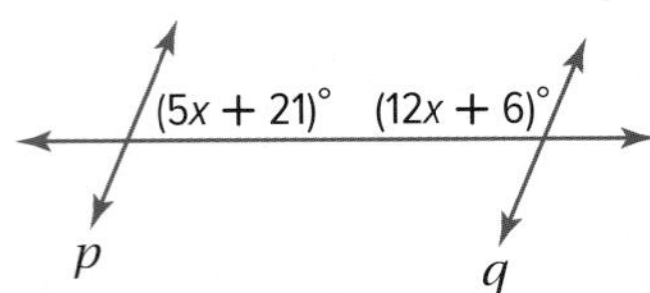

65. **LANDSCAPING** Find the measure needed for $m\angle ADC$ that will make $\overline{AB} \parallel \overline{CD}$ if $m\angle BAD = 45$.

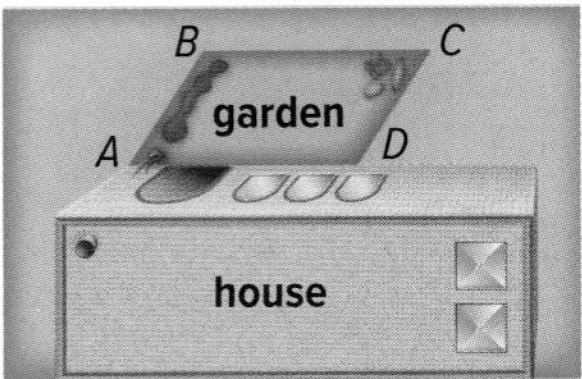

Example 12

Given the following information, determine which lines, if any, are parallel. State the postulate or theorem that justifies your answer.

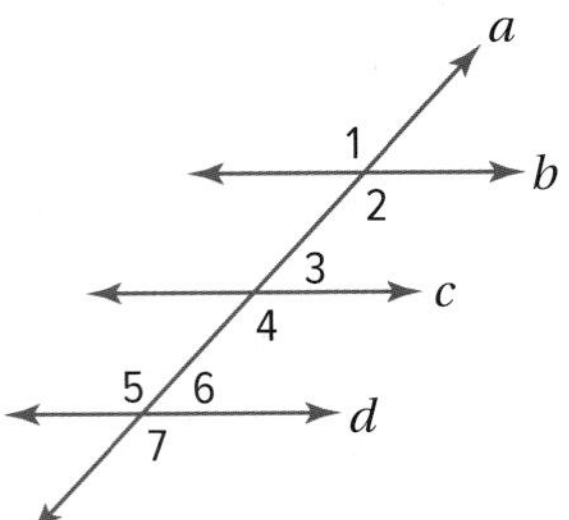

a. $\angle 1 \cong \angle 7$

$\angle 1$ and $\angle 7$ are alternate exterior angles of lines b and d.

Because $\angle 1 \cong \angle 7$, $b \parallel d$ by the Converse of the Alternate Exterior Angles Theorem.

b. $\angle 4 \cong \angle 5$

$\angle 4$ and $\angle 5$ are alternate interior angles of lines c and d.

Because $\angle 4 \cong \angle 5$, $c \parallel d$ by the Converse of the Alternate Interior Angles Theorem.

2-10 Perpendiculars and Distance

Copy each figure. Draw the segment that represents the distance indicated.

66. X to $\overline{VW}$

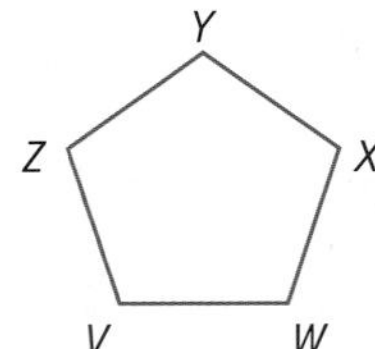

67. L to $\overline{JK}$

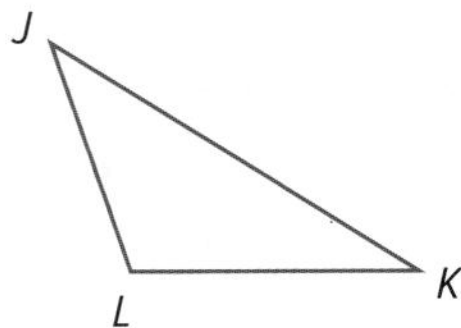

68. **HOME DÉCOR** Scott wants to hang two rows of framed pictures in parallel lines on his living room wall. He first spaces the nails on the wall in a line for the top row. Next, he hangs a weighted plumb line from each nail and measures an equal distance below each nail for the second row. Why does this ensure that the two rows of pictures will be parallel?

Example 13

Copy the figure. Draw the segment that represents the distance from point A to $\overline{CD}$.

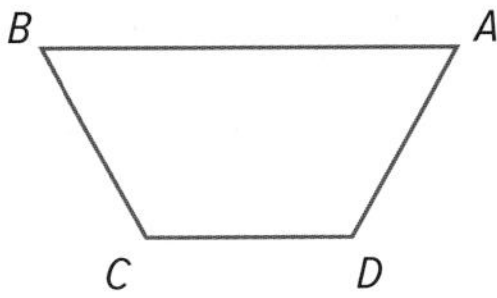

The distance from a line to a point not on the line is the length of the segment perpendicular to the line that passes through the point.

Extend $\overline{CD}$ and draw the segment perpendicular to $\overline{CD}$ from A.

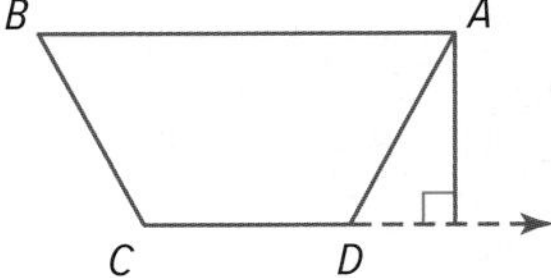

CHAPTER 2

Practice Test

Write a conjecture that describes the pattern in each sequence. Then use your conjecture to find the next item in the sequence.

1. 15, 30, 45, 60

2. 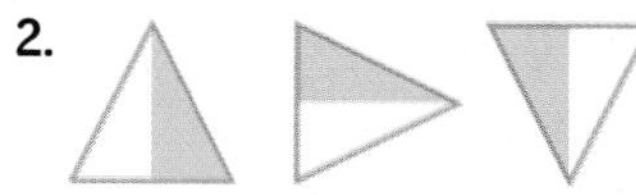

Use the following statements to write a compound statement for each conjunction or disjunction. Then find its truth value.

p: $5 < -3$

q: All vertical angles are congruent.

r: If $4x = 36$, then $x = 9$.

3. p and q

4. $(p \vee q) \wedge r$

5. **PROOF** Write a paragraph proof.

Given: $\overline{JK} \cong \overline{CB}$, $\overline{KL} \cong \overline{AB}$

Prove: $\overline{JL} \cong \overline{AC}$

6. Determine whether the stated conclusion is valid based on the given information. If not, write *invalid.* Explain your reasoning.

Given: If a lawyer passes the bar exam, then he or she can practice law. Candice passed the bar exam.

Conclusion: Candice can practice law.

Determine whether each statement is *always, sometimes,* or *never* true.

7. Two angles that are supplementary form a linear pair.

8. If B is between A and C, then $AC + AB = BC$.

9. If two lines intersect to form congruent adjacent angles, then the lines are perpendicular.

Find the measure of each numbered angle and name the theorems that justify your work.

10. $m\angle 1 = x$, $m\angle 2 = x - 6$

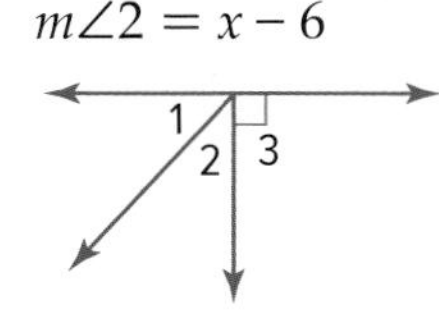

11. $m\angle 7 = 2x + 15$, $m\angle 8 = 3x$

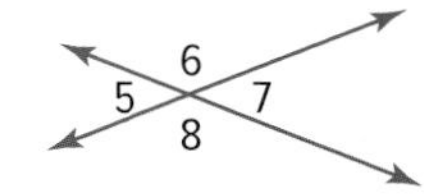

12. **MULTIPLE CHOICE** If a triangle has one obtuse angle, then it is an obtuse triangle.

Which of the following statements is the contrapositive of the conditional above?

A If a triangle is not obtuse, then it has one obtuse angle.

B If a triangle does not have one obtuse angle, then it is not an obtuse triangle.

C If a triangle is not obtuse, then it does not have one obtuse angle.

D If a triangle is obtuse, then it has one obtuse angle.

In the figure, $m\angle 8 = 96$ and $m\angle 12 = 42$. Find the measure of each angle. Tell which postulate(s) or theorem(s) you used.

13. $\angle 9$

14. $\angle 11$

15. $\angle 6$

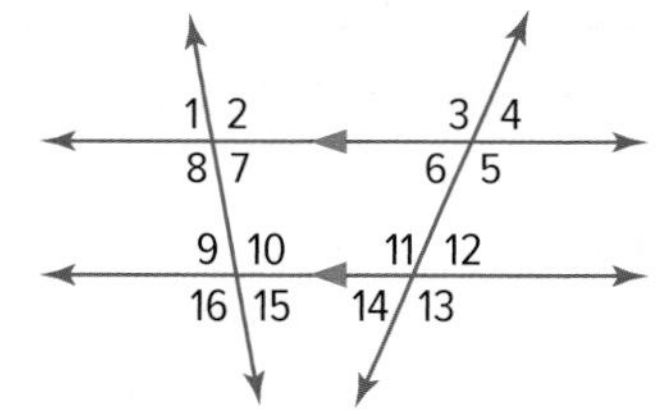

Determine the slope of the line that contains the given points.

16. $A(0, 6)$, $B(4, 0)$

17. $E(5, 4)$, $F(8, 1)$ -1

Write an equation in slope-intercept form for each line described.

18. passes through $(0, 7)$, parallel to $y = 4x - 19$

19. passes through $(-8, 1)$, perpendicular to $y = 2x - 17$

20. Find x so that $a \parallel b$. Identify the postulate or theorem you used.

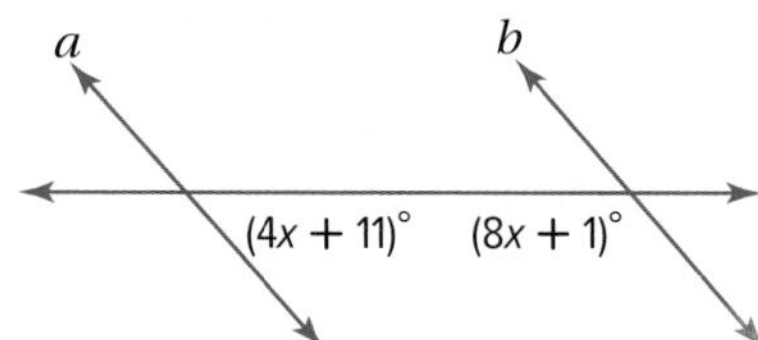

21. Line ℓ contains points $(-4, 2)$ and $(3, -5)$. Find the distance from line ℓ to point $P(1, 2)$.

22. Write an equation in slope-intercept form for the perpendicular bisector of $\overline{AB}$ where $A(-2, 4)$ and $B(4, 6)$.

CHAPTER 2

Preparing for Assessment

Performance Task

Provide a clear solution to each part of the task. Be sure to show all of your work, include all relevant drawings, and justify your answers.

EVENT PLANNING Members of a charitable organization and city council members participate in a planning committee for an annual benefit race and vendor fair to raise money for a cause that improves life in the city.

Part A

The planning committee is deciding on the best place to hold the event. A large park might work, but there are too many participants for them to race along the same paths. The park is a large rectangle, as shown to the right. The planning committee proposes that the participants begin the race at corner *A* of the park but half of the runners would run East to corner *B* and half of the runners would run South to corner *D*. The finish line would be at corner *C*.

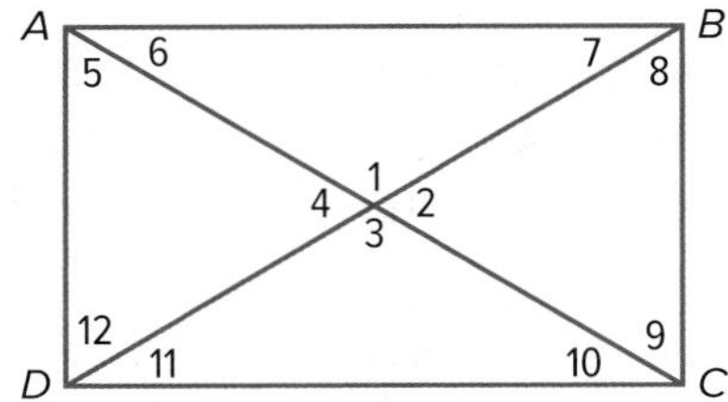

1. **Critique** Write a proof that shows that each group of participants would have to run the same distance to complete the race.

Part B

The vendor fair will be set up along the diagonal paths that run from corners *A* to *C* and from corners *B* to *D*. The vendors will purchase a location along the paths to set up their booths. The most expensive vendor locations will be at points of intersection. The price will correspond to the size of the angle, because the larger angles will provide the vendor with more space.

2. **Perseverance** Identify all sets of congruent angles in the park.
3. Identify all pairs of complementary angles.
4. Identify all pairs of supplementary angles.

Part C

The following year, even more participants sign up. The planning committee decides that the park will not be large enough to accommodate the race participants, so they the race onto the city streets. They plan the route on a grid. Each unit is one square city block, which has a length of one-eighth mile. The origin represents the starting and ending point, which is also where the first aid team will be in case of injuries. The planning committee wants to ensure that the first aid team is within one-half mile of all points on the route.

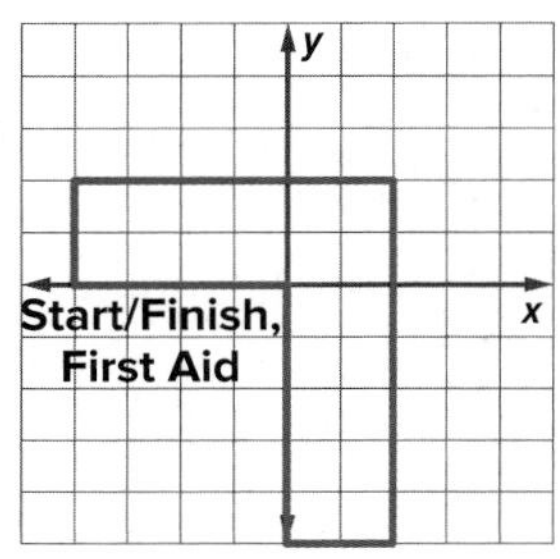

5. Determine whether the route satisfies this constraint. Explain why or why not.

Part D

The proposed route has a hill at the very beginning of the race. The planning committee wants to use the slope of the hill to advertise the event to athletes who might want to participate. A local engineer provides a map with the horizontal length and height of the hill.

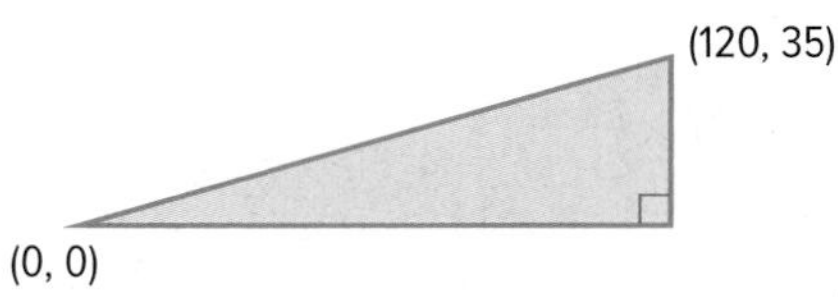

6. What is the slop of the hill?

Test-Taking Strategy

Example

Read the problem. Identify what you need to know. Then use the information in the problem to solve.

In a school of 292 students, 94 participate in sports, 122 participate in academic clubs, and 31 participate in both. How many students at the school do not participate in sports or academic clubs?

A 95 **C** 122

B 107 **D** 138

Step 1 **What is the problem asking you to solve?**
I need to find the number of students who do not participate in clubs or sports.

Step 2 **Can a Venn diagram help you solve the problem? If so, how?**
Yes. Draw overlapping circles to show the number of students who participate in sports and the number of students who participate in clubs. The overlapping region shows the number of students who participate in both.

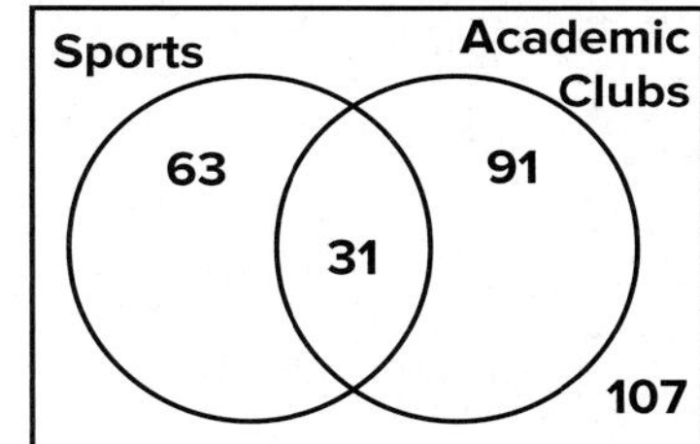

Step 3 **How can you check that the Venn diagram is reasonable? What is the correct answer?**
I should make sure the totals represented by each region of the completed diagram match the given information in the problem.

The correct answer is B.

Test-Taking Tip

Venn Diagrams
Solving geometry problems frequently requires the use of logical reasoning. You can use a Venn diagram to help you solve logical-reasoning problems on standardized tests.

Apply the Strategy

Read the question. Identify what you need to know. Then use the information in the problem to solve.

A random sample of 195 people were asked whether they had seen a comedy movie or an action movie in the past week. Of those surveyed, 76 had seen an action movie, 24 had seen both types of movies, and 88 had seen neither. How many people had seen a comedy movie in the past week?

A 7 **C** 55

B 31 **D** 107

Answer the questions.

a. What is the problem asking you to solve?

b. Can a Venn diagram help you solve the problem? If so, how?

c. What is the correct answer?

CHAPTER 2

Preparing for Assessment

Cumulative Assessment

Read each question. Then fill in the correct answer on the answer document provided by your teacher or on a sheet of paper.

1. Which of the following conditional statements has a true converse?

- ○ **A** If two angles are vertical angles, then the angles are congruent.
- ○ **B** Supplementary angles are angles with measures that have a sum of 180°.
- ○ **C** All angles are geometric figures.
- ○ **D** If an angle is a right angle, then it is not an acute angle.

2. Which of the following statements is an example of a theorem? Select all that apply.

- ☐ **A** If $a = b$, then $b = a$.
- ☐ **B** If $a = b$ and $c \neq 0$, then $\frac{a}{c} = \frac{b}{c}$.
- ☐ **C** If $a + 5 = b + 6$, then $a = b + 1$.
- ☐ **D** If triangle ABC has angle measures $a = 90$ and $b = 60$, then $c = 30$.

3. What is the measure of $\angle CED$ in degrees?

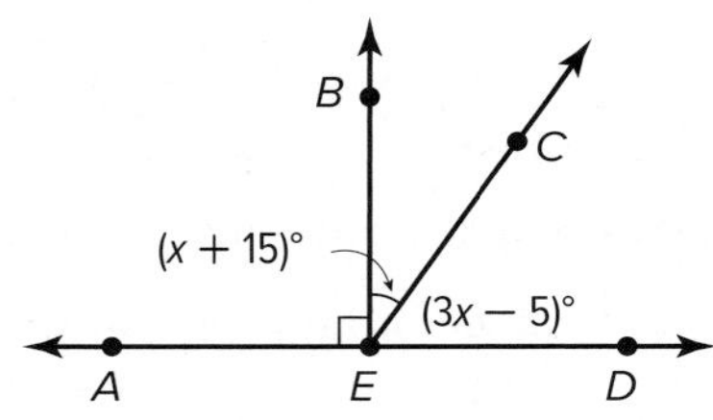

4. Identify the pairs of congruent angles. Select all that apply.

- ☐ **A** $\angle 1$ and $\angle 2$
- ☐ **B** $\angle 1$ and $\angle 5$
- ☐ **C** $\angle 2$ and $\angle 4$
- ☐ **D** $\angle 2$ and $\angle 7$
- ☐ **E** $\angle 4$ and $\angle 5$
- ☐ **F** $\angle 4$ and $\angle 6$

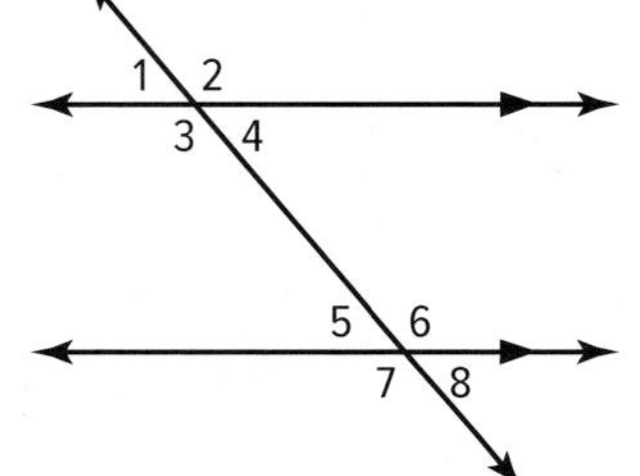

5. The table shows three statements that Carmina wrote in her notebook.

p	*Line segment* is a defined term.
q	*Plane* is an undefined term.
r	*Collinear* is a defined term.

Which of the following compound statements is true?

- ○ **A** p and q
- ○ **B** p and $\sim r$
- ○ **C** $\sim q$ or $\sim r$
- ○ **D** $\sim q$ and r

6. Which of the following statements about the figure is false?

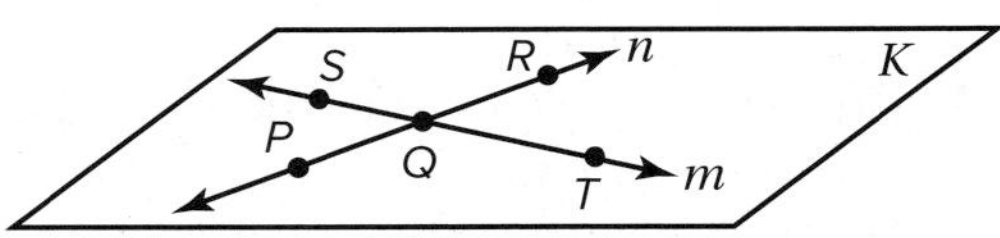

- ○ **A** Points P, Q, and R are collinear.
- ○ **B** Because point Q is between points P and R, $PQ + QR = PQ$.
- ○ **C** Plane K is a flat surface made up of points that extends infinitely in all directions.
- ○ **D** Lines m and n intersect only at point Q.

7. Points P, Q, and R are collinear, and Q lies between P and R.

$PQ = 2x + 5$ and $PR = 4x + 10$.

a. Find an expression in terms of x for QR.

b. What can you conclude about point Q?

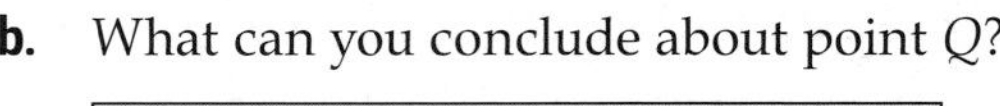

Test-Taking Tip

Question 5 An *and* statement is true only when both statements that form it are true. An *or* statement is true if at least one of the statements that form it is true.

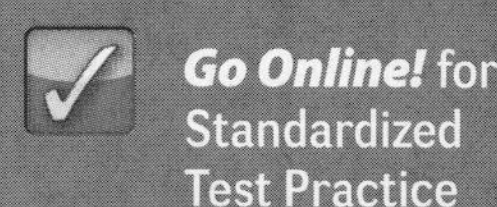

Go Online! for Standardized Test Practice

8. Tyrell was asked to write the following proof.

Given: $\overline{KL} \cong \overline{KM}$

Prove: $\overline{JK} + \overline{KM} = \overline{JL}$

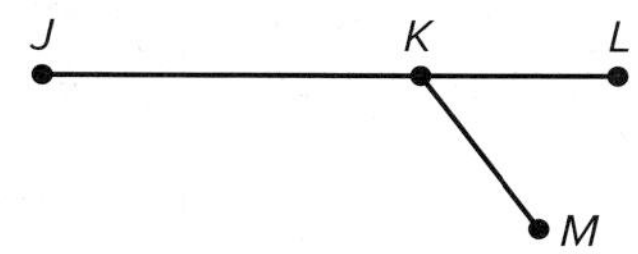

Which of the following should Tyrell use as a reason in his proof?

- ◯ **A** Subtraction Property of Equality
- ◯ **B** Definition of midpoint
- ◯ **C** Segment Addition Postulate
- ◯ **D** Transitive Property of Congruence

9. Joshua made a sequence of figures using toothpicks, as shown.

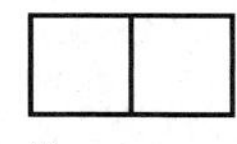

1 square 4 toothpicks	2 squares 7 toothpicks	3 squares 10 toothpicks

Which of the following is the best conjecture Joshua can make based on the figures?

- ◯ **A** A row of n squares consists of $4n$ toothpicks.
- ◯ **B** A row of n squares consists of $3n + 1$ toothpicks.
- ◯ **C** A row of n squares consists of $4n - 1$ toothpicks.
- ◯ **D** A row of n squares consists of $n + 3$ toothpicks.

10. $\angle M$ and $\angle N$ are vertical angles. $m\angle M = 6x - 2$ and $m\angle N = 4x + 38$. What is the measure of $\angle M$ in degrees?

11. Use the diagram to complete the proof.

Given: $\overline{AB} \perp \overline{CD}$

$\angle AOD$ is bisected by EF.

Prove: $\angle BOG = 15°$

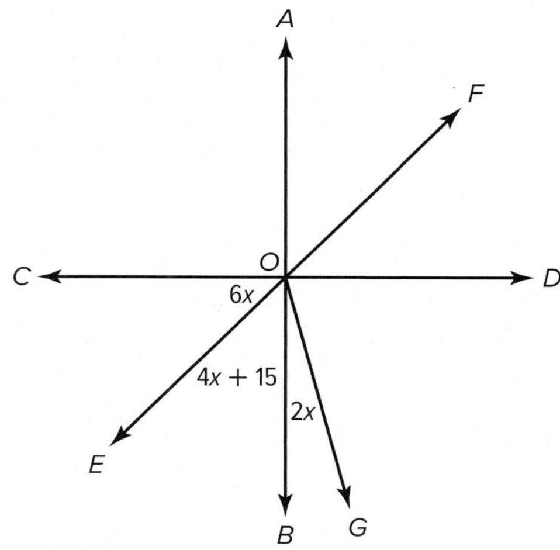

	Statement		Reason
	$AB \perp CD$		Given
	$\angle AOD$ is bisected by EF.		Given
	$m\angle AOD = m\angle COB = 90°$	**a.**	
b.			Definition of a bisector
c.	$m\angle COE = m\angle FOD$; $m\angle EOB =$		Vertical angles theorem
d.	$6x +$ $\quad = 90$		Definition of a complementary angle
	$10x + 15 = 90$	**e.**	
f.		**g.**	
h.	$x =$		Division Property of Equality
	$\angle BOG = 2x = 2(7.5) = 15$	**i.**	

12. Write the equation of the line that is perpendicular to the line $y = \frac{1}{4}x - 9$ and passes through the point (0, 2).

Need Extra Help?

If you missed Question...	1	2	3	4	5	6	7	8	9	10	11	12
Go to Lesson...	2-2	2-5	2-6	2-7	2-2	2-4	2-5	2-5	2-1	2-6	2-6	2-9

CHAPTER 3
Rigid Transformations and Symmetry

THEN

You created logical arguments and identified line relationships.

NOW

In this chapter, you will:

- Name and draw figures that have been reflected, translated, rotated, or dilated.
- Recognize and draw compositions of transformations.
- Identify symmetry in two- and three-dimensional figures.

WHY

PHOTOGRAPHY Photographers use reflections, rotations, and symmetry to make photographs interesting and visually appealing.

Use the Mathematical Practices to complete the activity.

1. Use Tools Take a picture that shows a reflection or search the Internet for such an image.

2. Sense-Making What do you already know about reflection? Can you identify the line of reflection in the image you took or selected?

3. Apply Math Use the tools in the Geometer's Sketchpad to draw the primary objects in your photograph. Use the transformations tool to reflect your drawing. Move the line of reflection and discuss the results with the class.

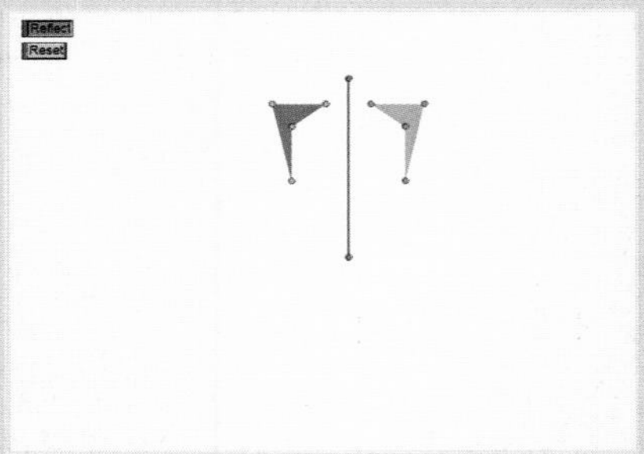

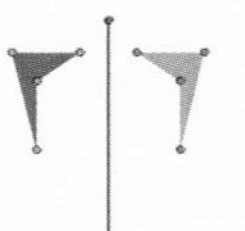

Michael DeYoung/Blend Images

Go Online to Guide Your Learning

Explore & Explain

Transformations Tool

Use the **Transformations** tool to practice working with reflections in the coordinate plane in Lesson 3-1.

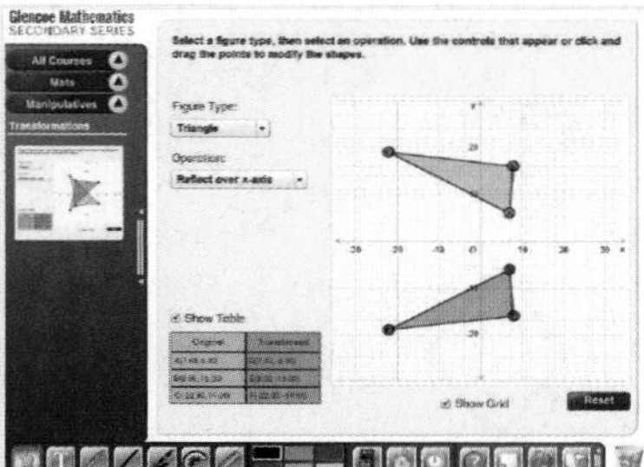

Geometry Tools

Use the **Geometry Tools** to explore the effects of transformations on geometric figures in Lesson 3-4.

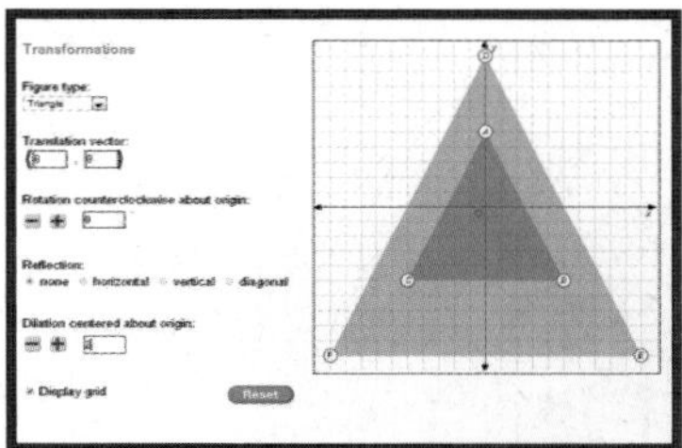

eBook

Interactive Student Guide

Before starting the chapter, answer the **Chapter Focus** preview questions. Check your answers as you complete each lesson. At the end of the chapter, try the **Performance Task**.

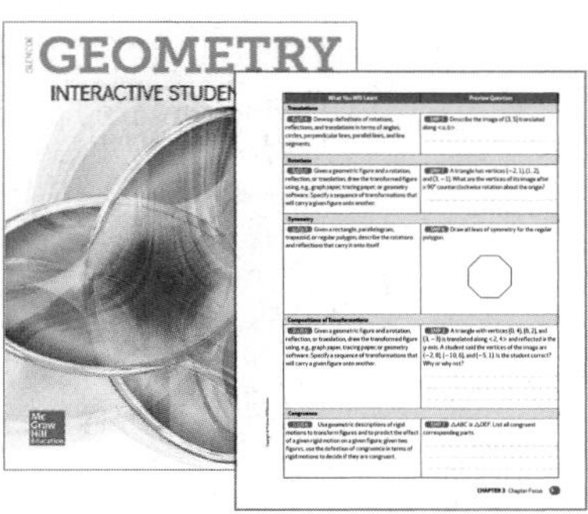

Organize

Foldables

Make this **Transformations and Symmetry** foldable to help you organize your notes. Begin with three sheets of notebook paper, folded in half. Fold each paper lengthwise to form a pocket, glue the sheets side-by-side, and label each pocket.

Collaborate

Chapter Project

In the **Graphic Design** project, you will use what you have learned about transformation of polygons to complete a project.

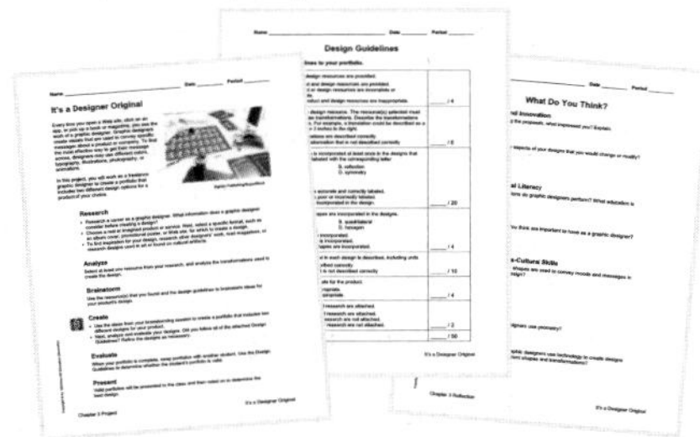

Focus

LEARNSMART®

Need help studying? **Complete the Congruence, Proof, and Constructions** domain in LearnSmart to review for the chapter test.

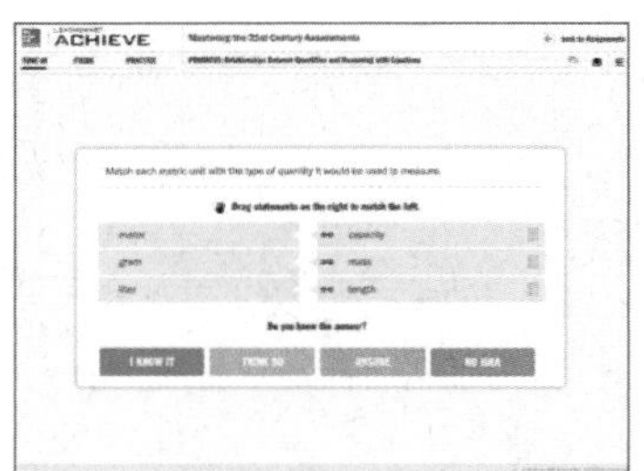

ALEKS®

You can use the **Similarities and Transformations** topic in ALEKS to explore what you know about transformations and what you are ready to learn.*

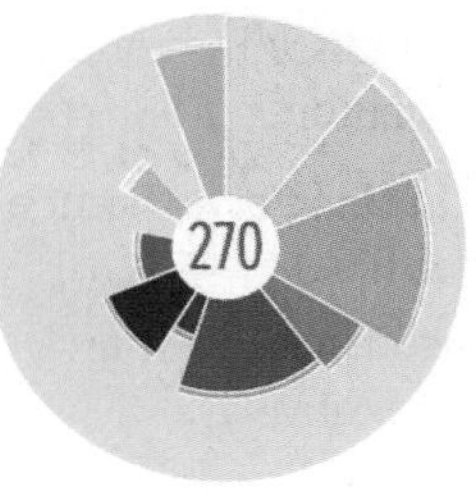

* Ask your teacher if this is part of your program.

Go Online! for Vocabulary Review Games and key vocabulary in 13 languages

Get Ready for the Chapter

Connecting Concepts

Concept Check

Identify the type of congruence transformation shown as a *reflection*, *translation*, or *rotation*.

1. Figure A to Figure B
2. Figure D to Figure A
3. Figure A to Figure C

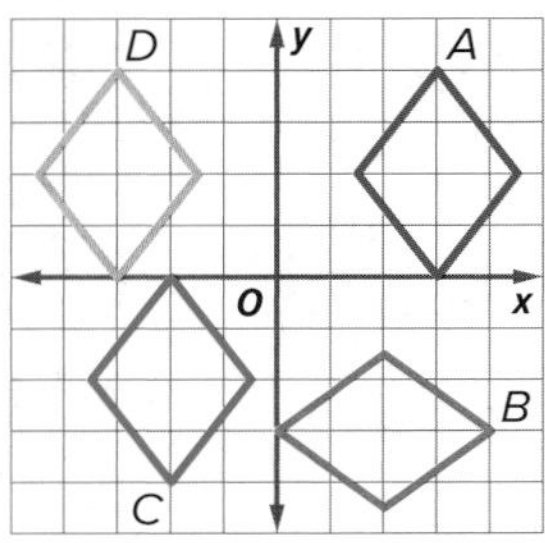

Find the coordinates of the image after performing the indicated transformation on each figure.

4. $\triangle JKL$; reflection in y-axis
5. $\square ABCD$; translation along vector $\langle 3, -2 \rangle$
6. If a drummer in the marching band moves from position (1, 4) to position (5, 1), how would you write the component form of the vector to describe his movement?

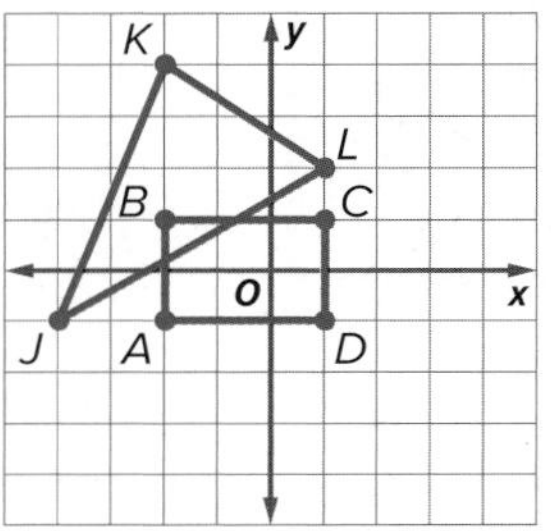

Performance Task Review

You can use the concepts and skills in the chapter to solve problems in a real-world setting. Understanding coordinate grids will help you finish the Performance Task at the end of the chapter.

MP In this Performance Task you will:

- reason abstractly and quantitatively
- use appropriate tools strategically

New Vocabulary

English		Español
line of reflection	p. 221	línea de reflexión
center of rotation	p. 240	centro de rotación
angle of rotation	p. 240	ángulo de rotación
composition of transformations	p. 249	composición de transformaciones
symmetry	p. 259	símetria
reflectional symmetry	p. 259	símetria de reflectional
line of symmetry	p. 259	eje de símetria

Review Vocabulary

reflection reflexión a transformation representing a flip of the figure over a point, line or plane

rotation rotación a transformation that turns every point of a preimage through a specified angle and direction about a fixed point

translation traslación a transformation that moves all points of a figure the same distance in the same direction

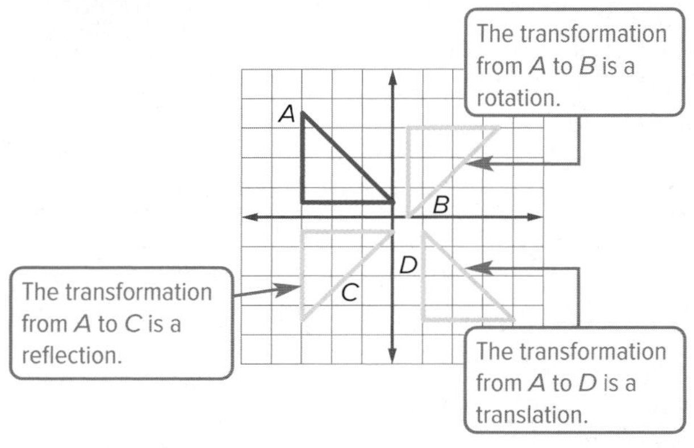

EXPLORE 3-1

Geometry Lab

Reflections

Recall that a reflection is a flip over a line called the line of reflection.

Mathematical Practices
7 Look for and make use of structure.

A reflection is a rigid transformation that preserves congruence.

In this activity, you will use tracing paper to explore reflections.

Activity 1 Reflect a Figure in the *y*-Axis

Given quadrilateral *ABCD*.

Step 1 Trace the figure onto a piece of tracing paper.

Step 2 Use the tracing paper to reflect the figure in the *y*-axis.

Step 3 Record the coordinates of the vertices of the image and preimage in a table like the one shown.

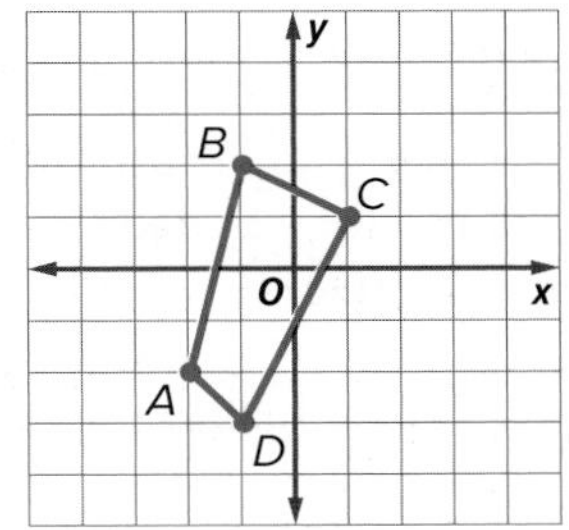

	X	Y		X	Y
A	−2	−2	**A′**		
B	−1	2	**B′**		
C	1	1	**C′**		
D	−1	−3	**D′**		

1A. What pattern do you observe between the coordinates of the preimage and image?

1B. Devise a rule that could be used to find the coordinates of the image of a figure reflected in the *y*-axis.

Activity 2 Reflect a Figure in the *x*-Axis

Given quadrilateral *JKLM*.

Step 1 Trace the figure onto a piece of tracing paper.

Step 2 Use the tracing paper to reflect the figure in the *x*-axis.

Step 3 Record the coordinates of the vertices of the image and preimage in a table like the one shown.

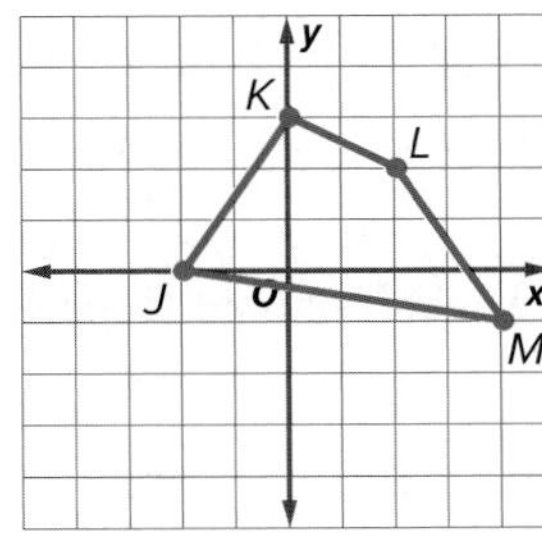

	X	Y		X	Y
J	−2	0	**J′**		
K	0	3	**K′**		
L	2	2	**L′**		
M	4	−1	**M′**		

2A. What pattern do you observe between the coordinates of the preimage and image?

2B. Devise a rule that could be used to find the coordinates of the image of a figure reflected in the *x*-axis.

(continued on the next page)

EXPLORE 3-1

Geometry Lab

Reflections *Continued*

Rigid motion transformations can be used to show that two figures are congruent by the **principle of superposition**. This states that two figures are congruent if and only if there is a rigid motion or a series of rigid motions that maps one figure exactly onto the other.

You can use tracing paper to determine whether two figures on the coordinate plane are congruent. If the figures are congruent, you can also determine and describe the rigid motion that maps one figure onto the other.

Activity 3 Determine Congruency

Triangle *ABC* has vertices *A*(−4, 2), *B*(−5, −3), and *C*(3, 1), and triangle *DEF* has vertices *D*(2, −4), *E*(−3, −5), and *F*(1, 3). Determine whether △*ABC* is congruent to △*DEF*.

Step 1 Trace △*ABC* onto a piece of tracing paper.

Step 2 Move the tracing paper around the coordinate plane using rigid motions to determine if △*ABC* is congruent to △*DEF*.

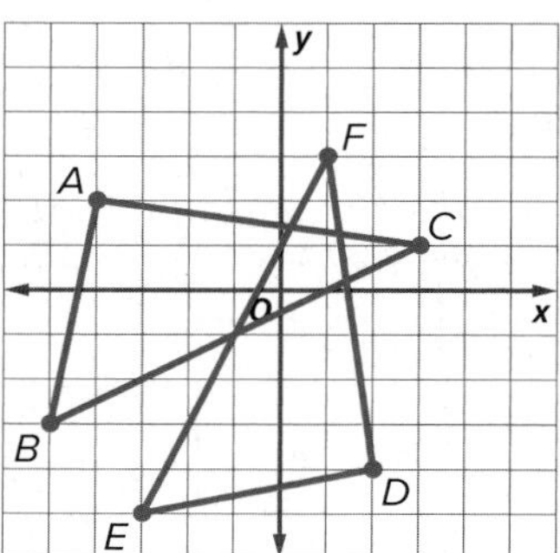

3A. Is △*ABC* congruent to △*DEF*? If so, what rigid motion maps △*ABC* onto △*DEF*?

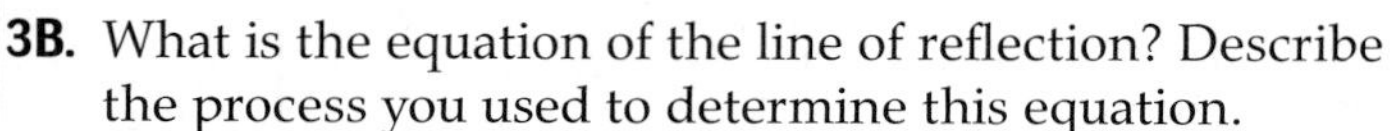

3B. What is the equation of the line of reflection? Describe the process you used to determine this equation.

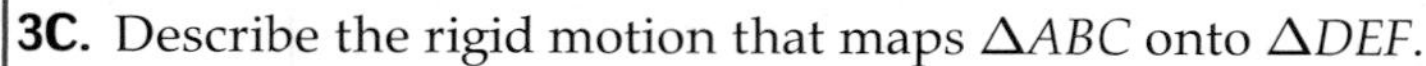

3C. Describe the rigid motion that maps △*ABC* onto △*DEF*.

Exercises

Find the coordinates of the image of each figure under the given reflection.

1. *A*(4, 2), *B*(3, 2), *C*(−1, 1); *y*-axis

2. *J*(−4, 0), *K*(−2, 4), *L*(3, −1); *x*-axis

Determine whether each pair of figures is congruent. Explain.

3.

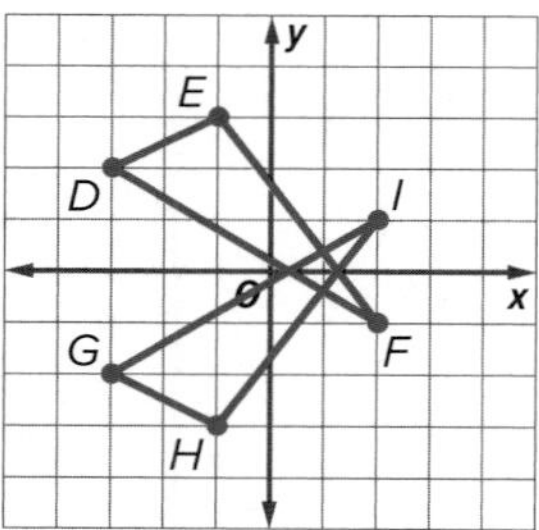

4.

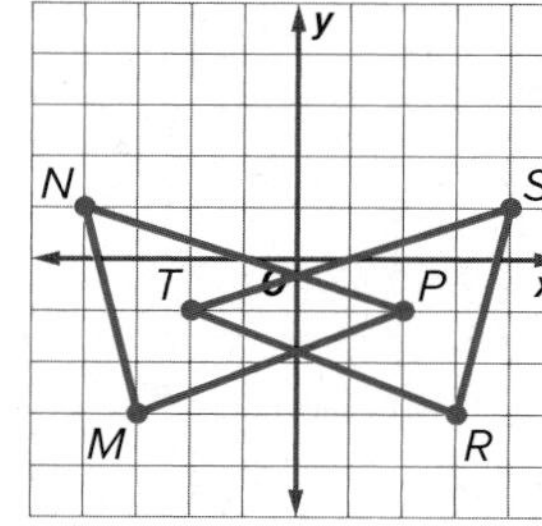

LESSON 1
Reflections

∴Then	∴Now	∴Why?
• You identified reflections and verified them as congruence transformations.	• **1** Given a geometric figure and a reflection, draw the transformed figure. **2** Describe the effects of reflections on the coordinate plane.	• Notice in this water reflection that the distance a point lies above the water line appears the same as the distance its image lies below the water.

New Vocabulary
line of reflection

Mathematical Practices

5 Use appropriate tools strategically.

1 Draw Reflections In Lesson 1-7, you learned that a reflection or *flip* is a transformation in a line called the **line of reflection**. Each point of the preimage and its corresponding point on the image are the same distance from this line.

Key Concept Reflection in a Line

A reflection in a line is a function that maps a point to its image such that

- if the point is on the line, then the image and preimage are the same point, or
- if the point is not on the line, the line is the perpendicular bisector of the segment joining the two points.

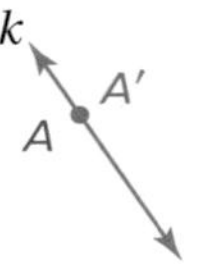

A is on line *k*.

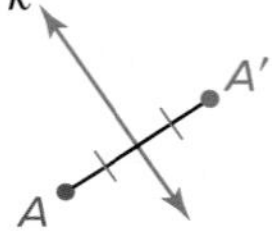

A is not on line *k*.

A′, *A″*, *A‴*, and so on, name corresponding points for one or more transformations.

To reflect a polygon in a line, reflect each of the polygon's vertices. Then connect these vertices to form the reflected image.

Example 1 Reflect a Figure in a Line

Copy the figure and the given line of reflection. Then draw the reflected image in this line using a ruler.

Step 1 Draw a line through each vertex that is perpendicular to line k.

Step 2 Measure the distance from point A to line k. Then locate A' the same distance from line k on the opposite side.

Step 3 Repeat Step 2 to locate points B' and C'. Then connect vertices A', B', and C' to form the reflected image.

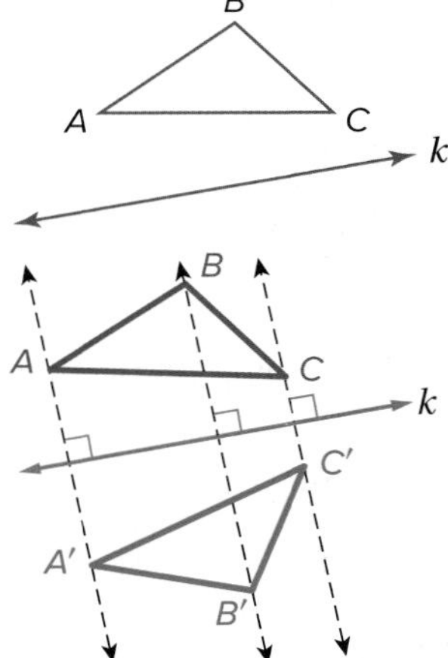

Guided Practice

1A.

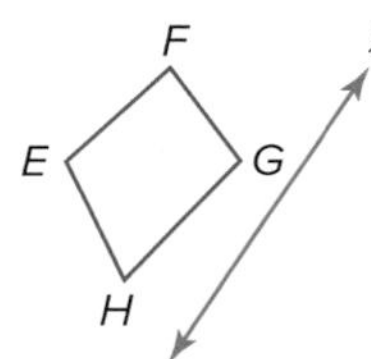

1B.

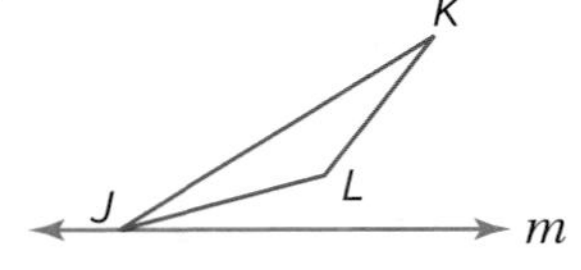

1C.

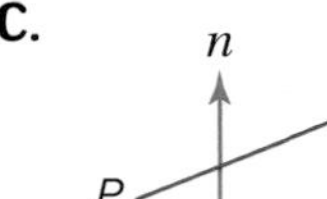

Thomas Hinsche/Getty Images

Recall that a reflection is a *congruence transformation* or *isometry*. That means the preimage and the image are congruent figures: they have the same size and shape.

Real-World Career

Photographer
Photographers take photos for a variety of reasons such as journalism, art, to record an event, or for scientific purposes. In some photography fields such as photojournalism and scientific photography, a bachelor's degree is required. For others, such as portrait photography, technical proficiency is the only requirement.

Real-World Example 2 Minimize Distance by Using a Reflection

SHOPPING Suppose you are going to shop in Store B, return to your car, and then go to Store G. Where along line *s* of parking spaces should you park to minimize the distance you will walk?

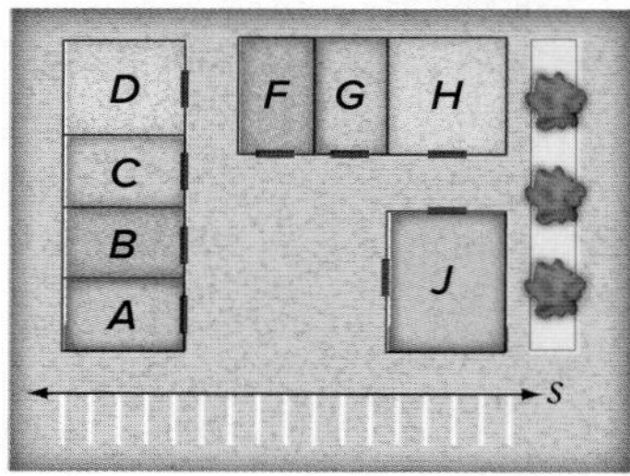

Understand You are asked to locate a point P on line s such that $BP + PG$ has the least possible value.

Plan If B, P, and G were collinear, the shortest distance between the three points would be the straight line connecting B to G. Use the reflection of point B in line s to find the location for point P.

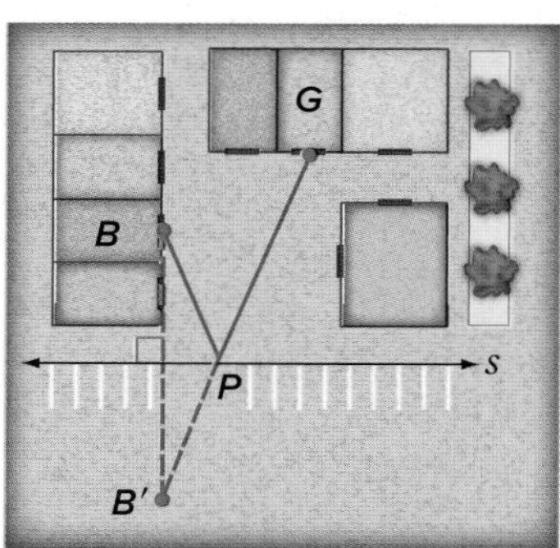

Solve Find B' by reflecting B across the line s. Draw $\overline{B'G}$. Locate P at the intersection of line s and $\overline{B'G}$.

Check $\overline{BP} \cong \overline{B'P}$, as long as P is located along the line of reflection s. You should park at the point of intersection between $\overline{B'G}$ and s to minimize the distance you will walk.

The shortest distance between 2 points is a straight line. If you park at some other location along s, the path from B' to P to G is no longer a straight line. Thus, parking anywhere else along s will result in a longer walk.

Guided Practice

2. **TICKET SALES** Joy wants to select a good location to sell tickets for a dance. Locate point P such that the distance someone would have to walk from Hallway A, to point P on the wall, and then to their next class in Hallway B is minimized.

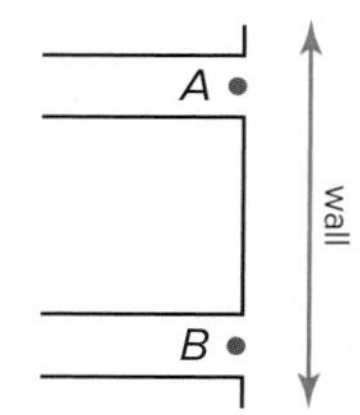

2 Describe Reflections in the Coordinate Plane

Reflections can also be performed in the coordinate plane by using the techniques presented in Example 3.

Study Tip

MP Sense-Making Reflections, like all isometries, preserve distance, angle measure, betweenness of points, and collinearity. Because a reflection is a congruence transformation, a preimage and its image are congruent; the orientation of a preimage and its image, however, are reversed.

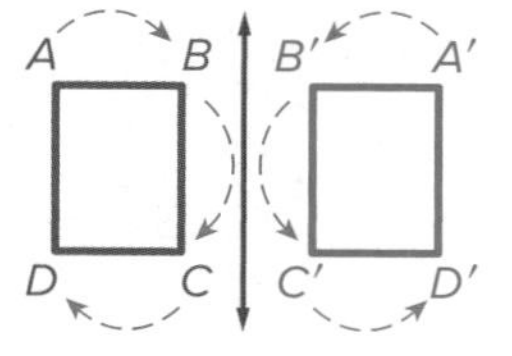

Example 3 Reflect a Figure in a Horizontal or Vertical Line

Triangle *JKL* has vertices $J(0, 3)$, $K(-2, -1)$, and $L(-6, 1)$. Graph $\triangle JKL$ and its image in the line $x = -4$.

Find a corresponding point for each vertex so that a vertex and its image are equidistant from the line $x = -4$.

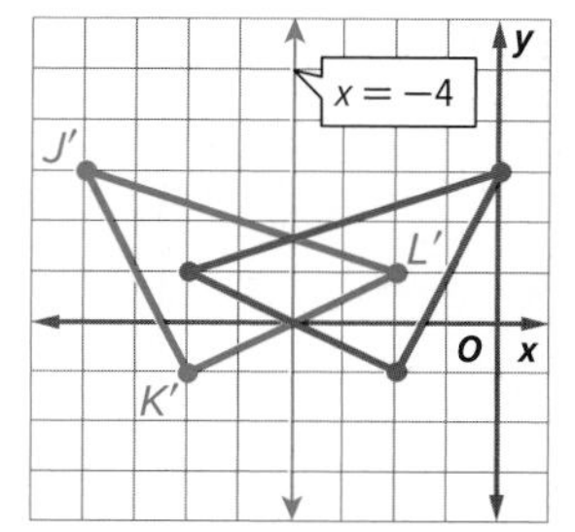

EpicStockMedia/Getty Images

Guided Practice

Trapezoid *RSTV* has vertices $R(-1, 1)$, $S(4, 1)$, $T(4, -1)$, and $V(-1, -3)$. Graph trapezoid *RSTV* and its image in the given line.

3A. $y = -3$

3B. $x = 2$

When the line of reflection is the x- or y-axis, you can use the following rule.

Key Concept Reflection in the x- or y-axis

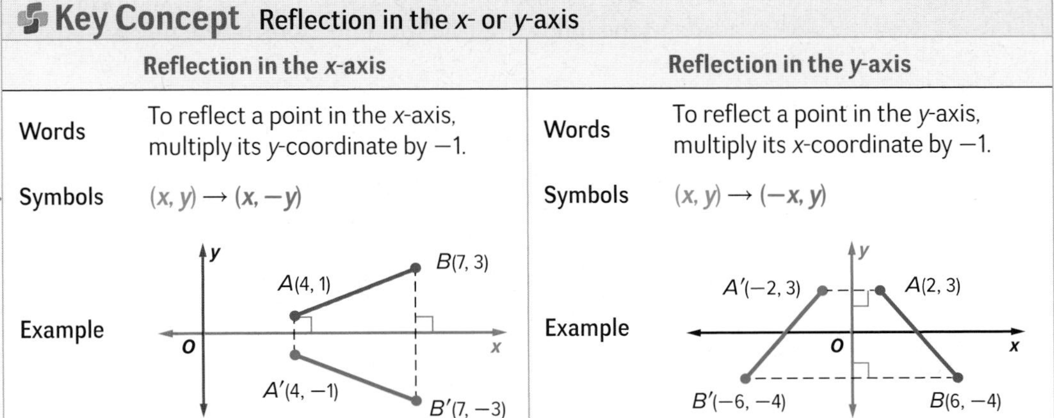

	Reflection in the x-axis		Reflection in the y-axis
Words	To reflect a point in the x-axis, multiply its y-coordinate by -1.	Words	To reflect a point in the y-axis, multiply its x-coordinate by -1.
Symbols	$(x, y) \rightarrow (x, -y)$	Symbols	$(x, y) \rightarrow (-x, y)$
Example		Example	

Reading Math

Coordinate Notation The expression $P(a, b) \rightarrow P'(a, -b)$ can be read as *point P with coordinates a and b is mapped to new location P prime with coordinates a and negative b.*

Example 4 Reflect a Figure in the x- or y-axis

a. Graph $\triangle ABC$ with vertices $A(-5, 3)$, $B(2, 0)$, and $C(1, 2)$ and its reflection in the x-axis.

Multiply the y-coordinate of each vertex by -1.

$(x, y) \rightarrow (x, -y)$

$A(-5, 3) \rightarrow A'(-5, -3)$

$B(2, 0) \rightarrow B'(2, 0)$

$C(1, 2) \rightarrow C'(1, -2)$

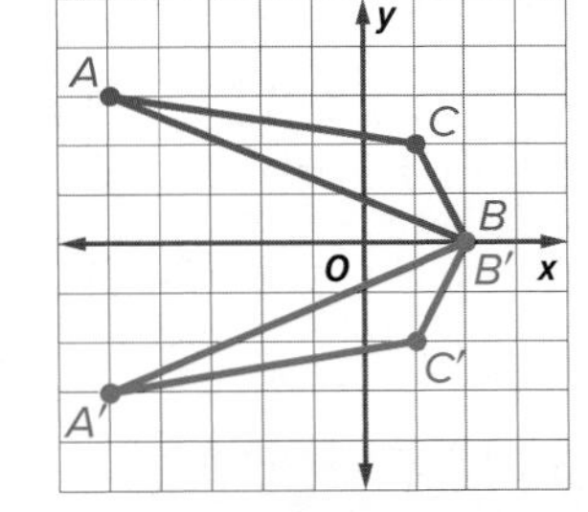

Study Tip

Invariant Points In Example 4a, point B is called an *invariant point* because it maps onto itself. Only points that lie on the line of reflection are invariant under a reflection.

b. Describe the transformation shown using coordinate notation.

The points in the preimage map to the points in the image as described below.

$P(-4, 1) \rightarrow P'(4, 1)$

$Q(2, 3) \rightarrow Q'(-2, 3)$

$R(2, -1) \rightarrow R'(-2, -1)$

$S(-4, -3) \rightarrow S'(4, -3)$

$(x, y) \rightarrow (-x, y)$

The transformation is a reflection in the y-axis.

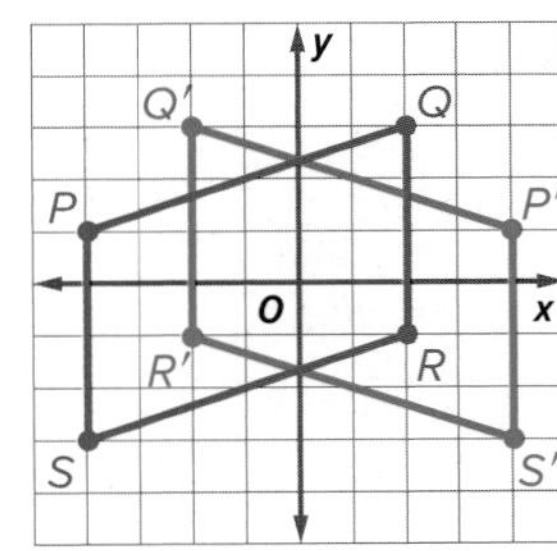

Guided Practice

4A. Given rectangles with vertices $E(-4, -1)$, $F(2, 2)$, $G(3, 0)$, and $H(-3, -3)$ and $E'(-4, 1)$, $F'(2, -2)$, $G'(3, 0)$, and $H'(-3, 3)$, describe the transformation using coordinate notation.

4B. Graph $\triangle JKL$ with vertices $J(3, 2)$, $K(2, -2)$, and $L(4, -5)$ and its reflection in the y-axis.

You can also reflect a figure in the line $y = x$.

Review Vocabulary

Perpendicular Lines Two nonvertical lines are perpendicular if and only if the product of their slopes is −1.

The slope of $y = x$ is 1. In the graph shown, $\overline{CC'}$ is perpendicular to $y = x$, so its slope is −1. From $C(-3, 2)$, move right 2.5 units and down 2.5 units to reach $y = x$. From this point on $y = x$, move right 2.5 units and down 2.5 units to locate $C'(2, -3)$. Using a similar method, the image of $D(-3, -1)$ is found to be $D'(-1, -3)$.

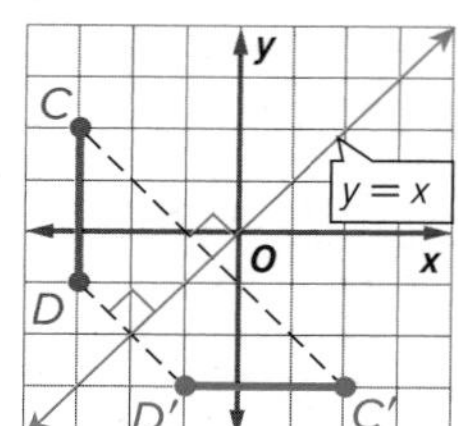

Comparing the coordinates of these and other examples leads to the following rule for reflections in the line $y = x$.

Key Concept Reflection in Line $y = x$

Words To reflect a point in the line $y = x$, interchange the x- and y-coordinates.

Symbols $(x, y) \rightarrow (y, x)$

Example

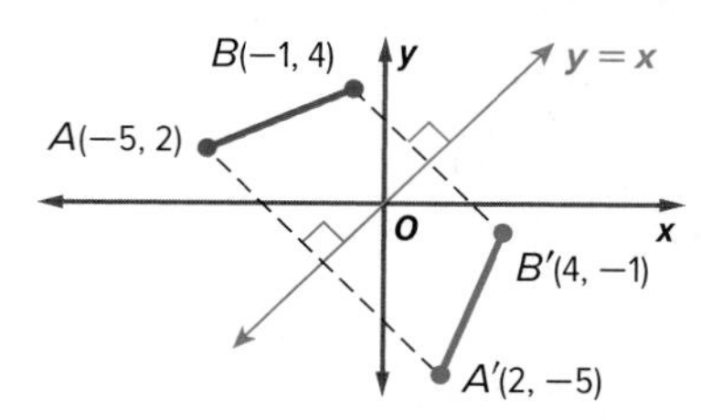

Example 5 Reflect a Figure in the Line $y = x$

Quadrilateral $JKLM$ has vertices $J(2, 2)$, $K(4, 1)$, $L(3, -3)$, and $M(0, -4)$. Graph $JKLM$ and its image $J'K'L'M'$ in the line $y = x$.

Interchange the x- and y-coordinates of each vertex.

(x, y)	→	(y, x)
$J(2, 2)$	→	$J'(2, 2)$
$K(4, 1)$	→	$K'(1, 4)$
$L(3, -3)$	→	$L'(-3, 3)$
$M(0, -4)$	→	$M'(-4, 0)$

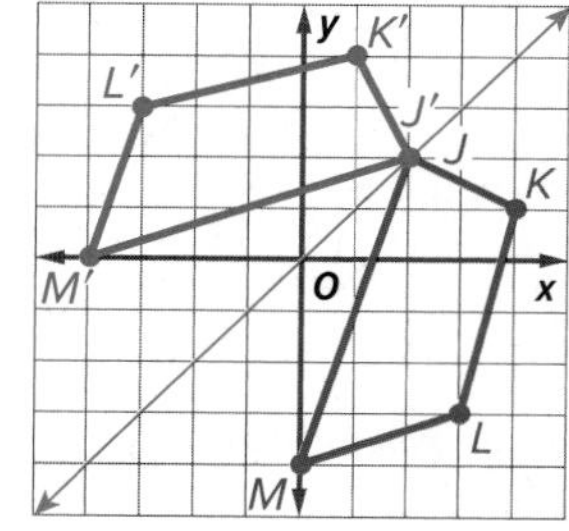

Study Tip

Preimage and Image In this book, the preimage will always be blue and the image will always be green.

Go Online!

Investigate reflections by using the **Geometry Tools** in ConnectED.

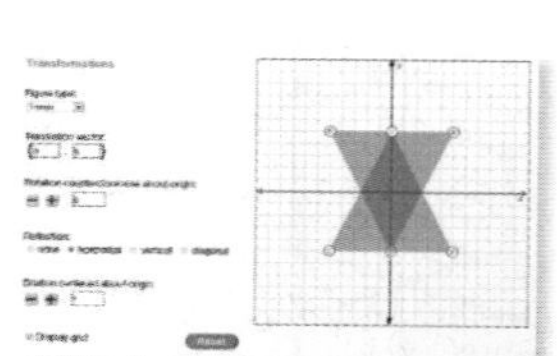

Guided Practice

5. $\triangle BCD$ has vertices $B(-3, 3)$, $C(1, 4)$, and $D(-2, -4)$. Graph $\triangle BCD$ and its image in the line $y = x$.

Concept Summary Reflection in the Coordinate Plane

Reflection in the x-axis	Reflection in the y-axis	Reflection in the line $y = x$
$P(x, y)$; $P'(x, -y)$	$P(x, y)$; $P'(-x, y)$	$P(x, y)$; $P'(y, x)$
$(x, y) \rightarrow (x, -y)$	$(x, y) \rightarrow (-x, y)$	$(x, y) \rightarrow (y, x)$

Check Your Understanding

= Step-by-Step Solutions begin on page R13.

Example 1 **Copy the figure and the given line of reflection. Then draw the reflected image in this line using a ruler.**

1. 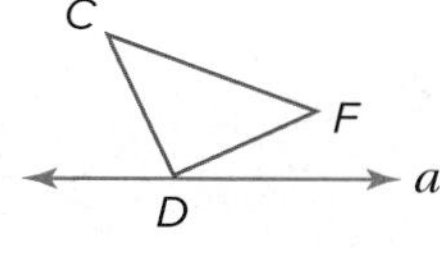

2. 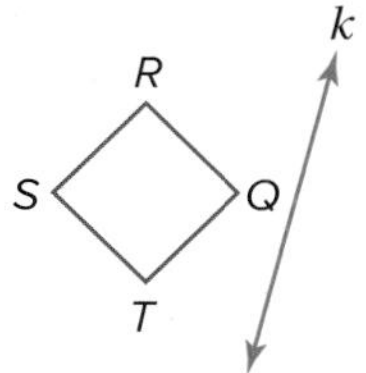

3.
X
W
Y
f
Z

Example 2 **4. SPORTING EVENTS** Toru is waiting at a café for a friend to bring him a ticket to a sold-out sporting event. At what point P along the street should the friend try to stop his car to minimize the distance Toru will have to walk from the café, to the car, and then to the arena entrance? Draw a diagram.

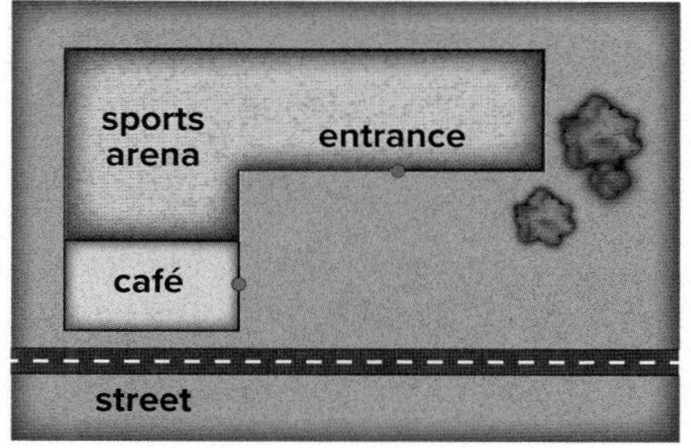

Example 3 **Graph $\triangle ABC$ and its image in the given line.**

5. $y = -2$ **6.** $x = 3$

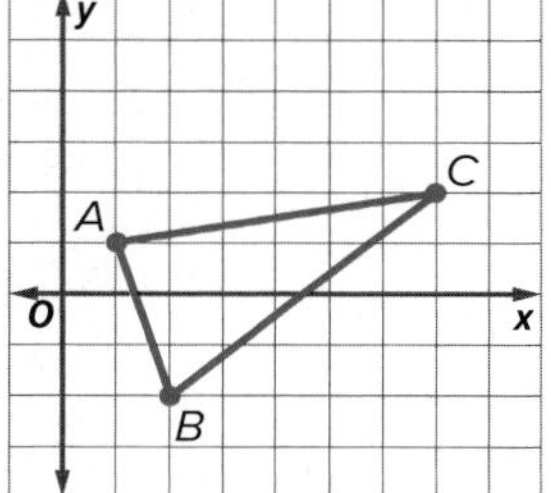

Examples 4–5 **Graph each figure and its image under the given reflection.**

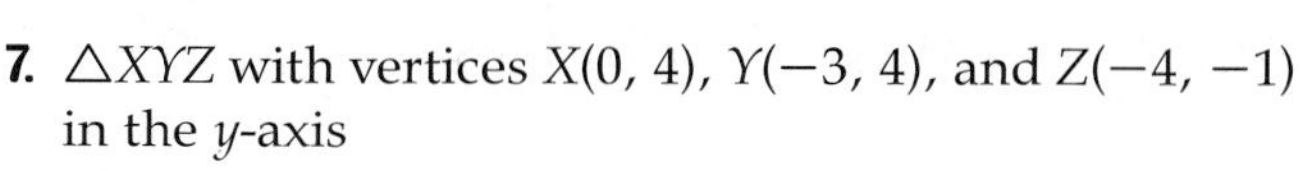
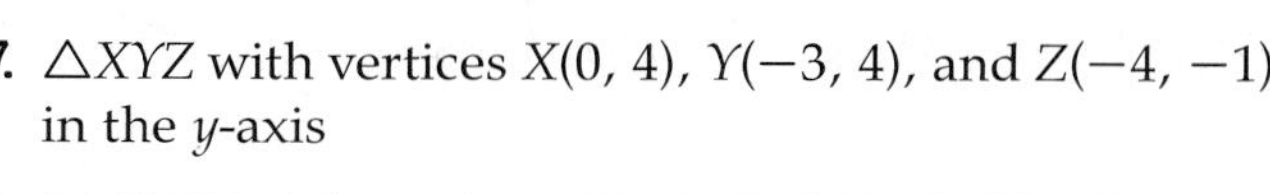

7. $\triangle XYZ$ with vertices $X(0, 4)$, $Y(-3, 4)$, and $Z(-4, -1)$ in the y-axis

8. $\square QRST$ with vertices $Q(-1, 4)$, $R(4, 4)$, $S(3, 1)$, and $T(-2, 1)$ in the x-axis

9. Given quadrilaterals $J(-3, 1)$, $K(-1, 3)$, $L(1, 3)$, and $M(-3, -1)$ and its image $J'(1, -3)$, $K'(3, -1)$, $L'(3, 1)$, and $M'(-1, -3)$, describe the transformation using coordinate notation.

Practice and Problem Solving

Extra Practice is on page R3.

Example 1 **MP TOOLS** Copy the figure and the given line of reflection. Then draw the reflected image in this line using a ruler.

10. 

11. 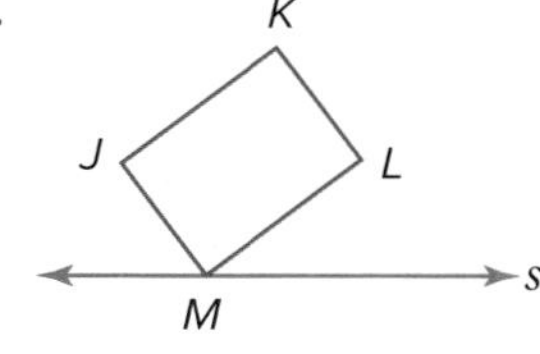

12. 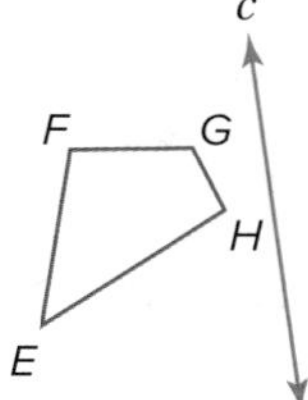

13. 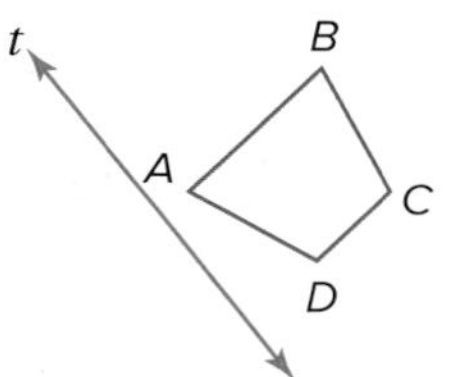

14. 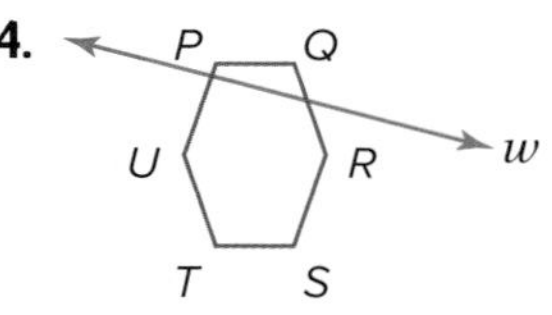

15. 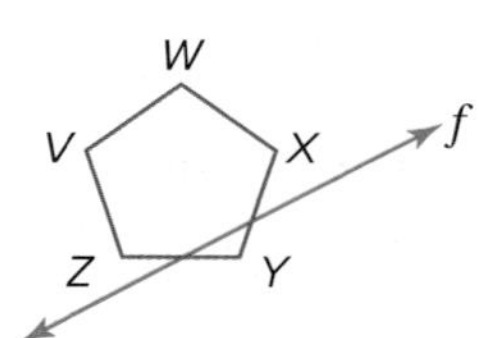

Example 2 **SPORTS When a ball is rolled or struck without spin against a wall, it bounces off the wall and travels in a ray that is the reflected image of the path of the ball if it had gone straight through the wall. Use this information in Exercises 16 and 17.**

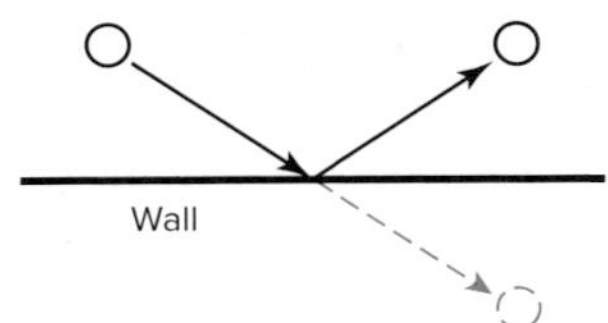

16. **BILLIARDS** Tadeo is playing billiards. He wants to strike the eight ball with the cue ball so that the eight ball bounces off the rail and rolls into the indicated pocket. If the eight ball moves with no spin, draw a diagram showing the exact point P along the right rail where the eight ball should hit after being struck by the cue ball.

17. **INDOOR SOCCER** Abby is playing indoor soccer, and she wants to hit the ball to her teammate at point C, but must avoid an opposing player at point B. She decides to hit the ball at a point A so that it bounces off the side wall. Draw a diagram that shows the exact point along the top wall toward which Abby should aim.

Example 3 **Graph each figure and its image in the given line.**

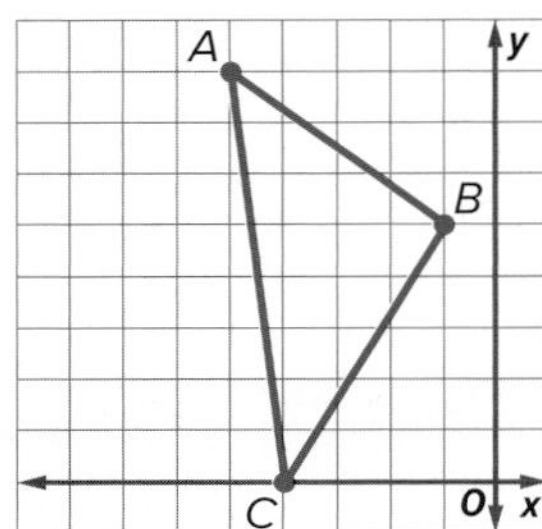

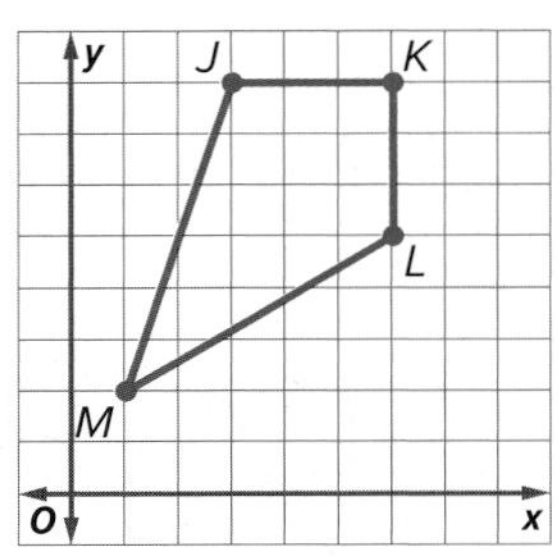

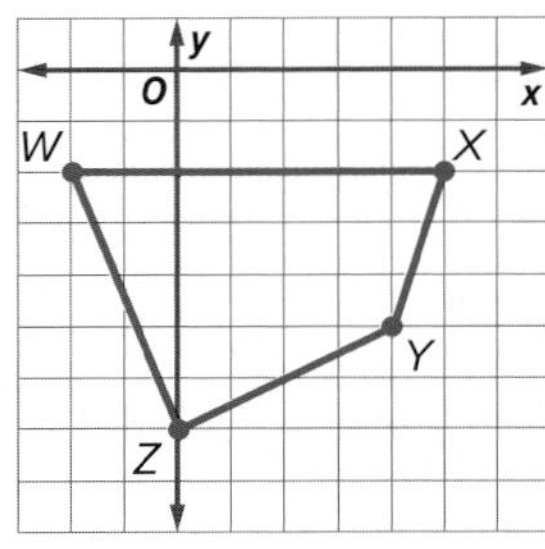

18. $\triangle ABC$; $y = 3$

19. $\triangle ABC$; $x = -1$

20. $JKLM$; $x = 1$

21. $JKLM$; $y = 4$

22. $WXYZ$; $y = -4$

23. $WXYZ$; $x = -2$

Examples 4–5 MP **STRUCTURE Graph each figure and its image under the given reflection.**

24. rectangle $ABCD$ with vertices $A(-5, 2)$, $B(1, 2)$, $C(1, -1)$, and $D(-5, -1)$ in the line $y = -2$

25. square $JKLM$ with vertices $J(-4, 6)$, $K(0, 6)$, $L(0, 2)$, and $M(-4, 2)$ in the y-axis

26. $\triangle FGH$ with vertices $F(-3, 2)$, $G(-4, -1)$, and $H(-6, -1)$ in the line $y = x$

27. ▱$WXYZ$ with vertices $W(2, 3)$, $X(7, 3)$, $Y(6, -1)$, and $Z(1, -1)$ in the x-axis

Given each figure and image, describe the transformation using coordinate notation.

28. $P(-1, 4)$, $Q(2, 4)$, $R(1, -1)$, $S(-1, -1)$ and $P'(1, 4)$, $Q'(-2, 4)$, $R'(-1, -1)$, $S'(1, -1)$

29. $S(-3, -2)$, $T(-2, 3)$, $U(2, 2)$ and $S'(-2, -3)$, $T'(3, -2)$, $U'(2, 2)$

Each figure shows a preimage and its reflected image in some line. Copy each figure and draw the line of reflection.

30.

31.

32.

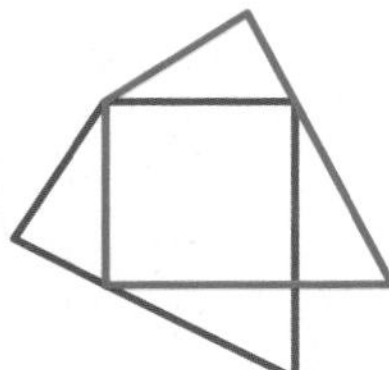

CONSTRUCTION To construct the reflection of a figure in a line using only a compass and a straightedge, you can use:

- the construction of a line perpendicular to a given line through a point not on the line (Lesson 1-5), and
- the construction of a segment congruent to a given segment (Lesson 1-2).

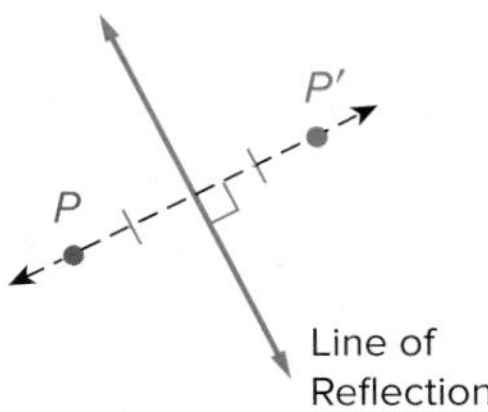

MP TOOLS Copy each figure and the given line of reflection. Then construct the reflected image.

33.

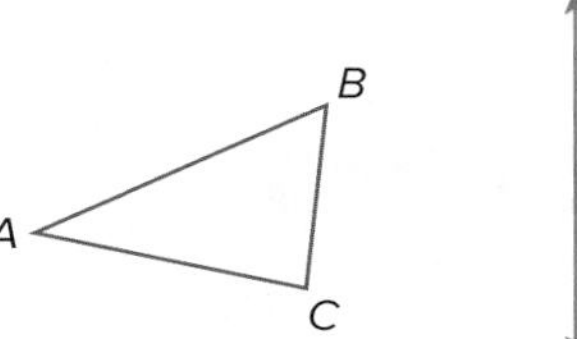

34.

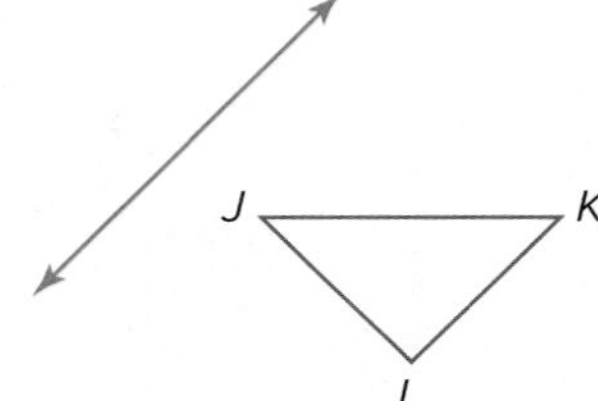

35 **ARCHITECTURE** Refer to the photo of the building at the right.

a. Identify the line of reflection.

b. Identify the preimage/image pairs. Assume that architectural features on the left-hand side of the picture represent the preimages.

ALGEBRA Graph the line $y = 2x - 3$ and its reflected image in the given line. What is the equation of the reflected image?

36. x-axis

37. y-axis

38. $y = x$

39. Reflect $\triangle CDE$ shown below in the line $y = 3x$.

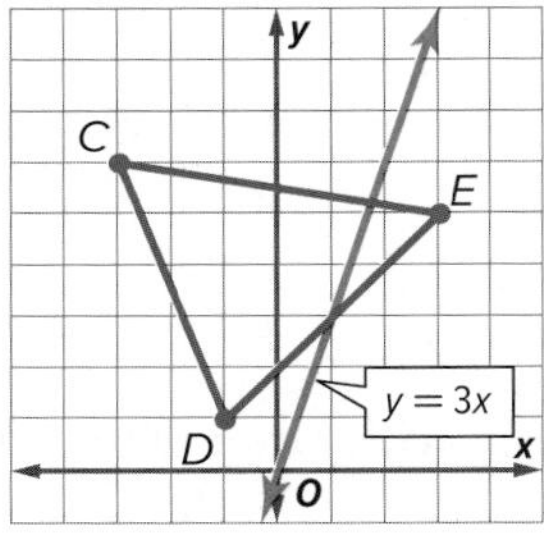

40. Relocate vertex C so that $ABCDE$ is convex, and all sides remain the same length.

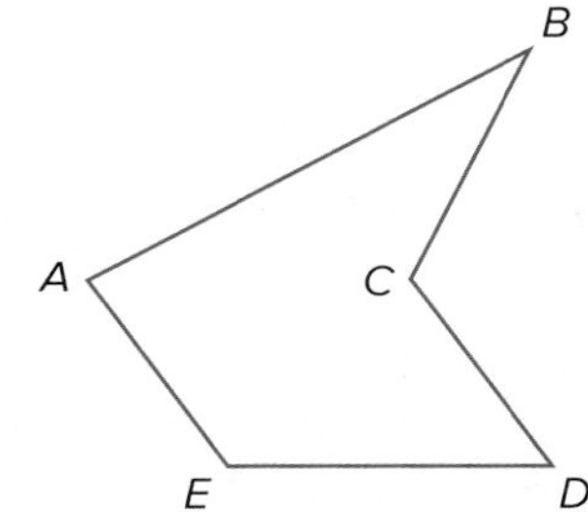

ALGEBRA Graph the reflection of each function in the given line. Then write the equation of the reflected image.

41. x-axis

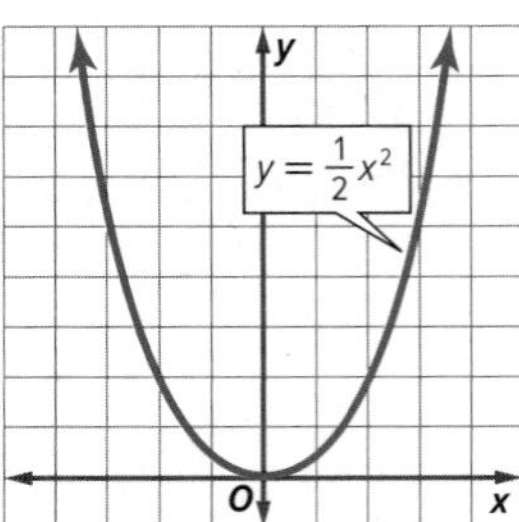

42. y-axis

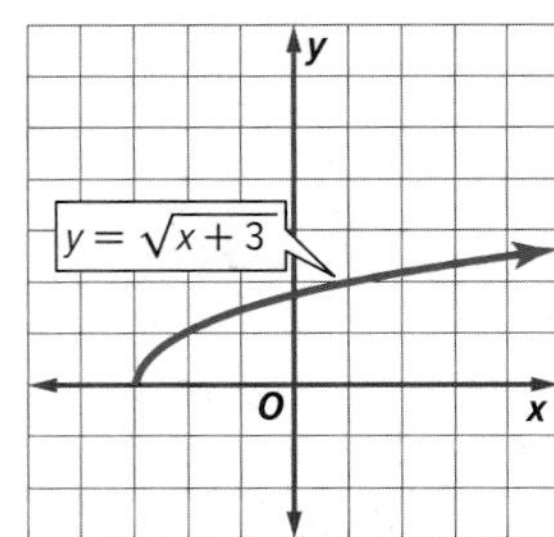

43. x-axis

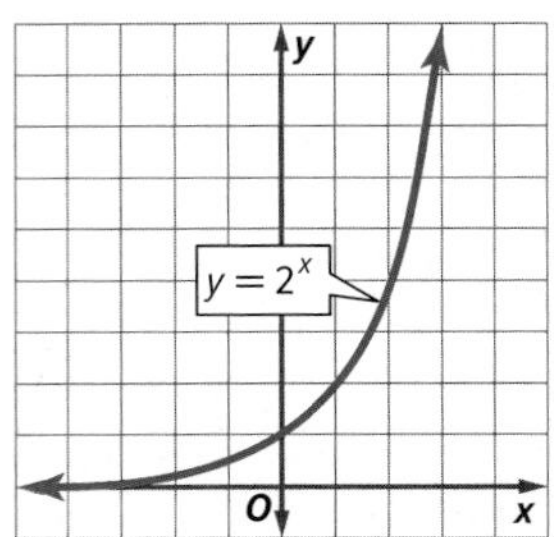

44. MULTIPLE REPRESENTATIONS In this problem, you will investigate a reflection in the origin.

a. Geometric Draw $\triangle ABC$ in the coordinate plane so that each vertex is a whole-number ordered pair.

b. Graphical Locate each reflected point A', B', and C' so that the reflected point, the original point, and the origin are collinear, and both the original point and the reflected point are equidistant from the origin.

c. Tabular Copy and complete the table below.

	$\triangle ABC$		$\triangle A'B'C'$	
Coordinates	A		A'	
	B		B'	
	C		C'	

d. Verbal Make a conjecture about the relationship between corresponding vertices of a figure reflected in the origin.

H.O.T. Problems Use Higher-Order Thinking Skills

45. ERROR ANALYSIS Jamil and Ashley are finding the coordinates of the image of (2, 3) after a reflection in the x-axis. Is either of them correct? Explain.

Jamil	Ashley
C′(2, −3)	C′(−2, 3)

46. WRITING IN MATH Describe how to reflect a figure not on the coordinate plane across a line.

47. CHALLENGE A point in the second quadrant with coordinates $(-a, b)$ is reflected in the x-axis. If the reflected point is then reflected in the line $y = -x$, what are the final coordinates of the image?

48. OPEN-ENDED Draw a polygon on the coordinate plane that when reflected in the x-axis looks exactly like the original figure.

49. CHALLENGE When $A(4, 3)$ is reflected in a line, its image is $A'(-1, 0)$. Find the equation of the line of reflection. Explain your reasoning.

50. MP **PRECISION** The image of a point reflected in a line is *always, sometimes,* or *never* located on the other side of the line of reflection.

51. WRITING IN MATH Suppose points P, Q, and R are collinear, with point Q between points P and R. Describe a plan for a proof that the reflection of points P, Q, and R in a line preserves collinearity and betweenness of points.

52. Quadrilateral $ABCD$ has vertices $A(1, 2)$, $B(3, 6)$, $C(7, 6)$, and $D(7, 0)$. What are the coordinates of C' under the reflection $(x, y) \rightarrow (-x, y)$? MP 7

- ○ A $(-7, 6)$
- ○ B $(7, -6)$
- ○ C $(6, 7)$
- ○ D $(-7, -6)$

53. Amira drew $\triangle JKL$ as shown. Then she accurately drew the image of $\triangle JKL$ under the reflection $(x, y) \rightarrow (y, x)$ and labeled the image $\triangle J'K'L'$.

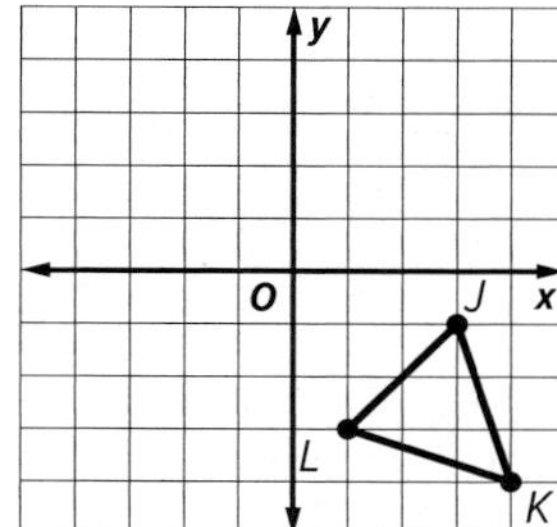

Which of the following statements about $\triangle J'K'L'$ must be true? MP 6

- ○ A $\triangle J'K'L'$ lies entirely in Quadrant I.
- ○ B $\triangle J'K'L'$ lies entirely in Quadrant II.
- ○ C $\triangle J'K'L'$ lies entirely in Quadrant III.
- ○ D $\triangle J'K'L'$ coincides with $\triangle JKL$.

54. Jennifer draws a line and a point P not on the line. Then she accurately draws the reflection image of point P, which she labels P'. She finds that the distance from point P to the line of reflection is 4 centimeters. What is the length of $\overline{PP'}$? MP 6

- ○ A 2 cm
- ○ B 4 cm
- ○ C 8 cm
- ○ D 16 cm

55. What is the x-coordinate of the image of the point $(3, -2)$ after a reflection in the line $y = x$? MP 1

x-coordinate = ☐

56. Jamal draws a rectangle with vertices $J(-2, -2)$, $K(1, -2)$, and $L(-1, 2)$, and then its reflection at $J'(2, -2)$, $K'(-1, -2)$, and $L'(1, 2)$. Describe the transformation using coordinate notation. MP 1

57. The vertices of $\triangle DEF$ are $D(3, 3)$, $E(2, 0)$, and $F(6, 1)$. After a reflection in the line $y = x$, which of the following points lie in the interior of $\triangle D'E'F'$? MP 2

- ○ A $(-1, 3)$
- ○ B $(1, 3)$
- ○ C $(3, -1)$
- ○ D $(-1, -3)$

58. Kyle performs a reflection of a figure and finds that the x-coordinate of each vertex every point on the x-axis is mapped to itself. Which of the following could be Kyle's reflection? MP 2, 7

- ○ A $(x, y) \rightarrow (x, -y)$
- ○ B $(x, y) \rightarrow (y, x)$
- ○ C $(x, y) \rightarrow (-x, y)$
- ○ D $(x, y) \rightarrow (x + 4, y)$

59. **MULTI-STEP** Carlos is using a coordinate plane to design a mural. He starts by drawing $\triangle QRS$ as shown. MP 1, 4

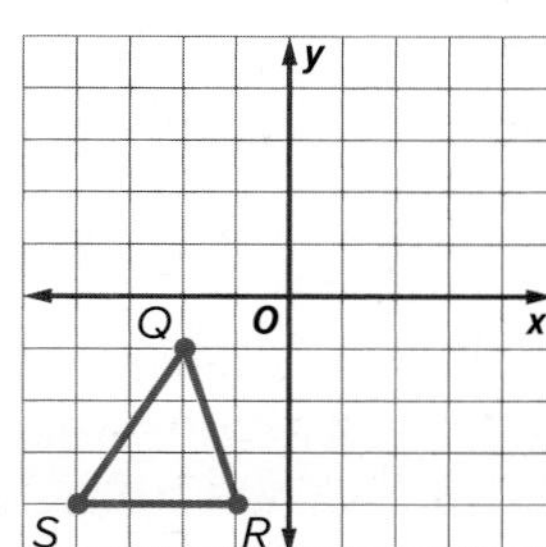

a. Carlos decides to reflect the triangle in the x-axis. In which quadrant does the reflection image lie?

b. Carlos also reflects the triangle so that the image of point S is $S'(4, -4)$. Write a coordinate rule for the reflection Carlos used.

c. Describe a reflection Carlos can use if he wants the image of $\triangle QRS$ to intersect the x-axis.

EXPLORE 3-2

Geometry Lab
Translations

Mathematical Practices
MP 7 Look for and make use of structure.

Recall that a translation is a slide that moves all points of the original figure the same distance in the same direction.

You can use tracing paper to explore translations.

Activity 1 Translate a Figure on the Coordinate Plane with Tracing Paper

Work cooperatively.

Given the preimage $\triangle DEF$ with vertices $D(1, -4)$, $E(1, -1)$, and $F(4, 0)$, find the image $\triangle D'E'F'$ translated along $<-4, 4>$.

Step 1 Trace $\triangle DEF$ onto the tracing paper.

Step 2 Slide the tracing paper along the vector $<-4, 4>$.

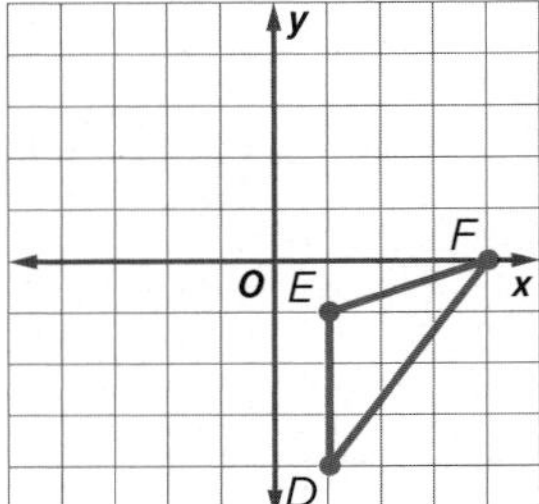

1a. Find the coordinates of the vertices of the image. Record your findings in a table like the one shown.

Coordinates of Vertices of Image	
D'	(____, ____)
E'	(____, ____)
F'	(____, ____)

b. Use the graph to determine the distance from each vertex to the corresponding image. Record your findings in a table like the one shown.

Distance from Vertex to Image		
DD' ____ unit(s)	EE' ____ unit(s)	FF' ____ unit(s)

c. What do you observe about the distances from each vertex to the image?

d. Use the graph to determine the slope of each segment from the vertex to the corresponding image.

Record your findings in a table like the one shown.

Slope		
DD' ____	EE' ____	FF' ____

e. What do you observe about the slopes from each vertex to the line of reflection?

f. What can you conclude about the relationship between $\triangle DEF$ and $\triangle D'E'F'$?

g. Explain how this activity demonstrates that a translation is a rigid transformation.

Recall that rigid motion transformations can be used to show that two figures are congruent by the **principle of superposition**. This states that two figures are congruent if and only if there is a rigid motion or a series of rigid motions that maps one figure exactly onto the other.

You can use tracing paper to determine whether two figures on the coordinate plane are congruent. If the figures are congruent, you can also determine and describe the rigid motion that maps one figure onto the other.

Activity 2 Determine Congruence on the Coordinate Plane

Triangle JKL has vertices $J(-3, -2)$, $K(-2, 1)$, and $L(-1, -1)$, and triangle QRS has vertices $Q(0, 2)$, $R(1, 5)$, and $S(2, 3)$. Determine whether $\triangle JKL$ is congruent to $\triangle QRS$.

Step 1 Trace $\triangle JKL$ onto a piece of tracing paper.

Step 2 Move the tracing paper around the coordinate plane using rigid motions to determine if $\triangle JKL$ is congruent to $\triangle QRS$.

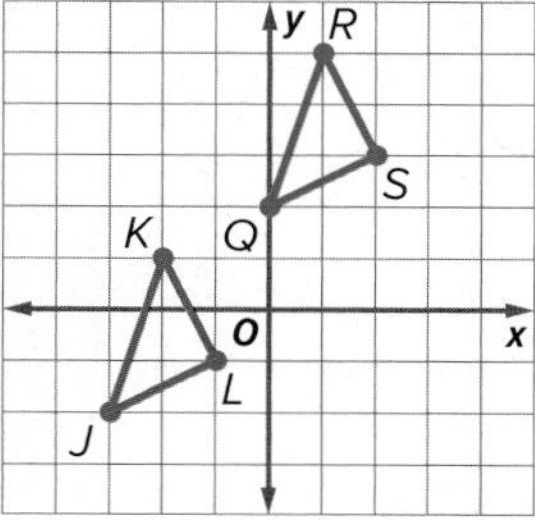

2A. Is $\triangle JKL$ congruent to $\triangle QRS$? If so, what rigid motion maps $\triangle JKL$ onto $\triangle QRS$?

2B. What is the component form of the translation vector? Describe the process you used.

2C. Describe the rigid motion that maps $\triangle JKL$ onto $\triangle DEF$.

Analyze the Results

Work cooperatively.

1. Use tracing paper to translate the shape along the vector $\langle 3, 2 \rangle$. Write the new coordinates of the vertices.

2. **ARGUMENTS** Katie says that when you translate in the y direction, you add, and when you translate in the x direction, you subtract. Is she correct? Explain.

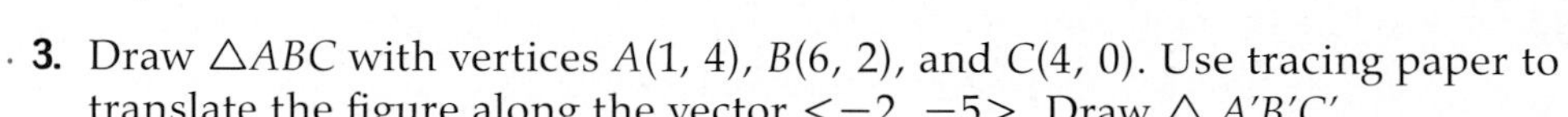

3. Draw $\triangle ABC$ with vertices $A(1, 4)$, $B(6, 2)$, and $C(4, 0)$. Use tracing paper to translate the figure along the vector $\langle -2, -5 \rangle$. Draw $\triangle A'B'C'$.

4. What are the coordinates of $A'B'C'$?

5. Use your tracing paper to trace $A'B'C'$ along the vector $\langle 6, -6 \rangle$. What are the coordinates of the new image, $A''B''C''$?

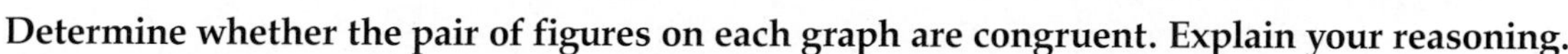

Determine whether the pair of figures on each graph are congruent. Explain your reasoning.

6. 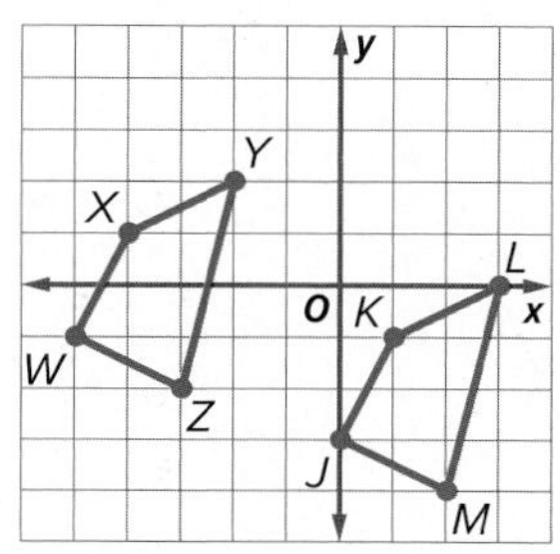

7. 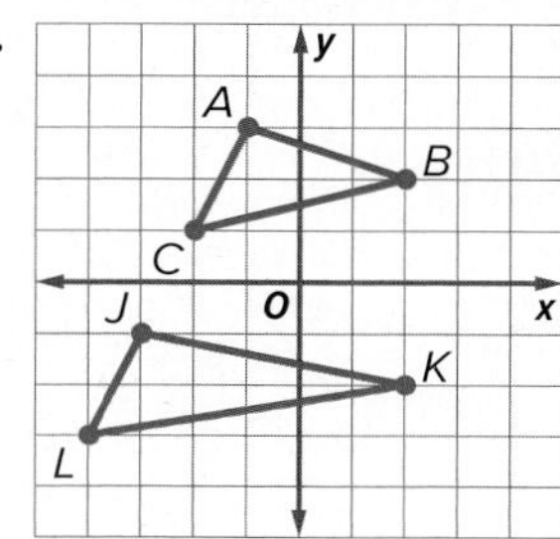

LESSON 2

Translations

Then

- You found the magnitude and direction of vectors.

Now

1 Draw translations.

2 Draw translations in the coordinate plane.

Why?

- Stop-motion animation is a technique in which an object is moved by very small amounts between individually photographed frames. When the series of frames is played as a continuous sequence, the result is the illusion of movement.

New Vocabulary

translation vector

Mathematical Practices

5 Use appropriate tools strategically.

7 Look for and make use of structure.

1 Draw Translations Recall that a translation or *slide* is a transformation that moves all points of a figure the same distance in the same direction. Since vectors can be used to describe both distance and direction, vectors can be used to define translations.

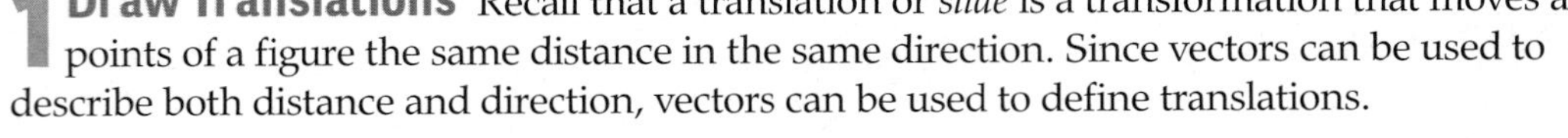

Key Concept Translation

A translation is a function that maps each point to its image along a vector, called the **translation vector**, such that

- each segment joining a point and its image has the same length as the vector, and
- this segment is also parallel to the vector.

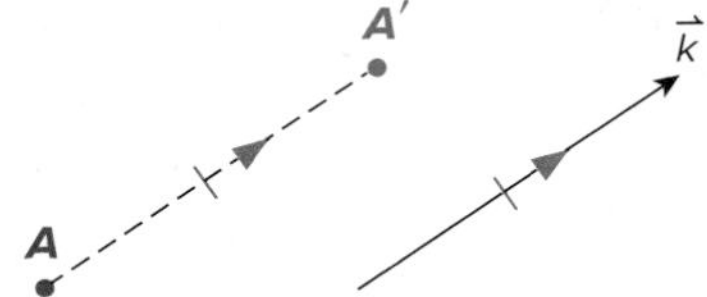

Point A' is a translation of point A along translation vector $\vec{k}$.

Example 1 Draw a Translation

Copy the figure and the given translation vector. Then draw the translation of the figure along the translation vector.

Step 1 Draw a line through each vertex parallel to vector $\vec{w}$.

Step 2 Measure the length of vector $\vec{w}$. Locate point X' by marking off this distance along the line through vertex X, starting at X and in the same direction as the vector.

Step 3 Repeat Step 2 to locate points Y' and Z'. Then connect vertices X', Y', and Z' to form the translated image.

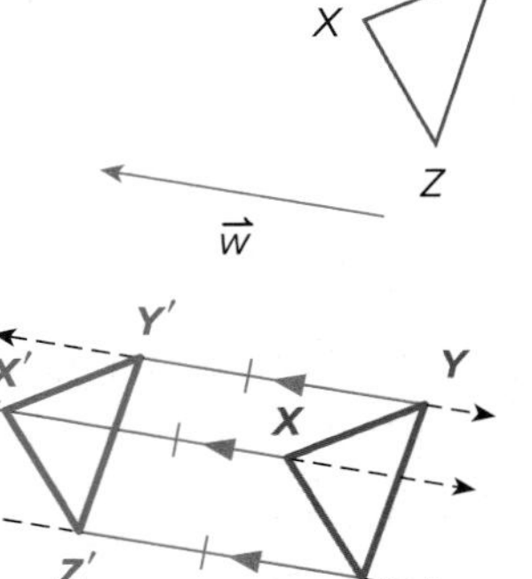

Guided Practice

1A.

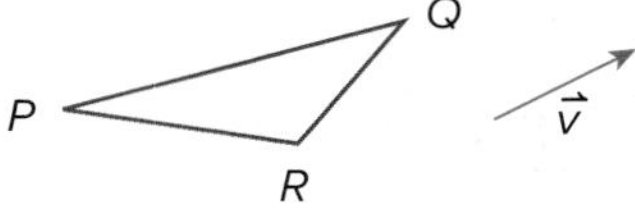

1B.

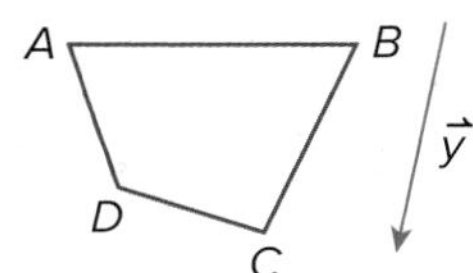

2 Draw Translations in the Coordinate Plane

Recall that a vector in the coordinate plane can be written as $\langle a, b \rangle$, where a represents the horizontal change and b is the vertical change from the vector's tip to its tail. $\overrightarrow{CD}$ is represented by $\langle 2, -4 \rangle$.

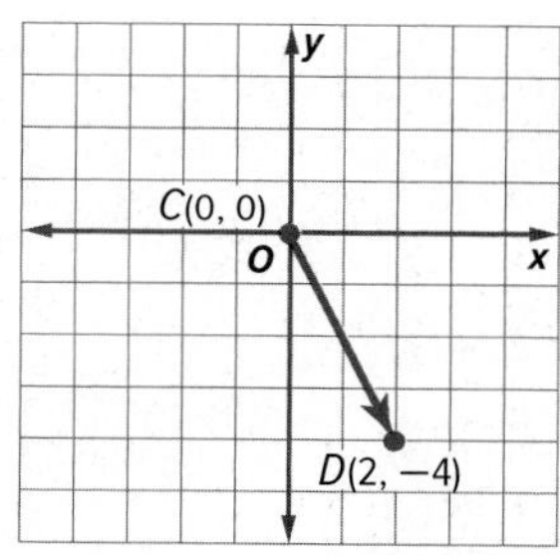

Written in this form, called the component form, a vector can be used to translate a figure in the coordinate plane.

> **Reading Math**
>
> **Horizontal and Vertical Translations** When the translation vector is of the form $\langle a, 0 \rangle$, the translation is horizontal only. When the translation vector is of the form $\langle 0, b \rangle$, the translation is vertical only.

Key Concept Translation in the Coordinate Plane

Words To translate a point along vector $\langle a, b \rangle$, add a to the x-coordinate and b to the y-coordinate.

Symbols $(x, y) \rightarrow (x + a, y + b)$

Example The image of $P(-2, 3)$ translated along vector $\langle 7, 4 \rangle$ is $P'(5, 7)$.

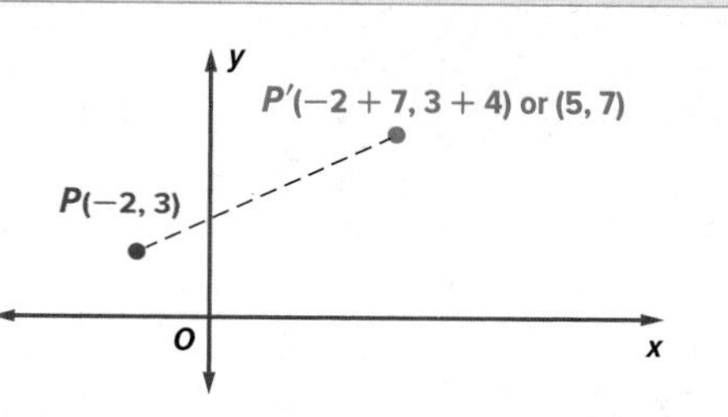

A translation is another type of congruence transformation or isometry.

Example 2 Translations in the Coordinate Plane

Graph each figure and its image along the given vector.

a. $\triangle EFG$ with vertices $E(-7, -1)$, $F(-4, -4)$, and $G(-3, -1)$; $\langle 2, 5 \rangle$

The vector indicates a translation 2 units right and 5 units up.

$(x, y) \rightarrow (x + 2, y + 5)$

$E(-7, -1) \rightarrow E'(-5, 4)$

$F(-4, -4) \rightarrow F'(-2, 1)$

$G(-3, -1) \rightarrow G'(-1, 4)$

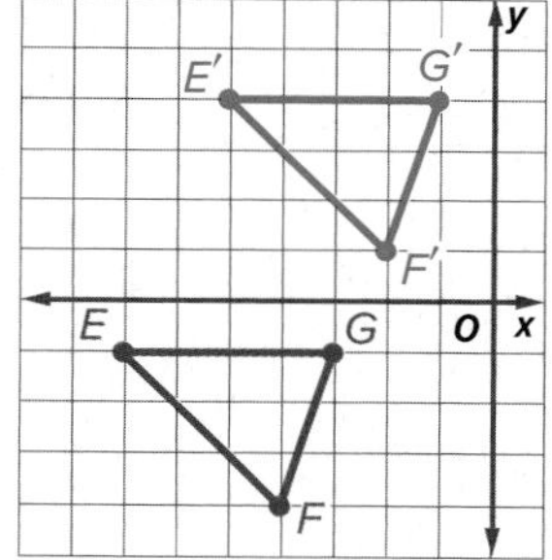

b. square $JKLM$ with vertices $J(3, 4)$, $K(5, 2)$, $L(7, 4)$, and $M(5, 6)$; $\langle -3, -4 \rangle$

The vector indicates a translation 3 units left and 4 units down.

$(x, y) \rightarrow (x + (-3), y + (-4))$

$J(3, 4) \rightarrow J'(0, 0)$

$K(5, 2) \rightarrow K'(2, -2)$

$L(7, 4) \rightarrow L'(4, 0)$

$M(5, 6) \rightarrow M'(2, 2)$

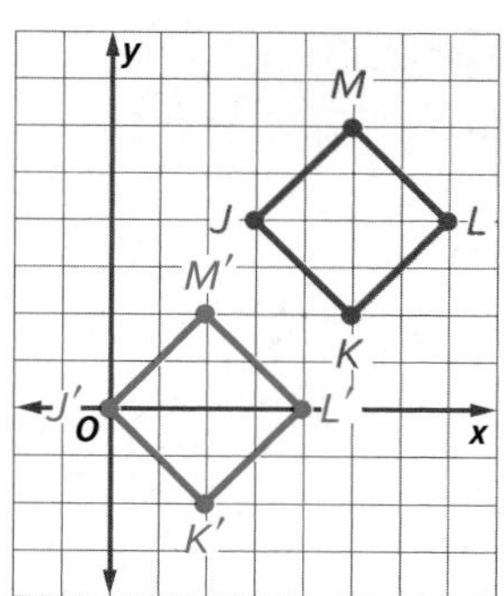

> **Math History Link**
>
> **Felix Klein** (1849–1925) Klein's definition of geometry as the study of the properties of a space that remain invariant under a group of transformations allowed for the inclusion of both Euclidean and non-Euclidean geometry.

Guided Practice

2A. $\triangle ABC$ with vertices $A(2, 6)$, $B(1, 1)$, and $C(7, 5)$; $\langle -4, -1 \rangle$

2B. quadrilateral $QRST$ with vertices $Q(-8, -2)$, $R(-9, -5)$, $S(-4, -7)$, and $T(-4, -2)$; $\langle 7, 1 \rangle$

Real-World Link

Marching bands often make use of a series of formations that can include geometric shapes. Usually, each band member has an assigned position in each formation. *Floating* is the movement of a group of members together without changing the shape or size of their formation.

Real-World Example 3 Describing Translations

MARCHING BAND In one part of a marching band's performance, a line of trumpet players starts at position 1, marches to position 2, and then to position 3. Each unit on the graph represents one step.

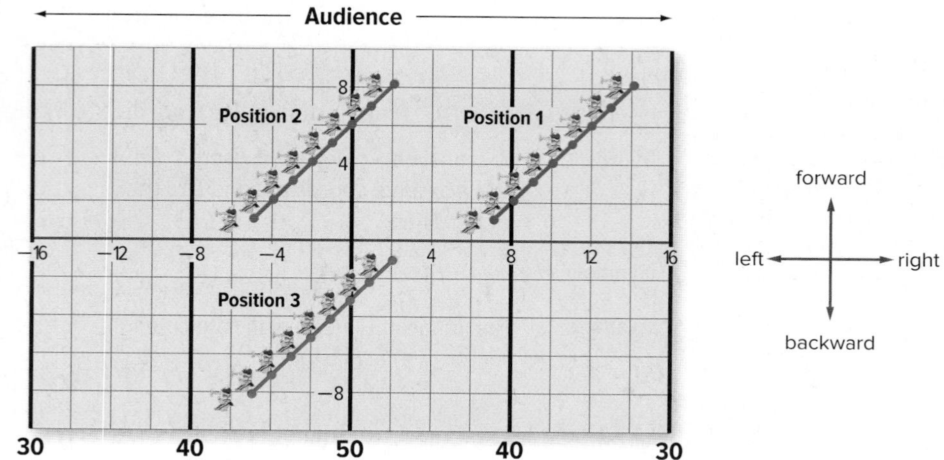

a. Describe the translation of the trumpet line from position 1 to position 2 in coordinate notation and in words.

One point on the line in position 1 is (14, 8). In position 2, this point moves to (2, 8). Use the translation function $(x, y) \rightarrow (x + a, y + b)$ to write and solve equations to find a and b.

$(14 + a, 8 + b)$ or $(2, 8)$

$14 + a = 2$ $\quad\quad$ $8 + b = 8$

$a = -12$ $\quad\quad$ $b = 0$

coordinate notation: $(x, y) \rightarrow (x + (-12), y + 0)$

So, the trumpet line is translated 12 steps *left* but no steps forward or backward from position 1 to position 2.

b. Describe the translation of the line from position 1 to position 3 using a translation vector.

$(14 + a, 8 + b)$ or $(2, -1)$

$14 + a = 2$ $\quad\quad$ $8 + b = -1$

$a = -12$ $\quad\quad$ $b = -9$

translation vector: $\langle -12, -9 \rangle$

Go Online!

What is the effect of a translation on a figure? Investigate by using the **Geometry Tools** in ConnectED. Discuss your findings with a partner. Ask for clarification as you need it.

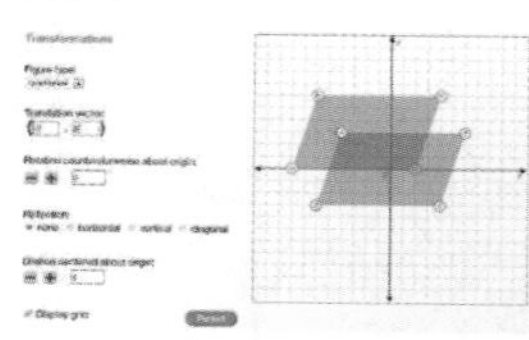

Guided Practice

3. ANIMATION A coin is filmed using stop-motion animation so that it appears to move.

A. Describe the translation from A to B in coordinate notation and in words.

B. Describe the translation from A to C using a translation vector.

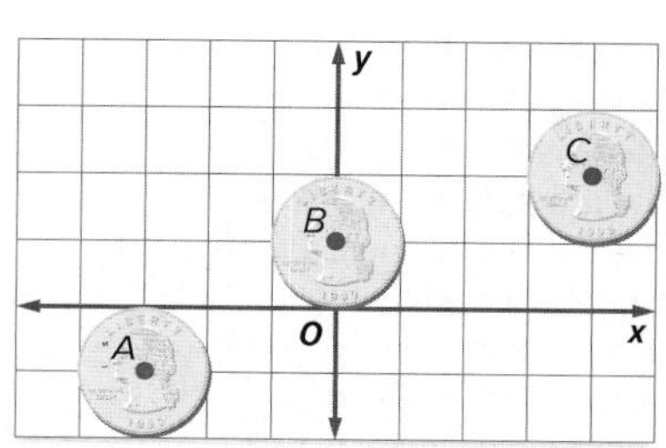

Check Your Understanding

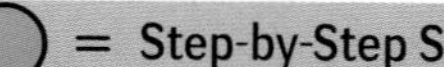
= Step-by-Step Solutions begin on page R13.

Example 1 **Copy the figure and the given translation vector. Then draw the translation of the figure along the translation vector.**

1.
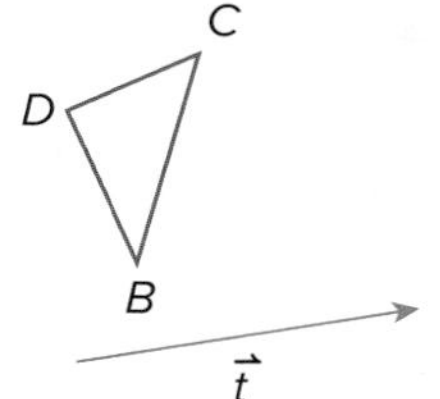

2.
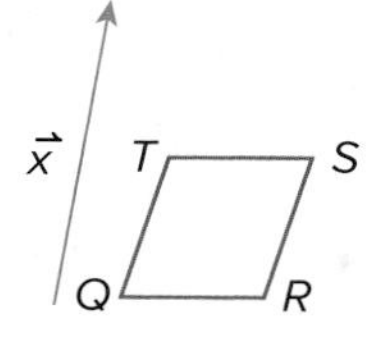

3.
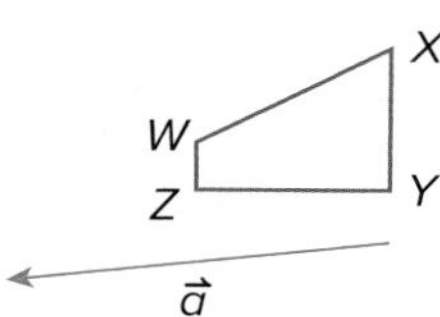

Example 2 **Graph each figure and its image along the given vector.**

4. trapezoid $JKLM$ with vertices $J(2, 4)$, $K(1, 1)$, $L(5, 1)$ and $M(4, 4)$; $\langle 7, 1 \rangle$

5. $\triangle DFG$ with vertices $D(-8, 8)$, $F(-10, 4)$, and $G(-7, 6)$; $\langle 5, -2 \rangle$

6. parallelogram $WXYZ$ with vertices $W(-6, -5)$, $X(-2, -5)$, $Y(-1, -8)$, and $Z(-5, -8)$; $\langle -1, 4 \rangle$

Example 3

7. **VIDEO GAMES** The object of the video game shown is to manipulate the colored tiles left or right as they fall from the top of the screen to completely fill each row without leaving empty spaces. If the starting position of the tile piece at the top of the screen is (x, y), use coordinate notation to describe the translation that will fill the indicated row.

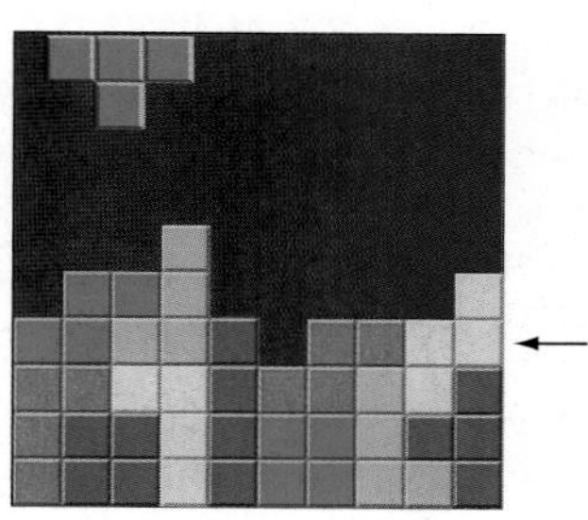

Practice and Problem Solving

Extra Practice is on page R3.

Example 1 **TOOLS** Copy the figure and the given translation vector. Then draw the translation of the figure along the translation vector.

8.
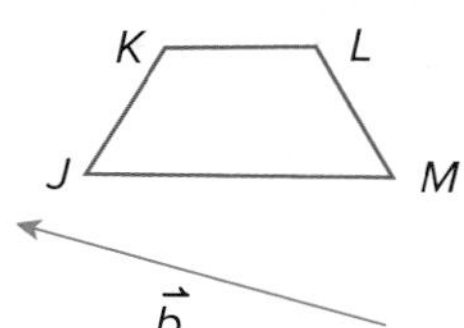

9.
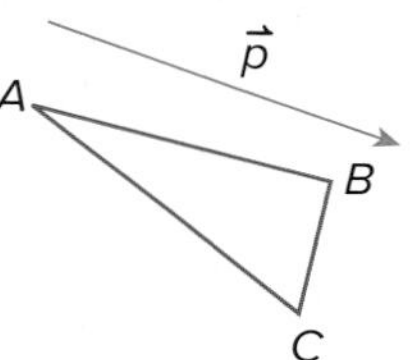

10.
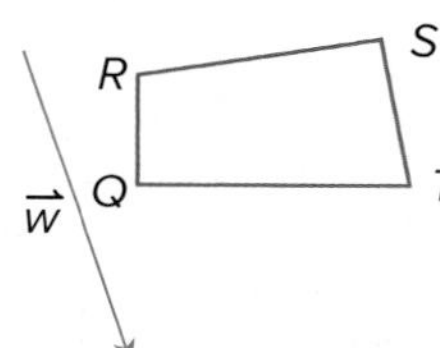

11.
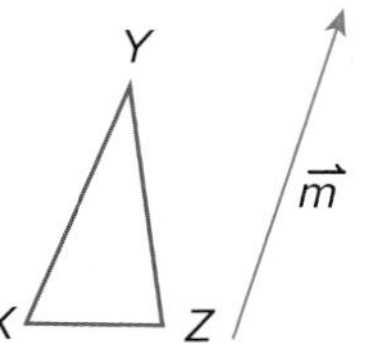

12.
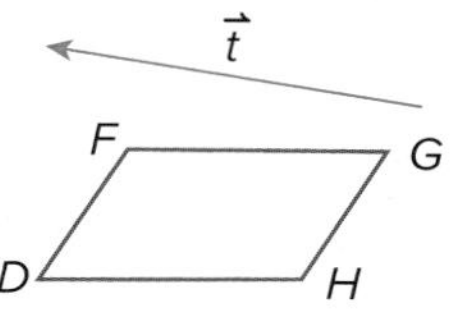

13.
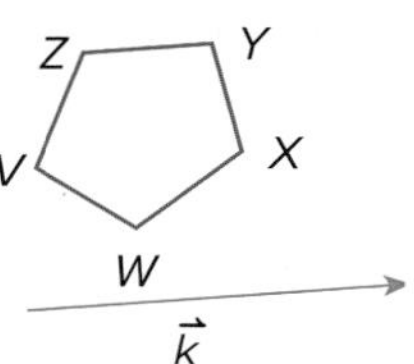

Example 2 **Graph each figure and its image along the given vector.**

14. $\triangle ABC$ with vertices $A(1, 6)$, $B(3, 2)$, and $C(4, 7)$; $\langle 4, -1 \rangle$

15. $\triangle MNP$ with vertices $M(4, -5)$, $N(5, -8)$, and $P(8, -6)$; $\langle -2, 5 \rangle$

16. rectangle $QRST$ with vertices $Q(-8, 4)$, $R(-8, 2)$, $S(-3, 2)$, and $T(-3, 4)$; $\langle 2, 3 \rangle$

17. quadrilateral $FGHJ$ with vertices $F(-4, -2)$, $G(-1, -1)$, $H(0, -4)$, and $J(-3, -6)$; $\langle -5, -2 \rangle$

18. $\square WXYZ$ with vertices $W(-3, -1)$, $X(1, -1)$, $Y(2, -4)$, and $Z(-2, -4)$; $\langle -3, 4 \rangle$

19. trapezoid $JKLM$ with vertices $J(-4, -2)$, $K(-1, -2)$, $L(0, -5)$, and $M(-5, -5)$; $\langle 6, 5 \rangle$

Example 3

20. MP **MODELING** Brittany's neighborhood is shown on the grid at the right.

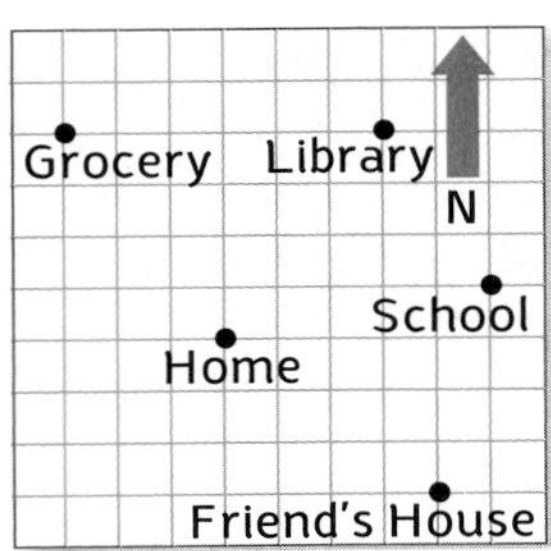

a. If she leaves home and travels 4 blocks north and 3 blocks east, what is her new location?

b. Use words to describe two possible translations that will take Brittany home from school.

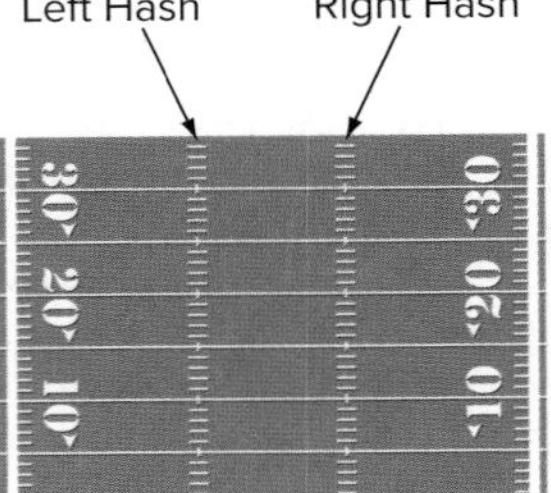

21 **FOOTBALL** A wide receiver starts from his 15-yard line on the right hash mark and runs a route that takes him 12 yards to the left and down field for a gain of 17 yards. Write a translation vector to describe the receiver's route.

22. **MULTI-STEP** Each type of chess piece has a rule for how it can be moved. A knight can move two squares horizontally and then one square vertically, or two squares vertically and one square horizontally. The bishop can only move diagonally. When a chess piece moves to a space that is occupied by an opponent's piece, it is *attacking* that piece.

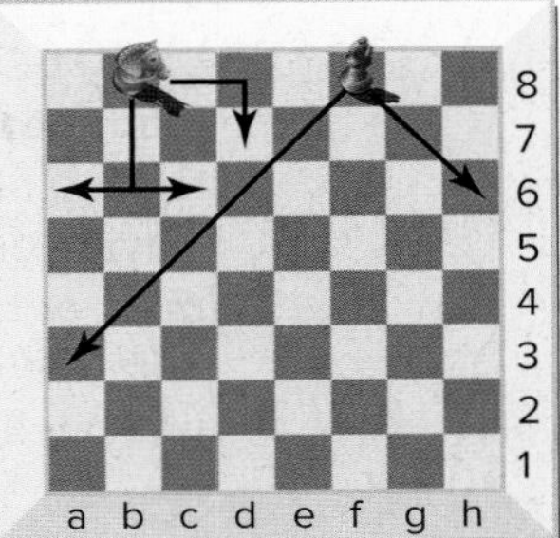

a. What is the maximum number of knights that could be placed on an 8-by-8 chessboard so that none of the pieces can attack each other? What is the maximum number of bishops?

b. Explain your solution process.

Write each translation vector to map the green image to the blue preimge.

23.

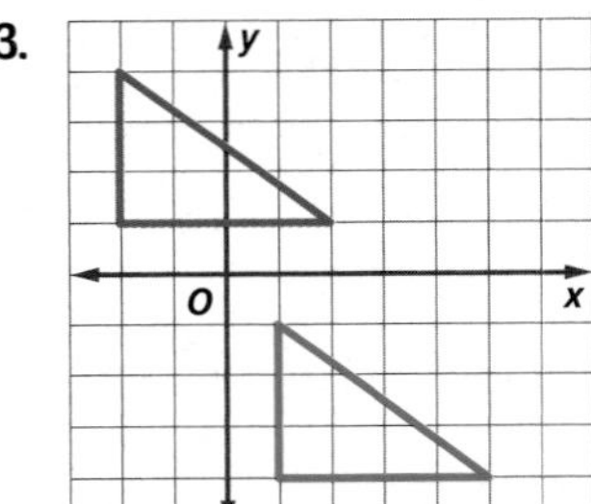

24.

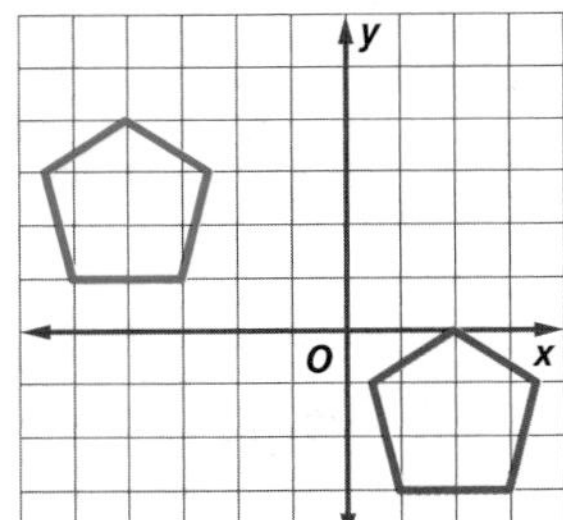

25. **CONCERTS** Dexter's family buys tickets every year for a concert. Last year they were in seats C3, C4, C5, and C6. This year, they will be in seats D16, D17, D18, and D19. Write a translation in words and using vector notation that can be used to describe the change in their seating.

D 1 2 3 4 5 6 7 8 9 10 11 12 13 | 14 15 16 17 18 19 20 21 22 23 24 25 26 D
C 1 2 3 4 5 6 7 8 9 10 11 12 13 | aisle | 14 15 16 17 18 19 20 21 22 23 24 25 26 C
B 1 2 3 4 5 6 7 8 9 10 11 12 13 | 14 15 16 17 18 19 20 21 22 23 24 25 26 B

MP SENSE-MAKING Graph the translation of each function along the given vector. Then write the equation of the translated image.

26. $\langle 4, 1 \rangle$

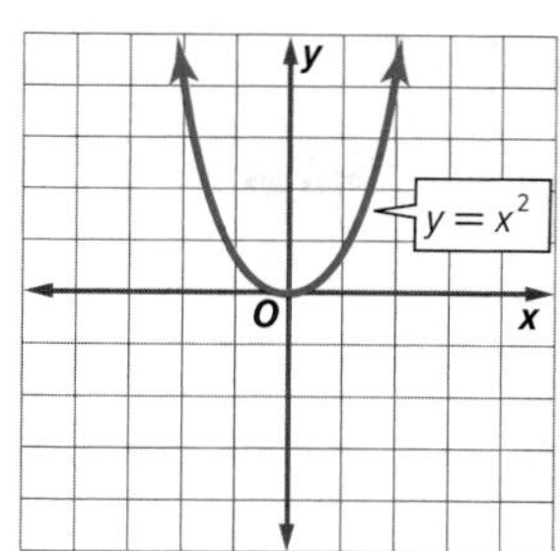

27 $\langle -2, 0 \rangle$

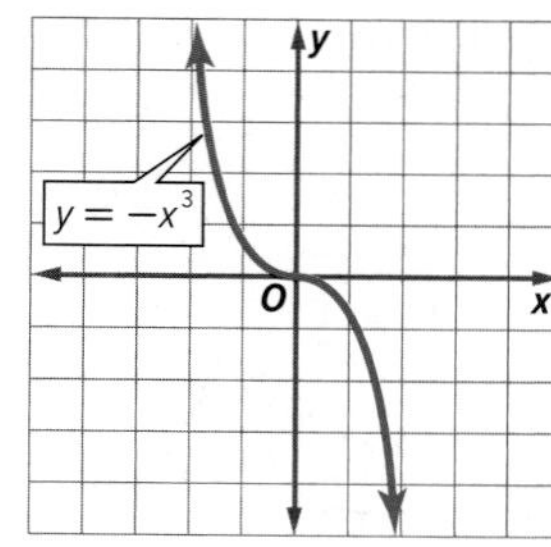

28. ROLLER COASTERS The length of the roller coaster track from the top of a hill to the bottom of the hill is 125 feet and the vertical height is 75 feet. If the position at the top of the hill is (x, y), use coordinate notation to describe the translation to the bottom of the hill. Round to the nearest foot.

29. MULTIPLE REPRESENTATIONS In this problem, you will investigate reflections in a pair of parallel lines.

a. Geometric On patty paper, draw $\triangle ABC$ and a pair of vertical lines ℓ and m. Reflect $\triangle ABC$ in line ℓ by folding the patty paper. Then reflect $\triangle A'B'C'$, in line m. Label the final image $\triangle A''B''C''$.

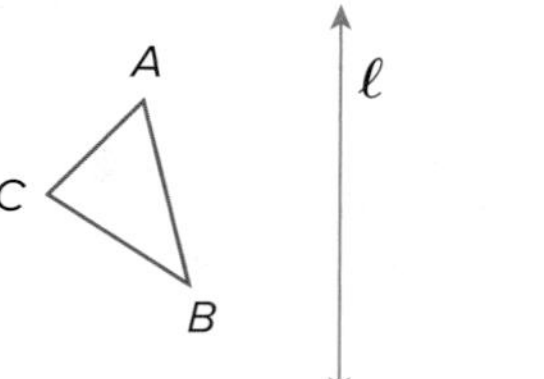

b. Geometric Draw two different triangles and label them $\triangle DEF$ and $\triangle JKL$. Repeat the process in part **a** for $\triangle DEF$ reflected in vertical lines n and p and $\triangle JKL$ reflected in vertical lines q and r.

c. Tabular Copy and complete the table below.

Distance Between Corresponding Points (cm)		Distance Between Vertical Lines (cm)	
A and A'', B and B'', C and C''		ℓ and m	
D and D'', E and E'', F and F''		n and p	
J and J'', K and K'', L and L''		q and r	

d. Verbal Describe the result of two reflections in two vertical lines using one transformation.

H.O.T. Problems Use Higher-Order Thinking Skills

30. MP REASONING Determine a rule to find the final image of a point that is translated along $\langle x + a, y + b \rangle$ and then $\langle x + c, y + d \rangle$.

31. CHALLENGE A line $y = mx + b$ is translated using the vector $\langle a, b \rangle$. Write the equation of the translated line. What is the value of the y-intercept?

32. OPEN-ENDED Draw a figure on the coordinate plane so that the figure has the same orientation after it is reflected in the line $y = 1$. Explain what must be true in order for this to occur.

33. WRITING IN MATH Compare and contrast coordinate notation and vector notation for translations.

34. WRITING IN MATH Recall from Lesson 3-1 that an invariant point maps onto itself. Can invariant points occur with translations? Explain why or why not.

35. On a grid, $\triangle PQR$ has vertices $P(2, 1)$, $Q(3, -1)$, and $R(-1, -2)$, as shown. MP 1, 5

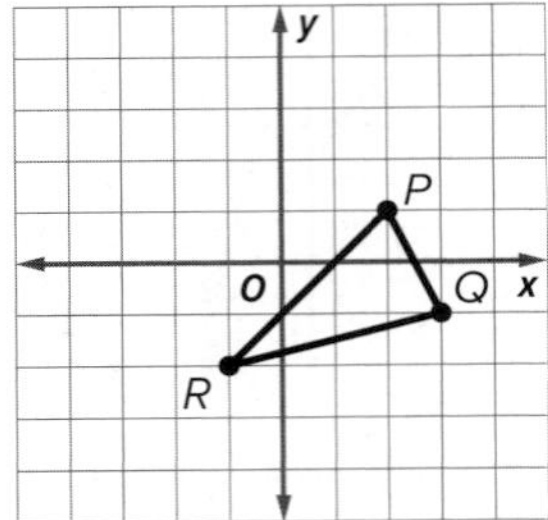

a. $\triangle PQR$ is translated along $(x, y) \rightarrow (x - 2, y + 3)$. Find the coordinates of the vertices of $\triangle P'Q'R'$.

$P' = ($ ______ $)$
$Q' = ($ ______ $)$
$R' = ($ ______ $)$

b. Which point lies in the interior of $\triangle P'Q'R'$?

- **A** $(0, -1)$
- **B** $(2, -4)$
- **C** $(-2, -4)$
- **D** $(-2, 2)$

36. Which of the following is the image of the line $y = x - 1$ under the translation $(x, y) \rightarrow (x - 1, y + 1)$? MP 2, 4

- **A** $y = x$
- **B** $y = x - 3$
- **C** $y = x + 1$
- **D** $y = x - 1$

37. Which of the following vectors will translate a point with the coordinates $(3, -5)$ up and to the left on a coordinate plane? MP 2

- **A** $\langle -3, 0 \rangle$
- **B** $\langle 3, 3 \rangle$
- **C** $\langle -3, 3 \rangle$
- **D** $\langle 0, 3 \rangle$

38. Given a triangle with vertices $A(-7, -1)$, $B(-2, -3)$, and $C(2, 4)$, which of the following represents the coordinates of the vertices of the image after a translation of the triangle 5 units right and 3 units up? MP 2

- **A** $A(-4, 4)$, $B(1, 2)$, $C(5, 9)$
- **B** $A(-2, 2)$, $B(3, 0)$, $C(7, 7)$
- **C** $A(-12, 2)$, $B(-7, 0)$, $C(-3, 7)$
- **D** $A(-2, -4)$, $B(3, -6)$, $C(7, 1)$

39. Which of the following statements is true for both the translation $(x, y) \rightarrow (x + 2, y)$ and the translation $(x, y) \rightarrow (x, y + 2)$? MP 2, 4

- **A** A point and its image have the same y-coordinate.
- **B** Points on the x-axis are mapped to other points on the x-axis.
- **C** The translation vector is $\langle 0, 2 \rangle$.
- **D** A point in Quadrant I is mapped to another point in Quadrant I.

40. What is the distance between any point and its image under the translation $(x, y) \rightarrow (x + 4, y - 6)$? Round to the nearest tenth. MP 2, 6

______

41. Under which of the following translations does the image of rectangle $JKLM$ not overlap rectangle $JKLM$? MP 1, 2, 7

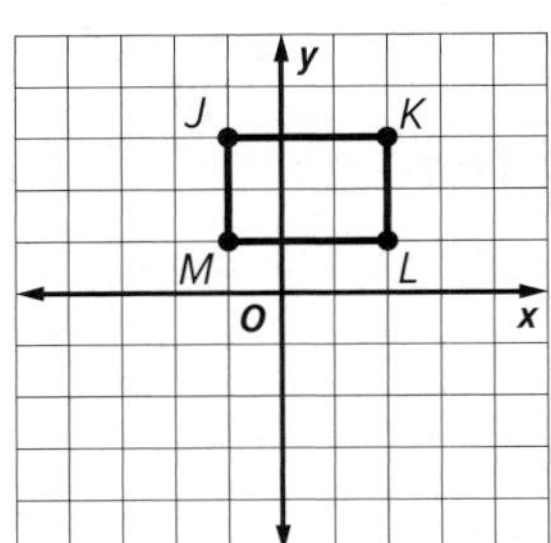

- **A** $(x, y) \rightarrow (x + 3, y + 2)$
- **B** $(x, y) \rightarrow (x - 1.5, y - 1.5)$
- **C** $(x, y) \rightarrow (x, y - 1.9)$
- **D** $(x, y) \rightarrow (x + 3.1, y)$
- **E** $(x, y) \rightarrow (x + 2.8, y - 1)$

EXPLORE 3-3

Geometry Lab

Rotations

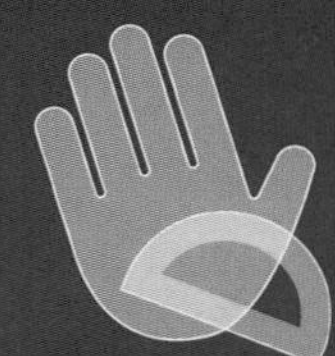

A **rotation** is a type of transformation that moves a figure about a fixed point, or center of rotation, through a specific angle and in a specific direction. In this activity you will use tracing paper to explore the properties of rotations. Or if you prefer, use similar steps to explore with dynamic geometry software.

Mathematical Practices

MP 5 Use appropriate tools strategically.

Activity Explore Rotations

Work cooperatively.

Step 1 On a piece of tracing paper, draw quadrilateral *ABCD* and a point *P*.

Step 2 On another piece of tracing paper, trace quadrilateral *ABCD* and point *P*. Label the new quadrilateral *A′B′C′D′* and the new point *P*.

Step 3 Position the tracing paper so that both points *P* coincide. Rotate the paper so that *ABCD* and *A′B′C′D′* do not overlap. Tape the two pieces of tracing paper together.

Step 4 Measure the distance between *A*, *B*, *C*, and *D* to point *P*. Repeat for quadrilateral *A′B′C′D′*. Then copy and complete the table below.

Quadrilateral	Length			
ABCD	AP	BP	CP	DP
A′B′C′D′	A′P′	B′P′	C′P′	D′P′

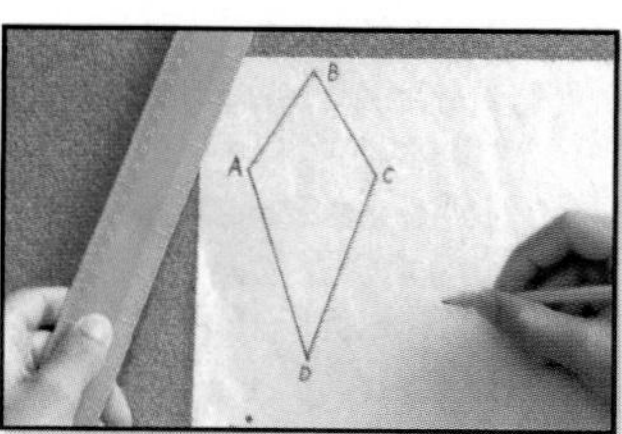

Step 1

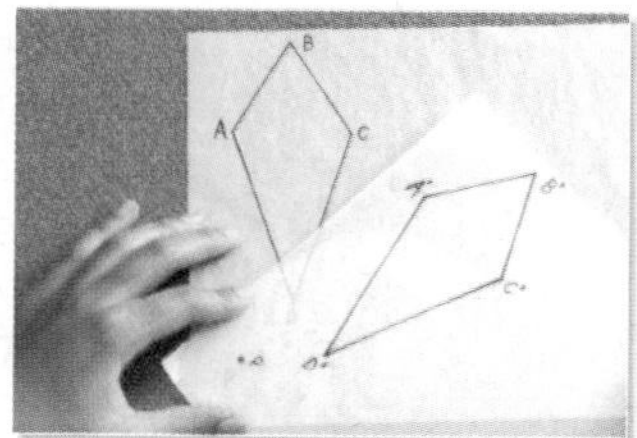

Step 2 and 3

Exercises

Work cooperatively.

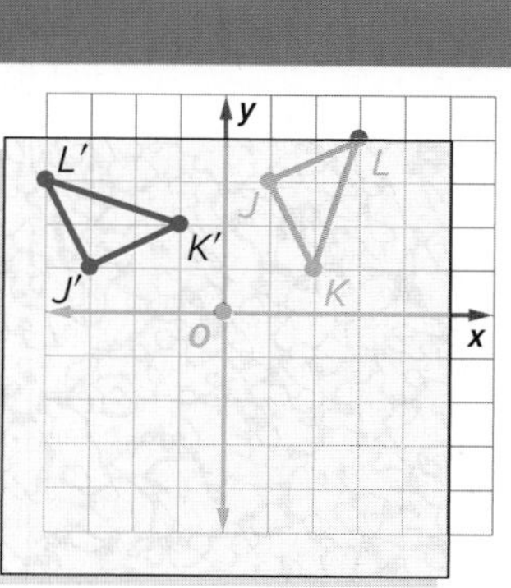

1. Graph $\triangle JKL$ with vertices $J(1, 3)$, $K(2, 1)$, and $L(3, 4)$ on a coordinate plane, and then trace on tracing paper.
   a. Use a protractor to rotate each vertex 90° counterclockwise about the origin as shown. What are the vertices of the image?
   b. Rotate $\triangle JKL$ 180° about the origin. What are the vertices of the image?
   c. Use the Distance Formula to find the distance from points *J*, *K*, and *L* to the origin. Repeat for *J′K′L′* and *J″K″L″*.
2. **WRITING IN MATH** If you rotate point (4, 2) 90° and 180° about the origin, how do the *x*- and *y*-coordinates change?
3. **MAKE A PREDICTION** What are the new coordinates of a point (*x*, *y*) that is rotated 270°?
4. **MAKE A CONJECTURE** Make a conjecture about the distances from the center of rotation *P* to each corresponding vertex of *ABCD* and *A′B′C′D′*.

Ed-Imaging

LESSON 3

Rotations

Then

- You identified rotations and verified them as congruence transformations.

Now

1. Given a geometric figure and a rotation, draw the transformed figure.
2. Describe the effects of rotations on the coordinate plane.

Why?

- Windmills convert the wind's energy into electricity through the rotation of turbine blades.

New Vocabulary

center of rotation
angle of rotation

Mathematical Practices

5 Use appropriate tools strategically.

7 Look for and make use of structure.

1 Draw Rotations

In Lesson 1-7, you learned that a rotation or turn moves every point of a preimage through a specified angle and direction about a fixed point.

Key Concept Rotation

A rotation about a fixed point, called the **center of rotation**, through an angle of $x°$ is a function that maps a point to its image such that

- if the point is the center of rotation, then the image and preimage are the same point, or
- if the point is not the center of rotation, then the image and preimage are the same distance from the center of rotation and the measure of the **angle of rotation** formed by the preimage, center of rotation, and image points is x.

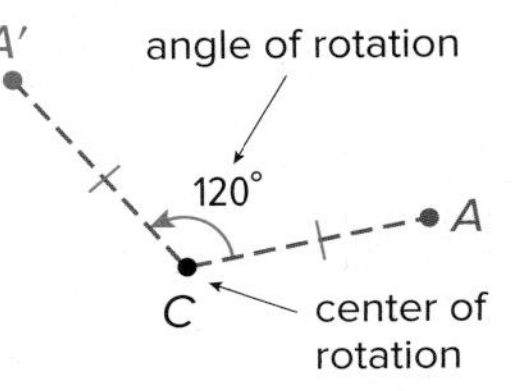

A' is the image of A after a 120° rotation about point C.

The direction of a rotation can be either clockwise or counterclockwise. Assume that all rotations are counterclockwise unless stated otherwise.

clockwise

counterclockwise

Example 1 Draw a Rotation

Copy $\triangle ABC$ and point K. Then use a protractor and ruler to draw a 140° rotation of $\triangle ABC$ about point K.

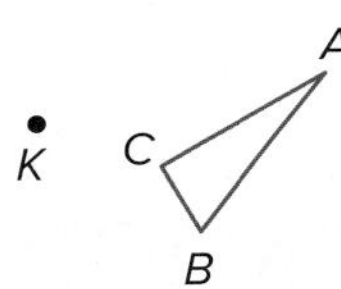
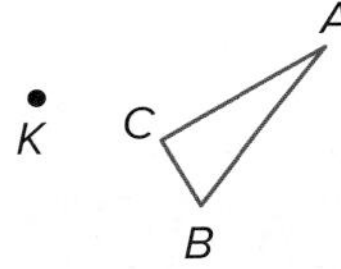

Step 1 Draw a segment from A to K.

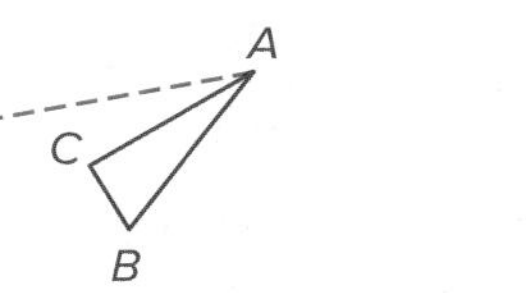

Step 2 Draw a 140° angle using $\overline{KA}$.

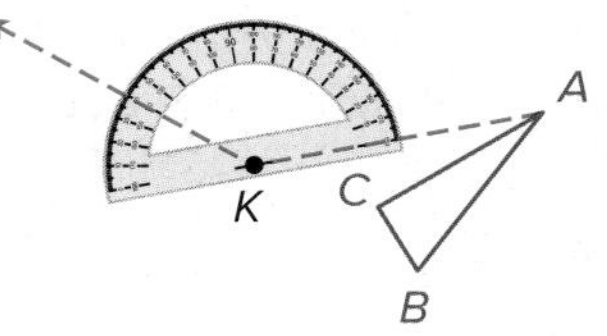

Step 3 Use a ruler to draw A' such that $KA' = KA$.

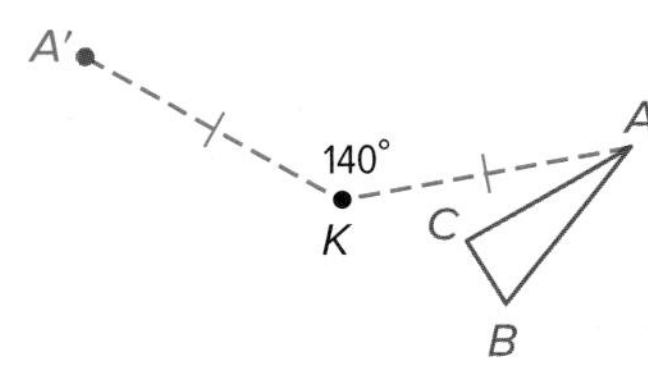

Step 4 Repeat Steps 1–3 for vertices B and C and draw $\triangle A'B'C'$.

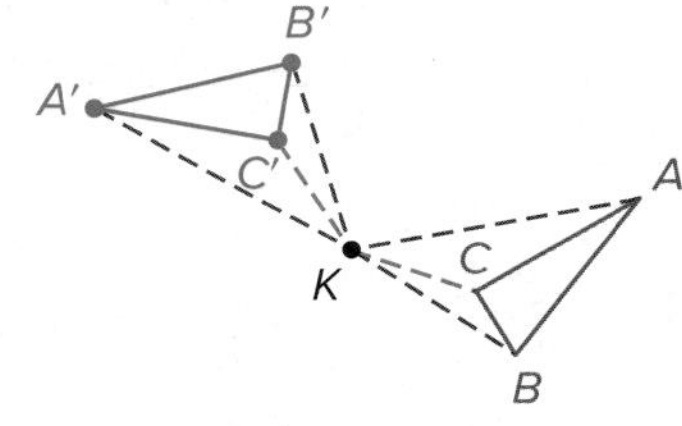

Guided Practice

Copy each figure and point K. Then use a protractor and ruler to draw a rotation of the figure the given number of degrees about K.

1A. 65°

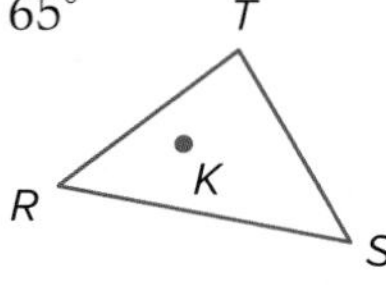

1B. 170°

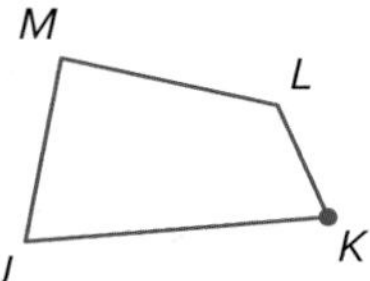

> **Study Tip**
>
> **Rotation direction** Counterclockwise refers to the direction opposite that of the hands of a clock. Clockwise rotations can be designated by a negative measure. For example, a rotation of −90° about the origin is a rotation 90° clockwise about the origin.

2 Describe the Effects of Rotations

Rotations are functions that map points of the plane to other points of the plane. You can describe rotations using the following rules.

Key Concept Rotations in the Coordinate Plane

90° Rotation To rotate a point 90° counterclockwise about the origin, multiply the y-coordinate by −1 and then interchange the x- and y-coordinates. **Symbols** $(x, y) \rightarrow (-y, x)$	**Example** 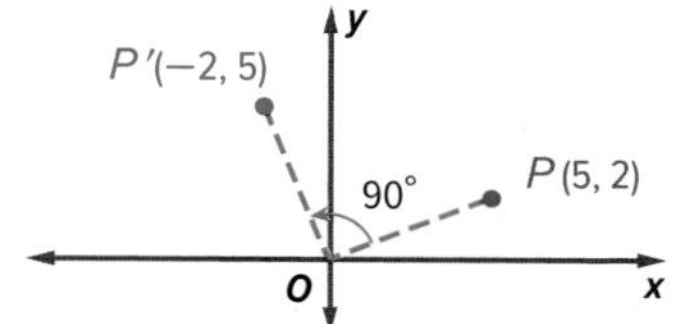
180° Rotation To rotate a point 180° counterclockwise about the origin, multiply the x- and y-coordinates by −1. **Symbols** $(x, y) \rightarrow (-x, -y)$	**Example** 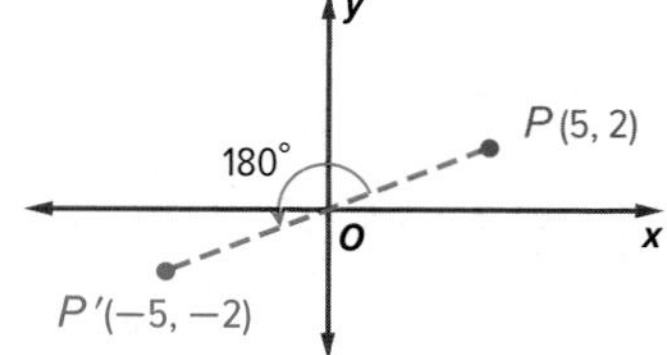
270° Rotation To rotate a point 270° counterclockwise about the origin, multiply the x-coordinate by −1 and then interchange the x- and y-coordinates. **Symbols** $(x, y) \rightarrow (y, -x)$	**Example** 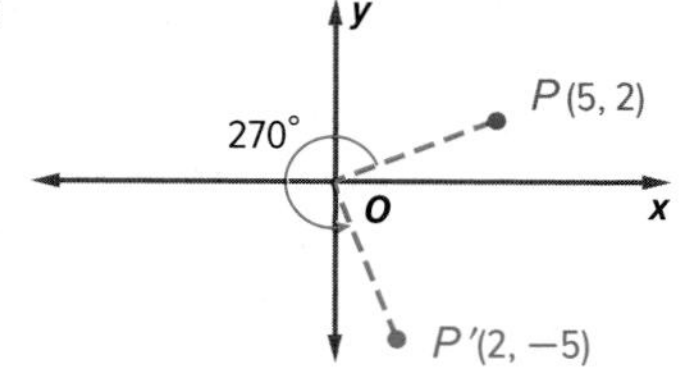

> **Study Tip**
>
> **360° Rotation** A rotation of 360° about a point returns a figure to its original position. That is, the image under a 360° rotation is equal to the preimage.

Example 2 Rotate a Figure About the Origin

Triangle PQR has vertices $P(1, 1)$, $Q(4, 5)$, and $R(5, 1)$. Graph $\triangle PQR$ and its image after a rotation 90° about the origin.

Multiply the y-coordinate of each vertex by −1 and interchange.

$(x, y) \rightarrow (-y, x)$

$P(1, 1) \rightarrow P'(-1, 1)$

$Q(4, 5) \rightarrow Q'(-5, 4)$

$R(5, 1) \rightarrow R'(-1, 5)$

Graph $\triangle PQR$ and its image $\triangle P'Q'R'$.

Guided Practice

2. Parallelogram $FGHJ$ has vertices $F(2, 1)$, $G(7, 1)$, $H(6, -3)$, and $J(1, -3)$. Graph $FGHJ$ and its image after a rotation 180° about the origin.

When combined with translations, the rules for rotations can be used to rotate figures about points other than the origin.

Go Online! 

Investigate rotations by using the **Geometry Tools** in ConnectED.

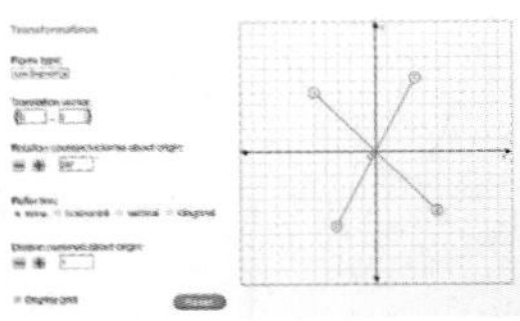

Example 3 Rotate About a Point Other Than the Origin

Triangle *ABC* has vertices *A*(−8, 5), *B*(−6, 9), and *C*(−3, 6). Graph △*ABC* after a rotation 180° about the point (−5, 3).

Step 1 Make a prediction.

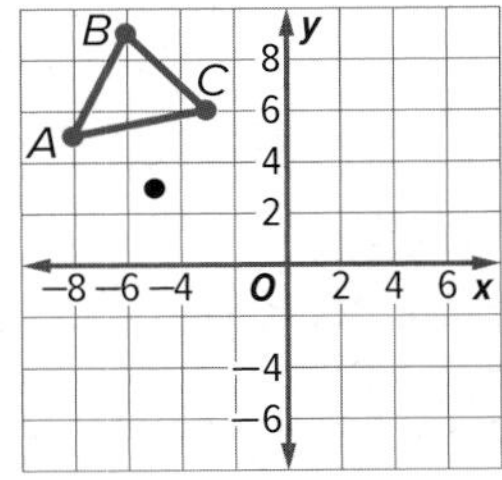

Graph △*ABC* and predict where the image of △*ABC* will be on the coordinate plane after the given rotation.

Prediction: The image of △*ABC* after a 180° rotation about the point (−5, 3) will be a triangle in the third quadrant.

Step 2 Map the center of rotation to the origin.

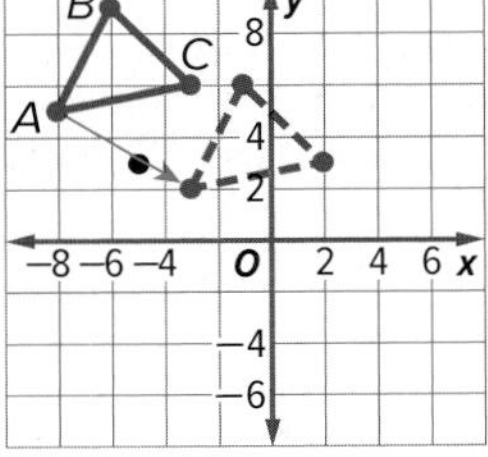

To map the center of rotation to the origin, translate the center of rotation along the vector <5, −3>. Then translate the vertices of △*ABC* along the same vector.

$(x, y) \rightarrow (x + 5, y - 3)$

$A(-8, 5) \rightarrow (-3, 2)$

$B(-6, 9) \rightarrow (-1, 6)$

$C(-3, 6) \rightarrow (2, 3)$

Step 3 Rotate 180° about the origin.

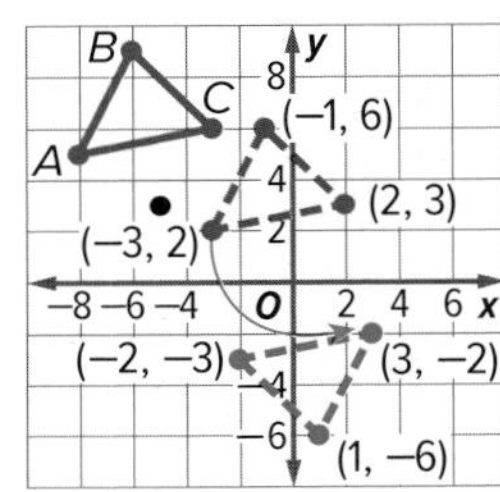

$(x, y) \rightarrow (-x, -y)$

$(-3, 2) \rightarrow (3, -2)$

$(-1, 6) \rightarrow (1, -6)$

$(2, 3) \rightarrow (-2, -3)$

Watch Out!

Multiple Transformations When you rotate a figure about a point other than the origin, be sure to perform the translation, rotation, and translation in the correct order.

Step 4 Map the center of rotation to its original location.

To map the center of rotation to its original location, translate the center of rotation along the vector <−5, 3>. Then translate the vertices of the rotated triangle along the same vector.

$(x, y) \rightarrow (x - 5, y + 3)$

$(3, -2) \rightarrow A'(-2, 1)$

$(1, -6) \rightarrow B'(-4, -3)$

$(-2, -3) \rightarrow C'(-7, 0)$

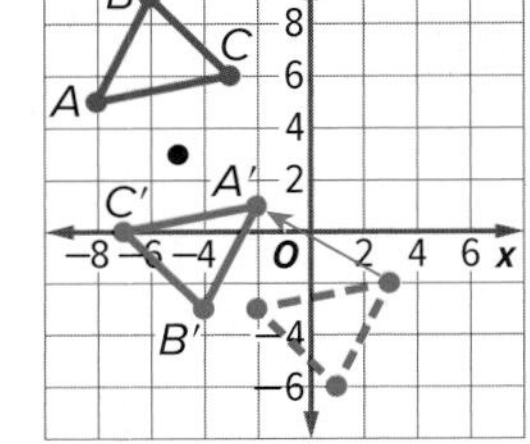

CHECK The image matches the prediction because the majority of the image is in the third quadrant.

Guided Practice

3. Triangle *RST* has vertices *R*(1, −3), *S*(3, −4), and *T*(1, −7). Graph △*RST* after a rotation 90° about the point (2, −2).

Check Your Understanding

◯ = Step-by-Step Solutions begin on page R13.

Example 1

Copy each polygon and point K. Then use a protractor and ruler to draw the specified rotation of each figure about point K.

1. 45°

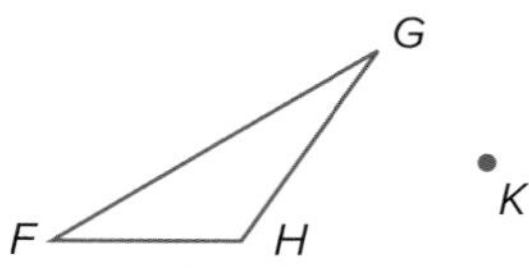

2. 120°

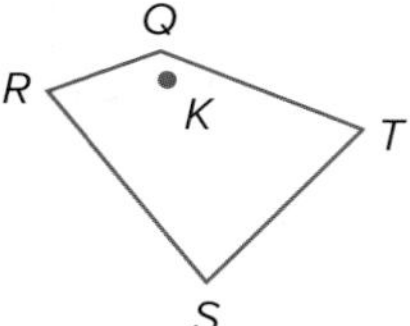

Example 2

3. Triangle DFG has vertices $D(-2, 6)$, $F(2, 8)$, and $G(2, 3)$. Graph $\triangle DFG$ and its image after a rotation 180° about the origin.

Example 3

Graph each figure and its image after the specified rotation.

4. $\triangle ABC$ has vertices $A(3, 2)$, $B(5, 1)$, and $C(3, -1)$; 90° about the point $(1, 1)$
5. trapezoid $JKLM$ has vertices $J(-3, 3)$, $K(0, 3)$, $L(1, 1)$, and $M(-3, 1)$; 180° about the point $(1, -1)$
6. $\triangle DEF$ has vertices $D(-4, 1)$, $E(-2, 4)$, and $F(-2, 1)$; 270° about the point $(-3, -2)$

Practice and Problem Solving

Extra Practice is on page R3.

Example 1

MP TOOLS Copy each polygon and point K. Then use a protractor and ruler to draw the specified rotation of each figure about point K.

7. 90°

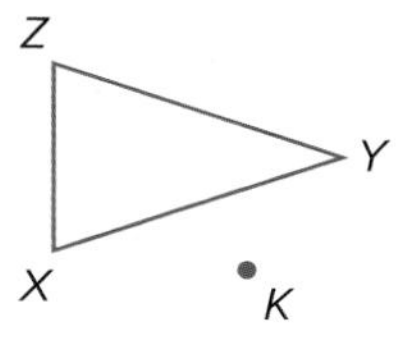

8. 15°

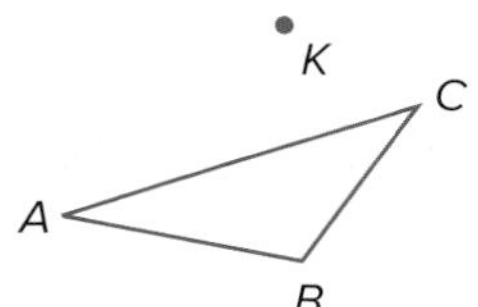

9. 145°

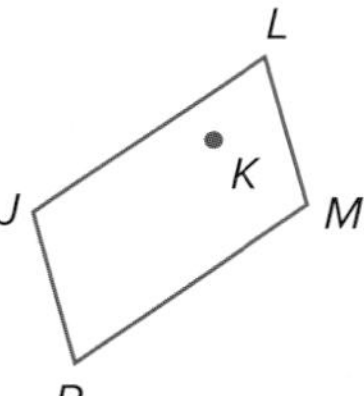

10. 30°

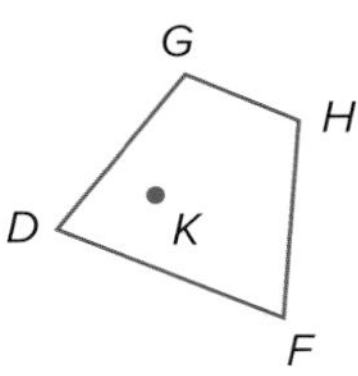

11. 260°

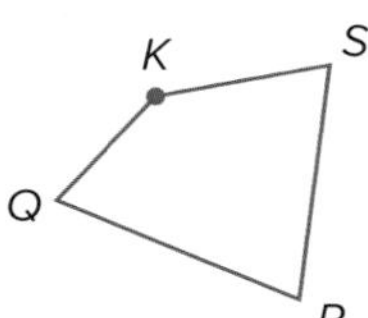

12. 50°

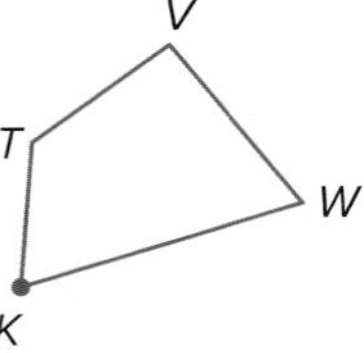

PINWHEELS Find the angle of rotation to the nearest tenth of a degree that maps P onto P'. Explain your reasoning.

13.

14.

15.

(l)BRIAN ELLIOTT/Alamy, (c)JupiterImages/Thinkstock/Alamy, (r)Pick and Mix Images/Alamy

Examples 2–3 **Graph each figure and its image after the specified rotation.**

16. $\triangle JKL$ has vertices $J(2, 6)$, $K(5, 2)$, and $L(7, 5)$; 90° about the origin

17. rhombus $WXYZ$ has vertices $W(-3, 4)$, $X(0, 7)$, $Y(3, 4)$, and $Z(0, 1)$; 90° about the origin

18. $\triangle FGH$ has vertices $F(2, 4)$, $G(5, 6)$, and $H(7, 2)$; 180° about the origin

19. $\triangle PQR$ has vertices $P(1, 2)$, $Q(1, -1)$, and $R(-2, -1)$; 180° about the point $(2, -2)$

20. trapezoid $ABCD$ has vertices $A(-3, 4)$, $B(-1, 3)$, $C(-1, 0)$, and $D(-3, -1)$; 270° about the point $(2, 1)$

21. $\triangle JKL$ has vertices $J(-1, -1)$, $K(3, -1)$, and $L(-1, -3)$; 90° about the point $(-3, -1)$

22. **WEATHER** A weathervane is used to indicate the direction of the wind. If the vane is pointing northeast and rotates 270°, what is the new wind direction?

23. MP **MODELING** The photograph of the Grande Roue, or Big Wheel, at the right, appears blurred because of the camera's shutter speed—the length of time the camera's shutter was open. The diameter of the wheel is 60 meters.

a. Estimate the angle of rotation in the photo. (*Hint:* Use points A and A'.)

b. If the Ferris wheel makes one revolution per minute, use your estimate from part **a** to estimate the camera's shutter speed.

Alan Schein/Alamy

Each figure shows a preimage and its image after a rotation about point P. Copy each figure, locate point P, and find the angle of rotation.

24.

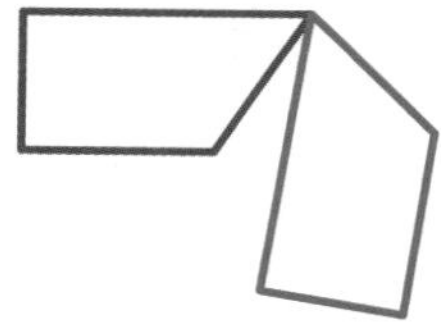

25.

ALGEBRA **Give the equation of the line $y = -x - 2$ after a rotation about the origin through the given angle. Then describe the relationship between the equations of the image and preimage.**

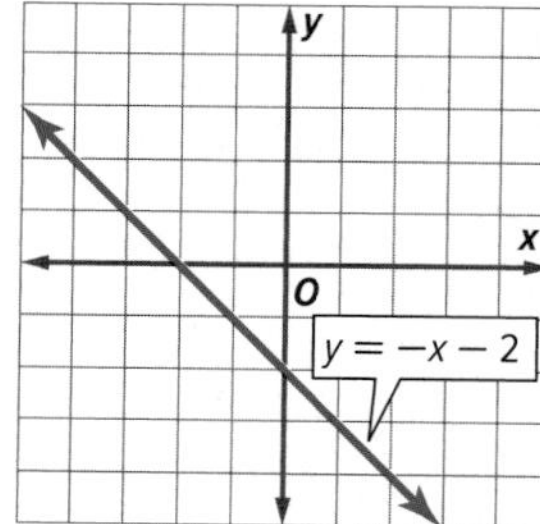

26. 90°

27. 180°

28. 270°

29. 360°

ALGEBRA **Rotate the line the specified number of degrees about the x- and y-intercepts and find the equation of the resulting image.**

30. $y = x - 5$; 90°

31. $y = 2x + 4$; 180°

32. $y = 3x - 2$; 270°

33 **RIDES** An amusement park ride consists of four circular cars. The ride rotates at a rate of 0.25 revolution per second. In addition, each car rotates 0.5 revolution per second. If Jane is positioned at point P when the ride begins, what coordinates describe her position after 31 seconds?

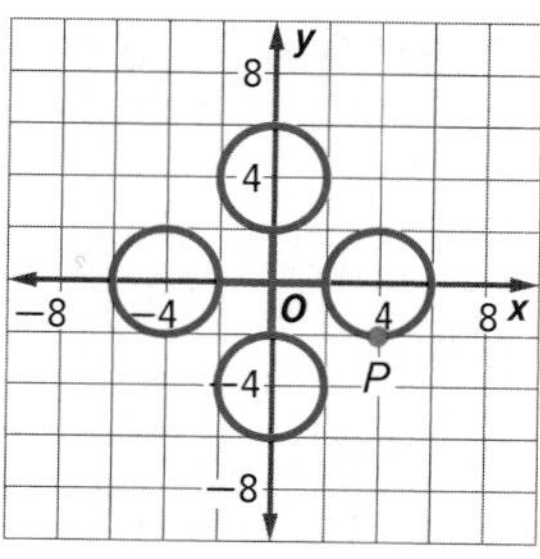

34. **MULTI-STEP** Refer to the information given in Exercise 23. Carlos is trying to duplicate the process used to capture the image of the Grande Roue using the Texas Star, which has a height of 212 feet and makes 1.5 revolutions per minute. The Grande Roue makes one revolution per minute.

a. What shutter speed should he use?

b. Explain your solution process.

35. **MULTIPLE REPRESENTATIONS** In this problem, you will investigate reflections over a pair of intersecting lines.

a. **Geometric** On a coordinate plane, draw a triangle and a pair of intersecting lines. Label the triangle *ABC* and the lines ℓ and m. Reflect $\triangle ABC$ in the line ℓ. Then reflect $\triangle A'B'C'$ in the line m. Label the final image $A''B''C''$.

b. **Geometric** Repeat the process in part **a** two more times in two different quadrants. Label the second triangle *DEF* and reflect it in intersecting lines n and p. Label the third triangle *MNP* and reflect it in intersecting lines q and r.

c. **Tabular** Measure the angle of rotation of each triangle about the point of intersection of the two lines. Copy and complete the table below.

Angle of Rotation Between Figures		Angle Between Intersecting Lines	
$\triangle ABC$ and $\triangle A''B''C''$		l and m	
$\triangle DEF$ and $\triangle D''E''F''$		n and p	
$\triangle MNP$ and $\triangle M''N''P''$		q and r	

d. **Verbal** Make a conjecture about the angle of rotation of a figure about the intersection of two lines after the figure is reflected in both lines.

H.O.T. Problems Use Higher-Order Thinking Skills

36. **WRITING IN MATH** Are collinearity and betweenness of points maintained under rotation? Explain.

37. **CHALLENGE** Point *C* has coordinates *C*(5, 5). The image of this point after a rotation of 100° about a certain point is *C'*(−5, 7.5). Use construction to estimate the coordinates of the center of this rotation. Explain.

38. **OPEN-ENDED** Draw a figure on the coordinate plane. Describe a nonzero rotation that maps the image onto the preimage with no change in orientation.

39. MP **CONSTRUCT ARUGMENTS** Is the reflection of a figure in the *x*-axis equivalent to the rotation of that same figure 180° about the origin? Explain.

40. **WRITING IN MATH** Do invariant points *sometimes*, *always*, or *never* occur in a rotation? Explain your reasoning.

41. $\triangle GHJ$ has vertices $G(1, 3)$, $H(4, 3)$, and $J(2, 0)$. $\triangle G'H'J'$ is the image of $\triangle GHJ$ under the rotation $(x, y) \rightarrow (-y, x)$. Which of the following is a true statement about $\triangle G'H'J'$? MP 6

- ○ **A** $\triangle G'H'J'$ lies entirely in Quadrant III.
- ○ **B** $\triangle G'H'J$ intersects the positive y-axis.
- ○ **C** $\triangle G'H'J$ intersects the x-axis.
- ○ **D** $\triangle G'H'J$ overlaps $\triangle GHJ$.

42. Chloe is using a coordinate plane to create a geometric pattern for an art project. She starts by drawing $\triangle RST$, as shown. Next, she wants to rotate the triangle so that its image intersects the x-axis.

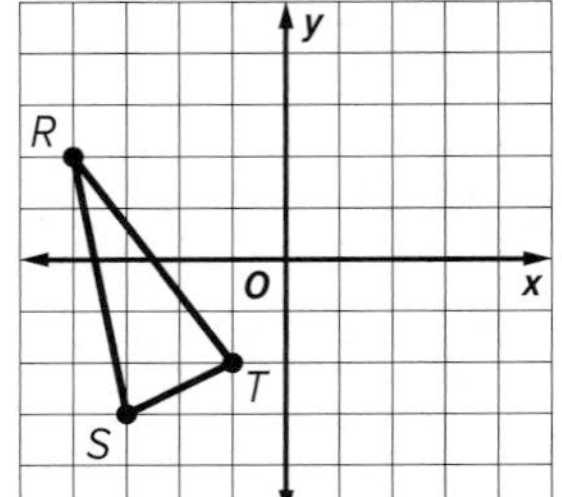

Which of the following transformations could Chloe use? MP 4

- ○ **A** $(x, y) \rightarrow (y, -x)$
- ○ **B** $(x, y) \rightarrow (x, -y)$
- ○ **C** $(x, y) \rightarrow (x + 5, y)$
- ○ **D** $(x, y) \rightarrow (-x, -y)$

43. Which of the following points has an image in Quadrant III under the rotation $(x, y) \rightarrow (y, -x)$? MP 2

- ○ **A** $(2, -1)$
- ○ **B** $(3, 2)$
- ○ **C** $(-1, -4)$
- ○ **D** $(-2, 3)$

44. Milo graphs the line $x = 3$. Then he graphs the image of the line using the rotation $(x, y) \rightarrow (-y, x)$. Which of the following points lies on the image of the line? MP 7

- ○ **A** $(-3, 2)$
- ○ **B** $(4, 3)$
- ○ **C** $(-1, -3)$
- ○ **D** $(2, -3)$

45. Malik is using the figure shown below to write a definition of a rotation of $x°$ about point P. Which of the following will be part of the definition? MP 6

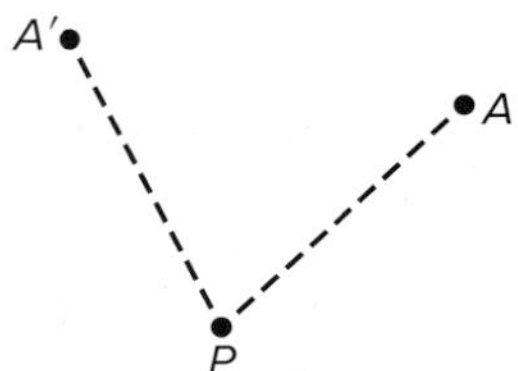

- ☐ **A** $AP = A'P$
- ☐ **B** Point A is translated along a vector to point A'.
- ☐ **C** $m\angle P = x$
- ☐ **D** The image of point P is point P.
- ☐ **E** The image of point A is point A'.

46. **MULTI-STEP** A landscaper uses a coordinate plane to design flower beds for a mall. She places a triangular flower bed on the coordinate plane with vertices at $(1, 3)$, $(4, 5)$, and $(5, 3)$. MP 4

a. The landscaper decides to rotate the bed 180° about the point $(3, 1)$. Predict the quadrant in which the image of the bed will lie. Explain how you made the prediction.

b. Graph the bed and its image after the rotation.

c. The landscaper places a rectangular bed on the coordinate plane with vertices at $(-3, 1)$, $(-1, 1)$, $(-1, -2)$, and $(-3, -2)$. What rotation should she use to reposition the bed so that its vertices are $(-3, 1)$, $(-3, 3)$, $(0, 3)$, and $(0, 1)$?

CHAPTER 3

Mid-Chapter Quiz

Lessons 3-1 through 3-3

Copy the figure and the given line of reflection. Then draw the reflected image in this line using a ruler. (Lesson 3-1)

1.

2.

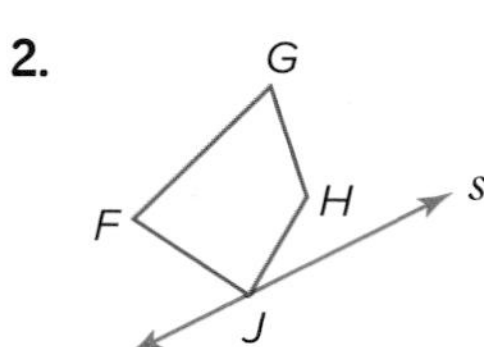

Graph each figure and its image after the specified reflection. (Lesson 3-1)

3. △*FGH* has vertices *F* (−4, 3), *G* (−2, 0), and *H* (−1, 4); in the *y*-axis

4. rhombus *QRST* has vertices *Q* (2, 1), *R*(4, 3), *S*(6, 1), and *T*(4, −1); in the *x*-axis

5. CLUBS The drama club is selling candy during the intermission of a school play.

a. Locate point *P* along the wall to represent the candy table so that people coming from either door *A* or door *B* would walk the same distance to the table. (Lesson 3-1)

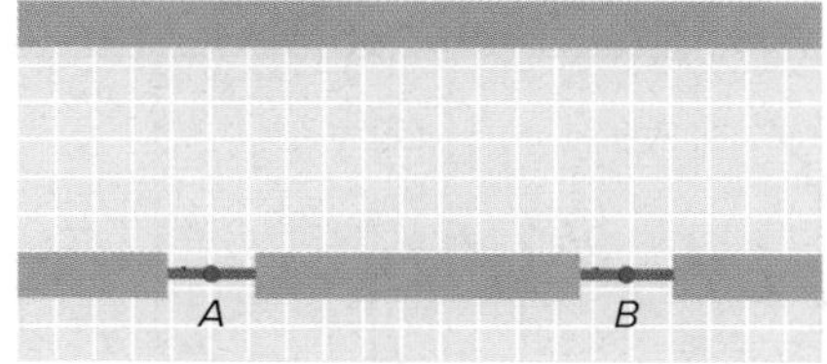

b.  What mathematical practice did you use to solve this problem?

Graph each figure and its image after the specified translation. (Lesson 3-2)

6. △*ABC* with vertices *A*(0, 0), *B*(2, 1), *C*(1, −3); ⟨3, −1⟩

7. rectangle *JKLM* has vertices *J*(−4, 2), *K*(−4, −2), *L*(−1, −2), and *M*(−1, 2); ⟨5, −3⟩

Copy the figure and the given translation vector. Then draw the translation of the figure along the translation vector. (Lesson 3-2)

8.

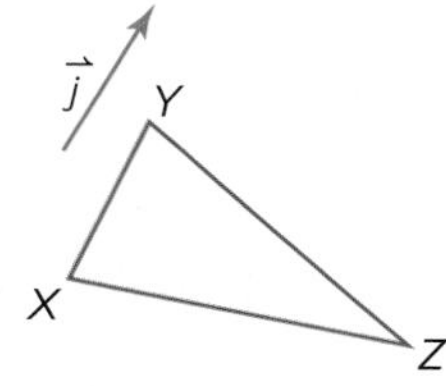

9.

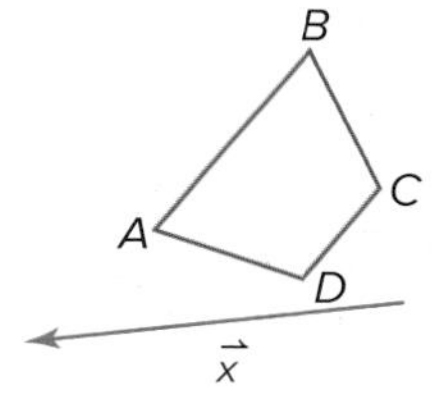

10. COMICS Alex is making a comic. He uses graph paper to make sure the dimensions of his drawings are accurate. If he draws a coordinate plane with two flies as shown below, what vector represents the movement from fly 1 to fly 2? (Lesson 3-2)

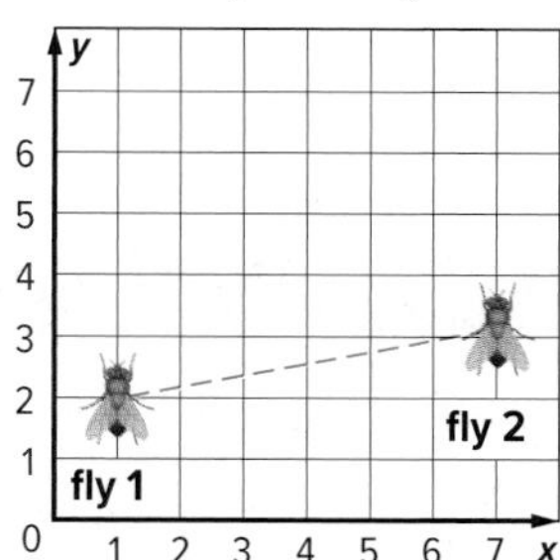

Copy each polygon and point *R*. Then use a protractor and ruler to draw the specified rotation of each figure about point *R*. (Lesson 3-3)

11. 45°

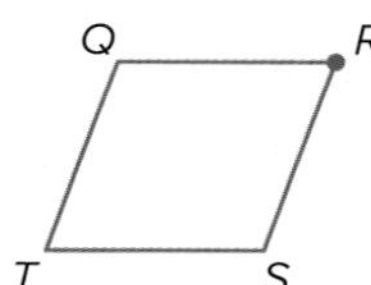

12. 60°

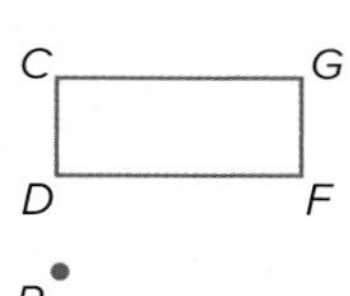

13. MULTIPLE CHOICE What is the image of point *M* after a rotation of 90° about the origin? (Lesson 3-3)

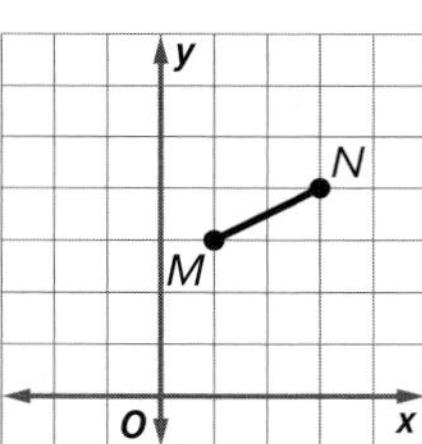

A (−3, 1)

B (−3, −1)

C (−1, −3)

D (3, 1)

Graph each figure and its image after the specified rotation. (Lesson 3-3)

14. △*RST* has vertices *R*(−3, 0), *S*(−1, −4), and *T*(0, −1); 90°

15. square *JKLM* has vertices *J*(−1, 2), *K*(−1, −2), *L*(3, −2), and *M*(3, 2); 180°

EXPLORE 3-4

Geometry Software Lab

Compositions of Transformations

In this lab, you will use Geometer's Sketchpad to explore the effects of performing multiple transformations on a figure.

Mathematical Practices
MP **5** Use appropriate tools strategically.

Activity

Work cooperatively. Reflect a figure in two vertical lines.

Step 1 Use the line segment tool to construct a triangle with one vertex pointing to the left so that you can easily see changes as you perform transformations. Label the triangle *ABC*.

Step 2 Insert and label a line *m* to the right of $\triangle ABC$. Insert a point so that the distance from the point to line *m* is greater than the width of $\triangle ABC$. Draw the line parallel to line *m* through the point and label the new line *r*.

Step 3 Select line *m* and choose **Mark Mirror** from the **Transform** menu. Select all sides and vertices of $\triangle ABC$ and choose **Reflect** from the **Transform** menu.

Step 4 Repeat the process you used in Step 3 to reflect the new image in line *r*.

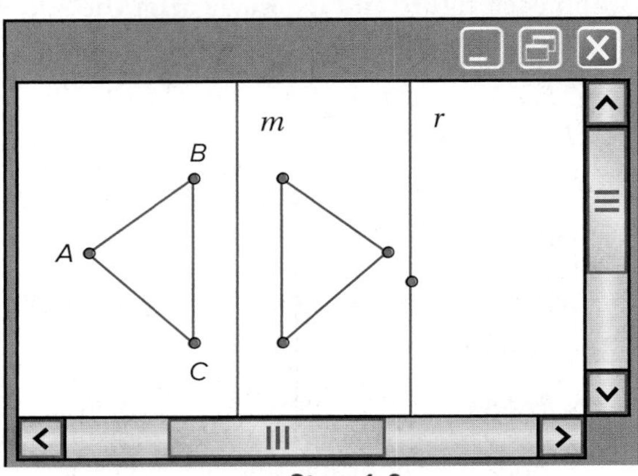

Steps 1–3

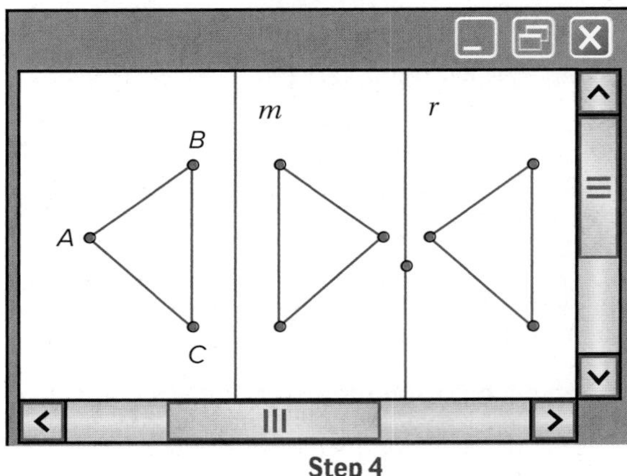

Step 4

Analyze the Results Work cooperatively.

1. How are the original figure and the final figure related?
2. What single transformation could be used to produce the final figure?
3. If you move line *m*, what happens? if you move line *r*?
4. **MAKE A CONJECTURE** If you reflected the figure in a third line, what single transformation could be used to produce the final figure? Explain your reasoning.
5. Repeat the activity for a pair of perpendicular lines. What single transformation could be used to produce the same final figure?
6. If you reflected the figure from Exercise 5 in a third line perpendicular to the second line, is there a single transformation that could be used to produce the final figure? Explain your reasoning.

LESSON 4

Compositions of Transformations

Then

- You drew reflections, translations, and rotations.

Now

1. Draw glide reflections and other compositions of isometries in the coordinate plane.
2. Draw compositions of reflections in parallel and intersecting lines.

Why?

- The pattern of footprints left in the sand after a person walks along the edge of a beach illustrates the composition of two different transformations—translations and reflections.

New Vocabulary

composition of transformations

glide reflection

Mathematical Practices

1 Make sense of problems and persevere in solving them.

2 Reason abstractly and quantitatively.

3 Construct viable arguments and critique the reasoning of others.

4 Model with mathematics.

1 Glide Reflections

When a transformation is applied to a figure and then another transformation is applied to its image, the result is called a **composition of transformations**. A glide reflection is one type of composition of transformations.

Key Concept Glide Reflection

A **glide reflection** is the composition of a translation followed by a reflection in a line parallel to the translation vector.

Example

The glide reflection shown is the composition of a translation along $\vec{w}$ followed by a reflection in line ℓ.

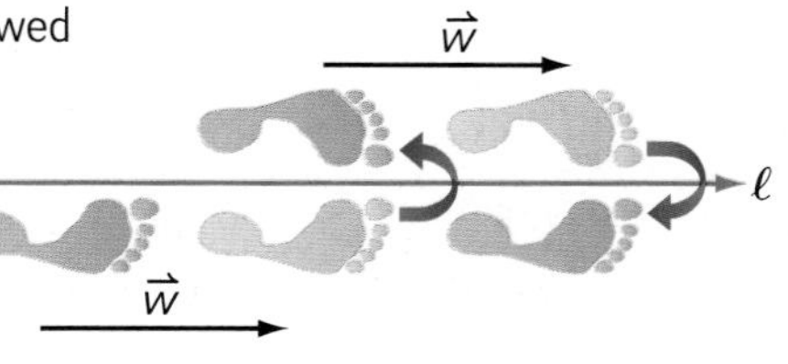

Example 1 Graph a Glide Reflection

Triangle *JKL* has vertices *J*(6, −1), *K*(10, −2), and *L*(5, −3). Graph △*JKL* and its image after a translation along ⟨0, 4⟩ and a reflection in the *y*-axis.

Step 1 translation along ⟨0, 4⟩

(x, y)	→	$(x, y + 4)$
$J(6, -1)$	→	$J'(6, 3)$
$K(10, -2)$	→	$K'(10, 2)$
$L(5, -3)$	→	$L'(5, 1)$

Step 2 reflection in the *y*-axis

(x, y)	→	$(-x, y)$
$J'(6, 3)$	→	$J''(-6, 3)$
$K'(10, 2)$	→	$K''(-10, 2)$
$L'(5, 1)$	→	$L''(-5, 1)$

Step 3 Graph △*JKL* and its image △*J″K″L″*.

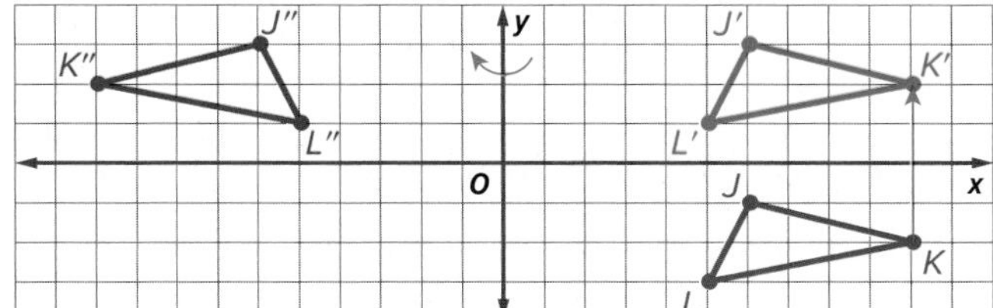

Guided Pracacice

Triangle *PQR* has vertices *P*(1, 1), *Q*(2, 5), and *R*(4, 2). Graph △*PQR* and its image after the indicated glide reflection.

1A. Translation: along ⟨−2, 0⟩
Reflection: in *x*-axis

1B. Translation: along ⟨−3, −3⟩
Reflection: in $y = x$

In Example 1, $\triangle JKL \cong \triangle J'K'L'$ and $\triangle J'K'L' \cong \triangle J''K''L''$. By the Transitive Property of Congruence, $\triangle JKL \cong \triangle J''K''L''$. This suggests the following theorem.

Theorem 3.1 Composition of Isometries

The composition of two (or more) isometries is an isometry.

You will prove one case of Theorem 3.1 in Exercise 30.

Study Tips

MP Sense-Making Glide reflections, reflections, translations, and rotations are the only four *rigid motions* or isometries in a plane.

So, the composition of two or more isometries—reflections, translations, or rotations—results in an image that is congruent to its preimage.

Example 2 Graph Other Compositions of Isometries

The endpoints of $\overline{CD}$ are $C(-7, 1)$ and $D(-3, 2)$. Graph $\overline{CD}$ and its image after a reflection in the x-axis and a rotation 90° about the origin.

Step 1 reflection in the x-axis

$(x, y) \rightarrow (x, -y)$

$C(-7, 1) \rightarrow C'(-7, -1)$

$D(-3, 2) \rightarrow D'(-3, -2)$

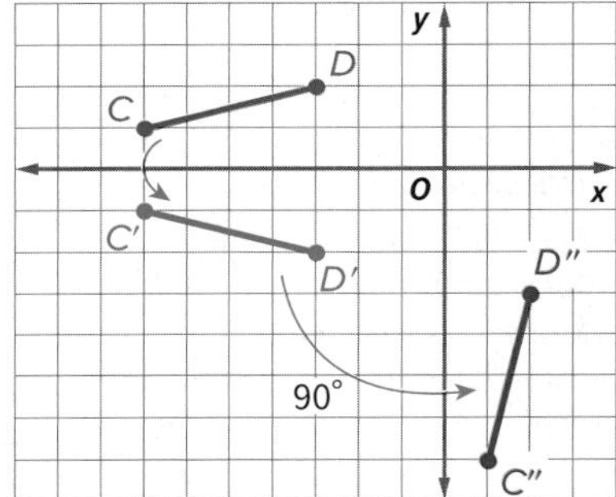

Step 2 rotation 90° about origin

$(x, y) \rightarrow (-y, x)$

$C'(-7, -1) \rightarrow C''(1, -7)$

$D'(-3, -2) \rightarrow D''(2, -3)$

Reading Math

Double Primes Double primes are used to indicate that a vertex is the image of a second transformation.

Step 3 Graph $\overline{CD}$ and its image $\overline{C''D''}$.

Guided Practice

Triangle ABC has vertices $A(-6, -2)$, $B(-5, -5)$, and $C(-2, -1)$. Graph $\triangle ABC$ and its image after the composition of transformations in the order listed.

2A. Translation: along $\langle 3, -1 \rangle$
Reflection: in y-axis

2B. Rotation: 180° about origin
Translation: along $\langle -2, 4 \rangle$

2 Compositions of Two Reflections

The composition of two reflections in parallel lines is the same as a translation.

Theorem 3.2 Reflections in Parallel Lines

The composition of two reflections in parallel lines can be described by a translation vector that is

- perpendicular to the two lines, and
- twice the distance between the two lines.

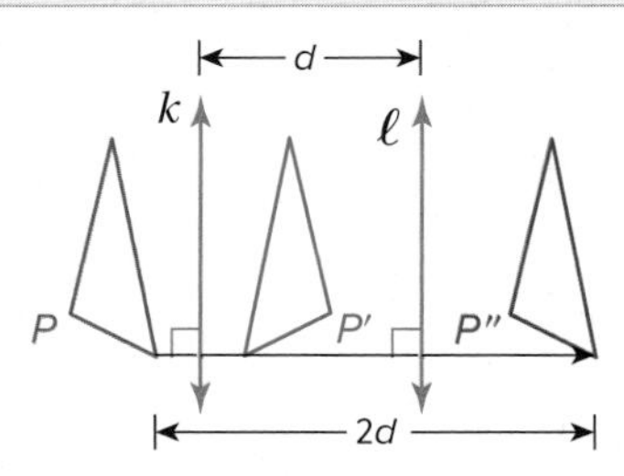

You will prove Theorem 3.2 in Exercise 34.

The composition of two reflections in intersecting lines is the same as a rotation.

Theorem 3.3 Reflections in Intersecting Lines

The composition of two reflections in intersecting lines can be described by a rotation

- about the point where the lines intersect and
- through an angle that is twice the measure of the acute or right angle formed by the lines.

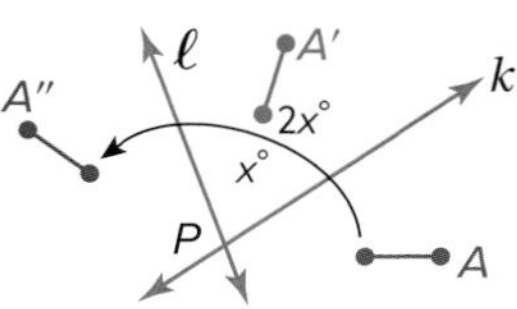

You will prove Theorem 3.3 in Exercise 35.

Example 3 Reflect a Figure in Two Lines

Copy and reflect figure A in line m and then line p. Then describe a single transformation that maps A onto A''.

a.

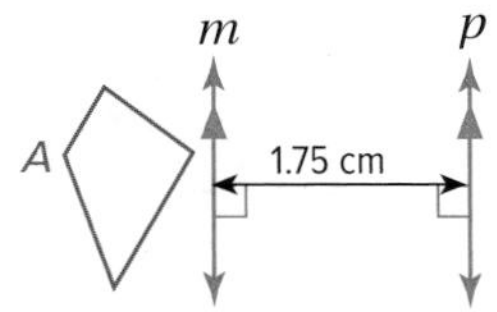

Step 1 Reflect A in line m.

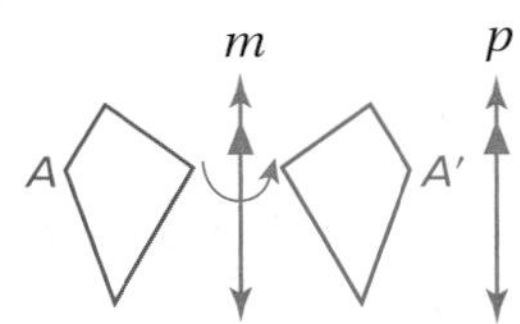

Step 2 Reflect A' in line p.

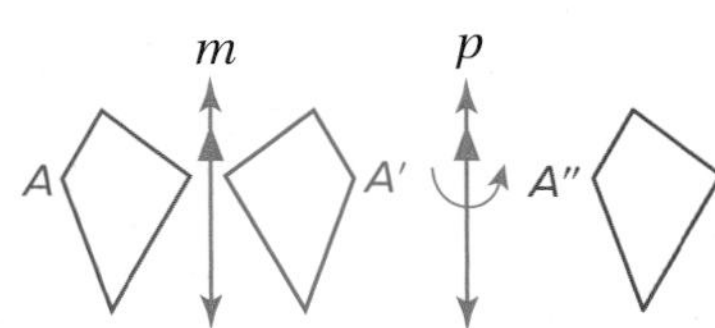

By Theorem 3.2, the composition of two reflections in parallel vertical lines m and p is equivalent to a horizontal translation right $2 \cdot 1.75$ or 3.5 centimeters.

b.

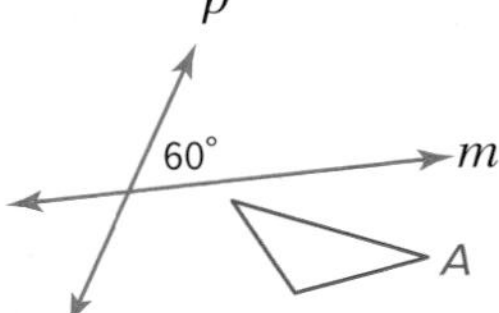

Step 1

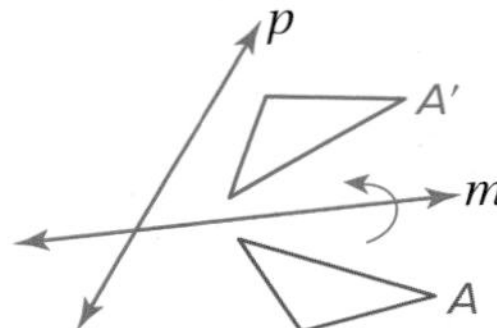

Step 2

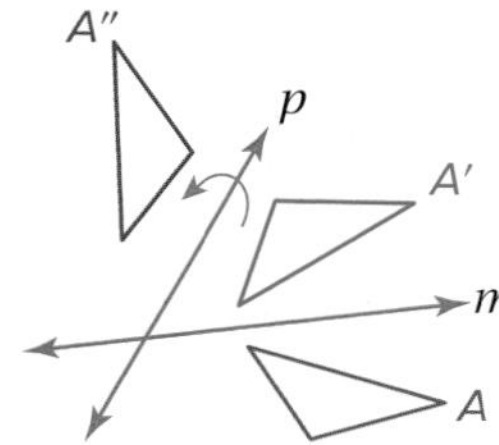

By Theorem 3.3, the composition of two reflections in intersecting lines m and p is equivalent to a $2 \cdot 60°$ or $120°$ counterclockwise rotation about the point where lines m and p intersect.

> **Watch Out**
>
> **Order of Composition** Be sure to compose two transformations according to the order in which they are given.

Guided Practice

Copy and reflect figure B in line n and then line q. Then describe a single transformation that maps B onto B''.

3A.

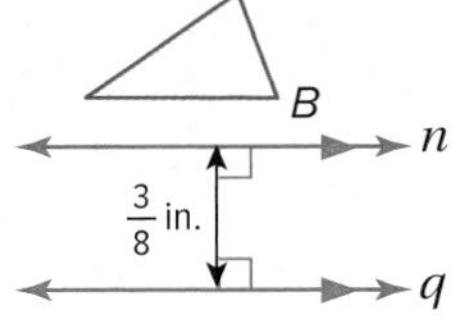

3B.

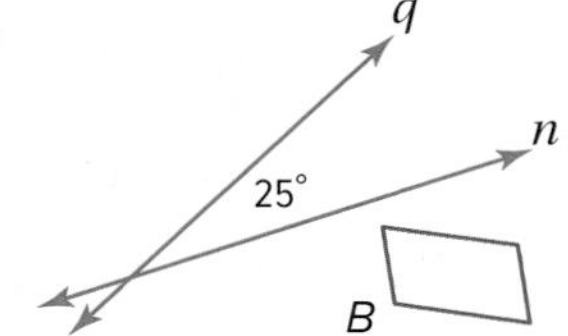

Example 4 Find the Preimage

Determine the preimage given the image and the composition of transformations.

a. translation along $\vec{k}$, reflection in line j

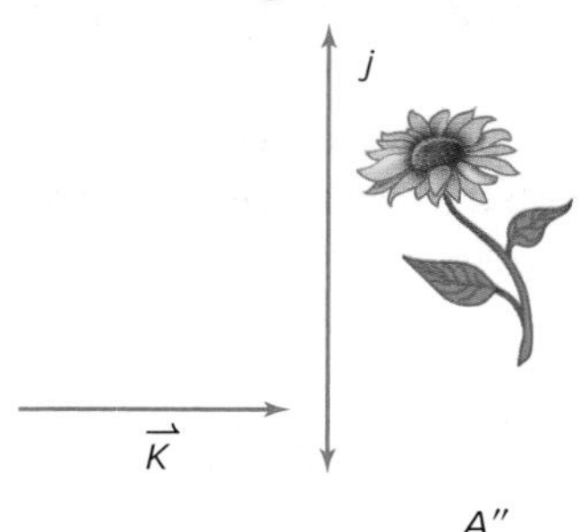

The figure shown is the image after a translation and a reflection. Work backwards to find A' and then the preimage, A.

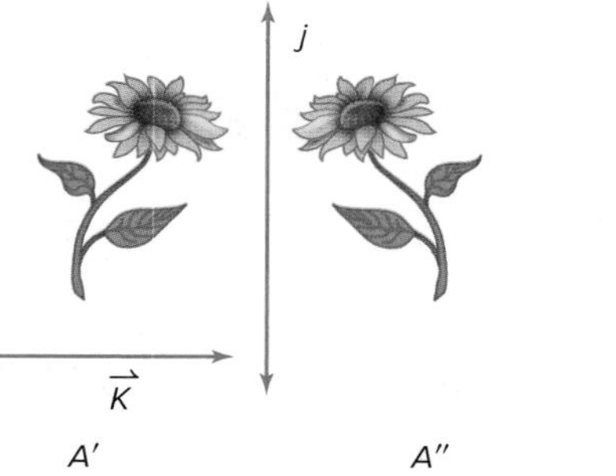

A'' is the reflection of A' in the line.

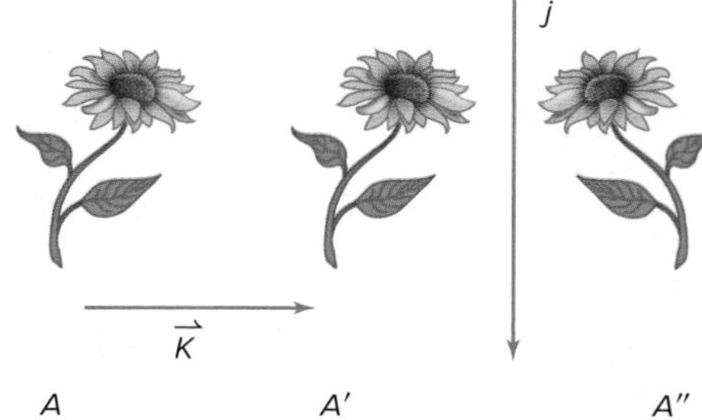

A' is the translated image of A.

b. reflection in x-axis, reflection in y-axis, $A''(-1, -1)$, $B''(-2, -4)$, $C''(-4, -4)$, and $D''(-5, -1)$

First graph $A''B''C''D''$. Then reflect the figure in the y-axis first to find $A'B'C'D'$. Then reflect in the x-axis. $ABCD$ has coordinates $A(1, 1)$, $B(2, 4)$, $C(4, 4)$, and $D(5, 1)$.

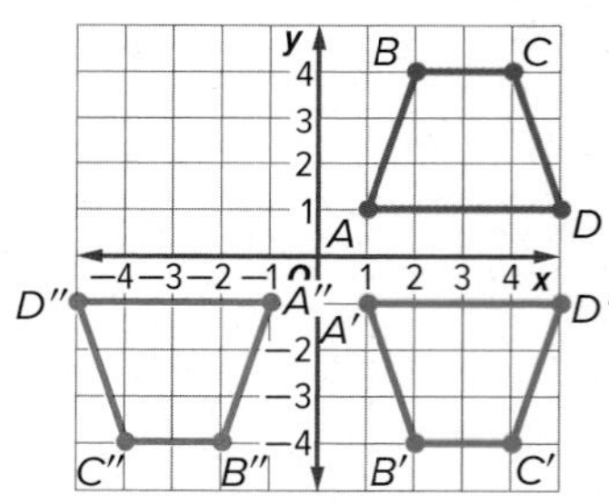

Guided Practice

A. translation along $\vec{j}$, reflection in line ℓ

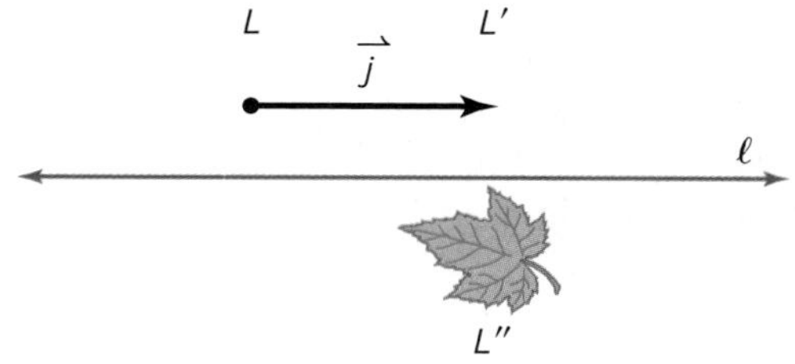

B. rotation 90° about origin, reflection x-axis

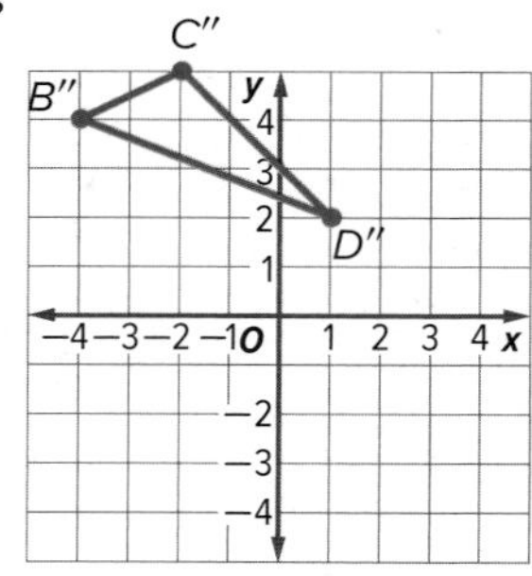

Concept Summary Compositions of Translations

Glide Reflection	Translation	Rotation
the composition of a reflection and a translation	the composition of two reflections in parallel lines	the composition of two reflections in intersecting lines

Real-World Link

In carpets, border patterns result when any of several basic transformations are repeated in one direction. There are seven possible combinations: translations, horizontal reflections, vertical reflections, vertical followed by horizontal reflections, glide reflections, rotations, and reflections followed by glide reflections.

Source: The Textile Museum

Real-World Example 5 Identify Transformations

PATTERNS Identify the preimage in each pattern. Then identify the sequence of transformations that will carry the preimage onto the image(s).

a.

The pattern is created by successive translations of the first four potted plants, which is the preimage. This pattern can be created by combining two reflections in lines m and p as shown. Notice that line m goes through the center of the preimage.

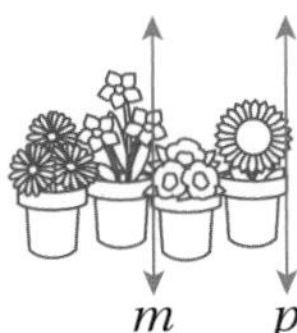

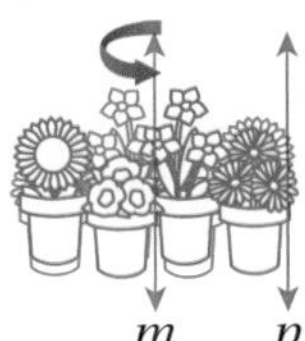

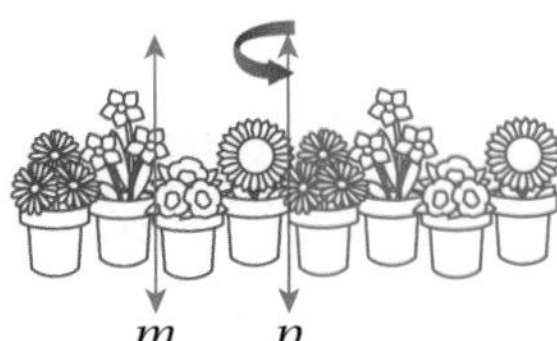

b.

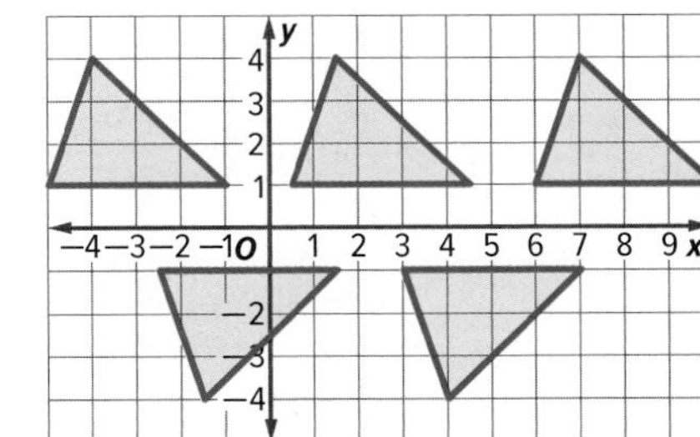

The pattern is created by glide reflection. Using the triangle in Quadrant IV as the preimage, this pattern can be continued by combining a translation along vector $\vec{v}$ with a reflection in the x-axis as shown.

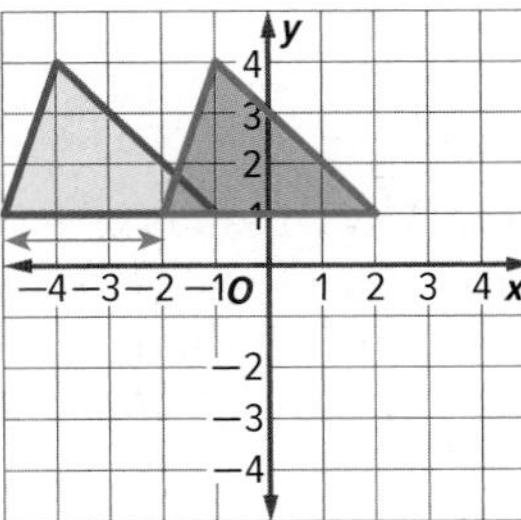

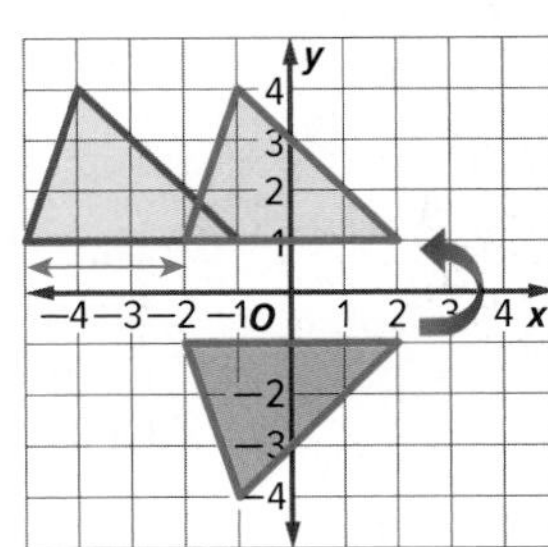

Guided Practice

5. **CARPET PATTERNS** Identify the preimage in each pattern. Then identify the sequence of transformations that will carry the preimage onto the image(s).

A.

B.

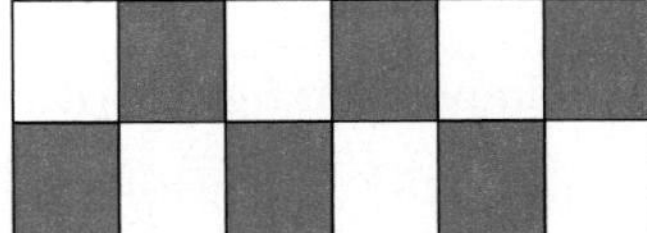

Martin Child/Photodisc/Getty Images

Check Your Understanding

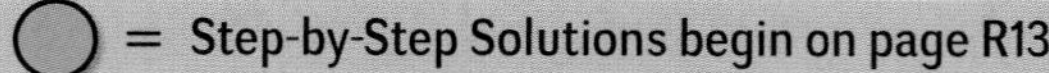

Example 1 **Triangle *CDE* has vertices *C*(−5, −1), *D*(−2, −5), and *E*(−1, −1). Graph △*CDE* and its image after the indicated glide reflection.**

1. Translation: along ⟨4, 0⟩
   Reflection: in *x*-axis

2. Translation: along ⟨0, 6⟩
   Reflection: in *y*-axis

Example 2

3. The endpoints of $\overline{JK}$ are *J*(2, 5) and *K*(6, 5). Graph $\overline{JK}$ and its image after a reflection in the *x*-axis and a rotation 90° about the origin.

Example 3 **Copy and reflect figure *S* in line *m* and then line *p*. Then describe a single transformation that maps *S* onto *S*″.**

4.

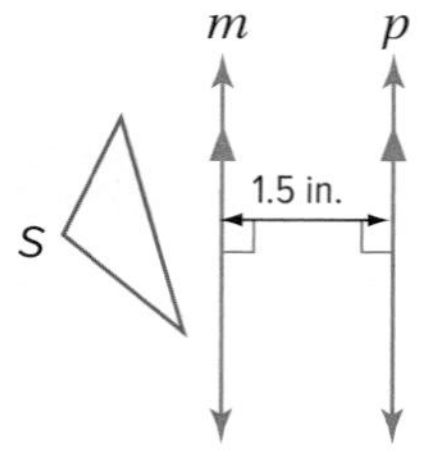

m p
1.5 in.
S

5.

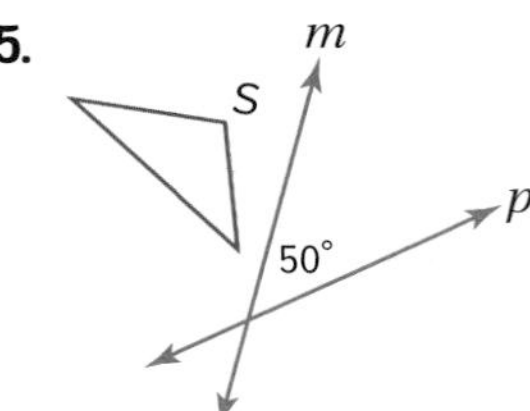

Example 4 **Determine the preimage given the image and the composition of transformations.**

6. reflection in line *j*, translation along $\vec{n}$

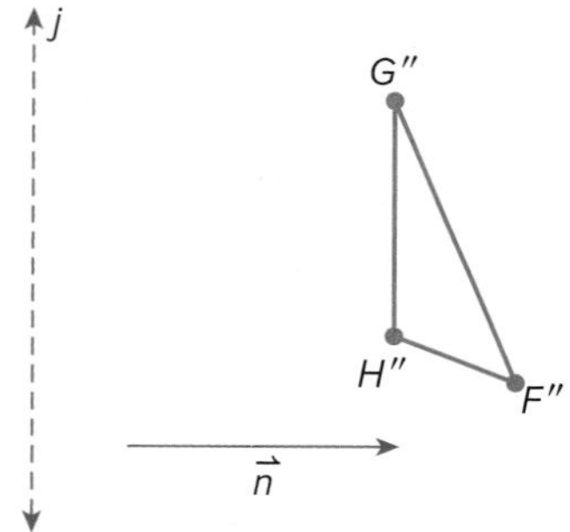

7. rotation 180° about origin, translation 4 units down

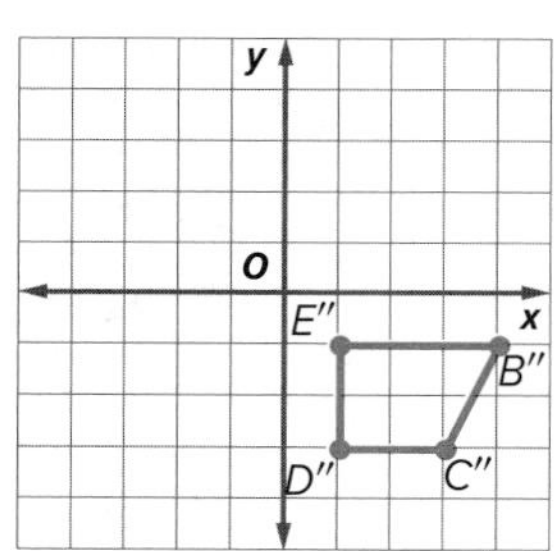

Example 5

8. **TILE PATTERNS** Viviana is creating a pattern for the top of a table with tiles in the shape of isosceles triangles. Identify the sequence of transformations that was used to transform the white triangle to the blue triangle.

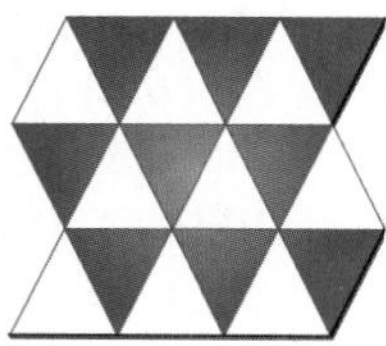

Practice and Problem Solving

Extra Practice is on page R3.

Example 1 **Graph each figure with the given vertices and its image after the indicated glide reflection.**

9. △*RST*: *R*(1, −4), *S*(6, −4), *T*(5, −1)
   Translation: along ⟨2, 0⟩
   Reflection: in *x*-axis

10. △*JKL*: *J*(1, 3), *K*(5, 0), *L*(7, 4)
    Translation: along ⟨−3, 0⟩
    Reflection: in *x*-axis

11. △*XYZ*: *X*(−7, 2), *Y*(−5, 6), *Z*(−2, 4)
    Translation: along ⟨0, −1⟩
    Reflection: in *y*-axis

12. △*ABC*: *A*(2, 3), *B*(4, 7), *C*(7, 2)
    Translation: along ⟨0, 4⟩
    Reflection: in *y*-axis

Example 2 **SENSE-MAKING** Graph each figure with the given vertices and its image after the indicated composition of transformations.

13. $\overline{WX}$: $W(-4, 6)$ and $X(-4, 1)$
Reflection: in x-axis
Rotation: 90° about origin

14. $\overline{AB}$: $A(-3, 2)$ and $B(3, 8)$
Rotation: 90° about origin
Translation: along $\langle 4, 4 \rangle$

15. $\overline{FG}$: $F(1, 1)$ and $G(6, 7)$
Reflection: in x-axis
Rotation: 180° about origin

16. $\overline{RS}$: $R(2, -1)$ and $S(6, -5)$
Translation: along $\langle -2, -2 \rangle$
Reflection: in y-axis

Example 3 **Copy and reflect figure D in line m and then line p. Then describe a single transformation that maps D onto D''.**

17.

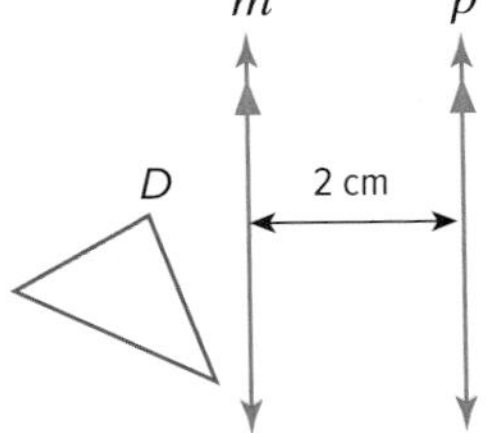

18.

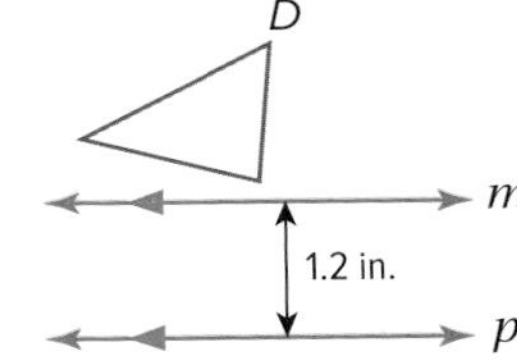

19.

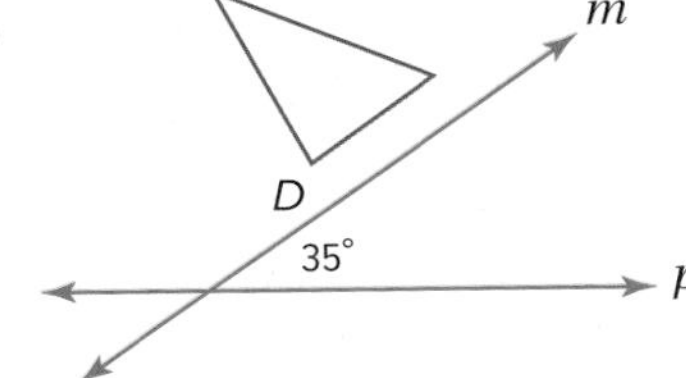

20.

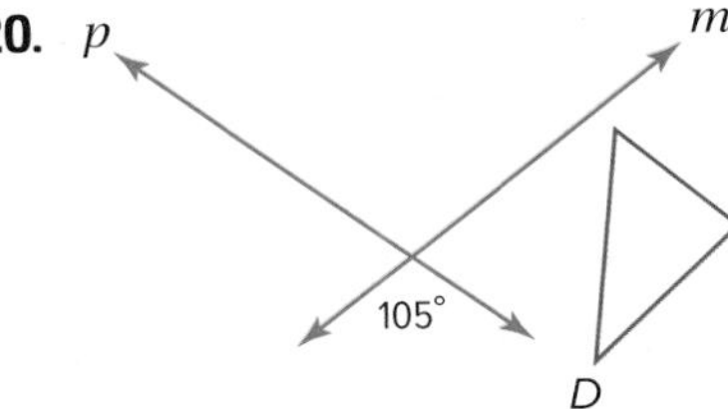

Example 4 **Determine the preimage given the image and the composition of transformations.**

21. rotation 180° about A, translation along $\vec{k}$

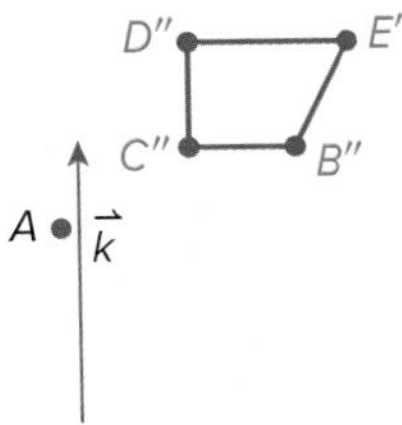

22. reflection in line $x = y$, rotation 90° about the origin

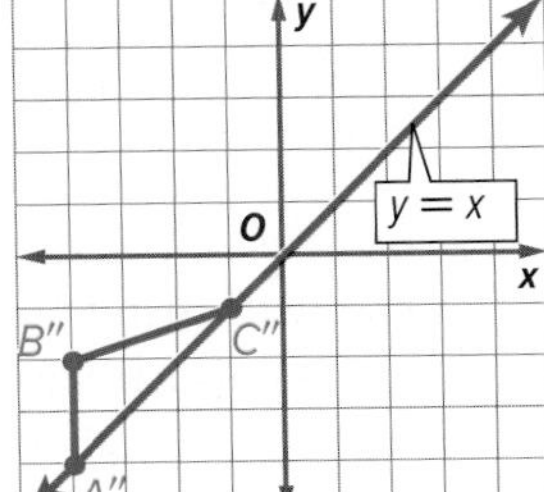

Example 5 **MODELING** **Identify the sequence of transformations that will create the outlined kimono fabric pattern.**

23.

24.

25.

26. **MP MODELING** Elizabeth has airbrushed the pattern shown onto her skateboard. What combination of transformations did she use to create the pattern?

ALGEBRA Graph each figure and its image after the indicated transformations.

27. Rotation: 90° about the origin
Reflection: in x-axis

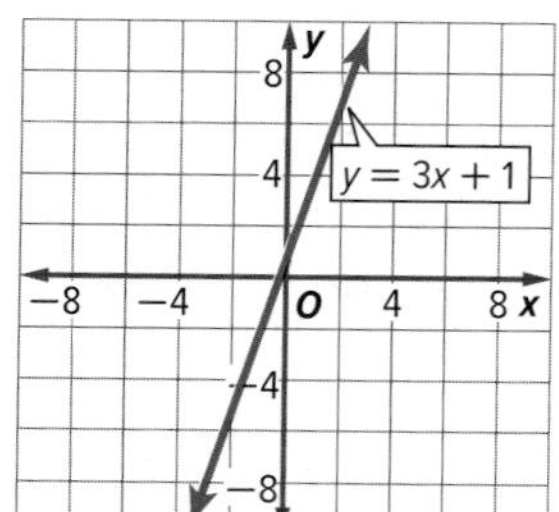

28. Reflection: in x-axis
Reflection: in y-axis

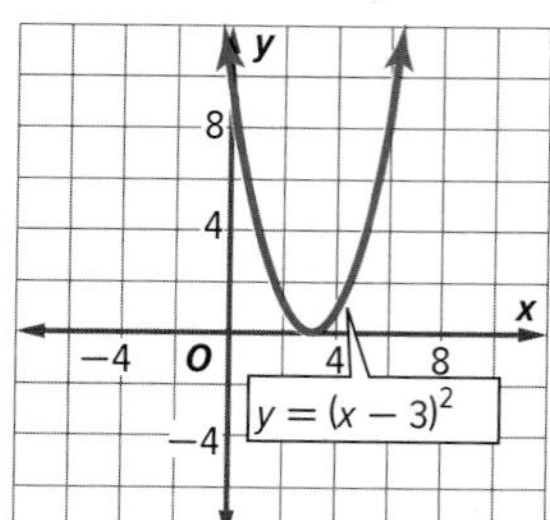

29. $\triangle ABC$ is reflected in the x-axis and then rotated 180° about the origin. $\triangle A''B''C''$ has vertices $A''(3, 1)$, $B''(2, 3)$, and $C''(1, 0)$. Find the coordinates of $\triangle ABC$.

30. **PROOF** Write a paragraph proof for one case of the Composition of Isometries Theorem.

Given: A translation along $\langle a, b\rangle$ maps X to X' and Y to Y'. A reflection in z maps X' to X'' and Y' to Y''.

Prove: $\overline{XY} \cong \overline{X''Y''}$

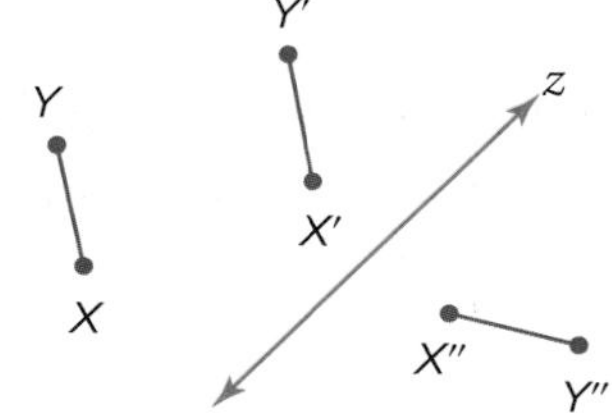

31. **KNITTING** Tonisha is knitting a scarf using the tumbling blocks pattern shown at the right. Describe the transformations combined to transform the red figure to the blue figure.

Identify the sequence of transformations that will carry the preimage to the final image.

32.

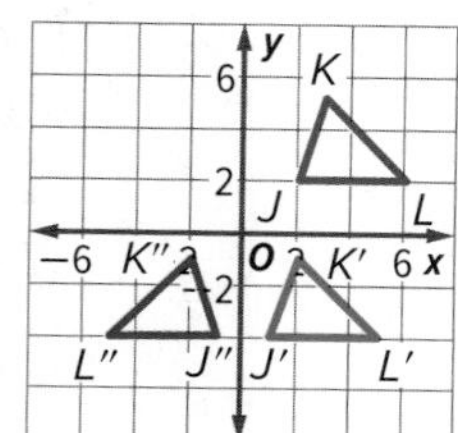

33.

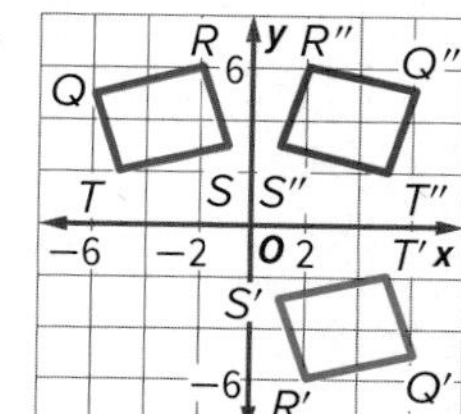

34. PROOF Write a two-column proof of Theorem 3.2.

Given: A reflection in line p maps $\overline{BC}$ to $\overline{B'C'}$.
A reflection in line q maps $\overline{B'C'}$ to $\overline{B''C''}$.
$p \parallel q$, $AD = x$

Prove: **a.** $\overline{BB''} \perp p$, $\overline{BB''} \perp q$

b. $BB'' = 2x$

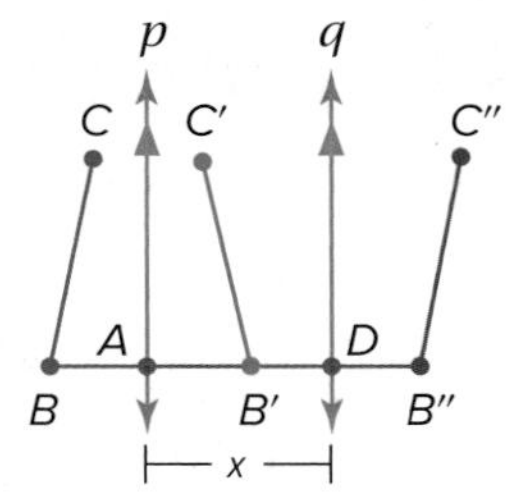

35. PROOF Write a paragraph proof of Theorem 3.3.

Given: Lines ℓ and m intersect at point P.
A is any point not on ℓ or m.

Prove: **a.** If you reflect point A in m, and then reflect its image A' in ℓ, A'' is the image of A after a rotation about point P.

b. $m\angle APA'' = 2(m\angle SPR)$

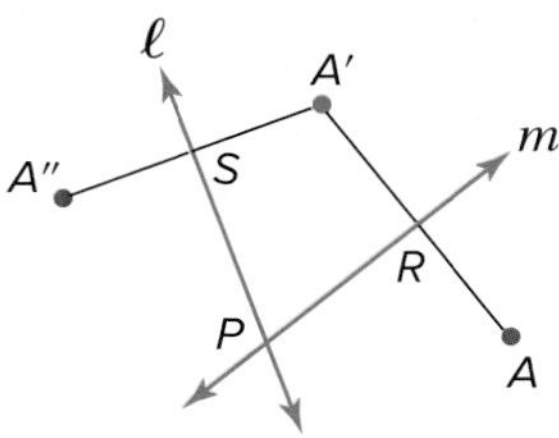

H.O.T. Problems Use Higher-Order Thinking Skills

36. ERROR ANALYSIS Daniel and Lolita are translating $\triangle XYZ$ along $\langle 2, 2\rangle$ and reflecting it in the line $y = 2$. Daniel says that the transformation is a glide reflection. Lolita disagrees and says that the transformation is a composition of transformations. Is either of them correct? Explain your reasoning.

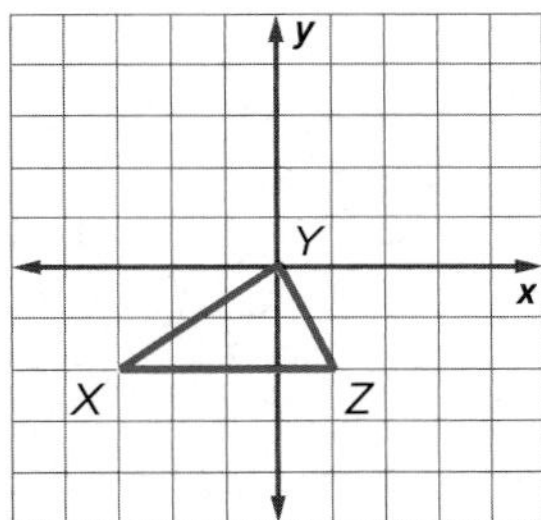

37. WRITING IN MATH Do any points remain invariant under glide reflections? under compositions of transformations? Explain.

38. CHALLENGE If $PQRS$ is translated along $\langle 3, -2\rangle$, reflected in $y = -1$, and rotated 90° about the origin, what are the coordinates of $P'''Q'''R'''S'''$?

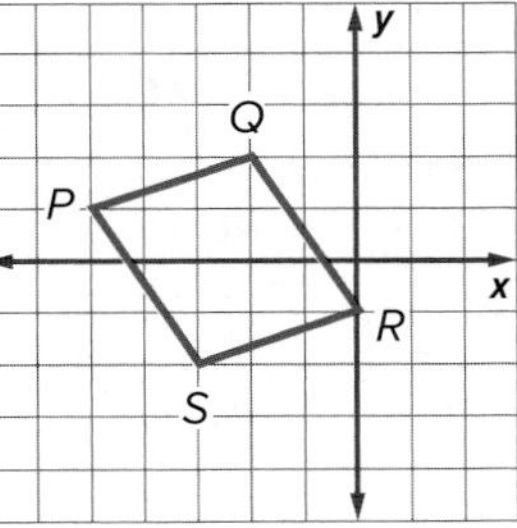

39. MP CONSTRUCT ARGUMENTS If an image is to be reflected in the line $y = x$ and the x-axis, does the order of the reflections affect the final image? Explain.

40. OPEN-ENDED Identify a glide reflection or sequence of transformations that can be used to transform $\triangle ABC$ to $\triangle DEF$.

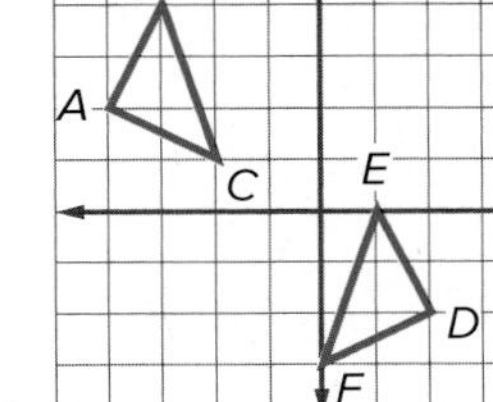

41 MP ARGUMENTS When two rotations are performed on a single image, does the order of the rotations *sometimes*, *always*, or *never* affect the location of the final image? Explain.

42. WRITING IN MATH Compare and contrast glide reflections and compositions of transformations.

43. Which composition of transformations maps $\overline{JK}$ to $\overline{LM}$? MP 2

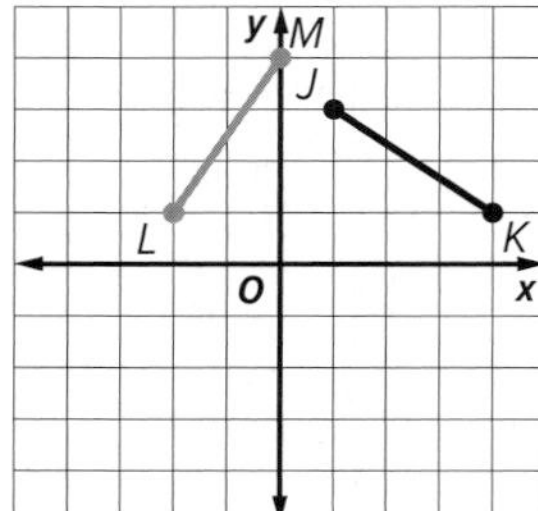

- A rotation 90° about the origin and translation along ⟨0, −1⟩
- B reflection in y-axis and translation along ⟨−1, −2⟩
- C translation along ⟨0, −1⟩ and rotation 90° about the origin
- D translation along ⟨−1, 1⟩ and reflection in y-axis

44. Miguel draws $\triangle ABC$ as shown. Then he rotates the triangle 180° about the origin and reflects the image in the y-axis.

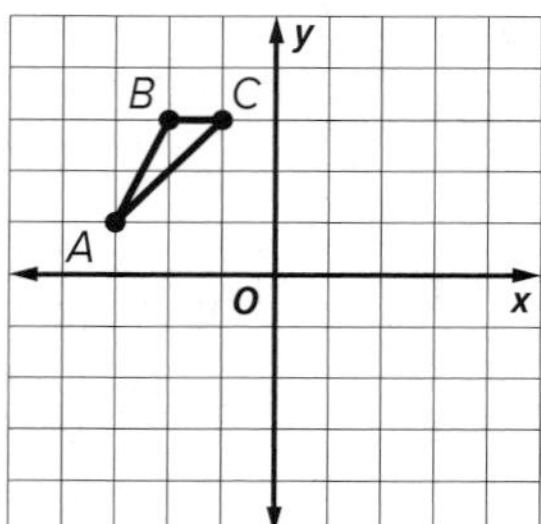

Which of the following points is a vertex of the final image of the triangle? MP 1

- A (−1, −3)
- B (3, 1)
- C (1, −3)
- D (3, −2)
- E (−2, −2)

45. MULT-STEP Jada is creating a design on a sheet of graph paper by drawing and transforming a triangle. She draws $\triangle PQR$ with vertices $P(-4, 1)$, $Q(-4, -3)$, and $R(-2, -2)$ as shown. MP 1

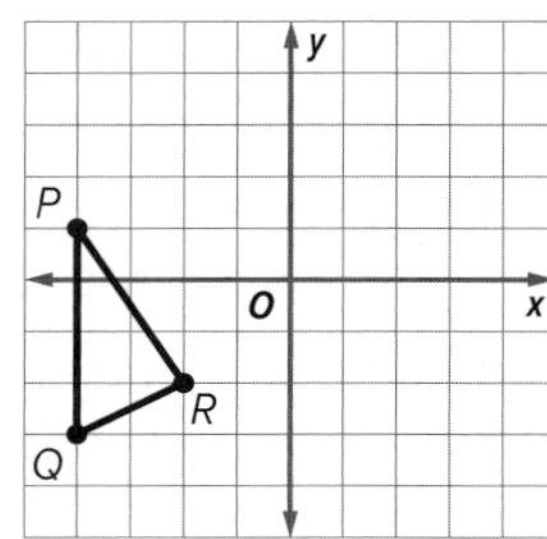

a. Jada first translates $\triangle PQR$ along ⟨5, 2⟩. Given the translation $(x, y) \rightarrow (x + 5, y + 2)$, what are the coordinates of the vertices of $\triangle P'Q'R'$?

$P' =$ ( )

$Q' =$ ( )

$R' =$ ( )

Copy the original image onto a sheet of graph paper, and then draw the image $\triangle P'Q'R'$.

b. Jada next reflects the image $\triangle P'Q'R'$ in the x-axis to get the final image, $\triangle P''Q''R''$. Draw the final image on a sheet of graph paper. What are the coordinates of the vertices of $\triangle P''Q''R''$?

$P'' =$ ( )

$Q'' =$ ( )

$R'' =$ ( )

c. Which of the following points lies in the interior of $\triangle P''Q''R''$?

- A (2, 0)
- B (−2, 0)
- C (2, 3)
- D (2, −3)
- E (3, −2)
- F (3, 2)

LESSON 5
Symmetry

Then

- You drew reflections and rotations of figures.

Now

1. Use line symmetry to describe the reflections that carry a figure onto itself.
2. Use rotational symmetry to describe the rotations that carry a figure onto itself.

Why?

- In the animal kingdom, the symmetry of an animal's body is often an indication of the animal's complexity. Animals displaying line symmetry, such as insects, are usually more complex life forms than those displaying rotational symmetry, like a jellyfish.

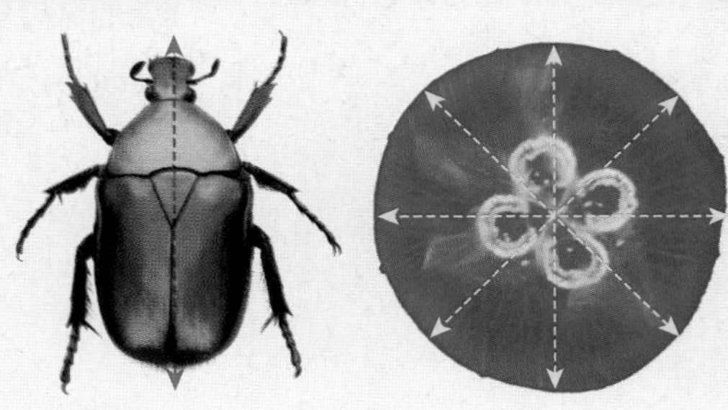

New Vocabulary

symmetry
line symmetry
line of symmetry
rotational symmetry
center of symmetry
order of symmetry
magnitude of symmetry

Mathematical Practices

4 Model with mathematics.
6 Attend to precision.
8 Look for and express regularity in repeated reasoning.

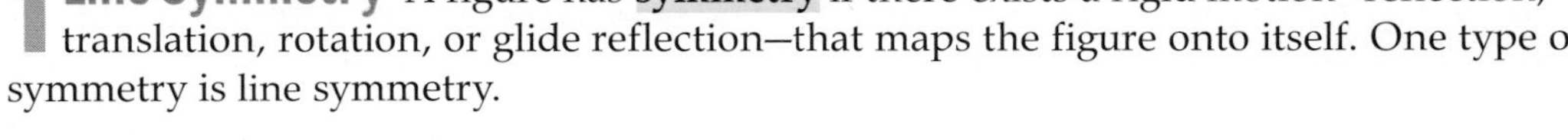

1 Line Symmetry

A figure has **symmetry** if there exists a rigid motion—reflection, translation, rotation, or glide reflection—that maps the figure onto itself. One type of symmetry is line symmetry.

Key Concept Line Symmetry

A figure in the plane has **line symmetry** (or *reflectional symmetry*) if the figure can be mapped onto itself by a reflection in a line, called a **line of symmetry** (or *axis of symmetry*).

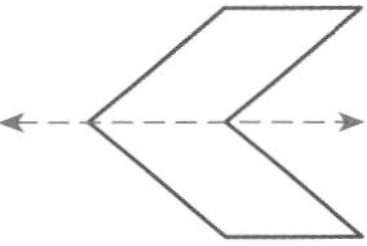

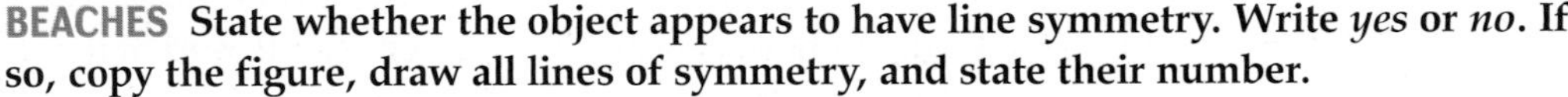

Real-World Example 1 Identify Line Symmetry

BEACHES **State whether the object appears to have line symmetry. Write *yes* or *no*. If so, copy the figure, draw all lines of symmetry, and state their number.**

a.

Yes; the crab has one line of symmetry.

b.

Yes; the starfish has five lines of symmetry.

c.

No; there is no line in which the oyster shell can be reflected so that it maps onto itself.

Guided Practice

State whether the figure has line symmetry. Write *yes* or *no*. If so, copy the figure, draw all lines of symmetry, and state their number.

1A.

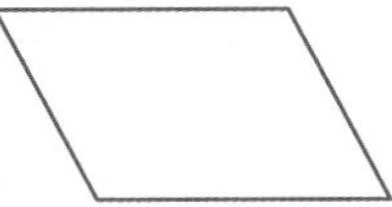

1B.

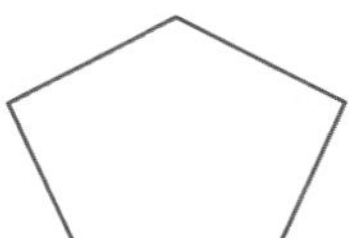

1C.

(l to r, t to b) Brand X Pictures/PunchStock, Jo Bradford/Green Island Art Studios/Moment/Getty Images, Stockbyte/age footstock (bc, br) Siede Preis/Photodisc/Getty Images

2 Rotational Symmetry

While line symmetry is based on reflections, rotational symmetry is based on rotations.

Key Concept Rotational Symmetry

A figure in the plane has **rotational symmetry** (or *radial symmetry*) if the figure can be mapped onto itself by a rotation between 0° and 360° about the center of the figure, called the **center of symmetry** (or *point of symmetry*).

Examples The figure below has rotational symmetry because a rotation of 90°, 180°, or 270° maps the figure onto itself.

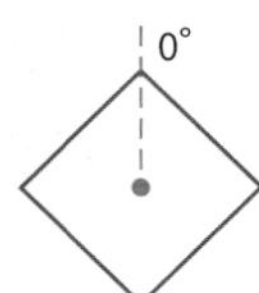

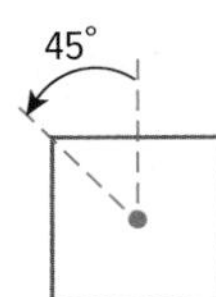

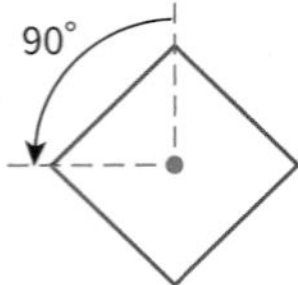

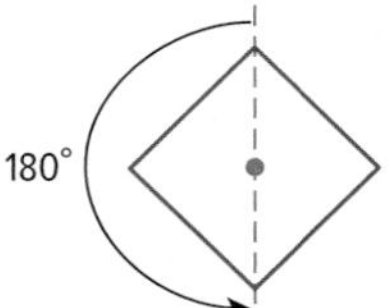

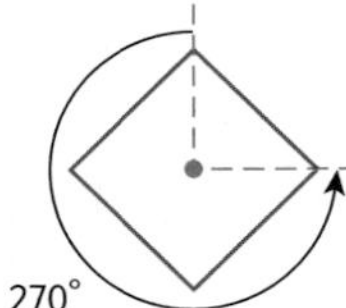

The number of times a figure maps onto itself as it rotates from 0° to 360° is called the **order of symmetry**. The **magnitude of symmetry** (or angle of rotation) is the smallest angle through which a figure can be rotated so that it maps onto itself.

$$\text{magnitude} = 360° \div \text{order}$$

The figure above has rotational symmetry of order 4 and magnitude 90°.

Go Online!

In Personal Tutor videos for this lesson, teachers describe how to identify symmetry. Watch with a partner, then try describing how to identify symmetry for them. Have them ask questions to help your understanding.

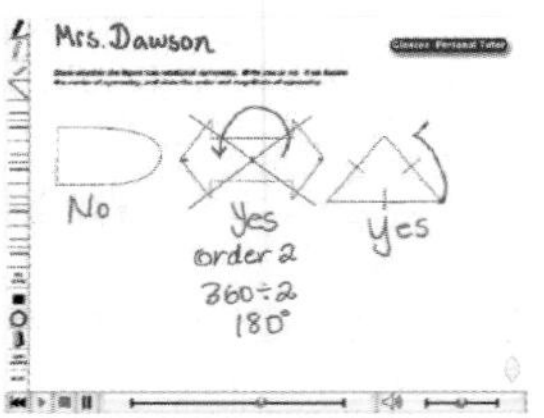

Example 2 Identify Rotational Symmetry

State whether the figure has rotational symmetry. Write *yes* or *no*. If so, copy the figure, locate the center of symmetry, and state the order and magnitude of symmetry.

a.

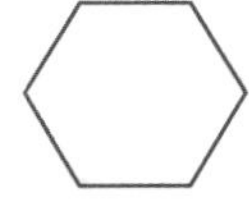

Yes; the regular hexagon has order 6 rotational symmetry and magnitude 360° ÷ 6 or 60°. The center is the intersection of the diagonals.

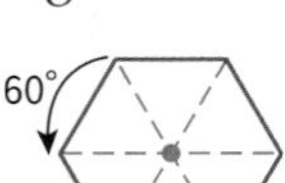

b.

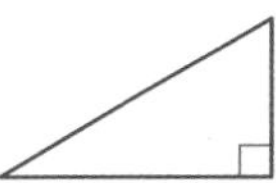

No; no rotation between 0° and 360° maps the right triangle onto itself.

c.

Yes; the figure has order 2 rotational symmetry and magnitude 360° ÷ 2 or 180°. The center is the intersection of the diagonals.

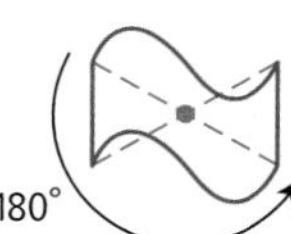

Study Tip

Point Symmetry A figure has *point symmetry* if the figure can be mapped onto itself by a rotation of 180°. This playing card exhibits point symmetry. It looks the same right-side up as upside down.

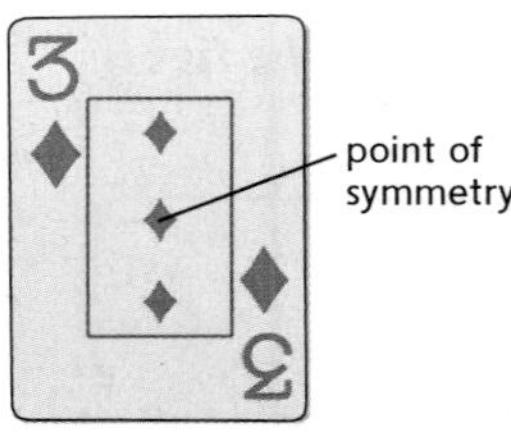

Guided Practice

FLOWERS State whether the flower appears to have rotational symmetry. Write *yes* or *no*. If so, copy the flower, locate the center of symmetry, and state the order and magnitude of symmetry.

2A.

2B.

2C.

(l to r) Jupiterimages/Thinkstock/Alamy, Frahaus/E+/Getty Images, eli_asenova/E+/Getty Images

When you are given a figure on a coordinate plane, you can use what you know about reflections and rotations on the coordinate plane to describe the rigid motions that map the figure onto itself.

Example 3 Describe Reflections and Rotations

State whether the figure has line symmetry and/or rotational symmetry. If so, describe the reflections and/or rotations that map the figure onto itself.

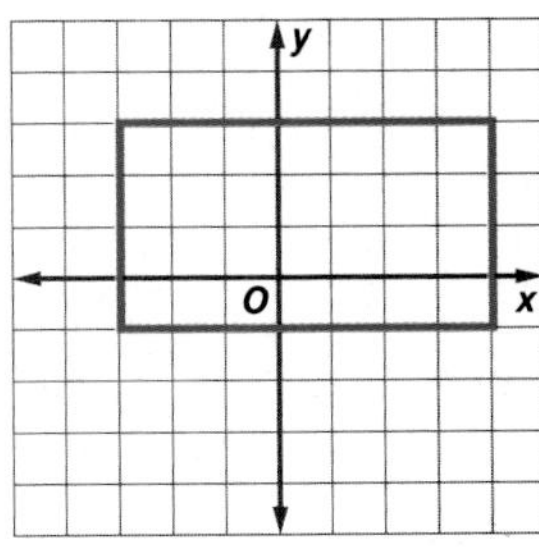

Step 1 Determine line symmetry.

The figure is a rectangle, so there are two lines of symmetry. Write the equation for each line of symmetry.

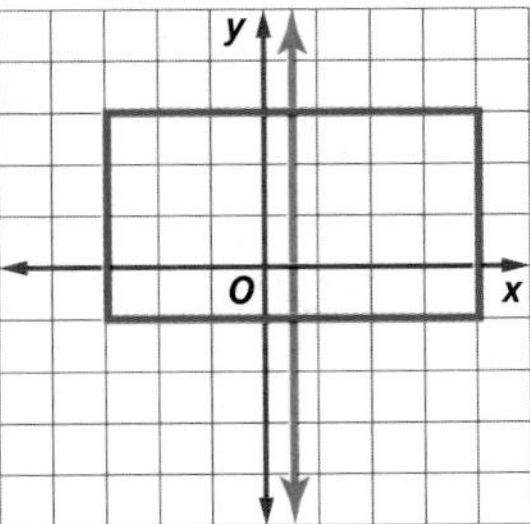

Line of symmetry: $x = 0.5$

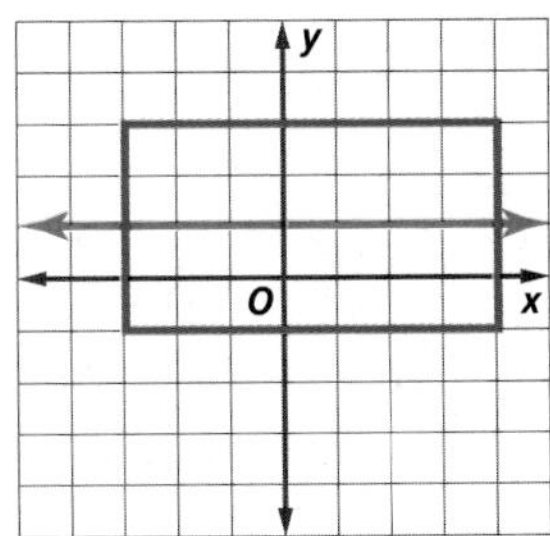

Line of symmetry: $y = 1$

Watch Out!

Rotations A rotation of 360° will map any figure onto itself. For this reason, rotations of 360° are not included when describing the rotational symmetry of a figure.

Step 2 Determine rotational symmetry.

The rectangle has order 2 rotational symmetry and magnitude $360° \div 2 = 180°$.

Draw the diagonals to find the center of symmetry.

The center of symmetry is (0.5, 1).

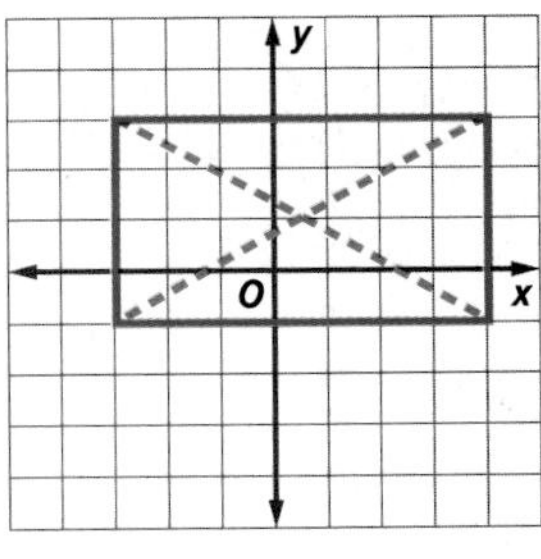

Step 3 Describe the reflections and rotations that map the figure onto itself.

The reflection in the line $x = 0.5$ maps the rectangle onto itself.

The reflection in the line $y = 1$ maps the rectangle onto itself.

The rotation of 180° around the point (0.5, 1) maps the rectangle onto itself.

Guided Practice

3A.

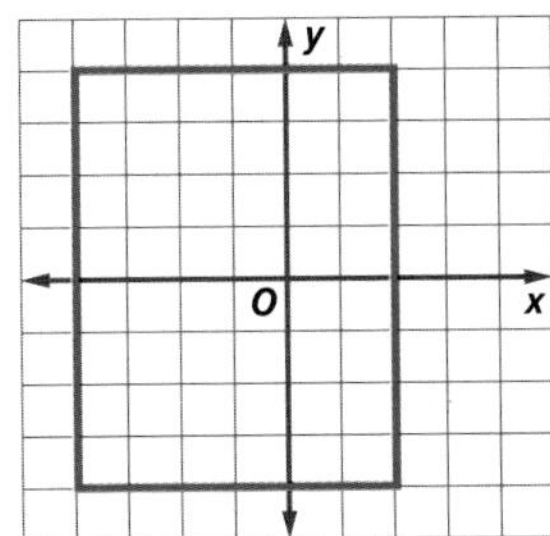

3B.

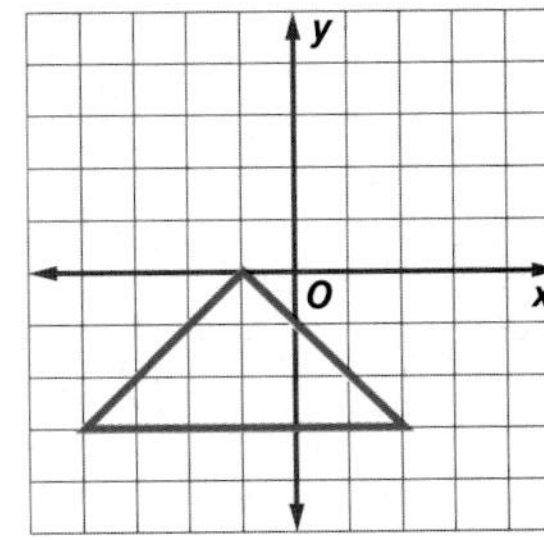

3C.

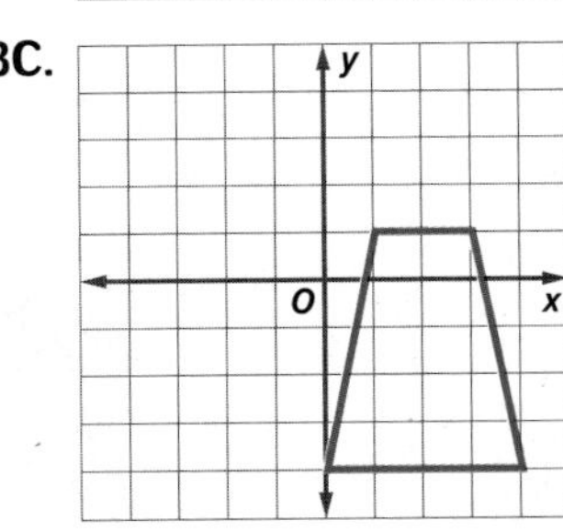

3D.

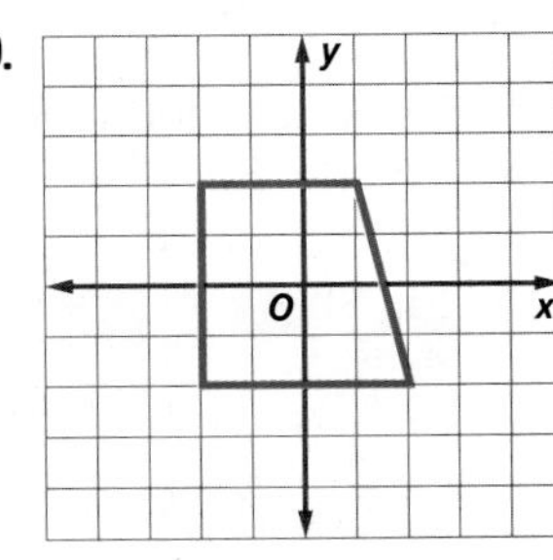

Check Your Understanding

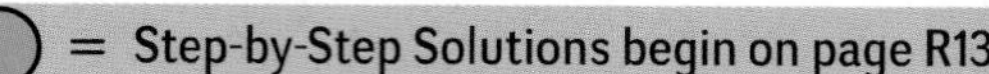

Example 1 **State whether the figure appears to have line symmetry. Write *yes* or *no*. If so, copy the figure, draw all lines of symmetry, and state their number.**

1. 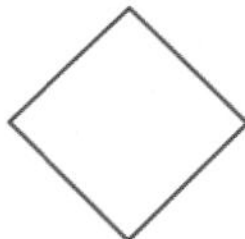

2. 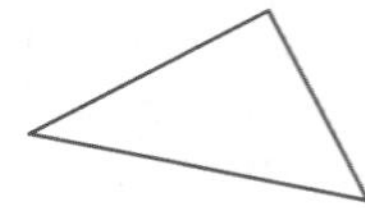

3. 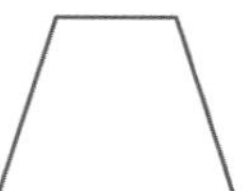

Example 2 **State whether the figure has rotational symmetry. Write *yes* or *no*. If so, copy the figure, locate the center of symmetry, and state the order and magnitude of symmetry.**

4. 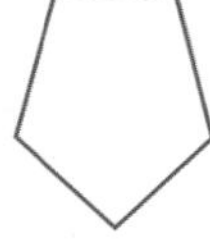

5. 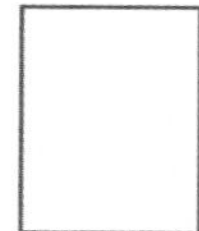

6. 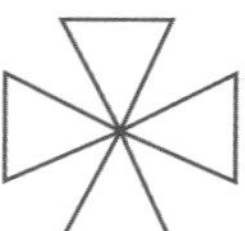

Example 3 **State whether the figure has line symmetry and/or rotational symmetry. If so, describe the reflections and/or rotations that map the figure onto itself.**

7. 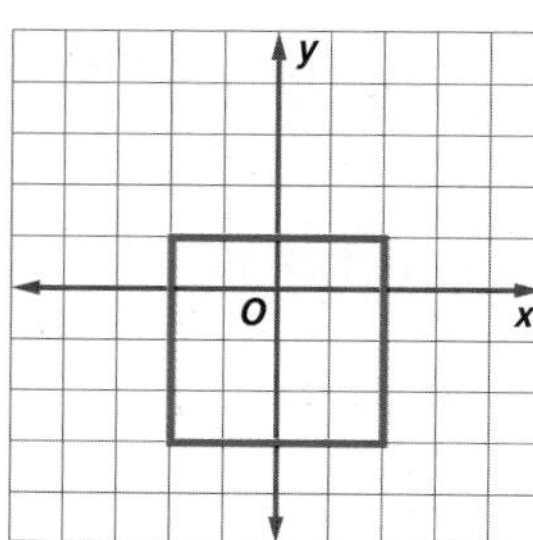

8. 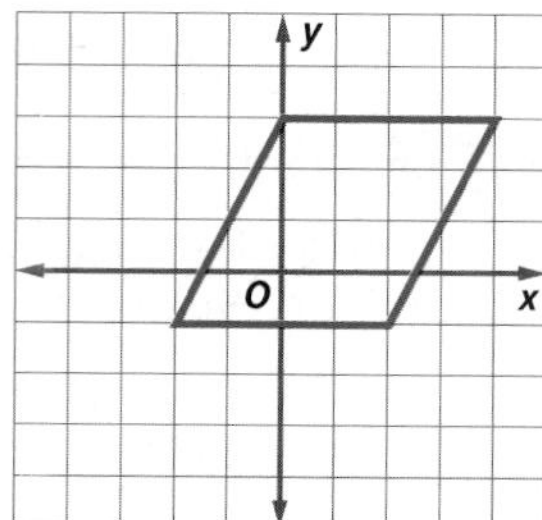

Practice and Problem Solving

Extra Practice is on page R3.

Example 1 **REGULARITY** State whether the figure appears to have line symmetry. Write *yes* or *no*. If so, copy the figure, draw all lines of symmetry, and state their number.

9. 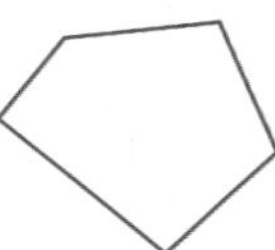

10. 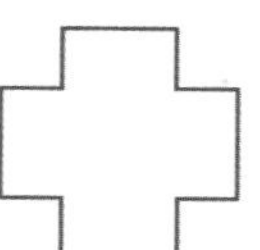

11. 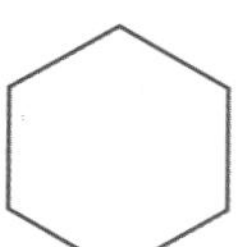

12. 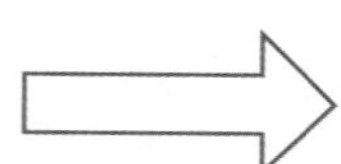

13. 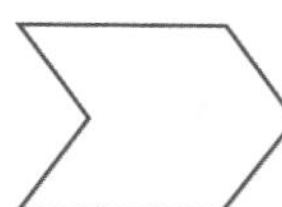

14. 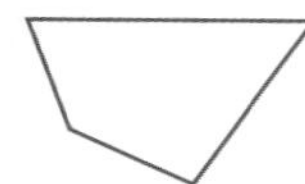

FLAGS State whether each flag design appears to have line symmetry. Write *yes* or *no*. If so, copy the flag, draw all lines of symmetry, and state their number.

15. 

16. 

17. 

Example 2 **State whether the figure has rotational symmetry. Write *yes* or *no*. If so, copy the figure, locate the center of symmetry, and state the order and magnitude of symmetry.**

18. 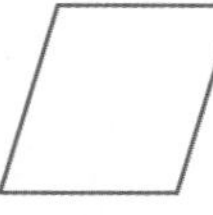

19. 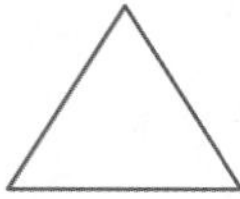

20. 

21. 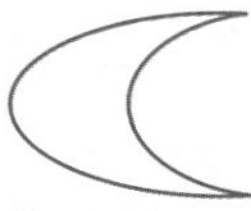

22. 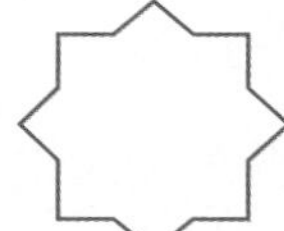

23. 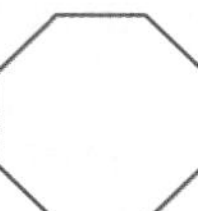

WHEELS State whether each wheel cover appears to have rotational symmetry. Write *yes* or *no*. If so, state the order and magnitude of symmetry.

24. 

25. 

26. 

Example 3 **State whether the figure has line symmetry and/or rotational symmetry. If so, describe the reflections and/or rotations that map the figure onto itself.**

27. 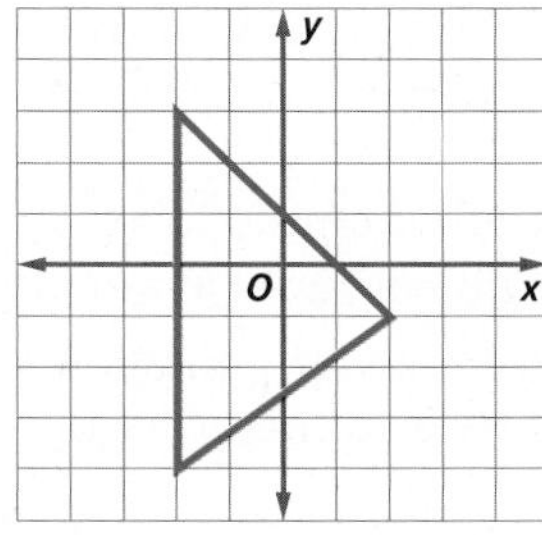

28. 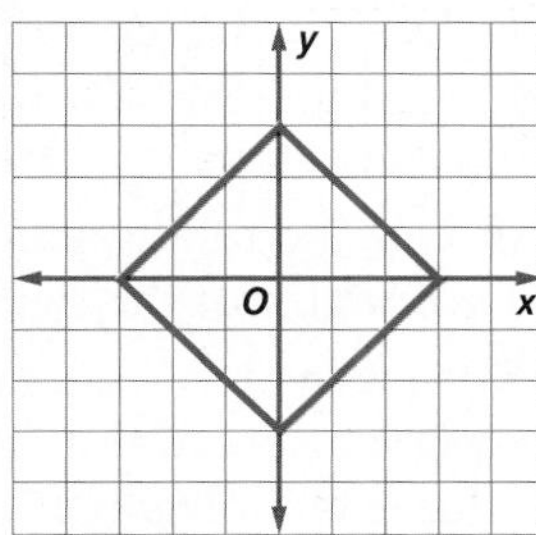

29. 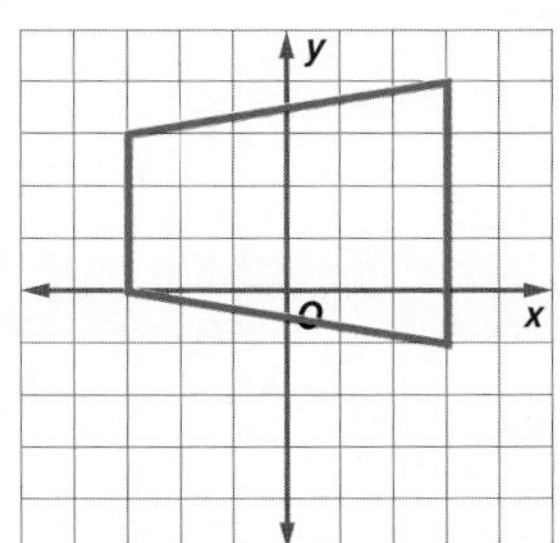

30. 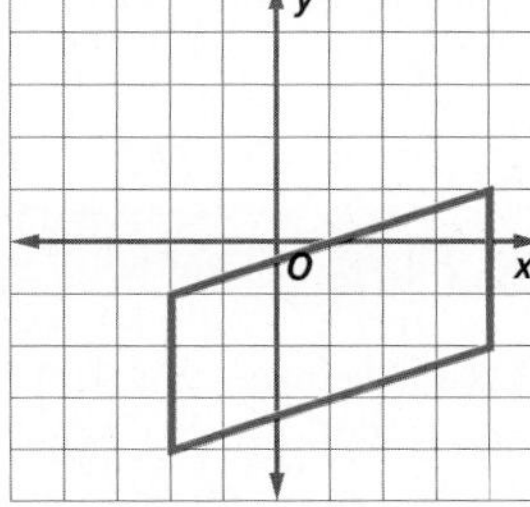

31. **MP MODELING** Symmetry is an important component of photography. Photographers often use reflection in water to create symmetry in photos. The photo at the right is a long exposure shot of the Eiffel tower reflected in a pool.

a. Describe the line symmetry created by the photo.

b. Is there rotational symmetry in the photo? Explain your reasoning.

COORDINATE GEOMETRY Determine whether the figure with the given vertices has *line* symmetry and/or *rotational* symmetry.

32. $R(-3, 3)$, $S(-3, -3)$, $T(3, 3)$

33. $A(-4, 0)$, $B(0, 4)$, $C(4, 0)$, $D(0, -4)$

34. $F(0, -4)$, $G(-3, -2)$, $H(-3, 2)$, $J(0, 4)$, $K(3, 2)$, $L(3, -2)$

35. $W(-2, 3)$, $X(-3, -3)$, $Y(3, -3)$, $Z(2, 3)$

ALGEBRA Graph the function and determine whether the graph has *line* and/or *rotational* symmetry. If so, state the order and magnitude of symmetry, and write the equations of any lines of symmetry.

36. $y = x$

37. $y = x^2 + 1$

38. $y = -x^3$

39. Refer to the rectangle on the coordinate plane.

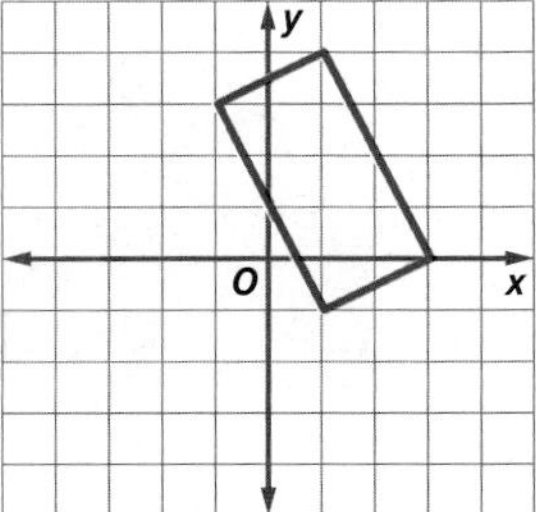

a. What are the equations of the lines of symmetry of the rectangle?

b. What happens to the equations of the lines of symmetry when the rectangle is rotated 90° counterclockwise around its center of symmetry? Explain.

40. **MULTIPLE REPRESENTATIONS** In this problem, you will use dynamic geometric software to investigate line and rotational symmetry in regular polygons.

a. Geometric Use The Geometer's Sketchpad to draw an equilateral triangle. Use the reflection tool under the transformation menu to investigate and determine all possible lines of symmetry. Then record their number.

b. Geometric Use the rotation tool under the transformation menu to investigate the rotational symmetry of the figure in part **a**. Then record its order of symmetry.

c. Tabular Repeat the process in parts **a** and **b** for a square, regular pentagon, and regular hexagon. Record the number of lines of symmetry and the order of symmetry for each polygon.

d. Verbal Make a conjecture about the number of lines of symmetry and the order of symmetry for a regular polygon with n sides.

H.O.T. Problems Use Higher-Order Thinking Skills

41. **ERROR ANALYSIS** Jaime says that Figure A has only line symmetry, and Jewel says that Figure A has only rotational symmetry. Is either of them correct? Explain your reasoning.

Figure A

42. **CHALLENGE** A quadrilateral in the coordinate plane has exactly two lines of symmetry, $y = x - 1$ and $y = -x + 2$. Find possible vertices for the figure. Graph the figure and the lines of symmetry.

43. **REASONING** A figure has infinitely many lines of symmetry. What is the figure? Explain.

44. **OPEN-ENDED** Draw a figure with line symmetry but not rotational symmetry. Explain.

45. **WRITING IN MATH** How are line symmetry and rotational symmetry related?

46. Sasha owns a tile store. For each tile in her store, she calculates the sum of the number of lines of symmetry and the order of symmetry, and then she enters this value into a database. Which value should she enter in the database for the tile shown here? MP 4

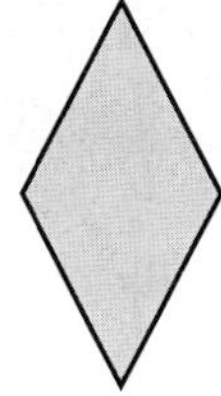
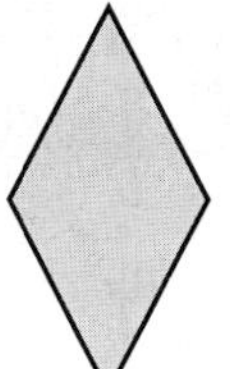

- A 2
- B 3
- C 4
- D 8

47. Patrick drew a figure that has line symmetry but not rotational symmetry. Which of the following could be the figure that Patrick drew?
 MP 2

- A 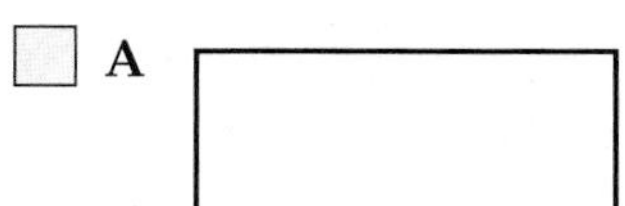
- B 
- C 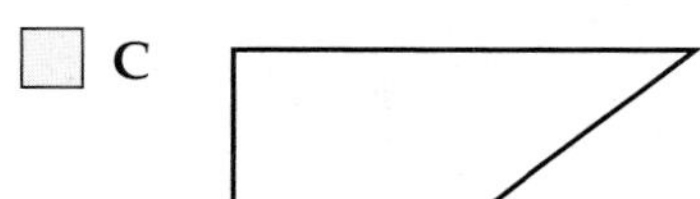
- D 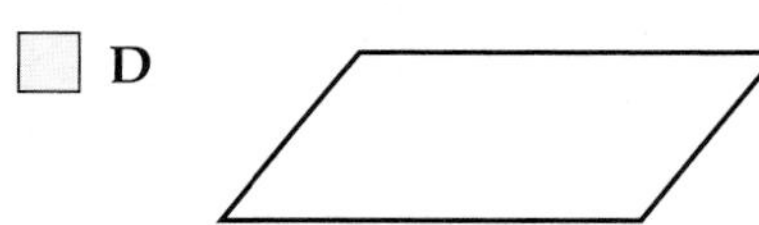
- E 

48. Which of the following figures may have exactly one line of symmetry and no rotational symmetry? MP 2

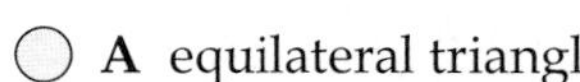

- A equilateral triangle
- B equiangular triangle
- C isosceles triangle
- D scalene triangle

49. Camryn plotted the points $P(-2, 2)$, $Q(-1, 4)$, and $R(0, 2)$. Which of the following additional points can she plot so that the resulting quadrilateral $PQRS$ has line symmetry but not rotational symmetry? MP 2

- A $S(-1, 0)$
- B $S(-1, -2)$
- C $S(0, 0)$
- D $S(-2, -1)$

50. What is the order of symmetry for the figure below? MP 8

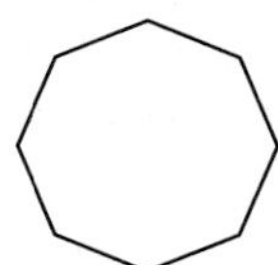

51. **MULTI-STEP** Roberto is a graphic designer. He is using a coordinate plane to design a new logo for a client. He starts by drawing the lines of symmetry shown. MP 4

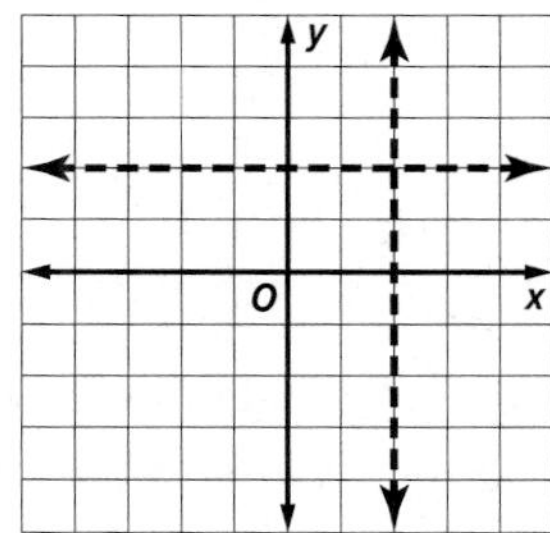

a. The logo will be based on a rectangle that has the given lines of symmetry. One of the vertices of the rectangle is $(-3, 1)$. What are the other vertices of the rectangle?

b. What is the order of symmetry for this rectangle?

c. Suppose Roberto removes the vertex at $(-3, 1)$ and decides that the logo should be based on a square instead of a rectangle. Describe one way that he can assign coordinates to the vertices of the square so that the given lines are still lines of symmetry.

52. What is the magnitude of symmetry for a regular polygon that has 12 sides? MP 8

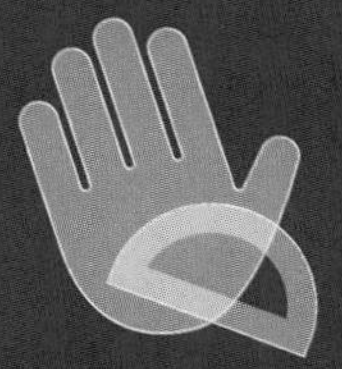

EXTEND 3-5

Geometry Lab

Exploring Constructions with a Reflective Device

A reflective device is a tool made of semitransparent plastic that reflects objects. It works best if you lay it on a flat surface in a well-lit room. You can use a reflective device to explore transformations of geometric objects.

Mathematical Practices

MP **5** Use appropriate tools strategically.

Activity 1 Reflect a Triangle

Work cooperatively. Use a reflective device to reflect $\triangle ABC$ in w. Label the reflection $\triangle A'B'C'$.

Step 1 Draw $\triangle ABC$ and the line of reflection w.

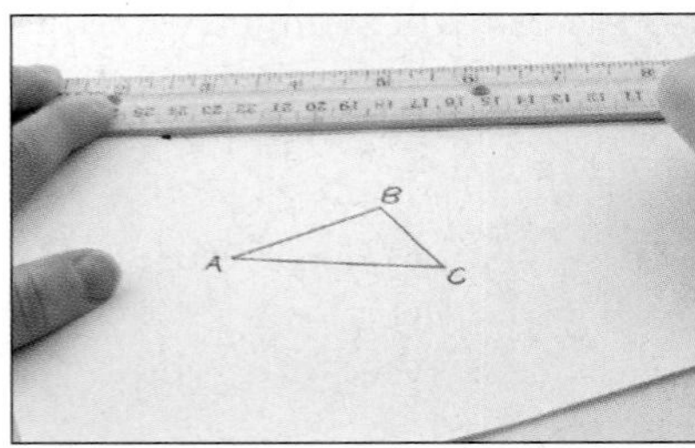

Step 2 With the reflective device on line w, draw points for the vertices of the reflection.

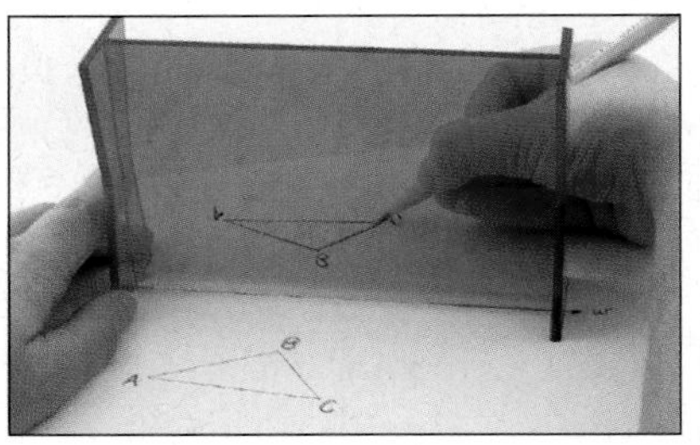

Step 3 Use a straightedge to connect the points to form $\triangle A'B'C'$.

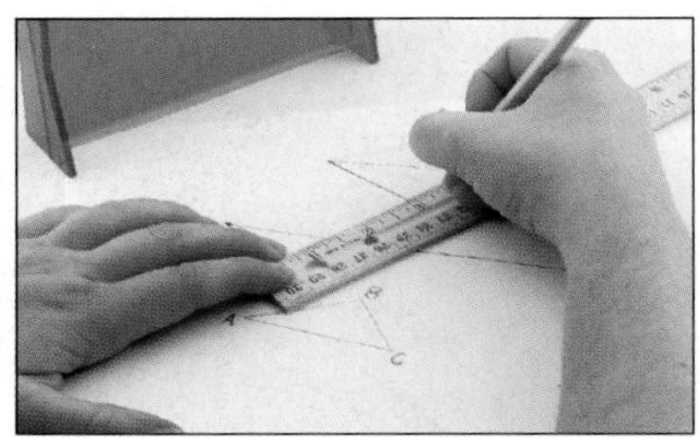

We have used a compass, straightedge, string, and paper folding to make geometric constructions. You can also use a reflective device for constructions.

Activity 2 Construct Lines of Symmetry

Work cooperatively. Use a reflective device to construct the lines of symmetry for a regular hexagon.

Step 1 Draw a regular hexagon. Place the reflective device on the shape and move it until one half of the shape matches the reflection of the other half. Draw the line of symmetry.

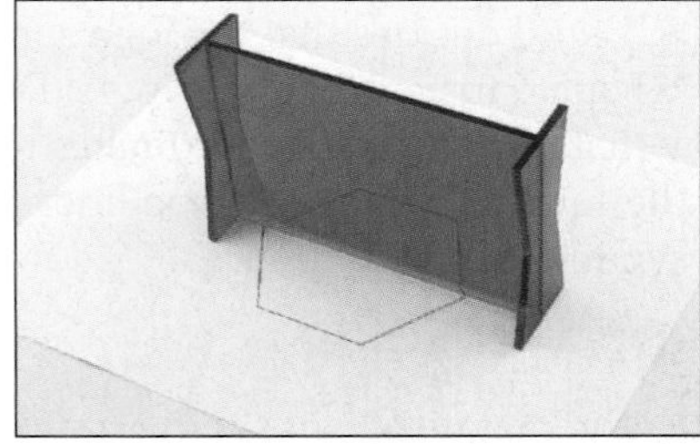

Step 2 Repeat Step 1 until you have found all the lines of symmetry.

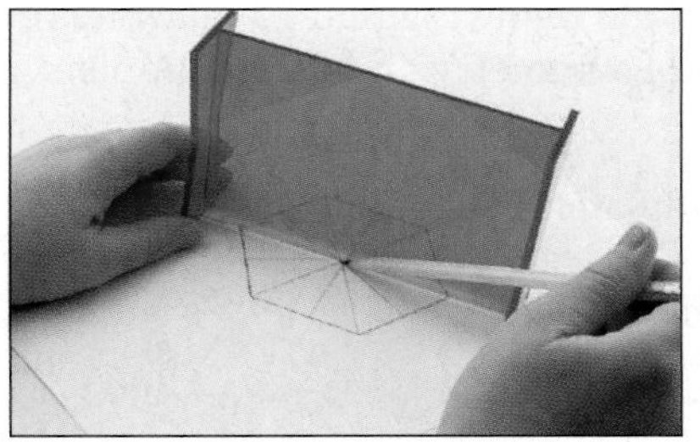

Activity 3 Construct a Parallel line

Work cooperatively. Use a reflective device to reflect line ℓ to line m that is parallel and passes through point P.

Step 1

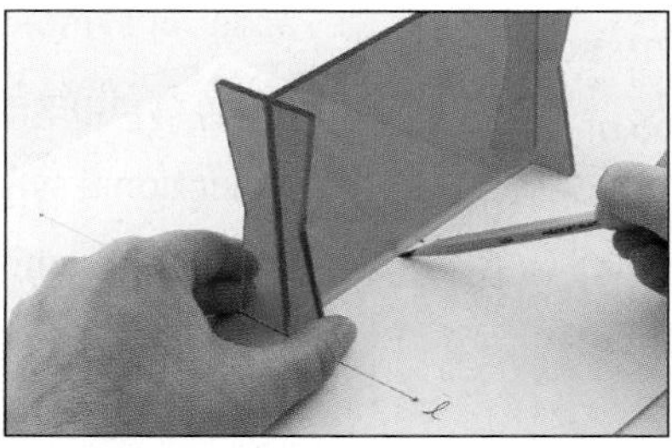

Draw line ℓ and point P. Place a short side of the reflective device on line ℓ and the long side on point P. Draw a line. This line is perpendicular to ℓ through P.

Step 2

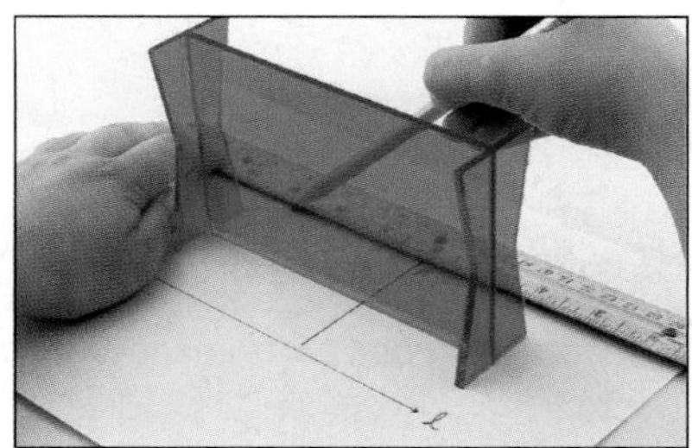

Place the reflective device so that the perpendicular line coincides with itself and the reflection of line ℓ passes through point P. Use a straightedge to draw the parallel line m through P.

You can use a reflective device to construct perpendicular bisectors of a triangle. In Chapter 5, you'll also see how to construct perpendicular bisectors with paper folding.

Activity 4 Construct Perpendicular Bisectors

Work cooperatively. Use a reflective device to find the circumcenter of $\triangle ABC$.

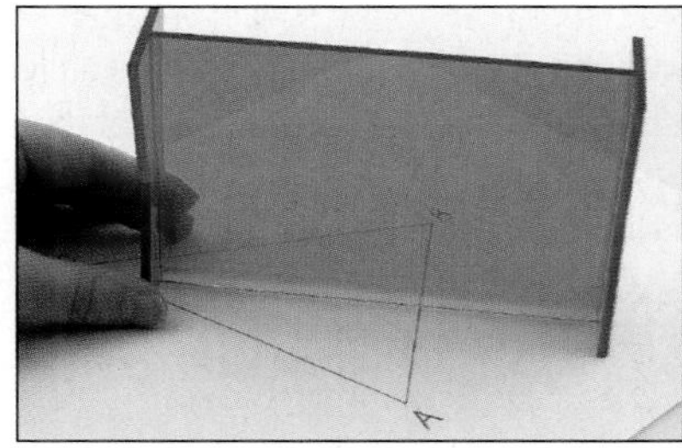

Step 1 Draw $\triangle ABC$. Place the reflective device between A and B and adjust it until A and B coincide. Draw the line of symmetry.

Step 2 Repeat Step 1 for sides $\overline{AC}$ and $\overline{BC}$. Then place a point at the intersection of the three perpendicular bisectors. This is the circumcenter of the triangle.

Model and Analyze

Work cooperatively.

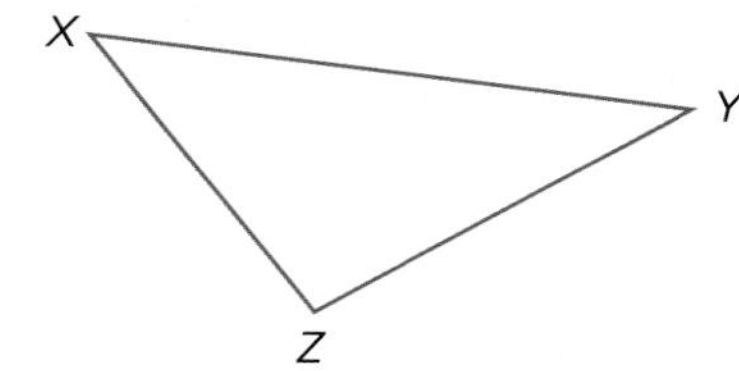

1. How do you know that the steps in Activity 4 give the actual perpendicular bisector and the circumcenter of $\triangle ABC$?
2. Construct the angle bisectors and find the incenter of $\triangle XYZ$.
3. Describe how you used the reflective device for the construction.

CHAPTER 3

Study Guide and Review

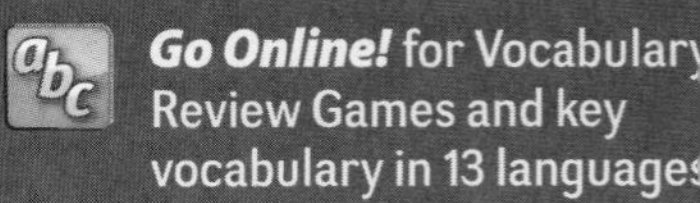

Study Guide

Key Concepts

Reflections (Lesson 3-1)

- A reflection is a transformation representing a flip of a figure over a point, line, or plane.

Translations (Lesson 3-2)

- A translation is a transformation that moves all points of a figure the same distance in the same direction.
- A translation maps each point to its image along a translation vector.

Rotations (Lesson 3-3)

- A rotation turns each point in a figure through the same angle about a fixed point.

Compositions of Transformations (Lesson 3-4)

- A translation can be represented as a composition of reflections in parallel lines and a rotation can be represented as a composition of reflections in intersecting lines.

Symmetry (Lesson 3-5)

- The line of symmetry in a figure is a line where the figure could be folded in half so that the two halves match exactly.
- The number of times a figure maps onto itself as it rotates from 0° to 360° is called the order of symmetry.
- The magnitude of symmetry is the smallest angle through which a figure can be rotated so that it maps onto itself.

FOLDABLES® Study Organizer

Use your Foldable to review the chapter. Working with a partner can be helpful. Ask for clarification of concepts as needed.

Key Vocabulary

angle of rotation (p. 240)
axis symmetry (p. 261)
center of rotation (p. 240)
composition of transformations (p. 249)
glide reflection (p. 249)
line of symmetry (p. 259)
magnitude of symmetry (p. 260)
order of symmetry (p. 260)
plane symmetry (p. 261)
reflectional symmetry (p. 259)
rotational symmetry (p. 260)
symmetry (p. 259)

Vocabulary Check

Choose the term that best completes each sentence.

1. When a transformation is applied to a figure, and then another transformation is applied to its image, this is a(n) (composition of transformations, order of symmetries).
2. If a figure is folded across a straight line and the halves match exactly, the fold line is called the (line of reflection, line of symmetry).
3. A (dilation, glide reflection) enlarges or reduces a figure proportionally.
4. The number of times a figure maps onto itself as it rotates from 0° to 360° is called the (magnitude of symmetry, order of symmetry).
5. A (line of reflection, translation vector) is the same distance from each point of a figure and its image.
6. A figure has (a center of rotation, symmetry) if it can be mapped onto itself by a rigid motion.

Concept Check

Choose the term that best completes each sentence.

7. A glide reflection includes both a reflection and a ________.
8. To rotate a point ________ degrees counterclockwise about the origin, multiply the y-coordinate by -1 and then interchange the x- and y-coordinates.
9. A figure has ________ symmetry if the figure can be mapped onto itself by a rotation between 0° and 360° about the center of the figure.

Lesson-by-Lesson Review 

3-1 Reflections

Graph each figure and its image under the given reflection.

10. rectangle *ABCD* with *A*(2, −4), *B*(4, −6), *C*(7, −3), and *D*(5, −1) in the *x*-axis
11. triangle *XYZ* with *X*(−1, 1), *Y*(−1, −2), and *Z*(3, −3) in the *y*-axis
12. quadrilateral *QRST* with *Q*(−4, −1), *R*(−1, 2), *S*(2, 2), and *T*(0, −4) in the line $y = x$
13. **ART** Anita is making the two-piece sculpture shown for a memorial garden. In her design, one piece of the sculpture is a reflection of the other, to be placed beside a sidewalk that would be located along the line of reflection. Copy the figures and draw the line of reflection.

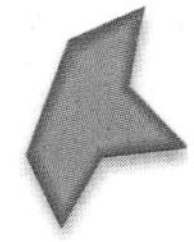

Example 1

Graph △*JKL* with vertices *J*(1, 4), *K*(2, 1), and *L*(6, 2) and its reflected image in the *x*-axis.

Multiply the *y*-coordinate of each vertex by −1.

(*x*, *y*)	→	(*x*, −*y*)
J(1, 4)	→	*J*′(1, −4)
K(2, 1)	→	*K*′(2, −1)
L(6, 2)	→	*L*′(6, −2)

Graph △*JKL* and its image △*J*′*K*′*L*′.

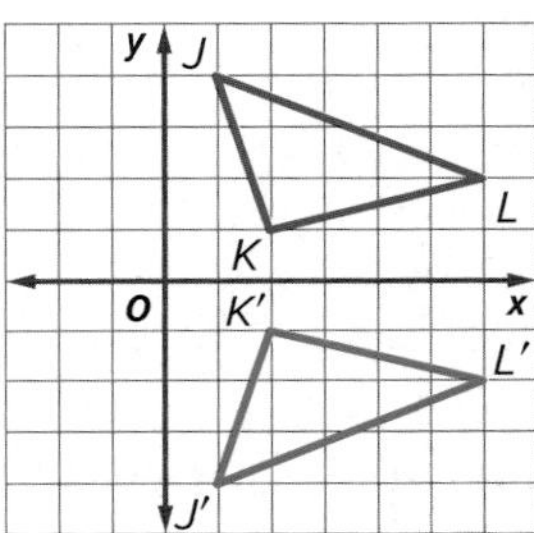

3-2 Translations

14. Graph △*ABC* with vertices *A*(0, −1), *B*(2, 0), *C*(3, −3) and its image along ⟨−5, 4⟩.
15. Copy the figure and the given translation vector. Then draw the translation of the figure along the translation vector.

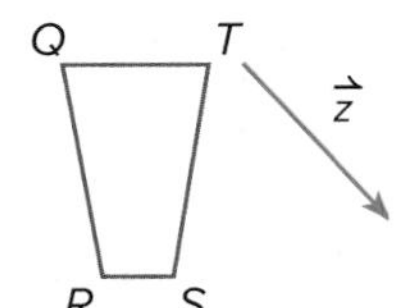

16. **DANCE** Five dancers are positioned onstage as shown. Dancers *B*, *F*, and *C* move along ⟨0, −2⟩, while dancer *A* moves along ⟨5, −1⟩. Draw the dancers' final positions.

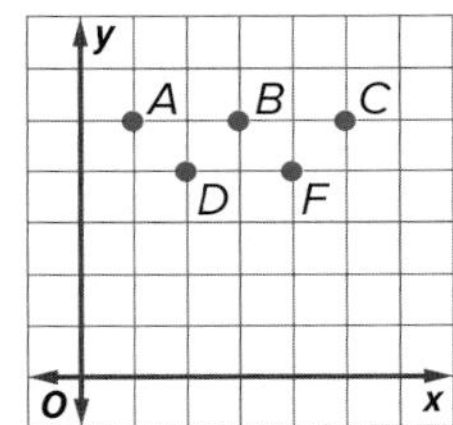

Example 2

Graph △*XYZ* with vertices *X*(2, 2), *Y*(5, 5), *Z*(5, 3) and its image along ⟨−3, −5⟩.

The vector indicates a translation 3 units left and 5 units down.

(*x*, *y*)	→	(*x* − 3, *y* − 5)
X(2, 2)	→	*X*′(−1, −3)
Y(5, 5)	→	*Y*′(2, 0)
Z(5, 3)	→	*Z*′(2, −2)

Graph △*XYZ* and its image △*X*′*Y*′*Z*′.

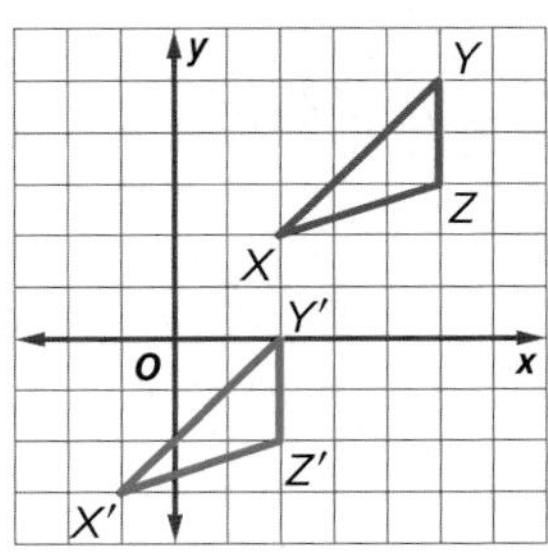

Study Guide and Review *Continued*

3-3 Rotations

Copy each polygon and point P. Then use a protractor and ruler to draw the specified rotation about point P.

17. 50°

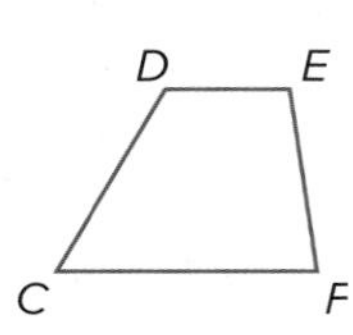

18. 90°

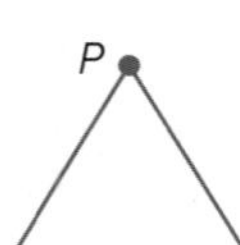

19. 130°

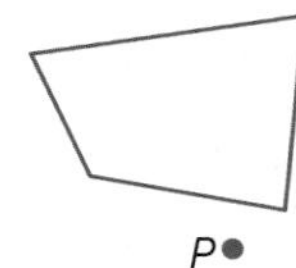

Graph each figure and its image after the specified rotation about the origin.

20. $\triangle MNO$ with vertices $M(-2, 2)$, $N(0, -2)$, $O(1, 0)$; 180°

21. $\triangle DGF$ with vertices $D(1, 2)$, $G(2, 3)$, $F(1, 3)$; 90°

Write a rule to describe each transformation.

22.

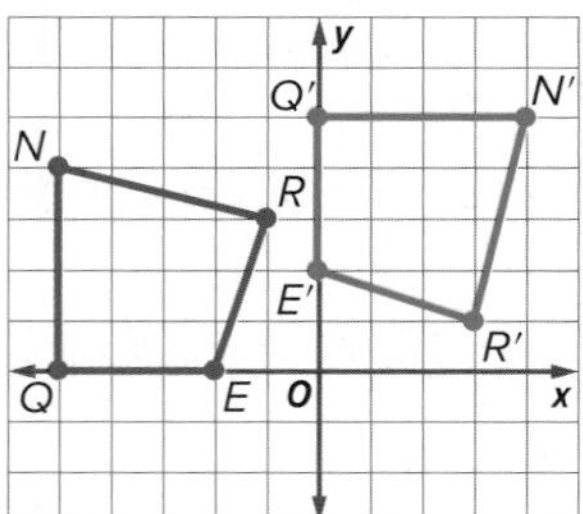

23.

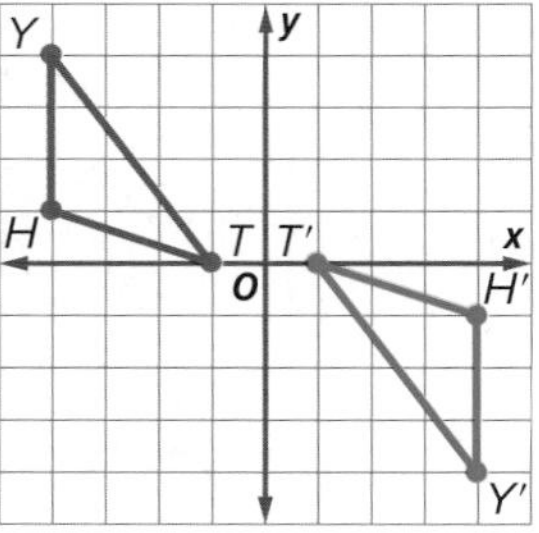

Example 3

Triangle ABC has vertices $A(-4, 0)$, $B(-3, 4)$, and $C(-1, 1)$. Graph $\triangle ABC$ and its image after a rotation 270° about the origin.

One method to solve this is to combine a 180° rotation with a 90° rotation. Multiply the x- and y-coordinates of each vertex by −1.

(x, y)	→	$(-x, -y)$
$A(-4, 0)$	→	$A'(4, 0)$
$B(-3, 4)$	→	$B'(3, -4)$
$C(-1, 1)$	→	$C'(1, -1)$

Multiply the y-coordinate of each vertex by −1 and interchange.

$(-x, -y)$	→	$(-y, x)$
$A'(4, 0)$	→	$A''(0, 4)$
$B'(3, -4)$	→	$B''(4, 3)$
$C'(1, -1)$	→	$C''(1, 1)$

Graph $\triangle ABC$ and its image $\triangle A''B''C''$.

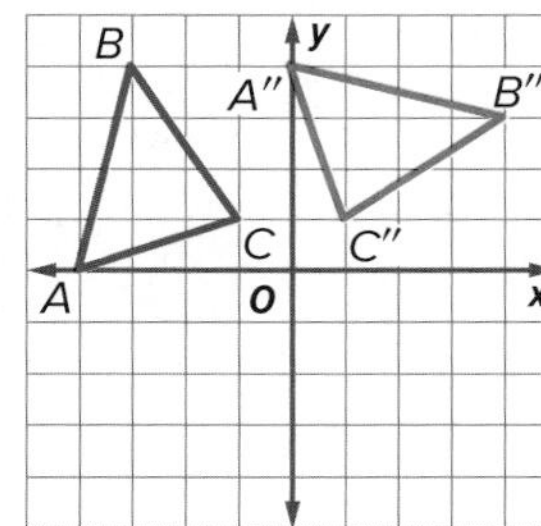

3-4 Compositions of Transformations

Graph each figure with the given vertices and its image after the indicated transformation.

24. $\overline{CD}$: $C(3, 2)$ and $D(1, 4)$
    Reflection: in $y = x$
    Rotation: 270° about the origin.

25. $\overline{GH}$: $G(-2, -3)$ and $H(1, 1)$
    Translation: along $\langle 4, 2\rangle$
    Reflection: in the x-axis

26. **PATTERNS** Jeremy is creating a pattern for the border of a poster using a stencil. Describe the transformation combination that he used to create the pattern below.

27. Copy and reflect figure T in line ℓ and then line m. Then describe a single transformation that maps T onto T''.

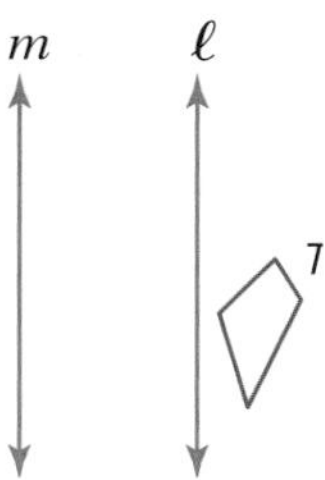

Example 4

The endpoints of $\overline{RS}$ are $R(4, 3)$ and $S(1, 1)$. Graph $\overline{RS}$ and its image after a translation along $\langle -5, -1\rangle$ and a rotation 180° about the origin.

Step 1 translation along $\langle -5, -1\rangle$

(x, y)	→	$(x - 5, y - 1)$
$R(4, 3)$	→	$R'(-1, 2)$
$S(1, 1)$	→	$S'(-4, 0)$

Step 2 rotation 180° about origin

(x, y)	→	$(-x, -y)$
$R'(-1, 2)$	→	$R''(1, -2)$
$S'(-4, 0)$	→	$S''(4, 0)$

Step 3 Graph $\overline{RS}$ and its image $\overline{R''S''}$.

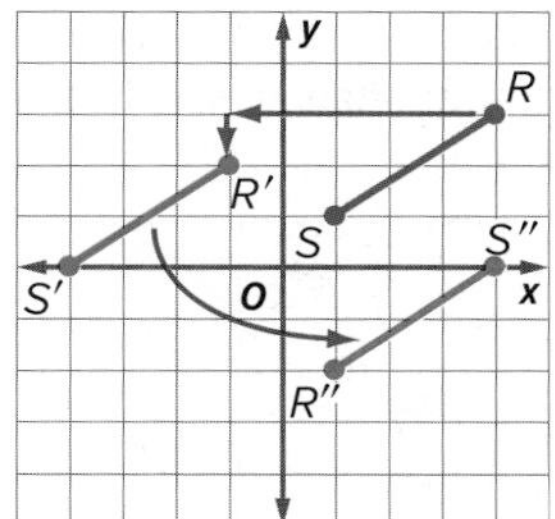

Study Guide and Review *Continued*

3-5 Symmetry

State whether each figure appears to have reflectional symmetry. Write *yes* or *no*. If so, copy the figure, draw all lines of symmetry, and state their number.

28. 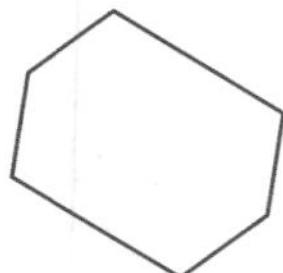

29. 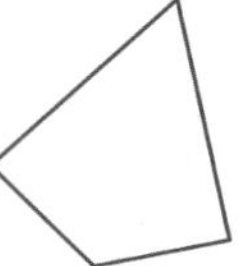

State whether each figure has rotational symmetry. Write *yes* or *no*. If so, copy the figure, locate the center of symmetry, and state the order and magnitude of symmetry.

30. 

31. 

32. **KNITTING** Amy is creating a pattern for a scarf she is knitting for her friend. How many lines of symmetry are there in the pattern?

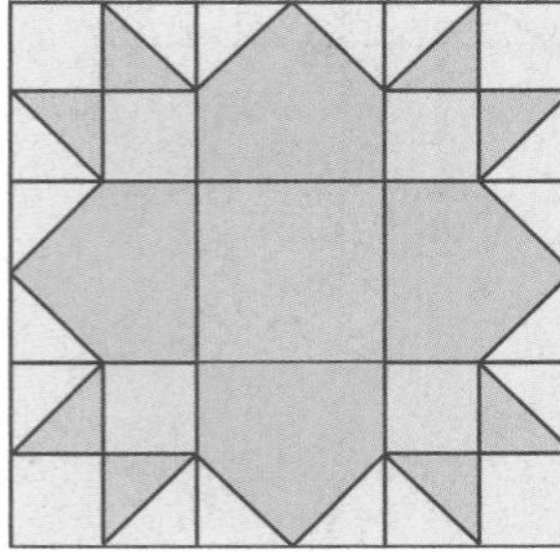

Example 5

State whether each figure appears to have reflectional symmetry. Write *yes* or *no*. If so, draw all lines of symmetry, and state their number.

a. Square

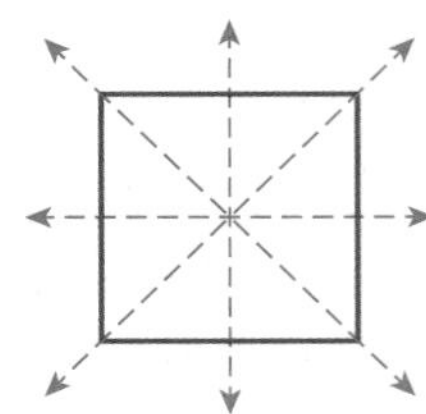

Yes, 4

b. The letter H

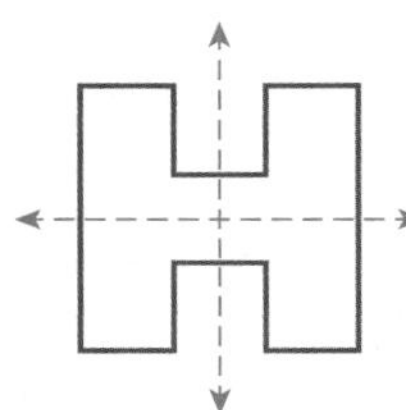

Yes, 2

Practice Test

Copy the figure and the given line of reflection. Then draw the reflected image in this line using a ruler.

1. 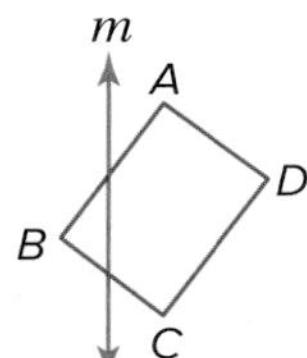

2. 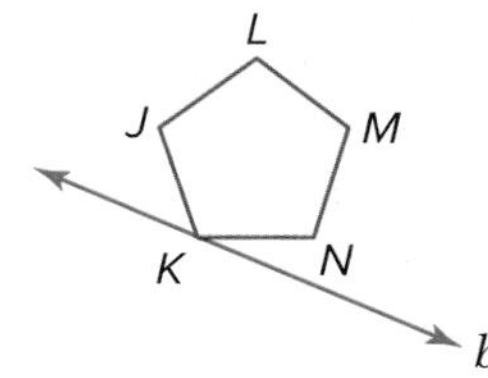

3. Point J has coordinates $(-1, 4)$. What are the coordinates of point J', the image of J after a rotation of 180* about the origin?

4. Point P has coordinates $(-1, 2)$. Its image P' has coordinates $(-5, -5)$ after a translation. Find the coordinates of the point $(4, 3)$ after the same translation.

5. A figure is translated along $\langle 3, -3 \rangle$. What translation will move the image back to its original position?

6. **PARKS** Isabel is on a ride at an amusement park that slides the rider to the right, and then rotates counterclockwise about its own center 60° every 2 seconds. How many seconds pass before Isabel completes one full rotation?

State whether each figure has *plane* symmetry, *axis* symmetry, *both*, or *neither*.

7. 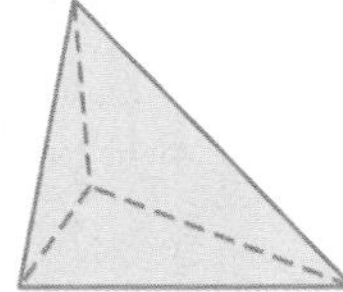

8. 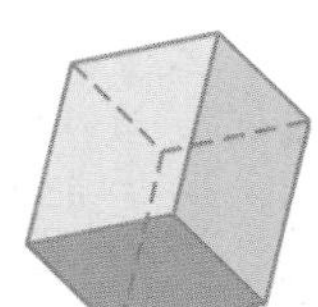

Graph each figure and its image under the given transformation.

9. $\square FGHJ$ with vertices $F(-1, -1)$, $G(-2, -4)$, $H(1, -4)$, and $J(2, -1)$ in the x-axis

10. $\triangle ABC$ with vertices $A(0, -1)$, $B(2, 0)$, $C(3, -3)$; $\langle -5, 4 \rangle$

11. quadrilateral $WXYZ$ with vertices $W(2, 3)$, $X(1, 1)$, $Y(3, 0)$, and $Z(5, 2)$; 180° about the origin

Tell whether each image is result of a translation, reflection, or roation of the preimage.

12. preimage: $\triangle ABC$ with vertices $A(1, 2)$, $B(0, -5)$, $C(6, -3)$
image: $\triangle A'B'C'$ with vertices $A(-1, 2)$, $B(0, -5)$, $C(-6, -3)$

13. preimage: quadrilateral $LMNP$ with vertices $L(-8, 0)$, $M(-8, 4)$, $N(0, 0)$, and $P(0, 4)$
image: quadrilateral $L'M'N'P'$ with vertices $L(0, 8)$, $M(4, 8)$, $N(0, 0)$, and $P(4, 0)$

Copy the figure and the given translation vector. Then draw the translation of the figure along the translation vector.

14. 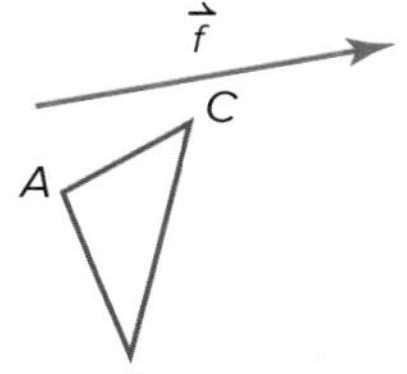

15. 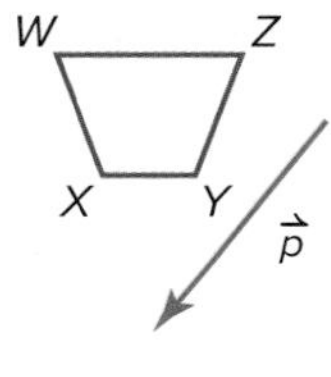

16. **ART** Below is an artist's rendition of what Stonehenge, a famous archaeological site in England, would have looked like before the stones fell or were removed. What is the order and magnitude of symmetry for the outer ring?

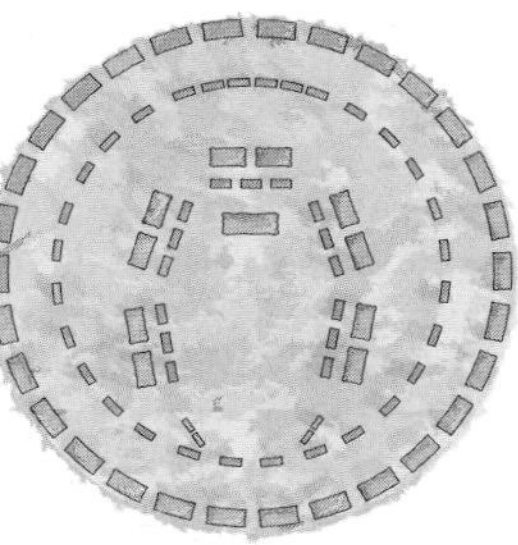

17. **MULTIPLE CHOICE** Which transformation or combination of transformations best represents the figure shown below?

A reflection
B glide reflection
C rotation

Preparing for Assessment

Performance Task

Provide a clear solution to each part of the task. Be sure to show all of your work, include all relevant drawings, and justify your answers.

DESIGN The board members on city council are holding a contest for the best design of a city park. The park will occupy a city block. Plans should be submitted using a coordinate grid to represent the area. Mark is a design student who decides to enter the contest. He is planning to use what he knows about transformations in his plans.

Part A

1. The contest rules state that every entry must include 2 water fountains, a drainage pond with a fountain, and a picnic area. Can Mark use rigid motion transformations to create his plan? Explain.

Part B

2. Mark's 1st draft is shown at the right. The drainage pond is placed in the center of the park with the fountain at the origin point. Mark placed a water fountain at (2, −2). He plans to place the second one using a reflection in the y-axis and a rotation of 180° around the center fountain. What are the coordinates of the second water fountain?

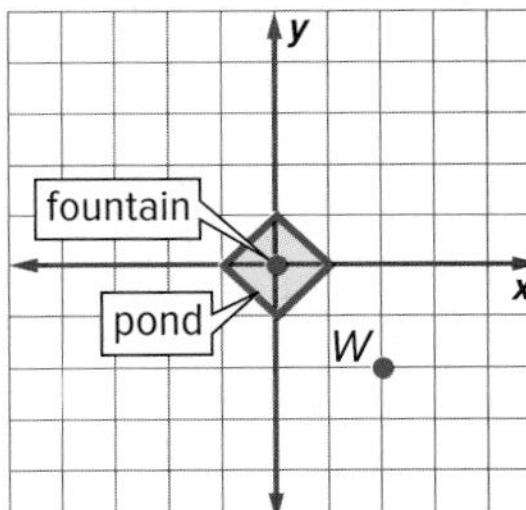

3. Describe two different single transformations that Mark could have used to locate the second water fountain.

Part C

Mark added two picnic tables to his design draft.

4. Describe a sequence of transformations that would map picnic table 1 to picnic table 2. Explain your solution process.

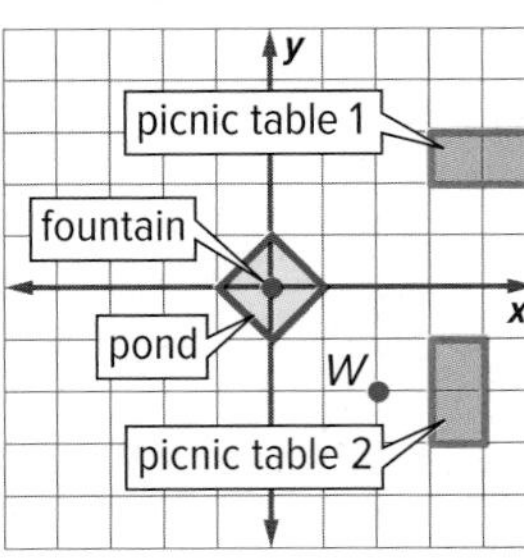

5. Mark wants to add a 3rd picnic table to Quadrant III. Describe the transformation(s) Mark could use to place this table.

6. Identify examples of line and rotational symmetry in Mark's design.

Test-Taking Strategy

Example

Read the problem. Identify what you need to know. Then use the information given to solve the problem.

The vertices of $\triangle LMN$ are $L(5, 6)$, $M(2, 0)$, and $N(-8, 8)$. After the figure is translated, the vertices have coordinates, in random order, $(-2, 0)$, $(1, 6)$, and $(-12, 8)$. Which of the following best describes the translation?

A $(x, y) \rightarrow (x, y - 4)$

B $(x, y) \rightarrow (x - 4, y)$

C $(x, y) \rightarrow (x, y + 4)$

D $(x, y) \rightarrow (x + 4, y)$

Step 1 **What in the problem gives you a clue that you should work backwards to solve?**
The problem gives vertices of a triangle *before* and *after* a translation.

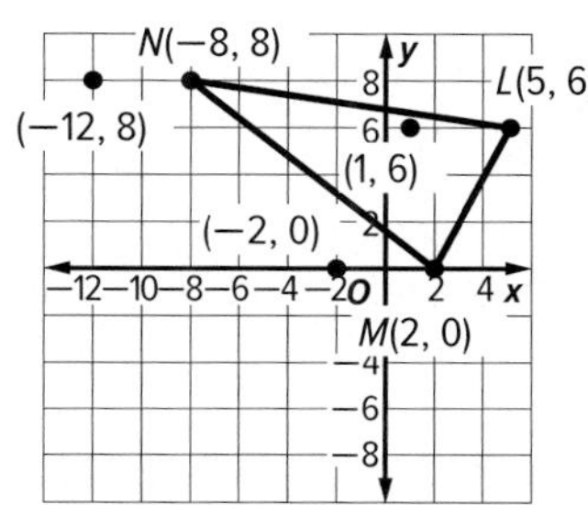

Step 2 **What steps will you take to solve the problem?**
I will plot and label the original points on a coordinate plane. Then I'll plot the points after the translation. Then I'll compare the two and see what translation must have been performed to get from one to the other.

Preimage	$(x, y) \rightarrow (x - 4, y)$	Image
$L(5, 6)$	$(x, y) \rightarrow (5 - 4, 6)$	$(1, 6)$
$M(2, 0)$	$(x, y) \rightarrow (2 - 4, 0)$	$(-2, 0)$
$N(-8, 8)$	$(x, y) \rightarrow (-8 - 4, 8)$	$(-12, 8)$

Step 3 **What is the best way to check your answer?**
Work in the other direction. I can start with the original points, perform the translation in the answer I found, and see if it results in the "after" points given in the problem.

The correct answer is B.

Test-Taking Tip

Work Backwards
Most problems require finding the end result. However, in some problems, the end result is given and the problem asks for an intermediate step. To solve problems like this, you must work backwards.

Apply the Strategy

Read the problem. Identify what you need to know. Then use the information given to solve the problem.

The coordinate grid shows the final image when a point was rotated 90° clockwise about the origin and shifted 7 units right. What were the original coordinates?

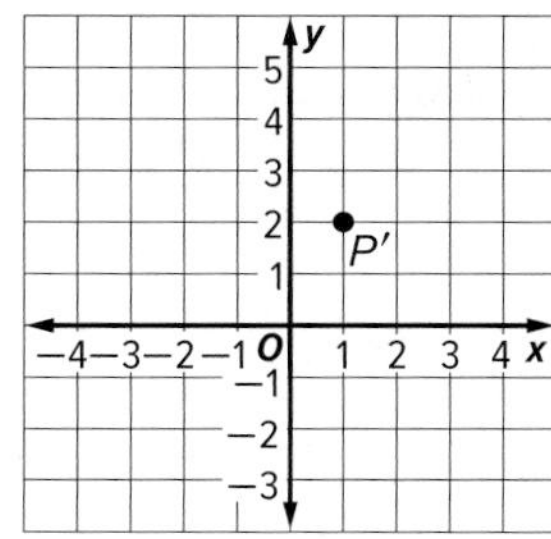

A $(-2, 1)$

B $(2, 6)$

C $(2, -1)$

D $(-2, -6)$

Answer the questions below.

a. What clues do you see that indicate you should work backwards?

b. What steps will you take to solve?

c. What is the best way to check your answer?

d. What is the correct answer?

CHAPTER 3

Preparing for Assessment

Cumulative Review

Read each question. Then fill in the correct answer on the answer document provided by your teacher or on a sheet of paper.

1. $\triangle PQR$ has vertices $P(-1, -2)$, $Q(1, -3)$, and $R(-2, -4)$. $\triangle P'Q'R'$ is the image of $\triangle PQR$ under the rotation $(x, y) \rightarrow (y, -x)$. Which of the following is a true statement?

○ **A** $\triangle P'Q'R'$ intersects the positive x-axis.

○ **B** $\triangle P'Q'R'$ lies entirely in Quadrant II.

○ **C** $\triangle P'Q'R'$ intersects the y-axis.

○ **D** $\triangle P'Q'R'$ lies to the left of the y-axis.

2. Point A is located on a coordinate plane at $(4, 7)$ and point B at $(4, -3)$. These points are first reflected in the y-axis, then translated left 2 units to create points C and D. If points C and D are connected with points A and B to form a rectangle, what is the resulting area of the figure in square units?

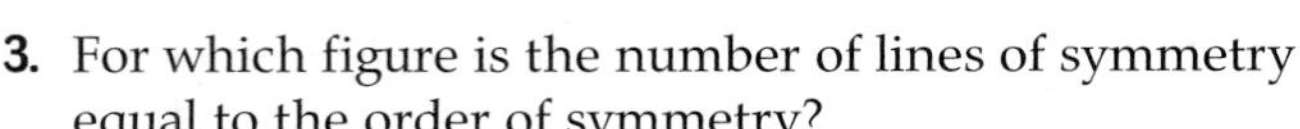

3. For which figure is the number of lines of symmetry equal to the order of symmetry?

○ **A**

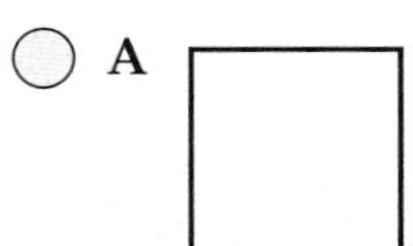

○ **B**

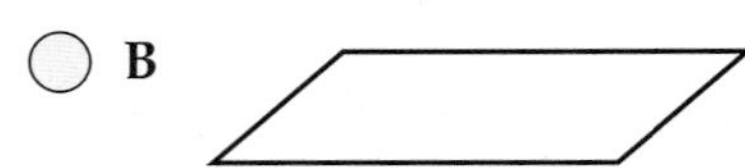

○ **C**

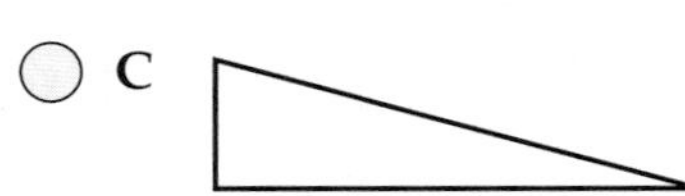

○ **D**

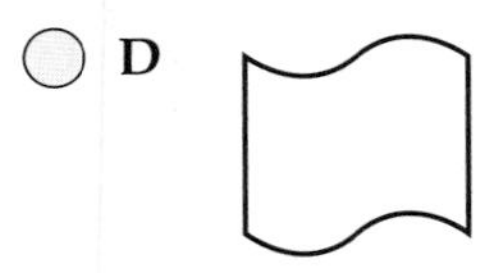

4. Which of the following is the image of the line $y = 3x + 2$ under the reflection $(x, y) \rightarrow (x, -y)$?

○ **A** $y = -3x - 2$

○ **B** $y = -3x + 2$

○ **C** $y = \frac{1}{3}x - \frac{2}{3}$

○ **D** $y = -\frac{1}{3}x - 2$

5. Carlos draws $\triangle PQR$ as shown. Then he translates the triangle along $\langle -4, 1 \rangle$ and rotates the image 180° about the origin.

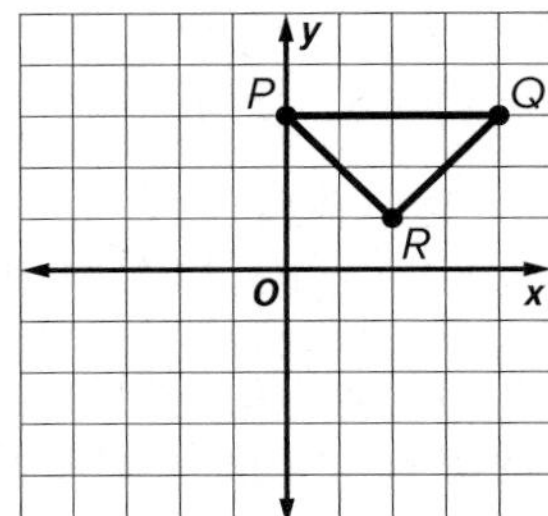

Which of the following points lies in the interior of the final image of the triangle?

A $(-6, -1)$

B $(0, -4)$

C $(2, -3)$

D $(-3, -2)$

6. Triangle DFG has coordinates $D(-5, -1)$, $F(4, 3)$, and $G(1, -3)$. If the triangle is reflected in the line $y = x$, what are the coordinates of image triangle $D'F'G'$?

Test-Taking Tip

Question 5 To check your answer, use the image you drew and perform the transformations in reverse to see if it results in the original figure.

7. Jermaine plots the point $P(4, -2)$. Then he translates the point using the translation $(x, y) \rightarrow (x - 3, y + 2)$ and labels the image P'. Finally, he draws $\overline{PP'}$. What is the length of $\overline{PP'}$ to the nearest tenth?

8. Rima is designing a logo for a website. She starts by drawing $\triangle STU$ with vertices $S(0, 2)$, $T(2, 2)$, and $U(2,0)$. Then she reflects the triangle in the x-axis. Which is a true statement about the final image?

   ☐ **A** The point (0, 2) lies on the resulting figure.

   ☐ **B** One of its vertices has coordinates $(-4, 4)$.

   ☐ **C** It never intersects $\triangle STU$.

   ☐ **D** It is congruent to $\triangle STU$.

   ☐ **E** It is similar to $\triangle STU$.

9. Amanda draws a circle on a coordinate plane. She translates it 4 units up, reflects it across the x-axis, and then rotates it 90° counterclockwise about the origin of the plane, which results in the figure shown below. What was the center of the original circle she drew? Write your answer in the form (x, y).

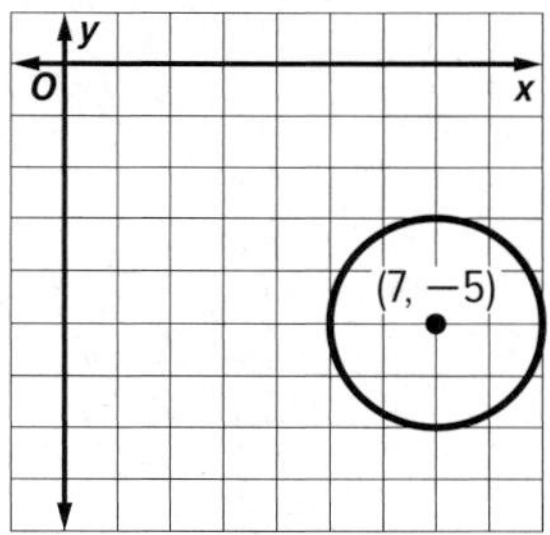

10. The equation of line m is $y = 2$. Line n is the image of line m under the reflection $(x, y) \rightarrow (y, x)$. Which of the following is a true statement about the lines?

    ○ **A** Line m is parallel to line n.

    ○ **B** Line m and line n coincide.

    ○ **C** Line m and line n intersect at the origin.

    ○ **D** Line m and line n are perpendicular.

11. The Elmwood High School debate team wants to base their team logo on a pentagon that has line symmetry but not rotational symmetry. Which of the following figures could the team use?

    ○ **A**

    ○ **C**

    ○ **B**

    ○ **D**

12. Marcus draws the figure below on a coordinate plane. First, he reflects it in the y-axis. Then he reflects it in the x-axis. Finally, he rotates it 90° counterclockwise about the origin. Which of the following transformations could he have performed which would have resulted in the same figure? Select all that apply.

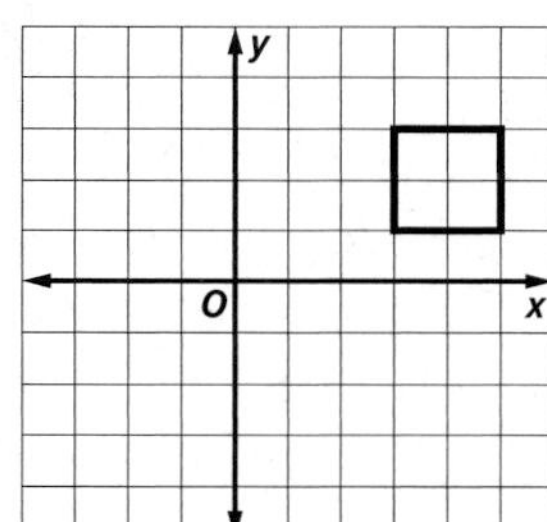

   ☐ **A** rotation 90° clockwise

   ☐ **B** translation down 2 units

   ☐ **C** reflection in the x-axis

   ☐ **D** translation down 6 units and left 2 units

   ☐ **E** reflection in the x-axis and rotation about the bottom-left vertex

Need Extra Help?

If you missed Question...	1	2	3	4	5	6	7	8	9	10	11	12
Go to Lesson...	3-3	3-2	3-5	3-4	3-1	3-2	3-2	3-1	3-3	3-4	3-5	3-1

CHAPTER 4

Triangles and Congruence

THEN

You learned about segments and angles, and discovered relationships between their measures.

NOW

In this chapter you will:

- Apply special relationships about the interior and exterior angles.
- Identify corresponding parts of congruent triangles and prove triangles congruent.
- Apply the definition of congruence in terms of rigid motion to triangles.

WHY

FITNESS Triangles are used to add strength to many structures, including fitness equipment such as bike frames.

Use the Mathematical Practices to complete the activity.

Apply Math Look for two triangles in the photo that appear to be the same size and shape. Which rigid motion transformation can be used to show that the triangles are congruent?

Model with Mathematics Use the Transformations tool in ConnectED to model these triangles.

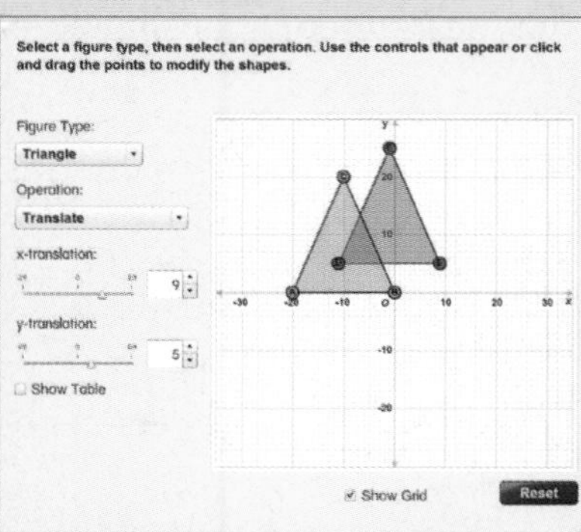

Construct an Argument Work with your classmates to develop an argument to show that these two triangles are congruent.

John Kelly/Media Bakery

Go Online to Guide Your Learning

Explore & Explain

Geometry Tools

Investigate using the **Triangle Relationships** tool. How does changing the length of one side or measure of one angle affect congruence?

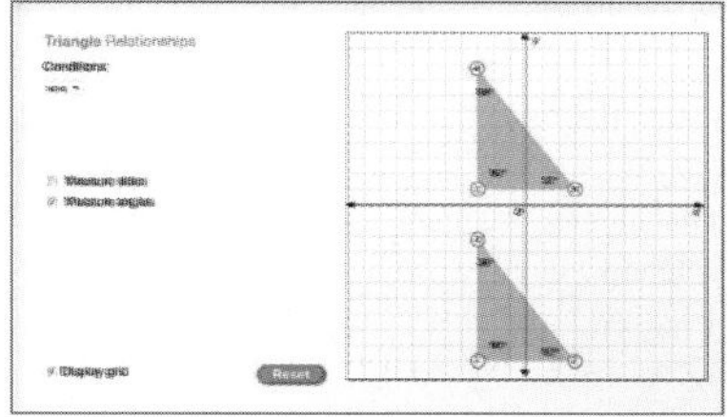

The Geometer's Sketchpad

Visualize and explore triangle congruence using the Triangle Congruence sketch in Lesson 4-3.

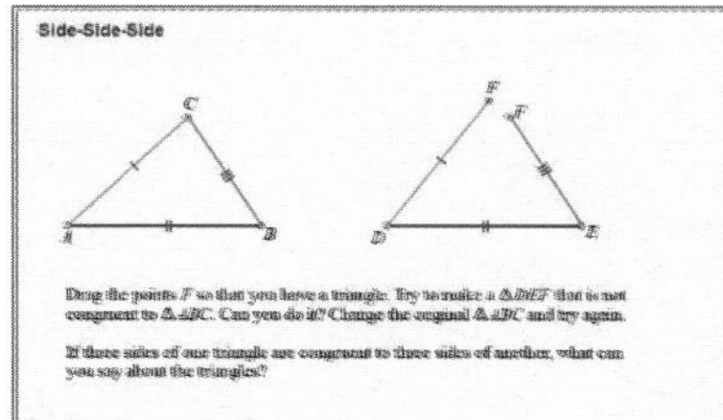

eBook

Interactive Student Guide

Before starting the chapter, answer the **Chapter Focus** preview questions. Check your answers as you complete each lesson. At the end of the chapter, try the **Performance Task**.

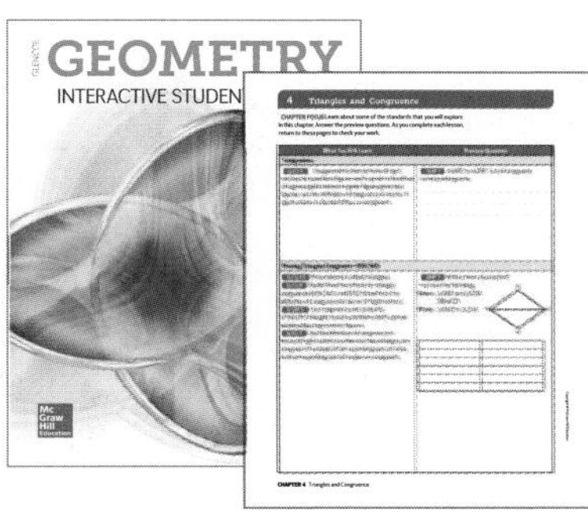

Organize

Foldables

Get organized! Create a **Triangle Congruence Foldable** before you start the chapter to arrange your notes on triangles.

Collaborate

Chapter Project

In the **Make Your Mark** project, you will apply what you learn about congruence transformations to a project that addresses global awareness.

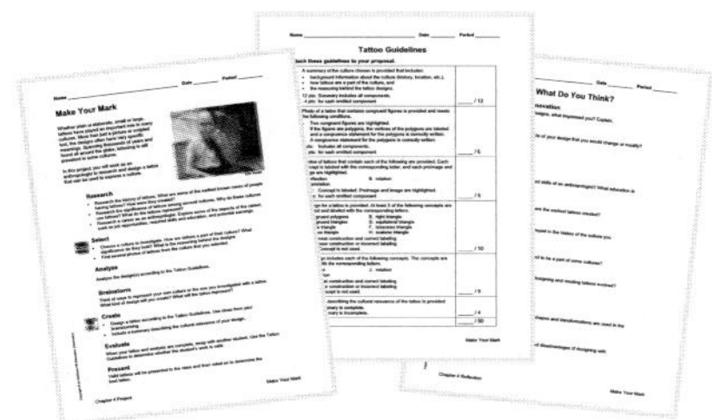

Focus

LEARNSMART®

Need help studying? Complete the **Congruence, Proof, and Constructions** domains in LearnSmart to review for the chapter test.

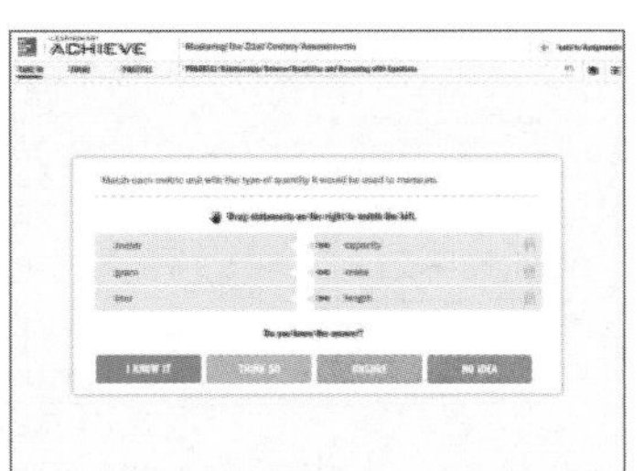

ALEKS®

You can use the **Triangles and Lines and Angles** topic in ALEKS to find out what you know about triangles and what you are ready to learn.*

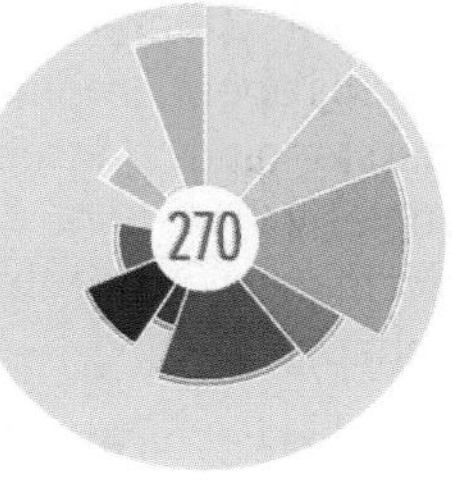

* Ask your teacher if this is part of your program.

Get Ready for the Chapter

Go Online! for Vocabulary Review Games and key vocabulary in 13 languages.

Connecting Concepts

Concept Check

Review the concepts used in this chapter by answering each question below.

1. $\angle A$ is a right angle, $\angle B$ is an acute angle, and $\angle C$ is an obtuse angle. List the angles in order from least measure to greatest measure.
2. Two adjacent angles form an obtuse angle. How many of the adjacent angles can be right angles? Explain.
3. Three angles that share a vertex form a line. Give possible measures for each of the three angles.
4. Two parallel lines are intersected by a transversal. Explain whether it is possible for the same-side interior angles to be complementary.

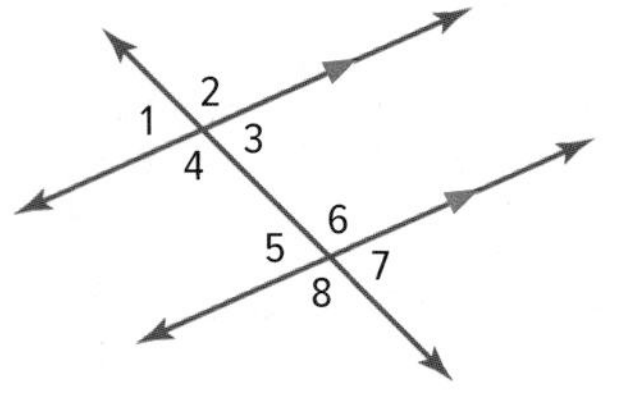

5. Two parallel lines are intersected by a transversal. Explain whether it is possible for the eight angles formed to all be congruent.
6. Explain how to find the distance between two points in the coordinate plane that have the same x-coordinate.
7. A right triangle is drawn in the coordinate plane so that its legs are parallel to the axes. Describe two ways to find the length of the hypotenuse.

Performance Task Preview

You can use the concepts and skills in this chapter to solve problems about graphic design. Understanding congruent triangles will help you finish the Performance Task at the end of the chapter.

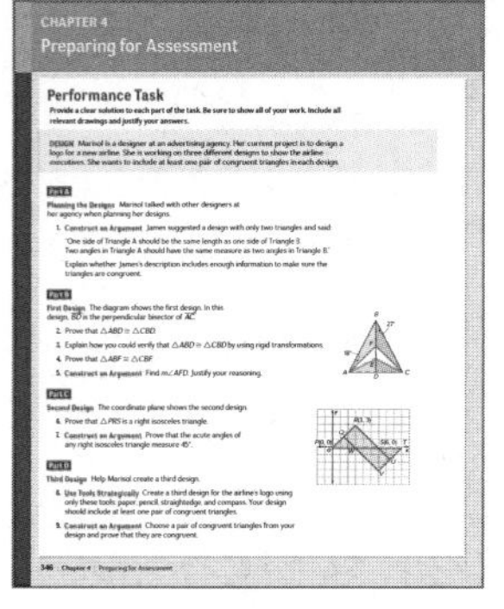

In this Performance Task you will:

- construct an argument
- use tools strategically

New Vocabulary

English		Español
auxiliary line	p. 282	linea auxiliar
exterior angles	p. 284	ángulos externos
remote interior angles	p. 284	ángulos internos no adyacentes
corollary	p. 285	corolario
congruent polygons	p. 291	polígonos congruentes
corresponding parts	p. 291	partes correspondientes
included angle	p. 302	ángulo incluido
included side	p. 311	lado incluido
base angle	p. 325	ángulo de la base
legs of an isosceles triangle	p. 325	catetos de un triángulo isósceles
vertex angle	p. 325	ángulo de vértice

Review Vocabulary

alternate interior angles ángulos alternos internos nonadjacent interior angles that lie on opposite sides of a transversal

consecutive interior angles ángulos internos consecutivos interior angles that lie on the same side of a transversal

corresponding angles ángulos correspondientes angles that lie on the same side of a transversal and the same side of two lines intersected by the transversal

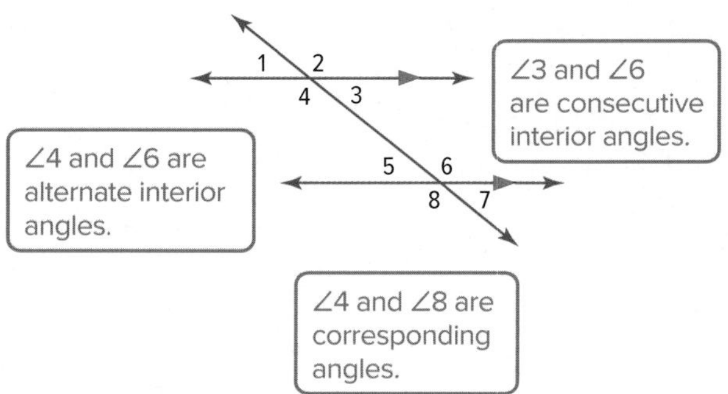

EXPLORE 4-1

Geometry Lab

Angles of Triangles

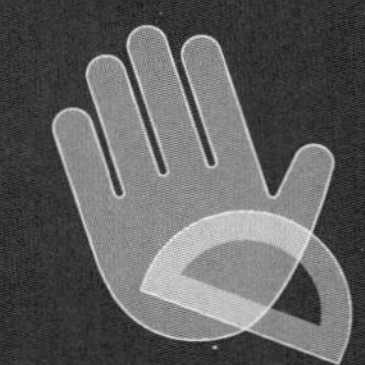

In this lab, you will find special relationships among the angles of a triangle.

Mathematical Practices
 **4** Model with mathematics.
5 Use appropriate tools strategically.

Activity 1 Interior Angles of a Triangle

Step 1

Draw and cut out several different triangles. Label the vertices A, B, and C.

Step 2

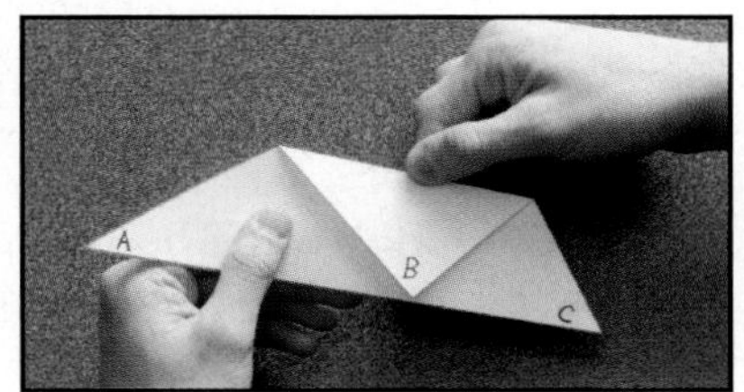

For each triangle, fold vertex B down so that the fold line is parallel to $\overline{AC}$. Relabel as vertex B.

Step 3

Then fold vertices A and C so that they meet vertex B. Relabel as vertices A and C.

Analyze the Results Work cooperatively.

1. Angles A, B, and C are called *interior angles* of triangle ABC. What type of figure do these three angles form when joined together in Step 3?
2. **Make a conjecture** about the sum of the measures of the interior angles of a triangle.

Activity 2 Exterior Angles of a Triangle

Step 1

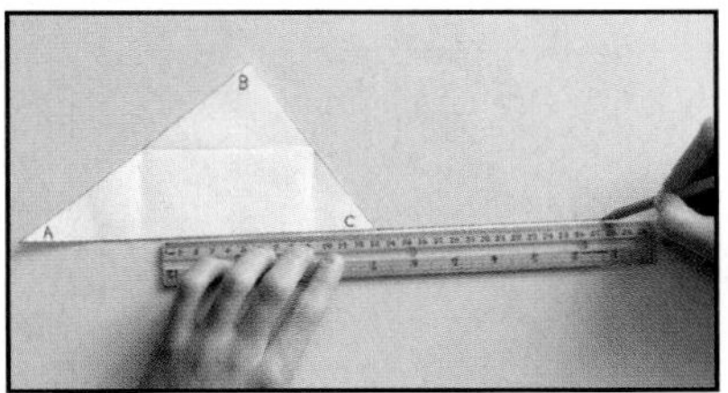

Unfold each triangle from Activity 1 and place each on a separate piece of paper. Extend AC as shown.

Step 2

For each triangle, tear off $\angle A$ and $\angle B$.

Step 3

Arrange $\angle A$ and $\angle B$ so that they fill the angle adjacent to $\angle C$ as shown.

Model and Analyze the Results Work cooperatively.

3. The angle adjacent to $\angle C$ is called an *exterior angle* of triangle ABC. **Make a conjecture** about the relationship among $\angle A$, $\angle B$, and the exterior angle at C.
4. Repeat the steps in Activity 2 for the exterior angles of $\angle A$ and $\angle B$ in each triangle.
5. **Make a conjecture** about the measure of an exterior angle and the sum of the measures of its nonadjacent interior angles.

LESSON 1

Angles of Triangles

Then

- You classified triangles by their side or angle measures.

Now

1. Apply the Triangle Angle-Sum Theorem.
2. Apply Exterior Angle Theorem.

Why?

- Each year, NASA sponsors the RASC-AL Exploration Robo-Ops Competition in which students design and build a robot.

One test of a robot's movements is to program it to move in a triangular path. The sum of the measures of the pivot angles through which the robot must turn will always be the same.

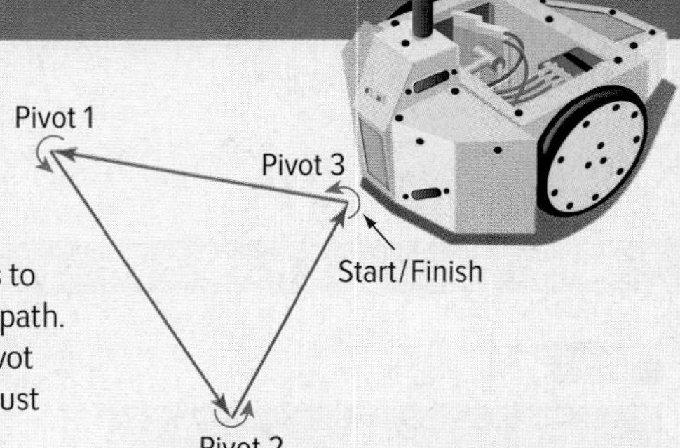

New Vocabulary

auxiliary line
exterior angle
remote interior angles
flow proof
corollary

Mathematical Practices

1 Make sense of problems and persevere in solving them.

3 Construct viable arguments and critique the reasoning of others.

8 Look for and express regularity in repeated reasoning.

1 Triangle Angle-Sum Theorem

The Triangle Angle-Sum Theorem gives the relationship among the interior angle measures of any triangle.

Theorem 4.1 Triangle Angle-Sum Theorem

Words The sum of the measures of the angles of a triangle is 180.

Example $m\angle A + m\angle B + m\angle C = 180$

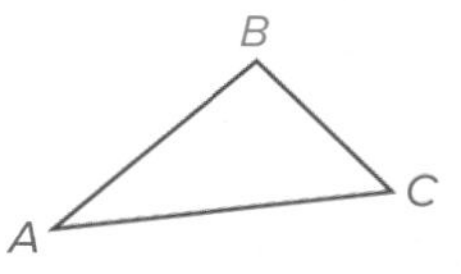

The proof of the Triangle Angle-Sum Theorem requires the use of an auxiliary line. An **auxiliary line** is an extra line or segment drawn in a figure to help analyze geometric relationships. As with any statement in a proof, you must justify any properties of an auxiliary line that you have drawn.

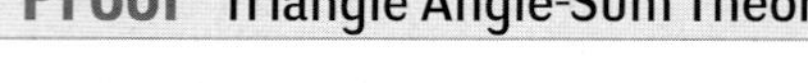

Proof Triangle Angle-Sum Theorem

Given: $\triangle ABC$

Prove: $m\angle 1 + m\angle 2 + m\angle 3 = 180$

Proof:

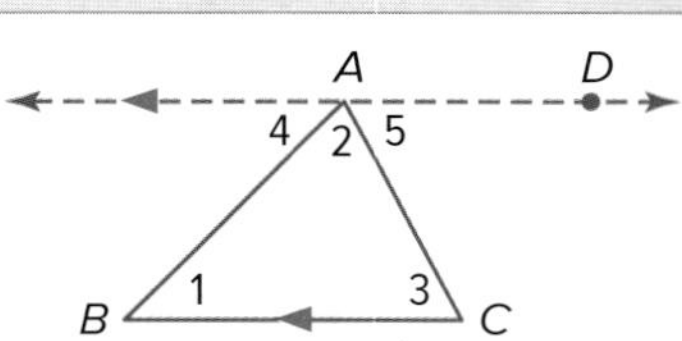

Statements	Reasons
1. $\triangle ABC$	1. Given
2. Draw $\overleftrightarrow{AD}$ through A parallel to $\overline{BC}$.	2. Parallel Postulate
3. $\angle 4$ and $\angle BAD$ form a linear pair.	3. Def. of a linear pair
4. $\angle 4$ and $\angle BAD$ are supplementary.	4. If 2 ∠s form a linear pair, they are supplementary.
5. $m\angle 4 + m\angle BAD = 180$	5. Def. of suppl. ∠s
6. $m\angle BAD = m\angle 2 + m\angle 5$	6. Angle Addition Postulate
7. $m\angle 4 + m\angle 2 + m\angle 5 = 180$	7. Substitution
8. $\angle 4 \cong \angle 1$, $\angle 5 \cong \angle 3$	8. Alt. Int. ∠s Theorem
9. $m\angle 4 = m\angle 1$, $m\angle 5 = m\angle 3$	9. Def. of ≅ ∠s
10. $m\angle 1 + m\angle 2 + m\angle 3 = 180$	10. Substitution

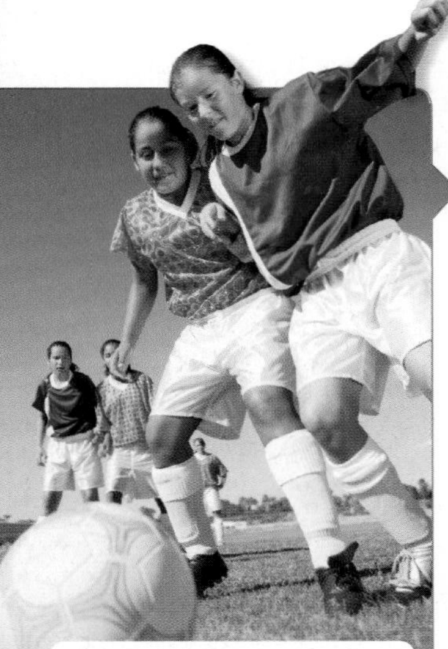

The Triangle Angle-Sum Theorem can be used to determine the measure of the third angle of a triangle when the other two angle measures are known.

Real-World Example 1 Use the Triangle Angle-Sum Theorem

SOCCER The diagram shows the path of the ball in a passing drill created by four friends. Find the measure of each numbered angle.

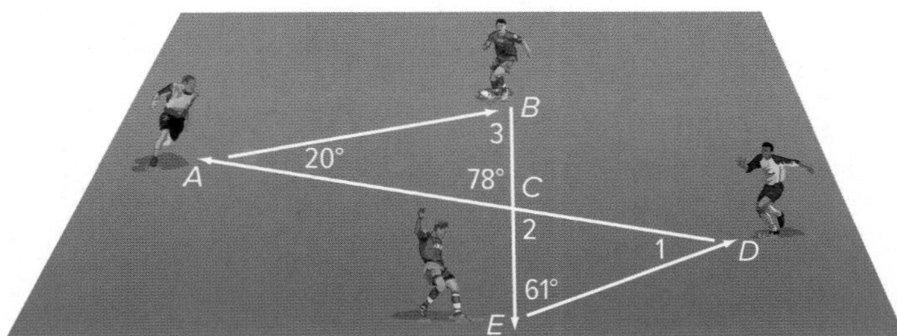

Understand You know the measures of two angles of one triangle and one measure of another. You also know that $\angle ACB$ and $\angle 2$ are vertical angles.

Plan Find $m\angle 3$ using the Triangle Angle-Sum Theorem. Use the Vertical Angles Theorem to find $m\angle 2$. Then you will have enough information to find the measure of $\angle 1$ in $\triangle CDE$.

Solve

$m\angle 3 + m\angle BAC + m\angle ACB = 180$ — Triangle Angle-Sum Theorem

$m\angle 3 + 20 + 78 = 180$ — Substitution

$m\angle 3 + 98 = 180$ — Simplify.

$m\angle 3 = 82$ — Subtract 98 from each side.

$\angle ACB$ and $\angle 2$ are congruent vertical angles. So, $m\angle 2 = 78$.

Use $m\angle 2$ and $\angle CED$ of $\triangle CDE$ to find $m\angle 1$.

$m\angle 1 + m\angle 2 + m\angle CED = 180$ — Triangle Angle-Sum Theorem

$m\angle 1 + 78 + 61 = 180$ — Substitution

$m\angle 1 + 139 = 180$ — Simplify.

$m\angle 1 = 41$ — Subtract 139 from each side.

Check The sums of the measures of the angles of $\triangle ABC$ and $\triangle CDE$ should be 180.

$\triangle ABC$: $m\angle 3 + m\angle BAC + m\angle ACB = 82 + 20 + 78$ or 180 ✓

$\triangle CDE$: $m\angle 1 + m\angle 2 + m\angle CED = 41 + 78 + 61$ or 180 ✓

By identifying each part of the problem, this complex problem could be separated into three manageable pieces. The properties of triangles were used to check the reasonableness of the answers found.

Guided Practice

Find the measures of each numbered angle.

1A.

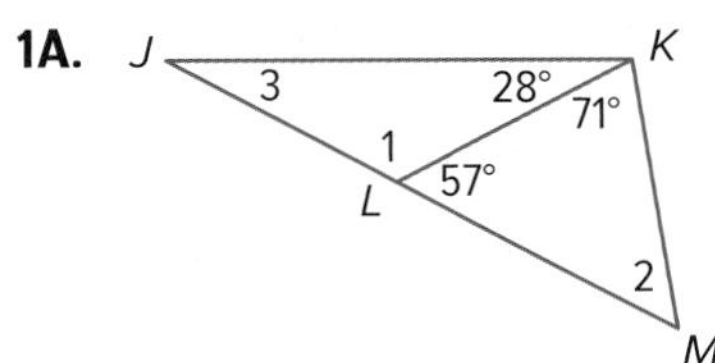

1B.

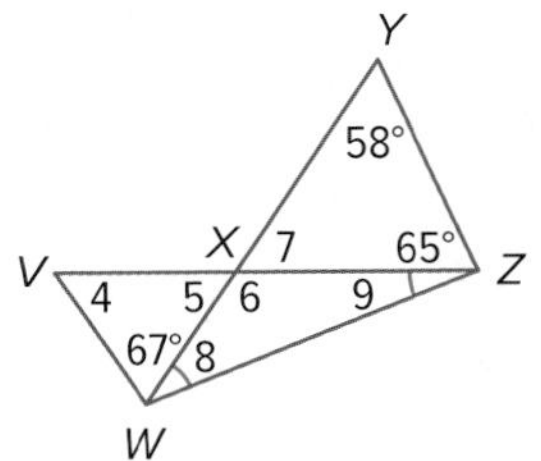

Real-World Link

The pass-and-move soccer drill incorporates several fundamental aspects of passing. All passes in this drill are made in a triangle, which is the basis of all ball movement. Additionally, the players are forced to move immediately after passing the ball.

2 Exterior Angle Theorem

In addition to its three interior angles, a triangle can have **exterior angles** formed by one side of the triangle and the extension of an adjacent side. Each exterior angle of a triangle has two **remote interior angles** that are not adjacent to the exterior angle.

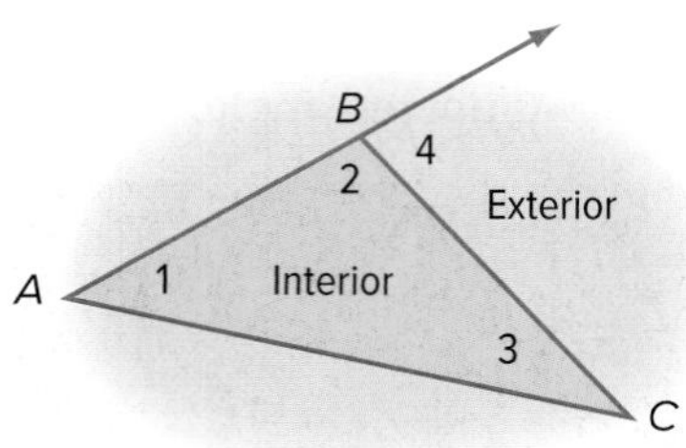

$\angle 4$ is an exterior angle of $\triangle ABC$. Its two remote interior angles are $\angle 1$ and $\angle 3$.

Theorem 4.2 Exterior Angle Theorem

The measure of an exterior angle of a triangle is equal to the sum of the measures of the two remote interior angles.

Example $m\angle A + m\angle B = m\angle 1$

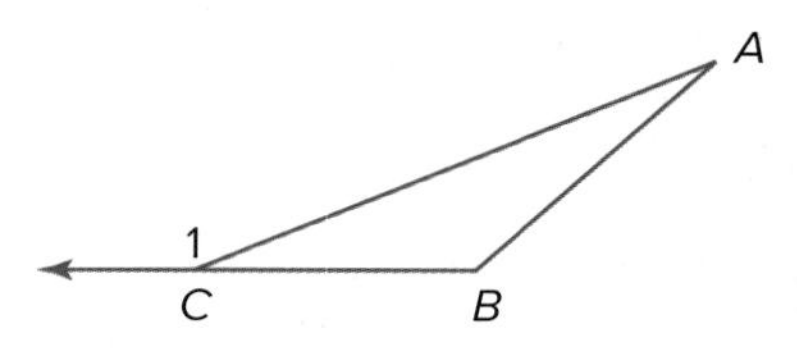

Reading Math

Flowchart Proof A flow proof is sometimes called a *flowchart* proof.

Flow Proofs Flow proofs can be written vertically or horizontally.

A **flow proof** uses statements written in boxes and arrows to show the logical progression of an argument. The reason justifying each statement is written below the box. You can use a flow proof to prove the Exterior Angle Theorem.

Proof Exterior Angle Theorem

Given: $\triangle ABC$

Prove: $m\angle A + m\angle B = m\angle 1$

Flow Proof:

$\triangle ABC$
Given
↓
$m\angle A + m\angle B + m\angle 2 = 180$
Triangle Angle-Sum Theorem

$\angle 2$ and $\angle 1$ form a linear pair.
Definition of a linear pair
↓
$\angle 2$ and $\angle 1$ are supplementary.
If 2 ∠s form a linear pair, they are supplementary.
↓
$m\angle 2 + m\angle 1 = 180$
Definition of supplementary

↘ ↙

$m\angle A + m\angle B + m\angle 2 = m\angle 2 + m\angle 1$
Substitution
↓
$m\angle A + m\angle B = m\angle 1$
Subtraction Property of Equality

Go Online!

The Triangle Angle-Sum and Exterior Angle Theorems are widely used in problems arising in everyday life, society, and the workplace. If you need extra practice, look for **Worksheets** in ConnectED.

The Exterior Angle Theorem can also be used to find missing measures.

Real-World Example 2 Use the Exterior Angle Theorem

FITNESS Find the measure of $\angle JKL$ in the Triangle Pose shown.

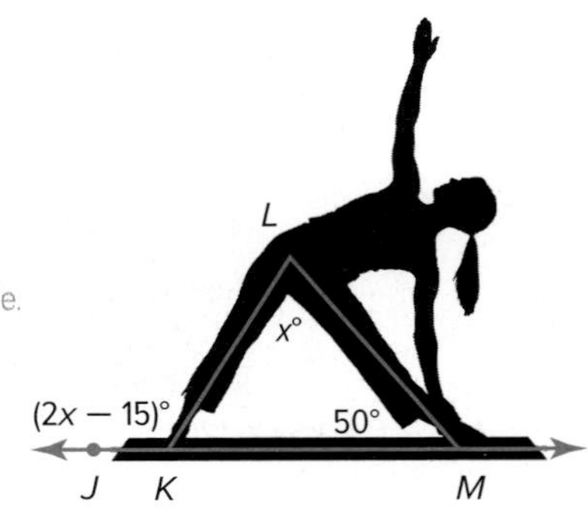

$m\angle KLM + m\angle LMK = m\angle JKL$	Exterior Angle Theorem
$x + 50 = 2x - 15$	Substitution
$50 = x - 15$	Subtract x from each side.
$65 = x$	Add 15 to each side.

So, $m\angle JKL = 2(65) - 15$ or 115.

Guided Practice

2. CLOSET ORGANIZING Tanya mounts the shelving bracket shown to the wall of her closet. What is the measure of $\angle 1$, the angle that the bracket makes with the wall?

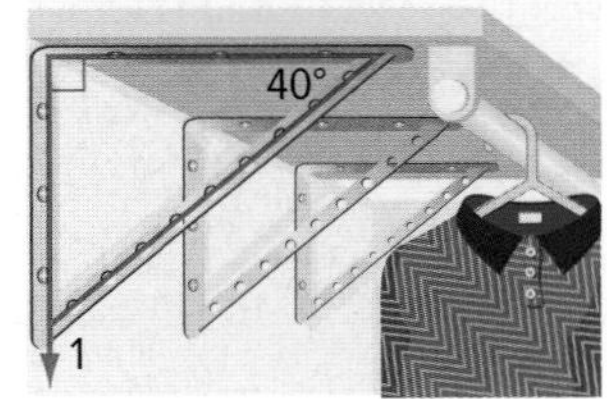

Real-World Career

Personal Trainer
Personal trainers instruct and motivate individuals in exercise activities. They demonstrate various exercises and help clients improve their exercise techniques. Personal trainers must obtain certification in the fitness field.

A **corollary** is a theorem with a proof that follows as a direct result of another theorem. As with a theorem, a corollary can be used as a reason in a proof. The corollaries below follow directly from the Triangle Angle-Sum Theorem.

Corollaries Triangle Angle-Sum Corollaries

4.1 The acute angles of a right triangle are complementary.

Abbreviation: *Acute ∠s of a rt. △ are comp.*

Example: If $\angle C$ is a right angle, then $\angle A$ and $\angle B$ are complementary.

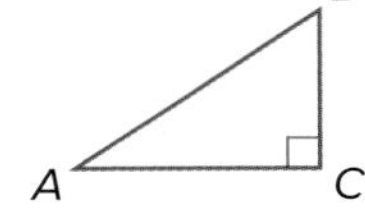

4.2 There can be at most one right or obtuse angle in a triangle.

Example: If $\angle L$ is a right or an obtuse angle, then $\angle J$ and $\angle K$ must be acute angles.

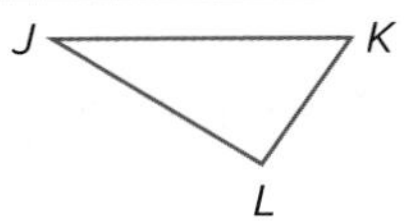

You will prove Corollaries 4.1 and 4.2 in Exercises 34 and 35.

Example 3 Find Angle Measures in Right Triangles

Find the measures of each numbered angle.

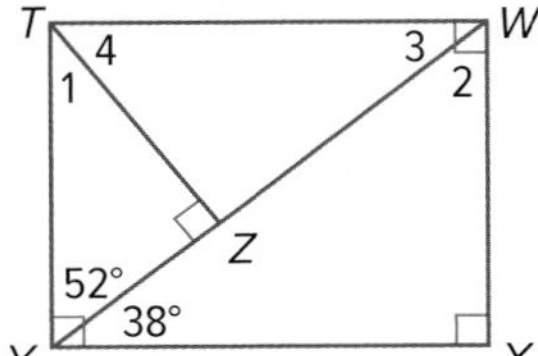

$m\angle 1 + m\angle TYZ = 90$	Acute ∠s of a rt. △ are comp.
$m\angle 1 + 52 = 90$	Substitution
$m\angle 1 = 38$	Subtract 52 from each side.

Guided Practice

3A. $\angle 2$ **3B.** $\angle 3$ **3C.** $\angle 4$

Study Tip

Check for Reasonableness When you are solving for the measure of one or more angles of a triangle, always check to make sure that the sum of the angles measures is 180.

Check Your Understanding

◯ = Step-by-Step Solutions begin on page R13.

Example 1 Find the measures of each numbered angle.

1.

2.

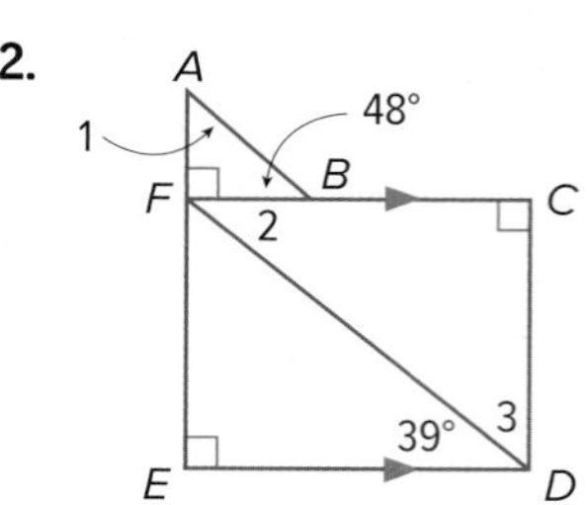

Example 2 Find each measure.

3. $m\angle 2$

4. $m\angle MPQ$

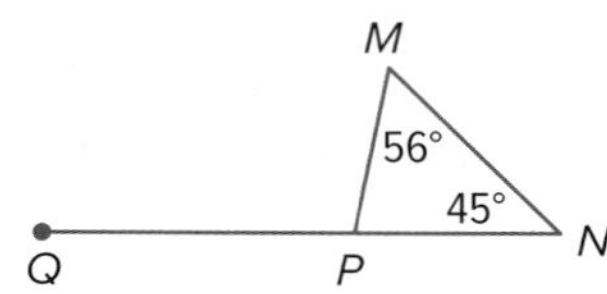

DECK CHAIRS The brace of this deck chair forms a triangle with the rest of the chair's frame as shown. If $m\angle 1 = 95$ and $m\angle 3 = 55$, find each measure.

5. $m\angle 4$
6. $m\angle 6$
7. $m\angle 2$
8. $m\angle 5$

Example 3

MP REGULARITY Find each measure.

9. $m\angle 1$

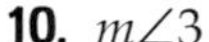

10. $m\angle 3$
11. $m\angle 2$

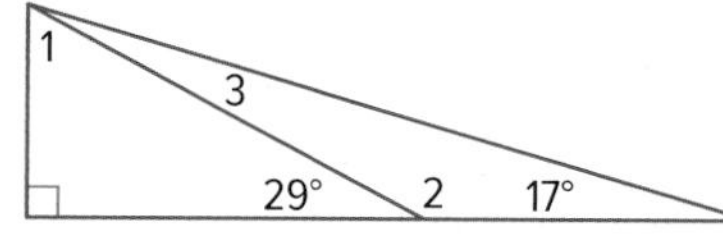

Practice and Problem Solving

Extra Practice is on page R4.

Example 1 Find the measure of each numbered angle.

12.

13.

14.

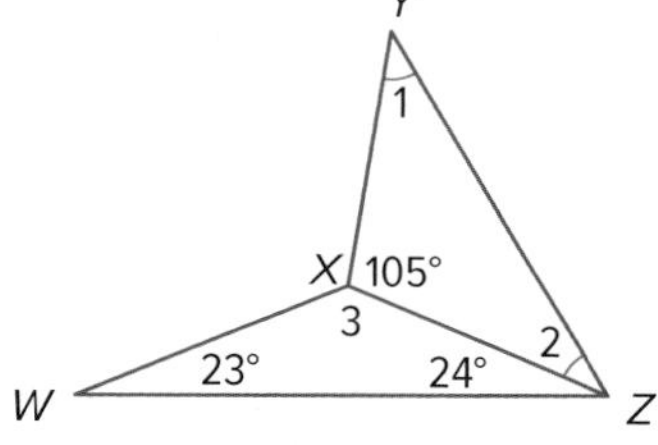

15.

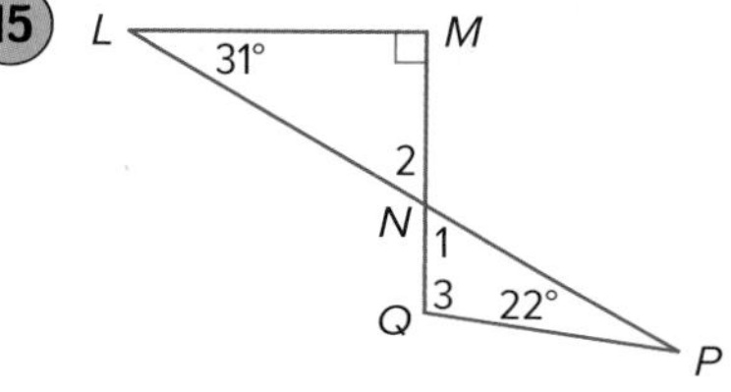

16. **AIRPLANES** The path of an airplane can be modeled using two sides of a triangle as shown. The distance covered during the plane's ascent is equal to the distance covered during its descent.

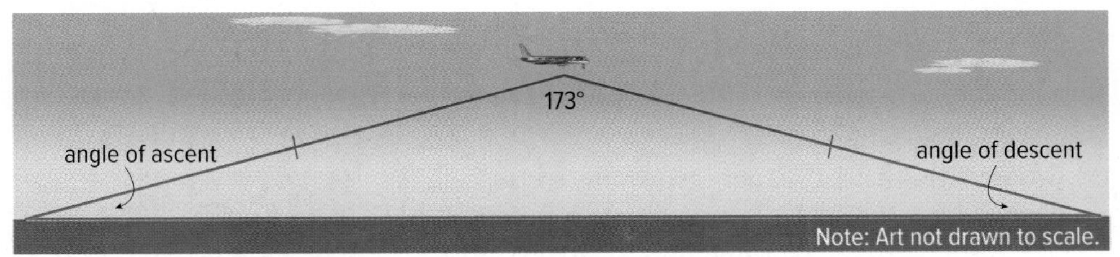

a. Classify the model using its sides and angles.

b. The angles of ascent and descent are congruent. Find their measures.

Example 2 **Find each measure.**

17. $m\angle 1$

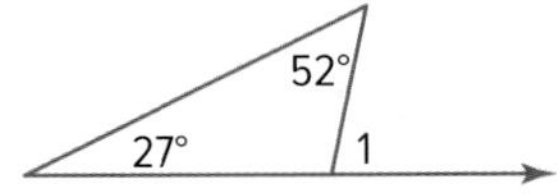

18. $m\angle 3$

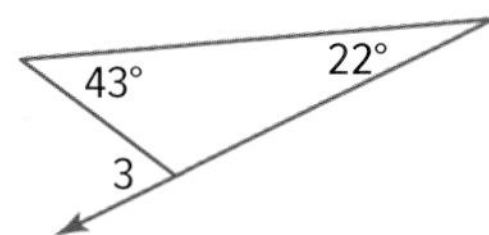

19. $m\angle 2$

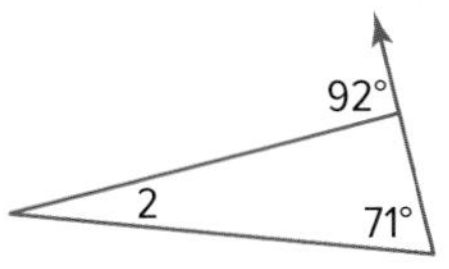

20. $m\angle 4$

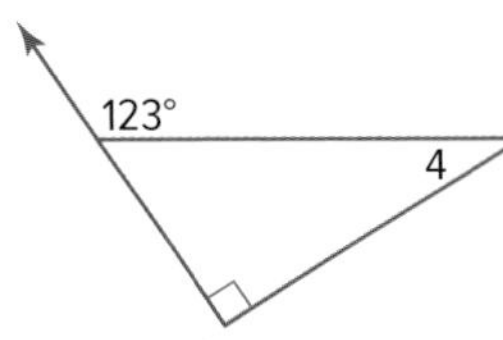

21. $m\angle ABC$

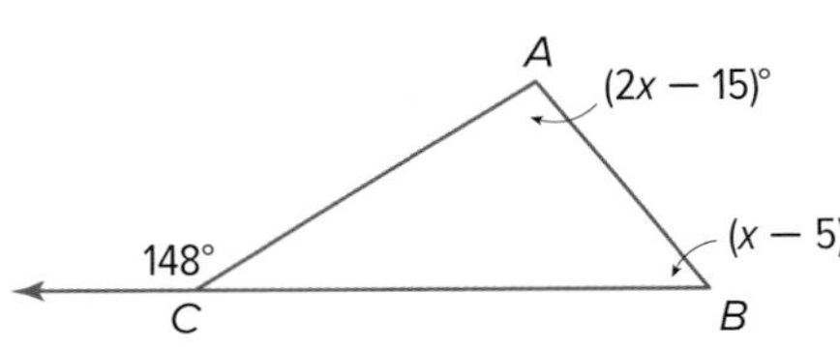

22. $m\angle JKL$

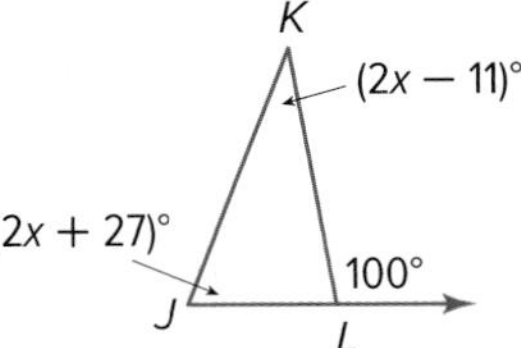

Example 3 23. **WHEELCHAIR RAMP** Suppose the wheelchair ramp shown makes a 12° angle with the ground. What is the measure of the angle the ramp makes with the van door?

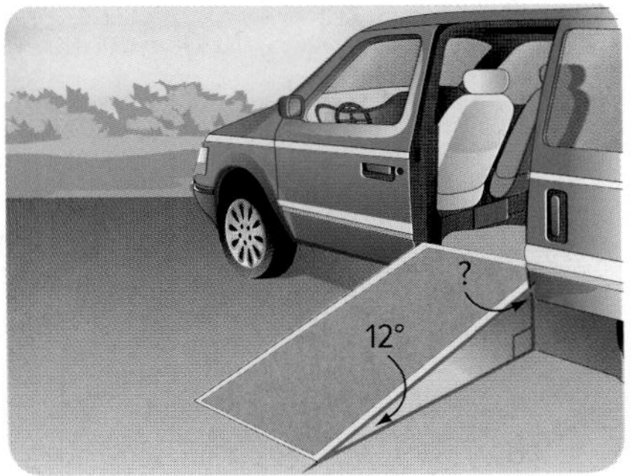

MP REGULARITY **Find each measure.**

24. $m\angle 1$

25. $m\angle 2$

26. $m\angle 3$

27. $m\angle 4$

28. $m\angle 5$

29. $m\angle 6$

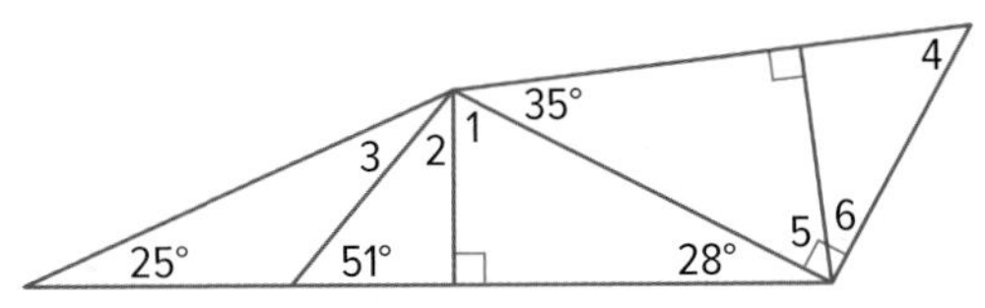

ALGEBRA **Find the value of x. Then find the measure of each angle.**

30. 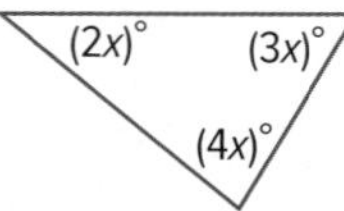

31. 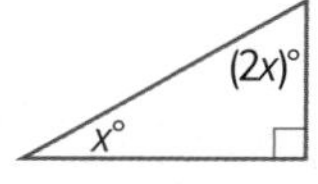

32. 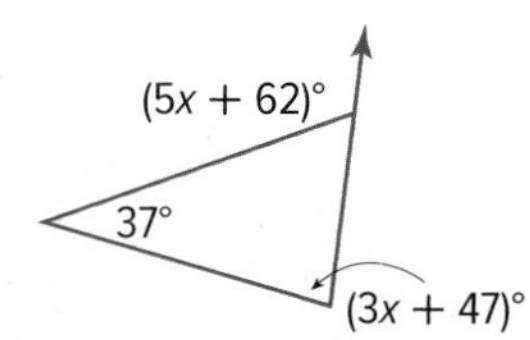

33. **GARDENING** A landscaper is forming an isosceles triangle in a flower bed using chrysanthemums. She wants $m\angle A$ to be three times the measure of $\angle B$ and $\angle C$. What should the measure of each angle be?

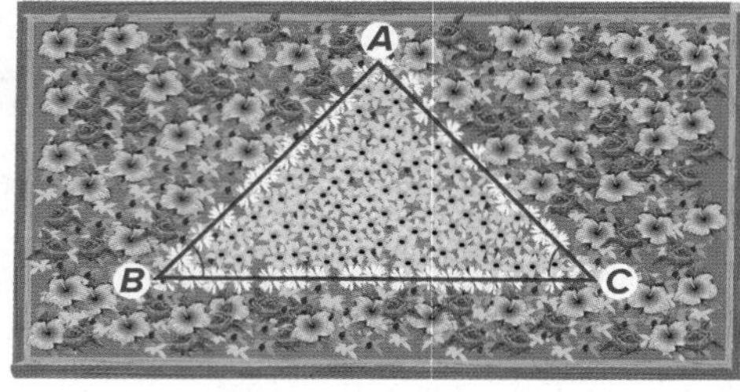

PROOF **Write the specified type of proof.**

34. flow proof of Corollary 4.1

35. paragraph proof of Corollary 4.2

MP **REGULARITY** **Find the measure of each numbered angle.**

36. 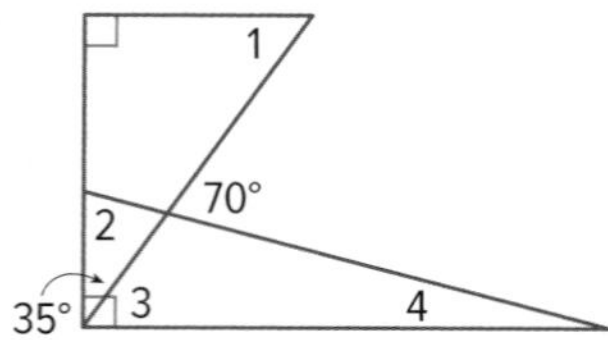

37. 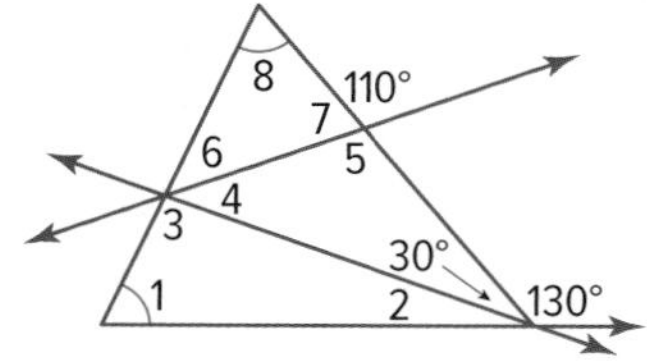

38. **ALGEBRA** Classify the triangle shown by its angles. Explain your reasoning.

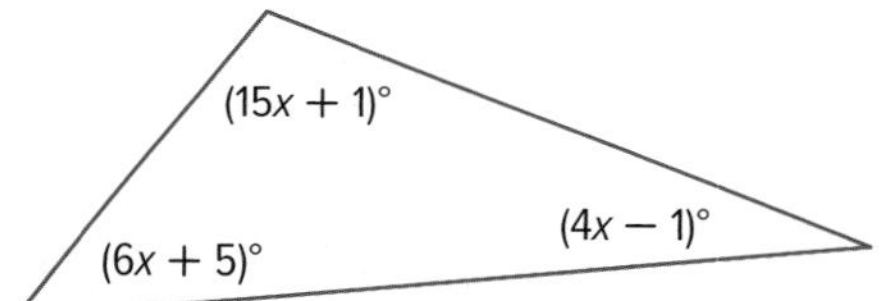

39. **ALGEBRA** The measure of the larger acute angle in a right triangle is two degrees less than three times the measure of the smaller acute angle. Find the measure of each angle.

40. Determine whether the following statement is *true* or *false*. If false, give a counterexample. If true, give an argument to support your conclusion.

If the sum of two acute angles of a triangle is greater than 90, then the triangle is acute.

41. **ALGEBRA** In $\triangle XYZ$, $m\angle X = 157$, $m\angle Y = y$, and $m\angle Z = z$. Write an inequality to describe the possible measures of $\angle Z$. Explain your reasoning.

42. **CARS** Refer to the photo at the right.
    a. Find $m\angle 1$ and $m\angle 2$.
    b. If the support for the hood were shorter than the one shown, how would $m\angle 1$ change? Explain.
    c. If the support for the hood were shorter than the one shown, how would $m\angle 2$ change? Explain.

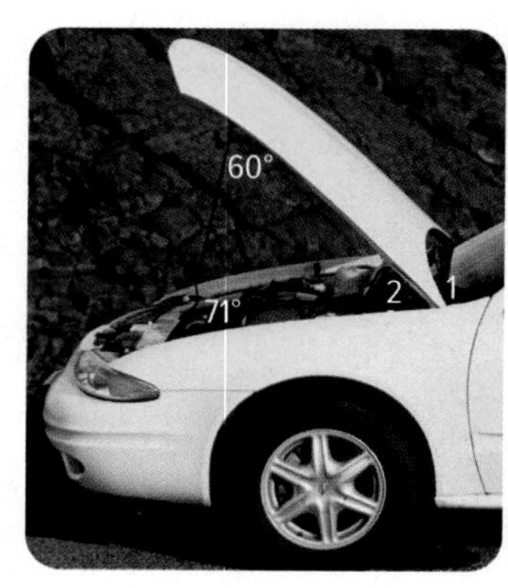

PROOF **Write the specified type of proof.**

43. two-column proof

Given: *RSTUV* is a pentagon.

Prove: $m\angle S + m\angle STU + m\angle TUV + m\angle V + m\angle VRS = 540$

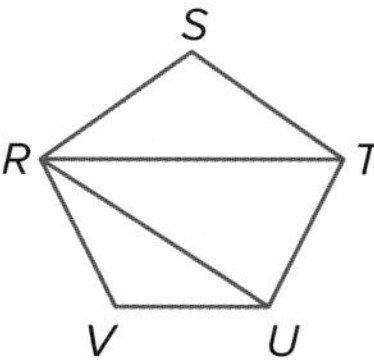

44. flow proof

Given: $\angle 3 \cong \angle 5$

Prove: $m\angle 1 + m\angle 2 = m\angle 6 + m\angle 7$

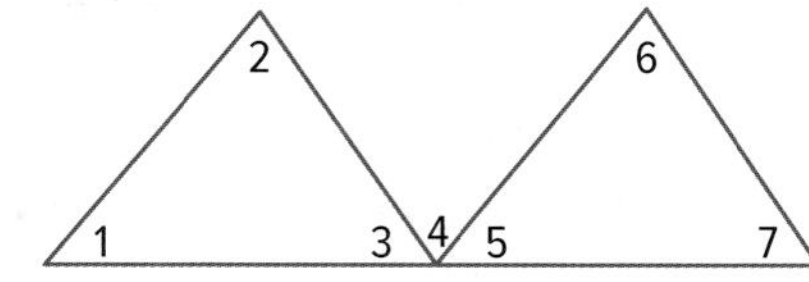

45. **MULTIPLE REPRESENTATIONS** In this problem, you will explore the sum of the measures of the exterior angles of a triangle.

   a. **Geometric** Draw five different triangles, extending the sides and labeling the angles as shown. Be sure to include at least one obtuse, one right, and one acute triangle.

   b. **Tabular** Measure the exterior angles of each triangle. Record the measures for each triangle and the sum of these measures in a table.

   c. **Verbal** Make a conjecture about the sum of the exterior angles of a triangle. State your conjecture using words.

   d. **Algebraic** State the conjecture you wrote in part **c** algebraically.

   e. **Analytical** Write a paragraph proof of your conjecture.

H.O.T. Problems Use Higher-Order Thinking Skills

46. **CRITIQUE** Curtis measured and labeled the angles of the triangle as shown. Arnoldo says that at least one of his measures is incorrect. Explain in at least two different ways how Arnoldo knows that this is true.

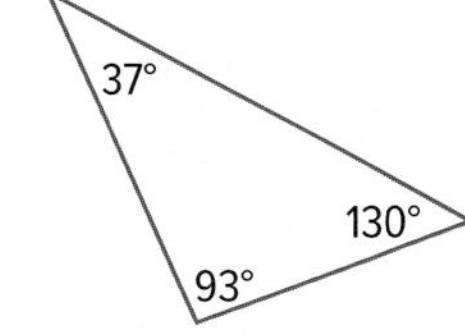

47. **WRITING IN MATH** Explain how you could use the image to the right to prove that the sum of the measures of the interior angles of a triangle is 180.

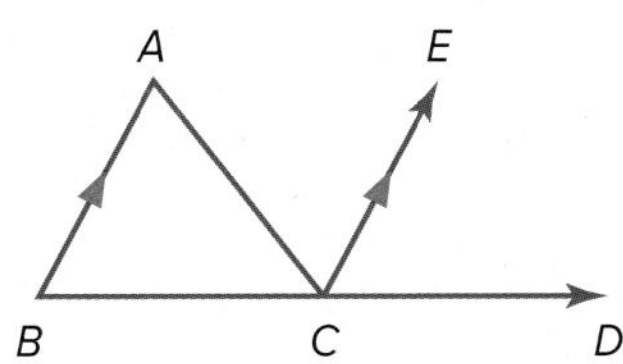

48. **OPEN-ENDED** Construct a right triangle and measure one of the acute angles. Find the measure of the second acute angle using calculation and explain your method. Confirm your result using a protractor.

49. **CHALLENGE** Find the values of y and z in the figure at the right.

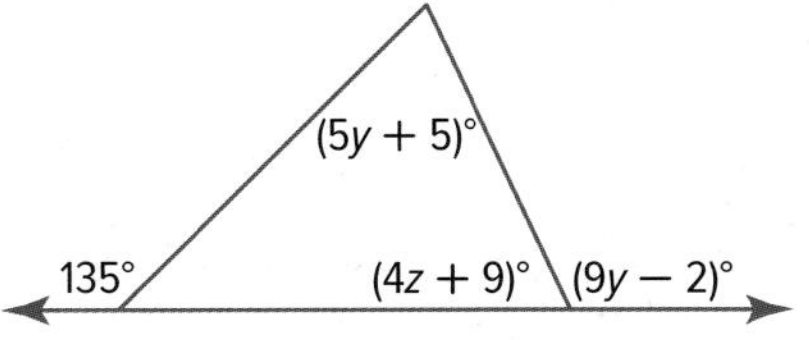

50. **REASONING** If an exterior angle adjacent to $\angle A$ is acute, is $\triangle ABC$ acute, right, obtuse, or can its classification not be determined? Explain your reasoning.

51. **WRITING IN MATH** Explain why a triangle cannot have an obtuse, acute, and a right exterior angle.

52. Which is the value of x? MP 1, 8

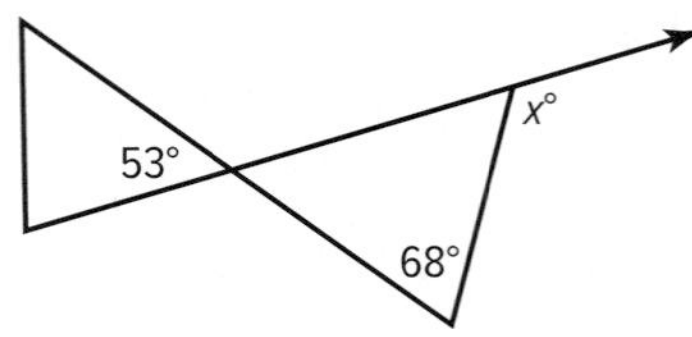

- A 53
- B 59
- C 68
- D 121

53. Find the measure of each numbered angle. MP 1, 8

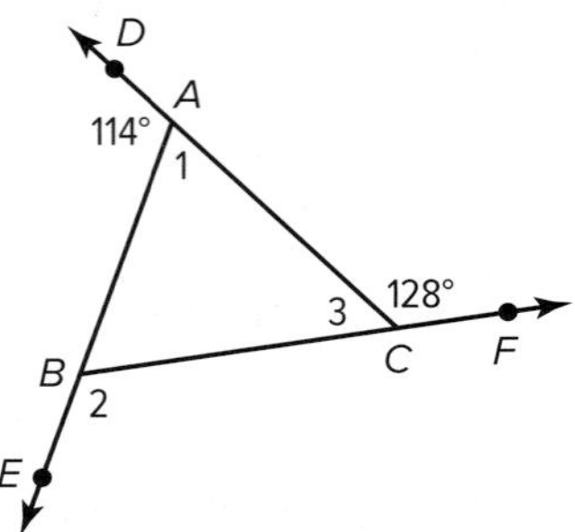

a. $m\angle 1 =$ ☐°

b. $m\angle 2 =$ ☐°

c. $m\angle 3 =$ ☐°

54. Which of the following statements are true? Select all that apply. MP 2

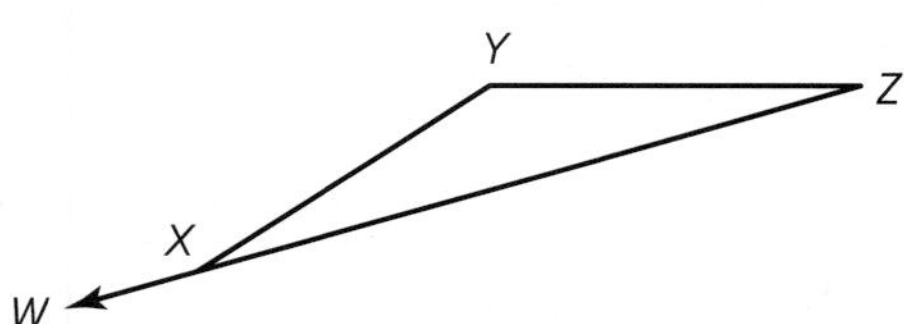

- ☐ A $m\angle Y + m\angle Z + m\angle YXZ = 180^\circ$
- ☐ B $m\angle YXW = 180^\circ - m\angle YXZ$
- ☐ C $m\angle YXZ = m\angle Y + m\angle Z$
- ☐ D $m\angle YXW = m\angle Y + m\angle Z$
- ☐ E $m\angle Z = m\angle YXZ$

55. In the figure, $m\angle B = 75$. The measure of $\angle A$ is half the measure of $\angle C$. What is the measure of $\angle C$? MP 1, 2

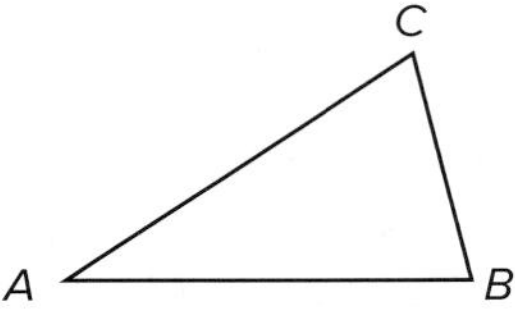

°

56. In a right triangle, the measure of one acute angle is 3.5 times the measure of the other acute angle. What is the measure of the larger acute angle? MP 1, 8

°

57. Complete the proof. MP 3

Given: $\triangle DEF$ is equiangular.

Prove: $m\angle D = m\angle E = m\angle F = 60^\circ$

Statements	Reasons
1. $\triangle DEF$ is equiangular.	1. ☐
2. $m\angle D = m\angle E = m\angle F$	2. Definition of equiangular
3. ☐	3. Triangle Angle Sum Theorem
4. $m\angle D + m\angle D + m\angle D = 180^\circ$ $m\angle E + m\angle E + m\angle E = 180^\circ$ $m\angle F + m\angle F + m\angle F = 180^\circ$	4. Substitution Property
5. $3 \cdot m\angle D = 180^\circ$ $3 \cdot m\angle E = 180^\circ$ $3 \cdot m\angle F = 180^\circ$	5. Simplify
6. $m\angle D = 60^\circ$ $m\angle E = 60^\circ$ $m\angle F = 60^\circ$	6. ☐

58. **MULTI-STEP** The measures of the angles of a triangle are in the ratio 2 : 5 : 8. MP 1, 6

a. Find the measures of the angles.

b. Classify the triangle as *acute, right,* or *obtuse*.

LESSON 2
Congruent Triangles

Then	Now	Why?
• You identified and used congruent angles.	• **1** Name and use corresponding parts of congruent polygons. **2** Prove triangles congruent using the definition of congruence.	• As an antitheft device, many manufacturers make car stereos with removable faceplates. The shape and size of the faceplate and of the space where it fits must be exactly the same for the faceplate to properly attach to the car's dashboard.

New Vocabulary

principle of superposition
congruent polygons
corresponding parts

Mathematical Practices

3 Construct viable arguments and critique the reasoning of others.

1 Congruence and Corresponding Parts Recall that if a geometric figure can be mapped exactly to another geometric figure, they are congruent. This is called the **principle of superposition**. In two **congruent polygons**, all of the parts of one polygon are congruent to the **corresponding parts** or matching parts of the other polygon. These corresponding parts include *corresponding angles* and *corresponding sides*. Polygon congruence can also be defined in terms of rigid motions such as reflections, translations, and rotations.

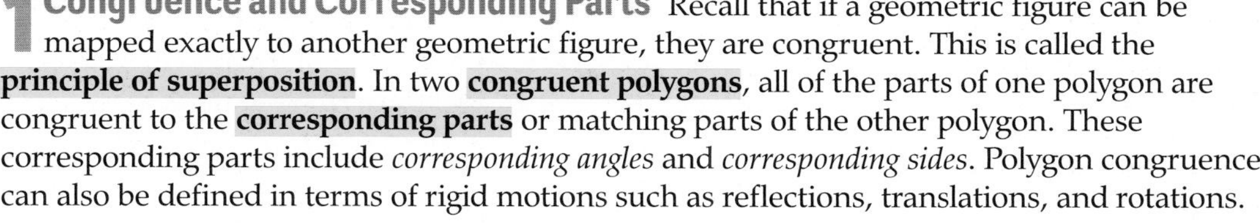

Key Concept Definition of Congruent Polygons

Words Two polygons are congruent if and only if their corresponding parts are congruent.

Two polygons are congruent if and only if a rigid motion or a series of rigid motions maps one polygon exactly onto the other.

Example Corresponding Angles

$\angle A \cong \angle H$ $\quad$ $\angle B \cong \angle J$ $\quad$ $\angle C \cong \angle K$

Corresponding Sides

$\overline{AB} \cong \overline{HJ}$ $\quad$ $\overline{BC} \cong \overline{JK}$ $\quad$ $\overline{AC} \cong \overline{HK}$

Congruence Statement

$\triangle ABC \cong \triangle HJK$

Model

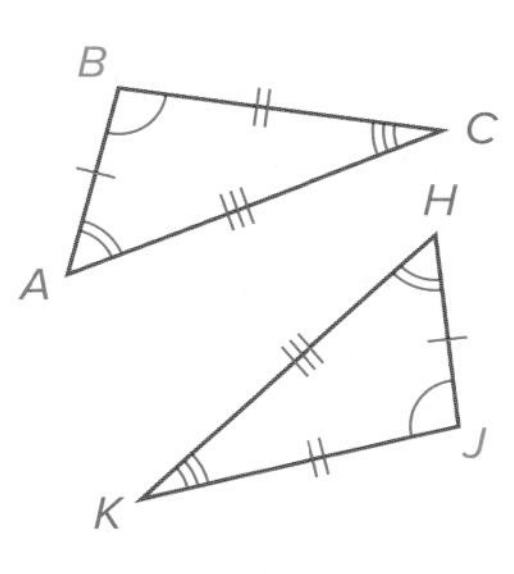

Valid congruence statements list corresponding vertices in the same order.

Example 1 Proving Polygons Congruent

Show that the polygons are congruent by using rigid motions and by identifying all congruent corresponding parts. Then write a congruence statement.

A
B
C
Y
X
Z

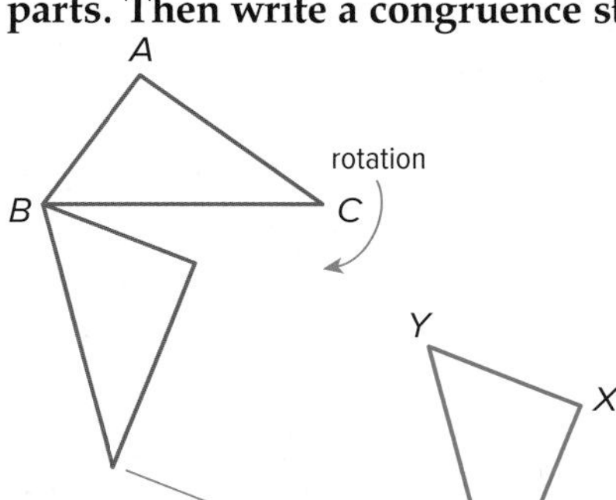

$\triangle ABC$ can be mapped exactly onto $\triangle XYZ$ by a combination of a rotation and a translation.

Therefore, the triangles are congruent.

Math History Link

Johann Carl Friedrich Gauss (1777–1855) Gauss developed the congruence symbol to show that two sides of an equation were the same even if they weren't equal. He made many advances in math and physics, including a proof of the fundamental theorem of algebra.

Source: The Granger Collection, New York

Now use corresponding parts.

Angles: $\angle A \cong \angle X$
$\angle B \cong \angle Y$
$\angle C \cong \angle Z$

Sides: $\overline{AB} \cong \overline{XY}$
$\overline{BC} \cong \overline{YZ}$
$\overline{AC} \cong \overline{XZ}$

All corresponding parts of the two polygons are congruent. Therefore, $\triangle ABC \cong \triangle XYZ$.

Guided Practice

1A.

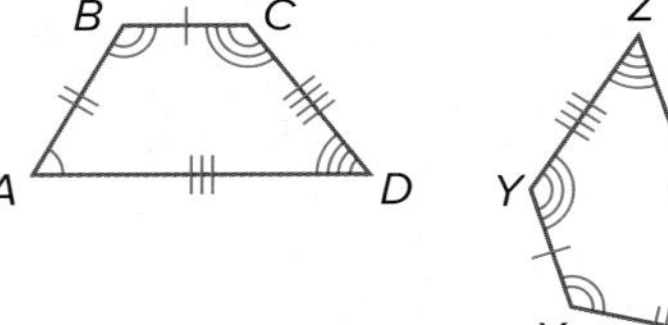

1B.

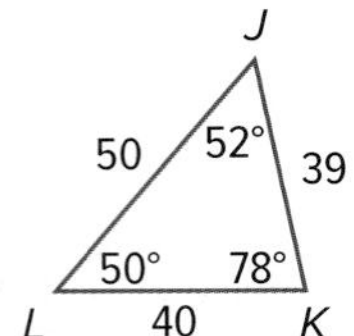

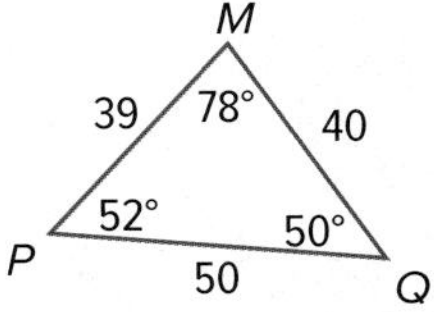

The phrase "if and only if" in the congruent polygon definition means that both the conditional and its converse are true. So, if two polygons are congruent, then their corresponding parts are congruent. For triangles, we say *Corresponding parts of congruent triangles are congruent,* or CPCTC.

Study Tip

MP Structure You can use a congruence statement to help you correctly identify corresponding sides.

$\triangle ABC \cong \triangle DFE$
$\overline{BC} \cong \overline{FE}$

Example 2 Use Corresponding Parts of Congruent Triangles

In the diagram, $\triangle ABC \cong \triangle DFE$. Find the values of x and y.

$\angle F \cong \angle B$	CPCTC
$m\angle F = m\angle B$	Definition of congruence
$8y - 5 = 99$	Substitution
$8y = 104$	Add 5 to each side.
$y = 13$	Divide each side by 8.
$\overline{FE} \cong \overline{BC}$	CPCTC
$FE = BC$	Definition of congruence
$2y + x = 38.4$	Substitution
$2(13) + x = 38.4$	Substitution
$26 + x = 38.4$	Simplify.
$x = 12.4$	Subtract 26 from each side.

Guided Practice

2. In the diagram, $\triangle RSV \cong \triangle TVS$. Find the values of x and y.

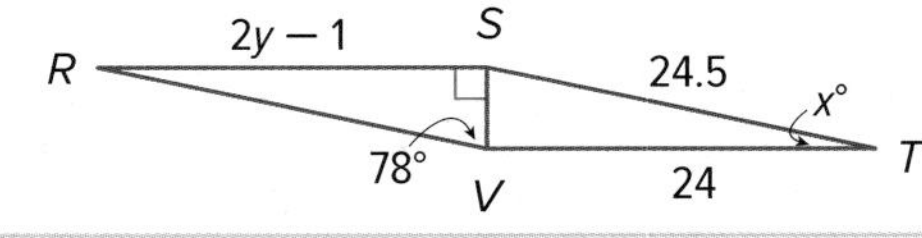

2 Prove Triangles Congruent

The Triangle Angle-Sum Theorem you learned in Lesson 4-1 leads to another theorem about the angles in two triangles.

Theorem 4.3 Third Angles Theorem

Words: If two angles of one triangle are congruent to two angles of a second triangle, then the third angles of the triangles are congruent.

Example: If $\angle C \cong \angle K$ and $\angle B \cong \angle J$, then $\angle A \cong \angle L$.

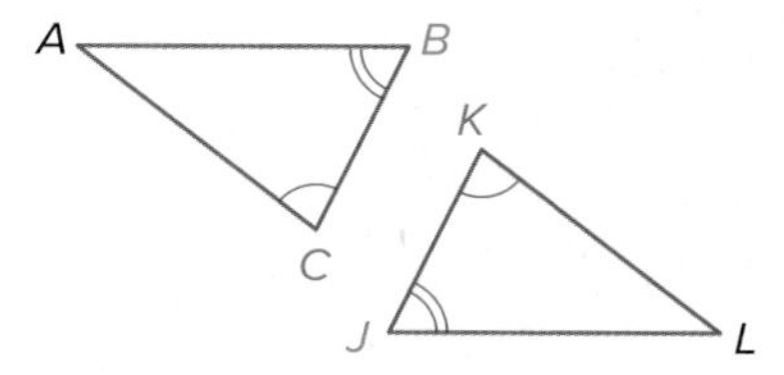

You will prove this theorem in Exercise 21.

Real-World Example 3 Use the Third Angles Theorem

PARTY PLANNING The planners of the Senior Banquet decide to fold the dinner napkins using the Triangle Pocket Fold so that they can place a small gift in the pocket. If $\angle NPQ \cong \angle RST$, and $m\angle NPQ = 40$, find $m\angle SRT$.

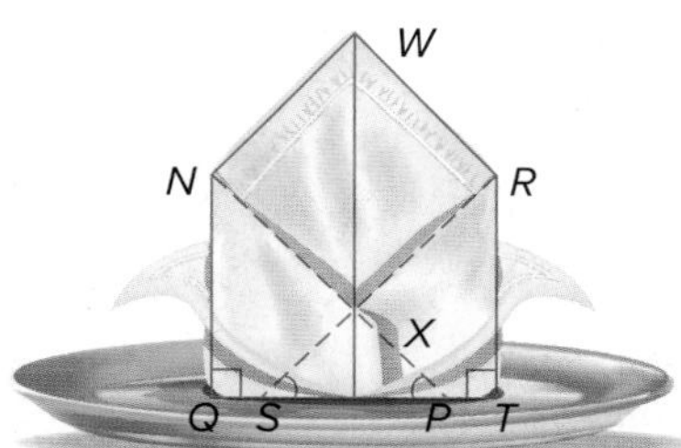

$\angle NPQ \cong \angle RST$, and because all right angles are congruent, $\angle NQP \cong \angle RTS$. So by the Third Angles Theorem, $\angle QNP \cong \angle SRT$. By the definition of congruence, $m\angle QNP = m\angle TRS$.

$m\angle QNP + m\angle NPQ = 90$	The acute angles of a right triangle are complementary.
$m\angle QNP + 40 = 90$	Substitution
$m\angle QNP = 50$	Subtract 40 from each side.

By substitution, $m\angle SRT = m\angle QNP$ or 50.

Guided Practice

3. In the diagram above, if $\angle WNX \cong \angle WRX$, $\overline{WX}$ bisects $\angle NXR$, $m\angle WNX = 88$, and $m\angle NXW = 49$, find $m\angle NWR$. Explain your reasoning.

Real-World Link

Using some basic skills with napkin folding can add an elegant touch to any party. Many of the folds use triangles.

Example 4 Prove that Two Triangles Are Congruent

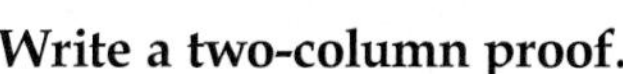

Write a two-column proof.

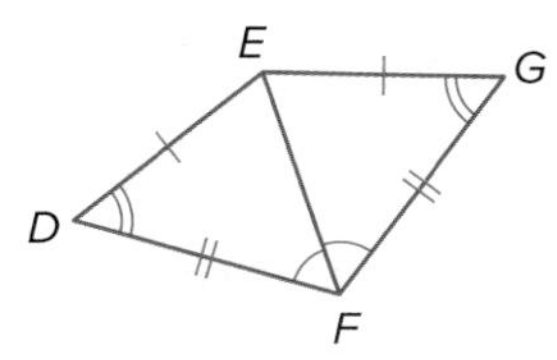

Given: $\overline{DE} \cong \overline{GE}$, $\overline{DF} \cong \overline{GF}$, $\angle D \cong \angle G$, $\angle DFE \cong \angle GFE$

Prove: $\triangle DEF \cong \triangle GEF$

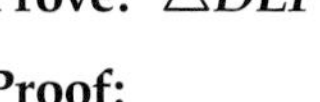

Proof:

Statements	Reasons
1. $\overline{DE} \cong \overline{GE}$, $\overline{DF} \cong \overline{GF}$	**1.** Given
2. $\overline{EF} \cong \overline{EF}$	**2.** Reflexive Property of Congruence
3. $\angle D \cong \angle G$, $\angle DFE \cong \angle GFE$	**3.** Given
4. $\angle DEF \cong \angle GEF$	**4.** Third Angles Theorem
5. $\triangle DEF \cong \triangle GEF$	**5.** Definition of congruent polygons

Study Tip

MP Construct Arguments When two triangles share a common side, use the Reflexive Property of Congruence to establish that the common side is congruent to itself.

Go Online!

Investigate when two triangles are congruent by using the **Geometry Tools** in ConnectED.

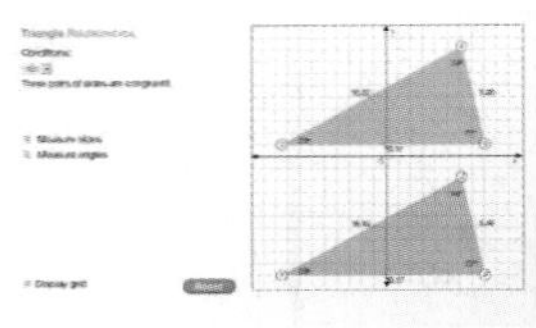

Guided Practice

4. Write a two-column proof.

Given: $\angle J \cong \angle P$, $\overline{JK} \cong \overline{PM}$, $\overline{JL} \cong \overline{PL}$, and L bisects $\overline{KM}$.

Prove: $\triangle JLK \cong \triangle PLM$

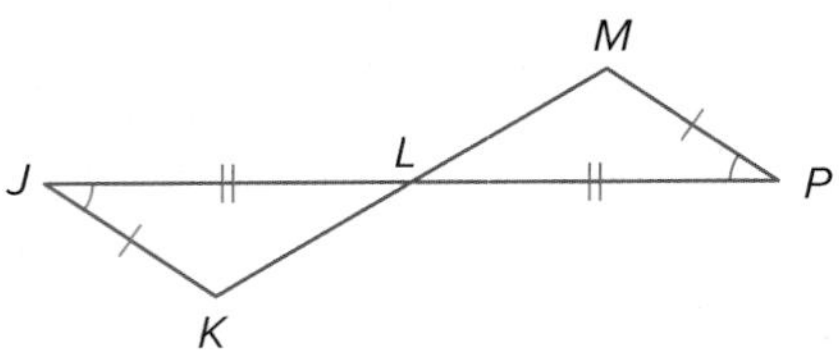

Like congruence of segments and angles, congruence of triangles is reflexive, symmetric, and transitive.

Theorem 4.4 Properties of Triangle Congruence

Reflexive Property of Triangle Congruence

$\triangle ABC \cong \triangle ABC$

Symmetric Property of Triangle Congruence

If $\triangle ABC \cong \triangle EFG$, then $\triangle EFG \cong \triangle ABC$.

Transitive Property of Triangle Congruence

If $\triangle ABC \cong \triangle EFG$ and $\triangle EFG \cong \triangle JKL$, then $\triangle ABC \cong \triangle JKL$.

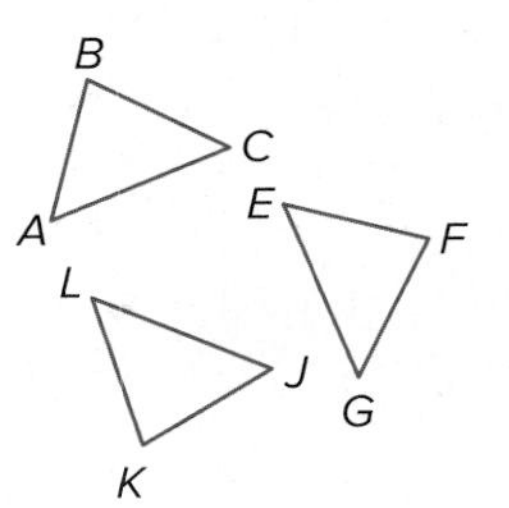

You will prove the reflexive, symmetric, and transitive parts of Theorem 4.4 in Exercises 27, 22, and 26, respectively.

Check Your Understanding

 = Step-by-Step Solutions begin on page R13.

Example 1 **Show that the polygons are congruent by using rigid motions and by identifying all congruent corresponding parts. Then write a congruence statement.**

1.

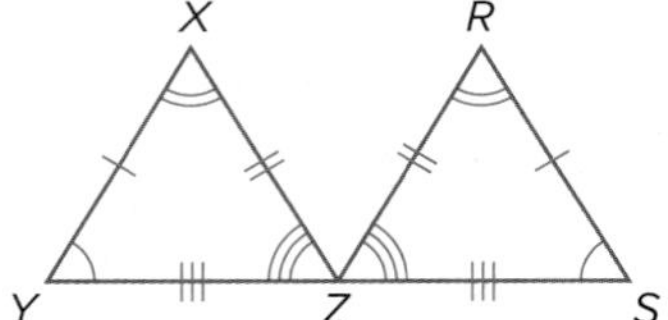

2.

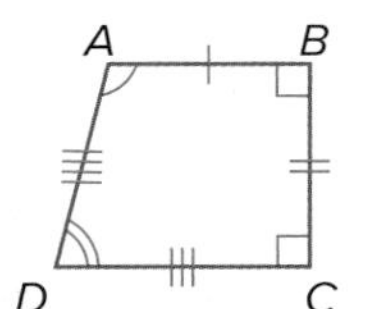

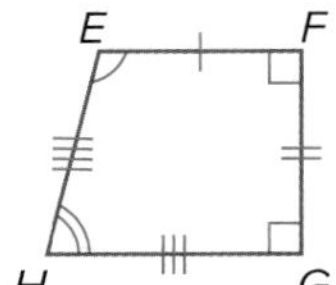

3. TOOLS Sareeta is changing the tire on her bike and the nut securing the tire looks like the one shown. Which of the sockets below should she use with her wrench to remove the tire? Explain your reasoning.

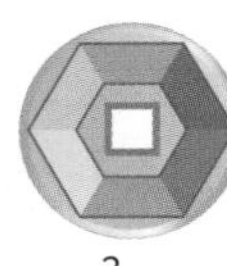

$\frac{3}{8}$ in.

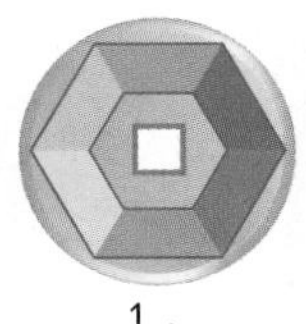

$\frac{1}{2}$ in.

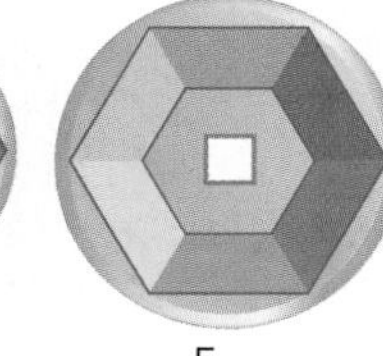

$\frac{5}{8}$ in.

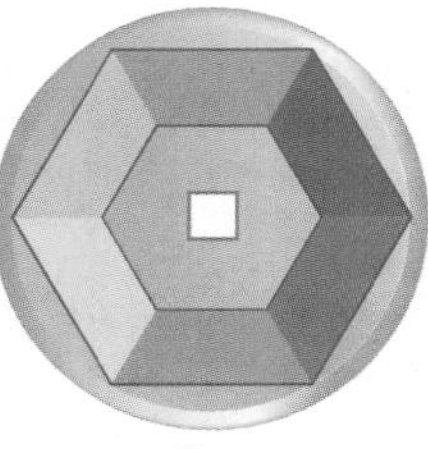

$\frac{3}{4}$ in.

Example 2 **In the figure, $\triangle LMN \cong \triangle QRS$.**

4. Find x.

5 Find y.

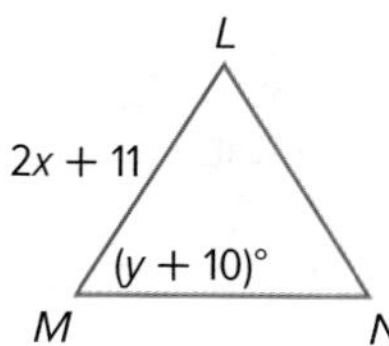

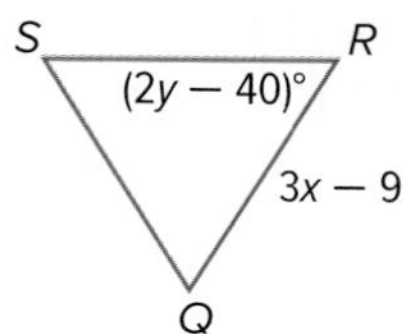

Example 3 **MP REGULARITY** **Find x. Explain your reasoning.**

6.

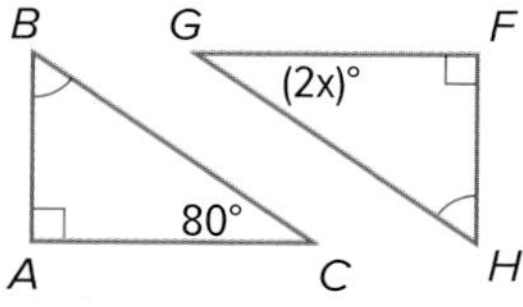

7.

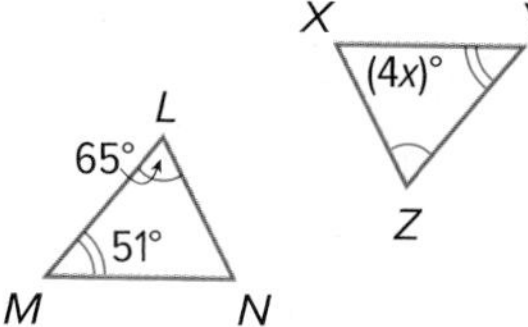

Example 4 **8. PROOF** Write a paragraph proof.

Given: $\angle WXZ \cong \angle YXZ$, $\angle XZW \cong \angle XZY$, $\overline{WX} \cong \overline{YX}$, $\overline{WZ} \cong \overline{YZ}$

Prove: $\triangle WXZ \cong \triangle YXZ$

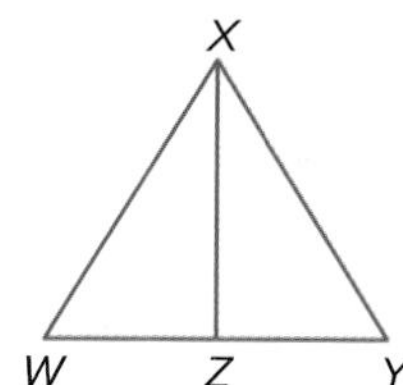

Practice and Problem Solving

Extra Practice is on page R4.

Example 1 **Show that the polygons are congruent by using rigid motions and by identifying all congruent corresponding parts. Then write a congruence statement.**

9.

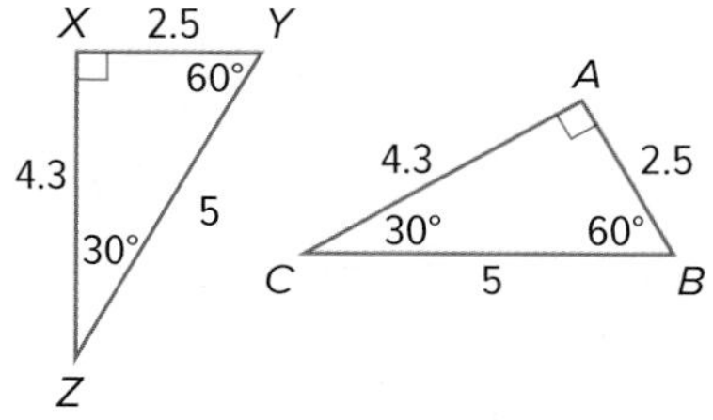

10.

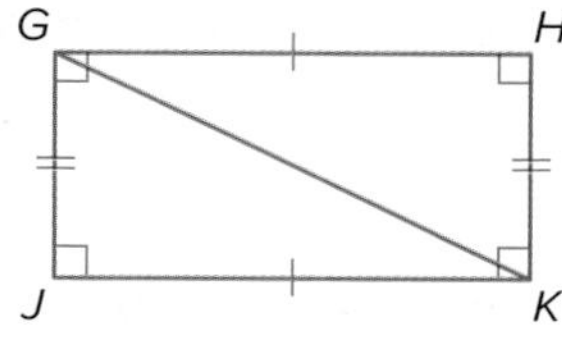

11.

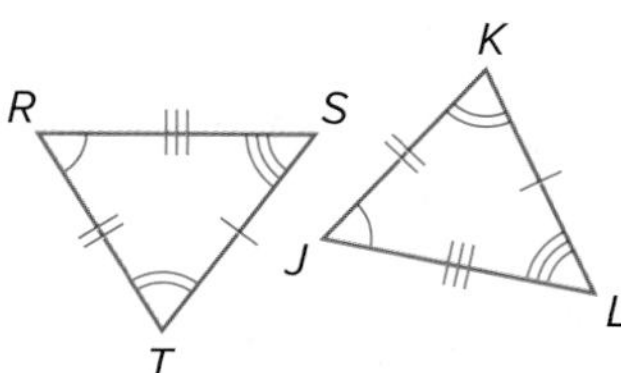

12.

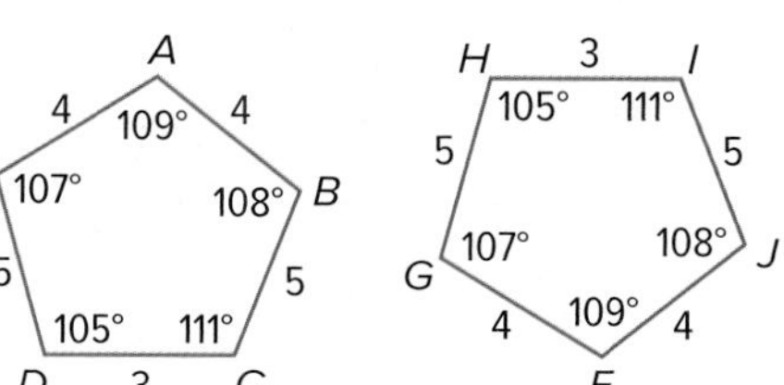

Example 2 **Polygon $BCDE \cong$ polygon $RSTU$. Find each value.**

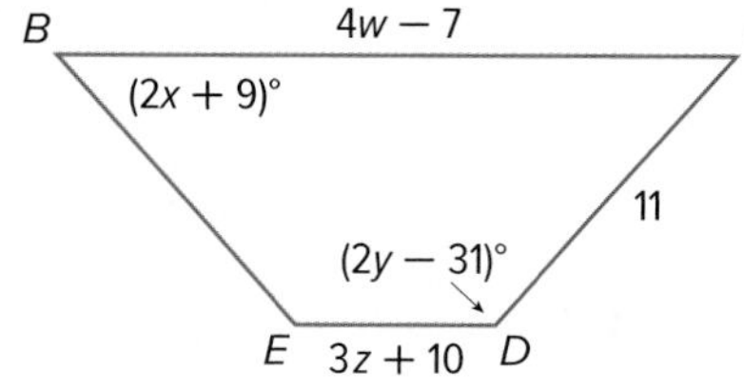

13. x

14. y

15 z

16. w

17. **SAILING** To ensure that sailboat races are fair, the boats and their sails are required to be the same size and shape.

a. Write a congruence statement relating the triangles in the photo.

b. Name three pairs of congruent segments.

c. Name three pairs of congruent angles.

Example 3 **Find x and y.**

18. 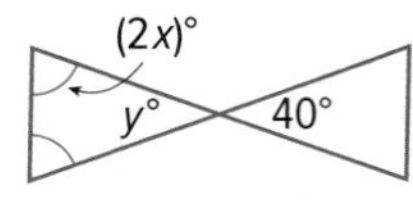

19. 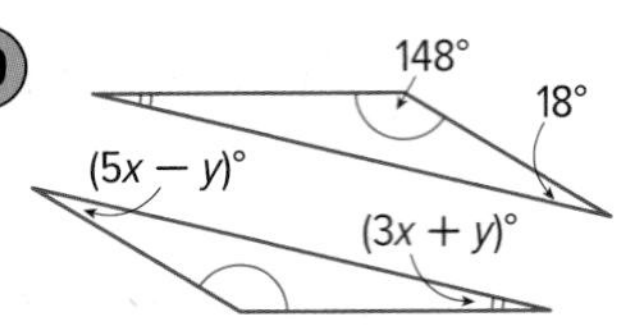

20. 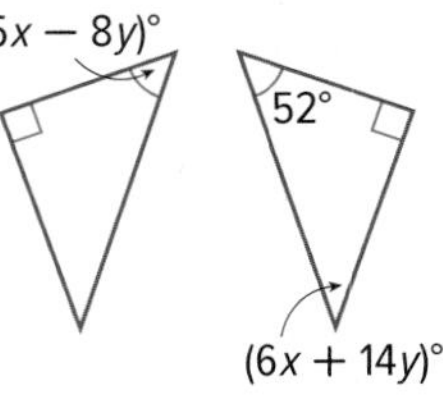

Example 4 21. **PROOF** Write a two-column proof of Theorem 4.3.

22. **PROOF** Put the statements used to prove the theorem below in the correct order. Provide the reasons for each statement.

Congruence of triangles is symmetric. (Theorem 4.4)

Given: $\triangle RST \cong \triangle XYZ$

Prove: $\triangle XYZ \cong \triangle RST$

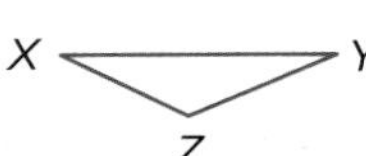

Proof:

$\angle X \cong \angle R$, $\angle Y \cong \angle S$, $\angle Z \cong \angle T$, $\overline{XY} \cong \overline{RS}$, $\overline{YZ} \cong \overline{ST}$, $\overline{XZ} \cong \overline{RT}$	$\angle R \cong \angle X$, $\angle S \cong \angle Y$, $\angle T \cong \angle Z$, $\overline{RS} \cong \overline{XY}$, $\overline{ST} \cong \overline{YZ}$, $\overline{RT} \cong \overline{XZ}$	$\triangle RST \cong \triangle XYZ$	$\triangle XYZ \cong \triangle RST$
?	?	?	?

CONSTRUCT ARGUMENTS **Write a two-column proof.**

23. **Given:** $\overline{BD}$ bisects $\angle B$.
$\overline{BD} \perp \overline{AC}$

Prove: $\angle A \cong \angle C$

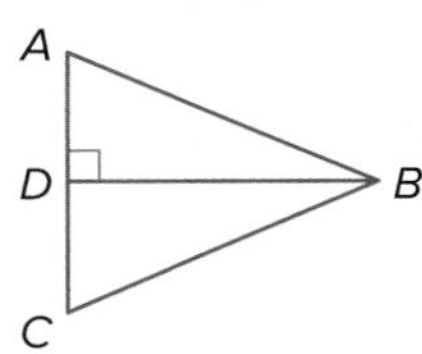

24. **Given:** $\angle P \cong \angle T$, $\angle S \cong \angle Q$
$\overline{TR} \cong \overline{PR}$, $\overline{RP} \cong \overline{RQ}$,
$\overline{RT} \cong \overline{RS}$
$\overline{PQ} \cong \overline{TS}$

Prove: $\triangle PRQ \cong \triangle TRS$

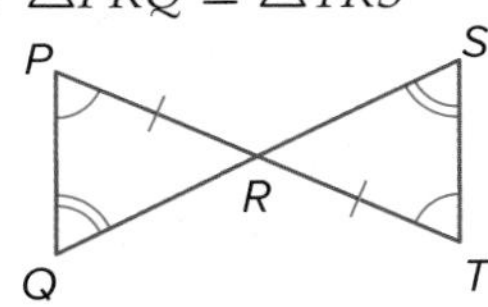

25. **SCRAPBOOKING** Lanie is using a flower-shaped corner decoration punch for a scrapbook she is working on. If she punches the corners of two pages as shown, what property guarantees that the punched designs are congruent? Explain.

PROOF **Write the specified type of proof of the indicated part of Theorem 4.4.**

26. Congruence of triangles is transitive. (paragraph proof)

27. Congruence of triangles is reflexive. (flow proof)

ALGEBRA **Draw and label a figure to represent the congruent triangles. Then find x and y.**

28. $\triangle ABC \cong \triangle DEF$, $AB = 7$, $BC = 9$, $AC = 11 + x$, $DF = 3x + 5$, and $DE = 2y - 5$

29. $\triangle LMN \cong \triangle RST$, $m\angle L = 49$, $m\angle M = 10y$, $m\angle S = 70$, and $m\angle T = 4x + 9$

30. $\triangle JKL \cong \triangle MNP$, $JK = 12$, $LJ = 5$, $PM = 2x - 3$, $m\angle L = 67$, $m\angle K = y + 4$ and $m\angle N = 2y - 15$

31. **PENNANTS** Marren is decorating an area of 100 square feet for the pep rally. She is using a string of pennants that are congruent isosceles triangles.

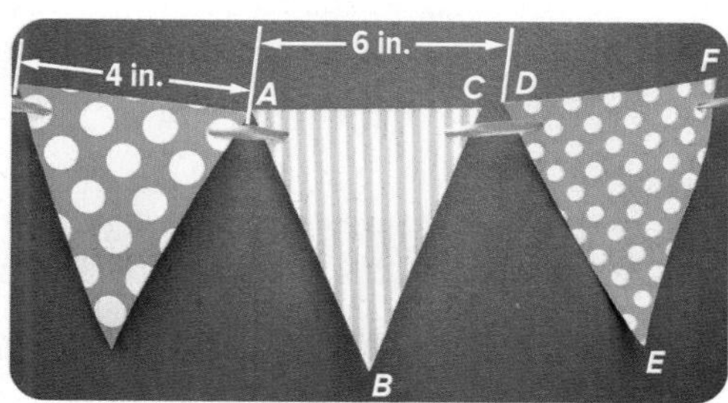

a. List seven pairs of congruent segments in the photo.

b. If the area decorated is a square, how long will the pennant string need to be?

c. How many pennants will be on the string?

32. **MP** **SENSE-MAKING** In the photo of New York City's Chrysler Building at the right, $\overline{TS} \cong \overline{ZY}$, $\overline{XY} \cong \overline{RS}$, $\overline{TR} \cong \overline{ZX}$, $\angle X \cong \angle R$, $\angle T \cong \angle Z$, $\angle Y \cong \angle S$, and $\triangle HGF \cong \triangle LKJ$.

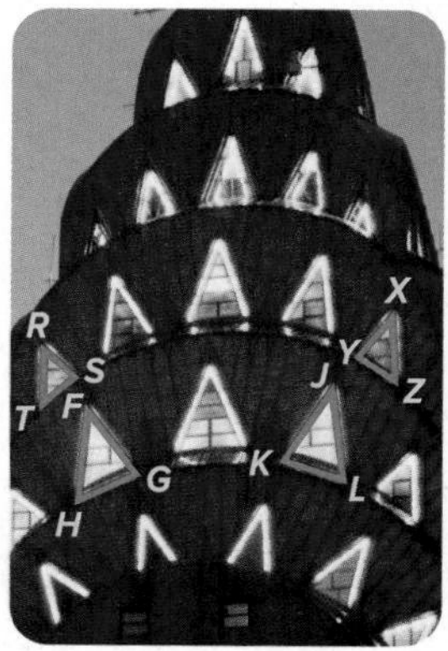

a. Which triangle, if any, is congruent to $\triangle YXZ$? Explain your reasoning.

b. Which side(s) are congruent to $\overline{JL}$? Explain your reasoning.

c. Which angle(s) are congruent to $\angle G$? Explain your reasoning.

33. **MULTIPLE REPRESENTATIONS** In this problem, you will explore criteria for triangle congruence. Select a tool such as compass and straightedge or dynamic geometry software.

a. **Geometric** Draw two triangles that meet each of the following criteria.
- the sides of the first triangle are congruent to the corresponding sides of the second triangle
- two of the sides and the included angle are congruent
- two angles and the included side are congruent
- two angles and a nonincluded side are congruent
- two sides and a nonincluded angle are congruent
- the angles of one are congruent to the corresponding angles of the second triangle

b. **Verbal** Analyze the triangles drawn in part **a** by drawing additional triangles or transforming those you drew. Which criteria resulted in congruent triangles? Which criteria did not result in congruent triangles?

c. **Verbal** Make a conjecture about the criteria to determine whether two triangles are congruent.

34. **PATTERNS** The pattern shown is created using regular polygons.

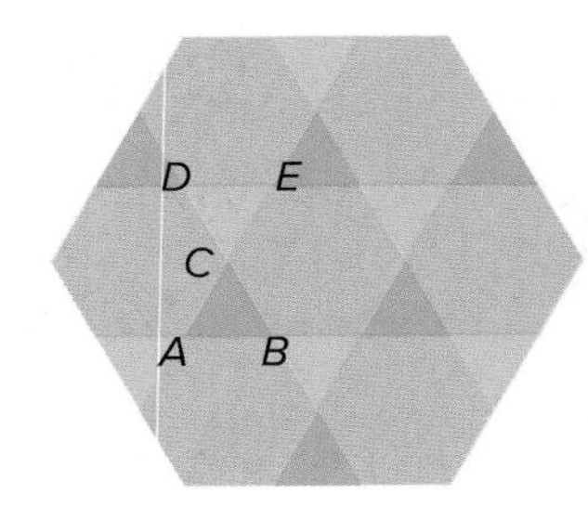

a. What two polygons are used to create the pattern?
b. Name a pair of congruent triangles.
c. Name a pair of corresponding angles.
d. If $CB = 2$ inches, what is AE? Explain.
e. What is the measure of $\angle D$? Explain.

35. **FITNESS** A fitness instructor is starting a new aerobics class using fitness hoops. She wants to confirm that all of the hoops are the same size. What measure(s) can she use to prove that all of the hoops are congruent? Explain your reasoning.

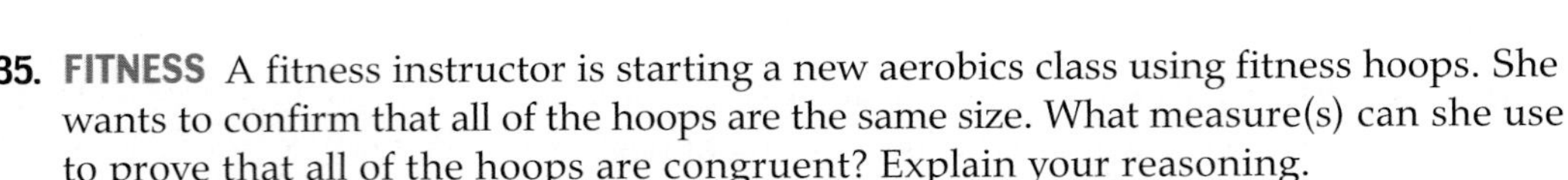

H.O.T. Problems Use Higher-Order Thinking Skills

36. **WRITING IN MATH** Explain why the order of the vertices is important when naming congruent triangles. Give an example to support your answer.

37. **ERROR ANALYSIS** Jasmine and Will are evaluating the congruent figures below. Jasmine says that $\triangle CAB \cong \triangle ZYX$ and Will says that $\triangle ABC \cong \triangle YXZ$. Is either of them correct? Explain.

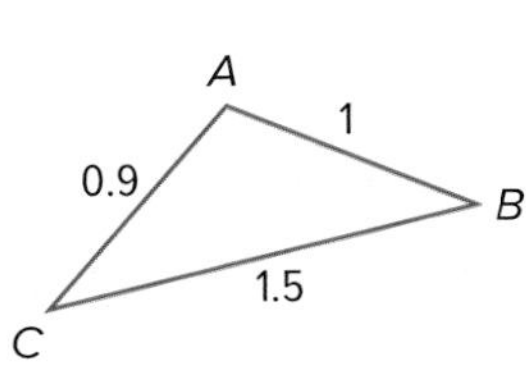

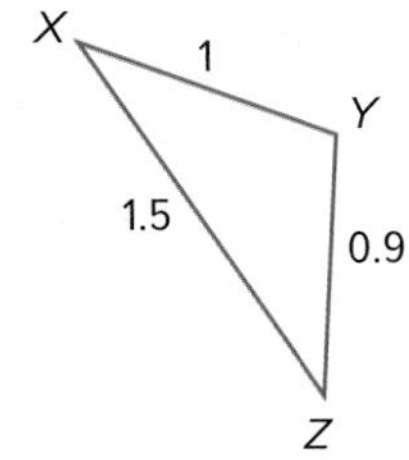

38. **WRITE A QUESTION** A classmate is using the Third Angles Theorem to show that if two corresponding pairs of the angles of two triangles are congruent, then the third pair is also congruent. Write a question to help him decide if he can use the same strategy for quadrilaterals.

39. MP **PERSEVERANCE** Find x and y if $\triangle PQS \cong \triangle RQS$.

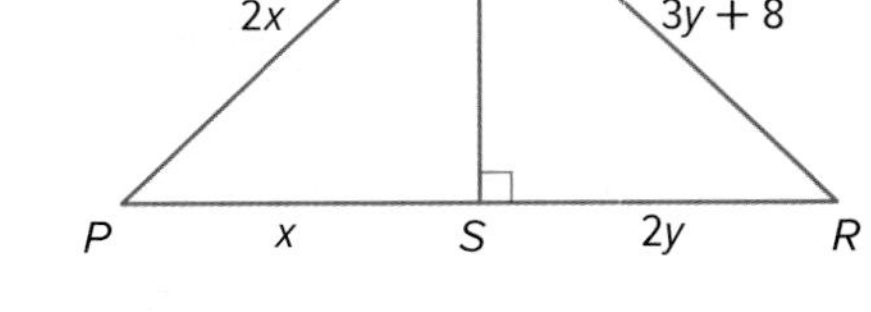

MP **CRITIQUE ARGUMENTS** Determine whether each statement is *true* or *false*. If false, give a counterexample. If true, explain your reasoning.

40. Two triangles with two pairs of congruent corresponding angles and three pairs of congruent corresponding sides are congruent.

41. Two triangles with three pairs of corresponding congruent angles are congruent.

42. MP **SENSE-MAKING** Write a paragraph proof to prove polygon $ABED \cong$ polygon $FEBC$.

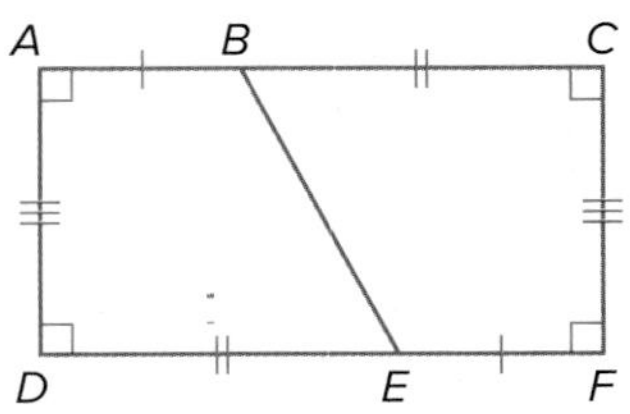

43. **WRITING IN MATH** Determine whether the following statement is *always, sometimes,* or *never* true. Explain your reasoning.

Equilateral triangles are congruent.

44. A cement path is placed as shown in a square region of a park. If the triangular grassy areas along both sides of the path are congruent, what is the perimeter of the path? MP 1

- A 40 ft
- B 48 ft
- C 60 ft
- D 105 ft

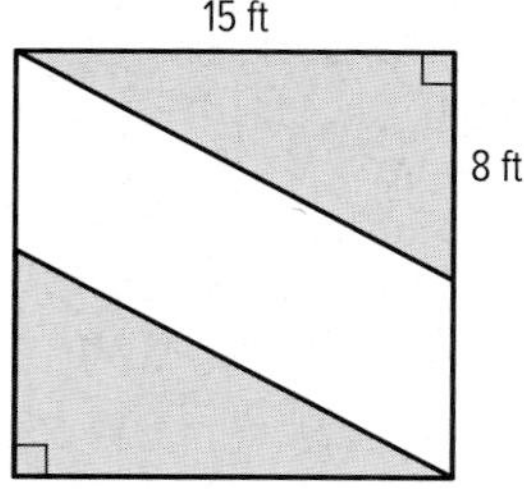

45. $\triangle QRS \cong \triangle TUQ$

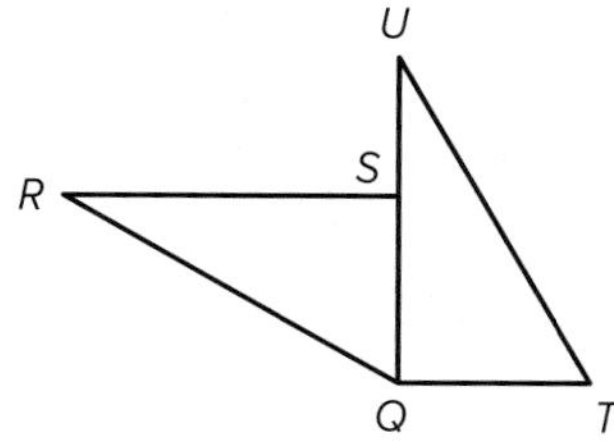

Which statement is not necessarily true? MP 1

- A $\overline{RS} \cong \overline{UQ}$
- B $\overline{SQ} \cong \overline{QT}$
- C $\angle T \cong \angle R$
- D $\angle RSQ \cong \angle UQT$
- E $\angle RQS \cong \angle UTQ$

46. The opposite sides of quadrilateral $WXYZ$ are congruent. Prove that $\triangle WXZ \cong \triangle YZX$. MP 3

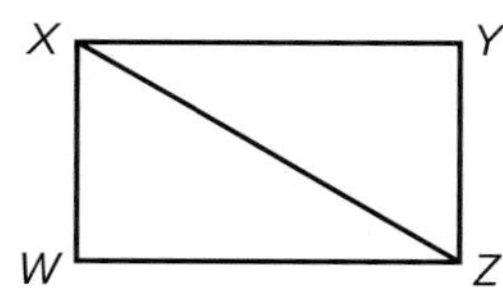

47. Use the following information to find the values of x and y. $\triangle TUV \cong \triangle HJK$, $TU = 14$, $UV = 18$, $TV = 4y + 1$, $JK = 2x - 4$, and $HK = 6y - 5$. MP 1, 3

48. A regular hexagon is divided into six congruent triangles.

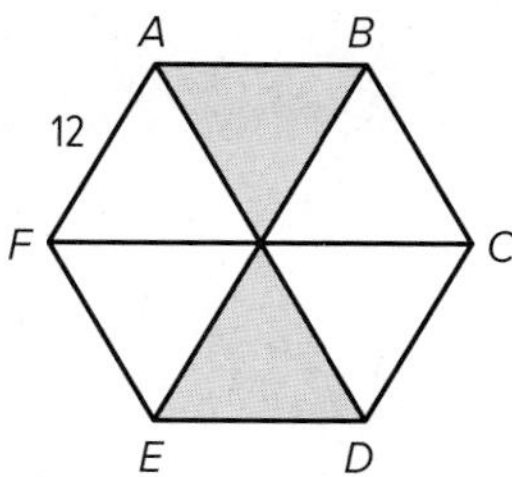

Which of the following best approximates the area of the shaded region? MP 1

- A 62.35 square units
- B 124.71 square units
- C 249.41 square units
- D 374.12 square units

49. **MULTI-STEP** The graph shows $\triangle FGH$ and $\triangle JKL$. MP 3

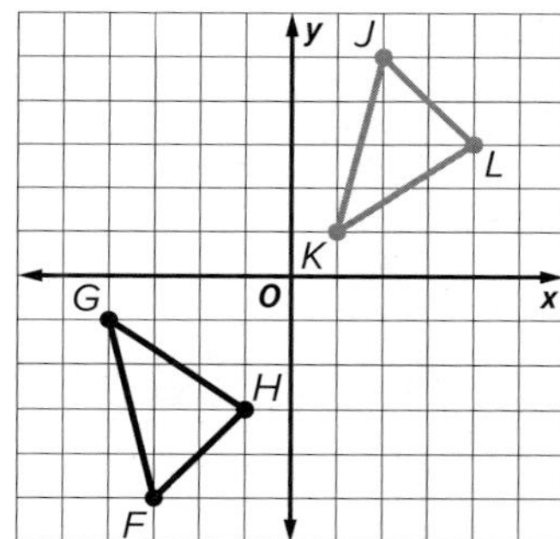

a. Describe a set of rigid motions that could be performed on $\triangle FGH$ to prove that it is congruent to $\triangle JKL$.

b. Perform the set of rigid motions, and list the coordinates of the vertices of the image of $\triangle FGH$ at each stage.

c. Given that $m\angle F = 59°$ and $m\angle G = 42°$, what is the measure of $\angle L$? Explain your reasoning.

50. $\triangle ABC \cong \triangle DEF$. $AB = 4x - 2$, $BC = 10$, $DE = 18$, $EF = 3y + 1$, and $DF = 2x + 4$. MP 3

a. Find the values of x and y.

b. Find DF.

LESSON 3

Proving Triangles Congruent–SSS, SAS

Then	Now	Why?
• You proved triangles congruent using the definition of congruence.	• **1** Use the SSS Postulate to test for triangle congruence. **2** Use the SAS Postulate to test for triangle congruence.	• An A-frame sandwich board is a convenient way to display information. Not only does it fold flat for easy storage, but with each sidearm locked into place, the frame is extremely sturdy. With the sidearms the same length and positioned the same distance from the top on either side, the open frame forms two congruent triangles.

New Vocabulary
included angle

Mathematical Practices

1 Make sense of problems and persevere in solving them.

3 Construct viable arguments and critique the reasoning of others.

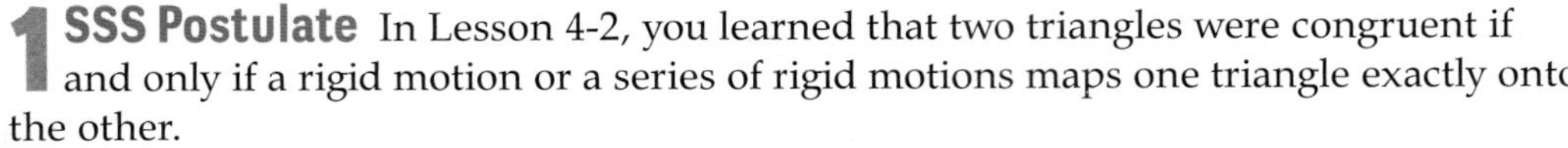

1 SSS Postulate In Lesson 4-2, you learned that two triangles were congruent if and only if a rigid motion or a series of rigid motions maps one triangle exactly onto the other.

The sandwich board demonstrates that if two triangles have the same three side lengths, then they are congruent. This is expressed in the postulate below.

Postulate 4.1 Side-Side-Side (SSS) Congruence

If three sides of one triangle are congruent to three sides of a second triangle, then the triangles are congruent.

Example If **S**ide $\overline{AB} \cong \overline{DE}$,
Side $\overline{BC} \cong \overline{EF}$, and
Side $\overline{AC} \cong \overline{DF}$,
then $\triangle ABC \cong \triangle DEF$.

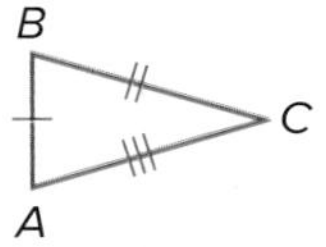

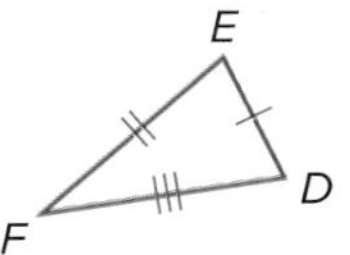

Example 1 Use SSS to Prove Triangles Congruent

Write a flow proof.

Given: $\overline{GH} \cong \overline{KJ}$, $\overline{HL} \cong \overline{JL}$, and L is the midpoint of $\overline{GK}$.

Prove: $\triangle GHL \cong \triangle KJL$

Flow Proof:

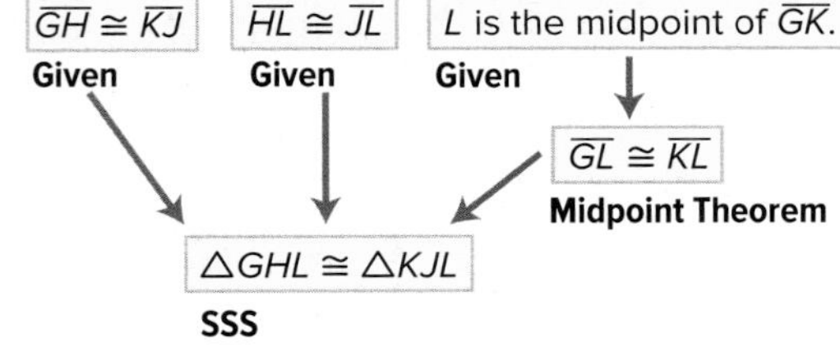

Guided Practice

1. Write a flow proof.

Given: $\triangle QRS$ is isosceles with $\overline{QR} \cong \overline{SR}$.
$\overline{RT}$ bisects $\overline{QS}$ at point T.

Prove: $\triangle QRT \cong \triangle SRT$

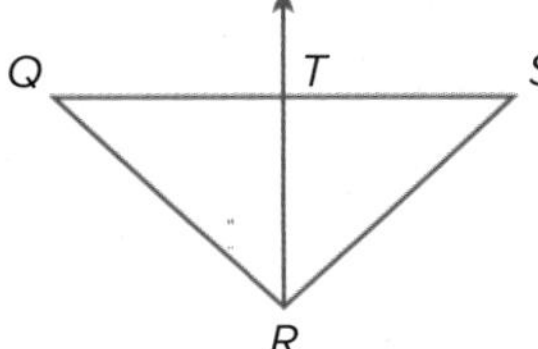

Example 2 SSS on the Coordinate Plane

Triangle *ABC* has vertices *A*(1, 1), *B*(0, 3), and *C*(2, 5). Triangle *EFG* has vertices *E*(1, −1), *F*(2, −5), and *G*(4, −4).

a. Graph both triangles on the same coordinate plane.

b. Use your graph to make a conjecture as to whether the triangles are congruent. Explain your reasoning using rigid motions.

c. Write a logical argument using coordinate geometry to support the conjecture you made in part b.

Read the Item

You are asked to do three things in this problem. In part **a**, you are to graph $\triangle ABC$ and $\triangle EFG$ on the same coordinate plane. In part **b**, you should make a conjecture that $\triangle ABC \cong \triangle EFG$ or $\triangle ABC \ncong \triangle EFG$ based on your graph. Finally, in part **c**, you are asked to prove your conjecture.

Study Tip

MP Tools When you are solving problems using the coordinate plane, remember to use tools like the Distance, Midpoint, and Slope Formulas to solve problems and to check your solutions.

Solve the Item

a.

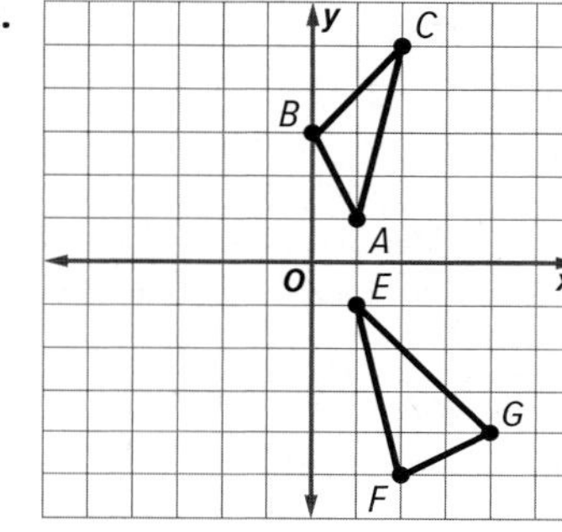

b. From the graph, there does not appear to be a combination of rigid motions that will map one triangle onto the other, so we can conjecture that they are not congruent.

c. Use the Distance Formula to show that not all corresponding sides have the same measure.

$AB = \sqrt{(0-1)^2 + (3-1)^2}$

$= \sqrt{1+4}$ or $\sqrt{5}$

$BC = \sqrt{(2-0)^2 + (5-3)^2}$

$= \sqrt{4+4}$ or $\sqrt{8}$

$AC = \sqrt{(2-1)^2 + (5-1)^2}$

$= \sqrt{1+16}$ or $\sqrt{17}$

$EF = \sqrt{(2-1)^2 + [-5-(-1)]^2}$

$= \sqrt{1+16}$ or $\sqrt{17}$

$FG = \sqrt{(4-2)^2 + [-4-(-5)]^2}$

$= \sqrt{4+1}$ or $\sqrt{5}$

$EG = \sqrt{(4-1)^2 + [-4-(-1)]^2}$

$= \sqrt{9+9}$ or $\sqrt{18}$

While $AB = FG$ and $AC = EF$, $BC \neq EG$. Since SSS congruence is not met, $\triangle ABC \ncong \triangle EFG$.

Go Online! 

$\triangle ABC \ncong \triangle EFG$ is read as *triangle ABC is not congruent to triangle EFG.* To hear pronunciations of expressions, log into your **eStudent Edition**. Ask a partner or your teacher for clarification as you need it.

Guided Practice

2. Triangle *JKL* has vertices *J*(2, 5), *K*(1, 1), and *L*(5, 2). Triangle *NPQ* has vertices *N*(−3, 0), *P*(−7, 1), and *Q*(−4, 4).

a. Graph both triangles on the same coordinate plane.

b. Use your graph to make a conjecture as to whether the triangles are congruent. Explain your reasoning.

c. Write a logical argument using coordinate geometry to support the conjecture you made in part **b**.

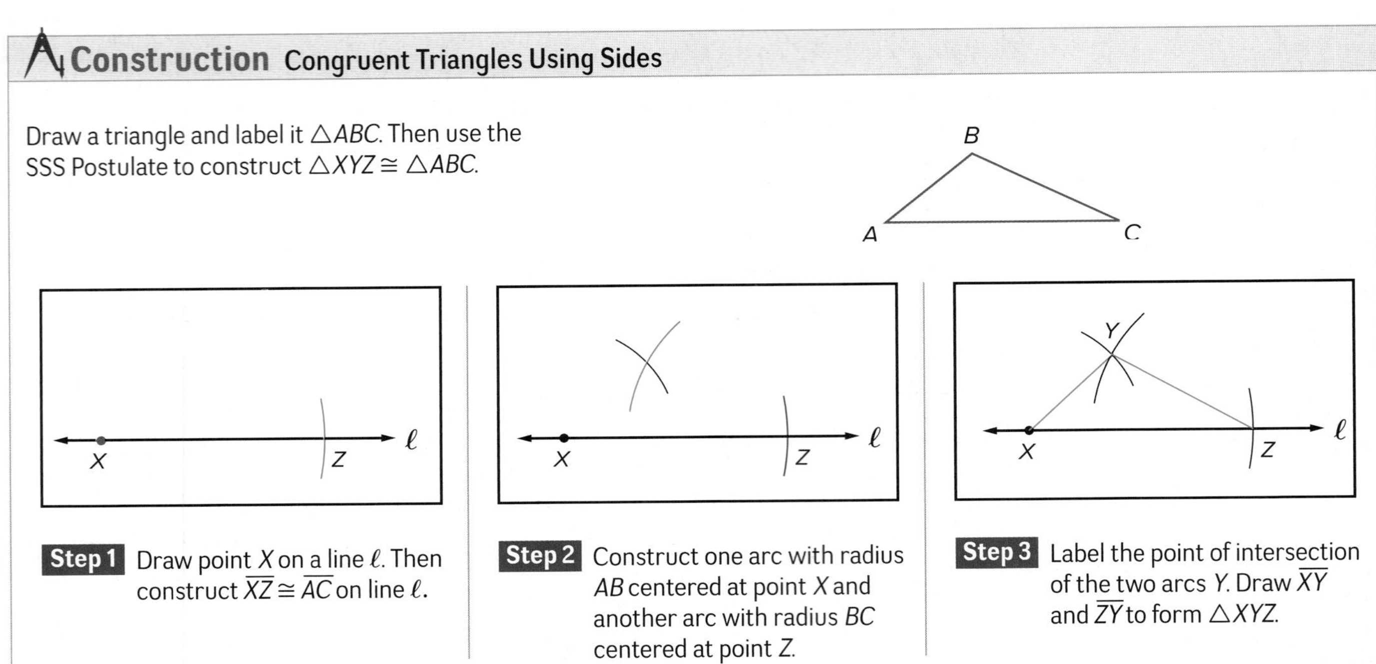

2 SAS Postulate

The angle formed by two adjacent sides of a polygon is called an **included angle**. Consider included angle *JKL* formed by the hands on the first clock shown below. Any time the hands form an angle with the same measure, the distance between the ends of the hands $\overline{JL}$ and $\overline{PR}$ will be the same.

$\triangle PKR \cong \triangle JKL$

Any two triangles formed using the same side lengths and included angle measure will be congruent. This illustrates the following postulate.

Study Tip

Side-Side-Angle The measures of two sides and a nonincluded angle are not sufficient to prove two triangles congruent.

Postulate 4.2 Side-Angle-Side (SAS) Congruence

Words If two sides and the included angle of one triangle are congruent to two sides and the included angle of a second triangle, then the triangles are congruent.

Example If **S**ide $\overline{AB} \cong \overline{DE}$,
Angle $\angle B \cong \angle E$, and
Side $\overline{BC} \cong \overline{EF}$,
then $\triangle ABC \cong \triangle DEF$.

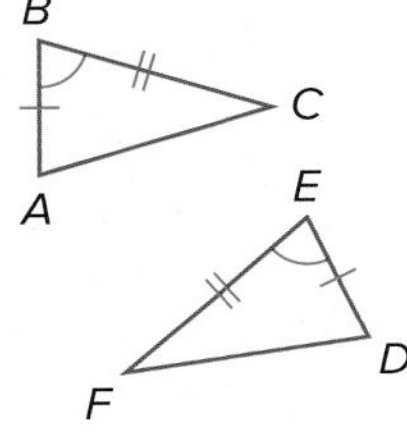

Real-World Example 3 Use SAS to Prove Triangles are Congruent

LIGHTING **The scaffolding for stage lighting shown appears to be made up of congruent triangles. If $\overline{WX} \cong \overline{YZ}$ and $\overline{WX} \parallel \overline{ZY}$, write a two-column proof to prove that $\triangle WXZ \cong \triangle YZX$.**

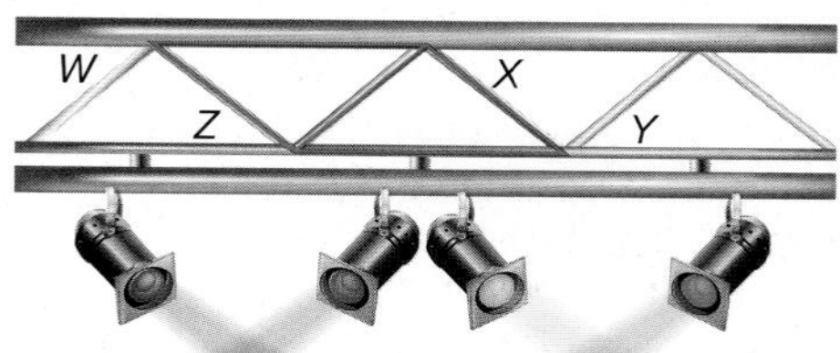

Proof:

Statements	Reasons
1. $\overline{WX} \cong \overline{YZ}$	1. Given
2. $\overline{WX} \parallel \overline{ZY}$	2. Given
3. $\angle WXZ \cong \angle XZY$	3. Alternate Interior Angles Theorem
4. $\overline{XZ} \cong \overline{ZX}$	4. Reflexive Property of Congruence
5. $\triangle WXZ \cong \triangle YZX$	5. SAS

Real-World Career

Lighting Technicians In the motion picture industry, gaffers—or lighting technicians—place the lighting required for a film. Gaffers make sure the angles the lights form are in the correct positions. They may have college or technical school degrees, or they may have completed a formal training program.

Guided Practice

3. **EXTREME SPORTS** The wings of the hang glider shown appear to be congruent triangles. If $\overline{FG} \cong \overline{GH}$ and $\overline{JG}$ bisects $\angle FGH$, prove that $\triangle FGJ \cong \triangle HGJ$.

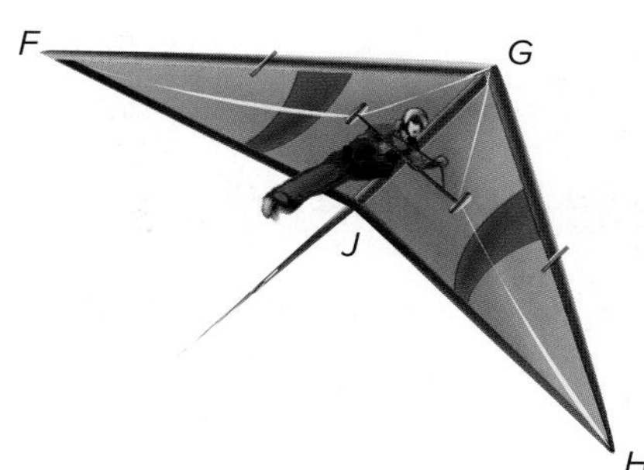

Construction Congruent Triangles Using Two Sides and the Included Angle

Draw a triangle and label it $\triangle ABC$. Then use the SAS Postulate to construct $\triangle RST \cong \triangle ABC$.

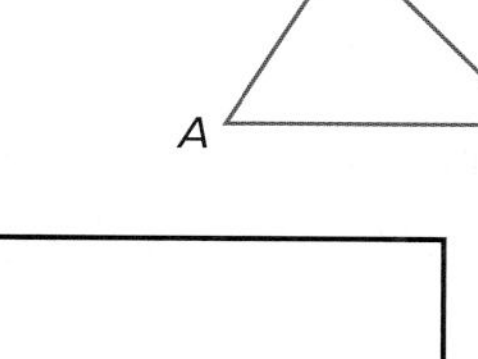

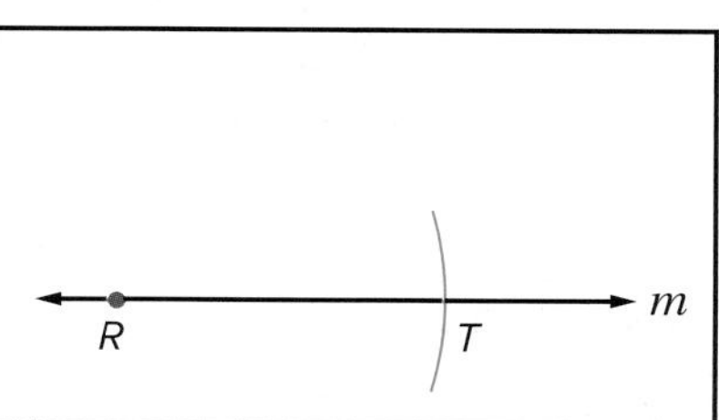

Step 1 Draw point R on a line m. Then construct $\overline{RT} \cong \overline{AC}$ on line m.

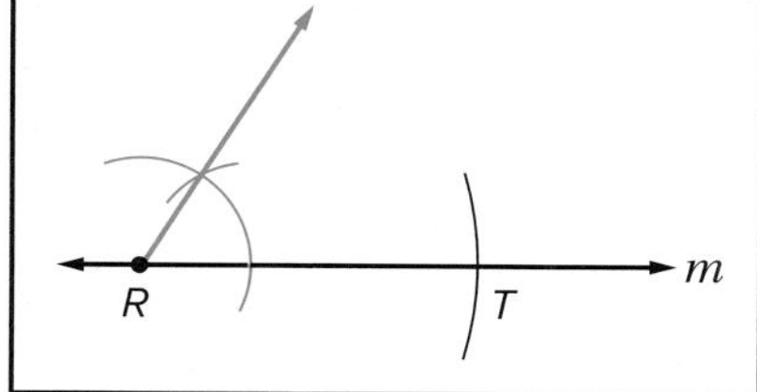

Step 2 Construct $\angle R \cong \angle A$ using $\overrightarrow{RT}$ as a side of the angle and point R.

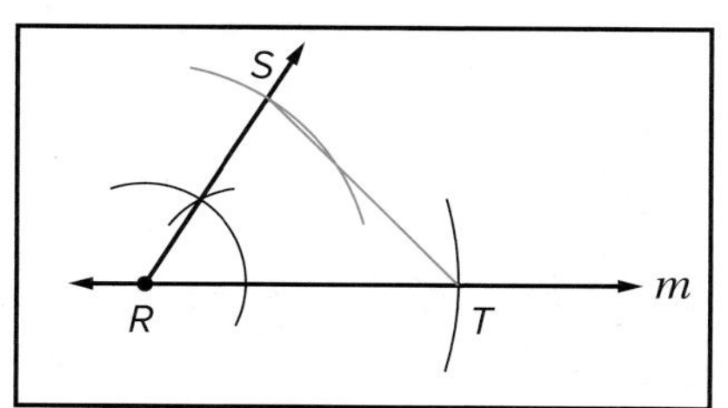

Step 3 Construct $\overline{RS} \cong \overline{AB}$. Then draw $\overline{ST}$ to form $\triangle RST$.

Example 4 Use SAS or SSS in Proofs

Write a paragraph proof.

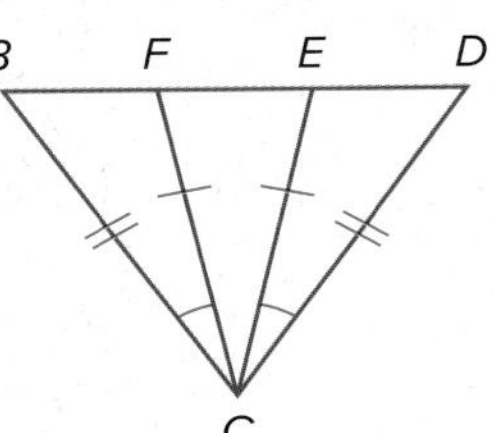

Given: $\overline{BC} \cong \overline{DC}$, $\angle BCF \cong \angle DCE$, $\overline{FC} \cong \overline{EC}$

Prove: $\angle CFD \cong \angle CEB$

Proof:

Since $\overline{BC} \cong \overline{DC}$, $\angle BCF \cong \angle DCE$, and $\overline{FC} \cong \overline{EC}$, then $\triangle BCF \cong \triangle DCE$ by SAS. By CPCTC, $\angle CFB \cong \angle CED$. $\angle CFD$ forms a linear pair with $\angle CFB$, and $\angle CEB$ forms a linear pair with $\angle CED$. By the Congruent Supplements Theorem, $\angle CFD$ is supplementary to $\angle CFB$ and $\angle CEB$ is supplementary to $\angle CED$. Since angles supplementary to the same angle or congruent angles are congruent, $\angle CFD \cong \angle CEB$.

Study Tip

MP Sense-Making When triangles overlap, it can be helpful to draw each triangle separately and label the congruent parts. In Example 4, the figure could have been separated as shown.

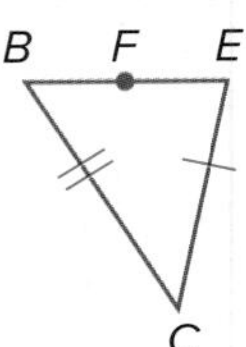

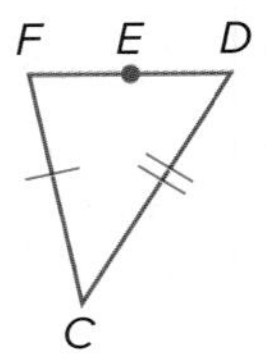

Guided Practice

4. Write a two-column proof.

Given: $\overline{MN} \cong \overline{PN}$, $\overline{LM} \cong \overline{LP}$

Prove: $\angle LNM \cong \angle LNP$

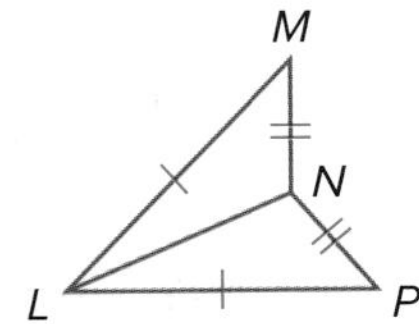

Check Your Understanding

◯ = Step-by-Step Solutions begin on page R13.

Example 1

1. OPTICAL ILLUSION The figure shown is a pattern formed using four large congruent squares and four small congruent squares.

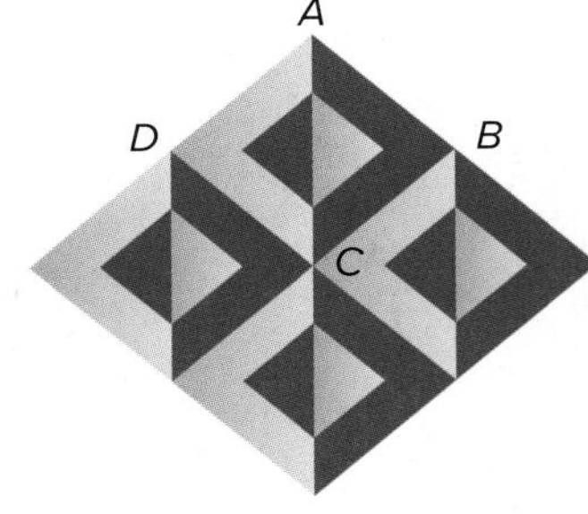

a. How many different-sized triangles are used to create the illusion?

b. Use the Side-Side-Side Congruence Postulate to prove that $\triangle ABC \cong \triangle CDA$.

c. What is the relationship between $\overleftrightarrow{AB}$ and $\overleftrightarrow{CD}$? Explain your reasoning.

Example 2

2. MULTIPLE CHOICE Triangle ABC has vertices $A(-3, -5)$, $B(-1, -1)$, and $C(-1, -5)$. Triangle XYZ has vertices $X(5, -5)$, $Y(3, -1)$, and $Z(3, -5)$. Which congruence statement describes the relationship between $\triangle ABC$ and $\triangle XYZ$?

A $\triangle ABC$ and $\triangle XYZ$ are both right triangles, so $\triangle ABC \cong \triangle XYZ$.

B Since $AC = XZ$, $CB = ZY$, and $BA = YX$, $\triangle ABC \cong \triangle XYZ$ by SSS congruence.

C Since $\triangle ABC$ and $\triangle XYZ$ are both right triangles, and $CB = ZY$ and $BA = YX$, $\triangle ABC \cong \triangle XYZ$ by SSS congruence.

D Since $AC = XZ$, $AB = ZY$, and $BC = YX$, $\triangle ABC \cong \triangle XYZ$ by SSS congruence.

Example 3

(3) EXERCISE In the exercise diagram, if $\overline{LP} \cong \overline{NO}$, $\angle LPM \cong \angle NOM$, and $\triangle MOP$ is equilateral, write a paragraph proof to show that $\triangle LMP \cong \triangle NMO$.

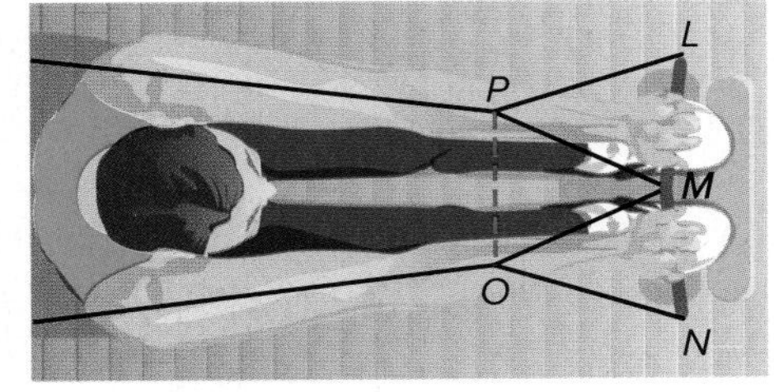

Example 4 **4. Write a two-column proof.**

Given: $\overline{BA} \cong \overline{DC}$, $\angle BAC \cong \angle DCA$
Prove: $\overline{BC} \cong \overline{DA}$

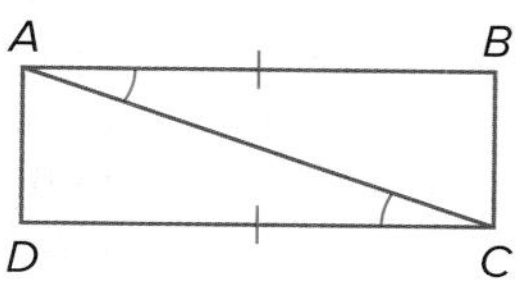

Practice and Problem Solving

Extra Practice is on page R4.

Example 1 **PROOF** **Write the specified type of proof.**

5. paragraph proof

Given: $\overline{QR} \cong \overline{SR}$, $\overline{ST} \cong \overline{QT}$
Prove: $\triangle QRT \cong \triangle SRT$

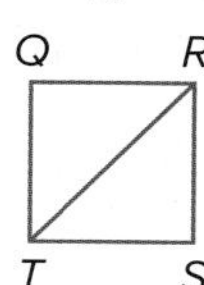

6. two-column proof

Given: $\overline{AB} \cong \overline{ED}$, $\overline{CA} \cong \overline{CE}$; $\overline{AC}$ bisects $\overline{BD}$.
Prove: $\triangle ABC \cong \triangle EDC$

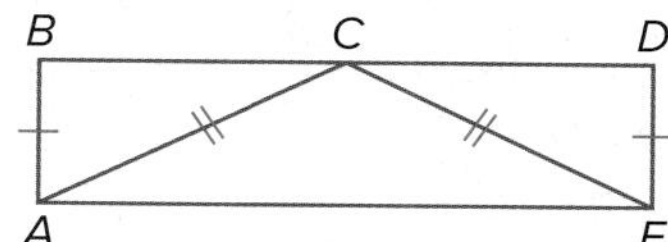

7. BRIDGES The Fred Hartman Bridge in Harris County is the longest cable-stayed bridge in Texas, spanning 2.6 miles across Houston Ship Channel. It is supported using steel cables suspended from two concrete supports. If the supports are the same height above the roadway and perpendicular to the roadway, and the topmost cables meet at a point midway between the supports, prove that the two triangles shown in the photo are congruent.

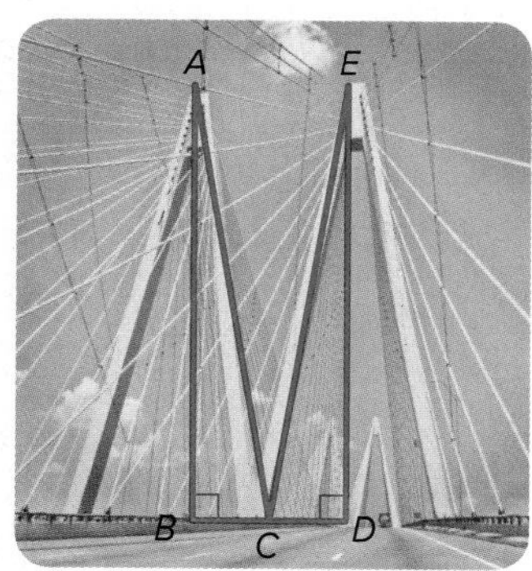

Example 2 **SENSE-MAKING** **Determine whether $\triangle MNO \cong \triangle QRS$. Explain using rigid motions.**

8. $M(2, 5)$, $N(5, 2)$, $O(1, 1)$, $Q(-4, 4)$, $R(-7, 1)$, $S(-3, 0)$

9. $M(0, -1)$, $N(-1, -4)$, $O(-4, -3)$, $Q(3, -3)$, $R(4, -4)$, $S(3, 3)$

10. $M(0, -3)$, $N(1, 4)$, $O(3, 1)$, $Q(4, -1)$, $R(6, 1)$, $S(9, -1)$

11. $M(4, 7)$, $N(5, 4)$, $O(2, 3)$, $Q(2, 5)$, $R(3, 2)$, $S(0, 1)$

Example 3 **PROOF** **Write the specified type of proof.**

12. two-column proof

Given: $\overline{BD} \perp \overline{AC}$, $\overline{BD}$ bisects $\overline{AC}$.
Prove: $\triangle ABD \cong \triangle CBD$

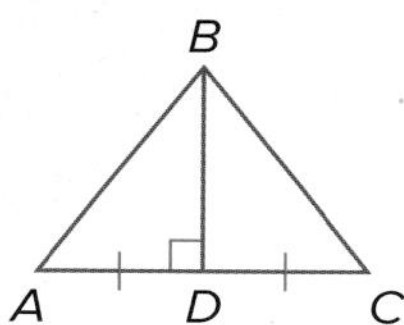

13. paragraph proof

Given: R is the midpoint of $\overline{QS}$ and $\overline{PT}$.
Prove: $\triangle PRQ \cong \triangle TRS$

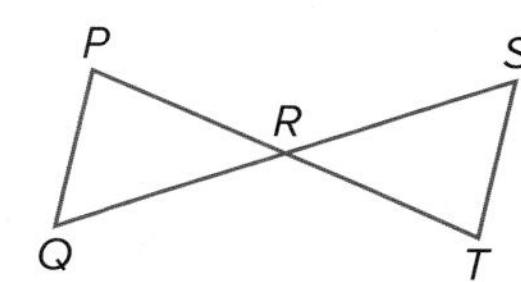

Example 4 **PROOF** **Write the specified type of proof.**

14. flow proof

Given: $\overline{JM} \cong \overline{NK}$; L is the midpoint of $\overline{JN}$ and $\overline{KM}$.

Prove: $\angle MJL \cong \angle KNL$

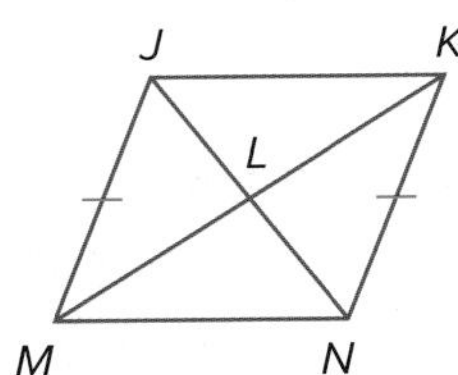

15. paragraph proof

Given: $\triangle XYZ$ is equilateral. $\overline{WY}$ bisects $\angle XYZ$.

Prove: $\overline{XW} \cong \overline{ZW}$

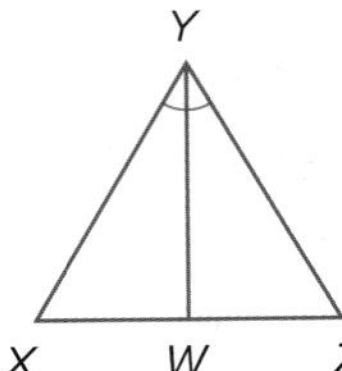

CONSTRUCT ARGUMENTS **Determine which postulate can be used to prove that the triangles are congruent. If it is not possible to prove congruence, write *not possible*. If it is possible, describe the rigid motions that map one triangle onto the other.**

16. 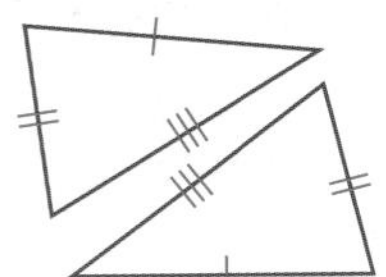

17. 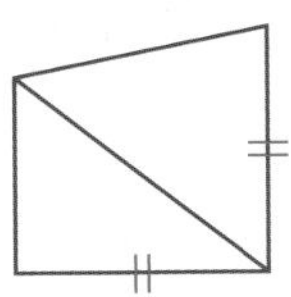

18. 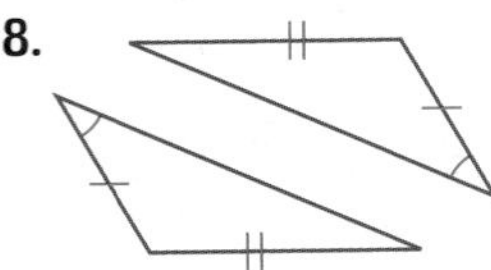

19. 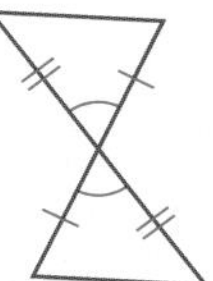

20. **SIGNS** Refer to the diagram at the right.

a. Identify the three-dimensional figure represented by the wet floor sign.

b. If $\overline{AB} \cong \overline{AD}$ and $\overline{CB} \cong \overline{DC}$, prove that $\triangle ACB \cong \triangle ACD$.

c. Why do the triangles not look congruent in the diagram?

PROOF **Write a flow proof.**

21. **Given:** $\overline{MJ} \cong \overline{ML}$; K is the midpoint of $\overline{JL}$.

Prove: $\triangle MJK \cong \triangle MLK$

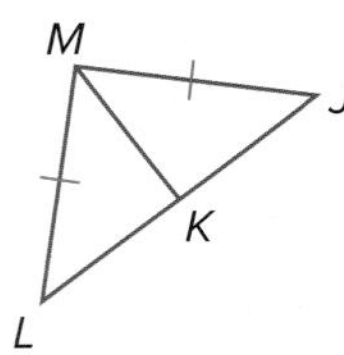

22. **Given:** $\triangle TPQ \cong \triangle SPR$

$\angle TQR \cong \angle SRQ$

Prove: $\triangle TQR \cong \triangle SRQ$

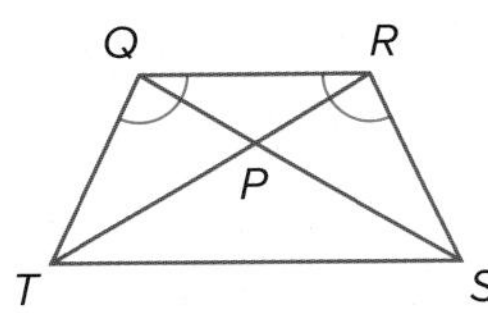

23. **SOFTBALL** Use the diagram of a fast-pitch softball diamond shown. Let F = first base, S = second base, T = third base, P = pitching point, and R = home plate.

a. Write a two-column proof to prove that the distance from first base to third base is the same as the distance from home plate to second base.

b. Write a two-column proof to prove that the angle formed between second base, home plate, and third base is the same as the angle formed between second base, home plate, and first base.

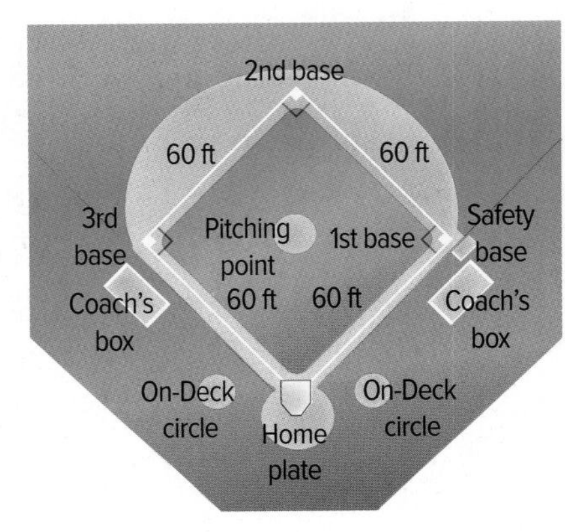

PROOF **Write a two-column proof.**

24. **Given:** $\overline{YX} \cong \overline{WZ}$, $\overline{YX} \parallel \overline{ZW}$

Prove: $\triangle YXZ \cong \triangle WZX$

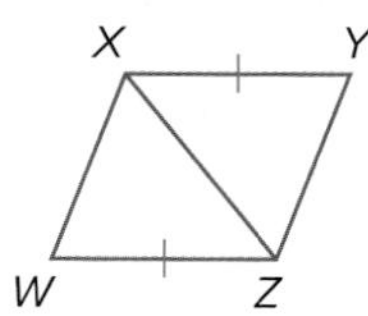

25. **Given:** $\triangle EAB \cong \triangle DCB$

Prove: $\triangle EAD \cong \triangle DCE$

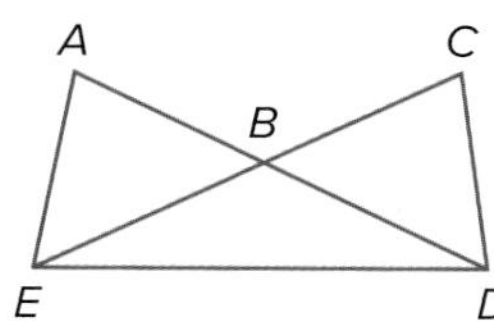

26. **CONSTRUCT ARGUMENTS** **Write a paragraph proof.**

Given: $\overline{HL} \cong \overline{HM}$, $\overline{PM} \cong \overline{KL}$, $\overline{PG} \cong \overline{KJ}$, $\overline{GH} \cong \overline{JH}$

Prove: $\angle G \cong \angle J$

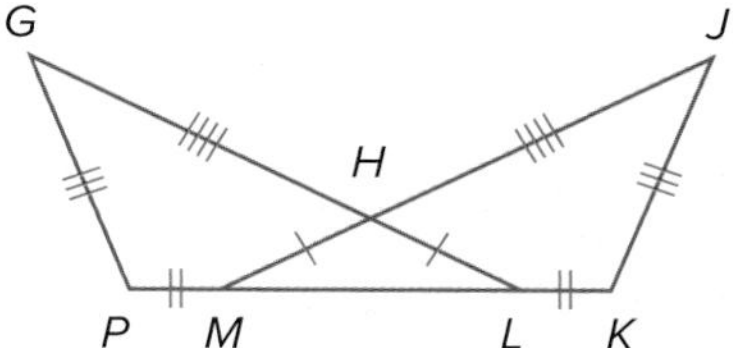

H.O.T. Problems Use Higher-Order Thinking Skills

REASONING **Describe the steps in an activity to illustrate the indicated criterion for triangle congruence for $\triangle ABC$ and $\triangle XYZ$. Then explain how this criterion follows from the principle of superposition.**

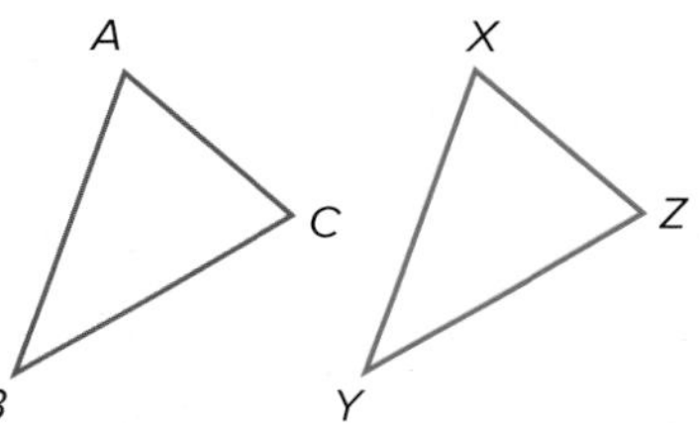

27. SSS

28. SAS

29. **CHALLENGE** **Refer to the graph shown.**

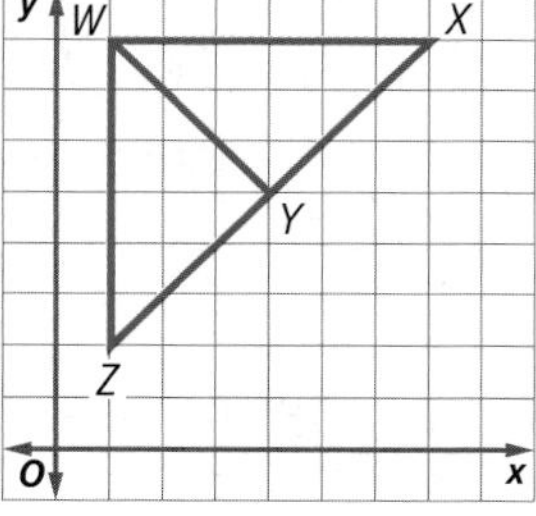

a. Describe two methods you could use to prove that $\triangle WYZ$ is congruent to $\triangle WYX$. You may not use a ruler or a protractor. Which method do you think is more efficient? Explain.

b. Are $\triangle WYZ$ and $\triangle WYX$ congruent? Explain your reasoning.

30. **REASONING** Determine whether the following statement is *true* or *false*. If true, explain your reasoning. If *false*, provide a counterexample.

If the congruent sides in one isosceles triangle have the same measure as the congruent sides in another isosceles triangle, then the triangles are congruent.

31. **ERROR ANALYSIS** Bonnie says that $\triangle PQR \cong \triangle XYZ$ by SAS. Shada disagrees. She says that there is not enough information to prove that the two triangles are congruent. Is either of them correct? Explain.

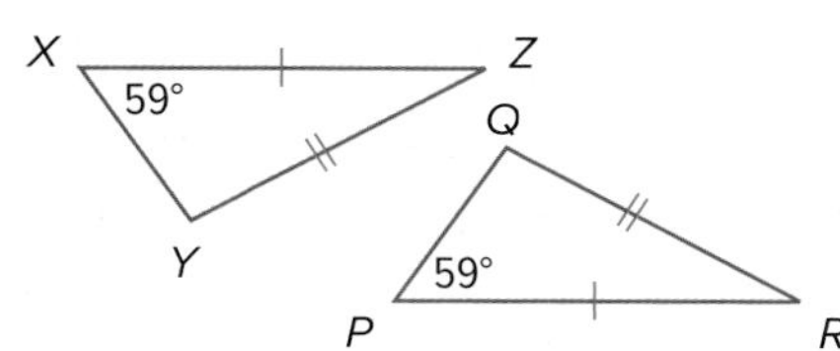

32. **OPEN-ENDED** Use a straightedge to draw obtuse triangle ABC. Then construct $\triangle XYZ$ so that it is congruent to $\triangle ABC$ using either SSS or SAS. Justify your construction mathematically and verify it using measurement.

33. **WRITING IN MATH** Two pairs of corresponding sides of two right triangles are congruent. Are the triangles congruent? Explain your reasoning.

34. Which triangles best show a counterexample to the statement below? MP 3

> If the hypotenuse of one right triangle is congruent to the hypotenuse of another, then the triangles are congruent.

○ **A**

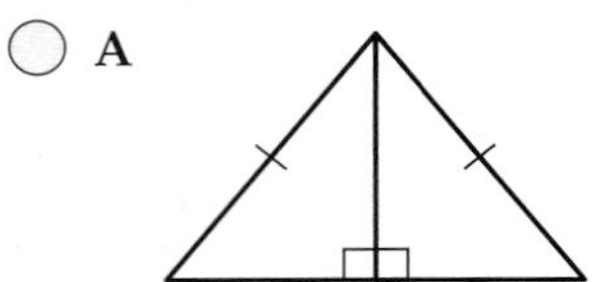

○ **B**

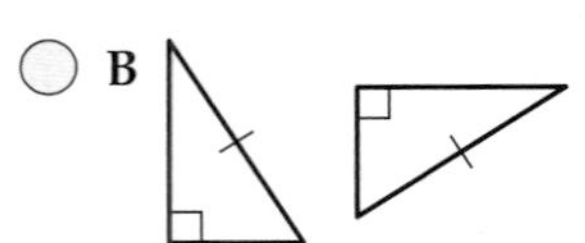

○ **C**

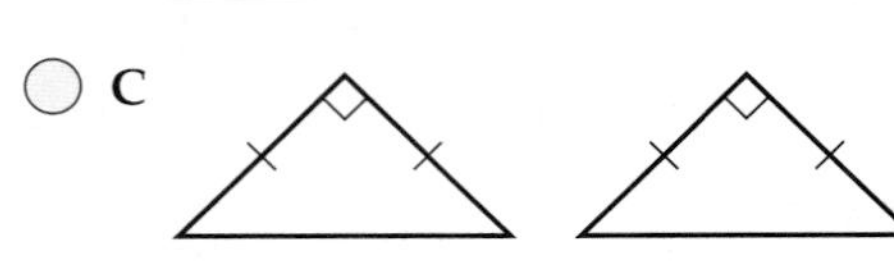

○ **D**

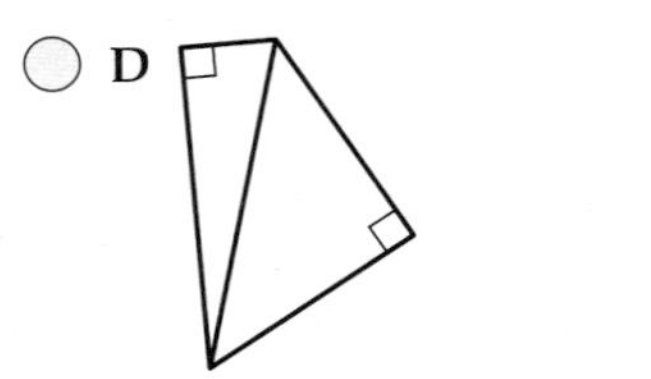

35. **MULTI-STEP** Refer to the diagram showing $\triangle ABC$ and $\triangle DCB$. MP 1, 3

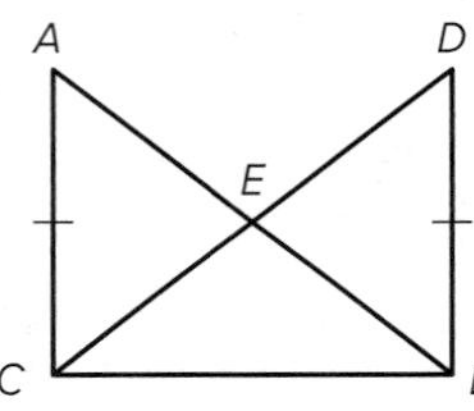

a. Name one pair of segments that must be congruent to prove that the triangles are congruent.

b. Name one pair of angles that must be congruent to be able to prove that the triangles are congruent.

c. Assume the angle pair named in part **b** is congruent. Describe the rigid motion(s) that maps one triangle onto the other.

d. Assume the angle pair named in part **b** is congruent. What postulate can be used to prove that the triangles are congruent?

e. Write a paragraph proof to show $\triangle ABC \cong \triangle DCB$ given $\angle ABC \cong \angle DBC$.

36. $\triangle ABC \cong \triangle DEF$. Point D is located at $(11, 4)$. Point F is located at $(7, 0)$. Find one possible location for point E. MP 1

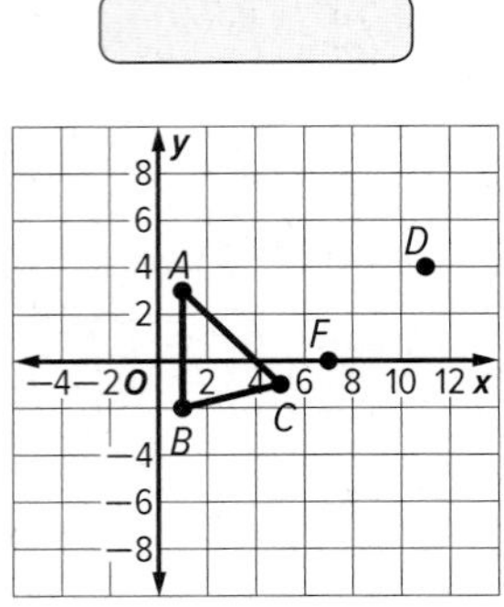

37. Identify the type of congruence transformation shown. MP 2

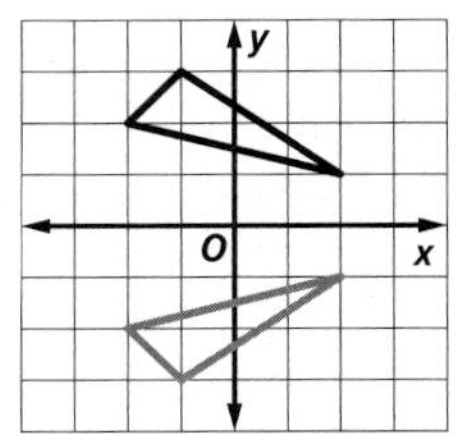

○ **A** translation

○ **B** rotation about the origin

○ **C** reflection in the x-axis

○ **D** dilation

38. Select the triangles that can be proven congruent by SSS or SAS. MP 2

☐ **A**

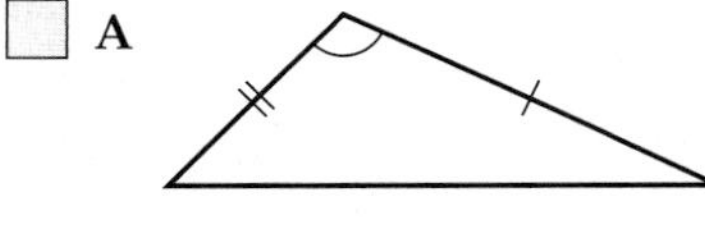

☐ **B**

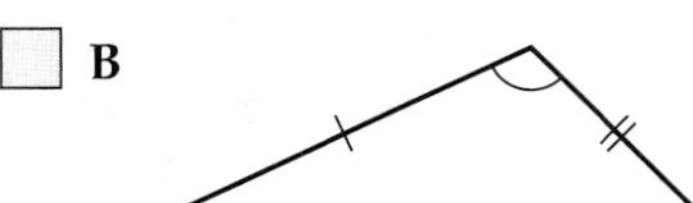

☐ **C**

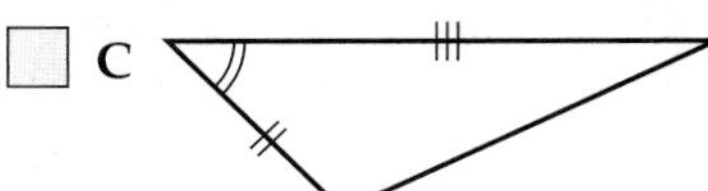

☐ **D**

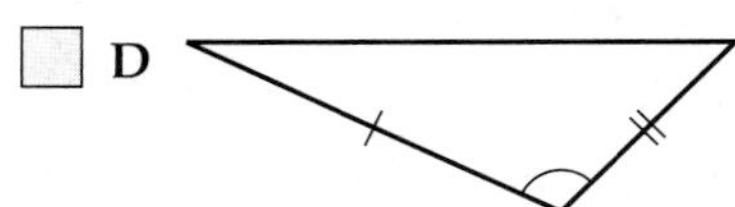

EXTEND 4-3

Geometry Lab

Proving Constructions

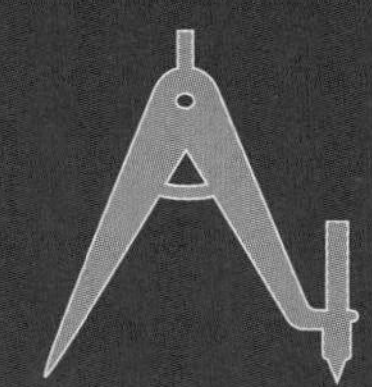

When you perform a construction using a straightedge and compass, you assume that segments constructed using the same compass setting are congruent. You can use this information, along with definitions, postulates, and theorems, to prove constructions.

Mathematical Practices 3, 5

Activity

Work cooperatively. Follow the steps below to bisect an angle. Then prove the construction.

Step 1

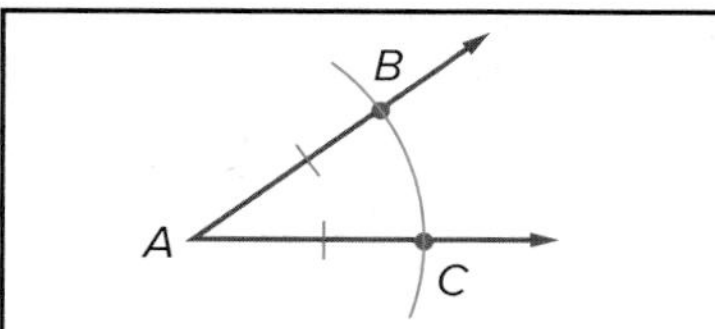

Draw any angle with vertex A. Place the compass point at A and draw an arc that intersects both sides of $\angle A$. Label the points B and C. Mark the congruent segments.

Step 2

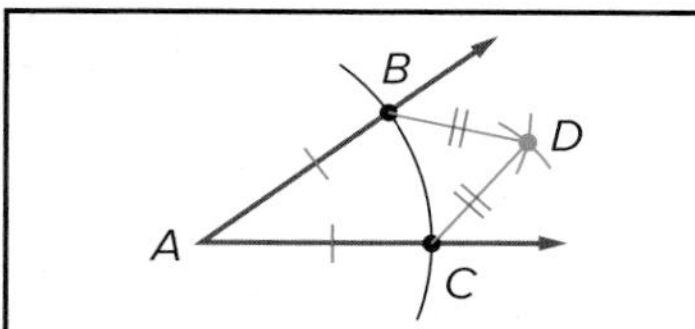

With the compass point at B, draw an arc in the interior of $\angle A$. With the same radius, draw an arc from C intersecting the first arc at D. Draw the segments $\overline{BD}$ and $\overline{CD}$. Mark the congruent segments.

Step 3

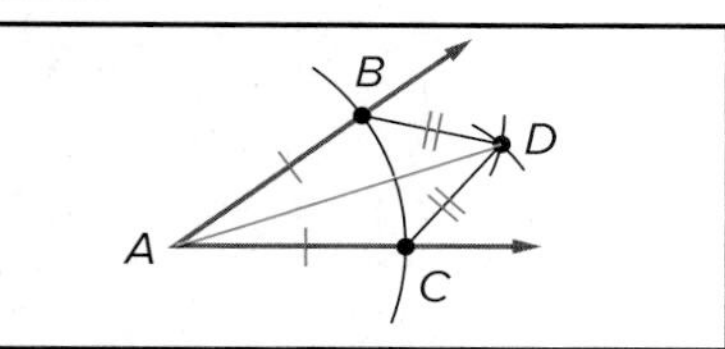

Draw $\overline{AD}$.

Given: Description of steps and diagram of construction

Prove: $\overline{AD}$ bisects $\angle BAC$.

Proof:

Statements	Reasons
1. $\overline{AB} \cong \overline{AC}$	1. The same compass setting was used from point A to construct points B and C.
2. $\overline{BD} \cong \overline{CD}$	2. The same compass setting was used from points B and C to construct point D.
3. $\overline{AD} \cong \overline{AD}$	3. Reflexive Property
4. $\triangle ABD \cong \triangle ACD$	4. SSS Postulate
5. $\angle BAD \cong \angle CAD$	5. CPCTC
6. $\overline{AD}$ bisects $\angle BAC$.	6. Definition of angle bisector

Exercises

Work cooperatively.

1. Construct a line parallel to a given line through a given point. Write a two-column proof of your construction.
2. Construct an equilateral triangle. Write a paragraph proof of your construction.
3. **CHALLENGE** Construct the bisector of a segment that is also perpendicular to the segment and write a two-column proof of your construction. (*Hint:* You will need to use more than one pair of congruent triangles.)

CHAPTER 4

Mid-Chapter Quiz

Lessons 4-1 through 4-3

Find the measure of each numbered angle. (Lesson 4-1)

1. $m\angle 1$
2. $m\angle 2$
3. $m\angle 3$

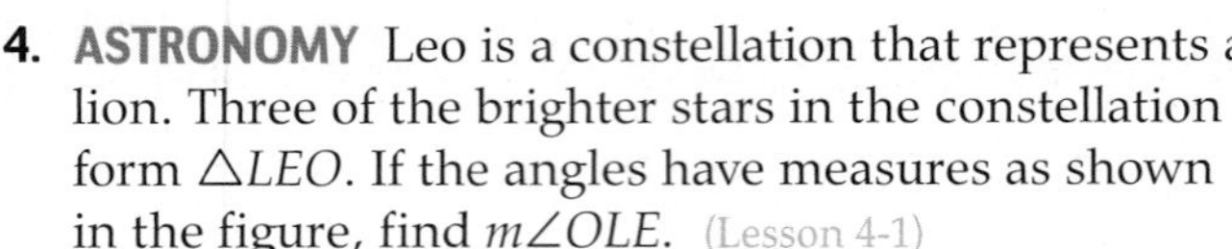

4. **ASTRONOMY** Leo is a constellation that represents a lion. Three of the brighter stars in the constellation form $\triangle LEO$. If the angles have measures as shown in the figure, find $m\angle OLE$. (Lesson 4-1)

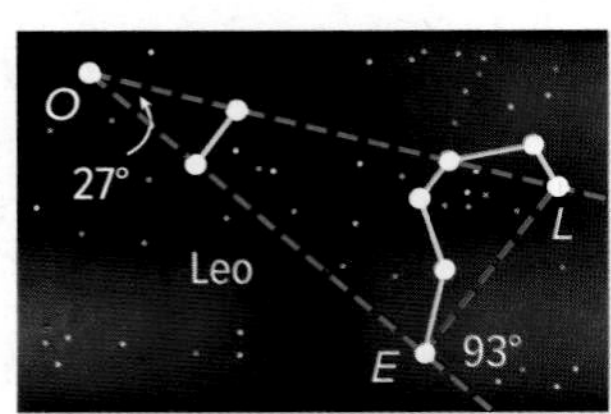

Find the measure of each numbered angle. (Lesson 4-1)

5. $m\angle 4$
6. $m\angle 5$
7. $m\angle 6$
8. $m\angle 7$

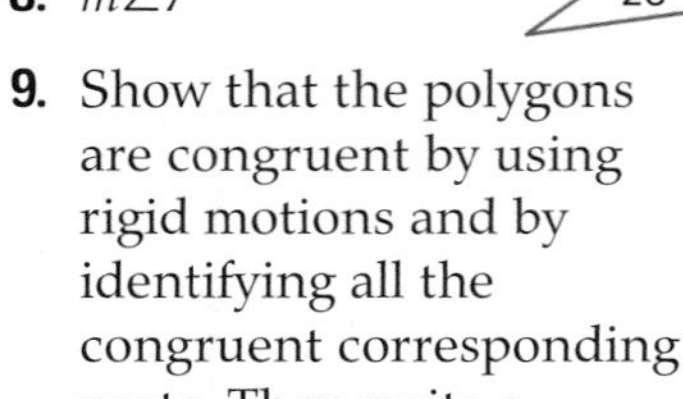

9. Show that the polygons are congruent by using rigid motions and by identifying all the congruent corresponding parts. Then write a congruence statement.

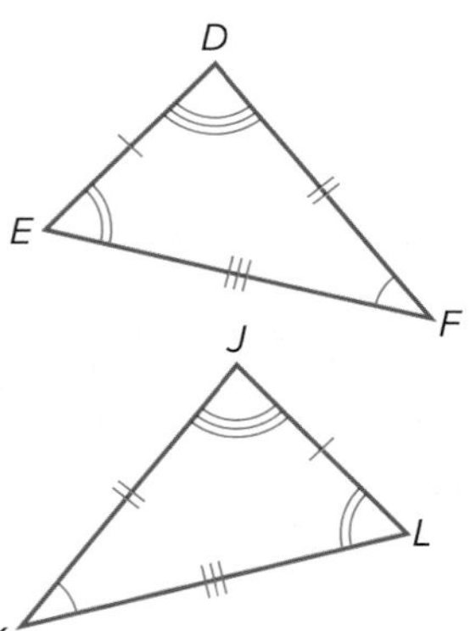

In the diagram, $\triangle RST \cong \triangle ABC$. (Lesson 4-2)

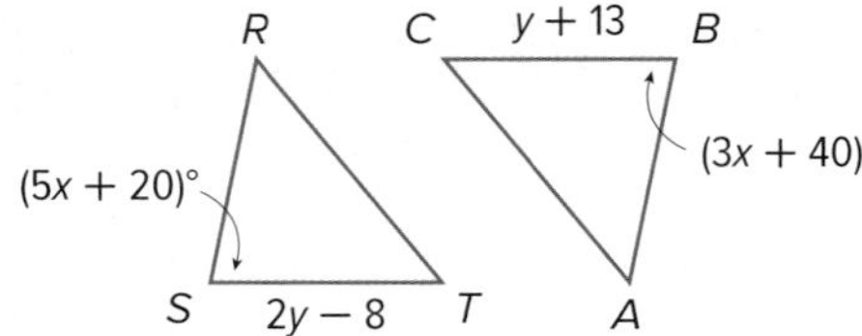

10. Find x.
11. Find y.
12. Find RT if $AC = 4x + 2$.
13. Find $m\angle BAC$ if $m\angle SRT = (2y + 8)°$.

14. **ARCHITECTURE** The diagram shows an A-frame house with various points labeled. Assume that segments and angles that appear to be congruent in the diagram are congruent. Indicate which triangles are congruent. (Lesson 4-2)

15. **MULTIPLE CHOICE** Determine which statement is true given that $\triangle CBX \cong \triangle SML$. (Lesson 4-2)

A $\overline{MO} \cong \overline{SL}$

B $\overline{XC} \cong \overline{ML}$

C $\angle X \cong \angle S$

D $\angle XCB \cong \angle LSM$

16. **BRIDGES** A bridge truss is shown in the diagram, where $\overline{AC} \perp \overline{BD}$ and B is the midpoint of $\overline{AC}$.

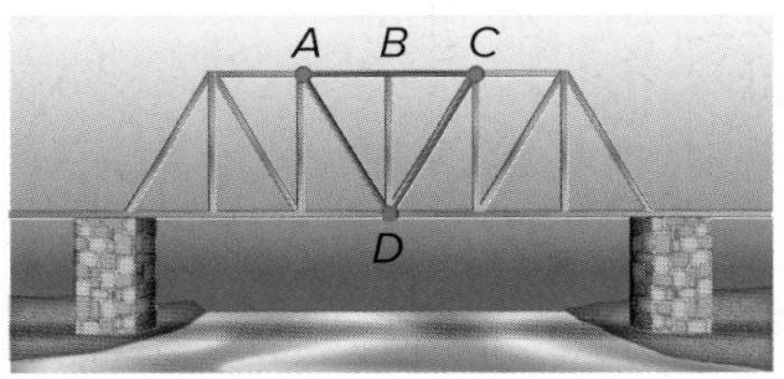

a. What method can be used to prove that $\triangle ABD \cong \triangle CBD$? (Lesson 4-3)

b. MP Which mathematical practice did you use to solve this problem?

Determine whether $\triangle PQR \cong \triangle XYZ$. If so, tell what rigid motion maps one onto the other. (Lesson 4-3)

17. $P(3, -5)$, $Q(11, 0)$, $R(1, 6)$, $X(5, 1)$, $Y(13, 6)$, $Z(3, 12)$
18. $P(-3, -3)$, $Q(-5, 1)$, $R(-2, 6)$, $X(2, -6)$, $Y(3, 3)$, $Z(5, -1)$
19. $P(8, 1)$, $Q(-7, -15)$, $R(9, -6)$, $X(5, 11)$, $Y(-10, -5)$, $Z(6, 4)$

20. **Write a two-column proof.** (Lesson 4-3)

Given: $\triangle LMN$ is isosceles with $\overline{LM} \cong \overline{NM}$, and $\overline{MO}$ bisects $\angle LMN$.

Prove: $\triangle MLO \cong \triangle MNO$

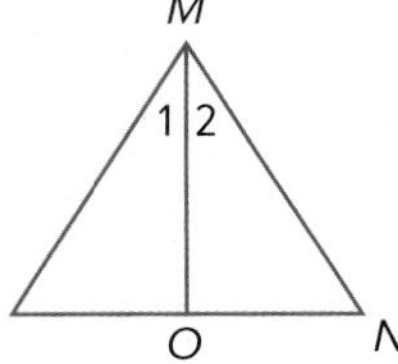

LESSON 4

Proving Triangles Congruent–ASA, AAS

Then

- You proved triangles congruent using SSS and SAS.

Now

1. Use the ASA congruence criterion to prove triangles congruent.
2. Use the AAS congruence criterion to prove triangles congruent.

Why?

- Competitive rowing, also called crew, involves two or more people who sit facing the stern of the boat, with each rower pulling one or two oars. In high school competitions, a race—called a regatta—usually requires a body of water that is more than 1500 meters long. Congruent triangles can be used to measure distances that are not easily measured directly, like the length of a regatta course.

New Vocabulary

included side

Mathematical Practices

3 Construct viable arguments and critique the reasoning of others.

1 ASA Postulate

An **included side** is the side located between two consecutive angles of a polygon. In $\triangle ABC$ at the right, $\overline{AC}$ is the included side between $\angle A$ and $\angle C$.

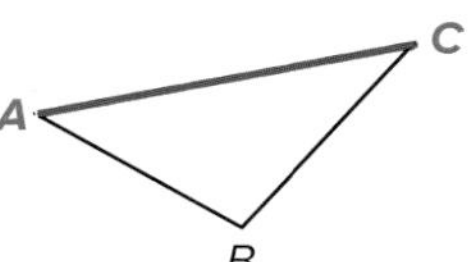

Postulate 4.3 Angle-Side-Angle (ASA) Congruence

If two angles and the included side of one triangle are congruent to two angles and the included side of another triangle, then the triangles are congruent.

Example If **A**ngle $\angle A \cong \angle D$,
Side $\overline{AB} \cong \overline{DE}$, and
Angle $\angle B \cong \angle E$,
then $\triangle ABC \cong \triangle DEF$.

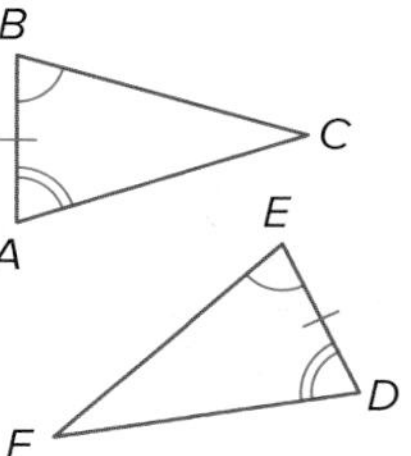

Construction Congruent Triangles Using Two Angles and Included Side

Draw a triangle and label it $\triangle ABC$. Then use the ASA Postulate to construct $\triangle XYZ \cong \triangle ABC$.

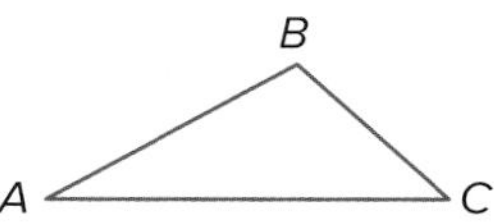

Step 1

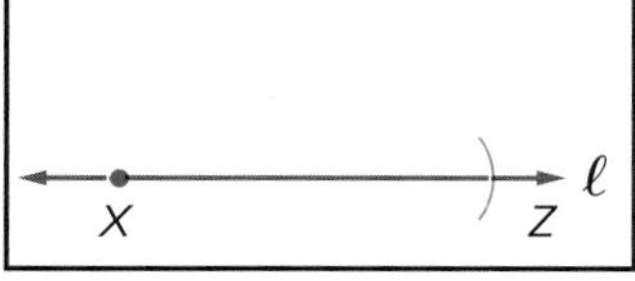

Draw a line ℓ and select a point X. Construct $\overline{XZ}$ such that $\overline{XZ} \cong \overline{AC}$.

Step 2

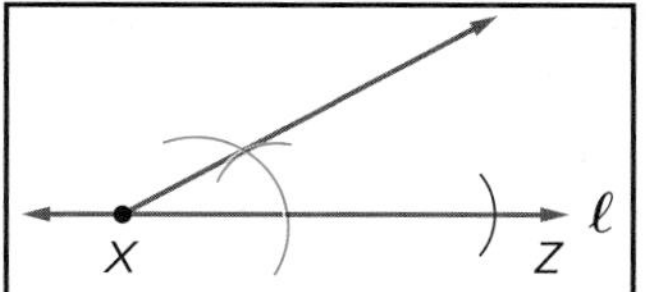

Construct an angle congruent to $\angle A$ at X using $\overleftrightarrow{XZ}$ as a side of the angle.

Step 3

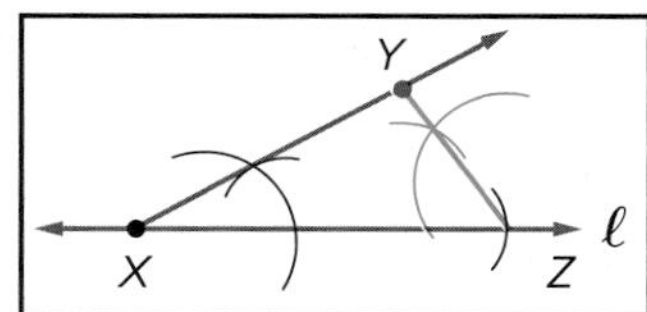

Construct an angle congruent to $\angle C$ at Z using $\overleftrightarrow{XZ}$ as a side of the angle. Label the point where the new sides of the angles meet as Y.

Example 1 Use ASA to Prove Triangles Congruent

Write a two-column proof.

Given: $\overline{QS}$ bisects $\angle PQR$.
$\angle PSQ \cong \angle RSQ$

Prove: $\triangle PQS \cong \triangle RQS$

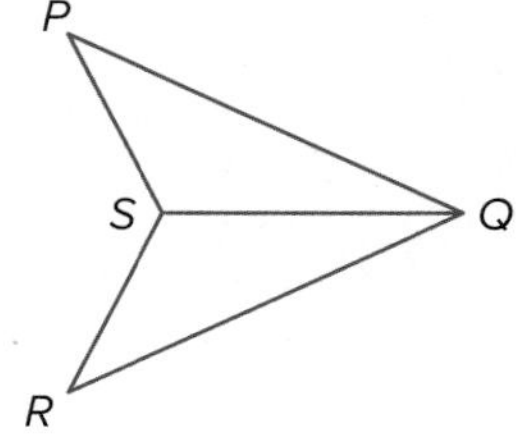

Proof:

Statements	Reasons
1. $\overline{QS}$ bisects $\angle PQR$. $\angle PSQ \cong \angle RSQ$	**1.** Given
2. $\angle PQS \cong \angle RQS$	**2.** Definition of Angle Bisector
3. $\overline{QS} \cong \overline{QS}$	**3.** Reflexive Property of Congruence
4. $\triangle PQS \cong \triangle RQS$	**4.** ASA

Guided Practice

1. Write a flow proof.
Given: $\overline{ZX}$ bisects $\angle WZY$; $\overline{XZ}$ bisects $\angle YXW$.
Prove: $\triangle WXZ \cong \triangle YXZ$

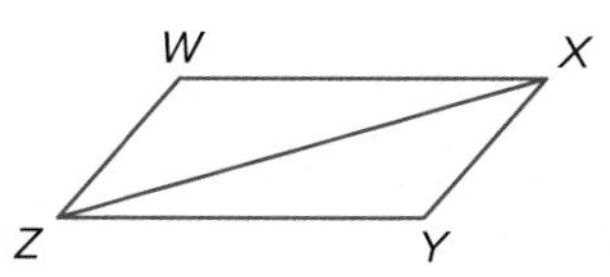

2 **AAS Theorem** The congruence of two angles and a nonincluded side are also sufficient to prove two triangles congruent. This congruence relationship is a theorem because it can be proved using the Third Angles Theorem.

Go Online! 

Is knowing that two angles and one pair of corresponding sides of two triangles are congruent sufficient to prove that the triangles are congruent? Investigate by using the **Geometry Tools** in ConnectED.

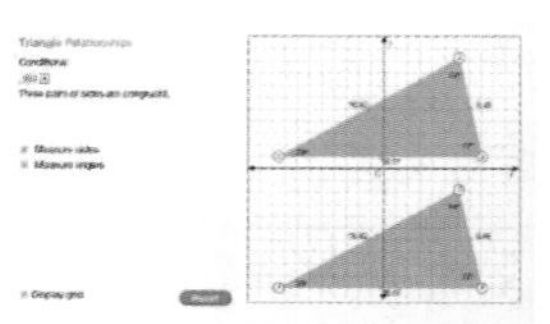

Theorem 4.5 Angle-Angle-Side (AAS) Congruence

If two angles and the nonincluded side of one triangle are congruent to the corresponding two angles and side of a second triangle, then the two triangles are congruent.

Example If **Angle** $\angle A \cong \angle D$,
Angle $\angle B \cong \angle E$, and
Side $\overline{BC} \cong \overline{EF}$,
then $\triangle ABC \cong \triangle DEF$.

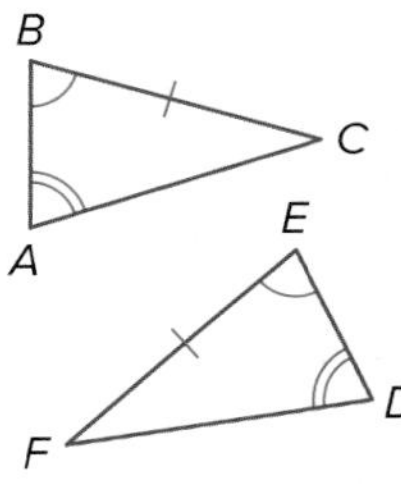

Proof Angle-Angle-Side Theorem

Given: $\angle L \cong \angle Q$, $\angle M \cong \angle R$, $\overline{MN} \cong \overline{RS}$

Prove: $\triangle LMN \cong \triangle QRS$

Proof:

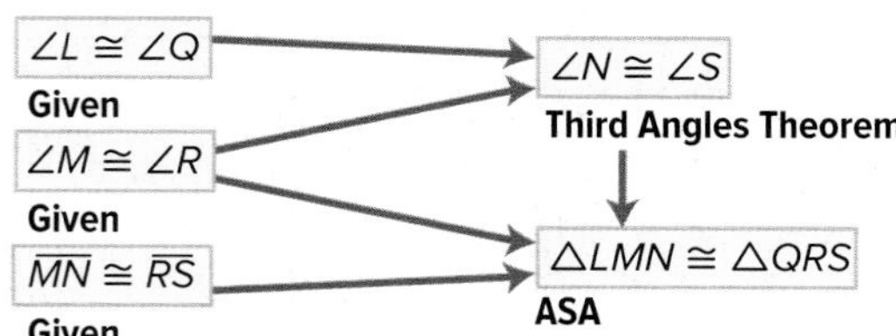

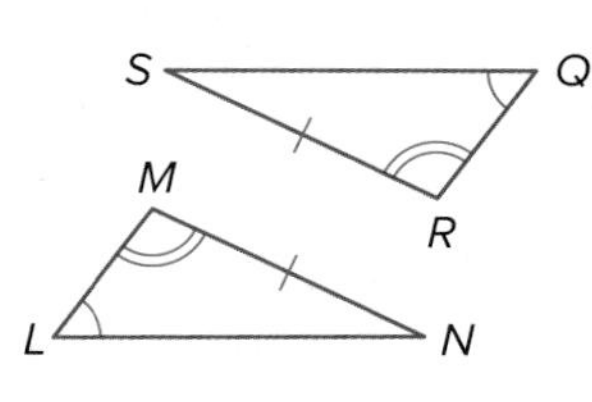

Example 2 Use AAS to Prove Triangles Congruent

Write a paragraph proof.

Given: $\angle DAC \cong \angle BEC$
$\overline{DC} \cong \overline{BC}$

Prove: $\triangle ACD \cong \triangle ECB$

A B F E D C

Proof: We are given that $\angle DAC \cong \angle BEC$ and $\overline{DC} \cong \overline{BC}$. $\angle C \cong \angle C$ by the Reflexive Property. By AAS, $\triangle ACD \cong \triangle ECB$.

Guided Practice

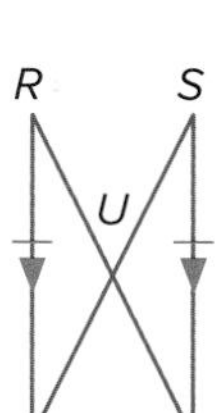

2. Write a flow proof.
Given: $\overline{RQ} \cong \overline{ST}$ and $\overline{RQ} \parallel \overline{ST}$
Prove: $\triangle RUQ \cong \triangle TUS$

You can use congruent triangles to measure distances that are difficult to measure directly.

Real-World Example 3 Apply Triangle Congruence

COMMUNITY SERVICE Jeremias is working with a community service group to build a bridge across a creek at a local park. The bridge will span the creek between points *C* and *B*. Jeremias located a fixed point *D* to use as a reference point so that the segments have the relationships shown. *A* is the midpoint of $\overline{CD}$ and *DE* is 15 feet. How long does the bridge need to be?

In order to determine the length of $\overline{CB}$, we must first prove that the two triangles Jeremias has created are congruent.

- Because $\overline{CD}$ is perpendicular to both $\overline{CB}$ and $\overline{DE}$, the segments form right angles as shown on the diagram.
- All right angles are congruent, so $\angle BCA \cong \angle EDA$.
- Point *A* is the midpoint of $\overline{CD}$, so $\overline{CA} \cong \overline{AD}$.
- $\angle BAC$ and $\angle EAD$ are vertical angles, so they are congruent.

Therefore, by ASA, $\triangle BAC \cong \triangle EAD$.

Since $\triangle BAC \cong \triangle EAD$, $\overline{DE} \cong \overline{CB}$ by CPCTC. Since the measure of $\overline{DE}$ is 15 feet, the measure of $\overline{CB}$ is also 15 feet. Therefore, the bridge needs to be 15 feet long.

Study Tip

Construct Arguments In Example 3, $\angle B$ and $\angle E$ are congruent by the Third Angles Theorem. Congruence of all three corresponding angles is not sufficient, however, to prove two triangles congruent.

Guided Practice

3. In the sign scaffold shown at the right, $\overline{BC} \perp \overline{AC}$ and $\overline{DE} \perp \overline{CE}$. $\angle BAC \cong \angle DCE$, and $\overline{AB} \cong \overline{CD}$. Write a paragraph proof to show that $\overline{BC} \cong \overline{DE}$.

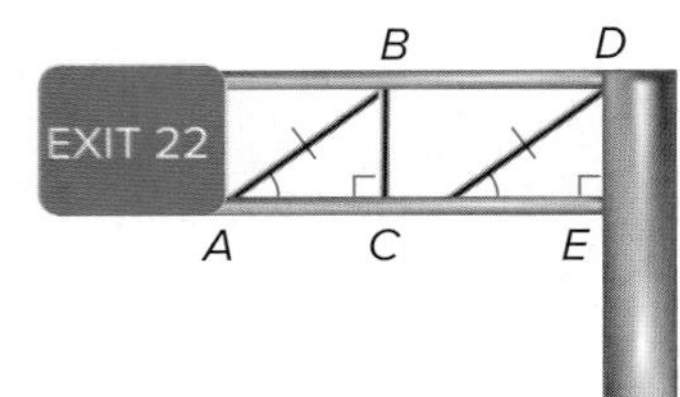

You have learned several methods for proving triangle congruence.

Concept Summary Proving Triangles Congruent

SSS	SAS	ASA	AAS
Three pairs of corresponding sides are congruent.	Two pairs of corresponding sides and their included angles are congruent.	Two pairs of corresponding angles and their included sides are congruent.	Two pairs of corresponding angles and the corresponding nonincluded sides are congruent.

Check Your Understanding

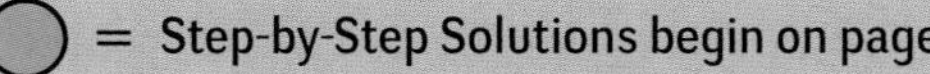

Example 1 **PROOF Write the specified type of proof.**

1. two-column proof

   Given: $\overline{CB}$ bisects $\angle ABD$ and $\angle ACD$.

   Prove: $\triangle ABC \cong \triangle DCB$

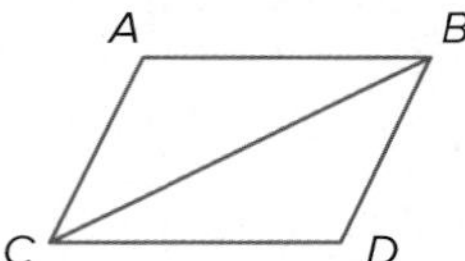

2. flow proof

   Given: $\overline{JK} \parallel \overline{LM}$, $\overline{JL} \parallel \overline{KM}$

   Prove: $\triangle JML \cong \triangle MJK$

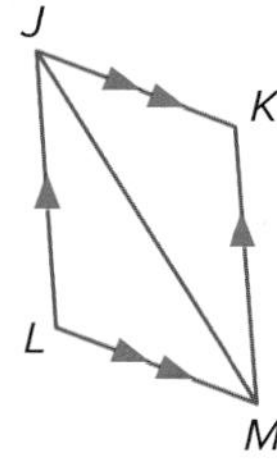

Example 2

3. paragraph proof

   Given: $\angle K \cong \angle M$, $\overline{JK} \cong \overline{JM}$, $\overline{JL}$ bisects $\angle KLM$.

   Prove: $\triangle JKL \cong \triangle JML$

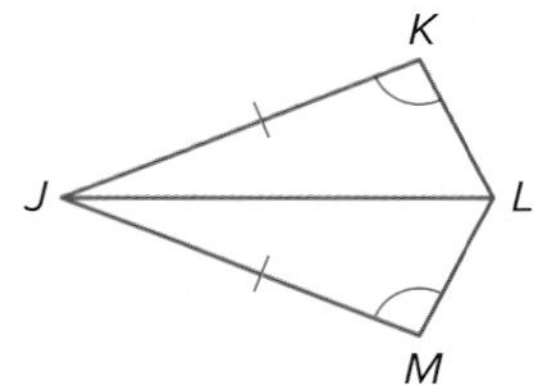

4. two-column proof

   Given: $\overline{GH} \parallel \overline{FJ}$

   $m\angle G = m\angle J = 90$

   Prove: $\triangle HJF \cong \triangle FGH$

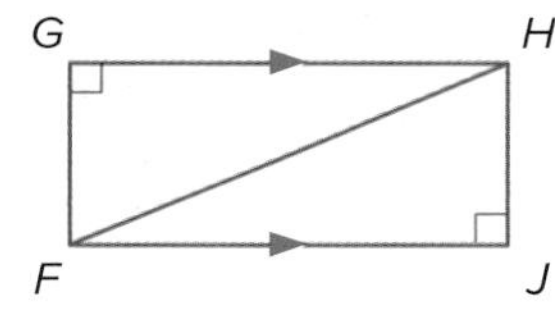

Example 3

5 **BRIDGE BUILDING** A surveyor needs to find the distance from point A to point B across a canyon. She places a stake at A, and a coworker places a stake at B on the other side of the canyon. The surveyor then locates C on the same side of the canyon as A such that $\overline{CA} \perp \overline{AB}$. A fourth stake is placed at E, the midpoint of $\overline{CA}$. Finally, a stake is placed at D such that $\overline{CD} \perp \overline{CA}$ and D, E, and B are sited as lying along the same line.

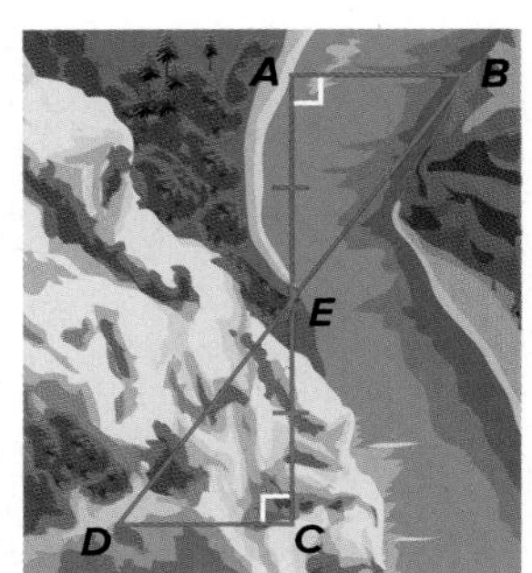

a. Explain how the surveyor can use the triangles formed to find AB.

b. If $AC = 1300$ meters, $DC = 550$ meters, and $DE = 851.5$ meters, what is AB? Explain your reasoning.

Practice and Problem Solving

Extra Practice is on page R4.

Example 1

PROOF **Write a paragraph proof.**

6. **Given:** $\overline{CE}$ bisects $\angle BED$; $\angle BCE$ and $\angle ECD$ are right angles.

Prove: $\triangle ECB \cong \triangle ECD$

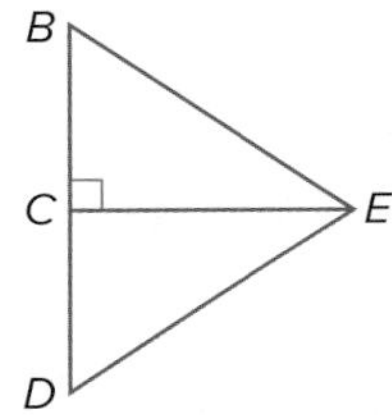

7. **Given:** $\angle W \cong \angle Y$, $\overline{WZ} \cong \overline{YZ}$, $\overline{XZ}$ bisects $\angle WZY$.

Prove: $\triangle XWZ \cong \triangle XYZ$

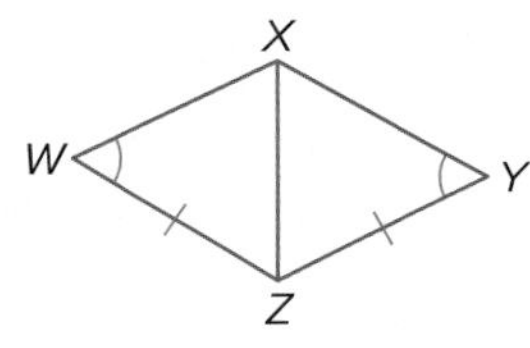

8. **TOYS** The object of the toy shown is to make the two spheres meet and strike each other repeatedly on one side of the wand and then again on the other side. If $\angle JKL \cong \angle MLK$ and $\angle JLK \cong \angle MKL$, prove that $\overline{JK} \cong \overline{ML}$.

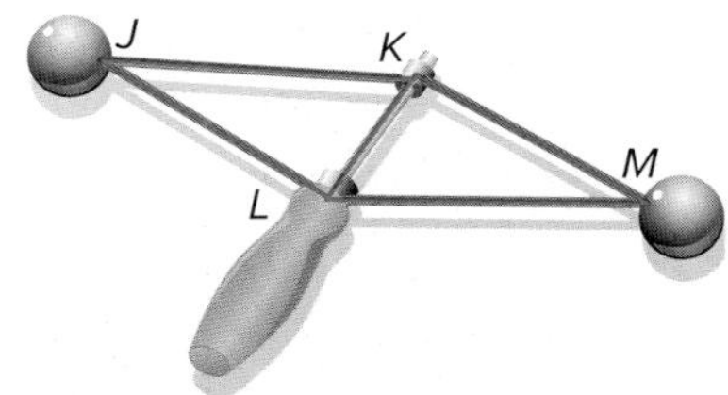

Example 2

PROOF **Write a two-column proof.**

9 **Given:** V is the midpoint of $\overline{YW}$; $\overline{UY} \parallel \overline{XW}$.

Prove: $\triangle UVY \cong \triangle XVW$

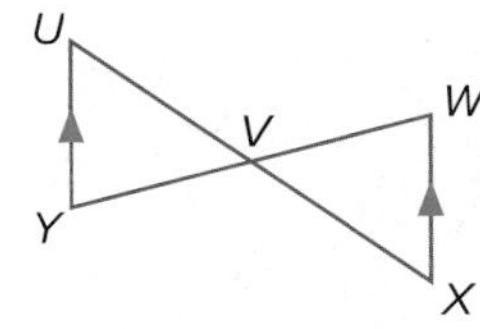

10. **Given:** $\overline{MS} \cong \overline{RQ}$, $\overline{MS} \parallel \overline{RQ}$

Prove: $\triangle MSP \cong \triangle RQP$

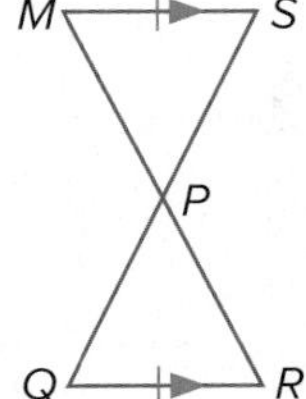

11. MP **CONSTRUCT ARGUMENTS** Write a flow proof.

Given: $\angle A$ and $\angle C$ are right angles.

$\angle ABE \cong \angle CBD$, $\overline{AE} \cong \overline{CD}$

Prove: $\overline{BE} \cong \overline{BD}$

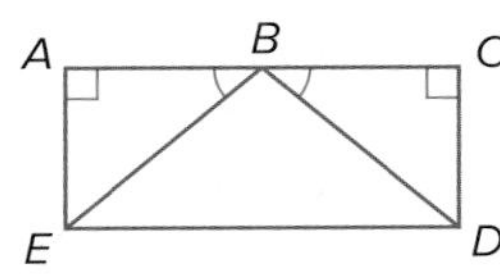

12. **PROOF** Write a flow proof.

Given: $\overline{KM}$ bisects $\angle JML$; $\angle J \cong \angle L$.

Prove: $\overline{JM} \cong \overline{LM}$

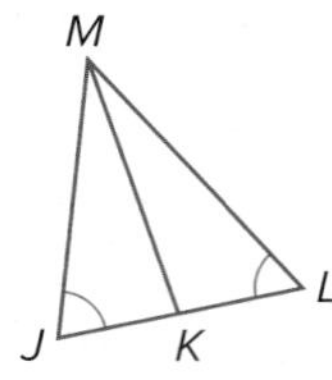

Example 3

13. **PERSEVERANCE** A high school wants to hold a 1500-meter regatta on Lake Powell but is unsure if the lake is long enough. To measure the distance across the lake, the crew members locate the vertices of the triangles below and find the measures of the lengths of $\triangle HJK$ as shown below.

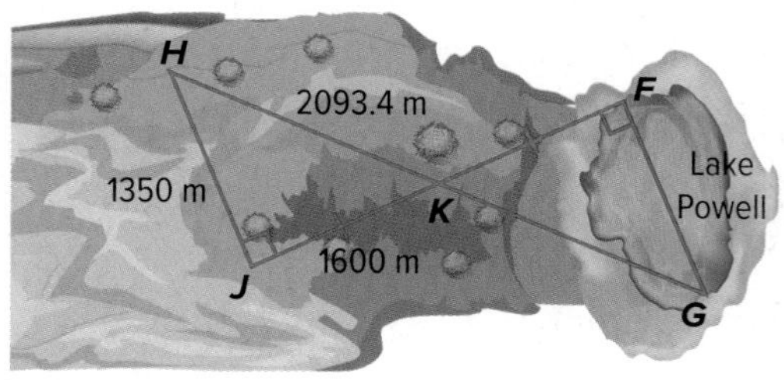

a. Explain how the crew team can use the triangles formed to estimate the distance FG across the lake.

b. Using the measures given, is the lake long enough for the team to use as the location for their regatta? Explain your reasoning.

ALGEBRA Find the value of the variable that yields congruent triangles.

14. $\triangle BCD \cong \triangle WXY$

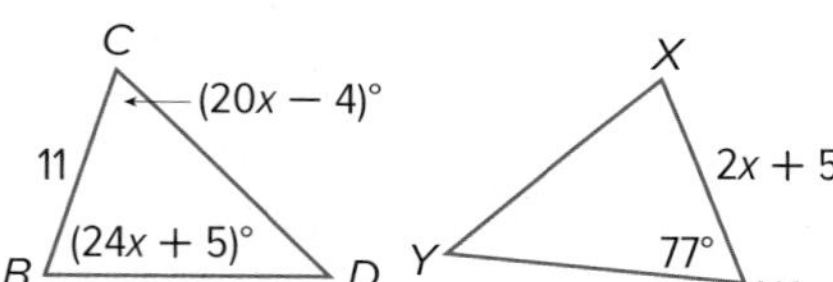

15. $\triangle MHJ \cong \triangle PQJ$

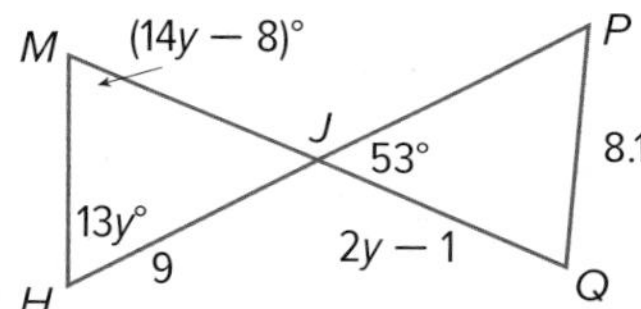

16. **THEATER DESIGN** The trusses of the roof of the outdoor theater shown below appear to be several different pairs of congruent triangles. Assume that trusses that appear to lie on the same line actually lie on the same line.

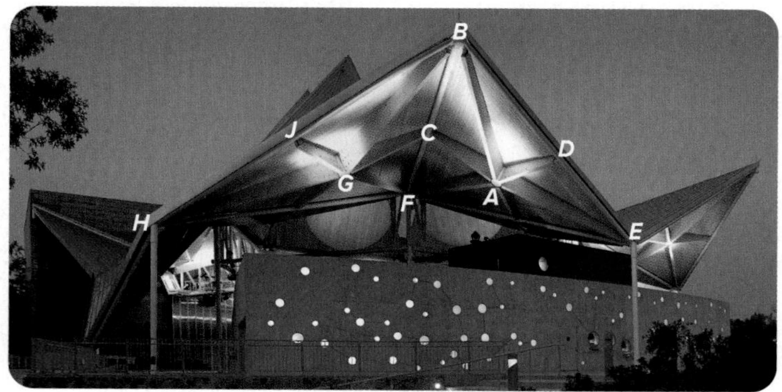

a. If $\overline{AB}$ bisects $\angle CBD$ and $\angle CAD$, prove that $\triangle ABC \cong \triangle ABD$.

b. If $\triangle ABC \cong \triangle ABD$ and $\angle FCA \cong \angle EDA$, prove that $\triangle CAF \cong \triangle DAE$.

c. If $\overline{HB} \cong \overline{EB}$, $\angle BHG \cong \angle BEA$, $\angle HGJ \cong \angle EAD$, and $\angle JGB \cong \angle DAB$, prove that $\triangle BHG \cong \triangle BEA$.

PROOF **Write a paragraph proof.**

17. **Given:** $\overline{AE} \perp \overline{DE}$, $\overline{EA} \perp \overline{AB}$, C is the midpoint of $\overline{AE}$.

Prove: $\overline{CD} \cong \overline{CB}$

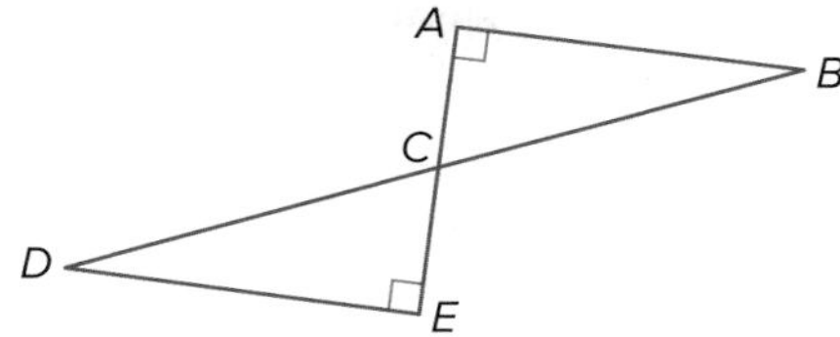

18. **Given:** $\angle F \cong \angle J$, $\overline{FH} \parallel \overline{GJ}$

Prove: $\overline{FH} \cong \overline{JG}$

PROOF **Write a two-column proof.**

19. **Given:** $\angle K \cong \angle M$, $\overline{KP} \perp \overline{PR}$, $\overline{MR} \perp \overline{PR}$

Prove: $\angle KPL \cong \angle MRL$

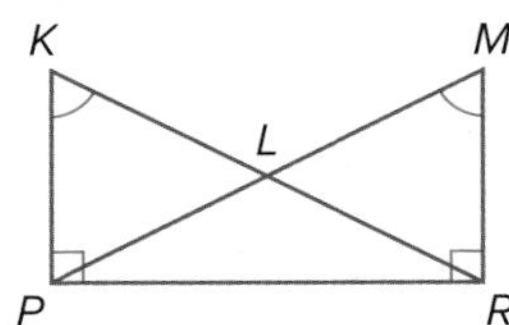

20. **Given:** $\overline{QR} \cong \overline{SR} \cong \overline{WR} \cong \overline{VR}$

Prove: $\overline{QT} \cong \overline{WU}$

21 **FITNESS** The seat tube of a bicycle forms a triangle with each seat and chain stay as shown. If each seat stay makes a 44° angle with its corresponding chain stay and each chain stay makes a 68° angle with the seat tube, show that the two seat stays are the same length.

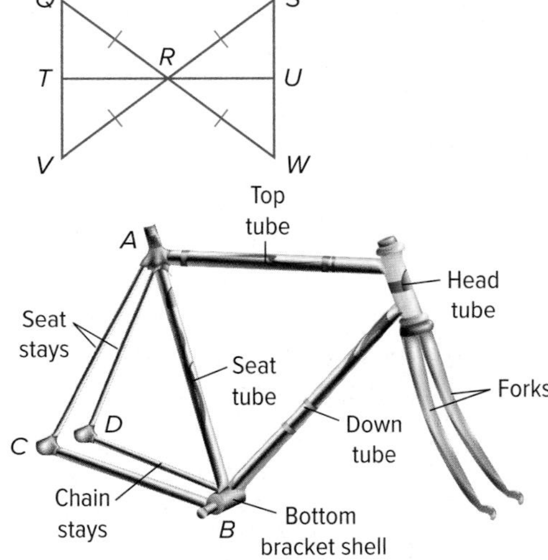

H.O.T. Problems Use Higher-Order Thinking Skills

22. **OPEN-ENDED** Draw and label two triangles that could be proved congruent by ASA.

23. **MP** **REASONING** Make a conjecture about each geometric relationship using the given construction.

a. Construct a triangle with two congruent angles. Make a conjecture about the sides opposite the constructed angles.

b. Construct a triangle with two congruent sides. Make a conjecture about the angles opposite the congruent sides.

24. **MP** **CONSTRUCT ARGUMENTS** Find a counterexample to show why SSA (Side-Side-Angle) cannot be used to prove the congruence of two triangles.

25. **MP** **REASONING** Describe the steps in an activity to illustrate the ASA criterion for triangle congruence for $\triangle ABC$ and $\triangle XYZ$. Then explain how this criterion follows from the principle of superposition.

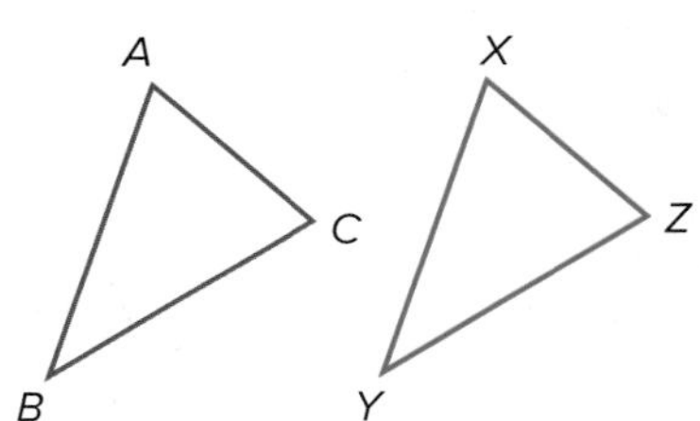

26. **WRITING IN MATH** How do you know which method (SSS, SAS, etc.) to use when proving triangle congruence? Use a chart to explain your reasoning.

27. MULTI-STEP Given: $\angle J \cong \angle M$ and $\angle K \cong \angle N$ MP 3

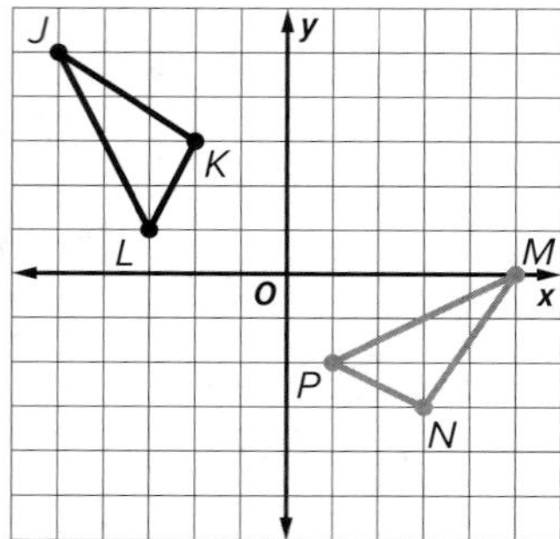

a. Prove that $\triangle JKL \cong \triangle MNP$ using ASA or AAS.

b. Describe a set of rigid motions that could be performed on $\triangle JKL$ to prove that it is congruent to $\triangle MNP$.

c. Perform the set of rigid motions, and list the coordinates of the vertices of the image of $\triangle JKL$ at each stage.

28. $\triangle EGH$ is an equilateral triangle and $\angle EDG \cong \angle EFH$. What is the perimeter of $\triangle EGH$? MP 1, 2

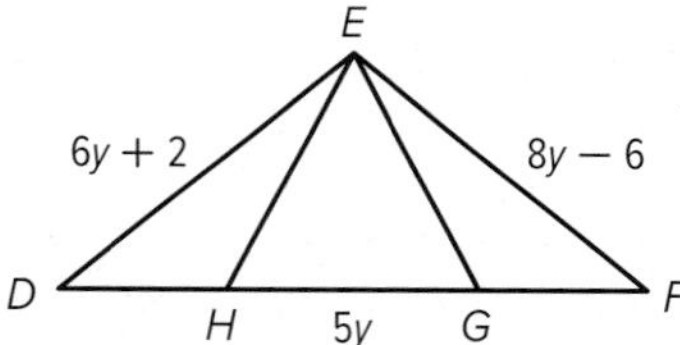

- A 20 units
- B 26 units
- C 30 units
- D 60 units
- E 72 units
- F 112 units

29. Given: $\overline{BC}$ is perpendicular to $\overline{AD}$; $\angle A \cong \angle D$

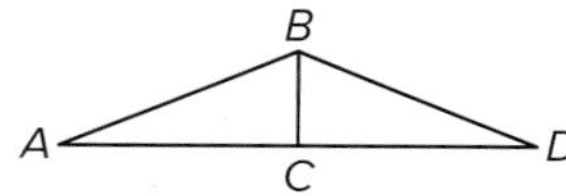

Prove that $\triangle ABC \cong \triangle DBC$. MP 3

30. Which *additional* information could be used to prove that $\triangle QRU \cong \triangle TRS$? Select all that apply. MP 3

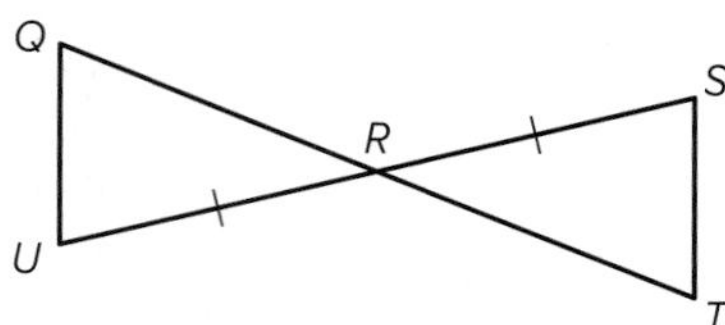

- ☐ A $\overline{QR} \cong \overline{TR}$
- ☐ B $\overline{QU} \cong \overline{TS}$
- ☐ C $\overline{US} \cong 2\overline{RS}$
- ☐ D $\angle U \cong \angle S$
- ☐ E $\angle QRU \cong \angle TRS$
- ☐ F R is the midpoint of $\overline{US}$.

31. Given that $\triangle ACD \cong \triangle CAB$ and $\overline{AD} \parallel \overline{BC}$, what is the measure, in degrees, of $\angle CBA$? MP 2

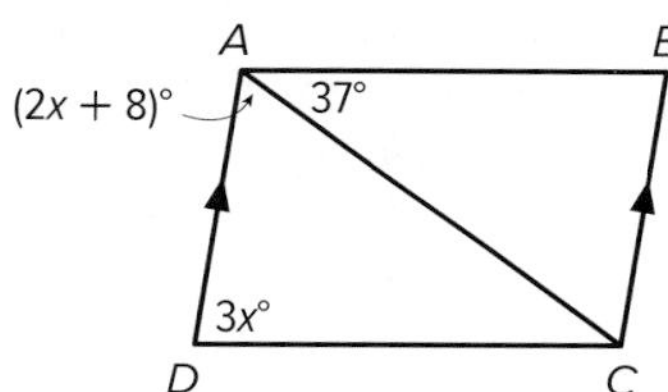

32. Given: $\overline{AC}$ and $\overline{DE}$ intersect at B.
$\overline{AD} \cong \overline{EC}$
$\angle 1 \cong \angle 2$

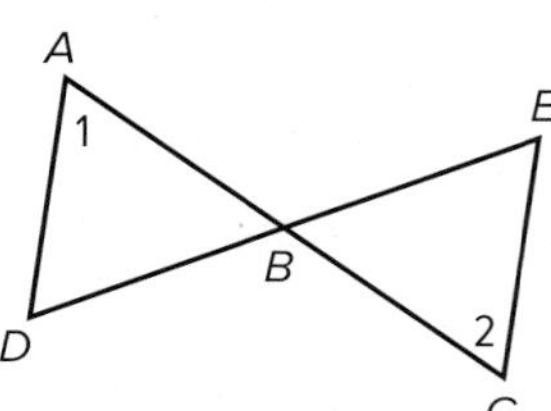

Prove that $\triangle ABD \cong \triangle CBE$. MP 3

33. Name the third pair of congruent parts needed to prove that $\triangle ABC \cong \triangle XYZ$ using ASA.
$\angle A \cong \angle X$
$\angle B \cong \angle Y$ MP 1

LESSON 5

Proving Right Triangles Congruent

Then

- You proved triangles congruent using SSS, SAS, ASA, and AAS.

Now

- 1 Use the right triangle congruence theorems to prove relationships in geometric figures.

Why?

- Many structures like bridges have support beams in the shape of triangles. Making sure the angles and lengths are correct is critical to the integrity of the structure.

Mathematical Practices

3 Construct viable arguments and critique the reasoning of others.

1 Right Triangle Congruence Theorems

There are four triangle congruence theorems that apply to right triangles. You can use these theorems to prove triangles congruent.

Theorem Right Triangle Congruence

Theorem 4.6 Leg-Leg Congruence
If the legs of one right triangle are congruent to the corresponding legs of another right triangle, then the triangles are congruent.

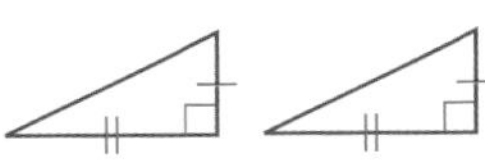

Abbreviation LL

Theorem 4.7 Hypotenuse-Angle Congruence
If the hypotenuse and acute angle of one right triangle are congruent to the hypotenuse and corresponding acute angle of another right triangle, then the two triangles are congruent.

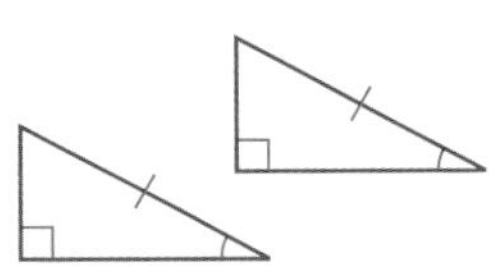

Abbreviation HA

Theorem 4.8 Leg-Angle Congruence
If one leg and an acute angle of one right triangle are congruent to the corresponding leg and acute angle of another right triangle, then the triangles are congruent.

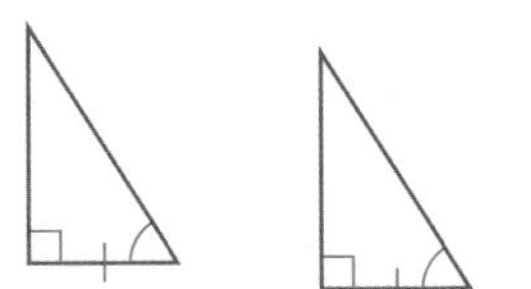

Abbreviation LA

Theorem 4.9 Hypotenuse-Leg Congruence
If the hypotenuse and a leg of one right triangle are congruent to the hypotenuse and corresponding leg of another right triangle, then the triangles are congruent.

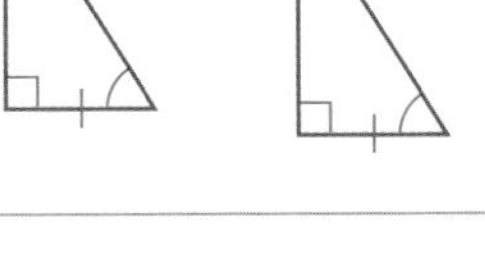

Abbreviation HL

You will prove Theorems 4.6, 4.7, 4.8, and 4.9 in Exercises 15–18.

Example 1 Right Triangle Congruence

Determine whether each pair of triangles is congruent. If yes, include the theorem or postulate that applies and describe the series of rigid motions that map one triangle onto the other.

a.

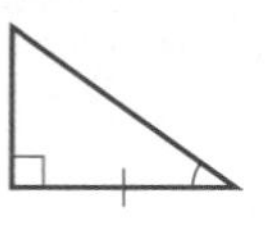

One pair of corresponding legs is congruent. One pair of corresponding acute angles is congruent. So, the triangles are congruent by LA.

The first triangle is mapped onto the other by a counterclockwise rotation of 90° and a translation.

b.

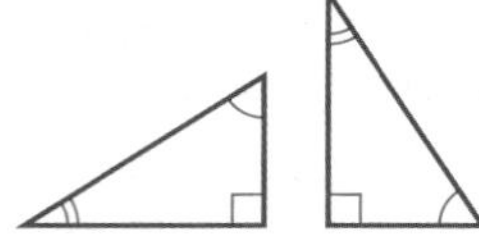

There is not enough information to determine triangle congruence.

At least one pair of corresponding sides must be congruent in order to show that these triangles are congruent.

Guided Practice

1.

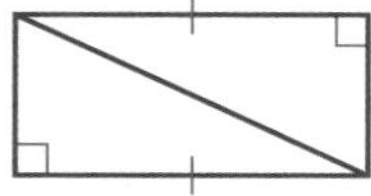

Example 2 Prove Right Triangle Congruence

Write the specified type of proof.

a. flow proof

Given: $\triangle PQR$ and $\triangle PSR$ are right triangles with right angles $\angle Q$ and $\angle S$ and $\angle PRQ \cong \angle PRS$.

Prove: $\triangle PQR \cong \triangle PSR$

Proof:

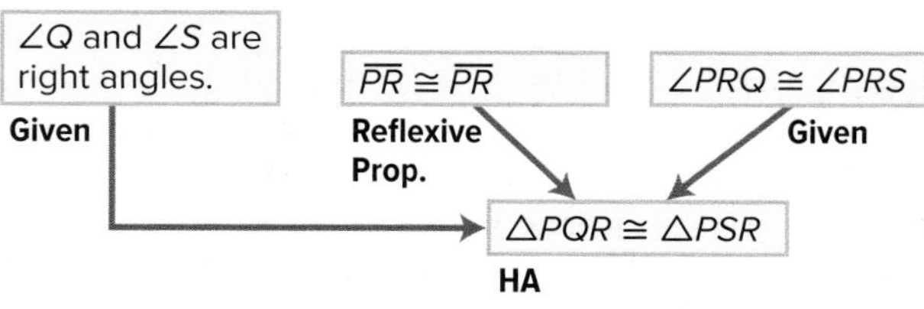

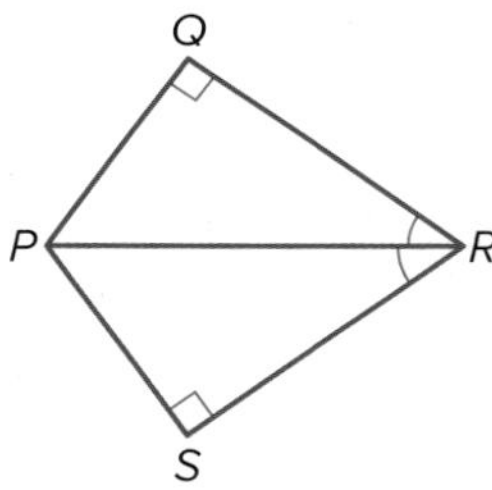

b. two-column proof

Given: $\angle V$ and $\angle X$ are right angles. W is the midpoint of $\overline{VX}$.

Prove: $\triangle UVW \cong \triangle YXW$

Proof:

Statements	Reasons
1. $\angle V$ and $\angle X$ are right angles.	**1.** Given
2. $\triangle UVW$ and $\triangle YXW$ are right triangles.	**2.** Definition of right triangles
3. $\angle UWV \cong \angle YWX$	**3.** Vertical Angles Thm
4. W is the midpoint of $\overline{VX}$.	**4.** Given
5. $\overline{VW} \cong \overline{WX}$	**5.** Midpoint Thm
6. $\triangle UVW \cong \triangle YXW$	**6.** LA

Watch Out

When you use the LA Theorem, be sure that the congruent legs and angles are corresponding.

Guided Practice

2. paragraph proof

Given: $\angle A$ and $\angle C$ are right angles.
$\overline{AB} \parallel \overline{DC}$

Prove: $\overline{BC} \parallel \overline{AD}$

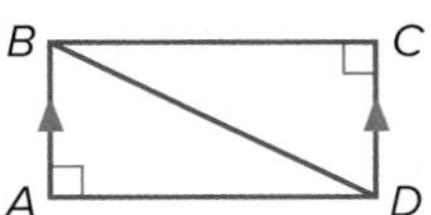

Real-World Example 3 Prove Right Triangles Congruent

ARCHITECTURE In the diagram of the cable-stayed bridge, the tower support is perpendicular to the bridge deck. The base of the tower support is located at the midpoint of the segment connecting the ends of the cables.

Given: $\overline{KL} \perp \overline{JM}$
L is the midpoint of $\overline{JM}$.

Prove: $\triangle JKL \cong \triangle MKL$

Proof:

Statements	Reasons
1. $\overline{KL} \perp \overline{JM}$	1. Given
2. $\angle KLJ$ and $\angle KLM$ are rt. ∠s	2. Def. of $\perp$ lines
3. $\triangle JKL$ and $\triangle MKL$ are rt. △s	3. Def. of rt. △
4. L is the midpoint of $\overline{JM}$.	4. Given
5. $\overline{JL} \cong \overline{LM}$	5. Midpoint Theorem
6. $\overline{KL} \cong \overline{KL}$	6. Reflexive Property
7. $\triangle JKL \cong \triangle MKL$	7. LL

Guided Practice

3. **ART** Quilters use triangles to make patterns that look like pinwheels. Write a paragraph proof.

Given: $\angle E$ and $\angle B$ are right angles.
C is the midpoint of $\overline{AD}$.

Prove: $\triangle ABC \cong \triangle DEC$

Check Your Understanding

◯ = Step-by-Step Solutions begin on page R13.

Example 1 **Determine whether each pair of triangles is congruent. If yes, include the theorem or postulate that applies and describe the series of rigid motions that map one triangle onto the other.**

1.

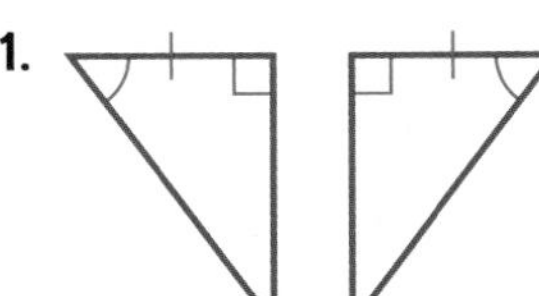

2.

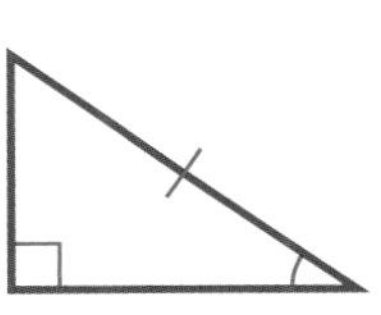

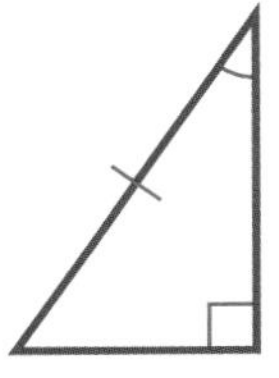

Examples 2–3 **PROOF** **Write a two-column proof.**

3 **Given:** $\overline{FH} \perp \overline{EG}$

$\overline{FH}$ bisects $\overline{EG}$.

Prove: $\triangle EFH \cong \triangle GFH$

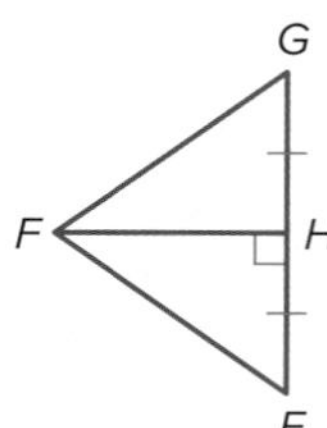

4. **Given:** $\overline{AD} \cong \overline{BC}$

$\angle A$ and $\angle C$ are right $\angle s$.

Prove: $\triangle ABD \cong \triangle CDB$

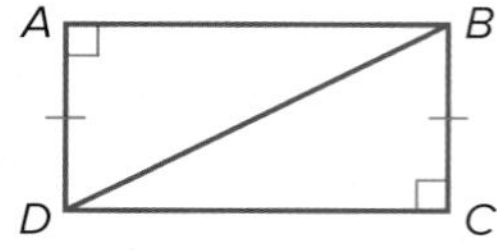

Practice and Problem Solving

Extra Practice is on page R4.

Example 1 **Determine whether each pair of triangles is congruent. If yes, include the theorem or postulate that applies and describe the series of rigid motions that map one triangle onto the other.**

5.

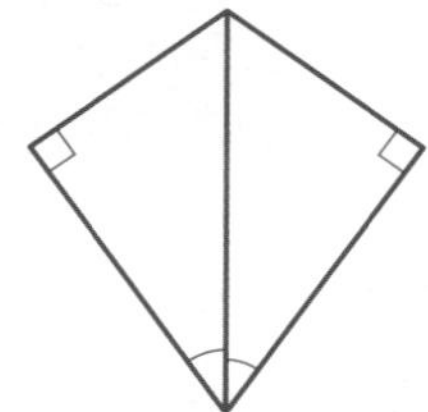

6.

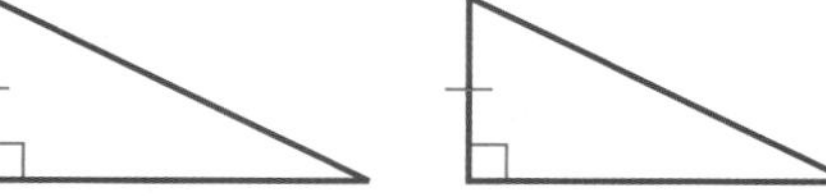

Examples 2–3 **PROOF** **Write a two-column proof.**

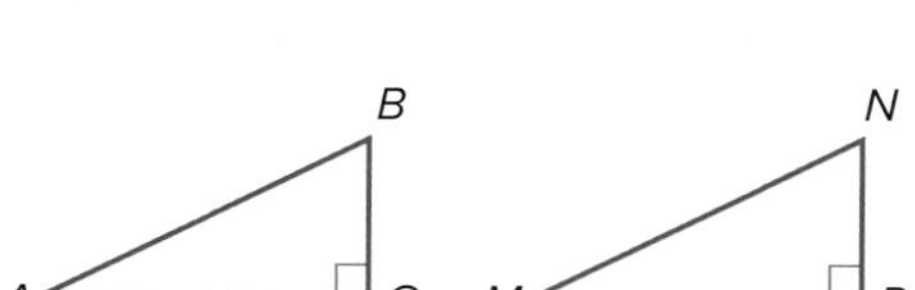

7. **Given:** $\overline{AB} \perp \overline{BC}$, $\overline{DC} \perp \overline{BC}$

$\overline{AC} \cong \overline{BD}$

Prove: $\overline{AB} \cong \overline{DC}$

8. **Given:** $\overline{AB} \parallel \overline{DC}$, $\overline{AB} \perp \overline{BC}$

E is the midpoint of $\overline{AC}$ and $\overline{BD}$.

Prove: $\overline{AC} \cong \overline{DB}$

9. **Given:** $\overline{AB} \perp \overline{BC}$, $\overline{DC} \perp \overline{BC}$

$\overline{AB} \cong \overline{DC}$

Prove: $\angle A \cong \angle D$

Which pairs of corresponding parts need to be congruent to prove $\triangle ABC \cong \triangle MNP$ using the indicated theorem?

10. LA

11. LL

12. HA

13. HL

14. Write a two-column proof.

Given: $\angle B$ and $\angle D$ are right angles.

$\overline{AD} \parallel \overline{BC}$

Prove: $\triangle ABC \cong \triangle CDA$

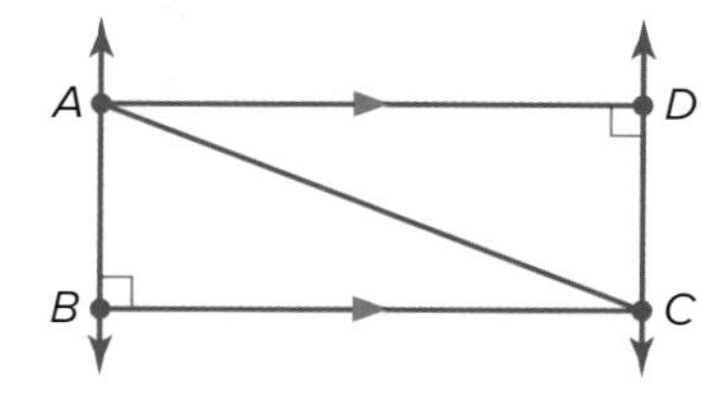

PROOF **Write the specified type of proof.**

15. two-column proof for the Hypotenuse-Leg Theorem (Hint: Use the Pythagorean Theorem.)

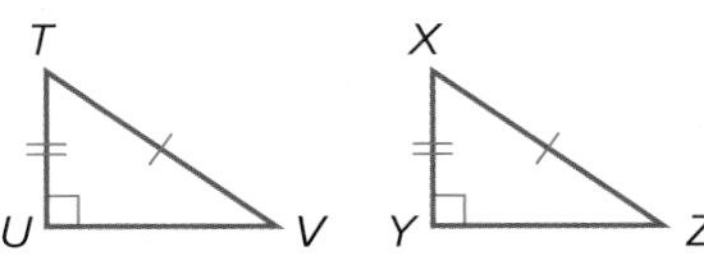

16. paragraph proof for the Leg-Leg Theorem

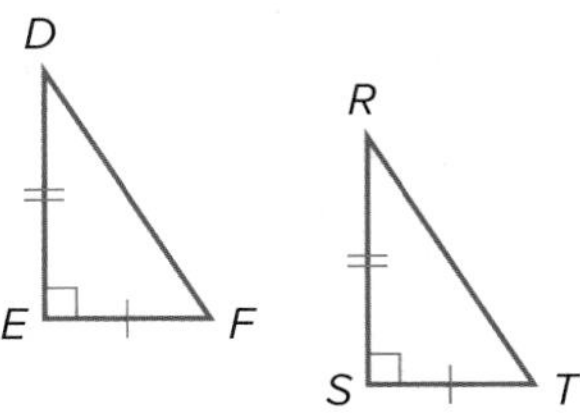

17. two column proof for the Leg-Angle Theorem (Hint: There are two possible cases.)

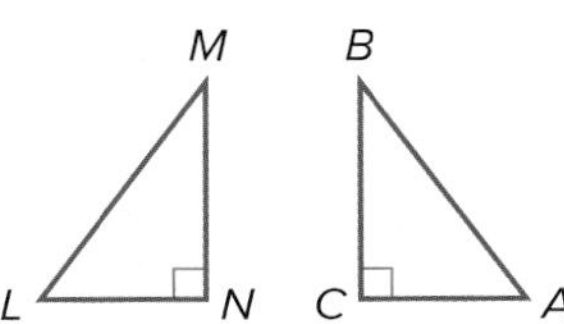

18. paragraph proof for the Hypotenuse-Angle Theorem

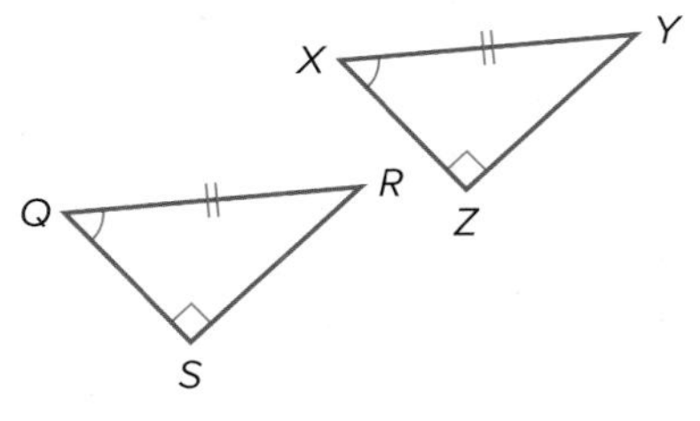

19. **CONSTRUCT ARGUMENTS** Craig and his brother are painting a house. The brothers' ladders are each the same length. If they both place the bottom of their ladders an equal distance away from the house, will each ladder reach to the same height on the house? Justify your argument.

H.O.T. Problems Use Higher-Order Thinking Skills

20. **CRITIQUE ARGUMENTS** Carlos sees two right triangles that have congruent hypotenuses and concludes that they are congruent by HA. Is he correct? If not, show a counterexample.

21. **CHALLENGE** Triangle *ABC* is equilateral. Segment *AD* cuts the triangle in half, making two congruent right triangles with acute angles of 30° and 60°. Write a paragraph proof to show that if the shorter leg of a 30-60-90 triangle is equal to x, then the hypotenuse is $2x$ and the longer leg is $x\sqrt{3}$.

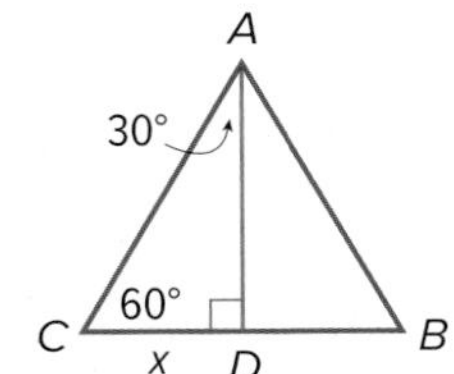

22. **REASONING** Suppose that an acute angle and one leg of a right triangle are congruent to an acute angle and one leg of another right triangle. Are the triangles congruent? Explain.

23. **OPEN-ENDED** Angle C of $\triangle ABC$ is a right angle, and $\angle F$ of $\triangle DEF$ is a right angle. The hypotenuses of both triangles are congruent. What additional piece(s) of information would be necessary to conclude that the two triangles are congruent?

24. **WRITING IN MATH** Explain how the Pythagorean Theorem could be used to confirm LL and HL.

25. What congruence rule can be used to prove $\triangle ABC \cong \triangle DEF$? MP 2

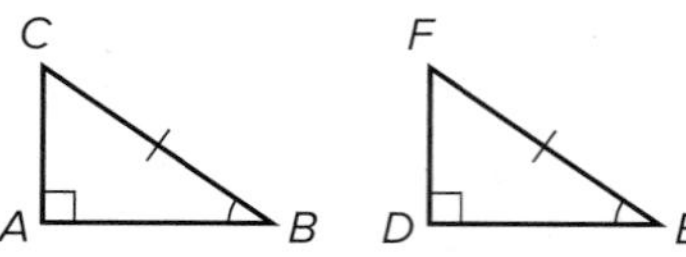

- ◯ **A** HL
- ◯ **B** HA
- ◯ **C** LA
- ◯ **D** The triangles might not be congruent.

26. In $\triangle TUV$, $\angle T$ is a right angle, and in $\triangle XYZ$, $\angle X$ is a right angle. To prove that $\triangle TUV \cong \triangle XYZ$ by LL, what pieces of information do you need to know? MP 2

- ◯ **A** $\overline{TU} \cong \overline{XY}$ and $\overline{UV} \cong \overline{YZ}$
- ◯ **B** $\overline{UV} \cong \overline{YZ}$ and $\overline{TV} \cong \overline{XZ}$
- ◯ **C** $\overline{TU} \cong \overline{XY}$ and $\overline{TV} \cong \overline{XZ}$
- ◯ **D** $\angle U \cong \angle X$ and $\overline{TU} \cong \overline{XY}$

27. In $\triangle ABC$, $\angle A$ is a right angle. In $\triangle DEF$, $\angle D$ is a right angle. For each given statement, which theorem can be used to prove congruence? If not enough information is provided, write *not enough information*. MP 2, 3

a. $\angle B \cong \angle E$; $\angle C \cong \angle F$

b. $\angle B \cong \angle E$; $\overline{BC} \cong \overline{EF}$

c. $\overline{BC} \cong \overline{EF}$, $\overline{BA} \cong \overline{ED}$

d. $\angle B \cong \angle E$, $\overline{AC} \cong \overline{DF}$

e. $\overline{BA} \cong \overline{ED}$, $\overline{AC} \cong \overline{DF}$

28. Use the information in the diagram. MP 2, 3

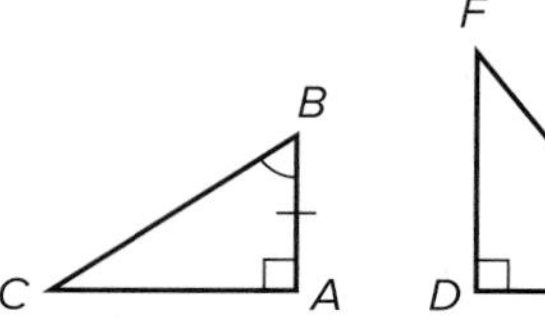

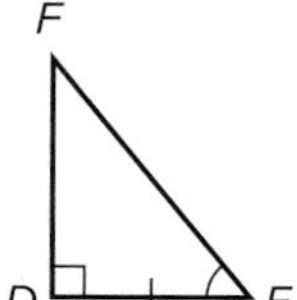

a. Are the triangles congruent? Explain.

b. Describe a series rigid motions that can map $\triangle ABC$ onto $\triangle DEF$.

29. Are the triangles congruent? Explain. MP 2, 3

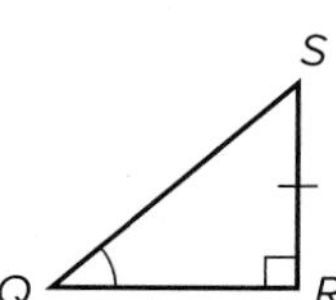

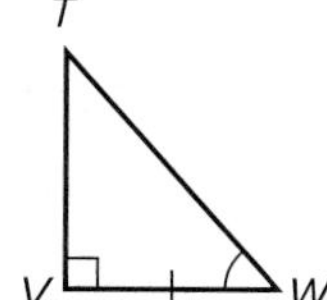

30. **MULTI-STEP** $\triangle WXY$ is an equilateral triangle. The distance from Y to Z is 42 feet. MP 3

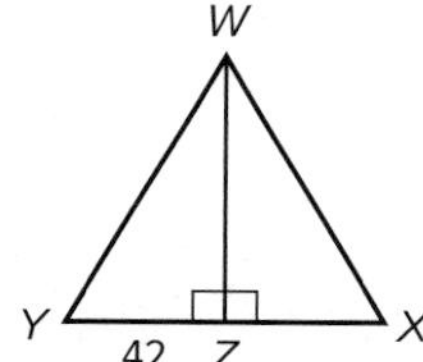

a. Write a paragraph proof to show $\triangle WYZ \cong \triangle WXZ$.

b. Find the length of $\overline{WX}$.

c. Find the length of $\overline{WZ}$.

d. Find the area of $\triangle WXY$.

31. The hypotenuses of two right triangles are congruent and have lengths $21 - 2x$ and $3x + 1$. The corresponding legs have lengths 5 and $x + 1$. Determine whether the triangles are congruent Justify your answer. MP 2

LESSON 6

Isosceles and Equilateral Triangles

Then

- You identified isosceles and equilateral triangles.

Now

1. Use properties of isosceles triangles.
2. Use properties of equilateral triangles.

Why?

- Engineers design roller coasters with triangular reinforcements between the tracks. This design lends support and stability. The triangle supports shown in the photo are isosceles triangles.

New Vocabulary

legs of an isosceles triangle
vertex angle
base angles

Mathematical Practices

2 Reason abstractly and quantitatively.
3 Construct viable arguments and critique the reasoning of others.

1 Properties of Isosceles Triangles

Recall that isosceles triangles have at least two congruent sides. The parts of an isosceles triangle have special names.

The two congruent sides are called the **legs of an isosceles triangle**, and the angle with sides that are the legs is called the **vertex angle**. The side of the triangle opposite the vertex angle is called the *base*. The two angles formed by the base and the congruent sides are called the **base angles**.

$\angle 1$ is the vertex angle.

$\angle 2$ and $\angle 3$ are the base angles.

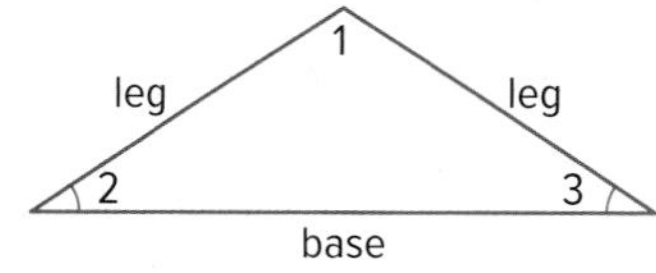

Theorems Isosceles Triangle

4.10 Isosceles Triangle Theorem
If two sides of a triangle are congruent, then the angles opposite those sides are congruent.

Example If $\overline{AC} \cong \overline{BC}$, then $\angle 2 \cong \angle 1$.

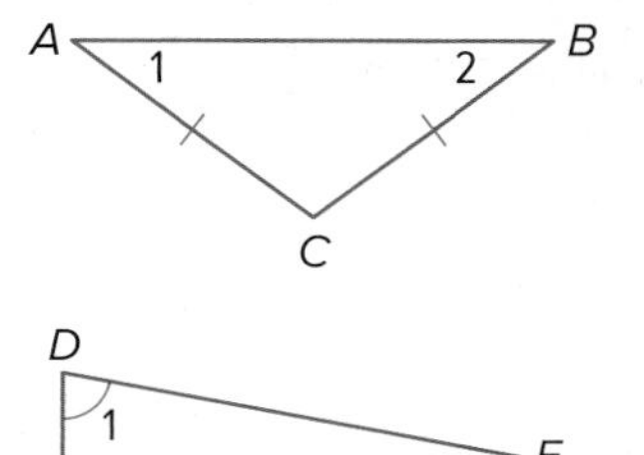

4.11 Converse of Isosceles Triangle Theorem
If two angles of a triangle are congruent, then the sides opposite those angles are congruent.

Example If $\angle 1 \cong \angle 2$, then $\overline{FE} \cong \overline{DE}$.

You will prove Theorem 4.11 in Exercise 37.

Example 1 Congruent Segments and Angles

a. Name two unmarked congruent angles.

$\angle ACB$ is opposite $\overline{AB}$ and $\angle B$ is opposite $\overline{AC}$, so $\angle ACB \cong \angle B$.

b. Name two unmarked congruent segments.

$\overline{AD}$ is opposite $\angle ACD$ and $\overline{AC}$ is opposite $\angle D$, so $\overline{AD} \cong \overline{AC}$.

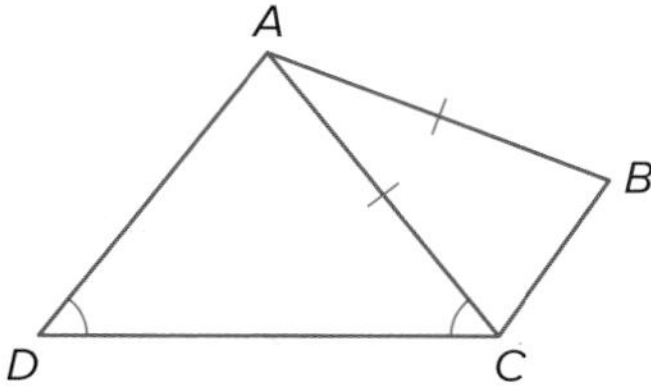

Guided Practice

1A. Name two unmarked congruent angles.

1B. Name two unmarked congruent segments.

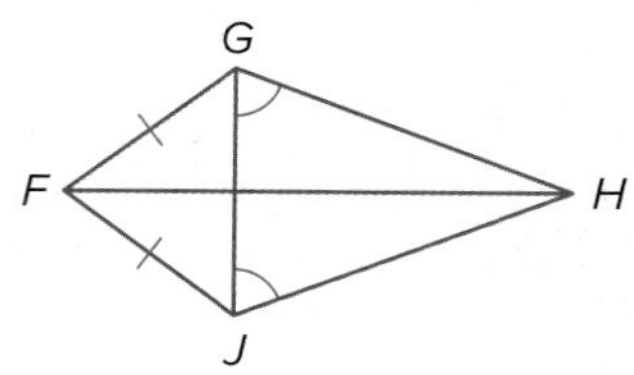

To prove the Isosceles Triangle Theorem, draw an auxiliary line and use the two triangles formed.

Proof Isosceles Triangle Theorem

Given: $\triangle LMP$; $\overline{LM} \cong \overline{LP}$

Prove: $\angle M \cong \angle P$

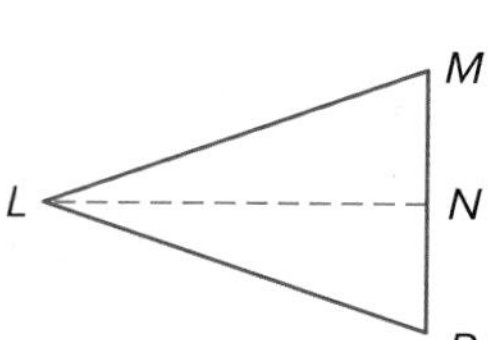

Proof:

Statements	Reasons
1. Let N be the midpoint of $\overline{MP}$.	1. Every segment has exactly one midpoint.
2. Draw an auxiliary segment $\overline{LN}$.	2. Two points determine a line.
3. $\overline{MN} \cong \overline{PN}$	3. Midpoint Theorem
4. $\overline{LN} \cong \overline{LN}$	4. Reflexive Property of Congruence
5. $\overline{LM} \cong \overline{LP}$	5. Given
6. $\triangle LMN \cong \triangle LPN$	6. SSS
7. $\angle M \cong \angle P$	7. CPCTC

Go Online! 

Recall that an equilateral triangle is a triangle with three congruent sides. **Vocabulary** is critically important to learning geometry. Find all the terms with animations, English pronunciations, and translations into 13 languages in the eGlossary in ConnectED.

2 Properties of Equilateral Triangles

The Isosceles Triangle Theorem leads to two corollaries about the angles of an equilateral triangle.

Corollaries Equilateral Triangle

4.3 A triangle is equilateral if and only if it is equiangular.

Example If $\angle A \cong \angle B \cong \angle C$, then $\overline{AB} \cong \overline{BC} \cong \overline{CA}$.

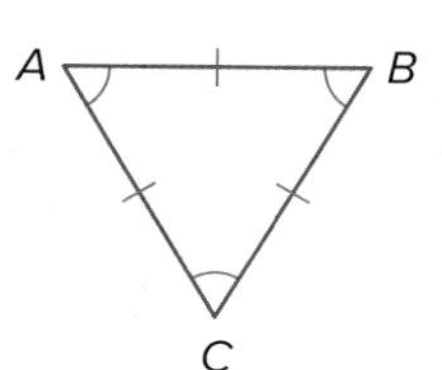

4.4 Each angle of an equilateral triangle measures 60.

Example If $\overline{DE} \cong \overline{EF} \cong \overline{FD}$, then $m\angle D = m\angle E = m\angle F = 60$.

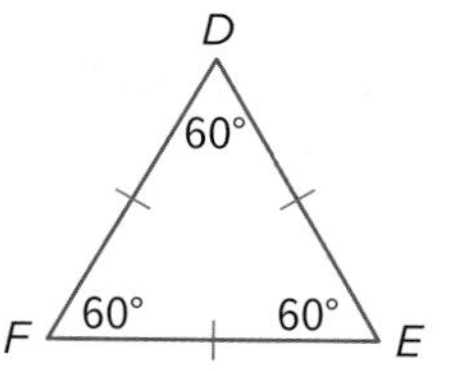

You will prove Corollaries 4.3 and 4.4 in Exercises 35 and 36.

Example 2 Find Missing Measures

Find each measure.

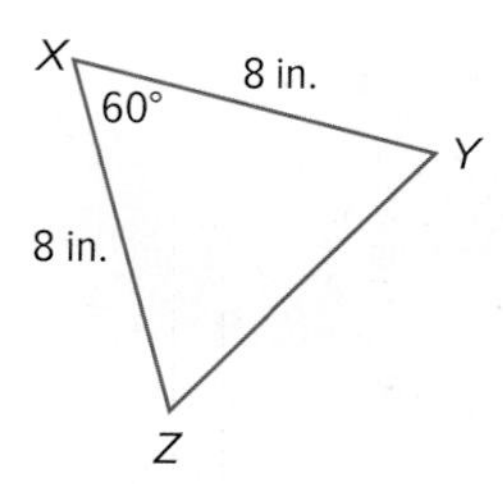

a. $m\angle Y$

Because $XY = XZ$, $\overline{XY} \cong \overline{XZ}$. By the Isosceles Triangle Theorem, base angles Z and Y are congruent, so $m\angle Z = m\angle Y$. Use the Triangle Angle-Sum Theorem to write and solve an equation to find $m\angle Y$.

$m\angle X + m\angle Y + m\angle Z = 180$	Triangle Angle-Sum Theorem
$60 + m\angle Y + m\angle Y = 180$	$m\angle X = 60$, $m\angle Z = m\angle Y$
$60 + 2(m\angle Y) = 180$	Simplify.
$2(m\angle Y) = 120$	Subtract 60 from each side.
$m\angle Y = 60$	Divide each side by 2.

b. YZ

$m\angle Z = m\angle Y$, so $m\angle Z = 60$ by substitution. Because $m\angle X = 60$, all three angles measure 60, so the triangle is equiangular. Because an equiangular triangle is also equilateral, $XY = XZ = ZY$. Because $XY = 8$ inches, $YZ = 8$ inches by substitution.

Study Tip

Isosceles Triangles As you discovered in Example 2, any isosceles triangle that has one 60° angle must be an equilateral triangle.

Guided Practice

2A. $m\angle M$ **2B.** PN

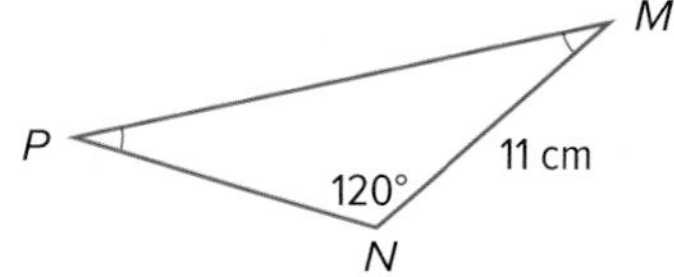

You can use the properties of equilateral triangles and algebra to find missing values.

Example 3 Find Missing Values

ALGEBRA Find the value of each variable.

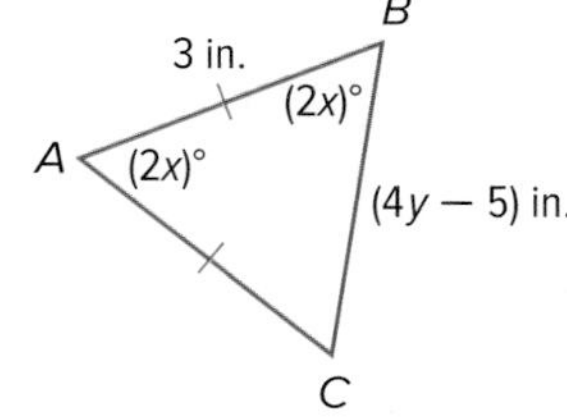

Since $\angle B = \angle A$, $\overline{AC} \cong \overline{BC}$ by the Converse of the Isosceles Triangle Theorem. All of the sides of the triangle are congruent, so the triangle is equilateral. Each angle of an equilateral triangle measures 60°, so $2x = 60$ and $x = 30$.

The triangle is equilateral, so all of the sides are congruent, and the lengths of all of the sides are equal.

$AB = BC$	Definition of equilateral triangle
$3 = 4y - 5$	Substitution
$8 = 4y$	Add 5 to each side.
$2 = y$	Divide each side by 4.

Guided Practice

3. Find the value of each variable.

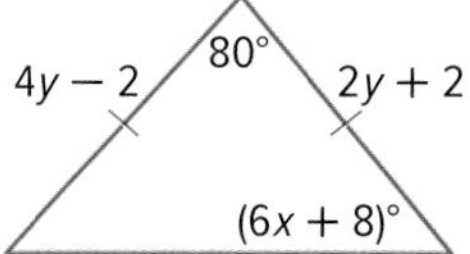

Real-World Link

Biosphere II is the largest totally enclosed ecosystem ever built, covering 3.14 acres in Oracle, Arizona. The controlled-environment facility is 91 feet at its highest point, and it has 6500 windows that enclose a volume of 7.2 million cubic feet.

Source: University of Arizona

Real-World Example 4 Apply Triangle Congruence

ENVIRONMENT Refer to the photo of Biosphere II at the right. $\triangle ACE$ is an equilateral triangle. F is the midpoint of $\overline{AE}$, D is the midpoint of $\overline{EC}$, and B is the midpoint of $\overline{CA}$. Prove that $\triangle FBD$ is also equilateral.

Given: $\triangle ACE$ is equilateral. F is the midpoint of $\overline{AE}$, D is the midpoint of $\overline{EC}$, and B is the midpoint of $\overline{CA}$.

Prove: $\triangle FBD$ is equilateral.

Proof:

Statements	Reasons
1. $\triangle ACE$ is equilateral.	1. Given
2. F is the midpoint of $\overline{AE}$, D is the midpoint of $\overline{EC}$, and B is the midpoint of $\overline{CA}$.	2. Given
3. $m\angle A = 60$, $m\angle C = 60$, $m\angle E = 60$	3. Each angle of an equilateral triangle measures 60.
4. $\angle A \cong \angle C \cong \angle E$	4. Definition of congruence and substitution
5. $\overline{AE} \cong \overline{EC} \cong \overline{CA}$	5. Definition of equilateral triangle
6. $AE = EC = CA$	6. Definition of congruence
7. $\overline{AF} \cong \overline{FE}$, $\overline{ED} \cong \overline{DC}$, $\overline{CB} \cong \overline{BA}$	7. Midpoint Theorem
8. $AF = FE$, $ED = DC$, $CB = BA$	8. Definition of congruence
9. $AF + FE = AE$, $ED + DC = EC$, $CB + BA = CA$	9. Segment Addition Postulate
10. $AF + AF = AE$, $FE + FE = AE$, $ED + ED = EC$, $DC + DC = EC$, $CB + CB = CA$, $BA + BA = CA$	10. Substitution Property
11. $2AF = AE$, $2FE = AE$, $2ED = EC$, $2DC = EC$, $2CB = CA$, $2BA = CA$	11. Addition Property
12. $2AF = AE$, $2FE = AE$, $2ED = AE$, $2DC = AE$, $2CB = AE$, $2BA = AE$	12. Substitution Property
13. $2AF = 2ED = 2CB$, $2FE = 2DC = 2BA$	13. Transitive Property
14. $AF = ED = CB$, $FE = DC = BA$	14. Division Property
15. $\overline{AF} \cong \overline{ED} \cong \overline{CB}$, $\overline{FE} \cong \overline{DC} \cong \overline{BA}$	15. Definition of congruence
16. $\triangle AFB \cong \triangle EDF \cong \triangle CBD$	16. SAS
17. $\overline{DF} \cong \overline{FB} \cong \overline{BD}$	17. CPCTC
18. $\triangle FBD$ is equilateral.	18. Definition of equilateral triangle

Guided Practice

4. Given that $\triangle ACE$ is equilateral, $\overline{FB} \parallel \overline{EC}$, $\overline{FD} \parallel \overline{BC}$, $\overline{BD} \parallel \overline{EF}$, and D is the midpoint of $\overline{EC}$, prove that $\triangle FED \cong \triangle BDC$.

Check Your Understanding

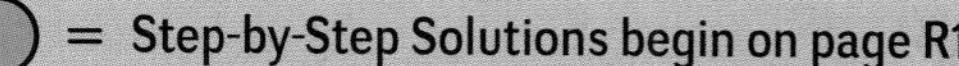

Example 1 **Refer to the figure at the right.**

1. If $\overline{AB} \cong \overline{CB}$, name two congruent angles.
2. If $\angle EAC \cong \angle ECA$, name two congruent segments.

Example 2 **Find each measure.**

3. FH

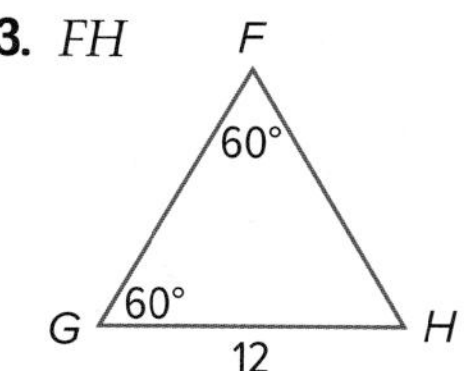

4. $m\angle MRP$

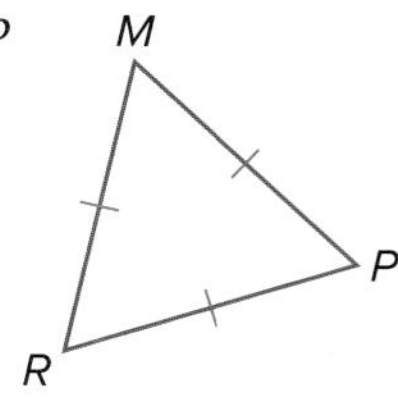

Example 3 **SENSE-MAKING** **Find the value of each variable.**

5.

6.

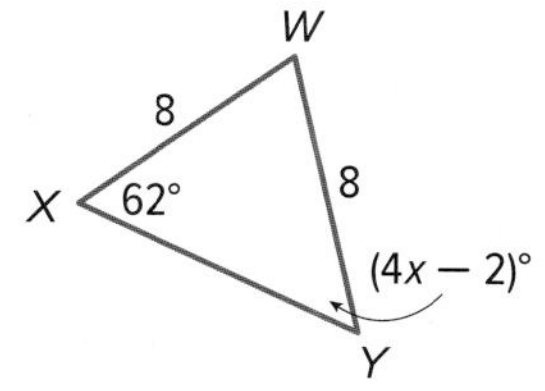

Example 4 7. **PROOF** **Write a two-column proof.**

Given: $\triangle ABC$ is isosceles; $\overline{EB}$ bisects $\angle ABC$.

Prove: $\triangle ABE \cong \triangle CBE$

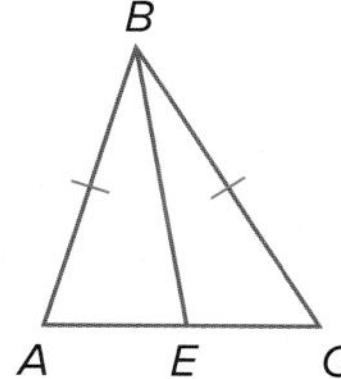

8. **ROLLER COASTERS** The roller coaster track shown in the photo on page 285 appears to be composed of congruent triangles. A portion of the track is shown.

   a. If $\overline{QR}$ and $\overline{ST}$ are perpendicular to $\overline{QT}$, $\triangle VSR$ is isosceles with base $\overline{SR}$, and $\overline{QT} \parallel \overline{SR}$, prove that $\triangle RQV \cong \triangle STV$.

   b. If $VR = 2.5$ meters and $QR = 2$ meters, find the distance between $\overline{QR}$ and $\overline{ST}$. Explain your reasoning.

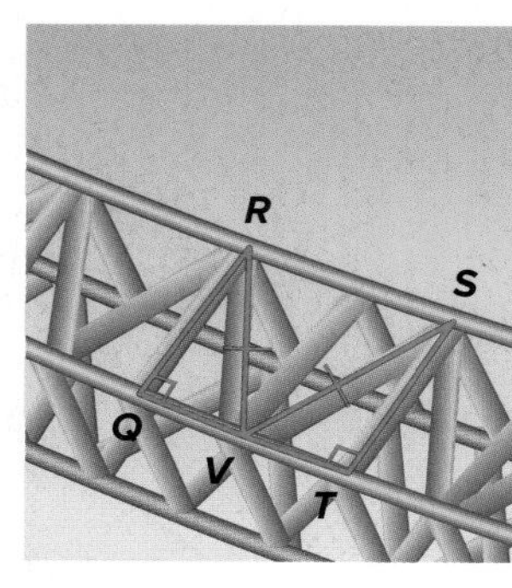

Practice and Problem Solving

Extra Practice is on page R4.

Example 1 **Refer to the figure at the right.**

9. If $\overline{AB} \cong \overline{AE}$, name two congruent angles.
10. If $\angle ABF \cong \angle AFB$, name two congruent segments.
11. If $\overline{CA} \cong \overline{DA}$, name two congruent angles.
12. If $\angle DAE \cong \angle DEA$, name two congruent segments.
13. If $\angle BCF \cong \angle BFC$, name two congruent segments.
14. If $\overline{FA} \cong \overline{AH}$, name two congruent angles.

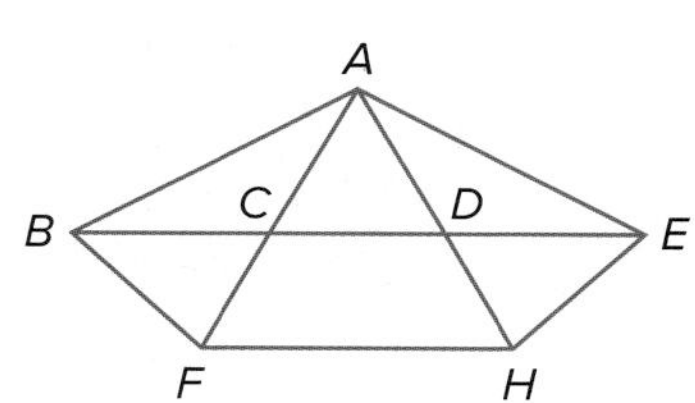

Example 2 **Find each measure.**

15. $m\angle BAC$

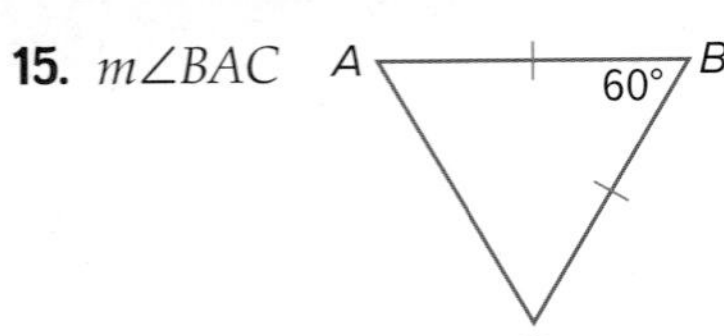

16. $m\angle SRT$

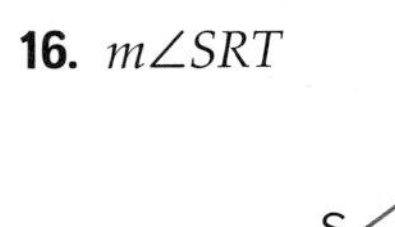
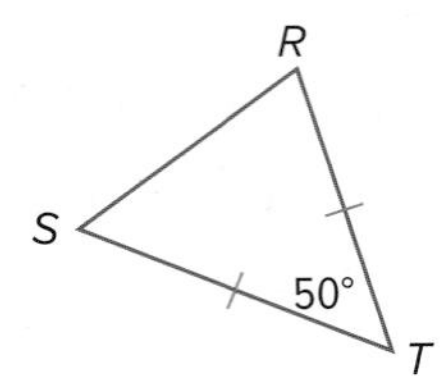

17. TR

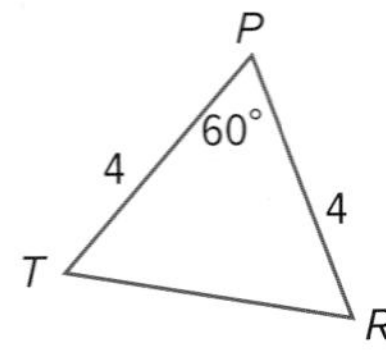

18. CB

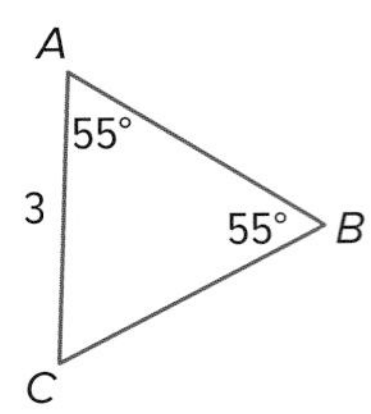

Example 3 **REGULARITY** **Find the value of each variable.**

19.

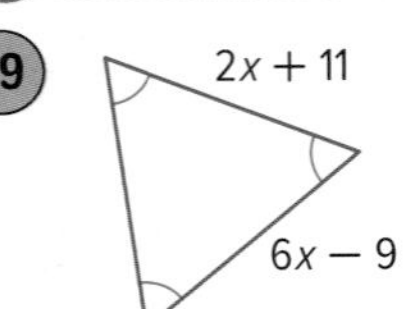

20.

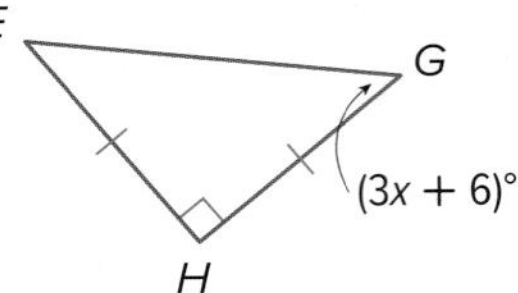

21.

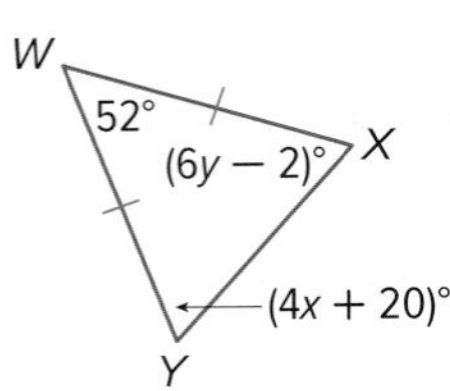

22.

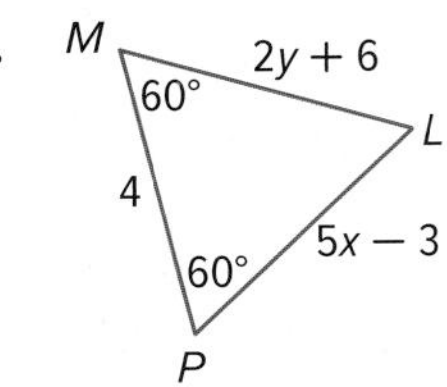

Example 4 **PROOF** **Write a paragraph proof.**

23. **Given:** $\triangle HJM$ is isosceles, and $\triangle HKL$ is equilateral. $\angle JKH$ and $\angle HKL$ are supplementary and $\angle HLK$ and $\angle MLH$ are supplementary.

Prove: $\angle JHK \cong \angle MHL$

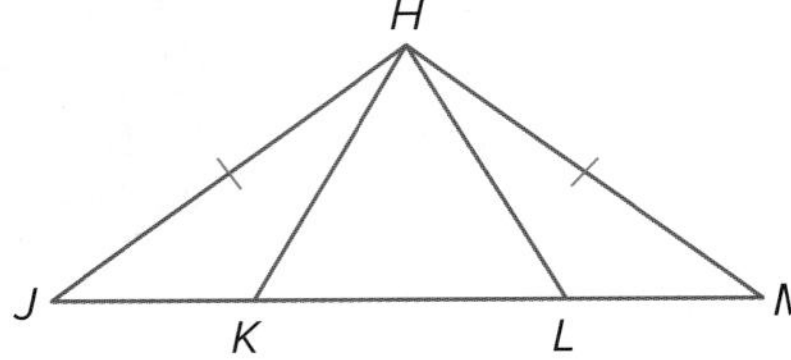

24. **Given:** $\overline{XY} \cong \overline{XZ}$
W is the midpoint of $\overline{XY}$.
Q is the midpoint of $\overline{XZ}$.
Prove: $\overline{WZ} \cong \overline{QY}$

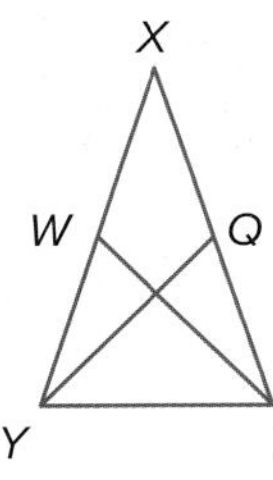

25. **BABYSITTING** While babysitting her neighbor's children, Elisa observes that the supports on either side of a park swing set form two sets of triangles. Using a jump rope to measure, Elisa is able to determine that $\overline{AB} \cong \overline{AC}$, but $\overline{BC} \not\cong \overline{AB}$.

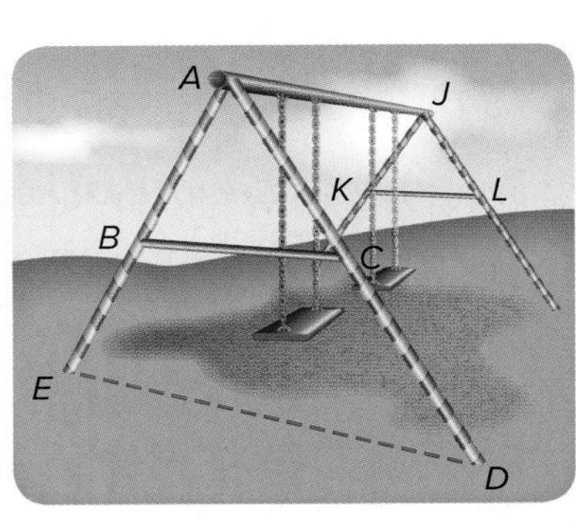

a. Elisa estimates $m\angle BAC$ to be 50. Based on this estimate, what is $m\angle ABC$? Explain.

b. If $\overline{BE} \cong \overline{CD}$, show that $\triangle AED$ is isosceles.

c. If $\overline{BC} \parallel \overline{ED}$ and $\overline{ED} \cong \overline{AD}$, show that $\triangle AED$ is equilateral.

d. If $\triangle JKL$ is isosceles, what is the minimum information needed to prove that $\triangle ABC \cong \triangle JLK$? Explain your reasoning.

26. **CHIMNEYS** In the picture, $\overline{BD} \perp \overline{AC}$ and $\triangle ABC$ is an isosceles triangle with base $\overline{AC}$. Show that the chimney of the house, represented by $\overline{BD}$, bisects the angle formed by the sloped sides of the roof, $\angle ABC$.

27. **CONSTRUCTION** Construct three different isosceles right triangles. Explain your method. Then verify your constructions using measurement and mathematics.

28. **PROOF** Based on your construction in Exercise 27, make and prove a conjecture about the relationship between the base angles of an isosceles right triangle.

MP **REGULARITY** **Find each measure.**

29 $m\angle CAD$

30. $m\angle ACD$

31. $m\angle ACB$

32. $m\angle ABC$

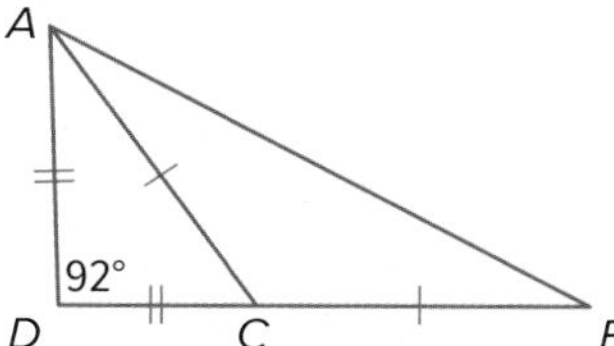

33. **FITNESS** In the diagram, the rider will use his bike to hop across the tops of each of the concrete solids shown. If each triangle is isosceles with vertex angles G, H, and J, and $\overline{BG} \cong \overline{HC}$, $\overline{HD} \cong \overline{JF}$, $\angle G \cong \angle H$, and $\angle H \cong \angle J$, show that the distance from B to F is three times the distance from D to F.

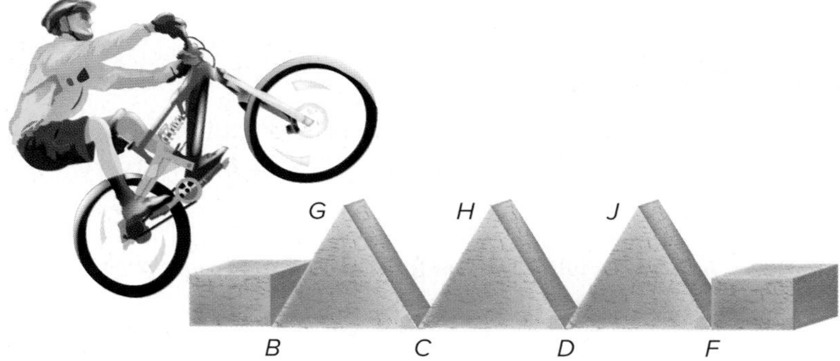

34. **Given:** $\triangle XWV$ is isosceles; $\overline{ZY} \perp \overline{YV}$.

Prove: $\angle X$ and $\angle YZV$ are complementary.

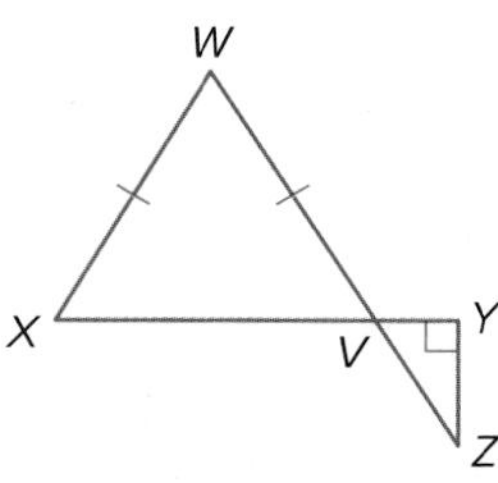

PROOF **Write a two-column proof of each corollary or theorem.**

35. Corollary 4.3

36. Corollary 4.4

37. Theorem 4.11

Find the value of each variable.

38.

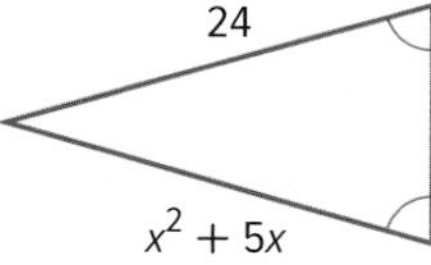

39.

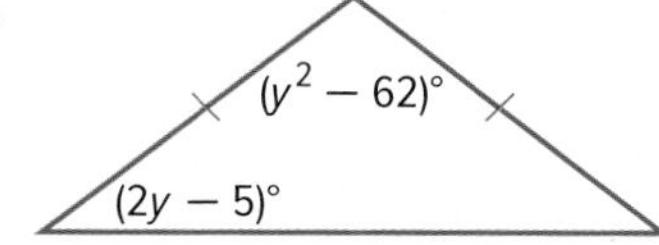

GAMES **Use the diagram of a game timer shown to find each measure.**

40. $m\angle LPM$

41 $m\angle LMP$

42. $m\angle JLK$

43. $m\angle JKL$

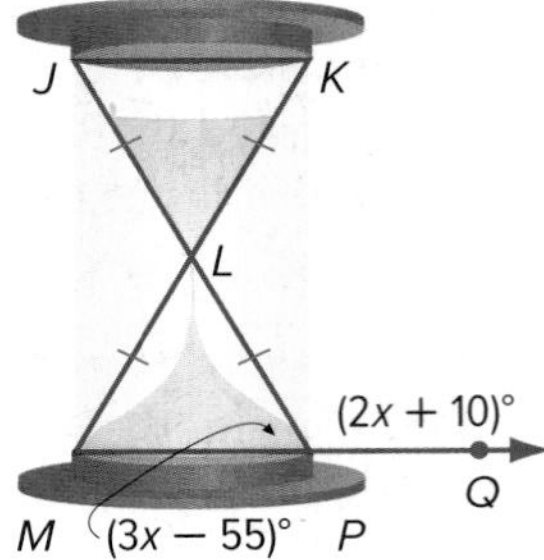

44. MULTIPLE REPRESENTATIONS In this problem, you will explore possible measures of the interior angles of an isosceles triangle given the measure of one exterior angle.

a. **Geometric** Use a ruler and a protractor to draw three different isosceles triangles, extending one of the sides adjacent to the vertex angle and to one of the base angles, and labeling as shown.

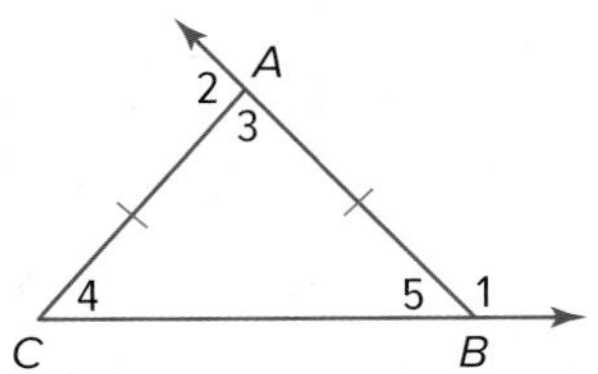

b. **Tabular** Use a protractor to measure and record $m\angle 1$ for each triangle. Use $m\angle 1$ to calculate the measures of $\angle 3$, $\angle 4$, and $\angle 5$. Then find and record $m\angle 2$ and use it to calculate these same measures. Organize your results in two tables.

c. **Verbal** Explain how you used $m\angle 1$ to find the measures of $\angle 3$, $\angle 4$, and $\angle 5$. Then explain how you used $m\angle 2$ to find these same measures.

d. **Algebraic** If $m\angle 1 = x$, write an expression for the measures of $\angle 3$, $\angle 4$, and $\angle 5$. Likewise, if $m\angle 2 = x$, write an expression for these same angle measures.

H.O.T. Problems Use Higher-Order Thinking Skills

45. CHALLENGE In the figure at the right, if $\triangle WJZ$ is equilateral and $\angle ZWP \cong \angle WJM \cong \angle JZL$, prove that $\overline{WP} \cong \overline{ZL} \cong \overline{JM}$.

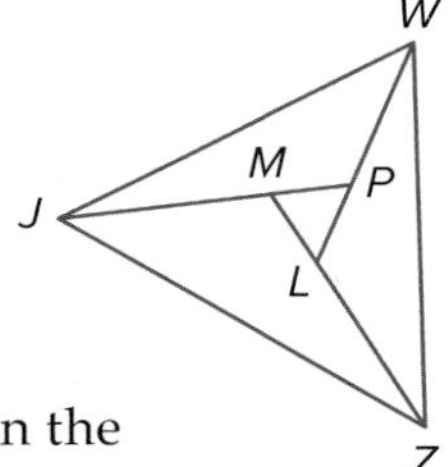

MP PRECISION **Determine whether the following statements are *sometimes*, *always*, or *never* true. Explain.**

46. If the measure of the vertex angle of an isosceles triangle is an integer, then the measure of each base angle is an integer.

47. If the measures of the base angles of an isosceles triangle are integers, then the measure of its vertex angle is odd.

48. ERROR ANALYSIS Alexis and Miguela are finding $m\angle G$ in the figure shown. Alexis says that $m\angle G = 35$, while Miguela says that $m\angle G = 60$. Is either of them correct? Explain your reasoning.

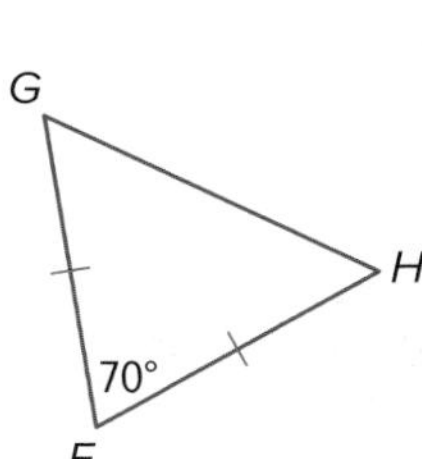

49. OPEN-ENDED If possible, draw an isosceles triangle with base angles that are obtuse. If it is not possible, explain why not.

50. MP **REASONING** In isosceles $\triangle ABC$, $m\angle B = 90$. Draw the triangle. Indicate the congruent sides and label each angle with its measure.

51. **WRITING IN MATH** How can triangle classifications help you prove triangle congruence?

52. **MULTI-STEP** $\triangle AED$ and $\triangle BEC$ are congruent and isosceles. MP 1, 2

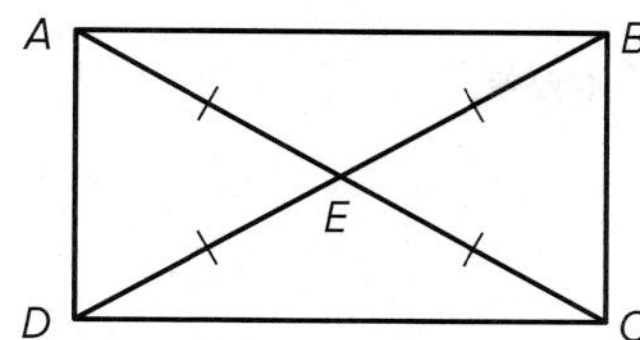

a. Which statement is not necessarily true from the information given?

- ◯ **A** $\triangle AEB \cong \triangle DEC$
- ◯ **B** $\overline{AD} \cong \overline{BE}$
- ◯ **C** $\angle CAD \cong \angle DBC$
- ◯ **D** $\overline{AB} \cong \overline{DC}$

b. If $m\angle EAB = 40$, what is the measure of $\angle DCE$? ▭°

c. If $m\angle ADE = 50$, what is the measure of $\angle BEC$? ▭°

d. What measure would $\angle AEB$ need to be so that $\overline{AB} \cong \overline{EC}$? Explain.

53. What is $m\angle FEG$? MP 2, 8

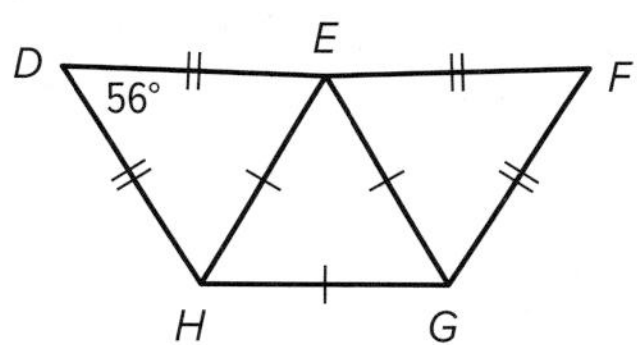

- ◯ **A** 34
- ◯ **B** 56
- ◯ **C** 60
- ◯ **D** 62
- ◯ **E** 124

54. A jeweler bends wire to make charms in the shape of isosceles triangles. The base of the triangle measures 2 centimeters and another side measures 3.5 centimeters. How many charms can be made from a piece of wire that is 63 centimeters long? MP 2

▭

55. In the figure below, $\triangle NDG \cong \triangle LGD$. MP 2, 8

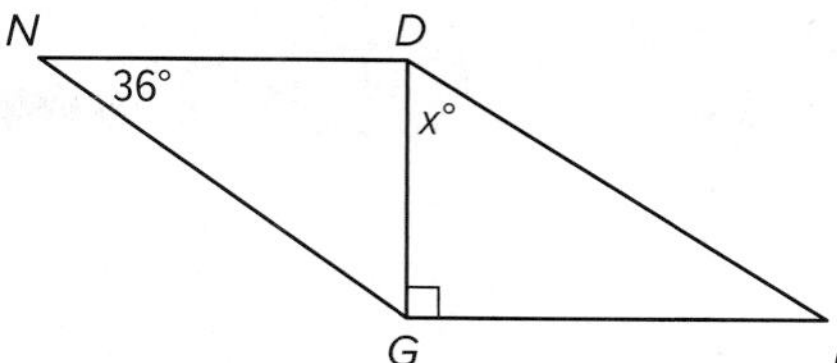

a. What is $m\angle L$? ▭°

b. What is the value of x? ▭

c. What is $m\angle DGN$? ▭°

56. An equilateral triangle has a perimeter of 216 inches. What is the length of one side of the triangle? MP 2

▭

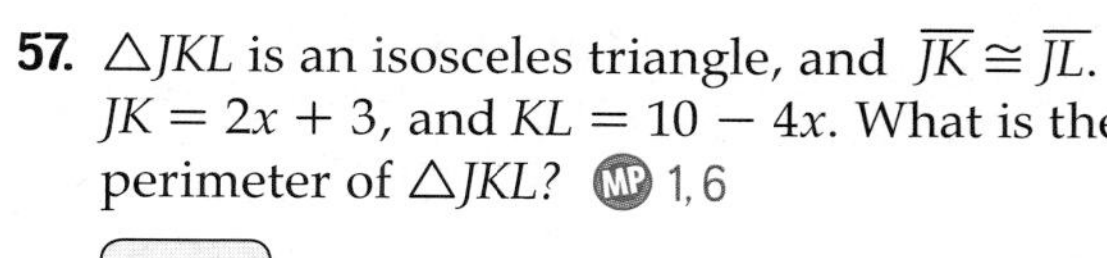

57. $\triangle JKL$ is an isosceles triangle, and $\overline{JK} \cong \overline{JL}$. $JK = 2x + 3$, and $KL = 10 - 4x$. What is the perimeter of $\triangle JKL$? MP 1, 6

▭

58. $\triangle ABC$ is an isosceles triangle with base angles measuring $(5x + 5)^\circ$ and $(7x - 19)^\circ$. Find the value of x. MP 6

▭

59. $\triangle QRV$ is an equilateral triangle. What additional information would be enough to prove $\triangle VRT$ is an equilateral triangle? MP 1, 2

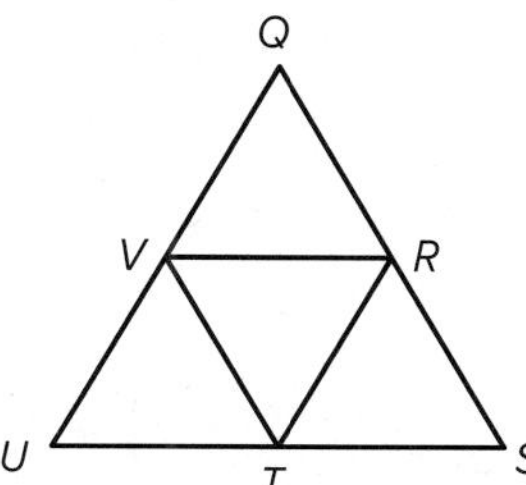

- ◯ **A** $\overline{TS} \cong \overline{RT}$
- ◯ **B** $m\angle VRT = 60$
- ◯ **C** $\triangle QRV \cong \triangle VTU$
- ◯ **D** $\triangle VRT \cong \triangle RST$

LESSON 7

Triangles and Coordinate Proof

Then

- You used coordinate geometry to prove triangle congruence

Now

1. Position and label triangles for use in coordinate proofs.
2. Write coordinate proofs.

Why?

- A global positioning system (GPS) receives transmissions from satellites that allow the exact location of a car to be determined. The information can be used with navigation software to provide driving directions.

New Vocabulary

coordinate proof

Mathematical Practices

3 Construct viable arguments and critique the reasoning of others.

6 Attend to precision

7 Look for and make use of structure.

1 Position and Label Triangles

As with global positioning systems, knowing the coordinates of a figure in a coordinate plane allows you to explore its properties and draw conclusions about it. **Coordinate proofs** use figures in the coordinate plane and algebra to prove geometric concepts. The first step in a coordinate proof is placing the figure on the coordinate plane.

Example 1 Position and Label a Triangle

Position and label right triangle MNP on the coordinate plane so that leg $\overline{MN}$ is a units long and leg $\overline{NP}$ is b units long.

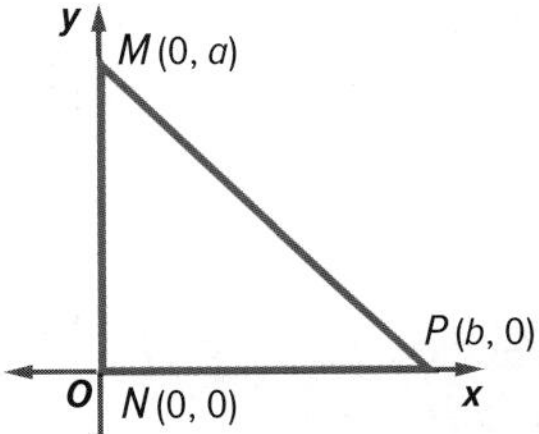

- The length(s) of the side(s) that are along the axes will be easier to determine than the length(s) of side(s) that are not along an axis. Since this is a right triangle, two sides can be located on an axis.
- Placing the right angle of the triangle, $\angle N$, at the origin will allow the two legs to be along the x- and y-axes.
- Position the triangle in the first quadrant.
- Since M is on the y-axis, its x-coordinate is 0. Its y-coordinate is a because the leg is a units long.
- Since P is on the x-axis, its y-coordinate is 0. Its x-coordinate is b because the leg is b units long.

Guided Practice

1. Position and label isosceles triangle JKL on the coordinate plane so that its base $\overline{JL}$ is a units long, vertex K is on the y-axis, and the height of the triangle is b units.

Key Concept Placing Triangles on Coordinate Plane

Step 1 Use the origin as a vertex or center of the triangle.

Step 2 Place at least one side of a triangle on an axis.

Step 3 Keep the triangle within the first quadrant if possible.

Step 4 Use coordinates that make computations as simple as possible.

Example 2 Identify Missing Coordinates

Name the missing coordinates of isosceles triangle XYZ.

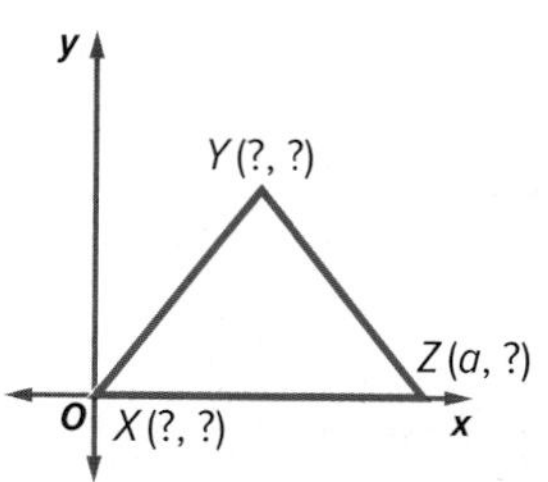

Vertex X is positioned at the origin; its coordinates are (0, 0).

Vertex Z is on the x-axis, so its y-coordinate is 0. The coordinates of vertex Z are $(a, 0)$.

$\triangle XYZ$ is isosceles, so using a vertical segment from Y to the x-axis and the Hypotenuse-Leg Theorem shows that the x-coordinate of Y is halfway between 0 and a or $\frac{a}{2}$. We cannot write the y-coordinate in terms of a, so call it b. The coordinates of point Y are $\left(\frac{a}{2}, b\right)$.

Study Tip

Structure The intersection of the x- and y-axes forms a right angle, so it is a convenient place to locate the right angle of a figure such as a right triangle.

Guided Practice

2. Name the missing coordinates of isosceles right triangle ABC.

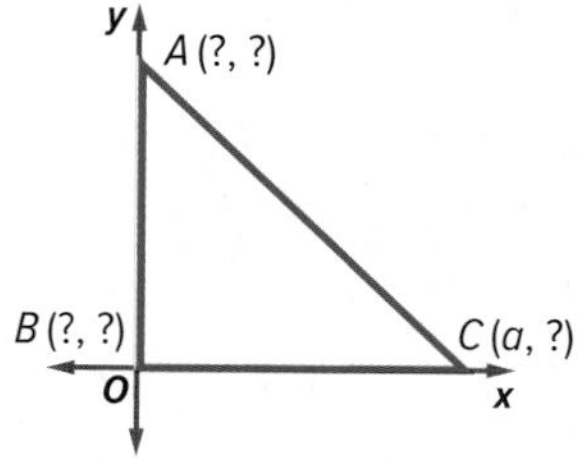

2 Write Coordinate Proofs

After a triangle is placed on the coordinate plane and labeled, we can use coordinate proofs to verify properties and to prove theorems.

Go Online!

Log into ConnectED to watch an **Animation** demonstrating how to place figures on the coordinate plane for coordinate proofs.

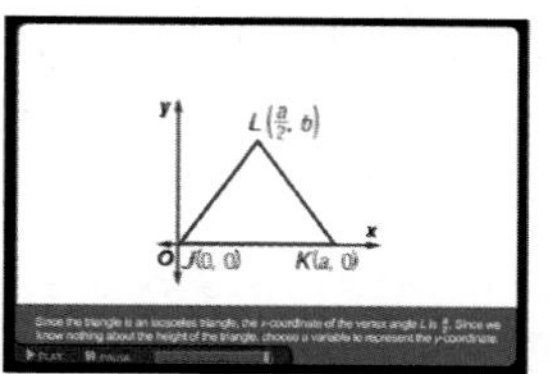

Example 3 Write a Coordinate Proof

Write a coordinate proof to show that a line segment joining the midpoints of two sides of a triangle is parallel to the third side.

Place a vertex at the origin and label it A. Use coordinates that are multiples of 2 because the Midpoint Formula involves dividing the sum of the coordinates by 2.

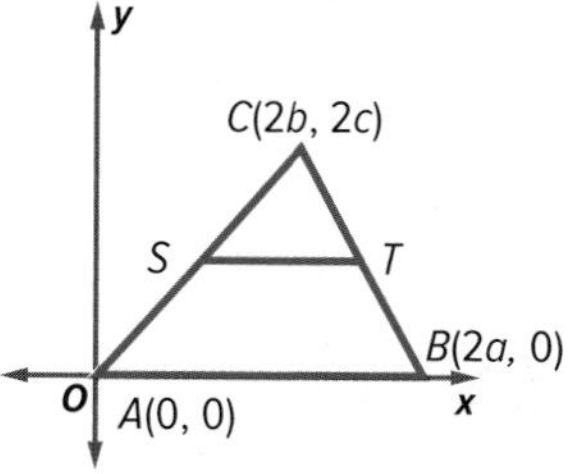

Given: $\triangle ABC$
S is the midpoint of $\overline{AC}$.
T is the midpoint of $\overline{BC}$.

Prove: $\overline{ST} \parallel \overline{AB}$

Proof:

By the Midpoint Formula, the coordinates of S are $\left(\frac{2b+0}{2}, \frac{2c+0}{2}\right)$ or (b, c) and the coordinates of T are $\left(\frac{2a+2b}{2}, \frac{0+2c}{2}\right)$ or $(a+b, c)$.

By the Slope Formula, the slope of $\overline{ST}$ is $\frac{c-c}{a+b-b}$ or 0 and the slope of $\overline{AB}$ is $\frac{0-0}{2a-0}$ or 0.

Since $\overline{ST}$ and $\overline{AB}$ have the same slope, $\overline{ST} \parallel \overline{AB}$.

Guided Practice

3. Write a coordinate proof to show that $\triangle ABX \cong \triangle CDX$.

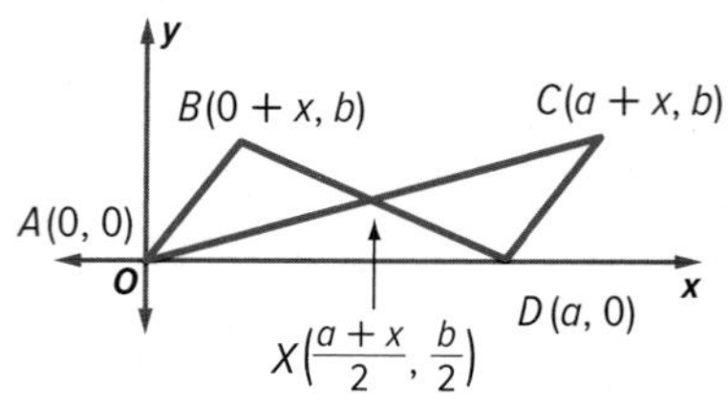

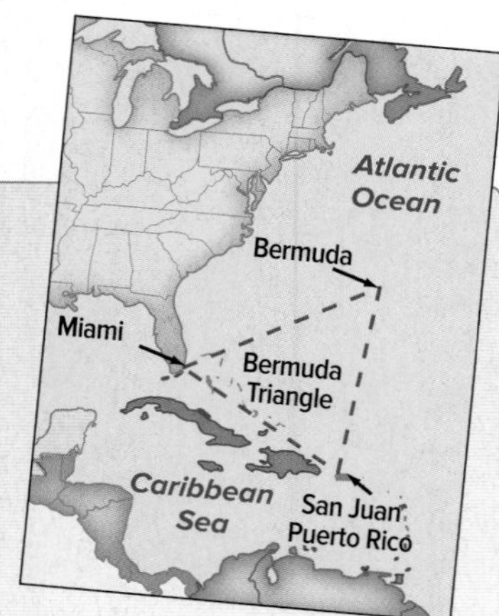

Real-World Link

More than 50 ships and 20 airplanes have mysteriously disappeared from a section of the North Atlantic Ocean off of North America commonly referred to as the Bermuda Triangle.

Source: *Encyclopaedia Britannica*

The techniques used for coordinate proofs can be used to solve real-world problems.

Real-World Example 4 Classify Triangles

GEOGRAPHY The Bermuda Triangle is a region formed by Miami, Florida; San Juan, Puerto Rico; and Bermuda. The approximate coordinates of each location, respectively, are 25.8°N 80.27°W, 18.48°N 66.12°W, and 33.37°N 64.68°W. Write a coordinate proof to prove that the Bermuda Triangle is scalene.

The first step is to label the coordinates of each location. Let M represent Miami, B represent Bermuda, and P represent Puerto Rico.

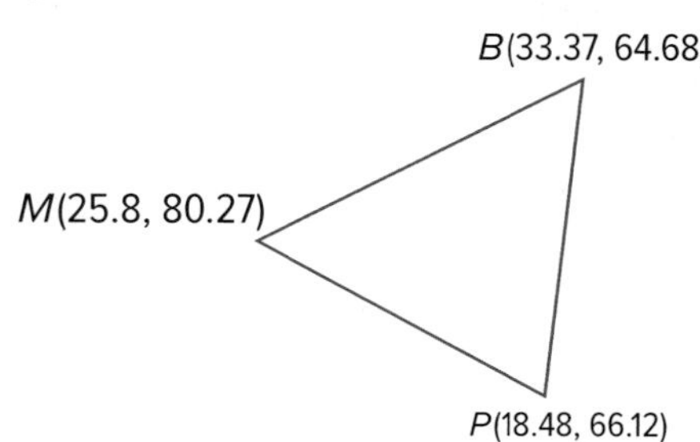

If no two sides of $\triangle MPB$ are congruent, then the Bermuda Triangle is scalene. Use the Distance Formula and a calculator to find the distance between each location.

$$MB = \sqrt{(33.37 - 25.8)^2 + (64.68 - 80.27)^2}$$

$$\approx 17.33$$

$$MP = \sqrt{(25.8 - 18.48)^2 + (80.27 - 66.12)^2}$$

$$\approx 15.93$$

$$PB = \sqrt{(33.37 - 18.48)^2 + (64.68 - 66.12)^2}$$

$$\approx 14.96$$

Because each side is a different length, $\triangle MPB$ is scalene. Therefore, the Bermuda Triangle is scalene.

Guided Practice

4. **GEOGRAPHY** In 2006, a group of art museums collaborated to form the West Texas Triangle to promote their collections. This region is formed by the cities of Odessa, Albany, and San Angelo. The approximate coordinates of each location, respectively, are 31.9°N 102.3°W, 32.7°N 99.3°W, and 31.4°N 100.5°W. Write a coordinate proof to prove that the West Texas Triangle is approximately isosceles.

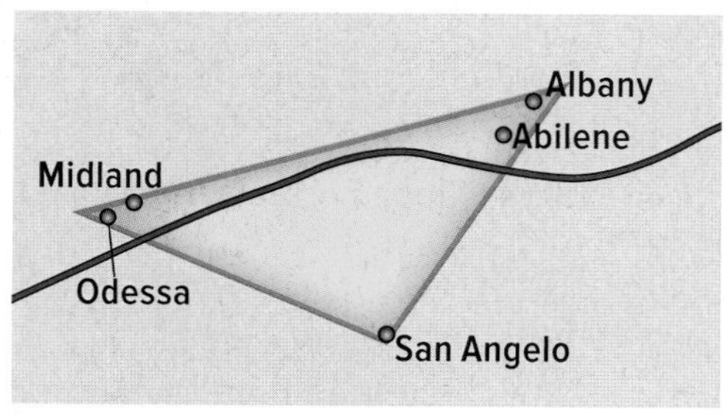

Check Your Understanding

◯ = Step-by-Step Solutions begin on page R13.

Go Online! for a Self-Check Quiz

Example 1 **Position and label each triangle on the coordinate plane.**

1. right $\triangle ABC$ with legs $\overline{AC}$ and $\overline{AB}$ so that $\overline{AC}$ is $2a$ units long and leg $\overline{AB}$ is $2b$ units long
2. isosceles $\triangle FGH$ with base $\overline{FG}$ that is $2a$ units long

Example 2 **Name the missing coordinate(s) of each triangle.**

3. 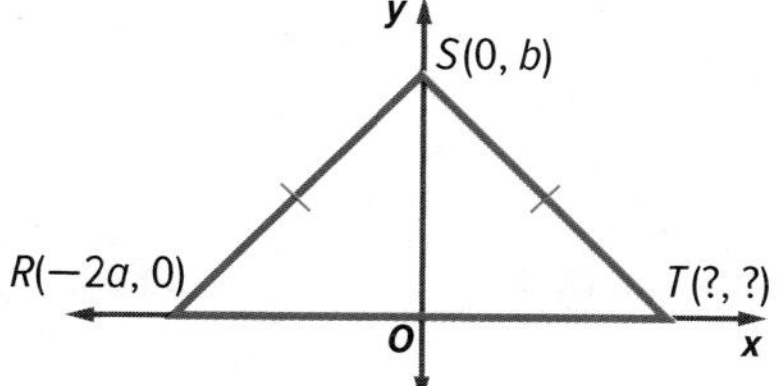

4. 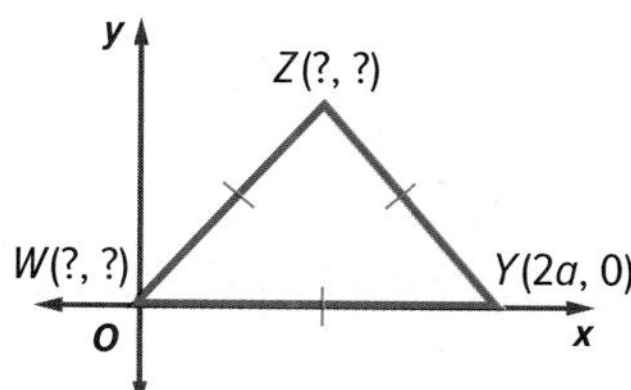

Example 3 5. **MP CONSTRUCT ARGUMENTS** Write a coordinate proof to show that $\triangle FGH \cong \triangle FDC$.

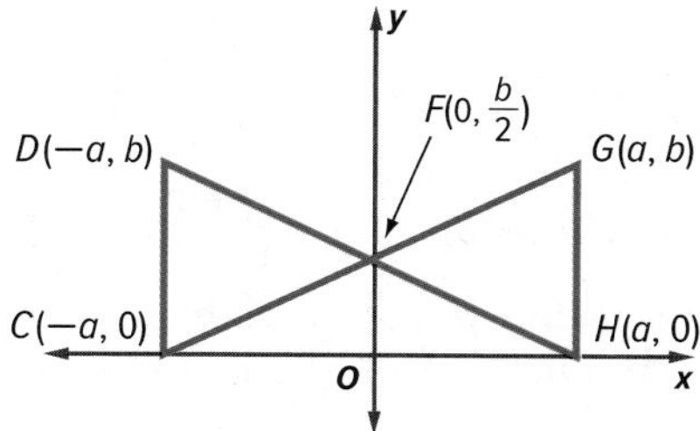

Example 4 6. **FLAGS** Write a coordinate proof to prove that the large triangle in the center of the flag is isosceles. The dimensions of the flag are 3 feet by 5 feet and point B of the triangle bisects the right side of the flag.

Practice and Problem Solving

Extra Practice is on page R4.

Example 1 **Position and label each triangle on the coordinate plane.**

7. isosceles $\triangle ABC$ with base $\overline{AB}$ that is a units long
8. right $\triangle XYZ$ with hypotenuse $\overline{YZ}$, the length of $\overline{XY}$ is b units long, and the length of $\overline{XZ}$ is three times the length of $\overline{XY}$
9. isosceles right $\triangle RST$ with hypotenuse $\overline{RS}$ and legs $3a$ units long
10. right $\triangle JKL$ with legs $\overline{JK}$ and $\overline{KL}$ so that $\overline{JK}$ is a units long and leg $\overline{KL}$ is $4b$ units long
11. equilateral $\triangle GHJ$ with sides $\frac{1}{2}a$ units long
12. equilateral $\triangle DEF$ with sides $4b$ units long

Example 2 **Name the missing coordinate(s) of each triangle.**

13. 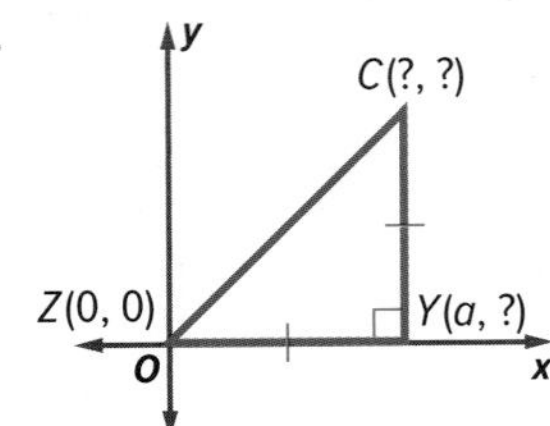

14. 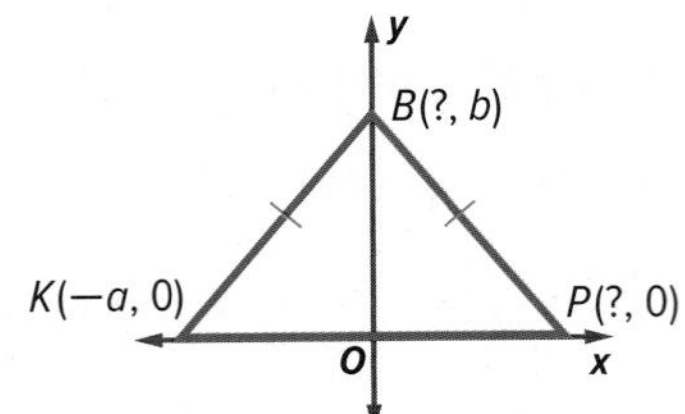

15. 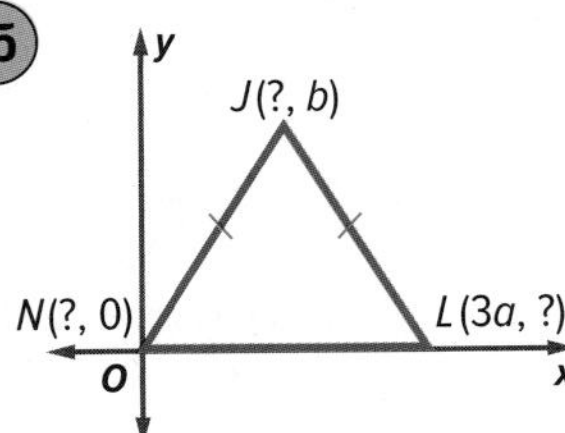

16. 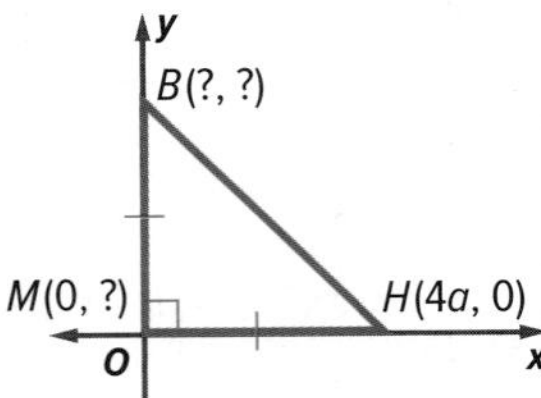

17. 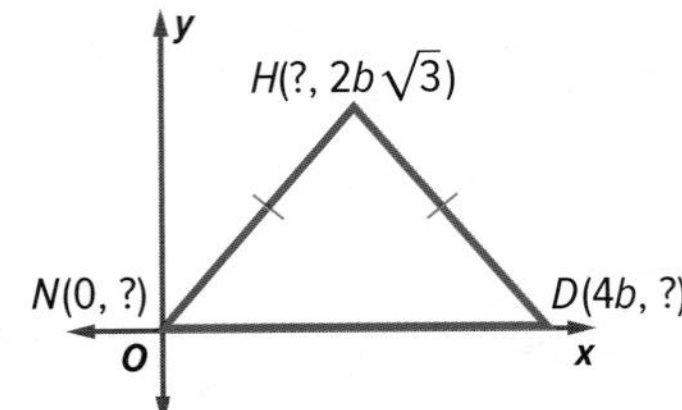

18. 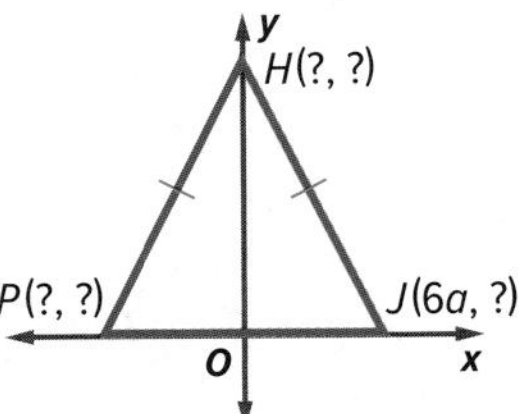

Example 3 **CONSTRUCT ARGUMENTS** **Write a coordinate proof for each statement.**

19. The segments joining the base vertices to the midpoints of the legs of an isosceles triangle are congruent.

20. The three segments joining the midpoints of the sides of an isosceles triangle form another isosceles triangle.

Example 4 **PROOF** **Write a coordinate proof for each statement.**

21. The measure of the segment that joins the vertex of the right angle in a right triangle to the midpoint of the hypotenuse is one-half the measure of the hypotenuse.

22. If a line segment joins the midpoints of two sides of a triangle, then its length is equal to one half the length of the third side.

23. **RESEARCH TRIANGLE** The cities of Raleigh, Durham, and Chapel Hill, North Carolina, form what is known as the Research Triangle. The approximate latitude and longitude of Raleigh are 35.82°N 78.64°W, of Durham are 35.99°N 78.91°W, and of Chapel Hill are 35.92°N 79.04°W. Show that the triangle formed by these three cities is scalene.

24. **PARTY PLANNING** Three friends live in houses with backyards adjacent to a neighborhood bike path. They decide to have a round-robin party using their three homes, inviting their friends to start at one house and then move to each of the other two. If one friend's house is centered at the origin, then the location of the other homes are (5, 12) and (13, 0). Write a coordinate proof to prove that the triangle formed by these three homes is isosceles.

Draw $\triangle XYZ$ and find the slope of each side of the triangle. Determine whether the triangle is a right triangle. Explain.

25. $X(0, 0)$, $Y(2h, 2h)$, $Z(4h, 0)$

26. $X(0, 0)$, $Y(1, h)$, $Z(2h, 0)$

27. **MULTI-STEP** Two high school clubs have gone camping. Club A pitches their tent 25 miles north of the ranger's station. Club B wants to set up their tent so that it is 9 miles north of the ranger's station and forms a right triangle with the ranger's station and Club A's tent.

a. Where should Club B set up camp?

b. What assumptions did you make? Write a coordinate proof to prove that the figure formed is a right triangle.

28. **PROOF** Write a coordinate proof to prove that $\triangle ABC$ is an isosceles triangle if the vertices are $A(0, 0)$, $B(a, b)$, and $C(2a, 0)$.

29 **WATERCRAFT** Three personal watercraft vehicles launch from the same dock. The first vehicle leaves the dock traveling due northeast, while the second vehicle travels due northwest. Meanwhile, the third vehicle leaves the dock traveling due north.

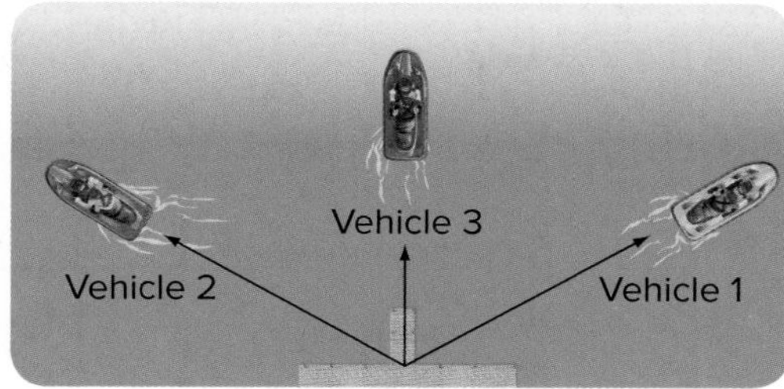

The first and second vehicles stop about 300 yards from the dock, while the third stops about 212 yards from the dock.

a. If the dock is located at (0, 0), sketch a graph to represent this situation. What is the equation of the line along which the first vehicle lies? What is the equation of the line along which the second vehicle lies? Explain your reasoning.

b. Write a coordinate proof to prove that the dock, the first vehicle, and the second vehicle form an isosceles right triangle.

c. Find the coordinates of the locations of all three watercrafts. Explain your reasoning.

d. Write a coordinate proof to prove that the positions of all three watercrafts are approximately collinear and that the third watercraft is at the midpoint between the other two.

H.O.T. Problems Use Higher-Order Thinking Skills

30. MP **REASONING** The midpoints of the sides of a triangle are located at $(a, 0)$, $(2a, b)$, and (a, b). If one vertex is located at the origin, what are the coordinates of the other vertices? Explain your reasoning.

MP **REASONING** **Find the coordinates of point L so $\triangle JKL$ is the indicated type of triangle. Point J has coordinates (0, 0) and point K has coordinates $(2a, 2b)$.**

31. scalene triangle **32.** right triangle **33.** isosceles triangle

34. **OPEN-ENDED** Draw an isosceles right triangle on the coordinate plane so that the midpoint of its hypotenuse is the origin. Label the coordinates of each vertex.

35. MP **CONSTRUCT ARGUMENTS** Use a coordinate proof to show that if you add n units to each x-coordinate of the vertices of a triangle and m to each y-coordinate, the resulting figure is congruent to the original triangle.

36. MP **REASONING** A triangle has vertex coordinates (0, 0) and $(a, 0)$. If the coordinates of the third vertex are in terms of a, and the triangle is isosceles, identify the coordinates and position the triangle on the coordinate plane.

37. **WRITING IN MATH** Explain why following each guideline below for placing a triangle on the coordinate plane is helpful in proving coordinate proofs.

a. Use the origin as a vertex of the triangle.

b. Place at least one side of the triangle on the x- or y-axis.

c. Keep the triangle within the first quadrant if possible.

38. $\triangle ACD$ is an equilateral triangle.

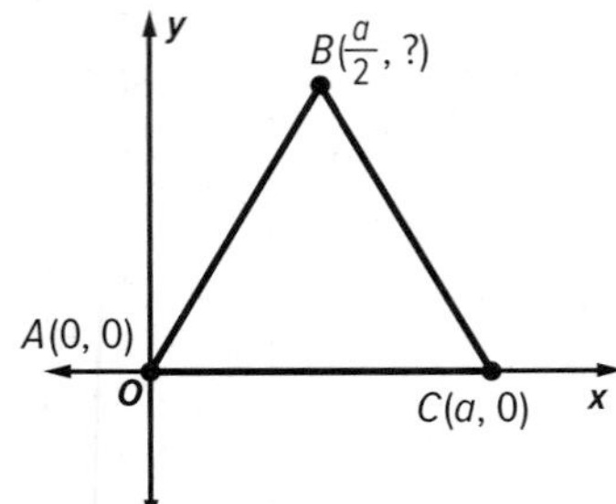

What is the y-coordinate of point B? MP 2, 6

- ○ **A** $\frac{a}{2}$
- ○ **B** $\frac{\sqrt{3}a}{2}$
- ○ **C** a
- ○ **D** $\frac{3a}{2}$

39. What are the coordinates of point R in the triangle? MP 2, 6

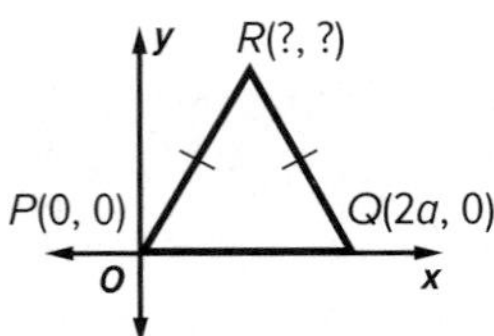

- ○ **A** $\left(\frac{a}{2}, a\right)$
- ○ **B** (a, b)
- ○ **C** $\left(\frac{b}{2}, a\right)$
- ○ **D** $\left(\frac{a}{2}, \frac{b}{2}\right)$
- ○ **E** $\left(\frac{b}{2}, \frac{a}{2}\right)$

40. A right triangle has vertices at (0, 0) and $(a, 0)$. Which coordinates could be the coordinates of third vertex? Select all that apply. MP 2, 6

- ☐ **A** $(0, a)$
- ☐ **B** $(-a, 0)$
- ☐ **C** (a, a)
- ☐ **D** $(0, -a)$
- ☐ **E** $(-a, a)$
- ☐ **F** $(a, -a)$
- ☐ **G** $(-a, -a)$
- ☐ **H** (a, b), where $b \neq a$

41. The figure shown models a ramp with a height of 12 inches. MP 2

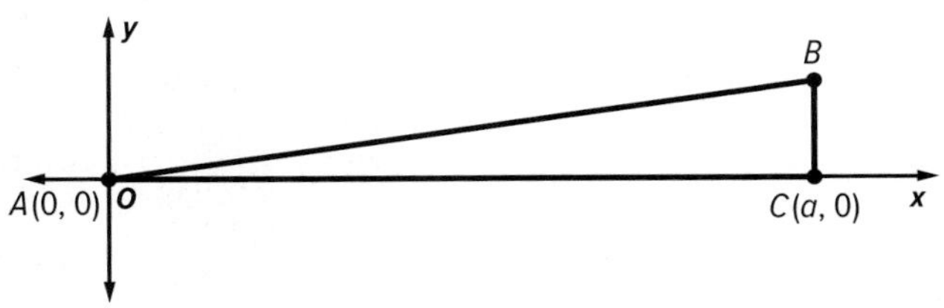

What is the slope of the ramp?

- ○ **A** 0
- ○ **B** $\frac{12}{a}$
- ○ **C** $\frac{a}{12}$
- ○ **D** 12
- ○ **E** $12a$

42. **MULTI-STEP** $\triangle ACD$ is an isosceles right triangle. Point B is the midpoint of $\overline{AC}$. MP 1, 2, 6

a. What are the coordinates of point C? ☐

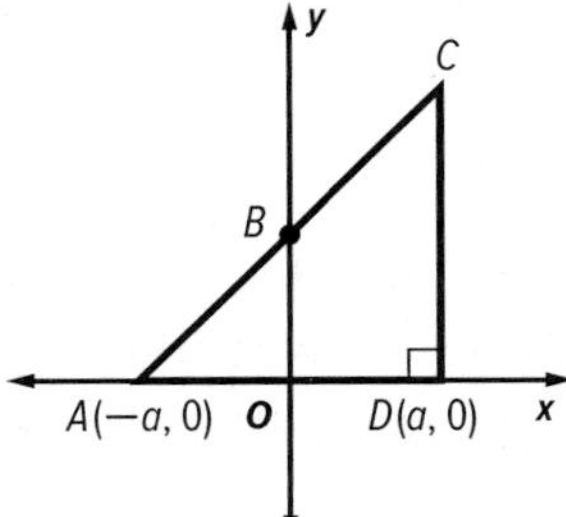

b. What are the coordinates of point B?

- ○ **A** $(0, a)$
- ○ **B** $\left(0, \frac{a}{2}\right)$
- ○ **C** $(0, \sqrt{2}a)$
- ○ **D** $(0, 2\sqrt{2}a)$

c. What is the slope of $\overline{AC}$? ☐

d. What is the length of $\overline{AC}$? Explain.

Study Guide and Review

Study Guide

Key Concepts

Angles of Triangles (Lesson 4-1)

- The measure of an exterior angle is equal to the sum of its two remote interior angles.

Congruent Triangles (Lesson 4-2 through 4-5)

- In a congruence transformation, the position of the image and preimage may differ, but the two figures remain congruent.
- SSS: If all of the corresponding sides of two triangles are congruent, then the triangles are congruent.
- SAS: If two pairs of corresponding sides of two triangles and the included angles are congruent, then the triangles are congruent.
- ASA: If two pairs of corresponding angles of two triangles and the included sides are congruent, then the triangles are congruent.
- AAS: If two pairs of corresponding angles of two triangles are congruent, and a corresponding pair of nonincluded sides is congruent, then the triangles are congruent.
- LL: If the legs of a right triangle are congruent to the corresponding parts of another right triangle, then the triangles are congruent.
- HA: If the hypotenuse and an acute angle are congruent to the corresponding parts of another right triangle, then the triangles are congruent.
- LA: If one leg and the adjacent acute angle are congruent to the corresponding parts of another right triangle, then the triangles are congruent.
- HL: If the hypotenuse and one leg of a right triangle are congruent to the corresponding parts of another right triangle, then the triangles are congruent.

Isosceles and Equilateral Triangles (Lesson 4-6)

- The base angles of an isosceles triangle are congruent and a triangle is equilateral if it is equiangular.

Coordinate Proofs (Lesson 4-7)

- Coordinate proofs use algebra to prove geometric concepts.

Study Organizer

Use your Foldable to review the chapter. Working with a partner can be helpful. Ask for clarification of concepts as needed.

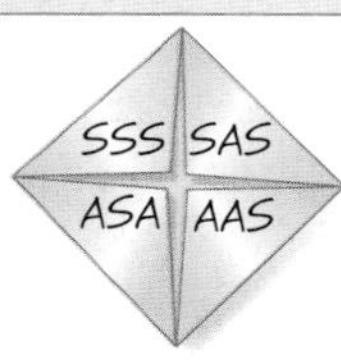

Key Vocabulary

auxiliary line (p. 282)
base angles (p. 325)
congruent polygons (p. 291)
coordinate proof (p. 334)
corollary (p. 285)
corresponding parts (p. 291)
exterior angle (p. 284)
flow proof (p. 284)
included angle (p. 302)
included side (p. 311)
remote interior angles (p. 284)
vertex angle (p. 325)

Vocabulary Check

State whether each sentence is *true* or *false*. If *false*, replace the underlined word or phrase to make a true sentence.

1. The <u>vertex</u> angles of an isosceles triangle are congruent.
2. An <u>included</u> side is the side located between two consecutive angles of a polygon.
3. The three types of <u>congruence transformations</u> are rotation, reflection, and translation.
4. A <u>rotation</u> moves all points of a figure the same distance and in the same direction.
5. A <u>flow proof</u> uses figures in the coordinate plane and algebra to prove geometric concepts.
6. The measure of an <u>exterior angle</u> of a triangle is equal to the sum of the measures of its two remote interior angles.
7. To use the <u>HL Theorem</u> to prove two right triangles are congruent, you must know that the hypotenuse and an acute angle of one right triangle are congruent to the corresponding parts of another right triangle.
8. An equilateral triangle is a special case of <u>isosceles</u> triangle.

Concept Check

9. Explain whether the SAS Postulate can be used to prove two right triangles congruent.
10. How are using the SSS Postulate and SAS Postulate different when proving triangles congruent?
11. How are the HA and AAS Theorems similar?

Lesson-by-Lesson Review 

4-1 Angles of Triangles

Find the measure of each numbered angle.

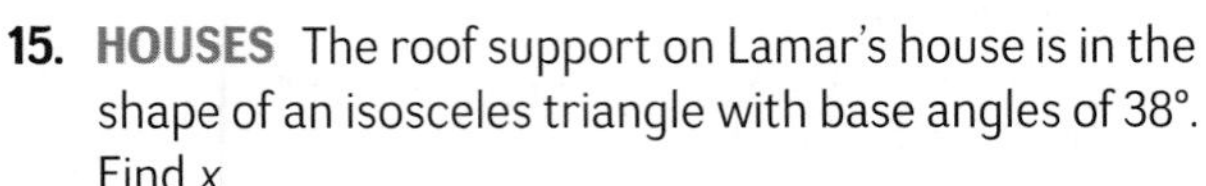

12. ∠1
13. ∠2
14. ∠3
15. **HOUSES** The roof support on Lamar's house is in the shape of an isosceles triangle with base angles of 38°. Find *x*.

Example 1

Find the measure of each numbered angle.

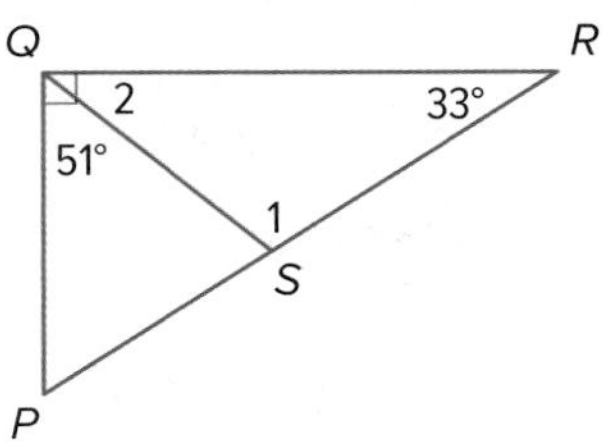

$$m\angle 2 + m\angle PQS = 90$$
$$m\angle 2 + 51 = 90 \quad \text{Substitution}$$
$$m\angle 2 = 39 \quad \text{Subtract 51 from each side.}$$

$$m\angle 1 + m\angle 2 + 33 = 180 \quad \text{Triangle Angle-Sum Theorem}$$
$$m\angle 1 + 39 + 33 = 180 \quad \text{Substitution}$$
$$m\angle 1 + 72 = 180 \quad \text{Simplify.}$$
$$m\angle 1 = 108 \quad \text{Subtract.}$$

4-2 Congruent Triangles

Show that the polygons are congruent by identifying all congruent corresponding parts. Then write a congruence statement.

16. 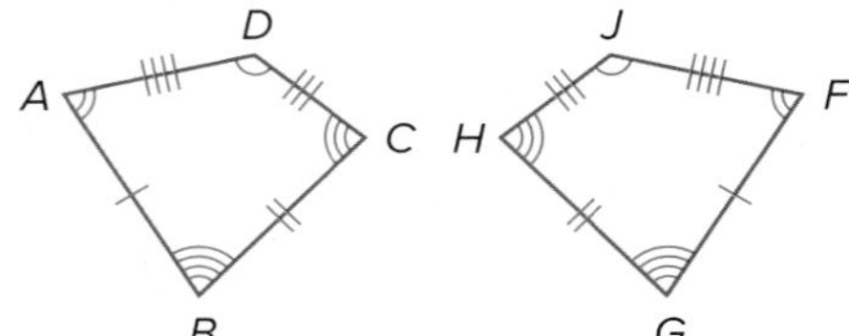

17. 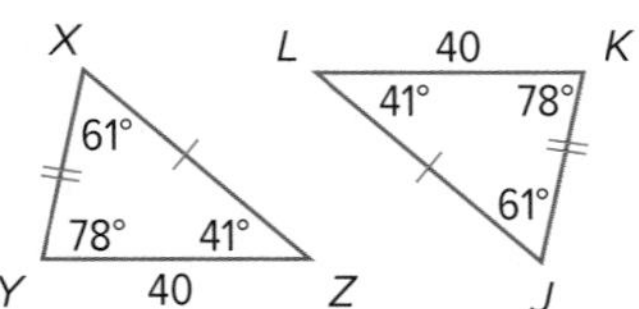

18. **MOSAIC TILING** A section of a mosaic tiling is shown. Name the triangles that appear to be congruent.

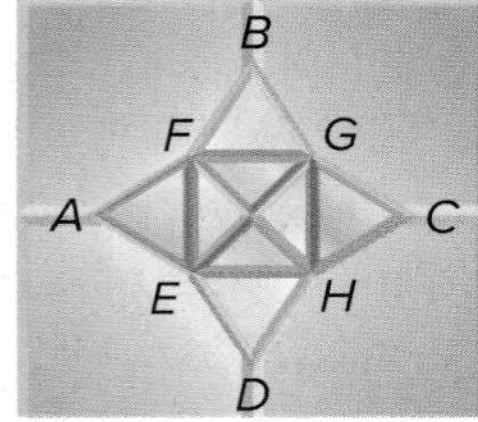

Example 2

Show that the polygons are congruent by identifying all the congruent corresponding parts. Then write a congruence statement.

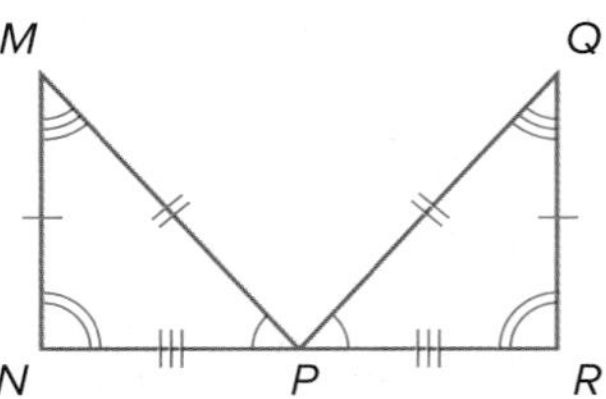

$\triangle MNP$ can be mapped exactly onto $\triangle QRP$ by reflecting it horiztonally.

Angles: $\angle N \cong \angle R$, $\angle M \cong \angle Q$, $\angle MPN \cong \angle QPR$

Sides: $\overline{MN} \cong \overline{QR}$, $\overline{MP} \cong \overline{QP}$, $\overline{NP} \cong \overline{RP}$

All corresponding parts of the two triangles are congruent. Therefore, $\triangle MNP \cong \triangle QRP$.

4-3 Proving Triangles Congruent—SSS, SAS

Determine whether $\triangle ABC \cong \triangle XYZ$. Explain.

19. $A(5, 2)$, $B(1, 5)$, $C(0, 0)$, $X(-3, 3)$, $Y(-7, 6)$, $Z(-8, 1)$

20. $A(3, -1)$, $B(3, 7)$, $C(7, 7)$, $X(-7, 0)$, $Y(-7, 4)$, $Z(1, 4)$

Determine which postulate can be used to prove that the triangles are congruent. If it is not possible to prove that they are congruent, write *not possible*.

21.

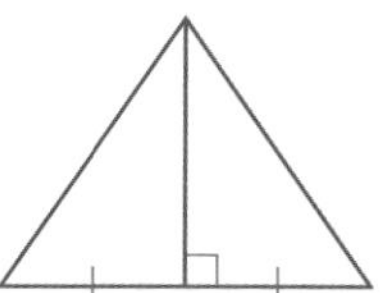

22.

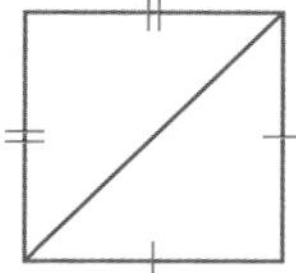

Example 3

Write a two-column proof.

Given: $\triangle KPL$ is equilateral.

$\overline{JP} \cong \overline{MP}$,

$\angle JPK \cong \angle MPL$

Prove: $\triangle JPK \cong \triangle MPL$

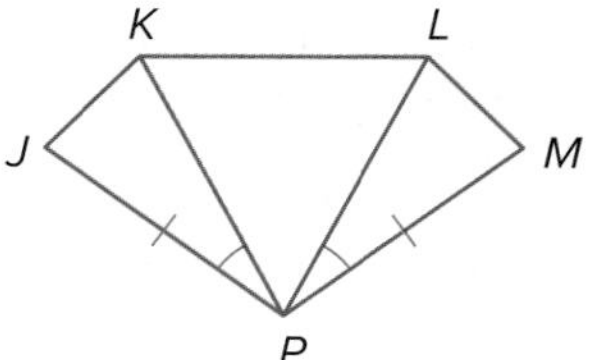

Statements	Reasons
1. $\triangle KPL$ is equilateral.	**1.** Given
2. $\overline{PK} \cong \overline{PL}$	**2.** Def. of equilateral $\triangle$
3. $\overline{JP} \cong \overline{MP}$	**3.** Given
4. $\angle JPK \cong \angle MPL$	**4.** Given
5. $\triangle JPK \cong \triangle MPL$	**5.** SAS

4-4 Proving Triangles Congruent—ASA, AAS

23. Write a two-column proof.

Given: $\overline{AB} \parallel \overline{DC}$, $\overline{AB} \cong \overline{DC}$

Prove: $\triangle ABE \cong \triangle CDE$

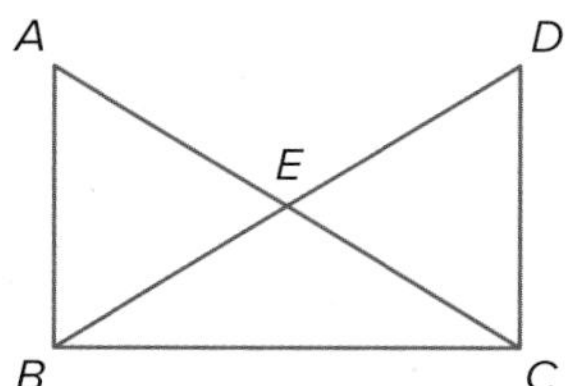

24. KITES Denise's kite is shown in the figure at the right. Describe a rigid motion that maps $\triangle WXY$ onto $\triangle WZY$. Given that $\overline{WY}$ bisects both $\angle XWZ$ and $\angle XYZ$, write a two-column proof to prove that $\triangle WXY \cong \triangle WZY$.

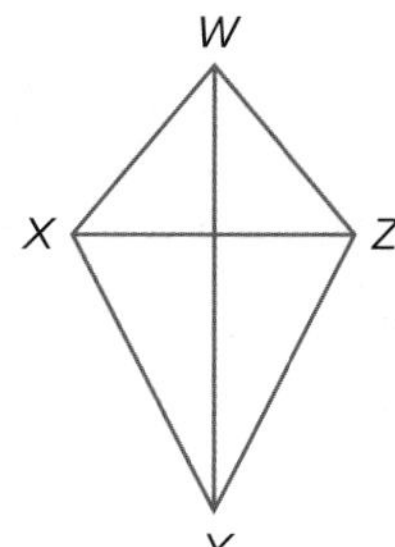

Example 4

Write a flow proof.

Given: $\overline{PQ}$ bisects $\angle RPS$.

$\angle R \cong \angle S$

Prove: $\triangle RPQ \cong \triangle SPQ$

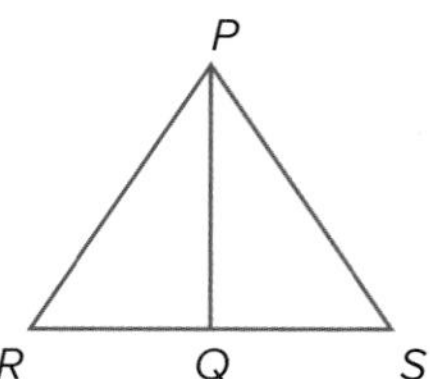

Flow Proof:

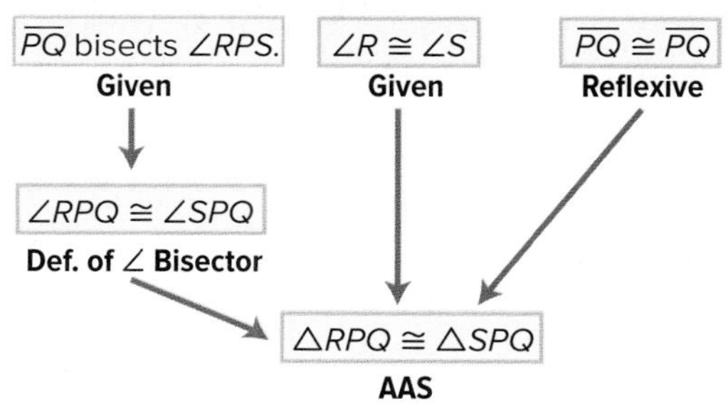

Study Guide and Review *Continued*

4-5 Proving Right Triangles Congruent

Determine whether each pair of triangles is congruent. If yes, include the theorem or postulate that applies and describe the series of rigid motions that map one triangle onto the other.

25.

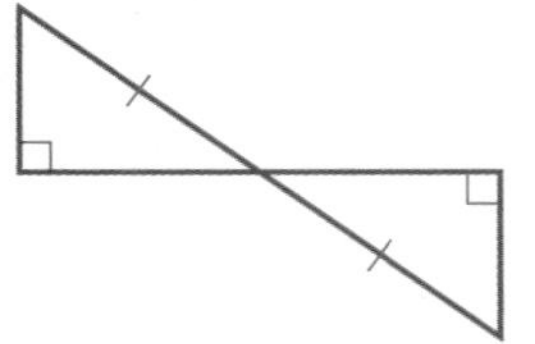

26.

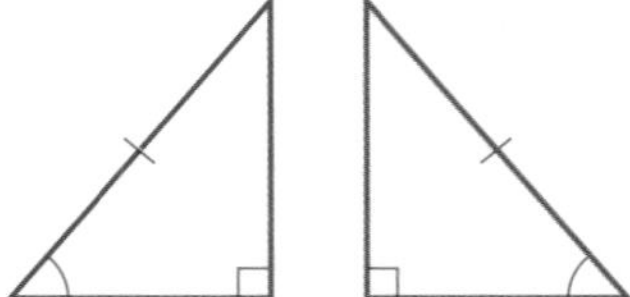

Example 5

Determine whether each pair of triangles is congruent. If yes, include the theorem or postulate that applies and describe the series of rigid motions that map one triangle onto the other.

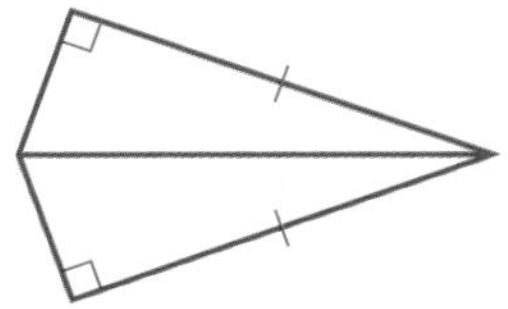

The hypotenuses are congruent and one pair of legs is congruent. So, the triangles are congruent by HL. One triangle is mapped onto the other by a reflection across the hypotenuse.

4-6 Isosceles and Equilateral Triangles

27. PAINTING Pam is painting using a wooden easel. The support bar on the easel forms an isosceles triangle with the two front supports. According to the figure shown, what are the measures of the base angles of the triangle?

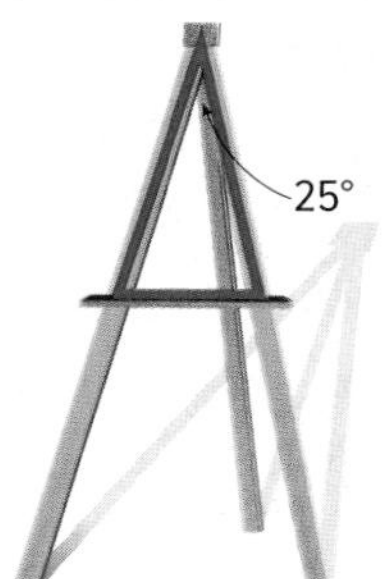

Example 6

Find AB.

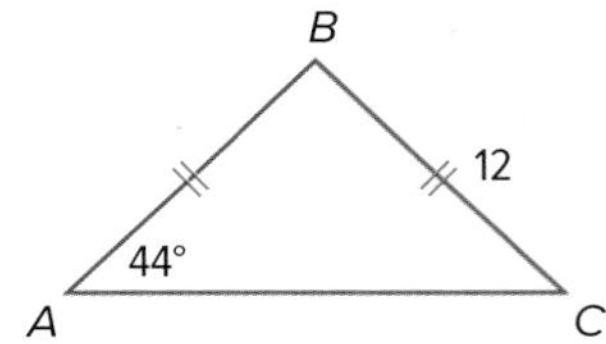

$AB = BC$, so $\triangle ABC$ is isosceles. Since $BC = 12$, $AB = 12$ by substitution.

4-7 Triangles and Coordinate Proof

Position and label each triangle on the coordinate plane.

28. right $\triangle MNO$ with right angle at point M and legs of lengths a and $2a$.

29. isosceles $\triangle WXY$ with height h and base $\overline{WY}$ with length $2a$.

30. right isosceles $\triangle ABC$ with right angle at point B and legs of length n.

Example 7

Position and label equilateral $\triangle XYZ$ with sides of $2a$.

- Place one vertex at the origin.
- Place one side of the triangle along the x-axis.
- Place the third vertex above the midpoint of $\overline{XY}$.

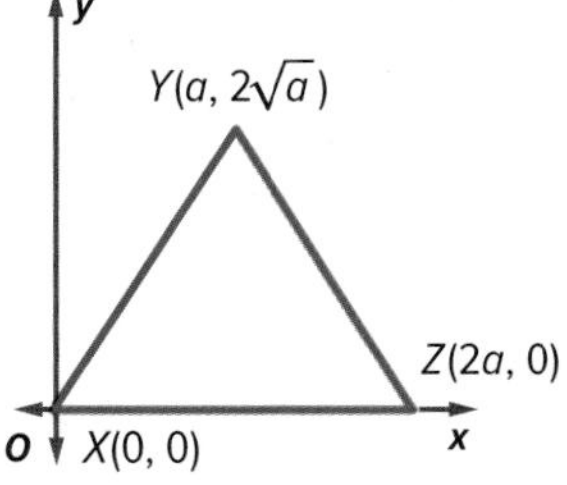

Practice Test

1. Show that the polygons are congruent by using rigid motions and by identifying all the congruent corresponding parts. Then write a congruence statement.

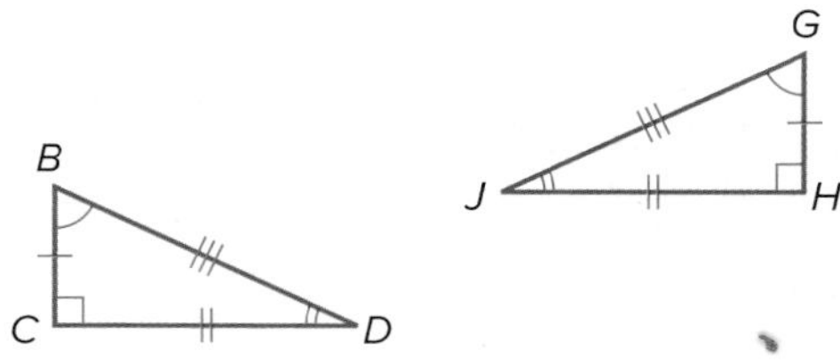

Find the measure of each numbered angle.

2. $\angle 1$ 3. $\angle 2$

4. $\angle 3$ 5. $\angle 4$

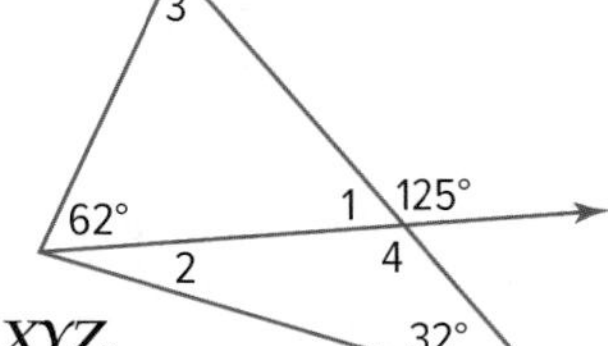

In the diagram, $\triangle RST \cong \triangle XYZ$.

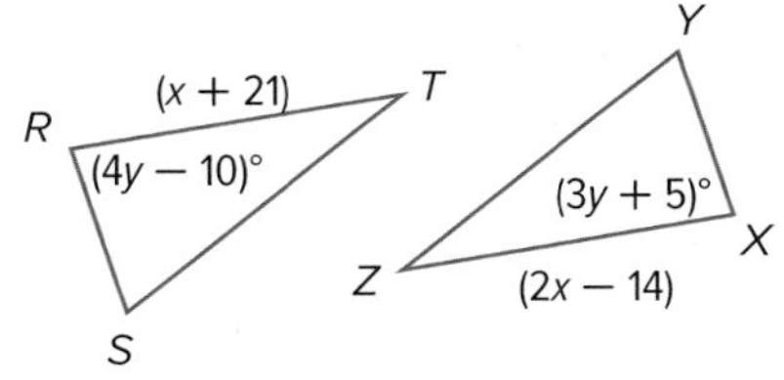

6. Find x.

7. Find y.

8. **PROOF** Write a flow proof.

Given: $\overline{XY} \parallel \overline{WZ}$ and $\overline{XW} \parallel \overline{YZ}$
Prove: $\triangle XWZ \cong \triangle ZYX$

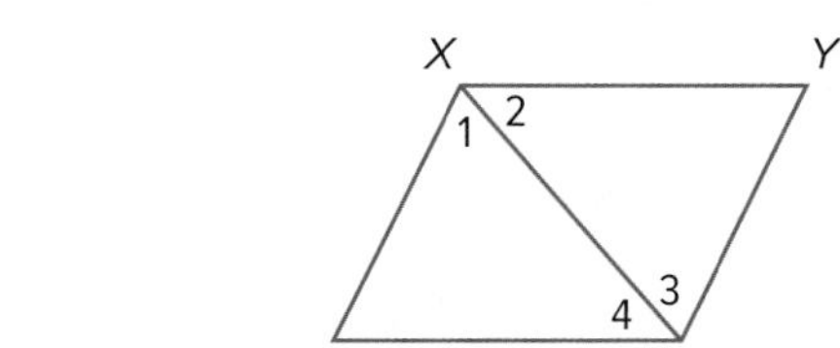

9. **MULTIPLE CHOICE** Find x.

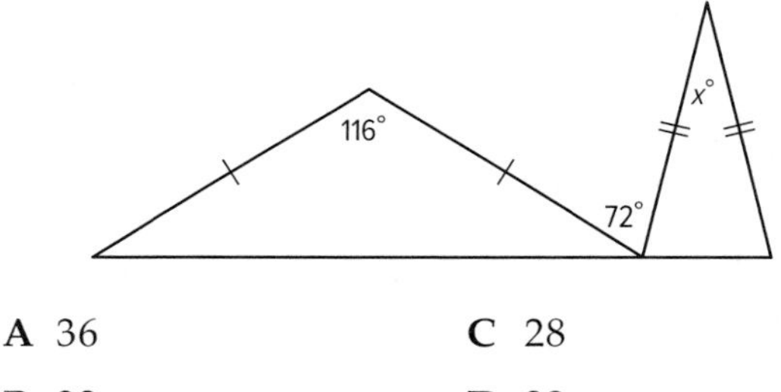

A 36 **C** 28

B 32 **D** 22

10. Determine whether $\triangle TJD \cong \triangle SEK$ given $T(-4, -2)$, $J(0, 5)$, $D(1, -1)$, $S(-1, 3)$, $E(3, 10)$, and $K(4, 4)$. Explain.

Determine which postulate or theorem can be used to prove each pair of triangles congruent. If it is not possible to prove them congruent, write *not possible*. If it is possible, describe the rigid motions that map one triangle onto the other.

11. 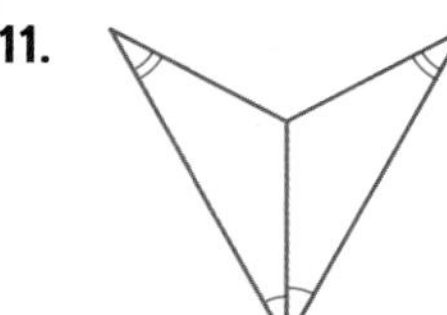

12. 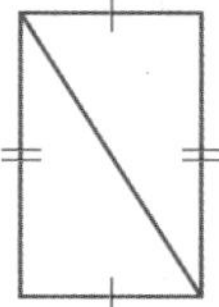

13. 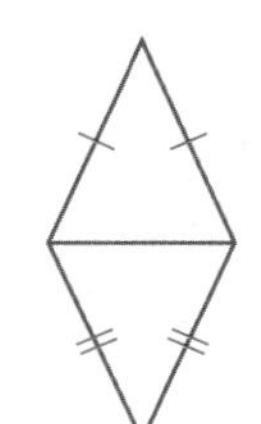

14. 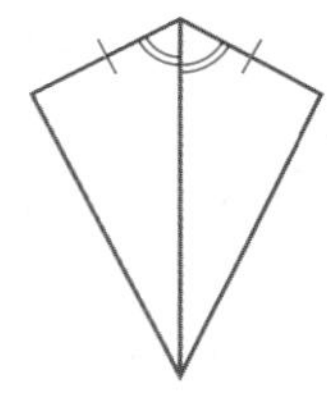

15. **LANDSCAPING** Angie has laid out a design for a garden consisting of two triangular areas as shown below. The points are $A(0, 0)$, $B(0, 5)$, $C(3, 5)$, $D(6, 5)$, and $E(6, 0)$. Name the type of congruence transformation for the preimage $\triangle ABC$ to $\triangle EDC$.

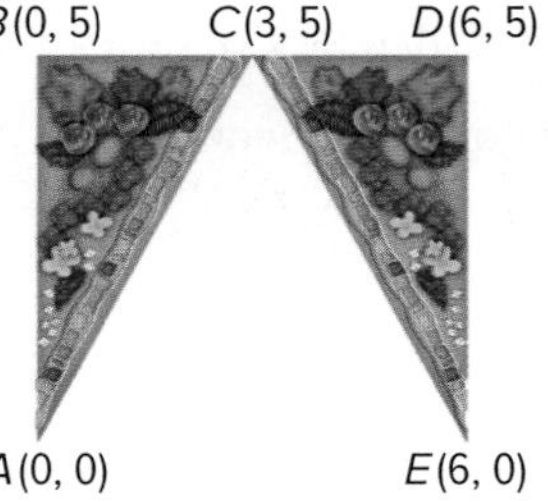

16. Naomi cuts out isosceles triangle RST. She folds the triangle so that $\overline{RS}$ lies on $\overline{TS}$ and makes a crease. Then she unfolds the triangle. Naomi claims that the two smaller triangles formed by the crease are congruent by HL. Do you agree? Explain.

17. **PROOF** $\triangle ABC$ is a right isosceles triangle with hypotenuse $\overline{AB}$. M is the midpoint of $\overline{AB}$. Write a coordinate proof to show that $\overline{CM}$ is perpendicular to $\overline{AB}$.

Preparing for Assessment

Performance Task

Provide a clear solution to each part of the task. Be sure to show all of your work. Include all relevant drawings and justify your answers.

DESIGN Marisol is a designer at an advertising agency. Her current project is to design a logo for a new airline. She is working on three different designs to show the airline executives. She wants to include at least one pair of congruent triangles in each design.

Part A

Planning the Designs Marisol talked with other designers at her agency when planning her designs.

1. **Construct an Argument** James suggested a design with only two triangles and said:

   "One side of Triangle A should be the same length as one side of Triangle B.
   Two angles in Triangle A should have the same measure as two angles in Triangle B."

   Explain whether James's description includes enough information to make sure the triangles are congruent.

Part B

First Design The diagram shows the first design. In this design, $\overline{BD}$ is the perpendicular bisector of $\overline{AC}$.

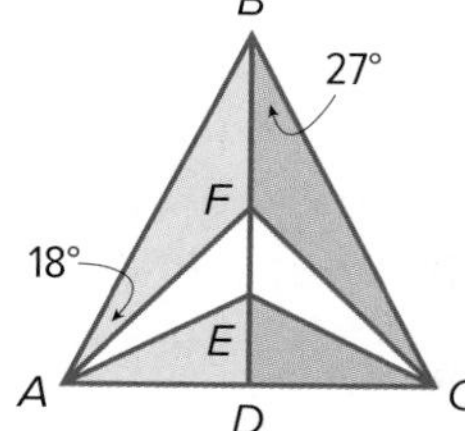

2. Write a two-column proof to show that $\triangle ABD \cong \triangle CBD$.
3. Explain how you could verify that $\triangle ABD \cong \triangle CBD$ by using rigid transformations.
4. Prove that $\triangle ABF \cong \triangle CBF$.
5. **Construct an Argument** Find $m\angle AFD$. Justify your reasoning.

Part C

Second Design The coordinate plane shows the second design.

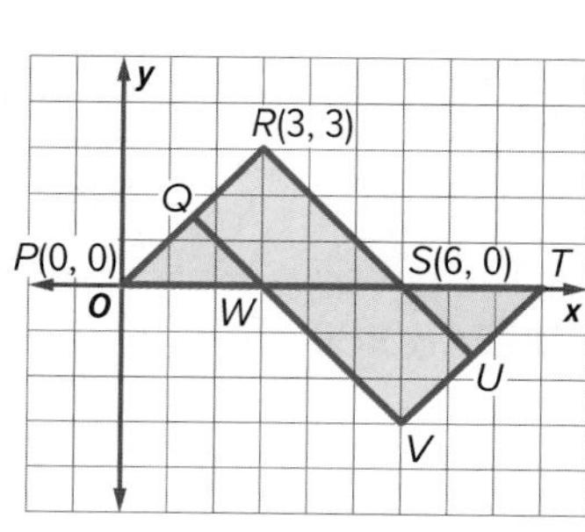

6. Prove that $\triangle PRS$ is a right isosceles triangle.
7. **Construct an Argument** Prove that the acute angles of any right isosceles triangle each measure 45°.

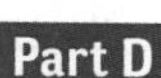

Part D

Third Design Help Marisol create a third design.

8. **Use Tools Strategically** Create a third design for the airline's logo using only these tools: paper, pencil, straightedge, and compass. Your design should include at least one pair of congruent triangles.
9. **Construct an Argument** Choose a pair of congruent triangles from your design and prove that they are congruent.

Test-Taking Strategy

Example

Read the problem. Identify what you need to know. Then use the information in the problem to solve.

Given: $\triangle AQT$ and $\triangle KML$ with right angles $\angle A$ and $\angle K$ and $\angle A \cong \angle K$ Determine which set of true statements is not sufficient to show congruence by using rigid motions.

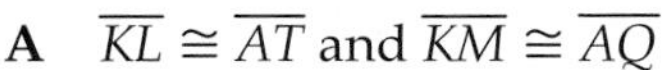

A $\overline{KL} \cong \overline{AT}$ and $\overline{KM} \cong \overline{AQ}$

B $\overline{LM} \cong \overline{TQ}$

C $\angle Q \cong \angle M$, $\angle A \cong \angle K$, and $\overline{AQ} \cong \overline{AM}$

D $\angle T \cong \angle L$ and $\overline{AQ} \cong \overline{AM}$

Step 1 **What do you need to find?**
the condition that is not sufficient to show congruence by using rigid motions

Step 2 **Is there enough information given to solve the problem?**
You need to choose the additional information that is needed to show congruence.

Step 3 **What information, if any, is not needed to solve the problem?**
You do not need to know that $\angle A \cong \angle K$ if you already know that $\angle A$ and $\angle K$ are right angles.

Step 4 **Are there any obvious wrong answers? If so, which one(s)? Explain.**
No. You need to analyze each answer choice in the context of the diagram.

Step 5 **What is the correct answer?**
B; another pair of angles or sides must be congruent. The condition in choice B is not sufficient to show congruence.

Test-Taking Tip

Reading Math Problems
The first step to solving any math problem is to read the problem. When reading a math problem to get the information you need to solve, it is helpful to use special reading strategies.

Apply the Strategy

Read the problem. Identify what you need to know. Then use the information in the problem to solve.

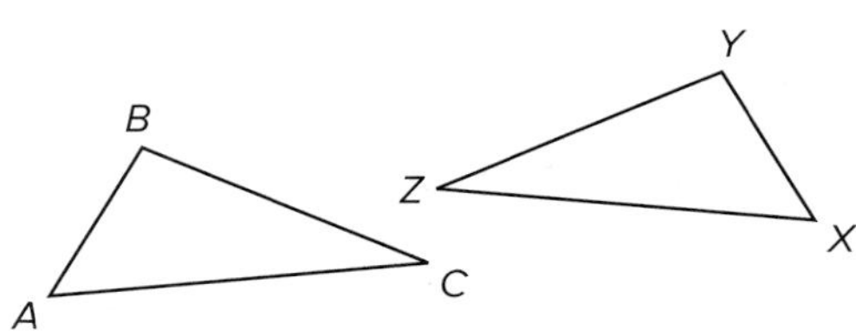

In the figure, $\angle C \cong \angle Z$ and $\overline{AC} \cong \overline{XZ}$. What additional information is *not* sufficient to prove that $\triangle ABC \cong \triangle XYZ$?

A $\overline{BC} \cong \overline{YZ}$ **C** $\angle B \cong \angle Y$

B $\overline{AB} \cong \overline{XY}$ **D** $\angle A \cong \angle X$

Answer the questions below.

a. What do you need to find?

b. Is there enough information given to solve the problem?

c. What information, if any, is not needed to solve the problem?

d. Are there any obvious wrong answers? If so, which one(s)? Explain.

e. What is the correct answer?

CHAPTER 4

Preparing for Assessment

Cumulative Review

Read each question. Then fill in the correct answer on the answer document provided by your teacher or on a sheet of paper.

1. $\triangle ABC \cong \triangle PQR$. Find the measure of $\angle B$.

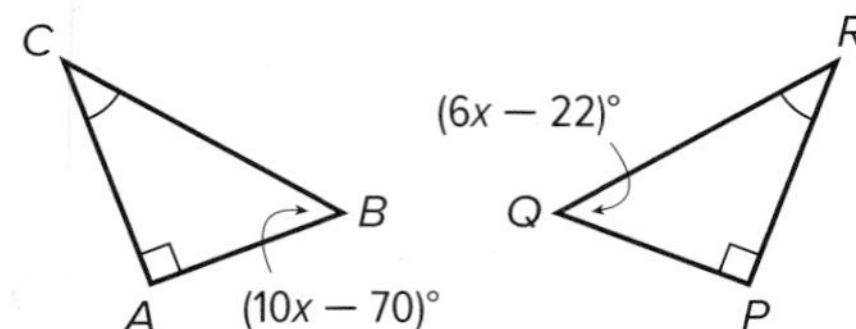

○ **A** 12

○ **B** 23

○ **C** 40

○ **D** 50

2. $\triangle KLM$ is mapped onto $\triangle RST$ by a rigid motion.

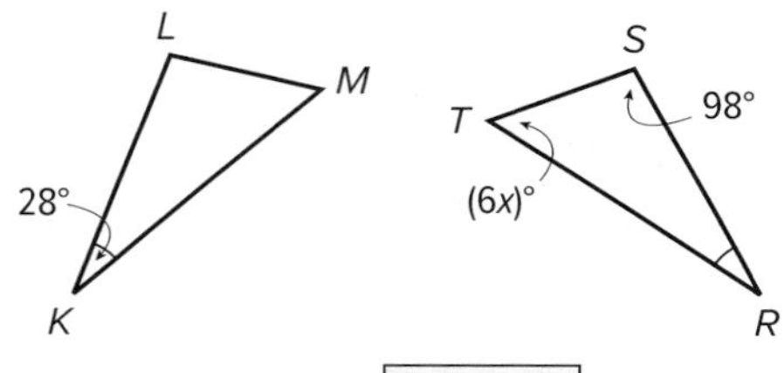

a. Find the value of x. ☐

b. Find $m\angle M$. ☐

3. In $\triangle ACD$, $\overline{AE} \cong \overline{DE}$. Which postulates or theorems can be used to prove $\triangle ACE \cong \triangle DCE$?

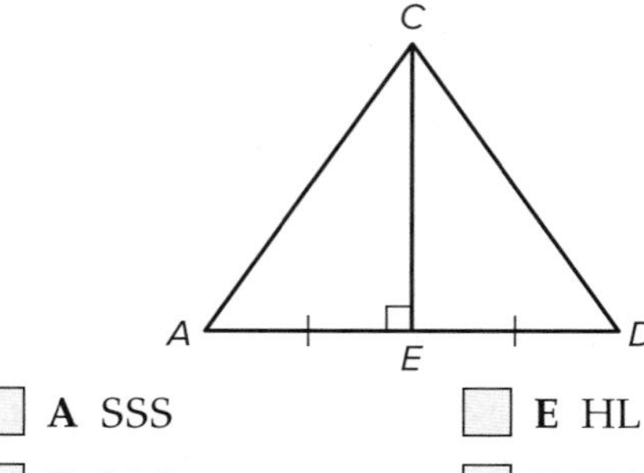

☐ **A** SSS	☐ **E** HL
☐ **B** SAS	☐ **F** HA
☐ **C** ASA	☐ **G** LL
☐ **D** AAS	☐ **H** LA

4. What is $m\angle RSQ$ in the figure below?

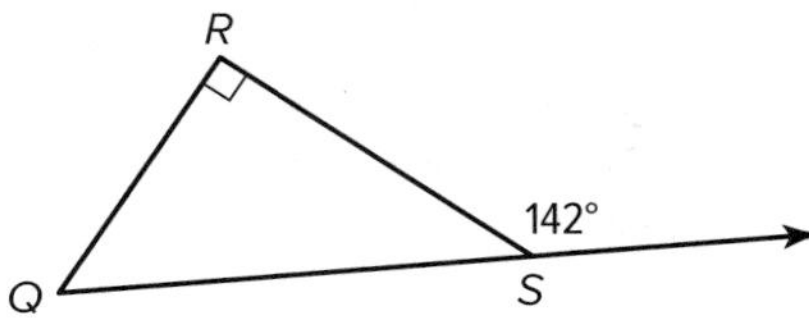

○ **A** 38

○ **B** 42

○ **C** 45

○ **D** 52

5. The three angles of a triangle measure $(3x - 21)°$, $(4x + 2)°$, and $(x + 15)°$. What is the measure of the smallest angle in degrees?

6. MULTI-STEP A throw rug is made from four right triangles.

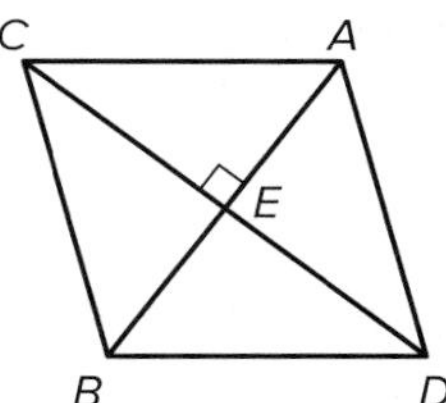

a. Name one pair of angles that must be congruent to prove $\triangle ACE \cong \triangle BCE$.

b. Assume the angle pair named in part **a** is congruent. Describe the rigid motion(s) that maps one triangle onto the other.

c. Assume the angle pair named in part **a** is congruent. What postulate or theorem can be used to prove the triangles are congruent?

d. Assume the throw rug is made up of four congruent right triangles. Describe the rigid motion(s) that maps $\triangle BCE$ onto $\triangle ADE$.

e. MP Which mathematical practice did you use to solve this problem?

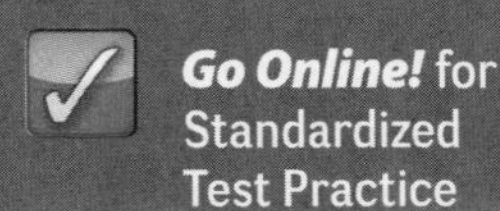

7. Which theorem or postulate can be used to show $\triangle GBE \cong \triangle ABD$?

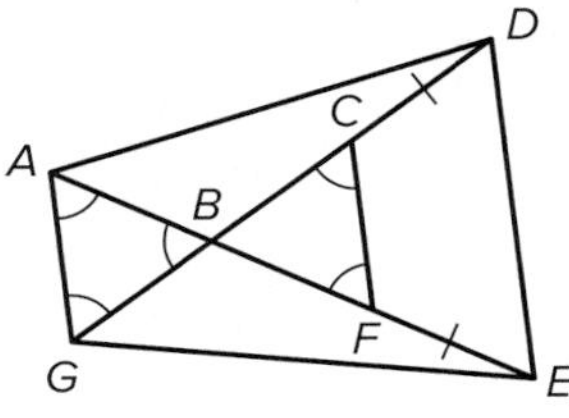

A SSS

B ASA

C AAS

D SAS

8. Select the triangles that can be proven congruent by SSS or SAS. Choose all that apply.

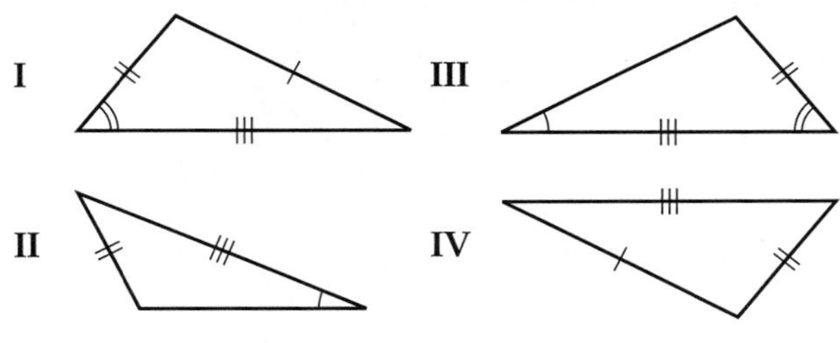

A I and IV

B II and III

C I and III

D II and IV

9. Right triangle $\triangle HJK$ has right angle J at the origin and vertex K at $(2a, 0)$. Give the coordinates of vertex H if the legs have lengths $2a$ and $3a$.

10. In the figure below, $\triangle AFE \cong \triangle DHE \cong \triangle BFG \cong \triangle CHG$.

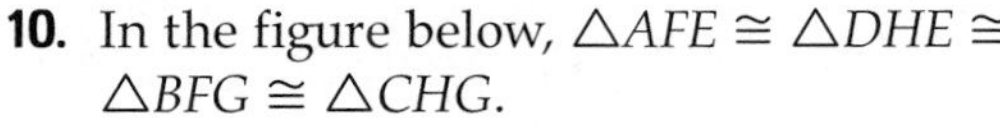

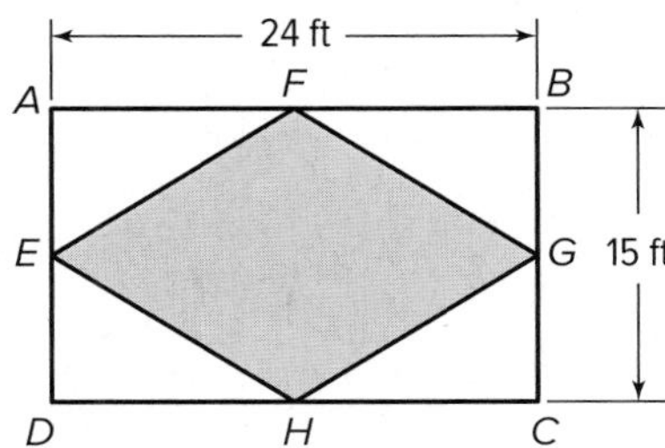

Find the area of the shaded region in square feet.

11. What type of transformation is shown?

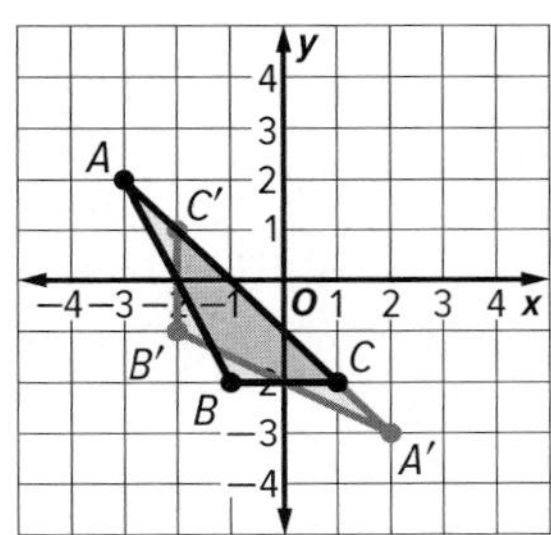

A reflection in the x-axis

B rotation

C translation

D reflection in the y-axis

12. In the figure, $m\angle 2 = 40$. Find $m\angle 1$.

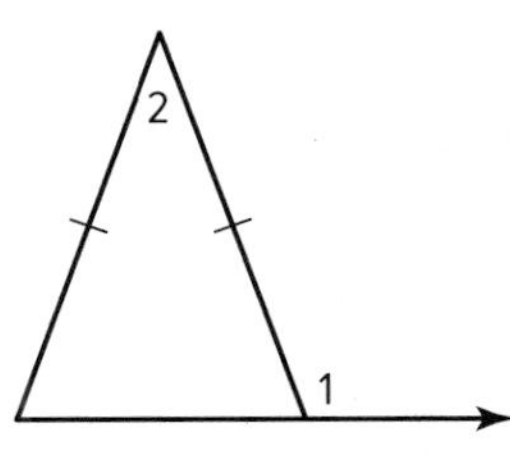

A 40

B 70

C 110

D 140

Need Extra Help?

If you missed Question...	1	2	3	4	5	6	7	8	9	10	11	12
Go to Lesson...	4-2	4-2	4-5	4-1	4-1	4-4	4-3	4-3	4-7	4-2	3-1	4-6

CHAPTER 5

Relationships in Triangles

THEN

You learned how to classify triangles.

NOW

In this chapter, you will:

- Learn about special segments and points related to triangles.
- Learn about relationships between the sides and angles of triangles.
- Learn to write indirect proofs.

WHY

SKATE PARK Triangle relationships are used to find and compare angle measures and distances. To create a skate park, planners use these relationships to ensure that, for example, momentum gained going down one ramp will carry a person up another. They seek efficiency and balance for the skateboarders.

Use the Mathematical Practices to complete the activity.

1. Sense-Making What elements of geometry, in addition to triangle relationships, do planners of a skate park need to consider?

2. Use Tools Use the Internet to search for "skate park designs" or related terms. What are important considerations when planning the ramps and inclines of a skate park?

3. Apply Math Use The Geometer's Sketchpad in ConnectED to construct at least three components of a skate park that would allow a skateboarder to move from one to another without needing to apply more force.

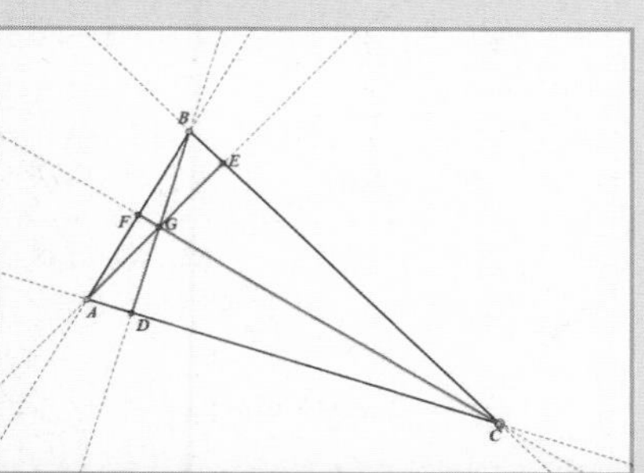

Go Online to Guide Your Learning

Explore & Explain

2D Tool and Ruler

Use the **2D** tool and ruler to enhance your understanding of the triangle inequality in Lesson 5-5.

Triangle Special Segments

Use the Triangle Special Segments tool to demonstrate the theorems presented in Lesson 5-1 and to explore centroids and orthocenters of triangles discussed in Lesson 5-2.

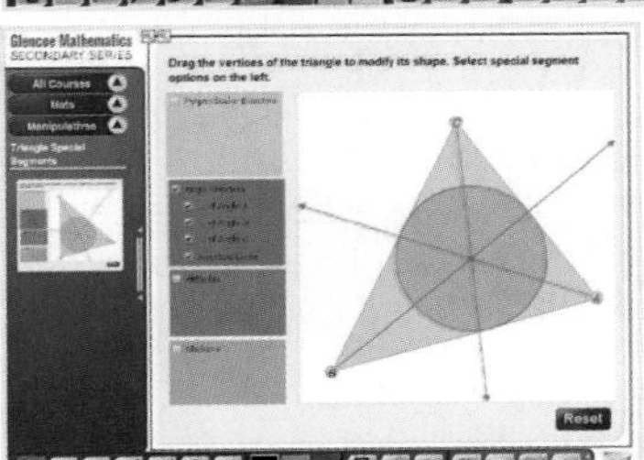

eBook

Interactive Student Guide

Before starting the chapter, answer the **Chapter Focus** preview questions. Check your answers as you complete each lesson. At the end of the chapter, try the **Performance Task**.

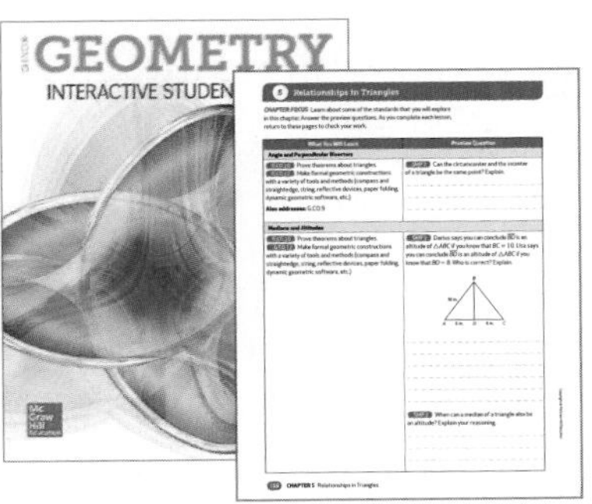

Organize

Foldables

Relationships in Triangles Make this Foldable to help you organize your Chapter 5 notes about relationships in triangles. Begin with seven sheets of grid paper stacked. Fold the top right corner to the bottom edge, fold the rectangular part in half, and staple at the fold. Label each sheet with a lesson number and the rectangular tab with the chapter title.

Collaborate

Chapter Project

In the **Architecture: Triangular Design** project, you will use what you have learned about special segments of triangles to complete a project.

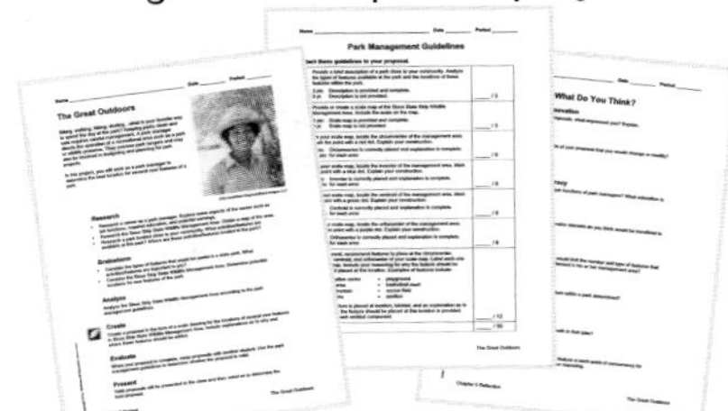

Focus

LEARNSMART

Need help studying? Complete the **Congruence, Proof, and Constructions** domain in LearnSmart to review for the chapter test.

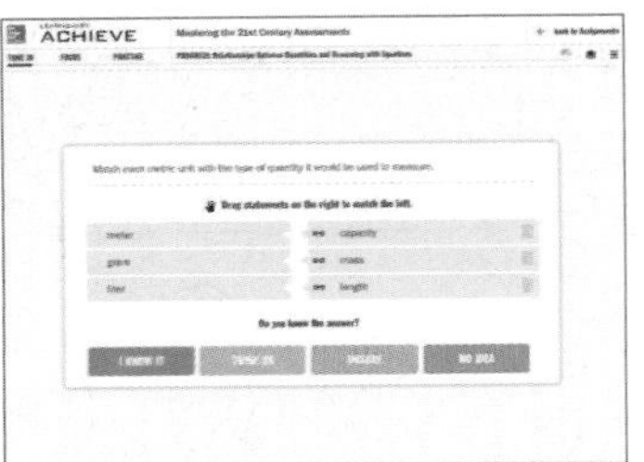

ALEKS

You can use the **Triangles** topic in ALEKS to explore what you know about relationships in triangles and what you are ready to learn.*

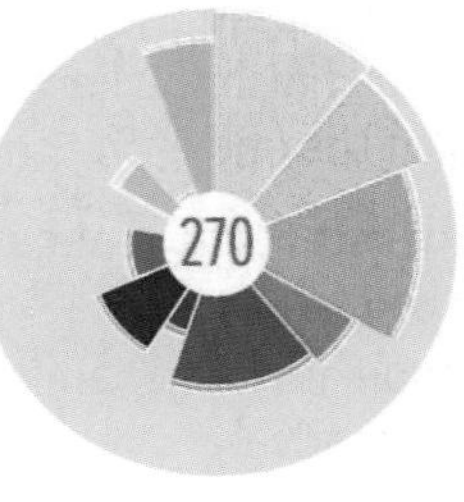

* Ask your teacher if this is part of your program.

Go Online! for Vocabulary Review Games and key vocabulary in 13 languages.

Get Ready for the Chapter

Connecting Concepts

Concept Check

Review the concepts used in this chapter by answering the questions below.

1. What do you know about $m\angle B$ and $m\angle C$?
2. What do you know about the length of $\overline{BC}$?
3. What type of triangle is shown?
4. What is the Isosceles Triangle Theorem?
5. If two angles of a triangle are congruent, what do you know about the sides opposite those angles?
6. If K is the midpoint of $\overline{JL}$, how are the points J, K, and L related?
7. If K is the midpoint of $\overline{JL}$, what do you know about JK and KL?
8. Nina added 15 songs to her digital media player. She now has more than 120 songs. How many songs were originally on the player?

Performance Task Preview

You can use the concepts and skills in the chapter to solve problems in a real-world setting. Understanding relationships in triangles will help you finish the Performance Task at the end of the chapter.

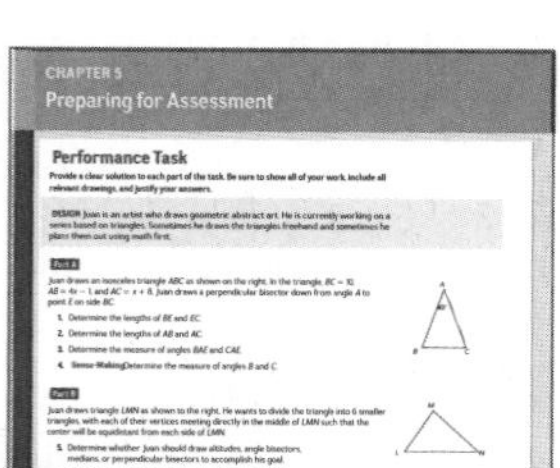
CHAPTER 5
Preparing for Assessment
Performance Task

In this Performance Task you will:

- reason abstractly and quantitatively
- make sense of problems and persevere in solving them

New Vocabulary

English		Español
perpendicular bisector	p. 354	mediatriz
concurrent lines	p. 355	rectas concurrentes
point of concurrency	p. 355	punto de concurrencia
circumcenter	p. 355	circuncentro
incenter	p. 358	incentro
median	p. 365	mediana
centroid	p. 365	baricentro
altitude	p. 367	altura
orthocenter	p. 367	ortocentro
indirect reasoning	p. 385	razonamiento indirecto
indirect proof	p. 385	demostración indirecta
proof by contradiction	p. 385	demostración por contradicción

Review Vocabulary

angle bisector bisectriz de un ángulo **a ray that divides an angle into two congruent angles** (Lesson 1-4)

midpoint punto medio **the point on a segment exactly halfway between the endpoints of the segment** (Lesson 1-3)

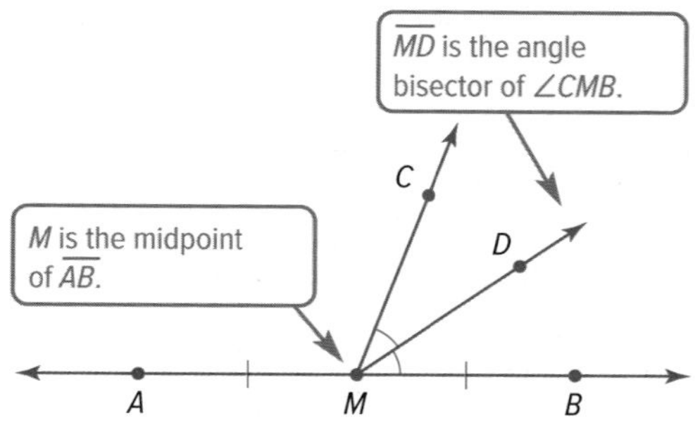

EXPLORE 5-1

Geometry Lab

Constructing Bisectors

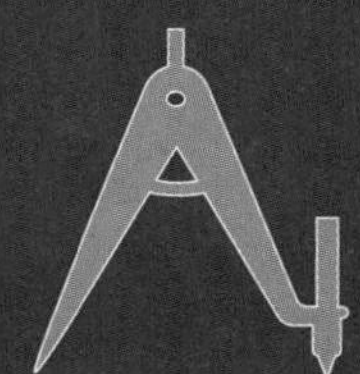

Paper folding can be used to construct special segments in triangles.

Mathematical Practices

MP **5** Use appropriate tools strategically.

Construction 1 Perpendicular Bisector

Work cooperatively. Construct a perpendicular bisector of the side of a triangle.

Step 1 Draw and cut $\triangle MPQ$.

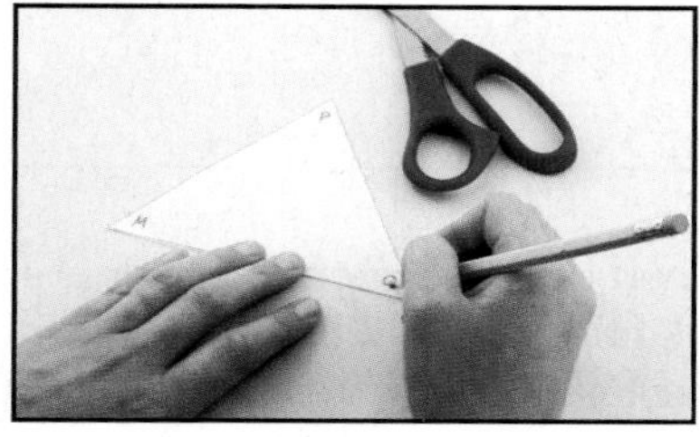

Step 2 Fold the triangle in half so M and Q touch.

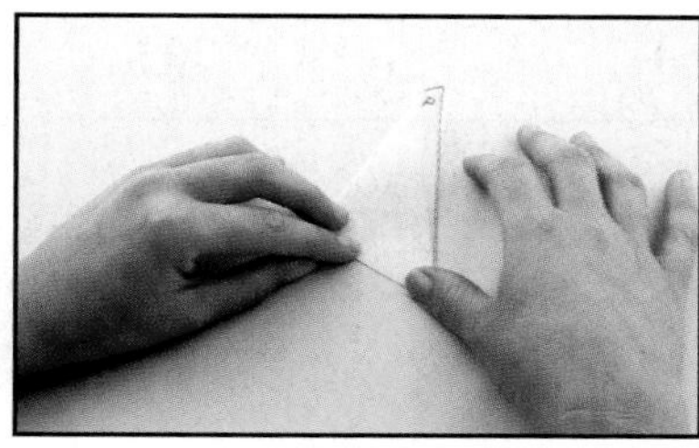

Step 3 Draw $\overline{AB}$ along the fold.

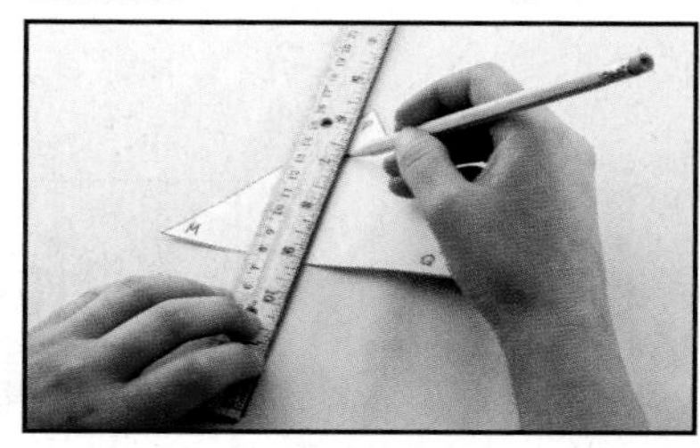

$\overline{AB}$ is the perpendicular bisector of $\overline{MQ}$.

An angle bisector in a triangle is a line containing a vertex of the triangle and bisecting that angle.

Construction 2 Angle Bisector

Work cooperatively. Construct an angle bisector of a triangle.

Step 1

Draw and cut out $\triangle ABC$.

Step 2

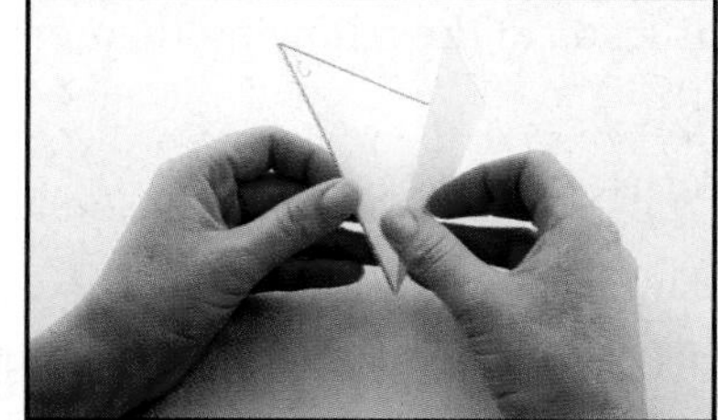

Fold in half through A aligning $\overline{AC}$ and $\overline{AB}$.

Step 3

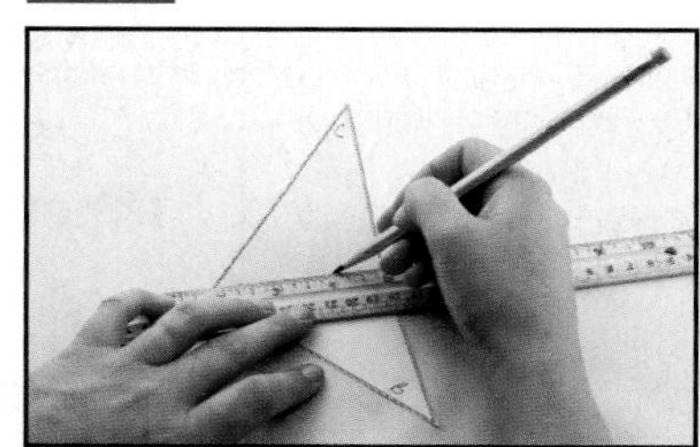

Draw $\overline{AL}$ along the fold. $\overline{AL}$ is an angle bisector of $\triangle ABC$.

Model and Analyze **Work cooperatively.**

1. Construct the perpendicular bisectors of the other two sides of $\triangle MPQ$. Construct the angle bisectors of the other two angles of $\triangle ABC$. Make a conjecture about the intersection of angle bisectors and the intersection of perpendicular bisectors in other triangles.
2. Refer to the construction of a segment bisector on page 28. How does the construction of a segment bisector relate to the construction of a perpendicular bisector of a segment?

Repeat the two constructions for each type of triangle.

3. acute
4. obtuse
5. right

LESSON 1

Bisectors of Triangles

Then

- You used segment and angle bisectors.

Now

1. Identify and use perpendicular bisectors in triangles.
2. Identify and use angle bisectors in triangles.

Why?

- Creating a work triangle in a kitchen can make food preparation more efficient by cutting down on the number of steps you have to take. To locate the point that is equidistant from the sink, stove, and refrigerator, you can use the perpendicular bisectors of the triangle.

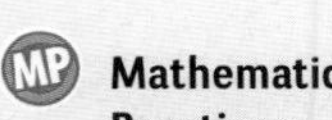

New Vocabulary

perpendicular bisector
concurrent lines
point of concurrency
circumcenter
incenter

Mathematical Practices

1 Make sense of problems and persevere in solving them.

3 Construct viable arguments and critique the reasoning of others.

7 Look for and make use of structure.

1 Perpendicular Bisectors

In Lesson 1-3, you learned that a segment bisector is any segment, line, or plane that intersects a segment at its midpoint. If a bisector is also perpendicular to the segment, it is called a **perpendicular bisector**.

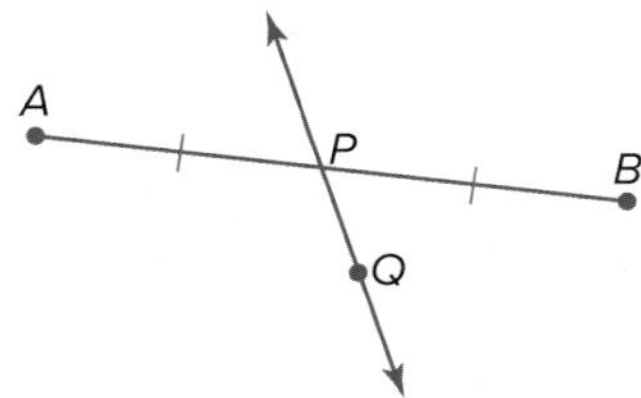

$\overleftrightarrow{PQ}$ is a bisector of $\overline{AB}$.

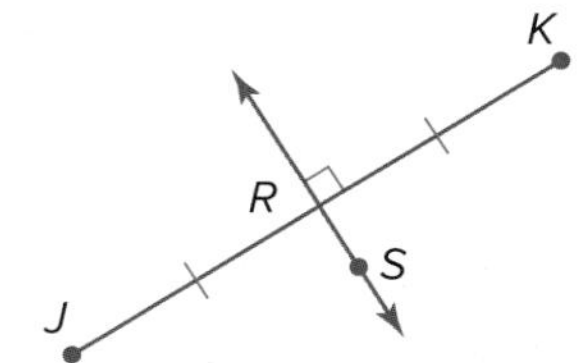

$\overleftrightarrow{RS}$ is a perpendicular bisector of $\overline{JK}$.

Recall that a *locus* is a set of points that satisfies a particular condition. The perpendicular bisector of a segment is the locus of points in a plane equidistant from the endpoints of the segment. This leads to the following theorems.

Theorems Perpendicular Bisectors

5.1 Perpendicular Bisector Theorem

If a point is on the perpendicular bisector of a segment, then it is equidistant from the endpoints of the segment.

Example: If $\overline{CD}$ is a $\perp$ bisector of $\overline{AB}$, then $AC = BC$.

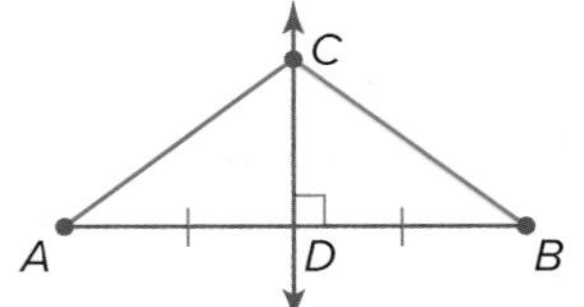

5.2 Converse of the Perpendicular Bisector Theorem

If a point is equidistant from the endpoints of a segment, then it is on the perpendicular bisector of the segment.

Example: If $AE = BE$, then E lies on $\overline{CD}$, the $\perp$ bisector of $\overline{AB}$.

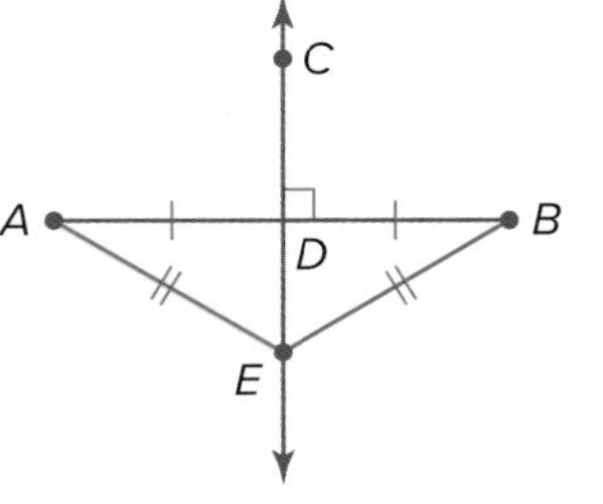

You will prove Theorems 5.1 and 5.2 in Exercises 39 and 37, respectively.

Eric Hernandez/Getty Images

Study Tip

Perpendicular Bisectors The perpendicular bisector of a side of a triangle does not necessarily pass through a vertex of the triangle. For example, in $\triangle XYZ$ below, the perpendicular bisector of $\overline{XY}$ does not pass through point Z.

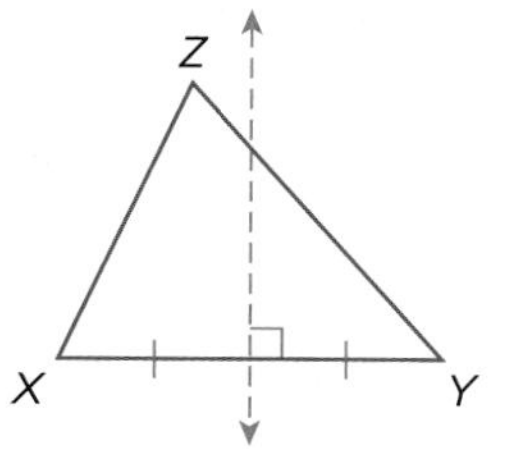

Go Online!

Investigate the bisectors of a triangle using the **Geometry Tools** in ConnectED.

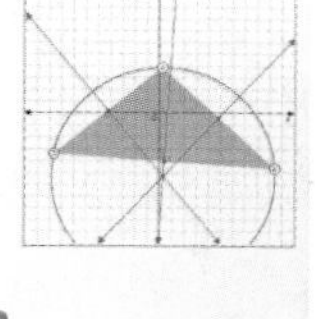

Example 1 Use the Perpendicular Bisector Theorems

Find each measure.

a. AB

From the information in the diagram, we know that $\overleftrightarrow{CA}$ is the perpendicular bisector of $\overline{BD}$.

$AB = AD$ Perpendicular Bisector Theorem

$AB = 4.1$ Substitution

b. WY

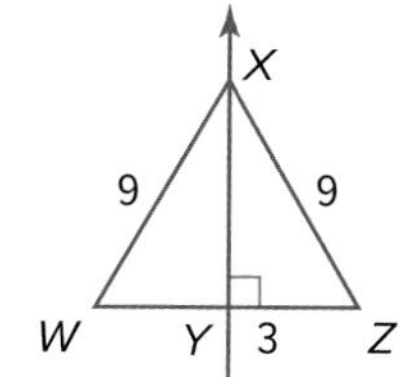

Since $WX = ZX$ and $\overleftrightarrow{XY} \perp \overline{WZ}$, $\overleftrightarrow{XY}$ is the perpendicular bisector of $\overline{WZ}$ by the Converse of the Perpendicular Bisector Theorem. By the definition of segment bisector, $WY = YZ$. Since $YZ = 3$, $WY = 3$.

c. RT

$\overleftrightarrow{SR}$ is the perpendicular bisector of $\overline{QT}$.

$RT = RQ$ Perpendicular Bisector Theorem

$4x - 7 = 2x + 3$ Substitution

$2x - 7 = 3$ Subtract $2x$ from each side.

$2x = 10$ Add 7 to each side.

$x = 5$ Divide each side by 2.

So $RT = 4(5) - 7$ or 13.

Guided Practice

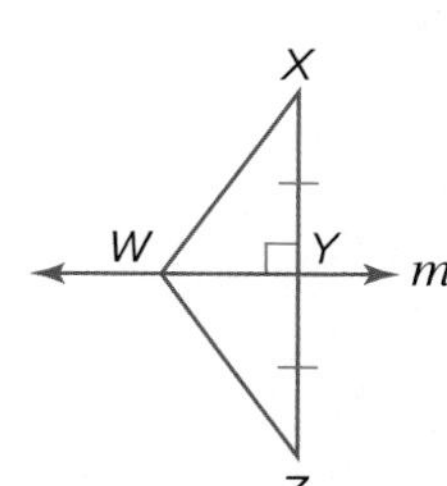

1A. If $WX = 25.3$, $YZ = 22.4$, and $WZ = 25.3$, find XY.

1B. If m is the perpendicular bisector of XZ and $WZ = 14.9$, find WX.

1C. If m is the perpendicular bisector of XZ, $WX = 4a - 15$, and $WZ = a + 12$, find WX.

When three or more lines intersect at a common point, the lines are called **concurrent lines**. The point where concurrent lines intersect is called the **point of concurrency**.

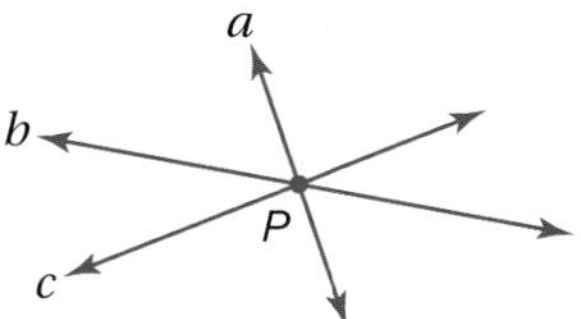

Lines a, b, and c are concurrent at P.

A triangle has three sides, so it also has three perpendicular bisectors. These bisectors are concurrent lines. The point of concurrency of the perpendicular bisectors is called the **circumcenter** of the triangle.

Theorem 5.3 Circumcenter Theorem

Words The perpendicular bisectors of a triangle intersect at a point called the *circumcenter* that is equidistant from the vertices of the triangle.

Example If P is the circumcenter of $\triangle ABC$, then $PB = PA = PC$.

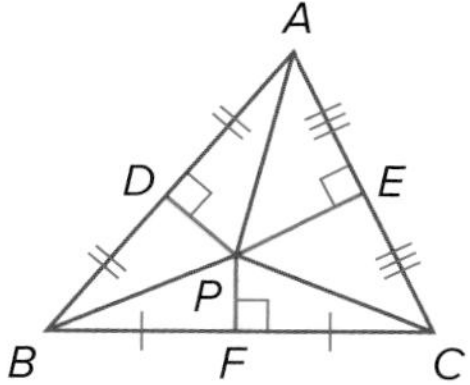

Reading Math ELL

Circum– The prefix *circum–* means *about* or *around*. You know that the *circumference* of a circle is the distance around it. The circumcenter is the center of a circle around a triangle that contains the vertices of the triangle.

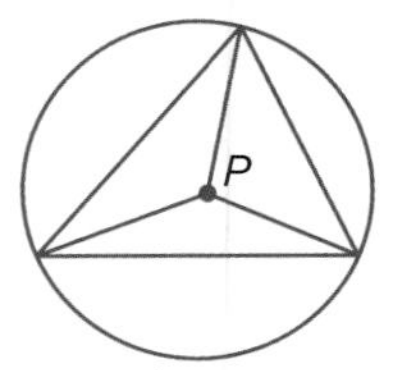

The circumcenter can be on the interior, exterior, or side of a triangle.

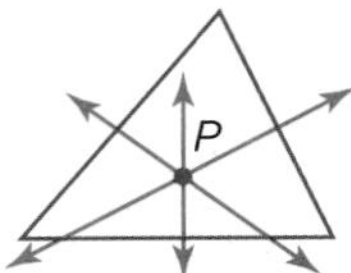

acute triangle

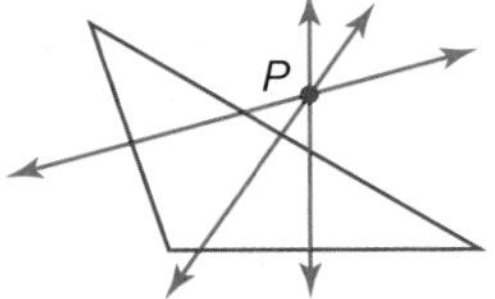

obtuse triangle

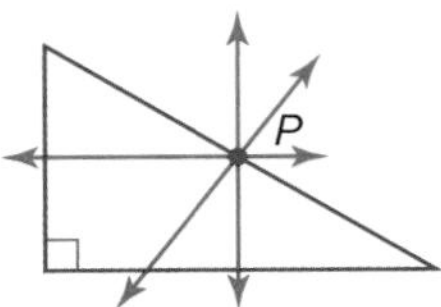

right triangle

Proof Circumcenter Theorem

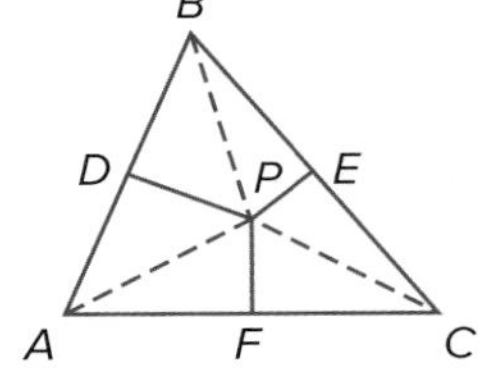

Given: $\overline{PD}$, $\overline{PF}$, and $\overline{PE}$ are perpendicular bisectors of $\overline{AB}$, $\overline{AC}$, and $\overline{BC}$, respectively.

Prove: $AP = CP = BP$

Paragraph Proof:

Since P lies on the perpendicular bisector of $\overline{AC}$, it is equidistant from A and C. By the definition of equidistant, $AP = CP$. The perpendicular bisector of $\overline{BC}$ also contains P. Thus, $CP = BP$. By the Transitive Property of Equality, $AP = BP$. Thus, $AP = CP = BP$.

Real-World Example 2 Use the Circumcenter Theorem

INTERIOR DESIGN A stove S, sink K, and refrigerator R are positioned in a kitchen as shown. Find the location for the center of an island work station so that it is the same distance from these three points.

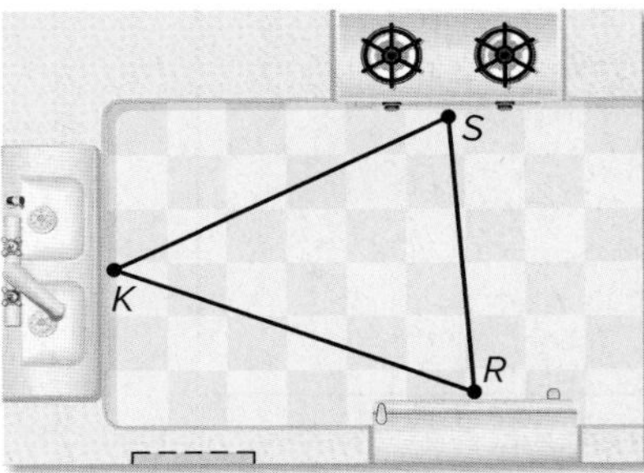

By the Circumcenter Theorem, a point equidistant from three points is found by using the perpendicular bisectors of the triangle formed by those points.

Copy $\triangle SKR$, and use a ruler and protractor to draw the perpendicular bisectors. The location for the center of the island is C, the circumcenter of $\triangle SKR$.

Guided Practice

2. To water his triangular garden, Alex needs to place a sprinkler equidistant from each vertex. Where should Alex place the sprinkler?

Real-World Link

Some basic rules of thumb for a kitchen work triangle are that the sides of the triangle should be no greater than 9 feet and no less than 4 feet. Also, the perimeter of the triangle should be no more than 26 feet and no less than 12 feet.

Source: Merillat

2 Angle Bisectors

Recall from Lesson 1-4 that an angle bisector divides an angle into two congruent angles. The angle bisector can be a line, segment, or ray.

The bisector of an angle can be described as the locus of points in the interior of the angle equidistant from the sides of the angle. This description leads to the following theorems.

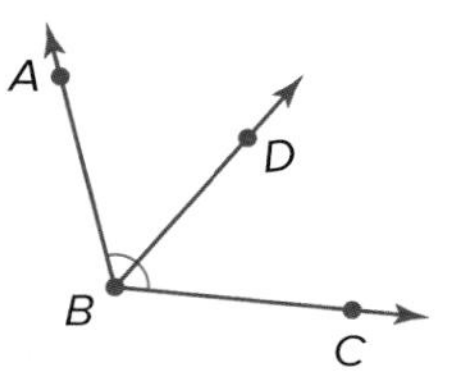

$\overrightarrow{BD}$ is the angle bisector of $\angle ABC$.

George Doyle/Stockbyte/Getty Images

Theorems Angle Bisectors

5.4 Angle Bisector Theorem

If a point is on the bisector of an angle, then it is equidistant from the sides of the angle.

Example: If $\overrightarrow{BF}$ bisects $\angle DBE$, $\overline{FD} \perp \overrightarrow{BD}$, and $\overline{FE} \perp \overrightarrow{BE}$, then $DF = FE$.

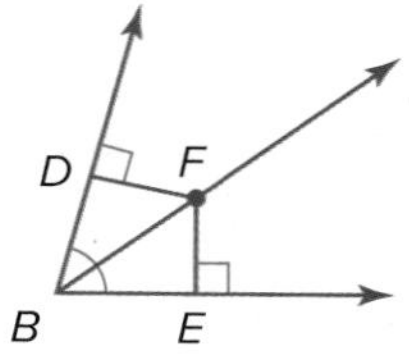

5.5 Converse of the Angle Bisector Theorem

If a point in the interior of an angle is equidistant from the sides of the angle, then it is on the bisector of the angle.

Example: If $\overline{FD} \perp \overrightarrow{BD}$, $\overline{FE} \perp \overrightarrow{BE}$, and $DF = FE$, then $\overrightarrow{BF}$ bisects $\angle DBE$.

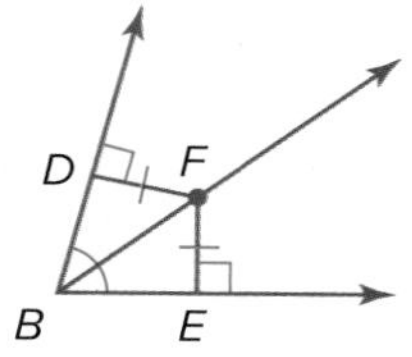

You will prove Theorems 5.4 and 5.5 in Exercises 43 and 40.

Example 3 Use the Angle Bisector Theorems

Find each measure.

a. XY

$XY = XW$	Angle Bisector Theorem
$XY = 7$	Substitution

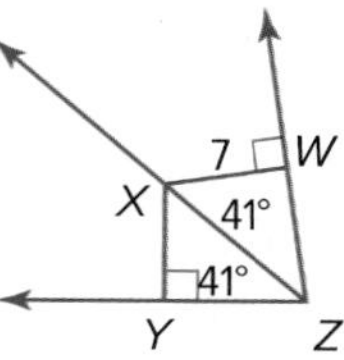

Study Tip

Angle Bisector For part **b**, only knowing that $JL = LM$ would not be enough information to conclude that $\overrightarrow{KL}$ bisects $\angle JKM$.

b. $m\angle JKL$

Since $\overline{LJ} \perp \overrightarrow{KJ}$, $\overline{LM} \perp \overrightarrow{KM}$, $\overline{LJ} \cong \overline{LM}$, L is equidistant from the sides of $\angle JKM$. By the Converse of the Angle Bisector Theorem, $\overrightarrow{KL}$ bisects $\angle JKM$.

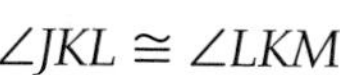

$\angle JKL \cong \angle LKM$	Definition of angle bisector
$m\angle JKL = m\angle LKM$	Definition of congruent angles
$m\angle JKL = 37$	Substitution

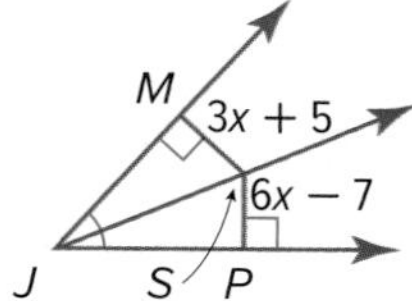

c. SP

$SP = SM$	Angle Bisector Theorem
$6x - 7 = 3x + 5$	Substitution
$3x - 7 = 5$	Subtract $3x$ from each side.
$3x = 12$	Add 7 to each side.
$x = 4$	Divide each side by 3.

So, $SP = 6(4) - 7$ or 17

M
3x + 5
6x − 7
J
S
P

Guided Practice

3A. If $m\angle BAC = 38$, $BC = 5$, and $DC = 5$, find $m\angle DAC$.

3B. If $m\angle BAC = 40$, $m\angle DAC = 40$, and $DC = 10$, find BC.

3C. If $\overrightarrow{AC}$ bisects $\angle DAB$, $BC = 4x + 8$, and $DC = 9x - 7$, find BC.

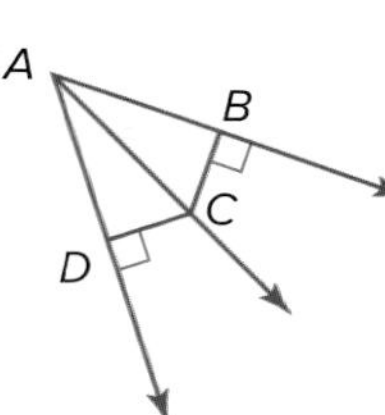

Reading Math

MP Sense-Making The incenter is the center of a circle that intersects each side of the triangle at one point. For this reason, the incenter always lies in the interior of a triangle.

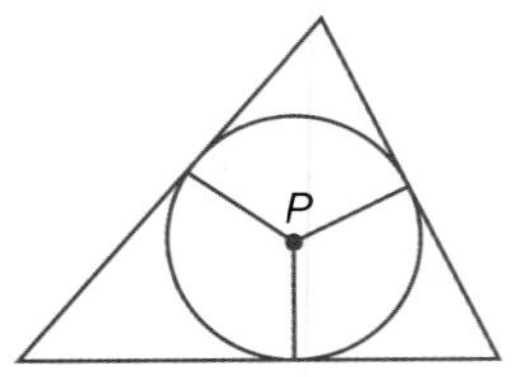

Similar to perpendicular bisectors, since a triangle has three angles, it also has three angle bisectors. The angle bisectors of a triangle are concurrent, and their point of concurrency is called the **incenter** of a triangle.

Theorem 5.6 Incenter Theorem

Words The angle bisectors of a triangle intersect at a point called the *incenter* that is equidistant from the sides of the triangle.

Example If P is the incenter of $\triangle ABC$, then $PD = PE = PF$.

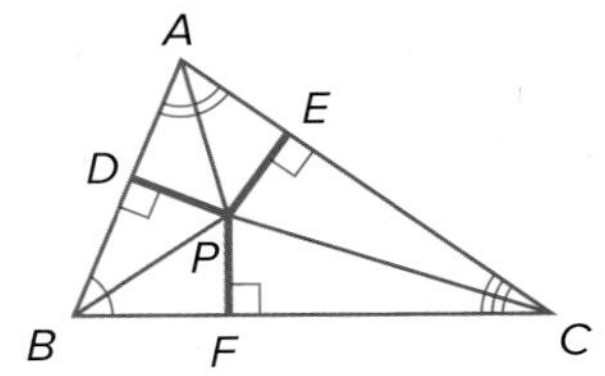

You will prove Theorem 5.6 in Exercise 38.

Example 4 Use the Incenter Theorem

If J is the incenter of $\triangle ABC$, find each measure.

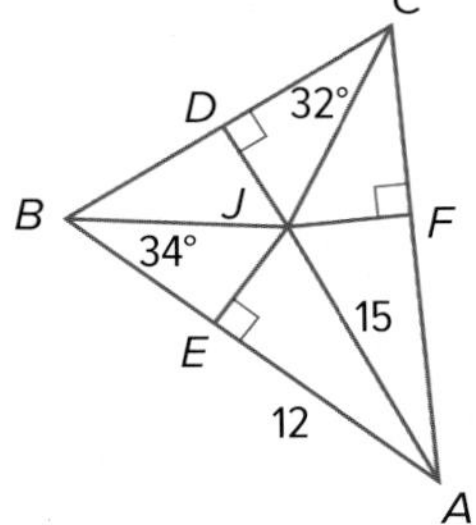

a. JF

By the Incenter Theorem, since J is equidistant from the sides of $\triangle ABC$, $JF = JE$. Find JF by using the Pythagorean Theorem.

$a^2 + b^2 = c^2$	Pythagorean Theorem
$JE^2 + 12^2 = 15^2$	Substitution
$JE^2 + 144 = 225$	$12^2 = 144$ and $15^2 = 225$.
$JE^2 = 81$	Subtract 144 from each side.
$JE = \pm 9$	Take the square root of each side.

Since length cannot be negative, use only the positive square root, 9. Since $JE = JF$, $JF = 9$.

b. $m\angle JAC$

Since $\overrightarrow{BJ}$ bisects $\angle CBE$, $m\angle CBE = 2m\angle JBE$. So $m\angle CBE = 2(34)$ or 68. Likewise, $m\angle DCF = 2m\angle DCJ$, so $m\angle DCF = 2(32)$ or 64.

$m\angle CBE + m\angle DCF + m\angle FAE = 180$	Triangle Angle-Sum Theorem
$68 + 64 + m\angle FAE = 180$	$m\angle CBE = 68$, $m\angle DCF = 64$
$132 + m\angle FAE = 180$	Simplify.
$m\angle FAE = 48$	Subtract 132 from each side.

Since $\overrightarrow{AJ}$ bisects $\angle FAE$, $2m\angle JAC = m\angle FAE$. This means that $m\angle JAC = \frac{1}{2}m\angle FAE$, so $m\angle JAC = \frac{1}{2}(48)$ or 24.

Guided Practice

If P is the incenter of $\triangle XYZ$, find each measure.

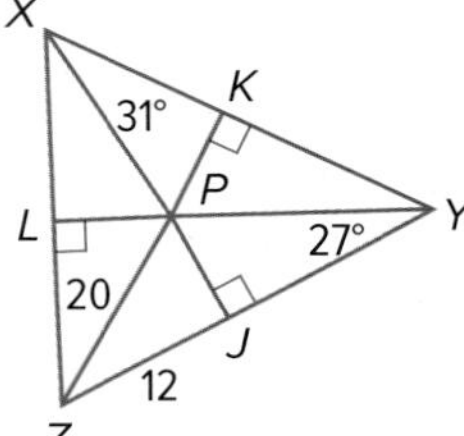

4A. PK

4B. $m\angle LZP$

Check Your Understanding

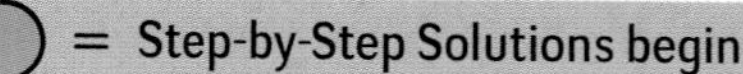
= Step-by-Step Solutions begin on page R13.

Example 1 **Find each measure.**

1. XW

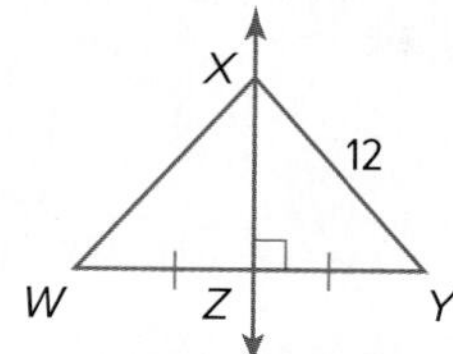

2. AC

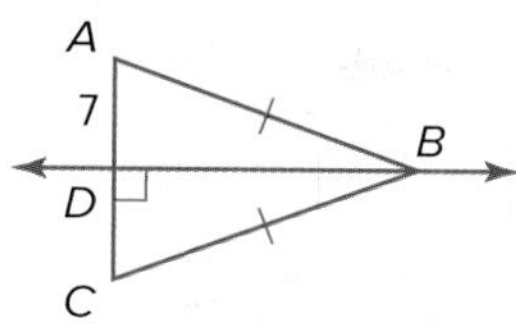

3. LP

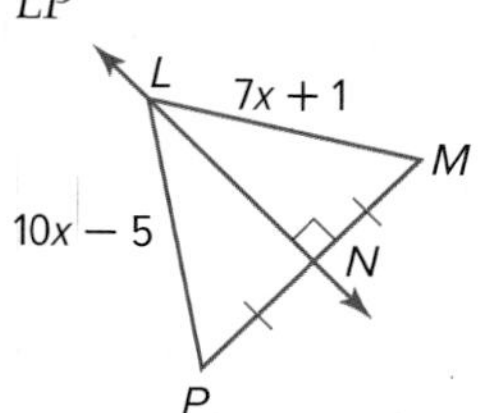

Example 2

4. ADVERTISING Four friends are passing out flyers at a mall food court. Three of them take as many flyers as they can and position themselves as shown. The fourth one keeps the supply of additional flyers. Copy the positions of points A, B, and C. Then position the fourth friend at D so that she is the same distance from each of the other three friends.

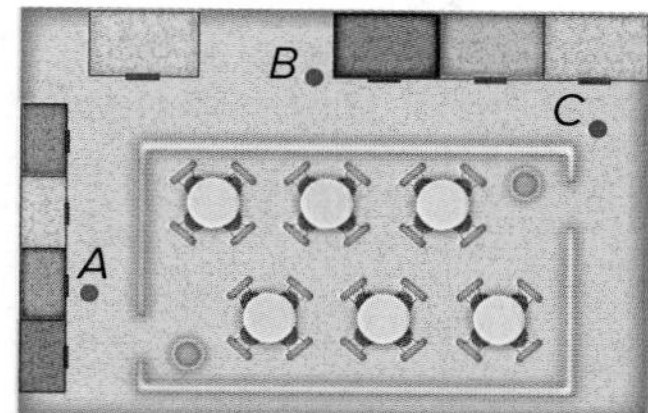

Example 3 **Find each measure.**

5. CP

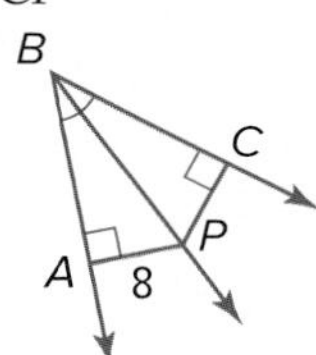

6. $m\angle WYZ$

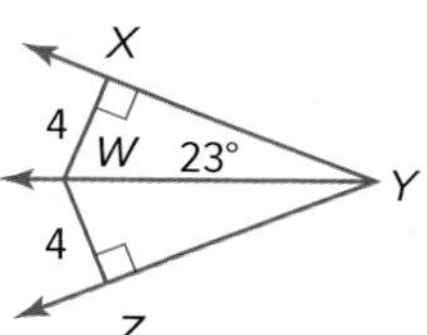

7. QM

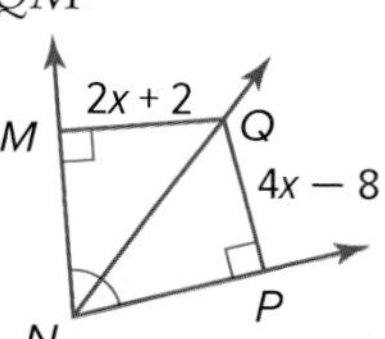

Example 4

8. MP **SENSE-MAKING** If Q is the incenter of $\triangle JLN$, find JQ.

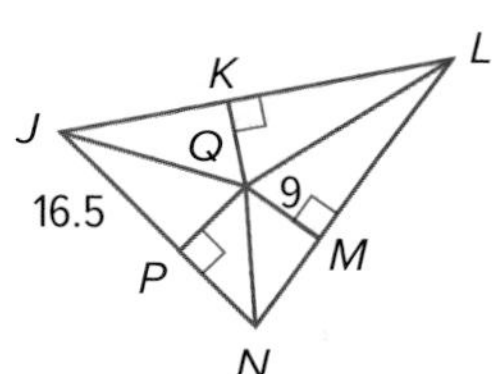

Practice and Problem Solving

Extra Practice is on page R5.

Example 1 **Find each measure.**

9. NP

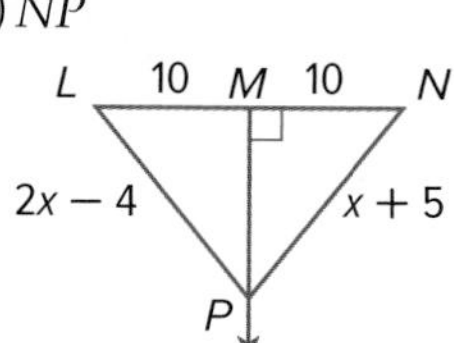

10. PS

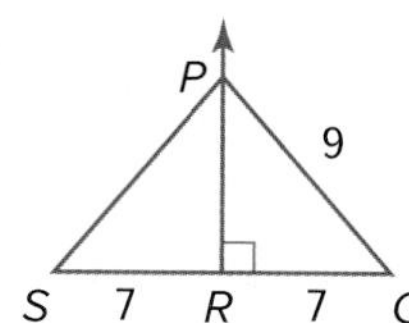

11. KL

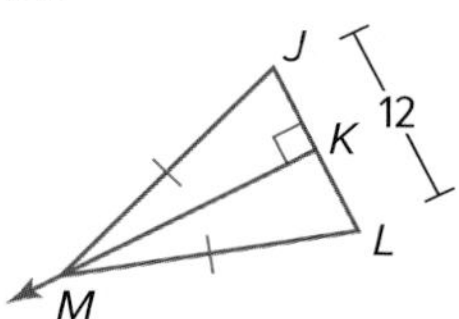

12. EG

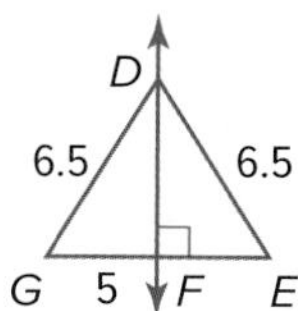

13. CD

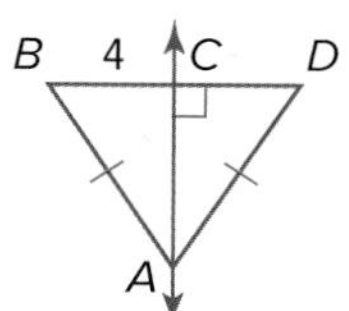

14. SW

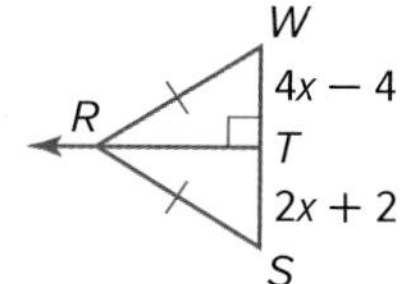

Example 2

15. **STATE FAIR** The state fair has set up the location of the midway, livestock competition, and food vendors. The fair planners decided that they want to locate the portable restrooms the same distance from each location. Copy the positions of points *M*, *L*, and *F*. Then find the location for the restrooms and label it *R*.

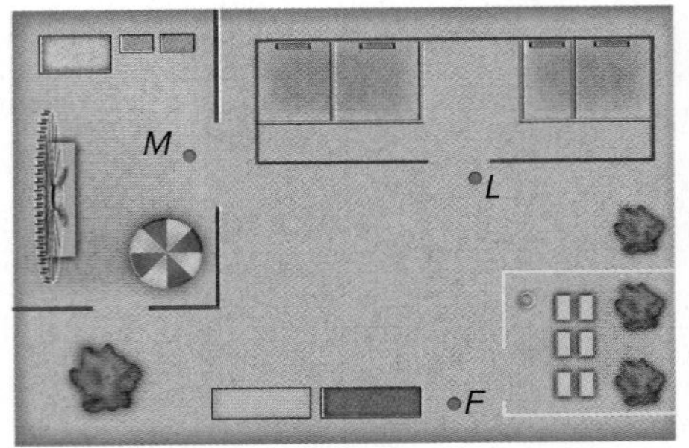

16. **SCHOOL** A school system has built an elementary, middle, and high school at the locations shown in the diagram. Copy the positions of points *E*, *M*, and *H*. Then find the location for the bus yard *B* that will service these schools so that it is the same distance from each school.

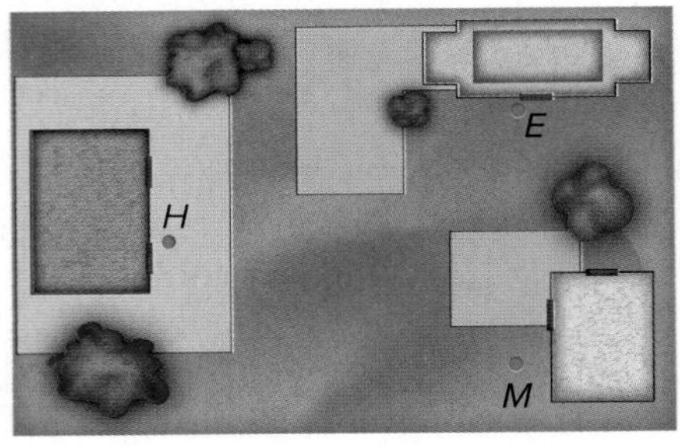

Point *D* is the circumcenter of $\triangle ABC$. List any segment(s) congruent to each segment.

17. $\overline{AD}$

18. $\overline{BF}$

19. $\overline{AH}$

20. $\overline{DC}$

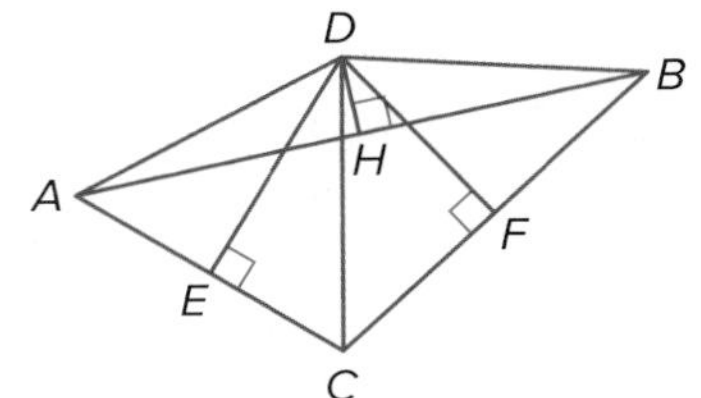

Example 3

Find each measure.

21. *AF*

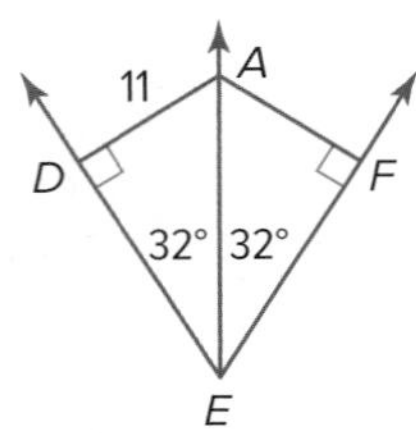

22. $m\angle DBA$

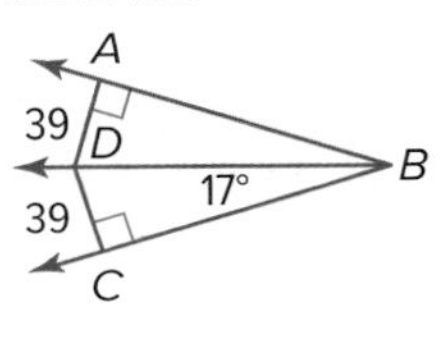

23. $m\angle PNM$

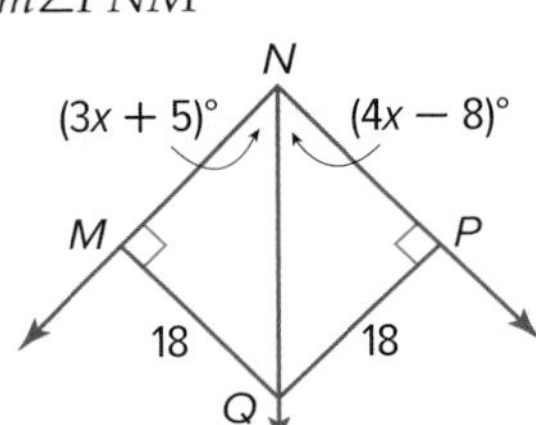

24. *XA*

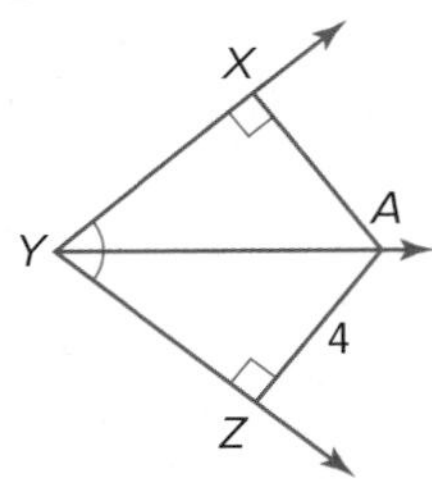

25. $m\angle PQS$

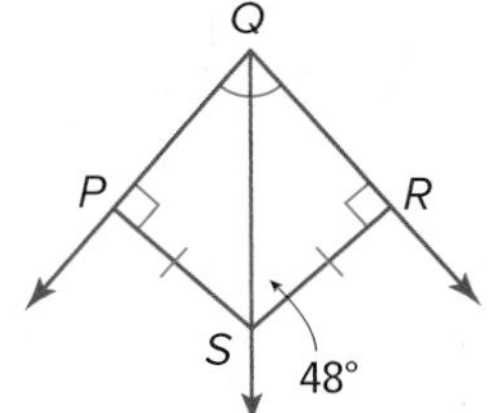

26. *PN*

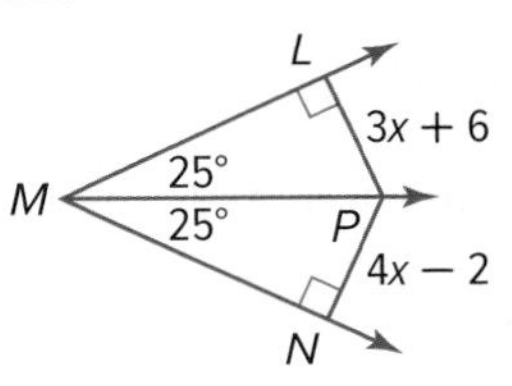

Example 4

MP STRUCTURE If *P* is the incenter of $\triangle AEC$, find each measure.

27. *PB*

28. *DE*

29. $m\angle DAC$

30. $m\angle DEP$

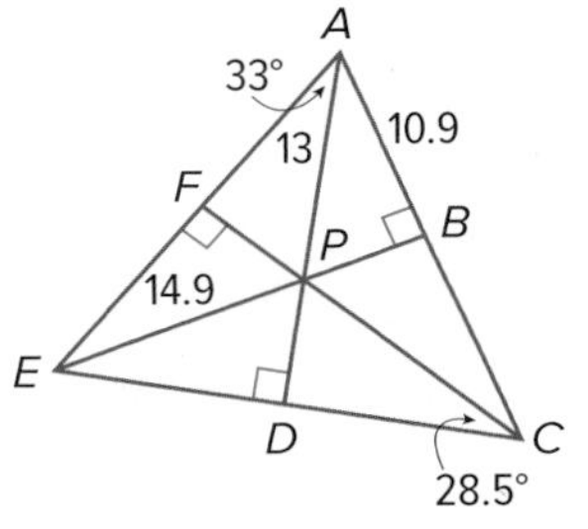

31 **INTERIOR DESIGN** You want to place a centerpiece on a corner table so that it is located the same distance from each edge of the table. Make a sketch to show where you should place the centerpiece. Explain your reasoning.

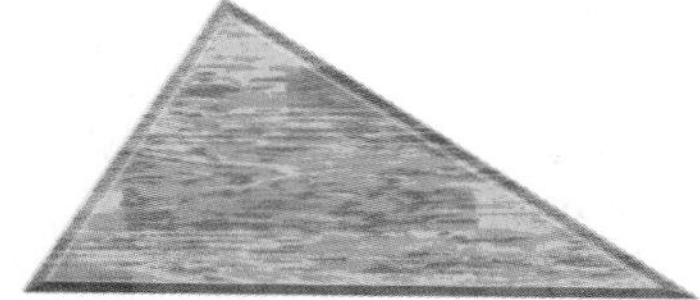

Determine whether there is enough information given in each diagram to find the value of x. Explain your reasoning.

32.

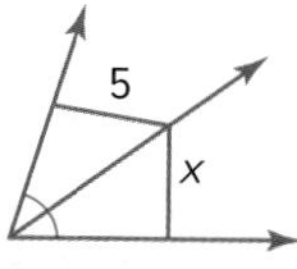

33.

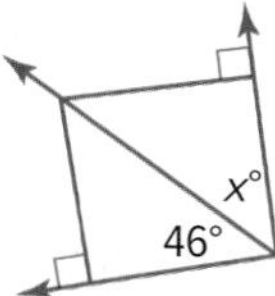

34.

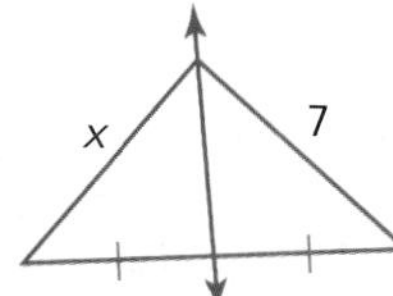

35.

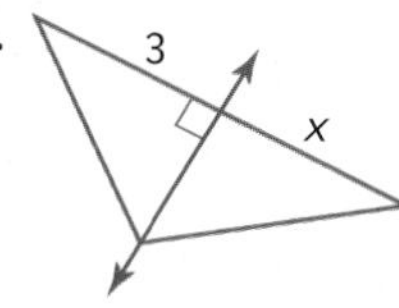

36. **SOCCER** A soccer player P is approaching the opposing team's goal as shown in the diagram. To make the goal, the player must kick the ball between the goalposts at L and R. The goalkeeper faces the kicker. He then tries to stand so that if he needs to dive to stop a shot, he is as far from the left-hand side of the shot angle as the right-hand side.

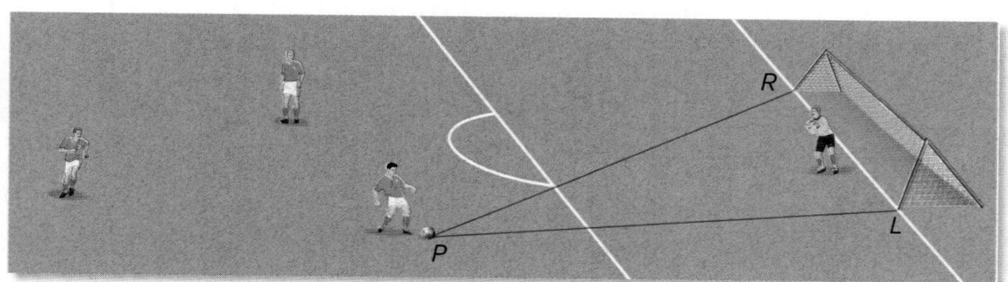

a. Describe where the goalkeeper should stand. Explain your reasoning.

b. Copy $\triangle PRL$. Use a compass and a straightedge to locate a point G where the goalkeeper should stand.

c. If the ball is kicked so it follows the path from P to R, construct the shortest path the goalkeeper should take to block the shot. Explain your reasoning.

PROOF **Write a two-column proof.**

37. Theorem 5.2

Given: $\overline{CA} \cong \overline{CB}, \overline{AD} \cong \overline{BD}$

Prove: C and D are on the perpendicular bisector of $\overline{AB}$.

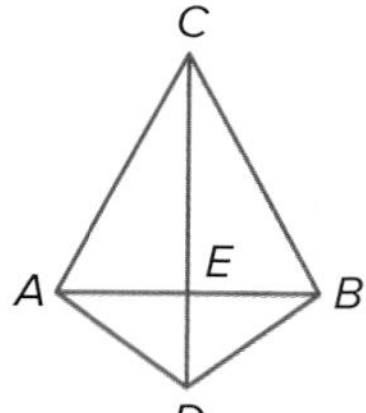

38. Theorem 5.6

Given: $\triangle ABC$, angle bisectors $\overline{AD}, \overline{BE}$, and $\overline{CF}$

$\overline{KP} \perp \overline{AB}, \overline{KQ} \perp \overline{BC}$,

$\overline{KR} \perp \overline{AC}$

Prove: $KP = KQ = KR$

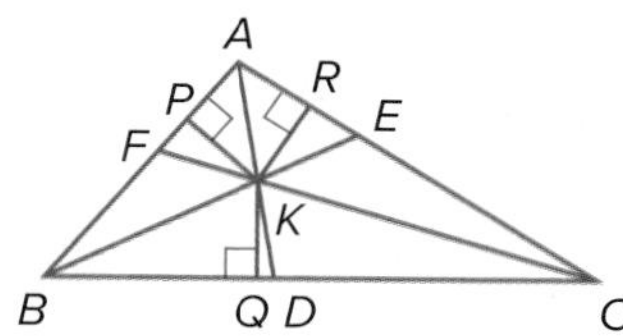

MP **CONSTRUCT ARGUMENTS** **Write a paragraph proof of each theorem.**

39. Theorem 5.1

40. Theorem 5.5

COORDINATE GEOMETRY **Write an equation in slope-intercept form for the perpendicular bisector of the segment with the given endpoints. Justify your answer.**

41. $A(-3, 1)$ and $B(4, 3)$

42. $C(-4, 5)$ and $D(2, -2)$

43. **PROOF** Write a two-column proof of Theorem 5.4.

44. **MULTI-STEP** Mykia is designing a logo for her school. The logo will be sewn onto a 3-foot by 5-foot piece of white fabric. Her logo consists of a circle inside an isosceles triangle with an height of 5 feet. Mykia wants the circle to be as large as possible so she can put a picture of the school's mascot in the center of the circle.

    a. Create a design for the logo, including the center of the circle. If Mykia wants the triangle to be red fabric and the circle to be blue fabric, how can she determine how much fabric of each color to buy?

    b. Describe and evaluate your solution process.

    c. What assumptions did you make?

COORDINATE GEOMETRY **Find the coordinates of the circumcenter of the triangle with the given vertices. Explain.**

45. $A(0, 0)$, $B(0, 6)$, $C(10, 0)$

46. $J(5, 0)$, $K(5, -8)$, $L(0, 0)$

47. **LOCUS** Consider $\overline{CD}$. Describe the set of all points in space that are equidistant from C and D.

H.O.T. Problems Use Higher-Order Thinking Skills

48. **ERROR ANALYSIS** Claudio says that from the information supplied in the diagram, he can conclude that K is on the perpendicular bisector of $\overline{LM}$. Caitlyn disagrees. Is either of them correct? Explain your reasoning.

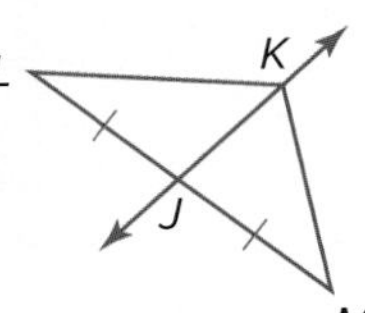

49. **OPEN-ENDED** Draw a triangle with an incenter located inside the triangle but a circumcenter located outside. Justify your drawing by using a straightedge and a compass to find both points of concurrency.

MP **CONSTRUCT ARGUMENTS** **Determine whether each statement is *sometimes*, *always*, or *never* true. Justify your reasoning using a counterexample or proof.**

50. The angle bisectors of a triangle intersect at a point that is equidistant from the vertices of the triangle.

51. In an isosceles triangle, the perpendicular bisector of the base is also the angle bisector of the opposite vertex.

MP **CONSTRUCT ARGUMENTS** **Write a two-column proof for each of the following.**

52. **Given:** Plane Y is a perpendicular bisector of $\overline{DC}$.
Prove: $\angle ADB \cong \angle ACB$

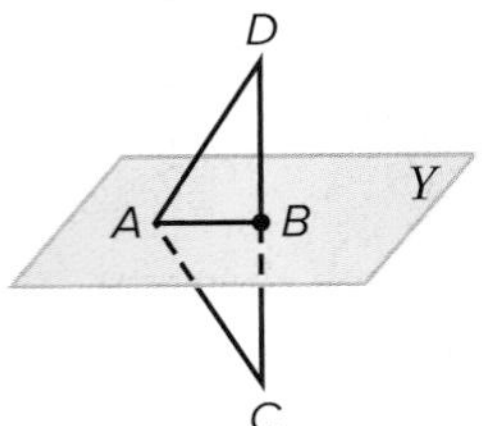

53. **Given:** Plane Z is an angle bisector of $\angle KJH$, $\overline{KJ} \cong \overline{HJ}$
Prove: $\overline{MH} \cong \overline{MK}$

54. **WRITING IN MATH** Compare and contrast the perpendicular bisectors and angle bisectors of a triangle. How are they alike? How are they different? Be sure to compare their points of concurrency.

55. Stephanie is designing a kite using the figure. What is the length of $\overline{PR}$? MP 2

- A 4
- B 5
- C 14
- D 28

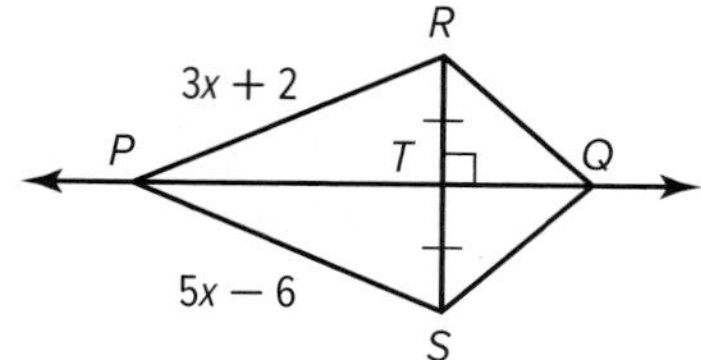

56. A triangular-shaped garden is shown. Can a fountain be placed at the circumcenter and still be inside the garden?

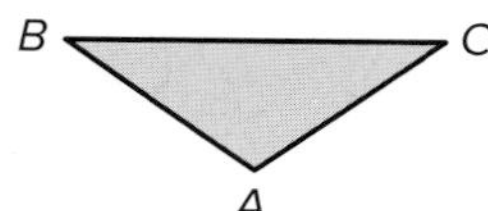

MP 1, 4

57. In the figure, point P is the circumcenter of $\triangle JKL$.

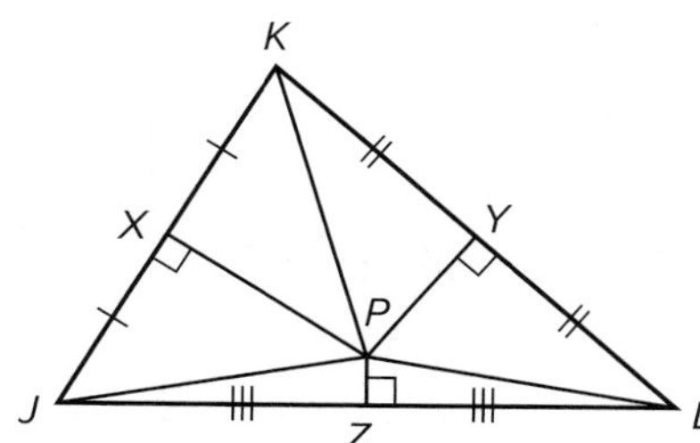

Which of the following statements must be true? MP 1

I. $PX = PY$

II. $PJ = PK$

III. $PK = PL$

- A I only
- B II only
- C III only
- D II and III only
- E I, II, and III

58. $\overleftrightarrow{BD}$ is the perpendicular bisector of $\overline{AC}$.

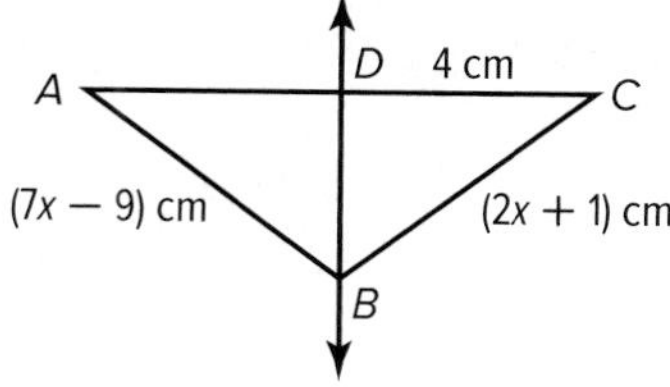

What is the perimeter of $\triangle ABC$? MP 2

- A 5 cm
- B 9 cm
- C 14 cm
- D 18 cm

59. **MULTI-STEP** Given: $MQ = 0.9$, $MP = 2.9$, and $MN = 1.8$ MP 2

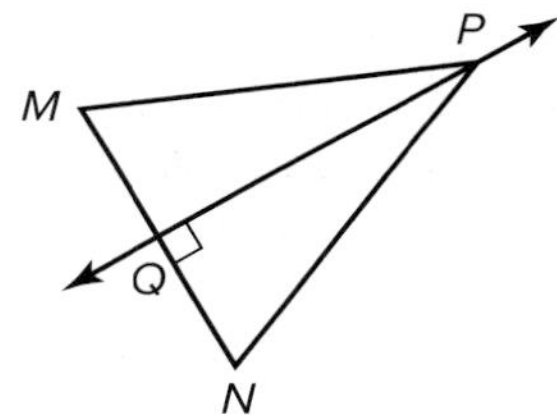

a. What is the value of NQ?

b. What is the relationship of $\triangle MPQ$ and $\triangle NPQ$? Explain.

c. What is the perimeter of $\triangle MPN$?

- A 2.9
- B 5.6
- C 7.6
- D 9.4

60. Determine whether there is enough information given in the diagram to find the value of x. Explain your reasoning. MP 2, 3

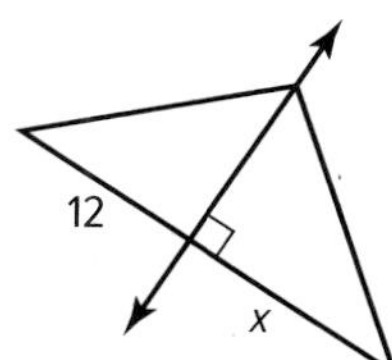

EXPLORE 5-2

Geometry Lab

Constructing Medians and Altitudes

A *median* of a triangle is a segment with endpoints that are a vertex and the midpoint of the side opposite that vertex. You can use the construction for the midpoint of a segment to construct a median.

Mathematical Practices
MP **5 Use appropriate tools strategically.**

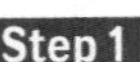

Construction 1 Median of a Triangle

Step 1

Wrap the end of string around a pencil. Place the thumbtack on vertex D and then on vertex E to draw intersecting arcs above and below $\overline{DE}$. Label the points of intersection R and S.

Step 2

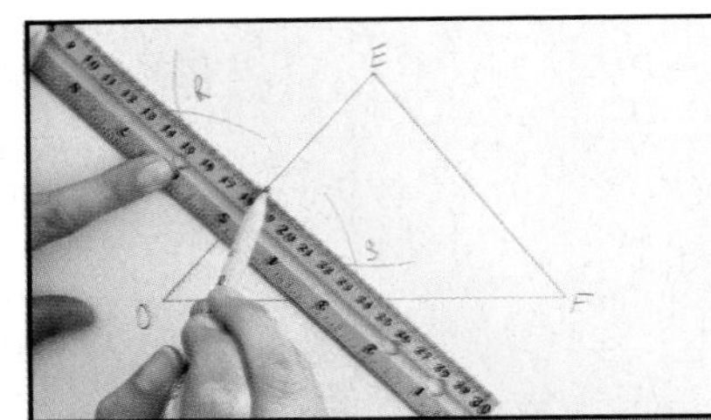

Use a straightedge to find the point where $\overline{RS}$ intersects $\overline{DE}$. Label the point M. This is the midpoint of $\overline{DE}$.

Step 3

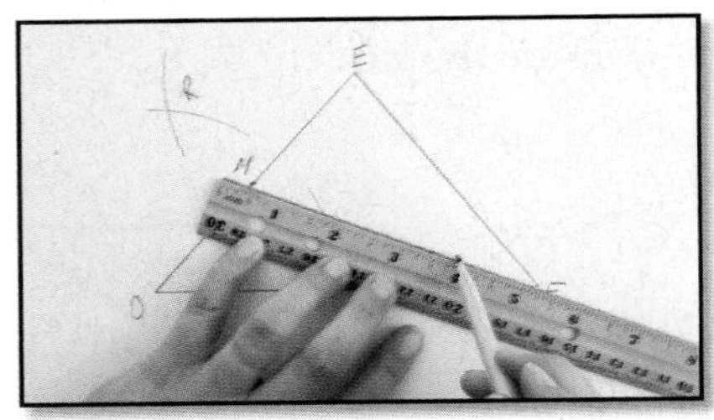

Draw a line through F and M. $\overline{FM}$ is a median of $\triangle DEF$.

An *altitude* of a triangle is a segment from a vertex of the triangle to the opposite side and is perpendicular to the opposite side.

Construction 2 Altitude of a Triangle

Step 1

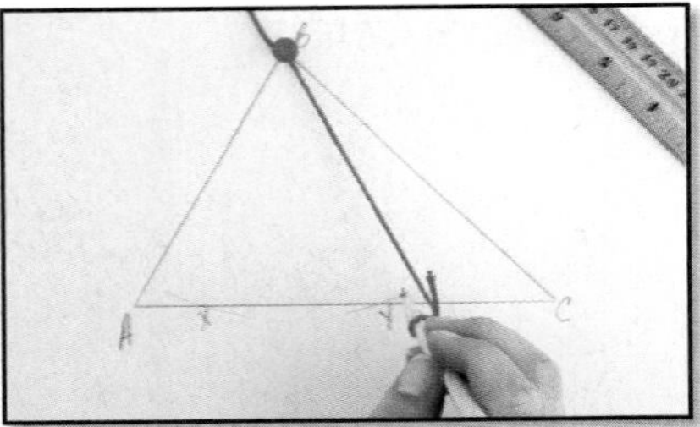

Place the thumbtack on vertex B and draw two arcs intersecting $\overline{AC}$. Label the points where the arcs intersect the sides as X and Y.

Step 2

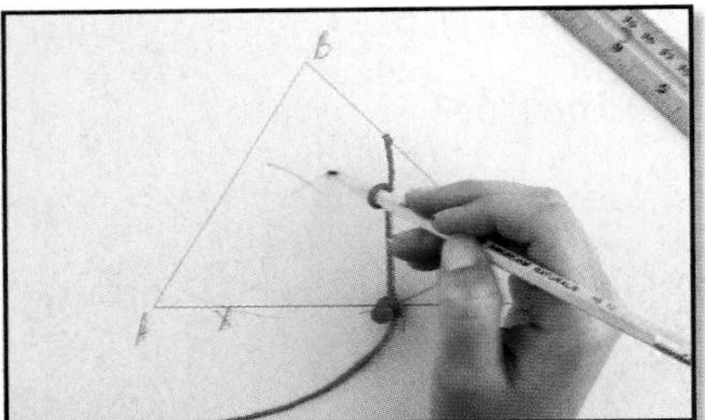

Adjust the length of the string so it is greater than $\frac{1}{2}XY$. Place the tack on X. Draw an arc above $\overline{AC}$ and draw an arc from Y. Label the points of intersection of the arcs H.

Step 3

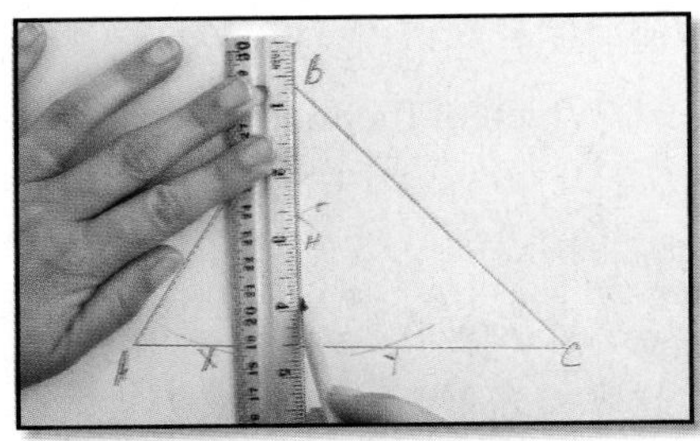

Use a straightedge to draw $\overleftrightarrow{BH}$. Label the point where $\overleftrightarrow{BH}$ intersects $\overline{AC}$ as D. $\overline{BD}$ is an altitude of $\triangle ABC$ and is perpendicular to $\overline{AC}$.

Model and Analyze Work cooperatively.

1. Construct the medians of the other two sides of $\triangle DEF$. What do you notice about the medians of a triangle?
2. Construct the altitudes to the other two sides of $\triangle ABC$. What do you observe?

LESSON 2

Medians and Altitudes of Triangles

Then

- You identified and used perpendicular and angle bisectors in triangles.

Now

1. Identify and use medians in triangles.
2. Identify and use altitudes in triangles.

Why?

- The balancing bird is a kinetic or a moving sculpture that uses the principles of balancing and equilibrium. The center of gravity is a special point on the object. It is the point where the mass of the body is perfectly balanced. The toy balances and rotates freely.

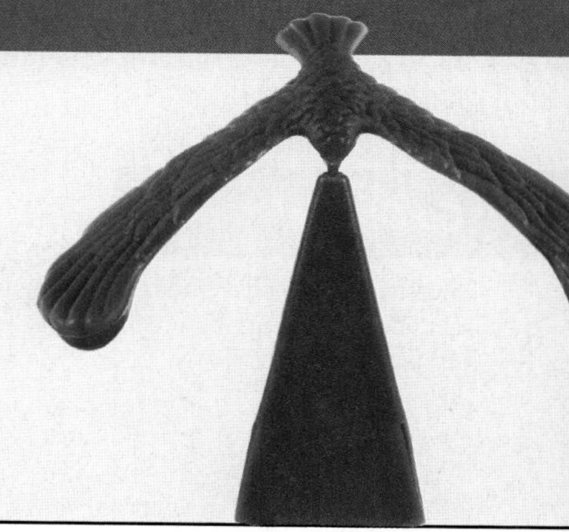

New Vocabulary
median
centroid
altitude
orthocenter

Mathematical Practices

3 Construct viable arguments and critique the reasoning of others.

6 Attend to precision.

1 Medians

A **median** of a triangle is a segment with endpoints being a vertex of a triangle and the midpoint of the opposite side.

Every triangle has three medians that are concurrent. The point of concurrency of the medians of a triangle is called the **centroid** and is always inside the triangle.

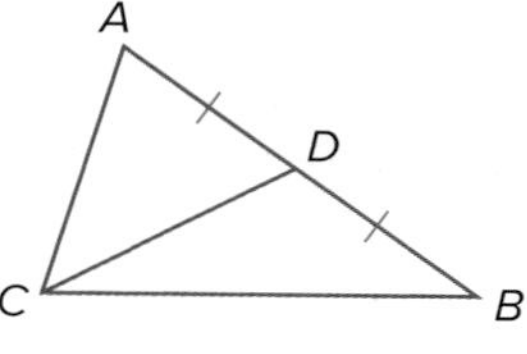

$\overline{CD}$ is a median of $\triangle ABC$.

Theorem 5.7 Centroid Theorem

The medians of a triangle intersect at a point called the centroid that is two thirds of the distance from each vertex to the midpoint of the opposite side.

Examples If P is the centroid of $\triangle ABC$, then
$AP = \frac{2}{3}AK$, $BP = \frac{2}{3}BL$, and $CP = \frac{2}{3}CJ$.

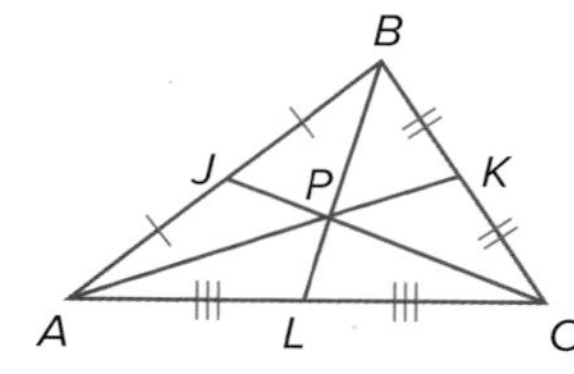

You will prove Theorem 5.7 in Exercise 36.

Example 1 Use the Centroid Theorem

In $\triangle ABC$, Q is the centroid and $BE = 9$. Find BQ and QE.

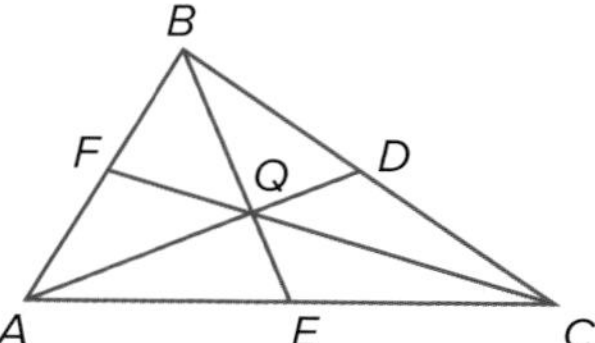

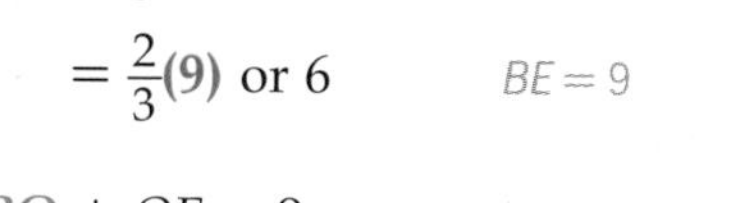

$BQ = \frac{2}{3}BE$	Centroid Theorem
$= \frac{2}{3}(9)$ or 6	$BE = 9$

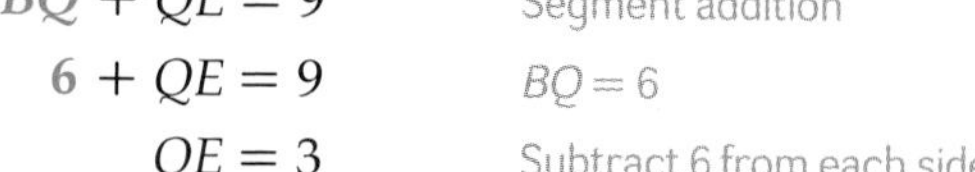

$BQ + QE = 9$	Segment addition
$6 + QE = 9$	$BQ = 6$
$QE = 3$	Subtract 6 from each side.

Guided Practice **In $\triangle ABC$ above, $FC = 15$. Find each length.**

1A. FQ

1B. QC

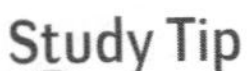

Study Tip

MP Reasoning In Example 2, you can also use number sense to find KP. Since $KP = \frac{2}{3}KT$, $PT = \frac{1}{3}KT$ and $KP = 2PT$. Therefore, if $PT = 2$, then $KP = 2(2)$ or 4.

Go Online!

Investigate the medians and altitudes of a triangle by using the **Geometry Tools** in ConnectED.

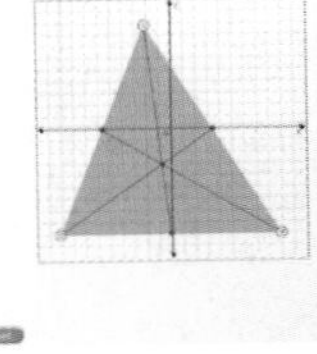

Example 2 Use the Centroid Theorem

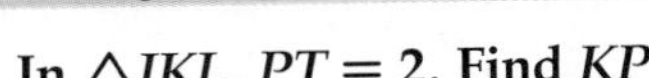

In $\triangle JKL$, $PT = 2$. Find KP.

Since $\overline{JR} \cong \overline{RK}$, R is the midpoint of $\overline{JK}$ and $\overline{LR}$ is a median of $\triangle JKL$. Likewise, S and T are the midpoints of $\overline{KL}$ and $\overline{LJ}$, respectively, so $\overline{JS}$ and $\overline{KT}$ are also medians of $\triangle JKL$. Therefore, point P is the centroid of $\triangle JKL$.

$KP = \frac{2}{3}KT$	Centroid Theorem
$KP = \frac{2}{3}(KP + PT)$	Segment addition and substitution
$KP = \frac{2}{3}(KP + 2)$	$PT = 2$
$KP = \frac{2}{3}KP + \frac{4}{3}$	Distributive Property
$\frac{1}{3}KP = \frac{4}{3}$	Subtract $\frac{2}{3}KP$ from each side.
$KP = 4$	Multiply each side by 3.

Guided Practice

In $\triangle JKL$ above, $RP = 3.5$ and $JP = 9$. Find each measure.

2A. PL

2B. PS

All polygons have a balance point or centroid. The centroid is also the balancing point or *center of gravity* for a triangular region. The center of gravity is the point at which the region is stable under the influence of gravity.

Real-World Example 3 Find the Centroid on Coordinate Plane

PERFORMANCE ART A performance artist plans to balance triangular pieces of metal during her next act. When one such triangle is placed on the coordinate plane, its vertices are located at (1, 10), (5, 0), and (9, 5). What are the coordinates of the point where the artist should support the triangle so that it will balance?

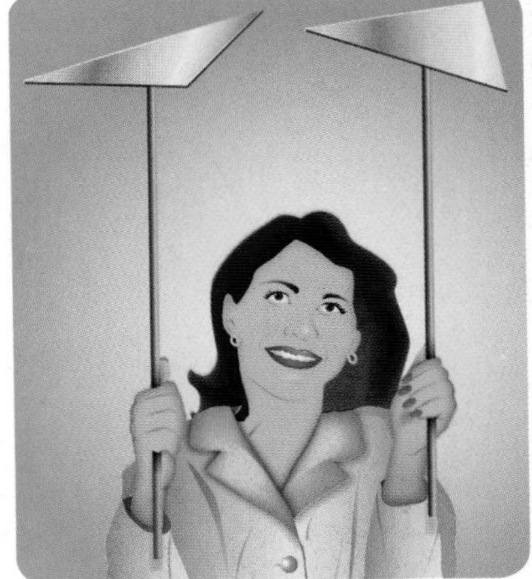

Understand You need to find the centroid of the triangle with the given coordinates. This is the point at which the triangle will balance.

Plan Graph and label the triangle with vertices $A(1, 10)$, $B(5, 0)$, and $C(9, 5)$. Since the centroid is the point of concurrency of the medians of a triangle, use the Midpoint Theorem to find the midpoint of one of the sides of the triangle. The centroid is two-thirds the distance from the opposite vertex to that midpoint.

Math History Link

Pierre de Fermat (1601–1665) Another triangle center is the Fermat point, which minimizes the sum of the distances from the three vertices. Fermat is one of the best-known mathematicians for writing proofs.

Solve Graph $\triangle ABC$.

Find the midpoint D of side $\overline{AB}$ with endpoints $A(1, 10)$ and $B(5, 0)$.

$$D\left(\frac{1+5}{2}, \frac{10+0}{2}\right) = D(3, 5)$$

Graph point D. Notice that $\overline{DC}$ is a horizontal line. The distance from $D(3, 5)$ to $C(9, 5)$ is $9 - 3$ or 6 units.

If P is the centroid of $\triangle ABC$, then $PC = \frac{2}{3}DC$. So the centroid is $\frac{2}{3}(6)$ or 4 units to the left of C. The coordinates of P are $(9 - 4, 5)$ or $(5, 5)$.

The performer should balance the triangle at the point (5, 5).

Check Use a different median to check your answer. The midpoint F of side $\overline{AC}$ is $F\left(\frac{1+9}{2}, \frac{10+5}{2}\right)$ or $F(5, 7.5)$. $\overline{BF}$ is a vertical line, so the distance from B to F is $7.5 - 0$ or 7.5. $\overline{PB} = \frac{2}{3}(7.5)$ or 5, so P is 5 units up from B. The coordinates of P are $(5, 0 + 5)$ or $(5, 5)$. ✓

Any side could have been used to find the centroid. Using all three sides is a simple way to check the solution. Point P appears to be in the middle of the triangle, so the answer is reasonable.

Guided Practice

3. A second triangle has vertices at (0, 4), (6, 11.5), and (12, 1). What are the coordinates of the point where the artist should support the triangle so that it will balance? Explain your reasoning.

2 Altitudes

An **altitude** of a triangle is a segment from a vertex to the line containing the opposite side and perpendicular to the line containing that side. An altitude can lie in the interior, exterior, or on the side of a triangle.

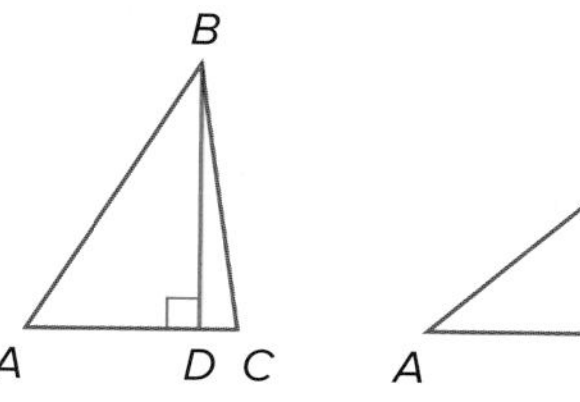

$\overline{BD}$ is an altitude from B to $\overline{AC}$.

Reading Math

Height of a Triangle The length of an altitude is known as the *height* of the triangle. The height of a triangle is used to calculate the triangle's area.

Every triangle has three altitudes. If extended, the altitudes of a triangle intersect in a common point.

Key Concept Orthocenter

The lines containing the altitudes of a triangle are concurrent, intersecting at a point called the **orthocenter**.

Example The lines containing altitudes $\overline{AF}$, $\overline{CD}$, and $\overline{BG}$ intersect at P, the orthocenter of $\triangle ABC$.

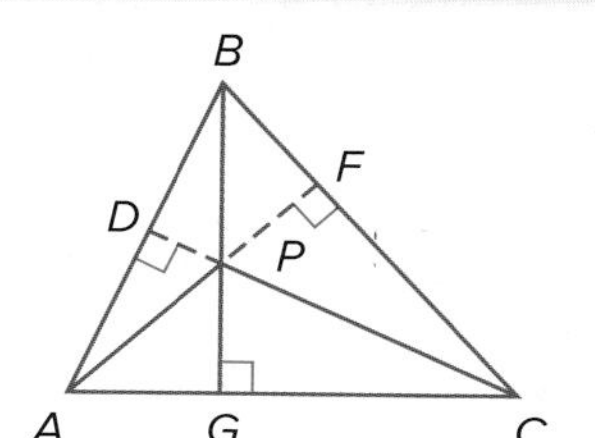

Oldtime/Alamy

Example 4 Find the Orthocenter on a Coordinate Plane

COORDINATE GEOMETRY **The vertices of $\triangle FGH$ are $F(-2, 4)$, $G(4, 4)$, and $H(1, -2)$. Find the coordinates of the orthocenter of $\triangle FGH$.**

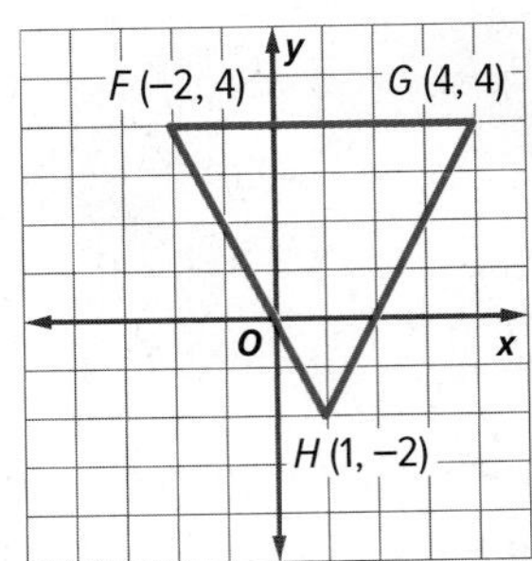

Step 1 Graph $\triangle FGH$. To find the orthocenter, find the point where two of the three altitudes intersect.

Step 2 Find an equation of the altitude from F to $\overline{GH}$. The slope of $\overline{GH}$ is $\frac{4-(-2)}{4-1}$ or 2, so the slope of the altitude, which is perpendicular to $\overline{GH}$, is $-\frac{1}{2}$.

$y - y_1 = m(x - x_1)$ Point-slope form

$y - 4 = -\frac{1}{2}[x - (-2)]$ $m = -\frac{1}{2}$ and $(x_1, y_1) = F(-2, 4)$.

$y - 4 = -\frac{1}{2}(x + 2)$ Simplify.

$y - 4 = -\frac{1}{2}x - 1$ Distributive Property

$y = -\frac{1}{2}x + 3$ Add 4 to each side.

Find an equation of the altitude from G to $\overline{FH}$. The slope of $\overline{FH}$ is $\frac{-2-4}{1-(-2)}$ or -2, so the slope of the altitude is $\frac{1}{2}$.

$y - y_1 = m(x - x_1)$ Point-slope form

$y - 4 = \frac{1}{2}(x - 4)$ $m = \frac{1}{2}$ and $(x_1, y_1) = G(4, 4)$

$y - 4 = \frac{1}{2}x - 2$ Distributive Property

$y = \frac{1}{2}x + 2$ Add 4 to each side.

Step 3 Solve the resulting system of equations $\begin{cases} y = -\frac{1}{2}x + 3 \\ y = \frac{1}{2}x + 2 \end{cases}$ to find the point of intersection of the altitudes.

Adding the two equations to eliminate x results in $2y = 5$ or $y = \frac{5}{2}$.

$y = \frac{1}{2}x + 2$ Equation of altitude from G

$\frac{5}{2} = \frac{1}{2}x + 2$ $y = \frac{5}{2}$

$\frac{1}{2} = \frac{1}{2}x$ Subtract $\frac{4}{2}$ or 2 from each side.

$1 = x$ Multiply each side by 2.

The coordinates of the orthocenter of $\triangle FGH$ are $\left(1, \frac{5}{2}\right)$ or $\left(1, 2\frac{1}{2}\right)$.

Study Tip

MP Reasoning Use the corner of a sheet of paper to draw the altitudes of each side of the triangle.

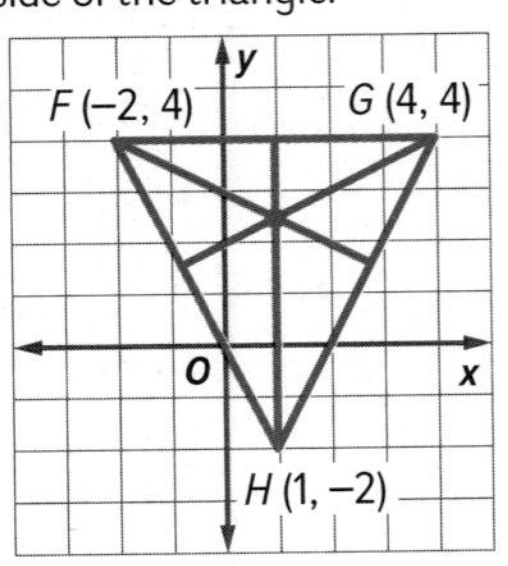

The intersection is located at approximately $\left(1, 2\frac{1}{2}\right)$, so the answer is reasonable.

Guided Practice

4. Find the coordinates of the orthocenter of $\triangle ABC$ graphed at the right.

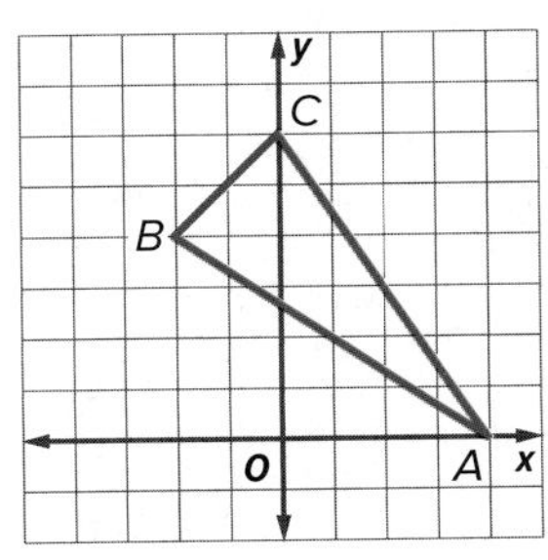

Concept Summary Special Segments and Points in Triangles

Name	Example	Point of Concurrency	Special Property	Example
perpendicular bisector		circumcenter	The circumcenter P of $\triangle ABC$ is equidistant from each vertex.	
angle bisector		incenter	The incenter Q of $\triangle ABC$ is equidistant from each side of the triangle.	
median		centroid	The centroid R of $\triangle ABC$ is two thirds of the distance from each vertex to the midpoint of the opposite side.	
altitude		orthocenter	The lines containing the altitudes of $\triangle ABC$ are concurrent at the orthocenter S.	

Check Your Understanding

 = Step-by-Step Solutions begin on page R13.

Examples 1–2 **In $\triangle ACE$, P is the centroid, $PF = 6$, and $AD = 15$. Find each measure.**

 PC

2. AP

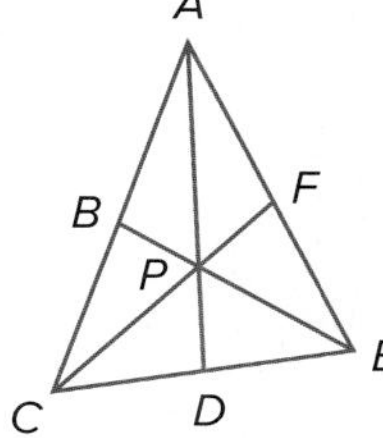

Example 3

3. **INTERIOR DESIGN** An interior designer is creating a custom coffee table for a client. The top of the table is a glass triangle that needs to balance on a single support. If the coordinates of the vertices of the triangle are (3, 6), (5, 2), and (7, 10), at what point should the support be placed?

Example 4

4. **COORDINATE GEOMETRY** Find the coordinates of the orthocenter of $\triangle ABC$ with vertices $A(-3, 3)$, $B(-1, 7)$, and $C(3, 3)$.

Examples 1–2 **In $\triangle SZU$, $UJ = 9$, $VJ = 3$, and $ZT = 18$. Find each length.**

5. YJ

6. SJ

7. YU

8. SV

9. JT

10. ZJ

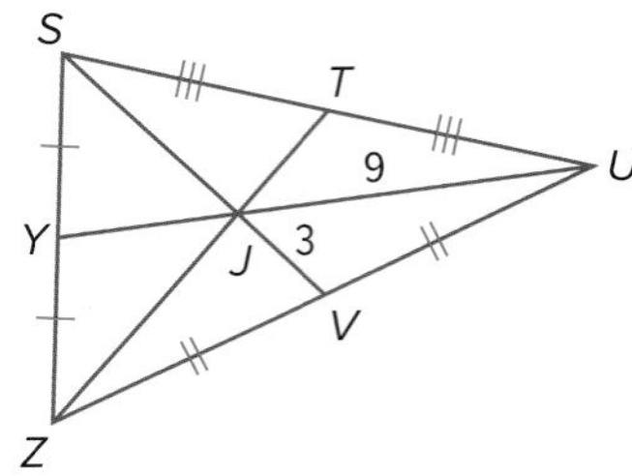

Example 3 **COORDINATE GEOMETRY** Find the coordinates of the centroid of each triangle with the given vertices.

11. $A(-1, 11)$, $B(3, 1)$, $C(7, 6)$

12. $X(5, 7)$, $Y(9, -3)$, $Z(13, 2)$

13 **INTERIOR DESIGN** Emilia made a collage with pictures of her friends. She wants to hang the collage from the ceiling in her room so that it is parallel to the ceiling. A diagram of the collage is shown in the graph at the right. At what point should she place the string?

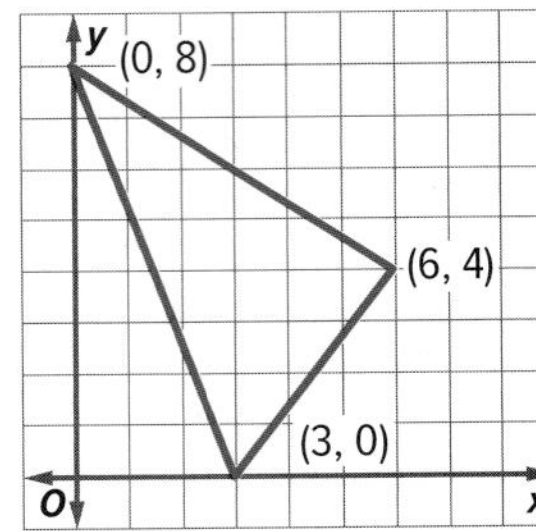

Example 4 **COORDINATE GEOMETRY** Find the coordinates of the orthocenter of each triangle with the given vertices.

14. $J(3, -2)$, $K(5, 6)$, $L(9, -2)$

15. $R(-4, 8)$, $S(-1, 5)$, $T(5, 5)$

Identify each segment $\overline{BD}$ as a(n) altitude, median, or perpendicular bisector.

16.

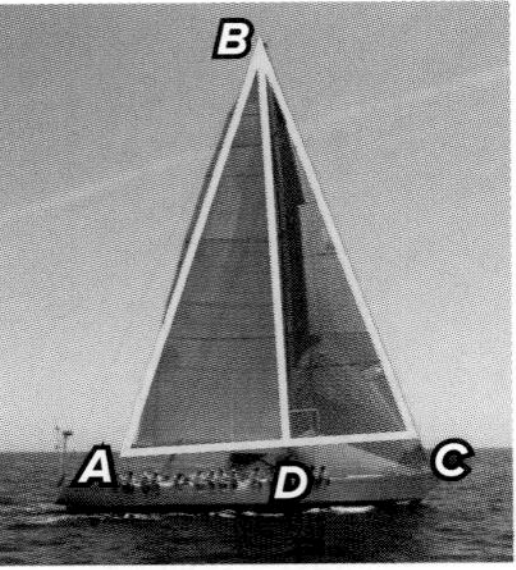

17.

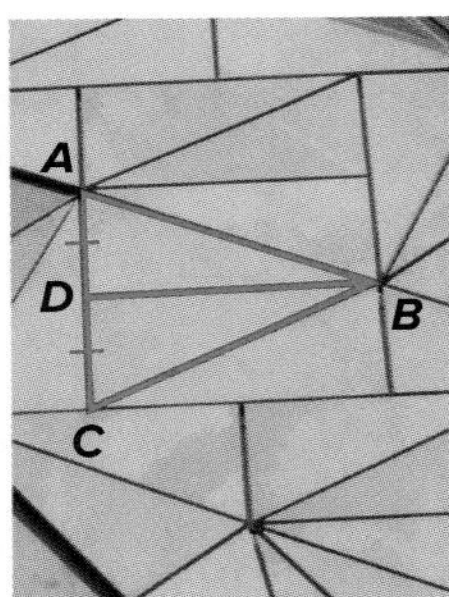

18.

19.

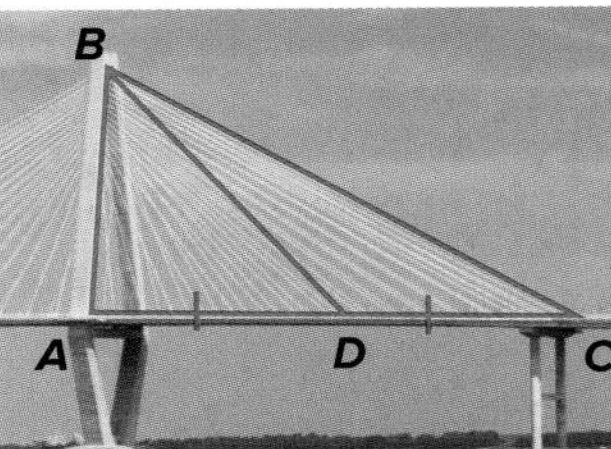

20. **MP** **SENSE-MAKING** In the figure at the right, if J, P, and L are the midpoints of $\overline{KH}$, $\overline{HM}$, and $\overline{MK}$, respectively, find x, y, and z.

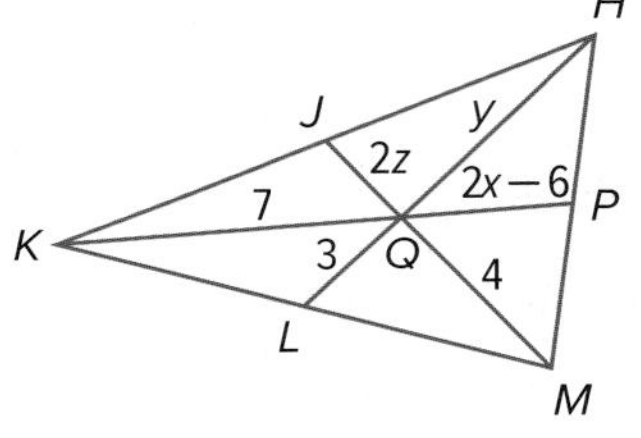

Find the value of x for $\triangle RST$ with medians $\overline{RM}$, $\overline{SL}$ and $\overline{TK}$, and centroid J.

21. $SL = x(JL)$

22. $JT = x(TK)$

23. $JM = x(RJ)$

ALGEBRA Use the figure at the right.

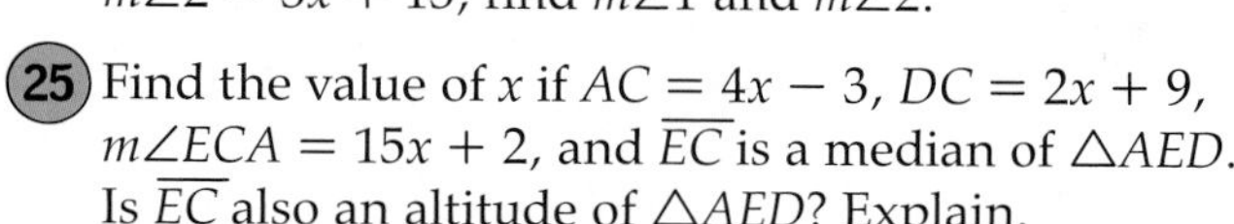

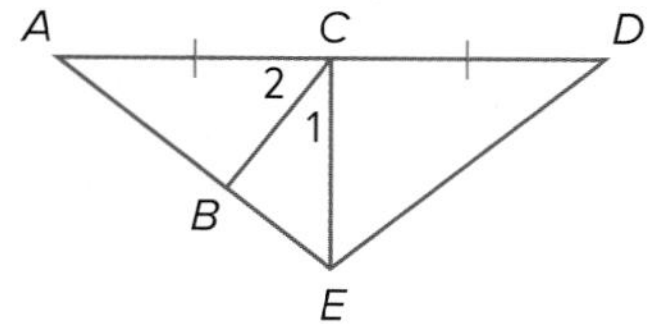

24. If $\overline{EC}$ is an altitude of $\triangle AED$, $m\angle 1 = 2x + 7$, and $m\angle 2 = 3x + 13$, find $m\angle 1$ and $m\angle 2$.

25. Find the value of x if $AC = 4x - 3$, $DC = 2x + 9$, $m\angle ECA = 15x + 2$, and $\overline{EC}$ is a median of $\triangle AED$. Is $\overline{EC}$ also an altitude of $\triangle AED$? Explain.

26. **GAMES** The game board shown is shaped like an equilateral triangle and has indentations for game pieces. The game's objective is to remove pegs by jumping over them until there is only one peg left. Copy the game board's outline and determine which of the points of concurrency the blue peg represents: *circumcenter, incenter, centroid,* or *orthocenter*. Explain your reasoning.

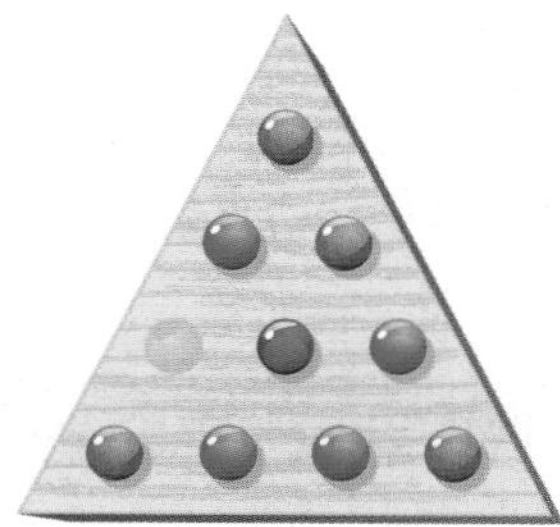

MP CONSTRUCT ARGUMENTS Use the given information to determine whether $\overline{LM}$ is a *perpendicular bisector, median,* and/or an *altitude* of $\triangle JKL$.

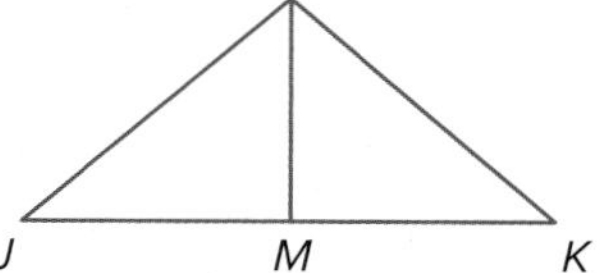

27. $\overline{LM} \perp \overline{JK}$

28. $\triangle JLM \cong \triangle KLM$

29. $\overline{JM} \cong \overline{KM}$

30. $\overline{LM} \perp \overline{JK}$ and $\overline{JL} \cong \overline{KL}$

31. **PROOF** Write a paragraph proof.

Given: $\triangle XYZ$ is isosceles.
$\overline{WY}$ bisects $\angle Y$.

Prove: $\overline{WY}$ is a median.

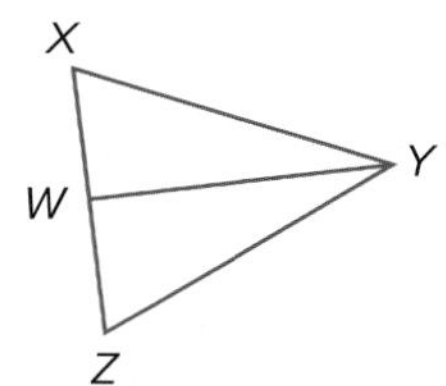

32. **PROOF** Write an algebraic proof.

Given: $\triangle XYZ$ with medians $\overline{XR}$, $\overline{YS}$, $\overline{ZQ}$

Prove: $\frac{XP}{PR} = 2$

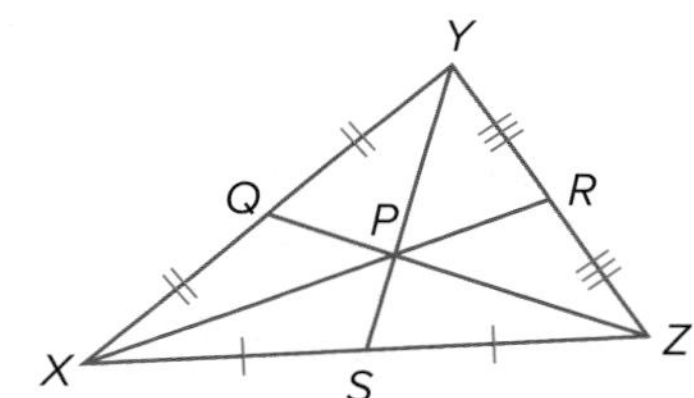

33. **MULTIPLE REPRESENTATIONS** In this problem, you will investigate the location of the points of concurrency for any equilateral triangle.

a. Concrete Construct an equilateral triangle. Use tracing paper to make three additional copies of the triangle. Construct the circumcenter on the first triangle, the incenter on the second triangle, the centroid on the third triangle, and the orthocenter on the remaining triangle.

b. Verbal Make a conjecture about the relationships among the four points of concurrency of any equilateral triangle.

c. Graphical Position an equilateral triangle and its circumcenter, incenter, centroid, and orthocenter on the coordinate plane using variable coordinates. Determine the coordinates of each point of concurrency.

ALGEBRA In $\triangle JLP$, $m\angle JMP = 3x - 6$, $JK = 3y - 2$, and $LK = 5y - 8$.

34. If $\overline{JM}$ is an altitude of $\triangle JLP$, find x.

35. Find LK if $\overline{PK}$ is a median.

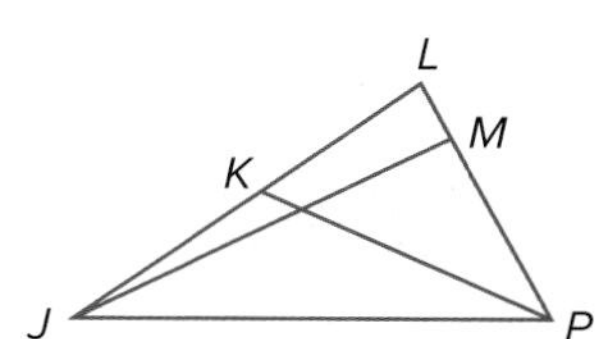

36. **PROOF** Write a coordinate proof to prove the Centroid Theorem.

Given: $\triangle ABC$, medians $\overline{AR}$, $\overline{BS}$, and $\overline{CQ}$

Prove: The medians intersect at point P and P is two thirds of the distance from each vertex to the midpoint of the opposite side.

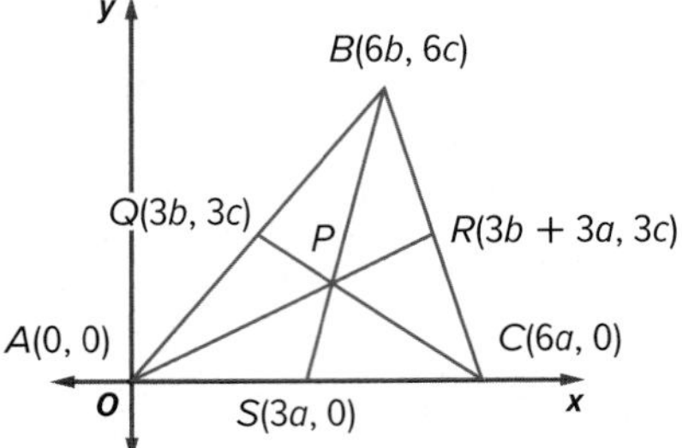

(*Hint*: First, find the equations of the lines containing the medians. Then find the coordinates of point P and show that all three medians intersect at point P. Next, use the Distance Formula and multiplication to show $AP = \frac{2}{3}AR$, $BP = \frac{2}{3}BS$, and $CP = \frac{2}{3}CQ$.)

H.O.T. Problems Use Higher-Order Thinking Skills

37. **ERROR ANALYSIS** Based on the figure at the right, Luke says that $\frac{2}{3}AP = AD$. Kareem disagrees. Is either of them correct? Explain your reasoning.

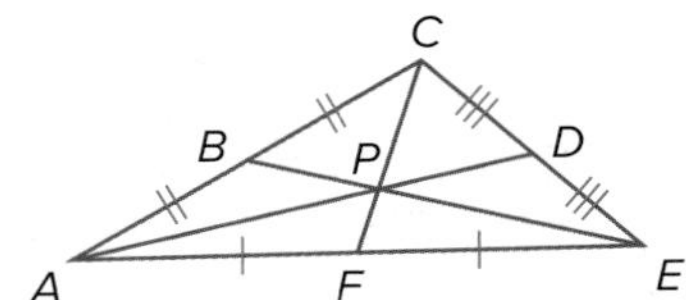

38. **CONSTRUCT ARGUMENTS** Determine whether the following statement is *true* or *false*. If true, explain your reasoning. If false, provide a counterexample.

The orthocenter of a right triangle is always located at the vertex of the right angle.

39. **STRUCTURE** $\triangle ABC$ has vertices $A(-3, 3)$, $B(2, 5)$, and $C(4, -3)$. What are the coordinates of the centroid of $\triangle ABC$? Explain the process you used to reach your conclusion.

40. **WRITING IN MATH** Compare and contrast the perpendicular bisectors, medians, and altitudes of a triangle.

41. **CHALLENGE** In the figure at the right, segments $\overline{AD}$ and $\overline{CE}$ are medians of $\triangle ACB$, $\overline{AD} \perp \overline{CE}$, $AB = 10$, and $CE = 9$. Find CA.

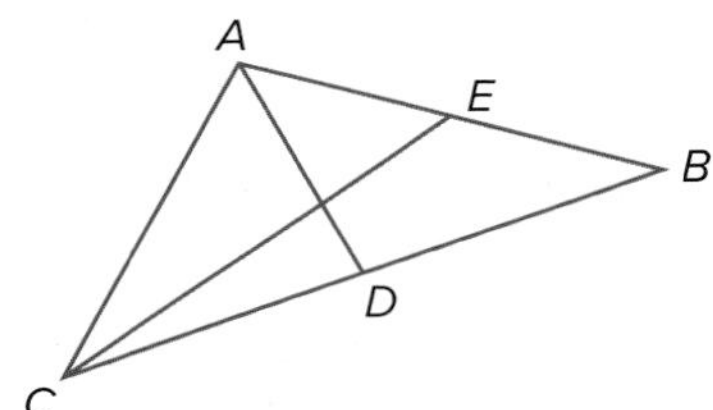

42. **STRUCTURE** In this problem, you will investigate the relationships among three points of concurrency in a triangle.

a. Draw an acute triangle and find the circumcenter, centroid, and orthocenter.

b. Draw an obtuse triangle and find the circumcenter, centroid, and orthocenter.

c. Draw a right triangle and find the circumcenter, centroid, and orthocenter.

d. Make a conjecture about the relationships among the circumcenter, centroid, and orthocenter.

43. **WRITING IN MATH** Use area to explain why the centroid of a triangle is its center of gravity. Then use this explanation to describe the location for the balancing point for a rectangle.

44. A local park is in the shape of a triangle with three straight paths, as shown. A fountain F is located at the centroid of the park, and $PF = 60$ meters. What is the length of $\overline{FL}$? MP 2

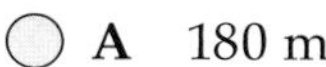

○ **A** 180 m

○ **B** 120 m

○ **C** 60 m

○ **D** 24 m

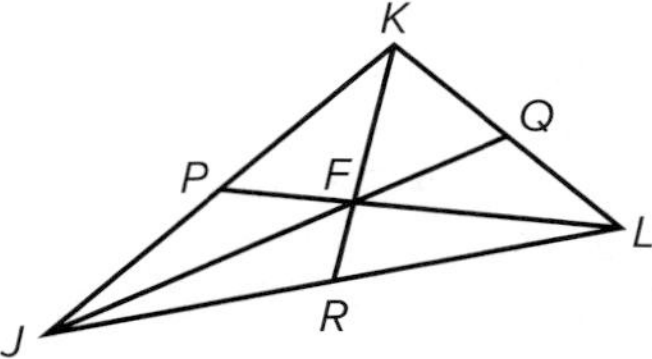

45. Peter draws $\triangle ABC$ and identifies a point that lies outside of $\triangle ABC$ as the intersection of the medians of the triangle. Could Peter be correct? MP 3

46. Kendall drew $\triangle XYZ$ and then located point C as shown.

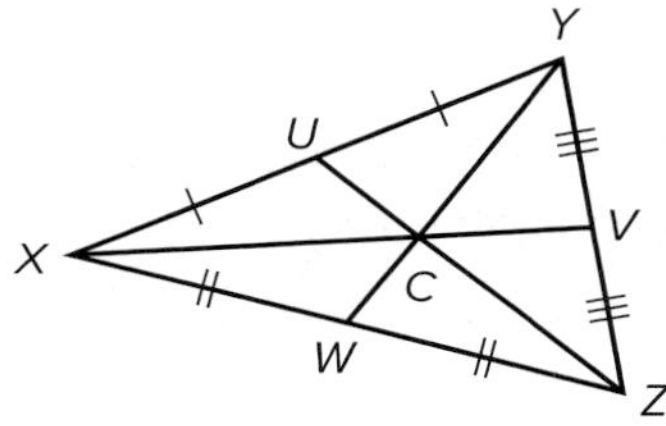

Which of the following statements about Kendall's triangle must be true? MP 2, 7

○ **A** Point C is equidistant from the vertices of $\triangle XYZ$.

○ **B** Point C is equidistant from the sides of $\triangle XYZ$.

○ **C** The triangle will balance at point C.

○ **D** The length of $\overline{CV}$ is equal to the length of $\overline{CW}$.

47. The vertices of $\triangle ABC$ are $A(-5, 1)$, $B(1, 7)$, and $C(1, -5)$. What are the coordinates of the centroid of the triangle? MP 1

○ **A** $(-5, 1)$

○ **B** $(-3, 1)$

○ **C** $(-1, 1)$

○ **D** $(1, 1)$

48. Point Q is the centroid of $\triangle RST$.

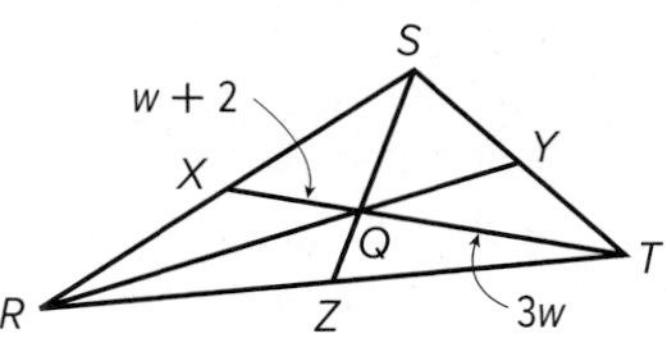

What is the length of $\overline{QX}$? MP 2, 6

○ **A** 3

○ **B** 4

○ **C** 6

○ **D** 12

49. Which type of segments intersect in a triangle to form an orthocenter? MP 2

○ **A** medians

○ **B** altitudes

○ **C** angle bisectors

○ **D** perpendicular bisectors

50. Which point(s) ALWAYS lie inside a triangle? Check all that apply. MP 2

☐ **A** incenter

☐ **B** centroid

☐ **C** orthocenter

☐ **D** circumcenter

51. MULTI-STEP $\triangle ABC$ is an equilateral triangle with vertices at $A(-3, 0)$, $B(3, 0)$, and $C(0, 3\sqrt{3})$. MP 3, 6

a. What are the coordinates of the centroid?

b. What are the coordinates of the incenter?

c. What are the coordinates of the circumcenter?

d. What are the coordinates of the orthocenter?

e. What is the relationship between all these points? Explain.

LESSON 3

Inequalities in One Triangle

Then

- You found the relationship between the angle measures of a triangle.

Now

1. Recognize and apply properties of inequalities to the measures of the angles of a triangle.
2. Recognize and apply properties of inequalities to the relationships between the angles and sides of a triangle.

Why?

- To create the appearance of depth in a room, interior designers use a technique called *triangulation*. A basic example of this technique is the placement of an end table on each side of a sofa with a painting over the sofa.

  The measures of the base angles of the triangle should be less than the measure of the other angle.

Mathematical Practices

1 Make sense of problems and persevere in solving them.

3 Construct viable arguments and critique the reasoning of others.

8 Look for and express regularity in repeated reasoning.

1 Angle Inequalities

In algebra, you learned about the inequality relationship between two real numbers. This relationship is often used in proofs.

Key Concept Definition of Inequality

Words	For any real numbers a and b, $a > b$ if and only if there is a positive number c such that $a = b + c$.
Example	If $5 = 2 + 3$, then $5 > 2$ and $5 > 3$.

The table below lists some of the properties of inequalities you studied in algebra.

Key Concept Properties of Inequality for Real Numbers

The following properties are true for any real numbers a, b, and c.

Comparison Property of Inequality	$a < b$, $a = b$, or $a > b$
Transitive Property of Inequality	**1.** If $a < b$ and $b < c$, then $a < c$. **2.** If $a > b$ and $b > c$, then $a > c$.
Addition Property of Inequality	**1.** If $a > b$, then $a + c > b + c$. **2.** If $a < b$, then $a + c < b + c$.
Subtraction Property of Inequality	**1.** If $a > b$, then $a - c > b - c$. **2.** If $a < b$, then $a - c < b - c$.

The definition of inequality and the properties of inequalities can be applied to the measures of angles and segments, since these are real numbers. Consider $\angle 1$, $\angle 2$, and $\angle 3$ in the figure shown.

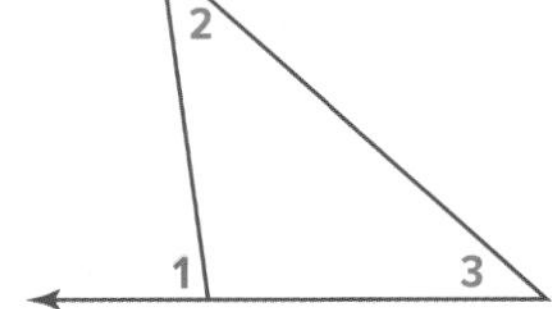

By the Exterior Angle Theorem, you know that $m\angle 1 = m\angle 2 + m\angle 3$.

Since the angle measures are positive numbers, we can also say that

$$m\angle 1 > m\angle 2 \quad \text{and} \quad m\angle 1 > m\angle 3$$

by the definition of inequality. This result suggests the following theorem.

> **Review Vocabulary**
>
> **Remote Interior Angle** Each exterior angle of a triangle has two *remote interior angles* that are not adjacent to the exterior angle.

Theorem 5.8 Exterior Angle Inequality

The measure of an exterior angle of a triangle is greater than the measure of either of its corresponding remote interior angles.

Example: $m\angle 1 > m\angle A$
$m\angle 1 > m\angle B$

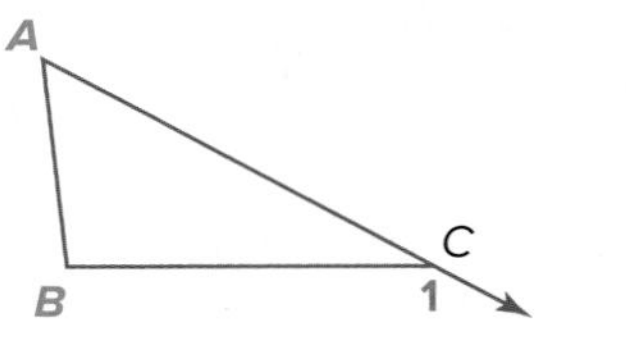

The proof of Theorem 5.8 is in Lesson 5-4.

Example 1 Use the Exterior Angle Inequality Theorem

Use the Exterior Angle Inequality Theorem to list all of the angles that satisfy the stated condition.

a. measures less than $m\angle 7$

$\angle 7$ is an exterior angle to $\triangle KML$, with $\angle 4$ and $\angle 5$ as corresponding remote interior angles. By the Exterior Angle Inequality Theorem, $m\angle 7 > m\angle 4$ and $m\angle 7 > m\angle 5$.

$\angle 7$ is also an exterior angle to $\triangle JKL$, with $\angle 1$ and $\angle JKL$ as corresponding remote interior angles. So, $m\angle 7 > m\angle 1$ and $m\angle 7 > m\angle JKL$. Since $m\angle JKL = m\angle 2 + m\angle 4$, by substitution $m\angle 7 > m\angle 2 + m\angle 4$. Therefore, $m\angle 7 > m\angle 2$.

So, the angles with measures less than $m\angle 7$ are $\angle 1$, $\angle 2$, $\angle 4$, $\angle 5$.

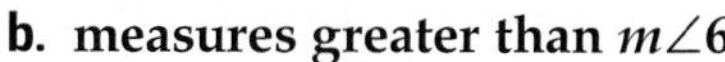

b. measures greater than $m\angle 6$

$\angle 3$ is an exterior angle to $\triangle KLM$. So by the Exterior Angle Inequality Theorem, $m\angle 3 > m\angle 6$. Because $\angle 8$ is an exterior angle to $\triangle JKL$, $m\angle 8 > m\angle 6$. Thus, the measures of $\angle 3$ and $\angle 8$ are greater than $m\angle 6$.

Guided Practice

1A. measures less than $m\angle 1$

1B. measures greater than $m\angle 8$

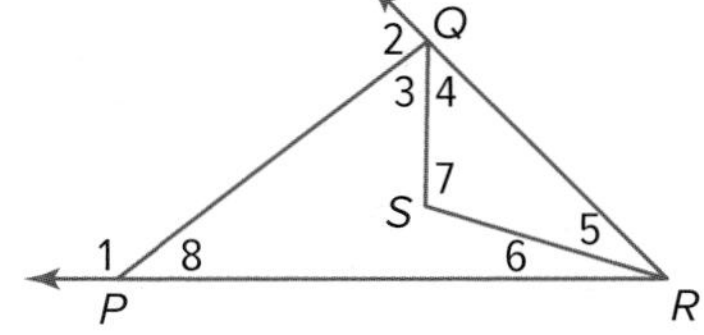

> **Watch Out!**
>
> **Identifying Side Opposite** Be careful to correctly identify the side opposite an angle. The sides that form the angle cannot be the sides opposite the angle.

2 Angle-Side Inequalities

In Lesson 4-6, you learned that if two sides of a triangle are congruent, or the triangle is isosceles, then the angles opposite those sides are congruent. What relationship exists if the sides are not congruent? Examine the longest and shortest sides and smallest and largest angles of a scalene obtuse triangle.

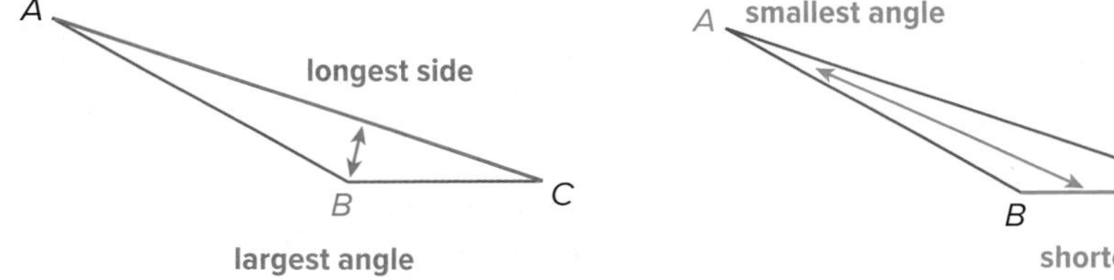

Notice that the longest side and largest angle of $\triangle ABC$ are opposite each other. Likewise, the shortest side and smallest angle are opposite each other.

Watch Out!

Symbols for Angles and Inequalities The symbol for angle (∠) looks similar to the symbol for less than (<), especially when handwritten. Be careful to write the symbols correctly in situations where both are used.

The side-angle relationships in an obtuse scalene triangle are true for all triangles, and are stated using inequalities in the theorems below.

Theorems Angle-Side Relationships in Triangles

5.9 If one side of a triangle is longer than another side, then the angle opposite the longer side has a greater measure than the angle opposite the shorter side.

Example: $XY > YZ$, so $m\angle Z > m\angle X$.

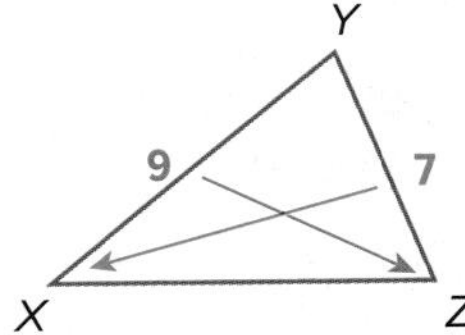

5.10 If one angle of a triangle has a greater measure than another angle, then the side opposite the greater angle is longer than the side opposite the lesser angle.

Example: $m\angle J > m\angle K$, so $KL > JL$.

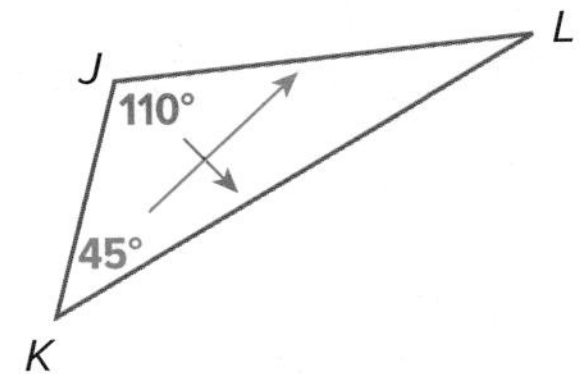

Proof Theorem 5.9

Given: $\triangle ABC$, $AB > BC$

Prove: $m\angle BCA > m\angle A$

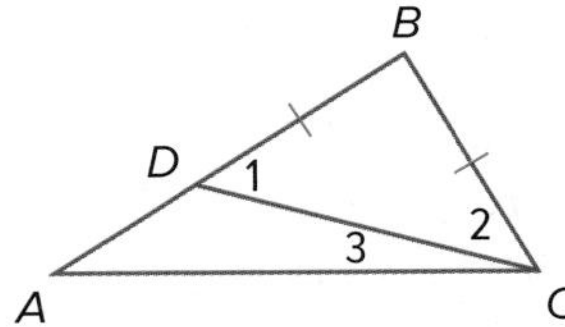

Proof:

Because $AB > BC$ in the given $\triangle ABC$, there exists a point D on $\overline{AB}$ such that $BD = BC$. Draw $\overline{CD}$ to form isosceles $\triangle BCD$. By the Isosceles Triangle Theorem, $\angle 1 \cong \angle 2$, so $m\angle 1 = m\angle 2$ by the definition of congruent angles.

By the Angle Addition Postulate, $m\angle BCA = m\angle 2 + m\angle 3$, so $m\angle BCA > m\angle 2$ by the definition of inequality. By substitution, $m\angle BCA > m\angle 1$.

By the Exterior Angle Inequality Theorem, $m\angle 1 > m\angle A$. Therefore, because $m\angle BCA > m\angle 1$ and $m\angle 1 > m\angle A$, by the Transitive Property of Inequality, $m\angle BCA > m\angle A$.

You will prove Theorem 5.10 in Lesson 5-4, Exercise 31.

Go Online! 

In Personal Tutor videos for this lesson, teachers describe how to solve problems involving inequalities in triangles. Watch with a partner then try describing how to solve a problem for them. Have them ask questions to help your understanding.

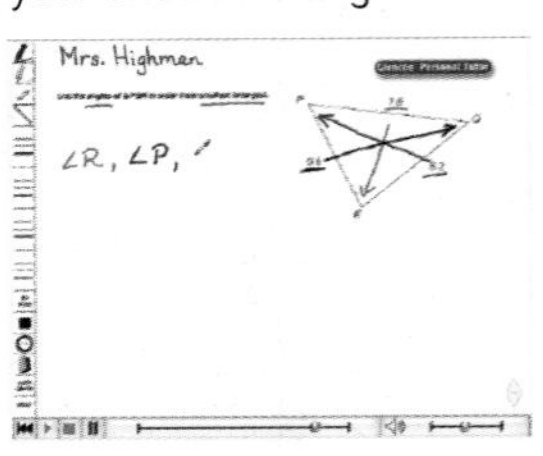

Example 2 Order Triangle Angle Measures

List the angles of $\triangle PQR$ in order from smallest to largest.

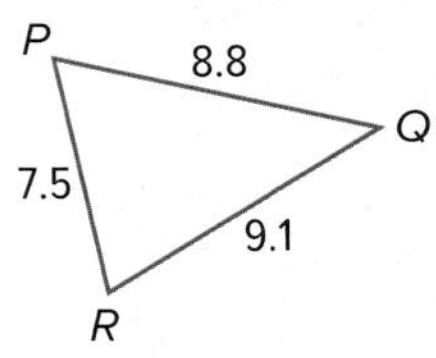

The sides from shortest to longest are $\overline{PR}$, $\overline{PQ}$, and $\overline{QR}$. The angles opposite these sides are $\angle Q$, $\angle R$, and $\angle P$, respectively. So the angles from smallest to largest are $\angle Q$, $\angle R$, and $\angle P$.

Guided Practice

2. List the angles and sides of $\triangle ABC$ in order from smallest to largest.

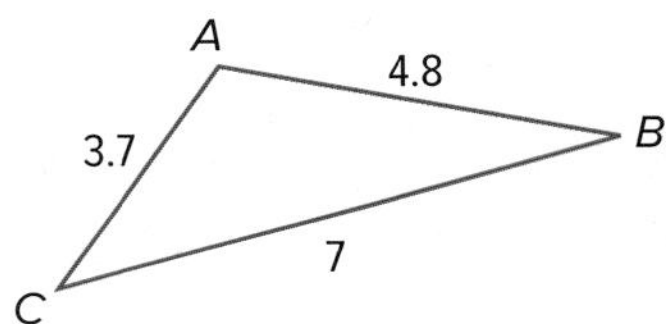

Example 3 Order Triangle Side Lengths

List the sides of $\triangle FGH$ in order from shortest to longest.

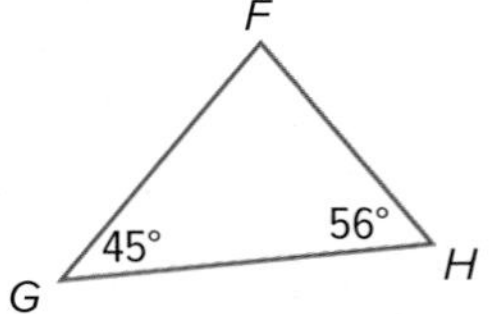

First find the missing angle measure using the Triangle Angle-Sum Theorem.

$m\angle F = 180 - (45 + 56)$ or 79

So, the angles from smallest to largest are $\angle G$, $\angle H$, and $\angle F$. The sides opposite these angles are $\overline{FH}$, $\overline{FG}$, and $\overline{GH}$, respectively. So, the sides from shortest to longest are $\overline{FH}$, $\overline{FG}$, and $\overline{GH}$.

Guided Practice

3. List the angles and sides of $\triangle WXY$ in order from smallest to largest.

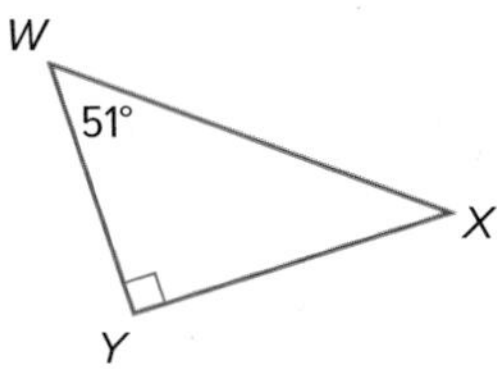

You can use angle-side relationships in triangles to solve real-world problems.

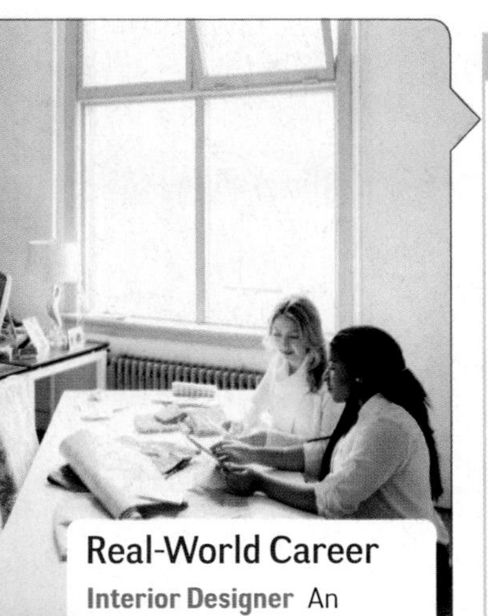

Real-World Career

Interior Designer An interior designer decorates a space so that it is visually pleasing and comfortable for people to live or work in. Designers must know color and paint theory, lighting design, and space planning. A bachelor's degree is recommended for entry-level positions. Graduates usually enter a 1- to 3-year apprenticeship before taking a licensing exam.

Hero Images/Getty Images

Real-World Example 4 Angle-Side Relationships

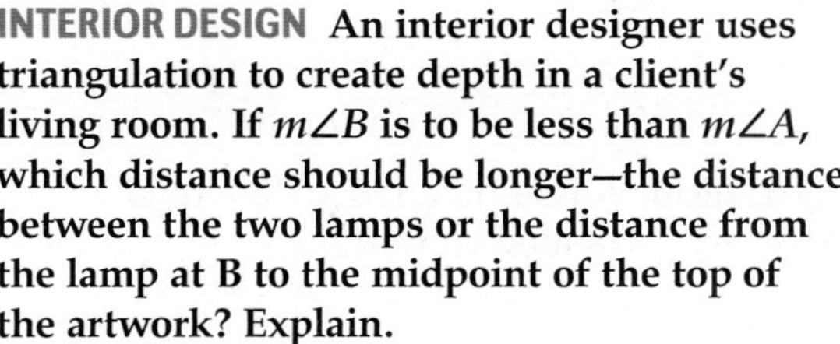

INTERIOR DESIGN An interior designer uses triangulation to create depth in a client's living room. If $m\angle B$ is to be less than $m\angle A$, which distance should be longer—the distance between the two lamps or the distance from the lamp at B to the midpoint of the top of the artwork? Explain.

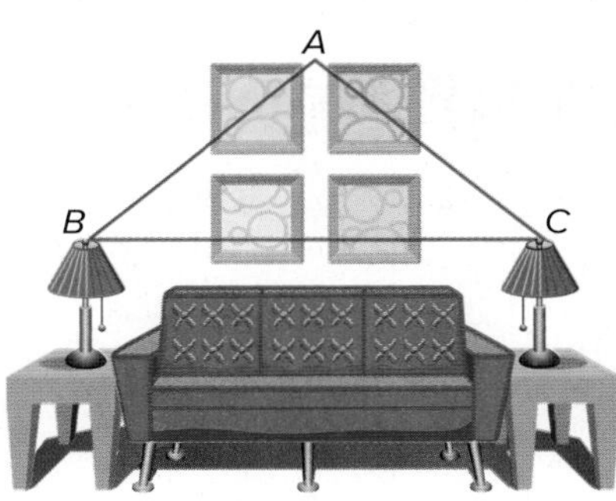

According to Theorem 5.10, in order for $m\angle B < m\angle A$, the length of the side opposite $\angle B$ must be less than the length of the side opposite $\angle A$. Since $\overline{AC}$ is opposite $\angle B$, and $\overline{BC}$ is opposite $\angle A$, then $AC < BC$ and $BC > AC$. So BC, the distance between the lamps, must be greater than the distance from the lamp at B to the midpoint of the top of the artwork.

Guided Practice

4. **LIFEGUARDING** During lifeguard training, an instructor simulates a person in distress so that trainees can practice their rescue skills. If the instructor, Trainee 1, and Trainee 2 are located in the positions shown on the diagram, which of the two trainees is closest to the instructor?

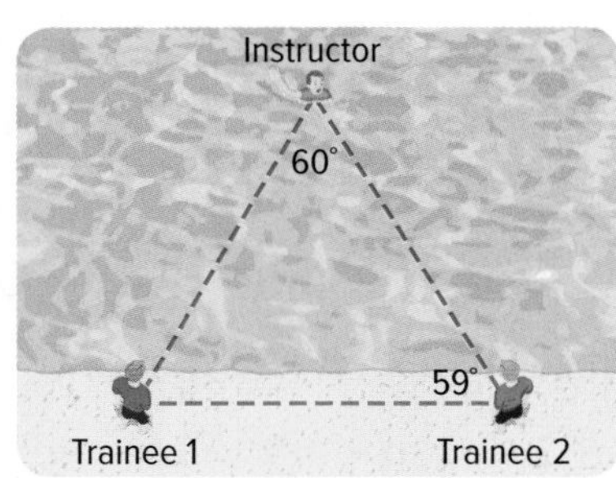

Check Your Understanding

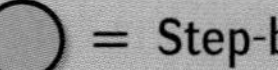 = Step-by-Step Solutions begin on page R13.

Example 1 **Use the Exterior Angle Inequality Theorem to list all of the angles that satisfy the stated condition.**

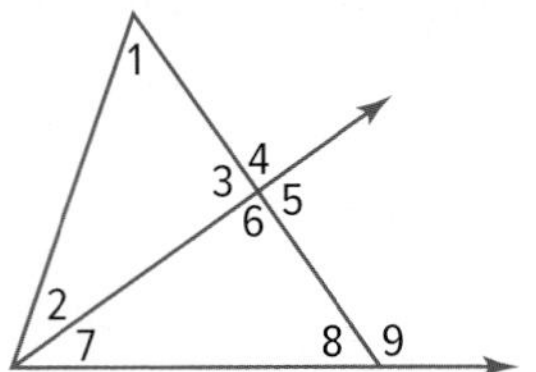

1. measures less than $m\angle 4$
2. measures greater than $m\angle 7$
3. measures greater than $m\angle 2$
4. measures less than $m\angle 9$

Examples 2–3 **List the angles and sides of each triangle in order from smallest to largest.**

5. 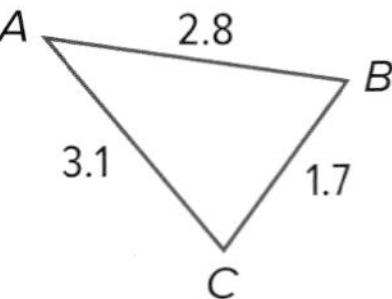

6. 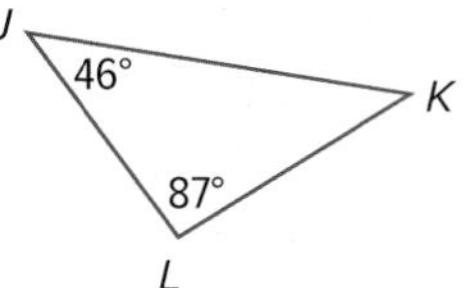

Example 4  7. **HANG GLIDING** The supports on a hang glider form triangles like the one shown. Which is longer—the support represented by $\overline{AC}$ or the support represented by $\overline{BC}$? Explain your reasoning.

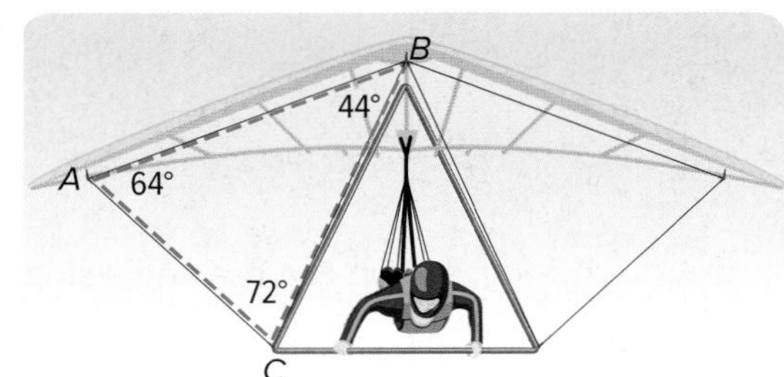

Practice and Problem Solving

Extra Practice is on page R5.

Example 1 **SENSE-MAKING** **Use the Exterior Angle Inequality Theorem to list all of the angles that satisfy the stated condition.**

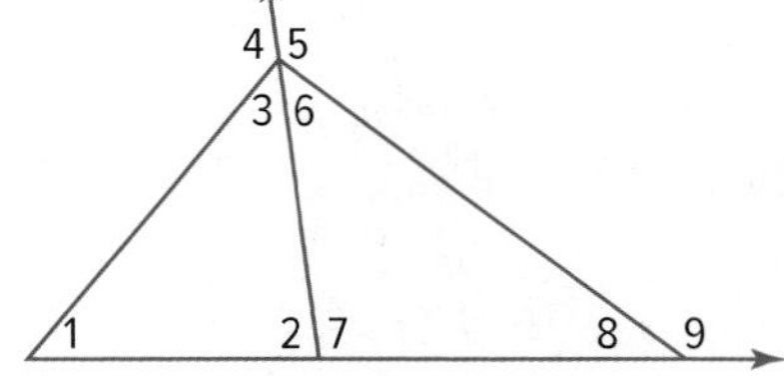

8. measures greater than $m\angle 2$
9. measures less than $m\angle 4$
10. measures less than $m\angle 5$
11. measures less than $m\angle 9$
12. measures greater than $m\angle 8$
13. measures greater than $m\angle 7$

Examples 2–3 **List the angles and sides of each triangle in order from smallest to largest.**

14. 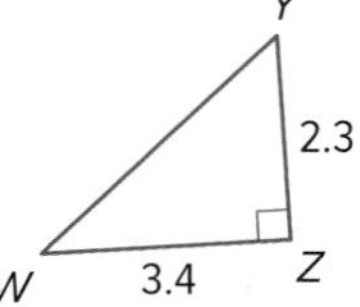

15. 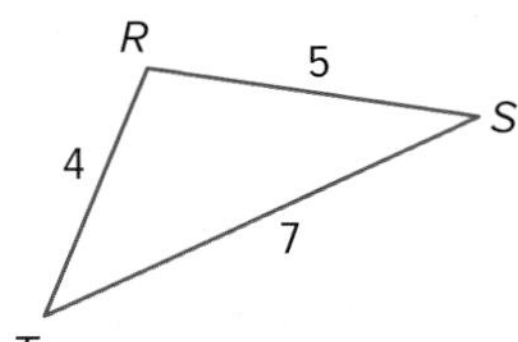

16. 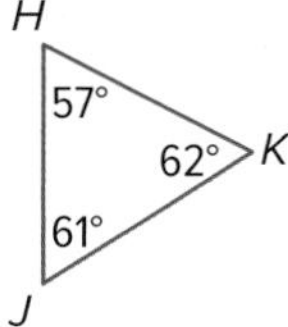

17. 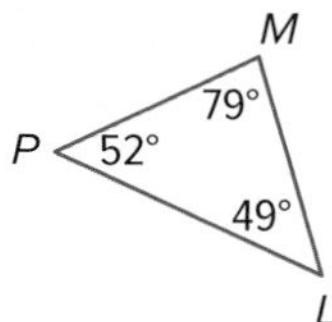

18. 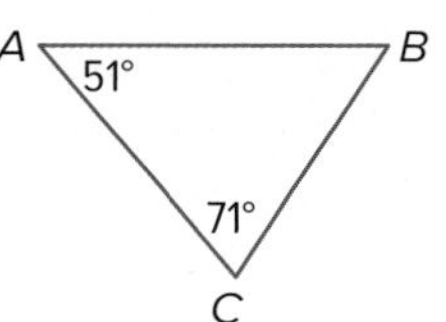

19. 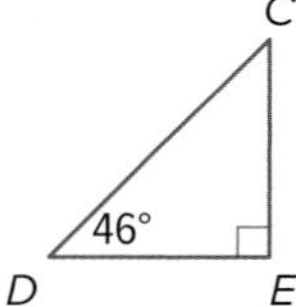

Example 4

20. SPORTS Ben, Gilberto, and Hannah are playing Ultimate. Hannah is trying to decide if she should pass to Ben or Gilberto. Which player should she choose in order to have the shorter passing distance? Explain your reasoning.

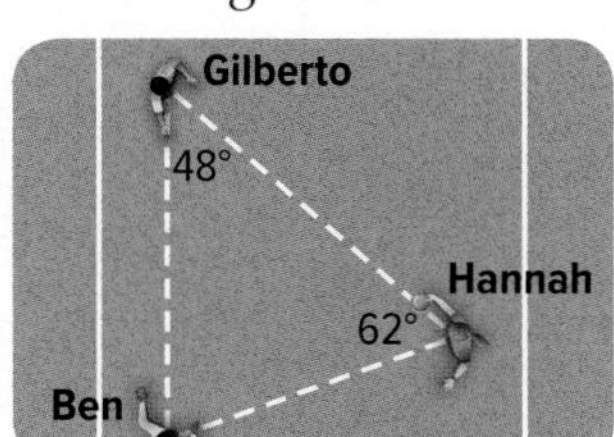

21. RAMPS The wedge below represents a bike ramp. Which is longer, the length of the ramp $\overline{XZ}$ or the length of the top surface of the ramp $\overline{YZ}$? Explain your reasoning using Theorem 5.9.

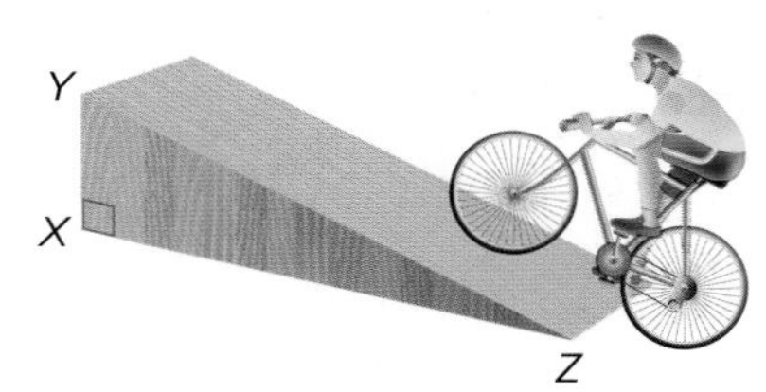

List the angles and sides of each triangle in order from smallest to largest.

22.

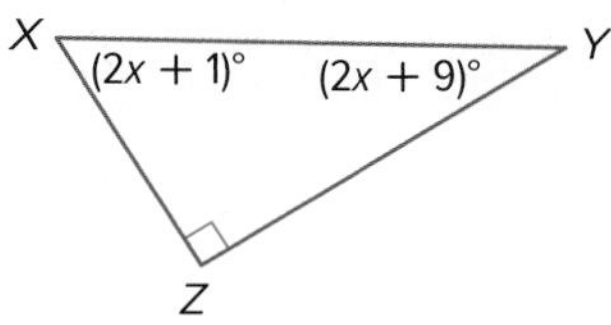

23.

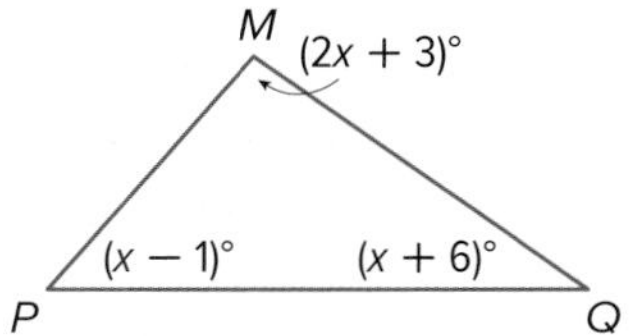

Use the figure at the right to determine which angle has the greatest measure.

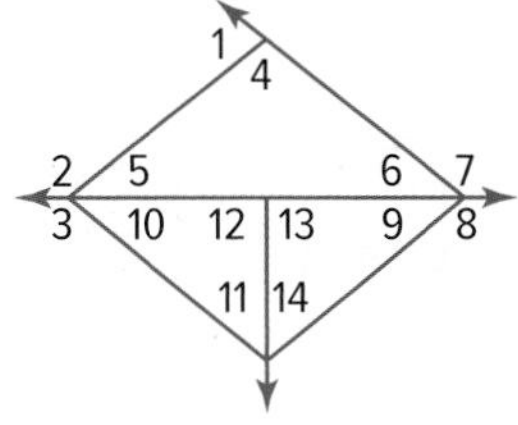

24. $\angle 1, \angle 5, \angle 6$ **25.** $\angle 2, \angle 4, \angle 6$

26. $\angle 7, \angle 4, \angle 5$ **27.** $\angle 3, \angle 11, \angle 12$

28. $\angle 3, \angle 9, \angle 14$ **29.** $\angle 8, \angle 10, \angle 11$

MP SENSE-MAKING **Use the figure at the right to determine the relationship between the measures of the given angles.**

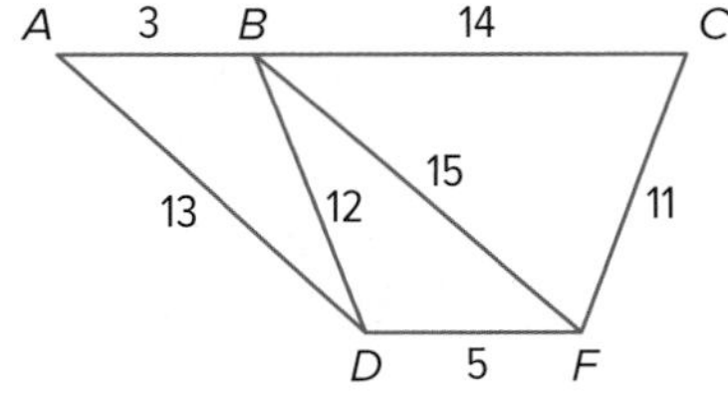

30. $\angle ABD, \angle BDA$ **31.** $\angle BCF, \angle CFB$

32. $\angle BFD, \angle BDF$ **33.** $\angle DBF, \angle BFD$

Use the figure at the right to determine the relationship between the given lengths.

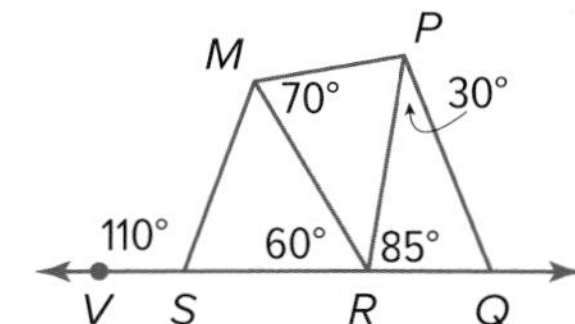

34. SM, MR **35.** RP, MP

36. RQ, PQ **37.** RM, RQ

38. HIKING Justin and his family are hiking around a lake as shown in the diagram at the right. Order the angles of the triangle formed by their path from largest to smallest.

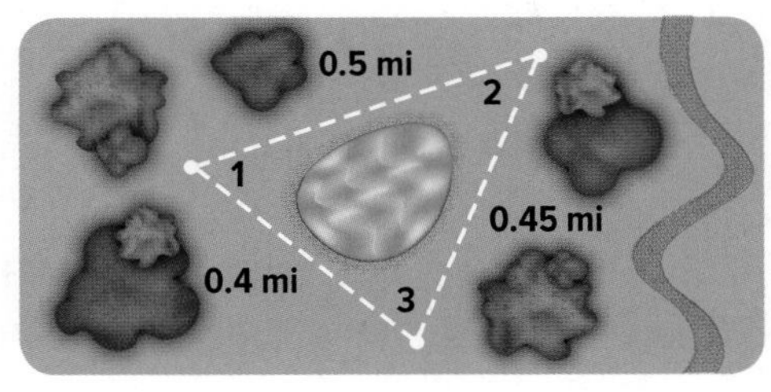

COORDINATE GEOMETRY List the angles of each triangle with the given vertices in order from smallest to largest. Justify your answer.

39 $A(-4, 6), B(-2, 1), C(5, 6)$

40. $X(-3, -2), Y(3, 2), Z(-3, -6)$

41. List the side lengths of the triangles in the figure from shortest to longest. Explain your reasoning.

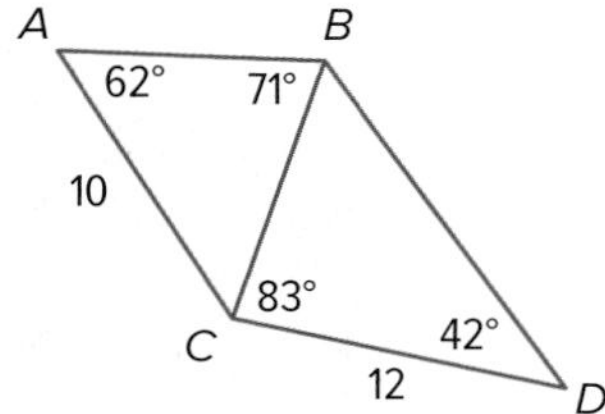

42. MULTIPLE REPRESENTATIONS In this problem, you will explore the relationship between the sides of a triangle.

a. **Geometric** Draw three triangles, including one acute, one obtuse, and one right triangle. Label the vertices of each triangle *A*, *B*, and *C*.

b. **Tabular** Measure the length of each side of the three triangles. Then copy and complete the table.

Triangle	*AB*	*BC*	*AB* + *BC*	*CA*
Acute				
Obtuse				
Right				

c. **Tabular** Create two additional tables like the one above, finding the sum of *BC* and *CA* in one table and the sum of *AB* and *CA* in the other.

d. **Algebraic** Write an inequality for each of the tables you created relating the measure of the sum of two of the sides to the measure of the third side of a triangle.

e. **Verbal** Make a conjecture about the relationship between the measure of the sum of two sides of a triangle and the measure of the third side.

H.O.T. Problems Use Higher-Order Thinking Skills

43. WRITING IN MATH Analyze the information given in the diagram and explain why the markings must be incorrect.

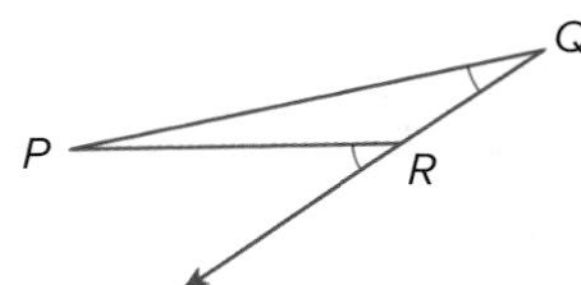

44. MP **CONSTRUCT ARGUMENTS** Using only a ruler, draw $\triangle ABC$ such that $m\angle A > m\angle B > m\angle C$. Justify your drawing.

45. MP **REASONING** Give a possible measure for $\overline{AB}$ in $\triangle ABC$ shown. Explain your reasoning.

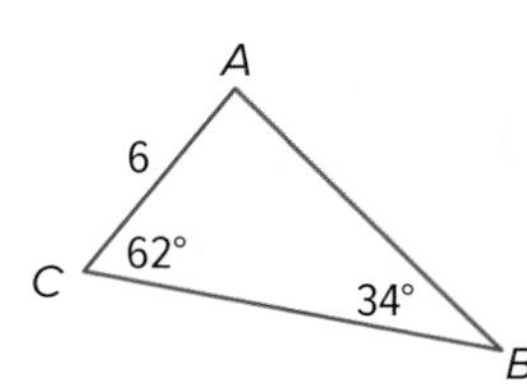

46. MP **CONSTRUCT ARGUMENTS** Is the base of an isosceles triangle *always*, *sometimes*, or *never* the longest side of the triangle? Explain.

47. MP **STRUCTURE** Use the side lengths in the figure to list the numbered angles in order from smallest to largest given that $m\angle 2 = m\angle 5$. Explain your reasoning.

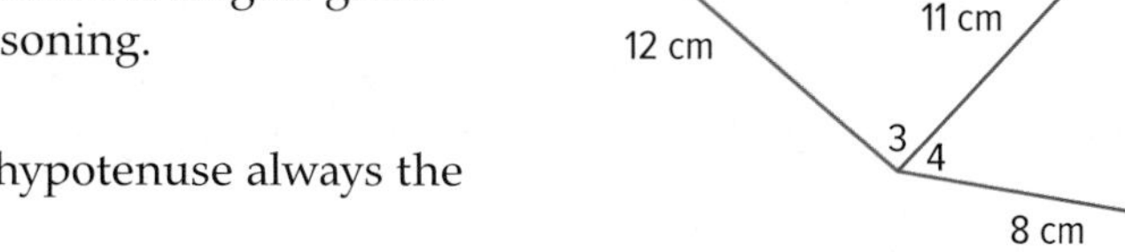

48. **WRITING IN MATH** Why is the hypotenuse always the longest side of a triangle?

49. Juan used geometry software to draw $\triangle MNP$ as shown. Which of the following is a true statement about $\triangle MNP$? MP 2, 7

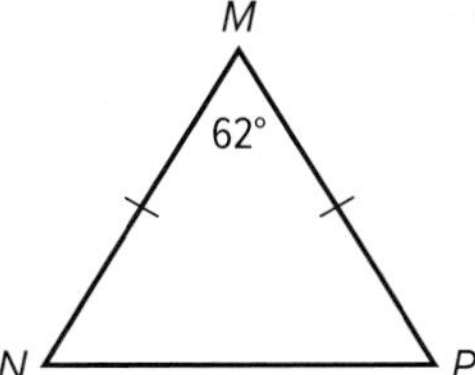

- ◯ **A** $\triangle MNP$ is equilateral.
- ◯ **B** $\overline{NP}$ is the longest side of $\triangle MNP$.
- ◯ **C** $\overline{NP}$ is the shortest side of $\triangle MNP$.
- ◯ **D** $\angle M$ is the smallest angle of $\triangle MNP$.

50. Given $\triangle ABC$ and $AB > BC$, prove that $m\angle BCA > m\angle A$. MP 3

51. Mr. Chen asked his students to write as many inequalities as possible using the angle relationships in the figure.

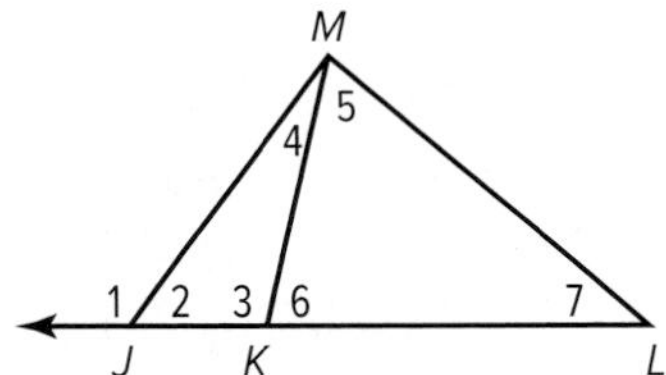

Which of the following inequalities cannot be justified using the information in the figure? MP 3

- ◯ **A** $m\angle 1 > m\angle 7$
- ◯ **B** $m\angle 5 > m\angle 4$
- ◯ **C** $m\angle 3 > m\angle 5$
- ◯ **D** $m\angle 3 > m\angle 7$

52. Two angles of a triangle have measures 45° and 92°. What type of triangle is it? MP 2

- ◯ **A** obtuse scalene
- ◯ **B** acute scalene
- ◯ **C** obtuse isosceles
- ◯ **D** acute isosceles

53. Katrina is designing a divided pen for rabbits with the angle measures shown.

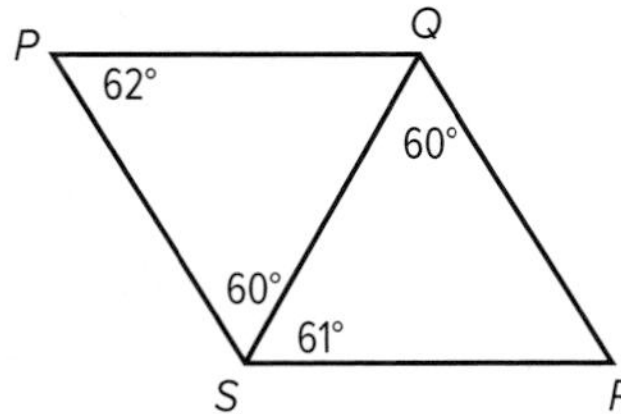

Which of the following is the shortest side of the rabbit pen? MP 1

- ◯ **A** $\overline{SQ}$
- ◯ **B** $\overline{QR}$
- ◯ **C** $\overline{RS}$
- ◯ **D** $\overline{SP}$

54. In $\triangle ABC$, $AB = 7.2$ cm, $BC = 6.1$ cm, and $AC = 3.5$ cm. Which of the following lists shows the angles of $\triangle ABC$ in order from smallest to largest? MP 2

- ◯ **A** $\angle B, \angle C, \angle A$
- ◯ **B** $\angle A, \angle B, \angle C$
- ◯ **C** $\angle C, \angle A, \angle B$
- ◯ **D** $\angle B, \angle A, \angle C$

55. **MULTI-STEP** Use the Exterior Angle Inequality Theorem answer each question. MP 2, 6

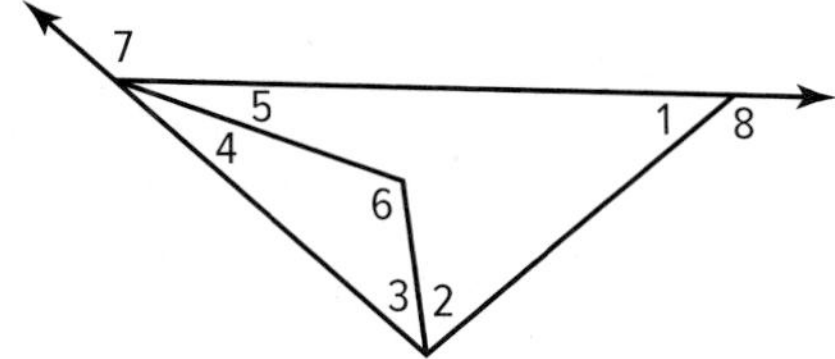

a. measures less than $\angle 8$.

b. measures greater than $\angle 1$.

c. If $m\angle 8 = 116°$ and $\angle 2 \cong \angle 3 \cong \angle 4 \cong \angle 5$, what is the $m\angle 6$? Explain.

CHAPTER 5
Mid-Chapter Quiz
Lessons 5-1 through 5-3

Find each measure. (Lesson 5-1)

1. AB

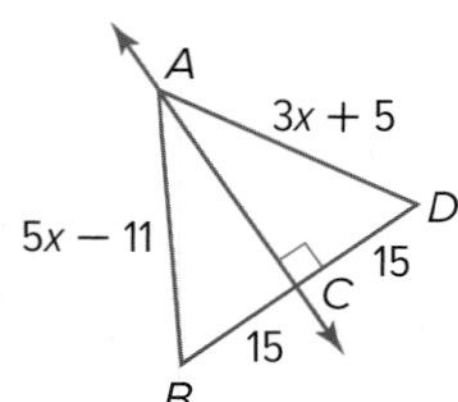

2. JL

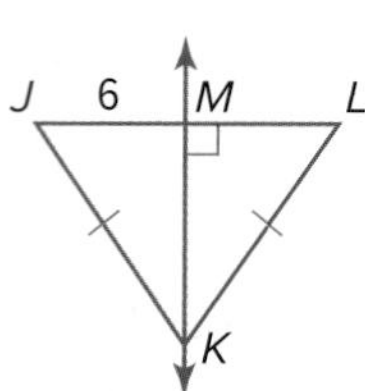

3. **CAMP** Camp Onawatchi ends with a game of capture the flag. The starting locations of three teams are shown in the diagram below, with the flag at a point equidistant from each team's base. (Lesson 5-1)

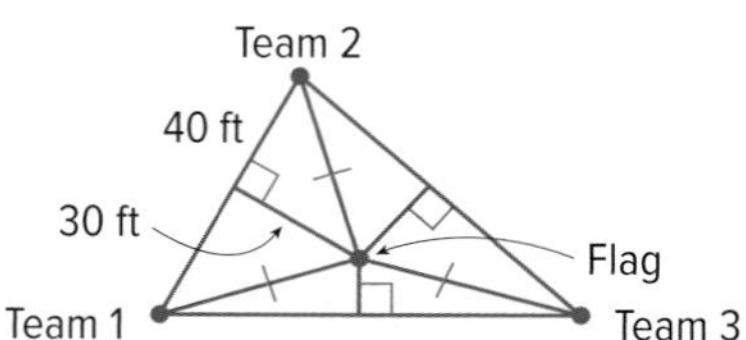

a. How far from each base is the flag in feet?

b. MP What mathematical practice did you use to solve this problem?

4. $\angle MNP$

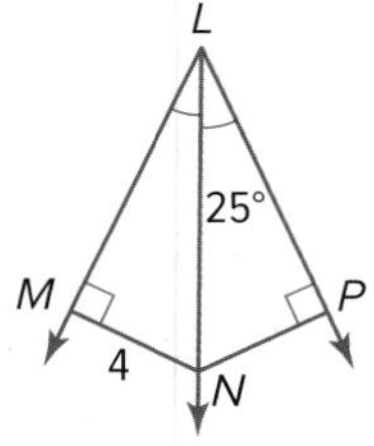

5. XY

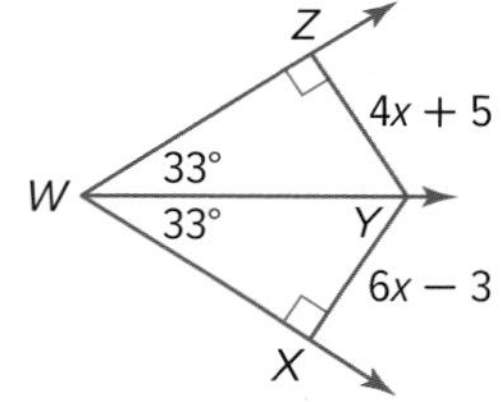

In $\triangle RST$, Z is the centroid and $RZ = 18$. Find each length. (Lesson 5-2)

6. ZV

7. SZ

8. SR

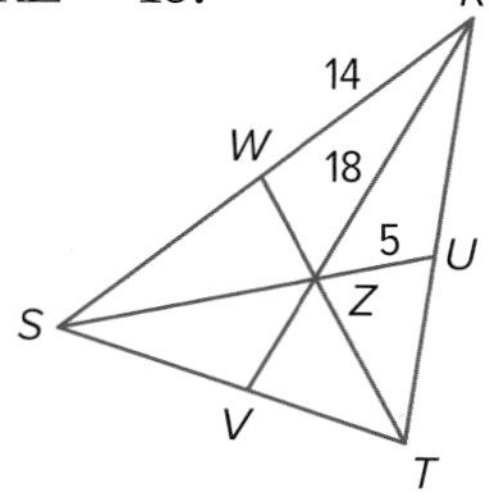

COORDINATE GEOMETRY Find the coordinates of the centroid of each triangle with the given vertices. (Lesson 5-2)

9. $A(1, 7)$, $B(4, 2)$, $C(7, 7)$

10. $X(-11, 0)$, $Y(-11, -8)$, $Z(-1, -4)$

11. $R(-6, 4)$, $S(-2, -2)$, $T(2, 4)$

12. $J(-5, 5)$, $K(-5, -1)$, $L(1, 2)$

13. **ARCHITECTURE** An architect is designing a high school building. Describe how to position the central office so that it is at the intersection of each hallway connected to the three entrances to the school. (Lesson 5-2)

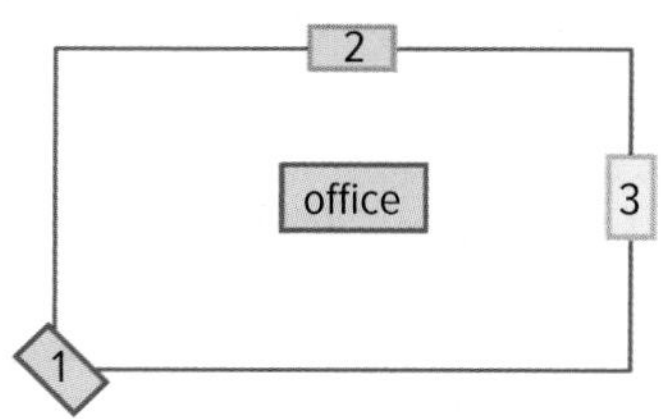

List the angles and sides of each triangle in order from smallest to largest. (Lesson 5-3)

14.

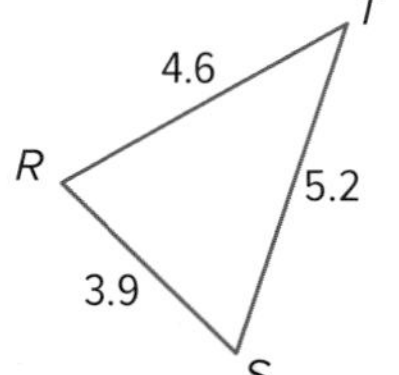

15.

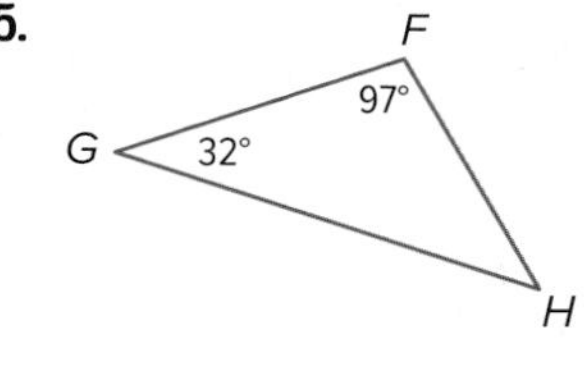

16. **VACATION** Kailey plans to fly over the route marked on the map of Hawaii below. (Lesson 5-3)

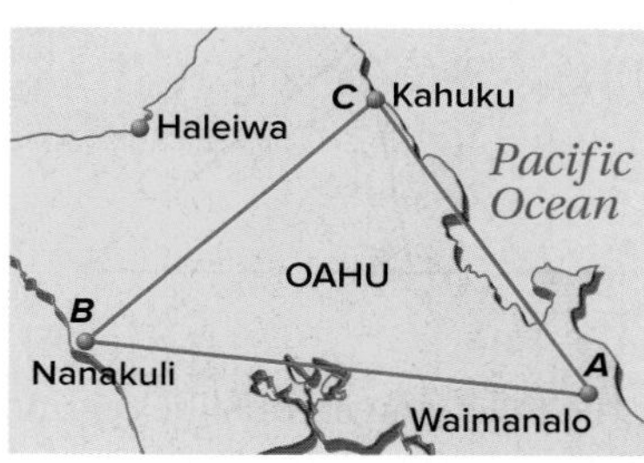

a. If $m\angle A = 2 + m\angle B$ and $m\angle C = 2(m\angle B) - 14$, what are the measures of the three angles?

b. What are the lengths of Kailey's trip in order of least to greatest?

c. The length of the entire trip is about 68 miles. The middle leg is 11 miles greater than one-half the length of the shortest leg. The longest leg is 12 miles greater than three-fourths of the shortest leg. What are the lengths of the legs of the trip?

Use the Exterior Angle Inequality Theorem to list all of the angles that satisfy the stated condition. (Lesson 5-3)

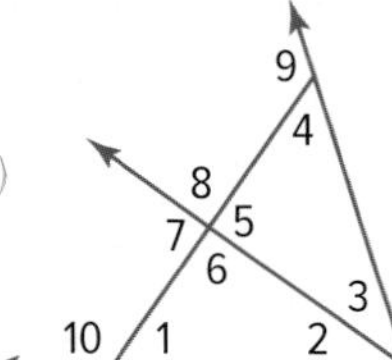

17. measures less than $m\angle 8$

18. measures greater than $m\angle 3$

19. measures less than $m\angle 10$

EXPLORE 5-4

Geometry Lab

Matrix Logic

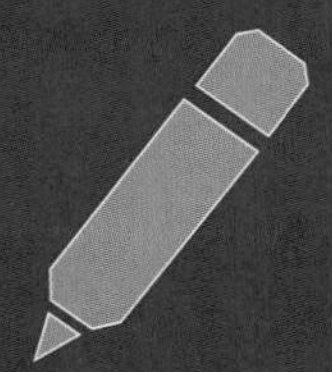

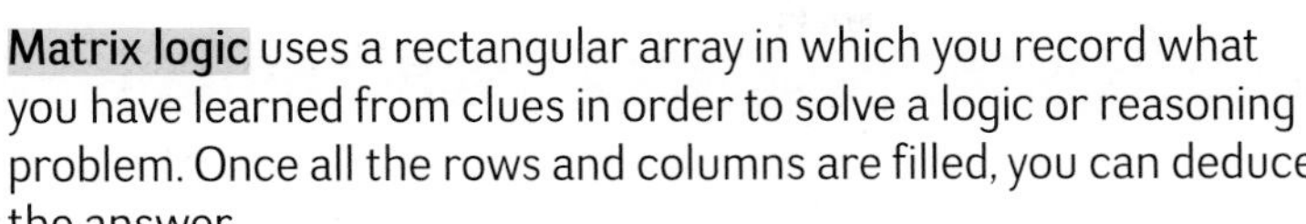

Matrix logic uses a rectangular array in which you record what you have learned from clues in order to solve a logic or reasoning problem. Once all the rows and columns are filled, you can deduce the answer.

Activity

FOOD Matt, Abby, Javier, Corey, and Keisha go to an Italian restaurant. Each orders his or her favorite dish: ravioli, pizza, lasagna, manicotti, or spaghetti. Javier loves ravioli, but Matt does not like pasta dishes. Abby does not like lasagna or manicotti. Corey's favorite dish does not end in the letter i. What does each person order?

Step 1 Create an appropriate matrix.

Use a 5 × 5 matrix that includes each person's name as the header for each row and their possible favorite foods as the header for each column.

		Favorite Dish				
		ravioli	pizza	lasagna	manicotti	spaghetti
Name	Matt	×	✓	×	×	×
	Abby	×	×			
	Javier	✓	×	×	×	×
	Corey	×	×			
	Keisha	×	×			

Step 2 Use each clue and logical reasoning to fill in the matrix.

- Since Javier loves ravioli, place a ✓ in Javier's row under ravioli and an × in every other cell in his row. Since only one person likes each dish, you can place an × in every other cell in the ravioli column.
- Since Matt does not like pasta, you know that Matt cannot like manicotti, ravioli, lasagna, or spaghetti, which are all pasta dishes. Therefore, Matt must like pizza. Place a ✓ in Matt's row under pizza. Place an × in every other cell in Matt's row and in every other cell in the pizza column.
- Since Abby does not like lasagna or manicotti, place an × in Abby's row under lasagna and manicotti. This leaves only spaghetti without an × for Abby's row. Therefore, you can conclude that Abby must like spaghetti. Place a ✓ in that cell and an × in every other cell in the spaghetti column.
- From the matrix, you can see that Corey's favorite dish must be either lasagna or manicotti. However, since Corey's favorite dish does not end in the letter i, you can conclude that Corey must like lasagna. In Corey's row, place a under lasagna and an × under manicotti.
- This leaves only one empty cell in Keisha's row, so you can conclude that her favorite dish is manicotti.

		Favorite Dish				
		ravioli	pizza	lasagna	manicotti	spaghetti
Name	Matt	×	✓	×	×	×
	Abby	×	×	×	×	✓
	Javier	✓	×	×	×	×
	Corey	×	×			×
	Keisha	×	×			×

		Favorite Dish				
		ravioli	pizza	lasagna	manicotti	spaghetti
Name	Matt	×	✓	×	×	×
	Abby	×	×	×	×	✓
	Javier	✓	×	×	×	×
	Corey	×	×	✓	×	×
	Keisha	×	×	×	✓	×

Step 3 Use your matrix to state the answer to the problem.

From the matrix you can state that Matt orders pizza, Abby orders spaghetti, Javier orders ravioli, Corey orders lasagna, and Keisha orders manicotti.

(continued on the next page)

EXPLORE 5-4

Geometry Lab

Matrix Logic *Continued*

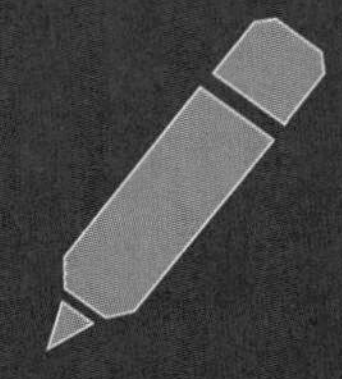

Exercises

Work cooperatively. Use a matrix to solve each problem.

1. **SPORTS** Trey, Nathan, Parker, and Chen attend the same school. Each participates in a different school sport: basketball, football, track, or tennis. Use the following clues to determine in which sport each student participates.
   - Nathan does not like track or basketball.
   - Trey does not participate in football or tennis.
   - Parker prefers an indoor winter sport.
   - Chen scored four touchdowns in the final game of the season.

2. **FAMILY** The Martin family has five children. Use the following clues to determine in what order the children were born.
   - Grace is older than Hannah.
   - Thomas is younger than Sarah.
   - Hannah is older than Thomas and Samuel.
   - Samuel is older than Thomas.
   - Sarah is older than Grace.

3. **PETS** Alejandra, Tamika, and Emily went to a pet store. Each girl chose a different pet to adopt: a dog, a rabbit, or a cat. Each girl named her pet Sweet Pea, Zuzu, or Roscoe. Use the following clues and the matrix shown to determine the animal each girl adopted and what name she gave her pet.
   - The girl who adopted a dog did not name it Sweet Pea.
   - Tamika's pet, who she named Zuzu, is not the type of animal that hops.
   - Roscoe, who is not a cat, was adopted by Emily.
   - The rabbit was not adopted by Alejandra.

		Pet			Pet Name		
		dog	rabbit	cat	Sweet Pea	Zuzu	Roscoe
Names	Alejandra						
	Tamika						
	Emily						
Pet Name	Sweet Pea						
	Zuzu						
	Roscoe						

4. **GEOMETRY** Kasa, Marcus, and Jason each drew a triangle, no two of which share the same side or angle classification. Use the following clues to determine what type of triangle each person has drawn.
   - Kasa did not draw an equilateral triangle.
   - Marcus' triangle has one angle that measures 25 and another that measures 65.
   - Jason drew a triangle with at least one pair of congruent sides.
   - The obtuse triangle has two congruent angles.

LESSON 4

Indirect Proof

Then

- You wrote paragraph, two-column, and flow proofs.

Now

1. Write indirect algebraic proofs.
2. Write indirect geometric proofs.

Why?

- Matthew: "I'm almost positive Friday is not a teacher work day, but I can't prove it."

  Kim: "Let's assume that Friday *is* a teacher work day. What day is our next Geometry test?"

  Ana: "Hmmm . . . according to the syllabus, it's this Friday. But we don't have tests on teacher work days—we're not in school."

  Jamal: "Exactly—so that proves it! This Friday can't be a teacher work day."

New Vocabulary

indirect reasoning
indirect proof
proof by contradiction

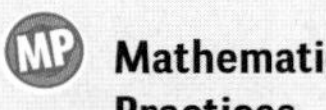

Mathematical Practices

3 Construct viable arguments and critique the reasoning of others.
2 Reason abstractly and quantitatively.

1 Indirect Algebraic Proof

The proofs you have written have been *direct proofs*—you started with a true hypothesis and proved that the conclusion was true. In the example above, the students used **indirect reasoning**, by assuming that a conclusion was false and then showing that this assumption led to a contradiction.

In an **indirect proof** or **proof by contradiction**, you temporarily assume that what you are trying to prove is false. By showing this assumption to be logically impossible, you prove your assumption false and the original conclusion true. Sometimes this is called *proof by negation.*

Key Concept How to Write an Indirect Proof

Step 1 Identify the conclusion you are asked to prove. Make the assumption that this conclusion is false by assuming that the opposite is true.

Step 2 Use logical reasoning to show that this assumption leads to a contradiction of the hypothesis, or some other fact, such as a definition, postulate, theorem, or corollary.

Step 3 Point out that since the assumption leads to a contradiction, the original conclusion, what you were asked to prove, must be true.

Example 1 State the Assumption for Starting an Indirect Proof

State the assumption necessary to start an indirect proof of each statement.

a. If 6 is a factor of n, then 2 is a factor of n.

The conclusion of the conditional statement is 2 *is a factor of* n. The negation of the conclusion is 2 *is not a factor of* n.

b. $\angle 3$ is an obtuse angle.

If $\angle 3$ *is an obtuse angle* is false, then $\angle 3$ *is not an obtuse angle* must be true.

Guided Practice

1A. $x > 5$

1B. $\triangle XYZ$ is an equilateral triangle.

Flying Colours Ltd/Photodisc/Getty Images

Indirect proofs can be used to prove algebraic concepts.

Example 2 Write an Indirect Algebraic Proof

Write an indirect proof to show that if $-3x + 4 > 16$, then $x < -4$.

Given: $-3x + 4 > 16$

Prove: $x < -4$

Step 1 **Indirect Proof:**
The negation of $x < -4$ is $x \geq -4$. So, assume that $x > -4$ or $x = -4$ is true.

Step 2 Make a table with several possibilities for x assuming $x > -4$ or $x = -4$.

x	-4	-3	-2	-1	0
$-3x + 4$	16	13	10	7	4

When $x > -4$, $-3x + 4 < 16$ and when $x = -4$, $-3x + 4 = 16$.

Step 3 In both cases, the assumption leads to the contradiction of the given information that $-3x + 4 > 16$. Therefore, the assumption that $x \geq -4$ must be false, so the original conclusion that $x < -4$ must be true.

Reading Math

contradiction Everyday use - an inconsistency or discrepancy; Math meaning - a principle of logic stating that an assumption cannot be *A* and the opposite of *A* at the same time. For example, the sky cannot be both blue and not blue at the same time. A coin flip cannot be both heads and tails at the same time.

Guided Practice

Write an indirect proof of each statement.

2A. If $7x > 56$, then $x > 8$.

2B. If $-c$ is positive, then c is negative.

Indirect reasoning and proof can be used in everyday situations.

Real-World Link

\$100–\$500 the range in price of a girl's prom dress

\$75–\$125 the range in cost for a tuxedo rental

around \$150 the cost of a fancy dinner for two

\$100–\$200 the range in cost of prom tickets per couple

Real-World Example 3 Indirect Algebraic Proof

PROM COSTS Javier asked his friend Christopher the cost of his meal and his date's meal when he went to dinner for prom. Christopher could not remember the individual costs, but he did remember that the total bill, not including tip, was over \$60. Use indirect reasoning to show that at least one of the meals cost more than \$30.

Let the cost of one meal be x and the cost of the other meal be y.

Step 1 **Given:** $x + y > 60$

Prove: $x > 30$ or $y > 30$

Indirect Proof:
Assume that $x \leq 30$ and $y \leq 30$.

Step 2 If $x \leq 30$ and $y \leq 30$, then $x + y \leq 30 + 30$ or $x + y \leq 60$. This is a contradiction because we know that $x + y > 60$.

Step 3 Because the assumption that $x \leq 30$ and $y \leq 30$ leads to a contradiction of a known fact, the assumption must be false. Therefore, the conclusion that $x > 30$ or $y > 30$ must be true. Thus, at least one of the meals had to cost more than \$30.

GuidedPractice

3. **TRAVEL** Cleavon traveled over 360 miles on his trip, making just two stops. Use indirect reasoning to prove that he traveled more than 120 miles on one leg of his trip.

Stockbyte/Getty Images

Indirect proofs are often used to prove concepts in number theory. In such proofs, it is helpful to remember that you can represent an even number with the expression $2k$ and an odd number with the expression $2k + 1$ for any integer k.

Example 4 Indirect Proofs in Number Theory

Write an indirect proof to show that if $x + 2$ is an even integer, then x is an even integer.

Step 1 **Given:** $x + 2$ is an even integer.

Prove: x is an even integer.

Indirect Proof:
Assume that x is an odd integer. This means that $x = 2k + 1$ for some integer k.

Step 2 $x + 2 = (2k + 1) + 2$ Substitution of assumption

$= (2k + 2) + 1$ Commutative Property

$= 2(k + 1) + 1$ Distributive Property

Now determine whether $2(k + 1) + 1$ is an even or odd integer. Since k is an integer, $k + 1$ is also an integer. Let m represent the integer $k + 1$.

$2(k + 1) + 1 = 2m + 1$ Substitution

So, $x + 2$ can be represented by $2m + 1$, where m is an integer. But this representation means that $x + 2$ is an odd integer, which contradicts the given statement that $x + 2$ is an even integer.

Step 3 Since the assumption that x is an odd integer leads to a contradiction of the given statement, the original conclusion that x is an even integer must be true.

Watch Out!

MP Construct Arguments Proof by contradiction and using a counterexample are not the same. A counterexample helps you disprove a conjecture. It cannot be used to prove a conjecture.

Guided Practice

4. Write an indirect proof to show that if the square of an integer is odd, then the integer is odd.

2 Indirect Proof with Geometry

Indirect reasoning can be used to prove statements in geometry, such as the Exterior Angle Inequality Theorem.

Example 5 Geometry Proof

If an angle is an exterior angle of a triangle, prove that its measure is greater than the measure of either of its corresponding remote interior angles.

A, 1, B, 2, C, 3, 4

Step 1 Draw a diagram of this situation. Then identify what you are given and what you are asked to prove.

Given: $\angle 4$ is an exterior angle of $\triangle ABC$.

Prove: $m\angle 4 > m\angle 1$ and $m\angle 4 > m\angle 2$.

Indirect Proof:
Assume that $m\angle 4 \ngtr m\angle 1$ or $m\angle 4 \ngtr m\angle 2$.
In other words, $m\angle 4 \leq m\angle 1$ or $m\angle 4 \leq m\angle 2$.

(continued on the next page)

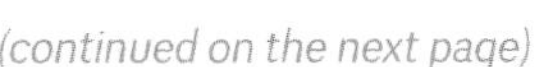

Step 2 You need only show that the assumption $m\angle 4 \leq m\angle 1$ leads to a contradiction. The argument for $m\angle 4 \leq m\angle 2$ follows the same reasoning.

$m\angle 4 \leq m\angle 1$ means that either $m\angle 4 = m\angle 1$ or $m\angle 4 < m\angle 1$.

Case 1 $m\angle 4 = m\angle 1$

$m\angle 4 = m\angle 1 + m\angle 2$ — Exterior Angle Theorem

$m\angle 4 = m\angle 4 + m\angle 2$ — Substitution

$0 = m\angle 2$ — Subtract $m\angle 4$ from each side.

This contradicts the fact that the measure of an angle is greater than 0, so $m\angle 4 \neq m\angle 1$.

Case 2 $m\angle 4 < m\angle 1$

By the Exterior Angle Theorem, $m\angle 4 = m\angle 1 + m\angle 2$. Because angle measures are positive, the definition of inequality implies that $m\angle 4 > m\angle 1$. This contradicts the assumption that $m\angle 4 < m\angle 1$.

Step 3 In both cases, the assumption leads to the contradiction of a theorem or definition. Therefore, the original conclusion that $m\angle 4 > m\angle 1$ and $m\angle 4 > m\angle 2$ must be true.

Study Tip

MP Reasoning Remember that the contradiction in an indirect proof is not always of the given information or the assumption. It can be of a known fact or definition, such as in Case 1 of Example 5—the measure of an angle must be greater than 0.

Guided Practice

5. Write an indirect proof.

Given: $\overline{MO} \cong \overline{ON}$, $\overline{MP} \not\cong \overline{NP}$

Prove: $\angle MOP \not\cong \angle NOP$

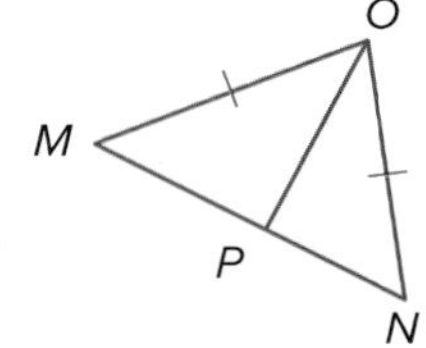

Check Your Understanding

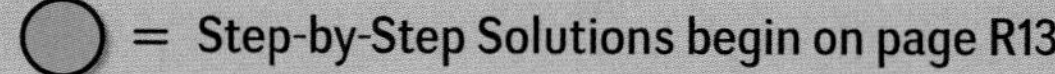

Example 1 **State the assumption you would make to start an indirect proof of each statement.**

1. $\overline{AB} \cong \overline{CD}$

2. $\triangle XYZ$ is a scalene triangle.

3. If $4x < 24$, then $x < 6$.

4. $\angle A$ is not a right angle.

Example 2 **Write an indirect proof of each statement.**

5. If $2x + 3 < 7$, then $x < 2$.

6. If $3x - 4 > 8$, then $x > 4$.

Example 3

7. LACROSSE Christina scored 13 points for her high school lacrosse team during the last six games. Prove that her average points per game was less than 3.

Example 4

8. Write an indirect proof to show that if $5x - 2$ is an odd integer, then x is an odd integer.

Example 5 **Write an indirect proof of each statement.**

9. The hypotenuse of a right triangle is the longest side.

10. If two angles are supplementary, then they both cannot be obtuse angles.

Practice and Problem Solving

Extra Practice is on page R5.

Example 1 **State the assumption you would make to start an indirect proof of each statement.**

11. If $2x > 16$, then $x > 8$.

12. $\angle 1$ and $\angle 2$ are not supplementary angles.

13. If two lines have the same slope, the lines are parallel.

14. If the consecutive interior angles formed by two lines and a transversal are supplementary, the lines are parallel.

15. If a triangle is not equilateral, the triangle is not equiangular.

16. An odd number is not divisible by 2.

Example 2 **Write an indirect proof of each statement.**

17. If $2x - 7 > -11$, then $x > -2$.

18. If $5x + 12 < -33$, then $x < -9$.

19. If $-3x + 4 < 7$, then $x > -1$.

20. If $-2x - 6 > 12$, then $x < -9$.

Example 3

21. **COMPUTER GAMES** Kwan-Yong bought two computer games for just over \$80 before tax. A few weeks later, his friend asked how much each game cost. Kwan-Yong could not remember the individual prices. Use indirect reasoning to show that at least one of the games cost more than \$40.

22. **FUNDRAISING** Jamila's school is having a Fall Carnival to raise money for a local charity. The cost of an adult ticket to the carnival is \$6 and the cost of a child's ticket is \$2.50. If 375 total tickets were sold and the profit was more than \$1460, prove that at least 150 adult tickets were sold.

Examples 4–5 MP **CONSTRUCT ARGUMENTS Write an indirect proof of each statement.**

23. **Given:** xy is an odd integer.
Prove: x and y are both odd integers.

24. **Given:** n^2 is even.
Prove: n^2 is divisible by 4.

25. **Given:** x is an odd number.
Prove: x is not divisible by 4.

26. **Given:** xy is an even integer.
Prove: x or y is an even integer.

27. **Given:** $XZ > YZ$
Prove: $\angle X \not\cong \angle Y$

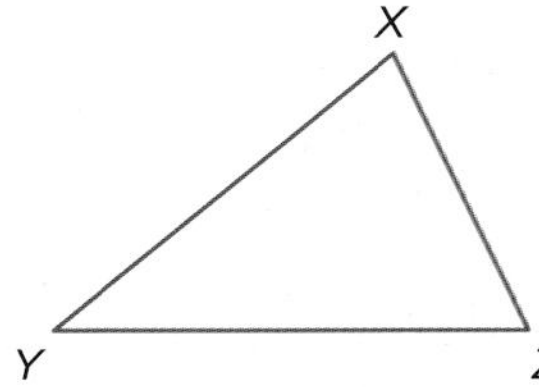

28. **Given:** $\triangle ABC$ is equilateral.
Prove: $\triangle ABC$ is equiangular.

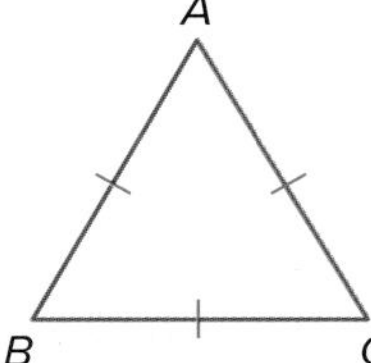

29. In an isosceles triangle neither of the base angles can be a right angle.

30. A triangle can have only one right angle.

31. Write an indirect proof for Theorem 5.10.

32. Write an indirect proof to show that if $\frac{1}{b} < 0$, then b is negative.

33. **BASKETBALL** In basketball, there are three possible ways to score three points in a single possession. A player can make a basket from behind the three-point line, a player may be fouled while scoring a two-point shot and be allowed to shoot one free throw, or a player may be fouled behind the three-point line and be allowed to shoot three free throws. When Katsu left to get in the concession line, the score was 28 home team to 26 visiting team. When she returned, the score was 28 home team to 29 visiting team. Katsu concluded that a player on the visiting team had made a three-point basket. Prove or disprove her assumption using an indirect proof.

34. **GAMES** A computer game involves a knight on a quest for treasure. At the end of the journey, the knight approaches the two doors shown below.

A servant tells the knight that one of the signs is true and the other is false. Use indirect reasoning to determine which door the knight should choose. Explain your reasoning.

35. **SURVEYS** Luisa's local library conducted an online poll of teens to find out what activities teens participate in to preserve the environment. The results of the poll are shown in the graph.

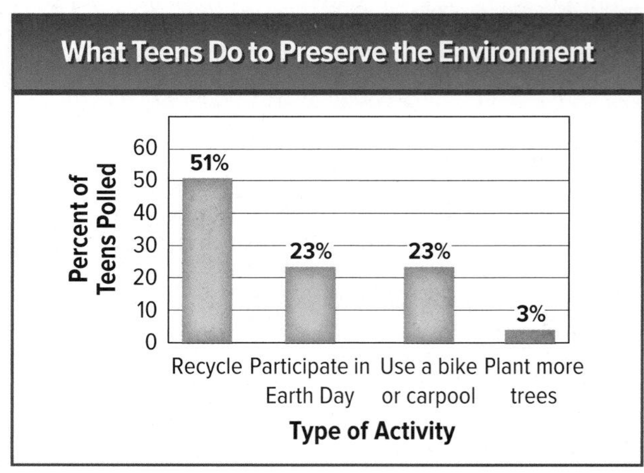

a. Prove: *More than half of teens polled said that they recycle to preserve the environment.*

b. If 400 teens were polled, verify that 92 said that they participate in Earth Day.

36. **MP REASONING** James, Hector, and Mandy all have different color cars. Only one of the statements below is true. Use indirect reasoning to determine which statement is true. Explain.

(1) James has a red car.

(2) Hector does not have a red car.

(3) Mandy does not have a blue car.

Determine whether each statement about the shortest distance between a point and a line or plane can be proved using a direct or indirect proof. Then write a proof of each statement.

37. **Given:** $\overline{AB} \perp$ line p

Prove: $\overline{AB}$ is the shortest segment from A to line p.

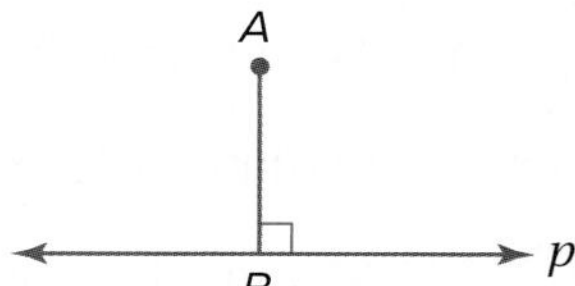

38. **Given:** $\overline{PQ} \perp$ plane M

Prove: $\overline{PQ}$ is the shortest segment from P to plane M.

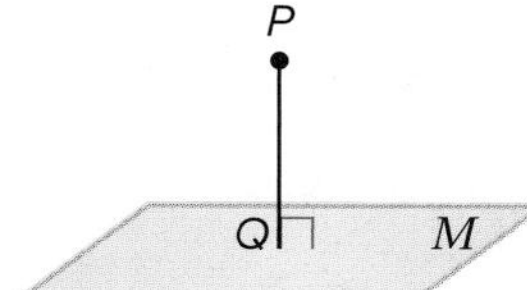

39. **NUMBER THEORY** In this problem, you will make and prove a conjecture about a number theory relationship.

a. Write an expression for *the sum of the cube of a number and three.*

b. Create a table that includes the value of the expression for 10 different values of n. Include both odd and even values of n.

c. Write a conjecture about n when the value of the expression is even.

d. Write an indirect proof of your conjecture.

H.O.T. Problems Use Higher-Order Thinking Skills

40. **WRITING IN MATH** Explain the procedure for writing an indirect proof.

41. **OPEN-ENDED** Write a statement that can be proven using indirect proof. Include the indirect proof of your statement.

42. MP **CONSTRUCT ARGUMENTS** If x is a rational number, then it can be represented by the quotient $\frac{a}{b}$ for some integers a and b, if $b \neq 0$. An irrational number cannot be represented by the quotient of two integers. Write an indirect proof to show that the product of a nonzero rational number and an irrational number is an irrational number.

43. **ERROR ANALYSIS** Amber and Raquel are trying to verify the following statement using indirect proof. Is either of them correct? Explain your reasoning.

If the sum of two numbers is even, then the numbers are even.

Amber	Raquel
The statement is true. If one of the numbers is even and the other number is zero, then the sum is even. Since the hypothesis is true even when the conclusion is false, the statement is true.	The statement is true. If the two numbers are odd, then the sum is even. Since the hypothesis is true when the conclusion is false, the statement is true.

44. **WRITING IN MATH** Refer to Exercise 8. Write the contrapositive of the statement and write a direct proof of the contrapositive. How are the direct proof of the contrapositive of the statement and the indirect proof of the statement related?

45. Miyuki is asked to prove that a right triangle cannot have an obtuse angle. She draws $\triangle RST$ and then assumes that $\angle T$ is an obtuse angle. She shows that this leads to a contradiction.

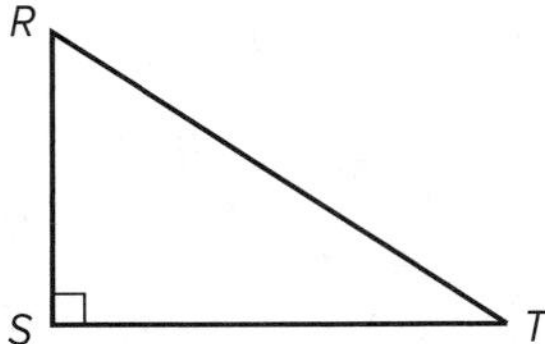

Which of the following statements could Miyuki use in her indirect proof? MP 3

- A The base angles of an isosceles triangle are congruent.
- B Each angle of an equilateral triangle measures 60°.
- C The sum of the measures of the angles in a triangle is 180°.
- D If two parallel lines are cut by a transversal, then the pairs of alternate interior angles are congruent.
- E Vertical angles are congruent.

46. Tomas wants to write an indirect proof that a scalene triangle cannot have two congruent angles. He draws a scalene $\triangle JKL$ and assumes that $\angle J \cong \angle K$.

Which of the following statements could Tomas use in his proof? MP 3

- A Each angle of an equilateral triangle measures 60°.
- B If two angles of a triangle are congruent, then the sides opposite them are congruent.
- C The acute angles of a right triangle are complementary.
- D The measure of an exterior angle of a triangle is equal to the sum of the measures of the two remote interior angles.

47. Colin is given a figure of $\triangle PQR$ that has some of the angle measures given. He is asked to write an indirect proof that $m\angle P < 45$. What assumption should Colin make to start his indirect proof? MP 3

- A Assume that $m\angle P < 45$.
- B Assume that $m\angle P \geq 45$.
- C Assume that $m\angle P = 45$.
- D Assume that $m\angle P > 45$.

48. Lisa wants to write an indirect proof that two angles of a triangle cannot be supplementary. She draws $\triangle ABC$ and assumes that $\angle A$ and $\angle B$ are supplementary.

Which of the following statements could Lisa use in her proof? MP 3

- A The acute angles of a right triangle are complementary.
- B If two parallel lines are cut by a transversal, then each pair of consecutive interior angles is supplementary.
- C If two sides of a triangle are congruent, then the angles opposite them are congruent.
- D The sum of the measures of the angles in a triangle is 180°.

49. **MULTI-STEP** Andrew wants to write an indirect proof that $\angle 1$ is an acute angle. MP 3

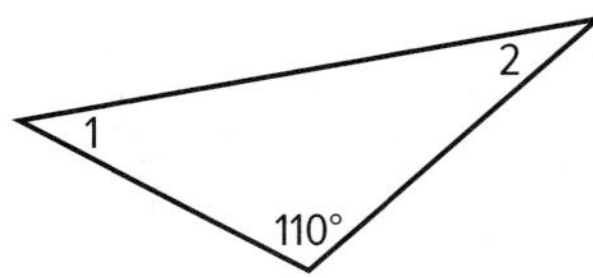

a. What assumption should Andrew make to start his indirect proof?

b. Which theorem about triangles should Andrew use as part of his proof?

c. Andrew also needs to write an indirect proof that $\angle 2$ is not a right angle. Show how to write this proof.

EXPLORE 5-5

Graphing Technology Lab

The Triangle Inequality

You can use the Cabri™ Jr. application on a TI-83/84 Plus graphing calculator to discover properties of triangles.

Mathematical Practices

MP **5** Use the appropriate tools strategically.

Activity

Work cooperatively. Construct a triangle. Observe the relationship between the sum of the lengths of two sides and the length of the other side.

Step 1 Construct a triangle using the triangle tool on the **F2** menu. Then use the **Alph-Num** tool on the **F5** menu to label the vertices as *A*, *B*, and *C*.

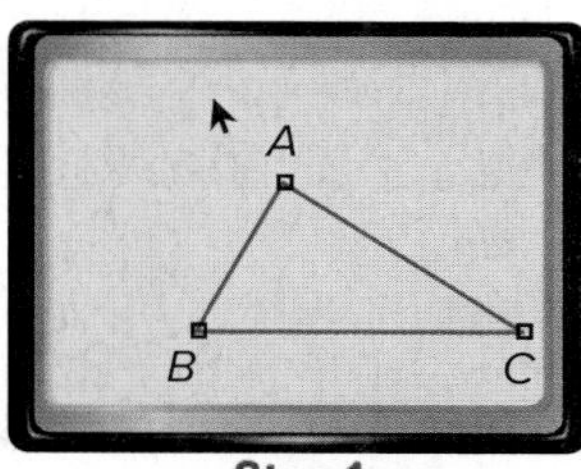

Step 1

Step 2 Access the **distance & length** tool, shown as **D. & Length**, under **Measure** on the **F5** menu. Use the tool to measure each side of the triangle.

Step 3 Display $AB + BC$, $AB + CA$, and $BC + CA$ by using the **Calculate** tool on the **F5** menu. Label the measures.

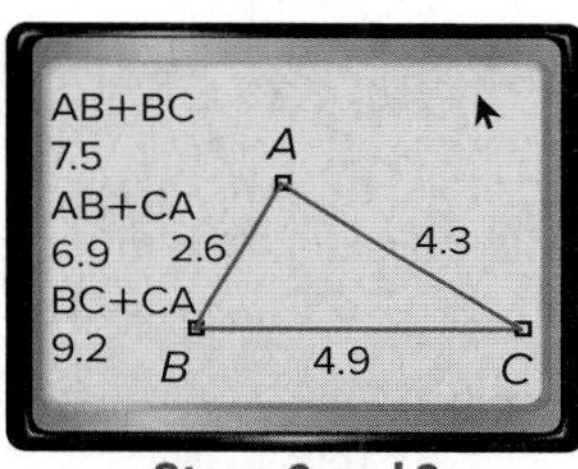

Steps 2 and 3

Step 4 Click and drag the vertices to change the shape of the triangle.

Analyze the Results Work cooperatively.

1. Replace each ● with <, >, or = to make a true statement.
   $AB + BC$ ● CA $\quad$ $AB + CA$ ● BC $\quad$ $BC + CA$ ● AB
2. Click and drag the vertices to change the shape of the triangle. Then review your answers to Exercise 1. What do you observe?
3. Click on point *A* and drag it to lie on line *BC*. What do you observe about *AB*, *BC*, and *CA*? Are *A*, *B*, and *C* the vertices of a triangle? Explain.
4. **Make a conjecture** about the sum of the lengths of two sides of a triangle and the length of the third side.
5. Do the measurements and observations you made in the Activity and in Exercises 1–3 constitute a proof of the conjecture you made in Exercise 4? Explain.
6. Replace each ● with <, >, or = to make a true statement.
   $|AB - BC|$ ● CA $\quad$ $|AB - CA|$ ● BC $\quad$ $|BC - CA|$ ● AB

   Then click and drag the vertices to change the shape of the triangle and review your answers. What do you observe?
7. How could you use your observations to determine the possible lengths of the third side of a triangle if you are given the lengths of the other two sides?

LESSON 5

The Triangle Inequality

Then

- You recognized and applied properties of inequalities to the relationships between the angles and sides of a triangle.

Now

1. Use the Triangle Inequality Theorem to identify possible triangles.
2. Prove triangle relationships using the Triangle Inequality Theorem.

Why?

- On a home improvement show, a designer wants to use scrap pieces of cording from another sewing project to decorate the triangular throw pillows that she and the homeowner have made. To minimize waste, she wants to use the scraps without cutting them. She selects three scraps at random and tries to form a triangle. Two such attempts are shown.

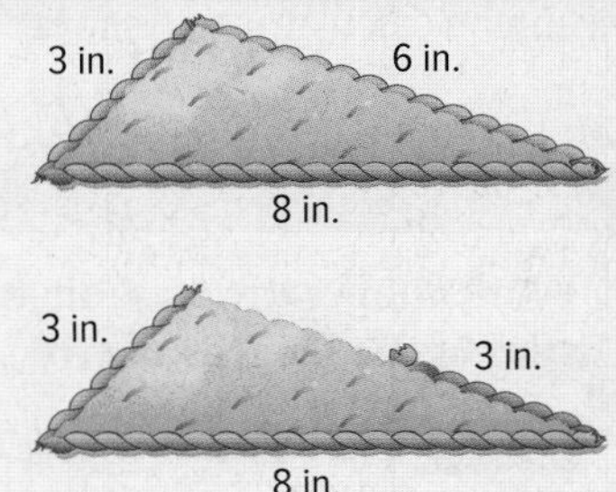

Mathematical Practices

1 Make sense of problems and persevere in solving them.

2 Reason abstractly and quantitatively.

4 Model with mathematics.

1 The Triangle Inequality

While a triangle is formed by three segments, a special relationship must exist among the lengths of the segments in order for them to form a triangle.

Theorem 5.11 Triangle Inequality Theorem

The sum of the lengths of any two sides of a triangle must be greater than the length of the third side.

Examples $PQ + QR > PR$
$QR + PR > PQ$
$PR + PQ > QR$

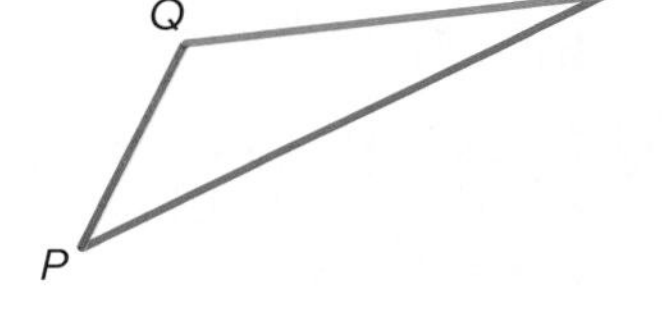

You will prove Theorem 5.11 in Exercise 23.

To show that it is not possible to form a triangle with three side lengths, you need only show that one of the three triangle inequalities is not true.

Example 1 Identify Possible Triangles Given Side Lengths

Is it possible to form a triangle with the given side lengths? If not, explain why not.

a. 8 in., 15 in., 17 in.

Check each inequality.

$8 + 15 \overset{?}{>} 17$, $23 > 17$ ✓

$8 + 17 \overset{?}{>} 15$, $25 > 15$ ✓

$15 + 17 \overset{?}{>} 8$, $32 > 8$ ✓

Since the sum of each pair of side lengths is greater than the third side length, sides with lengths 8, 15, and 17 inches will form a triangle.

b. 6 m, 8 m, 14 m

$6 + 8 \overset{?}{>} 14$

$14 \not> 14$ ✗

Since the sum of one pair of side lengths is not greater than the third side length, sides with lengths 6, 8, and 14 meters will not form a triangle.

Guided Practice

1A. 15 yd, 16 yd, 30 yd

1B. 2 ft, 8 ft, 11 ft

When the lengths of two sides of a triangle are known, the third side can be any length in a range of values. You can use the Triangle Inequality Theorem to determine the range of possible lengths for the third side.

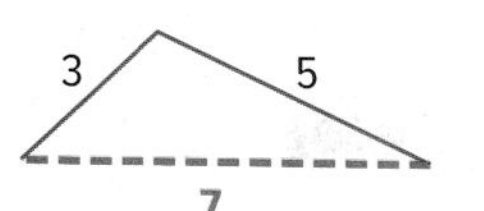

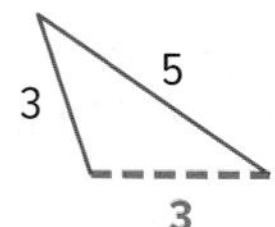

Example 2 Find Possible Side Lengths

Go Online!

Discover relationships among the measures of the sides and angles in a triangle with a **Geometer's Sketchpad®** sketch in ConnectED.

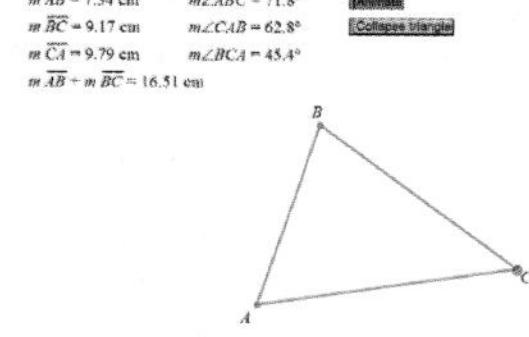

If the measures of two sides of a triangle are 3 feet and 7 feet, which is the *least* possible whole number measure for the third side?

A 3 ft **B** 4 ft **C** 5 ft **D** 10 ft

Read the Item

You need to determine which value is the least possible measure for the third side of a triangle with sides that measure 3 feet and 7 feet.

Solve the Item

To determine the least possible measure from the choices given, first determine the range of possible measures for the third side.

Draw a diagram and let x represent the length of the third side.

3 ft
7 ft
x ft

Next, set up and solve each of the three triangle inequalities.

$3 + 7 > x$	$3 + x > 7$	$x + 7 > 3$
$10 > x$ or $x < 10$	$x > 4$	$x > -4$

Notice that $x > -4$ is always true for any whole number measure for x. Combining the two remaining inequalities, the range of values that fit both inequalities is $x > 4$ and $x < 10$, which can be written as $4 < x < 10$.

Reading Math

Multiple Inequality Symbols The compound inequality $4 < x < 10$ is read *x is between 4 and 10.*

The least whole number value between 4 and 10 is 5. So the correct answer is choice C.

Guided Practice

2. Which of the following could *not* be the value of n?

A 7 **C** 13

B 10 **D** 22

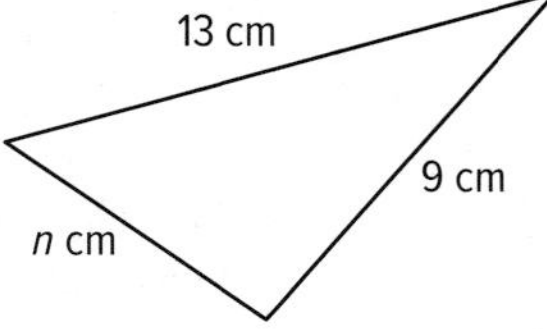

2 Proofs Using the Triangle Inequality Theorem

You can use the Triangle Inequality Theorem as a reason in proofs.

Real-World Example 3 Proof Using Triangle Inequality Theorem

TRAVEL The distance from Colorado Springs, Colorado, to Abilene, Texas, is the same as the distance from Colorado Springs to Tulsa, Oklahoma. Prove that a direct flight from Colorado Springs to Tulsa through Lincoln, Nebraska, is a greater distance than a nonstop flight from Colorado Springs to Abilene.

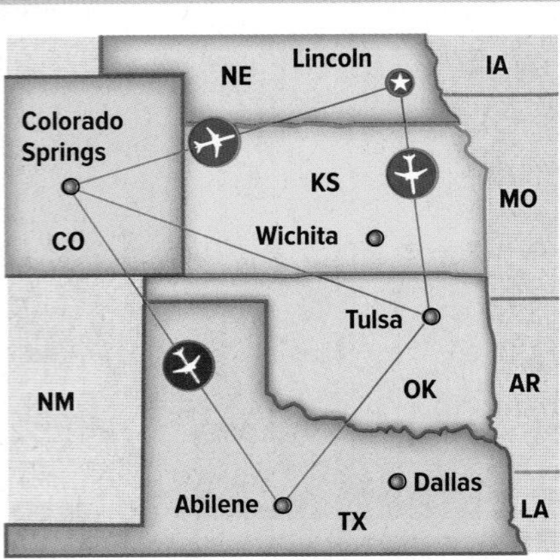

Real-World Link

A direct flight is not the same as a nonstop flight. For a direct flight, passengers do not change planes, but the plane may make one or more stops before continuing to its final destination.

Draw a simpler diagram of the situation and label the diagram. Draw in side $\overline{LT}$ to form $\triangle CTL$.

Given: $CA = CT$

Prove: $CL + LT > CA$

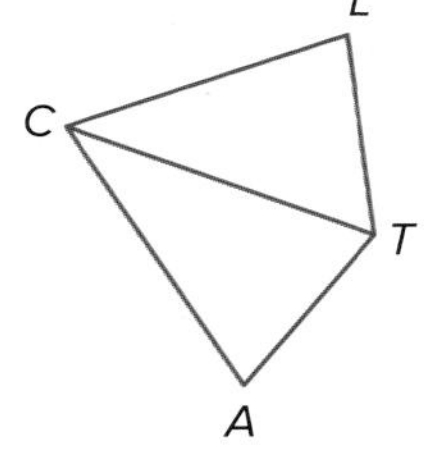

Proof:

Statements	Reasons
1. $CA = CT$	**1.** Given
2. $CL + LT > CT$	**2.** Triangle Inequality Theorem
3. $CL + LT > CA$	**3.** Substitution

Guided Practice

3. Write a two-column proof.

Given: $GL = LK$

Prove: $JH + GH > JK$

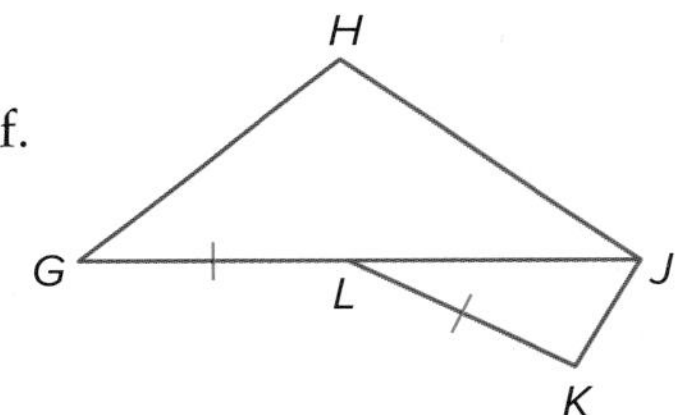

Check Your Understanding

◯ = Step-by-Step Solutions begin on page R13.

Example 1 **Is it possible to form a triangle with the given side lengths? If not, explain why not.**

(1) 5 cm, 7 cm, 10 cm **2.** 3 in., 4 in., 8 in. **3.** 6 m, 14 m, 10 m

Example 2

4. MULTIPLE CHOICE If the measures of two sides of a triangle are 5 yards and 9 yards, what is the least possible measure of the third side if the measure is an integer?

A 4 yd **B** 5 yd **C** 6 yd **D** 14 yd

Example 3

5. PROOF Write a two-column proof.

Given: $\overline{XW} \cong \overline{YW}$

Prove: $YZ + ZW > XW$

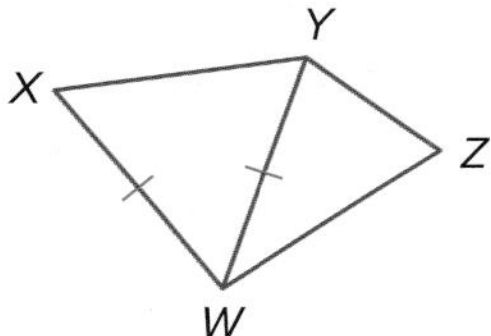

Ed Boettcher/Corbis Premium Collection/Alamy

Practice and Problem Solving

Extra Practice is on page R5.

Example 1 **Is it possible to form a triangle with the given side lengths? If not, explain why not.**

6. 4 ft, 9 ft, 15 ft

7. 11 mm, 21 mm, 16 mm

8. 9.9 cm, 1.1 cm, 8.2 cm

9. 2.1 in., 4.2 in., 7.9 in.

10. $2\frac{1}{2}$ m, $1\frac{3}{4}$ m, $5\frac{1}{8}$ m

11. $1\frac{1}{5}$ km, $4\frac{1}{2}$ km, $3\frac{3}{4}$ km

Example 2 **Find the range for the measure of the third side of a triangle given the measures of two sides.**

12. 4 ft, 8 ft

13. 5 m, 11 m

14. 2.7 cm, 4.2 cm

15. 3.8 in., 9.2 in.

16. $\frac{1}{2}$ km, $3\frac{1}{4}$ km

17. $2\frac{1}{3}$ yd, $7\frac{2}{3}$ yd

Example 3 **PROOF** Write a two-column proof.

18. Given: $\angle BCD \cong \angle CDB$

Prove: $AB + AD > BC$

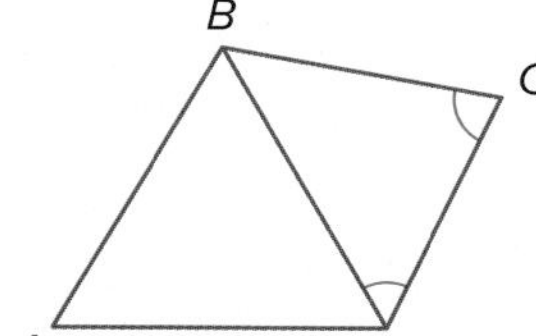

19. Given: $\overline{JL} \cong \overline{LM}$

Prove: $KJ + KL > LM$

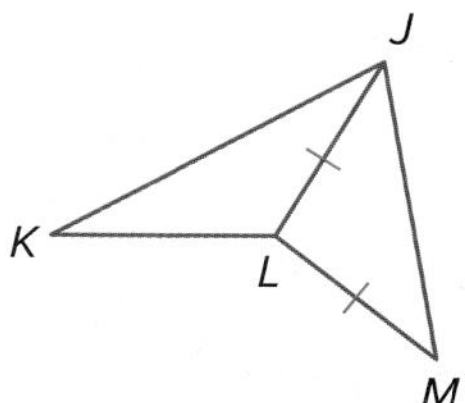

MP SENSE-MAKING Determine the possible values of x.

20.

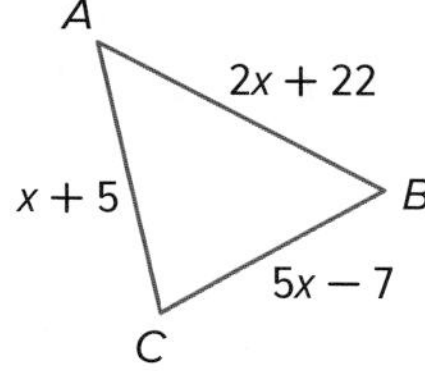

21.

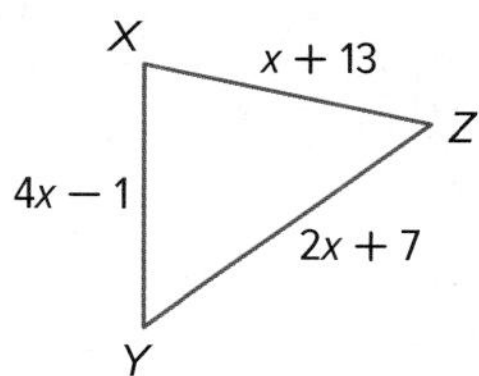

22. TRAVEL Keyan wants to take the most efficient route from his hotel to the hockey game at The Sportsplex. He can either take Highway 521 or take Highway 3 and Route 11 from his hotel to the arena.

a. Which of these two possible routes is the shorter? Explain your reasoning.

b. Suppose Keyan always drives the speed limit. If he chooses to take Highway 521, he is on it for 30 miles and the speed limit is 40 miles per hour. If he chooses to take the other route, the speed limit is 60 miles per hour on both roads and he is on Highway 3 for 22 miles and Route 11 for 25 miles. Which route is faster? Explain your reasoning.

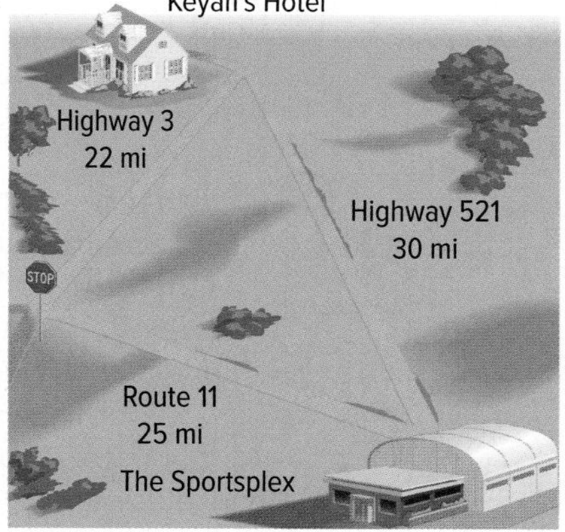

23. PROOF Write a two-column proof.

Given: $\triangle ABC$

Prove: $AC + BC > AB$ (Triangle Inequality Theorem)

(*Hint*: Draw auxiliary segment $\overline{CD}$, so that C is between B and D and $\overline{CD} \cong \overline{AC}$.)

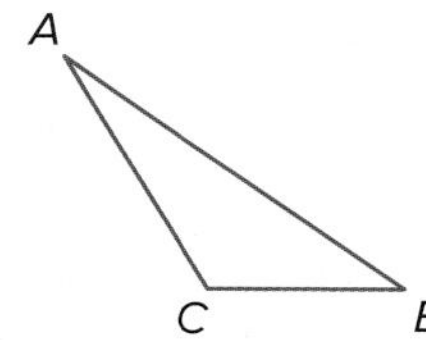

24. **MULTI-STEP** Toya rides her bike 3680 feet down her street, Meadow Court, turns left onto Holly Lane, and rides 2740 feet to get to the park. One day she decides to try to find a shorter path to the park by cutting straight through the field that is next to the two roads. However, if she decides to cut through the field, she must walk instead of riding her bike.

a. If Toya is able to ride her bike 3 times as fast as she can walk, how long should the path through the field be for it to be a shortcut? How much time would Toya save by taking the shortcut, if she can ride her bike 9 miles per hour?

b. Describe your solution process.

c. What assumptions did you make?

Find the range of possible measures of x if each set of expressions represents measures of the sides of a triangle.

25. x, 4, 6

26. 8, x, 12

27. $x + 1$, 5, 7

28. $x - 2$, 10, 12

29. $x + 2$, $x + 4$, $x + 6$

30. x, $2x + 1$, $x + 4$

31. **DRAMA CLUB** Anthony and Catherine are working on a ramp up to the stage for the drama club's next production. Anthony's sketch of the ramp is shown below. Catherine is concerned about the measurements and thinks they should recheck the measures before they start cutting the wood. Is Catherine's concern valid? Explain your reasoning.

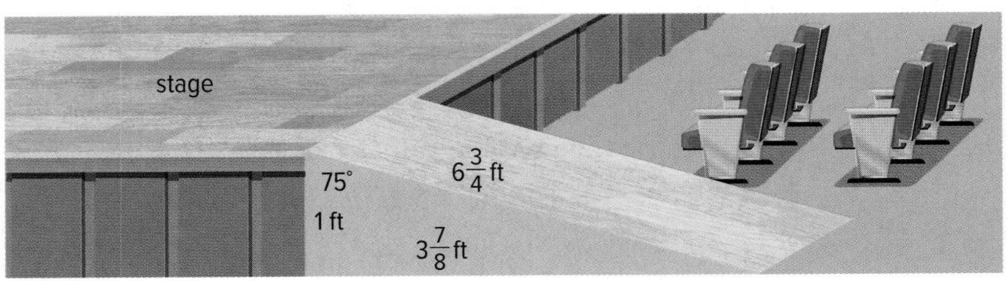

32. MP **MODELING** Aisha is riding her bike to the park and can take one of two routes. The most direct route from her house is to take Main Street, but it is safer to take Route 3 and then turn right onto Clay Road as shown. The additional distance she will travel if she takes Route 3 to Clay Road is between how many miles?

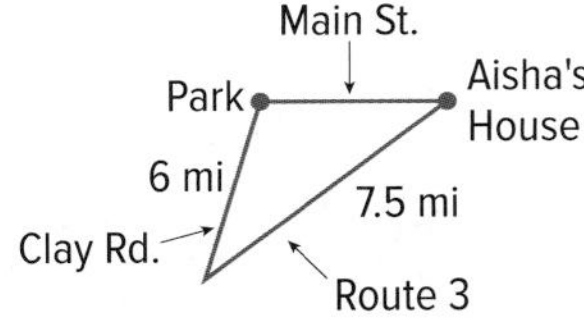

33 **DESIGN** Carlota designed an awning that she and her friends could take to the beach. Carlota decides to cover the top of the awning with material that will drape 6 inches over the front. What length of material should she buy to use with her design so that it covers the top of the awning, including the drape, when the supports are open as far as possible? Assume that the width of the material is sufficient to cover the awning.

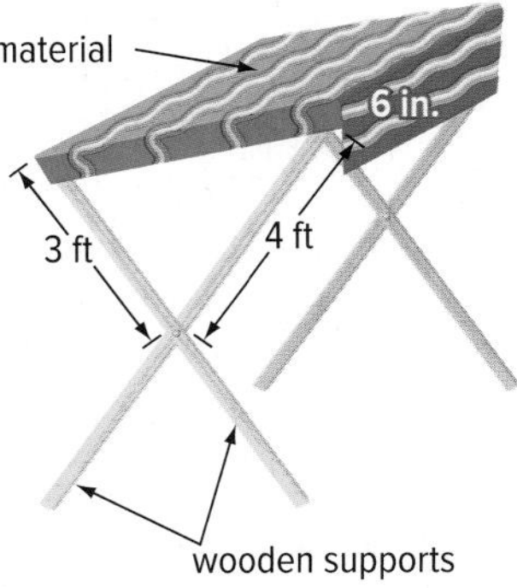

ESTIMATION Without using a calculator, determine if it is possible to form a triangle with the given side lengths. Explain.

34. $\sqrt{8}$ ft, $\sqrt{2}$ ft, $\sqrt{35}$ ft

35. $\sqrt{99}$ yd, $\sqrt{48}$ yd, $\sqrt{65}$ yd

36. $\sqrt{3}$ m, $\sqrt{15}$ m, $\sqrt{24}$ m

37. $\sqrt{122}$ in., $\sqrt{5}$ in., $\sqrt{26}$ in.

MP REASONING Determine whether the given coordinates are the vertices of a triangle. Explain.

38. $X(1, -3)$, $Y(6, 1)$, $Z(2, 2)$

39. $F(-4, 3)$, $G(3, -3)$, $H(4, 6)$

40. $J(-7, -1)$, $K(9, -5)$, $L(21, -8)$

41. $Q(2, 6)$, $R(6, 5)$, $S(1, 2)$

42. MULTIPLE REPRESENTATIONS In this problem, you will use inequalities to make comparisons between the sides and angles of two triangles.

a. Geometric Draw three pairs of triangles that have two pairs of congruent sides and one pair of sides that is not congruent. Mark each pair of congruent sides. Label each triangle pair ABC and DEF, where $\overline{AB} \cong \overline{DE}$ and $\overline{AC} \cong \overline{DF}$.

b. Tabular Copy the table below. Measure and record the values of BC, $m\angle A$, EF, and $m\angle D$ for each triangle pair.

Triangle Pair	BC	$m\angle A$	EF	$m\angle D$
1				
2				
3				

c. Verbal Make a conjecture about the relationship between the angles opposite the noncongruent sides of a pair of triangles that have two pairs of congruent legs.

H.O.T. Problems Use Higher-Order Thinking Skills

43. MP REASONING What is the range of possible perimeters for figure $ABCDE$ if $AC = 7$ and $DC = 9$? Explain your reasoning.

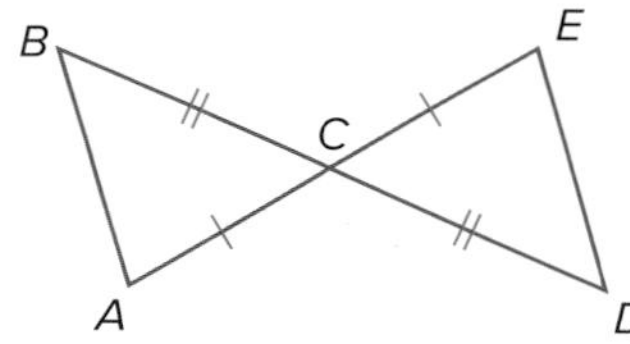

44. CHALLENGE What is the range of lengths of each leg of an isosceles triangle if the measure of the base is 6 inches? Explain.

45. WRITING IN MATH What can you tell about a triangle when given three side lengths? Include at least two items.

46. CHALLENGE The sides of an isosceles triangle are whole numbers, and its perimeter is 30 units. What is the probability that the triangle is equilateral?

47. OPEN-ENDED The length of one side of a triangle is 2 inches. Draw a triangle in which the 2-inch side is the shortest side and one in which the 2-inch side is the longest side. Include side and angle measures on your drawing.

48. WRITING IN MATH Suppose your house is $\frac{3}{4}$ mile from a park and the park is 1.5 miles from a shopping center.

a. If your house, the park, and the shopping center are noncollinear, what do you know about the distance from your house to the shopping center? Explain your reasoning.

b. If the three locations are collinear, what do you know about the distance from your house to the shopping center? Explain your reasoning.

Preparing for Assessment

49. Which of the following could not be the length of $\overline{MN}$? MP 7

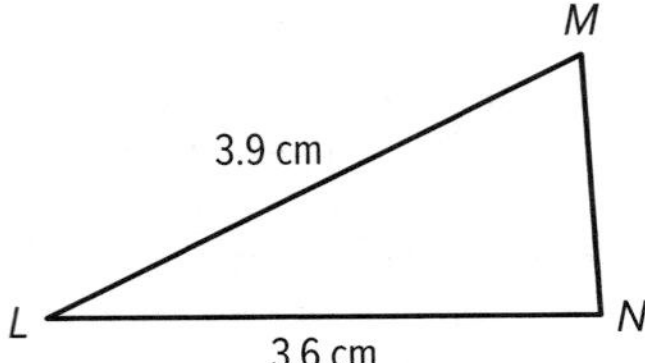

- ○ A 0.4 cm
- ○ B 3.3 cm
- ○ C 6.2 cm
- ○ D 7.5 cm

50. Three towns are connected by straight roads as shown in the figure. Nick is driving from Crestfield to Seaview.

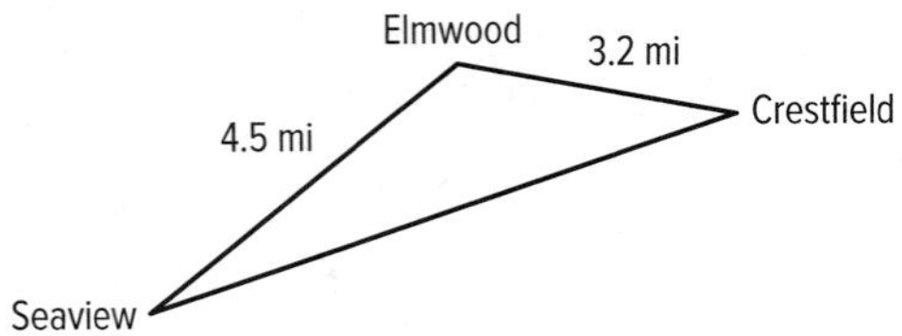

What is the greatest whole number of miles Nick may have to drive? MP 4

- ○ A 2 mi
- ○ B 7 mi
- ○ C 8 mi
- ○ D 15 mi

51. A mountain is 2.8 miles at the base and has one slope that is 1.6 miles. Which is the best description for the length of the other slope? MP 4

- ○ A between 1.2 mi and 4.4 mi
- ○ B between 1.6 mi and 2.8 mi
- ○ C less than 4.4 mi
- ○ D greater than 1.2 mi

52. In $\triangle ABC$, $AB = 6.5$ m and $BC = 8.1$ m. What is the greatest possible whole-number perimeter of $\triangle ABC$, in meters? MP 7

perimeter = ☐

53. The figure shows five straight hiking trails. Selena plans to hike from point T to point R.

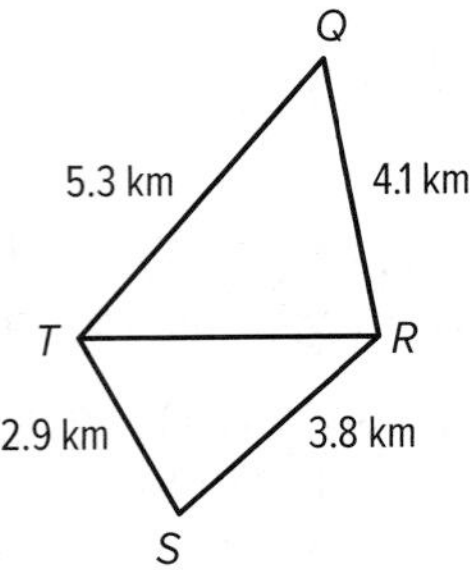

Which of the following are possible distances Selena might hike, assuming she completes the trail? MP 4

- ☐ A 1.2 km
- ☐ B 4.2 km
- ☐ C 5.1 km
- ☐ D 6.7 km
- ☐ E 9.1 km
- ☐ F 9.4 km

54. **MULTI-STEP** Naomi has some wooden dowels that she wants to glue together to form a triangular picture frame. The lengths of some of the dowels are shown in the table. MP 4

Dowel	Length (cm)
P	18.5
Q	20.6
R	31.1
S	40.8

a. Can Naomi make the frame using dowels Q, R, and S? Explain.

b. Can Naomi make the frame using dowels P, Q, and S? Explain.

c. Naomi decides to use dowels Q and R and a third dowel that is not shown in the table. What is the greatest possible whole-number length for the third dowel, in centimeters?

d. How many different frames can Naomi make with the dowels in the table?

LESSON 6

Inequalities in Two Triangles

Then

- You used inequalities to make comparisons in one triangle.

Now

- 1 Apply the Hinge Theorem or its converse to make comparisons in two triangles.
- 2 Prove triangle relationships using the Hinge Theorem or its converse.

Why?

- **A car jack is used to lift a car. The jack shown below is one of the simplest still in use today. Notice that as the jack is lowered, the legs of isosceles $\triangle ABC$ remain congruent, but the included angle A widens and $\overline{BC}$, the side opposite $\angle A$, lengthens.**

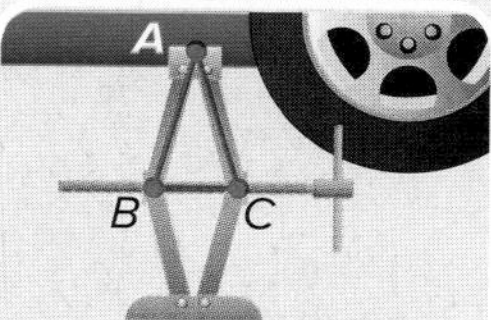

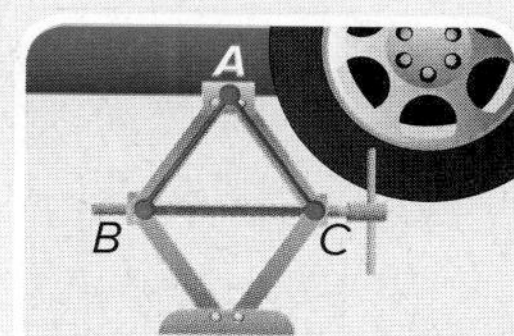

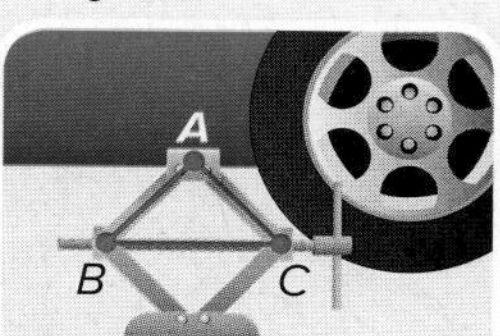

MP Mathematical Practices

3 Construct viable arguments and critique the reasoning of others.

1 Make sense of problems and persevere in solving them.

1 Hinge Theorem

The observation in the example above is true of any type of triangle and illustrates the following theorems.

Theorems Inequalities in Two Triangles

5.12 Hinge Theorem If two sides of a triangle are congruent to two sides of another triangle, and the included angle of the first is larger than the included angle of the second triangle, then the third side of the first triangle is longer than the third side of the second triangle.

Example: If $\overline{AB} \cong \overline{FG}$, $\overline{AC} \cong \overline{FH}$, and $m\angle A > m\angle F$, then $BC > GH$.

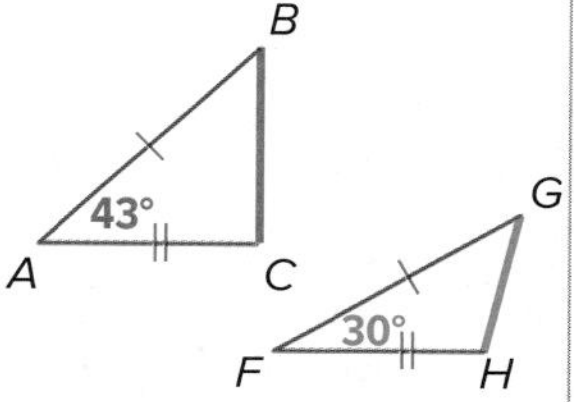

5.13 Converse of the Hinge Theorem If two sides of a triangle are congruent to two sides of another triangle, and the third side in the first is longer than the third side in the second triangle, then the included angle measure of the first triangle is greater than the included angle measure in the second triangle.

Example: If $\overline{JL} \cong \overline{PR}$, $\overline{KL} \cong \overline{QR}$, and $PQ > JK$, then $m\angle R > m\angle L$.

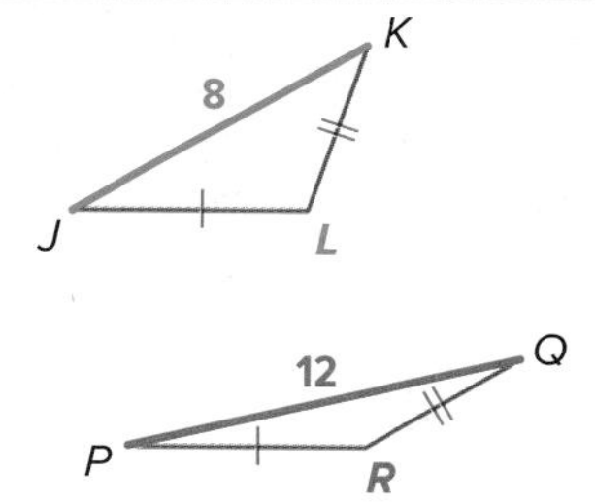

The proof of Theorem 5.12 is on p. 402. You will prove Theorem 5.13 in Exercise 28.

Example 1 Use the Hinge Theorem and its Converse

Compare the given measures.

a. WX **and** XY

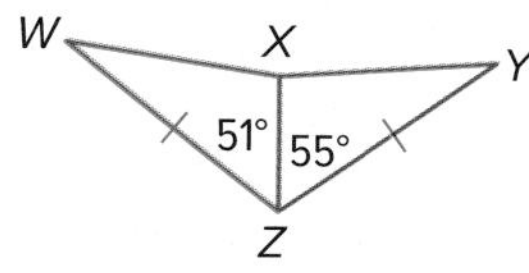

In $\triangle WXZ$ and $\triangle YXZ$, $\overline{WZ} \cong \overline{YZ}$, $\overline{XZ} \cong \overline{XZ}$, and $\angle YZX > \angle WZX$. By the Hinge Theorem, $m\angle WZX < m\angle YZX$, so $WX < XY$.

b. $m\angle FCD$ **and** $m\angle BFC$

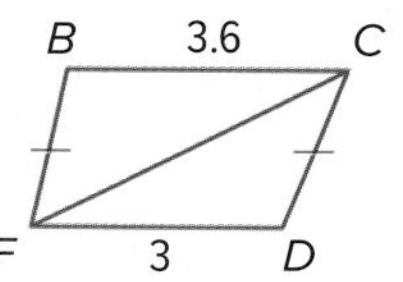

In $\triangle BCF$ and $\triangle DFC$, $\overline{BF} \cong \overline{DC}$, $\overline{FC} \cong \overline{CF}$, and $BC > FD$. By the Converse of the Hinge Theorem, $\angle BFC > \angle DCF$.

Go Online!

Have a question? Send a message to your teacher in ConnectED.

Guided Practice

Compare the given measures.

1A. JK and MQ

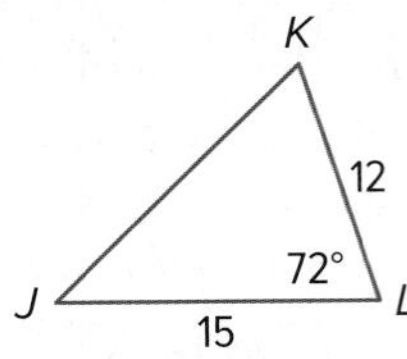

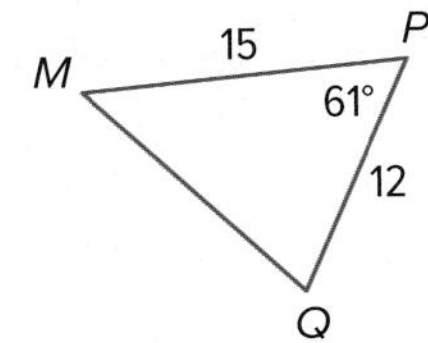

1B. $m\angle SRT$ and $m\angle VRT$

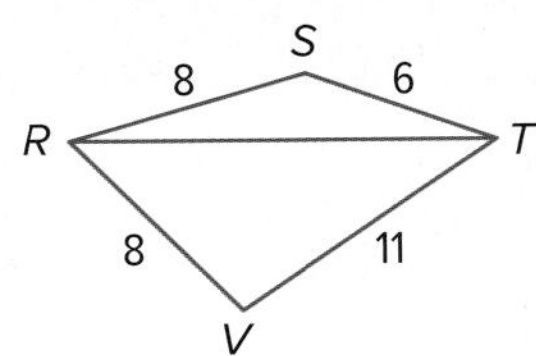

Proof Hinge Theorem

Given: $\triangle ABC$ and $\triangle DEF$, $\overline{AC} \cong \overline{DF}$, $\overline{BC} \cong \overline{EF}$ $m\angle F > m\angle C$

Prove: $DE > AB$

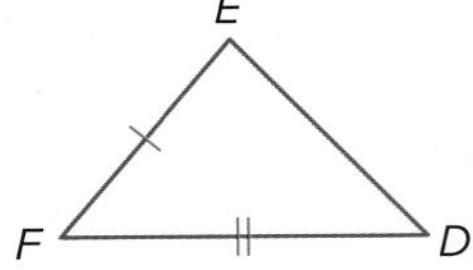

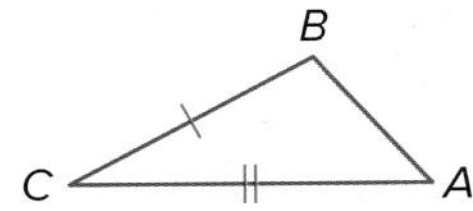

Proof:

We are given that $\overline{AC} \cong \overline{DF}$ and $\overline{BC} \cong \overline{EF}$. We also know that $m\angle F > m\angle C$.

Draw auxiliary ray FP such that $m\angle DFP = m\angle C$ and that $\overline{PF} \cong \overline{BC}$. This leads to two cases.

Case 1 P lies on $\overline{DE}$.

Then $\triangle FPD \cong \triangle CBA$ by SAS. Thus, $PD = BA$ by CPCTC and the definition of congruent segments.

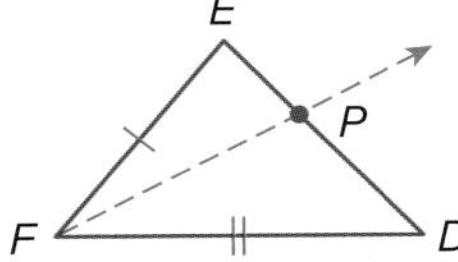

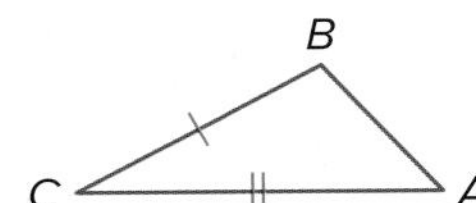

By the Segment Addition Postulate, $DE = EP + PD$. Also, $DE > PD$ by the definition of inequality. Therefore, $DE > AB$ by substitution.

Case 2 P does not lie on $\overline{DE}$.

Then let the intersection of $\overline{FP}$ and $\overline{ED}$ be point T, and draw another auxiliary segment $\overline{FQ}$ such that Q is on $\overline{DE}$ and $\angle EFQ \cong \angle QFP$. Then draw auxiliary segments $\overline{PD}$ and $\overline{PQ}$.

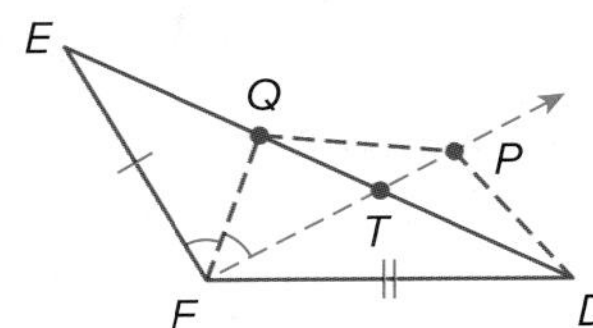

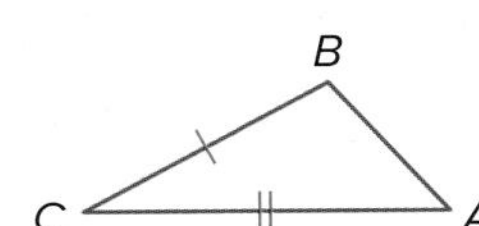

Because $\overline{FP} \cong \overline{BC}$ and $\overline{BC} \cong \overline{EF}$, we have $\overline{FP} \cong \overline{EF}$ by the Transitive Property. Also $\overline{QF}$ is congruent to itself by the Reflexive Property. Thus, $\triangle EFQ \cong \triangle PFQ$ by SAS. By CPCTC, $\overline{EQ} \cong \overline{PQ}$ or $EQ = PQ$. Also, $\triangle FPD \cong \triangle CBA$ by SAS. So, $\overline{PD} \cong \overline{BA}$ by CPCTC and $PD = BA$.

In $\triangle QPD$, $QD + PQ > PD$ by the Triangle Inequality Theorem. By substitution, $QD + EQ > \text{PD}$. Since $ED = QD + EQ$ by the Segment Addition Postulate, $ED > PD$. Using substitution, $ED > BA$ or $DE > AB$.

Study Tip

MP Sense-Making The Hinge Theorem is also called the SAS Inequality Theorem. The Converse of the Hinge Theorem is also called the SSS Inequality Theorem.

You can use the Hinge Theorem to solve real-world problems.

Real-World Example 2 Use the Hinge Theorem

PAINTBALL Two paintball teams leave from the same base camp. Team A goes 7.5 miles due west and then turns 35° north of west and goes 5 miles. Team B goes 7.5 miles due east and then turns 40° north of east and goes 5 miles. At this point, which team is farther from the base camp? Explain your reasoning.

Analyze Using the sets of directions given in the problem, you need to determine which paintball team is farther from base camp. A turn of 35 north of west is correctly interpreted as shown.

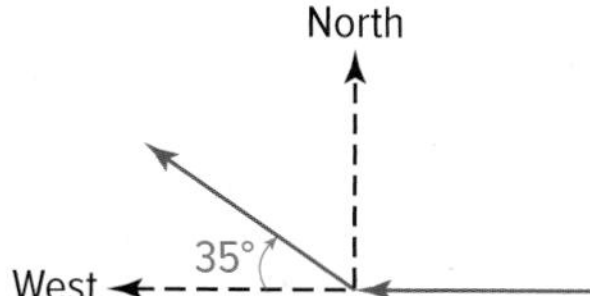

Plan Draw a diagram of the situation.

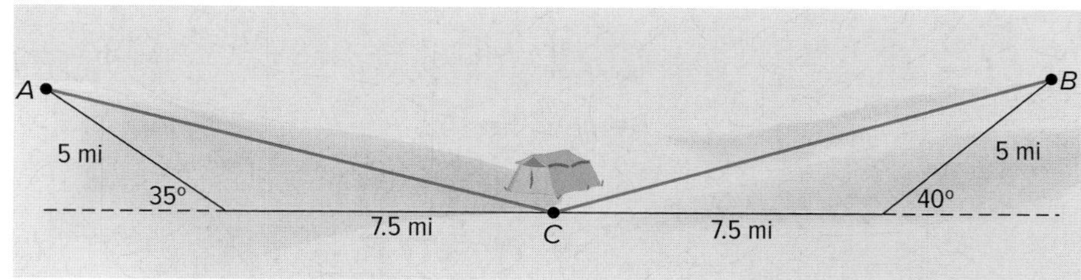

The paths taken by each team and the straight-line distance back to the camp form two triangles. Each team goes 7.5 miles and then turns and goes 5 miles.

Use linear pairs to find the measures of the included angles. Then apply the Hinge Theorem to compare the distance each team is from base camp.

Solve The included angle for the path made by Team A measures 180 − 35 or 145. The included angle for the path made by Team B is 180 − 40 or 140.

Since 145 > 140, $AC > BC$ by the Hinge Theorem. So Team A is farther from the base camp.

Check Team B turned 5° more than Team A did back toward base camp, so they should be closer to the base camp that Team A. Thus, Team A should be farther from the base camp.

Problem-Solving Tip

Draw a Diagram Draw a diagram to help you see and correctly interpret a problem that has been described in words.

Guided Practice

2A. SKIING Two groups of skiers leave from the same lodge. Group A goes 4 miles due east and then turns 70° north of east and goes 3 miles. Group B goes 4 miles due west and then turns 75° north of west and goes 3 miles. At this point, which group is *farther* from the lodge? Explain your reasoning.

2B. SKIING In problem 2A, suppose Group A instead went 4 miles west and then turned 45° north of west and traveled 3 miles. Which group would be *closer* to the lodge? Explain your reasoning.

When the included angle of one triangle is greater than the included angle in a second triangle, the Converse of the Hinge Theorem is used.

Study Tip

Using Additional Facts When finding a range for the possible values for *x*, you may need to use one of the following facts.

- The measure of any angle is always greater than 0 and less than 180.
- The measure of any segment is always greater than 0.

Example 3 Apply Algebra to the Relationships in Triangles

ALGEBRA Find the range of possible values for x.

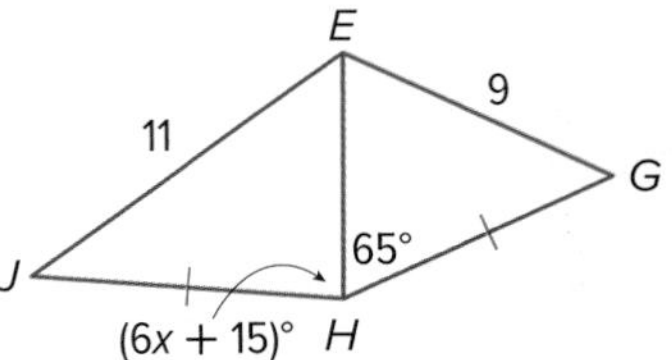

Step 1 From the diagram, we know that $\overline{JH} \cong \overline{GH}$, $\overline{EH} \cong \overline{EH}$, and $JE > EG$.

$m\angle JHE > m\angle EHG$	Converse of the Hinge Theorem
$6x + 15 > 65$	Substitution
$x > 8\frac{1}{3}$	Solve for x.

Step 2 Use the fact that the measure of any angle in a triangle is less than 180 to write a second inequality.

$m\angle JHE < 180$	
$6x + 15 < 180$	Substitution
$x < 27.5$	Solve for x.

Step 3 Write $x > 8\frac{1}{3}$ and $x < 27.5$ as the compound inequality $8\frac{1}{3} < x < 27.5$.

Guided Practice

3. Find the range of possible values for x.

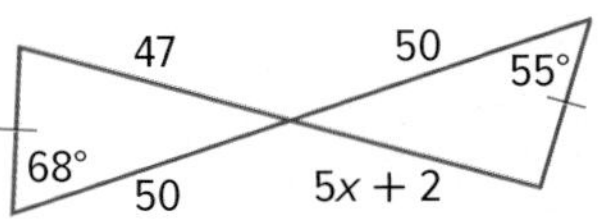

2 Prove Relationships In Two Triangles

You can use the Hinge Theorem and its converse to prove relationships in two triangles.

Example 4 Prove Triangle Relationships Using the Hinge Theorem

Write a two-column proof.

Given: $\overline{AB} \cong \overline{AD}$

Prove: $EB > ED$

Proof:

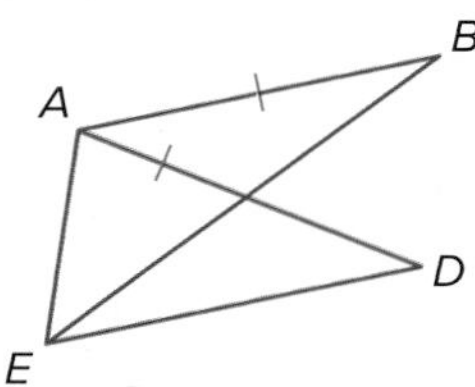

Statements	Reasons
1. $\overline{AB} \cong \overline{AD}$	**1.** Given
2. $\overline{AE} \cong \overline{AE}$	**2.** Reflexive Property
3. $m\angle EAB = m\angle EAD + m\angle DAB$	**3.** Angle Addition Postulate
4. $m\angle EAB > m\angle EAD$	**4.** Definition of Inequality
5. $EB > ED$	**5.** Hinge Theorem

Guided Practice

4. Write a two-column proof.

Given: $\overline{RQ} \cong \overline{ST}$

Prove: $RS > TQ$

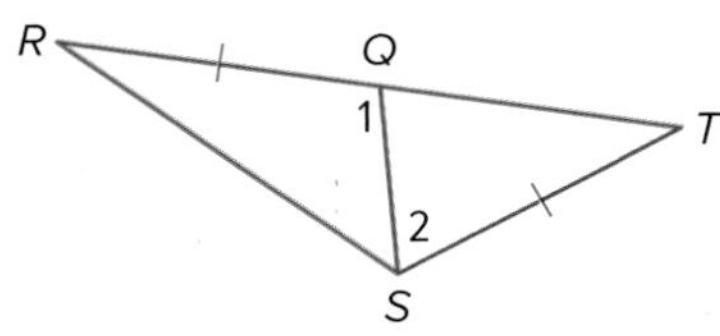

Example 5 Prove Relationships Using Converse of the Hinge Theorem

Write a flow proof.

Given: T is the midpoint of $\overline{ZX}$.
$\overline{ST} \cong \overline{WT}$
$SZ > WX$

Prove: $m\angle XTR > m\angle ZTY$

Flow Proof:

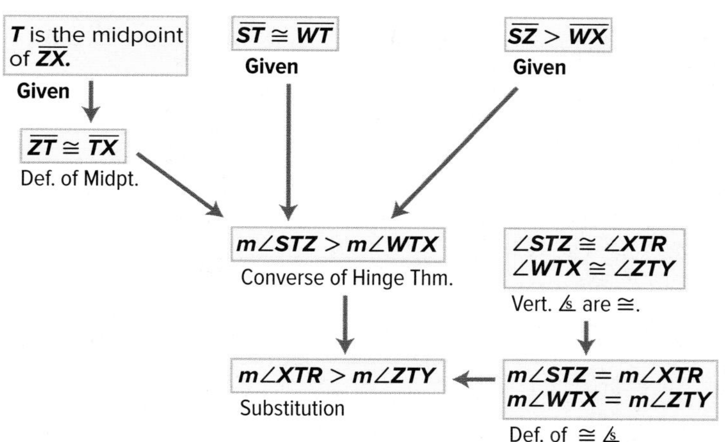

Guided Practice

5. Write a two-column proof.

Given: $\overline{NK}$ is a median of $\triangle JMN$.
$JN > NM$

Prove: $m\angle 1 > m\angle 2$

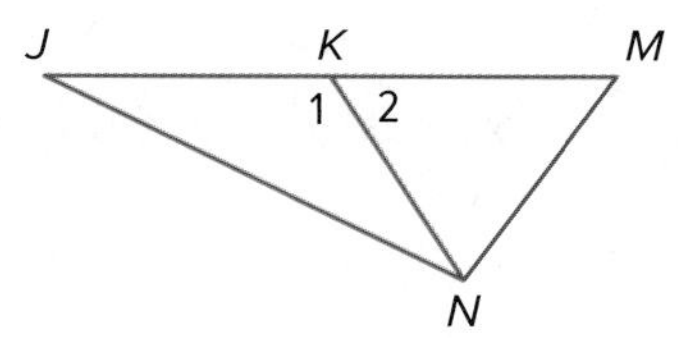

Check Your Understanding

 = Step-by-Step Solutions begin on page R13.

Example 1 **Compare the given measures.**

1. $m\angle ACB$ and $m\angle GDE$

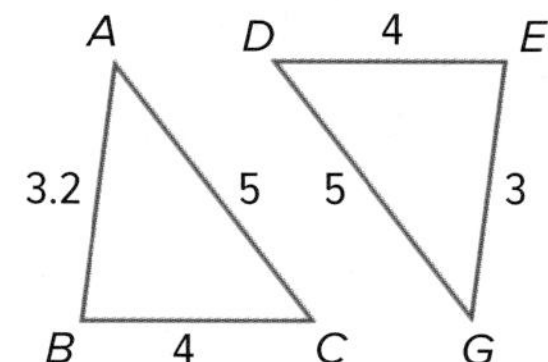

2. JL and KM

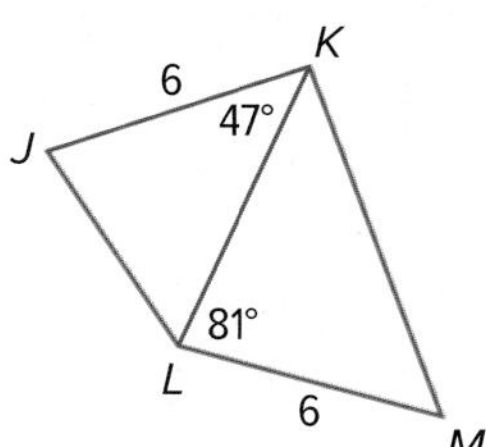

3 QT and ST

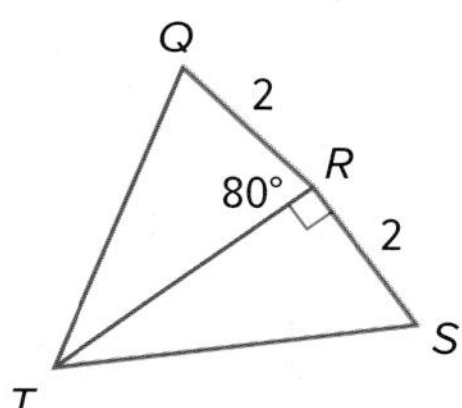

4. $m\angle XWZ$ and $m\angle YZW$

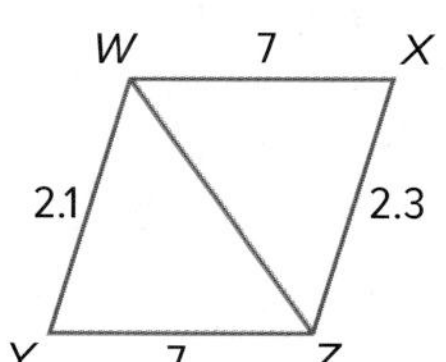

Example 2

5. **SWINGS** The position of the swing changes based on how hard the swing is pushed.

   a. Which pairs of segments are congruent?

   b. Is the measure of $\angle A$ or the measure of $\angle D$ greater? Explain.

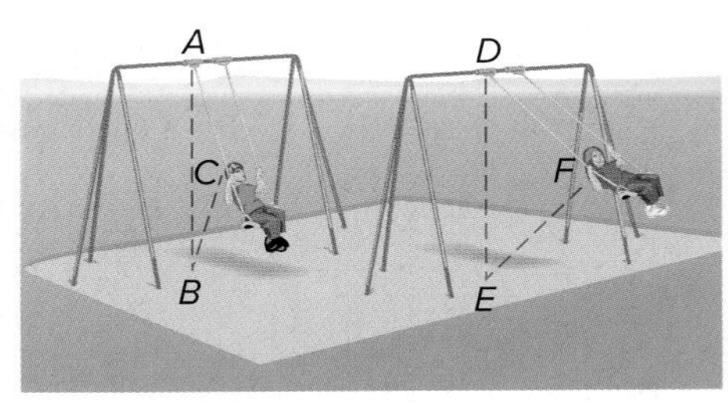

Example 3

Find the range of possible values for x.

6. 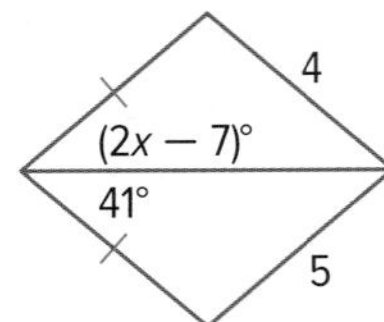

7. 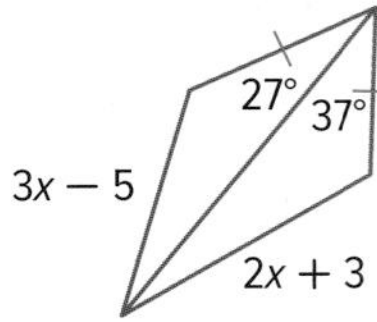

Examples 4–5

MP **CONSTRUCT ARGUMENTS** **Write a two-column proof.**

8. **Given:** $\triangle YZX$
   $\overline{YZ} \cong \overline{XW}$

   Prove: $ZX > YW$

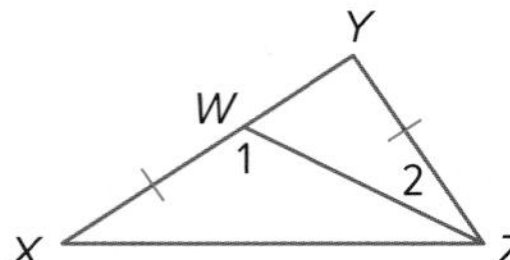

9. **Given:** $\overline{AD} \cong \overline{CB}$
   $DC < AB$

   Prove: $m\angle CBD < m\angle ADB$

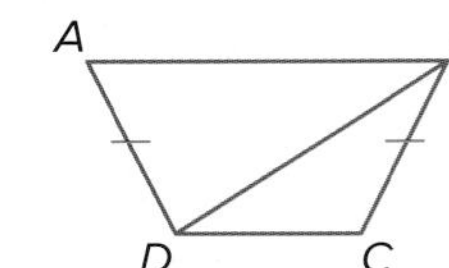

Practice and Problem Solving

Extra Practice is on page R5.

Example 1

Compare the given measures.

10. $m\angle BAC$ and $m\angle DGE$

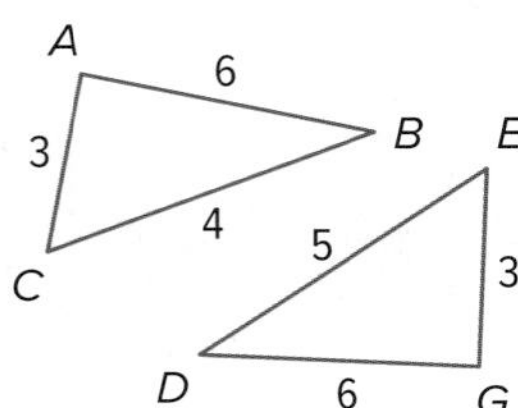

11. $m\angle MLP$ and $m\angle TSR$

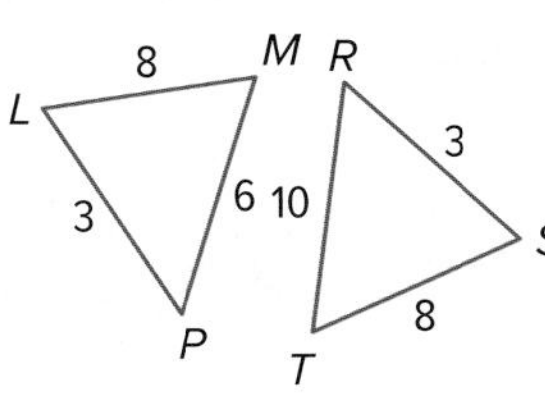

12. SR and XY

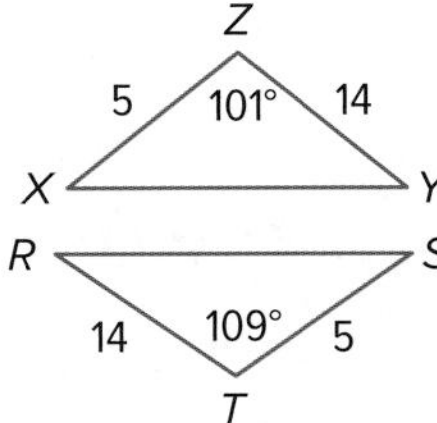

13. $m\angle TUW$ and $m\angle VUW$

T, U, 11, W, 12, V

14. PS and SR

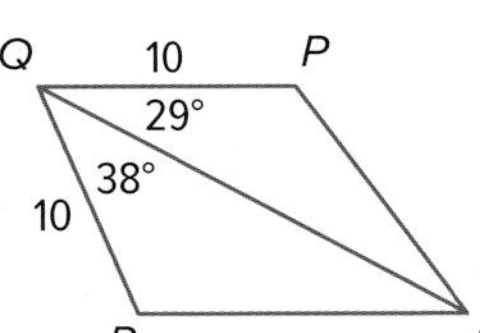

15. JK and HJ

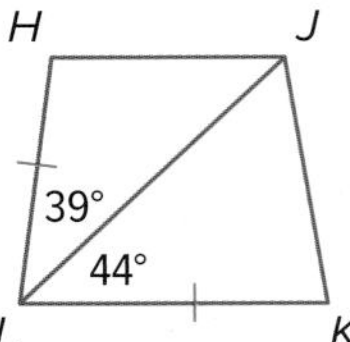

Example 2

16. **CAMPING** Pedro and Joel are camping in a national park. One morning, Pedro decides to hike to the waterfall. He leaves camp and goes 5 miles east then turns 15° south of east and goes 2 more miles. Joel leaves the camp and travels 5 miles west, then turns 35° north of west and goes 2 miles to the lake for a swim.

    a. When they reach their destinations, who is closer to the camp? Explain your reasoning. Include a diagram.

    b. Suppose instead of turning 35° north of west, Joel turned 10° south of west. Who would then be farther from the camp? Explain your reasoning. Include a diagram.

Example 3 **Find the range of possible values for x.**

17.

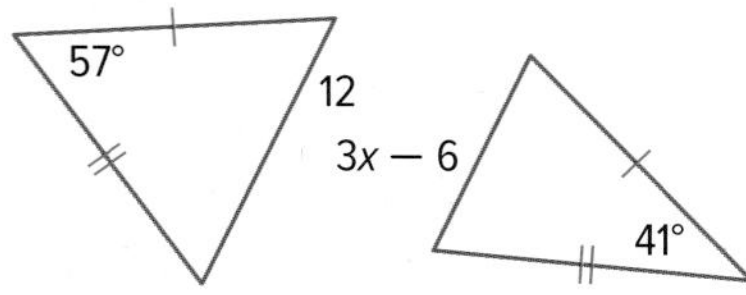

18.

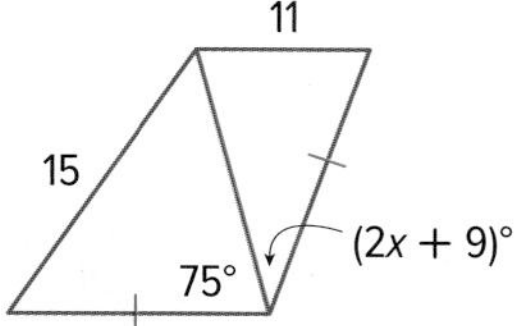

19

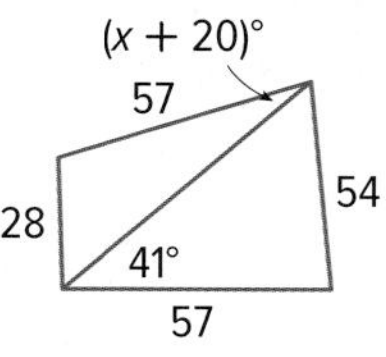

20.

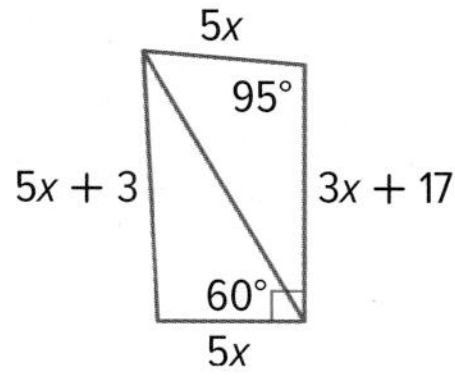

21. CRANES In the diagram, a crane is shown lifting an object to two different heights. The length of the crane's arm is fixed, and $\overline{MP} \cong \overline{RT}$. Is $\overline{MN}$ or $\overline{RS}$ shorter? Explain your reasoning.

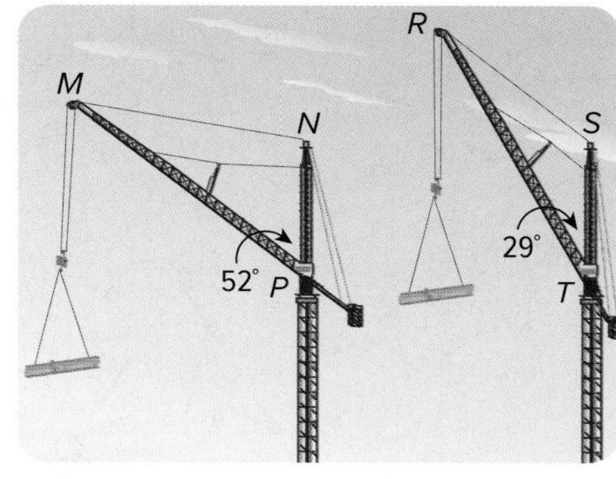

22. LOCKERS Neva and Shawn both have their lockers open as shown in the diagram. Whose locker hinge forms a larger angle? Explain your reasoning.

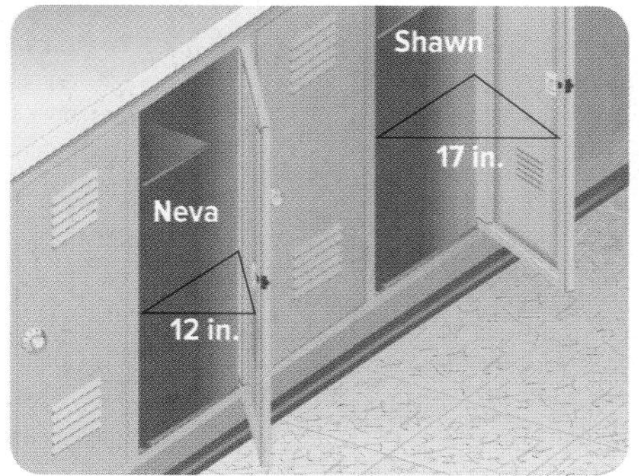

Examples 4–5 **MP CONSTRUCT ARGUMENTS** **Write a two-column proof.**

23. Given: $\overline{LK} \cong \overline{JK}$, $\overline{RL} \cong \overline{RJ}$
K is the midpoint of $\overline{QS}$.
$m\angle SKL > m\angle QKJ$

Prove: $RS > QR$

24. Given: $\overline{VR} \cong \overline{RT}$, $\overline{WV} \cong \overline{WT}$
$m\angle SRV > m\angle QRT$
R is the midpoint of $\overline{SQ}$.

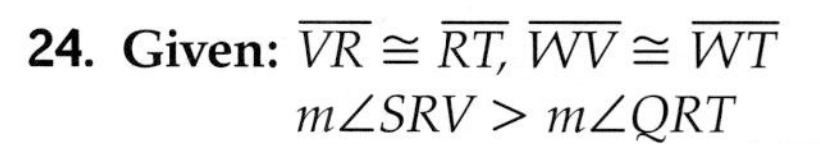

Prove: $WS > WQ$

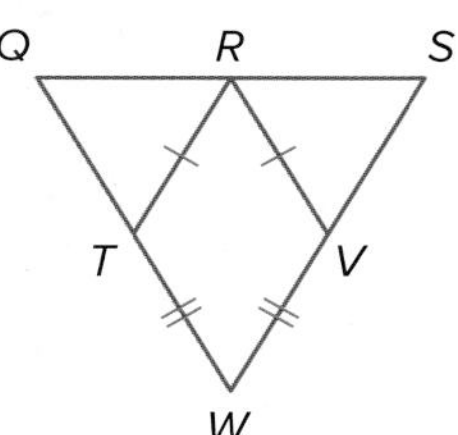

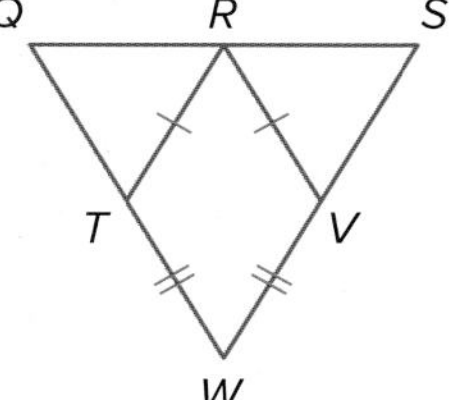

25. Given: $\overline{XU} \cong \overline{VW}$, $VW > XW$
$\overline{XU} \parallel \overline{VW}$

Prove: $m\angle XZU > m\angle UZV$

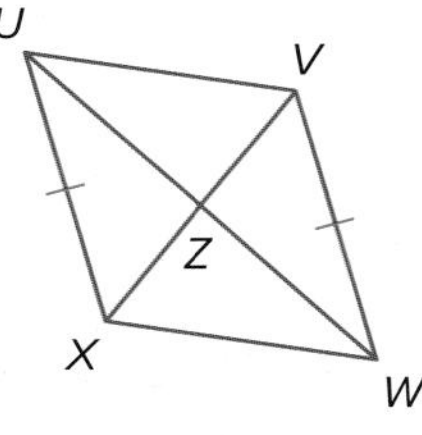

26. Given: $\overline{AF} \cong \overline{DJ}$, $\overline{FC} \cong \overline{JB}$
$AB > DC$

Prove: $m\angle AFC > m\angle DJB$

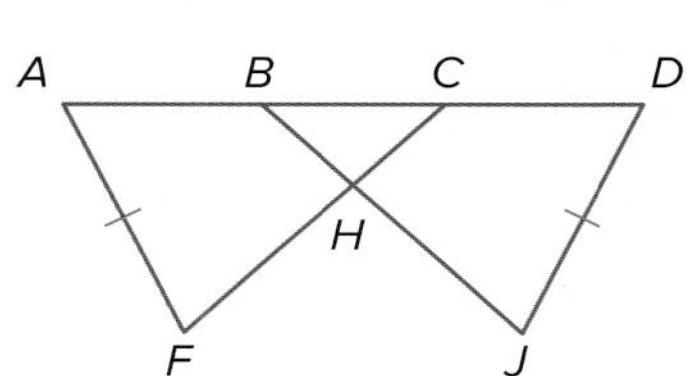

27. **EXERCISE** Anica is doing knee-supported bicep curls as part of her strength training.

a. Is the distance from Anica's fist to her shoulder greater in Position 1 or Position 2? Justify your answer using measurement.

b. Is the measure of the angle formed by Anica's elbow greater in Position 1 or Position 2? Explain your reasoning.

28. **PROOF** Use an indirect proof to prove the Converse of the Hinge Theorem (Theorem 5.13).

Given: $\overline{RS} \cong \overline{UW}$
$\overline{ST} \cong \overline{WV}$
$RT > UV$

Prove: $m\angle S > m\angle W$

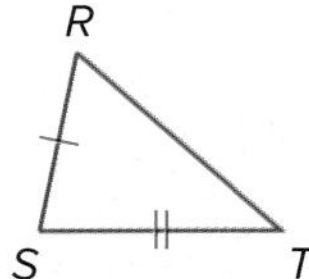

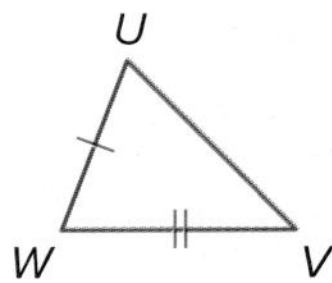

29. **PROOF** If $\overline{PR} \cong \overline{PQ}$ and $SQ > SR$, write a two-column proof to prove $m\angle 1 < m\angle 2$.

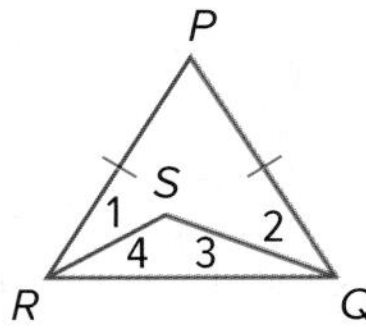

30. **SCAVENGER HUNT** Stephanie, Mario, Lee, and Luther are participating in a scavenger hunt as part of a geography lesson. Their map shows that the next clue is 50 feet due east and then 75 feet 35° east of north starting from the fountain in the school courtyard. When they get ready to turn and go 75 feet 35° east of north, they disagree about which way to go, so they split up and take the paths shown in the diagram below.

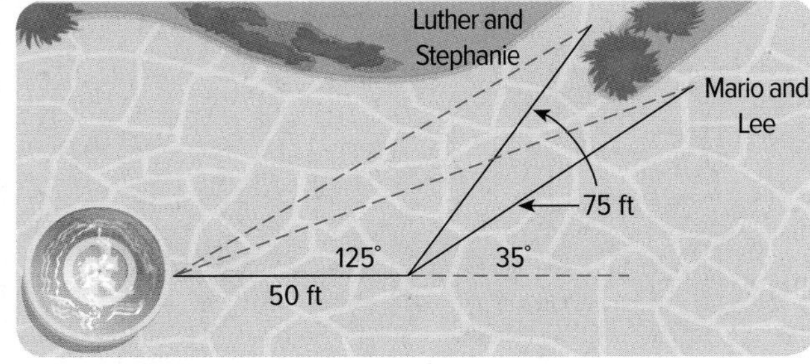

a. Which pair chose the correct path? Explain your reasoning.

b. Which pair is closest to the fountain when they stop? Explain your reasoning.

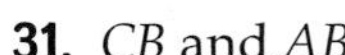

SENSE-MAKING Use the figure at the right to write an inequality relating the given pair of angle or segment measures.

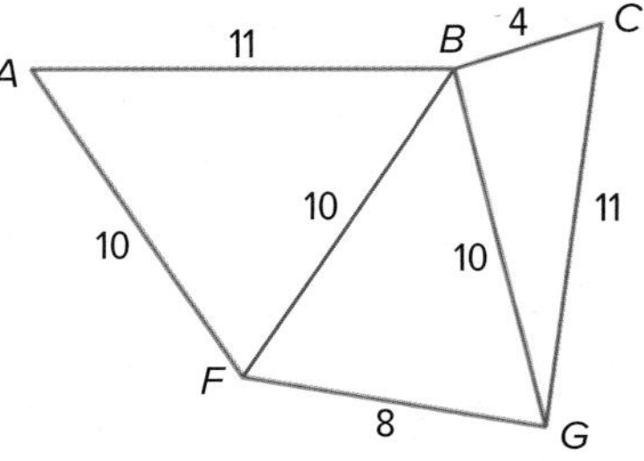

31. CB and AB

32. $m\angle FBG$ and $m\angle BFA$

33. $m\angle BGC$ and $m\angle FBA$

Use the figure at the right to write an inequality relating the given pair of angles or segment measures.

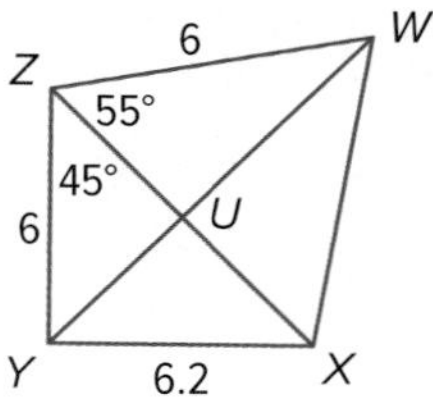

34. $m\angle ZUY$ and $m\angle ZUW$

35. WU and YU

36. WX and XY

37. MULTIPLE REPRESENTATIONS In this problem, you will investigate properties of polygons.

a. Geometric Draw a three-sided, a four-sided, and a five-sided polygon. Label the three-sided polygon ABC, the four-sided polygon $FGHJ$, and the five-sided polygon $PQRST$. Use a protractor to measure and label each angle.

b. Tabular Copy and complete the table below.

Number of sides	Angle Measures				Sum of Angles
3	$m\angle A$		$m\angle C$		
	$m\angle B$				
4	$m\angle F$		$m\angle H$		
	$m\angle G$		$m\angle J$		
5	$m\angle P$		$m\angle S$		
	$m\angle Q$		$m\angle T$		
	$m\angle R$				

c. Verbal Make a conjecture about the relationship between the number of sides of a polygon and the sum of the measures of the angles of the polygon.

d. Logical What type of reasoning did you use in part **c**? Explain.

e. Algebraic Write an algebraic expression for the sum of the measures of the angles for a polygon with n sides.

H.O.T. Problems Use Higher-Order Thinking Skills

38. MP **CONSTRUCT ARGUMENTS** If $m\angle LJN > m\angle KJL$, $\overline{KJ} \cong \overline{JN}$, and $\overline{JN} \perp \overline{NL}$, which is greater, $m\angle LKN$ or $m\angle LNK$? Explain your reasoning.

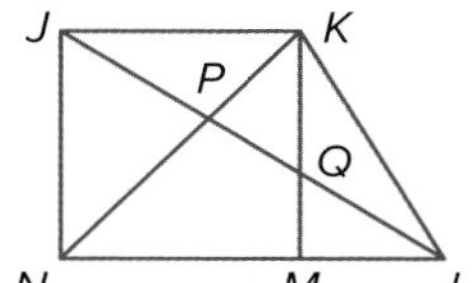

39. MP **REASONING** Give a real-world example of an object that uses a hinge. Draw two sketches in which the hinge on your object is adjusted to two different positions. Use your sketches to explain why Theorem 5.13 is called the Hinge Theorem.

40. MP **CONSTRUCT ARGUMENTS** Given $\triangle RST$ with median $\overline{RQ}$, if RT is greater than or equal to RS, what are the possible classifications of $\triangle RQT$? Explain your reasoning.

41. MP **REASONING** If $\overline{BD}$ is a median and $AB < BC$, then $\angle BDC$ is *always*, *sometimes*, or *never* an acute angle. Explain.

42. WRITING IN MATH Compare and contrast the Hinge Theorem to the SAS Postulate for triangle congruence.

43. Which of the following is a correct inequality based on the information given in the figure?
MP 7

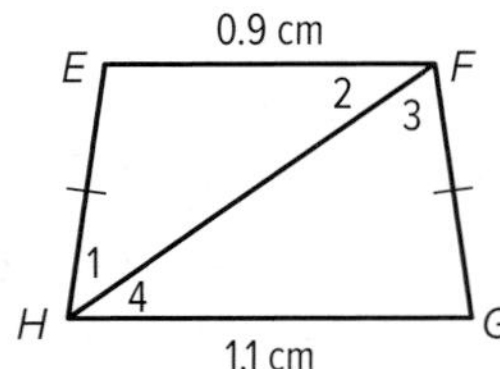

○ **A** $m\angle 3 > m\angle 1$

○ **B** $m\angle 2 > m\angle 3$

○ **C** $m\angle 4 > m\angle 1$

○ **D** $m\angle 1 > m\angle 3$

44. In the figure, $\overline{RS} \cong \overline{RU}$.

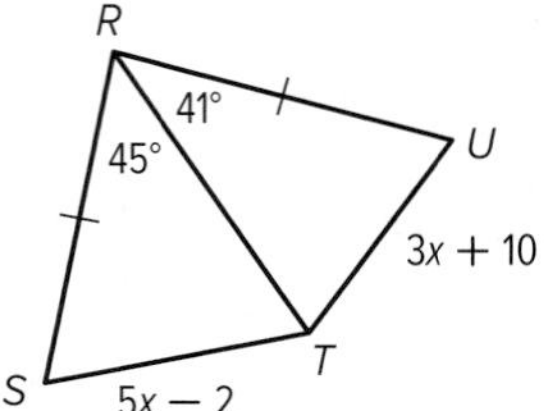

Which of the following inequalities best describes the possible values of x? MP 7

○ **A** $x > 4$

○ **B** $x < 6$

○ **C** $x > 6$

○ **D** $x > 1$

45. In $\triangle ABC$, $\triangle JKL$, and $\triangle MNP$, $\overline{AB} \cong \overline{JK} \cong \overline{MN}$, $\overline{BC} \cong \overline{KL} \cong \overline{NP}$, and $m\angle B < m\angle K < m\angle N$. Also, $AC = 3.5$ centimeters and $MP = 4.7$ centimeters.

Which of the following is a possible length of $\overline{JL}$?
MP 7

○ **A** 3.5 cm

○ **B** 4.2 cm

○ **C** 5.1 cm

○ **D** 8.2 cm

46. Carlos is comparing the plans for two different bicycle frames. The plans are shown below.

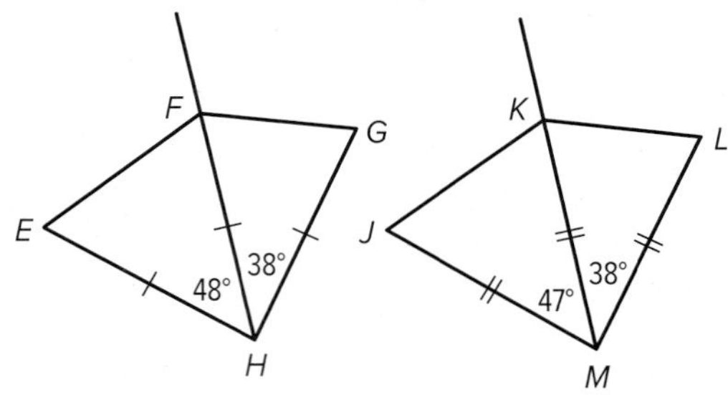

Which of the following statements about the bicycle frames must be true? MP 4

○ **A** $\overline{FG} \cong \overline{KL}$

○ **B** $\triangle FGH \cong \triangle KLM$

○ **C** $EF > JK$

○ **D** $JK > KL$

47. Eric and Heather are each taking a group of campers hiking in the woods. Eric's group leaves camp and goes 2 miles east, then turns 20° south of east and goes 4 more miles. Heather's group leaves camp and travels 2 miles west, then turns 30° north of west and goes 4 more miles. How many degrees south of east would Eric have needed to turn for his group and Heather's group to be the same distance from camp after the two legs of the hike?
MP 4

48. **MULTI-STEP** A set designer is making a sketch of a large spider web that will be painted on the rear wall of a theater. Part of the sketch is shown below.
MP 4

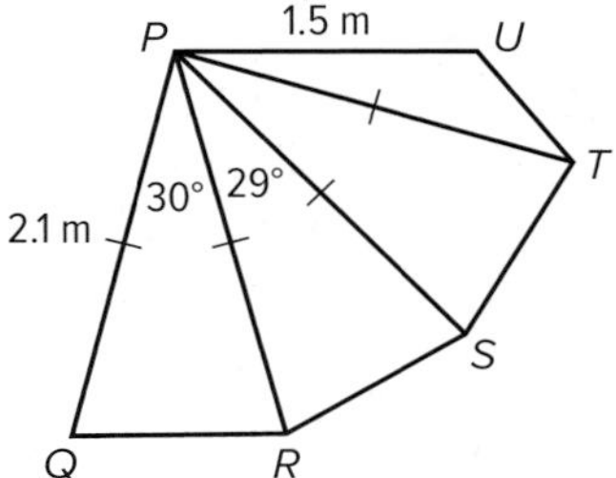

a. What conclusion, if any, can the designer make about $\overline{QR}$ and $\overline{RS}$? Explain.

b. The designer makes the sketch so that $RS > TS$. What conclusion, if any, can she make about $m\angle TPS$? Explain.

c. The designer states that $UT < TS$ by the Hinge Theorem. Do you agree or disagree? Explain.

CHAPTER 5

Study Guide and Review

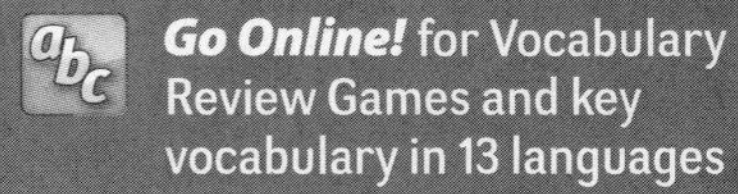

Study Guide

Key Concepts

Special Segments in Triangles (Lessons 5-1 and 5-2)

- The special segments of triangles are perpendicular bisectors, angle bisectors, medians, and altitudes.
- The intersection points of each of the special segments of a triangle are called the points of concurrency.
- The points of concurrency for a triangle are the circumcenter, incenter, centroid, and orthocenter.
- The centroid is two thirds of the distance from each vertex to the midpoint of the opposite side.

Indirect Proof (Lesson 5-4)

- Writing an Indirect Proof:
  1. Assume that the conclusion is false.
  2. Show that this assumption leads to a contradiction.
  3. Since the false conclusion leads to an incorrect statement, the original conclusion must be true.

Triangle Inequalities (Lessons 5-3, 5-5, and 5-6)

- The largest angle in a triangle is opposite the longest side, and the smallest angle is opposite the shortest side.
- The sum of the lengths of any two sides of a triangle is greater than the length of the third side.
- **Hinge Theorem** (SAS Inequality): In two triangles, if two sides are congruent, then the measure of the included angle determines which triangle has the longer third side.
- **Converse of Hinge Theorem** (SSS Inequality): In two triangles, if two corresponding sides of each triangle are congruent, then the length of the third side determines which triangle has the included angle with the greater measure.

FOLDABLES® Study Organizer

Use your Foldable to review the chapter. Working with a partner can be helpful. Ask for clarification of concepts as needed.

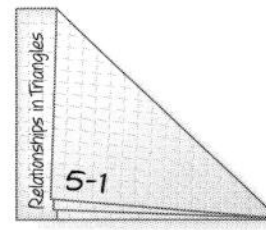

Key Vocabulary

altitude (p. 367)
centroid (p. 365)
circumcenter (p. 355)
concurrent lines (p. 355)
incenter (p. 358)
indirect proof (p. 385)
indirect reasoning (p. 385)
median (p. 365)
orthocenter (p. 367)
perpendicular bisector (p. 354)
point of concurrency (p. 355)
proof by contradiction (p. 385)

Vocabulary Check

State whether each sentence is *true* or *false*. If *false*, replace the underlined term to make a true sentence.

1. The altitudes of a triangle intersect at the centroid.
2. The point of concurrency of the medians of a triangle is called the incenter.
3. The circumcenter of a triangle is equidistant from the vertices of the triangle.
4. To find the centroid of a triangle, first construct the angle bisectors.
5. The perpendicular bisectors of a triangle are concurrent lines.
6. A proof by contradiction uses indirect reasoning.
7. A median of a triangle connects the midpoint of one side of the triangle to the midpoint of another side of the triangle.
8. The incenter is the point at which the angle bisectors of a triangle intersect.

Concept Check

9. Explain how to write a proof by contradiction.
10. Explain how to locate the largest angle in a scalene triangle. Then explain when a triangle does not have one largest angle.

Study Guide and Review *Continued*

Lesson-by-Lesson Review 

5-1 Bisectors of Triangles

11. Find EG if G is the incenter of $\triangle ABC$.

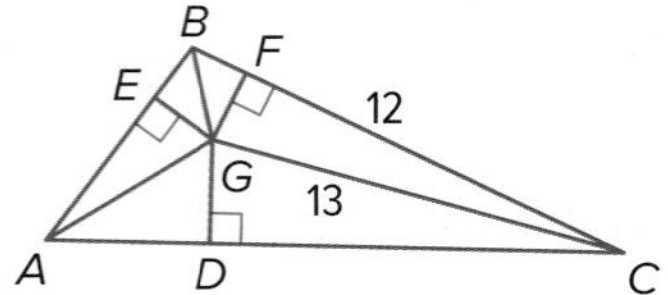

Find each measure.

12. RS

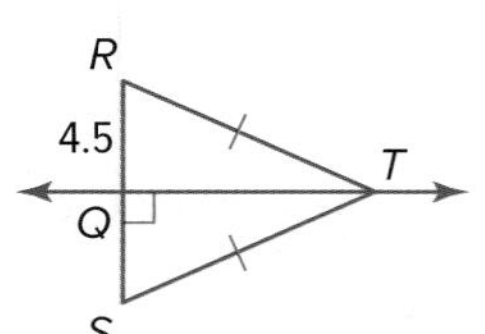

13. XZ

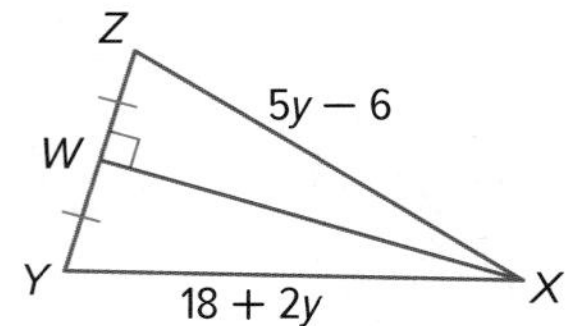

14. BASEBALL Jackson, Trevor, and Scott are warming up before a baseball game. One of their warm-up drills requires three players to form a triangle, with one player in the middle. Where should the fourth player stand so that he is the same distance from the other three players?

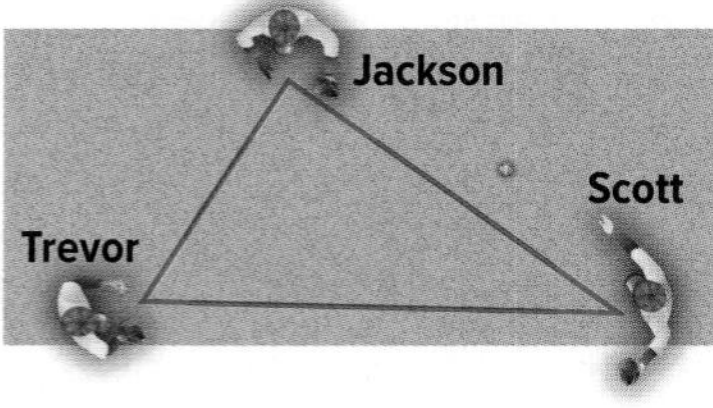

Example 1

Find each measure if Q is the incenter of $\triangle JKL$.

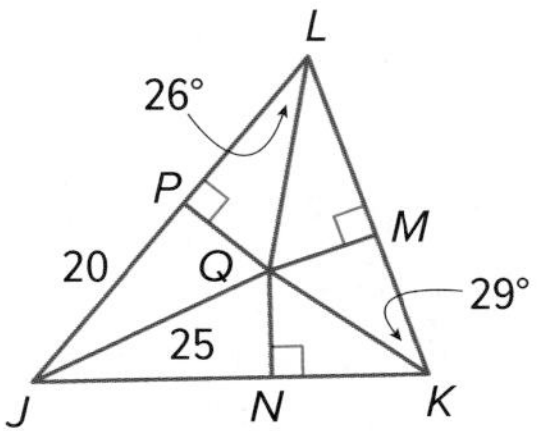

a. $\angle QJK$

$m\angle KLP + m\angle MKN + m\angle NJP = 180$ — $\triangle$ Sum Theorem

$2(26) + 2(29) + m\angle NJP = 180$ — Substitution

$110 + m\angle NJP = 180$ — Simplify.

$m\angle NJP = 70$ — Subtract.

Because $\overrightarrow{JQ}$ bisects $\angle NJP$, $2m\angle QJK = m\angle NJP$. So, $m\angle QJK = \frac{1}{2}m\angle NJP$, so $m\angle QJK = \frac{1}{2}(70)$ or 35.

b. QP

$a^2 + b^2 = c^2$ — Pythagorean Theorem

$(QP)^2 + 20^2 = 25^2$ — Substitution

$(QP)^2 + 400 = 625$ — $20^2 = 400$ and $25^2 = 625$

$(QP)^2 = 225$ — Subtract.

$QP = 15$ — Simplify.

5-2 Medians and Altitudes of Triangles

15. The vertices of $\triangle DEF$ are $D(0, 0)$, $E(0, 7)$, and $F(6, 3)$. Find the coordinates of the orthocenter of $\triangle DEF$.

16. PROM Georgia is on the prom committee. She wants to hang a dozen congruent triangles from the ceiling so that they are parallel to the floor. She sketched out one triangle on a coordinate plane with coordinates (0, 4), (3, 8), and (6, 0). If each triangle is to be hung by one chain, what are the coordinates of the point where the chain should attach to the triangle?

Example 2

In $\triangle EDF$, T is the centroid and $FT = 12$. Find TQ.

$FT = \frac{2}{3}FQ$

$FT = \frac{2}{3}(FT + TQ)$

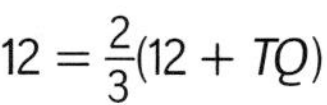

$12 = \frac{2}{3}(12 + TQ)$ — $FT = 12$

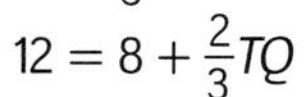

$12 = 8 + \frac{2}{3}TQ$ — Distributive Property

$4 = \frac{2}{3}TQ$ — Subtract.

$6 = TQ$ — Multiply.

5-3 Inequalities in One Triangle

List the angles and sides of each triangle in order from smallest to largest.

17.

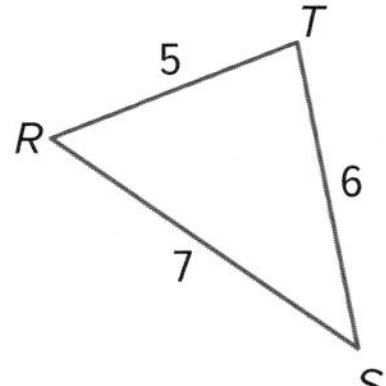

18.

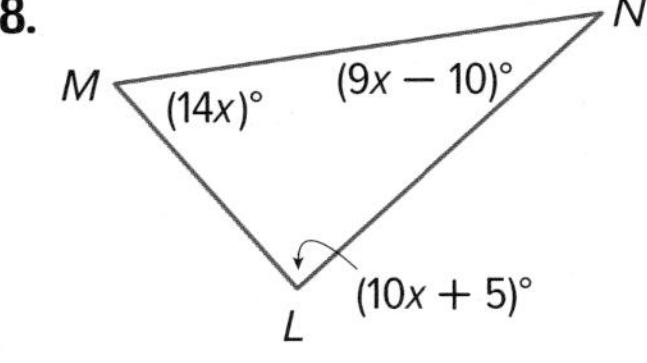

19. NEIGHBORHOODS Anna, Sarah, and Irene live at the intersections of the three roads that make the triangle shown. If the girls want to spend the afternoon together, is it a shorter path for Anna to stop and get Sarah and go on to Irene's house, or for Sarah to stop and get Irene and then go on to Anna's house?

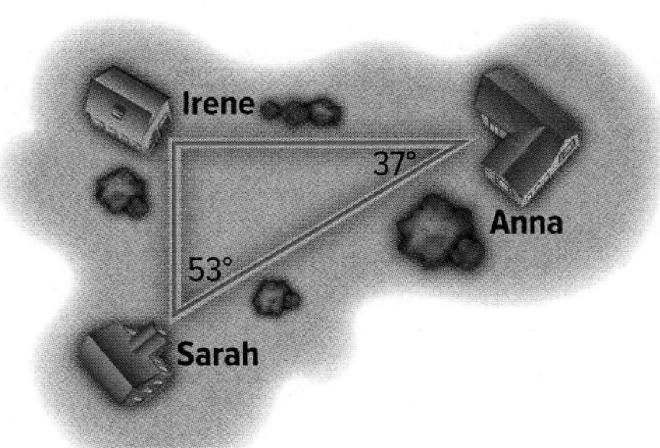

Example 3

List the angles and sides of $\triangle ABC$ in order from smallest to largest.

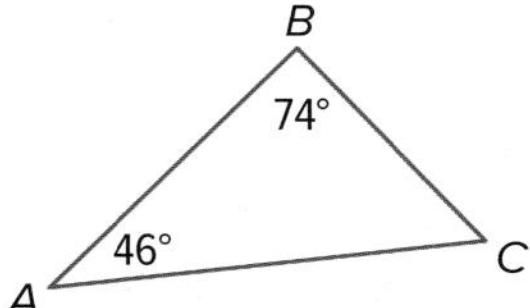

a. First, find the missing angle measure using the Triangle Angle-Sum Theorem.

$m\angle C = 180 - (46 + 74)$ or 60

So, the angles from smallest to largest are $\angle A$, $\angle C$, and $\angle B$.

b. The sides from shortest to longest are $\overline{BC}$, $\overline{AB}$, and $\overline{AC}$.

5-4 Indirect Proof

State the assumption you would make to start an indirect proof of each statement.

20. $m\angle A \geq m\angle B$

21. $\triangle FGH \cong \triangle MNO$

22. $\triangle KLM$ is a right triangle.

23. If $3y < 12$, then $y < 4$.

24. Write an indirect proof to show that if two angles are complementary, neither angle is a right angle.

25. CONCESSIONS Isaac purchased two items at a concession stand and spent over \$10. Use indirect reasoning to show that at least one of the items he purchased was over \$5.

Example 4

State the assumption necessary to start an indirect proof of each statement.

a. $\overline{XY} \not\cong \overline{JK}$

$\overline{XY} \cong \overline{JK}$

b. If $3x < 18$, then $x < 6$.

The conclusion of the conditional statement is $x < 6$. The negation of the conclusion is $x \geq 6$.

c. $\angle 2$ is an acute angle.

If *$\angle 2$ is an acute angle* is false, then *$\angle 2$ is not an acute angle* must be true. This means that *$\angle 2$ is an obtuse or right angle* must be true.

CHAPTER 5

Study Guide and Review *Continued*

5-5 The Triangle Inequality

Is it possible to form a triangle with the given lengths? If not, explain why not.

26. 5, 6, 9

27. 3, 4, 8

Find the range for the measure of the third side of a triangle given the measure of two sides.

28. 5 ft, 7 ft

29. 10.5 cm, 4 cm

30. BIKES Leonard rides his bike to visit Josh. Since High Street is closed, he has to travel 2 miles down Main Street and turn to travel 3 miles farther on 5th Street. If the three streets form a triangle with Leonard and Josh's house as two of the vertices, find the range of the possible distance between Leonard and Josh's houses when traveling straight down High Street.

Example 5

Is it possible to form a triangle with the lengths 7, 10, and 9 feet? If not, explain why not.

Check each inequality.

$7 + 10 > 9$	$7 + 9 > 10$	$10 + 9 > 7$
$17 > 9$ ✓	$16 > 10$ ✓	$19 > 7$ ✓

Because the sum of each pair of side lengths is greater than the third side length, sides with lengths 7, 10, and 9 feet will form a triangle.

5-6 Inequalities of Two Triangles

Compare the given measures.

31. $m\angle ABC$, $m\angle DEF$

32. QT and RS

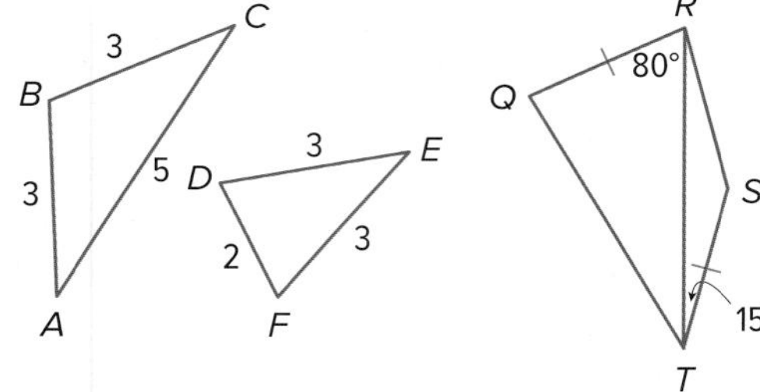

33. BOATING Rose and Connor each row across a pond heading to the same point. Neither of them has rowed a boat before, so they both go off course as shown in the diagram. After two minutes, they have each traveled 50 yards. Who is closer to their destination?

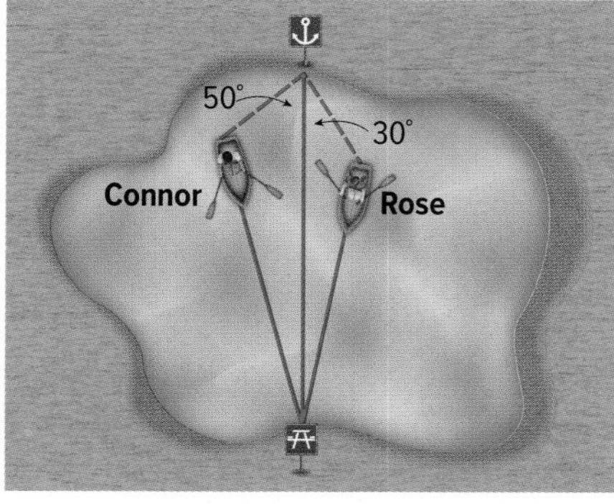

Example 6

Compare the given measures.

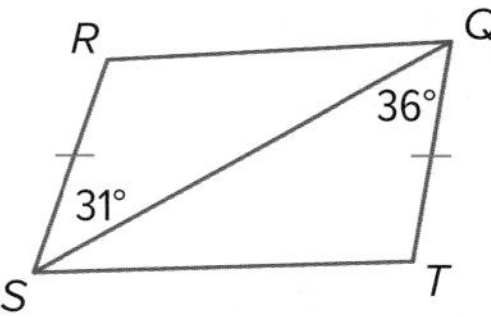

a. ***RQ* and *ST***

In $\triangle QRS$ and $\triangle STQ$, $\overline{RS} \cong \overline{TQ}$, $\overline{QS} \cong \overline{QS}$, and $\angle SQT > \angle RSQ$. By the Hinge Theorem, $m\angle SQT < m\angle RSQ$, so $RQ < ST$.

b. measure of $\angle KML$ and $\angle KMJ$

In $\triangle JKM$ and $\triangle LKM$, $\overline{JM} \cong \overline{LM}$, $\overline{KM} \cong \overline{KM}$, and $LK > JK$. By the Converse of the Hinge Theorem, $\angle KML > \angle KMJ$.

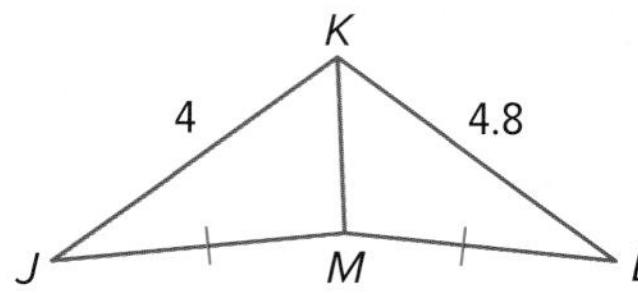

CHAPTER 5

Practice Test

1. **GARDENS** Maggie wants to plant a circular flower bed within a triangular area set off by three pathways. Which point of concurrency related to triangles would she use for the center of the largest circle that would fit inside the triangle?

In $\triangle CDF$, K is the centroid and $DK = 16$. Find each length.

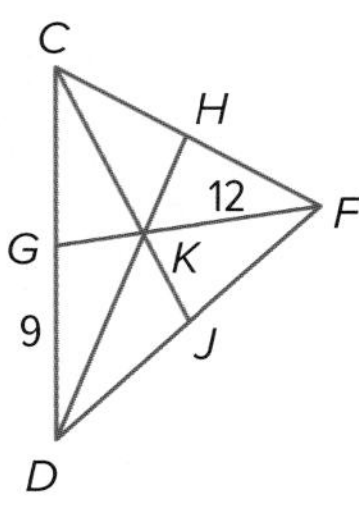

2. KH
3. CD
4. FG

5. **PROOF** Write an indirect proof.

   Given: $5x + 7 \geq 52$

   Prove: $x \geq 9$

Find each measure.

6. $m\angle TQR$

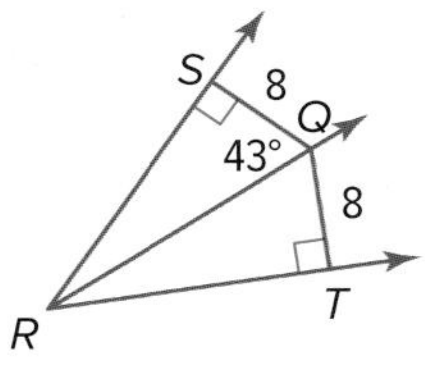

7. XZ

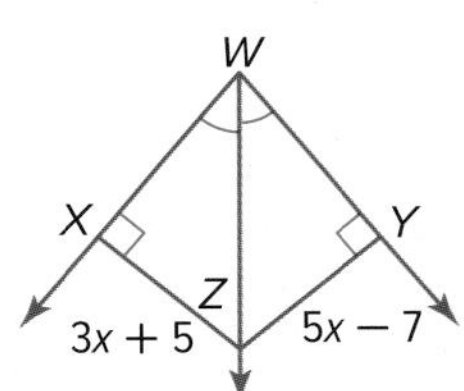

8. **GEOGRAPHY** The distance from Tonopah to Round Mountain is equal to the distance from Tonopah to Warm Springs. The distance from Tonopah to Hawthorne is the same as the distance from Tonopah to Beatty. Determine which distance is greater, Round Mountain to Hawthorne or Warm Springs to Beatty.

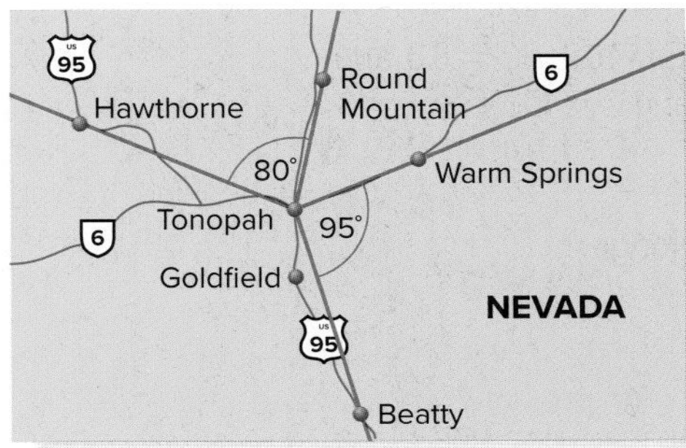

9. **MULTIPLE CHOICE** If the measures of two sides of a triangle are 3.1 feet and 4.6 feet, which is the *least* possible whole number measure for the third side?

   A 1.6 ft
   B 2 ft
   C 7.5 ft
   D 8 ft

Point H is the incenter of $\triangle ABC$. Find each measure.

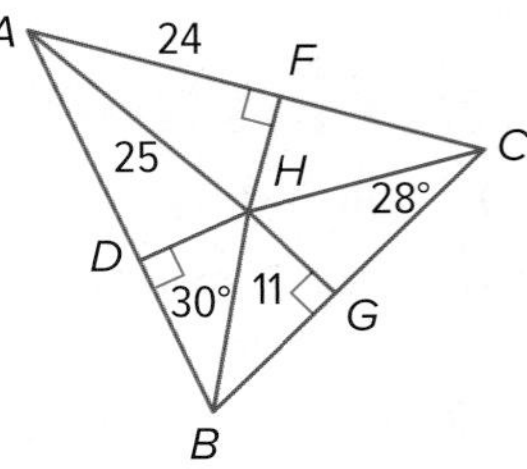

10. DH
11. BD
12. $m\angle HAC$
13. $m\angle DHG$

14. **MULTIPLE CHOICE** If the lengths of two sides of a triangle are 5 and 11, what is the range of possible lengths for the third side?

    A $6 < x < 10$
    B $5 < x < 11$
    C $6 < x < 16$
    D $x < 5$ or $x > 11$

Compare the given measures.

15. AB and BC

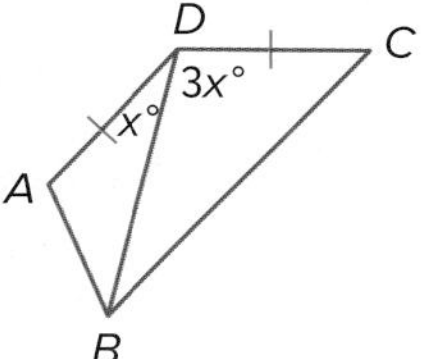

16. $\angle RST$ and $\angle JKL$

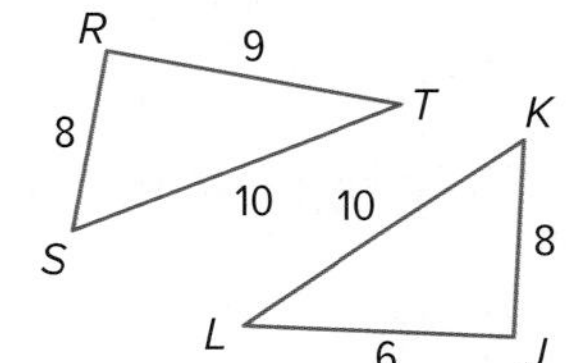

State the assumption necessary to start an indirect proof of each statement.

17. If 8 is a factor of n, then 4 is a factor of n.
18. $m\angle M > m\angle N$
19. If $3a + 7 \leq 28$, then $a \leq 7$.

Use the figure to determine which angle has the greatest measure.

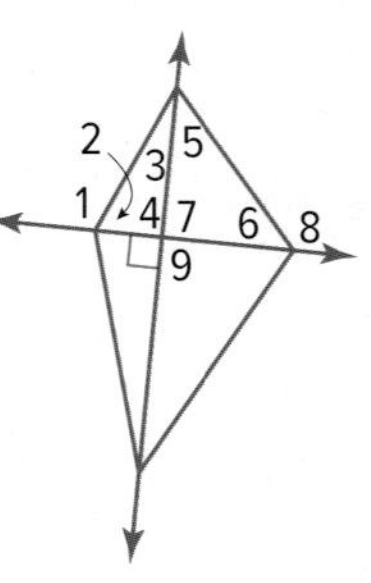

20. $\angle 1, \angle 5, \angle 6$
21. $\angle 9, \angle 8, \angle 3$
22. $\angle 4, \angle 3, \angle 2$

23. **PROOF** Write a two-column proof.

    Given: $\overline{RQ}$ bisects $\angle SRT$.

    Prove: $m\angle SQR > m\angle SRQ$

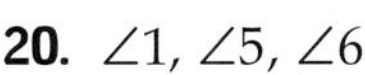

Find the range for the measure of the third side of a triangle given the measures of the two sides.

24. 10 ft, 16 ft
25. 23 m, 39 m

Preparing for Assessment

Performance Task

Provide a clear solution to each part of the task. Be sure to show all of your work, include all relevant drawings, and justify your answers.

DESIGN Juan is an artist who draws geometric abstract art. He is currently working on a series based on triangles. Sometimes he draws the triangles freehand and sometimes he plans them out using math first.

Part A

Juan draws an isosceles triangle *ABC* as shown on the right. In the triangle, $BC = 10$, $AB = 4x - 1$, and $AC = x + 8$. Juan draws a perpendicular bisector down from angle *A* to point *E* on side *BC*.

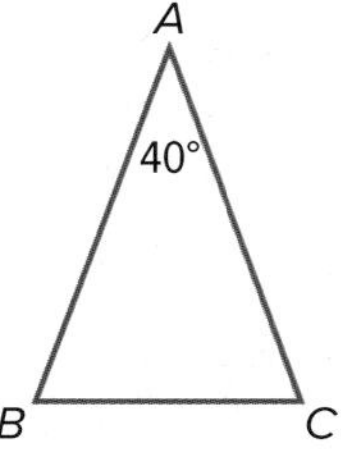

1. Determine the lengths of *BE* and *EC*.
2. Determine the lengths of *AB* and *AC*.
3. Determine the measure of angles *BAE* and *CAE*.
4. **Sense-Making** Determine the measure of angles *B* and *C*.

Part B

Juan draws triangle *LMN* as shown to the right. He wants to divide the triangle into 6 smaller triangles such that each triangle has one vertex that lies directly in the middle of *LMN*. He wants these vertices to be equidistant from each side of *LMN*.

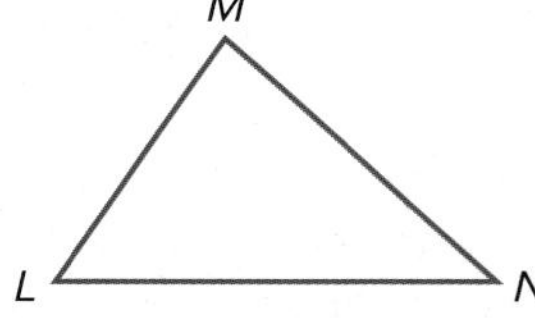

5. Determine whether Juan should draw altitudes, angle bisectors, medians, or perpendicular bisectors to accomplish his goal.
6. Explain where the point of currency would be relative to the incenter if Juan used each of the four segments listed above.

Part C

Juan draws two triangles. Triangle *ABC* has side measures as follows: $AB = 10.2$, $BC = 8.8$, and $AC = 7.6$. Triangle *XYZ* has angle measures as follows: $X = 42°$, $Y = 36°$, and $Z = 102°$.

7. List the angles of triangle *ABC* in order from smallest to greatest.
8. **Reasoning** List the sides of triangle *XYZ* in order from shortest to longest.

Part D

Juan is planning out a drawing. He is deciding between triangles with the following side lengths: 8, 12, 15; 9, 10, 21; and 6, 8, 15.

9. Determine which of the lengths listed could form a triangle.
10. For any sets of side lengths that could not form a triangle, determine what the minimum or maximum length of the longest side must be, given the lengths of the shorter two sides, to form a triangle.

Test-Taking Strategy

Example

Read the problem. Identify what you need to know. Then use the information in the problem to solve.

What is the measure of $\angle KLM$?

A 32

B 44

C 78

D 94

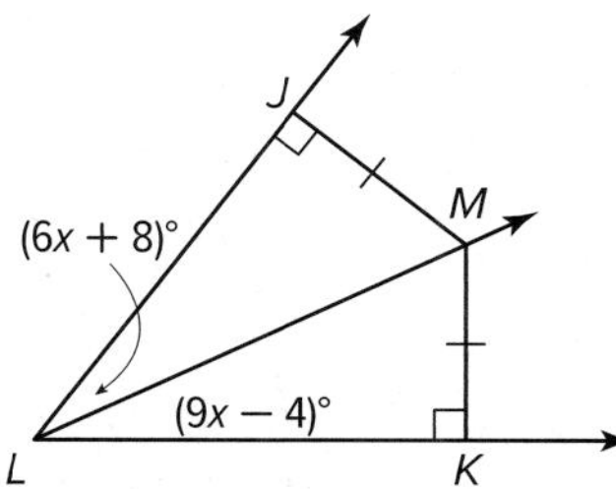

Step 1 **What are you being asked to find? Can you eliminate any unreasonable answers using the given information?**
I need to find the measure of $\angle KLM$. Because $\triangle KLM$ is a right triangle and $\angle KLM$ is not the right angle, I know the answer must be less than 90.

Step 2 **Are any answer choices clearly incorrect?**
Choice D is clearly incorrect, because it is greater than 90.

Step 3 **Of the choices remaining, which one is correct?**
By the Angle Bisector Theorem, $\angle KLM$ must be congruent to $\angle JLM$.
So, $6x + 8 = 9x - 4$, and $x = 4$. Substituting $x = 4$ into the expression for $\angle KLM$, I find that the measure of $\angle KLM$ is 32.
The correct answer is A.

> **Test-Taking Tip**
>
> **Eliminating Unreasonable Answers**
> You can eliminate unreasonable answers to determine the correct answer when solving multiple choice test items.

Apply the Strategy

Read the problem. Identify what you need to know. Then use the information in the problem to solve.

If two of the exterior angles of a triangle measure 123° and 147°, then which of the following statements is true?

A The triangle is a right triangle.

B The triangle is an acute triangle.

C The triangle is an obtuse triangle.

D The triangle is an equilateral triangle.

Answer the questions below.

a. What are you being asked to find? Can you eliminate any unreasonable answers using the given information?

b. Are any answer choices clearly incorrect?

c. Of the choices remaining, which one is correct?

CHAPTER 5

Preparing for Assessment

Cumulative Review

Read each question. Then fill in the correct answer on the answer document provided by your teacher or on a sheet of paper.

1. A triangular community garden has three straight paths, as shown in the figure. There is a birdbath B at the centroid of the garden.

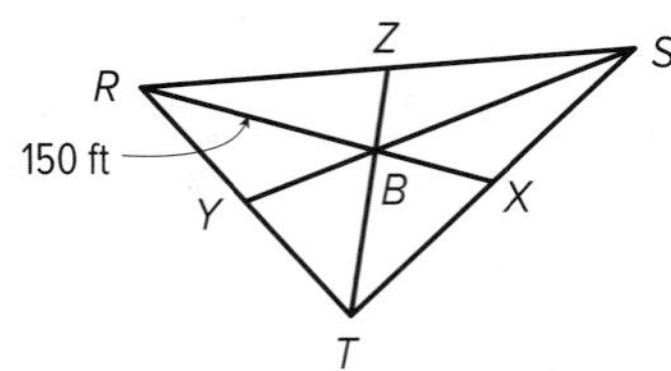

What is the length, in feet, of $\overline{BX}$?

2. In Sara's town, the library, the gym, and the train station form the vertices of a triangle. A straight road connects each pair of buildings. The distance from the library to the gym is 6.7 kilometers, and the distance from the gym to the train station is 4.9 kilometers. What is the greatest whole number of kilometers Sara might travel as she drives from the library to the train station?

Test-Taking Tip

Question 2 To get started, draw a diagram of the given information and consider how you can apply the Triangle Inequality theorem.

3. Which of the following are not undefined terms? Select all that apply.

- [ ] **A** point
- [ ] **B** line
- [ ] **C** plane
- [ ] **D** line segment
- [ ] **E** congruent segments

4. In the figure, $\overline{QR} \cong \overline{QT}$.

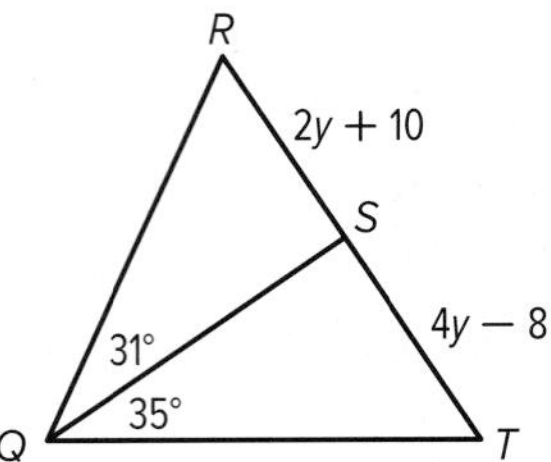

Which of the following inequalities best describes the possible values of y?

- **A** $y < 9$
- **B** $y > 3$
- **C** $y > 1$
- **D** $y > 9$

5. Dina is painting lines in a parking lot, as shown.

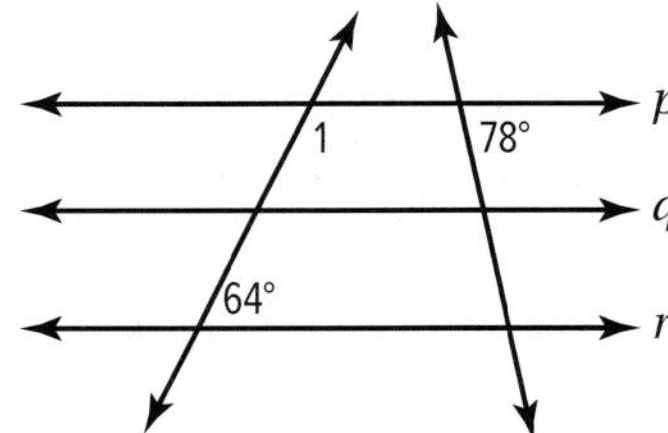

In order to ensure that line p is parallel to line r, what should be the measure, in degrees, of $\angle 1$?

6. Which of the sets of segments could be arranged to form a triangle? Select all that apply.

- [ ] **A** 3, 4, 7
- [ ] **B** 5, 12, 13
- [ ] **C** 7, 9, 11
- [ ] **D** 10, 16, 30
- [ ] **E** 12, 18, 21

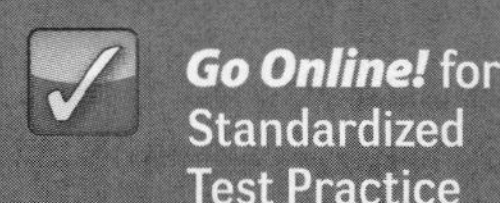

7. In the figure, $\overleftrightarrow{CD}$ is the perpendicular bisector of $\overline{AB}$.

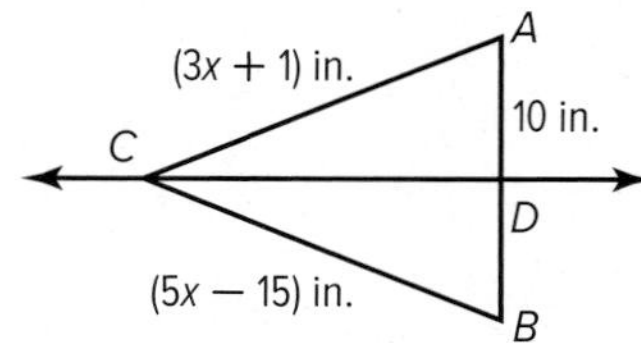

What is the length of the shortest side of $\triangle ABC$?

- A 8 in.
- B 10 in.
- C 20 in.
- D 25 in.

8. In $\triangle MNP$, $MP = 30$ mm, $MN = 28$ mm, and $NP = 23$ mm. List the angles of $\triangle MNP$ in order from smallest to largest.

9. Which of the following conditional statements has a true converse?

- A If you live in Dallas, then you live in Texas.
- B If $m\angle P < 90$, then $\angle P$ is an acute angle.
- C If an animal is a dog, then it is not a cat.
- D If today is Monday, then tomorrow is Saturday.

10. Match the name of a special triangle segment with its point of concurrency.

altitude	centroid
angle bisector	circumcenter
median	incenter
perpendicular bisector	orthocenter

11. Point C is the centroid of $\triangle RST$ and $SV = 24$ inches.

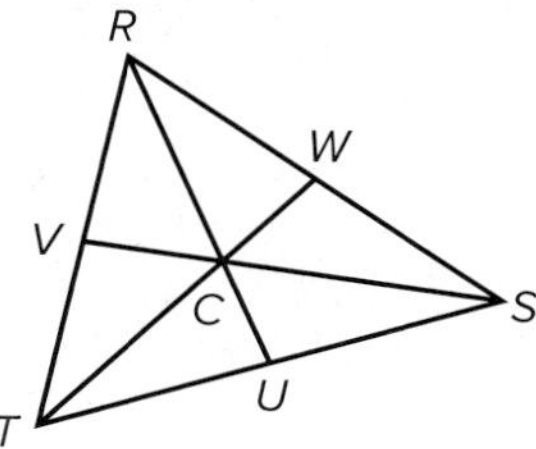

What is the length of $\overline{CV}$?

- A 8 in.
- B 12 in.
- C 16 in.
- D 36 in.

12. Alberto is joining wooden rods to make a triangular picture frame. The figure shows the two rods he has already joined together.

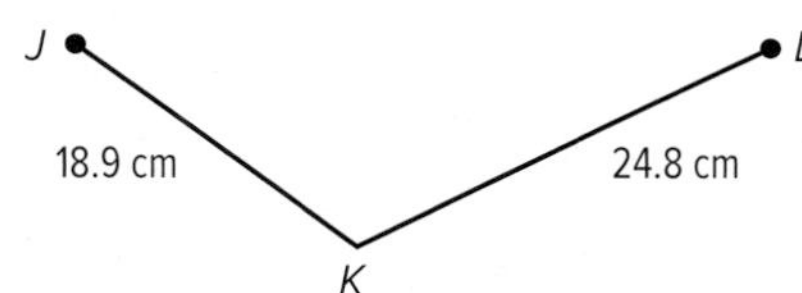

What is the greatest possible whole-number length of the rod that will join points J and L?

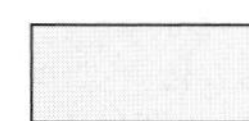

13. In $\triangle PQR$, $\overline{QS}$ is the perpendicular bisector of $\overline{PR}$. Given that $PQ = 9$ and $PS = 6$, what is the perimeter of $\triangle PQR$?

- A 15
- B 21
- C 24
- D 30

Need Extra Help?

If you missed Question...	1	2	3	4	5	6	7	8	9	10	11	12	13
Go to Lesson...	5-2	5-5	1-1	5-6	3-5	5-5	5-1	5-3	2-3	5-1	5-2	5-5	5-1

CHAPTER 6

Quadrilaterals

THEN

You classified polygons. You recognized and applied properties of polygons.

NOW

In this chapter, you will:

- Find and use the sum of the measures of the interior and exterior angles of a polygon.
- Recognize and apply properties of quadrilaterals.
- Compare quadrilaterals.

WHY

FUN AND GAMES Many sports and games are conducted on courts, fields, or boards. Geometric properties of shapes can be used to find the specific angle measures and side lengths in game equipment, playing fields, and game boards.

Use the Mathematical Practices to complete the activity.

1. Sense-Making What is the shape of the playing field or board for your favorite sport or game? What are the dimensions?

2. Use Tools Use the Quadrilateral Geometry Tool to draw an example of the playing field or board.

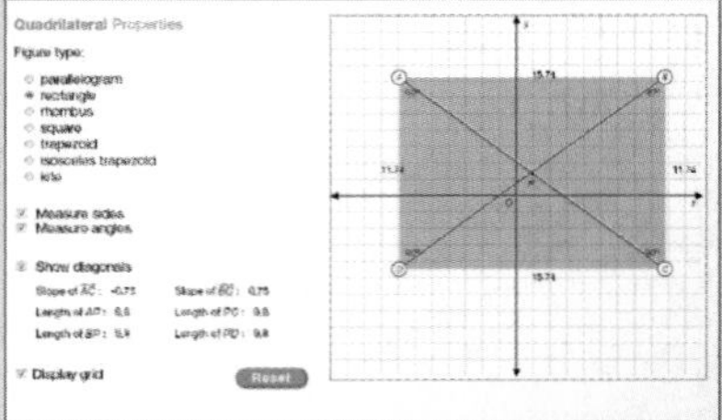

3. Apply Math Is the playing field or board a quadrilateral? Is it a rectangle, rhombus, square, or trapezoid?

4. Discuss How would you create your own game board or playing field for a sport and make sure that it was correct?

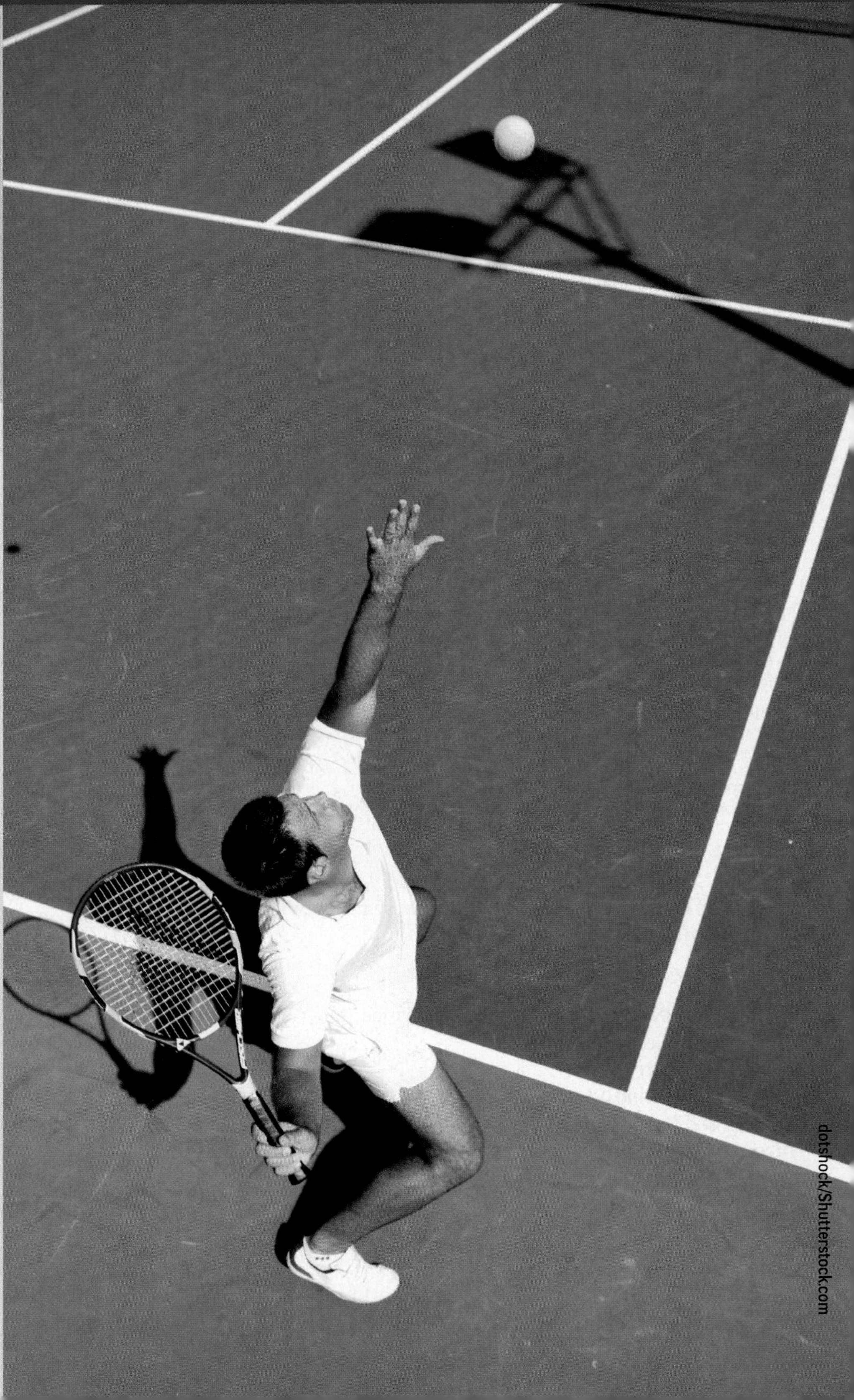

dotshock/Shutterstock.com

Go Online to Guide Your Learning

Explore & Explain

2-D Tool

Use the **2-D** and **Line Segment** tools to investigate the theorem presented in Lesson 6-1 and to explore the properties of trapezoids and kites discussed in Lesson 6-6. The **2-D** tool can also be used to create, change, and investigate quadrilaterals throughout the chapter.

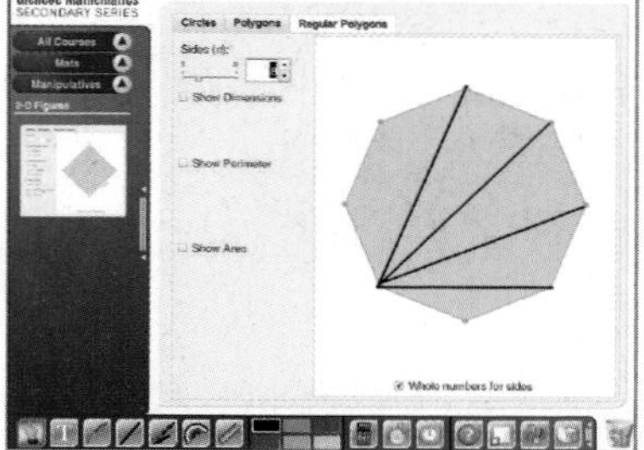

The Geometer's Sketchpad

Use **The Geometer's Sketchpad** to illustrate how the sum of the measures of the interior angles in a convex polygon is determined and to explore parallelograms, rectangles, rhombi, and isosceles trapezoids and the properties of these quadrilaterals.

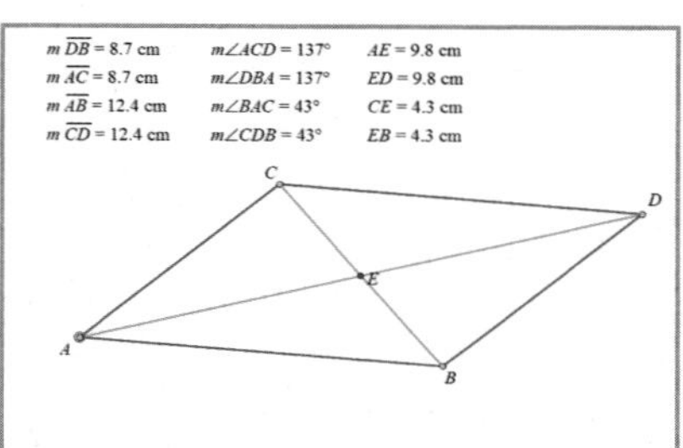

eBook Interactive Student Guide

Before starting the chapter, answer the **Chapter Focus** preview questions. Check your answers as you complete each lesson. At the end of the chapter, try the **Performance Task**.

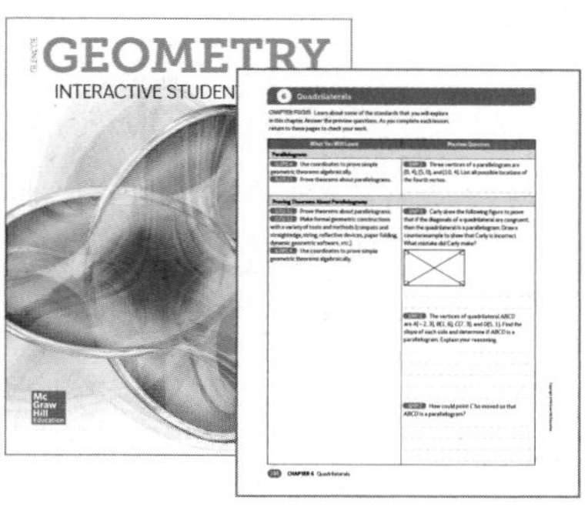

Organize

Foldables

Get organized! Create a **Quadrilaterals Foldable** before you start this chapter to help you organize your notes about quadrilaterals.

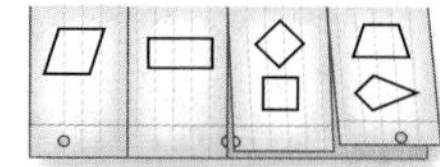

Collaborate

Chapter Project

In the **Fashion Forward** project, you will use what you have learned about quadrilaterals to complete a project that uses entrepreneurial skills.

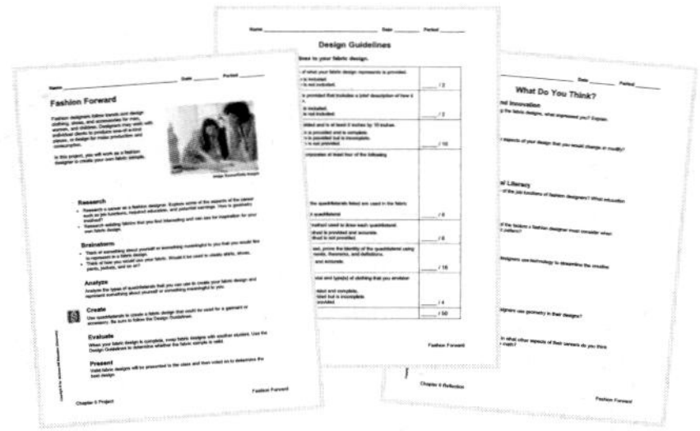

Focus

LEARNSMART®

Need help studying? Complete the **Congruence, Proof, and Construction** and the **Connecting Algebra and Geometry Through Coordinates** domains in LearnSmart to review for the chapter test.

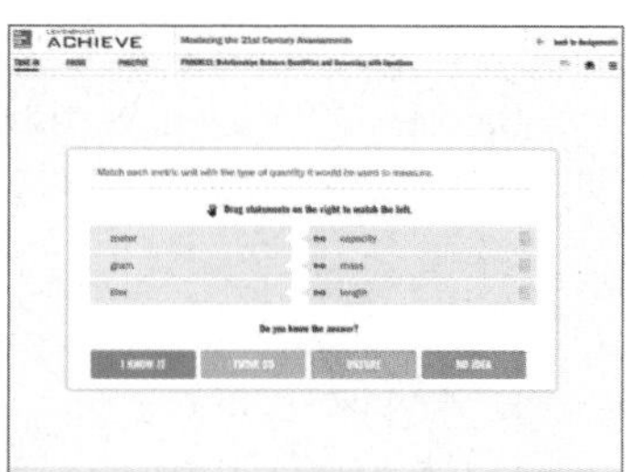

ALEKS®

You can use the **Polygons and Circles** topic in ALEKS to explore what you know about relationships in triangles and what you are ready to learn.*

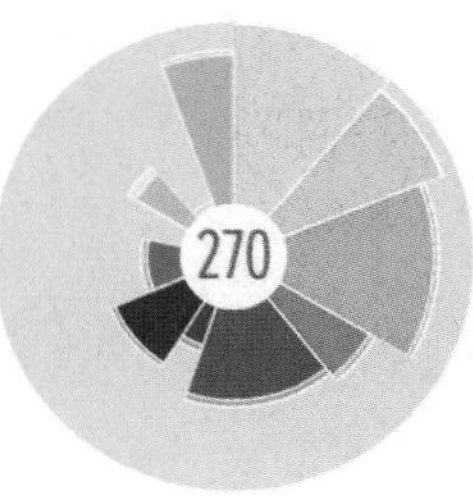

* Ask your teacher if this is part of your program.

Go Online! for Vocabulary Review Games and key vocabulary in 13 languages.

Get Ready for the Chapter

Connecting Concepts

Concept Check

Review the concepts used in this chapter by answering the questions below.

For Exercises 1–4, refer to the figure.

1. What theorem could be used to determine $m\angle 1$?
2. Write an equation to find $m\angle 1$.
3. What theorem could be used to determine $m\angle 2$?
4. Write an equation to find $m\angle 2$.

For Exercises 5–7, refer to $\triangle XYZ$.

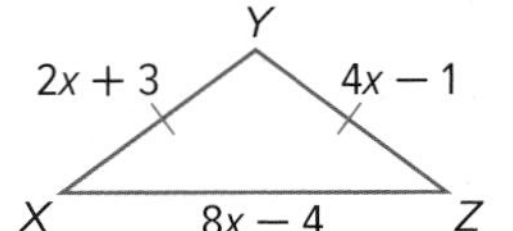

5. What do you know about XY and YZ?
6. How can you use what you know about XY and YZ to determine the value of x?
7. Knowing the value of x, how can you determine the length of each side of the triangle shown?

Performance Task Preview

You can use the concepts and skills in the chapter to help design a set for a play. Understanding quadrilaterals will help you finish the Performance Task at the end of the chapter.

In this Performance Task you will:

- make sense of problems
- reason abstractly and quantitatively

New Vocabulary

English		Español
diagonal	p. 423	diagonal
parallelogram	p. 433	paralelogramo
rectangle	p. 453	rectángulo
rhombus	p. 460	rombo
square	p. 461	cuadrado
trapezoid	p. 469	trapecio
base	p. 469	base
legs of a trapezoid	p. 469	catetos de un trapecio
base angle	p. 469	ángulo de la base
isosceles trapezoid	p. 469	trapecio isósceles
midsegment of a trapezoid	p. 471	segmento medio de un trapecio
kite	p. 472	cometa

Review Vocabulary

exterior angle ángulo externo an angle formed by one side of a triangle and the extension of another side

remote interior angle ángulos internos no adyacentes the angles of a triangle that are not adjacent to a given exterior angle

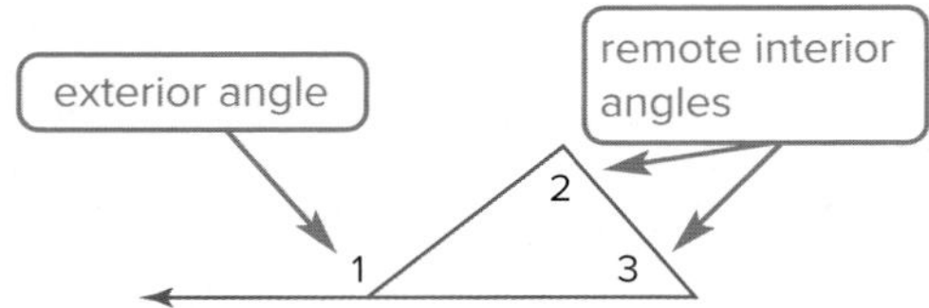

slope pendiente for a (nonvertical) line containing two points (x_1, y_1) and (x_2, y_2), the number m given by the formula $m = \frac{(y_2 - y_1)}{(x_2 - x_1)}$ where $x_2 \neq x_1$

LESSON 1

Angles of Polygons

Then

- You named and classified polygons.

Now

1. Find and use the sum of the measures of the interior angles of a polygon.
2. Find and use the sum of the measures of the exterior angles of a polygon.

Why?

- To create their honeycombs, young worker honeybees excrete flecks of wax that are carefully molded by other bees to form hexagonal cells. The cells are less than 0.1 millimeter thick, but they support almost 25 times their own weight. The cell walls all stand at exactly the same angle to one another. This angle is the measure of the interior angle of a regular hexagon.

 **New Vocabulary**

diagonal

 **Mathematical Practices**

3 Construct viable arguments and critique the reasoning of others.
4 Model with mathematics.
7 Look for and make use of structure.

1 Polygon Interior Angles Sum

A **diagonal** of a polygon is a segment that connects any two nonconsecutive vertices.

The vertices of polygon *PQRST* that are not consecutive with vertex *P* are vertices *R* and *S*. Therefore, polygon *PQRST* has two diagonals from vertex *P*, $\overline{PR}$ and $\overline{PS}$. Notice that the diagonals from vertex *P* separate the polygon into three triangles.

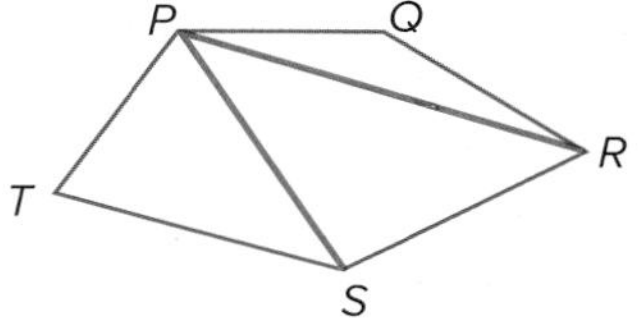

The sum of the angle measures of a polygon is the sum of the angle measures of the triangles formed by drawing all the possible diagonals from one vertex.

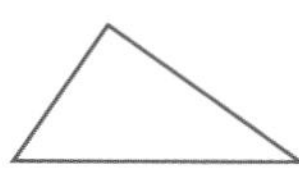

Triangle

Quadrilateral

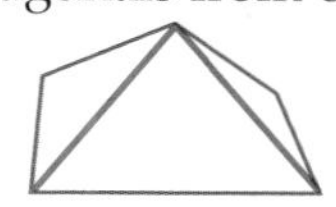

Pentagon

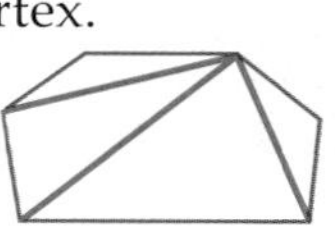

Hexagon

Since the sum of the angle measures of a triangle is 180, we can make a table and look for a pattern to find the sum of the angle measures for any convex polygon.

Polygon	Number of Sides	Number of Triangles	Sum of Interior Angle Measures
triangle	3	1	(1)180 or 180
quadrilateral	4	2	(2)180 or 360
pentagon	5	3	(3)180 or 540
hexagon	6	4	(4)180 or 720
n-gon	*n*	$n - 2$	$(n - 2)180$

This leads to the following theorem.

Theorem 6.1 Polygon Interior Angles Sum

The sum of the interior angle measures of an *n*-sided convex polygon is $(n - 2) \cdot 180$.

Example $m\angle A + m\angle B + m\angle C + m\angle D + m\angle E = (5 - 2) \cdot 180$

$= 540$

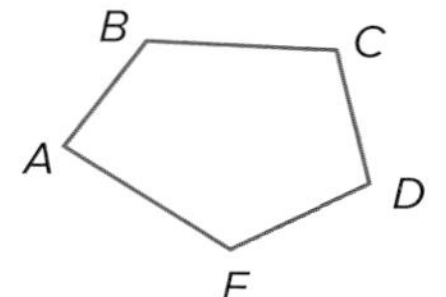

You will prove Theorem 6.1 for octagons in Exercise 42.

You can use the Polygon Interior Angles Sum Theorem to find the sum of the interior angles of a polygon and to find missing measures in polygons.

Study Tip

Naming Polygons Remember, a polygon with n-sides is an n-gon, but several polygons have special names.

Number of Sides	Polygon
3	triangle
4	quadrilateral
5	pentagon
6	hexagon
7	heptagon
8	octagon
9	nonagon
10	decagon
11	hendecagon
12	dodecagon
n	n-gon

Go Online!

See how the sum of the angle measures in a triangle is related to the sum of the angle measures in other polygons with a **Geometer's Sketchpad®** sketch in ConnectED.

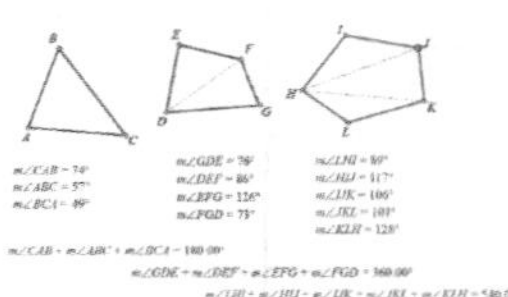

Example 1 Find the Interior Angles Sum of a Polygon

a. Find the sum of the measures of the interior angles of a convex heptagon.

A heptagon has seven sides. Use the Polygon Interior Angles Sum Theorem to find the sum of its interior angle measures.

$(n - 2) \cdot 180 = (7 - 2) \cdot 180$ $n = 7$

$= 5 \cdot 180$ or 900 Simplify.

The sum of the measures is 900.

CHECK Draw a convex polygon with seven sides. Use a protractor to measure each angle to the nearest degree. Then find the sum of these measures.

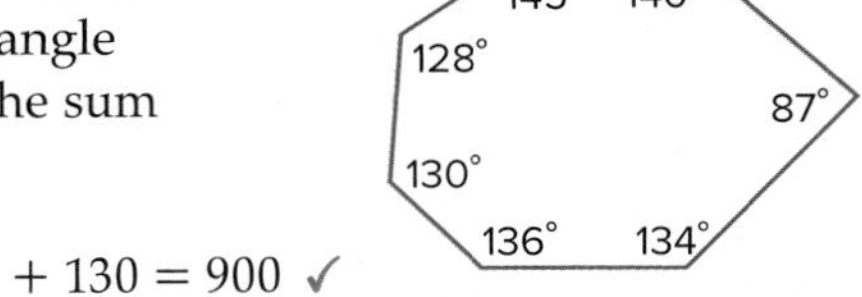

$128 + 145 + 140 + 87 + 134 + 136 + 130 = 900$ ✓

b. ALGEBRA Find the measure of each interior angle of quadrilateral $ABCD$.

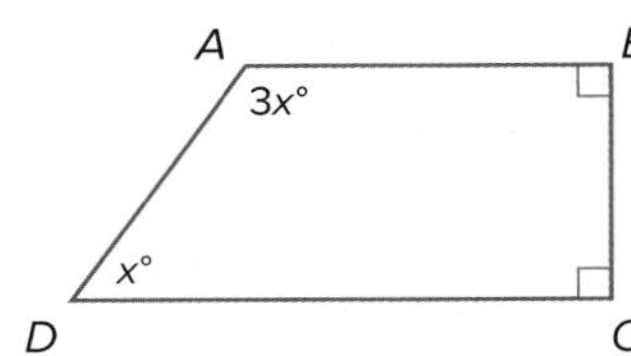

Step 1 Find x.

Since there are 4 angles, the sum of the interior angle measures is $(4 - 2) \cdot 180$ or 360.

$360 = m\angle A + m\angle B + m\angle C + m\angle D$ Sum of interior angle measures

$360 = 3x + 90 + 90 + x$ Substitution

$360 = 4x + 180$ Combine like terms.

$180 = 4x$ Subtract 180 from each side.

$45 = x$ Divide each side by 4.

Step 2 Use the value of x to find the measure of each angle.

$m\angle A = 3x = 3(45)$ or 135 $m\angle B = 90$ $m\angle C = 90$ $m\angle D = x = 45$

Guided Practice

1A. Find the sum of the measures of the interior angles of a convex octagon.

1B. Find the measure of each interior angle of pentagon $HJKLM$ shown.

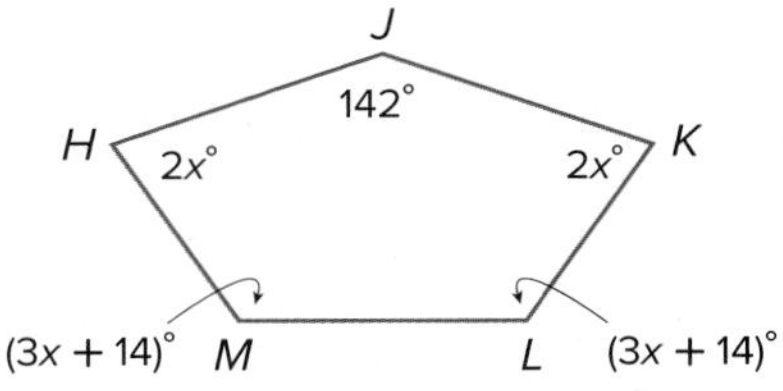

Review Vocabulary

regular polygon
a convex polygon in which all of the sides are congruent and all of the angles are congruent

Real-World Example 2 Interior Angle Measures of a Regular Polygon

TENTS The poles for a tent form the vertices of a regular hexagon. When the poles are properly positioned, what is the measure of the angle formed at a corner of the tent?

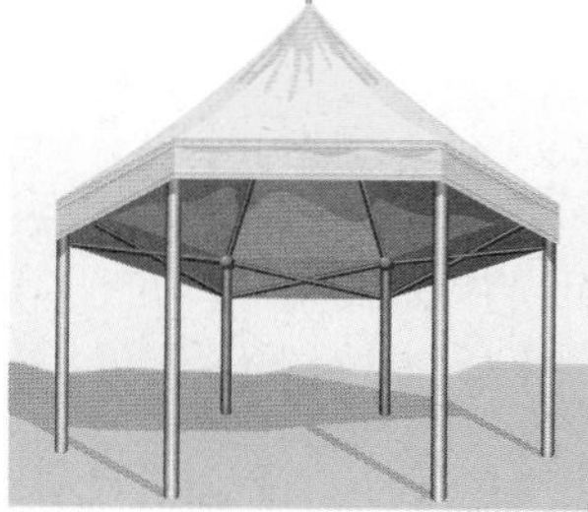

Understand Draw a diagram of the situation.

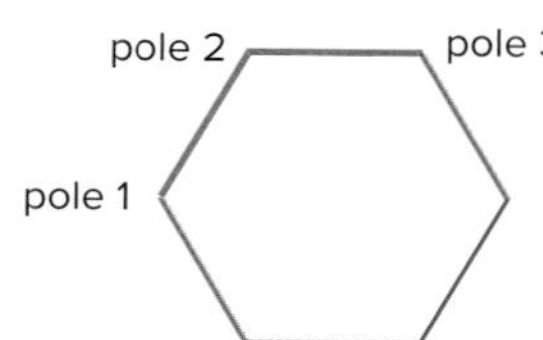

The measure of the angle formed at a corner of the tent is an interior angle of a regular hexagon.

Plan Use the Polygon Interior Angles Sum Theorem to find the sum of the measures of the angles. Since the angles of a regular polygon are congruent, divide this sum by the number of angles to find the measure of each interior angle.

Solve **Step 1** Find the sum of the interior angle measures.

$$(n - 2) \cdot 180 = (6 - 2) \cdot 180 \quad n = 6$$

$$= 4 \cdot 180 \text{ or } 720 \quad \text{Simplify.}$$

Step 2 Find the measure of one interior angle.

$$\frac{\text{sum of interior angle measures}}{\text{number of congruent angles}} = \frac{720}{6} \quad \text{Substitution}$$

$$= 120 \quad \text{Divide.}$$

The angle at a corner of the tent measures 120.

Check To verify that this measure is correct, use a ruler and a protractor to draw a regular hexagon using 120 as the measure of each interior angle. The last side drawn should connect with the beginning point of the first segment drawn.

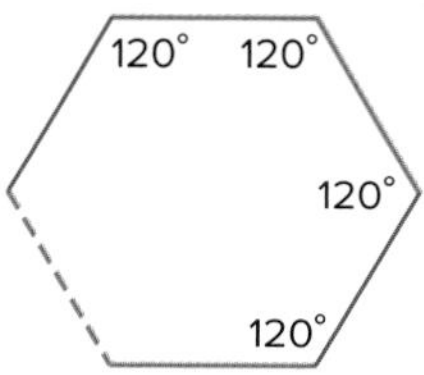

The diagram of a regular hexagon models the tent in the problem. Diagrams can help you visualize three-dimensional problems to make them easier to solve.

Guided Practice

2A. COINS Find the measure of each interior angle of the regular hendecagon that appears on the face of a Susan B. Anthony one-dollar coin.

2B. HOT TUBS A certain company makes hot tubs in a variety of different shapes. Find the measure of each interior angle of the regular nonagon model.

Real-World Link

Susan B. Anthony was a leader of the women's suffrage movement in the late 1800s, which eventually led to the Nineteenth Amendment giving women the right to vote. In 1979, the Susan B. Anthony one-dollar coin was first minted, making her the first woman to be depicted on U.S. currency.

Source: *Encyclopaedia Britannica*

Given the interior angle measure of a regular polygon, you can also use the Polygon Interior Angles Sum Theorem to find a polygon's number of sides.

Example 3 Find Number of Sides Given Interior Angle Measure

The measure of an interior angle of a regular polygon is 135. Find the number of sides in the polygon.

Let n = the number of sides in the polygon. Since all angles of a regular polygon are congruent, the sum of the interior angle measures is $135n$. By the Polygon Interior Angles Sum Theorem, the sum of the interior angle measures can also be expressed as $(n - 2) \cdot 180$.

$135n = (n - 2) \cdot 180$	Write an equation.
$135n = 180n - 360$	Distributive Property
$-45n = -360$	Subtract $180n$ from each side.
$n = 8$	Divide each side by -45.

The polygon has 8 sides.

Guided Practice

3. The measure of an interior angle of a regular polygon is 144. Find the number of sides in the polygon.

Review Vocabulary

exterior angle an angle formed by one side of a polygon and the extension of another side

2 Polygon Exterior Angles Sum

Does a relationship exist between the number of sides of a convex polygon and the sum of its exterior angle measures? Examine the polygons below in which an exterior angle has been measured at each vertex.

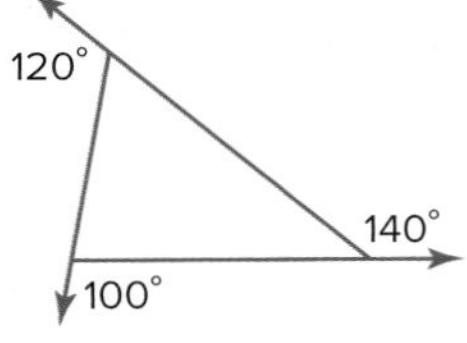

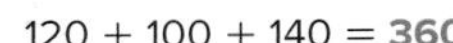

120 + 100 + 140 = **360**

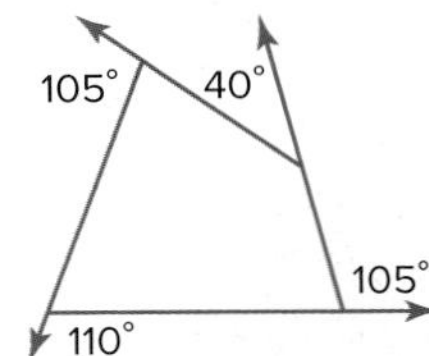

105 + 110 + 105 + 40 = **360**

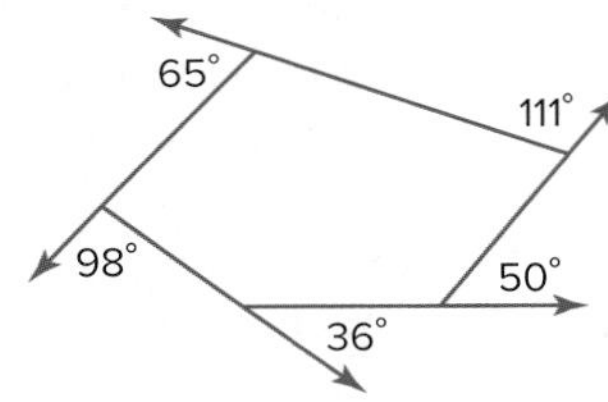

65 + 98 + 36 + 50 + 111= **360**

Notice that the sum of the exterior angle measures in each case is 360. This suggests the following theorem.

Theorem 6.2 Polygon Exterior Angles Sum

The sum of the exterior angle measures of a convex polygon, one angle at each vertex, is 360.

Example
$m\angle 1 + m\angle 2 + m\angle 3 + m\angle 4 + m\angle 5 + m\angle 6 = 360$

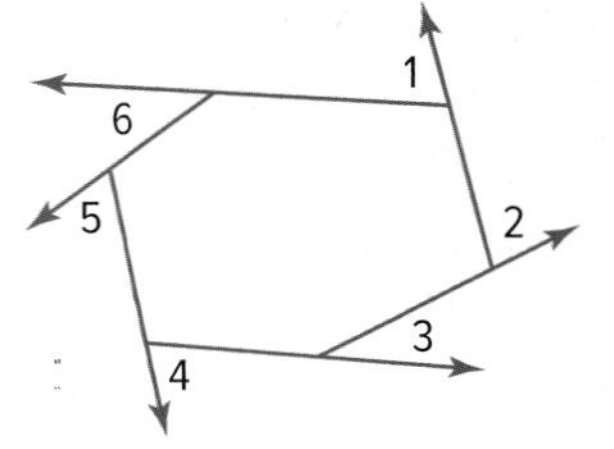

You will prove Theorem 6.2 in Exercise 43.

Example 4 Find Exterior Angle Measures of a Polygon

a. ALGEBRA Find the value of x in the diagram.

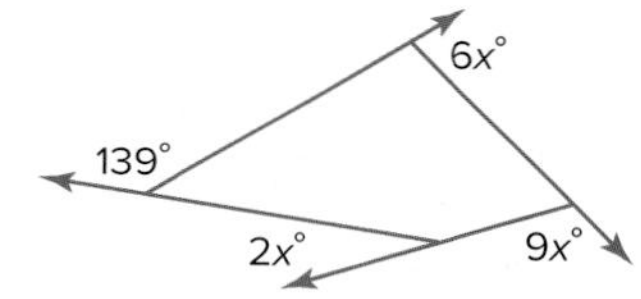

Use the Polygon Exterior Angles Sum Theorem to write an equation. Then solve for x.

$$(2x - 5) + 5x + 2x + (6x - 5) + (3x + 10) = 360$$

$$(2x + 5x + 2x + 6x + 3x) + [-5 + (-5) + 10] = 360$$

$$18x = 360$$

$$x = \frac{360}{18} \text{ or } 20$$

Study Tip

MP Perseverance To find the measure of each exterior angle of a regular polygon, you can find the measure of each interior angle and subtract this measure from 180, since an exterior angle and its corresponding interior angle are supplementary.

b. Find the measure of each exterior angle of a regular nonagon.

A regular nonagon has 9 congruent sides and 9 congruent interior angles. The exterior angles are also congruent, since angles supplementary to congruent angles are congruent. Let n = the measure of each exterior angle and write and solve an equation.

$9n = 360$ Polygon Exterior Angles Sum Theorem

$n = 40$ Divide each side by 9.

The measure of each exterior angle of a regular nonagon is 40.

Guided Practice

4A. Find the value of x in the diagram.

4B. Find the measure of each exterior angle of a regular dodecagon.

Check Your Understanding

◯ = Step-by-Step Solutions begin on page R13.

Go Online! for a Self-Check Quiz

Example 1 **Find the sum of the measures of the interior angles of each convex polygon.**

1. decagon

2. pentagon

Find the measure of each interior angle.

3.

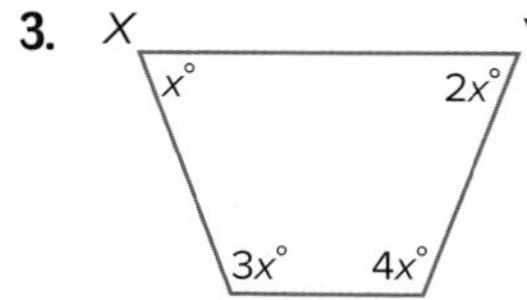

4.

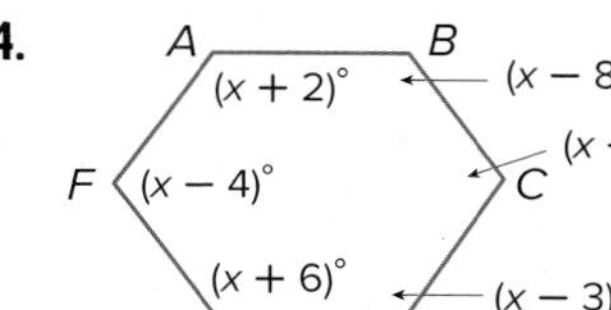

Example 2 **5** **AMUSEMENT** The Wonder Wheel at Coney Island in Brooklyn, New York, is a regular polygon with 16 sides. What is the measure of each interior angle of the polygon?

Example 3 **The measure of an interior angle of a regular polygon is given. Find the number of sides in the polygon.**

6. 150

7. 170

Example 4 **Find the value of x in each diagram.**

8.

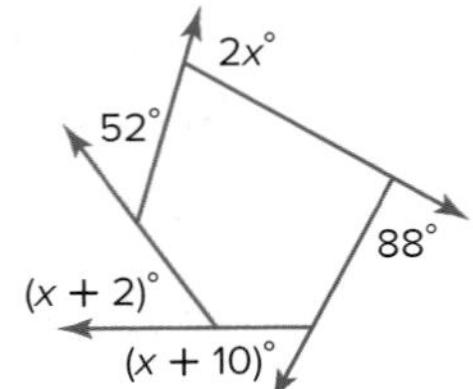

9.

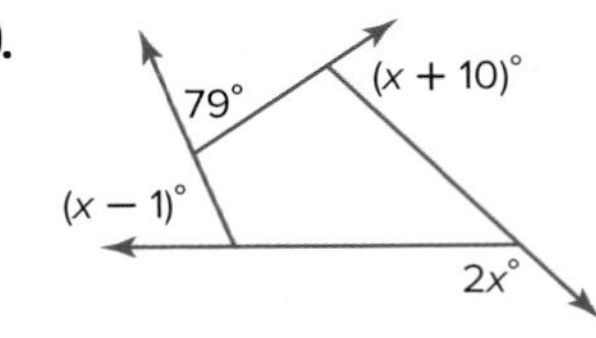

Find the measure of each exterior angle of each regular polygon.

10. quadrilateral

11. octagon

Practice and Problem Solving

Extra Practice is on page R6.

Find the sum of the measures of the interior angles of each convex polygon.

12. dodecagon

13. 20-gon

14. 29-gon

15. 32-gon

Find the measure of each interior angle.

16.

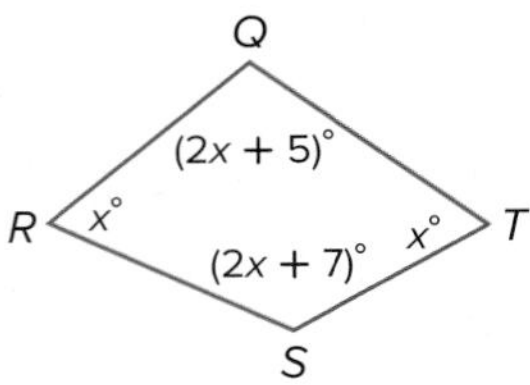

17

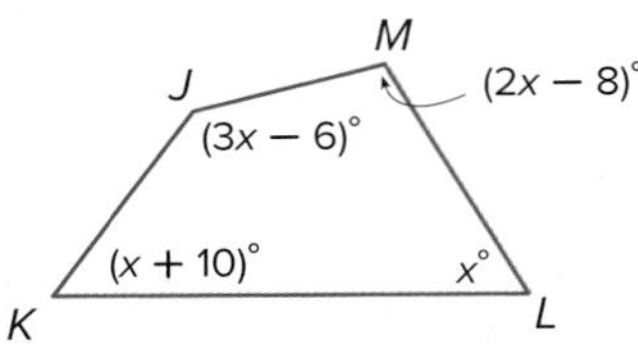

18.

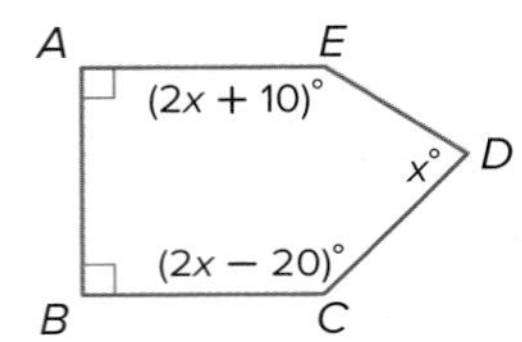

19.

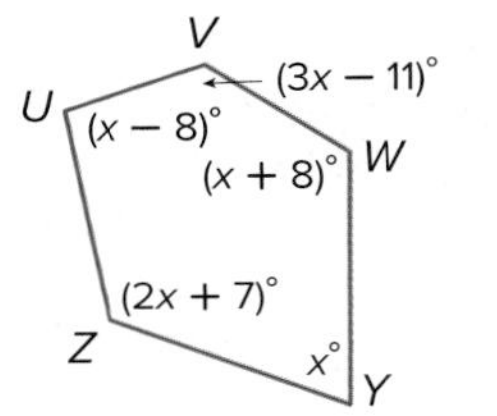

Example 1 **20. BASEBALL** In baseball, home plate is a pentagon. The dimensions of home plate are shown. What is the sum of the measures of the interior angles of home plate?

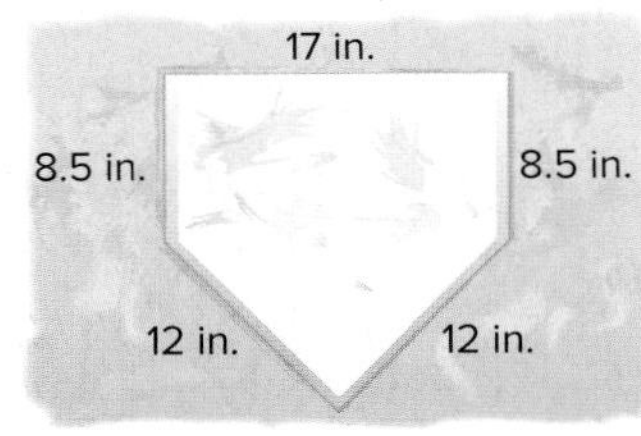

Example 2 **Find the measure of each interior angle of each regular polygon.**

21. dodecagon

22. pentagon

23. decagon

24. nonagon

25. MP **MODELING** Hexagonal chess is played on a regular hexagonal board comprised of 92 small hexagons in three colors. The chess pieces are arranged so that a player can move any piece at the start of a game.

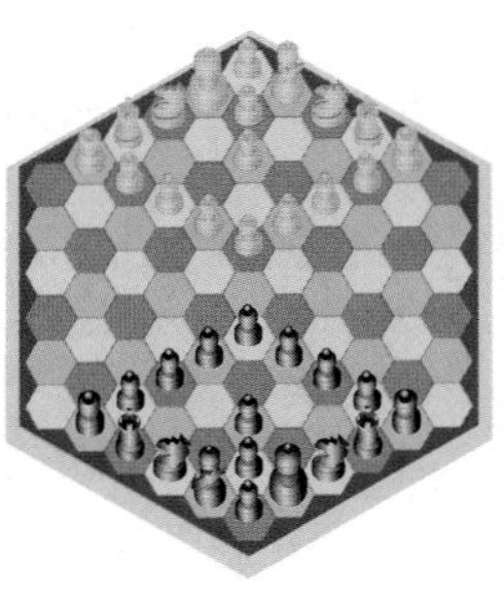

a. What is the sum of the measures of the interior angles of the chess board?

b. Does each interior angle have the same measure? If so, give the measure. Explain your reasoning.

Example 3 **The measure of an interior angle of a regular polygon is given. Find the number of sides in the polygon.**

26. 60

27. 90

28. 120

29. 156

Example 4 **Find the value of x in each diagram.**

30.

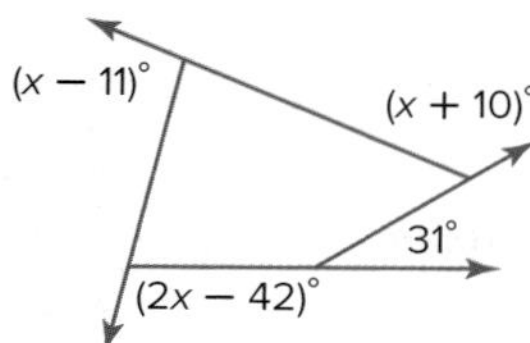

31.

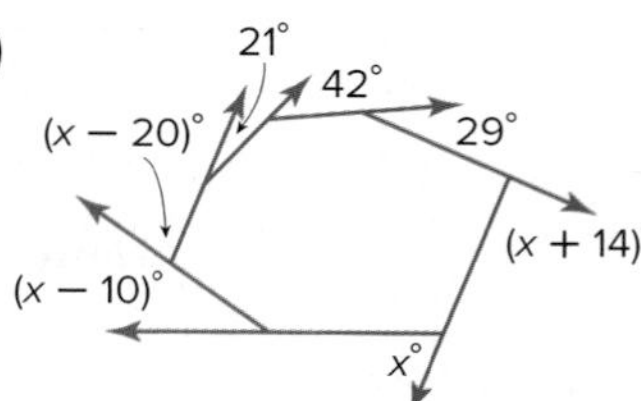

32.

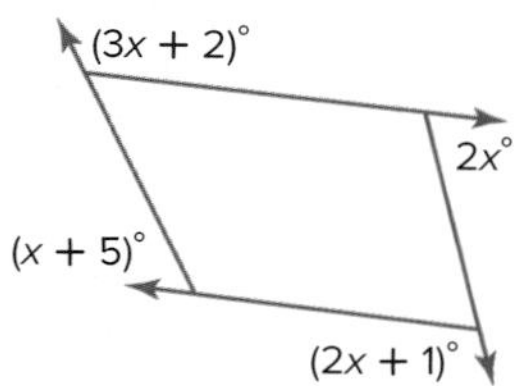

33.

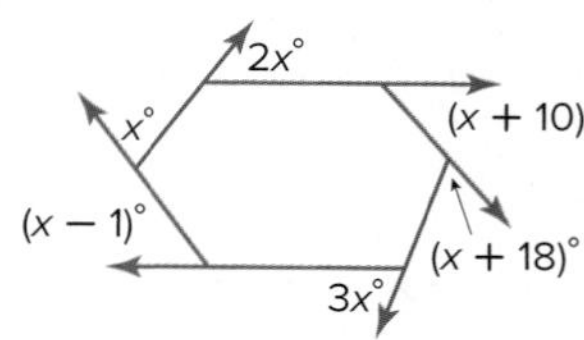

Find the measure of each exterior angle of each regular polygon.

34. decagon **35.** pentagon **36.** hexagon **37.** 15-gon

38. **COLOR GUARD** During the halftime performance for a football game, the color guard is planning a new formation in which seven members stand around a central point and stretch their flag to the person immediately to their left as shown.

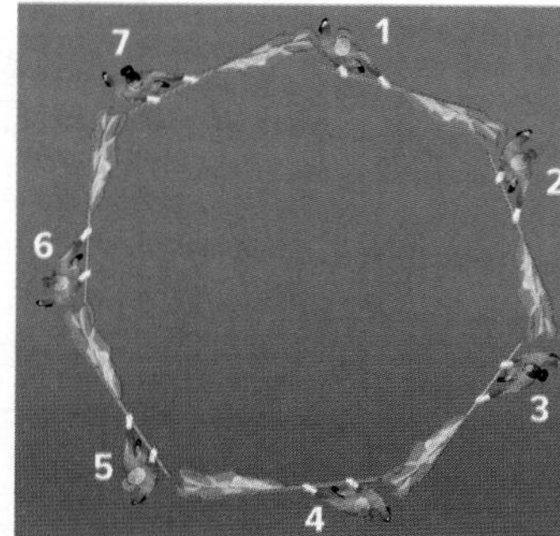

a. What is the measure of each exterior angle of the formation?

b. If the perimeter of the formation is 38.5 feet, how long is each flag?

Find the measures of an exterior angle and an interior angle given the number of sides of each regular polygon. Round to the nearest tenth, if necessary.

39. 7 **40.** 13 **41.** 14

42. **PROOF** Write a paragraph proof to prove the Polygon Interior Angles Sum Theorem for octagons.

43. **PROOF** Use algebra to prove the Polygon Exterior Angles Sum Theorem.

44. **MULTI-STEP** Mary is building a storage shed from wood panels in the shape of a regular polygon. Each rectangular wall panel is 7 feet wide and 8 feet tall. She wants the shed to have the most walls possible so she can hang tools on them. The panels are \$0.89 per square foot. The brackets used to join two panels together cost \$6.50. Mary's budget for the wall material is \$500.

a. How many walls will Mary be able to build without exceeding her budget? What is the shape of her storage shed?

b. Describe your solution process.

c. What assumptions did you make?

ALGEBRA Find the measure of each interior angle.

45. decagon, in which the measures of the interior angles are $x + 5$, $x + 10$, $x + 20$, $x + 30$, $x + 35$, $x + 40$, $x + 60$, $x + 70$, $x + 80$, and $x + 90$

46. polygon $ABCDE$, in which the measures of the interior angles are $6x$, $4x + 13$, $x + 9$, $2x - 8$, and $4x - 1$

47 **THEATER** The drama club would like to build a theater in the round, so the audience can be seated on all sides of the stage, for its next production.

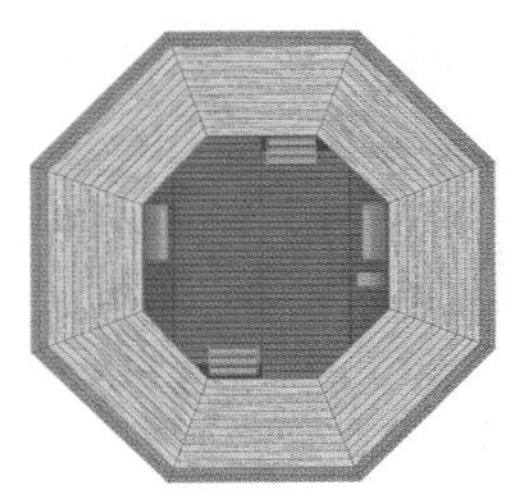

a. The stage will be a regular octagon with a total perimeter of 60 feet. To what length should each board be cut to form the sides of the stage?

b. At what angle should each board be cut so that they will fit together as shown? Explain your reasoning.

48. **MULTIPLE REPRESENTATIONS** In this problem, you will explore angle and side relationships in special quadrilaterals.

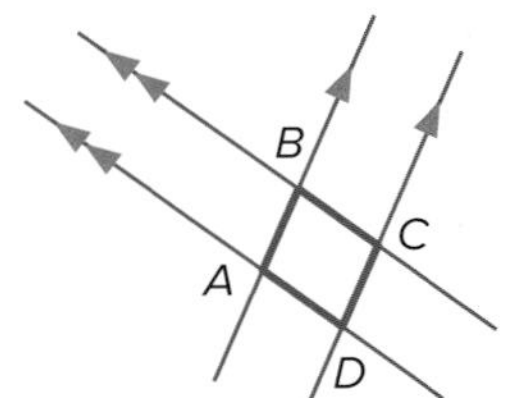

a. **Geometric** Draw two pairs of parallel lines that intersect like the ones shown. Label the quadrilateral formed by *ABCD*. Repeat these steps to form two additional quadrilaterals, *FGHJ* and *QRST*.

b. **Tabular** Copy and complete the table below.

Quadrilateral	Lengths and Measures							
ABCD	$m\angle A$		$m\angle B$		$m\angle C$		$m\angle D$	
	AB		*BC*		*CD*		*DA*	
FGHJ	$m\angle F$		$m\angle G$		$m\angle H$		$m\angle J$	
	FG		*GH*		*HJ*		*JF*	
QRST	$m\angle Q$		$m\angle R$		$m\angle S$		$m\angle T$	
	QR		*RS*		*ST*		*TQ*	

c. **Verbal** Make a conjecture about the relationship between the angles opposite each other in a quadrilateral formed by two pairs of parallel lines.

d. **Verbal** Make a conjecture about the relationship between two consecutive angles in a quadrilateral formed by two pairs of parallel lines.

e. **Verbal** Make a conjecture about the relationship between the sides opposite each other in a quadrilateral formed by two pairs of parallel lines.

H.O.T. Problems Use Higher-Order Thinking Skills

49. **ERROR ANALYSIS** Marcus says that the sum of the exterior angles of a decagon is greater than that of a heptagon because a decagon has more sides. Liam says that the sum of the exterior angles for both polygons is the same. Is either of them correct? Explain your reasoning.

50. **CHALLENGE** Find the values of *a*, *b*, and *c* if *QRSTVX* is a regular hexagon. Justify your answer.

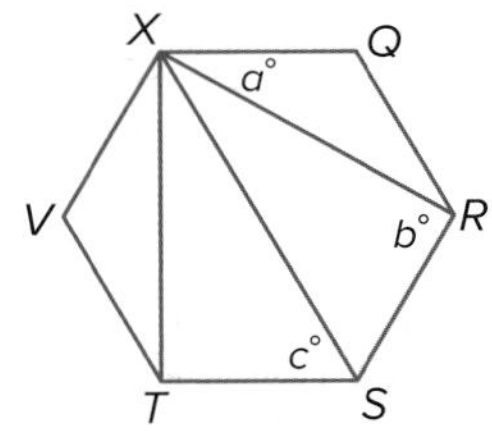

51. MP **CONSTRUCT ARGUMENTS** If two sides of a regular hexagon are extended to meet at a point in the exterior of the polygon, will the triangle formed *always, sometimes,* or *never* be equilateral? Justify your answer.

52. **OPEN-ENDED** Sketch a polygon and find the sum of its interior angles. How many sides does a polygon with twice this interior angles sum have? Justify your answer.

53. **WRITING IN MATH** Explain how triangles are related to the Polygon Interior Angles Sum Theorem.

54. Find the measure of angle A. MP 2

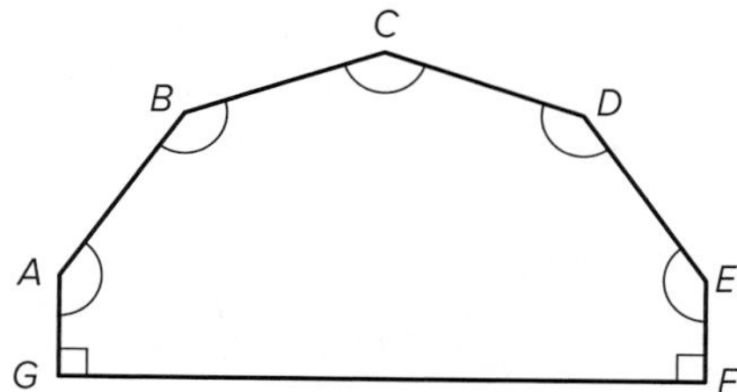

◯ **A** 51.4

◯ **B** 108

◯ **C** 128.6

◯ **D** 144

55. Polygons $ABCEGJ$ and $ABDFH$ are both regular polygons.

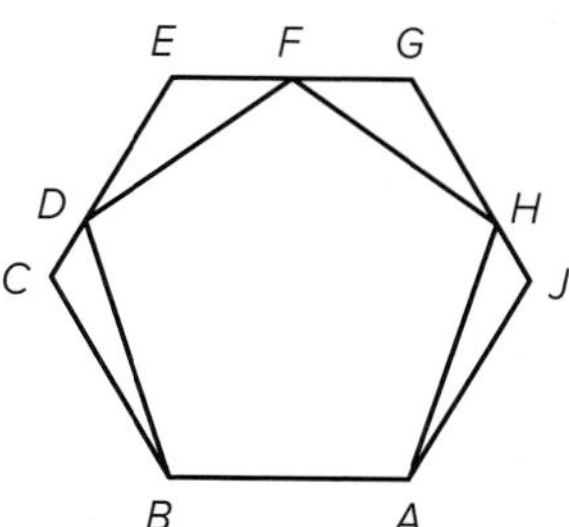

What is the measure of angle CBD? MP 2

◯ **A** 12

◯ **B** 18

◯ **C** 108

◯ **D** 120

56. The measures of the angles of a triangle are $x°$, $2x°$, and $(x - 20)°$. Which of the following are measures of interior angles in the triangle? Select all that apply. MP 2

☐ 15

☐ 30

☐ 50

☐ 80

☐ 100

☐ 180

57. The figure below shows a regular hexagon and a regular pentagon.

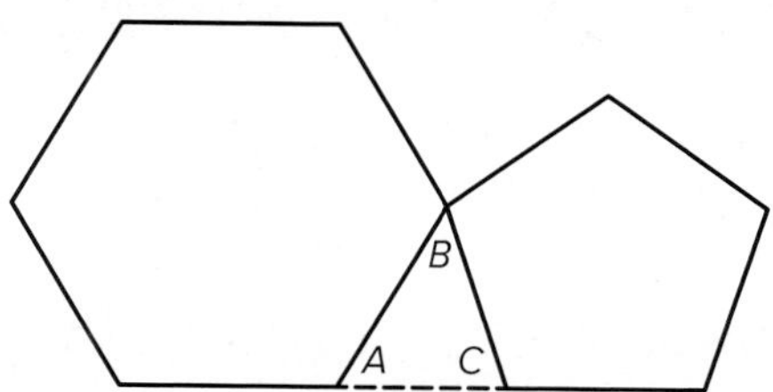

Find the measure of angle ABC. MP 2

◯ **A** 12

◯ **B** 48

◯ **C** 84

◯ **D** 132

58. **MULTI-STEP** Use pentagon $ABCDE$ to answer the questions. MP 2

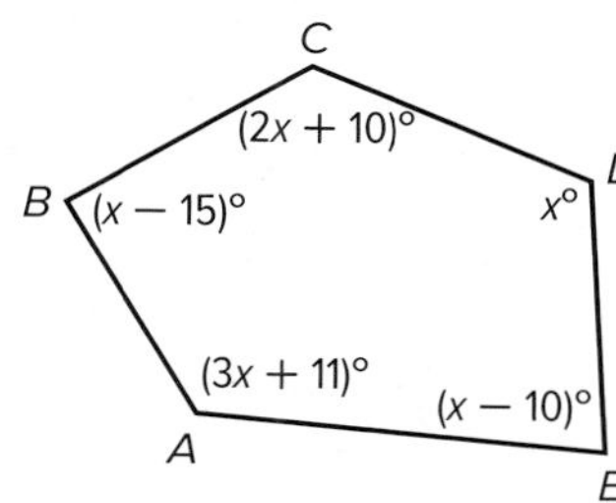

a. How many triangles will this polygon divide into? ☐

b. What is the value of x? ☐

c. Find the measure of each interior angle.

$m\angle A$ = ☐°

$m\angle B$ = ☐°

$m\angle C$ = ☐°

$m\angle D$ = ☐°

$m\angle E$ = ☐°

d. What mathematical practice did you use to solve this problem?

59. The sum of the measures of the interior angles of a decagon is ☐°. MP 2

60. The sum of the measures of the exterior angles of a decagon is ☐°. MP 2

EXTEND 6-1

Spreadsheet Lab

Angles of Polygons

It is possible to find the interior and exterior measurements along with the sum of the interior angles of any regular polygon with *n* number of sides by using a spreadsheet.

Mathematical Practices
MP **5** Use appropriate tools strategically.

Activity

Work cooperatively. Design a spreadsheet using the following steps.

- Label the columns as shown in the spreadsheet below.
- Enter the digits 3–10 in the first column.
- The number of triangles in a polygon is 2 fewer than the number of sides. Write a formula in Cell B2 to subtract 2 from each number in column A.
- Enter a formula in Cell C2 so the spreadsheet will calculate the sum of the measures of the interior angles. Remember that the formula is $S = (n - 2)180$.
- Continue to enter formulas so that the indicated computation is performed. Then, copy each formula through Row 9. The final spreadsheet will appear as below.

Polygons and Angles

	A	B	C	D	E	F
1	Number of Sides	Number of Triangles	Sum of Measures of Interior Angles	Measure of Each Interior Angle	Measure of Each Exterior Angle	Measures of Exterior Angles
2	3	1	180	60	120	360
3	4	2	360	90	90	360
4	5	3	540	108	72	360
5	6	4	720	120	60	360
6	7	5	900	128.57	51.43	360
7	8	6	1080	135	45	360
8	9	7	1260	140	40	360
9	10	8	1440	144	36	360

Sheet 1 | Sheet 2 | Sheet 3

Exercises

Work cooperatively.

1. Write the formula to find the measure of each interior angle in the polygon.
2. Write the formula to find the sum of the measures of the exterior angles.
3. What is the measure of each interior angle if the number of sides is 1? 2?
4. Is it possible to have values of 1 and 2 for the number of sides? Explain.

For Exercises 5–8, use the spreadsheet.

5. How many triangles are in a polygon with 17 sides?
6. Find the measure of an exterior angle of a regular polygon with 16 sides.
7. Find the measure of an interior angle of a regular polygon with 115 sides.
8. If the measure of the exterior angles is 0, find the measure of the interior angles. Is this possible? Explain.

LESSON 2

Parallelograms

Then

- You classified polygons with four sides as quadrilaterals.

Now

1. Recognize and apply properties of the sides and angles of parallelograms.
2. Recognize and apply properties of the diagonals of parallelograms.

Why?

- The arm of the basketball goal shown can be adjusted to a height of 10 feet or 5 feet. Notice that as the height is adjusted, each pair of opposite sides of the quadrilateral formed by the arms remains parallel.

New Vocabulary
parallelogram

Mathematical Practices

3 Construct viable arguments and critique the reasoning of others.
4 Model with mathematics.
6 Attend to precision.
8 Look for and express regularity in repeated reasoning.

1 Sides and Angles of Parallelograms

A **parallelogram** is a quadrilateral with both pairs of opposite sides parallel. To name a parallelogram, use the symbol ▱. In ▱$ABCD$, $\overline{BC} \parallel \overline{AD}$ and $\overline{AB} \parallel \overline{DC}$ by definition.

Other properties of parallelograms are given in the theorems below.

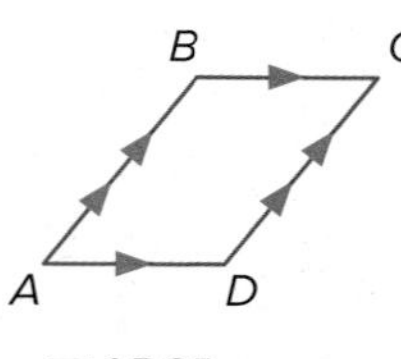

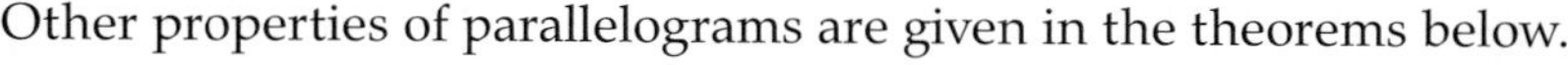

Theorems Properties of Parallelograms

6.3 If a quadrilateral is a parallelogram, then its opposite sides are congruent.

Abbreviation *Opp. sides of a ▱ are ≅.*

Example If $JKLM$ is a parallelogram, then $\overline{JK} \cong \overline{ML}$ and $\overline{JM} \cong \overline{KL}$.

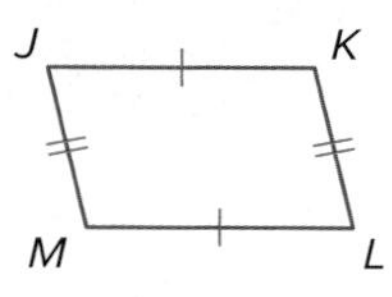

6.4 If a quadrilateral is a parallelogram, then its opposite angles are congruent.

Abbreviation *Opp. ∠s of a ▱ are ≅.*

Example If $JKLM$ is a parallelogram, then $\angle J \cong \angle L$ and $\angle K \cong \angle M$.

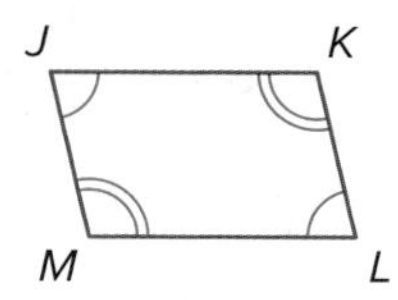

6.5 If a quadrilateral is a parallelogram, then its consecutive angles are supplementary.

Abbreviation *Cons. ∠s in a ▱ are supplementary.*

Example If $JKLM$ is a parallelogram, then $x + y = 180$.

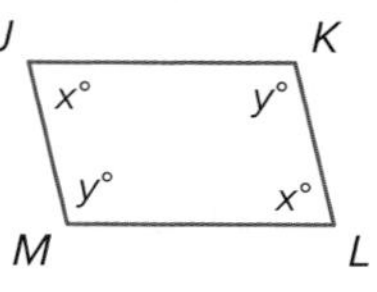

6.6 If a parallelogram has one right angle, then it has four right angles.

Abbreviation *If a ▱ has 1 rt. ∠, it has 4 rt. ∠s.*

Example In ▱$JKLM$, if $\angle J$ is a right angle, then $\angle K$, $\angle L$, and $\angle M$ are also right angles.

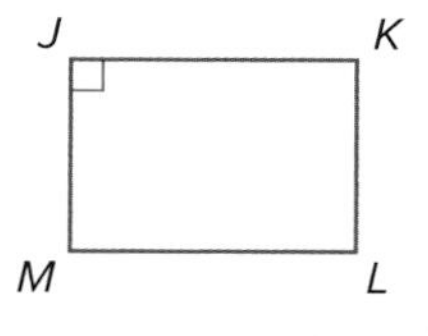

You will prove Theorems 6.3, 6.5, and 6.6 in Exercises 28, 26, and 7, respectively.

Proof Theorem 6.4

Write a two-column proof of Theorem 6.4.

Given: ▱$FGHJ$

Prove: $\angle F \cong \angle H$, $\angle J \cong \angle G$

Proof:

Statements	Reasons
1. ▱$FGHJ$	1. Given
2. $\overline{FG} \parallel \overline{JH}$; $\overline{FJ} \parallel \overline{GH}$	2. Definition of parallelogram
3. $\angle F$ and $\angle J$ are supplementary. $\angle J$ and $\angle H$ are supplementary. $\angle H$ and $\angle G$ are supplementary.	3. If parallel lines are cut by a transversal, consecutive interior angles are supplementary.
4. $\angle F \cong \angle H$, $\angle J \cong \angle G$	4. Supplements of the same angles are congruent.

Study Tip

Including a Figure Theorems are presented in general terms. In a proof, you must include a drawing so that you can refer to segments and angles specifically.

Real-World Example 1 Use Properties of Parallelograms

BASKETBALL In ▱$ABCD$, suppose $m\angle A = 55$, $AB = 2.5$ feet, and $BC = 1$ foot. Find each measure.

a. DC

$DC = AB$ — Opp. sides of a ▱ are ≅.

$= 2.5$ ft — Substitution

b. $m\angle B$

$m\angle B + m\angle A = 180$ — Cons. ∡ in a ▱ are supplementary.

$m\angle B + 55 = 180$ — Substitution

$m\angle B = 125$ — Subtract 55 from each side.

c. $m\angle C$

$m\angle C = m\angle A$ — Opp. ∡ of a ▱ are ≅.

$= 55$ — Substitution

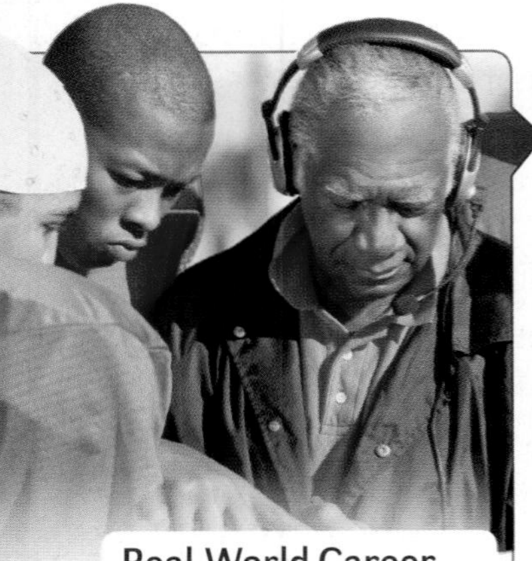

Real-World Career

Coach Coaches organize amateur and professional atheletes, teaching them the fundamentals of a sport. They manage teams during both practice sessions and competitions. Additional tasks may include selecting and issuing sports equiment, materials, and supplies. Head coaches at public secondary schools usually have a bachelor's degree.

Guided Practice

1. MIRRORS The wall-mounted mirror shown uses parallelograms that change shape as the arm is extended. In ▱$JKLM$, suppose $m\angle J = 47$. Find each measure.

A. $m\angle L$ **B.** $m\angle M$

C. Suppose the arm was extended farther so that $m\angle J = 90$. What would be the measure of each of the other angles? Justify your answer.

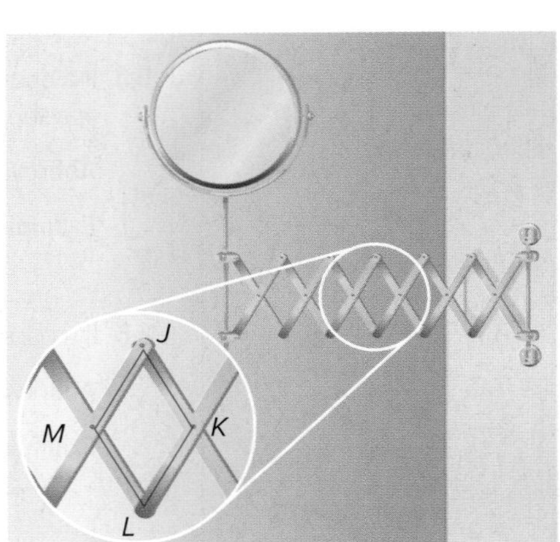

moodboard/SuperStock

Diagonals of Parallelograms
The diagonals of a parallelogram have special properties as well.

Go Online!

Investigate the attributes of parallelograms by using the **Geometry Tools** in ConnectED.

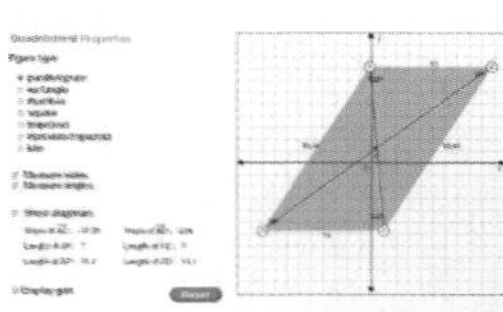

Theorems Diagonals of Parallelograms

6.7 If a quadrilateral is a parallelogram, then its diagonals bisect each other.

Abbreviation *Diag. of a ▱ bisect each other.*

Example If *ABCD* is a parallelogram, then $\overline{AP} \cong \overline{PC}$ and $\overline{DP} \cong \overline{PB}$.

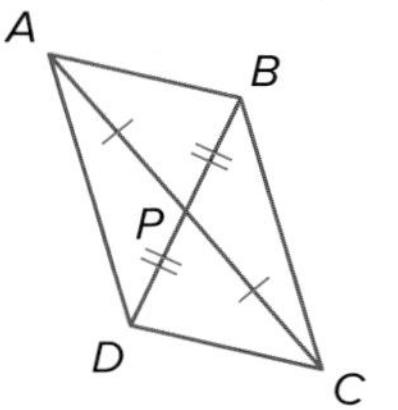

6.8 If a quadrilateral is a parallelogram, then each diagonal separates the parallelogram into two congruent triangles.

Abbreviation *Diag. separates a ▱ into 2 ≅ △s.*

Example If *ABCD* is a parallelogram, then $\triangle ABD \cong \triangle CDB$.

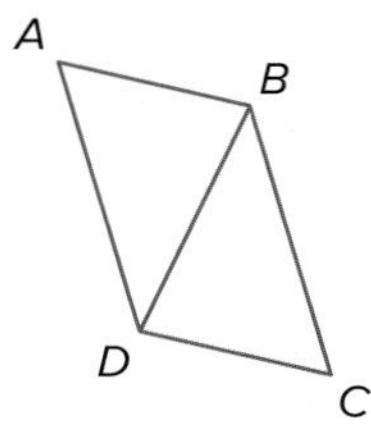

You will prove Theorems 6.7 and 6.8 in Exercises 29 and 27, respectively.

Example 2 Use Properties of Parallelograms and Algebra

ALGEBRA If *QRST* is a parallelogram, find the value of the indicated variable.

a. x

$\overline{QT} \cong \overline{RS}$	Opp. sides of a ▱ are ≅.
$QT = RS$	Definition of congruence
$5x = 27$	Substitution
$x = 5.4$	Divide each side by 5.

b. y

$\overline{TP} \cong \overline{PR}$	Diag. of a ▱ bisect each other.
$TP = PR$	Definition of congruence
$2y - 5 = y + 4$	Substitution
$y = 9$	Subtract y and add 5 to each side.

c. z

$\triangle TQS \cong \triangle RSQ$	Diag. separates a ▱ into 2 ≅ △s.
$\angle QST \cong \angle SQR$	CPCTC
$m\angle QST = m\angle SQR$	Definition of congruence
$3z = 33$	Substitution
$z = 11$	Divide each side by 3.

Study Tip

Congruent Triangles A parallelogram with two diagonals divides the figure into two pairs of congruent triangles.

Guided Practice

Find the value of each variable in the given parallelogram.

2A.

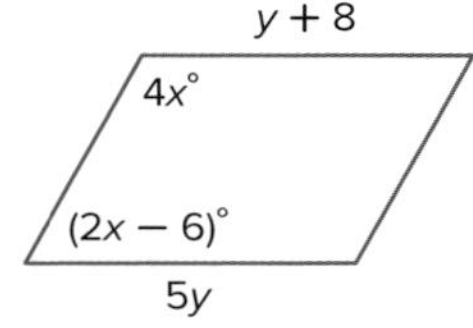

2B.

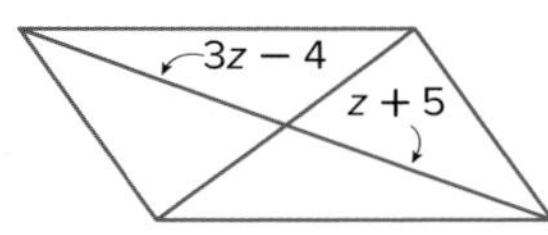

You can use Theorem 6.7 to determine the coordinates of the intersection of the diagonals of a parallelogram on a coordinate plane given the coordinates of the vertices.

Example 3 Parallelograms and Coordinate Geometry

COORDINATE GEOMETRY Determine the coordinates of the intersection of the diagonals of ▱*FGHJ* with vertices *F*(−2, 4), *G*(3, 5), *H*(2, −3), and *J*(−3, −4).

Since the diagonals of a parallelogram bisect each other, their intersection point is the midpoint of $\overline{FH}$ and $\overline{GJ}$. Find the midpoint of $\overline{FH}$ with endpoints (−2, 4) and (2, −3).

$$\left(\frac{x_1 + x_2}{2}, \frac{y_1 + y_2}{2}\right) = \left(\frac{-2 + 2}{2}, \frac{4 + (-3)}{2}\right) \quad \text{Midpoint Formula}$$

$$= (0, 0.5) \quad \text{Simplify.}$$

The coordinates of the intersection of the diagonals of ▱*FGHJ* are (0, 0.5).

CHECK Find the midpoint of $\overline{GJ}$ with endpoints (3, 5) and (−3, −4).

$$\left(\frac{3 + (-3)}{2}, \frac{5 + (-4)}{2}\right) = (0, 0.5) \checkmark$$

Study Tip

MP Regularity Graph the parallelogram in Example 3 and the point of intersection of the diagonals you found. Draw the diagonals. The point of intersection appears to be correct.

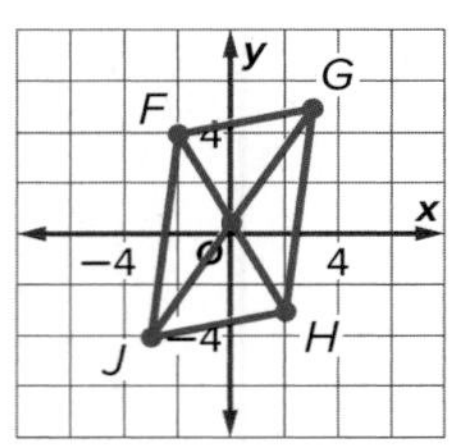

Guided Practice

3. **COORDINATE GEOMETRY** Determine the coordinates of the intersection of the diagonals of *RSTU* with vertices *R*(−8, −2), *S*(−6, 7), *T*(6, 7), and *U*(4, −2).

You can use the properties of parallelograms and their diagonals to write proofs.

Example 4 Proofs Using the Properties of Parallelograms

Write a paragraph proof.

Given: ▱*ABDG*, $\overline{AF} \cong \overline{CF}$

Prove: $\angle BDG \cong \angle C$

Proof:

We are given *ABDG* is a parallelogram. Because opposite angles in a parallelogram are congruent, $\angle BDG \cong \angle A$. We are also given that $\overline{AF} \cong \overline{CF}$. By the Isosceles Triangle Theorem, $\angle A \cong \angle C$. So, by the Transitive Property of Congruence, $\angle BDG \cong \angle C$.

Guided Practice

4. Write a two-column proof.

Given: ▱*HJKP* and ▱*PKLM*

Prove: $\overline{HJ} \cong \overline{ML}$

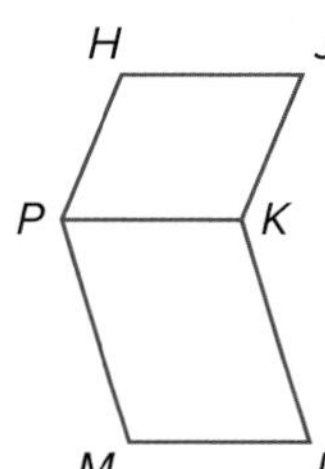

Check Your Understanding

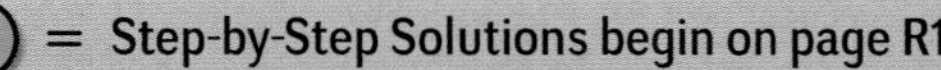

Example 1

1. **NAVIGATION** To chart a course, sailors use a *parallel ruler.* One edge of the ruler is placed along the line representing the direction of the course to be taken. Then the other ruler is moved until its edge reaches the compass rose printed on the chart. Reading the compass determines which direction to travel. The rulers and the crossbars of the tool form ▱$MNPQ$.

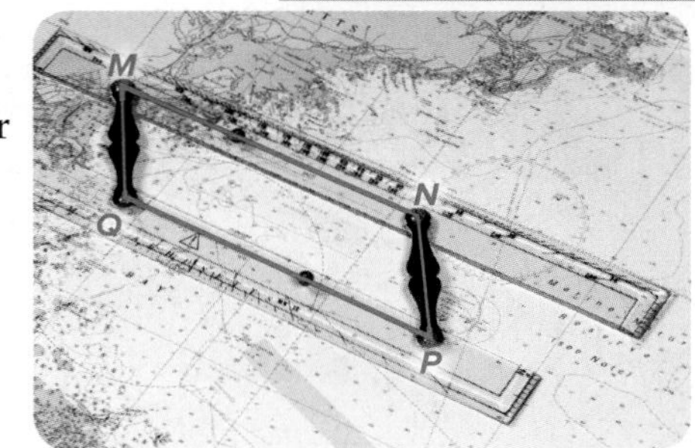

a. If $m\angle NMQ = 32$, find $m\angle MNP$.

b. If $m\angle MQP = 125$, find $m\angle MNP$.

c. If $MQ = 4$, what is NP?

Example 2

ALGEBRA Find the value of each variable in each parallelogram.

2.

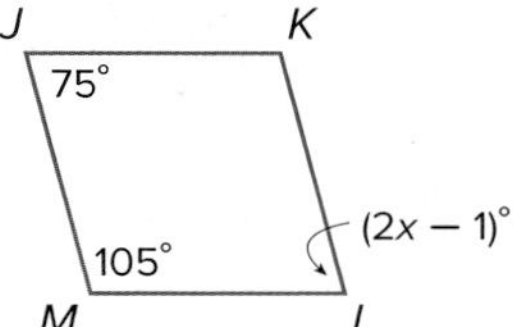

3.

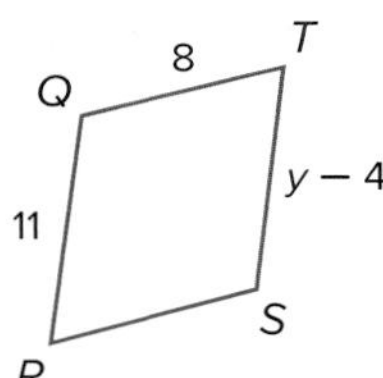

4.

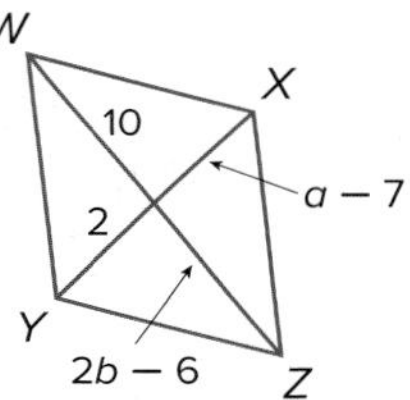

5.

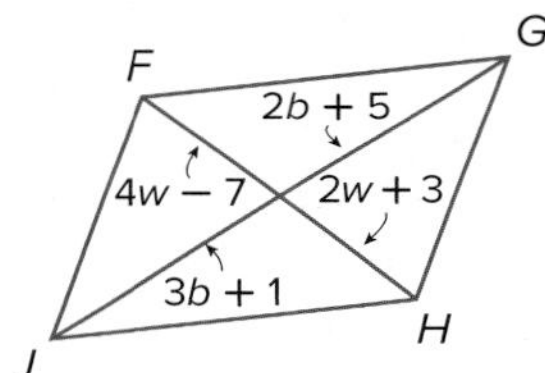

Example 3

6. **COORDINATE GEOMETRY** Determine the coordinates of the intersection of the diagonals of ▱$ABCD$ with vertices $A(-4, 6)$, $B(5, 6)$, $C(4, -2)$, and $D(-5, -2)$.

Example 4

MP CONSTRUCT ARGUMENTS Write the indicated type of proof.

7. paragraph

Given: ▱$ABCD$, $\angle A$ is a right angle.
Prove: $\angle B$, $\angle C$, and $\angle D$ are right angles. (Theorem 6.6)

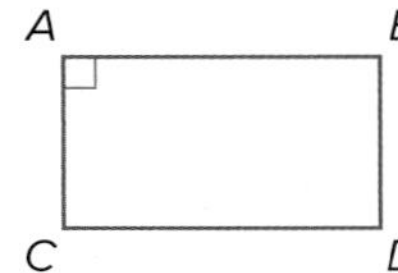

8. two-column

Given: $ABCH$ and $DCGF$ are parallelograms.
Prove: $\angle A \cong \angle F$

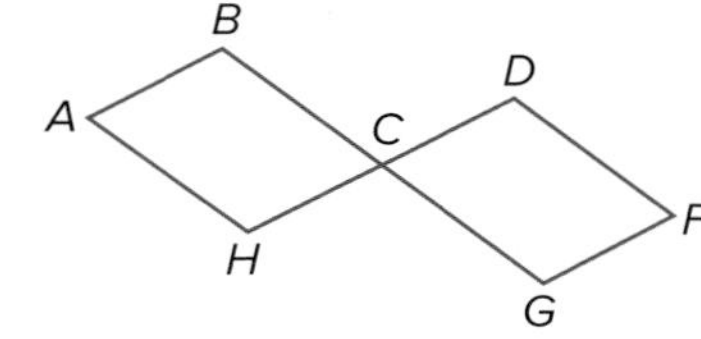

Practice and Problem Solving

Extra Practice is on page R6.

Example 1

Use ▱$PQRS$ to find each measure.

9. $m\angle R$

10. QR

11. QP

12. $m\angle S$

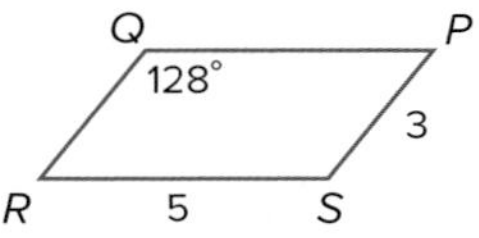

13. **HOME DECOR** The slats on Venetian blinds are designed to remain parallel in order to direct the path of light coming in a window. In $\square FGHJ$, $FJ = \frac{3}{4}$ inch, $FG = 1$ inch, and $m\angle JHG = 62$. Find each measure.

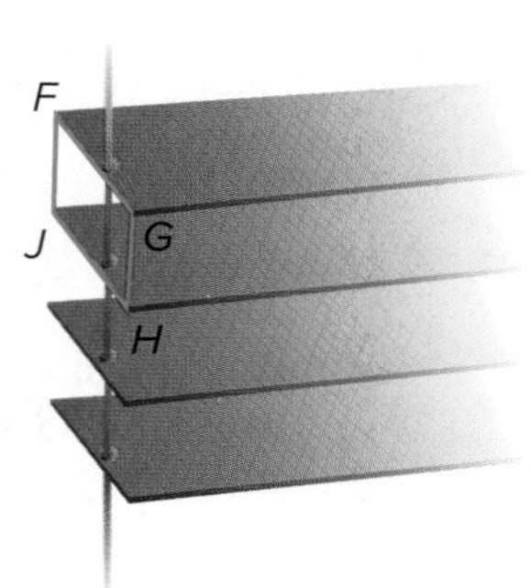

a. JH

b. GH

c. $m\angle JFG$

d. $m\angle FJH$

14. **MP MODELING** Wesley is a member of the kennel club in his area. His club uses accordion fencing like the section shown at the right to block out areas at dog shows.

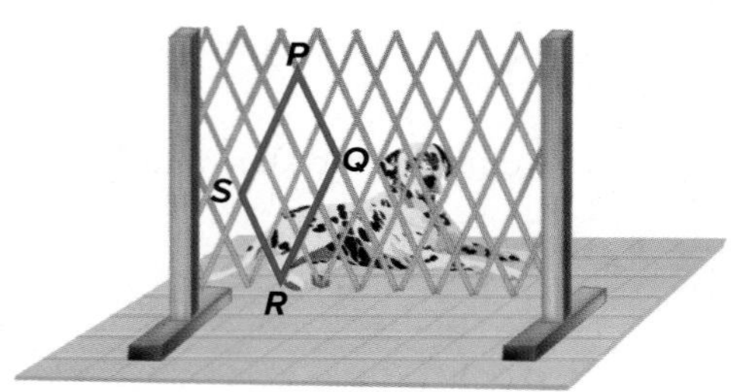

a. Identify two pairs of congruent segments.

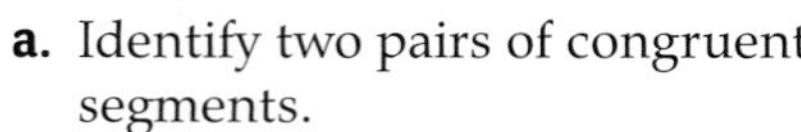

b. Identify two pairs of supplementary angles.

Example 2

ALGEBRA **Find the value of each variable in each parallelogram.**

15.

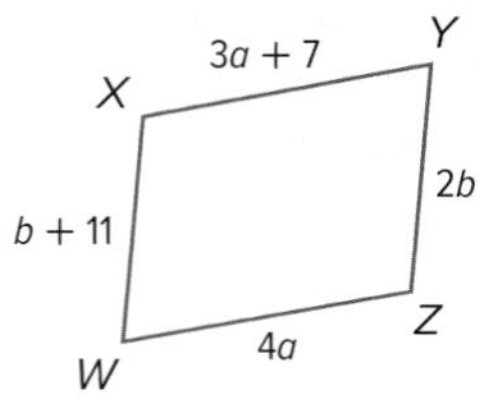

16.

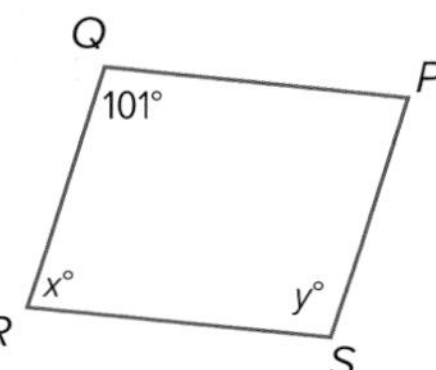

17.

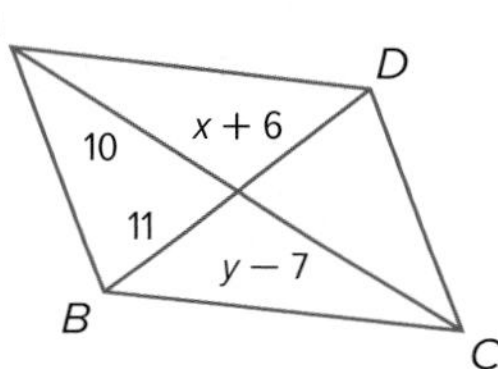

18.

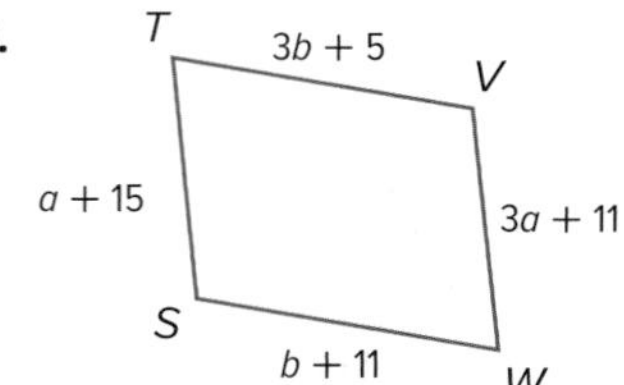

19.

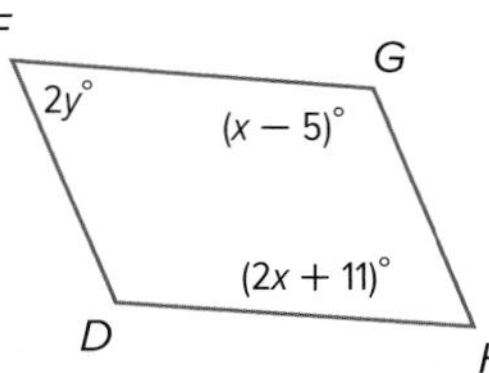

20.

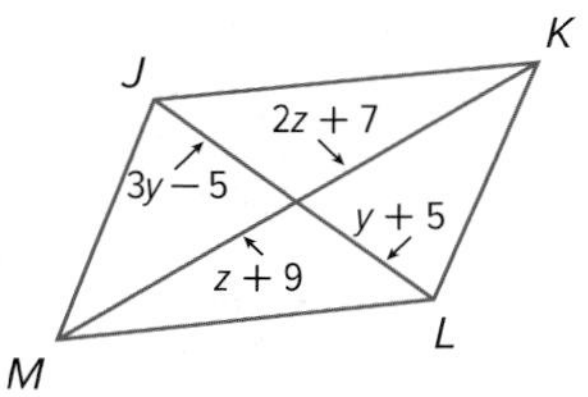

Example 3

COORDINATE GEOMETRY **Find the coordinates of the intersection of the diagonals of $\square WXYZ$ with the given vertices.**

21. $W(-1, 7)$, $X(8, 7)$, $Y(6, -2)$, $Z(-3, -2)$

22. $W(-4, 5)$, $X(5, 7)$, $Y(4, -2)$, $Z(-5, -4)$

Example 4

PROOF **Write a two-column proof.**

23. Given: $WXTV$ and $ZYVT$ are parallelograms.
Prove: $\overline{WX} \cong \overline{ZY}$

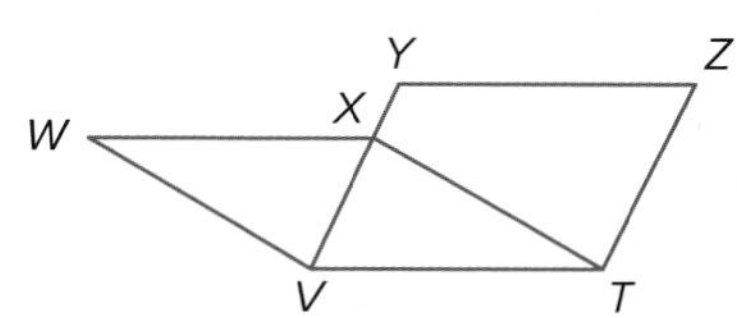

24. Given: $\square BDHA$, $\overline{CA} \cong \overline{CG}$
Prove: $\angle BDH \cong \angle G$

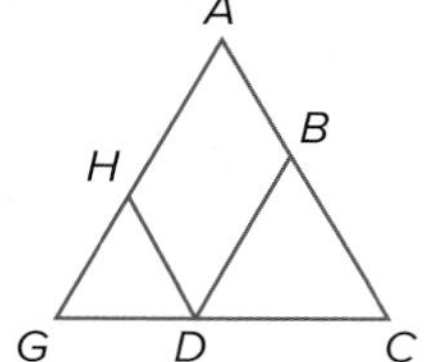

25. FLAGS Refer to the Alabama state flag at the right.

Given: $\triangle ACD \cong \triangle CAB$

Prove: $\overline{DP} \cong \overline{PB}$

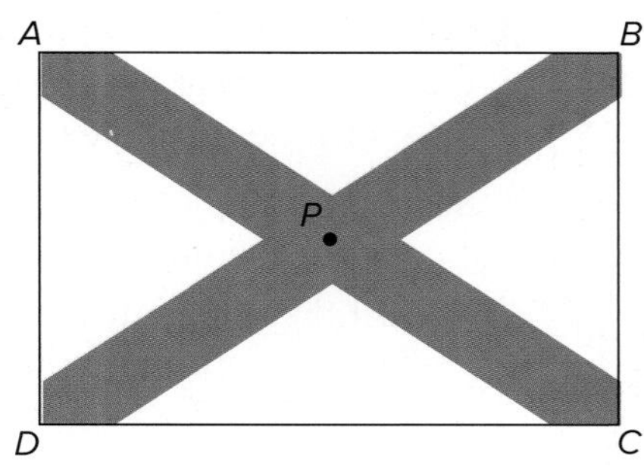

CONSTRUCT ARGUMENTS Write the indicated type of proof.

26. two-column
Given: ▱*GKLM*
Prove: $\angle G$ and $\angle K$, $\angle K$ and $\angle L$, $\angle L$ and $\angle M$, and $\angle M$ and $\angle G$ are supplementary. (Theorem 6.5)

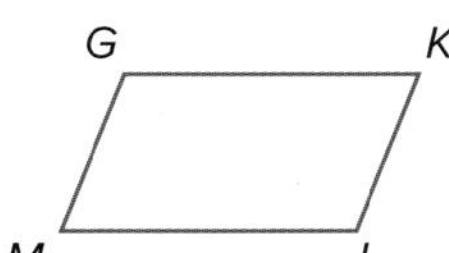

27. two-column
Given: ▱*WXYZ*
Prove: $\triangle WXZ \cong \triangle YZX$ (Theorem 6.8)

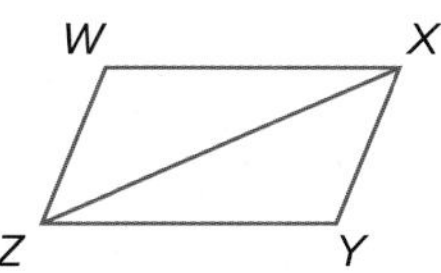

28. two-column
Given: ▱*PQRS*
Prove: $\overline{PQ} \cong \overline{RS}$, $\overline{QR} \cong \overline{SP}$ (Theorem 6.3)

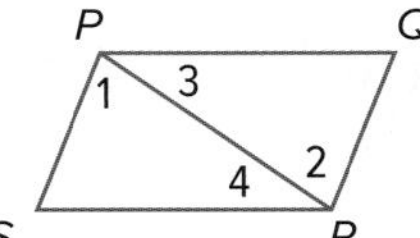

29. paragraph
Given: ▱*ACDE* is a parallelogram.
Prove: $\overline{EC}$ bisects $\overline{AD}$. (Theorem 6.7)

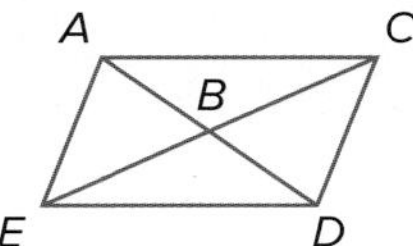

30. COORDINATE GEOMETRY Use the graph shown.

a. Use the Distance Formula to determine if the diagonals of *JKLM* bisect each other. Explain.

b. Determine whether the diagonals are congruent. Explain.

c. Use slopes to determine if the consecutive sides are perpendicular. Explain.

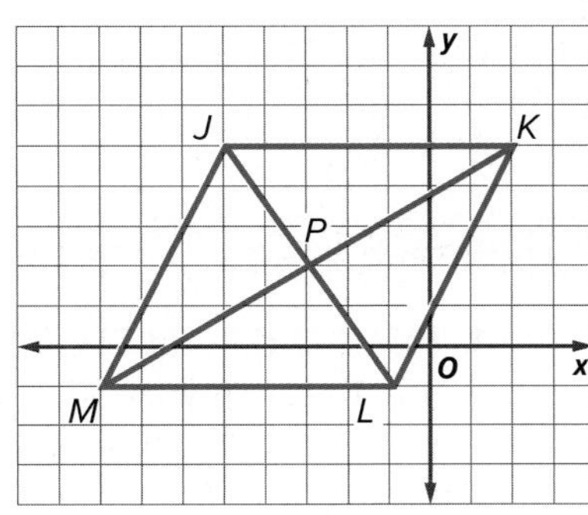

ALGEBRA Use ▱*ABCD* to find each measure or value.

31. x

32. y

33 $m\angle AFB$

34. $m\angle DAC$

35. $m\angle ACD$

36. $m\angle DAB$

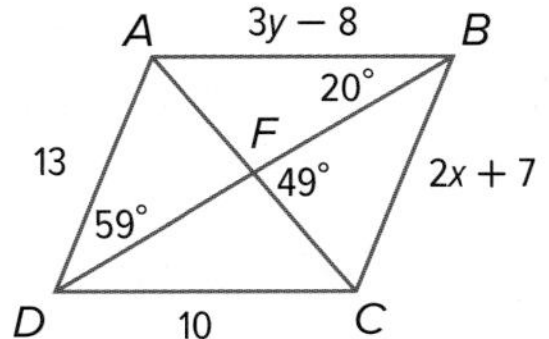

37. COORDINATE GEOMETRY ▱*ABCD* has vertices $A(-3, 5)$, $B(1, 2)$, and $C(3, -4)$. Determine the coordinates of vertex D if it is located in Quadrant III. Explain.

38. MECHANICS Scissor lifts are variable elevation work platforms. One is shown at the right. In the diagram, *ABCD* and *DEFG* are congruent parallelograms.

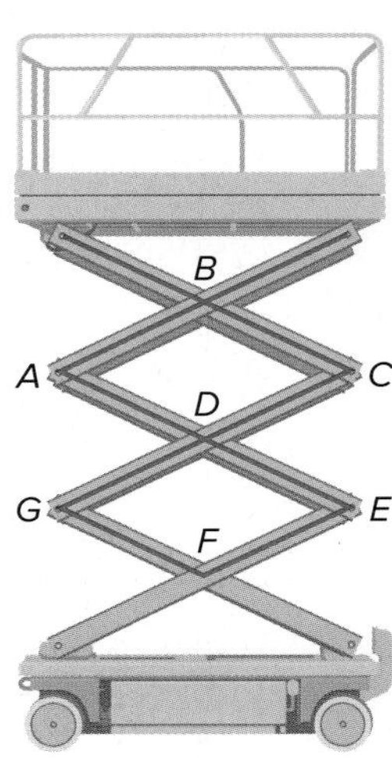

a. List the angle(s) congruent to $\angle A$. Explain your reasoning.

b. List the segment(s) congruent to $\overline{BC}$. Explain your reasoning.

c. List the angle(s) supplementary to $\angle C$. Explain your reasoning.

PROOF **Write a two-column proof.**

39 **Given:** $\square YWVZ$, $\overline{VX} \perp \overline{WY}$, $\overline{YU} \perp \overline{VZ}$
Prove: $\triangle YUZ \cong \triangle VXW$

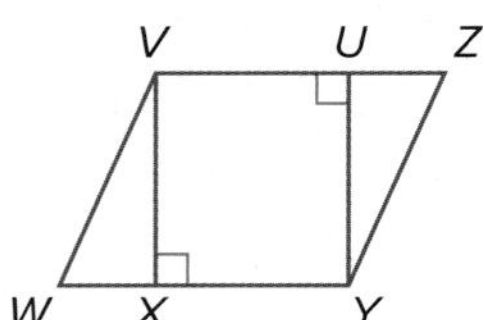

40. MULTIPLE REPRESENTATIONS In this problem, you will explore tests for parallelograms.

a. **Geometric** Draw three pairs of segments that are both congruent and parallel and connect the endpoints to form quadrilaterals. Label one quadrilateral *ABCD*, one *MNOP*, and one *WXYZ*. Measure and label the sides and angles of the quadrilaterals.

b. **Tabular** Copy and complete the table below.

Quadrilateral	Opposite Sides Congruent?	Opposite Angles Congruent?	Parallelogram
ABCD			
MNOP			
WXYZ			

c. **Verbal** Make a conjecture about quadrilaterals with one pair of segments that are both congruent and parallel.

H.O.T. Problems Use Higher-Order Thinking Skills

41. CHALLENGE *ABCD* is a parallelogram with side lengths as indicated in the figure at the right. The perimeter of *ABCD* is 22. Find *AB*.

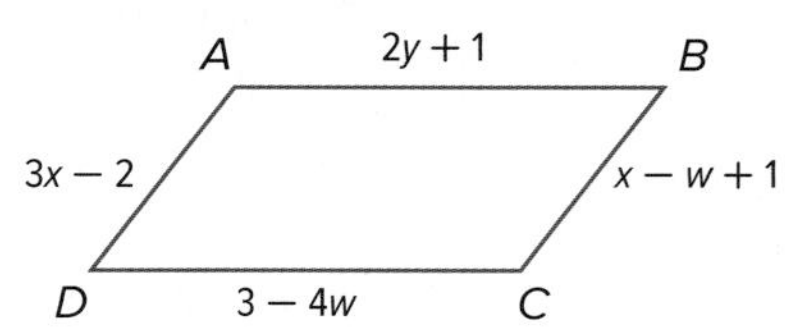

42. WRITING IN MATH Explain why parallelograms are *always* quadrilaterals, but quadrilaterals are *sometimes* parallelograms.

43. OPEN-ENDED Provide a counterexample to show that parallelograms are not always congruent if their corresponding sides are congruent.

44. MP REASONING Find $m\angle 1$ and $m\angle 10$ in the figure at the right. Explain.

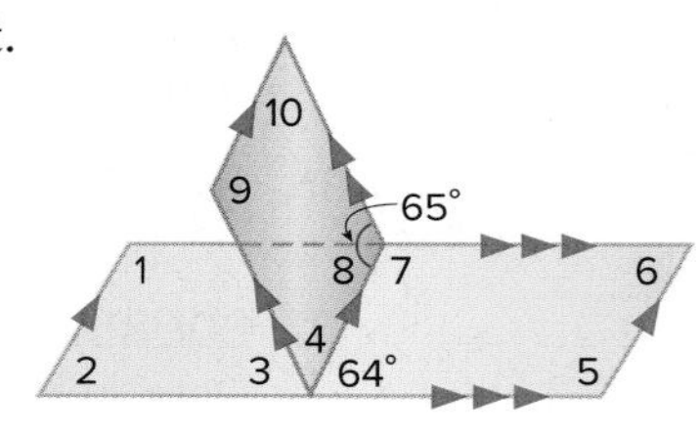

45. WRITING IN MATH Summarize the properties of the sides, angles, and diagonals of a parallelogram.

Preparing for Assessment

46. The park near Karla's home is shaped like a parallelogram, as shown in the figure. Side $\overline{JK}$ is 3 times as long as side $\overline{KL}$, and the perimeter of the park is 1200 feet.

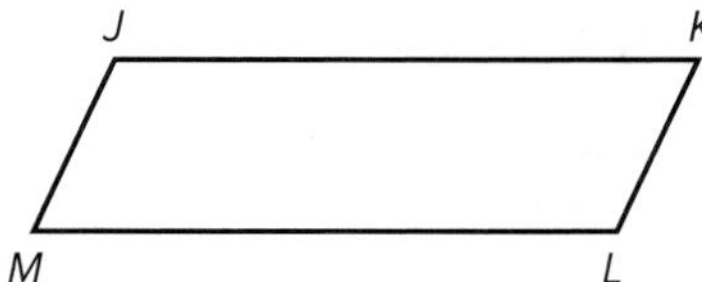

What is the length of $\overline{KL}$? MP 2

- ○ **A** 450 ft
- ○ **B** 400 ft
- ○ **C** 300 ft
- ○ **D** 150 ft

47. In parallelogram $ABCD$, $\angle A$ and $\angle B$ are consecutive angles. The measure of $\angle A$ is 40 more than the measure of $\angle B$. What is the measure of $\angle A$? MP 2

- ○ **A** 65
- ○ **B** 70
- ○ **C** 110
- ○ **D** 140

48. In the figure, quadrilateral $MNPQ$ is a parallelogram. Which of the following statements about the figure must be true? MP 3

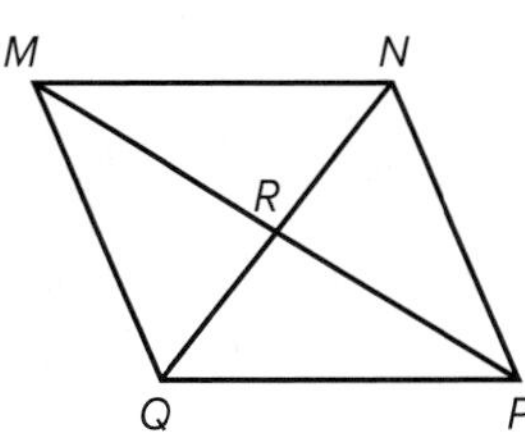

I. $\angle MNP \cong \angle PQM$

II. $\overline{MR} \cong \overline{PR}$

III. $\overline{MQ} \cong \overline{MN}$

- ○ **A** I only
- ○ **B** II only
- ○ **C** III only
- ○ **D** I and II only
- ○ **E** I, II, and III

49. In parallelogram $ABCD$, the measure of $\angle A$ is 55°, the measure of $\angle C$ is $(4x + 11)°$. MP 2

a. What is the value of x? ▭

b. What is the measure of $\angle B$? ▭°

50. What is the x-coordinate of the intersection of the diagonals of a parallelogram with vertices $A(-5, 0)$, $B(3, 4)$, $C(6, 3)$, and $D(-2, -1)$? MP 6

▭

51. **MULTI-STEP** Quadrilateral $PQRS$ is a parallelogram. MP 3

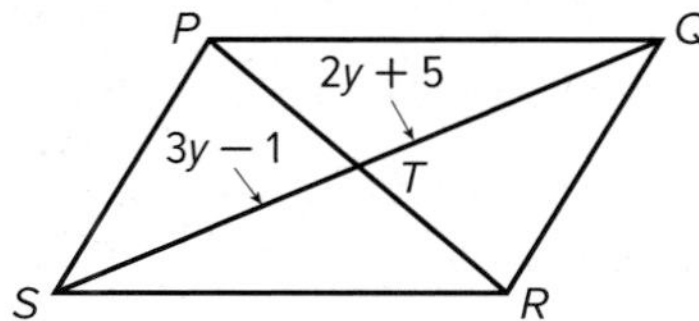

a. Which Theorem defines a relationship between $\overline{PR}$ and $\overline{SQ}$? ▭

b. Find SQ. ▭

c. Prove $\triangle PST \cong \triangle QRT$.

d. What mathematical practice did you use to solve this problem?

52. The coordinates of a quadrilateral are $J(-2, 4)$, $K(-1, -1)$, $L(9, 1)$, and $M(8, 6)$. Which statements are true? Select all that apply. MP 6

- ☐ **A** $\overline{JM} \cong \overline{LM}$
- ☐ **B** $\overline{JL} \cong \overline{KM}$
- ☐ **C** $\triangle JML \cong \triangle LKJ$
- ☐ **D** $\triangle JMK \cong \triangle LKM$
- ☐ **E** $JKLM$ is a rhombus.

53. In parallelogram $EFGH$, the measure of $\angle E$ is $(4x + 18)°$ and the measure of $\angle G$ is $(2x + 54)°$. What must be true about the parallelogram? MP 1

- ○ **A** $EFGH$ is a rhombus.
- ○ **B** $EFGH$ is a trapezoid.
- ○ **C** The measure of $\angle E$ is 18°.
- ○ **D** The measure of $\angle F$ is 90°.

EXPLORE 6-3

Geometry Software Lab

Parallelograms

You can use The Geometer's Sketchpad to discover properties of parallelograms.

Mathematical Practices

MP **5** Use appropriate tools strategically.

Activity

Construct a quadrilateral with two diagonals that bisect each other.

Step 1 Construct a segment using the segment tool. Label the endpoints A and C.

Step 2 Construct the midpoint of $\overline{AC}$. Label it P.

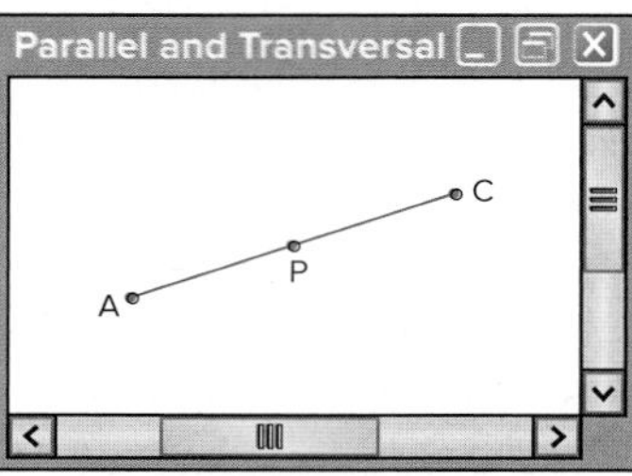

Steps 1 and 2

Step 3 Use the circle tool to draw a circle centered at P.

Step 4 Use the segment tool to draw a diameter of the circle. Label the endpoints B and D.

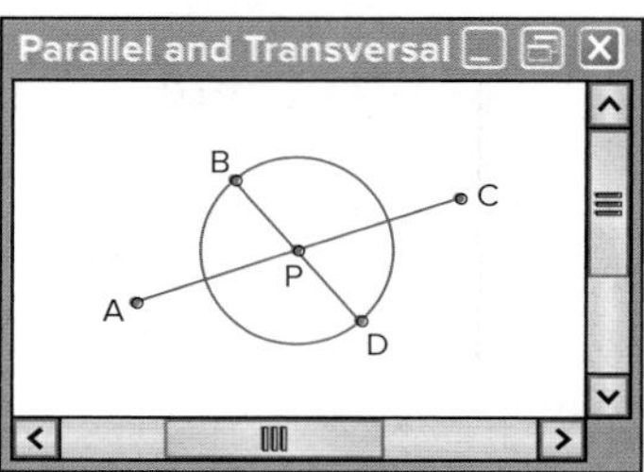

Steps 3 and 4

Step 5 Draw quadrilateral $ABCD$. Hide the circle.

Step 6 **Make a conjecture** about the slopes of each side of the quadrilateral.

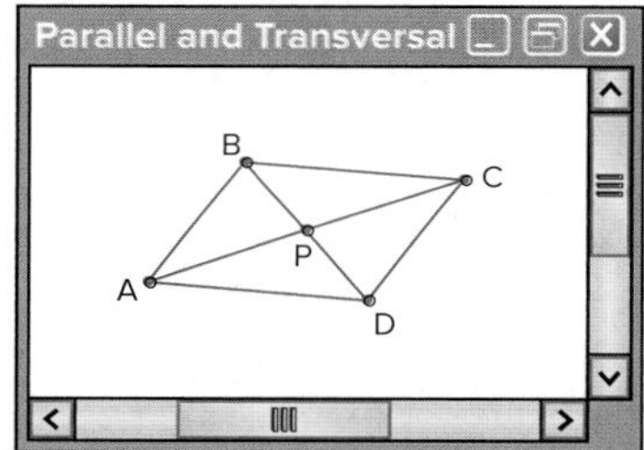

Step 5 and 6

Analyze the Results

Work cooperatively.

1. What is the relationship between $\overline{AC}$ and $\overline{BD}$? Explain how you know.
2. What do you observe about the slopes of opposite sides of the quadrilateral? What type of quadrilateral is $ABCD$? Explain.
3. Click on point D and drag it to change the shape of $ABDC$. What do you observe?
4. Make a conjecture about a quadrilateral with a pair of diagonals that bisect each other.
5. Construct a quadrilateral with both pairs of opposite sides congruent. Then analyze the slopes of the sides of the quadrilateral. Make a conjecture based on your observations.

LESSON 3

Tests for Parallelograms

Then

- You recognized and applied properties of parallelograms.

Now

1. Recognize the conditions that ensure a quadrilateral is a parallelogram.
2. Prove that a set of points forms a parallelogram in the coordinate plane.

Why?

- Lexi and Rosalinda cut strips of bulletin board paper at an angle to form the hallway display shown. Their friends asked them how they cut the strips so that their sides were parallel without using a protractor.

Rosalinda explained that since the left and right sides of the paper were parallel, she only needed to make sure that the sides were cut to the same length to guarantee that a strip would form a parallelogram.

Mathematical Practices

3 Construct viable arguments and critique the reasoning of others.

2 Reason abstractly and quantitatively.

1 Conditions for Parallelograms

If a quadrilateral has each pair of opposite sides parallel, it is a parallelogram by definition.

This is not the only test, however, that can be used to determine if a quadrilateral is a parallelogram.

Theorems Conditions for Parallelograms

6.9 If both pairs of opposite sides of a quadrilateral are congruent, then the quadrilateral is a parallelogram.

Abbreviation *If both pairs of opp. sides are ≅, then quad. is a ▱.*

Example If $\overline{AB} \cong \overline{DC}$ and $\overline{AD} \cong \overline{BC}$, then $ABCD$ is a parallelogram.

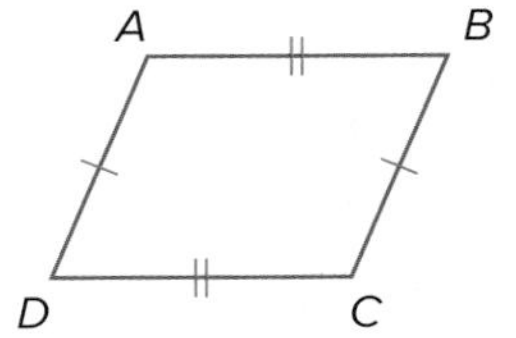

6.10 If both pairs of opposite angles of a quadrilateral are congruent, then the quadrilateral is a parallelogram.

Abbreviation *If both pairs of opp. ∠s are ≅, then quad. is a ▱.*

Example If $\angle A \cong \angle C$ and $\angle B \cong \angle D$, then $ABCD$ is a parallelogram.

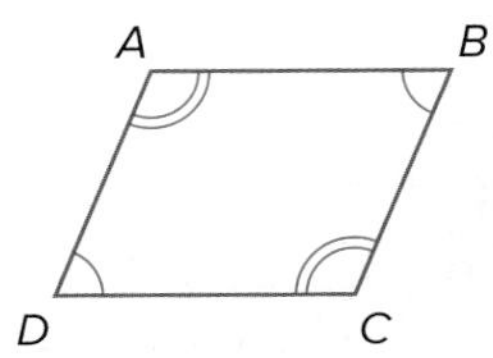

6.11 If the diagonals of a quadrilateral bisect each other, then the quadrilateral is a parallelogram.

Abbreviation *If diag. bisect each other, then quad. is a ▱.*

Example If $\overline{AC}$ and $\overline{DB}$ bisect each other, then $ABCD$ is a parallelogram.

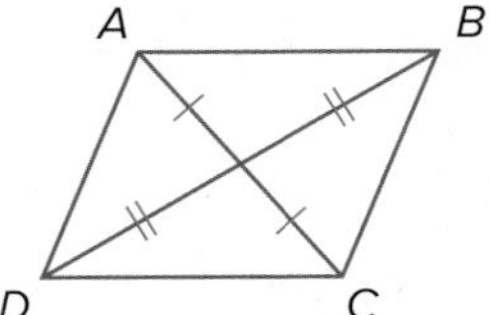

6.12 If one pair of opposite sides of a quadrilateral is both parallel and congruent, then the quadrilateral is a parallelogram.

Abbreviation *If one pair of opp. sides is ≅ and ||, then the quad. is a ▱.*

Example If $\overline{AB} \parallel \overline{DC}$ and $\overline{AB} \cong \overline{DC}$, then $ABCD$ is a parallelogram.

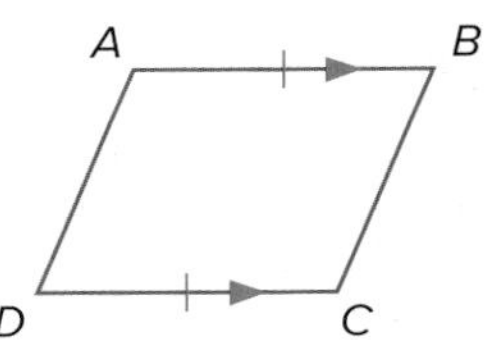

You will prove Theorems 6.10, 6.11, and 6.12 in Exercises 30, 32, and 33, respectively.

Proof Theorem 6.9

Write a paragraph proof of Theorem 6.9.

Given: $\overline{WX} \cong \overline{ZY}$, $\overline{WZ} \cong \overline{XY}$

Prove: $WXYZ$ is a parallelogram.

Paragraph Proof:

Two points determine a line, so we can draw auxiliary line $\overline{ZX}$ to form $\triangle ZWX$ and $\triangle XYZ$. We are given that $\overline{WX} \cong \overline{ZY}$ and $\overline{WZ} \cong \overline{XY}$. Also, $\overline{ZX} \cong \overline{XZ}$ by the Reflexive Property of Congruence. So $\triangle ZWX \cong \triangle XYZ$ by SSS. By CPCTC, $\angle WXZ \cong \angle YZX$ and $\angle WZX \cong \angle YXZ$. This means that $\overline{WX} \parallel \overline{ZY}$ and $\overline{WZ} \parallel \overline{XY}$ by the Alternate Interior Angles Converse. Opposite sides of $WXYZ$ are parallel, so by definition $WXYZ$ is a parallelogram.

Example 1 Identify Parallelograms

Determine whether the quadrilateral is a parallelogram. Justify your answer.

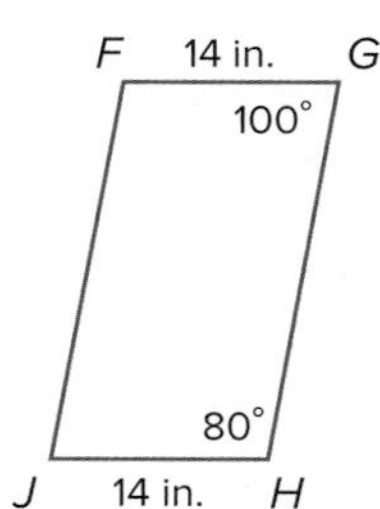

Opposite sides $\overline{FG}$ and $\overline{JH}$ are congruent because they have the same measure. Also, since $\angle FGH$ and $\angle GHJ$ are supplementary consecutive interior angles, $\overline{FG} \parallel \overline{JH}$. Therefore, by Theorem 6.12, $FGHJ$ is a parallelogram.

Guided Practice

1A. 12 cm, 5 cm, 5 cm, 12 cm

1B. 85°, 85°

You can use the conditions of parallelograms to prove relationships in real-world situations.

Real-World Example 2 Use Parallelograms to Prove Relationships

FISHING The diagram shows a side view of the tackle box at the left. In the diagram, $PQ = RS$ and $PR = QS$. Explain why the upper and middle trays remain parallel no matter to what height the trays are raised or lowered.

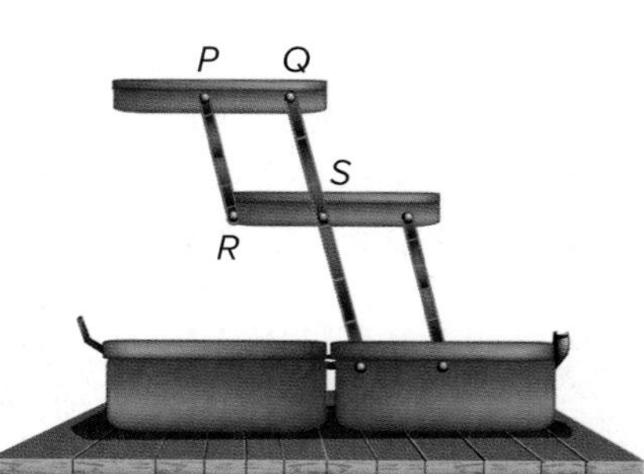

Because both pairs of opposite sides of quadrilateral $PQSR$ are congruent, $PQRS$ is a parallelogram by Theorem 6.9. By the definition of a parallelogram, opposite sides are parallel, so $\overline{PQ} \parallel \overline{RS}$. Therefore, no matter the vertical position of the trays, they will always remain parallel.

Guided Practice

2. BANNERS In the example at the beginning of the lesson, explain why the cuts made by Lexi and Rosalinda are parallel.

Real-World Link

A 2- or 3-cantilever tackle box is often used to organize lures and other fishing supplies. The trays lift up and away so that all items in the box are easily accessible.

You can also use the conditions of parallelograms along with algebra to find missing values that make a quadrilateral a parallelogram.

Watch Out!

MP Sense-Making In Example 3, if x is 4, then y must be 2.5 in order for $FGHJ$ to be a parallelogram. In other words, if x is 4 and y is 1, then $FGHJ$ is not a parallelogram.

Example 3 Use Parallelograms and Algebra to Find Values

If $FK = 3x - 1$, $KG = 4y + 3$, $JK = 6y - 2$, and $KH = 2x + 3$, find x and y so that the quadrilateral is a parallelogram.

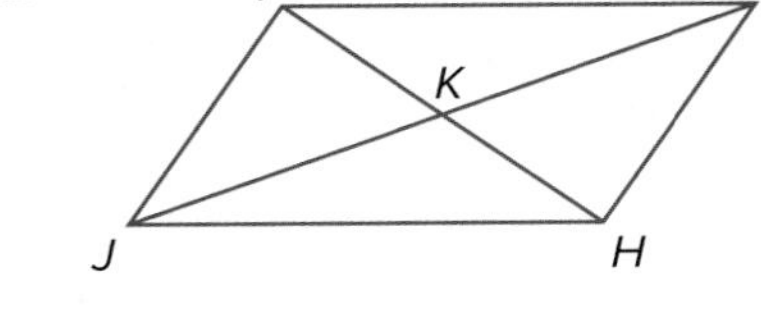

By Theorem 6.11, if the diagonals of a quadrilateral bisect each other, then it is a parallelogram. So find x such that $\overline{FK} \cong \overline{KH}$ and y such that $\overline{JK} \cong \overline{KG}$.

$FK = KH$	Definition of ≅
$3x - 1 = 2x + 3$	Substitution
$x - 1 = 3$	Subtract $2x$ from each side.
$x = 4$	Add 1 to each side.
$JK = KG$	Definition of ≅
$6y - 2 = 4y + 3$	Substitution
$2y - 2 = 3$	Subtract $4y$ from each side.
$2y = 5$	Add 2 to each side.
$y = 2.5$	Divide each side by 2.

So, when x is 4 and y is 2.5, quadrilateral $FGHJ$ is a parallelogram.

Guided Practice

Find x and y so that each quadrilateral is a parallelogram.

3A.

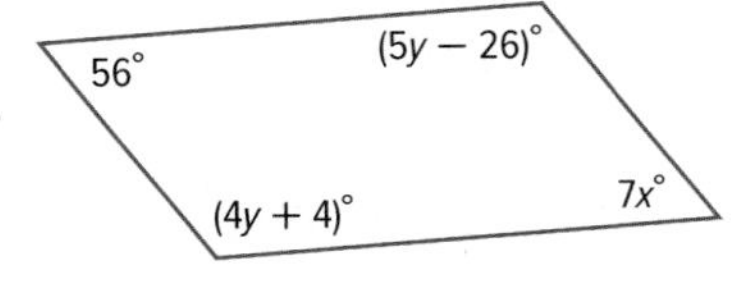

3B.

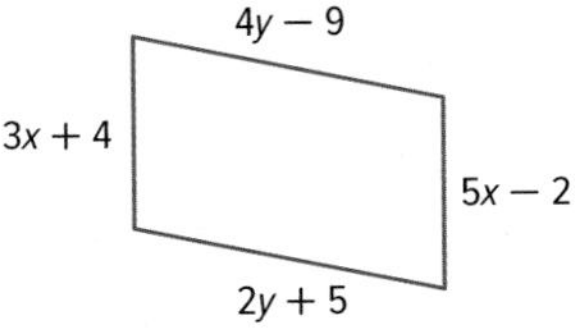

Go Online!

How can you determine whether a quadrilateral is a parallelogram? Investigate by using the **Geometry Tools** in ConnectED. Discuss your findings with a partner. Ask for clarification as you need it.

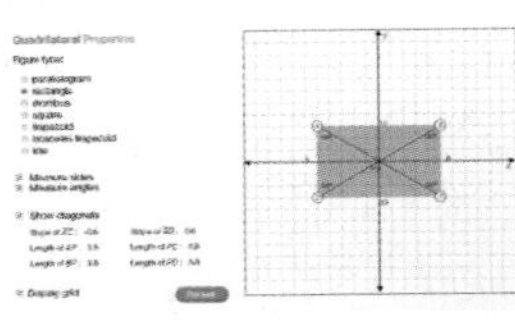

You have learned the conditions of parallelograms. The following list summarizes how to use the conditions to prove a quadrilateral is a parallelogram.

Concept Summary

Prove that a Quadrilateral Is a Parallelogram

- Show that both pairs of opposite sides are parallel. (Definition)
- Show that both pairs of opposite sides are congruent. (Theorem 6.9)
- Show that both pairs of opposite angles are congruent. (Theorem 6.10)
- Show that the diagonals bisect each other. (Theorem 6.11)
- Show that a pair of opposite sides is both parallel and congruent. (Theorem 6.12)

Study Tip

MP Sense-Making To show that a quadrilateral is a parallelogram, you can also use the Midpoint Formula. If the midpoint of each diagonal is the same point, then the diagonals bisect each other.

2 Parallelograms on the Coordinate Plane

We can use the Distance, Slope, and Midpoint Formulas to determine whether a quadrilateral in the coordinate plane is a parallelogram.

Example 4 Parallelograms and Coordinate Geometry

COORDINATE GEOMETRY Graph quadrilateral *KLMN* with vertices *K*(2, 3), *L*(8, 4), *M*(7, −2), and *N*(1, −3). Determine whether the quadrilateral is a parallelogram. Justify your answer using the Slope Formula.

If the opposite sides of a quadrilateral are parallel, then it is a parallelogram.

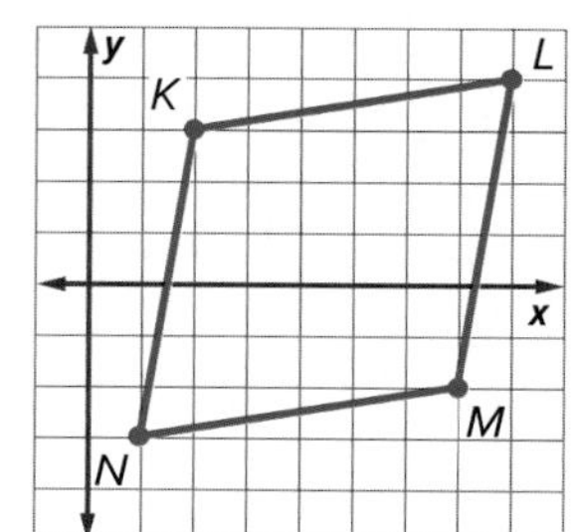

slope of $\overline{KL} = \frac{4-3}{8-2}$ or $\frac{1}{6}$

slope of $\overline{NM} = \frac{-2-(-3)}{7-1}$ or $\frac{1}{6}$

slope of $\overline{KN} = \frac{-3-3}{1-2} = \frac{-6}{-1}$ or 6

slope of $\overline{LM} = \frac{-2-4}{7-8} = \frac{-6}{-1}$ or 6

Because opposite sides have the same slope, $\overline{KL} \parallel \overline{NM}$ and $\overline{KN} \parallel \overline{LM}$. Therefore, *KLMN* is a parallelogram by definition.

Guided Practice

Determine whether the quadrilateral is a parallelogram. Justify your answer using the given formula.

4A. *A*(3, 3), *B*(8, 2), *C*(6, −1), *D*(1, 0); Distance Formula

4B. *F*(−2, 4), *G*(4, 2), *H*(4, −2), *J*(−2, −1); Midpoint Formula

In Chapter 4, you learned that variable coordinates can be assigned to the vertices of triangles. Then the Distance, Slope, and Midpoint Formulas were used to write coordinate proofs of theorems. The same can be done with quadrilaterals.

Review Vocabulary

coordinate proof a proof that uses figures in the coordinate plane and algebra to prove geometric concepts

Example 5 Parallelograms and Coordinate Proofs

Write a coordinate proof for the following statement.

If one pair of opposite sides of a quadrilateral is both parallel and congruent, then the quadrilateral is a parallelogram.

Step 1 Position quadrilateral *ABCD* on the coordinate plane such that $\overline{AB} \parallel \overline{DC}$ and $\overline{AB} \cong \overline{DC}$.

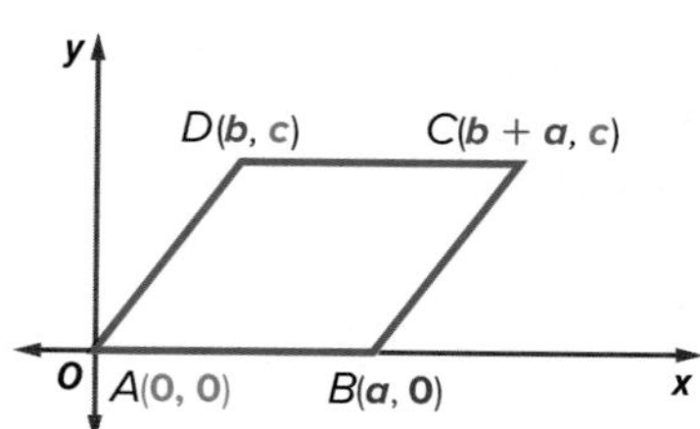

- Begin by placing the vertex *A* at the origin.
- Let $\overline{AB}$ have a length of *a* units. Then *B* has coordinates **(*a*, 0)**.
- Because horizontal segments are parallel, position the endpoints of $\overline{DC}$ so that they have the same *y*-coordinate, ***c***.
- So that the distance from *D* to *C* is also *a* units, let the *x*-coordinate of *D* be ***b*** and of *C* be ***b* + *a***.

Math History Link

René Descartes
(1596–1650)
René Descartes was a French mathematician who was the first to use a coordinate grid. It has been said that he first thought of locating a point on a plane with a pair of numbers when he was watching a fly on the ceiling, but this is a myth.

Step 2 Use your figure to write a proof.

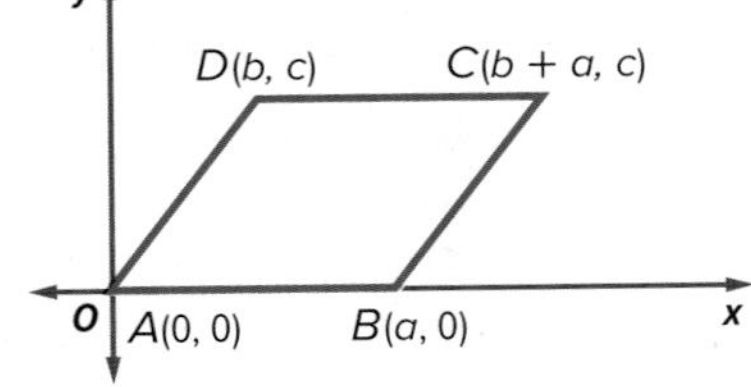

Given: quadrilateral $ABCD$, $\overline{AB} \parallel \overline{DC}$, $\overline{AB} \cong \overline{DC}$

Prove: $ABCD$ is a parallelogram.

Coordinate Proof:

By definition, a quadrilateral is a parallelogram if opposite sides are parallel. We are given that $\overline{AB} \parallel \overline{DC}$, so we need only show that $\overline{AD} \parallel \overline{BC}$.

Use the Slope Formula.

slope of $\overline{AD} = \frac{c-0}{b-0} = \frac{c}{b}$ slope of $\overline{BC} = \frac{c-0}{b+a-a} = \frac{c}{b}$

Because $\overline{AD}$ and $\overline{BC}$ have the same slope, $\overline{AD} \parallel \overline{BC}$. So quadrilateral $ABCD$ is a parallelogram because opposite sides are parallel.

Guided Practice

5. Write a coordinate proof of this statement: *If a quadrilateral is a parallelogram, then opposite sides are congruent.*

Check Your Understanding

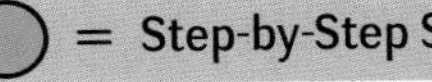

= Step-by-Step Solutions begin on page R13.

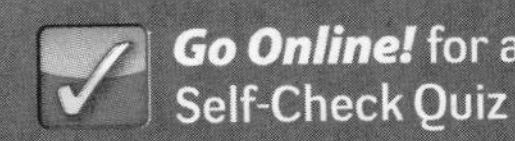

Example 1 **Determine whether each quadrilateral is a parallelogram. Justify your answer.**

1.

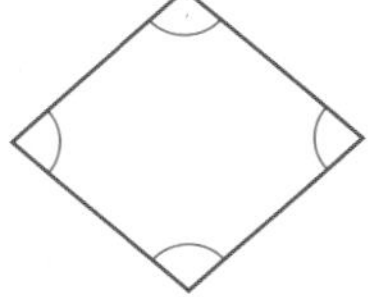

2.

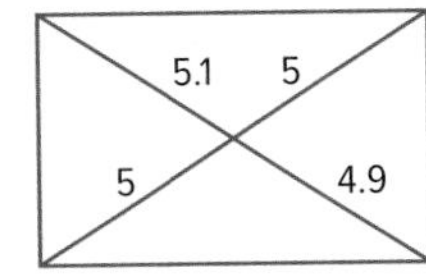

Example 2

3. KITES Charmaine is building the kite shown below. She wants to be sure that the string around her frame forms a parallelogram before she secures the material to it. How can she use the measures of the wooden portion of the frame to prove that the string forms a parallelogram? Explain your reasoning.

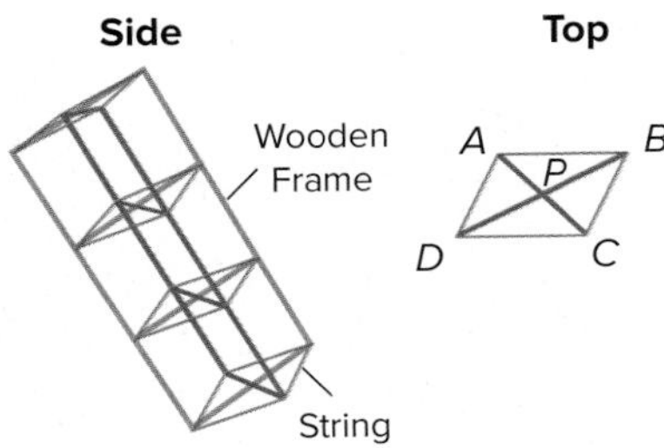

Example 3 **ALGEBRA Find x and y so that the quadrilateral is a parallelogram.**

4.

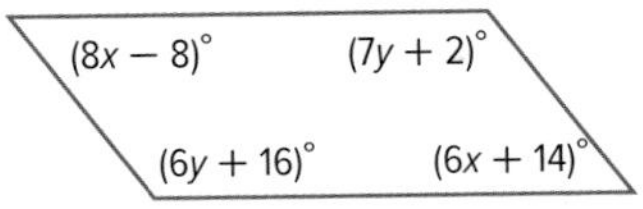

(figure: $(8x-8)°$, $(7y+2)°$, $(6y+16)°$, $(6x+14)°$)

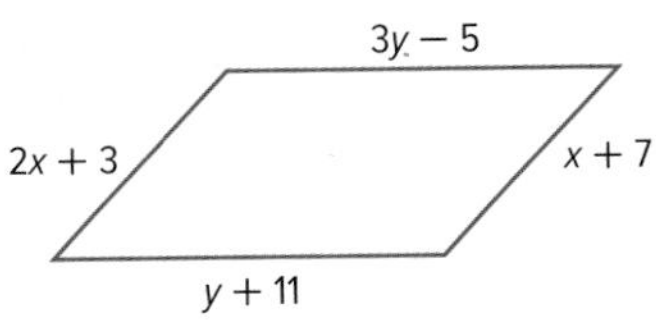

Example 4 **COORDINATE GEOMETRY** **Graph each quadrilateral with the given vertices. Determine whether the figure is a parallelogram. Justify your answer with the method indicated.**

6. $A(-2, 4)$, $B(5, 4)$, $C(8, -1)$, $D(-1, -1)$; Slope Formula

7. $W(-5, 4)$, $X(3, 4)$, $Y(1, -3)$, $Z(-7, -3)$; Midpoint Formula

Example 5 8. Write a coordinate proof for the statement: *If a quadrilateral is a parallelogram, then its diagonals bisect each other.*

Practice and Problem Solving

Extra Practice is on page R6.

Example 1 **MP CONSTRUCT ARGUMENTS** **Determine whether each quadrilateral is a parallelogram. Justify your answer.**

9. 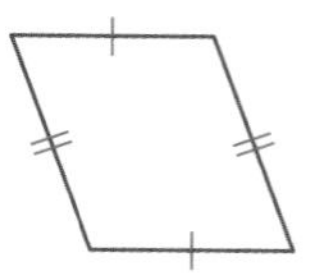

10. 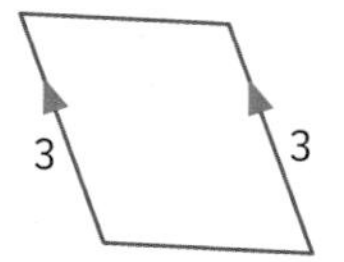

11. 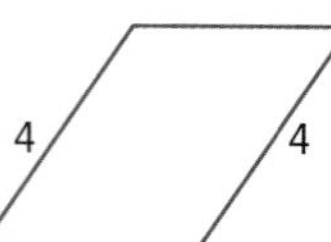

12. 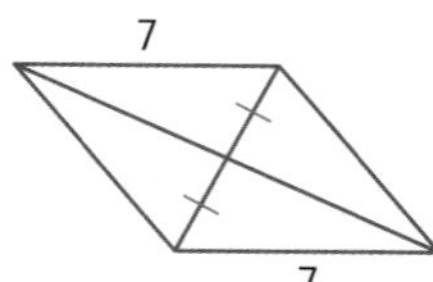

13. 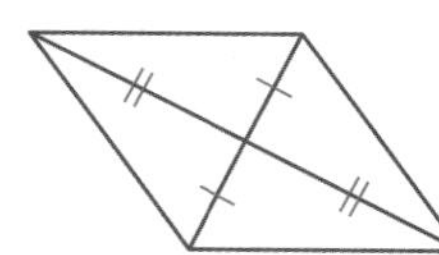

14. 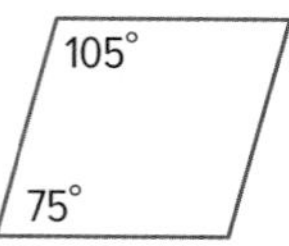

Example 2 15. **PROOF** If $ACDH$ is a parallelogram, B is the midpoint of $\overline{AC}$, and F is the midpoint of $\overline{HD}$, write a flow proof to prove that $ABFH$ is a parallelogram.

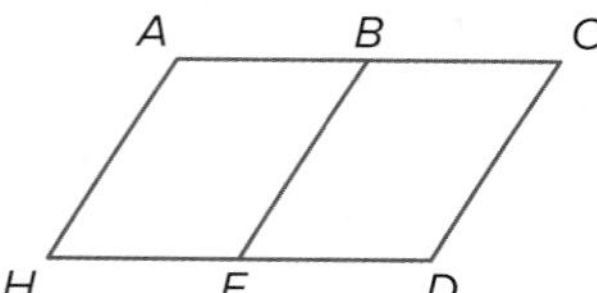

16. **PROOF** If $WXYZ$ is a parallelogram, $\angle W \cong \angle X$, and M is the midpoint of $\overline{WX}$, write a paragraph proof to prove that ZMY is an isosceles triangle.

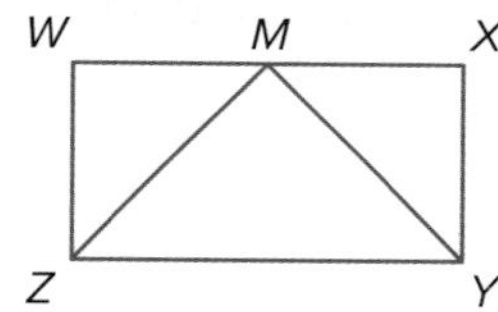

17. **REPAIR** Parallelogram lifts are used to elevate large vehicles for maintenance. In the diagram, $ABEF$ and $BCDE$ are parallelograms. Write a two-column proof to show that $ACDF$ is also a parallelogram.

Example 3 **ALGEBRA** **Find x and y so that the quadrilateral is a parallelogram.**

18. 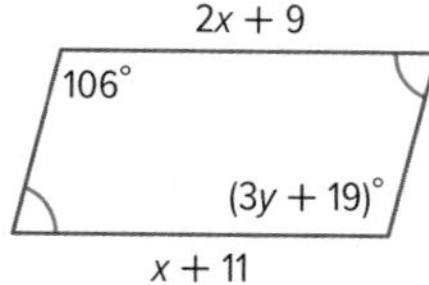

19. 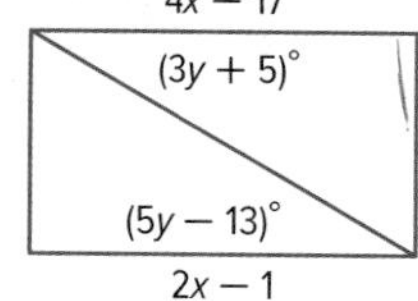

20. $\left(\frac{1}{4}x\right)^\circ$ $(4x - 8)^\circ$ $(8y - 12)^\circ$ $(y - 8)^\circ$

ALGEBRA **Find x and y so that the quadrilateral is a parallelogram.**

21.

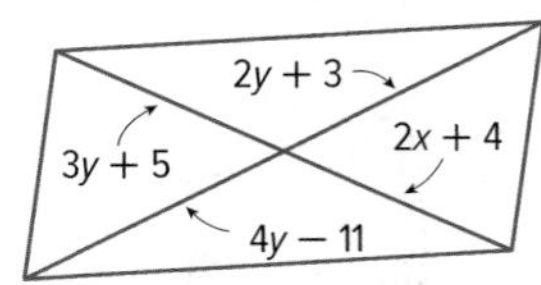

22.

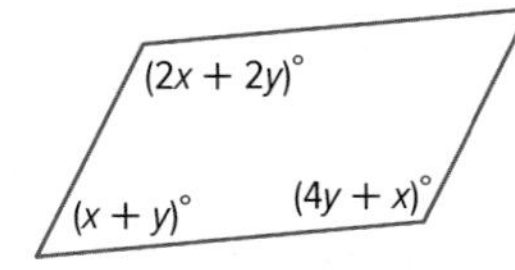

23.

$2x + 4y$

21 $\quad$ $3x + 3y$

$6y + \frac{1}{2}x$

Example 4 **COORDINATE GEOMETRY** Graph each quadrilateral with the given vertices. Determine whether the figure is a parallelogram. Justify your answer with the method indicated.

24. $A(-3, 4)$, $B(4, 5)$, $C(5, -1)$, $D(-2, -2)$; Slope Formula

25. $J(-4, -4)$, $K(-3, 1)$, $L(4, 3)$, $M(3, -3)$; Distance Formula

26. $V(3, 5)$, $W(1, -2)$, $X(-6, 2)$, $Y(-4, 7)$; Slope Formula

27. $Q(2, -4)$, $R(4, 3)$, $S(-3, 6)$, $T(-5, -1)$; Distance and Slope Formulas

Example 5 **28.** Write a coordinate proof for the statement: *If both pairs of opposite sides of a quadrilateral are congruent, then the quadrilateral is a parallelogram.*

29. Write a coordinate proof for the statement: *If a parallelogram has one right angle, it has four right angles.*

30. **PROOF** Write a paragraph proof of Theorem 6.10.

31 **PANTOGRAPH** A pantograph is a device that can be used to copy an object and either enlarge or reduce it based on the dimensions of the pantograph.

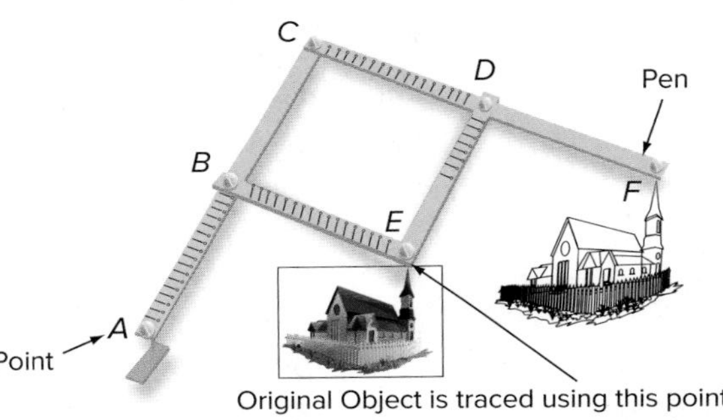

a. If $\overline{AC} \cong \overline{CF}$, $\overline{AB} \cong \overline{CD} \cong \overline{BE}$, and $\overline{DF} \cong \overline{DE}$, write a paragraph proof to show that $\overline{BE} \parallel \overline{CD}$.

b. The scale of the copied object is the ratio of CF to BE. If AB is 12 inches, DF is 8 inches, and the width of the original object is 5.5 inches, what is the width of the copy?

PROOF **Write a two-column proof.**

32. Theorem 6.11

33. Theorem 6.12

34. **CONSTRUCTION** Explain how you can use Theorem 6.11 to construct a parallelogram. Then construct a parallelogram using your method.

MP **REASONING** **Name the missing coordinates for each parallelogram.**

35

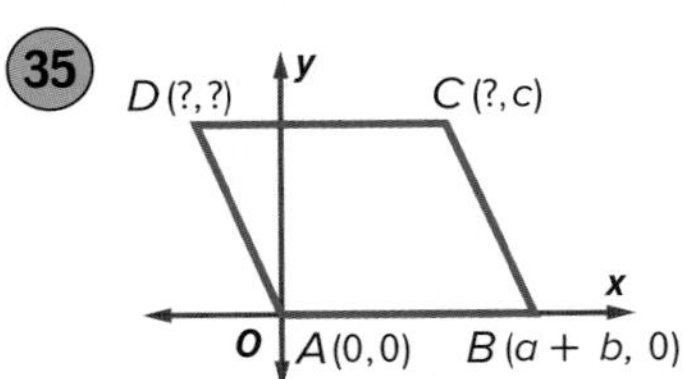

36.

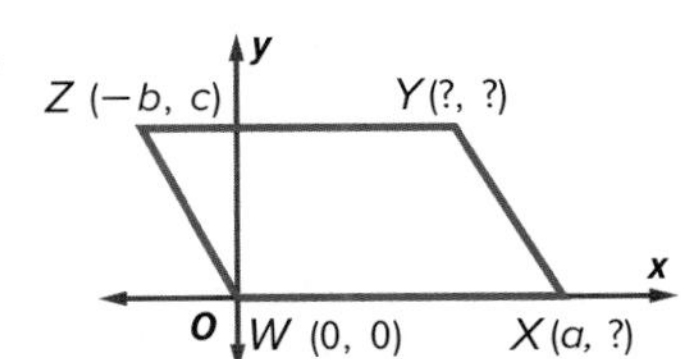

37. **SERVICE** While replacing a handrail, a contractor uses a carpenter's square to confirm that the vertical supports are perpendicular to the top step and the ground, respectively. How can the contractor prove that the two handrails are parallel using the fewest measurements? Assume that the top step and the ground are both level.

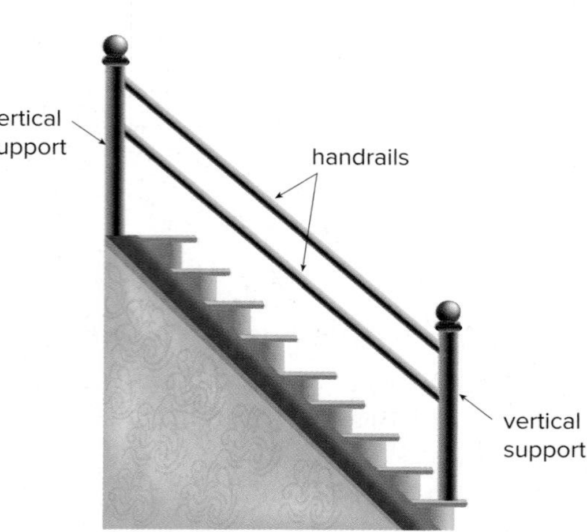

38. **PROOF** Write a coordinate proof to prove that the segments joining the midpoints of the sides of any quadrilateral form a parallelogram.

39. **MULTIPLE REPRESENTATIONS** In this problem, you will explore the properties of rectangles. A rectangle is a quadrilateral with four right angles.

a. **Geometric** Draw three rectangles with varying lengths and widths. Label one rectangle *ABCD*, one *MNOP*, and one *WXYZ*. Draw the two diagonals for each rectangle.

b. **Tabular** Measure the diagonals of each rectangle, and complete the table at the right.

c. **Verbal** Write a conjecture about the diagonals of a rectangle.

Rectangle	Side	Length
ABCD	$\overline{AC}$	
	$\overline{BD}$	
MNOP	$\overline{MO}$	
	$\overline{NP}$	
WXYZ	$\overline{WY}$	
	$\overline{XZ}$	

H.O.T. Problems Use Higher-Order Thinking Skills

40. **CHALLENGE** The diagonals of a parallelogram meet at the point (0, 1). One vertex of the parallelogram is located at (2, 4), and a second vertex is located at (3, 1). Find the locations of the remaining vertices.

41. **WRITING IN MATH** Compare and contrast Theorem 6.9 and Theorem 6.3.

42. MP **CONSTRUCT ARGUMENTS** If two parallelograms have four congruent corresponding angles, are the parallelograms *sometimes*, *always*, or *never* congruent? Justify your answer.

43. **OPEN-ENDED** Position and label a parallelogram on the coordinate plane differently than shown in either Example 5, Exercise 35, or Exercise 36.

44. **CHALLENGE** If *ABCD* is a parallelogram and $\overline{AJ} \cong \overline{KC}$, show that quadrilateral *JBKD* is a parallelogram.

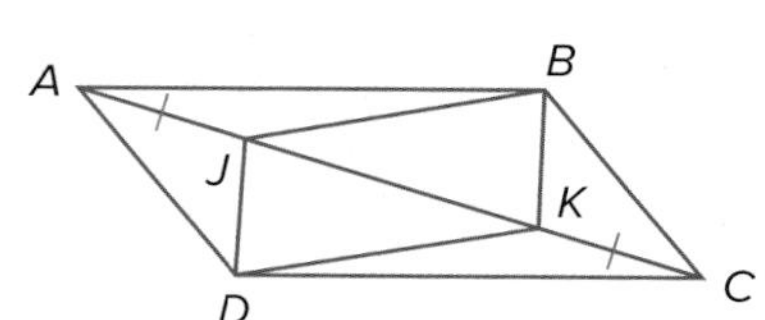

45. **WRITING IN MATH** How can you prove that a quadrilateral is a parallelogram?

46. Amir is using software to design a geometric pattern for a quilt. He starts by making quadrilateral *PQRS*, as shown. He wants to know if the quadrilateral he made is a parallelogram.

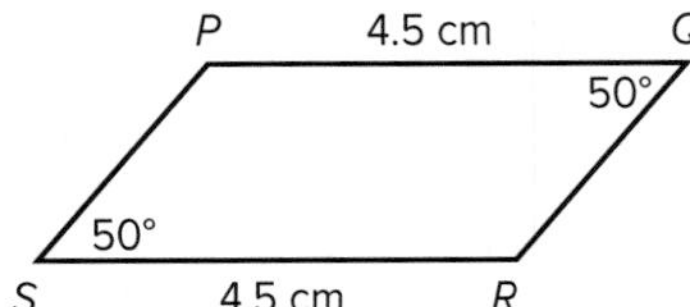

Which of the following best describes quadrilateral *PQRS*? MP 3, 4

◯ **A** It is a parallelogram, because a pair of opposite sides are congruent.

◯ **B** It is a parallelogram, because a pair of opposite angles are congruent.

◯ **C** It is a parallelogram, because a pair of opposite sides are both parallel and congruent.

◯ **D** There is not enough information to conclude that the quadrilateral is a parallelogram.

47. Quadrilateral *ABCD* has vertices *A*(0, 1), *B*(1, 4), *C*(5, 5), and *D*(4, 2). Kalina uses the slope formula to show that $\overline{BC}$ is parallel to $\overline{AD}$. Which of the following can she use as her next step to show that *ABCD* is a parallelogram? MP 2, 6

I. Use the Slope Formula to show that $\overline{AB}$ is parallel to $\overline{CD}$.

II. Use the Distance Formula to show that $\overline{BC}$ is congruent to $\overline{AD}$.

III. Use the Distance Formula to show that $\overline{AB}$ is congruent to $\overline{CD}$.

◯ **A** I only

◯ **B** II only

◯ **C** I and II only

◯ **D** III only

◯ **E** I, II, and III

48. In quadrilateral *WXYZ*, $WX = 4x - 15$, $XY = 4x + 20$, $YZ = 3x + 5$, and $ZW = 6x - 20$. For what value of x is quadrilateral *WXYZ* a parallelogram? MP 2, 3

49. **MULTI-STEP** Use the figure to answer the questions. MP 2, 3, 6

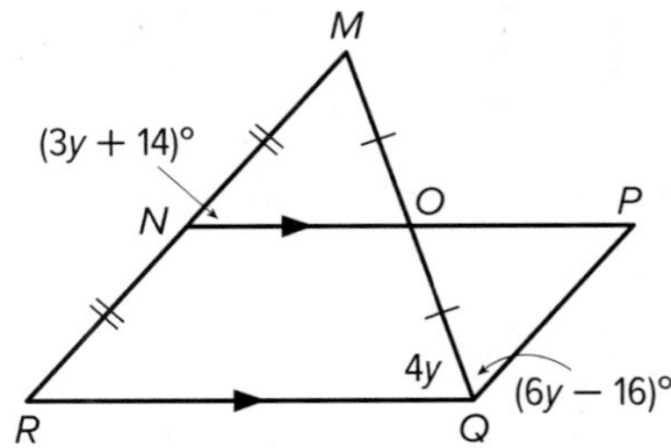

a. What is $m\angle MNO$?

b. What is $m\angle OQP$?

c. What is $m\angle MON$?

d. What is $m\angle OPQ$?

e. Is $\overline{MR} \parallel \overline{PQ}$? Explain.

f. Is *NPQR* a parallelogram? Explain.

g. What mathematical practice did you use to solve this problem?

50. Three of the vertices of a parallelogram are at (3, −4), (−1, −2), and (6, 1). Which point could be the location of the fourth vertex? Select all that apply. MP 2, 5

☐ **A** (10, −1)

☐ **B** (3, 7)

☐ **C** (2, 3)

☐ **D** (−4, −7)

☐ **E** (−6, −1)

51. Which are ways to determine that a quadrilateral is a parallelogram? Select all that apply. MP 2

☐ **A** The quadrilateral has two pairs of congruent angles.

☐ **B** The quadrilateral has both pairs of opposite sides congruent.

☐ **C** The quadrilateral has one pair of opposite sides that are parallel and congruent.

☐ **D** The quadrilateral has diagonals that are different lengths.

☐ **E** The quadrilateral has both pairs of opposite angles congruent.

Mid-Chapter Quiz

Lessons 6-1 through 6-3

Find the sum of the measures of the interior angles of each convex polygon. (Lesson 6-1)

1. pentagon
2. heptagon
3. 18-gon
4. 23-gon

Find the measure of each interior angle. (Lesson 6-1)

5.

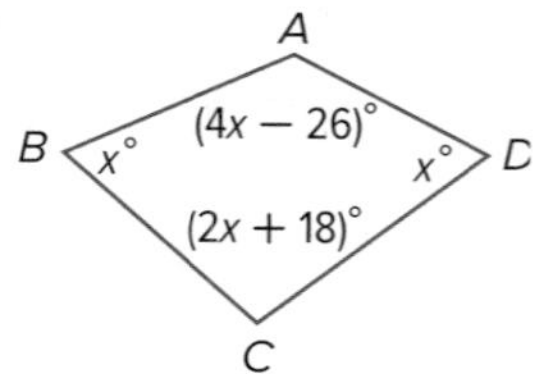

6.

Q
R
P
S
$(2x - 16)°$
$2x°$
$x°$
$(x + 10)°$

The sum of the measures of the interior angles of a regular polygon is given. Find the number of sides in the polygon. (Lesson 6-1)

7. 720
8. 1260
9. 1800
10. 4500

Find the value of x in each diagram. (Lesson 6-1)

11.

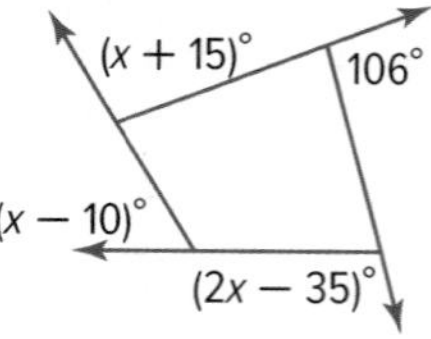

12.

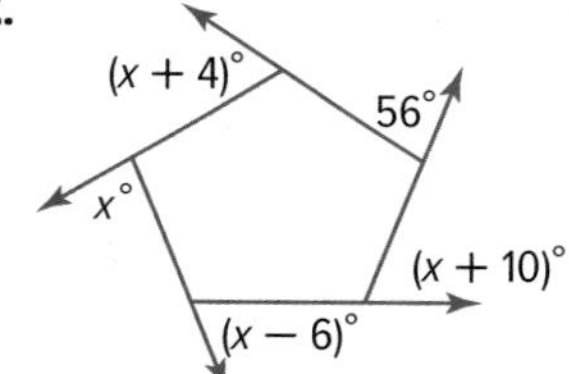

Use ▱$WXYZ$ to find each measure. (Lesson 6-2)

13. $m\angle WZY$
14. WZ
15. $m\angle XYZ$

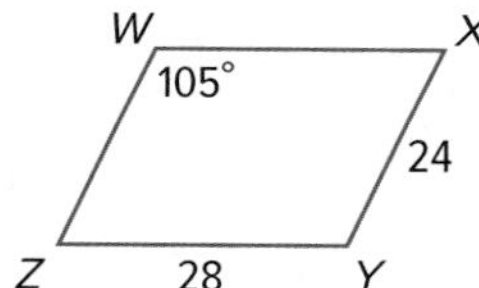

16. **DESIGN** Describe two ways to ensure that the pieces of the design at the right would fit properly together. (Lesson 6-2)

ALGEBRA Find the value of each variable in each parallelogram. (Lesson 6-2)

17.

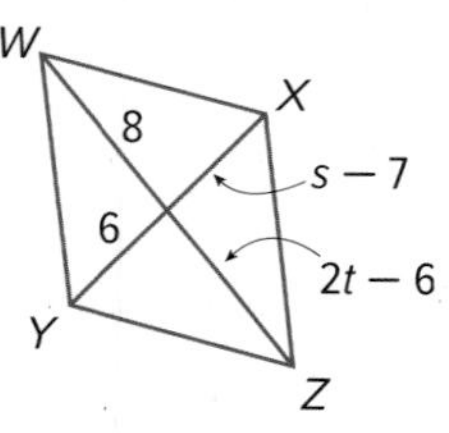

18.

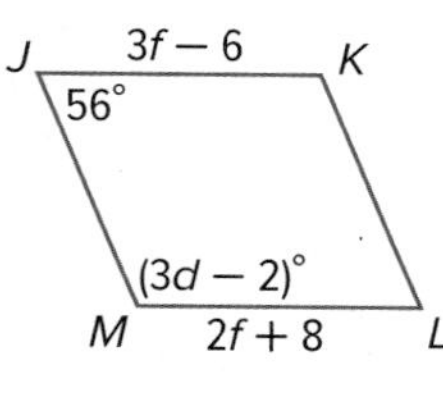

19. **PROOF** Write a two-column proof. (Lesson 6-2)

Given: ▱$GFBA$ and ▱$HACD$

Prove: $\angle F \cong \angle D$

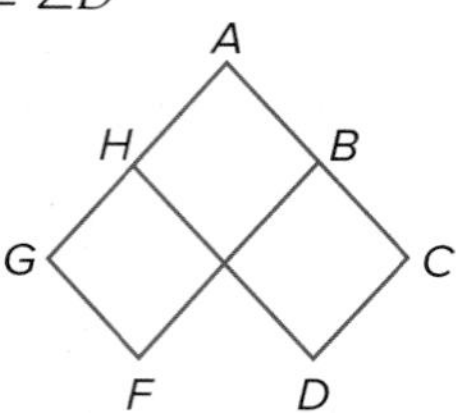

Find x and y so that each quadrilateral is a parallelogram. (Lesson 6-3)

20.

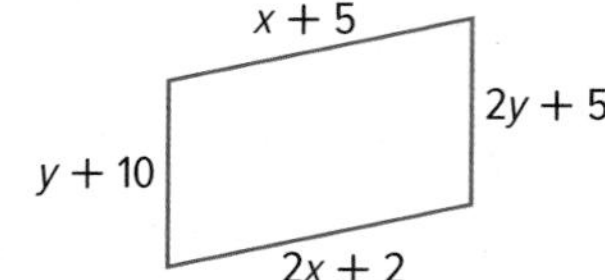

21.

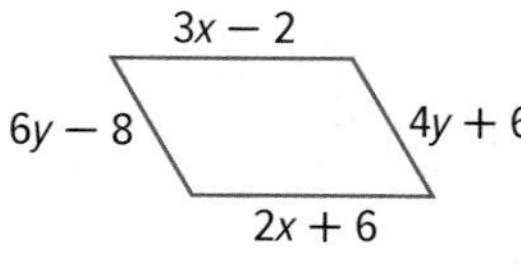

22. **MUSIC** This keyboard stand with legs joined at the midpoints always remains parallel to the floor. (Lesson 6-3)

a. Explain why this occurs.

b. MP What mathematical practice did you use to solve this problem?

23. **MULTIPLE CHOICE** Which of the following quadrilaterals is not a parallelogram? (Lesson 6-3)

A

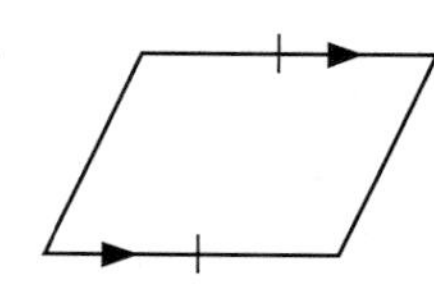

C

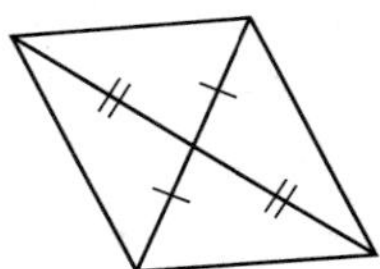

B

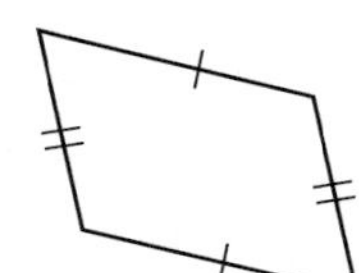

D

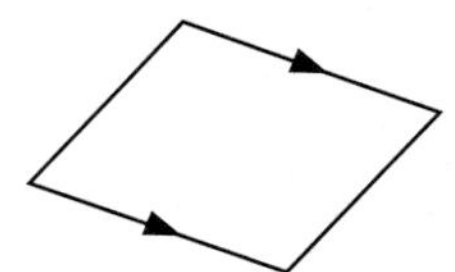

COORDINATE GEOMETRY Determine whether the figure is a parallelogram. Justify your answer with the method indicated. (Lesson 6-3)

24. $A(-6, -5)$, $B(-1, -4)$, $C(0, -1)$, $D(-5, -2)$; Distance Formula

25. $Q(-5, 2)$, $R(-3, -6)$, $S(2, 2)$, $T(-1, 6)$; Slope Formula

LESSON 4

Special Parallelograms: Rectangles

Then

- You used properties of parallelograms and determined whether quadrilaterals were parallelograms.

Now

1. Recognize and apply properties of rectangles.
2. Determine whether parallelograms are rectangles.

Why?

- Leonardo is in charge of set design for a school play. He needs to use paint to create the appearance of a doorway on a lightweight solid wall. The doorway is to be a rectangle 36 inches wide and 80 inches tall. How can Leonardo be sure that he paints a rectangle?

New Vocabulary

rectangle

Mathematical Practices

3 Construct viable arguments and critique the reasoning of others.

5 Use appropriate tools strategically.

1 Properties of Rectangles

A **rectangle** is a parallelogram with four right angles. By definition, a rectangle has the following properties.

- All four angles are right angles.
- Opposite sides are parallel and congruent.
- Opposite angles are congruent.
- Consecutive angles are supplementary.
- Diagonals bisect each other.

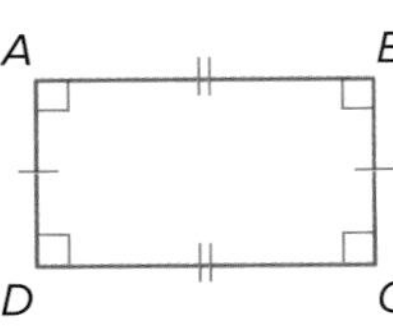

Rectangle $ABCD$

In addition, the diagonals of a rectangle are congruent.

Theorem 6.13 Diagonals of a Rectangle

If a parallelogram is a rectangle, then its diagonals are congruent.

Abbreviation *If a ▱ is a rectangle, diag. are ≅.*

Example If ▱$JKLM$ is a rectangle, then $\overline{JL} \cong \overline{MK}$.

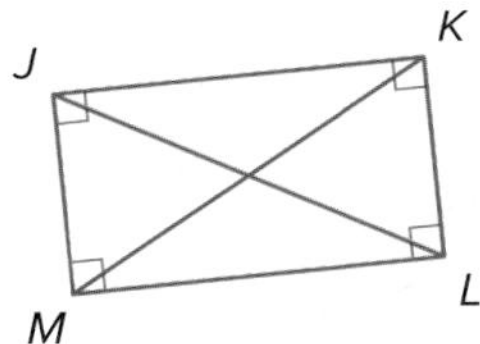

You will prove Theorem 6.13 in Exercise 33.

Real-World Example 1 Use Properties of Rectangles

PARKS The Parks and Wildlife organization plans to add two diagonal walking paths, as shown, to one of its rectangular parks. If $PS = 180$ meters and $PR = 200$ meters, find QT.

P Q T S R

$\overline{QS} \cong \overline{PR}$	If a ▱ is a rectangle, diag. are ≅.
$QS = PR$	Definition of congruence
$QS = 200$	Substitution

Because $PQRS$ is a rectangle, it is a parallelogram. The diagonals of a parallelogram bisect each other, so $QT = ST$.

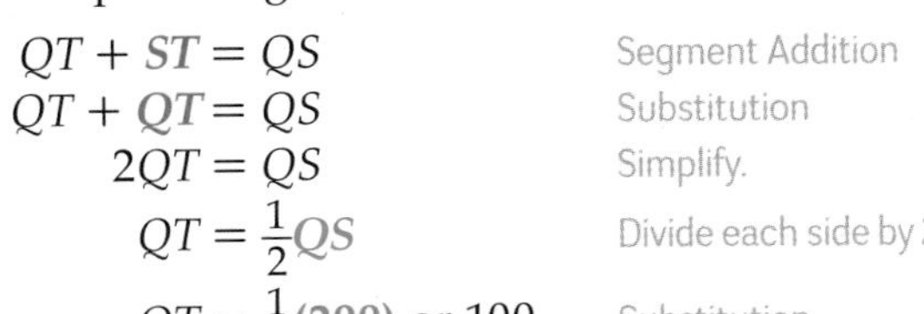

$QT + ST = QS$	Segment Addition
$QT + QT = QS$	Substitution
$2QT = QS$	Simplify.
$QT = \frac{1}{2}QS$	Divide each side by 2.
$QT = \frac{1}{2}(200)$ or 100	Substitution

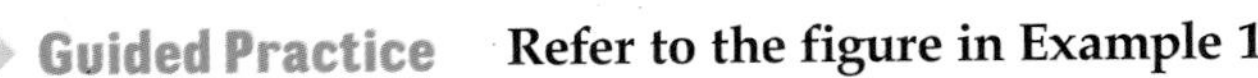

Guided Practice **Refer to the figure in Example 1.**

1A. If $TS = 120$ meters, find PR.

1B. If $m\angle PRS = 64$, find $m\angle SQR$.

You can use the properties of rectangles along with algebra to find missing values.

Example 2 Use Properties of Rectangles and Algebra

ALGEBRA Quadrilateral *JKLM* is a rectangle. If $m\angle KJL = 2x + 4$ and $m\angle JLK = 7x + 5$, find x.

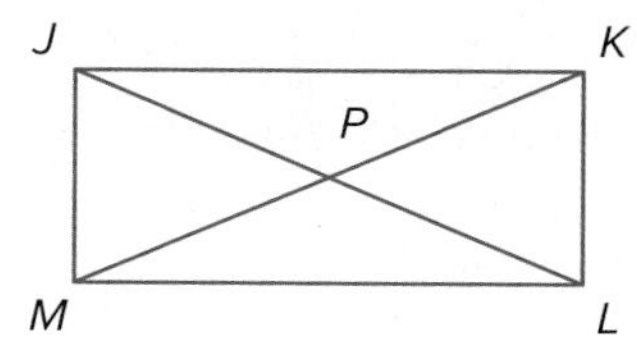

Because *JKLM* is a rectangle, it has four right angles. So, $m\angle MLK = 90$. Because a rectangle is a parallelogram, opposite sides are parallel. Alternate interior angles of parallel lines are congruent, so $\angle JLM \cong \angle KJL$ and $m\angle JLM = m\angle KJL$.

$m\angle JLM + m\angle JLK = 90$	Angle Addition
$m\angle KJL + m\angle JLK = 90$	Substitution
$2x + 4 + 7x + 5 = 90$	Substitution
$9x + 9 = 90$	Add like terms.
$9x = 81$	Subtract 9 from each side.
$x = 9$	Divide each side by 9.

Guided Practice

2. Refer to the figure in Example 2. If $JP = 3y - 5$ and $MK = 5y + 1$, find y.

Study Tip

Right Angles Recall from Theorem 6.6 that if a parallelogram has one right angle, then it has four right angles.

2 Prove that Parallelograms are Rectangles

The converse of Theorem 6.13 is also true.

Theorem 6.14 Diagonals of a Rectangle

If the diagonals of a parallelogram are congruent, then the parallelogram is a rectangle.

Abbreviation *If diag. of a ▱ are ≅, then ▱ is a rectangle.*

Example If $\overline{WY} \cong \overline{XZ}$ in ▱*WXYZ*, then ▱*WXYZ* is a rectangle.

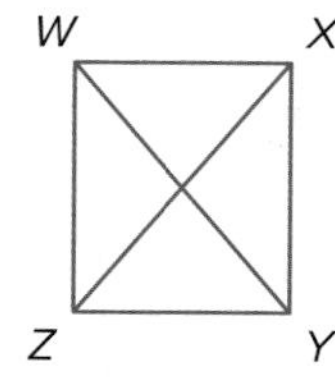

You will prove Theorem 6.14 in Exercise 34.

Real-World Example 3 Proving Rectangle Relationships

DODGEBALL A community recreation center has created an outdoor dodgeball playing field. To be sure that it meets the ideal playing field requirements, they measure the sides of the field and its diagonals. If $AB = 60$ feet, $BC = 30$ feet, $CD = 60$ feet, $AD = 30$ feet, $AC = 67$ feet, and $BD = 67$ feet, explain how the recreation center can be sure that the playing field is rectangular.

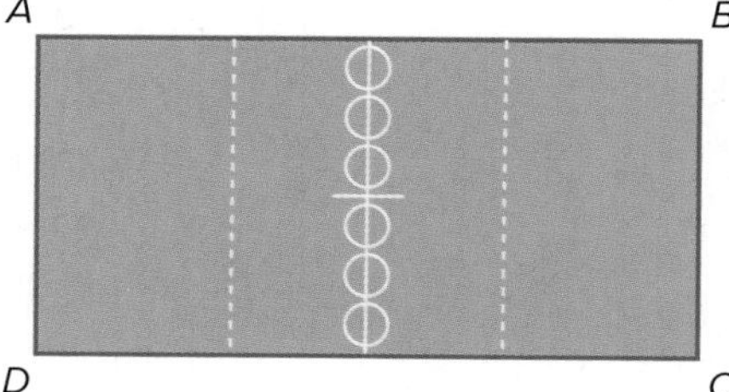

Because $AB = CD$, $BC = AD$, and $AC = BD$, $\overline{AB} \cong \overline{CD}$, $\overline{BC} \cong \overline{AD}$, and $\overline{AC} \cong \overline{BD}$. Because $\overline{AB} \cong \overline{CD}$ and $\overline{BC} \cong \overline{AD}$, *ABCD* is a parallelogram. Because $\overline{AC}$ and $\overline{BD}$ are congruent diagonals in ▱*ABCD*, ▱*ABCD* is a rectangle.

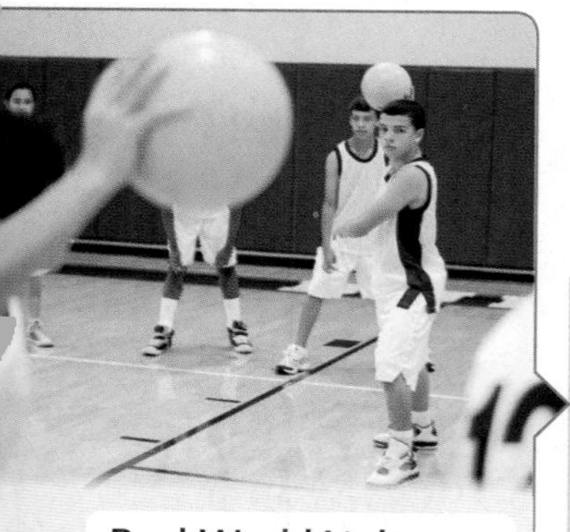

Real-World Link

The game of dodgeball is played on a rectangular playing field ideally 60 feet long and 30 feet wide. The field is divided into two equal sections by a center-line and attack-lines that are 3 meters (9.8 feet) from, and parallel to, the centerline.

Source: National Amateur Dodgeball Assoc.

Real-World Link

The Mosaic Youth Theater in Detroit, Michigan, is a professional performing arts training program for young people ages 12 to 18. Students are involved in all aspects of performances, including set and lighting design, set construction, stage management, sound, and costumes.

Guided Practice

3. SET DESIGN Refer to the beginning of the lesson. Leonardo measures the sides of his figure and confirms that they have the desired measures as shown. Using a carpenter's square, he also confirms that the measure of the bottom left corner of the figure is a right angle. Can he conclude that the figure is a rectangle? Explain.

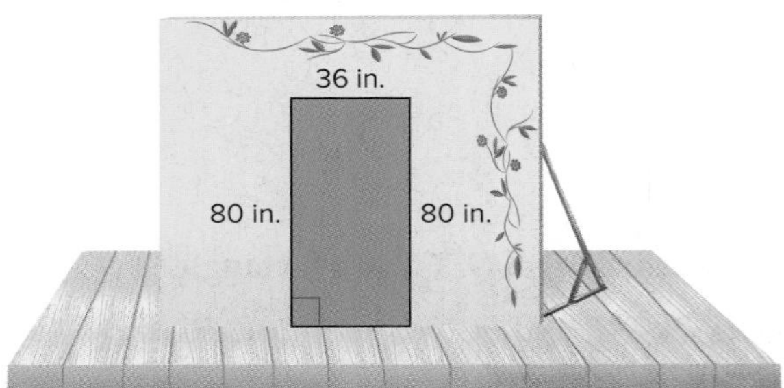

You can also use the properties of rectangles to prove that a quadrilateral positioned on a coordinate plane is a rectangle given the coordinates of the vertices.

Example 4 Rectangles and Coordinate Geometry

COORDINATE GEOMETRY Quadrilateral *PQRS* has vertices *P*(−5, 3), *Q*(1, −1), *R*(−1, −4), and *S*(−7, 0). Determine whether *PQRS* is a rectangle by using the Distance Formula.

Step 1 Use the Distance Formula to determine whether *PQRS* is a parallelogram by determining if opposite sides are congruent.

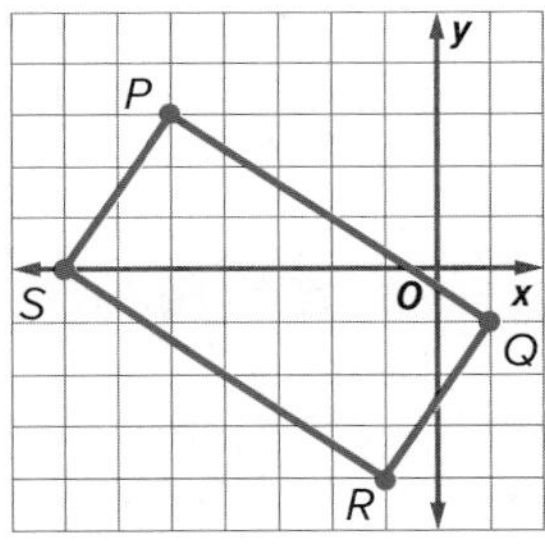

$PQ = \sqrt{(-5-1)^2 + [3-(-1)]^2}$ or $\sqrt{52}$

$RS = \sqrt{[-1-(-7)]^2 + (-4-0)^2}$ or $\sqrt{52}$

$PS = \sqrt{[-5-(-7)]^2 + (3-0)^2}$ or $\sqrt{13}$

$QR = \sqrt{[1-(-1)]^2 + [-1-(-4)]^2}$ or $\sqrt{13}$

Because opposite sides of the quadrilateral have the same measure, they are congruent. So, quadrilateral *PQRS* is a parallelogram.

Step 2 Determine whether the diagonals of ▱*PQRS* are congruent.

$PR = \sqrt{[-5-(-1)]^2 + [3-(-4)]^2}$ or $\sqrt{65}$

$QS = \sqrt{[1-(-7)]^2 + (-1-0)^2}$ or $\sqrt{65}$

Because the diagonals have the same measure, they are congruent. So, ▱*PQRS* is a rectangle.

Study Tip

MP Sense-Making A rectangle is a parallelogram, but a parallelogram is not necessarily a rectangle.

Guided Practice

4. Quadrilateral *JKLM* has vertices *J*(−10, 2), *K*(−8, −6), *L*(5, −3), and *M*(2, 5). Determine whether *JKLM* is a rectangle using the Slope Formula.

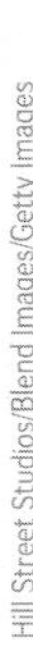

Check Your Understanding

○ = Step-by-Step Solutions begin on page R13.

Example 1 **FARMING An X-brace on a rectangular barn door is both decorative and functional. It helps to prevent the door from warping over time. If $ST = 3\frac{13}{16}$ feet, $PS = 7$ feet, and $m\angle PTQ = 67$, find each measure.**

1. QR
2. SQ
3. $m\angle TQR$
4. $m\angle TSR$

Example 2 **ALGEBRA Quadrilateral $DEFG$ is a rectangle.**

5. If $FD = 3x - 7$ and $EG = x + 5$, find EG.
6. If $m\angle EFD = 2x - 3$ and $m\angle DFG = x + 12$, find $m\angle EFD$.

Example 3

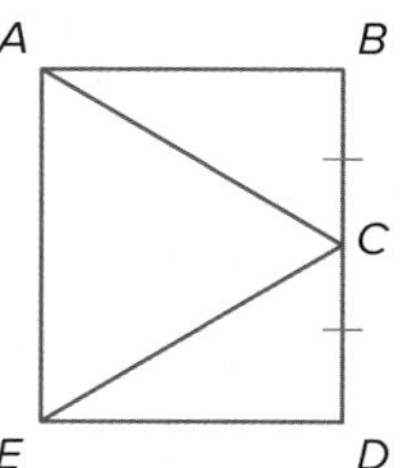

7. **PROOF** If $ABDE$ is a rectangle and $\overline{BC} \cong \overline{DC}$, prove that $\overline{AC} \cong \overline{EC}$.

Example 4 **COORDINATE GEOMETRY Graph each quadrilateral with the given vertices. Determine whether the figure is a rectangle. Justify your answer using the indicated formula.**

8. $W(-4, 3)$, $X(1, 5)$, $Y(3, 1)$, $Z(-2, -2)$; Slope Formula
9. $A(4, 3)$, $B(4, -2)$, $C(-4, -2)$, $D(-4, 3)$; Distance Formula

Practice and Problem Solving

Extra Practice is on page R6.

Example 1 **FENCING X-braces are also used to provide support in rectangular fencing. If $AB = 6$ feet, $AD = 2$ feet, and $m\angle DAE = 65$, find each measure.**

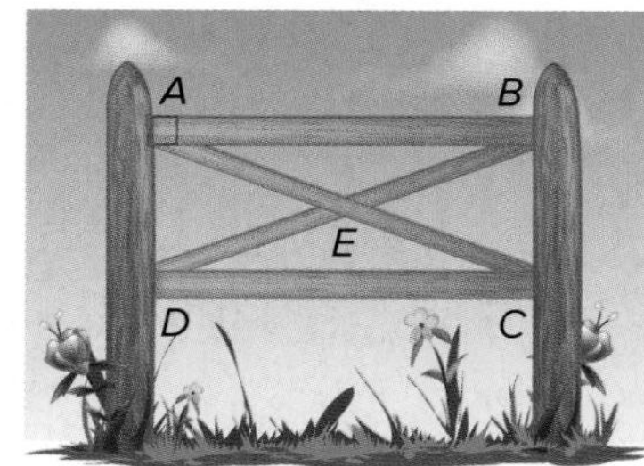

10. BC
11. DB
12. $m\angle CEB$
13. $m\angle EDC$

Example 2 **MP REGULARITY Quadrilateral $WXYZ$ is a rectangle.**

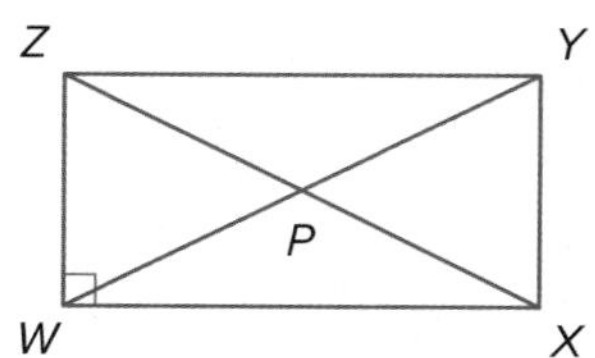

14. If $ZY = 2x + 3$ and $WX = x + 4$, find WX.
15. If $PY = 3x - 5$ and $WP = 2x + 11$, find ZP.
16. If $m\angle ZYW = 2x - 7$ and $m\angle WYX = 2x + 5$, find $m\angle ZYW$.
17. If $ZP = 4x - 9$ and $PY = 2x + 5$, find ZX.
18. If $m\angle XZY = 3x + 6$ and $m\angle XZW = 5x - 12$, find $m\angle YXZ$.
19. If $m\angle ZXW = x - 11$ and $m\angle WZX = x - 9$, find $m\angle ZXY$.

Example 3 **PROOF** **Write a two-column proof.**

20. **Given:** $ABCD$ is a rectangle.
Prove: $\triangle ADC \cong \triangle BCD$

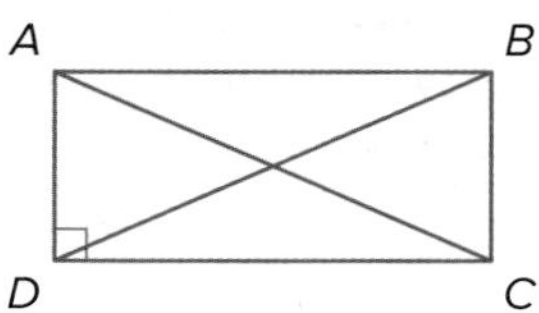

21. **Given:** $QTVW$ is a rectangle.
$\overline{QR} \cong \overline{ST}$
Prove: $\triangle SWQ \cong \triangle RVT$

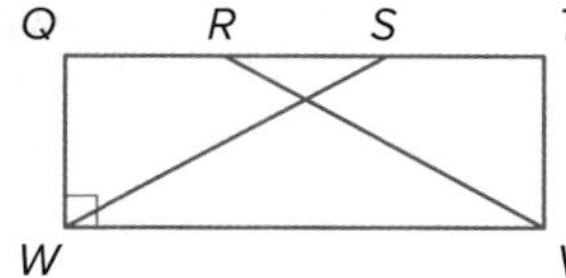

Example 4 **COORDINATE GEOMETRY** **Graph each quadrilateral with the given vertices. Determine whether the figure is a rectangle. Justify your answer using the indicated formula.**

22. $W(-2, 4)$, $X(5, 5)$, $Y(6, -2)$, $Z(-1, -3)$; Slope Formula

23. $J(3, 3)$, $K(-5, 2)$, $L(-4, -4)$, $M(4, -3)$; Distance Formula

24. $Q(-2, 2)$, $R(0, -2)$, $S(6, 1)$, $T(4, 5)$; Distance Formula

25. $G(1, 8)$, $H(-7, 7)$, $J(-6, 1)$, $K(2, 2)$; Slope Formula

Quadrilateral $ABCD$ is a rectangle. Find each measure if $m\angle 2 = 40$.

26. $m\angle 1$ **27.** $m\angle 7$ **28.** $m\angle 3$

 **29** $m\angle 5$ **30.** $m\angle 6$ **31.** $m\angle 8$

32. **MODELING** Jody is building a new bookshelf using wood and metal supports like the one shown. To what length should she cut the metal supports in order for the bookshelf to be *square*, which means that the angles formed by the shelves and the vertical supports are all right angles? Explain your reasoning.

PROOF **Write a two-column proof.**

33. Theorem 6.13 **34.** Theorem 6.14

PROOF **Write a paragraph proof of each statement.**

35. If a parallelogram has one right angle, then it is a rectangle.

36. If a quadrilateral has four right angles, then it is a rectangle.

37. **CONSTRUCTION** Construct a rectangle using the construction for congruent segments and the construction for a line perpendicular to another line through a point on the line. Justify each step of the construction.

38. **SPORTS** The end zone of a football field is 160 feet wide and 30 feet long. Kyle is responsible for painting the field. He has finished the end zone. Explain how Kyle can confirm that the end zone is the regulation size and be sure that it is also a rectangle using only a tape measure.

ALGEBRA **Quadrilateral $WXYZ$ is a rectangle.**

39. If $XW = 3$, $WZ = 4$, and $XZ = b$, find YW.

40. If $XZ = 2c$, $ZY = 6$, and $XY = 8$, find WY.

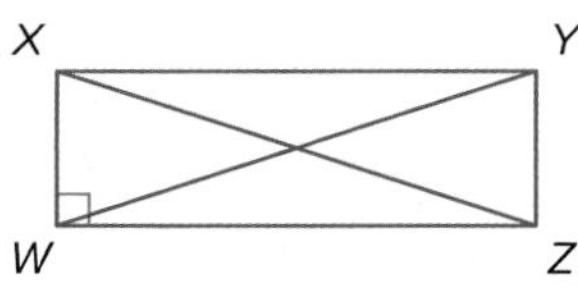

41. **MULTI-STEP** Samantha's rectangular vegetable garden is 40 feet by 56 feet. She is going to plant another rectangular garden that is 43 feet wide. She wants to buy a hose that can stretch over the length of the diagonals of either garden. The Garden Center sells hoses in 50-, 75-, and 100-foot lengths. Samantha wants to buy the smallest hose possible that will still be able to stretch the length of the diagonal of either garden.

a. If the distance from each vegetable garden to the water faucet is 5 feet, find the maximum length the other vegetable garden could be and still require Samantha to only buy the same length of hose as her existing vegetable garden.

b. Describe your solution process.

PROOF Write a coordinate proof of each statement.

42. The diagonals of a rectangle are congruent.

43. If the diagonals of a parallelogram are congruent, then it is a rectangle.

44. **MULTIPLE REPRESENTATIONS** In the problem, you will explore properties of other special parallelograms.

a. **Geometric** Draw three parallelograms, each with all four sides congruent. Label one parallelogram *ABCD*, one *MNOP*, and one *WXYZ*. Draw the two diagonals of each parallelogram and label the intersections *R*.

b. **Tabular** Use a protractor to measure the appropriate angles and complete the table below.

Parallelogram	ABCD		MNOP		WXYZ	
Angle	$\angle ARB$	$\angle BRC$	$\angle MRN$	$\angle NRO$	$\angle WRX$	$\angle XRY$
Angle Measure						

c. **Verbal** Make a conjecture about the diagonals of a parallelogram with four congruent sides.

H.O.T. Problems Use Higher-Order Thinking Skills

45. **CHALLENGE** In rectangle *ABCD*, $m\angle EAB = 4x + 6$, $m\angle DEC = 10 - 11y$, and $m\angle EBC = 60$. Find the values of x and y.

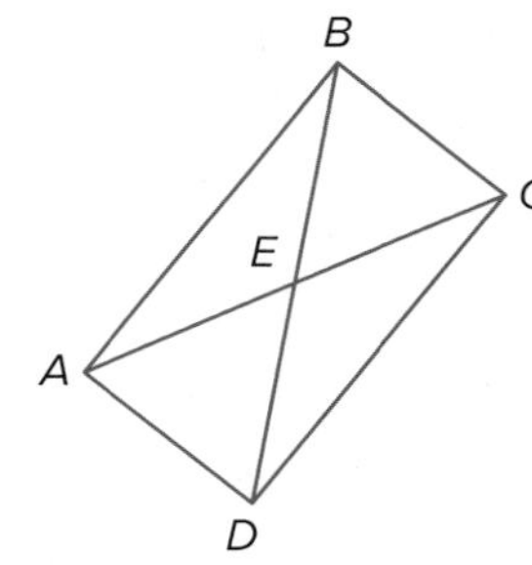

46. **CRITIQUE ARGUMENTS** Parker says that any two congruent acute triangles can be arranged to make a rectangle. Tamika says that only two congruent right triangles can be arranged to make a rectangle. Is either of them correct? Explain your reasoning.

47. **REASONING** In the diagram at the right, lines n, p, q, and r are parallel and lines ℓ and m are parallel. How many rectangles are formed by the intersecting lines?

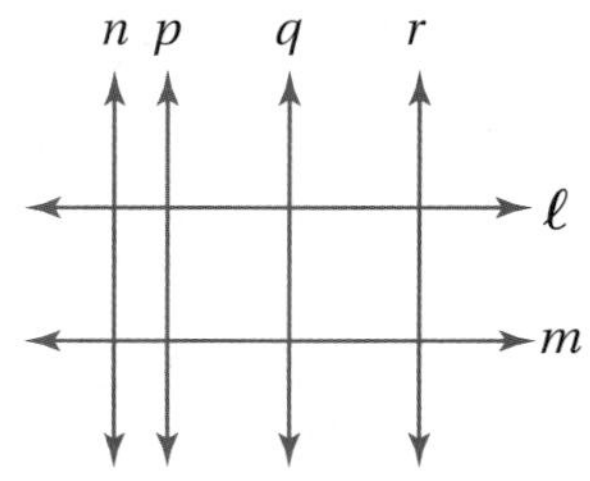

48. **OPEN-ENDED** Write the equations of four lines having intersections that form the vertices of a rectangle. Verify your answer using coordinate geometry.

49. **WRITING IN MATH** Why are all rectangles parallelograms, but all parallelograms are not rectangles? Explain.

50. **MULTI-STEP** Elena and Joshua use chalk to draw a race course for remote-controlled cars, as shown. They draw the course so that $RSTU$ is a rectangle with $RS = 50$ feet and $RT = 60$ feet. During the first race, the cars travel from R to S to V and back to R. How far do the cars travel? MP 4

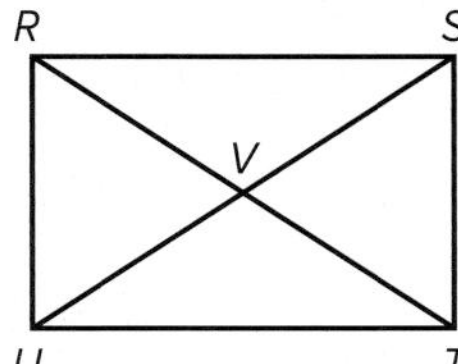

- A 170 ft
- B 110 ft
- C 80 ft
- D 60 ft

Quadrilateral $ABCD$ is a rectangle and $x = 16$. Find each measure.

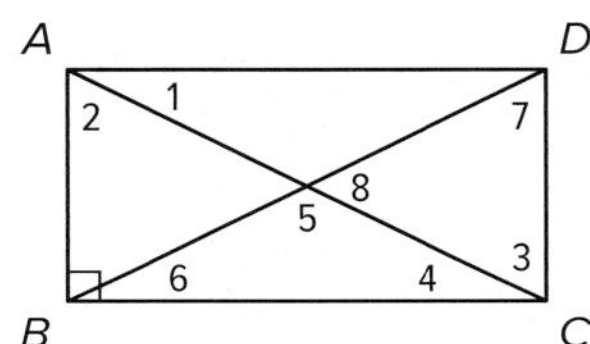

51. $m\angle 1 = x + 3$ MP 2

- A 16°
- B 19°
- C 64°
- D 71°

52. $m\angle 2 = 4x + 7$ MP 2

- A 16°
- B 19°
- C 64°
- D 71°

53. $m\angle 3$ MP 2

- A 16°
- B 19°
- C 64°
- D 71°

54. $m\angle 5$ MP 2

- A 19°
- B 38°
- C 71°
- D 142°

55. $m\angle 6$ MP 2

- A 16°
- B 19°
- C 64°
- D 71°

56. $m\angle 8$ MP 2

- A 16°
- B 19°
- C 38°
- D 71°

57. Quadrilateral $GHJK$ is a rectangle.

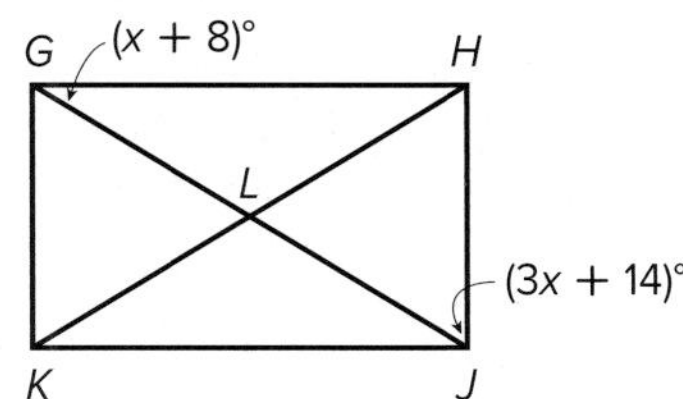

What is the measure of $\angle HJG$? MP 2

- A 3
- B 5
- C 17
- D 25
- E 65

58. Li Ming drew parallelogram $ABCD$ so that A, B, C, and D are consecutive vertices. Which of the following will guarantee that $ABCD$ is a rectangle? MP 6

- A $\overline{AB} \cong \overline{CD}$
- B $\overline{AD} \cong \overline{DC}$
- C $\overline{AC} \cong \overline{BD}$
- D $\overline{BC} \cong \overline{CD}$

59. Marciano plots points J, K, L, and M on a coordinate plane. He shows that quadrilateral $JKLM$ is a parallelogram. Which of the following can he do to prove that $JKLM$ is a rectangle? MP 3

- A Use the Distance Formula to show that $JL = MK$.
- B Use the Midpoint Formula to show that $\overline{JL}$ and $\overline{MK}$ bisect each other.
- C Use the Slope Formula to show that $\overline{JK}$ and $\overline{ML}$ are parallel.
- D Use the Distance Formula to show that $JK = ML$.

60. In rectangle $PQRS$, $PQ = x + 50$, $PR = 2x + 10$, and $SQ = 4x - 130$. What is the length of $\overline{PR}$? MP 1

- A 70
- B 90
- C 120
- D 150

LESSON 5

Special Parallelograms: Rhombi, Squares

Then

- You determined whether quadrilaterals were parallelograms and/or rectangles.

Now

1. Recognize and apply the properties of rhombi and squares.
2. Determine whether quadrilaterals are rectangles, rhombi, or squares.

Why?

- The netting used for goals in such sports as soccer, basketball, and hockey is made out of rhombus-shaped nylon mesh. Similarly shaped netting is used for tennis and volleybal nets. A rhombus and a square are both types of equilateral parallelograms.

New Vocabulary
rhombus
square

Mathematical Practices

3 Construct viable arguments and critque the reasoning of others.
2 Reason abstractly and quantitatively.

1 Properties of Rhombi and Squares

A **rhombus** is a parallelogram with all four sides congruent. A rhombus has all the properties of a parallelogram and the two additional characteristics described in the theorems below.

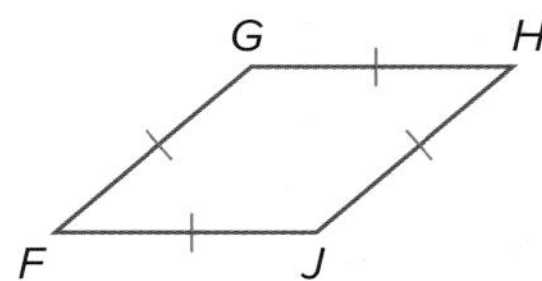

Theorems Diagonals of a Rhombus

6.15 If a parallelogram is a rhombus, then its diagonals are perpendicular.

Example If ▱$ABCD$ is a rhombus, then $\overline{AC} \perp \overline{BD}$.

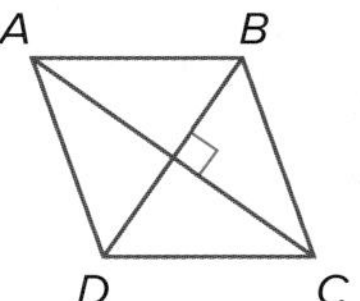

6.16 If a parallelogram is a rhombus, then each diagonal bisects a pair of opposite angles.

Example If ▱$NPQR$ is a rhombus, then $\angle 1 \cong \angle 2$, $\angle 3 \cong \angle 4$, $\angle 5 \cong \angle 6$, and $\angle 7 \cong \angle 8$.

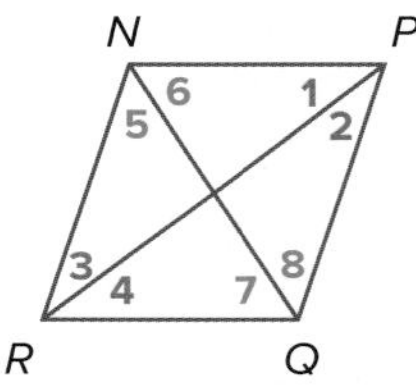

You will prove Theorem 6.16 in Exercise 34.

Proof Theorem 6.15

Given: $ABCD$ is a rhombus.

Prove: $\overline{AC} \perp \overline{BD}$

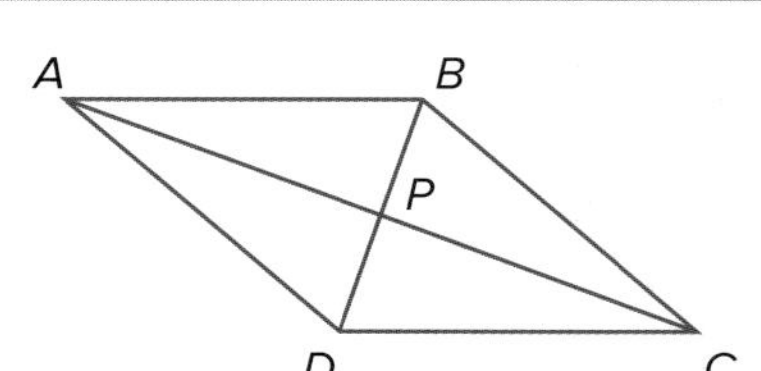

Paragraph Proof:

Because $ABCD$ is a rhombus, by definition $\overline{AB} \cong \overline{BC}$. A rhombus is a parallelogram and the diagonals of a parallelogram bisect each other, so $\overline{BD}$ bisects $\overline{AC}$ at P. Thus, $\overline{AP} \cong \overline{PC}$. $\overline{BP} \cong \overline{BP}$ by the Reflexive Property. So, $\triangle APB \cong \triangle CPB$ by SSS. $\angle APB \cong \angle CPB$ by CPCTC. $\angle APB$ and $\angle CPB$ also form a linear pair. Two congruent angles that form a linear pair are right angles. $\angle APB$ is a right angle, so $\overline{AC} \perp \overline{BD}$ by the definition of perpendicular lines.

Reading Math

Rhombi The plural form of rhombus is *rhombi*, pronounced ROM-bye.

Example 1 Use Properties of a Rhombus

The diagonals of rhombus $FGHJ$ intersect at K. Use the given information to find each measure or value.

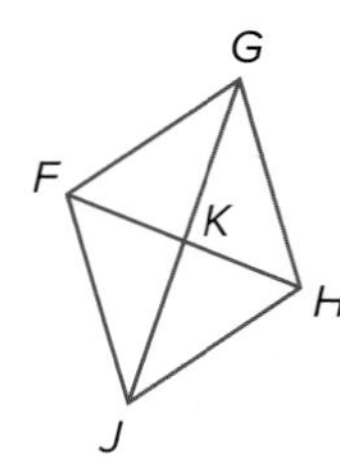

a. If $m\angle FJH = 82$, find $m\angle KHJ$.

Because $FGHJ$ is a rhombus, diagonal $\overline{JG}$ bisects $\angle FJH$. Therefore, $m\angle KJH = \frac{1}{2}m\angle FJH$. So $m\angle KJH = \frac{1}{2}(82)$ or 41. Because the diagonals of a rhombus are perpendicular, $m\angle JKH = 90$ by the definition of perpendicular lines.

$m\angle KJH + m\angle JKH + m\angle KHJ = 180$	Triangle Angle-Sum Theorem
$41 + 90 + m\angle KHJ = 180$	Substitution
$131 + m\angle KHJ = 180$	Simplify.
$m\angle KHJ = 49$	Subtract 131 from each side.

b. ALGEBRA If $GH = x + 9$ and $JH = 5x - 2$, find x.

$\overline{GH} \cong \overline{JH}$	By definition, all sides of a rhombus are congruent.
$GH = JH$	Definition of congruence
$x + 9 = 5x - 2$	Substitution
$9 = 4x - 2$	Subtract x from each side.
$11 = 4x$	Add 2 to each side.
$2.75 = x$	Divide each side by 4.

Go Online!

What special relationships exist in rhombi and squares? Investigate by using the **Geometry Tools** in ConnectED.

Guided Practice

Refer to rhombus $FGHJ$ above.

1A. If $FK = 5$ and $FG = 13$, find KJ.

1B. ALGEBRA If $m\angle JFK = 6y + 7$ and $m\angle KFG = 9y - 5$, find y.

A **square** is a parallelogram with four congruent sides and four right angles. Recall that a parallelogram with four right angles is a rectangle, and a parallelogram with four congruent sides is a rhombus. Therefore, a parallelogram that is both a rectangle and a rhombus is also a square.

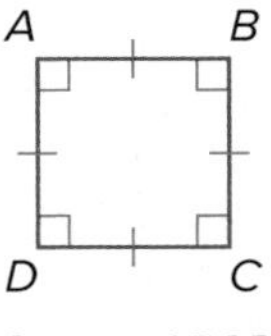

Square $ABCD$

The Venn diagram summarizes the relationships among parallelograms, rhombi, rectangles, and squares.

Concept Summary Parallelograms

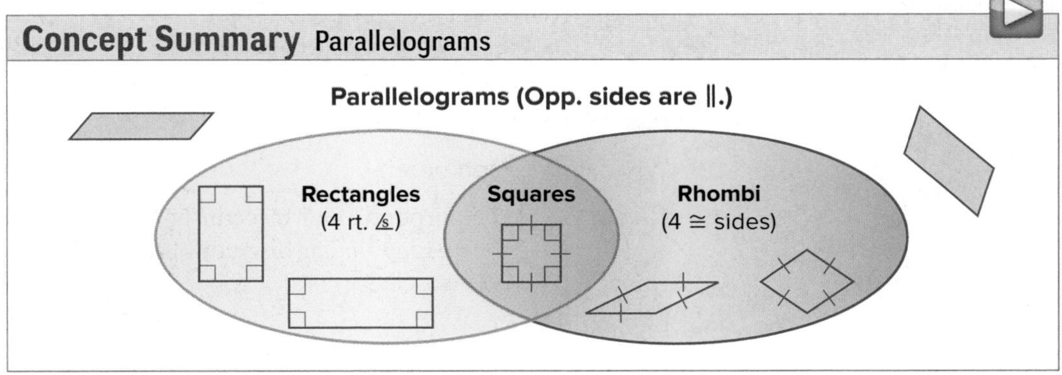

All of the properties of parallelograms, rectangles, and rhombi apply to squares. For example, the diagonals of a square bisect each other (parallelogram), are congruent (rectangle), and are perpendicular (rhombus).

2 Prove that Quadrilaterals are Rhombi or Squares

The theorems below provide conditions for rhombi and squares.

> **Study Tip**
>
> **Common Misconception** Theorems 6.17, 6.18, and 6.19 apply only if you already know that a quadrilateral is a parallelogram.

Theorems Conditions for Rhombi and Squares

6.17 If the diagonals of a parallelogram are perpendicular, then the parallelogram is a rhombus. (Converse of Theorem. 6.15)

Example If $\overline{JL} \perp \overline{KM}$, then $\square JKLM$ is a rhombus.

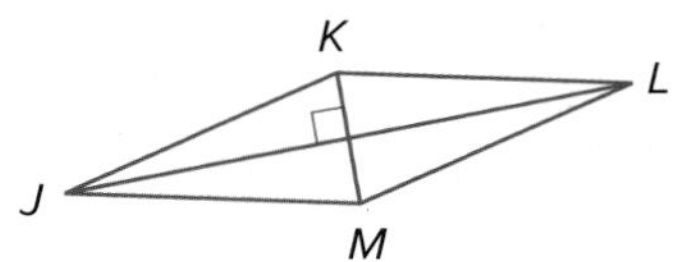

6.18 If one diagonal of a parallelogram bisects a pair of opposite angles, then the parallelogram is a rhombus. (Converse of Theorem. 6.16)

Example If $\angle 1 \cong \angle 2$ and $\angle 3 \cong \angle 4$, or $\angle 5 \cong \angle 6$ and $\angle 7 \cong \angle 8$, then $\square WXYZ$ is a rhombus.

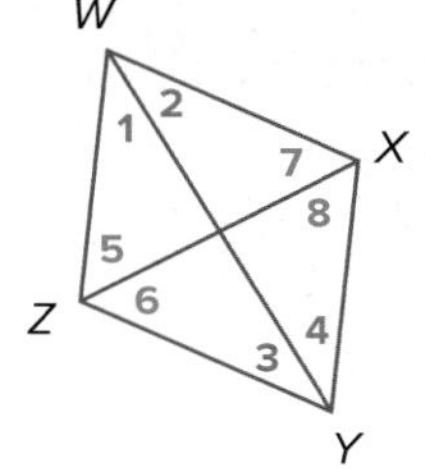

6.19 If one pair of consecutive sides of a parallelogram are congruent, the parallelogram is a rhombus.

Example If $\overline{AB} \cong \overline{BC}$, then $\square ABCD$ is a rhombus.

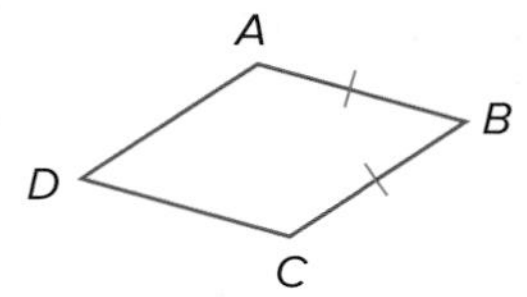

6.20 If a quadrilateral is both a rectangle and a rhombus, then it is a square.

You will prove Theorems 6.17–6.20 in Exercises 35–38, respectively.

You can use the properties of rhombi and squares to write proofs.

Example 2 Proofs Using Properties of Rhombi and Squares

Write a paragraph proof.

Given: *JKLM* is a parallelogram.

$\triangle JKL$ is isosceles.

Prove: *JKLM* is a rhombus.

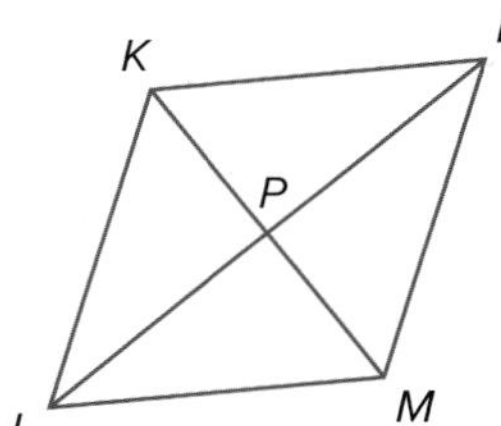

Paragraph Proof:

Because it is given that $\triangle JKL$ is isosceles, $\overline{KL} \cong \overline{JK}$ by definition. These are consecutive sides of the given parallelogram *JKLM*. So, by Theorem 6.19, *JKLM* is a rhombus.

> **Study Tip**
>
> **Congruent Triangles** Because a rhombus has four congruent sides, one diagonal separates the rhombus into two congruent isosceles triangles. Drawing two diagonals separates the rhombus into four congruent right triangles.

Guided Practice

2. Write a paragraph proof.

Given: $\overline{SQ}$ is the perpendicular bisector of $\overline{PR}$.
$\overline{PR}$ is the perpendicular bisector of $\overline{SQ}$.
$\triangle RMS$ is isosceles.

Prove: *PQRS* is a square.

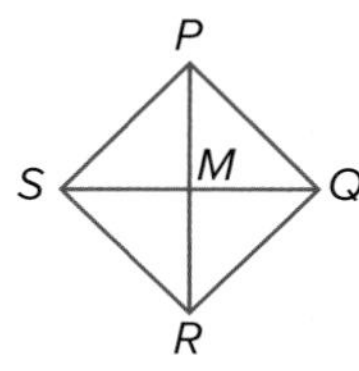

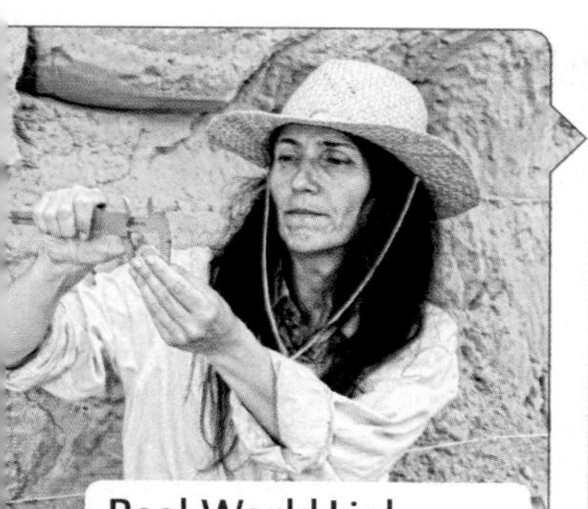

Real-World Link

Archaeology is the study of artifacts that provide information about human life and activities in the past. Since humans only began writing about 5000 years ago, information from periods before that time must be gathered from the objects that archaeologists locate.

Source: Encyclopaedia Britannica

Real-World Example 3 Use Conditions for Rhombi and Squares

ARCHAEOLOGY The key to the successful excavation of an archaeological site is accurate mapping. How can archaeologists be sure that the region they have marked off is a 1-meter by 1-meter square?

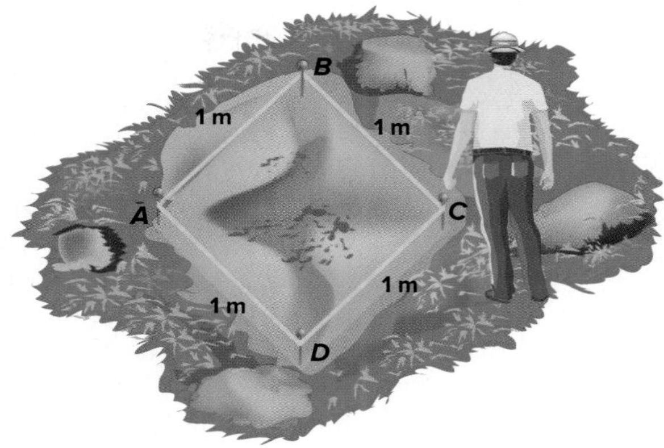

Each side of quadrilateral *ABCD* measures 1 meter. Because opposite sides are congruent, *ABCD* is a parallelogram. Because consecutive sides of ▱*ABCD* are congruent, it is a rhombus. If the archaeologists can show that ▱*ABCD* is also a rectangle, then by Theorem 6.20, ▱*ABCD* is a square.

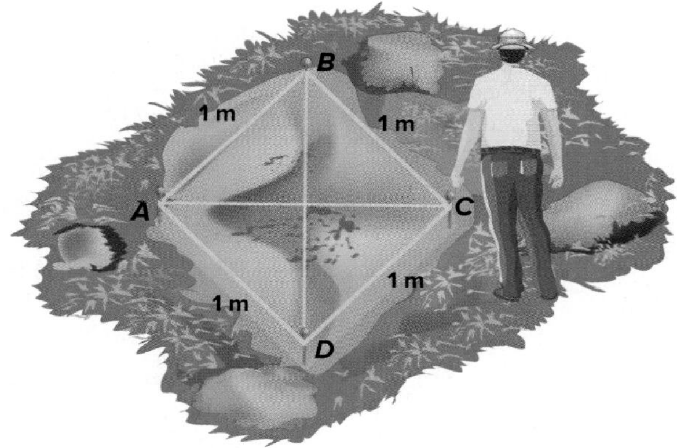

If the diagonals of a parallelogram are congruent, then the parallelogram is a rectangle. So if the archaeologists measure the length of string needed to form each diagonal and find that these lengths are equal, then *ABCD* is a square.

Guided Practice

3. QUILTING Kathy is designing a quilt with blocks like the one shown.

A. If she marks the diagonals of each yellow piece and determines that each pair of diagonals is perpendicular, can she conclude that each yellow piece is a rhombus? Explain.

B. If all four angles of the green piece have the same measure and the bottom and left sides have the same measure, can she conclude that the green piece is a square? Explain.

In Chapter 4, you used coordinate geometry to classify triangles. Coordinate geometry can also be used to classify quadrilaterals.

Example 4 Classify Quadrilaterals Using Coordinate Geometry

COORDINATE GEOMETRY **Determine whether ▱$JKLM$ with vertices $J(-7, -2)$, $K(0, 4)$, $L(9, 2)$, and $M(2, -4)$ is a *rhombus*, a *rectangle*, or a *square*. List all that apply. Explain.**

Problem-Solving Tip

MP Sense-Making When analyzing a figure using coordinate geometry, graph the figure to help formulate a conjecture and also to help check the reasonableness of the answer you obtain algebraically.

Understand Plot and connect the vertices on a coordinate plane.

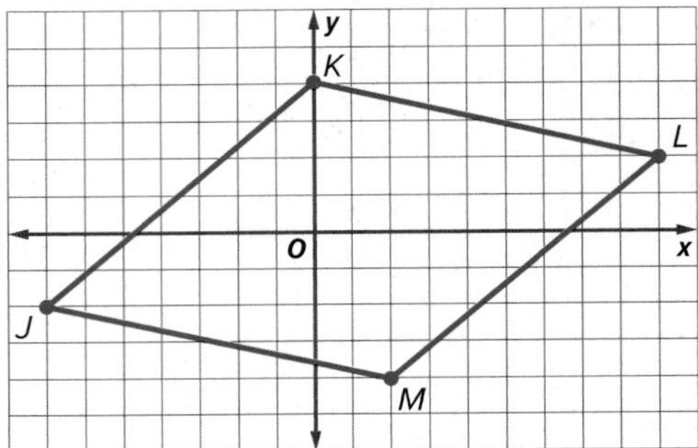

It appears from the graph that the parallelogram has four congruent sides, but no right angles. So, it appears that the figure is a rhombus, but not a square or a rectangle.

Plan If the diagonals of the parallelogram are congruent, then it is a rectangle. If they are perpendicular, then it is a rhombus. If they are both congruent and perpendicular, the parallelogram is a rectangle, a rhombus, and a square.

Solve **Step 1** Use the Distance Formula to compare the diagonal lengths.

$KM = \sqrt{(2-0)^2 + (-4-4)^2} = \sqrt{68}$ or $2\sqrt{17}$

$JL = \sqrt{[9-(-7)]^2 + [2-(-2)]^2} = \sqrt{272}$ or $4\sqrt{17}$

Since $2\sqrt{17} \neq 4\sqrt{17}$, the diagonals are not congruent. So, ▱$JKLM$ is *not* a rectangle. Since the figure is not a rectangle, it also *cannot* be a square.

Study Tip

Square and Rhombus A square is a rhombus, but a rhombus is not necessarily a square.

Step 2 Use the Slope Formula to determine whether the diagonals are perpendicular.

slope of $\overline{KM} = \frac{-4-4}{2-0} = \frac{-8}{2}$ or -4

slope of $\overline{JL} = \frac{2-(-2)}{9-(-7)} = \frac{4}{16}$ or $\frac{1}{4}$

Since the product of the slopes of the diagonals is -1, the diagonals are perpendicular, so ▱$JKLM$ is a rhombus.

Check $JK = \sqrt{[4-(-2)]^2 + [0-(-7)]^2}$ or $\sqrt{85}$

$KL = \sqrt{(9-0)^2 + (2-4)^2}$ or $\sqrt{85}$

So, ▱$JKLM$ is a rhombus by Theorem 6.20.

Since the slope of $\overline{JK} = \frac{4-(-2)}{0-(-7)}$ or $\frac{6}{7}$, the slope of $\overline{KL} = \frac{2-4}{9-0}$ or $-\frac{2}{9}$, and the product of these slopes is not -1, consecutive sides $\overline{JK}$ and $\overline{KL}$ are not perpendicular. Therefore, $\angle JKL$ is not a right angle. So ▱$JKLM$ is not a rectangle or a square.

Guided Practice

4. **Given** $J(5, 0)$, $K(8, -11)$, $L(-3, -14)$, $M(-6, -3)$, determine whether parallelogram $JKLM$ is a *rhombus*, a *rectangle*, or a *square*. List all that apply. Explain.

Check Your Understanding

Example 1 **ALGEBRA** **Quadrilateral *ABCD* is a rhombus. Find each value or measure.**

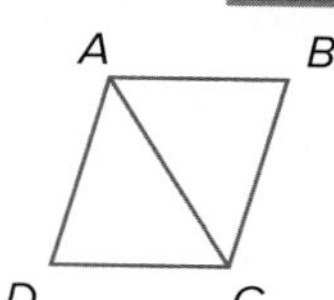

1. If $m\angle BCD = 64$, find $m\angle BAC$.
2. If $AB = 2x + 3$ and $BC = x + 7$, find CD.

Examples 2–3

3. **PROOF** Write a two-column proof to prove that if $ABCD$ is a rhombus with diagonal $\overline{DB}$, then $\overline{AP} \cong \overline{CP}$.

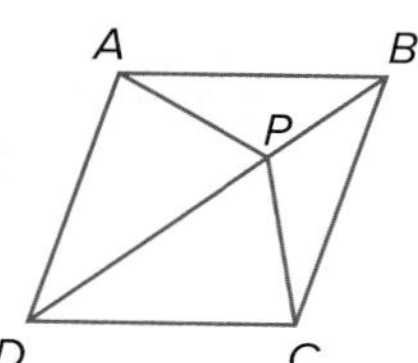

4. **GAMES** The checkerboard below is made up of 64 congruent black and red squares. Use this information to prove that the board itself is a square.

Example 4 **COORDINATE GEOMETRY** Given each set of vertices, determine whether ▱*QRST* is a ***rhombus***, a ***rectangle***, or a ***square***. List all that apply. Explain.

5. $Q(1, 2), R(-2, -1), S(1, -4), T(4, -1)$
6. $Q(-2, -1), R(-1, 2), S(4, 1), T(3, -2)$

Practice and Problem Solving

Extra Practice is on page R6.

Example 1 **ALGEBRA** **Quadrilateral *ABCD* is a rhombus. Find each value or measure.**

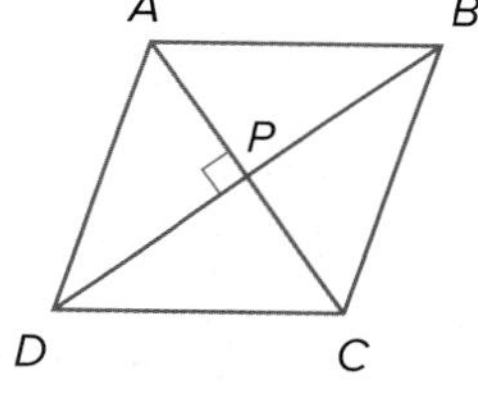

7. If $AB = 14$, find BC.
8. If $m\angle BCD = 54$, find $m\angle BAC$.
9. If $AP = 3x - 1$ and $PC = x + 9$, find AC.
10. If $DB = 2x - 4$ and $PB = 2x - 9$, find PD.
11. If $m\angle ABC = 2x - 7$ and $m\angle BCD = 2x + 3$, find $m\angle DAB$.
12. If $m\angle DPC = 3x - 15$, find x.

Example 2 **MP CONSTRUCT ARGUMENTS** **Write a two-column proof.**

13. **Given:** $\overline{WZ} \parallel \overline{XY}$, $\overline{WX} \parallel \overline{ZY}$
$\overline{WZ} \cong \overline{ZY}$
Prove: $WXYZ$ is a rhombus.

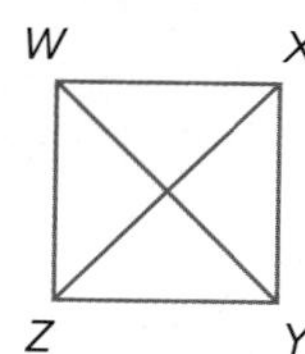

14. **Given:** $QRST$ is a parallelogram.
$\overline{TR} \cong \overline{QS}$, $m\angle QPR = 90$
Prove: $QRST$ is a square.

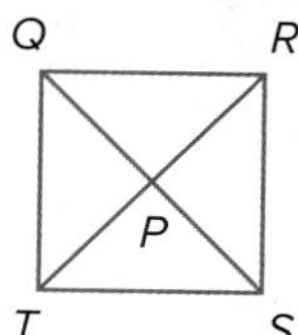

15. **Given:** $ABCD$ is a parallelogram.
$\triangle ABC \cong \triangle ADC$
Prove: $ABCD$ is a rhombus.

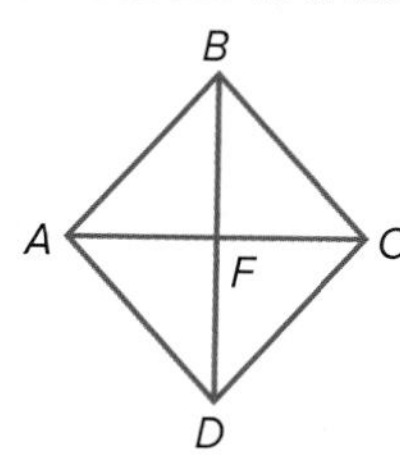

16. **Given:** $ACDH$ and $BCDF$ are parallelograms; $\overline{BF} \cong \overline{AB}$.
Prove: $ABFH$ is a rhombus.

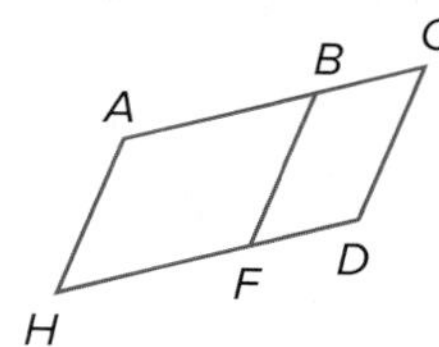

Example 3

17. **ROADWAYS** Main Street and High Street intersect as shown in the diagram. Each of the crosswalks is the same length. Classify the quadrilateral formed by the crosswalks. Explain your reasoning.

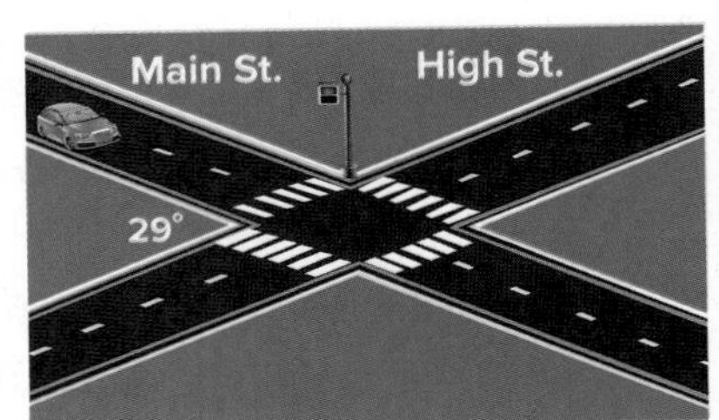

18. **MP MODELING** A landscaper has staked out the area for a square garden as shown. She has confirmed that each side of the quadrilateral formed by the stakes is congruent and that the diagonals are perpendicular. Is this information enough for the landscaper to be sure that the garden is a square? Explain your reasoning.

Example 4

COORDINATE GEOMETRY Given each set of vertices, determine whether ▱*JKLM* is a *rhombus*, a *rectangle*, or a *square*. List all that apply. Explain.

19. $J(-4, -1)$, $K(1, -1)$, $L(4, 3)$, $M(-1, 3)$

20. $J(-3, -2)$, $K(2, -2)$, $L(5, 2)$, $M(0, 2)$

21. $J(-2, -1)$, $K(-4, 3)$, $L(1, 5)$, $M(3, 1)$

22. $J(-1, 1)$, $K(4, 1)$, $L(4, 6)$, $M(-1, 6)$

***ABCD* is a rhombus. If $PB = 12$, $AB = 15$, and $m\angle ABD = 24$, find each measure.**

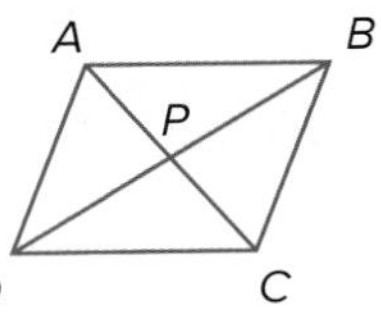

23 AP

24. CP

25. $m\angle BDA$

26. $m\angle ACB$

***WXYZ* is a square. If $WT = 3$, find each measure.**

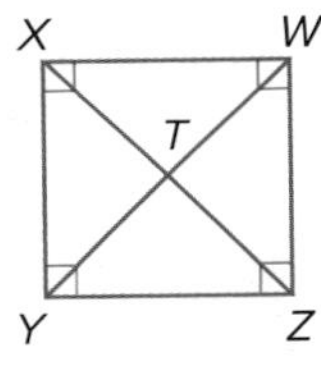

27. ZX

28. XY

29. $m\angle WTZ$

30. $m\angle WYX$

Classify each quadrilateral.

31.

32.

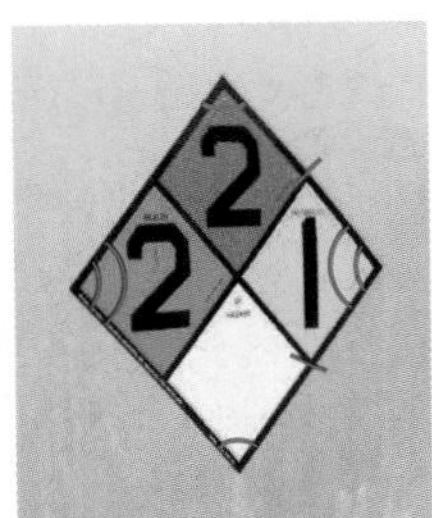

33.

PROOF Write a paragraph proof.

34. Theorem 6.16

35. Theorem 6.17

36. Theorem 6.18

37. Theorem 6.19

38. Theorem 6.20

CONSTRUCTION Use diagonals to construct each figure. Justify each construction.

39. rhombus

40. square

PROOF Write a coordinate proof of each statement.

41. The diagonals of a square are perpendicular.

42. The segments joining the midpoints of the sides of a rectangle form a rhombus.

43. **DESIGN** The tile pattern below consists of regular octagons and quadrilaterals. Classify the quadrilaterals in the pattern and explain your reasoning.

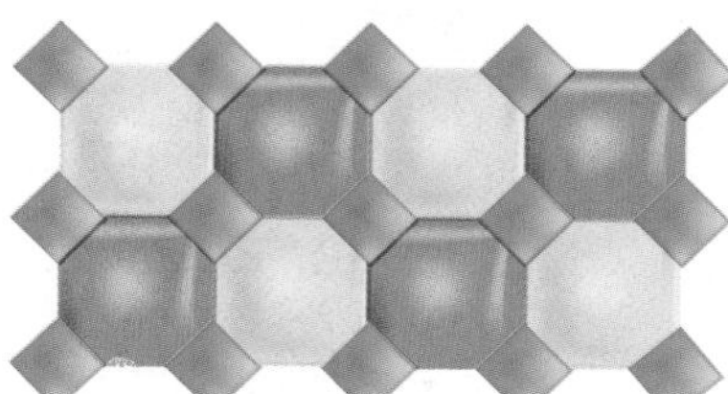

44. **REPAIR** The window pane shown needs to be replaced. What are the dimensions of the replacement pane?

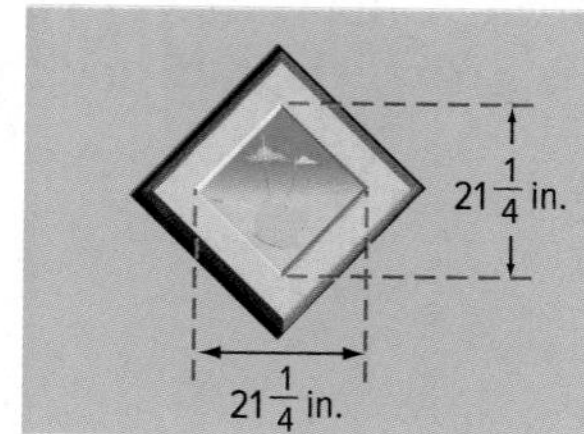

45. **MULTIPLE REPRESENTATIONS** In this problem, you will explore the properties of kites, which are quadrilaterals with exactly two distinct pairs of adjacent congruent sides.

a. **Geometric** Draw three kites with varying side lengths. Label one kite *ABCD*, one *PQRS*, and one *WXYZ*. Then draw the diagonals of each kite, labeling the point of intersection *N* for each kite.

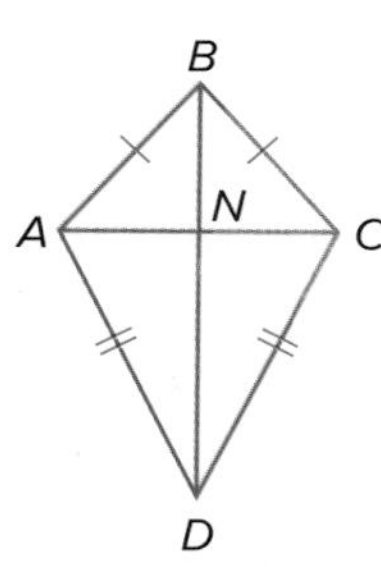

Kite *ABCD*

b. **Tabular** Measure the distance from *N* to each vertex. Record your results in a table like the one shown.

Figure	Distance from *N* to Each Vertex Along Shorter Diagonal		Distance from *N* to Each Vertex Along Longer Diagonal	
ABCD				
PQRS				
WXYZ				

c. **Verbal** Make a conjecture about the diagonals of a kite.

H.O.T. Problems Use Higher-Order Thinking Skills

46. **ERROR ANALYSIS** In parallelogram *PQRS*, $\overline{PR} \cong \overline{QS}$. Lola thinks that the parallelogram is a square, and Xavier thinks that it is a rhombus. Is either of them correct? Explain your reasoning.

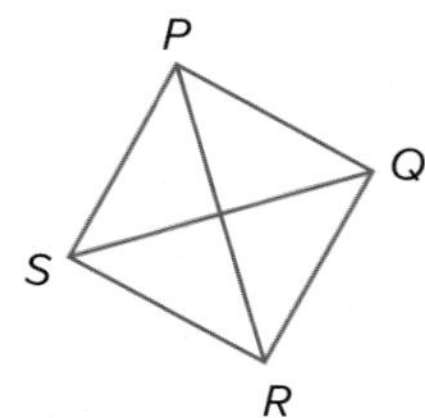

47. MP **REASONING** Determine whether the statement is *true* or *false*. Then write the converse, inverse, and contrapositive of the statement and determine the truth value of each. Explain your reasoning.

If a quadrilateral is a square, then it is a rectangle.

48. **CHALLENGE** The area of square *ABCD* is 36 square units and the area of $\triangle EBF$ is 20 square units. If $\overline{EB} \perp \overline{BF}$ and $AE = 2$, find the length of $\overline{CF}$.

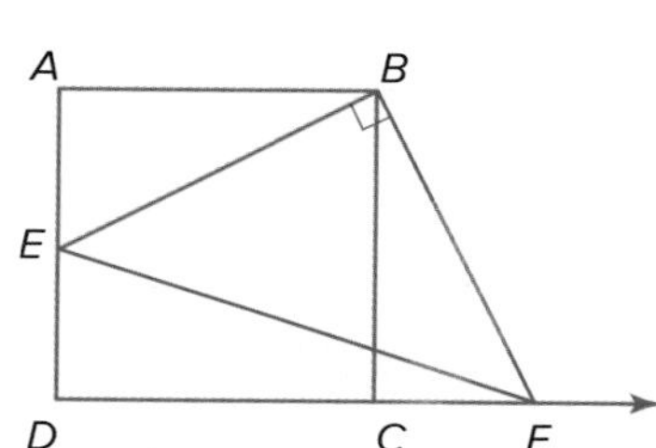

49. **OPEN-ENDED** Find the vertices of a square with diagonals that are contained in the lines $y = x$ and $y = -x + 6$. Justify your reasoning.

50. **WRITING IN MATH** Compare all of the properties of the following quadrilaterals: parallelograms, rectangles, rhombi, and squares.

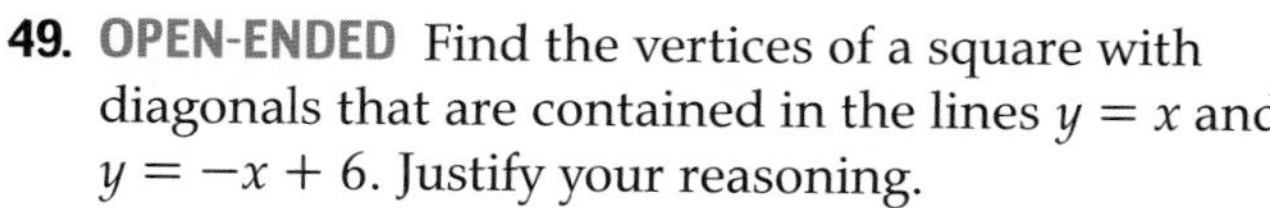

51. Julia is designing a pair of earrings. The figure shows one of the earrings. Julia knows that quadrilateral *JKLM* is a parallelogram and that $m\angle KLN = 54$.

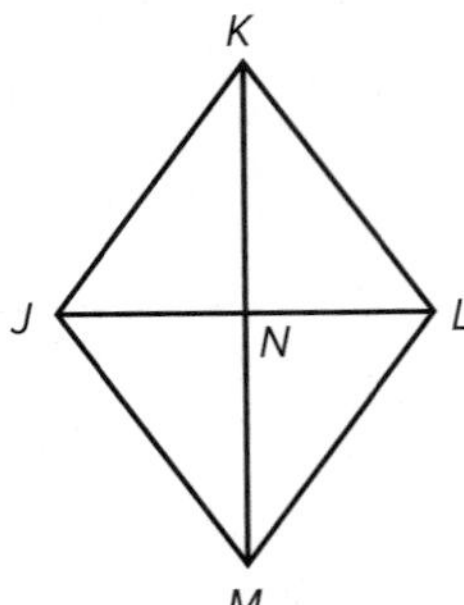

What should the measure of $\angle LKN$ be in order for the earring to be a rhombus? MP 4

- ○ **A** 36
- ○ **B** 54
- ○ **C** 90
- ○ **D** 108

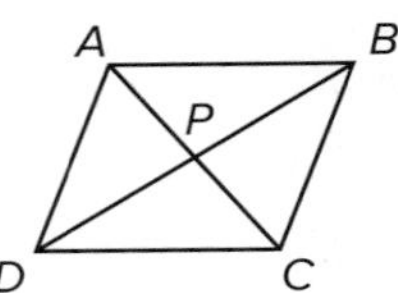

***ABCD* is a rhombus. If $PB = 15$, $AB = 25$ and $m\angle ABD = 33$, find each measure.**

52. $CP =$ ▭ MP 2

- ○ **A** 16
- ○ **B** 18
- ○ **C** 20
- ○ **D** 24

53. $AP =$ ▭ MP 2

- ○ **A** 16
- ○ **B** 18
- ○ **C** 20
- ○ **D** 24

54. $m\angle ACB =$ ▭ MP 2

- ○ **A** 33
- ○ **B** 57
- ○ **C** 63
- ○ **D** 90

55. $m\angle BDA =$ ▭ MP 2

- ○ **A** 33
- ○ **B** 57
- ○ **C** 63
- ○ **D** 90

56. **MULTI-STEP** The vertices of a parallelogram are $P(-1, 7)$, $Q(5, 5)$, $R(3, -1)$, and $S(-3, 1)$. Which of the following terms apply to the parallelogram? MP 3

I. square

II. rectangle

III. rhombus

- ○ **A** I only
- ○ **B** II only
- ○ **C** III only
- ○ **D** II and III only
- ○ **E** I, II, and III

57. The diagonals of rhombus *ABCD* are $\overline{AC}$ and $\overline{BD}$. Which of the following is not necessarily true? MP 2, 6

- ○ **A** $\overline{AC} \perp \overline{BD}$
- ○ **B** $\overline{AC} \cong \overline{BD}$
- ○ **C** $\overline{AC}$ and $\overline{BD}$ bisect each other.
- ○ **D** $\overline{AC}$ bisects $\angle A$.

58. A supermarket sells trays of vegetables in the shape of a rhombus. The diagonals of the tray form four compartments for the vegetables as shown.

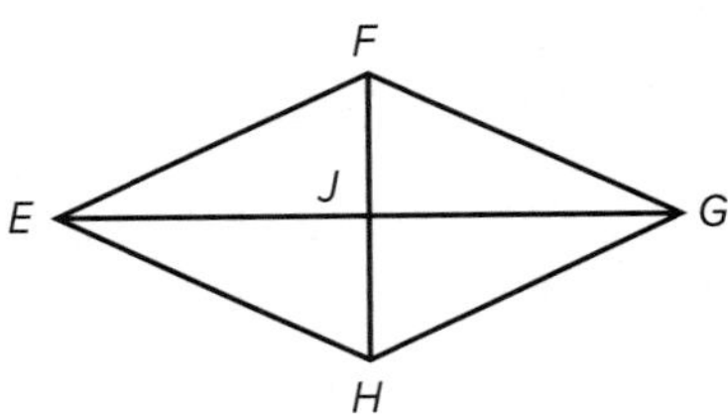

Given that $m\angle EFG = 130$, what is $m\angle HGJ$? MP 4

- ○ **A** 130
- ○ **B** 65
- ○ **C** 50
- ○ **D** 25

59. *QRST* is a parallelogram and $x = 10$. What is $m\angle QRS$? MP 2

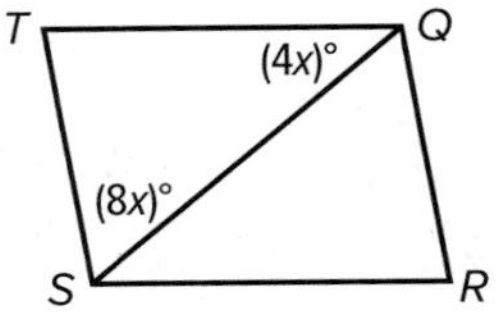

LESSON 6

Trapezoids and Kites

Then	Now	Why?
You used properties of special parallelograms.	1 Apply properties of trapezoids. 2 Apply properties of kites.	In gymnastics, vaulting boxes made out of high compression foam are used as spotting platforms, vaulting horses, and steps. The left and right side of each section is a *trapezoid*.

New Vocabulary
trapezoid
bases
legs of a trapezoid
base angles
isosceles trapezoid
midsegment of a trapezoid
kite

Mathematical Practices

1 Make sense of problems and persevere in solving them.

2 Reason abstractly and quantitatively.

1 Properties of Trapezoids

A **trapezoid** is a quadrilateral with exactly one pair of parallel sides. The parallel sides are called **bases**. The nonparallel sides are called **legs**. The **base angles** are formed by the base and one of the legs. In trapezoid $ABCD$, $\angle A$ and $\angle B$ are one pair of base angles and $\angle C$ and $\angle D$ are the other pair. If the legs of a trapezoid are congruent, then it is an **isosceles trapezoid**.

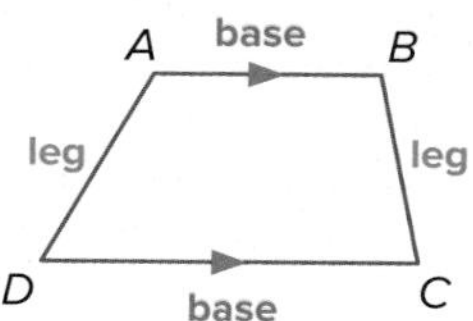

Theorems Isosceles Trapezoids

6.21 If a trapezoid is isosceles, then each pair of base angles is congruent.

Example If trapezoid $FGHJ$ is isosceles, then $\angle G \cong \angle H$ and $\angle F \cong \angle J$.

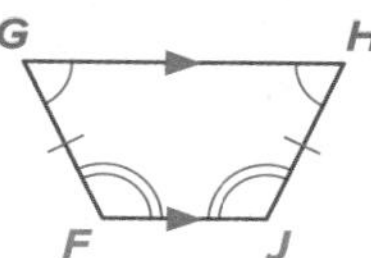

6.22 If a trapezoid has one pair of congruent base angles, then it is an isosceles trapezoid.

Example If $\angle L \cong \angle M$, then trapezoid $KLMP$ is isosceles.

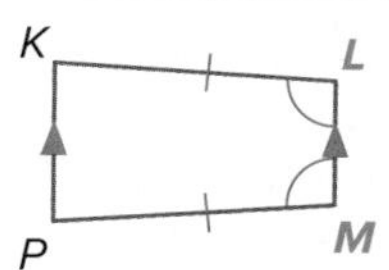

6.23 A trapezoid is isosceles if and only if its diagonals are congruent.

Example If trapezoid $QRST$ is isosceles, then $\overline{QS} \cong \overline{RT}$. Likewise, if $\overline{QS} \cong \overline{RT}$, then trapezoid $QRST$ is isosceles.

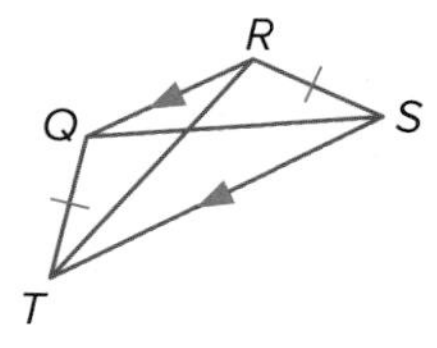

You will prove Theorem 6.21, Theorem 6.22, and the other part of Theorem 6.23 in Exercises 28, 29, and 30.

Proof Part of Theorem 6.23

Given: $ABCD$ is an isosceles trapezoid.

Prove: $\overline{AC} \cong \overline{BD}$

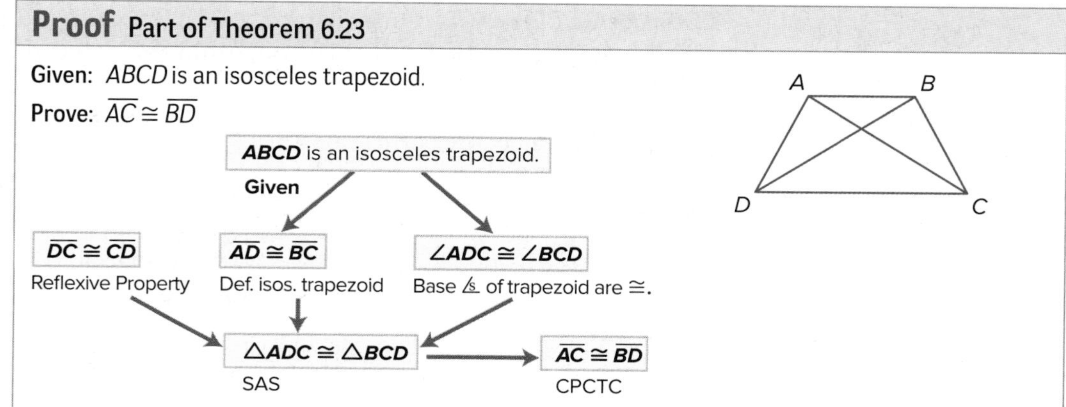

Real-World Example 1 Use Properties of Isosceles Trapezoids

MUSIC The speaker shown is an isosceles trapezoid. If $m\angle FJH = 85$, $FL = 8$ inches, and $JG = 19$ inches, find each measure.

a. $m\angle FGH$

Because $FGHJ$ is an isosceles trapezoid, $\angle FJH$ and $\angle GHJ$ are congruent base angles. So, $m\angle GHJ = m\angle FJH = 85$.

Because $FGHJ$ is a trapezoid, $\overline{FG} \parallel \overline{JH}$.

$m\angle FGH + m\angle GHJ = 180$	Consecutive Interior Angles Theorem
$m\angle FGH + 85 = 180$	Substitution
$m\angle FGH = 95$	Subtract 85 from each side.

b. LH

Because $FGHJ$ is an isosceles trapezoid, diagonals $\overline{FH}$ and $\overline{JG}$ are congruent.

$FH = JG$	Definition of congruent
$FL + LH = JG$	Segment Addition
$8 + LH = 19$	Substitution
$LH = 11$ cm	Subtract 8 from each side.

Real-World Link

Speakers are amplifiers that intensify sound waves so that they are audible to the unaided ear. Amplifiers exist in devices such as televisions, stereos, and computers.

Source: How Stuff Works

Guided Practice

1. **CAFETERIA TRAYS** To save space at a square table, cafeteria trays often incorporate trapezoids into their design. If $WXYZ$ is an isosceles trapezoid and $m\angle YZW = 45$, $WV = 15$ centimeters, and $VY = 10$ centimeters, find each measure.

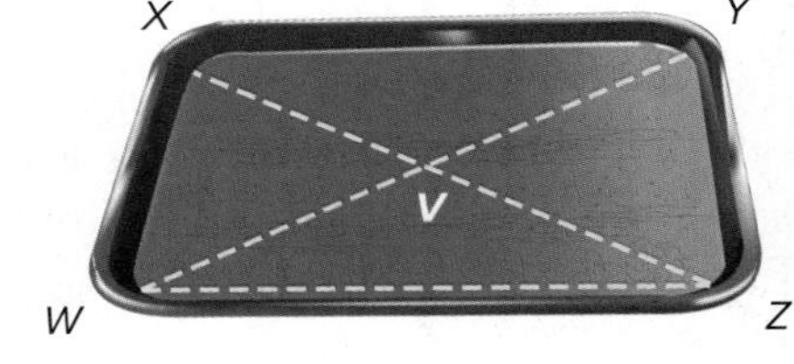

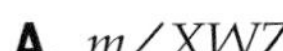

A. $m\angle XWZ$ **B.** $m\angle WXY$

C. XZ **D.** XV

Study Tip

MP Sense-Making The base angles of a trapezoid are only congruent if the trapezoid is isosceles.

You can use coordinate geometry to determine whether a trapezoid is an isosceles trapezoid.

Example 2 Isosceles Trapezoids and Coordinate Geometry

COORDINATE GEOMETRY Quadrilateral $ABCD$ has vertices $A(-3, 4)$, $B(2, 5)$, $C(3, 3)$, and $D(-1, 0)$. Show that $ABCD$ is a trapezoid and determine whether it is an isosceles trapezoid.

Graph and connect the vertices of $ABCD$.

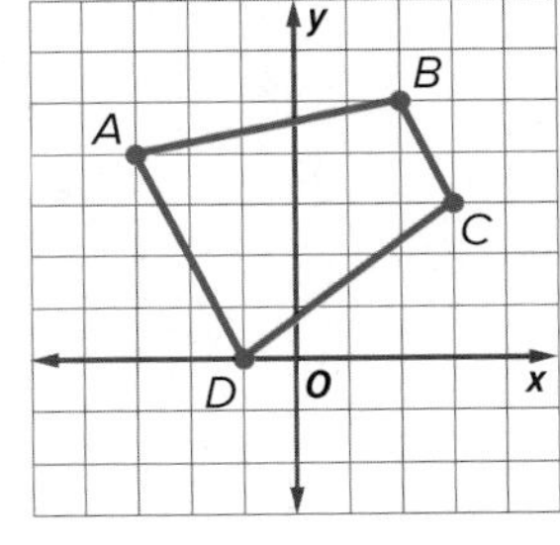

Step 1 Use the Slope Formula to compare the slopes of opposite sides $\overline{BC}$ and $\overline{AD}$ and of opposite sides $\overline{AB}$ and $\overline{DC}$. A quadrilateral is a trapezoid if exactly one pair of opposite sides is parallel.

Opposite sides $\overline{BC}$ and $\overline{AD}$:

slope of $\overline{BC} = 3 - \frac{5}{3} - 2 = -\frac{2}{1}$ or -2

slope of $\overline{AD} = \frac{0 - 4}{-1 - (-3)} = \frac{-4}{2}$ or -2

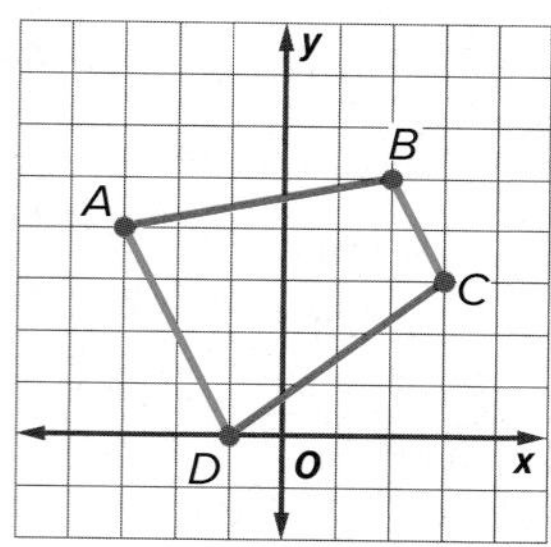

Because the slopes of $\overline{BC}$ and $\overline{AD}$ are equal, $\overline{BC} \parallel \overline{AD}$.

Opposite sides $\overline{AB}$ and $\overline{DC}$:

slope of $\overline{AB} = \frac{5 - 4}{2 - (-3)} = \frac{1}{5}$

slope of $\overline{DC} = \frac{0 - 3}{-1 - 3} = \frac{-3}{-4}$ or $\frac{3}{4}$

Reading Math

Symbols Recall that the symbol $\nparallel$ means is *not parallel to.*

Because the slopes of $\overline{AB}$ and $\overline{DC}$ are *not* equal, $\overline{BC} \nparallel \overline{AD}$. Because quadrilateral *ABCD* has only one pair of opposite sides that are parallel, quadrilateral *ABCD* is a trapezoid.

Step 2 Use the Distance Formula to compare the lengths of legs $\overline{AB}$ and $\overline{DC}$. A trapezoid is isosceles if its legs are congruent.

$AB = \sqrt{(-3 - 2)^2 + (4 - 5)^2}$ or $\sqrt{26}$

$DC = \sqrt{(-1 - 3)^2 + (0 - 3)^2} = \sqrt{25}$ or 5

Because $AB \neq DC$, legs $\overline{AB}$ and $\overline{DC}$ are *not* congruent. Therefore, trapezoid *ABCD* is not isosceles.

Guided Practice

2. Quadrilateral *QRST* has vertices $Q(-8, -4)$, $R(0, 8)$, $S(6, 8)$, and $T(-6, -10)$. Show that *QRST* is a trapezoid and determine whether *QRST* is an isosceles trapezoid.

Reading Math

Midsegment A midsegment of a trapezoid can also be called a *median.*

The **midsegment of a trapezoid** is the segment that connects the midpoints of the legs of the trapezoid.

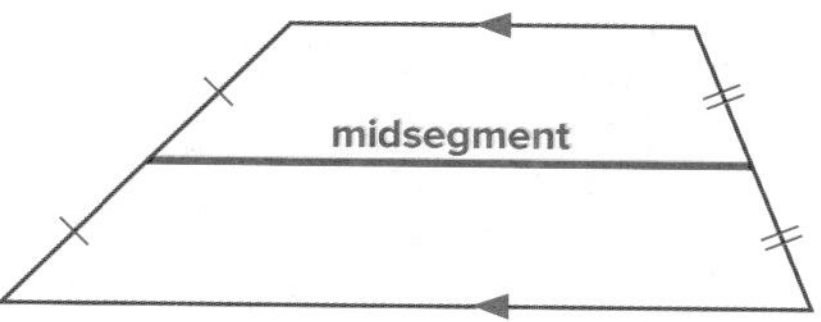

The theorem below relates the midsegment and the bases of a trapezoid.

Theorem 6.24 Trapezoid Midsegment Theorem

The midsegment of a trapezoid is parallel to each base and its measure is one half the sum of the lengths of the bases.

Example If $\overline{BE}$ is the midsegment of trapezoid *ACDF*, then $\overline{AF} \parallel \overline{BE}$, $\overline{CD} \parallel \overline{BE}$, and $BE = \frac{1}{2}(AF + CD)$.

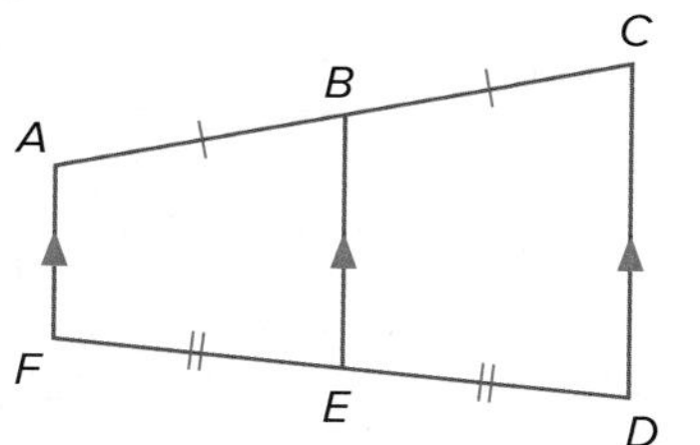

You will prove Theorem 6.24 in Exercise 33.

Example 3 Midsegment of a Trapezoid

In the figure, $\overline{LH}$ is the midsegment of trapezoid $FGJK$. What is the value of x?

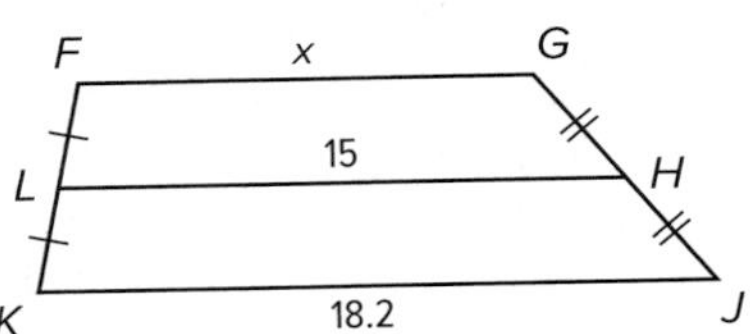

Note: The figure is not drawn to scale.

Read the Item

You are given the measure of the midsegment of a trapezoid and the measure of one of its bases. You are asked to find the measure of the other base.

Solve the Item

$LH = \frac{1}{2}(FG + KJ)$	Trapezoid Midsegment Theorem
$15 = \frac{1}{2}(x + 18.2)$	Substitution
$30 = x + 18.2$	Multiply each side by 2.
$11.8 = x$	Subtract 18.2 from each side.

Go Online!

Investigate the attributes of trapezoids and kites by using the **Geometry Tools** in ConnectED.

Guided Practice

3A. Trapezoid $ABCD$ is shown below. If $\overline{FG}$ is parallel to $\overline{AD}$, what is the x-coordinate of point G?

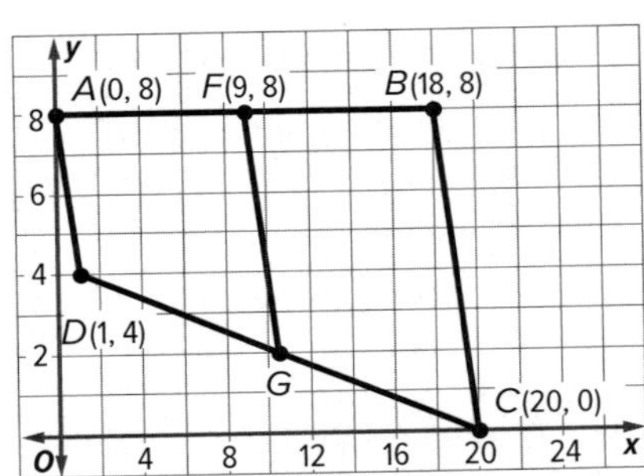

3B. In the figure, $\overline{MN}$ is the midpoint of trapezoid $QRST$. Write the equation for the midpoint, and then solve for the value of x.

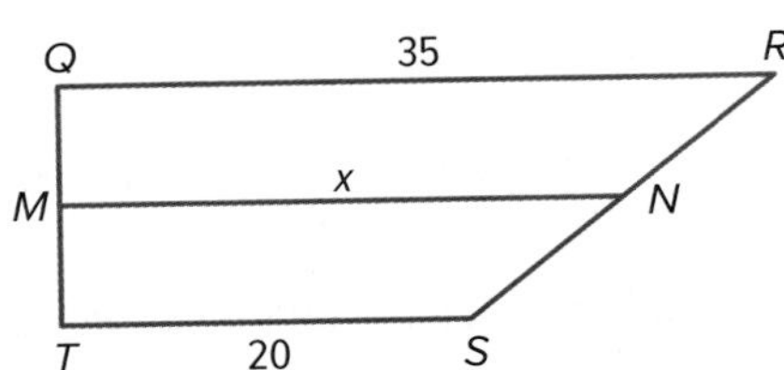

2 Properties of Kites

A **kite** is a quadrilateral with exactly two pairs of consecutive congruent sides. Unlike a parallelogram, the opposite sides of a kite are not congruent or parallel.

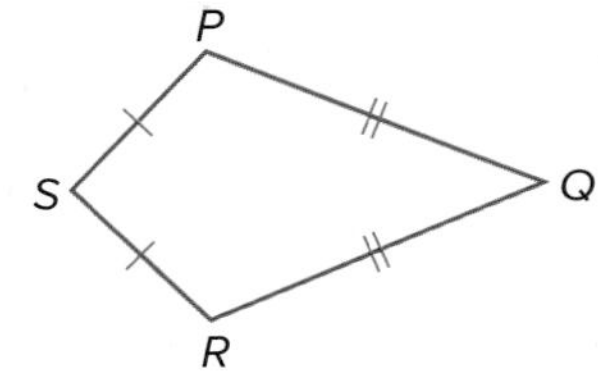

Study Tip

Kites The congruent angles of a kite are included by the noncongruent adjacent sides.

Theorems Kites

6.25 If a quadrilateral is a kite, then its diagonals are perpendicular.

Example If quadrilateral *ABCD* is a kite, then $\overline{AC} \perp \overline{BD}$.

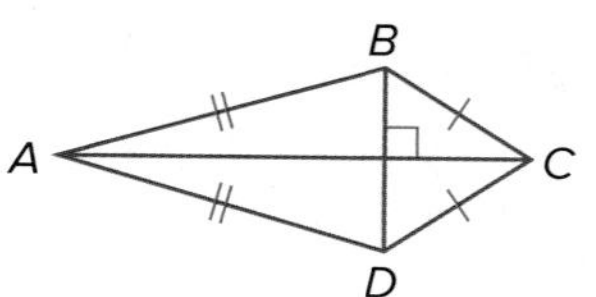

6.26 If a quadrilateral is a kite, then exactly one pair of opposite angles is congruent.

Example If quadrilateral *JKLM* is a kite, $\overline{JK} \cong \overline{KL}$, and $\overline{JM} \cong \overline{LM}$, then $\angle J \cong \angle L$ and $\angle K \ncong \angle M$.

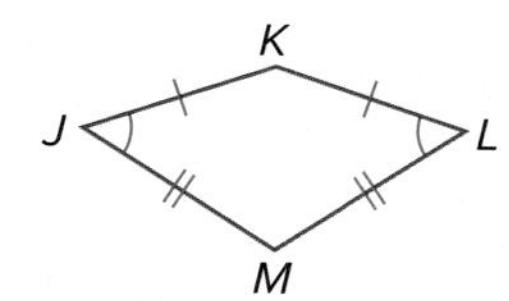

You will prove Theorems 6.25 and 6.26 in Exercises 31 and 32, respectively.

You can use the theorems above, the Pythagorean Theorem, and the Polygon Interior Angles Sum Theorem to find missing measures in kites.

Real-World Link

The fastest recorded speed of a kite is over 120 miles per hour. The record for the highest single kite flown is 12,471 feet.

Source: Borealis Kites

Example 4 Use Properties of Kites

a. If *FGHJ* is a kite, find $m\angle GFJ$.

Because a kite can only have one pair of opposite congruent angles and $\angle G \ncong \angle J$, then $\angle F \cong \angle H$. So, $m\angle F = m\angle H$. Write and solve an equation to find $m\angle F$.

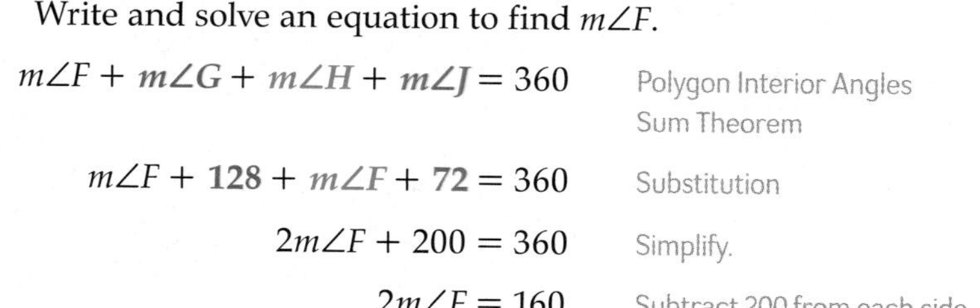

$m\angle F + m\angle G + m\angle H + m\angle J = 360$ Polygon Interior Angles Sum Theorem

$m\angle F + 128 + m\angle F + 72 = 360$ Substitution

$2m\angle F + 200 = 360$ Simplify.

$2m\angle F = 160$ Subtract 200 from each side.

$m\angle F = 80$ Divide each side by 2.

b. If *WXYZ* is a kite, find *ZY*.

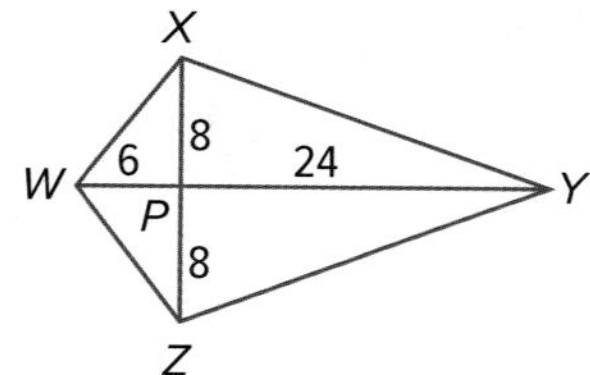

Because the diagonals of a kite are perpendicular, they divide *WXYZ* into four right triangles. Use the Pythagorean Theorem to find *ZY*, the length of the hypotenuse of right $\triangle YPZ$.

$PZ^2 + PY^2 = ZY^2$ Pythagorean Theorem

$8^2 + 24^2 = ZY^2$ Substitution

$640 = ZY^2$ Simplify.

$\sqrt{640} = ZY$ Take the square root of each side.

$8\sqrt{10} = ZY$ Simplify.

Guided Practice

4A. If $m\angle BAD = 38$ and $m\angle BCD = 50$, find $m\angle ADC$.

4B. If $BT = 5$ and $TC = 8$, find *CD*.

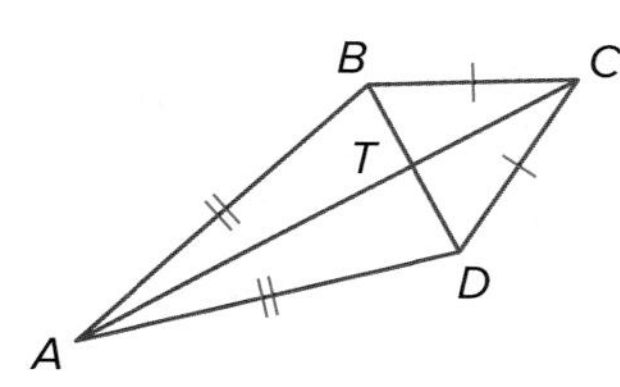

Check Your Understanding

◯ = Step-by-Step Solutions begin on page R13.

Go Online! for a Self-Check Quiz

Example 1 **Find each measure.**

1. $m\angle D$

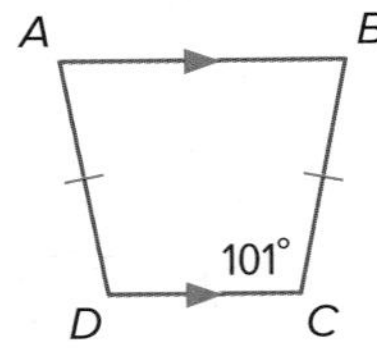

2. WT, if $ZX = 20$ and $TY = 15$

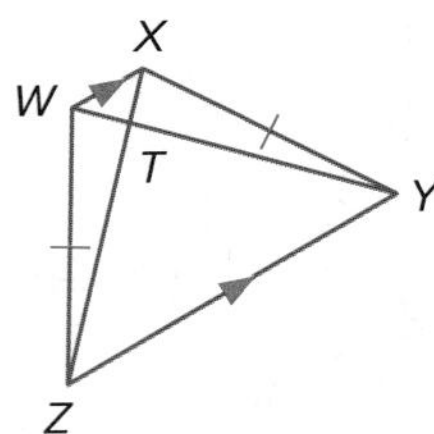

Example 2 COORDINATE GEOMETRY **Quadrilateral $ABCD$ has vertices $A(-4, -1)$, $B(-2, 3)$, $C(3, 3)$, and $D(5, -1)$.**

3. Verify that $ABCD$ is a trapezoid.
4. Determine whether $ABCD$ is an isosceles trapezoid. Explain.

Example 3 5. In the figure at the right, $\overline{YZ}$ is the midsegment of trapezoid $TWRV$. Determine the value of x.

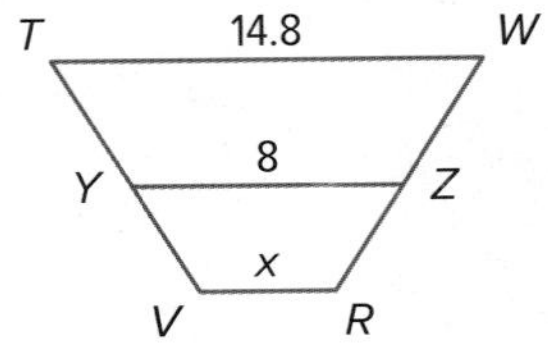

Example 4 MP SENSE-MAKING **If $ABCD$ is a kite, find each measure.**

6. AB

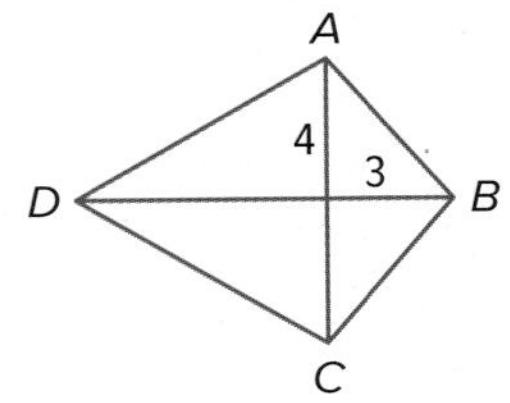

7. $m\angle C$

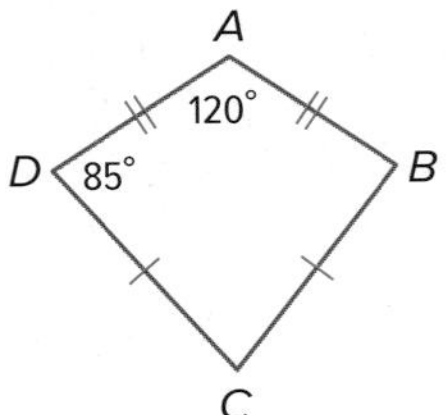

Practice and Problem Solving

Extra Practice is on page R6.

Example 1 **Find each measure.**

8. $m\angle K$

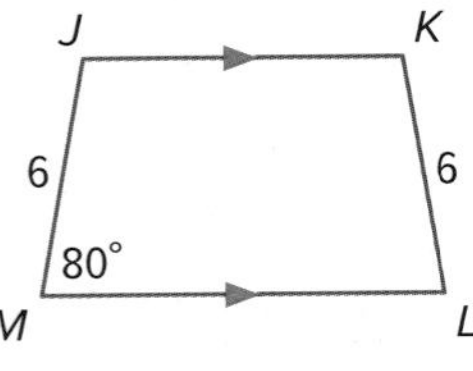

9. $m\angle Q$

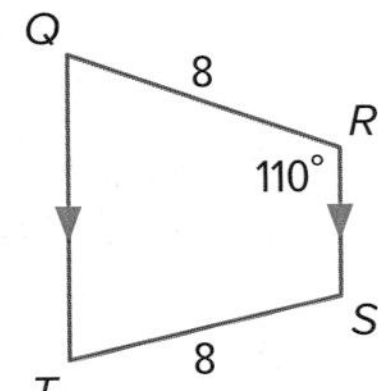

10. JL, if $KP = 4$ and $PM = 7$

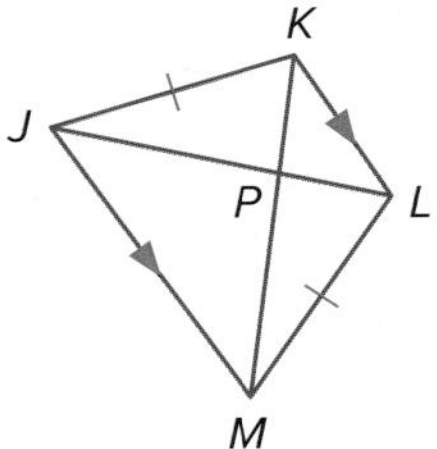

11. PW, if $XZ = 18$ and $PY = 3$

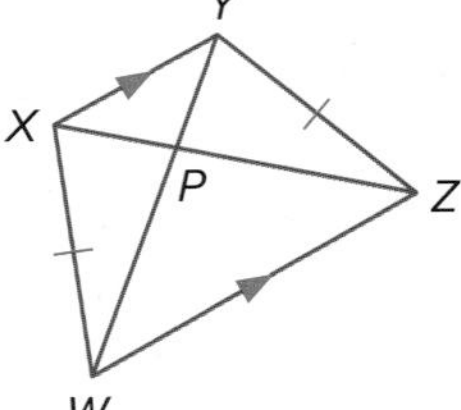

Example 2 COORDINATE GEOMETRY **For each quadrilateral with the given vertices, verify that the quadrilateral is a trapezoid and determine whether the figure is an isosceles trapezoid.**

12. $A(-2, 5)$, $B(-3, 1)$, $C(6, 1)$, $D(3, 5)$
13. $J(-4, -6)$, $K(6, 2)$, $L(1, 3)$, $M(-4, -1)$
14. $Q(2, 5)$, $R(-2, 1)$, $S(-1, -6)$, $T(9, 4)$
15. $W(-5, -1)$, $X(-2, 2)$, $Y(3, 1)$, $Z(5, -3)$

Example 3 **For trapezoid *QRTU*, *V* and *S* are midpoints of the legs.**

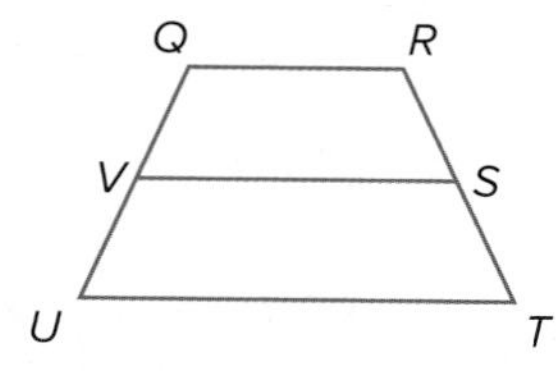

16. If $QR = 12$ and $UT = 22$, find VS.
17. If $QR = 4$ and $UT = 16$, find VS.
18. If $VS = 9$ and $UT = 12$, find QR.
19. If $TU = 26$ and $SV = 17$, find QR.
20. If $QR = 2$ and $VS = 7$, find UT.
21. If $RQ = 5$ and $VS = 11$, find UT.
22. **DESIGN** Juana is designing a window box. She wants the end of the box to be a trapezoid with the dimensions shown. If she wants to put a shelf in the middle for the plants to rest on, about how wide should she make the shelf?

23. **MUSIC** The keys of the xylophone shown form a trapezoid. If the length of the lower pitched C is 6 inches long, and the higher pitched D is 1.8 inches long, how long is the G key?

Example 4 **SENSE-MAKING If *WXYZ* is a kite, find each measure.**

24. YZ

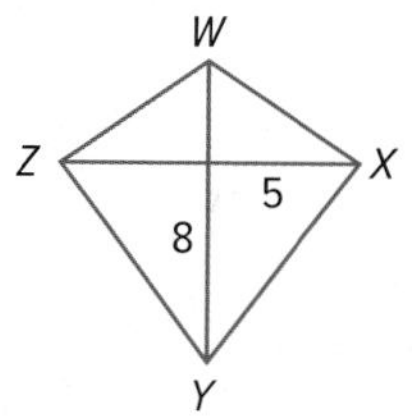

25. WP

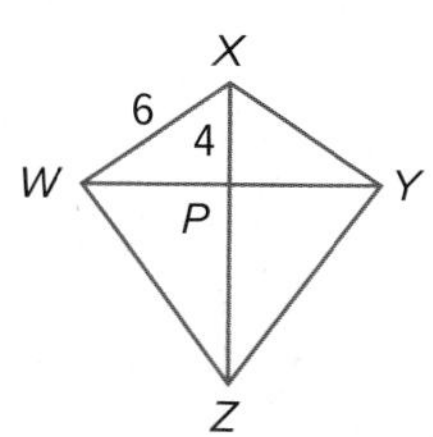

26. $m\angle X$

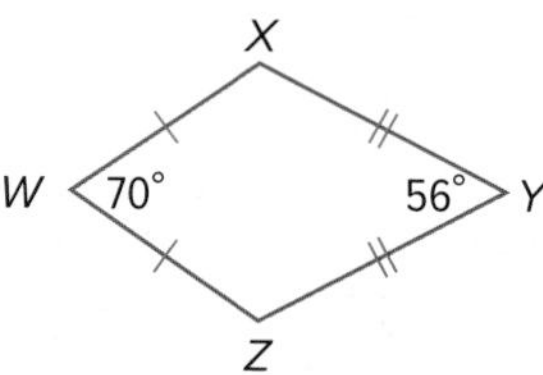

27. $m\angle Z$

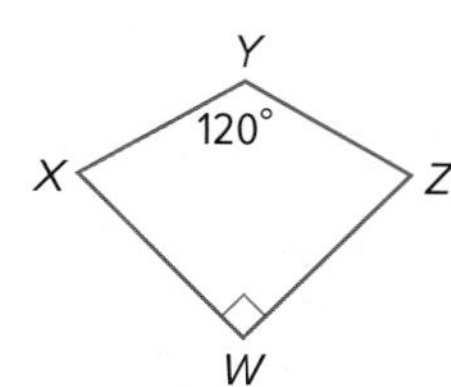

PROOF **Write a paragraph proof for each theorem.**

28. Theorem 6.21
29. Theorem 6.22
30. Theorem 6.23
31. Theorem 6.25
32. Theorem 6.26

33. **PROOF** Write a coordinate proof for Theorem 6.24.

34. **COORDINATE GEOMETRY** Refer to quadrilateral $ABCD$.

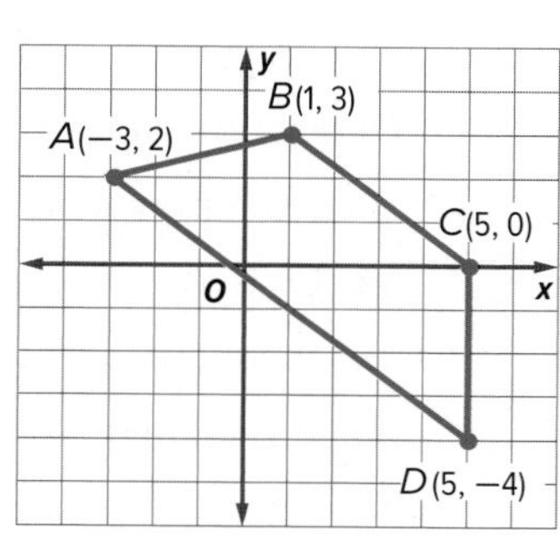

  a. Determine whether the figure is a trapezoid. If so, is it isosceles? Explain.

  b. Is the midsegment contained in the line with equation $y = -x + 1$? Justify your answer.

  c. Find the length of the midsegment.

ALGEBRA ***ABCD*** **is a trapezoid.**

35. If $AC = 3x - 7$ and $BD = 2x + 8$, find the value of x so that $ABCD$ is isosceles.

36. If $m\angle ABC = 4x + 11$ and $m\angle DAB = 2x + 33$, find the value of x so that $ABCD$ is isosceles.

SPORTS **The end of the batting cage shown is an isosceles trapezoid. If $PT = 12$ feet, $ST = 28$ feet, and $m\angle PQR = 110$, find each measure.**

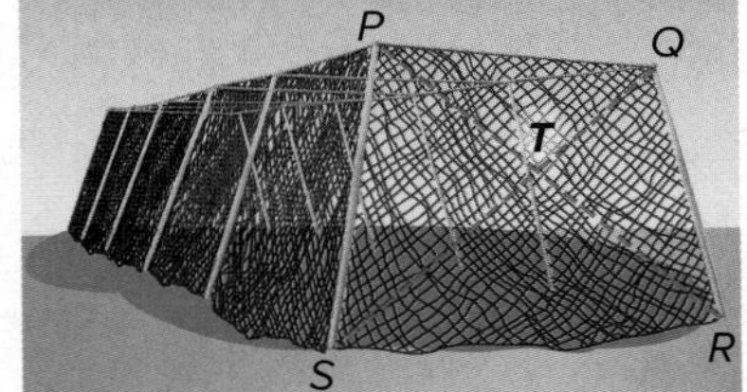

37. TR

38. SQ

39. $m\angle QRS$

40. $m\angle QPS$

ALGEBRA **For trapezoid** ***QRST***, ***M*** **and** ***P*** **are midpoints of the legs.**

41. If $QR = 16$, $PM = 12$, and $TS = 4x$, find x.

42. If $TS = 2x$, $PM = 20$, and $QR = 6x$, find x.

43. If $PM = 2x$, $QR = 3x$, and $TS = 10$, find PM.

44. If $TS = 2x + 2$, $QR = 5x + 3$, and $PM = 13$, find TS.

SHOPPING **The side of the shopping bag shown is an isosceles trapezoid. If $EC = 9$ inches, $DB = 19$ inches, $m\angle ABE = 40$, and $m\angle EBC = 35$, find each measure.**

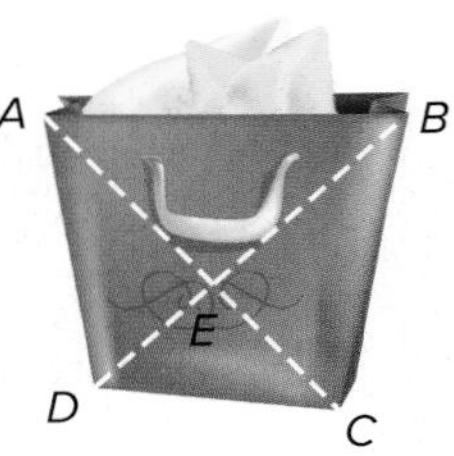

45. AE

46. AC

47. $m\angle BCD$

48. $m\angle EDC$

ALGEBRA ***WXYZ*** **is a kite.**

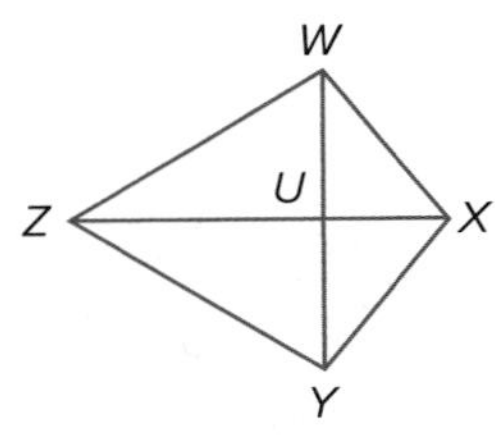

49. If $m\angle WXY = 120$, $m\angle WZY = 4x$, and $m\angle ZWX = 10x$, find $m\angle ZYX$.

50. If $m\angle WXY = 13x + 24$, $m\angle WZY = 35$, and $m\angle ZWX = 13x + 14$, find $m\angle ZYX$.

MP CONSTRUCT ARGUMENTS **Write a two-column proof.**

51. Given: $ABCD$ is an isosceles trapezoid.

Prove: $\angle DAC \cong \angle CBD$

52. Given: $\overline{WZ} \cong \overline{ZV}$, $\overline{XY}$ bisects $\overline{WZ}$ and $\overline{ZV}$, and $\angle W \cong \angle ZXY$.

Prove: $WXYV$ is an isosceles trapezoid.

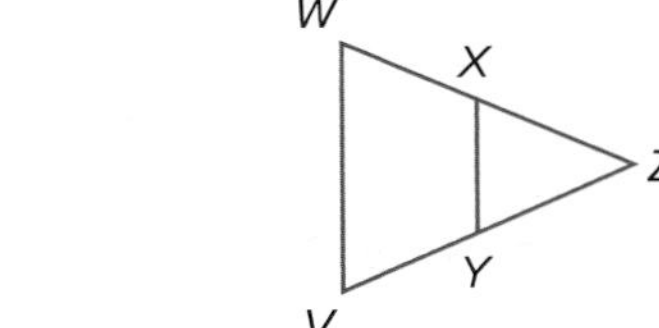
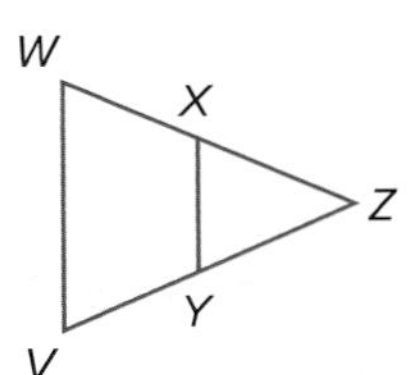

Determine whether each statement is *always*, *sometimes*, or *never* true. Explain.

53. The opposite angles of a trapezoid are supplementary.

54. One pair of opposite sides are parallel in a kite.

55. A square is a rhombus.

56. A rectangle is a square.

57. A parallelogram is a rectangle.

58. **KITES** Refer to the kite at the right. Using the properties of kites, write a two-column proof to show that $\triangle MNR$ is congruent to $\triangle PNR$.

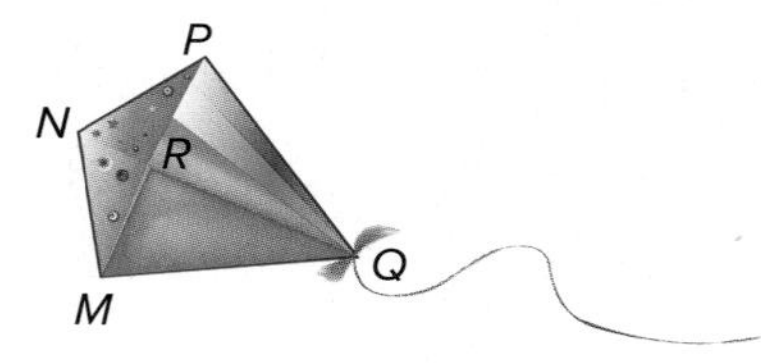

59. **VENN DIAGRAM** Create a Venn diagram that incorporates all quadrilaterals, including trapezoids, isosceles trapezoids, kites, and quadrilaterals that cannot be classified as anything other than quadrilaterals.

COORDINATE GEOMETRY Determine whether each figure is a *trapezoid*, a *parallelogram*, a *square*, a *rhombus*, or a *quadrilateral* given the coordinates of the vertices. Choose the most specific term. Explain.

60. $A(-1, 4)$, $B(2, 6)$, $C(3, 3)$, $D(0, 1)$

61. $W(-3, 4)$, $X(3, 4)$, $Y(5, 3)$, $Z(-5, 1)$

62. **MULTIPLE REPRESENTATIONS** In this problem, you will explore proportions in kites.

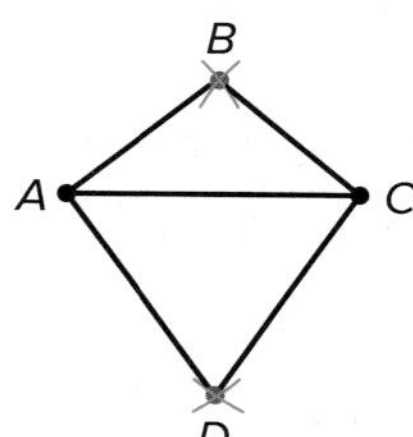

a. **Geometric** Draw a segment. Construct a noncongruent segment that perpendicularly bisects the first segment. Connect the endpoints of the segments to form a quadrilateral *ABCD*. Repeat the process two times. Name the additional quadrilaterals *PQRS* and *WXYZ*.

b. **Tabular** Copy and complete the table below.

Figure	Side	Length	Side	Length	Side	Length	Side	Length
ABCD	AB		BC		CD		DA	
PQRS	PQ		QR		RS		SP	
WXYZ	WX		XY		YZ		ZW	

c. **Verbal** Make a conjecture about a quadrilateral in which the diagonals are perpendicular, exactly one diagonal is bisected, and the diagonals are not congruent.

PROOF Write a coordinate proof of each statement.

63. The diagonals of an isosceles trapezoid are congruent.

64. The median of an isosceles trapezoid is parallel to the bases.

H.O.T. Problems Use Higher-Order Thinking Skills

65. **ERROR ANALYSIS** Bedagi and Belinda are trying to determine $m\angle A$ in kite *ABCD* shown. Is either of them correct? Explain.

Bedagi

$m\angle A = 45$

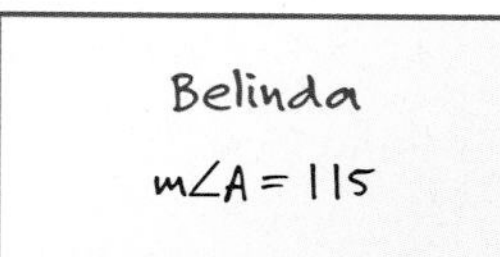

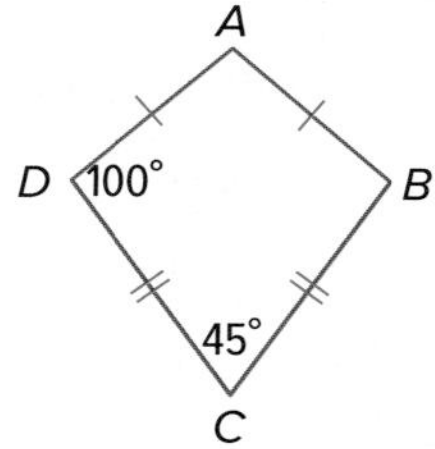

66. **CHALLENGE** If the parallel sides of a trapezoid are contained by the lines $y = x + 4$ and $y = x - 8$, what equation represents the line contained by the midsegment?

67. **MP CONSTRUCT ARGUMENTS** Is it *sometimes*, *always*, or *never* true that a square is also a kite? Explain.

68. **OPEN-ENDED** Sketch two noncongruent trapezoids *ABCD* and *FGHJ* in which $\overline{AC} \cong \overline{FH}$ and $\overline{BD} \cong \overline{GJ}$.

69. **WRITING IN MATH** Describe the properties a quadrilateral must possess in order for the quadrilateral to be classified as a trapezoid, an isosceles trapezoid, or a kite. Compare the properties of all three quadrilaterals.

70. Quadrilateral $PQRS$ has vertices $P(-4, 1)$, $Q(-1, 2)$, $R(2, -2)$, and $S(-4, -4)$ as shown in the graph. Which of the following best describes the quadrilateral? MP 2

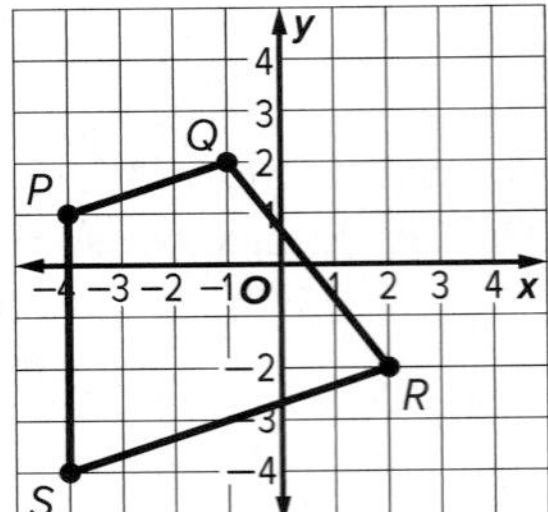

- ○ **A** isosceles trapezoid
- ○ **B** kite
- ○ **C** parallelogram
- ○ **D** trapezoid

71. Trapezoid $MNPQ$ has vertices $M(-1, 3)$, $N(1, 4)$, $P(3, 2)$, and $Q(-3, -1)$. What is the length of the midsegment of the trapezoid? MP 2

- ○ **A** $\sqrt{45}$
- ○ **B** $\sqrt{20}$
- ○ **C** $\sqrt{8}$
- ○ **D** $\sqrt{5}$

72. Quadrilateral $WXYZ$ has vertices $W(-2, 1)$, $X(1, 2)$, $Y(4, -2)$, and $Z(-1, -2)$. Which of the following terms best describes the quadrilateral? MP 2

- ○ **A** kite
- ○ **B** parallelogram
- ○ **C** rhombus
- ○ **D** trapezoid

73. Quadrilateral $ABCD$ has vertices $A(-4, -1)$, $B(1, 3)$, $C(6, 2)$, and $D(-4, -6)$. Which of the following best describes the quadrilateral? MP 2

- ○ **A** kite
- ○ **B** parallelogram
- ○ **C** trapezoid
- ○ **D** isosceles trapezoid

74. Two sides of quadrilateral $FGHJ$ are shown.

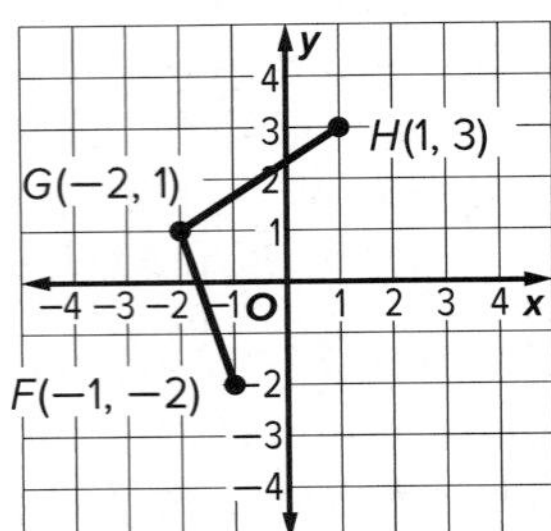

For which coordinates of vertex J is the resulting quadrilateral not a trapezoid? MP 2

- ○ **A** $(4, -6)$
- ○ **B** $(3, -3)$
- ○ **C** $(5, 2)$
- ○ **D** $(2, 0)$
- ○ **E** $(8, 4)$

75. MULTI-STEP Use the figure to answer the questions. M is the midpoint of $\overline{AD}$, N is the midpoint of $\overline{BC}$. MP 1, 2

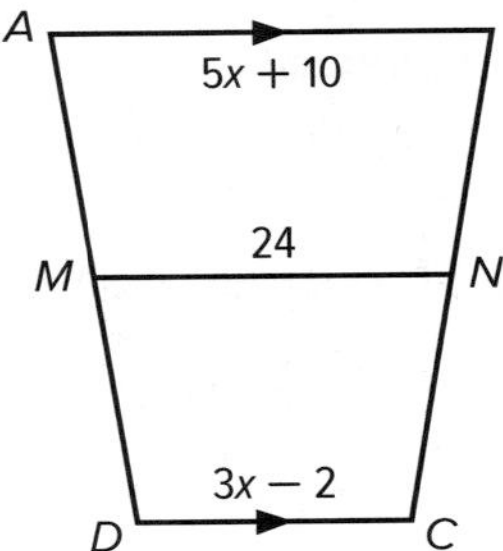

a. What is the shape of $ABCD$? ☐

b. What is the value of x? ☐

c. Find the measure of $\overline{AB}$. ☐

d. Find the measure of $\overline{CD}$. ☐

e. Which statements must be true? Select all that apply.

- ☐ **A** $\overline{AB} \parallel \overline{CD}$
- ☐ **B** $\overline{AB} \parallel \overline{MN}$
- ☐ **C** $\overline{BN} \cong \overline{MD}$
- ☐ **D** $\overline{AD} \cong \overline{BC}$
- ☐ **E** $\overline{AM} \cong \overline{MD}$

f. What mathematical practice did you use to solve this problem?

Study Guide and Review

Study Guide

Key Concepts

Angles of Polygons (Lesson 6-1)

- The sum of the measures of the interior angles of a polygon is given by the formula $S = (n - 2)180$.
- The sum of the measures of the exterior angles of a convex polygon is 360.

Properties of Parallelograms (Lessons 6-2 and 6-3)

- Opposite sides are congruent and parallel.
- Opposite angles are congruent.
- Consecutive angles are supplementary.
- If a parallelogram has one right angle, it has four right angles.
- Diagonals bisect each other.

Properties of Rectangles, Rhombi, Squares, and Trapezoids (Lesson 6-4 through 6-6)

- A rectangle has all the properties of a parallelogram. Diagonals are congruent and bisect each other. All four angles are right angles.
- A rhombus has all the properties of a parallelogram. All sides are congruent. Diagonals are perpendicular. Each diagonal bisects a pair of opposite angles.
- A square has all the properties of a parallelogram, a rectangle, and a rhombus.
- In an isosceles trapezoid, both pairs of base angles are congruent and the diagonals are congruent.

FOLDABLES® Study Organizer

Use your Foldable to review the chapter. Working with a partner can be helpful. Ask for clarification of concepts as needed.

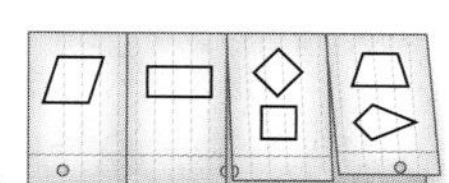

Key Vocabulary

- **base** (p. 469)
- **base angle** (p. 469)
- **diagonal** (p. 423)
- **isosceles trapezoid** (p. 469)
- **kite** (p. 472)
- **legs** (p. 469)
- **midsegment of a trapezoid** (p. 471)
- **parallelogram** (p. 433)
- **rectangle** (p. 453)
- **rhombus** (p. 460)
- **square** (p. 461)
- **trapezoid** (p. 469)

Vocabulary Check

State whether each sentence is *true* or *false*. If *false*, replace the underlined word or phrase to make a true sentence.

1. <u>No</u> angles in an isosceles trapezoid are congruent.
2. If a parallelogram is a <u>rectangle</u>, then the diagonals are congruent.
3. A <u>midsegment of a trapezoid</u> is a segment that connects any two nonconsecutive vertices.
4. The base of a trapezoid is one of the <u>parallel</u> sides.
5. The diagonals of a <u>rhombus</u> are perpendicular.
6. The <u>diagonal</u> of a trapezoid is the segment that connects the midpoints of the legs.
7. A rectangle <u>is not always</u> a parallelogram.
8. A quadrilateral with only one set of parallel sides is a <u>parallelogram</u>.

Concept Check

9. Explain how to prove that a given parallelogram is a square.
10. Explain how to determine the number of sides in a regular polygon given the measure of one of its interior angles.

Lesson-by-Lesson Review 

6-1 Angles of Polygons

Find the sum of the measures of the interior angles of each convex polygon.

11. decagon

12. 15-gon

13. **SNOWFLAKES** The snowflake decoration at the right suggests a regular hexagon. Find the sum of the measures of the interior angles of the hexagon.

The measure of an interior angle of a regular polygon is given. Find the number of sides in the polygon.

14. 135

15. ≈ 166.15

Example 1

Find the sum of the measures of the interior angles of a convex 22-gon.

$m = (n - 2)180$	Write an equation.
$= (22 - 2)180$	Substitution
$= 20 \cdot 180$	Subtract.
$= 3600$	Multiply.

Example 2

The measure of an interior angle of a regular polygon is 157.5. Find the number of sides in the polygon.

$157.5n = (n - 2)180$	Write an equation.
$157.5n = 180n - 360$	Distributive Property
$-22.5n = -360$	Subtract.
$n = 16$	Divide.

The polygon has 16 sides.

6-2 Parallelograms

Use ▱$ABCD$ to find each measure.

16. $m\angle ADC$

17. AD

18. AB

19. $m\angle BCD$

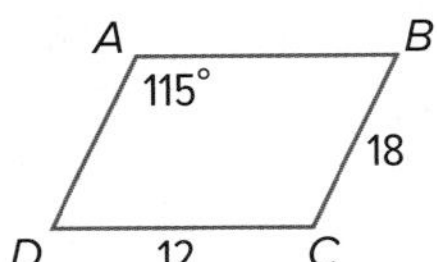

ALGEBRA Find the value of each variable in each parallelogram.

20.

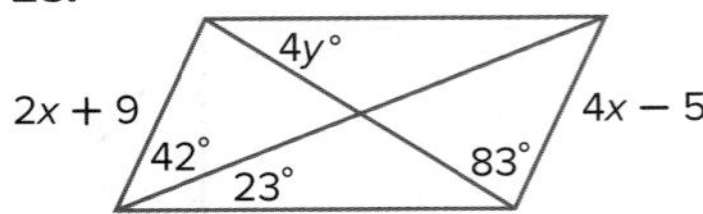

21.

2y + 19
(2x + 41)°
115°
3y + 13

22. **DESIGN** What type of information is needed to determine whether the shapes that make up the stained glass window below are parallelograms?

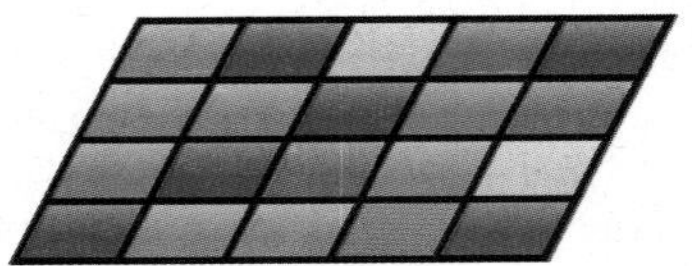

Example 3

ALGEBRA If *KLMN* is a parallelogram, find the value of the indicated variable.

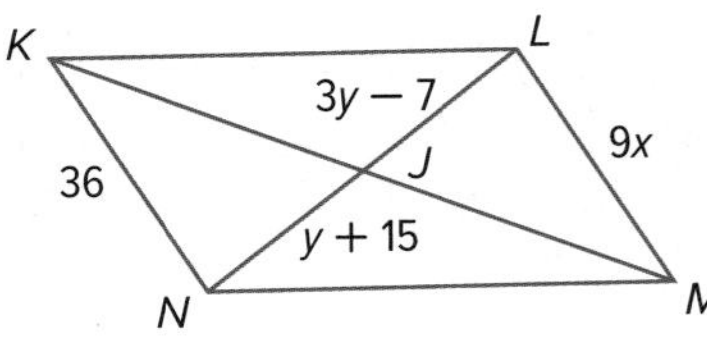

a. x

$\overline{KN} \cong \overline{LM}$	Opp. sides of a ▱ are ≅.
$KN = LM$	Definition of congruence
$36 = 9x$	Substitution
$4 = x$	Divide.

b. y

$\overline{NJ} \cong \overline{JL}$	Diag. of a ▱ bisect each other.
$NJ = JL$	Definition of congruence
$y + 15 = 3y - 7$	Substitution
$-2y = -22$	Subtract.
$y = 11$	Divide.

6-3 Tests for Parallelograms

Determine whether each quadrilateral is a parallelogram. Justify your answer.

23. 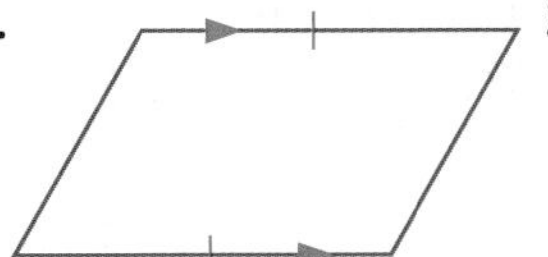

24. 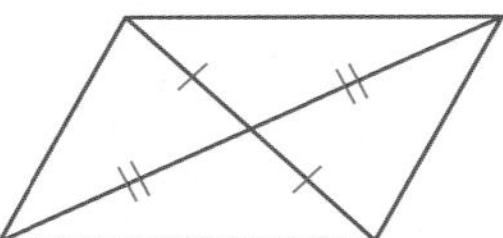

25. PROOF Write a two-column proof.

Given: ▱$ABCD$, $\overline{AE} \cong \overline{CF}$

Prove: Quadrilateral $EBFD$ is a parallelogram.

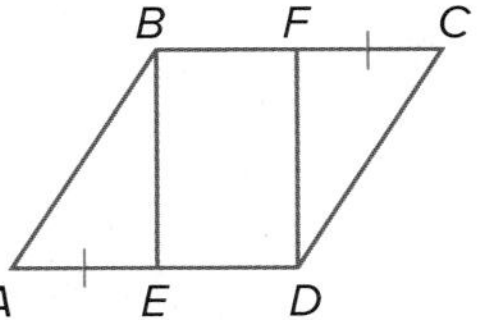

ALGEBRA Find x and y so that the quadrilateral is a parallelogram.

26. 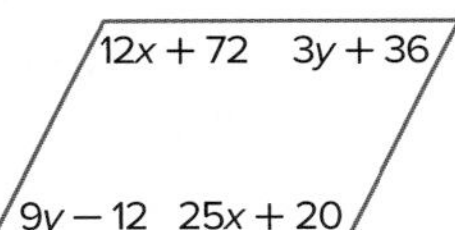

27. 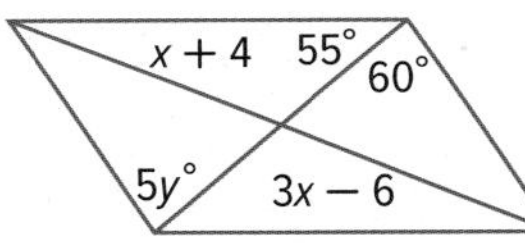

Example 4

If $TP = 4x + 2$, $QP = 2y - 6$, $PS = 5y - 12$, and $PR = 6x - 4$, find x and y so that the quadrilateral is a parallelogram.

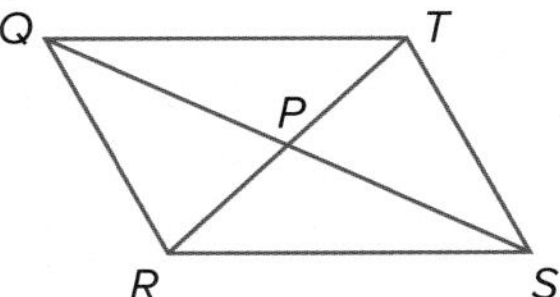

Find x such that $\overline{TP} \cong \overline{PR}$ and y such that $\overline{QP} \cong \overline{PS}$.

$TP = PR$	Definition of ≅
$4x + 2 = 6x - 4$	Substitution
$-2x = -6$	Subtract.
$x = 3$	Divide.

$QP = PS$	Definition of ≅
$2y - 6 = 5y - 12$	Substitution
$-3y = -6$	Subtract.
$y = 2$	Divide.

6-4 Special Parallelograms: Rectangles

28. PARKING The lines of the parking space shown below are parallel. How wide is the space (in inches)?

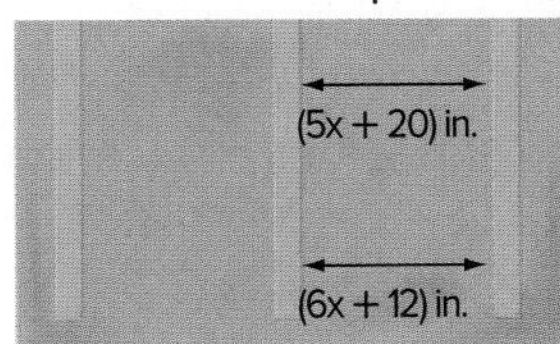

ALGEBRA Quadrilateral $EFGH$ is a rectangle.

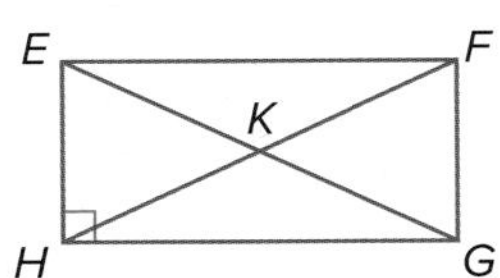

29. If $m\angle FEG = 57$, find $m\angle GEH$.

30. If $m\angle HGE = 13$, find $m\angle FGE$.

31. If $FK = 32$ feet, find EG.

32. Find $m\angle HEF + m\angle EFG$.

33. If $EF = 4x - 6$ and $HG = x + 3$, find EF.

Example 5

ALGEBRA Quadrilateral $ABCD$ is a rectangle. If $m\angle ADB = 4x + 8$ and $m\angle DBA = 6x + 12$, find x.

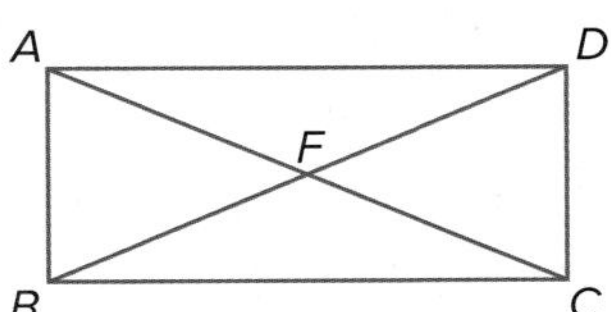

$ABCD$ is a rectangle, so $m\angle ABC = 90$. Because the opposite sides of a rectangle are parallel, and the alternate interior angles of parallel lines are congruent, $\angle DBC \cong \angle ADB$ and $m\angle DBC = m\angle ADB$.

$m\angle DBC + m\angle DBA = 90$	Angle Addition
$m\angle ADB + m\angle DBA = 90$	Substitution
$4x + 8 + 6x + 12 = 90$	Substitution
$10x + 20 = 90$	Add.
$10x = 70$	Subtract.
$x = 7$	Divide.

Study Guide and Review *Continued*

6-5 Special Parallelograms: Rhombi and Squares

ALGEBRA *ABCD* is a rhombus. If $EB = 9$, $AB = 12$ and $m\angle ABD = 55$, find each measure.

34. AE

35. $m\angle BDA$

36. CE

37. $m\angle ACB$

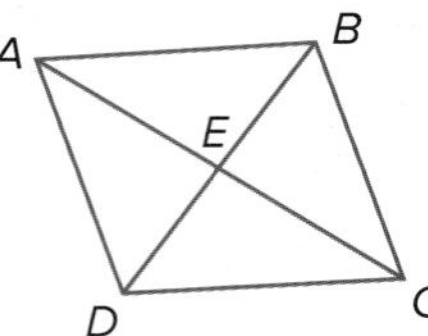

38. LOGOS A car company uses the symbol shown at the right for their logo. If the inside space of the logo is a rhombus, what is the length of $\overline{FJ}$?

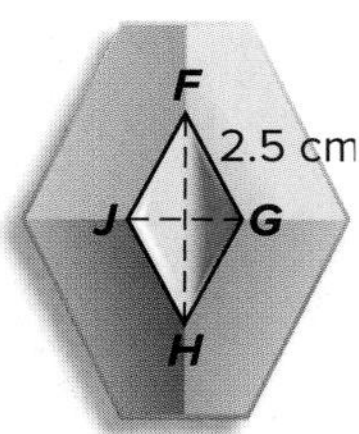

COORDINATE GEOMETRY Given each set of vertices, determine whether ▱*QRST* is a *rhombus*, a *rectangle*, or a *square*. List all that apply. Explain.

39. $Q(12, 0)$, $R(6, -6)$, $S(0, 0)$, $T(6, 6)$

40. $Q(-2, 4)$, $R(5, 6)$, $S(12, 4)$, $T(5, 2)$

Example 6

The diagonals of rhombus *QRST* intersect at *P*. Use the information to find each measure or value.

a. ALGEBRA If $QT = x + 7$ and $TS = 2x - 9$, find x.

$\overline{QT} \cong \overline{TS}$	Def. of rhombus
$QT = TS$	Def. of congruence
$x + 7 = 2x - 9$	Substitution
$-x = -16$	Subtract.
$x = 16$	Divide.

b. If $m\angle QTS = 76$, find $m\angle TSP$.

$\overline{TR}$ bisects $\angle QTS$. Therefore, $m\angle PTS = \frac{1}{2}m\angle QTS$. So $m\angle PTS = \frac{1}{2}(76)$ or 38. Because the diagonals of a rhombus are perpendicular, $m\angle TPS = 90$.

$m\angle PTS + m\angle TPS + m\angle TSP = 180$	△ Sum Thm.
$38 + 90 + m\angle TSP = 180$	Substitution
$128 + m\angle TSP = 180$	Add.
$m\angle TSP = 52$	Subtract.

6-6 Trapezoids and Kites

Find each measure.

41. GH

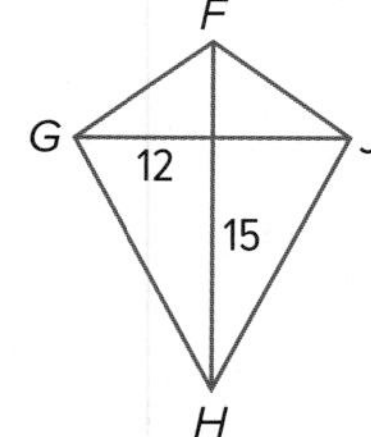

42. $m\angle Z$

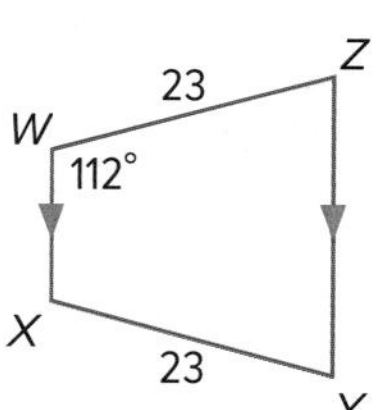

43. DESIGN Renee designed the square tile as an art project.

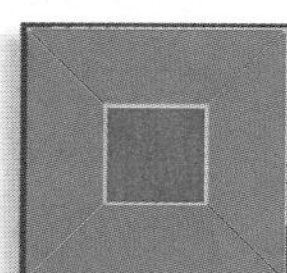

a. Describe a way to determine if the trapezoids in the design are isosceles.

b. If the perimeter of the tile is 48 inches and the perimeter of the red square is 16 inches, what is the perimeter of one of the trapezoids?

Example 7

If *QRST* is a kite, find $m\angle RST$.

Because $\angle Q \cong \angle S$, $m\angle Q = m\angle S$. Write and solve an equation to find $m\angle S$.

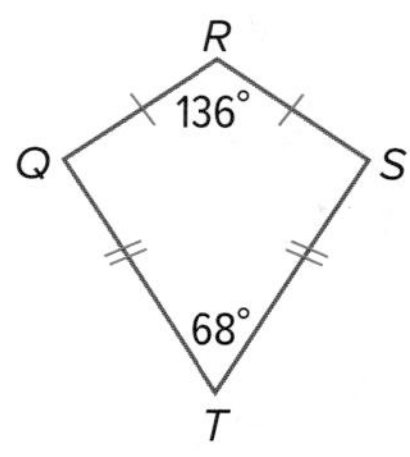

$m\angle Q + m\angle R + m\angle S + m\angle T = 360$	Polygon Int. ∠s Sum Thm
$m\angle Q + 136 + m\angle S + 68 = 360$	Substitution
$2m\angle S + 204 = 360$	Simplify.
$2m\angle S = 156$	Subtract.
$m\angle S = 78$	Divide.

CHAPTER 6
Practice Test

Find the sum of the measures of the interior angles of each convex polygon.

1. hexagon
2. 16-gon

3. **ART** Jen is making a frame to stretch a canvas over for a painting. She nailed four pieces of wood together at what she believes will be the four vertices of a square.
   a. How can she be sure that the canvas will be a square?
   b. If the canvas has the dimensions shown below, what are the missing measures?

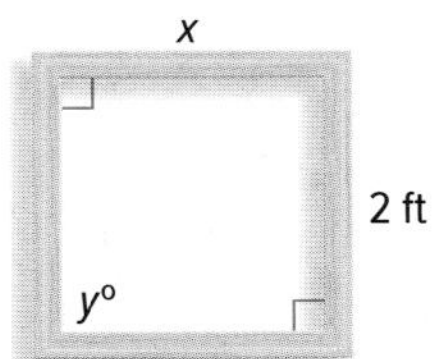

Quadrilateral *ABCD* is an isosceles trapezoid.

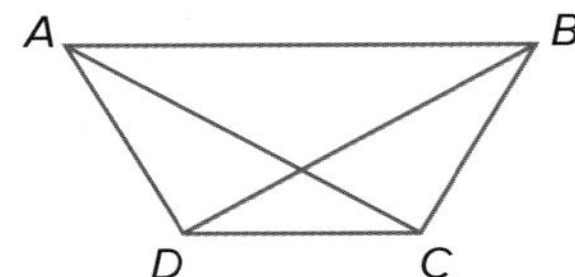

4. Which angle is congruent to $\angle BCD$?
5. Which side is parallel to $\overline{AB}$?
6. Which segment is congruent to $\overline{AC}$?

The measure of the interior angles of a regular polygon is given. Find the number of sides in the polygon.

7. 900
8. 1980
9. 2880
10. 5400

11. **MULTIPLE CHOICE** If *QRST* is a parallelogram, what is the value of *x*?

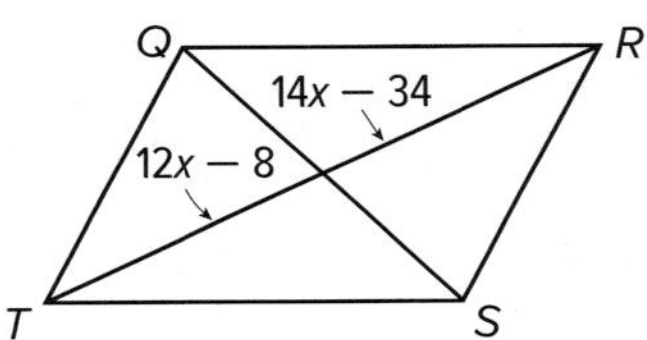

A 11
B 12
C 13
D 14

If *CDFG* is a kite, find each measure.

12. *GF*
13. $m\angle D$

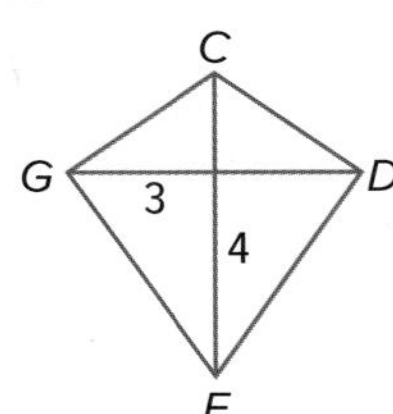

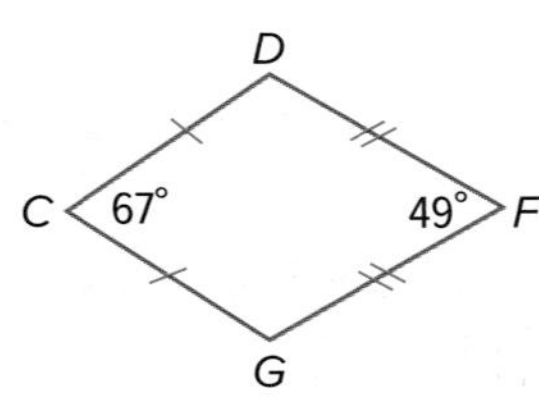

ALGEBRA Quadrilateral *MNOP* is a rhombus. Find each value or measure.

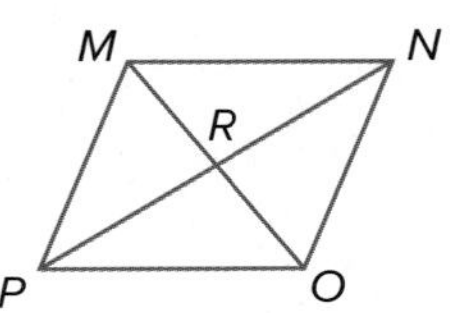

14. $m\angle MRN$
15. If $PR = 12$, find *RN*.
16. If $m\angle PON = 124$, find $m\angle POM$.

17. **CONSTRUCTION** The Smiths are building an addition to their house. Mrs. Smith is cutting an opening for a new window. If she measures to see that the opposite sides are congruent and that the diagonal measures are congruent, can Mrs. Smith be sure that the window opening is rectangular? Explain.

Use ▱*JKLM* to find each measure.

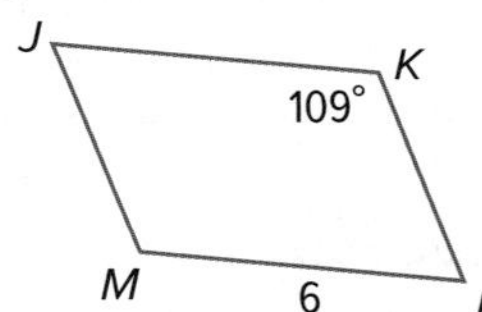

18. $m\angle JML$
19. *JK*
20. $m\angle KLM$

ALGEBRA Quadrilateral *DEFG* is a rectangle.

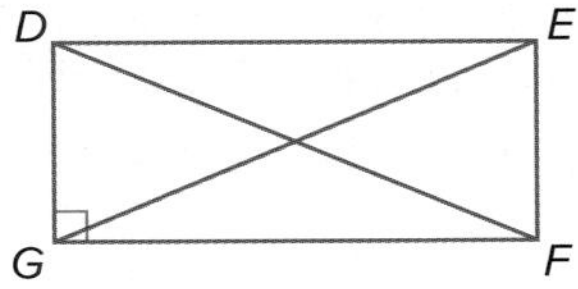

21. If $DF = 2(x + 5) - 7$ and $EG = 3(x - 2)$, find *EG*.
22. If $m\angle EDF = 5x - 3$ and $m\angle DFG = 3x + 7$, find $m\angle EDF$.
23. If $DE = 14 + 2x$ and $GF = 4(x - 3) + 6$, find *GF*.

Determine whether each quadrilateral is a parallelogram. Justify your answer.

24. 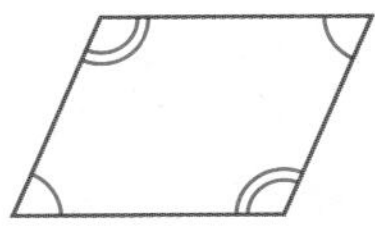

25. 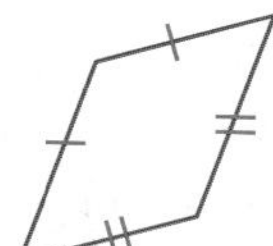

Preparing for Assessment

Performance Task

Provide a clear solution to each part of the task. Be sure to show all of your work, include all relevant drawings, and justify your answers.

DESIGN A high school shop class is working with the drama club to create a set for the school's annual play.

Part A

First, the shop class made six rectangular backdrops. They want to make sure the backdrops are perfect rectangles. The shop teacher verifies that opposite sides are the same length, but he doesn't have the proper tools to measure the angles.

1. **Reasoning** Explain another way the teacher can determine whether the shapes are rectangles.

Part B

The drama club wants the shop class to modify some items from last year's play. One set piece is a regular hexagon and one is a regular octagon.

2. They want to divide the octagon into evenly-sized triangles. Determine the smallest number of evenly-sized triangles into which the octagon can be divided.
3. Determine each of the interior angle measures of the triangles that come from the octagon.
4. **Sense-Making** For the hexagon, the drama club wants a piece added to the hexagon as shown. Determine the interior angle measures of the right triangle that will fit next to the hexagon.

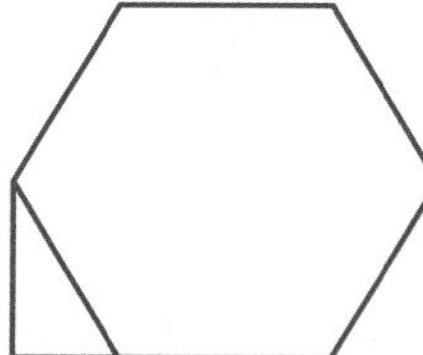

Part C

The drama club needs two backdrops shaped like parallelograms that fit together as shown. The shop class makes three quadrilaterals. The first quadrilateral has angle measures 40°, 120°, 40°, and 120°. The second quadrilateral has angle measures 37°, 143°, 37°, and 143°. The third quadrilateral has angle measures 30°, 150°, 45°, and 135°.

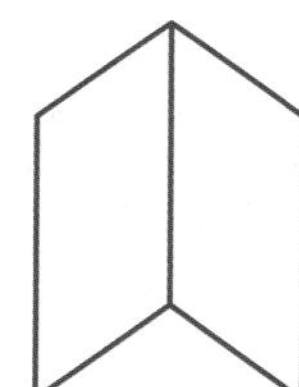

5. Determine which quadrilaterals, if any, are parallelograms.
6. If you determined any of the shapes were *not* parallelograms, explain why they are not.

Part D

The drama club has a set piece shaped like an isosceles trapezoid with base lengths 8 feet and 14 feet. They need the shop class to cut the trapezoid along the midsegment.

7. Determine the lengths of the bases of the two trapezoids made by cutting the original trapezoid.

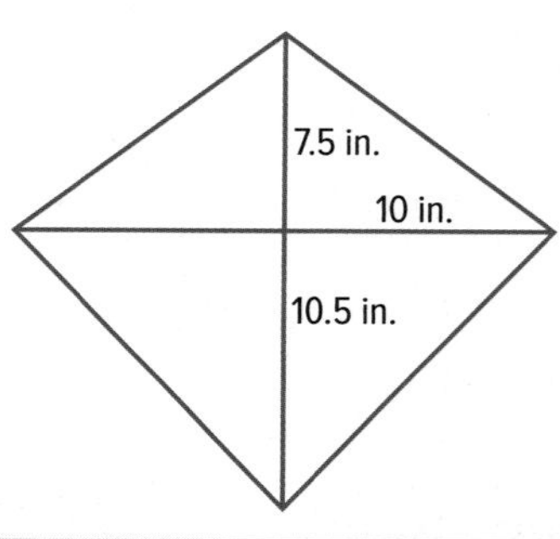

Part E

The drama club also asked for a wooden kite shape as shown. The shop class plans to use the letter T as the frame of the kite and add the sides of the kite, as shown.

8. Determine the perimeter of the kite.

Test-Taking Strategy

Example

Read the problem. Identify what you need to know. Then use the information in the problem to solve it.

$\overline{RS}$ is the midsegment of trapezoid $MNOP$. What is the length of $\overline{RS}$?

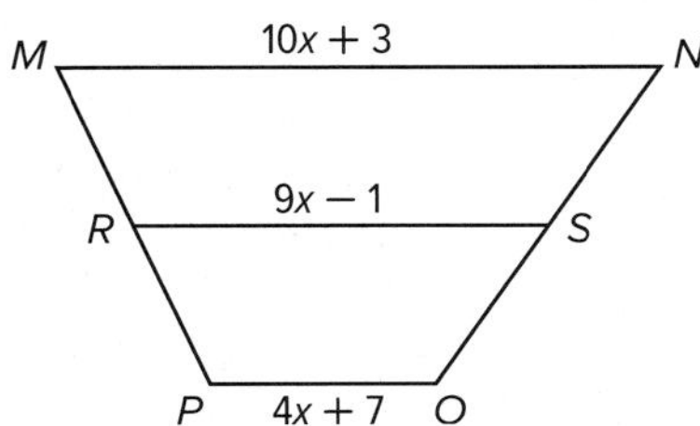

A 14 units

B 19 units

C 23 units

D 26 units

Step 1 **What do you need to find?**
the length of midsegment $\overline{RS}$

Step 2 **Is there a definition or property that you can use?**
Yes; the midsegment of a trapezoid is one half the sum of the lengths of the bases.

Step 3 **Can you set up an equation to help find the unknown?**
Yes; $\frac{1}{2}[(10x + 3) + (4x + 7)] = 9x - 1$

Step 4 **Is the solution to the equation the answer to the problem?**
No; the value of x must be substituted into $RS = 9x - 1$.

Step 5 **What is the correct answer?** D

Test-Taking Tip

Applying Definitions and Properties Many geometry problems require the application of definitions and properties in order to solve them. Definitions and properties can be used to solve for unknowns in a problem. Look for key words such as "sum," "perimeter," "angle measure," or "quadrilateral" to determine whether a definition or property can be used to help solve problems.

Apply the Strategy

Read the problem. Identify what you need to know. Then use the information in the problem to solve. Show your work.

If $\overline{AB} \parallel \overline{DC}$, find x.

A 32.5

B 65

C 105

D 115

Answer the questions below.

1. What do you need to find?
2. Is there a definition or property that you can use?
3. Can you set up an equation to help find the unknown?
4. Is the solution to the equation the answer to the problem?
5. What is the correct answer?

CHAPTER 6

Preparing for Assessment

Cumulative Review

Read each question. Then fill in the correct answer on the answer document provided by your teacher or on a sheet of paper.

1. Jayden uses software to draw a regular polygon. He finds that the measure of each exterior angle is 20.

   a. How many sides does his polygon have?

   ☐ sides

   b. MP What mathematical practice did you use to solve this problem?

2. The figure shows the back of a rectangular bookcase. $QN = (x + 11)$ inches and $MP = (2x - 14)$ inches.

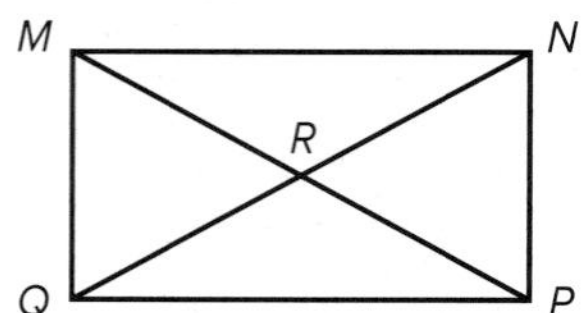

   What is the length of $\overline{QR}$?

   ○ A 12.5 in. ○ C 25 in.

   ○ B 18 in. ○ D 36 in.

3. $\overline{RS}$ is the perpendicular bisector of $\overline{TU}$.

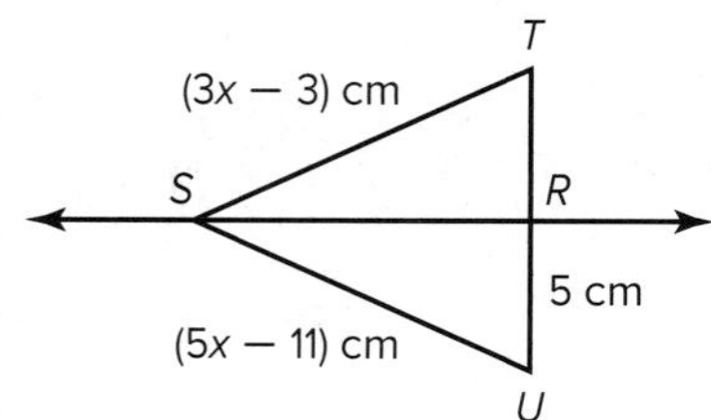

   What is the perimeter of $\triangle STU$?

   ○ A 28 cm ○ C 18 cm

   ○ B 23 cm ○ D 14 cm

4. A rectangular prism has a volume of 72.8 meters. The length of the prism is 5.2 meters and the height of the prism is 4 meters. What is the width of the prism in meters?

   ☐

5. Shantelle designed this logo for a jewelry store. She knows that quadrilateral *EFGH* is a parallelogram.

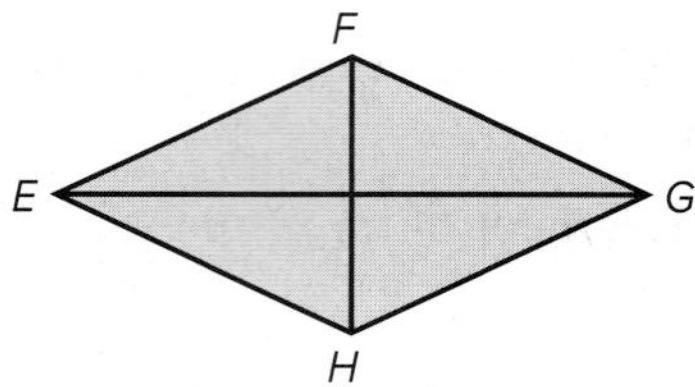

   Which of the following would let Shantelle conclude that *EFGH* is a rhombus?

   I. $\overline{FH} \cong \overline{EG}$

   II. $\overline{FH} \perp \overline{EG}$

   III. $\overline{FH}$ and $\overline{EG}$ bisect each other.

   ○ A I only

   ○ B II only

   ○ C III only

   ○ D II and III only

6. Trapezoid *PQRS* has vertices $P(-5, 2)$, $Q(-1, 4)$, $R(2, 2)$, and $S(4, -4)$. What is the slope of the trapezoid's midsegment?

   ○ A -3 ○ C $-\frac{2}{3}$

   ○ B $-\frac{3}{2}$ ○ D $\frac{1}{2}$

Test-Taking Tip

Question 6 Check the reasonableness of your answer by plotting the vertices and graphing the trapezoid.

7. In $\triangle KLM$, $m\angle K = 43°$, $m\angle L = 59°$, and $m\angle M = 78°$. Which of the following lists shows the sides of $\triangle KLM$ in order from shortest to longest?

   ○ A $\overline{KL}, \overline{MK}, \overline{LM}$

   ○ B $\overline{LM}, \overline{KL}, \overline{MK}$

   ○ C $\overline{MK}, \overline{LM}, \overline{KL}$

   ○ D $\overline{LM}, \overline{MK}, \overline{KL}$

Go Online! for Standardized Test Practice

8. What is the equation of the line that passes through the point $P(1, 0)$ and is parallel to the line with equation $-2x + y = 3$?

A $y = 2x - 2$

B $y = -\frac{1}{2}x + \frac{1}{2}$

C $y = -2x + 2$

D $y = \frac{1}{2}x - \frac{1}{2}$

9. Quadrilateral *JKLM* is a parallelogram.

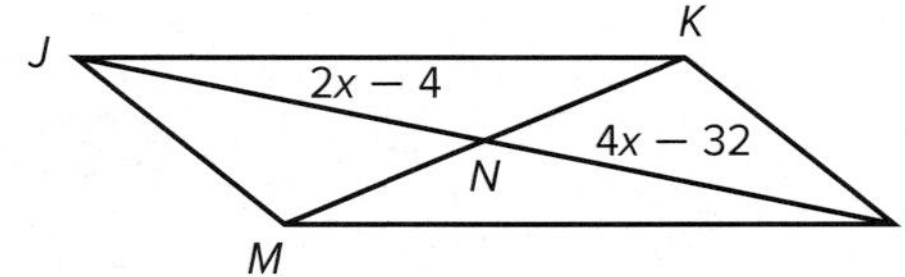

What is the length of diagonal $\overline{JL}$?

A 14

B 24

C 28

D 48

10. In the figure, $\overline{BD} \perp \overline{AC}$.

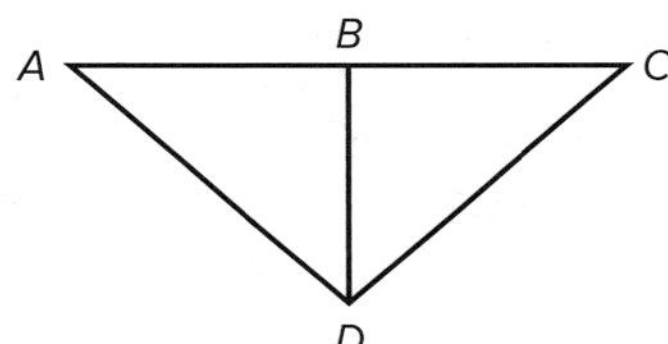

Which additional piece of information is needed to prove $\triangle ABD \cong \triangle CBD$ by the SAS Congruence Postulate?

A $\overline{BD}$ bisects $\angle ADC$.

B $\overline{AD} \cong \overline{CD}$

C $\overline{BD}$ is the perpendicular bisector of $\overline{AC}$.

D $\angle A \cong \angle C$

11. The slope of the line through points $P(-2, -1)$ and $Q(1, y)$ is 2. What is the value of y?

A 7

B 5

C $\frac{1}{2}$

D -7

12. What value of x guarantees that quadrilateral *ABCD* is a parallelogram?

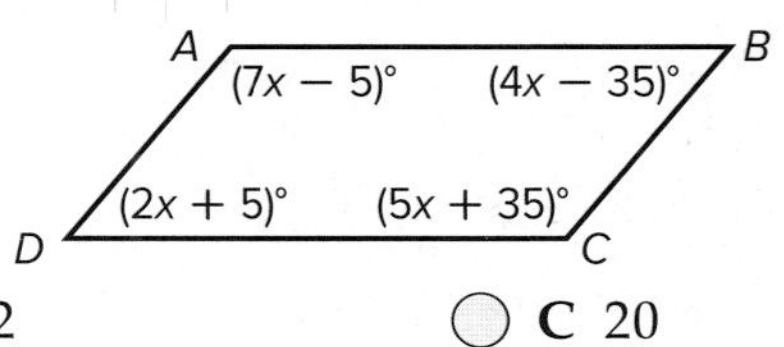

A 2

B 15

C 20

D 45

13. What is the measure of the smallest interior angle of quadrilateral *PQRS*?

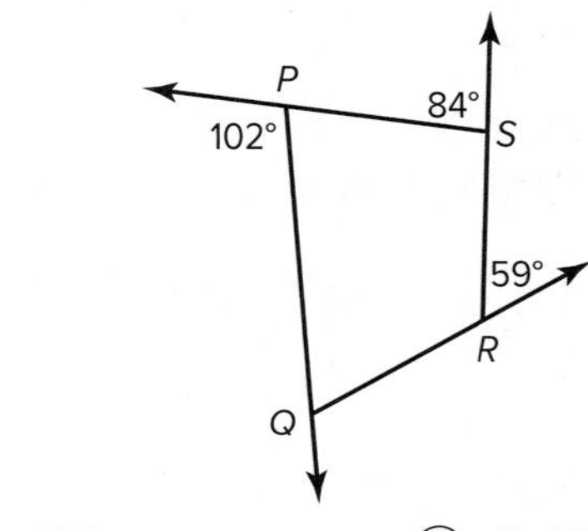

A 121

B 78

C 65

D 59

14. Antonio is building a gate for his garden fence. The gate has two diagonal braces, as shown.

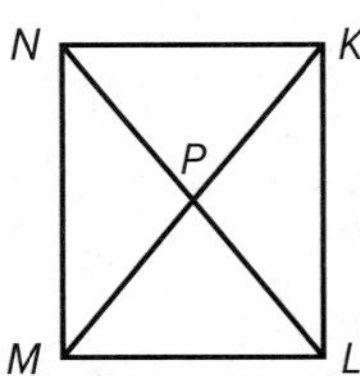

Select all steps Antonio should take to ensure that the gate is a rectangle.

A Check that $NL = MK$.

B Check that $\angle P$ is a right angle.

C Check that $KL = LM = MN = NK$.

D Check that $\overline{NL}$ and $\overline{MK}$ bisect each other.

Need Extra Help?

If you missed Question...	1	2	3	4	5	6	7	8	9	10	11	12	13	14
Go to Lesson...	6-1	6-4	5-1	1-8	6-5	6-6	5-3	2-8	6-2	4-3	2-8	6-3	6-1	6-4

CHAPTER 7

Similarity

THEN

You learned about angles of polygons, including quadrilaterals, and applied them to real-world applications.

NOW

In this chapter you will:

- Identify dilations.
- Identify similar polygons.
- Identify AA, SSS, and SAS similarity and the parts of similar triangles.
- Use parallel lines and proportional parts.

MP WHY

SPORTS Triangles can be used in sports to describe the path of a ball, such as a bounce pass from one person to another.

Use the Mathematical Practices to complete the activity.

1. Sense-Making What sports involve a bounce pass? Can you think of sports where the ball is struck, follows a path to the ground or playing surface, and then bounces up?

2. Apply Math Write a problem involving a bounce path and two triangles.

3. Modeling Use the Geometry Tool in ConnectED to model the triangles in the problem you created.

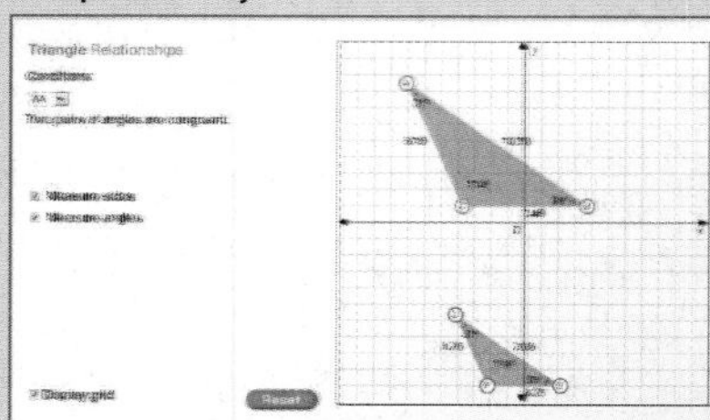

4. Reasoning What can you do to make your problem more realistic? What other forces are at play in the real world when a ball is bounced or struck?

Go Online to Guide Your Learning

Explore & Explain

Transformation Tool

Use the **Transformation** tool to explore similarity transformations discussed in Lesson 7-2.

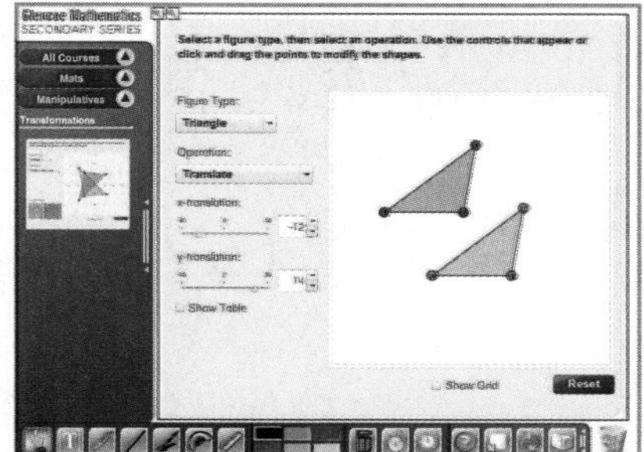

The Geometer's Sketchpad

Use the **The Geometer's Sketchpad** to discover principles of similarity and develop the definition of similar polygons, to explore the results of cutting through a triangle with a line parallel to a side, and to explore how to use similar triangles to find a distance that cannot be measured directly.

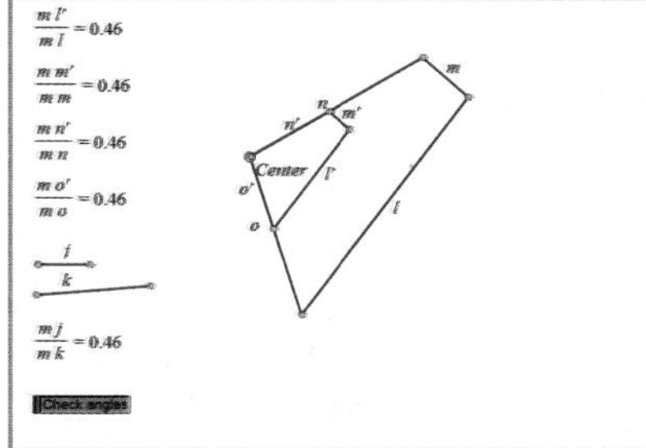

eBook

Interactive Student Guide

Before starting the chapter, answer the **Chapter Focus** preview questions. Check your answers as you complete each lesson. At the end of the chapter, try the **Performance Task**.

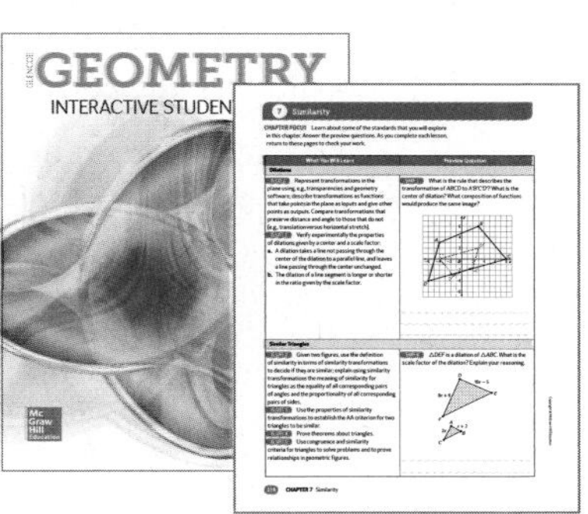

Organize

Foldables

Get organized! Create this Similarity Foldable before you start the chapter to help you organize your notes about dilations, similar polygons, similar triangles, parallel lines, and proportional parts.

Collaborate

Chapter Project

In the **Model Makers** project, you will use what you have learned about similarities and proportional parts to complete a business project.

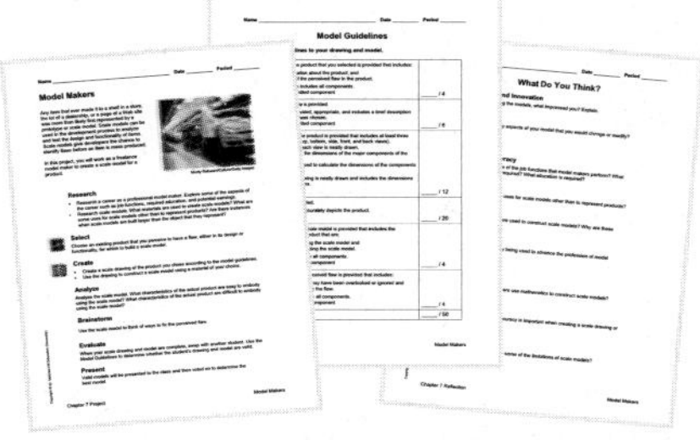

Focus

LEARNSMART®

Need help studying? Complete the **Similarity, Proof, and Trigonometry** domain in LearnSmart to review for the chapter test.

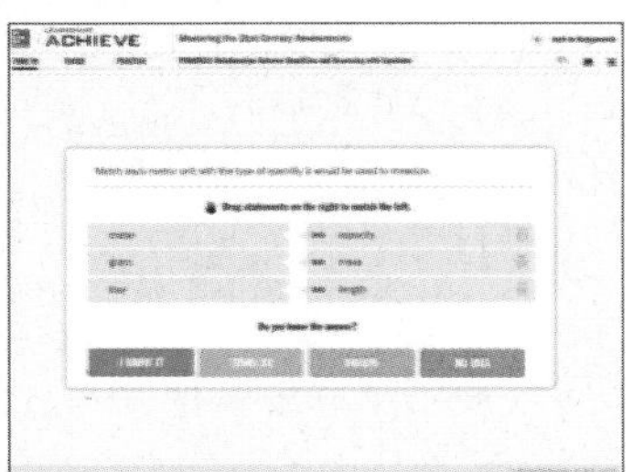

ALEKS®

You can use the **Similarities and Transformations** topic in ALEKS to explore what you know about relationships in triangles and what you are ready to learn.*

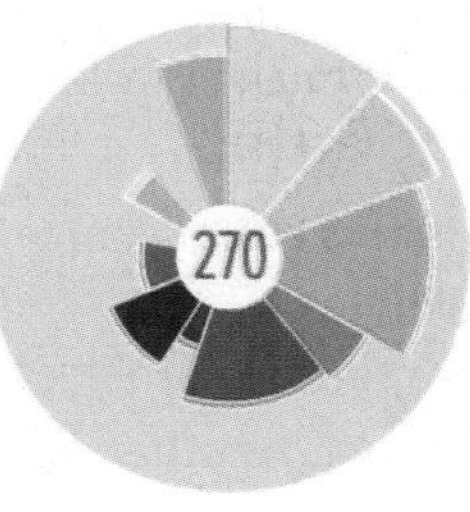

* Ask your teacher if this is part of your program.

Go Online! for Vocabulary Review Games and key vocabulary in 13 languages.

Get Ready for the Chapter

Connecting Concepts

Concept Check

Review the concepts used in this chapter by answering the questions below.

1. Given the equation $\frac{x+9}{2} = \frac{3x-1}{8}$, what would be the first step to solve?
2. Given the equation $3(4x - 3) = 5(2x + 11)$, what rule would you apply to begin solving?

In the figure, $\overrightarrow{QP}$ and $\overrightarrow{QR}$ are opposite rays and $\overrightarrow{QT}$ bisects $\angle SQR$.

3. How can you express $m\angle SQR$ in terms of $m\angle TQR$?

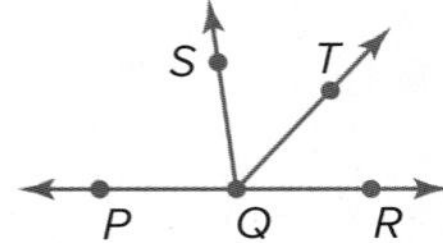

4. Justify your answer.
5. If $m\angle SQR = 6x + 8$ and $m\angle TQR = 4x - 14$, what mathematical method would you apply first to solve the equation?
6. What do you know about $m\angle SQT$ in relation to $m\angle SQR$? How?
7. Knowing the value of x, how can you determine $m\angle SQT$?

Performance Task Review

You can use the concepts and skills in the chapter to design projects for a carpentry business. Understanding similarities will help you finish the Performance Task at the end of the chapter.

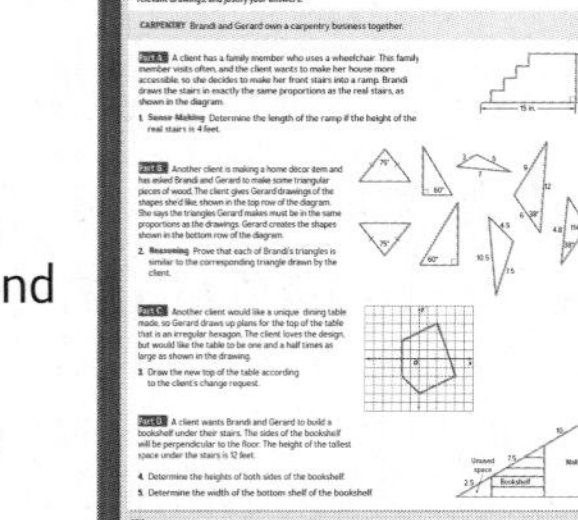

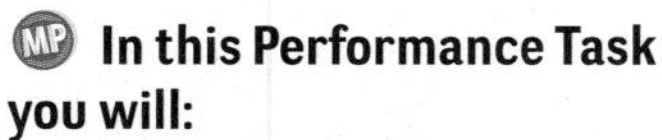

In this Performance Task you will:

- make sense of problems and persevere in solving them
- model with mathematics
- attend to precision

New Vocabulary

English		Español
dilation	p. 492	homotecia
similar polygons	p. 502	polígonos semejantes
similarity transformation	p. 502	transformación de semejanza
scale factor	p. 504	factor de escala
midsegment of a triangle	p. 535	segmento medio de un triángulo

Review Vocabulary

altitude altura a segment drawn from a vertex of a triangle perpendicular to the line containing the other side

angle bisector bisectriz de un ángulo a ray that divides an angle into two congruent angles

median mediana a segment drawn from a vertex of a triangle to the midpoint of the opposite side

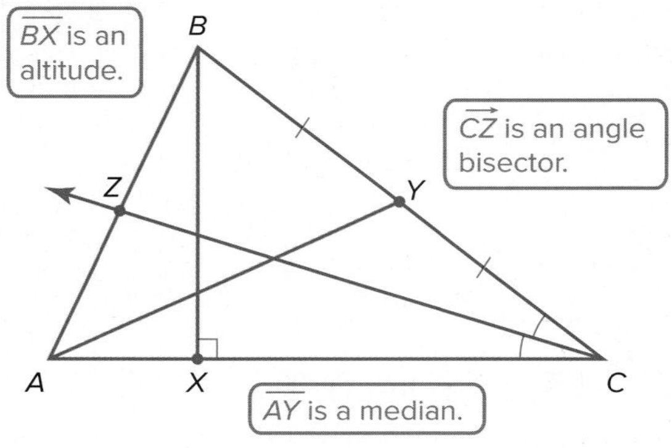

EXPLORE 7-1

Graphing Technology Lab

Dilations

You can use TI-Nspire Technology to explore properties of dilations.

Mathematical Practices
MP 5 Use appropriate tools strategically.

Activity 1 Dilation of a Triangle

Work cooperatively. Dilate a triangle by a scale factor of 1.5.

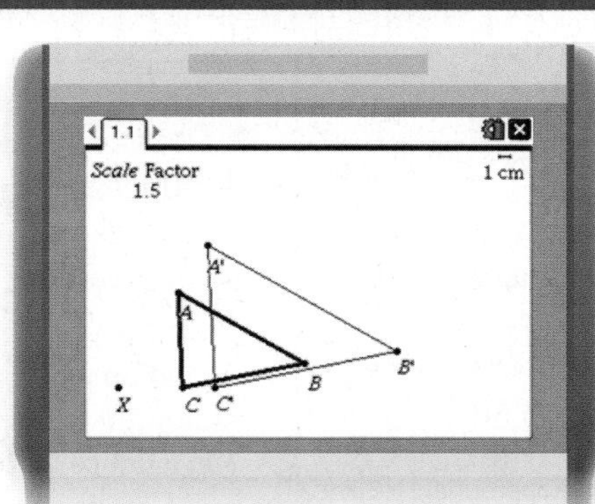

Step 1 Add a new **Geometry** page. Then, from the **Points & Lines** menu, use the **Point** tool to add a point and label it *X*.

Step 2 From the **Shapes** menu, select **Triangle** and specify three points. Label the points *A*, *B*, and *C*.

Step 3 From the **Actions** menu, use the **Text** tool to separately add the text *Scale Factor* and *1.5* to the page.

Step 4 From the **Transformation** menu, select **Dilation**. Then select point *X*, $\triangle ABC$, and the text *1.5*.

Step 5 Label the points on the image *A′*, *B′*, and *C′*.

Analyze the Results **Work cooperatively.**

1. Using the **Slope** tool on the **Measurement** menu, describe the effect of the dilation on $\overline{AB}$. That is, how are the lines through $\overline{AB}$ and $\overline{A'B'}$ related?
2. What is the effect of the dilation on the line passing through side $\overline{CA}$?
3. What is the effect of the dilation on the line passing through side $\overline{CB}$?

Activity 2 Dilation of a Polygon

Work cooperatively. Dilate a polygon by a scale factor of −0.5.

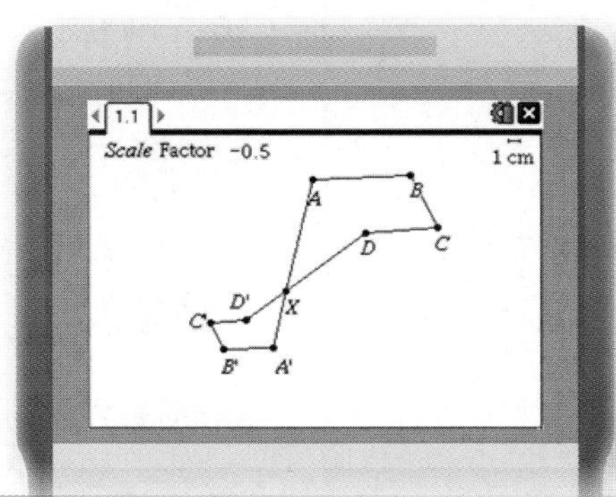

Step 1 Add a new **Geometry** page and draw polygon *ABCDX* as shown. Add the text *Scale Factor* and *−0.5* to the page.

Step 2 From the **Transformation** menu, select **Dilation**. Then select point *X*, polygon *ABCDX*, and the text *−0.5*.

Step 3 Label the points on the image *A′*, *B′*, *C′*, and *D′*.

Model and Analyze **Work cooperatively.**

4. Analyze the effect of the dilation in Activity 2 on sides that contain the center of the dilation.

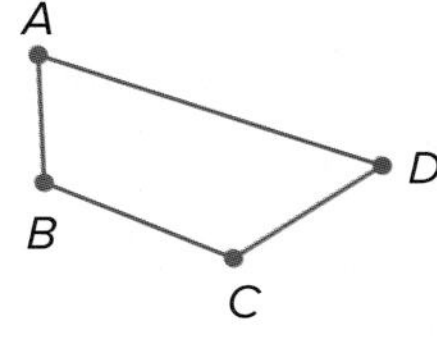

5. Analyze the effect of a dilation of trapezoid *ABCD* shown with a scale factor of 0.75 and the center of the dilation at *A*.
6. **MAKE A CONJECTURE** Describe the effect of a dilation on segments that pass through the center of a dilation and segments that do not pass through the center of a dilation.

LESSON 1

Dilations

Then

- You identified dilations and verified them as similarity transformations.

Now

1. Draw dilations.
2. Draw dilations in the coordinate plane.

Why?

- Charles can resize his photos before uploading them to a social networking site. Scaling down the size or enlarging the size of the original photo is an example of a *dilation*.

New Vocabulary
dilation

Mathematical Practices

1 Make sense of problems and persevere in solving them.

5 Use appropriate tools strategically.

1 Draw Dilations

A **dilation** or *scaling* is a similarity transformation that enlarges or reduces a figure proportionally with respect to a *center* point and a *scale* factor.

Key Concept Dilation

A dilation with center C and positive scale factor k, $k \neq 1$, is a function that maps a point P in a figure to its image such that

- if point P and C coincide, then the image and preimage are the same point, or
- if point P is not the center of dilation, then P' lies on $\overrightarrow{CP}$ and $CP' = k(CP)$.

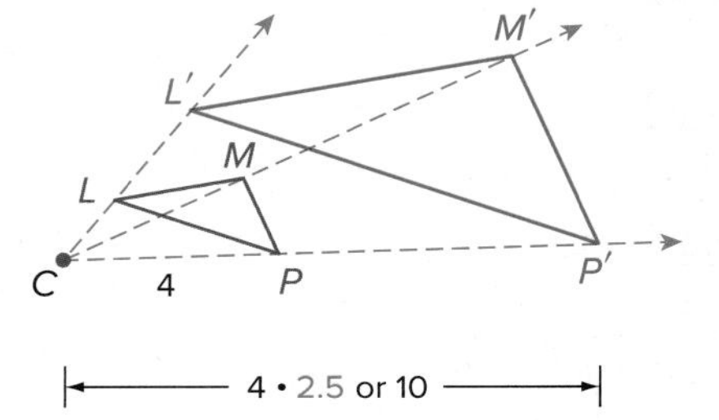

$\triangle L'M'P'$ is the image of $\triangle LMP$ under a dilation with center C and scale factor 2.5.

Example 1 Draw a Dilation

Copy $\triangle ABC$ and point D. Then use a ruler to draw the image of $\triangle ABC$ under a dilation with center D and scale factor $\frac{1}{2}$.

Step 1 Draw rays from D though each vertex.

Step 2 Locate A' on $\overrightarrow{DA}$ such that $DA' = \frac{1}{2}DA$.

Step 3 Locate B' on $\overrightarrow{DB}$ and C' on $\overrightarrow{DC}$ in the same way. Then draw $\triangle A'B'C'$.

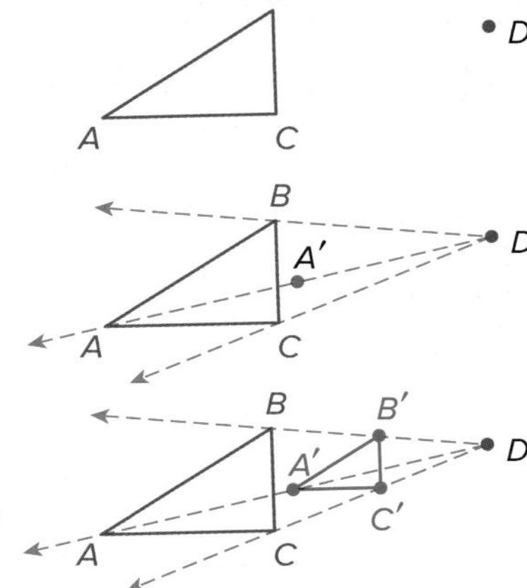

Guided Practice

Copy the figure and point J. Then use a ruler to draw the image of the figure under a dilation with center J and the scale factor k indicated.

1A. $k = \frac{3}{2}$

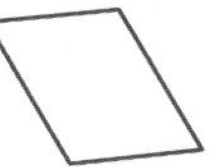

1B. $k = 0.75$

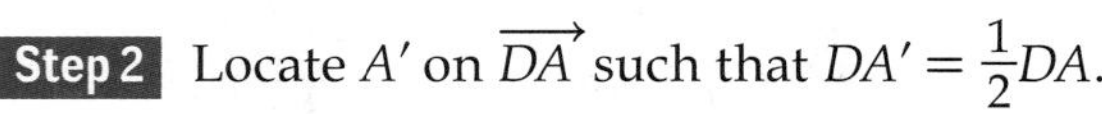

Geber86/E+/Getty Images

If $k > 1$, then the dilation is an *enlargement*. If $0 < k < 1$, then the dilation is a *reduction*. A dilation with a scale factor of 1 is called an *isometry dilation*. It produces an image that coincides with the preimage. The two figures are congruent.

Real-World Example 2 Find the Scale Factor of a Dilation

PHOTOGRAPHY **To create different-sized prints, you can adjust the distance between a film negative and the enlarged print by using a photographic enlarger. Suppose the distance between the light source C and the negative is 45 millimeters (CP). To what distance PP' should you adjust the enlarger to create a 22.75-centimeter wide print ($X'Y'$) from a 35-millimeter wide negative (XY)?**

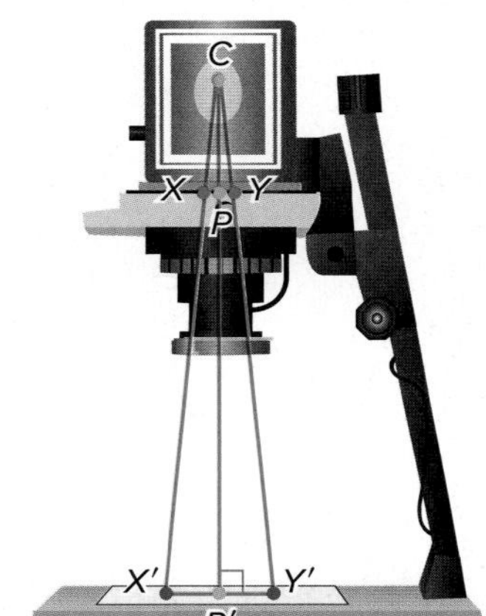

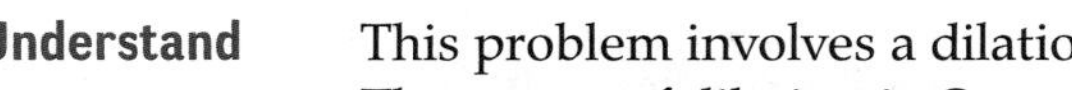

Understand This problem involves a dilation. The center of dilation is C, $XY = 35$ mm, $X'Y' = 22.75$ cm or 227.5 mm, and $CP = 45$ mm. You are asked to find PP'.

Plan Find the scale factor of the dilation from the preimage XY to the image $X'Y'$. Use the scale factor to find CP' and then use CP and CP' to find PP'.

Solve The scale factor k of the enlargement is the ratio of a length on the image to a corresponding length on the preimage.

$k = \frac{\text{image length}}{\text{preimage length}} = \frac{X'Y'}{XY}$ Scale factor of image; image $= X'Y'$, preimage $= XY$

$= \frac{227.5}{35}$ or 6.5 Divide.

Use this scale factor of 6.5 to find CP'.

$CP' = k(CP)$ Definition of dilation

$= 6.5(45)$ $k = 6.5$ and $CP = 45$

$= 292.5$ Multiply.

Use CP' and CP to find PP'.

$CP + PP' = CP'$ Segment addition

$45 + PP' = 292.5$ $CP = 45$ and $CP' = 292.5$

$PP' = 247.5$ Subtract 45 from each side.

So the enlarger should be adjusted so that the distance from the negative to the enlarged print (PP') is 247.5 millimeters or 24.75 centimeters.

Check $k = \frac{\text{distance from } C \text{ to } P'}{\text{distance from } C \text{ to } P}$

$= \frac{CP'}{CP} = \frac{292.5}{45}$ or 6.5 ✓

Because the dilation is an enlargement, the scale factor should be greater than 1. Because $6.5 > 1$, the scale factor found is reasonable.

Problem-Solving Tip

MP Perseverance To prevent calculations errors, estimate the answer to a problem before solving. In Example 2, estimate the scale factor of the dilation to be about $\frac{240}{40}$ or 6. Then CP' would be about $6 \cdot 50$ or 300 and PP' about $300 - 50$ or 250 millimeters, which is 25 centimeters. A measure of 24.75 centimeters is close to this estimate, so the answer is reasonable.

Guided Practice

2. Determine whether the dilation from Figure Q to Q' is an *enlargement* or a *reduction*. Then find the scale factor of the dilation and x.

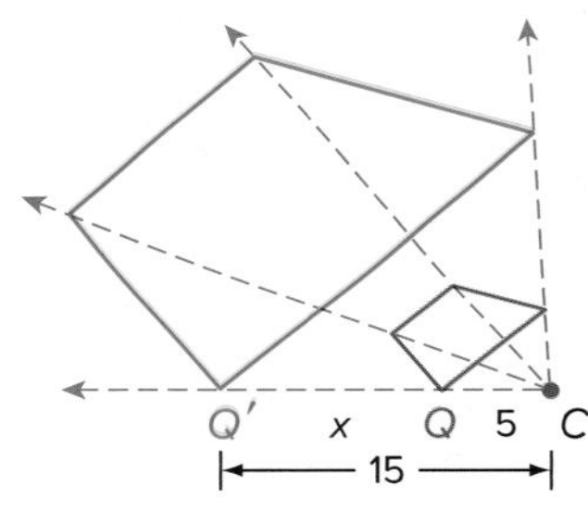

2 Dilations in the Coordinate Plane

You can use the following rules to find the image of a figure after a dilation centered at the origin.

Study Tip

Negative Scale Factors Dilations can also have negative scale factors. You will investigate this type of dilation in Exercise 40.

Key Concept Dilations in the Coordinate Plane

Words To find the coordinates of an image after a dilation centered at the origin, multiply the x- and y-coordinates of each point on the preimage by the scale factor of the dilation, k.

Symbols $(x, y) \rightarrow (kx, ky)$

Example

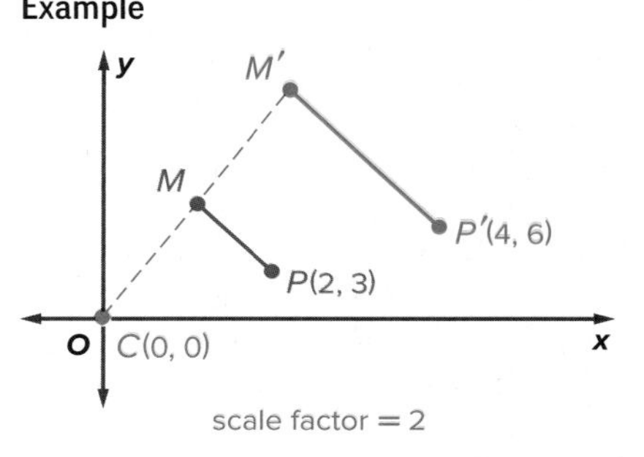

Go Online!

Investigate dilations by using the **Geometry Tools** in ConnectED.

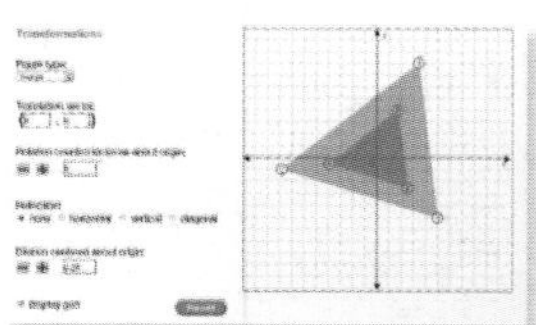

Example 3 Dilations in the Coordinate Plane

Graph the image of each polygon with the given vertices after a dilation at the indicated center with the given scale factor.

a. $J(-2, 4)$, $K(-2, -2)$, $L(-4, -2)$, $M(-4, 2)$; **origin;** $k = 2.5$

The distance from (0, 0) to J' should be 2.5 times longer than the distance from (0, 0) to J. To find the coordinates of the image, multiply the x- and y-coordinates of each vertex by the scale factor, 2.5.

$(x, y) \rightarrow (2.5x, 2.5y)$

$J(-2, 4) \rightarrow J'(-5, 10)$

$K(-2, -2) \rightarrow K'(-5, -5)$

$L(-4, -2) \rightarrow L'(-10, -5)$

$M(-4, 2) \rightarrow M'(-10, 5)$

Graph $JKLM$ and its image $J'K'L'M'$.

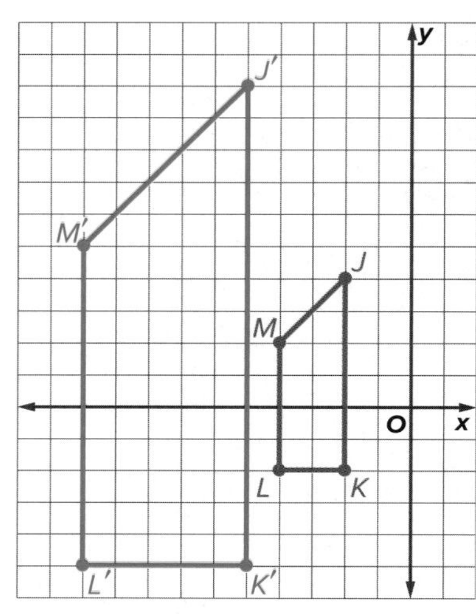

b. $B(-1, -1)$, $C(-4, -1)$, $D(-4, -3)$, $E(-1, -3)$; $(1, 1)$; $k = 2$

First, graph $BCDE$ and label the center of dilation A.

Plot the distance from A to B. Then double this distance to locate B'. B is 2 units left and 2 units down from A. So B' must be 4 units to the left and 4 units down from A. Continue on in this manner to locate each vertex of the image.

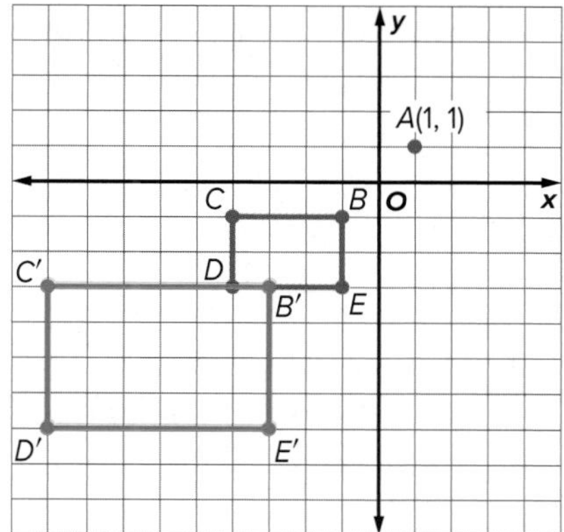

Guided Practice

3A. $Q(0, 6)$, $R(-6, -3)$, $S(6, -3)$; origin; $k = \frac{1}{3}$

3B. $F(-5, -3)$, $G(-6, -1)$, $H(-1, -1)$, $J(-2, -3)$; $(1, -1)$; $k = 1$

Example 4 Compositions of Dilations

Graph $QRST$ and its images after a composition of dilations centered at the origin with scale factor 2 and scale factor 1.5, given $Q(1, 1)$, $R(1, 3)$, $S(3, 3)$, and $T(4, 1)$.

Step 1 Multiply the x- and y-coordinates of each vertex by the scale factor of the first dilation.

$(x, y) \rightarrow (2x, 2y)$

$Q(1, 1) \rightarrow Q'(2, 2)$

$R(1, 3) \rightarrow R'(2, 6)$

$S(3, 3) \rightarrow S'(6, 6)$

$T(4, 1) \rightarrow T'(8, 2)$

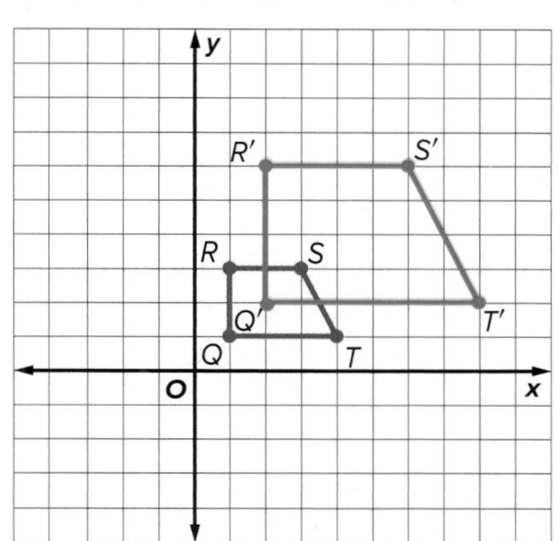

Step 2 Multiply the x- and y-coordinates of $Q'R'S'T'$ by the scale factor of the second dilation.

$(x, y) \rightarrow (1.5x, 1.5y)$

$Q'(2, 2) \rightarrow Q''(3, 3)$

$R'(2, 6) \rightarrow R''(3, 9)$

$S'(6, 6) \rightarrow S''(9, 9)$

$T'(8, 2) \rightarrow T''(12, 3)$

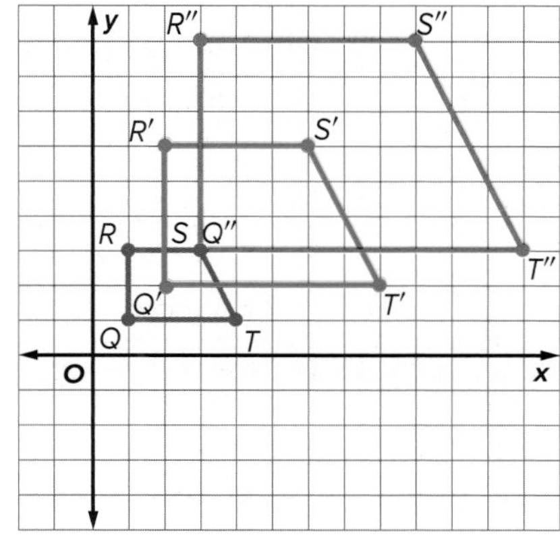

Guided Practice

Graph each figure and its images after the indicated transformations with the given center.

4A. $A(-5, 2)$, $B(-1, 8)$, $C(2, 4)$; dilation: 2; dilation: $\frac{1}{3}$; center $(0, 0)$

4B. $J(-2, 1)$, $K(-2, 6)$, $L(-4, 6)$, $M(-4, 1)$; dilation: $\frac{1}{2}$, reflection: x-axis, center $(1, 2)$

Check Your Understanding

◯ = Step-by-Step Solutions begin on page R13.

Example 1 **Copy the figure and point M. Then use a ruler to draw the image of the figure under a dilation with center M and the scale factor k indicated.**

1. $k = \frac{1}{4}$

2. $k = 2$

Example 2 **3** Determine whether the dilation from Figure B to B' is an *enlargement* or a *reduction*. Then find the scale factor of the dilation and x.

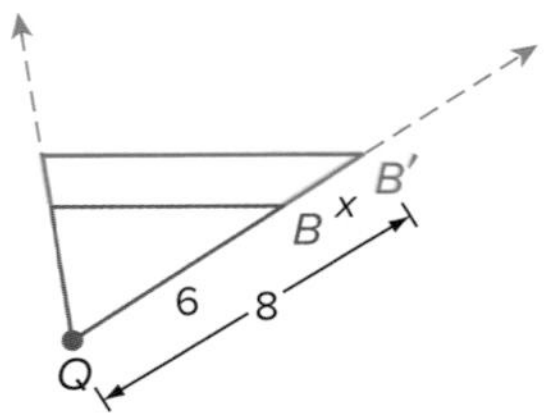

4. **BIOLOGY** Under a microscope, a single-celled organism 200 microns in length appears to be 50 millimeters long. If 1 millimeter = 1000 microns, what magnification setting (scale factor) was used? Explain your reasoning.

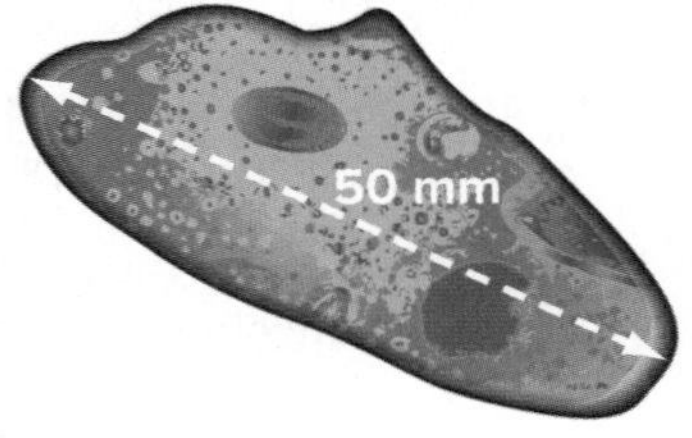

Example 3 **Graph the image of each polygon with the given vertices after a dilation at the indicated center with the given scale factor.**

5. $W(0, 0)$, $X(6, 6)$, $Y(6, 0)$; origin; $k = 1.5$

6. $Q(-4, 4)$, $R(-4, -4)$, $S(4, -4)$, $T(4, 4)$; origin; $k = \frac{1}{2}$

7. $A(-1, 4)$, $B(2, 4)$, $C(3, 2)$, $D(-2, 2)$; $(-4, -3)$; $k = 2$

8. $J(-2, 0)$, $K(2, 4)$, $L(8, 0)$, $M(2, -4)$; $(-2, 0)$; $k = \frac{3}{4}$

Example 4 **Graph each figure and its image after the indicated transformations with the given center.**

9. $C(-7, -3)$, $D(-7, 6)$, $E(1, 3)$; dilation: $\frac{1}{2}$; dilation: 4; center $(0, 0)$

10. $W(4, 4)$, $X(1, -3)$, $Y(6, -6)$, $Z(13, 4)$; dilation: $\frac{1}{3}$, reflection: y-axis, center $(1, 1)$

Practice and Problem Solving

Extra Practice is on page R7.

Example 1 **MP TOOLS** Copy the figure and point S. Then use a ruler to draw the image of the figure under a dilation with center **S** and the scale factor **k** indicated.

11. $k = \frac{5}{2}$

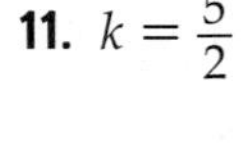

12. $k = 3$

13. $k = 0.8$

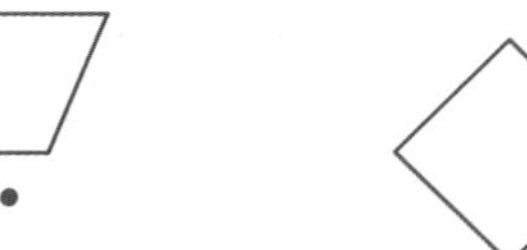

14. $k = \frac{1}{3}$

Example 2 **Determine whether the dilation from figure W to W' is an *enlargement* or a *reduction*. Then find the scale factor of the dilation and x.**

15.

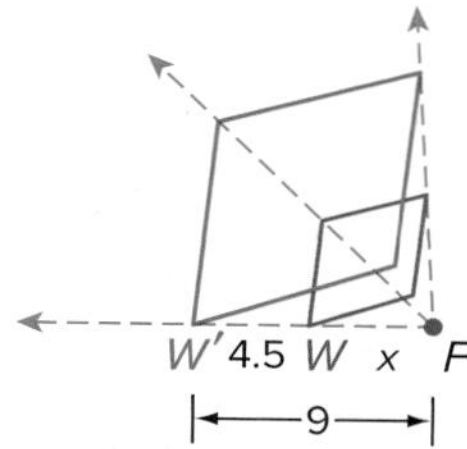

16.

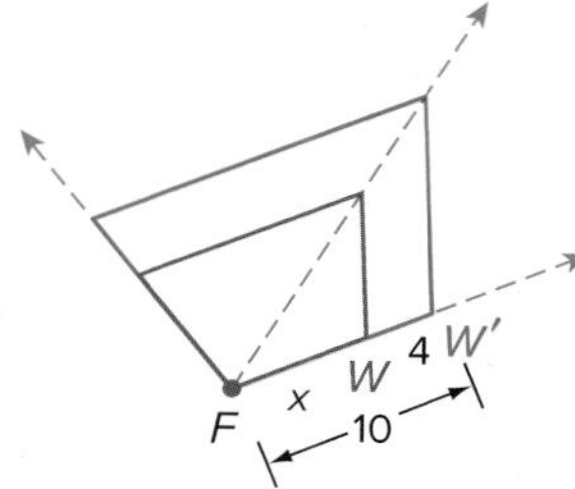

17.

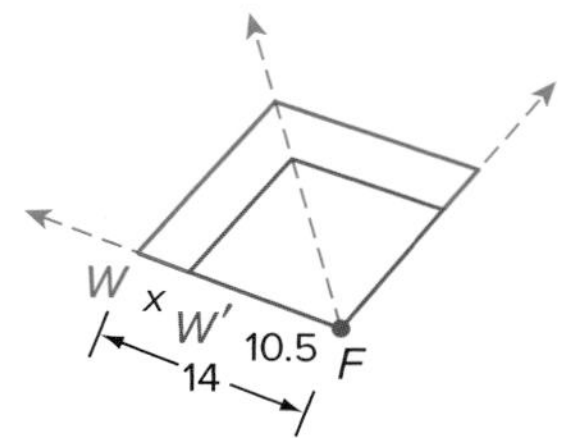

18.

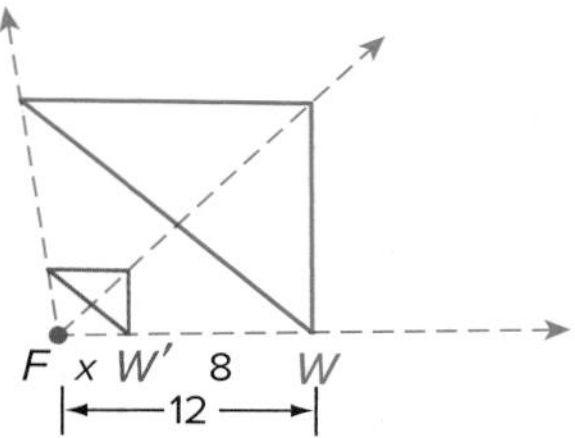

INSECTS When viewed under a microscope, each insect has the measurement given on the picture. Given the actual measure of each insect, what magnification was used? Explain your reasoning.

19.

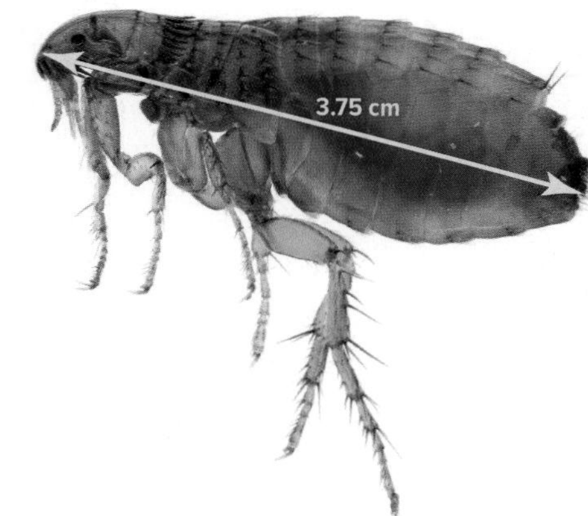

20.

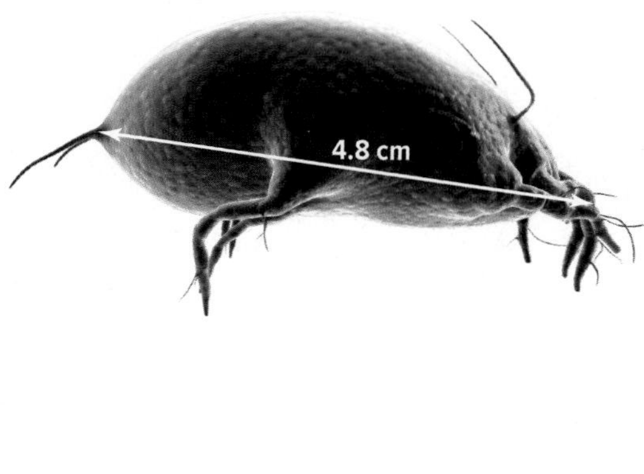

Example 3 MP **SENSE-MAKING** Graph the image of each polygon with the given vertices after a dilation at the indicated center with the given scale factor.

21. $J(-8, 0)$, $K(-4, 4)$, $L(-2, 0)$; origin; $k = 0.5$

22. $S(0, 0)$, $T(-4, 0)$, $V(-8, -8)$; origin; $k = 1.25$

23. $A(9, 9)$, $B(3, 3)$, $C(6, 0)$; origin; $k = \frac{1}{3}$

24. $D(4, 4)$, $F(0, 0)$, $G(8, 0)$; origin; $k = 0.75$

25. $M(-2, 0)$, $P(0, 2)$, $Q(2, 0)$, $R(0, -2)$; $(-4, -4)$; $k = 2.5$

26. $W(2, 2)$, $X(2, 0)$, $Y(0, 1)$, $Z(1, 2)$; $(4, -2)$; $k = 3$

Example 4 **Graph each figure and its images after the indicated transformations with the given center.**

27. dilation: $\frac{1}{3}$; dilation: 2; center (0, 0)

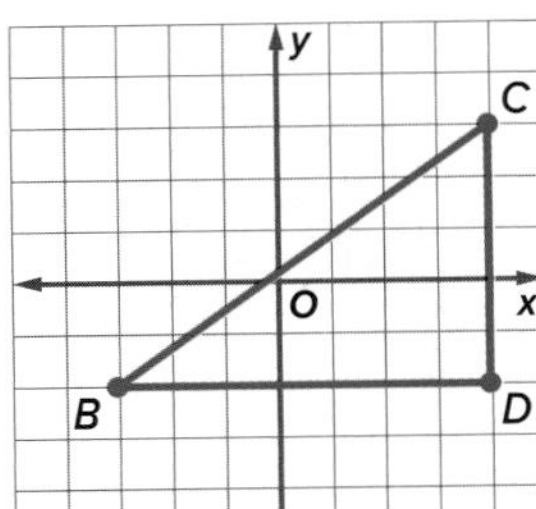

28. dilation: $\frac{1}{2}$; dilation: 2; center (0, 0)

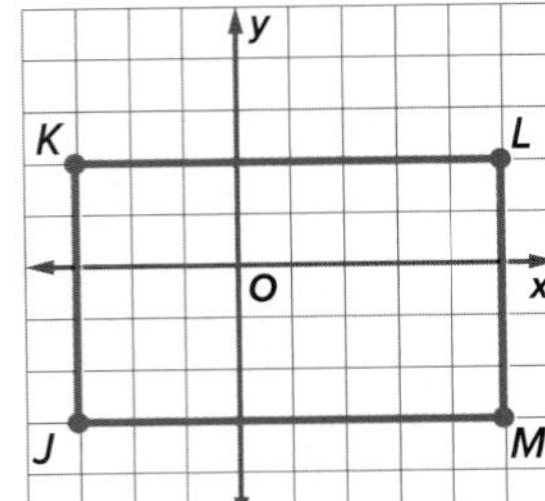

29. dilation: 2; dilation: $\frac{3}{2}$; center (1, 1)

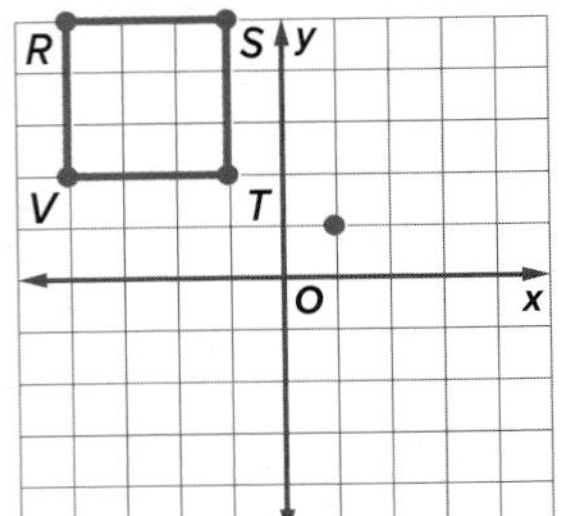

30. dilation: $\frac{1}{2}$; dilation: 3; center (−2, 0)

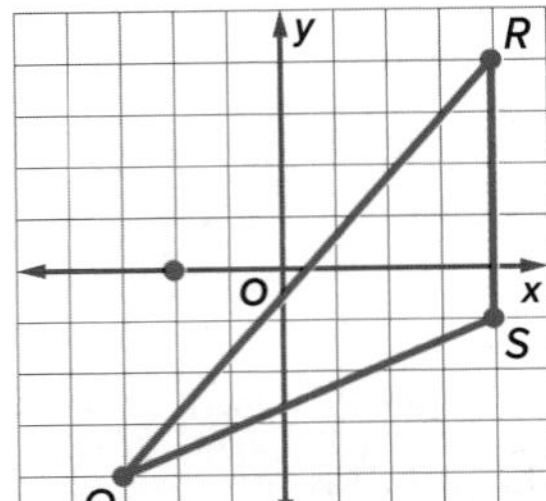

31. dilation: $\frac{1}{2}$; reflection: y-axis; center (0, 0)

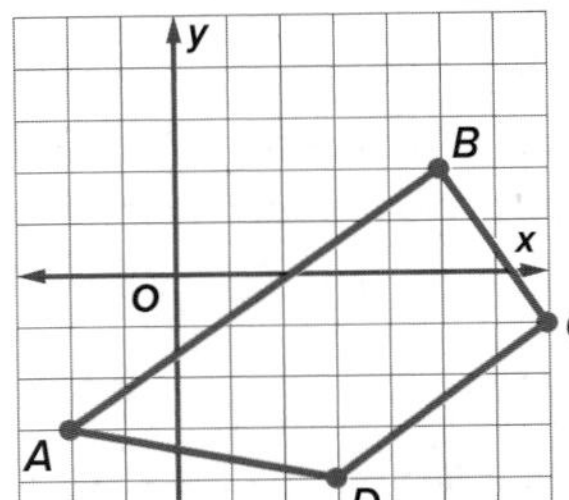

32. dilation: 3; rotation: 180°; center (−2, −2)

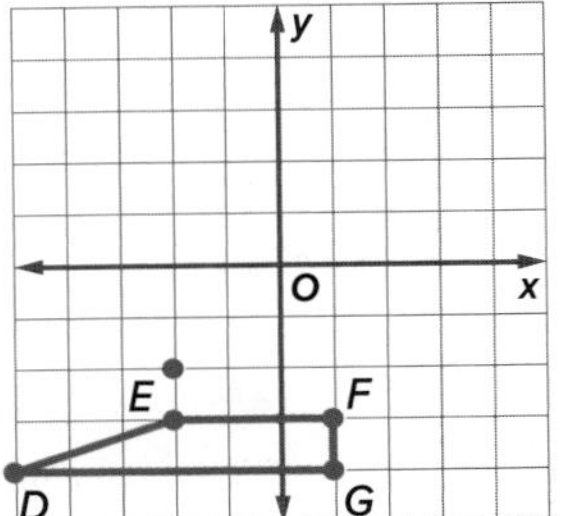

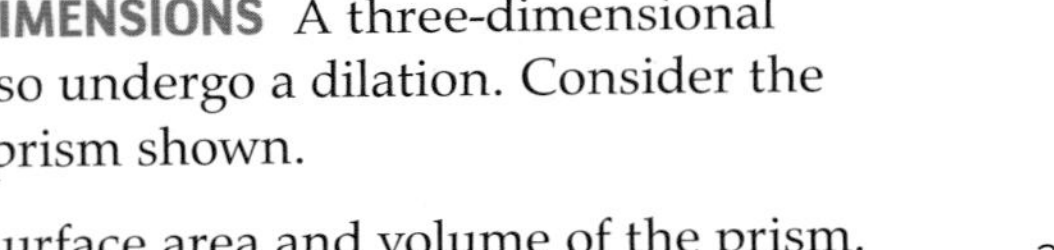

33 **CHANGING DIMENSIONS** A three-dimensional figure can also undergo a dilation. Consider the rectangular prism shown.

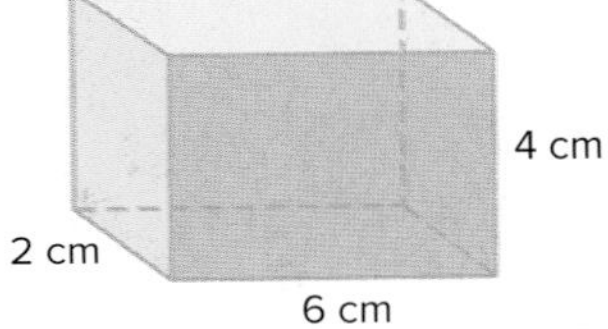

a. Find the surface area and volume of the prism.

b. Find the surface area and volume of the prism after a dilation with a scale factor of 2.

c. Find the surface area and volume of the prism after a dilation with a scale factor of $\frac{1}{2}$.

d. How many times as great is the surface area and volume of the image as the preimage after each dilation?

e. Make a conjecture as to the effect a dilation with a positive scale factor r would have on the surface area and volume of a prism.

34. **PHOTOGRAPHY AND ART** To make a scale drawing of a photograph, students overlay a $\frac{1}{4}$-inch grid on a 5-inch by 7-inch high contrast photo, overlay a $\frac{1}{2}$-inch grid on a 10-inch by 14-inch piece of drawing paper, and then sketch the image in each square of the photo to the corresponding square on the drawing paper.

a. What is the scale factor of the dilation?

b. To create an image that is 10 times as large as the original, what size grids are needed?

c. What would be the area of a grid drawing of a 5-inch by 7-inch photo that used 2-inch grids?

35. **COORDINATE GEOMETRY** Refer to the graph of $FGHJ$.

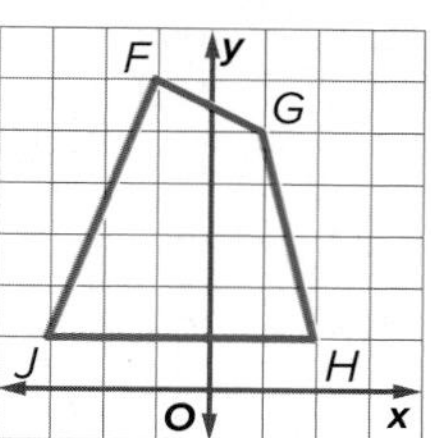

a. Dilate $FGHJ$ by a scale factor of $\frac{1}{2}$ centered at the origin, and then reflect the dilated image in the y-axis.

b. Complete the composition of transformations in part **a** in reverse order.

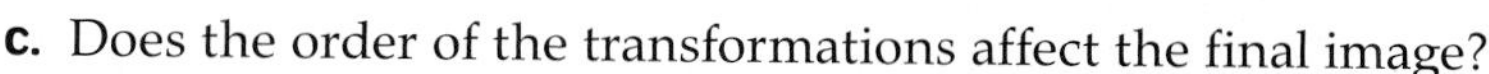

c. Does the order of the transformations affect the final image?

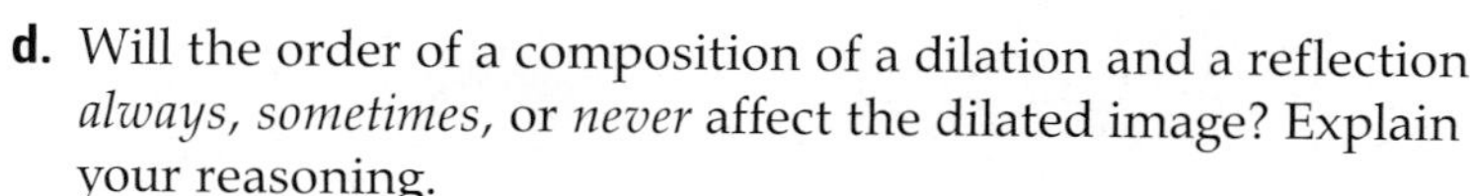

d. Will the order of a composition of a dilation and a reflection *always, sometimes,* or *never* affect the dilated image? Explain your reasoning.

36. MP **PERSEVERANCE** Refer to the graph of $\triangle DEF$.

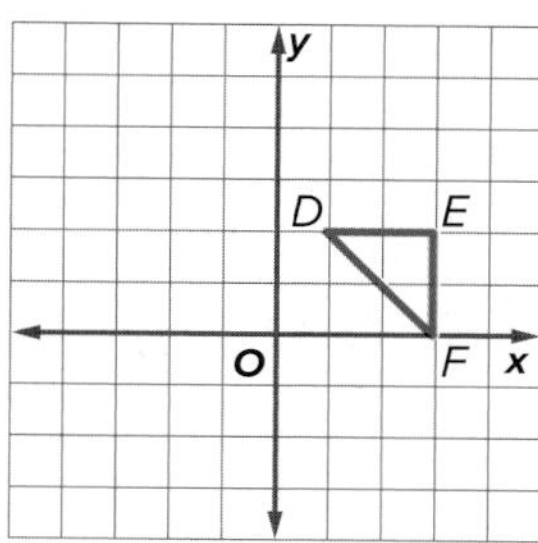

a. Graph the dilation of $\triangle DEF$ centered at point D with a scale factor of 3.

b. Describe the dilation as a composition of transformations including a dilation with a scale factor of 3 centered at the origin.

c. If a figure is dilated by a scale factor of 3 with a center of dilation (x, y), what composition of transformations, including a dilation with a scale factor of 3 centered at the origin, will produce the same final image?

37. **HEALTH** A coronary artery may be dilated with a balloon catheter as shown. The cross section of the middle of the balloon is a circle.

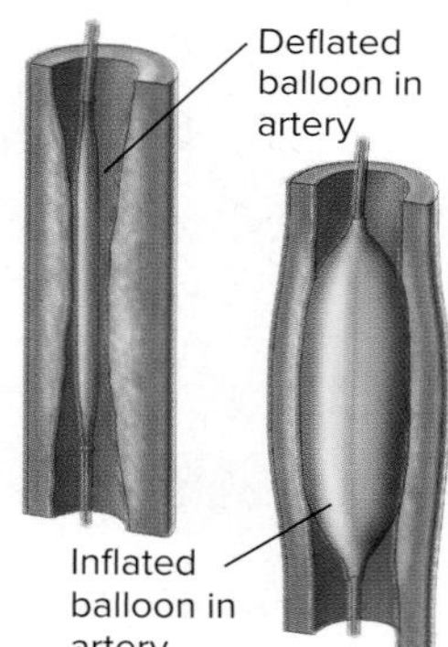

a. A surgeon inflates a balloon catheter in a patient's coronary artery, dilating the balloon from a diameter of 1.5 millimeters to 2 millimeters. Find the scale factor of this dilation.

b. Find the cross-sectional area of the balloon before and after the dilation.

Each figure shows a preimage and its image after a dilation centered at point P. Copy each figure, locate point P, and estimate the scale factor.

38.

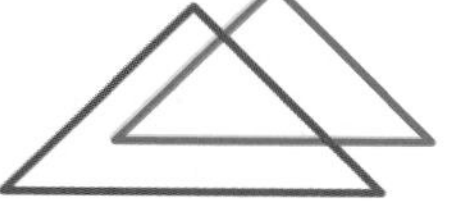

39.

40. **MULTIPLE REPRESENTATIONS** In this problem, you will investigate dilations centered at the origin with negative scale factors.

a. **Geometric** Draw $\triangle ABC$ with points $A(-2, 0)$, $B(2, -4)$, and $C(4, 2)$. Then draw the image of $\triangle ABC$ after a dilation centered at the origin with a scale factor of -2. Repeat the dilation with scale factors of $-\frac{1}{2}$ and -3. Record the coordinates for each dilation.

b. **Verbal** Make a conjecture about the function relationship for a dilation centered at the origin with a negative scale factor.

c. **Analytical** Write the function rule for a dilation centered at the origin with a scale factor of $-k$.

d. **Verbal** Describe a dilation centered at the origin with a negative scale factor as a composition of transformations.

41 **PHOTOGRAPHY** Becca took this photo for the school newspaper. The original picture was too large to fit in the space she had within the article, so she used her computer to reduce the width and height to fit the space exactly. Why did the newspaper's editor reject her image? What are the dimensions of the photo she resubmits? Explain.

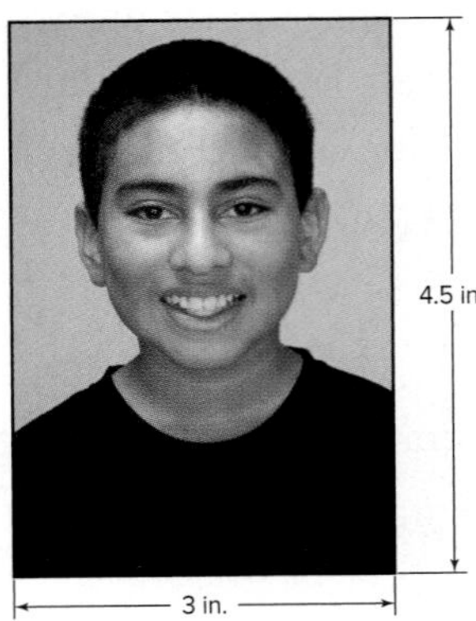

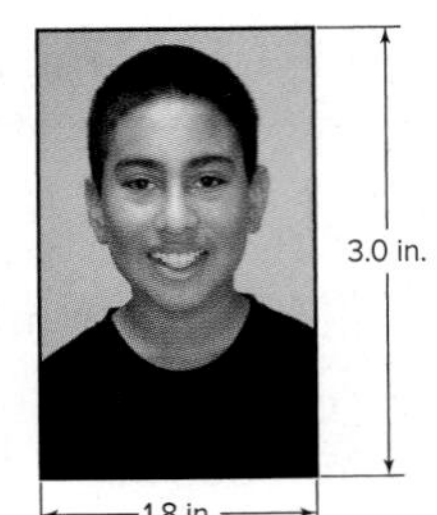

42. Describe the dilation from AB to $A'B'$ and $A'B'$ to $A''B''$ in the triangles shown.

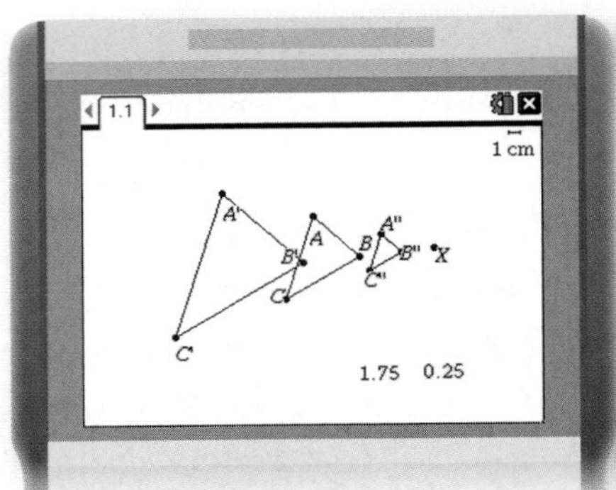

***WXYZ* has vertices *W*(6, 2), *X*(3, 7), *Y*(−1, 4), and *Z*(4, −2).**

43. Graph $WXYZ$ and find the perimeter of the figure.

44. Graph the image of $WXYZ$ after a dilation of $\frac{1}{2}$ centered at the origin. Find the perimeter of the dilated image and compare it to the perimeter of $WXYZ$.

H.O.T. Problems Use Higher-Order Thinking Skills

45. **CHALLENGE** Find the equation for the dilated image of the line $y = 4x - 2$ if the dilation is centered at the origin with a scale factor of 1.5.

46. **WRITING IN MATH** Are parallel lines (parallelism) and collinear points (collinearity) preserved under all transformations? Explain.

47. MP **CONSTRUCT ARGUMENTS** Determine whether invariant points are *sometimes*, *always*, or *never* maintained for the transformations described below. If so, describe the invariant point(s). If not, explain why invariant points are not possible.

a. dilation of $ABCD$ with scale factor 1

b. rotation of $\overline{AB}$ 74° about B

c. reflection of $\triangle MNP$ in the x-axis

d. translation of $PQRS$ along $\langle 7, 3 \rangle$

e. dilation of $\triangle XYZ$ centered at the origin with scale factor 2

48. **OPEN-ENDED** Graph a triangle. Dilate the triangle so that its area is four times the area of the original triangle. State the scale factor and center of your dilation.

49. **WRITING IN MATH** Can you use transformations to create congruent figures, similar figures, and equal figures? Explain.

50. Point A has coordinates $A(2, 1)$. What is the distance between point A and its final image after a reflection in the y-axis and a dilation centered at the origin with a scale factor of 2? Round to the nearest tenth. MP 1,7

51. Zariah drew $\triangle JKL$ on a coordinate plane and then applied a composition of two transformations to produce $\triangle MNP$.

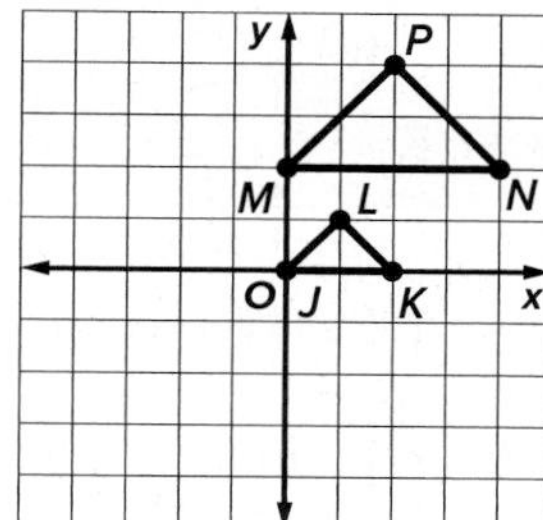

Which statement appears to be true? MP 1,7

- A $\triangle JKL \cong \triangle MNP$
- B The transformations included a dilation with a scale factor of 2.
- C Zariah used only rigid motions.
- D The area of $\triangle MNP$ is 2 times the area of $\triangle JKL$.

52. Theo plots the point $A(1, 3)$. Then he translates it along $\langle -2, -1 \rangle$ and dilates the image using a dilation centered at the origin with a scale factor of 3. What are the coordinates of the final image? MP 7

- A (2, 5)
- B (−1, 2)
- C (1, 8)
- D (−6, −3)
- E (−3, 6)

53. $\overline{JK}$ has endpoints $J(2, 4)$ and $K(6, 2)$. Maria dilates the segment using a dilation centered at the origin with a scale factor of $\frac{1}{2}$ and then reflects the image in the x-axis. What is the midpoint of the final image? MP 7

- A (2, −1.5)
- B (−2, 1.5)
- C (2, 1.5)
- D (1.5, 2)

54. **MULTI-STEP** Point P is $(-4, 3)$. It is translated along $\langle 7, -1 \rangle$ and then dilated by a factor of 3.5. MP 6

a. Which formula is used to find the coordinates of the translated point in Step 1?

- A $(x + 7, y + 7)$
- B $(x + 7, y - 1)$
- C $(x + 1, y - 7)$
- D $(x + 7, y + 1)$

b. Which formula is used to find the coordinates of the dilated point in Step 2?

- A $(x + 3.5, y + 3.5)$
- B $(x - 3.5, y - 3.5)$
- C $(3.5x, 3.5y)$
- D $\left(\frac{x}{3.5}, \frac{y}{3.5}\right)$

c. What are the coordinates of the final point?

- A (10.5, 7)
- B (−14, 10.5)
- C (3, 2)
- D (7, 10.5)

d. What are the coordinates of point $Q(9, -3)$ after these transformations?

- A (16, −4)
- B (31.5, −10.5)
- C (−3, 9)
- D (56, −7)

e. The final image of a point under these transformations is (14, 3.5). What are the coordinates of the point?

- A (4, 1)
- B (−3, 2)
- C (7, 2.5)
- D (1, 2)

LESSON 2

Similar Polygons

Then

- You drew the image of a figure after a dilation.

Now

1. Use the definition of similarity to identify similar polygons.
2. Solve problems by using the properties of similar polygons.

Why?

- People often customize their computer desktops using photos, centering the images at their original size or stretching them to fit the screen. This second method distorts the image because the original and new images are not geometrically similar.

New Vocabulary

similar polygons
similarity transformation
scale factor

Mathematical Practices

1 Make sense of problems and persevere in solving them.
6 Attend to precision.

1 Identify Similar Polygons

Two polygons are **similar polygons** if one can be obtained from the other by a dilation or by a dilation with one or more rigid transformations. A dilation is a type of similarity transformation. A **similarity transformation** occurs when a figure and its image have the same shape but not necessarily the same size.

Example 1 Determine Whether Polygons Are Similar

Determine whether the given polygons are similar. Explain.

a. ***JKLM* and *PQRS***

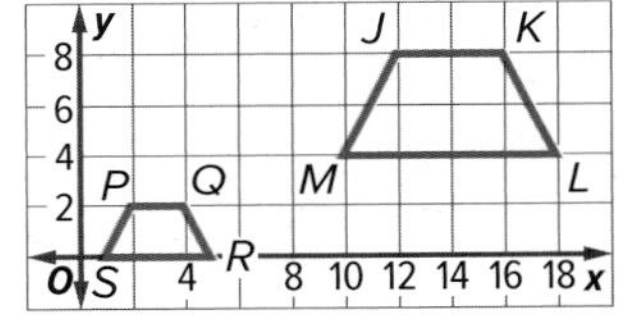

First dilate polygon *JKLM* so that its image is the same size as polygon *PQRS*.

Dilate polygon *JKLM* using a dilation centered at the origin with scale factor 0.5.

The image of polygon *JKLM* is polygon *J′K′L′M′*.

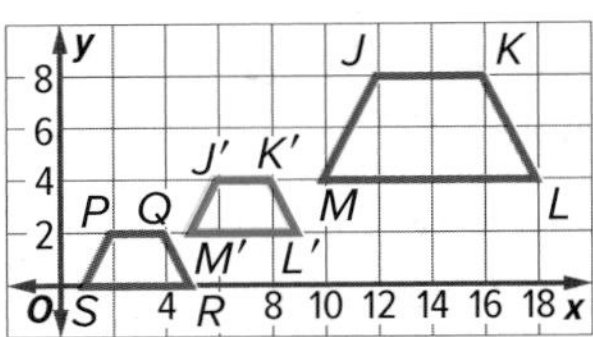

Now translate polygon *J′K′L′M′* so that vertex *J′* maps to vertex *P*.

A translation along the vector $(-4, -2)$ maps polygon *J′K′L′M′* to polygon *PQRS*.

Polygon *JKLM* and polygon *PQRS* are similar because one polygon can be obtained from the other polygon by a similarity transformation followed by a rigid transformation.

b. ***ABCD* and *EFGH***

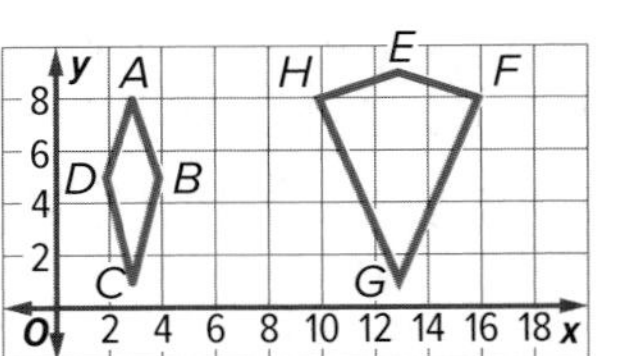

Polygon *ABCD* and polygon *EFGH* do not have the same shape.

The polygons are not similar because there is no similarity transformation and no combination of similarity transformation and rigid transformations that will map one polygon onto the other.

Guided Practice

Determine whether the given polygons are similar. Explain.

1A. $ABCD$ and $PQRS$

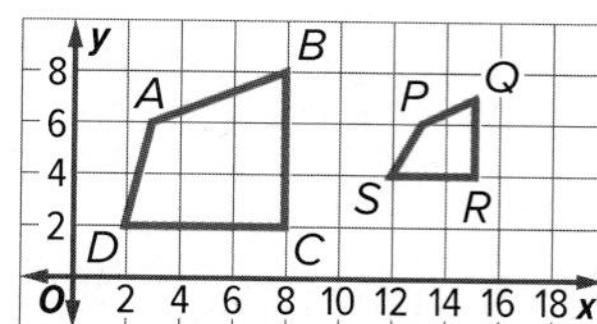

1B. $\triangle JKL$ and $\triangle XYZ$

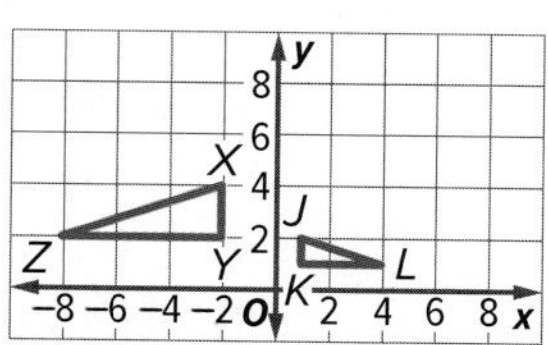

2 Use Properties of Similar Polygons

When two polygons are similar, there are important relationships among the polygons' angles and sides.

Key Concept Similar Polygons

Two polygons are similar if and only if their corresponding angles are congruent and corresponding side lengths are proportional.

Example In the diagram below, $ABCD$ is similar to $WXYZ$.

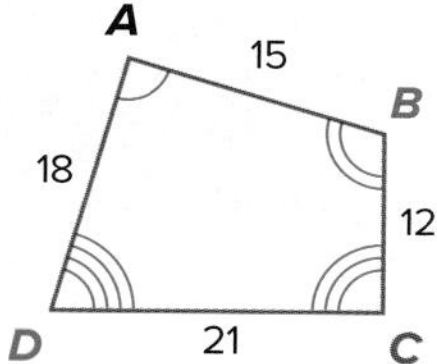

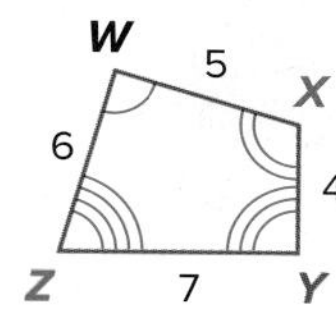

Corresponding angles

$\angle A \cong \angle W$, $\angle B \cong \angle X$, $\angle C \cong \angle Y$, and $\angle D \cong \angle Z$

Corresponding sides

$$\frac{AB}{WX} = \frac{BC}{XY} = \frac{CD}{YZ} = \frac{DA}{ZW} = \frac{3}{1}$$

Symbols $ABCD \sim WXYZ$

As with congruence statements, the order of vertices in a similarity statement like $ABCD \sim WXYZ$ is important. It identifies the corresponding angles and sides.

Watch Out!

Proportions There are many equivalent ways to write a proportion. When you check your answer to a problem like Example 2, be sure to consider different ways to write the proportion.

Example 2 Use a Similarity Statement

If $\triangle FGH \sim \triangle JKL$, list all pairs of congruent angles, and write a proportion that relates the corresponding sides.

Use the similarity statement.

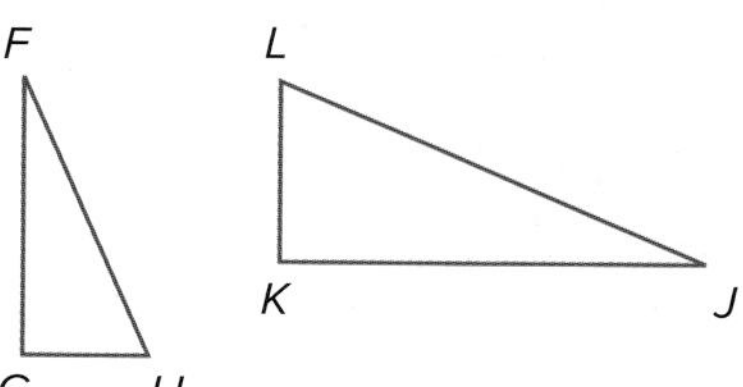

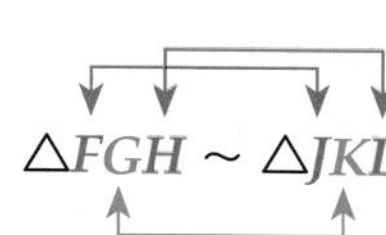

Congruent angles: $\angle F \cong \angle J$, $\angle G \cong \angle K$, $\angle H \cong \angle L$

Proportion: $\frac{FG}{JK} = \frac{GH}{KL} = \frac{HF}{LJ}$

Guided Practice

2. In the diagram, $NPQR \sim UVST$. List all pairs of congruent angles, and write a proportion that relates the corresponding sides.

N P S T R Q V U

Go Online! 

Discover the principles of similarity with a **Geometer's Sketchpad®** sketch in ConnectED. Discuss your findings with a partner. Ask for clarification as you need it.

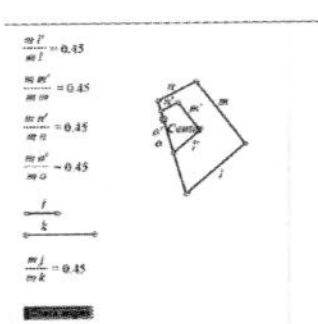

The ratio of the lengths of the corresponding sides of two similar polygons is called the **scale factor**. The scale factor depends on the order of comparison.

In the diagram, $\triangle ABC \sim \triangle XYZ$.

The scale factor from $\triangle XYZ$ to $\triangle ABC$ is $\frac{6}{3}$ or 2.

The scale factor from $\triangle ABC$ to $\triangle XYZ$ is $\frac{3}{6}$ or $\frac{1}{2}$.

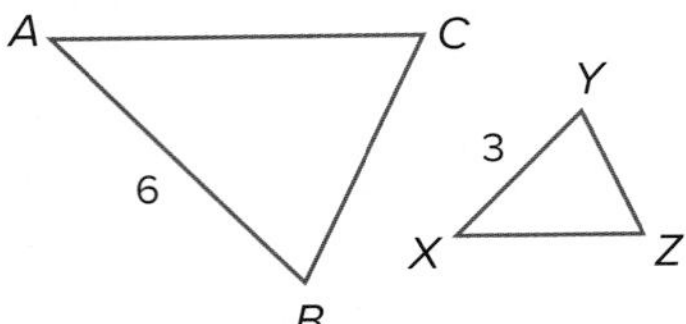

Real-World Example 3 Identify Similar Polygons

PHOTO EDITING Kuma wants to use the rectangular photo shown as the background for her computer's desktop, but she needs to resize it. Determine whether the following rectangular images are similar. If so, write the similarity statement and scale factor. Explain your reasoning.

a. 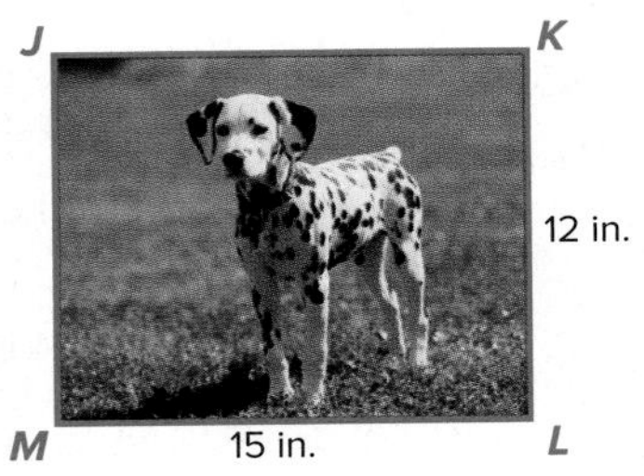

b.

a. Step 1 Compare corresponding angles.

Because all angles of a rectangle are right angles and right angles are congruent, corresponding angles are congruent.

Step 2 Compare corresponding sides.

$\frac{DC}{HG} = \frac{10}{14}$ or $\frac{5}{7}$ $\quad \frac{BC}{FG} = \frac{8}{12}$ or $\frac{2}{3}$ $\quad \frac{5}{7} \neq \frac{2}{3}$

Because corresponding sides are not proportional, $ABCD \nsim EFGH$. So the photos are not similar.

Reading Math

Similarity Symbol The symbol $\nsim$ is read as *is not similar to.*

b. Step 1 Because $ABCD$ and $JKLM$ are both rectangles, corresponding angles are congruent.

Step 2 Compare corresponding sides.

$\frac{DC}{ML} = \frac{10}{15}$ or $\frac{2}{3}$ $\quad \frac{BC}{KL} = \frac{8}{12}$ or $\frac{2}{3}$ $\quad \frac{2}{3} = \frac{2}{3}$

Because corresponding sides are proportional, $ABCD \sim JKLM$. So the rectangles are similar with a scale factor from $JKLM$ to $ABCD$ of $\frac{2}{3}$.

Guided Practice

3. Determine whether the triangles shown are similar. If so, write the similarity statement and scale factor. Explain your reasoning.

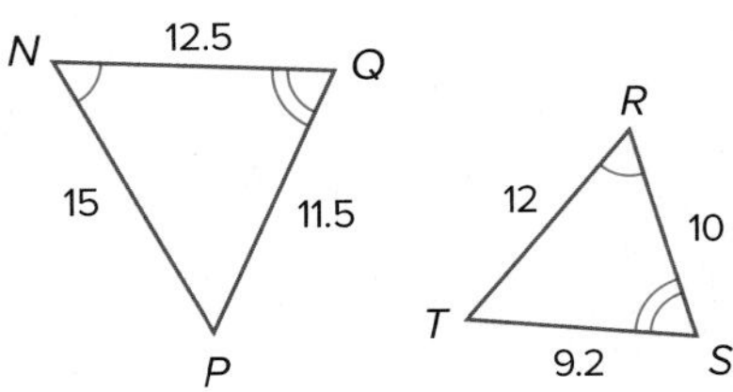

Example 4 Use Similar Figures to Find Missing Measures

In the diagram, $ACDF \sim VWYZ$.

a. Find x.

Use the corresponding side lengths to write a proportion.

$\frac{CD}{WY} = \frac{DF}{YZ}$	Similarity proportion
$\frac{9}{6} = \frac{x}{10}$	$CD = 9$, $WY = 6$, $DF = x$, $YZ = 10$
$9(10) = 6(x)$	Cross Products Property
$90 = 6x$	Multiply.
$15 = x$	Divide each side by 6.

b. Find y.

Use the corresponding side lengths to write a proportion.

$\frac{CD}{WY} = \frac{FA}{ZV}$	Similarity proportion
$\frac{9}{6} = \frac{12}{3y - 1}$	$CD = 9$, $WY = 6$, $FA = 12$, $ZV = 3y - 1$
$9(3y - 1) = 6(12)$	Cross Products Property
$27y - 9 = 72$	Multiply.
$27y = 81$	Add 9 to each side.
$y = 3$	Divide each side by 27.

A C 12 9 F x D

Y 10 Z 6 3y − 1 W V

Guided Practice

Find the value of each variable if $\triangle JLM \sim \triangle QST$.

4A. x

4B. y

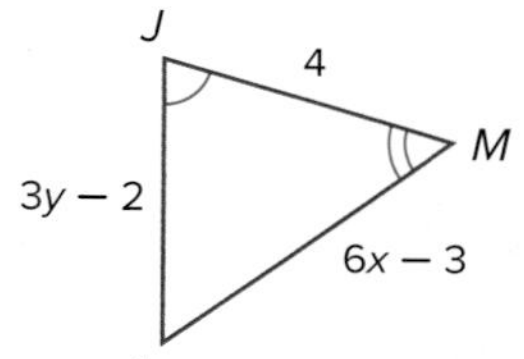

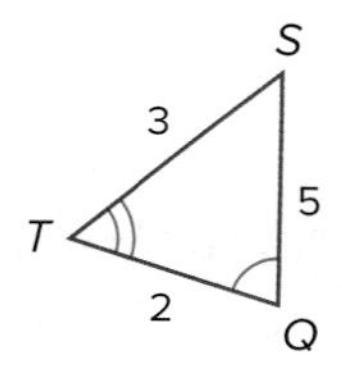

Study Tip

MP Sense-Making When only two congruent angles of a triangle are given, remember that you can use the Third Angles Theorem to establish that the remaining corresponding angles are also congruent.

In similar polygons, the ratio of any two corresponding lengths is proportional to the scale factor between them. This leads to the following theorem about the perimeters of two similar polygons.

Theorem 7.1 Perimeters of Similar Polygons

If two polygons are similar, then their perimeters are proportional to the scale factor between them.

Example If $ABCD \sim JKLM$, then

$$\frac{AB + BC + CD + DA}{JK + KL + LM + MJ} = \frac{AB}{JK} = \frac{BC}{KL} = \frac{CD}{LM} = \frac{DA}{MJ}$$

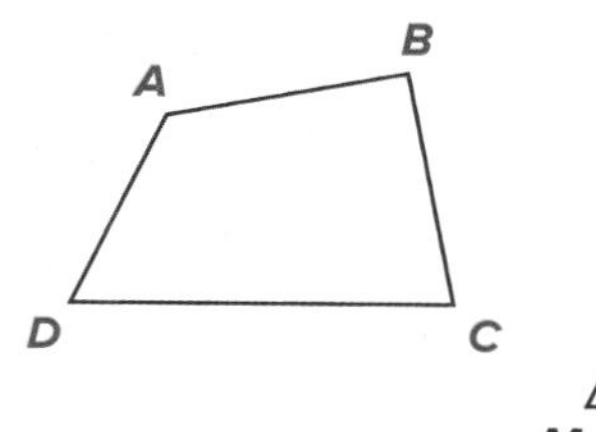

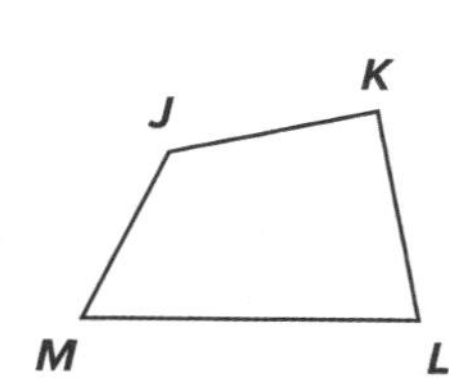

Example 5 Use a Scale Factor to Find Perimeter

If $ABCDE \sim PQRST$, find the scale factor of $ABCDE$ to $PQRST$ and the perimeter of each polygon.

The scale factor from $PQRST$ to $ABCDE$ is $\frac{CD}{RS}$ or $\frac{4}{3}$.

Because $\overline{BC} \cong \overline{AB}$ and $\overline{AE} \cong \overline{CD}$, the perimeter of $ABCDE$ is $8 + 8 + 4 + 6 + 4$ or 30.

Use the perimeter of $ABCDE$ and the scale factor to write a proportion. Let x represent the perimeter of $PQRST$.

$$\frac{4}{3} = \frac{\text{perimeter of } ABCDE}{\text{perimeter of } PQRST} \quad \text{Theorem 7.1}$$

$$\frac{4}{3} = \frac{30}{x} \quad \text{Substitution}$$

$$(3)(30) = 4x \quad \text{Cross Products Property}$$

$$22.5 = x \quad \text{Solve.}$$

So, the perimeter of $PQRST$ is 22.5.

Watch Out!

Perimeter Remember that perimeter is the distance around a figure. Be sure to find the sum of all side lengths when finding the perimeter of a polygon. You may need to use other markings or geometric principles to find the length of unmarked sides.

Guided Practice

5. If $MNPQ \sim XYZW$, find the scale factor from $XYZW$ to $MNPQ$ and the perimeter of each polygon.

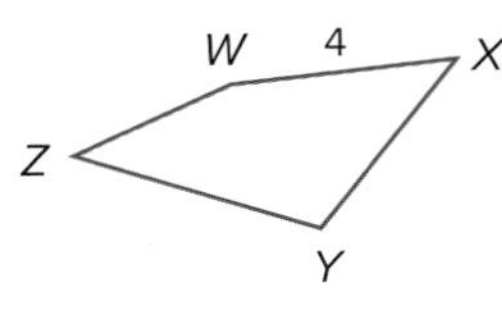

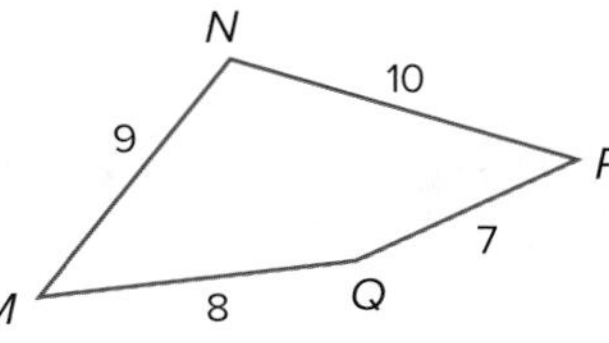

Check Your Understanding

◯ = Step-by-Step Solutions begin on page R13.

Example 1 **Determine whether the given polygons are similar. Explain.**

1. $DEFG$ and $JKLM$

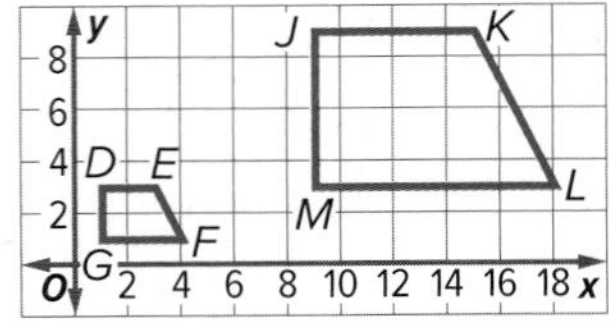

2. $ABCD$ and $RSTU$

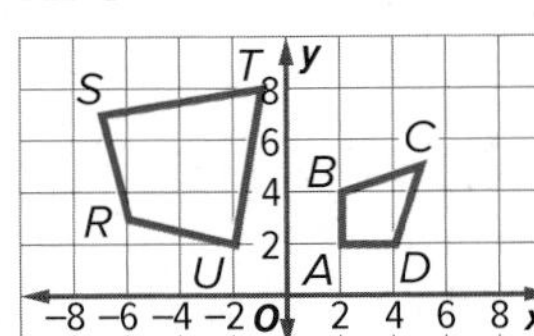

Example 2 **List all pairs of congruent angles, and write a proportion that relates the corresponding sides for each pair of similar polygons.**

3. $\triangle ABC \sim \triangle ZYX$

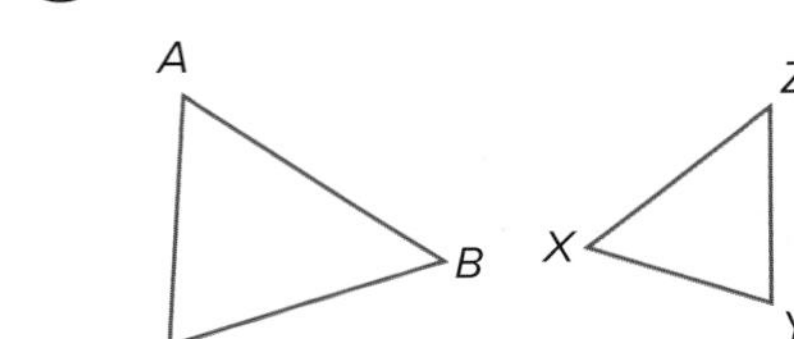

4. $JKLM \sim TSRQ$

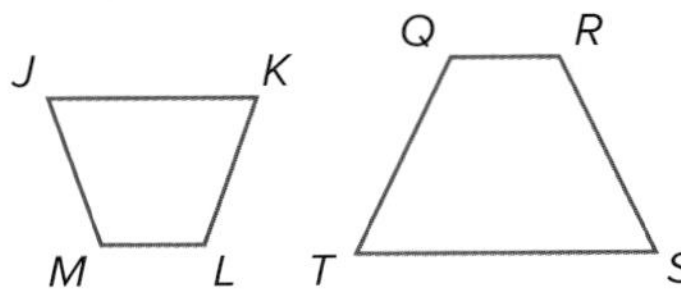

Example 3 **Determine whether each pair of figures is similar. If so, write the similarity statement and scale factor. If not, explain your reasoning.**

5.

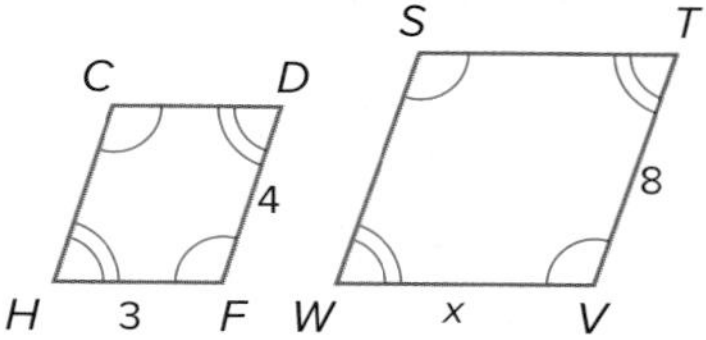

6.

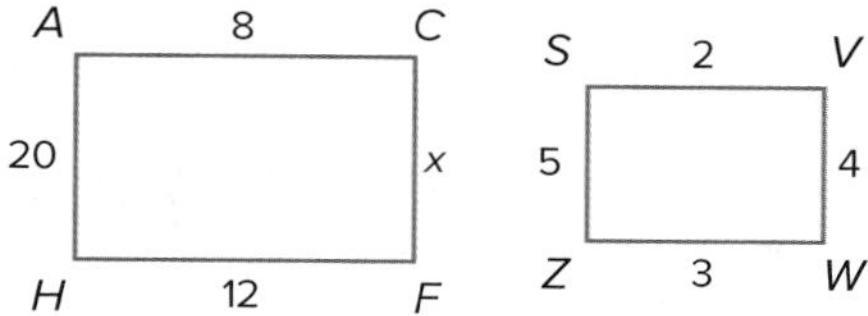

Example 4 **Each pair of polygons is similar. Find the value of x.**

7. C D H 3 F 4 S T W x V 8

8. A 8 C 20 x H 12 F S 2 V 5 4 Z 3 W

Example 5 **9. DESIGN** On the blueprint of the apartment shown, the balcony measures 1 inch wide by 1.75 inches long. If the actual length of the balcony is 7 feet, what is the perimeter of the balcony?

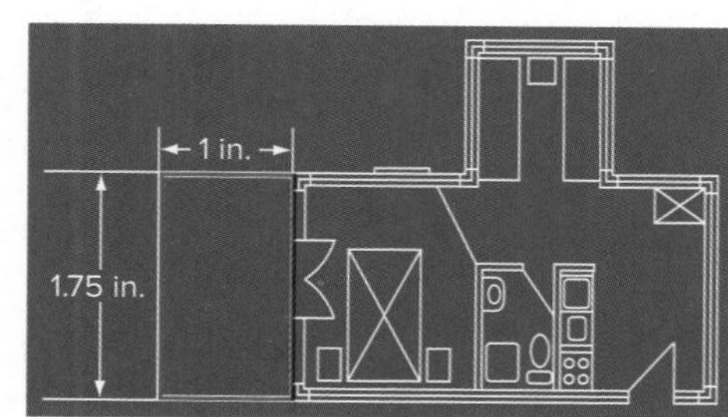

Practice and Problem Solving

Extra Practice is on page R7.

Example 1 **Determine whether the given polygons are similar. Explain.**

10. $\triangle GHJ$ and $\triangle KLM$

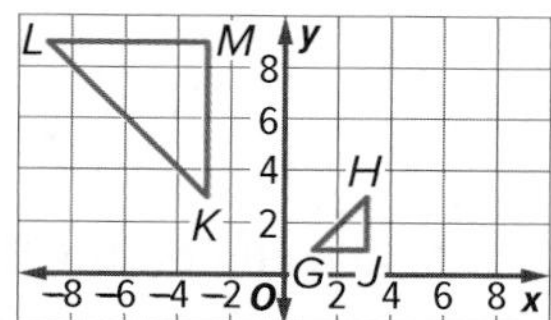

11. $PQRS$ and $WXYZ$

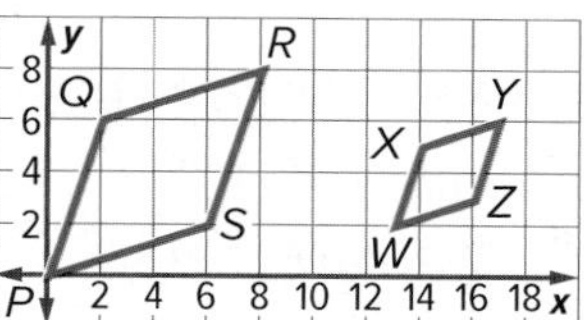

Example 2 **List all pairs of congruent angles, and write a proportion that relates the corresponding sides for each pair of similar polygons.**

12. $\triangle CHF \sim \triangle YWS$

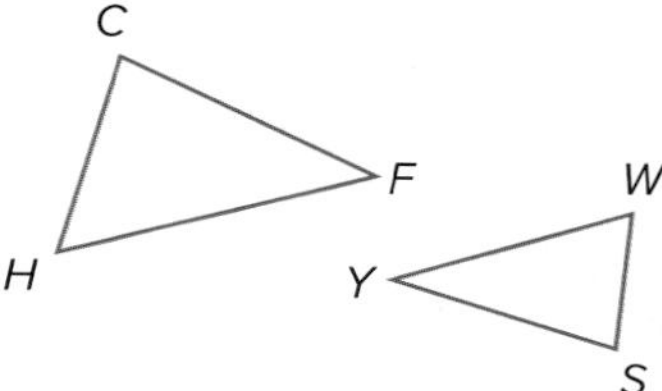

13. $JHFM \sim PQST$

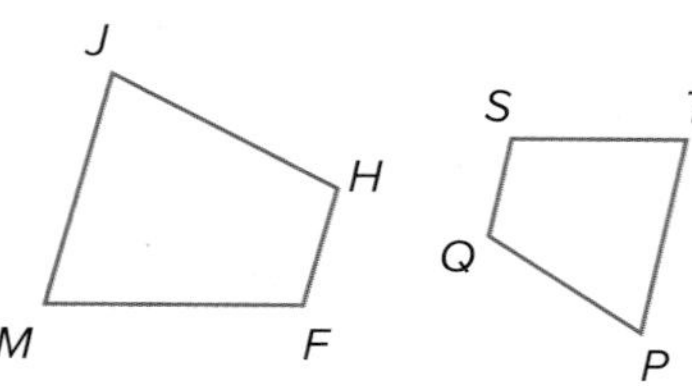

Example 3 **CONSTRUCT ARGUMENTS** **Determine whether each pair of figures is similar. If so, write the similarity statement and scale factor. If not, explain your reasoning.**

14.

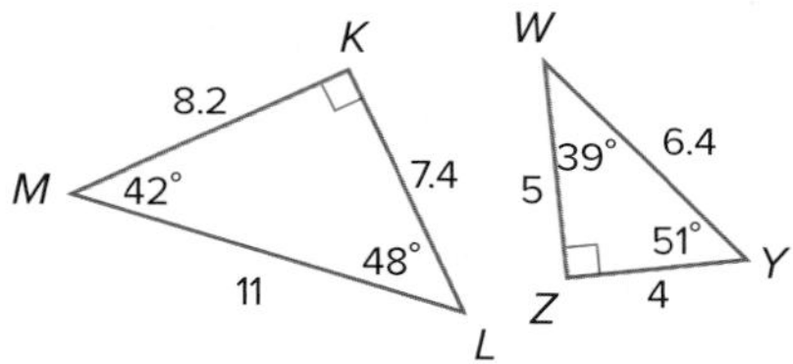

15

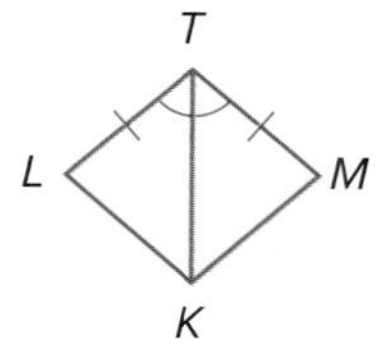

16. **GAMES** The dimensions of a hockey rink are 200 feet by 85 feet. Are the hockey rink and the air hockey table shown similar? Explain your reasoning.

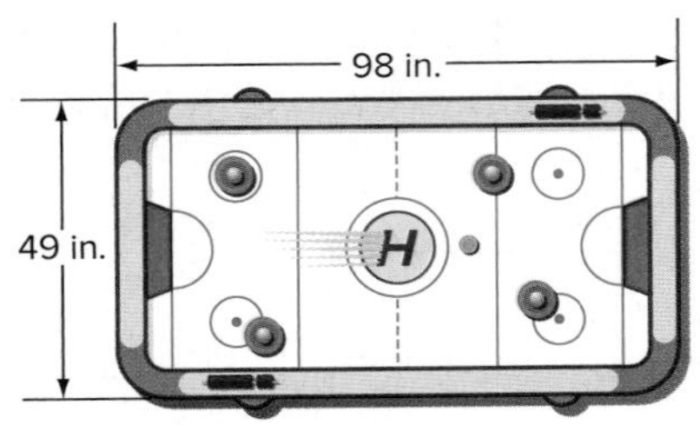

17. **COMPUTERS** The dimensions of a 17-inch flat panel computer screen are approximately $13\frac{1}{4}$ by $10\frac{3}{4}$ inches. The dimensions of a 19-inch flat panel computer screen are approximately $14\frac{1}{2}$ by 12 inches. To the nearest tenth, are the computer screens similar? Explain your reasoning.

Example 4

MP REGULARITY Each pair of polygons is similar. Find the value of x.

18.

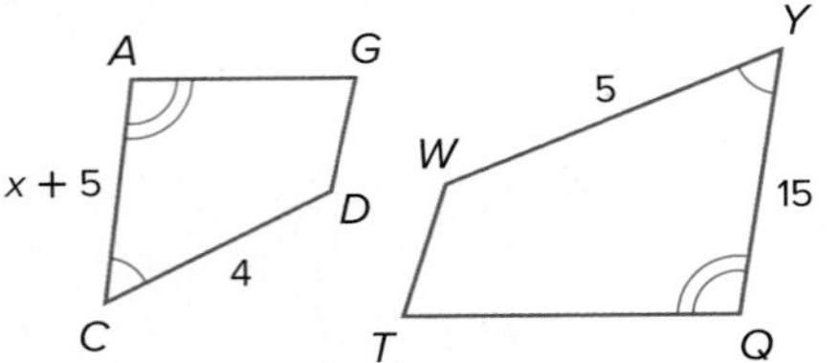

19.

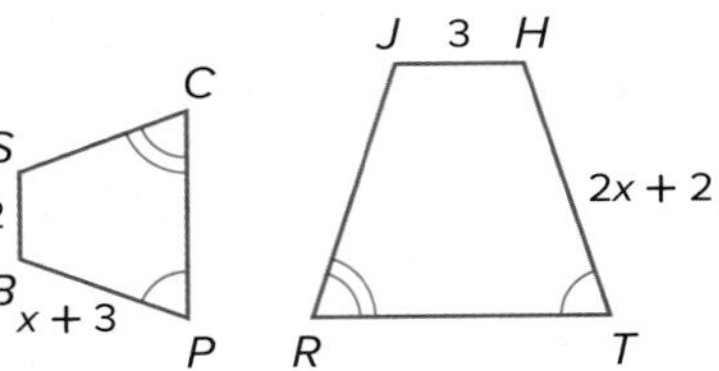

20.

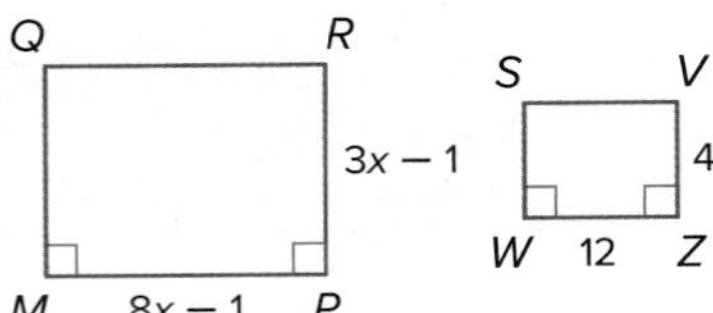

21.

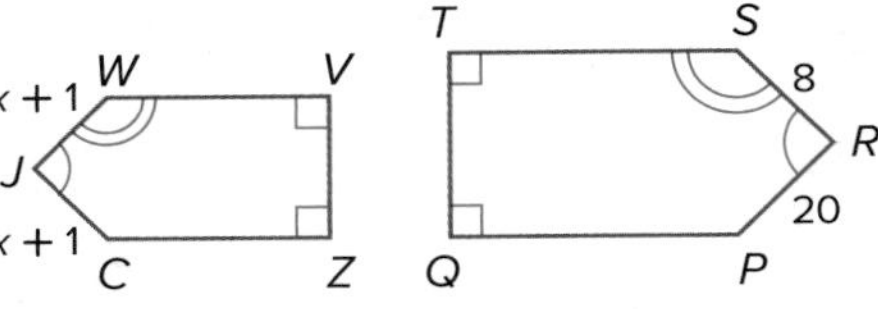

Example 5

22. Rectangle $ABCD$ has a width of 8 yards and a length of 20 yards. Rectangle $QRST$, which is similar to rectangle $ABCD$, has a length of 40 yards. Find the scale factor of rectangle $ABCD$ to rectangle $QRST$ and the perimeter of each rectangle.

Find the perimeter of the given triangle.

23. $\triangle DEF$, if $\triangle ABC \sim \triangle DEF$, $AB = 5$, $BC = 6$, $AC = 7$, and and $DE = 3$

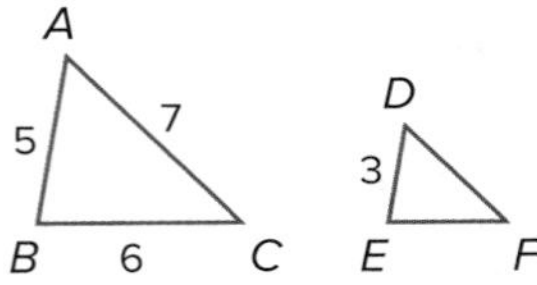

24. $\triangle WZX$, if $\triangle WZX \sim \triangle SRT$, $ST = 6$, $WX = 5$, and the perimeter of $\triangle SRT = 15$

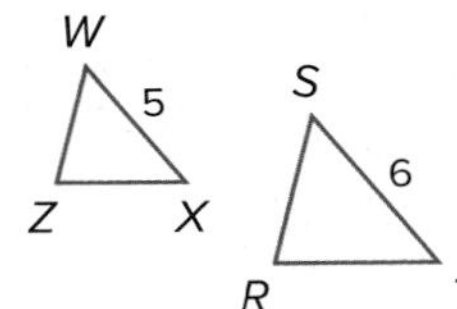

25. $\triangle CBH$, if $\triangle CBH \sim \triangle FEH$, $ADEG$ is a parallelogram, $CH = 7$, $FH = 10$, $FE = 11$, and $EH = 6$

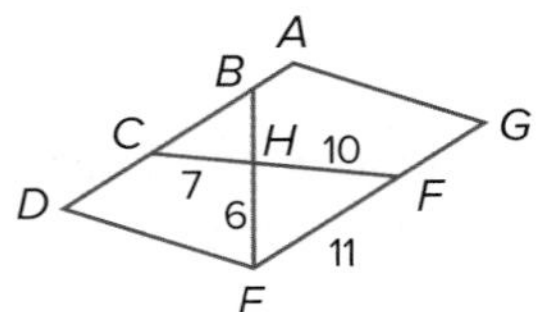

26. $\triangle DEF$, if $\triangle DEF \sim \triangle CBF$, perimeter of $\triangle CBF = 27$, $DF = 6$, $FC = 8$

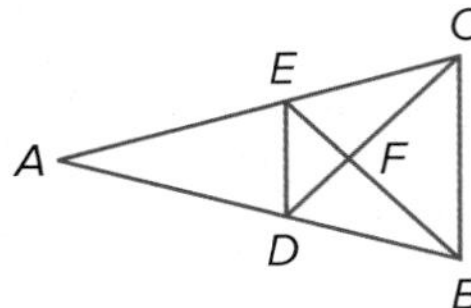

27. Two similar rectangles have a scale factor of 2 : 4. The perimeter of the large rectangle is 80 meters. Find the perimeter of the small rectangle.

28. Two similar rectangles have a scale factor of 3 : 2. The perimeter of the small rectangle is 50 feet. Find the perimeter of the large rectangle.

29. **PROOF** Write a paragraph proof of Theorem 7.1.

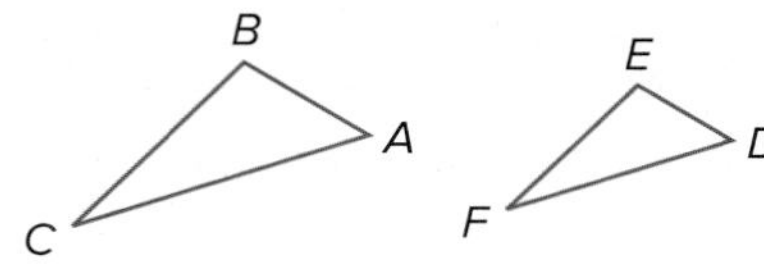

Given: $\triangle ABC \sim \triangle DEF$ and $\frac{AB}{DE} = \frac{m}{n}$

Prove: $\frac{\text{perimeter of } \triangle ABC}{\text{perimeter of } \triangle DEF} = \frac{m}{n}$

30. **PHOTOS** You are enlarging the photo shown at the right for your school yearbook. If the dimensions of the original photo are $2\frac{1}{3}$ inches by $1\frac{2}{3}$ inches and the scale factor of the old photo to the new photo is 2:3, what are the dimensions of the new photo?

31. **CHANGING DIMENSIONS** Rectangle *QRST* is similar to rectangle *JKLM* with sides in a ratio of 4:1.

   a. What is the ratio of the areas of the two rectangles?

   b. Suppose the dimension of each rectangle is tripled. What is the new ratio of the sides of the rectangles?

   c. What is the ratio of the areas of these larger rectangles?

   d. Suppose only one pair of corresponding dimensions of each rectangle is doubled. What is the new ratio of the sides of the rectangles?

32. **CHANGING DIMENSIONS** In the figure shown, $\triangle FGH \sim \triangle XYZ$.

   a. Show that the perimeters of $\triangle FGH$ and $\triangle XYZ$ have the same ratio as their corresponding sides.

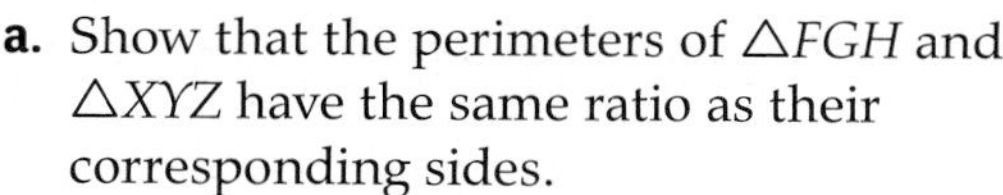

   b. If 6 units are added to the lengths of each side, are the new triangles similar? Explain.

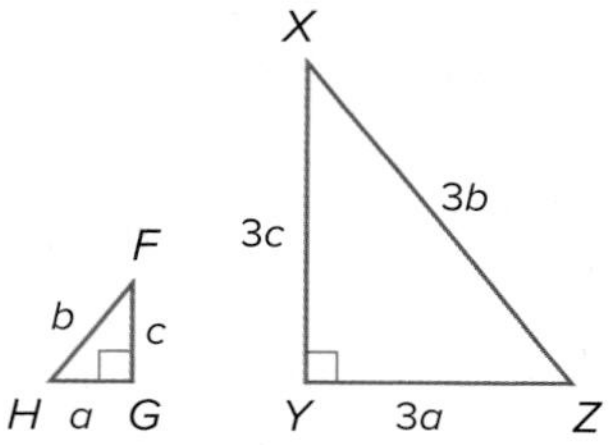

33. **MULTIPLE REPRESENTATIONS** In this problem, you will investigate similarity in squares.

   a. **Geometric** Draw three different-sized squares. Label them *ABCD*, *PQRS*, and *WXYZ*. Measure and label each square with its side length.

   b. **Tabular** Calculate and record in a table the ratios of corresponding sides for each pair of squares: *ABCD* and *PQRS*, *PQRS* and *WXYZ*, and *WXYZ* and *ABCD*. Is each pair of squares similar?

   c. **Verbal** Make a conjecture about the similarity of all squares.

H.O.T. Problems   Use Higher-Order Thinking Skills

34. **CHALLENGE** For what value(s) of *x* is $BEFA \sim EDCB$?

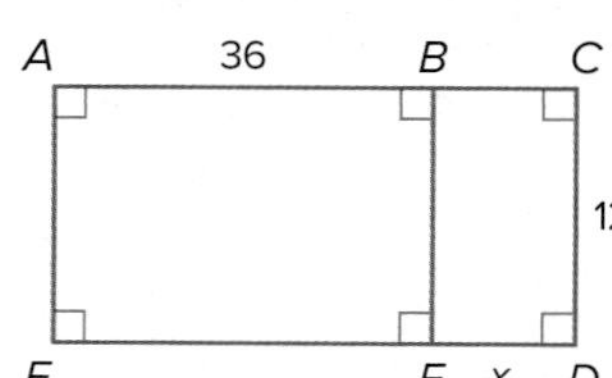

35. **OPEN-ENDED** Find a counterexample for the following statement.

   All rectangles are similar.

36. **MP REASONING** Draw two regular pentagons of different sizes. Are the pentagons similar? Will any two regular polygons with the same number of sides be similar? Explain.

37. **WRITING IN MATH** How can you describe the relationship between two figures?

38. In the figure, $JKLM \sim PQRS$. What is the perimeter of $PQRS$? MP 1

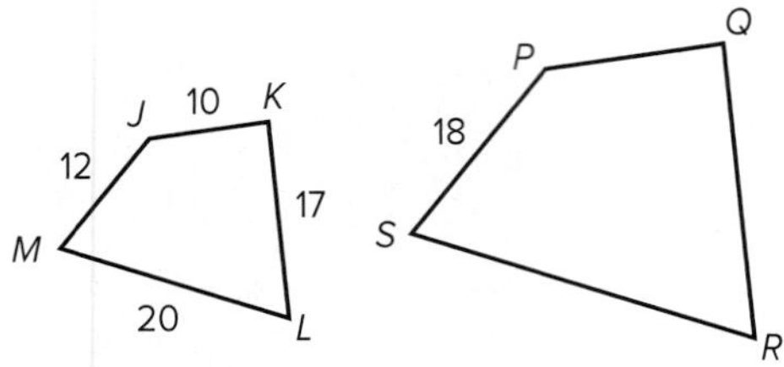

- A 18
- B 39.3
- C 59
- D 88.5

39. Braden drew two rectangles, $RSTU$ and $VWXY$, so that $RSTU \sim VWXY$. The ratio of the perimeter of $RSTU$ to the perimeter of $VWXY$ is $\frac{3}{4}$. Given that the length of $RSTU$ is 24 and the width of $RSTU$ is 12, what is the length of $VWXY$? MP 1

- A 9
- B 16
- C 18
- D 32

40. In the figure, $\triangle ABC \sim \triangle DEF$.

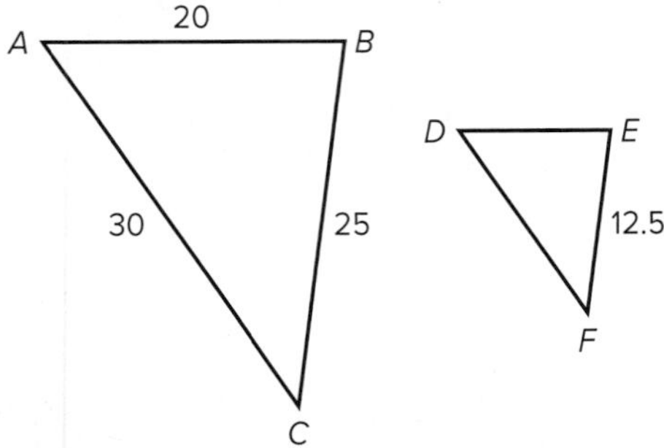

What is the perimeter of $\triangle DEF$? MP 1

perimeter = ☐

41. Two similar rectangles have a scale factor of 3 : 5. The perimeter of the larger rectangle is 65 meters. What is the perimeter in meters of the smaller rectangle? MP 1

perimeter = ☐

42. An architect is designing two triangular support structures for a roof. The support structures must be similar. The architect designs them so that $\triangle KLM \sim \triangle PQR$, as shown. She wants the ratio of the perimeter of $\triangle KLM$ to the perimeter of $\triangle PQR$ to be $\frac{3}{2}$.

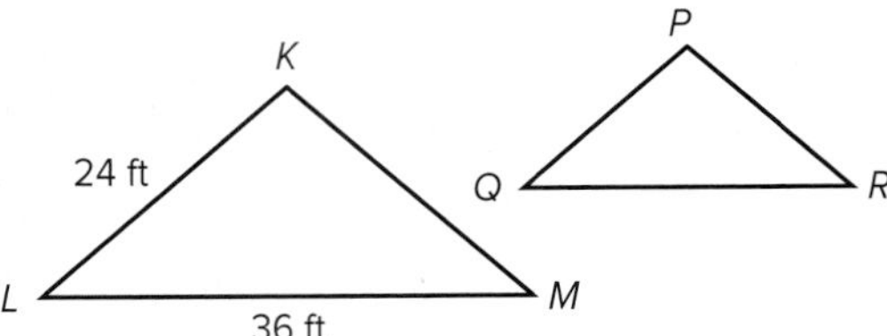

What length should the architect use for $\overline{QR}$? MP 4

- A 54 ft
- B 36 ft
- C 24 ft
- D 16 ft

43. **MULTI-STEP** Yasmina uses a coordinate plane to design jewelry to sell at a craft fair. The figure shows the design for a pair of earrings that will be based on polygon $DEFG$. MP 4

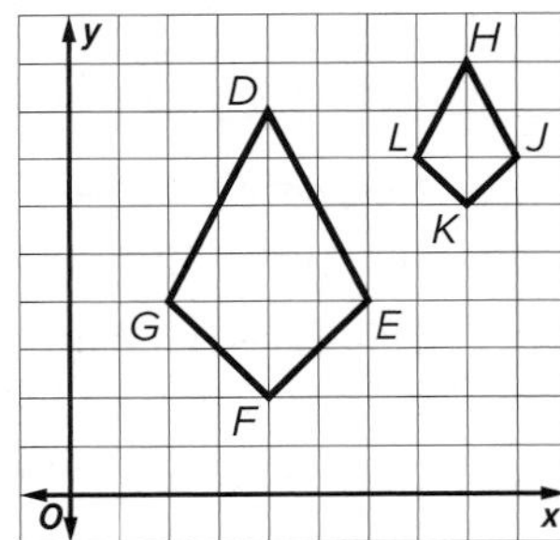

a. Yasmina decides to base the earrings on polygon $HJKL$ instead. Use transformations to explain why polygons $DEFG$ and $HJKL$ are similar.

b. Write a similarity statement for the polygons.

c. Write a proportion that shows how the sides of the polygons are related.

d. Suppose the perimeter of polygon $DEFG$ is 6.2 centimeters. What is the perimeter of polygon $HJKL$? Explain how you know.

LESSON 3

Similar Triangles: AA Similarity

Then	Now	Why?
● You used the properties of similarity to compare polygons.	● 1 Use the AA similarity criterion to prove triangles similar. 2 Solve problems by using the properties of similar triangles.	● Julian wants to draw a similar version of his lacrosse club's logo on a poster. He first draws a line at the bottom of the poster. Next, he uses a cutout of the original triangle to copy the two bottom angles. Finally, he extends the non-common sides of the two angles.

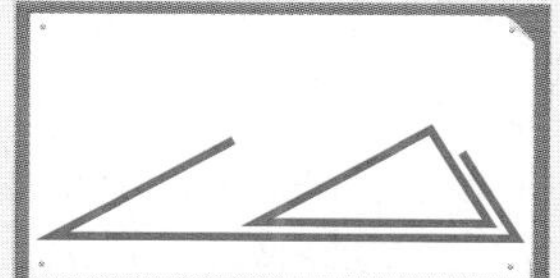

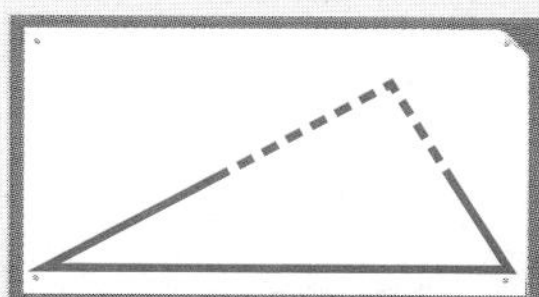

New Vocabulary
similar triangles

Mathematical Practices

3 Construct viable arguments and critique the reasoning of others.

4 Model with mathematics.

1 Use the AA Similarity Criterion

Two triangles are **similar triangles** if there exists a composition of similarity transformations and rigid transformations that maps one triangle onto the other triangle. You can prove that two triangles are similar by showing that all corresponding angles are congruent and all corresponding sides are proportional. There are also shortcuts to proving triangle similarity.

Consider $\triangle ABC$ and $\triangle DEF$ in which $\angle A \cong \angle D$ and $\angle B \cong \angle E$. To prove that the triangles are similar, we must show that there is a similarity transformation or a combination of a similarity transformation and one or more rigid transformations that maps one triangle onto the other.

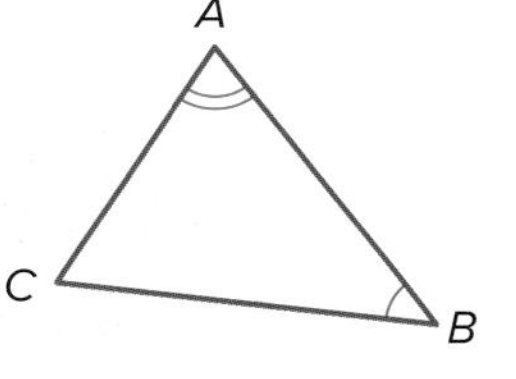

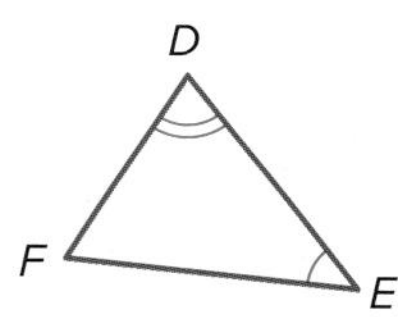

The Third Angles Theorem allows us to conclude that $\angle C \cong \angle F$.

To analyze the sides of the triangles, translate $\triangle DEF$ so that $\angle E'$ coincides with $\angle B$.

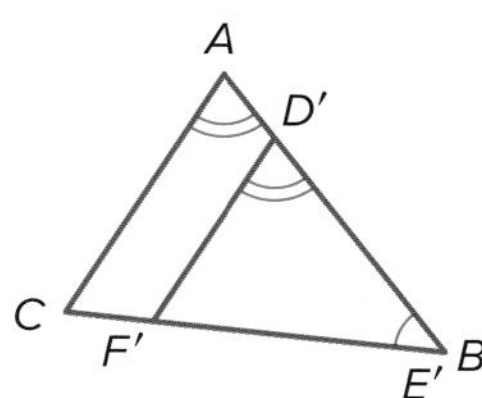

Dilations preserve angle measures, so $\angle D \cong \angle D'$, $\angle E \cong \angle E'$, and $\angle F \cong \angle F'$. By substitution, $\angle A \cong \angle D'$, $\angle B \cong \angle E'$, and $\angle C \cong \angle F'$. Notice that because $\angle B \cong \angle E'$, D' lies on $\overline{AB}$ and F' lies on $\overline{CE}$. Suppose we perform a dilation of $\triangle D'E'F'$ with respect to the center B. We would like to move D' to A. The scale factor which will accomplish this is $\frac{AB}{D'B}$.

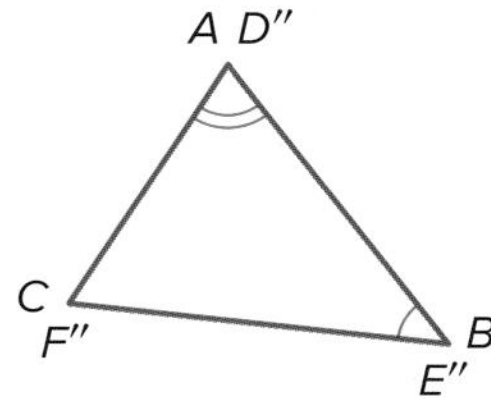

Go Online!

How do you know whether two triangles are similar? Investigate by using the **Geometry Tools** in ConnectED.

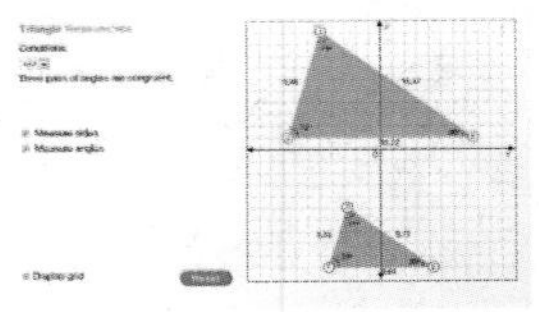

To check that F maps to C, recall that $\angle A \cong \angle D$. Angles are preserved by dilations and so this means that $\overline{D'F'}$ must map to $\overline{AC}$. This is proved by contradiction. If the dilation mapped F' to a point other than C, then there would be segments from A to two different points on $\overleftrightarrow{BC}$ that form the same angle with $\overline{AB}$. This is not possible, so $\overline{D'F'}$ must map to $\overline{AC}$.

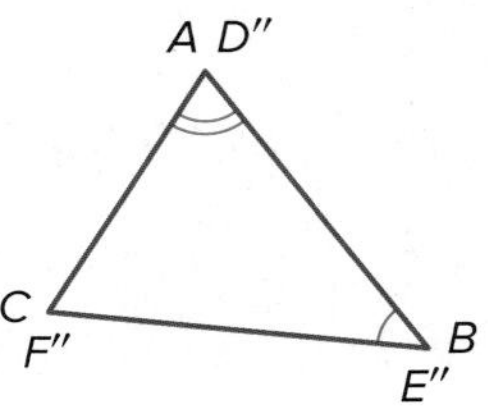

Because we have found a combination of a similarity transformation and a rigid transformation that maps $\triangle DEF$ onto $\triangle ABC$, the two triangles are similar.

This demonstrates the Angle-Angle Similarity Postulate.

Postulate 7.1 Angle-Angle (AA) Similarity

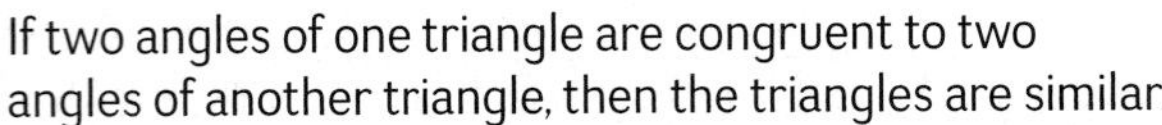

If two angles of one triangle are congruent to two angles of another triangle, then the triangles are similar.

Example If $\angle A \cong \angle F$ and $\angle B \cong \angle G$, then $\triangle ABC \sim \triangle FGH$.

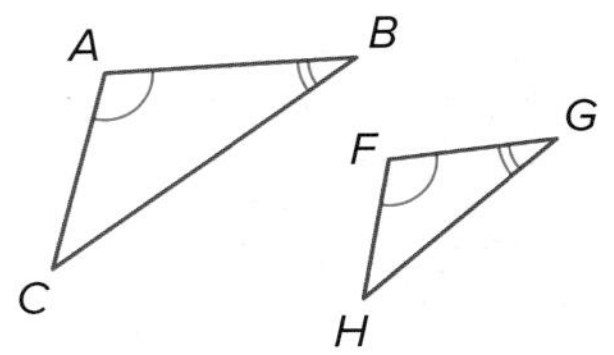

You can use Angle-Angle Similarity to determine that two triangles are similar. To determine which angles of two triangles correspond, begin by comparing the angles with the greatest measures, then the angles with the next greatest measures, and finish by comparing the angles with the least measures.

Example 1 Use the AA Similarity Postulate

Determine whether the triangles are similar. If so, write a similarity statement. Explain your reasoning.

a.

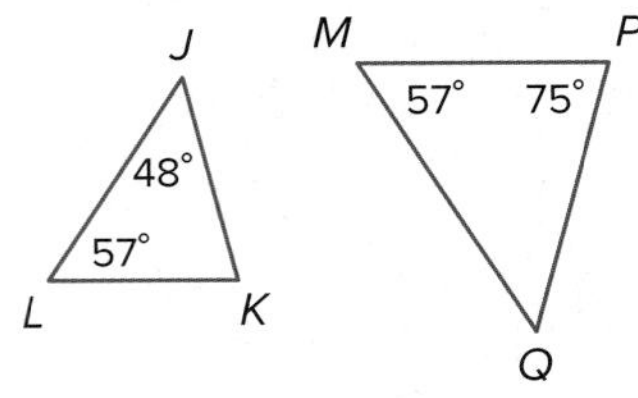

Since $m\angle L = m\angle M$, $\angle L \cong \angle M$. By the Triangle Sum Theorem, $57 + 48 + m\angle K = 180$, so $m\angle K = 75$. Since $m\angle P = 75$, $\angle K \cong \angle P$. So, $\triangle LJK \sim \triangle MQP$ by AA Similarity.

b.

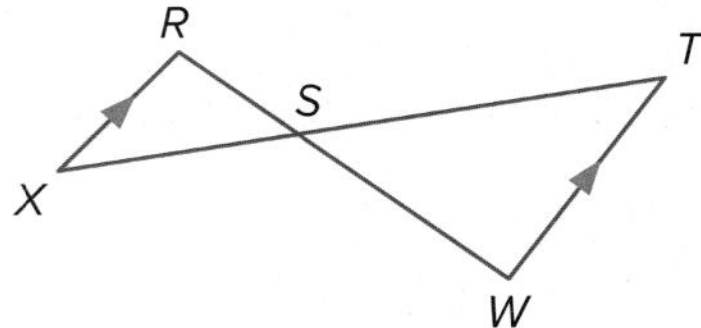

$\angle RSX \cong \angle WST$ by the Vertical Angles Theorem. Since $\overline{RX} \parallel \overline{TW}$, $\angle R \cong \angle W$. So, $\triangle RSX \sim \triangle WST$ by AA Similarity.

> **Study Tip**
> **MP Sense-Making** It is helpful to redraw similar triangles so that the corresponding side lengths have the same orientation.

c.

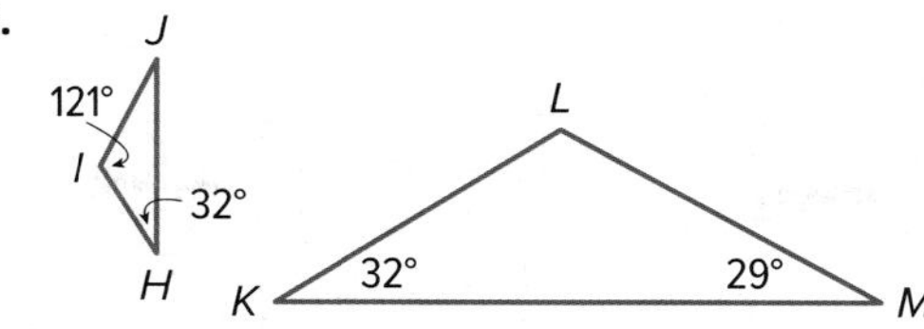

$180° - (121° + 32°) = 27^0$

Two angles not are congruent.

So, $\triangle HIJ \nsim \triangle KLM$.

d.

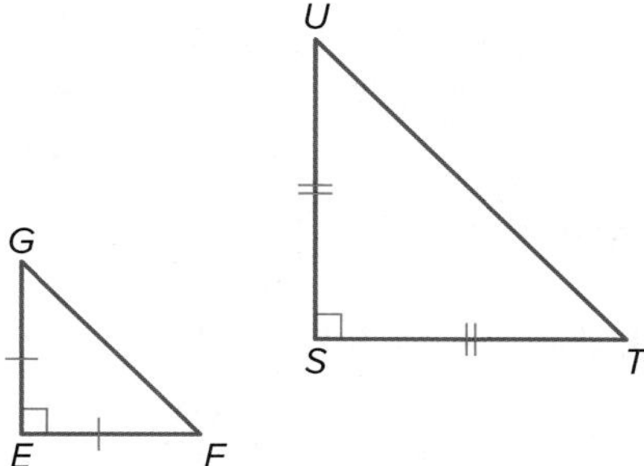

By the Isosceles Triangle Theorem, $\angle F \cong \angle G$.

$\frac{180° - 90°}{2} = 45°$.

Because both of these triangles are Isosceles Right triangles, we know that the unknown angles are 45°. Two pairs of angles are congruent. $\triangle EFG \sim \triangle STU$ by the AA Similarity Theorem.

1A.

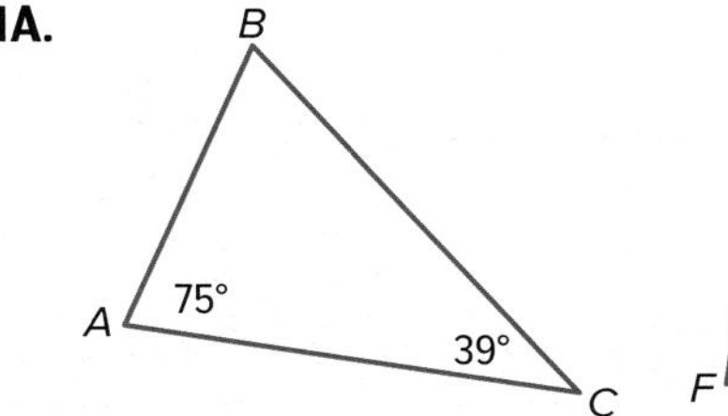

1B.

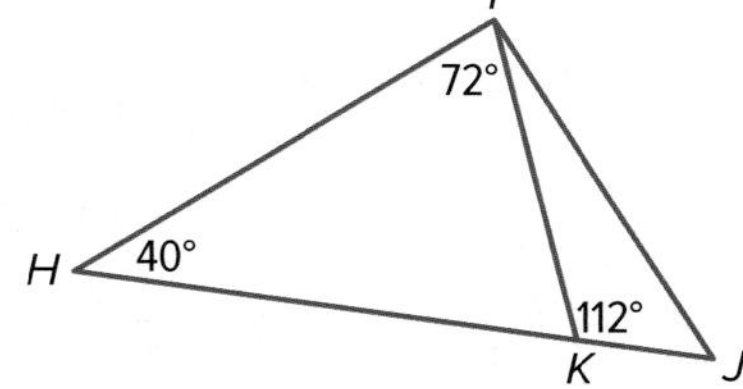

You can decide whether the information you have is sufficient to prove that two triangles are similar.

Example 2 Identify Sufficient Conditions

> **Study Tip**
> **Identifying Nonexamples** Sometimes test questions require you to find a nonexample, as in this case. You must check each option until you find a valid nonexample. If you would like to check your answer, confirm that each additional option is correct.

In the figure, $\angle LNM$ is a right angle. Which of the following would not be sufficient to prove that $\triangle LNM \sim \triangle MNO$?

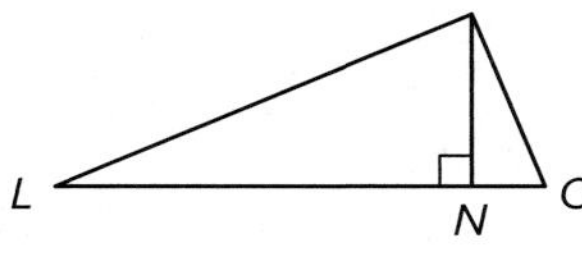

A $\angle LMN \cong \angle MON$

B $\angle LNM \cong \angle MNO$

C $m\angle MLN = 28°$and $m\angle OMN = 28°$

D $m\angle LMN = 52°$and $m\angle MON = 52°$

$\angle LNM$ and$\angle MNO$ form a straight angle, and $\angle LNM$ is a right angle, so $\angle MNO$ is a also a right angle. All right angles are congruent, so $\angle LNM \cong \angle MNO$. Check each answer choice until you find one that does not supply a sufficient additional condition to prove that $\triangle LNM \sim \triangle MNO$.

Choices A, C, and D: You only need to identify one more pair of congruent angles to use AA. Each of these choices has at least one more congruent pair of angles.

Choice B: This choice restates the pair of right angles you know are congruent. It does not give sufficient information to prove the triangles are similar.

Guided Practice

2. If $\triangle JKL$ and $\triangle FGH$ are two triangles such that $\angle J \cong \angle F$, which of the following would be sufficient information to prove that the triangles are congruent?

F $\angle G \cong \angle F$ **G** $\angle G \cong \angle L$ **H** $\angle G \cong \angle K$ **J** $\angle G \cong \angle F$

2 Use Similar Triangles

Use Similar Triangles Like the congruence of triangles, similarity of triangles is reflexive, symmetric, and transitive.

Theorem 7.2 Properties of Similarity

Reflexive Property of Similarity	$\triangle ABC \sim \triangle ABC$
Symmetric Property of Similarity	If $\triangle ABC \sim \triangle DEF$, then $\triangle DEF \sim \triangle ABC$.
Transitive Property of Similarity	If $\triangle ABC \sim \triangle DEF$, and $\triangle DEF \sim \triangle XYZ$, then $\triangle ABC \sim \triangle XYZ$.

You will prove Theorem 7.2 in Exercise 25.

Example 3 Parts of Similar Triangles

Find *BE* and *AD*.

Since $\overline{BE} \parallel \overline{CD}$, $\angle ABE$ $\angle BCD$, and $\angle AEB \cong \angle EDC$ because they are corresponding angles. By AA Similarity, $\triangle ABE \sim \triangle ACD$.

Study Tip

Proportions An additional proportion that is true for Example 3 is $\frac{AC}{CD} = \frac{AB}{BE}$.

$\frac{AB}{AC} = \frac{BE}{CD}$	Definition of Similar Polygons
$\frac{3}{5} = \frac{x}{3.5}$	$AC = 5$, $CD = 3.5$, $AB = 3$, $BE = x$
$3.5 \cdot 3 = 5 \cdot x$	Cross Products Property
$2.1 = x$	

BE is 2.1.

$\frac{AC}{AB} = \frac{AD}{AE}$	Definition of Similar Polygons
$\frac{5}{3} = \frac{y+3}{y}$	$AC = 5$, $AB = 3$, $AD = y + 3$, $AE = y$
$5 \cdot y = 3(y + 3)$	Cross Products Property
$5y = 3y + 9$	Distributive Property
$2y = 9$	Subtract $3y$ from each side.
$y = 4.5$	AD is $y + 3$ or 7.5.

Guided Practice

Find each measure.

3A. *QP* and *MP*

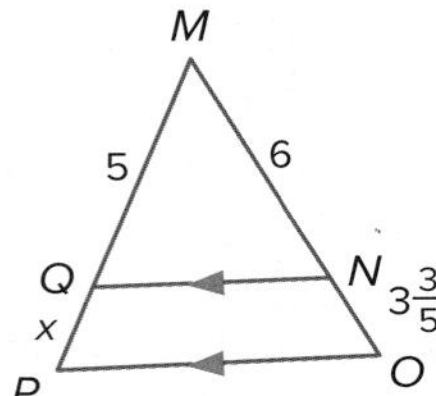

3B. *WR* and *RT*

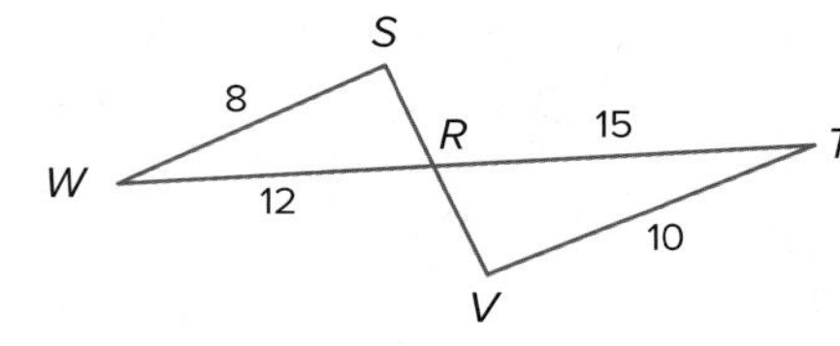

When you can represent real-world situations with similar triangles, you can use proportions and the properties of the triangles to find measurements. This is called indirect measurement.

Real-World Example 4 Indirect Measurement

ROLLER COASTERS Hallie is estimating the height of the Superman roller coaster in Mitchellville, Maryland. She is 5 feet 3 inches tall and her shadow is 3 feet long. If the length of the shadow of the roller coaster is 40 feet, how tall is the roller coaster?

Understand Make a sketch of the situation. Hallie's height of 5 feet 3 inches is equivalent to 5.25 feet.

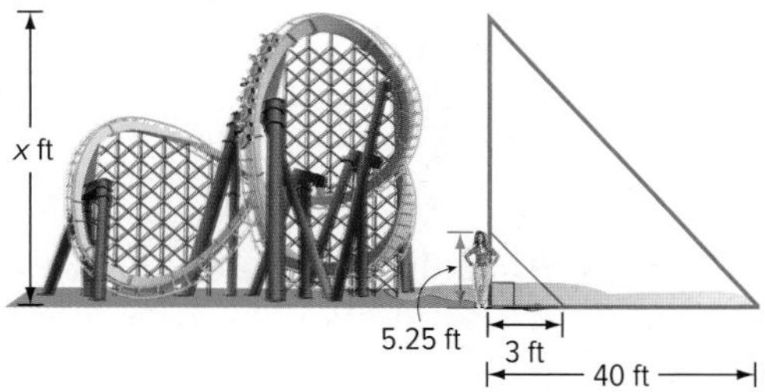

Plan In shadow problems, you can assume that the angles formed by the Sun's rays with any two objects are congruent and that the two objects form the sides of two right triangles.

Because two pairs of angles are congruent, the right triangles are similar by the AA Similarity Postulate. Write a proportion.

$$\frac{\text{Hallie's height}}{\text{coaster's height}} = \frac{\text{Hallie's shadow length}}{\text{coaster's shadow length}}$$

Solve Using the proportion, substitute the known values and let x represent roller coaster's height.

$\frac{5.25}{x} = \frac{3}{40}$ Substitution

$3 \cdot x = 40(5.25)$ Cross Products Property

$3x = 210$ Simplify.

$x = 70$ Divide each side by 3.

The roller coaster is 70 feet tall.

Check The roller coaster's shadow length is $\frac{40 \text{ ft}}{3 \text{ ft}}$ or about 13.3 times Hallie's shadow length. Check to see that the roller coaster's height is about 13.3 times Hallie's height. $\frac{70 \text{ ft}}{5.25 \text{ ft}} \approx 13.3$ ✓

Problem-Solving Tip

Reasonable Answers When you have solved a problem, check your answer for reasonableness. In this example, Hallie's shadow is a little more than half her height. The coaster's shadow is also a little more than half of the height. Therefore, the answer is reasonable.

Guided Practice

4. BUILDINGS Adam is standing next to the Palmetto Building in Columbia, South Carolina. He is 6 feet tall and the length of his shadow is 9 feet. If the length of the shadow of the building is 322.5 feet, how tall is the building?

Check Your Understanding

○ = Step-by-Step Solutions begin on page R13.

Go Online! for a Self-Check Quiz

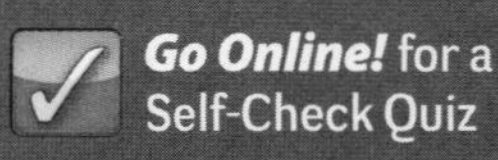

Examples 1 **Determine whether the triangles are similar, using the AA Similarity Theorem. If so, write a similarity statement.**

1.

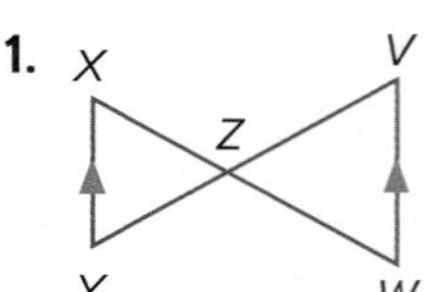

2.

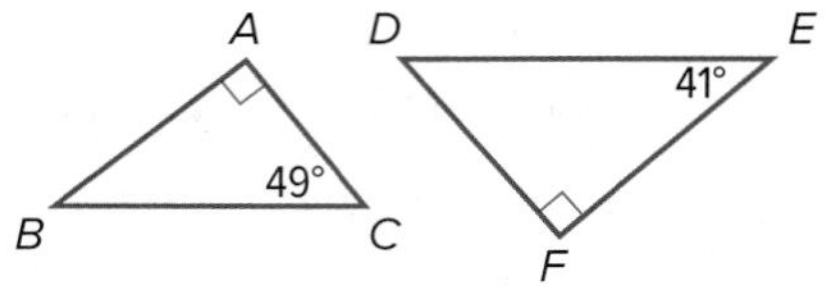

3.

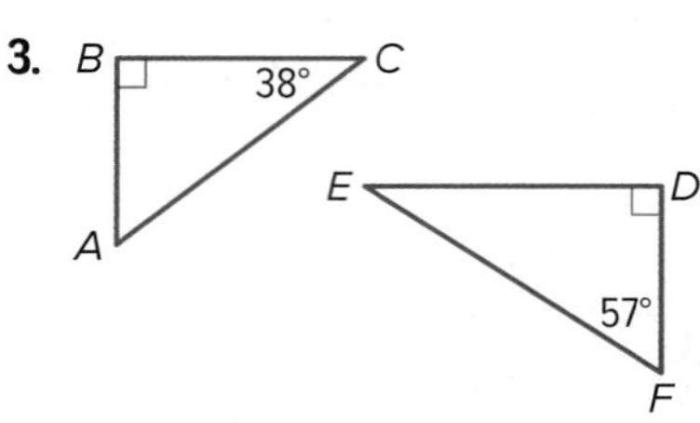

4.

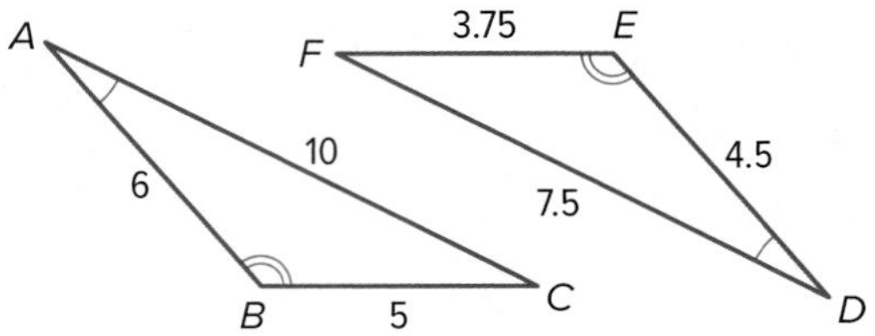

Example 2 **5. MULTIPLE CHOICE** In the figure, $\overline{AB}$ intersects $\overline{DE}$ at point C. Which additional information would be enough to prove that $\triangle ADC \sim \triangle BEC$?

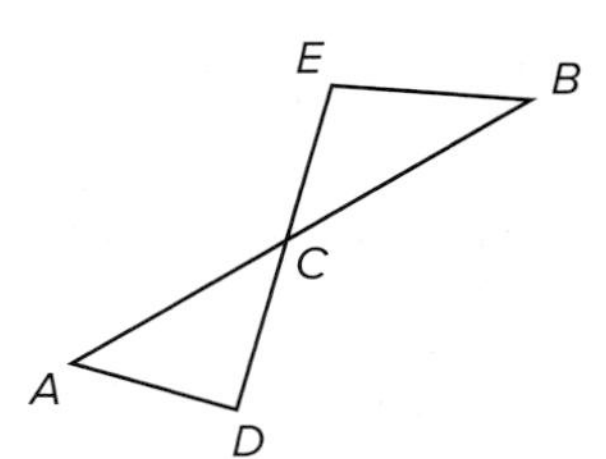

A $\angle DAC$ and $\angle ECB$ are congruent.

B $\overline{AC}$ and $\overline{BC}$ are congruent.

C $\overline{AD}$ and $\overline{EB}$ are parallel.

D $\angle CBE$ is a right angle.

Example 3 **MP STRUCTURE Identify the similar triangles. Find each measure.**

6. XZ

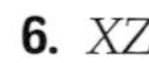

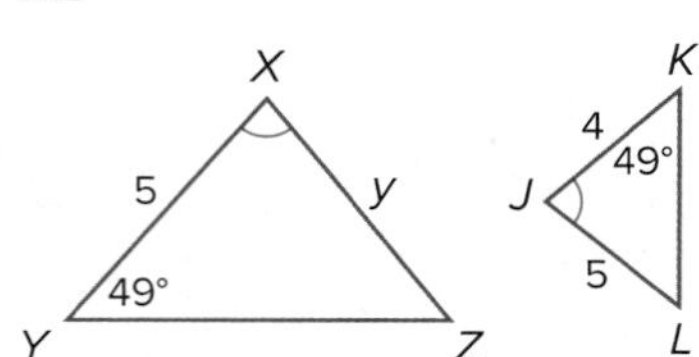

7. VS

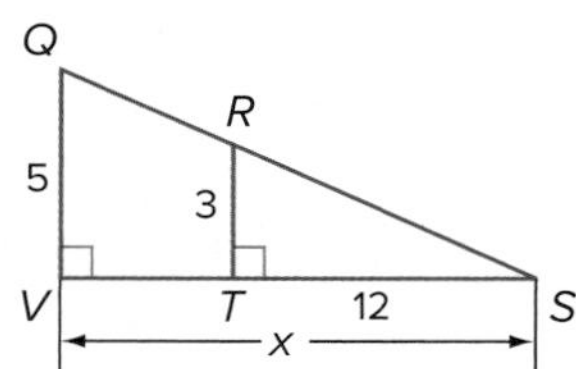

Example 4 **8. COMMUNICATION** A cell phone tower casts a 100-foot shadow. At the same time, a 4-foot, 6-inch post near the tower casts a shadow of 3 feet 4 inches. Find the height of the tower.

Practice and Problem Solving

Extra Practice is on page R7.

Examples 1, 2 **Determine whether the triangles are similar. If so, write a similarity statement. Explain your reasoning.**

9. $\triangle ACE$, $\triangle BCD$

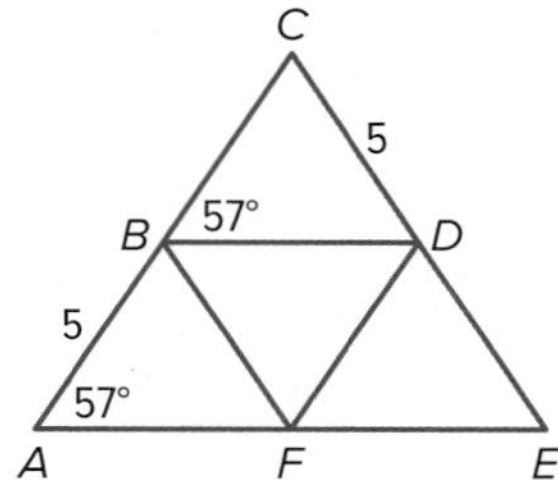

10.

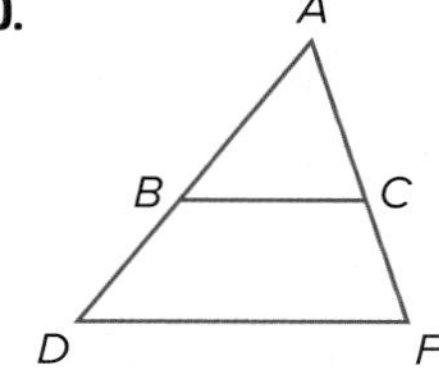

11

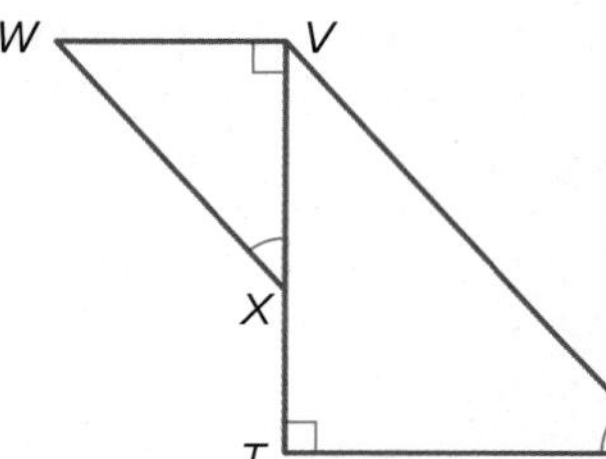

Examples 1–3 **Determine whether the triangles are similar. If so, write a similarity statement. Explain your reasoning.**

12.

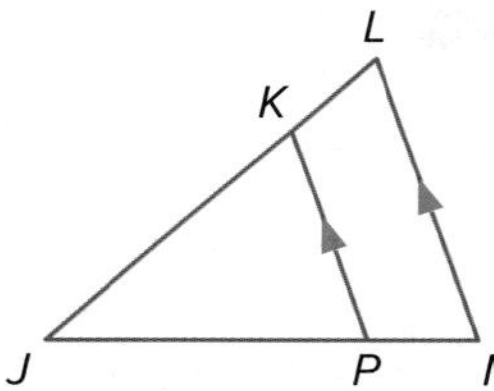

13.

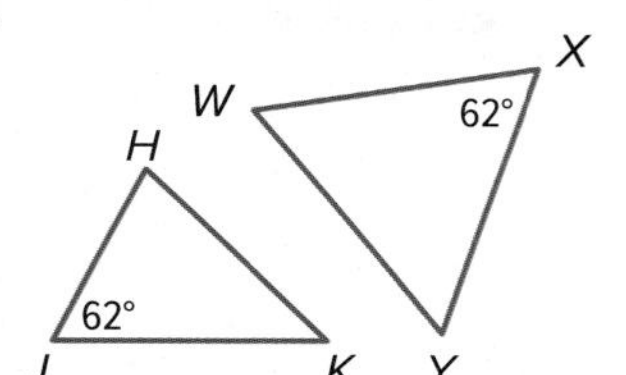

14.

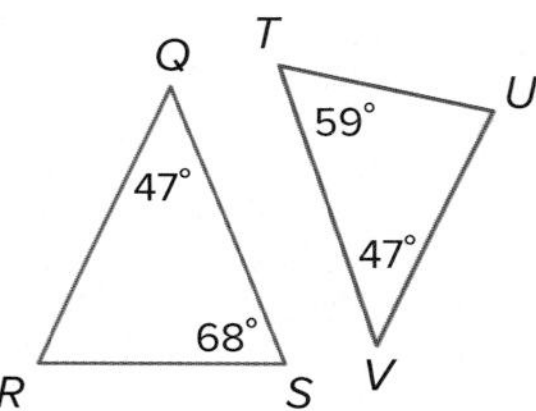

15. **MODELING** Scientists are studying eyes and how light is refracted by the lens of the eyes to create two similar triangles. Prove that the triangles are similar using similarity transformations and the relationship between the angles and sides of the triangles.

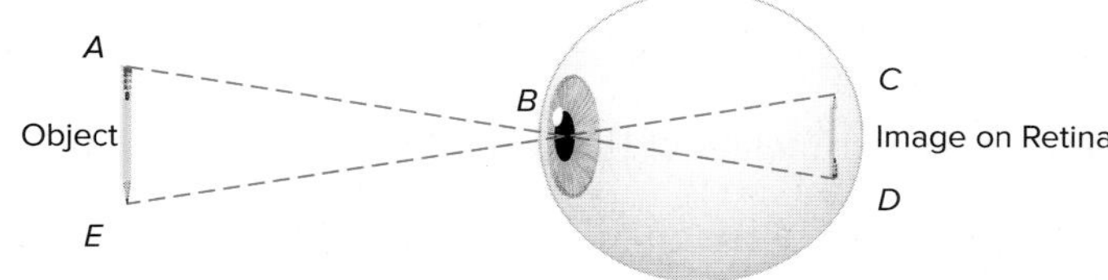

Example 4 **ALGEBRA Identify the similar triangles. Then find each measure.**

16. EG

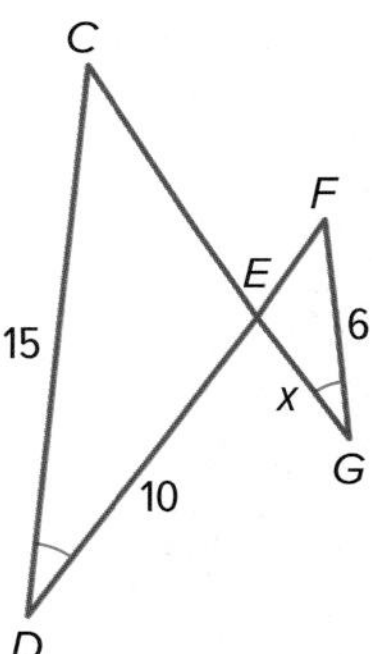

17. ST

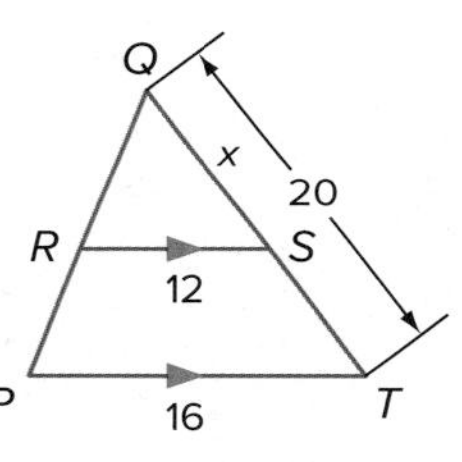

18. WZ, UZ

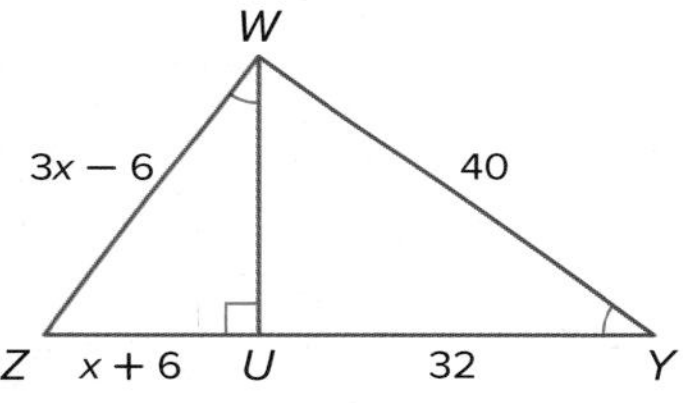

19. HJ, HK

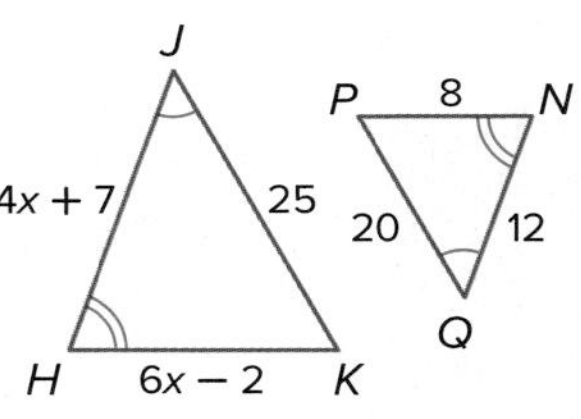

20. EB

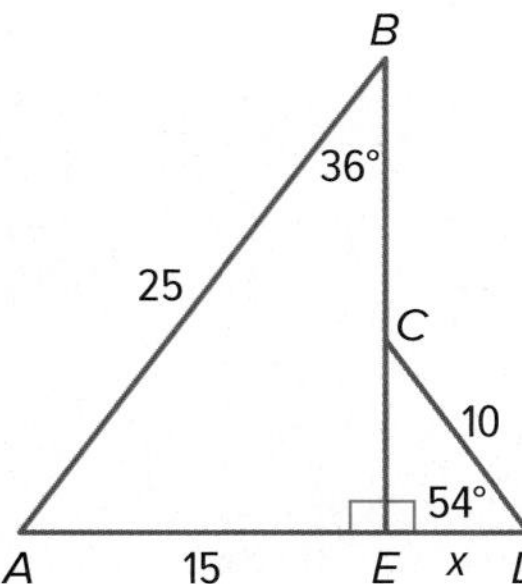

21. GD, DH

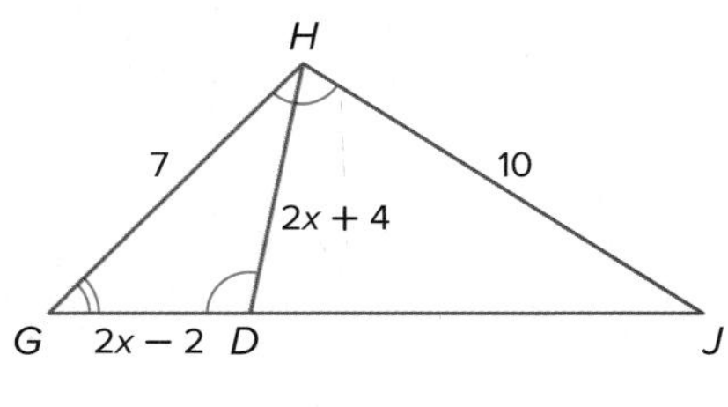

Example 5 **22.** **STATUES** Mei is standing next to a statue in the park. If Mei is 5 feet tall, her shadow is 3 feet long, and the statue's shadow is $10\frac{1}{2}$ feet long, how tall is the statue?

23. **SPORTS** When Alonzo, who is 5′11″ tall, stands next to a basketball goal, his shadow is 2′ long, and the basketball goal's shadow is 4′4″ long. About how tall is the basketball goal?

24. **BIRDWATCHING** Taylor sees the nest of a rare bird near the top of a tree. He wants to report its position to the local conservation group. Taylor is 6 feet tall and casts a 3.5-foot shadow. The tree with the nest casts a shadow 6 feet long. About how far above the ground is the nest?

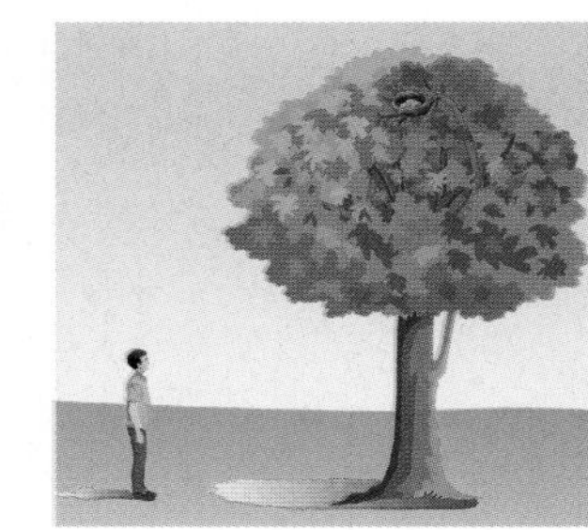

25. **PROOF** Write a two-column proof for each part of Theorem 7.2.

PROOF **Write a two-column proof.**

26. Given: *ABCD* is a trapezoid.

Prove: $\frac{DP}{PB} = \frac{CP}{PA}$

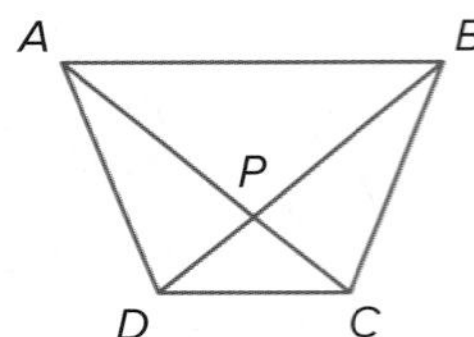

27. Given: $\triangle XYZ$ and $\triangle ABC$ are right triangles; $\angle Z \cong \angle C$.

Prove: $\triangle YXZ \sim \triangle BAC$

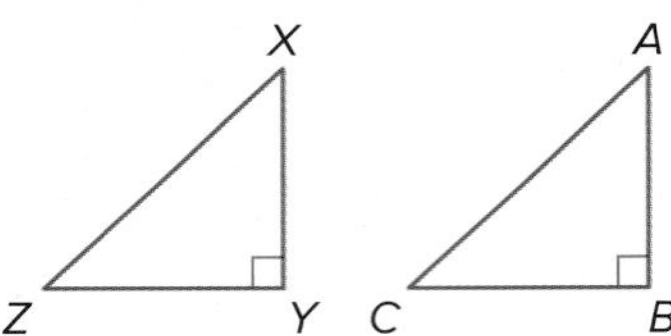

COORDINATE GEOMETRY $\triangle XYZ$ **and** $\triangle WYV$ **have vertices** $X(-1, -9)$, $Y(5, 3)$, $Z(-1, 6)$, $W(1, -5)$, **and** $V(1, 5)$.

28. Graph the triangles, and prove that $\triangle XYZ \sim \triangle WYV$.

29. Find the ratio of the perimeters of the two triangles.

30. **MP MODELING** When Luis's dad threw a bounce pass to him, the angles formed by the basketball's path were congruent. The ball landed $\frac{2}{3}$ of the way between them before it bounced back up. If Luis's dad released the ball 40 inches above the floor, at what height did Luis catch the ball?

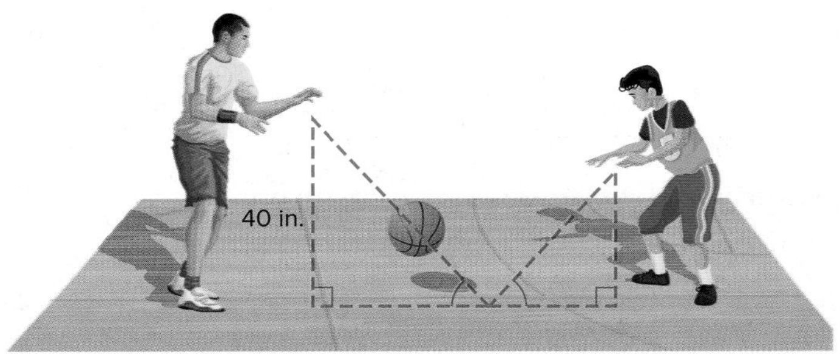

31. **BILLIARDS** When a ball is deflected off a smooth surface, the angles formed by the path are congruent. Booker hit the orange ball and it followed the path from *A* to *B* to *C* as shown below. What was the total distance traveled by the ball from the time Booker hit it until it came to rest at the end of the table?

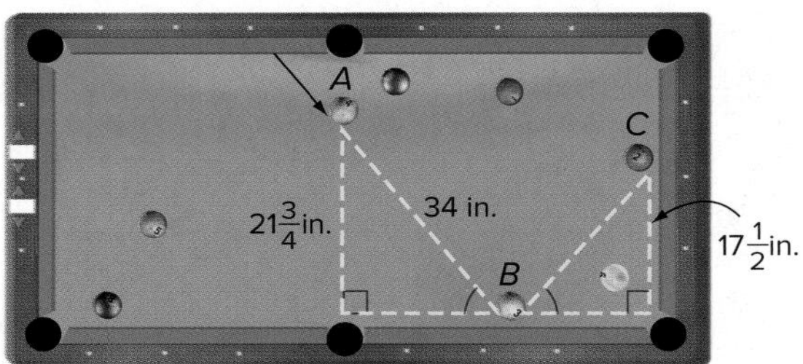

32. If possible, describe a similarity transformation or a combination of a similarity transformation and a rigid transformation that could be used to demonstrate that the triangles are similar.

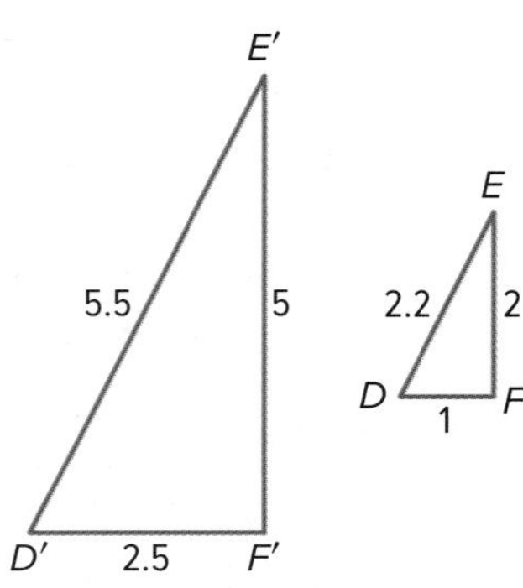

33. CHANGING DIMENSIONS Assume that $\triangle ABC \sim \triangle JKL$.

a. If the lengths of the sides of $\triangle JKL$ are half the lengths of the corresponding sides of $\triangle ABC$, and the perimeter of $\triangle ABC$ is 40 inches, what is the perimeter of $\triangle JKL$? How is the perimeter related to the scale factor from $\triangle ABC$ to $\triangle JKL$?

b. If the lengths of the sides of $\triangle ABC$ are three times the lengths of the corresponding sides of $\triangle JKL$, and the perimeter of $\triangle ABC$ is 21 inches, what is the perimeter of $\triangle JKL$? How is the perimeter related to the scale factor from $\triangle ABC$ to $\triangle JKL$?

34. MEDICINE Certain medical treatments involve laser beams that contact and penetrate the skin. Refer to the diagram at the right.

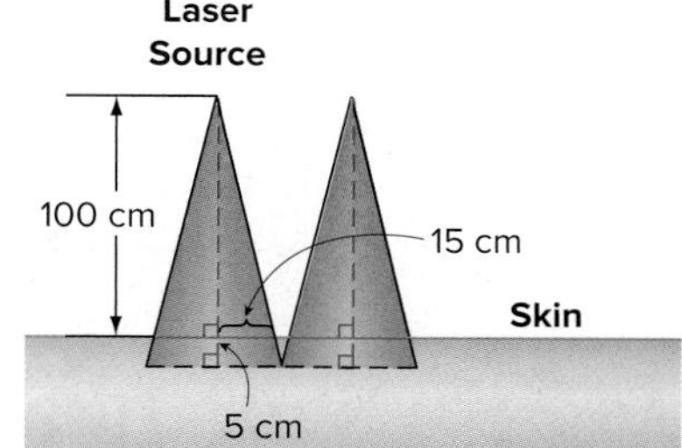

a. What needs to be true for the triangles formed by the edges of the laser beams and the surface of the skin to be similar to triangles formed by the edges of the laser beams and the treated area beneath the skin?

b. What needs to be true for the triangles formed by the edges of the two laser beams and the surface of the skin to be similar to each other?

c. If all of the triangles are similar, how far apart should the laser sources be placed to ensure that the areas treated by each source do not overlap or leave an area untreated?

35. MULTIPLE REPRESENTATIONS In this problem, you will explore proportional parts of triangles.

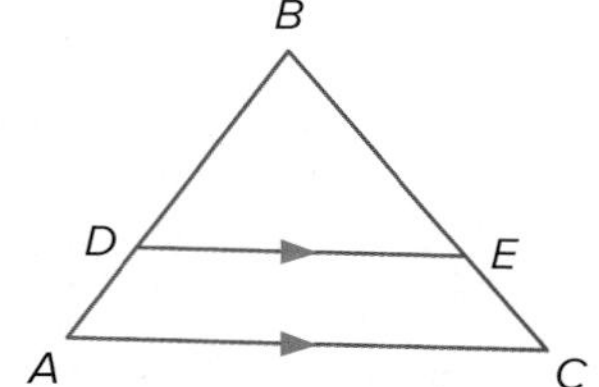

a. **Geometric** Draw $\triangle ABC$ with $\overline{DE}$ parallel to $\overline{AC}$ as shown at the right.

b. **Tabular** Measure and record the lengths AD, DB, CE, and EB and the ratios $\frac{AD}{DB}$ and $\frac{CE}{EB}$ in a table.

c. **Verbal** Make a conjecture about the segments created by a line parallel to one side of a triangle and intersecting the other two sides.

H.O.T. Problems Use Higher-Order Thinking Skills

36. WRITING IN MATH Explain why the AA Postulate requires that only two pairs of corresponding angles rather than three are shown to be congruent in order to prove that two triangles are similar.

37. CHALLENGE $\overline{YW}$ is an altitude of $\triangle XYZ$. Can you prove that $\triangle WXY$ and $\triangle WYZ$ are similar triangles? Explain.

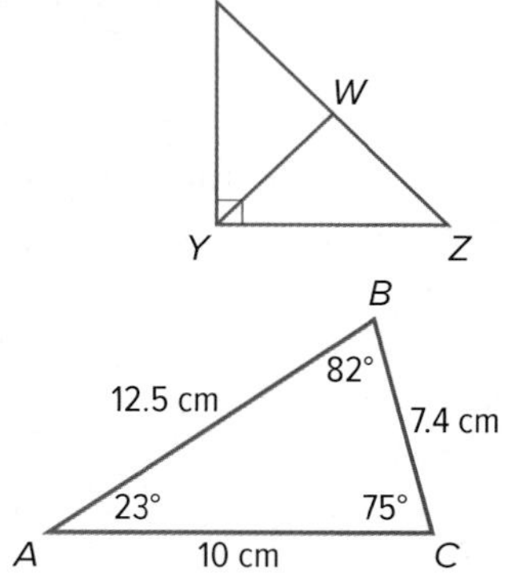

38. MP **REASONING** In a pair of similar triangles, the sides of one triangle measure 3, 3.25, and 4.75 4.5 units, The sides of the second triangle measure $x - 0.46$, x, and $x + 2.76$ units. Find the value of x.

39. OPEN-ENDED Draw a triangle that is similar to $\triangle ABC$ shown. Explain how you know your triangle is similar to $\triangle ABC$.

40. The ladder of a slide is perpendicular to the ground. The slide slopes downward at an angle of 58 degrees. What angles cannot be used to prove that $\triangle JKL \sim \triangle JMN$ by AA Similarity? MP 4

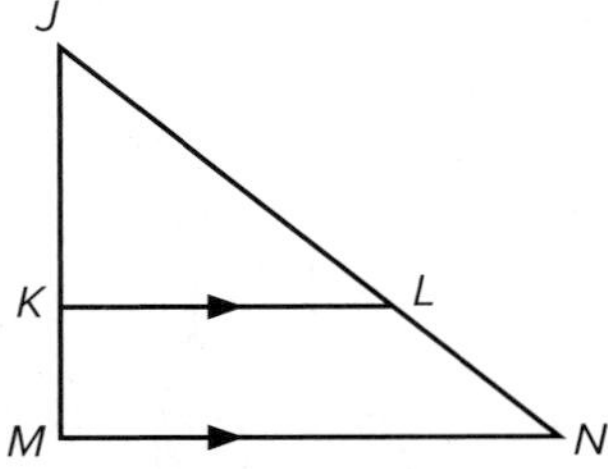

- A $\angle KJL \cong \angle MJN$ and $\angle JKL \cong \angle JMN$
- B $\angle KLJ \cong \angle MNJ$ and $\angle JKL \cong \angle JMN$
- C $\angle KLJ \cong \angle NLK$ and $\angle JKL \cong \angle MKL$
- D $\angle KLJ \cong \angle MNJ$ and $\angle KJL \cong \angle MJN$

41. $\triangle JKL \sim \triangle PQR$, the measure of angle P is 34°, and the measure of angle R is 72°. The measure of angle K is given by the expression $(7x + 32)°$. What is the value of x? MP 1

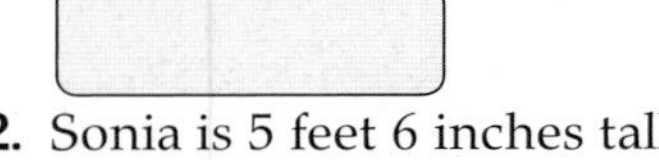

42. Sonia is 5 feet 6 inches tall. She wants to estimate the height of a palm tree near her school. She measures her shadow and the palm tree's shadow and then sets up the sketch shown below.

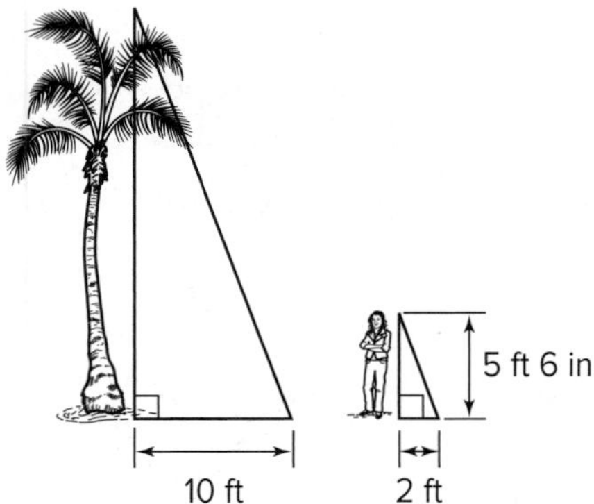

Which of the following is the best estimate of the height of the palm tree? MP 4

- A 28 ft
- B 27.5 ft
- C 13.5 ft
- D 3.6 ft

43. In the figure below, $\overline{EB} \parallel \overline{DC}$. MP 3

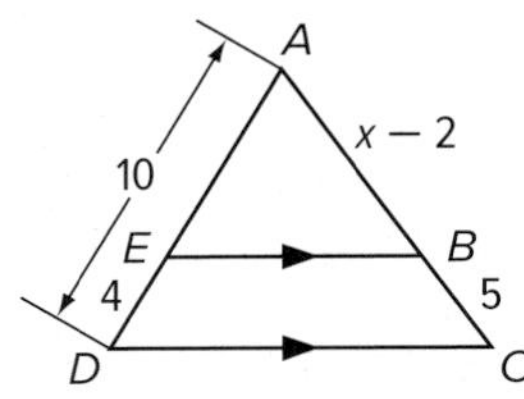

a. Which proportion correctly represents the situation?

- A $\frac{x-2}{6} = \frac{4}{10}$
- B $\frac{6}{x-2} = \frac{4}{10}$
- C $\frac{x-2}{6} = \frac{4}{5}$
- D $\frac{6}{x-2} = \frac{4}{5}$

b. Find the value of x and the measure of $\overline{AB}$.

- A $x = 4.4$, $AB = 2.4$
- B $x = 6.8$, $AB = 4.8$
- C $x = 9.5$, $AB = 7.5$
- D $x = 17$, $AB = 15$

44. MULTI-STEP These two triangles are similar and $\angle F \cong \angle R$. MP 3

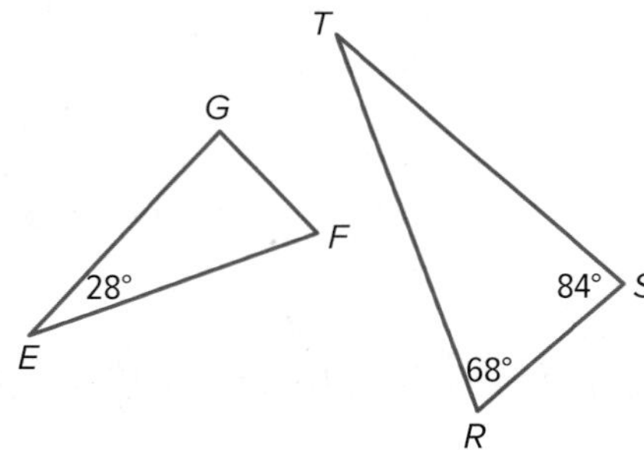

a What is the measure of angle F?

- A 28°
- B 68°
- C 74°
- D 96°

b Which of the following are true? Select all that apply.

- A $\angle G \cong \angle T$
- B $\angle T \cong \angle E$
- C $\triangle EFG \sim \triangle RST$
- D $\triangle EFG \sim \triangle TRS$
- E $\triangle FEG$ is a dilation of $\triangle RTS$.
- F $\triangle RTS$ is a dilation of $\triangle FEG$.

LESSON 4

Similar Triangles: SSS and SAS Similarity

Then	Now	Why?
• You used the AA Similarity Postulate to prove triangles are similar.	• **1** Use the SSS similarity criterion to prove triangles are similar. **2** Use the SAS similarity criterion to prove triangles are similar.	• The sports store where Miguel works sells 2-person tents and larger 6-person tents with ends shaped like isosceles triangles. Miguel sets up a display with a 2-person tent so that the angle formed at the top of the tent between the two equal-length sides is 65°. If he sets up a 6-person tent so that the angle formed at the top of the tent between the two equal-length sides is 65° are the triangles in the tents similar?

Mathematical Practice Standards

1 Make sense of problems and persevere in solving them.

2 Reason abstractly and quantitatively.

3 Construct viable arguments and critique the reasoning of others.

4 Model with mathematics.

1 SSS Similarity

You can use the AA Similarity Postulate to prove other statements about triangle similarity. The SSS Similarity Theorem states that two triangles are similar if the ratio of the lengths of each pair of corresponding sides is the same.

Theorem 7.3 SSS Similarity

Side-Side-Side (SSS) Similarity
If the corresponding side lengths of two triangles are proportional, then the triangles are similar.

Example If $\frac{JK}{MP} = \frac{KL}{PQ} = \frac{LJ}{QM}$, then $\triangle JKL \sim \triangle MPQ$.

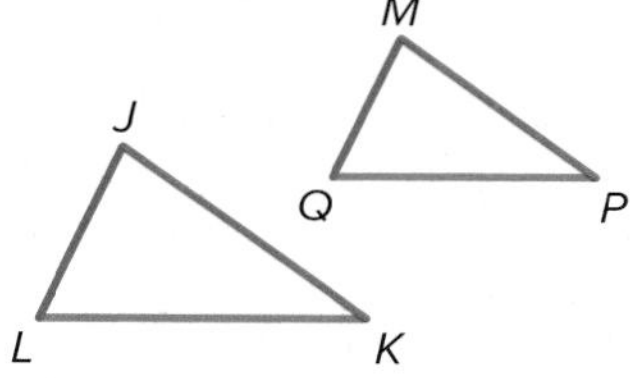

Proof Theorem 7.3

Given: $\frac{AB}{FG} = \frac{BC}{GH} = \frac{AC}{FH}$

Prove: $\triangle ABC \sim \triangle FGH$

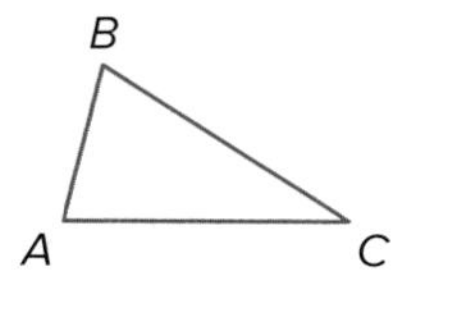

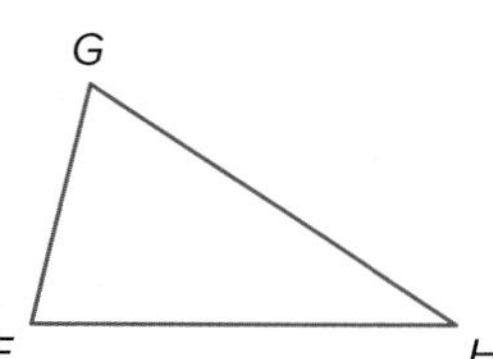

Paragraph Proof:

Locate J on $\overline{FG}$ so that $JG = AB$. Draw $\overline{JK}$ so that $\overline{JK} \parallel \overline{FH}$. Label $\angle GJK$ as $\angle 1$.

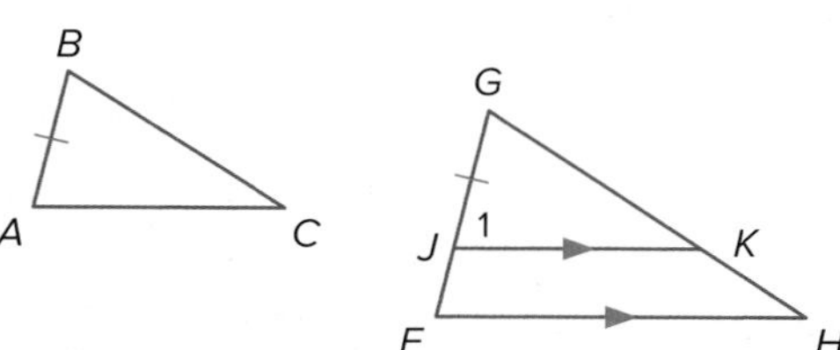

Because $\angle G \cong \angle G$ by the Reflexive Property and $\angle 1 \cong \angle F$ by the Corresponding Angles Postulate, $\triangle GJK \sim \triangle GFH$ by the AA Similarity Postulate.

By the definition of similar polygons, $\frac{JG}{FG} = \frac{GK}{GH} = \frac{JK}{FH}$. By substitution, $\frac{AB}{FG} = \frac{GK}{GH} = \frac{JK}{FH}$.

(continued on the next page)

Go Online! 

How do you know whether two triangles are similar? Investigate by using the **Geometry Tools** in ConnectED.

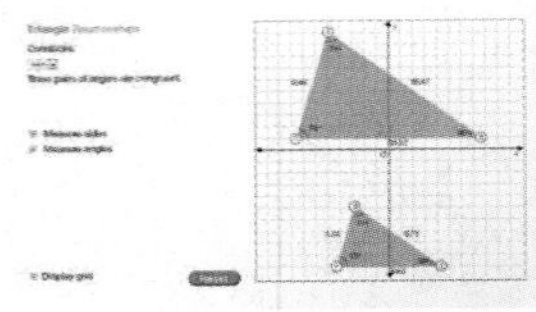

Because we are also given that $\frac{AB}{FG} = \frac{BC}{GH} = \frac{AC}{FH}$, we can say that $\frac{GK}{GH} = \frac{BC}{GH}$ and $\frac{JK}{FH} = \frac{AC}{FH}$. This means that $GK = BC$ and $JK = AC$, so $\overline{GK} \cong \overline{BC}$ and $\overline{JK} \cong \overline{AC}$.

By SSS, $\triangle ABC \cong \triangle JGK$.

By CPCTC, $\angle B \cong \angle G$ and $\angle A \cong \angle 1$. Because $\angle 1 \cong \angle F$, $\angle A \cong \angle F$ by the Transitive Property. By AA Similarity, $\triangle ABC \sim \triangle FGH$.

Example 1 Use the SSS Similarity Theorem

Determine whether the triangles are similar. If so, write a similarity statement. Explain your reasoning.

a. P 6 Q, 8, 5, R, 12.5, 20, T 15 S

$\frac{PR}{SR} = \frac{8}{20}$ or $\frac{2}{5}$, $\frac{PQ}{ST} = \frac{6}{15}$ or $\frac{2}{5}$, and $\frac{QR}{TR} = \frac{5}{12.5} = \frac{50}{125}$ or $\frac{2}{5}$. So, $\triangle PQR \sim \triangle STR$ by the SSS Similarity Theorem.

b.

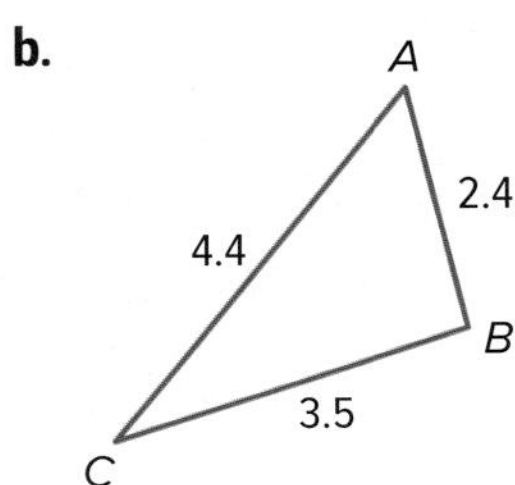

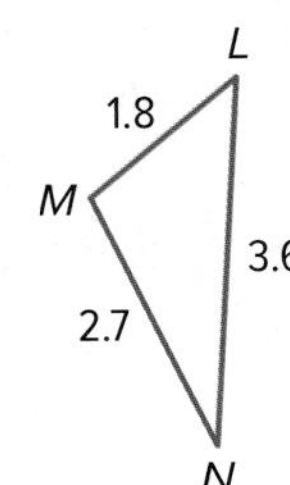

$\frac{AB}{LM} = \frac{2.4}{1.8}$ or $\frac{4}{3}$, $\frac{BC}{MN} = \frac{3.5}{2.7}$ or $\frac{35}{27}$, $\frac{AC}{LN} = \frac{4.4}{3.6}$ or $\frac{11}{9}$.

The sides are not proportional, so $\triangle ABC \nsim \triangle LMN$.

Guided Practice

1A.

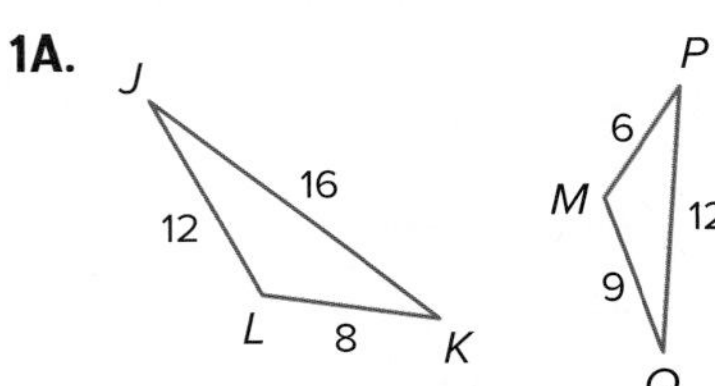

1B.

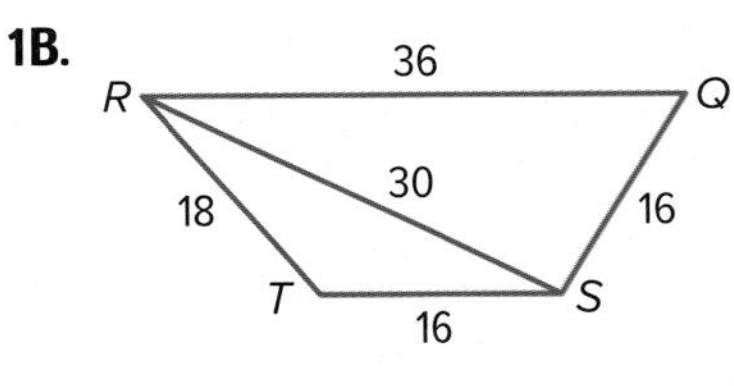

The SSS Similarity Theorem allows you to analyze triangles on the coordinate plane.

Example 2 Analyze Triangles on the Coordinate Plane

Determine whether $\triangle ABC$ with vertices $A(-2, 1)$, $B(3, -1)$, and $C(1, 4)$ is similar to $\triangle XYZ$ with vertices $X(-2, 4)$, $Y(2, -6)$, and $Z(-8, -2)$. Explain your reasoning.

Use the Distance Formula to find the measures of the sides.

$AB = \sqrt{(-2-(3))^2 + (-1-(1))^2} = \sqrt{(-5)^2 + (-2)^2}$ or $\sqrt{29}$

$BC = \sqrt{(3-1)^2 + (-1-4)^2} = \sqrt{(2)^2 + ((-5))^2}$ or $\sqrt{29}$

$AC = \sqrt{(-2-1)^2 + (1-4)^2} = \sqrt{(-3)^2 + (-3)^2}$ or $3\sqrt{2}$

$XY = \sqrt{(-2-2)^2 + (4-(-6))^2} = \sqrt{(-4)^2 + 10^2}$ or $2\sqrt{29}$

$YZ = \sqrt{(2-(-8))^2 + (-6-(-2))^2} = \sqrt{(10)^2 + (-4)^2}$ or $2\sqrt{29}$

$XZ = \sqrt{(-2-(-8))^2 + (4-(-2))^2} = \sqrt{6^2 + 6^2}$ or $6\sqrt{2}$

$\frac{AB}{XY} = \frac{\sqrt{29}}{2\sqrt{29}}$ or $\frac{1}{2}$ $\quad \frac{BC}{YZ} = \frac{\sqrt{29}}{2\sqrt{29}}$ or $\frac{1}{2}$ $\quad \frac{AC}{XZ} = \frac{3\sqrt{2}}{6\sqrt{2}}$ or $\frac{1}{2}$

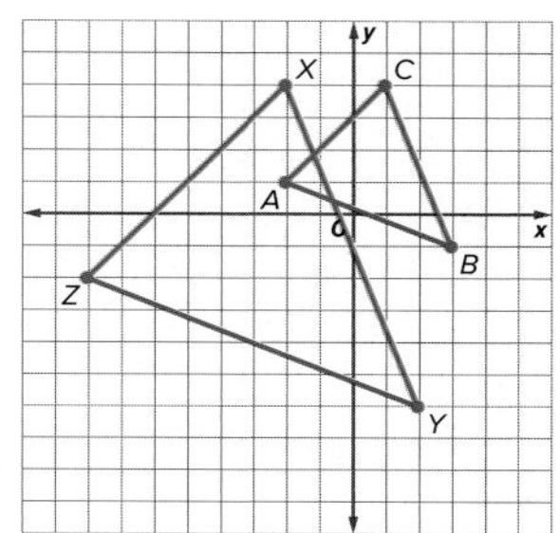

Because the side measures are proportional, $\triangle ABC$ is similar to $\triangle XYZ$ by SSS Similarity.

Guided Practice

Determine whether the triangles are similar.

2A. $\triangle MNO$ with $M(5, 2)$, $N(3, -8)$, and $O(0, -2)$ and $\triangle PQR$ with $P(11, 5)$, $Q(7, -15)$, and $R(1, -3)$

2B. $\triangle FGH$ with $F(1, 10)$, $G(3, -5)$, and $H(7, 5)$ and $\triangle JKL$ with $J(2, 7)$, $K(3, -1)$, and $L(5, 4)$

2 SAS Similarity

The SAS Similarity Theorem states that two triangles are similar if the ratio of the lengths of two pairs of corresponding sides is the same and their included angles are congruent.

Theorem 7.4 SAS Similarity

Side-Angle-Side (SAS) Similarity
If the lengths of two sides of one triangle are proportional to the lengths of two corresponding sides of another triangle and the included angles are congruent, then the triangles are similar.

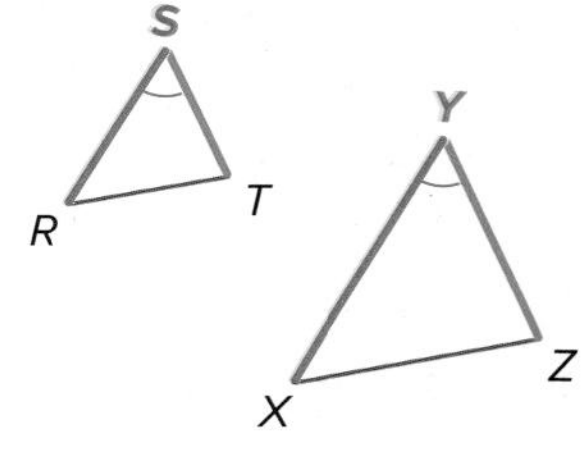

Example If $\frac{RS}{XY} = \frac{ST}{YZ}$ and $\angle S \cong \angle Y$, then $\triangle RST \sim \triangle XYZ$.

You will prove Theorem 7.4 in Exercise 28.

Example 3 Use the SAS Similarity Theorem

Determine whether the triangles are similar. If so, write a similarity statement. Explain your reasoning.

a.

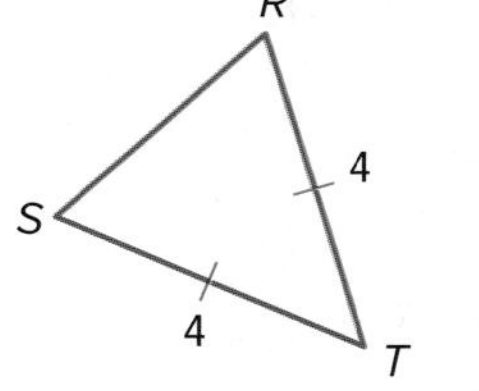

$\frac{RT}{UW} = \frac{4}{2.5}$ or 1.6 $\qquad \frac{VW}{ST} = \frac{4}{2.5}$ or 1.6

The lengths of the two pairs of sides are proportional. However we don't know that the lengths of the third pair of sides are proportional or that the included angles are congruent. There is not enough information to prove that the triangles are congruent.

b.

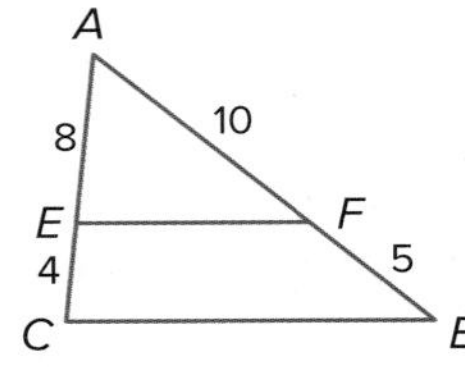

By the Reflexive Property, $\angle A \cong \angle A$.

$\frac{AF}{AB} = \frac{10}{10 + 5} = \frac{10}{15}$ or $\frac{2}{3}$ and $\frac{AE}{AC} = \frac{8}{8 + 4} = \frac{8}{12}$ or $\frac{2}{3}$.

Because the lengths of the sides that include $\angle A$ are proportional, $\triangle AEF \sim \triangle ACB$ by the SAS Similarity Theorem.

Guided Practice

3A.

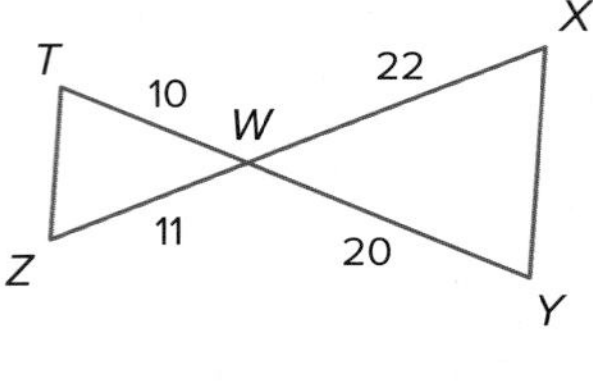

3B.

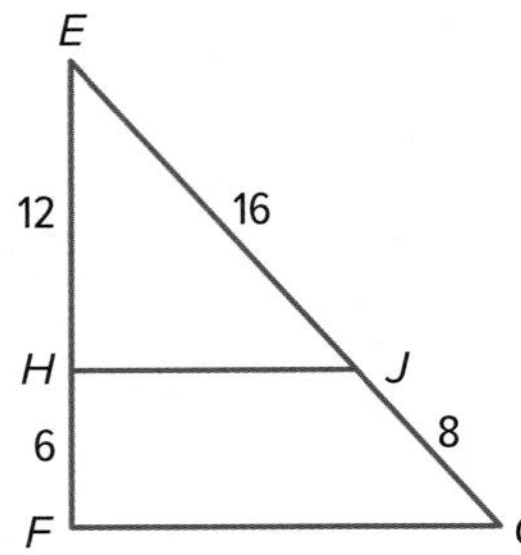

You can also use the SSS Similarity Theorem to analyze triangles on the coordinate plane.

Example 4 Analyze Triangles on the Coordinate Plane

Determine whether $\triangle ABC$ with vertices $A(-2, 7)$, $B(-2, -8)$, and $C(4, 4)$ is similar to $\triangle DEC$ with vertices $D(0, 6)$, and $E(0, -4)$. Explain your reasoning.

$\angle ACB$ $\angle DCE$ by the Reflexive Property. Use the distance formula to find the measures of the sides for which $\angle C$ is the included angle.

$AC = \sqrt{(-2 - 4)^2 + (7 - 4)^2} = \sqrt{(-6)^2 + 3^2}$ or $3\sqrt{5}$

$BC = \sqrt{(-2 - 4)^2 + (-8 - 4)^2} = \sqrt{(-6)^2 + (-(12))^2}$ or $6\sqrt{5}$

$DC = \sqrt{(0 - 4)^2 + (6 - 4)^2} = \sqrt{(-4)^2 + 2^2}$ or $2\sqrt{5}$

$EC = \sqrt{(0 - 4)^2 + (-4 - 4)^2} = \sqrt{(-4)^2 + (-(8))^2}$ or $4\sqrt{5}$

$\frac{BC}{EC} = \frac{6\sqrt{5}}{4\sqrt{5}}$ or 1.5 $\qquad$ $\frac{AC}{DC} = \frac{3\sqrt{5}}{2\sqrt{5}}$ or 1.5

Because two pairs of side measures are proportional and the included angles are congruent, $\triangle ABC$ is similar to $\triangle DEC$ by SAS Similarity.

Study Tip

Using Diagrams Marking a diagram as you identify the corresponding congruent and proportional parts of triangles can help you verify that the similarity criteria are met.

Guided Practice

Determine whether the triangles are similar.

4A. $\triangle PQR$ with $P(1, 5)$, $Q(7, -2)$, and $R(3, -3)$ and $\triangle STR$ with $S(2, 5)$, and $T(6, -2)$

4B. $\triangle EFG$ with $E(-1, 1)$, $F(2, -5)$, and $G(4, 2)$ and $\triangle ABG$ with $A(-6, 0)$, and $B(0, -12)$

Once you have established that two triangles are similar, you can write and solve a proportion to find an unknown measurement.

Real-World Example 5 Indirect Measurement

ECOLOGY A park volunteer is doing an inventory of the trees in the park, noting the species and height of each tree. He stands so that his shadow coincides with the tree's shadow. He is 5 feet 9 inches tall and his shadow is 12 feet long. If the length of the tree's shadow is 17.2 feet, how tall is the tree?

Understand Make a sketch of the situation. 5 feet 9 inches is equivalent to 5.75 feet. If the tree's shadow is 17.2 feet, the length of the tree's shadow behind the volunteer is 5.2 feet.

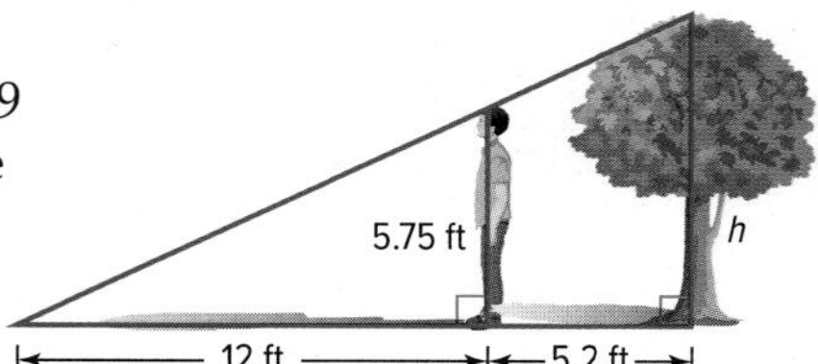

> **Study Tip**
> **MP Structure** Proportions must have corresponding parts such as height to length. This demonstrates the corresponding sides that are proportional.

Plan In shadow problems, you can assume that the angles formed by the Sun's rays with any two objects are congruent and that the objects form the sides of two right triangles. Because two corresponding pairs of sides are proportional, the triangles are similar by the SAS Similarity Theorem.

$$\frac{\text{volunteer's height}}{\text{tree's height}} = \frac{\text{volunteer's shadow length}}{\text{tree's shadow length}}$$

Solve Substitute the known values and let h = tree's height.

$$\frac{5.75}{h} = \frac{12}{17.2}$$

$$12 \cdot h = (5.75)(17.2)$$

$$12h = 98.9$$

$$h \approx 8.2$$

The tree is about 8.2 feet tall.

Check The tree's shadow length is $\frac{17.2}{12} \approx 1.4$ times the volunteer's shadow length. Check to see that the tree's height is about 1.4 times the volunteer's height. $\frac{8.2}{5.75} \approx 1.4$.

Guided Practice

5. **MEASUREMENT** Connor is standing next to the flagpole at his school and wants to use his shadow to help measure the pole. The length of his shadow is 8 feet when the flagpole's shadow is 50 feet. How high is the flagpole to the nearest foot?

In addition to the definition of similar triangles, you now know three methods for proving that triangles are similar.

Concept Summary Triangle Similarity

AA Similarity Postulate

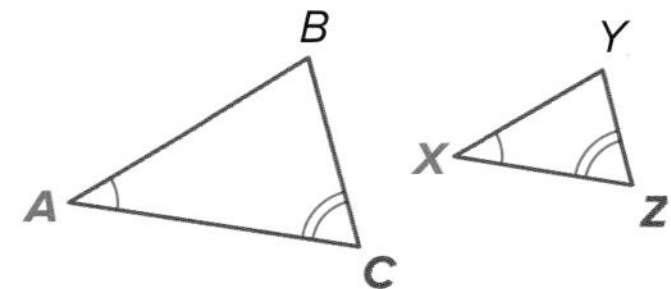

If $\angle A \cong \angle X$ and $\angle C \cong \angle Z$, then $\triangle ABC \sim \triangle XYZ$.

SSS Similarity Theorem

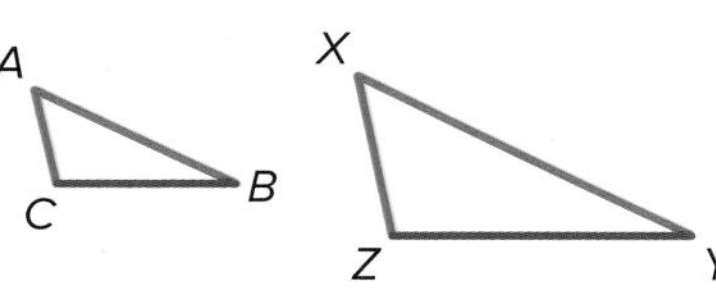

If $\frac{AB}{XY} = \frac{BC}{YZ} = \frac{CA}{ZX}$, then $\triangle ABC \sim \triangle XYZ$.

SAS Similarity Theorem

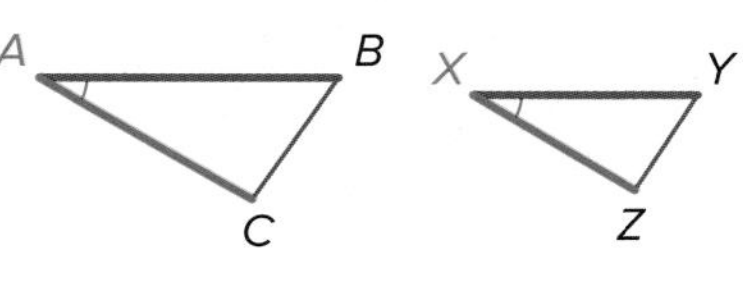

If $\angle A \cong \angle X$ and $\frac{AB}{XY} = \frac{CA}{ZX}$, then $\triangle ABC \sim \triangle XYZ$.

Check Your Understanding

= Step-by-Step Solutions begin on page R13.

Go Online! for a Self-Check Quiz

Examples 1, 3 **Determine whether the triangles are similar. If so, write a similarity statement. Explain your reasoning.**

1. 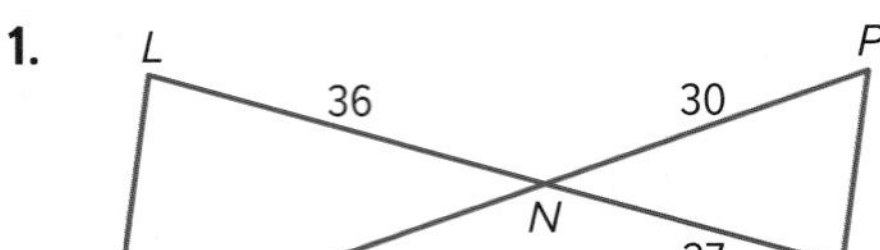

2. 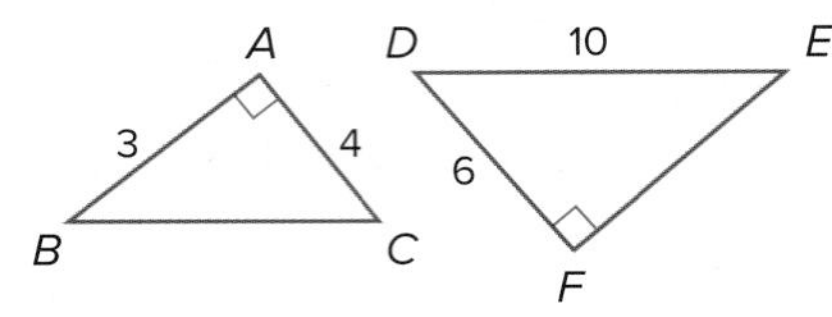

3. 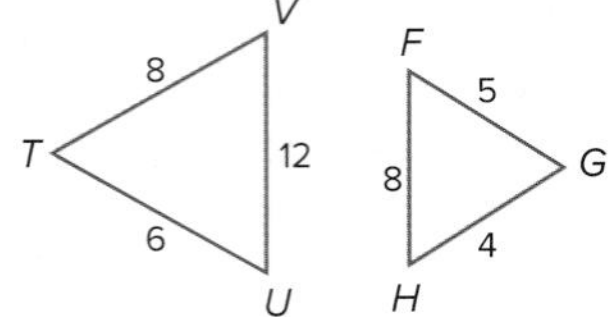

4. 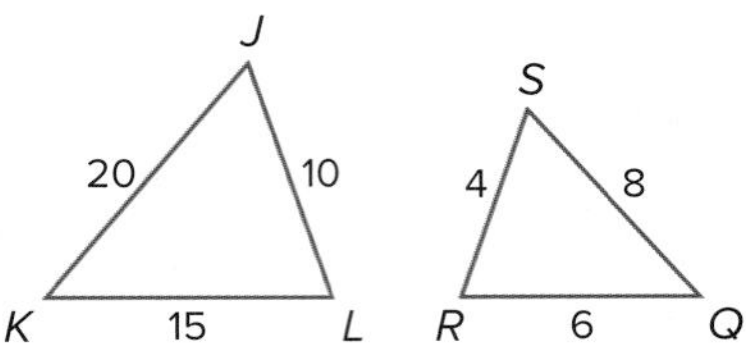

Examples 2, 4 **Determine whether the triangles are similar.**

5. $\triangle LMN$ with $L(10, -2)$, $M(-2, 4)$, and $N(6, -4)$ and $\triangle PQR$ with $P(-1, 5)$, $Q(2, -1)$, and $R(-3, 3)$

6. $\triangle ABC$ with $A(2, -2)$, $B(5, -4)$, and $C(-3, -3)$ and $\triangle ADE$ with $D(8, -6)$, and $E(-8, -4)$

7. In the figure, $\frac{AB}{DE} = \frac{BC}{EF}$. Determine whether the given information is sufficient to prove that $\triangle ABC \sim \triangle DEF$. Explain.

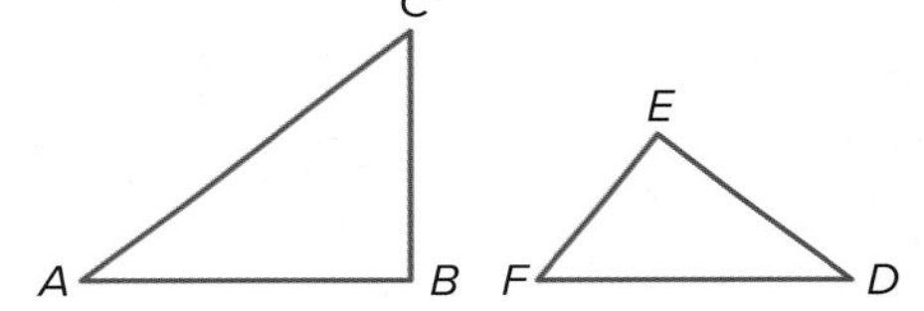

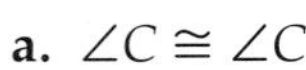

a. $\angle C \cong \angle C$

b. $\frac{AB}{DE} = \frac{AC}{DF}$

c. $\angle B \cong \angle E$

d. $\angle A \cong \angle D$

Example 5

8. **SPORTS** The goalpost at the North High School football field casts a shadow 50 feet long. The distance from the post to the shadow of the crossbar is 16 feet. If the crossbar is 10 feet above the ground, how tall is the goalpost?

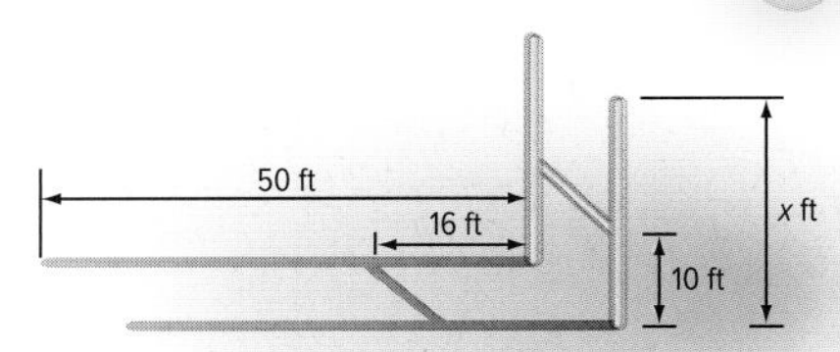

Practice and Problem Solving

Extra Practice is on page R7.

Examples 1, 3 **Determine whether the triangles are similar. If so, write a similarity statement. If not, what would be sufficient to prove the triangles similar? Explain your reasoning.**

9. 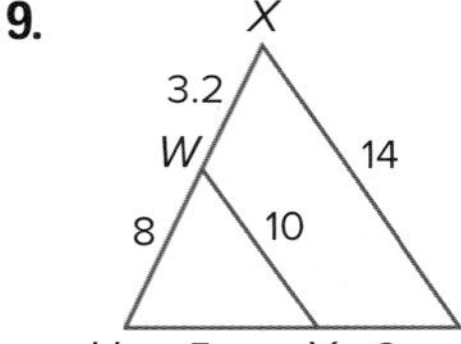

10. 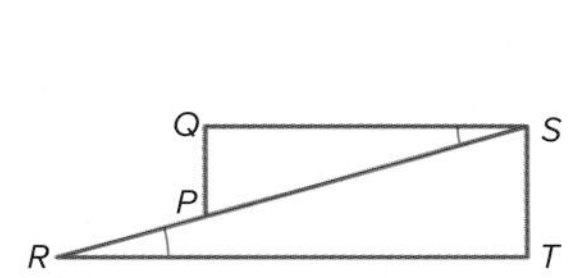

11. 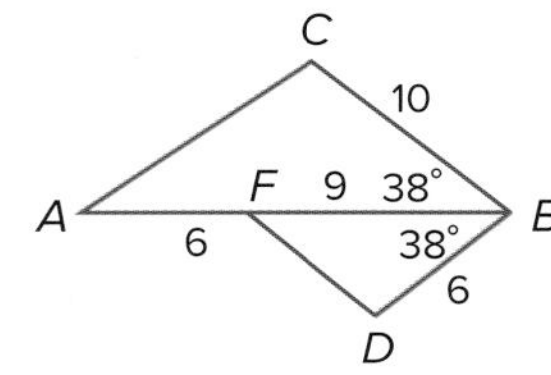

Examples 1, 3 **Determine whether the triangles are similar. If so, write a similarity statement. If not, what would be sufficient to prove the triangles similar? Explain your reasoning.**

12.

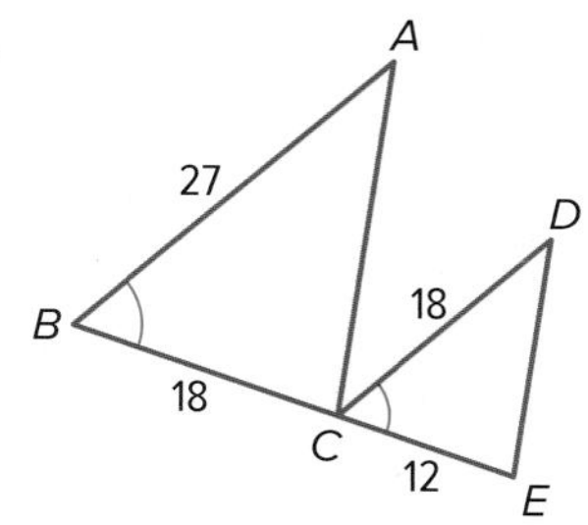

13.

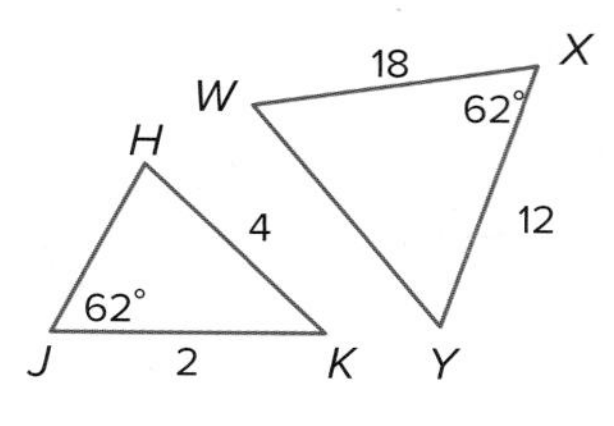

14.

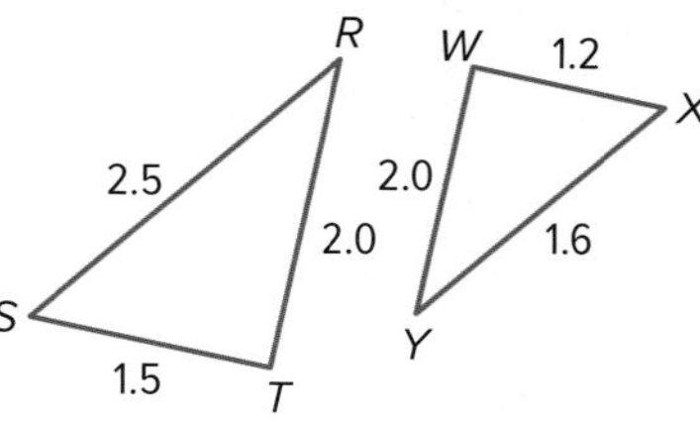

15.

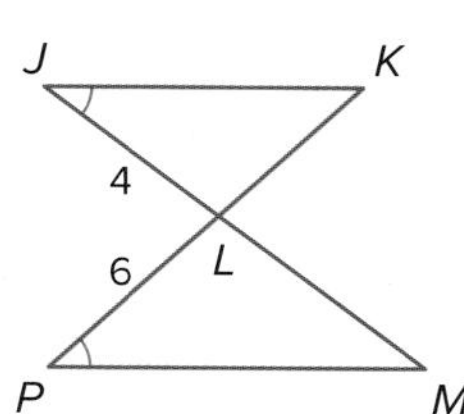

16.

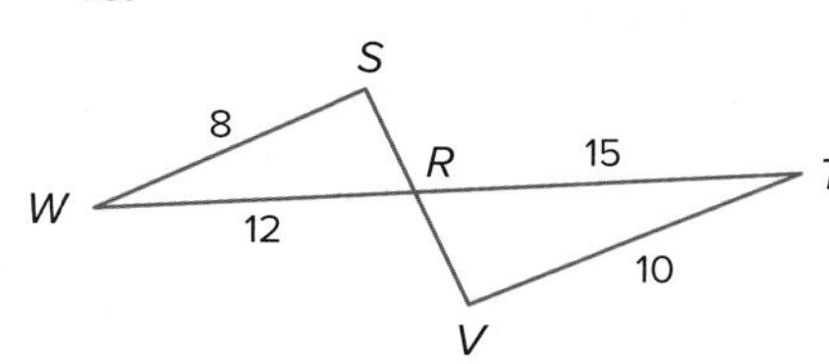

17.

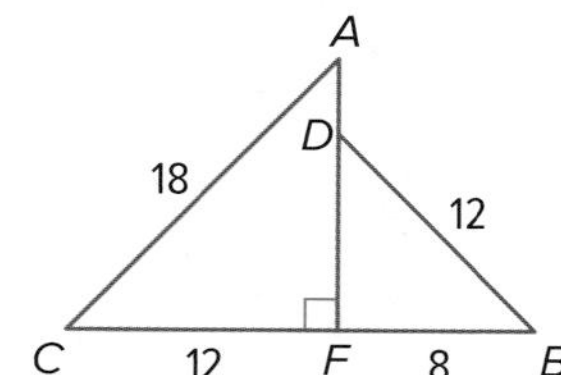

Determine whether the triangles are similar.

Example 2

18. $\triangle LMN$ with $L(-6, -2)$, $M(2, 4)$, and $N(8, -4)$ and $\triangle PQR$ with $P(3, 1)$, $Q(-1, -2)$, and $R(-2, 2)$

19. $\triangle DEF$ with $D(3, -1)$, $E(-1, 4)$, and $F(2, -3)$ and $\triangle GHI$ with $G(-3, 9)$, $H(12, -3)$, and $I(-9, 6)$

20. $\triangle RST$ with $R(1.5, -1.0)$, $S(-3.5, 1.0)$, and $T(3.0, -2.5)$ and $\triangle UVW$ with $U(-2.0, 3.0)$, $V(2.0, 6.0)$, and $W(-5.0, 5.0)$

21. $\triangle JKL$ with $J(1.2, -2.0)$, $K(-3.2, 4.0)$, and $L(3.6, -2.4)$ and $\triangle MNO$ with $M(-0.3, 0.5)$, $N(0.8, -1.0)$, and $O(0.9, -0.6)$

Example 4

22. $\triangle ABC$ with $A(-2, 4)$, $B(3, -4)$, and $C(1, -6)$ and $\triangle ADE$ with $D(8, -12)$, and $E(4, -16)$

23. $\triangle FGH$ with $F(5, 5)$, $G(20, -7)$, and $H(5, -10)$ and $\triangle FJK$ with $J(10, 1)$, and $K(5, 0)$

24. $\triangle BCD$ with $B(1.7, -1.0)$, $C(3.3, 0.6)$, and $D(-1.5, -1.8)$ and $\triangle BFG$ with $F(2.1, -0.6)$, and $G(0.9, -1.2)$

25. $\triangle MNP$ with $M(2.5, -1.2)$, $N(2.5, -0.4)$, and $P(3.1, -2.3)$ and $\triangle MRT$ with $R(2.5, 0.4)$, and $T(4.3, -4.5)$

Example 5

26. FORESTRY A hypsometer, as shown, can be used to estimate the height of a tree. Bartolo looks through the straw to the top of the tree and obtains the readings given. Find the height of the tree.

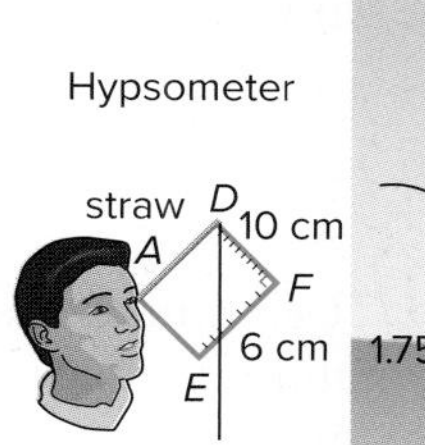

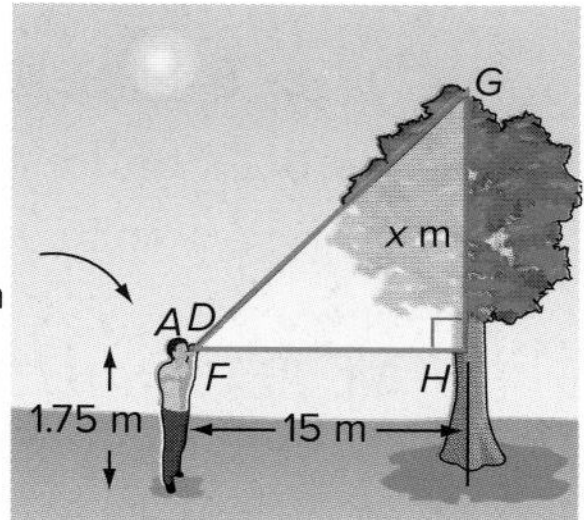

27. Given: $\triangle XYZ$ and $\triangle ABC$ are right triangles; $\frac{XY}{AB} = \frac{YZ}{BC}$

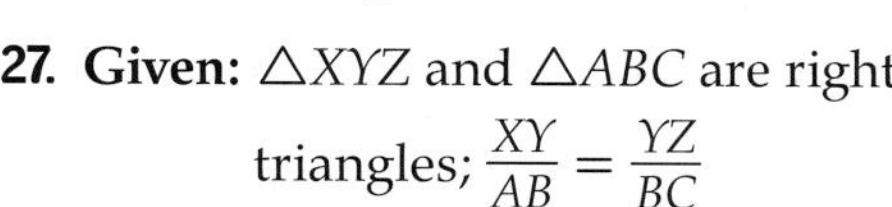

Prove: $\triangle YXZ \sim \triangle BAC$

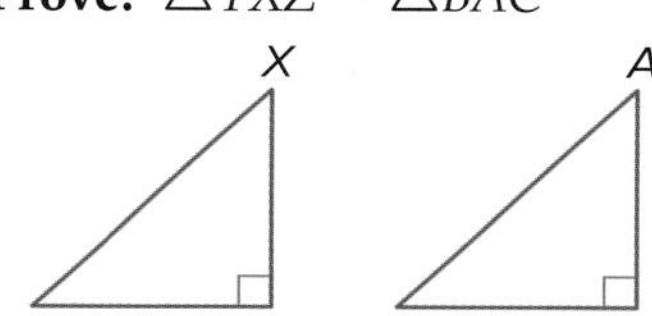

28. PROOF Write a two-column proof for Theorem 7.4.

MP STRUCTURE Identify the similar triangles. Find each measure.

29. KL

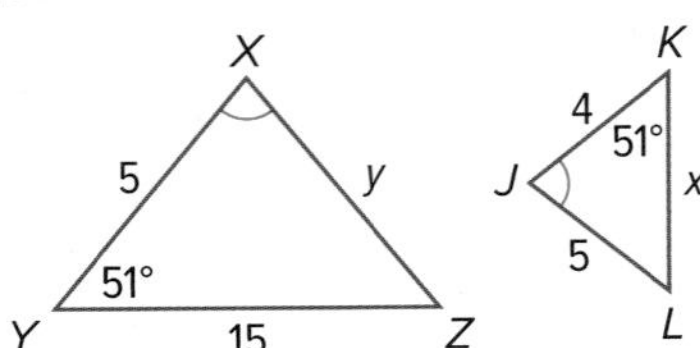

30. GH

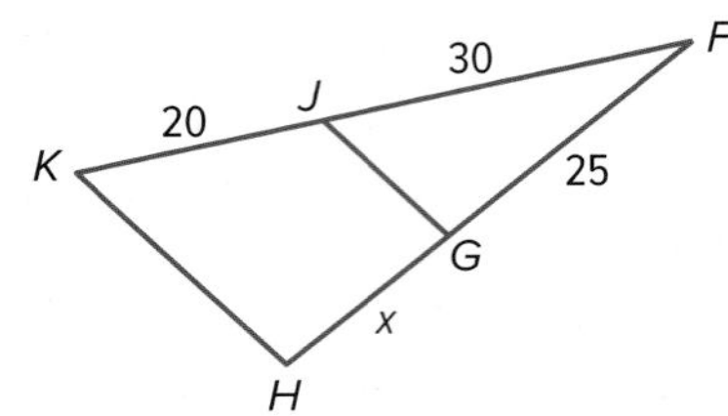

PROOF Write a two-column proof.

31. Given: $\triangle ABC$ and $\triangle DEF$ are right triangles; $DE = \frac{2}{3}AB$, $EF = \frac{2}{3}BC$

Prove: $\frac{DF}{AC} = \frac{DE}{AB}$

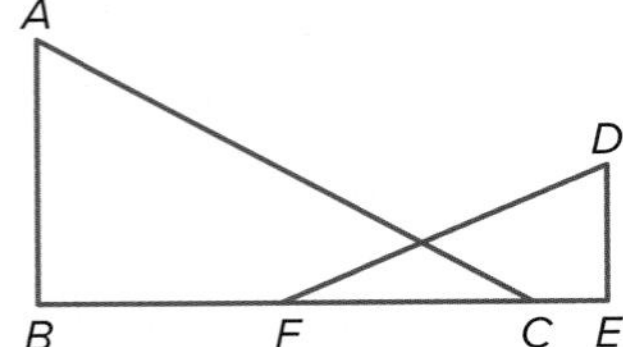

32. RIDES The Power Tower ride at Cedar Point amusement park is casting a shadow that is about 76 feet long. Addy is 5 feet 7 inches tall and her shadow is about 17 inches long.

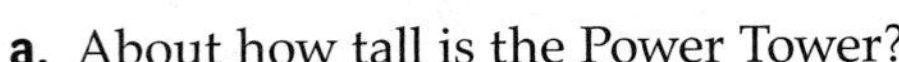

a. About how tall is the Power Tower?

b. If the shadow of the Valravn roller coaster is about 56.5 feet long, about how tall is the Valravn?

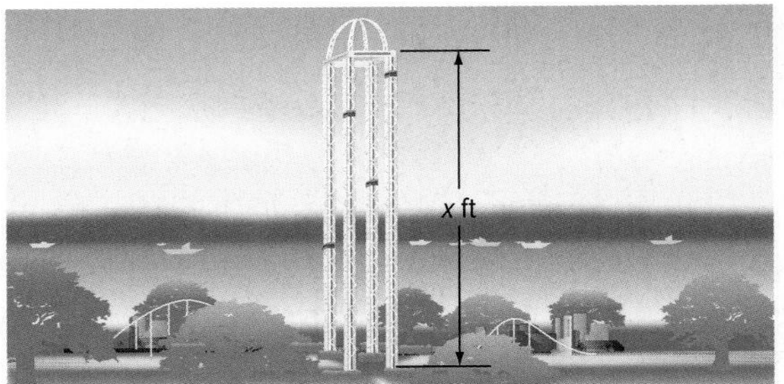

33. $\triangle WXY$ has coordinates $W(0, 0)$, $X(1, 4)$, and $Y(4, 2)$.

a. Graph the triangle.

b. $\triangle WXY$ is dilated in the coordinate plane with a scale factor of 2 and center $(0, 0)$. What are the coordinates of the image triangle, $\triangle W'X'Y'$?

c. Use SSS Similarity to prove that $\triangle WXY \sim \triangle W'X'Y'$.

d. Find the ratio of the perimeters of the two triangles. Justify your reasoning.

34. REASONING Use similar triangles to show that the slope of the line through any two points on that line is constant. That is, if points A, B, A' and B' are on line ℓ, use similar triangles to show that the slope of the line from A to B is equal to the slope of the line from A' to B'.

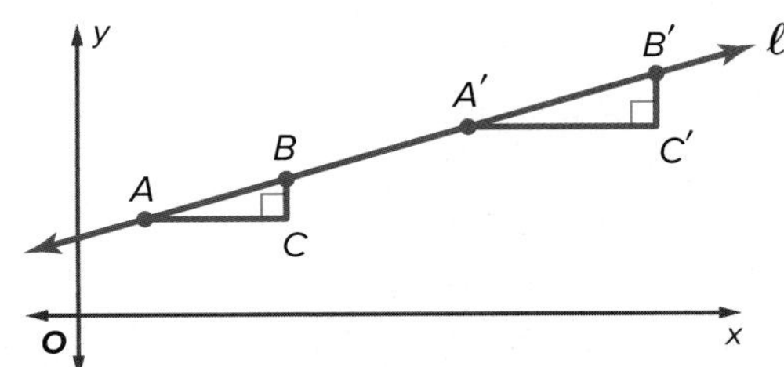

35. Assume that $\triangle ABC \sim \triangle JKL$.

a. If the lengths of the sides of $\triangle JKL$ are half the length of the sides of $\triangle ABC$, and the area of $\triangle ABC$ is 40 square inches, what is the area of $\triangle JKL$? How is the area related to the scale factor of $\triangle ABC$ to $\triangle JKL$?

b. If the lengths of the sides of $\triangle ABC$ are three times the length of the sides of $\triangle JKL$, and the area of $\triangle ABC$ is 63 square inches, what is the area of $\triangle JKL$? How is the area related to the scale factor of $\triangle ABC$ to $\triangle JKL$?

36. **ARCHITECTURAL DESIGN** A-frame houses feature steeply-angled sides that begin at or near the foundation line and meet at the top in the shape of the letter A. Jessica is designing an A-frame house that will be 40 feet tall and the base of the house will be 60 feet long. She will build a second floor balcony around the outside of the house, 15 feet above the ground. The left side of the house will be 50 feet long and the balcony will intersect the side 18.75 feet from the bottom. The height of the house bisects the base of the house and the balcony.

a. Draw a diagram to model the side of the house.

b. Calculate the total length of the balcony.

37. **MULTIPLE REPRESENTATION** In this problem, you will explore proportional parts of parallel lines.

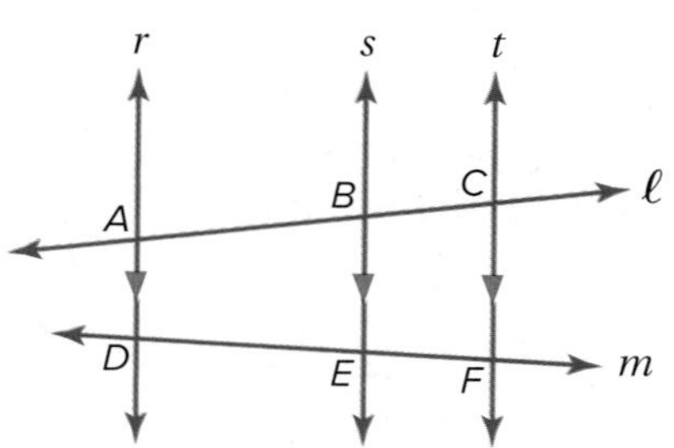

a. **Geometric** Draw parallel lines r, s, and t with transversals l and m intersecting all three lines, as shown at the right.

b. **Tabular** Measure and record the lengths AB, BC, DE, EF, and the ratios $\frac{AB}{BC}$ and $\frac{DE}{EF}$.

c. **Verbal** Make a conjecture about the segments created by transversals intersecting parallel lines.

H.O.T. Problems Use Higher-Order Thinking Skills

38. **WRITING IN MATH** Compare and contrast the AA Similarity Postulate, the SSS Similarity Theorem, and the SAS Similarity Theorem.

39. **REASONING** Explain how you know $\triangle EFG \sim \triangle DBC$. Find EG and DB.

40. **REASONING** A triangle has sides that measure 3 inches, 4 inches, and 5 inches. Each side length is increased by x inches. Is the new triangle similar to the original triangle? Justify your reasoning.

41. **OPEN-ENDED** Draw a triangle that is similar to $\triangle ABC$ shown. Explain how you know that it is similar.

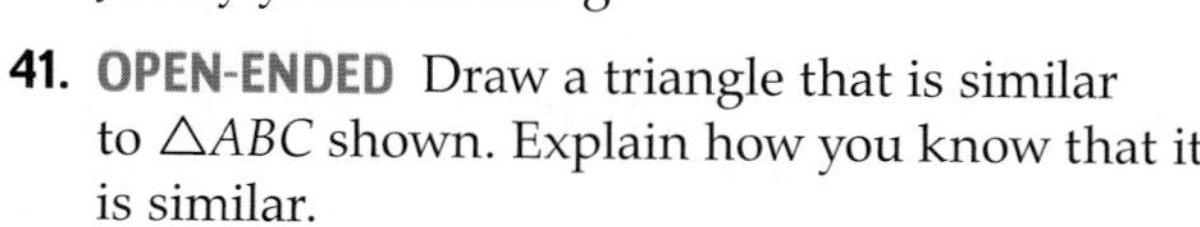

42. **WRITING IN MATH** How can you choose an appropriate scale for a model or drawing?

43. **REASONING** Explain how to determine whether $\triangle KLM \sim \triangle OPN$. Then, if possible, find KL and OP.

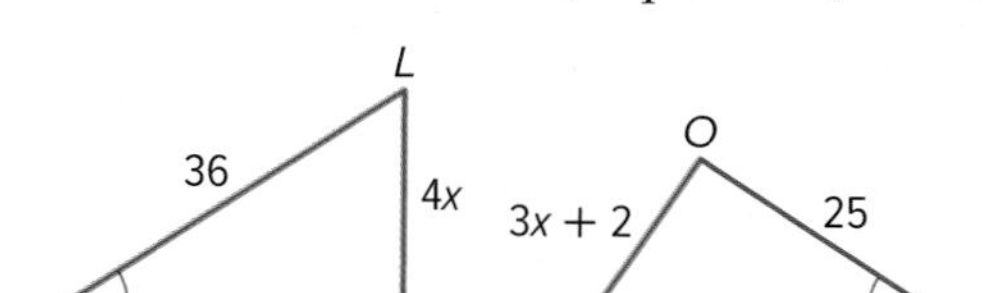

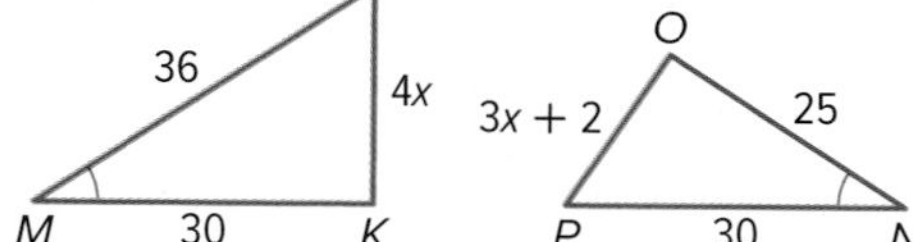

44. **OPEN-ENDED** Create a real-world problem that uses similar triangles to find the height of a tall object.

Preparing for Assessment

45. MULTI-STEP

a. In the figure, $\frac{AB}{DE} = \frac{BC}{EF}$. Which additional information would be sufficient to prove that the triangles are similar? Select all that apply. MP 2

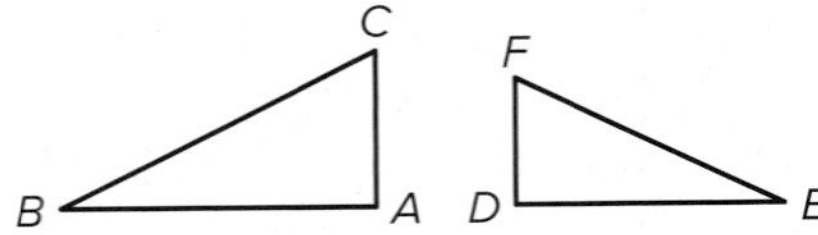

- ☐ **A** $\angle C \cong \angle F$
- ☐ **B** $\frac{AB}{DE} = \frac{BC}{EF}$
- ☐ **C** $\angle B \cong \angle E$
- ☐ **D** $\angle A \cong \angle D$

b. If the information you chose in part **a** is true, which of the following is a similarity statement for the two triangles? MP 2

- ○ **A** $\triangle ABC \sim \triangle FED$
- ○ **B** $\triangle ABC \sim \triangle EDF$
- ○ **C** $\triangle ABC \sim \triangle DFE$
- ○ **D** $\triangle ABC \sim \triangle DEF$

c. Which of the following statements can be proved true because they are corresponding parts of similar triangles? Select all that apply. MP 3

- ☐ **A** $\angle C \cong \angle D$
- ☐ **B** $\frac{AC}{FD} = \frac{BC}{EF}$
- ☐ **C** $\frac{AB}{DF} = \frac{BC}{EF}$
- ☐ **D** $\angle B \cong \angle E$
- ☐ **E** $\angle A \cong \angle D$
- ☐ **F** $\frac{AB}{EF} = \frac{BC}{DE}$

46. In the figure, $VT = 8\sqrt{3}$ and $RT = 10\sqrt{3}$. Determine whether the triangles are similar. If so, write a similarity statement. If not, what additional information would be sufficient to prove the triangles similar? Explain your reasoning. MP 2

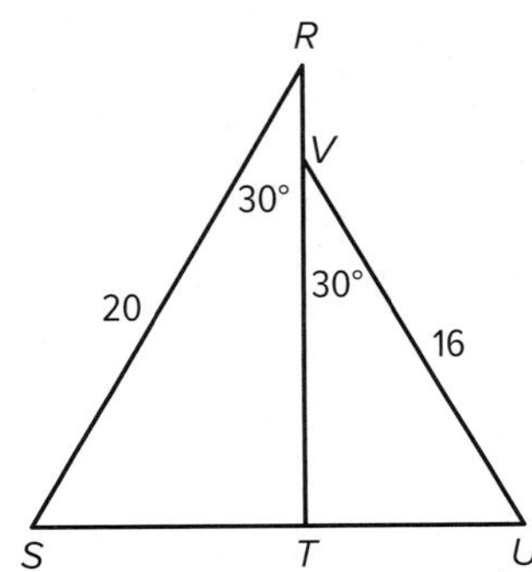

47. In the figure, $\angle P \cong \angle S$.

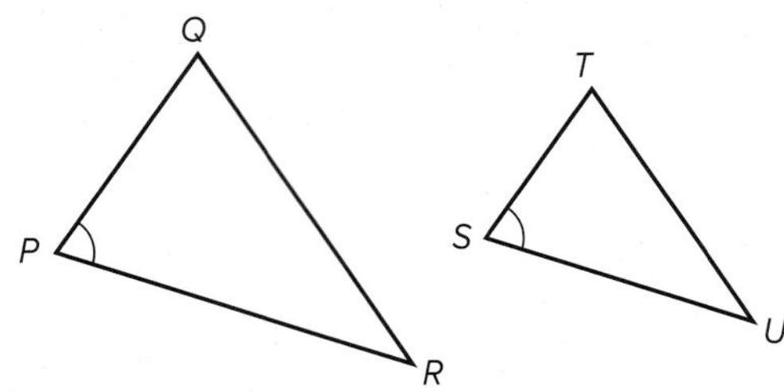

Which of the following additional pieces of information would allow you to prove that $\triangle PQR \sim \triangle STU$? MP 2

I. $\angle Q \cong \angle T$

II. $PQ = 8$, $ST = 6$, $PR = 12$, $SU = 9$

III. $\overline{QR} \cong \overline{TU}$

- ○ **A** I only
- ○ **B** II only
- ○ **C** III only
- ○ **D** I and II only
- ○ **E** I, II, and III

48. Michelle's dog is 35 centimeters tall. She wants to estimate the height of the street sign on the corner. She measures her dog's shadow and the street sign's shadow and then draws the sketch shown below.

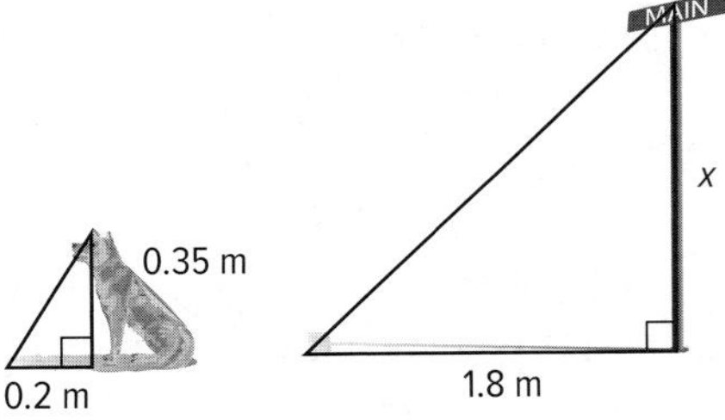

Which of the following is the best estimate of the height of the street sign? MP 4

- ○ **A** 3.6 m
- ○ **B** 3.15 m
- ○ **C** 2.0 m
- ○ **D** 1.03 m

EXTEND 7-4

Geometry Lab

Proofs of Perpendicular and Parallel Lines

You have learned that two straight lines that are neither horizontal nor vertical are perpendicular if and only if the product of their slopes is −1. In this activity, you will use similar triangles to prove the first half of this theorem: if two straight lines are perpendicular, then the products of their slopes is −1.

Mathematical Practices

MP 3 Construct viable arguments and critique the reasoning of others.

Activity 1 Perpendicular Lines

Given: **Slope of $\overleftrightarrow{AC} = m_1$, slope of $\overleftrightarrow{CE} = m_2$, and $\overleftrightarrow{AC} \perp \overleftrightarrow{CE}$**

Prove: $m_1 m_2 = -1$

Step 1 On a coordinate plane, construct $\overleftrightarrow{AC} \perp \overleftrightarrow{CE}$ and transversal $\overleftrightarrow{BD}$ parallel to the x-axis through C. Then construct a right triangle $\triangle ABC$ such that $\overline{AC}$ is the hypotenuse and a right triangle $\triangle EDC$ such that $\overline{CE}$ is the hypotenuse. The legs of both triangles should be parallel to the x-and y-axes, as shown.

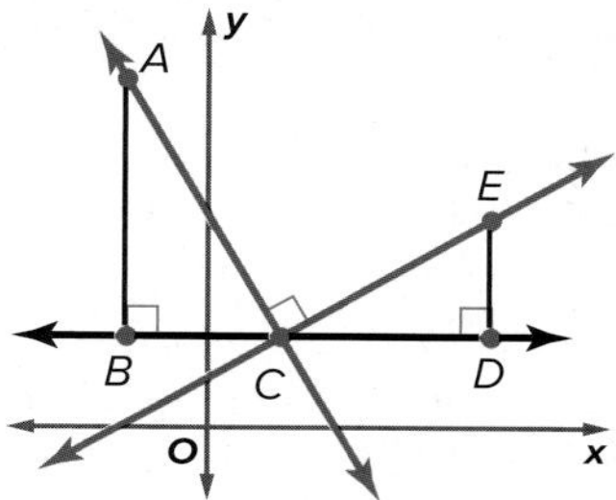

Step 2 Find the slopes of $\overleftrightarrow{AC}$ and $\overleftrightarrow{CE}$.

Slope of $\overleftrightarrow{AC}$

$m_1 = \frac{\text{rise}}{\text{run}}$ Slope Formula

$= \frac{-AB}{BC}$ or $-\frac{AB}{BC}$ rise $= -AB$, run $= BC$

Slope of $\overleftrightarrow{CE}$

$m_2 = \frac{\text{rise}}{\text{run}}$ Slope Formula

$= \frac{DE}{CD}$ rise $= DE$, run $= CD$

Step 3 Show that $\triangle ABC \sim \triangle CDE$.

Because $\triangle ACB$ is a right triangle with right angle B, $\angle BAC$ is complementary to $\angle ACB$. It is given that $\overleftrightarrow{AC} \perp \overleftrightarrow{CE}$, so we know that $\angle ACE$ is a right angle.

By construction, $\angle BCD$ is a straight angle. So, $\angle ECD$ is complementary to $\angle ACB$. Because angles complementary to the same angle are congruent, $\angle BAC \cong \angle ECD$.

Because right angles are congruent, $\angle B \cong \angle D$. Therefore, by AA Similarity, $\triangle ABC \sim \triangle CDE$.

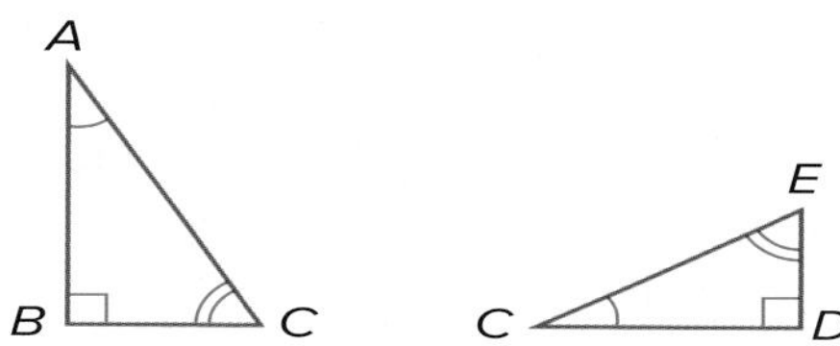

Step 4 Use the fact that $\triangle ABC \sim \triangle CDE$ to show that $m_1 m_2 = -1$.

Because $m_1 = -\frac{AB}{BC}$ and $m_2 = \frac{DE}{CD}$, $m_1 m_2 = \left(-\frac{AB}{BC}\right)\left(\frac{DE}{CD}\right)$. Because two similar polygons have proportional sides, $\frac{AB}{BC} = \frac{CD}{DE}$. Therefore, by substitution, $m_1 m_2 = \left(-\frac{CD}{DE}\right)\left(\frac{DE}{CD}\right)$ or −1.

Geometry Lab

Proofs of Perpendicular and Parallel Lines *Continued*

Model **Work cooperatively.**

1. **PROOF** Use the diagram from Activity 1 to prove the second half of the theorem.

   Given: Slope of $\overleftrightarrow{CE} = m_1$, slope of $\overleftrightarrow{AC} = m_2$, and $m_1 m_2 = -1$; $\triangle ABC$ is a right triangle with right angle B. $\triangle CDE$ is a right triangle with right angle D.

   Prove: $\overleftrightarrow{CE} \perp \overleftrightarrow{AC}$

You can also use similar triangles to prove that two distinct nonvertical lines are parallel if and only if they have the same slope. In this activity, you will prove the first half of the biconditional.

Activity 1 Parallel Lines

Given: Slope of $\overleftrightarrow{FG} = m_1$, slope of $\overleftrightarrow{JK} = m_2$, and $m_1 = m_2$. $\triangle FHG$ is a right triangle with right angle H. $\triangle JLK$ is a right triangle with right angle L.

Prove: $\overleftrightarrow{FG} \parallel \overleftrightarrow{JK}$

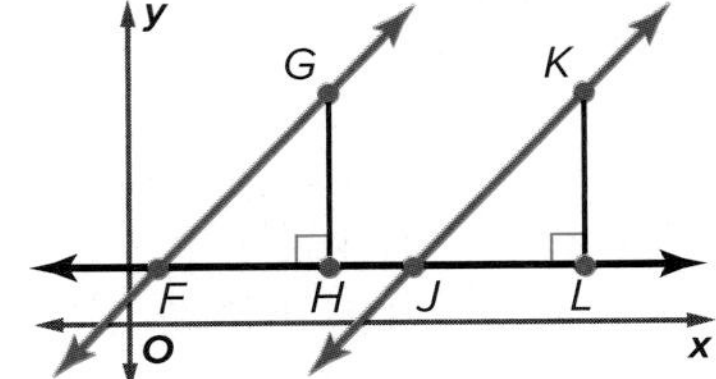

Step 1 On a coordinate plane, construct $\overleftrightarrow{FG}$ and $\overleftrightarrow{JK}$, right $\triangle FHG$, and right $\triangle JLK$. Then draw horizontal transversal $\overleftrightarrow{FL}$, as shown.

Step 2 Find the slopes of $\overleftrightarrow{FG}$ and $\overleftrightarrow{JK}$.

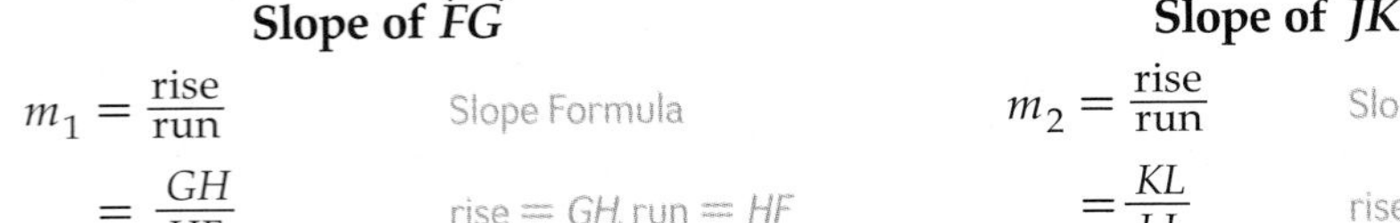

Slope of $\overleftrightarrow{FG}$

$m_1 = \frac{\text{rise}}{\text{run}}$ Slope Formula

$= \frac{GH}{HF}$ rise = GH, run = HF

Slope of $\overleftrightarrow{JK}$

$m_2 = \frac{\text{rise}}{\text{run}}$ Slope Formula

$= \frac{KL}{LJ}$ rise = KL, run = LJ

Step 3 Show that $\triangle FHG \sim \triangle JLK$.

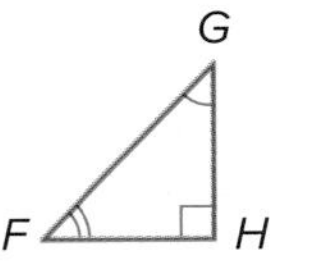

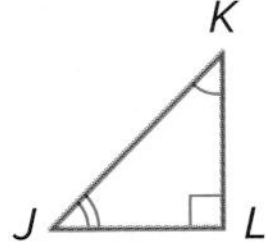

It is given that $m_1 = m_2$. By substitution, $\frac{GH}{HF} = \frac{KL}{LJ}$. This ratio can be rewritten as $\frac{GH}{KL} = \frac{HF}{LJ}$, so two pairs of corresponding sides are proportional. Because $\angle H$ and $\angle L$ are right angles, $\angle H \cong \angle L$.

Therefore, by SAS Similarity, $\triangle FHG \sim \triangle JLK$.

Step 4 Use the fact that $\triangle FHG \sim \triangle JLK$ to prove that $\overleftrightarrow{FG} \parallel \overleftrightarrow{JK}$.

Corresponding angles in similar triangles are congruent, so $\angle GFH \cong \angle KJL$. Notice that these are corresponding angles formed when $\overleftrightarrow{FG}$ and $\overleftrightarrow{JK}$ are cut by transversal $\overleftrightarrow{FL}$. Thus, by the Converse of the Corresponding Angles Postulate, $\overleftrightarrow{FG} \parallel \overleftrightarrow{JK}$.

Model **Work cooperatively.**

2. **PROOF** Use the diagram from Activity 2 to prove the second half of the theorem.

   Given: Slope of $\overleftrightarrow{FG} = m_1$, slope of $\overleftrightarrow{JK} = m_2$, and $\overleftrightarrow{FG} \parallel \overleftrightarrow{JK}$

   Prove: $m_1 = m_2$

CHAPTER 7

Mid-Chapter Quiz

Lessons 7-1 through 7-4

Dilate the figure with the given vertices after a dilation at the indicated center with the given scale factor. Name the coordinates of the image. (Lesson 7-1)

1. $A(1, 4)$, $B(6, 8)$, $C(4, 9)$; center $(3, 6)$; $k = 0.5$
2. $P(-3, 2)$, $Q(-3, 6)$, $R(-6, 1)$; center $(0, 0)$; $k = 3$
3. $X(-1, -1)$, $Y(3, 3)$, $Z(2, -2)$; center $(1, -1)$; $k = 2$
4. **BUSINESS** A graphic designer has designed a 40 centimeter by 40 centimeter logo for a business with a circle with a radius of 5 centimeters at the center. They now need to reduce the logo to a 2 centimeter by 2 centimeter size to fit on business cards. (Lesson 7-1)
   a. What will the scale factor of the dilation be?
   b. What will the diameter of the circle be on the business card?
5. **BIOLOGY** Mary studies red blood cells under a microscope to diagnose leukemia. The microscope is attached to a large screen on the wall where the picture is enlarged. If a red blood cell image is 15.04 centimeters on the screen and 0.00000752 meters in real life, how many times has the microscope amplified the image? (Lesson 7-1)

Each pair of polygons is similar. Find the value of x. (Lesson 7-2)

6. 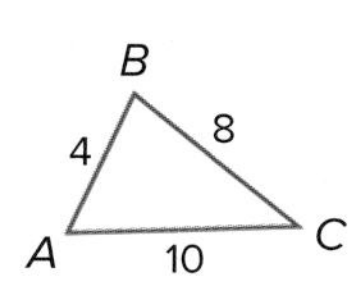

Q
9
18
P
x
R

7. 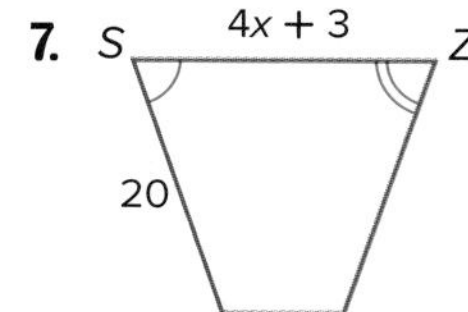

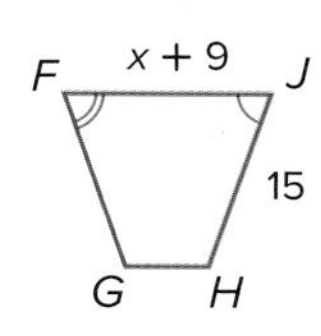

8. **MULTIPLE CHOICE** Two similar polygons have a scale factor of 3 : 5. The perimeter of the larger polygon is 120 feet. What is the perimeter of the smaller polygon? (Lesson 7-2)

   A 68 ft
   B 72 ft
   C 192 ft
   D 200 ft

Determine whether the triangles are similar. If so, write a similarity statement. If not, what would be sufficient to prove the triangles similar? Explain your reasoning. (Lesson 7-3)

9. 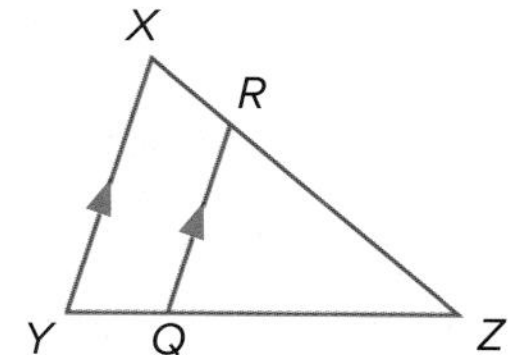

10. 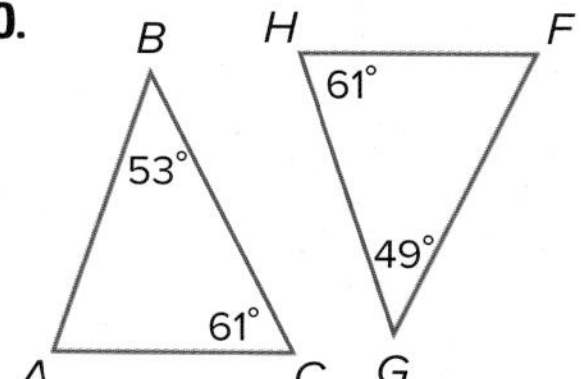

ALGEBRA Identify the similar triangles. Find the value of x in each case. (Lesson 7-3)

11. SR

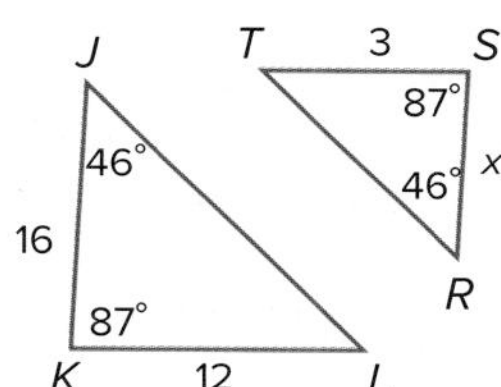

12. AF

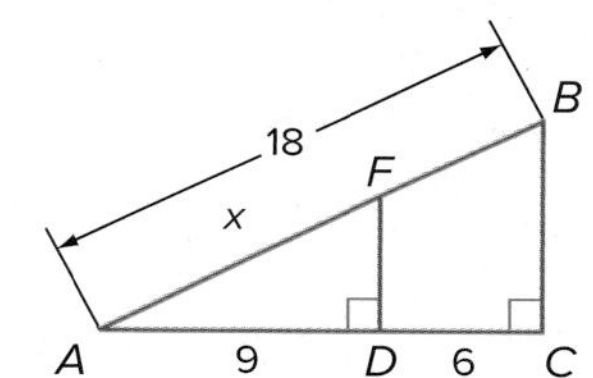

13. Is $\triangle TWZ$ similar to $\triangle YWX$? Explain your reasoning. (Lesson 7-4)

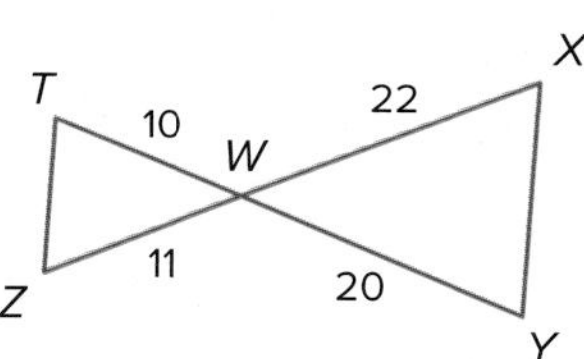

14. **TENNIS** Justin is playing tennis. When serving, he stands 12 feet away from the net, which is 3 feet tall. The ball is served from a height of 7.5 feet. Justin thinks the ball travels about 21.4 feet before it hits the ground 8 feet from the net on the opposite side. (Lesson 7-4)
    a. How far does the ball travel before it reaches the net? Round your answer to the nearest tenth if necessary.
    b. What assumptions did you make to solve for the distance the ball travels?
    c. What mathematical practice did you use to solve this problem?

LESSON 5

Parallel Lines and Proportional Parts

::Then

- You used proportions to solve problems between similar triangles.

::Now

- 1 Use proportional parts within triangles.
- 2 Use proportional parts with parallel lines.

::Why?

- Photographers have many techniques at their disposal that can be used to add interest to a photograph. One such technique is the use of a vanishing point perspective, in which an image with parallel lines, such as train tracks, is photographed so that the lines appear to converge at a point on the horizon.

 **New Vocabulary**

midsegment of a triangle

 **Mathematical Practices**

1 Make sense of problems and persevere in solving them.

3 Construct viable arguments and critique the reasoning of others.

1 Proportional Parts Within Triangles

When a triangle contains a line that is parallel to one of its sides, the two triangles formed can be proved similar using the Angle-Angle Similarity Postulate. Because the triangles are similar, their sides are proportional.

Theorem 7.5 Triangle Proportionality Theorem

If a line is parallel to one side of a triangle and intersects the other two sides, then it divides the sides into segments of proportional lengths.

Example If $\overline{BE} \parallel \overline{CD}$, then $\frac{AB}{BC} = \frac{AE}{ED}$.

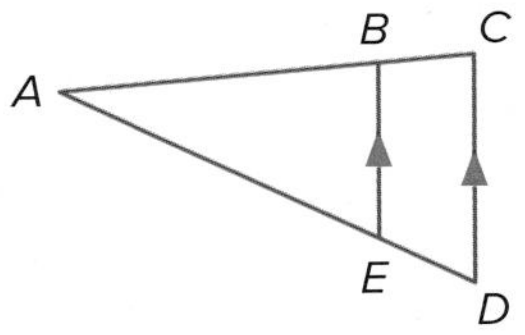

You will prove Theorem 7.5 in Exercise 30.

Example 1 Find the Length of a Side

In $\triangle PQR$, $\overline{ST} \parallel \overline{RQ}$. If $PT = 7.5$, $TQ = 3$, and $SR = 2.5$, find PS.

Use the Triangle Proportionality Theorem.

$\frac{PS}{SR} = \frac{PT}{TQ}$ Triangle Proportionality Theorem

$\frac{PS}{2.5} = \frac{7.5}{3}$ Substitute.

$PS \cdot 3 = (2.5)(7.5)$ Cross Products Property

$3PS = 18.75$ Multiply.

$PS = 6.25$ Divide each side by 3.

P, S, T, Q, R

Guided Practice

1. If $PS = 12.5$, $SR = 5$, and $PT = 15$, find TQ.

flickr/Getty Images

Math History Link

Galileo Galilei (1564–1642) Galileo was born in Pisa, Italy. He studied philosophy, astronomy, and mathematics. Galileo made essential contributions to all three disciplines.

Source: *Encyclopaedia Britannica*

The converse of Theorem 7.5 can be proven using the proportional parts of a triangle.

Theorem 7.6 Converse of Triangle Proportionality Theorem

If a line intersects two sides of a triangle and separates the sides into proportional corresponding segments, then the line is parallel to the third side of the triangle.

Example If $\frac{AE}{EB} = \frac{CD}{DB}$, then $\overline{AC} \parallel \overline{ED}$.

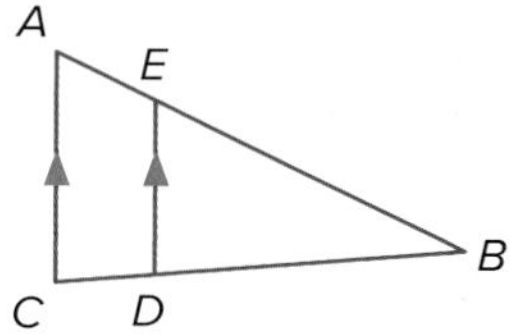

You will prove Theorem 7.6 in Exercise 31.

Example 2 Determine if Lines Are Parallel

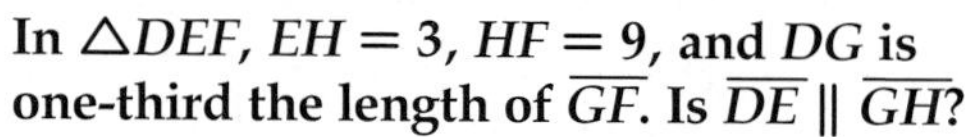

In $\triangle DEF$, $EH = 3$, $HF = 9$, and DG is one-third the length of $\overline{GF}$. Is $\overline{DE} \parallel \overline{GH}$?

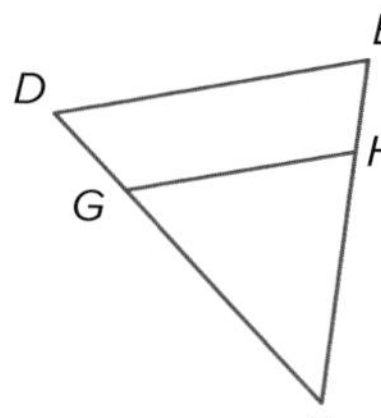

Using the converse of the Triangle Proportionality Theorem, in order to show that $\overline{DE} \parallel \overline{GH}$, we must show that $\frac{DG}{GF} = \frac{EH}{HF}$.

Find and simplify each ratio. Let $DG = x$. Because DG is one-third of GF, $GF = 3x$.

$\frac{DG}{GF} = \frac{x}{3x}$ or $\frac{1}{3}$ $\qquad$ $\frac{EH}{HF} = \frac{3}{9}$ or $\frac{1}{3}$

Because $\frac{1}{3} = \frac{1}{3}$, the sides are proportional, so $\overline{DE} \parallel \overline{GH}$.

Guided Practice

2. DG is half the length of $\overline{GF}$, $EH = 6$, and $HF = 10$. Is $\overline{DE} \parallel \overline{GH}$?

Study Tip ELL

MP Sense-Making The three midsegments of a triangle form the *midsegment triangle.*

A **midsegment of a triangle** is a segment with endpoints that are the midpoints of two sides of the triangle. Every triangle has three midsegments. The midsegments of $\triangle ABC$ are $\overline{RP}$, $\overline{PQ}$, and $\overline{RQ}$.

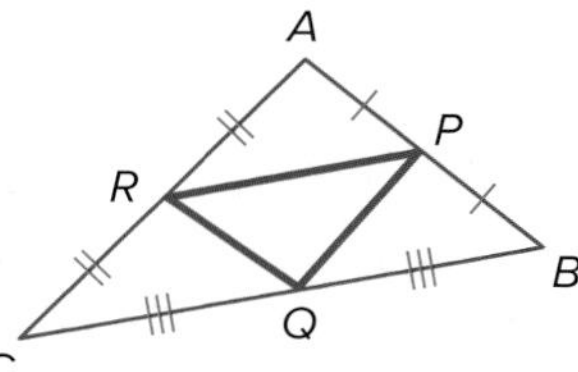

A special case of the Triangle Proportionality Theorem is the Triangle Midsegment Theorem. The Triangle Midsegment Theorem is similar to the Trapezoid Midsegment Theorem, which states that the midsegment of a trapezoid is parallel to the bases and its length is one half the sum of the measures of the bases.

Theorem 7.7 Triangle Midsegment Theorem

A midsegment of a triangle is parallel to one side of the triangle, and its length is one half the length of that side.

Example If J and K are midpoints of $\overline{FH}$ and $\overline{HG}$, respectively, then $\overline{JK} \parallel \overline{FG}$ and $JK = \frac{1}{2}FG$.

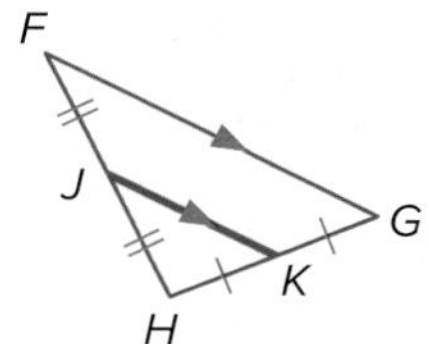

You will prove Theorem 7.7 in Exercise 32.

Example 3 Use the Triangle Midsegment Theorem

In the figure, $\overline{XY}$ and $\overline{XZ}$ are midsegments of $\triangle RST$. Find each measure.

a. XZ

$XZ = \frac{1}{2}RT$ Triangle Midsegment Theorem

$XZ = \frac{1}{2}(13)$ Substitution

$XZ = 6.5$ Simplify.

b. ST

$XY = \frac{1}{2}ST$ Triangle Midsegment Theorem

$7 = \frac{1}{2}ST$ Substitution

$14 = ST$ Multiply each side by 2.

c. $m\angle RYX$

By the Triangle Midsegment Theorem, $\overline{XZ} \parallel \overline{RT}$.

$\angle RYX \cong \angle YXZ$ Alternate Interior Angles Theorem

$m\angle RYX = m\angle YXZ$ Definition of congruence

$m\angle RYX = 124$ Substitution

Guided Practice

Find each measure.

3A. DE

3B. DB

3C. $m\angle FED$

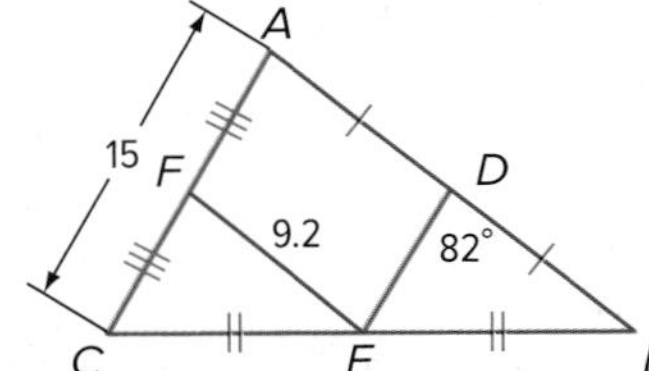

2 Proportional Parts with Parallel Lines

Another special case of the Triangle Proportionality Theorem involves three or more parallel lines cut by two transversals. Notice that if transversals a and b are extended, they form triangles with the parallel lines.

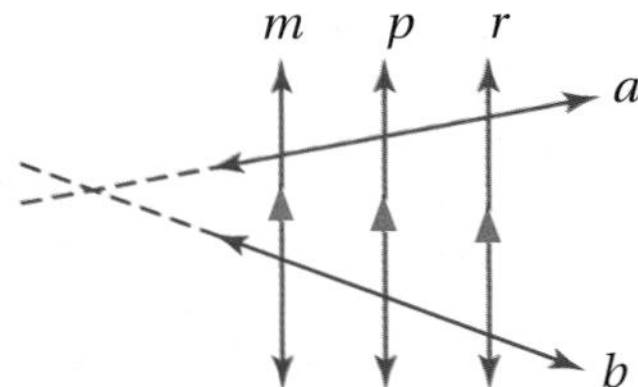

Study Tip

Other Proportions Two other proportions can be written for the example in Corollary 7.1.

$\frac{AB}{EF} = \frac{BC}{FG}$ and $\frac{AC}{BC} = \frac{EG}{FG}$

Corollary 7.1 Proportional Parts of Parallel Lines

If three or more parallel lines intersect two transversals, then they cut off the transversals proportionally.

Example If $\overline{AE} \parallel \overline{BF} \parallel \overline{CG}$, then $\frac{AB}{BC} = \frac{EF}{FG}$.

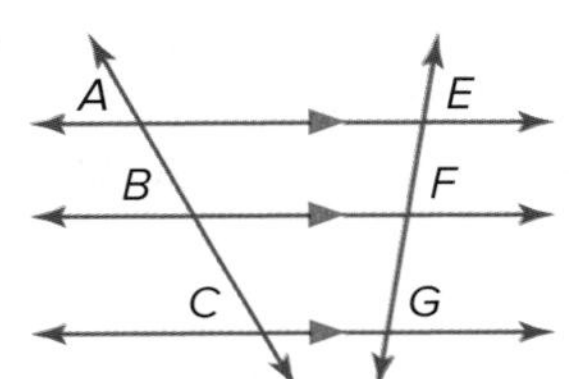

You will prove Corollary 7.1 in Exercise 28.

Real-World Link

To make a two-dimensional drawing appear three-dimensional, an artist provides several perceptual cues.

- *size* - faraway items look smaller
- *clarity* - closer objects appear more in focus
- *detail* - nearby objects have texture, while distant ones are roughly outlined

Source: Center for Media Literacy

Real-World Example 4 Use Proportional Segments of Transversals

ART Megan is drawing a hallway in one-point perspective. She uses the guidelines shown to draw two windows on the left wall. If segments $\overline{AD}$, $\overline{BC}$, $\overline{WZ}$, and $\overline{XY}$ are all parallel, $AB = 8$ centimeters, $DC = 9$ centimeters, and $ZY = 5$ centimeters, find WX.

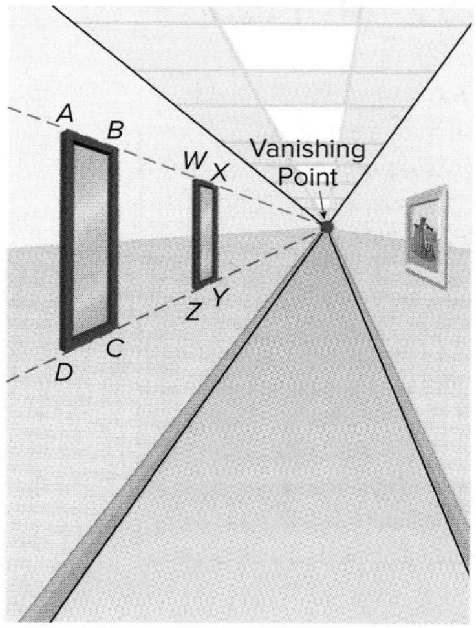

By Corollary 7.1, if $\overline{AD} \parallel \overline{BC} \parallel \overline{WZ} \parallel \overline{XY}$, then $\frac{AB}{WX} = \frac{DC}{ZY}$.

$\frac{AB}{WX} = \frac{DC}{ZY}$ Corollary 7.1

$\frac{8}{WX} = \frac{9}{5}$ Substitute.

$WX \cdot 9 = 8 \cdot 5$ Cross Products Property

$9WX = 40$ Simplify.

$WX = \frac{40}{9}$ Divide each side by 4.

The distance between W and X should be $\frac{40}{9}$ or about 4.4 centimeters.

CHECK The ratio of DC to ZY is 9 to 5, which is about 10 to 5 or 2 to 1. The ratio of AB to WX is 8 to 4.4 or about 8 to 4 or 2 to 1 as well, so the answer is reasonable. ✓

Guided Practice

4. REAL ESTATE *Frontage* is the measurement of a property's boundary that runs along the side of a particular feature such as a street, lake, ocean, or river. Find the ocean frontage for Lot A to the nearest tenth of a yard.

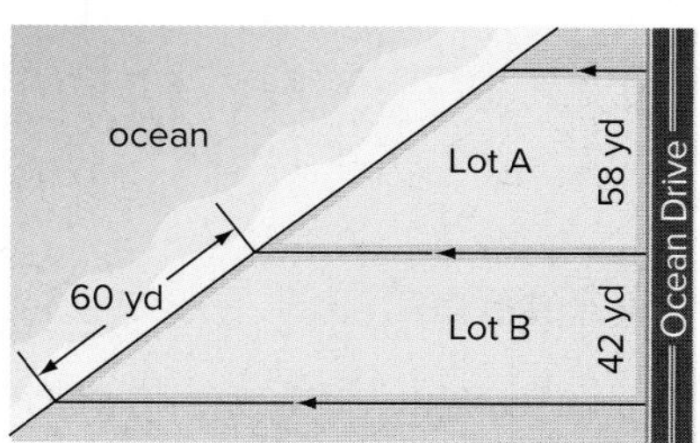

If the scale factor of the proportional segments is 1, then they separate the transversals into congruent parts.

Corollary 7.2 Congruent Parts of Parallel Lines

If three or more parallel lines cut off congruent segments on one transversal, then they cut off congruent segments on every transversal.

Example If $\overline{AE} \parallel \overline{BF} \parallel \overline{CG}$, and $\overline{AB} \cong \overline{BC}$, then $\overline{EF} \cong \overline{FG}$.

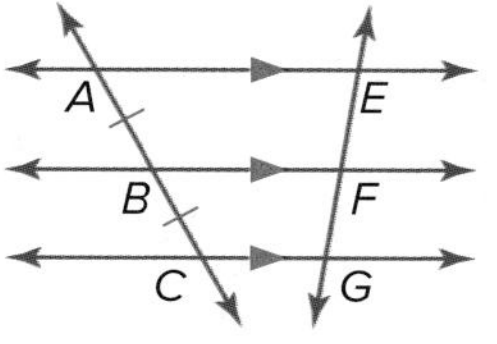

You will prove Corollary 7.2 in Exercise 29.

Go Online!

Log into ConnectED to watch an **Animation** of the trisecting a segment construction. Pause and continue to follow along using your tools.

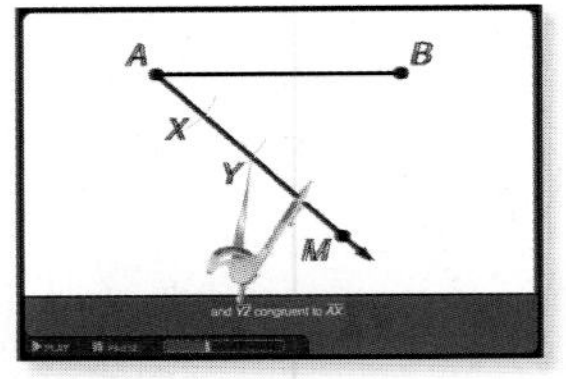

Real-World Example 5 Use Congruent Segments of Transversals

ALGEBRA Find x and y.

Because $\overleftrightarrow{JM} \parallel \overleftrightarrow{KP} \parallel \overleftrightarrow{LQ}$ and $\overline{MP} \cong \overline{PQ}$, then $\overline{JK} \cong \overline{KL}$ by Corollary 7.2.

$JK = KL$	Definition of congruence
$6x - 5 = 4x + 3$	Substitution
$2x - 5 = 3$	Subtract $4x$ from each side.
$2x = 8$	Add 5 to each side.
$x = 4$	Divide each side by 2.
$MP = PQ$	Definition of congruence
$3y + 8 = 5y - 7$	Substitution
$8 = 2y - 7$	Subtract $3y$ from each side.
$15 = 2y$	Add 7 to each side.
$7.5 = y$	Divide each side by 2.

Guided Practice

5A.

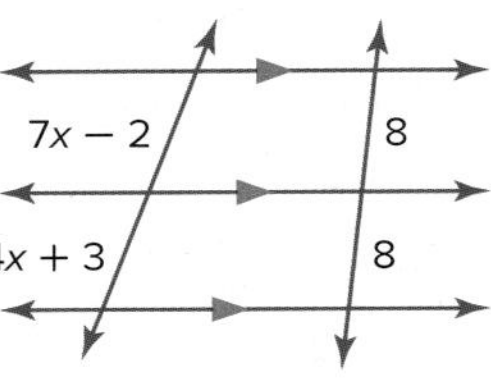

5B.

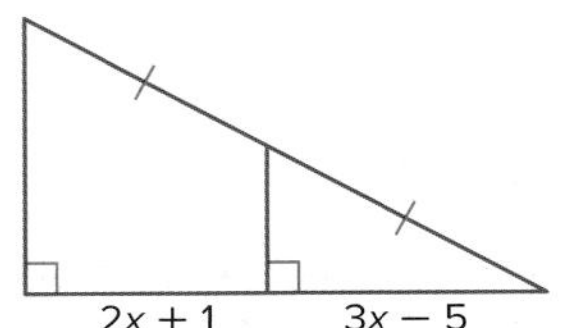

It is possible to separate a segment into two congruent parts by constructing the perpendicular bisector of a segment. However, a segment cannot be separated into three congruent parts by constructing perpendicular bisectors. To do this, you must use parallel lines and Corollary 7.2.

Construction Trisect a Segment

Draw a segment $\overline{AB}$. Then use Corollary 7.2 to trisect $\overline{AB}$.

Step 1 Draw $\overline{AC}$. Then with the compass at A, mark off an arc that intersects $\overline{AC}$ at X.

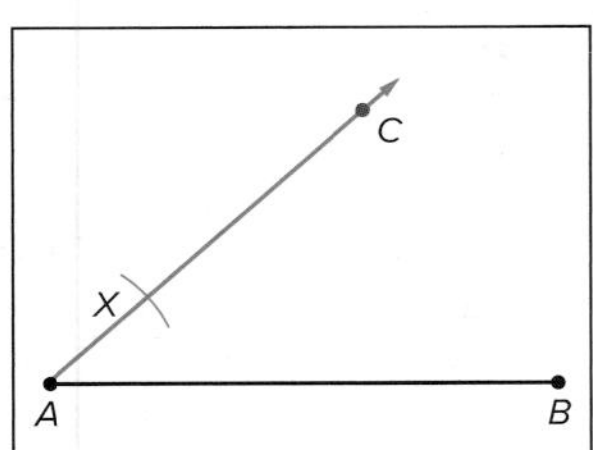

Step 2 Use the same compass setting to mark off Y and Z such that $\overline{AX} \cong \overline{XY} \cong \overline{YZ}$. Then draw ZB.

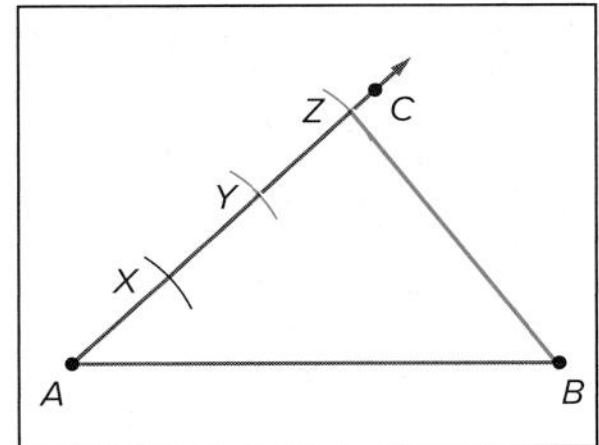

Step 3 Construct lines through Y and X that are parallel to $\overline{ZB}$. Label the intersection points on $\overline{AB}$ as J and K.

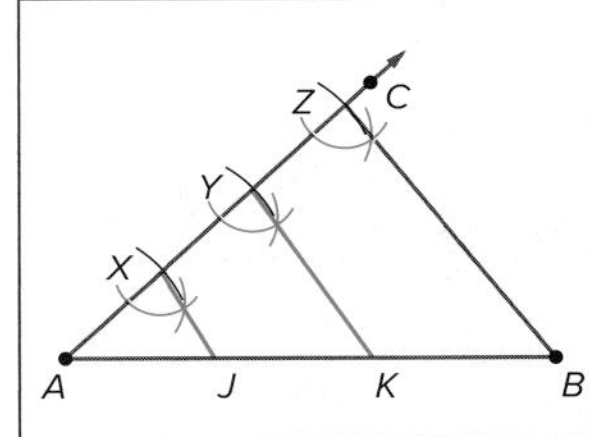

Conclusion: Because parallel lines cut off congruent segments on transversals, $\overline{AJ} \cong \overline{JK} \cong \overline{KB}$.

Check Your Understanding

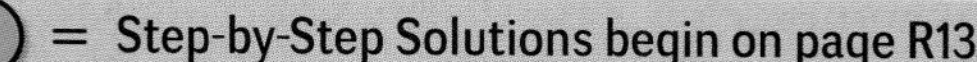

Example 1

1. If $XM = 4$, $XN = 6$, and $NZ = 9$, find XY.
2. If $XN = 6$, $XM = 2$, and $XY = 10$, find NZ.

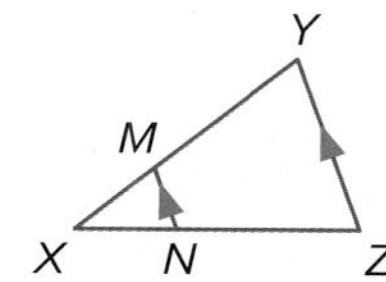

Example 2

3. In $\triangle ABC$, $BC = 15$, $BE = 6$, $DC = 12$, and $AD = 8$. Determine whether $\overline{DE} \parallel \overline{AB}$. Justify your answer.

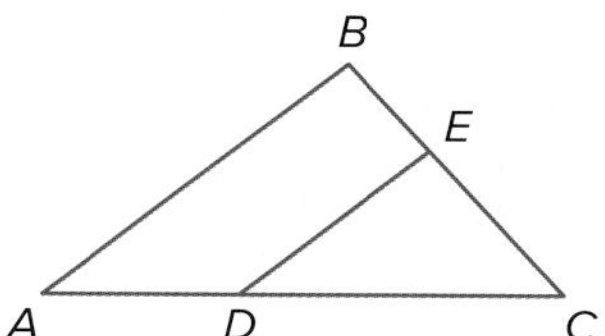

4. In $\triangle JKL$, $JK = 15$, $JM = 5$, $LK = 13$, and $PK = 9$. Determine whether $\overline{JL} \parallel \overline{MP}$. Justify your answer.

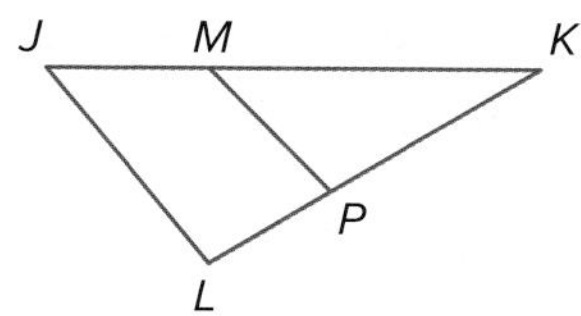

Example 3

$\overline{JH}$ is a midsegment of $\triangle KLM$. Find the value of x.

5.

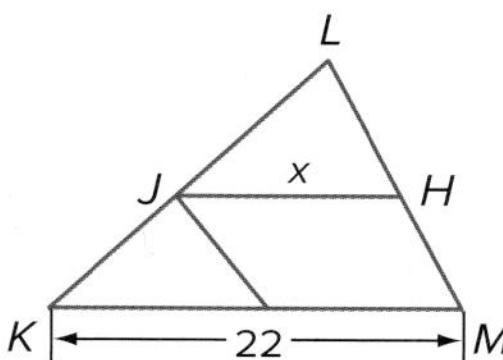

6.

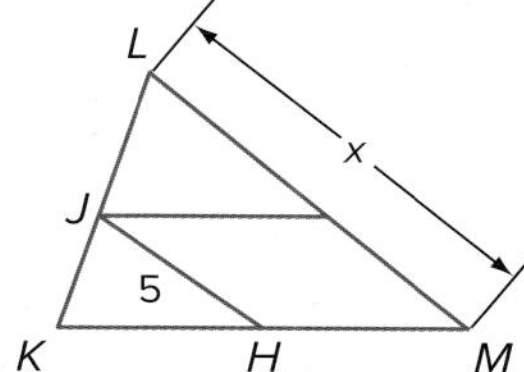

Example 4

7. **MAPS** Refer to the map at the right. 3rd Avenue and 5th Avenue are parallel. If the distance from 3rd Avenue to City Mall along State Street is 3201 feet, find the distance between 5th Avenue and City Mall along Union Street. Round to the nearest tenth.

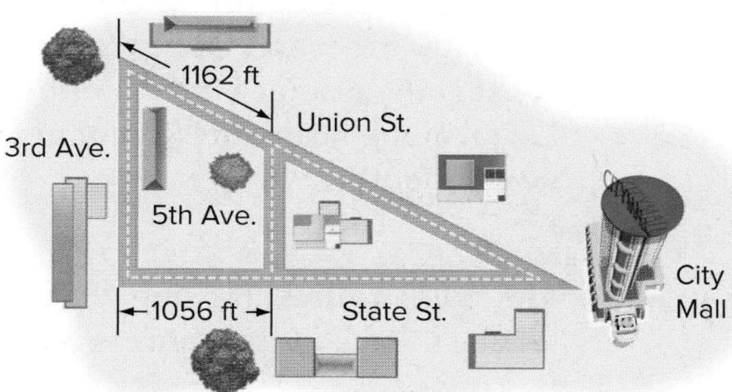

Example 5

ALGEBRA Find x and y.

8.

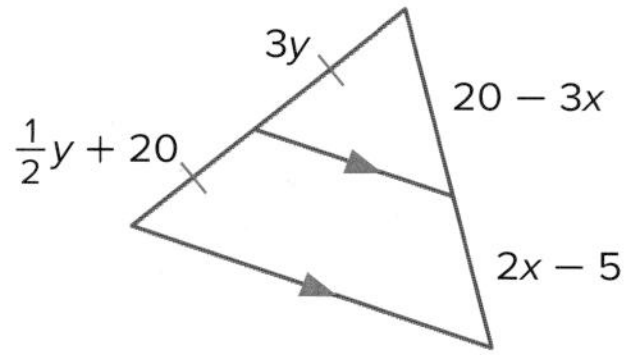

9.

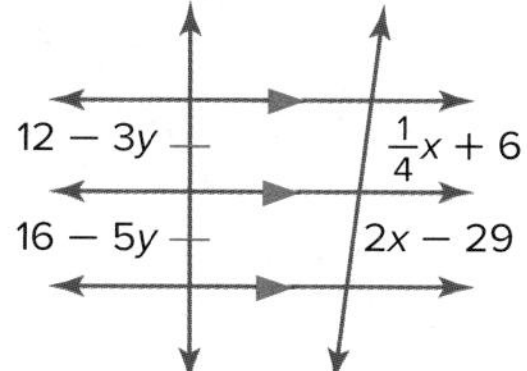

Practice and Problem Solving

Extra Practice is on page R7.

Example 1

10. If $AB = 6$, $BC = 4$, and $AE = 9$, find ED.
11. If $AB = 12$, $AC = 16$, and $ED = 5$, find AE.
12. If $AC = 14$, $BC = 8$, and $AD = 21$, find ED.
13. If $AD = 27$, $AB = 8$, and $AE = 12$, find BC.

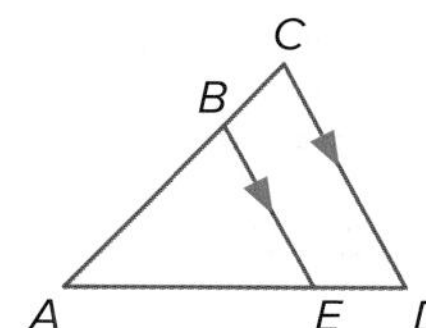

Example 2 **Determine whether $\overline{VY} \parallel \overline{ZW}$. Justify your answer.**

14. $ZX = 18$, $ZV = 6$, $WX = 24$, and $YX = 16$

15. $VX = 7.5$, $ZX = 24$, $WY = 27.5$, and $WX = 40$

16. $ZV = 8$, $VX = 2$, and $YX = \frac{1}{2}WY$

17. $WX = 31$, $YX = 21$, and $ZX = 4ZV$

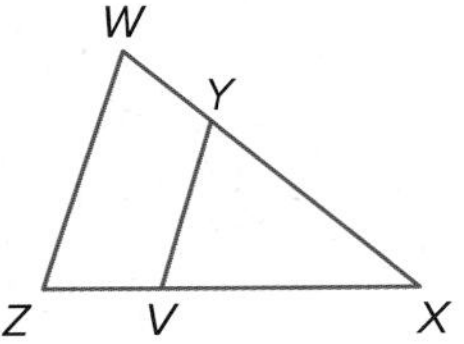

Example 3 **$\overline{JH}$, $\overline{JP}$, and $\overline{PH}$ are midsegments of $\triangle KLM$. Find the value of x.**

18.

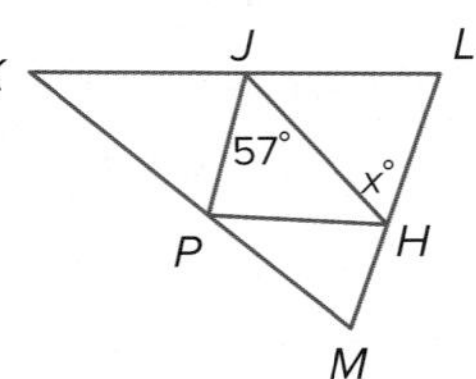

19

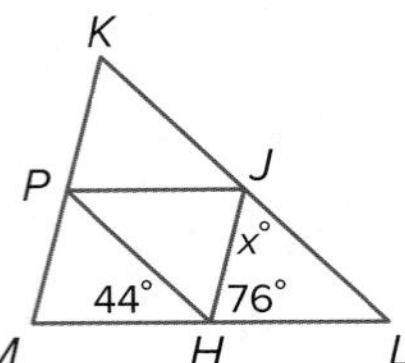

20.

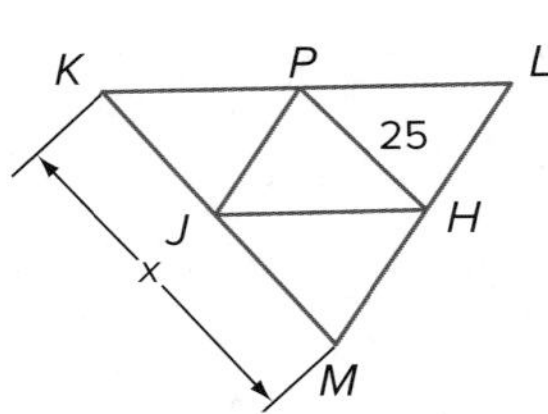

21.

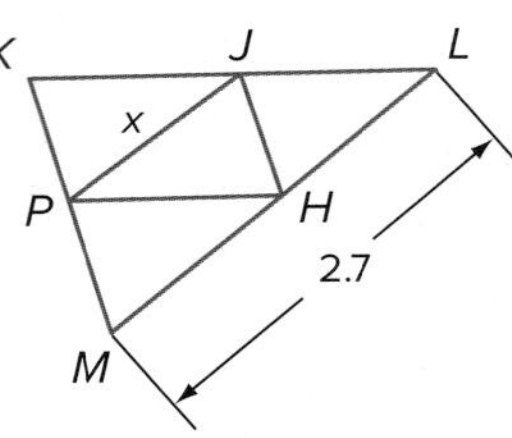

Example 4 **22.** **MODELING** In Charleston, South Carolina, Logan Street is parallel to both King Street and Smith Street between Beaufain Street and Queen Street. What is the distance from Smith to Logan along Beaufain? Round to the nearest foot.

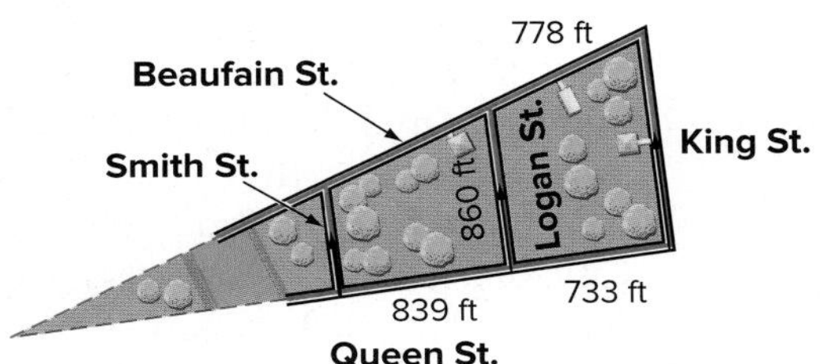

23. **ART** Tonisha drew the line of dancers shown below for her perspective project in art class. Each of the dancers is parallel. Find the lower distance between the first two dancers.

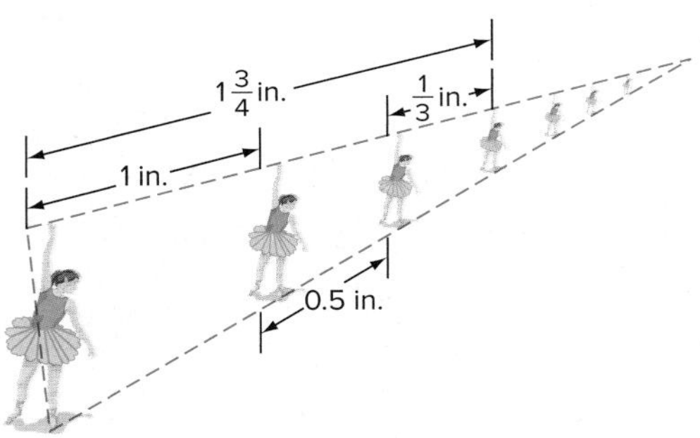

Example 5 **ALGEBRA Find x and y.**

24.

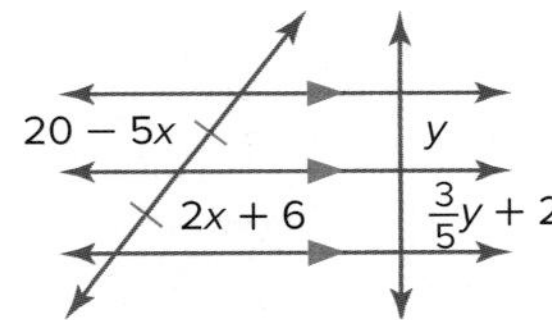

25.

ALGEBRA **Find x and y.**

26.

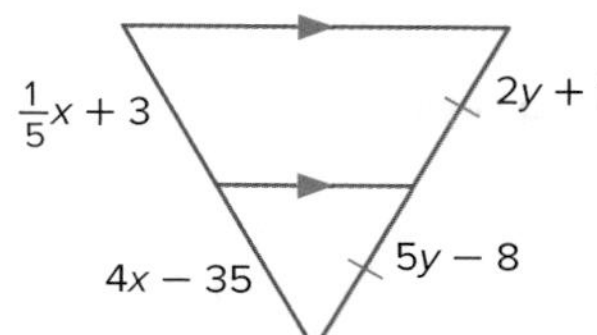

27.

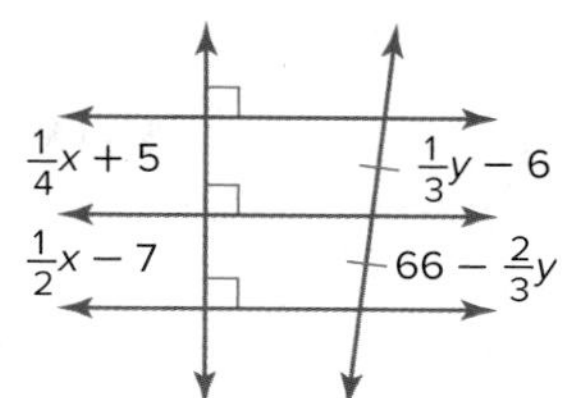

CONSTRUCT ARGUMENTS **Write a paragraph proof.**

28. Corollary 7.1

29. Corollary 7.2

30. Theorem 7.5

CONSTRUCT ARGUMENTS **Write a two-column proof.**

31. Theorem 7.6

32. Theorem 7.7

Refer to $\triangle QRS$.

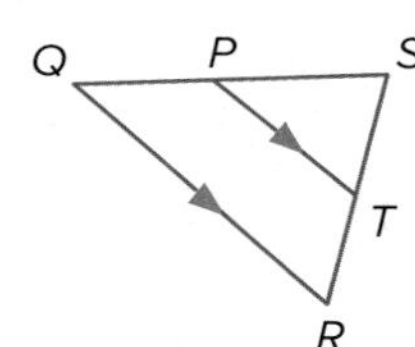

33. If $ST = 8$, $TR = 4$, and $PT = 6$, find QR.

34. If $SP = 4$, $PT = 6$, and $QR = 12$, find SQ.

35 If $CE = t - 2$, $EB = t + 1$, $CD = 2$, and $CA = 10$, find t and CE.

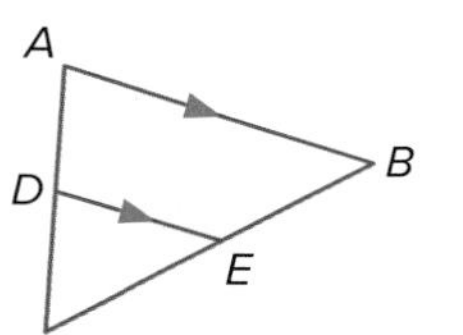

36. If $WX = 7$, $WY = a$, $WV = 6$, and $VZ = a - 9$, find WY.

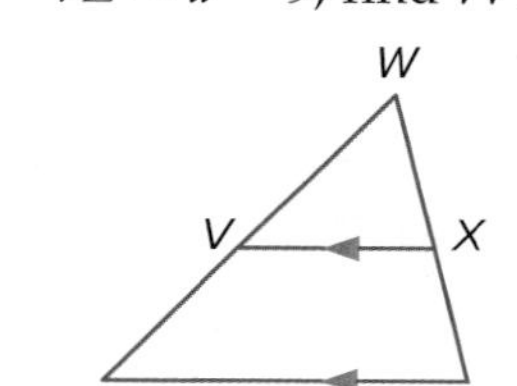

37. If $QR = 2$, $XW = 12$, $QW = 15$, and $ST = 5$, find RS and WV.

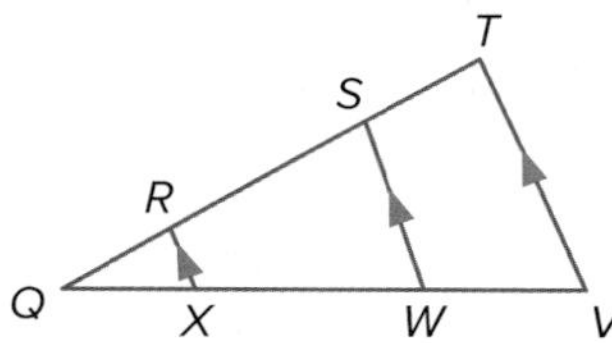

38. If $LK = 4$, $MP = 3$, $PQ = 6$, $KJ = 2$, $RS = 6$, and $LP = 2$, find ML, QR, QK, and JH.

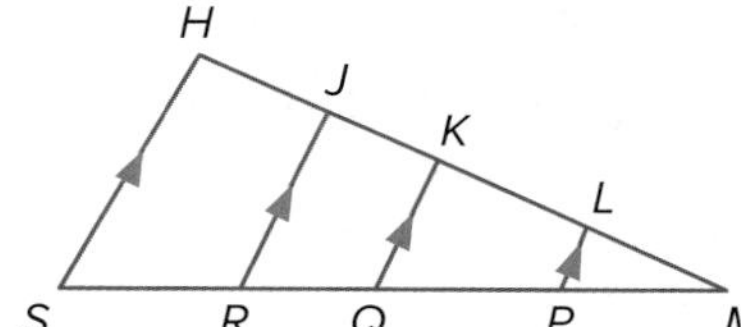

39. **DRAWING TOOLS** The golden ratio is found widely in art, architecture, and nature. The golden ratio is 1.618:1. A Fibonacci gauge is made of two pairs of wooden segments that are pinned so $\overline{AE} \parallel \overline{BG}$ and $\overline{FD} \parallel \overline{AC}$. As the sides of the gauge are moved, the ratio of EC to ED is always the golden ratio. Explain why this gauge works.

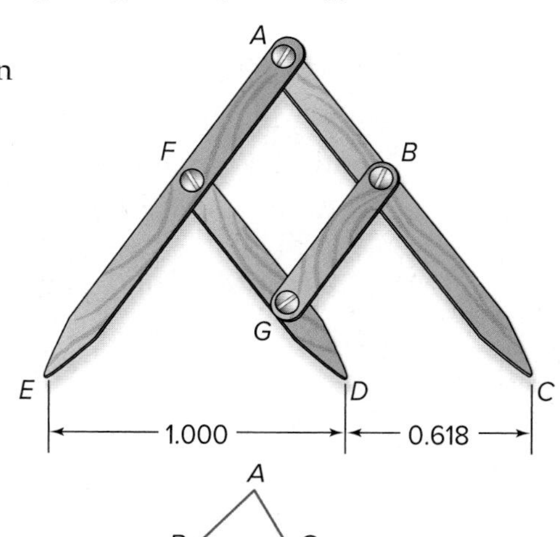

Determine the value of x so that $\overline{BC} \parallel \overline{DF}$.

A
B
C
D
F

40. $AB = x + 5$, $BD = 12$, $AC = 3x + 1$, and $CF = 15$

41. $AC = 15$, $BD = 3x - 2$, $CF = 3x + 2$, and $AB = 12$

42. **COORDINATE GEOMETRY** $\triangle ABC$ has vertices $A(-8, 7)$, $B(0, 1)$, and $C(7, 5)$. Draw $\triangle ABC$. Determine the coordinates of the midsegment of $\triangle ABC$ that is parallel to $\overline{BC}$. Justify your answer.

43. **HOUSES** Refer to the diagram of the gable at the right. Each piece of siding is a uniform width. Find the lengths of $\overline{FG}$, $\overline{EH}$, and $\overline{DJ}$.

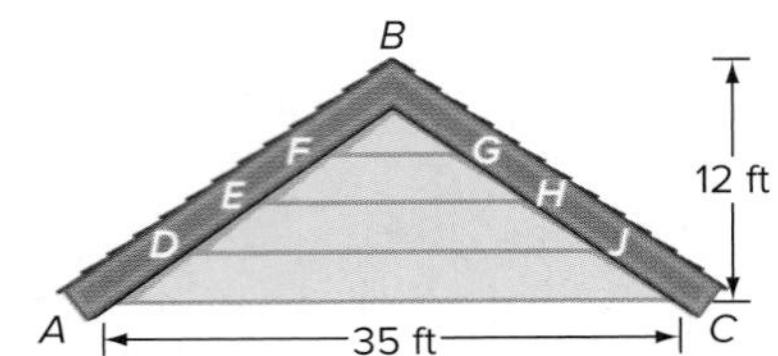

CONSTRUCTIONS **Construct each segment as directed.**

44. a segment separated into five congruent segments

45. a segment separated into two segments in which their lengths have a ratio of 1 to 3

46. a segment 3 inches long, separated into four congruent segments

47. **MULTIPLE REPRESENTATIONS** In this problem, you will explore angle bisectors and proportions.

  a. **Geometric** Draw three triangles: one acute, one right, and one obtuse. Label one triangle ABC and draw angle bisector $\overrightarrow{BD}$. Label the second MNP with angle bisector $\overrightarrow{NQ}$ and the third WXY with angle bisector $\overrightarrow{XZ}$.

  b. **Tabular** Copy and complete the table at the right with the appropriate values.

  c. **Verbal** Make a conjecture about the segments of a triangle created by an angle bisector.

Triangle	Length		Ratio	
ABC	AD		$\frac{AD}{CD}$	
	CD			
	AB		$\frac{AB}{CB}$	
	CB			
MNP	MQ		$\frac{MQ}{PQ}$	
	PQ			
	MN		$\frac{MN}{PN}$	
	PN			
WXY	WZ		$\frac{WZ}{YZ}$	
	YZ			
	WX		$\frac{WX}{YX}$	
	YX			

H.O.T. Problems Use Higher-Order Thinking Skills

48. **MP CRITIQUE ARGUMENTS** Jacob and Sebastian are finding the value of x in $\triangle JHL$. Jacob says that MP is one half of JL, so x is 4.5. Sebastian says that JL is one half of MP, so x is 18. Is either of them correct? Explain.

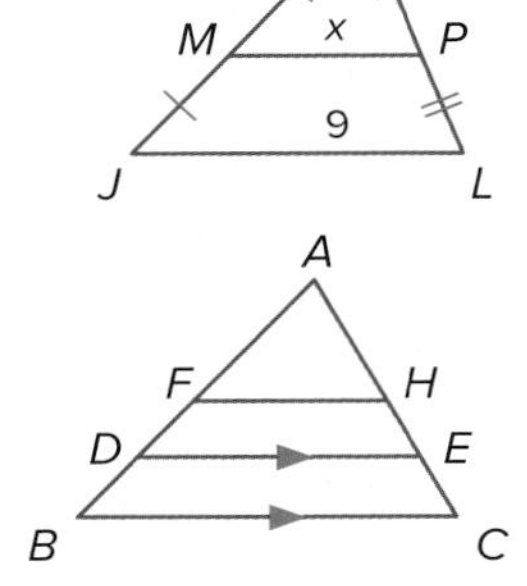

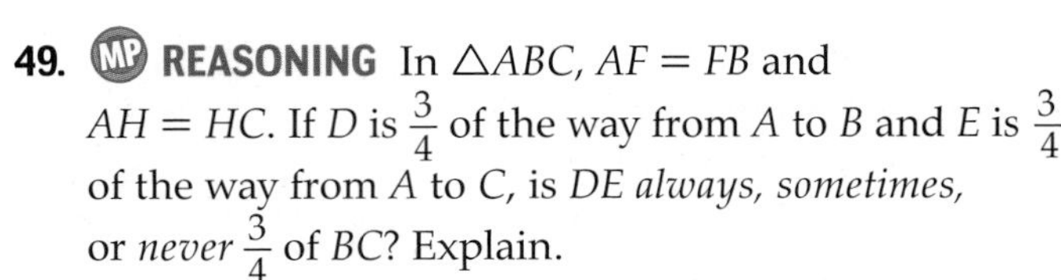

49. **MP REASONING** In $\triangle ABC$, $AF = FB$ and $AH = HC$. If D is $\frac{3}{4}$ of the way from A to B and E is $\frac{3}{4}$ of the way from A to C, is DE *always, sometimes,* or *never* $\frac{3}{4}$ of BC? Explain.

50. **CHALLENGE** Write a two-column proof.

Given: $AB = 4$, $BC = 4$, and $CD = DE$

Prove: $\overline{BD} \parallel \overline{AE}$

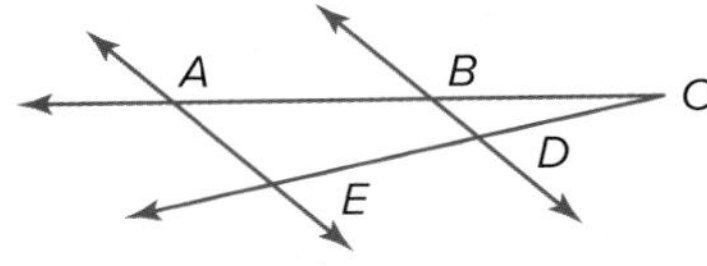

51. **OPEN-ENDED** Draw three segments, a, b, and c, of all different lengths. Draw a fourth segment, d, such that $\frac{a}{b} = \frac{c}{d}$.

52. **WRITING IN MATH** Compare the Triangle Proportionality Theorem and the Triangle Midsegment Theorem.

53. $\overline{RS}$ is a midsegment of $\triangle XYZ$.

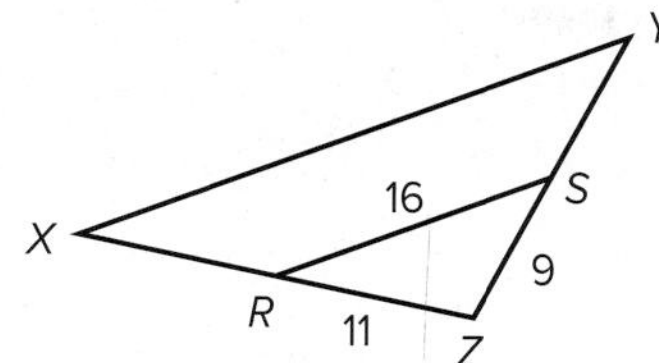

What is the perimeter of $\triangle XYZ$? MP 6

- A 88
- B 72
- C 56
- D 52
- E 36

54. In $\triangle ABC$, P is the midpoint of $\overline{AB}$ and Q is the midpoint of $\overline{BC}$. Which of the following statements is not necessarily true? MP 3

- A $\overline{PQ} \parallel \overline{AC}$
- B $PQ = \frac{1}{2}AC$
- C $BQ = QC$
- D $BP = BQ$

55. In $\triangle FGH$, $\overleftrightarrow{JK}$ is parallel to $\overline{GH}$.

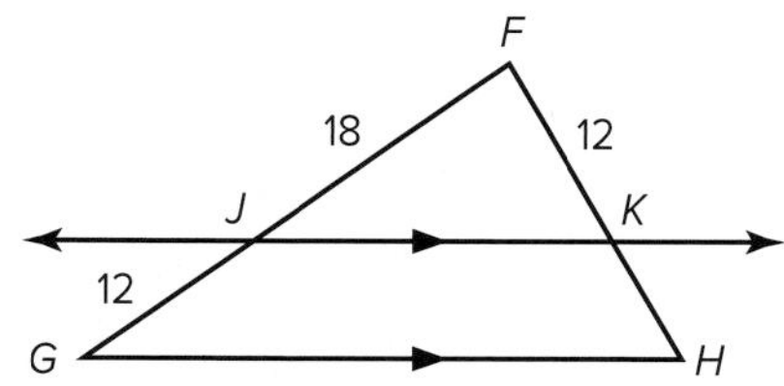

What is the length of $\overline{FH}$? MP 6

- A 8
- B 18
- C 20
- D 24

56. In $\triangle RST$, X is the midpoint of $\overline{RS}$ and Y is the midpoint of $\overline{RT}$. Given that $XY = 3z + 2$ and $ST = 7z - 1$, what is the value of z? MP 2, 3

57. If the vertices of $\triangle JKL$ are (0, 0), (0, 10), and (10, 10), what is the area of $\triangle JKL$ in square units? MP 1

58. **MULTI-STEP** First, Second, and Third Streets are parallel in a town. Waller Avenue is perpendicular to these streets, and Shaw Road runs at an angle to them. What is the distance along Waller Avenue from First Street to Third Street? MP 1, 6

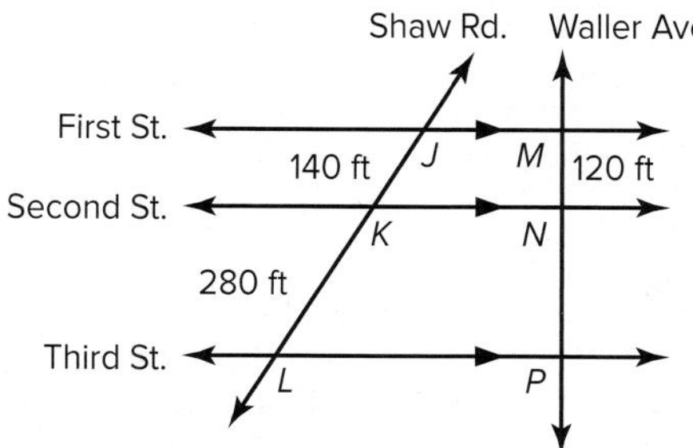

a. Which theorem can you use to solve this problem?

b. Which proportion is correct, and is helpful for solving this problem?

- A $\frac{KN}{JM} = \frac{KL}{JK}$
- B $\frac{JK}{JM} = \frac{KN}{MN}$
- C $\frac{JK}{KL} = \frac{MN}{NP}$
- D $\frac{KN}{NP} = \frac{KL}{LP}$

c. What is NP?

d. What is the distance from First Street to Third Street?

59. In the figure, $\overleftrightarrow{BC} \parallel \overleftrightarrow{DE}$. $AD = 10$; $AB = 6$ and $BC = 4$. Find DE. MP 6

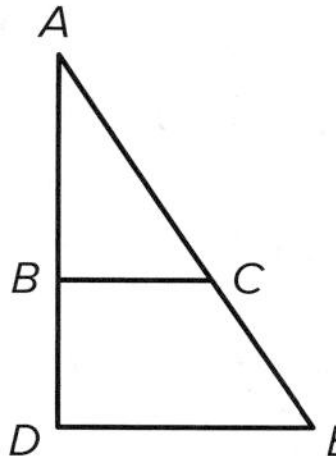

LESSON 6

Parts of Similar Triangles

Then

- You learned that corresponding sides of similar polygons are proportional.

Now

1. Recognize and use proportional relationships of corresponding angle bisectors, altitudes, and medians of similar triangles.
2. Use the Triangle Bisector Theorem.

Why?

- The "Rule of Thumb" uses the average ratio of a person's arm length to the distance between his or her eyes and the altitudes of similar triangles to estimate the distance between a person and an object of approximately known width.

Mathematical Practices

1 Make sense of problems and persevere in solving them.

3 Construct viable arguments and critique the reasoning of others.

1 Special Segments of Similar Triangles

You learned in Lesson 7-2 that the corresponding side lengths of similar polygons, such as triangles, are proportional. This concept can be extended to other segments in triangles.

Theorems Special Segments of Similar Triangles

7.8 If two triangles are similar, the lengths of corresponding altitudes are proportional to the lengths of corresponding sides.

Abbreviation ~$\triangle$s *have corr. altitudes proportional to corr. sides.*

Example If $\triangle ABC \sim \triangle FGH$, then $\frac{AD}{FJ} = \frac{AB}{FG}$.

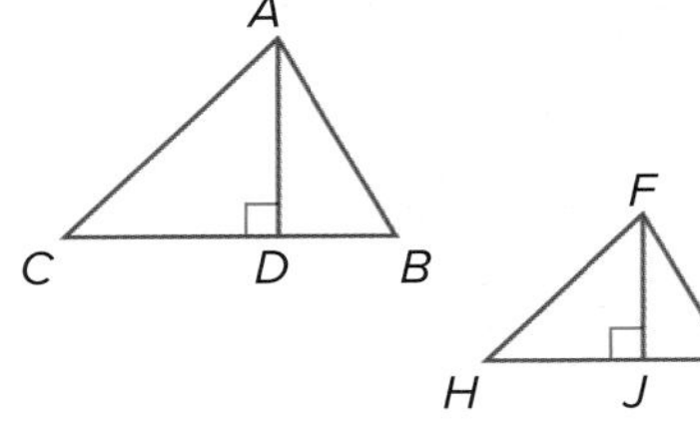

7.9 If two triangles are similar, the lengths of corresponding angle bisectors are proportional to the lengths of corresponding sides.

Abbreviation ~$\triangle$s *have corr. $\angle$ bisectors proportional to corr. sides.*

Example If $\triangle KLM \sim \triangle QRS$, then $\frac{LP}{RT} = \frac{LM}{RS}$.

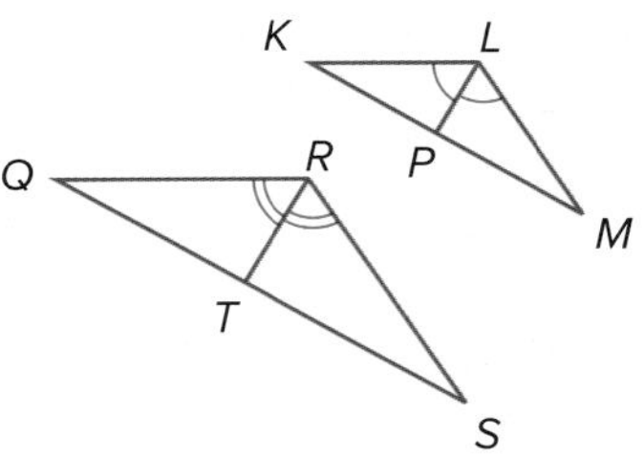

7.10 If two triangles are similar, the lengths of corresponding medians are proportional to the lengths of corresponding sides.

Abbreviation ~$\triangle$s *have corr. medians proportional to corr. sides.*

Example If $\triangle ABC \sim \triangle WXY$, then $\frac{CD}{YZ} = \frac{AB}{WX}$.

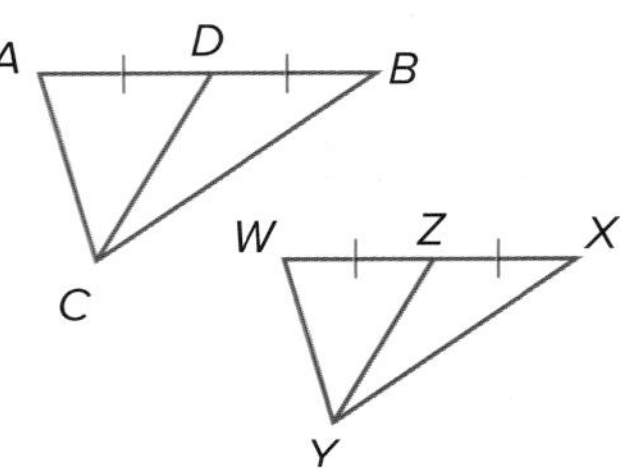

You will prove Theorems 7.9 and 7.10 in Exercises 18 and 19, respectively.

Real-World Career

Athletic Trainer Athletic trainers help prevent and treat sports injuries. They ensure that protective equipment is used properly and that people understand safe practices that prevent injury. An athletic trainer must have a bachelor's degree to be certified. Most also have master's degrees. Refer to Exercise 29.

Proof Theorem 7.8

Given: $\triangle FGH \sim \triangle KLM$
$\overline{FJ}$ and $\overline{KP}$ are altitudes.

Prove: $\frac{FJ}{KP} = \frac{HF}{MK}$

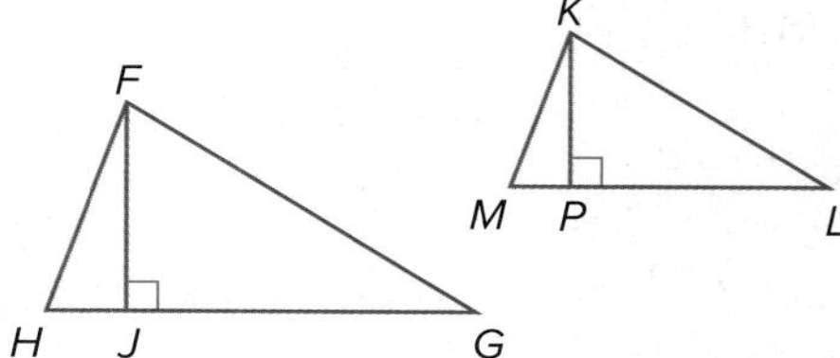

Paragraph Proof:
Because $\triangle FGH \sim \triangle KLM$, $\angle H \cong \angle M$. $\angle FJH \cong \angle KPM$ because they are both right angles created by the altitudes drawn to the opposite sides and all right angles are congruent.

Thus $\triangle HFJ \sim \triangle MKP$ by AA Similarity. So $\frac{FJ}{KP} = \frac{HF}{MK}$ by the definition of similar polygons.

Because the corresponding altitudes are chosen at random, we need not prove Theorem 7.8 for every pair of altitudes.

You can use special segments in similar triangles to find missing measures.

Example 1 Use Special Segments in Similar Triangles

In the figure, $\triangle ABC \sim \triangle FDG$. Find the value of x.

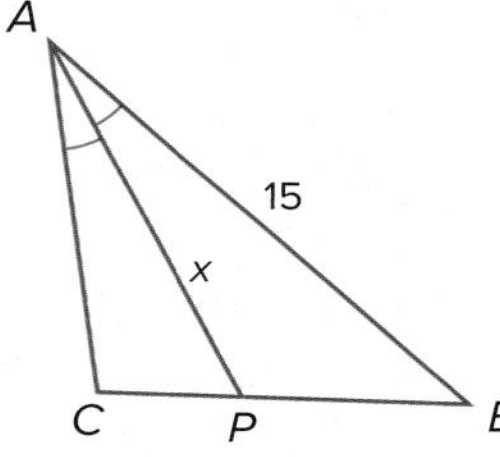

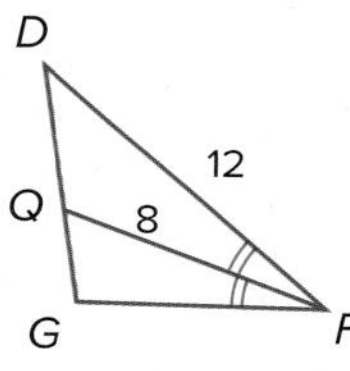

$\overline{AP}$ and $\overline{FQ}$ are corresponding angle bisectors and $\overline{AB}$ and $\overline{FD}$ are corresponding sides of similar triangles ABC and FDG.

$\frac{AP}{FQ} = \frac{AB}{FD}$	$\sim\triangle$s have corr. $\angle$ bisectors proportional to the corr. sides.
$\frac{x}{8} = \frac{15}{12}$	Substitution
$8 \cdot 15 = x \cdot 12$	Cross Products Property
$120 = 12x$	Simplify.
$10 = x$	Divide each side by 12.

Study Tip

MP Sense-Making Example 1 could also have been solved by first finding the scale factor between $\triangle ABC$ and $\triangle FDG$. The ratio of the angle bisector in $\triangle ABC$ to the angle bisector in $\triangle FDG$ would then be equal to this scale factor.

Guided Practice

Find the value of x.

1A.

1B.

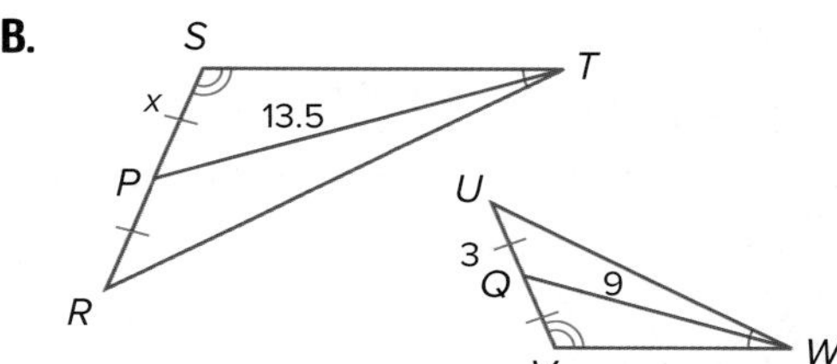

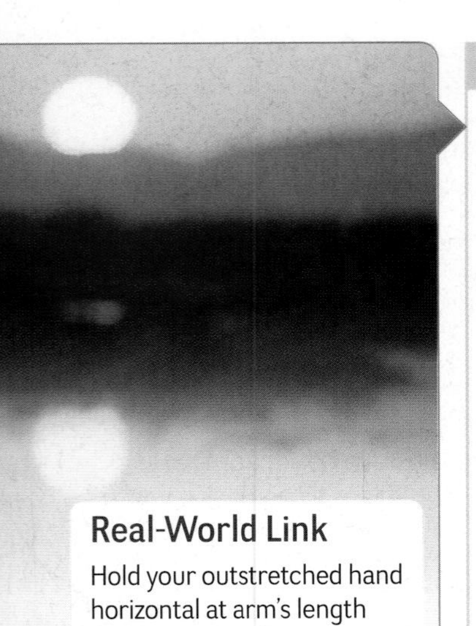

Real-World Link

Hold your outstretched hand horizontal at arm's length with your palm facing you; for each hand width the sun is above the horizon, there is one remaining hour of sunlight.

Source: Sail Island Channels

Go Online!

Watch **Personal Tutor** videos to hear descriptions of using similar triangles to solve problems. Try describing how to solve a problem for a partner. Have them ask you questions to help your understanding.

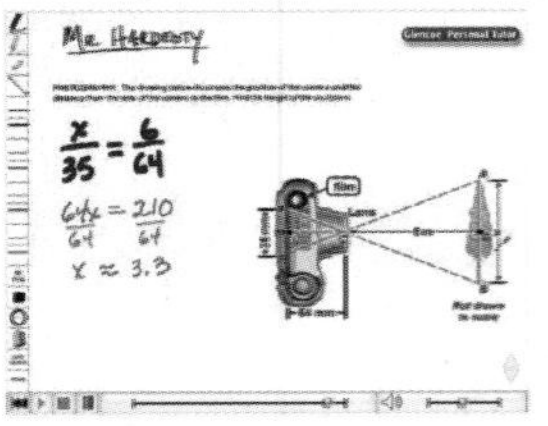

Real-World Example 2 Use Similar Triangles to Solve Problems

ESTIMATING DISTANCES **Liliana holds her arm straight out in front of her with her elbow straight and her thumb pointing up. Closing one eye, she aligns one edge of her thumb with a car she is sighting. Next she switches eyes without moving her head or her arm. The car appears to jump 4 car widths. If Liliana's arm is about 10 times longer than the distance between her eyes, and the car is about 5.5 feet wide, estimate the distance from Liliana's thumb to the car.**

Understand Make a diagram labeling the given distances and the distance you need to find as x. Label the vertices of the triangles formed.

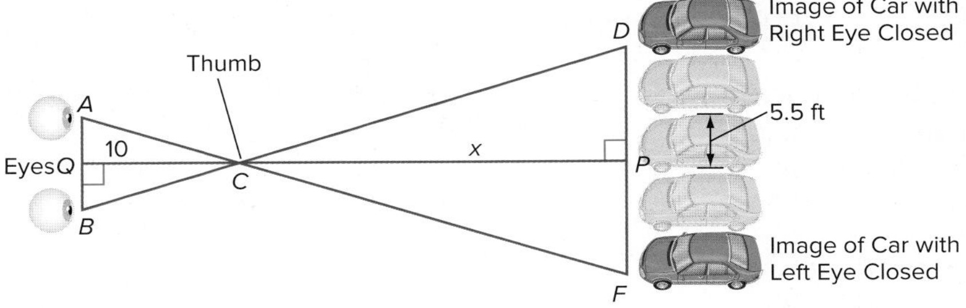

Note: Not drawn to scale.

We assume that if Liliana's thumb is straight out in front of her, then $\overline{QC}$ is an altitude of $\triangle ABC$. Likewise, $\overline{PC}$ is the corresponding altitude. We assume that $\overline{AB} \parallel \overline{DF}$.

Plan Because $\overline{AB} \parallel \overline{DF}$, $\angle BAC \cong \angle DFC$ and $\angle CBA \cong \angle CDF$ by the Alternate Interior Angles Theorem. Therefore $\triangle ABC \sim \triangle FDC$ by AA Similarity. Write a proportion and solve for x.

Solve

$\frac{PC}{QC} = \frac{AB}{DF}$	Theorem 7.8
$\frac{10}{x} = \frac{1}{5.5 \cdot 4}$	Substitution
$\frac{10}{x} = \frac{1}{22}$	Simplify.
$10 \cdot 22 = x \cdot 1$	Cross Products Property
$220 = x$	Simplify.

So the estimated distance to the car is 220 feet.

Check The ratio of Liliana's arm length to the width between her eyes is 10 to 1. The ratio of the distance to the car to the distance the image of the car jumped is 220 to 22 or 10 to 1. ✓

Creating a diagram allows us to compare the corresponding angles and thus determine that the triangles are similar.

Guided Practice

2. Suppose Liliana stands at the back of her classroom and sights a clock on the wall at the front of the room. If the clock is 30 centimeters wide and appears to move 3 clock widths when she switches eyes, estimate the distance from Liliana's thumb to the clock.

2 Triangle Angle Bisector Theorem

An angle bisector of a triangle also divides the side opposite the angle proportionally.

Theorem 7.11 Triangle Angle Bisector

An angle bisector in a triangle separates the opposite side into two segments that are proportional to the lengths of the other two sides.

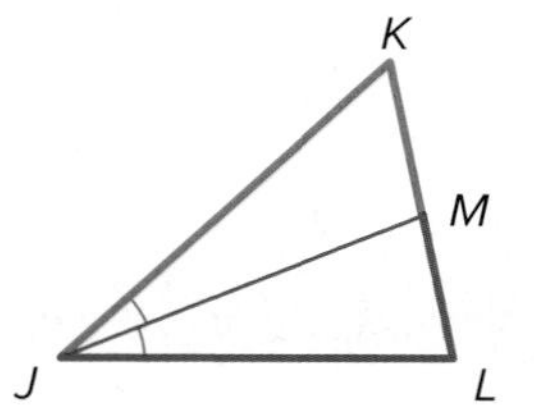

Example If $\overline{JM}$ is an angle bisector of $\triangle JKL$,

then $\frac{KM}{LM} = \frac{KJ}{LJ}$. ← segments with vertex K ← segments with vertex L

You will prove Theorem 7.11 in Exercise 25.

Example 3 Use the Triangle Angle Bisector Theorem

Find x.

Because $\overline{RT}$ is an angle bisector of $\triangle QRS$, you can use the Triangle Angle Bisector Theorem to write a proportion.

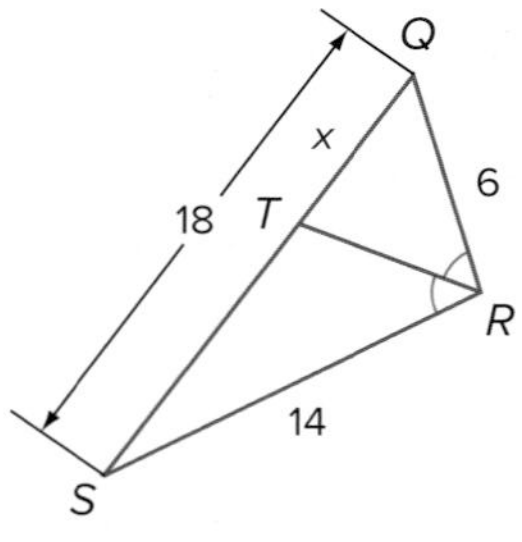

$\frac{QT}{ST} = \frac{QR}{SR}$	Triangle Angle Bisector Theorem
$\frac{x}{18 - x} = \frac{6}{14}$	Substitution
$(18 - x)(6) = x \cdot 14$	Cross Products Property
$108 - 6x = 14x$	Simplify.
$108 = 20x$	Add $6x$ to each side.
$5.4 = x$	Divide each side by 20.

Guided Practice

Find the value of x.

3A.

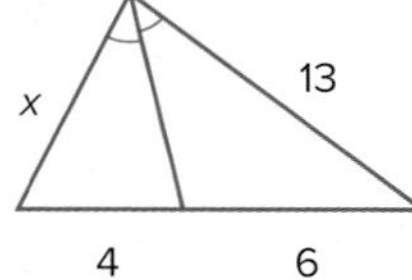

3B.

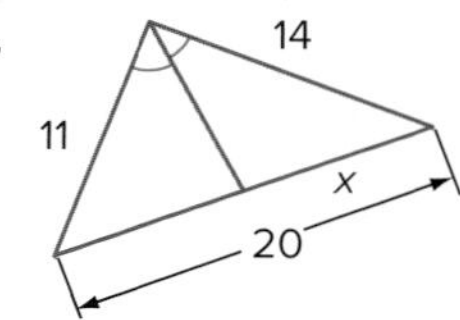

Check Your Understanding

 = Step-by-Step Solutions begin on page R13.

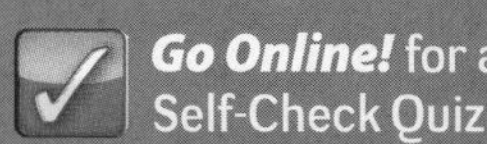

Example 1 Find x.

1

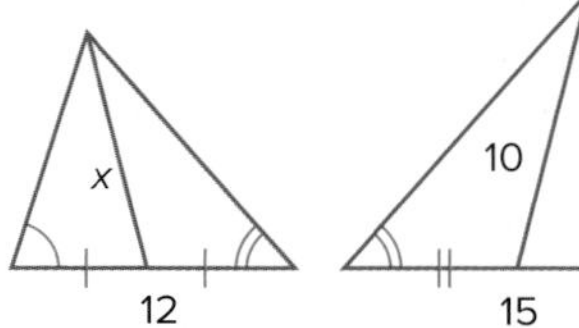

2.

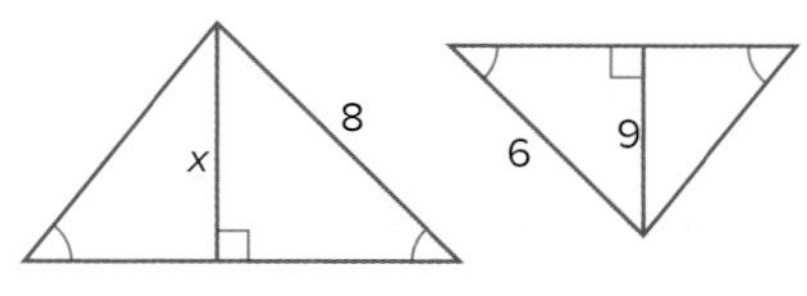

Example 2

3. VISION A cat that is 10 inches tall forms a retinal image that is 7 millimeters tall. If $\triangle ABE \sim \triangle DBC$ and the distance from the pupil to the retina is 25 millimeters, how far away from your pupil is the cat?

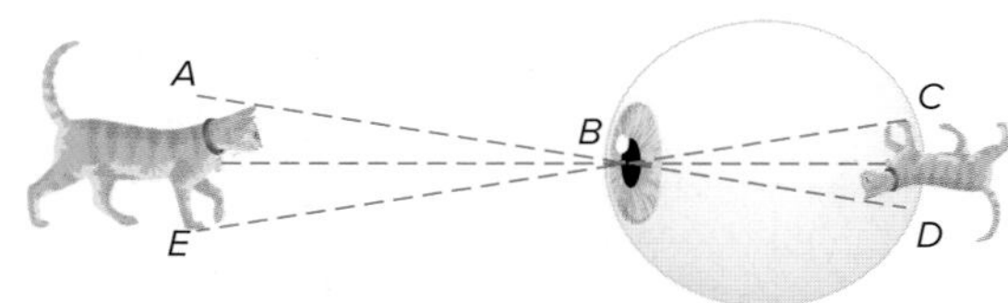

Example 3 **Find the value of each variable.**

4. 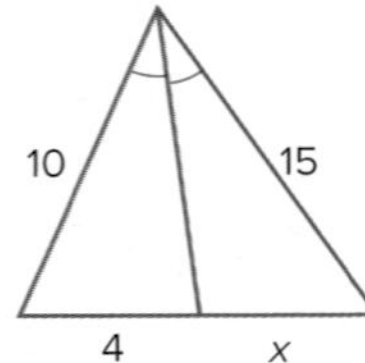

5. 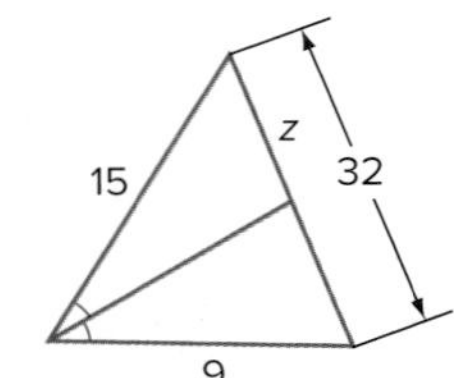

Practice and Problem Solving

Extra Practice is on page R7.

Example 1 **Find x.**

6. 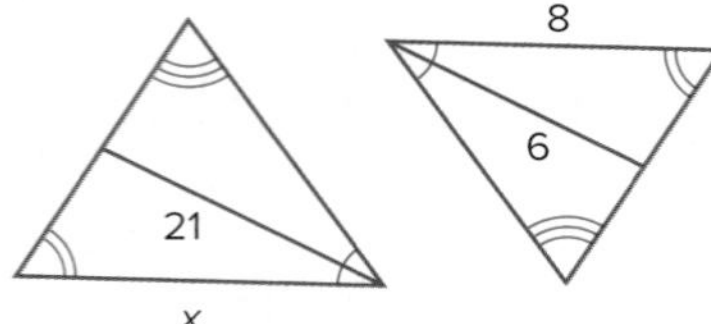

7. 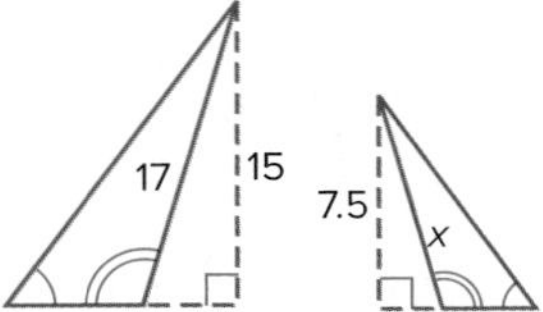

8. 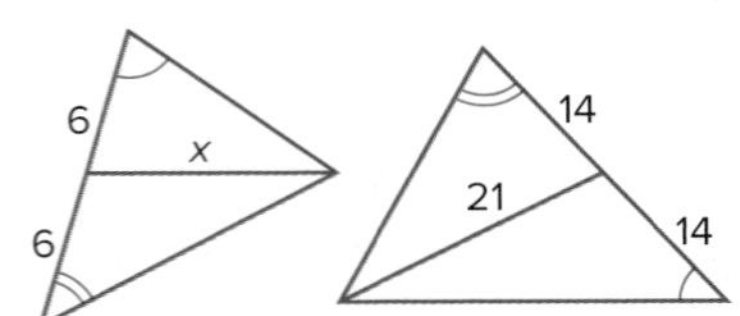

9. 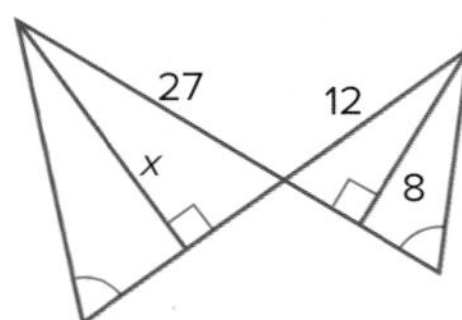

Example 2 10. **ROADWAYS** The intersection of the two roads shown forms two similar triangles. If AC is 382 feet, MP is 248 feet, and the gas station is 50 feet from the intersection, how far from the intersection is the bank?

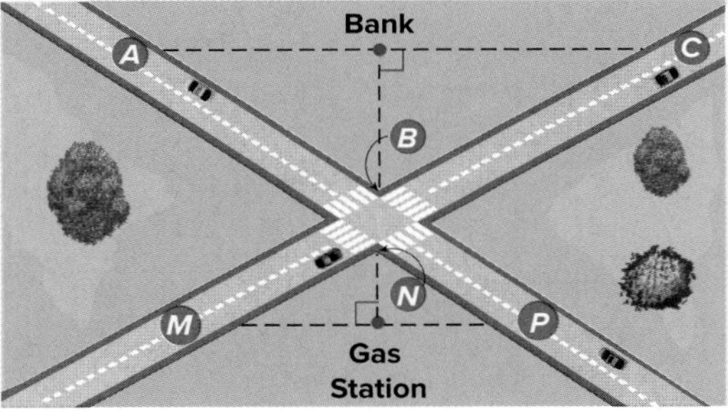

Example 3 **SENSE-MAKING Find the value of each variable.**

11. 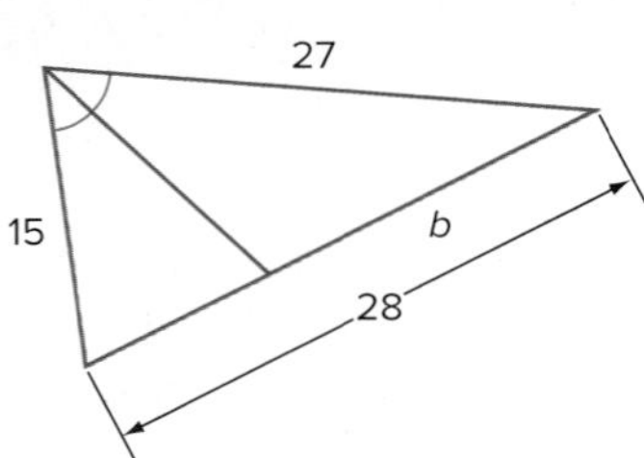

12. 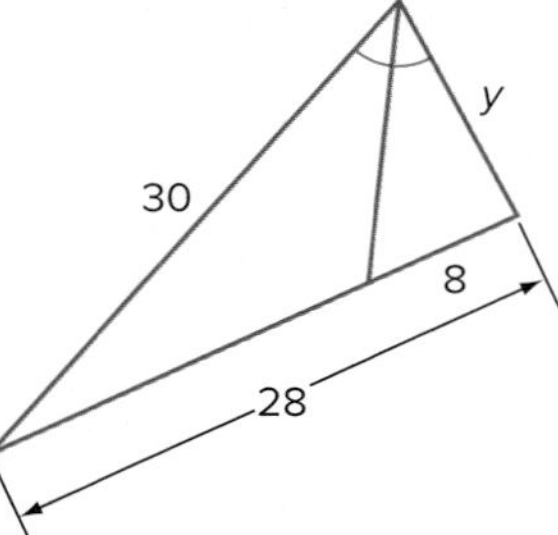

13. 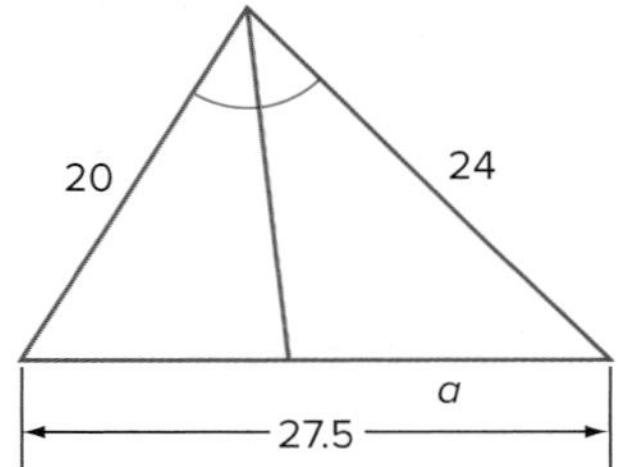

14. 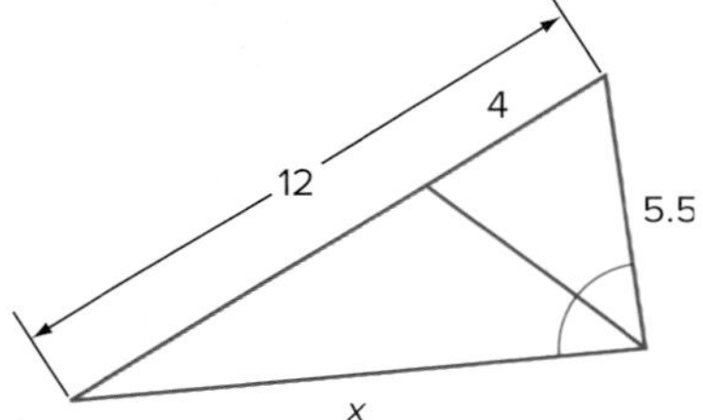

15. **ALGEBRA** If $\overline{AB}$ and $\overline{JK}$ are altitudes, $\triangle DAC \sim \triangle MJL$, $AB = 9$, $AD = 4x - 8$, $JK = 21$, and $JM = 5x + 3$, find x.

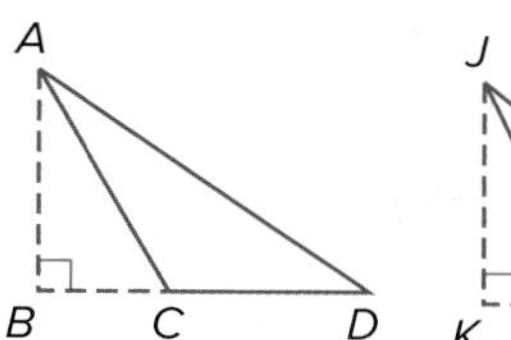

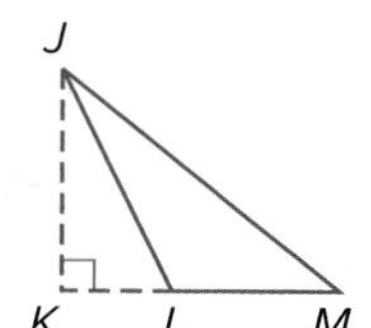

16. **ALGEBRA** If $\overline{NQ}$ and $\overline{VX}$ are medians, $\triangle PNR \sim \triangle WVY$, $NQ = 8$, $PR = 12$, $WY = 7x - 1$, and $VX = 4x + 2$, find x.

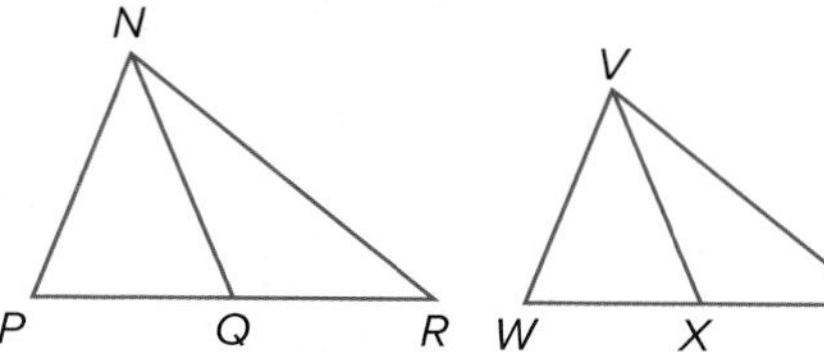

17. If $\triangle SRY \sim \triangle WXQ$, $\overline{RT}$ is an altitude of $\triangle SRY$, $\overline{XV}$ is an altitude of $\triangle WXQ$, $RT = 5$, $RQ = 4$, $QY = 6$, and $YX = 2$, find XV.

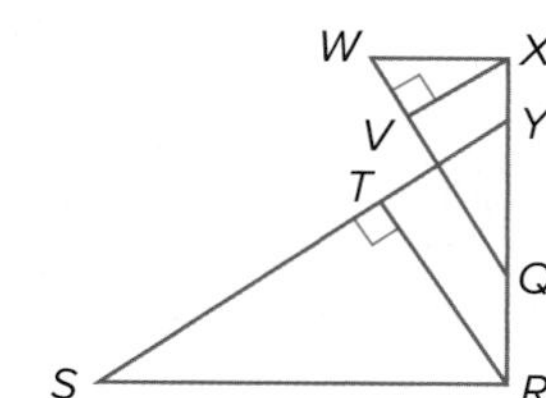

18. **PROOF** Write a paragraph proof of Theorem 7.9.

19. **PROOF** Write a two-column proof of Theorem 7.10.

ALGEBRA **Find x.**

20.

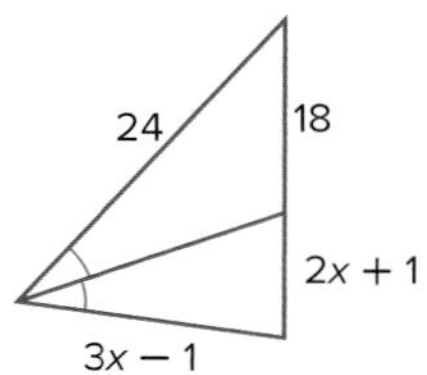

21.

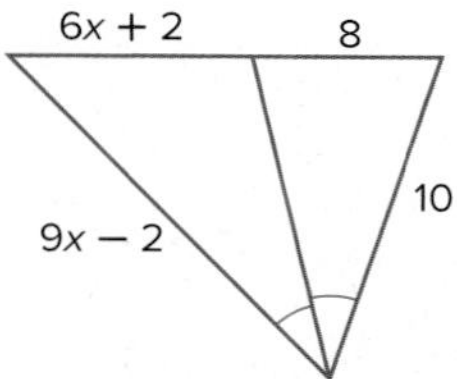

22.

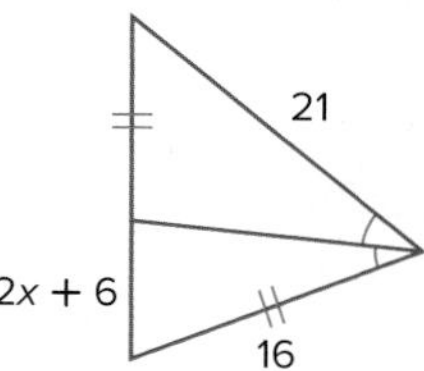

23.

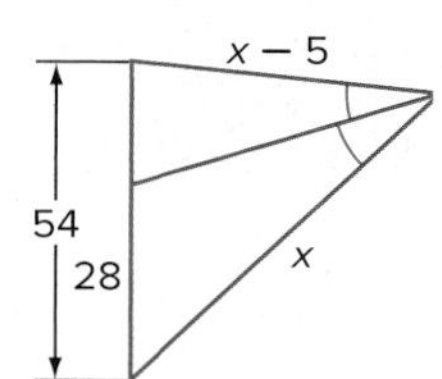

24. **SPORTS** Consider the triangle formed by the path between a batter, center fielder, and right fielder as shown. If the batter gets a hit that bisects the triangle at $\angle B$, is the center fielder or the right fielder closer to the ball? Explain your reasoning.

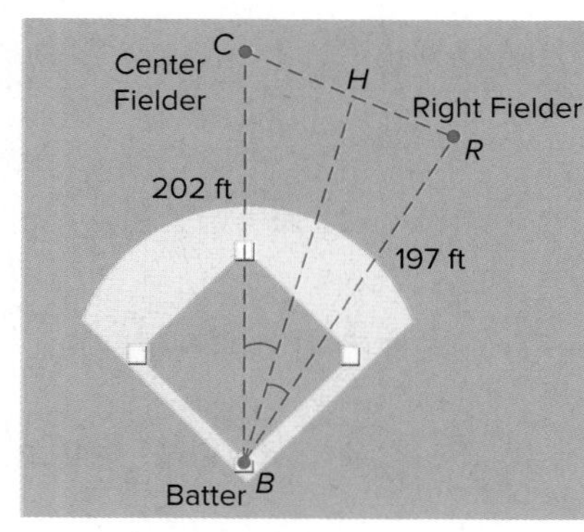

CONSTRUCT ARGUMENTS **Write a two-column proof.**

25. Theorem 7.11

Given: $\overline{CD}$ bisects $\angle ACB$.
By construction, $\overline{AE} \parallel \overline{CD}$.

Prove: $\frac{AD}{DB} = \frac{AC}{BC}$

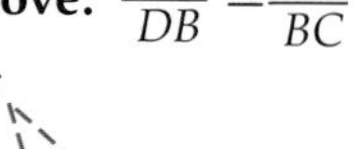

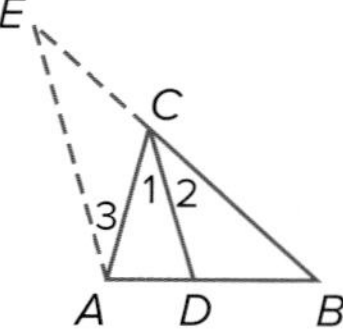

26. **Given:** $\angle H$ is a right angle.
L, K, and M are midpoints.

Prove: $\angle LKM$ is a right angle.

PROOF **Write a two-column proof.**

27. **Given:** $\triangle QTS \sim \triangle XWZ$, $\overline{TR}$ and $\overline{WY}$ are angle bisectors.

Prove: $\frac{TR}{WY} = \frac{QT}{XW}$

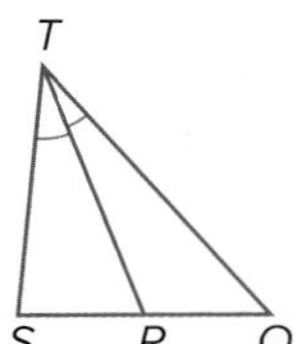

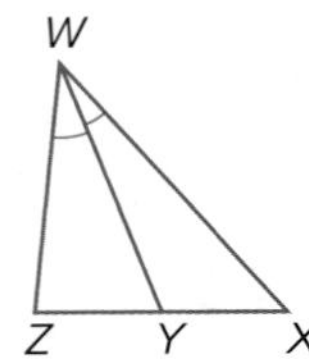

28. **Given:** $\overline{FD} \parallel \overline{BC}$, $\overline{BF} \parallel \overline{CD}$, $\overline{AC}$ bisects $\angle C$.

Prove: $\frac{DE}{EC} = \frac{BA}{AC}$

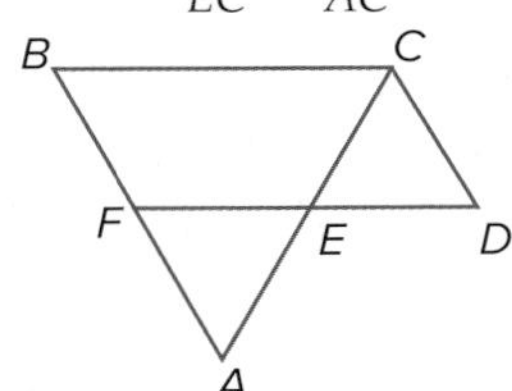

29. **SPORTS** During football practice, Trevor threw a pass to Ricardo as shown below. If Eli is farther from Trevor when he completes the pass to Ricardo and Craig and Eli move at the same speed, who will reach Ricardo to tackle him first?

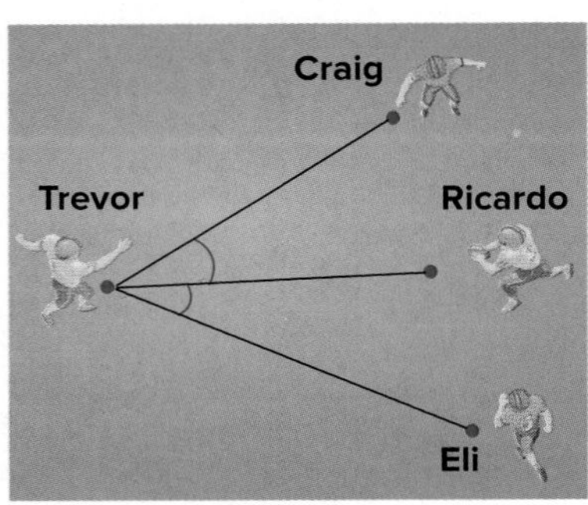

30. **SHELVING** In the bookshelf shown, the distance between each shelf is 13 inches and $\overline{AK}$ is a median of $\triangle ABC$. If EF is $3\frac{1}{3}$ inches, what is BK?

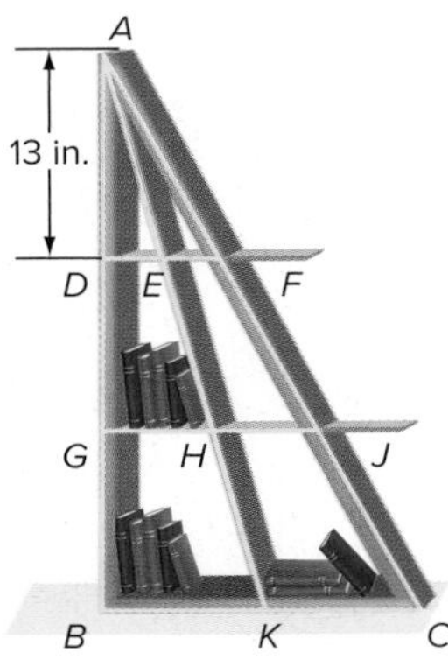

H.O.T. Problems Use Higher-Order Thinking Skills

31. **ERROR ANALYSIS** Chun and Traci are determining the value of x in the figure. Chun says to find x, solve the proportion $\frac{5}{8} = \frac{15}{x}$, but Traci says to find x, the proportion $\frac{5}{x} = \frac{8}{15}$ should be solved. Is either of them correct? Explain.

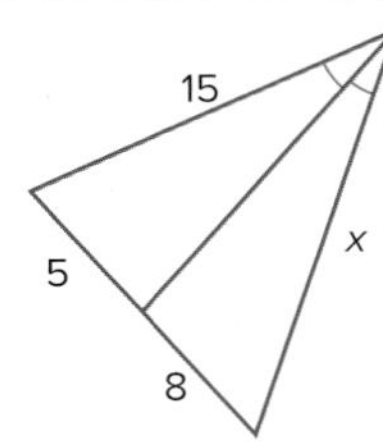

32. **MP CONSTRUCT ARGUMENTS** Find a counterexample to the following statement. Explain.

If the measure of an altitude and side of a triangle are proportional to the corresponding altitude and corresponding side of another triangle, then the triangles are similar.

33. **CHALLENGE** The perimeter of $\triangle PQR$ is 94 units. $\overline{QS}$ bisects $\angle PQR$. Find PS and RS.

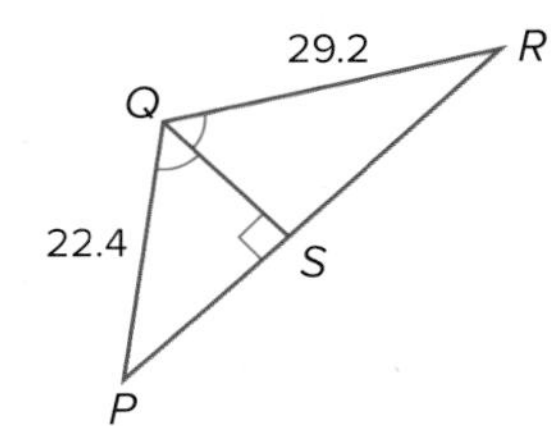

34. **OPEN-ENDED** Draw two triangles so that the measures of corresponding medians and a corresponding side are proportional, but the triangles are not similar.

35. **WRITING IN MATH** Compare and contrast Theorem 7.9 and the Triangle Angle Bisector Theorem.

36. $\overline{MQ}$ is an angle bisector of $\triangle MNP$, as shown.

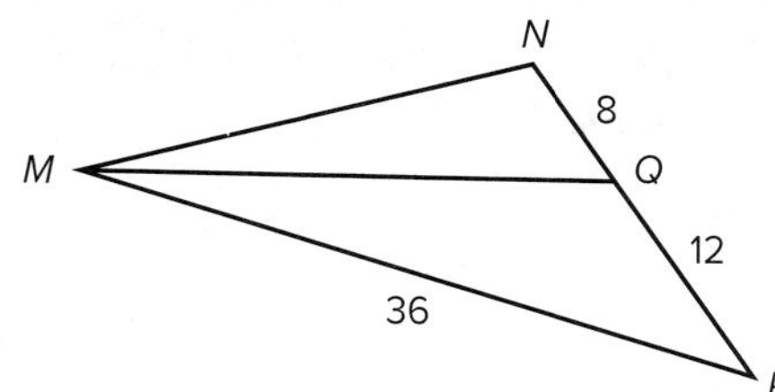

What is the perimeter of $\triangle MNP$? MP 1, 6

- A 24
- B 56
- C 80
- D 88

37. If $\triangle ABC \sim \triangle QRS$, what is the value of x? MP 1, 5

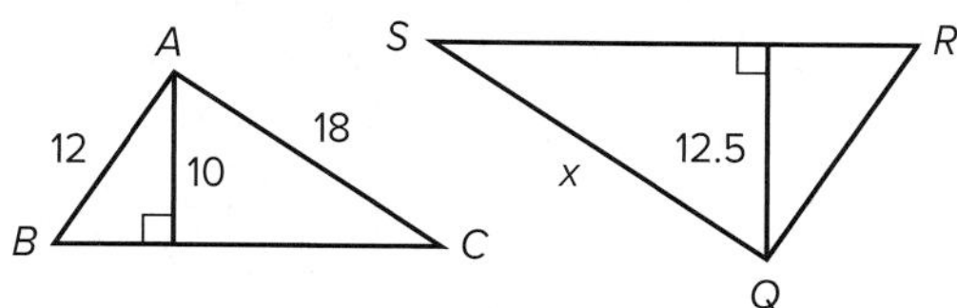

- A 14.4
- B 15
- C 20.5
- D 22.5

38. In $\triangle JKL$, $\overline{LJ}$ is 1.5 times as long as $\overline{LM}$. MP 2

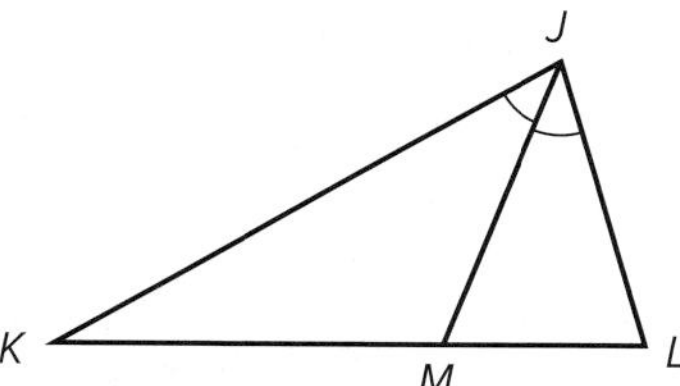

Which of the following statements must be true?

I. $\overline{KJ}$ is 1.5 times as long as $\overline{KM}$.

II. $\overline{KM}$ is 1.5 times as long as $\overline{LM}$.

III. $\overline{KJ}$ is 1.5 times as long as $\overline{LJ}$.

- A I only
- B II only
- C III only
- D II and III only
- E I, II, and III

39. The measure of one of the base angles of an isosceles triangle is one third of the measure of the vertex angle. What is the measure of one of the base angles? MP 1

40. MULTI-STEP In the figure, $\triangle EFG \sim \triangle HJK$ and $\overline{EX}$ and $\overline{HY}$ are medians. MP 1, 2, 6

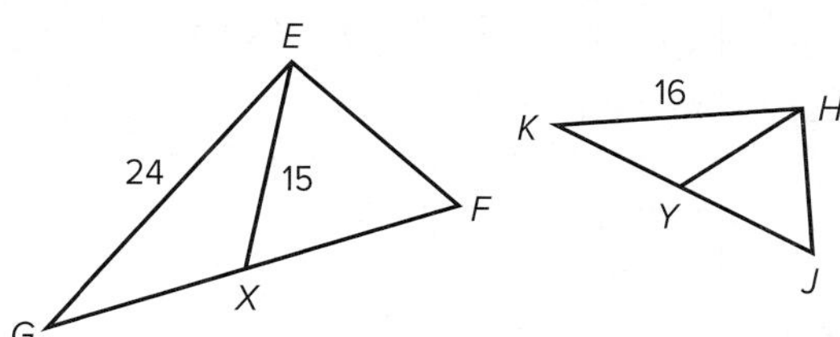

a. Which proportion could be used to find the length of $\overline{HY}$?

- A $\frac{EG}{GX} = \frac{KH}{KY}$
- B $\frac{EG}{HY} = \frac{EX}{KH}$
- C $\frac{EG}{EX} = \frac{HY}{KH}$
- D $\frac{EG}{EX} = \frac{KH}{HY}$

b. Write three sets of similar triangles found in the diagram.

c. What is the length of $\overline{HY}$?

- A 8
- B 10
- C 16
- D 24

d. The perimeter of $\triangle EGX$ is 51. What is the length of $\overline{KY}$?

- A 12
- B 8
- C 10
- D 6

EXTEND 7-6

Geometry Lab

Fractals

A **fractal** is a geometric figure that is created using iteration. **Iteration** is a process of repeating the same operation over and over again. Fractals are **self-similar**, which means that the smaller details of the shape have the same geometric characteristics as the original form.

Mathematical Practices

MP 8 Look for and express regularity in repeated reasoning.

Activity 1

Stage 0 Draw an equilateral triangle on isometric dot paper in which each side is 8 units long.

Stage 1 Connect the midpoints of the sides to form another triangle. Shade the center triangle.

Stage 2 Repeat the process using the three unshaded triangles. Connect the midpoints of the sides to form three other triangles.

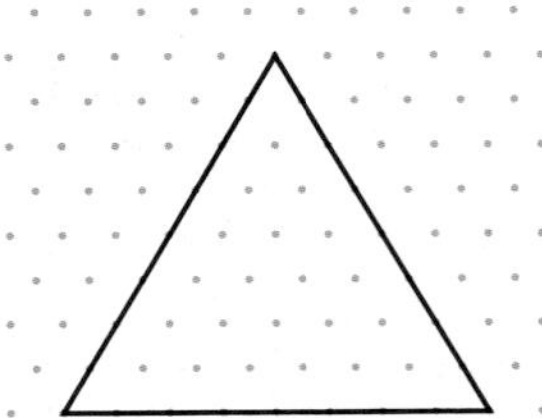

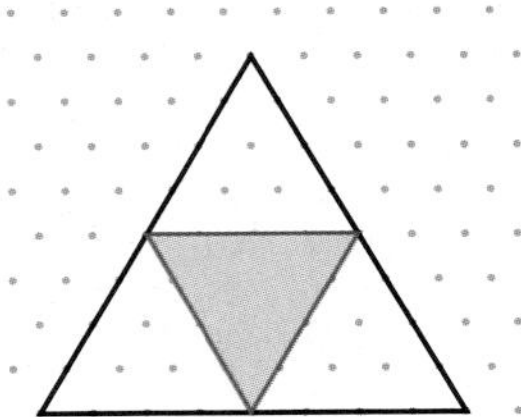

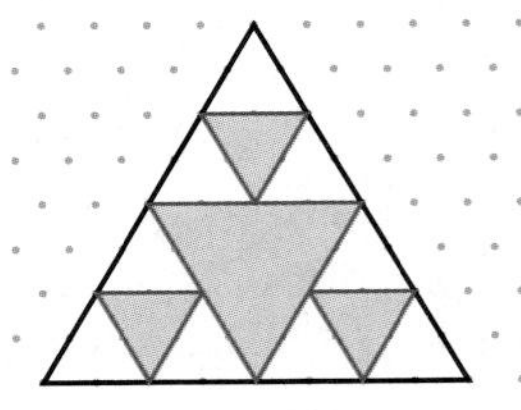

If you repeat this process indefinitely, the figure that results is called the Sierpinski Triangle.

Analyze the Results Work cooperatively.

1. If you continue the process, how many unshaded triangles will you have at Stage 3?
2. What is the perimeter of an unshaded triangle in Stage 4?
3. If you continue the process indefinitely, what will happen to the perimeters of the unshaded triangles?
4. **CHALLENGE** Complete the proof below.

   Given: $\triangle KAP$ is equilateral. D, F, M, B, C, and E are midpoints of $\overline{KA}$, $\overline{AP}$, $\overline{PK}$, $\overline{DA}$, $\overline{AF}$, and $\overline{FD}$, respectively.

   Prove: $\triangle BAC \sim \triangle KAP$

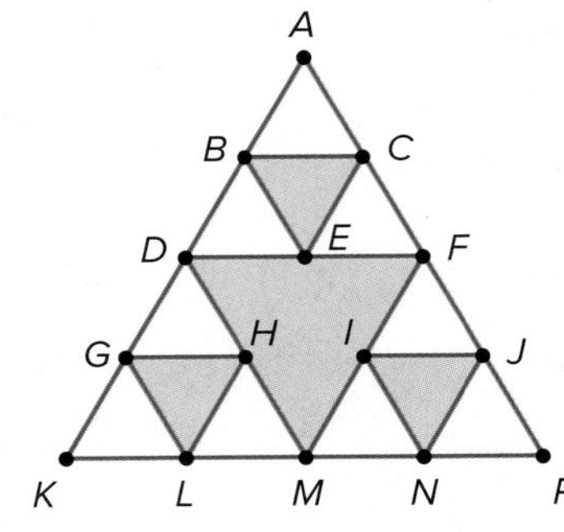

5. A *fractal tree* can be drawn by making two new branches from the endpoint of each original branch, each one third as long as the previous branch.

   a. Draw Stages 3 and 4 of a fractal tree. How many total branches do you have in Stages 1 through 4? (Do not count the stems.)

   b. Write an expression to predict the number of branches at each stage.

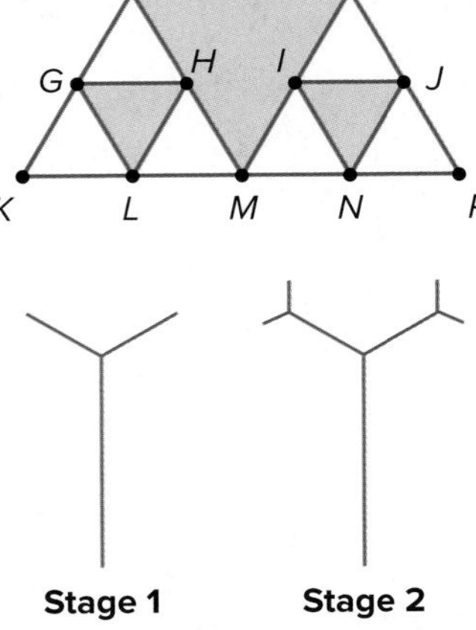

CHAPTER 7
Study Guide and Review

Study Guide

Key Concepts

Dilations (Lesson 7-1)

- Dilations enlarge or reduce figures proportionally.

Similar Polygons and Triangles (Lessons 7-2, 7-3, and 7-4)

- Two polygons are similar if and only if their corresponding angles are congruent and the measures of their corresponding sides are proportional.
- Two triangles are similar if:

  AA: Two angles of one triangle are congruent to two angles of the other triangle.

  SSS: The measures of the corresponding sides of the two triangles are proportional.

  SAS: The measures of two sides of one triangle are proportional to the measures of two corresponding sides of another triangle and their included angles are congruent.

Parallel Lines and Proportional Parts (Lesson 7-5)

- If a line is parallel to one side of a triangle and intersects the other two sides in two distinct points, then it separates these sides into segments of proportional length.
- A midsegment of a triangle is parallel to one side of the triangle and its length is one half the length of that side.

Parts of Similar Triangles (Lesson 7-6)

- Two triangles are similar when each of the following are proportional in measure:
  - their perimeters
  - their corresponding altitudes
  - their corresponding angle bisectors
  - their corresponding medians

Use your Foldable to review the chapter. Working with a partner can be helpful. Ask for clarification of concepts as needed.

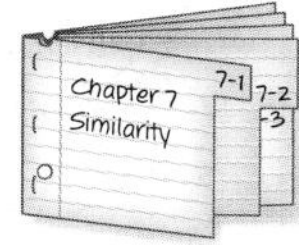

Key Vocabulary

dilation (p. 511)

enlargement (p. 511)

midsegment of a triangle (p. 535)

reduction (p. 511)

scale factor (p. 504)

similar polygons (p. 503)

similarity transformation (p. 511)

Vocabulary Check

Choose the letter of the word or phrase that best completes each statement.

a. scale factor
b. AA Similarity Postulate
c. SSS Similarity Theorem
d. SAS Similarity Theorem
e. Symmetric Property of Similarity
f. midsegment
g. dilation
h. enlargement
i. reduction
j. Transitive Property of Similarity

1. A(n) ___?___ of a triangle has endpoints that are the midpoints of two sides of the triangle.
2. If $\angle A \cong \angle X$ and $\angle C \cong \angle Z$, then $\triangle ABC \sim \triangle XYZ$ by the ___?___.
3. A(n) ___?___ is an example of a similarity transformation.
4. The ratio of the lengths of two corresponding sides of two similar polygons is the ___?___.
5. A dilation with a scale factor of $\frac{2}{5}$ will result in a(n) ___?___.
6. If $\angle A \cong \angle X$ and $\frac{AB}{XY} = \frac{AC}{XZ}$, then $\triangle ABC \sim \triangle XYZ$ by the ___?___.
7. If $\triangle ABC \sim \triangle DEF$ and $\triangle DEF \sim \triangle XYZ$, then $\triangle ABC \sim \triangle XYZ$ by the ___?___.

Concept Check

8. Explain whether two triangles must be similar if two sides of one triangle are proportional to the corresponding sides of the other triangle and an angle of one triangle is congruent to an angle of the other triangle.
9. A line segment connects the midpoints of two sides of a triangle. What conclusions that can be made about this line segment?

Study Guide and Review *Continued*

Lesson-by-Lesson Review 

7-1 Dilations

10. Copy the figure and point S. Then use a ruler to draw the image of the figure under a dilation with center S and scale factor $r = 1.25$.

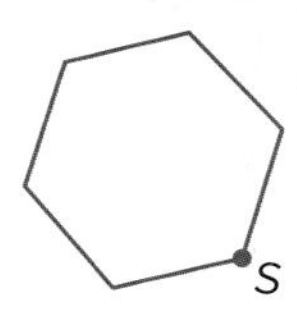

11. Determine whether the dilation from figure W to W' is an *enlargement* or a *reduction*. Then find the scale factor of the dilation and x.

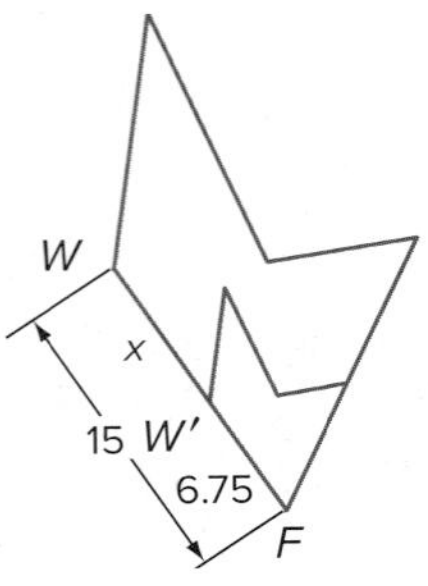

Example 1

Square $ABCD$ has vertices $A(0, 0)$, $B(0, 8)$, $C(8, 8)$, and $D(8, 0)$. Find the image of $ABCD$ after a dilation centered at the origin with a scale factor of 0.5.

Multiply the x- and y-coordinates of each vertex by the scale factor, 0.5.

$(x, y) \rightarrow (0.5x, 0.5y)$

$A(0, 0) \rightarrow A'(0, 0)$

$B(0, 8) \rightarrow B'(0, 4)$

$C(8, 8) \rightarrow C'(4, 4)$

$D(8, 0) \rightarrow D'(4, 0)$

Graph $ABCD$ and its image $A'B'C'D'$.

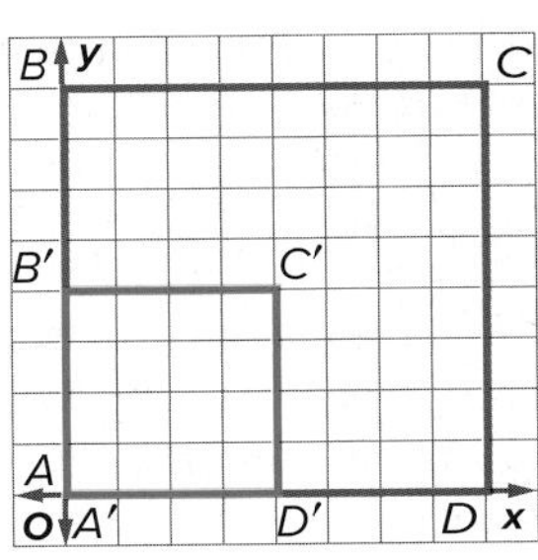

7-2 Similar Polygons

Determine whether each pair of figures is similar. If so, write the similarity statement and scale factor. If not, explain your reasoning.

12.

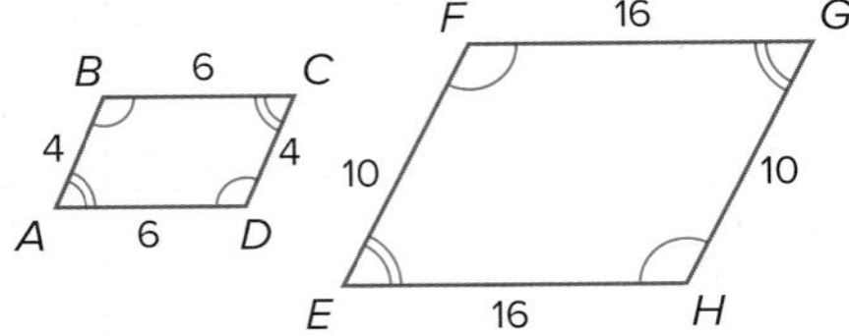

13.

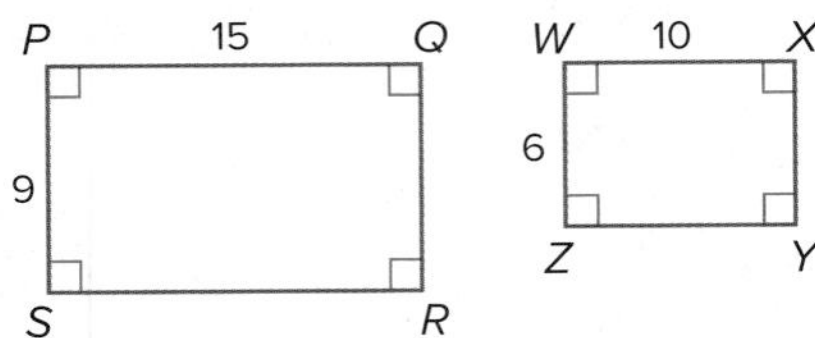

14. PHOTOS If the dimensions of a photo are 2 inches by 3 inches and the dimensions of a poster are 8 inches by 12 inches, are the photo and poster similar? Explain.

Example 2

Determine whether the pair of triangles is similar. If so, write the similarity statement and scale factor. If not, explain your reasoning.

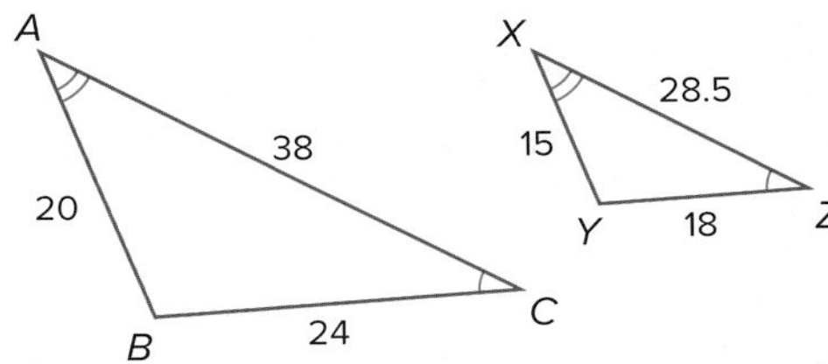

$\angle A \cong \angle X$ and $\angle C \cong \angle Z$, so by the Third Angle Theorem, $\angle B \cong \angle Y$. All of the corresponding angles are therefore congruent.

Similar polygons must also have proportional side lengths. Check the ratios of corresponding side lengths.

$\frac{AB}{XY} = \frac{20}{15}$ or $\frac{4}{3}$ $\quad \frac{BC}{YZ} = \frac{24}{18}$ or $\frac{4}{3}$ $\quad \frac{AC}{XZ} = \frac{38}{28.5}$ or $\frac{4}{3}$

Because corresponding sides are proportional, $\triangle ABC \sim \triangle XYZ$. So, the triangles are similar with a scale factor of $\frac{4}{3}$.

7-3 Similar Triangles: AA Similarity

Determine whether the triangles are similar. If so, write a similarity statement. Explain your reasoning.

15.

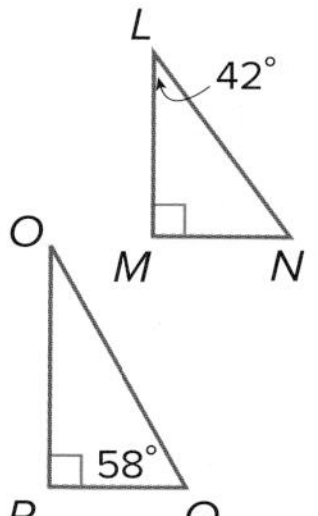

16.

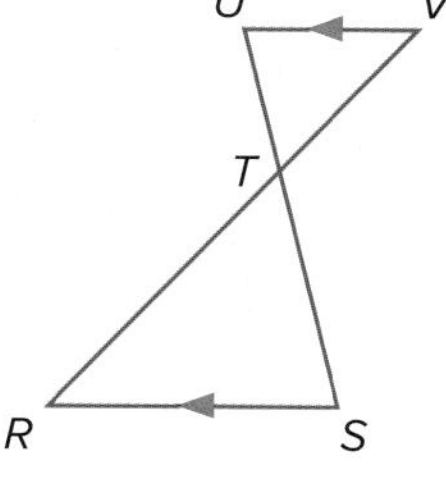

17. **TREES** To estimate the height of a tree, Dave stands in the shadow of the tree so that his shadow and the tree's shadow end at the same point. Dave is 6 feet 4 inches tall and his shadow is 15 feet long. If he is standing 66 feet away from the tree, what is the height of the tree?

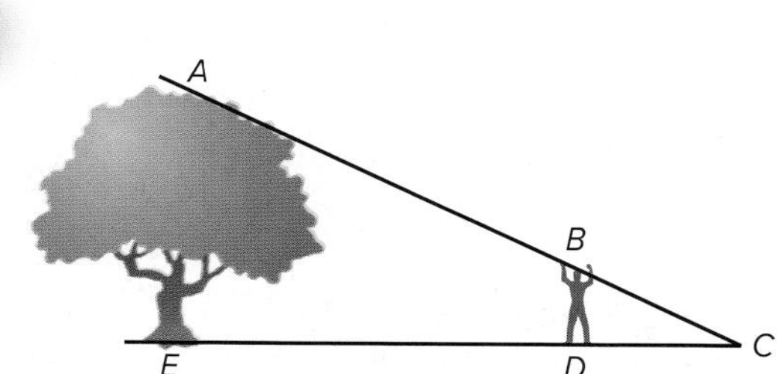

Example 3

Determine whether triangles *DEF* and *JKF* are similar. If so, write a similarity statement. Explain your reasoning.

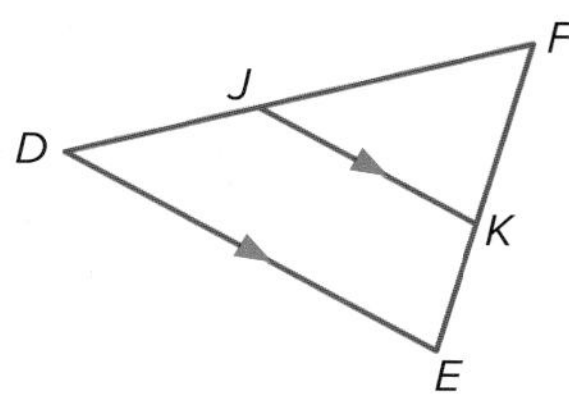

$\angle F \cong \angle F$ by the Reflexive Property.

Because $\overline{DE} \parallel \overline{JK}$, $\angle FJK \cong \angle FDE$ by the Corresponding Angles Postulate.

Because two angles of one triangle are congruent to two angles of the other triangle, $\triangle DEF \sim \triangle JKF$ by AA Similarity.

7-4 Similar Triangles: SSS and SAS Similarity

Determine whether the triangles are similar. If so, write a similarity statement. Explain your reasoning.

18.

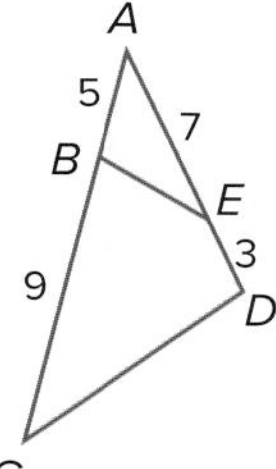

19.

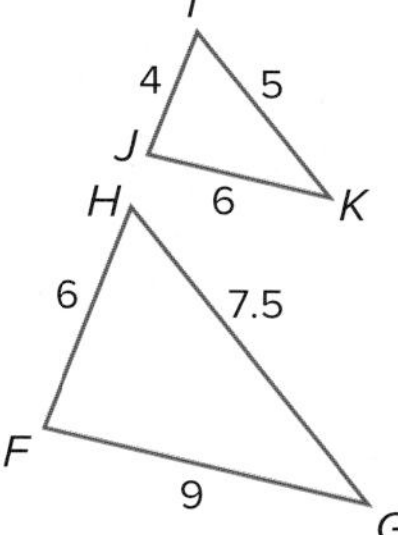

Example 4

Determine whether triangles *WZX* and *XZY* are similar. If so, write a similarity statement. Explain your reasoning.

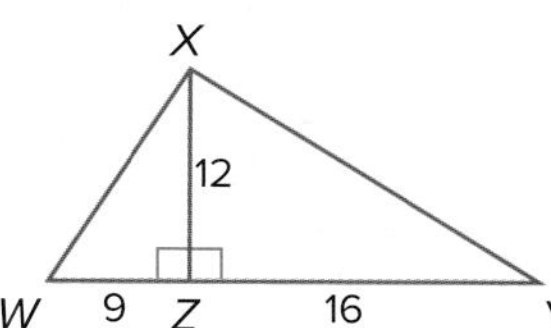

$\angle WZX \cong \angle XZY$ because they are both right angles. Now compare the ratios of the legs of the right triangles.

$$\frac{WZ}{XZ} = \frac{9}{12} = \frac{3}{4} \qquad \frac{XZ}{YZ} = \frac{12}{16} = \frac{3}{4}$$

Because two pairs of sides are proportional with the included angles congruent, $\triangle WZX \sim \triangle XZY$ by SAS Similarity.

7-5 Parallel Lines and Proportional Parts

Find x.

20.

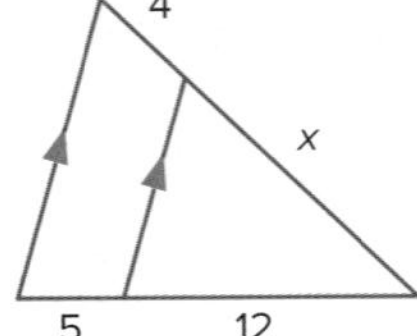

21.

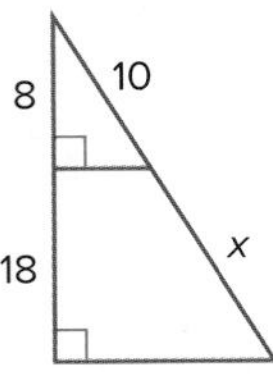

22. STREETS Find the distance along Broadway between 37th Street and 36th Street.

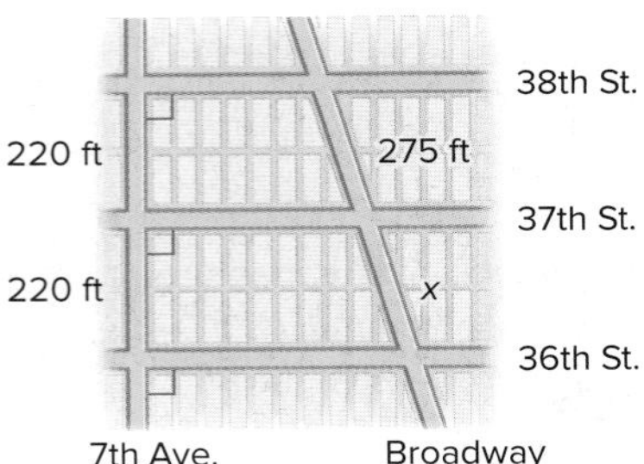

Example 5

ALGEBRA Find x and y.

$FK = KG$

$3x + 7 = 4x - 1$

$-x = -8$

$x = 8$

$FJ = JH$	Definition of congruence
$y + 12 = 2y - 5$	Substitution
$-y = -17$	Subtract.
$y = 17$	Simplify.

7-6 Parts of Similar Triangles

Find the value of each variable.

23.

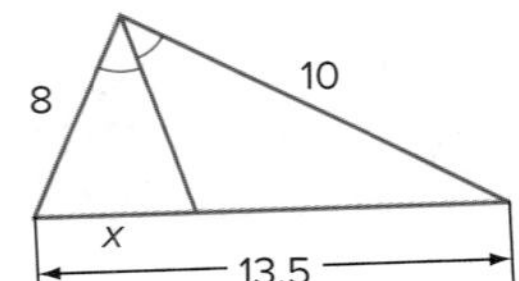

24.

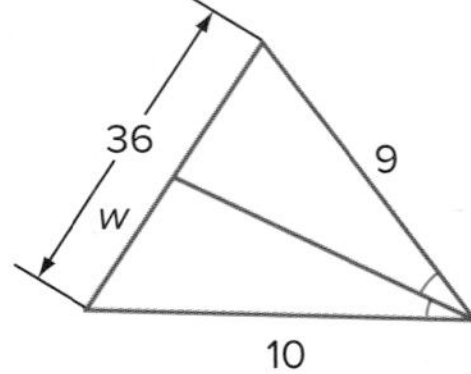

25. MAPS The scale given on a map of the state of Missouri indicates that 3 inches represents 50 miles. The cities of St. Louis, Springfield, and Kansas City form a triangle. If the measurements of the lengths of the sides of this triangle on the map are 15 inches, 10 inches, and 13 inches, find the perimeter of the actual triangle formed by these cities to the nearest mile.

Example 6

Find x.

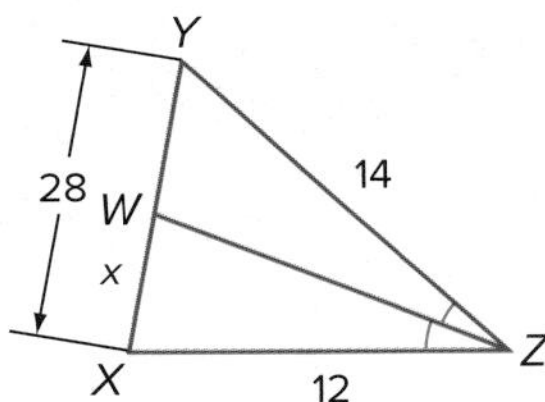

Use the Triangle Angle Bisector Theorem to write a proportion.

$\frac{WX}{YW} = \frac{XZ}{YZ}$	Triangle Angle Bisector Thm.
$\frac{x}{28 - x} = \frac{12}{14}$	Substitution
$(28 - x)(12) = x \cdot 14$	Cross Products Property
$336 - 12x = 14x$	Simplify.
$336 = 26x$	Add.
$12.9 = x$	Simplify.

CHAPTER 7

Practice Test

Copy the figure and point M. Then use a ruler to draw the image of the figure under a dilation with center M and the scale factor r indicated.

1. $r = 1.5$

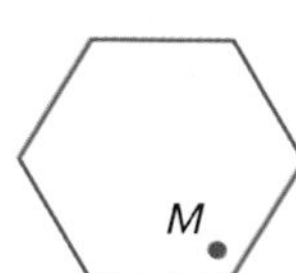

2. $r = \frac{1}{3}$

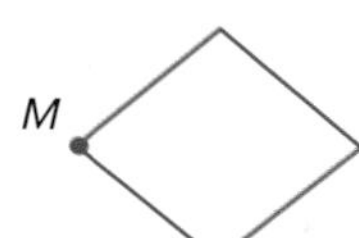

Determine whether each pair of figures is similar. If so, write the similarity statement and scale factor. If not, explain your reasoning.

3.

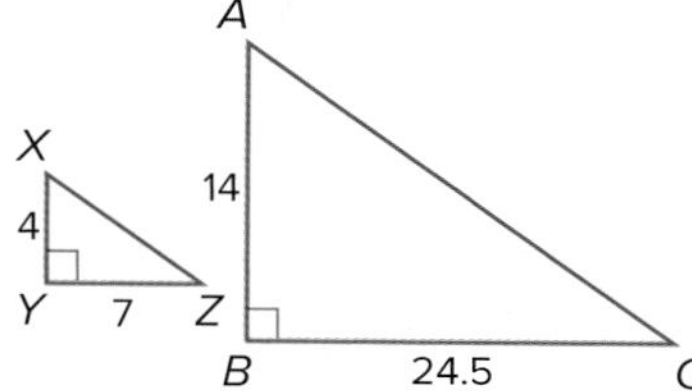

4.

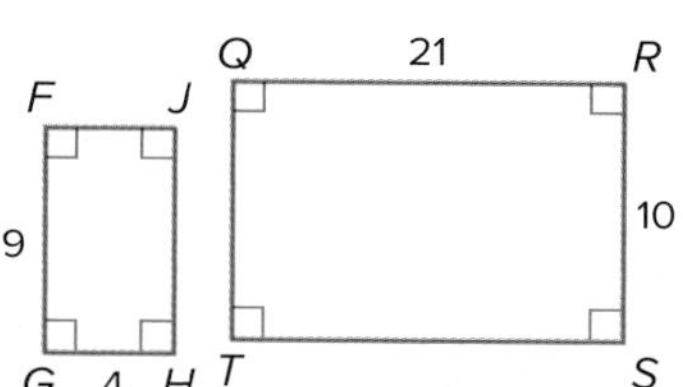

ALGEBRA Find x and y. Round to the nearest tenth if necessary.

5.

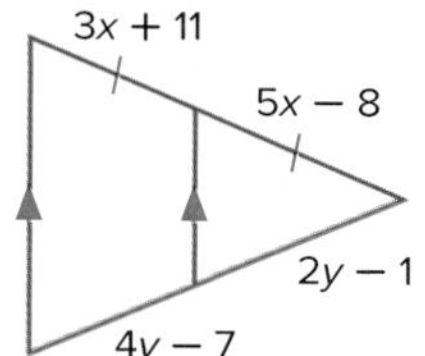

6.

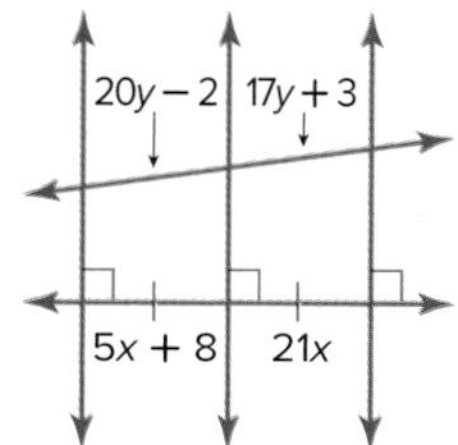

7. **ALGEBRA** Equilateral $\triangle MNP$ has perimeter $12a + 18b$. $\overline{QR}$ is a midsegment. What is QR?

8. **ALGEBRA** Right isosceles $\triangle ABC$ has hypotenuse with length h. $\overline{DE}$ is a midsegment with length $4x$ that is not parallel to the hypotenuse. What is the perimeter of $\triangle ABC$?

9. Determine whether the triangles are similar. If so, write a similarity statement. Explain your thinking.

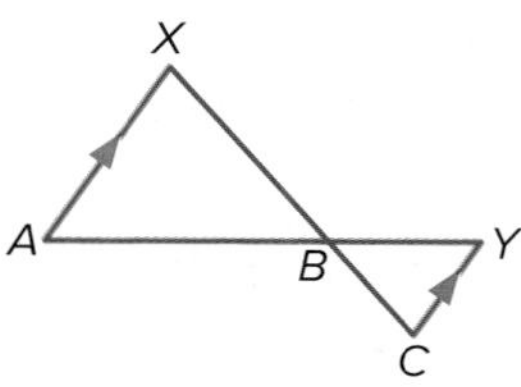

Find x.

10.

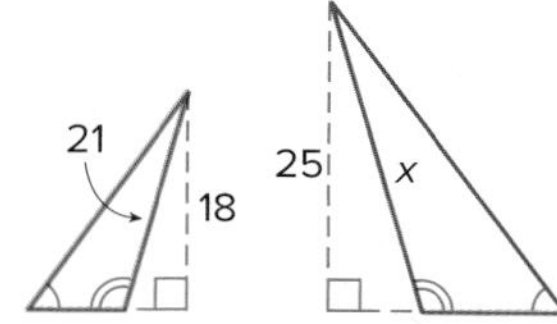

11.

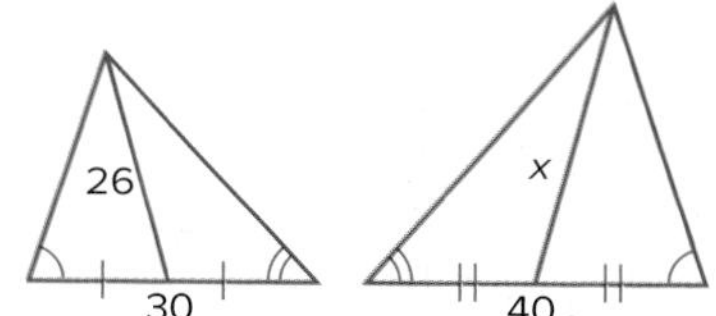

Determine whether the dilation from A to B is an *enlargement* or a *reduction*. Then find the scale factor of the dilation.

12.

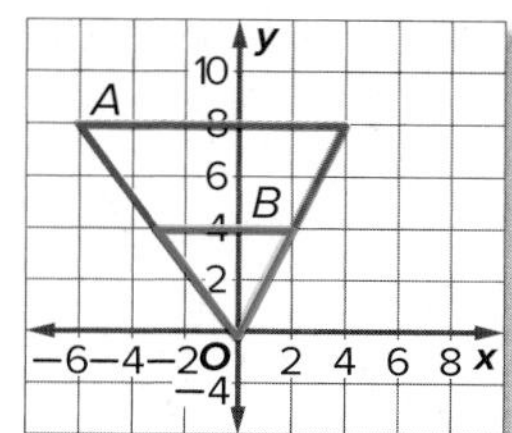

13.

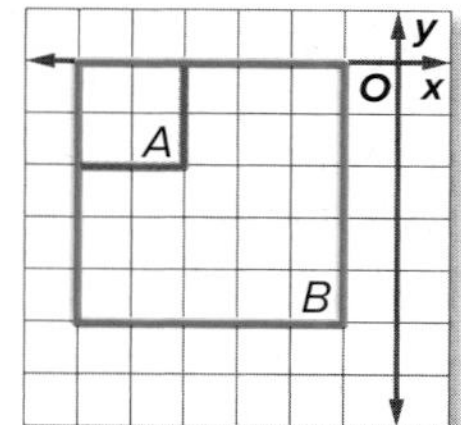

14. **ALGEBRA** Identify the similar triangles. Find WZ and UZ.

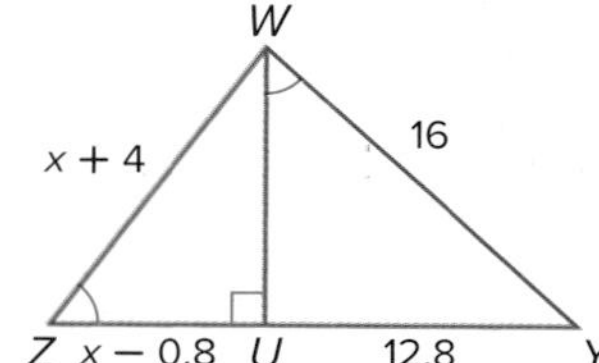

Preparing for Assessment

Performance Task

Provide a clear solution to each part of the task. Be sure to show all of your work, include all relevant drawings, and justify your answers.

CARPENTRY Brandi and Gerard own a carpentry business together.

Part A A client has a family member who uses a wheelchair. This family member visits often, and the client wants to make her house more accessible, so she decides to make her front stairs into a ramp. Brandi draws the stairs in exactly the same proportions as the real stairs, as shown in the diagram.

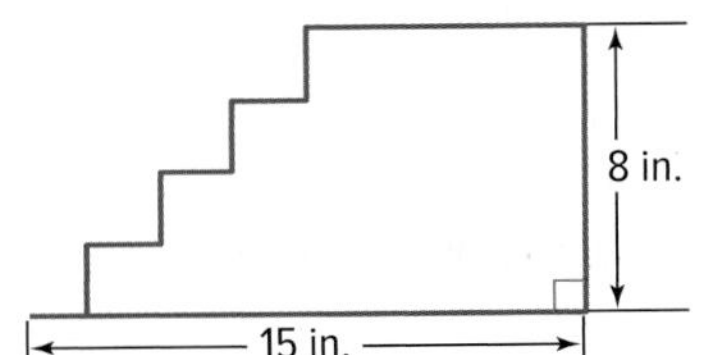

1. **Sense-Making** Determine the length of the ramp if the height of the real stairs is 4 feet.

Part B Another client is making a home décor item and has asked Brandi and Gerard to make some triangular pieces of wood. The client gives Gerard drawings of the shapes she'd like, shown in the top row of the diagram. She says the triangles Gerard makes must be in the same proportions as the drawings. Gerard creates the shapes shown in the bottom row of the diagram.

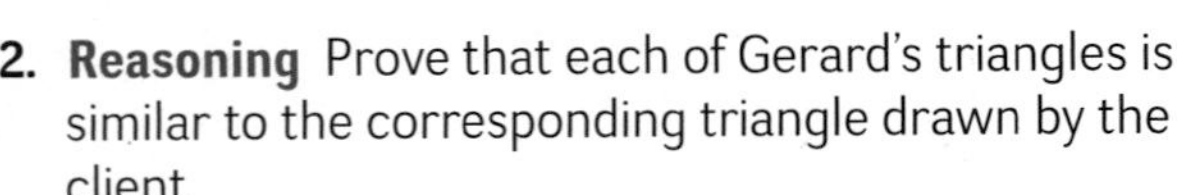

2. **Reasoning** Prove that each of Gerard's triangles is similar to the corresponding triangle drawn by the client.

Part C Another client would like a unique dining table made, so Brandi draws up plans for the top of the table that is an irregular hexagon. The client loves the design, but would like the table to be one and a half times as large as shown in the drawing.

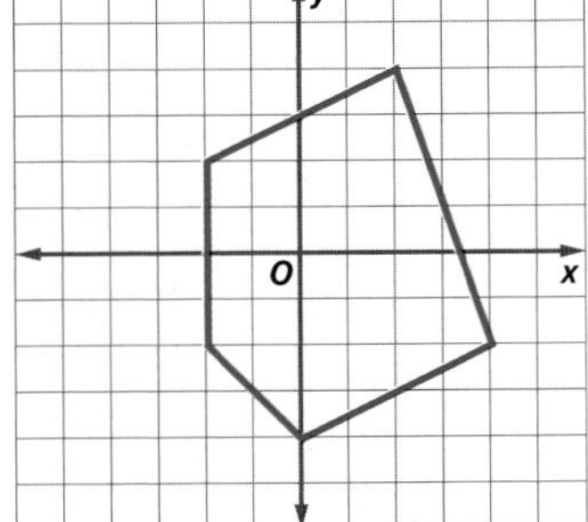

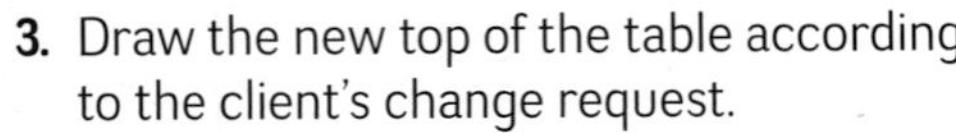

3. Draw the new top of the table according to the client's change request.

Part D A client wants Brandi and Gerard to build a bookshelf under their stairs. The sides of the bookshelf will be perpendicular to the floor. The height of the tallest space under the stairs is 12 feet.

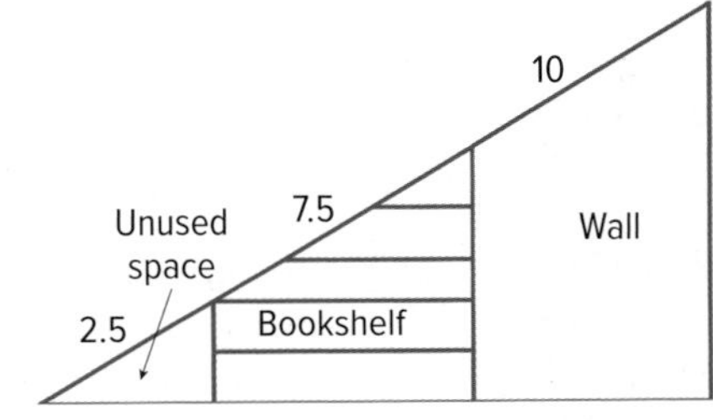

4. Determine the heights of both sides of the bookshelf.
5. Determine the width of the bottom shelf of the bookshelf.

Test-Taking Strategy

Example

Read the problem. Identify what you need to know. Then use the information in the problem to solve.

In the triangle shown, $\angle MQN \cong \angle RQS$. Which of the following would not be sufficient to prove that $\triangle QMN \sim \triangle QRS$?

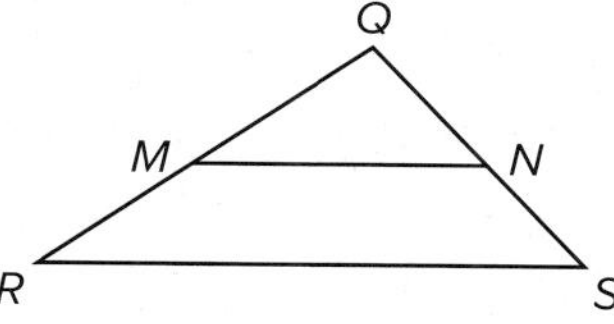

A $\angle QMN \cong \angle QRS$

B $\overline{MN} \parallel \overline{RS}$

C $\overline{QN} \cong \overline{NS}$

D $\frac{QM}{QR} = \frac{QN}{QS}$

Step 1 **What is the problem statement in the question? Are there any keywords that indicate this is a nonexample type question?**
The problem statement is that triangle *QMN* is similar to triangle *QRS*. The word "not" tells me this is a nonexample question.

Step 2 **Are any answer choices clearly incorrect? Are any in the wrong format? Do any have the wrong units?**
No, in this case no answers can be immediately eliminated because all of them appear to be the kind of statement that might prove similarity.

Step 3 **Which answer choice will you test first?**
The answer choices are not in any particular order, so I will start with choice A.

The correct answer is C.

Test-Taking Tip

Identifying Nonexamples
Multiple choice items sometimes ask you to determine which of the given answer choices is a nonexample. These types of problems require a different approach when solving them.

Apply the Strategy

Read the problem. Identify what you need to know. Then use the information in the problem to solve.

Consider the figure below.

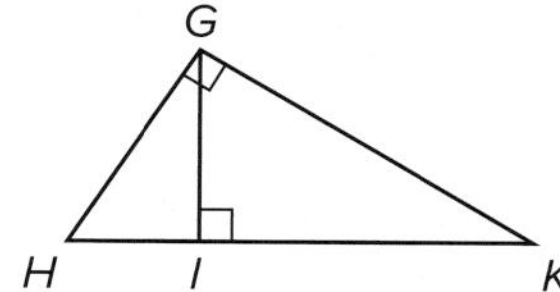

Which of the following is not sufficient to prove that $\triangle GIK \sim \triangle HIG$?

A $\angle GKI \cong \angle HGI$

B $\frac{HI}{GI} = \frac{GI}{IK}$

C $\frac{GH}{GI} = \frac{GK}{HK}$

D $\angle IGK \cong \angle IHG$

Answer the questions below.

a. What is the problem statement in the question? Are there any keywords that indicate this is a nonexample type question?

b. Are any answer choices clearly incorrect?

c. Which answer choice will you test first?

d. What is the correct answer?

CHAPTER 7
Preparing for Assessment
Cumulative Review

Read each question. Then fill in the correct answer on the answer document provided by your teacher or on a sheet of paper.

1. Colin cuts out a triangle and locates the centroid C as shown in the figure.

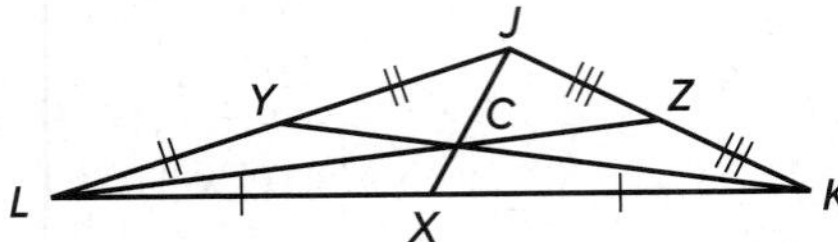

Given that $CX = 6$ inches, what is the length of $\overline{JX}$?

○ **A** 4 in.
○ **B** 9 in.
○ **C** 12 in.
○ **D** 18 in.

2. The figure shows one part of the support structure of a bridge. $\overline{RS}$ is parallel to $\overline{UV}$.

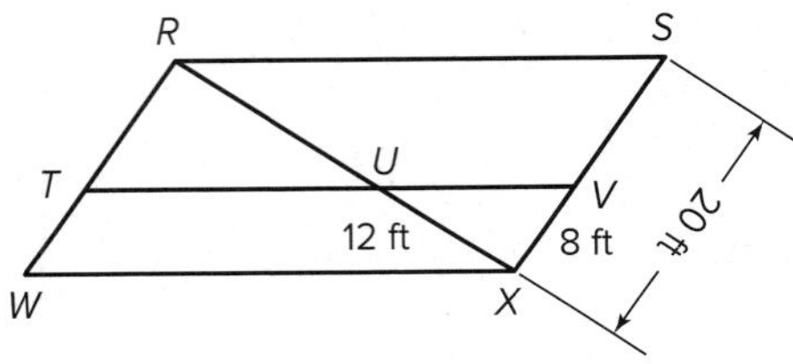

What is the length of $\overline{RU}$?

○ **A** 4.8 ft
○ **B** 18 ft
○ **C** 24 ft
○ **D** 30 ft

3. Quadrilateral $ABCD$ is a parallelogram. What should be the value of x in order to guarantee that quadrilateral $ABCD$ is a rhombus?

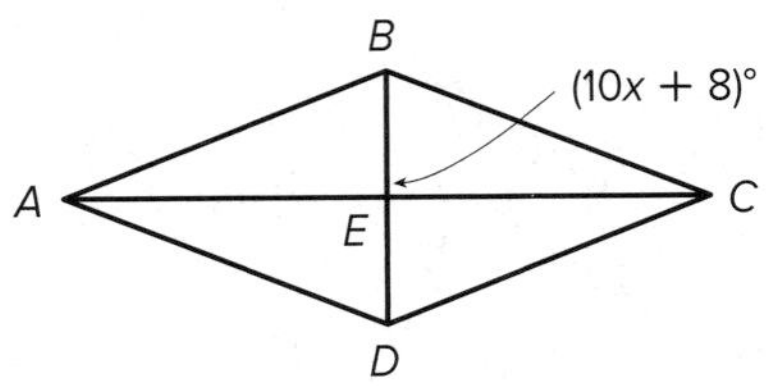

[ ]

Test-Taking Tip

Question 3 Ask yourself what must be true about the diagonals of a parallelogram in order for the parallelogram to be a rhombus.

4. Dennis drew $\triangle PQR$ and then dilated it to create $\triangle XYZ$, as shown.

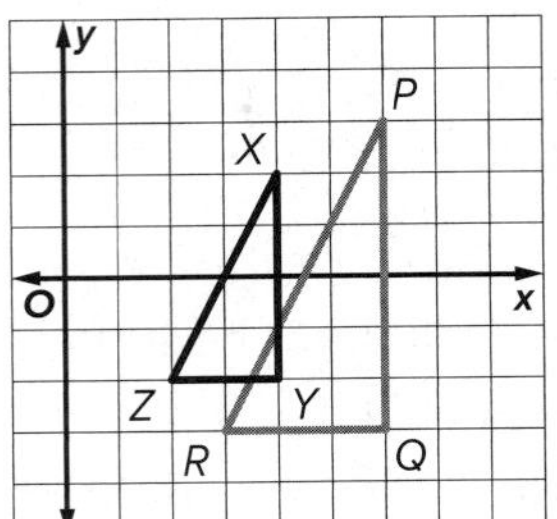

Select all of the true statements.

☐ **A** The scale factor of the dilation is less than 1.
☐ **B** The scale factor of the dilation is $\frac{2}{3}$.
☐ **C** The scale factor of the dilation is $\frac{3}{2}$.
☐ **D** $\triangle PQR \cong \triangle XYZ$
☐ **E** $\triangle PQR \sim \triangle XYZ$

5. What is the equation of the line that passes through (3, 4) and is perpendicular to the line $3x + y = 3$?

○ **A** $y = \frac{1}{3}x + 3$
○ **B** $y = -\frac{1}{3}x + 5$
○ **C** $y = -3x + 13$
○ **D** $y = 3x - 5$

6. In the figure, $ABCD \sim PQRS$. What is the perimeter of $ABCD$?

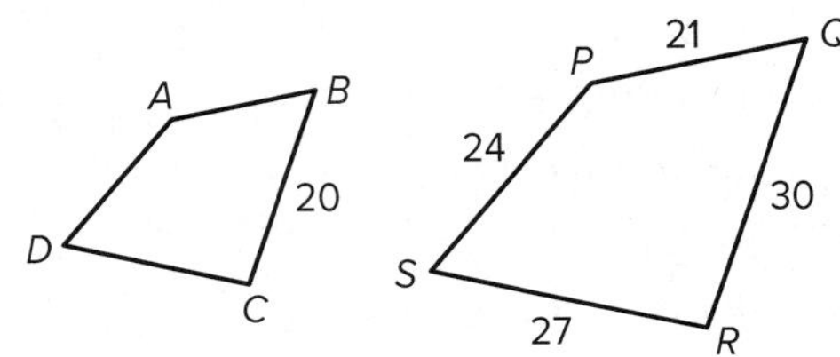

[ ]

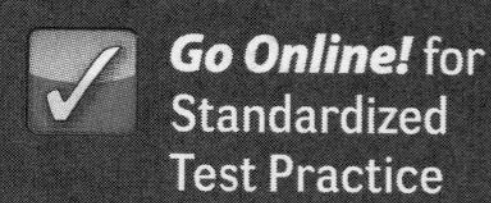

7. In $\triangle JKL$, $\overleftrightarrow{MN}$ is parallel to $\overline{KJ}$.

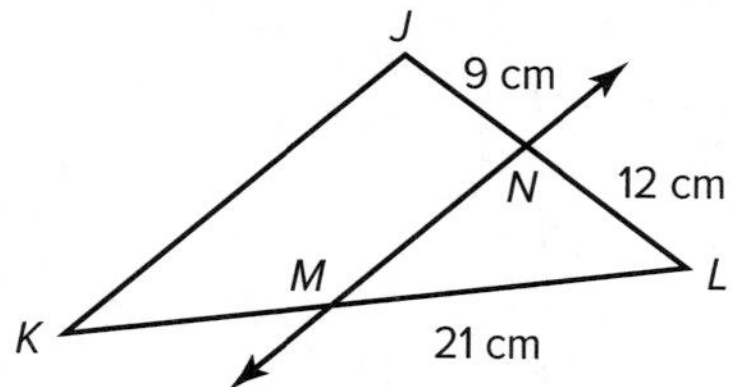

What is the length of $\overline{KL}$?

- ◯ **A** 15.75 cm
- ◯ **B** 36.75 cm
- ◯ **C** 39 cm
- ◯ **D** 49 cm

8. In $\triangle PQR$, the measure of $\angle P$ is 3 times the measure of $\angle Q$.

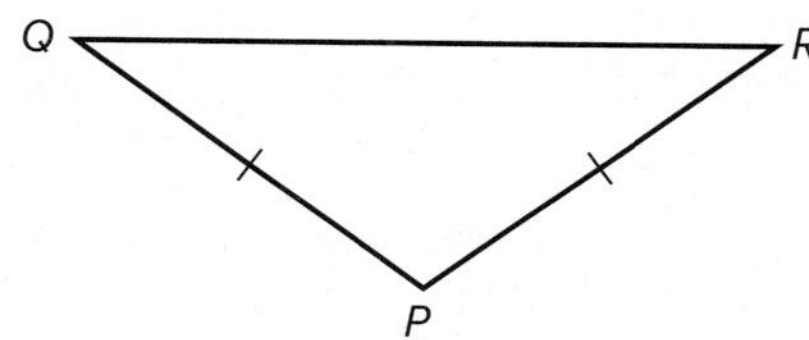

What is the measure of $\angle R$?

- ◯ **A** 108
- ◯ **B** 60
- ◯ **C** 45
- ◯ **D** 36

9. What is the value of y?

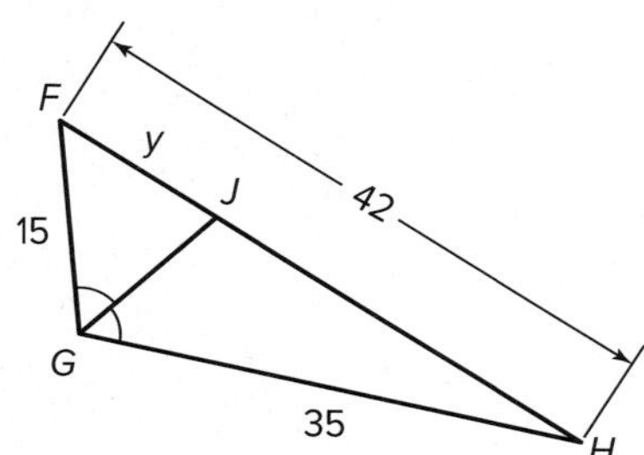

- ◯ **A** 12.6
- ◯ **B** 18
- ◯ **C** 22
- ◯ **D** 29.4

10. The figure shows the distances along several straight roads in Isabella's town. She drives the road directly from the library to city hall.

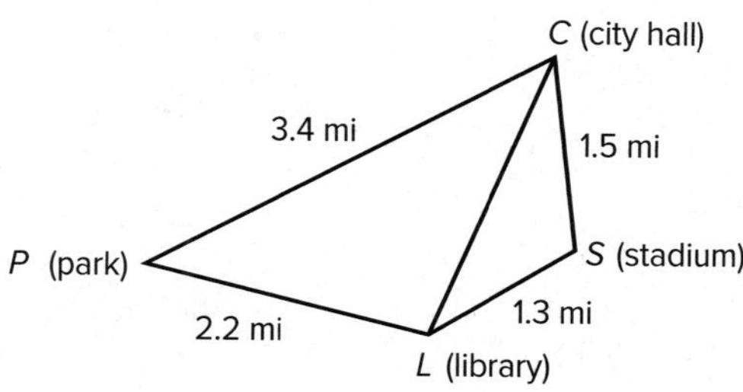

a. Which of the following is a possible distance Isabella might drive?

- ◯ **A** 0.9 mi
- ◯ **B** 1.2 mi
- ◯ **C** 1.5 mi
- ◯ **D** 2.8 mi

b. MP What mathematical practice did you use to solve this problem?

11. Which additional piece of information, if any, is sufficient to prove that $\triangle MNP \sim \triangle QRP$?

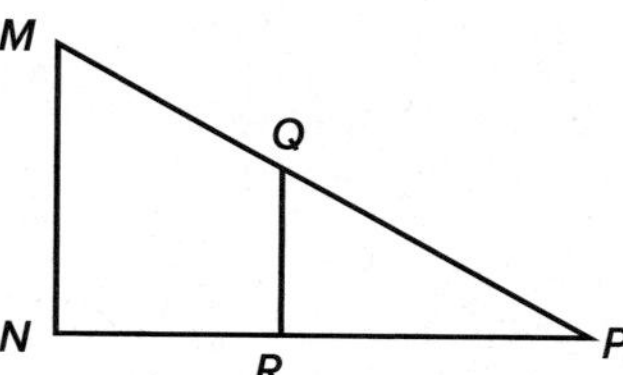

- ◯ **A** Point Q is the midpoint of $\overline{MP}$.
- ◯ **B** $\frac{MN}{MP} = \frac{RP}{QR}$
- ◯ **C** $\overline{MN}$ is parallel to $\overline{QR}$.
- ◯ **D** No additional information is needed.

12. $\overline{JK}$ is a midsegment of $\triangle RST$.

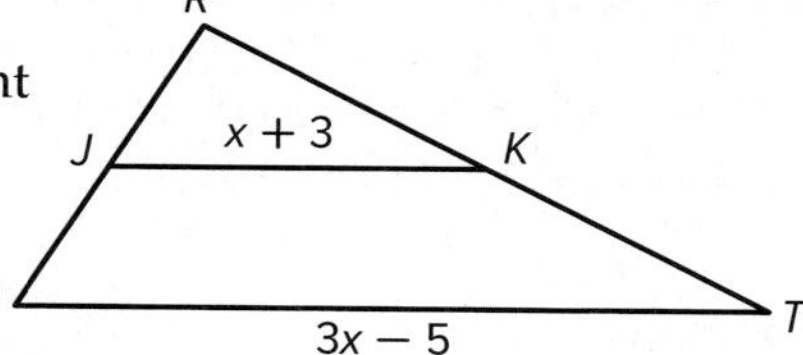

What is the length of $\overline{JK}$?

- ◯ **A** 28
- ◯ **B** 14
- ◯ **C** 11
- ◯ **D** 7

Need Extra Help?

If you missed Question...	1	2	3	4	5	6	7	8	9	10	11	12
Go to Lesson...	5-2	7-3	6-5	7-2	2-8	7-2	7-5	4-6	7-6	5-5	7-3	7-5

CHAPTER 8

Right Triangles and Trigonometry

THEN

You solved proportions.

NOW

In this chapter, you will:

- Use the Pythagorean Theorem.
- Use properties of special right triangles.
- Use trigonometry to find missing measures of triangles.

WHY

EVENT PLANNING Properties of triangles and trigonometry can be used to determine the dimensions of decorations and supplies.

Use the Mathematical Practices to complete the activity.

1. Sense-Making How could right triangles be used to plan and build decorations for a special event?

2. Apply Math Write a problem using a right triangle to determine the dimensions of a decoration (i.e., a stand-up prop) or the length of a rope that suspends a decoration and is anchored to the ground, the wall, or other points.

3. Modeling Use the Ruler Grids mat and the protractor and line segment tools in ConnectED to model the right triangle for the problem you created, and determine side lengths.

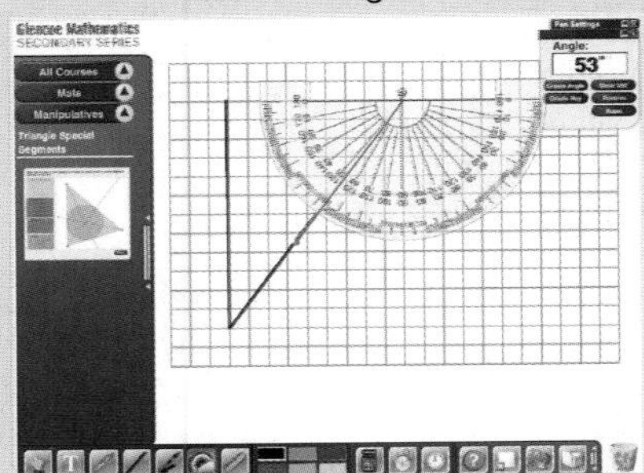

4. Discuss In what other ways could you use the properties of right triangles to plan for and place decorations?

RHIMAGE/Shutterstock.com

Go Online to Guide Your Learning

Explore & Explain

Triangle Special Segments

Use the **Triangle Special Segments** tool with the protractor and ruler to explore the theorems discussed in Lesson 8-1 and to enhance understanding of other concepts contained in this chapter.

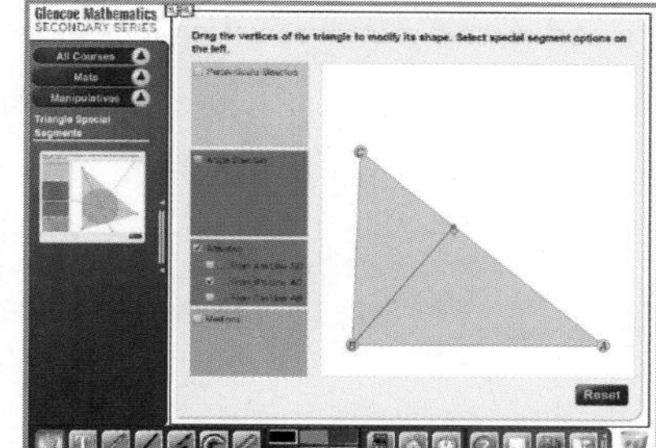

The Geometer's Sketchpad

Use the **The Geometer's Sketchpad** to provide a way to use squares and their areas to investigate the Pythagorean Theorem and to explore 30-60-90 triangles and the relationship among the side lengths and among areas. The Geometer's Sketchpad can also illustrate how to use the Law of Sines and the Law of Cosines, as discussed in Lessons 8-6 and 8-7.

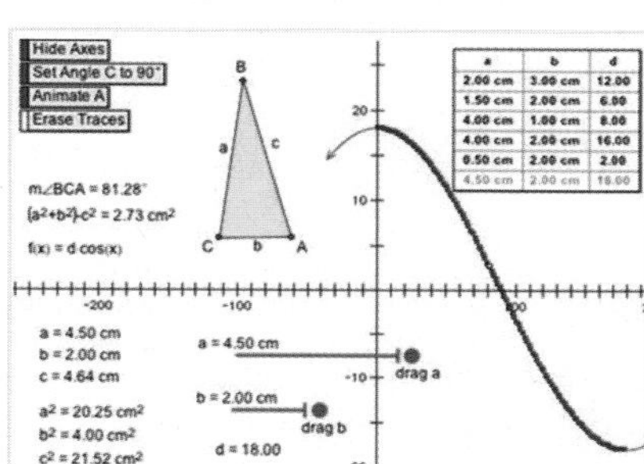

eBook

Interactive Student Guide

Before starting the chapter, answer the **Chapter Focus** preview questions. Check your answers as you complete each lesson. At the end of the chapter, try the **Performance Task**.

Organize

Foldables

Get organized! Create this **Right Triangles and Trigonometry Foldable** before you begin this chapter to help you organize your notes about right angles and trigonometry.

Collaborate

Chapter Project

In the **Surveyors** project, you will use what you have learned about right triangles and trigonometry to complete a business literacy project.

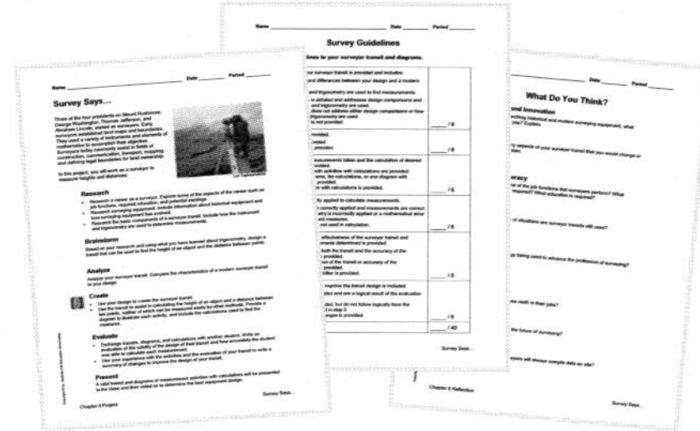

Focus

LEARNSMART

Need help studying? Complete the **Similarity, Proof, and Trigonometry** domain in LearnSmart to review for the chapter test.

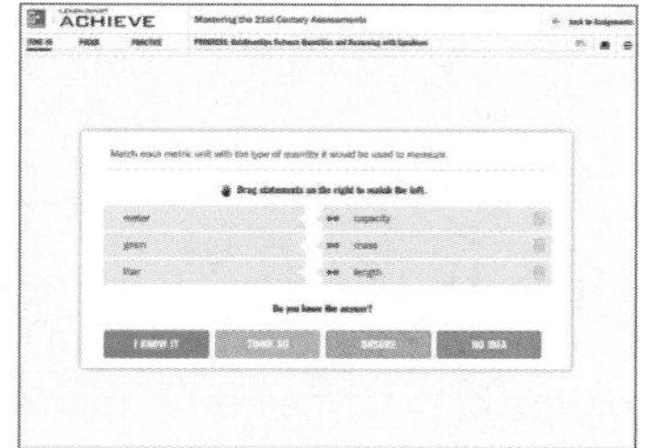

ALEKS

You can use the **Triangles** topic in ALEKS to explore what you know about right triangles and trigonometry and what you are ready to learn.*

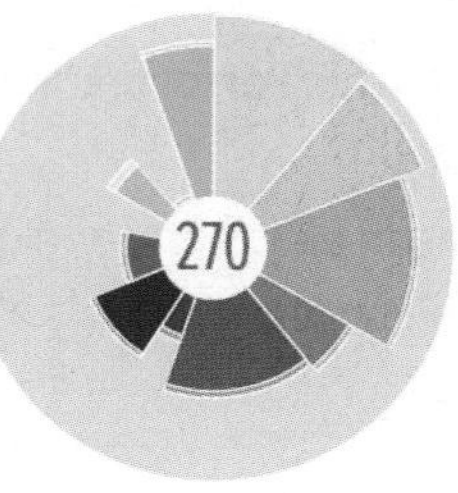

* Ask your teacher if this is part of your program.

Get Ready for the Chapter

Go Online! for Vocabulary Review Games and key vocabulary in 13 languages.

Connecting Concepts

Concept Check

Review the concepts used in this chapter by answering the questions below.

1. What is the first step to simplify $\frac{6}{\sqrt{3}}$?
2. Simplify $\frac{6\sqrt{3}}{3}$.
3. Describe the Pythagorean Theorem.
4. Given the right triangle shown, what equation could you write to solve for x?

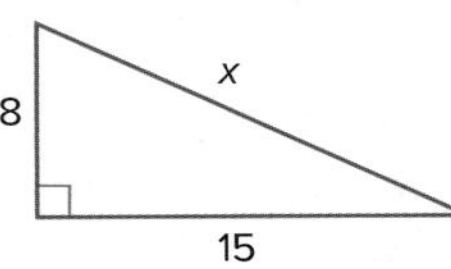

5. Describe how to solve the equation $289 = x^2$ for x?
6. Anna is making a banner out of 4 congruent triangles as shown. She is planning to trim the perimeter with blue trim. Write an equation to solve for x.

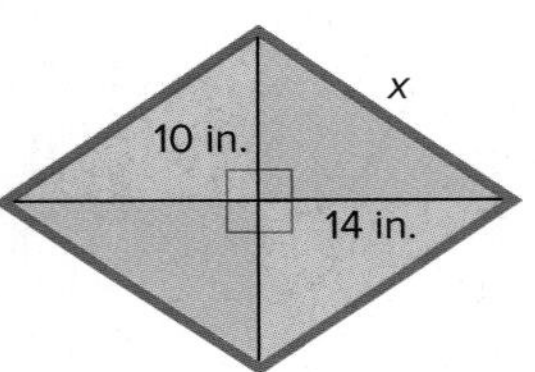

7. How much blue trim will Anna need in total?
8. Describe the steps needed to graph a line segment with endpoints $A(-4, 2)$ and $B(3, -2)$.

Performance Task Review

You can use the concepts and skills in the chapter to solve problems related to navigation. Understanding right triangles and trigonometry will help you finish the Performance Task at the end of the chapter.

In this Performance Task you will:

- reason abstractly and quantitatively
- use appropriate tools strategically

New Vocabulary

English		Español
geometric mean	p. 565	media geométrica
Pythagorean triple	p. 576	triplete pitágorico
trigonometry	p. 596	trigonométria
trigonometric ratio	p. 596	razón trigonométrica
sine	p. 596	seno
cosine	p. 596	coseno
tangent	p. 596	tangente
inverse sine	p. 599	inverso del seno
inverse cosine	p. 599	inverso del coseno
inverse tangent	p. 599	inverse del tangente
angle of elevation	p. 608	ángulo de elevación
angle of depression	p. 608	ángulo de depresión
Law of Sines	p. 617	ley de los senos
solving a triangle	p. 617	resolver un triángulo
ambiguous case of the Law of Sines	p. 618	caso ambiguo de la ley de los senos
Law of Cosines	p. 624	ley do los cosenos

Review Vocabulary

altitude altura a segment drawn from a vertex of a triangle perpendicular to the line containing the other side

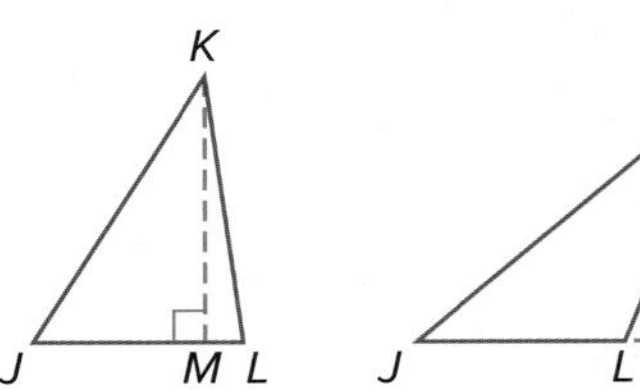

$\overline{KM}$ is an altitude of $\triangle JKL$.

Pythagorean Theorem Teorema de Pitágoras If a and b are the measures of the legs of a right triangle and c is the measure of the hypotenuse, then $a^2 + b^2 = c^2$.

LESSON 1

Geometric Mean

Then	Now	Why?
You used proportional relationships of corresponding angle bisectors, altitudes, and medians of similar triangles.	**1** Find the geometric mean between two numbers. **2** Solve problems involving relationships between parts of a right triangle and the altitude to its hypotenuse.	Photographing very tall or very wide objects can be challenging. It can be difficult to include the entire object in your shot without distorting the image. If your camera is set for a vertical viewing angle of 90° and you know the height of the object you wish to photograph, you can use the geometric mean of the distance from the top of the object to your camera level and the distance from the bottom of the object to camera level.

New Vocabulary
geometric mean

Mathematical Practices

7 Look for and make use of structure.

3 Construct viable arguments and critique the reasoning of others.

1 Geometric Mean

When the means of a proportion are the same number, that number is called the geometric mean of the extremes. The **geometric mean** between two numbers is the positive square root of their product.

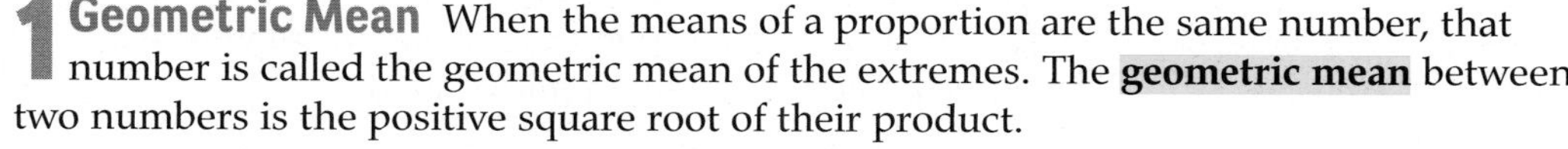

$$\text{extreme} \rightarrow \frac{a}{x} = \frac{x}{b} \leftarrow \text{mean}$$
mean → x, b ← extreme

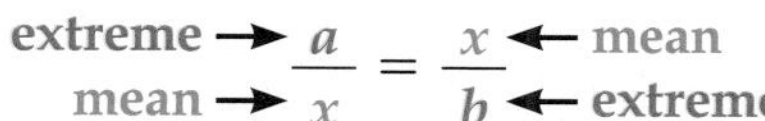

Key Concept Geometric Mean

Words	The geometric mean of two positive numbers a and b is the number x such that $\frac{a}{x} = \frac{x}{b}$. So, $x^2 = ab$ and $x = \sqrt{ab}$.
Example	The geometric mean of $a = 9$ and $b = 4$ is 6, because $6 = \sqrt{9 \cdot 4}$.

Example 1 Geometric Mean

Find the geometric mean between 8 and 10.

$x = \sqrt{ab}$	Definition of geometric mean
$= \sqrt{8 \cdot 10}$	$a = 8$ and $b = 10$
$= \sqrt{(4 \cdot 2) \cdot (2 \cdot 5)}$	Factor.
$= \sqrt{16 \cdot 5}$	Associative Property
$= 4\sqrt{5}$	Simplify.

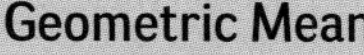

The geometric mean between 8 and 10 is $4\sqrt{5}$ or about 8.9.

Guided Practice

Find the geometric mean between each pair of numbers.

1A. 5 and 45

1B. 12 and 15

2 Geometric Means in Right Triangles

In a right triangle, an altitude drawn from the vertex of the right angle to the hypotenuse forms two additional right triangles. These three right triangles share a special relationship.

Review Vocabulary

altitude (of a triangle) a segment from a vertex to the line containing the opposite side and perpendicular to the line containing that side

Theorem 8.1

If the altitude is drawn to the hypotenuse of a right triangle, then the two triangles formed are similar to the original triangle and to each other.

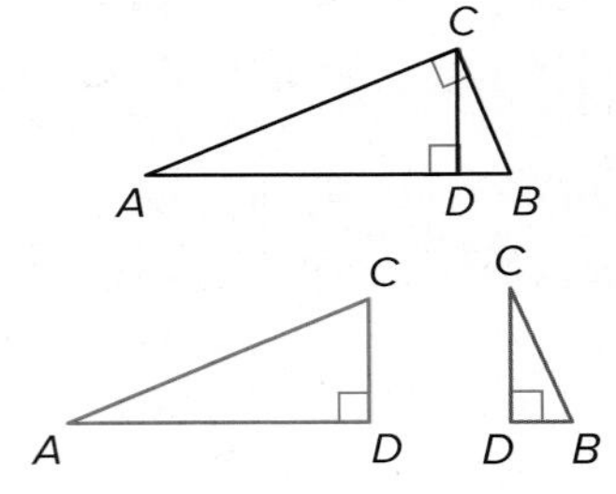

Example If $\overline{CD}$ is the altitude to hypotenuse $\overline{AB}$ of right $\triangle ABC$, then $\triangle ACD \sim \triangle ABC$, $\triangle CBD \sim \triangle ABC$, and $\triangle ACD \sim \triangle CBD$.

You will prove Theorem 8.1 in Exercise 39.

Example 2 Identify Similar Right Triangles

Write a similarity statement identifying the three similar right triangles in the figure.

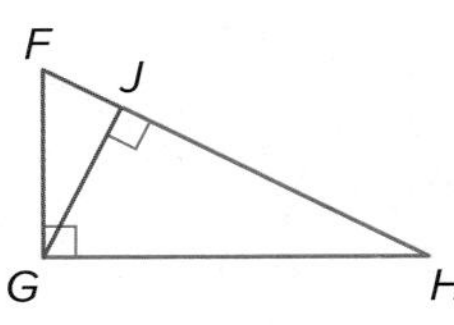

Separate the triangle into two triangles along the altitude. Then sketch the three triangles, reorienting the smaller ones so that their corresponding angles and sides are in the same positions as the original triangle.

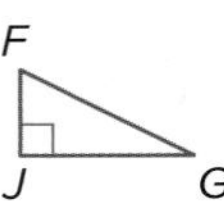

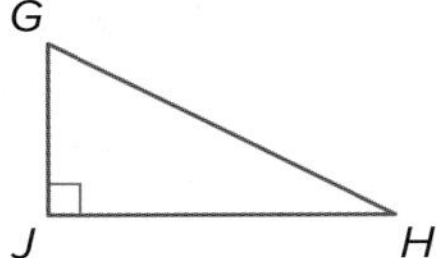

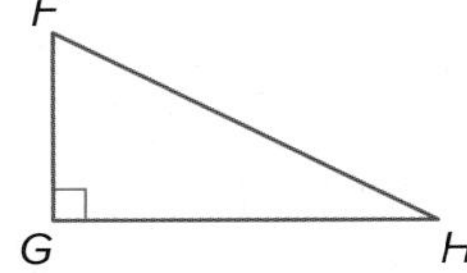

So by Theorem 8.1, $\triangle FJG \sim \triangle GJH \sim \triangle FGH$.

Study Tip

MP Sense-Making To reorient the right triangles in Example 2, first match up the right angles. Then match up the shorter sides.

Guided Practice

2A.

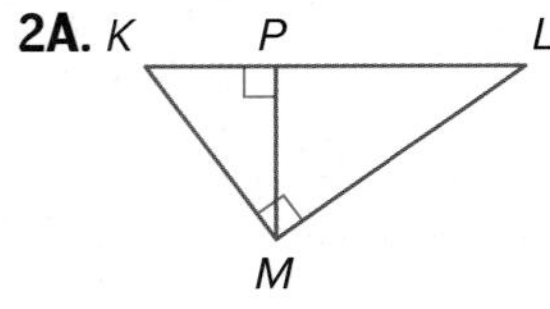

2B.

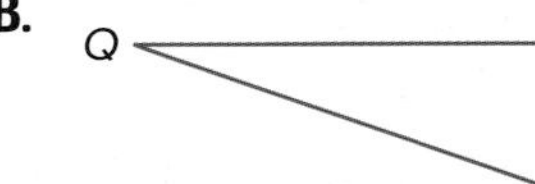

From Theorem 8.1, you know that altitude $\overline{CD}$ drawn to the hypotenuse of right triangle ABC forms three similar triangles: $\triangle ACB \sim \triangle ADC \sim \triangle CDB$. By the definition of similar polygons, you can write the following proportions comparing the side lengths of these triangles.

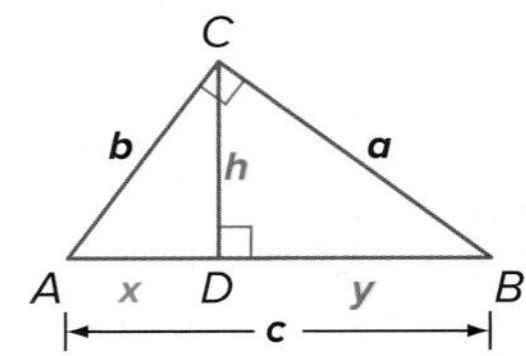

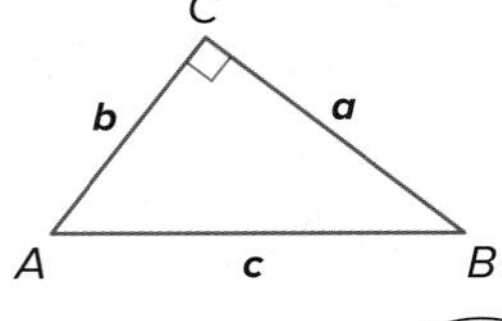

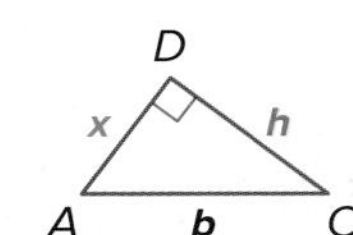

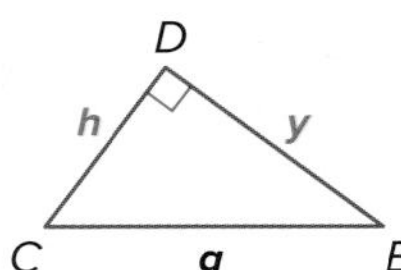

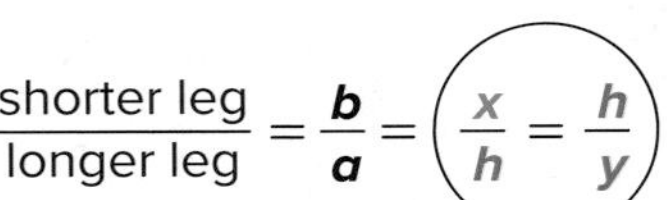

$\frac{\text{shorter leg}}{\text{longer leg}} = \frac{b}{a} = \frac{x}{h} = \frac{h}{y}$

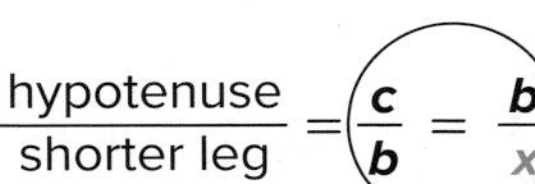

$\frac{\text{hypotenuse}}{\text{shorter leg}} = \frac{c}{b} = \frac{b}{x} = \frac{a}{h}$

$\frac{\text{hypotenuse}}{\text{longer leg}} = \frac{c}{a} = \frac{b}{h} = \frac{a}{y}$

Notice that the circled relationships involve geometric means. This leads to the theorems at the top of the next page.

Theorems Right Triangle Geometric Mean Theorems

8.2 Geometric Mean (Altitude) Theorem The altitude drawn to the hypotenuse of a right triangle separates the hypotenuse into two segments. The length of this altitude is the geometric mean between the lengths of these two segments.

Example If $\overline{CD}$ is the altitude to hypotenuse $\overline{AB}$ of right $\triangle ABC$, then $\frac{x}{h} = \frac{h}{y}$ or $h = \sqrt{xy}$.

8.3 Geometric Mean (Leg) Theorem The altitude drawn to the hypotenuse of a right triangle separates the hypotenuse into two segments. The length of a leg of this triangle is the geometric mean between the length of the hypotenuse and the segment of the hypotenuse adjacent to that leg.

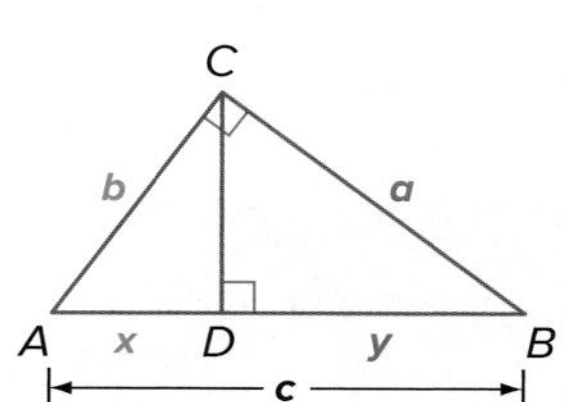

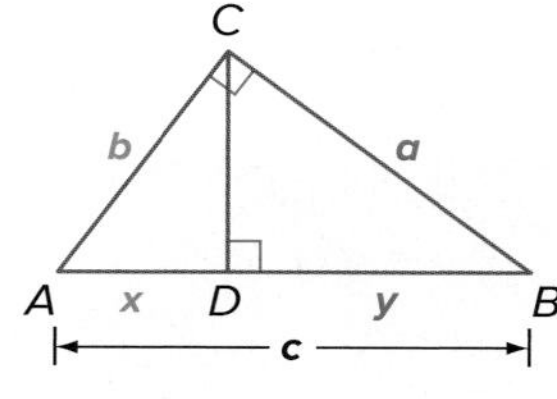

Example If $\overline{CD}$ is the altitude to hypotenuse $\overline{AB}$ of right $\triangle ABC$, then $\frac{c}{b} = \frac{b}{x}$ or $b = \sqrt{xc}$ and $\frac{c}{a} = \frac{a}{y}$ or $a = \sqrt{yc}$.

You will prove Theorems 8.2 and 8.3 in Exercises 40 and 41, respectively.

Go Online!

In Example 3, the value of *x* could also be found by solving the proportion $\frac{5}{x} = \frac{x}{20}$. Search proportions in ConnectED to find resources to review how to set up and solve proportions.

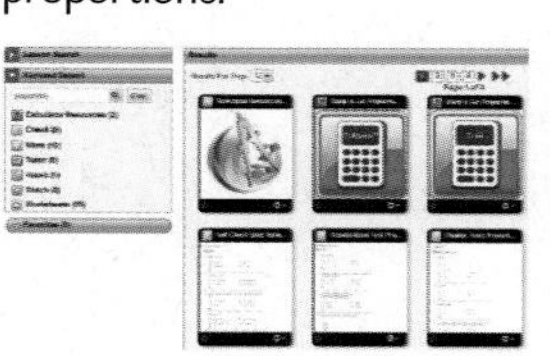

Example 3 Use Geometric Mean with Right Triangles

Find *x*, *y*, and *z*.

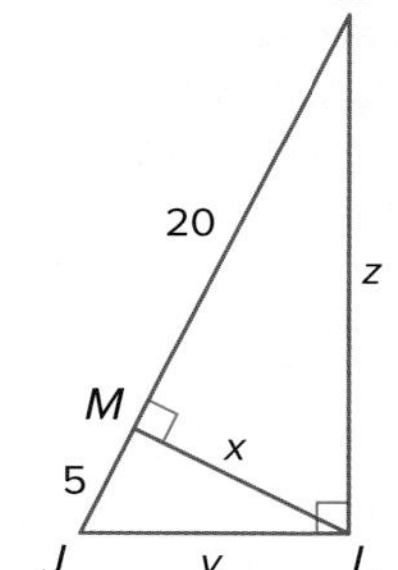

Since *x* is the measure of the altitude drawn to the hypotenuse of right $\triangle JKL$, *x* is the geometric mean of the lengths of the two segments that make up the hypotenuse, *JM* and *MK*.

$x = \sqrt{JM \cdot MK}$	Geometric Mean (Altitude) Theorem
$= \sqrt{5 \cdot 20}$	Substitution
$= \sqrt{100}$ or 10	Simplify.

Since *y* is the measure of leg $\overline{JL}$, *y* is the geometric mean of *JM*, the measure of the segment adjacent to this leg, and *JK*, the measure of the hypotenuse.

$y = \sqrt{JM \cdot JK}$	Geometric Mean (Leg) Theorem
$= \sqrt{5 \cdot (20 + 5)}$	Substitution
$= \sqrt{125}$ or about 11.2	Use a calculator to simplify.

Since *z* is the measure of leg $\overline{KL}$, *z* is the geometric mean of *MK*, the measure of the segment adjacent to $\overline{KL}$, and *JK*, the measure of the hypotenuse.

$z = \sqrt{MK \cdot JK}$	Geometric Mean (Leg) Theorem
$= \sqrt{20 \cdot (20 + 5)}$	Substitution
$= \sqrt{500}$ or about 22.4	Use a calculator to simplify.

Guided Practice

Find x, y, and z.

3A.

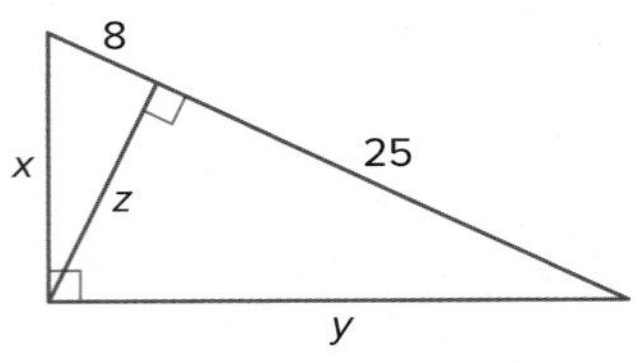

3B.

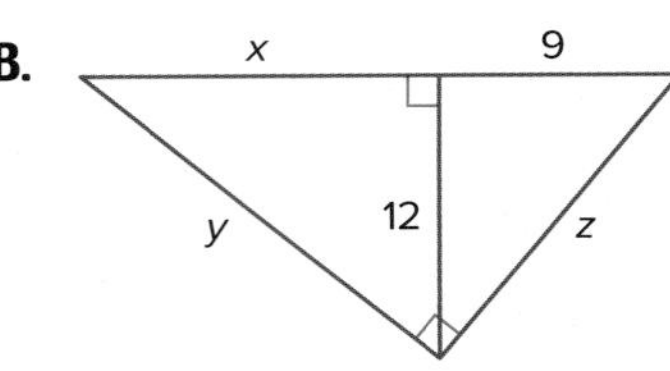

You can use geometric mean to measure height indirectly.

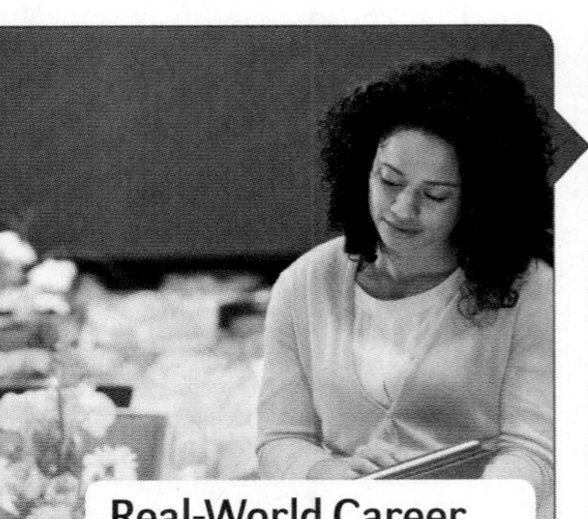

Real-World Career

Event Planner
Event planners organize events including choosing a location, arranging for food, and scheduling entertainment. They also coordinate services like transportation and photography.

Most of the skills required for event planning are acquired through on-the-job experience.

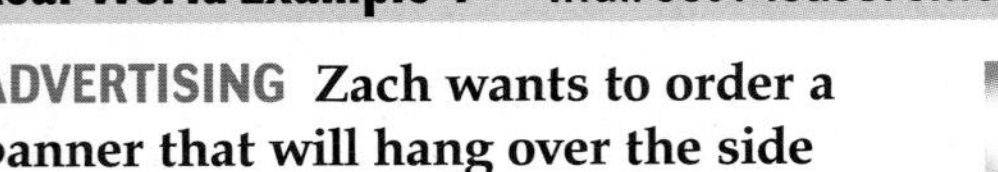

Real-World Example 4 Indirect Measurement

ADVERTISING Zach wants to order a banner that will hang over the side of his high school baseball stadium grandstand and reach the ground.

To find this height, he uses a cardboard square to line up the top and bottom of the grandstand. He measures his distance from the grandstand and from the ground to his eye level. Find the height of the grandstand to the nearest foot.

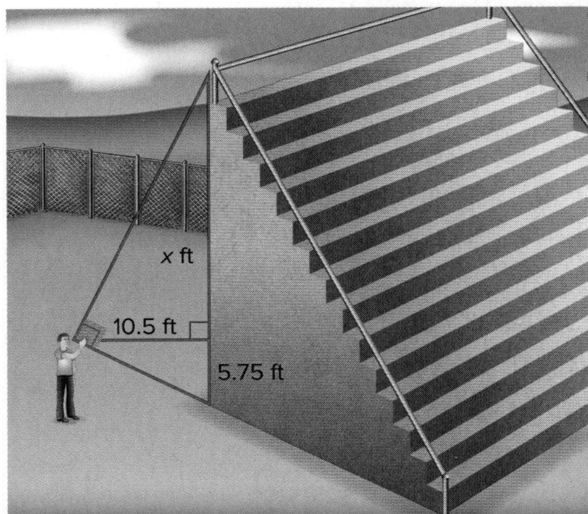

Note: Not drawn to scale.

The distance from Zach to the grandstand is the altitude to the hypotenuse of a right triangle. The length of this altitude is the geometric mean of the two segments that make up the hypotenuse. The shorter segment has the measure of 5.75 feet. Let the unknown measure be x feet.

$10.5 = \sqrt{5.75 \cdot x}$ Geometric Mean (Altitude) Theorem

$110.25 = 5.75x$ Square each side.

$19.17 \approx x$ Divide each side by 5.75.

The height of the grandstand is the total length of the hypotenuse, $5.75 + 19.17$, or about 25 feet.

Guided Practice

4. SPORTS A community center needs to estimate the cost of installing a rock climbing wall by estimating the height of the wall. Sue holds a book up to her eyes so that the top and bottom of the wall are in line with the bottom edge and binding of the cover. If her eye level is 5 feet above the ground and she stands 11 feet from the wall, how high is the wall? Draw a diagram and explain your reasoning.

Check Your Understanding

 = Step-by-Step Solutions begin on page R13.

Example 1 **Find the geometric mean between each pair of numbers.**

1. 5 and 20 **2.** 36 and 4 **3.** 40 and 15

Example 2 **4.** Write a similarity statement identifying the three similar triangles in the figure.

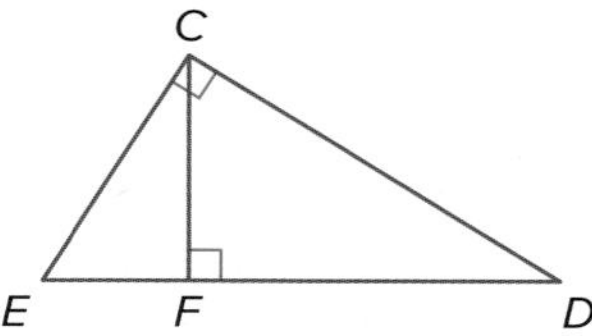

Example 3 **Find x, y, and z.**

5.

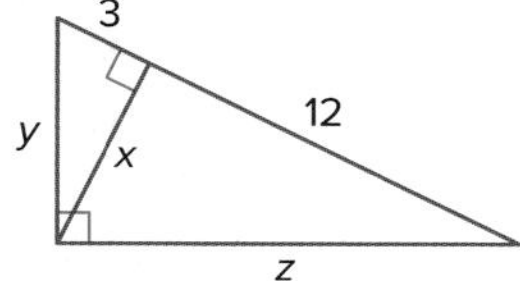

6.

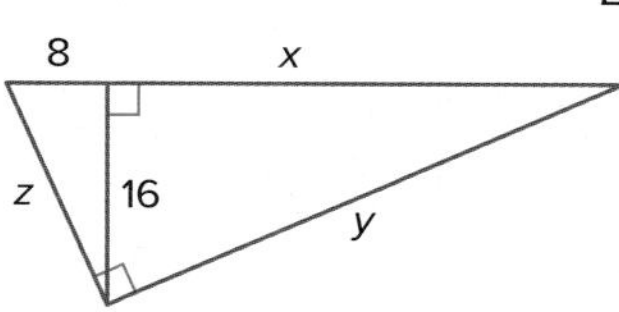

Example 4 **7.** **MODELING** Corey is visiting the Jefferson Memorial with his family. He wants to estimate the height of the statue of Thomas Jefferson. Corey stands so that his line of vision to the top and base of the statue form a right angle as shown in the diagram. About how tall is the statue?

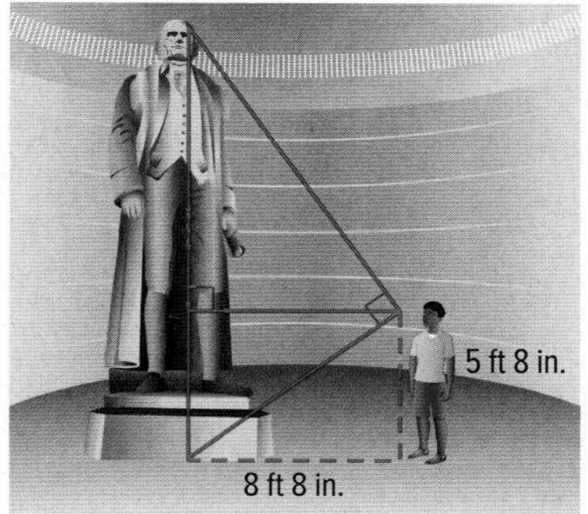

Note: Not drawn to scale.

Practice and Problem Solving

Extra Practice is on page R8.

Example 1 **Find the geometric mean between each pair of numbers.**

8. 81 and 4 **9.** 25 and 16 **10.** 20 and 25

11. 36 and 24 **12.** 12 and 2.4 **13.** 18 and 1.5

Example 2 **Write a similarity statement identifying the three similar triangles in the figure.**

14.

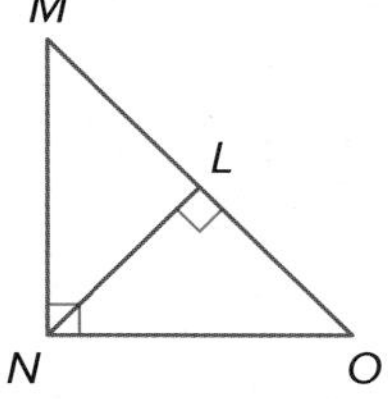

15.

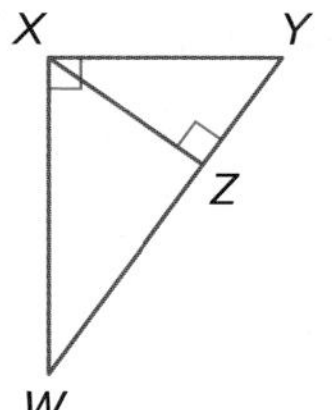

16.

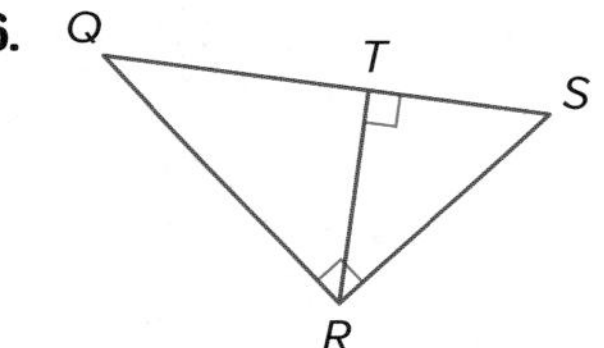

17.

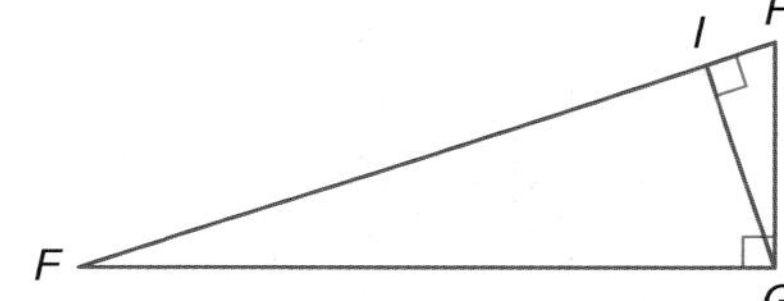

Example 3 **Find x, y, and z.**

18.

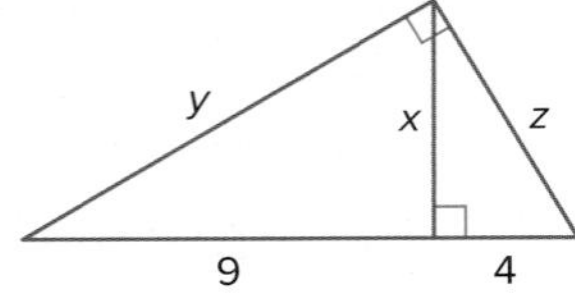

19.

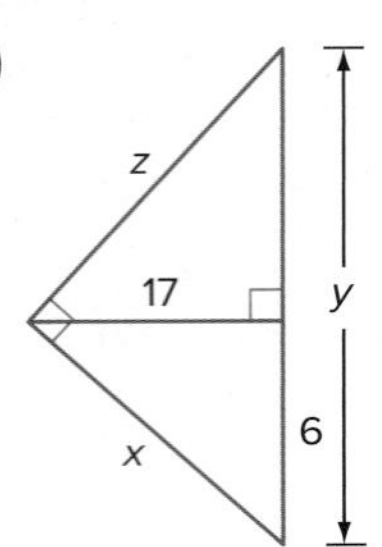

20.

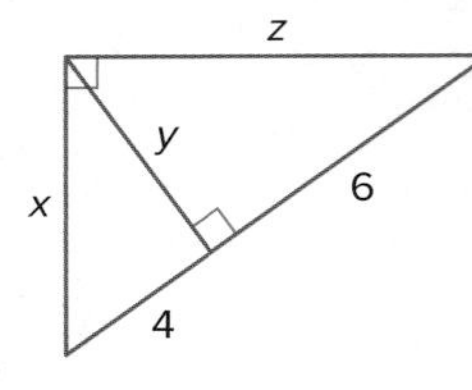

21.

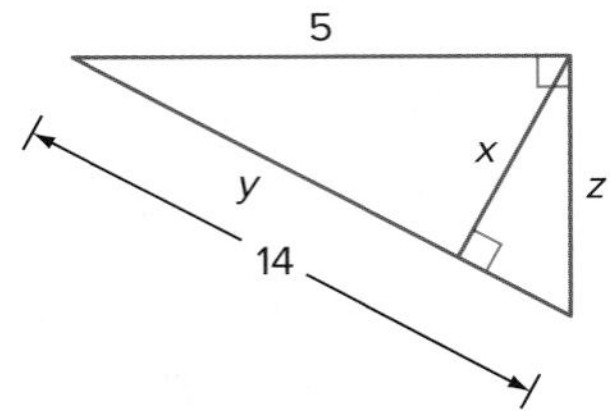

22.

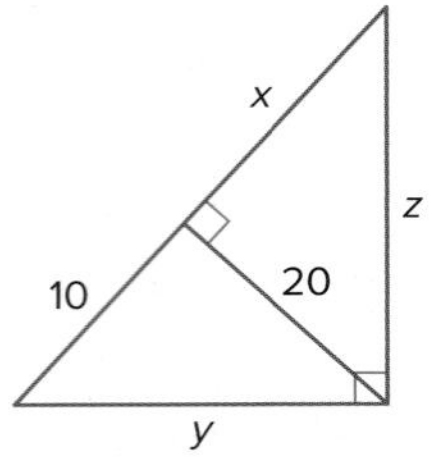

23.

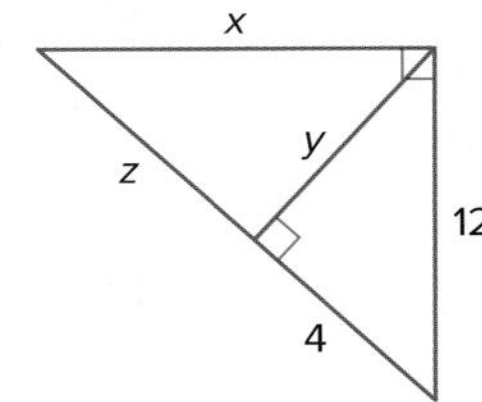

Example 4

24. **MP MODELING** Evelina is hanging silver stars from the gym ceiling using string for the homecoming dance. She wants the ends of the strings where the stars will be attached to be 7 feet from the floor. Use the diagram to determine how long she should make the strings.

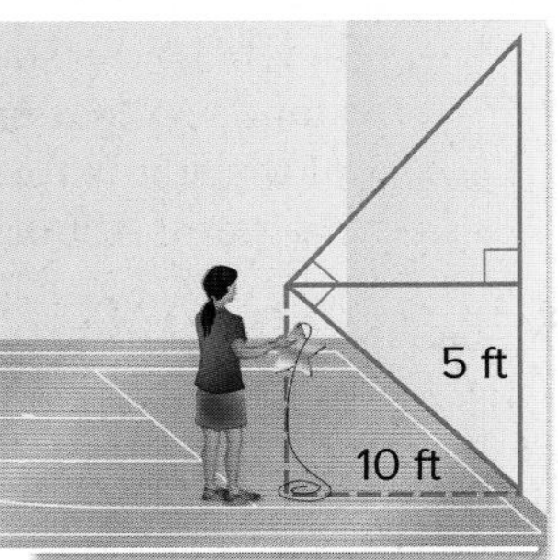

Note: Not drawn to scale.

25. **MP MODELING** Makayla is using a book to sight the top of a waterfall. Her eye level is 5 feet from the ground and she is a horizontal distance of 28 feet from the waterfall. Find the height of the waterfall to the nearest tenth of a foot.

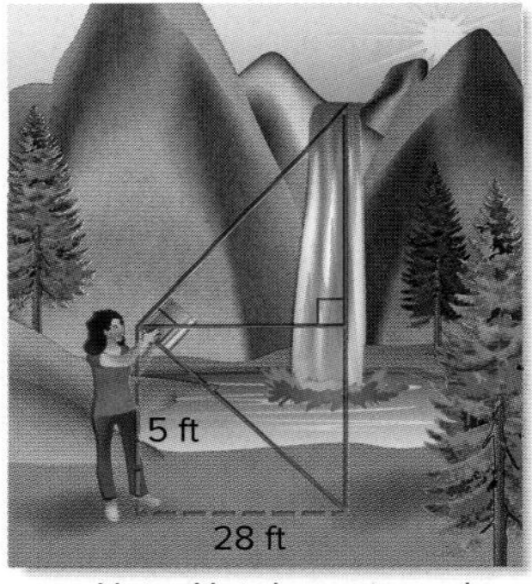

Note: Not drawn to scale.

Find the geometric mean between each pair of numbers.

26. $\frac{1}{5}$ and 60

27. $\frac{3\sqrt{2}}{7}$ and $\frac{5\sqrt{2}}{7}$

28. $\frac{3\sqrt{5}}{4}$ and $\frac{5\sqrt{5}}{4}$

Find x, y, and z.

29.

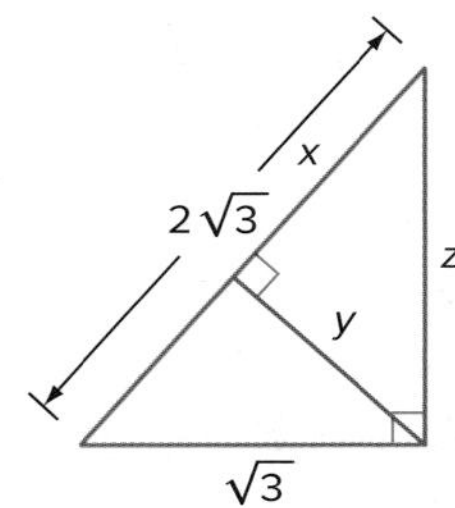

30.

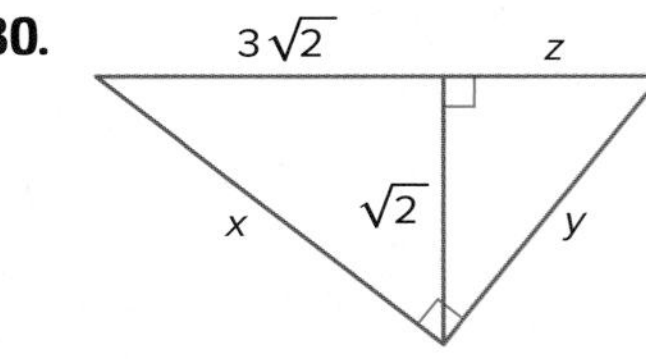

31. **ALGEBRA** The geometric mean of a number and four times the number is 22. What is the number?

Use similar triangles to find the value of *x*.

32. 

33. 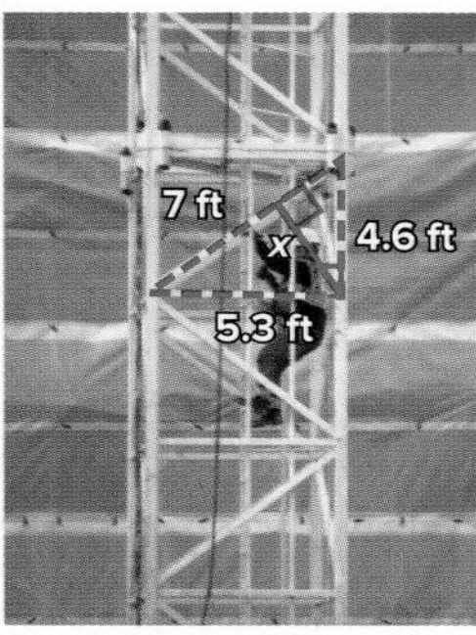

34. 

ALGEBRA Find the value of the variable.

35. 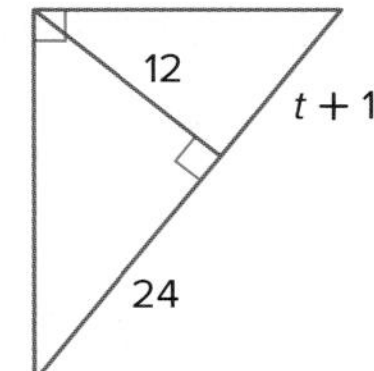

36. 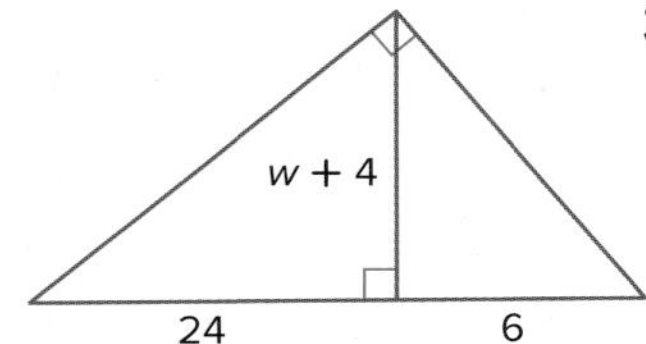

37. 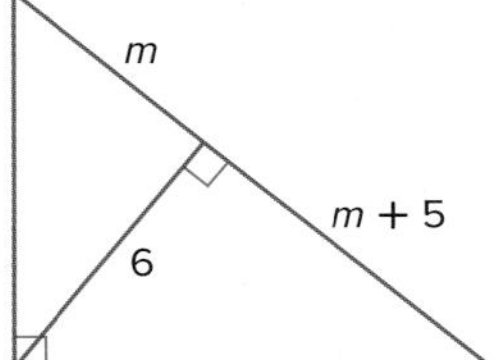

38. **CONSTRUCTION** A room-in-attic truss is a truss design that provides support while leaving an area that can be enclosed as living space. In the diagram, $\angle BCA$ and $\angle EGB$ are right angles, $\triangle BEF$ is isosceles, $\overline{CD}$ is an altitude of $\triangle ABC$, and $\overline{EG}$ is an altitude of $\triangle BEF$. If $DB = 5$ feet, $CD = 6$ feet 4 inches, $BF = 10$ feet 10 inches, and $EG = 4$ feet 6 inches, what is AE?

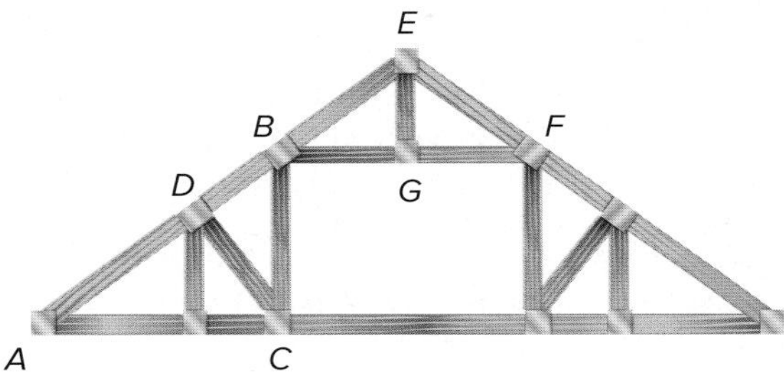

MP CONSTRUCT ARGUMENTS **Write a proof for each theorem.**

39. Theorem 8.1

40. Theorem 8.2

41. Theorem 8.3

42. **TRUCKS** In photography, the angle formed by the top of the subject, the camera, and the bottom of the subject is called the viewing angle, as shown at the right. Natalie is taking a picture of Bigfoot #5, which is 15 feet 6 inches tall. She sets her camera on a tripod that is 5 feet above ground level. The vertical viewing angle of her camera is set for 90°.

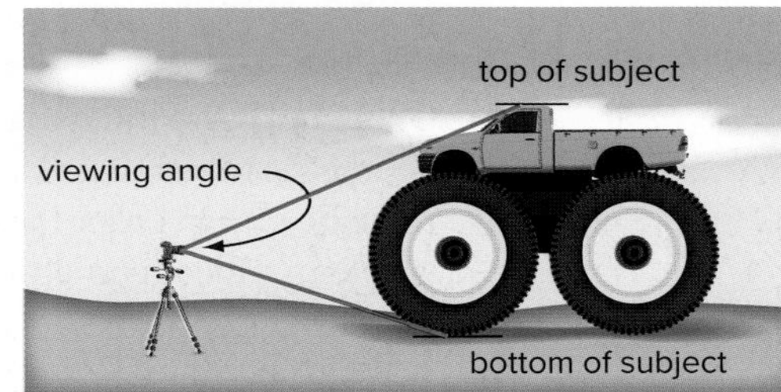

a. Sketch a diagram of this situation.

b. How far away from the truck should Natalie stand so that she perfectly frames the entire height of the truck in her shot?

43 **FINANCE** The average rate of return on an investment over two years is the geometric mean of the two annual returns. If an investment returns 12% one year and 7% the next year, what is the average rate of return on this investment over the two-year period?

44. PROOF Derive the Pythagorean Theorem using the figure at the right and the Geometric Mean (Leg) Theorem.

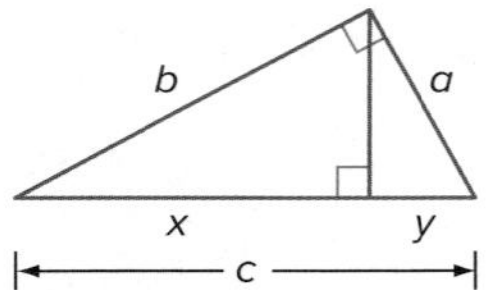

Determine whether each statement is *always*, *sometimes*, or *never* true. Explain your reasoning.

45 The geometric mean for consecutive positive integers is the mean of the two numbers.

46. The geometric mean for two perfect squares is a positive integer.

47. The geometric mean for two positive integers is another integer.

48. MULTIPLE REPRESENTATIONS In this problem, you will investigate geometric mean.

a. **Tabular** Copy and complete the table of five ordered pairs (x, y) such that $\sqrt{xy} = 8$.

x	y	$\sqrt{xy}$
		8
		8
		8
		8
		8

b. **Graphical** Graph the ordered pairs from your table in a scatter plot.

c. **Verbal** Make a conjecture as to the type of graph that would be formed if you connected the points from your scatter plot. Do you think the graph of any set of ordered pairs that results in the same geometric mean would have the same general shape? Explain your reasoning.

H.O.T. Problems Use Higher-Order Thinking Skills

49. ERROR ANALYSIS Aiden and Tia are finding the value of x in the triangle shown. Is either of them correct? Explain your reasoning.

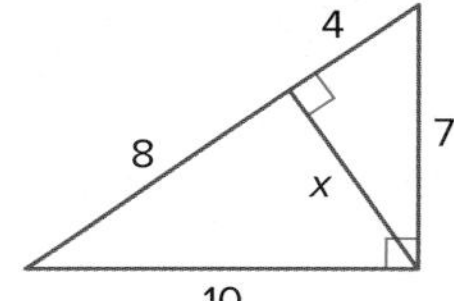

Aiden	Tia
$\frac{4}{x} = \frac{x}{7}$	$\frac{4}{x} = \frac{x}{10}$
$x \approx 5.3$	$x \approx 6.3$

50. CHALLENGE Refer to the figure at the right. Find x, y, and z.

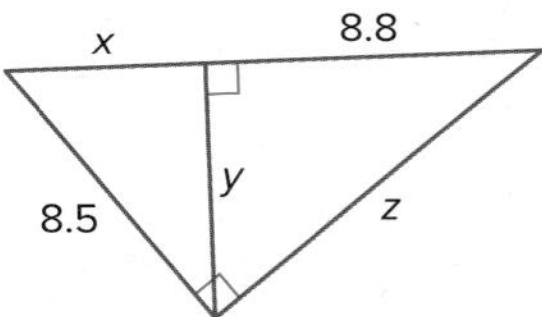

51. OPEN-ENDED Find two pairs of whole numbers with a geometric mean that is also a whole number. What condition must be met in order for a pair of numbers to produce a whole-number geometric mean?

52. MP REASONING Refer to the figure at the right. The orthocenter of $\triangle ABC$ is located 6.4 units from point D. Find BC.

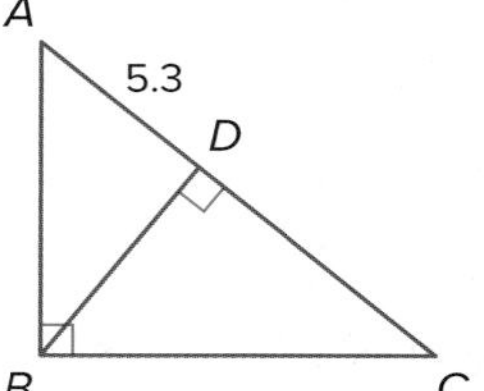

53. WRITING IN MATH Compare and contrast the arithmetic and geometric means of two numbers. When will the two means be equal? Justify your reasoning.

54. The figure shows the paths in a botanical garden.

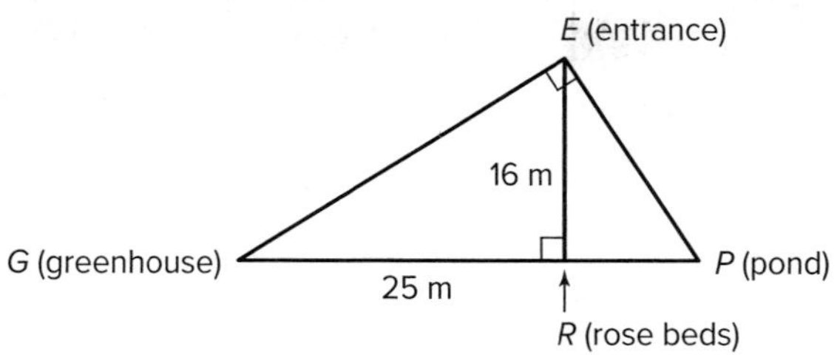

What is the distance from the greenhouse to the pond? MP 4

- A 10.24 m
- B 35.24 m
- C 39.06 m
- D 64.06 m

55. What is the area of $\triangle JKL$? MP 2

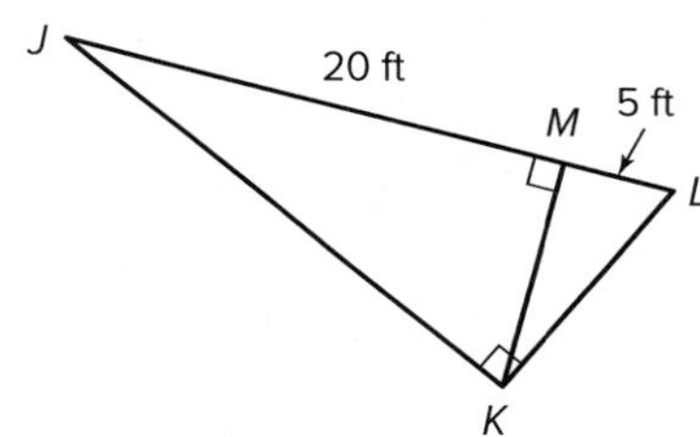

- A 10 ft^2
- B 50 ft^2
- C 125 ft^2
- D 250 ft^2

56. Which of the following is the best estimate of the length of $\overline{AB}$? MP 2

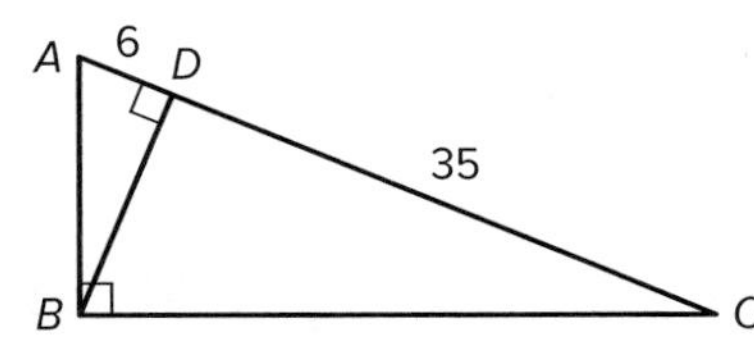

- A 37.9
- B 15.7
- C 14.5
- D 8.9

57. Zachary wants to estimate the height of a cliff. He stands 15 feet from the base of the cliff.

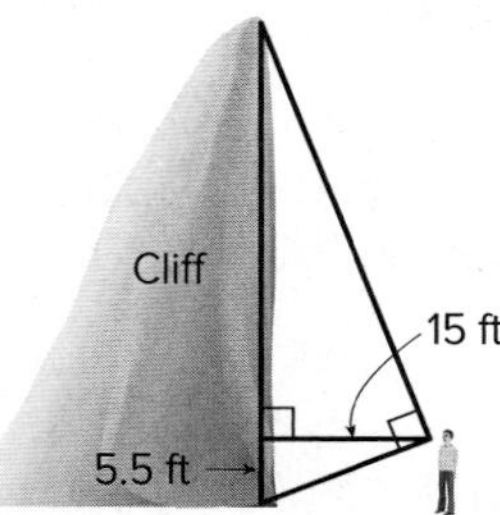

If his eye level is at 5.5 feet, which of the following is closest to the height of the cliff? MP 4

- A 10.6 ft
- B 17.5ft
- C 40.9 ft
- D 46.4 ft

58. In the figure, $\overline{AD}$ is perpendicular to $\overline{BC}$, and $\overline{AB}$ is perpendicular to $\overline{AC}$. What is BC? MP 3

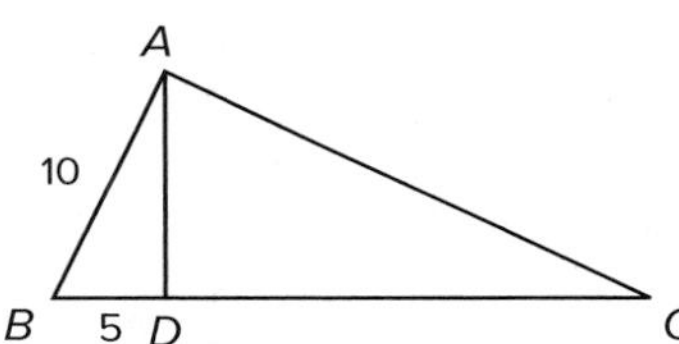

- A $5\sqrt{3}$
- B 15
- C $10\sqrt{3}$
- D 20

59. MULTI-STEP An altitude is drawn to the hypotenuse of a right triangle, separating the hypotenuse into two segments. The segments of the hypotenuse are in the ratio 4 : 9. MP 1,7

a. What is the ratio of the length of the shorter segment of the hypotenuse to the length of the altitude?

b. Suppose the length of the altitude is 42 inches. What are the lengths of the hypotenuse segments?

EXPLORE 8-2

Geometry Lab
Proofs Without Words

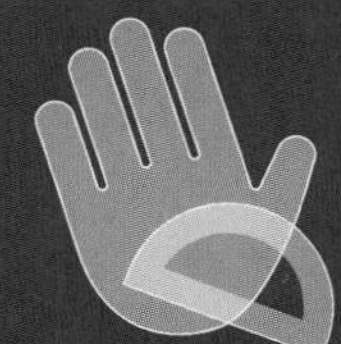

In Chapter 1, you learned that the Pythagorean Theorem relates the measures of the legs and the hypotenuse of a right triangle. You can prove the Pythagorean Theorem by using diagrams without words.

Mathematical Practices
MP 4 Model with mathematics

Activity

Prove the Pythagorean Theorem by using paper and algebra.

Step 1

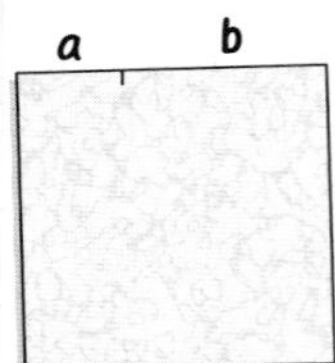

On a piece of tracing paper, mark one side a and b as shown above.

Step 2

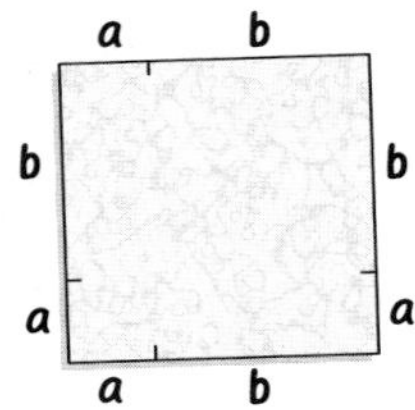

Copy these measures on each of the other sides.

Step 3

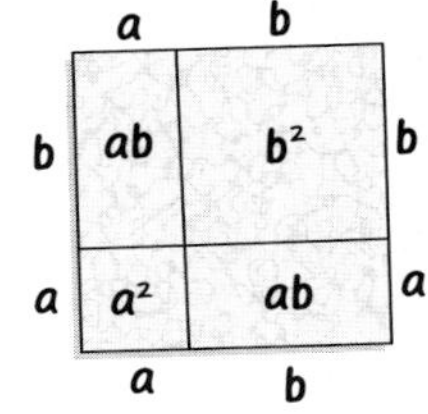

Fold the paper into four sections and label the area of each section.

Step 4

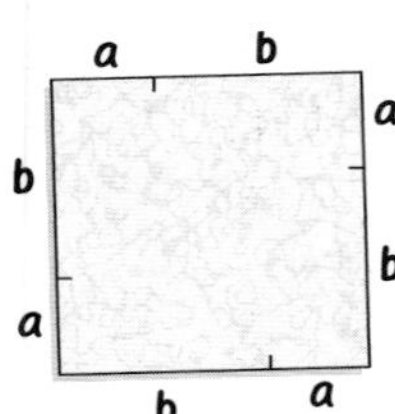

On another piece of tracing paper, mark each side a and b as shown above.

Step 5

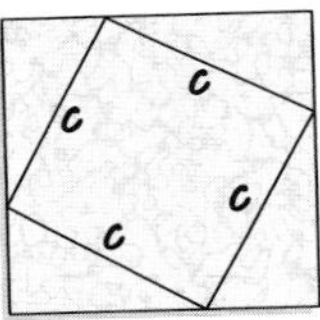

Connect the marks using a straightedge. Let c represent the length of each hypotenuse.

Step 6

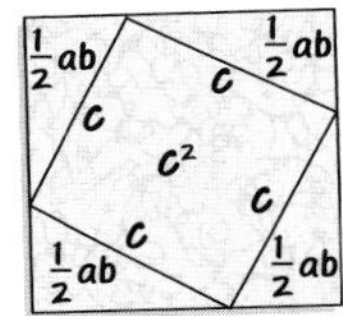

Label the area of each triangle $\frac{1}{2}ab$ and the area of each square c^2.

Step 7 Place the squares side by side and color the corresponding regions that have the same area. For example, $ab = \frac{1}{2}ab + \frac{1}{2}ab$. The parts that are not shaded tell us that $a^2 + b^2 = c^2$.

Analyze the Results **Work cooperatively.**

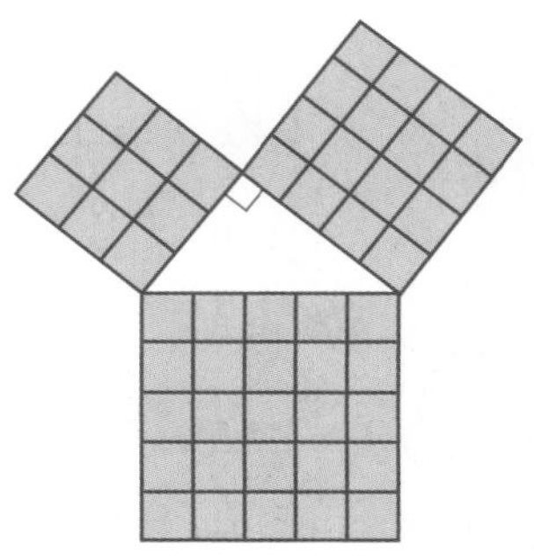

1. Use a ruler to measure a, b, and c. Do these measures confirm that $a^2 + b^2 = c^2$?
2. Repeat the activity with different a and b values. What do you notice?
3. **WRITING IN MATH** Explain why the diagram at the right is an illustration of the Pythagorean Theorem.
4. **CHALLENGE** Draw a geometric diagram to show that for any positive numbers a and b, $a + b > \sqrt{a^2 + b^2}$. Explain.

LESSON 2
The Pythagorean Theorem and Its Converse

Then

- You used the Pythagorean Theorem to develop the Distance Formula.

Now

1. Use the Pythagorean Theorem.
2. Use the Converse of the Pythagorean Theorem.

Why?

- Tether lines are used to steady an inflatable snowman. Suppose you know the height at which the tether lines are attached to the snowman and how far away you want to anchor the tether in the ground. You can use the converse of the Pythagorean Theorem to adjust the lengths of the tethers to keep the snowman perpendicular to the ground.

New Vocabulary
Pythagorean triple

Mathematical Practices

1 Make sense of problems and persevere in solving them.

4 Model with mathematics.

1 The Pythagorean Theorem

The Pythagorean Theorem is perhaps one of the most famous theorems in mathematics. It relates the lengths of the hypotenuse (side opposite the right angle) and legs (sides adjacent to the right angle) of a right triangle.

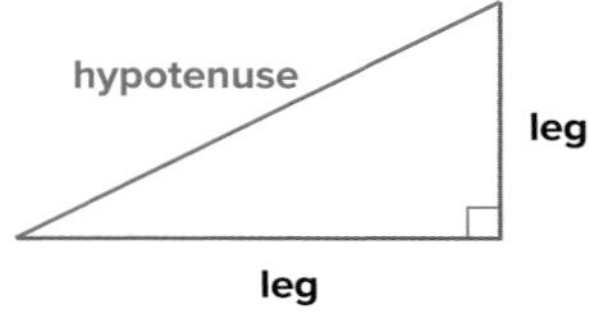

Theorem 8.4 Pythagorean Theorem

Words In a right triangle, the sum of the squares of the lengths of the legs is equal to the square of the length of the hypotenuse.

Symbols If $\triangle ABC$ is a right triangle with right angle C, then $a^2 + b^2 = c^2$.

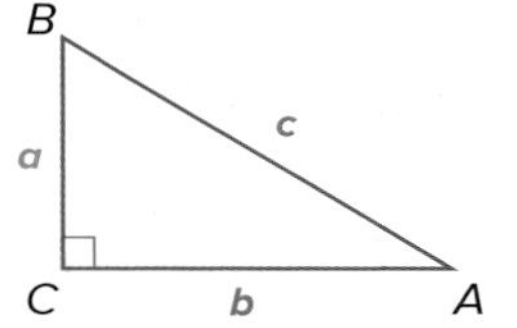

The geometric mean can be used to prove the Pythagorean Theorem.

Proof Pythagorean Theorem

Given: $\triangle ABC$ with right angle at C

Prove: $a^2 + b^2 = c^2$

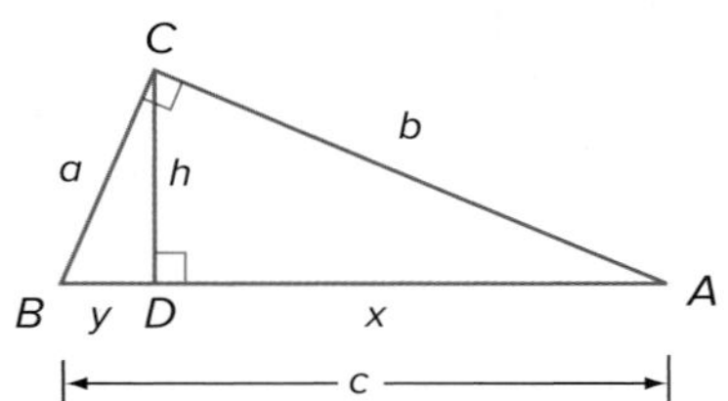

Proof:

Draw right triangle ABC so C is the right angle. Then draw the altitude from C to $\overline{AB}$. Let $AB = c$, $AC = b$, $BC = a$, $AD = x$, $DB = y$, and $CD = h$. Two geometric means now exist.

$\frac{c}{a} = \frac{a}{y}$ and $\frac{c}{b} = \frac{b}{x}$ Geometric Mean (Leg) Theorem

$a^2 = cy$ $\quad b^2 = cx$ Cross products

$a^2 + b^2 = cy + cx$ Add the equations.

$a^2 + b^2 = c(y + x)$ Factor.

$a^2 + b^2 = c \cdot c$ Because $c = y + x$, substitute c for $(y + x)$.

$a^2 + b^2 = c^2$ Simplify.

You can use the Pythagorean Theorem to find the measure of any side of a right triangle given the lengths of the other two sides.

Example 1 **Find Missing Measures Using the Pythagorean Theorem**

Find x.

a.

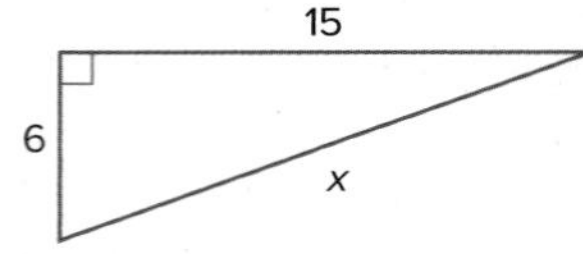

The side opposite the right angle is the hypotenuse, so $c = x$.

$a^2 + b^2 = c^2$	Pythagorean Theorem
$6^2 + 15^2 = x^2$	$a = 6$ and $b = 15$
$261 = x^2$	Simplify.
$\sqrt{261} = x$	Take the positive square root of each side.
$3\sqrt{29} = x$	Simplify.

Study Tip

Positive Square Root When finding the length of a side using the Pythagorean Theorem, use only the positive and not the negative square root, since length cannot be negative.

b.

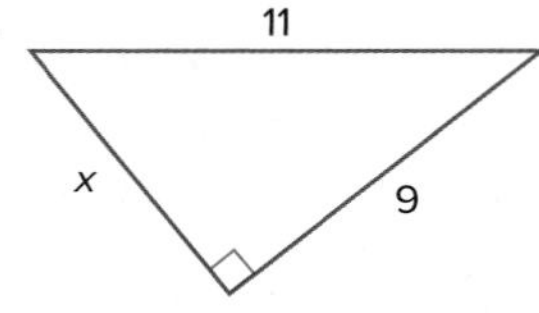

The hypotenuse is 11, so $c = 11$.

$a^2 + b^2 = c^2$	Pythagorean Theorem
$x^2 + 9^2 = 11^2$	$a = x$ and $b = 9$
$x^2 + 81 = 121$	Simplify.
$x^2 = 40$	Subtract 81 from each side.
$x = \sqrt{40}$ or $2\sqrt{10}$	Take the positive square root of each side and simplify.

Guided Practice

1A.

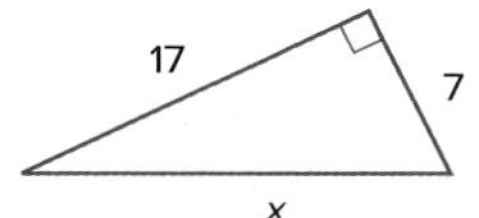

1B.

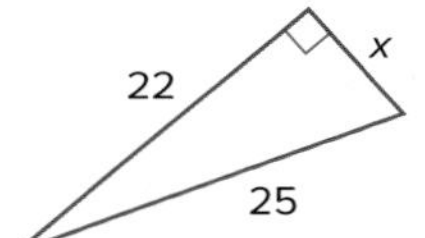

A **Pythagorean triple** is a set of three nonzero whole numbers a, b, and c, such that $a^2 + b^2 = c^2$. One common Pythagorean triple is 3, 4, 5; that is, the sides of a right triangle are in the ratio 3:4:5. The most common Pythagorean triples are shown below in the first row. The triples below these are found by multiplying each number in the triple by the same factor.

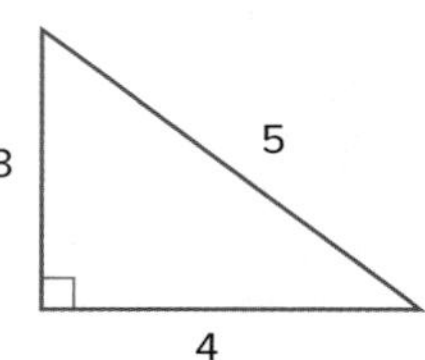

Study Tip

Pythagorean Triples If the measures of the sides of any right triangle are not whole numbers, the measures do not form a Pythagorean triple.

Key Concept **Common Pythagorean Triples**

3, 4, 5	5, 12, 13	8, 15, 17	7, 24, 25
6, 8, 10	10, 24, 26	16, 30, 34	14, 48, 50
9, 12, 15	15, 36, 39	24, 45, 51	21, 72, 75
$3x, 4x, 5x$	$5x, 12x, 13x$	$8x, 15x, 17x$	$7x, 24x, 25x$

The largest number in each triple is the length of the hypotenuse.

Example 2 Use a Pythagorean Triple

Reading Math

3-4-5 A right triangle with side lengths 3, 4, and 5 is called a *3-4-5 right triangle*.

Use a Pythagorean triple to find x. Explain your reasoning.

Notice that 15 and 12 are both multiples of 3, because $15 = 3 \cdot 5$ and $12 = 3 \cdot 4$. Because 3, 4, 5 is a Pythagorean triple, the missing leg length x is $3 \cdot 3$ or 9.

CHECK $12^2 + 9^2 \stackrel{?}{=} 15^2$ Pythagorean Theorem

$225 = 225$ ✓ Simplify.

Guided Practice

2A.

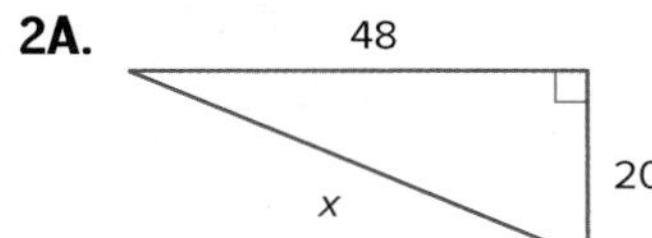

2B.

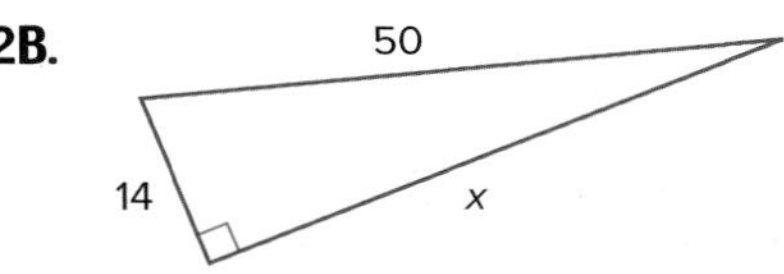

The Pythagorean Theorem can be used to solve many real-world problems.

Example 3 Use the Pythagorean Theorem

Damon is locked out of his house. The only open window is on the second floor, which is 12 feet above the ground. He needs to borrow a ladder from his neighbor. If he must place the ladder 5 feet from the house to avoid some bushes, what length of ladder does Damon need?

A 7 feet **C** 13 feet

B 11 feet **D** 17 feet

Note: Not drawn to scale.

Read the Item

The distance the ladder is from the house, the height the ladder reaches, and the length of the ladder itself make up the lengths of the sides of a right triangle. You need to find the length of the ladder, which is the hypotenuse.

Study Tip

MP Sense-Making Because the hypotenuse of a right triangle is always the longest side, the length of the ladder in Example 3 must be greater than 5 or 12 feet. Because 7 and 11 feet are both less than 12 feet, choices A and B can be eliminated.

Solve the Item

Method 1 Use a Pythagorean triple.

The lengths of the legs are 5 and 12. 5, 12, 13 is a Pythagorean triple, so the length of the ladder is 13 feet.

Method 2 Use the Pythagorean Theorem.

Let x represent the length of the ladder.

$5^2 + 12^2 = x^2$ Pythagorean Theorem

$169 = x^2$ Simplify.

$\sqrt{169} = x$ Take the positive square root of each side.

$13 = x$ Simplify.

So, the answer is choice C.

Go Online! 

The Pythagorean Theorem and its converse are widely used in problems arising in everyday life, society, and the workplace. Ask your teacher a questions about it by sending a message in ConnectED.

Guided Practice

3. The distance from the base of a ladder to a wall that it leans against should be at least one fourth of the ladder's total length. What is the maximum distance x up the wall that a 20-foot ladder will reach, to the nearest tenth?

   A 12 feet **C** 20.6 feet

   B 19.4 feet **D** 30.6 feet

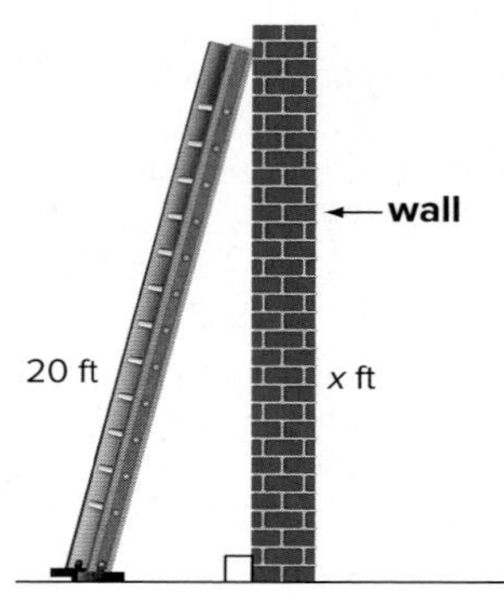

2 Converse of the Pythagorean Theorem

Converse of the Pythagorean Theorem The converse of the Pythagorean Theorem also holds. You can use this theorem to help you determine whether a triangle is a right triangle given the measures of all three sides.

Theorem 8.5 Converse of the Pythagorean Theorem

Words If the sum of the squares of the lengths of the shortest sides of a triangle is equal to the square of the length of the longest side, then the triangle is a right triangle.

Symbols If $a^2 + b^2 = c^2$, then $\triangle ABC$ is a right triangle.

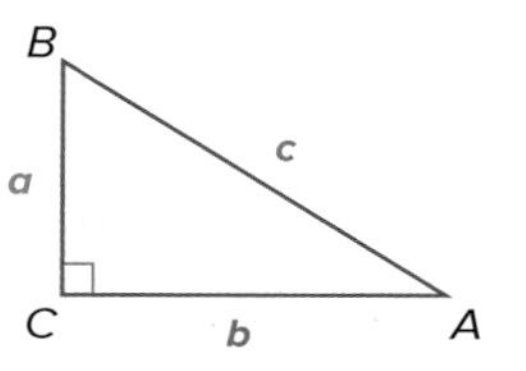

You will prove Theorem 8.5 in Exercise 35.

Study Tip

Determining the Longest Side If the measures of any of the sides of a triangle are expressed as radicals, you may wish to use a calculator to determine which length is the longest.

You can also use side lengths to classify a triangle as acute or obtuse.

Theorem Pythagorean Inequality Theorems

8.6 If the square of the length of the longest side of a triangle is less than the sum of the squares of the lengths of the other two sides, then the triangle is an acute triangle.

Symbols If $c^2 < a^2 + b^2$, then $\triangle ABC$ is acute.

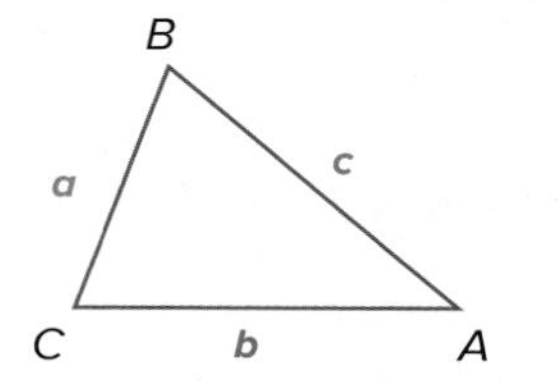

8.7 If the square of the length of the longest side of a triangle is greater than the sum of the squares of the lengths of the other two sides, then the triangle is an obtuse triangle.

Symbols If $c^2 > a^2 + b^2$, then $\triangle ABC$ is obtuse.

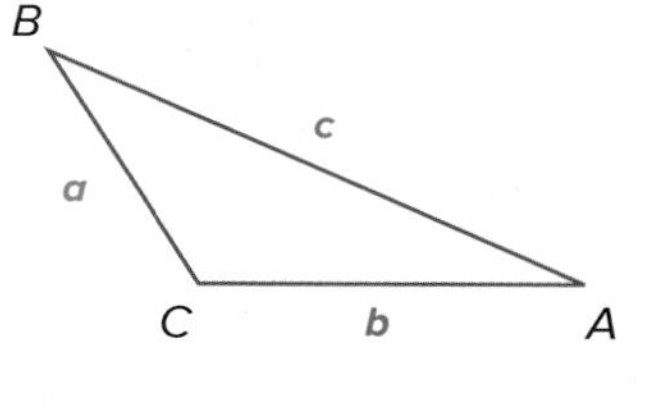

You will prove Theorems 8.6 and 8.7 in Exercises 36 and 37, respectively.

Example 4 Classify Triangles

Determine whether each set of numbers can be the measures of the sides of a triangle. If so, classify the triangle as *acute*, *right*, or *obtuse*. Justify your answer.

a. 7, 14, 16

Step 1 Determine whether the measures can form a triangle using the Triangle Inequality Theorem.

$7 + 14 > 16$ ✓ $\quad 14 + 16 > 7$ ✓ $\quad 7 + 16 > 14$ ✓

The side lengths 7, 14, and 16 can form a triangle.

Step 2 Classify the triangle by comparing the square of the longest side to the sum of the squares of the other two sides.

$c^2 \stackrel{?}{=} a^2 + b^2$ Compare c^2 and $a^2 + b^2$.

$16^2 \stackrel{?}{=} 7^2 + 14^2$ Substitution

$256 > 245$ Simplify and compare.

Because $c^2 > a^2 + b^2$, the triangle is obtuse.

b. 9, 40, 41

Step 1 Determine whether the measures can form a triangle.

$9 + 40 > 41$ ✓ $\quad 40 + 41 > 9$ ✓ $\quad 9 + 41 > 40$ ✓

The side lengths 9, 40, and 41 can form a triangle.

Step 2 Classify the triangle.

$c^2 \stackrel{?}{=} a^2 + b^2$ Compare c^2 and $a^2 + b^2$.

$41^2 \stackrel{?}{=} 9^2 + 40^2$ Substitution

$1681 = 1681$ Simplify and compare.

Because $c^2 = a^2 + b^2$, the triangle is a right triangle.

Guided Practice

4A. 11, 60, 61 **4B.** $2\sqrt{3}, 4\sqrt{2}, 3\sqrt{5}$ **4C.** 6.2, 13.8, 20

Review Vocabulary

Triangle Inequality Theorem The sum of the lengths of any two sides of a triangle must be greater than the length of the third side.

Check Your Understanding

 = Step-by-Step Solutions begin on page R13.

Example 1 Find x.

1.

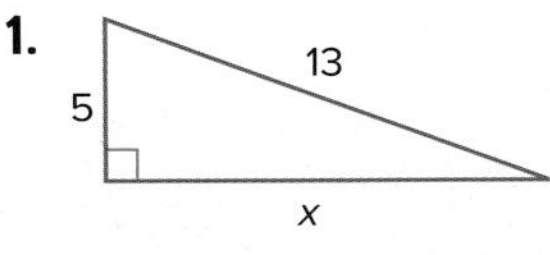

2.

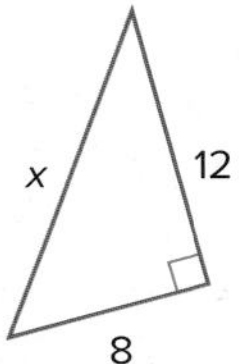

3.

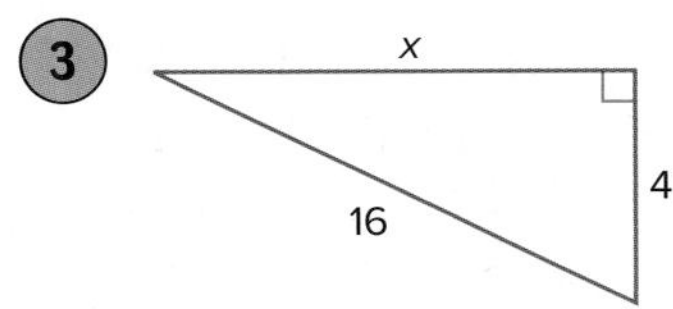

Example 2

4. Use a Pythagorean triple to find x. Explain your reasoning.

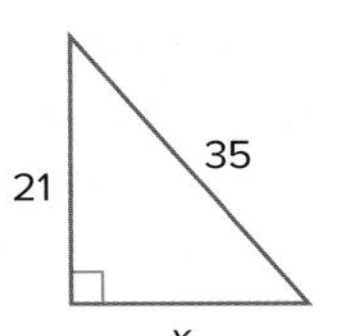

Example 3

5. **MULTIPLE CHOICE** The mainsail of a boat is shown. What is the length, in feet, of $\overline{LN}$?

A 52.5
B 65
C 72.5
D 75

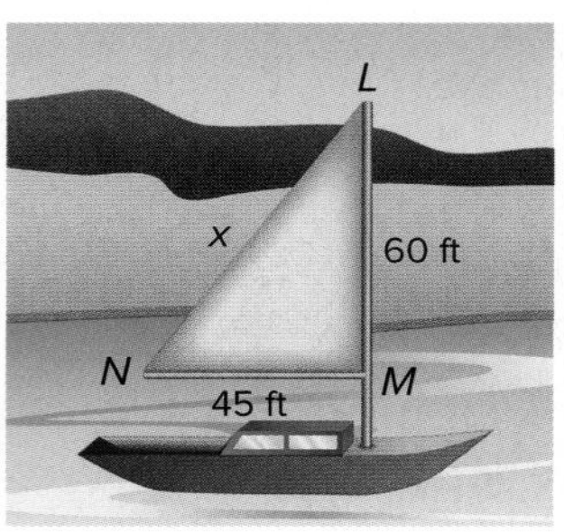

Example 4

Determine whether each set of numbers can be the measures of the sides of a triangle. If so, classify the triangle as *acute*, *obtuse*, or *right*. Justify your answer.

6. 15, 36, 39
7. 16, 18, 26
8. 15, 20, 24

Practice and Problem Solving

Extra Practice is on page R8.

Example 1

Find x.

9. 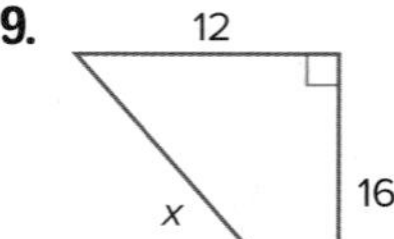

10. 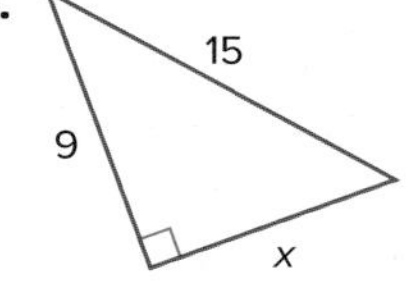

11. 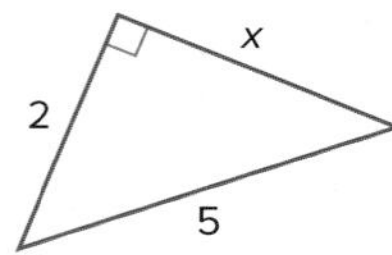

12. 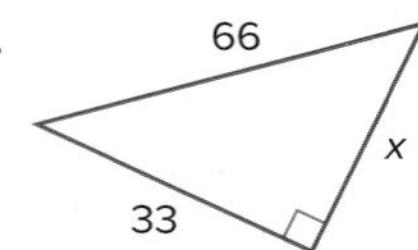

13. 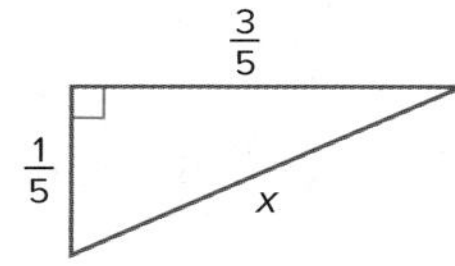

14. 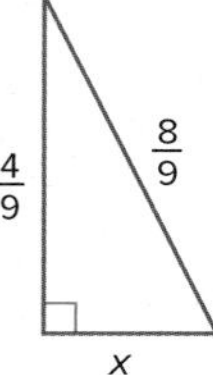

Example 2

PERSEVERANCE Use a Pythagorean Triple to find x.

15. 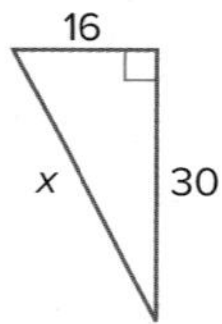

16. 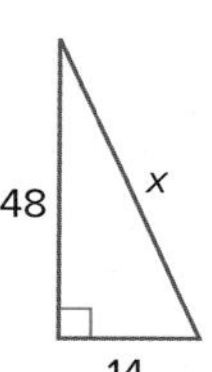

17. 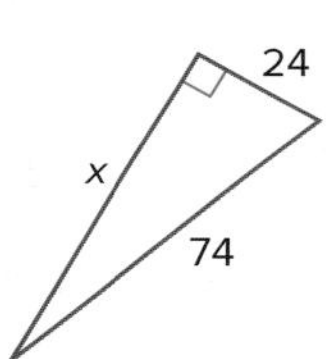

18. 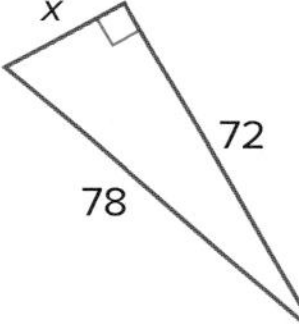

Example 3

19. **BASKETBALL** The support for a basketball goal forms a right triangle as shown. What is the length x of the horizontal portion of the support?

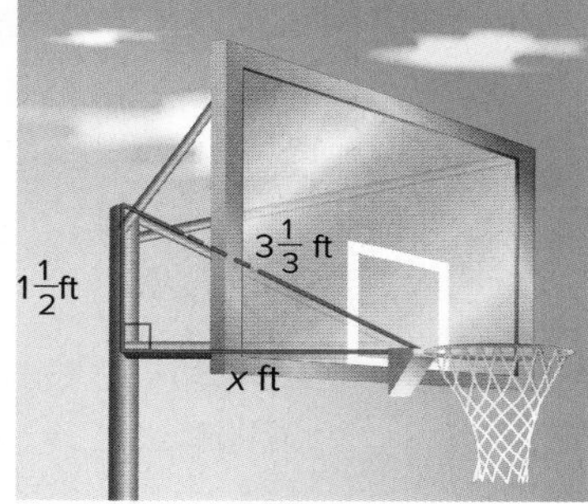

20. **DRIVING** The street that Khaliah usually uses to get to school is under construction. She has been taking the detour shown. If the construction starts at the point where Khaliah leaves her normal route and ends at the point where she reenters her normal route, about how long is the stretch of road under construction?

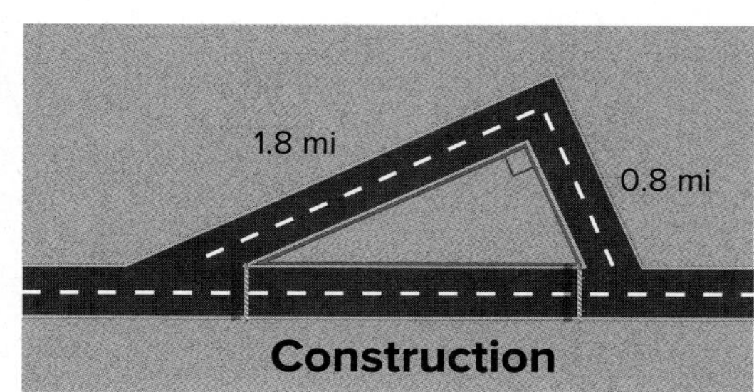

Example 4 **Determine whether each set of numbers can be the measures of the sides of a triangle. If so, classify the triangle as *acute*, *obtuse*, or *right*. Justify your answer.**

21. 7, 15, 21
22. 10, 12, 23
23. 4.5, 20, 20.5
24. 44, 46, 91
25. 4.2, 6.4, 7.6
26. 4, 12, 14

Find x.

27.
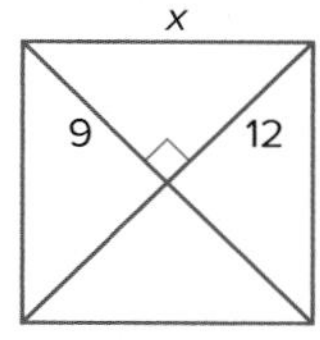

28.
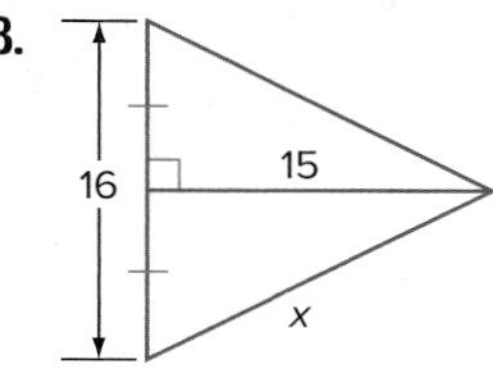

29.
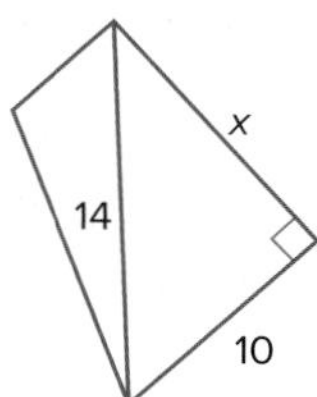

COORDINATE GEOMETRY **Determine whether $\triangle XYZ$ is an *acute*, *right*, or *obtuse* triangle for the given vertices. Explain.**

30. $X(-3, -2)$, $Y(-1, 0)$, $Z(0, -1)$
31. $X(-7, -3)$, $Y(-2, -5)$, $Z(-4, -1)$
32. $X(1, 2)$, $Y(4, 6)$, $Z(6, 6)$
33. $X(3, 1)$, $Y(3, 7)$, $Z(11, 1)$

34. **JOGGING** Brett jogs in the park three times a week. Usually, he takes a $\frac{3}{4}$-mile path that cuts through the park. Today, the path is closed, so he is taking the orange route shown. How much farther will he jog on his alternate route than he would have if he had followed his normal path?

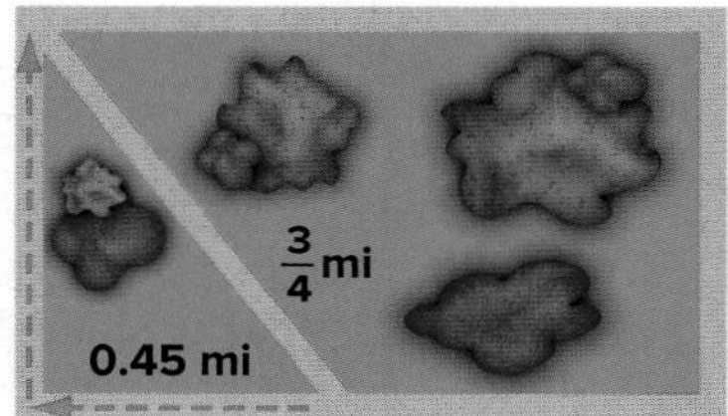

35. **PROOF** Write a paragraph proof of Theorem 8.5.

PROOF Write a two-column proof for each theorem.

36. Theorem 8.6
37. Theorem 8.7

SENSE-MAKING **Find the perimeter and area of each figure.**

38.
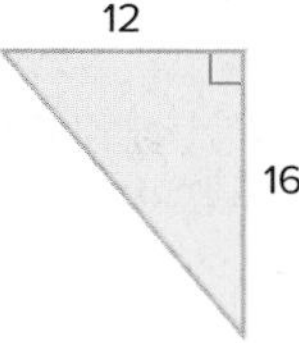

39.
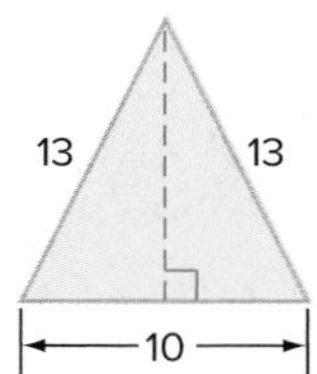

40.
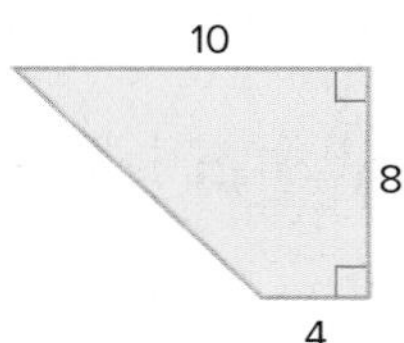

41. **ALGEBRA** The sides of a triangle have lengths x, $x + 5$, and 25. If the length of the longest side is 25, what value of x makes the triangle a right triangle?

42. **ALGEBRA** The sides of a triangle have lengths $2x$, 8, and 12. If the length of the longest side is $2x$, what values of x make the triangle acute?

43. **TELEVISION** The screen aspect ratio, or the ratio of the width to the height, of a high-definition television is 16:9. The size of a television is given by the diagonal distance across the screen. If an HDTV is 41 inches wide, what is its screen size?

44. **PLAYGROUND** According to the *Handbook for Public Playground Safety,* the ratio of the vertical distance to the horizontal distance covered by a slide should not be more than about 4 to 7. If the horizontal distance allotted in a slide design is 14 feet, approximately how long should the slide be?

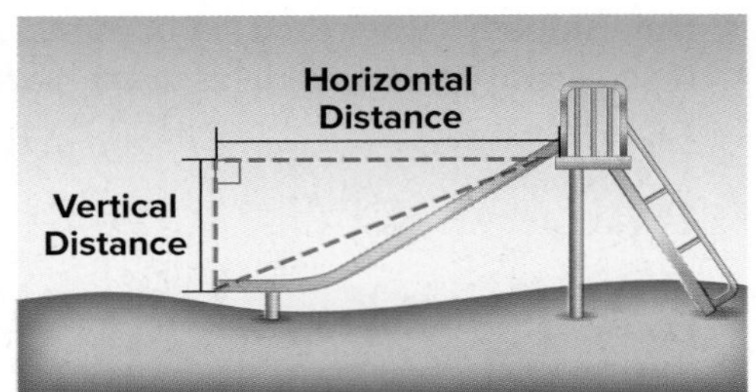

Find x.

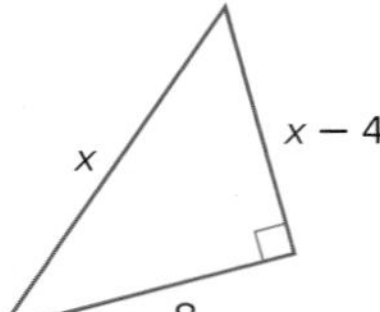

46.

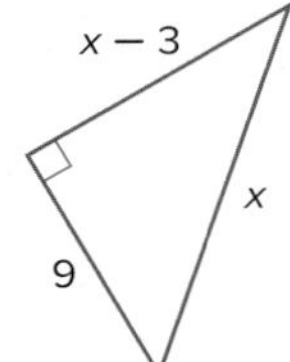

47.

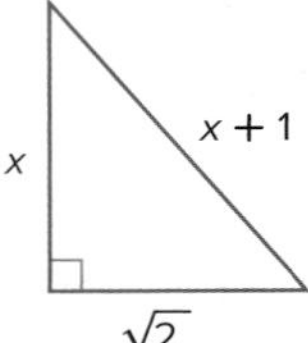

48. **MULTIPLE REPRESENTATIONS** In this problem, you will investigate special right triangles.

a. **Geometric** Draw three different isosceles right triangles that have whole-number side lengths. Label the triangles *ABC, MNP,* and *XYZ* with the right angle located at vertices *A, M,* and *X,* respectively. Label the leg lengths of each side, and find the exact length of the hypotenuse.

b. **Tabular** Copy and complete the table below.

Triangle	Length				Ratio	
ABC	*BC*		*AB*		$\frac{BC}{AB}$	
MNP	*NP*		*MN*		$\frac{NP}{MN}$	
XYZ	*YZ*		*XY*		$\frac{YZ}{XY}$	

c. **Verbal** Make a conjecture about the ratio of the hypotenuse to a leg of an isosceles right triangle.

H.O.T. Problems Use Higher-Order Thinking Skills

49. **CHALLENGE** Find the value of x in the figure at the right.

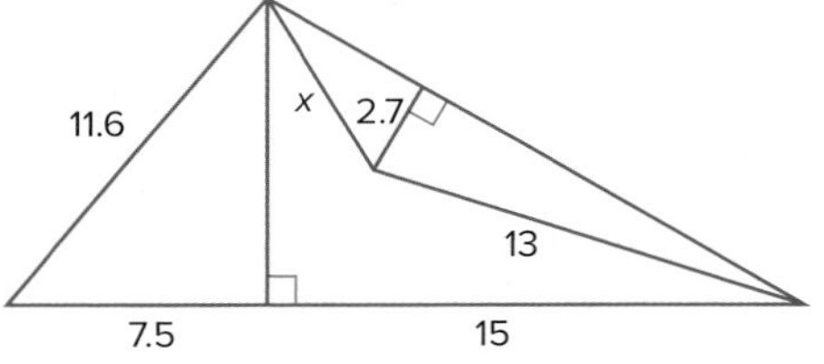

50. **ARGUMENTS** *True* or *false*? Any two right triangles with the same hypotenuse have the same area. Explain your reasoning.

51. **OPEN-ENDED** Draw a right triangle with side lengths that form a Pythagorean triple. If you double the length of each side, is the resulting triangle *acute, right,* or *obtuse*? if you halve the length of each side? Explain.

52. **WRITING IN MATH** Research *incommensurable magnitudes* and describe how this phrase relates to the use of irrational numbers in geometry. Include one example of an irrational number used in geometry.

53. Jessica has three wooden rods with the lengths shown below. She wants to place them together, if possible, to create a triangular picture frame.

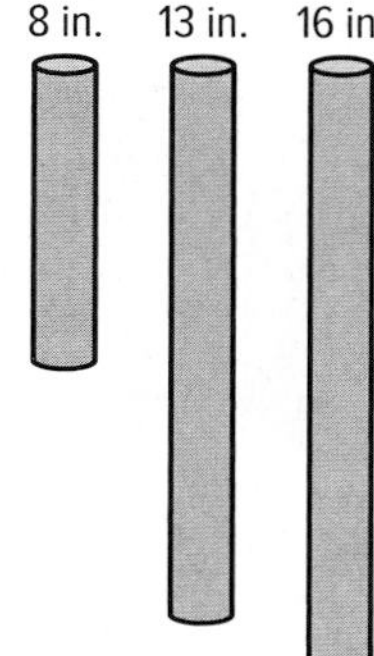

Which of the following is the best description of the frame Jessica can create? MP 1, 4

- ○ **A** The frame will be an acute triangle.
- ○ **B** The frame will be a right triangle.
- ○ **C** The frame will be an obtuse triangle.
- ○ **D** The rods cannot be placed together to form a triangle.

54. Dontrell's school has a rectangular lawn with the dimensions shown. Students often walk along the diagonal of the lawn as a shortcut.

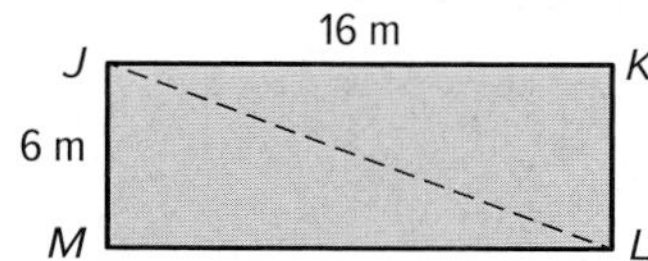

To the nearest tenth of a meter, how much shorter is it to walk directly from J to L rather than from J to K to L? MP 1, 4

- ○ **A** 4.9 m
- ○ **B** 9.4 m
- ○ **C** 10.0 m
- ○ **D** 17.1 m

55. A square park has a diagonal walkway from one corner to another. If the walkway is 120 meters long, what is the approximate length of each side of the park in meters? MP 1, 4 ☐

56. Lin wants to build a triangular vegetable bed with sides that are 7 feet, 8 feet, and 10 feet long. Which of the following best describes the vegetable bed? MP 1, 4

- ○ **A** It will be an acute triangle.
- ○ **B** It will be a right triangle.
- ○ **C** It will be an obtuse triangle.
- ○ **D** Lin cannot form a triangle with these side lengths.

57. Which of the following is the best estimate of the perimeter of $\triangle RST$? MP 1, 6

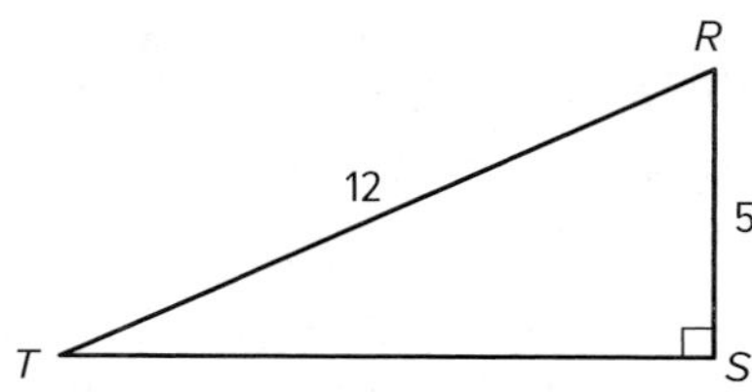

- ○ **A** 30.0
- ○ **B** 27.9
- ○ **C** 17.0
- ○ **D** 10.9

58. **MULTI-STEP** A gardener wishes to make a triangular garden. He has fence segments of length 8 feet, 14 feet, 15 feet, 17 feet, and 20 feet. MP 1, 4

a. What combination of fence lengths will make a right triangle? ☐

b. What combination of fence lengths will make an acute triangle? ☐

c. What combinations of fence lengths will make an obtuse triangle? ☐

59. A rhombus has a perimeter of 20 units. The length of one diagonal for the rhombus is 8. MP 1

a. Find the side length of the rhombus. ☐

b. Find the length of the other diagonal. ☐

EXTEND 8-2

Geometry Lab

Coordinates in Space

You have used ordered pairs of two coordinates to describe the location of a point on the coordinate plane. Because space has three dimensions, a point requires three numbers, or coordinates, to describe its location in space.

Mathematical Practices

4 Model with mathematics

A point in space is represented by an **ordered triple** of real numbers (x, y, z). In the figure at the right, the ordered triple (2, 3, 6) locates point P. Notice that a rectangular prism is used to show perspective.

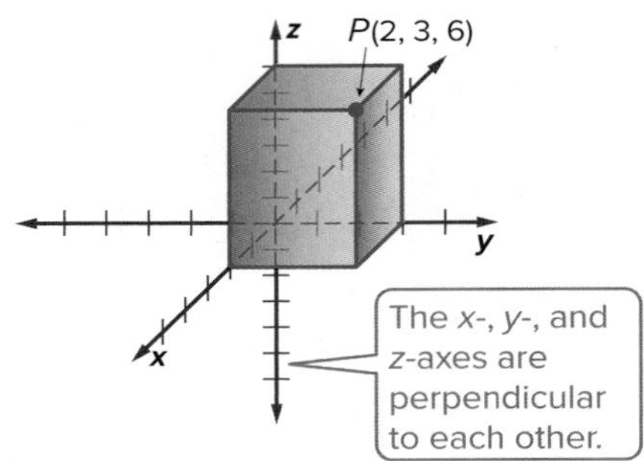

Activity 1 Graph a Rectangular Solid

Work cooperatively. Graph a rectangular solid that has two vertices, $L(4, -5, 2)$ and the origin. Label the coordinates of each vertex.

Step 1 Plot the x-coordinate first. Draw a segment from the origin 4 units in the positive direction.

Step 2 To plot the y-coordinate, draw a segment five units in the negative direction.

Step 3 Next, to plot the z-coordinate, draw a segment two units long in the positive direction.

Step 4 Label the coordinate L.

Step 5 Draw the rectangular prism and label each vertex: $L(4, -5, 2)$, $K(0, -5, 2)$, $J(0, 0, 2)$, $M(4, 0, 2)$ $Q(4, -5, 0)$, $P(0, -5, 0)$, $N(0, 0, 0)$, and $R(4, 0, 0)$.

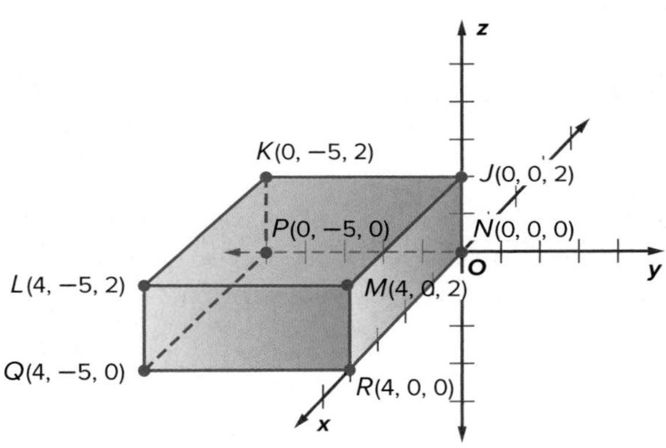

Finding the distance between points and the midpoint of a segment in space is similar to finding distance and a midpoint in the coordinate plane.

Key Concept Distance and Midpoint Formulas in Space

If A has coordinates $A(x_1, y_1, z_1)$ and B has coordinates $B(x_2, y_2, z_2)$, then

$$AB = \sqrt{(x_2 - x_1)^2 + (y_2 - y_1)^2 + (z_2 - z_1)^2}.$$

The midpoint M of $\overline{AB}$ has coordinates

$$M\left(\frac{x_1 + x_2}{2}, \frac{y_1 + y_2}{2}, \frac{z_1 + z_2}{2}\right).$$

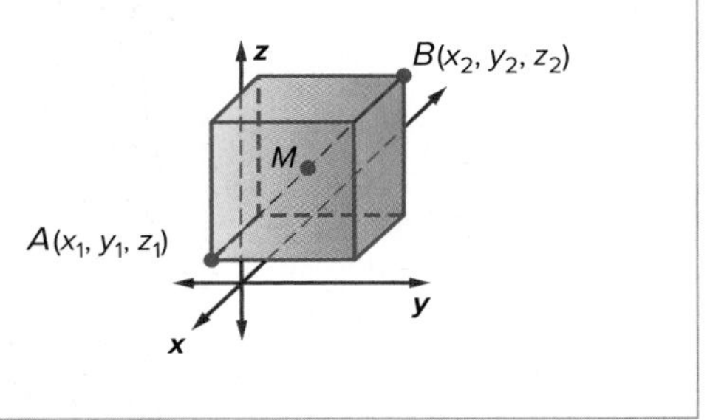

Activity 2 Distance and Midpoint Formulas in Space

Work cooperatively. Consider $J(2, 4, 9)$ and $K(-4, -5, 11)$.

a. Find JK.

$$JK = \sqrt{(x_2 - x_1)^2 + (y_2 - y_1)^2 + (z_2 - z_1)^2} \quad \text{Distance Formula in Space}$$
$$= \sqrt{(-4 - 2)^2 + (-5 - 4)^2 + (11 - 9)^2} \quad \text{Substitution}$$
$$= \sqrt{121} \quad \text{Simplify.}$$
$$= 11 \quad \text{Use a calculator.}$$

b. Determine the coordinates of the midpoint M of $\overline{JK}$.

$$M = \left(\frac{x_1 + x_2}{2}, \frac{y_1 + y_2}{2}, \frac{z_1 + z_2}{2}\right) \quad \text{Midpoint Formula in Space}$$
$$= \left(\frac{2 + (-4)}{2}, \frac{4 + (-5)}{2}, \frac{9 + 11}{2}\right) \quad \text{Substitution}$$
$$= \left(-1, -\frac{1}{2}, 10\right) \quad \text{Simplify.}$$

Exercises

Work cooperatively. Graph a rectangular solid that contains the given point and the origin as vertices. Label the coordinates of each vertex.

1. $A(2, 1, 5)$
2. $P(-1, 4, 2)$
3. $C(-2, 2, 2)$
4. $R(3, -4, 1)$
5. $P(4, 6, -3)$
6. $G(4, 1, -3)$
7. $K(-2, -4, -4)$
8. $W(-1, -3, -6)$
9. $W(3, 3, 4)$

Determine the distance between each pair of points. Then determine the coordinates of the midpoint M of the segment joining the pair of points.

10. $D(0, 0, 0)$ and $E(1, 5, 7)$
11. $G(-3, -4, 6)$ and $H(5, -3, -5)$
12. $K(2, 2, 0)$ and $L(-2, -2, 0)$
13. $P(-2, -5, 8)$ and $Q(3, -2, -1)$
14. $A(4, 7, 9)$ and $B(-3, 8, -8)$
15. $W(-12, 8, 10)$ and $Z(-4, 1, -2)$
16. $F\left(\frac{3}{5}, 0, \frac{4}{5}\right)$ and $G(0, 3, 0)$
17. $G(1, -1, 6)$ and $H\left(\frac{1}{5}, -\frac{2}{5}, 2\right)$
18. $B(\sqrt{3}, 2, 2\sqrt{2})$ and $C(-2\sqrt{3}, 4, 4\sqrt{2})$
19. $S(6\sqrt{3}, 4, 4\sqrt{2})$ and $T(4\sqrt{3}, 5, \sqrt{2})$

20. PROOF Write a coordinate proof of the Distance Formula in Space.

Given: A has coordinates $A(x_1, y_1, z_1)$, and B has coordinates $B(x_2, y_2, z_2)$.

Prove: $AB = \sqrt{(x_2 - x_1)^2 + (y_2 - y_1)^2 + (z_2 - z_1)^2}$

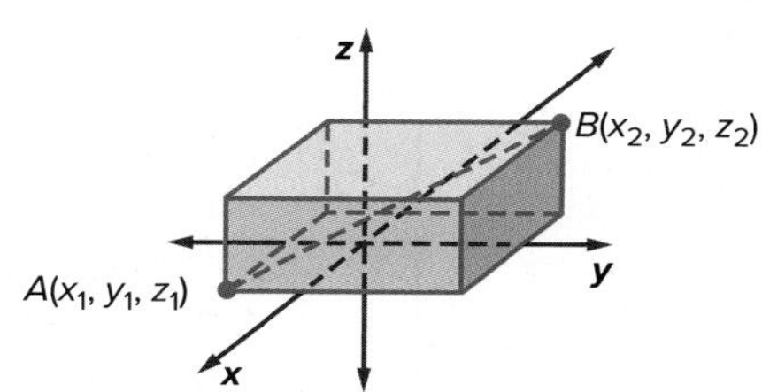

21. WRITING IN MATH Compare and contrast the Distance and Midpoint Formulas on the coordinate plane and in three-dimensional coordinate space.

LESSON 3

Special Right Triangles

Then

- You used properties of isosceles and equilateral triangles.

Now

1. Use the properties of 45°-45°-90° triangles.
2. Use the properties of 30°-60°-90° triangles.

Why?

- As part of a packet for students attending a regional student council meeting, Lyndsay orders triangular highlighters. She wants to buy rectangular boxes for the highlighters and other items, but she is concerned that the highlighters will not fit in the box she has chosen. If she knows the length of a side of the highlighter, Lyndsay can use the properties of special right triangles to determine if it will fit in the box.

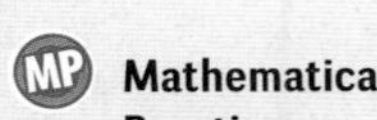

Mathematical Practices

1 Make sense of problems and persevere in solving them.

7 Look for and make use of structure.

1 Properties of 45°-45°-90° Triangles

The diagonal of a square forms two congruent isosceles right triangles. Because the base angles of an isosceles triangle are congruent, the measure of each acute angle is $90 \div 2$ or 45. Such a triangle is also known as a 45°-45°-90° triangle.

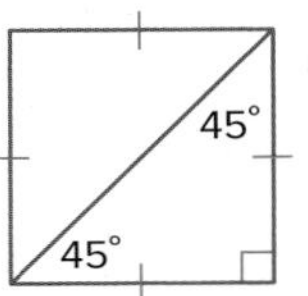

You can use the Pythagorean Theorem to find a relationship among the side lengths of a 45°-45°-90° right triangle.

$\ell^2 + \ell^2 = h^2$	Pythagorean Theorem
$2\ell^2 = h^2$	Simplify.
$\sqrt{2\ell^2} = \sqrt{h^2}$	Take the positive square root of each side.
$\ell\sqrt{2} = h$	Simplify.

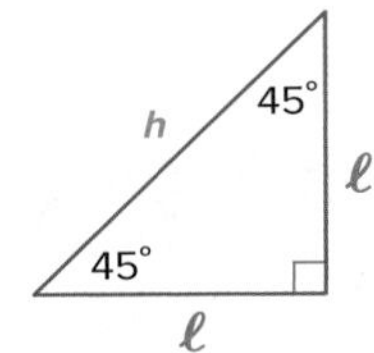

This algebraic proof verifies the following theorem.

Theorem 8.8 45°-45°-90° Triangle Theorem

In a 45°-45°-90° triangle, the legs ℓ are congruent and the length of the hypotenuse h is $\sqrt{2}$ times the length of a leg.

Symbols In a 45°-45°-90° triangle, $\ell = \ell$ and $h = \ell\sqrt{2}$.

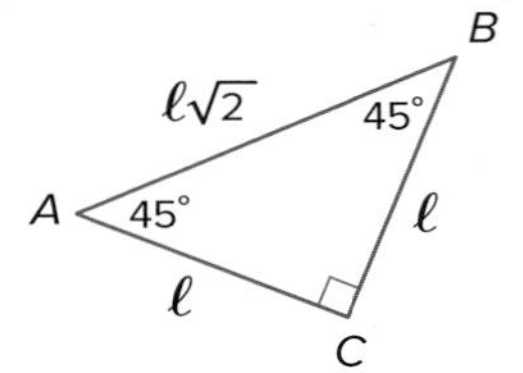

Example 1 Find the Hypotenuse Length in a 45°-45°-90° Triangle

Find x.

a.

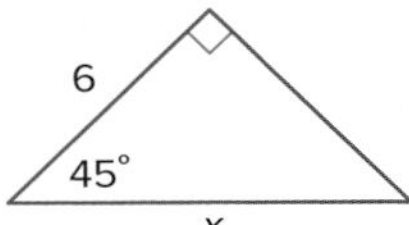

The acute angles of a right triangle are complementary, so the measure of the third angle is $90 - 45$ or 45. Because this is a 45°-45°-90° triangle, use Theorem 8.8.

$h = \ell\sqrt{2}$	Theorem 8.8
$x = 6\sqrt{2}$	Substitution

b.

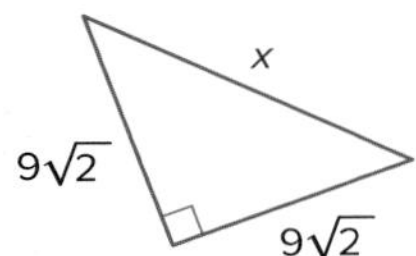

The legs of this right triangle have the same measure, so it is isosceles. Since this is a 45°-45°-90° triangle, use Theorem 8.8.

$h = \ell\sqrt{2}$	Theorem 8.8
$x = 9\sqrt{2} \cdot \sqrt{2}$	Substitution
$x = 9 \cdot 2$ or 18	$\sqrt{2} \cdot \sqrt{2} = 2$

Guided Practice

Find x.

1A.

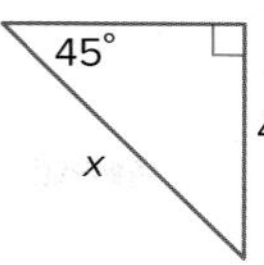

1B.

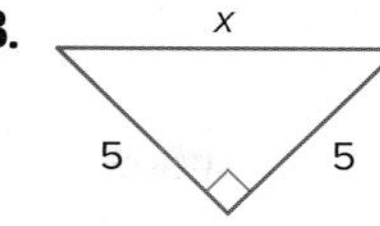

1C.

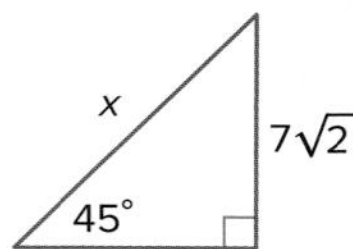

You can also work backward using Theorem 8.8 to find the lengths of the legs of a 45°-45°-90° triangle given the length of its hypotenuse.

Example 2 Find the Leg Lengths in a 45°-45°-90° Triangle

Find x.

The legs of this right triangle have the same measure, x, so it is a 45°-45°-90° triangle. Use Theorem 8.8 to find x.

$h = \ell\sqrt{2}$ — 45°-45°-90° Triangle Theorem

$12 = x\sqrt{2}$ — Substitution

$\frac{12}{\sqrt{2}} = x$ — Divide each side by $\sqrt{2}$.

$\frac{12}{\sqrt{2}} \cdot \frac{\sqrt{2}}{\sqrt{2}} = x$ — Rationalize the denominator.

$\frac{12\sqrt{2}}{2} = x$ — Multiply.

$6\sqrt{2} = x$ — Simplify.

Review Vocabulary

rationalizing the denominator a method used to eliminate radicals from the denominator of a fraction

Guided Practice

2A.

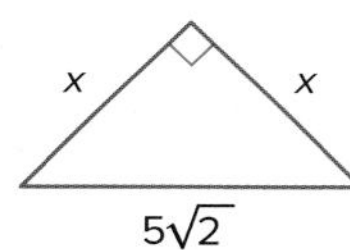

2B.

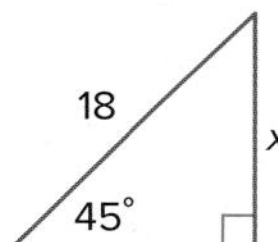

2 Properties of 30°-60°-90° Triangles

A 30°-60°-90° triangle is another *special* right triangle or right triangle with side lengths that share a special relationship. You can use an equilateral triangle to find this relationship.

When an altitude is drawn from any vertex of an equilateral triangle, two congruent 30°-60°-90° triangles are formed. In the figure shown, $\triangle ABD \cong \triangle CBD$, so $\overline{AD} \cong \overline{CD}$. If $AD = x$, then $CD = x$ and $AC = 2x$. Because $\triangle ABC$ is equilateral, $AB = 2x$ and $BC = 2x$.

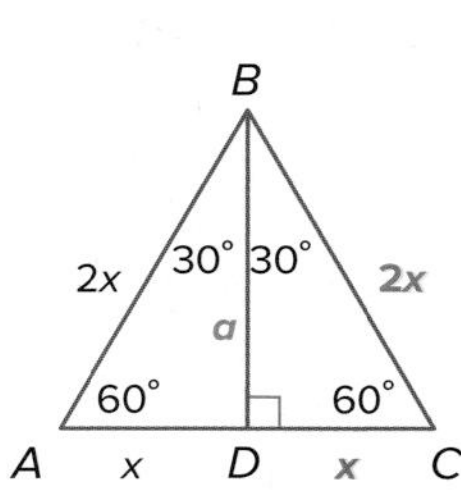

Study Tip

MP Sense-Making Notice that an altitude of an isosceles triangle is also a median of the triangle. In the figure at the right, $\overline{BD}$ bisects $\overline{AC}$.

Use the Pythagorean Theorem to find a, the length of the altitude $\overline{BD}$, which is also the longer leg of $\triangle BDC$.

$a^2 + x^2 = (2x)^2$ — Pythagorean Theorem

$a^2 + x^2 = 4x^2$ — Simplify.

$a^2 = 3x^2$ — Subtract x^2 from each side.

$a = \sqrt{3x^2}$ — Take the positive square root of each side.

$a = x\sqrt{3}$ — Simplify.

> **Study Tip**
> **Use Ratios** The lengths of the sides of a 30°-60°-90° triangle are in a ratio of 1 to $\sqrt{3}$ to 2 or 1 : $\sqrt{3}$: 2.

This algebraic proof verifies the following theorem.

Theorem 8.9 30°-60°-90° Triangle Theorem

In a 30°-60°-90° triangle, the length of the hypotenuse h is 2 times the length of the shorter leg s, and the length of the longer leg ℓ is $\sqrt{3}$ times the length of the shorter leg.

Symbols In a 30°-60°-90° triangle, $h = 2s$ and $\ell = s\sqrt{3}$.

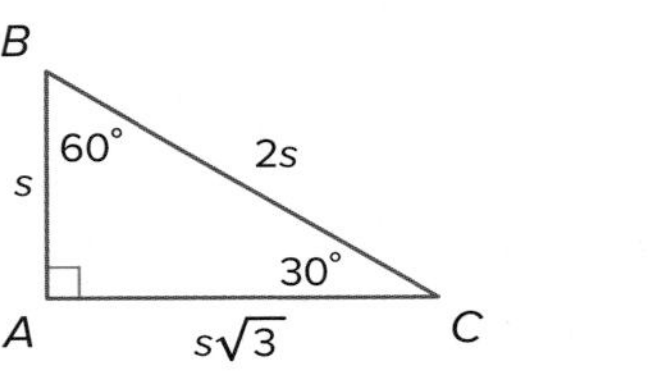

Remember, the shortest side of a triangle is opposite the smallest angle. So the shorter leg in a 30°-60°-90° triangle is opposite the 30° angle, and the longer leg is opposite the 60° angle.

> **Go Online!** 
> Discover the special relationships among the lengths of the sides of 30°-60° right triangles and why they hold with a **Geometer's Sketchpad®** sketch in ConnectED. Discuss your findings with a partner. Ask for clarification as you need it.
>
> 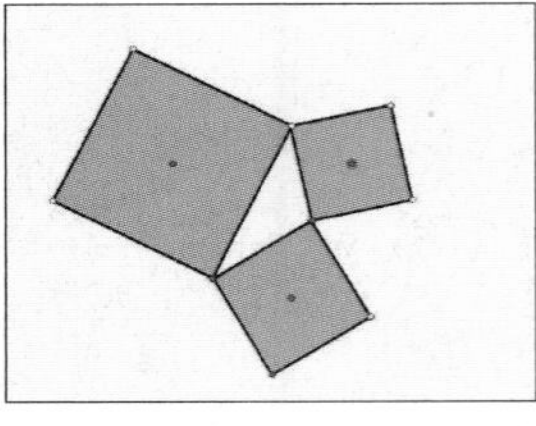

Example 3 Find Lengths in a 30°-60°-90° Triangle

Find x and y.

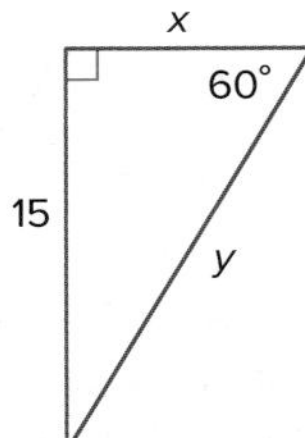

The acute angles of a right triangle are complementary, so the measure of the third angle in this triangle is 90 − 60 or 30. This is a 30°-60°-90° triangle.

Use Theorem 8.9 to find x, the length of the shorter side.

$\ell = s\sqrt{3}$	Theorem 8.9
$15 = x\sqrt{3}$	Substitution
$\frac{15}{\sqrt{3}} = x$	Divide each side by $\sqrt{3}$.
$\frac{15}{\sqrt{3}} \cdot \frac{\sqrt{3}}{\sqrt{3}} = x$	Rationalize the denominator.
$\frac{15\sqrt{3}}{\sqrt{3} \cdot \sqrt{3}} = x$	Multiply.
$\frac{15\sqrt{3}}{3} = x$	$\sqrt{3} \cdot \sqrt{3} = 3$
$5\sqrt{3} = x$	Simplify.

Now use Theorem 8.9 to find y, the length of the hypotenuse.

$h = 2s$	Theorem 8.9
$y = 2(5\sqrt{3})$ or $10\sqrt{3}$	Substitution

Guided Practice

Find x and y.

3A.

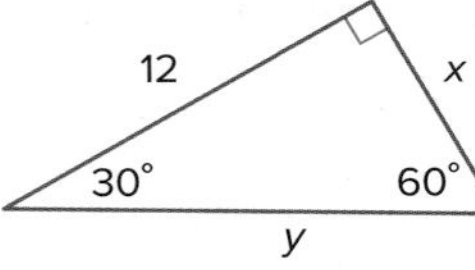

3B.

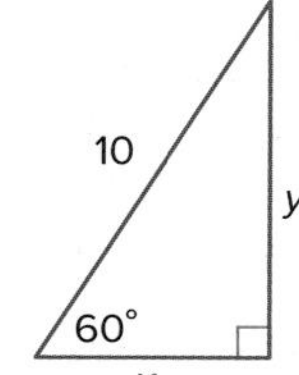

3C.

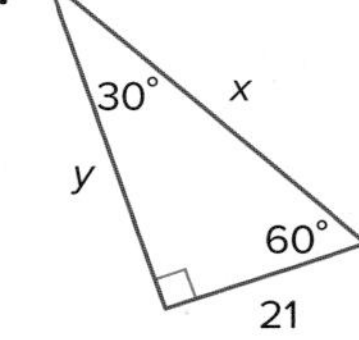

Brand X Pictures/Punchstock

Real-World Example 4 Use Properties of Special Right Triangles

INVENTIONS A company makes crayons that "do not roll off tables" by shaping them as triangular prisms with equilateral bases. Sixteen of these crayons fit into a box shaped like a triangular prism that is $1\frac{1}{2}$ inches wide. The crayons stand on end in the box and the base of the box is equilateral. What are the dimensions of each crayon?

Understand You know that 16 crayons with equilateral triangular bases fit into a prism. You need to find the base length and height of each crayon.

Plan Guess and check to determine the arrangement of 16 crayons that would stack to fill the box. Find the width of one crayon and use the 30°-60°-90° Triangle Theorem to find its altitude.

Solve Make a guess that 4 equilateral crayons will fit across the base of the box. A sketch shows that the total number of crayons it takes to fill the box using 4 crayons across the base is 16. ✓

The width of the box is $1\frac{1}{2}$ inches, so the width of one crayon is $1\frac{1}{2} \div 4$ or $\frac{3}{8}$ inch.

Draw an equilateral triangle representing one crayon. Its altitude forms the longer leg of two 30°-60°-90° triangles. Use Theorem 8.9 to find the approximate length of the altitude a.

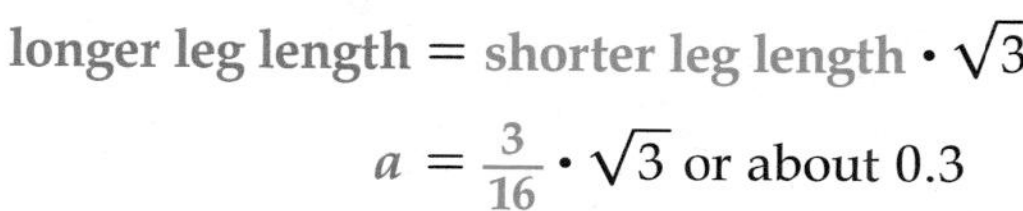

longer leg length = shorter leg length $\cdot \sqrt{3}$

$$a = \frac{3}{16} \cdot \sqrt{3} \text{ or about } 0.3$$

Each crayon is $\frac{3}{8}$ or about 0.4 inch by about 0.3 inch.

Check Find the height of the box using the 30°-60°-90° Triangle Theorem. Then divide by four, since the box is four crayons high. The result is a crayon height of about 0.3 inch. ✓

Problem-Solving Tip

Guess and Check In Example 4, suppose your first guess had been that the box was 5 crayons wide.

The sketch of this possibility reveals that this leads to a stack of 25, not 16 crayons.

Guided Practice

4. FURNITURE The top of the aquarium coffee table shown is an isosceles right triangle. The table's longest side, $\overline{AC}$, measures 107 centimeters. What is the distance from vertex B to side $\overline{AC}$? What are the lengths of the other two sides?

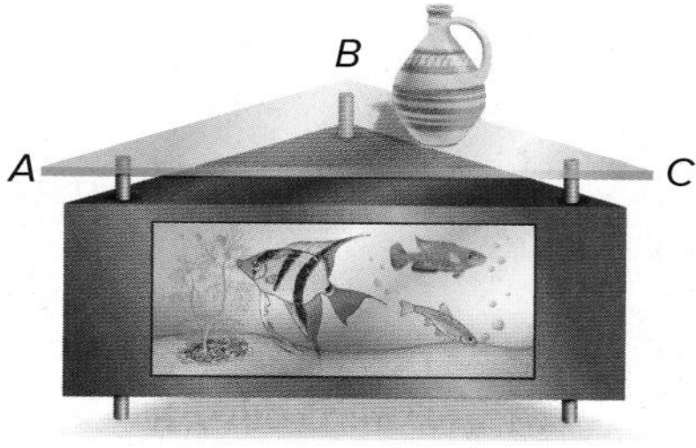

Check Your Understanding

 = Step-by-Step Solutions begin on page R13.

Examples 1-2 Find x.

1. 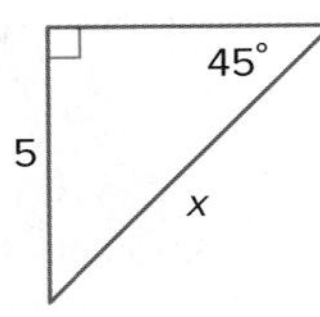

2. 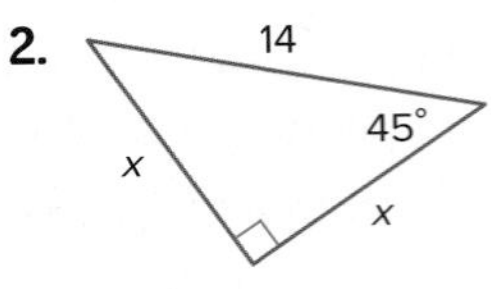

3. 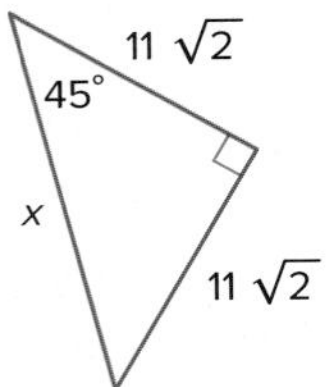

Example 3 Find x and y.

4. 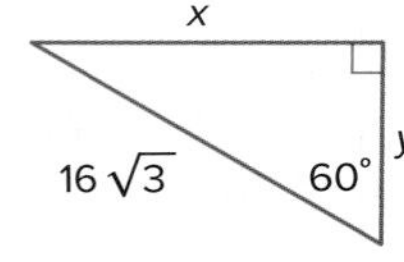

5. 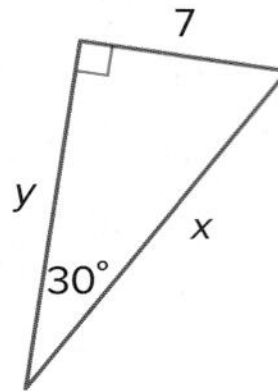

6. 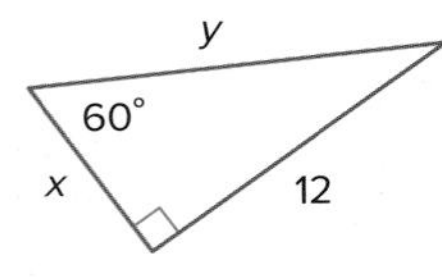

Example 4

7. **ART** Paulo is mailing an engraved plaque that is $3\frac{1}{4}$ inches high to the winner of a chess tournament. He has a mailer that is a triangular prism with 4-inch equilateral triangle bases as shown in the diagram. Will the plaque fit through the opening of the mailer? Explain.

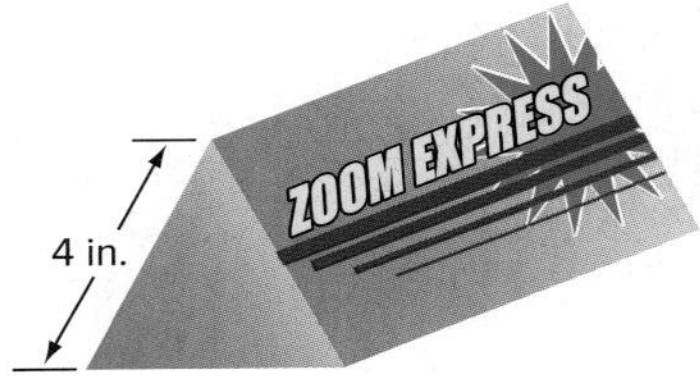

Practice and Problem Solving

Extra Practice is on page R8.

Examples 1-2  **SENSE-MAKING** Find x.

8. 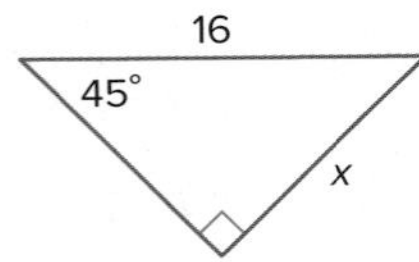

9. 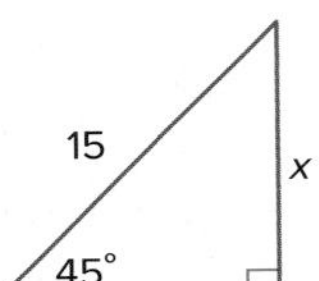

10. 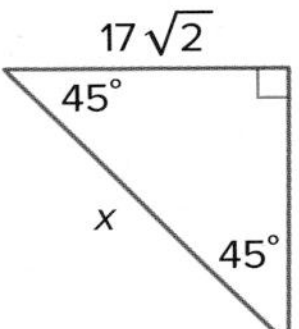

11. 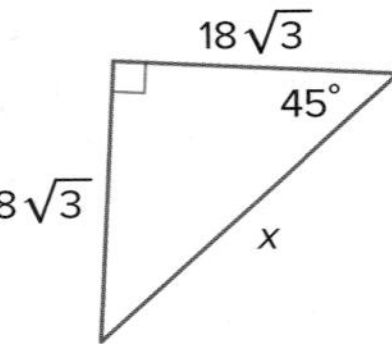

12. 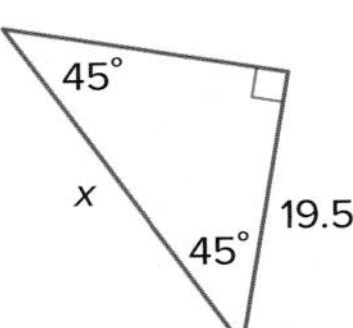

13. 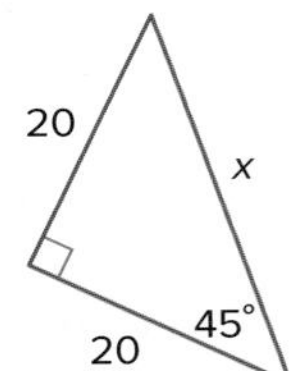

14. If a 45°-45°-90° triangle has a hypotenuse length of 9, find the leg length.

15. Determine the length of the leg of a 45°-45°-90° triangle with a hypotenuse length of 11.

16. What is the length of the hypotenuse of a 45°-45°-90° triangle if the leg length is 6 centimeters?

17. Find the length of the hypotenuse of a 45°-45°-90° triangle with a leg length of 8 centimeters.

Example 3 **Find x and y.**

18.

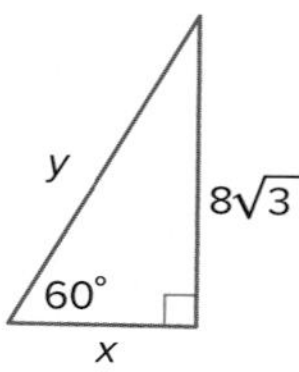

19.

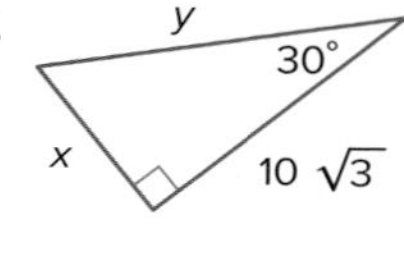

20.

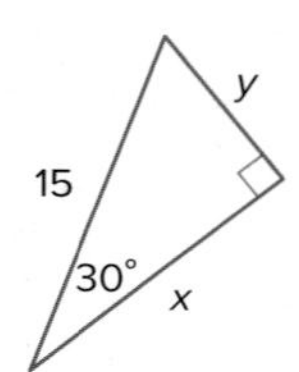

21.

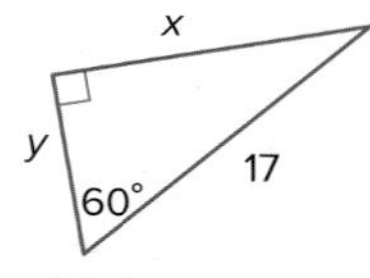

22.

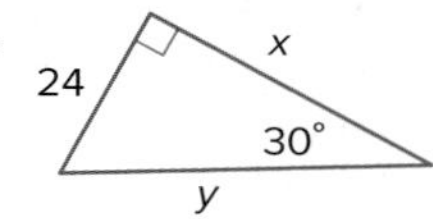

23.

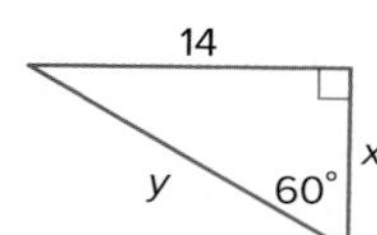

24. An equilateral triangle has an altitude length of 18 feet. Determine the length of a side of the triangle.

25. Find the length of the side of an equilateral triangle that has an altitude length of 24 feet.

Example 4 **26.** MP **MODELING** Refer to the beginning of the lesson. Each highlighter is an equilateral triangle with 9-centimeter sides. Will the highlighter fit in a 10-centimeter by 7-centimeter rectangular box? Explain.

27. **EVENT PLANNING** Grace is having a party, and she wants to decorate the gable of the house as shown. The gable is an isosceles right triangle and she knows that the height of the gable is 8 feet. What length of lights will she need to cover the gable below the roof line?

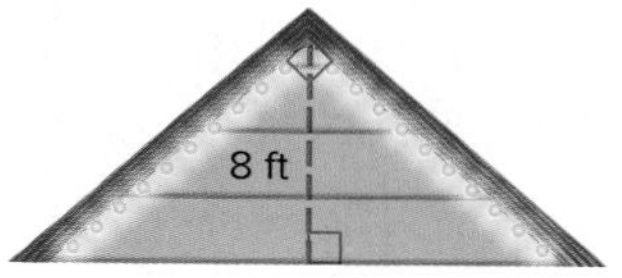

Find x and y.

28.

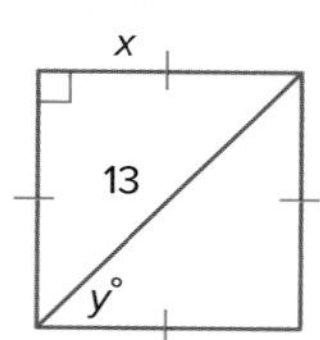

29.

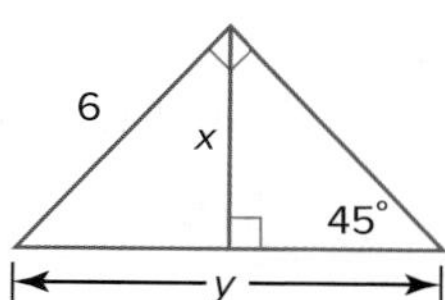

30.

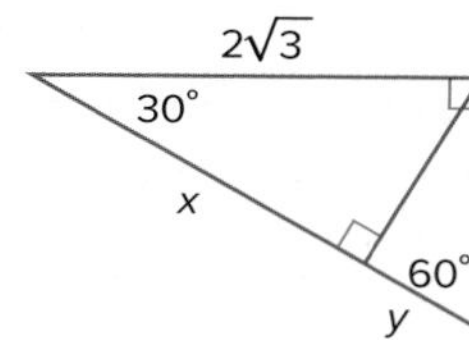

31.

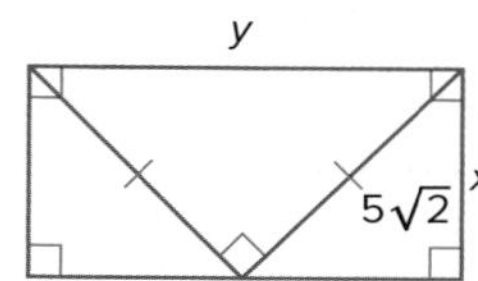

32.

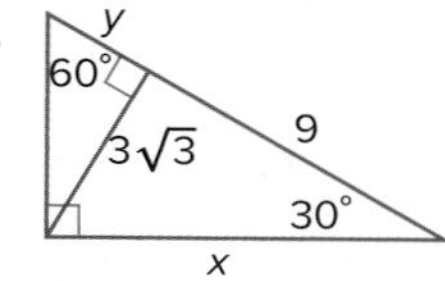

33.

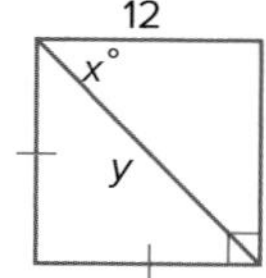

34. **QUILTS** The quilt block shown is made up of a square and four isosceles right triangles. What is the value of x? What is the side length of the entire quilt block?

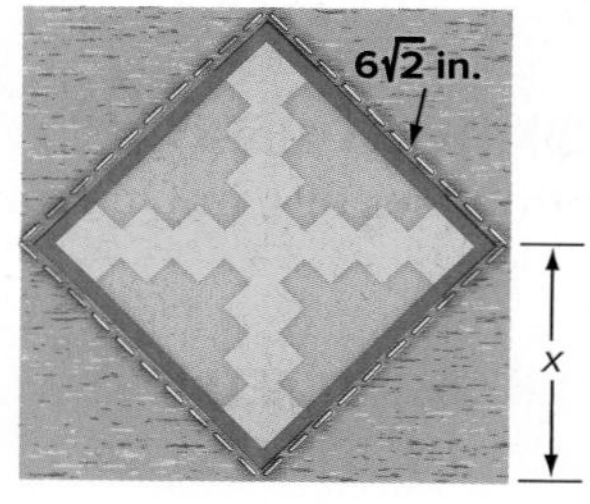

35. **ZIP LINE** Suppose a zip line is anchored in one corner of a course shaped like a rectangular prism. The other end is anchored in the opposite corner as shown. If the zip line makes a 60° angle with post $\overline{AF}$, find the zip line's length, AD.

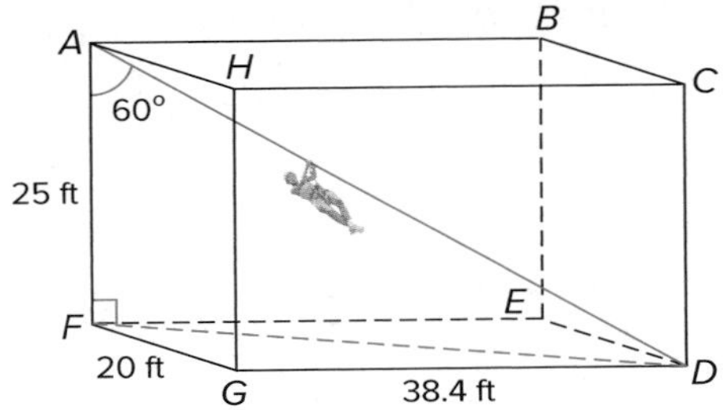

36. **GAMES** Kei is building a bean bag toss for the school carnival. He is using a 2-foot back support that is perpendicular to the ground 2 feet from the front of the board. He also wants to use a support that is perpendicular to the board as shown in the diagram. How long should he make the support?

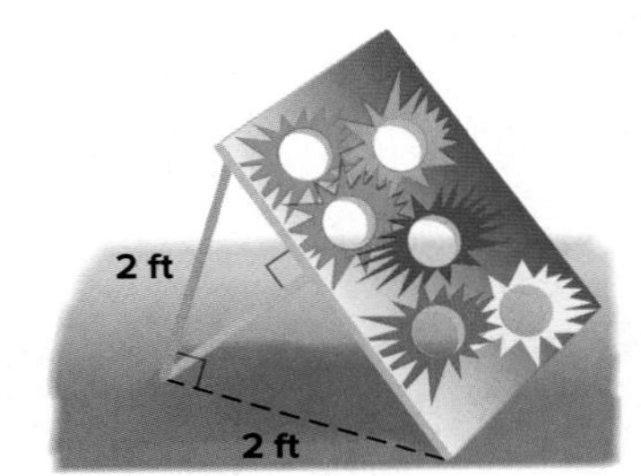

37. Find x, y, and z.

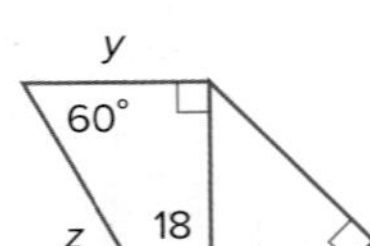

38. Each triangle in the figure is a 45°-45°-90° triangle. Find x.

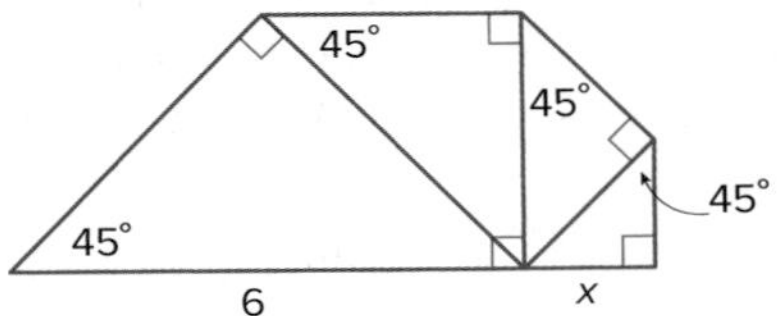

39. **MP MODELING** The dump truck shown has a 15-foot bed length. What is the height of the bed h when angle x is 30°? 45°? 60°?

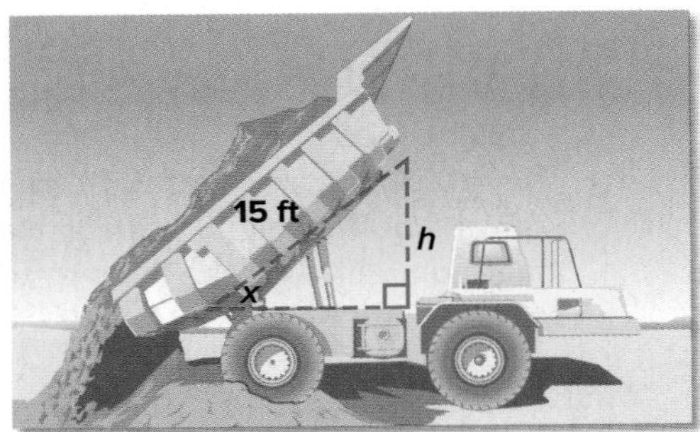

40. Find x, y, and z, and the perimeter of trapezoid $PQRS$.

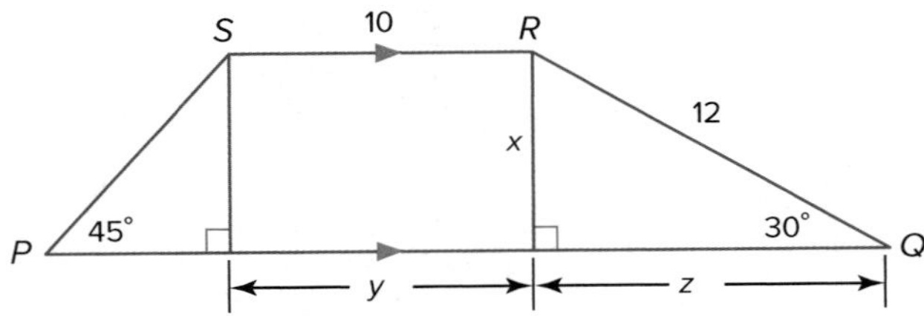

41. **COORDINATE GEOMETRY** $\triangle XYZ$ is a 45°-45°-90° triangle with right angle Z. Find the coordinates of X in Quadrant I for $Y(-1, 2)$ and $Z(6, 2)$.

42. **COORDINATE GEOMETRY** $\triangle EFG$ is a 30°-60°-90° triangle with $m\angle F = 90$. Find the coordinates of E in Quadrant III for $F(-3, -4)$ and $G(-3, 2)$. $\overline{FG}$ is the longer leg.

43. **COORDINATE GEOMETRY** $\triangle JKL$ is a 45°-45°-90° triangle with right angle K. Find the coordinates of L in Quadrant IV for $J(-3, 5)$ and $K(-3, -2)$.

44. EVENT PLANNING Eva has reserved a gazebo at a local park for a party. She wants to be sure that there will be enough space for her 12 guests to be in the gazebo at the same time. She wants to allow 8 square feet of area for each guest. If the floor of the gazebo is a regular hexagon and each side is 7 feet, will there be enough room for Eva and her friends? Explain. (*Hint:* Use the Polygon Interior Angle Sum Theorem and the properties of special right triangles.)

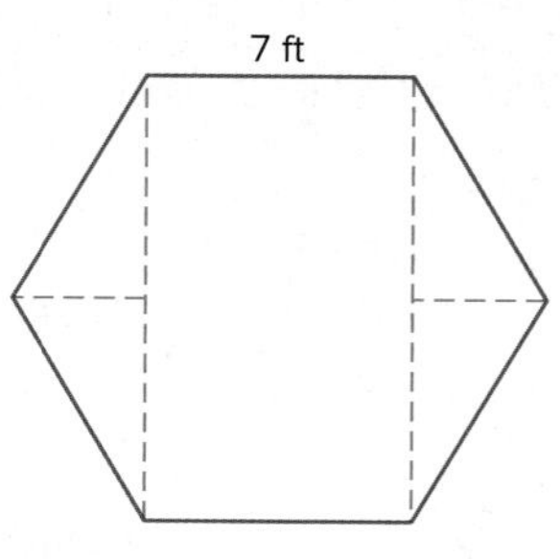

45 MULTIPLE REPRESENTATIONS In this problem, you will investigate ratios in right triangles. Use a protractor and straightedge or dynamic geometry software.

a. **Geometric** Draw three similar right triangles with a 50° angle. Label one triangle *ABC* where angle *A* is the right angle and *B* is the 50° angle. Label a second triangle *MNP* where *M* is the right angle and *N* is the 50° angle. Label the third triangle *XYZ* where *X* is the right angle and *Y* is the 50° angle.

b. **Tabular** Copy and complete the table below.

Triangle	Length				Ratio	
ABC	*AC*		*BC*		$\frac{BC}{AC}$	
MNP	*MP*		*NP*		$\frac{NP}{MP}$	
XYZ	*XZ*		*YZ*		$\frac{YZ}{XZ}$	

c. **Verbal** Make a conjecture about the ratio of the leg opposite the 50° angle to the hypotenuse in any right triangle with an angle measuring 50°.

H.O.T. Problems Use Higher-Order Thinking Skills

46. CRITIQUE Carmen and Audrey want to find *x* in the triangle shown. Is either of them correct? Explain.

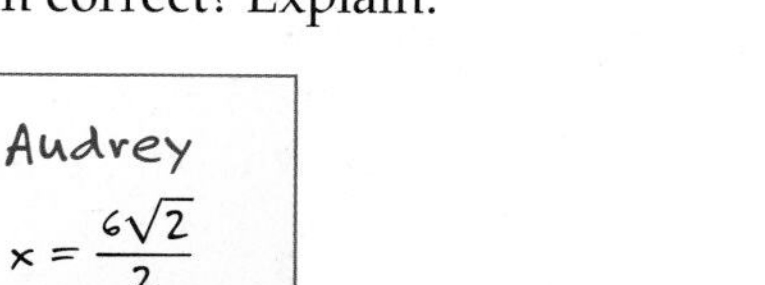

Carmen	Audrey
$x = \frac{6\sqrt{3}}{2}$	$x = \frac{6\sqrt{2}}{2}$
$x = 3\sqrt{3}$	$x = 3\sqrt{2}$

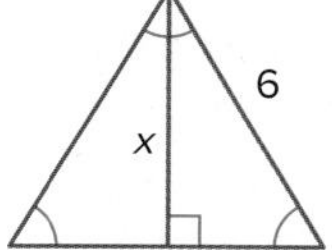

47. OPEN-ENDED Draw a rectangle that has a diagonal twice as long as its width. Then write an equation to find the length of the rectangle.

48. CHALLENGE Find the perimeter of quadrilateral *ABCD*.

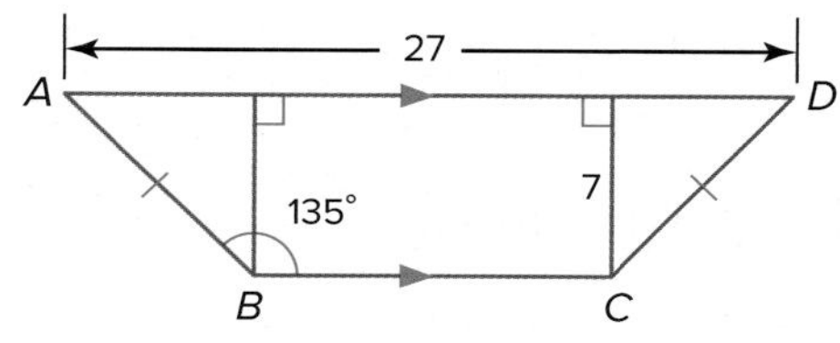

49. REASONING The ratio of the measure of the angles of a triangle is 1:2:3. The length of the shortest side is 8. What is the perimeter of the triangle?

50. WRITING IN MATH Why are some right triangles considered *special?*

51. A yield sign approximates an equilateral triangle with sides that are 36 inches long. Which of these is the best estimate of the height of the sign? MP 4

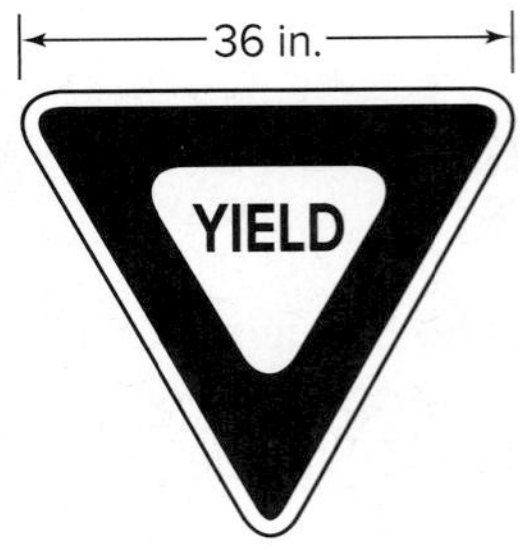

- ◯ **A** 18.0 in.
- ◯ **B** 25.5 in.
- ◯ **C** 31.2 in.
- ◯ **D** 62.4 in.

52. The area of an equilateral triangle is $8\sqrt{3}$ square units. Find the length of one side of the triangle. MP 8

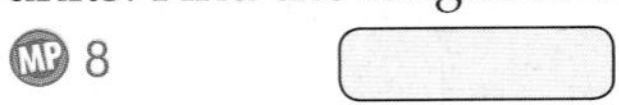

53. The diagonal of a square measures 10 units. Find the perimeter of the square. MP 8

54. In a stained-glass window, each colored pane of glass is separated by a metal strip. The window itself is also surrounded by metal strips. Hailey is making a square stained-glass window as shown.

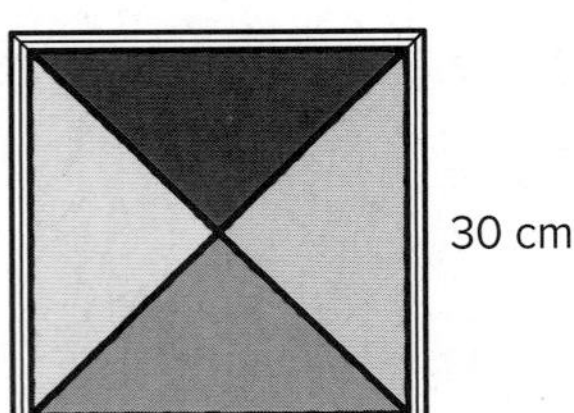

Which of these represents the total length of the metal strips Hailey will need? MP 8

- ◯ **A** $30 + 60\sqrt{2}$ cm
- ◯ **B** 120 cm
- ◯ **C** $120 + 30\sqrt{2}$ cm
- ◯ **D** $120 + 60\sqrt{2}$ cm
- ◯ **E** $120 + 60\sqrt{3}$ cm

55. What is the value of x? MP 8

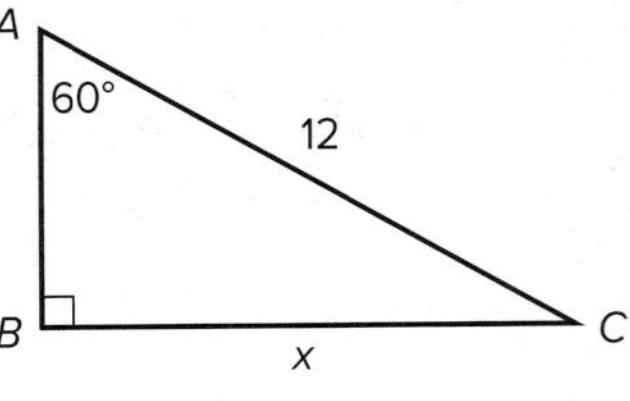

- ◯ **A** 6
- ◯ **B** $4\sqrt{3}$
- ◯ **C** $6\sqrt{3}$
- ◯ **D** $12\sqrt{3}$

56. What is the length of $\overline{KJ}$ in the figure below? MP 8

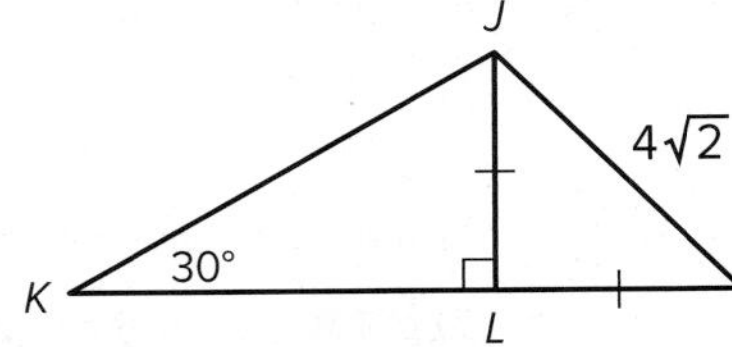

- ◯ **A** $8\sqrt{2}$
- ◯ **B** 8
- ◯ **C** $4\sqrt{3}$
- ◯ **D** 4

57. **MULTI-STEP** A piece of wire measures 24 units. MP 1, 8

a. The wire is first bent to form a square. What is the length of a diagonal of the square?

b. Then, the wire is bent to form an equilateral triangle. Find the height of the triangle.

58. Find the exact value of y. MP 1

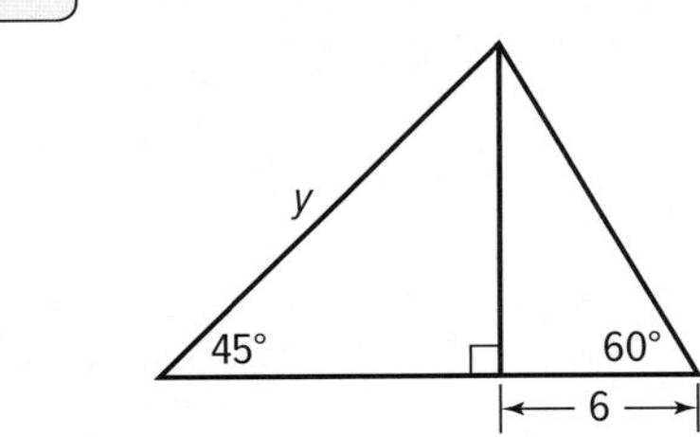

59. $m\angle BCA = 45°$ and $m\angle D = 30°$. If $BC = 6$, find AD. MP 1

A
B C D

EXPLORE 8-4

Graphing Technology Lab

Trigonometry

You have investigated patterns in the measures of special right triangles. *Trigonometry* is the study of the patterns in all right triangles. You can use the Cabri Jr. application on a graphing calculator to investigate these patterns.

Mathematical Practices

5 Use appropriate tools strategically.

Activity Investigate Trigonometric Ratios

Work cooperatively.

Step 1 Use the line tool on the **F2** menu to draw a horizontal line. Label the points on the line A and B.

Step 2 Press **F2** and choose the **Perpendicular** tool to create a perpendicular line through point B. Draw and label a point C on the perpendicular line.

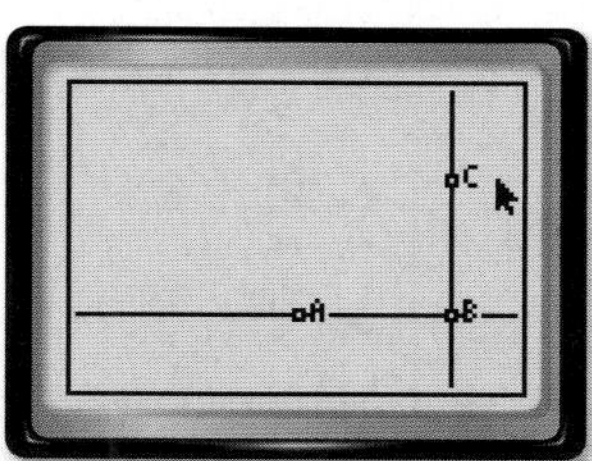

Steps 1 and 2

Step 3 Use the **Segment** tool on the **F2** menu to draw $\overline{AC}$.

Step 4 Find and label the measures of $\overline{BC}$ and $\overline{AC}$ using the **Distance** and **Length** tool under **Measure** on the **F5** menu. Use the **Angle** tool to find the measure of $\angle A$.

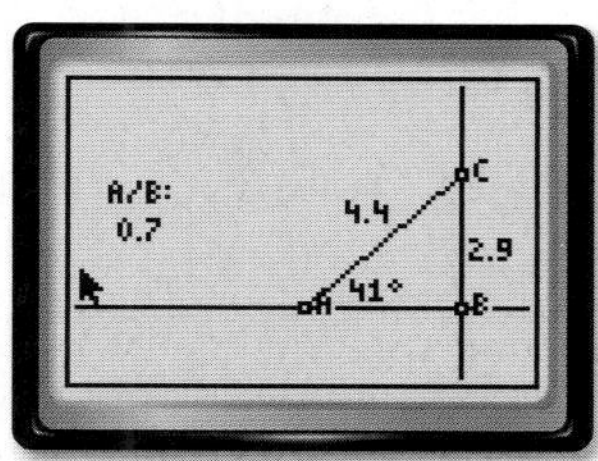

Steps 3 through 5

Step 5 Calculate and display the ratio $\frac{BC}{AC}$ using the **Calculate** tool on the **F5** menu. Label the ratio as A/B.

Step 6 Press CLEAR. Then use the arrow keys to move the cursor close to point B. When the arrow is clear, press and hold the ALPHA key. Drag B and observe the ratio.

Analyze the Results

Work cooperatively.

1. Discuss the effect on $\frac{BC}{AC}$ by dragging point B on $\overline{BC}$, $\overline{AC}$, and $\angle A$.
2. Use the calculate tool to find the ratios $\frac{AB}{AC}$ and $\frac{BC}{AB}$. Then drag B and observe the ratios.
3. **MAKE A CONJECTURE** The *sine, cosine*, and *tangent* functions are trigonometric functions based on angle measures. Make a note of $m\angle A$. Exit Cabri Jr. and use SIN, COS, and TAN on the calculator to find *sine, cosine* and *tangent* for $m\angle A$. Compare the results to the ratios you found in the activity. Make a conjecture about the definitions of sine, cosine, and tangent.

LESSON 4

Trigonometry

Then

- You used the Pythagorean Theorem to find missing lengths in right triangles.

Now

1. Find trigonometric ratios using right triangles.
2. Use trigonometric ratios to find angle measures in right triangles.

Why?

- The steepness of a hiking trail is often expressed as a percent of grade. The steepest part of Bright Angel Trail in the Grand Canyon National Park has about a 15.7% grade. This means that the trail rises or falls 15.7 feet over a horizontal distance of 100 feet. You can use trigonometric ratios to determine that this steepness is equivalent to an angle of about 9°.

New Vocabulary

trigonometry
trigonometric ratio
sine
cosine
tangent
inverse sine
inverse cosine
inverse tangent

Mathematical Practices

1 Make sense of problems and persevere in solving them.
2 Reason abstractly and quantitatively.
5 Use appropriate tools strategically.
8 Look for and express regularity in repeated reasoning.

1 Trigonometric Ratios

The word **trigonometry** comes from two Greek terms, *trigon*, meaning triangle, and *metron*, meaning measure. The study of trigonometry involves triangle measurement. A **trigonometric ratio** is a ratio of the lengths of two sides of a right triangle. One trigonometric ratio of $\triangle ABC$ is $\frac{AC}{AB}$.

By AA Similarity, a right triangle with a given acute angle measure is similar to every other right triangle with the same acute angle measure. So, trigonometric ratios are constant for a given angle measure.

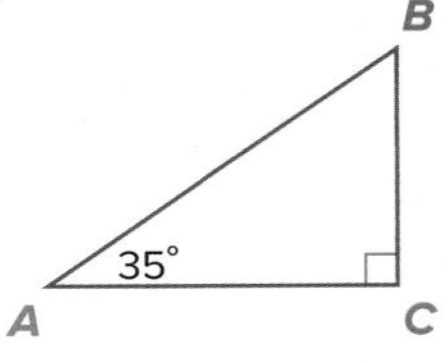

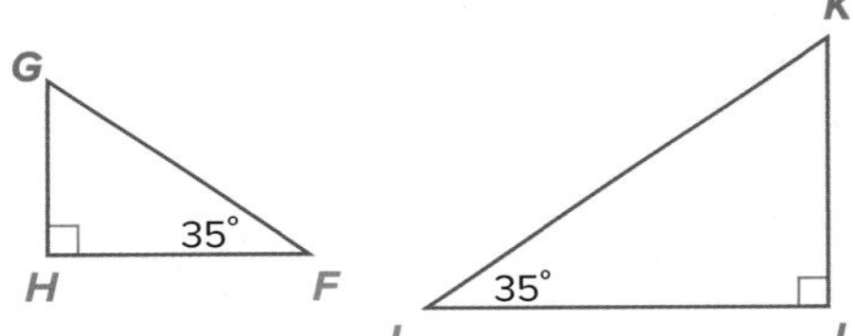

$\triangle ABC \sim \triangle FGH \sim \triangle JKL$, so $\frac{AC}{AB} = \frac{FH}{FG} = \frac{JL}{JK}$

The names of the three most common trigonometric ratios are given below.

Key Concept Trigonometric Ratios

Words	Symbols	
If $\triangle ABC$ is a right triangle with acute $\angle A$, then the **sine** of $\angle A$ (written sin A) is the ratio of the length of the leg opposite $\angle A$ (opp) to the length of the hypotenuse (hyp).	$\sin A = \frac{\text{opp}}{\text{hyp}}$ or $\frac{a}{c}$ $\sin B = \frac{\text{opp}}{\text{hyp}}$ or $\frac{b}{c}$	A, b, c, C, a, B
If $\triangle ABC$ is a right triangle with acute $\angle A$, then the **cosine** of $\angle A$ (written cos A) is the ratio of the length of the leg adjacent $\angle A$ (adj) to the length of the hypotenuse (hyp).	$\cos A = \frac{\text{adj}}{\text{hyp}}$ or $\frac{b}{c}$ $\cos B = \frac{\text{adj}}{\text{hyp}}$ or $\frac{a}{c}$	
If $\triangle ABC$ is a right triangle with acute $\angle A$, then the **tangent** of $\angle A$ (written tan A) is the ratio of the length of the leg opposite $\angle A$ (opp) to the length of the leg adjacent $\angle A$ (adj).	$\tan A = \frac{\text{opp}}{\text{adj}}$ or $\frac{a}{b}$ $\tan B = \frac{\text{opp}}{\text{adj}}$ or $\frac{b}{a}$	

Study Tip

Memorizing Trigonometric Ratios SOH-CAH-TOA is a mnemonic device for learning the ratios for sine, cosine, and tangent using the first letter of each word in the ratios.

$\sin A = \frac{\text{opp}}{\text{hyp}}$

$\cos A = \frac{\text{adj}}{\text{hyp}}$

$\tan A = \frac{\text{opp}}{\text{adj}}$

Example 1 Find Sine, Cosine, and Tangent Ratios

Express each ratio as a fraction and as a decimal to the nearest hundredth.

a. $\sin P$

$\sin P = \frac{\text{opp}}{\text{hyp}}$

$= \frac{15}{17}$ or about 0.88

b. $\cos P$

$\cos P = \frac{\text{adj}}{\text{hyp}}$

$= \frac{8}{17}$ or about 0.47

c. $\tan P$

$\tan P = \frac{\text{opp}}{\text{adj}}$

$= \frac{15}{8}$ or about 1.88

d. $\sin Q$

$\sin Q = \frac{\text{opp}}{\text{hyp}}$

$= \frac{8}{17}$ or about 0.47

e. $\cos Q$

$\cos Q = \frac{\text{adj}}{\text{hyp}}$

$= \frac{15}{17}$ or about 0.88

f. $\tan Q$

$\tan Q = \frac{\text{opp}}{\text{adj}}$

$= \frac{8}{15}$ or about 0.53

Guided Practice

1. Find $\sin J$, $\cos J$, $\tan J$, $\sin K$, $\cos K$, and $\tan K$. Express each ratio as a fraction and as a decimal to the nearest hundredth.

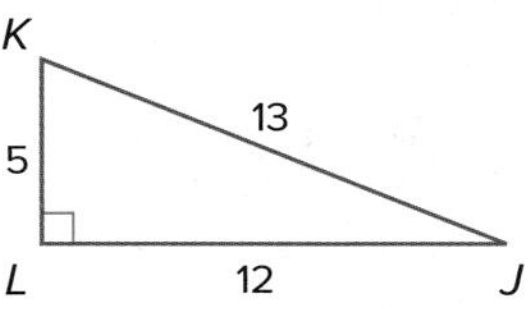

Special right triangles can be used to find the sine, cosine, and tangent of 30°, 60°, and 45° angles.

Example 2 Use Special Right Triangles to Find Trigonometric Ratios

Use a special right triangle to express the tangent of 30° as a fraction and as a decimal to the nearest hundredth.

Draw and label the side lengths of a 30°-60°-90° right triangle, with x as the length of the shorter leg.

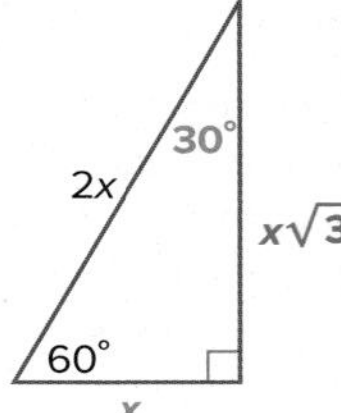

The side opposite the 30° angle has a measure of x.

The side adjacent to the 30° angle has a measure of $x\sqrt{3}$.

$\tan 30° = \frac{\text{opp}}{\text{adj}}$ Definition of tangent ratio

$= \frac{x}{x\sqrt{3}}$ Substitution

$= \frac{1}{\sqrt{3}} \cdot \frac{\sqrt{3}}{\sqrt{3}}$ Simplify and rationalize the denominator.

$= \frac{\sqrt{3}}{3}$ or about 0.58 Simplify and use a calculator.

Guided Practice

2. Use a special right triangle to express the cosine of 45° as a fraction and as a decimal to the nearest hundredth.

Real-World Link

The grade of a trail often changes many times. Average grade is the average of several consecutive running grades of a trail. Maximum grade is the smaller section of a trail that exceeds the trail's typical running grade. Trails often have maximum grades that are much steeper than the average running grade.

Source: Federal Highway Administration

Real-World Example 3 Estimate Measures Using Trigonometry

HIKING A certain part of a hiking trail slopes upward at about a 5° angle. After traveling a horizontal distance of 100 feet along this part of the trail, what would be the change in a hiker's vertical position? What distance has the hiker traveled along the path?

Let $m\angle A = 5$. The vertical change in the hiker's position is x, the measure of the leg opposite $\angle A$. The horizontal distance traveled is 100 feet, the measure of the leg adjacent to $\angle A$. Since the length of the leg opposite and the leg adjacent to a given angle are involved, write an equation using a tangent ratio.

$\tan A = \frac{\text{opp}}{\text{adj}}$ Definition of tangent ratio

$\tan 5° = \frac{x}{100}$ Substitution

$100 \cdot \tan 5° = x$ Multiply each side by 100.

Use a calculator to find x.

100 [TAN] 5 [ENTER] **8.748866353**

The hiker is about 8.75 feet higher than when he started walking.

The distance y traveled along the path is the length of the hypotenuse, so you can use a cosine ratio to find this distance.

$\cos A = \frac{\text{adj}}{\text{hyp}}$ Definition of cosine ratio

$\cos 5° = \frac{100}{y}$ Substitution

$y \cdot \cos 5° = 100$ Multiply each side by y.

$y = \frac{100}{\cos 5°}$ Divide each side by cos 5°.

Use a calculator to find y.

100 [÷] [COS] 5 [ENTER] **100.3819838**

The hiker has traveled a distance of about 100.38 feet along the path.

Study Tip

Graphing Calculator Be sure your graphing calculator is in degree mode rather than radian mode.

Guided Practice

Find x to the nearest hundredth.

3A. 18, x, 25°

3B.

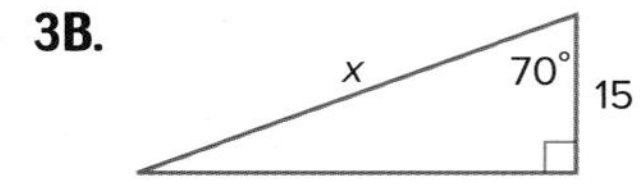

3C. ARCHITECTURE The front of the vacation cottage shown is an isosceles triangle. What is the height x of the cottage above its foundation? What is the length y of the roof? Explain your reasoning.

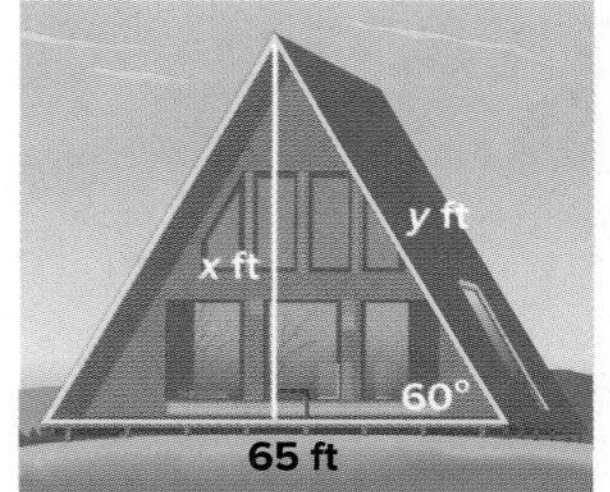

2 Use Inverse Trigonometric Ratios

In Example 2, you found that $\tan 30° \approx 0.58$. It follows that if the tangent of an acute angle is 0.58, then the angle measures approximately 30.

If you know the sine, cosine, or tangent of an acute angle, you can use a calculator to find the measure of the angle, which is the inverse of the trigonometric ratio.

> **Reading Math**
>
> **Inverse Trigonometric Ratios** The expression $\sin^{-1} x$ is read *the inverse sine of x* and is interpreted as the angle with sine x. Be careful not to confuse this notation with the notation for negative exponents—$\sin^{-1} x \neq \frac{1}{\sin x}$. **Instead, this notation is similar to the notation for an inverse function, $f^{-1}(x)$.**

Key Concept Inverse Trigonometric Ratios

Words	If $\angle A$ is an acute angle and the sine of A is x, then the **inverse sine** of x is the measure of $\angle A$.
Symbols	If $\sin A = x$, then $\sin^{-1} x = m\angle A$.
Words	If $\angle A$ is an acute angle and the cosine of A is x, then the **inverse cosine** of x is the measure of $\angle A$.
Symbols	If $\cos A = x$, then $\cos^{-1} x = m\angle A$.
Words	If $\angle A$ is an acute angle and the tangent of A is x, then the **inverse tangent** of x is the measure of $\angle A$.
Symbols	If $\tan A = x$, then $\tan^{-1} x = m\angle A$.

So if $\tan 30° \approx 0.58$, then $\tan^{-1} 0.58 \approx 30°$.

Example 4 Find Angle Measures Using Inverse Trigonometric Ratios

Use a calculator to find the measure of $\angle A$ to the nearest tenth.

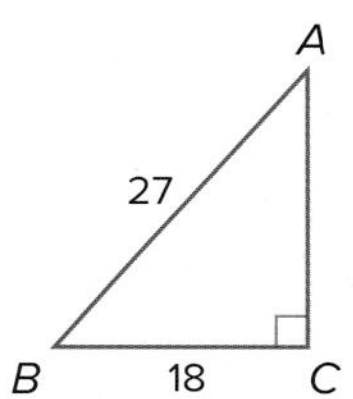

The measures given are those of the leg opposite $\angle A$ and the hypotenuse, so write an equation using the sine ratio.

$\sin A = \frac{18}{27}$ or $\frac{2}{3}$ $\qquad$ $\sin A = \frac{\text{opp}}{\text{hyp}}$

If $\sin A = \frac{2}{3}$, then $\sin^{-1} \frac{2}{3} = m\angle A$. Use a calculator.

KEYSTROKES: [2nd] [SIN $^{-1}$] [( ] 2 [÷] 3 [ )] [ENTER] 41.8103149

So, $m\angle A \approx 41.8°$.

> **Study Tip**
>
> **MP Tools** Use a graphing calculator. The second functions of the [SIN], [COS], and [TAN] keys are usually the inverses.

Guided Practice

Use a calculator to find the measure of $\angle A$ to the nearest tenth.

4A.

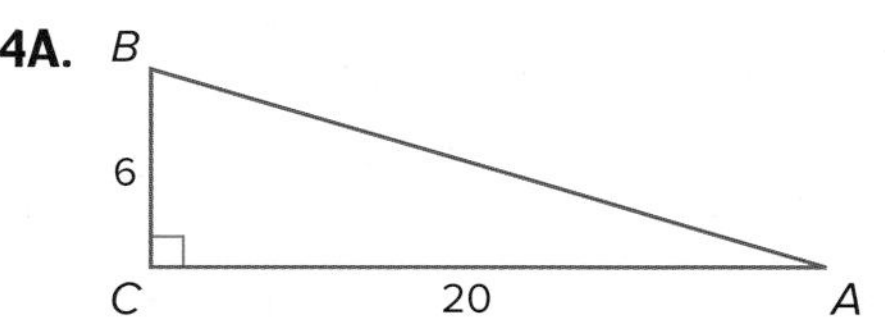

4B.

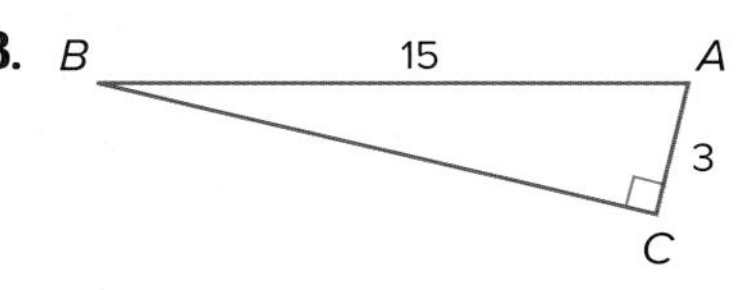

When you use given measures to find the unknown angle and side measures of a right triangle, this is known as *solving a right triangle*. To solve a right triangle, you need to know

- two side lengths or
- one side length and the measure of one acute angle.

Example 5 Solve a Right Triangle

Solve the right triangle. Round side measures to the nearest tenth and angle measures to the nearest degree.

Step 1 Find $m\angle X$ by using a tangent ratio.

$$\tan X = \frac{9}{5} \qquad \tan X = \frac{\text{opp}}{\text{adj}}$$

$$\tan^{-1}\frac{9}{5} = m\angle X \qquad \text{Definition of inverse tangent}$$

$$60.9453959 \approx m\angle X \qquad \text{Use a calculator.}$$

So, $m\angle X \approx 61$.

Step 2 Find $m\angle Y$ using Corollary 4.1, which states that the acute angles of a right triangle are complementary.

$$m\angle X + m\angle Y = 90 \qquad \text{Corollary 4.1}$$

$$61 + m\angle Y \approx 90 \qquad m\angle X \approx 61$$

$$m\angle Y \approx 29 \qquad \text{Subtract 61 from each side.}$$

So, $m\angle Y \approx 29$.

Step 3 Find XY by using the Pythagorean Theorem.

$$(XZ)^2 + (ZY)^2 = (XY)^2 \qquad \text{Pythagorean Theorem}$$

$$5^2 + 9^2 = (XY)^2 \qquad \text{Substitution}$$

$$106 = (XY)^2 \qquad \text{Simplify.}$$

$$\sqrt{106} = XY \qquad \text{Take the positive square root of each side.}$$

$$10.3 \approx XY \qquad \text{Use a calculator.}$$

So $XY \approx 10.3$.

Study Tip

Alternative Methods Right triangles can often be solved using different methods. In Example 5, $m\angle Y$ could have been found using a tangent ratio, and $m\angle X$ and a sine ratio could have been used to find XY.

WatchOut!

Approximation If using calculated measures to find other measures in a right triangle, be careful not to round values until the last step. So in the following equation, use $\tan^{-1}\frac{9}{5}$ instead of its approximate value, 61°.

$$XY = \frac{9}{\sin X} = \frac{9}{\sin\left(\tan^{-1}\frac{9}{5}\right)} \approx 10.3$$

Guided Practice

Solve each right triangle. Round side measures to the nearest tenth and angle measures to the nearest degree.

5A.

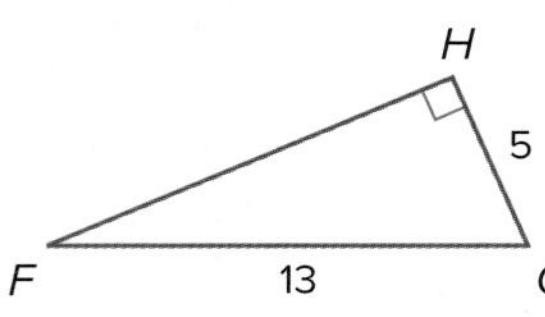

5B.

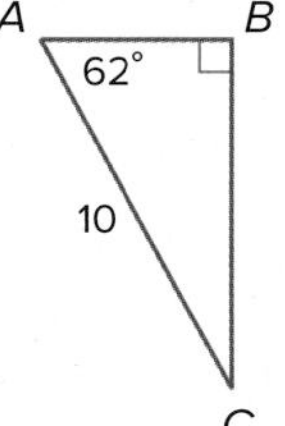

5C.

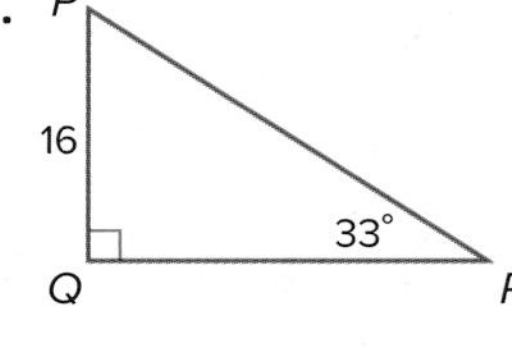

Check Your Understanding

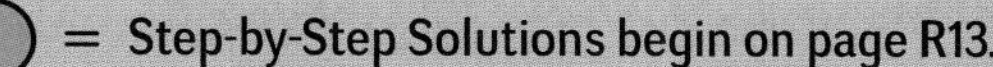

Example 1 **Express each ratio as a fraction and as a decimal to the nearest hundredth.**

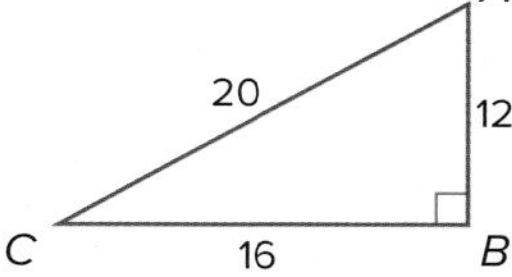

1. $\sin A$
2. $\tan C$
3. $\cos A$
4. $\tan A$
5. $\cos C$
6. $\sin C$

Example 2

7. Use a special right triangle to express $\sin 60°$ as a fraction and as a decimal to the nearest hundredth.

Example 3 **Find x. Round to the nearest hundredth.**

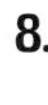

8. 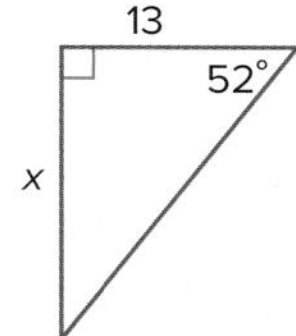

9. 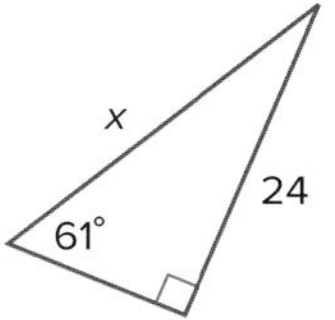

10. 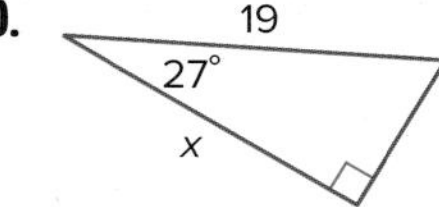

11. **SPORTS** David is building a bike ramp. He wants the angle that the ramp makes with the ground to be 20°. If the board he wants to use for his ramp is $3\frac{1}{2}$ feet long, about how tall will the ramp need to be at the highest point?

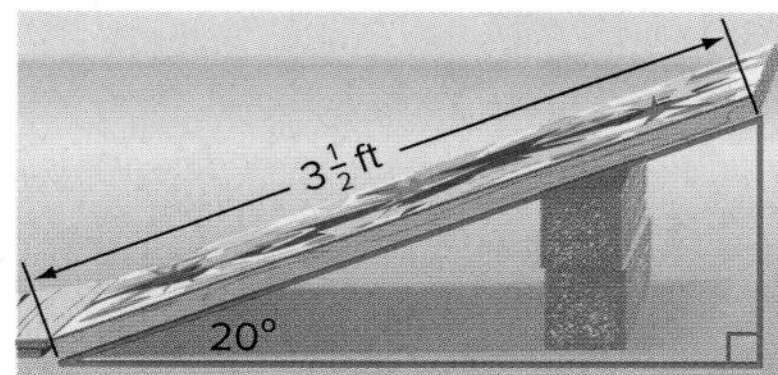

Example 4 **TOOLS Use a calculator to find the measure of $\angle Z$ to the nearest tenth.**

12. 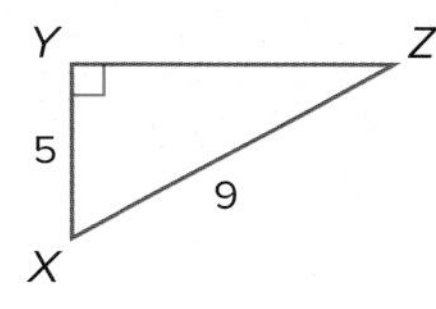

13. 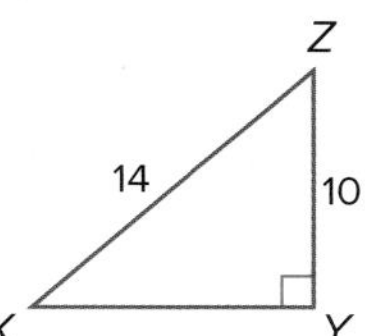

14. 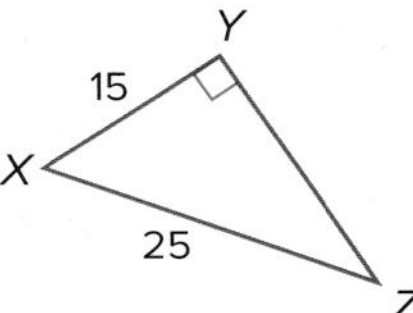

Example 5

15. Solve the right triangle. Round side measures to the nearest tenth and angle measures to the nearest degree.

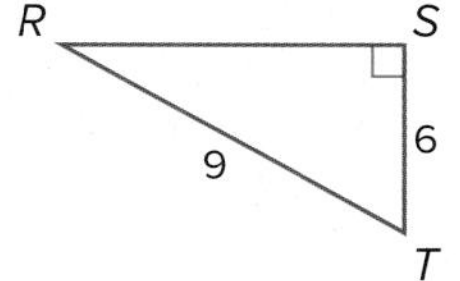

Practice and Problem Solving

Extra Practice is on page R8.

Example 1 **Find $\sin J$, $\cos J$, $\tan J$, $\sin L$, $\cos L$, and $\tan L$. Express each ratio as a fraction and as a decimal to the nearest hundredth.**

16. 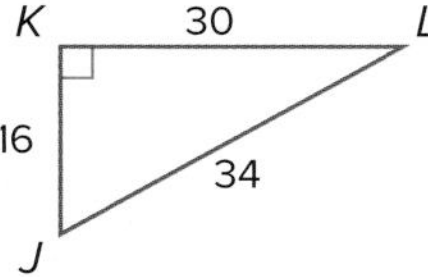

17. 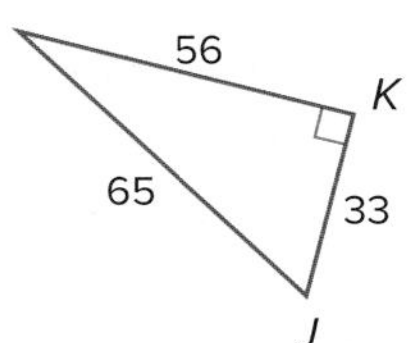

18. 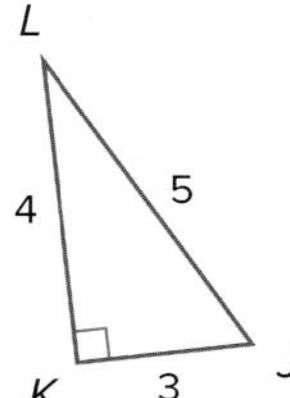

19. 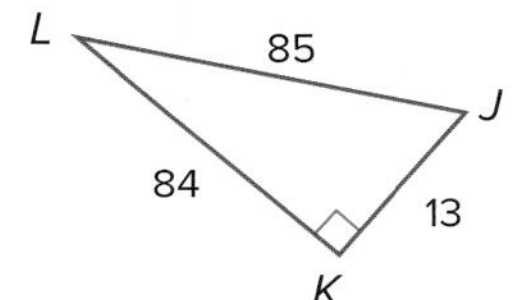

20. 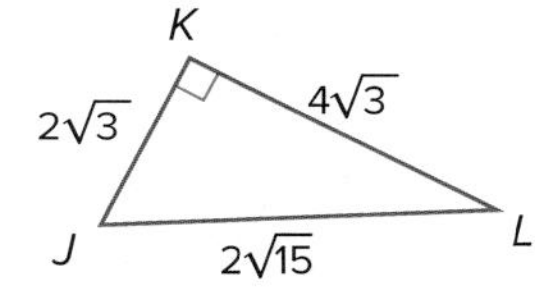

21. 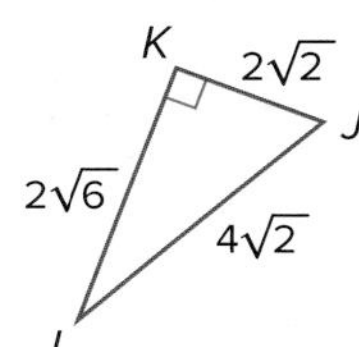

Example 2 **Use a special right triangle to express each trigonometric ratio as a fraction and as a decimal to the nearest hundredth.**

22. tan 60° **23.** cos 30° **24.** sin 45°

25. sin 30° **26.** tan 45° **27.** cos 60°

Example 3 **Find x. Round to the nearest tenth.**

28.
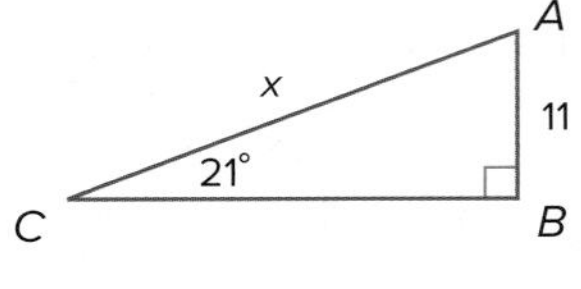

29.
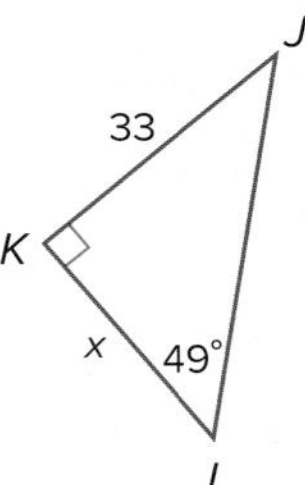

30.
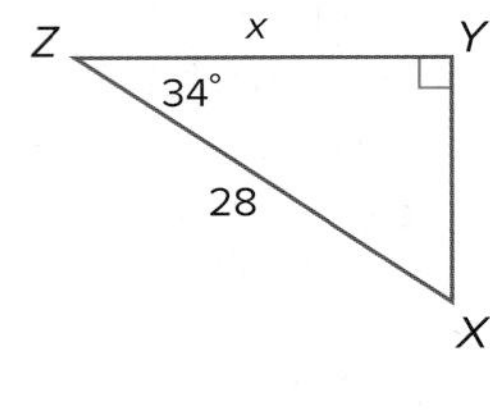

31.
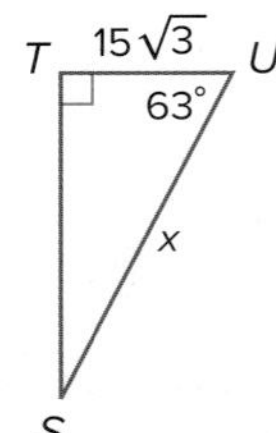

32.
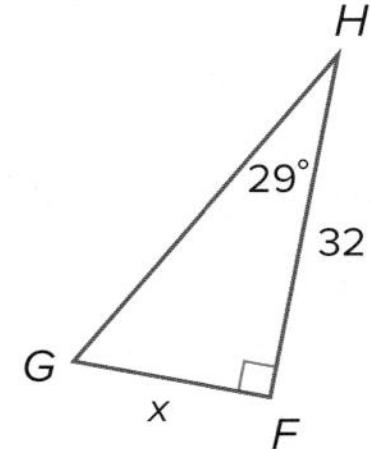

33.
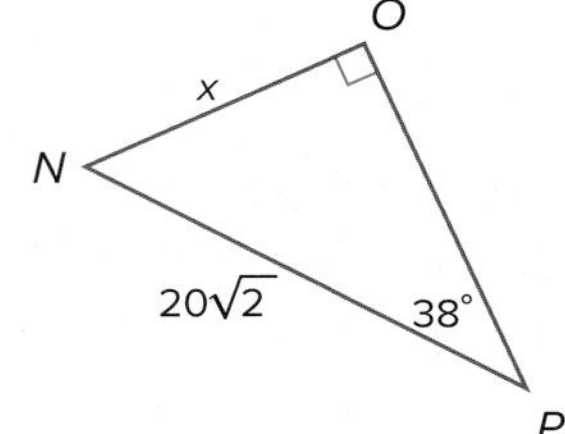

34. **GYMNASTICS** The springboard that Eric uses in his gymnastics class has 6-inch coils and forms an angle of 14.5° with the base. About how long is the springboard?

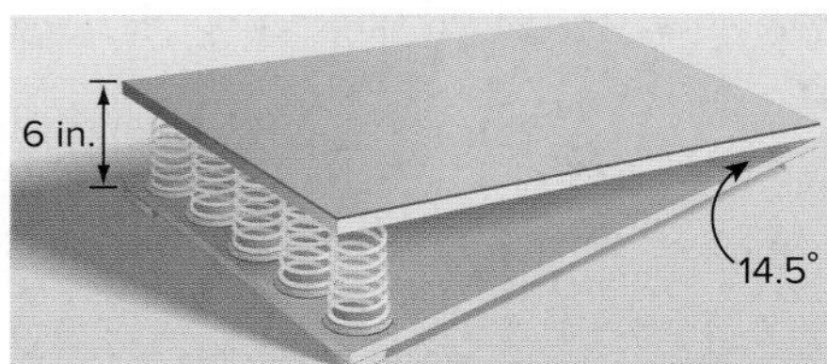

35 **ROLLER COASTERS** The angle of ascent of the first hill of a roller coaster is 55°. If the length of the track from the beginning of the ascent to the highest point is 98 feet, what is the height of the roller coaster when it reaches the top of the first hill?

Example 4 **MP TOOLS Use a calculator to find the measure of $\angle T$ to the nearest tenth.**

36.
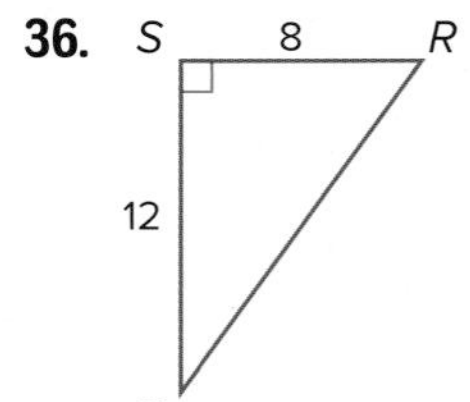

37.
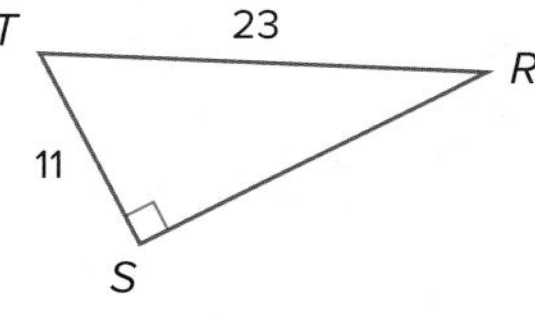

38.
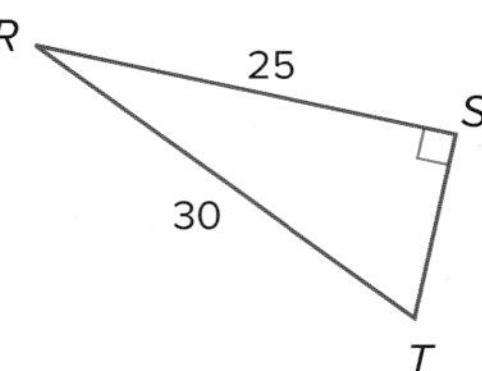

39.
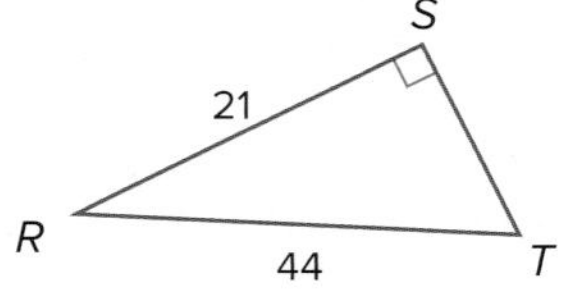

40.
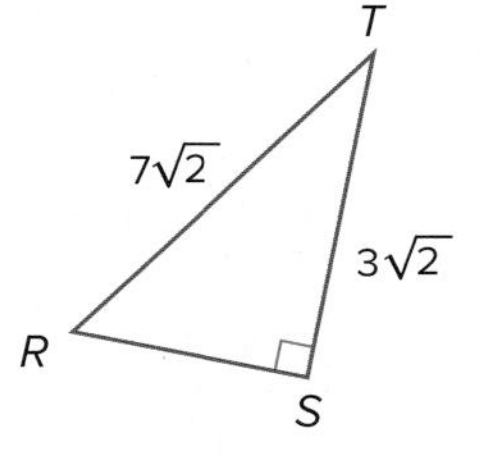

41.
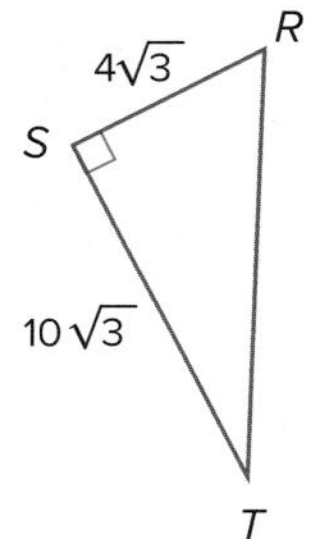

Example 5 **Solve each right triangle. Round side measures to the nearest tenth and angle measures to the nearest degree.**

42.

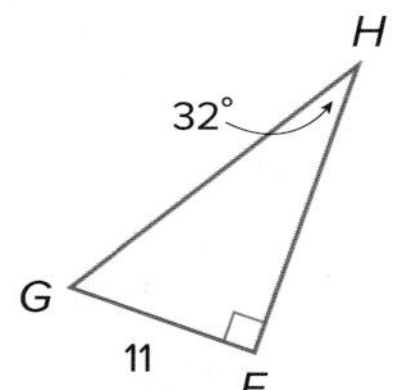

43.

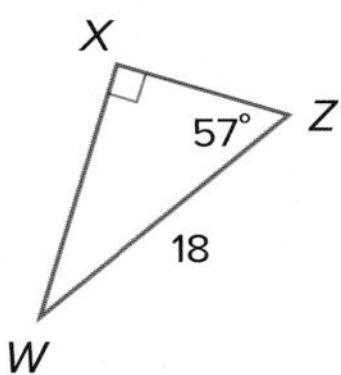

44.

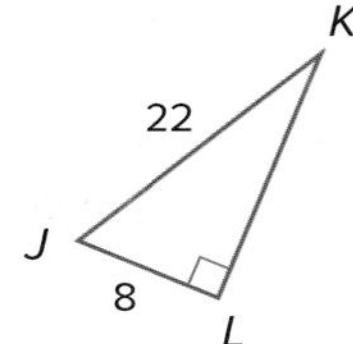

45.

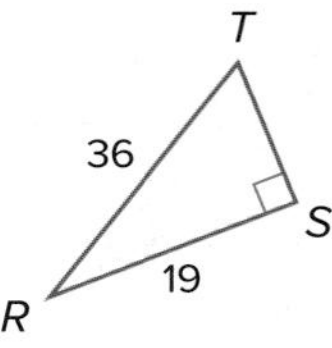

46. BACKPACKS Ramón has a rolling backpack that is $3\frac{3}{4}$ feet tall when the handle is extended. When he is pulling the backpack, Ramon's hand is 3 feet from the ground. What angle does his backpack make with the floor? Round to the nearest degree.

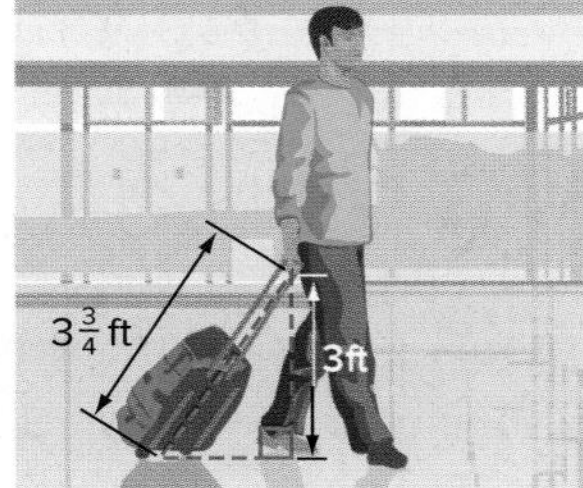

COORDINATE GEOMETRY **Find the measure of each angle to the nearest tenth of a degree using the Distance Formula and an inverse trigonometric ratio.**

47 $\angle K$ in right triangle JKL with vertices $J(-2, -3)$, $K(-7, -3)$, and $L(-2, 4)$

48. $\angle Y$ in right triangle XYZ with vertices $X(4, 1)$, $Y(-6, 3)$, and $Z(-2, 7)$

49. $\angle A$ in right triangle ABC with vertices $A(3, 1)$, $B(3, -3)$, and $C(8, -3)$

50. SCHOOL SPIRIT Hana is making a pennant for each of the 18 girls on her basketball team. She will use $\frac{1}{2}$-inch seam binding to finish the edges of the pennants.

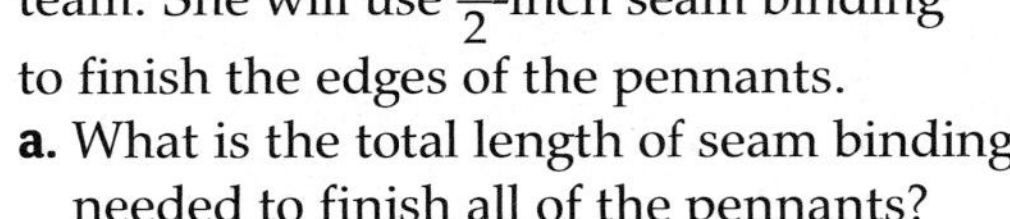

a. What is the total length of seam binding needed to finish all of the pennants?

b. If seam binding is sold in 3-yard packages at a cost of $1.79, how much will it cost?

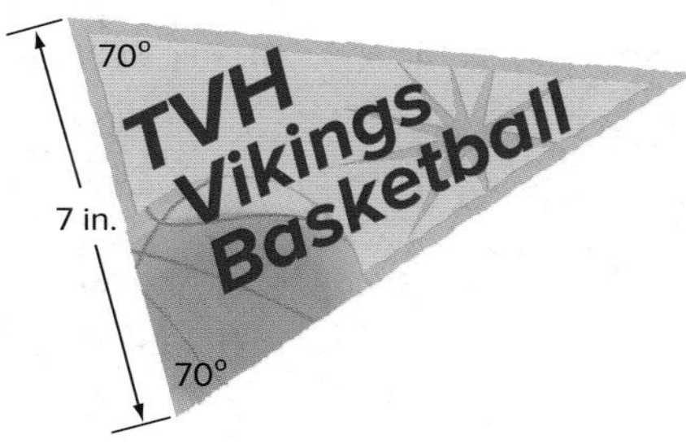

SENSE-MAKING **Find the perimeter and area of each triangle. Round to the nearest hundredth.**

51.

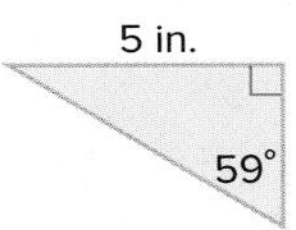

52.

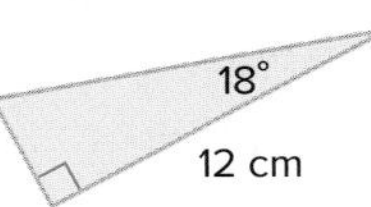

53.

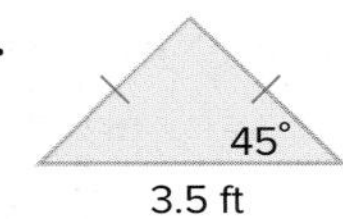

54. Find the tangent of the greater acute angle in a triangle with side lengths of 3, 4, and 5 centimeters.

55. Find the cosine of the smaller acute angle in a triangle with side lengths of 10, 24, and 26 inches.

56. ESTIMATION Ethan and Tariq want to estimate the area of the field that their team will use for soccer practice. They know that the field is rectangular, and they have paced off the width of the field as shown. They used the fence posts at the corners of the field to estimate that the angle between the length of the field and the diagonal is about 40°. If they assume that each of their steps is about 18 inches, what is the area of the practice field in square feet? Round to the nearest square foot.

40°

280 steps

Find x and y. Round to the nearest tenth.

57 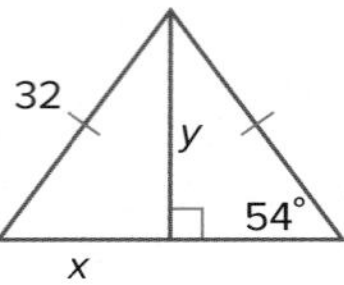

58. 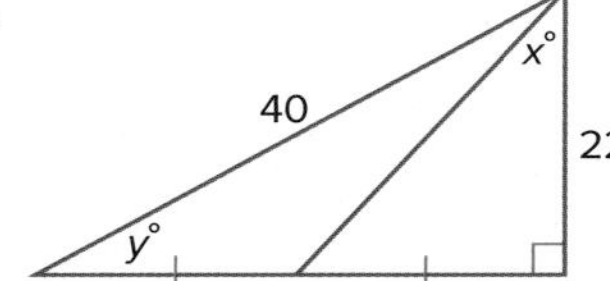

59. 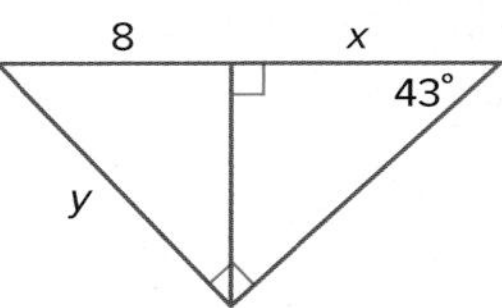

60. COORDINATE GEOMETRY Show that the slope of a line at 225° from the x-axis is equal to the tangent of 225°.

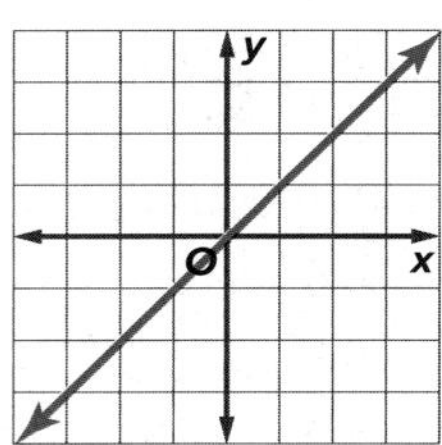

61. MULTIPLE REPRESENTATIONS In this problem, you will investigate an algebraic relationship between the sine and cosine ratios.

a. Geometric Draw three right triangles that are not similar to each other. Label the triangles ABC, MNP, and XYZ, with the right angles located at vertices B, N, and Y, respectively. Measure and label each side of the three triangles.

b. Tabular Copy and complete the table below.

Triangle	Trigonometric Ratios				Sum of Ratios Squared	
ABC	cos A		sin A		$(\cos A)^2 + (\sin A)^2 =$	
	cos C		sin C		$(\cos C)^2 + (\sin C)^2 =$	
MNP	cos M		sin M		$(\cos M)^2 + (\sin M)^2 =$	
	cos P		sin P		$(\cos P)^2 + (\sin P)^2 =$	
XYZ	cos X		sin X		$(\cos X)^2 + (\sin X)^2 =$	
	cos Z		sin Z		$(\cos Z)^2 + (\sin Z)^2 =$	

c. Verbal Make a conjecture about the sum of the squares of the cosine and sine of an acute angle of a right triangle.

d. Algebraic Express your conjecture algebraically for an angle X.

e. Analytical Show that your conjecture is valid for angle A in the figure at the right using the trigonometric functions and the Pythagorean Theorem.

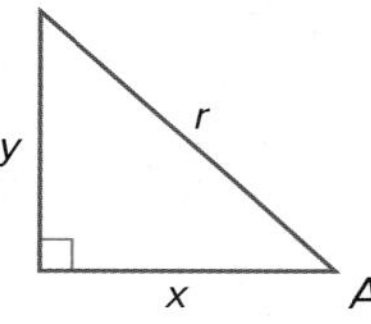

H.O.T. Problems Use Higher-Order Thinking Skills

62. CHALLENGE Solve $\triangle ABC$. Round to the nearest whole number.

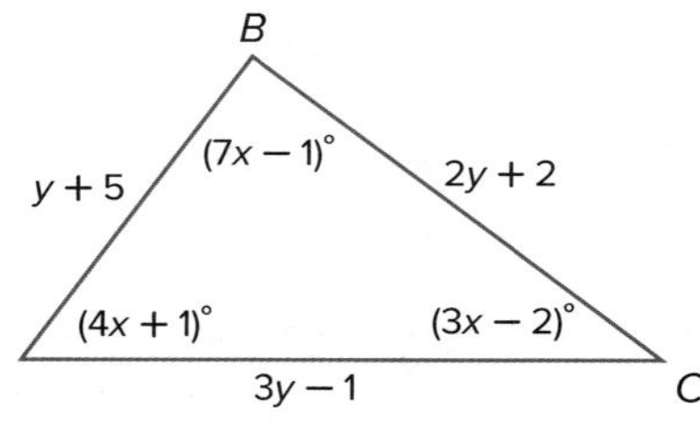

63. MP REASONING Are the values of sine and cosine for an acute angle of a right triangle always less than 1? Explain.

64. MP REASONING What is the relationship between the sine and cosine of complementary angles? Explain your reasoning and use the relationship to find cos 50 if sin 40 $\approx$ 0.64.

65. WRITING IN MATH Explain how you can use ratios of the side lengths to find the angle measures of the acute angles in a right triangle.

66. A vertical pole is supported by a guy wire, as shown in the figure. What is the height of the pole, in feet, to the nearest tenth of a foot? MP 4

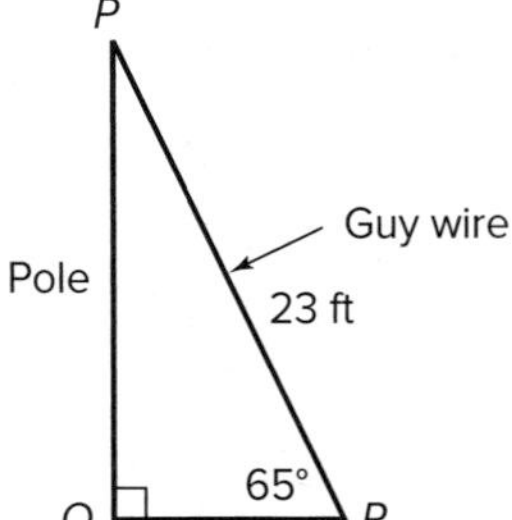

67. The legs of a right triangle are 5 and 6 units long. Find the measure of the second largest angle of the triangle. Round your answer to the nearest tenth. MP 1

68. In right triangle ABC, $\sin A = \frac{2}{5}$. Find $\sin(90° - A)$. MP 1

69. A bookshelf that hangs on a wall is supported by a bracket, as shown in the figure.

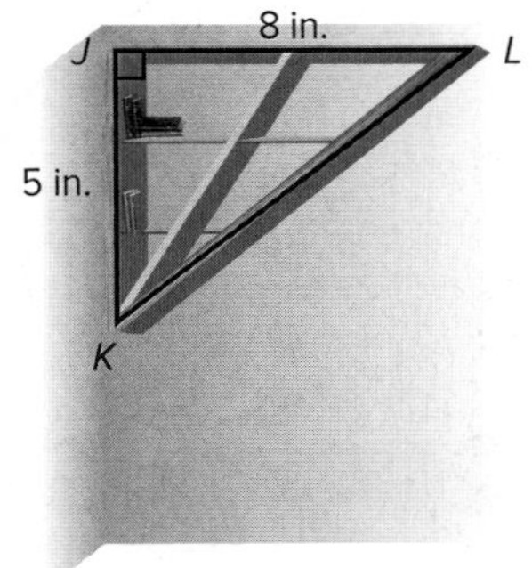

Which of the following is the best estimate of the measure of $\angle L$? MP 4

- A 32
- B 39
- C 51
- D 58

70. In $\triangle ABC$, $\angle B$ is a right angle and $m\angle A = 40$. Given that $AC = 15$, what is AB to the nearest tenth? MP 1

71. Erin wants to find the length of $\overline{PN}$ in $\triangle MNP$.

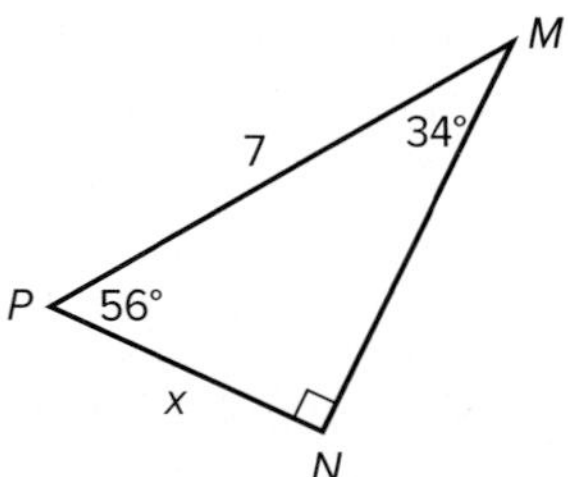

Which of the following equations can Erin use to find the length of $\overline{PN}$? MP 1

I. $\sin 34° = \frac{x}{7}$

II. $\cos 56° = \frac{x}{7}$

III. $\tan 56° = \frac{7}{x}$

- A I only
- B II only
- C I and II only
- D I and III only
- E I, II, and III

72. A five-meter-long ladder leans against a wall, with the top of the ladder being four meters above the ground. What is the approximate angle that the ladder makes with the ground? Round to the nearest degree. MP 1

73. MULTI-STEP Given an acute angle A where $\tan A = \frac{2}{5}$, find: MP 1

a. $\cos A$

b. $\sin A$

c. $\cos(90° - A)$

d. $\sin(90° - A)$

e. $\sin^2 A + \cos^2 A$

EXTEND 8-4

Graphing Technology Lab
Secant, Cosecant, and Cotangent

In the previous lesson, you used the trigonometric functions sine, cosine, and tangent to find angle relationships in right angles. In this activity, you will use the reciprocals of those functions, cosecant, secant, and cotangent, to explore angle and side relationships in right triangles.

Mathematical Practices

MP 5 Use appropriate tools strategically.

Key Concept Reciprocal Trigonometric Ratios

Words	Symbols
The **cosecant** of $\angle A$ (written csc A) is the reciprocal of sin A.	$\csc A = \frac{1}{\sin A}$ or $\frac{c}{a}$
The **secant** of $\angle A$ (written sec A) is the reciprocal of cos A.	$\sec A = \frac{1}{\cos A}$ or $\frac{c}{b}$
The **cotangent** of $\angle A$ (written cot A) is the reciprocal of tan A.	$\cot A = \frac{1}{\tan A}$ or $\frac{b}{a}$

Activity Find Trigonometric Values

Work cooperatively.

Step 1 Draw and label a right triangle with the dimensions shown at the right.

Step 2 Use your graphing calculator to find the values for sin A, cos A, and tan A.

Step 3 Next, find the value for csc A by dividing 1 by [SIN] A. Repeat step 3 to find sec A and cot A.

Step 4 Copy the table below and record your results. Next, find the value of each trigonometric function for angle C.

Angle	sin	cos	tan	csc	sec	cot
A						
C						

Exercises

Work cooperatively.

1. Find the values of the six trigonometric functions for a 45° angle in a 45° –45° –90° triangle with legs that are 4 cm.
2. In $\triangle FGH$, $\tan F = \frac{5}{12}$. Find cot F and sin F if $\angle G$ is a right angle.
3. Find the values of the six trigonometric functions for angle T in $\triangle RST$ if $m\angle R = 36°$. Round to the nearest hundredth.

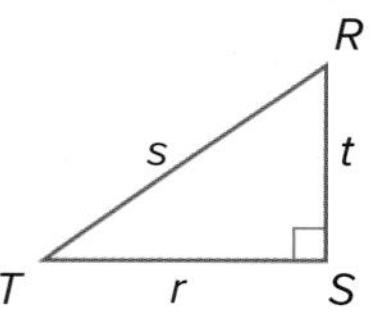

Mid-Chapter Quiz

Lessons 8-1 through 8-4

Find the geometric mean between each pair of numbers. (Lesson 8-1)

1. 12 and 3
2. 63 and 7
3. 45 and 20
4. 50 and 10

Write a similarity statement identifying the three similar triangles in each figure. (Lesson 8-1)

5. 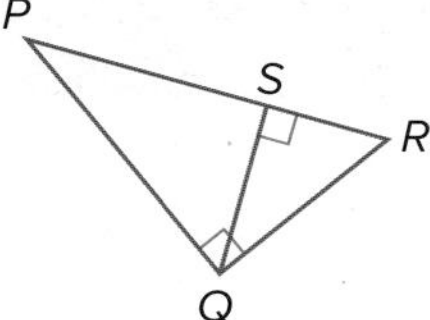

6. 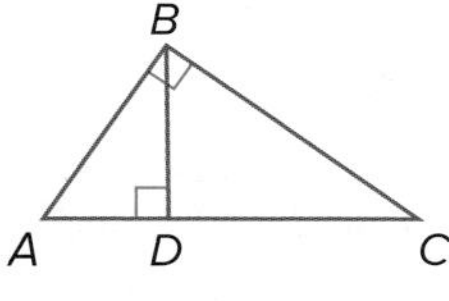

7. Find x, y, and z. (Lesson 8-1)

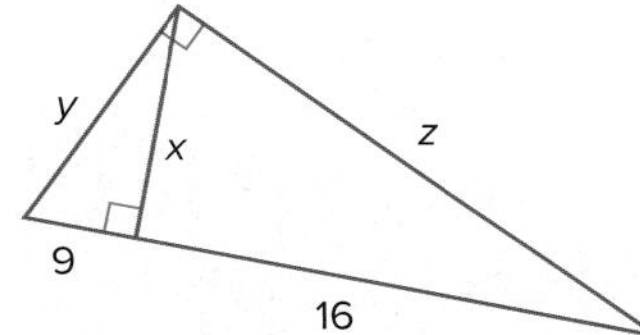

8. **PARKS** There is a small park in a corner made by two perpendicular streets. The park is 100 feet by 150 feet, with a diagonal path, as shown below. What is the length of path $\overline{AC}$? (Lesson 8-2)

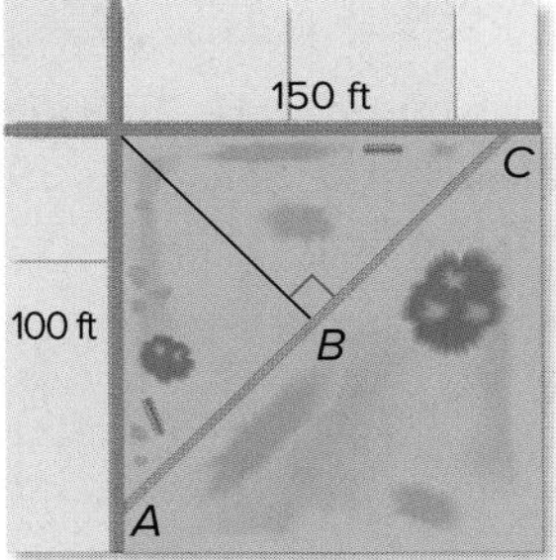

Find x. Round to the nearest hundredth. (Lesson 8-2)

9. 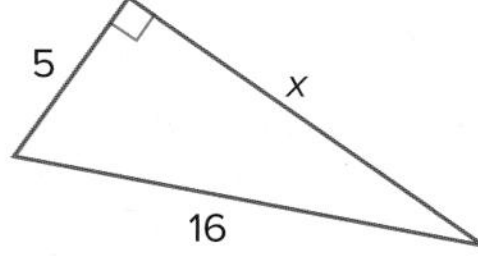

10. 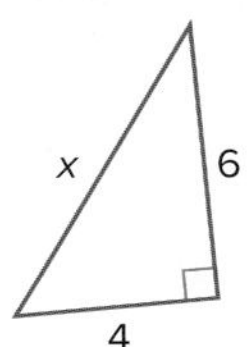

11. **MULTIPLE CHOICE** Which of the following sets of numbers is not a Pythagorean triple? (Lesson 8-2)

A 9, 12, 15
B 21, 72, 75
C 15, 36, 39
D 8, 13, 15

Find x. (Lesson 8-3)

12. 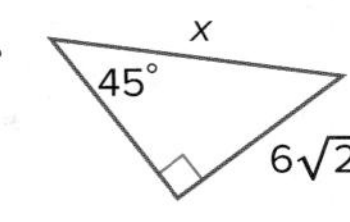

13. 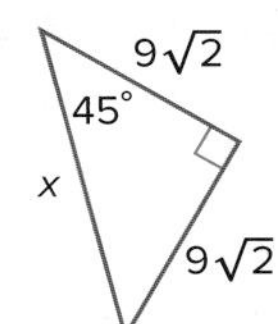

14. **DESIGN** Jamie designed a pinwheel to put in her garden. In the pinwheel, the blue triangles are congruent equilateral triangles, each with an altitude of 4 inches. The red triangles are congruent isosceles right triangles. The hypotenuse of a red triangle is congruent to a side of the blue triangle. (Lesson 8-3)

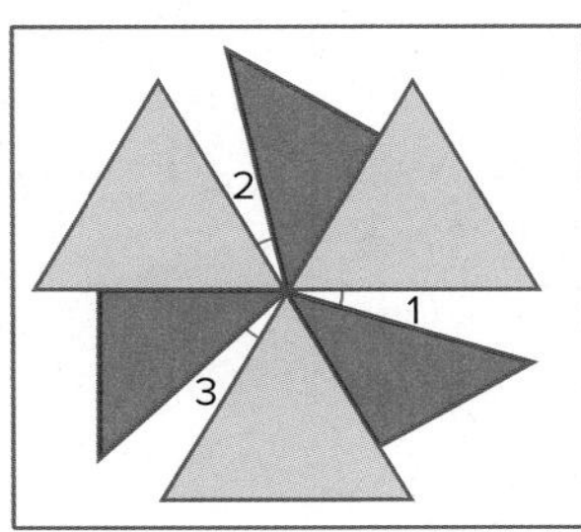

a. If angles 1, 2, and 3 are congruent, find the measure of each angle.
b. Find the perimeter of the pinwheel.
c. MP What mathematical practice did you use to solve this problem?

Find x and y. (Lesson 8-3)

15. 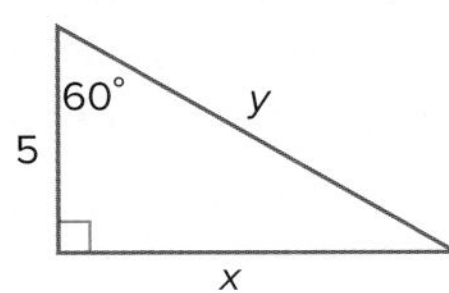

16. 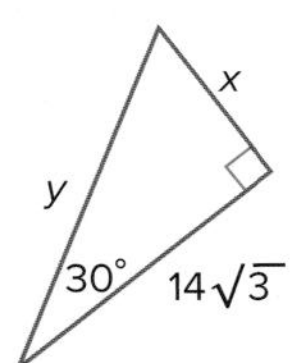

Express each ratio as a fraction and as a decimal to the nearest hundredth. (Lesson 8-4)

17. tan M
18. cos M
19. cos N
20. sin N

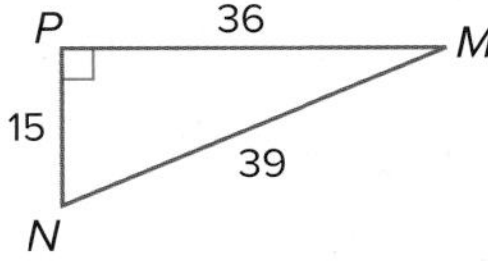

21. Solve the right triangle. Round angle measures to the nearest degree and side measures to the nearest tenth. (Lesson 8-4)

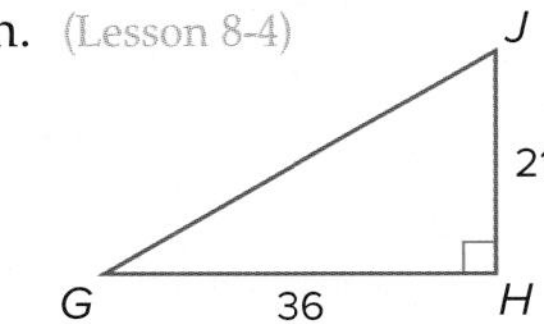

LESSON 5

Angles of Elevation and Depression

∴ Then	∴ Now	∴ Why?
● You used similar triangles to measure distances indirectly.	● **1** Solve problems involving angles of elevation and depression. **2** Use angles of elevation and depression to find the distance between two objects.	● To make a field goal, a kicker must kick the ball with enough force and at an appropriate angle of elevation to ensure that the ball will reach the goalpost at a level high enough to make it over the horizontal bar. This angle must change depending on the initial placement of the ball away from the base of the goalpost.

New Vocabulary

angle of elevation
angle of depression

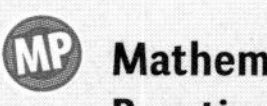

Mathematical Practices

4 Model with mathematics.

1 Make sense of problems and persevere in solving them.

1 Angles of Elevation and Depression An **angle of elevation** is the angle formed by a horizontal line and an observer's line of sight to an object above the horizontal line. An **angle of depression** is the angle formed by a horizontal line and an observer's line of sight to an object below the horizontal line.

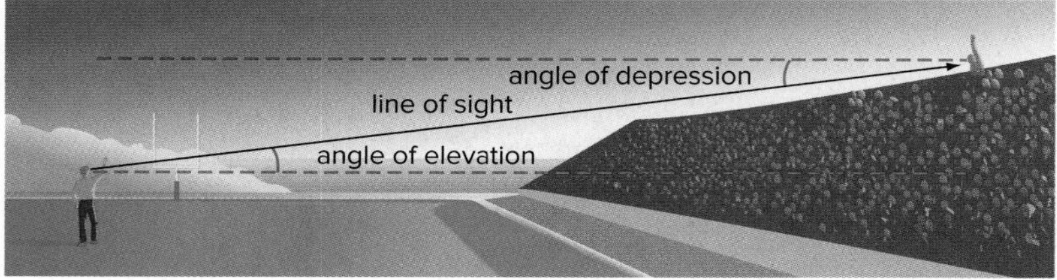

Horizontal lines are parallel, so the angle of elevation and the angle of depression in the diagram are congruent by the Alternate Interior Angles Theorem.

Example 1 Angle of Elevation

VACATION Leah wants to see a castle in an amusement park. She sights the top of the castle at an angle of elevation of 38°. She knows that the castle is 190 feet tall. If Leah is 5.5 feet tall, how far is she from the castle to the nearest foot?

Make a sketch to represent the situation.

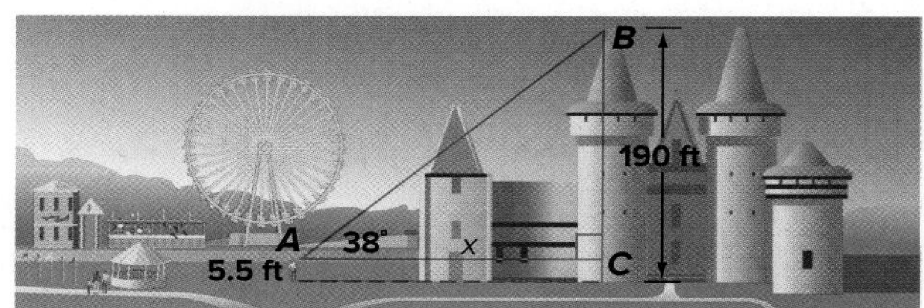

Because Leah is 5.5 feet tall, $BC = 190 - 5.5$ or 184.5 feet. Let x represent the distance from Leah to the castle, AC.

$\tan A = \frac{BC}{AC}$ — $\tan = \frac{\text{opposite}}{\text{adjacent}}$

$\tan 38° = \frac{184.5}{x}$ — $m\angle A = 38$, $BC = 184.5$, $AC = x$

$x = \frac{184.5}{\tan 38°}$ — Solve for x.

$x \approx 236.1$ — Use a calculator.

Leah is about 236 feet from the castle.

Juice Images/Alamy

Guided Practice

1. **FOOTBALL** The cross bar of a goalpost is 10 feet high. If a field goal attempt is made 25 yards from the base of the goalpost that clears the goal by 1 foot, what is the smallest angle of elevation at which the ball could have been kicked to the nearest degree?

Example 2 Angle of Depression

EMERGENCY A search and rescue team is airlifting people from the scene of a boating accident when they observe another person in need of help. If the angle of depression to this other person is 42° and the helicopter is 18 feet above the water, what is the horizontal distance from the rescuers to this person to the nearest foot?

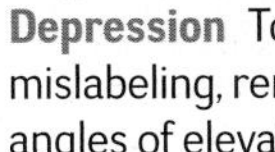

WatchOut!

Angles of Elevation and Depression To avoid mislabeling, remember that angles of elevation and depression are always formed with a horizontal line and never with a vertical line.

Make a sketch of the situation.

A, B, 42° angle of depression, 18 ft, D, x, 42°, C

Note: Art not drawn to scale.

Because $\overrightarrow{AB}$ and $\overline{DC}$ are parallel, $m\angle BAC = m\angle ACD$ by the Alternate Interior Angles Theorem.

Let x represent the horizontal distance from the rescuers to the person DC.

$\tan C = \frac{AD}{DC}$	$\tan = \frac{\text{opposite}}{\text{adjacent}}$
$\tan 42° = \frac{18}{x}$	$C = 42$, $AD = 18$, and $DC = x$
$x\tan 42° = 18$	Multiply each side by x.
$x = \frac{18}{\tan 42°}$	Divide each side by tan 42°.
$x \approx 20.0$	Use a calculator.

The horizontal distance from the rescuers to the person is 20.0 feet.

Math HistoryLink

Eratosthenes (276–194 B.C.) Eratosthenes was a mathematician who was born in Cyrene, which is now Libya. He used the angle of elevation of the Sun at noon in the cities of Alexandria and Syene (now Egypt) to measure the circumference of Earth.

Source: *Encyclopaedia Britannica*

Guided Practice

2. **LIFEGUARDING** A lifeguard is watching a beach from a line of sight 6 feet above the ground. She sees a swimmer at an angle of depression of 8°. How far away from the tower is the swimmer?

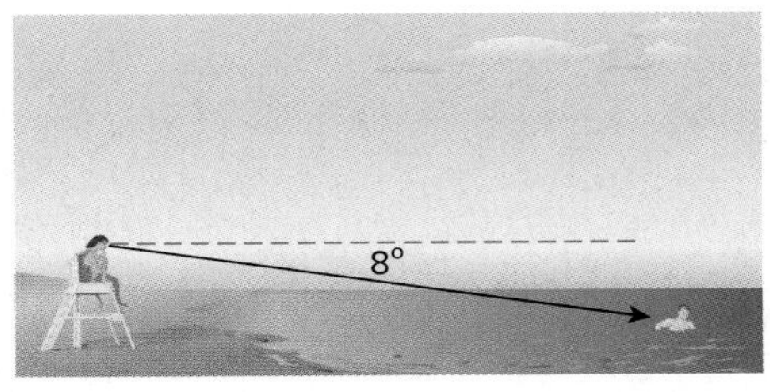

2 Two Angles of Elevation or Depression

Angles of elevation or depression to two different objects can be used to estimate the distance between those objects. Similarly, the angles from two different positions of observation to the same object can be used to estimate the object's height.

Real-World Link

In the United States, lumber volume is measured in board-feet, which is defined as a piece of wood containing 144 cubic inches. Woodland owners often estimate the lumber volume of trees they own to determine how many to cut and sell.

Source: The Ohio State University School of Natural Resources

Example 3 Use Two Angles of Elevation or Depression

TREE REMOVAL To estimate the height of a tree she wants removed, Mrs. Long sights the tree's top at a 70° angle of elevation. She then steps back 10 meters and sights the top at a 26° angle. If Mrs. Long's line of sight is 1.7 meters above the ground, how tall is the tree to the nearest meter?

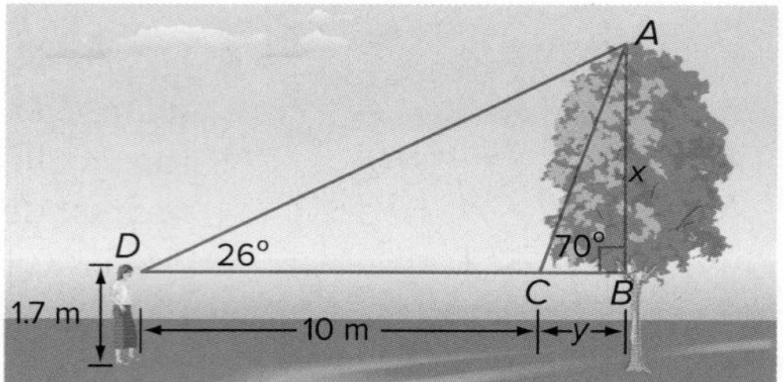

Understand $\triangle ABC$ and $\triangle ABD$ are right triangles. The height of the tree is the sum of Mrs. Long's height and AB.

Plan Since her initial distance from the tree is not given, write and solve a system of equations using both triangles. Let $AB = x$ and $CB = y$. So $DB = y + 10$ and the height of the tree is $x + 1.7$.

Solve Use $\triangle ABC$.

$$\tan 70^\circ = \frac{x}{y} \qquad \tan = \frac{\text{opposite}}{\text{adjacent}};\ m\angle ACB = 70$$

$$y \tan 70^\circ = x \qquad \text{Multiply each side by } y.$$

Use $\triangle ABD$.

$$\tan 26^\circ = \frac{x}{y + 10} \qquad \tan = \frac{\text{opposite}}{\text{adjacent}};\ m\angle D = 26$$

$$(y + 10) \tan 26^\circ = x \qquad \text{Multiply each side by } y + 10.$$

Substitute the value for x from $\triangle ABD$ in the equation for $\triangle ABC$ and solve for y.

$$y \tan 70^\circ = x$$

$$y \tan 70^\circ = (y + 10) \tan 26^\circ$$

$$y \tan 70^\circ = y \tan 26^\circ + 10 \tan 26^\circ$$

$$y \tan 70^\circ - y \tan 26^\circ = 10 \tan 26^\circ$$

$$y(\tan 70^\circ - \tan 26^\circ) = 10 \tan 26^\circ$$

$$y = \frac{10 \tan 26^\circ}{\tan 70^\circ - \tan 26^\circ}$$

Use a calculator to find that $y \approx 2.16$. Using the equation from $\triangle ABC$, $x = 2.16 \tan 70^\circ$ or about 5.9. The height of the tree is 5.9 + 1.7 or 7.6, which is about 8 meters.

Check Substitute the value for y in the equation from $\triangle ABD$.

$x = (2.16 + 10) \tan 26^\circ$ or about 5.9.

When using the angles of depression to two different objects to calculate the distance between them, it is important to remember that the two objects must lie in the same horizontal plane. Eight meters is reasonable for the tree height.

Guided Practice

3. **SKYSCRAPERS** Two buildings are sited from atop a 200-meter skyscraper. Building A is sited at a 35° angle of depression, while Building B is sighted at a 36° angle of depression. How far apart are the two buildings to the nearest meter?

Check Your Understanding

◯ = Step-by-Step Solutions begin on page R13.

Example 1

1. **BIKING** Lenora wants to build the bike ramp shown. Find the length of the base of the ramp.

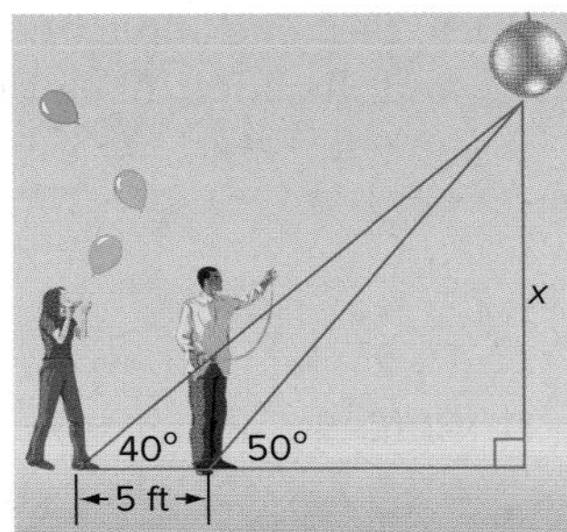

Example 2

2. **BASEBALL** A fan is seated in the upper deck of a stadium 200 feet away from home plate. If the angle of depression to the field is 62°, at what height is the fan sitting?

Example 3

3. **MP MODELING** Annabelle and Rich are setting up decorations for their school dance. Rich is standing 5 feet directly in front of Annabelle under a disco ball. If the angle of elevation from Annabelle to the ball is 40° and from Rich to the ball is 50°, how high is the disco ball?

Practice and Problem Solving

Extra Practice is found on page R8.

Example 1

4. **HOCKEY** A hockey player takes a shot 20 feet away from a 5-foot goal. If the puck travels at a 15° angle of elevation toward the center of the goal, will the player score?

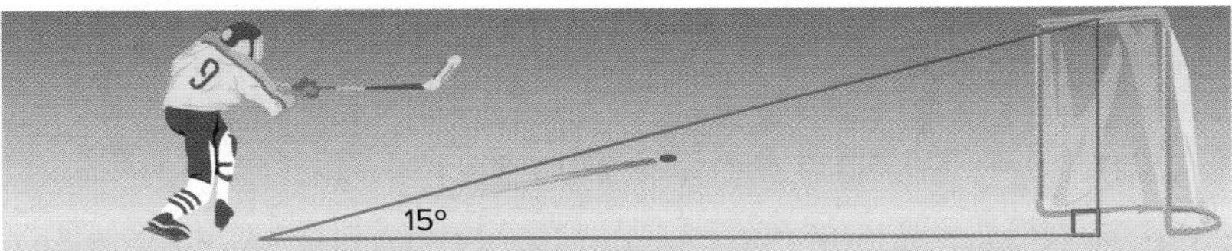

5. **MOUNTAINS** Find the angle of elevation to the peak of a mountain for an observer who is 155 meters from the mountain if the observer's eye is 1.5 meters above the ground and the mountain is 350 meters tall.

Example 2

6. **WATERPARK** Two water slides are 50 meters apart on level ground. From the top of the taller slide, you can see the top of the shorter slide at an angle of depression of 15°. If you know that the top of the other slide is approximately 15 meters above the ground, about how far above the ground are you? Round to the nearest tenth of a meter.

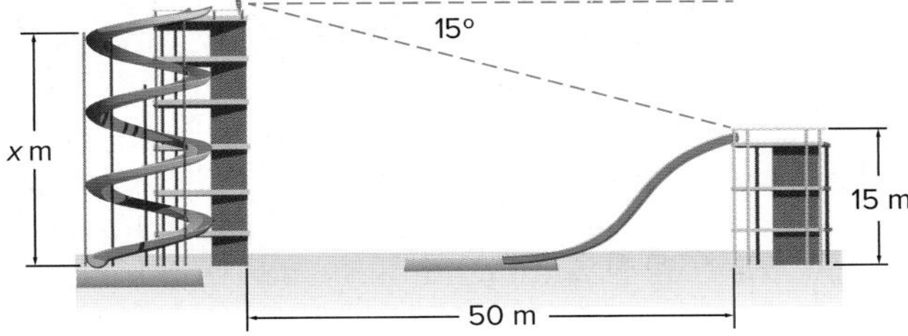

7. **AVIATION** Due to a storm, a pilot flying at an altitude of 528 feet has to land. If he has a horizontal distance of 2000 feet to land, at what angle of depression should he land?

Example 3

8. **PYRAMIDS** Miko and Tyler are visiting the Great Pyramid in Egypt. From where Miko is standing, the angle of elevation to the top of the pyramid is 48.6°. From Tyler's position, the angle of elevation is 50°. If they are standing 20 feet apart and they are each 5 feet 6 inches tall, how tall is the pyramid?

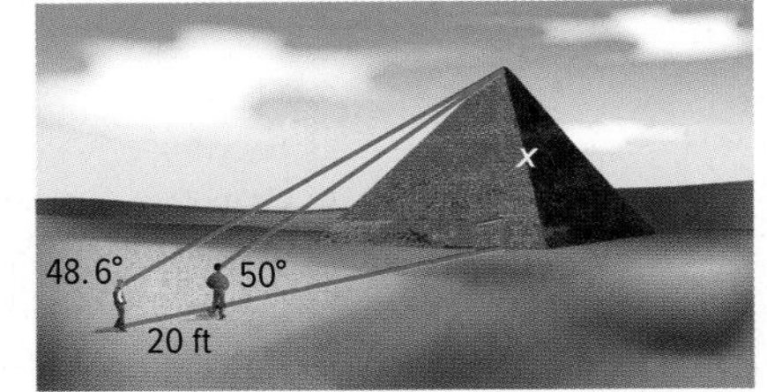

9 **DIVING** Austin is standing on the high dive at the local pool. Austin's line of sight is 3 feet above the diving board. Two of his friends are in the water as shown. If the angle of depression to one of his friends is 40°, and 30° to his other friend who is 5 feet beyond the first, how tall is the platform?

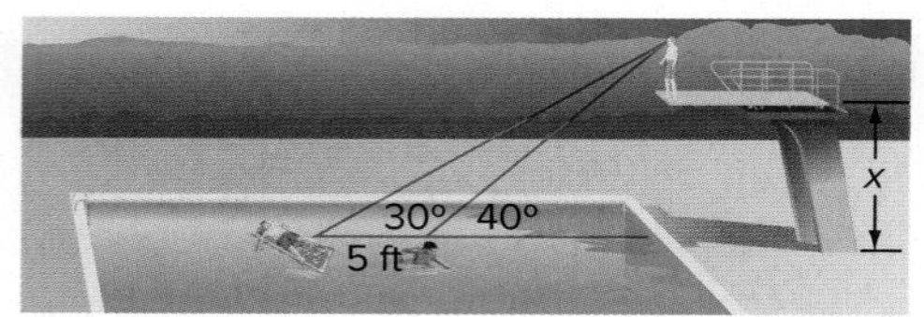

10. **BASKETBALL** Claire and Marisa are waiting to get a rebound during a basketball game. If the height of the basketball hoop is 10 feet, the angle of elevation between Claire and the goal is 35°, and the angle of elevation between Marisa and the goal is 25°, how far apart are they standing?

11. **RIVERS** Hugo is standing in the top of St. Louis' Gateway Arch, looking down on the Mississippi River. The angle of depression to the closer bank is 45° and the angle of depression to the farther bank is 18°. The arch is 630 feet tall. Estimate the width of the river at that point.

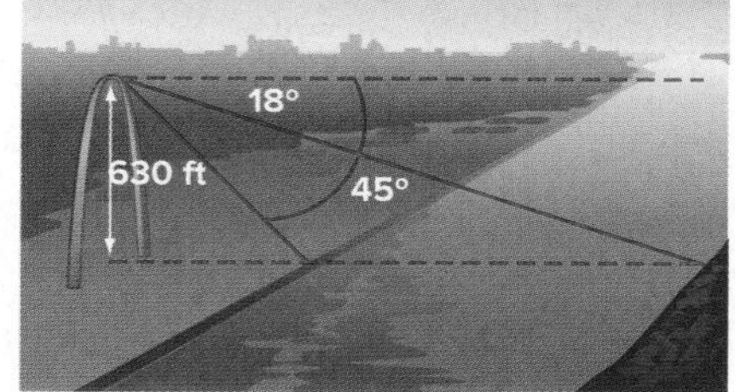

12. **MODELING** The Unzen Volcano in Japan has a magma reservoir located 15 kilometers beneath the Chijiwa Bay, located east of the volcano. A magma channel, which connects the reservoir to the volcano, rises at a 40° angle of elevation toward the volcano. What length of magma channel is below sea level?

13. **BRIDGES** Suppose you are standing in the middle of the platform of the Akashi Kaikyo Bridge. If the height from the top of the platform holding the suspension cables is 297 meters and the length from the platform to the center of the bridge is 995 meters, what is the angle of depression from the center of the bridge to the platform?

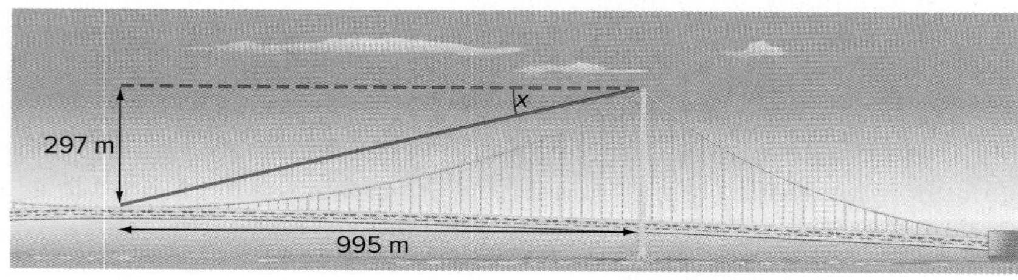

14. **LIGHTHOUSES** Little Gull Island Lighthouse shines a light from a height of 91 feet with a 6° angle of depression. Plum Island Lighthouse, 1800 feet away, shines a light from a height of 34 feet with a 2° angle of depression. Which light will reach a boat that sits exactly between Little Gull Island Lighthouse and Plum Island Lighthouse?

15. **MULTI-STEP** Tom is designing a passive solar home in El Paso, along the 31.8° latitude. His design will include an overhang to shield the windows from the Sun.

Day of the Year	Angle of Elevation of the Sun
Longest Day	90° + 23° − the latitude
Shortest Day	90° − 23° − the latitude

a. If the windows will be 12 feet high with the overhang beginning 2 feet above the windows, what length should he make the overhang?

b. Explain your solution process.

16. **MAINTENANCE** Two telephone repair workers arrive at a location to restore electricity after a power outage. One of the workers climbs up the telephone pole while the other worker stands 10 feet to the left of the pole. If the terminal box is located 30 feet above ground on the pole and the angle of elevation from the truck to the repair worker is 70°, how far is the worker on the ground standing from the truck?

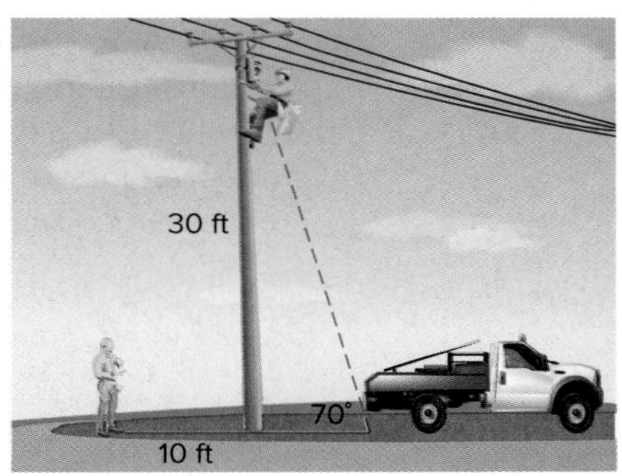

17. **PHOTOGRAPHY** A digital camera with a panoramic lens is described as having a view with an angle of elevation of 38°. If the camera is on a 3-foot tripod aimed directly at a 124-foot-tall monument, how far from the monument should you place the tripod to see the entire monument in your photograph?

18. **MP MODELING** As a part of their weather unit, Anoki's science class took a hot-air balloon ride. As they passed over a fenced field, the angle of depression of the closer side of the fence was 32°, and the angle of depression of the farther side of the fence was 27°. If the height of the balloon was 800 feet, estimate the width of the field.

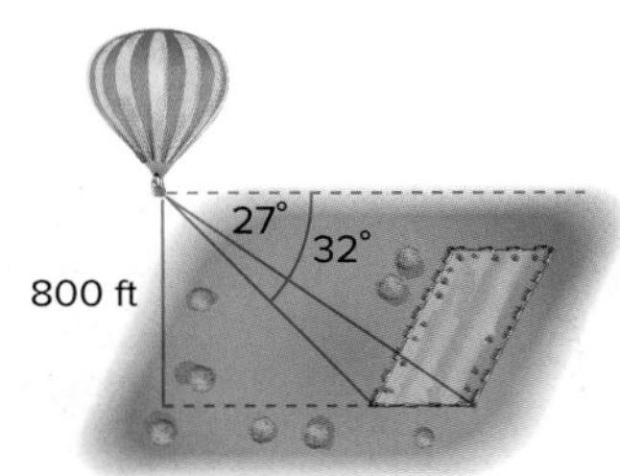

19. **MARATHONS** The Badwater Ultramarathon is a race that begins at the lowest point in California's Death Valley and ends on the highest point in the state, Mount Whitney. The race starts at a depth of 86 meters below sea level and the peak of Mount Whitney is 4421 meters above sea level.

a. Find the angle of elevation to the peak of Mount Whitney if the horizontal distance to its base is 1200 meters.

b. If the angle of depression to the start of the race in Death Valley is 38°, what is the horizontal distance from a point at sea level?

20. **AMUSEMENT PARKS** India, Enrique, and Trina went to an amusement park while visiting Japan. They went on a Ferris wheel that was 100 meters in diameter and on an 80-meter cliff-dropping slide.

a. When Enrique and Trina are at the topmost point on the Ferris wheel shown below, how far are they from India?

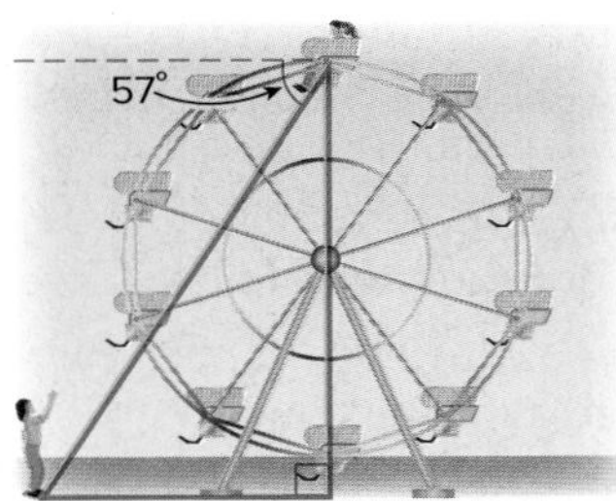

b. If the cliff-dropping ride has an angle of depression of 46°, how long is the slide?

21. **DARTS** Kelsey and José are throwing darts from a distance of 8.5 feet. The center of the bull's-eye on the dartboard is 5.7 feet from the floor. José throws from a height of 6 feet, and Kelsey throws from a height of 5 feet. What are the angles of elevation or depression from which each must throw to get a bull's-eye? Ignore other factors such as air resistance, velocity, and gravity.

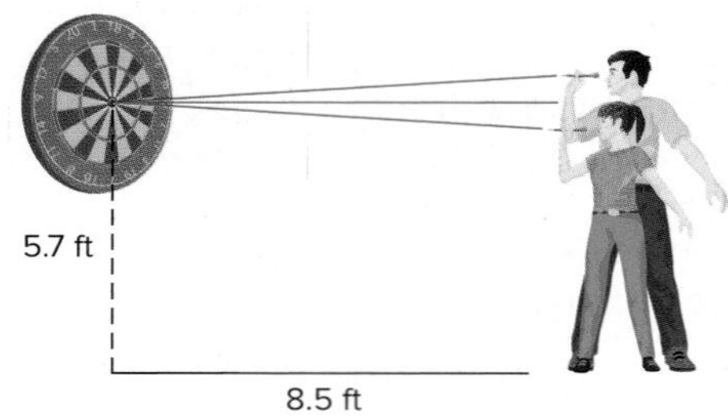

22. **MULTIPLE REPRESENTATIONS** In this problem, you will investigate relationships between the sides and angles of triangles.

a. **Geometric** Draw three triangles. Make one acute, one obtuse, and one right. Label one triangle ABC, a second MNP, and the third XYZ. Label the side lengths and angle measures of each triangle.

b. **Tabular** Copy and complete the table below.

Triangle	Ratios		
ABC	$\frac{\sin A}{BC} =$	$\frac{\sin B}{CA} =$	$\frac{\sin C}{AB} =$
MNP	$\frac{\sin M}{NP} =$	$\frac{\sin N}{PM} =$	$\frac{\sin P}{MN} =$
XYZ	$\frac{\sin X}{YZ} =$	$\frac{\sin Y}{ZX} =$	$\frac{\sin Z}{XY} =$

c. **Verbal** Make a conjecture about the ratio of the sine of an angle to the length of the leg opposite that angle for a given triangle.

H.O.T. Problems Use Higher-Order Thinking Skills

23. **ERROR ANALYSIS** Terrence and Rodrigo are trying to determine the relationship between angles of elevation and depression. Terrence says that if you are looking up at someone with an angle of elevation of 35°, then they are looking down at you with an angle of depression of 55°, which is the complement of 35°. Rodrigo disagrees and says that the other person would be looking down at you with an angle of depression equal to your angle of elevation, or 35°. Is either of them correct? Explain.

24. **CHALLENGE** Find the value of x. Round to the nearest tenth.

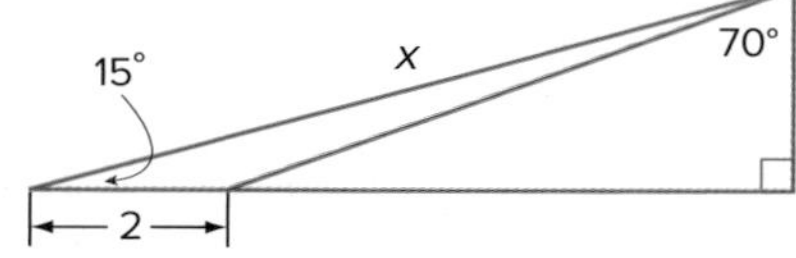

25. **MP** **REASONING** Classify the statement below as *true* or *false*. Explain.

As a person moves closer to an object he or she is sighting, the angle of elevation increases.

26. **OPEN-ENDED** A classmate finds the angle of elevation of an object, but she is trying to find the angle of depression. Write a question to help her solve the problem.

27. **WRITING IN MATH** Describe a way that you can estimate the height of an object without using trigonometry by choosing your angle of elevation. Explain your reasoning.

28. Leon wants to estimate the height of a building. Leon's eyes are 5.5 feet above ground. He stands 20 feet from the building and sights the top of the building, as shown in the figure. What is the building's height to the nearest tenth of a foot? MP 4

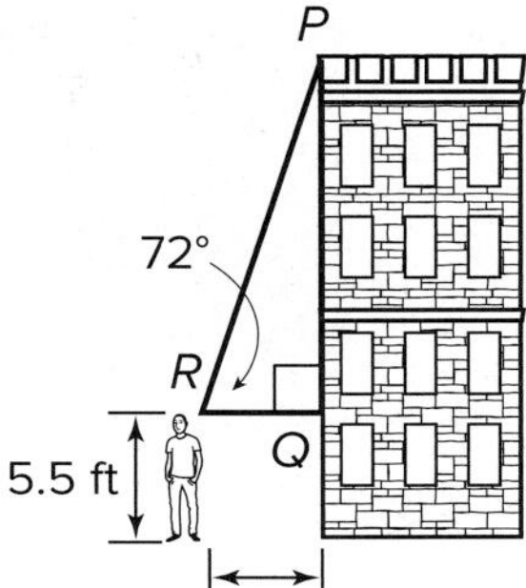

- A 11.8 ft
- B 24.5 ft
- C 61.6 ft
- D 67.1 ft

29. A tree is 70 feet tall. The angle of elevation of the Sun is 60°. Find the exact length of the tree's shadow. MP 4

30. The pilot of an airplane finds the angle of depression of an airport to be 16°. If the horizontal distance to the airport is 60,000 feet, find the altitude of the plane. MP 4

31. A passenger in a hot-air balloon spots a small fire on the ground. The angle of depression to the fire is 30°, and the height of the hot-air balloon is 150 feet. To the nearest foot, what is the horizontal distance from the hot-air balloon to the fire? MP 4

- A 75 ft
- B 87 ft
- C 130 ft
- D 260 ft
- E 300 ft

32. A searchlight is 6500 feet from a weather station. If the angle of elevation to the spot of light on the clouds above the station is 45°, how high is the cloud ceiling in feet? MP 4

33. The angle of depression from the top of a skyscraper to a fountain on the ground is 81°. The skyscraper is 421 meters tall. What is the distance from the base of the skyscraper to the fountain, in meters, rounded to the nearest tenth? MP 4

34. Ariela is standing at the top of a tower that is 50 feet tall. She spots a helicopter in the distance. The angle of elevation from Ariela to the helicopter is 21°, as shown. MP 4

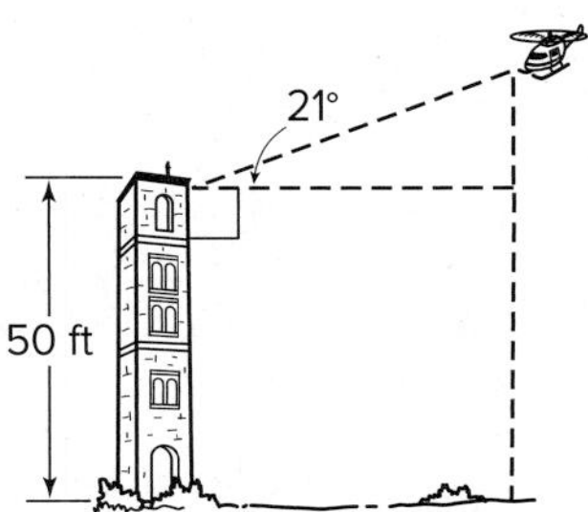

Which of the following best describes what Ariela should do to find the height of the helicopter?

- A Solve $\tan 21° = \frac{x}{50}$.
- B Solve $\tan 21° = \frac{50}{x}$.
- C Solve $\cos 21° = \frac{x}{50}$.
- D There is not enough information to find the height of the helicopter.

35. A ladder rests against a vertical wall. The base of the ladder is 28 ft from the wall. If the angle between the top of the ladder and the wall is 30°, find the length of the ladder. MP 4

36. **MULTI-STEP** A lamp post is anchored to the ground by a 400-ft wire. The angle the wire makes with the ground is θ, and the end of the wire is 200 ft from the base of the post. MP 4

a. Find θ.

The angle the wire makes with the ground is θ, and the end of the wire is x ft from the base of the post.

b. Express x in terms of a trigonometric function of θ.

LESSON 6

The Law of Sines

Then	Now	Why?
You found side lengths and angle measures of right triangles.	1 Find the area of a triangle using two sides and an included angle. 2 Use the Law of Sines to solve triangles.	Mars has hundreds of thousands of craters. These craters are named after famous scientists, science fiction authors, and towns on Earth. The craters named Wahoo, Wabash, and Naukan are shown in the figure. You can use trigonometry to find the distance between Wahoo and Naukan.

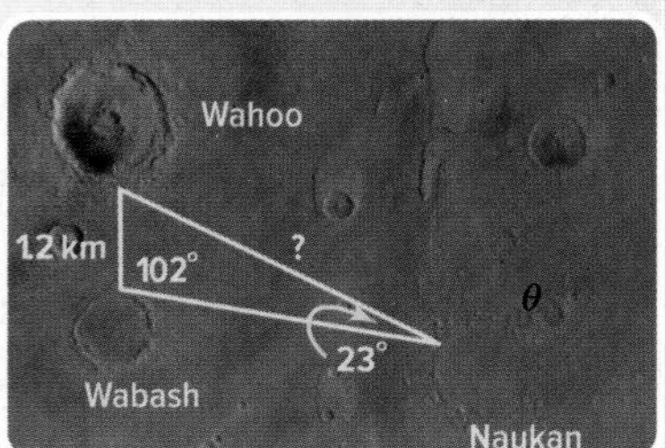

New Vocabulary

Law of Sines
solving a triangle
ambiguous case

MP Mathematical Practices

1 Make sense of problems and persevere in solving them.
4 Model with mathematics.
5 Use appropriate tools strategically.

1 Find the Area of a Triangle

In the triangle at the right, $\sin A = \frac{h}{c}$, or $h = c \sin A$.

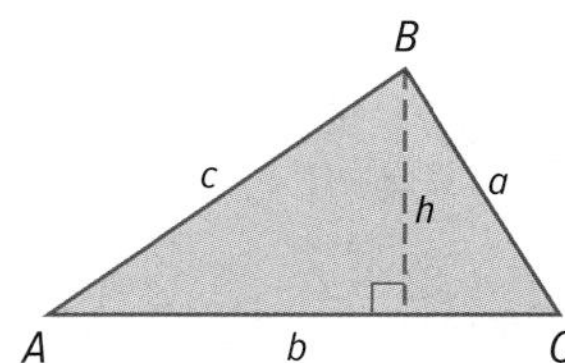

$\text{Area} = \frac{1}{2}bh$	Formula for area of a triangle
$\text{Area} = \frac{1}{2}b(c \sin A)$	Replace h with $c \sin A$.
$\text{Area} = \frac{1}{2}bc \sin A$	Simplify.

You can use this formula or two other formulas to find the area of a triangle if you know the lengths of two sides and the measure of the included angle.

Key Concept Area of a Triangle

Words The area of a triangle is one half the product of the lengths of two sides and the sine of their included angle.

Symbols $\text{Area} = \frac{1}{2}bc \sin A = \frac{1}{2}ac \sin B = \frac{1}{2}ab \sin C$

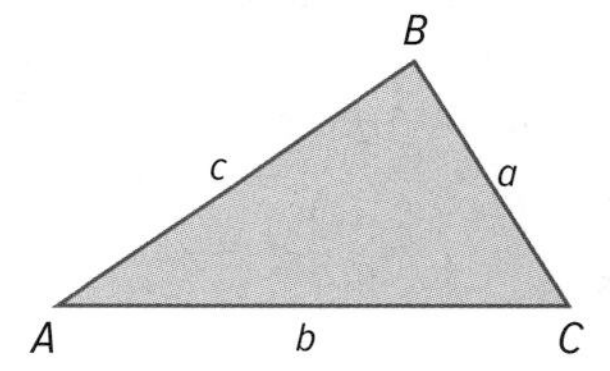

Example 1 Find the Area of a Triangle

Find the area of $\triangle ABC$ to the nearest tenth.

In $\triangle ABC$, $a = 8$, $b = 9$, and $C = 104°$.

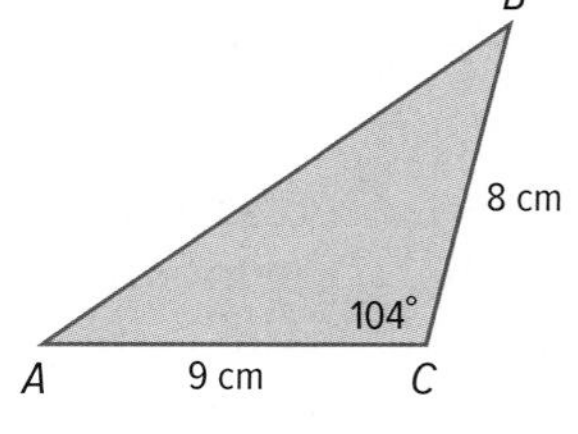

$\text{Area} = \frac{1}{2}ab \sin C$	Based on the known measures, use the third area formula.
$= \frac{1}{2}(8)(9) \sin 104°$	Substitution
$\approx 34.9 \text{ cm}^2$	Simplify.

MENTAL CHECK Round the sin 104° to sin 90° because the sin of 90° is 1.

$\frac{1}{2}(8)(9)\sin 90° = \frac{1}{2}(8)(9)(1) = 36$

This is close to the answer of 34.9 square centimeters.

Guided Practice

1. Find the area of $\triangle ABC$ to the nearest tenth if $A = 31°$, $b = 18$ meters, and $c = 22$ meters.

2 Use the Law of Sines to Solve Triangles

You can use the area formulas to derive the **Law of Sines**, which shows the relationships between side lengths of a triangle and the sines of the angles opposite them.

$\frac{1}{2}bc \sin A = \frac{1}{2}ac \sin B = \frac{1}{2}ab \sin C$	Set the area formulas equal to each other.
$bc \sin A = ac \sin B = ab \sin C$	Multiply each expression by 2.
$\frac{bc \sin A}{abc} = \frac{ac \sin B}{abc} = \frac{ab \sin C}{abc}$	Divide each expression by *abc*.
$\frac{\sin A}{a} = \frac{\sin B}{b} = \frac{\sin C}{c}$	Simplify.

Go Online! 

You have used the trigonometric ratios to find missing parts of right triangles. Explore how to find missing parts of oblique triangles using **The Geometer's Sketchpad®** activity in ConnectED.

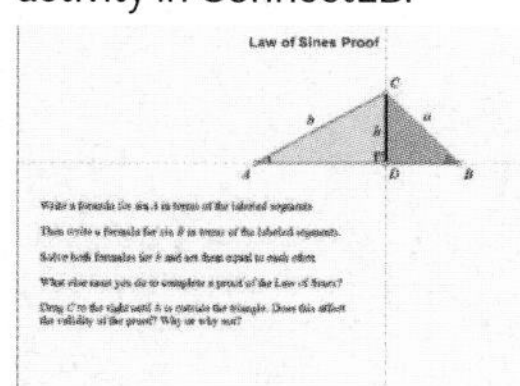

Key Concept Law of Sines

In $\triangle ABC$, if sides with lengths *a*, *b*, and *c* have opposite angles with measures *A*, *B*, and *C*, respectively, then the following is true.

$$\frac{\sin A}{a} = \frac{\sin B}{b} = \frac{\sin C}{c}$$

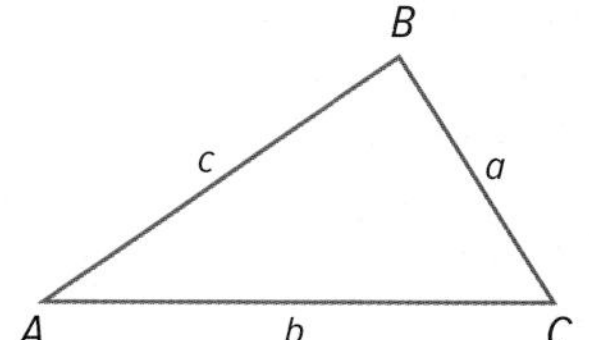

You can use the Law of Sines to solve a triangle if you know either of the following.

- the measures of two angles and any side (AAS or ASA)

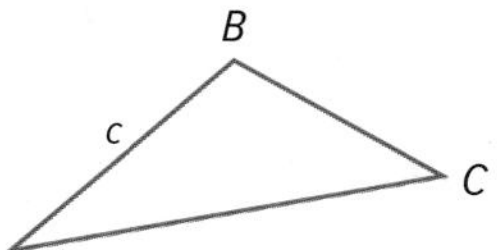

- the measures of two sides and the angle opposite one of the sides (SSA)

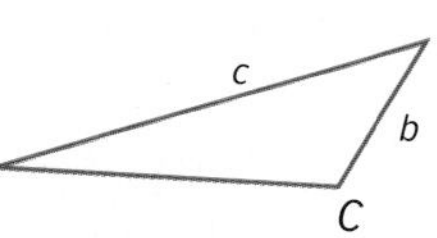

Using given measures to find all unknown side lengths and angle measures of a triangle is called **solving a triangle**.

Study Tip

The Law of Sines may also be written as $\frac{a}{\sin A} = \frac{b}{\sin B} = \frac{c}{\sin C}$. So, the expressions below could also be used to solve the triangle in Example 2.

- $\frac{a}{\sin 55°} = \frac{3}{\sin 80°}$
- $\frac{b}{\sin 45°} = \frac{3}{\sin 80°}$

Example 2 Solve a Triangle Given Two Angles and a Side

Solve $\triangle ABC$. Round to the nearest tenth if necessary.

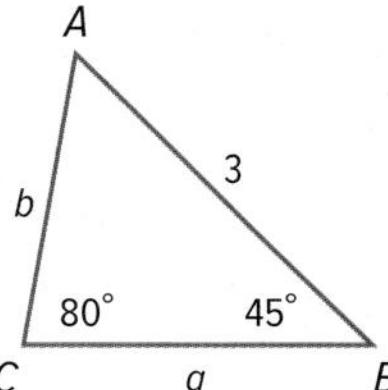

Step 1 Find the measure of the third angle.
$m\angle A = 180 - (80 + 45)$ or $55°$

Step 2 Use the Law of Sines to find side lengths *a* and *b*. Write an equation to find each variable.

$\frac{\sin A}{a} = \frac{\sin C}{c}$	Law of Sines	$\frac{\sin B}{b} = \frac{\sin C}{c}$
$\frac{\sin 55°}{a} = \frac{\sin 80°}{3}$	Substitution	$\frac{\sin 45°}{b} = \frac{\sin 80°}{3}$
$a = \frac{3 \sin 55°}{\sin 80°}$	Solve for each variable.	$b = \frac{3 \sin 45°}{\sin 80°}$
$a \approx 2.5$	Use a calculator.	$b \approx 2.2$

So, $A = 55°$, $a \approx 2.5$, and $b \approx 2.2$.

Guided Practice

2. Solve $\triangle NPQ$ if $P = 42°$, $Q = 65°$, and $n = 5$.

Study Tip

Two Solutions A situation in which two solutions for a triangle exist is called the *ambiguous case*.

If you are given the measures of two angles and a side, exactly one triangle is possible. However, if you are given the measures of two sides and the angle opposite one of them, zero, one, or two triangles may be possible. This is known as the **ambiguous case**. So, when solving a triangle using the SSA case, zero, one, or two solutions are possible.

Study Tip

A is Acute In the figures at the right, the altitude h is compared to a because h is the minimum distance from C to $\overline{AB}$ when A is acute.

$\sin A = \frac{\text{opp}}{\text{hyp}}$

$\sin A = \frac{h}{b}$

Key Concept Possible Triangles in SSA Case

Consider a triangle in which a, b, and $m\angle A$ are given.

$\angle A$ is Acute.

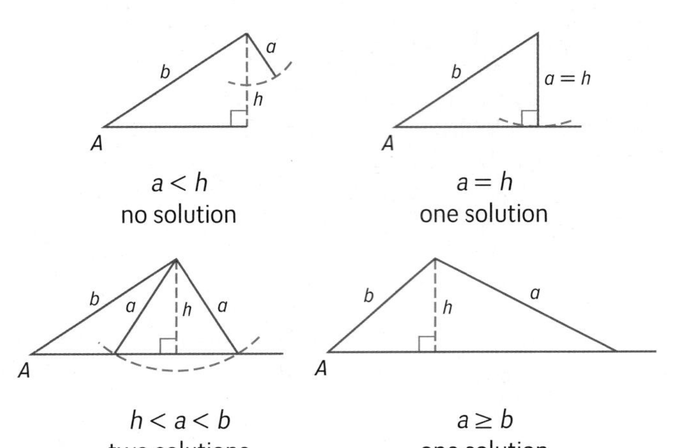

$\angle A$ is Right or Obtuse.

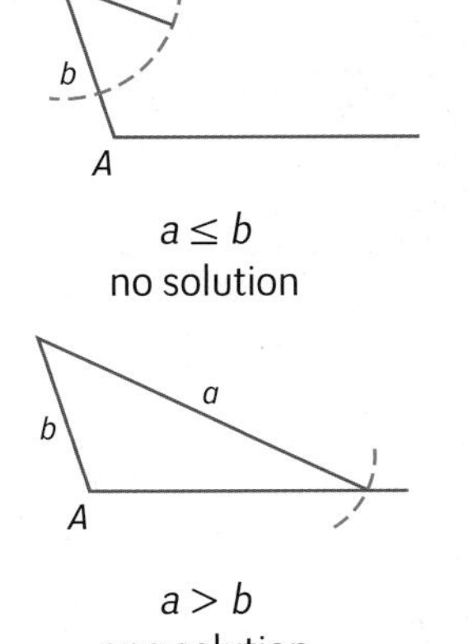

Because $\sin A = \frac{h}{b}$, you can use $h = b \sin A$ to find h in acute triangles.

Example 3 Solve a Triangle Given Two Sides and an Angle

Determine whether each triangle has *no* solution, *one* solution, or *two* solutions. Then solve the triangle. Round side lengths to the nearest tenth and angle measures to the nearest degree.

a. In $\triangle RST$, $R = 105°$, $r = 9$, and $s = 6$.

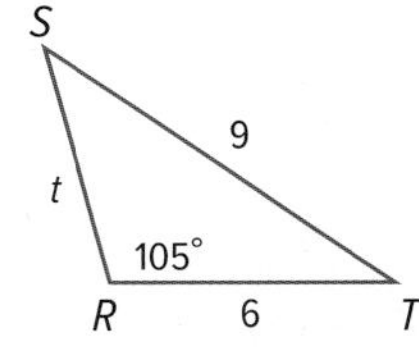

Because $\angle R$ is obtuse and $9 > 6$, you know that one solution exists.

Step 1 Use the Law of Sines to find $m\angle S$.

$\frac{\sin S}{6} = \frac{\sin 105°}{9}$ Law of Sines

$\sin S = \frac{6 \sin 105°}{9}$ Multiply each side by 6.

$\sin S \approx 0.6440$ Use a calculator.

$S \approx 40°$ Use the $\sin^{-1}$ function.

Step 2 Find $m\angle T$.

$m\angle T \approx 180 - (105 + 40)$ or $35°$

Step 3 Use the Law of Sines to find t.

$\frac{\sin 35°}{t} \approx \frac{\sin 105°}{9}$ Law of Sines

$t \approx \frac{9 \sin 35°}{\sin 105°}$ Solve for t.

$t \approx 5.3$ Use a calculator.

So, $S \approx 40°$, $T \approx 35°$, and $t \approx 5.3$.

b. In $\triangle ABC$, $A = 54°$, $a = 6$, and $b = 8$.

Because $\angle A$ is acute and $6 < 8$, find h and compare it to a.

$b \sin A = 8 \sin 54°$ — $b = 8$ and $A = 54°$

≈ 6.5 — Use a calculator.

Because $6 < 6.5$ or $a < h$, there is no solution.

c. In $\triangle ABC$, $A = 35°$, $a = 17$, and $b = 20$.

Because $\angle A$ is acute and $17 < 20$, find h and compare it to a.

$b \sin A = 20 \sin 35°$ — $b = 20$ and $A = 35°$

≈ 11.5 — Use a calculator.

Because $11.5 < 17 < 20$ or $h < a < b$, there are two solutions. So, there are two triangles to be solved

Case 1 $\angle B$ **is acute.**

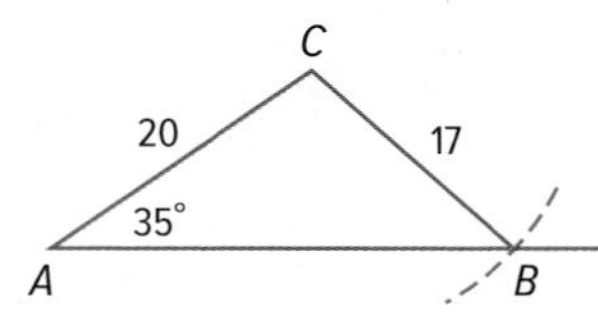

Step 1 Find $m\angle B$.

$\frac{\sin B}{20} = \frac{\sin 35°}{17}$ — Law of Sines

$\sin B = \frac{20 \sin 35°}{17}$ — Solve for sin B.

$\sin B \approx 0.6748$ — Use a calculator.

$B \approx 42°$ — Find $\sin^{-1} 0.6748$.

Step 2 Find $m\angle C$.

$m\angle C \approx 180 - (35 + 42)$ or $103°$

Step 3 Find c.

$\frac{\sin 103°}{c} = \frac{\sin 35°}{17}$ — Law of Sines

$c = \frac{17 \sin 103°}{\sin 35°}$ — Solve for c.

$c \approx 28.9$ — Simplify.

Case 2 $\angle B$ **is obtuse.**

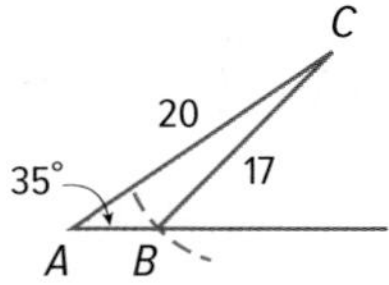

Step 1 Find $m\angle B$.

The sine function also has a positive value in Quadrant II. So, find an obtuse angle B for which $\sin B \approx 0.6748$.

$m\angle B \approx 180° - 42°$ or $138°$

Step 2 Find $m\angle C$.

$m\angle C \approx 180 - (35 + 138)$ or $7°$

Step 3 Find c.

$\frac{\sin 7°}{c} \approx \frac{\sin 35°}{17}$ — Law of Sines

$c \approx \frac{17 \sin 7°}{\sin 35°}$ — Solve for c.

$c \approx 3.6$ — Simplify.

So, one solution is $B \approx 42°$, $C \approx 103°$, and $c \approx 28.9$, and another solution is $B \approx 138°$, $C \approx 7°$, and $c \approx 3.6$.

> **Study Tip**
>
> **Reference Angle** In the triangle in Case 2, you are using the reference angle 42° to find the other value of B.

Guided Practice

Determine whether each triangle has *no* solution, *one* solution, or *two* solutions. Then solve the triangle. Round side lengths to the nearest tenth and angle measures to the nearest degree.

3A. In $\triangle RST$, $R = 95°$, $r = 10$, and $s = 12$.

3B. In $\triangle MNP$, $N = 32°$, $n = 7$, and $p = 4$.

3C. In $\triangle ABC$, $A = 47°$, $a = 15$, and $b = 18$.

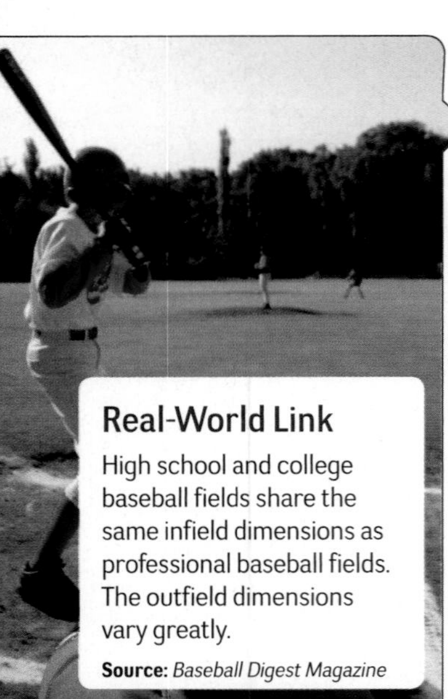

Real-World Link

High school and college baseball fields share the same infield dimensions as professional baseball fields. The outfield dimensions vary greatly.

Source: *Baseball Digest Magazine*

Real-World Example 4 Use the Law of Sines to Solve a Problem

BASEBALL A baseball is hit between the second and third bases and is caught at point *B*, as shown in the figure. How far away from second base was the ball caught?

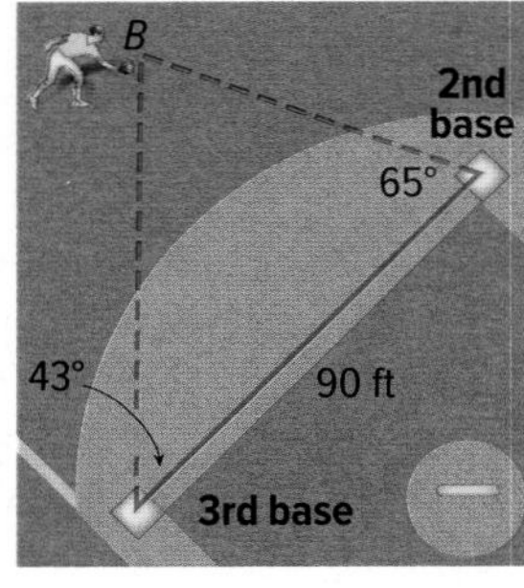

$\frac{\sin 72^\circ}{90} = \frac{\sin 43^\circ}{x}$ Law of Sines

$x \sin 72^\circ = 90 \sin 43^\circ$ Cross products

$x = \frac{90 \sin 43^\circ}{\sin 72^\circ}$ Solve for *x*.

$x \approx 64.5$ Use a calculator.

So, the distance is about 64.5 feet.

Guided Practice

4. How far away from third base was the ball caught?

Check Your Understanding

 = Step-by-Step Solutions begin on page R13.

Example 1 **Find the area of $\triangle ABC$ to the nearest tenth, if necessary.**

1. 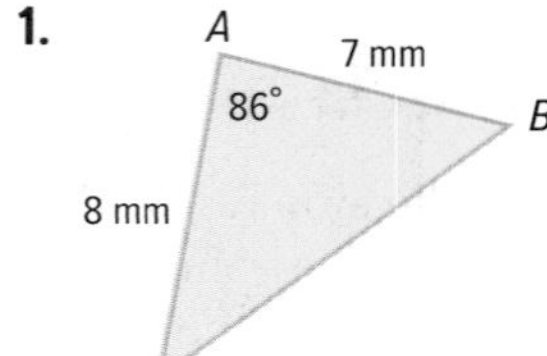

2. 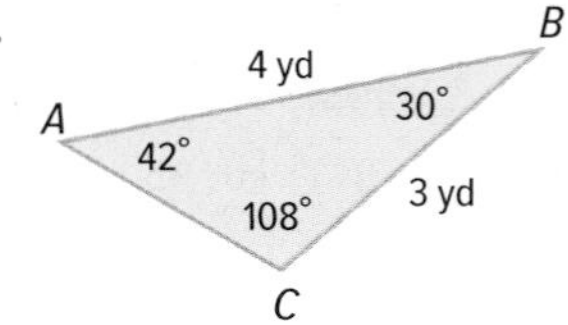

3 $A = 40^\circ$, $b = 11$ cm, $c = 6$ cm

4. $B = 103^\circ$, $a = 20$ in., $c = 18$ in.

Example 2 **Solve each triangle. Round side lengths to the nearest tenth and angle measures to the nearest degree.**

5. 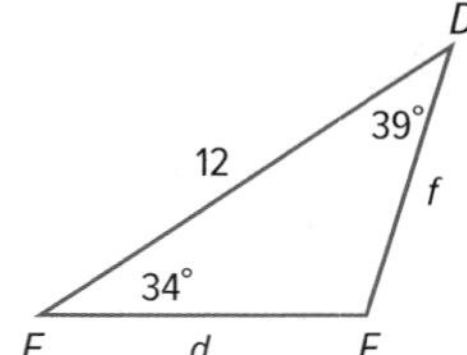

6. 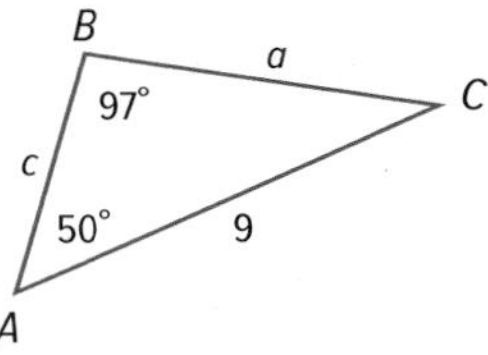

7. Solve $\triangle FGH$ if $G = 80^\circ$, $H = 40^\circ$, and $g = 14$.

Example 3 **MP PERSEVERANCE Determine whether each $\triangle ABC$ has *no* solution, *one* solution, or *two* solutions. Then solve the triangle. Round side lengths to the nearest tenth and angle measures to the nearest degree.**

8. $A = 95^\circ$, $a = 19$, $b = 12$

9. $A = 60^\circ$, $a = 15$, $b = 24$

10. $A = 34^\circ$, $a = 8$, $b = 13$

11. $A = 30^\circ$, $a = 3$, $b = 6$

Example 4 **12. SPACE** Refer to the beginning of the lesson. Find the distance between the Wahoo Crater and the Naukan Crater on Mars.

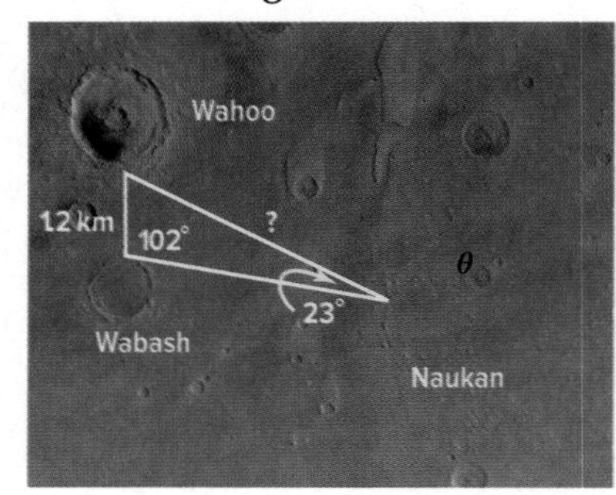

Practice and Problem Solving

Extra Practice is on page R8.

Example 1 **Find the area of $\triangle ABC$ to the nearest tenth.**

13. 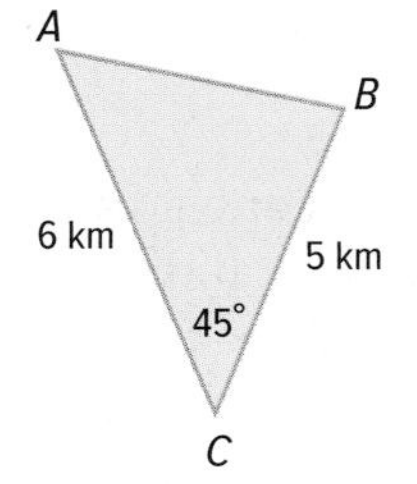

14. 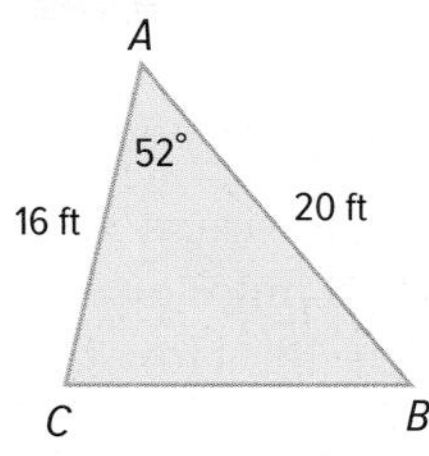

15. 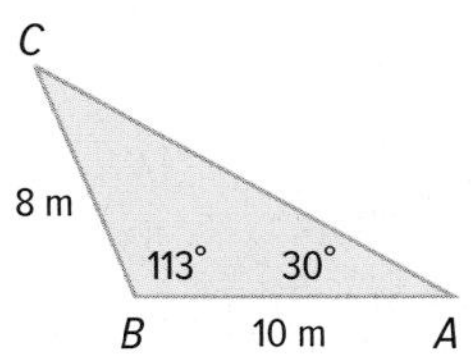

16. 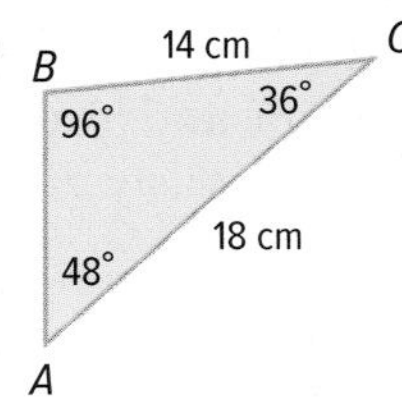

17. $C = 25°$, $a = 4$ ft, $b = 7$ ft
18. $A = 138°$, $b = 10$ in., $c = 20$ in.
19. $B = 92°$, $a = 14.5$ m, $c = 9$ m
20. $C = 116°$, $a = 2.7$ cm, $b = 4.6$ cm

Example 2 **MP REASONING Solve each triangle. Round side lengths to the nearest tenth and angle measures to the nearest degree.**

21. 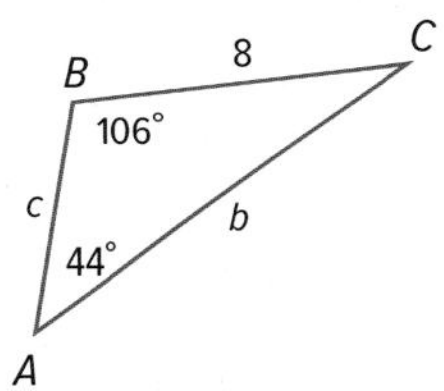

22. 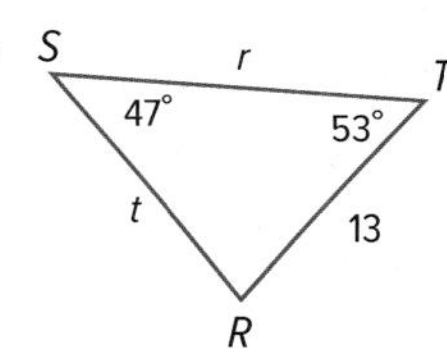

23. 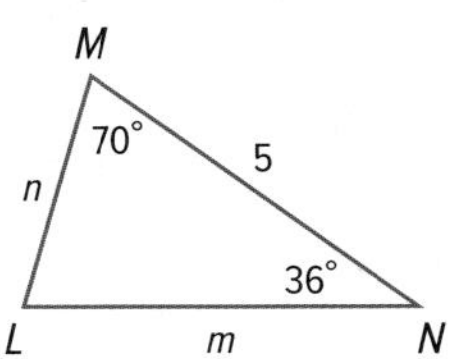

24. 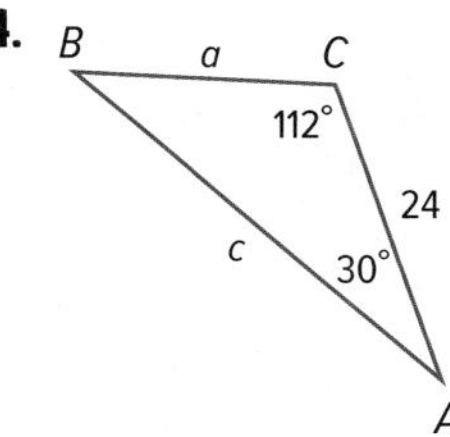

25. Solve $\triangle HJK$ if $H = 53°$, $J = 20°$, and $h = 31$.
26. Solve $\triangle NPQ$ if $P = 109°$, $Q = 57°$, and $n = 22$.
27. Solve $\triangle ABC$ if $A = 50°$, $a = 2.5$, and $C = 67°$.
28. Solve $\triangle ABC$ if $B = 18°$, $C = 142°$, and $b = 20$.

Example 3 **Determine whether each $\triangle ABC$ has *no* solution, *one* solution, or *two* solutions. Then solve the triangle. Round side lengths to the nearest tenth and angle measures to the nearest degree.**

29. $A = 100°$, $a = 7$, $b = 3$
30. $A = 75°$, $a = 14$, $b = 11$
31. $A = 38°$, $a = 21$, $b = 18$
32. $A = 52°$, $a = 9$, $b = 20$
33. $A = 42°$, $a = 5$, $b = 6$
34. $A = 44°$, $a = 14$, $b = 19$
35. $A = 131°$, $a = 15$, $b = 32$
36. $A = 30°$, $a = 17$, $b = 34$

Example 4 **GEOGRAPHY In Hawaii, the distance from Hilo to Kailua is 57 miles, and the distance from Hilo to Captain Cook is 55 miles.**

Kailua 57 mi Hilo
80° 55 mi
Captain Cook

37. What is the measure of the angle formed at Hilo?

38. What is the distance between Kailua and Captain Cook?

39. **TORNADOES** Tornado sirens A, B, and C form a triangular region in one area of a city. Sirens A and B are 8 miles apart. The angle formed at siren A is 112°, and the angle formed at siren B is 40°. How far apart are sirens B and C?

40. **MYSTERIES** The Bermuda Triangle is a region of the Atlantic Ocean between Bermuda, Miami, Florida, and San Juan, Puerto Rico. It is an area where ships and airplanes have been rumored to mysteriously disappear.

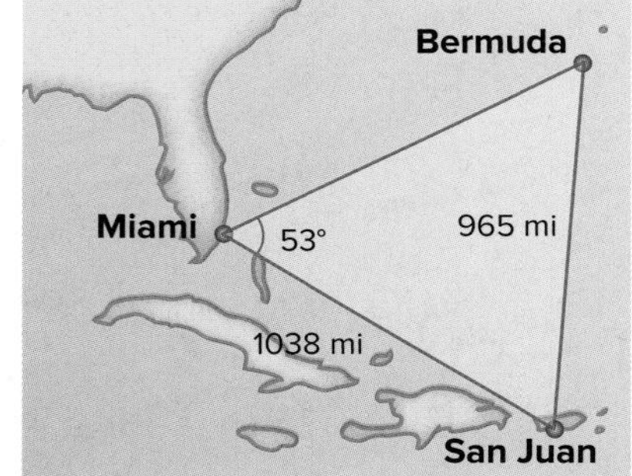

   a. What is the distance between Miami and Bermuda?

   b. What is the approximate area of the Bermuda Triangle?

41. **BICYCLING** One side of a triangular cycling path is 4 miles long. The angle opposite this side is 64°. Another angle formed by the triangular path measures 66°.

   a. Sketch a drawing of the situation. Label the missing sides a and b.

   b. Write equations that could be used to find the lengths of the missing sides.

   c. What is the perimeter of the path?

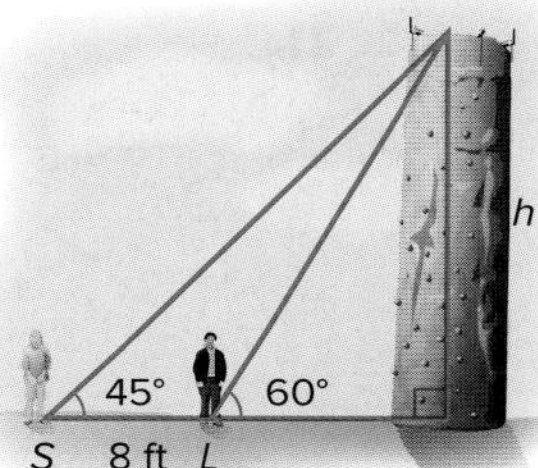

42. **ROCK CLIMBING** Savannah S and Leon L are standing 8 feet apart in front of a rock climbing wall, as shown at the right. What is the height of the wall? Round to the nearest tenth.

H.O.T. Problems Use Higher-Order Thinking Skills

43. **ERROR ANALYSIS** In $\triangle RST$, $R = 56°$, $r = 24$, and $t = 12$. Cameron and Gabriela are using the Law of Sines to find T. Is either of them correct? Explain your reasoning.

Cameron

$\frac{\sin T}{12} = \frac{\sin 56°}{24}$

$\sin T \approx 0.4145$

$T \approx 24.5°$

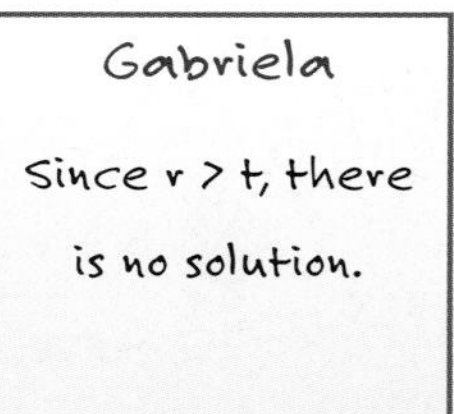

44. **OPEN ENDED** Create an application problem involving right triangles and the Law of Sines. Then solve your problem, drawing diagrams if necessary.

45. **CHALLENGE** Using the figure at the right, derive the formula Area $= \frac{1}{2}bc \sin A$.

46. **REASONING** Find the side lengths of two different triangles ABC that can be formed if $A = 55°$ and $C = 20°$.

47. **WRITING IN MATH** Use the Law of Sines to explain why a and b do not have unique values in the figure shown.

48. **OPEN ENDED** Given that $E = 62°$ and $d = 38$, find a value for e such that no triangle DEF can exist. Explain your reasoning.

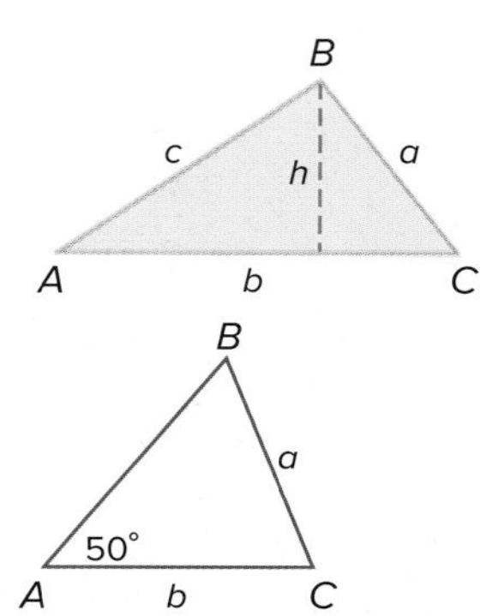

49. In $\triangle XYZ$, $XZ = 2$ centimeters, $XY = 6$ centimeters, and $\angle X = 70°$. What is the area of $\triangle XYZ$ to the nearest square centimeter? MP 1

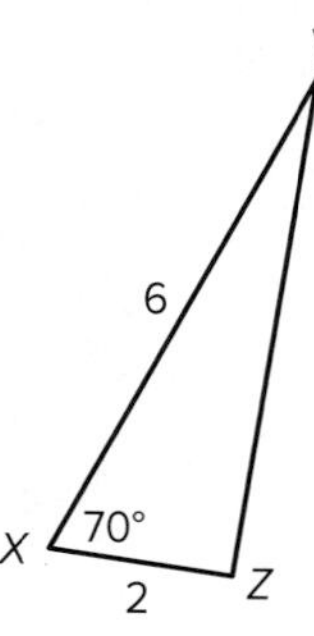

50. In scalene triangle $\triangle QRS$, angles Q, R, and S measure $x°$, $2x°$, and $3x°$, respectively. What is the ratio of side length s to side length r? MP 1

- ○ **A** $\frac{\sin 60°}{\sin 90°}$
- ○ **B** $\frac{\sin 90°}{\sin 60°}$
- ○ **C** $\frac{\sin 90°}{\sin 30°}$
- ○ **D** $\frac{\sin 30°}{\sin 60°}$

51. In $\triangle ABC$, $m\angle B = 70°$, $m\angle C = 42°$, and $c = 22$. Which expressions represent the distances a and b? Select all that apply. MP 1

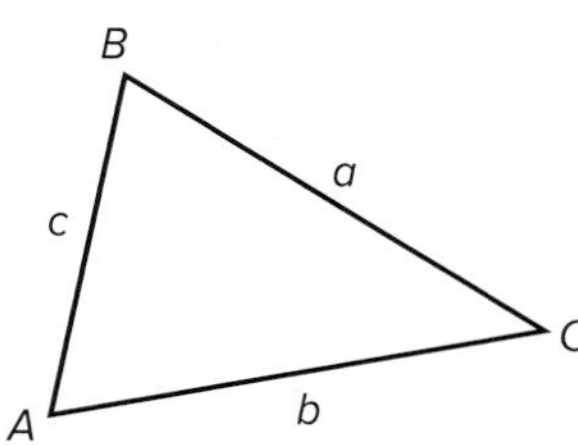

- ☐ **A** $\frac{22 \sin 42°}{\sin 70°}$
- ☐ **B** $\frac{22 \sin 70°}{\sin 68°}$
- ☐ **C** $\frac{22 \sin 68°}{\sin 42°}$
- ☐ **D** $\frac{22 \sin 68°}{\sin 70°}$
- ☐ **E** $\frac{22 \sin 42°}{\sin 68°}$
- ☐ **F** $\frac{22 \sin 70°}{\sin 42°}$

52. The diagram below shows the distance of two boats from shore. MP 1, 4

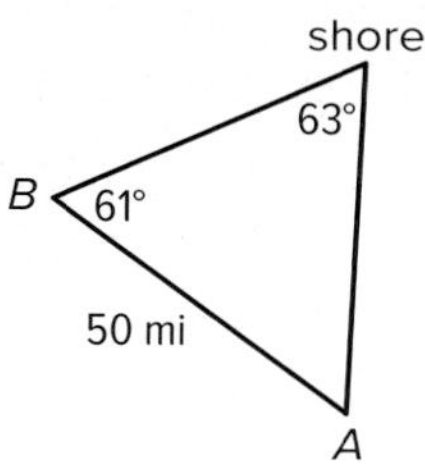

a. How far is boat A from the shore to the nearest tenth of a mile?

- ○ **A** 46.5 mi
- ○ **B** 49.1 mi
- ○ **C** 50.9 mi
- ○ **D** 53.7 mi

b. How far is boat B from the shore to the nearest tenth of a mile?

- ○ **A** 46.5 mi
- ○ **B** 49.1 mi
- ○ **C** 50.9 mi
- ○ **D** 53.7 mi

53. **MULTI-STEP** Juan looks up at the top of a building at an angle of 49°. George, who is 100 feet behind Juan, looks up at the top of a building at an angle of 37°. MP 1, 4

a. Draw a diagram of the scenario.

b. How far is George from the building to the nearest tenth of a foot?

c. How tall is the building to the nearest tenth of a foot?

54. The area of an acute triangle ABC is $20\sqrt{3}$. The length of $\overline{AB}$ is 8 and the length of $\overline{BC}$ is 10. MP 1

a. Which angle is determined by this information?

b. What is the measure of this angle?

LESSON 7
The Law of Cosines

::Then

- You solved triangles by using the Law of Sines.

::Now

1. Use the Law of Cosines to solve triangles.
2. Choose methods to solve triangles.

::Why?

- You can use trigonometry to find the distance from a ship used to lower a submersible into the ocean and a shipwreck spotted by the submersible on the ocean floor.

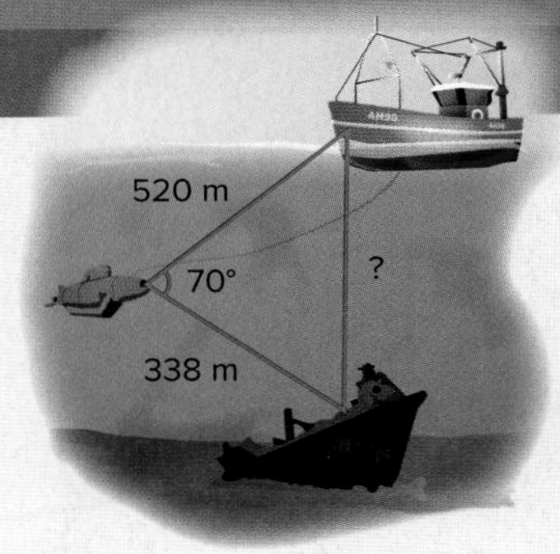

New Vocabulary
Law of Cosines

Mathematical Practices

1 Make sense of problems and persevere in solving them.
4 Model with mathematics.
5 Use appropriate tools strategically.
6 Attend to precision.

1 Use the Law of Cosines to Solve Triangles

You cannot use the Law of Sines to solve a triangle like the one shown above. You can use the **Law of Cosines** if:

- the measures of two sides and the included angle are known (SAS).
- the measures of three sides are known (SSS).

Key Concept Law of Cosines

In $\triangle ABC$, if sides with lengths a, b, and c are opposite angles with measures A, B, and C, respectively, then the following are true.

$$a^2 = b^2 + c^2 - 2bc \cos A$$
$$b^2 = a^2 + c^2 - 2ac \cos B$$
$$c^2 = a^2 + b^2 - 2ab \cos C$$

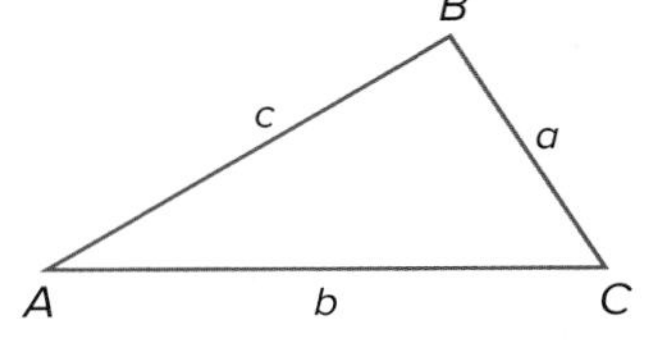

Example 1 Solve a Triangle Given Two Sides and the Included Angle

Solve $\triangle ABC$.

Step 1 Use the Law of Cosines to find the missing side length.

$b^2 = a^2 + c^2 - 2ac \cos B$ Law of Cosines
$b^2 = 7^2 + 5^2 - 2(7)(5) \cos 36°$ $a = 7, c = 5, B = 36°$
$b^2 \approx 17.4$ Use a calculator to simplify.
$b \approx 4.2$ Take the positive square root of each side.

Step 2 Use the Law of Sines to find a missing angle measure. Finding the smaller angle first can help avoid an error if the larger angle might be obtuse.

$\frac{\sin C}{5} \approx \frac{\sin 36°}{4.2}$ $\frac{\sin C}{c} = \frac{\sin B}{b}$
$\sin C \approx \frac{5 \sin 36°}{4.2}$ Multiply each side by 5.
$C \approx 44°$ Use the $\sin^{-1}$ function.

Step 3 Find the measure of the other angle. $m\angle A \approx 180° - (36° + 44°)$ or $100°$
So, $b \approx 4.2$, $A \approx 100°$, and $C \approx 44°$.

Guided Practice

1. Solve $\triangle FGH$ if $G = 82°$, $f = 6$, and $h = 4$.

Go Online! 

Watch the **Personal Tutor** describe how to solve trigonometry problems with a partner. Then try describing how to solve a problem for them. Have them ask questions to help your understanding.

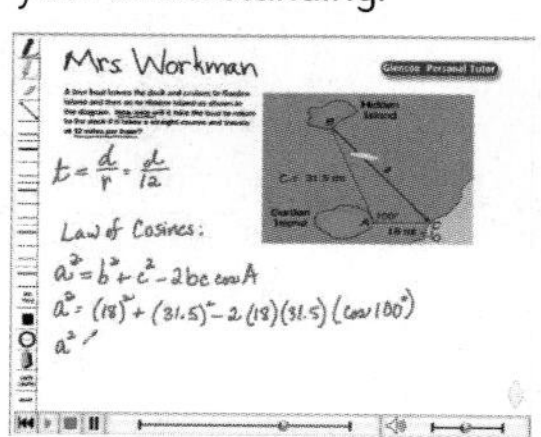

When you are only given the three side lengths of a triangle, you can solve it by using the Law of Cosines. The first step is to find the measure of the largest angle. This is done to ensure the other two angles are acute when using the Law of Sines.

Example 2 **Solve a Triangle Given Three Sides**

Solve △*ABC*.

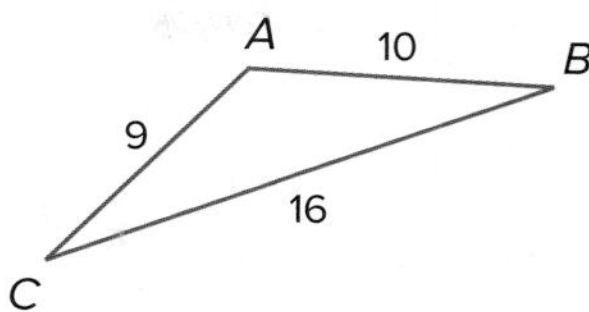

Step 1 Use the Law of Cosines to find the measure of the largest angle, $\angle A$.

$a^2 = b^2 + c^2 - 2bc \cos A$	Law of Cosines
$16^2 = 9^2 + 10^2 - 2(9)(10) \cos A$	$a = 16$, $b = 9$, and $c = 10$
$16^2 - 9^2 - 10^2 = -2(9)(10) \cos A$	Subtract 9^2 and 10^2 from each side.
$\frac{16^2 - 9^2 - 10^2}{-2(9)(10)} = \cos A$	Divide each side by $-2(9)(10)$.
$-0.4167 \approx \cos A$	Use a calculator to simplify.
$115° \approx A$	Use the $\cos^{-1}$ function.

Step 2 Use the Law of Sines to find the measure of $\angle B$.

$\frac{\sin B}{9} \approx \frac{\sin 115°}{16}$	$\frac{\sin B}{b} = \frac{\sin A}{a}$
$\sin B \approx \frac{9 \sin 115°}{16}$	Multiply each side by 9.
$\sin B \approx 0.5098$	Use a calculator.
$B \approx 31°$	Use the $\sin^{-1}$ function.

Step 3 Find the measure of $\angle C$.

$m\angle C \approx 180° - (115° + 31°)$ or about $34°$

So, $A \approx 115°$, $B \approx 31°$, and $C \approx 34°$.

Guided Practice

2. Solve △ABC if $a = 5$, $b = 11$, and $c = 8$.

Review Vocabulary

oblique a triangle that has no right angle

2 Choose a Method to Solve Triangles

You can use the Law of Sines and the Law of Cosines to solve problems involving oblique triangles. You need to know the measure of at least one side and any two other parts. If the triangle has a solution, you must decide whether to use the Law of Sines or the Law of Cosines to begin solving it.

Concept Summary **Solving Oblique Triangles**

Given	Begin by Using
two angles and any sides	Law of Sines
two sides and an angle opposite one of them	Law of Sines
two sides and their included angle	Law of Cosines
three sides	Law of Cosines

Real-World Example 3 Use the Law of Cosines

SCUBA DIVING A scuba diver looks up 20° and sees a turtle 9 feet away. She looks down 40° and sees a blue parrotfish 12 feet away. How far apart are the turtle and the blue parrotfish?

turtle B
scuba diver
A
9 ft
20°
40°
12 ft
a
blue parrotfish
C

Understand You know the angles formed by the scuba diver's line of sight and her distance from the sea creatures.

Plan Use the information to draw and label a diagram. Because two sides and the included angle of a triangle are given, you can use the Law of Cosines to solve the problem.

Solve

$a^2 = b^2 + c^2 - 2bc \cos A$	Law of Cosines
$a^2 = 12^2 + 9^2 - 2(12)(9) \cos 60$	$b = 12$, $c = 9$, and $A = 60$
$a^2 = 117$	Use a calculator.
$a \approx 10.8$	Find the positive value of a.

So, the turtle and the blue parrotfish are about 10.8 feet apart.

Check Using the Law of Sines, you can find that $B \approx 74°$ and $C \approx 46°$. Because $C < A < B$ and $c < a < b$, the solution is reasonable.

You can use the two small triangles to find the two parts of the total distance individually and then confirm that their sum is about 10.8.

Real-World Link

The record for the deepest seawater scuba dive was 1090 feet, made by a scuba diver in the Red Sea.

Source: *Guinness Book of World Records*

Guided Practice

3. **MARATHONS** Amelia ran 6 miles in one direction. She then turned 79° and ran 7 miles. At the end of the run, how far was Amelia from her starting point?

Check Your Understanding

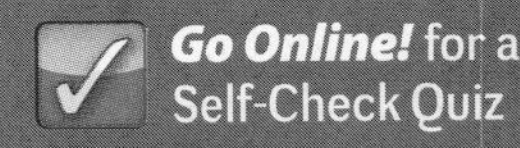

Go Online! for a Self-Check Quiz

Examples 1–2 **Solve each triangle. Round side lengths to the nearest tenth and angle measures to the nearest degree.**

1.

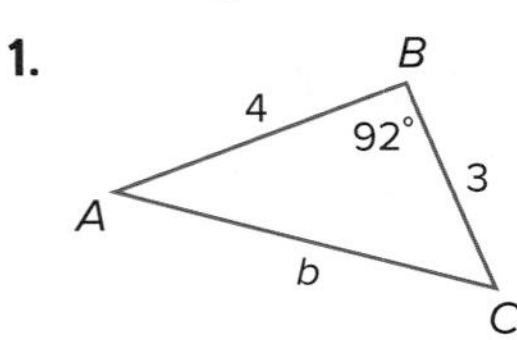

2.

A
10
B
14
20
C

3. $a = 5, b = 8, c = 12$

4. $B = 110°, a = 6, c = 3$

Example 3 **PRECISION Determine whether each triangle should be solved by beginning with the Law of *Sines* or the Law of *Cosines*. Then solve the triangle.**

5.

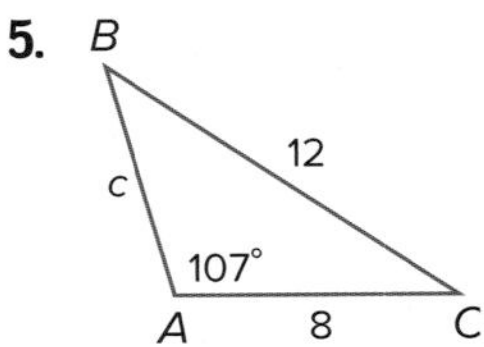

6.

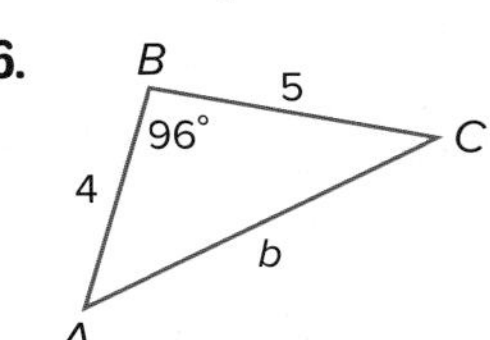

7. In $\triangle RST$, $R = 35°$, $s = 16$, and $t = 9$.

8. **FOOTBALL** In a football game, the quarterback is 20 yards from Receiver A. He turns 40° to see Receiver B, who is 16 yards away. How far apart are the two receivers?

Purestock/SuperStock

Practice and Problem Solving

Extra Practice is on page R8.

Examples 1–2 **Solve each triangle. Round side lengths to the nearest tenth and angle measures to the nearest degree.**

9.

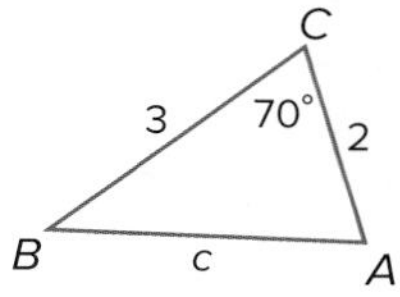

10.

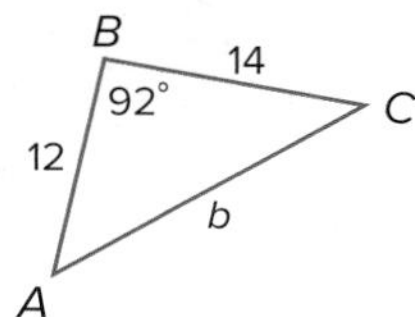

11.

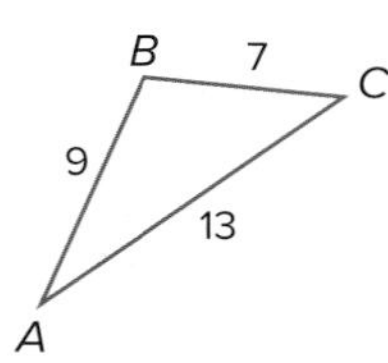

12.

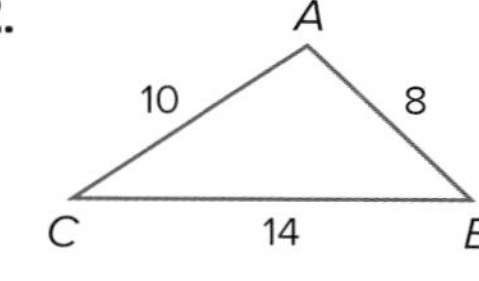

13. $A = 116°, b = 5, c = 3$

14. $C = 80°, a = 9, b = 2$

15. $f = 10, g = 11, h = 4$

16. $w = 20, x = 13, y = 12$

Example 3 **Determine whether each triangle should be solved by beginning with the Law of *Sines* or the Law of *Cosines*. Then solve the triangle.**

17.

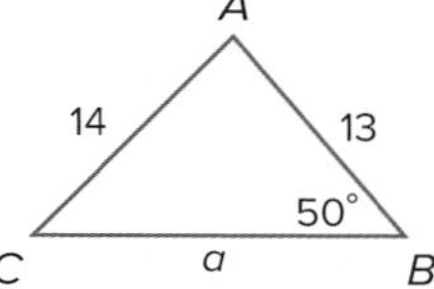

18.

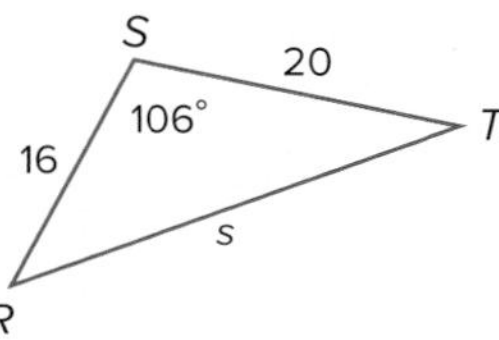

19.

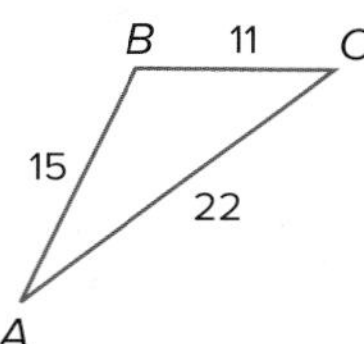

20.

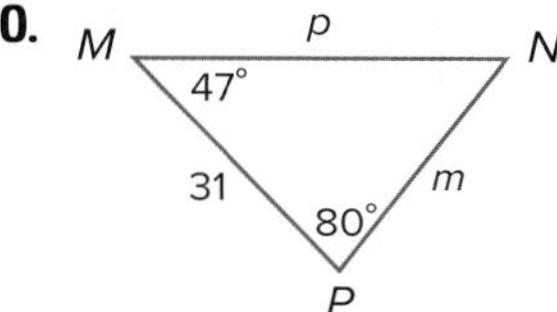

21. In $\triangle ABC$, $C = 84°$, $c = 7$, and $a = 2$.

22. In $\triangle HJK$, $h = 18$, $j = 10$, and $k = 23$.

23 **EXPLORATION** Find the distance between the ship and the shipwreck shown in the diagram. Round to the nearest tenth.

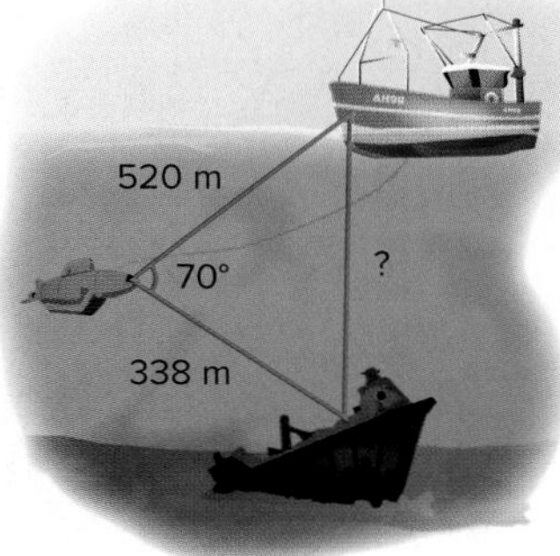

24. **GEOMETRY** A parallelogram has side lengths 8 centimeters and 12 centimeters. One angle between them measures 42°. To the nearest tenth, what is the length of the shorter diagonal?

25. **RACING** A triangular cross-country course has side lengths 1.8 kilometers, 2 kilometers, and 1.2 kilometers. What are the angles formed between each pair of sides?

26. MP **MODELING** A triangular plot of farmland measures 0.9 by 0.5 by 1.25 miles.

a. If the plot of land is fenced on the border, what will be the angles at which the fences of the three sides meet? Round to the nearest degree.

b. What is the area of the plot of land?

27. **LAND** Some land is in the shape of a triangle. The distances between each vertex of the triangle are 140 yards, 210 yards, and 300 yards, respectively. Use the Law of Cosines to find the area of the land to the nearest square yard.

28. **RIDES** Two bumper cars at an amusement park ride collide as shown below.

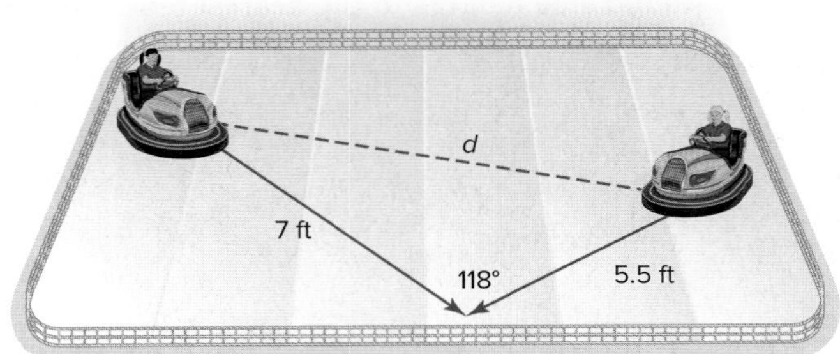

a. How far apart d were the two cars before they collided?

b. Before the collision, a third car was 10 feet from the blue car and 13 feet from the orange car. Describe the angles formed by the three cars before the collision.

29. **PICNICS** A triangular picnic area is 11 yards by 14 yards by 10 yards.

a. Sketch and label a drawing to represent the picnic area.

b. Describe how you could find the area of the picnic area.

c. What is the area? Round to the nearest tenth.

30. **WATERSPORTS** A person on a personal watercraft makes a trip from point A to point B to point C traveling 28 miles per hour. She then returns from point C back to her starting point traveling 35 miles per hour. How many minutes did the entire trip take? Round to the nearest tenth.

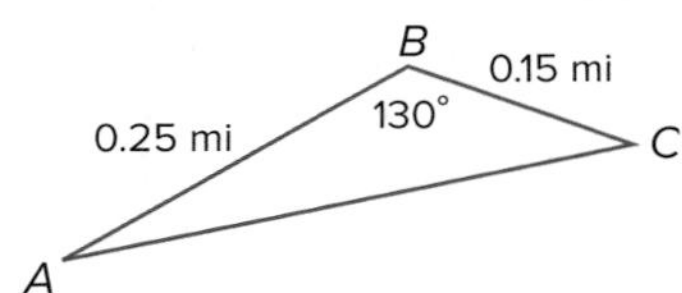

Solve each triangle. Round side lengths to the nearest tenth and angle measures to the nearest degree.

31.

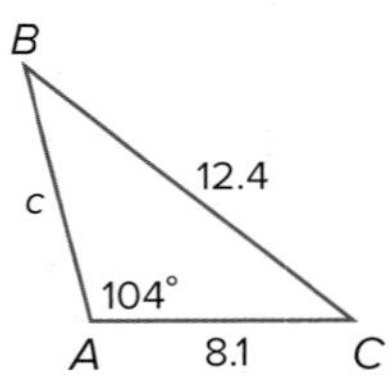

32.

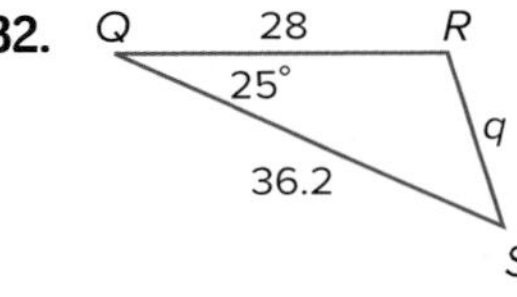

33.

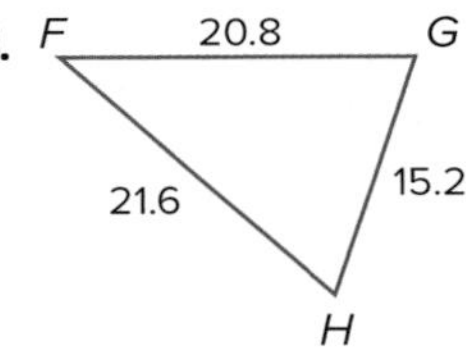

H.O.T. Problems Use Higher-Order Thinking Skills

34. **CHALLENGE** Use the figure and the Pythagorean Theorem to derive the Law of Cosines. Use the hints below.

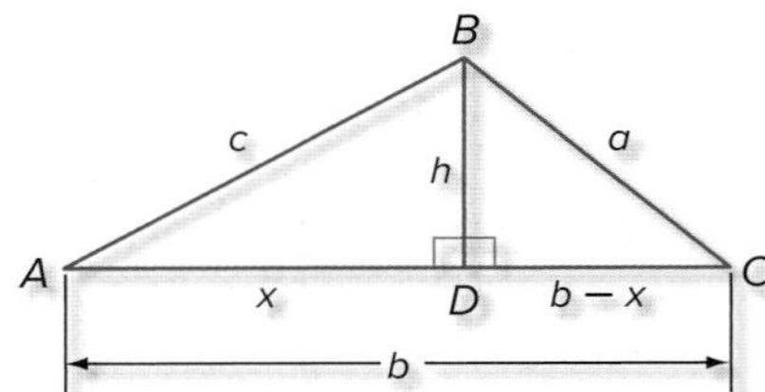

- First, use the Pythagorean Theorem for $\triangle DBC$.
- In $\triangle ADB$, $c^2 = x^2 + h^2$.
- $\cos A = \frac{x}{c}$

35. **MP CONSTRUCT ARGUMENTS** Three sides of a triangle measure 10.6 centimeters, 8 centimeters, and 14.5 centimeters. Explain how to find the measure of the largest angle. Then find the measure of the angle to the nearest degree.

36. **OPEN-ENDED** Create an application problem involving right triangles and the Law of Cosines. Then solve your problem, drawing diagrams if necessary.

37. **WRITING IN MATH** How do you know which method to use when solving a triangle?

38. A manufacturing company produces triangular-shaped support brackets. The side lengths of the brackets measure 6 feet, 8 feet, and 5 feet. What trigonometric relationship represents the measure of the largest interior angle, angle A, formed by the support brackets? MP 4

○ **A** $\cos A = -\frac{1}{20}$

○ **B** $\sin A = -\frac{1}{20}$

○ **C** $\sin A = -\frac{1}{2}$

○ **D** $\cos A = \frac{1}{20}$

39. In $\triangle ABC$, the given dimensions are in the diagram.

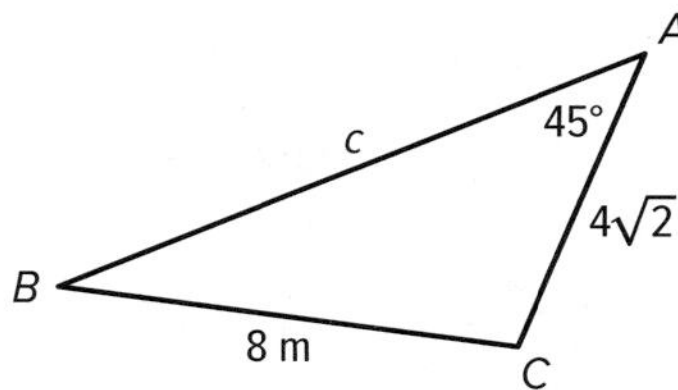

Which of the following trigonometric ratios in $\triangle ABC$ is equal to $\frac{1}{2}$? MP 1, 4

○ **A** $\sin B$

○ **B** $\cos B$

○ **C** $\sin C$

○ **D** $\cos C$

40. Select all of the correct expressions used in finding the missing lengths. MP 4, 5

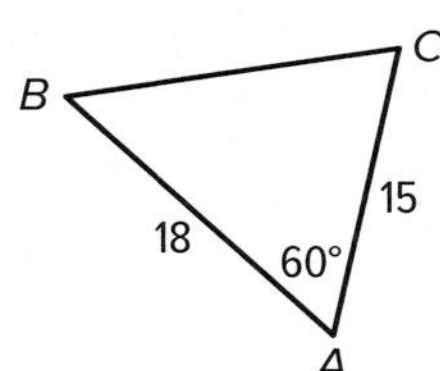

☐ **A** $a^2 = 15^2 + 15^2 - 2\,(15)\,(15)\cos 60°$

☐ **B** $a^2 = 15^2 + 18^2 - 2\,(15)\,(18)\cos 60°$

☐ **C** $C = \sin^{-1}\left(\frac{18 \sin 17°}{60}\right)$

☐ **D** $C = \sin^{-1}\left(\frac{18 \sin 60°}{17}\right)$

☐ **E** $B = \sin^{-1}\left(\frac{15 \sin 60°}{17}\right)$

41. A garden has sides of 8 meters, 11 meters, and 15 meters. What is the area of the garden to the nearest square meter? MP 1, 4

42. **MULTI-STEP** Carlotta paddled a kayak from a dock and traveled 4 miles east. She then steered the kayak 80° north of east and traveled another 2 miles before she anchored. MP 1, 4, 5

a. Draw a diagram of the scenario.

b. To the nearest tenth of a mile, find the distance from Carlotta's starting point to the point where she anchored the kayak.

c. What law did you use to find the distance? Explain your reasoning.

d. Find the measure of the two other angles of the triangle to the nearest degree.

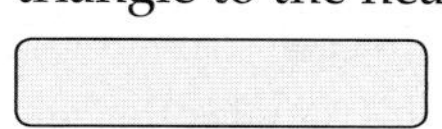

43. **MULTI-STEP** In $\triangle ABC$, $m\angle A = 50°$, $m\angle B = 35°$, and $a = 12$. MP 1

a. Find the measure of $\angle C$.

b. Find b to the nearest tenth.

c. Find c to the nearest tenth.

d. Find the perimeter of $\triangle ABC$

44. In an isosceles triangle, the vertex angle is 30° and the base is 12 cm long. Find the perimeter of the triangle to the nearest integer. MP 5

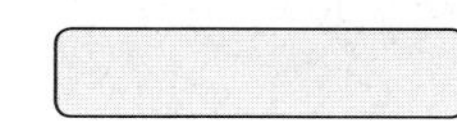

45. In a triangle, $\sin A$ is twice the value of $\sin B$. Given that $a = 4$, find the value of b. MP 5

Study Guide and Review

Go Online! for Vocabulary Review Games and key vocabulary in 13 languages

Study Guide

Key Concepts

Geometric Mean (Lesson 8-1)

- For two positive numbers a and b, the geometric mean is the positive number x where $a : x = x : b$ is true.

Pythagorean Theorem (Lesson 8-2)

- Let $\triangle ABC$ be a right triangle with right angle C. Then $a^2 + b^2 = c^2$.

Special Right Triangles (Lesson 8-3)

- The measures of the sides of a 45°-45°-90° triangle are x, x, and $x\sqrt{2}$.
- The measures of the sides of a 30°-60°-90° triangle are x, $2x$, and $x\sqrt{3}$.

Trigonometry (Lesson 8-4)

- $\sin A = \frac{\text{opposite leg}}{\text{hypotenuse}}$
- $\cos A = \frac{\text{adjacent leg}}{\text{hypotenuse}}$
- $\tan A = \frac{\text{opposite leg}}{\text{adjacent leg}}$

Angles of Elevation and Depression (Lesson 8-5)

- An angle of elevation is the angle formed by a horizontal line and the line of sight to an object above.
- An angle of depression is the angle formed by a horizontal line and the line of sight to an object below.

Laws of Sines and Cosines (Lessons 8-6 and 8-7)

Let $\triangle ABC$ be any triangle.

- Law of Sines: $\frac{\sin A}{a} = \frac{\sin B}{b} = \frac{\sin C}{c}$
- Law of Cosines: $a^2 = b^2 + c^2 - 2bc \cos A$
  $b^2 = a^2 + c^2 - 2ac \cos B$
  $c^2 = a^2 + b^2 - 2ab \cos C$
- Area $= \frac{1}{2}bc \sin A = \frac{1}{2}ac \sin B = \frac{1}{2}ab \sin C$

FOLDABLES® Study Organizer

Use your Foldable to review the chapter. Working with a partner can be helpful. Ask for clarification of concepts as needed.

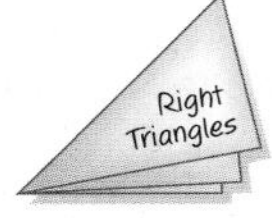

Key Vocabulary

ambiguous case (p. 608)
angle of depression (p. 608)
angle of elevation (p. 608)
cosine (p. 596)
geometric mean (p. 565)
inverse cosine (p. 599)
inverse sine (p. 599)
inverse tangent (p. 599)
Law of Cosines (p. 624)
Law of Sines (p. 616)
Pythagorean triple (p. 576)
sine (p. 596)
solving a triangle (p. 624)
tangent (p. 596)
trigonometric ratio (p. 596)
trigonometry (p. 596)

Vocabulary Check

State whether each sentence is *true* or *false*. If *false*, replace the underlined word or phrase to make a true sentence.

1. The arithmetic mean of two numbers is the positive square root of the product of the numbers.
2. A Pythagorean triple is a set of three nonzero whole numbers a, b, and c such that $a^2 + b^2 = c^2$.
3. To find the length of the hypotenuse of a right triangle, take the square root of the difference of the squares of the legs.
4. An angle of elevation is the angle formed by a horizontal line and an observer's line of sight to an object below the horizon.
5. The Law of Sines can be used to find an angle measure when given three side lengths.
6. A trigonometric ratio is a ratio of the lengths of two sides of a right triangle.

Concept Check

7. Explain the relationship between the angle of elevation and angle of depression between two objects.
8. Explain when it would be more useful to solve a triangle by beginning with the Law of Cosines than with the Law of Sines.

Lesson-by-Lesson Review 

8-1 Geometric Mean

Find the geometric mean between each pair of numbers.

9. 9 and 4

10. $\sqrt{20}$ and $\sqrt{80}$

11. $\frac{8\sqrt{2}}{3}$ and $\frac{4\sqrt{2}}{3}$

12. Find x, y, and z.

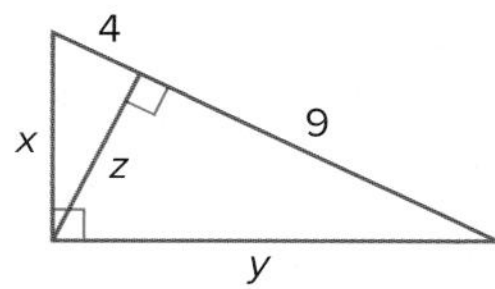

13. **DANCES** Mike is hanging a string of lights on his barn for a square dance. Using a book to sight the top and bottom of the barn, he can see he is 15 feet from the barn. If his eye level is 5 feet from the ground, how tall is the barn?

Example 1

Find the geometric mean between 10 and 15.

$x = \sqrt{ab}$	Definition of geometric mean
$= \sqrt{10 \cdot 15}$	$a = 10$ and $b = 15$
$= \sqrt{(5 \cdot 2) \cdot (3 \cdot 5)}$	Factor.
$= \sqrt{25 \cdot 6}$	Associative Property
$= 5\sqrt{6}$	Simplify.

8-2 The Pythagorean Theorem and Its Converse

Find x.

14. 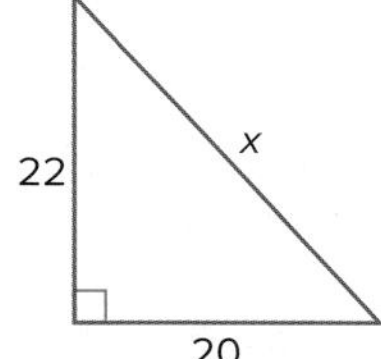

15. 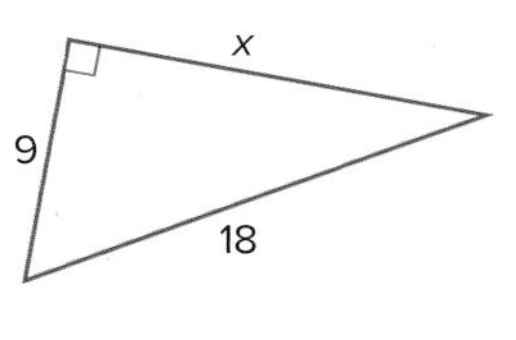

Determine whether each set of numbers can be the measures of the sides of a triangle. If so, classify the triangle as *acute*, *obtuse*, or *right*. Justify your answer.

16. 7, 24, 25

17. 13, 15, 16

18. 65, 72, 88

19. **SWIMMING** Alexi walks 27 meters south and 38 meters east to get around a lake. Her sister swims directly across the lake. How many meters to the nearest tenth did Alexi's sister save by swimming?

Example 2

Find x.

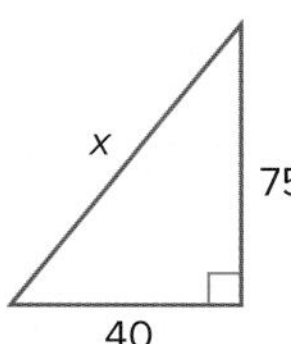

The side opposite the right angle is the hypotenuse, so $c = x$.

$a^2 + b^2 = c^2$	Pythagorean Theorem
$40^2 + 75^2 = x^2$	$a = 40$ and $b = 75$
$7225 = x^2$	Simplify.
$\sqrt{7225} = x$	Take the positive square root of each side.
$85 = x$	Simplify.

8-3 Special Right Triangles

Find *x* and *y*.

20.

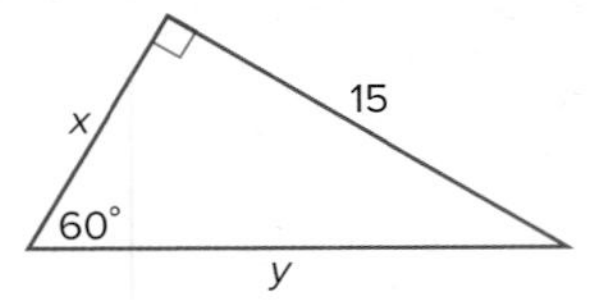

21.

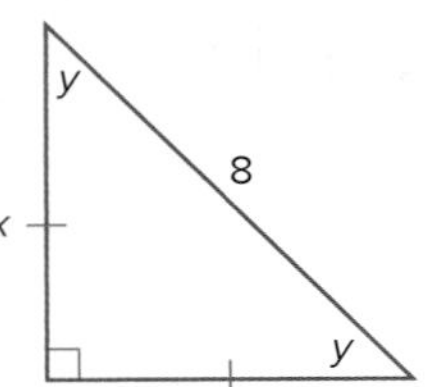

22. CLIMBING Jason is adding a climbing wall to his younger brother's swing-set. If he starts building 5 feet away from the existing structure and wants it to have a 60° angle, how long should the wall be?

Example 3

Find *x* and *y*.

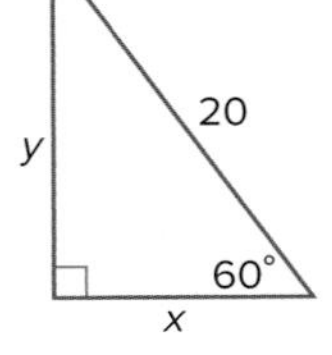

The measure of the third angle in this triangle is 90 − 60 or 30. This is a 30°-60°-90° triangle.

$h = 2s$	30°-60°-90° Triangle Theorem
$20 = 2x$	Substitute.
$10 = x$	Divide.

Now find *y*, the length of the longer leg.

$\ell = s\sqrt{3}$	30°-60°-90° Triangle Theorem
$y = 10\sqrt{3}$	Substitute.

8-4 Trigonometry

Express each ratio as a fraction and as a decimal to the nearest hundredth.

23. sin *A* **24.** tan *B*

25. sin *B* **26.** cos *A*

27. tan *A* **28.** cos *B*

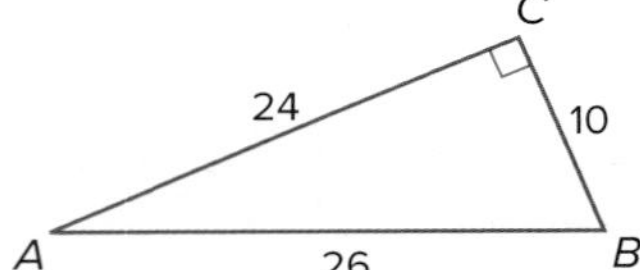

Find *x*.

29.

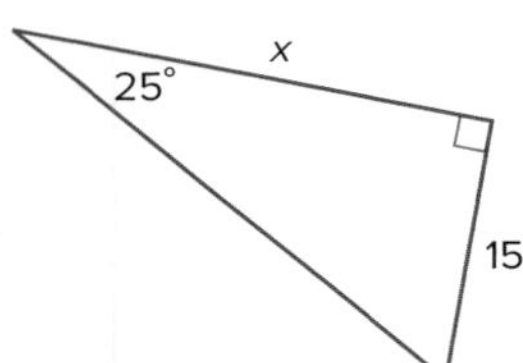

30.

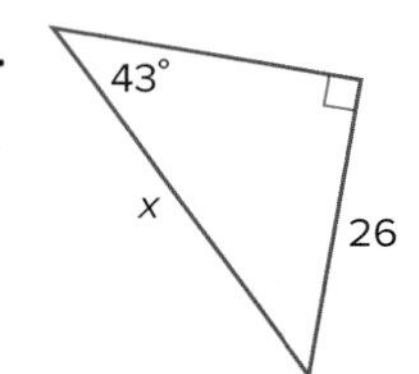

31. GARDENING Sofia wants to put a flower bed in the corner of her yard by laying a stone border that starts 3 feet from the corner of one fence and ends 6 feet from the corner of the other fence. Find the angles, *x* and *y*, the fence makes with the border.

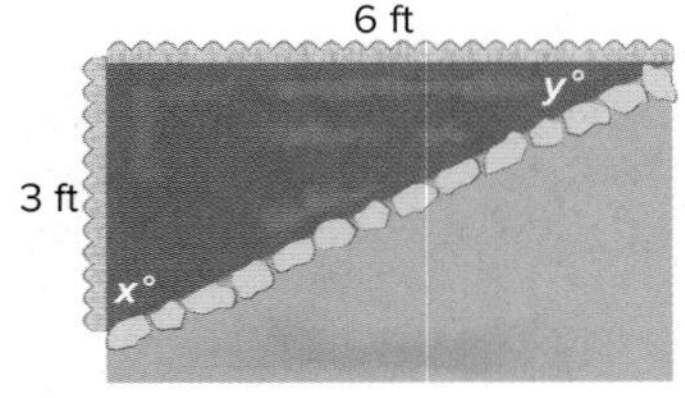

Example 4

Express each ratio as a fraction and as a decimal to the nearest hundredth.

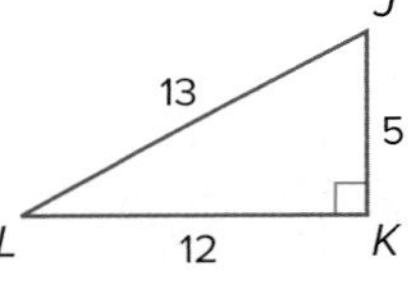

a. sin *L*

$\sin L = \frac{5}{13}$ or about 0.38 $\quad \sin L = \frac{\text{opp}}{\text{hyp}}$

b. cos *L*

$\cos L = \frac{12}{13}$ or about 0.92 $\quad \cos L = \frac{\text{adj}}{\text{hyp}}$

c. tan *L*

$\tan L = \frac{5}{12}$ or 0.42 $\quad \tan L = \frac{\text{opp}}{\text{adj}}$

8-5 Angles of Elevation and Depression

32. **JOBS** Tom delivers papers on a rural route from his car. If he throws a paper from a height of 4 feet, and it lands 15 feet from the car, at what angle of depression did he throw the paper to the nearest degree?

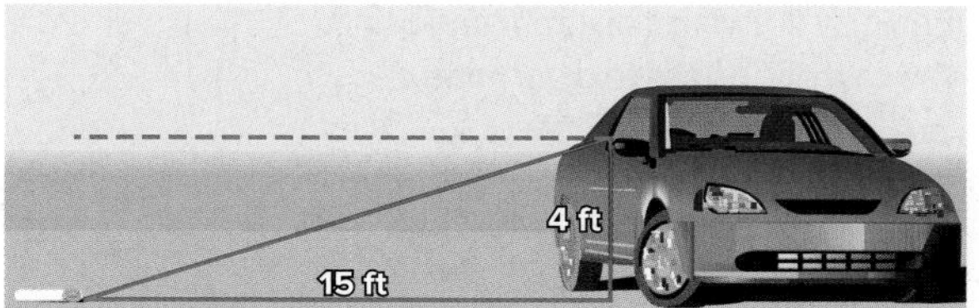

33. **TOWER** There is a cell phone tower in the field across from Jen's house. If Jen walks 50 feet from the tower and finds the angle of elevation from her position to the top of the tower to be 60°, how tall is the tower?

Example 5

Sarah's cat climbed up a tree. If she sights her cat at an angle of elevation of 40°, and her eyes are 5 feet off the ground, how high up from the ground is her cat?

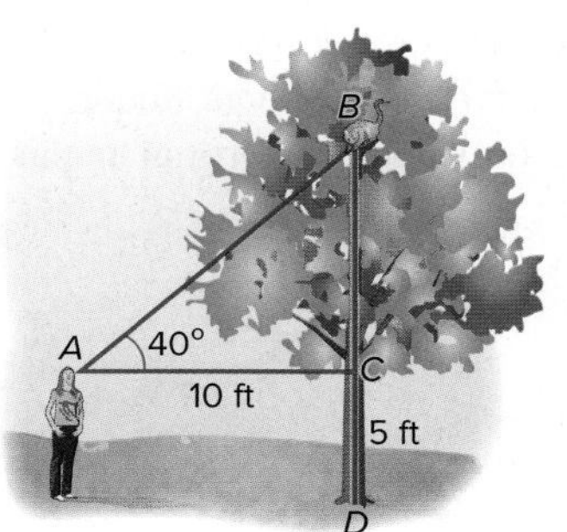

To find how far up the tree the cat is, find CB.

$\tan 40 = \frac{CB}{10}$ $\quad \tan = \frac{\text{opposite}}{\text{adjacent}}$

$10(\tan 40) = CB$ $\quad$ Multiply each side by 10.

$8.4 = CB$ $\quad$ Simplify.

Since Sarah's eyes are 5 feet from the ground, add 5 to 8.4. Sarah's cat is 13.4 feet up.

8-6 Law of Sines

Determine whether each triangle has *no* solution, *one* solution, or *two* solutions. Then solve each triangle. Round measures of sides to the nearest tenth and measures of angles to the nearest degree.

34. $C = 118°, c = 10, a = 4$
35. $A = 25°, a = 15, c = 18$
36. $A = 70°, a = 5, c = 16$
37. **BOAT** Kira and Mallory are standing on opposite sides of a river. How far is Kira from the boat? Round to the nearest tenth if necessary.

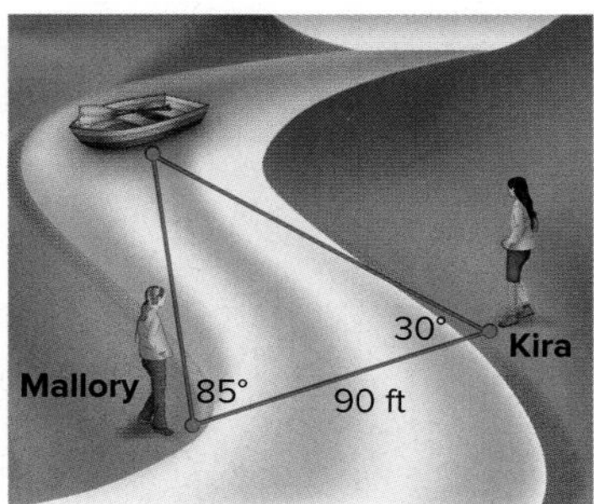

Example 6

Solve $\triangle ABC$.

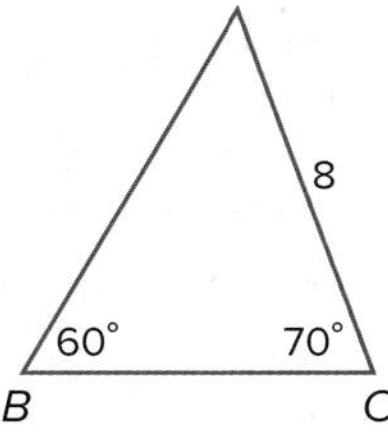

First, find the measure of the third angle.

$60° + 70° + a = 180°$

$A = 50°$

Now use the Law of Sines to find a and c. Write two equations, each with one variable.

$\frac{\sin B}{b} = \frac{\sin C}{c}$ $\qquad$ $\frac{\sin B}{b} = \frac{\sin A}{a}$

$\frac{\sin 60°}{8} = \frac{\sin 70°}{c}$ $\qquad$ $\frac{\sin 60°}{8} = \frac{\sin 50°}{a}$

$c = \frac{8 \sin 70°}{\sin 60°}$ $\qquad$ $a = \frac{8 \sin 50°}{\sin 60°}$

$c \approx 8.7$ $\qquad$ $a \approx 7.1$

Therefore, $A = 50°$, $c \approx 8.7$, and $a \approx 7.1$.

CHAPTER 8

Study Guide and Review *Continued*

8-7 Law of Cosines

Determine whether each triangle should be solved by beginning with the Law of *Sines* or Law of *Cosines*. Then solve each triangle. Round measures of sides to the nearest tenth and measures of angles to the nearest degree.

38.

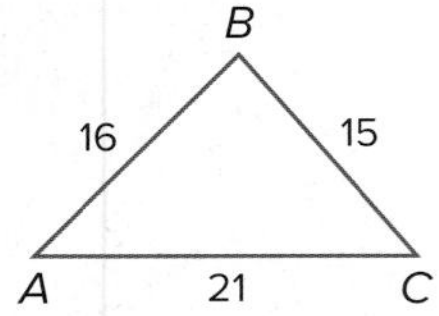

39.

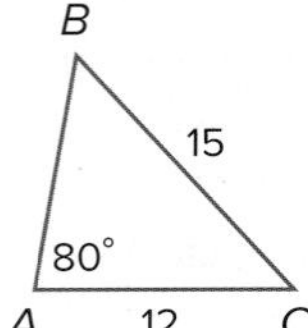

40. $C = 75°$, $a = 5$, $b = 7$

41. $A = 42°$, $a = 9$, $b = 13$

42. $b = 8.2$, $c = 15.4$, $A = 35°$

43. **FARMING** A farmer wants to fence a piece of his land. Two sides of the triangular field have lengths of 120 feet and 325 feet. The measure of the angle between those sides is 70°. How much fencing will the farmer need?

44. In a rhombus whose side length is 22 and the smaller angle is 55°, find the length of the shorter diagonal to the nearest tenth.

45. A triangular walking course has 2 sides of 240 feet and 360 feet, and the angle between these sides measures 38°. Find the length of the third side of the course to the nearest foot.

46. Three sides of a triangle measure 20 meters, 30 meters, and 40 meters. Find the largest angle of the triangle to the nearest degree.

Example 7

Solve $\triangle ABC$ for $C = 55°$, $b = 11$, and $a = 18$.

You are given the measure of two sides and the included angle. Begin by drawing a diagram and using the Law of Cosines to determine c.

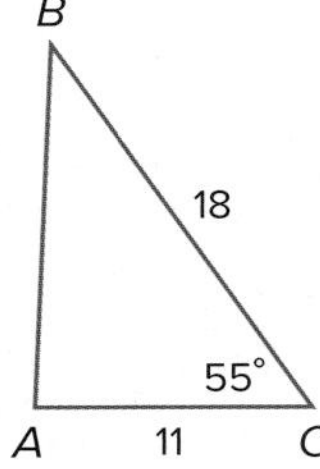

$c^2 = a^2 + b^2 - 2ab \cos C$

$c^2 = 182 + 112 - 2(18)(11) \cos 55°$

$c^2 \approx 217.9$

$c \approx 14.8$

Next, you can use the Law of Sines to find the measure of angle A.

$\frac{\sin A}{18} \approx \frac{\sin 55°}{14.8}$

$\sin A \approx \frac{18 \sin 55°}{14.8}$ or A is about 85.0°

The measure of the angle B is approximately $180 - (85.0 + 55)$ or 40.0°.

Therefore, $c \approx 14.8$, $A \approx 85.0°$, and $B \approx 40.0°$

Practice Test

Find the geometric mean between each pair of numbers.

1. 7 and 11
2. 12 and 9
3. 14 and 21
4. $4\sqrt{3}$ and $10\sqrt{3}$
5. Find x, y, and z.

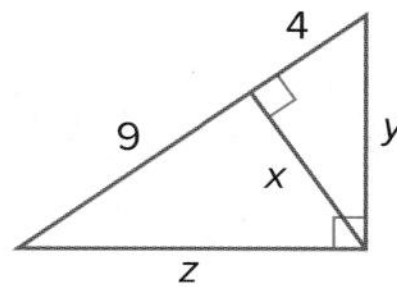

6. **FAIRS** Blake is setting up his tent at a Renaissance fair. If the tent is 8 feet tall and the tether can be staked no more than two feet from the tent, how long should the tether be?

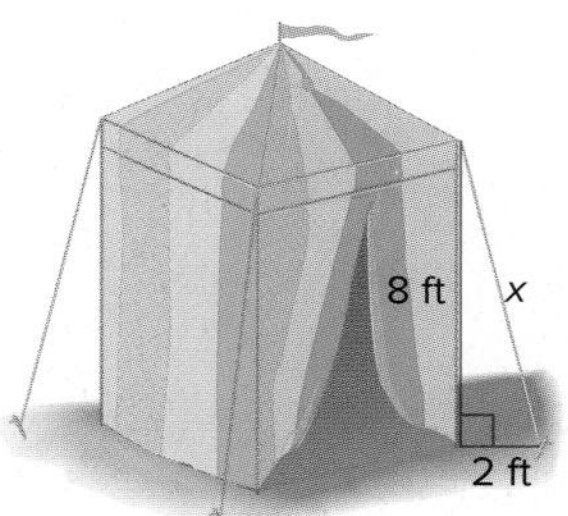

Use a calculator to find the measure of $\angle R$ to the nearest tenth.

7.

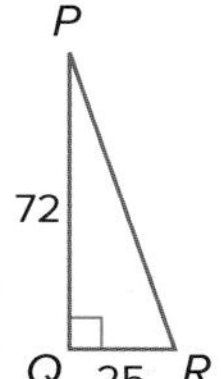

8.

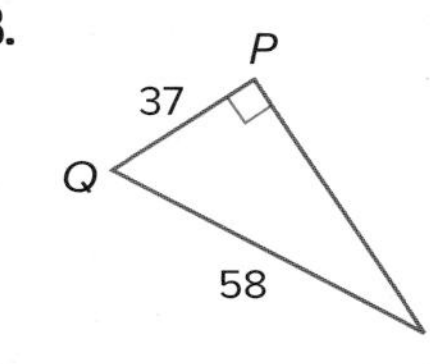

9. Find x and y.

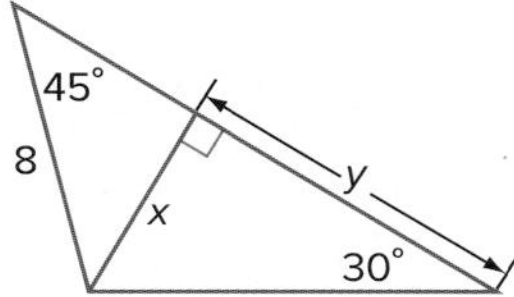

Express each ratio as a fraction and as a decimal to the nearest hundredth.

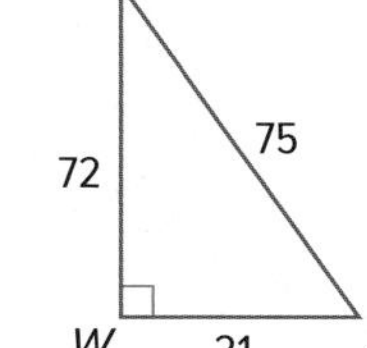

10. $\cos X$
11. $\tan X$
12. $\tan V$
13. $\sin V$

Find the area of $\triangle ABC$ to the nearest tenth.

14. $A = 48°$, $b = 7$ in., $c = 12$ in.
15. $C = 128°$, $a = 1.25$ m, $b = 16$ m

16. **SPACE** Anna is watching a space shuttle launch 6 miles from Cape Canaveral in Florida. When the angle of elevation from her viewing point to the shuttle is 80°, how high is the shuttle, if it is going straight up?

Determine whether each $\triangle ABC$ has *no* solution, *one* solution, or *two* solutions. Then solve the triangle. Round side lengths to the nearest tenth and angle measures to the nearest degree.

17. $A = 23°$, $a = 14$, $b = 11$
18. $A = 112°$, $a = 5$, $b = 9$

19. **MULTIPLE CHOICE** Which of the following is the length of the leg of a 45°-45°-90° triangle with a hypotenuse of 20?

A 10
B $10\sqrt{2}$
C 20
D $20\sqrt{2}$

Find x.

20.

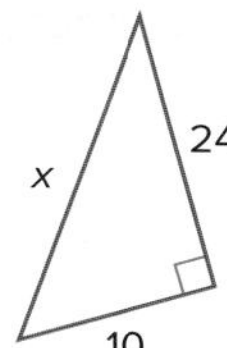

21.

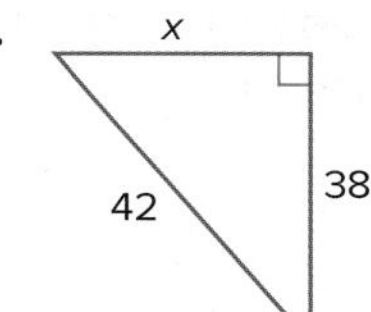

22. **WHALE WATCHING** Isaac is looking through binoculars on a whale watching trip when he notices a sea otter in the distance. If he is 20 feet above sea level in the boat, and the angle of depression is 30°, how far away from the boat is the otter to the nearest foot?

Determine whether each triangle should be solved by beginning with the Law of Sines or Law of Cosines. Then solve each triangle. Round side lengths to the nearest tenth and angle measures to the nearest degree.

23.

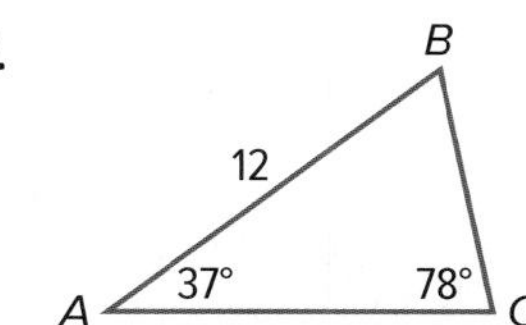

24.

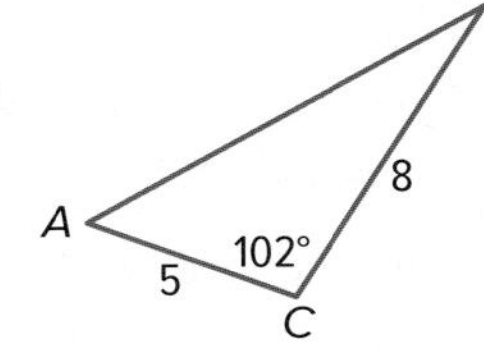

25. Solve $\triangle FGH$. Round to the nearest degree.

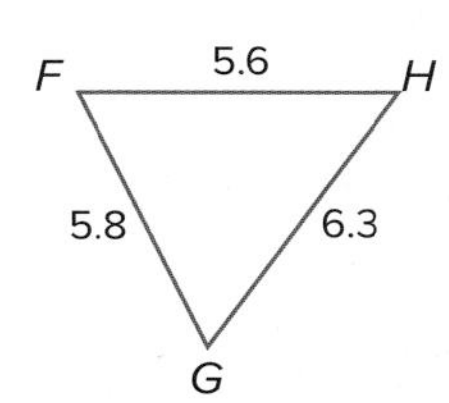

Preparing for Assessment

Performance Task

Provide a clear solution to each part of the task. Be sure to show all of your work. Include all relevant drawings and justify your answers.

NAVIGATION Jessika and her family are on vacation at the coast. They take a boat from a small island along a 10 mile path that is perpendicular to the horizontal shoreline of the mainland. A supply store is located 5 miles from the docks on the mainland shoreline.

Part A

1. What is the distance from the supply store to their boat on the island?

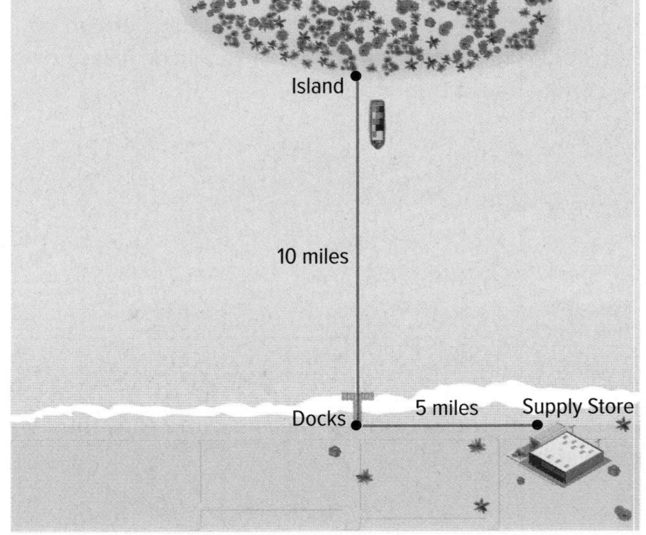

Part B

Jessika wants to find the angle formed by the shoreline and the segment from the supply store to her family's boat on the island.

2. **Reasoning** Write a trigonometric equation to find this angle.

3. Use a calculator to find this angle to the nearest tenth of a degree.

Part C

The docks are relocated to a location on the shoreline that is 8 miles farther away from the supply store.

4. How far does the boat have to travel now to get from the island to the new docks?

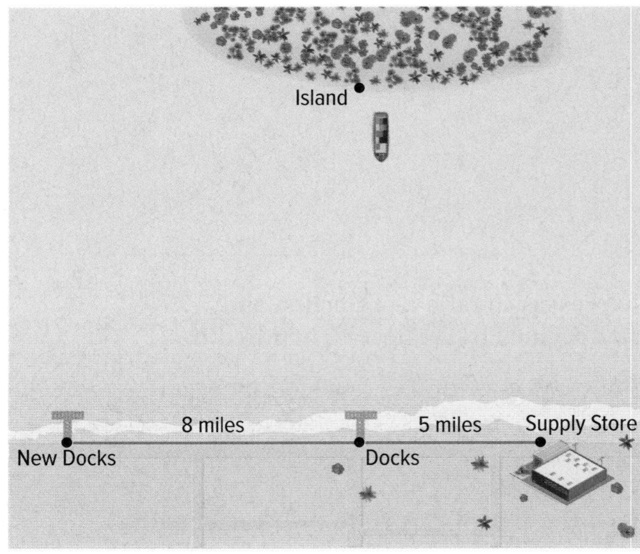

Part D

A bearing is an angle formed clockwise from due north. The island is north of the mainland.

5. **Reasoning** At what bearing does the boat need to travel from the island to reach the new docks?

Test-Taking Strategy

Example

Read the problem. Identify what you need to know. Then use the information in the problem to solve.

The ratio of the width to the height of a high-definition television is 16:9. This is also called the *aspect ratio* of the television. The size of a television is given in terms of the diagonal distance across the screen. If an HD television is 25.5 inches tall, what is its screen size?

d

A 48 inches **C** 51 inches

B 50 inches **D** 52 inches

Step 1 **What are you asked to find?**
The length of the diagonal of the television.

Step 2 **What formula can you use?**
The diagonal, height, and width form a right triangle, so use the Pythagorean Theorem to find the diagonal.

Step 3 **What is the correct answer?**

Find the width of the screen. Set up and solve a proportion using the aspect ratio 16:9.

$\frac{16}{9} = \frac{w}{25.5}$ ← width of the screen ← height of the screen

$9w = 408$ Cross Products Property

$w = 45\frac{1}{3}$ Divide each side by 9.

So, the width of the screen is $45\frac{1}{3}$ inches. Use the Pythagorean Theorem to solve for the diagonal distance. Let c = diagonal, a = height, and b = width.

$c^2 = a^2 + b^2$ Pythagorean Theorem

$c^2 = (25.5)^2 + \left(45\frac{1}{3}\right)^2$ Substitute for a and b.

$c \approx 52.01$ Simplify. Take the positive square root of both sides to solve for c.

The diagonal distance of the screen is about 52 inches. So, the answer is D.

Test-Taking Tip

Use a Formula Sometimes it is necessary to use a formula to solve problems on standardized tests. In some cases, you may even be given a formula sheet for reference.

Apply the Strategy

Read the problem. Identify what you need to know. Then use the information in the problem to solve.

Christine is flying a kite on the end of a taut string. The kite is 175 feet above the ground and is a horizontal distance of 130 feet from where Christine is standing. How much kite string has Christine let out? Round to the nearest foot.

A 204 ft **C** 225 ft

B 218 ft **D** 236

Answer the questions below.

a. What are you asked to find?

b. What formula can you use?

c. What is the correct answer?

CHAPTER 8

Preparing for Assessment

Cumulative Review

Read each question. Then fill in the correct answer on the answer document provided by your teacher or on a sheet of paper.

1. Jeffrey leans his skateboard against a wall. The skateboard is 30 inches long, and the base of the skateboard is 8 inches from the wall.

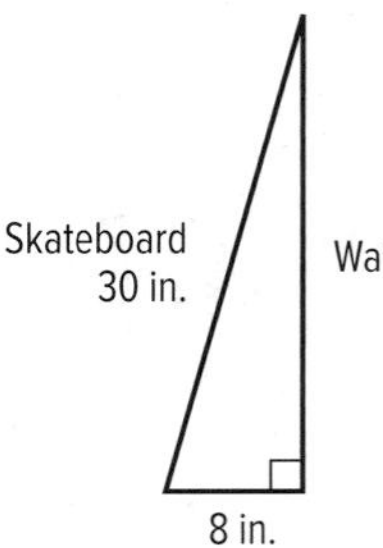

Which of the following is the best estimate of the measure of the angle the skateboard makes with the ground?

- A 14.9
- B 15.5
- C 74.5
- D 75.1

2. Marisol's new tablet is 9 inches long and has a diagonal of 11 inches. What is the perimeter of the tablet, in inches, to the nearest tenth of an inch?

Test-Taking Tip

Question 2 This problem requires two steps. First use the Pythagorean Theorem to find the width of the tablet. Then find the perimeter.

3. What should be the value of x in order for quadrilateral $ABCD$ to be a parallelogram?

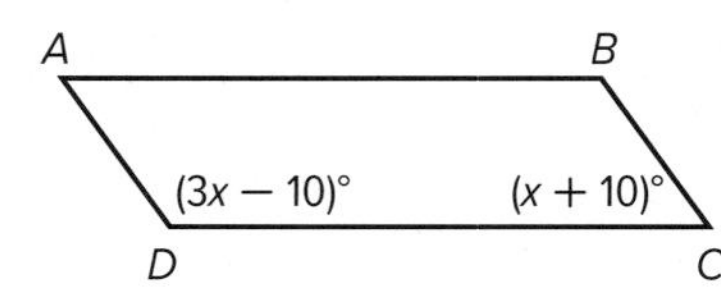

- A 125
- B 55
- C 45
- D 10

4. Sunny is designing a small bookcase with parallel shelves and slanted sides, as shown in the figure.

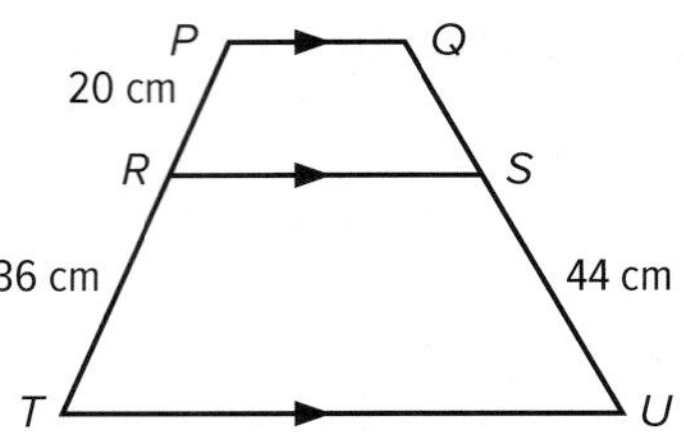

Which of the following is the best estimate of the length of $\overline{QU}$?

- A 24.4 cm
- B 68.4 cm
- C 28.0 cm
- D 79.2 cm

5. What is the area of $\triangle PQR$?

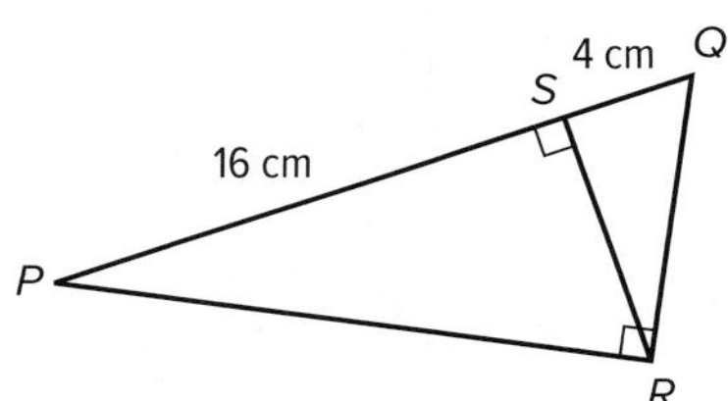

- A 160 m^2
- B 100 m^2
- C 80 m^2
- D 64 m^2

6. Forest rangers in two observation towers spot a fire in the distance. The towers are 500 feet apart. The rangers each measure the angle to the fire as shown in the figure.

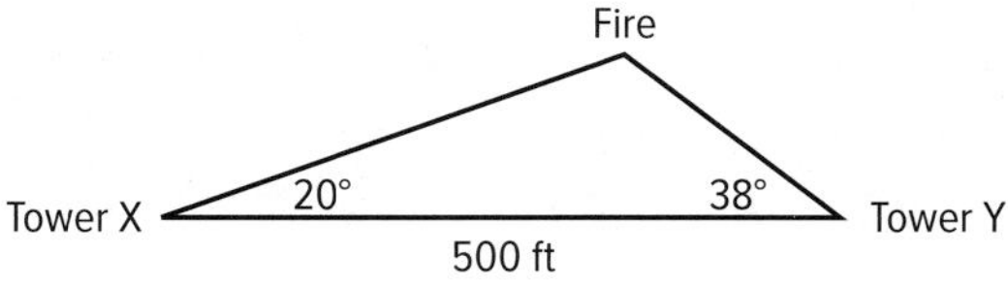

a. What is the distance from Tower X to the fire to the nearest foot?

______ ft

b. MP What mathematical practice did you use to solve this problem?

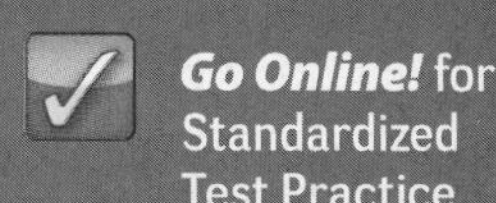

7. Kendrick draws and cuts out an equilateral triangle with sides 5 inches long. He folds the triangle in half, as shown. What is the value of x to the nearest tenth of an inch?

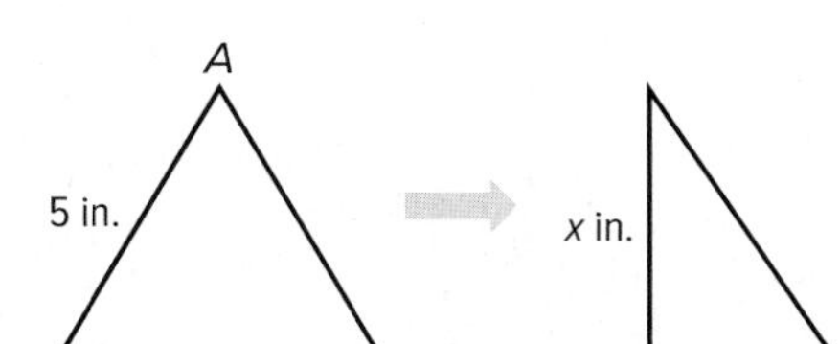

8. Line p is parallel to line q.

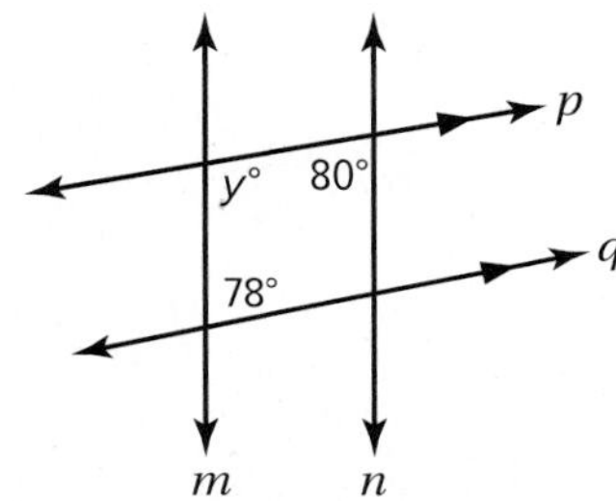

What is the value of y?

- **A** 78
- **B** 80
- **C** 100
- **D** 102

9. From the top of a 65-foot-tall building, Addie spots a motorcycle at street level. The angle of depression from Addie to the motorcycle is 38°. What is the horizontal distance from the base of the building to the motorcycle?

- **A** 105.6 ft
- **B** 83.2 ft
- **C** 51.2 ft
- **D** 50.8 ft

10. Find the measure of $\overline{PQ}$ if Q is the midpoint of PR, $PQ = 9x - 18$, and $QR = 3x + 36$.

11. Quadrilateral $JKLM$ is a parallelogram.

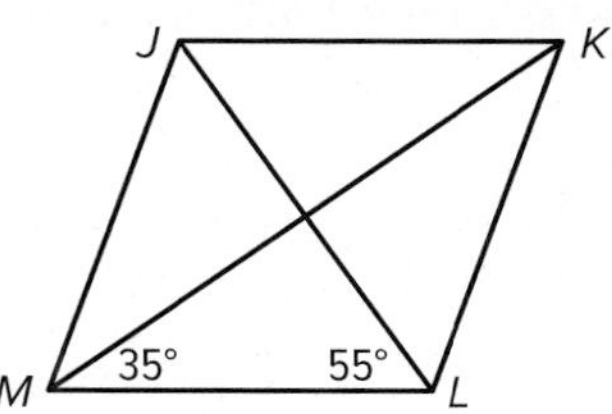

Which of the following is the best name for quadrilateral $JKLM$?

- **A** kite
- **B** rectangle
- **C** rhombus
- **D** square

12. Claire wants to cut three straws and place them together to form an obtuse triangle. Which of the following could be the lengths of the straws that Claire uses?

- **A** 6 in., 7 in., 8 in.
- **B** 6 in., 7 in., 10 in.
- **C** 6 in., 8 in., 10 in.
- **D** 6 in., 8 in., 14 in.

13. A park is a square with straight paths along the diagonals, as shown.

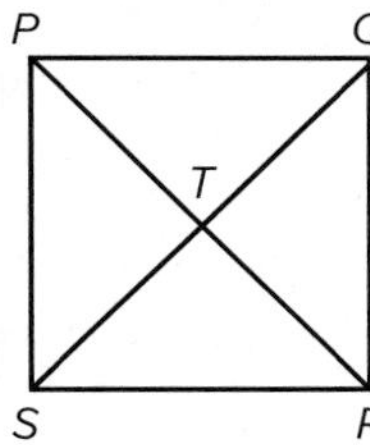

Select all statements about the path that must be true.

- **A** The length of $\overline{SR}$ is $\sqrt{2}$ times the length of $\overline{TR}$.
- **B** The length of $\overline{PS}$ is $\sqrt{2}$ times the length of $\overline{PR}$.
- **C** The length of $\overline{QS}$ is $\sqrt{2}$ times the length of $\overline{QR}$.
- **D** The length of $\overline{SR}$ is $\sqrt{3}$ times the length of $\overline{TR}$.
- **E** The length of $\overline{PS}$ is $\sqrt{3}$ times the length of $\overline{PR}$.

Need Extra Help?

If you missed Question...	1	2	3	4	5	6	7	8	9	10	11	12	13
Go to Lesson...	8-4	8-2	6-3	7-5	8-1	8-6	8-3	2-7	8-5	1-3	6-5	8-2	8-3

CHAPTER 9

Circles

THEN

You learned about special segments and angle relationships in triangles.

NOW

In this chapter, you will:

- Learn the relationships between central angles, arcs, and inscribed angles in a circle.
- Define and use secants and tangents.
- Use an equation to identify or describe a circle.

MP WHY

SCIENCE The actual shape of a rainbow is a complete circle. The portion of the circle that can be seen above the horizon is a special segment of a circle called an arc.

Use the Mathematical Practices to complete the activity.

1. Use Tools Use the Internet to learn more about the way in which a rainbow is created and why it is actually a full circle.

2. Reasoning If a rainbow is a full circle, why do you think you only see a portion of it? Can you think of a way that could be represented on a coordinate plane?

3. Modeling Use the Explore Circle tool to create a circle representative of a rainbow. Place the circle so that the *x*-axis represents the horizon as we see it and model the arc of the rainbow that we typically see.

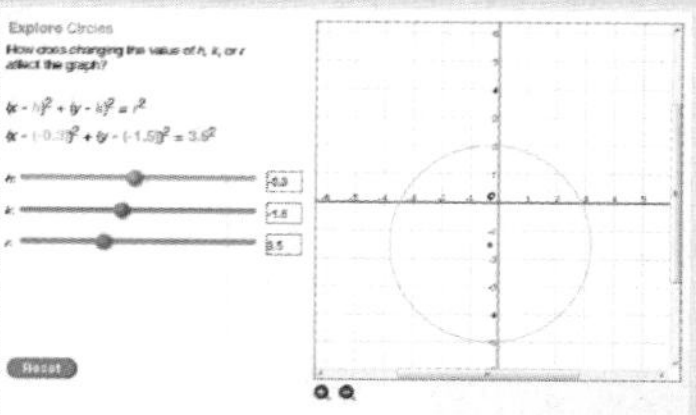

Go Online to Guide Your Learning

Explore & Explain

Triangle Special Segments

Use the **Triangle Special Segments** tool to construct inscribed and circumscribed circles discussed in Lesson 9-1.

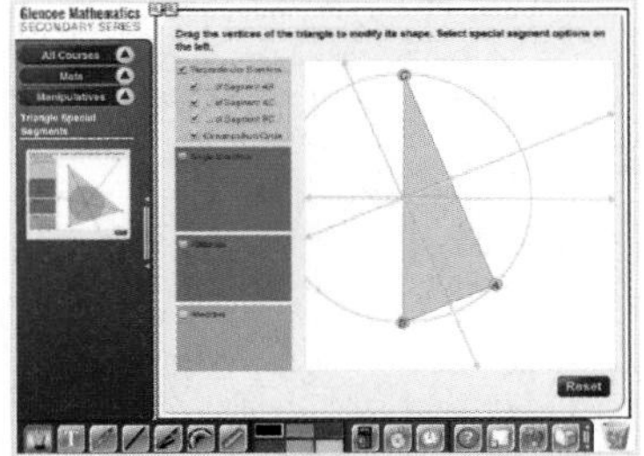

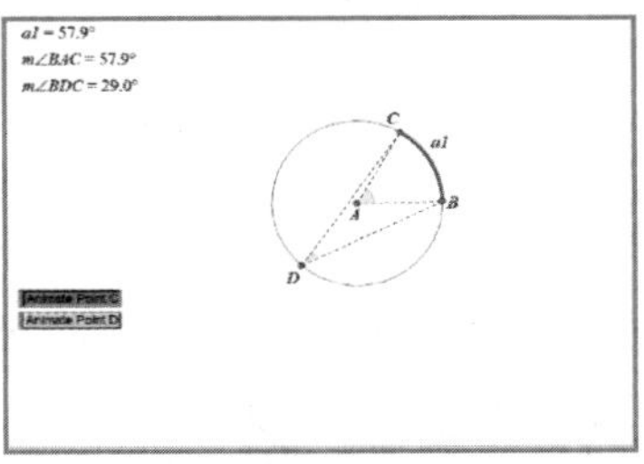

The Geometer's Sketchpad

Use the **The Geometer's Sketchpad** to explore the relationship between a circle's circumference and its diameter, the relationships in a circle between angles and the arcs they intercept, the properties of chords in a circle, and to construct two tangent segments to a circle and investigate their properties.

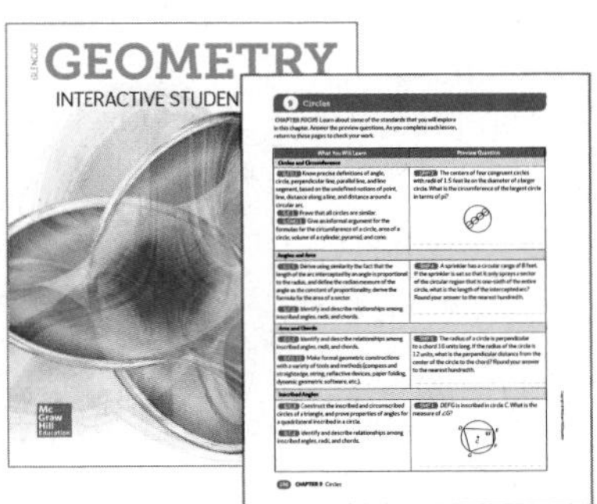

eBook

Interactive Student Guide

Before starting the chapter, answer the **Chapter Focus** preview questions. Check your answers as you complete each lesson. At the end of the chapter, try the **Performance Task**.

Organize

Foldables

Get organized! Before beginning this chapter, create this Foldable to help you organize your notes on circles.

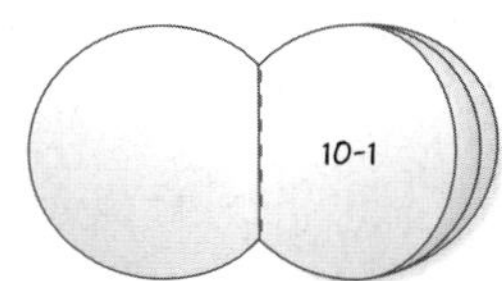

FOLDABLES®

Collaborate

Chapter Project

In the **Olympic Games** project, you will use what you have learned about circles to complete a project that addresses global awareness.

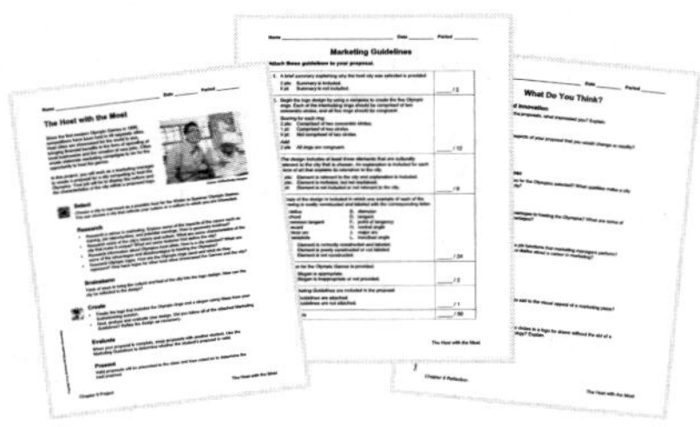

Focus

LEARNSMART®

Need help studying? Complete the **Connecting Algebra and Geometry Through Coordinates** domain in LearnSmart to review for the chapter test.

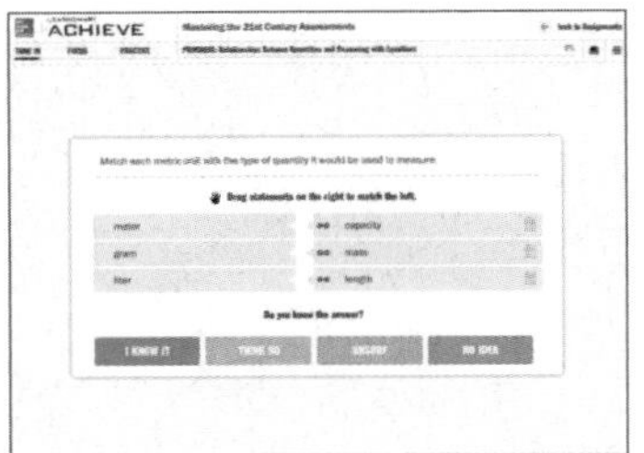

ALEKS®

You can use the **Polygons and Circles** topic in ALEKS to explore what you know about circles and what you are ready to learn.*

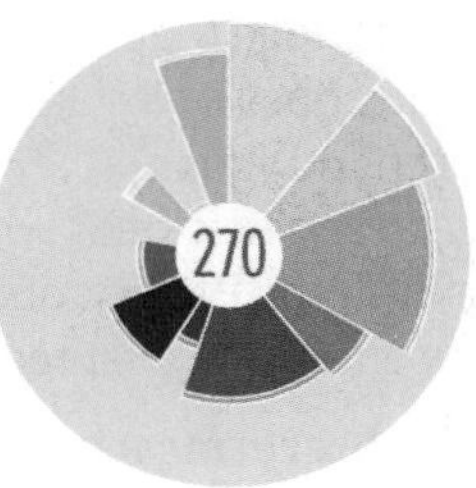

* Ask your teacher if this is part of your program.

Go Online! for Vocabulary Review Games and key vocabulary in 13 languages.

Get Ready for the Chapter

Connecting Concepts

Concept Check

Review the concepts used in this chapter by answering the questions below.

1. If you are asked to find 26% of a number, what is the first step you need to take?
2. You are asked to find 15% of 35. You convert 15% to 0.15. What is your next step?
3. Classify the triangle shown.
4. What fact about special right triangles can you use to solve for x?
5. What is another way you can solve for x?

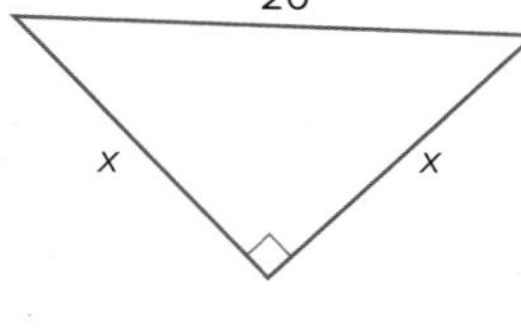

6. What formula would you use to solve $x^2 + 4x - 40 = 0$?
7. What process is used to derive the Quadratic Formula?
8. Given a rectangle that is 3 feet wide and 8 feet long, what equation would allow you to determine length of a diagonal that would cut the triangle into two right triangles?

Performance Task Preview

You can use the concepts and skills in the chapter to design a circular badge that circumscribes a circle around different polygons. Understanding circles will help you finish the Performance Task at the end of the chapter.

MP In this Performance Task you will:

- make sense of problems
- reason abstractly and quantitatively
- construct an argument

New Vocabulary

English		Español
circle	p. 643	círculo
center	p. 643	centro
radius	p. 643	radio
chord	p. 643	cuerda
diameter	p. 643	diámetro
circumference	p. 645	circunferencia
pi (π)	p. 645	pi (π)
inscribed	p. 646	inscrito
circumscribed	p. 646	circunscrito
central angle	p. 652	ángulo central
arc	p. 652	arco
inscribed angle	p. 669	ángulo inscrito
intercepted arc	p. 669	arco intersecado
tangent	p. 678	tangente
secant	p. 687	secante
focus	p. 703	foco
directrix	p. 703	directriz

Review Vocabulary

coplanar coplanar points that lie in the same plane

degree grado $\frac{1}{360}$ of the circular rotation about a point

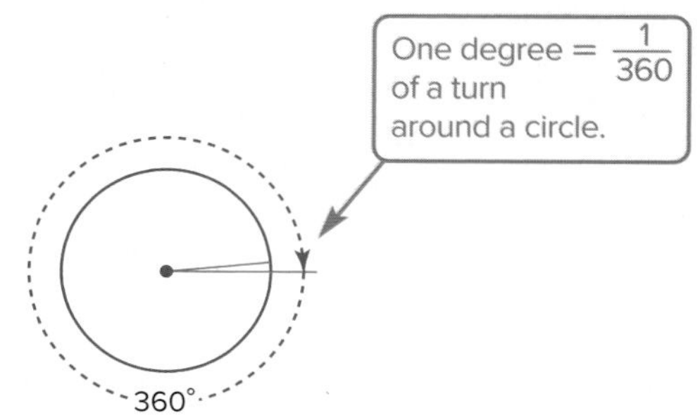

LESSON 1

Circles and Circumference

Then

- You identified and used parts of parallelograms.

Now

1. Identify and use parts of circles.
2. Solve problems involving the circumference of a circle.

Why?

- The maxAir ride shown speeds back and forth and rotates counterclockwise. At times, the riders are upside down 140 feet above the ground experiencing "airtime"—a feeling of weightlessness. The ride's width—or *diameter*—is 44 feet. You can find the distance that a rider travels in one rotation by using this measure.

New Vocabulary

circle
center
radius
chord
diameter
concentric circles
circumference
pi (π)
inscribed
circumscribed

Mathematical Practices

4 Model with mathematics.

1 Make sense of problems and persevere in solving them.

1 Segments in Circles

A **circle** is the locus or set of all points in a plane equidistant from a given point called the **center** of the circle.

Segments that intersect a circle have special names.

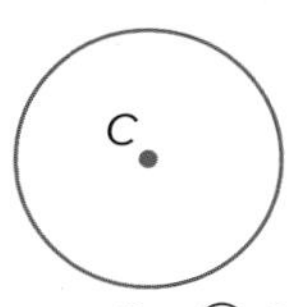

Circle C or $\odot C$

Key Concept Special Segments in a Circle

A **radius** (plural radii) is a segment with endpoints at the center and on the circle.

Examples $\overline{CD}$, $\overline{CE}$, and $\overline{CF}$ are radii of $\odot C$.

A **chord** is a segment with endpoints on the circle.

Examples $\overline{AB}$ and $\overline{DE}$ are chords of $\odot C$.

A **diameter** of a circle is a chord that passes through the center and is made up of collinear radii.

Example $\overline{DE}$ is a diameter of $\odot C$. Diameter $\overline{DE}$ is made up of collinear radii $\overline{CD}$ and $\overline{CE}$.

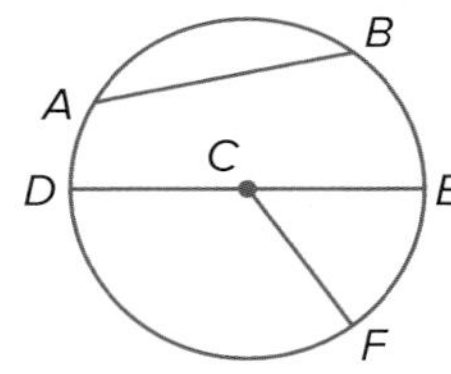

Example 1 Identify Segments in a Circle

a. Name the circle and identify a radius.

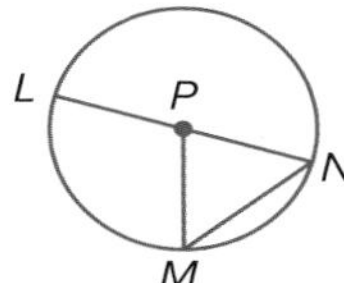

The circle has a center at P, so it is named circle P, or $\odot P$. Three radii are shown: $\overline{PL}$, $\overline{PN}$, and $\overline{PM}$.

b. Identify a chord and a diameter of the circle.

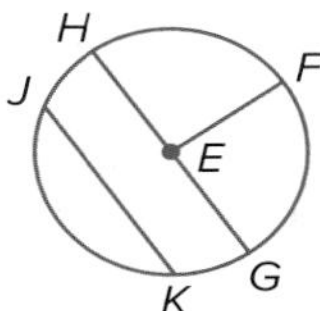

Two chords are shown: $\overline{JK}$ and $\overline{HG}$. $\overline{HG}$ goes through the center, so $\overline{HG}$ is a diameter.

Guided Practice

1. Name the circle, a radius, a chord, and a diameter of the circle.

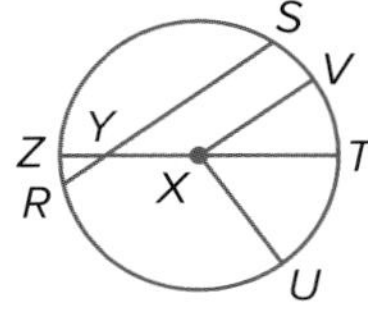

Reading Math

Precision The words *radius* and *diameter* are used to describe lengths as well as segments. Because a circle has many different radii and diameters, the phrases *the radius* and *the diameter* refer to lengths rather than segments.

By definition, the distance from the center of a circle to any point on the circle is always the same. Therefore, all radii r of a circle are congruent. Because a diameter d is composed of two radii, all diameters of a circle are also congruent.

Key Concept Radius and Diameter Relationships

If a circle has radius r and diameter d, the following relationships are true.

Radius Formula $r = \frac{d}{2}$ or $r = \frac{1}{2}d$

Diameter Formula $d = 2r$

Example 2 Find Radius and Diameter

If $QV = 8$ inches, what is the diameter of $\odot Q$?

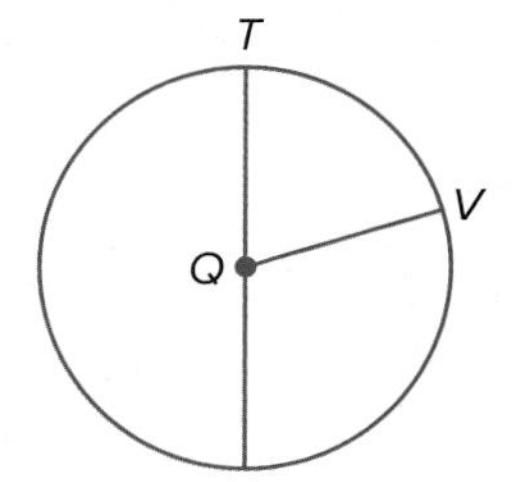

$d = 2r$ Diameter Formula

$= 2(8)$ or 16 Substitute and simplify.

The diameter of $\odot Q$ is 16 inches.

Guided Practice

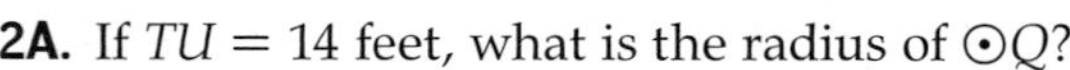

2A. If $TU = 14$ feet, what is the radius of $\odot Q$?

2B. If $QT = 11$ meters, what is QU?

As with other figures, pairs of circles can be congruent, similar, or share other special relationships.

Review Vocabulary

coplanar points that lie in the same plane

Key Concept Circle Pairs

Two circles are congruent if and only if they have congruent radii.

Example $\overline{GH} \cong \overline{JK}$, so $\odot G \cong \odot J$.

All circles are similar.

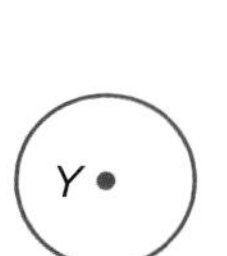

Example $\odot X \sim \odot Y$

Concentric circles are coplanar circles that have the same center.

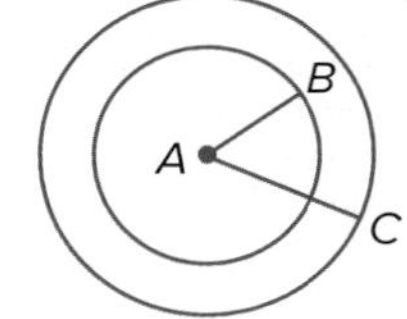

Example $\odot A$ with radius $\overline{AB}$ and $\odot A$ with radius $\overline{AC}$ are concentric.

You will prove that all circles are similar in Exercise 52.

Two circles can intersect in two different ways.

2 Points of Intersection	1 Point of Intersection	No Points of Intersection

Go Online! 

Circles and circumference are widely used in problems arising in everyday life, society, and the workplace. If you need extra practice, look for **Worksheets** in ConnectED.

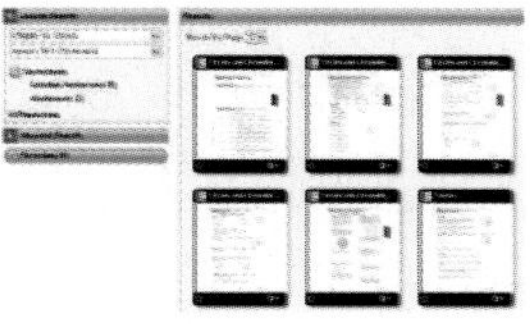

The segment connecting the centers of the two intersecting circles contains the radii of the two circles.

Example 3 Find Measures in Intersecting Circles

The diameter of $\odot S$ is 30 units, the diameter of $\odot R$ is 20 units, and $DS = 9$ units. Find CD.

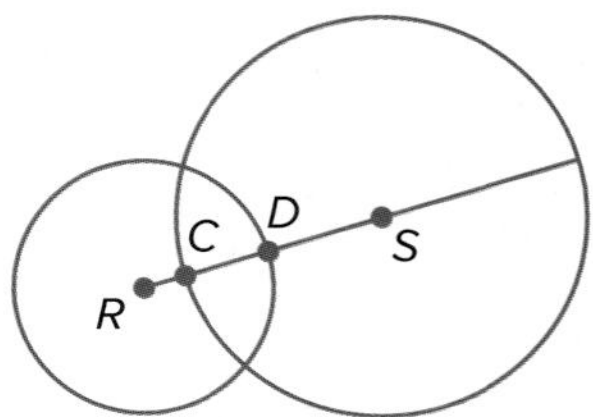

Because the diameter of $\odot S$ is 30, $CS = 15$.
$\overline{CD}$ is part of radius $\overline{CS}$.

$CD + DS = CS$	Segment Addition Postulate
$CD + 9 = 15$	Substitution
$CD = 6$	Subtract 9 from each side.

Guided Practice

3. Use the diagram above to find RC.

2 Circumference

The **circumference** of a circle is the distance around the circle. By definition, the ratio $\frac{C}{d}$ is an irrational number called **pi (π)**. Two formulas for circumference can be derived by using this definition.

$\frac{C}{d} = \pi$	Definition of pi
$C = \pi d$	Multiply each side by d.
$C = \pi(2r)$	$d = 2r$
$C = 2\pi r$	Simplify.

Key Concept Circumference

Words If a circle has diameter d or radius r, the circumference C equals the diameter times pi or twice the radius times pi.

Symbols $C = \pi d$ or $C = 2\pi r$

Real-World Example 4 Find Circumference

TENNIS Find the circumference of the helipad described at the left.

$C = \pi d$	Circumference Formula
$= \pi(79)$	Substitution
$= 79\pi$	Simplify.
≈ 248.19	Use a calculator.

The circumference of the helipad is 79π feet or about 248.19 feet.

Guided Practice

Find the circumference of each circle described. Round to the nearest hundredth.

4A. radius = 2.5 centimeters

4B. diameter = 16 feet

Real-World Link

The world's highest tennis court doubles as the helipad of the Burj Al Arab hotel in the United Arab Emirates. The helipad has a diameter of 79 feet and is nearly 700 feet high.

Source: Burj Al Arab, Emporis Buildings

These circumference formulas can also be used to determine the diameter and radius of a circle when the circumference is given.

Example 5 Find Diameter and Radius

Find the diameter and radius of a circle to the nearest hundredth if the circumference of the circle is 106.4 millimeters.

$C = \pi d$	Circumference Formula	$r = \frac{1}{2}d$	Radius Formula
$106.4 = \pi d$	Substitution	$\approx \frac{1}{2}(33.87)$	$d \approx 33.87$
$\frac{106.4}{\pi} = d$	Divide each side by π.	≈ 16.94 mm	Use a calculator.
$33.87 \text{ mm} \approx d$	Use a calculator.		

Study Tip

Levels of Accuracy Because π is irrational, its value cannot be given as a terminating decimal. Using a value of 3 for π provides a quick estimate in calculations. Using a value of 3.14 or $\frac{22}{7}$ provides a closer approximation. For the most accurate approximation, use the π key on a calculator. Unless stated otherwise, assume that in this text, a calculator with a π key was used to generate answers.

Guided Practice

5. Find the diameter and radius of a circle to the nearest hundredth if the circumference of the circle is 77.8 centimeters.

A polygon is **inscribed** in a circle if all of its vertices lie on the circle. A circle is **circumscribed** about a polygon if it contains all the vertices of the polygon.

- Quadrilateral *LMNP* is *inscribed in* ⊙*K*.

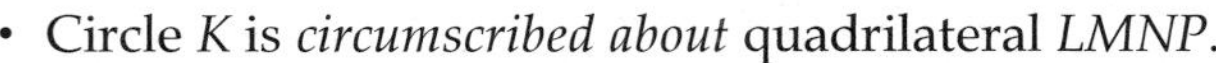

- Circle *K* is *circumscribed about* quadrilateral *LMNP*.

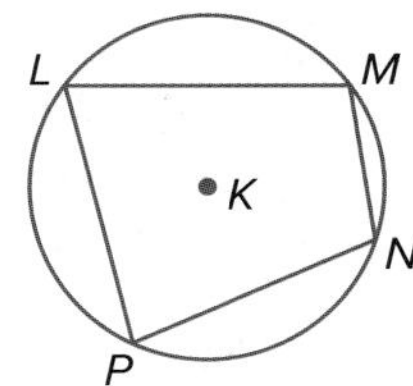

Example 6 Circumference of Circumscribed Polygons

A square with side length of 9 inches is inscribed in ⊙*J*. Find the exact circumference of ⊙*J*.

Study Tip

Circumcircle A *circumcircle* is a circle that passes through all of the vertices of a polygon.

You need to find the diameter of the circle and use it to calculate the circumference.

First, draw a diagram. The diagonal of the square is the diameter of the circle and the hypotenuse of a right triangle.

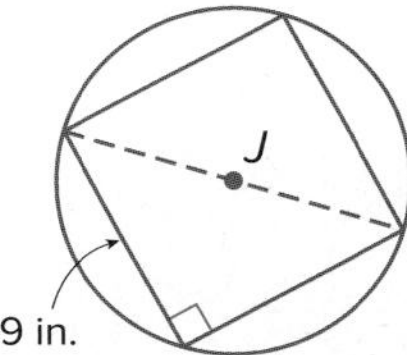

$a^2 + b^2 = c^2$	Pythagorean Theorem
$9^2 + 9^2 = c^2$	Substitution
$162 = c^2$	Simplify.
$9\sqrt{2} = c$	Take the positive square root of each side.

The diameter of the circle is $9\sqrt{2}$ inches.

Find the circumference in terms of π by substituting $9\sqrt{2}$ for d in $C = \pi d$. The exact circumference is $9\pi\sqrt{2}$ inches.

Guided Practice

Find the exact circumference of each circle by using the given polygon.

6A. inscribed right triangle with legs 7 meters and 3 meters long

6B. circumscribed square with side 10 feet long

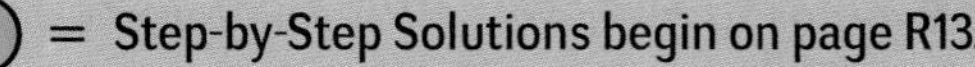

Check Your Understanding

= Step-by-Step Solutions begin on page R13.

Examples 1–2 **For Exercises 1–4, refer to $\odot N$.**

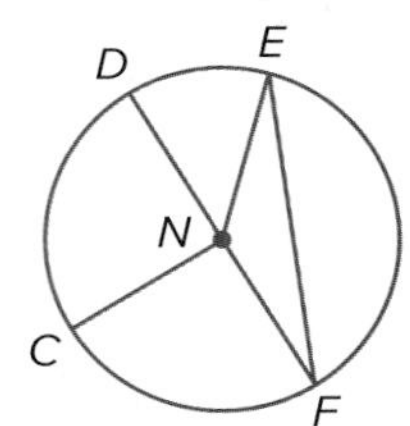

1. Name the circle.
2. Identify each.
   a. a chord b. a diameter c. a radius
3. If $CN = 8$ centimeters, find DN.
4. If $EN = 13$ feet, what is the diameter of the circle?

Example 3 **The diameters of $\odot A$, $\odot B$, and $\odot C$ are 8 inches, 18 inches, and 11 inches, respectively. Find each measure.**

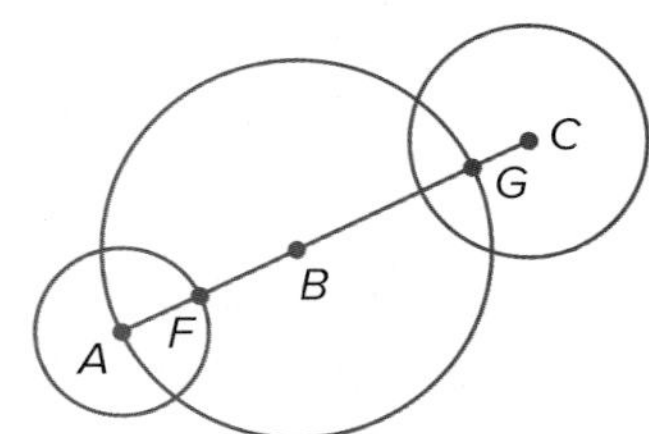

5. FG
6. FB

Example 4

7. **RIDES** A circular ride similar to the one described at the beginning of the lesson has a diameter of 44 feet. What are the radius and circumference of the ride? Round to the nearest hundredth, if necessary.

Example 5

8. **MP MODELING** The circumference of the circular swimming pool shown is about 56.5 feet. What are the diameter and radius of the pool? Round to the nearest hundredth.

Example 6

9. The right triangle shown is inscribed in $\odot D$. Find the exact circumference of $\odot D$.

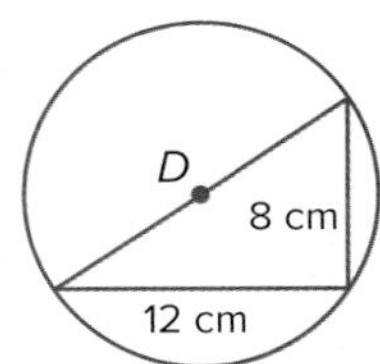

Practice and Problem Solving

Extra Practice is on page R9.

Examples 1–2 **For Exercises 10–13, refer to $\odot R$.**

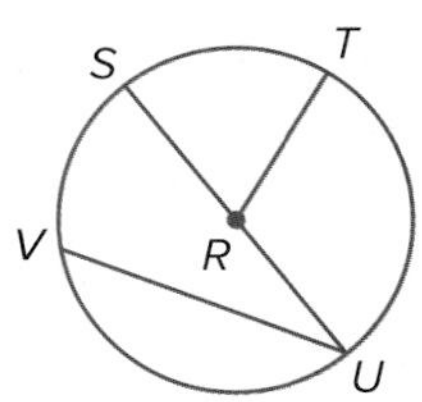

10. Name the center of the circle.
11. Identify a chord that is also a diameter.
12. Is $\overline{VU}$ a radius? Explain.
13. If $SU = 16.2$ centimeters, what is RT?

For Exercises 14–17, refer to $\odot F$.

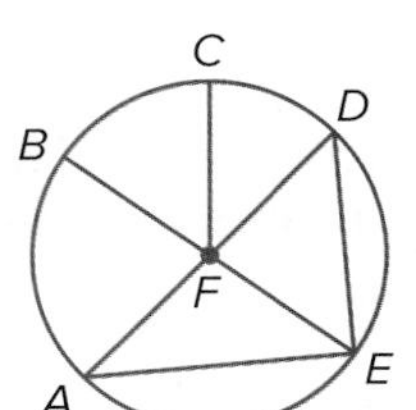

14. **Identify a chord that is not a diameter.**
15. If $CF = 14$ inches, what is the diameter of the circle?
16. Is $\overline{AF} \cong \overline{EF}$? Explain.
17. If $DA = 7.4$ centimeters, what is EF?

Example 3 **Circle J has a radius of 10 units, $\odot K$ has a radius of 8 units, and $BC = 5.4$ units. Find each measure.**

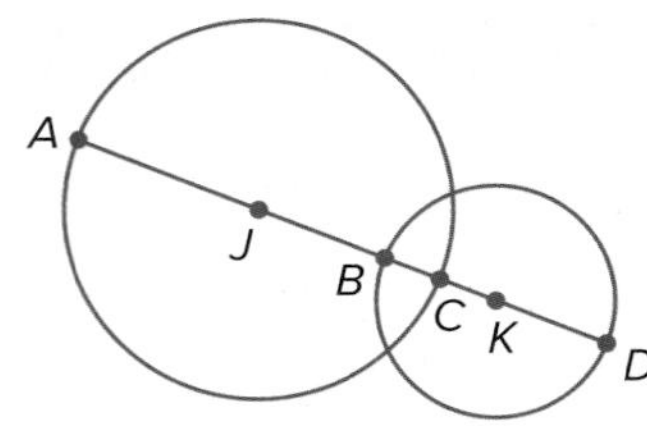

18. CK

19. AB

20. JK

21. AD

Example 4

22. **PIZZA** Find the radius and circumference of the pizza shown. Round to the nearest hundredth, if necessary.

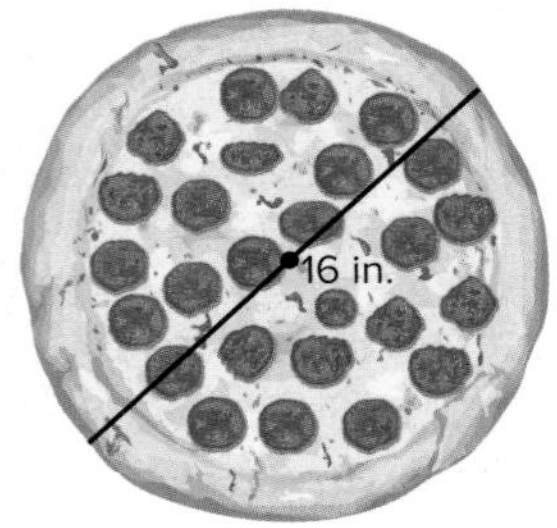

23. **BICYCLES** A bicycle has tires with a diameter of 26 inches. Find the radius and circumference of a tire. Round to the nearest hundredth, if necessary.

Example 5 **Find the diameter and radius of a circle with the given circumference. Round to the nearest hundredth.**

24. $C = 18$ in.

25. $C = 124$ ft

26. $C = 375.3$ cm

27. $C = 2608.25$ m

Example 6 MP **SENSE-MAKING** Find the exact circumference of each circle by using the given inscribed or circumscribed polygon.

28.

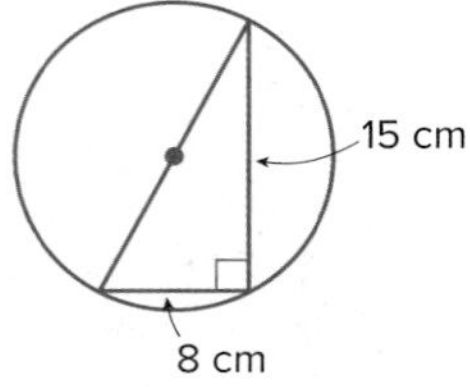

29.

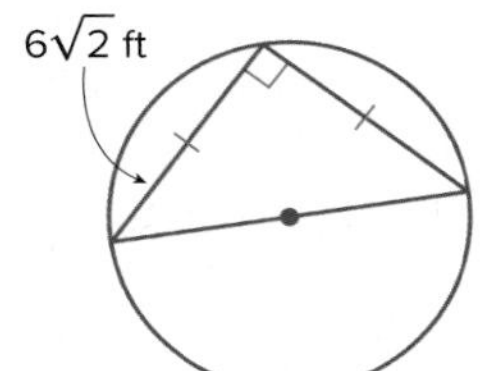

30.

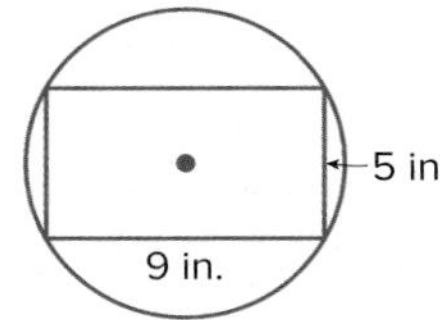

31.

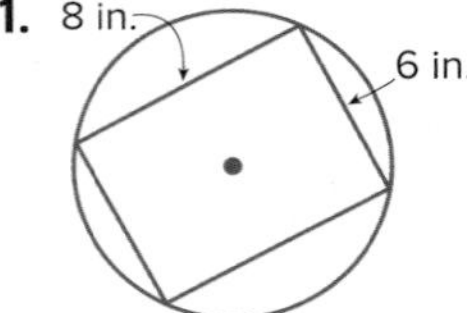

32.

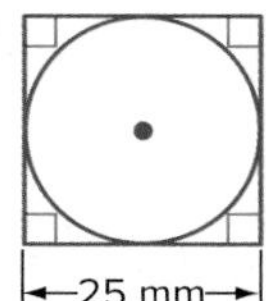

33.

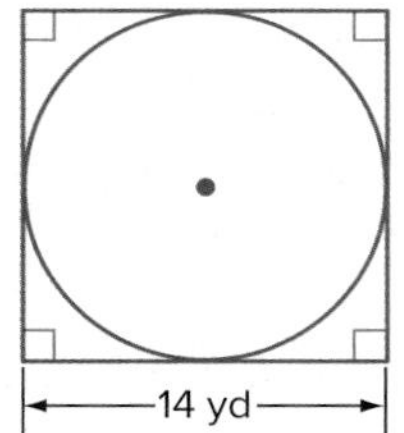

34. **DISC GOLF** Disc golf is similar to regular golf, except that a flying disc is used instead of a ball and clubs. For professional competitions, the maximum weight of a disc in grams is 8.3 times the diameter in centimeters. What is the maximum allowable weight for a disc with circumference 66.92 centimeters? Round to the nearest tenth.

35. **PATIOS** Mr. Martinez is going to build the patio shown.

    a. What is the patio's approximate circumference?

    b. If Mr. Martinez changes the plans so that the inner circle has a circumference of approximately 25 feet, what should the radius of the circle be to the nearest foot?

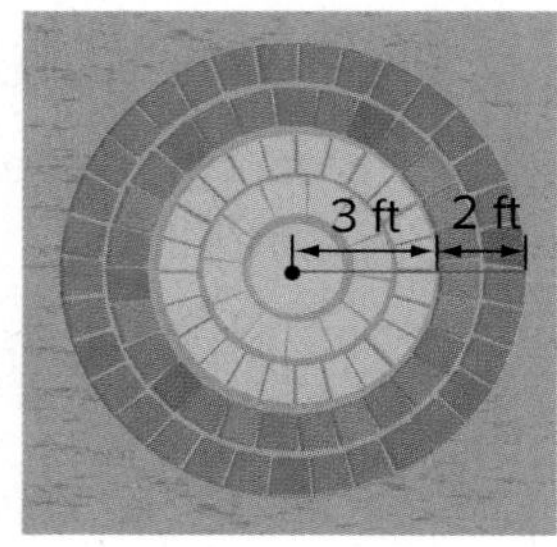

The radius, diameter, or circumference of a circle is given. Find each missing measure to the nearest hundredth.

36. $d = 8\frac{1}{2}$ in., $r =$ __?__, $C =$ __?__

37. $r = 11\frac{2}{5}$ ft, $d =$ __?__, $C =$ __?__

38. $C = 35x$ cm, $d =$ __?__, $r =$ __?__

39. $r = \frac{x}{8}$, $d =$ ____, $C =$ __?__

Determine whether the circles in the figures below appear to be *congruent, concentric,* or *neither.*

40.

41.

42.

43. **HISTORY** The *Indian Shell Ring* on Hilton Head Island approximates a circle. If each unit on the coordinate grid represents 25 feet, how far would someone have to walk to go completely around the ring? Round to the nearest tenth.

44. **MODELING** A brick path is being installed around a circular pond. The pond has a circumference of 68 feet. The outer edge of the path is going to be 4 feet from the pond all the way around. What is the approximate circumference of the path? Round to the nearest hundredth.

45. **MULTIPLE REPRESENTATIONS** In this problem, you will explore changing dimensions in circles.

a. **Geometric** Use a compass to draw three circles in which the scale factor from each circle to the next larger circle is 1 : 2.

b. **Tabular** Calculate the radius (to the nearest tenth) and circumference (to the nearest hundredth) of each circle. Record your results in a table.

c. **Verbal** Explain why these three circles are geometrically similar.

d. **Verbal** Make a conjecture about the ratio between the circumferences of two circles when the ratio between their radii is 2.

e. **Analytical** The scale factor from $\odot A$ to $\odot B$ is $\frac{b}{a}$. Write an equation relating the circumference (C_A) of $\odot A$ to the circumference (C_B) of $\odot B$.

f. **Numerical** If the scale factor from $\odot A$ to $\odot B$ is $\frac{1}{3}$, and the circumference of $\odot A$ is 12 inches, what is the circumference of $\odot B$?

46. **BUFFON'S NEEDLE** Measure the length ℓ of a needle (or toothpick) in centimeters. Next, draw a set of horizontal lines that are ℓ centimeters apart on a sheet of plain white paper.

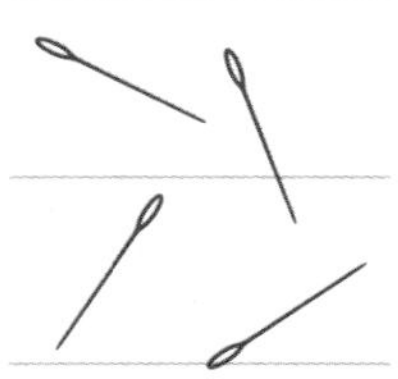

a. Drop the needle onto the paper. When the needle lands, record whether it touches one of the lines as a hit. Record the number of hits after 25, 50, and 100 drops.

b. Calculate the ratio of two times the total number of drops to the number of hits after 25, 50, and 100 drops.

c. How are the values you found in part **b** related to π?

47 **MAPS** The concentric circles on the map below show the areas that are 5, 10, 15, 20, 25, and 30 miles from downtown Phoenix.

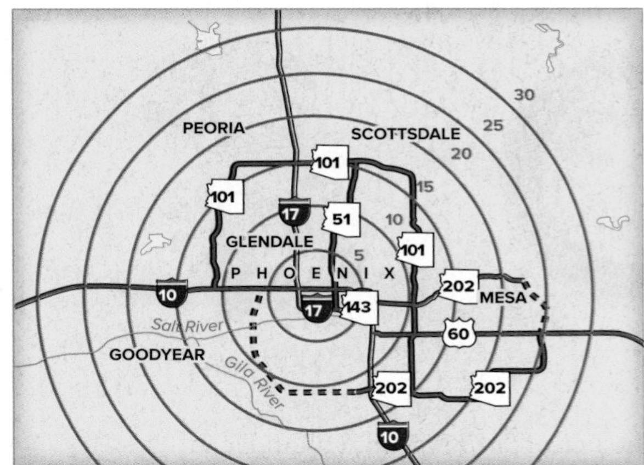

a. How much greater is the circumference of the outermost circle than the circumference of the center circle?

b. As the radii of the circles increase by 5 miles, by how much does the circumference increase?

H.O.T. Problems Use Higher-Order Thinking Skills

48. **WRITING IN MATH** How can we describe the relationships that exist between circles and lines?

49. **REASONING** In the figure, a circle with radius r is inscribed in a regular polygon and circumscribed about another.

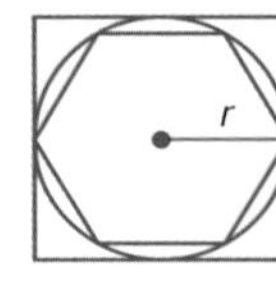

a. What are the perimeters of the circumscribed and inscribed polygons in terms of r? Explain.

b. Is the circumference C of the circle greater or less than the perimeter of the circumscribed polygon? the inscribed polygon? Write a compound inequality comparing C to these perimeters.

c. Rewrite the inequality from part **b** in terms of the diameter d of the circle and interpret its meaning.

d. As the number of sides of both the circumscribed and inscribed polygons increase, what will happen to the upper and lower limits of the inequality from part **c**, and what does this imply?

50. **CHALLENGE** The sum of the circumferences of circles H, J, and K shown at the right is 56π units. Find KJ.

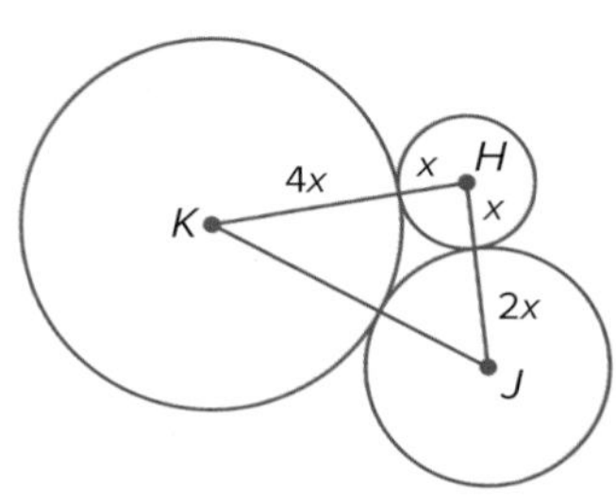

51. **REASONING** Is the distance from the center of a circle to a point in the interior of a circle *sometimes, always,* or *never* less than the radius of the circle? Explain.

52. **CONSTRUCT ARGUMENTS** Use the locus definition of a circle and dilations to prove that all circles are similar.

53. **CHALLENGE** In the figure, $\odot P$ is inscribed in equilateral triangle LMN. What is the circumference of $\odot P$?

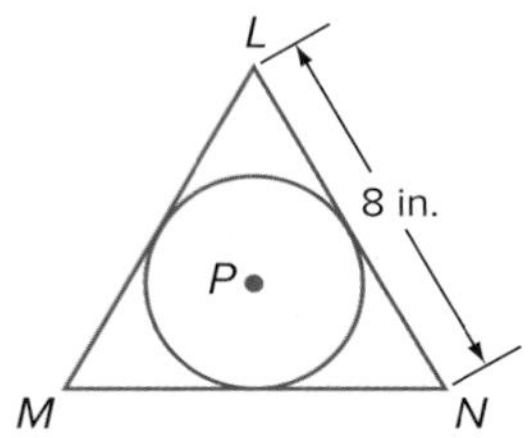

54. **WRITING IN MATH** Research and write about the history of pi and its importance to the study of geometry.

55. What is the circumference of $\odot C$ to the nearest hundredth? MP 1, 6

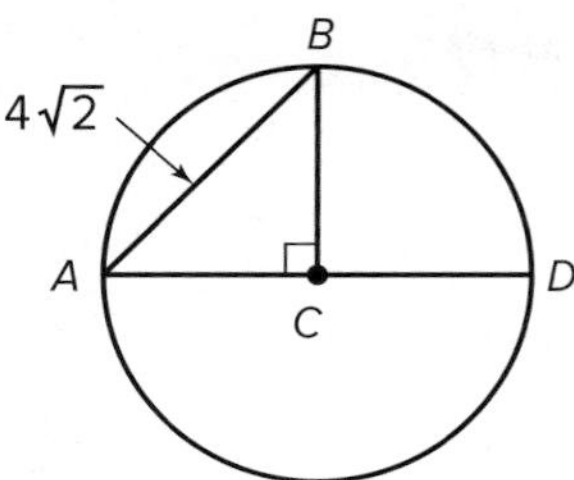

56. *MNPQ* is a square. The radius of $\odot R$ is 3.

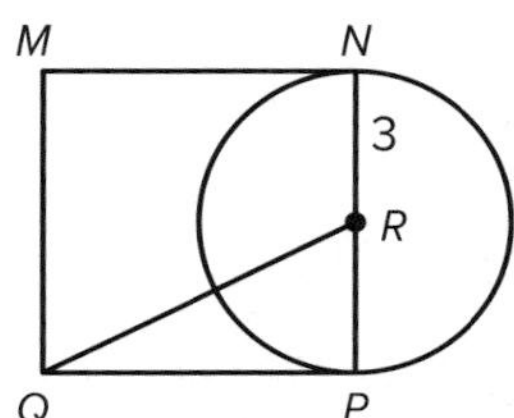

Which of the following is the length of $\overline{QR}$? MP 1, 6

- A 3
- B $3\sqrt{2}$
- C 6
- D $3\sqrt{5}$

57. In the figure, *C* is the midpoint of $\overline{AE}$, *B* is the midpoint of $\overline{AC}$, *D* is the midpoint of $\overline{CE}$, and $AE = 32$.

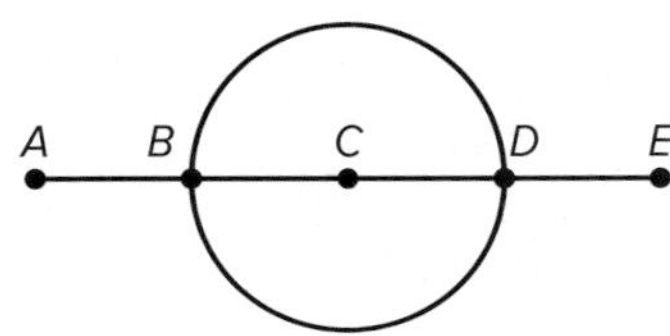

What is the circumference of $\odot C$ to the nearest hundredth? MP 1, 6

58. The right triangle $\triangle JKL$ is inscribed in $\odot M$, as shown.

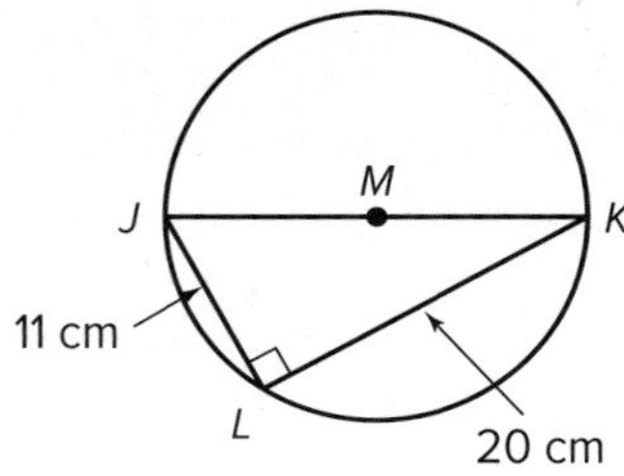

Which of the following is the best estimate of the circumference of $\odot M$? MP 1, 6

- A 35.9 cm
- B 62.8 cm
- C 71.7 cm
- D 143.4 cm
- E 409.2 cm

59. **MULTI-STEP** Square *JKLM* has an area of 9 square meters. A circle with center *M* passes through *J* and *L*. MP 1, 6

a. Name the circle.

b. Name a line segment that is a radius of the circle.

c. What is the circumference of the circle?

60. Triangle *DEF* is an equilateral triangle with side lengths of 2 inches. MP 1, 6

a. Explain how to find the diameter of circle *C*.

b. What is the circumference of circle *C*?

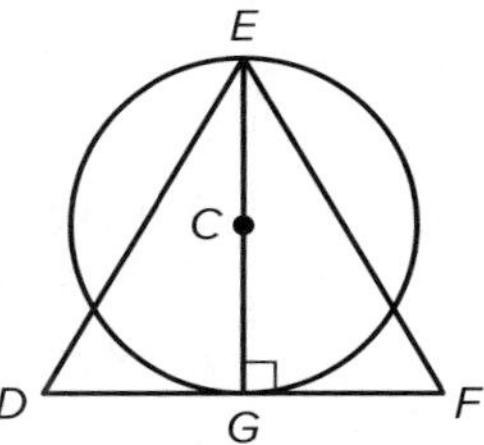

LESSON 2

Measuring Angles and Arcs

Then

- You measured angles and identified congruent angles.

Now

1. Identify central angles, major arcs, minor arcs, and semicircles, and find their measures.
2. Find arc lengths.

Why?

- The thirteen stars of the Betsy Ross flag are arranged equidistant from each other and from a fixed point. The distance between consecutive stars varies depending on the size of the flag, but the measure of the central angle formed by the center of the circle and any two consecutive stars is always the same.

New Vocabulary

central angle
arc
minor arc
major arc
semicircle
congruent arcs
adjacent arcs
arc length
radian measure

Mathematical Practices

6 Attend to precision.
4 Model with mathematics.

1 Angles and Arcs

A **central angle** of a circle is an angle with a vertex in the center of the circle. Its sides contain two radii of the circle. $\angle ABC$ is a central angle of $\odot B$.

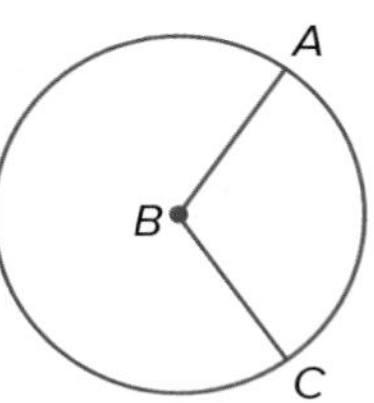

Recall from Lesson 1-4 that a *degree* is $\frac{1}{360}$ of the circular rotation about a point. This leads to the following relationship.

Key Concept Sum of Central Angles

Words The sum of the measures of the central angles of a circle with no interior points in common is 360.

Example $m\angle 1 + m\angle 2 + m\angle 3 = 360$

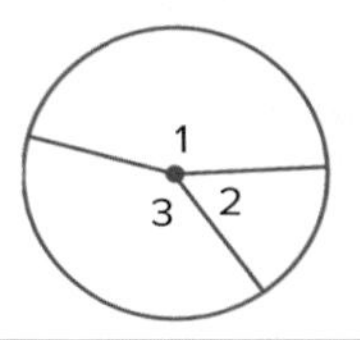

Example 1 Find Measures of Central Angles

Find the value of *x*.

$m\angle GFH + m\angle HFJ + m\angle GFJ = 360$	Sum of Central Angles
$130 + 90 + m\angle GFJ = 360$	Substitution
$220 + m\angle GFJ = 360$	Simplify.
$m\angle GFJ = 140$	Subtract 220 from each side.

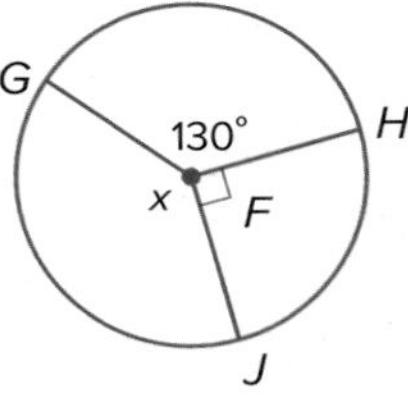

Guided Practice

1A.

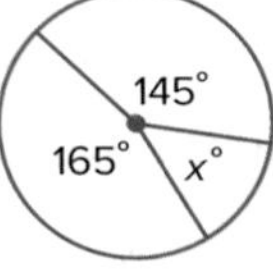

1B.

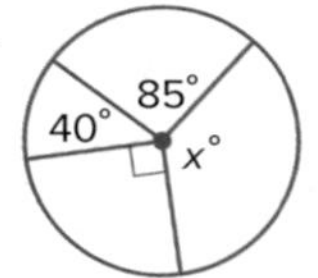

An **arc** is a portion of a circle defined by two endpoints. A central angle separates the circle into two arcs with measures related to the measure of the central angle.

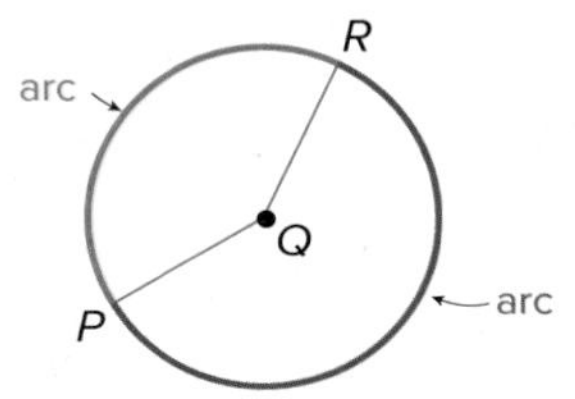

Study Tip

Naming Arcs Minor arcs are named by their endpoints. Major arcs and semicircles are named by their endpoints and another point on the arc that lies between these endpoints.

Key Concept Arcs and Arc Measure

Arc	Measure
A **minor arc** is the shortest arc connecting two endpoints on a circle.	The measure of a minor arc is less than 180 and equal to the measure of its related central angle. $m\widehat{AB} = m\angle ACB = x$
A **major arc** is the longest arc connecting two endpoints on a circle.	The measure of a major arc is greater than 180, and equal to 360 minus the measure of the minor arc with the same endpoints. $m\widehat{ADB} = 360 - m\widehat{AB} = 360 - x$
A **semicircle** is an arc with endpoints that lie on a diameter.	The measure of a semicircle is 180. $m\widehat{ADB} = 180$

Example 2 Classify Arcs and Find Arc Measures

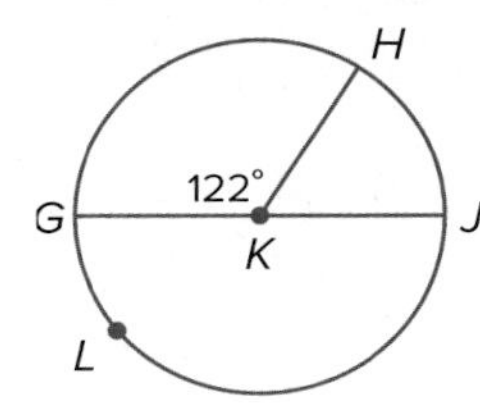

$\overline{GJ}$ is a diameter of $\odot K$. Identify each arc as a *major arc*, *minor arc*, or *semicircle*. Then find its measure.

a. $\widehat{GH}$

$\widehat{GH}$ is a minor arc, so $m\widehat{GH} = m\angle GKH$ or 122.

b. $\widehat{GLH}$

$\widehat{GLH}$ is a major arc that shares the same endpoints as minor arc $\widehat{GH}$.

$m\widehat{GHL} = 360 - m\widehat{GH}$

$= 360 - 122$ or 238

c. $\widehat{GLJ}$

$\widehat{GLJ}$ is a semicircle, so $m\widehat{GLJ} = 180$.

Guided Practice

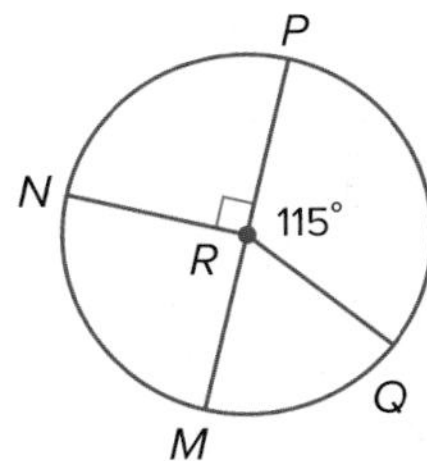

$\overline{PM}$ is a diameter of $\odot R$. Identify each arc as a *major arc*, *minor arc*, or *semicircle*. Then find its measure.

2A. $\widehat{MQ}$ **2B.** $\widehat{MNP}$ **2C.** $\widehat{MNQ}$

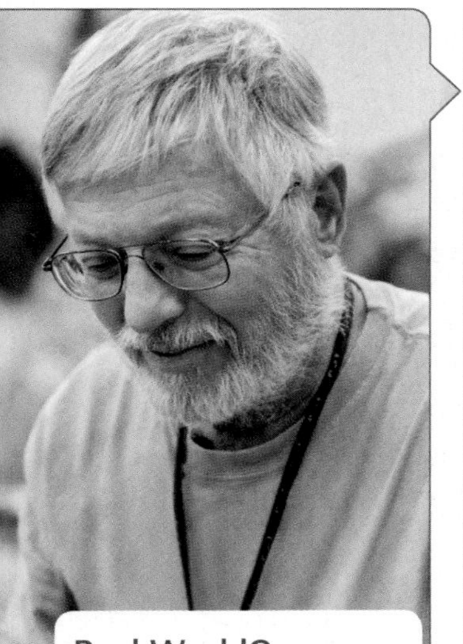

Real-WorldCareer

Historical Researcher Research in museums includes authentication, verification, and description of artifacts. Employment as a historical researcher requires a minimum of a bachelor's degree in history.

Congruent arcs are arcs in the same or congruent circles that have the same measure.

Theorem 9.1

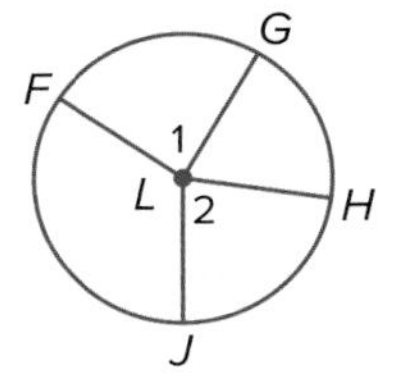

Words In the same circle or in congruent circles, two minor arcs are congruent if and only if their central angles are congruent.

Example If $\angle 1 \cong \angle 2$, then $\widehat{FG} \cong \widehat{HJ}$.
If $\widehat{FG} \cong \widehat{HJ}$, then $\angle 1 \cong \angle 2$.

You will prove Theorem 9.1 in Exercise 54.

Real-World Example 3 Find Arc Measures in Circle Graphs

SPORTS Refer to the circle graph. Find $m\widehat{CD}$.

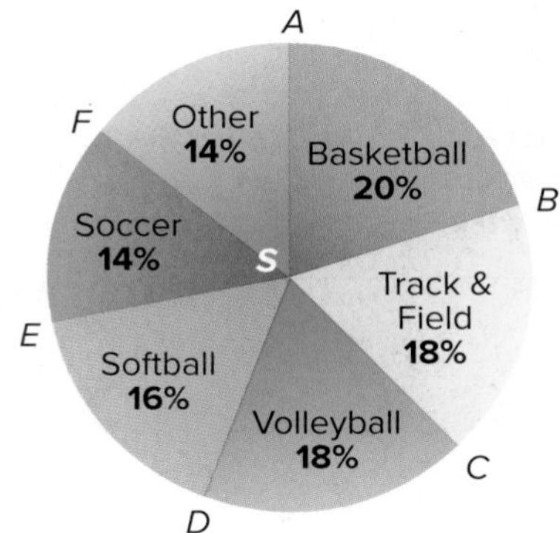

$\widehat{CD}$ is a minor arc. $m\widehat{CD} = m\angle CSD$

$\angle CSD$ represents 18% of the whole, or 18% of the circle.

$m\angle CSD = 0.18(360)$ Find 18% of 360.

$= 64.8$ Simplify.

Guided Practice

3A. $m\widehat{EF}$

3B. $m\widehat{FA}$

Adjacent arcs are arcs in a circle that have exactly one point in common. In $\odot M$, $\widehat{HJ}$ and $\widehat{JK}$ are adjacent arcs. As with adjacent angles, you can add the measures of adjacent arcs.

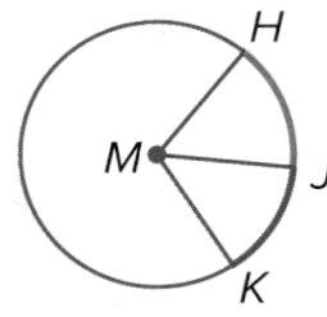

Postulate 9.1 Arc Addition Postulate

Words The measure of an arc formed by two adjacent arcs is the sum of the measures of the two arcs.

Example $m\widehat{XYZ} = m\widehat{XY} + m\widehat{YZ}$

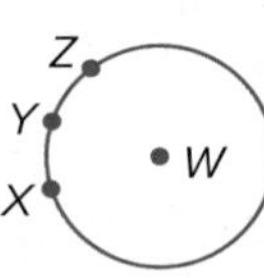

Math History Link

Euclid (c. 325–265 B.C.) The 13 books of Euclid's *Elements* are influential works of science. In them, geometry and other branches of mathematics are logically developed. Book 3 of *Elements* is devoted to circles, arcs, and angles.

Example 4 Use Arc Addition to Find Measures of Arcs

Find $m\widehat{AED}$ in $\odot F$.

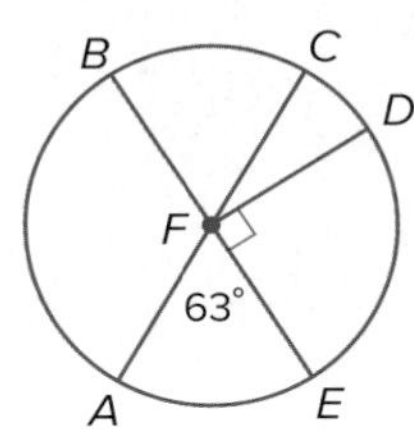

$m\widehat{AED} = m\widehat{AE} + m\widehat{ED}$ Arc Addition Postulate

$= m\angle AFE + m\angle EFD$ $m\widehat{AE} = m\angle AFE$, $m\widehat{ED} = m\angle EFD$

$= 63 + 90$ or 153 Substitution

Guided Practice

4A. $m\widehat{CE}$

4B. $m\widehat{ABD}$

Watch Out!

Arc Length The length of an arc is given in linear units, such as centimeters. The measure of an arc is given in degrees.

2 Arc Length

Arc length is the distance between the endpoints along an arc measured in linear units. Because an arc is a portion of a circle, its length is a fraction of the circumference.

Key Concept Arc Length

Words The ratio of the length of an arc ℓ to the circumference of the circle is equal to the ratio of the degree measure of the arc to 360.

Proportion $\frac{\ell}{2\pi r} = \frac{x}{360}$ or

Equation $\ell = \frac{x}{360} \cdot 2\pi r$

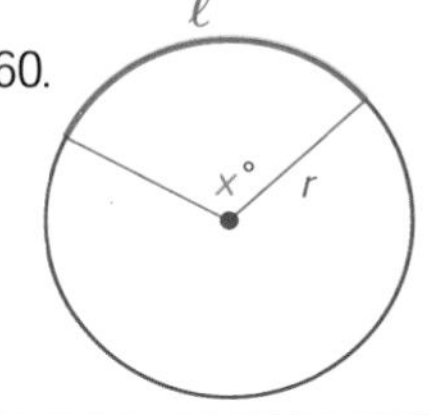

Example 5 Find Arc Length

Find the length of $\widehat{ZY}$. Round to the nearest hundredth.

a.

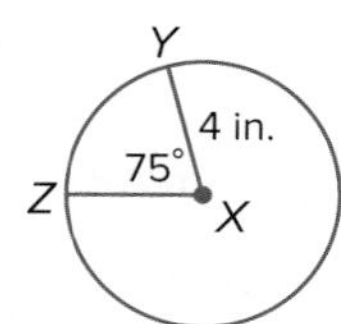

$\ell = \frac{x}{360} \cdot 2\pi r$ Arc Length Equation

$= \frac{75}{360} \cdot 2\pi(4)$ Substitution

≈ 5.24 in. Use a calculator.

b.

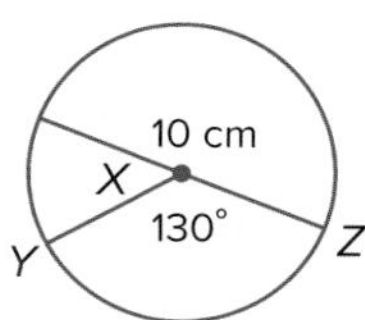

$\ell = \frac{x}{360} \cdot 2\pi r$ Arc Length Equation

$= \frac{130}{360} \cdot 2\pi(5)$ Substitution

≈ 11.34 cm Use a calculator.

> **Study Tip**
>
> **MP Sense-Making** The arc lengths in Examples 5a and 5b could also have been calculated using the arc length proportion $\frac{\ell}{2\pi r} = \frac{x}{360}$.

Guided Practice

Find the length of $\widehat{AB}$. Round to the nearest hundredth.

5A.

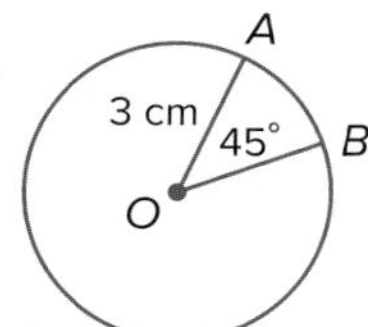

5B.

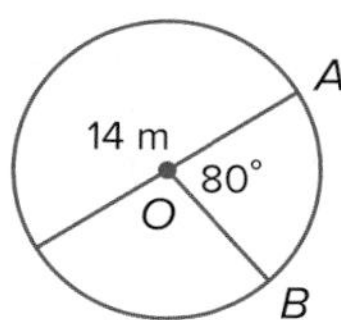

5C.

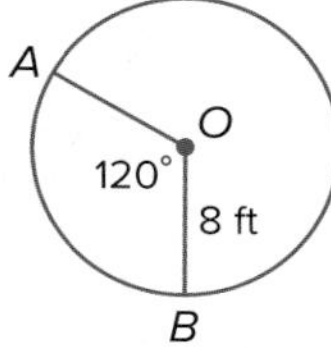

Angles can also be measured in units that are based on arc length. The **radian measure**, θ, of a central angle is the ratio of the arc length to the radius of the circle: $\theta = \frac{\ell}{r}$ radians.

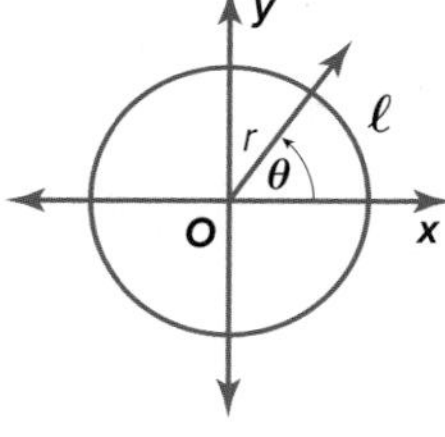

$\theta = \frac{\ell}{r}$ radians

> **Study Tip**
>
> **Radian Measure** The circumference of a circle is $2\pi r$. So, one complete revolution around a circle equals 2π radians.
>
> $\theta = \frac{2\pi r}{r}$
>
> $= 2\pi$ radians $= 360°$

Example 6 Find Arc Length Using Radian Measure

Find the length of $\widehat{ZY}$. Round to the nearest hundredth.

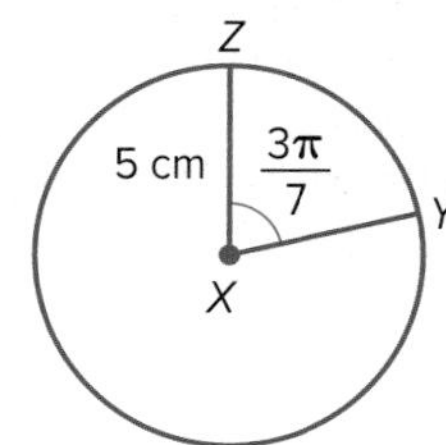

$\theta = \frac{\ell}{r}$ Arc Length Equation

$\frac{3}{7}\pi = \frac{\ell}{5}$ $\theta = \frac{3}{7}\pi, r = 5$

$5\left(\frac{3}{7}\pi\right) = \ell$ Multiply each side by 5.

$\frac{15}{7}\pi = \ell$ Simplify.

$6.73 \approx \ell$ Use a calculator.

Guided Practice

6A.

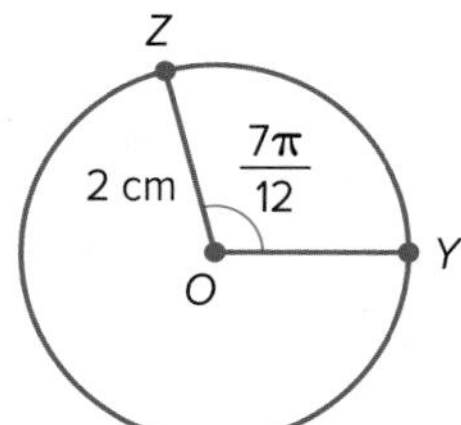

6B.

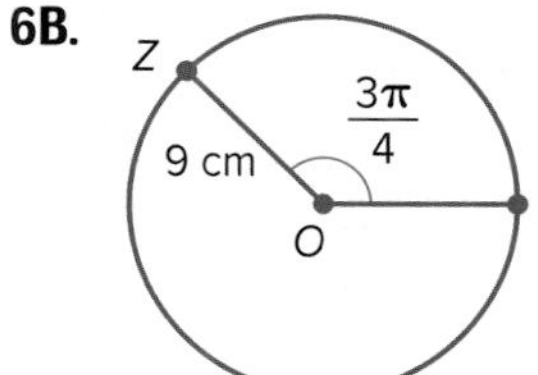

Check Your Understanding

Example 1 G.C.2

Find the value of x.

1. 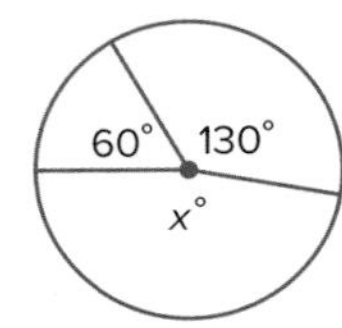

2. 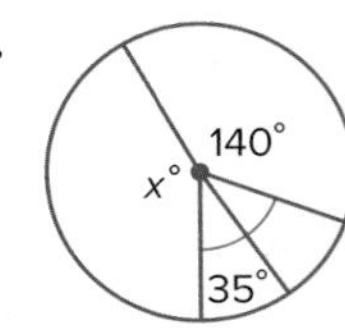

Example 2 G.C.2

PRECISION $\overline{HK}$ and $\overline{IG}$ are diameters of ⊙L. Identify each arc as a *major arc, minor arc,* or *semicircle*. Then find its measure.

3. $\widehat{IHJ}$
4. $\widehat{HI}$
5. $\widehat{HGK}$

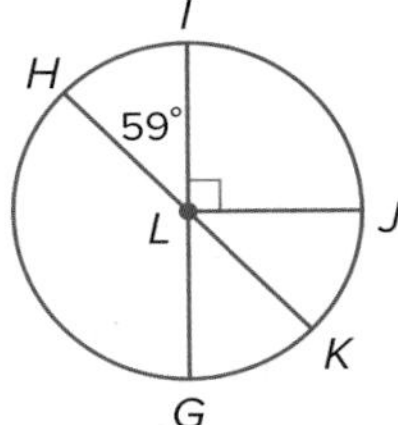

Example 3 G.C.5

6. **FITNESS** The graph shows the results of a survey taken by high school students regarding what activities they participate in after school.
   a. Find $m\widehat{AB}$.
   b. Find $m\widehat{BC}$.
   c. Describe the type of arc that the category Sports represents.

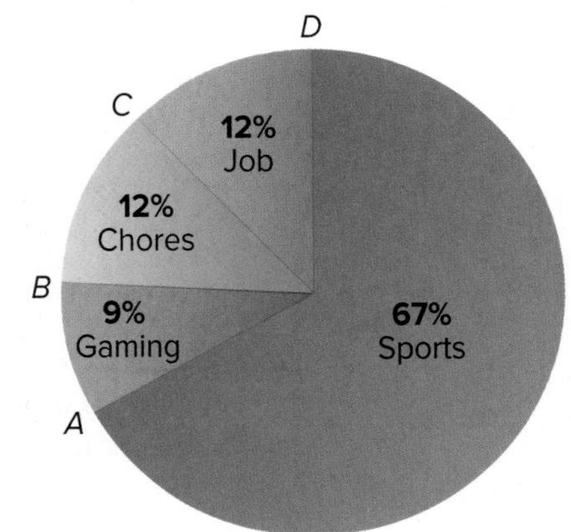

Example 4 G.C.2

$\overline{QS}$ is a diameter of ⊙V. Find each measure.

7. $m\widehat{STP}$
8. $m\widehat{QRT}$
9. $m\widehat{PQR}$

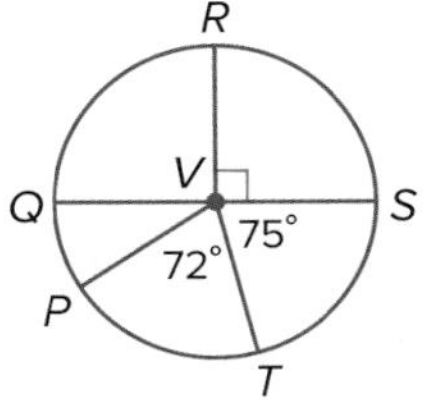

Examples 5-6 G.C.5

Find the length of $\widehat{JK}$. Round to the nearest hundredth.

10. 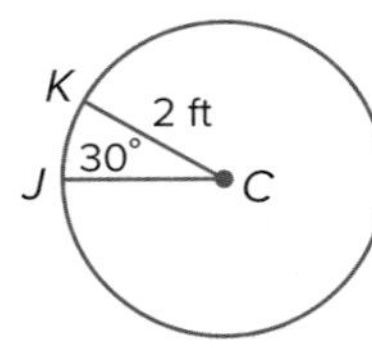

11. 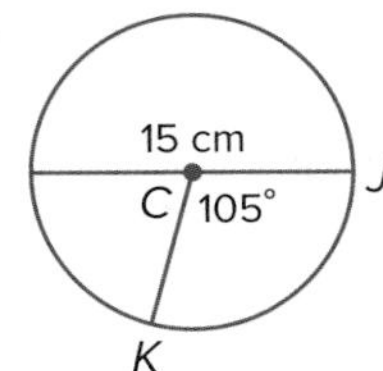

12. 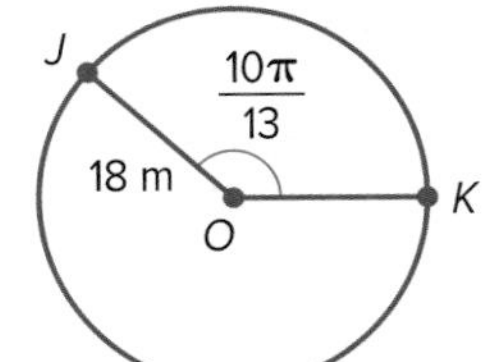

13. 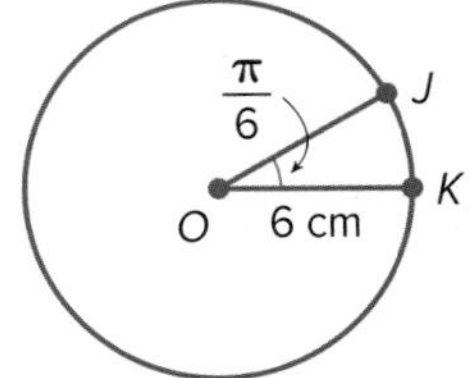

Practice and Problem Solving

Extra Practice is on page R9.

Example 1 G.C.2

Find the value of x.

14. 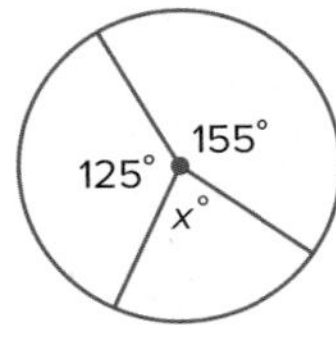

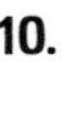 15. 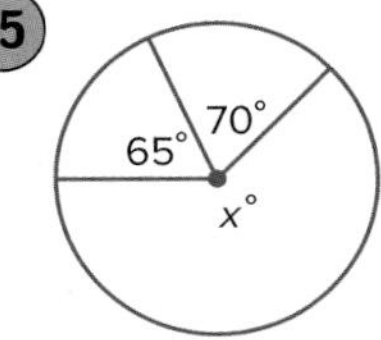

16. 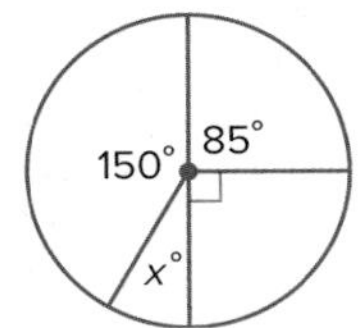

17. 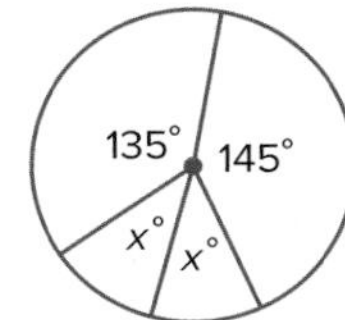

Example 2 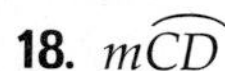 $\overline{AD}$ **and** $\overline{CG}$ **are diameters of** $\odot B$. **Identify each arc as a *major arc, minor arc,* or *semicircle.* Then find its measure.**

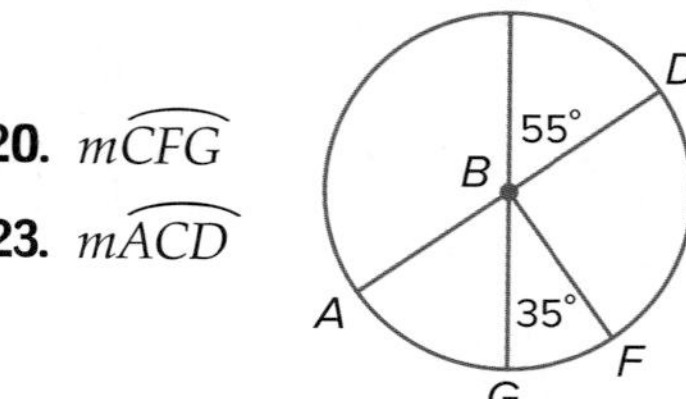

18. $m\widehat{CD}$
19. $m\widehat{AC}$
20. $m\widehat{CFG}$
21. $m\widehat{CGD}$
22. $m\widehat{GCF}$
23. $m\widehat{ACD}$
24. $m\widehat{AG}$
25. $m\widehat{ACF}$

Example 3

26. **SHOPPING** The graph shows the results of a survey in which teens were asked where the best place was to shop for clothes.

Best Places to Clothes Shop

None of these 9%
Online 9%
Vintage stores 4%
Flea markets 2%
Mall 76%

a. What would be the arc measures associated with the mall and vintage stores categories?

b. Describe the kinds of arcs associated with the category "Mall" and the category "None of these."

c. Are there any congruent arcs in this graph? Explain.

27. **MP MODELING** The table shows the distribution of endangered animals by species in a south central part of the United States.

a. If you were to construct a circle graph of this information, what would be the arc measures associated with the first two categories?

b. Describe the kind of arcs associated with the first and last category.

c. Are there any congruent arcs in this graph? Explain.

Endangered Animals in South Central U.S.	
mammals	25%
fish	28%
reptiles	21%
amphibians	11%
invertebrates	15%

Source: All About Wildlife

Example 2.4 **ENTERTAINMENT** **Use the Ferris wheel shown to find each measure.**

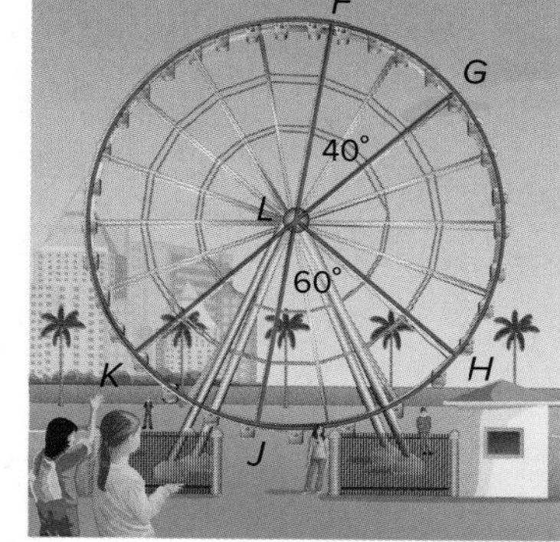

28. $m\widehat{FG}$
29. $m\widehat{JH}$
30. $m\widehat{JKF}$
31. $m\widehat{JFH}$
32. $m\widehat{GHF}$
33. $m\widehat{GHK}$
34. $m\widehat{HK}$
35. $m\widehat{JKG}$
36. $m\widehat{KFH}$
37. $m\widehat{HGF}$

Examples 5-6 **Use** $\odot P$ **to find the length of each arc. Round to the nearest hundredth.**

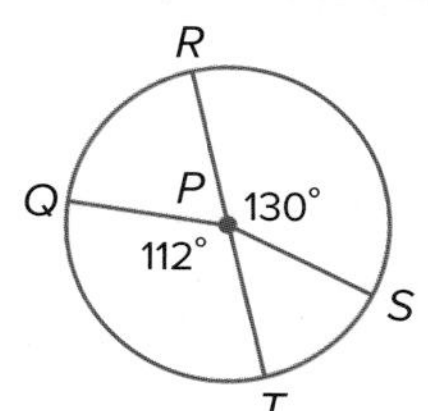

38. $\widehat{RS}$, if the radius is 2 inches
39. $\widehat{QT}$, if the diameter is 9 centimeters
40. $\widehat{QR}$, if $PS = 4$ millimeters
41. $\widehat{RS}$, if $RT = 15$ inches
42. $\widehat{QRS}$, if $RT = 11$ feet
43. $\widehat{RTS}$, if $PQ = 3$ meters

HISTORY The figure shows the stars in the Betsy Ross flag referenced at the beginning of the lesson.

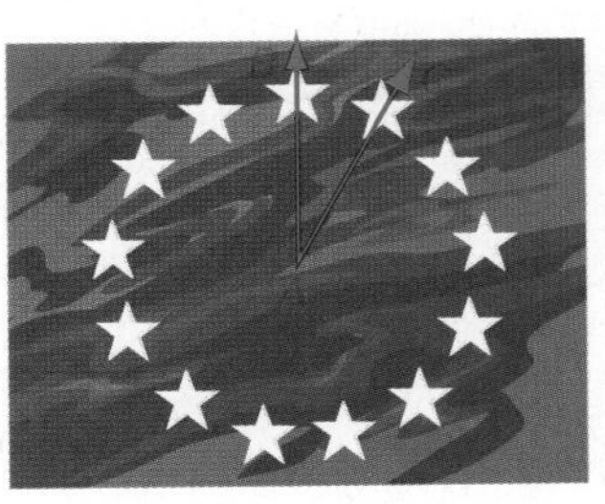

44. What is the measure of central angle A? Explain how you determined your answer.

45. If the diameter of the circle were doubled, what would be the effect on the arc length from the center of one star B to the next star C?

46. FARMS The *Pizza Farm* in Madera, California, is a circle divided into eight equal slices, as shown at the right. Each "slice" is used for growing or grazing pizza ingredients.

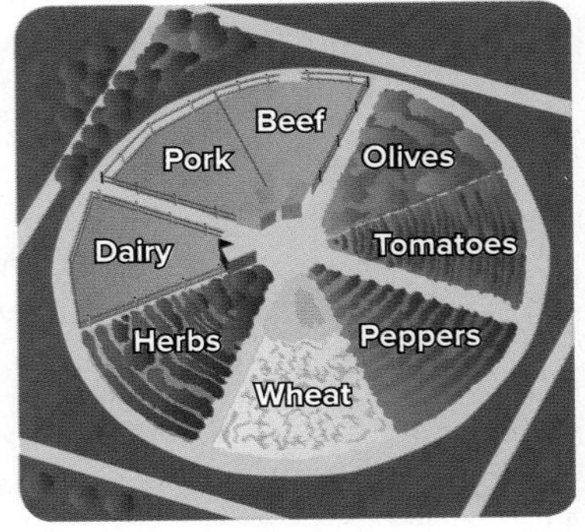

a. What is the total arc measure of the slices containing olives, tomatoes, and peppers?

b. The circle is 125 feet in diameter. What is the arc length of one slice? Round to the nearest hundredth.

MP REASONING Find each measure. Round each linear measure to the nearest hundredth and each arc measure to the nearest degree.

47. circumference of $\odot S$

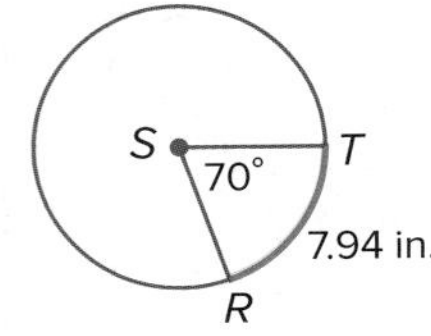

48. $m\widehat{CD}$

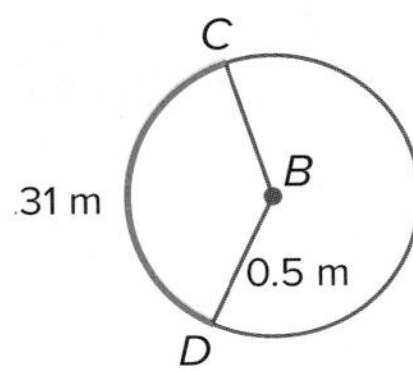

49. radius of $\odot K$

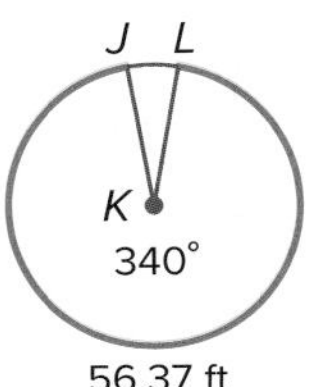

ALGEBRA In $\odot C$, $m\angle HCG = 2x$ and $m\angle HCD = 6x + 28$. Find each measure.

50. $m\widehat{EF}$

51. $m\widehat{HD}$

52. $m\widehat{HGF}$

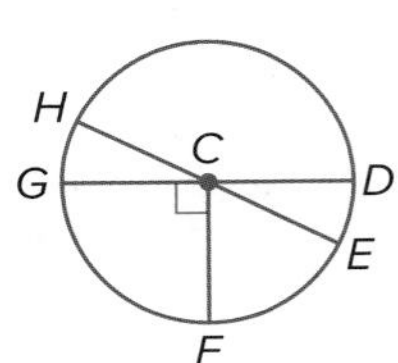

53 RIDES A pirate ship ride follows a semicircular path, as shown in the diagram.

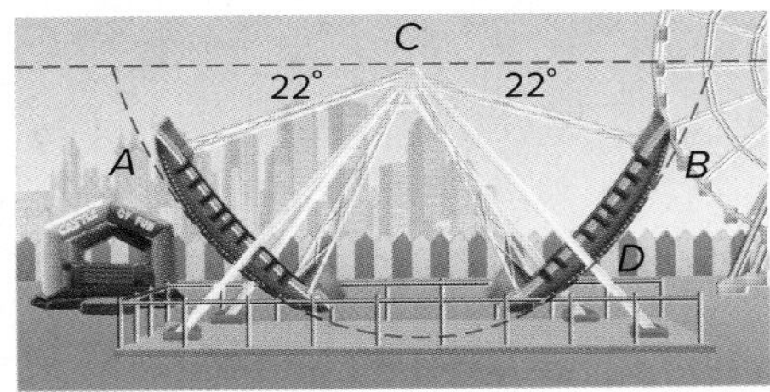

a. What is $m\widehat{AB}$?

b. If $CD = 62$ feet, what is the length of $\widehat{AB}$? Round to the nearest hundredth.

54. PROOF Write a two-column proof of Theorem 9.1.

Given: $\angle BAC \cong \angle DAE$

Prove: $\widehat{BC} \cong \widehat{DE}$

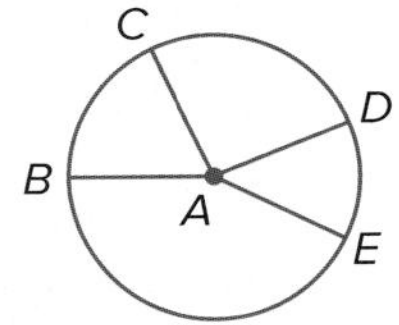

55. **COORDINATE GEOMETRY** In the graph, point M is located at the origin. Find each measure in $\odot M$. Round each linear measure to the nearest hundredth and each arc measure to the nearest tenth degree.

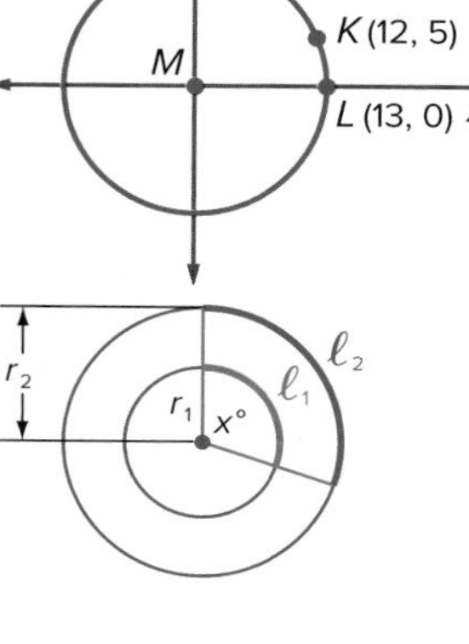

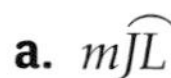

a. $m\widehat{JL}$ **b.** $m\widehat{KL}$ **c.** $m\widehat{JK}$
d. length of $\widehat{JL}$ **e.** length of $\widehat{JK}$

56. **ARC LENGTH AND RADIAN MEASURE** In this problem, you will use concentric circles to show that the length of the arc intercepted by a central angle of a circle is dependent on the circle's radius.

a. Compare the measures of arc ℓ_1 and arc ℓ_2. Then compare the lengths of arc ℓ_1 and arc ℓ_2. What do these two comparisons suggest?

b. Use similarity transformations (dilations) to explain why the length of an arc ℓ intercepted by a central angle of a circle is proportional to the circle's radius r. That is, explain why we can say that for this diagram, $\frac{\ell_1}{r_1} = \frac{\ell_2}{r_2}$.

c. Write expressions for the lengths of arcs ℓ_1 and ℓ_2. Use these expressions to identify the constant of proportionality k in $\ell = kr$.

d. The expression that you wrote for k in part **c** converts the degree measure of an angle to the *radian measure* of an angle. Use it to find the radian measure of an angle measuring 90°.

H.O.T. Problems Use Higher-Order Thinking Skills

57. **ERROR ANALYSIS** Brody says that $\widehat{WX}$ and $\widehat{YZ}$ are congruent since their central angles have the same measure. Selena says they are not congruent. Is either of them correct? Explain your reasoning.

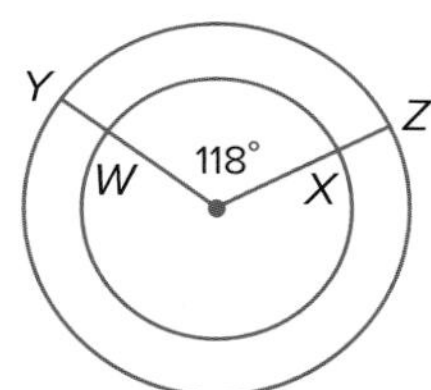

CONSTRUCT ARGUMENTS Determine whether each statement is *sometimes, always,* or *never* true. Explain your reasoning.

58. The measure of a minor arc is less than 180.

59. If a central angle is obtuse, its corresponding arc is a major arc.

60. The sum of the measures of adjacent arcs of a circle depends on the measure of the radius.

61. **CHALLENGE** The measures of $\widehat{LM}$, $\widehat{MN}$, and $\widehat{NL}$ are in the ratio 5 : 3 : 4. Find the measure of each arc.

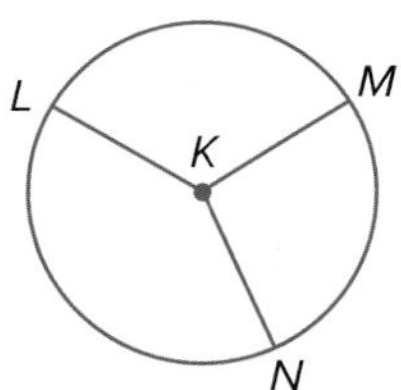

62. **OPEN-ENDED** Draw a circle and locate three points on the circle. Estimate the measures of the three nonoverlapping arcs that are formed. Then use a protractor to find the measure of each arc. Label your circle with the arc measures.

63. **CHALLENGE** The time shown on an analog clock is 8:10. What is the measure of the angle formed by the hands of the clock?

64. **WRITING IN MATH** Describe the three different types of arcs in a circle and the method for finding the measure of each one.

65. A model train runs on a circular track with a diameter of 4 meters, as shown. Which of the following is the best estimate of the distance the train travels as it moves from the station to the grain silo? MP 6

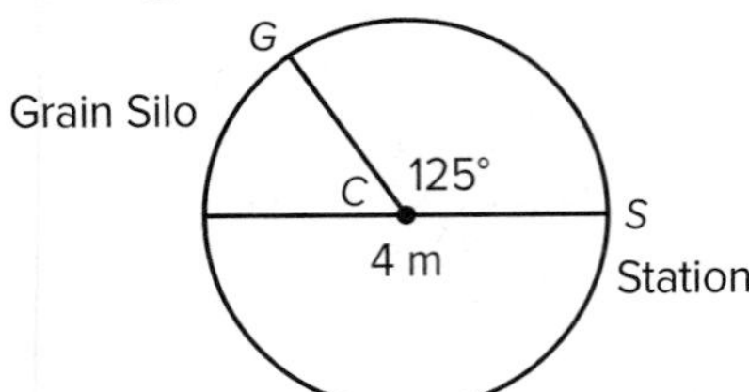

- ◯ **A** 12.6 m
- ◯ **B** 8.7 m
- ◯ **C** 4.4 m
- ◯ **D** 2.2 m

66. In $\odot J$, $\angle KJL \cong \angle LJM \cong \angle MJN$. Sofia wants to calculate the length of $\widehat{LN}$.

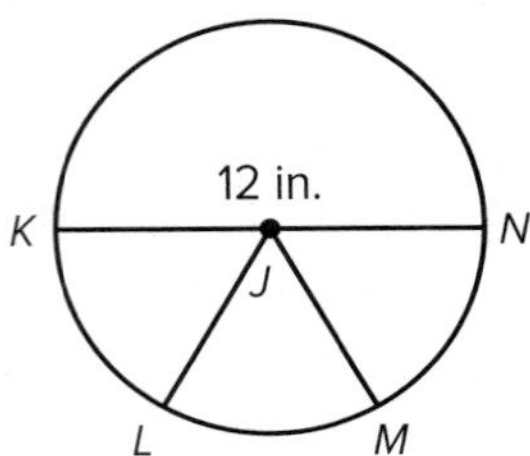

Which expression can Sofia use to find the required length? MP 6

- ◯ **A** $\frac{60}{360} \cdot 12\pi$
- ◯ **B** $\frac{60}{360} \cdot 24\pi$
- ◯ **C** $\frac{120}{360} \cdot 24\pi$
- ◯ **D** $\frac{120}{360} \cdot 12\pi$
- ◯ **E** $\frac{120}{360} \cdot 6\pi$

67. In $\odot B$, $m\angle LBM = 3x$ and $m\angle LBQ = 4x + 61$.

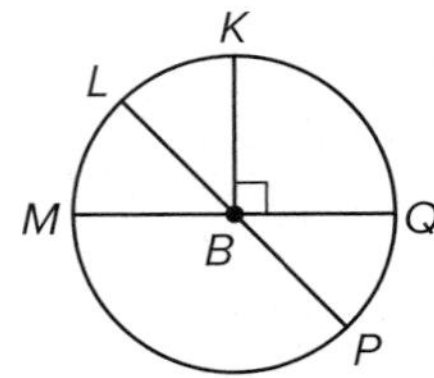

What is the measure of $\angle PBQ$? MP 6

[ ]

68. **MULTI-STEP** The table shows the polling numbers for an upcoming election MP 6

Distribution of Voters	
Candidate A	32%
Candidate B	30%
Candidate C	28%
Candidate D	5%
Candidate E	5%

a. If you were to construct a circle graph of this information, what would be the arc measures for candidates B and D?

b. Suppose candidates A and B are combined on the circle graph. What would be the arc measure? And what type of arc would this be?

c. Suppose candidates C, D, and E are combined on the circle graph. What would be the arc measure? And what type of arc would this be?

d. Are there any congruent arcs in this graph? Explain.

69. The minute hand of a clock is 3 inches long. Which of the following is the best estimate of the distance the tip of the hand moves as the time changes from 12:30 to 12:45? MP 6

- ◯ **A** 0.8 in.
- ◯ **B** 2.4 in.
- ◯ **C** 4.7 in.
- ◯ **D** 9.4 in.

70. In $\odot Q$, the length of $\widehat{ST}$ is 3π centimeters.

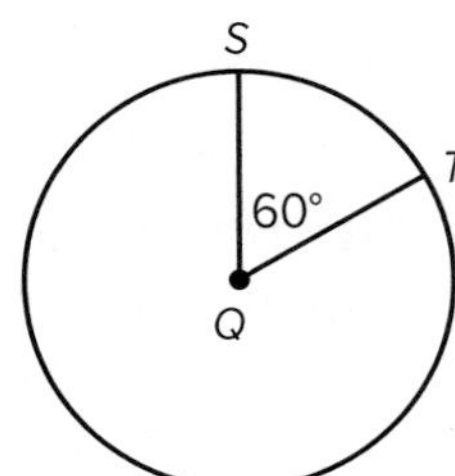

What is the radius of the circle? MP 2

- ◯ **A** 0.11 cm
- ◯ **B** 1.5 cm
- ◯ **C** 9 cm
- ◯ **D** 18 cm

LESSON 3

Arcs and Chords

Then

- You used the relationships between arcs and angles to find measures.

Now

1. Recognize and use relationships between arcs and chords.
2. Recognize and use relationships between arcs, chords, and diameters.

Why?

- Embroidery hoops are used in sewing, quilting, and cross-stitching, as well as for embroidering. The endpoints of the snowflake shown are both the endpoints of a chord and the endpoints of an arc.

Mathematical Practices

4 Model with mathematics.

3 Construct viable arguments and critique the reasoning of others.

1 Arcs and Chords

A *chord* is a segment with endpoints on a circle. If a chord is not a diameter, then its endpoints divide the circle into a major and a minor arc.

Theorem 9.2

Words In the same circle or in congruent circles, two minor arcs are congruent if and only if their corresponding chords are congruent.

Example $\widehat{FG} \cong \widehat{HJ}$ if and only if $\overline{FG} \cong \overline{HJ}$.

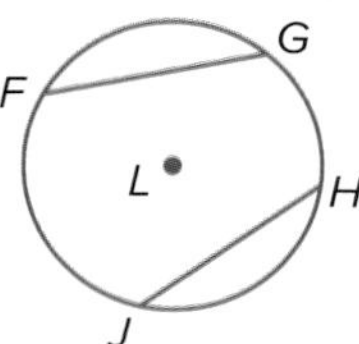

Proof Theorem 9.2 (part 1)

Given: $\odot P$; $\widehat{QR} \cong \widehat{ST}$

Prove: $\overline{QR} \cong \overline{ST}$

Proof:

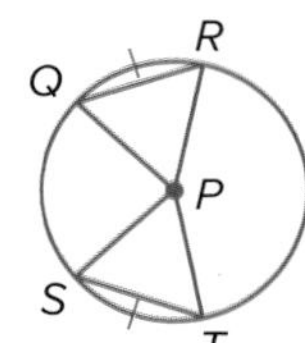

Statements	Reasons
1. $\odot P$, $\widehat{QR} \cong \widehat{ST}$	1. Given
2. $\angle QPR \cong \angle SPT$	2. If arcs are $\cong$, their corresponding central $\angle$s are $\cong$.
3. $\overline{QP} \cong \overline{PR} \cong \overline{SP} \cong \overline{PT}$	3. All radii of a circle are $\cong$.
4. $\triangle PQR \cong \triangle PST$	4. SAS
5. $\overline{QR} \cong \overline{ST}$	5. CPCTC

You will prove part 2 of Theorem 9.2 in Exercise 25.

Real-World Example 1 Use Congruent Chords to Find Arc Measure

CRAFTS In the embroidery hoop, $\overline{AB} \cong \overline{CD}$ and $m\widehat{AB} = 60$. Find $m\widehat{CD}$.

$\overline{AB}$ and $\overline{CD}$ are congruent chords, so the corresponding arcs $\widehat{AB}$ and $\widehat{CD}$ are congruent. $m\widehat{AB} = m\widehat{CD} = 60$

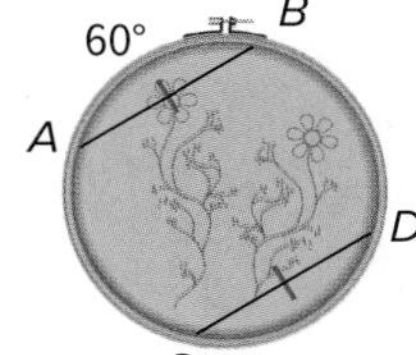

Guided Practice

1. If $m\widehat{AB} = 78$ in the embroidery hoop, find $m\widehat{CD}$.

Example 2 Use Congruent Arcs to Find Chord Lengths

ALGEBRA In the figures, $\odot J \cong \odot K$ and $\widehat{MN} \cong \widehat{PQ}$. Find PQ.

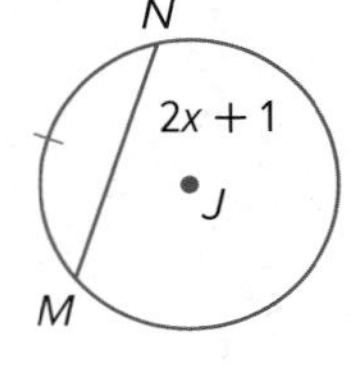

$\widehat{MN}$ and $\widehat{PQ}$ are congruent arcs in congruent circles, so the corresponding chords $\overline{MN}$ and $\overline{PQ}$ are congruent.

$MN = PQ$	Definition of congruent segments
$2x + 1 = 3x - 7$	Substitution
$8 = x$	Simplify.

So, $PQ = 3(8) - 7$ or 17.

Guided Practice

2. In $\odot W$, $\widehat{RS} \cong \widehat{TV}$. Find RS.

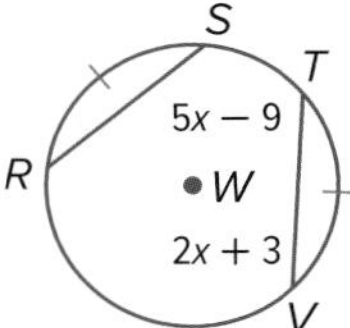

Study Tip

Arc Bisectors In the figure below, $\overline{FH}$ is an arc bisector of $\widehat{JG}$.

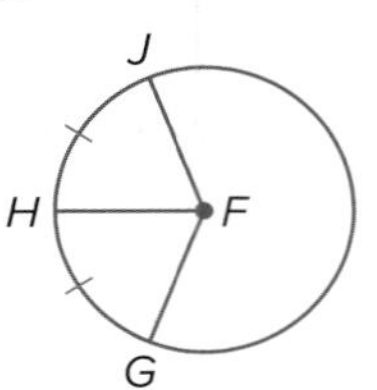

2 Bisecting Arcs and Chords

If a line, segment, or ray divides an arc into two congruent arcs, then it *bisects* the arc.

Theorems

9.3 If a diameter (or radius) of a circle is perpendicular to a chord, then it bisects the chord and its arc.

Example If diameter $\overline{AB}$ is perpendicular to chord $\overline{XY}$, then $\overline{XZ} \cong \overline{ZY}$ and $\widehat{XB} \cong \widehat{BY}$.

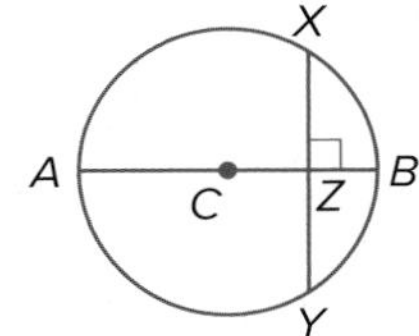

9.4 The perpendicular bisector of a chord is a diameter (or radius) of the circle.

Example If $\overline{AB}$ is a perpendicular bisector of chord $\overline{XY}$, then $\overline{AB}$ is a diameter of $\odot C$.

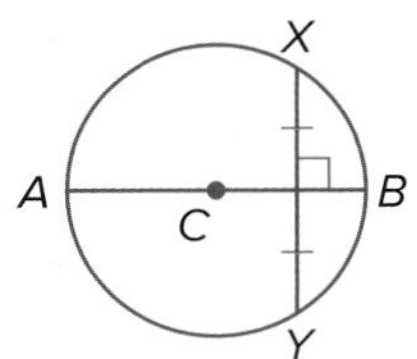

You will prove Theorems 9.3 and 9.4 in Exercises 26 and 28, respectively.

Example 3 Use a Radius Perpendicular to a Chord

In $\odot S$, $m\widehat{PQR} = 98$. Find $m\widehat{PQ}$.

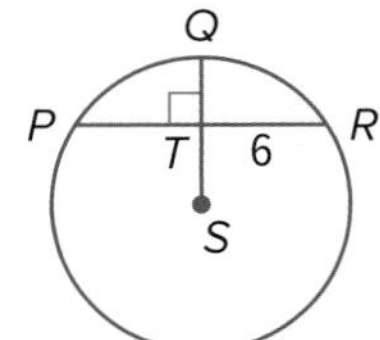

Radius $\overline{SQ}$ is perpendicular to chord $\overline{PR}$. So by Theorem 9.3, $\overline{SQ}$ bisects $\widehat{PQR}$. Therefore, $m\widehat{PQ} = m\widehat{QR}$. By substitution, $m\widehat{PQ} = \frac{98}{2}$ or 49.

Guided Practice

3. In $\odot S$, find PR.

Real-World Link

To make stained glass windows, glass is heated to a temperature of 2000 degrees, until it is the consistency of taffy. The colors are caused by the addition of metallic oxides.

Source: *Artistic Stained Glass by Regg*

Go Online!

You can add any known information to a figure to help you solve a problem. In Example 4 radius $\overline{JK}$ was drawn. Work with a partner to complete the Self-Check Quiz. Take turns describing how to solve each problem and any information added to each figure. Ask for clarification as you need it.

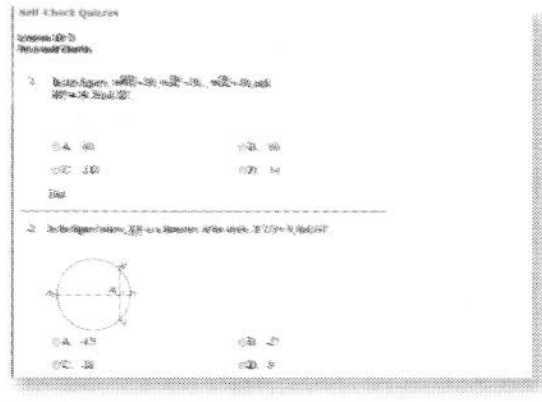

Real-World Example 4 Use a Diameter Perpendicular to a Chord

STAINED GLASS In the stained glass window, diameter $\overline{GH}$ is 30 inches long and chord $\overline{KM}$ is 22 inches long. Find JL.

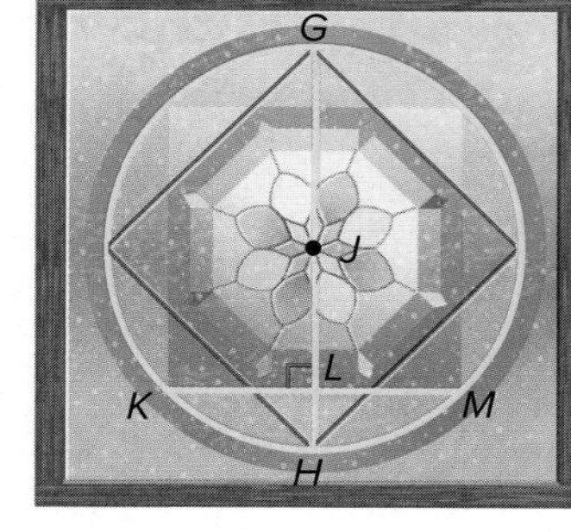

Step 1 Draw radius $\overline{JK}$.

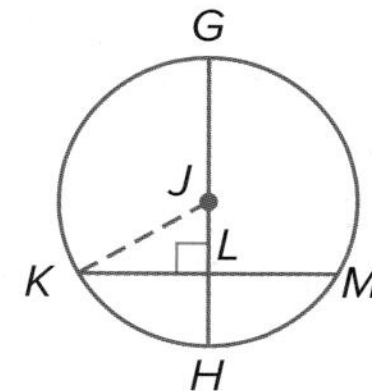

This forms right $\triangle JKL$.

Step 2 Find JK and KL.

Because $GH = 30$ inches, $JH = 15$ inches. All radii of a circle are congruent, so $JK = 15$ inches.

Because diameter $\overline{GH}$ is perpendicular to $\overline{KM}$, $\overline{GH}$ bisects chord $\overline{KM}$ by Theorem 9.3. So, $KL = \frac{1}{2}(22)$ or 11 inches.

Step 3 Use the Pythagorean Theorem to find JL.

$KL^2 + JL^2 = JK^2$	Pythagorean Theorem
$11^2 + JL^2 = 15^2$	$KL = 11$ and $JK = 15$
$121 + JL^2 = 225$	Simplify.
$JL^2 = 104$	Subtract 121 from each side.
$JL = \sqrt{104}$	Take the positive square root of each side.

So, JL is $\sqrt{104}$ or about 10.20 inches long.

Guided Practice

4. In $\odot R$, find TV. Round to the nearest hundredth.

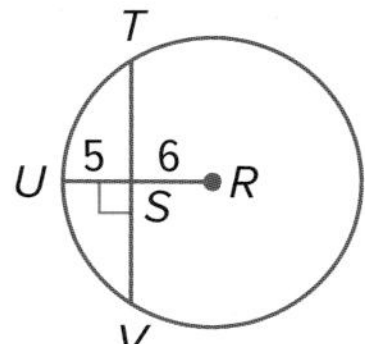

In addition to Theorem 9.2, you can use the following theorem to determine whether two chords in a circle are congruent.

Theorem 9.5

Words In the same circle or in congruent circles, two chords are congruent if and only if they are equidistant from the center.

Example $\overline{FG} \cong \overline{JH}$ if and only if $LX = LY$.

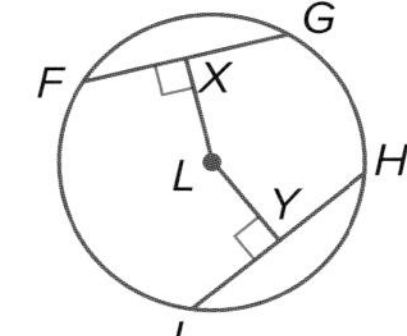

You will prove Theorem 9.5 in Exercises 29 and 30.

Example 5 Chords Equidistant from Center

ALGEBRA In $\odot A$, $WX = XY = 22$. Find AB.

Because chords $\overline{WX}$ and $\overline{XY}$ are congruent, they are equidistant from A. So, $AB = AC$.

$AB = AC$

$5x = 3x + 4$ Substitution

$x = 2$ Simplify.

So, $AB = 5(2)$ or 10.

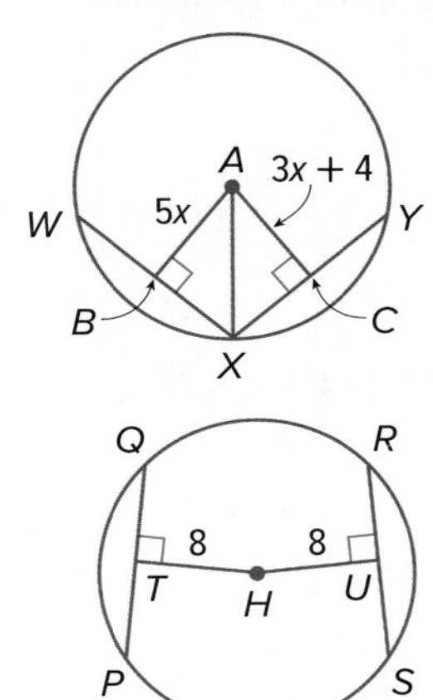

Guided Practice

5. In $\odot H$, $PQ = 3x - 4$ and $RS = 14$. Find x.

You can use Theorem 9.5 to find the point equidistant from three noncollinear points.

Construction Circle Through Three Noncollinear Points

Step 1

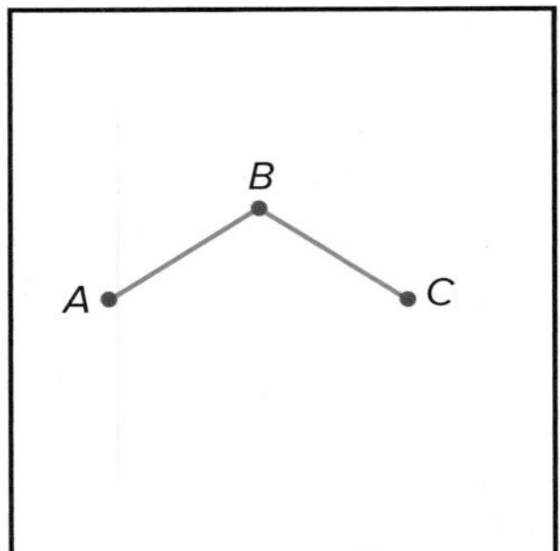

Draw three noncollinear points A, B, and C. Then draw segments $\overline{AB}$ and $\overline{BC}$.

Step 2

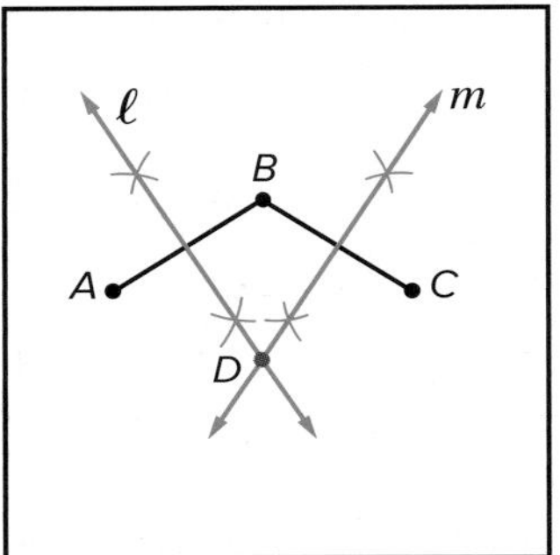

Construct the perpendicular bisectors ℓ and m of $\overline{AB}$ and $\overline{BC}$. Label the point of intersection D.

Step 3

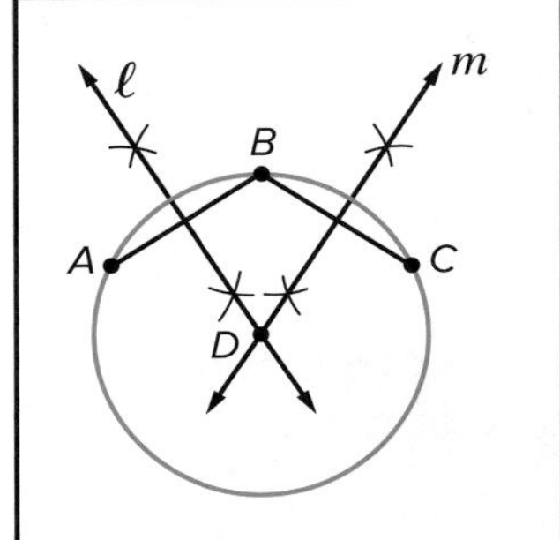

By Theorem 9.4, lines ℓ and m contain diameters of $\odot D$. With the compass at point D, draw a circle through points A, B, and C.

Check Your Understanding

 = Step-by-Step Solutions begin on page R13.

Examples 1–2 ALGEBRA Find the value of x.

1

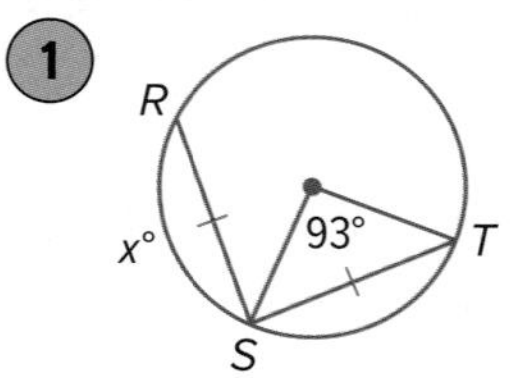

2.

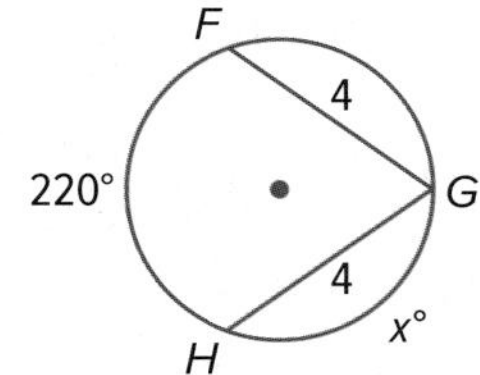

3.

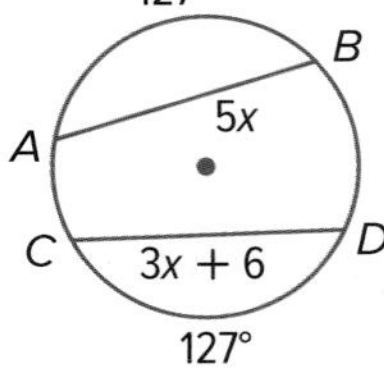

Examples 3–4 In $\odot P$, $JK = 10$ and $m\widehat{JLK} = 134$. Find each measure. Round to the nearest hundredth.

4. 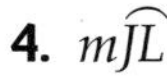 $m\widehat{JL}$

5. PQ

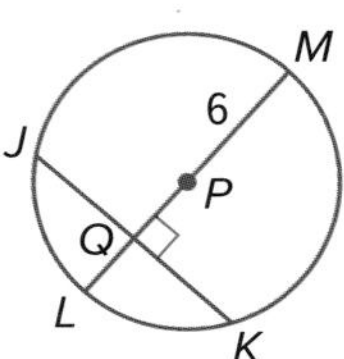

Example 5 **6.** In $\odot J$, $GH = 9$, $KL = 4x + 1$. Find x.

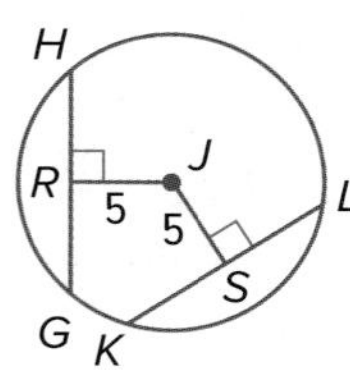

Practice and Problem Solving

Extra Practice is on page R9.

Examples 1–2 **ALGEBRA** Find the value of x.

7.

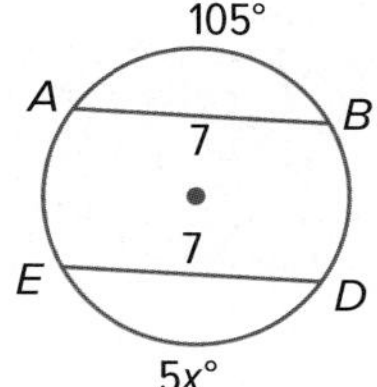

8.

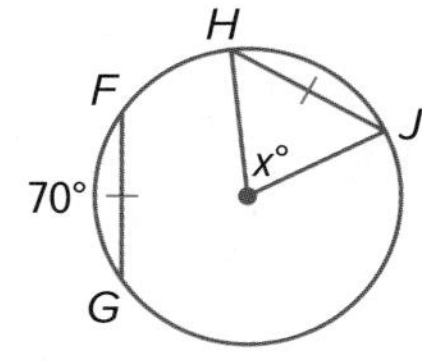

9.

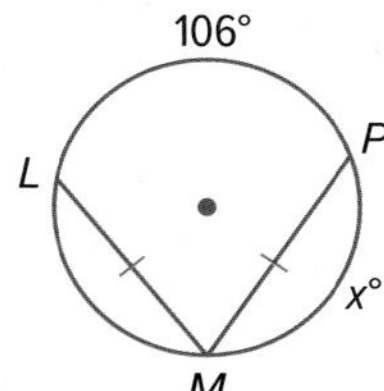

10.

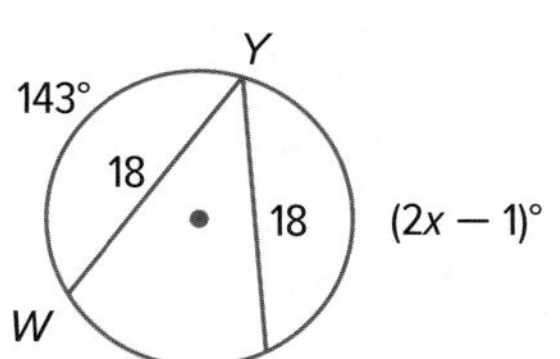

11.

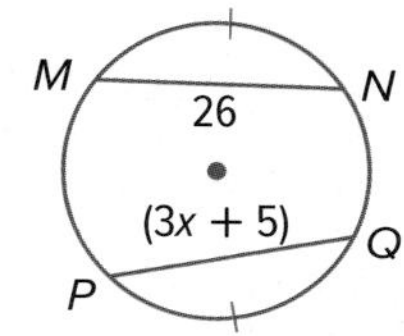

12.

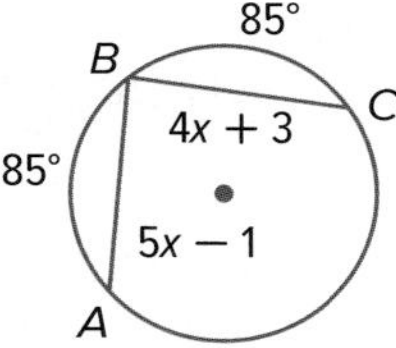

13 $\odot C \cong \odot D$

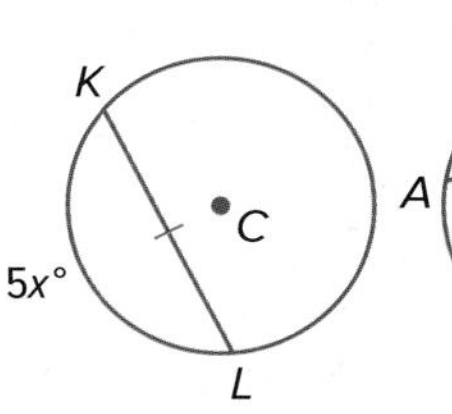
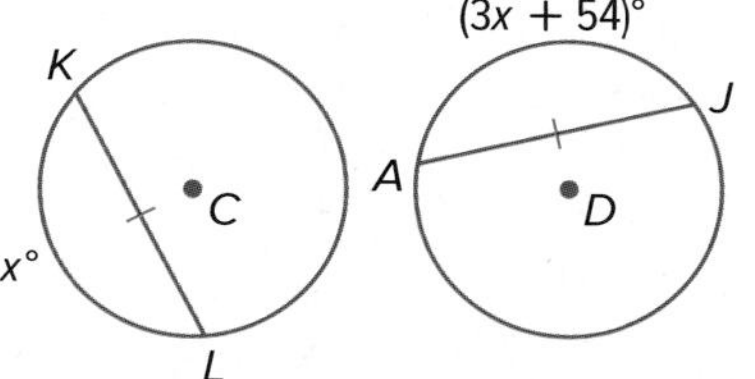

14. $\odot P \cong \odot Q$

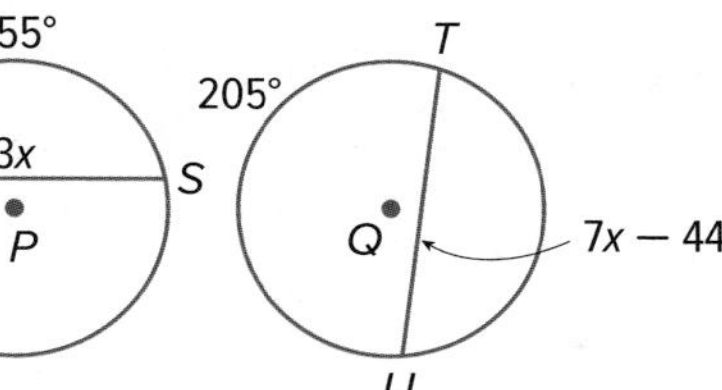

15. **MODELING** Angie is in a jewelry-making class at her local arts center. She wants to make a pair of triangular earrings from a metal circle. She knows that $\widehat{AC}$ is 115°. If she wants to cut two equal parts off so that $\widehat{AB} = \widehat{BC}$, what is x?

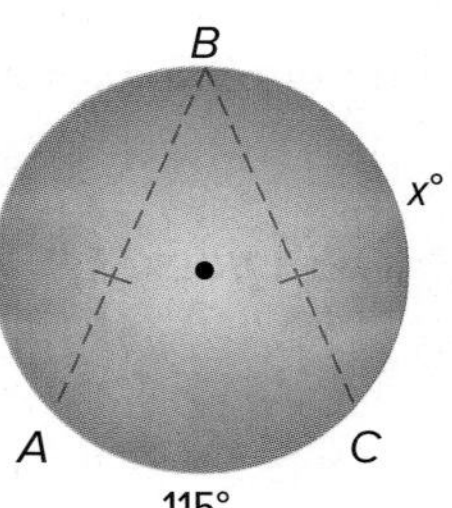

Examples 3–4 **In $\odot A$, the radius is 14 and $CD = 22$. Find each measure. Round to the nearest hundredth, if necessary.**

16. CE

17. EB

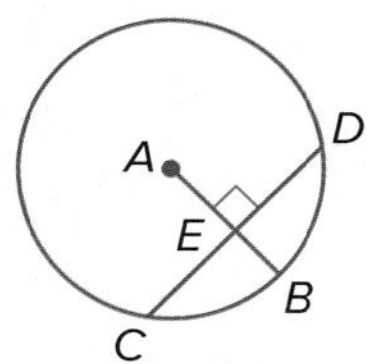

In $\odot H$, the diameter is 18, $LM = 12$, and $m\widehat{LM} = 84$. Find each measure. Round to the nearest hundredth, if necessary.

18. $m\widehat{LK}$

19. HP

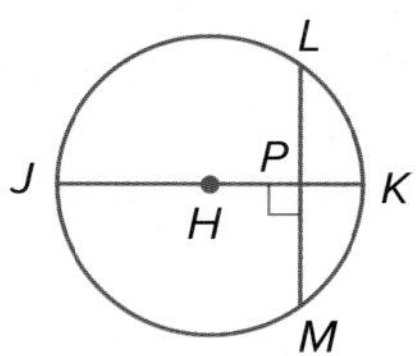

20. **SNOWBOARDING** The snowboarding rail shown is an arc of a circle in which $\overline{BD}$ is part of the diameter. If $\widehat{ABC}$ is about 32% of a complete circle, what is $m\widehat{AB}$?

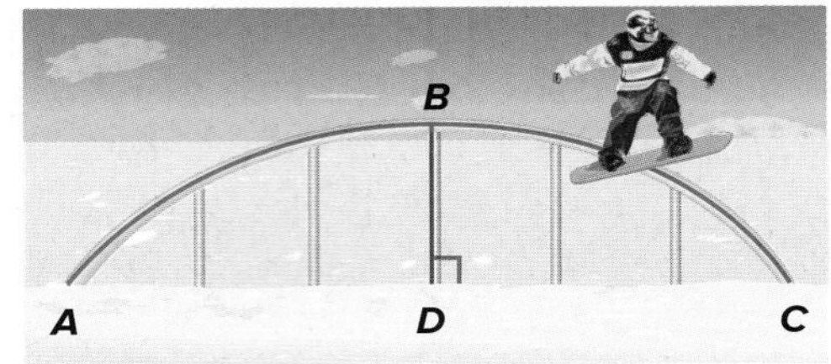

21. **ROADS** The curved road at the right is part of $\odot C$, which has a radius of 88 feet. What is AB? Round to the nearest tenth.

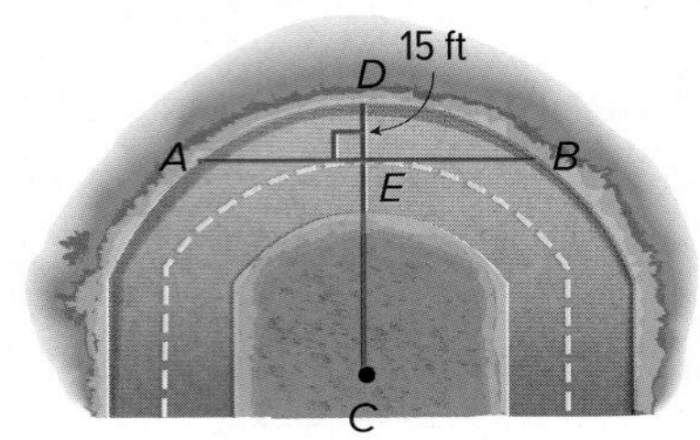

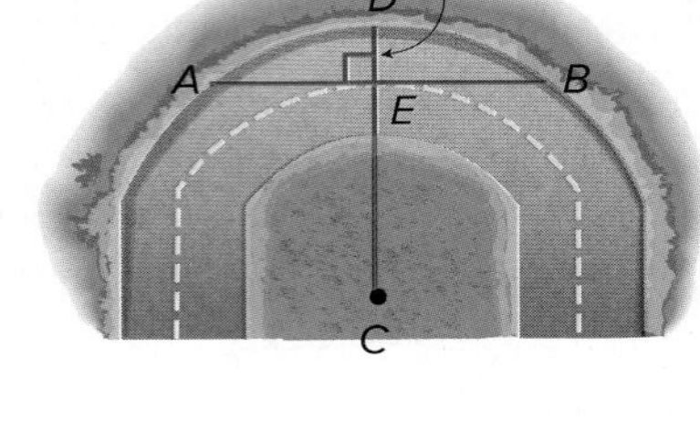

Example 5

22. **ALGEBRA** In $\odot F$, $\overline{AB} \cong \overline{BC}$, $DF = 3x - 7$, and $FE = x + 9$. What is x?

23. **ALGEBRA** In $\odot S$, $LM = 16$ and $PN = 4x$. What is x?

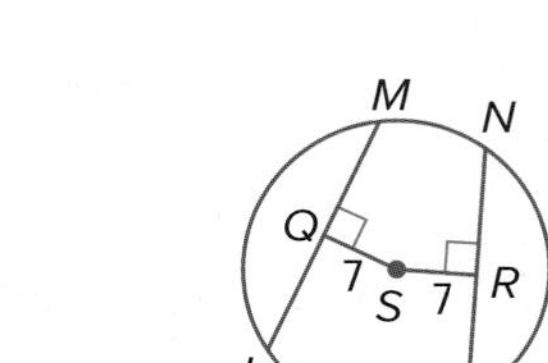

PROOF Write a two-column proof.

24. **Given:** $\odot P$, $\overline{KM} \perp \overline{JP}$

Prove: $\overline{JP}$ bisects $\overline{KM}$ and $\widehat{KM}$.

PROOF Write the specified type of proof.

25. paragraph proof of Theorem 9.2, part 2

Given: $\odot P$, $\overline{QR} \cong \overline{ST}$

Prove: $\widehat{QR} \cong \widehat{ST}$

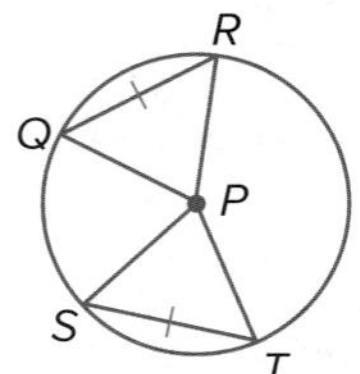

26. two-column proof of Theorem 9.3

Given: $\odot C$, $\overline{AB} \perp \overline{XY}$

Prove: $\overline{XZ} \cong \overline{YZ}$, $\widehat{XB} \cong \widehat{YB}$

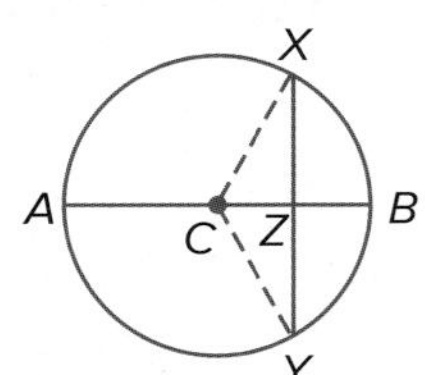

27. **DESIGN** Roberto is designing a logo for a friend's coffee shop according to the design at the right, where each chord is equal in length. What is the measure of each arc and the length of each chord?

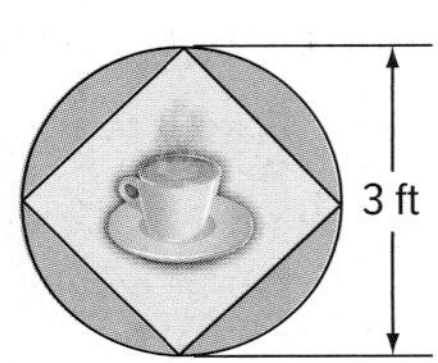

28. **MP CONSTRUCT ARGUMENTS** Write a two-column proof of Theorem 9.4.

MP CONSTRUCT ARGUMENTS Write a two-column proof of the indicated part of Theorem 9.5.

29. In a circle, if two chords are equidistant from the center, then they are congruent.

30. In a circle, if two chords are congruent, then they are equidistant from the center.

ALGEBRA Find the value of x.

31 $\overline{AB} \cong \overline{DF}$

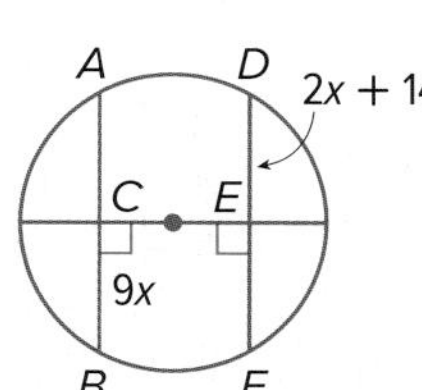

32. $\overline{GH} \cong \overline{KJ}$

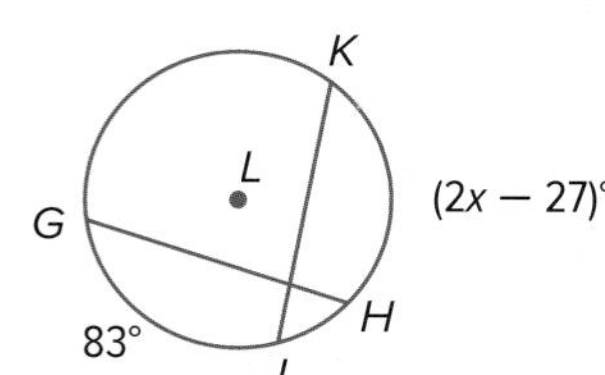

33. $\widehat{WTY} \cong \widehat{TWY}$

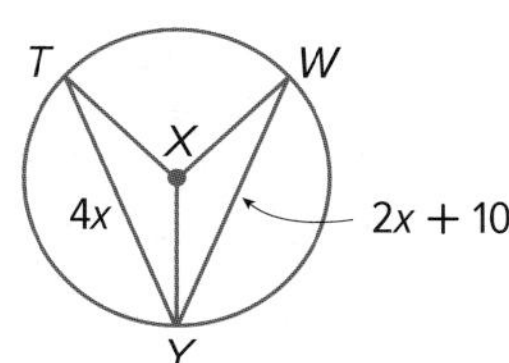

34. MULTI-STEP A retail store manager wants to set up a display of a new fashion line. There are three entrances into the store.

a. Where should the display be placed to get maximum exposure?

b. Describe your solution process including any assumptions made. Include a diagram with your response.

H.O.T. Problems Use Higher-Order Thinking Skills

35. CHALLENGE The common chord $\overline{AB}$ between $\odot P$ and $\odot Q$ is perpendicular to the segment connecting the centers of the circles. If $AB = 10$, what is the length of $\overline{PQ}$? Explain your reasoning.

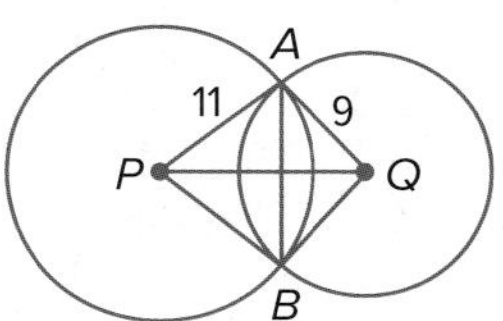

36. MP REASONING In a circle, $\overline{AB}$ is a diameter and $\overline{HG}$ is a chord that intersects $\overline{AB}$ at point X. Is it *sometimes*, *always*, or *never* true that $HX = GX$? Explain.

37. CHALLENGE Use a compass to draw a circle with chord $\overline{AB}$. Refer to this construction for the following problem.

Step 1 Construct $\overline{CD}$, the perpendicular bisector of $\overline{AB}$.

Step 2 Construct $\overline{FG}$, the perpendicular bisector of $\overline{CD}$. Label the point of intersection O.

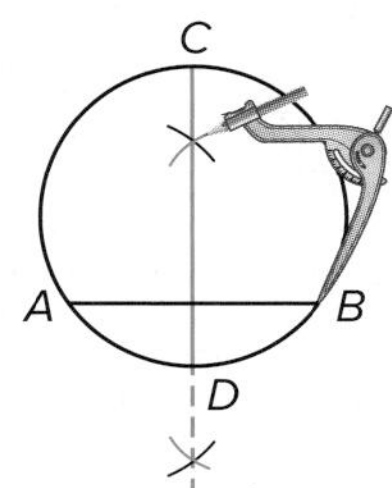

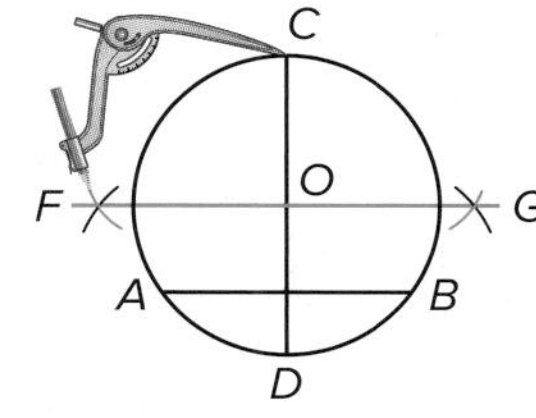

a. Use an indirect proof to show that $\overline{CD}$ passes through the center of the circle by assuming that the center of the circle is *not* on $\overline{CD}$.

b. Prove that O is the center of the circle.

38. OPEN-ENDED Construct a circle and draw a chord. Measure the chord and the distance that the chord is from the center. Find the length of the radius.

39. WRITING IN MATH If the measure of an arc in a circle is tripled, will the chord of the new arc be three times as long as the chord of the original arc? Explain your reasoning.

40. In the figure, $\odot G \cong \odot H$ and $\widehat{EF} \cong \widehat{KL}$. What is the length of $\overline{KL}$? MP 6

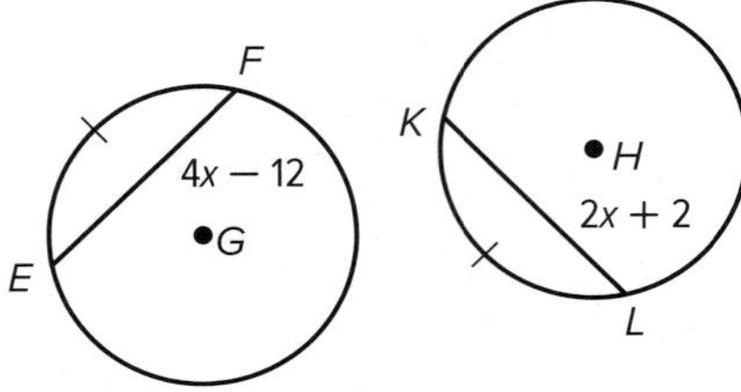

- A 32
- B 16
- C 12
- D 7

41. Which of the following is a valid conclusion that Matthew can make based on the figure? MP 2

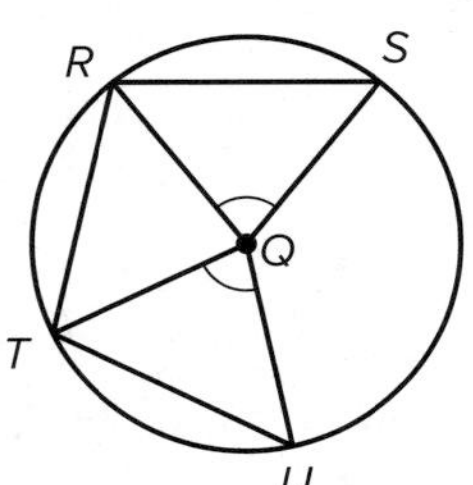

I. $\overline{RS} \cong \overline{TU}$

II. $\widehat{RT} \cong \widehat{TU}$

III. $\widehat{RS} \cong \widehat{TU}$

- A I only
- B III only
- C I and II only
- D I and III only
- E I, II, and III

42. If $CW = WF$ and $ED = 30$, what is DF? MP 6

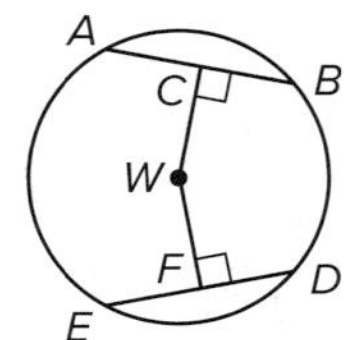

- A 60
- B 45
- C 30
- D 15

43. **MULTI-STEP** A chord $\overline{KM}$ is 24 feet and the diameter $\overline{GH}$ is 36 feet. Find the length of $\overline{JL}$. MP 2, 6

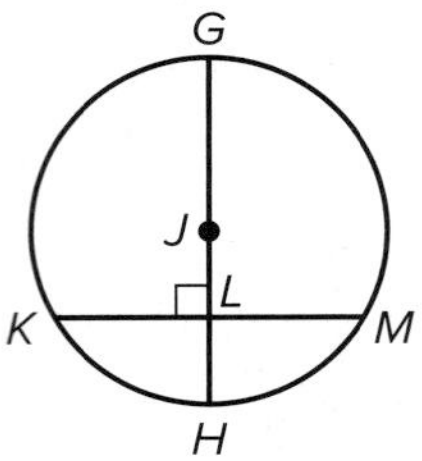

a. Name a triangle that includes $\overline{JL}$ as its side.

b. Find the lengths of the two sides of your triangle (other than $\overline{JL}$). Explain.

c. Use the Pythagorean Theorem to write an equation that includes JL as the variable.

d. Solve the equation and find JL.

44. In $\odot P$, $PS = PT = 12$, $HJ = 2x + 10$, and $KL = 4x - 8$.

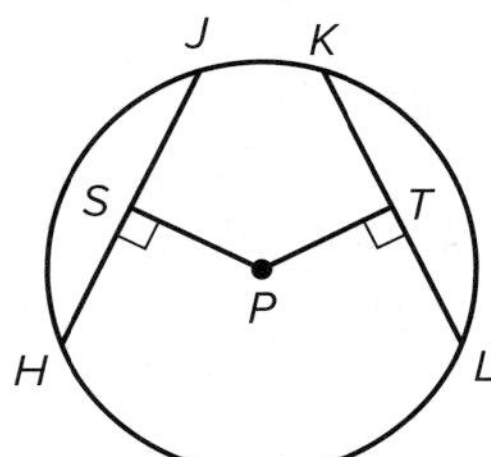

What is the length of $\overline{HS}$? MP 2

- A 6
- B 9
- C 14
- D 28

45. In $\odot C$, the diameter is 22 centimeters and $MN = 18$ centimeters.

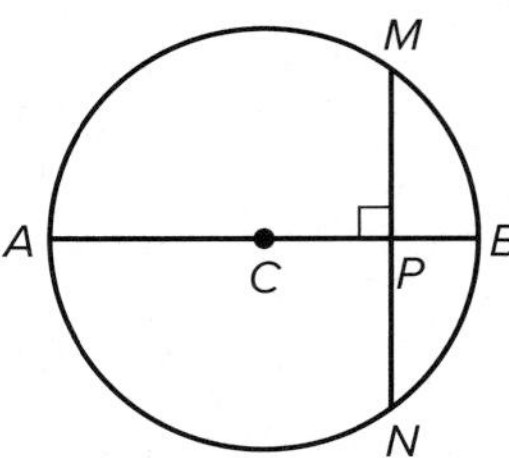

Which is the best estimate of the length of $\overline{CP}$? MP 6

- A 2.0 cm
- B 6.3 cm
- C 4.0 cm
- D 14.2 cm

LESSON 4

Inscribed Angles

Then	Now	Why?
You found measures of interior angles of polygons.	**1** Find measures of inscribed angles. **2** Find measures of angles of inscribed polygons.	The entrance to a school prom has a semicircular arch. Streamers are attached with one end at point *A* and the other end at point *B*. The middle of each streamer can then be attached to a different point *P* along the arch.

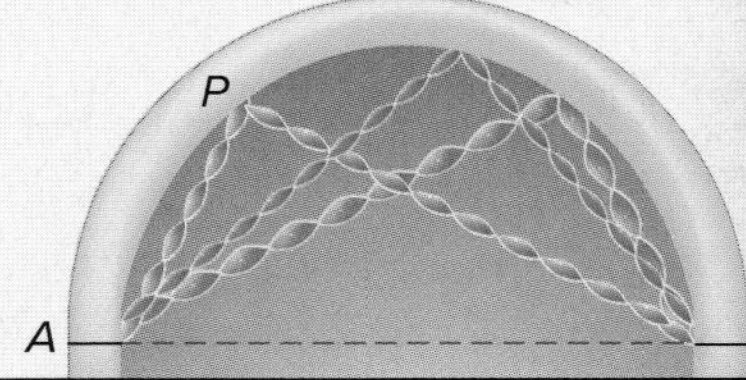

New Vocabulary
inscribed angle
intercepted arc

Mathematical Practices

3 Construct viable arguments and critique the reasoning of others.

7 Look for and make use of structure.

8 Look for and express regularity in repeated reasoning.

1 Inscribed Angles

Notice that the angle formed by each streamer appears to be congruent, no matter where point P is placed along the arch. An **inscribed angle** has a vertex on a circle and sides that contain chords of the circle. In $\odot C$, $\angle QRS$ is an inscribed angle.

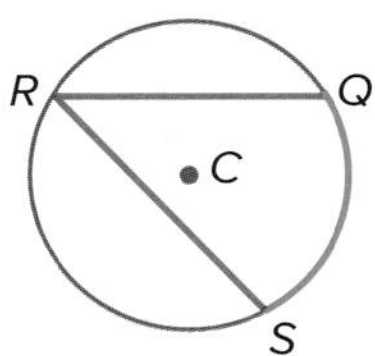

An **intercepted arc** has endpoints on the sides of an inscribed angle and lies in the interior of the inscribed angle. In $\odot C$, minor arc $\widehat{QS}$ is intercepted by $\angle QRS$.

There are three ways that an angle can be inscribed in a circle.

Case 1	Case 2	Case 3
P	P	P
Center *P* is on a side of the inscribed angle.	Center *P* is inside the inscribed angle.	The center *P* is in the exterior of the inscribed angle.

In Case 1, one side of the angle is a diameter of the circle.

For each of these cases, the following theorem holds true.

Theorem 9.6 Inscribed Angle Theorem

Words If an angle is inscribed in a circle, then the measure of the angle equals one half the measure of its intercepted arc.

Example $m\angle 1 = \frac{1}{2}m\widehat{AB}$ and $m\widehat{AB} = 2m\angle 1$

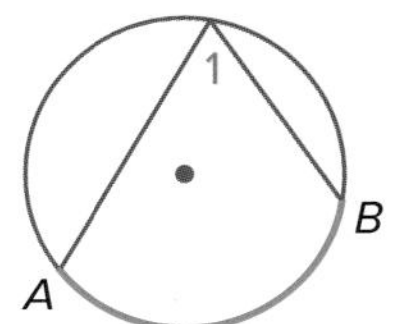

You will prove Cases 2 and 3 of the Inscribed Angle Theorem in Exercises 37 and 38.

moodboard/Alamy

Vocabulary Link

Inscribed

Everyday Use: written on or in a surface, such as inscribing the inside of a ring with an inscription

Math Use: touching only the sides (or interior) of another figure

Proof Inscribed Angle Theorem (Case 1)

Given: $\angle B$ is inscribed in $\odot P$.

Prove: $m\angle B = \frac{1}{2}m\widehat{AC}$

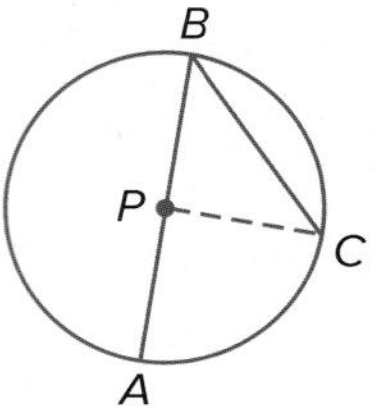

Proof:

Statements	Reasons
1. Draw an auxiliary radius $\overline{PC}$.	**1.** Two points determine a line.
2. $\overline{PB} \cong \overline{PC}$	**2.** All radii of a circle are ≅.
3. $\triangle PBC$ is isosceles.	**3.** Definition of isosceles triangle
4. $m\angle B = m\angle C$	**4.** Isosceles Triangle Theorem
5. $m\angle APC = m\angle B + m\angle C$	**5.** Exterior Angle Theorem
6. $m\angle APC = 2m\angle B$	**6.** Substitution (Steps 4, 5)
7. $m\widehat{AC} = m\angle APC$	**7.** Definition of arc measure
8. $m\widehat{AC} = 2m\angle B$	**8.** Substitution (Steps 6, 7)
9. $2m\angle B = m\widehat{AC}$	**9.** Symmetric Property of Equality
10. $m\angle B = \frac{1}{2}m\widehat{AC}$	**10.** Division Property of Equality

Example 1 Use Inscribed Angles to Find Measures

Find each measure.

a. $m\angle P$

$m\angle P = \frac{1}{2}m\widehat{MN}$

$= \frac{1}{2}(70)$ or 35

b. $m\widehat{PO}$

$m\widehat{PO} = 2m\angle N$

$= 2(56)$ or 112

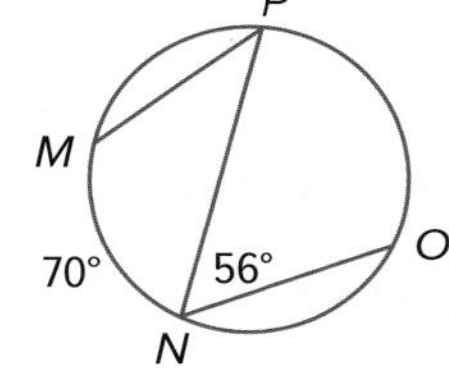

Guided Practice

1A. $m\widehat{CF}$

1B. $m\angle C$

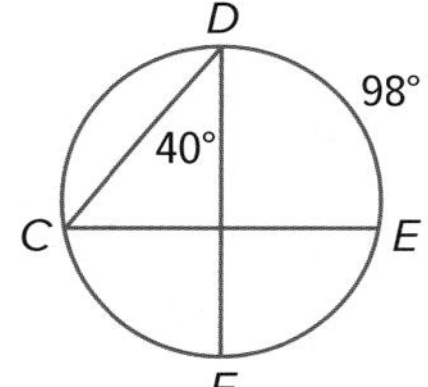

Two inscribed angles that intercept the same arc of a circle are related.

Theorem 9.7

Words If two inscribed angles of a circle intercept the same arc or congruent arcs, then the angles are congruent.

Example $\angle B$ and $\angle C$ both intercept $\widehat{AD}$. So, $\angle B \cong \angle C$.

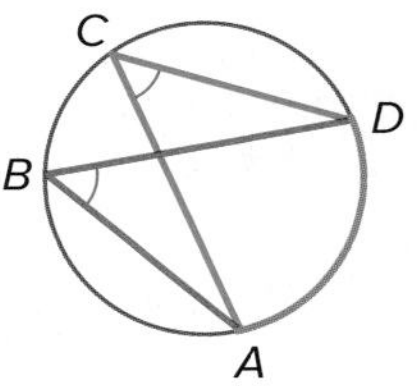

You will prove Theorem 9.7 in Exercise 39.

> **Study Tip**
> **Inscribed Polygons** Remember that for a polygon to be an inscribed polygon, *all* of its vertices must lie on the circle.

Example 2 Use Inscribed Angles to Find Measures

ALGEBRA Find $m\angle T$.

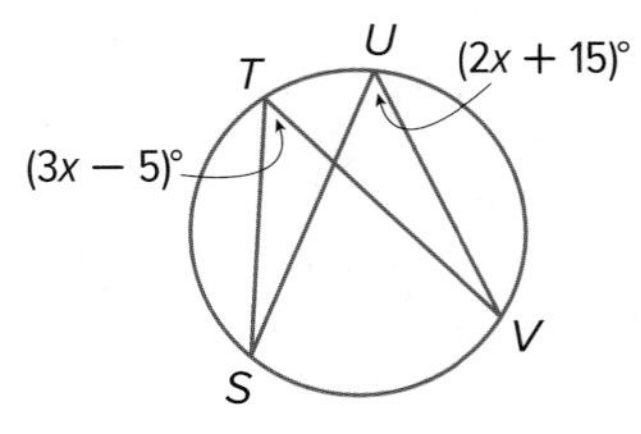

$\angle T \cong \angle U$	$\angle T$ and $\angle U$ both intercept $\overset{\frown}{SV}$.
$m\angle T = m\angle U$	Definition of congruent angles
$3x - 5 = 2x + 15$	Substitution
$x = 20$	Simplify.

So, $m\angle T = 3(20) - 5$ or 55.

Guided Practice

2. If $m\angle S = 3x$ and $m\angle V = (x + 16)$, find $m\angle S$.

Example 3 Use Inscribed Angles in Proofs

Write a two-column proof.

Given: $\overset{\frown}{JM} \cong \overset{\frown}{KL}$

Prove: $\triangle JMN \cong \triangle KLN$

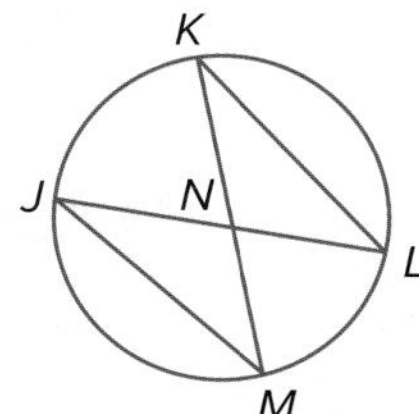

Proof:

Statements	Reasons
1. $\overset{\frown}{JM} \cong \overset{\frown}{KL}$	**1.** Given
2. $\overline{JM} \cong \overline{KL}$	**2.** If minor arcs are ≅, their corresponding chords are ≅.
3. $\angle M$ intercepts $\overset{\frown}{JK}$. $\angle L$ intercepts $\overset{\frown}{JK}$.	**3.** Definition of intercepted arc
4. $\angle M \cong \angle L$	**4.** Inscribed ∠s of same arc are ≅.
5. $\angle JNM \cong \angle KNL$	**5.** Vertical ∠s are ≅.
6. $\triangle JMN \cong \triangle KLN$	**6.** AAS

Guided Practice

3. Given: $\overset{\frown}{QR} \cong \overset{\frown}{ST}$, $\overset{\frown}{PQ} \cong \overset{\frown}{PT}$

Prove: $\triangle PQR \cong \triangle PTS$

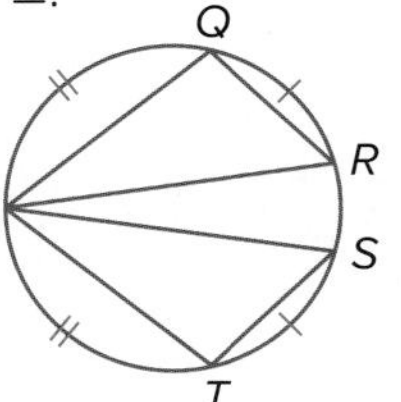

2 Angles of Inscribed Polygons

Triangles and quadrilaterals that are inscribed in circles have special properties.

> **Go Online!**
> 
> Tap into the power of your graphing calculator with Graphing Calculator Easy Files™. Practice vocabulary in English or Spanish with Lesson Vocabulary Review Files. Or review with a 5-Minute Check. Ask your teacher to assign them to you in ConnectED.

Theorem 9.8

Words An inscribed angle of a triangle intercepts a diameter or semicircle if and only if the angle is a right angle.

Example If $\overset{\frown}{FJH}$ is a semicircle, then $m\angle G = 90$. If $m\angle G = 90$, then $\overset{\frown}{FJH}$ is a semicircle and $\overline{FH}$ is a diameter.

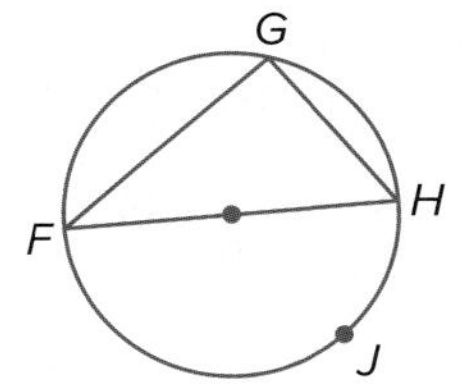

You will prove Theorem 9.8 in Exercise 40.

Example 4 Find Angle Measures in Inscribed Triangles

ALGEBRA Find $m\angle F$.

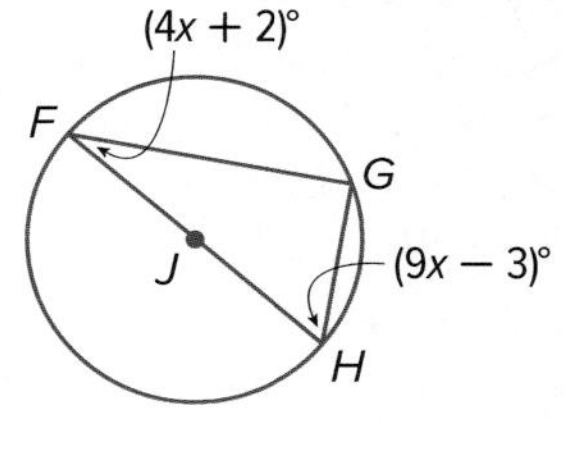

$\triangle FGH$ is a right triangle because $\angle G$ inscribes a semicircle.

$m\angle F + m\angle G + m\angle H = 180$	$\triangle$Angle-Sum Theorem
$(4x + 2) + 90 + (9x - 3) = 180$	Substitution
$13x + 89 = 180$	Simplify.
$13x = 91$	Subtract 89 from each side.
$x = 7$	Divide each side by 13.

So, $m\angle F = 4(7) + 2$ or 30.

Guided Practice

4. If $m\angle F = 7x + 2$ and $m\angle H = 17x - 8$, find x.

While many different types of triangles, including right triangles, can be inscribed in a circle, only certain quadrilaterals can be inscribed in a circle.

Study Tip

Construct Arguments Theorem 9.9 can be verified by considering that the arcs intercepted by opposite angles of an inscribed quadrilateral form a circle.

Theorem 9.9

Words If a quadrilateral is inscribed in a circle, then its opposite angles are supplementary.

Example If quadrilateral $KLMN$ is inscribed in $\odot A$, then $\angle L$ and $\angle N$ are supplementary and $\angle K$ and $\angle M$ are supplementary.

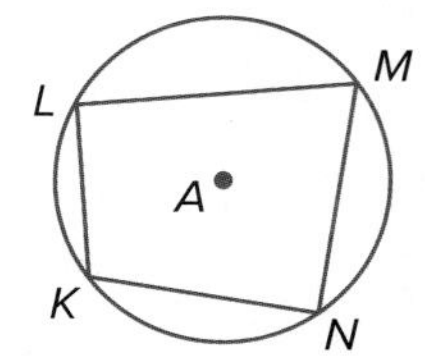

You will prove Theorem 9.9 in Exercise 31.

Real-World Example 5 Find Angle Measures

Real-World Link

Charms for jewelry first became popular during the age of the Egyptian Pharaohs. They were repopularized by Queen Victoria in the early twentieth century and by Louis Vuitton in 2001.

Source: *My Mother's Charms*

JEWELRY The necklace charm shown uses a quadrilateral inscribed in a circle. Find $m\angle A$ and $m\angle B$.

B
(2x − 30)°
A
C
x°
D

Because $ABCD$ is inscribed in a circle, opposite angles are supplementary.

$m\angle A + m\angle C = 180$	$m\angle B + m\angle D = 180$
$m\angle A + 90 = 180$	$(2x - 30) + x = 180$
$m\angle A = 90$	$3x - 30 = 180$
	$3x = 210$
	$x = 70$

So, $m\angle A = 90$ and $m\angle B = 2(70) - 30$ or 110.

Guided Practice

5. Quadrilateral $WXYZ$ is inscribed in $\odot V$. Find $m\angle X$ and $m\angle Y$.

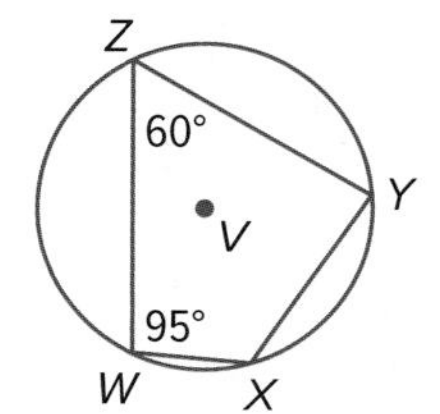

Zoonar RF/Getty Images

Check Your Understanding

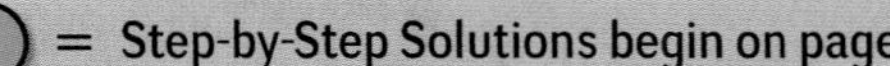
= Step-by-Step Solutions begin on page R13.

Example 1 **Find each measure.**

1. $m\angle B$

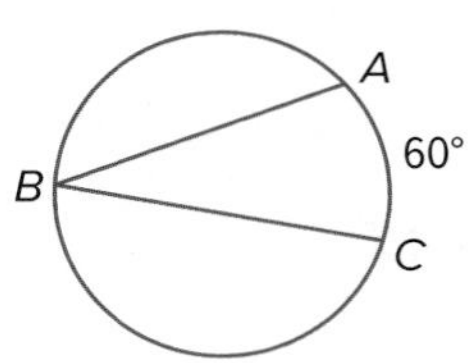

2. $m\widehat{RT}$

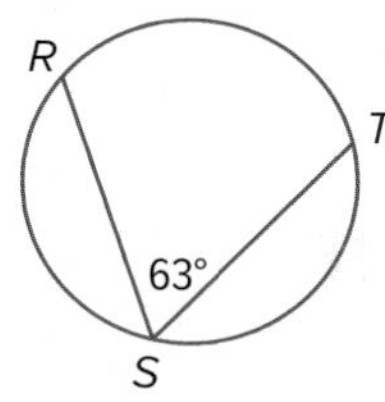

3. $m\widehat{WX}$

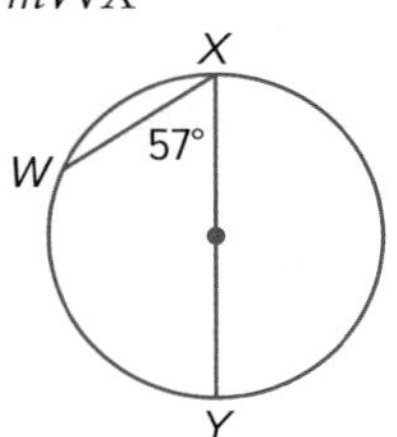

4. SCIENCE The diagram shows how light bends in a raindrop to make the colors of the rainbow. If $m\widehat{ST} = 144$, what is $m\angle R$?

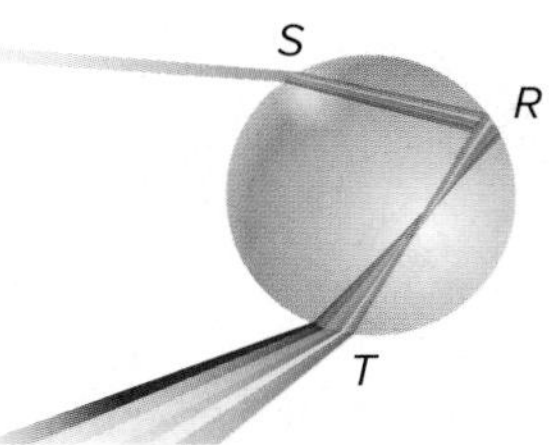

Example 2 **ALGEBRA** **Find each measure.**

5. $m\angle H$

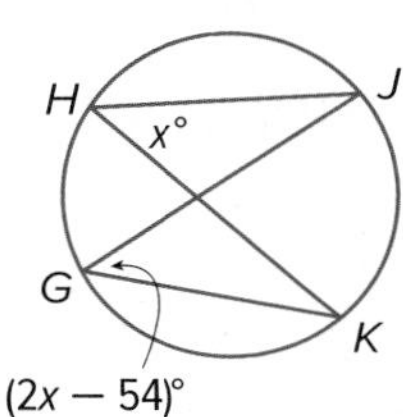

6. $m\angle B$

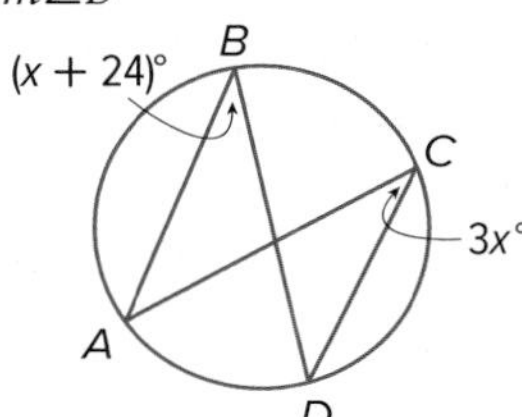

Example 3 **7. PROOF** **Write a two-column proof.**

Given: $\widehat{RS} \cong \widehat{UT}$

Prove: $\triangle RVS \cong \triangle UVT$

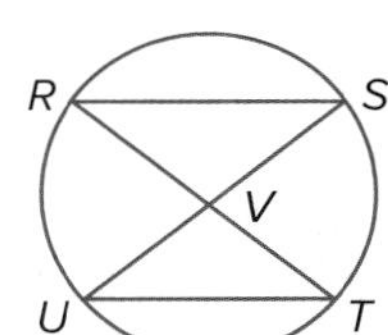

Examples 4–5 MP **STRUCTURE** **Find each value.**

8. $m\angle R$

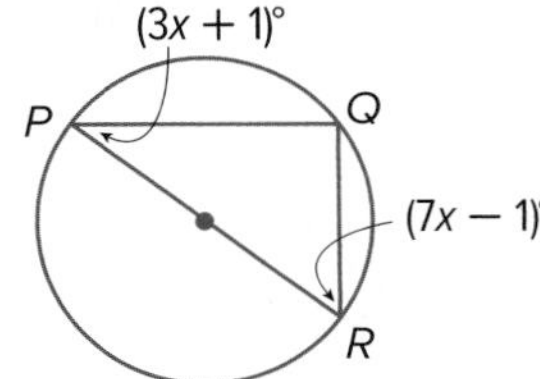

9. x

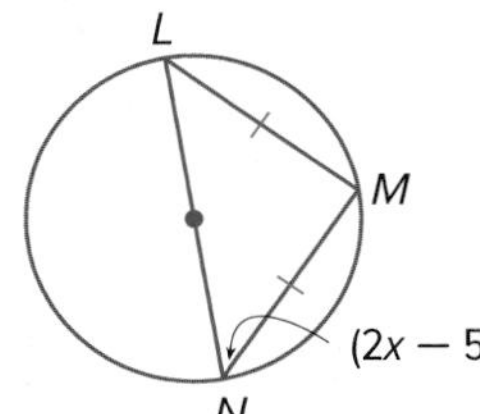

10. $m\angle C$ and $m\angle D$

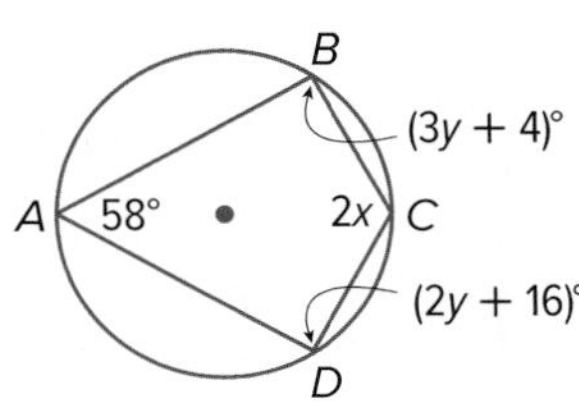

Practice and Problem Solving

Extra Practice is on page R9.

Example 1 **Find each measure.**

11. $m\widehat{DH}$

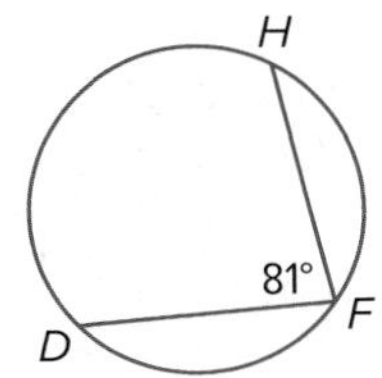

12. $m\angle K$

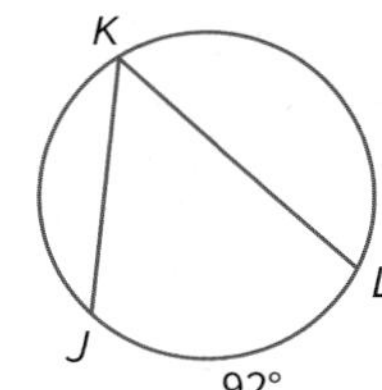

13 $m\angle P$

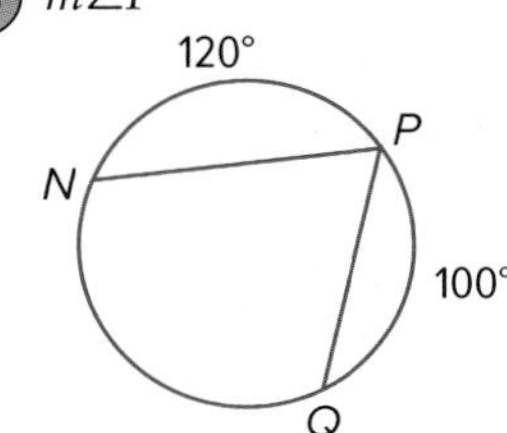

14. $m\widehat{AC}$

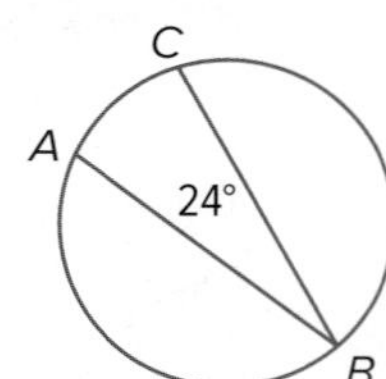

15. $m\widehat{GH}$

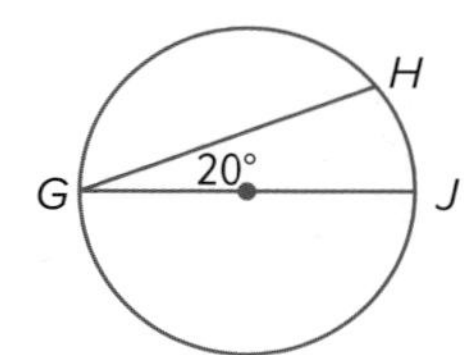

16. $m\angle S$

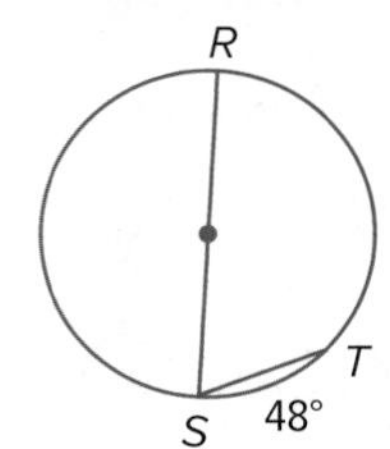

Example 2 **ALGEBRA** **Find each measure.**

17. $m\angle R$

18. $m\angle S$

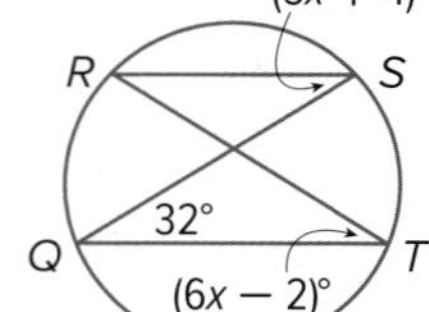

19. $m\angle A$

20. $m\angle C$

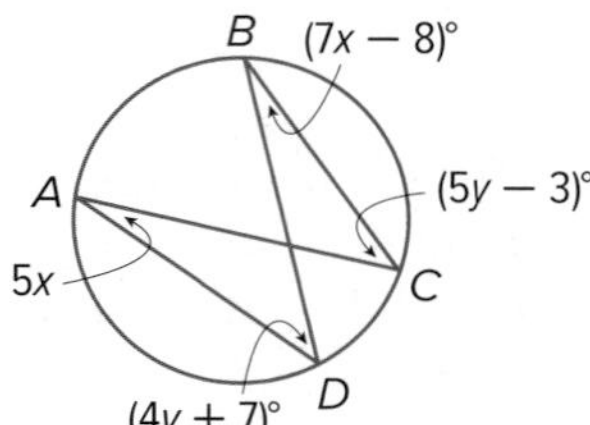

Example 3 **PROOF** **Write the specified type of proof.**

21. paragraph proof

Given: $m\angle T = \frac{1}{2}m\angle S$

Prove: $m\widehat{TUR} = 2m\widehat{URS}$

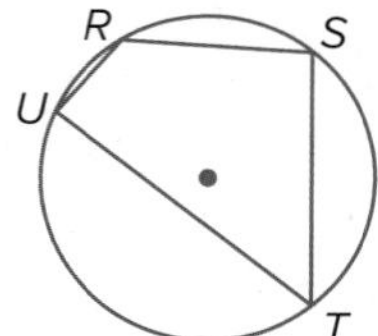

22. two-column proof

Given: $\odot C$

Prove: $\triangle KML \sim \triangle JMH$

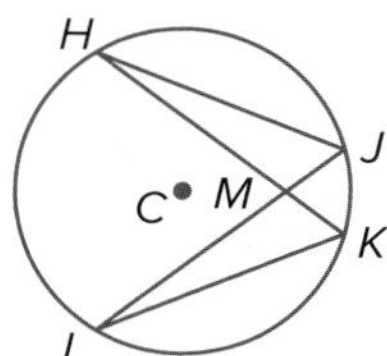

Example 4 **ALGEBRA** **Find each value.**

23. x

24. $m\angle T$

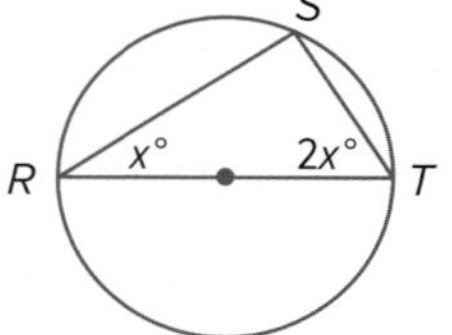

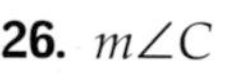

25. x

26. $m\angle C$

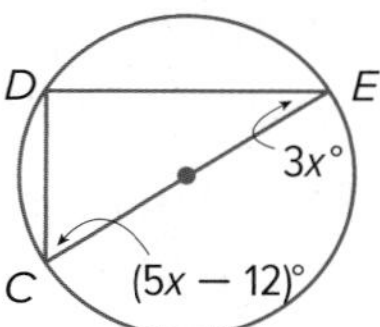

Example 5 **MP** **STRUCTURE** **Find each measure.**

27. $m\angle T$

28. $m\angle Z$

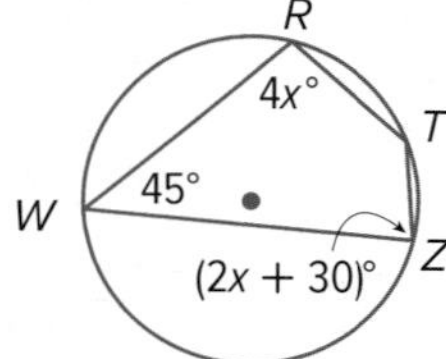

29. $m\angle H$

30. $m\angle G$

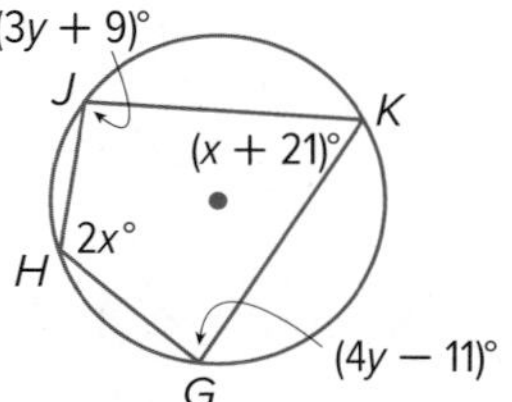

31. **PROOF** Write a paragraph proof for Theorem 9.9.

SIGNS **A stop sign in the shape of a regular octagon is inscribed in a circle. Find each measure.**

32. $m\widehat{NQ}$

33. $m\angle RLQ$

34. $m\angle LRQ$

35. $m\angle LSR$

36. ART Four different string art star patterns are shown. If all of the inscribed angles of each star shown are congruent, find the measure of each inscribed angle.

a. 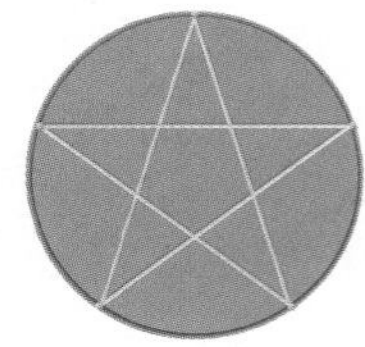
b. 
c. 
d. 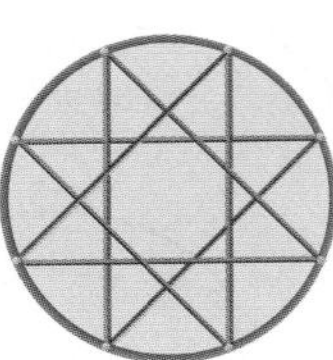

PROOF Write a two-column proof for each case of Theorem 9.6.

37. Case 2

Given: P lies inside $\angle ABC$. $\overline{BD}$ is a diameter.

Prove: $m\angle ABC = \frac{1}{2}m\widehat{AC}$

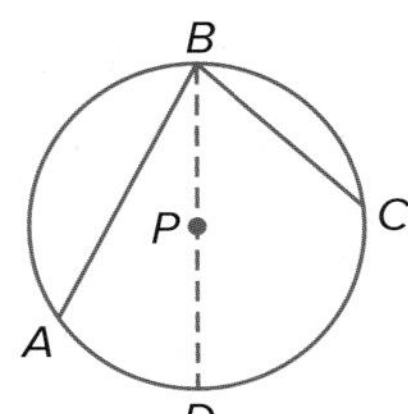

38. Case 3

Given: P lies outside $\angle ABC$. $\overline{BD}$ is a diameter.

Prove: $m\angle ABC = \frac{1}{2}m\widehat{AC}$

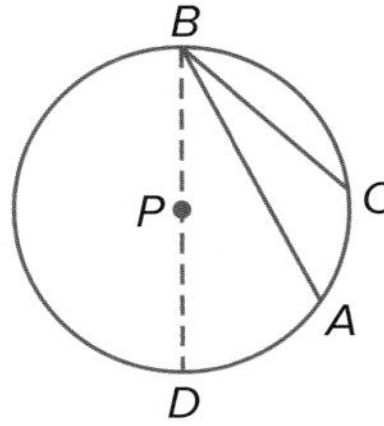

PROOF Write the specified proof for each theorem.

39. Theorem 9.7, two-column proof

40. Theorem 9.8, paragraph proof

41. MULTIPLE REPRESENTATIONS In this problem, you will investigate the relationship between the arcs of a circle that are cut by two parallel chords.

- **a. Geometric** Use a compass to draw a circle with parallel chords $\overline{AB}$ and $\overline{CD}$. Connect points A and D by drawing segment $\overline{AD}$.
- **b. Numerical** Use a protractor to find $m\angle A$ and $m\angle D$. Then determine $m\widehat{AC}$ and $m\widehat{BD}$. What is true about these arcs? Explain.
- **c. Verbal** Draw another circle and repeat parts **a** and **b**. Make a conjecture about arcs of a circle that are cut by two parallel chords.
- **d. Analytical** Use your conjecture to find $m\widehat{PR}$ and $m\widehat{QS}$ in the figure at the right. Verify by using inscribed angles to find the measures of the arcs.

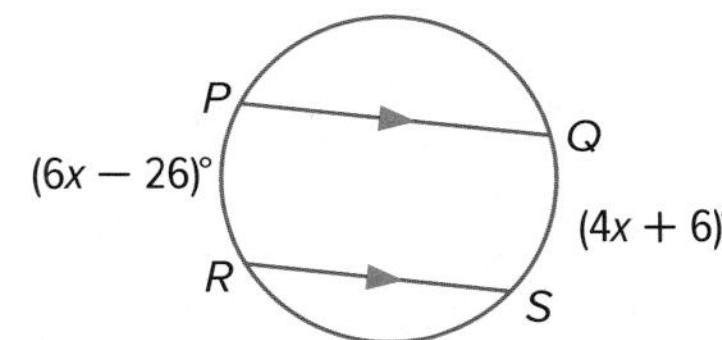

H.O.T. Problems Use Higher-Order Thinking Skills

MP CONSTRUCT ARGUMENTS Determine whether the quadrilateral can *always*, *sometimes*, or *never* be inscribed in a circle. Explain your reasoning.

42. square **43.** rectangle **44.** parallelogram **45.** rhombus **46.** kite

47. MP CHALLENGE A square is inscribed in a circle. What is the ratio of the area of the circle to the area of the square?

48. WRITING IN MATH A 45°-45°-90° right triangle is inscribed in a circle. If the radius of the circle is given, explain how to find the lengths of the right triangle's legs.

49. OPEN-ENDED Find and sketch a real-world logo with an inscribed polygon.

50. WRITING IN MATH Compare and contrast inscribed angles and central angles of a circle. If they intercept the same arc, how are they related?

Preparing for Assessment

51. Quadrilateral $ABCD$ is inscribed in $\odot K$. What is the measure of $\angle A$ in degrees? MP 6

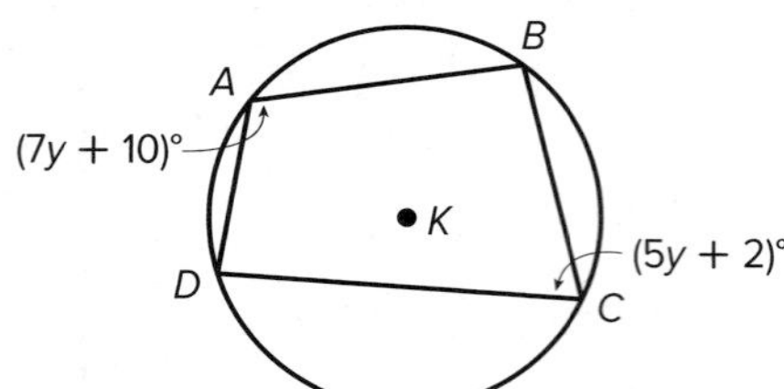

52. What is $m\widehat{SR}$? MP 6

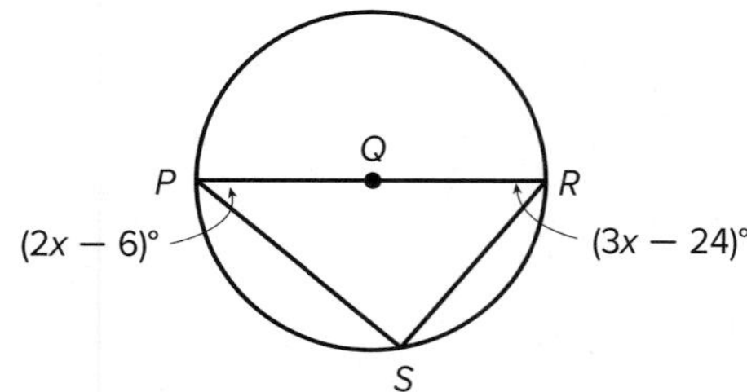

○ **A** 96

○ **B** 84

○ **C** 48

○ **D** 42

○ **E** 24

53. Quadrilateral $JKLM$ is inscribed in $\odot C$, as shown.

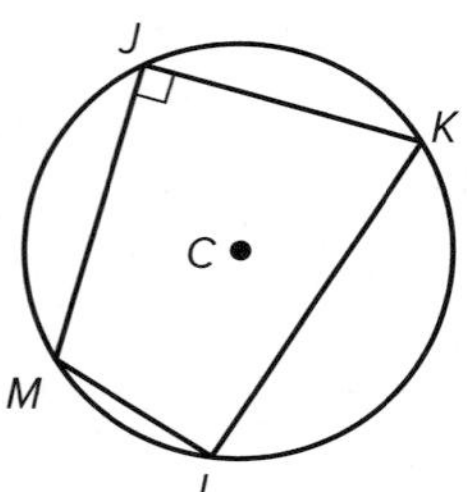

Which of the following statements must be true? MP 2

I. Quadrilateral $JKLM$ contains at least two right angles.

II. $\widehat{MLK}$ is a semicircle.

III. $\angle L$ and $\angle M$ are supplementary.

○ **A** I only

○ **B** II only

○ **C** II and III only

○ **D** I and II only

54. MULTI-STEP Look at the figure. MP 2, 3

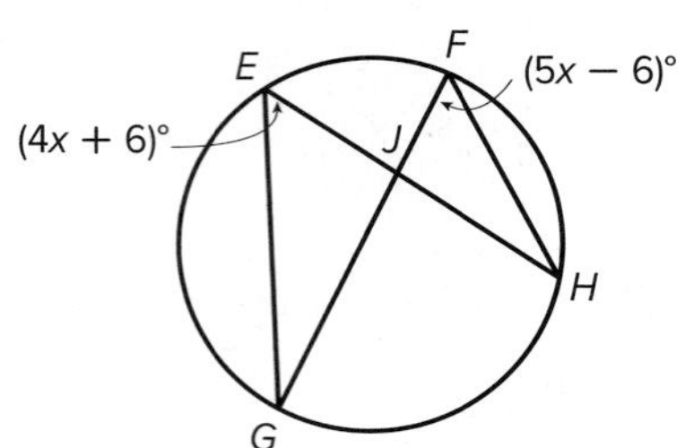

a. Which statements are true? Check all that apply.

☐ **A** $m\angle GFH = 2m\angle FEH$

☐ **B** $m\angle GEH = 2m\angle GFH$

☐ **C** $m\angle GEH = m\angle GFH$

☐ **D** $m\angle GEH = 11.2$

☐ **E** The measure of an inscribed angle is twice the measure of its intercepted arc.

☐ **F** The measure of an inscribed angle is half the measure of its intercepted arc.

☐ **G** The measure of an inscribed angle is equal to the measure of its intercepted arc.

☐ **H** All angles inscribed by the same arc have the same measure.

b. The solution for x below contains an error. Choose the step where the error occurs.

○ **A** $m\angle GEH = m\angle GFH$

○ **B** $4x + 6 = 5x - 6$

○ **C** $9x = 12$

○ **D** $x = 1.33$

c. What is the measure of $\angle GFH$?

○ **A** 22°　　○ **C** 54°

○ **B** 27°　　○ **D** 108°

d. Can you prove $\triangle EJG \cong \triangle FJH$? Explain.

e. Complete the following proof that shows $\triangle EJG \sim \triangle FJH$.

Statements	Reasons
1. $\angle GEH \cong \angle HFG$	**1.** Inscribed ∡ of same arc are ≅.
2. ☐	**2.** Inscribed ∡ of same arc are ≅.
3. $\triangle EJG \sim \triangle FJH$	**3.** AA Similarity

Mid-Chapter Quiz

Lessons 9-1 through 9-4

For Exercises 1–3, refer to ⊙*A*. (Lesson 9-1)

1. Name the circle.
2. Name a diameter.
3. Name a chord that is not a diameter.

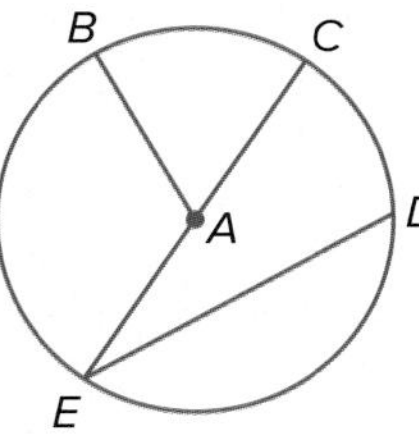

4. **BICYCLES** A bicycle has tires that are 24 inches in diameter. (Lesson 9-1)
   a. Find the circumference of one tire.
   b. How many inches does the tire travel after 100 rotations?
   c. 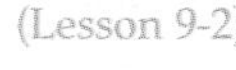 What mathematical practice did you use to solve this problem?

Find the radius and diameter of a circle with the given circumference. Round to the nearest hundredth. (Lesson 9-1)

5. $C = 23$ cm
6. $C = 78$ ft

7. **MULTIPLE CHOICE** What is the length of $\widehat{BC}$? (Lesson 9-2)
   - A 18°
   - B 2.20 cm
   - C 168°
   - D 30.79 cm

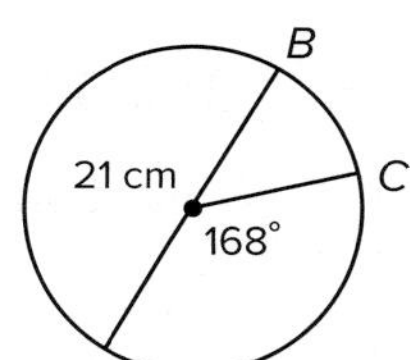

8. **GARDENS** The garden hose caddy shown has a diameter of 14.5 inches. (Lesson 9-2)

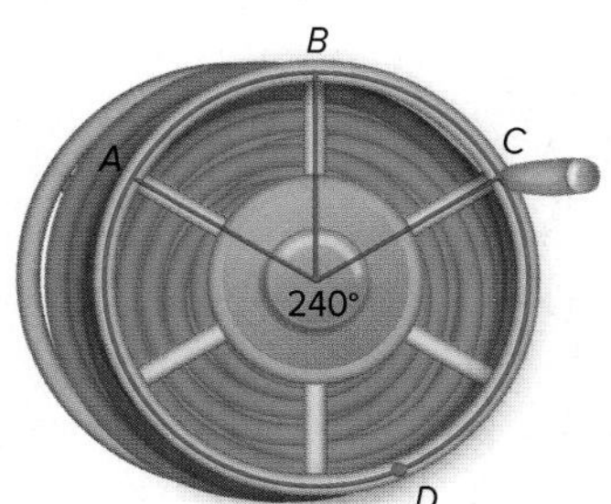

   a. Find $m\widehat{ADC}$.
   b. Find the length of $\widehat{ADC}$.

9. Find the value of x. (Lesson 9-3)

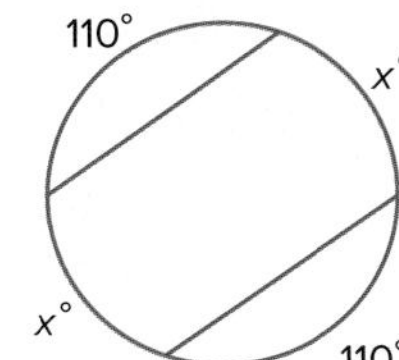

10. In ⊙*B*, $CE = 13.5$. Find BD. Round to the nearest hundredth. (Lesson 9-3)

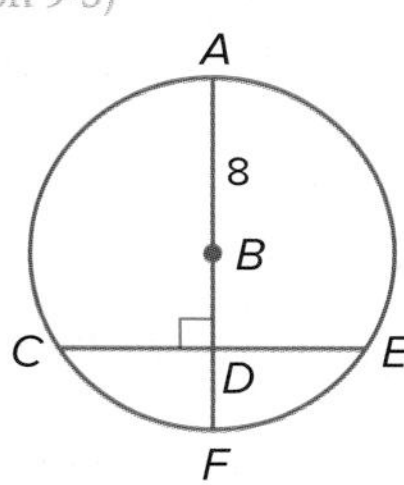

11. The two circles shown are congruent. Find x and the length of the chord. (Lesson 9-3)

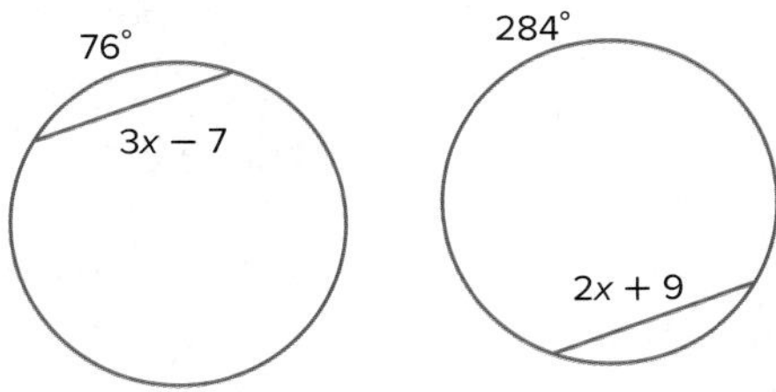

Find each measure. (Lesson 9-4)

12. $m\widehat{TU}$
13. $m\angle A$

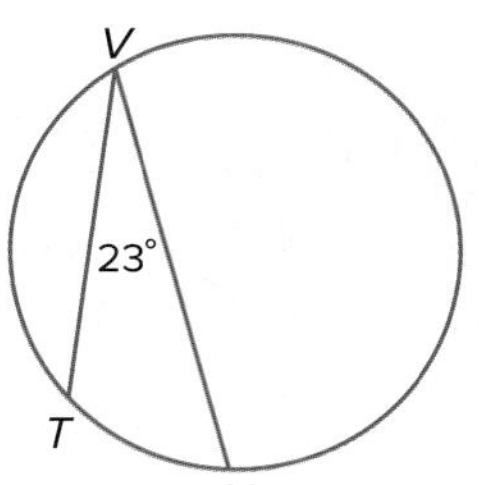

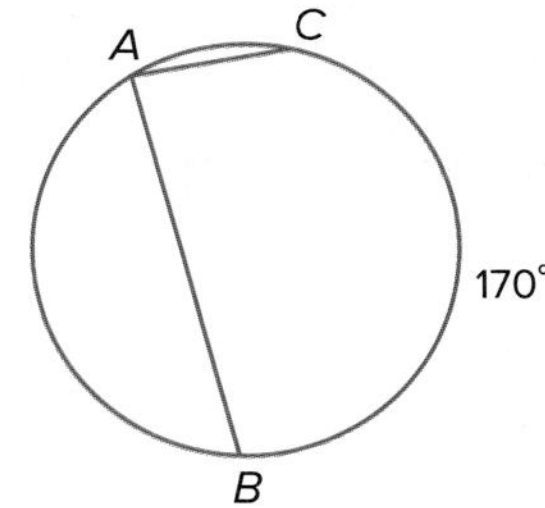

14. **MULTIPLE CHOICE** Find x. (Lesson 9-4)

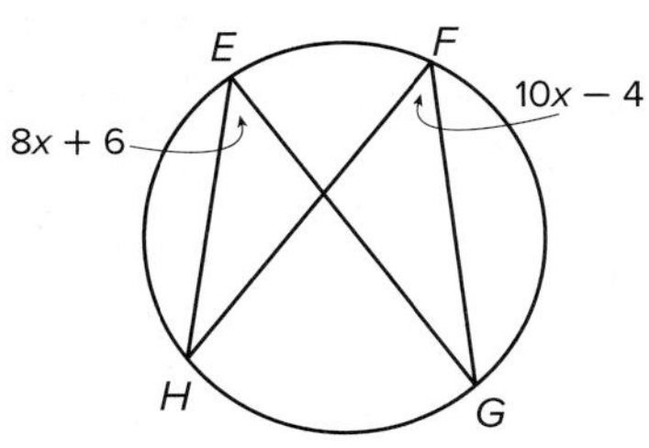

   - A 1.8
   - B 5
   - C 46
   - D 90

15. If a square with sides of 14 inches is inscribed in a circle, what is the diameter of the circle? (Lesson 9-4)

LESSON 5

Tangents

Then

- You used the Pythagorean Theorem to find side lengths of right triangles.

Now

1. Use properties of tangents.
2. Solve problems involving circumscribed polygons.

Why?

- The first bicycles were moved by pushing your feet on the ground. Modern bicycles use pedals, a chain, and gears. The chain loops around circular gears. The length of the chain between these gears is measured from the points of tangency.

New Vocabulary
tangent
point of tangency
common tangent

Mathematical Practices
1 Make sense of problems and persevere in solving them.
2 Reason abstractly and quantitatively.

1 Tangents

A **tangent** is a line in the same plane as a circle that intersects the circle in exactly one point, called the **point of tangency**. $\overleftrightarrow{AB}$ is tangent to $\odot C$ at point A. $\overline{AB}$ and $\overrightarrow{AB}$ are also called tangents.

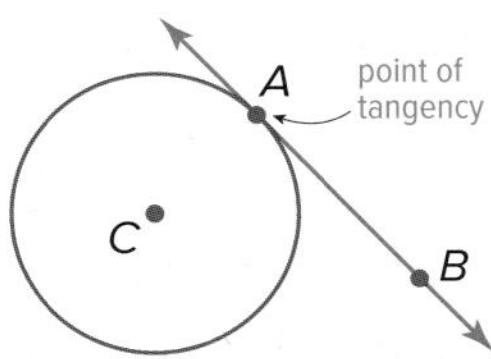

A **common tangent** is a line, ray, or segment that is tangent to two circles in the same plane. In each figure below, line ℓ is a common tangent of circles F and G.

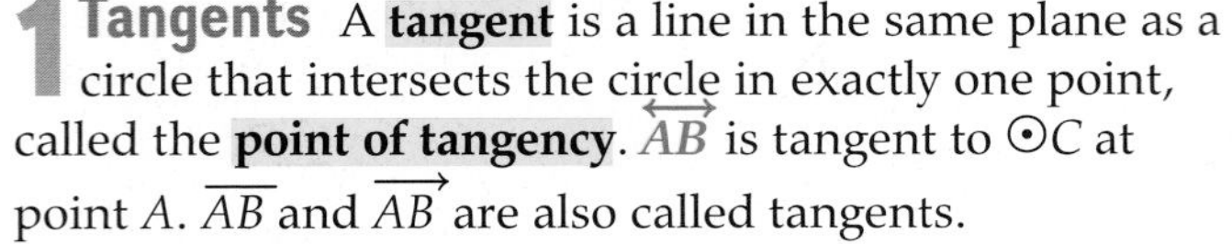

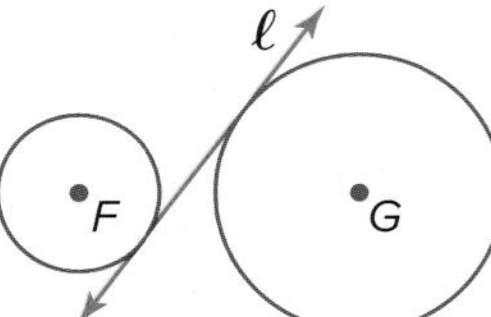

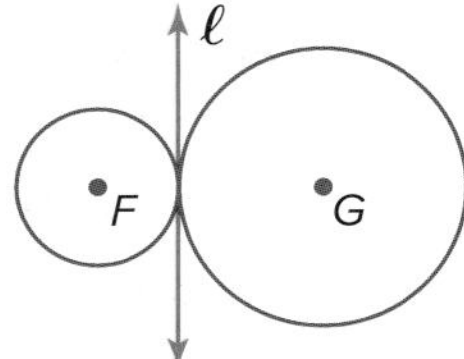

Example 1 Identify Common Tangents

Copy each figure and draw the common tangents. If no common tangent exists, state *no common tangent*.

a.

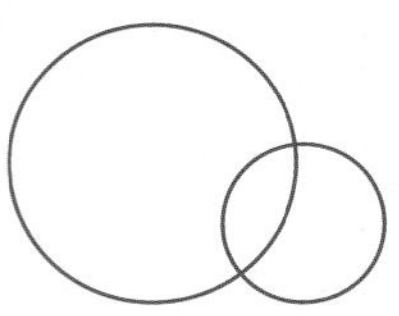

These circles have two common tangents.

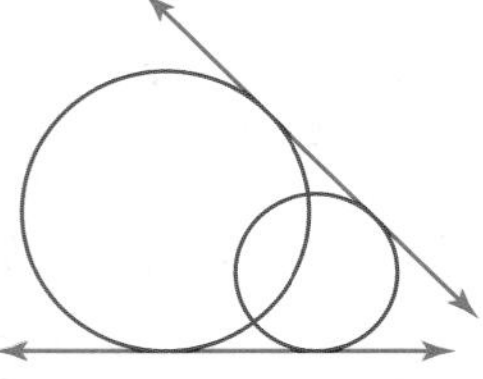

b.

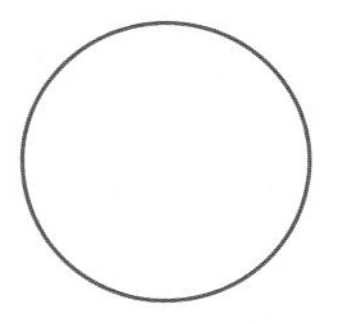

These circles have four common tangents.

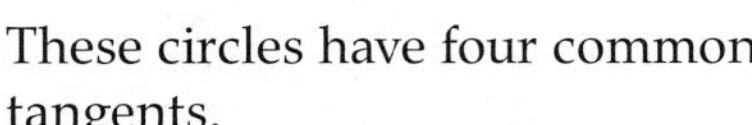

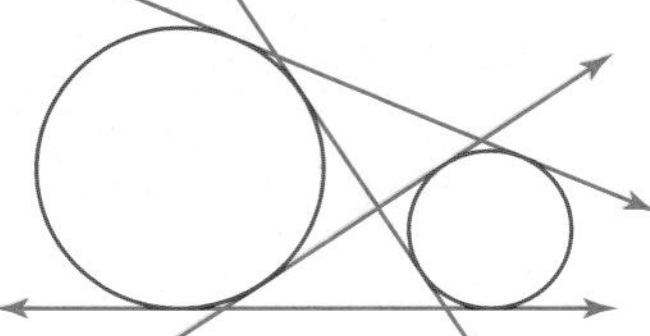

Guided Practice

1A.

1B.

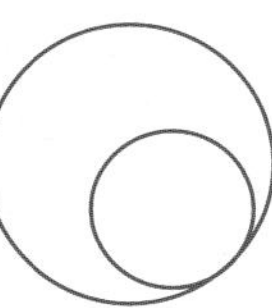

The shortest distance from a tangent to the center of a circle is the radius drawn to the point of tangency.

Theorem 9.10

Words In a plane, a line is tangent to a circle if and only if it is perpendicular to a radius drawn to the point of tangency.

Example Line ℓ is tangent to $\odot S$ if and only if $\ell \perp \overline{ST}$.

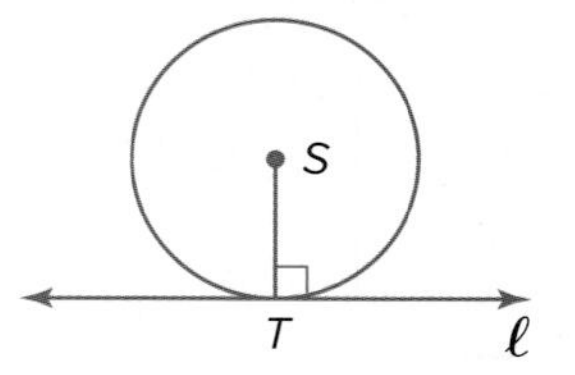

You will prove both parts of Theorem 9.10 in Exercises 32 and 33.

Example 2 Identify a Tangent

$\overline{JL}$ is a radius of $\odot J$. Determine whether $\overline{KL}$ is tangent to $\odot J$. Justify your answer.

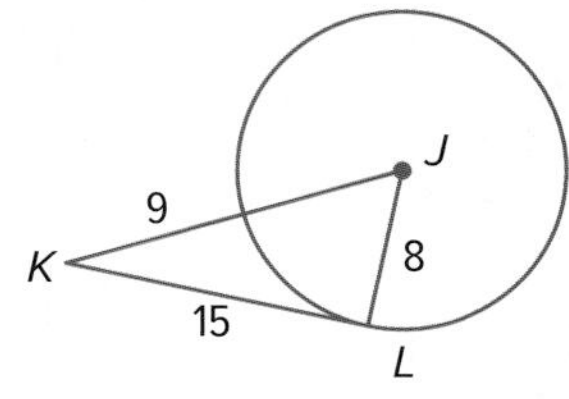

Test to see if $\triangle JKL$ is a right triangle.

$8^2 + 15^2 \stackrel{?}{=} (8 + 9)^2$ Pythagorean Theorem

$289 = 289$ ✓ Simplify.

$\triangle JKL$ is a right triangle with right angle JLK. So $\overline{KL}$ is perpendicular to radius $\overline{JL}$ at point L. Therefore, by Theorem 9.10, $\overline{KL}$ is tangent to $\odot J$.

Guided Practice

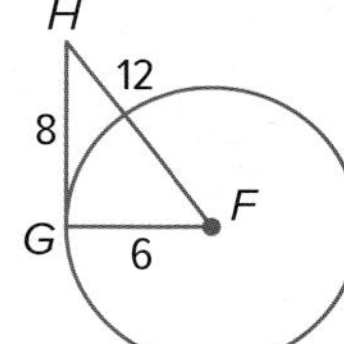

2. Determine whether $\overline{GH}$ is tangent to $\odot F$. Justify your answer.

You can also use Theorem 9.10 to identify missing values.

Example 3 Use a Tangent to Find Missing Values

$\overline{JH}$ is tangent to $\odot G$ at J. Find the value of x.

By Theorem 9.10, $\overline{JH} \perp \overline{GJ}$. So, $\triangle GHJ$ is a right triangle.

$GJ^2 + JH^2 = GH^2$ Pythagorean Theorem

$x^2 + 12^2 = (x + 8)^2$ $GJ = x$, $JH = 12$, and $GH = x + 8$

$x^2 + 144 = x^2 + 16x + 64$ Multiply.

$80 = 16x$ Simplify.

$5 = x$ Divide each side by 16.

Problem-Solving Tip

Solve a Simpler Problem You can use the *solve a simpler problem* strategy by sketching and labeling the right triangles without the circles. A drawing of the triangle in Example 3 is shown below.

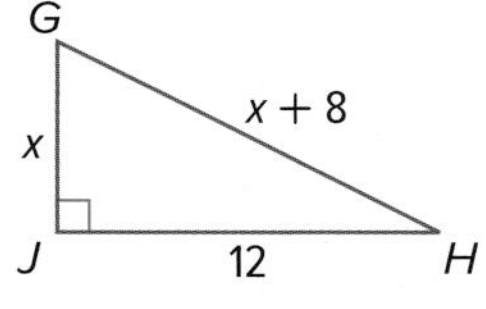

Guided Practice

Find the value of x. Assume that segments that appear to be tangent are tangent.

3A.

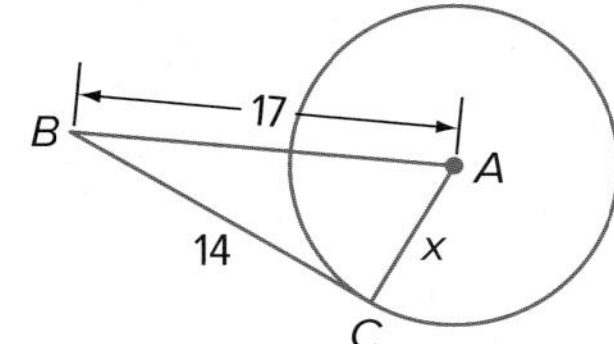

3B.

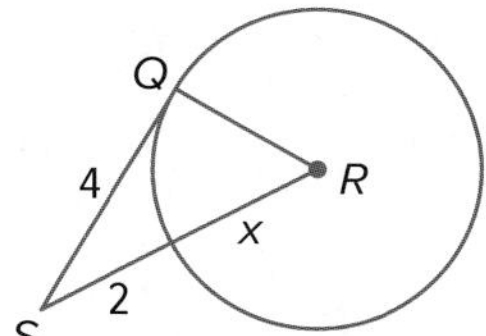

You can use Theorems 9.8 and 9.10 to construct a line tangent to a circle.

Construction Line Tangent to a Circle Through an External Point

Step 1 Use a compass to draw circle C and a point A outside circle C. Then draw $\overline{CA}$.

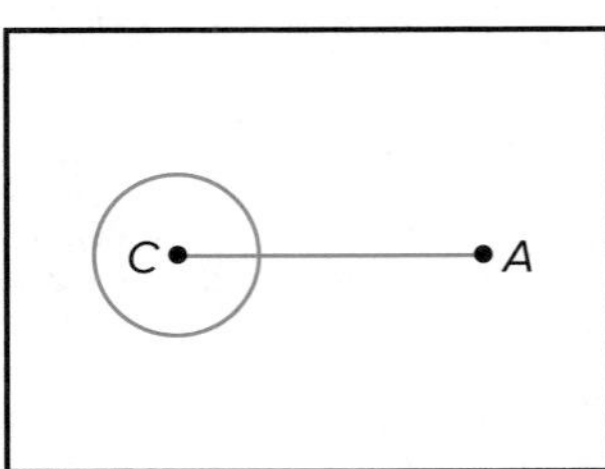

Step 2 Construct line ℓ, the perpendicular bisector of $\overline{CA}$. Label the point of intersection X.

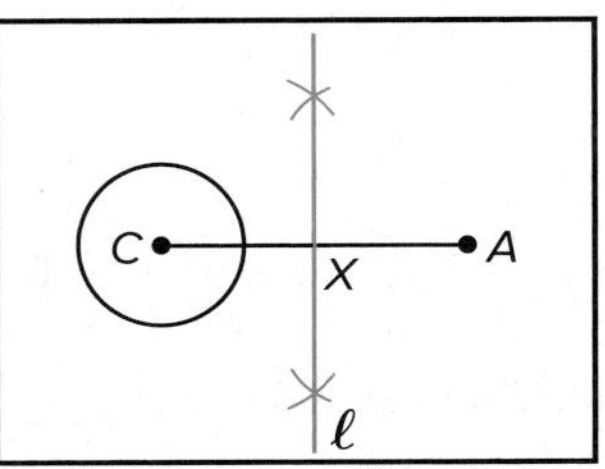

Step 3 Construct circle X with radius $\overline{XC}$. Label the points of intersection of the two circles D and E.

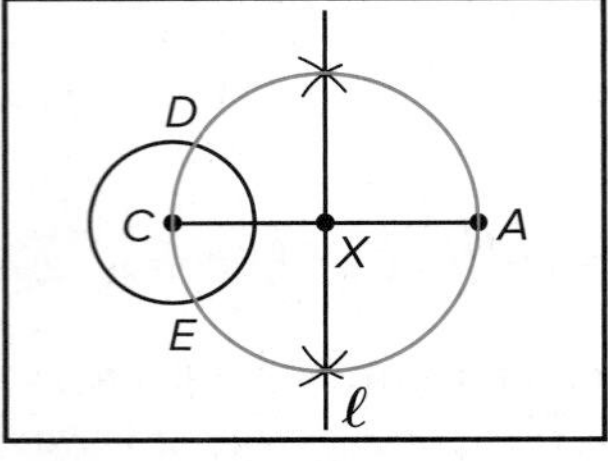

Step 4 Draw $\overleftrightarrow{AD}$ and $\overline{DC}$. $\triangle ADC$ is inscribed in a semicircle. So, $\angle ADC$ is a right angle and $\overleftrightarrow{AD}$ is tangent to $\odot C$.

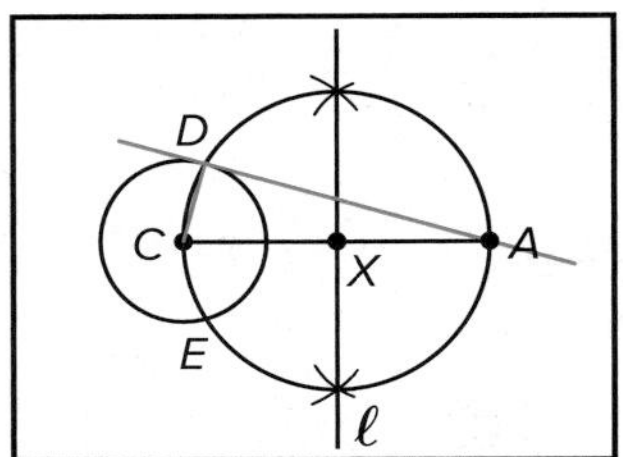

You will justify this construction in Exercise 36 and construct a line tangent to a circle through a point on the circle in Exercise 34.

More than one line can be tangent to the same circle.

Theorem 9.11

Words If two segments from the same exterior point are tangent to a circle, then they are congruent.

Example If $\overline{AB}$ and $\overline{CB}$ are tangent to $\odot D$, then $\overline{AB} \cong \overline{CB}$.

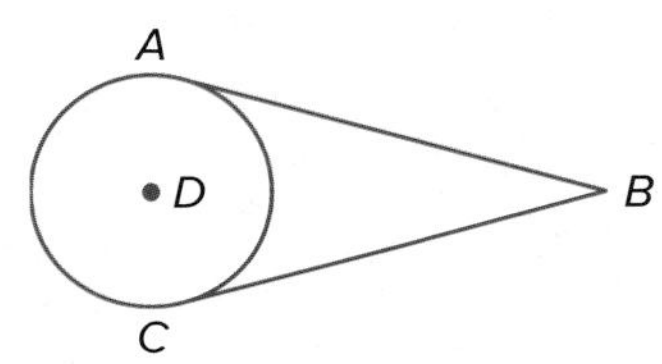

You will prove Theorem 9.11 in Exercise 28.

Example 4 Use Congruent Tangents to Find Measures

ALGEBRA $\overline{AB}$ and $\overline{CB}$ are tangent to $\odot D$. Find the value of x.

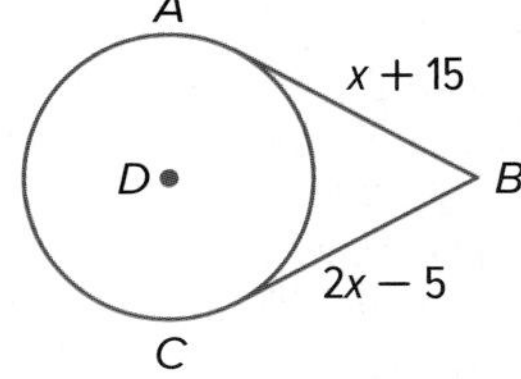

$AB = CB$	Tangents from the same exterior point are congruent.
$x + 15 = 2x - 5$	Substitution
$15 = x - 5$	Subtract x from each side.
$20 = x$	Add 5 to each side.

Guided Practice

ALGEBRA Find the value of x. Assume that segments that appear to be tangent are tangent.

4A.

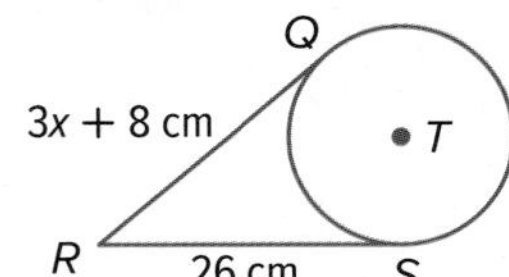

4B.

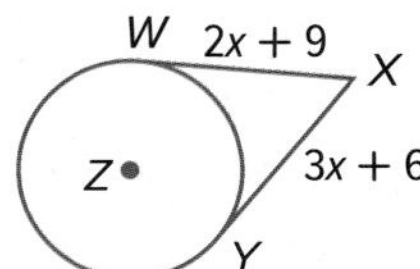

Go Online! 

Construct tangents and compare the lengths of two tangent segments from the common intersection point to the points of tangency with a **Geometer's Sketchpad®** sketch in ConnectED.

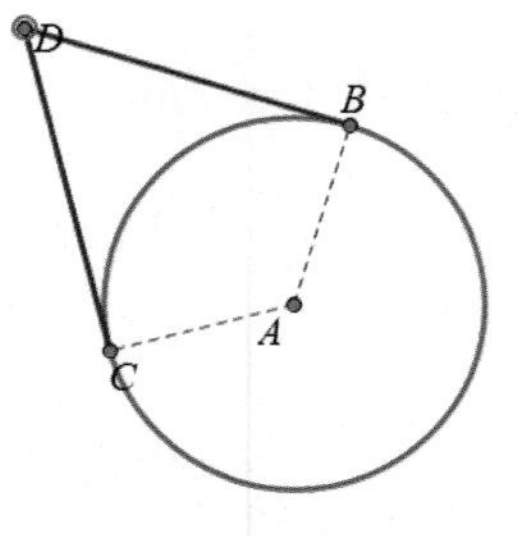

Watch Out

Identifying Circumscribed Polygons If a circle is tangent to one or more of the sides of a polygon, it does not mean that the polygon is circumscribed about the circle, as shown in the set of figures on the right.

2 Circumscribed Polygons

A polygon is circumscribed about a circle if every side of the polygon is tangent to the circle.

Circumscribed Polygons	Polygons Not Circumscribed

You can use Theorem 9.11 to find missing measures in circumscribed polygons.

Real-World Example 5 Find Measures in Circumscribed Polygons

GRAPHIC DESIGN A graphic designer is giving directions to create a larger version of the triangular logo shown. If $\triangle ABC$ is circumscribed about $\odot G$, find the perimeter of $\triangle ABC$.

Step 1 Find the missing measures.

Because $\triangle ABC$ is circumscribed about $\odot G$, $\overline{AE}$ and $\overline{AD}$ are tangent to $\odot G$, as are $\overline{BE}$, $\overline{BF}$, $\overline{CF}$, and $\overline{CD}$. Therefore, $\overline{AE} \cong \overline{AD}$, $\overline{BF} \cong \overline{BE}$, and $\overline{CF} \cong \overline{CD}$.

So, $AE = AD = 8$ feet, $BF = BE = 7$ feet.

By Segment Addition, $CF + FB = CB$, so $CF = CB - FB = 10 - 7$ or 3 feet. So, $CD = CF = 3$ feet.

Step 2 Find the perimeter of $\triangle ABC$.

$$\text{perimeter} = AE + EB + BC + CD + DA$$
$$= 8 + 7 + 10 + 3 + 8 \text{ or } 36$$

So, the perimeter of $\triangle ABC$ is 36 feet.

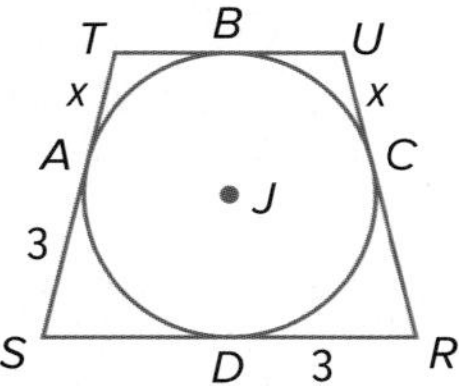

Guided Practice

5. Quadrilateral $RSTU$ is circumscribed about $\odot J$. If the perimeter is 18 units, find x.

Check Your Understanding

 = Step-by-Step Solutions begin on page R13.

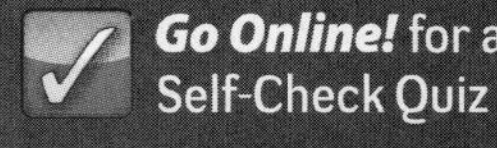

Example 1

1. Copy the figure shown, and draw the common tangents. If no common tangent exists, state *no common tangent*.

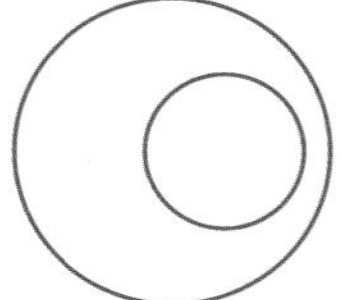

Example 2

Determine whether $\overline{FG}$ is tangent to $\odot E$. Justify your answer.

2.

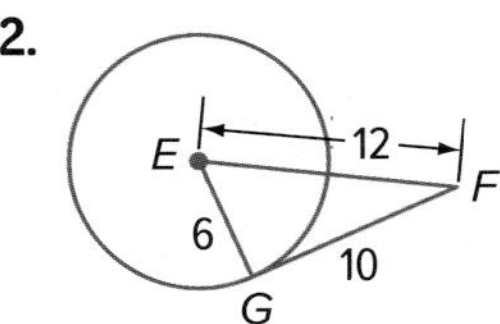

3

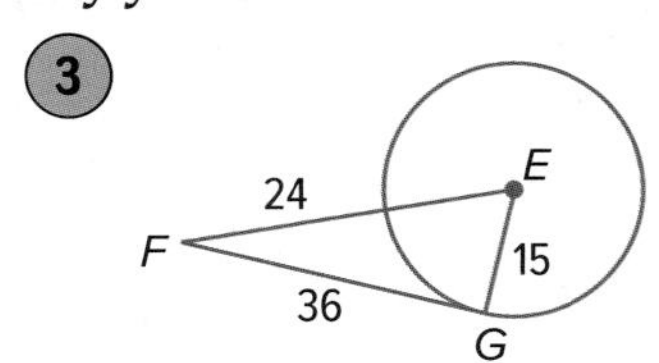

Examples 3–4 **Find x. Assume that segments that appear to be tangent are tangent.**

4.
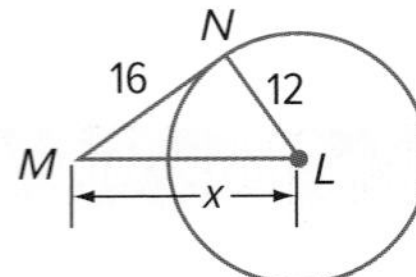

5.
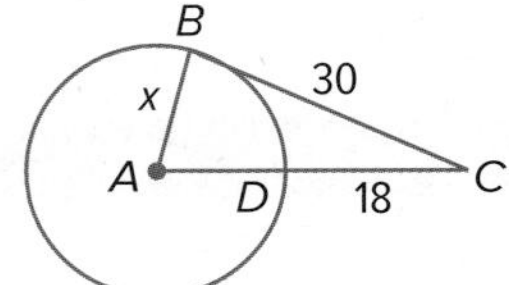

6.
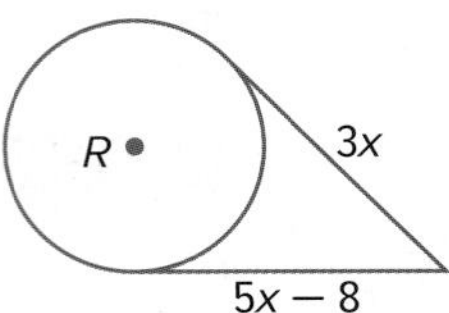

7. **CITY PLANNING** A landscape architect is designing a public park with two adjacent circular fountains. The walking paths are tangent to the fountains as shown. The lengths are given in feet. Find x and y.

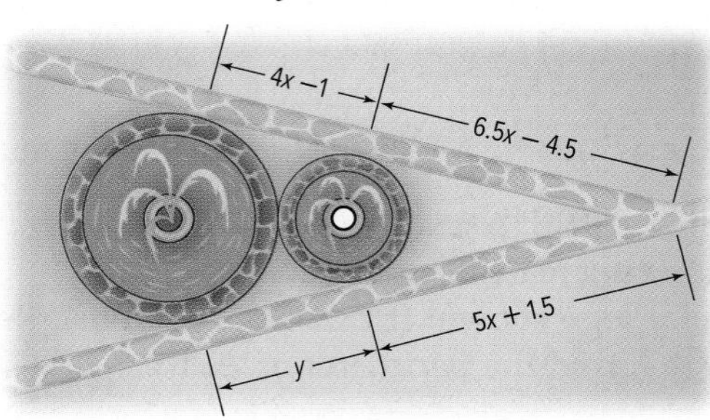

Example 5

8. **MP SENSE-MAKING** Triangle JKL is circumscribed about $\odot R$.

   a. Find x.

   b. Find the perimeter of $\triangle JKL$.

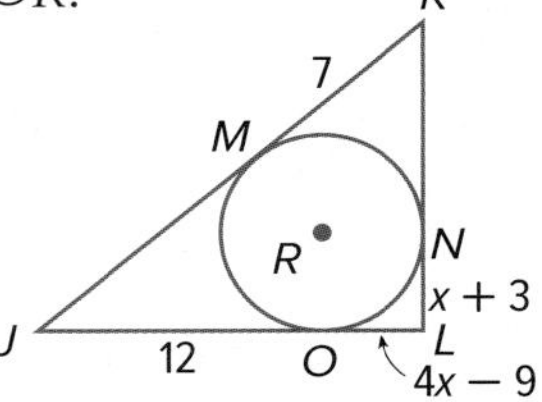

Practice and Problem Solving

Extra Practice is on page R9.

Example 1 **Copy each figure and draw the common tangents. If no common tangent exists, state *no common tangent*.**

9.

10.
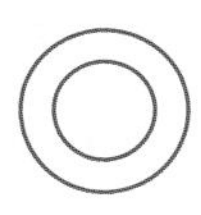

11.
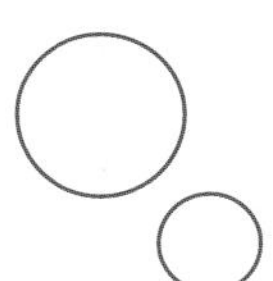

12.
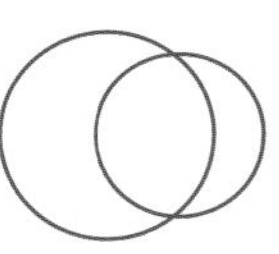

Example 2 **Determine whether each $\overline{XY}$ is tangent to the given circle. Justify your answer.**

13.
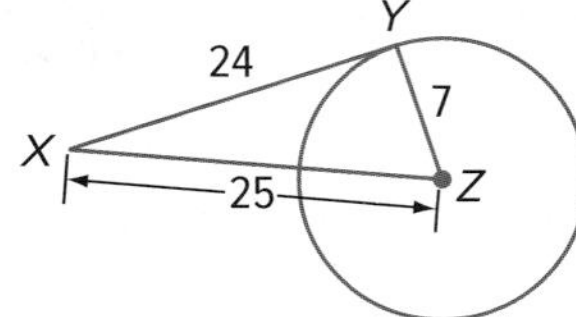

14.
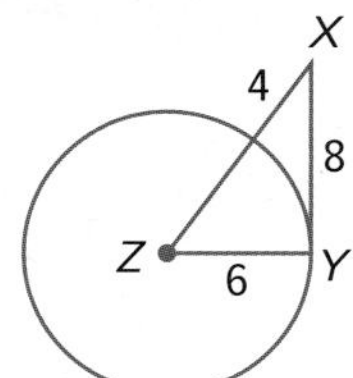

15.
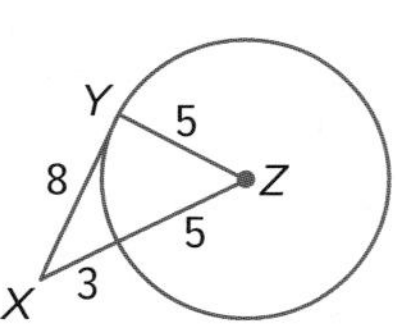

16.
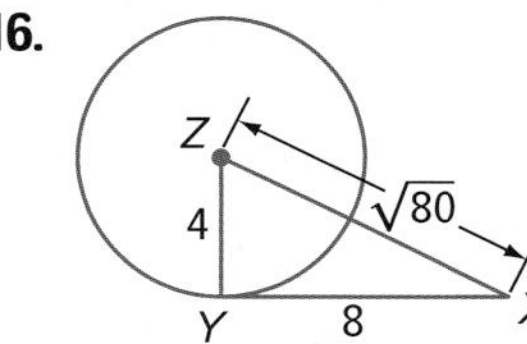

Examples 3–4 **Find x. Assume that segments that appear to be tangent are tangent. Round to the nearest tenth if necessary.**

17.

18.

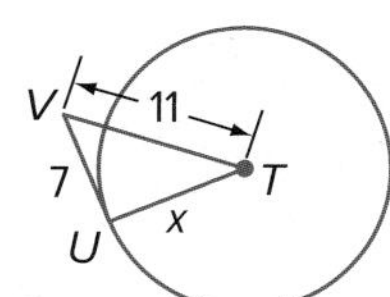

19.

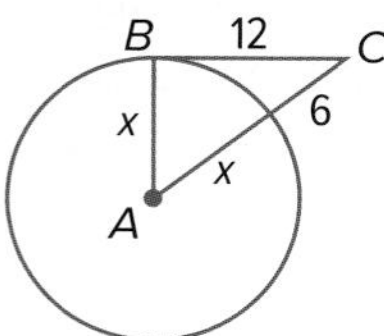

20.

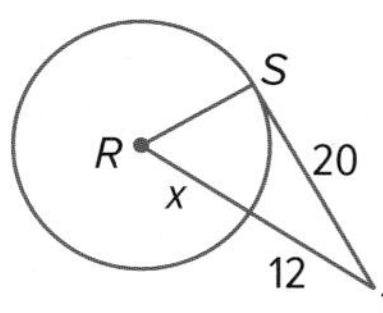

21.

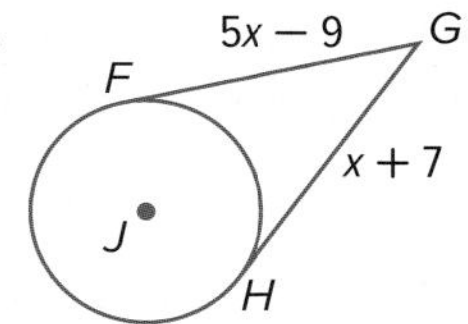

22.

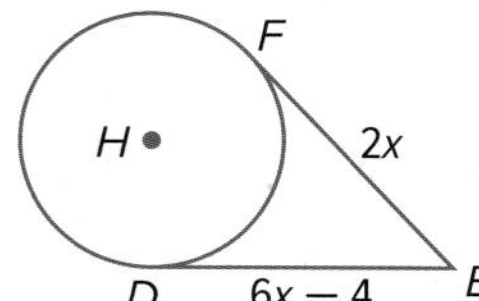

23. **ARBORS** In the arbor shown, $\overline{AC}$ and $\overline{BC}$ are tangents to $\odot D$. The radius of the circle is 26 inches and $EC = 20$ inches. Find each measure to the nearest hundredth.

a. AC

b. BC

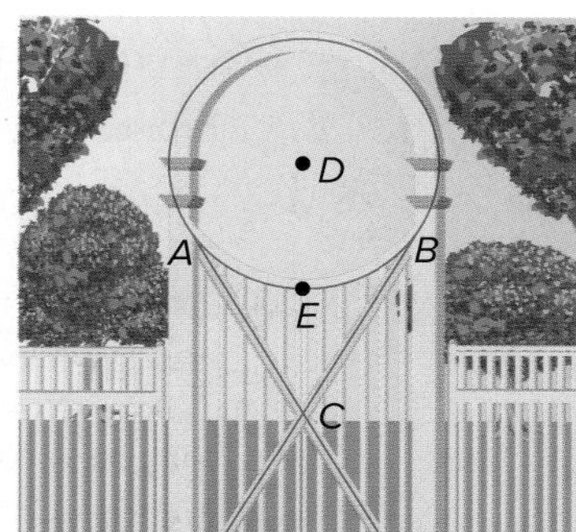

Example 5 **MP SENSE-MAKING** **Find the value of x. Then find the perimeter.**

24.

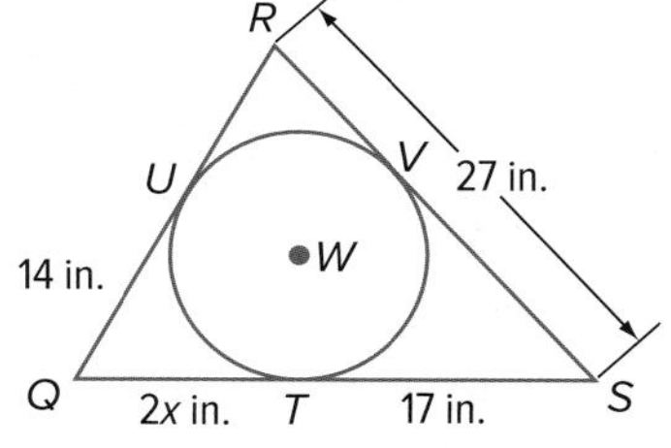

25.

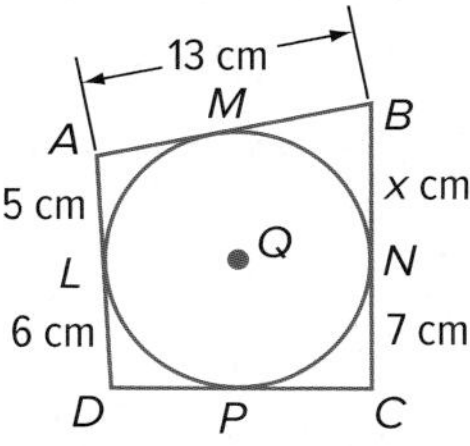

Find x to the nearest hundredth. Assume that segments that appear to be tangent are tangent.

26.

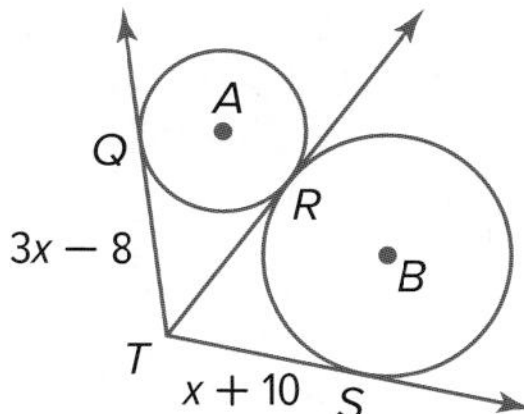

27.

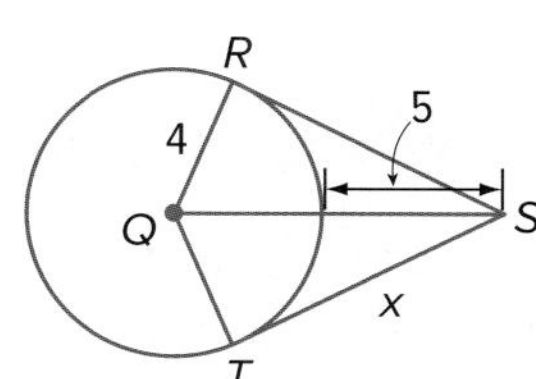

Write the specified type of proof.

28. Two-column proof of Theorem 9.11

Given: $\overline{AC}$ is tangent to $\odot H$ at C.
$\overline{AB}$ is tangent to $\odot H$ at B.

Prove: $\overline{AC} \cong \overline{AB}$

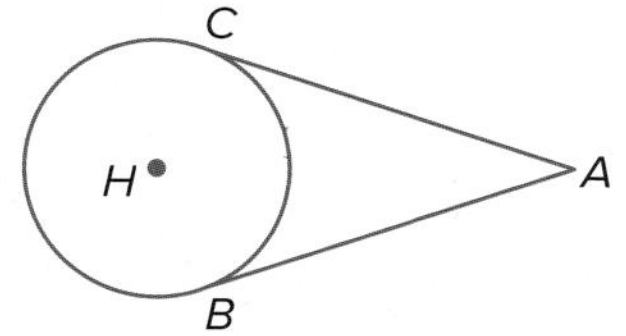

29. Two-column proof

Given: Quadrilateral $ABCD$ is circumscribed about $\odot P$.

Prove: $AB + CD = AD + BC$

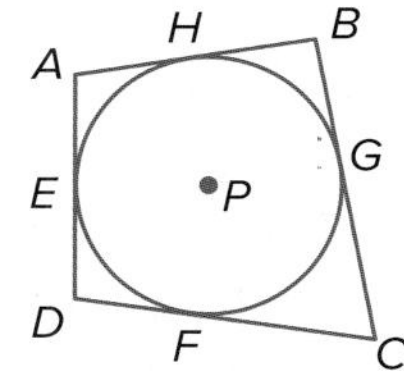

30. **SATELLITES** A satellite is 720 kilometers above Earth, which has a radius of 6360 kilometers. The region of Earth that is visible from the satellite is between the tangent lines $\overline{BA}$ and $\overline{BC}$. What is BA? Round to the nearest hundredth.

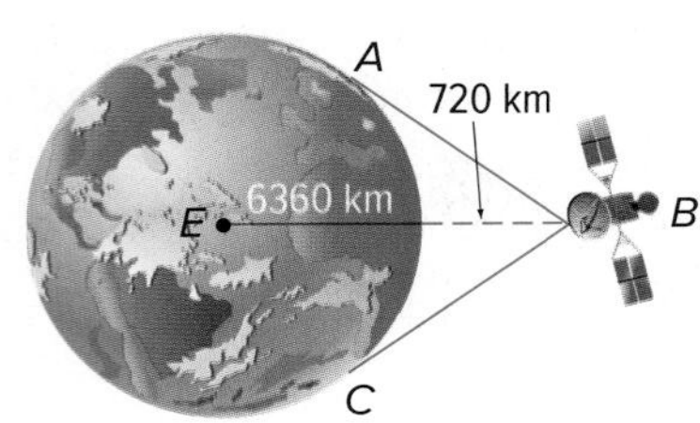

31. **SPACE TRASH** *Orbital debris* refers to materials from space missions that still orbit Earth. In 2007, a 1400-pound ammonia tank was discarded from a space mission. Suppose the tank has an altitude of 435 miles. What is the distance from the tank to the farthest point on Earth's surface from which the tank is visible? Assume that the radius of Earth is 4000 miles. Round to the nearest mile, and include a diagram of this situation with your answer.

32. **PROOF** Write an indirect proof to show that if a line is tangent to a circle, then it is perpendicular to a radius of the circle. (Part 1 of Theorem 9.10)

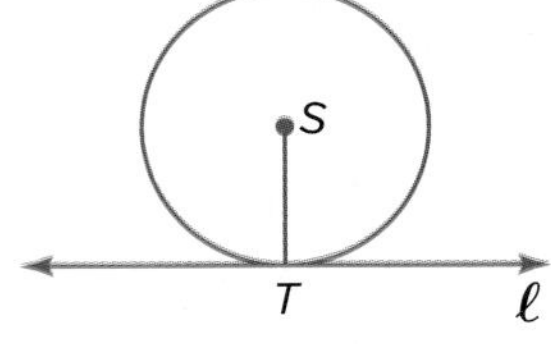

Given: ℓ is tangent to $\odot S$ at T; $\overline{ST}$ is a radius of $\odot S$.

Prove: $\ell \perp \overline{ST}$

(*Hint:* Assume ℓ is *not* $\perp$ to $\overline{ST}$.)

33. **PROOF** Write an indirect proof to show that if a line is perpendicular to the radius of a circle at its endpoint, then the line is a tangent of the circle. (Part 2 of Theorem 9.10)

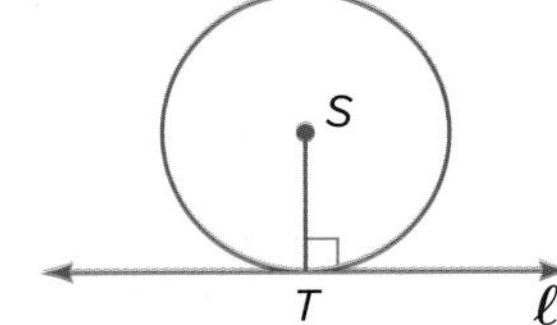

Given: $\ell \perp \overline{ST}$; $\overline{ST}$ is a radius of $\odot S$.

Prove: ℓ is tangent to $\odot S$.

(*Hint:* Assume ℓ is *not* tangent to $\odot S$.)

34. MP **TOOLS** Construct a line tangent to a circle through a point on the circle. Use a compass to draw $\odot A$. Choose a point P on the circle and draw $\overleftrightarrow{AP}$. Then construct a segment through point P perpendicular to $\overleftrightarrow{AP}$. Label the tangent line t. Explain and justify each step.

H.O.T. Problems Use Higher-Order Thinking Skills

35. **CHALLENGE** $\overline{PQ}$ is tangent to circles R and S in the diagram to the right. Find PQ. Explain your reasoning.

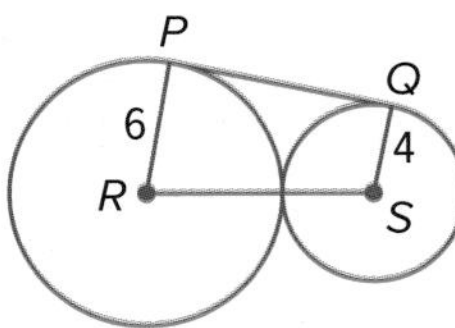

36. **WRITING IN MATH** Explain each step in the construction in page 680. Provide a justification for each step of the construction.

37. **OPEN-ENDED** Draw a circumscribed triangle and an inscribed triangle.

38. **REASONING** In the figure, $\overline{XY}$ and $\overline{XZ}$ are tangent to $\odot A$. $\overline{XZ}$ and $\overline{XW}$ are tangent to $\odot B$. Explain how segments $\overline{XY}$, $\overline{XZ}$, and $\overline{XW}$ can all be congruent if the circles have different radii.

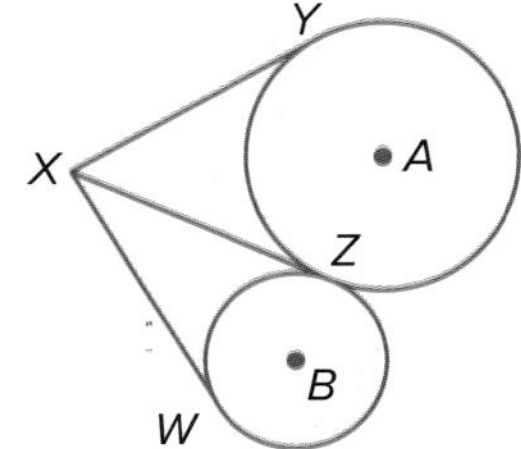

39. **WRITING IN MATH** Is it possible to draw a tangent from a point that is located anywhere outside, on, or inside a circle? Explain.

40. $\overline{MN}$ and $\overline{MP}$ are tangent to $\odot G$, as shown, with $MN = 2y - 2$, $MP = 4y - 32$, and $NP = y + 8$.

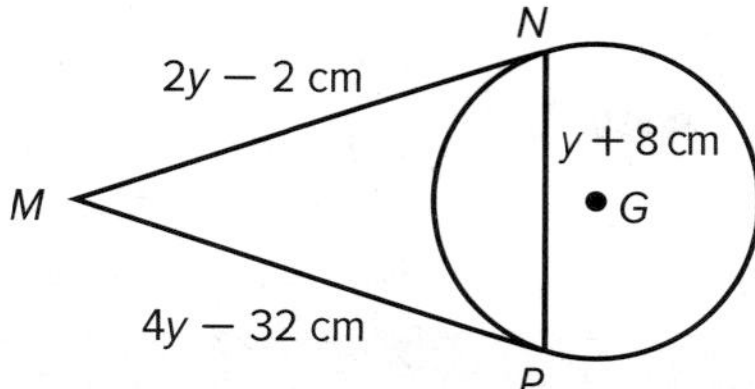

What is the perimeter of $\triangle MNP$? MP 1, 6

- A 15 cm
- B 28 cm
- C 44 cm
- D 56 cm
- E 79 cm

41. A square is inscribed in a circle with a radius of 6 inches. Find the length of each side of the square to the nearest tenth of an inch. MP 1, 6

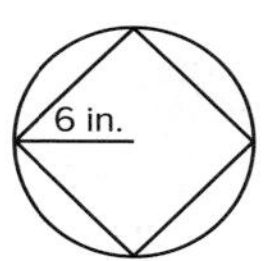

42. Circle C has a radius of 10. $\overline{KJ}$ is tangent to $\odot C$ at point K, and $LJ = 12$.

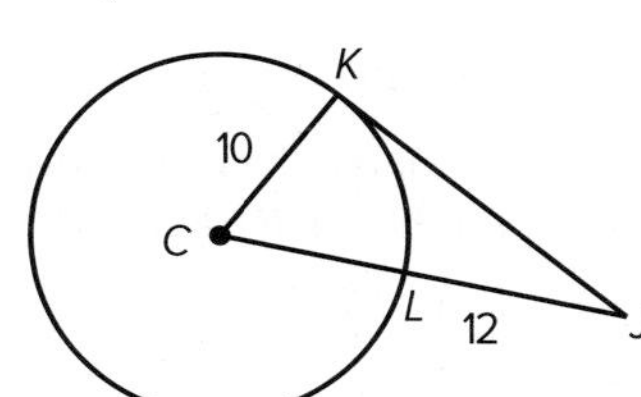

Which of the following is the best estimate of the length of $\overline{KJ}$? MP 2, 6

- A 22.0
- B 19.6
- C 12.0
- D 6.6

43. $\overline{PQ}$ is tangent to $\odot S$ at point Q, $m\angle PQS = (4y + 30)$, and $m\angle SPQ = (y + 51)$. What is $m\angle SPQ$? MP 2, 6

- A 15
- B 58
- C 66
- D 90

44. **MULTI-STEP** Look at the figure. $\overline{RS}$ is tangent to the circle at point R. MP 1, 2

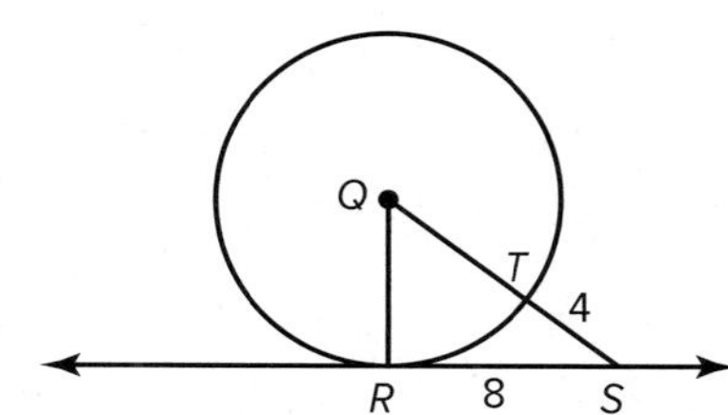

a. Which statements are true? Check all that apply.

- A $\overline{QR} \cong \overline{QT}$
- B $\overline{QR} \cong \overline{QS}$
- C $\overline{QT} \cong \overline{RS}$
- D $QR^2 + RS^2 = QS^2$
- E $QS^2 + QR^2 = RS^2$
- F $\overline{QT} + \overline{TS} = \overline{QS}$

b. Which fact is true only because $\overline{QR}$ is tangent to $\overline{RS}$?

- A $\overline{QR} \cong \overline{QT}$
- B $\overline{QR} \cong \overline{QS}$
- C $QR^2 + RS^2 = QS^2$
- D $QS^2 + QR^2 = RS^2$

c. What are the lengths of $\overline{QT}$ and $\overline{QS}$, respectively?

- A 4, 6
- B 8, 12
- C 4, 8
- D 6, 10

EXTEND 9-5

Geometry Lab

Inscribed and Circumscribed Circles

In this lab, you will perform constructions that involve inscribing or circumscribing a circle.

MP 5 Use appropriate tools strategically.

Activity 1 Construct a Circle Inscribed in a Triangle

Step 1

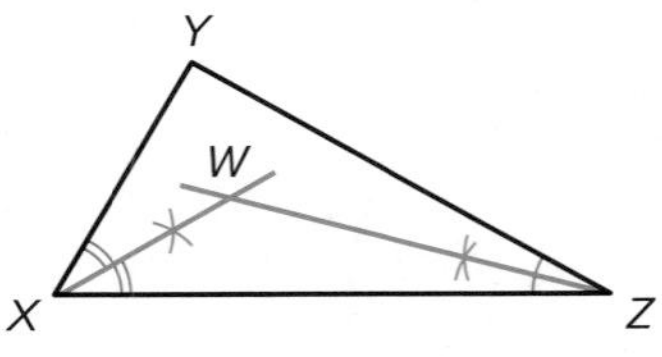

Draw a triangle XYZ and construct two angle bisectors of the triangle to locate the incenter W.

Step 2

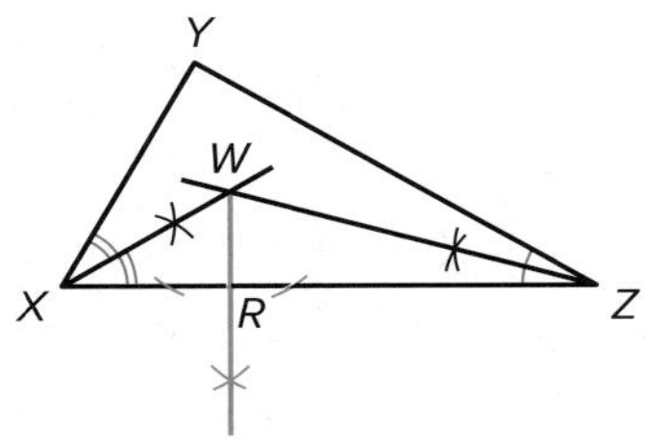

Construct a segment perpendicular to a side through the incenter. Label the intersection R.

Step 3

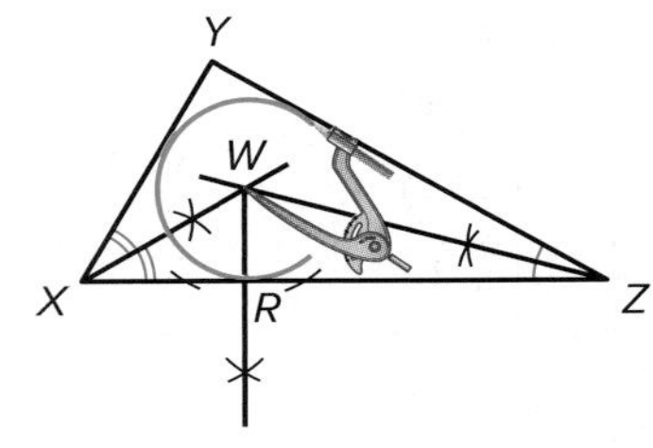

Set a compass of the length of $\overline{WR}$. Put the point of the compass on W and draw a circle with that radius.

Activity 2 Construct a Triangle Circumscribed About a Circle

Step 1

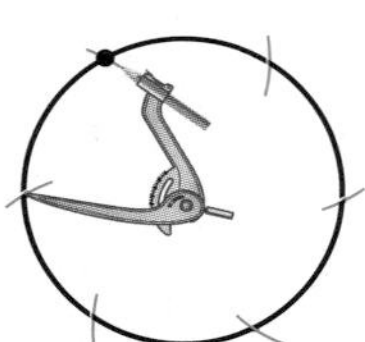

Construct a circle and draw a point. Use the same compass setting you used to construct the circle to construct an arc on the circle from the point. Continue as shown.

Step 2

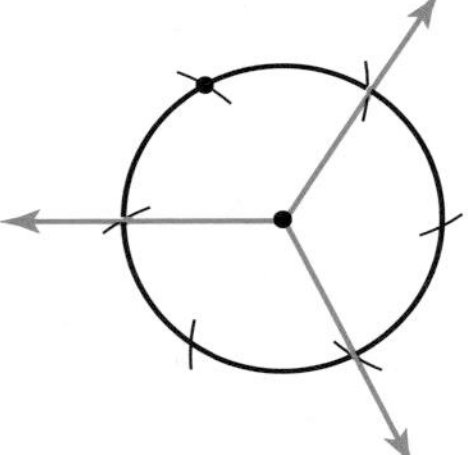

Draw rays from the center through every other arc.

Step 3

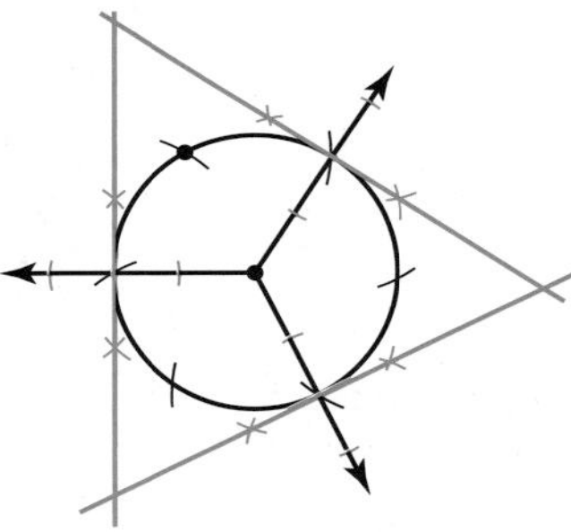

Construct a line perpendicular to each of the rays.

Model

1. Draw a right triangle and inscribe a circle in it.
2. Inscribe a regular hexagon in a circle. Then inscribe an equilateral triangle in a circle. (*Hint:* The first step of each construction is identical to Step 1 in Activity 2.)
3. Inscribe a square in a circle. Then circumscribe a square about a circle.
4. **CHALLENGE** Circumscribe a regular hexagon about a circle.

LESSON 6

Secants, Tangents, and Angle Measures

Then

- You found measures of segments formed by tangents to a circle.

Now

1. Find measures of angles formed by lines intersecting on or inside a circle.
2. Find measures of angles formed by lines intersecting outside the circle.

Why?

- An average person's field of vision is about 180°. Most cameras, including smartphone cameras, have a much narrower viewing angle of between 20° and 60°. Using specialized software, some smartphones can capture panoramic images that show a field of vision similar to or greater than that of the human eye. The viewing angle determines how much of a subject a camera can capture.

New Vocabulary

secant

Mathematical Practices

2 Reason abstractly and quantitatively.

3 Construct viable arguments and critique the reasoning of others.

1 Intersections On or Inside a Circle

A **secant** is a line that intersects a circle in exactly two points. Lines j and k are secants of $\odot C$.

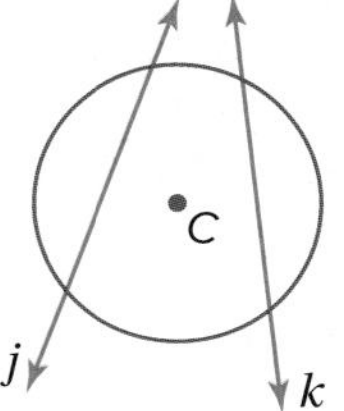

When two secants intersect inside a circle, the angles formed are related to the arcs they intercept.

Theorem 9.12

Words If two secants or chords intersect in the interior of a circle, then the measure of an angle formed is one half the *sum* of the measure of the arcs intercepted by the angle and its vertical angle.

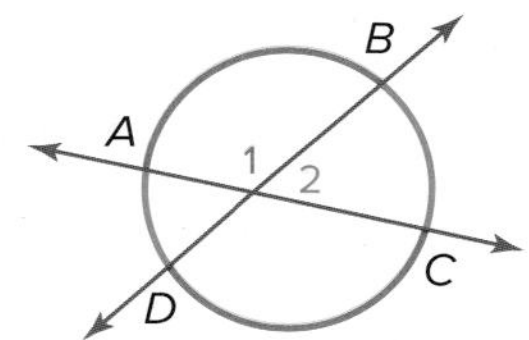

Example $m\angle 1 = \frac{1}{2}(m\widehat{AB} + m\widehat{CD})$ and $m\angle 2 = \frac{1}{2}(m\widehat{DA} + m\widehat{BC})$

Proof

Given: $\overleftrightarrow{HK}$ and $\overleftrightarrow{JL}$ intersect at M.

Prove: $m\angle 1 = \frac{1}{2}(m\widehat{JH} + m\widehat{LK})$

Proof:

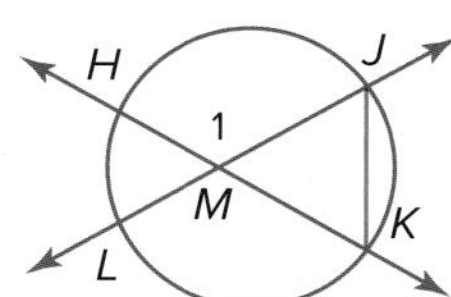

Statements	Reasons
1. $\overleftrightarrow{HK}$ and $\overleftrightarrow{JL}$ intersect at M.	1. Given
2. $m\angle 1 = m\angle MJK + m\angle MKJ$	2. Exterior Angle Theorem
3. $m\angle MJK = \frac{1}{2}m\widehat{LK}$, $m\angle MKJ = \frac{1}{2}m\widehat{JH}$	3. The measure of an inscribed $\angle$ equals half the measure of the intercepted arc.
4. $m\angle 1 = \frac{1}{2}m\widehat{LK} + \frac{1}{2}m\widehat{JH}$	4. Substitution
5. $m\angle 1 = \frac{1}{2}(m\widehat{JH} + m\widehat{LK})$	5. Distributive Property

Example 1 Use Intersecting Chords or Secants

Find x.

a.

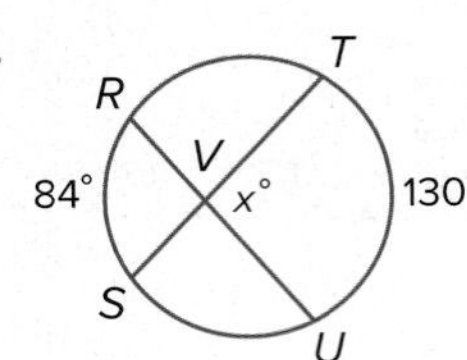

$$m\angle TVU = \frac{1}{2}(m\widehat{RS} + m\widehat{TU})$$ Theorem 9.12

$$x = \frac{1}{2}(84 + 130)$$ Substitution

$$= \frac{1}{2}(214) \text{ or } 107$$ Simplify.

Study Tip

Alternative Method In Example 1b, $m\angle DEB$ can also be found by first finding the sum of the measures of $\widehat{AC}$ and $\widehat{BD}$.

$m\widehat{AC} + m\widehat{BD}$

$= 360 - (m\widehat{AC} + m\widehat{CD})$

$= 360 - (143 + 75) = 142$

$m\angle DEB$

$= \frac{1}{2}(m\widehat{AC} + m\widehat{BD})$

$= \frac{1}{2}(142) = 71$

b.

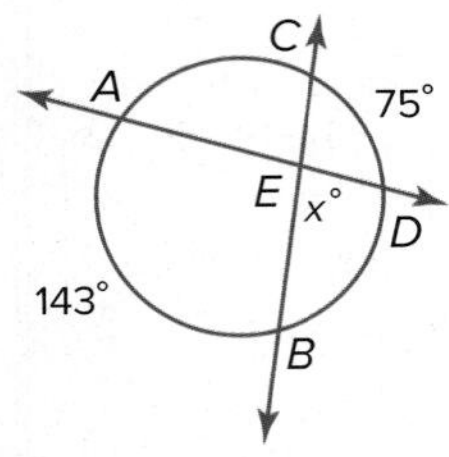

Step 1 Find $m\angle AEB$.

$$m\angle AEB = \frac{1}{2}(m\widehat{AB} + m\widehat{CD})$$ Theorem 9.12

$$= \frac{1}{2}(143 + 75)$$ Substitution

$$= \frac{1}{2}(218) \text{ or } 109$$ Simplify.

Step 2 Find x, the measure of $\angle DEB$.

$\angle AEB$ and $\angle DEB$ are supplementary angles.

So, $x = 180 - 109$ or 71.

c.

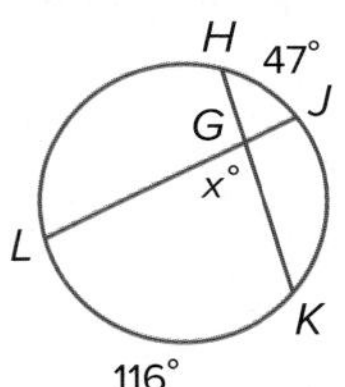

$$m\angle GLH = \frac{1}{2}(m\widehat{GH} + m\widehat{KJ})$$ Theorem 9.12

$$110 = \frac{1}{2}(x + 97)$$ Substitution

$$220 = x + 97$$ Multiply each side by 2.

$$123 = x$$ Subtract 97 from each side.

Guided Practice

Find the value for x in each diagram.

1A.

1B.

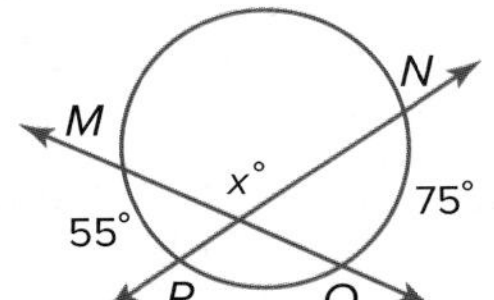

1C.

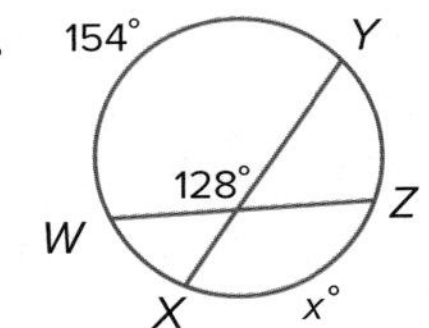

Recall that Theorem 9.6 states that the measure of an inscribed angle is half the measure of its intercepted arc. If one of the sides of this angle is tangent to the circle, this relationship still holds true.

Theorem 9.13

Words If a secant and a tangent intersect at the point of tangency, then the measure of each angle formed is one half the measure of its intercepted arc.

Example $m\angle 1 = \frac{1}{2}m\widehat{AB}$ and $m\angle 2 = \frac{1}{2}m\widehat{ACB}$

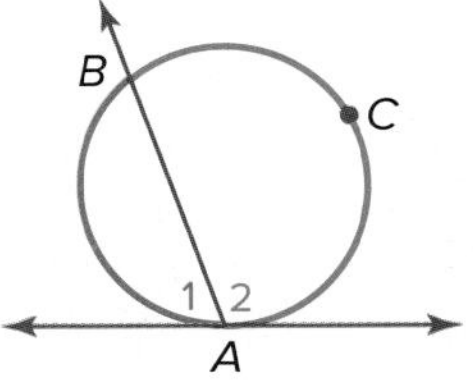

You will prove Theorem 9.13 in Exercise 33.

Go Online!

Follow along as you watch a **Personal Tutor** solve a problem involving secants and tangents.

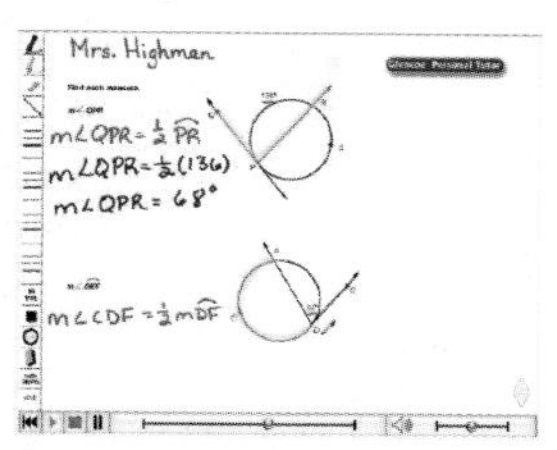

Example 2 Use Intersecting Secants and Tangents

Find each measure.

a. $m\angle QPR$

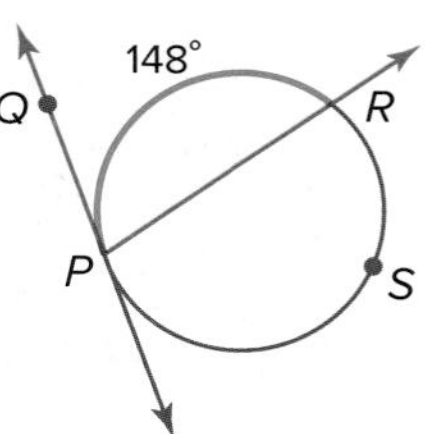

$m\angle QPR = \frac{1}{2}m\widehat{PR}$ Theorem 9.13

$= \frac{1}{2}(148)$ or 74 Substitute and simplify.

b. $m\widehat{DEF}$

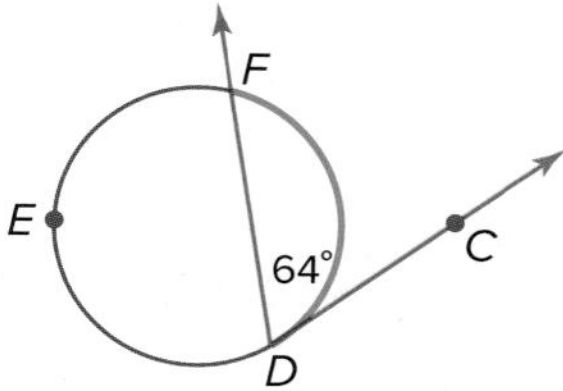

$m\angle CDF = \frac{1}{2}m\widehat{FD}$ Theorem 9.13

$64 = \frac{1}{2}m\widehat{FD}$ Substitution

$128 = m\widehat{FD}$ Multiply each side by 2.

$m\widehat{DEF} = 360 - m\widehat{FD} = 360 - 128$ or 232

Guided Practice

2A. Find $m\widehat{JLK}$.

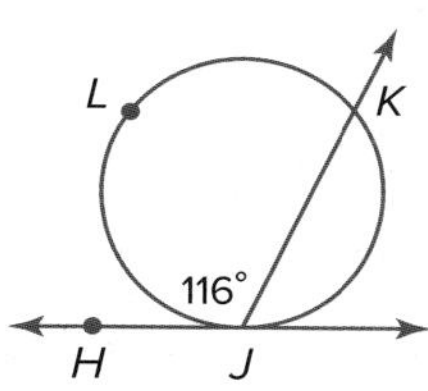

2B. Find $m\angle RQS$ if $m\widehat{QTS} = 238$.

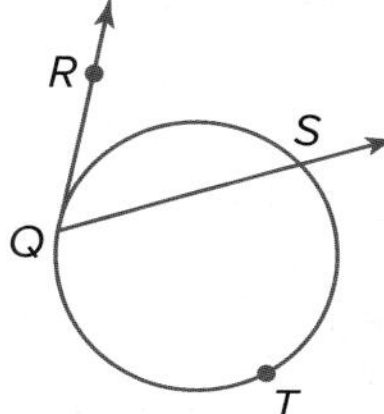

2 Intersections Outside a Circle

Secants and tangents can also meet outside a circle. The measure of the angle formed also involves half of the measures of the arcs they intercept.

Theorem 9.14

Words If two secants, a secant and a tangent, or two tangents intersect in the exterior of a circle, then the measure of the angle formed is one half the *difference* of the measures of the intercepted arcs.

Examples

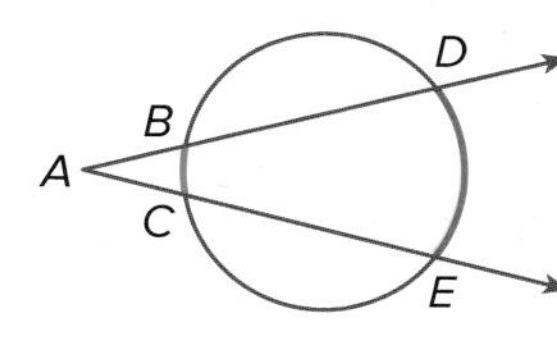

Two Secants

$m\angle A = \frac{1}{2}(m\widehat{DE} - m\widehat{BC})$

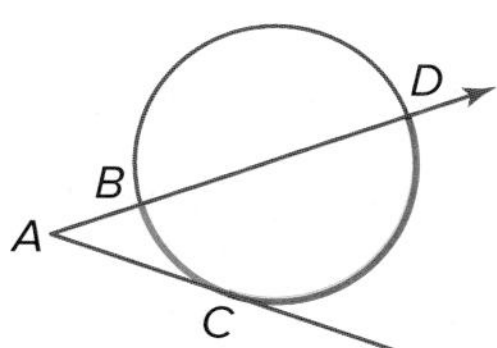

Secant-Tangent

$m\angle A = \frac{1}{2}(m\widehat{DC} - m\widehat{BC})$

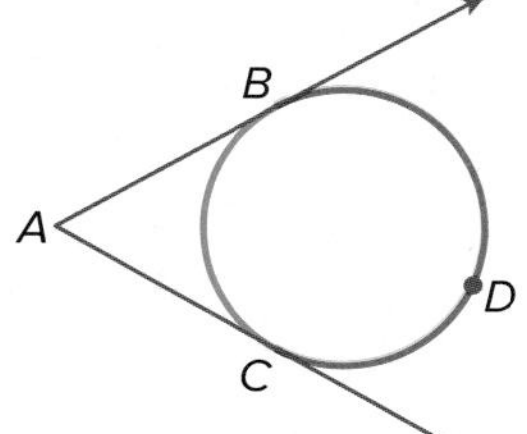

Two Tangents

$m\angle A = \frac{1}{2}(m\widehat{BDC} - m\widehat{BC})$

You will prove Theorem 9.14 in Exercises 30–32.

Study Tip

MP Sense-Making The measure of each $\angle A$ is half the absolute value of the difference of the arc measure. In this way, the order of the arc measures does not affect the outcome of the calculation.

Example 3 Use Tangents and Secants that Intersect Outside a Circle

Find each measure.

a. $m\angle L$

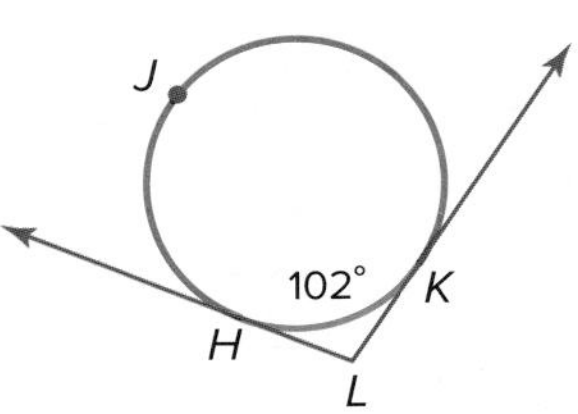

$m\angle L = \frac{1}{2}(m\widehat{HJK} - m\widehat{HK})$ Theorem 9.14

$= \frac{1}{2}(360 - 102) - 102$ Substitution

$= \frac{1}{2}(258 - 102) \text{ or } 78$ Simplify.

b. $m\widehat{CD}$

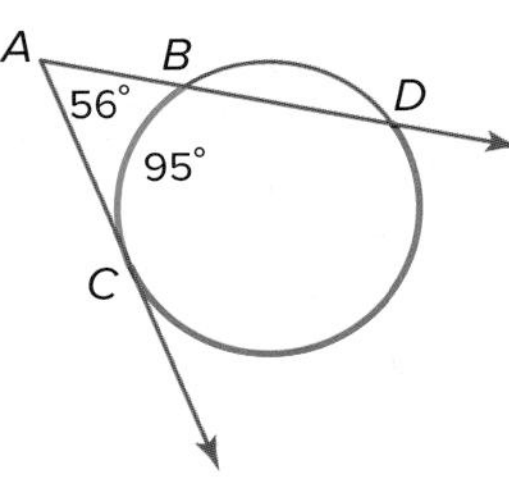

$m\angle A = \frac{1}{2}(m\widehat{CD} - m\widehat{BC})$ Theorem 9.14

$56 = \frac{1}{2}(m\widehat{CD} - 95)$ Substitution

$112 = m\widehat{CD} - 95$ Multiply each side by 2.

$207 = m\widehat{CD}$ Add 95 to each side.

Guided Practice

3A. $m\angle S$

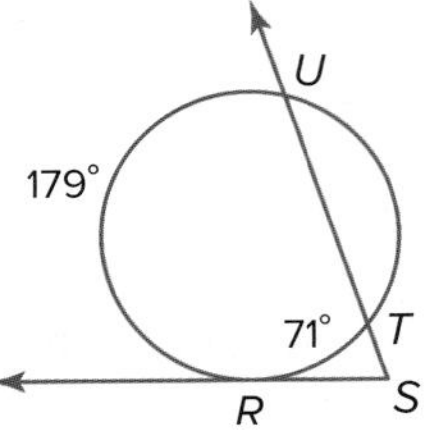

3B. $m\widehat{XZ}$

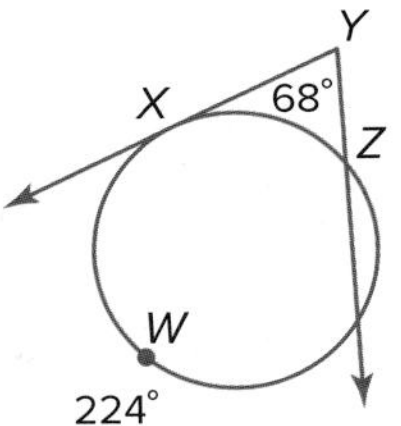

You can apply the properties of intersecting secants to solve real-world problems.

Real-World Example 4 Apply Properties of Intersecting Secants

SCIENCE The diagram shows the path of a light ray as it hits a drop of water. The ray is bent, or *refracted*, at points A, B, and C. If $m\widehat{AC} = 128$ and $m\widehat{XBY} = 84$, what is $m\angle D$?

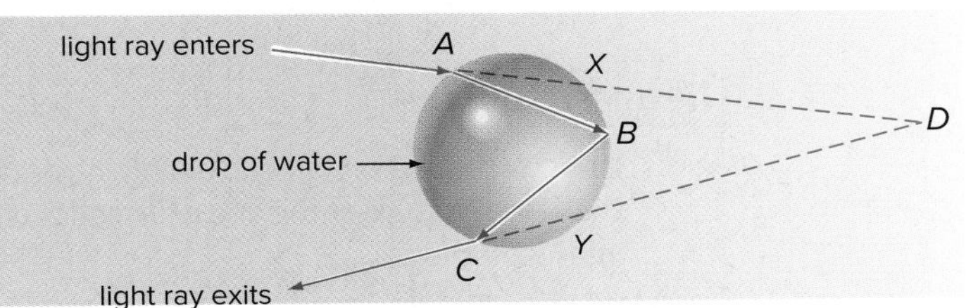

$m\angle D = \frac{1}{2}(m\widehat{AC} - m\widehat{XBY})$ Theorem 9.14

$= \frac{1}{2}(128 - 84)$ Substitution

$= \frac{1}{2}(44) \text{ or } 22$ Simplify.

Guided Practice

4. Find the value of x.

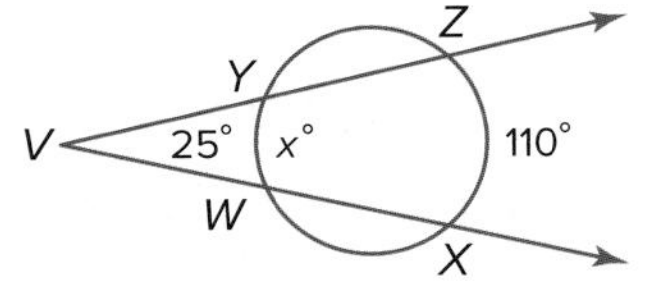

Real-World Link

There is a difference in the *index of refraction* between the two mediums such as air and glass. The index of refraction N is given by the equation $N = \frac{c}{V}$, where c is the speed of light and V is the velocity of light in that material.

Source: Microscopy Resource Center

Key Concept Circle and Angle Relationships

Vertex of Angle	Model(s)	Angle Measure
on the circle		one half the measure of the intercepted arc $m\angle 1 = \frac{1}{2}x$
inside the circle		one half the measure of the sum of the intercepted arc $m\angle 1 = \frac{1}{2}(x + y)$
outside the circle		one half the measure of the difference of the intercepted arcs $m\angle 1 = \frac{1}{2}(x - y)$

Check Your Understanding

 = Step-by-Step Solutions begin on page R13.

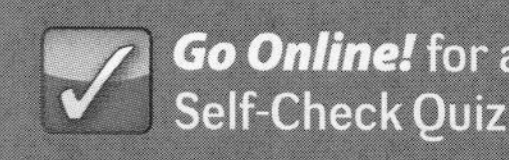

Examples 1–2 **Find each measure. Assume that segments that appear to be tangent are tangent.**

1. $m\angle 1$

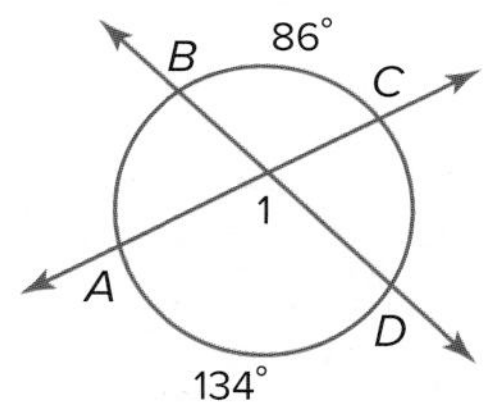

2. $m\widehat{TS}$

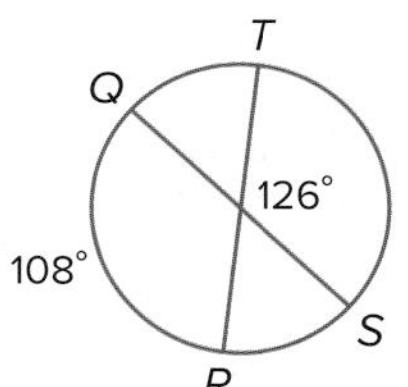

3. $m\angle 2$

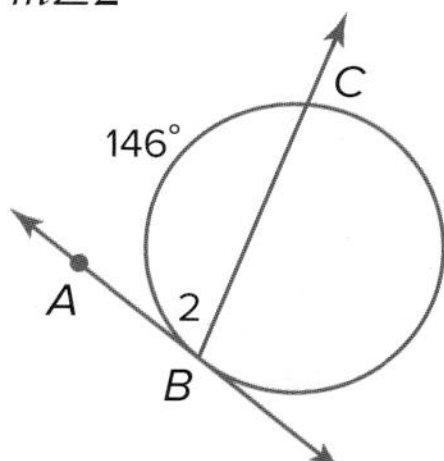

Examples 3–4 **4.** $m\angle H$

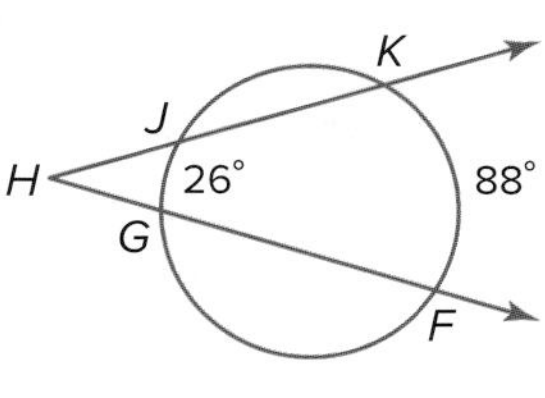

5. $m\widehat{QTS}$

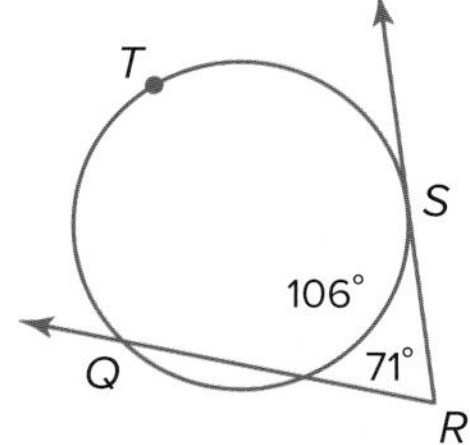

6. $m\widehat{LP}$

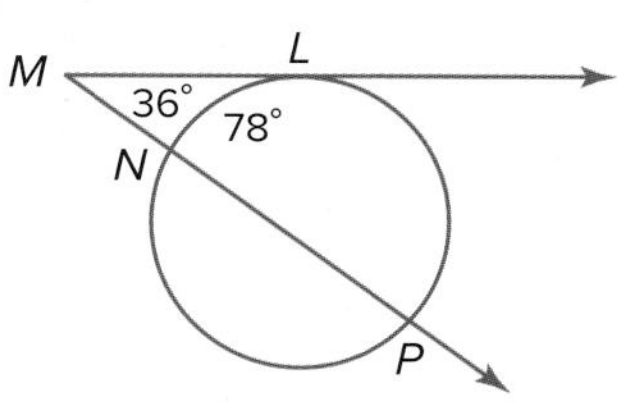

7. **STUNTS** A ramp is attached to the first of several barrels that have been strapped together for a circus motorcycle stunt as shown. What is the measure of the angle the ramp makes with the ground?

Practice and Problem Solving

Extra Practice is on page R9.

Examples 1–2 **Find each measure. Assume that segments that appear to be tangent are tangent.**

8. $m\angle 3$

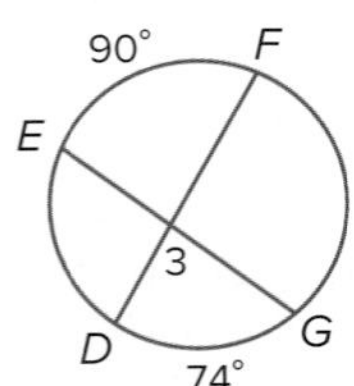

9. $m\angle 4$

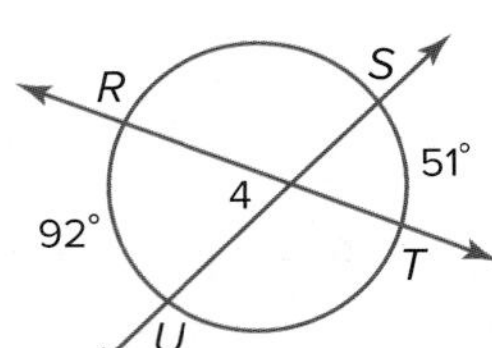

10. $m\angle JMK$

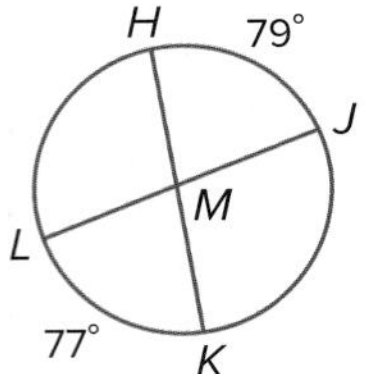

11. $m\widehat{RQ}$

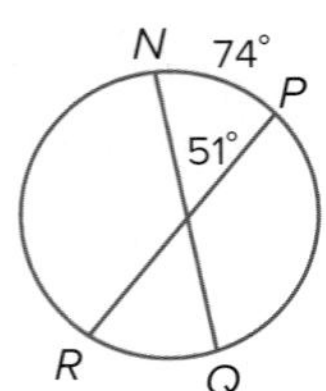

12. $m\angle K$

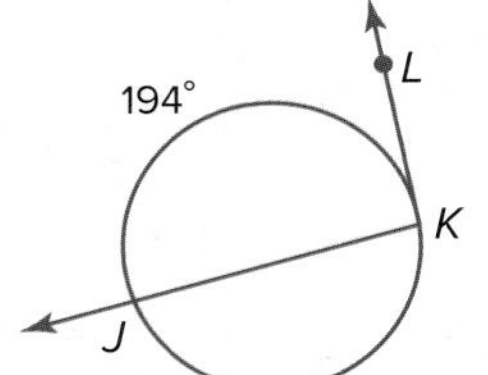

13. $m\widehat{PM}$

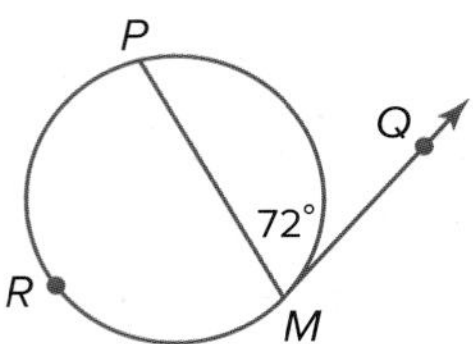

14. $m\angle ABD$

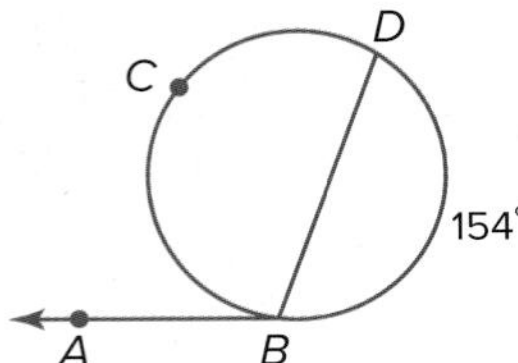

15. $m\angle DAB$

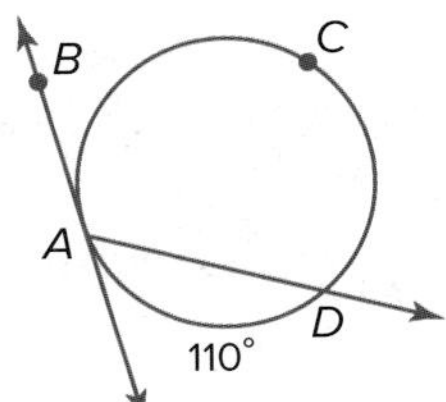

16. $m\widehat{GJF}$

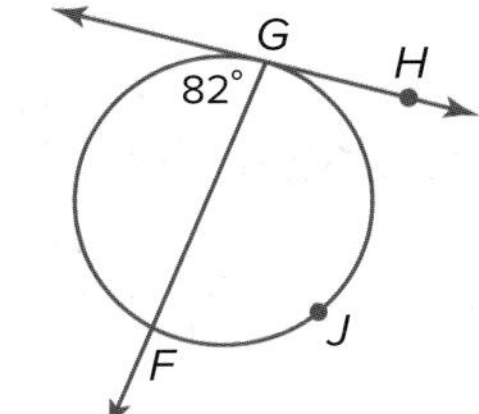

17. SPORTS The multi-sport field shown includes a softball field and a soccer field. If $m\widehat{ABC} = 200$, find each measure.

a. $m\angle ACE$

b. $m\angle ADC$

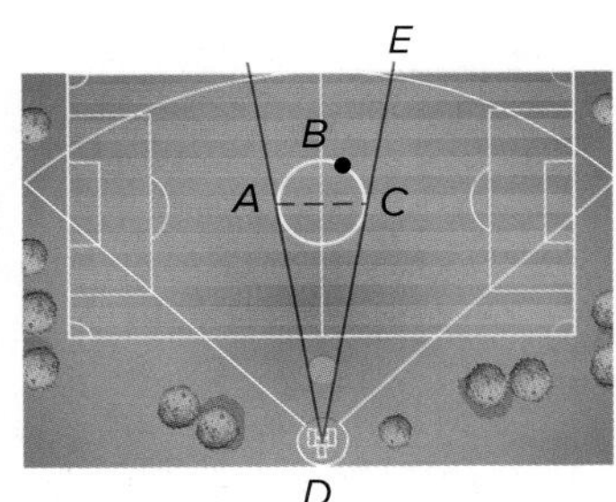

Examples 3–4 **STRUCTURE Find each measure.**

18. $m\angle A$

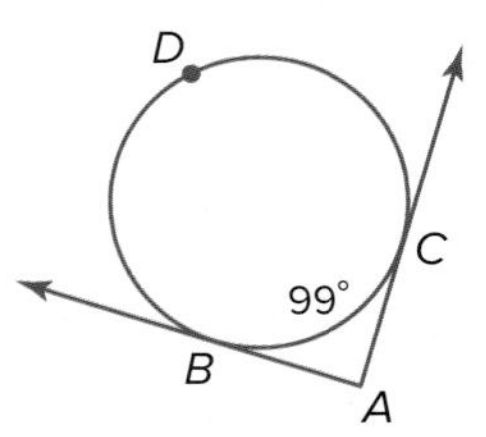

19. $m\angle W$

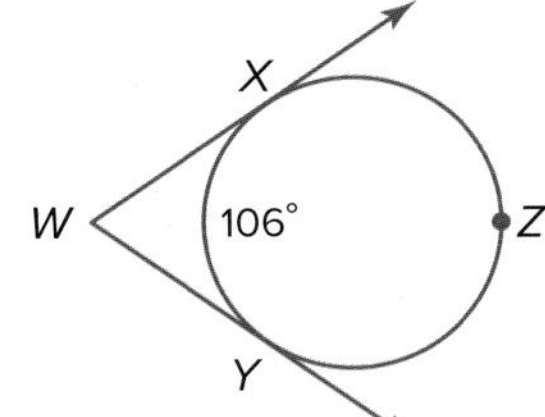

20. $m\widehat{JM}$

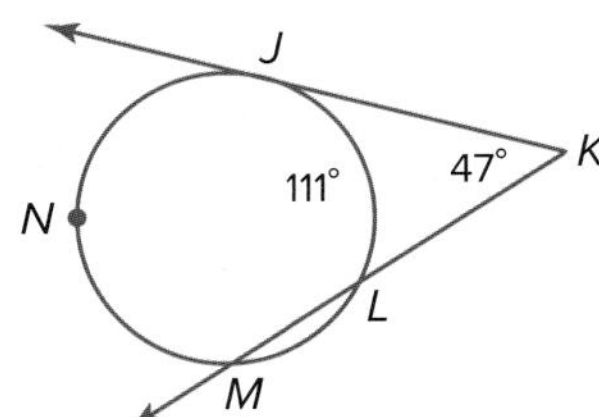

21. $m\widehat{XY}$

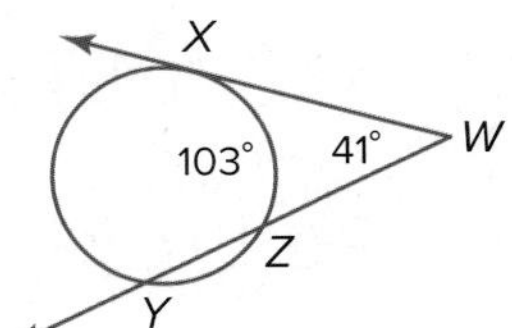

22. $m\angle R$

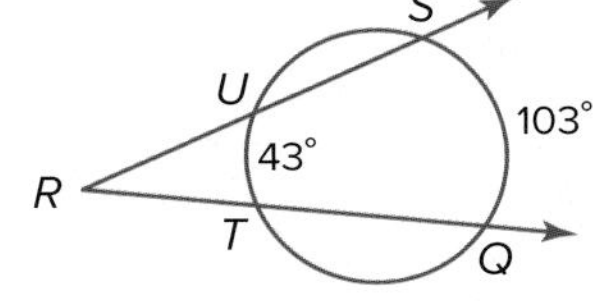

23. $m\widehat{SU}$

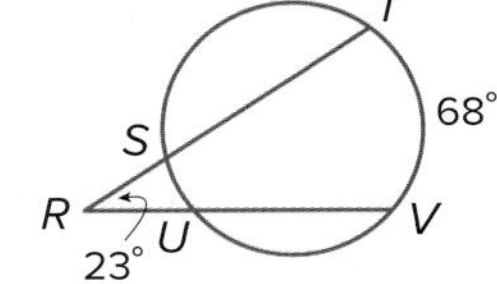

24. **JEWELRY** In the circular necklace shown, A and B are tangent points. If $x = 260$, what is y?

25. **SPACE** A satellite orbits above Earth's equator. Find x, the measure of the planet's arc, that is visible to the satellite.

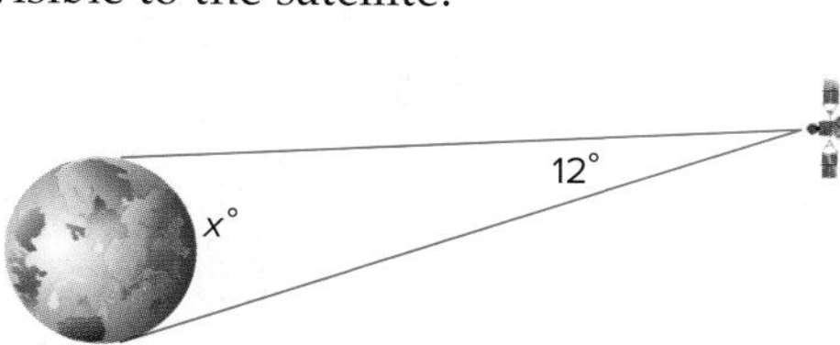

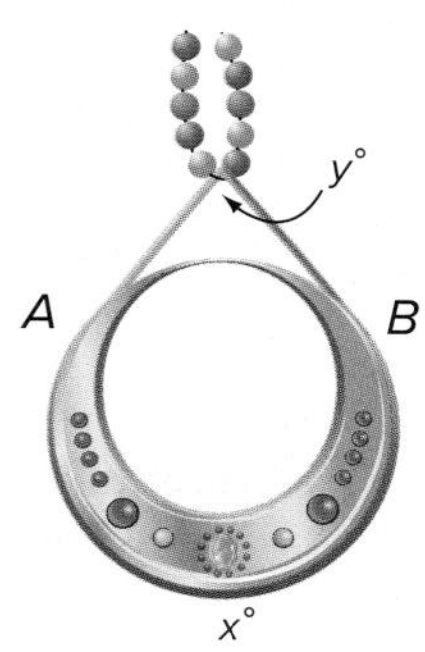

ALGEBRA **Find the value of x.**

26.

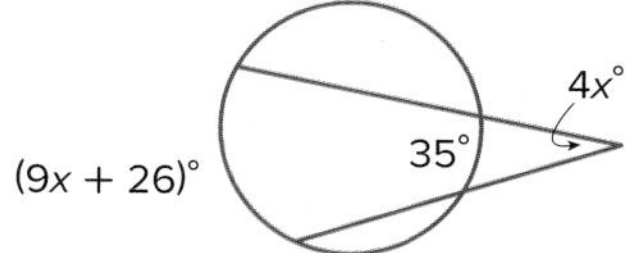

27

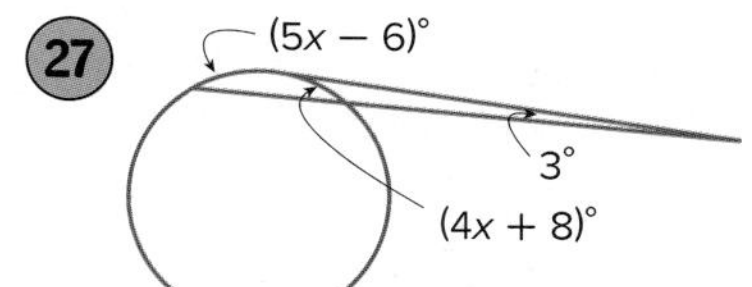

28.

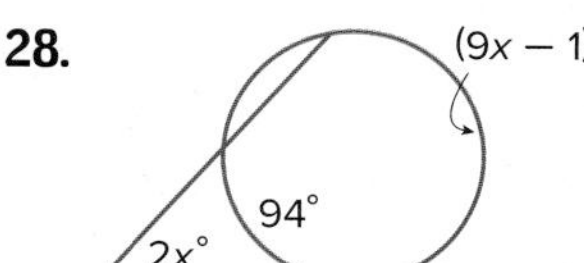

29. **PHOTOGRAPHY** A photographer frames a carousel in his camera shot as shown so that the lines of sight form tangents to the carousel.

a. If the camera's viewing angle is 35°, what is the arc measure of the carousel that appears in the shot?

b. If you want to capture an arc measure of 150° in the photograph, what viewing angle should be used?

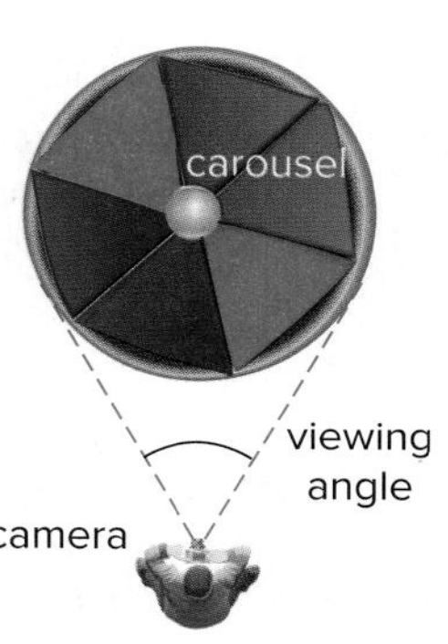

MP CONSTRUCT ARGUMENTS **For each case of Theorem 9.14, write a two-column proof.**

30. **Case 1**

Given: secants $\overrightarrow{AD}$ and $\overrightarrow{AE}$

Prove: $m\angle A = \frac{1}{2}(m\widehat{DE} - m\widehat{BC})$

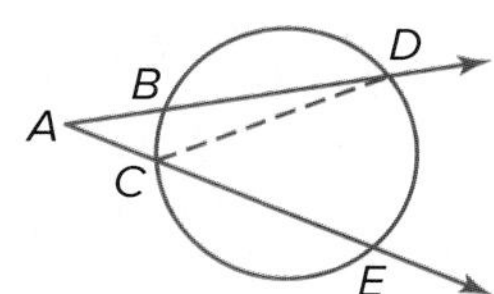

31. **Case 2**

Given: tangent $\overrightarrow{FM}$ and secant $\overrightarrow{FL}$

Prove: $m\angle F = \frac{1}{2}(m\widehat{LH} - m\widehat{GH})$

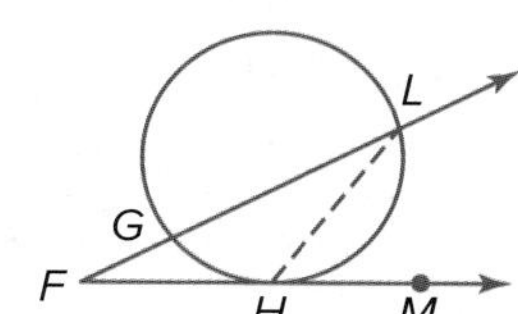

32. **Case 3**

Given: tangents $\overrightarrow{RS}$ and $\overrightarrow{RV}$

Prove: $m\angle R = \frac{1}{2}(m\widehat{SWT} - m\widehat{ST})$

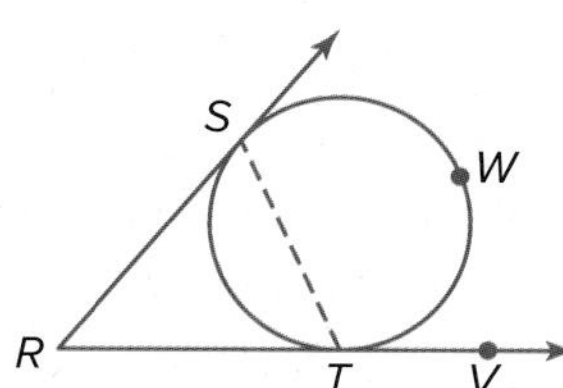

33. **PROOF** Write a paragraph proof of Theorem 9.13.

a. **Given:** $\overleftrightarrow{AB}$ is a tangent of $\odot O$.
$\overrightarrow{AC}$ is a secant of $\odot O$.
$\angle CAE$ is acute.

Prove: $m\angle CAE = \frac{1}{2}m\widehat{CA}$

b. Prove that if $\angle CAB$ is obtuse, $m\angle CAB = \frac{1}{2}m\widehat{CDA}$.

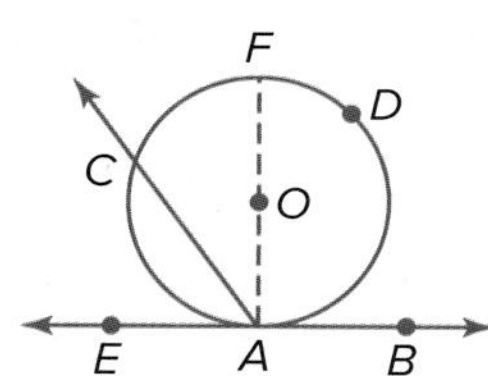

34. WALLPAPER In the wallpaper design shown, $\overline{BC}$ is a diameter of $\odot Q$. If $m\angle A = 26$ and $m\widehat{CE} = 67$, what is $m\widehat{DE}$?

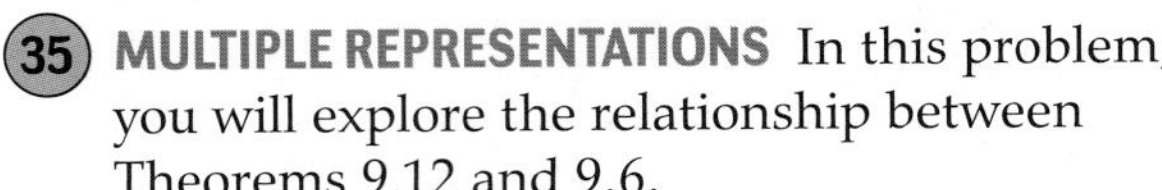

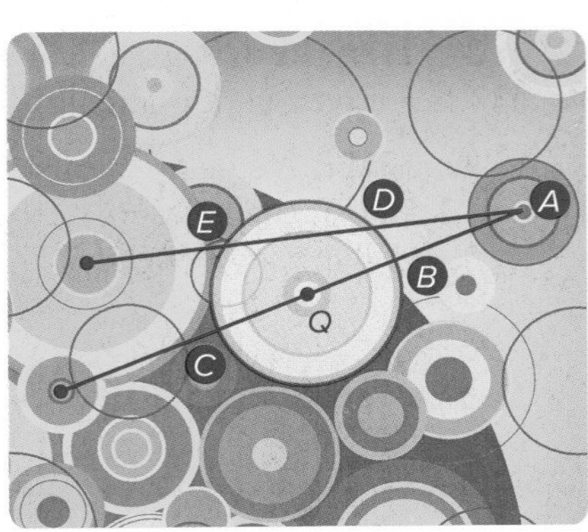

35) MULTIPLE REPRESENTATIONS In this problem, you will explore the relationship between Theorems 9.12 and 9.6.

a. **Geometric** Copy the figure shown. Then draw three successive figures in which the position of point D moves closer to point C, but points A, B, and C remain fixed.

b. **Tabular** Estimate the measure of $\widehat{CD}$ for each successive circle, recording the measures of $\widehat{AB}$ and $\widehat{CD}$ in a table. Then calculate and record the value of x for each circle.

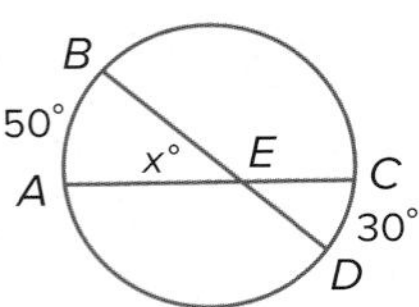

c. **Verbal** Describe the relationship between $m\widehat{AB}$ and the value of x as $m\widehat{CD}$ approaches zero. What type of angle does $\angle AEB$ become when $m\widehat{CD} = 0$?

d. **Analytical** Write an algebraic proof to show the relationship between Theorems 9.12 and 9.6 described in part **c**.

H.O.T. Problems Use Higher-Order Thinking Skills

36. WRITING IN MATH Explain how to find the measure of an angle formed by a secant and a tangent that intersect outside a circle.

37. CHALLENGE The circles below are concentric. What is x?

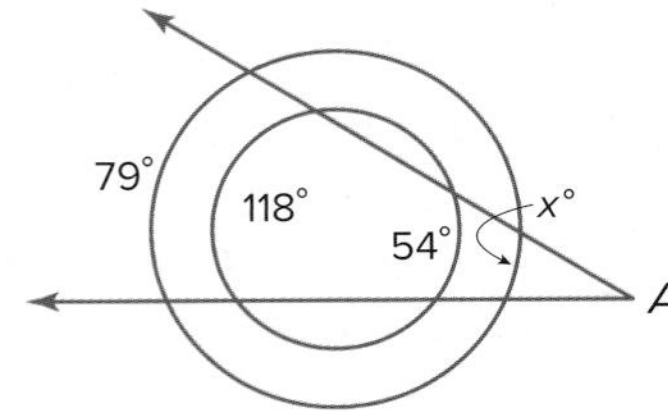

38. MP REASONING Isosceles $\triangle ABC$ is inscribed in $\odot D$. What can you conclude about $m\widehat{AB}$ and $m\widehat{BC}$? Explain.

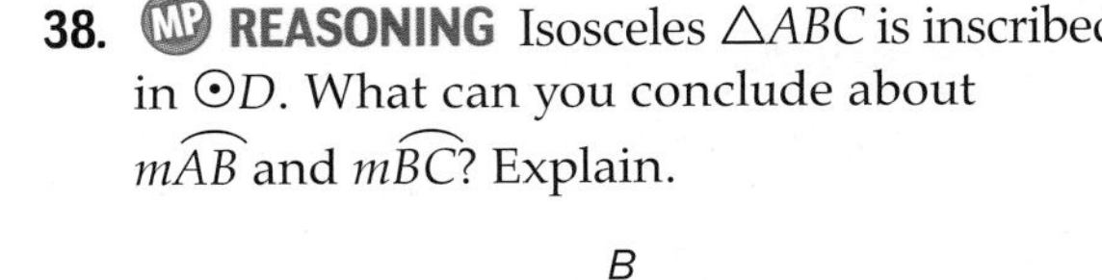

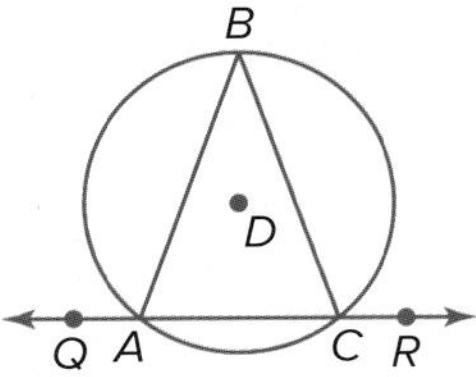

39. MP CONSTRUCT ARGUMENTS In the figure, $\overline{JK}$ is a diameter and $\overrightarrow{GH}$ is a tangent.

a. Describe the range of possible values for $m\angle G$. Explain your reasoning

b. If $m\angle G = 34$, find the measures of minor arcs J and KH. Explain your reasoning.

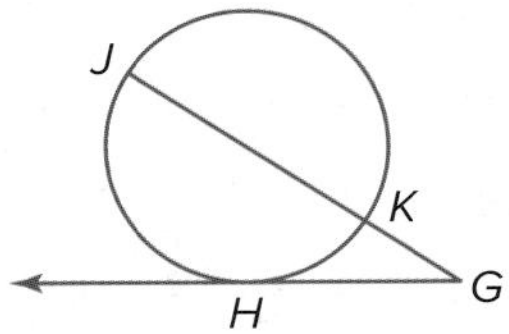

40. OPEN-ENDED Draw a circle and two tangents that intersect outside the circle. Use a protractor to measure the angle that is formed. Find the measures of the minor and major arcs formed. Explain your reasoning.

41. WRITING IN MATH A circle is inscribed within $\triangle PQR$. If $m\angle P = 50$ and $m\angle Q = 60$, describe how to find the measures of the three minor arcs formed by the points of tangency.

42. In the figure, $\overrightarrow{TW}$ bisects $\angle VTU$ and $m\widehat{TV} = 104$.

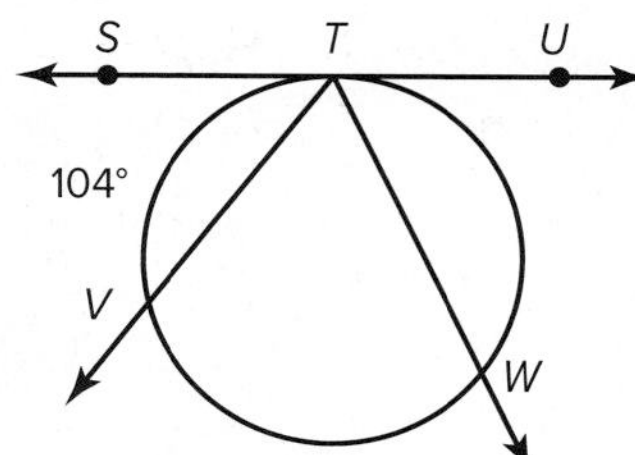

What is the measure of $\angle UTW$? MP 6

- ◯ **A** 38
- ◯ **B** 52
- ◯ **C** 64
- ◯ **D** 76
- ◯ **E** 128

43. If $m\angle AED = 95$ and $m\widehat{AD} = 120$, what is $m\angle BAC$? MP 6

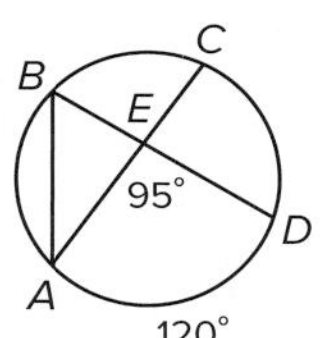

44. $\overline{MN}$ is tangent to $\odot C$ at point N.

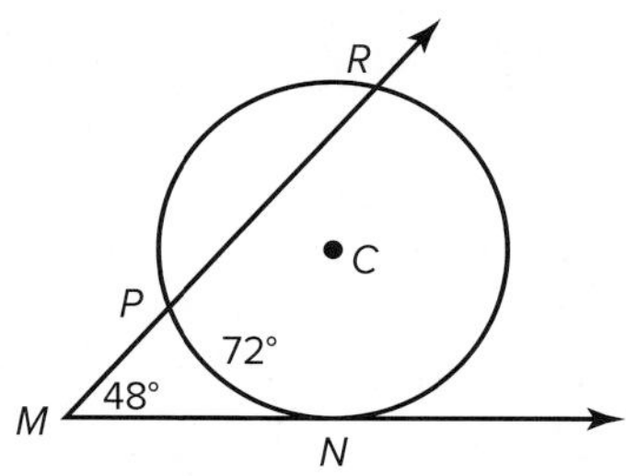

What is the measure of $\widehat{NR}$? MP 6

- ◯ **A** 24
- ◯ **B** 96
- ◯ **C** 120
- ◯ **D** 168

45. $\overline{GH}$ and $\overline{JK}$ are chords of a circle that intersect at point L. Given that $m\widehat{GJ} = 122$ and $m\widehat{HK} = 80$, what is $m\angle HLK$? MP 6

- ◯ **A** 101
- ◯ **B** 61
- ◯ **C** 40
- ◯ **D** 21

46. **MULTI-STEP** Look at the figure. MP 2, 3

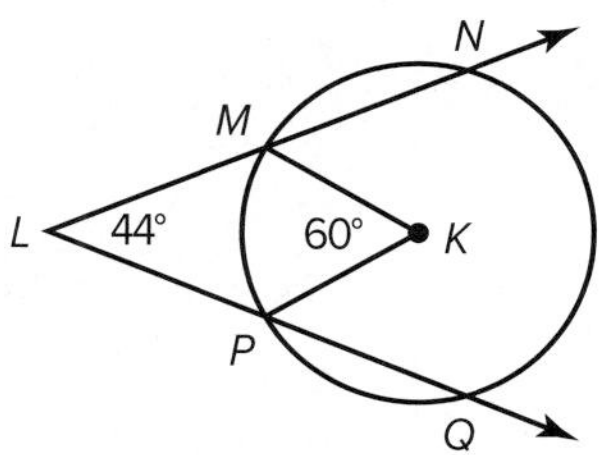

a. In the diagram, what is the arc measure of $\widehat{MP}$?

- ◯ **A** 30°
- ◯ **B** 45°
- ◯ **C** 60°
- ◯ **D** 120°

b. Which statements are true? Check all that apply.

- ☐ **A** $m\angle L = \frac{1}{2}(m\widehat{NQ} - m\widehat{MP})$
- ☐ **B** $44 = \frac{1}{2}(m\widehat{NQ} - 60)$
- ☐ **C** $88 = m\widehat{NQ} + 60$
- ☐ **D** $28 = m\widehat{NQ}$
- ☐ **E** The interior measure is equal to the arc measure.
- ☐ **F** An interior angle measure is always half the arc measure.

c. In the diagram, what is the arc measure of $\widehat{NQ}$?

- ◯ **A** 28°
- ◯ **B** 52°
- ◯ **C** 104°
- ◯ **D** 148°

LESSON 7

Equations of Circles

Then

- You wrote equations of lines using information about their graphs.

Now

1. Write the equation of a circle.
2. Graph a circle on the coordinate plane.

Why?

- Telecommunications towers emit radio signals that are used to transmit cellular calls. Each tower covers a circular area, and towers are arranged so that a signal is available at any location in the coverage area.

New Vocabulary

compound locus

Mathematical Practices

2 Reason abstractly and quantitatively.

7 Look for and make use of structure.

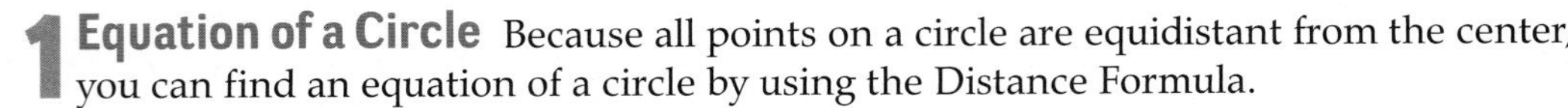

1 Equation of a Circle

Because all points on a circle are equidistant from the center, you can find an equation of a circle by using the Distance Formula.

Let (x, y) represent a point on a circle centered at the origin. Using the Pythagorean Theorem, $x^2 + y^2 = r^2$.

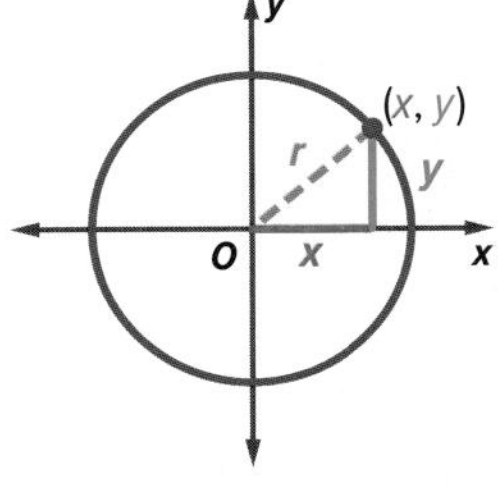

Now suppose that the center is not at the origin, but at the point (h, k). You can use the Distance Formula to develop an equation for the circle.

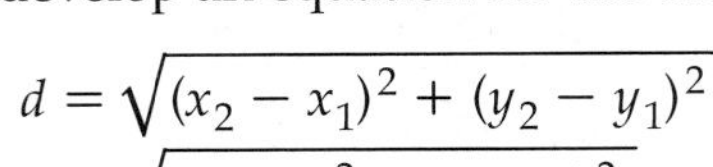

$d = \sqrt{(x_2 - x_1)^2 + (y_2 - y_1)^2}$	Distance Formula
$r = \sqrt{(x - h)^2 + (y - k)^2}$	$d = r, (x_1, y_1) = (h, k), (x_2, y_2) = (x, y)$
$r^2 = (x - h)^2 + (y - k)^2$	Square each side.

Key Concept Equation of a Circle in Standard Form

The standard form of the equation of a circle with center at (h, k) and radius r is $(x - h)^2 + (y - k)^2 = r^2$.

The standard form of the equation of a circle is also called the *center-radius* form.

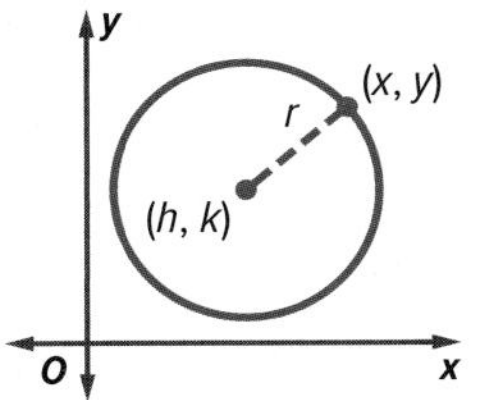

Example 1 Write an Equation Using the Center and Radius

Write the equation of each circle.

a. center at (1, −8), radius 7

$(x - h)^2 + (y - k)^2 = r^2$	Equation of a circle
$(x - 1)^2 + [y - (-8)]^2 = 7^2$	$(h, k) = (1, -8), r = 7$
$(x - 1)^2 + (y + 8)^2 = 49$	Simplify.

b. the circle graphed at the right

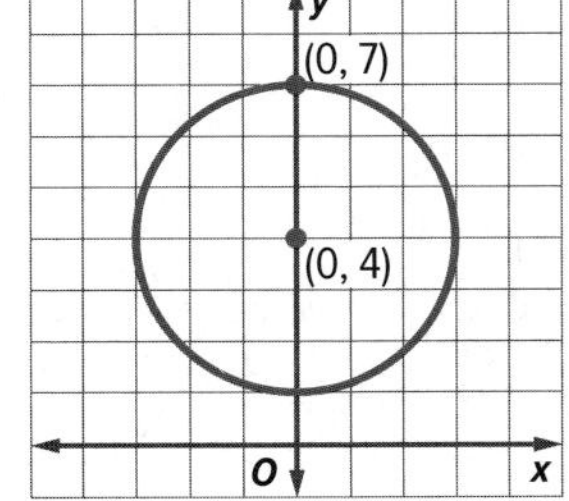

The center is at $(0, 4)$ and the radius is 3.

$(x - h)^2 + (y - k)^2 = r^2$	Equation of a circle
$(x - 0)^2 + (y - 4)^2 = 3^2$	$(h, k) = (0, 4), r = 3$
$x^2 + (y - 4)^2 = 9$	Simplify.

Guided Practice

1A. center at origin, radius $\sqrt{10}$

1B. center at $(4, -1)$, diameter 8

Tuan Tran/Getty Images

Example 2 Write an Equation Using the Center and a Point

Write the equation of the circle with center at (−2, 4), that passes through (−6, 7).

Step 1 Find the distance between the points to determine the radius.

$r = \sqrt{(x_2 - x_1)^2 + (y_2 - y_1)^2}$ Distance Formula

$= \sqrt{[-6 - (-2)]^2 + (7 - 4)^2}$ $(x_1, y_1) = (-2, 4)$ and $(x_2, y_2) = (-6, 7)$

$= \sqrt{25}$ or 5 Simplify.

Step 2 Write the equation using $h = -2$, $k = 4$, and $r = 5$.

$(x - h)^2 + (y - k)^2 = r^2$ Equation of a circle

$[x - (-2)]^2 + (y - 4)^2 = 5^2$ $h = -2$, $k = 4$, and $r = 5$

$(x + 2)^2 + (y - 4)^2 = 25$ Simplify.

Guided Practice

2. Write the equation of the circle with center at (−3, −5) that passes through (0, 0).

2 Graph Circles

You can use the equation of a circle to graph it on a coordinate plane. To do so, you may need to write the equation in standard form first.

Example 3 Graph a Circle

The equation of a circle is $x^2 + y^2 - 8x + 2y = -8$. State the coordinates of the center and the measure of the radius. Then graph the equation.

Write the equation in standard form by completing the square.

$x^2 + y^2 - 8x + 2y = -8$ Original equation

$x^2 - 8x + y^2 + 2y = -8$ Isolate and group like terms.

$x^2 - 8x + 16 + y^2 + 2y + 1 = -8 + 16 + 1$ Complete the squares.

$(x - 4)^2 + (y + 1)^2 = 9$ Factor and simplify.

$(x - 4)^2 + [y - (-1)]^2 = 3^2$ Write +1 as − (−1) and 9 as 3^2.

With the equation now in standard form, you can identify h, k, and r.

$(x - 4)^2 + [y - (-1)]^2 = 3^2$

$(x - h)^2 + (y - k)^2 = r^2$

So, $h = 4$, $k = -1$, and $r = 3$. The center is at (4, −1), and the radius is 3. Plot the center and four points that are 3 units from this point. Sketch the circle through these four points.

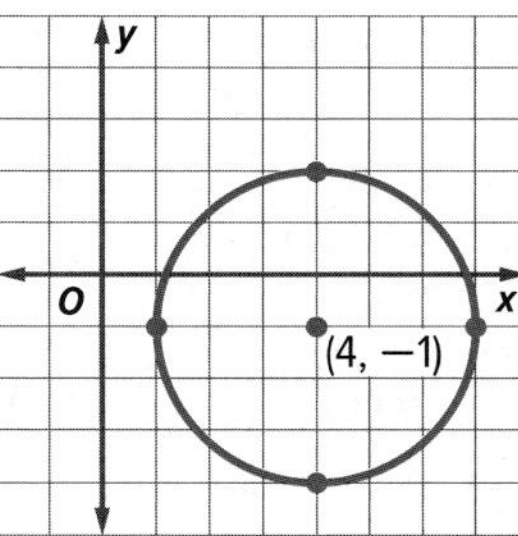

Go Online!

To complete the square for any quadratic expression of the form $x^2 + bx$, follow these steps.

Step 1 Find one half of b.

Step 2 Square the result in Step 1.

Step 3 Add the result of Step 2 to $x^2 + bx$.

Ask your teacher any questions you have about completing the square by sending a message in ConnectED.

Guided Practice

For each circle with the given equation, state the coordinates of the center and the measure of the radius. Then graph the equation.

3A. $x^2 + y^2 - 4 = 0$

3B. $x^2 + y^2 + 8x - 14y + 40 = 0$

Real-World Example 4 Use Three Points to Write an Equation

Real-World Link

About 1000 tornadoes are reported across the United States each year. The most violent tornadoes have wind speeds of 250 mph or more. Damage paths can be a mile wide and 50 miles long.

Source: National Oceanic & Atmospheric Administration

TORNADOES Three tornado sirens are placed strategically on a circle around a town so they can be heard by all. Write the equation of the circle on which they are placed if the coordinates of the sirens are $A(-8, 3)$, $B(-4, 7)$, and $C(-4, -1)$.

Understand You are given three points that lie on a circle.

Plan Graph $\triangle ABC$. Construct the perpendicular bisectors of two sides to locate the center of the circle. Then find the radius.

Use the center and radius to write an equation.

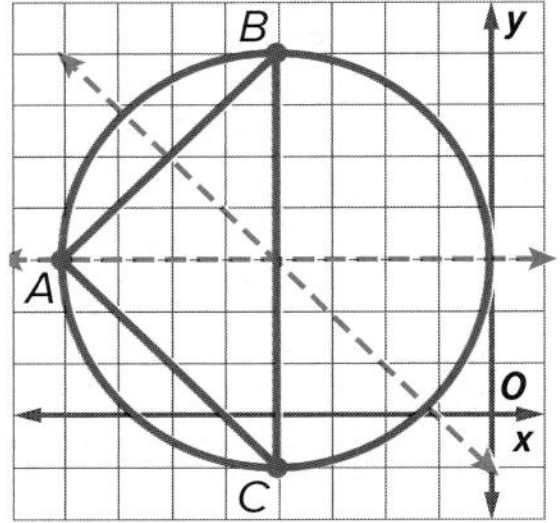

Solve The center appears to be at $(-4, 3)$.
The radius is 4. Write an equation.

$$(x - h)^2 + (y - k)^2 = r^2$$
$$[x - (-4)]^2 + (y - 3)^2 = 4^2$$
$$(x + 4)^2 + (y - 3)^2 = 16$$

Check Verify the center by finding the equations of the two bisectors and solving the system of equations. Verify the radius by finding the distance between the center and another point on the circle. ✓

Analyze the graph to find the relationship between the radius and center of the circle and the given information.

Guided Practice

4. Write an equation of a circle that contains $R(1, 2)$, $S(-3, 4)$, and $T(-5, 0)$.

A line can intersect a circle in at most two points.

Example 5 Intersections with Circles

Find the point(s) of intersection between $x^2 + y^2 = 4$ and $y = x$.

Graph these equations on the same coordinate plane. The points of intersection are solutions of both equations. You can estimate these points on the graph to be at about $(-1.4, -1.4)$ and $(1.4, 1.4)$. Use substitution to find the coordinates of these points algebraically.

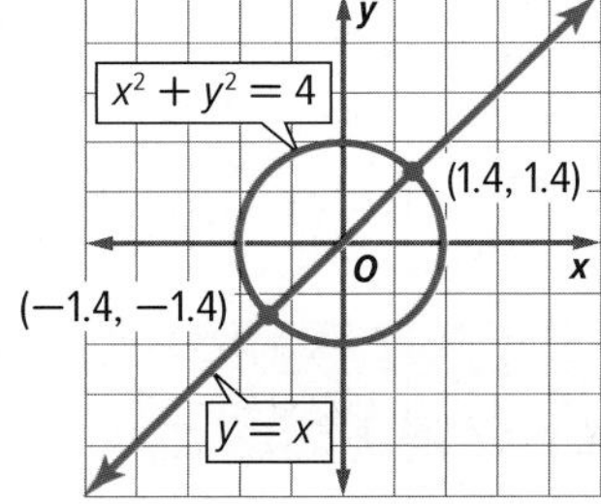

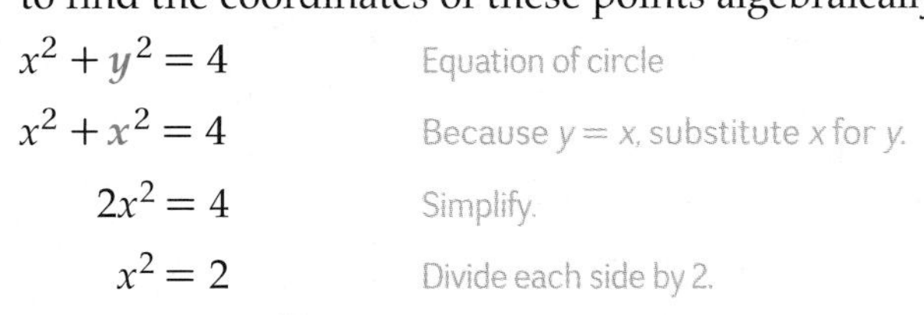

$x^2 + y^2 = 4$	Equation of circle
$x^2 + x^2 = 4$	Because $y = x$, substitute x for y.
$2x^2 = 4$	Simplify.
$x^2 = 2$	Divide each side by 2.
$x = \pm\sqrt{2}$	Take the square root of each side.

So $x = \sqrt{2}$ or $x = -\sqrt{2}$. Use the equation $y = x$ to find the corresponding y values.

$y = x$	Equation of line	$y = x$
$y = \sqrt{2}$	$x = \sqrt{2}$ or $x = -\sqrt{2}$	$y = -\sqrt{2}$

The points of intersection are located at $(\sqrt{2}, \sqrt{2})$ and $(-\sqrt{2}, -\sqrt{2})$ or at about $(-1.4, -1.4)$ and $(1.4, 1.4)$. Check these solutions in both of the original equations.

Study Tip

MP Sense-Making In addition to taking square roots, other quadratic techniques that you may need to apply in order to solve equations of the form $ax^2 + bx + c = 0$ include completing the square, factoring, and the Quadratic Formula, $x = \frac{-b \pm \sqrt{b^2 - 4ac}}{2a}$.

Guided Practice

5. Find the point(s) of intersection between $x^2 + y^2 = 8$ and $y = -x$.

Check Your Understanding

◯ = Step-by-Step Solutions begin on page R13.

Go Online! for a Self-Check Quiz

Examples 1–2 **Write the equation of each circle.**

1. center at (9, 0), radius 5
2. center at (3, 1), diameter 14
3. center at origin, passes through (2, 2)
4. center at (−5, 3), passes through (1, −4)
5. y, (4, 1), (2, 1), O, x
6. 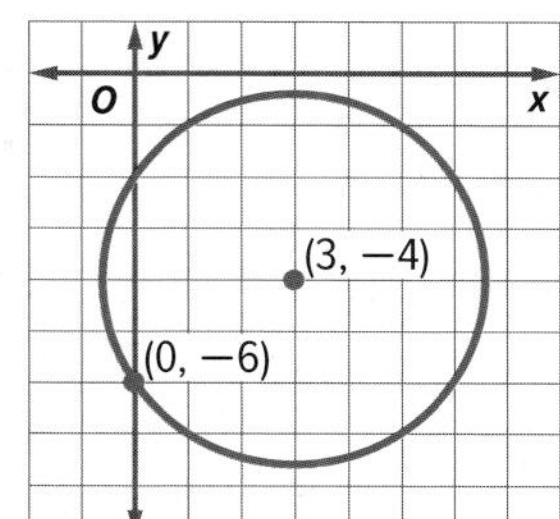

Example 3 **For each circle with the given equation, state the coordinates of the center and the measure of the radius. Then graph the equation.**

7. $x^2 - 6x + y^2 + 4y = 3$
8. $x^2 + (y + 1)^2 = 4$

Example 4

9. **RADIOS** Three radio towers are modeled by the points $R(4, 5)$, $S(8, 1)$, and $T(-4, 1)$. Determine the location of another tower equidistant from all three towers, and write an equation for the circle.

10. **COMMUNICATION** Three cell phone towers can be modeled by the points $X(6, 0)$, $Y(8, 4)$, and $Z(3, 9)$. Determine the location of another cell phone tower equidistant from the other three, and write an equation for the circle.

Example 5 **Find the point(s) of intersection, if any, between each circle and line with the equations given.**

11. $(x - 1)^2 + y^2 = 4$
    $y = x + 1$
12. $(x - 2)^2 + (y + 3)^2 = 18$
    $y = -2x - 2$

Practice and Problem Solving

Extra Practice is on page R9.

Examples 1–2 **STRUCTURE** **Write the equation of each circle.**

13. center at origin, radius 4
14. center at (6, 1), radius 7
15. center at (−2, 0), diameter 16
16. center at (8, −9), radius $\sqrt{11}$
17. center at (−3, 6), passes through (0, 6)
18. center at (1, −2), passes through (3, −4)
19. 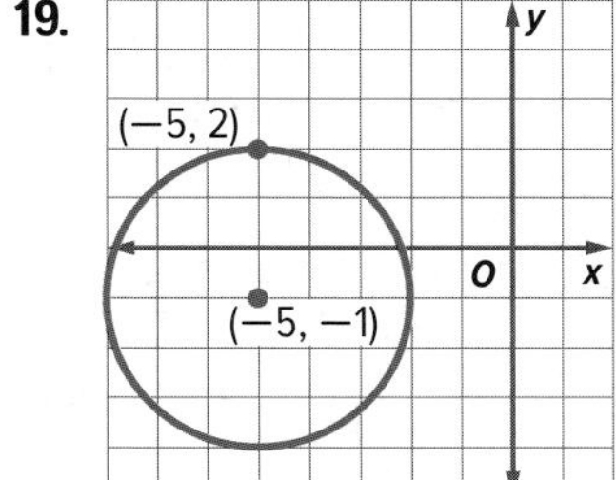

20. 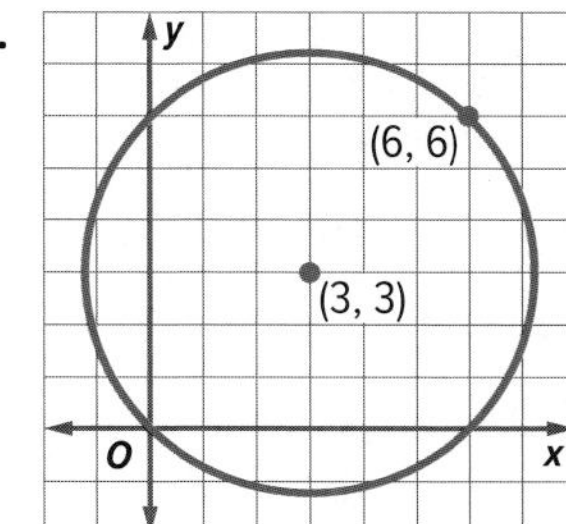

21. **WEATHER** A Doppler radar screen shows concentric rings around a storm. If the center of the radar screen is the origin and each ring is 15 miles farther from the center, what is the equation of the third ring?

22. **GARDENING** A sprinkler waters a circular area that has a diameter of 10 feet. The sprinkler is located 20 feet north of the house. If the house is located at the origin, what is the equation for the circle of area that is watered?

Example 3 **For each circle with the given equation, state the coordinates of the center and the measure of the radius. Then graph the equation.**

23. $x^2 + y^2 = 36$

24. $x^2 + y^2 - 4x - 2y = -1$

25 $x^2 + y^2 + 8x - 4y = -4$

26. $x^2 + y^2 - 16x = 0$

Example 4 **Write an equation of a circle that contains each set of points. Then graph the circle.**

27. $A(1, 6), B(5, 6), C(5, 0)$

28. $F(3, -3), G(3, 1), H(7, 1)$

Example 5 **Find the point(s) of intersection, if any, between each circle and line with the equations given.**

29. $x^2 + y^2 = 5$
$y = \frac{1}{2}x$

30. $x^2 + y^2 = 2$
$y = -x + 2$

31. $x^2 + (y + 2)^2 = 8$
$y = x - 2$

32. $(x + 3)^2 + y^2 = 25$
$y = -3x$

33. $x^2 + y^2 = 5$
$y = 3x$

34. $(x - 1)^2 + (y - 3)^2 = 4$
$y = -x$

Write the equation of each circle.

35. a circle with a diameter having endpoints at (0, 4) and (6, −4)

36. a circle with $d = 22$ and a center translated 13 units left and 6 units up from the origin

37. MP **MODELING** The higher a model rocket travels after it is launched, the larger the circle of possible landing sites becomes. Under normal wind conditions, the landing radius is three times the altitude of the rocket.

a. Write the equation of the landing circle for a rocket that travels 300 feet in the air. Assume the center of the circle is at the origin.

b. What would be the radius of the landing circle for a rocket that travels 1000 feet in the air? Assume the center of the circle is at the origin.

38. SKYDIVING Three of the skydivers in the circular formation shown have approximate coordinates of $G(13, -2)$, $H(-1, -2)$, and $J(6, -9)$.

a. What are the approximate coordinates of the center skydiver?

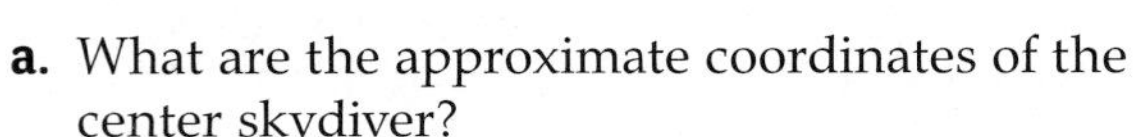

b. If each unit represents 1 foot, what is the diameter of the skydiving formation?

39. MULTI-STEP Consuela's favorite pizza place offers free delivery within a certain radius of the restaurant. The pizza place is 4 miles west and 5 miles north of Consuela's house. Her house is at the edge of their free-delivery radius.

a. Consuela rode her bike 1 mile south and 1 mile west to her friend's house. If they order pizza, will they get free delivery?

b. Describe your solution process.

40. INTERSECTIONS OF CIRCLES Graph $x^2 + y^2 = 4$ and $(x - 2)^2 + y^2 = 4$ on the same coordinate plane.

a. Estimate the point(s) of intersection between the two circles.

b. Solve $x^2 + y^2 = 4$ for y.

c. Substitute the value you found in part **b** into $(x - 2)^2 + y^2 = 4$ and solve for x.

d. Substitute the value you found in part **c** into $x^2 + y^2 = 4$ and solve for y.

e. Use your answers to parts **c** and **d** to write the coordinates of the points of intersection. Compare these coordinates to your estimate from part **a**.

f. Verify that the point(s) you found in part **d** lie on both circles.

41. Prove or disprove that the point $(1, 2\sqrt{2})$ lies on a circle centered at the origin and containing the point $(0, -3)$.

42. **MULTIPLE REPRESENTATIONS** A **compound locus** satisfies more than one distinct set of conditions. Consider the points A and B in the coordinate plane.

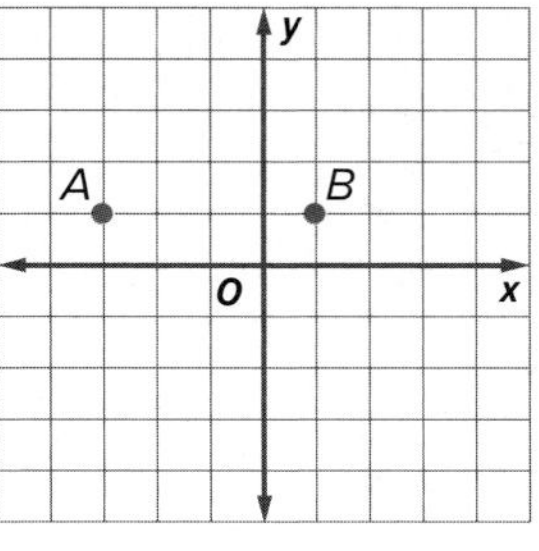

   a. Determine and graph the locus of points equidistant from A and B.
   b. Determine and graph the locus of all points in a plane that are a distance AB from B.
   c. Describe the locus of all points in a plane equidistant from a single point. Then describe the locus of all points that are both equidistant from A and B and are a distance of AB from B. Describe the graph of the compound locus.

Write an equation for each graph described. Then determine whether each point lies *on*, *inside*, or *outside* the circle.

43. a circle with center at the origin and radius 4 units

   a. $(4, 4)$ b. $(3, \sqrt{7})$ c. $(2, \sqrt{5})$ d. $(2\sqrt{3}, -2)$

44. a circle with center at $(2, 4)$ and radius 3 units

   a. $(2, 1)$ b. $(0, 4)$ c. $(2, 7)$ d. $(3, 1)$

45. A circle with a diameter of 12 has its center in the second quadrant. The lines $y = -4$ and $x = 1$ are tangent to the circle. Write an equation of the circle.

H.O.T. Problems Use Higher-Order Thinking Skills

46. **CHALLENGE** Write a coordinate proof to show that if an inscribed angle intercepts the diameter of a circle, as shown, the angle is a right angle.

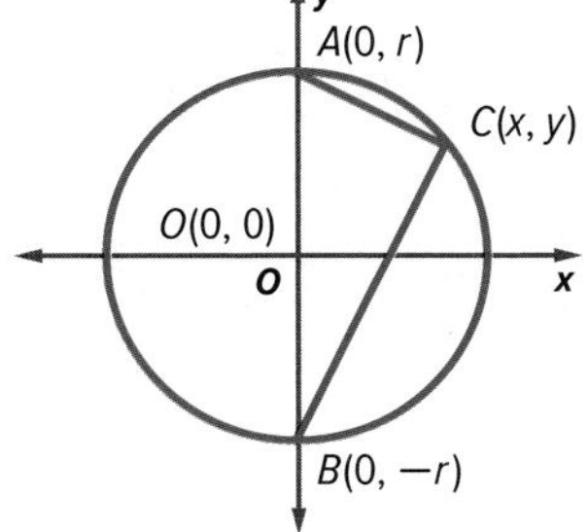

47. **MP REASONING** A circle has the equation $(x - 5)^2 + (y + 7)^2 = 16$. If the center of the circle is shifted 3 units right and 9 units up, what would be the equation of the new circle? Explain your reasoning.

48. **OPEN-ENDED** Graph three noncollinear points and connect them to form a triangle. Then construct the circle that circumscribes it.

49. **WRITING IN MATH** Seven new radio stations must be assigned broadcast frequencies. The stations are located at $A(9, 2)$, $B(8, 4)$, $C(8, 1)$, $D(6, 3)$, $E(4, 0)$, $F(3, 6)$, and $G(4, 5)$, where 1 unit = 50 miles.

   a. If stations that are more than 200 miles apart can share the same frequency, what is the least number of frequencies that can be assigned to these stations?
   b. Describe two different beginning approaches to solving this problem.
   c. Choose an approach, solve the problem, and explain your reasoning.

CHALLENGE **Find the coordinates of point P on $\overrightarrow{AB}$ that partitions the segment into the given ratio AP to PB.**

50. $A(0, 0)$, $B(3, 4)$, 2 to 3

51. $A(0, 0)$, $B(-8, 6)$, 4 to 1

52. **WRITING IN MATH** Describe the relationship between the Distance Formula, the equation of a circle centered at the origin, and the equation of a circle centered at a point not on the origin.

53. Ethan wrote the equation $(x + 1)^2 + (y + 2)^2 = 4$ for the circle shown in the figure. MP 3

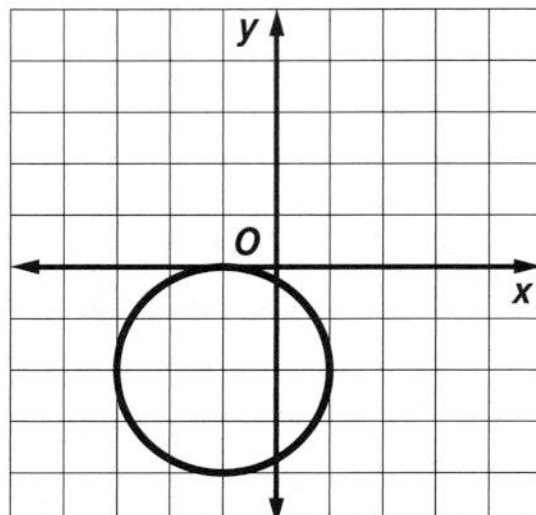

Which of the following is the best description of the equation Ethan wrote?

- ○ **A** His equation is correct.
- ○ **B** His equation shows an incorrect radius.
- ○ **C** His equation shows an incorrect center.
- ○ **D** His equation shows an incorrect radius and center.

54. The equation of a circle is $x^2 + y^2 + 4x - 8y = -16$. Which of the following is a true statement about the circle? MP 7

I. The radius of the circle is 4.

II. The center of the circle is $(2, -4)$.

III. The circle passes through $(0, 4)$.

- ○ **A** I only
- ○ **B** II only
- ○ **C** III only
- ○ **D** I and II only
- ○ **E** I, II, and III

55. What is the equation of the circle that passes through $(0, -2)$, $(8, -2)$, and $(4, -6)$? MP 4

- ○ **A** $(x - 4)^2 + (y + 2)^2 = 4$
- ○ **B** $(x + 4)^2 + (y - 2)^2 = 4$
- ○ **C** $(x - 4)^2 + (y + 2)^2 = 16$
- ○ **D** $(x + 4)^2 + (y - 2)^2 = 16$

56. MULTI-STEP Samantha is using a coordinate plane to design a video game. In the game, a character travels in a circle to pick up coins. Samantha places coins at $P(1, 2)$, $Q(4, -1)$, and $R(1, -4)$. Now she needs to determine the equation of a circle through those three points. MP 1, 4, 7

a. What two pieces of information does Samantha need in order to write the equation of a circle?

- ☐ **A** two points on the circle
- ☐ **B** the tangent equation
- ☐ **C** the center
- ☐ **D** the radius
- ☐ **E** a point in the interior of the circle
- ☐ **F** a point exterior to the circle

b. Describe a set of steps Samantha can take to find the equation of a circle through the points. Use a graph to support your description.

c. What are the coordinates of the center of the circle?

- ○ **A** $(0, 1)$
- ○ **B** $(1, -1)$
- ○ **C** $(-1, 0)$
- ○ **D** $(-1, -1)$

d. What is the radius of the circle?

- ○ **A** -1
- ○ **B** 1
- ○ **C** 3
- ○ **D** 9

e. What is the equation of the circle?

- ○ **A** $(x - 1)^2 - (y + 1)^2 = 9$
- ○ **B** $(x - 1)^2 - (y)^2 = 9$
- ○ **C** $(x - 1)^2 + (y)^2 = 9$
- ○ **D** $(x - 1)^2 + (y + 1)^2 = 9$

LESSON 8

Equations of Parabolas

Then

- You found the equation of a circle by using the Distance Formula.

Now

1. Write the equation of a parabola.
2. Graph parabolas on the coordinate plane.

Why?

- Parabolas are used to describe and model satellite dishes and parabolic telescopes. Incoming signals are concentrated at an antenna at the parabola's focus and then transmitted from there.

New Vocabulary

parabola
focus
directrix
axis of symmetry
vertex

Mathematical Practices

1 Make sense of problems and persevere in solving them.

2 Reason abstractly and quantitatively.

4 Model with mathematics.

7 Look for and make use of structure.

1 Write Equations of Parabolas

Geometrically, a **parabola** is a set of all points that are equidistant from a fixed point called the **focus** and a fixed line called the **directrix**. The line through the focus perpendicular to the directrix is the **axis of symmetry**. The point where the axis of symmetry intersects the parabola is the **vertex**.

Consider a parabola with focus $F(0, p)$ and directrix $y = -p$. The axis of symmetry is $x = 0$ and the vertex is $(0, 0)$.

Let $P(x, y)$ be any point on the parabola. Recall that the distance between a point and a line is the length of the segment perpendicular to the line through the point. So the distance from (x, y) to the directrix is PD for $D(x, -p)$. By the definition of a parabola, the distance from P to the focus, PF, and the distance from P to the directrix, PD, are equal.

$$PF = PD$$
$$\sqrt{(x - 0^2) + (y - p)^2} = \sqrt{(x - x)^2 + (y - (-p))^2}$$
$$\sqrt{x^2 + (y - p)^2} = \sqrt{(y + p)^2}$$
$$x^2 + (y - p)^2 = (y + p)^2$$
$$x^2 + y^2 - 2py + p^2 = y^2 + 2py + p^2$$
$$x^2 = 4yp$$

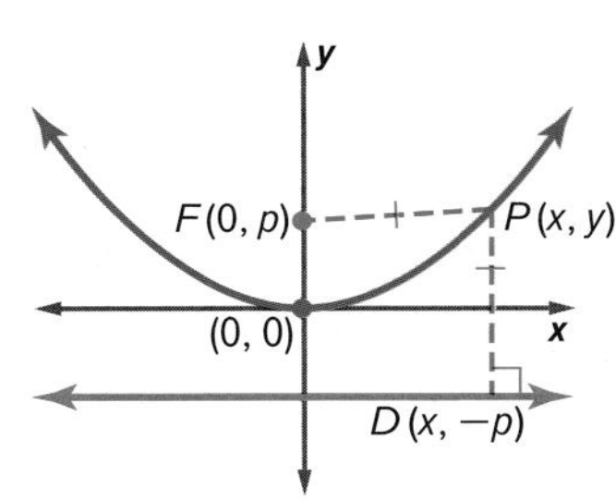

So, the equation of a parabola that has a vertical axis of symmetry is $x^2 = 4py$, or $\frac{1}{y = 4p}x^2$. Notice that when p is positive the parabola opens up. When p is negative, the parabola opens down.

The derivation of the equation of a parabola with a horizontal axis of symmetry is similar. Thus the equation of a parabola with vertex $(0, 0)$ that has a horizontal axis of symmetry is $y^2 = 4px$. When p is positive the parabola opens right. When p is negative, the parabola opens left.

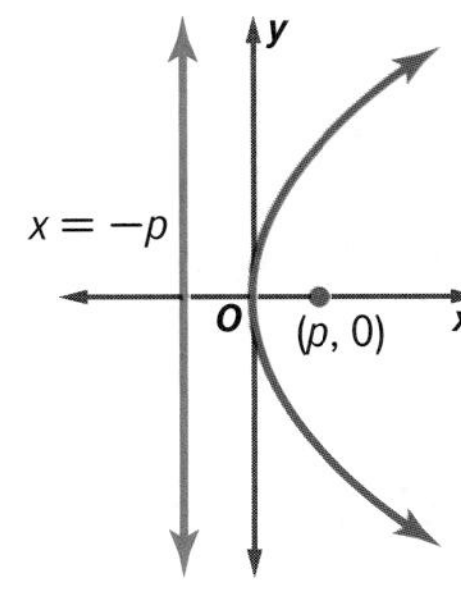

A parabola is translated in the plane in the same way as other geometric figures. Translating a parabola so that the vertex of the new parabola is (h, k) allows for standard forms of an equation of parabola with vertex at (h, k).

Key Concept Standard Form of Equations of Parabolas

Equation: $(x - h)^2 = 4p(y - k)$

Axis of Symmetry: vertical, $x = h$

Focus: $(h, k + p)$

Vertex: (h, k)

Directrix: $y = k - p$

Opens: up if $p > 0$, down if $p < 0$

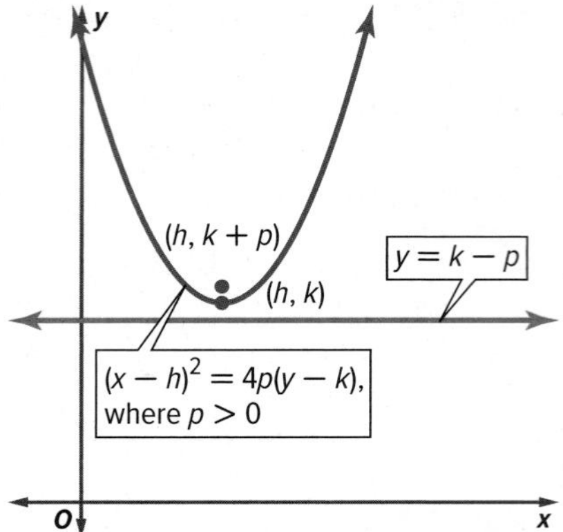

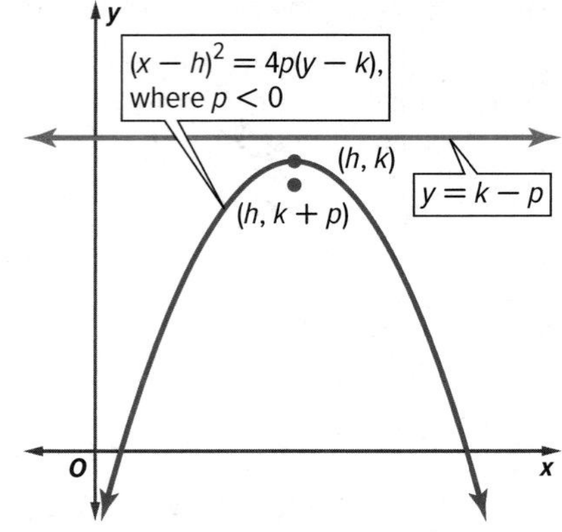

Equation: $(y - k)^2 = 4p(x - h)$

Axis of Symmetry: horizontal, $y = k$

Focus: $(h + p, k)$

Vertex: (h, k)

Directrix: $x = h - p$

Opens: right if $p > 0$, left if $p < 0$

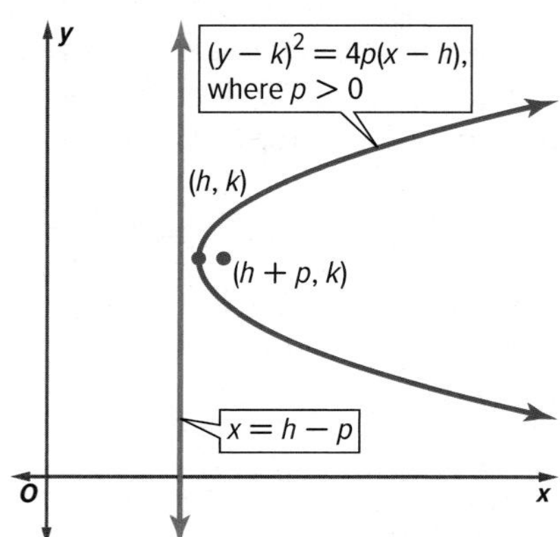

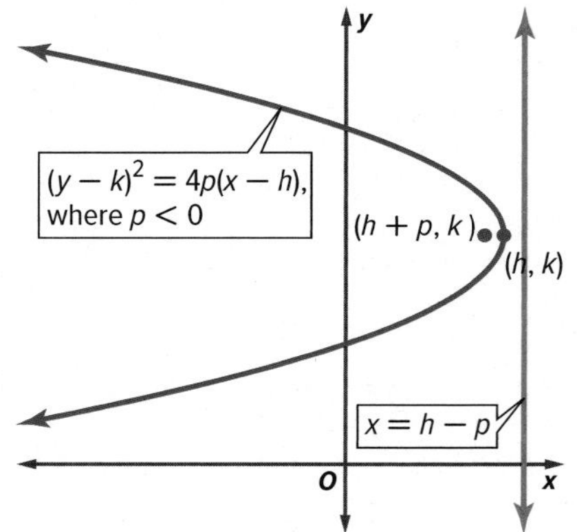

Example 1 Write the Equation of a Parabola

Write an equation for each parabola.

a. focus at (0, 1) and directrix $y = -1$

The directrix is horizontal, so the axis of symmetry is vertical. The equation is of the form $(x - h)^2 = 4p(y - k)$.

Use the focus and directrix to find the values of h, k, and p.

The focus is $(h, k + p) = (0, 1)$, so $h = 0$ and $k + p = 1$.

The directrix is $y = k - p$, so $k - p = -1$. Solve the system to find k and p.

$$\begin{aligned} k + p &= 1 \\ k - p &= -1 \\ \hline 2k &= 0 \\ k &= 0 \end{aligned} \qquad \begin{aligned} k + p &= 1 \\ 0 + p &= 1 \\ p &= 1 \end{aligned}$$

Study Tip

MP Structure The sign of p helps you determine which way the graph of the parabola opens. If $p < 0$, the graph will open down or to the left. If $p > 0$, the graph will open up or to the right.

Substitute to find the equation.

$(x - h)^2 = 4p(y - k)$

$(x - 0)^2 = 4(1)(y - 0)$

$x^2 = 4y$

The equation is $x^2 = 4y$ or $y = \frac{1}{4}x^2$.

b. vertex (−1, 4) and focus (−3, 4)

The vertex is (−1, 4), so $h = -1$ and $k = 4$.

The vertex and the focus lie on the axis of symmetry, so the equation of the axis of symmetry is $y = 4$. Thus, the axis of symmetry is horizontal and the equation is of the form $(y - k)^2 = 4p(x - h)$.

The focus is $(h + p, k)$. Use the x-coordinate to find the value of p.

$h + p = -3$

$-1 + p = -3$

$p = -2$

Write the equation.

$(y - k)^2 = 4p(x - h)$

$(y - 4)^2 = 4(-2)(x - (-1))$

$(y - 4)^2 = -8(x + 1)$

The equation of the parabola is $(y - 4)^2 = -8(x + 1)$.

Guided Practice

1A. focus at (0, −2) and directrix at $y = 2$

1B. vertex (−2, −1), focus (−4, −1)

Parabolic reflectors are often used in the real world to collect or send waves.

Real-World Example 2 Write an Equation of a Parabola

SATELLITES A parabolic satellite dish opens upward. The dish reflects the collected signals to the focus of the parabola. The distance between the vertex and the focus is 9.6 meters. Write an equation to represent a cross section of the dish.

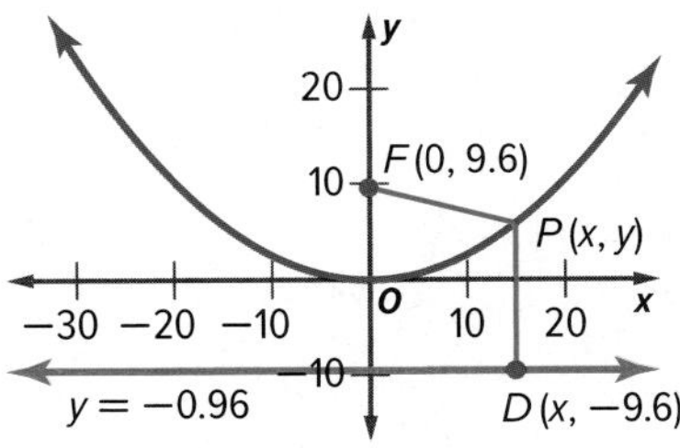

The parabola opens upward, so the axis of symmetry is vertical. The equation is of the form $(x - h)^2 = 4p(y - k)$.

Use (0, 0) as the vertex, so $h = 0$ and $k = 0$. The focus is 9.6 meters from the vertex, so $p = 9.6$ and the focus is (0, 9.6).

Substitute to find the equation.

$(x - h)^2 = 4p(y - k)$

$(x - 0)^2 = 4(9.6)(y - 0)$

$x^2 = 38.4y$

The equation is $x^2 = 38.4y$ or about $y = 0.026x^2$.

Guided Practice

2. **SEARCHLIGHTS** A searchlight uses parabolic reflectors to concentrate and project a beam of light. Suppose a reflector opens to the right and the distance between the vertex and the focus is 2.2 feet. Write an equation of the cross section of the reflector.

2 Graph Parabolas

You can analyze the equation of a parabola to find the key features of the parabola and to draw its graph.

Example 3 Analyze and Graph a Parabola

Graph each equation.

a. $x - 3 = -\frac{1}{16}(y + 1)^2$

Step 1 Write the equation in the standard form to identify h, k, and p.

$x - 3 = -\frac{1}{16}(y + 1)^2$	Original equation
$-16(x - 3) = (y + 1)^2$	Multiply each side by −16.
$(y + 1)^2 = -16(x - 3)$	Symmetric Property
$(y - (-1))^2 = 4(-4)(x - 3)$	$1 = -(-1)$ and $-16 = 4(-4)$

Thus $h = 3$, $k = -1$, and $p = -4$.

Step 2 Identify some key features of the graph. Then draw.

focus, $(h + p, k)$: $(3 + (-4), -1)$ or $(-1, -1)$

vertex, (h, k): $(3, -1)$

axis of symmetry, $y = k$: $y = -1$

directrix, $x = h - p$: $x = 3 - (-4)$ or $x = 7$

direction of opening: left since $p < 0$

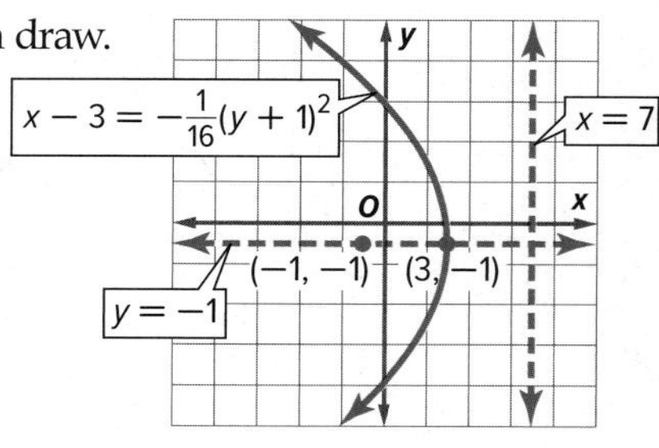

b. $y = -2(x + 4)^2 - 1$

Step 1

$y = -2(x + 4)^2 - 1$	Original equation
$y + 1 = -2(x + 4)^2$	Add 1 to each side.
$-\frac{1}{2}(y + 1) = (x + 4)^2$	Multiply each side by $-\frac{1}{2}$.
$(x + 4)^2 = -\frac{1}{2}(y + 1)$	Symmetric Property
$(x - (-4))^2 = 4\left(-\frac{1}{8}\right)(y - (-1))$	$-\frac{1}{2} = 4\left(-\frac{1}{8}\right)$, $1 = -(-1)$, $4 = -(-4)$

Thus $h = -4$, $k = -1$, and $p = -\frac{1}{8}$.

Step 2 **focus:** $\left(-4, -1 + \left(-\frac{1}{8}\right)\right)$ or $\left(-4, -\frac{9}{8}\right)$

vertex: $(-4, -1)$

axis of symmetry: $x = -4$

directrix: $y = -\frac{7}{8}$

direction of opening: down

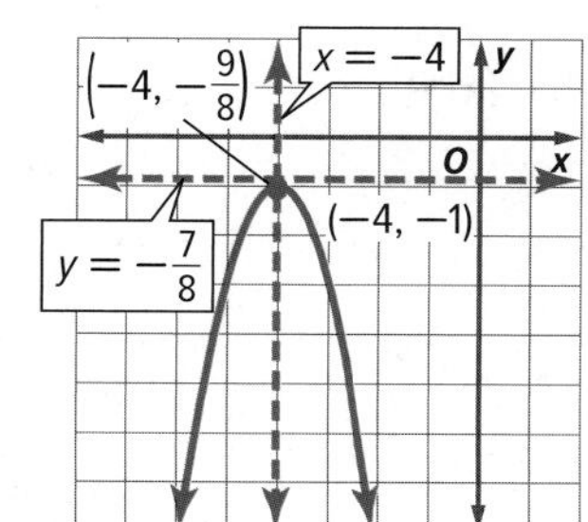

Guided Practice

3A. $x = \frac{1}{3}(y + 2)^2 + 7$

3B. $y = (x + 3)^2 - 13$

Check Your Understanding

◯ = Step-by-Step Solutions begin on page R13.

Example 1 **Write an equation for each parabola.**

1. focus (0, 6), directrix $y = -6$
2. focus (0, −8), directrix $y = 8$
3. focus (–2, 0), directrix $x = 2$
4. focus (3, 0), directrix $x = -3$
5. vertex $\left(-\frac{9}{16}, 1\right)$, focus $\left(-\frac{9}{16}, 3\right)$
6. vertex $\left(\frac{3}{8}, -2\right)$, directrix $x = -\frac{3}{8}$

Example 2

7. **MODELING** The parabolic reflector plate of a flashlight has its bulb located at the focus of the parabola. The distance between the vertex and the focus is 1.8 centimeters. Write an equation to represent the cross section of the reflector plate.

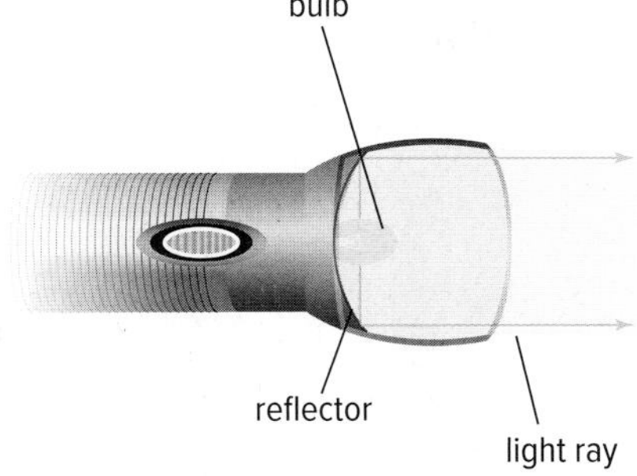

Example 3 **Graph each equation.**

8. $y = (x - 4)^2 - 6$
9. $y = 4(x + 5)^2 + 3$
10. $y = \frac{1}{8}x^2 + 2$
11. $x = -\frac{1}{4}(y - 4)^2 - 2$

Practice and Problem Solving

Extra Practice is on page R9.

Example 1 **Write an equation for each parabola.**

12. focus $\left(\frac{5}{2}, 0\right)$, directrix $x = -\frac{5}{2}$
13. focus $\left(0, -\frac{7}{4}\right)$, directrix $y = \frac{7}{4}$
14. focus (0, 8), directrix, $y = -8$
15. focus (0, −10), directrix, $y = 10$
16. focus (1, 4), directrix, $y = -4$
17. focus (2, −1), directrix, $y = 3$
18. focus (5, 2), directrix, $x = -1$
19. focus (4, −1), directrix, $x = 8$
20. vertex (0, 0), directrix, $y = -2$
21. vertex (0, 0), directrix, $x = 3$
22. vertex (4, 4), directrix, $y = 3$
23. vertex (1, −1), directrix, $y = 1$
24. vertex (5, 1), directrix, $x = 0$
25. vertex (−3, 1), directrix, $x = -4$
26. vertex (0, 6), directrix, $x = 2$
27. vertex (−7, 0), directrix, $x = -2$

Example 2

28. **CARS** The parabolic reflector plate of a car headlight has its bulb located at the focus of the parabola. The distance between the vertex and the focus is 6.5 centimeters.
    a. Write an equation of the cross section of the reflector plate.
    b. Graph the parabola.

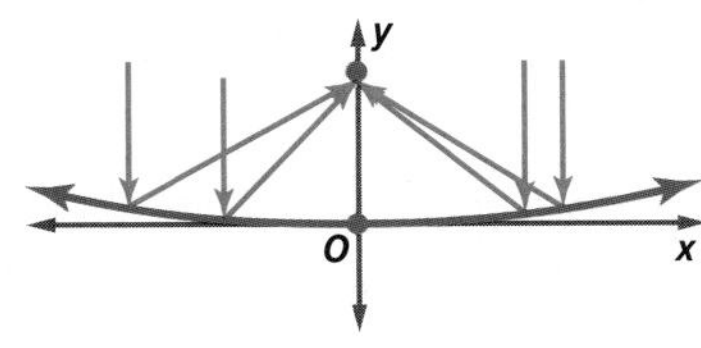

Example 3 **Graph each equation.**

29. $y = 2x^2$
30. $y = \frac{1}{3}x^2$
31. $y = -2x^2$
32. $x = \frac{1}{2}y^2$
33. $y = 2(x - 1)^2 - 4$
34. $x = -\frac{1}{32}y^2 - 6$
35. $y = \frac{1}{12}x^2 + 1$
36. $y = -2(x - 2)^2 + 3$
37. $x = -\frac{1}{16}(y + 5)^2 - 1$
38. $y = -\frac{1}{12}(x - 3)^2 + 5$
39. $x = \frac{1}{8}(y + 4)^2 - 4$
40. $x = -\frac{1}{16}(y - 4)^2 + 6$

Identify the focus and directrix of each parabola.

41. $\frac{1}{8}(x + 2)^2 - (y - 3) = 0$

42. $x - 4 = -\frac{1}{4}y^2$

43. $(y - 1)^2 - 24(x + 3) = 0$

44. SOLAR ENERGY Solar energy can be concentrated using parabolic reflecting plates. The collected energy is used to heat homes and produce electricity. Write an equation to represent the cross section of the parabolic plate shown.

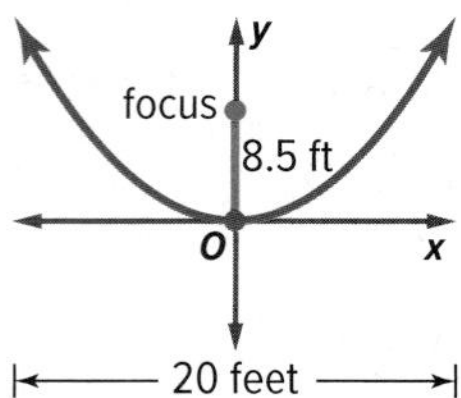

Write the equation of the parabola with the given directrix and with vertex at (0, 0).

45. $x = -\frac{3}{2}$

46. $y = \frac{1}{12}$

47. $y = -\frac{11}{6}$

48. $y = \frac{5}{12}$

49. $x = -4$

50. $x = \frac{1}{8}$

51. A parabola with vertex (0, 0) has equation $y^2 - 16x = 0$ and is shifted 3 units to the right and 2 units up. Write the equation of the new parabola. Then graph both parabolas on the same coordinate plane.

Write an equation of each parabola.

52. focus $(0, -3)$, directrix $y = 3$

53. focus $(0, 7)$, directrix $y = -7$

54. vertex $\left(-\frac{9}{5}, 1\right)$, focus $\left(-\frac{9}{5}, 3\right)$

55. vertex $\left(\frac{3}{4}, -2\right)$, directrix $x = -\frac{3}{4}$

Identify the coordinates of the vertex and focus, and the equation of the directrix of each parabola.

56. $y + 2 = \frac{1}{4}(x - 1)^2$

57. $x - 4 = 12(y + 2)^2$

58. $y = 8(x - 2)^2$

59. $x - 4 = (y + 6)^2$

60. $x + 1 = 4(y - 1)^2$

61. $y - 4 = x^2$

62. **MP CRITIQUE ARGUMENTS** Describe the error in the graph of each parabola.

a. $x^2 + 4y = 0$

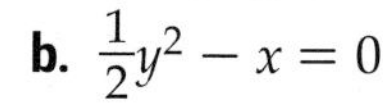

b. $\frac{1}{2}y^2 - x = 0$

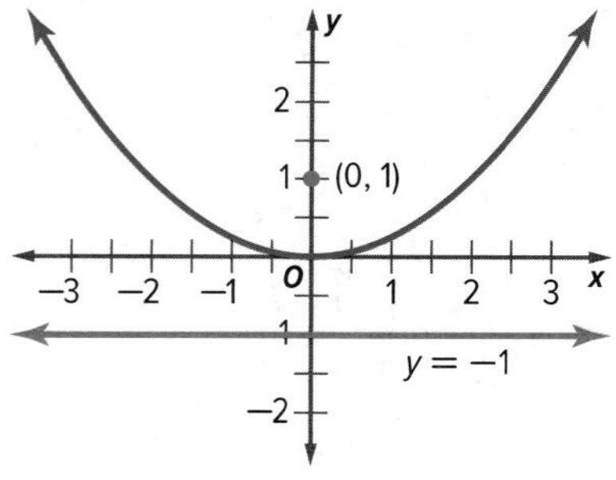

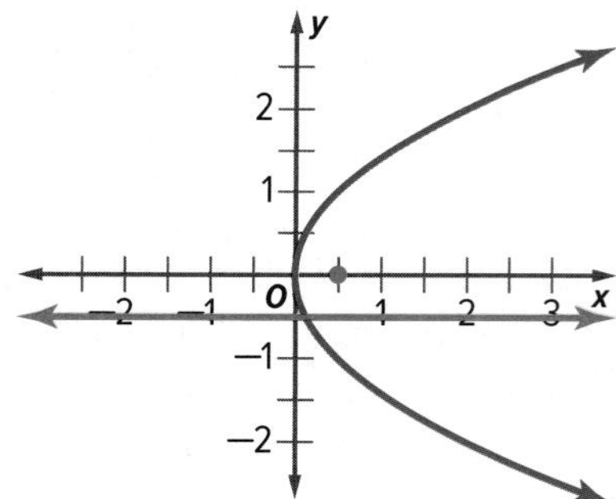

63. Graph $y = -x^2 + 4$ and $(x - 2)^2 = y$ on the same coordinate plane.

a. Estimate the point(s) of intersection of the two parabolas.

b. Substitute the expression $(x - 2)^2$ for y into $y = -x^2 + 4$ and solve for x.

c. Substitute the value(s) you found in part **b** into $y = -x^2 + 4$ and solve for y.

d. Use your answers to parts **b** and **c** to write the coordinates of the point(s) of intersection of the two graphs. Compare this to your estimates in part **a**.

e. Verify that the point(s) you found in part **d** lie on both parabolas.

64. Graph $y = \sqrt{x+2} - 1$ on a graphing calculator.

a. This graph is reflected in the line $y = -1$. Write the equation of the image.

b. Write an equation that represents the parabola formed by the original graph and its image.

65. TELEVISION A television station uses parabolic antennas to transmit their signals. An antenna used to transmit the signal has a focus 50 inches from the vertex.

a. Make two sketches of the antenna: one opening upwards and one opening to the right.

b. Use your sketches to write an equation for each sketch. Use the forms $x^2 = 4py$ and $y^2 = 4px$.

c. Does the equation you found in part **b** affect the depth of the antenna? Explain.

66. ASTRONOMY In astronomy, a radio telescope dish is shaped like a parabola and has special antennas to detect signals from space, known as the "radio universe." A radio telescope can be over 100 meters in diameter, but it can vary greatly in design, size, and configuration.

a. In a typical antenna, the distance between the focus and the vertex is 25 meters. What is an equation of the parabola? Assume that vertex is (0, 0).

b. Explain how a single dish antenna may work in a parabola-shaped telescope dish.

H.O.T. Problems Use Higher-Order Thinking Skills

67. WRITING IN MATH Explain how to find the distance from the focus to the directrix for the parabola $x = 4y^2$.

68. WRITING IN MATH The parabolic reflector plate of a bicycle head lamp has its bulb located at the focus of the parabola. Explain the possible advantages to using this design.

69. MP **REASONING** Explain how to find the equation of a parabola which has a vertex at $(3, -2)$ and a focus at $(6, -2)$.

70. WRITING IN MATH As the value of p increases, how does the width of the graph of $y = \frac{1}{4p}x^2$ change? Justify your reasoning.

71. CHALLENGE

a. What part of a parabola can be modeled by the equation $y = \sqrt{x}$?

b. What is the domain and range of the function in part **a**?

c. Explain how the equation $y = \sqrt{x}$ is related to a parabola of the form $y^2 = 4px$.

d. State a rule that you could use to help you use a graphing calculator to show the graph of $y^2 = 4px$.

72. MP **REASONING** How are a and p related in the same parabola expressed both as $y = ax^2$ and $x^2 = 4py$?

73. MP **REASONING** Predict how a change in the value of p in the equation $y^2 = 4px$ will affect the focus, directrix, and graph of $y^2 = 4px$. Then verify your prediction by graphing both in same coordinate plane.

74. WRITING IN MATH Explain the relationship between the Distance Formula, the equation of a parabola with vertex (0, 0), and the equation of a congruent parabola with a vertex not at the origin.

75. CHALLENGE Prove or disprove that the point $(-8, -4)$ lies on a parabola with vertex (0, 0) and containing the point $\left(-2\sqrt{2}, -\frac{\sqrt{2}}{8}\right)$.

Preparing for Assessment

76. Use the Distance Formula to derive the equation of a parabola with focus (0, 4) and directrix $y = -4$. MP 3

77. **MULTI-STEP** Consider the parabola with vertex (−4, 2) and focus (−4, 5).

a. Write the equation of the parabola. MP 1

b. Which of the following statements is true about the graph of the parabola? MP 1

- A The graph opens upward.
- B The graph opens downward.
- C The graph opens to the right.
- D The graph opens to the left.

c. Which of the following points are on the graph of the parabola? Select all that apply. MP 6

- A (2, 5)
- B $\left(-3, \frac{3}{4}\right)$
- C (0, 0)
- D (8, 14)
- E (−4, 2)
- F (6, 3)

d. Which of the following is the equation of the directrix of the parabola? MP 1

- A $y = -1$
- B $y = 1$
- C $x = 3$
- D $x = -1$

e. Describe how the vertex of the graph is translated from (0, 0). MP 1

78. Which of the following is the equation of a parabola with vertex (3, −5) and directrix $x = 7$? MP 1

- A $(y - 5)^2 = -16(x + 3)$
- B $(x - 3)^2 = -16(y + 5)$
- C $(y + 5)^2 = -16(x - 3)$
- D $(y + 5)^2 = 16(x - 3)$

79. Consider the parabola $(y - 1)^2 = -8(x + 5)$. MP 2

a. Identify the vertex.

b. Identify the focus.

c. Identify the directrix.

80. The graph of $x^2 = -2y$ is translated 1 unit down and 4 units left. Which of the following is the equation of the new parabola? MP 1

- A $(x + 1)^2 = 2(y - 4)$
- B $(x - 4)^2 = -2(y - 1)$
- C $(x + 4)^2 = -2(y - 1)$
- D $(x + 4)^2 = -2(y + 1)$

81. The parabolic reflector of a theatrical stage light has a bulb located at the focus of the reflector. If the focus of the parabola is 8.5 inches from the vertex, what are two possible equations of the parabola? MP 4

82. Write an equation of the parabola with focus (−4, −1) and directrix $y = 13$. MP 1

Study Guide and Review

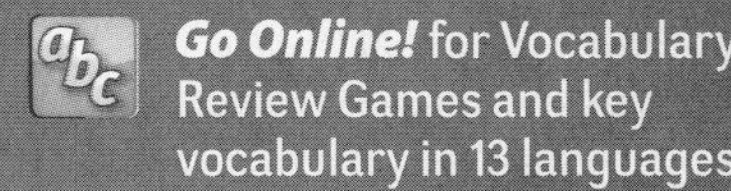

Study Guide

Key Concepts

Circles and Circumference (Lesson 9-1)

- The circumference of a circle is equal to πd or $2\pi r$.

Angles, Arcs, Chords, and Inscribed Angles
(Lessons 9-2 to 9-4)

- The sum of the measures of the central angles of a circle is 360°.
- The length of an arc is proportional to the length of the circumference.
- Diameters perpendicular to chords bisect chords and intercepted arcs.
- The measure of an inscribed angle is half the measure of its intercepted arc.

Tangents, Secants, and Angle Measures
(Lessons 9-5 and 9-6)

- A line that is tangent to a circle intersects the circle in exactly one point and is perpendicular to a radius.
- Two segments tangent to a circle from the same exterior point are congruent.
- The measure of an angle formed by two secant lines is half the positive difference of its intercepted arcs.
- The measure of an angle formed by a secant and tangent line is half its intercepted arc.

Equations of Circles and Parabolas
(Lessons 9-7 and 9-8)

- The equation of a circle with center (h, k) and radius r is $(x - h)^2 - (y - k)^2 = r^2$.
- A parabola is the locus of all points in a plane equidistant from a fixed point, called the focus, and a fixed line, called the directrix. The equation of a parabola can be found using the locus definition and the Distance Formula.

FOLDABLES® Study Organizer

Use your Foldable to review the chapter. Working with a partner can be helpful. Ask for clarification of concepts as needed.

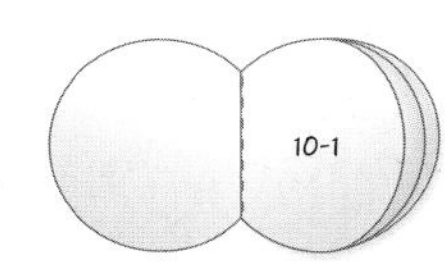

Key Vocabulary

adjacent arcs (p. 654)
arc (p. 652)
arc length (p. 654)
axis of symmetry (p. 703)
center (p. 643)
central angle (p. 652)
chord (p. 643)
circle (p. 643)
circumference (p. 645)
circumscribed (p. 646)
common tangent (p. 678)
concentric circles (p. 644)
congruent arcs (p. 653)
diameter (p. 643)
directrix (p. 703)
focus (p. 703)
inscribed (p. 646)
inscribed angle (p. 669)
intercepted arc (p. 669)
major arc (p. 653)
minor arc (p. 653)
parabola (p. 703)
pi (π) (p. 645)
point of tangency (p. 678)
radian measure (p. 655)
radius (p. 643)
secant (p. 687)
semicircle (p. 653)
tangent (p. 678)

Vocabulary Check

State whether each sentence is *true* or *false*. If *false*, replace the underlined word or phrase to make a true sentence.

1. Any segment with both endpoints on the circle is a radius of the circle.
2. A chord passing through the center of a circle is a diameter.
3. A central angle has the center as its vertex and its sides contain two radii of the circle.
4. An arc with a measure of less than 180° is a major arc.
5. An intercepted arc is an arc that has its endpoints on the sides of an inscribed angle and lies in the interior of the inscribed angle.
6. A common tangent is the point at which a line in the same plane as a circle intersects the circle.
7. Two circles are concentric circles if and only if they have congruent radii.

Concept Check

8. Explain the relationship between an inscribed angle and its intercepted arc.
9. Explain the difference between tangent and secant lines to a circle.

Study Guide and Review *Continued*

Lesson-by-Lesson Review 

9-1 Circles and Circumference

For Exercises 10–12, refer to ⊙*D*.

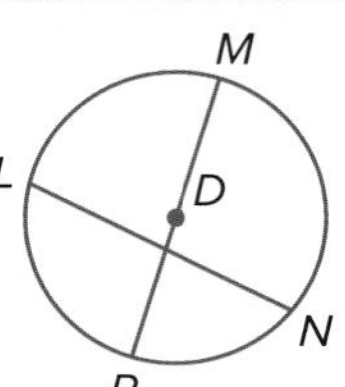

10. Name the circle.

11. Name a radius.

12. Name a chord that is not a diameter.

Find the diameter and radius of a circle with the given circumference. Round to the nearest hundredth.

13. $C = 43$ cm

14. $C = 26.7$ yd

15. $C = 108.5$ ft

16. $C = 225.9$ mm

Example 1

Find the circumference of ⊙*A*.

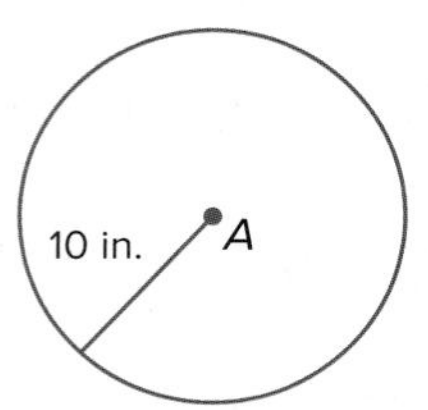

$C = 2\pi r$	Circumference formula
$= 2\pi(10)$	Substitution
≈ 62.83	Use a calculator.

The circumference of ⊙*A* is about 62.83 inches.

9-2 Measuring Angles and Arcs

Find the value of *x*.

17.

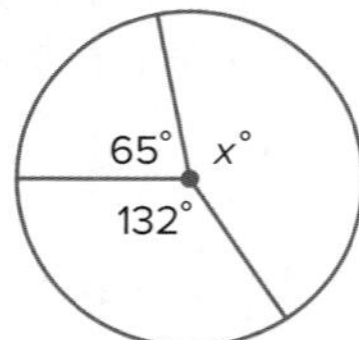

18.

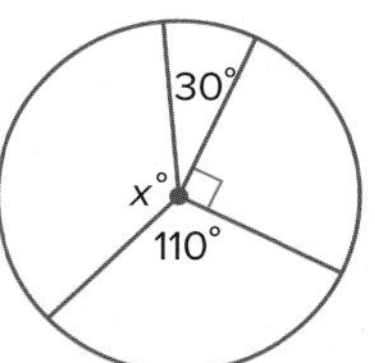

19. MOVIES The pie chart below represents the results of a survey taken by Mrs. Jameson regarding her students' favorite types of movies. Find each measure.

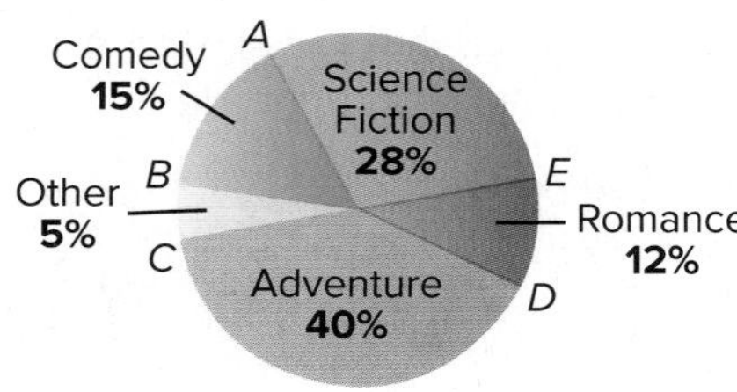

a. $m\widehat{AE}$

b. $m\widehat{BC}$

c. Describe the type of arc that the category Adventure represents.

Example 2

Find the value of *x*.

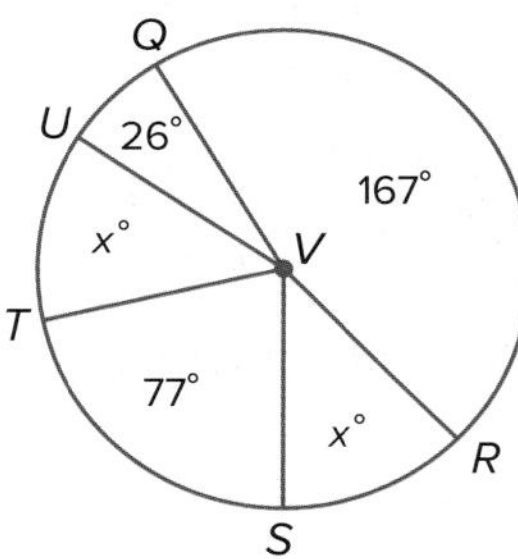

$m\angle QVR + m\angle RVS + m\angle SVT + m\angle TVU + m\angle UVQ = 360$	Sum of Central Angles
$167 + x + 77 + x + 26 = 360$	Substitution
$270 + 2x = 360$	Simplify.
$2x = 90$	Subtract.
$x = 45$	Divide.

9-3 Arcs and Chords

20. Find the value of x.

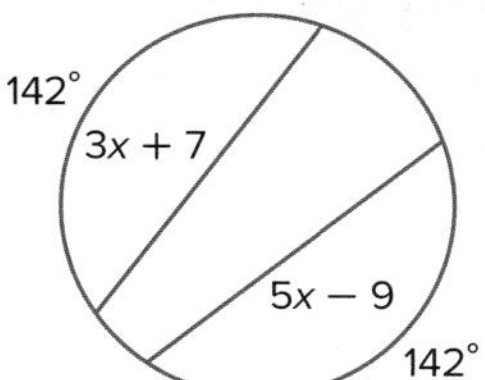

In $\odot K$, $MN = 16$ and $m\widehat{MN} = 98$. Find each measure. Round to the nearest hundredth.

21. $m\widehat{NJ}$ 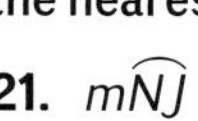

22. LN

23. GARDENING The top of the trellis shown is an arc of a circle in which $\overline{CD}$ is part of the diameter and $\overline{CD} \perp \overline{AB}$. If $\widehat{ACB}$ is about 28% of a complete circle, what is $m\widehat{CB}$?

Example 3

ALGEBRA In $\odot E$, $EG = EF$. Find AB.

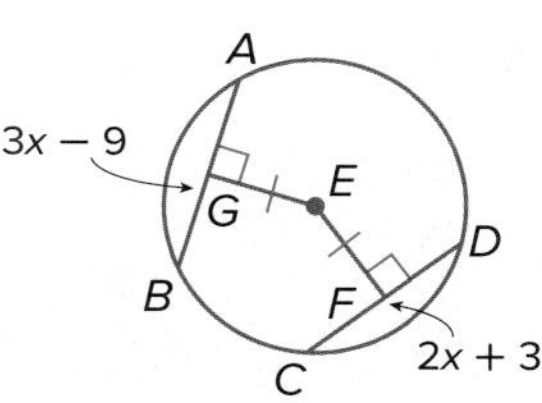

Because $\overline{EG}$ and $\overline{EF}$ are congruent, they are equidistant from E. So, $AB = CD$.

$AB = CD$	Theorem 10.5
$3x - 9 = 2x + 3$	Substitution
$3x = 2x + 12$	Add.
$x = 12$	Simplify.

So, $AB = 3(12) - 9$ or 27.

9-4 Inscribed Angles

Find each measure.

24. $m\angle 1$

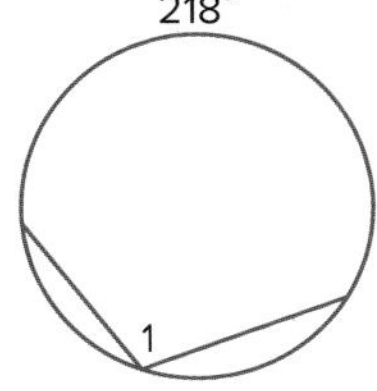

25. $m\widehat{GH}$

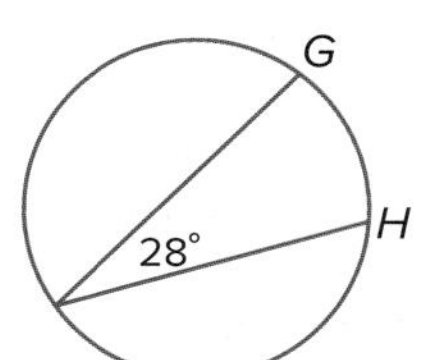

26. MARKETING In the logo at the right, $m\angle 1 = 42$. Find $m\angle 5$.

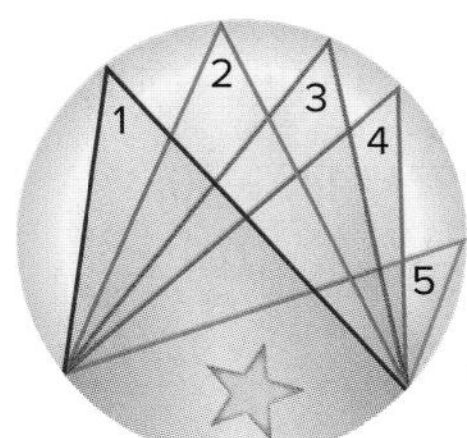

Example 4

Find $m\angle D$ and $m\angle B$.

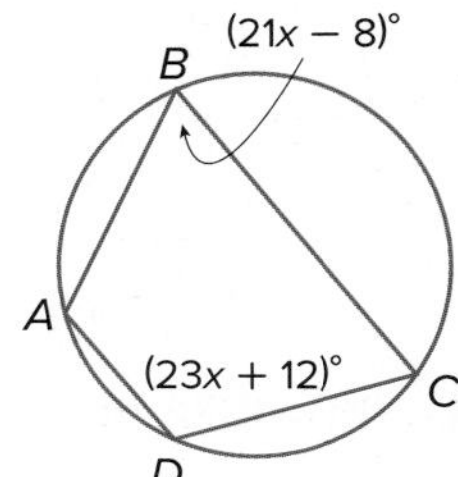

Because $ABCD$ is inscribed in a circle, opposite angles are supplementary.

$m\angle D + m\angle B = 180$	Definition of supplementary
$23x + 12 + 21x - 8 = 180$	Substitution
$44x + 4 = 180$	Simplify.
$44x = 176$	Subtract.
$x = 4$	Divide.

So, $m\angle D = 23(4) + 12$ or 104 and $m\angle B = 21(4) - 8$ or 76.

Study Guide and Review *Continued*

9-5 Tangents

27. SCIENCE FICTION In a story Todd is writing, instantaneous travel between a two-dimensional planet and its moon is possible when the time-traveler follows a tangent. Copy the figures below and draw all possible travel paths.

28. Find x and y. Assume that segments that appear to be tangent are tangent. Round to the nearest tenth if necessary.

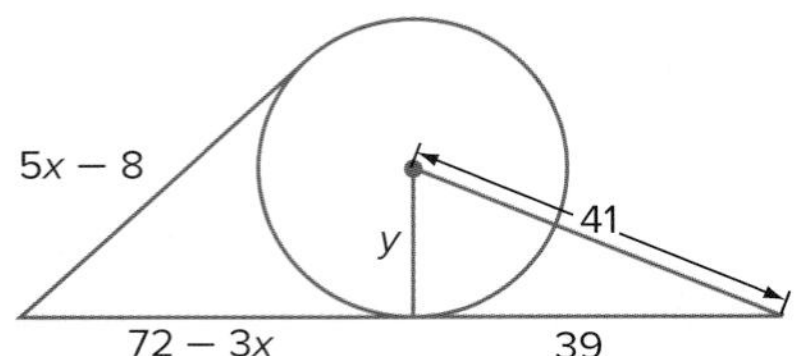

Example 5

In the figure, $\overline{KL}$ is tangent to ⊙M at K. Find the value of x.

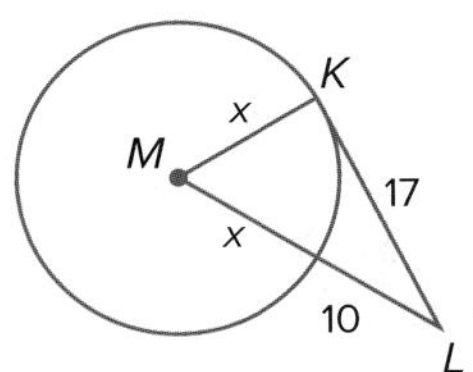

By Theorem 10.10, $\overline{MK} \perp \overline{KL}$. So, $\triangle MKL$ is a right triangle.

$KM^2 + KL^2 = ML^2$	Pythagorean Theorem
$x^2 + 17^2 = (x + 10)^2$	Substitution
$x^2 + 289 = x^2 + 20x + 100$	Multiply.
$289 = 20x + 100$	Simplify.
$189 = 20x$	Subtract.
$9.45 = x$	Divide.

9-6 Secants, Tangents, and Angle Measures

Find each measure. Assume that segments that appear to be tangent are tangent.

29. $m\angle 1$

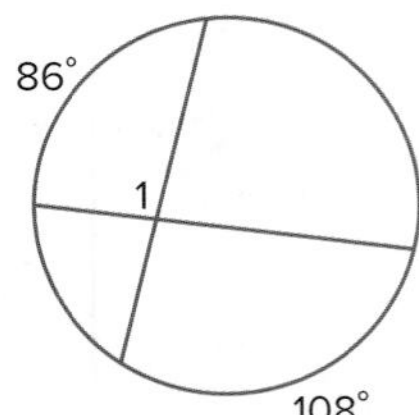

30. $m\widehat{AC}$

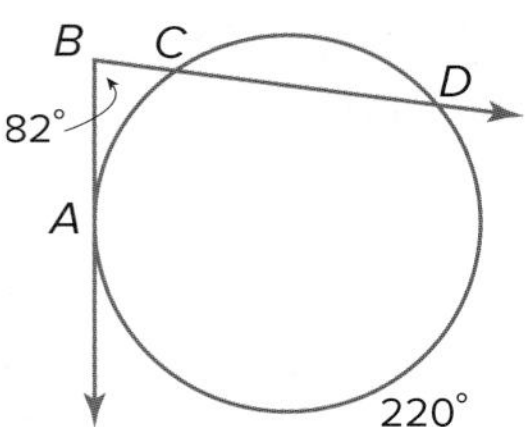

31. PHOTOGRAPHY Ahmed needs to take a close-up shot of an orange for his art class. He frames a shot of an orange as shown below, so that the lines of sight form tangents to the orange. If the measure of the camera's viewing angle is 34°, what is $m\widehat{ACB}$?

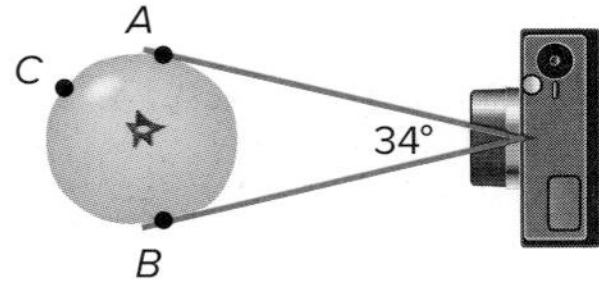

Example 6

Find the value of x.

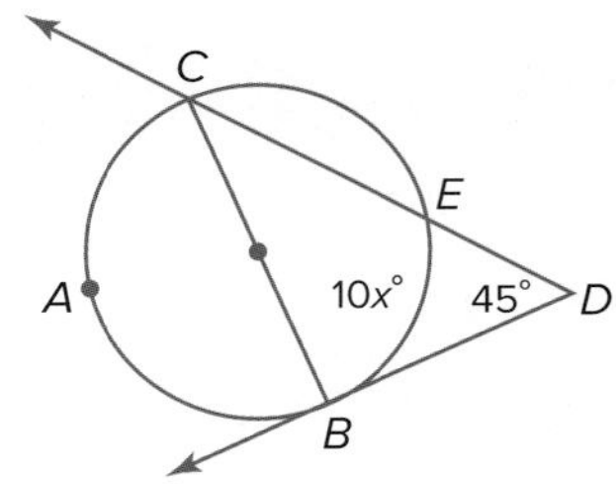

$\widehat{CAB}$ is a semicircle because $\overline{CB}$ is a diameter.

So, $m\widehat{CAB} = 180$.

$m\angle D = \frac{1}{2}(m\widehat{CAB} - m\widehat{EB})$	Theorem 10.14
$45 = \frac{1}{2}(180 - 10x)$	Substitution
$90 = 180 - 10x$	Multiply.
$-90 = -10x$	Subtract.
$9 = x$	Divide.

9-7 Equations of Circles

Write the equation of each circle.

32. center at $(-2, 4)$, radius 5

33. center at $(1, 2)$, diameter 14

Give the coordinates of the center and the measure of the radius of each circle.

34. $x^2 + y^2 = 16$

35. $(x - 1)^2 + (y + 5)^2 - 4 = 0$

36. $y^2 + (y + 4)^2 = 9$

37. $(x - 3)^2 + (y - 1)^2 + 7 = 16$

38. $(x - 5)^2 + (y + 1)^2 = 36$

39. $(y + 2)^2 = 49 - (x + 8)^2$

Find the point(s) of intersection between the circle and line.

40. $(x - 1)^2 + y^2 = 25; y = 5$

41. $(x - 3)^2 + (y + 2)^2 = 25; y = x$

42. $(x + 3)^2 + (y - 4)^2 = 9; y = -x + 4$

43. **FIREWOOD** In an outdoor training course, Kat learns a wood-chopping safety check that involves making a circle with her arm extended, to ensure she will not hit anything overhead as she chops. If her reach is 19 inches, the hatchet handle is 15 inches, and her shoulder is located at the origin, what is the equation of Kat's safety circle?

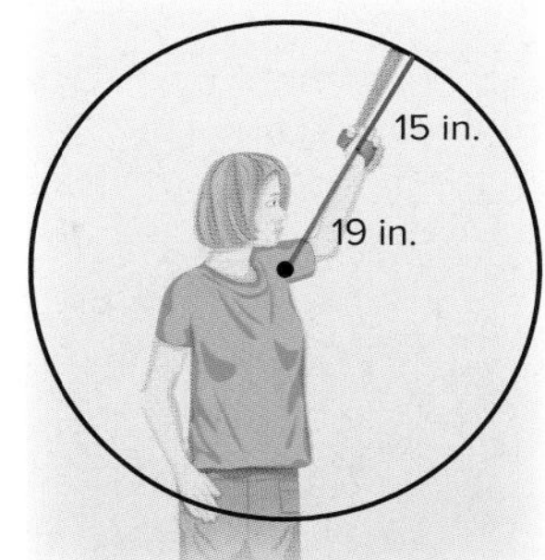

Example 7

Write the equation of the circle graphed below.

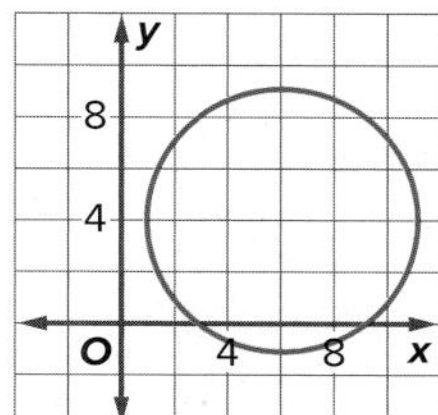

The center is at (6, 4) and the radius is 5.

$(x - h)^2 + (y - k)^2 = r^2$ Equation of a circle

$(x - 6)^2 + (y - 4)^2 = 5^2$ $(h, k) = (6, 4)$ and $r = 5$

$(x - 6)^2 + (y - 4)^2 = 25$ Simplify.

9-8 Equations of Parabolas

Write an equation of each parabola.

44. focus (0, 2), directrix $y = -2$

45. focus (4, 0), directrix $x = -4$

46. focus (0, −5), directrix $y = 5$

47. focus (4, −3), vertex (1, −3)

48. focus $\left(-2, -\frac{1}{2}\right)$, directrix $y = \frac{5}{2}$

Example 8

Write an equation of the parabola with focus at (0, 3) and directrix $y = -3$.

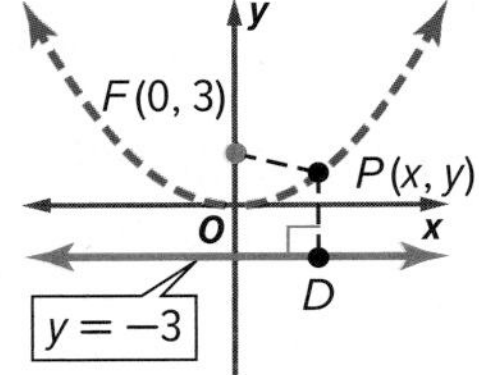

Step 1 Graph $F(0, 3)$ and $y = -3$.

Step 2 Label a point $D(x, -3)$ on $y = -3$ such that $\overline{PD}$ is perpendicular to the line $y = -3$.

Step 3 Use the Distance Formula to find PD and PF.

$$PD = \sqrt{(x - x)^2 + [y - (-3)]^2} = \sqrt{(y + 3)^2}$$

$$PF = \sqrt{(x - 0)^2 + (y - 3)^2} = \sqrt{x^2 + (y - 3)^2}$$

Step 4 Because $PD = PF$, set these expressions equal to each other.

$\sqrt{(y + 3)^2} = \sqrt{x^2 + (y - 3)^2}$ — $PD = PF$

$(y + 3)^2 = x^2 + (y - 3)^2$ — Square each side.

$y^2 + 6y + 9 = x^2 + y^2 - 6y + 9$ — Square binomials.

$12y = x^2$ or $y = \frac{1}{12}x^2$ — Subtract.

Practice Test

1. **POOLS** Amanda's family has a swimming pool that is 4 feet deep in their backyard. If the diameter of the pool is 25 feet, what is the circumference of the pool to the nearest foot?

2. Find the exact circumference of the circle below.

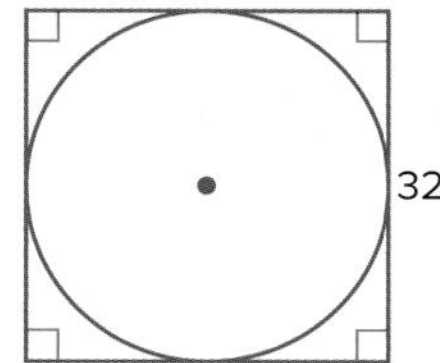

Find the value of *x*.

3.

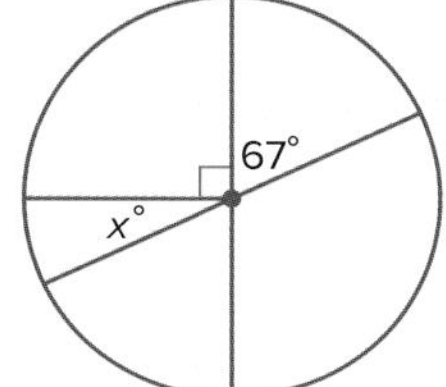

4.

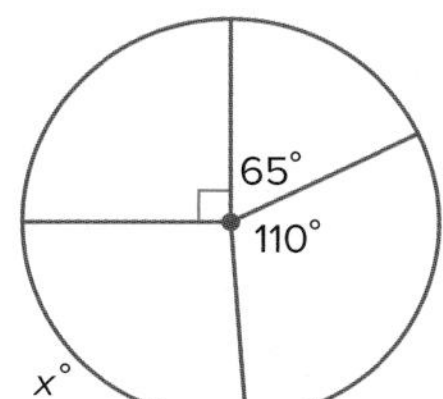

5.

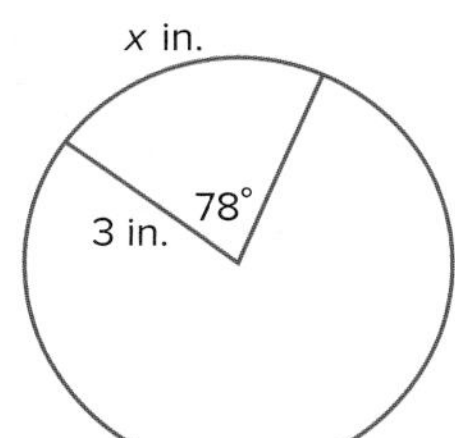

6.

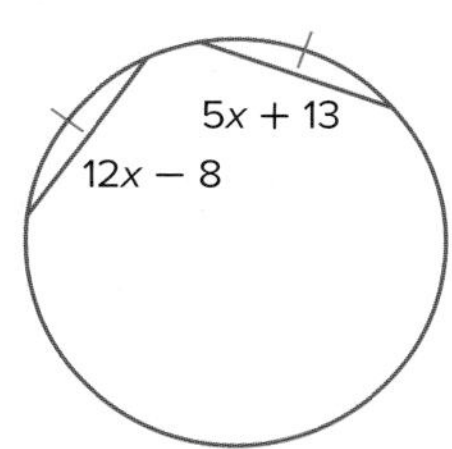

7. **MULTIPLE CHOICE** What is CD?

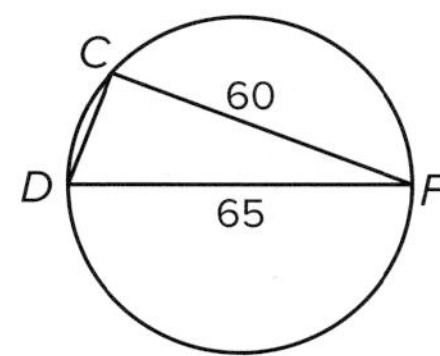

A 15
B 25
C 88.5
D Not enough information is given.

8. Find x if $\odot M \cong \odot N$.

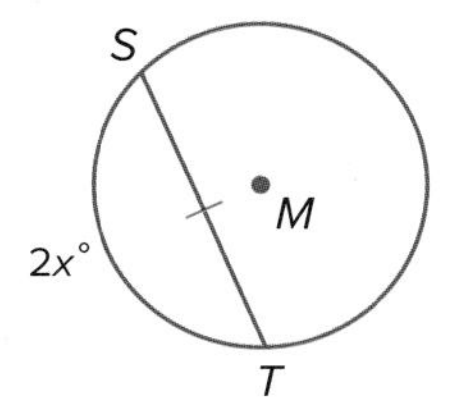

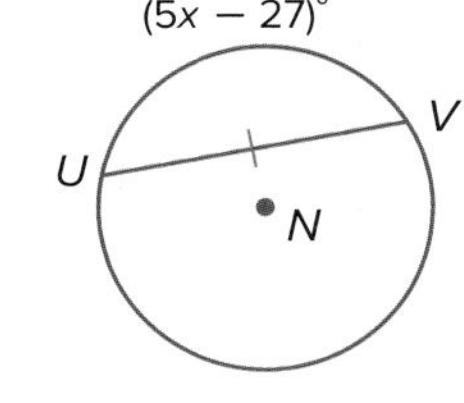

9. **MULTIPLE CHOICE** If $\overline{HK}$ is tangent to circle O, what is the radius of the circle?

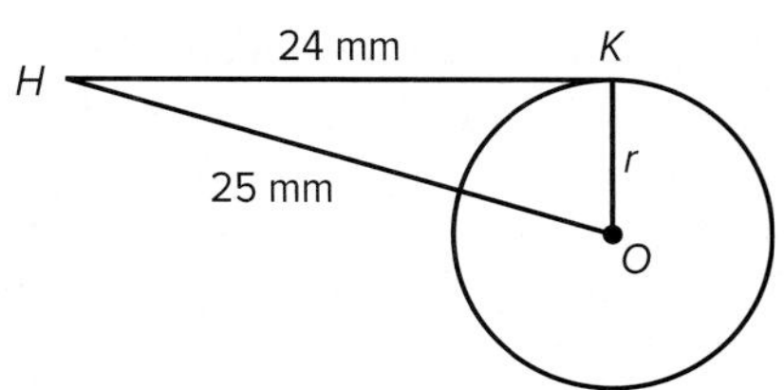

A 7 mm
B 8 mm
C 9 mm
D 10 mm

10. Determine whether $\overline{FG}$ is tangent to $\odot E$. Justify your answer.

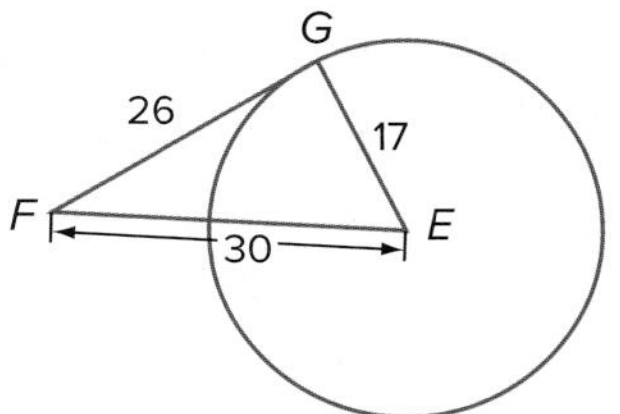

11. Find the perimeter of the triangle at the right. Assume that segments that appear to be tangent are tangent.

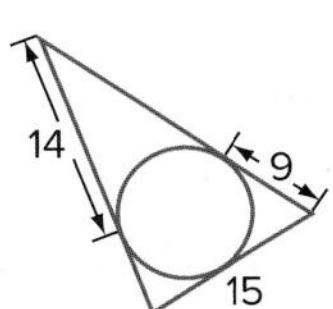

Find each measure.

12. $m\angle T$

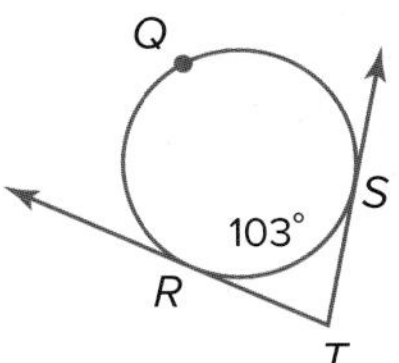

13. $m\angle AEB$

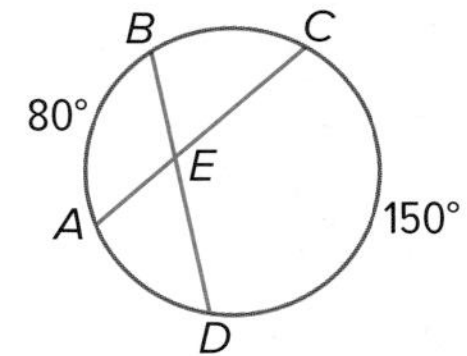

Write the equation of each circle.

14. center at (−2, 5), radius 4

15. center at (1, 0), diameter 12

Find an equation of the parabola with the focus and directrix given.

16. (0, 6), $y = -6$

17. $\left(-\frac{3}{4}, 0\right)$, $x = \frac{3}{4}$

CHAPTER 9

Preparing for Assessment

Performance Task

Provide a clear solution to each part of the task. Be sure to show all of your work. Include all relevant drawings and justify your answers.

BADGE DESIGN Jasmin is designing a circular badge where she circumscribes a circle around different polygons.

Part A

Sense-Making Jasmin is wondering if all triangles can be inscribed in a circle. She takes a particular triangle *ABC* where $m\angle A = 58°$; $m\angle B = 60°$, and $m\angle C = 62°$.

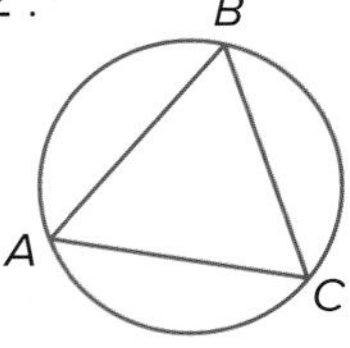

1. What are the measures of $\overset{\frown}{AB}$, $\overset{\frown}{CB}$, and $\overset{\frown}{AC}$?
2. Explain why a unique circle passes through the points *A*, *B*, and *C*.

Part B

Reasoning Next, Jasmin circumscribes a circle about quadrilateral *EFGH*.

3. Given $m\angle E = 130°$ and $m\angle F = 70°$, find $m\angle G$ and $m\angle H$.

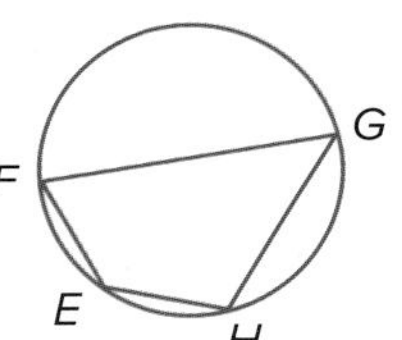

Part C

A quadrilateral inscribed in a circle can be referred to as a ***cyclic quadrilateral.*** Quadrilateral *EFGH* is an example of a cyclic quadrilateral.

4. State a property regarding pairs of angles in a cyclic quadrilateral.

Part D

Construct an Argument

5. Determine whether each of the following can be inscribed in a circle. Explain why or why not.
   - **a.** square
   - **b.** rhombus
   - **c.** rectangle
   - **d.** parallelogram
   - **e.** trapezoid
   - **f.** kite

Part E

A quadrilateral *ABCD* with vertices *A*(0, 0), *B*(−4, 3), *C*(−1, 7), and *D*(3, 4) is inscribed in a circle.

6. What type of special quadrilateral is *ABCD*?
7. Find an equation for the circle.

Test-Taking Strategy

Example

Read the problem. Identify what you need to know. Then use the given information to solve the problem.

Solve for x in the figure.

133°
$4x - 2$
$6x - 10$
133°

A 2 **C** 4

B 3 **D** 6

Step 1 **What parts of a circle are given in the diagram?**

two chords and two arc lengths

Step 2 **What properties can you use?**

Two chords are congruent if and only if their corresponding minor arcs are congruent.

Step 3 **How can you solve the problem?**

The minor arcs are congruent, so the chords are congruent. Write and solve an equation.

$4x - 2 = 6x - 10$ Definition of Congruent Segments

$-2x = -8$

$x = 4$ Simplify.

Step 4 **What is the answer?** C

Test-Taking Tip

Properties of Circles
When solving problems involving circles, start by identifying special properties and relationships among angles, arcs, and segments that intersect the circle. Then use those relationships to solve for unknown measures.

Apply the Strategy

Read each problem. Identify what you need to know. Then use the given information to solve the problem.

In the figure shown, what is $m\angle S$?

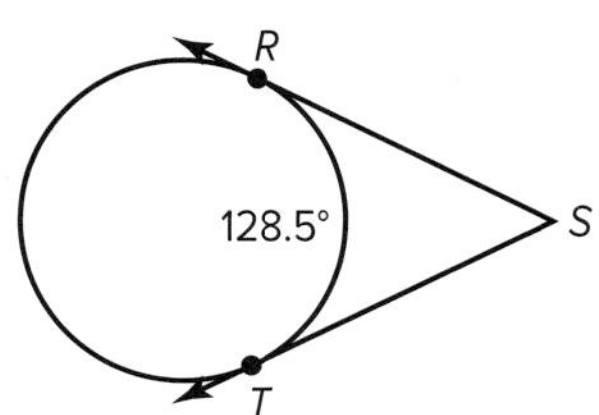

A 51.5 **C** 128.5

B 77 **D** 257

a. What parts of a circle are given in the diagram?

b. What properties can you use?

c. How can you solve the problem?

d. What is the answer?

CHAPTER 9

Preparing for Assessment

Cumulative Review

Read each question. Then fill in the correct answer on the answer document provided by your teacher or on a sheet of paper.

1. $\overline{RS}$ is tangent to circle Q at point R.

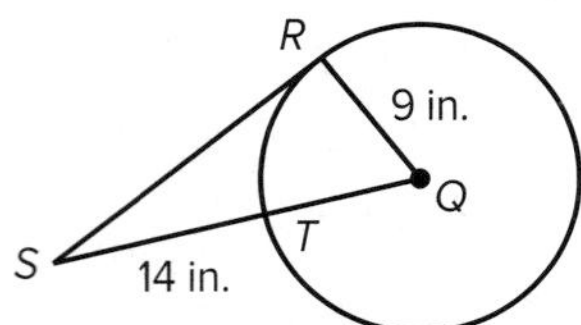

Which of the following is the best estimate of the perimeter of $\triangle QRS$?

- A 56.7 in.
- B 53.2 in.
- C 32.0 in.
- D 21.2 in.

2. Which of the following is the best description of the circle with equation $x^2 + y^2 - 6x + 4y + 9 = 0$?

- A The center is $(3, -2)$ and the radius is 2.
- B The center is $(-3, 2)$ and the radius is 2.
- C The center is $(3, -2)$ and the radius is 4.
- D The center is $(-3, 2)$ and the radius is 4.

3. Trapezoid $ABCD$ is shown below.

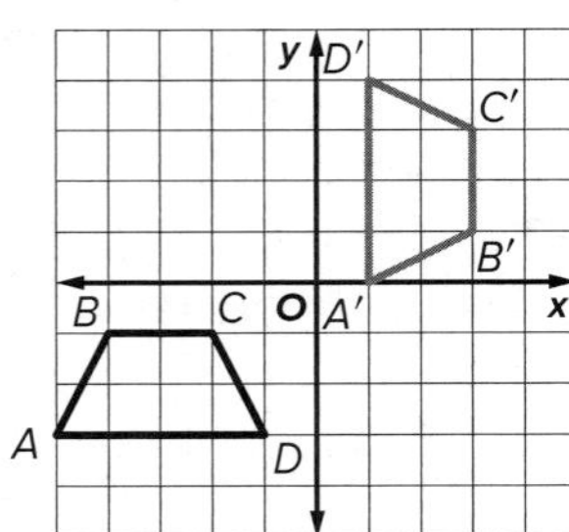

$ABCD$ is transformed to create a congruent image. What sequence of transformations created $A'B'C'D'$?

- A Two reflections and a translation
- B Two reflections and a rotation
- C Rotation, reflection, and translation
- D Rotation and reflection

4. Sean draws a triangle with vertices A, B, and C so that $AB = 11.6$ centimeters and $BC = 14.7$ centimeters. What is the greatest possible whole-number length of $\overline{AC}$, in centimeters?

5. A pendulum sweeps out an arc of a circle as shown.

a. Which of the following is the best estimate of the distance the tip of the pendulum travels as it moves from point R to point S and back to point R?

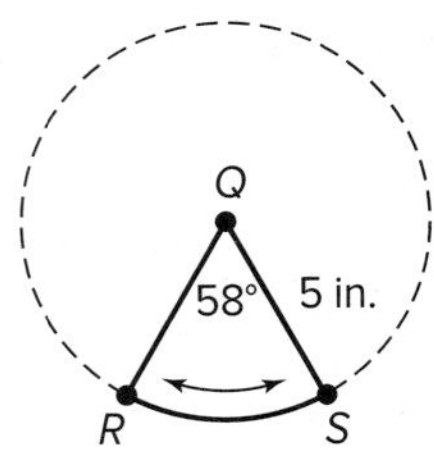

- A 5.1 in.
- B 10.1 in.
- C 52.7 in.
- D 390.0 in.

b. MP What mathematical practice did you use to solve this problem?

6. What is $m\angle KLM$?

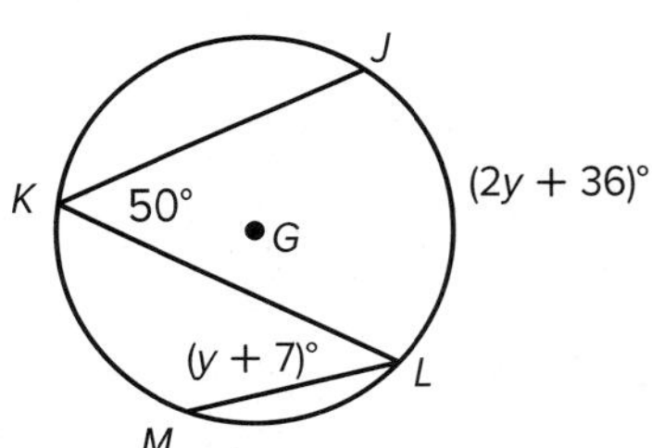

- A 50
- B 39
- C 32
- D 14

Test-Taking Tip

Question 6 There are two steps to solving this problem. First work with $\angle JKL$ and $\overset{\frown}{JL}$, then work with $\angle KLM$ once you know the value of y.

Go Online! for Standardized Test Practice

7. Square $ABCD$ is inscribed in $\odot Q$, as shown. What is the circumference of $\odot Q$, in meters? Round to the nearest tenth.

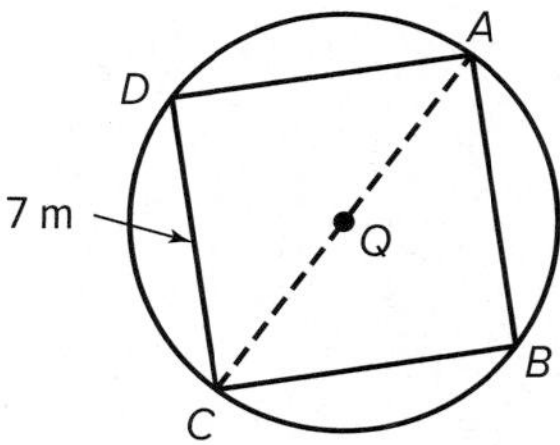

8. Melinda drew a figure with no line symmetry and no rotational symmetry. Select all terms that could describe the figure Melinda drew.

- ☐ **A** Isosceles triangle
- ☐ **B** Parallelogram
- ☐ **C** Regular pentagon
- ☐ **D** Scalene triangle
- ☐ **E** Trapezoid

9. The diameter of $\odot K$ is 22 centimeters and $\overline{PQ}$ is 18 centimeters long.

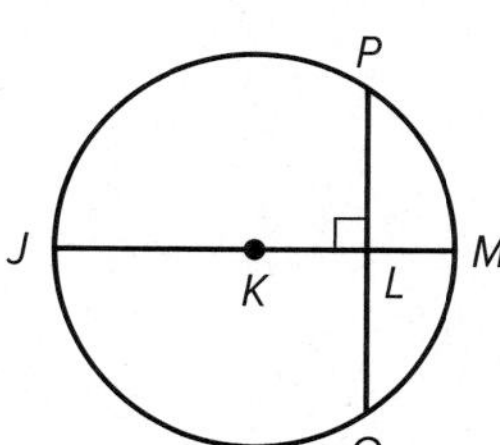

Which of the following is closest to the length of $\overline{KL}$?

○ **A** 4.0 cm ○ **C** 14.2 cm

○ **B** 6.3 cm ○ **D** 20.1 cm

10. What is the equation of the line through (2, 2) that is perpendicular to the line $2x + y = 1$?

○ **A** $y = \frac{1}{2}x + 1$ ○ **C** $y = -2x + 6$

○ **B** $y = -\frac{1}{2}x + 3$ ○ **D** $y = 2x - 2$

11. In the figure, $m\angle JKL = (2x + 2)$.

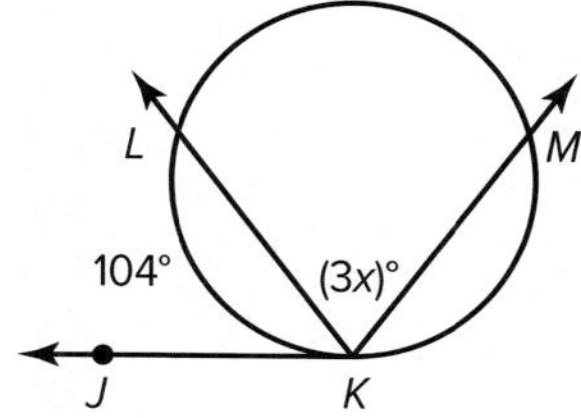

What is $m\widehat{LM}$?

○ **A** 25 ○ **C** 104

○ **B** 75 ○ **D** 150

12. Quadrilateral $JKLM$ is a rectangle.

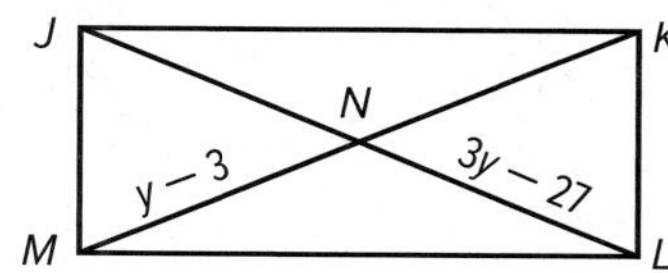

What is the length of the diagonal $\overline{JL}$?

○ **A** 24 ○ **C** 12

○ **B** 18 ○ **D** 9

13. Dinesh plotted the point $P(-1, 3)$. Then he applied the transformation $(x, y) \rightarrow (-y, x)$ followed by the transformation $(x, y) \rightarrow (-x, y)$. What is the distance between point P and its final image?

○ **A** $2\sqrt{2}$ ○ **C** $2\sqrt{5}$

○ **B** $4\sqrt{2}$ ○ **D** $2\sqrt{10}$

Need Extra Help?

If you missed Question...	1	2	3	4	5	6	7	8	9	10	11	12	13
Go to Lesson...	9-5	9-7	3-4	5-5	9-2	9-4	9-1	3-5	9-3	2-8	9-6	6-4	3-4

CHAPTER 10

Extending Area

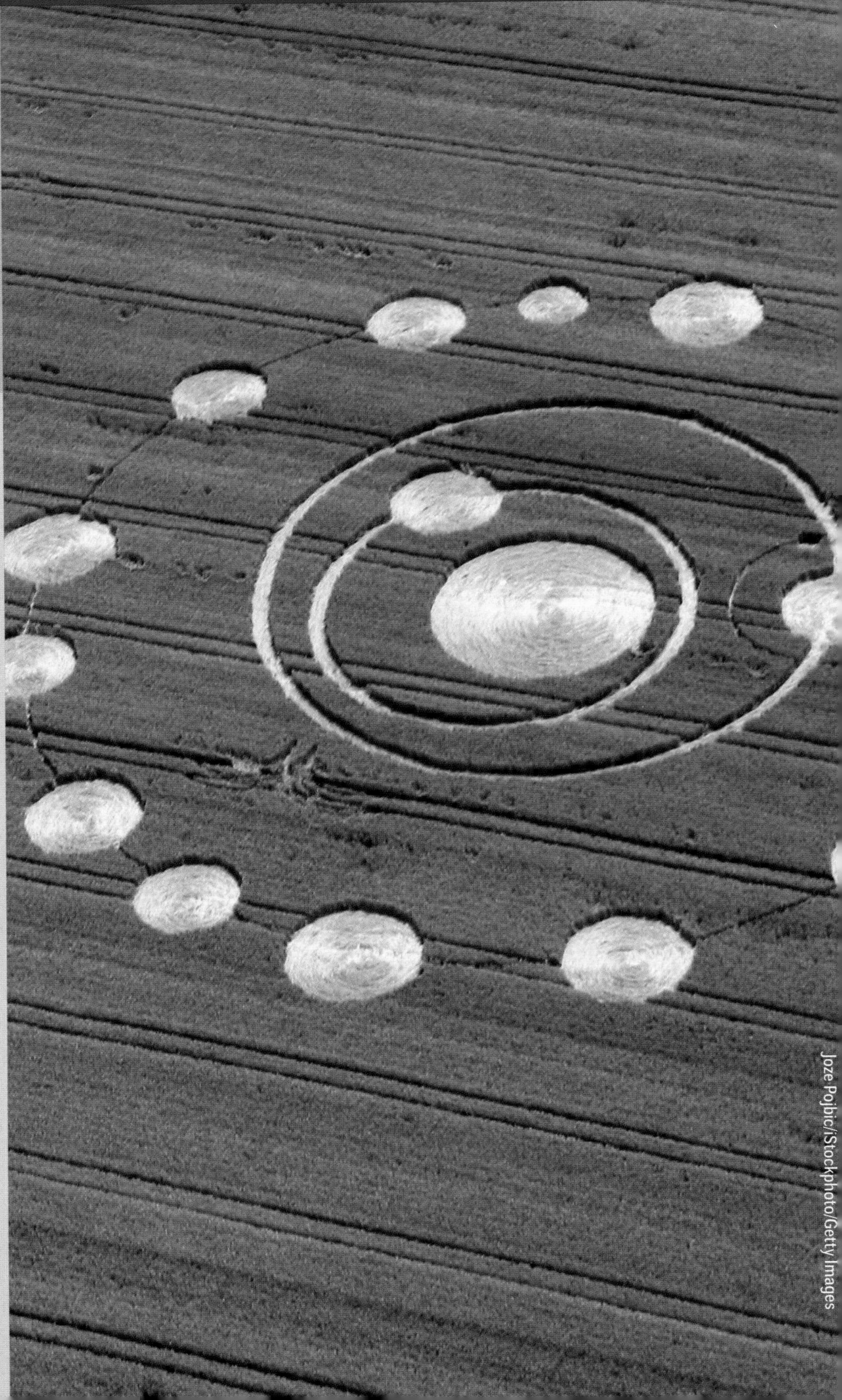

THEN

You learned about circles and angles within circles.

NOW

In this chapter, you will:

- Find areas of polygons.
- Solve problems involving areas and sectors of circles.
- Find scale factors using similar figures.

WHY

AGRICULTURE Center-pivot irrigation systems are credited with the formation of crop circles.

Use the Mathematical Practices to complete the activity.

1. Sense Making How might farmers use areas to determine the placement of irrigation systems?

2. Use Tools Use the Internet to learn more about center-pivot irrigation. What is a typical radius for one of these systems?

3. Reasoning If you know the radius of the center-pivot irrigation system, what else can you determine about the field it will irrigate?

4. Modeling Use the 2-D Figures tool to find the area irrigated by a typical center-pivot irrigation system.

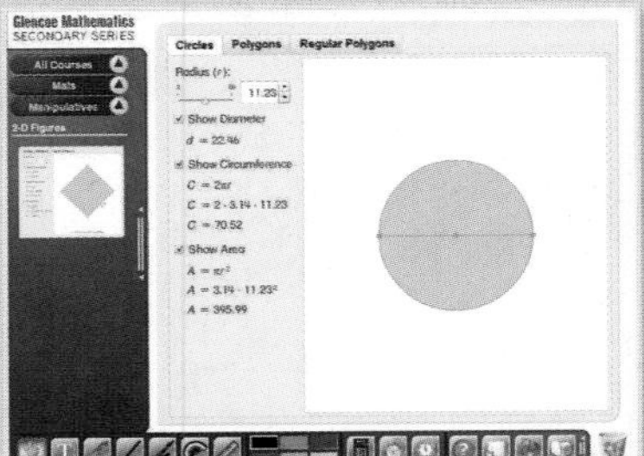

5. Discuss In what other ways would information about the area irrigated by the center-pivot system be useful?

Joze Pojbic/iStockphoto/Getty Images

Go Online to Guide Your Learning

Explore & Explain

3-D Figures

Use the **3-D Figures** tool find the surface areas of pyramids and cones, discussed in Lesson 10-6.

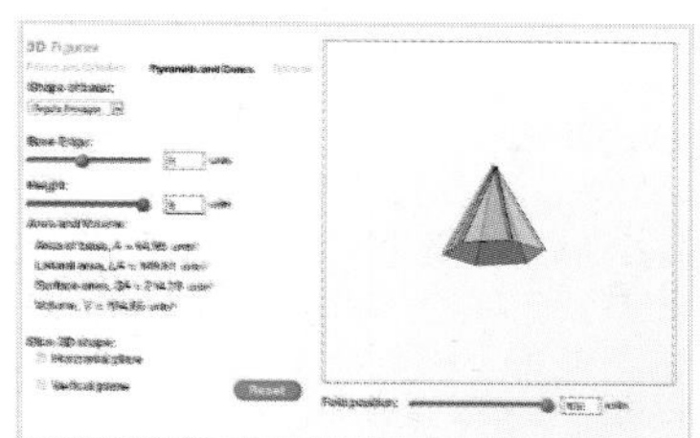

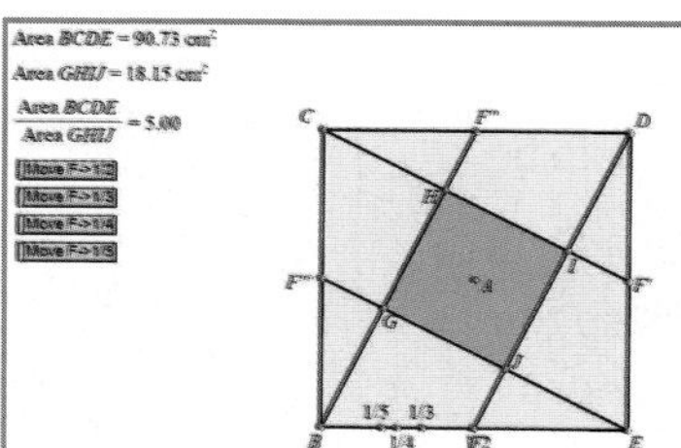

The Geometer's Sketchpad

Use **The Geometer's Sketchpad** to illustrate how to construct a square within a square and explore the relationship between them. You can also use The Geometer's Sketchpad to explore the relationships between the area of parallelograms and triangles and between the areas of similar figures.

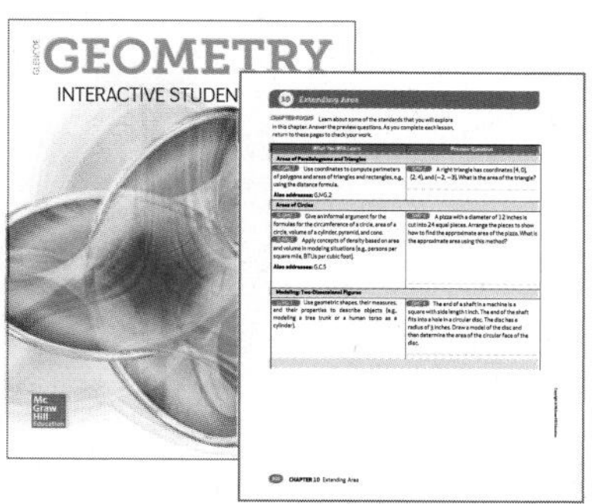

eBook

Interactive Student Guide

Before starting the chapter, answer the **Chapter Focus** preview questions. Check your answers as you complete each lesson. At the end of the chapter, try the **Performance Task**.

Organize

Foldables

Get organized! Before you begin this chapter, create this Foldable to help you organize your notes about areas of polygons and circles.

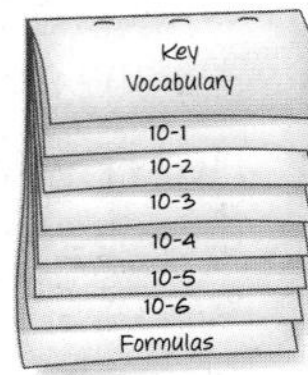

Collaborate

Chapter Project

In the **Real Estate** project, you will use what you have learned about areas of polygons and circles to complete a project that addresses business literacy.

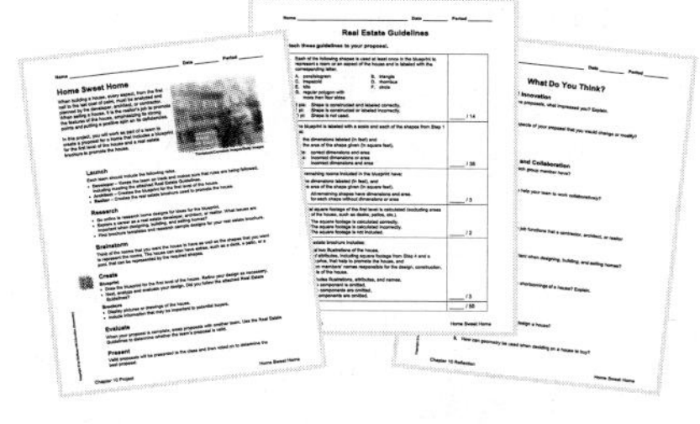

Focus

LEARNSMART®

Need help studying? Complete the **Circles** domain in LearnSmart to review for the chapter test.

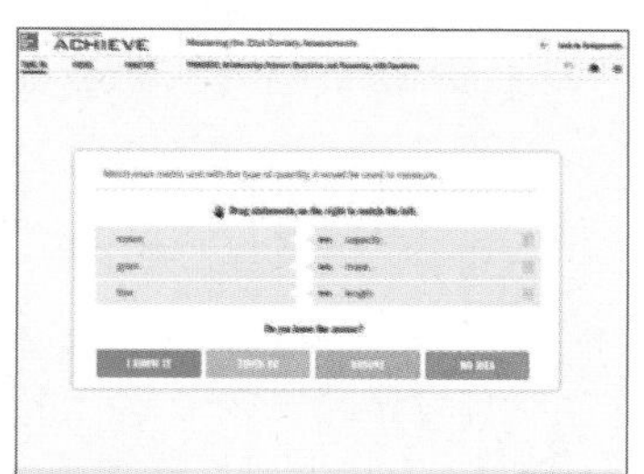

ALEKS®

You can use the **Polygons and Circles** and **Coordinate Geometry** topics in ALEKS to explore what you know about extending area and what you are ready to learn.*

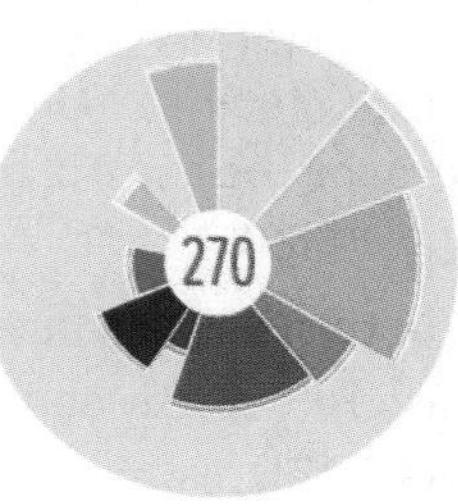

* Ask your teacher if this is part of your program.

Go Online! for Vocabulary Review Games and key vocabulary in 13 languages.

Get Ready for the Chapter

Connecting Concepts

Concept Check

Review the concepts used in this chapter by answering the questions below.

1. What is the equation to determine the area of a rectangle?
2. If you know the area of a rectangle and the length of one side of the rectangle, how could determine its width?
3. Given $a = 9$, $b = 10$, $c = 12$, and $d = 13$, what would be your first step in evaluating $\frac{1}{2}(ab + cd)$?
4. Given $A = 2(8 + 36)$, what would be the first step to solving the equation?
5. In the figure shown, what is the term for the side with length h?

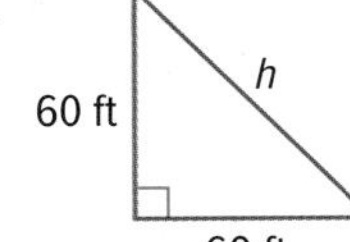

6. What type of triangle is this?
7. For this type of triangle, what is the value of the hypotenuse in terms of the length of the leg?
8. Find the value of h to the nearest hundredth.

Performance Task Review

You can use the concepts and skills in the chapter to solve problems in a real-world setting. Knowing how to determine the area of a polygon will help you finish the Performance Task at the end of the chapter in which you use your knowledge to determine the costs of projects in and around a home.

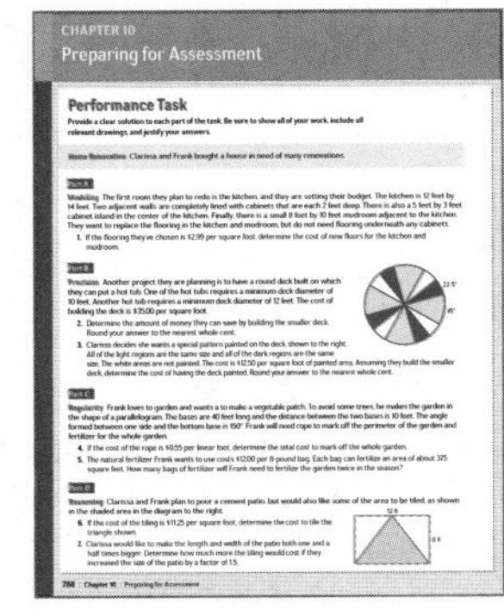
CHAPTER 10
Preparing for Assessment

Performance Task

In this Performance Task you will:

- reason abstractly and quantitatively
- model with mathematics
- attend to precision
- look for and express regularity in repeated reasoning

New Vocabulary

English		Español
base of a parallelogram	p. 725	base de un paralelogramo
base of a triangle	p. 727	base de un triángulo
height of a triangle	p. 727	altura de un triángulo
height of a trapezoid	p. 735	altura de un trapecio
sector of a circle	p. 744	sector de un círculo
segment of a circle	p. 748	segmento de un círculo
radius of a regular polygon	p. 752	radio de un polígono regular
central angle of a regular polygon	p. 752	ángulo central de un polígono regular
apothem	p. 752	apotema
composite figure	p. 754	figura compuesta
lateral face	p. 770	cara lateral
lateral edge	p. 770	arista lateral
altitude	p. 770	altura
lateral area	p. 770	área lateral
axis	p. 771	eje
regular pyramid	p. 773	pirámide regular
slant height	p. 773	altura oblicua
right cone	p. 775	cono recto
oblique cone	p. 775	cono oblicuo

Review Vocabulary

arc arco a part of a circle that is defined by two endpoints

central angle ángulo central an angle that intersects a circle in two points and has its vertex at the center of the circle

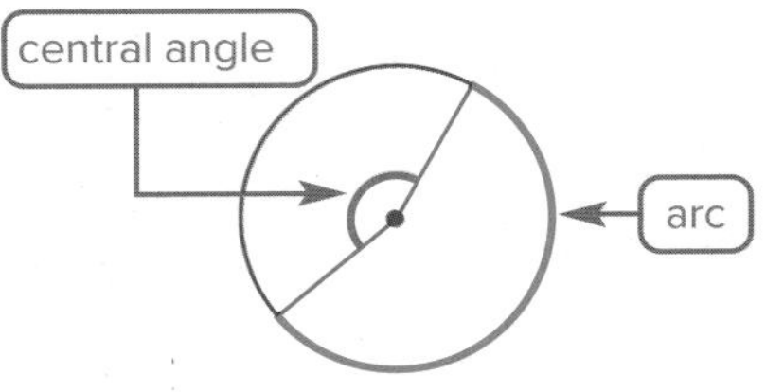

diagonal diagonal a segment that connects nonconsecutive vertices of a polygon

LESSON 1

Areas of Parallelograms and Triangles

Then

- You found areas of rectangles and squares.

Now

1. Find perimeters and areas of parallelograms.
2. Find perimeters and areas of triangles.

Why?

- A tangram is an ancient Chinese puzzle that can be rearranged to form different images, such as the animals shown. The area of the puzzle, before and after being rearranged, remains the same. It is the sum of all the areas of its pieces.

New Vocabulary

base of a parallelogram
height of a parallelogram
base of a triangle
height of a triangle

Mathematical Practices

1 Make sense of problems and persevere in solving them.
7 Look for and make use of structure.

1 Areas of Parallelograms

In Lesson 6-2, you learned that a *parallelogram* is a quadrilateral with both pairs of opposite sides parallel. Any side of a parallelogram can be called the **base of a parallelogram**. The **height of a parallelogram** is the perpendicular distance between any two parallel bases.

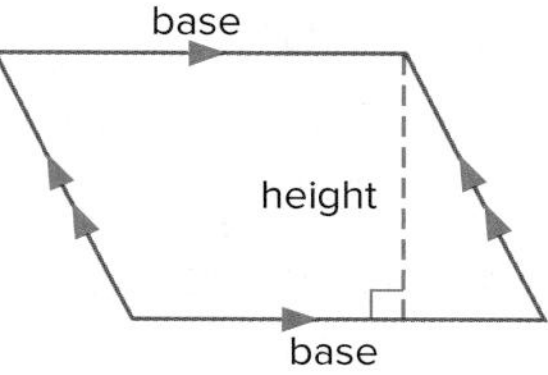

You can use the following postulate to develop the formula for the area of a parallelogram.

Postulate 10.1 Area Addition Postulate

The area of a region is the sum of the areas of its nonoverlapping parts.

In the figures below, a right triangle is cut off from one side of a parallelogram and translated to the other side as shown to form a rectangle with the same base and height.

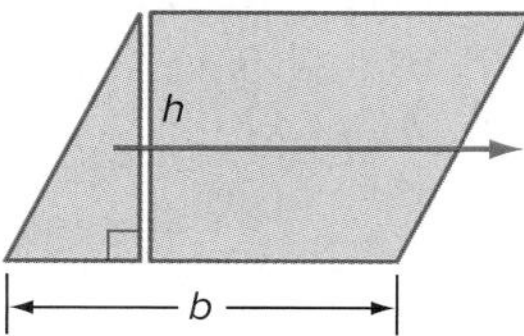

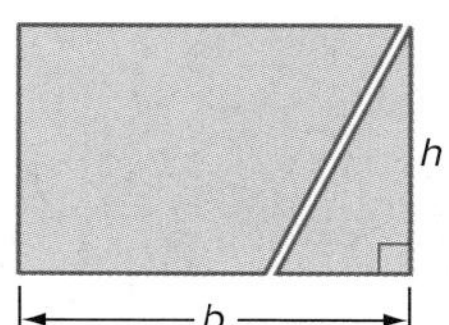

Recall from Lesson 1-6 that the area of a rectangle is the product of its base and height. By the Area Addition Postulate, a parallelogram with base b and height h has the same area as a rectangle with base b and height h.

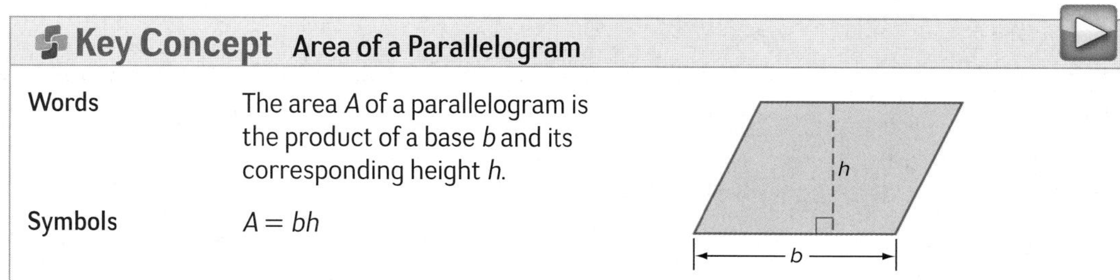

Key Concept Area of a Parallelogram

Words	The area A of a parallelogram is the product of a base b and its corresponding height h.	
Symbols	$A = bh$	

Example 1 Perimeter and Area of a Parallelogram

Find the perimeter and area of $\square ABCD$.

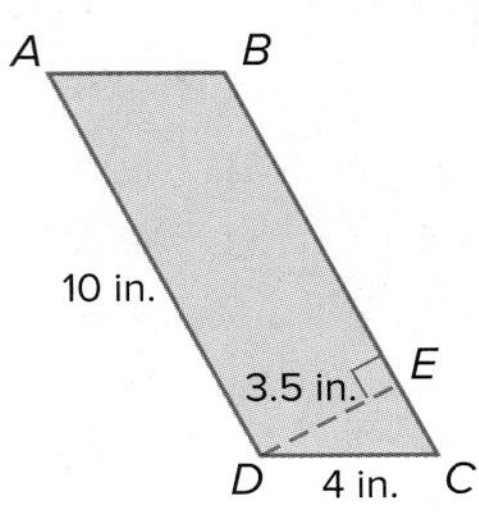

Perimeter

Because opposite sides of a parallelogram are congruent, $\overline{AB} \cong \overline{DC}$ and $\overline{BC} \cong \overline{AD}$. So $AB = 4$ inches and $BC = 10$ inches.

$$\text{Perimeter of } \square ABCD = AB + BC + DC + AD = 4 + 10 + 4 + 10 \text{ or } 28 \text{ in.}$$

Area

The height given, DE, is 3.5 inches. $\overline{BC}$ is the base, which measures 10 inches.

$A = bh$ — Area of a parallelogram

$= (10)(3.5)$ or 35 in^2 — $b = 10$ and $h = 3.5$

Study Tip

Heights of Figures The height of a figure can be measured by extending a base. In Example 1, the height of $\square ABCD$ that corresponds to base $\overline{DC}$ can be measured by extending $\overline{DC}$.

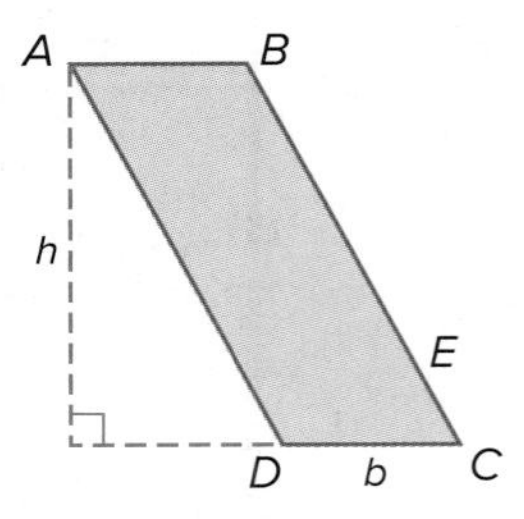

Guided Practice

Find the perimeter and area of each parallelogram.

1A.

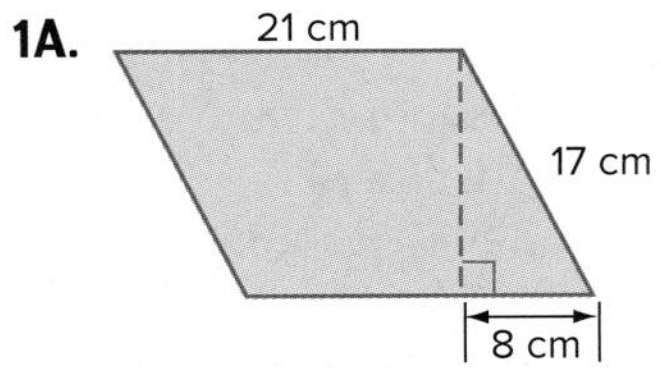

1B.

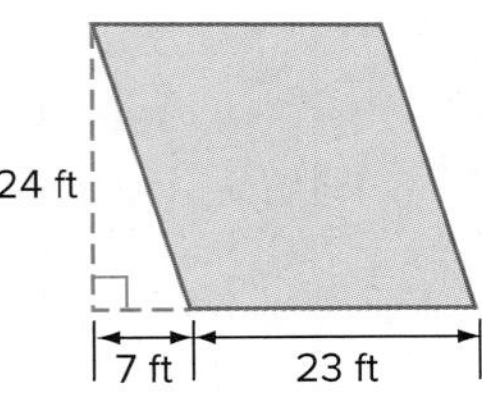

You may need to use trigonometry to find the area of a parallelogram.

Example 2 Area of a Parallelogram

Find the area of $\square EFGH$.

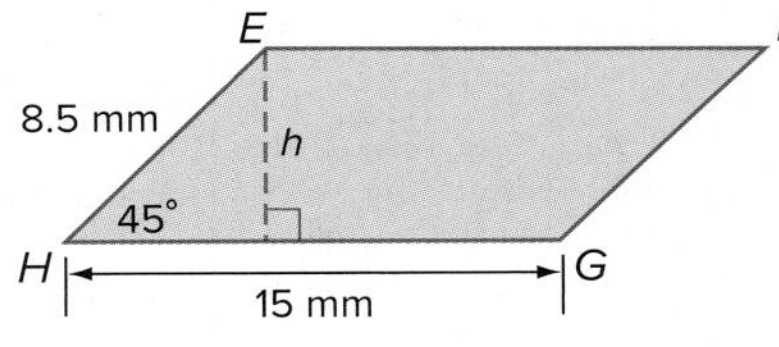

Step 1 Use a 45°-45°-90° triangle to find the height h of the parallelogram.

Recall that if the measure of the leg opposite the 45° angle is h, then the measure of the hypotenuse is $h\sqrt{2}$.

$h\sqrt{2} = 8.5$ — Substitute 8.5 for the measure of the hypotenuse.

$h = \frac{8.5}{\sqrt{2}}$ or about 6 mm — Divide each side by $\sqrt{2}$.

Step 2 Find the area.

$A = bh$ — Area of a parallelogram

$\approx (15)(6)$ or 90 mm^2 — $b = 15$ and $h \approx 6$

Watch Out

MP Precision Remember that perimeter is measured in linear units such as inches and centimeters. Area is measured in square units such as square feet and square millimeters.

Guided Practice

Find the area of each parallelogram. Round to the nearest tenth if necessary.

2A.

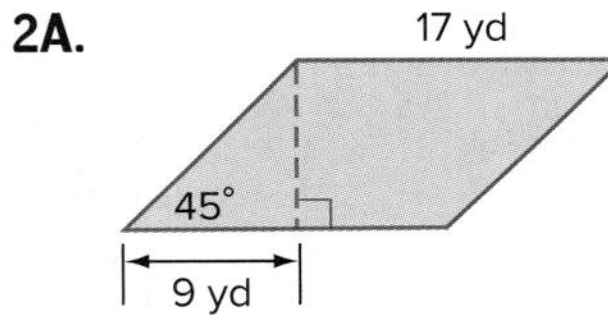

2B.

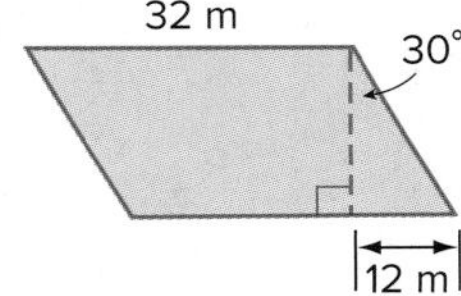

Review Vocabulary

altitude of a triangle
a segment from a vertex of a triangle to the line containing the opposite side and perpendicular to the line containing that side

2 Areas of Triangles

Like the base of a parallelogram, the **base of a triangle** can be any side. The **height of a triangle** is the length of an altitude drawn to a given base.

You can use the following postulate to develop the formula for the area of a triangle.

Postulate 10.2 Area Congruence Postulate

If two figures are congruent, then they have the same area.

In the figures below, a parallelogram is cut in half along a diagonal to form two congruent triangles with the same base and height.

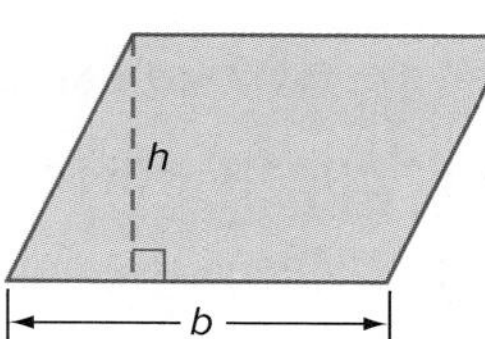

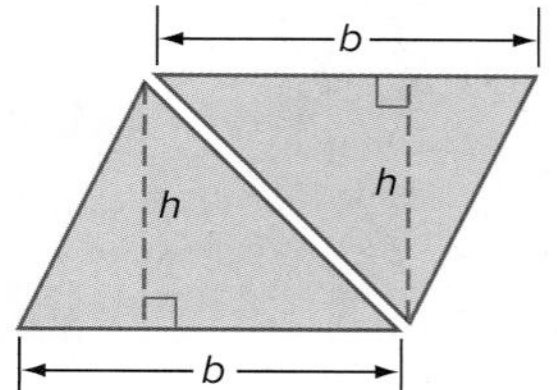

By the Area Congruence Postulate, the two congruent triangles have the same area. So, one triangle with base b and height h has half the area of a parallelogram with base b and height h.

Key Concept Area of a Triangle

Words The area A of a triangle is one half the product of a base b and its corresponding height h.

Symbols $A = \frac{1}{2}bh$ or $A = \frac{bh}{2}$

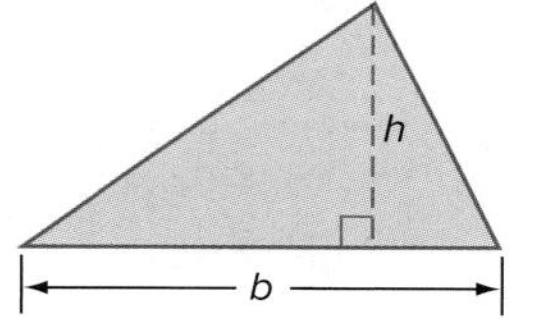

Real-World Example 3 Perimeter and Area of a Triangle

GARDENING D'Andre needs enough mulch to cover the triangular garden shown and enough paving stones to border it. If one bag of mulch covers 12 square feet and one paving stone provides a 4-inch border, how many bags of mulch and how many stones does he need to buy?

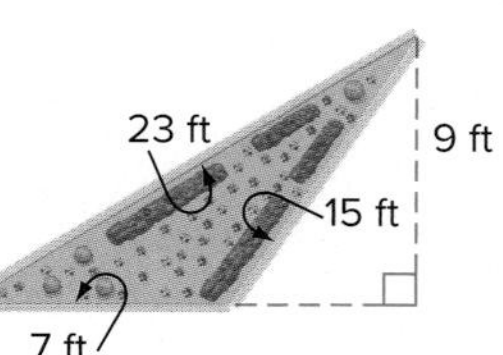

Step 1 Find the perimeter of the garden.

Perimeter of garden = 23 + 15 + 7 or 45 ft

Step 2 Find the area of the garden.

$A = \frac{1}{2}bh$ Area of a triangle

$= \frac{1}{2}(7)(9)$ or 31.5 ft^2 $b = 7$ and $h = 9$

Step 3 Use unit analysis to determine how many of each item are needed.

Bags of Mulch

$31.5 \text{ ft}^2 \cdot \frac{1 \text{ bag}}{12 \text{ ft}^2} = 2.625 \text{ bags}$

Paving Stones

$45 \text{ ft} \cdot \frac{12 \text{ in.}}{1 \text{ ft}} \cdot \frac{1 \text{ stone}}{4 \text{ in.}} = 135 \text{ stones}$

Round the number of bags up so there is enough mulch. He will need 3 bags of mulch and 135 paving stones.

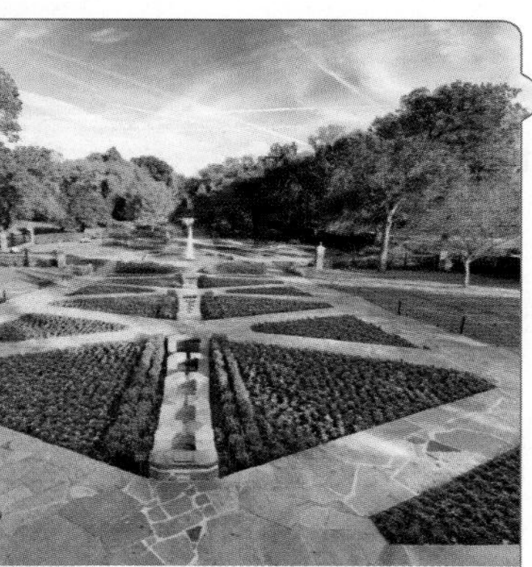

Real-World Link

Triangular gardens can serve as focal points in landscaping or simply result from intersecting walkways.

Guided Practice

Find the perimeter and area of each triangle.

3A. 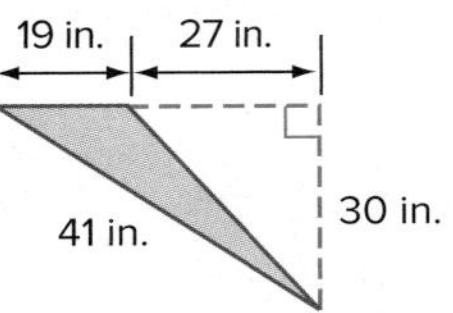

3B. 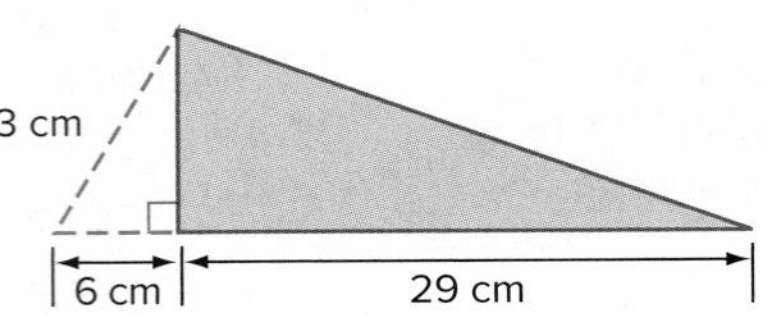

You can use the Distance Formula to find the perimeter of a polygon graphed on a coordinate plane.

Example 4 Perimeter and Area on the Coordinate Plane

Find the perimeter and area of $\triangle PQR$ with $P(-1, 3)$, $Q(-3, -1)$, and $R(4, -1)$.

Step 1 Find the perimeter of $\triangle PQR$.

Use the Distance Formula to find the length of each side.

$$PQ = \sqrt{(x_2 - x_1)^2 + (y_2 - y_1)^2}$$
$$= \sqrt{[-1 - (-3)]^2 + [3 - (-1)]^2}$$
$$= \sqrt{2^2 + 4^2} \text{ or } \sqrt{20}$$

$$QR = \sqrt{(x_2 - x_1)^2 + (y_2 - y_1)^2}$$
$$= \sqrt{[(-3) - 4]^2 + [(-1) - (-1)]^2}$$
$$= \sqrt{(-7)^2 + 0^2} \text{ or } 7$$

$$PR = \sqrt{(x_2 - x_1)^2 + (y_2 - y_1)^2}$$
$$= \sqrt{(-1 - 4)^2 + [3 - (-1)]^2}$$
$$= \sqrt{(-5)^2 + 4^2} \text{ or } \sqrt{41}$$

The perimeter of $\triangle PQR$ is $\sqrt{20} + 7 + \sqrt{41}$ or about 17.9 units.

Step 2 Find the area of $\triangle PQR$.

Using $\overline{QR}$ as the base, the height is the perpendicular distance from P to $\overline{QR}$. From the graph, the height is 4 units.

$A = \frac{1}{2}bh$ Area of a triangle

$= \frac{1}{2}(7)(4) \text{ or } 14$ Substitute and simplify.

The area of $\triangle PQR$ is 14 square units.

Guided Practice

4. Find the perimeter and area of $\triangle ABC$ with $A(6, -1)$, $B(1, -1)$, and $C(1, 6)$.

Check Your Understanding

= Step-by-Step Solutions begin on page R13.

Examples 1–3 **Find the perimeter and area of each parallelogram or triangle. Round to the nearest tenth if necessary.**

1. 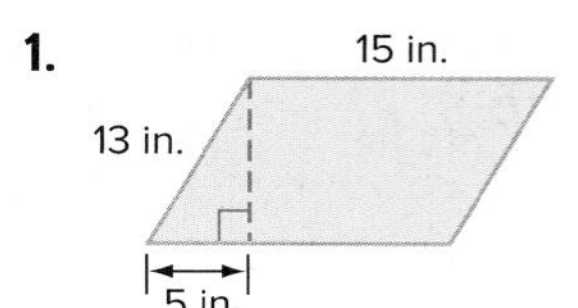

2. 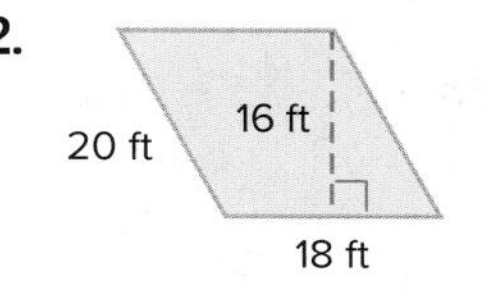

3. 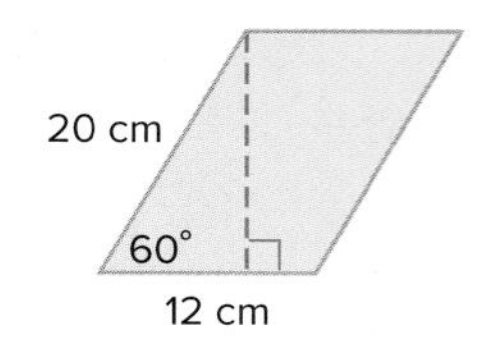

4. 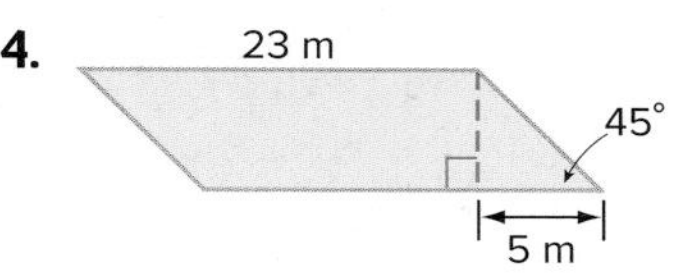

5. 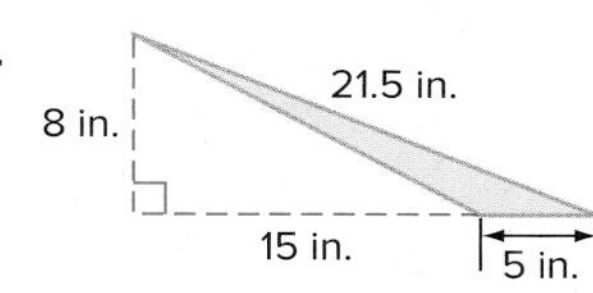

6. 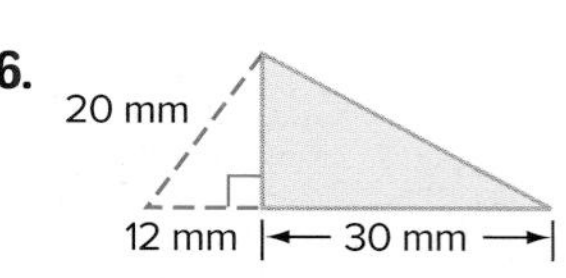

7. CRAFTS Marquez and Victoria are making pinwheels. Each pinwheel is composed of 4 triangles with the dimensions shown. Find the perimeter and area of one triangle.

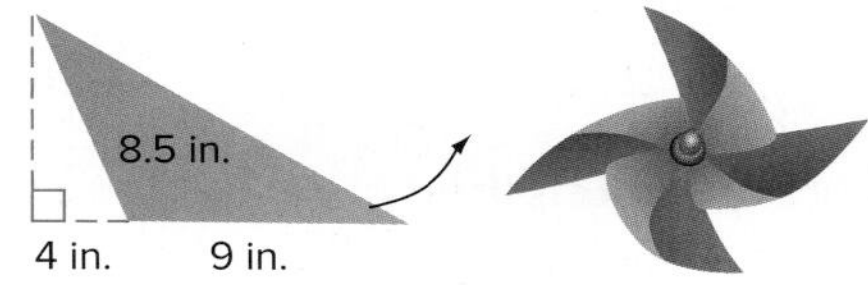

Example 4 **Find the perimeter and area of each triangle.**

8. $\triangle JKL$ with $J(5, 6)$, $K(-3, -1)$, and $L(-2, 6)$

9. $\triangle WXY$ with $W(7, 4)$, $X(-1, 5)$, and $Y(7, -1)$

Practice and Problem Solving

Extra Practice is on page R10.

Examples 1–3 **MP STRUCTURE Find the perimeter and area of each parallelogram or triangle. Round to the nearest tenth if necessary.**

10. 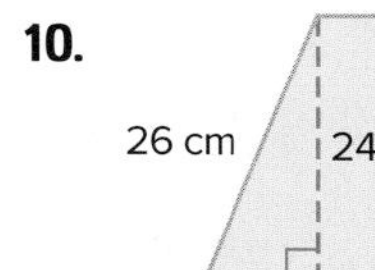

11 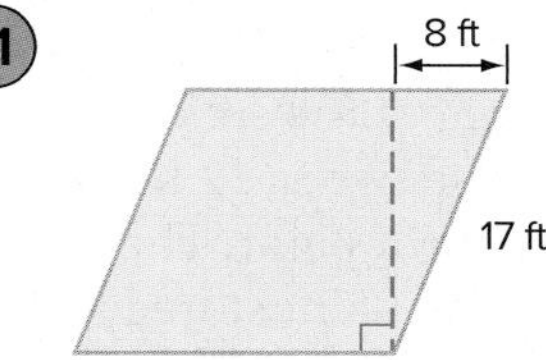

12. 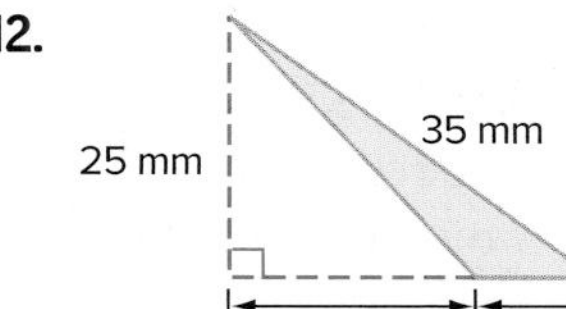

13. 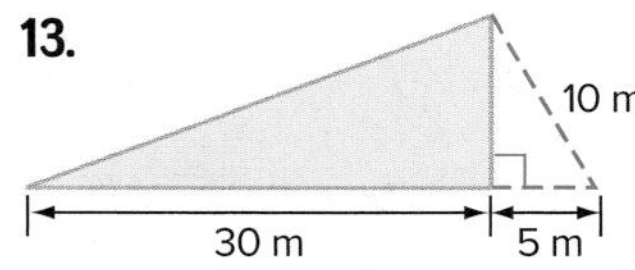

14. 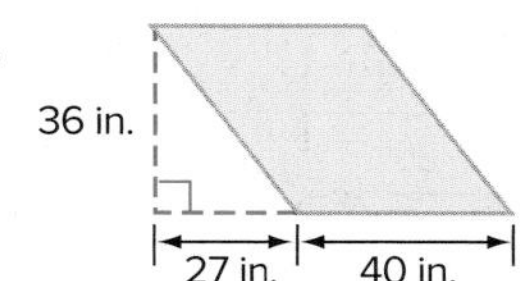

15. 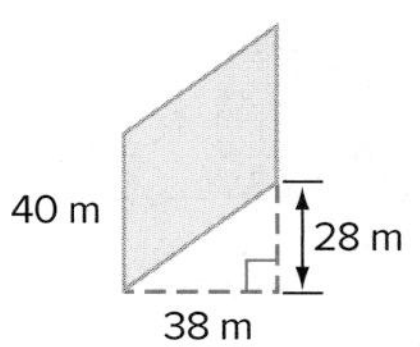

16. TANGRAMS The tangram shown is a 4-inch square.

a. Find the perimeter and area of the purple triangle. Round to the nearest tenth.

b. Find the perimeter and area of the blue parallelogram. Round to the nearest tenth.

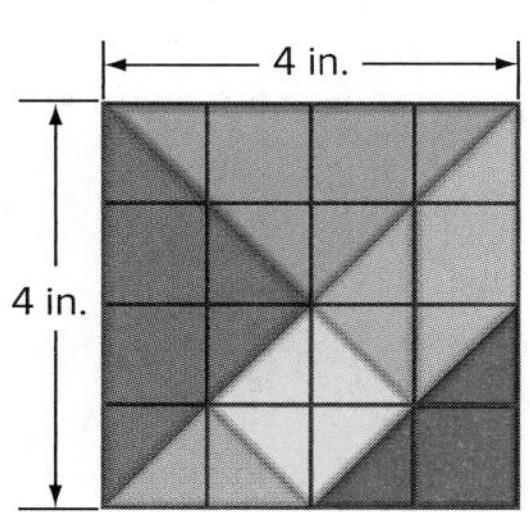

Example 2 **MP STRUCTURE Find the area of each parallelogram. Round to the nearest tenth if necessary.**

17.
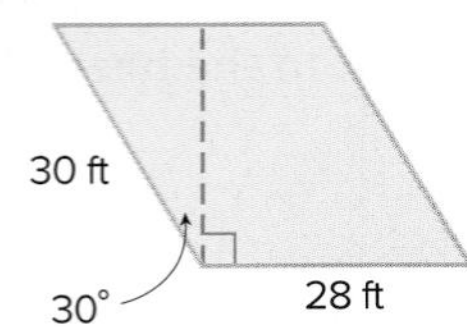

18.
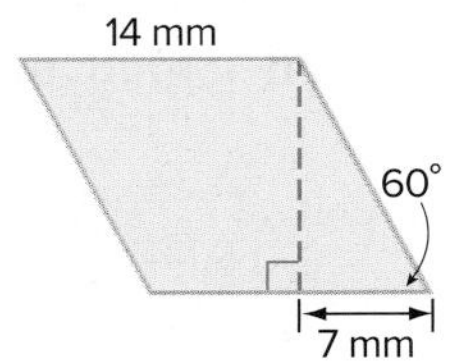

19.
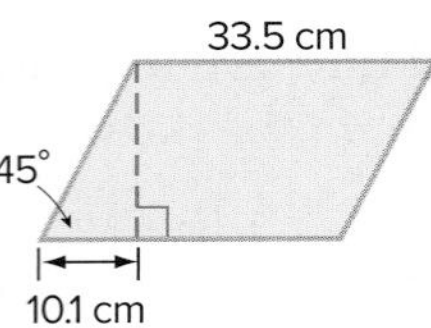

20.
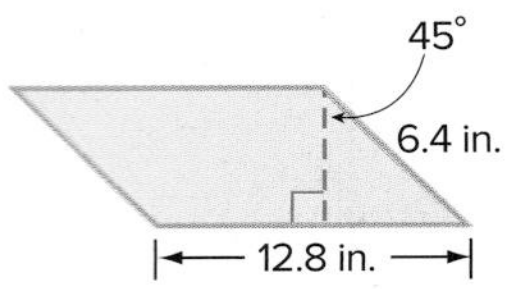

21.
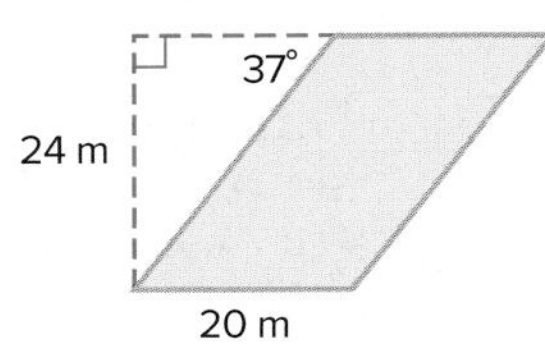

22.
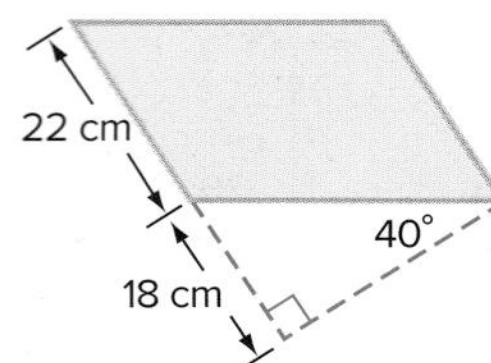

23. **WEATHER** Tornado watch areas are often shown on weather maps using parallelograms. What is the area of the region affected by the tornado watch shown? Round to the nearest square mile.

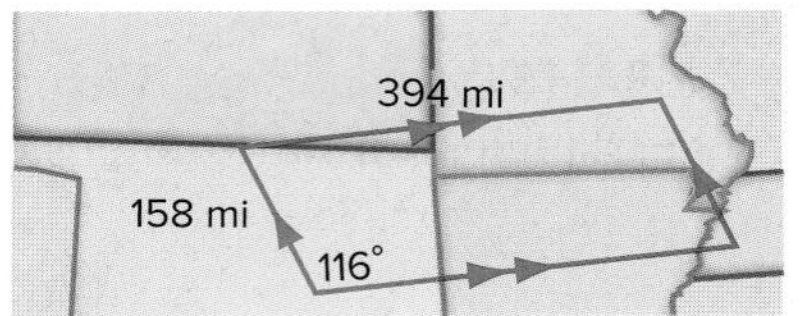

Example 4 **Find the perimeter and area of each figure.**

24. $\triangle DEF$ with $D(4, -4)$, $E(-5, 1)$, and $F(11, -4)$
25. $\triangle RST$ with $R(-8, -2)$, $S(-2, -2)$, and $T(-3, -7)$
26. $\triangle MNP$ with $M(0, 6)$, $N(-2, 8)$, and $P(-2, -1)$
27. $\square ABCD$ with $A(4, 7)$, $B(2, 1)$, $C(8, 1)$, and $D(10, 7)$

28. **FLAGS** Omar wants to make a replica of Guyana's national flag.

a. What is the area of the piece of fabric he will need for the red region? for the yellow region?

b. If the fabric costs $3.99 per square yard for each color and he buys exactly the amount of fabric he needs, how much will it cost to make the flag?

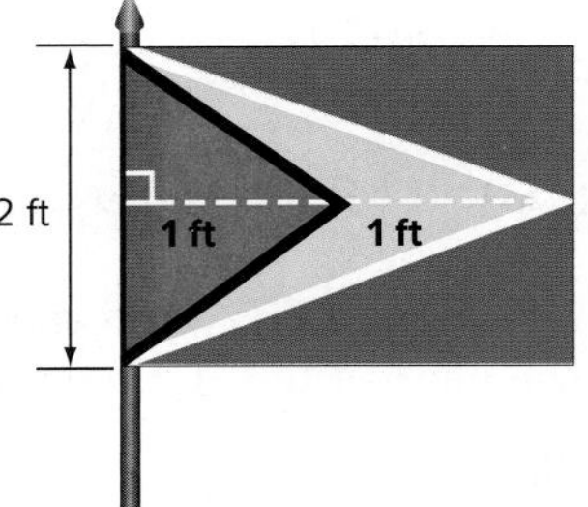

29. **MULTI-STEP** Madison is in charge of the set design for *Romeo and Juliet*. The backdrop shown is 12 feet wide and 20 feet tall and needs 3 coats of paint. The window is 4 feet wide and 1 foot high. The paint store has the following available. One quart of paint covers 87.5 square feet.

Size	8 oz	1 qt	1 gal
Cost ($)	3.75	14	30

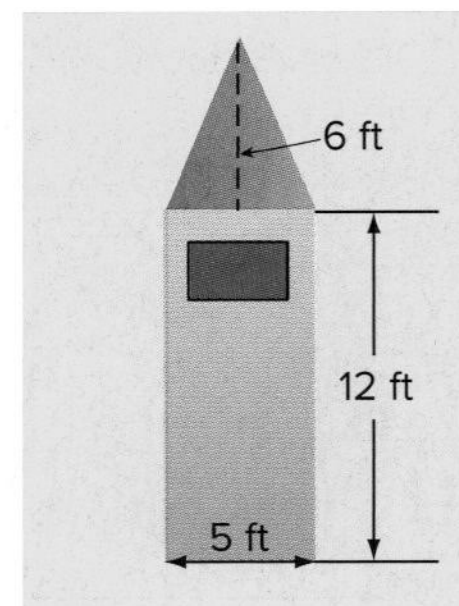

a. What should she buy to minimize cost?

b. Explain your solution process.

Find the perimeter and area of each figure. Round to the nearest hundredth, if necessary.

30.
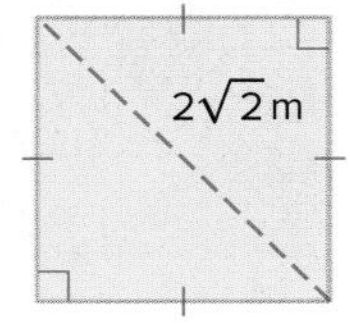

31.
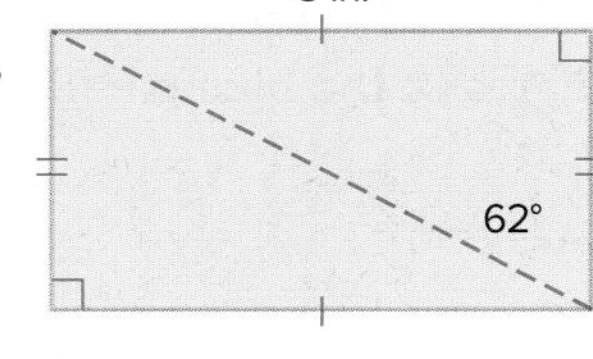

32.
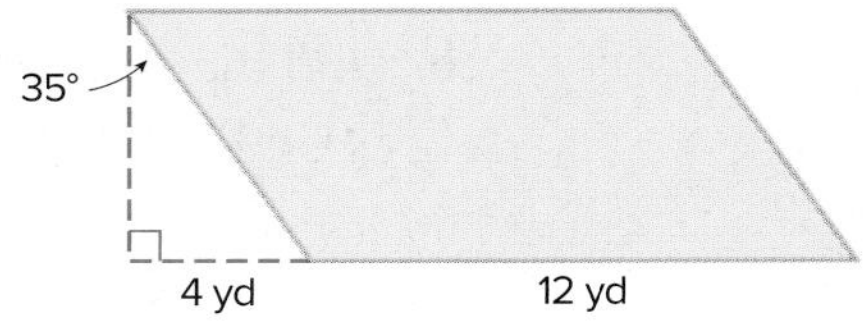

33. **ALGEBRA** The base of a triangle is twice its height. If the area of the triangle is 49 square feet, find its base and height.

34. **ALGEBRA** The height of a triangle is 3 meters less than its base. If the area of the triangle is 44 square meters, find its base and height.

35. **HERON'S FORMULA** Heron's Formula relates the lengths of the sides of a triangle to the area of the triangle. The formula is $A = \sqrt{s(s-a)(s-b)(s-c)}$, where s is the *semiperimeter*, or one half the perimeter of the triangle, and a, b, and c are the side lengths.

   a. Use Heron's Formula to find the area of a triangle with side lengths 7, 10, and 4.

   b. Show that the areas found for a 5-12-13 right triangle are the same using Heron's Formula and using the triangle area formula you learned earlier in this lesson.

36. **MULTIPLE REPRESENTATIONS** In this problem, you will investigate the relationship between the area and perimeter of a rectangle.

   a. **Algebraic** A rectangle has a perimeter of 12 units. If the length of the rectangle is x and the width of the rectangle is y, write equations for the perimeter and area of the rectangle.

   b. **Tabular** Tabulate all possible whole-number values for the length and width of the rectangle, and find the area for each pair.

   c. **Graphical** Graph the area of the rectangle with respect to its length.

   d. **Verbal** Describe how the area of the rectangle changes as its length changes.

   e. **Analytical** For what whole-number values of length and width will the area be greatest? least? Explain your reasoning.

H.O.T. Problems Use Higher-Order Thinking Skills

37. **CHALLENGE** Find the area of $\triangle ABC$ graphed at the right. Explain your method.

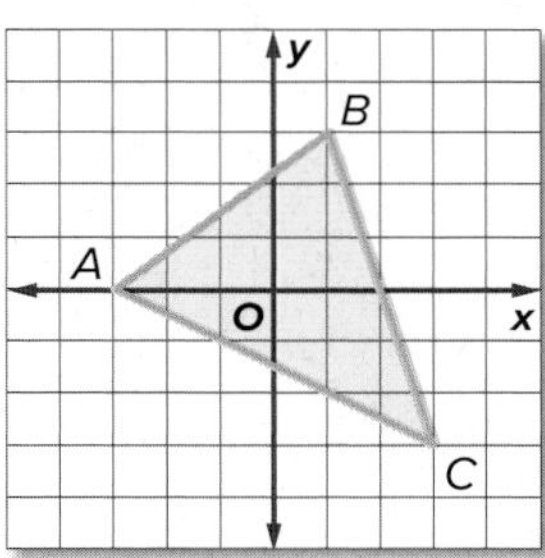

38. **MP CONSTRUCT ARGUMENTS** Will the perimeter of a nonrectangular parallelogram *always, sometimes,* or *never* be greater than the perimeter of a rectangle with the same area and the same height? Explain.

39. **WRITING IN MATH** Points J and L lie on line m. Point K lies on line p. If lines m and p are parallel, describe how the area of $\triangle JKL$ will change as K moves along line p.

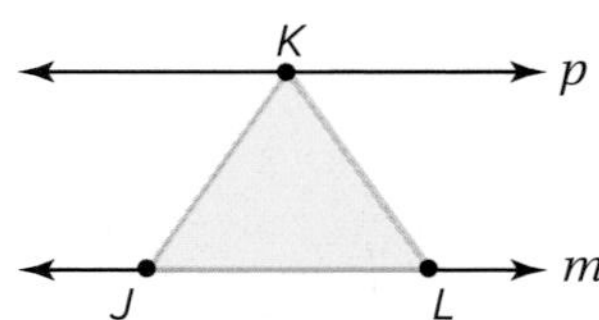

40. **OPEN-ENDED** The area of a polygon is 35 square units. The height is 7 units. Draw three different triangles and three different parallelograms that meet these requirements. Label the base and height on each.

41. **WRITING IN MATH** Describe two different ways you could use measurement to find the area of parallelogram $PQRS$.

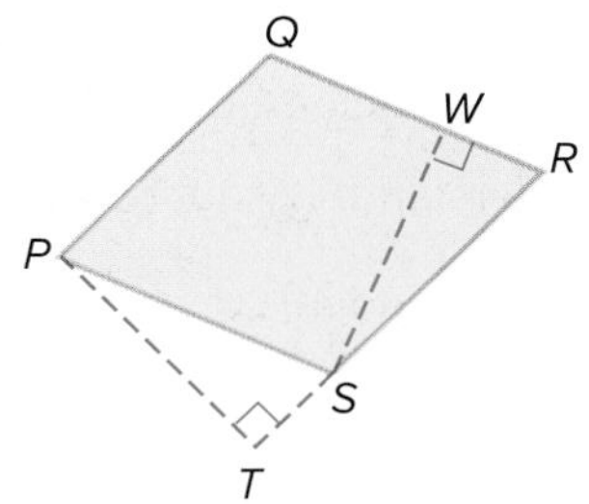

Consider $\triangle BCD$.

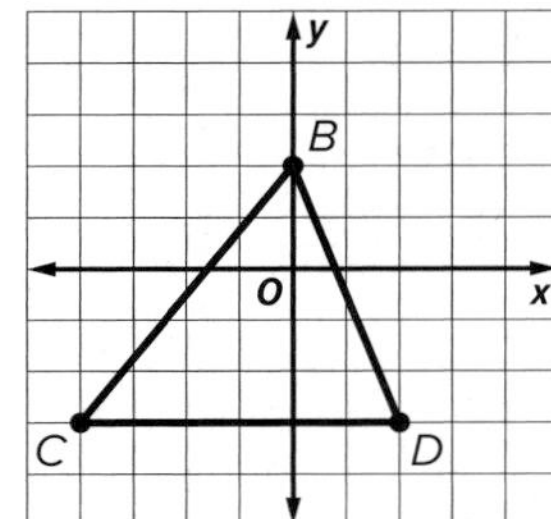

42. Find the perimeter of $\triangle BCD$. If necessary, round to the nearest tenth. MP 1, 7

- A 12.4 units
- B 17.8 units
- C 29.0 units
- D 35.6 units

43. Find the area of $\triangle BCD$. If necessary, round to the nearest tenth. MP 1, 7

- A 15 units²
- B 16.2 units²
- C 17.2 units²
- D 19.2 units²

44. Katie makes coasters by cutting pieces of cardboard into equilateral triangles with the dimensions shown.

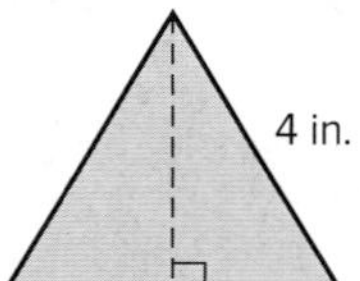

What is the area of each coaster? MP 4

- A $2\sqrt{3}$ in²
- B 4 in²
- C $4\sqrt{2}$ in²
- D $4\sqrt{3}$ in²

45. A parallelogram is in the shape of a rectangle. The parallelogram has a length of 7.5 meters and a width of 6.5 meters. What is the perimeter and area of the parallelogram? MP 1, 7

A 19 meters and 36 square meters

B 21 meters and 40 square meters

C 25 meters and 50 square meters

D 28 meters and 48.75 square meters

46. The area of parallelogram *STUV* is four times the area of parallelogram *MNPQ*.

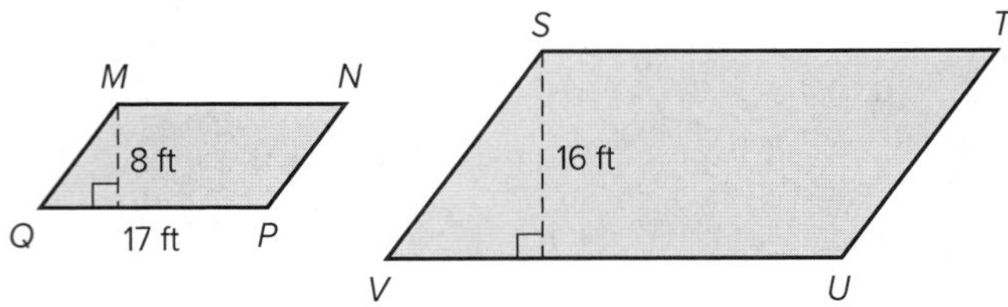

What is the length of $\overline{VU}$? MP 7

- A 68
- B 34
- C 32
- D 25

47. Ben uses software to draw a right triangle with a hypotenuse that has endpoints at (6, 0) and (0, 6). What is the area of the triangle in square units? MP 7

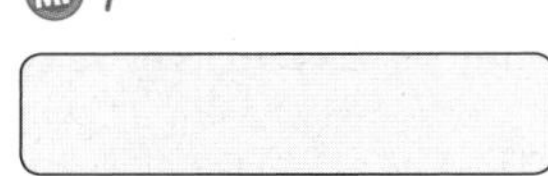

48. A raised garden is shaped like a parallelogram with two sides that meet at a 45° angle. The garden has an area of $84\sqrt{2}$ square feet and a base of 14 feet. What is the perimeter of the garden in feet? MP 1, 4

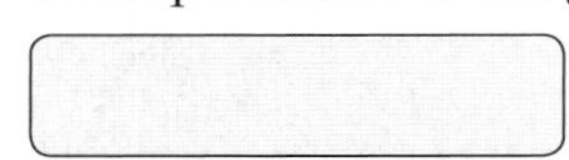

49. **MULTI-STEP** A parallelogram has a base that is six times the length of the height. MP 1, 2

a. What is an expression for the area of the parallelogram in terms of the height, *h*?

b. A triangle has the same area and height *h* as the parallelogram. What is an expression for the base of the triangle?

c. Find the area of the parallelogram if the height is 6.

d. Find the length of the base of a triangle that has the same area and a height of 6.

EXPLORE 10-2

Graphing Technology Lab

Areas of Trapezoids, Rhombi, and Kites

You can use the TI-Nspire Technology to explore special quadrilaterals.

Mathematical Practices

 5 Use appropriate tools strategically.

Activity 1

Work cooperatively.

Step 1 Open a new **Graphs** page. Select **Show Grid** from the **View** menu so that points can be placed at integer coordinates.

Step 2 Select **Line** from the **Points & Lines** menu, and draw a horizontal line.

Step 3 Select **Parallel** from the **Construction** menu to draw a line parallel to your original line through a point with the same *x*-coordinate as a point in Step 2.

Step 4 Place an additional point on the parallel line you just constructed using **Point on** from the **Points & Lines** menu. Label the four points as shown.

Step 5 From the **Shapes** menu, select **Polygon**, and draw a polygon using the four points you created. From the **Actions** menu, select **Attributes,** select the polygon, and increase the line thickness of the polygon.

Step 6 Display the area of the polygon using the **Area** tool from the **Measurement** menu. Move each of the points and observe the effect on the area.

Step 1:

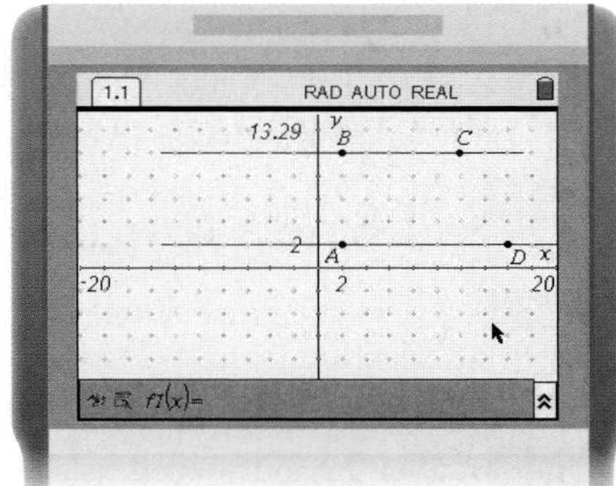

Step 5:

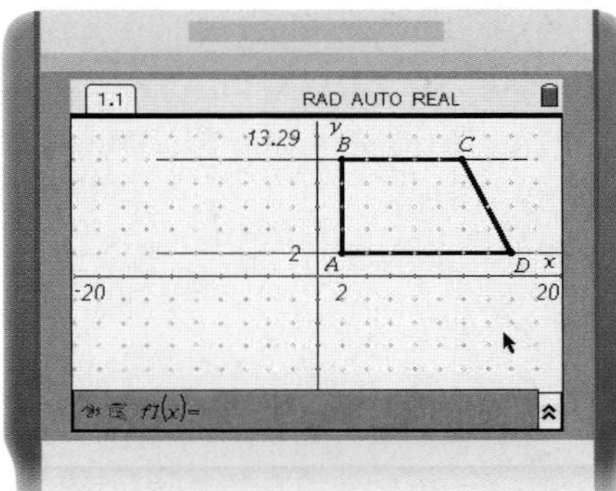

Step 6:

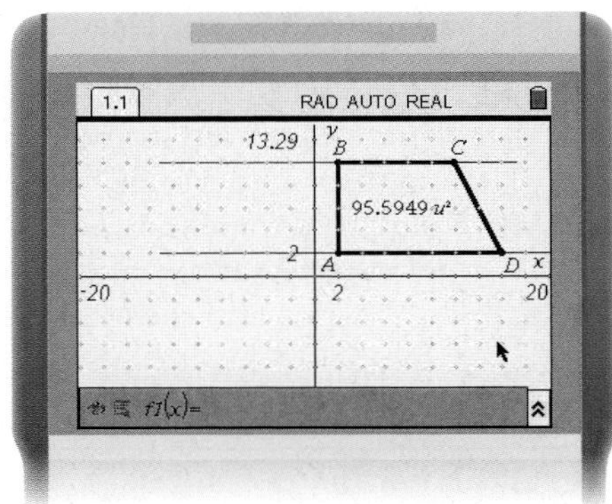

Analyze the Results

Work cooperatively

1. What type of quadrilateral is *ABCD*? Explain your reasoning.
2. **MAKE A CONJECTURE** Using the formulas you learned in Lesson 10-1, make a conjecture about the formula for the area of this type of quadrilateral if *BC* is b_1, *AD* is b_2, and *AB* is h. Explain.

(continued on the next page)

Graphing Technology Lab

Areas of Trapezoids, Rhombi, and Kites *Continued*

Activity 2

Work cooperatively.

Step 1 Open a new **Graphs** page. Select **Show Grid** from the **View** menu so that points can be placed at integer coordinates.

Step 1:

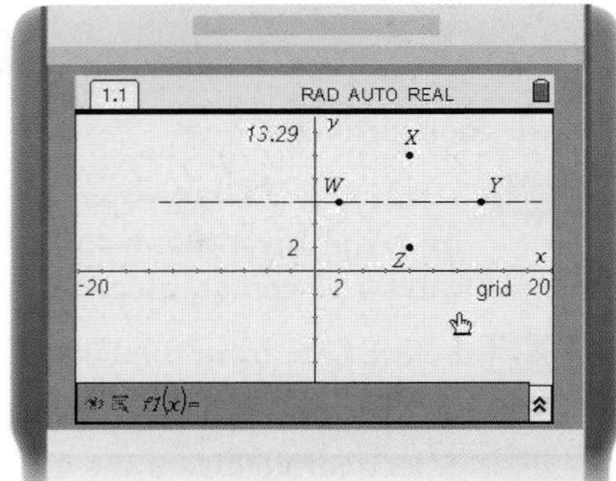

Step 2 Select **Line** from the **Points & Lines** menu, and draw a line.

Step 3 Place a point above the line by selecting **Point** from the **Points & Lines** menu.

Step 4 Reflect the point above the line by choosing **Reflection** from the **Transformation** menu, then select the point and then the line.

Step 5 Label the four points as shown.

Step 6 From the **Shapes** menu, select **Polygon**, and draw a polygon using points W, X, Y, and Z.

Step 6:

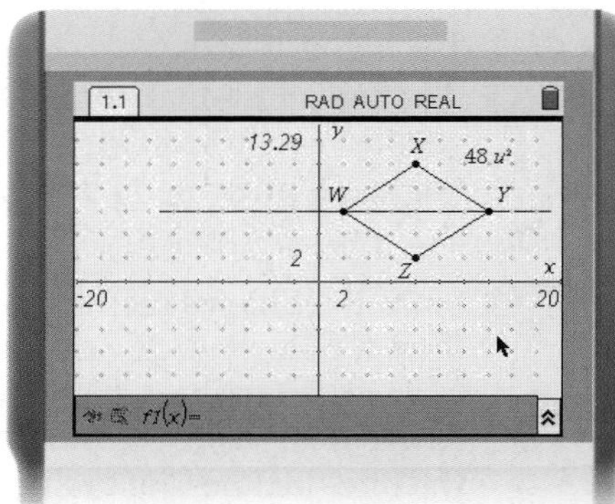

Step 7 Display the area of the polygon using the **Area** tool from the **Measurement** menu. Move points W, X, and Y, and observe the effect on the area.

Step 8 Select **Segment** from the **Points & Lines** menu to draw the diagonals of $WXYZ$.

Step 8:

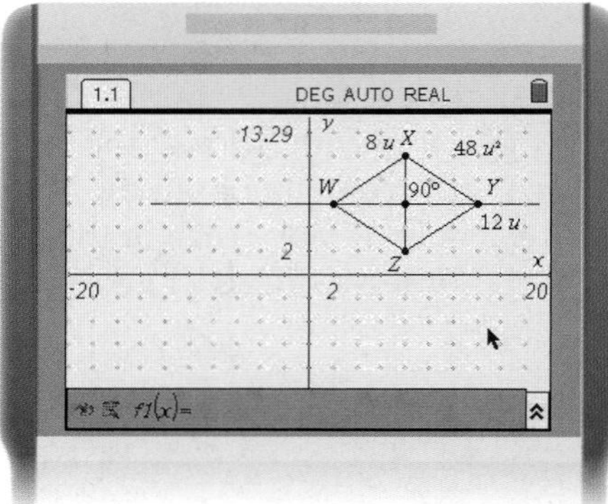

Step 9 Display the lengths of the diagonals using the **Length** tool from the **Measurement** menu, and display the angle between the diagonals using the **Angle** tool. Continue to move points W, X, and Y, and observe the effect on the area and the angle between the diagonals.

Analyze the Results

Work cooperatively

3. What type of quadrilateral is $WXYZ$? Explain your reasoning.
4. **MAKE A CONJECTURE** Using the formulas you learned in Lesson 10-1, develop a formula for the area of this type of quadrilateral. Let WY be d_1, and let XZ be d_2. Explain your reasoning.
5. **CHALLENGE** Construct a quadrilateral using two perpendicular lines and reflecting a point on each as you did in Step 4 of Activity 2. What type of quadrilateral is formed? Does the formula for the area you developed in Exercise 4 apply?

LESSON 2

Areas of Trapezoids, Rhombi, and Kites

Then

- You found areas of triangles and parallelograms.

Now

1. Find areas of trapezoids.
2. Find areas of rhombi and kites.

Why?

- Brianna has turned her hobby of making designer handbags and totes into a small business. Among her designs is a trapezoid-shaped handbag. To estimate the amount of material needed to produce each handbag, she needs to calculate the area of a trapezoid.

New Vocabulary
height of a trapezoid

Mathematical Practices

1 Make sense of problems and persevere in solving them.

7 Look for and make use of structure.

1 Areas of Trapezoids

In Lesson 6-6, you learned that a *trapezoid* is a quadrilateral with exactly one pair of parallel sides. These parallel sides are called *bases*. The **height of a trapezoid** is the perpendicular distance between its bases.

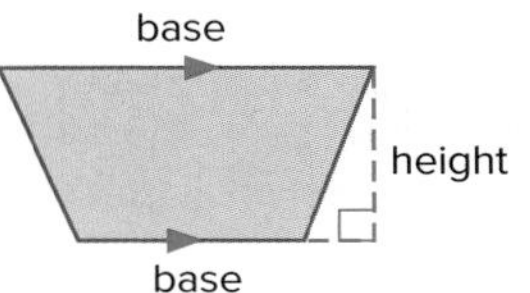

In the figure below, a translation and rotation of the first trapezoid results in two congruent trapezoids that fit together to form a parallelogram.

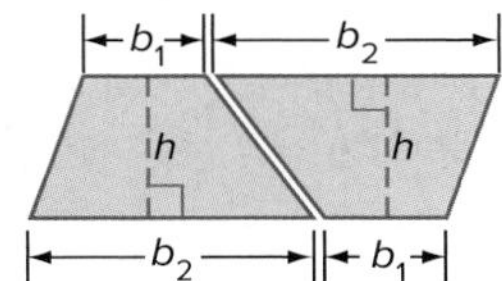

The area of the parallelogram is the product of the height h and the sum of the two bases, b_1 and b_2. The area of one trapezoid is one half the area of the parallelogram.

Key Concept Area of a Trapezoid

Words The area A of a trapezoid is one half the product of the height h and the sum of its bases, b_1 and b_2.

Symbols $A = \frac{1}{2}h(b_1 + b_2)$

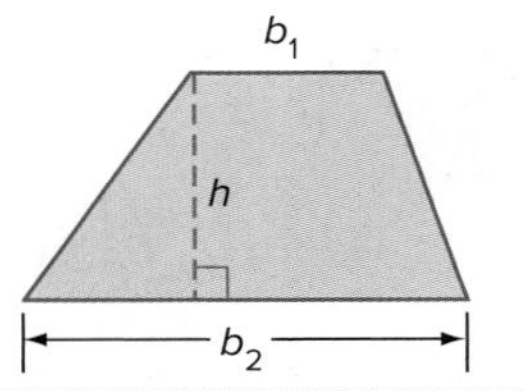

Real-World Example 1 Area of a Trapezoid

CRAFTS One of Brianna's trapezoid-shaped totes is shown. Find the amount of material used to make the side shown.

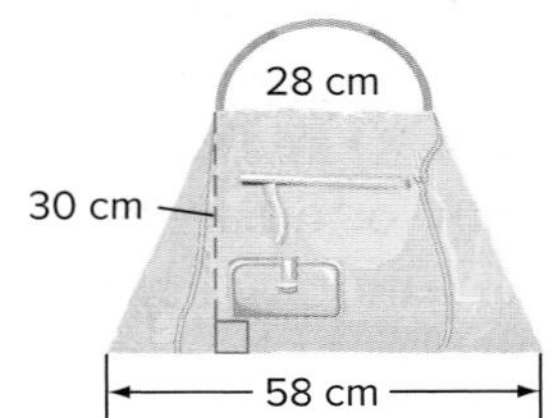

$A = \frac{1}{2}h(b_1 + b_2)$ Area of a trapezoid

$= \frac{1}{2}(30)(28 + 58)$ $h = 30, b_1 = 28, b_2 = 58$

$= 1290$ Simplify.

The tote requires 1290 square centimeters.

Guided Practice

1. **AUTOMOBILES** Find the area of glass used to make the windshield of a van shown at the right.

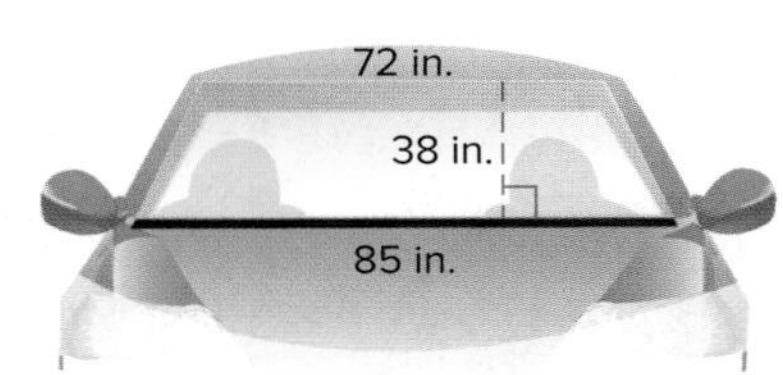

Juice Images/age fotostock

Example 2 Area of a Trapezoid

JEWELRY Emelia designed the pennant shown for her team. Find the area of the shaded portion of her team's pennant.

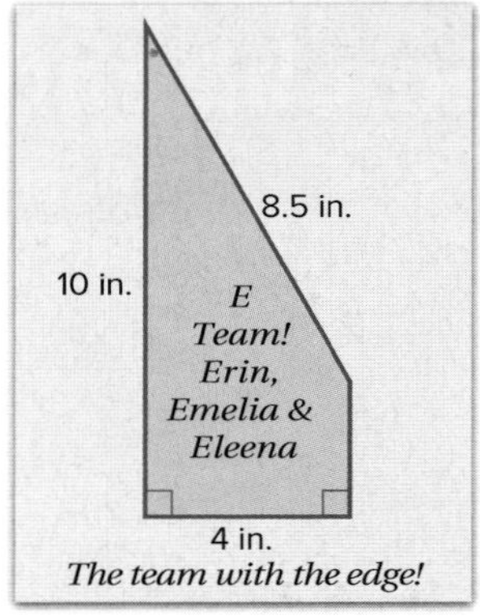

Read the Item

You are given a trapezoid with one base measuring 10 inches, a height of 4 inches, and a third side measuring 8.5 inches. To find the area of the trapezoid, first find the measure of the other base.

Solve the Item

Draw the segment shown to form a right triangle and a rectangle. The triangle has a hypotenuse of 8.5 inches and legs of 4 and ℓ inches. The rectangle has a length of 4 inches and a width of x inches.

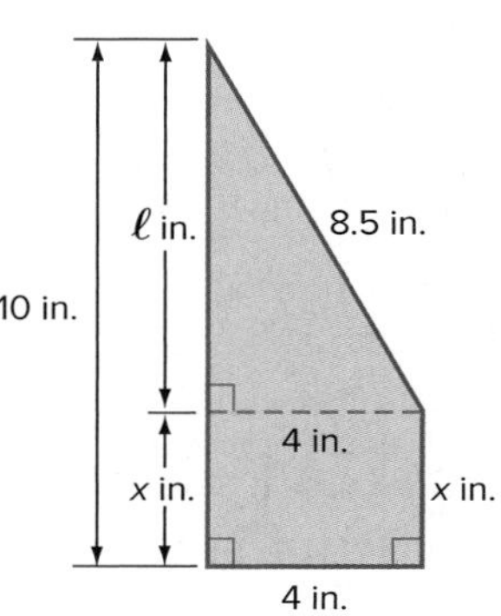

Use the Pythagorean Theorem to find ℓ.

$a^2 + b^2 = c^2$	Pythagorean Theorem
$\ell^2 + 4^2 = 8.5^2$	$a = \ell$, $b = 4$, and $c = 8.5$
$\ell^2 + 16 = 72.25$	Simplify.
$\ell^2 = 56.25$	Subtract 16 from each side.
$\ell = 7.5$	Take the positive square root of each side.

By Segment Addition, $\ell + x = 10$. So, $7.5 + x = 10$ and $x = 2.5$. The width of the rectangle is also the measure of the second base of the trapezoid.

$A = \frac{1}{2}h(b_1 + b_2)$	Area of a trapezoid
$= \frac{1}{2}(4)(10 + 2.5)$	$h = 4$, $b_1 = 10$, and $b_2 = 2.5$
$= 25$	Simplify.

So the pennant has an area of 25 square inches.

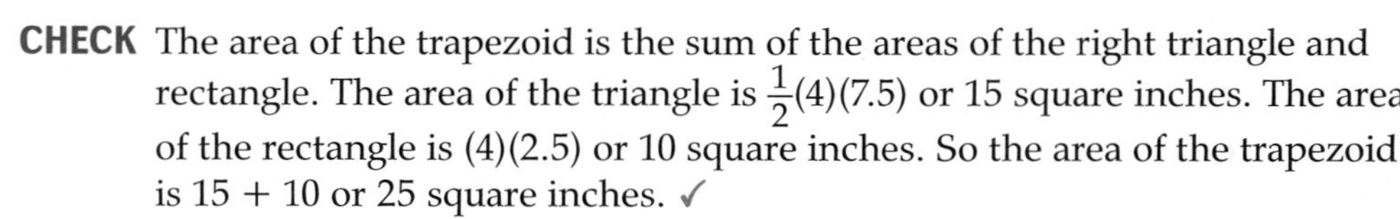

CHECK The area of the trapezoid is the sum of the areas of the right triangle and rectangle. The area of the triangle is $\frac{1}{2}(4)(7.5)$ or 15 square inches. The area of the rectangle is $(4)(2.5)$ or 10 square inches. So the area of the trapezoid is $15 + 10$ or 25 square inches. ✓

Study Tip

MP Modeling To solve some area problems, you need to draw in parallel and/or perpendicular lines to find information not provided.

Real-World Career

Craft Artist Craft artists create their art by hand to sell or exhibit. They work with a wide variety of materials including textiles, woods, metal, and ceramics.

Most artists receive some type of postsecondary training, and about 63% are self-employed. Craft artists make up about 3% of all artists.

Guided Practice

2. **JEWELRY** Owen designed the silver earrings shown that are shaped like isosceles trapezoids. What is the area of each earring?

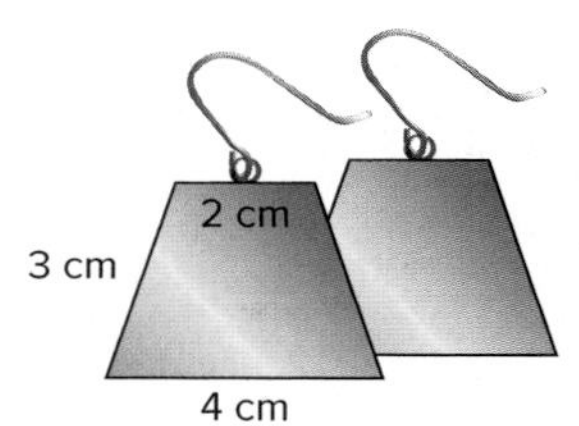

2 Areas of Rhombi and Kites

Recall from Lessons 6-5 and 6-6 that a *rhombus* is a parallelogram with all four sides congruent and a *kite* is a quadrilateral with exactly two pairs of consecutive congruent sides.

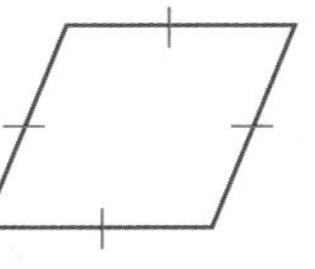

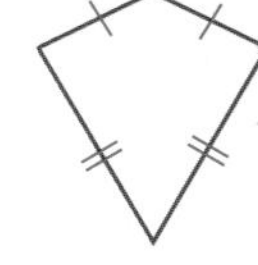

Review Vocabulary

diagonal a segment that connects any two nonconsecutive vertices in a polygoninverses.

The areas of rhombi and kites are related to the lengths of their diagonals.

Key Concept Area of a Rhombus or Kite

Words The area A of a rhombus or kite is one half the product of the lengths of its diagonals, d_1 and d_2.

Symbols $A = \frac{1}{2}d_1d_2$

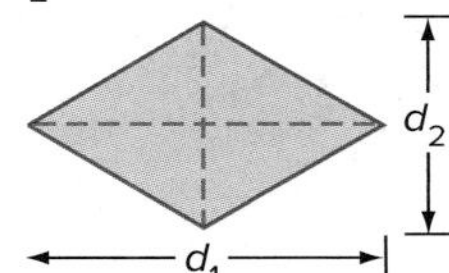

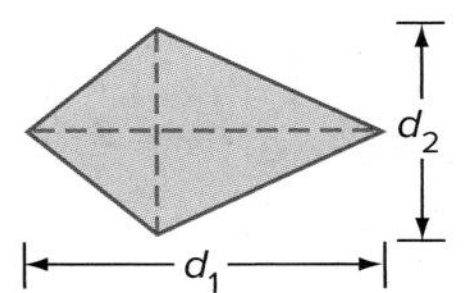

You will derive the formulas for the area of a kite and the area of a rhombus in Exercises 23 and 24.

Example 3 Area of a Rhombus and a Kite

Find the area of each rhombus or kite.

a.

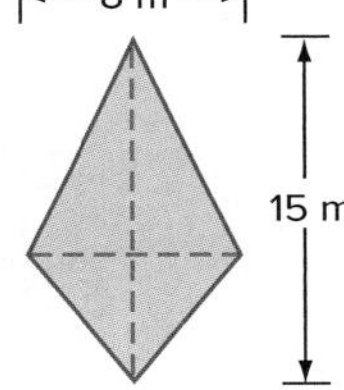

$A = \frac{1}{2}d_1d_2$ Area of a kite

$= \frac{1}{2}(8)(15)$ $d_1 = 8$ and $d_2 = 15$

$= 60 \text{ m}^2$ Simplify.

b.

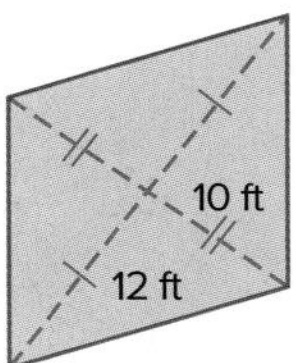

Step 1 Find the length of each diagonal.

Because the diagonals of a rhombus bisect each other, the lengths of the diagonals are 12 + 12 or 24 feet and 10 + 10 or 20 feet.

Step 2 Find the area of the rhombus.

$A = \frac{1}{2}d_1d_2$ Area of a rhombus

$= \frac{1}{2}(24)(20)$ $d_1 = 24$ and $d_2 = 20$

$= 240 \text{ ft}^2$ Simplify.

Guided Practice

Find the area of each rhombus or kite.

3A.

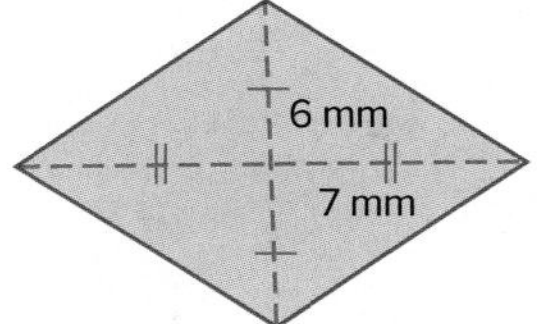

3B.

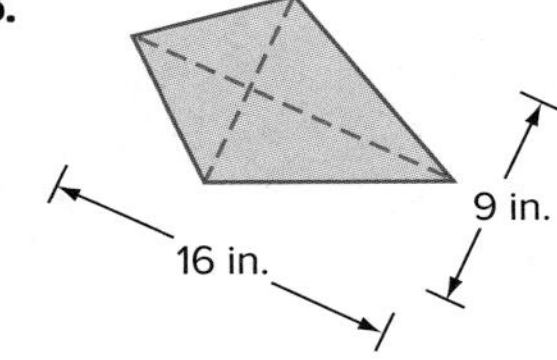

Math History Link

Heron of Alexandria (c. 10–70 A.D.) Heron was a mathematician and engineer in Roman Egypt. He developed a formula for finding the area of a triangle if the lengths of the sides are known.

You can use algebra to solve for unknown measures in trapezoids, rhombi, and kites.

Example 4 Use Area to Find Missing Measures

ALGEBRA One diagonal of a rhombus is twice as long as the other diagonal. If the area of the rhombus is 169 square millimeters, what are the lengths of the diagonals?

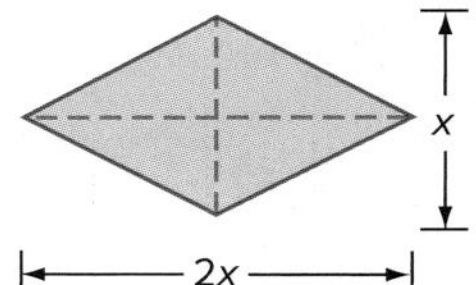

Step 1 Write an expression to represent each measure.

Let x represent the length of one diagonal. Then the length of the other diagonal is $2x$.

Step 2 Use the formula for the area of a rhombus to find x.

$A = \frac{1}{2}d_1d_2$	Area of a rhombus
$169 = \frac{1}{2}(x)(2x)$	$A = 169$, $d_1 = x$, and $d_2 = 2x$
$169 = x^2$	Simplify.
$13 = x$	Take the positive square root of each side.

So the lengths of the diagonals are 13 millimeters and 2(13) or 26 millimeters.

Go Online!

The area formulas in this Concept Summary are important to remember. Log into your **eStudent Edition** to bookmark this page.

Guided Practice

ALGEBRA Find x.

4A. $A = 92 \text{ in}^2$

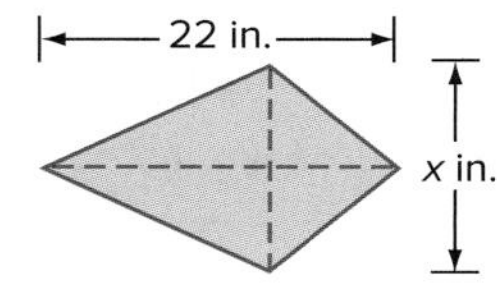

4B. $A = 177 \text{ cm}^2$

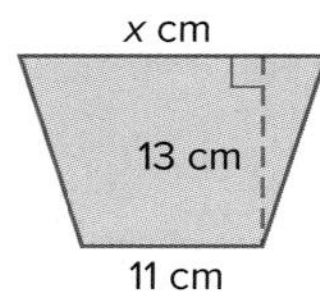

4C. ALGEBRA What is the area of the kite shown?

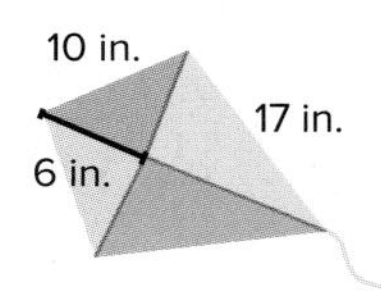

Concept Summary Areas of Polygons

Parallelogram	Triangles	Trapezoids	Rhombi and Kites
$A = bh$	$A = \frac{1}{2}bh$	$A = \frac{1}{2}h(b_1 + b_2)$	$A = \frac{1}{2}d_1d_2$

Check Your Understanding

◯ = Step-by-Step Solutions begin on page R13.

Go Online! for a Self-Check Quiz

Examples 1–3 **Find the area of each trapezoid, rhombus, or kite.**

1. 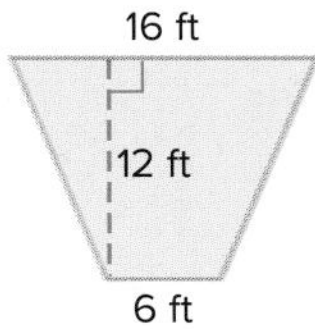

2. 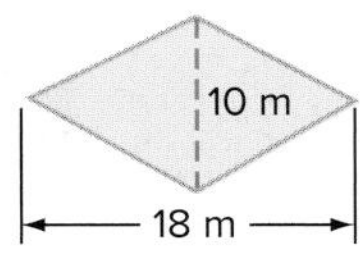

3. 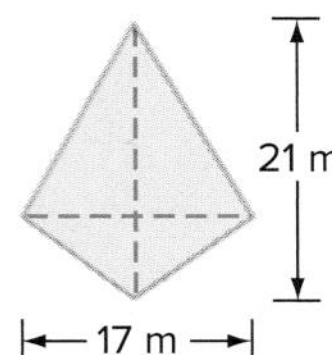

4. **PEP RALLY** Suki is designing posters for the Homecoming game. Her design is shown at the right. What is the area of the poster in square feet?

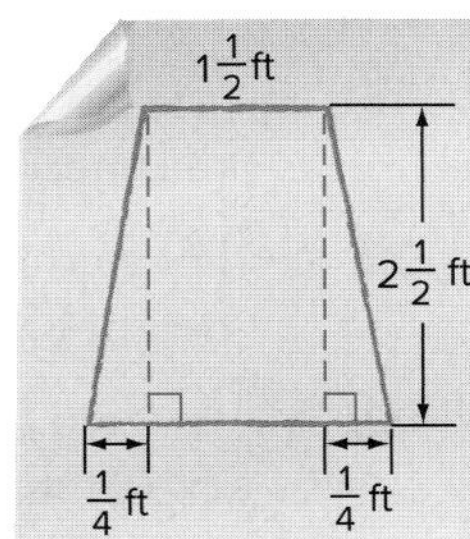

Example 4 **ALGEBRA** Find x.

5. $A = 78 \text{ cm}^2$

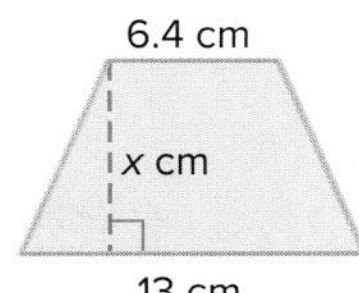

6. $A = 96 \text{ in}^2$

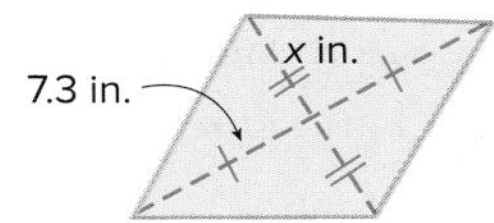

7. $A = 104 \text{ ft}^2$

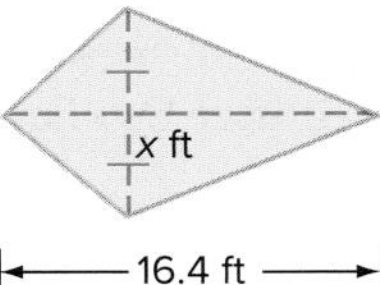

Practice and Problem Solving

Extra Practice is on page R10.

Examples 1–3 **STRUCTURE** **Find the area of each trapezoid, rhombus, or kite.**

8. 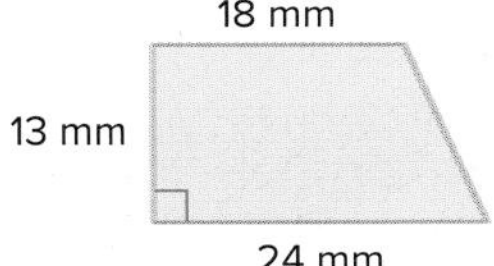

9. 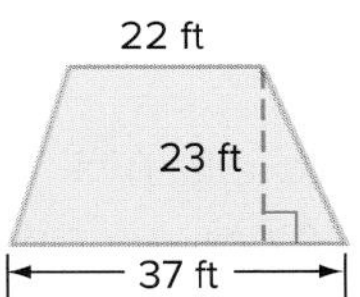

10. 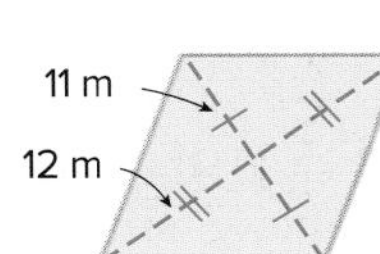

11. 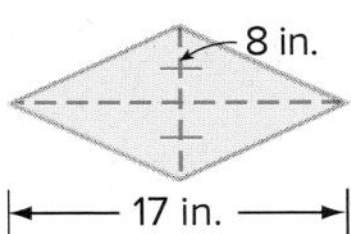

12. 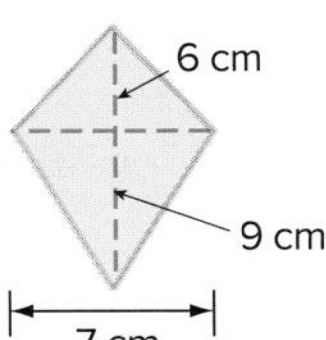

13. 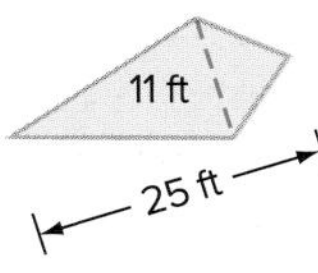

MICROSCOPES **Find the area of the identified portion of each magnified image. Assume that the identified portion is either a trapezoid, rhombus, or kite. Measures are provided in microns.**

14. human skin

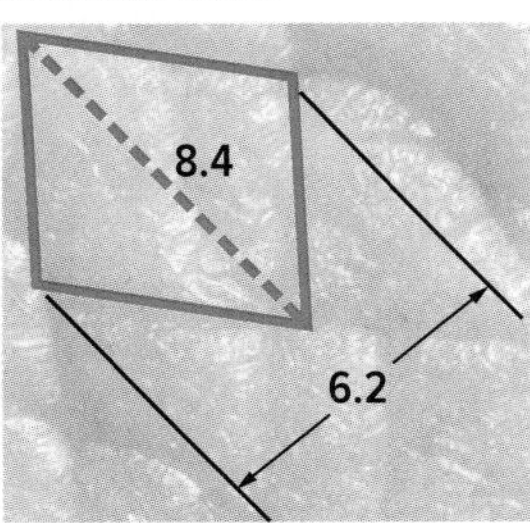

15. heartleaf plant

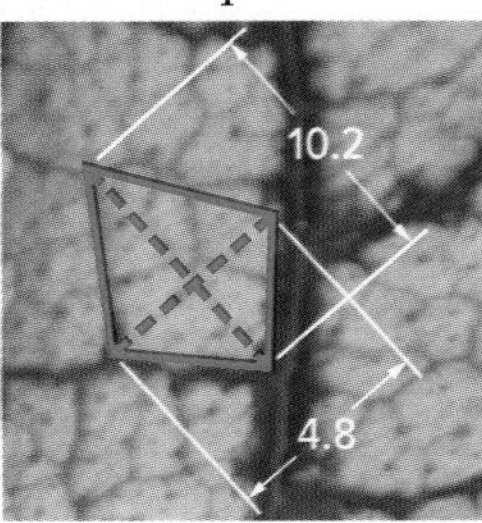

16. eye of a fly

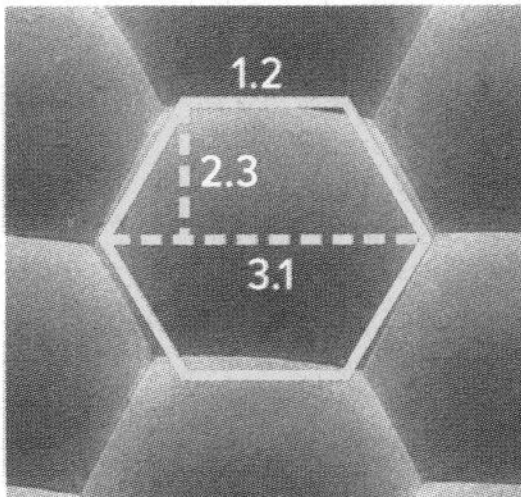

17. **JOBS** Jimmy works on his neighbors' yards after school to earn extra money to buy a car. He is going to plant grass seed in Mr. Troyer's yard. What is the area of the yard?

Example 4 **ALGEBRA Find each missing length.**

18. One diagonal of a kite is twice as long as the other diagonal. If the area of the kite is 240 square inches, what are the lengths of the diagonals?

19. The area of a rhombus is 168 square centimeters. If one diagonal is three times as long as the other, what are the lengths of the diagonals?

20. A trapezoid has base lengths of 12 and 14 feet with an area of 322 square feet. What is the height of the trapezoid?

21. A trapezoid has a height of 8 meters, a base length of 12 meters, and an area of 64 square meters. What is the length of the other base?

22. **HONORS** Estella has been asked to join an honor society at school. Before the first meeting, new members are asked to sand and stain the front side of a piece of wood in the shape of an isosceles trapezoid. What is the surface area that Estella will need to sand and stain?

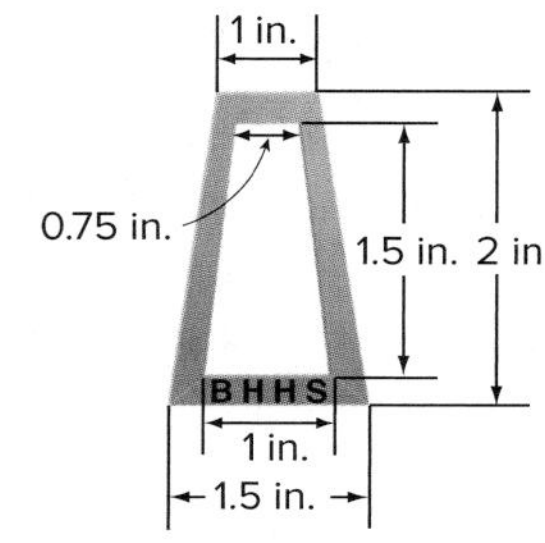

For each figure, provide a justification showing that $A = \frac{1}{2}d_1d_2$.

23.

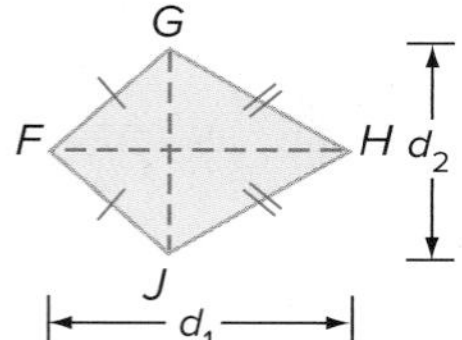

24.

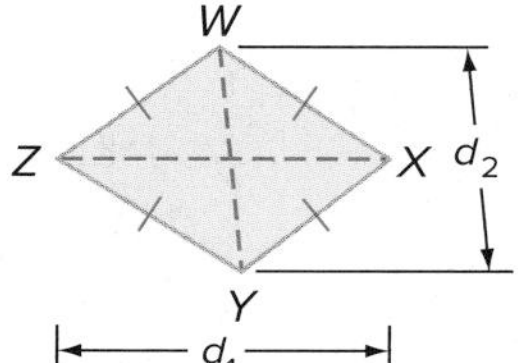

25. **CRAFTS** Ashanti is competing in a kite festival. The yellow, red, orange, green, and blue pieces of her kite design shown are congruent rhombi.

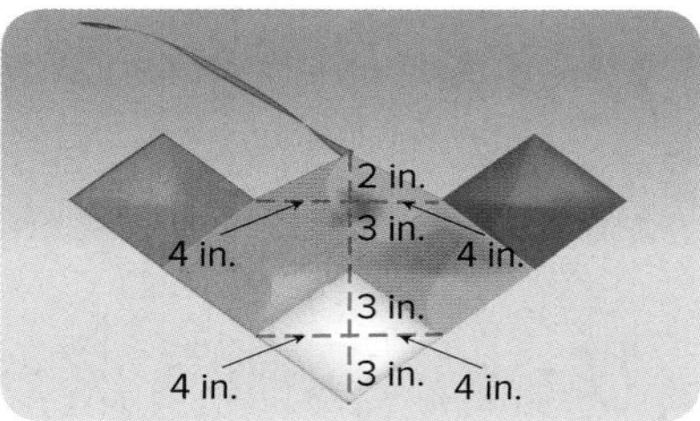

a. How much fabric of each color does she need to buy?

b. Ashanti wants the total area of her kite to be no greater than 200 square inches. Does her kite meet this requirement? Explain.

MP SENSE-MAKING Find the area of each quadrilateral with the given vertices.

26. $A(-8, 6)$, $B(-5, 8)$, $C(-2, 6)$, and $D(-5, 0)$

27. $W(3, 0)$, $X(0, 3)$, $Y(-3, 0)$, and $Z(0, -3)$

28. **METALS** When magnified in very powerful microscopes, some metals are composed of grains that have various polygonal shapes.

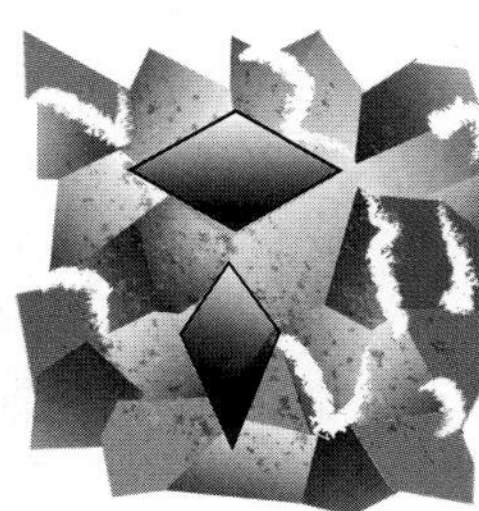

a. What is the area of figure 1 if the grain has a height of 4 microns and bases with lengths of 5 and 6 microns?

b. If figure 2 has perpendicular diagonal lengths of 3.8 microns and 4.9 microns, what is the area of the grain?

29. **PROOF** The figure at the right is a trapezoid that consists of two congruent right triangles and an isosceles right triangle. In 1876, James A. Garfield, the 20th president of the United States, discovered a proof of the Pythagorean Theorem using this diagram. Prove that $x^2 + y^2 = z^2$.

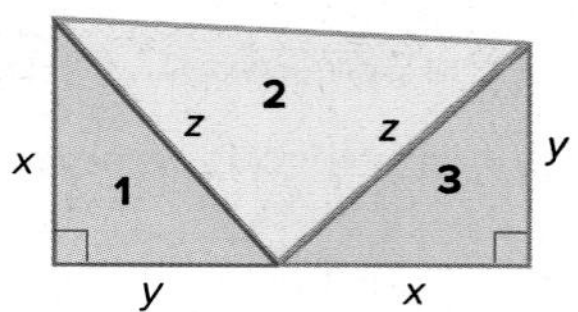

DIMENSIONAL ANALYSIS Find the perimeter and area of each figure in feet. Round to the nearest tenth, if necessary.

30.

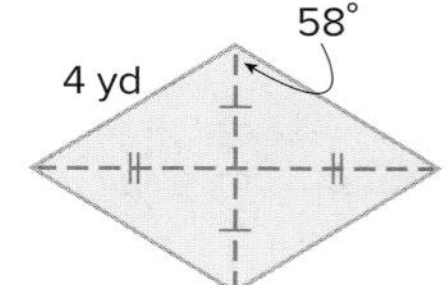

31.

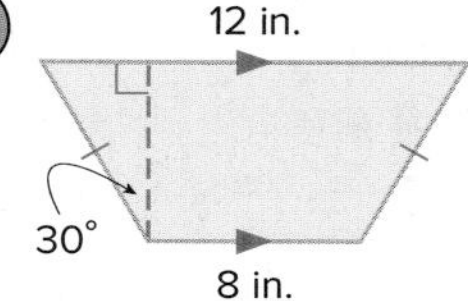

32.

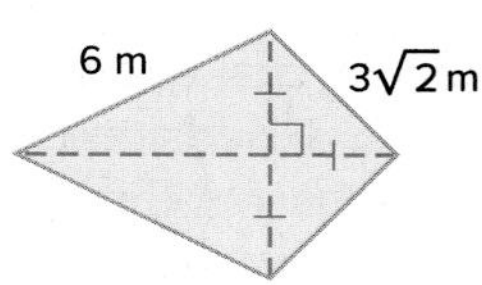

33. **MULTIPLE REPRESENTATIONS** In this problem, you will investigate perimeters of kites.

a. **Geometric** Draw a kite like the one shown if $x = 2$.

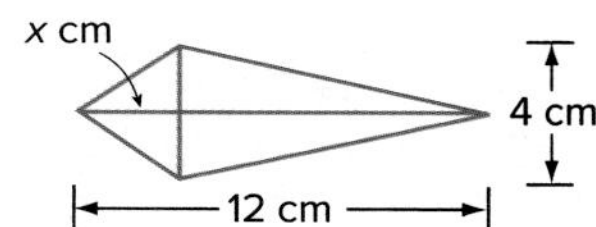

b. **Geometric** Repeat the process in part **a** for three x-values between 2 and 10 and for an x-value of 10. The overall length of the kite should remain 12 centimeters.

c. **Tabular** Measure and record in a table the perimeter of each kite, along with the x-value.

d. **Graphical** Graph the perimeter versus the x-value using the data from your table.

e. **Analytical** Make a conjecture about the value of x that will minimize the perimeter of the kite. What is the significance of this value?

H.O.T. Problems Use Higher-Order Thinking Skills

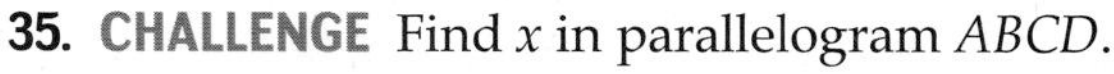

34. **CRITIQUE ARGUMENTS** Antonio and Madeline want to draw a trapezoid that has a height of 4 units and an area of 18 square units. Antonio says that only one trapezoid will meet the criteria. Madeline disagrees and thinks that she can draw several different trapezoids with a height of 4 units and an area of 18 square units. Is either of them correct? Explain your reasoning.

35. **CHALLENGE** Find x in parallelogram $ABCD$.

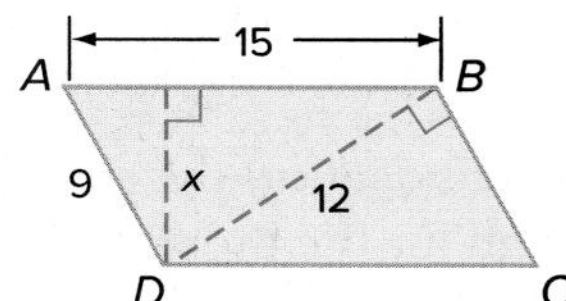

36. **OPEN-ENDED** Draw a kite and a rhombus with an area of 6 square inches. Label and justify your drawings.

37. **REASONING** If the areas of two rhombi are equal, are the perimeters *sometimes*, *always*, or *never* equal? Explain.

38. **WRITING IN MATH** How can you use trigonometry to find the area of a figure?

Preparing for Assessment

39. The Jays' team logo is a rhombus, as shown.

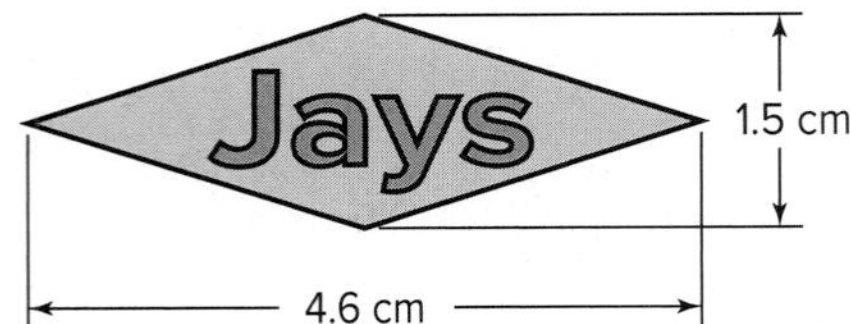

The team's manager makes an enlargement of the logo in which each dimension is 4 times larger than the version shown here. Which of the following best describes how the areas of the logos compare? MP 1, 4, 7

○ **A** The area of the enlargement is 2 times the area of the original.

○ **B** The area of the enlargement is 4 times the area of the original.

○ **C** The area of the enlargement is 8 times the area of the original.

○ **D** The area of the enlargement is 16 times the area of the original.

40. Li Mei drew the trapezoid shown below as the plan for a new concrete driveway apron. The contractor will charge \$9.50 per square foot to install the concrete. The homeowner will seal the concrete with sealer that costs \$40 for a container that covers 80 square feet.

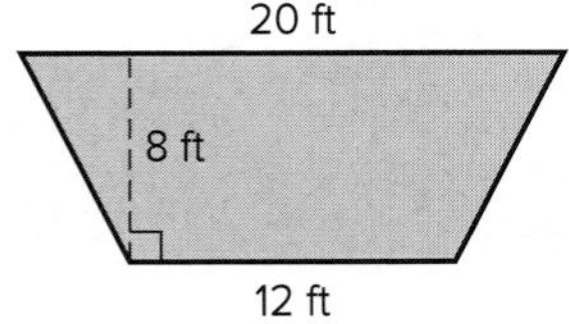

Which of the following statements about the driveway apron are true? Select all that apply. MP 1, 4

☐ **A** It will cost \$1216 for the concrete.

☐ **B** The homeowner will need four containers of sealer to seal the concrete.

☐ **C** A \$1300 budget for the concrete and sealer is enough for the project.

☐ **D** The area of concrete is 288 square feet.

41. One diagonal of a rhombus is four times as long as the other diagonal. If the area of the rhombus is 288 square meters, what are the lengths of the diagonals? MP 1, 7

42. The figure shows a kite that Alex created using geometry software.

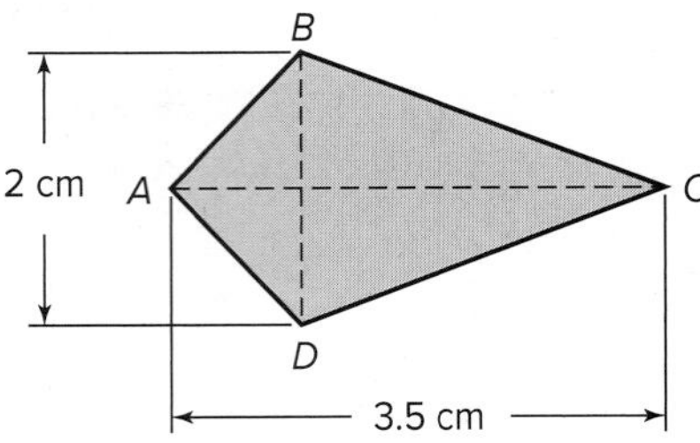

Which of the following will result in a kite with twice the area of *ABCD*? MP 1, 4, 7

I. Double the length of each diagonal.

II. Double the length of $\overline{BD}$.

III. Double the length of $\overline{AC}$.

○ **A** I only

○ **B** II only

○ **C** III only

○ **D** II and III

43. The sides of a soccer goal are in the form of a trapezoid. The bottom base of the trapezoid is 6.5 feet long, the top base of the trapezoid is 4.25 feet long, and the height of the trapezoid is 8 feet high. Find the area of one side of the soccer goal. MP 1, 4, 7

A 38 square feet

B 40 square feet

C 43 square feet

D 50 square feet

44. **MULTI-STEP** The area of a kite is 50 square inches. MP 1, 4, 7

a. If each diagonal is a whole number, how many distinct diagonal pairs have whole number lengths?

b. Which diagonal length, in inches, will guarantee that the kite is also a rhombus?

c. If the area of the kite is doubled, how many distinct diagonal pairs have whole number lengths?

LESSON 3

Areas of Circles and Sectors

Then

- You found the circumference of a circle.

Now

1. Find areas of circles.
2. Find areas of sectors of circles.

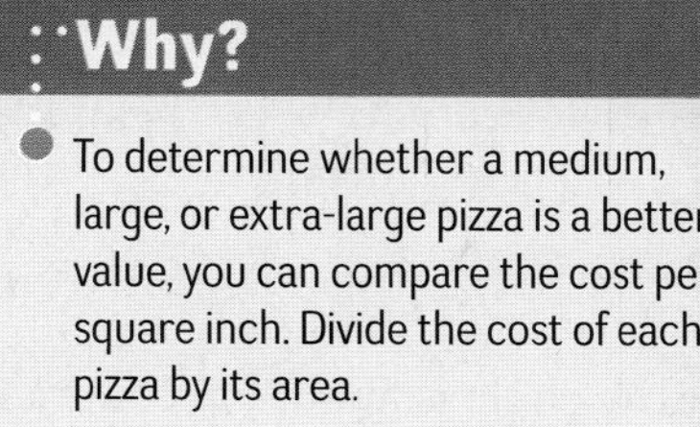

Why?

- To determine whether a medium, large, or extra-large pizza is a better value, you can compare the cost per square inch. Divide the cost of each pizza by its area.

New Vocabulary

sector of a circle
segment of a circle

Mathematical Practices

1 Make sense of problems and persevere in solving them.
6 Attend to precision.

1 Areas of Circles

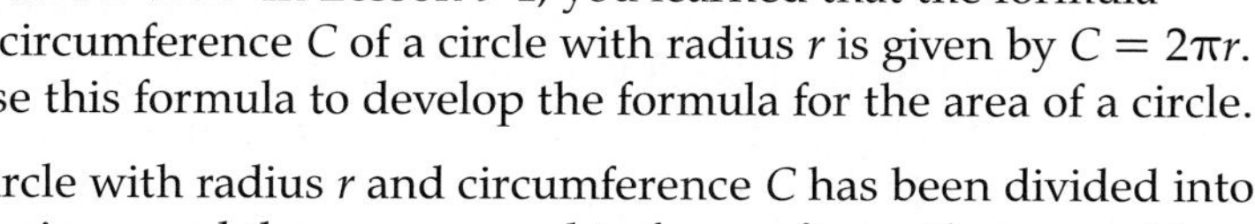

In Lesson 9-1, you learned that the formula for the circumference C of a circle with radius r is given by $C = 2\pi r$. You can use this formula to develop the formula for the area of a circle.

Below, a circle with radius r and circumference C has been divided into congruent pieces and then rearranged to form a figure that resembles a parallelogram.

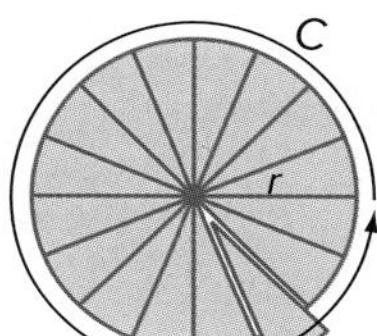

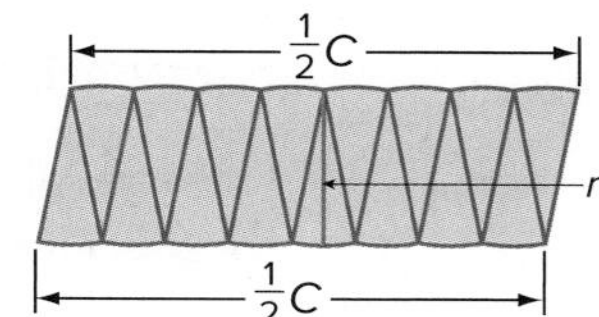

As the number of congruent pieces increases, the rearranged figure more closely approaches a parallelogram. The base of the parallelogram is $\frac{1}{2}C$ and the height is r, so its area is $\frac{1}{2}C \cdot r$. Because $C = 2\pi r$, the area of the parallelogram is also $\frac{1}{2}(2\pi r)r$ or πr^2.

Key Concept Area of a Circle

Words The area A of a circle is equal to π times the square of the radius r.

Symbols $A = \pi r^2$

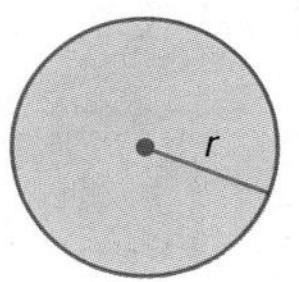

Real-World Example 1 Area of a Circle

SPORTS What is the area of the circular putting green shown to the nearest square foot?

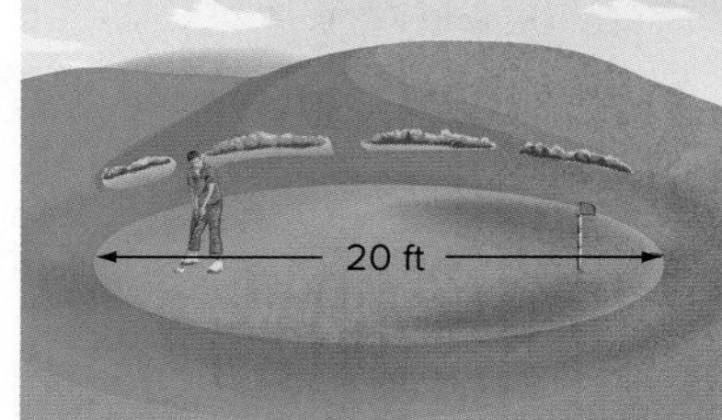

The diameter is 20 feet, so the radius is 10 feet.

$A = \pi r^2$	Area of a circle
$= \pi(10)^2$	$r = 10$
≈ 314	Use a calculator.

So, the area is about 314 square feet.

Guided Practice

1. **SPORTS** An archery target has a radius of 12 inches. What is the area of the target to the nearest square inch?

Example 2 Use the Area of a Circle to Find a Missing Measure

ALGEBRA Find the radius of a circle with an area of 95 square centimeters.

$A = \pi r^2$	Area of a circle
$95 = \pi r^2$	$A = 95$
$\frac{95}{\pi} = r^2$	Divide each side by π.
$5.5 \approx r$	Use a calculator. Take the positive square root of each side.

The radius of the circle is about 5.5 centimeters.

Guided Practice

2. **ALGEBRA** The area of a circle is 196π square yards. Find the diameter.

Review Vocabulary

central angle an angle with a vertex in the center of a circle and with sides that contain two radii of the circle

arc a portion of a circle defined by two endpoints

2 Areas of Sectors

A slice of a circular pizza is an example of a sector of a circle. A **sector of a circle** is a region of a circle bounded by a central angle and its intercepted major or minor arc. The formula for the area of a sector is similar to the formula for arc length.

Key Concept Area of a Sector

The ratio of the **area A of a sector** to the **area of the whole circle, πr^2**, is equal to the ratio of the **degree measure of the intercepted arc x** to 360.

Proportion: $\frac{A}{\pi r^2} = \frac{x}{360}$

Equation: $A = \frac{x}{360} \cdot \pi r^2$

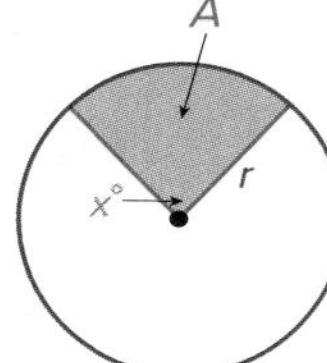

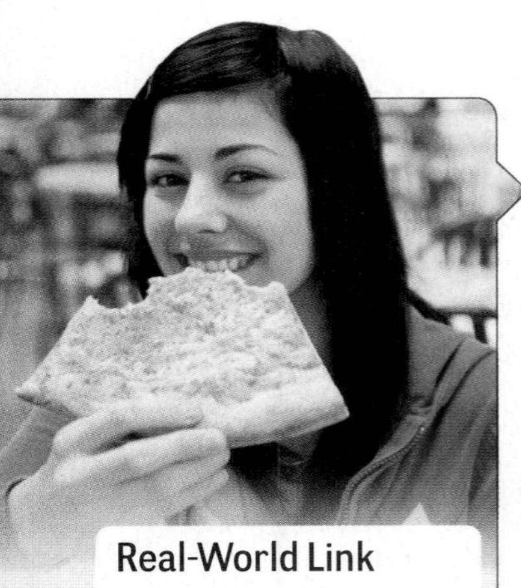

Real-World Link

About 3 billion pizzas are sold each year in the United States. That is equivalent to about 46 slices per person annually.

Source: Statistic Brain

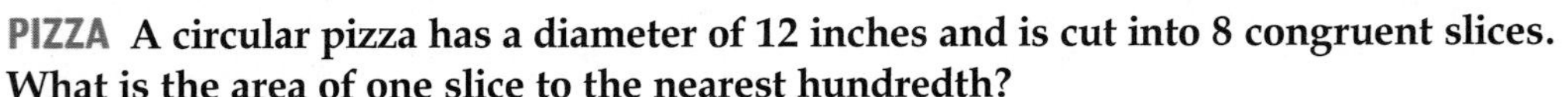

Real-World Example 3 Area of a Sector

PIZZA A circular pizza has a diameter of 12 inches and is cut into 8 congruent slices. What is the area of one slice to the nearest hundredth?

Step 1 Find the arc measure of a pizza slice.

Because the pizza is equally divided into 8 slices, each slice will have an arc measure of $360 \div 8$ or 45.

Step 2 Find the radius of the pizza. Use this measure to find the area of the sector, or slice.

The diameter is 12 inches, so the radius is 6 inches.

$A = \frac{x}{360} \cdot \pi r^2$	Area of a sector
$= \frac{45}{360} \cdot \pi(6)^2$	$x = 45$ and $r = 6$
≈ 14.14	Use a calculator.

So, the area of one slice of this pizza is about 14.14 square inches.

Go Online!

Watch Personal Tutor videos to hear descriptions of solving problems involving areas of circles and sectors. Try to describe how to solve a problem for a partner. Have them ask you questions to help your understanding.

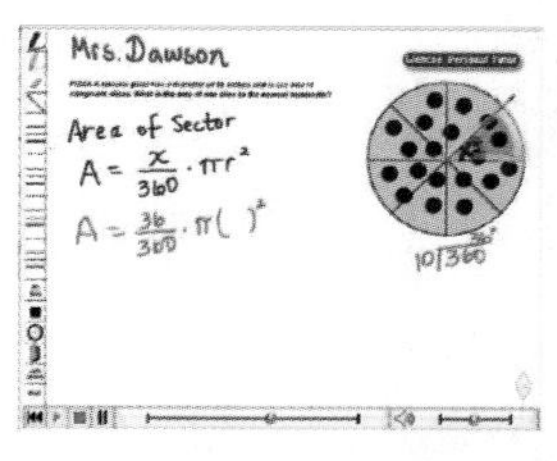

Guided Practice

Find the area of the shaded sector. Round to the nearest tenth.

3A.

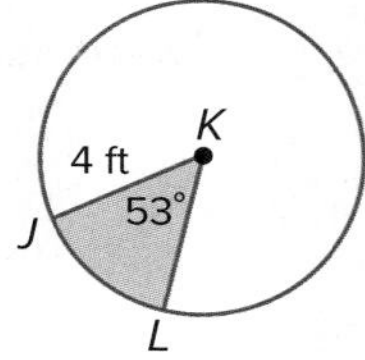

3B.

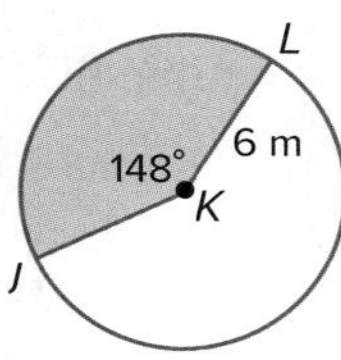

3C.

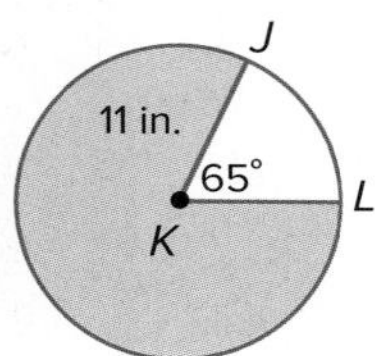

3D. **CRAFTS** The color wheel at the right is a tool that artists use to organize color schemes. If the diameter of the wheel is 10 inches and each of the 12 sections is congruent, find the approximate area covered by green hues.

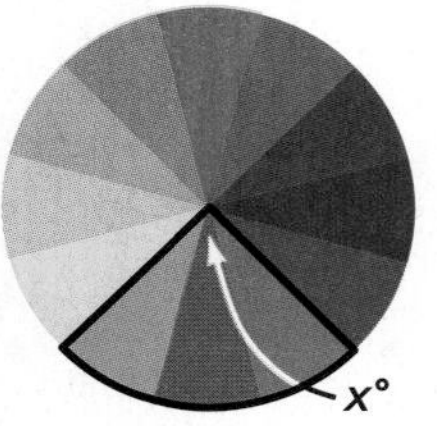

Check Your Understanding

 = Step-by-Step Solutions begin on page R13.

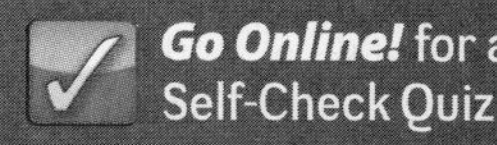

Example 1 **CONSTRUCTION** **Find the area of each circle. Round to the nearest tenth.**

1.

2.

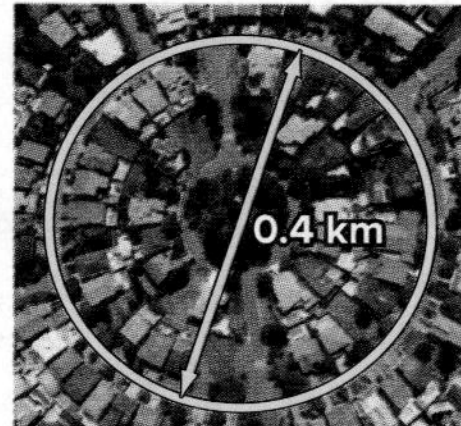

Example 2 **Find the indicated measure. Round to the nearest tenth.**

3. Find the diameter of a circle with an area of 74 square millimeters.

4. The area of a circle is 88 square inches. Find the radius.

Example 3 **Find the area of each shaded sector. Round to the nearest tenth.**

5.

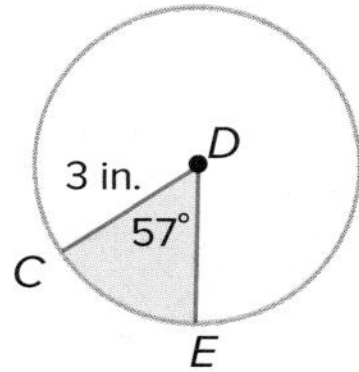

6.

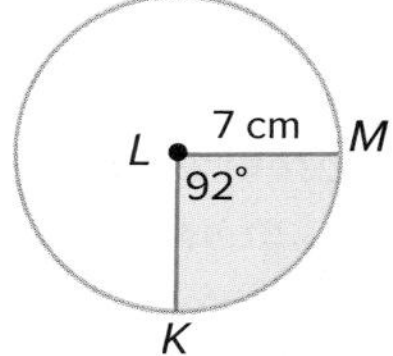

7. **BAKING** Chelsea is baking pies for a fundraiser at her school. She divides each 9-inch pie into 6 equal slices.

a. What is the area, in square inches, for each slice of pie?

b. If each slice costs $0.25 to make and she sells 8 pies at $1.25 for each slice, how much money will she raise?

Practice and Problem Solving

Extra Practice is on page R10.

Example 1

MODELING Find the area of each circle. Round to the nearest tenth.

8.
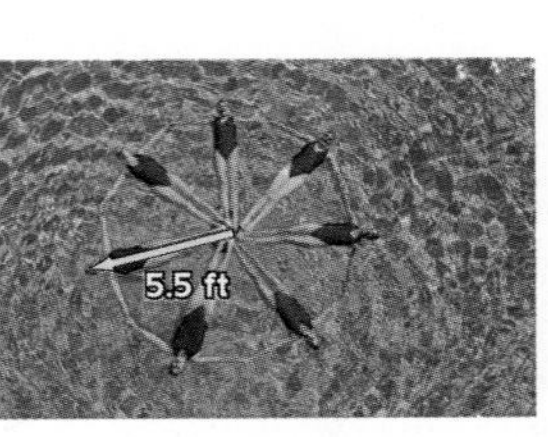

9.
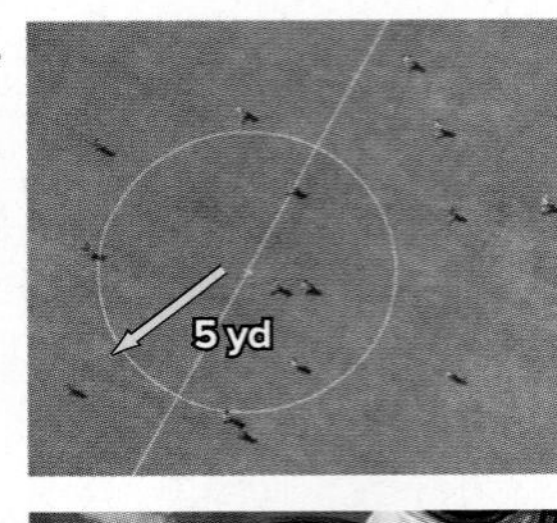

10.
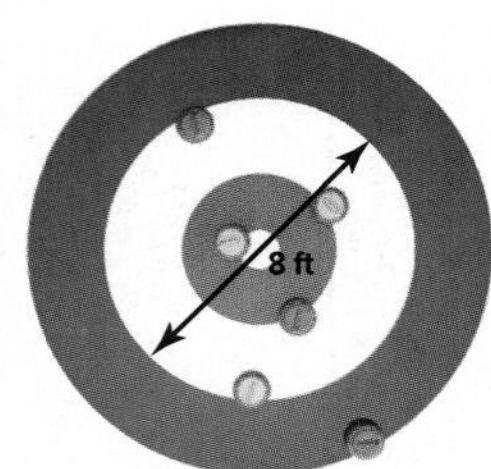

11.

12.
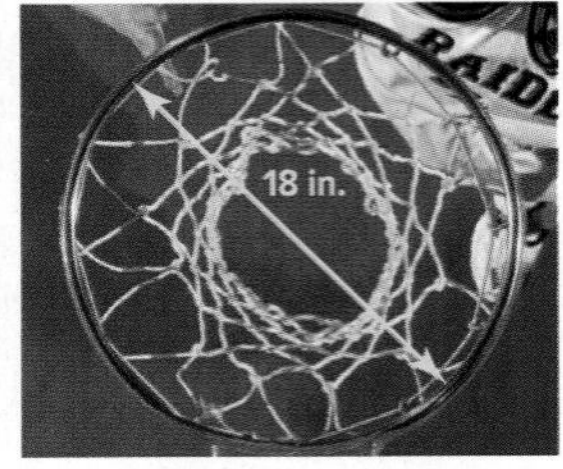

13.
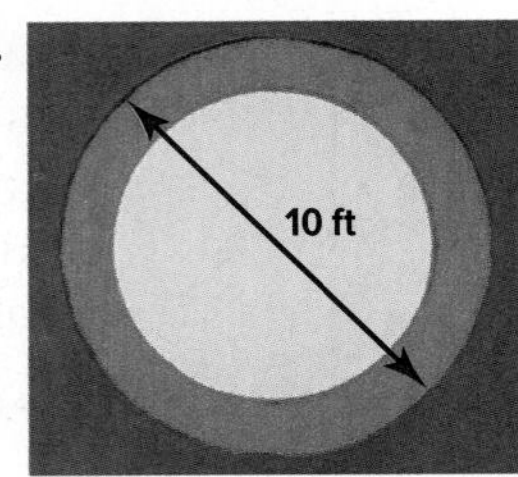

Example 2

Find the indicated measure. Round to the nearest tenth, if necessary.

14. The area of a circle is 68 square centimeters. Find the diameter.
15. Find the diameter of a circle with an area of 94 square millimeters.
16. The area of a circle is 112 square inches. Find the radius.
17. Find the radius of a circle with an area of 206 square feet.

Example 3

Find the area of each shaded sector. Round to the nearest tenth, if necessary.

18.
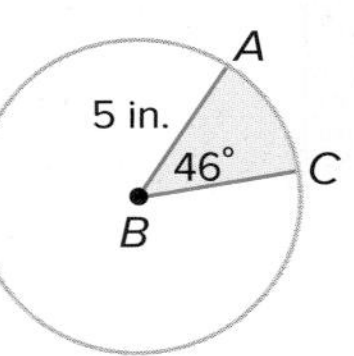

19.
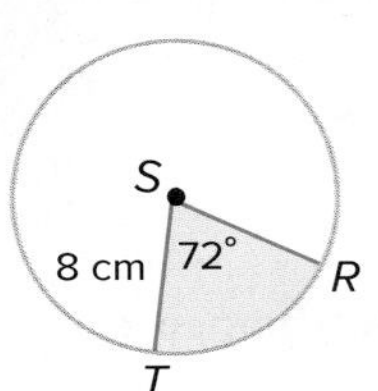

20.
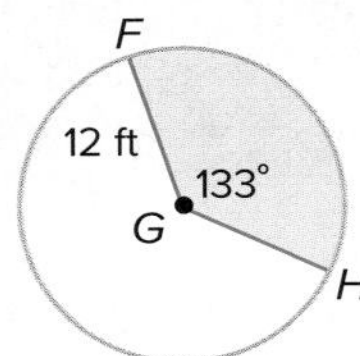

21.
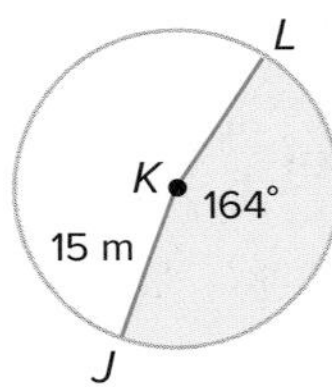

22.
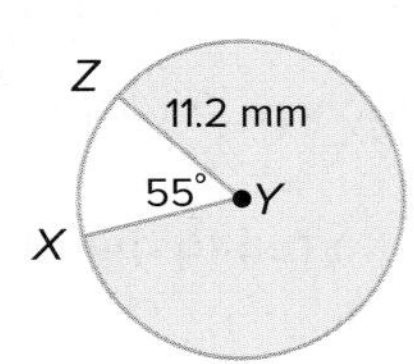

23.
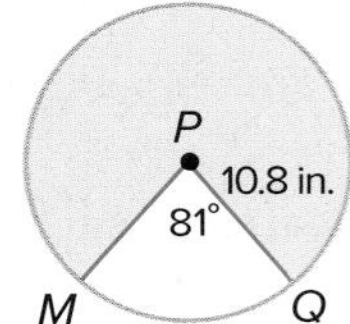

24. **MUSIC** The music preferences of students at Thomas Jefferson High are shown in the circle graph. Find the area of each sector and the degree measure of each intercepted arc if the radius of the circle is 1 unit.

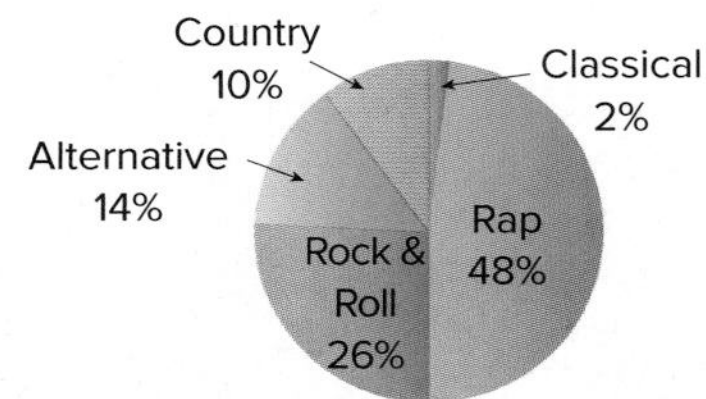

25. **JEWELRY** A jeweler makes a pair of earrings by cutting two 50° sectors from a silver disk.

a. Find the area of each sector.

b. If the weight of the silver disk is 2.3 grams, how many milligrams does the silver wedge for each earring weigh?

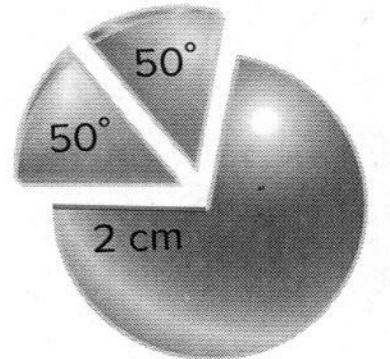

(tl) ©moodboard/Alamy; (tc) Stephan Zirwes/Getty Images; (tr) MileA/iStock/Getty Images; (bl) Kelly Redinger/Design Pics; (bc) PCN Photography/Alamy; (br) RunLenarun/Shutterstock.com

26. PROM Students voted on their favorite prom theme.

a. Create a circle graph with a diameter of 2 inches to represent these data.

b. Find the area of each theme's sector in your graph. Round to the nearest hundredth of an inch.

Theme	Percent
An Evening of Stars	11
Mardi Gras	32
Springtime in Paris	8
Night in Times Square	47
Undecided	2

SENSE-MAKING The area A of each shaded region is given. Find x.

27. $A = 66\text{ cm}^2$

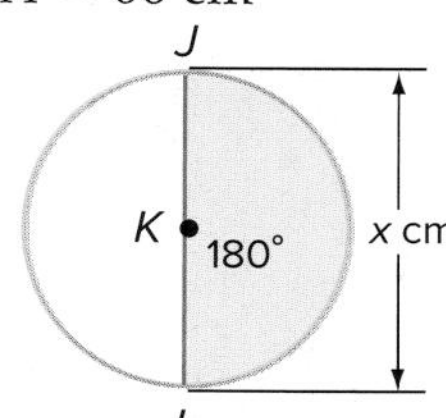

28. $A = 94\text{ in}^2$

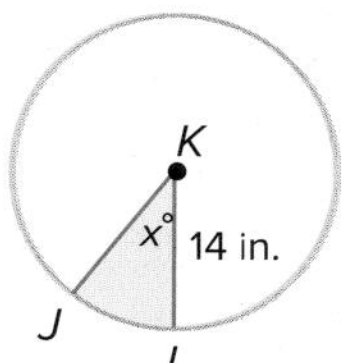

29. $A = 128\text{ ft}^2$

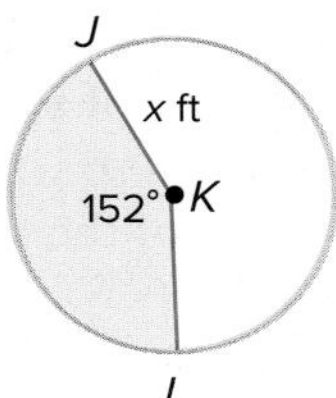

30. MULTI-STEP Luna is organizing a banquet for the Honor Society, and she needs 13 tablecloths for the round tables in the hall. The area of each table is approximately 29.27 square feet. She can rent tablecloths for $16 each or she can make them herself. Her local fabric store carries three different bolts of suitable fabric. The standard bolt is 60 inches wide and 100 yards long and costs $75. The wide bolt is 81 inches wide, 25 yards long, and costs $125. The extra-wide bolt is 90 inches wide, 25 yards long, and costs $150. Each tablecloth should cover the table with 9 inches of overhang.

a. How can Luna minimize the cost of the tablecloths?

b. Explain your reasoning.

c. What assumptions did you make?

31. TREES The age of a living tree can be determined by multiplying the diameter of the tree by its growth factor, or rate of growth.

a. What is the diameter of a live oak tree with a circumference of 36 feet?

b. If the growth factor of the live oak tree is 130, what is the age of the tree?

Find the area of the shaded region. Round to the nearest tenth.

32.

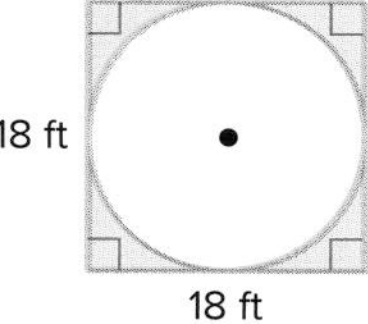

33.

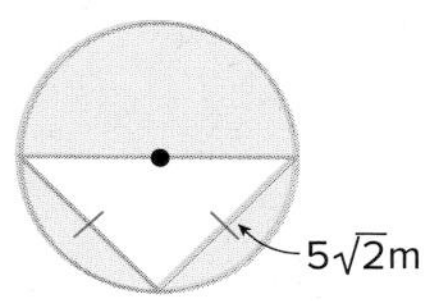

34.

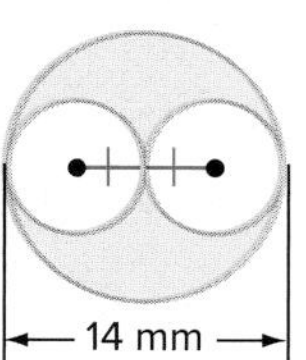

35.

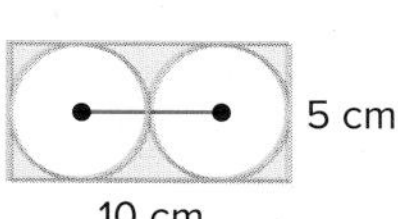

36.

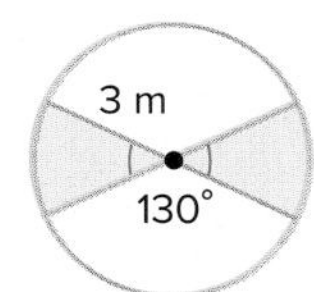

37.

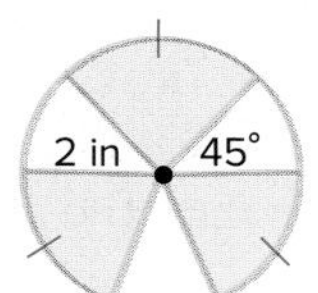

38. COORDINATE GEOMETRY What is the area of sector ABC shown on the graph?

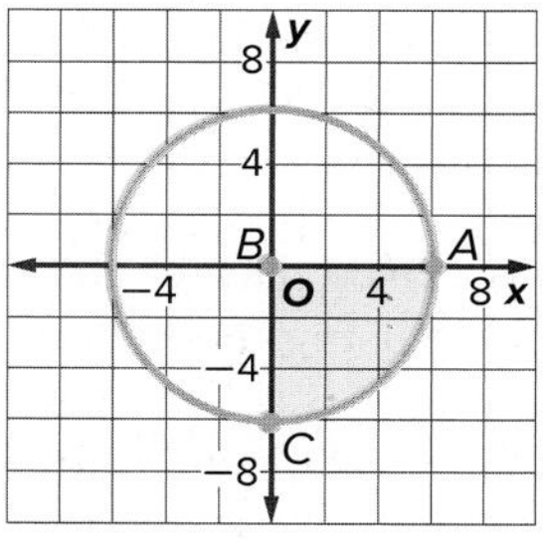

39. ALGEBRA The figure shown below is a sector of a circle. If the perimeter of the figure is 22 millimeters, find its area in square millimeters

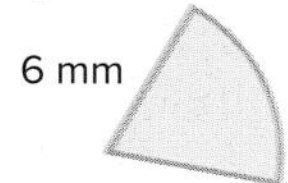

Find the area of each shaded region.

40.

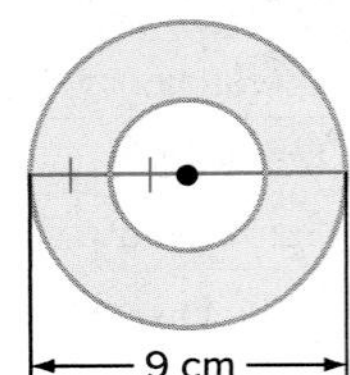

41

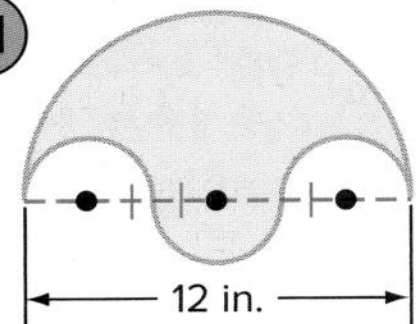

42.

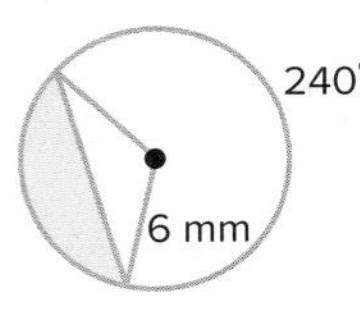

43. MULTIPLE REPRESENTATIONS In this problem, you will investigate segments of circles. A **segment of a circle** is the region bounded by an arc and a chord.

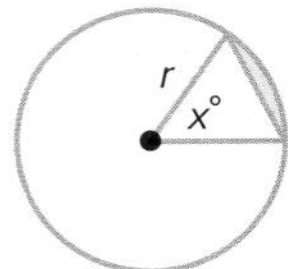

a. **Algebraic** Write an equation for the area A of a segment of a circle with a radius r and a central angle of $x°$. (*Hint*: Use trigonometry to find the base and height of the triangle.)

b. **Tabular** Calculate and record in a table ten values of A for x-values ranging from 10 to 90 if r is 12 inches. Round to the nearest tenth.

c. **Graphical** Graph the data from your table with the x-values on the horizontal axis and the A-values on the vertical axis.

d. **Analytical** Use your graph to predict the value of A when x is 63. Then use the formula you generated in part **a** to calculate the value of A when x is 63. How do the values compare?

H.O.T. Problems Use Higher-Order Thinking Skills

44. ERROR ANALYSIS Kristen and Chase want to find the area of the shaded region in the circle shown. Is either of them correct? Explain your reasoning.

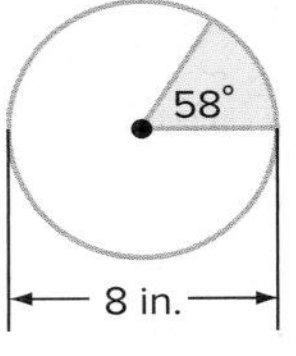

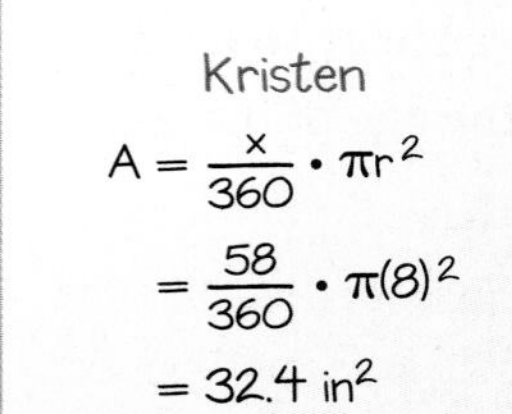

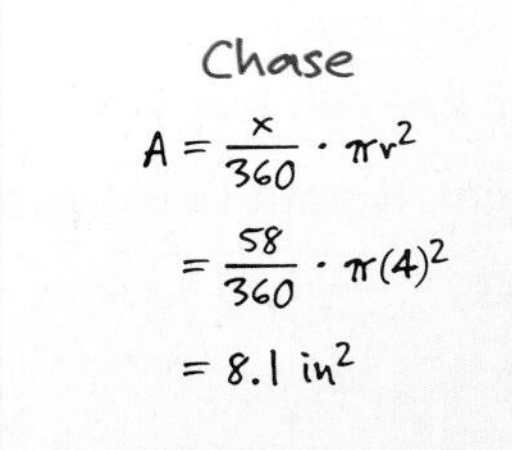

45. CHALLENGE Find the area of the shaded region. Round to the nearest tenth.

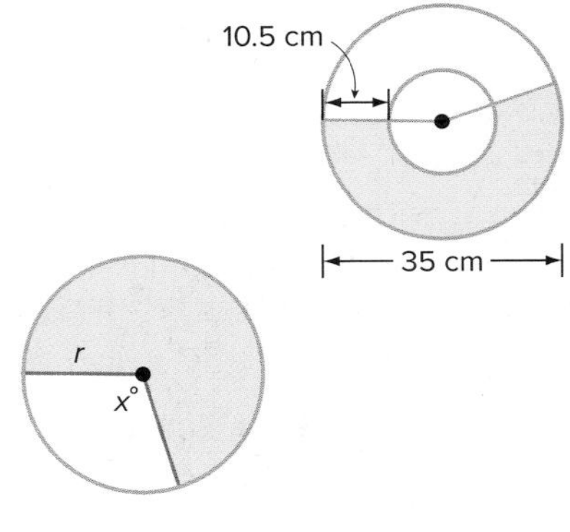

46. MP **CONSTRUCT ARGUMENTS** Refer to Exercise 43. Is the area of a sector of a circle *sometimes, always,* or *never* greater than the area of its corresponding segment?

47. WRITING IN MATH Describe two methods you could use to find the area of the shaded region of the circle. Which method do you think is more efficient? Explain your reasoning.

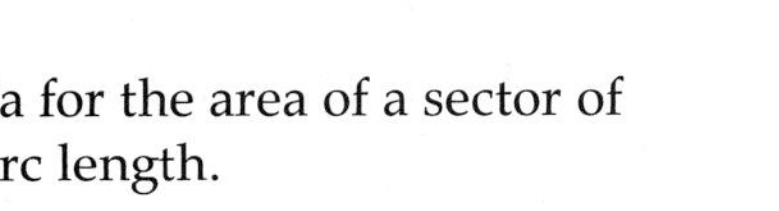

48. CHALLENGE Derive the formula for the area of a sector of a circle using the formula for arc length.

49. WRITING IN MATH If the radius of a circle doubles, will the measure of a sector of that circle double? Will it double if the arc measure of that sector doubles?

50. Visitors at a school carnival have a chance to toss a bean onto a circular tabletop that is divided into equal sectors, as shown.

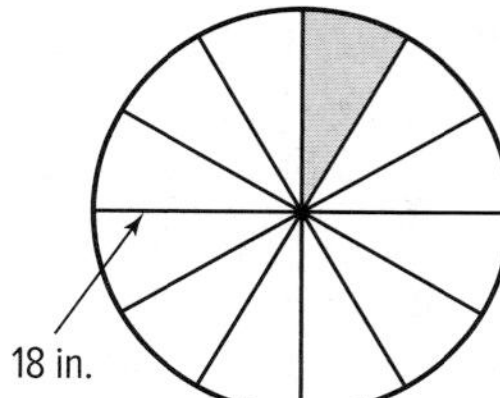

Visitors win a prize if the bean lands in the shaded sector. What is the area of this sector in square inches? Round to the nearest tenth. MP 1, 6

51. A lawn sprinkler sprays water 25 feet and moves back and forth through an angle of 150°.

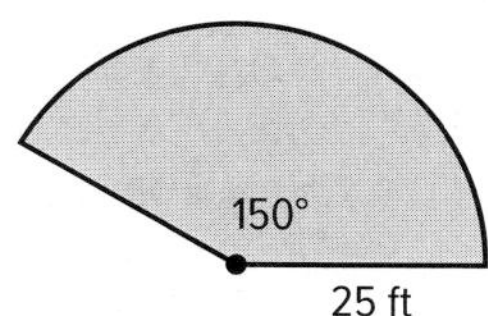

Which of the following is the best estimate of the area of the lawn that gets watered? MP 6

- **A** 65 ft^2
- **B** 818 ft^2
- **C** 1963 ft^2
- **D** 4712 ft^2

52. A sector of a circle has an intercepted arc that measures 120°. The area of the sector is 155.8 square centimeters. What is the radius of the circle in centimeters? Round to the nearest tenth.
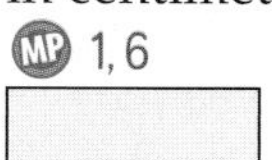 1, 6

53. One pizza with radius 9 inches is cut into 8 congruent sectors. Another pizza with the same radius is cut into 10 congruent sectors. How much more pizza, in square inches, is in a slice from the pizza cut into 8 sectors? MP 1, 6

- **A** 6.4
- **B** 25.4
- **C** 31.8
- **D** 57.2

54. Which expression represents the area of the shaded sector in square meters? MP 6

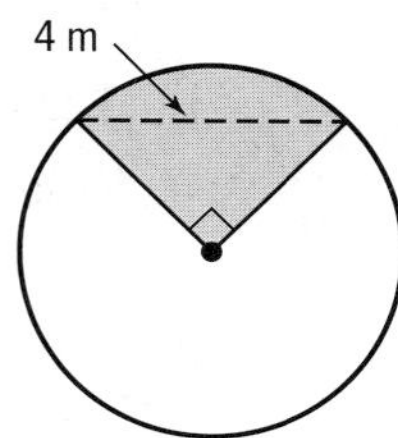

- **A** $\frac{\sqrt{2}}{2}\pi$
- **B** $\sqrt{2}\pi$
- **C** 2π
- **D** 4π

55. In $\odot C$, a sector has an area of 24π square inches. The radius of $\odot C$ is 12 inches. What is the measure, in degrees, of the arc that is intercepted by the sector? MP 1, 6

- **A** 360
- **B** 60π
- **C** 60
- **D** $\frac{180}{\pi}$

56. A circular pie has a diameter of 8 inches and is cut into 6 congruent slices. What is the area of one slice of pie? MP 1, 6

- **A** 6 square inches
- **B** 8.4 square inches
- **C** 21.6 square inches
- **D** 33.5 square inches

57. MULTI-STEP A regular hexagon, inscribed in a circle, is divided into 6 congruent triangles. The perimeter of the hexagon is 48 inches. MP 1, 6

a. What is the radius of the circle?

b. Find the area of each of the 6 sectors of the circle that have sides that coincide with sides of the congruent triangles. Round to the nearest tenth.

c. What is the area of one of the triangles? Round to the nearest tenth.

d. How much greater is the sector area than that of one of the triangles? Round to the nearest tenth.

CHAPTER 10

Mid-Chapter Quiz

Lessons 10-1 through 10-3

Find the perimeter and area of each parallelogram or triangle. Round to the nearest tenth if necessary. (Lesson 10-1)

1.
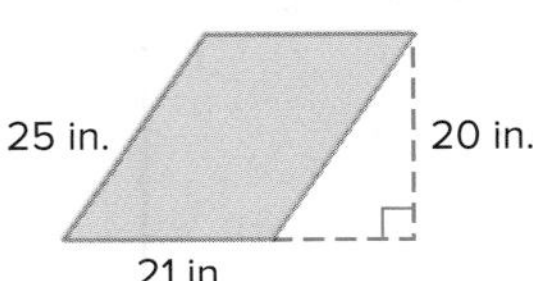

2.
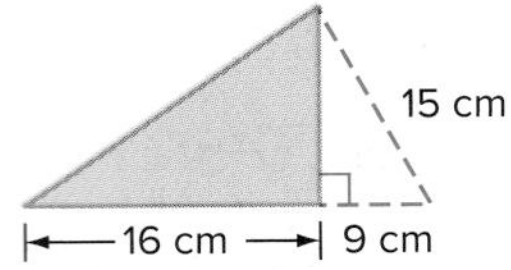

3.
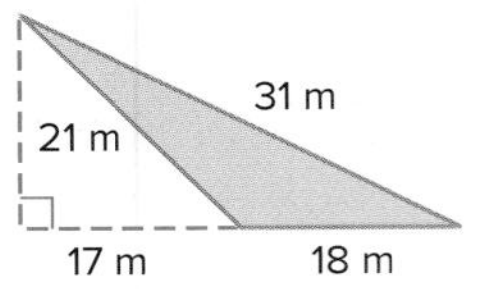

4.
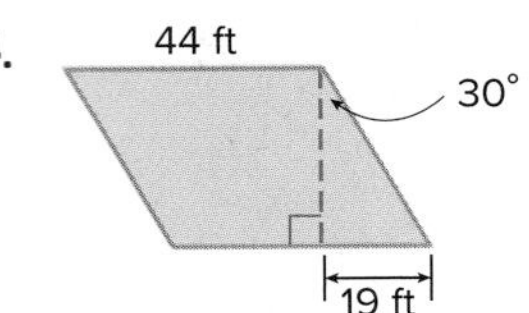

5. Find the perimeter and area of $\triangle FGH$ with $F(-3, 5)$, $G(-3, 10)$, and $H(0, 6)$. (Lesson 10-1)

6. **DESIGN** A plaque, as shown below, is made with a rhombus in the middle. (Lesson 10-2)

a. If the diagonals of the rhombus measure 7 inches and 9 inches, how much space is available for engraving text onto the award?

b. MP What mathematical practice did you use to solve this problem?

7. **MULTIPLE CHOICE** The area of a kite is 4 square feet. If the tail is to be 3 times longer than the kite's long diagonal, and the short diagonal measures 2 feet, how long should the kite's tail be? (Lesson 10-2)

A 4 feet
B 6 feet
C 7 feet
D 12 feet

Find the area of each trapezoid, rhombus, or kite. (Lesson 10-2)

8.
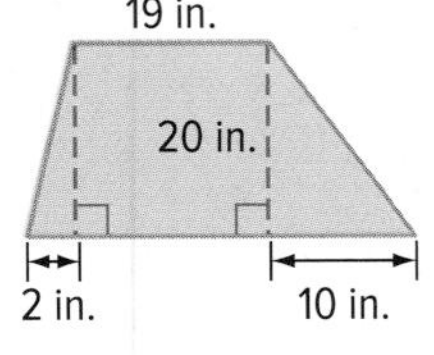

9.
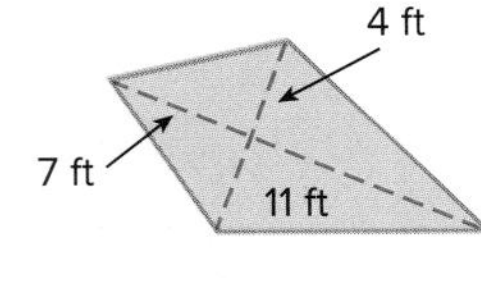

10.
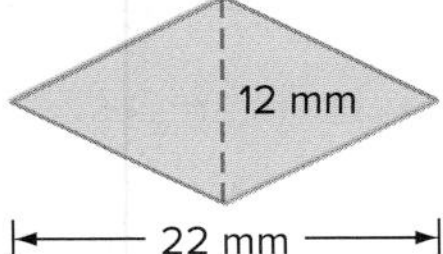

11.
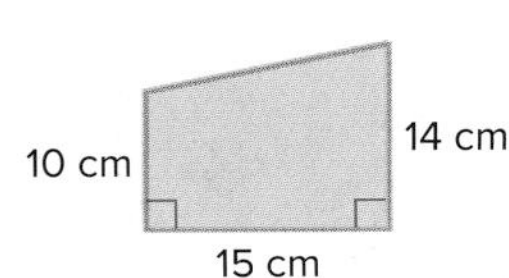

12. **ARCHAEOLOGY** The most predominant shape in Incan architecture is the trapezoid. The doorway pictured below is 3 feet wide at the top and 4 feet wide at the bottom. A person who is 5 feet 8 inches tall can barely pass through the doorway. How much fabric would be necessary to make a curtain for the doorway? (Lesson 10-2)

13. **ALGEBRA** A sector of a circle has a central angle measure of 30° and radius r. Write an expression for the perimeter of the sector in terms of r. (Lesson 10-3)

Find the area of each shaded sector. Round to the nearest tenth. (Lesson 10-3)

14.
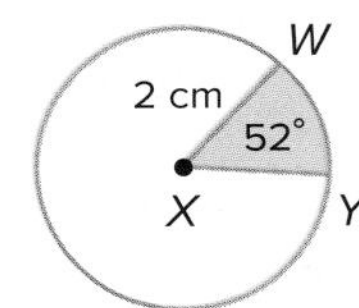

15.
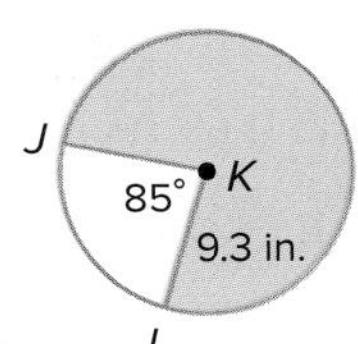

16.
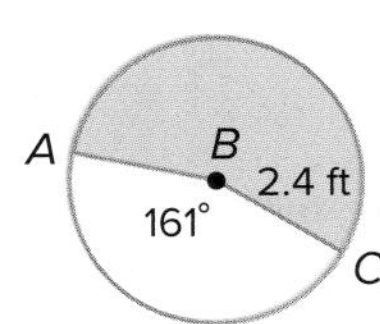

17.
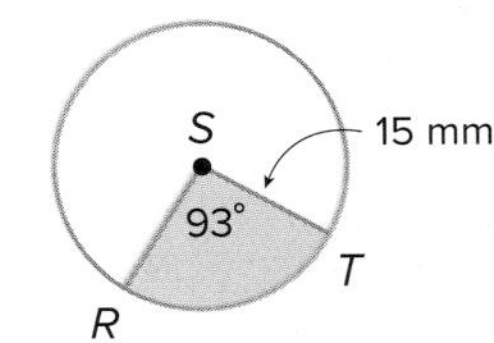

Find the indicated measure. Round to the nearest tenth. (Lesson 10-3)

18. The area of a circle is 52 square inches. Find the diameter.

19. Find the radius of a circle with an area of 104 square meters.

20. **FRUIT** The diameter of the orange slice shown is 9 centimeters. If each of the orange's 10 sections are congruent, find the approximate area covered by 8 sections. (Lesson 10-3)

EXPLORE 10-4

Geometry Lab

Investigating Areas of Regular Polygons

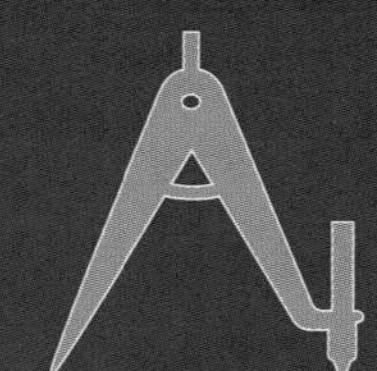

The point in the interior of a regular polygon that is equidistant from all of the vertices is the *center* of the polygon. A segment from the center that is perpendicular to a side of the polygon is an **apothem**.

Mathematical Practices

 5 Use appropriate tools strategically.

 7 Look for and make use of structure.

Activity

Work cooperatively.

Step 1 Copy regular pentagon $ABCDE$ and its center O.

Step 2 Draw the apothem from O to side $\overline{AB}$ by constructing the perpendicular bisector of $\overline{AB}$. Label the apothem measure as a. Label the measure of $\overline{AB}$ as s.

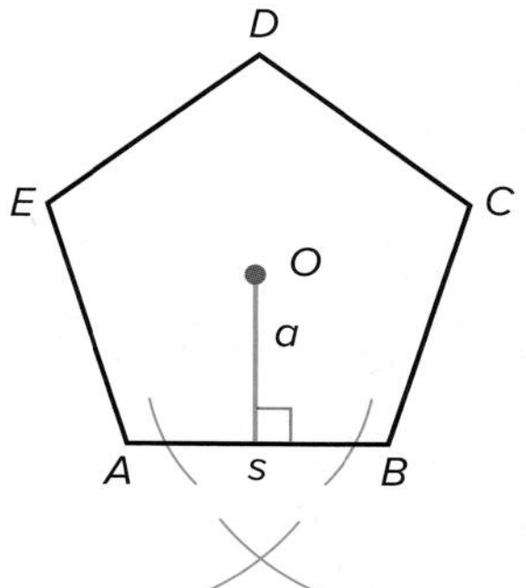

Step 3 Use a straightedge to draw $\overline{OA}$ and $\overline{OB}$.

Step 4 What measure in $\triangle AOB$ represents the base of the triangle? What measure represents the height?

Step 5 Find the area of $\triangle AOB$ in terms of s and a.

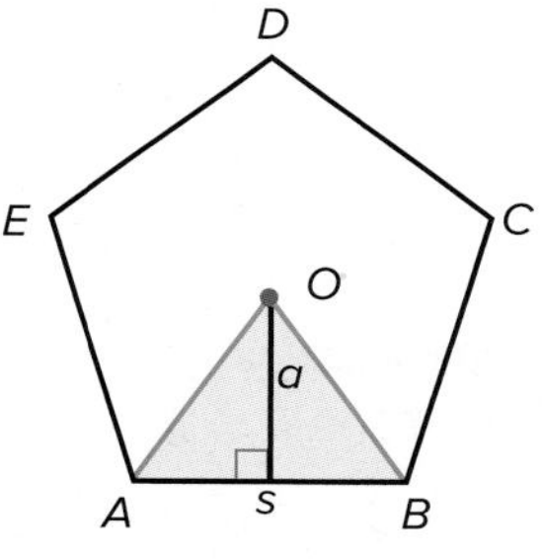

Step 6 Draw $\overline{OC}$, $\overline{OD}$, and $\overline{OE}$. What is true of the five small triangles formed?

Step 7 How do the areas of the five triangles compare?

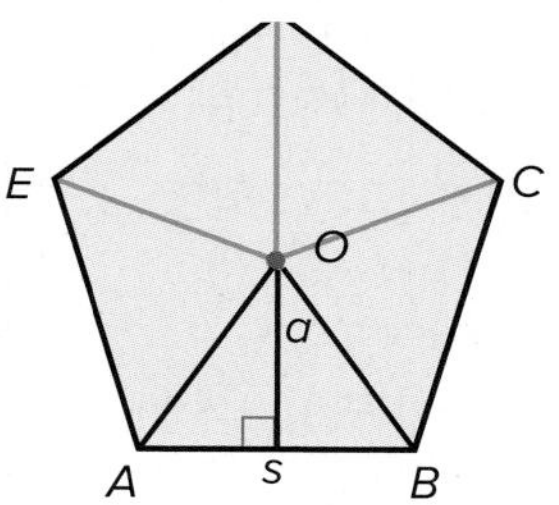

Analyze the Results

Work cooperatively

1. The area of a pentagon $ABCDE$ can be found by adding the areas of the given triangles that make up the pentagonal region.

   $A = \frac{1}{2}sa + \frac{1}{2}sa + \frac{1}{2}sa + \frac{1}{2}sa + \frac{1}{2}sa$

   $A = \frac{1}{2}(sa + sa + sa + sa + sa)$ or $\frac{1}{2}(5sa)$

   What does $5s$ represent?

2. Write a formula for the area of a pentagon in terms of perimeter P.

LESSON 4

Areas of Regular Polygons and Composite Figures

∴Then	∴Now	∴Why?
• You used inscribed and circumscribed figures and found the areas of circles.	• **1** Find areas of regular polygons. **2** Find areas of composite figures.	• The top of the table shown is a regular hexagon. Notice that the top is composed of six congruent triangular sections. To find the area of the table top, you can find the sum of the areas of the sections.

 **New Vocabulary**

center of a regular polygon
radius of a regular polygon
apothem
central angle of a regular polygon
composite figure

 **Mathematical Practices**

1 Make sense of problems and persevere in solving them.
6 Attend to precision.

1 Areas of Regular Polygons

In the figure, a regular pentagon is *inscribed* in ⊙P, and ⊙P is *circumscribed* about the pentagon. The **center of a regular polygon** and the **radius of a regular polygon** are also the center and the radius of its circumscribed circle.

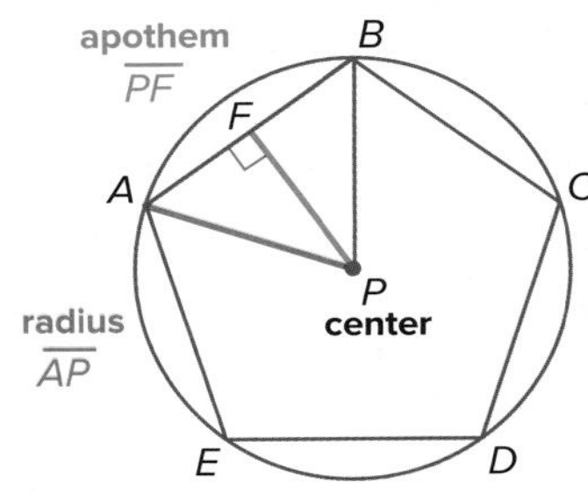

∠APB is a central angle of regular pentagon $ABCDE$.

A segment drawn from the center of a regular polygon perpendicular to a side of the polygon is called an **apothem**. Its length is the height of an isosceles triangle that has two radii as legs.

A **central angle of a regular polygon** has its vertex at the center of the polygon and its sides pass through consecutive vertices of the polygon. The measure of each central angle of a regular n-gon is $\frac{360}{n}$.

Example 1 Identify Segments and Angles in Regular Polygons

Square $FGHJ$ is inscribed in ⊙K. Identify the center, a radius, an apothem, and a central angle of the polygon. Then find the measure of a central angle.

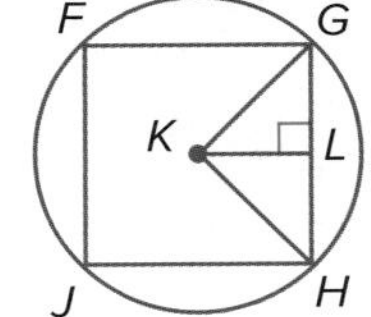

center: point K radius: $\overline{KG}$ or $\overline{KH}$

apothem: $\overline{KL}$ central angle: $\angle GKH$

A square is a regular polygon with 4 sides. Thus, the measure of each central angle of square $FGHJ$ is $\frac{360}{4}$ or 90.

Guided Practice

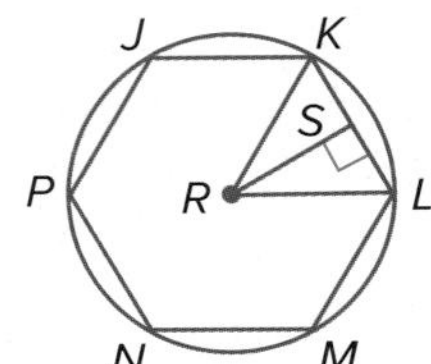

1. In the figure, regular hexagon $JKLMNP$ is inscribed in ⊙R. Identify the center, a radius, an apothem, and a central angle of the polygon. Then find the measure of a central angle.

You can find the area of any regular n-gon by dividing the polygon into congruent isosceles triangles. This strategy is sometimes called *decomposing the polygon into triangles.*

Real-World Example 2 Area of a Regular Polygon

Reading Math

Apothem Like the *radius* of a circle, the *apothem* of a polygon refers to the length of any apothem of the polygon.

ART Kang created the stained glass window shown. The window is a regular octagon with a side length of 15 inches and an apothem of 18.1 inches. What is the area covered by the window?

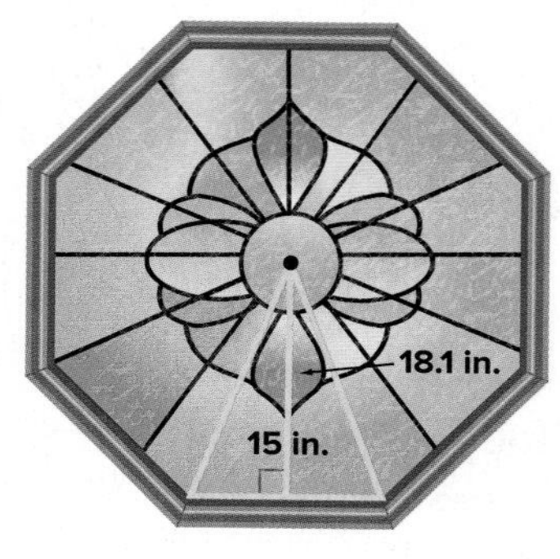

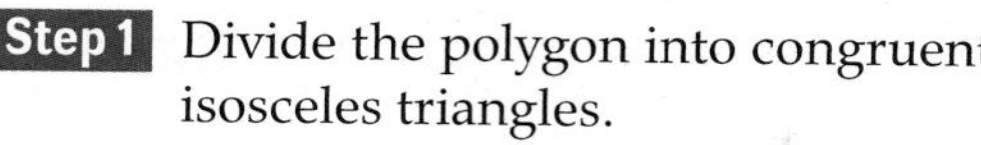

Step 1 Divide the polygon into congruent isosceles triangles.

Because the polygon has 8 sides, the polygon can be divided into 8 congruent isosceles triangles, each with a base of 15 inches and a height of 18.1 inches.

Step 2 Find the area of one triangle.

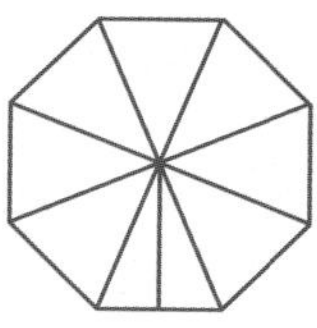

$A = \frac{1}{2}bh$	Area of a triangle
$= \frac{1}{2}(15)(18.1)$	$b = 15$ and $h = 18.1$
$= 135.75 \text{ in}^2$	Simplify.

Step 3 Multiply the area of one triangle by the total number of triangles.

Because there are 8 triangles, the area of the stained glass is 135.75 • 8 or 1086 square inches.

Guided Practice

2. HOT TUBS The cover of the hot tub shown is a regular pentagon. If the side length is 2.5 feet and the apothem is 1.7 feet, find the area of the lid to the nearest tenth.

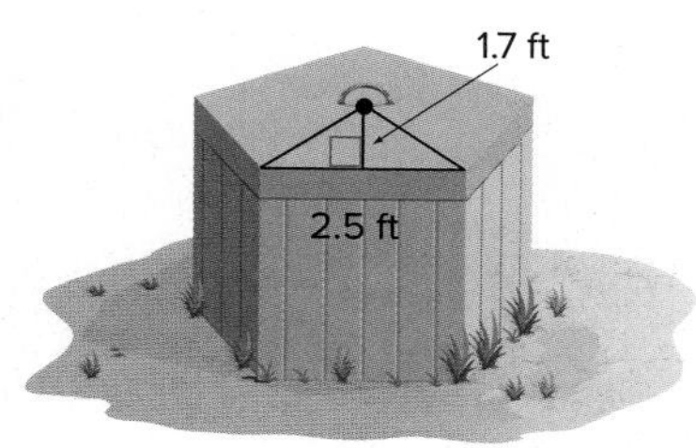

Watch Out!

Area of Regular Polygon this approach can only be applied to *regular* polygons.

From Example 2, we can develop a formula for the area of a regular n-gon with side length s and apothem a.

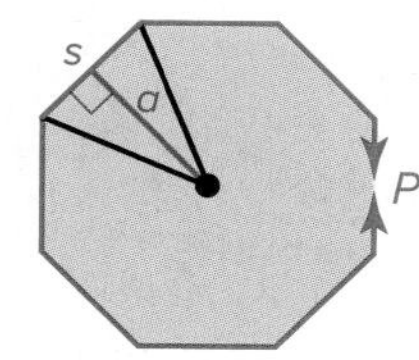

A = area of one triangle • number of triangles	
$= \frac{1}{2}$ • base • height • number of triangles	
$= \frac{1}{2} \cdot s \cdot a \cdot n$	Base of triangle is s and height is a. The number of triangles is n.
$= \frac{1}{2} \cdot a \cdot (n \cdot s)$	Commutative and Associative Properties
$= \frac{1}{2} \cdot a \cdot P$	The perimeter P of the polygon is $n \cdot s$.

Key Concept Area of a Regular Polygon

Words The area A of a regular n-gon with side length s is one half the product of the apothem a and perimeter P.

Symbols $A = \frac{1}{2}a(ns)$ or $A = \frac{1}{2}aP$.

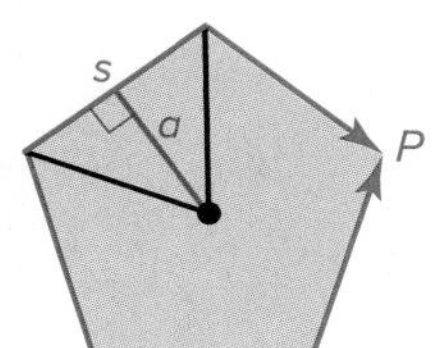

Example 3 Use the Formula for the Area of a Regular Polygon

Find the area of each regular polygon. Round to the nearest tenth.

a. regular hexagon

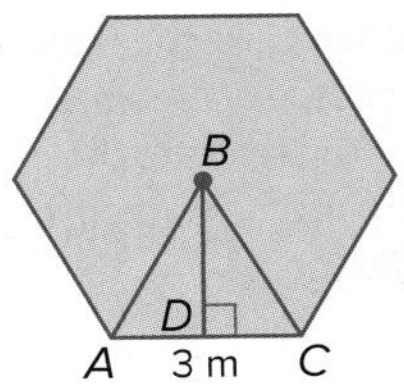

Step 1 Find the measure of a central angle.

A regular hexagon has 6 congruent central angles, so $m\angle ABC = \frac{360}{6}$ or 60.

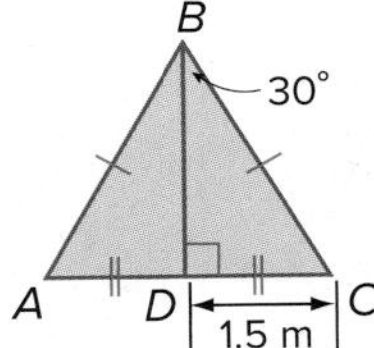

Step 2 Find the apothem.

Apothem $\overline{BD}$ is the height of isosceles $\triangle ABC$. It bisects $\angle ABC$, so $m\angle DBC = 30$. It also bisects $\overline{AC}$, so $DC = 1.5$ meters.

$\triangle BDC$ is a 30°–60°–90° triangle with a shorter leg that measures 1.5 meters, so $BD = 1.5\sqrt{3}$ meters.

Study Tip

MP Precision The altitude of an isosceles triangle from its vertex to its base is also an angle bisector and median of the triangle.

Step 3 Use the apothem and side length to find the area.

$A = \frac{1}{2}aP$	Area of a regular polygon
$= \frac{1}{2}(1.5\sqrt{3})(18)$	$a = 1.5\sqrt{3}$ and $P = 6(3)$ or 18
$\approx 23.4 \text{ m}^2$	Use a calculator.

b. regular pentagon

Step 1 A regular pentagon has 5 congruent central angles, so $m\angle QPR = \frac{360}{5}$ or 72.

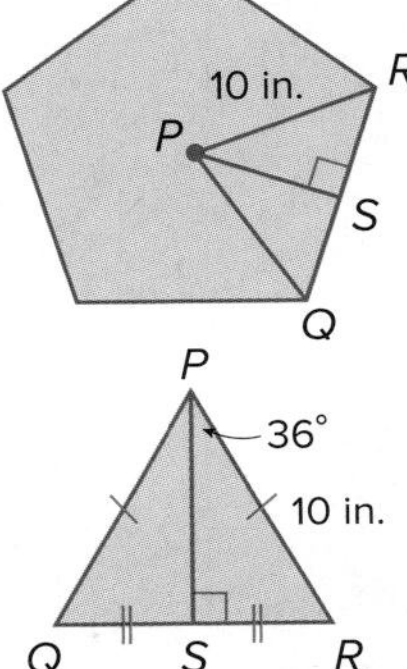

Step 2 Apothem $\overline{PS}$ is the height of isosceles $\triangle RPQ$. It bisects $\angle RPQ$, so $m\angle RPS = 36$. Use trigonometric ratios to find the side length and apothem of the polygon.

$\sin 36° = \frac{SR}{10}$ $\qquad$ $\cos 36° = \frac{PS}{10}$

$10 \sin 36° = SR$ $\qquad$ $10 \cos 36° = PS$

$QR = 2SR$ or $2(10 \sin 36°)$. So the pentagon's perimeter is $5 \cdot 2(10 \sin 36°)$ or $10(10 \sin 36°)$. The length of the apothem $\overline{PS}$ is $10 \cos 36°$.

Step 3 $A = \frac{1}{2}aP$	Area of a regular polygon
$= \frac{1}{2}(10 \cos 36°)[10(10 \sin 36°)]$	$a = 10 \cos 36°$, $P = 10(10 \sin 36°)$
$\approx 237.8 \text{ in}^2$	Use a calculator.

Guided Practice

3A.

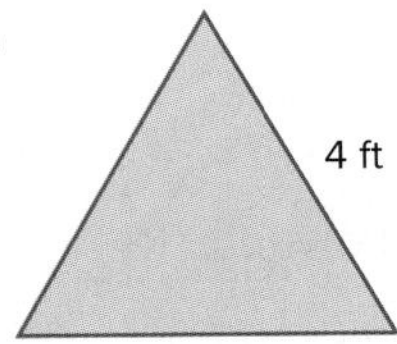

3B.

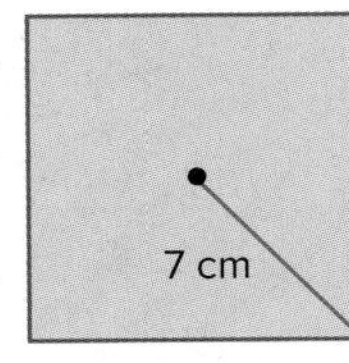

3C.

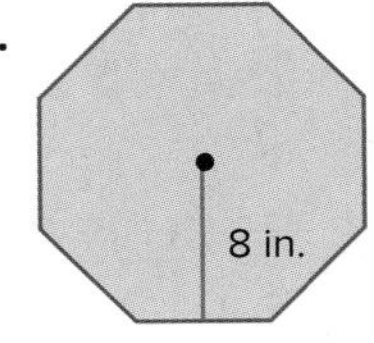

2 Areas of Composite Figures

A **composite figure** is a figure that can be separated into regions that are basic figures, such as triangles, rectangles, trapezoids, and circles. To find the area of a composite figure, find the area of each basic figure and then use the Area Addition Postulate.

Real-World Link

The first miniature golf course was built in Pinehurst, North Carolina, on a private estate owned by James Barber. There are currently between 5000 and 7500 miniature golf courses in the United States.

Source: Miniature Golf Association of the United States

Example 4 Find the Area of a Composite Figure by Adding

When viewed from above, the putting green at a miniature golf course is composed of a semicircle, trapezoid, and triangle. Which of the following best represents the area of carpet needed to cover the green?

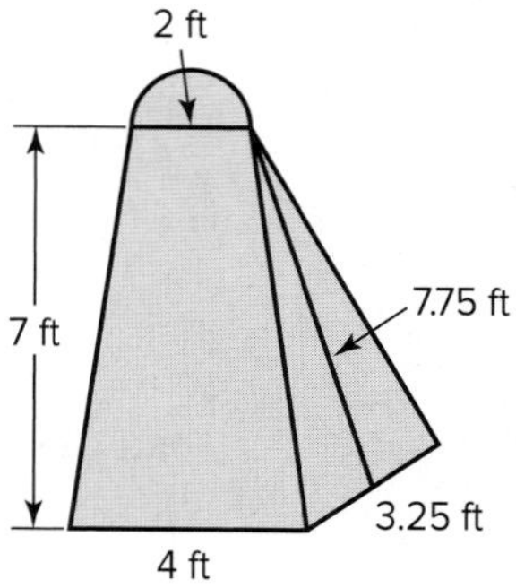

A 21 **B** 32 **C** 35 **D** 37

Read the Item

The area to be carpeted can be separated into a trapezoid with a height of 7 feet and bases of 2 feet and 4 feet, a triangle with a base of 3.25 feet and height of 7.75 feet, hypotenuse of 5.7 feet, and a semicircle with a diameter of 2 feet. Find the area of each figure separately and add to get the total area.

Solve the Item

Area of green = **area of trapezoid** + **area of triangle** + **area of semicircle**

$$= \frac{1}{2} \cdot h\,(b_1 + b_2) + \frac{1}{2} \cdot b \cdot h + \frac{1}{2}\pi r^2$$

$$\approx \frac{1}{2} \cdot 7 \cdot (2 + 4) + \frac{1}{2} \cdot 3.25 \cdot 7.75 + \frac{1}{2}\pi(1)^2$$

$$\approx 21 + 12.59 + 1.57 \text{ or about } 35.16 \text{ ft}^2$$

So, about 35 square feet of carpet is needed. The correct answer is C.

Guided Practice

The figure shown is composed of a semicircle, regular hexagon, and trapezoid. What is the area of the figure?

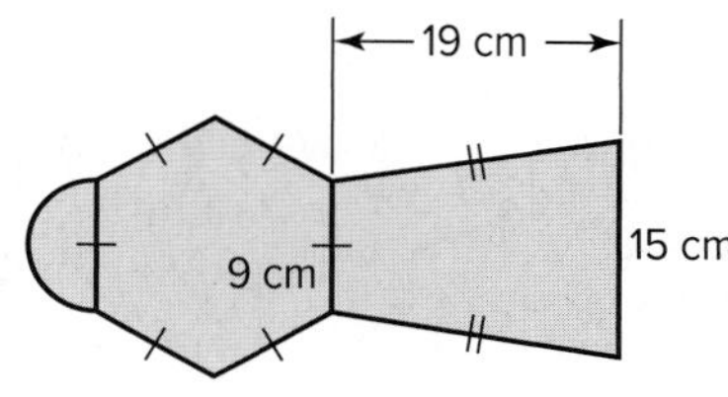

A 427.8 cm^2 **C** 454.3 cm^2

B 438.4 cm^2 **D** 470.2 cm^2

The areas of some figures can be found by subtracting the areas of basic figures.

Go Online!

Log into ConnectED to watch an **Animation** about areas of composite figures.

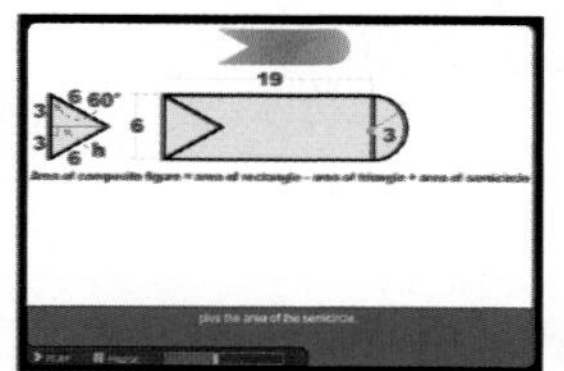

Example 5 Find the Area of a Composite Figure by Subtracting

Find the area of the figure. Round to the nearest tenth if necessary.

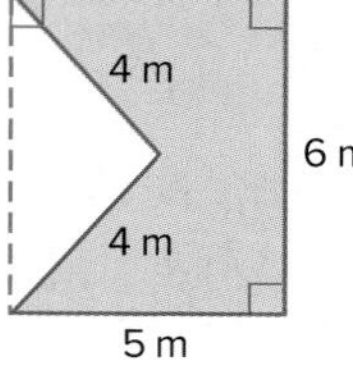

To find the area of the figure, subtract the area of the triangle from the area of the rectangle.

Using the Pythagorean Theorem, the height h of the triangle is $\sqrt{4^2 - 3^2}$ or $\sqrt{7}$ meters.

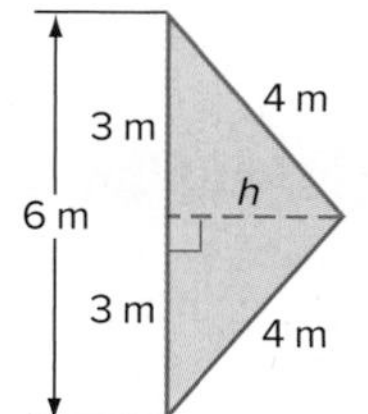

Area of figure = **Area of rectangle** − **Area of triangle**

$$= b \cdot h - \frac{1}{2}bh$$

$$= 5 \cdot 6 - \frac{1}{2}(6)(\sqrt{7})$$

$$\approx 30 - 7.9 \text{ or about } 22.1 \text{ m}^2$$

Guided Practice

5A.

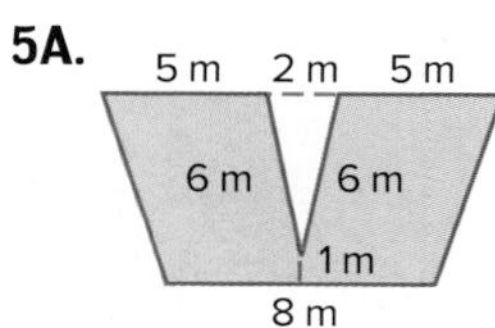

5B.

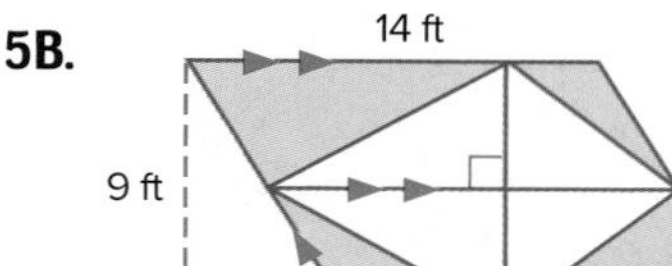

Check Your Understanding

○ = Step-by-Step Solutions begin on page R13.

Example 1

1. In the figure, regular heptagon *ABCDEFG* is inscribed in ⊙*P*. Identify the center, a radius, an apothem, and a central angle of the polygon. Then find the measure of a central angle.

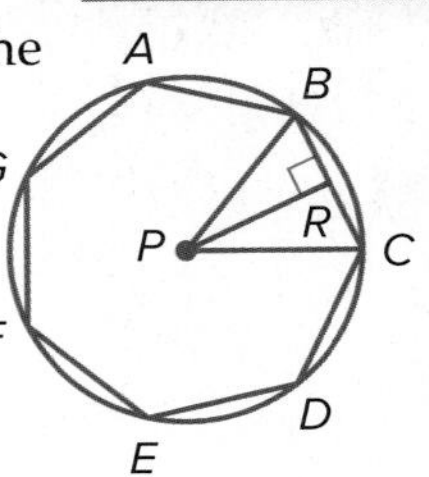

Examples 2–3 **Find the area of each regular polygon. Round to the nearest tenth.**

2. 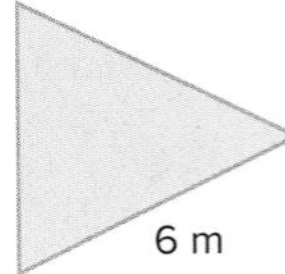

3. 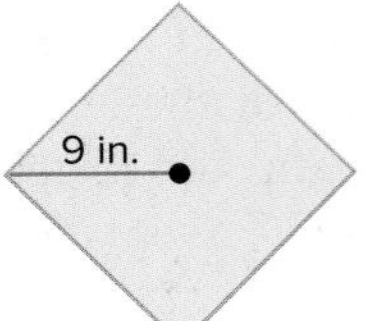

4. 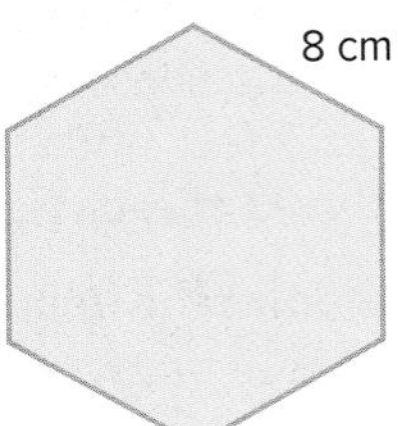

5. **POOLS** Kenton's job is to cover the community pool during fall and winter. Because the pool is in the shape of an octagon, he needs to find the area in order to have a custom cover made. If the pool has the dimensions shown at the right, what is the area of the pool?

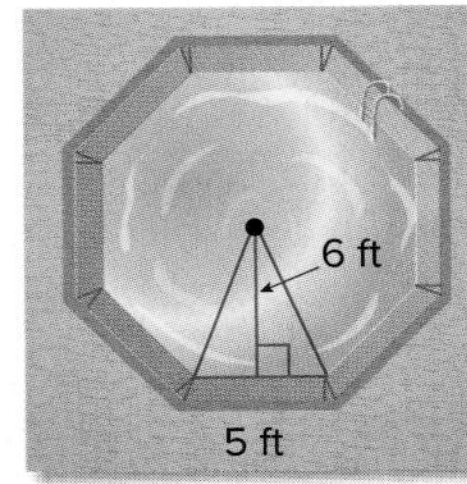

Example 4

6. **MULTIPLE CHOICE** The figure shown is composed of a regular hexagon and equilateral triangles. Which of the following best represents the area?

A 37.5 in^2

B $37.5\sqrt{3} \text{ in}^2$

C 75 in^2

D $75\sqrt{3} \text{ in}^2$

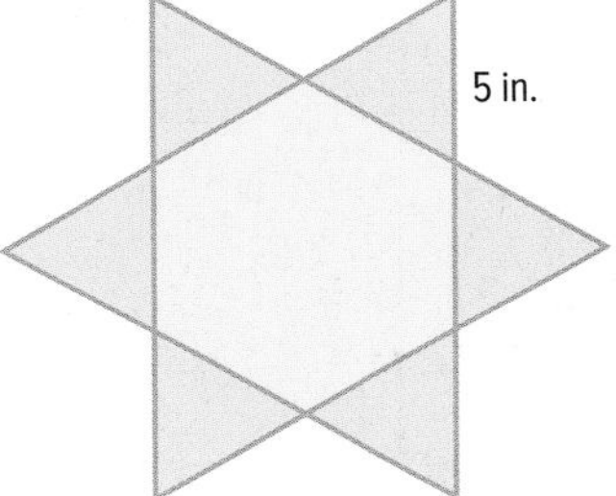

Example 5

7. **BASKETBALL** The basketball court in Jeff's school is painted as shown.

   a. What area of the court is blue? Round to the nearest square foot.

   b. What area of the court is red? Round to the nearest square foot.

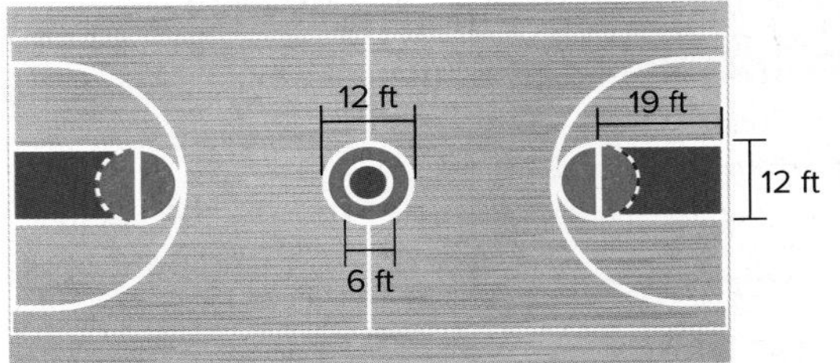

Note: Art not drawn to scale.

Practice and Problem Solving

Extra Practice is on page R10.

Example 1

MP SENSE-MAKING In each figure, a regular polygon is inscribed in a circle. Identify the center, a radius, an apothem, and a central angle of each polygon. Then find the measure of a central angle.

8. 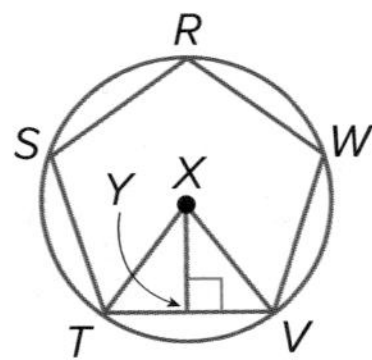

9. 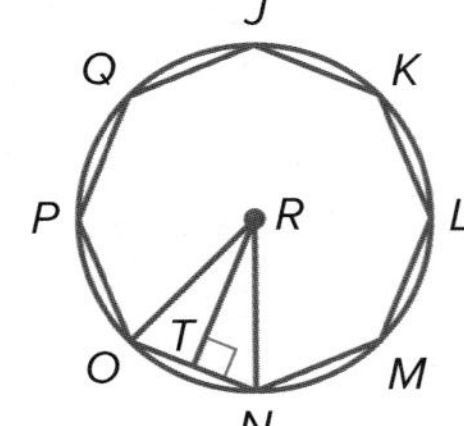

Examples 2–3 **Find the area of each regular polygon. Round to the nearest tenth.**

10.

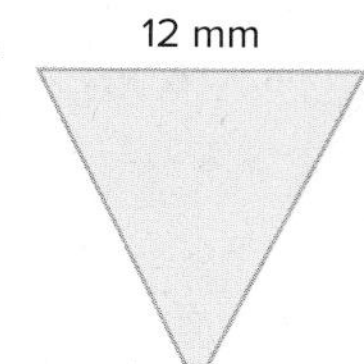

11.

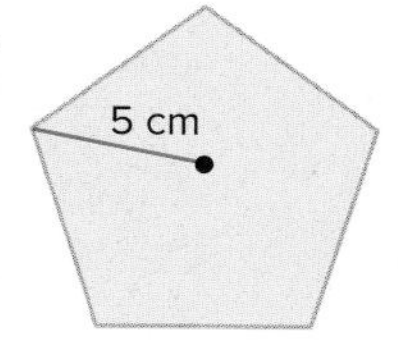

12.

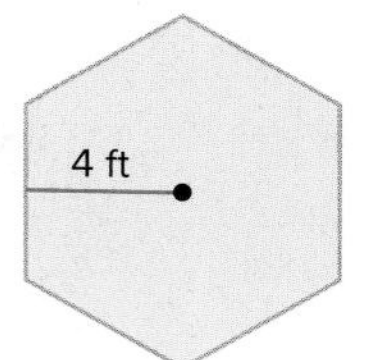

13.

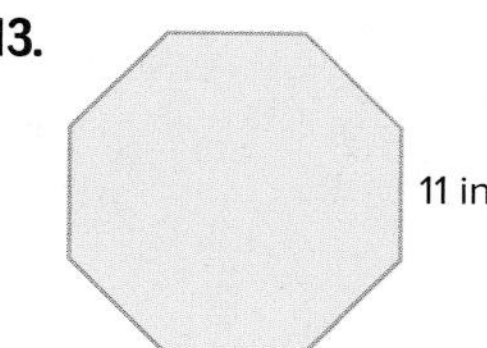

Example 4

14. CARPETING Ignacio's family is getting new carpet in their family room, and they want to determine how much the project will cost.

a. Use the floor plan shown to find the area to be carpeted.

b. If the carpet costs \$4.86 per square yard, how much will the project cost?

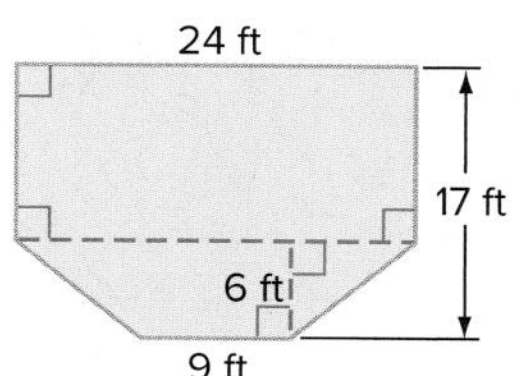

Examples 4–5 **SENSE-MAKING** **Find the area of each figure. Round to the nearest tenth if necessary.**

15

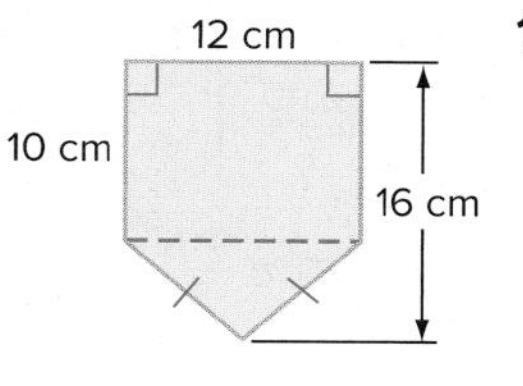

16.

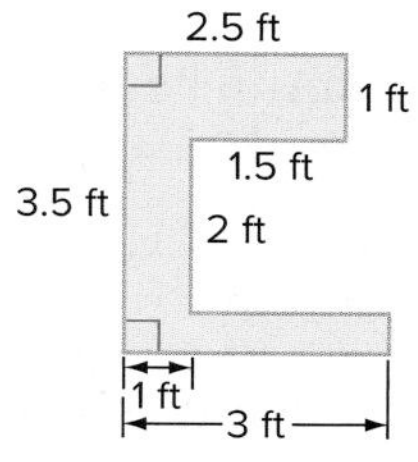

17.

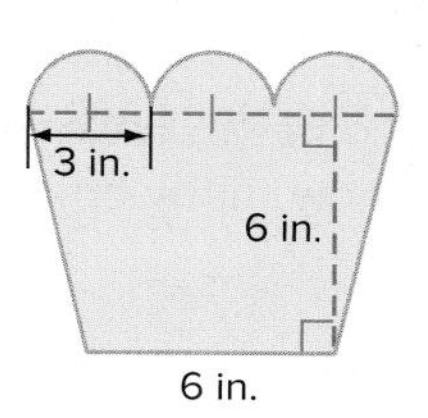

18.

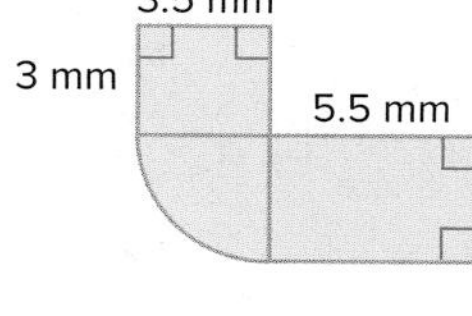

19. CRAFTS Latoya's greeting card company is making envelopes for a card from the pattern shown.

a. Find the perimeter and area of the pattern. Round to the nearest tenth.

b. If Latoya orders sheets of paper that are 2 feet by 4 feet, how many envelopes can she make per sheet?

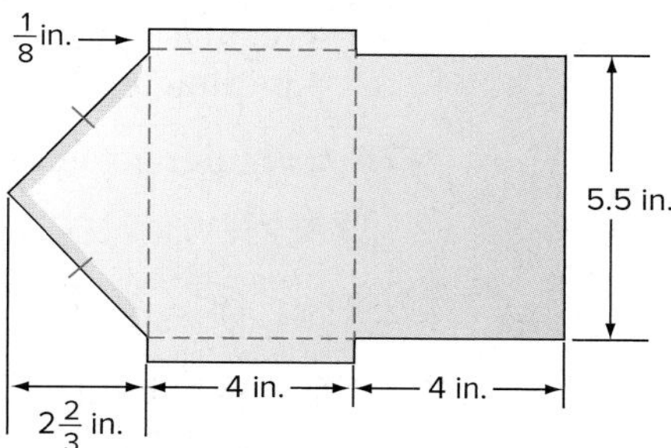

20. VOLUNTEERING James is making pinwheels at a summer camp. If they want to paint one side of each pinwheel, what is the approximate total area of 10 pinwheels?

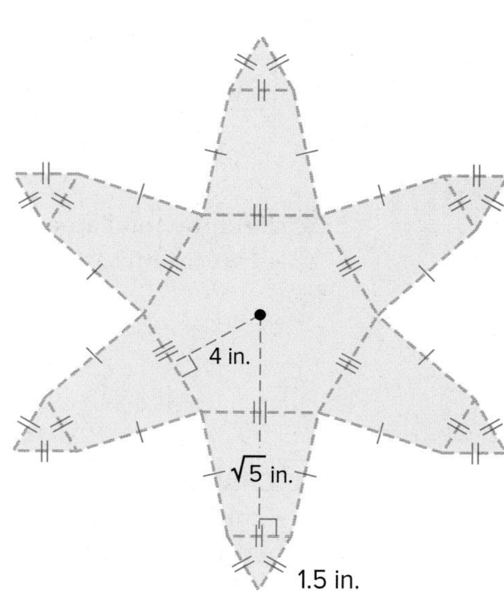

21. THEATRE Alison's drama club is planning on painting the amphitheater stage. Find the total area of the stage.

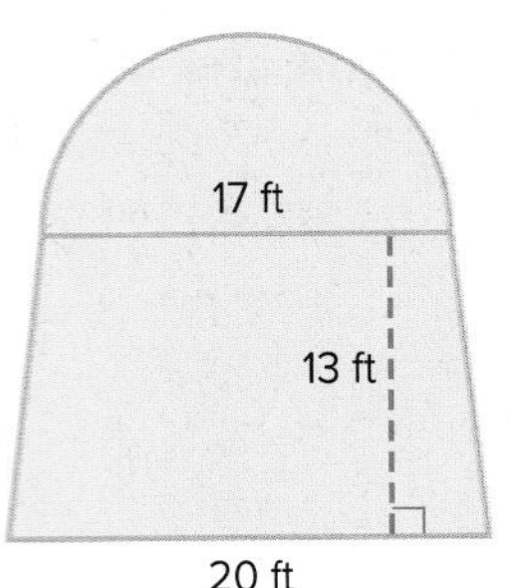

Find the area of each shaded region formed by each circle and regular polygon. Round to the nearest tenth.

22.

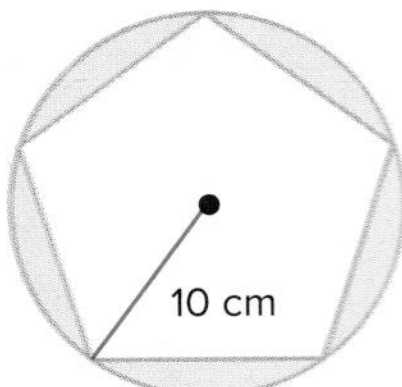

23.

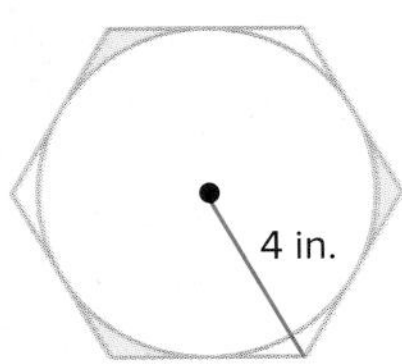

24.

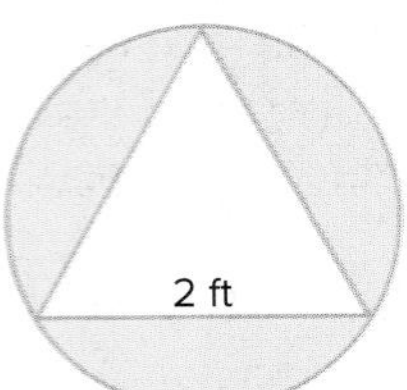

25. FLOORING JoAnn wants to lay 12″ × 12″ tile on her bathroom floor.

a. Find the area of the bathroom floor in her apartment floor plan.

b. If the tile comes in boxes of 15 and JoAnn buys no extra tile, how many boxes will she need?

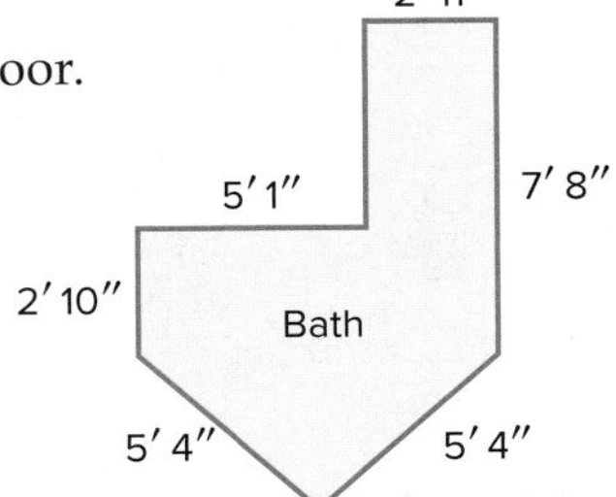

Find the perimeter and area of each figure. Round to the nearest tenth, if necessary.

26. **a regular hexagon with a side length of 12 centimeters**

27. **a regular pentagon circumscribed about a circle with a radius of 8 millimeters**

28. **a regular octagon inscribed in a circle with a radius of 5 inches**

MP **PERSEVERANCE** **Find the area of each shaded region. Round to the nearest tenth.**

29

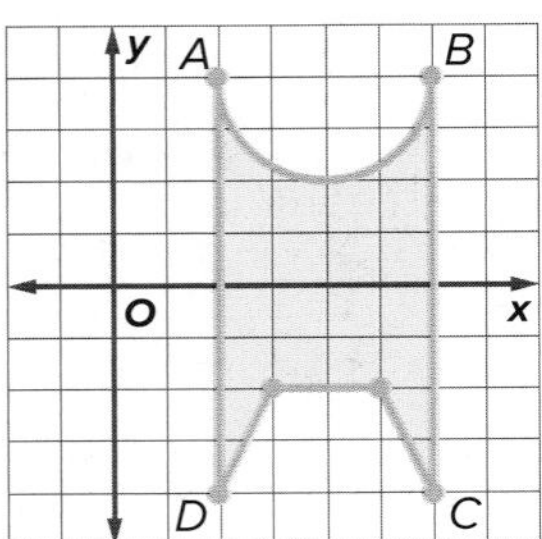

30.

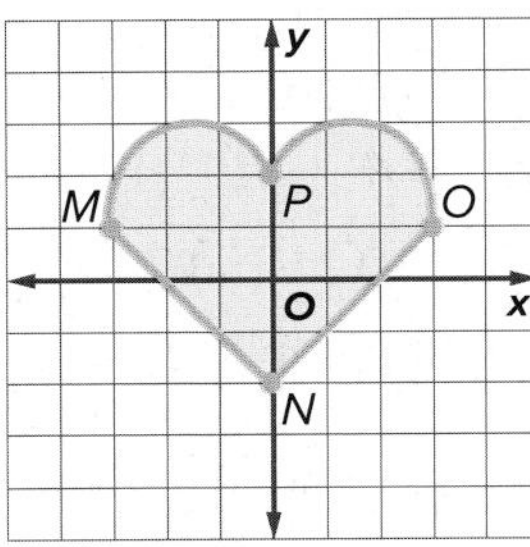

31.

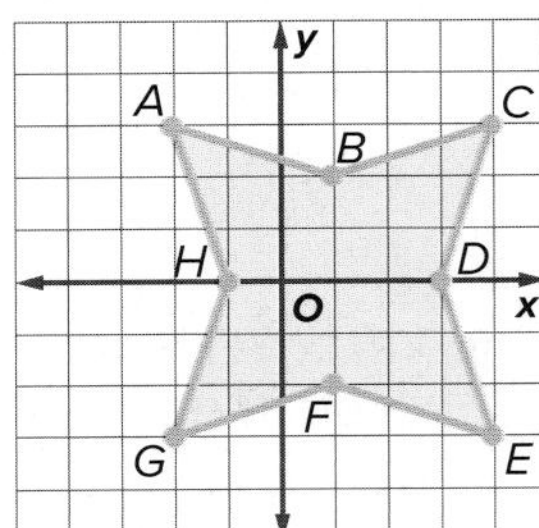

32. Find the total area of the shaded regions. Round to the nearest tenth.

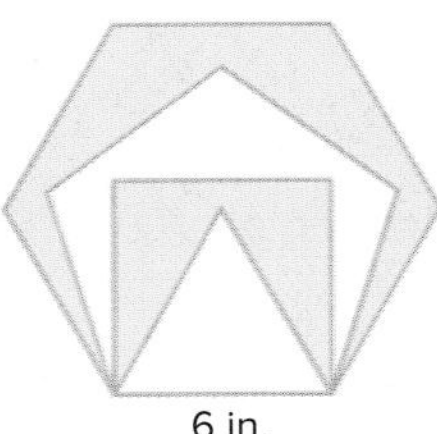

33. CHANGING DIMENSIONS Calculate the area of an equilateral triangle with a perimeter of 3 inches. Calculate the areas of a square, a regular pentagon, and a regular hexagon with perimeters of 3 inches. How does the area of a regular polygon with a fixed perimeter change as the number of sides increases?

34. **MULTIPLE REPRESENTATIONS** In this problem, you will investigate the areas of regular polygons inscribed in circles.

a. **Geometric** Draw a circle with a radius of 1 unit and inscribe a square. Repeat twice, inscribing a regular pentagon and hexagon.

b. **Algebraic** Use the inscribed regular polygons from part **a** to develop a formula for the area of an inscribed regular polygon in terms of angle measure x and number of sides n.

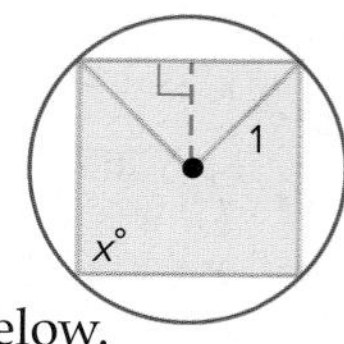

c. **Tabular** Use the formula you developed in part **b** to complete the table below. Round to the nearest hundredth.

Number of Sides, n	4	5	6	8	10	20	50	100
Interior Angle Measure, x								
Area of Inscribed Regular Polygon								

d. **Verbal** Make a conjecture about the area of an inscribed regular polygon with a radius of 1 unit as the number of sides increases.

H.O.T. Problems Use Higher-Order Thinking Skills

35. **ERROR ANALYSIS** Chloe and Flavio want to find the area of the hexagon shown. Is either of them correct? Explain your reasoning.

Chloe

$A = \frac{1}{2}Pa$

$= \frac{1}{2}(66)(9.5)$

$= 313.5 \text{ in}^2$

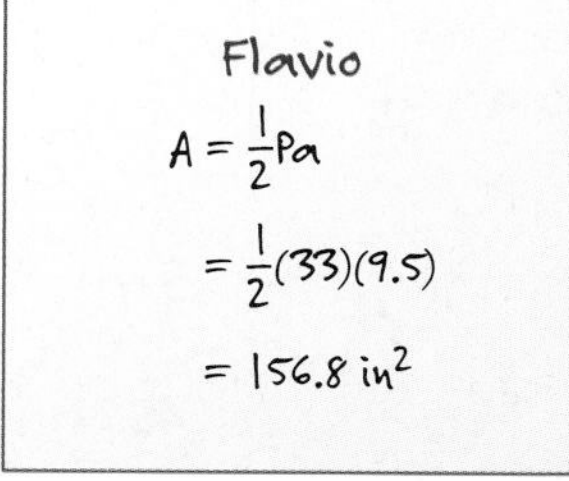

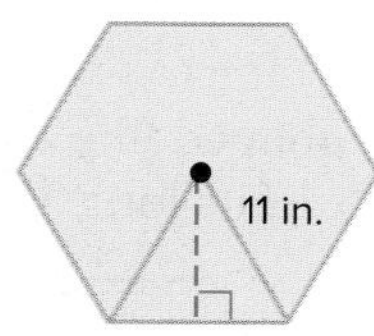

36. **SENSE-MAKING** Using the map of Nevada shown, estimate the area of the state. Explain your reasoning.

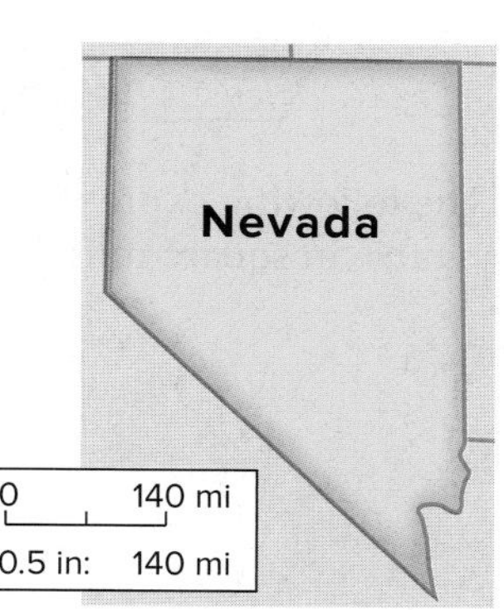

37. **OPEN-ENDED** Draw a pair of composite figures that have the same area. Make one composite figure out of a rectangle and a trapezoid, and make the other composite figure out of a triangle and a rectangle. Show the area of each basic figure.

38. **WRITING IN MATH** Consider the sequence of area diagrams shown.

a. What algebraic theorem do the diagrams prove? Explain your reasoning.

b. Create your own sequence of diagrams to prove a different algebraic theorem.

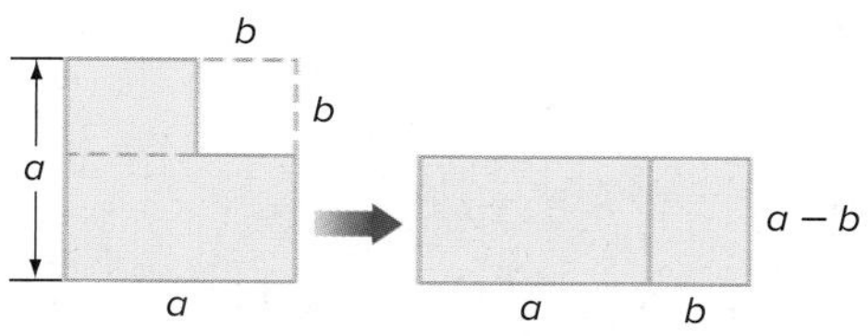

39. **WRITING IN MATH** How can you find the area of any figure?

40. Which of the following is the best estimate of the area of the concrete patio shown here? MP 1, 6

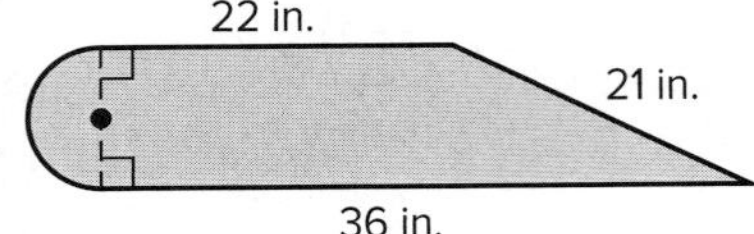

- ○ A 550 in^2
- ○ B 646 in^2
- ○ C 660 in^2
- ○ D 782 in^2
- ○ E 839 in^2

41. What is the area of a square with an apothem of 2 feet? MP 1, 6

- ○ A 16 ft^2
- ○ B 8 ft^2
- ○ C 4 ft^2
- ○ D 2 ft^2

42. A picnic table shaped like a regular hexagon has sides that are x units long.

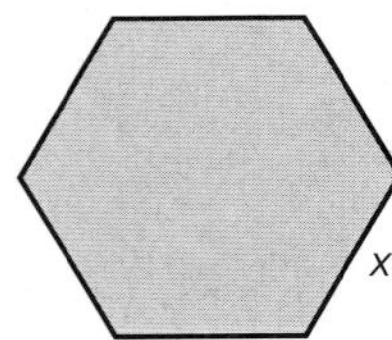

Which of the following expressions represents the area of the table in square units? MP 1, 6

- ○ A $\frac{3\sqrt{2}}{2}x^2$
- ○ B $\frac{\sqrt{3}}{4}x^2$
- ○ C $3\sqrt{3}\,x^2$
- ○ D $\frac{3\sqrt{3}}{2}x^2$

43. A stained glass panel is shaped like a regular pentagon has a side length of 7 inches. What is the area, to the nearest tenth? MP 1, 6

44. Find the area of the shaded figure in square inches. Round to the nearest tenth. MP 1, 6

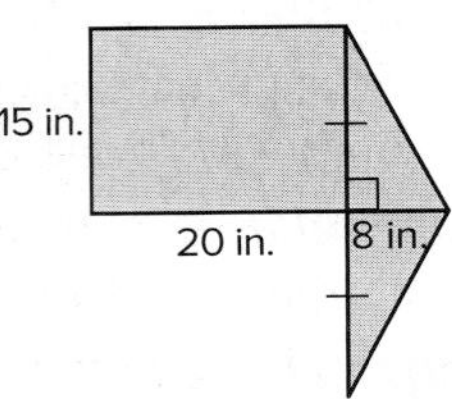

45. **MULTI-STEP** The dimensions of a patio are shown in the diagram. If the surface of the patio is to be painted, about how many square feet will be painted? MP 1, 6

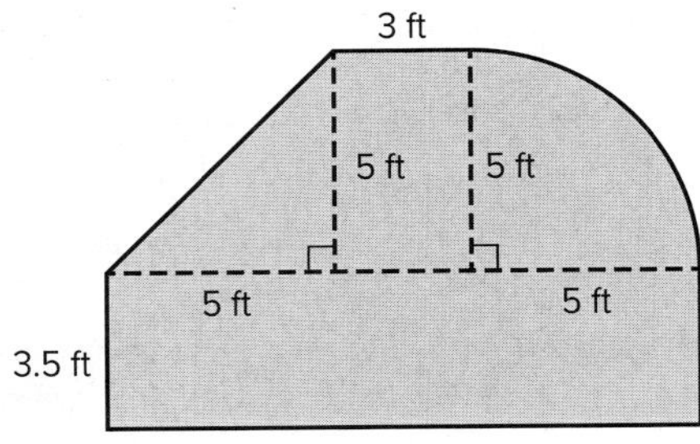

- ○ A 66.8 ft^2
- ○ B 92.6 ft^2
- ○ C 112.3 ft^2
- ○ D 151.5 ft^2

46. Find the area of the figure. Round to the nearest tenth. MP 1, 6

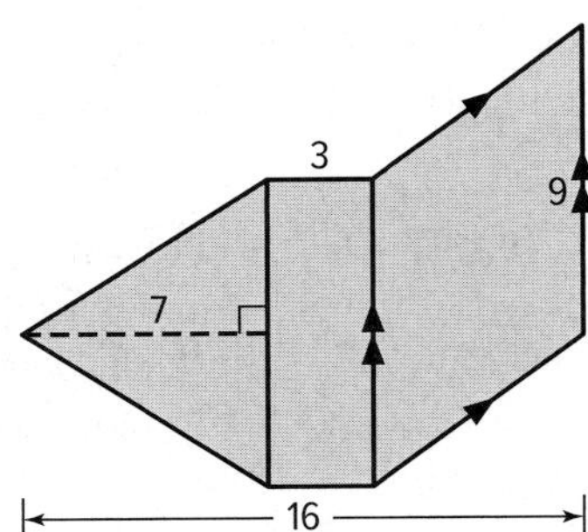

47. A circle is inscribed in a square. The diameter of the circle is 12 inches and is equal to the length of the sides of the square. If the circle is cut out of the square, what is the area of the remaining part of the square? Round your answer to the nearest tenth. MP 1, 6

EXTEND 10-4

Geometry Lab

Regular Polygons on the Coordinate Plane

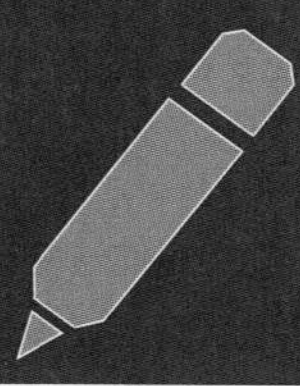

If you know the coordinates of two consecutive vertices of a regular polygon, you can use the Distance Formula to find the length of each side. For example, in the figure shown, the length of $\overline{AB}$ is $\sqrt{(3-1)^2+(1-4)^2}$ or $\sqrt{13}$. Using this measure, you can then find the perimeter and area of the figure using the techniques presented in Lesson 10-4.

You can also use the Distance Formula to find the perimeter and area of a regular polygon inscribed in a circle given the coordinates of the endpoints of a radius.

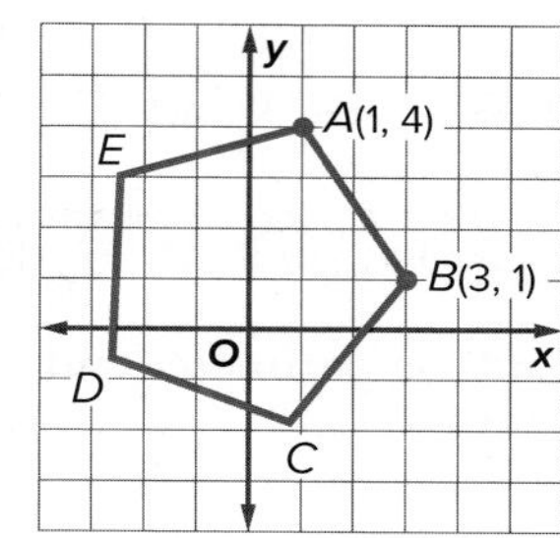

Mathematical Practices

8 Look for and express regularity in repeated reasoning.

Activity 1 Inscribed Polygon

Work cooperatively. Find the perimeter and area of octagon *ABCDEFGH*, which is inscribed in ⊙*O*. Round to the nearest tenth, if necessary.

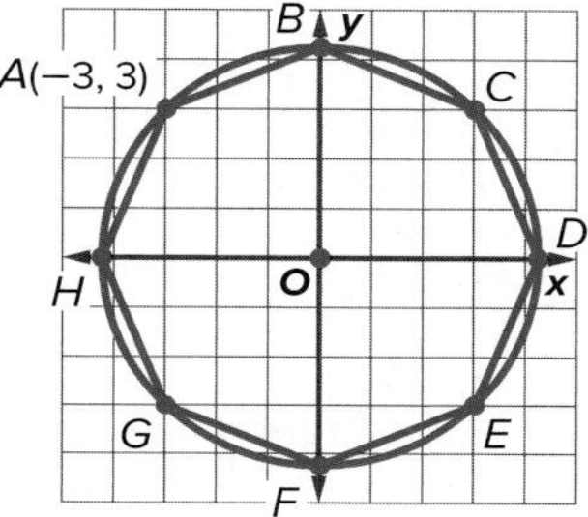

Step 1 Use the Distance Formula to find a radius of ⊙*O*.

$OA = \sqrt{(-3-0)^2+(3-0)^2}$ — $x_2=-3, x_1=0, y_2=3$, and $y_1=0$

$= \sqrt{18}$ or $3\sqrt{2}$ — Simplify.

Step 2 Find the perimeter and area.

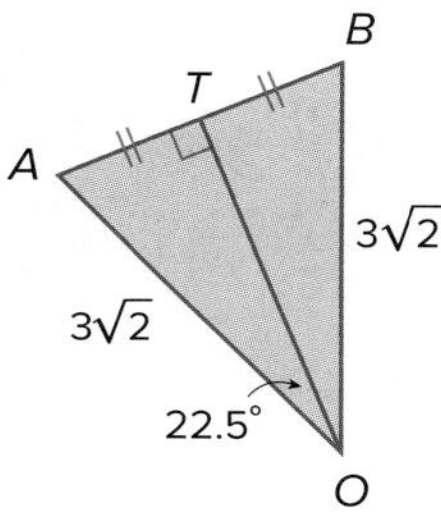

Because the octagon is inscribed in ⊙*O*, $\overline{OA}$ and $\overline{OB}$ are both radii of ⊙*O*. Therefore, $OA = OB = 3\sqrt{2}$. Let $\overline{OT}$ be an apothem of the octagon with length a. Then $\overline{OT}$ is also the height of isosceles $\triangle AOB$. Because the octagon is regular, $m\angle AOB$ is $360 \div 8$ or 45. Because $\overline{OT}$ bisects $\angle AOB$ and side $\overline{AB}$, $m\angle AOT = 45 \div 2$ or 22.5, and $AB = 2(AT)$.

Use trigonometric ratios to find a and AT.

$\cos 22.5° = \frac{a}{3\sqrt{2}}$ — $\cos\theta = \frac{\text{adj}}{\text{hyp}}$

$a = 3\sqrt{2}\cos 22.5°$ — Solve for a.

$\sin 22.5° = \frac{AT}{3\sqrt{2}}$ — $\sin\theta = \frac{\text{opp}}{\text{hyp}}$

$AT = 3\sqrt{2}\sin 22.5°$ — Solve for AT.

$AB = 2(AT)$, so $AB = 2(3\sqrt{2}\sin 22.5°)$ and the perimeter P of the octagon is $8(2)3\sqrt{2}\sin 22.5°$ or about 26.0 units. The area of the octagon is $\frac{1}{2}aP$, which is $\frac{1}{2}3\sqrt{2}\cos 22.5° \cdot 8(2)3\sqrt{2}\sin 22.5°$ or about 50.9 units^2.

EXTEND 10-4

Geometry Lab

Regular Polygons on the Coordinate Plane *Continued*

You can also use the Distance Formula to find the perimeter and area of a regular polygon circumscribed about a circle given the coordinates of the endpoints of a radius.

Activity 2 Circumscribed Polygon

Work cooperatively. Find the perimeter and area of hexagon *ABCDEF*, which is circumscribed about ⊙*Q*. Round to the nearest tenth, if necessary.

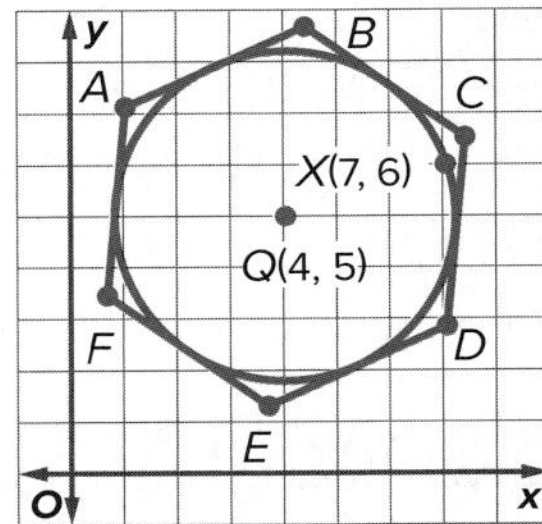

Step 1 Use the Distance Formula to find a radius of ⊙*Q*.

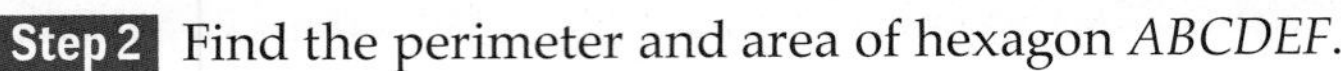

$QX = \sqrt{(7-4)^2 + (6-5)^2}$ or $\sqrt{10}$ $\quad x_2 = 7, x_1 = 4, y_2 = 6,$ and $y_1 = 5$

Step 2 Find the perimeter and area of hexagon *ABCDEF*.

Because the hexagon is circumscribed about ⊙*Q*, $\overline{AB}$ is tangent to the circle. Let the point of tangency be *T*. Since all radii of a circle are congruent, radius $\overline{QT}$ also measures $\sqrt{10}$. $\overline{QT}$ is an apothem of the hexagon, so $a = \sqrt{10}$.

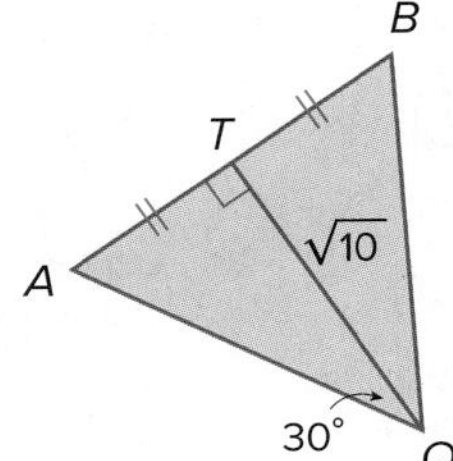

The apothem is also the height of isosceles △*AQB*. Because the hexagon is regular, $m\angle AQB$ is 360 ÷ 6 or 60. Because $\overline{QT}$ bisects ∠*AQB* and side $\overline{AB}$, $m\angle AQT = 60 \div 2$ or 30, and $AB = 2(AT)$. Use trigonometric ratios to find *AT*. Then find *AB*.

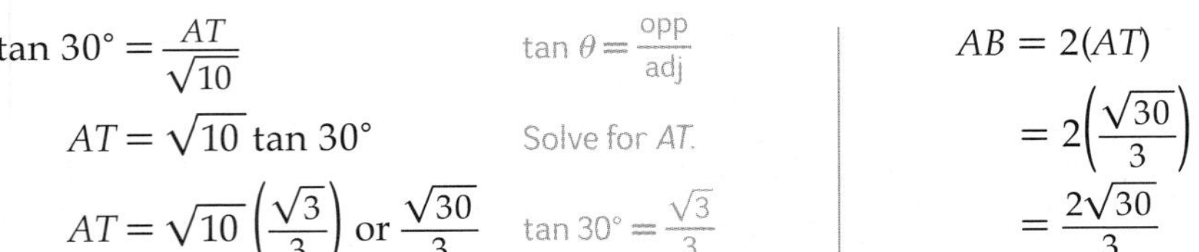

$\tan 30° = \frac{AT}{\sqrt{10}}$ $\quad \tan\theta = \frac{\text{opp}}{\text{adj}}$

$AT = \sqrt{10}\tan 30°$ $\quad$ Solve for *AT*.

$AT = \sqrt{10}\left(\frac{\sqrt{3}}{3}\right)$ or $\frac{\sqrt{30}}{3}$ $\quad \tan 30° = \frac{\sqrt{3}}{3}$

$AB = 2(AT)$

$= 2\left(\frac{\sqrt{30}}{3}\right)$

$= \frac{2\sqrt{30}}{3}$

The perimeter *P* of the hexagon is $6 \cdot \frac{2\sqrt{30}}{3}$ or $4\sqrt{30}$, which is about 21.9 units. The area of the hexagon is $\frac{1}{2}aP$, which is $\frac{1}{2}\sqrt{10}(4\sqrt{30})$ or about 34.6 units^2.

Exercises

Work cooperatively. Find the perimeter and area of each regular polygon with the given consecutive vertices. Round to the nearest tenth, if necessary.

1. pentagon *ABCDE*; *A*(1, 4), *B*(3, 1)
2. hexagon *ABCDEF*; *A*(−4, 2), *B*(0, 5)

Find the perimeter and area of each regular polygon inscribed in ⊙*O*, centered at the origin, and containing the given point. Round to the nearest tenth, if necessary.

3. pentagon *ABCDE*; *E*(−4, −1)
4. hexagon *ABCDEF*; *D*(4, −5)

Find the perimeter and area of each regular polygon circumscribed about ⊙*Q*, with the given center and point *X* on the circle. Round to the nearest tenth, if necessary.

5. pentagon *ABCDE*; *Q*(−2, 1); *X*(−1, 3)
6. octagon *ABCDEFGH*; *Q*(3, −1); *X*(1, −3)

LESSON 5

Area and Nonrigid Transformations

Then

- You used scale factors and proportions to solve problems involving the perimeters of similar figures.

Now

1. Find areas of similar figures by using scale factors.
2. Determine how changes in dimensions affect the areas of figures.

Why?

- Architecture firms often hire model makers to make scale models of projects that are used to sell their designs. Since the base of a model is geometrically similar to the base of the actual building it represents, their areas are related.

Mathematical Practices

1 Make sense of problems and persevere in solving them.
3 Construct viable arguments and critique the reasoning of others.
4 Model with mathematics.

1 Areas of Similar Figures

In Chapter 7 you learned that a nonrigid transformation changes the size but not the shape of a figure, producing a similar figure. In Lesson 7-2, you learned that if two polygons are similar, then their perimeters are proportional to the scale factor between them. The areas of two similar polygons share a different relationship.

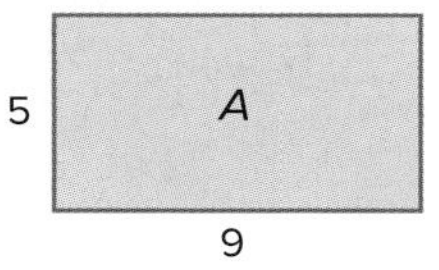

$$\frac{\text{perimeter of figure } B}{\text{perimeter of figure } A} = \frac{28k}{28} \text{ or } k$$

$$\frac{\text{area of figure } B}{\text{area of figure } A} = \frac{45k^2}{45} \text{ or } k^2$$

Theorem 10.1 Areas of Similar Polygons

Words If two polygons are similar, then their areas are proportional to the square of the scale factor between them.

Example If $ABCD \sim FGHJ$, then $\frac{\text{area of } FGHJ}{\text{area of } ABCD} = \left(\frac{FG}{AB}\right)^2$.

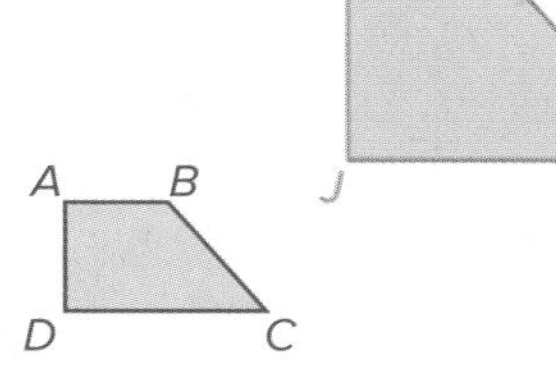

You will prove Theorem 10.1 for triangles in Exercise 22.

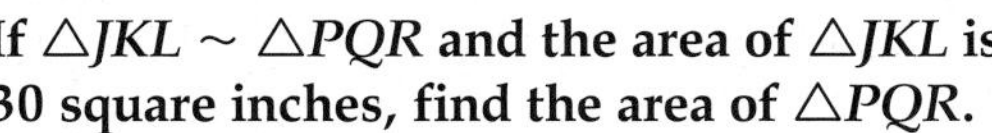

Example 1 Find Areas of Similar Polygons

If $\triangle JKL \sim \triangle PQR$ and the area of $\triangle JKL$ is 30 square inches, find the area of $\triangle PQR$.

The scale factor between $\triangle PQR$ and $\triangle JKL$ is $\frac{15}{12}$ or $\frac{5}{4}$, so the ratio of their areas is $\left(\frac{5}{4}\right)^2$.

$\frac{\text{area of } \triangle PQR}{\text{area of } \triangle JKL} = \left(\frac{5}{4}\right)^2$ Write a proportion.

$\frac{\text{area of } \triangle PQR}{30} = \frac{25}{16}$ Area of $\triangle JKL = 30$ and $\left(\frac{5}{4}\right)^2 = \frac{25}{16}$

$\text{area of } \triangle PQR = \frac{25}{16} \cdot 30$ Multiply each side by 30.

$\text{area of } \triangle PQR = 46.875$ Simplify.

So the area of $\triangle PQR$ is about 46.9 square inches.

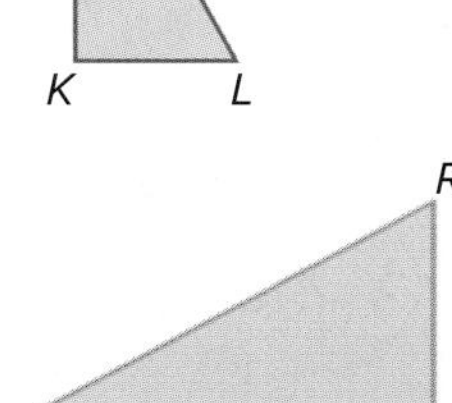

Lane Oatey/Blue Jean Images/Getty Images

Go Online!

Discover a relationship between the areas of similar figures with a **Geometer's Sketchpad®** sketch in ConnectED.

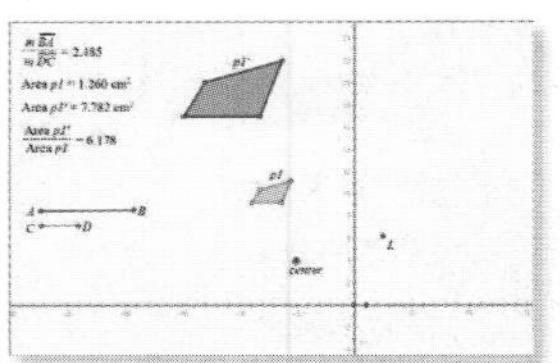

Guided Practice

For each pair of similar figures, find the area of the green figure.

1A.

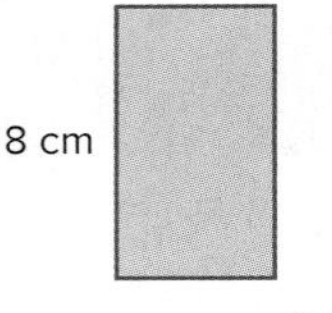

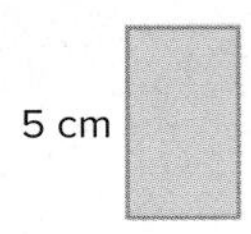

1B.

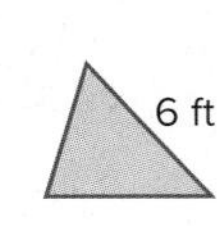

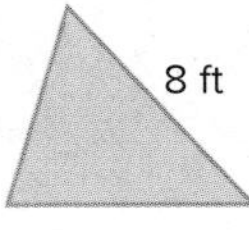

You can use the areas of similar figures to find the scale factor between them or a missing measure.

Example 2 Use Areas of Similar Figures

The area of ▱*ABCD* is 150 square meters. The area of ▱*FGHJ* is 54 square meters. If ▱*ABCD* ~ ▱*FGHJ*, find the scale factor of ▱*FGHJ* to ▱*ABCD* and the value of x.

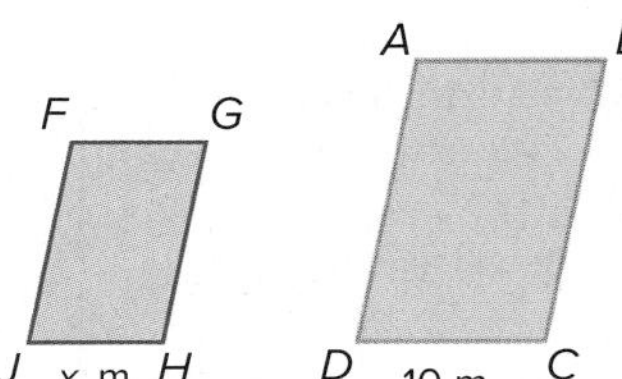

Let k be the scale factor between ▱*FGHJ* and ▱*ABCD*.

$\frac{\text{area of ▱}FGHJ}{\text{area of ▱}ABCD} = k^2$ Theorem 10.1

$\frac{54}{150} = k^2$ Substitution

$\frac{9}{25} = k^2$ Simplify.

$\frac{3}{5} = k$ Take the positive square root of each side.

So the scale factor of ▱*FGHJ* to ▱*ABCD* is $\frac{3}{5}$. Use this scale factor to find the value of x.

$\frac{JH}{DC} = k$ The ratio of corresponding lengths of similar polygons is equal to the scale factor between the polygons.

$\frac{x}{10} = \frac{3}{5}$ Substitution

$x = \frac{3}{5} \cdot 10$ or 6 Multiply each side by 10.

CHECK Confirm that $\frac{JH}{DC}$ is equal to the scale factor.

$\frac{JH}{DC} = \frac{6}{10} = \frac{3}{5}$ ✓

Watch Out!

Writing Ratios When finding the ratio of the area of Figure *A* to the area of Figure *B*, be sure to write your ratio as $\frac{\text{area of figure } A}{\text{area of figure } B}$.

Reading Math

Ratios Ratios can be written in different ways. For example, x to y, $x : y$, and $\frac{x}{y}$ are all representations of the ratio of x and y.

Guided Practice

For each pair of similar figures, use the given areas to find the scale factor of the blue to the green figure. Then find x.

2A.

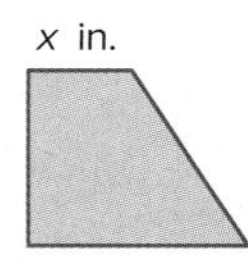

6 in.

$A = 72$ in²

2B.

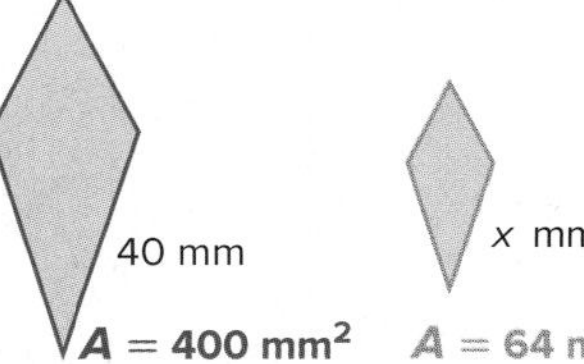

2 Dimensional Changes

When the dimensions of a figure are changed proportionally, the new figure is similar to the original figure. Changing the dimensions nonproportionally does not result in similar figures.

Real-World Example 3 Changing Dimensions

GARDENING Orlando and Mia each have 12-feet by 15-feet rectangular gardens which they plan to expand. Orlando's new garden will measure 18 feet by 22.5 feet while Mia's new garden will measure 18 feet by 18.75 feet. Describe how the changes in dimensions affect the areas of each garden.

Draw a diagram and label the measurements.

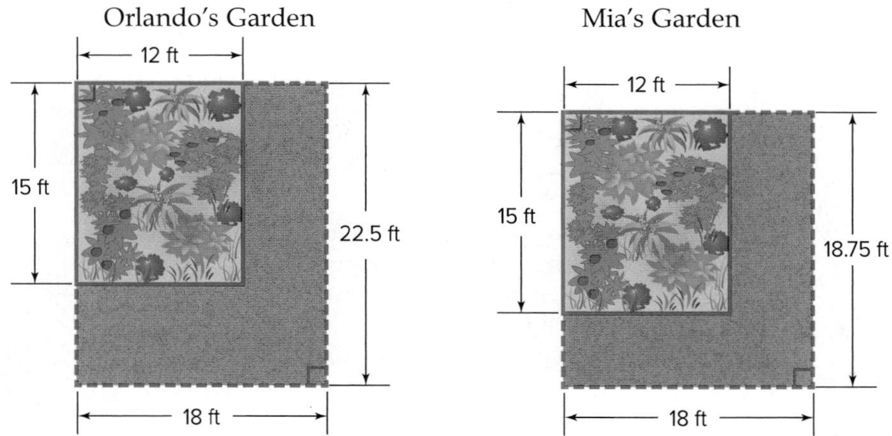

Next, compare the new dimensions to the original dimensions to determine whether the increases are proportional or nonproportional.

Orlando's Garden		Mia's Garden
$\frac{18 \text{ ft}}{12 \text{ ft}} = \frac{3}{2}$	$\frac{\text{new width}}{\text{original width}}$	$\frac{18 \text{ ft}}{12 \text{ ft}} = \frac{3}{2}$
$\frac{22.5 \text{ ft}}{15 \text{ ft}} = \frac{3}{2}$	$\frac{\text{new length}}{\text{original length}}$	$\frac{18.75 \text{ ft}}{15 \text{ ft}} = \frac{5}{4}$

Because each dimension of Orlando's garden increased by the same scale factor, this is a proportional dimension change. So the original garden and new garden are similar figures. By Theorem 10.1, the ratio of the areas is the square of the scale factor.

$$\frac{\text{new area}}{\text{original area}} = \frac{(18)(22.5)}{(12)(15)} = \frac{405}{180} = \frac{9}{4} = \left(\frac{3}{2}\right)^2$$

Because different scale factors were used for each dimension, Mia's garden is not changing proportionally. Compute the ratio of the new area to the original using the scale factors applied to each dimension.

$$\frac{\text{new area}}{\text{original area}} = \frac{(12 \cdot 1.5)(15 \cdot 1.25)}{(12)(15)} = 1.5 \cdot 1.25 \text{ or } 1.875$$

Notice that the ratio of the new area to the original area is the product of the scale factors used to enlarge the dimensions.

Orlando's garden is increasing proportionally so the ratio of the new area to the original area is the square of the scale factor. Mia's garden is not increasing proportionally. The ratio of the new area to the original area is the product of the scale factors.

Reading Math

Similar Circles Since all circles have the same shape, all circles are similar. Therefore, the areas of two circles are also related by the square of the scale factor between them.

Guided Practice

3. **CRAFTS** Miyoki is crocheting two circles. If the diameter of the smaller circle is about 8 centimeters and the diameter of the larger circle is about 12.6 centimeters, describe how the difference in dimensions affects the areas of the circles.

Check Your Understanding

◯ = Step-by-Step Solutions begin on page R13.

Example 1 For each pair of similar figures, find the area of the green figure.

1. 8 yd; 4 yd; $A = 36\text{ yd}^2$

2. 5 m; 7 m; $A = 40\text{ m}^2$

Example 2 For each pair of similar figures, use the given areas to find the scale factor from the blue to the green figure. Then find x.

3. x cm; 21 cm; $A = 875\text{ cm}^2$; $A = 315\text{ cm}^2$

4. 15 in.; x in.; $A = 153\text{ in}^2$; $A = 272\text{ in}^2$

Example 3

5. **MEMORIES** Zola has a picture frame that holds all of her school pictures. Each small opening is similar to the large opening in the center. If the center opening has an area of 33 square inches, what is the area of each small opening?

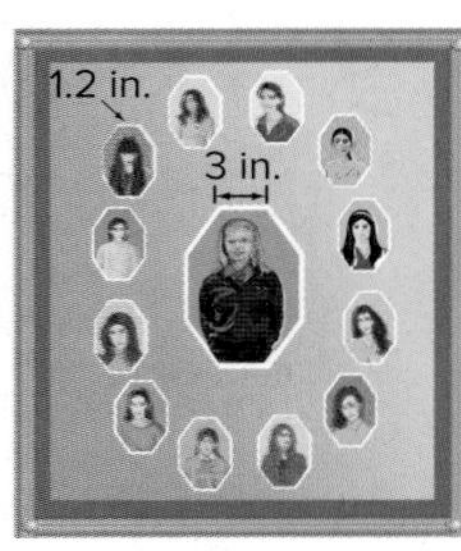

Practice and Problem Solving

Extra Practice is on page R10.

Example 1 For each pair of similar figures, find the area of the green figure.

6. 10 mm; 18 mm; $A = 25\text{ mm}^2$

7. 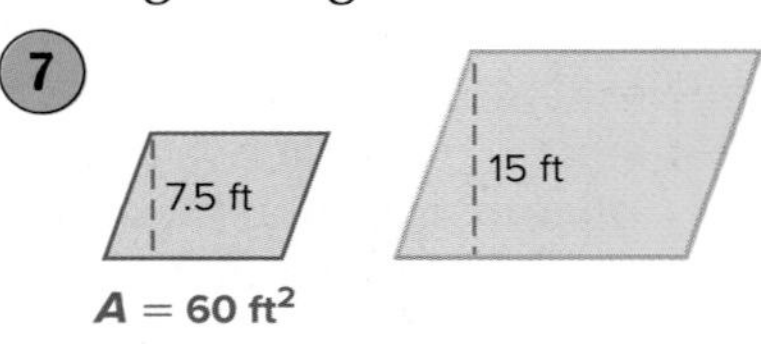

8. 28 in.; 15.4 in.; $A = 500\text{ in}^2$

9. 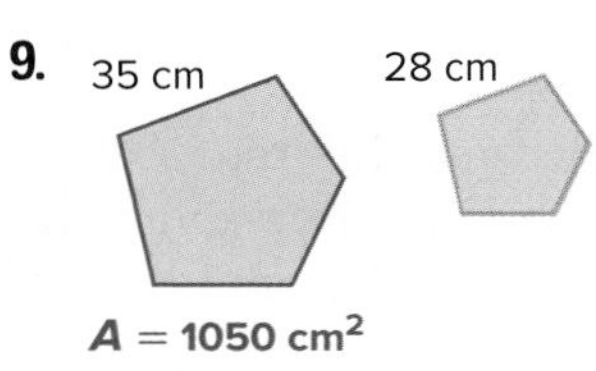

Example 2 **STRUCTURE** For each pair of similar figures, use the given areas to find the scale factor of the blue to the green figure. Then find x.

10. 12 m; x m; $A = 72\text{ m}^2$; $A = 50\text{ m}^2$

11. 14 in.; x in.; $A = 96\text{ in}^2$; $A = 150\text{ in}^2$

12. x ft; 14 ft; $A = 27\text{ ft}^2$; $A = 147\text{ ft}^2$

13. x cm; 24 cm; $A = 846\text{ cm}^2$; $A = 376\text{ cm}^2$

Example 3

14. **CRAFTS** Marina crafts unique trivets and other kitchenware. Her basic trivet design is an equilateral triangle with an area of about 3.9 square inches. She plans to make trivet A by increasing each side by $\frac{4}{3}$. To make trivet B, Marina will double the length of one side while keeping the height as measured from the doubled side the same as the basic trivet. What are the approximate areas of trivets A and B?

15. **CHANGING DIMENSIONS** A circle has a radius of 24 inches.
   a. If the area is doubled, how does the radius change?
   b. How does the radius change if the area is tripled?
   c. What is the change in the radius if the area is increased by a factor of x?

16. **CHANGING DIMENSIONS** A polygon has an area of 144 square meters.
   a. If the area is doubled, how does each side length change?
   b. How does each side length change if the area is tripled?
   c. What is the change in each side length if the area is increased by a factor of x?

17. **BAKING** Kaitlyn wants to use one of two regular hexagonal cake pans for a recipe she is making. The side length of the larger pan is 4.5 inches, and the area of the base of the smaller pan is 41.6 square inches.
   a. What is the side length of the smaller pan?
   b. The recipe that Kaitlyn is using calls for a circular cake pan with an 8-inch diameter. Which pan should she choose? Explain your reasoning.

18. MP **MODELING** Federico's family is putting hardwood floors in the two geometrically similar rooms shown. If the cost of flooring is constant and the flooring for the kitchen cost $2000, what will be the total flooring cost for the two rooms? Round to the nearest hundred dollars.

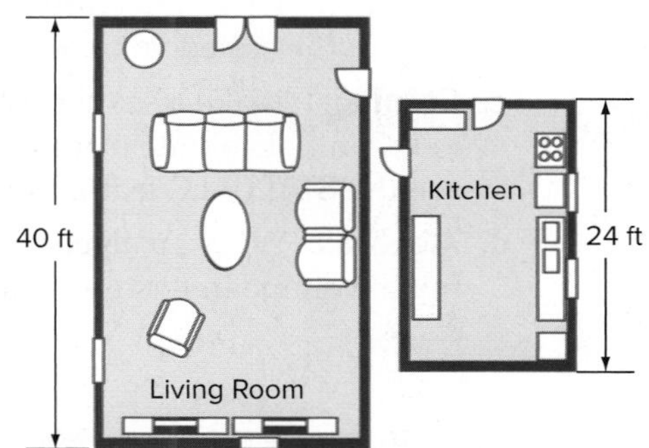

COORDINATE GEOMETRY Find the area of each figure. Use the segment length given to find the area of a similar polygon.

19. $J'L' = 3$

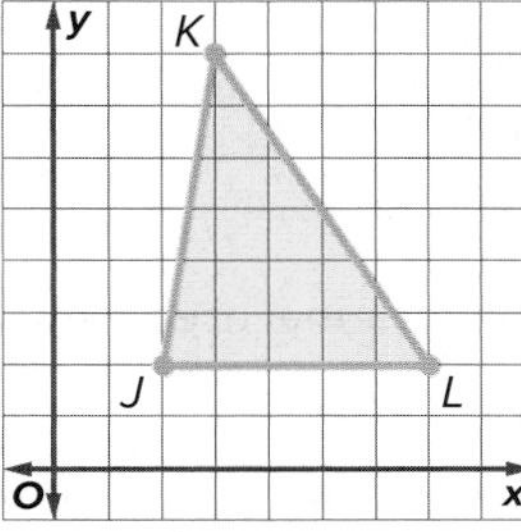

20. $W'X' = 8$

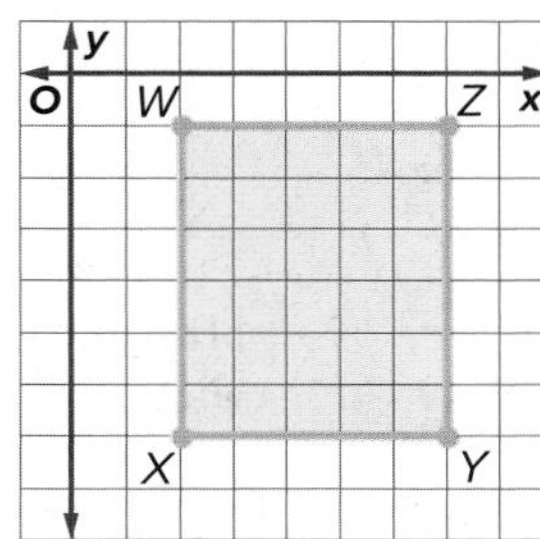

21. $B'C' = 5$

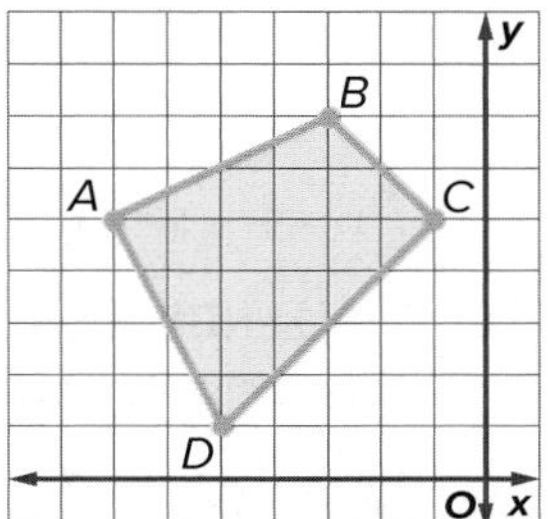

22. **PROOF** Write a paragraph proof.

Given: $\triangle ABC \sim \triangle XYZ$

Prove: $\frac{\text{area of } \triangle ABC}{\text{area of } \triangle XYZ} = \frac{a^2}{x^2}$

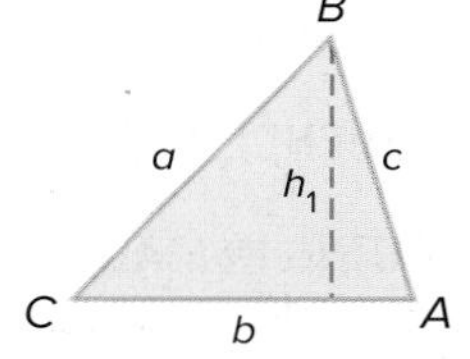

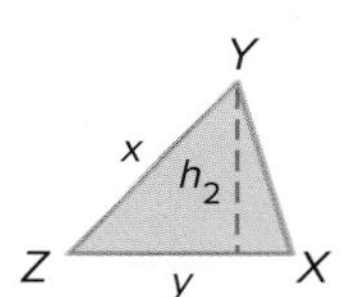

23 **STATISTICS** The graph shows the increase in high school tennis participation from 2000 to 2010.

a. Explain why the graph is misleading.

b. How could the graph be changed to more accurately represent the growth in high school tennis participation?

24. **MULTIPLE REPRESENTATIONS** In this problem, you will investigate changing dimensions proportionally in three-dimensional figures.

a. **Tabular** Copy and complete the table below for each scale factor of a rectangular prism that is 2 inches by 3 inches by 5 inches.

Scale Factor	Length (in.)	Width (in.)	Height (in.)	Volume (in^3)	Ratio of Scaled Volume to Initial Volume
1	3	2	5		
2					
3					
4					
5					
10					

b. **Verbal** Make a conjecture about the relationship between the scale factor and the ratio of the scaled volume to the initial volume.

c. **Graphical** Make a scatter plot of the scale factor and the ratio of the scaled volume to the initial volume using the **STAT PLOT** feature on your graphing calculator. Then use the **STAT CALC** feature to approximate the function represented by the graph.

d. **Algebraic** Write an algebraic expression for the ratio of the scaled volume to the initial volume in terms of scale factor k.

H.O.T. Problems Use Higher-Order Thinking Skills

25. MP **CRITIQUE ARGUMENTS** Violeta and Gavin are trying to come up with a formula that can be used to find the area of a circle with a radius r after it has been enlarged by a scale factor k. Is either of them correct? Explain your reasoning.

Violeta	Gavin
$A = k\pi r^2$	$A = \pi(r^2)^k$

26. **CHALLENGE** If you want the area of a polygon to be x% of its original area, by what scale factor should you multiply each side length?

27. MP **REASONING** A regular n-gon is enlarged, and the ratio of the area of the enlarged figure to the area of the original figure is R. Write an equation relating the perimeter of the enlarged figure to the perimeter of the original figure Q.

28. **OPEN-ENDED** Draw a pair of similar figures with areas that have a ratio of 4 : 1. Explain.

29. **WRITING IN MATH** Explain how to find the area of an enlarged polygon if you know the area of the original polygon and the scale factor of the enlargement.

30. In the figure, $\triangle PQR \sim \triangle STU$. The area of $\triangle PQR$ is 50 square centimeters and the area of $\triangle STU$ is 32 square centimeters.

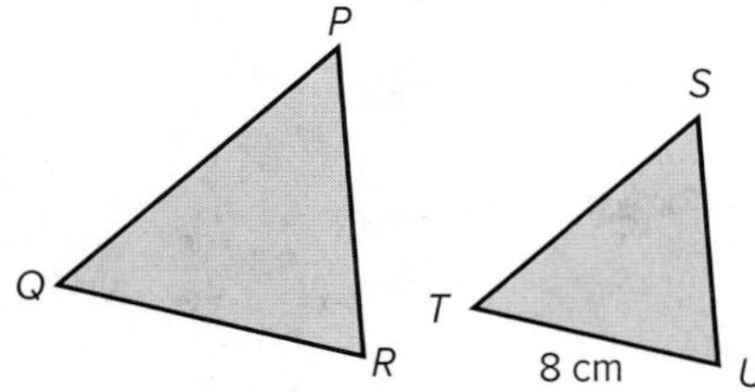

What is the length of $\overline{QR}$? MP 1

- A 5.12 cm
- B 6.4 cm
- C 10 cm
- D 12.5 cm
- E 26 cm

31. Connor drew the trapezoids below so that trapezoid *ABCD* ~ trapezoid *WXYZ*. The area of trapezoid *ABCD* is 55 square feet. MP 2

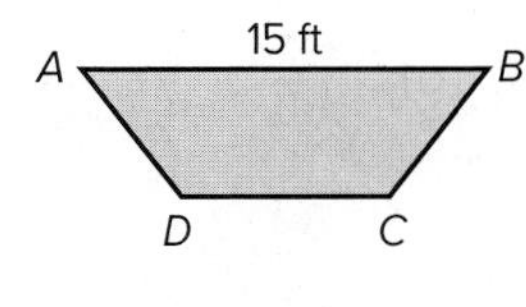

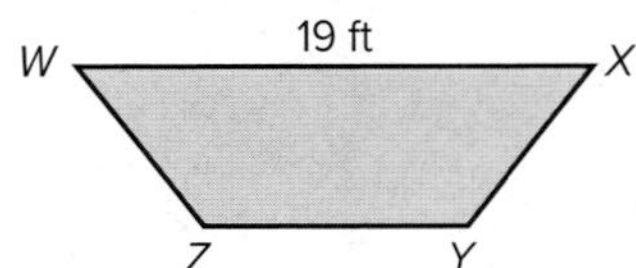

Which of the following is the best estimate of the area of trapezoid *WXYZ*?

- A 88 ft²
- B 70 ft²
- C 43 ft²
- D 34 ft²

32. If the area of a circle of radius r is 154 cm², what is the area of a circle of radius $4r$? MP 6

33. Two parallelograms are similar. Parallelogram *A* has an area of 48 square feet and has a base of 16 feet. Parallelogram *B* has an area of 27 square feet and has a base of x feet. What is the scale factor from parallelogram *A* to parallelogram *B* and was is the value of x? MP 1

34. One right $\triangle ABC$ has an area of 36 square inches. A similar right $\triangle DEF$ has an area of 4 square inches. One leg of $\triangle ABC$ is 6 inches. Find the length of the corresponding leg of $\triangle DEF$. Explain your work. MP 2

35. **MULTI-STEP** Alisha uses geometry software to draw ▱*FGHJ*, as shown. According to the software, the area of ▱*FGHJ* is 200 square millimeters. Alisha also uses the software to create ▱*KLMN* so that ▱*FGHJ* ~ ▱*KLMN*. MP 2

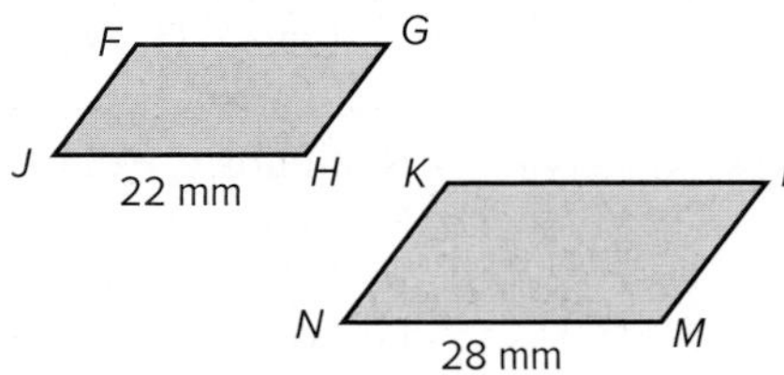

a. Write and simplify a ratio to show the scale factor between ▱*KLMN* and ▱*FGHJ*.

b. Write a proportion to show the ratio of the areas of the two parallelograms.

c. Solve the proportion to find the area of *KLMN*. Give the correct units.

36. Triangles *ABC* and *DEF* are similar. The area of triangle *ABC* is 15 square inches. The height of triangle *ABC* is 5 inches. The height of triangle *DEF* is 13 inches.

a. What is the scale factor from triangle *DEF* to triangle *ABC*? MP 1

b. What is the area of triangle *DEF*? MP 1

- A 26 square inches
- B 86 square inches
- C 101 square inches
- D 106 square inches
- E 156 square inches

LESSON 6

Surface Area

Then

- Find the perimeters and areas of similar figures by using scale factors.

Now

1. Find surface areas of prisms and cylinders.
2. Find surface areas of pyramids and cones.

Why?

- There are many styles of fish tanks that are cylinders and prisms. The style of tank like the one on the right allows people to walk around the tank to find the best view as they observe the fish.

New Vocabulary

lateral face
lateral edge
base edge
altitude
height
lateral area
axis
regular pyramid
slant height
composite solid

Mathematical Practices

1 Make sense of problems and preserve in solving them.
2 Reason abstractly and quantitatively.
6 Attend to precision.

1 Prisms and Cylinders

In a solid figure, faces that are not bases are called **lateral faces**. Lateral faces intersect each other at the **lateral edges**, which are parallel and congruent. The lateral faces intersect the base at the **base edges**. The **altitude** is a perpendicular segment that joins the planes of the bases. The **height** is the length of the altitude.

Recall that a prism is a polyhedron with two parallel congruent bases. In a right prism, the lateral edges are altitudes and the lateral faces are rectangles. In an oblique prism, the lateral edges are not perpendicular to the bases. At least one lateral face is not a rectangle.

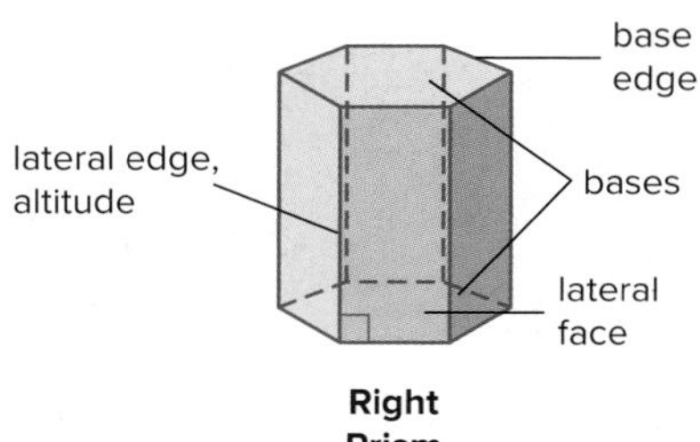

Right Prism

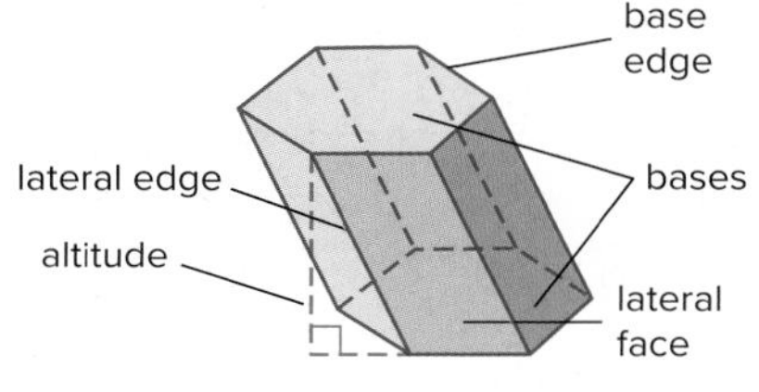

Oblique Prism

The **lateral area** L of a prism is the sum of the areas of the lateral faces. The net at the right shows how to find the lateral area of a prism.

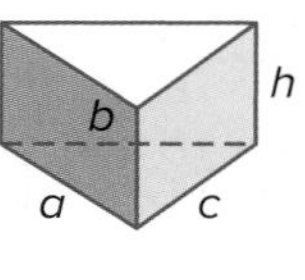

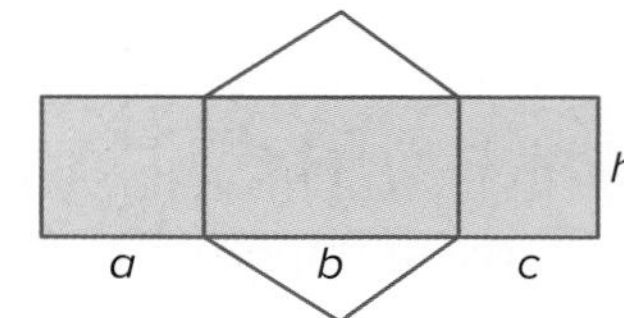

$L = a(h) + b(h) + c(h)$ Sum of the areas of the lateral faces

$= (a + b + c)(h)$ Distributive Property

$= Ph$ $P = a + b + c$

The surface area of a prism is the sum of the lateral area and the areas of the bases.

Key Concepts Lateral and Surface Area of a Prism

Words: The lateral area L of a right prism is the product of the perimeter of the base P and the height of the prism h.	**Words:** The surface area S of a right prism is the sum of the lateral area and twice the area of a base B.
Symbols: $L = Ph$	**Symbols:** $S = L + 2B$ or $S = Ph + 2B$

From this point on, you can assume that solids in the text are right solids. If a solid is oblique, it will be clearly stated.

WatchOut!

Right Prisms The bases of a right prism are congruent, but the lateral faces are not always congruent.

Example 1 Surface Area of a Prism

Find the surface area of the prism. Round to the nearest tenth.

Step 1 Find the missing side length of the base.

$c^2 = 6^2 + 5^2$ Pythagorean Theorem

$c^2 = 61$ Simplify.

$c \approx 7.8$ Take the positive square root of each side.

Step 2 Find the surface area.

$S = Ph + 2B$ Surface area of a prism

$\approx (5 + 6 + 7.8)(7) + 2\left(\frac{1}{2}\right)(5)(6)$ Substitution

≈ 161.6 Simplify.

The surface area of the prism is approximately 161.6 square centimeters.

Guided Practice

Find the surface area of each prism.

1A. A rectangular prism has a base with dimensions of 9 feet by 4 feet and a height of 6 feet.

1B. A regular hexagonal prism with side length of 5 centimeters for the base and a height of 12 centimeters.

Real-World Example 2 Use the Surface Area of a Prism

PACKAGING The Marketing Department of a shoe manufacturer is designing what will be printed on the boxes. Each box is a rectangular prism with a base measuring 6 inches by 14 inches and a height of 5 inches. What is the surface area of the shoebox available for printing?

$S = Ph + 2B$ Surface area of a prism

$= (2(6) + 2(14))(5) + 2(6)(14)$ Substitution

$= 368$ Simplify.

The surface area of the shoebox is 368 square inches.

Guided Practice

2. FOOD A piece of cheese in the shape of a triangular prism is shrink wrapped for sale. The base is a right triangle with legs 3 centimeters and 4 centimeters. The height of the piece is 4 centimeters. What is the surface area of the shrink wrap?

The **axis** of a cylinder is the segment with endpoints that are centers of the circular bases. If the axis is also an altitude, then the cylinder is a right cylinder. If not then the cylinder is oblique.

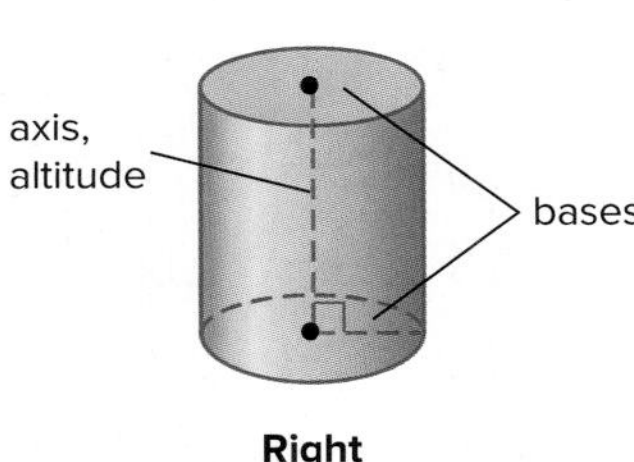

Right Cylinder

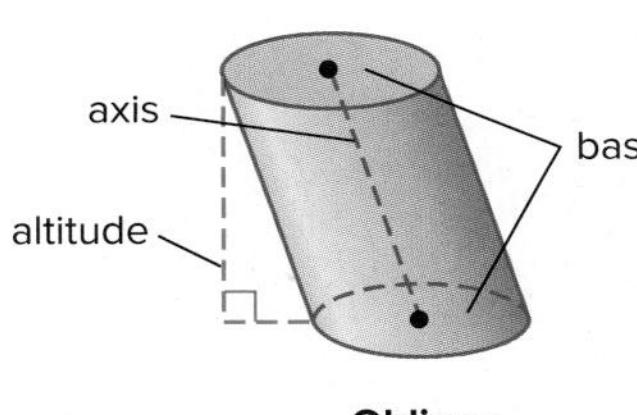

Oblique Cylinder

The lateral area of a right cylinder is the area of the curved surface. Like a right prism, the lateral area L equals Ph. Because the base is a circle, the perimeter is the circumference of the circle C.

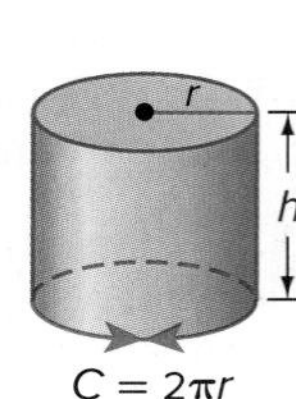

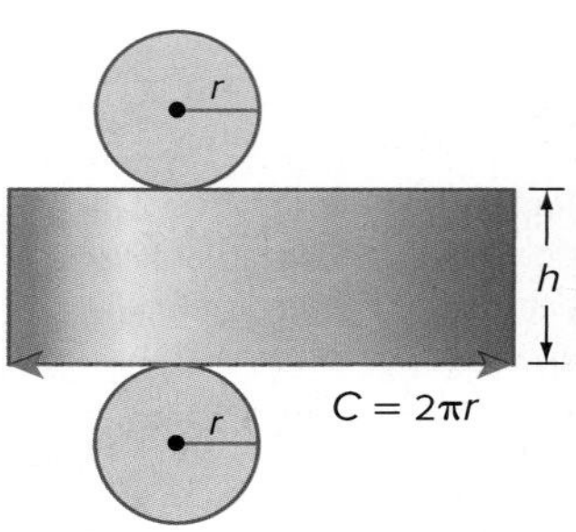

So, the lateral area is $Ch = 2\pi rh$.

The surface area is the sum of the lateral area and the areas of the bases.

Key Concepts Lateral and Surface Area of a Cylinder

Words: The lateral area L of a right cylinder with radius r is the product of the circumference of the base C and the height h of the cylinder.	**Words:** The surface area S of a right cylinder is the sum of the lateral area and twice the area of a base B.
Symbols: $L = Ch$ or $L = 2\pi rh$.	**Symbols:** $S = L + 2B$, $S = Ch + 2B$, or $S = 2\pi rh + 2\pi r^2$.

StudyTip

Formulas The formula for the surface area, $S = L + 2B$, is the same for a right prism and a right cylinder.

Example 3 Surface Area of a Cylinder

Find the surface area of the cylinder. Round to the nearest tenth.

$S = 2\pi rh + 2\pi r^2$ Surface area of a cylinder

$= 2\pi(7.5)(18) + 2\pi(7.5)^2$ Replace r with 7.5 and h with 18.

≈ 1201.6 Simplify.

15 mm

18 mm

The surface area of the cylinder is about 1201.7 square millimeters.

StudyTip

Estimation Before finding the lateral area of a cylinder, use mental math to estimate. To estimate, multiply the diameter by 3 (to approximate π) and then by the height of the cylinder.

Guided Practice

3A. The radius is 5 inches and the height is 9 inches.

3B. The diameter is 6 centimeters and the height is 4.8 centimeters.

You can use surface area formulas to solve real-world problems.

Real-World Example 4 Use the Surface Area of a Cylinder

CRAFTS Sheree used the rectangular piece of felt shown at the right to cover the curved surface of her cylindrical pencil holder. The circumference corresponds to the side measuring 12.6 inches. What is the surface area of the entire pencil holder to the nearest tenth?

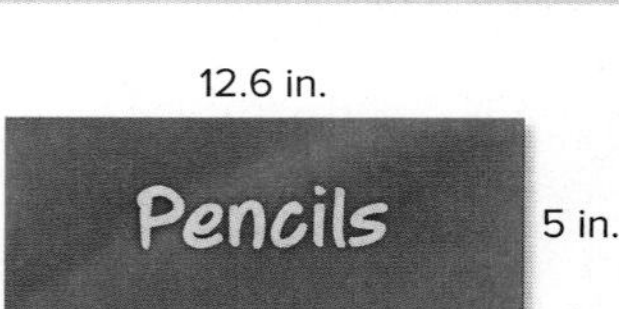

Step 1 Find the radius.

$C = 2\pi r$ Circumference of a circle

$12.6 = 2\pi r$ Replace C with 12.6.

$4.013 \approx 2r$ Divide both sides by π.

$2.0 \approx r$ Divide both sides by 2.

The radius of the pencil holder is about 2 inches.

Step 2 Find the surface area.

$S = L + 2B$	Surface area of a cylinder
$\approx 12.6 \cdot 5 + 2\pi(2.0)^2$	Replace L with $12.6 \cdot 5$ and r with 2.0.
≈ 88.1	Simplify.

The surface area of the pencil holder is about 88.1 square inches.

Guided Practice

9. **MANUFACTURING** A cylindrical juice can has a rectangular label with a length of 20 centimeters, which corresponds to the circumference, and height of 8 centimeters. What is the surface area of material needed to manufacture the can?

2 Pyramids and Cones

The *lateral faces* of a pyramid intersect at a common point called a *vertex*. Two lateral faces intersect at a *lateral edge*. A lateral face and the base intersect at a *base edge*. The *altitude* is the segment from the vertex perpendicular to the base.

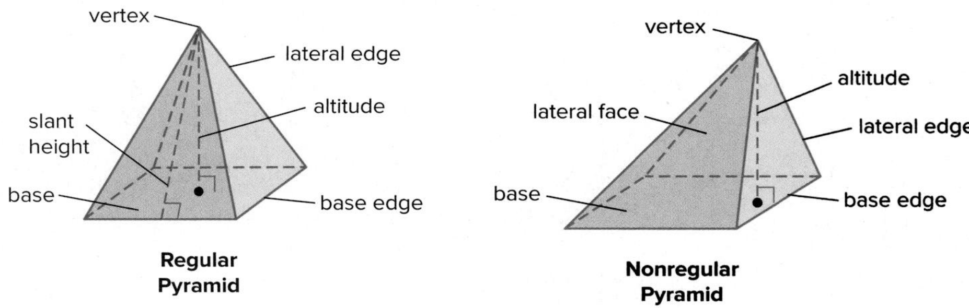

A **regular pyramid** has a base that is a regular polygon and the altitude has an endpoint at the center of the base. All the lateral edges are congruent and all the lateral faces are congruent isosceles triangles. The height of each lateral face is called the **slant height** ℓ of the pyramid.

The lateral area L of a regular pentagonal pyramid is the sum of the areas of the lateral faces, all of which are congruent triangles, as shown in the net at the right.

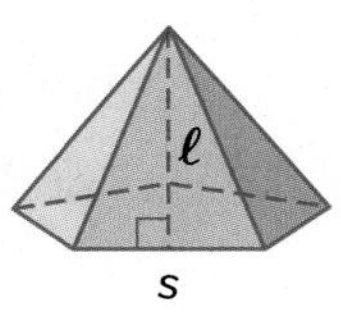

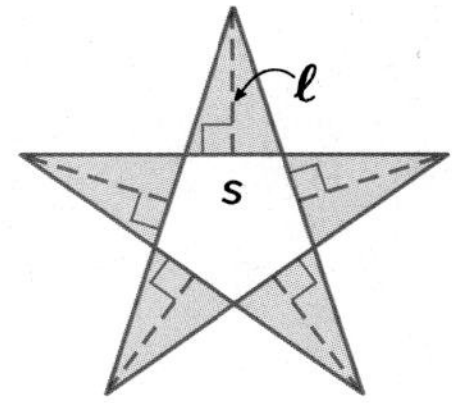

$L = \frac{1}{2}s\ell + \frac{1}{2}s\ell + \frac{1}{2}s\ell + \frac{1}{2}s\ell + \frac{1}{2}s\ell$	Sum of the areas of the lateral faces
$= \frac{1}{2}\ell(s + s + s + s + s)$	Distributive Property
$= \frac{1}{2}P\ell$	$P = s + s + s + s + s$

The surface area of a pyramid is the sum of the lateral area and the area of the base.

StudyTip

Making Connections The surface area of a pyramid equals $L + B$, not $L + 2B$, because a pyramid has only one base.

Key Concepts Lateral and Surface Area Formulas for a Pyramid

Words: The lateral area L of a regular pyramid is the product of one half the perimeter P of the base and the slant height ℓ.	**Words:** The surface area S of a right pyramid is the sum of the lateral area and twice the area of a base B.
Symbols: $L = \frac{1}{2}P\ell$	**Symbols:** $S = L + B$, or $S = \frac{1}{2}P\ell + B$

Example 5 Surface Area of a Regular Pyramid

Find the surface area of the regular pyramid. Round to the nearest tenth.

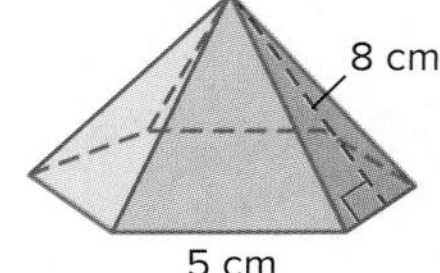

Step 1 Find the perimeter of the base.
$P = 6 \cdot 5$ or 30 centimeters

Step 2 Find the length of the apothem and the area of the base.

A central angle of the hexagon is $\frac{360°}{6}$ or 60°, so the angle formed in the triangle at the right is 30°.

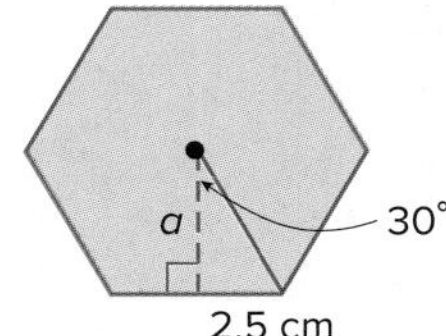

$\tan 30° = \frac{2.5}{a}$ — Write a trigonometric ratio to find the apothem a.

$a = \frac{2.5}{\tan 30°}$ — Solve for a.

≈ 4.3 — Use a calculator.

$A = \frac{1}{2}Pa$ — Area of a regular polygon

$\approx \frac{1}{2}(30)(4.3)$ — Replace P with 30 and a with 4.3.

≈ 64.5 — Multiply.

So, the area of the base B is about 64.5 square centimeters.

Step 3 Find the surface area of the pyramid.

$S = \frac{1}{2}P\ell + B$ — Surface area of a regular pyramid

$= \frac{1}{2}(30)(8) + 64.5$ — Substitute.

≈ 184.5 — Simplify.

The surface area of the pyramid is about 184.5 square centimeters.

Review Vocabulary

Trigonometric Ratios

$\sin A = \frac{opp}{hyp}$

$\cos A = \frac{adj}{hyp}$

$\tan A = \frac{opp}{adj}$

Guided Practice

Find the surface area of each pyramid. Round to the nearest tenth.

5A. an equilateral triangular pyramid with base edge 6 feet and a slant height of 8 feet

5B. a square pyramid with a base edge of 12 cm and a height of 16 cm

Real-World Example 6 Use the Surface Area of a Pyramid

MODELS Alison plans to paint a model of the Great Pyramid. What is the surface area if the model is a square pyramid with slant height of 10 inches and a base edge of 12 inches?

$S = \frac{1}{2}P\ell + B$ — Surface area of a regular pyramid

$= \frac{1}{2}(48)10 + 144$ — Replace P with 48, ℓ with 10, and B with 144.

$= 384$ — Simplify.

The surface area of the pyramid is 384 square inches.

Guided Practice

6. The height of another square pyramid model is 7 centimeters, and the base edge is 8 centimeters. What is the surface area of this model?

Recall that a cone has a circular base and a vertex. The axis of a cone is the segment with endpoints at the vertex and the center of the base. If the axis is also an altitude, then the cone is a *right cone*. If the axis is not the altitude, then the cone is an *oblique cone*.

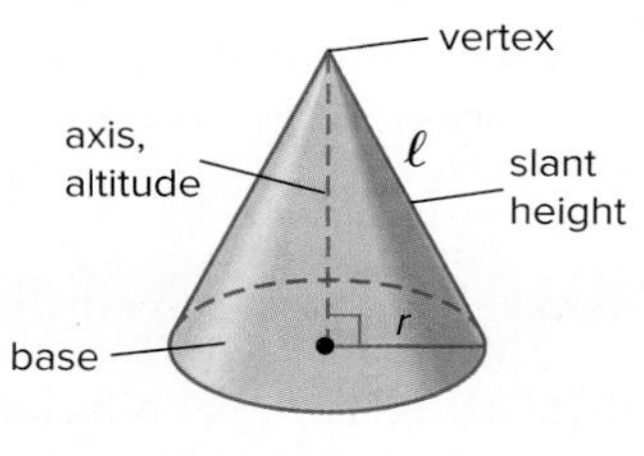

Right Cone

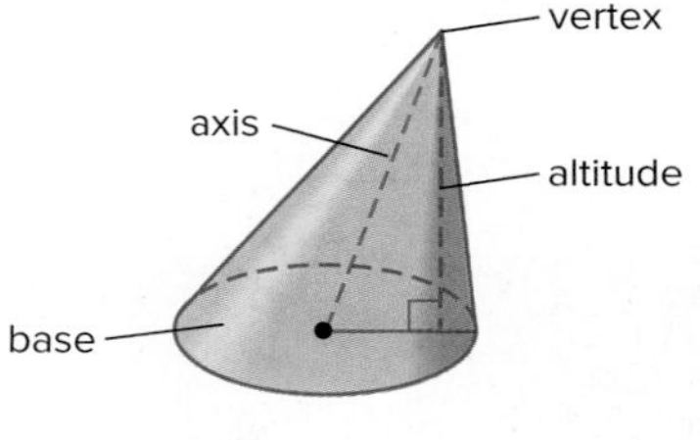

Oblique Cone

The net for a cone is shown at the right. The circle with radius r is the base of the cone. It has a circumference of $2\pi r$ and an area of πr^2. The sector with radius ℓ is the lateral surface of the cone. Its arc measure is $2\pi r$. You can use a proportion to find its area.

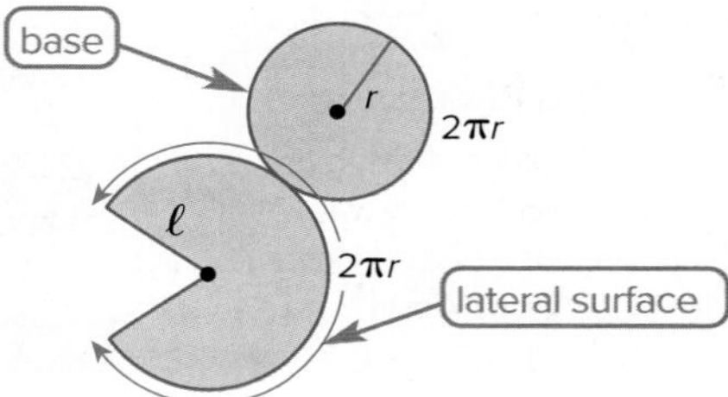

$$\frac{\text{area of sector}}{\text{area of circle}} = \frac{\text{measure of arc}}{\text{circumference of circle}}$$

$$\frac{\text{area of sector}}{\pi\ell^2} = \frac{2\pi r}{2\pi\ell}$$

$$\text{area of sector} = \pi\ell^2\left(\frac{2\pi r}{2\pi\ell}\right) = \pi r\ell$$

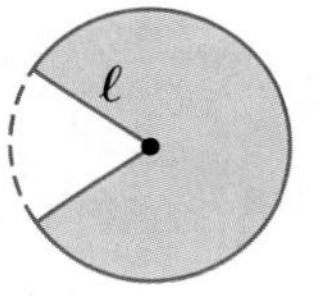

sector

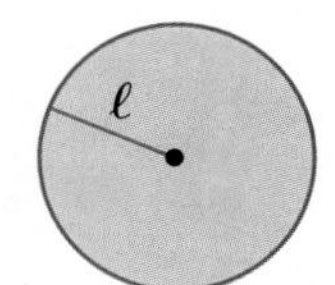

circle that contains the sector

The surface area of a cone is the sum of the lateral and the area of the base.

Key Concepts Lateral and Surface Area of a Cone

Words: The lateral area L of a right circular cone is the product of π, the radius r, and the slant height ℓ.	**Words:** The surface area S of a right circular cone is the sum of the lateral area and the area of a base.
Symbols: $L = \pi r\ell$.	**Symbols:** $S = L + B$ or $S = \pi r\ell + \pi r^2$.

Study Tip

MP Sense Making Like a pyramid, the lateral area L of a right circular cone equals $\frac{1}{2}P\ell$ where perimeter P is the circumference C of the base.

$L = \frac{1}{2}P\ell$

$= \frac{1}{2}C\ell$

$= \frac{1}{2}(2\pi r)\ell$

$= \pi r\ell$

Example 7 Surface Area of a Cone

Find the surface area of the cone. Round to the nearest tenth.

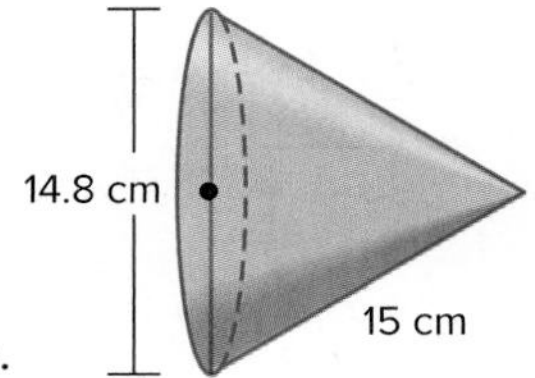

$S = \pi r\ell + \pi r^2$ Surface area of a cone

$= \pi(7.4)(15) + \pi(7.4)^2$ $r = 7.4$ and $\ell = 15$

≈ 520.8 Use a calculator.

The surface area of the cone is about 520.8 square centimeters.

Guided Practice

Find the surface area of each cone. Round to the nearest tenth.

7A. The radius is 0.8 millimeters with a slant height of 2.2 millimeters.

7B. The diameter is 12 inches with a slant height of 7 inches.

A **composite solid** is a three-dimensional figure that is composed of simpler figures. To find the lateral area or surface area of a composite figure, analyze each of the simpler figures contained in the composite figure.

Real-World Link

The construction of the town wall of Tallinn, Estonia, was begun in 1310. It was added to UNESCO's list of World Heritage Sites in 1997.

Real-World Example 8 Find the Lateral Area of a Composite Solid

ARCHITECTURE **A tower similar to the one shown at the left has a height of 16 feet and a radius of 12 feet. The cylindrical base is 10 feet tall. Find the lateral area of the tower.**

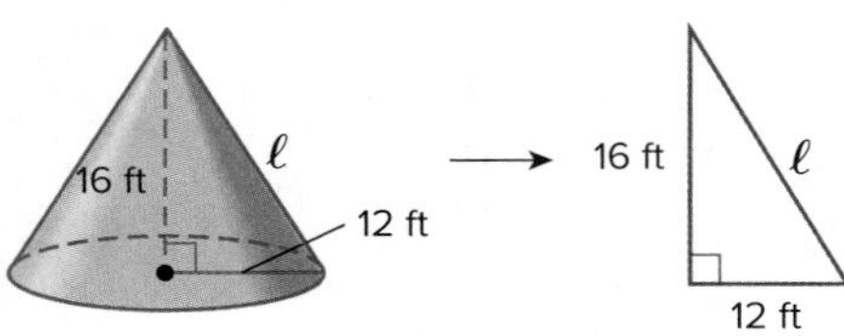

Step 1 Find the lateral area of the cone.

Use the Pythagorean Theorem to find the slant height.

$\ell^2 = 12^2 + 16^2$

$\ell^2 = 400$ Pythagorean Theorem

$\ell = 20$ Take the positive square root of each side.

Find the lateral area.

$L = \pi r \ell$ Lateral area of the cone

$= \pi(12)(20)$

≈ 754 $r = 12$ and $h = 20$

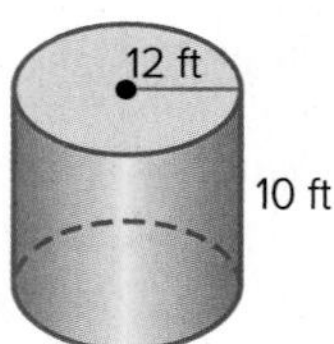

Step 2 Find the lateral area of the cylindrical base.

$L = 2\pi rh$ Lateral area of the cylinder

$= 2\pi(12)(10)$

≈ 754 $r = 12$ and $h = 10$

Step 3 Add the lateral areas.

$L_{cone} + L_{cylinder} = 754 + 754$

$= 1508$

The lateral area of the structure is about 1508 square feet.

Guided Practice

8. **OBELISK** An obelisk is a composite solid made of a square pyramid atop a rectangular prism. If the obelisk has a total height of 50 feet, a side length of 6 feet, and the pyramid is 8 feet tall, find the lateral area of the obelisk to the nearest tenth.

Check Your Understanding

◯ = Step-by-Step Solutions begin on page R13.

Examples 1, 3, 5, 7

Find the surface area of each solid. Round to the nearest tenth if necessary.

1.
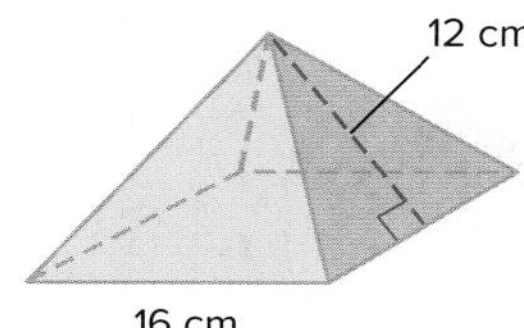

2.
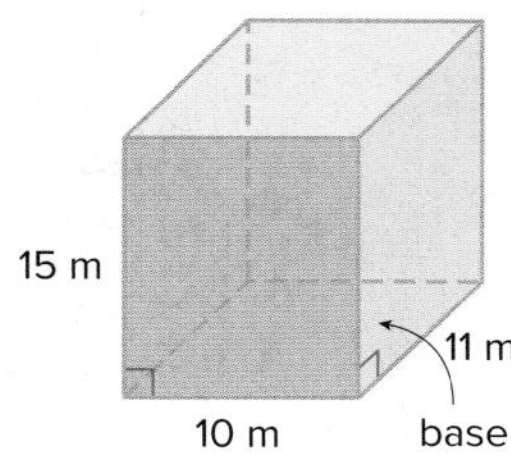

3.
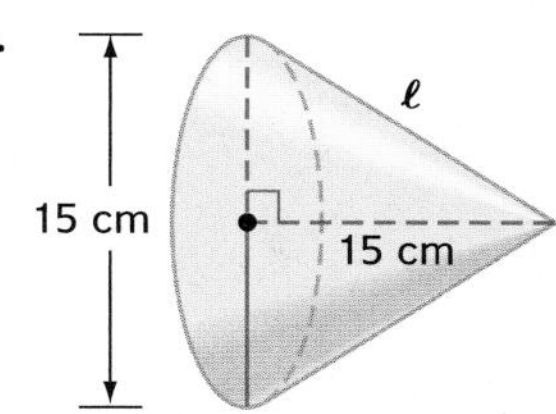

4.
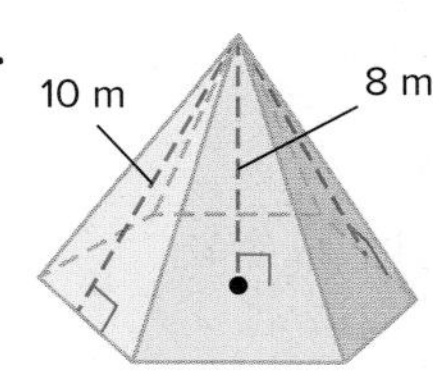

5.
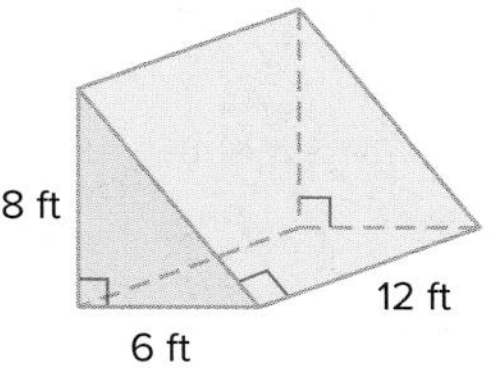

6.
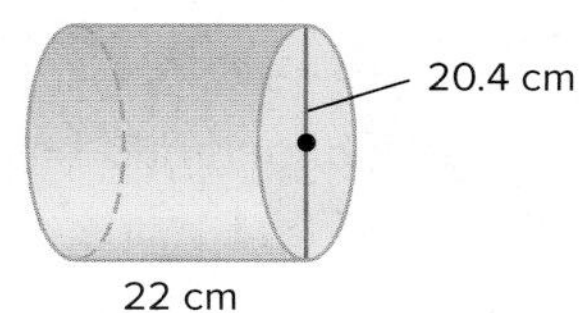

Examples 2, 4, 6, 8

7. **CARS** Evan is buying new tire rims that are 14 inches in diameter and 6 inches wide. Determine the surface area of each rim. Round to the nearest tenth.

8. **PATIO STONES** A patio stone has a rectangular base that is 3 inches by 8 inches and a height of 4 inches. What is the surface area of the stone?

9. **ROOFING** A pyramid shaped roof has a square base that is 30 feet wide and a slant height of 14 feet. How much roofing material is needed to cover the roof?

10. **TANKS** A storage tank is shown at the right. Round to the nearest tenth.
    a. Find the lateral area of the cylinder.
    b. Find the lateral area of the cone.
    c. Find the total lateral area of the tank.

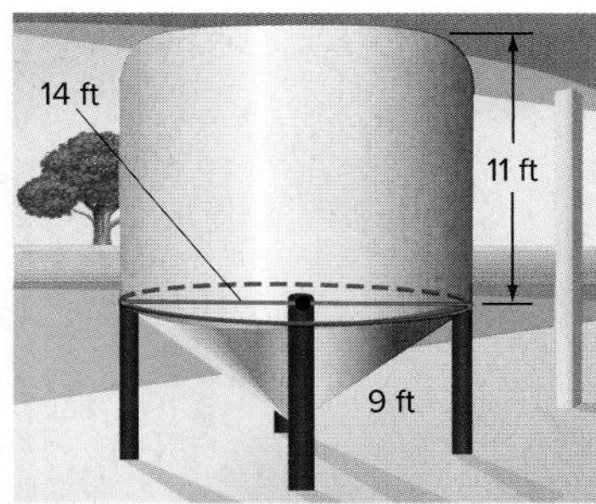

Practice and Problem Solving

Extra Practice is found on page R10.

Examples 1, 3, 5, 7

Find the surface area of each solid. Round to the nearest tenth if necessary.

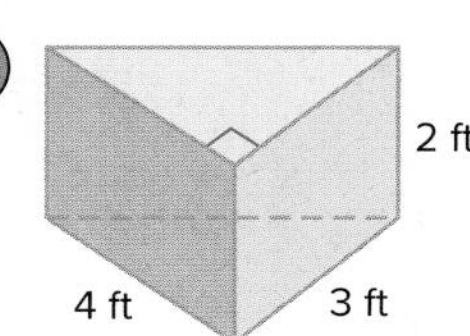

12.
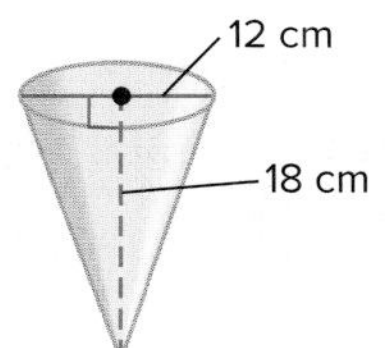

Examples 1, 3, 5, 7 **Find the surface area of each solid. Round to the nearest tenth if necessary.**

13.

1.1 cm
3.6 cm

14.

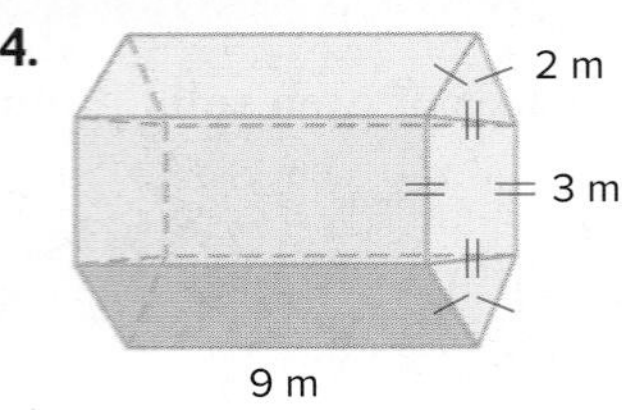

15.

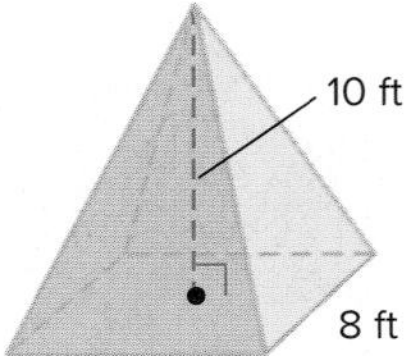

16.

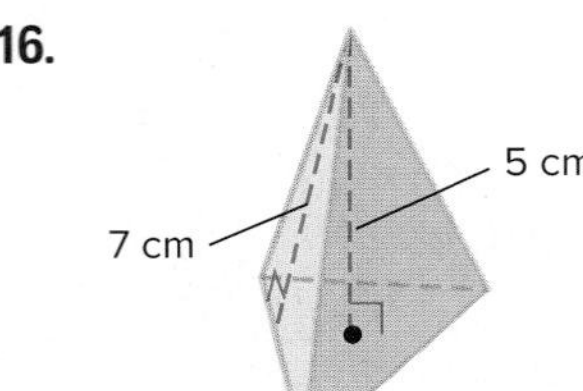

17. A cone has a diameter of 3.4 centimeters and the slant height is 6.5 centimeters.

18. A rectangular prism has $\ell = 25$ centimeters, $w = 18$ centimeters, and $h = 12$ centimeters.

19. A regular hexagonal pyramid has a base edge of 6 millimeters and a slant height of 9 millimeters.

20. A right triangular prism has $h = 6$ inches and a base with legs 9 inches and 12 inches long.

21. A cylinder has a diameter of 8 inches and a height of 6.2 inches.

22. A square pyramid has an altitude of 12 inches and a slant height of 18 inches.

23. A cylinder has a radius of 3 millimeters and a height of 15 millimeters.

24. A cone has an altitude of 5 feet and a slant height of 9.5 feet.

Examples 2, 4, 6, 8 **25.** Find the lateral area of the tent to the nearest tenth.

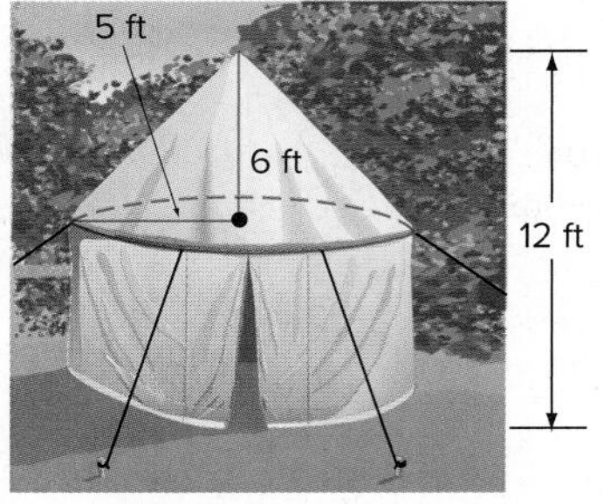

26. Find the lateral area of the dog house with a 12-inch square cut out of one face for the door.

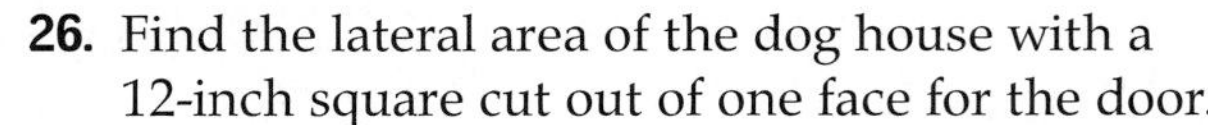

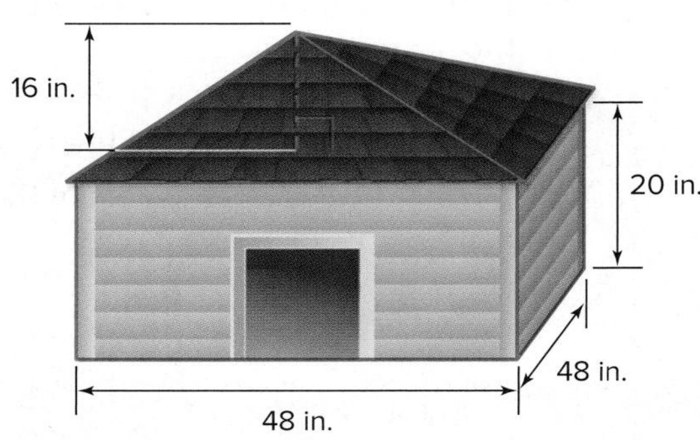

Find the surface area of each composite solid. Round to the nearest tenth if necessary.

27.

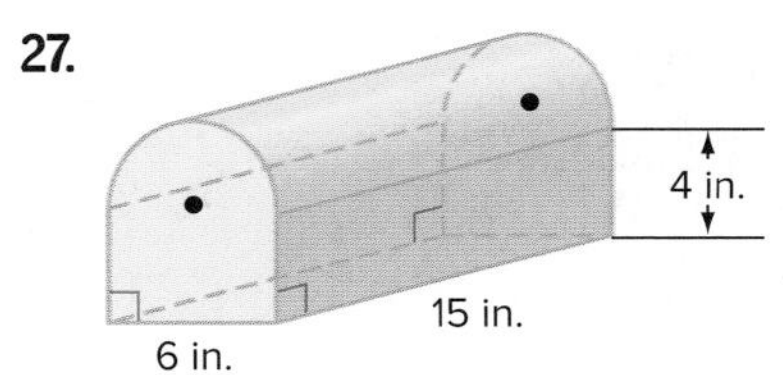

28.

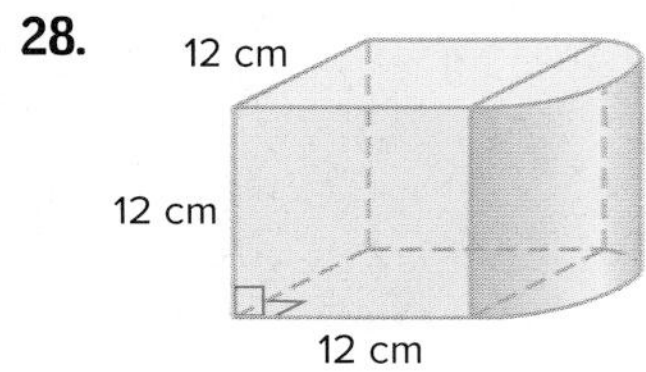

29. **MOUNTAINS** A conical mountain has a radius of 1.6 kilometers and a height of 0.5 kilometer. What is the lateral area of the mountain?

30. **AQUARIUMS** The Tower Aquarium in Henley Beach, Australia, is the world's largest cylindrical aquarium. It reaches a height of over 40 meters and is 36 meters in diameter. Visitors ascend through a column of water as they ride a split-level glass elevator up seven floors through the center of the aquarium. What is the approximate lateral area of the outside of the aquarium?

31. **HISTORY** Archaeologists recently discovered a 1500-year-old pyramid in Mexico City. The square pyramid measures 165 yards on each base edge and once stood 20 yards tall. What was the original lateral area of the pyramid?

32. **MONUMENTS** A *monolith* mysteriously appeared overnight at Seattle, Washington's Manguson Park. A hollow rectangular prism, the monolith was 9 feet tall, 4 feet wide, and 1 foot deep.

    a. Find the area in square feet of the structure's surfaces that lie above the ground.

    b. Use dimensional analysis to find the area in square yards.

33. **TEPEES** The dimensions of two canvas tepees are shown in the table below. Including the floors, approximately how much more canvas is used to make Tepee B than Tepee A.

Tepee	Diameter (ft)	Height (ft)
A	14	6
B	20	9

34. **DESIGN** A mailer needs to hold a poster that is almost 38 inches long and has a maximum rolled diameter of 6 inches.

    a. Design a mailer that is a triangular prism. Sketch the mailer and its net.

    b. Suppose you want to minimize the surface area of the mailer. What would be the dimensions of the mailer and its surface area?

35. **MULTI-STEP** Hector is designing a glass greenhouse for a city park. He has a 40-foot by 20-foot rectangular plot available. He wants the roof to be a triangular prism in which the center of the roof is 4 feet higher than the edges. The glass costs $25 per square foot, and Hector cannot spend more than $60,000 on glass.

    a. What is the maximum height that Hector should make the edge of the roof?

    b. Describe your solution process.

    c. What assumptions did you make?

36. **PETS** A *frustum* is the part of a solid that remains after the top portion has been cut by a plane parallel to the base. The ferret tent shown at the right is a frustum of a regular pyramid.

    a. Describe the faces of the solid.

    b. Find the surface area of the frustum formed by the tent.

    c. Another pet tent is made by cutting the top half of a pyramid with a height of 12 centimeters, slant height of 20 centimeters, and square base with side length of 32 centimeters. Find the surface area of the frustum.

37. The three-dimensional box needs to have a clear coating painted on all six faces. What is the approximate surface area of the box?

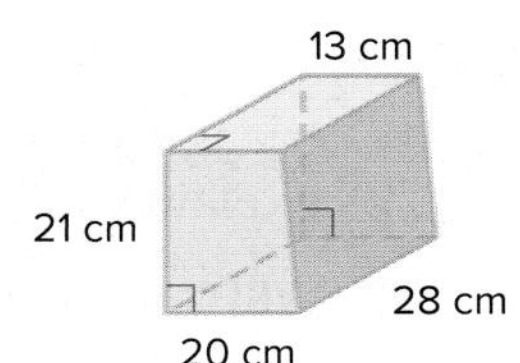

Find the surface area of each solid. Round to the nearest tenth.

38.

15 mm

32°

39.

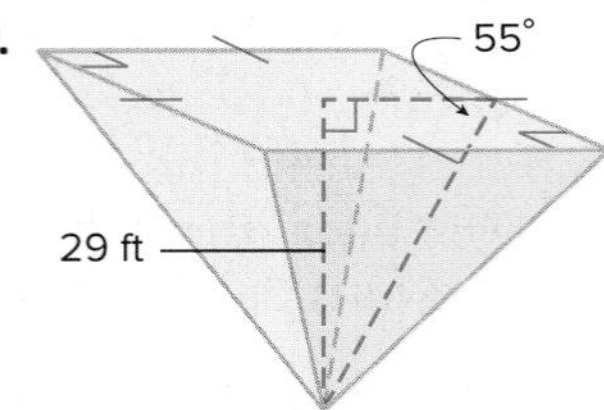

40. CONSTRUCTION A road roller is a construction vehicle with smooth and heavy rollers used for compacting roads and pavement. One of these rollers has a diameter of 48 inches and is 36 inches in length. What is the area covered by the roller in two full turns?

41. SUNCATCHERS Abby makes suncatchers of glass to sell at art shows. One style of suncatcher is a right hexagonal prism with a height of 9 centimeters and each base edge of 4 centimeters. What is the surface area of each suncatcher? (*Hint:* First, find the length of the apothem of the base.)

H.O.T. Problems Use Higher-Order Thinking Skills

42. WRITING IN MATH A square-based prism and a triangular prism are the same height. The base of the triangular prism is an equilateral triangle, with an altitude equal in length of the side of the square. Compare the lateral areas of the prisms.

43. MP REASONING A cone and a square pyramid have the same surface area. If the areas of their bases are also equal, do they have the same slant height as well? Explain.

44. MP CRITIQUE ARGUMENTS Montell and Derek are finding the surface area of a cylinder with a height of 5 centimeters and a radius of 6 centimeters. Is either of them correct? Explain your answer.

Montell	Derek
$S = \pi(6)^2 + \pi(6)(5)$	$S = 2\pi(6)^2 + 2\pi(6)(5)$
$= 36\pi + 30\pi$	$= 72\pi + 60\pi$
$= 66\pi$ cm^2	$= 132\pi$ cm^2

45. MP REASONING Classify the following statement as *sometimes, always,* or *never* true. Justify your reasoning.

The surface area of a cone of radius r and height h is less than the surface area of a cylinder of radius r and height h.

46. MP ARGUMENTS Determine whether the following is *true* or *false*. Explain your reasoning.

A regular polygonal pyramid and a cone both have height h units and base perimeter P units. Therefore, they have the same total surface area.

47. MP REASONING A right prism has a height of h units and a base that is an equilateral triangle of side ℓ units. Find the general formula for the total surface area of the prism. Explain your reasoning.

48. WRITING IN MATH Describe how to find the surface area of a regular polygonal pyramid with an n-gon base, height h units, and an apothem of a units.

49. Olivia makes a cylinder by bending the cardboard rectangle shown below so that the 8-centimeter sides join to form the lateral area of a cylinder. Then, she cuts out two cardboard circles to form the bases and attaches these to the lateral face.

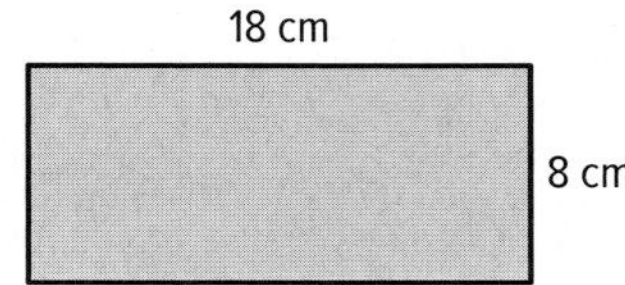

Which of the following is the best estimate of the surface area of the cylinder Olivia makes? MP 1

○ **A** 26 cm^2

○ **B** 52 cm^2

○ **C** 144 cm^2

○ **D** 196 cm^2

50. A cylindrical can has a circumference of 16π inches and a height of 20 inches. What is the surface area of the can in square inches? Round to the nearest tenth. MP 1, 6

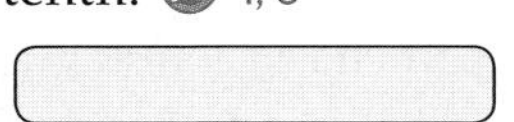

51. DeMarco is wrapping presents for a party. Each present is in a box shaped like a rectangular prism with the dimensions shown here. DeMarco plans to wrap 8 of the boxes.

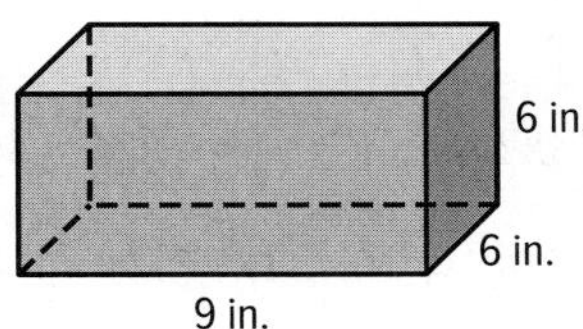

Which of the following is the best estimate for the least amount of wrapping paper DeMarco will need to buy? MP 1

○ **A** 1728 in^2

○ **B** 2016 in^2

○ **C** 2304 in^2

○ **D** 2592 in^2

52. The top of a gazebo in a park is the shape of a regular pentagonal pyramid. Each side of the pentagon is 10 feet long. If the slant height of the roof is about 6.9 feet, what is the lateral area of the roof to the nearest tenth? MP 1, 4, 6

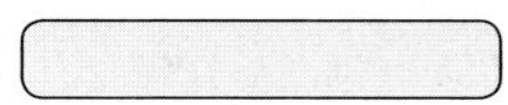

53. A model of a cone is used to demonstrate a new filter with top. To the nearest square millimeter, what is the surface area of the cone? MP 1, 6

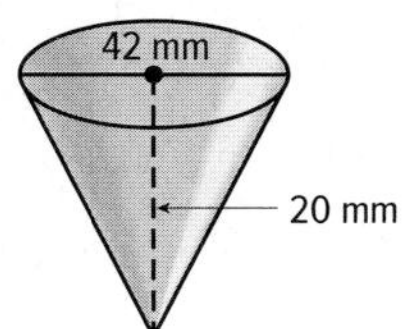

○ **A** 2705 mm^2

○ **B** 3299 mm^2

○ **C** 8820 mm^2

○ **D** 9368 mm^2

54. **MULTI-STEP** There are two separate buildings next to each other. The first is in the shape of a square prism. The dimensions of the base are 100 feet by 100 feet and the height of the building is 22 feet. The second building is in the shape of a square pyramid with the same dimensions. MP 1, 6

a. What is the surface area of the first building?

○ **A** 19,600 ft^2

○ **B** 28,800 ft^2

○ **C** 31,500 ft^2

○ **D** 48,000 ft^2

b. What is the surface area of the second building?

○ **A** 16,420 ft^2

○ **B** 18,720 ft^2

○ **C** 20,925 ft^2

○ **D** 38,000 ft^2

c. Why are the surface areas of the buildings different even though the dimensions are the same?

CHAPTER 10

Study Guide and Review

Go Online! for Vocabulary Review Games and key vocabulary in 13 languages

Study Guide

Key Concepts

Areas of Parallelograms and Triangles (Lesson 10-1)

- Area of a parallelogram: $A = bh$
- Area of a triangle: $A = \frac{1}{2}bh$ or $A = \frac{bh}{2}$

Areas of Trapezoids, Rhombi, and Kites (Lesson 10-2)

- Area of a trapezoid: $A = \frac{1}{2}h(b_1 + b_2)$
- Area of a rhombus or kite: $A = \frac{1}{2}d_1d_2$

Areas of Circles and Sectors (Lesson 10-3)

- Area of a circle: $A = \pi r^2$
- The ratio of the area A of a sector to the area of the whole circle, πr^2, is equal to the ratio of the degree measure of the intercepted arc x to 360.
  Proportion: $\frac{A}{\pi r^2} = \frac{x}{360}$ Equation: $A = \frac{x}{360} \cdot \pi r^2$

Areas of Regular Polygons and Composite Figures (Lesson 10-4)

- The area A of a regular n-gon with side length s is one half the product of the apothem a and perimeter P.
  $A = \frac{1}{2}a(ns)$ or $A = \frac{1}{2}aP$

Area and Nonrigid Transformations (Lesson 10-5)

- If two polygons are similar, then their areas are proportional to the square of the scale factor between them.
  If $ABCD \sim FGHJ$, then $\frac{\text{area of } FGHJ}{\text{area of } ABCD} = \left(\frac{FG}{AB}\right)^2$.

Surface Area (Lesson 10-6)

- Lateral surface area of a right prism: $L = Ph$
- Lateral surface area of a right cylinder: $L = 2\pi rh$
- Lateral surface area of a pyramid: $L = \frac{1}{2}p\ell$
- Lateral surface area of a right cone: $L = \pi r\ell$

FOLDABLES® Study Organizer

Use your Foldable to review the chapter. Working with a partner can be helpful. Ask for clarification of concepts as needed.

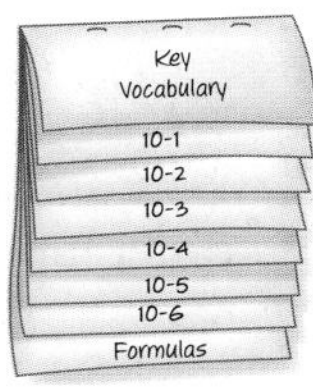

Key Vocabulary

altitude (p. 770)
apothem (p. 752)
axis (p. 771)
base edges (p. 770)
base of a parallelogram (p. 725)
base of a triangle (p. 727)
composite figure (p. 754)
height of a parallelogram (p. 725)
height of a solid (p. 770)
height of a trapezoid (p. 735)
height of a triangle (p. 727)
lateral area (p. 770)
lateral edge (p. 770)
lateral faces (p. 770)
oblique cone (p. 775)
regular pyramid (p. 775)
right cone (p. 775)
sector of a circle (p. 744)
slant height (p. 773)

Vocabulary Check

State whether each sentence is *true* or *false*. If *false*, replace the underlined term to make a true sentence.

1. The center of a trapezoid is the perpendicular distance between the bases.
2. A slice of pizza is a sector of a circle.
3. The center of a regular polygon is the distance from the middle to the circle circumscribed around the polygon.
4. The segment from the center of a square to the corner can be called the radius of the square.
5. A segment drawn perpendicular to a side of a regular polygon is called an apothem of the polygon.
6. The measure of each radial angle of a regular n-gon is $\frac{360}{n}$.
7. The slant height is the height of each lateral face of a pyramid or cone.
8. The height of a triangle is the length of an altitude drawn to a given base.

Concept Check

9. Explain how the areas of two similar polygons are related.
10. Explain how to determine the center of a regular polygon.

Lesson-by-Lesson Review 

10-1 Areas of Parallelograms and Triangles

Find the perimeter and area of each parallelogram or triangle. Round to the nearest tenth if necessary.

11.

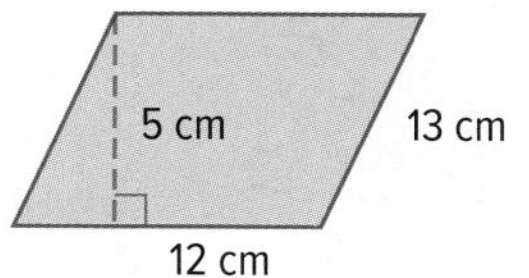

12.

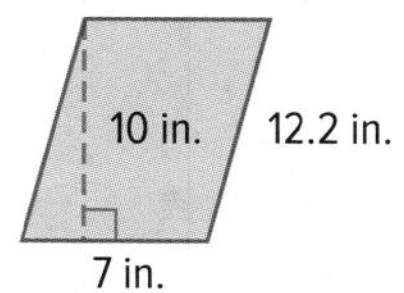

13.

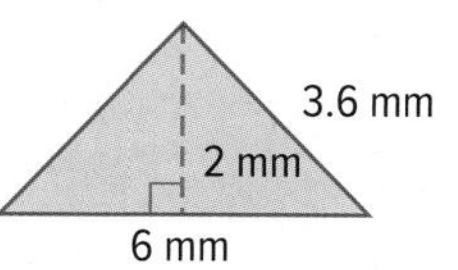

14.

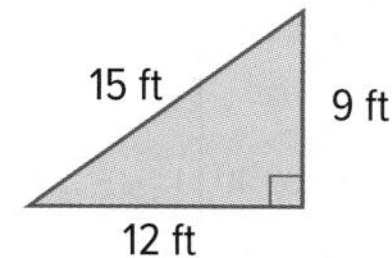

15. **PAINTING** Two of the walls of an attic in an A-frame house are triangular, each with a height of 12 feet and a width of 22 feet. How much paint is needed to paint one end of the attic?

Example 1

Find the perimeter and area of ▱$JKLM$.

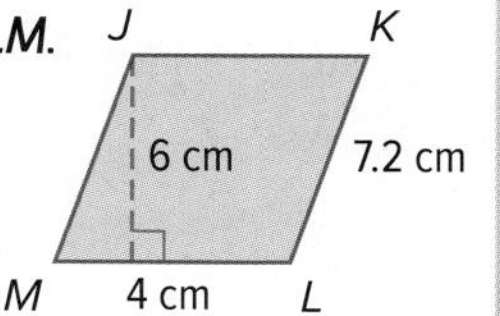

Perimeter

Perimeter of ▱$JKLM = JK + KL + LM + JM$

$= 4 + 7.2 + 4 + 7.2$ or 22.4 cm

Area

$A = bh$ Area of a parallelogram

$= (4)(6)$ or 24 cm^2 $\quad b = 4$ and $h = 6$

10-2 Areas of Trapezoids, Rhombi, and Kites

Find the area of each trapezoid, rhombus, or kite.

16.

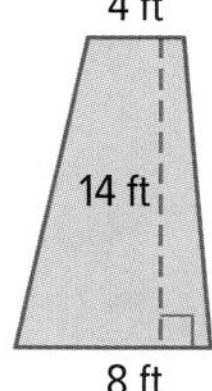

17.

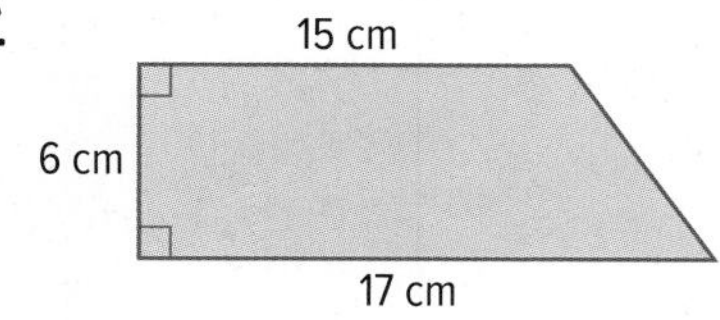

18.

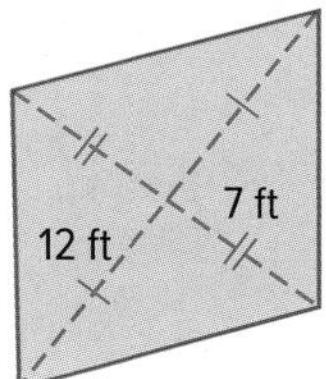

19.

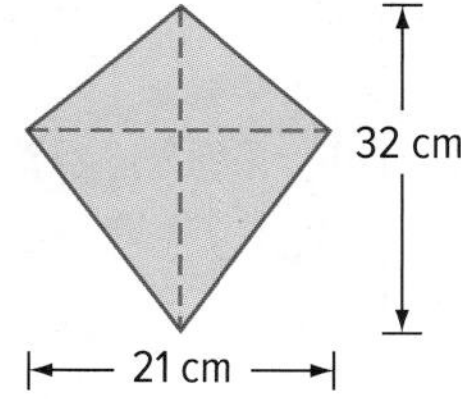

20. **KITES** Team Dragon's kite is 4 feet long and 3 feet across. How much fabric does it take to make their kite?

Example 2

Find the area of each rhombus or kite.

a.

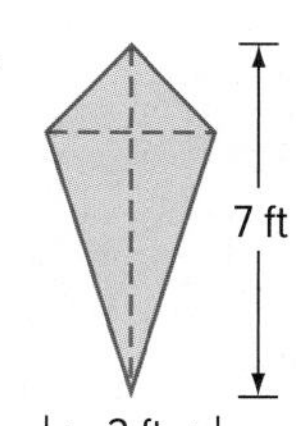

$A = \frac{1}{2}d_1d_2$ Area of a kite

$= \frac{1}{2}(7)(3)$ $\quad d_1 = 7$ and $d_2 = 3$

$= 10.5\text{ ft}^2$ Simplify.

b.

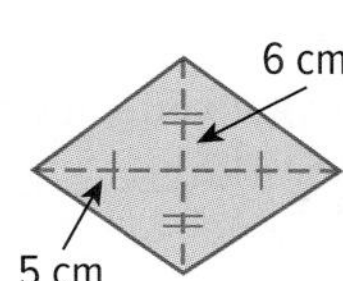

Since the diagonals of a rhombus bisect each other, the lengths of the diagonals are 6 + 6 or 12 centimeters and 5 + 5 or 10 centimeters.

$A = \frac{1}{2}d_1d_2$ Area of a rhombus

$= \frac{1}{2}(10)(12)$ $\quad d_1 = 10$ and $d_2 = 12$

$= 60\text{ cm}^2$ Simplify.

Study Guide and Review *Continued*

10-3 Areas of Circles and Sectors

Find the area of each shaded sector. Round to the nearest tenth.

21. 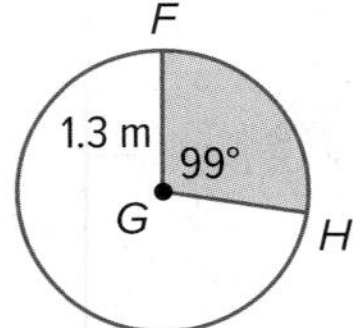

22. 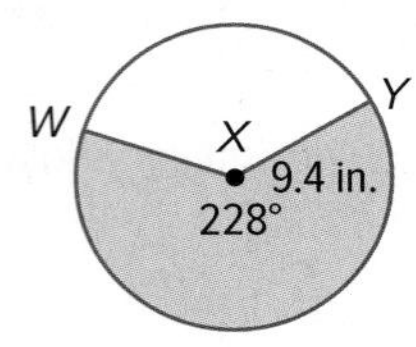

23. **BICYCLES** A bicycle tire decoration covers $\frac{1}{9}$ of the circle formed by the tire. If the tire has a diameter of 26 inches, what is the area of the decoration?

24. **PIZZA** Charlie and Kris ordered a 16-inch pizza and cut the pizza into 12 slices.

    a. If Charlie ate 3 pieces, what area of the pizza did he eat?

    b. If Kris ate 2 pieces, what area of the pizza did she eat?

    c. What is the area of leftover pizza?

Example 3

Find the area of the shaded sector. Round to the nearest tenth.

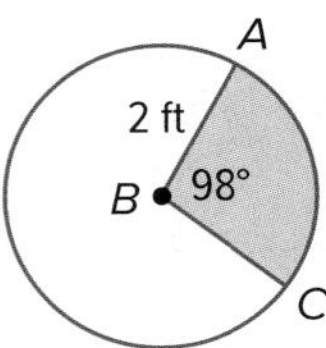

$A = \frac{x}{360} \cdot \pi r^2$ Area of a sector

$= \frac{98}{360} \cdot \pi(2)^2$ Substitution

$\approx 3.4 \text{ ft}^2$ Simplify.

10-4 Areas of Regular Polygons and Composite Figures

Find the area of each regular polygon or composite figure. Round to the nearest tenth.

25. 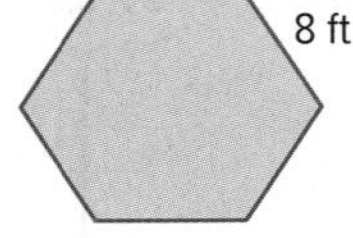

26. 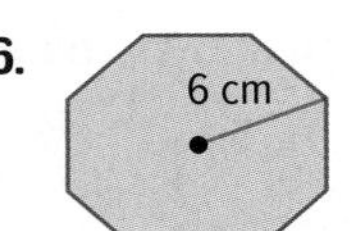

27. 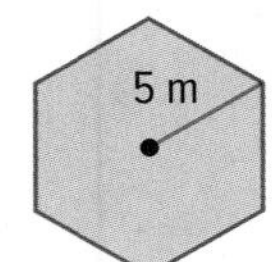

28. 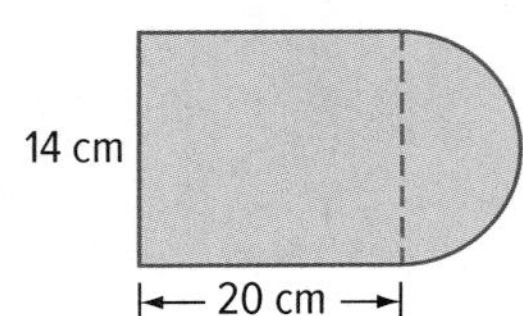

29. **JEWELRY** What is the area of the pendant shown below? Round to the nearest hundredth.

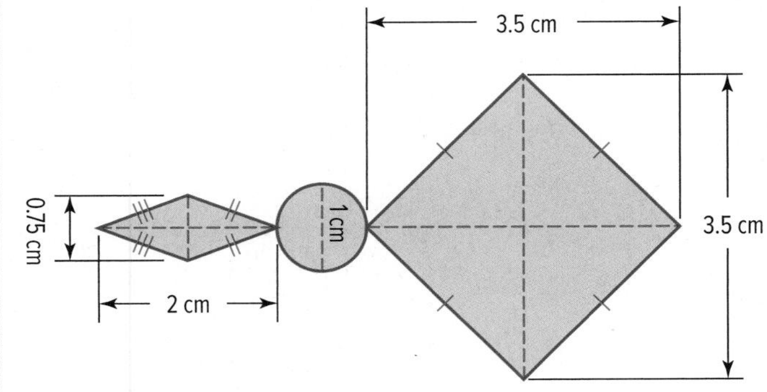

Example 4

Find the area of the figure.

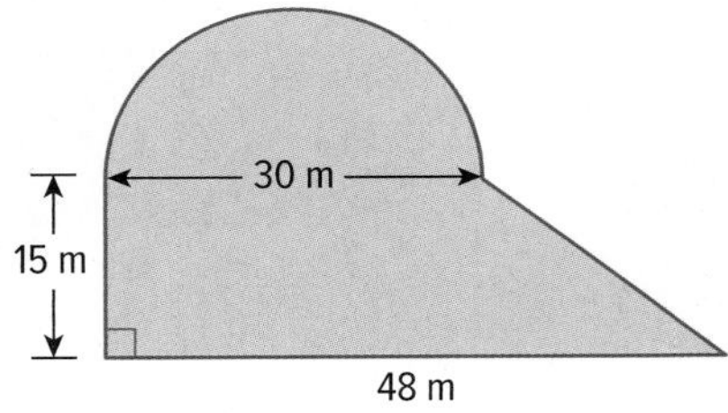

The composite shape is made up of a semicircle and a trapezoid.

Area = Area of semicircle + Area of trapezoid

$= \frac{180}{360} \cdot \pi \cdot r^2 + \frac{1}{2} \cdot h \cdot (b_1 + b_1)$

$\approx \frac{180}{360} \cdot \pi \cdot 15^2 + \frac{1}{2} \cdot 15 \cdot (30 + 48)$

$\approx 112.5\pi + 585$ or about 938.4 m^2

For each pair of similar figures, use the given areas to find the scale factor from the blue to the green figure. Then find x.

30.

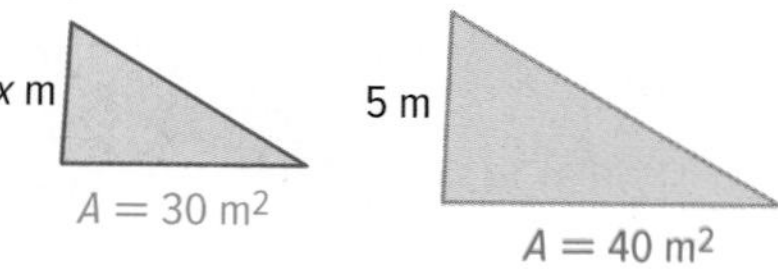

$A = 30\text{ m}^2$ $\quad A = 40\text{ m}^2$

31.

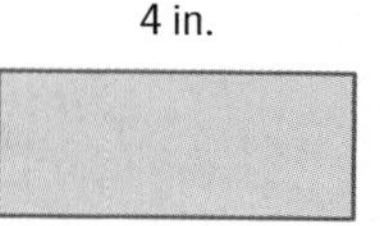

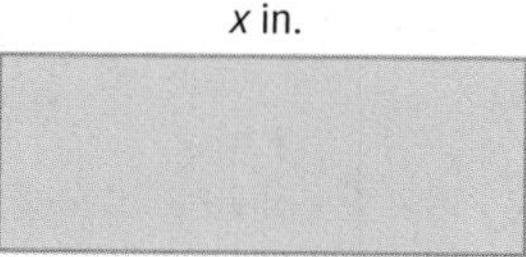

$A = 8\text{ in}^2$ $\quad A = 32\text{ in}^2$

32.

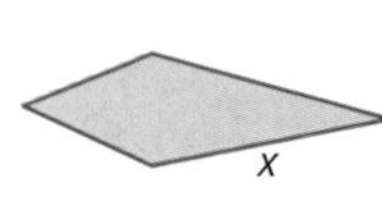

18 cm

$A = 525\text{ cm}^2$ $\quad A = 1575\text{ cm}^2$

COORDINATE GEOMETRY **Find the area of each figure. Use the segment length given to find the area of a similar polygon.**

33. $R'S' = 3$

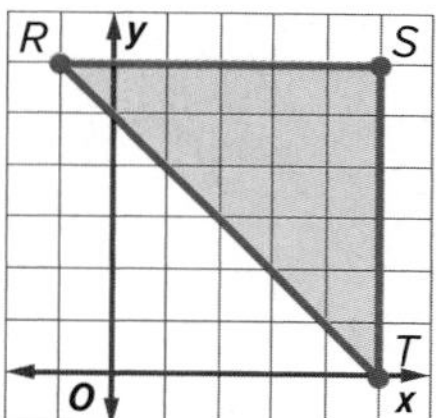

34. $K'L' = 15$

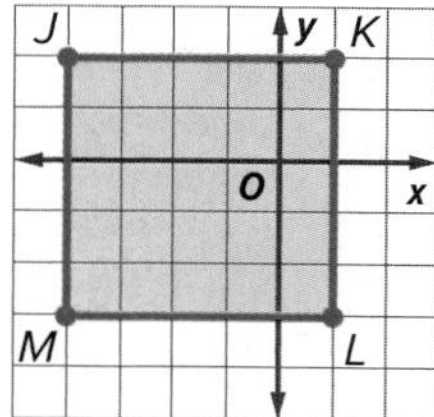

35. **LAND OWNERSHIP** Joshua's land is 600 square miles. He purchases an additional plot that is half the length and one-fourth the width of his original plot. What is the new area of his land?

Example 5

For the pair of similar figures, use the given areas to find the scale factor from the blue to the green figure. Then find x.

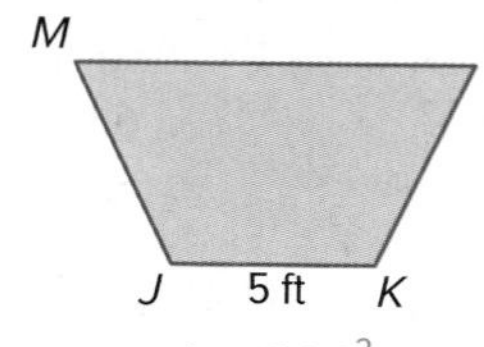

Q x T

R S

$A = 138\text{ ft}^2$ $\quad A = 5.52\text{ ft}^2$

Let k be the scale factor between trapezoid $JKLM$ and trapezoid $QRST$.

$\frac{\text{Area of trapezoid } JKLM}{\text{Area of trapezoid } QRST} = k^2$ Theorem 10.1

$\frac{138}{5.52} = k^2$ Substitution

$5 = k$ Take the positive square root of each side.

So, the scale factor from trapezoid $JKLM$ to trapezoid $QRST$ is 5. Use this scale factor to find the value of x.

$\frac{JK}{QT} = k$ The ratio of corresponding lengths of similar polygons is equal to the scale factor between the polygons.

$\frac{5}{x} = 5$ Substitution

$1 = x$ Simplify.

CHAPTER 10
Study Guide and Review *Continued*

10-6 Surface Area

Find the lateral area and surface area of each prism. Round to the nearest tenth if necessary.

36. 3 cm, 11 cm, 2 cm

37. 8 ft, 3 ft, 7 ft

Find the lateral area and surface area of each cylinder. Round to the nearest tenth.

38.

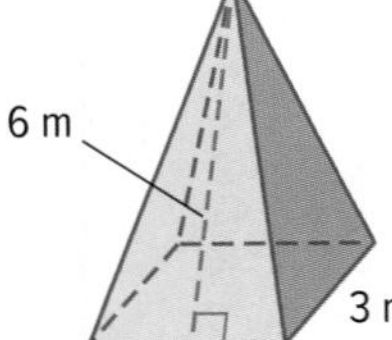

39.

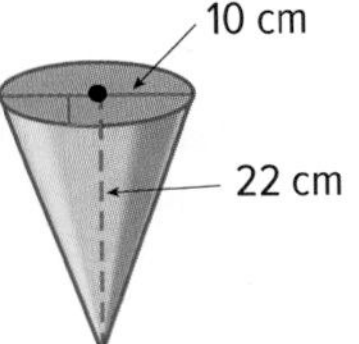

Find the lateral area and the surface area of each figure. Round to the nearest tenth.

40. 6 m, 3 m

41. 10 cm, 22 cm

Example 6

Find the surface area of the rectangular prism.

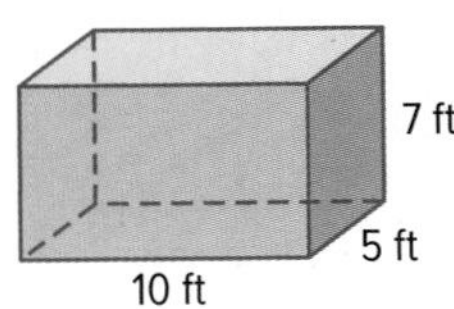

Use the 10-foot by 5-foot rectangle as the base.

$S = Ph + 2B$	Surface area of a prism
$= (2 \cdot 10 + 2 \cdot 5)(7) + 2(10 \cdot 5)$	Substitution
$= 310$	Simplify.

The surface area is 310 square feet.

Example 7

Find the surface area of the square pyramid. Round to the nearest tenth.

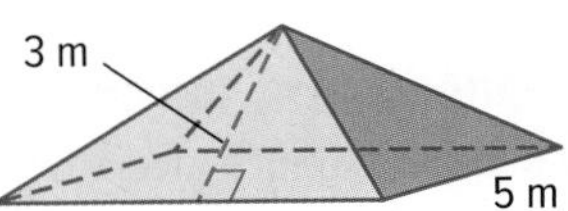

$S = \frac{1}{2}P\ell + B$	Surface area of a regular pyramid
$= \frac{1}{2}(4 \cdot 5)3 + 5 \cdot 5$	$P = 4 \cdot 5$ or 20, $\ell = 3$, $B = 5 \cdot 5$
$= 55$	Simplify.

The surface area is 55 square feet.

CHAPTER 10

Practice Test

Find the area and perimeter of each figure. Round to the nearest tenth if necessary.

1.
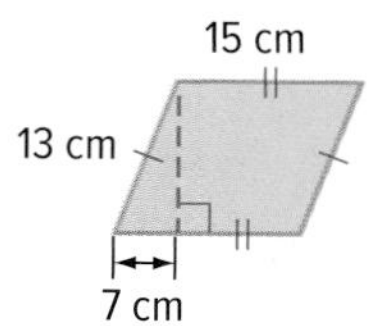

2.
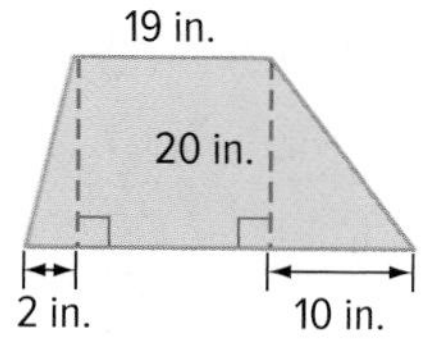

3.
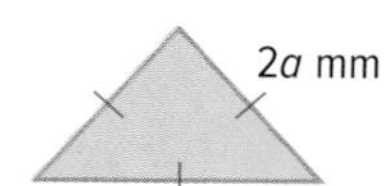

4.
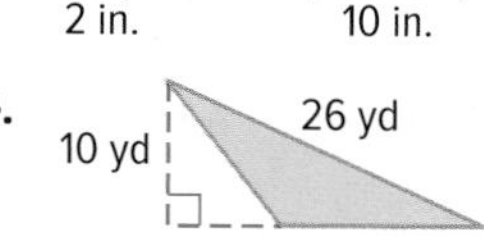

5. **ARCHAELOGY** The tile pattern shown was used in Pompeii for paving. If the diagonals of each rhombus are 2 and 3 inches, what area makes up each "cube" in the pattern?

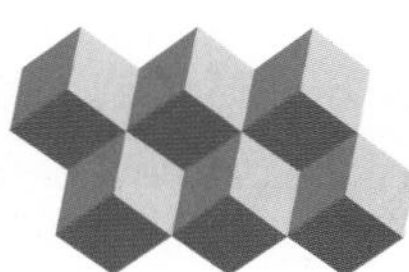

Find the area of each figure. Round to the nearest tenth if necessary.

6.
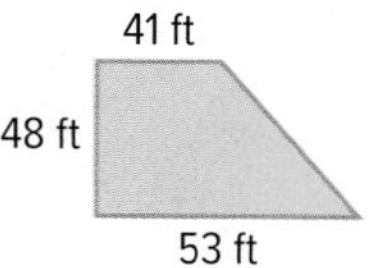

7.
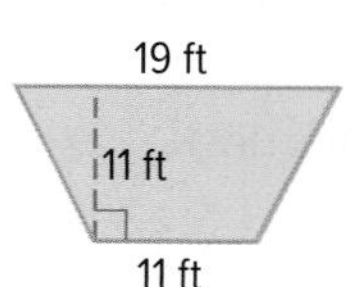

8.
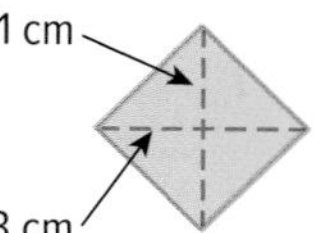

9.
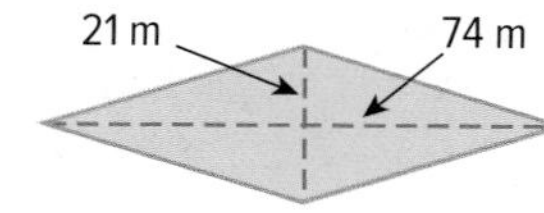

10. **GEMOLOGY** A gem is cut in a kite shape. It is 6.2 millimeters wide at its widest point and 5 millimeters long. What is the area?

11. **ALGEBRA** The area of a triangle is 16 square units. The base of the triangle is $x + 4$ and the height is x. Find x.

12. **ASTRONOMY** A large planetarium in the shape of a dome is being built. When it is complete, the base of the dome will have a circumference of 870 meters. How many square meters of land were required for this planetarium?

Find the area of each circle or sector. Round to the nearest tenth.

13.
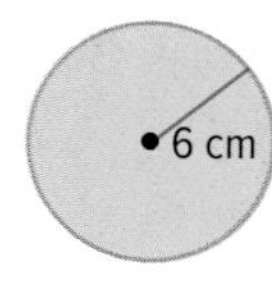

14.
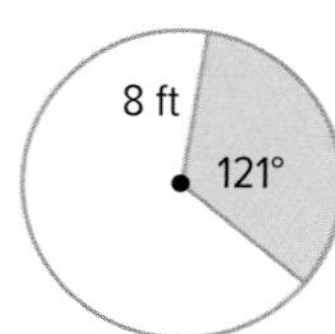

15. **MURALS** An artisan is creating a circular street mural for an art festival. The mural is going to be 50 feet wide. One sector of the mural spans 38°. What is the area of this sector to the nearest square foot?

Find the perimeter and area of each figure. Round to the nearest tenth if necessary.

16.
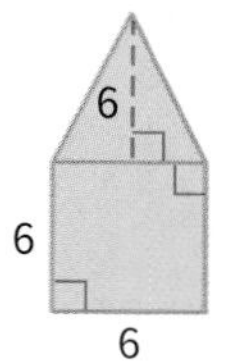

17.
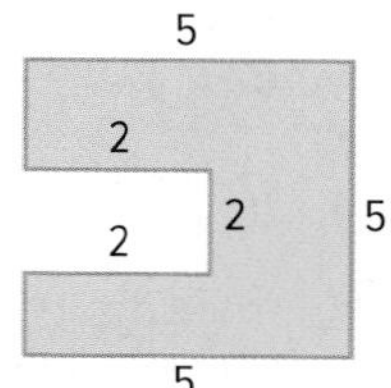

18. **FLOORING** Brian's service project is to build a tree house in the city park. Find the shaded area in the tree house. Round to the nearest tenth.

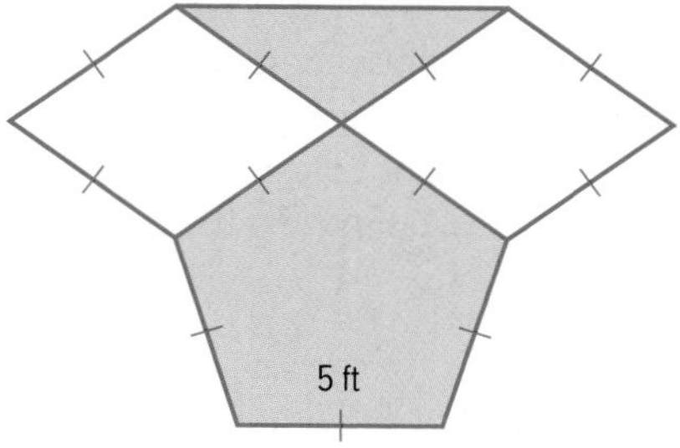

19. Find the lateral area and surface area of the tent model. Round to the nearest tenth if necessary.

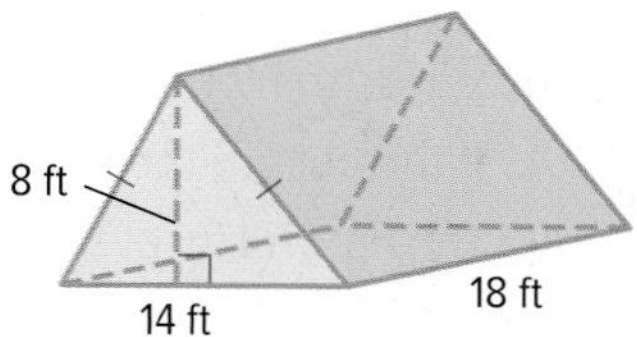

20. **BEEHIVE** Estimate the lateral area and surface area of the Turkish beehive room. Round to the nearest tenth if necessary.

CHAPTER 10

Preparing for Assessment

Performance Task

Provide a clear solution to each part of the task. Be sure to show all of your work, include all relevant drawings, and justify your answers.

Home Renovation Clarissa and Frank bought a house in need of many renovations.

Part A

Modeling The first room they plan to redo is the kitchen, and they are setting their budget. The kitchen is 12 feet by 14 feet. Two adjacent walls are completely lined with cabinets that are each 2 feet deep. There is also a 5 feet by 3 feet cabinet island in the center of the kitchen. Finally, there is a small 8 feet by 10 feet mudroom adjacent to the kitchen. They want to replace the flooring in the kitchen and mudroom, but do not need flooring underneath any cabinets.

1. If the flooring they've chosen is $2.99 per square foot, determine the cost of new floors for the kitchen and mudroom.

Part B

Precision Another project they are planning is to have a round deck built on which they can put a hot tub. One of the hot tubs requires a minimum deck diameter of 10 feet. Another hot tub requires a minimum deck diameter of 12 feet. The cost of building the deck is $35.00 per square foot.

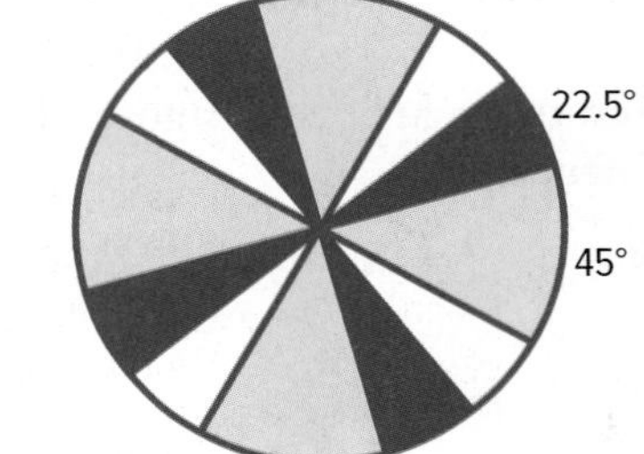

2. Determine the amount of money they can save by building the smaller deck. Round your answer to the nearest whole cent.
3. Clarissa decides she wants a special pattern painted on the deck, shown to the right. All of the light regions are the same size and all of the dark regions are the same size. The white areas are not painted. The cost is $12.50 per square foot of painted area. Assuming they build the smaller deck, determine the cost of having the deck painted. Round your answer to the nearest whole cent.

Part C

Regularity Frank loves to garden and wants a to make a vegetable patch. To avoid some trees, he makes the garden in the shape of a parallelogram. The bases are 40 feet long and the distance between the two bases is 10 feet. The angle formed between one side and the bottom base is 150°. Frank will need rope to mark off the perimeter of the garden and fertilizer for the whole garden.

4. If the cost of the rope is $0.55 per linear foot, determine the total cost to mark off the whole garden.
5. The natural fertilizer Frank wants to use costs $12.00 per 8-pound bag. Each bag can fertilize an area of about 375 square feet. How many bags of fertilizer will Frank need to fertilize the garden twice in the season?

Part D

Reasoning Clarissa and Frank plan to pour a cement patio, but would also like some of the area to be tiled, as shown in the shaded area in the diagram to the right.

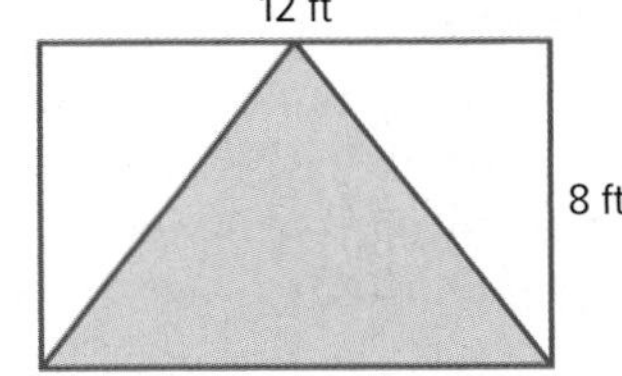

6. If the cost of the tiling is $11.25 per square foot, determine the cost to tile the triangle shown.
7. Clarissa would like to make the length and width of the patio both one and a half times bigger. Determine how much more the tiling would cost if they increased the size of the patio by a factor of 1.5.

Test-Taking Strategy

Example

Read the problem. Identify what you need to know. Then use the information in the problem to solve.

What is the area of the triangle? Round your answer to the nearest tenth.

A 112.5 m^2 **C** 152.5 m^2

B 172.5 m^2 **D** 195.5 m^2

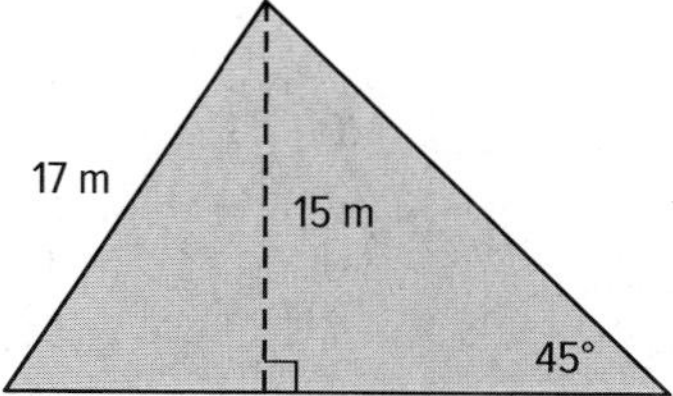

Step 1 **What are you being asked to solve? What information is given?**

I need to find the area of the triangle. The problem gives the height of the triangle, one side length, and one angle measure.

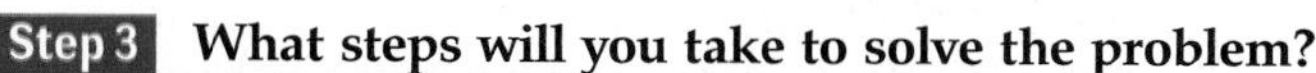

Step 2 **Are there any intermediate steps that need to be completed before you can solve the problem?**
To find the area, I need to find the length of the triangle's base.

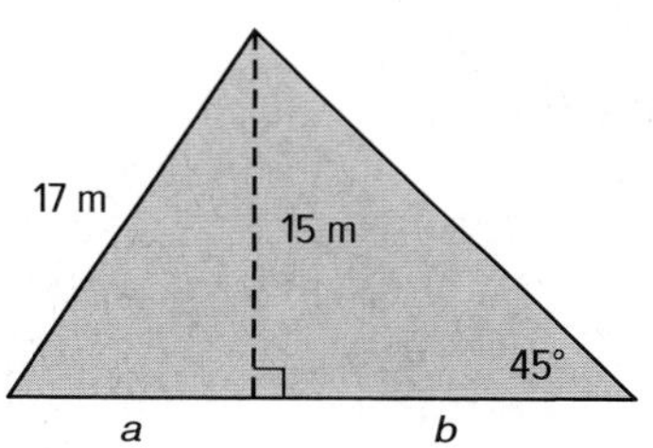

Step 3 **What steps will you take to solve the problem?**
I'll use the Pythagorean Theorem of find a and the special rules of right triangles to find b. Adding these together will give me the length of the base. Then I can use the formula for area of a triangle.

Step 4 **What is the correct answer?** The correct answer is B.

Test-Taking Tip

Multi-Step Problems When solving multi-step problems, decide what you're being asked to sove, list the information given, and choose a strategy to solve the problem.

Apply the Strategy

Read the problem. Identify what you need to know. Then use the information in the problem to solve.

Which of the following best represents the area of the figure shown?

A 350 in^2 **C** 460 in^2

B 410 in^2 **D** 470 in^2

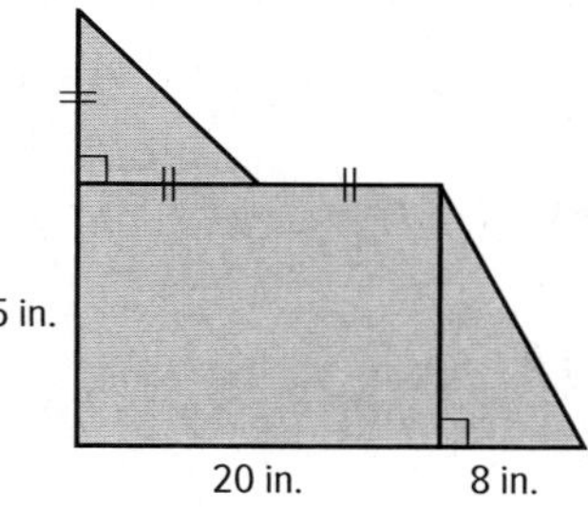

Answer the questions below.

a. What are you being asked to solve? What information is given?

b. Are there any intermediate steps that need to be completed before you can solve the problem?

c. What steps will you take to solve the problem?

d. What is the correct answer?

CHAPTER 10

Preparing for Assessment

Cumulative Review

Read each question. Then fill in the correct answer on the answer document provided by your teacher or on a sheet of paper.

1. In the figure, $\triangle JKL \sim \triangle MNP$. The area of $\triangle JKL$ is 324 square meters and the area of $\triangle MNP$ is 1764 square meters.

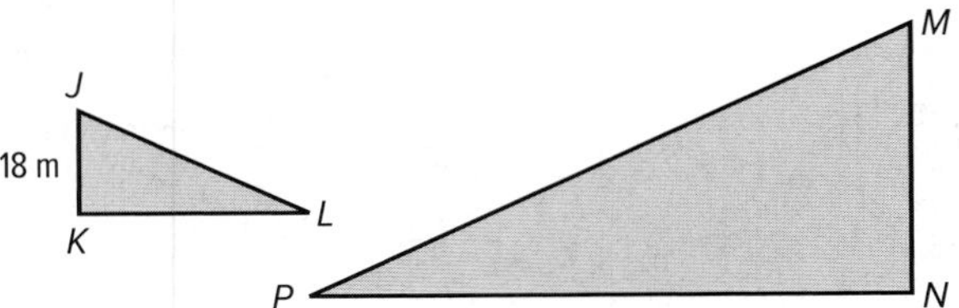

Which of the following is closest to the length of $\overline{MN}$?

- A 3 m
- B 8 m
- C 42 m
- D 98 m

2. Which of the following are true about the circle with equation $x^2 + y^2 - 6x + 6y = -14$?

- A The circle lies entirely in Quadrant IV.
- B The radius of the circle is 4.
- C The circle intersects both axes.
- D The center of the circle is $(3, -3)$.
- E The area of the circle is 4π.

3. Which of the following is the best estimate of the area of the regular octagon shown below?

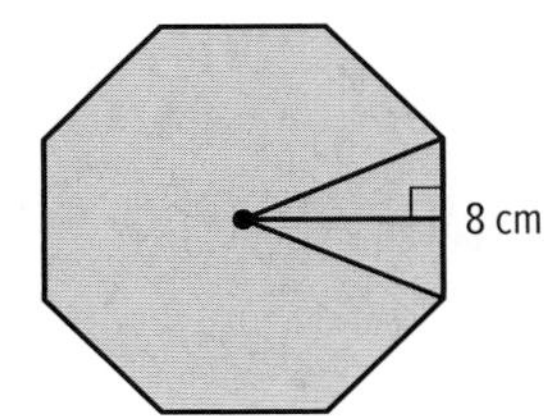

- A 618 cm^2
- B 309 cm^2
- C 128 cm^2
- D 39 cm^2

4. Malia wants to put a fence around the three sides of a right-triangular plot in her garden. She measures one side length and one acute angle, as shown in the figure.

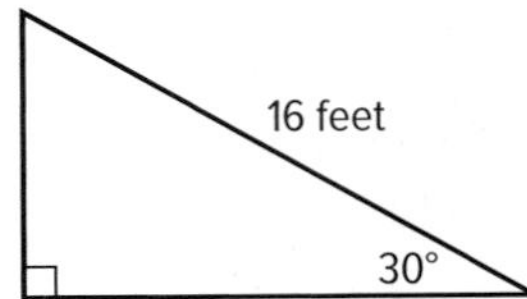

To the nearest foot, how many feet of fence does Malia need?

5. A circular pizza has a diameter of 18 inches. Charles slices the pizza into 12 equal sectors. What is the area of each sector, in square inches? Round to the nearest tenth.

Test-Taking Tip

Question 5 There are two ways to solve this problem. You can determine the measure of the intercepted arc for each sector and then use the formula for the area of a sector, or you can find the area of the whole pizza and divide by 12.

6. Kento enlarges $\triangle RST$ so that each dimension is 4 times the corresponding dimension shown in the figure.

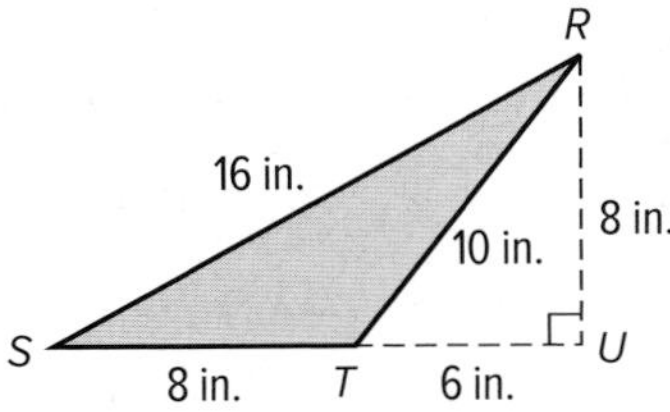

What is the area- of the enlarged triangle?

- A 128 in^2
- B 512 in^2
- C 640 in^2
- D 896 in^2

7. In parallelogram $ABCD$, side AB is 20 inches, side BC is 24 inches, and angle DAB is 60°. What are the area and perimeter of the parallelogram $ABCD$?

- A 88 in; 207.9 in^2
- B 44 in; 357.5 in^2
- C 88 in; 415.7 in^2
- D 44 in; 408.6 in^2

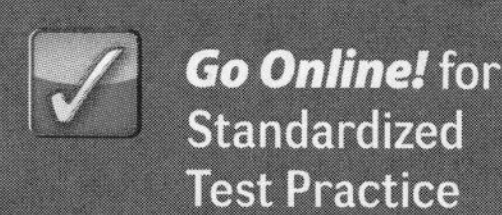

8. Which of the following is the best estimate of the area of the composite figure shown here?

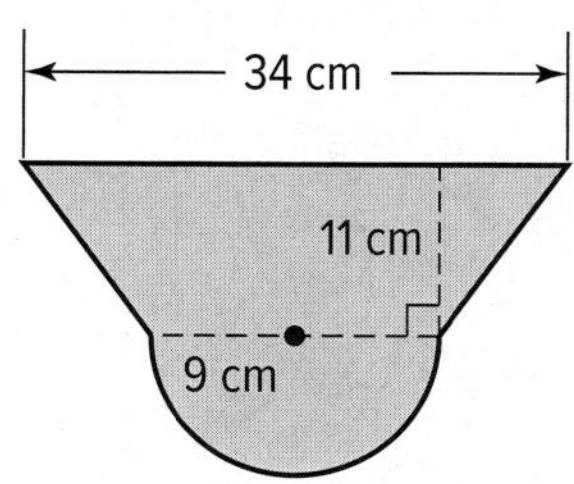

- A $540\ cm^2$
- B $413\ cm^2$
- C $364\ cm^2$
- D $286\ cm^2$

9. What is the surface area of the pyramid shown?

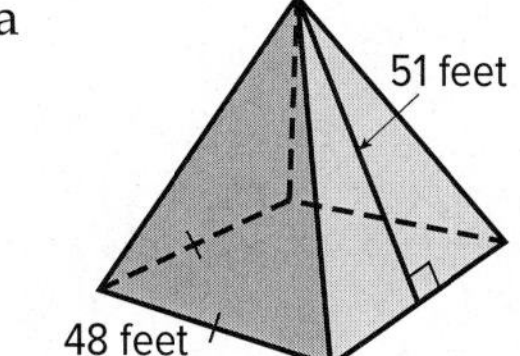

10. Dayana drew the trapezoid shown here.

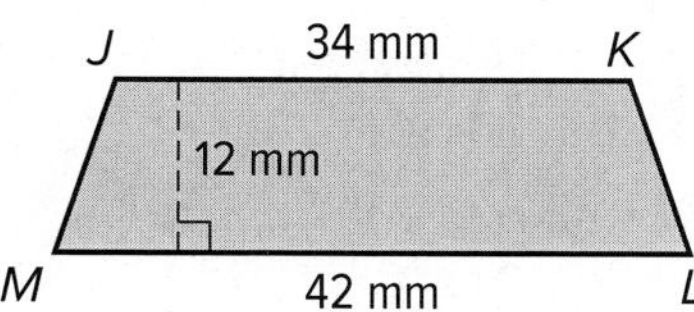

Which of the following will allow Dayana to create a new trapezoid with half the area of trapezoid *JKLM*? Select all that apply.

- A Draw a trapezoid in which each base is half as long as in trapezoid *JKLM*.
- B Draw a trapezoid in which the height is half as long as in trapezoid *JKLM*.
- C Draw a trapezoid in which only one base is half as long as in trapezoid *JKLM*.
- D Draw a trapezoid in which the height and only one base is half as long as in trapezoid *JKLM*.
- E Draw a trapezoid in which every dimension is half the corresponding dimension in trapezoid *JKLM*.

11. A rhombus has a longer diagonal of 18 centimeters and a perimeter of 45 centimeters. What is the area of the rhombus?

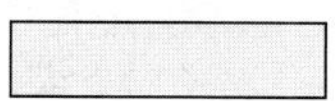

12. Marcello is making a cone-shaped base for a lamp. He goes online to order a piece of metal that is a sector of a circle, as shown. According to the website, the area of the sector is 180 square inches.

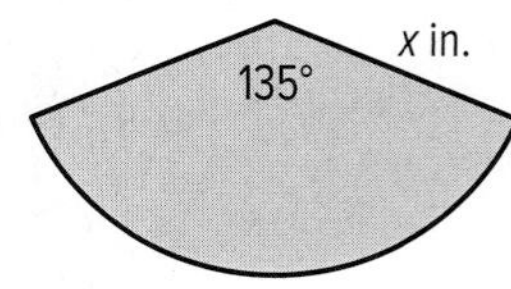

Which of the following is the best estimate of the value of x?

- A 4.6
- B 7.6
- C 12.4
- D 76.4

13. What is the surface area of the figure shown? Round your answer to the nearest whole number.

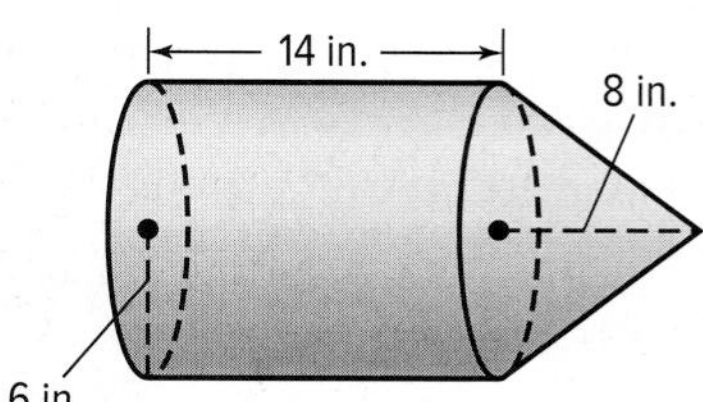

14. A kite has diagonals measuring 29 cm and 23 cm. What is the area of the kite?

- A $52\ cm^2$
- B $333.5\ cm^2$
- C $667\ cm^2$
- D $1334\ cm^2$

Need Extra Help?

If you missed Question...	1	2	3	4	5	6	7	8	9	10	11	12	13	14
Go to Lesson...	10-1	10-3	10-4	10-1	10-3	10-5	10-1	10-4	10-6	10-5	10-2	10-3	10-6	10-2

CHAPTER 11

Extending Volume

THEN

You identified three-dimensional figures and calculated the surface areas and volumes for some common solids.

NOW

In this chapter, you will:

- Find lateral areas, surface areas, and volumes of various three-dimensional figures.
- Investigate Euclidean and spherical geometries.
- Use properties of similar solids.

MP WHY

ARCHITECTURE Architects use shapes to create designs that are interesting and functional. A striking example of this is Biosphere 2.

Use the Mathematical Practices to complete the activity.

1. Use Tools Use the Internet to learn more about the Biosphere 2 project.

2. Reasoning Why was it important for architects to know the surface areas of each biome?

3. Apply Math Choose one of the sections of Biosphere 2 and estimate its total surface area.

4. Modeling Use the 2-D Figures tool to find the area of the shapes that make up the section you selected.

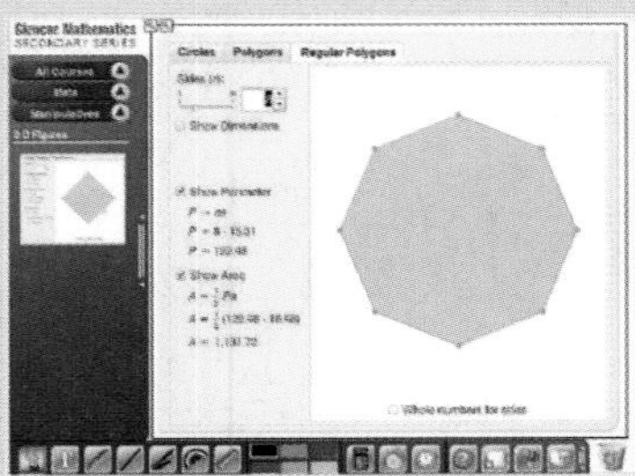

5. Discuss Why would knowing the volumes of the sections in Biosphere 2 be important?

Norman Pogson/Alamy

Go Online to Guide Your Learning

Explore & Explain

3-D Figures

Use the **3-D Figures** tool to explore the volume of different pyramids and cones, as discussed in Lesson 11-3, or to find the volume and surface area of a sphere, as discussed in Lesson 11-4.

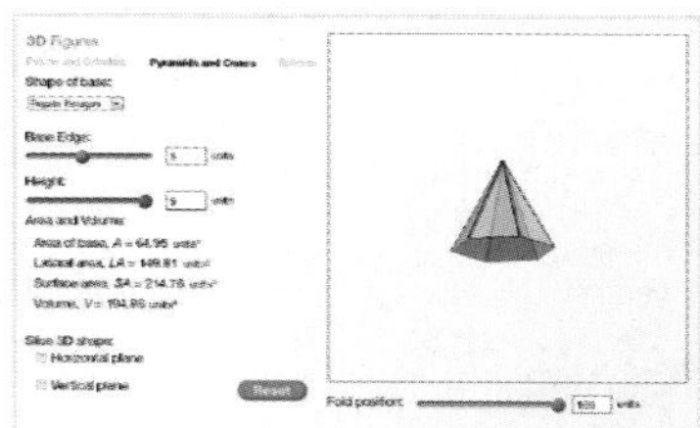

The Geometer's Sketchpad

Use **The Geometer's Sketchpad** to explore symmetries of regular polygons in Lesson 11-1 and the relationships between areas of similar figures in Lesson 11-6.

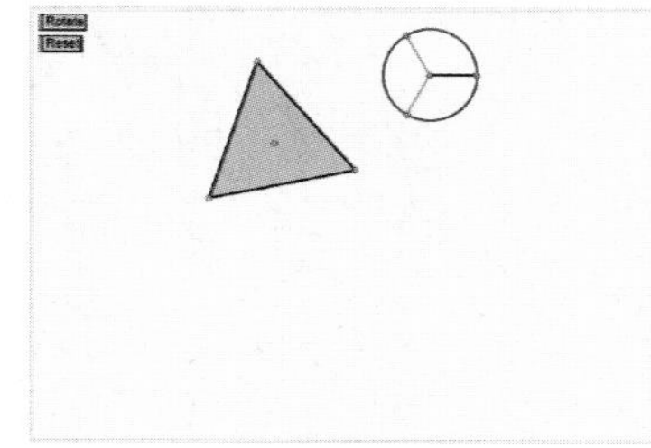

eBook Interactive Student Guide

Before starting the chapter, answer the **Chapter Focus** preview questions. Check your answers as you complete each lesson. At the end of the chapter, try the **Performance Task**.

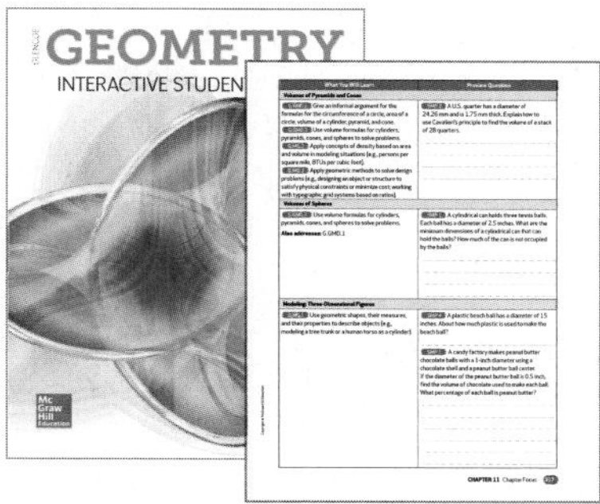

Organize

Foldables

Get organized! Create a **Surface Area and Volume Foldable** before you start the chapter to help you organize your notes about surface area and volume.

Collaborate

Chapter Project

In the **Vacation Resorts** project, you will use what you have learned about surface areas and volumes to complete a project that addresses business literacy.

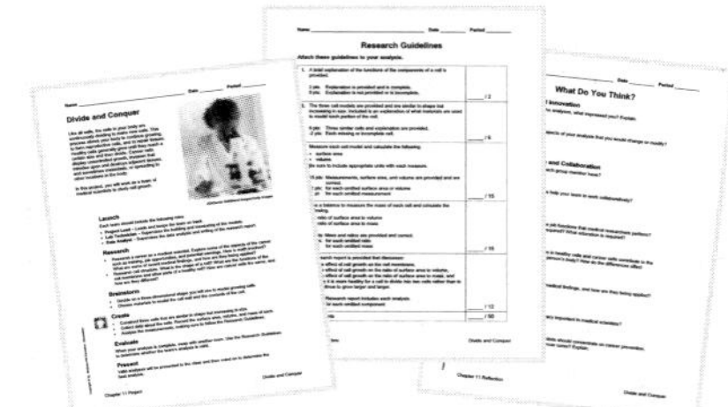

Focus

LEARNSMART

Need help studying? Complete the **Extend to Three Dimensions** domain in LearnSmart to review for the chapter test.

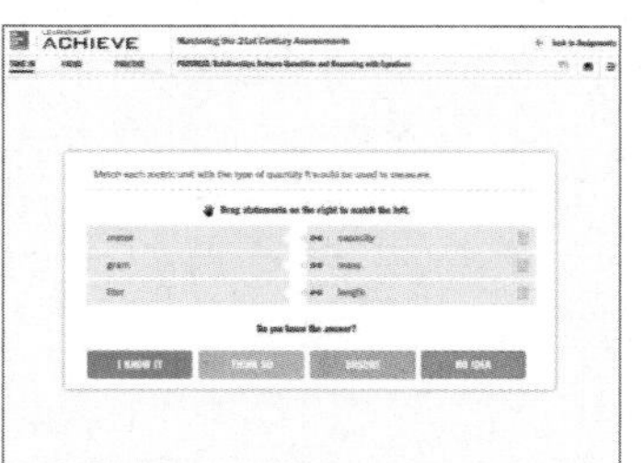

ALEKS

You can use the **Volumes and Surface Areas** topic in ALEKS to explore what you know about extending area and what you are ready to learn.*

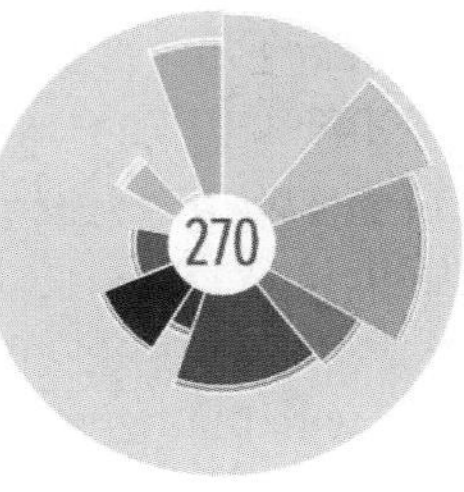

* Ask your teacher if this is part of your program.

Go Online! for Vocabulary Review Games and key vocabulary in 13 languages.

Get Ready for the Chapter

Connecting Concepts

Concept Check

Review the concepts used in this chapter by answering the questions below.

1. **CRAFTS** Michelle wants to cover a kite frame with fabric. The length of one diagonal is 16 inches and the other diagonal is 22 inches.
   a. Describe how to find the area of a kite.
   b. Find the area of the surface of the kite.

Find the area of each figure.

2. 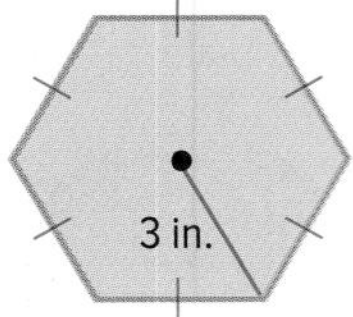

3. 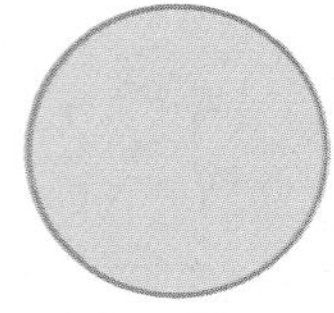

4. What is the equation to determine the area of a trapezoid?
5. Given $8^2 + 7^2 = c^2$, what would be the first step to solve the equation?
6. Given $c^2 = 133$, how would you solve the equation for c?

Performance Task Review

You can use the concepts and skills in the chapter to solve problems for a candle and soap company. Knowing how to determine surface areas and volumes will help you finish the Performance Task at the end of the chapter.

CHAPTER 11 Preparing for Assessment

Performance Task

In this Performance Task you will:

- make sense of problems and persevere in solving them
- model with mathematics
- attend to precision

New Vocabulary

English		Español
cross section	p. 797	sección transversal
solid of revolution	p. 798	sólido de revolución
great circle	p. 819	círculo mayor
poles	p. 819	postes
hemisphere	p. 819	hemisferio
Euclidean geometry	p. 827	geometría euclidiana
spherical geometry	p. 827	geometría esférica
non-Euclidean geometry	p. 828	geometría no euclidiana
similar solid	p. 834	sólidos semejantes
congruent solid	p. 834	sólidos congruentes
population density	p. 841	densidad demográfica
density	p. 842	densidad

Review Vocabulary

regular polyhedron poliedro regular a polyhedron in which all of the faces are regular congruent polygons

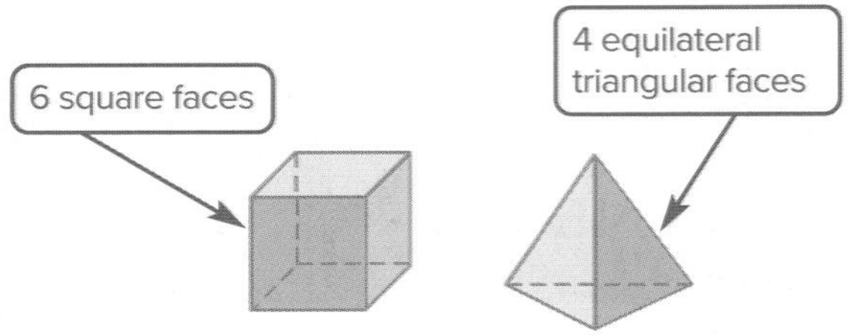

EXPLORE 11-1

Geometry Lab

Solids Formed by Translation

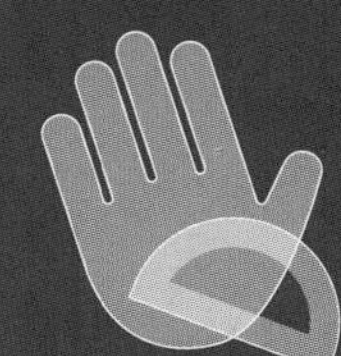

Mathematical Practices
MP 4 Model with mathematics.

We can relate some three-dimensional solids to two-dimensional figures with which we are already familiar. Some three-dimensional solids can be formed by translating a two-dimensional figure along a vector.

A **right solid** has base(s) that are perpendicular to the edges connecting them or connecting the base and the vertex of the solid. Some right solids are formed by translating a two-dimensional figure along a vector that is perpendicular to the plane in which the figure lies.

Activity 1

Identify and sketch the solid formed by translating a horizontal rectangle vertically.

To help visualize the solid formed, let a playing card represent the rectangle and lay it flat on a table so that it is horizontal. To show the translation of the rectangle vertically, stack other cards neatly, one by one, on top of the first.

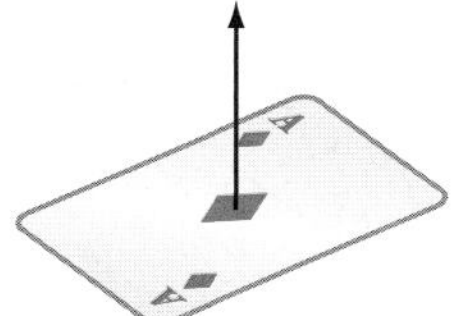

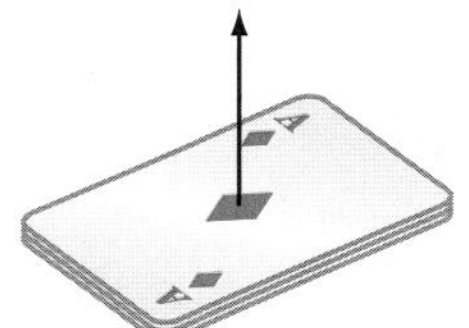

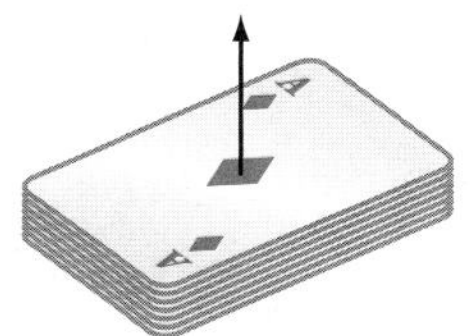

Notice that the solid formed is a right rectangular prism, which has a rectangular base, a translated copy of this base on the opposite side parallel to the base, and four congruent edges connecting the two congruent rectangles. These edges are parallel to each other but perpendicular to the bases. A sketch of the figure is shown.

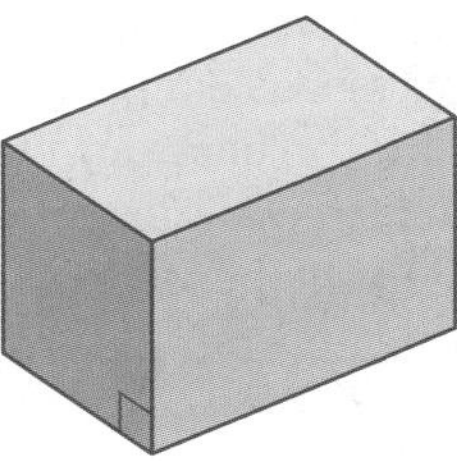

Model and Analyze

1. Use congruent triangular tangram pieces to identify and sketch the solid formed by translating a horizontal triangle vertically.

2. Use the coins from a roll of quarters to identify and sketch the solid formed by translating a horizontal circle vertically.

Identify and sketch the solid formed by translating a vertical two-dimensional figure horizontally.

3. rectangle

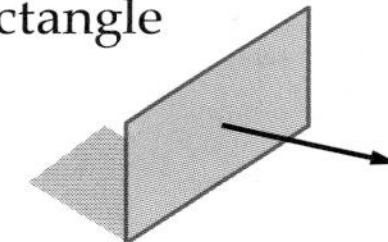

4. triangle

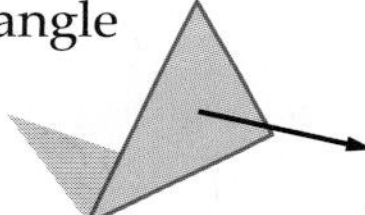

5. circle

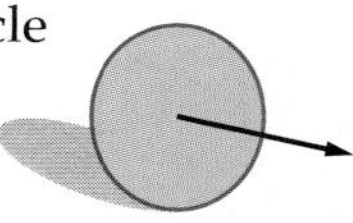

6. MP **REASONING** Are the solids formed in Exercises 3, 4, and 5 right solids? Explain your reasoning.

Solids Formed by Translation *Continued*

An **oblique solid** has base(s) that are not perpendicular to the edges connecting the two bases or vertex. An oblique solid can be formed by translating a two-dimensional figure along an oblique vector that is neither parallel nor perpendicular to the plane in which the two-dimensional figure lies.

Activity 2

Identify and sketch the solid formed by translating a horizontal rectangle along an oblique vector.

Let a playing card represent the rectangle. Lay it flat on a table so that it is horizontal. To show the translation of the rectangle along an oblique line, stack other cards one by one on top of the first so that the cards are shifted from the center of the previous card the same amount each time.

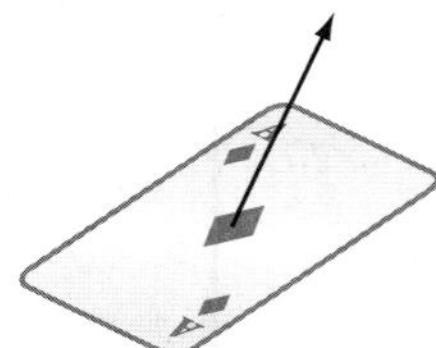

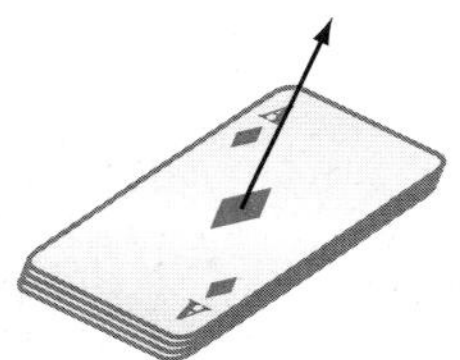

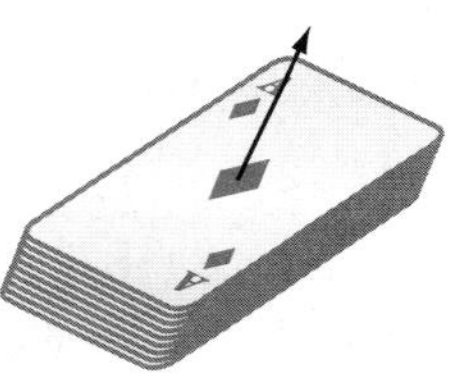

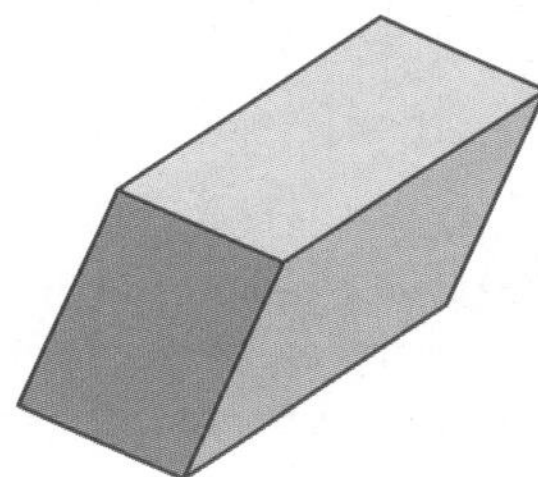

The solid formed is an oblique rectangular prism, which has a rectangular base, a translated copy of this base on the opposite side parallel to the base, and four congruent edges connecting the two congruent rectangles. These edges are parallel to each other but oblique to the bases. A sketch of the figure is shown.

Model and Analyze

Identify and sketch the solid formed by translating each vertical two-dimensional figure along an oblique vector. Use concrete models if needed.

7. triangle

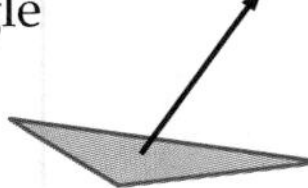

8. circle

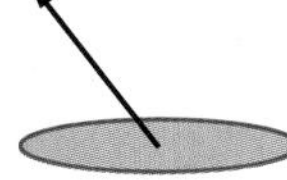

Identify each solid as *right*, *oblique*, or *neither*.

9.

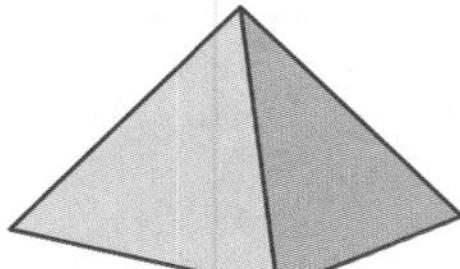

10.

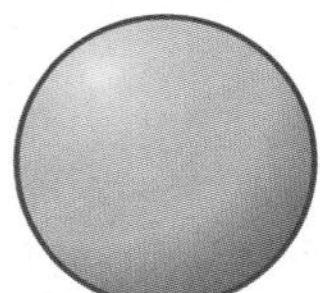

11.

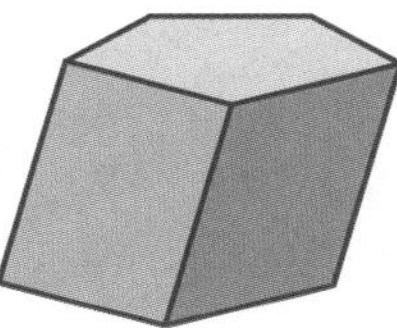

12. MP **REASONING** Can a pyramid with a square base be formed by translating the base vertically? Explain your reasoning.

LESSON 1

Cross Sections and Solids of Revolution

Then

- You explored area of two-dimensional figures and surface area of three-dimensional solids.

Now

1. Identify cross sections of three-dimensional solids.
2. Identify three-dimensional objects generated by rotations of two-dimensional figures.

Why?

- Video game developers use technology to make the gaming environments appear three-dimensional. As players move within the various video game worlds, objects are realistically shown in 3-D from different perspectives.

New Vocabulary

cross section
solid of revolution

Mathematical Practices

2 Reason abstractly and quantitatively.
4 Model with mathematics.

1 Investigate Cross Sections

A **cross section** is the intersection of a solid and a plane. The shape of the cross section depends on the angle of the plane.

Example 1 Identify Cross Sections of Solids

Describe each cross section.

a.

When a plane parallel to the base of the pyramid intersects the pyramid, the cross section is a square. When an angled plane passes through opposite faces of the pyramid, the cross section is a trapezoid. When a plane perpendicular to the base of the pyramid passes through the vertex of the pyramid, the cross section is a triangle.

b. **sphere** **cone** **cylinder**

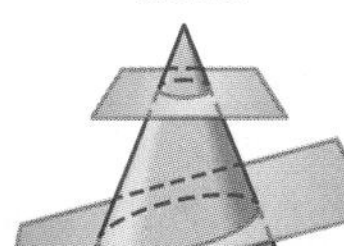

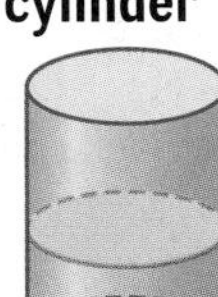

Every cross section of a sphere is a circle. When a plane intersects a cone or cylinder parallel to the base, then the cross section is a circle. When a plane intersects a cone at an angle relative to the base and does not intersect the vertex or base of the cone, then the cross section is an ellipse.

Guided Practice

1. Determine the shape of each cross section formed by the intersection of two planes with the solid. One plane is parallel to the base of each solid, and one plane is perpendicular to the base of each solid.

A.

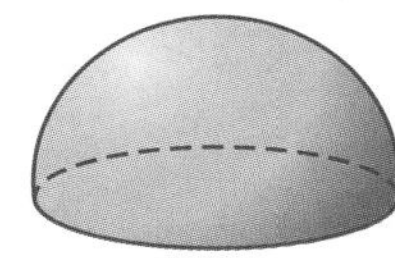

B.

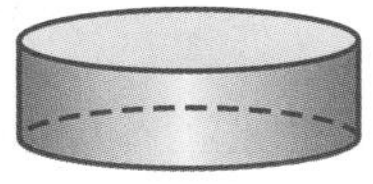

C.

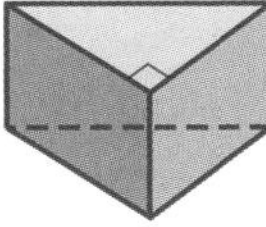

Hero Images/Getty Images

2 Investigate Solids of Revolution

A **solid of revolution** is a solid figure obtained by rotating a plane figure or curve around an axis. The shape of the solid of revolution depends on the location of the axis and the shape of the plane figure or curve being rotated.

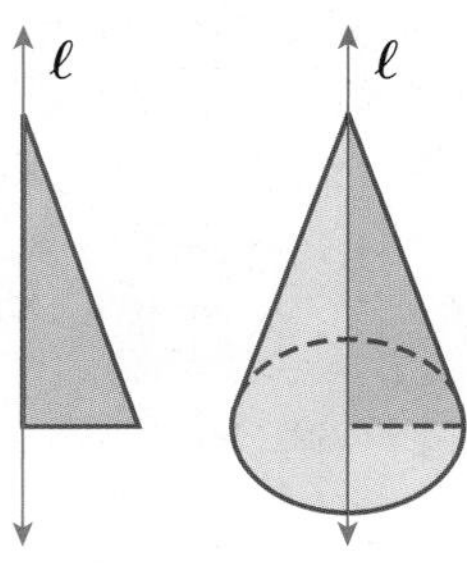

Study Tip

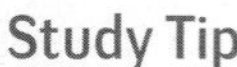

Tools You can use many different tools to help you visualize the solids of revolution. Straws or dowel rods could be used to represent the axis of rotations. Card stock or heavy construction paper can be attached to the straws or dowel rods to represent the two dimensional figures. You could also sketch the figure using dynamic geometry software or graph paper.

Example 2 Identify Solids of Revolution

Describe the three-dimensional solid generated by rotating the circle around the given axis.

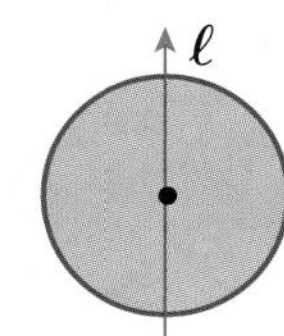

As the circle rotates about the axis, a sphere is formed.

Guided Practice

Describe the three-dimensional solid generated by rotating each two-dimensional shape around the given axis.

2A. Right triangle

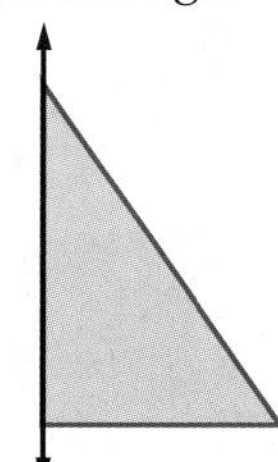

2B. Square

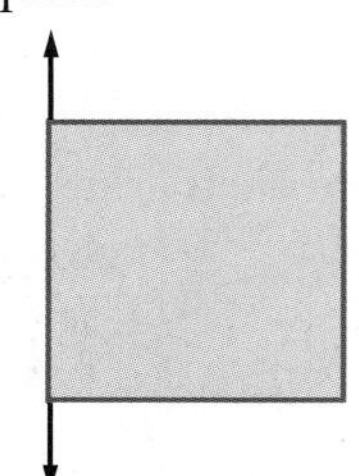

2C. Semicircle

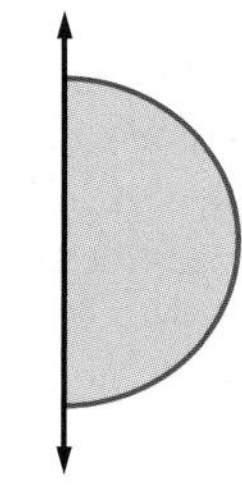

Check Your Understanding

 = Step-by-Step Solutions begin on page R13.

Example 1 **Determine the shape of each cross section formed by the intersection of the described plane with the solid.**

1. plane perpendicular to the base

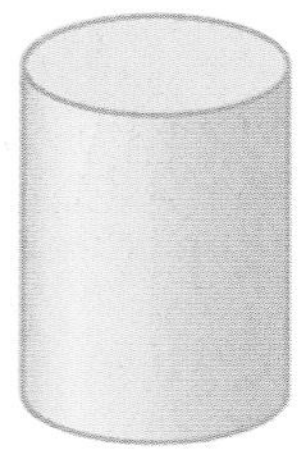

2. plane parallel to the base

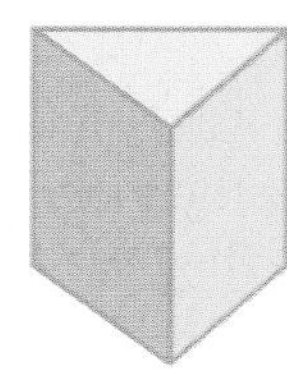

3. plane at an angle relative to the base through opposite faces

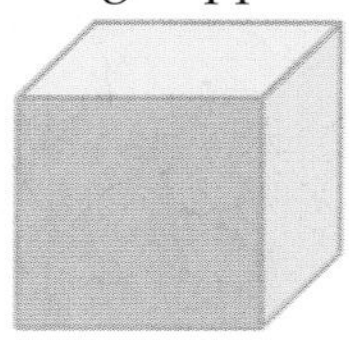

Example 2 **Describe the three-dimensional solid generated by rotating each two-dimensional shape around the given axis.**

4. rectangle

5. circle

6. triangle

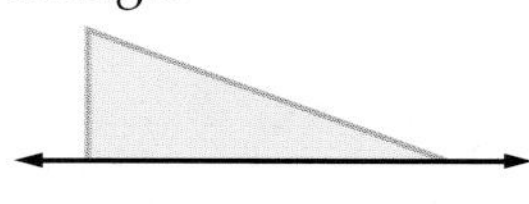

Practice and Problem Solving

Extra Practice is found on page R11.

Example 1 **Determine the shape of each cross section formed by the intersection of the described plane with the solid.**

7. plane at an angle relative to the bases that does not intersect either base

8. plane at an angle relative to the base that intersects the base

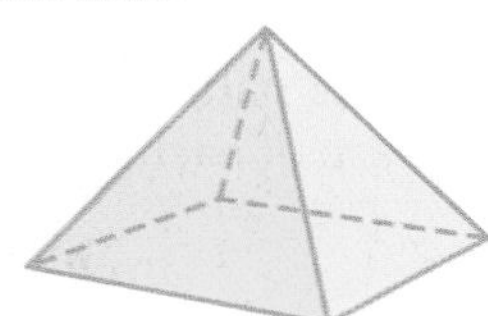

9. angled plane that intersects the sphere

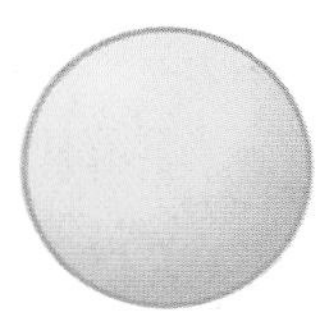

Example 2 **Describe the three-dimensional solid generated by rotating each two-dimensional shape around the given axis.**

10.

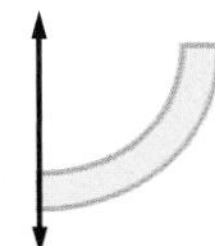

11. rectangle

12. exponential function

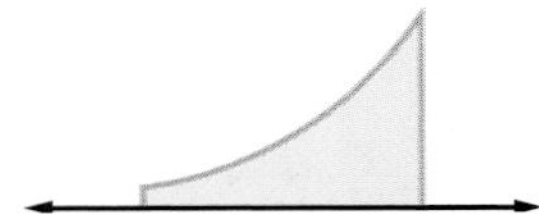

13. FOOD Describe how the cheese at the right can be sliced so that the slices form each shape.

a. rectangle

b. triangle

c. trapezoid

Describe each cross section.

14.

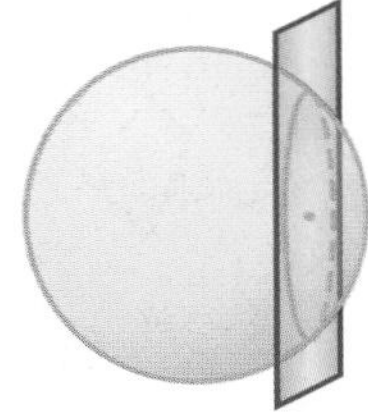

15.

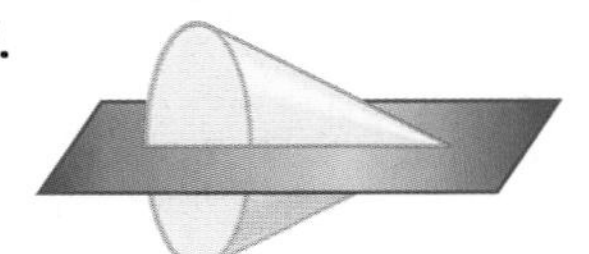

16. If a plane intersects with a cube at a vertex of the cube, what is the shape of the cross section? Explain your answer.

17. UFO Tanya has a model of a UFO. Sketch a two-dimensional figure that could be rotated around an axis to produce a three-dimensional solid similar to the model.

18. DESIGN Describe how you could create a tube with a length of 10 inches, a diameter of 2 inches, and a thickness of $\frac{1}{4}$ inch by rotating a 2-D figure around an axis. Make a sketch and label it.

19. POTTERY A potter creates three-dimensional objects by shaping the clay as it spins on a potter's wheel. Describe the line or curve that could be rotated around a vertical axis to produce the vase shown.

20. COOKIES Michelle is making cookies with a cylindrical roll of cookie dough. Describe how she can cut the cookie dough to make each shape.

a. circle **b.** longest rectangle **c.** oval **d.** shorter rectangle

21. **ART** A piece of clay in the shape of a rectangular prism is cut in half as shown at the right.

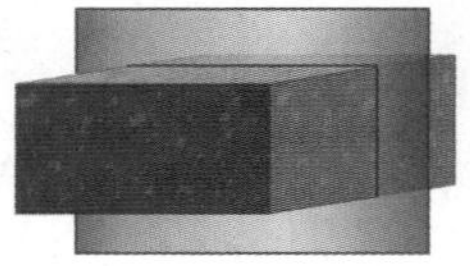

a. Describe the shape of the cross section.

b. Describe how the clay could be cut to make the cross section a triangle.

22. **EARTH SCIENCE** Crystals are solids in which the atoms are arranged in regular geometrical patterns. Sketch a cross section formed by a plane parallel to the base of each crystal.

a. tetragonal

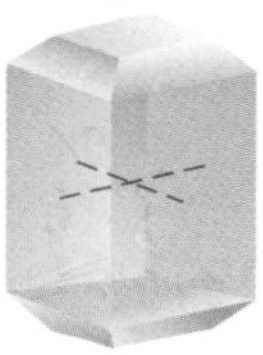

b. hexagonal

c. monoclinic

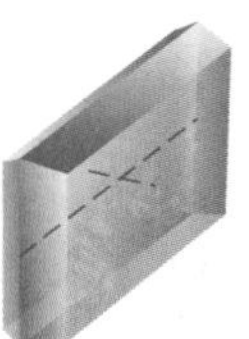

23. Which of the following chess pieces can be created by rotating a two-dimensional figure around a vertical axis?

H.O.T. Problems Use Higher-Order Thinking Skills

24. **MP CRITIQUE ARGUMENTS** Ellen says that if you slice a sphere at an angle, you get an elliptical cross section. Is she correct? Explain your answer.

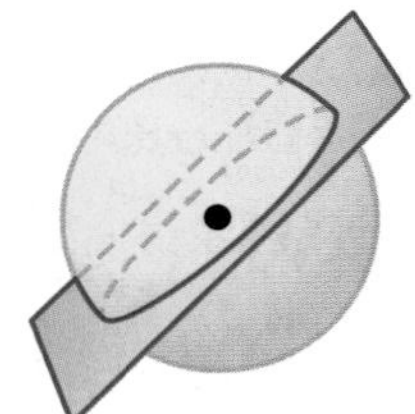

25. **MP REASONING** If you slice off the top of a cone, you are left with a **truncated cone.** What two-dimensional figure could be rotated around the axis to produce a truncated cone? Name and sketch the figure.

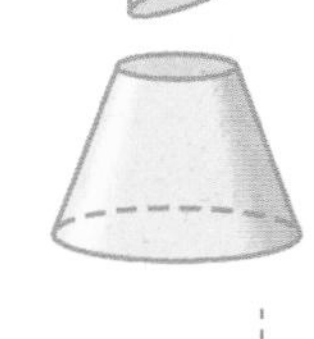

26. **CHALLENGE** If you slice a cone parallel to the base, the cross section is a circle. If the plane cuts at an angle through both sides of the cone, the cross section is an ellipse. What if the plane cuts at an angle through the side of the cone and through the base of the cone? How are such cross sections different from a circle or an ellipse? Research **conic sections** and describe each cross section of a cone in terms of the features of each conic section.

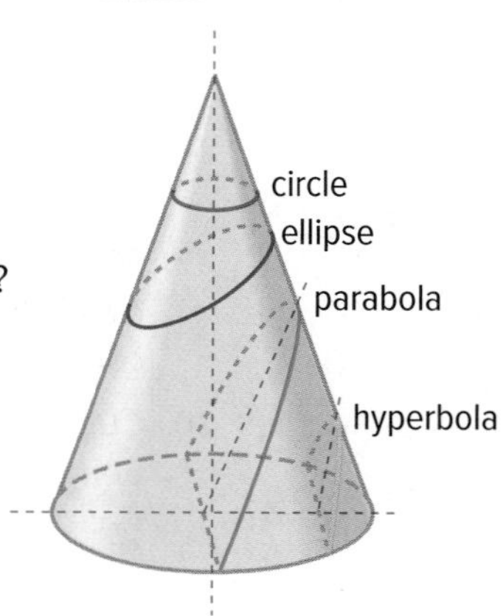

Determine the shape of each cross section. MP 7

27. 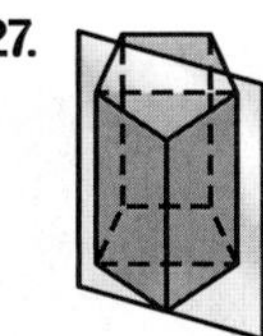

28. 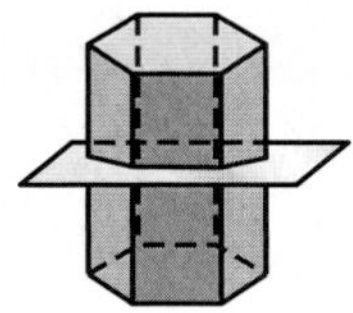

29. 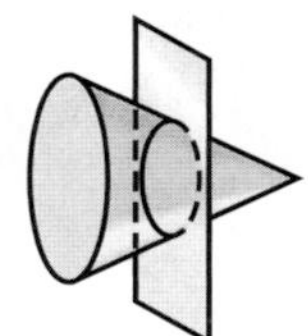

30. 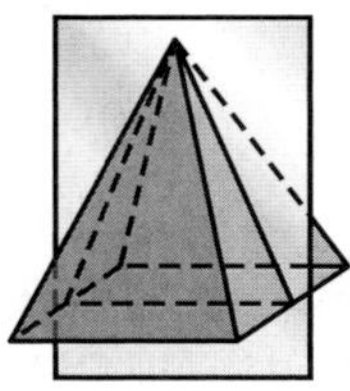

31. Which of the following three-dimensional solids can be generated by rotating a two-dimensional figure around an axis? MP 4, 7

- ○ A pyramid
- ○ B banana
- ○ C cube
- ○ D egg

32. Which of the following shapes could not be a cross section of the prism? MP 2

- ○ A octagon
- ○ B pentagon
- ○ C rectangle
- ○ D triangle

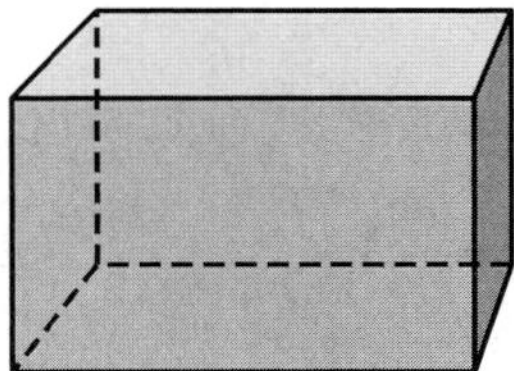

33. Eduardo has a piece of wood in the shape of a square pyramid. The sides of the base are 6 inches long. Eduardo cuts the pyramid with a single straight cut parallel to the base. Which of the following statements about the cross section must be true? MP 2

- ○ A The cross section is a square with an area less than 36 in^2.
- ○ B The cross section is a square with an area of 36 in^2.
- ○ C The cross section is a triangle with an area less than 36 in^2.
- ○ D The cross section is a triangle with an area of 36 in^2.

34. What shape is generated by rotating a rectangle around an axis parallel to one of its sides? MP 7

35. Describe how to generate a torus, or donut shape, with an outside radius of 6 inches. MP 2, 4

36. Describe how to generate a tube with the following dimensions: MP 2, 4

length = 20 feet

diameter = 8 inches

thickness = $\frac{1}{2}$ inch

37. Describe two ways to intersect a solid with a plane to produce a cross section that is a trapezoid. MP 2, 7

38. **MULTI-STEP** The figure below is a square pyramid with a height of 24 centimeters and a base length of 24 centimeters. MP 1, 2, 4

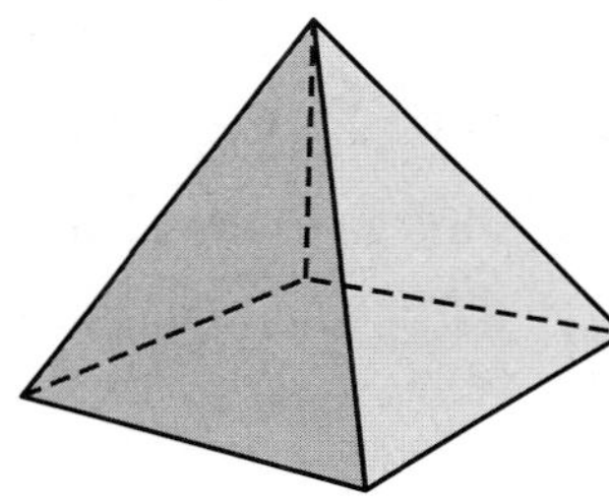

a. The pyramid is intersected by a plane perpendicular to the base of the pyramid and through the vertex. Determine the shape of the cross section.

b. What is the area of the cross section?

c. A plane parallel to the base of the pyramid intersects the pyramid at a height of 12 centimeters. Determine the shape of the cross section.

d. What is the area of the cross section?

e. A third plane intersects the pyramid at an angle. It goes through one face of the pyramid at a height of 12 centimeters, and it intersects the edge of the opposite side of the base. Determine the shape of the cross section.

LESSON 2

Volumes of Prisms and Cylinders

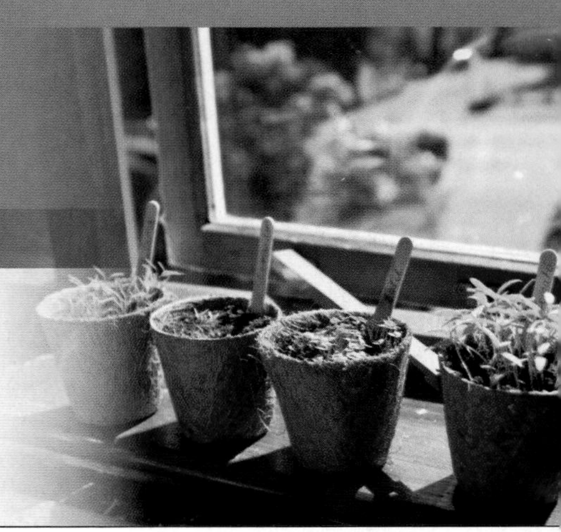

∴Then	∴Now	∴Why?
● You found surface areas of prisms and cylinders.	● **1** Find volumes of prisms. **2** Find volumes of cylinder.	● Planters come in a variety of shapes and sizes. You can approximate the amount of soil needed to fill a planter by finding the volume of the three-dimensional figure that it most resembles.

Mathematical Practices

1 Make sense of problems and persevere in solving them.

7 Look for and make use of structure.

1 Volume of Prisms

Recall that the volume of a solid is the measure of the amount of space the solid encloses. Volume is measured in cubic units.

The rectangular prism at the right has 6 • 4 or 24 cubic units in the bottom layer. Because there are two layers, the total volume is 24 • 2 or 48 cubic units.

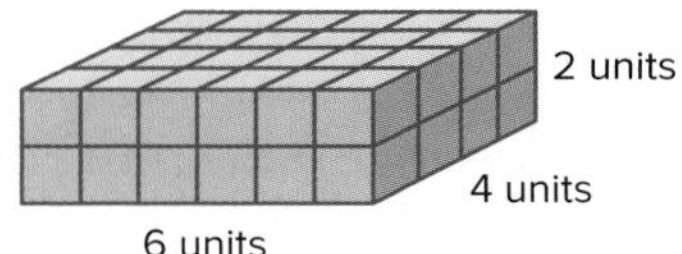

Key Concept Volume of a Prism

Words The volume V of a prism is $V = Bh$, where B is the area of a base and h is the height of the prism.

Symbols $V = Bh$

Model

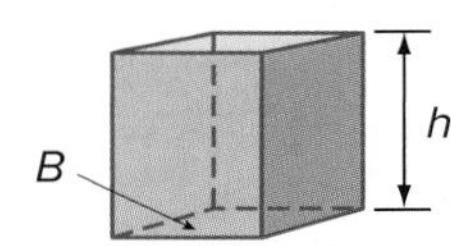

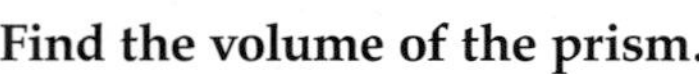

Example 1 Volume of a Prism

Find the volume of the prism.

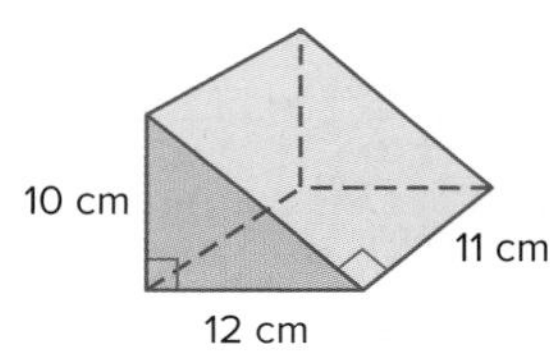

Step 1 Find the area of the base B.

$B = \frac{1}{2}bh$ Area of a triangle

$= \frac{1}{2}(12)(10)$ or 60 $b = 12$ and $h = 10$

Step 2 Find the volume of the prism.

$V = Bh$ Volume of a prism

$= 60(11)$ or 660 $B = 60$ and $h = 11$

The volume of the prism is 660 cubic centimeters.

Guided Practice

1A.

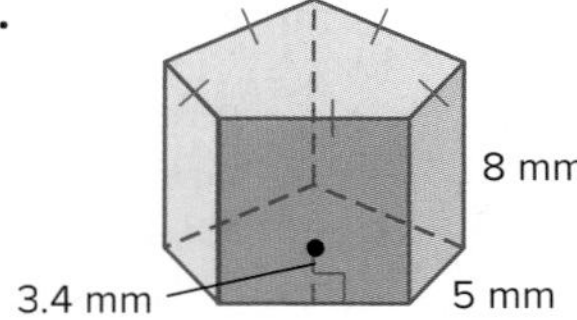

1B.

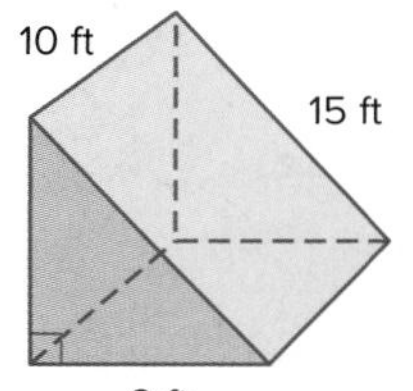

s0ulsurfing - Jason Swain/Getty Images

Go Online!

Investigate volumes of prisms and cylinders by using the **Geometry Tools** in ConnectED.

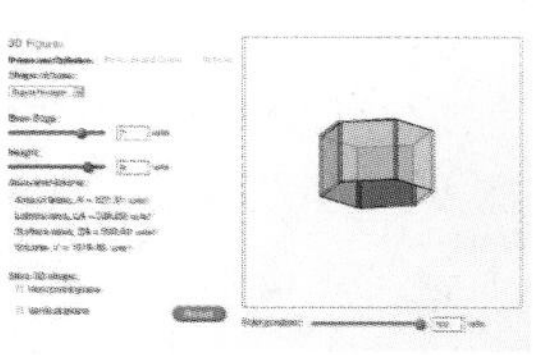

2 Volume of Cylinders

Like a prism, the volume of a cylinder can be thought of as consisting of layers. For a cylinder, these layers are congruent circular discs, similar to the coins in the roll shown. If we interpret the area of the base as the volume of a one-unit-high layer and the height of the cylinder as the number of layers, then the volume of the cylinder is equal to the volume of a layer times the number of layers or the area of the base times the height.

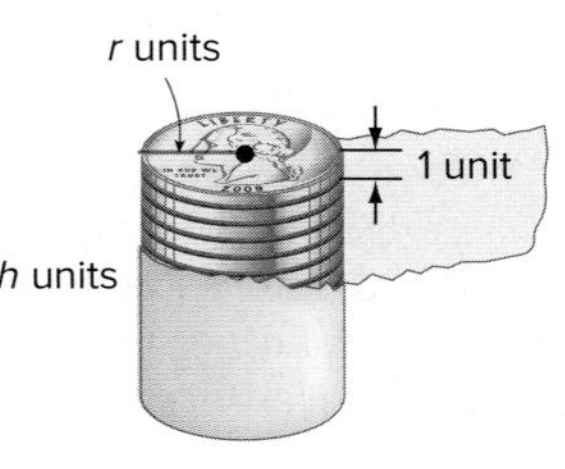

Key Concept Volume of a Cylinder

Words The volume V of a cylinder is $V = Bh$ or $V = \pi r^2 h$, where B is the area of the base, h is the height of the cylinder, and r is the radius of the base.

Symbols $V = Bh$ or $V = \pi r^2 h$

Model

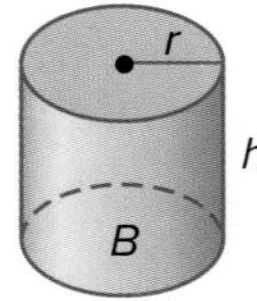

Example 2 Volume of a Cylinder

Find the volume of the cylinder at the right.

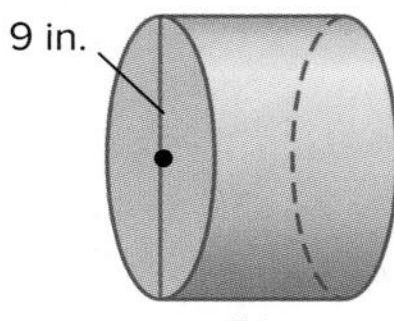

Estimate: $V \approx 3 \cdot 5^2 \cdot 5$ or 375 in^3

$V = \pi r^2 h$	Volume of a cylinder
$= \pi(4.5)^2(5)$	$r = 4.5$ and $h = 5$
≈ 318.1	Use a calculator.

The volume of the cylinder is about 318.1 cubic inches. This is fairly close to the estimate, so the answer is reasonable.

Guided Practice

2. Find the volume of a cylinder with a radius of 3 centimeters and a height of 8 centimeters. Round to the nearest tenth.

The first group of books at the right represents a right prism. The second group represents an oblique prism. Both groups have the same number of books. If all the books are the same size, then the volume of both groups is the same.

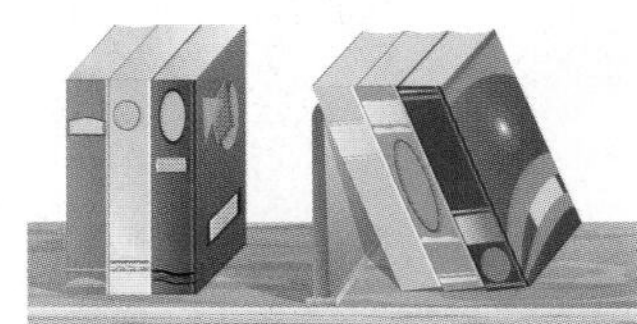

This demonstrates the following principle, which applies to all solids.

Watch Out

Cross-Sectional Area For solids with the same height to have the same volume, their cross sections must have the same area. The cross sections of the different solids do not have to be congruent polygons.

Key Concept Cavalieri's Principle

Words If two solids have the same height h and the same cross-sectional area B at every level, then they have the same volume.

Models

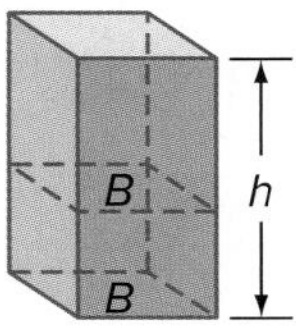

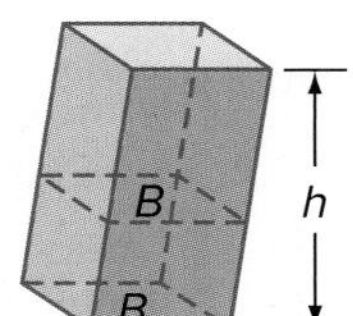

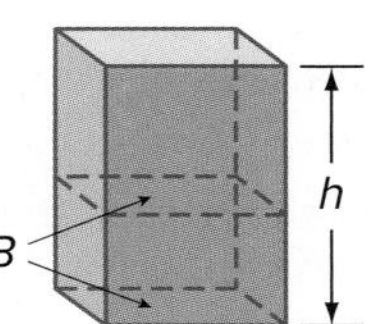

These prisms all have a volume of Bh.

> **Problem-Solving Tips**
>
> **MP Modeling** When solving problems involving volume of solids, one way to help you visualize the problem is to make a model of the solid.

Example 3 Volume of an Oblique Solid

Find the volume of an oblique hexagonal prism if the height is 6.4 centimeters and the base area is 17.3 square centimeters.

$V = Bh$ Volume of a prism

$= 17.3(6.4)$ $B = 17.3$ and $h = 6.4$

$= 110.72$ Simplify.

The volume is 110.72 cubic centimeters.

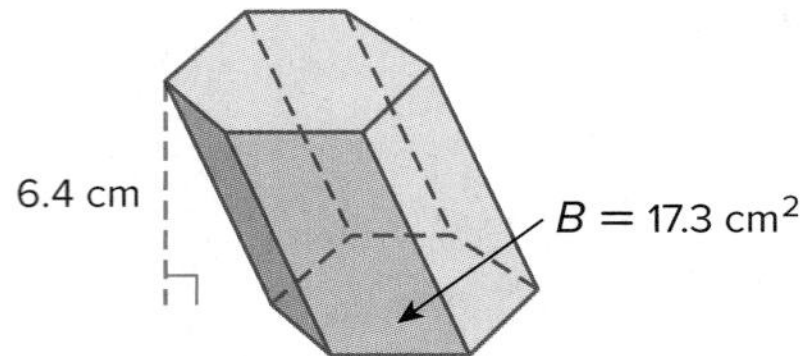

Guided Practice

3. Find the volume of an oblique cylinder that has a radius of 5 feet and a height of 3 feet. Round to the nearest tenth.

Example 4 Comparing Volumes of Solids

A manufacturer packages its products in cardboard boxes that have the same length and width, but different heights. If the volume of Box B is 150 cubic inches greater than the volume of Box A, what is the length of each box?

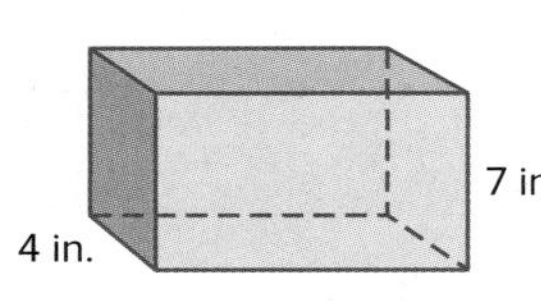

Box A

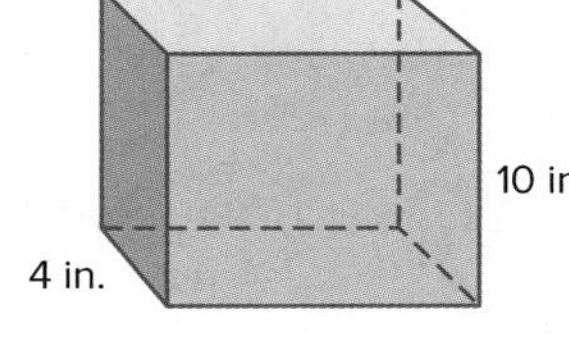

Box B

A 10 in. **B** $11\frac{1}{2}$ in. **C** 12 in. **D** $12\frac{1}{2}$ in.

Read the Item

You know two dimensions of each solid and that the difference between their volumes is 150 cubic inches.

Solve the Item

Volume of Box B − Volume of Box A = 150 Write an equation.

$4\ell \cdot 10 - 4\ell \cdot 7 = 150$ Use $V = Bh$.

$12\ell = 150$ Simplify.

$\ell = 12\frac{1}{2}$ Divide each side by 12.

The length of each box is $12\frac{1}{2}$ inches. The correct answer is D.

> **Real-World Career**
>
> **Architectural Engineer** An architectural engineer applies the technical skills of engineering to the design, construction, operation, maintenance, and renovation of buildings.
>
> Architectural engineers are required to have a bachelor's degree in engineering along with specialized coursework. Refer to Exercise 37.

Guided Practice

4. The containers at the right are filled with popcorn. About how many times as much popcorn does the larger container hold?

   F 1.6 times as much
   G 2.5 times as much
   H 3.3 times as much
   J 5.0 times as much

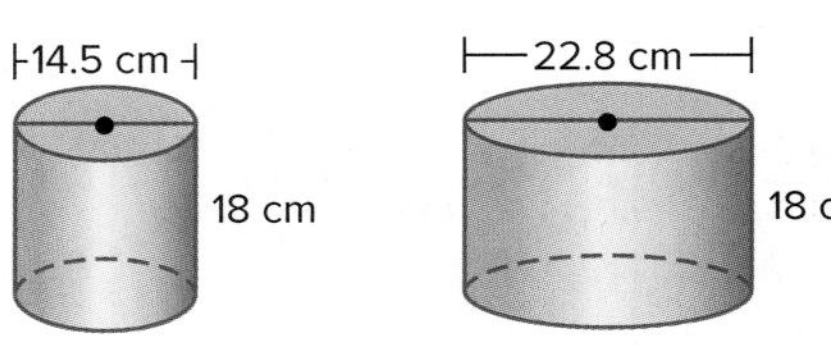

Check Your Understanding

◯ = Step-by-Step Solutions begin on page R13.

Examples 1 and 3 **Find the volume of each prism.**

1.

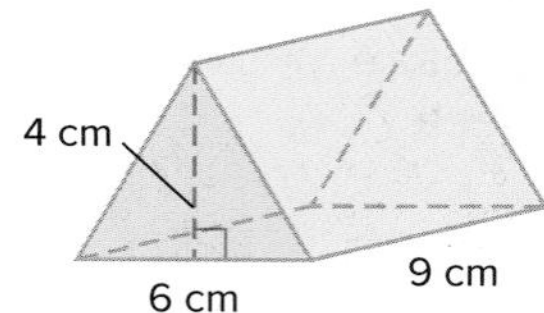

2.

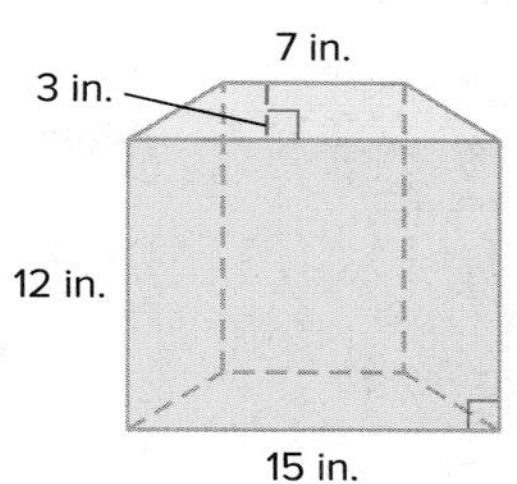

3. the oblique rectangular prism shown at the right

4. an oblique pentagonal prism with a base area of 42 square centimeters and a height of 5.2 centimeters

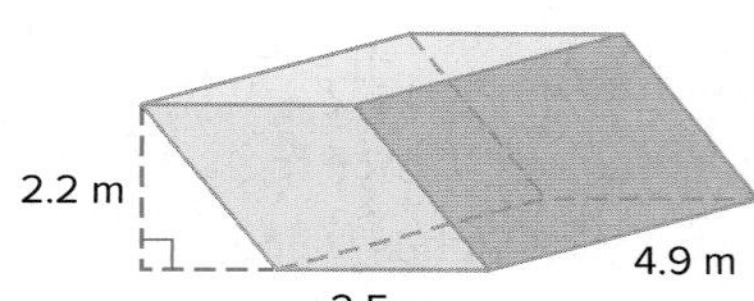

Examples 2–3 **Find the volume of each cylinder. Round to the nearest tenth.**

5.

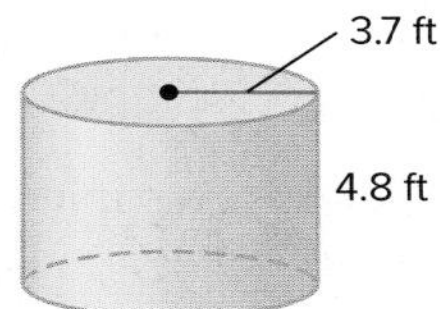

6.

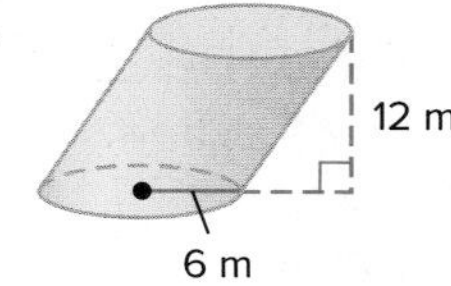

7. a cylinder with a diameter of 16 centimeters and a height of 5.1 centimeters

8. a cylinder with a radius of 4.2 inches and a height of 7.4 inches

Example 4

9. **MULTIPLE CHOICE** A rectangular lap pool measures 80 feet long by 20 feet wide. If it needs to be filled to four feet deep and each cubic foot holds 7.5 gallons, how many gallons will it take to fill the lap pool?

A 4000 **B** 6400 **C** 30,000 **D** 48,000

Practice and Problem Solving

Extra Practice is on page R11.

Examples 1 and 3

STRUCTURE **Find the volume of each prism.**

10.

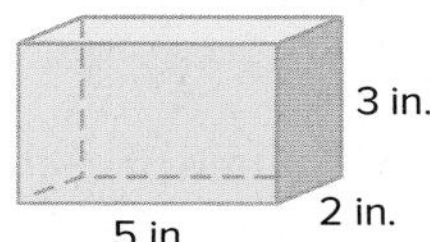

11.

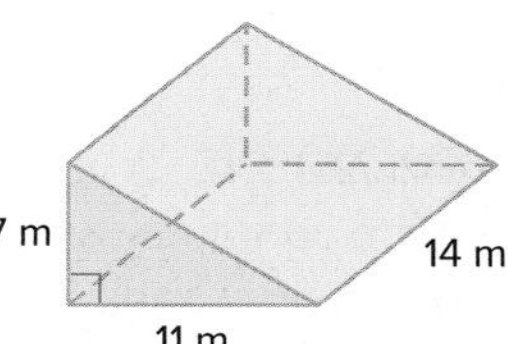

12.

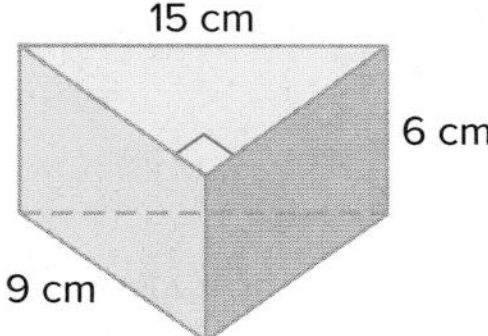

13.

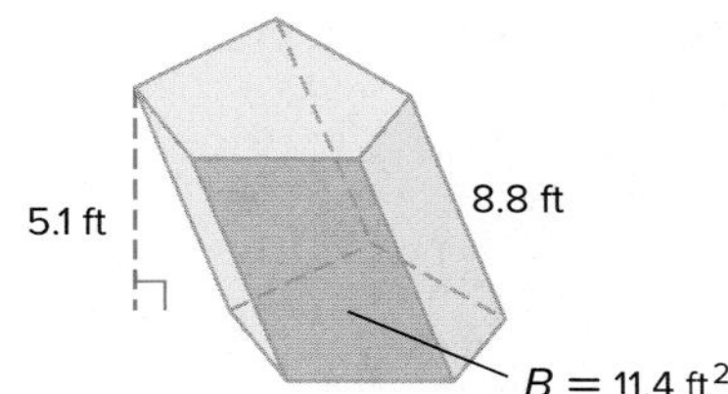

14. an oblique hexagonal prism with a height of 15 centimeters and a base area of 136 square centimeters

15. a square prism with a base edge of 9.5 inches and a height of 17 inches

Examples 2–3 **MP STRUCTURE** **Find the volume of each cylinder. Round to the nearest tenth.**

16.

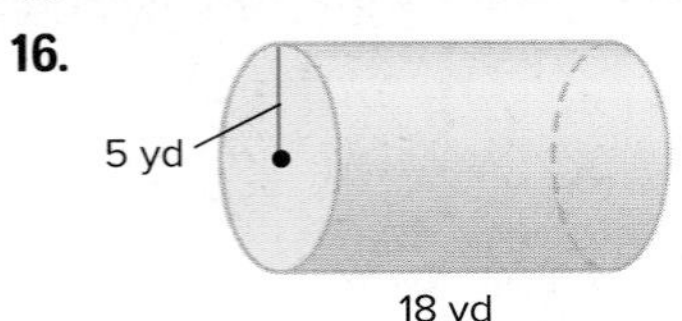

17.

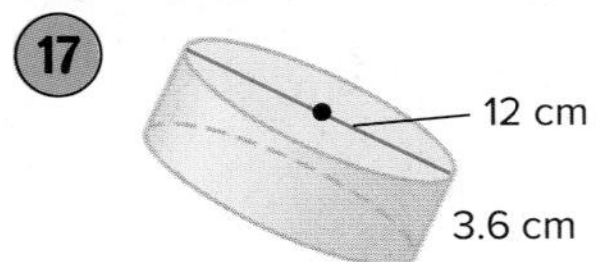

18.

19.

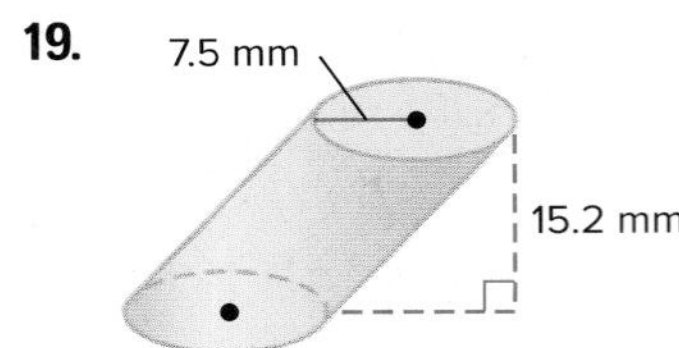

Example 4

20. PLANTER A planter is in the shape of a rectangular prism 18 inches long, $14\frac{1}{2}$ inches wide, and 12 inches high. What is the volume of potting soil in the planter if it is filled to $1\frac{1}{2}$ inches below the top?

21. SHIPPING The box shown is being used to ship two cylindrical candles. What is the volume of the empty space in the box?

22. CHANGING DIMENSIONS A cylinder has a radius of 5 centimeters and a height of 8 centimeters. Describe how each change affects the volume and surface area of the cylinder.

a. The height is tripled.

b. The radius is tripled.

c. Both the radius and the height are tripled.

d. The dimensions are exchanged.

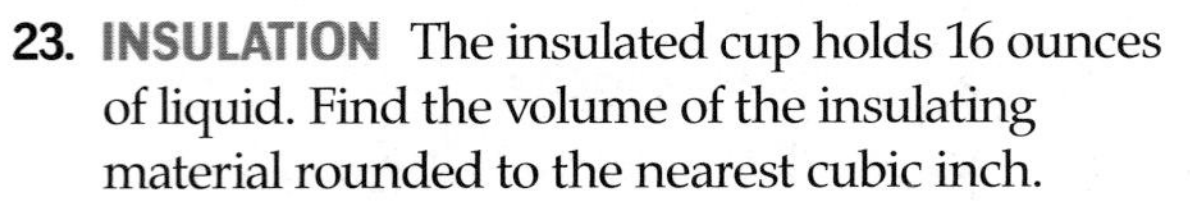

23. INSULATION The insulated cup holds 16 ounces of liquid. Find the volume of the insulating material rounded to the nearest cubic inch.

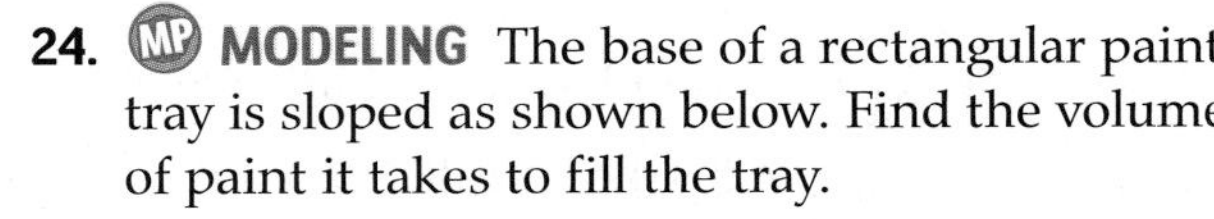

24. MP MODELING The base of a rectangular paint tray is sloped as shown below. Find the volume of paint it takes to fill the tray.

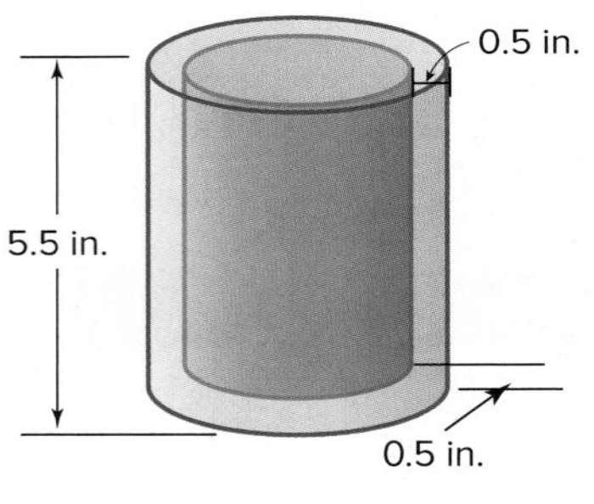

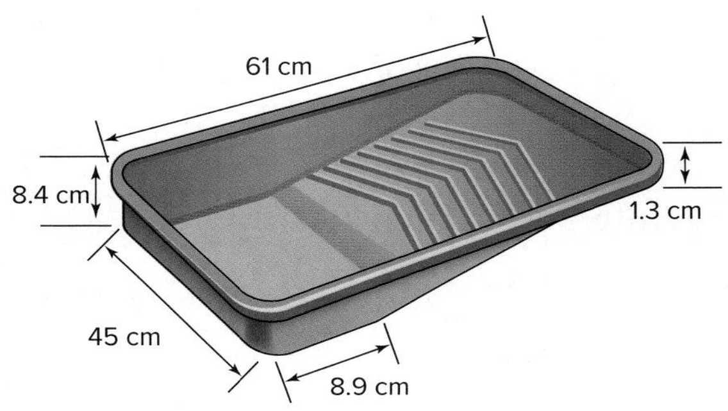

25. CHANGING DIMENSIONS A cereal company wants to increase the volume of each rectangular prism container by 25% without changing the base. Find the height of the new container if the original had a base of 8 inches by 2 inches and a height of 12 inches. What would the height be if the surface area of the container increased by 25%?

Find the volume of each composite solid. Round to the nearest tenth if necessary.

26.

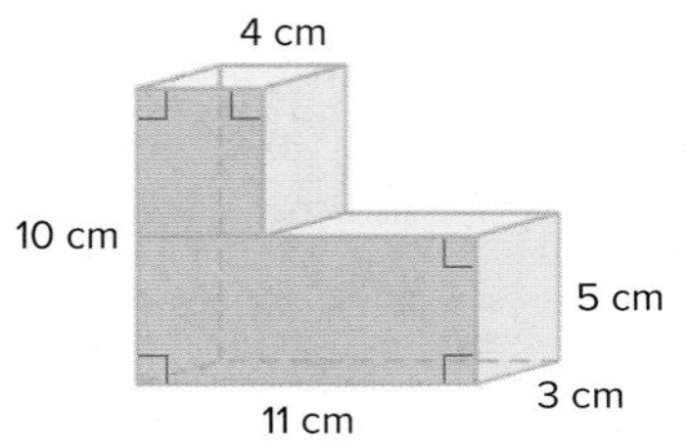

27.

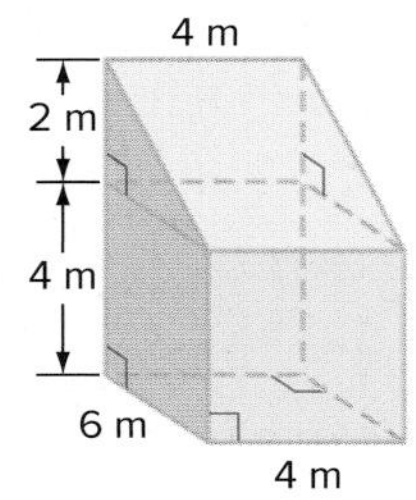

28.

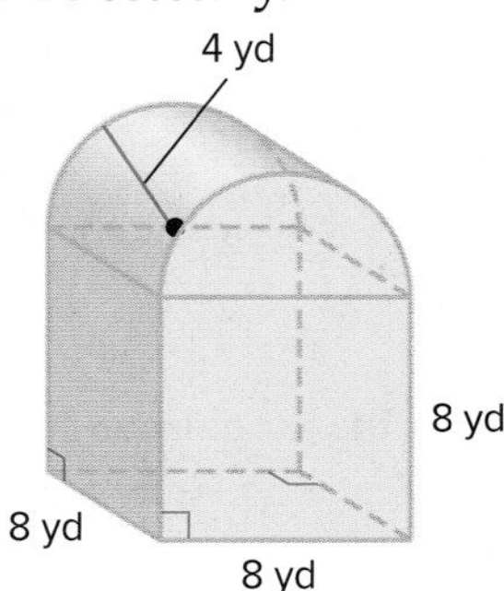

29. **FOOD** A cylindrical can of baked potato chips has a height of 27 centimeters and a radius of 4 centimeters. A new can is advertised as being 30% larger than the regular can. If both cans have the same radius, what is the height of the larger can?

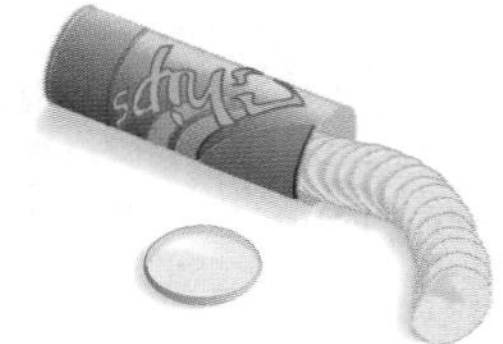

Find each measure to the nearest tenth.

30. A cylindrical can has a volume of 363 cubic centimeters. The diameter of the can is 9 centimeters. What is the height?

31. A cylinder has a surface area of 144π square inches and a height of 6 inches. What is the volume?

32. A rectangular prism has a surface area of 432 square inches, a height of 6 inches, and a width of 12 inches. What is the volume?

Find the volume of the solid formed by each net.

33.

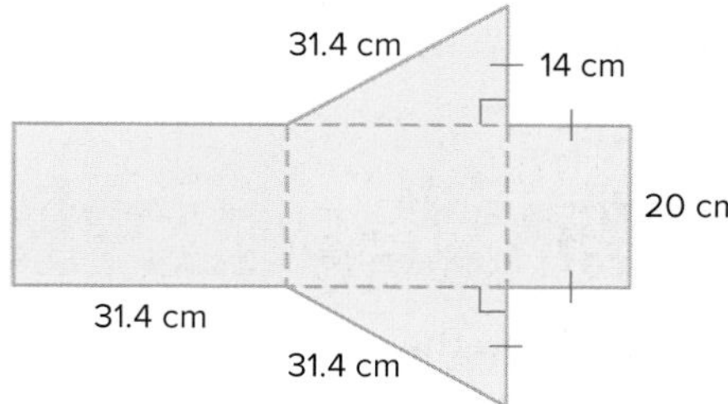

34.

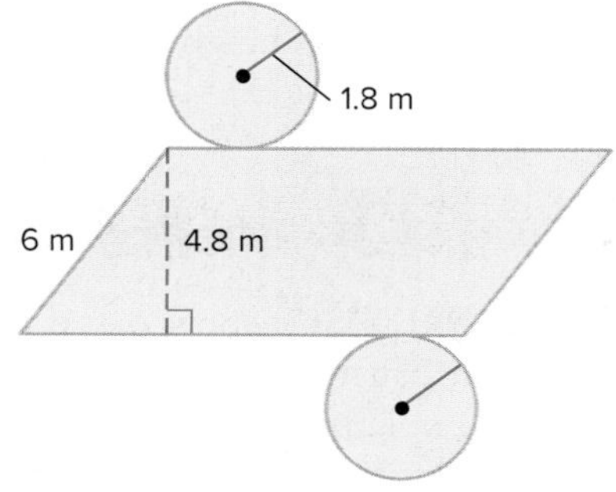

35. **SOIL** A soil scientist wants to determine the bulk density of a potting soil to assess how well a specific plant will grow in it. The density of the soil sample is the ratio of its weight to its volume.

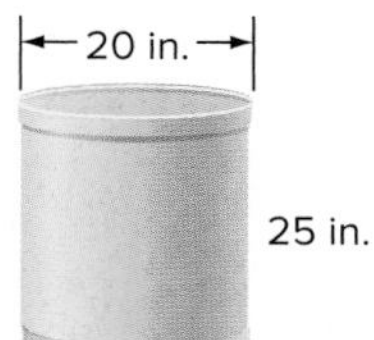

a. If the weight of the container with the soil is 20 pounds and the weight of the container alone is 5 pounds, what is the soil's bulk density?

b. Assuming that all other factors are favorable, how well should a plant grow in this soil if a bulk density of 0.0018 pound per square inch is desirable for root growth? Explain.

c. If a bag of this soil holds 2.5 cubic feet, what is its weight in pounds?

36. **DESIGN** Sketch and label (in inches) three different designs for a dry ingredient measuring cup that holds 1 cup. Be sure to include the dimensions in each drawing. (1 cup $\approx$ 14.4375 in^3)

37. MP **MODELING** A cylindrical stainless steel column is used to hide a ventilation system in a new building. According to the specifications, the diameter of the column can be between 30 centimeters and 95 centimeters. The height is to be 500 centimeters. What is the difference in volume between the largest and smallest possible column? Round to the nearest tenth.

38. **MULTI-STEP** Ryann is planning to build a sand castle. She wants her castle to be 4 feet high, 4 feet wide, and 6 feet deep. She has asked her brother Jack to bring sand over to her building site. They each have a bucket that is 8 inches in diameter and 16 inches tall. Each trip takes Jack about 30 seconds.

a. After how long will Ryann have all of the sand she could possibly need to complete her castle?

b. Describe your solution process.

c. What assumptions did you make?

39. Find the volume of the regular pentagonal prism at the right by dividing it into five equal triangular prisms. Describe the base area and height of each triangular prism.

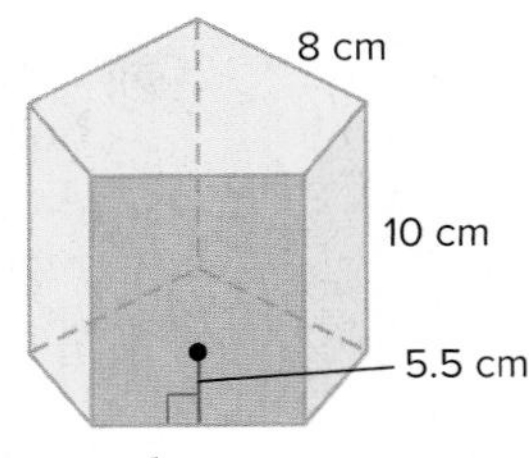

40. PATIOS Mr. Thomas is planning to remove an old patio and install a new rectangular concrete patio 20 feet long, 12 feet wide, and 4 inches thick. One contractor bid \$2225 for the project. A second contractor bid \$500 per cubic yard for the new patio and \$700 for removal of the old patio. Which is the less expensive option? Explain.

41. MULTIPLE REPRESENTATIONS In this problem, you will investigate cylinders.

a. Geometric Draw a right cylinder and an oblique cylinder with a height of 10 meters and a diameter of 6 meters.

b. Verbal A square prism has a height of 10 meters and a base edge of 6 meters. Is its volume greater than, less than, or equal to the volume of the cylinder? Explain.

c. Analytical Describe which change affects the volume of the cylinder more: multiplying the height by x or multiplying the radius by x. Explain.

H.O.T. Problems Use Higher-Order Thinking Skills

42. ERROR ANALYSIS Francisco and Valerie each calculated the volume of an equilateral triangular prism with an apothem of 4 units and height of 5 units. Is either of them correct? Explain your reasoning.

Francisco

$$V = Bh$$
$$= \frac{1}{2}aP \cdot h$$
$$= \frac{1}{2}(4)(24\sqrt{3}) \cdot 5$$
$$= 240\sqrt{3} \text{ cubic units}$$

Valerie

$$V = Bh$$
$$= \frac{\sqrt{3}}{2}s^2 \cdot h$$
$$= \frac{\sqrt{3}}{2}(4\sqrt{3})^2 \cdot 5$$
$$= 120\sqrt{3} \text{ cubic units}$$

43. CHALLENGE The cylindrical can below is used to fill a container with liquid. It takes three full cans to fill the container. Describe possible dimensions of the container if it is each of the following shapes.

2 in.
5 in.

a. rectangular prism

b. square prism

c. triangular prism with a right triangle as the base

44. WRITING IN MATH Write a helpful response to the following question posted on an Internet gardening forum.

I am new to gardening. The nursery will deliver a truckload of soil, which they say is 4 yards. I know that a yard is 3 feet, but what is a yard of soil? How do I know what to order?

45. OPEN-ENDED Draw and label a prism that has a volume of 50 cubic centimeters.

46. MP CONSTRUCT ARGUMENTS Determine whether the following statement is true or false. Explain.

Two cylinders with the same height and the same lateral area must have the same volume.

47. WRITING IN MATH How are the volume formulas for prisms and cylinders similar? How are they different?

48. The rectangular prism shown here has a square base and a volume of 132.3 cubic inches.

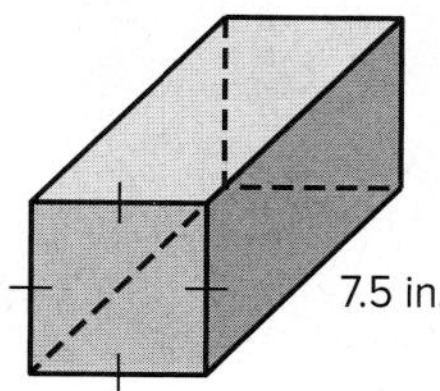

What is the perimeter of the base?
MP 1, 7

- ◯ **A** 30 in.
- ◯ **B** 17.64 in.
- ◯ **C** 16.8 in.
- ◯ **D** 4.2 in.

49. An aquarium is a rectangular prism that is 20 inches long, 1 foot wide, and 15 inches tall. Denise fills the aquarium using a container that holds 400 cubic inches of water. Assuming she always fills the container completely, how many times will Denise need to pour water from the container into the aquarium? MP 1, 7

50. Scott adds sand to the cylindrical container shown below so that the surface of the sand is 2 inches below the top of the container.

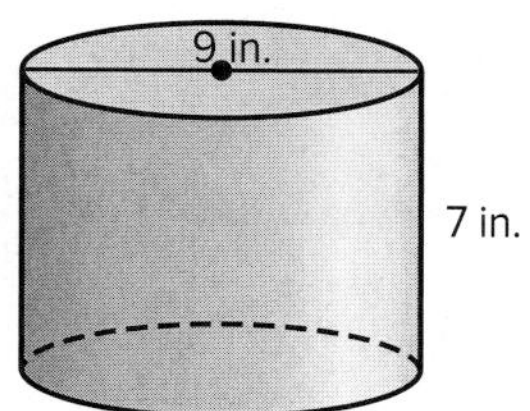

Which of the following is the best estimate of the volume of the sand in the container?
MP 1, 7

- ◯ **A** 127 in^3
- ◯ **B** 269 in^3
- ◯ **C** 318 in^3
- ◯ **D** 445 in^3
- ◯ **E** 1272 in^3

51. A cylindrical tank used for oil storage has a height that is half the length of its radius. If the volume of the tank is 1,122,360 cubic feet, what is the tank's radius in feet? Round to the nearest tenth. MP 1, 7

52. The cylindrical can of juice shown here has a volume of 300 cubic centimeters. What is the diameter of the can in centimeters? Round to the nearest tenth. MP 1, 7

- ◯ **A** 6.2 cm
- ◯ **C** 9.8 cm
- ◯ **B** 3.1 cm
- ◯ **D** 30 cm

53. A red cube has an edge length of 2 inches. A blue cube has an edge length that is double that of the red cube. What is the volume of the blue cube?
MP 1, 7

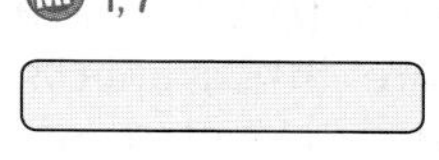

54. **MULTI-STEP** Kara has a cylindrical pillar candle that is 4 inches in diameter and 9 inches tall. She melts the candle and pours all of the wax into a square mold that is 4 inches on each side. MP 1, 7

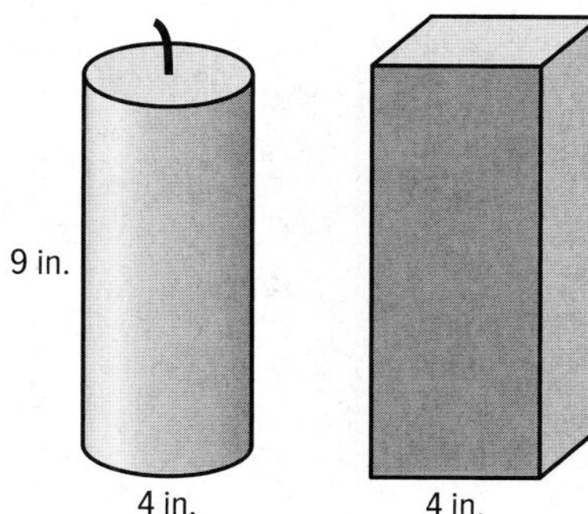

a. To the nearest cubic inch, what is the volume of the pillar candle?

b. Write an equation that makes the volume of the pillar candle equal to the volume of a square candle with 4-inch sides and an unknown height.

c. Solve the equation to find the height of the square candle.

d. If Kara wanted to make the square candle the same height as the cylindrical candle (9 inches), how much more wax would she need?

LESSON 3

Volumes of Pyramids and Cones

∴Then	∴Now	∴Why?
● You found surface areas of pyramids and cones.	● **1** Find volumes of pyramids. **2** Find volumes of cones.	● Marta is studying crystals that grow on rock formations. For a project, she is making a clay model of a crystal with a shape that is a composite of two congruent rectangular pyramids. The base of each pyramid will be 1 by 1.5 inches, and the total height will be 4 inches. Why is determining the volume of the model helpful in this situation?

Mathematical Practices

1 Make sense of problems and persevere in solving them.

7 Look for and make use of structure.

1 Volume of Pyramids

A triangular prism can be separated into three triangular pyramids as shown. Because all faces of a triangular pyramid are triangles, any face can be considered a base of the pyramid.

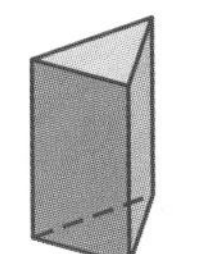

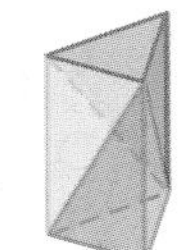

The yellow and orange pyramids have base area B_1 and height h_1. Therefore, by Cavalieri's Principle, they have the same volume. Likewise, the yellow and green pyramids have base area B_2 and height h_2, so they have the same volume.

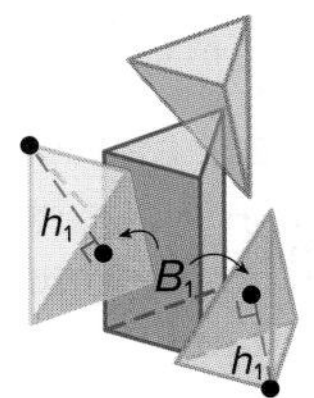

Because the orange and green pyramids have the same volume as the yellow pyramid, it follows that the volumes of all three pyramids are the same. Therefore, each pyramid has one third the volume of the prism with the same base area and height. This is true for a pyramid with any shape base.

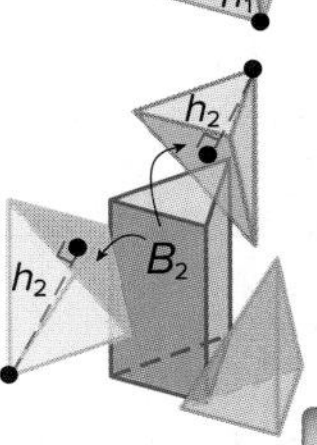

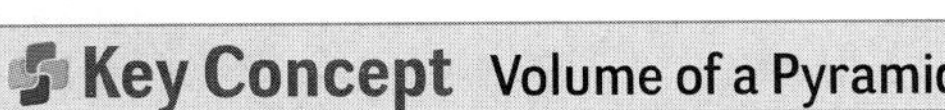

Key Concept Volume of a Pyramid

Words The volume of a pyramid is $V = \frac{1}{3}Bh$, where B is the area of the base and h is the height of the pyramid.

Symbols $V = \frac{1}{3}Bh$

Models

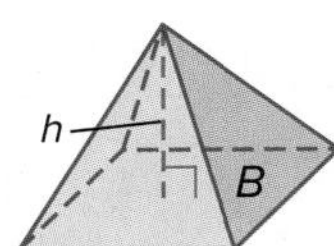

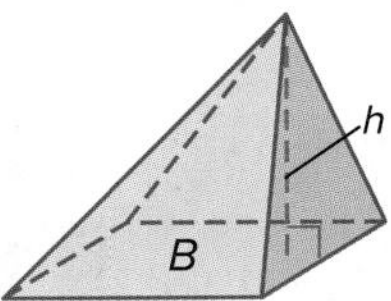

Example 1 Volume of a Pyramid

Find the volume of the pyramid.

$V = \frac{1}{3}Bh$ Volume of a pyramid

$= \frac{1}{3}(9.5 \cdot 8)(9)$ $B = 9.5 \cdot 8$ and $h = 9$

$= 228$ Simplify.

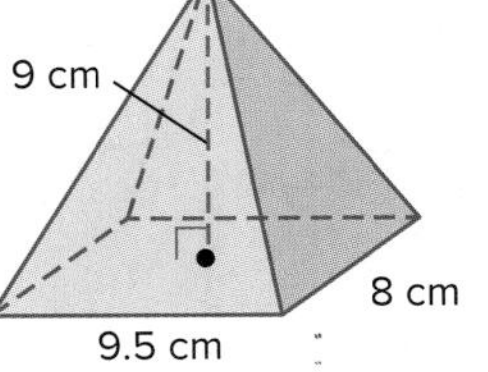

The volume of the pyramid is 228 cubic centimeters.

Go Online!

Watch an animation and answer questions to see how the volumes of pyramids and cones apply to architecture.

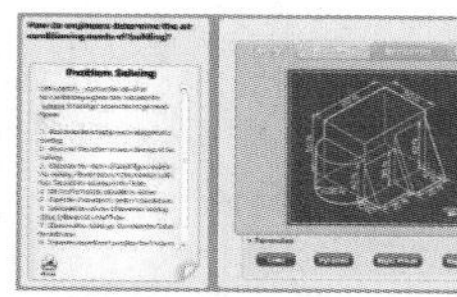

Guided Practice

1A.

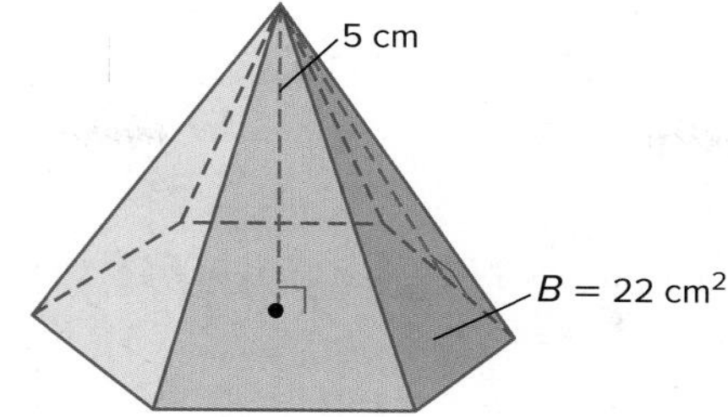

1B.

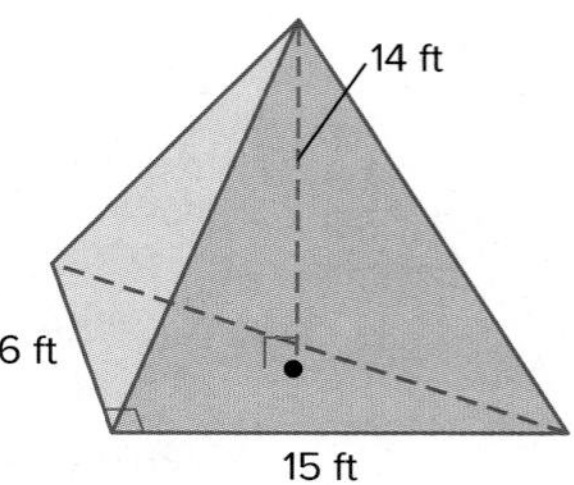

2 Volume of Cones

The pyramid and prism shown have the same base area B and height h as the cylinder and cone. Because the volume of the pyramid is one third the volume of the prism, then by Cavalieri's Principle, the volume of the cone must be one third the volume of the cylinder.

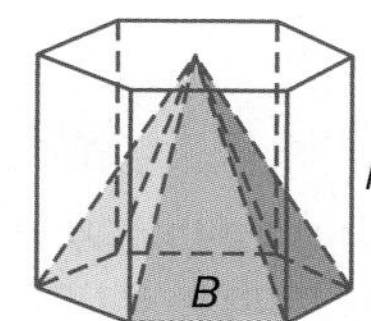

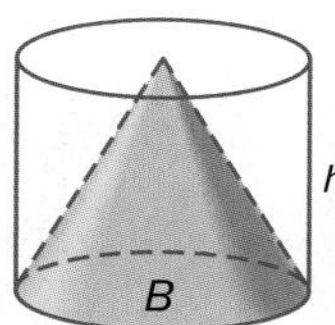

Watch Out!

Volumes of Cones The formula for the surface area of a cone only applies to right cones. However, the formula for volume applies to oblique cones as well as right cones.

Key Concept Volume of a Cone

Words The volume of a circular cone is $V = \frac{1}{3}Bh$, or $V = \frac{1}{3}\pi r^2 h$, where B is the area of the base, h is the height of the cone, and r is the radius of the base.

Models

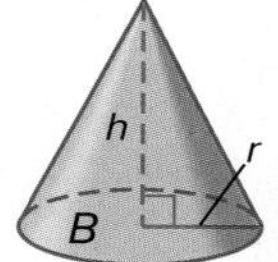

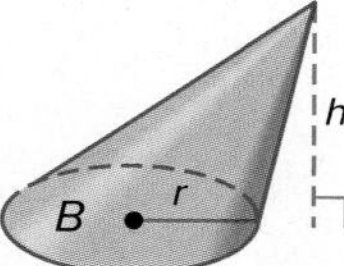

Symbols $V = \frac{1}{3}Bh$ or $V = \frac{1}{3}\pi r^2 h$

Example 2 Volume of a Cone

a. Find the volume of the cone. Round to the nearest tenth.

$V = \frac{1}{3}\pi r^2 h$ Volume of a cone

$\approx \frac{1}{3}\pi(3.2)^2(5.8)$ $r = 3.2$ and $h = 5.8$

≈ 62.2 Use a calculator.

The volume of the cone is approximately 62.2 cubic meters.

b. Find the volume of the cone. Round to the nearest tenth.

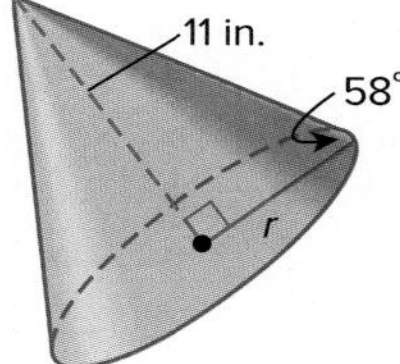

Step 1 Use trigonometry to find the radius.

$\tan 58° = \frac{11}{r}$ $\tan\theta = \frac{\text{opp}}{\text{adj}}$

$r = \frac{11}{\tan 58°}$ Solve for r.

$r \approx 6.9$ Use a calculator.

Step 2 Find the volume.

$V = \frac{1}{3}\pi r^2 h$ — Volume of a cone

$\approx \frac{1}{3}\pi(6.9)^2(11)$ — $r \approx 6.9$ and $h = 11$

≈ 548.4 — Use a calculator.

The volume of the cone is approximately 548.4 cubic inches.

Guided Practice

2A. 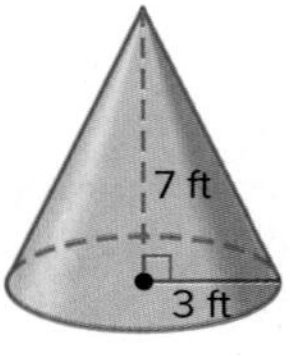

2B. 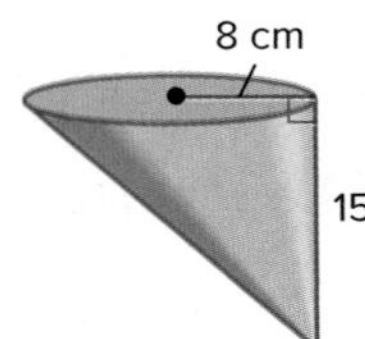

2C. 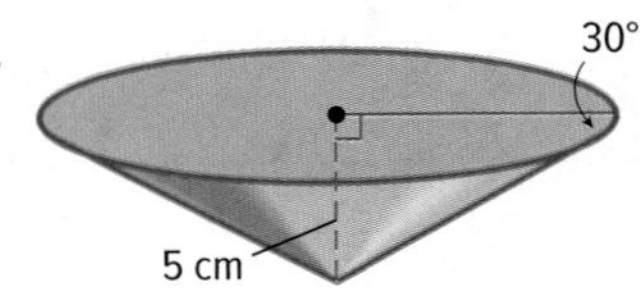

Real-World Link

The Washington Monument is the largest masonry structure in the world. By law, no other building in D.C. is allowed to be taller than the 555-foot-tall structure.

Source: Enchanted Learning

Real-World Example 3 Find Real-World Volumes

ARCHITECTURE At the top of the Washington Monument is a square pyramid, called a *pyramidion*. This pyramid has a height of 55.5 feet with base edges of approximately 34.5 feet. What is the volume of the pyramidion? Round to the nearest tenth.

Sketch and label the pyramid.

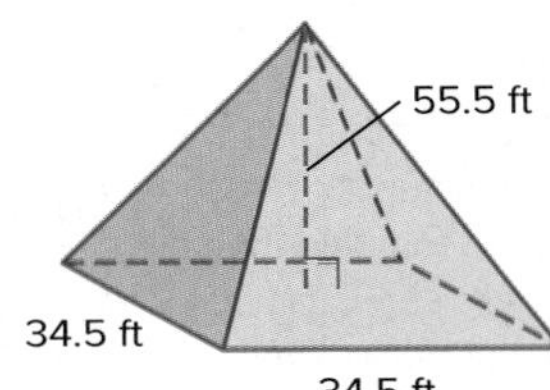

$V = \frac{1}{3}Bh$ — Volume of a pyramid

$= \frac{1}{3}(34.5 \cdot 34.5)(55.5)$ — $B = 34.5 \cdot 34.5$, $h = 55.5$

$\approx 22{,}019.6$ — Simplify.

The volume of the pyramidion atop the Washington Monument is about 22,019.6 cubic feet.

Guided Practice

3. **ARCHAEOLOGY** A pyramidion that was discovered in Saqqara, Egypt, in 1992 has a rectangular base 53 centimeters by 37 centimeters. It is 46 centimeters high. What is the volume of this pyramidion? Round to the nearest tenth.

The formulas for the volumes of solids are summarized below.

Concept Summary Volumes of Solids

Solid	prism	cylinder	pyramid	cone
Model	(h, B)	(r, h, B)	(h, B)	(h, r, B)
Volume	$V = Bh$	$V = Bh$ or $V = \pi r^2 h$	$V = \frac{1}{3}Bh$	$V = \frac{1}{3}Bh$ or $V = \frac{1}{3}\pi r^2 h$

Jupiterimages/Brand X/Alamy

Check Your Understanding

 = Step-by-Step Solutions begin on page R13.

Example 1 **Find the volume of each pyramid.**

1. 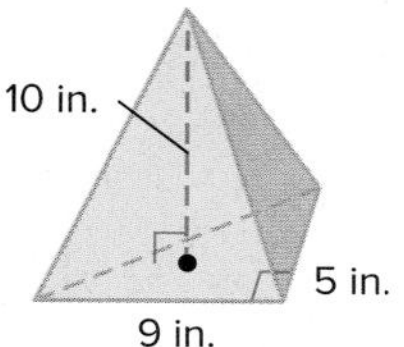

2. 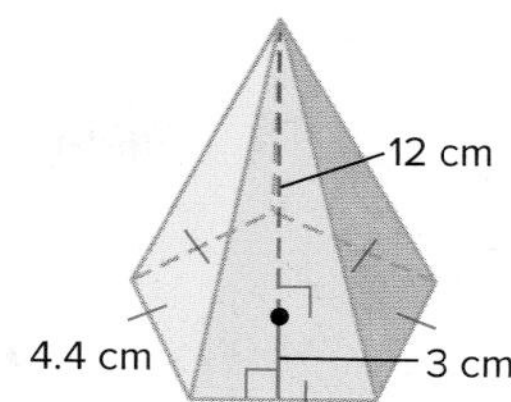

3. a rectangular pyramid with a height of 5.2 meters and a base 8 meters by 4.5 meters
4. a square pyramid with a height of 14 meters and a base with 8-meter side lengths

Example 2 **Find the volume of each cone. Round to the nearest tenth.**

5. 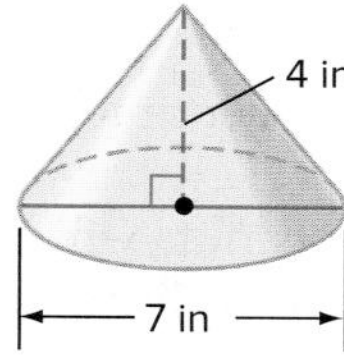

6. 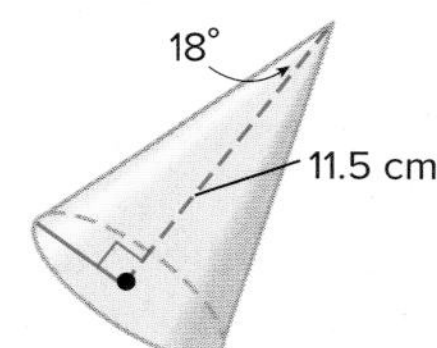

7. an oblique cone with a height of 10.5 millimeters and a radius of 1.6 millimeters
8. a cone with a slant height of 25 meters and a radius of 15 meters

Example 3

9. **HUTS** The Caddo Indians lived in tall cone-shaped grass huts made of wooden pole frames with long prairie grasses dried and threaded through the poles in layers. These houses normally measured 40 feet tall. Suppose the diameter is also 40 feet. What is the volume inside the hut?

Practice and Problem Solving

Extra Practice is on page R11.

Example 1 MP **STRUCTURE** Find the volume of each pyramid. Round to the nearest tenth.

10. 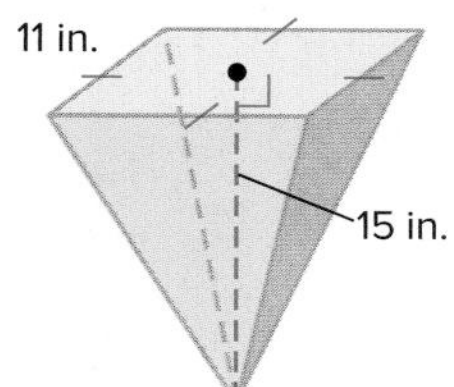

11. 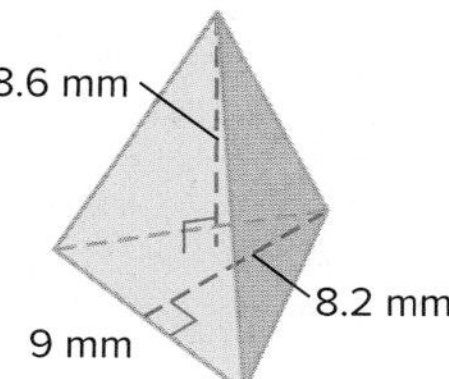

12. 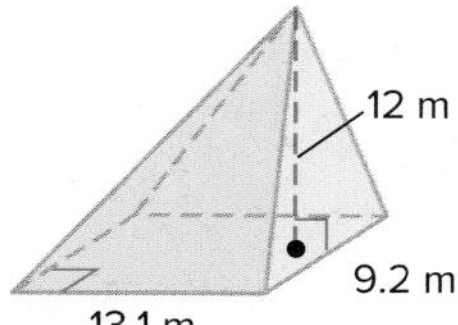

13. 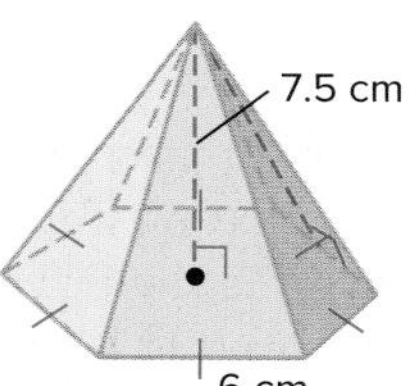

14. a pentagonal pyramid with a base area of 590 square feet and an altitude of 7 feet
15. a triangular pyramid with a height of 4.8 centimeters and a right triangle base with a leg 5 centimeters and hypotenuse 10.2 centimeters
16. A triangular pyramid with a right triangle base with a leg 8 centimeters and hypotenuse 10 centimeters has a volume of 144 cubic centimeters. Find the height.

Example 2 **Find the volume of each cone. Round to the nearest tenth.**

17.

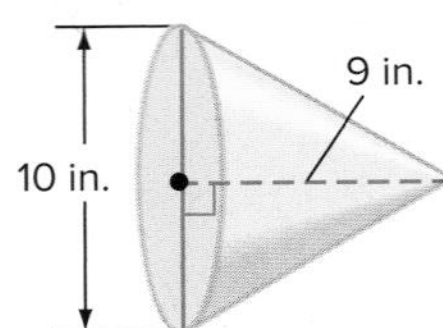

18.

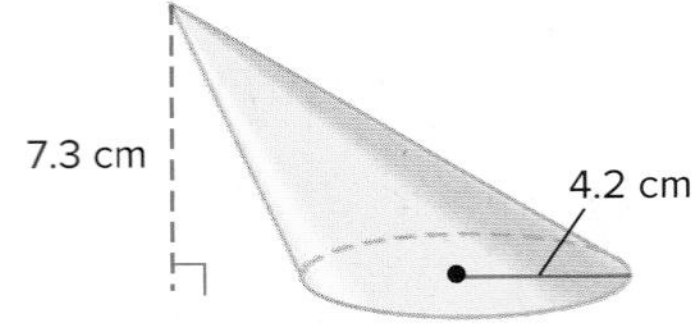

19.

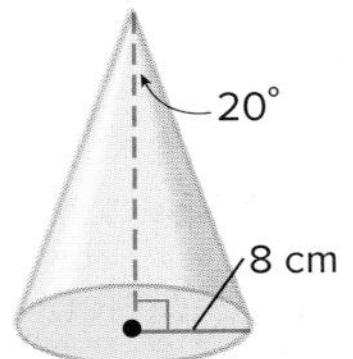

20.

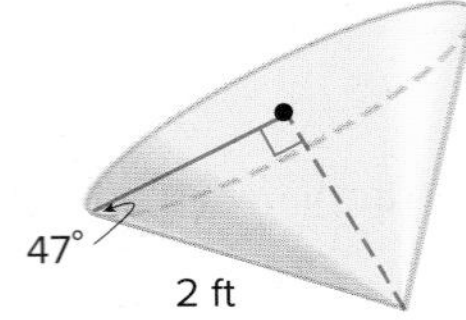

21. an oblique cone with a diameter of 16 inches and an altitude of 16 inches

22. a right cone with a slant height of 5.6 centimeters and a radius of 1 centimeter

Example 3 **23** **SNACKS** Approximately how many cubic centimeters of roasted peanuts will completely fill a paper cone that is 14 centimeters high and has a base diameter of 8 centimeters? Round to the nearest tenth.

24. MP **MODELING** A pyramid-shaped building in Memphis, Tennessee is approximately 350 feet tall, and its square base is 600 feet wide. Find the volume of this pyramid.

25. **GARDENING** The greenhouse at the right is a regular octagonal pyramid with a height of 5 feet. The base has side lengths of 2 feet. What is the volume of the greenhouse?

Find the volume of each solid. Round to the nearest tenth.

26.

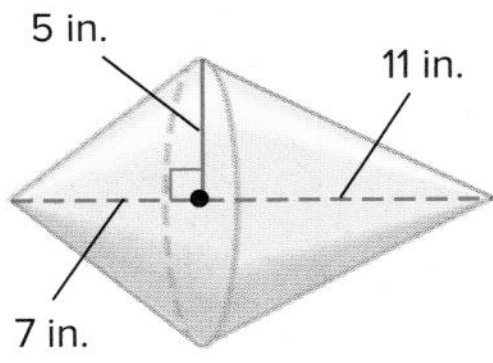

27.

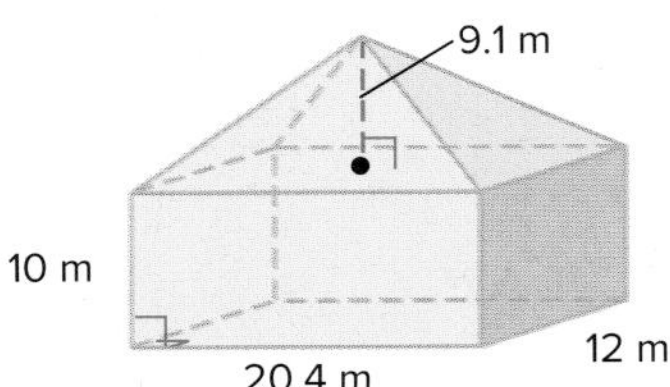

28.

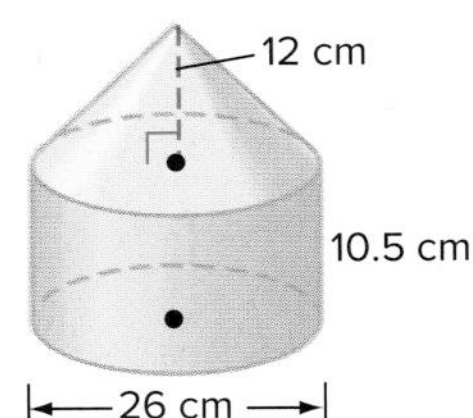

29. **HEATING** Sam is building an art studio in her backyard. To buy a heating unit for the space, she needs to determine the BTUs (British Thermal Units) required to heat the building. For new construction with good insulation, there should be 2 BTUs per cubic foot. What size unit does Sam need to purchase?

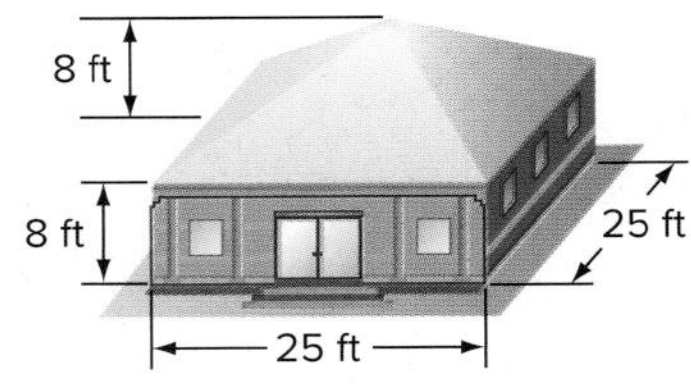

30. **SCIENCE** Refer to page 810. Determine the volume of the crystal model that Marta is making. Explain why knowing the volume is helpful in this situation.

31. **CHANGING DIMENSIONS** A cone has a radius of 4 centimeters and a height of 9 centimeters. Describe how each change affects the volume of the cone.

a. The height is doubled.

b. The radius is doubled.

c. Both the radius and the height are doubled.

Find each measure. Round to the nearest tenth if necessary.

32. A square pyramid has a volume of 862.5 cubic centimeters and a height of 11.5 centimeters. Find the side length of the base.

33. The volume of a cone is 196π cubic inches and the height is 12 inches. What is the diameter?

34. The lateral area of a cone is 71.6 square millimeters and the slant height is 6 millimeters. What is the volume of the cone?

35. **MULTIPLE REPRESENTATIONS** In this problem, you will investigate rectangular pyramids.

a. Geometric Draw two pyramids with different bases that have a height of 10 centimeters and a base area of 24 square centimeters.

b. Verbal What is true about the volumes of the two pyramids? Explain.

c. Analytical Explain how multiplying the base area and/or the height of the pyramid by 5 affects the volume of the pyramid.

H.O.T. Problems Use Higher-Order Thinking Skills

36. **MP CONSTRUCT ARGUMENTS** Determine whether the following statement is *always*, *sometimes*, or *never* true. Justify your reasoning.

The volume of a cone with radius r and height h equals the volume of a prism with height h.

37. **ERROR ANALYSIS** Alexandra and Cornelio are calculating the volume of the cone at the right. Is either of them correct? Explain your answer.

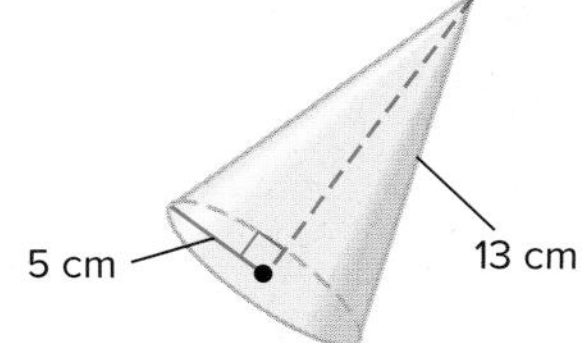

Alexandra	Cornelio
$V = \frac{1}{3}Bh$	$5^2 + 12^2 = 13^2$
$= \frac{1}{3}\pi(5^2)(13)$	$V = \frac{1}{3}Bh$
$\approx 340.3 \text{ cm}^3$	$= \frac{1}{3}\pi(5^2)(12)$
	$\approx 314.2 \text{ cm}^3$

38. **CHALLENGE** A cone has a volume of 568 cubic centimeters. What is the volume of a cylinder that has the same radius and height as the cone? Explain your reasoning.

39. **MP REASONING** Give an example of a pyramid and a prism that have the same base and the same volume. Explain your reasoning.

40. **WRITING IN MATH** Compare and contrast finding volumes of pyramids and cones with finding volumes of prisms and cylinders.

41. Cullen is buying a tent that is in the shape of a rectangular pyramid.

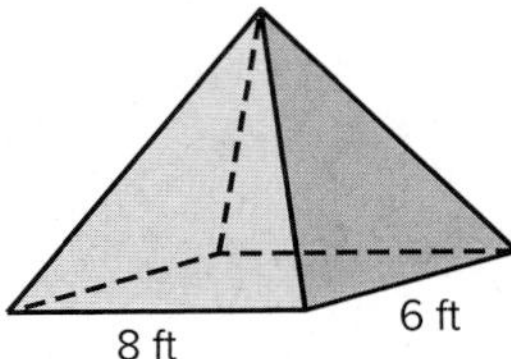

If the tent holds 88 cubic feet of air, how tall is the tent? MP 1, 7

- A $1\frac{5}{6}$ ft
- B $\frac{11}{18}$ ft
- C $5\frac{1}{2}$ ft
- D 10 ft

42. Tara has a cylindrical candle mold that is 8 inches high with a diameter of 3 inches. She would like to melt votive candles and reuse the wax. Each votive candle is a cylinder with a height of 1.5 inches and a radius of 0.5 inch. How many votive candles are needed to fill the candle mold? MP 1, 7

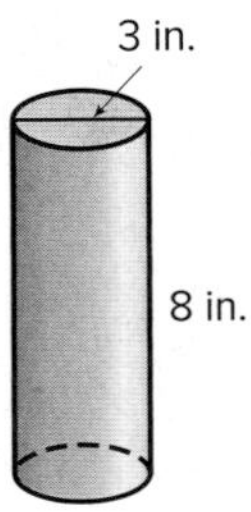

- A 12
- B 48
- C 57
- D 192
- E 226

43. A right circular cone has a height of 10 centimeters and a volume of 32 cubic centimeters. What is the radius of the cone? MP 1, 7

- A 0.5 cm
- B 1.75 cm
- C 2.25 cm
- D 2.75 cm

44. A conical sand toy has the dimensions shown. How many cubic centimeters of sand will it hold when it is filled to the top? MP 1, 7

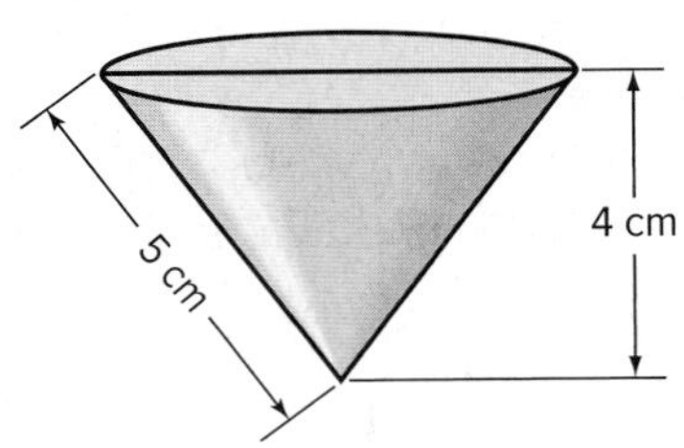

- A 12π
- B 15π
- C $\frac{80}{3}\pi$
- D $\frac{100}{3}\pi$

45. MULTI-STEP The figure shows a cone with a cylindrical hole cut out. What is the volume of this solid? MP 1, 7

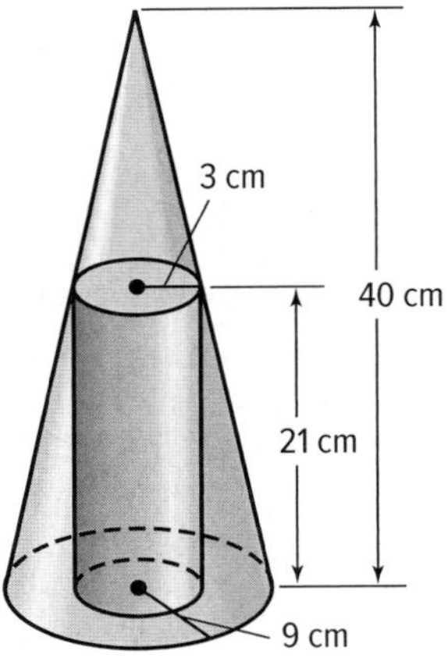

46. What is the height of the pyramid if the volume is 210 cubic feet and the base is 70 square feet? MP 1, 7

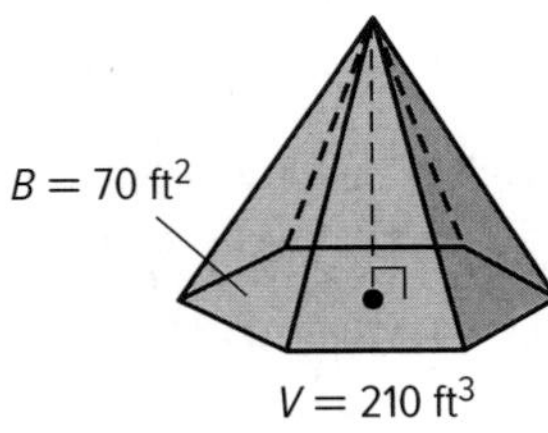

CHAPTER 11
Mid-Chapter Quiz
Lessons 11-1 through 11-3

Sketch each cross section formed by the intersection of the plane and solid. (Lesson 11-1)

1.

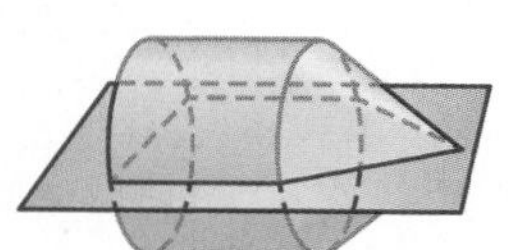

2.

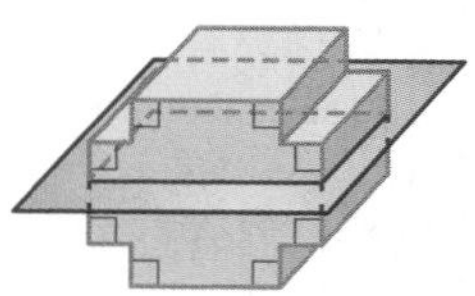

Describe the shape of the cross section. (Lesson 11-1)

3.

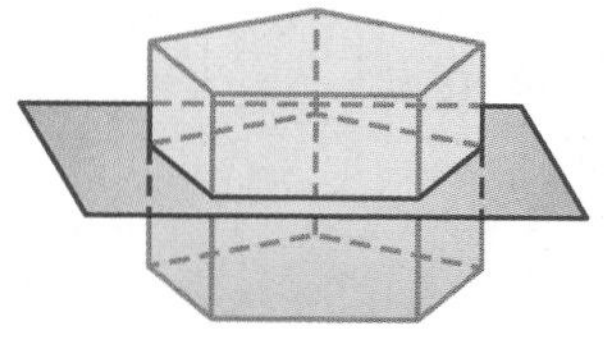

4.

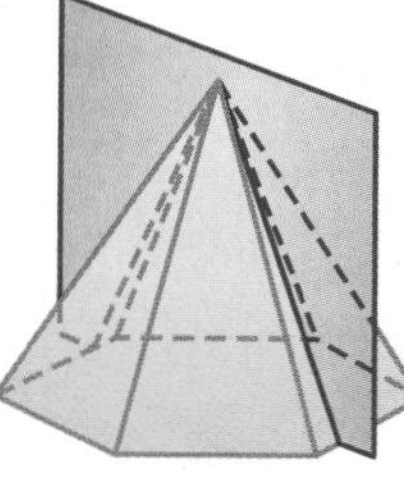

Identify the three-dimensional shape generated by rotating each two-dimensional shape around the given axis. (Lesson 11-1)

5.

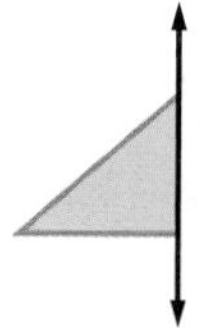

6.

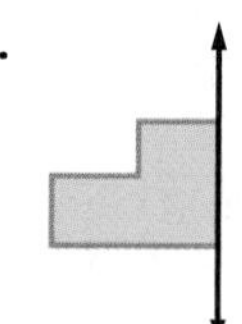

Find the volume of each prism or cylinder. Round to the nearest tenth if necessary. (Lesson 11-2)

7.

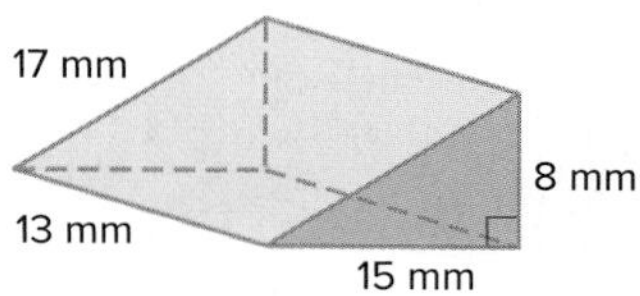

8.

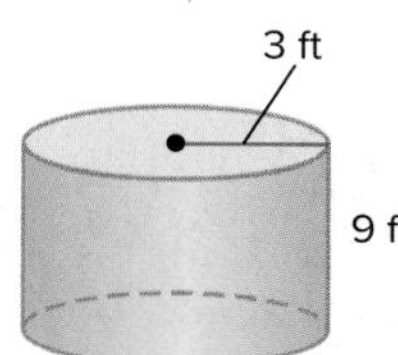

9.

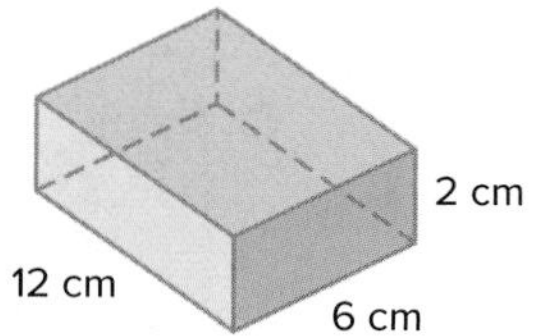

10.

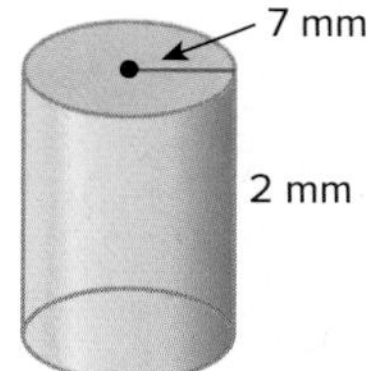

11.

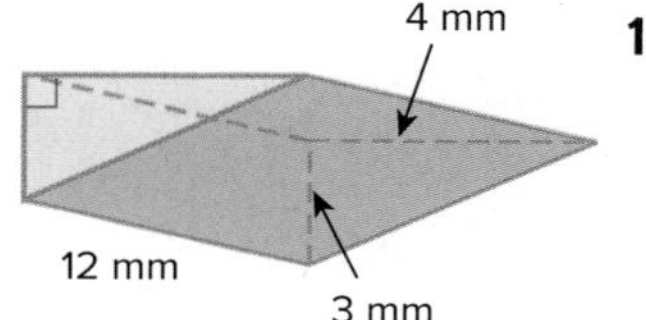

12.

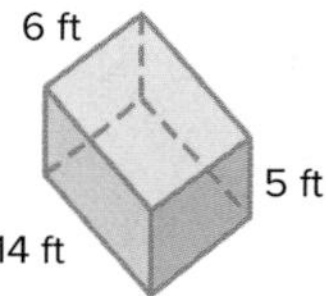

13. METEOROLOGY The TIROS weather satellites were a series of weather satellites that carried television and infrared cameras and were covered by solar cells. The cylinder-shaped body of a TIROS had a diameter of 42 inches and a height of 19 inches. (Lesson 11-2)

a. What was the volume available for carrying instruments and cameras? Round to the nearest tenth.

b. MP What mathematical practice did you use to solve this problem?

Find the volume of each pyramid or cone. Round to the nearest tenth if necessary. (Lesson 11-3)

14.

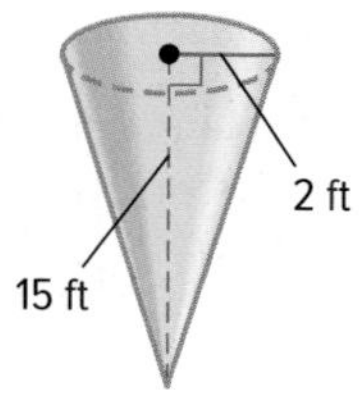

15.

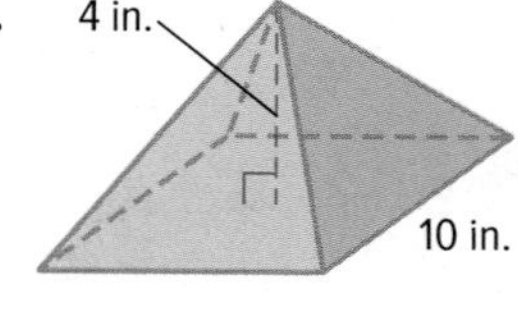

16.

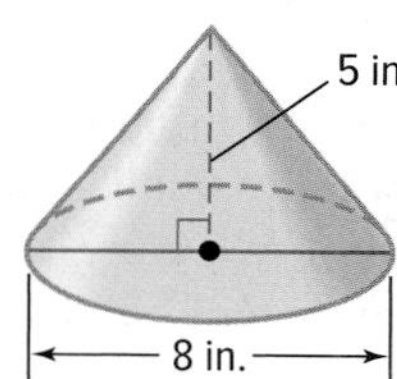

17.

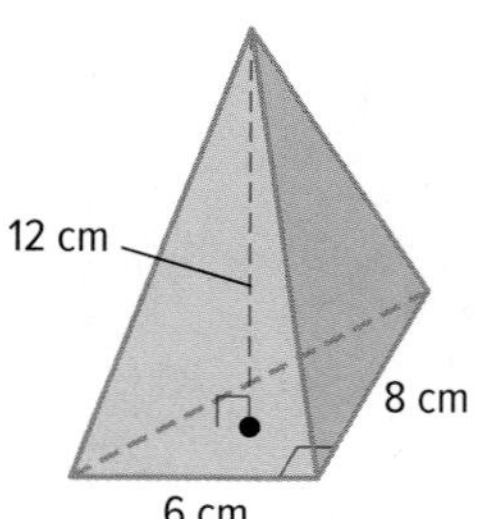

18.

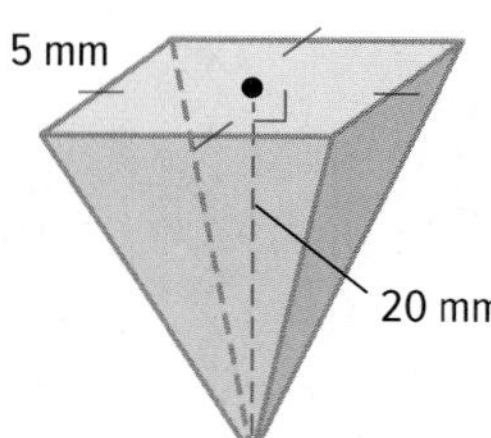

19.

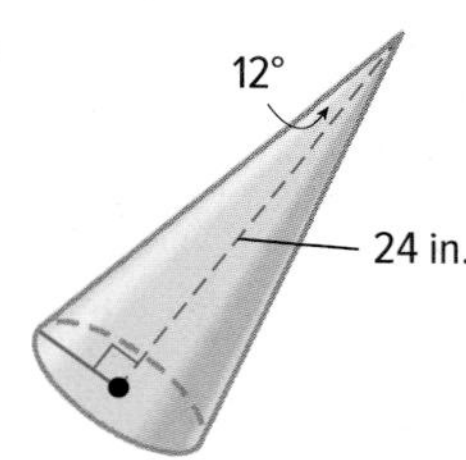

20. PLAYHOUSE Mary Anne is helping her dad make a playhouse for her little sister. The playhouse is a square pyramid on top of a cube. The pyramid's base is 5 feet and its height is 6 feet. Find the playhouse's volume in cubic feet. (Lesson 11-3)

21. COLLECTIONS Soledad collects unique salt and pepper shakers. She inherited a pair of shakers in the shape of regular square pyramids. Each edge of the base measures 3 centimeters and the height is 4 centimeters. Find the volume of one shaker. (Lesson 11-3)

LESSON 4

Spheres

∵Then	∵Now	∵Why?
• You found surface areas of prisms and cylinders.	**1** Find surface areas of spheres. **2** Find volumes of spheres.	• When you blow bubbles, soapy liquid surrounds a volume of air. Because of surface tension, the liquid maintains a shape that minimizes the surface area surrounding the air. The shape that minimizes surface area per unit of volume is a sphere.

New Vocabulary

great circle
pole
hemisphere

MP Mathematical Practices

1 Make sense of problems and persevere in solving them.

6 Attend to precision.

1 Surface Area of Spheres

Recall that a *sphere* is the locus of all points in space that are a given distance from a given point called the *center* of the sphere.

- A *radius* of a sphere is a segment from the center to a point on the sphere.
- A *chord* of a sphere is a segment that connects any two points on the sphere.
- A *diameter* of a sphere is a chord that contains the center.
- A *tangent* to a sphere is a line that intersects the sphere in exactly one point.

To develop a formula for the surface area of a sphere, consider a tennis ball. The covering of this sphere is comprised of two congruent dumbell-shaped pieces, each of which can be approximated by two congruent circles with radii equal to that of the sphere. So, the entire covering consists of approximately four congruent circles. The sum of these areas approximates the surface area of the sphere.

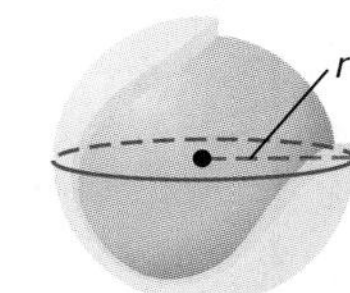

$S \approx 4A$ — Sum of circles with area A

$\approx 4(\pi r^2)$ or $4\pi r^2$ — $A = \pi r^2$

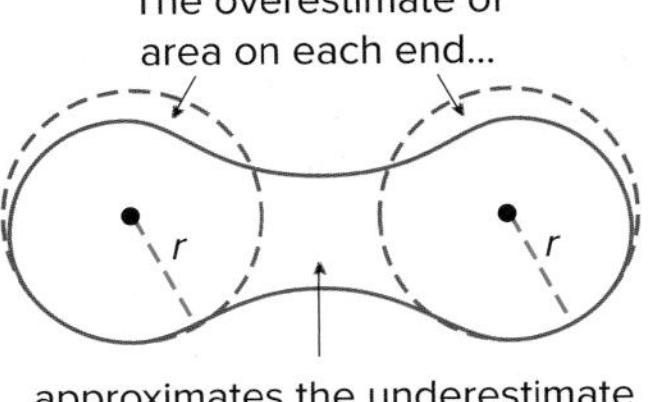

While its derivation is beyond the scope of this course, the exact formula is in fact $S = 4\pi r^2$.

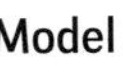

Key Concept Surface Area of a Sphere

Words The surface area S of a sphere is $S = 4\pi r^2$, where r is the radius.

Symbols $S = 4\pi r^2$

Model

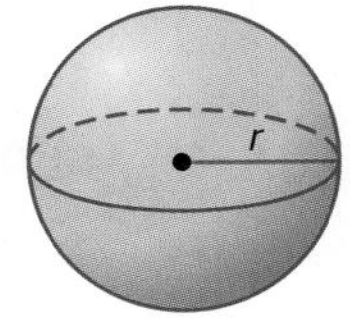

Example 1 Surface Area of a Sphere

Find the surface area of the sphere to the nearest tenth.

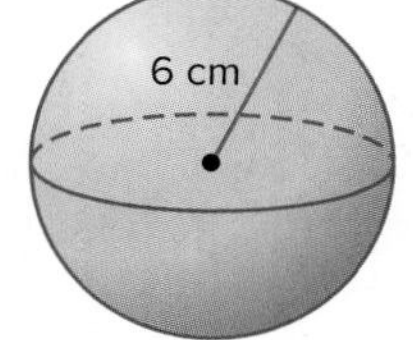

$S = 4\pi r^2$ Surface area of a sphere

$= 4\pi(6)^2$ Replace r with 6.

≈ 452.4 Use a calculator.

The surface area is about 452.4 square centimeters.

Guided Practice

1A.

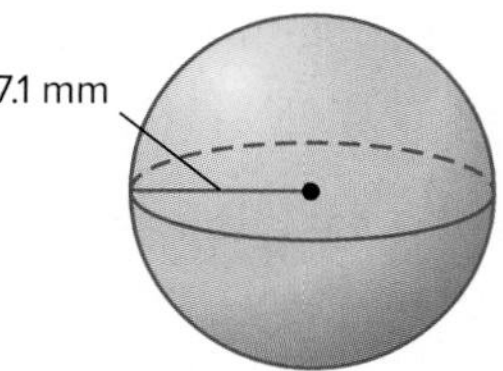

1B.

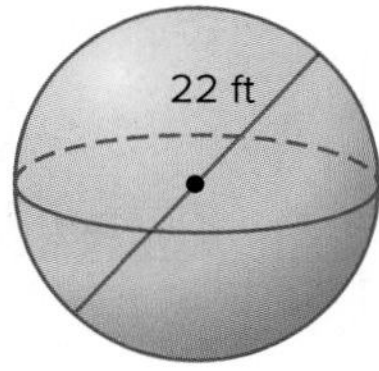

Recall that lateral area is defined as the sum of the area of the lateral faces of a solid. Because a sphere has no lateral faces, the surface area of a sphere is calculated.

Study Tip

Great Circles A sphere has an infinite number of great circles.

A plane can intersect a sphere in a point or in a circle. If the circle contains the center of the sphere, the intersection is called a **great circle**. The endpoints of a diameter of a great circle are called **poles**.

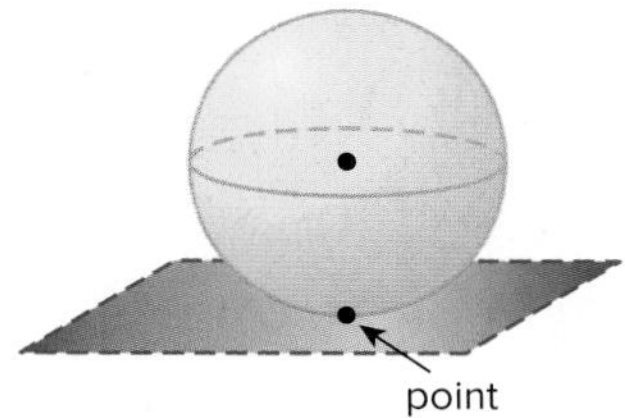

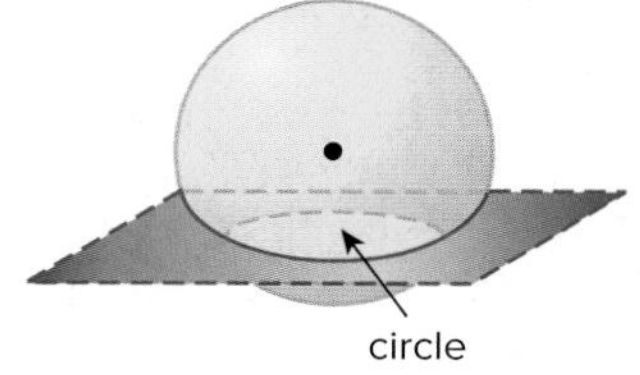

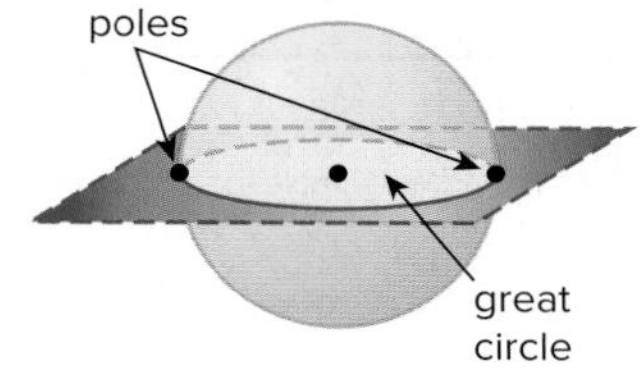

Because a great circle has the same center as the sphere and its radii are also radii of the sphere, it is the largest circle that can be drawn on a sphere. A great circle separates a sphere into two congruent halves, called **hemispheres**.

Example 2 Use Great Circles to Find Surface Area

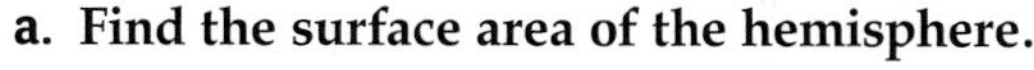

WatchOut!

Area of Hemisphere When finding the surface area of a hemisphere, do not forget to include the area of the great circle.

a. Find the surface area of the hemisphere.

Find half the area of a sphere with a radius of 2.8 centimeters. Then add the area of the great circle.

2.8 cm

$S = \frac{1}{2}(4\pi r^2) + \pi r^2$ Surface area of a hemisphere

$= \frac{1}{2}[4\pi(2.8)^2] + \pi(2.8)^2$ Replace r with 2.8.

$\approx 73.9 \text{ cm}^2$ Use a calculator.

b. Find the surface area of a sphere if the circumference of the great circle is 5π meters.

First, find the radius. The circumference of a great circle is $2\pi r$. So, $2\pi r = 5\pi$ or $r = 2.5$.

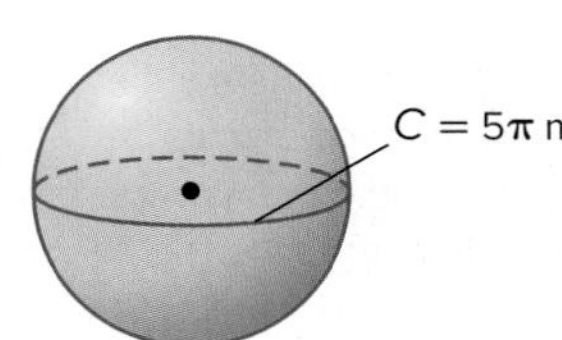

$S = 4\pi r^2$ Surface area of a sphere

$= 4\pi(2.5)^2$ Replace r with 2.5.

$\approx 78.5 \text{ m}^2$ Use a calculator.

c. Find the surface area of a sphere if the area of the great circle is approximately 130 square inches.

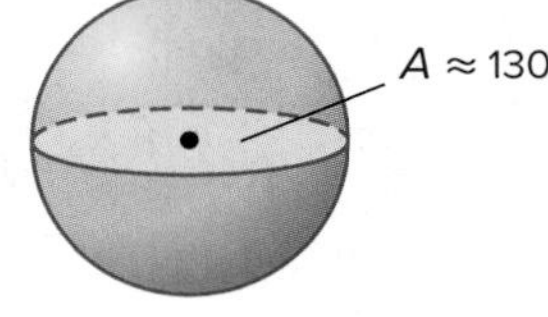

First, find the radius. The area of a great circle is πr^2. So, $\pi r^2 = 130$ or $r \approx 6.4$.

$S = 4\pi r^2$	Surface area of a sphere
$\approx 4\pi(6.4)^2$ or about 514.7 in^2	Replace r with 6.4. Use a calculator.

Guided Practice

Find the surface area of each figure. Round to the nearest tenth if necessary.

2A. sphere: circumference of great circle $= 16.2\pi$ ft

2B. hemisphere: area of great circle $\approx 94 \text{ mm}^2$

2C. hemisphere: circumference of great circle $= 36\pi$ cm

Go Online!

How do you find the surface area and volume of a sphere? Investigate by using the **Geometry Tools** in ConnectED. Discuss your findings with a partner. Ask for clarification as you need it.

2 Volume of Spheres

Suppose a sphere with radius r contains infinitely many pyramids with vertices at the center of the sphere. Each pyramid has height r and base area B. The sum of the volumes of all the pyramids equals the volume of the sphere.

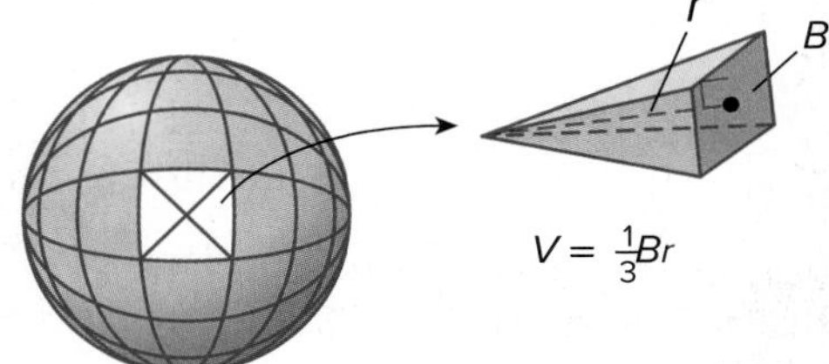

$V = \frac{1}{3}B_1r + \frac{1}{3}B_2r + \ldots + \frac{1}{3}B_nr$	Sum of volumes of pyramids
$= \frac{1}{3}r(B_1 + B_2 + \ldots + B_n)$	Distributive Property
$= \frac{1}{3}r(4\pi r^2)$	The sum of the pyramid base areas equals the surface area of the sphere.
$= \frac{4}{3}\pi r^3$	Simplify.

Study Tip

MP Modeling When solving problems involving volumes of solids, it is helpful to draw and label a diagram when no diagram is provided.

Key Concept Volume of a Sphere

Words The volume V of a sphere is $V = \frac{4}{3}\pi r^3$, where r is the radius of the sphere.

Symbols $V = \frac{4}{3}\pi r^3$

Model

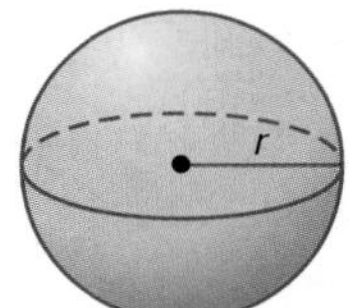

Example 3 Volumes of Spheres and Hemispheres

Find the volume of each sphere or hemisphere. Round to the nearest tenth.

a. a hemisphere with a radius of 6 meters

Estimate: $V \approx \frac{1}{2} \cdot \frac{4}{3} \cdot 3 \cdot 6^3$ or 432 m^3

$V = \frac{1}{2}\left(\frac{4}{3}\pi r^3\right)$	Volume of a hemisphere
$= \frac{2}{3}\pi(6)^3$ or about 452.4 m^3	Replace r with 6. Use a calculator.

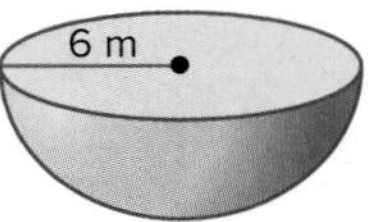

The volume of the hemisphere is about 452.4 cubic meters. This is close to the estimate, so the answer is reasonable.

Study Tip

MP Precision Remember to use the correct units when giving your answers. As with other solids, the surface area of a sphere is measured in square units, and volume is measured in cubic units.

b. a sphere with a great circle circumference of 18π centimeters

Step 1 Find the radius of the sphere.

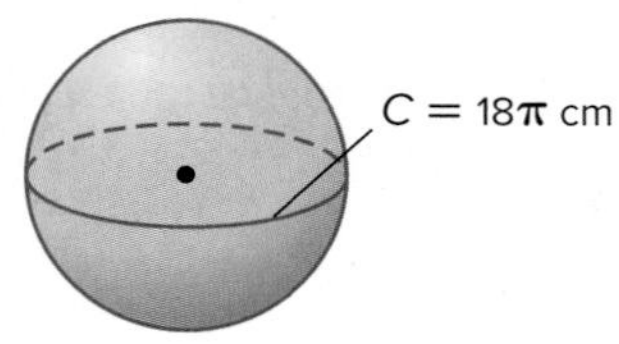

$C = 2\pi r$ Circumference of a circle

$18\pi = 2\pi r$ Replace C with 18π.

$r = 9$ Solve for r.

Step 2 Find the volume.

$V = \frac{4}{3}\pi r^3$ Volume of a sphere

$= \frac{4}{3}\pi(9)^3$ or about 3053.6 cm^3 Replace r with 9. Use a calculator.

Guided Practice

3A. sphere: diameter = 7.4 in.

3B. hemisphere: area of great circle ≈ 249 mm^2

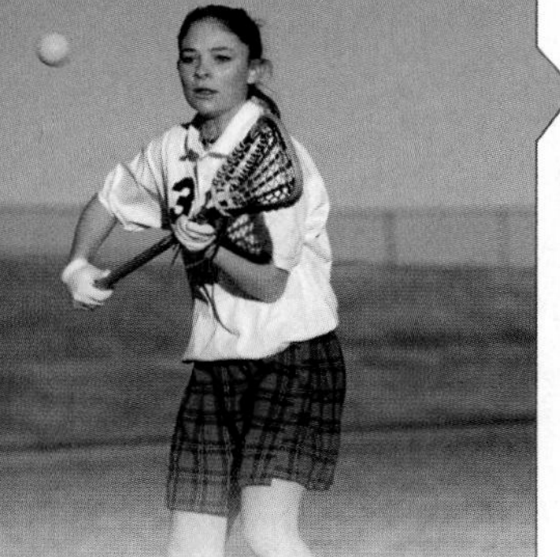

Real-World Link

Starting with the 2014 season, all lacrosse balls must meet the standard specifications. Balls are to weigh between 140 and 147 grams. Rebound is tested as well. They should bounce 70% from the falling point.

Source: NCAA

Real-World Example 4 Solve Problems Involving Solids

LACROSSE A regulation lacrosse ball is made of solid rubber. It takes up approximately 8.65 cubic inches. What is the circumference of the lacrosse ball? Round to the nearest tenth.

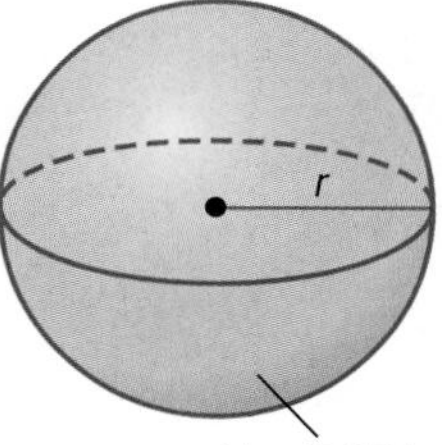

Understand Assume that the lacrosse ball is a sphere. You know that the volume is approximately 8.65 cubic inches. The circumference of the lacrosse ball is the circumference of the great circle.

Plan First use the volume formula to find the radius. Then find the circumference of the great circle.

Solve

$V = \frac{4}{3}\pi r^3$ Volume of a sphere

$8.65 \approx \frac{4}{3}\pi r^3$ Replace V with 8.65.

$2.06 \approx r^3$ Divide each side by $\frac{4}{3}\pi$.

Use a calculator to find $\sqrt[3]{2.06}$.

The radius of the sphere is approximately 1.3 inches. So, the circumference is $2\pi r = 2\pi(1.3)$ or approximately 8.2 inches.

Check You can work backward to check the solution.

If $C \approx 8.2$, then $r \approx 1.3$. If $r \approx 1.3$, then $V \approx \frac{4}{3}\pi \cdot 1.3^3$ or about 8.65 cubic inches. The solution is correct. ✓

Assuming that the lacrosse ball was a sphere allows use of the formula for the volume of a sphere to find a solution that closely approximates the actual circumference.

Guided Practice

4. SNOW CONES Ren makes a snow cone in a conical cup that is 3.5 inches tall with an opening 2.5 inches wide. He fills the cup to the top and then adds a hemispherical scoop that exactly covers the top of the cone. What is the volume of ice used to make the snow cone? Round to the nearest tenth.

Check Your Understanding

◯ = Step-by-Step Solutions begin on page R13.

Examples 1–2 **Find the surface area of each sphere or hemisphere. Round to the nearest tenth.**

1. 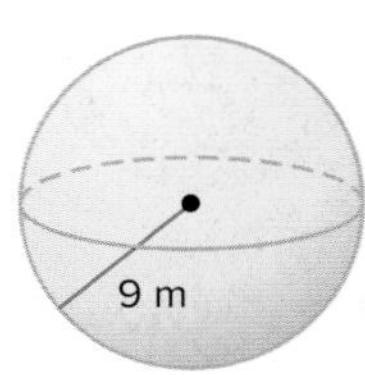

2. 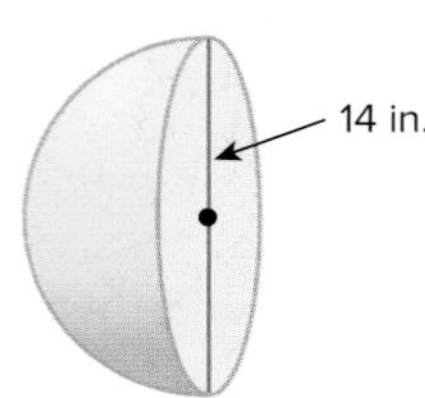

3. sphere: area of great circle = 36π yd^2
4. hemisphere: circumference of great circle ≈ 26 cm

Example 3 **Find the volume of each sphere or hemisphere. Round to the nearest tenth.**

5. sphere: radius = 10 ft
6. hemisphere: diameter = 16 cm
7. hemisphere: circumference of great circle = 24π m
8. sphere: area of great circle = 55π in^2

Example 4

9. **BASKETBALL** Basketballs used in professional games must have a circumference of $29\frac{1}{2}$ inches. What is the surface area of a basketball used in a professional game?

Practice and Problem Solving

Extra Practice is on page R11.

Examples 1–2 **Find the surface area of each sphere or hemisphere. Round to the nearest tenth.**

10. 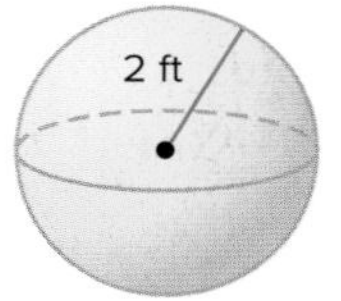

11. 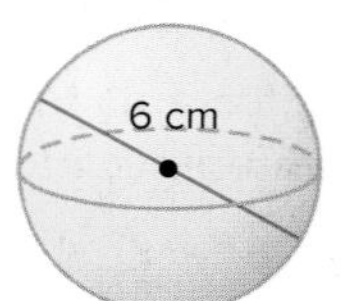

12. 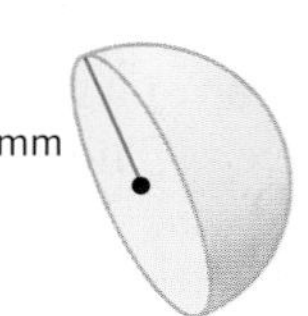

13. 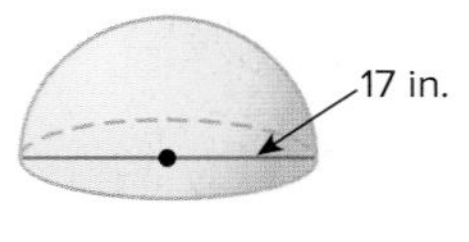

14. sphere: circumference of great circle = 2π cm
15. sphere: area of great circle ≈ 32 ft^2
16. hemisphere: area of great circle ≈ 40 in^2
17. hemisphere: circumference of great circle = 15π mm

Example 3 **MP PRECISION** Find the volume of each sphere or hemisphere. Round to the nearest tenth.

18. 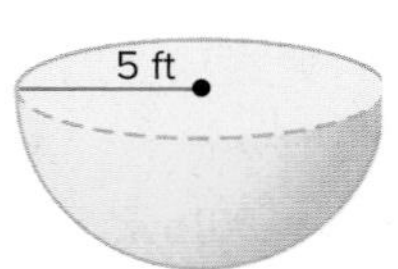

19. 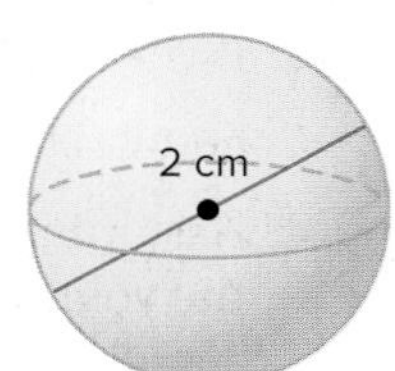

20. sphere: radius = 1.4 yd
21. hemisphere: diameter = 21.8 cm
22. sphere: area of great circle = 49π m^2
23. sphere: circumference of great circle ≈ 22 in.
24. hemisphere: circumference of great circle ≈ 18 ft
25. hemisphere: area of great circle ≈ 35 m^2

Example 4

26. **FISH** A *puffer fish* is able to "puff up" when threatened by gulping water and inflating its body. The puffer fish at the right is approximately a sphere with a diameter of 5 inches. Its surface area when inflated is about 1.5 times its normal surface area. What is the surface area of the fish when it is *not* puffed up?

27. **ARCHITECTURE** The Reunion Tower in Dallas, Texas, is topped by a spherical dome that has a surface area of approximately $13{,}924\pi$ square feet. What is the volume of the dome? Round to the nearest tenth.

28. **TREE HOUSE** The spherical tree house, or *tree sphere,* shown at the right has a diameter of 10.5 feet. Its volume is 1.8 times the volume of the first tree sphere that was built. What was the diameter of the first tree sphere? Round to the nearest foot.

MP **PERSEVERANCE** Find the surface area and the volume of each solid. Round to the nearest tenth.

29.

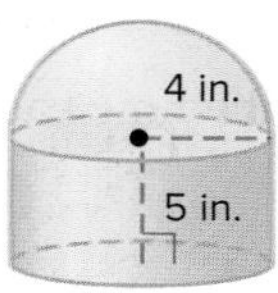

30.

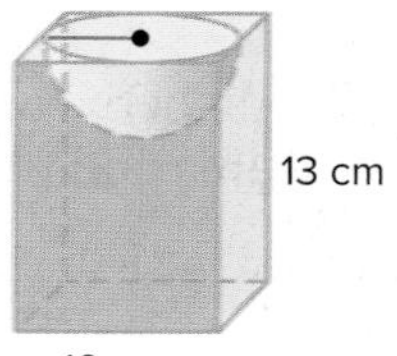

31. **TOYS** The spinning top at the right is a composite of a cone and a hemisphere.

  a. Find the surface area and the volume of the top. Round to the nearest tenth.

  b. If the manufacturer of the top makes another model with dimensions that are one half of the dimensions of this top, what are its surface area and volume?

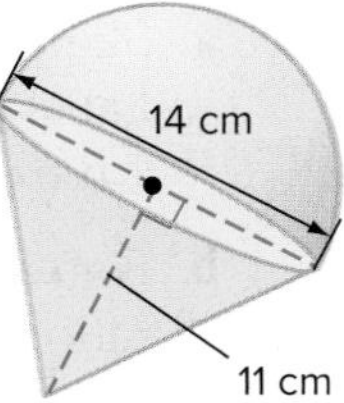

32. **BALLOONS** A spherical helium-filled balloon with a diameter of 30 centimeters can lift a 14-gram object. Find the size of a balloon that could lift a person who weighs 65 kilograms. Round to the nearest tenth.

Use sphere *S* to name each of the following.

33. a chord

34. a radius

35. a diameter

36. a tangent

37. a great circle

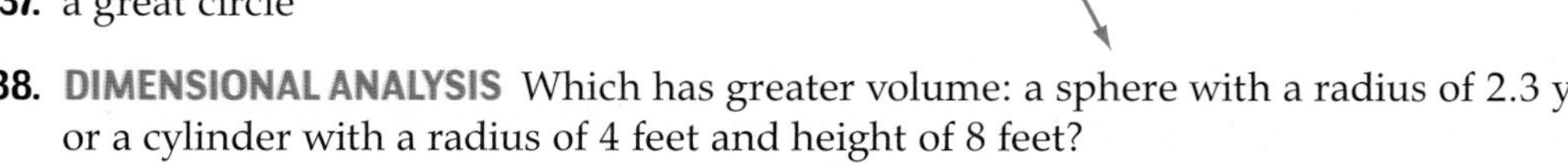

38. **DIMENSIONAL ANALYSIS** Which has greater volume: a sphere with a radius of 2.3 yards or a cylinder with a radius of 4 feet and height of 8 feet?

39. **INFORMAL PROOF** A sphere with radius r can be thought of as being made up of a large number of discs or thin cylinders. Consider the disc shown that is x units above or below the center of the sphere. Also consider a cylinder with radius r and height $2r$ that is hollowed out by two cones of height and radius r.

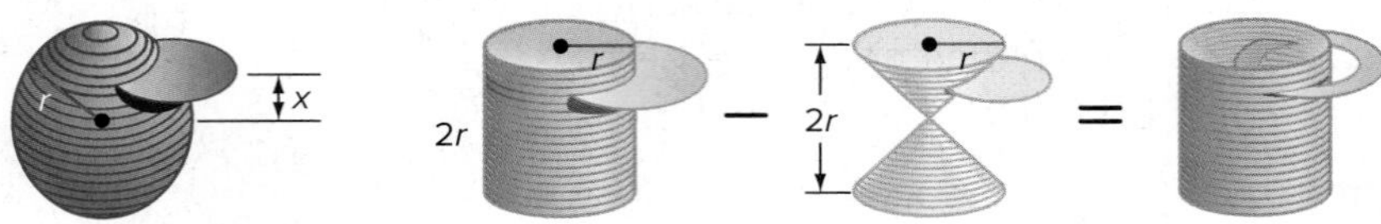

a. Find the radius of the disc from the sphere in terms of its distance x above the sphere's center. (*Hint:* Use the Pythagorean Theorem.)

b. If the disc from the sphere has a thickness of y units, find its volume in terms of x and y.

c. Show that this volume is the same as that of the hollowed-out disc with thickness of y units that is x units above the center of the cylinder and cone.

d. Since the expressions for the discs at the same height are the same, what guarantees that the hollowed-out cylinder and sphere have the same volume?

e. Use the formulas for the volumes of a cylinder and a cone to derive the formula for the volume of the hollowed-out cylinder and thus, the sphere.

MP **PERSEVERANCE** Describe the number and types of planes that produce reflectional symmetry in each solid. Then describe the angles of rotation that produce rotational symmetry in each solid.

40. sphere

41. hemisphere

CHANGING DIMENSIONS A sphere has a radius of 12 centimeters. Describe how each change affects the surface area and the volume of the sphere.

42. The radius is multiplied by 4.

43. The radius is divided by 3.

44. **DESIGN** A standard juice box holds 8 fluid ounces.

a. Sketch designs for three different juice containers that will each hold 8 fluid ounces. Label dimensions in centimeters. At least one container should be cylindrical. (*Hint:* $1 \text{ fl oz} \approx 29.57353 \text{ cm}^3$)

b. For each container in part **a**, calculate the surface area to volume (cm^2 per fl oz) ratio. Use these ratios to decide which of your containers can be made for the lowest materials cost. What shape container would minimize this ratio, and would this container be the cheapest to produce? Explain your reasoning.

H.O.T. Problems Use Higher-Order Thinking Skills

45. MP **MODELING** A cube has a volume of 216 cubic inches. Find the volume of a sphere that is circumscribed about the cube. Round to the nearest tenth.

46. **ERROR ANALYSIS** Diego says that the lateral area of a basketball is 29.5 square inches. Rodney says that you cannot calculate the lateral area of a sphere. Who is correct? Explain your reasoning.

47. **CHALLENGE** Sketch a sphere showing two examples of great circles. Sketch another sphere showing two examples of circles formed by planes intersecting the sphere that are *not* great circles.

48. **WRITING IN MATH** Write a ratio comparing the volume of a sphere with radius r to the volume of a cylinder with radius r and height $2r$. Then describe what the ratio means.

49. An artist is planning an exhibit in a hemispherical building with a diameter of 40 feet. The artist wants to cover the inside of the hemisphere (floor and ceiling) with black paint. Which of the following is the best estimate of the surface area that will be painted? MP 4

- A 2513 ft^2
- B 3770 ft^2
- C 5027 ft^2
- D 15,080 ft^2
- E 16,755 ft^2

50. What is the circumference of a sphere that has a volume of 288π cubic meters? MP 7

- A 144π m
- B 36π m
- C 12π m
- D 6π m

51. Adam has a spherical candle with a radius of 2 inches. He melts the candle completely and pours the soft wax into a mold in the shape of a rectangular prism. The dimensions of the mold are shown here.

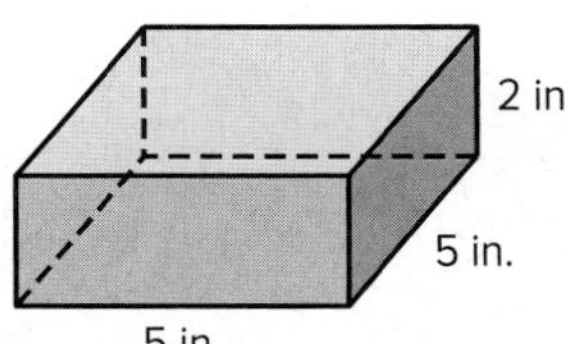

Which of the following best describes how the melted wax will fill the mold? MP 4

- A The wax will fill slightly more than $\frac{1}{3}$ of the mold.
- B The wax will fill the mold more than halfway.
- C The wax will fill the mold almost perfectly.
- D The wax will fill the mold, and there will be a lot of leftover wax.

52. A spherical disco ball has a surface area of 169π square centimeters. MP 7

a. Find the radius of the ball.

b. Find the volume of the ball.

53. MULTI-STEP Alice bought two spherical ornaments with radii of 2 centimeters and 4 centimeters. Find the following. MP 1

a. the volume of the first ornament

b. the volume of the second ornament

c. the surface area of the first ornament

d. the surface area of the second ornament

e. the ratio of the volume of the first ornament to the volume of the second ornament

f. the ratio of the surface area of the first ornament to the surface area of the second ornament

54. A ball has a diameter of 18 centimeters. A cylinder holds the ball exactly. What is the volume of the cylinder? MP 1

55. MULTI-STEP The length of the equator on a globe is 44 inches. Answer the following questions, rounding to the nearest tenth. MP 1

a. What is the diameter of the globe?

b. What is the radius of the globe?

c. What is the surface area of the globe?

d. What is the volume of the globe?

EXTEND 11-4

Geometry Lab

Locus and Spheres

Spheres are defined in terms of a locus of points in space. The definition of a sphere is the set of all points that are a given distance from a given point.

Mathematical Practices

MP 1 Make sense of problems and persevere in solving them.
6 Attend to precision.

Activity 1 Locus of Points a Given Distance from Endpoints

Work cooperatively. Find the locus of all points that are equidistant from a segment.

Collect the Data

- Draw a given line segment with endpoints *J* and *K*.

J K

- Create a set of points that are equidistant from the segment.

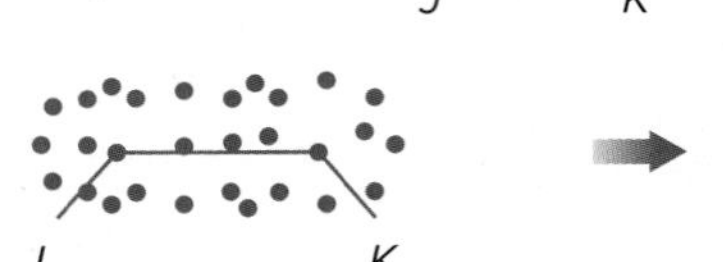

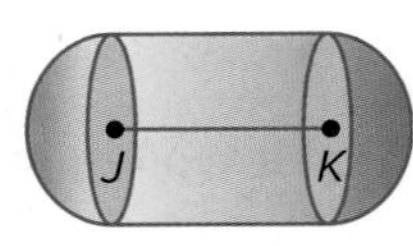

Analyze

1. Draw a figure and describe the locus of points in space that are 8 units from a segment that is 30 units long.

2. What three-dimensional shapes form the figure?

3. What are the radii and diameters of each hemisphere?

4. What are the diameter and the height of the cylinder?

Activity 2 Spheres That Intersect

Work cooperatively. Find the locus of all points that are equidistant from the centers of two intersecting spheres with the same radius.

Collect the Data

- Draw a line segment.
- Draw congruent overlapping spheres, with the centers at the endpoints of the given line segment.

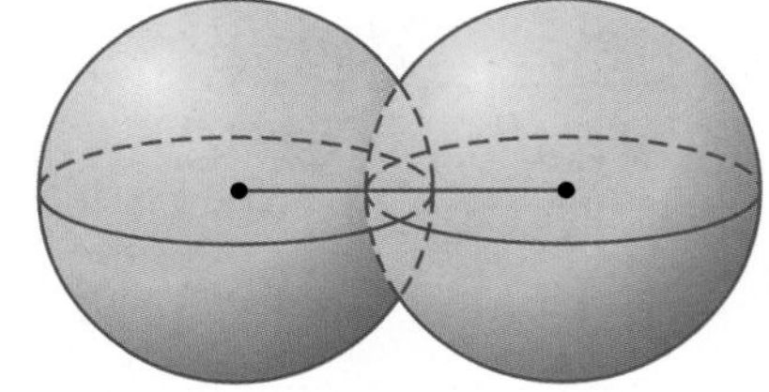

Analyze

5. What is the shape of the intersection of the upper hemispheres?

6. Can this be described as a locus of points in space or on a plane? Explain.

7. Describe this intersection as a locus.

8. **FIREWORKS** What is the locus of points that describes how particles from a fireworks explosion will disperse in an explosion at 400 feet above ground level if the expected distance a particle could travel is 200 feet?

LESSON 5

Spherical Geometry

Then

- You identified basic properties of spheres.

Now

1. Describe sets of points on a sphere.
2. Compare and contrast Euclidean and spherical geometries.

Why?

- Because Earth has a curved instead of a flat surface, the shortest path between two points on Earth is described by an arc of a great circle instead of a straight line.

New Vocabulary

Euclidean geometry
spherical geometry
non-Euclidean geometry

Mathematical Practices

3 Construct viable arguments and critique the reasoning of others.

1 Geometry on a Sphere

In this text, we have studied **Euclidean geometry**, either in the plane or in space. In plane Euclidean geometry, a *plane* is a flat surface made up of points that extend infinitely in all directions. In **spherical geometry**, or geometry on a sphere, a plane is the surface of a sphere.

Lines are also defined differently in spherical geometry.

Key Concept Lines in Plane and Spherical Geometry

Plane Euclidean Geometry	Spherical Geometry
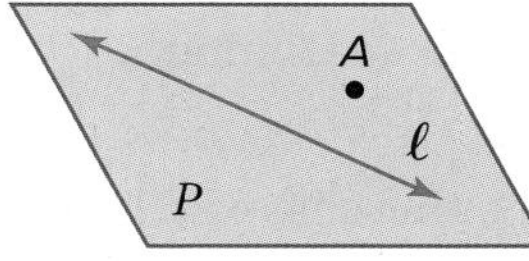	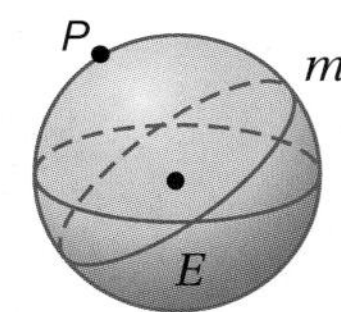
Plane *P* contains line ℓ and point *A* not on line ℓ.	Sphere *E* contains great circle *m* and point *p* not on *m*. Great circle *m* is a line on sphere *E*.

Example 1 Describe Sets of Points on a Sphere

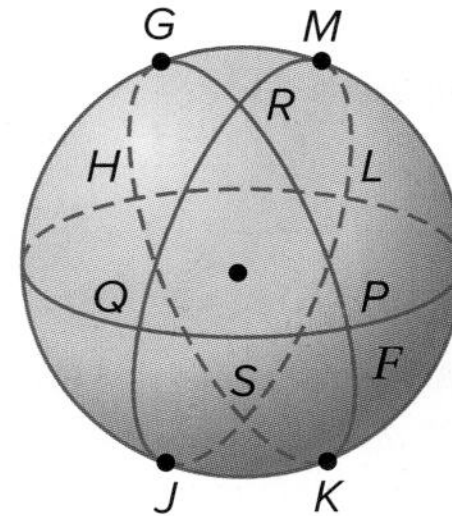

Name each of the following on sphere *F*.

a. two lines containing point *R*

$\overleftrightarrow{GP}$ and $\overleftrightarrow{MQ}$ are lines on sphere *F* that contain point *R*.

b. a segment containing point *K*

$\overline{PS}$ is a segment on sphere *F* that contains point *K*.

c. a triangle

$\triangle RQP$ is a triangle on sphere *F*.

Guided Practice

Name each of the following on sphere *F* above.

1A. two lines containing point *P*

1B. a segment containing point *Q*

1C. a triangle

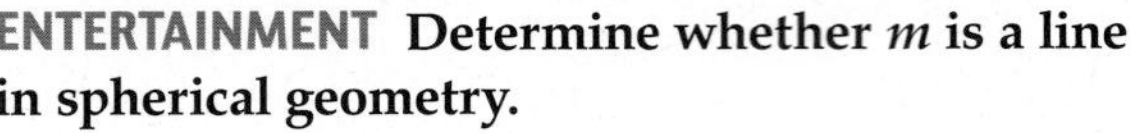

Real-World Example 2 Identify Lines in Spherical Geometry

ENTERTAINMENT Determine whether m is a line in spherical geometry.

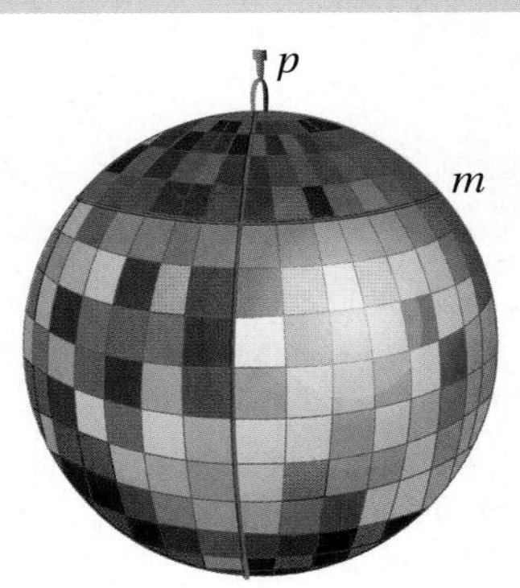

Notice that m does not go through two poles of the sphere. Therefore m is not a great circle and so not a line in spherical geometry.

Guided Practice

2. Determine whether p on the mirror ball shown is a line in spherical geometry.

> **Math History Link**
>
> **Georg F.B. Riemann (1826–1866)** Spherical geometry is sometimes called *Riemann geometry*, after Georg Riemann, a German mathematician responsible for the Riemannian Postulate, which states that through a point not on a line, there are no lines parallel to the given line.

2 Comparing Euclidean and Spherical Geometries

While some postulates and properties of Euclidean geometry are true in spherical geometry, others are not, or they are true only under certain circumstances. A **non-Euclidean geometry** is a geometry in which at least one of the postulates from Euclidean geometry fails.

Example 3 Compare Plane Euclidean and Spherical Geometries

> **Study Tip**
>
> **MP Sense-Making** Notice in Example 3b that the Parallel Postulate does not hold true on a sphere. Lines, or great circles, cannot be parallel in spherical geometry. Therefore, spherical geometry is non-Euclidean.

> **Study Tip**
>
> **Angle Measures** In Exercise 22, you will show that this is true for all triangles on a sphere in Spherical Geometry.

Determine whether the following postulate or property of plane Euclidean geometry has a corresponding statement in spherical geometry. If so, write the corresponding statement. If not, explain your reasoning.

a. Through any two points, there is exactly one line.

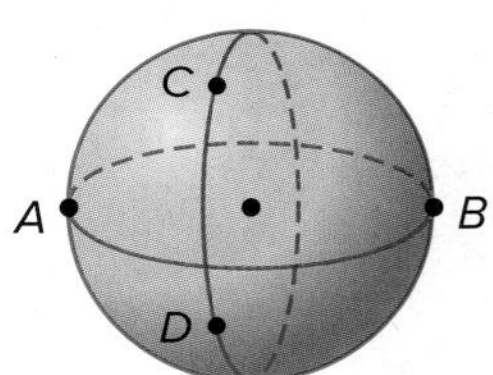

In the figure, there is more than one great circle (line) through polar points A and B. However, there is only one great circle through nonpolar points C and D. Therefore, a corresponding statement is that through any two nonpolar points, there is exactly one great circle (line).

b. If given a line and a point not on the line, there exists exactly one line through the point that is parallel to the given line.

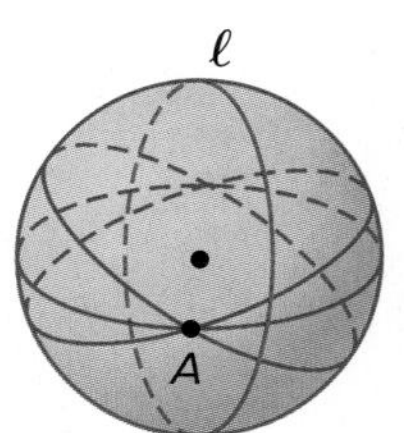

In the figure, notice that every great circle (line) containing point A will intersect line ℓ. Thus there exists no great circle through point A that is parallel to line ℓ.

c. The sum of the measures of a triangle is 180.

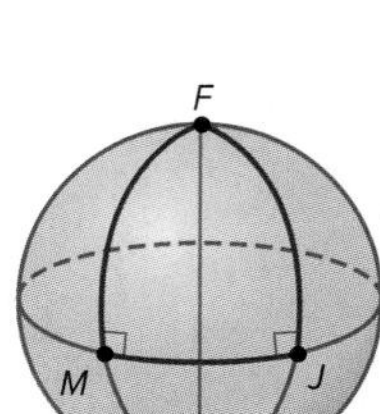

In the figure, $\angle M$ and $\angle J$ are both right angles. Thus, the sum of the measures of this triangle is greater than 180.

Guided Practice

3A. A line segment is the shortest path between two points.

3B. Through any two points, there is exactly one segment.

Check Your Understanding

○ = Step-by-Step Solutions begin on page R13.

Example 1 **Name each of the following on sphere B.**

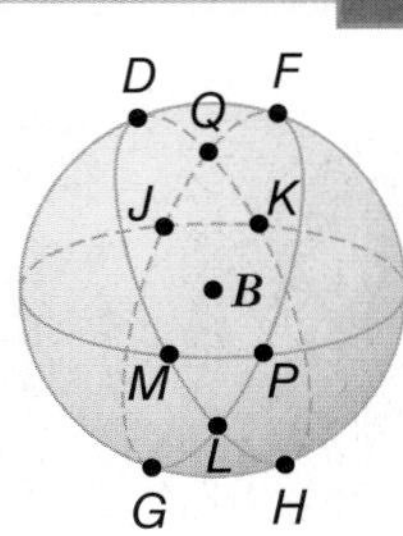

1. two lines containing point Q
2. a segment containing point L
3. a triangle
4. two segments on the same great circle

Example 2 **SPORTS** **Determine whether X on each of the spheres shown is a line in spherical geometry.**

5.

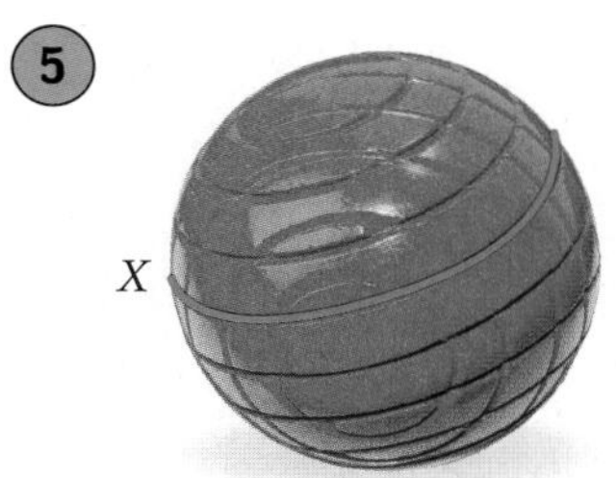

6.

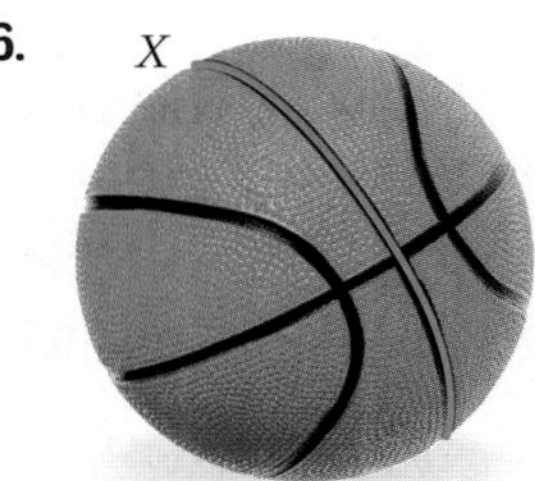

Example 3 **MP PERSEVERANCE** **Determine whether the following postulate or property of plane Euclidean geometry has a corresponding statement in spherical geometry. If so, write the corresponding statement. If not, explain your reasoning.**

7. The points on any line or line segment can be put into one-to-one correspondence with real numbers.
8. Perpendicular lines intersect at one point.

Practice and Problem Solving

Extra Practice is on page R11.

Example 1 **Name two lines containing point M, a segment containing point S, and a triangle in each of the following spheres.**

9.

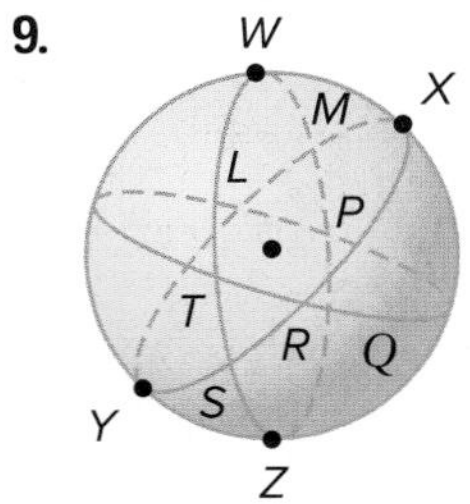

10.

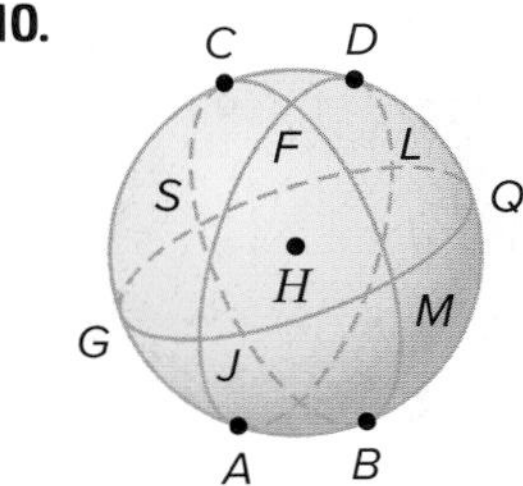

11. **SOCCER** Name each of the following on the soccer ball shown.
    a. two lines containing point B
    b. a segment containing point F
    c. a triangle
    d. a segment containing point C
    e. a line
    f. two lines containing point A

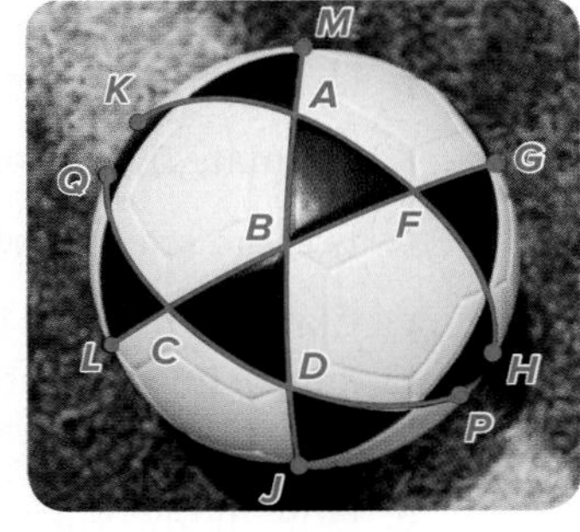

Example 2 **ARCHITECTURE** Determine whether w on each of the spheres shown is a line in spherical geometry.

12.

13.

14. **MODELING** Lines of latitude and longitude are used to describe positions on the Earth's surface. By convention, lines of longitude divide Earth vertically, while lines of latitude divide it horizontally.

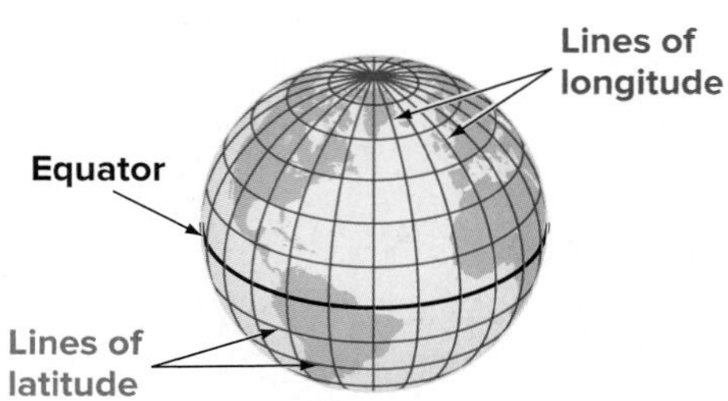

a. Are lines of longitude great circles? Explain.

b. Are lines of latitude great circles? Explain.

Example 3 **Determine whether the following postulate or property of plane Euclidean geometry has a corresponding statement in spherical geometry. If so, write the corresponding statement. If not, explain your reasoning.**

15 A line goes on infinitely in two directions.

16. Perpendicular lines form four 90° angles.

17. If three points are collinear, exactly one is between the other two.

18. If M is the midpoint of $\overline{AB}$, then $\overline{AM} \cong \overline{MB}$.

On a sphere, there are two distances that can be measured between two points. Use each figure and the information given to determine the distance between points J and K on each sphere. Round to the nearest tenth. Justify your answer.

19.

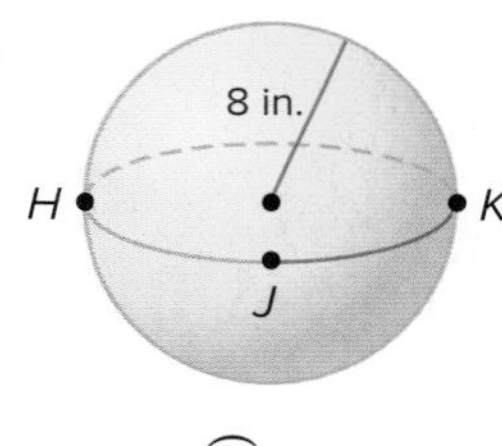

$m\widehat{JK} = 100$

20.

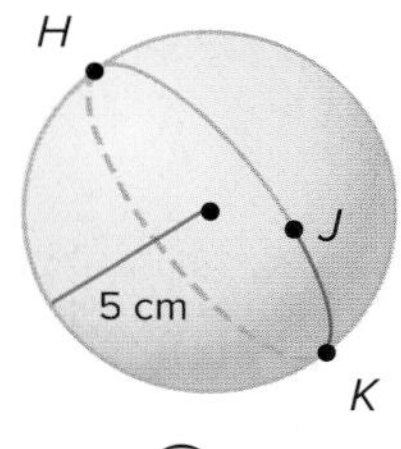

$m\widehat{JK} = 60$

21. **GEOGRAPHY** The location of Phoenix, Arizona, is 112° W longitude, 33.4° N latitude, and the location of Helena, Montana, is 112° W longitude, 46.6° N latitude. West indicates the location in terms of the prime meridian, and north indicates the location in terms of the equator. The mean radius of Earth is about 3960 miles.

a. Estimate the distance between Phoenix and Helena. Explain your reasoning.

b. Is there another way to express the distance between these two cities? Explain.

c. Can the distance between Washington, D.C., and Lisbon, Portugal, which lie on approximately the same lines of latitude, be calculated in the same way? Explain your reasoning.

d. How many other locations are there that are the same distance from Phoenix as Wichita is? Explain.

22. **MULTIPLE REPRESENTATIONS** In this problem, you will investigate triangles in spherical geometry.

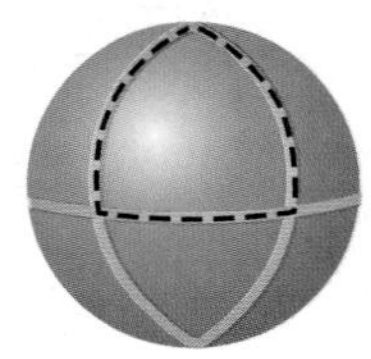

a. **Concrete** Use masking tape on a ball to mark three great circles. At least one of the three great circles should go through different poles than the other two. The great circles will form a triangle. Use a protractor to estimate the measure of each angle of the triangle.

b. **Tabular** Tabulate the measure of each angle of the triangle formed. Remove the tape and repeat the process two times so that you have tabulated the measure of three different triangles. Record the sum of the measures of each triangle.

c. **Verbal** Make a conjecture about the sum of the measures of a triangle in spherical geometry.

23. **QUADRILATERALS** Consider quadrilateral *ABCD* on sphere *P*. It has four sides with $\overline{DC} \perp \overline{CB}$, $\overline{AB} \perp \overline{CB}$, and $\overline{DC} \cong \overline{AB}$.

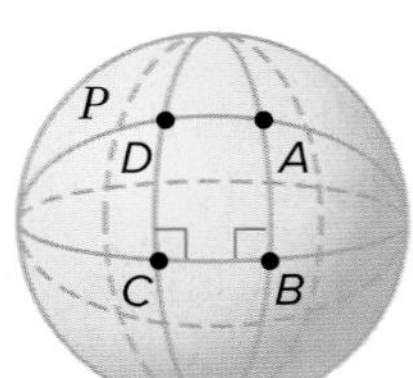

a. Is $\overline{CD} \perp \overline{DA}$? Explain your reasoning.

b. How does *DA* compare to *CB*?

c. Can a rectangle, as defined in Euclidean geometry, exist in non-Euclidean geometry? Explain your reasoning.

H.O.T. Problems Use Higher-Order Thinking Skills

24. **WRITING IN MATH** Compare and contrast Euclidean and spherical geometries. Be sure to include a discussion of planes and lines, including parallel and perpendicular lines, in both geometries.

25. **CHALLENGE** Geometries can be defined on curved surfaces other than spheres. Another type of non-Euclidean geometry is *hyperbolic geometry*. This geometry is defined on a curved saddle-like surface. Compare the sum of the angle measures of a triangle in hyperbolic, spherical, and Euclidean geometries.

Triangle in plane geometry

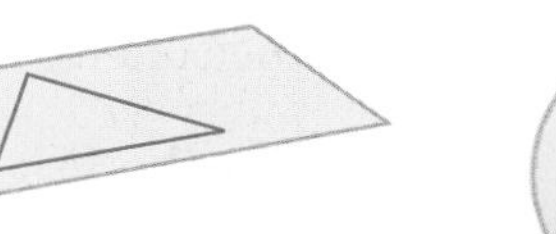

Triangle in spherical geometry

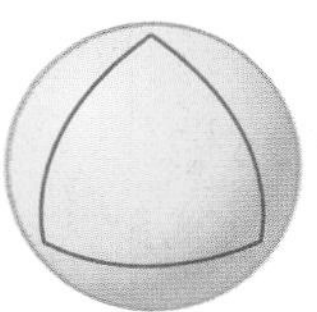

Triangle in hyperbolic geometry

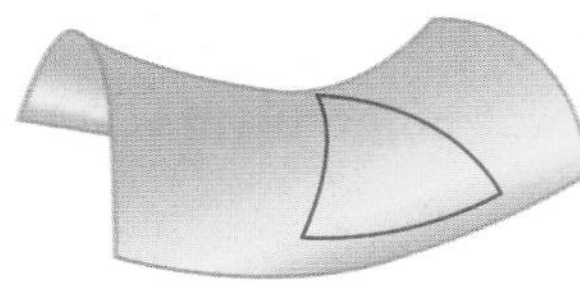

26. **CHALLENGE** Sketch a sphere with three points so that two of the points lie on a great circle and two of the points do not lie on a great circle.

27. **MP CONSTRUCT ARGUMENTS** A *small circle* of a sphere intersects at least two points, but does not go through opposite poles. Points *A* and *B* lie on a small circle of sphere *Q*. Will two small circles *sometimes*, *always*, or *never* be parallel? Draw a sketch and explain your reasoning.

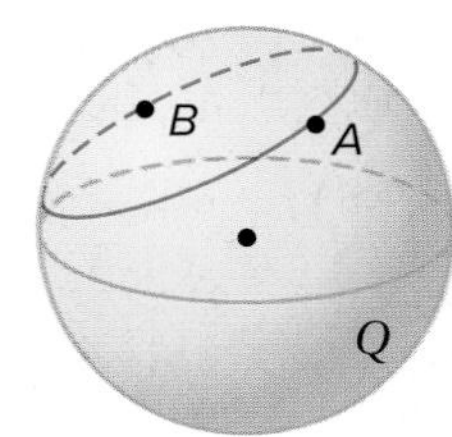

28. **WRITING IN MATH** Do similar or congruent triangles exist in spherical geometry? Explain your reasoning.

29. **MP REASONING** Is the statement *Spherical geometry is a subset of Euclidean geometry* true or false? Explain your reasoning.

30. **MP CONSTRUCT ARGUMENTS** Two planes are equidistant from the center of a sphere and intersect the sphere. What is true of the circles? Are they lines in spherical geometry?

Preparing for Assessment

31. Use spherical geometry to determine which of the following is a true statement about the figure. MP 8

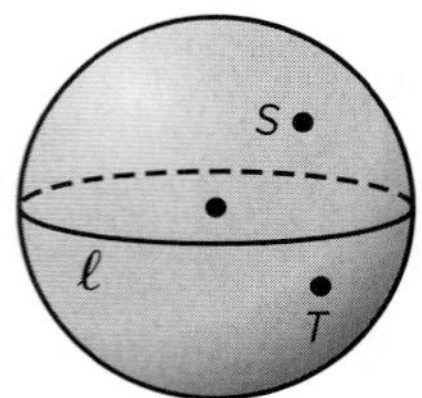

- ◯ **A** All lines through point S pass through point T.
- ◯ **B** All lines through point S intersect line ℓ.
- ◯ **C** All lines through point T are parallel to line ℓ.
- ◯ **D** There is exactly one line segment that has points S and T as endpoints.
- ◯ **E** There is a line through point S and a line through point T that do not intersect.

32. Jared wrote a statement in his math journal. His teacher read the statement and said, "That statement is true in Euclidean geometry but not in spherical geometry." Which of the following could be the statement Jared wrote? MP 8

- ◯ **A** Through any point, there is at least one line.
- ◯ **B** Given any two points, you can draw a line segment that has the points as endpoints.
- ◯ **C** Any three noncollinear points can be used as the vertices of a triangle.
- ◯ **D** If two lines intersect, then they intersect at exactly one point.

33. James is trying to shoot three balls of different sizes consecutively through a hoop with a diameter of 18 inches. The balls have volumes of 850π, 900π, and 950π cubic inches. How many of the balls will fit through the hoop? MP 1

34. Use spherical geometry and the figure shown here to determine which of the following statements is false. MP 8

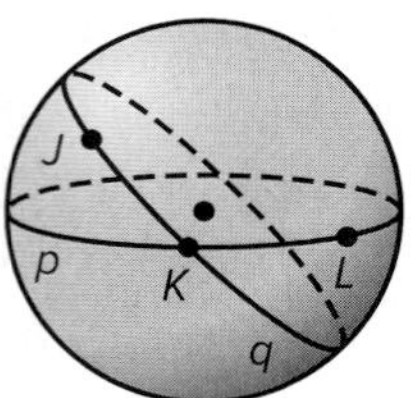

- ◯ **A** The figure shows two lines that intersect at point K.
- ◯ **B** There is exactly one line through points J and K.
- ◯ **C** It is possible to draw a line through point J that is parallel to $\overleftrightarrow{KL}$.
- ◯ **D** Every line through point L intersects $\overleftrightarrow{KL}$.

35. Which statement about spheres is not true? MP 8

- ◯ **A** There is only one great circle through any two points on a sphere that are not poles of the sphere.
- ◯ **B** A great circle is the intersection of a sphere and a plane that goes through the center of the sphere.
- ◯ **C** The shortest path between two points on a sphere is an arc of a great circle.
- ◯ **D** Two lines that are perpendicular to the same line on a sphere are parallel to each other.

36. **MULTI-STEP** Consider a sphere with a radius of 4 units. MP 1

a. Calculate the length of a great circle in a sphere with radius 4.

b. Calculate the length of a small circle in the same sphere if the center of the small circle is one unit away from the center of the sphere.

c. If a small circle on the same sphere has a length of 4π, how many units is its center from the center of the sphere?

EXTEND 11-5

Geometry Lab

Navigational Coordinates

A grid system of imaginary lines on Earth is used for locating places and navigation. Imaginary vertical lines drawn around Earth through the North and South Poles are called **meridians**, and they determine the measure of **longitude**. Imaginary horizontal lines parallel to the equator are called **parallels**, and they determine the measure of **latitude**.

The basic units for measurements are degrees, minutes, and seconds.
1 degree (°) = 60 minutes (′), and 60 minutes = 60 seconds (″).

Mathematical Practices

MP **5** Use appropriate tools strategically.

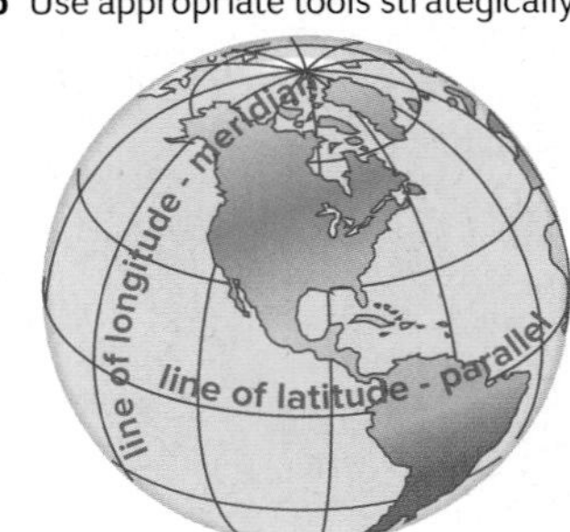

	Location of 0°	Direction	Maximum Degrees
Latitude (parallels)	equator	In northern hemisphere, all are degrees north. In southern hemisphere, all are degrees south.	90° at each pole
Longitude (meridians)	Prime Meridian through Greenwich, England	In eastern hemisphere, all are degrees east. In western hemisphere, all are degrees west.	180° at international dateline

Activity Investigate Latitude and Longitude

Work cooperatively. The table shows the latitude and longitude of three cities.

City	Latitude	Longitude
A	37°59′N	84°28′W
B	34°55′S	138°36′E
C	64°4′N	21°58′W

1. In which hemisphere is each city located?
2. Use a globe or map to name each city.
3. Earth is approximately a sphere with a radius of 3960 miles. The equator and all meridians are great circles. The circumference of a great circle is equal to the length of the equator or any meridian. Find the length of a great circle on Earth in miles.
4. Notice that the distance between each line of latitude is about the same. The distance from the equator to the North Pole is $\frac{1}{4}$ of the circumference of Earth, and each degree of latitude is $\frac{1}{90}$ of that distance. Estimate the distance between one pair of latitude lines in miles.

Analyze

Work cooperatively. The table shows the latitude and longitude of three cities.

City	Latitude	Longitude
F	1°28′S	48°29′W
G	13°45′N	100°30′E
H	41°17′S	174°47′E

5. Name the hemisphere in which each city is located.
6. MP **TOOLS** Use a globe or map to name each city.
7. Find the approximate distance between meridians at latitude of about 22° N. The direct distance between the two cities at the right is about 1646 miles.

Calcutta, India	22°34′N	88°24′E
Hong Kong, China	22°20′N	114°11′E

LESSON 6

Volume and Nonrigid Transformations

Then

- You compared surface areas and volumes of spheres.

Now

1. Identify scale factor by using dilation.
2. Find surface areas and volumes of similar solids by using scale factors.

Why?

- The gemstones at the right are cut in exactly the same shape, but their sizes are different. Their shapes are similar.

New Vocabulary
similar solid
congruent solid

Mathematical Practices
8 Look for and express regularity in repeated reasoning.

1 Dilation in Three Dimensions

In Chapter 7, you learned that a nonrigid transformation changes the size but not the shape of a figure, producing a similar figure. **Similar solids** have exactly the same shape but not necessarily the same size. All spheres are similar and all cubes are similar. In similar solids, the corresponding linear measures, such as height and radius, have equal ratios. The common ratio is called the *scale factor*. **Congruent solids** have exactly the same shape and the same size. Congruent solids are similar solids that have a scale factor of 1:1.

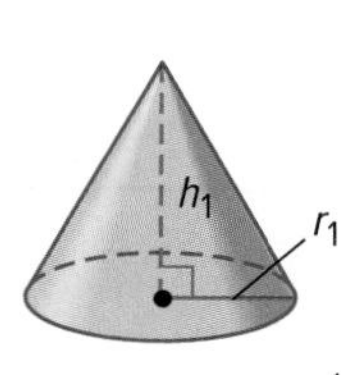

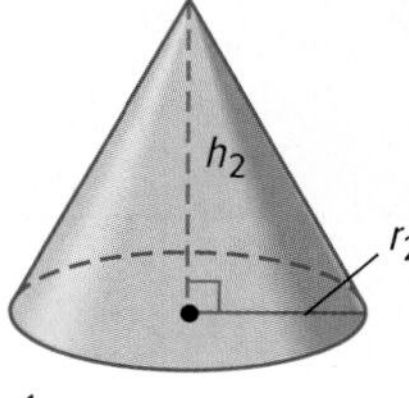

$$\frac{h^1}{h^2} = \frac{r^1}{r^2}$$

Example 1 Identify Similar and Congruent Solids

Determine whether the square pyramids are *similar*, *congruent*, or *neither*. If the pyramids are similar, state the scale factor.

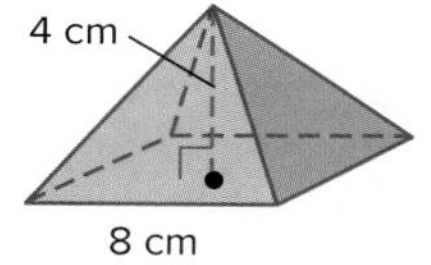

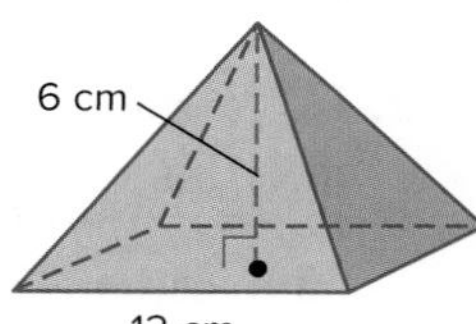

ratio of heights: $\frac{4}{6} = \frac{2}{3}$

ratio of base edges: $\frac{8}{12} = \frac{2}{3}$

The ratios of the corresponding measures are equal, so the pyramids are similar. The scale factor is 2:3. Because the scale factor is not 1:1, the solids are not congruent.

Guided Practice

1A.

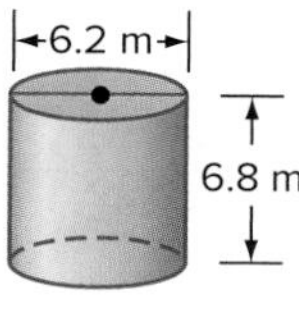

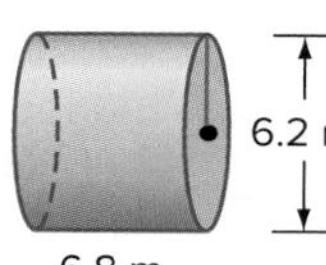

1B.

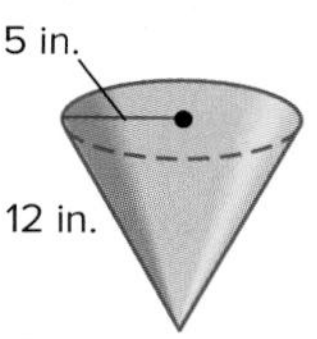

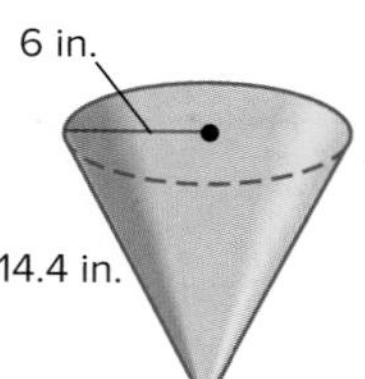

2 Properties of Similar Solids

The cubes at the right are similar solids with a scale factor of 3 : 2.

ratio of surface areas: 54 : 24 or 9 : 4
ratio of volumes: 27 : 8

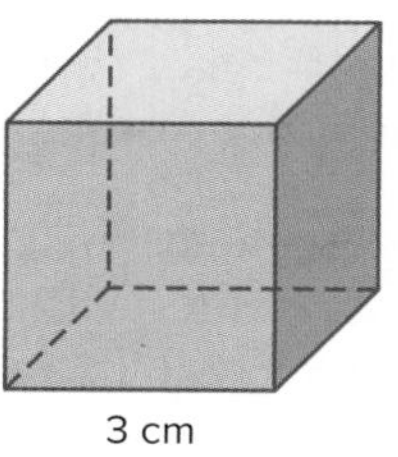

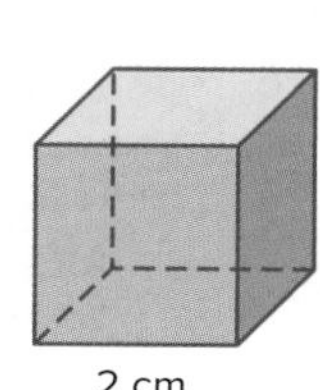

Burazin/Photographer's Choice RF/Getty Images

Notice that the ratio of surface areas, $9:4$, can be written as $3^2:2^2$. The ratio of volumes, $27:8$, can be written as $3^3:2^3$. This suggests the following theorem.

> **Study Tip**
> **Similar Solids and Area** If two solids are similar, then the ratio of any corresponding areas is $a^2:b^2$.

Theorem 11.1

Words If two similar solids have a scale factor of $a:b$, then the surface areas have a ratio of $a^2:b^2$, and the volumes have a ratio of $a^3:b^3$.

Models

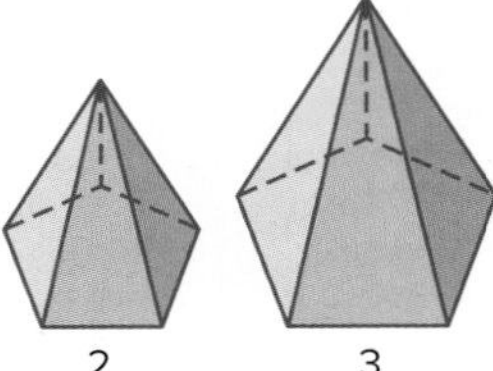

Example

scale factor	2:3
ratio of surface area	4:9
ratio of volumes	8:27

Figures must be similar for Theorem 11.1 to apply.

Example 2 Use Similar Solids to Solve Problems

The edge of a small cube measures 4 centimeters. If the scale factor between the small cube and a larger cube is $\frac{1}{3}$, what is the surface area of the larger cube?

First, find the surface area of the small cube.

$S = 6s$ Surface area of a cube

$= 6(4)^2$ Substitute.

$= 96$ Simplify.

The surface area of the small cube is 96 cm^2.

$\frac{\text{surface area of large cube}}{\text{surface area of small cube}} = \frac{3^2}{1^2}$ Use proportion.

$\frac{\text{surface area of large cube}}{96} = \frac{9}{1}$ Substitute.

surface area of large cube $= 9 \cdot 96$ or 864 Simplify.

The surface area of the large cube is 864 square centimeters.

> **Go Online!**
> Watch Personal Tutor videos to hear descriptions of solving problems involving congruent and similar solids. Try describing how to solve a problem for a partner. Have them ask you questions to help your understanding.

Guided Practice

2. Two similar rectangular prisms with square bases have surface areas of 98 square centimeters and 18 square centimeters. If one base edge of the larger rectangular prism measures 9 centimeters, what is the perimeter of one base of the smaller prism?

Example 3 Find the Volume of Similar Solids by Using Scale Factor

Rectangular prism A and rectangular prism B are similar. Find the volume of rectangular prism B.

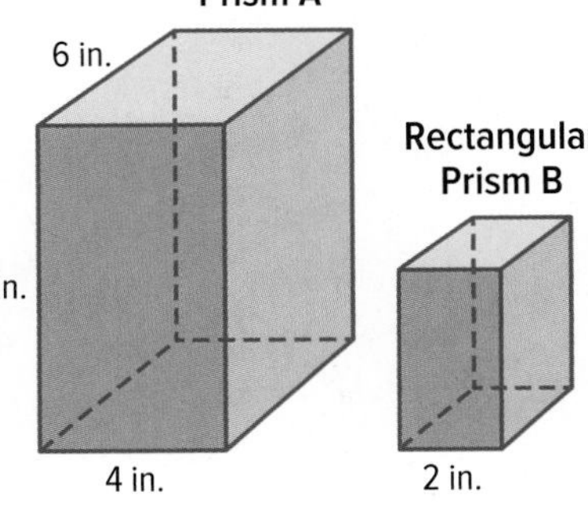

First find the scale factor.

$\frac{\text{length of the rectangular prism A}}{\text{length of the rectangular prism B}} = \frac{4}{2}$ or 2.

The scale factor is 2.

Find the volume of rectangular prism A.

Volume of rectangular prism

A = length × width × height

$V_A = 4 \times 6 \times 8$ Substitute.

$= 192$ Simplify.

The volume of rectangular prism A is 192 in^3.

> **Study Tip**
> **Similar Solids and Volume** If two solids are similar, then the ratio of any corresponding volume is $a^3:b^3$.

$$\frac{\text{Volume of rectangular prism A}}{\text{Volume of rectangular prism B}} = \frac{a^3}{b^3}$$ Write formulas.

$$\frac{192}{\text{Volume of rectangular prism B}} = \frac{2^3}{1^3}$$ Substitute.

$$\text{Volume of rectangular prism B} = \frac{192}{8}$$ Use proportion.

$$= 24$$ Simplify.

The volume of rectangular prism B is 24 in^3.

Guided Practice

3. Two circular cylinders are similar. The ratio of the radii is 3:2. The volume of the smaller cylinder is 60 cubic inches. Find the volume of the larger cylinder.

Real-World Example 4 Use Similar Solids to Find Unknown Values

CONTAINERS The containers at the right are similar cylinders. Find the height h of the smaller container.

h

10 in.

$V = 270\pi$ in^3

$V = 640\pi$ in^3

Understand You know the height of the larger container and the volumes of both containers.

Plan Use Theorem 11.1 to write a ratio comparing the volumes. Then find the scale factor and use it to find h.

Solve

$$\frac{\text{volume of small container}}{\text{volume of large container}} = \frac{270\pi}{640\pi}$$ Write a ratio comparing volumes.

$$= \frac{27}{64}$$ Simplify.

$$= \frac{3^3}{4^3}$$ The scale factor is 3:4.

Ratio of heights → $\frac{h}{10} = \frac{3}{4}$ ← Scale factor

$h \cdot 4 = 10 \cdot 3$ Find the cross products.

$h = 7.5$ Solve for h.

Check Because $\frac{7.5}{10} = 0.75 = \frac{3}{4}$, the solution is correct. ✓

For the two cylinders to be similar, all corresponding measurements must have the same scale factor.

Study Tip

MP Precision After finding the missing measure of similar solids, work backward to check your solutions.

Guided Practice

4. PUZZLES The world's largest puzzle cube has an edge length of 1.57 meters. The ratio of edge length of the largest puzzle cube to a regular puzzle cube is approximately $\frac{27.5}{1}$. Find the surface area of a regular puzzle cube rounded to the nearest hundredth.

Check Your Understanding

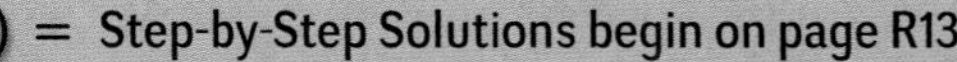

Example 1 **Determine whether each pair of solids is *similar*, *congruent*, or *neither*. If the solids are similar, state the scale factor.**

1.

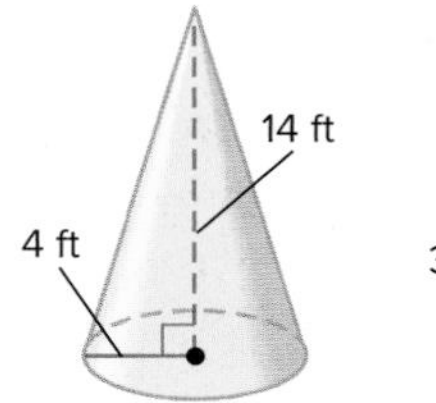

2.

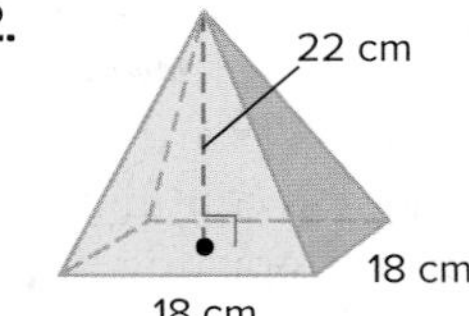

Example 2 **3.** Two similar cylinders have radii of 15 inches and 6 inches. If the surface area of the first cylinder is 2592 square inches, what is the surface area of the second cylinder?

Example 3 **4.** Two similar rectangular prisms have surface areas of 83.2 square feet and 20.8 square feet, respectively. If the volume of the first prism is 46.5 cubic feet, what is the volume of the second prism rounded to the nearest tenth?

Example 4 **5.** **EXERCISE BALLS** A company sells two different sizes of exercise balls. The ratio of the diameters is 15 : 11. If the diameter of the smaller ball is 55 centimeters, what is the volume of the larger ball? Round to the nearest tenth.

Practice and Problem Solving

Extra Practice is on page R11.

Example 1 **MP REGULARITY Determine whether each pair of solids is *similar*, *congruent*, or *neither*. If the solids are similar, state the scale factor.**

6.

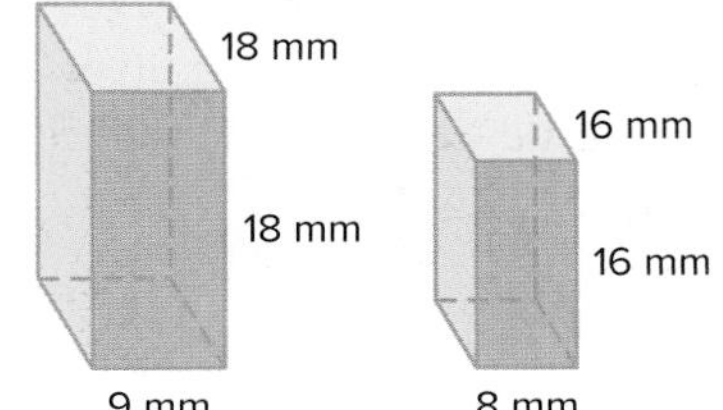

7.

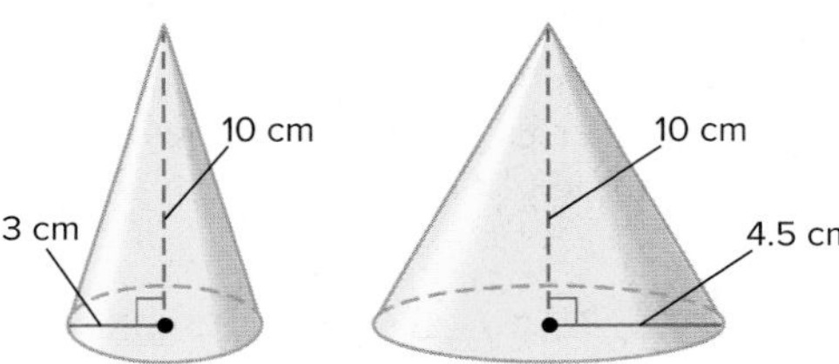

8.

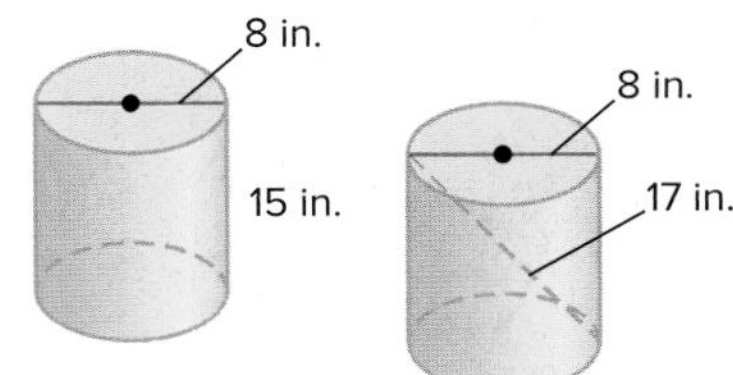

9.

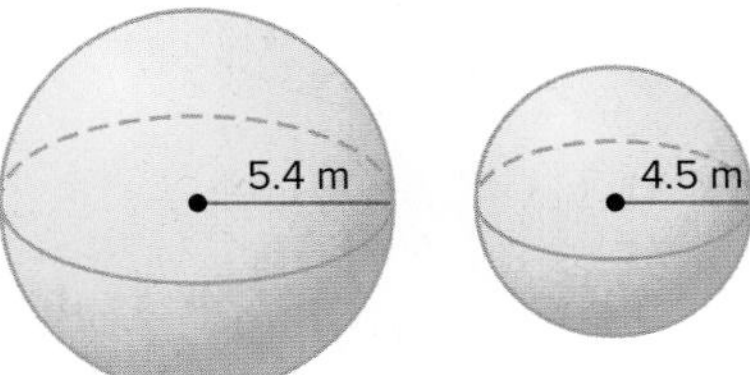

Example 2 **10.** Two similar pyramids have slant heights of 6 inches and 12 inches. If the volume of the large pyramid is 548 cubic inches, what is the volume of the small pyramid?

Example 3 **11** Two similar cylinders have heights of 35 meters and 25 meters. The volume of the shorter cylinder is 125π. What is the volume of the taller cylinder?

12. Two spheres have surface areas of 100π square centimeters and 16π square centimeters. What is the ratio of the volume of the large sphere to the volume of the small sphere?

13. Two similar hexagonal prisms have heights of 15 feet and 3 feet, respectively. If the volume of the first hexagonal prism is 250 cubic feet, what is the volume of the second hexagonal prism?

Example 4 **14.** **ICE CREAM** Two similar ice cream containers have volumes of 270π and 640π cubic inches. If the height of the larger cylinder is 10 inches, what is the area of the base of the smaller cylinder in terms of π?

15. **FOOD** A small cylindrical can of tuna has a radius of 4 centimeters and a height of 3.8 centimeters. A larger and similar can of tuna has a radius of 5.2 centimeters.

   a. What is the scale factor of the cylinders?

   b. What is the volume of the larger can? Round to the nearest tenth.

16. **SUITCASES** Two suitcases are similar rectangular prisms. The smaller suitcase is 68 centimeters long, 47 centimeters wide, and 27 centimeters deep. The larger suitcase is 85 centimeters long.

   a. What is the scale factor of the prisms?

   b. What is the volume of the larger suitcase? Round to the nearest tenth.

17. **CLASS RINGS** The 12-foot replica of the Aggie Ring at Haynes Ring Plaza on the Texas A&M campus weighs about 6500 pounds. Suppose that the ring was based on a ring that was 0.75 inch high, what is the scale factor?

18. The pyramids shown are congruent.

   a. What is the perimeter of the base of pyramid A?

   b. What is the area of the base of pyramid B?

   c. What is the volume of pyramid B?

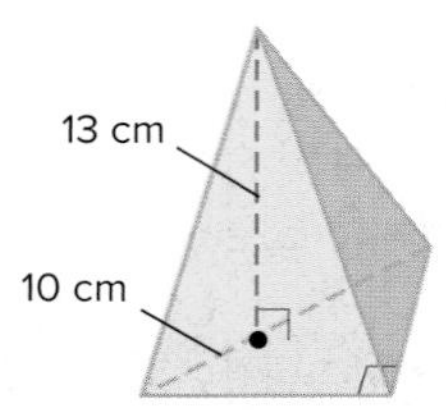

Pyramid A

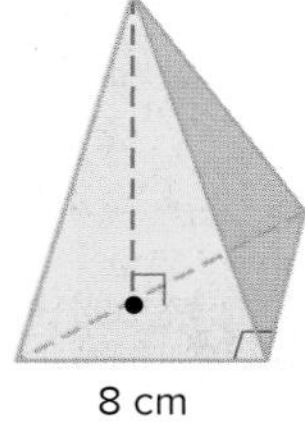

Pyramid B

19. **TECHNOLOGY** Jalissa and Mateo each have the same type of digital media player but in different colors. The players are congruent rectangular prisms. The volume of Jalissa's player is 4.92 cubic inches, the width is 2.4 inches, and the depth is 0.5 inch. What is the height of Mateo's player?

MP MODELING Each pair of solids below is similar.

20. What is the surface area of the smaller solid shown below?

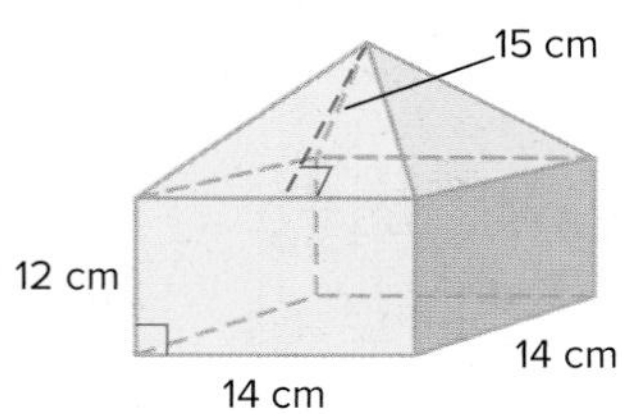

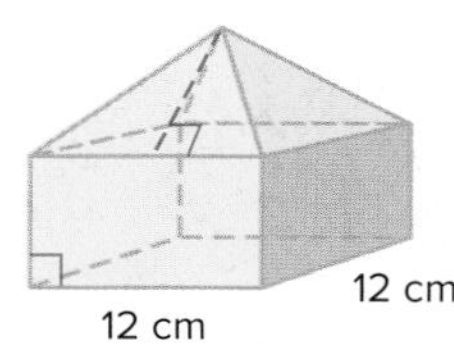

21. What is the volume of the larger solid shown below?

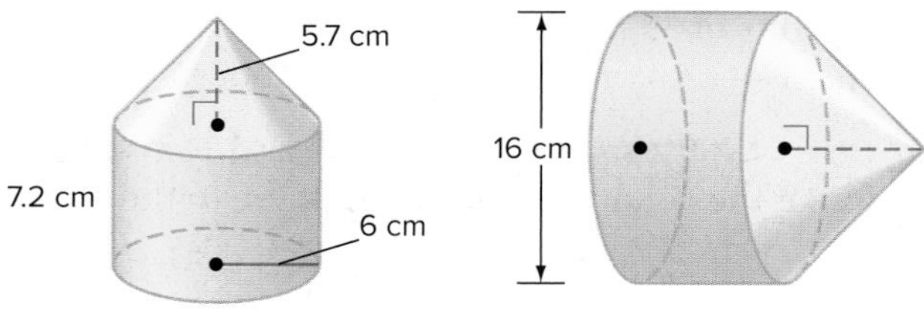

22. **DIMENSIONAL ANALYSIS** Two cylinders are similar. The height of the first cylinder is 23 centimeters, and the height of the other cylinder is 8 inches. If the volume of the first cylinder is 552π cubic centimeters, what is the volume of the other cylinder? Use $2.54 \text{ cm} = 1 \text{ in.}$

23. **DIMENSIONAL ANALYSIS** One sphere has a radius of 10 feet. The volume of a second sphere is 0.9 cubic meter. Use 2.54 centimeters = 1 inch to determine the scale factor from the first sphere to the second.

24. **ALGEBRA** Two similar cones have volumes of 343π cubic centimeters and 512π cubic centimeters. The height of each cone is equal to 3 times its radius. Find the radius and height of both cones.

25. **TENTS** Two tents are in the shape of hemispheres with circular floors. The ratio of their floor areas is 9:12.25. If the diameter of the smaller tent is 6 feet, what is the volume of the larger tent? Round to the nearest tenth.

26. **MODELING** In this problem, you will investigate similarity. The heights of two similar cylinders are in the ratio 2 to 3. The lateral area of the larger cylinder is 162π square centimeters, and the diameter of the smaller cylinder is 8 centimeters.

    a. **Verbal** What is the height of the larger cylinder? Explain your method.

    b. **Geometric** Sketch and label the two cylinders.

    c. **Analytical** How many times as great is the volume of the larger cylinder as the volume of the smaller cylinder?

H.O.T. Problems Use Higher-Order Thinking Skills

27. **MP CRITIQUE ARGUMENTS** Cylinder X has a diameter of 20 centimeters and a height of 11 centimeters. Cylinder Y has a radius of 30 centimeters and is similar to Cylinder X. Did Laura or Paloma correctly find the height of Cylinder Y? Explain your reasoning.

Laura	Paloma
Cylinder X: radius 10, height 11 Cylinder Y: radius 30, height a $\frac{10}{30} = \frac{11}{a}$, so $a = 33$.	Cylinder X: diameter 20, height 11 Cylinder Y: diameter 20, height a $\frac{20}{20} = \frac{11}{a}$, so $a = 11$.

28. **CHALLENGE** The ratio of the volume of Cylinder A to the volume of Cylinder B is 1:5. Cylinder A is similar to Cylinder C with a scale factor of 1:2, and Cylinder B is similar to Cylinder D with a scale factor of 1:3. What is the ratio of the volume of Cylinder C to the volume of Cylinder D? Explain your reasoning.

29. **WRITING IN MATH** Explain how the surface areas and volumes of the similar prisms shown at the right are related.

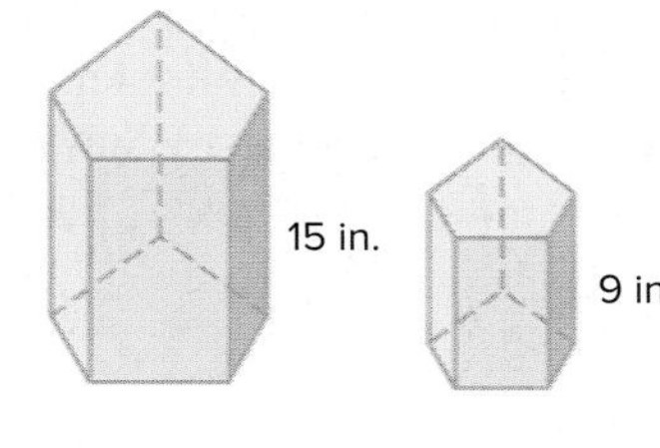

30. **OPEN-ENDED** Describe two nonsimilar triangular pyramids with similar bases.

31. **MP PERSEVERANCE** Plane P is parallel to the base of cone C, and the volume of the cone above the plane is $\frac{1}{8}$ of the volume of cone C. Find the height of cone C.

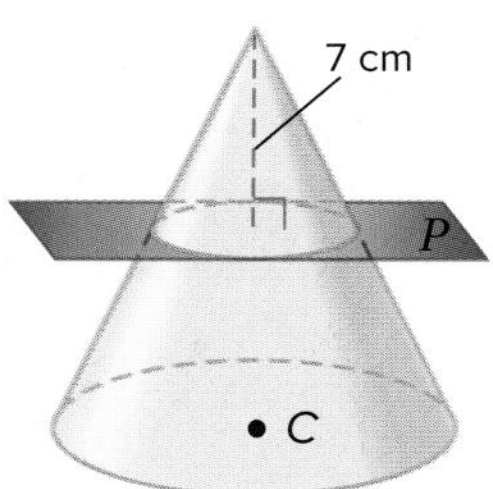

32. **WRITING IN MATH** Explain why all spheres are similar.

33. Two similar rectangular prisms have volumes of 60 cubic meters and 480 cubic meters. The surface area of the smaller prism is 94 square meters. What is the surface area of the larger prism? MP 3

- A 188 m^2
- B 376 m^2
- C 514 m^2
- D 752 m^2

34. Two beach balls have diameters of 8 inches and 20 inches. Which of the following statements must be true? MP 2

- A The ratio of the radii is 2:5.
- B The ratio of the surface areas is 8:125.
- C The ratio of the volumes is 4:25.
- D The ratio of the circumferences is 4:25.

35. The two cones in the figure are similar. The volume of the large cone is 18,000 cubic inches. What is the volume of the small cone? MP 2

- A $18{,}000 \text{ in}^3$
- B $83{,}333 \text{ in}^3$
- C 3888 in^3
- D 6480 in^3

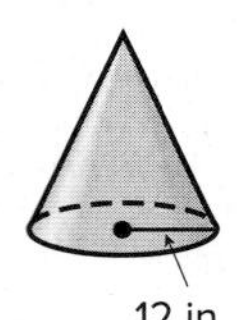

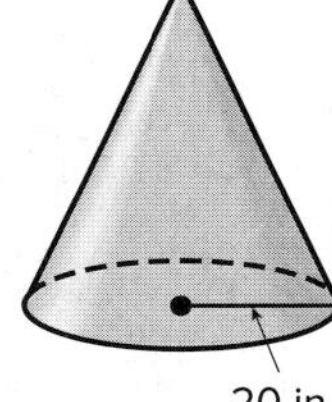

36. There are two square pyramids. One square pyramid has a height of 9 centimeters and base edges measuring 20 centimeters. The second square pyramid has a height of 4.5 centimeters and base edges measuring 10 centimeters. What is the scale factor for the two square pyramids? MP 2

- A 1 to 1
- B 2 to 1
- C 2 to 2
- D 3 to 1
- E 3 to 2

37. The two cylinders shown here are similar. The volume of the small cylinder is 16 cubic feet, and the volume of the large cylinder is 54 cubic feet.

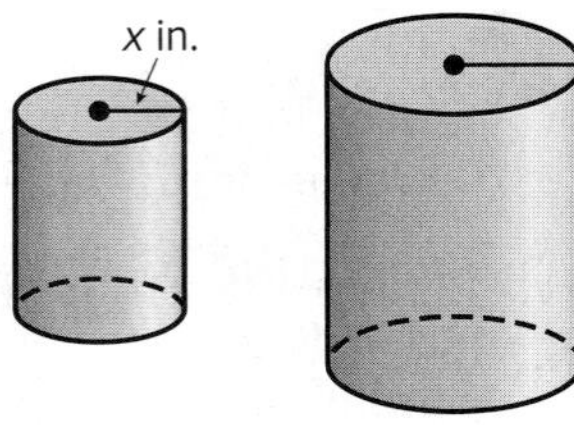

Which expression represents the radius of the large cylinder, in feet? MP 1

- A $\frac{8}{27}x$
- B $\frac{2}{3}x$
- C $\frac{1}{8}x$
- D $\frac{9}{4}x$
- E $\frac{27}{8}x$

38. Two spheres have radii of 20π meters and 6π meters. How many times larger is the surface area of the large sphere compared to the surface area of the small sphere? Round to the nearest tenth. MP 2

39. **MULTI-STEP** Triangular prism A and triangular prism B are similar. The scale factor of prism A to prism B is $\frac{2}{5}$. MP 2

a. The height of the triangular base of prism A is 4.8 feet. Find the height of the triangular base of prism B.

b. If the length of a side of prism A is 6 feet, what is the length of the corresponding side of prism B?

c. If prism B has a surface area of 80 square feet, what is the surface area of prism A?

d. If the volume of prism A is 32 cubic feet, what is the volume of prism B?

LESSON 7

Applying Measurements

Then

- You found areas of geometric figures and volumes of solids.

Now

1. Solve real-world problems involving density by using area.
2. Solve real-world problems involving density by using volume.

Why?

- Cities are more crowded than other areas. In a city, there are more people that live and work in an area of a given size. This means that the population density of a city is higher than other areas.

New Vocabulary

density
population density

Mathematical Practices

1 Make sense of problems and persevere in solving them.
4 Model with mathematics.
5 Use appropriate tools strategically.

1 Solve Density Problems Using Area

Density is a measure of the quantity of some physical property per unit of length, area, or volume. One example of density is **population density**, which is the measurement of population per unit of area. Population density is calculated for states, major cities, or other areas, based on data collected from the U.S. Census.

Key Concept Density Based on Area

Words Density is the ratio of objects to area.

Symbols $\text{density} = \dfrac{\text{number of objects}}{\text{area}}$

Real-World Example 1 Find Population Densities

CITIES Use the data in the table to find each of the following.

City	Population	Area (mi^2)
Boston, MA	617,594	48
Chicago, IL	2,695,598	228
Los Angeles, CA	?	469
New York, NY	8,175,133	?

a. Find the population density of Boston to the nearest tenth.

$$\text{Population density} = \frac{\text{population}}{\text{land area}}$$
$$= \frac{617{,}594}{48}$$
$$\approx 12{,}866.5 \text{ persons/mi}^2$$

b. Find the population of Los Angeles, given that the population density is 8086.6 persons per square mile.

Let p represent the population of Los Angeles.

$$\text{Population density} = \frac{\text{population}}{\text{land area}}$$
$$8086.6 = \frac{p}{469}$$
$$469(8086.6) = p$$
$$3{,}792{,}615 \approx p$$

Guided Practice

1A. Find the population density of Chicago to the nearest tenth.

1B. Find the area of New York, given that the population density is 26,980.6 persons per square mile.

Real-World Example 2 Apply Population Density

NATIONAL PARKS Looking at a proposal to establish a new campground at Yellowstone National Park, there is a concern about the number of wolves in the area. At last report, there were 98 wolves in the park. The new campground will be accepted if there are fewer than 2 wolves in the area that the campground would occupy. Use the data in the table to determine whether the campground can be established.

Location	Area
Entire park	3472 mi^2
Proposed campground	10 acres

Step 1 Find the population density of wolves in the park.

$$\frac{\text{population}}{\text{land area}} = \frac{2 \text{ wolves}}{3472 \text{ mi}^2}$$
$$\approx 0.028 \text{ wolves/mi}^2$$

Step 2 Find the population of wolves in the proposed campground. Convert the size of the campground from acres to square miles.

$$1 \text{ acre} = 0.0015625 \text{ mi}^2$$
$$10 \text{ acres} = 0.015625 \text{ mi}^2$$

The potential number of wolves in the proposed site is $0.015625 \times 0.028 = 0.0004375$ wolf.

Because 0.0004375 is less than 2, the proposed campground can be established.

Watch Out!

Dimensional Analysis In some cases, it may be necessary to use dimensional analysis to convert between units of measurement. To convert from one unit to another, use a numerical quantity known as a conversion factor.

Guided Practice

2. There are about 1.25 million alligators in Florida, which has an area of about 67,555 square miles. The state wants to build a new park with an area of 75 square miles, but the park cannot be built if there are more than 1000 alligators in the area. Should the state build the park? Explain.

2 Solve Density Problems Using Volume

Density is the measure of the quantity of some physical property per unit of length, area, or volume. If two objects have the same volume but different masses, the object with the greater mass will be denser. Density is often measured in grams per cubic centimeter (g/cm^3) or kilograms per cubic meter (kg/m^3).

Key Concept Density Based on Volume

Words Density is the ratio of mass to volume.

Symbols $\text{density} = \frac{\text{mass}}{\text{volume}}$

The figure shows the densities of some common materials.

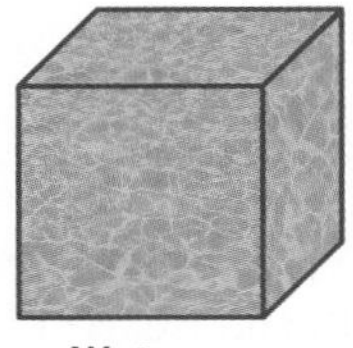

Water
1.0 g/cm^3

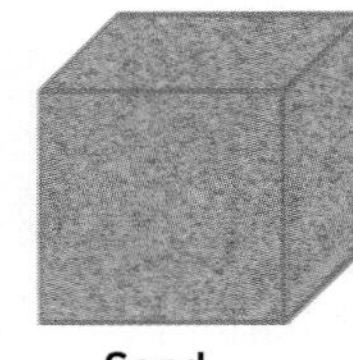

Sand
1.6 g/cm^3

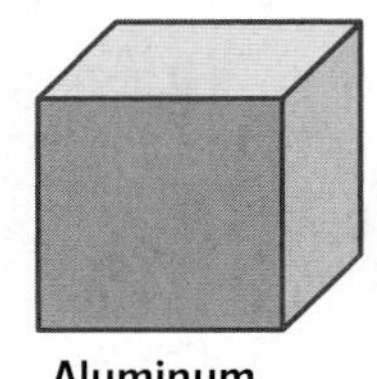

Aluminum
2.7 g/cm^3

Lead
11.4 g/cm^3

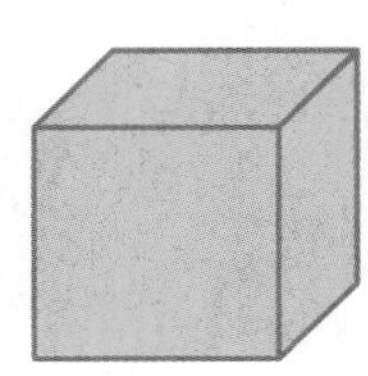

Gold
19.3 g/cm^3

Example 3 Find the Density of a Solid

The mass of the cube shown here is 275 grams. Find the density of the cube to the nearest tenth.

Step 1 Find the volume of the cube.

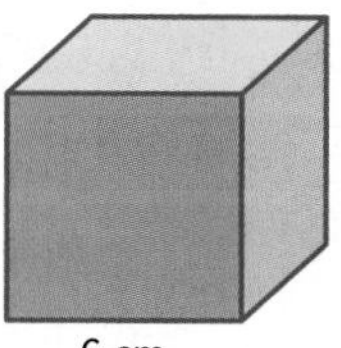

$V = s^3$ Volume of a cube

$= 6^3$ or 216 cm^3 Substitution, $s = 6$

Step 2 Find the density of the cube.

$$\text{density} = \frac{\text{mass}}{\text{volume}}$$

$$= \frac{275}{216} \approx 1.3\text{ g/cm}^3$$

Guided Practice

3. Find the density of a rectangular prism that is 6 inches by 2 inches by 4 inches with a mass of 16 ounces. Round to the nearest tenth.

When you calculate and compare densities in real-world problems, be sure you are using units consistently. You may need to convert units before calculating densities.

Real-World Example 4 Compare Densities

DRIVEWAYS **The Ramirez family wants to redo the driveway of their home. They have to choose between using asphalt or concrete, and they must make sure that whichever material they use has a density of less than 50 pounds per cubic foot because if the driveway is too heavy, it will sink. The driveway is 192 inches wide, 32 feet long, and 3 inches deep, and it will require about 9.5 tons of concrete or 5700 pounds of asphalt. Which material should the Ramirez family use?**

Step 1 Convert units in order to solve the problem using feet and pounds.

Convert 192 inches to feet: $\frac{192}{12} = 16$ ft Convert 3 inches to feet: $\frac{3}{12} = 0.25$ ft

Convert 9.5 tons to pounds: $9.5 \times 2000 = 19{,}000$ lb

Step 2 Find the volume of the driveway.

$V = \ell \times w \times h$

$= 32 \times 16 \times 0.25 = 128\text{ ft}^3$

Step 3 Calculate the density of each material.

$$\text{density of asphalt} = \frac{\text{mass}}{\text{volume}} = \frac{5700}{128} \approx 44.5\text{ lb/ft}^3$$

$$\text{density of concrete} = \frac{\text{mass}}{\text{volume}} = \frac{19{,}000}{128} \approx 148.4\text{ lb/ft}^3$$

The Ramirez family should use asphalt.

Study Tip

MP Sense-Making Before starting to solve the problem, look at all of the given units and decide which units will be most convenient to use for comparing densities.

Guided Practice

4. Two blocks are each 110 centimeters long, 100 millimeters wide, and 6 centimeters high. Block A is made of zinc, and its mass is 4.7 kilograms. Block B is made of iron, and its mass is 5200 grams. Which block has a greater density? Explain.

Check Your Understanding

◯ = Step-by-Step Solutions begin on page R13.

Example 1

1. A rectangular pen with length of 32 feet and width of 13 feet holds 56 rabbits. Find the population density of rabbits to the nearest hundredth.

2. There are 13,000 flowers in a square field. The population density of the flowers is 0.9 flower per square foot. Find the side length of the field to the nearest foot.

Example 2

3. **OWLS** The population density of the burrowing owl in Cape Coral, Florida, is 8.3 pairs per square mile. A new golf course is planned for a 2.4-square mile site where the owl population is estimated to be 17 pairs. Would Lee County approve the proposed club if their policy is to decline expansion when the estimated population density of owls is below the average density? Explain.

Example 3

4. A rectangular prism has a length of 15 centimeters, a width of 4.2 centimeters, and a height of 2 centimeters. The mass of the prism is 755 grams. Find the density of the prism to the nearest tenth.

5. A cube has an edge that is 12 feet long. The mass of the cube is 18,000 pounds. Find the density of the cube to the nearest tenth.

Example 4

6. **ART** A cylindrical sculpture has a radius of 50 millimeters and height of 15 centimeters. The mass of the sculpture is 3.7 kilograms. A sculpture in the shape of a rectangular prism has a length of 62 millimeters, width of 40 millimeters, and height of 12 centimeters. Its mass is 1250 grams. Which sculpture has a greater density? Explain.

Practice and Problem Solving

Extra Practice is found on page R11.

Example 1

Find the missing value in each row of the table. When necessary, round populations and areas to the nearest whole number and densities to the nearest tenth.

	Country	Population	Area (km^2)	Density (persons/km^2)
7.	Australia	?	7,682,300	2.5
8.	Japan	126,550,976	?	374
9.	Monaco	31,693	2	?
10.	Mongolia	?	1,553,556	1.7
11.	United States	275,562,673	9,147,593	?

Example 2

12. **CATTLE RANCHING** The Jackson family owns a ranch that covers 49,747 acres (1 acre = 0.0015625 square mile), but they do not use the entire ranch for cows. According to the USDA, ranches are required to have a minimum of 1.5 acres per cow.

    a. The Jackson family has 26,667 cows on the ranch, with a population density of 348.6 cows per square mile. How many square miles do they use for cows?

    b. If the family used the entire ranch for cows, what would be the maximum number of cows that they could have? What would the population density be in cows per square mile?

Example 3

13. A sphere has a radius of 14 centimeters and a mass of 36,249 grams. Find the density of the sphere to the nearest tenth.

14. A large metal cube measures 5 meters on each edge. Its mass is 537.5 kilograms. What is the density of the cube?

Example 4

15. **WOOD** Will has two blocks of wood. Block A is 2 feet long, 4 inches wide, and 2 inches tall. Block B is 1.5 feet long, 0.5 feet wide, and 3 inches tall. Block A has a mass of 3.8 pounds, and Block B has a mass of 120 ounces. Which block of wood has a greater density? Explain.

The table shows the densities of some common materials. Use the table and the given information to determine the most likely material that each solid is made of.

Material	Density (g/cm^3)
Cardboard	0.7
Chalk	2.5
Copper	8.9
Iron	7.2
Rubber	1.5

16. A rectangular prism is 12 cm long, 8.2 cm wide, and 3.3 cm tall. Its mass is 2.89 kg.

17. A sphere has a radius of 44 mm. Its mass is 535 kg.

18. A cylinder has a radius of 67 mm and a height of 12 cm. Its mass is 4230 g.

19. A cube has edges that are 0.1 m long. Its mass is 0.7 kg.

Each unit represents 1 kilometer on the coordinate plane. Use the given population to find the population density of the region shown. Round to the nearest tenth.

20. population: 2358.

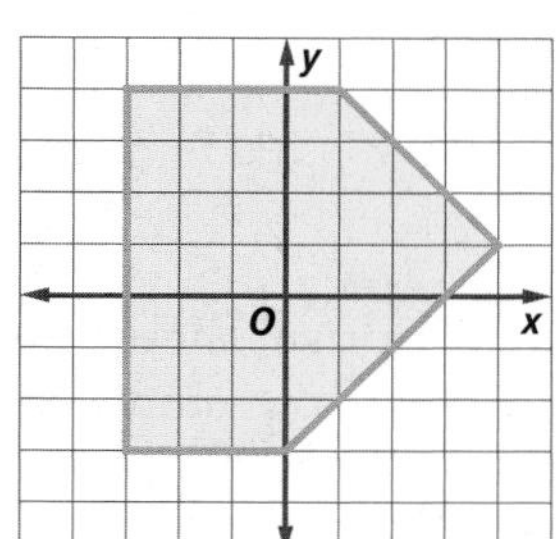

21 population: 55,323

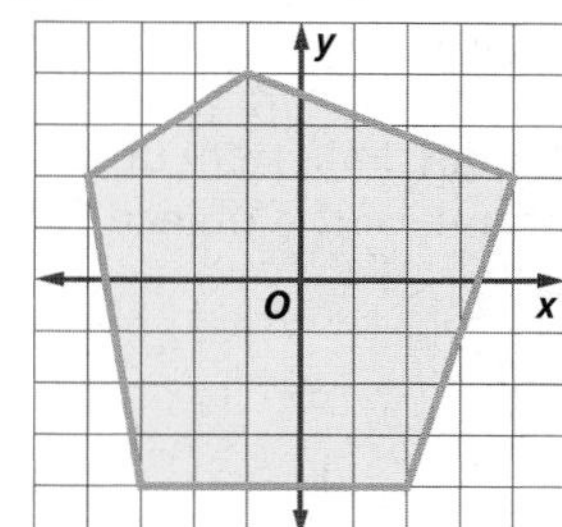

22. **EXERCISE** Saanvi uses a foam exercise roller to help her stretch after workouts. The more dense the roller, the more intense the stretching regimen is. The mass of the roller she is currently using is 436 grams.

a. What is the density of the foam roller?

b. If Saanvi uses an exercise roller that has a density of about 0.096 gram per cubic centimeter and is the same shape as the original roller, what is the mass of her new roller rounded to the nearest whole gram?

H.O.T. Problems Use Higher-Order Thinking Skills

23. **ERROR ANALYSIS** India has a population of 1.282 billion people and an area of 3.29 million square kilometers. Jane says the population density of India is 0.003 persons per square kilometer. Is Jane correct? Explain why or why not.

24. MP **REASONING** A cylinder and a sphere have the same mass. The height of the cylinder is equal to its radius. The radius of the sphere is equal to the radius of the cylinder. Which solid has a greater density? Explain.

25. **CHALLENGE** A rectangular prism has a length of 5 meters, a width of x meters, and a height of $(x + 2)$ meters. Its density is 1 kilogram per cubic meter, and its mass is 120 cubic meters. Find the value of x.

26. MP **CRITIQUE ARGUMENTS** An object will float in water if its density is less than 1 gram per cubic centimeter. Will a block of wood with a mass of 10 kilograms and dimensions 30 centimeters by 40 centimeters by 8 centimeters float in water? Explain.

27. **WRITING IN MATH** A student said that steel is heavier than plastic. Use density to explain what the student meant.

28. There are 64,288,000 gaming system owners in the United States. The area of the United States is 3,794,083 square miles. What is the population density of gaming system owners? MP 1

- ○ A 0.1 owner/mi^2
- ○ B 1.7 owners/mi^2
- ○ C 15 owners/mi^2
- ○ D 16.9 owners/mi^2
- ○ E 17.3 owners/mi^2

29. An object has a density of 4.6 grams per cubic centimeter. What is the volume of the object if its mass is 2773.8 grams? MP 1, 2

- ○ A 0.002 cm^3
- ○ B 603 cm^3
- ○ C 2769.2 cm^3
- ○ D 12,759.5 cm^3

30. Five pounds of flour completely fills a container that measures $6\frac{3}{4}$ inches by $6\frac{3}{4}$ inches by 6 inches. What is the density of the flour in ounces per cubic inch? MP 1, 2

- ○ A 0.02 oz/in^3
- ○ B 0.29 oz/in^3
- ○ C 3.42 oz/in^3
- ○ D 54.68 oz/in^3

31. Kenisha is going to plant 1000 seeds in a flower bed. For which of the following shapes of the flower bed will the density of seeds be greater than 50 seeds per square foot? MP 1, 4

- ☐ A a rectangle 6 ft long and 3 ft wide
- ☐ B a circle with a radius of 3 ft
- ☐ C a square with sides 5 ft long
- ☐ D a rectangle 5 ft long and 4 ft wide
- ☐ E a circle with a diameter of 2 ft
- ☐ F a square with sides 24 in. long
- ☐ G a circle with a radius of 1.5 ft

32. The population density of polar bears in a region of Canada is 5.1 bears per 1000 square kilometers. Which of the following is the best estimate of the number of polar bears that would be found in a national park within this region if the national park has an area of 26,500 square kilometers? MP 1, 4

- ○ A 5
- ○ B 135
- ○ C 5196
- ○ D 135,150
- ○ E 175,221

33. The rectangular prism shown here has a mass of m grams. Which expression can be used to find the density of the rectangular prism in grams per cubic centimeter? MP 2

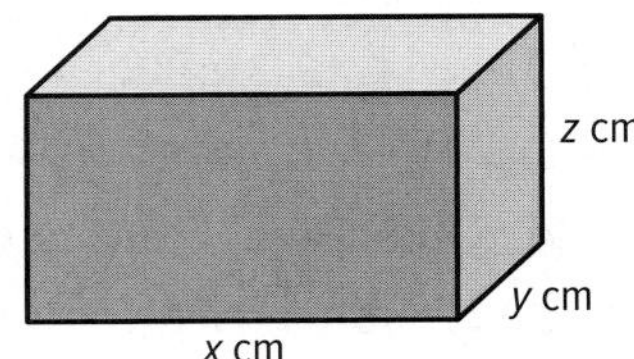

- ○ A $\frac{2xy + 2xz + 2yz}{m}$
- ○ B $\frac{m}{2xy + 2xz + 2yz}$
- ○ C $\frac{xyz}{m}$
- ○ D $\frac{m}{xyz}$

34. **MULTI-STEP** A cylinder has a diameter of 12 feet and a height of 18 feet. The mass of the cylinder is 6.2 tons. MP 1, 3

a. What is the density of the cylinder in pounds per cubic foot? Round to the nearest tenth.

b. Suppose the mass of the cylinder increases while its dimensions stay the same. How does this affect the density of the cylinder? Justify your answer.

c. What should the mass of the cylinder be, in tons, if the dimensions stay the same and the density changes to 10 pounds per cubic foot? Round to the nearest tenth.

Study Guide and Review

Study Guide

Key Concepts

Cross Sections and Solids of Revolution (Lesson 11-1)

- A cross section is the intersection of a solid and a plane.
- Rotations of two-dimensional objects produce three-dimensional objects.

Volumes of Prisms and Cylinders (Lesson 11-2)

- Volume of prism or cylinder: $V = Bh$

Volumes of Pyramids and Cones (Lesson 11-3)

- Volume of a pyramid: $V = \frac{1}{3}Bh$
- Volume of a cone: $V = \frac{1}{3}\pi r^2 h$

Spheres (Lesson 11-4)

- Surface area of a sphere: $S = 4\pi r^2$
- Volume of a sphere: $V = \frac{4}{3}\pi r^3$

Spherical Geometry (Lesson 11-5)

- Spherical geometry is geometry on a sphere.

Volume and Nonrigid Transformations (Lesson 11-6)

- If two solids are similar, then their volumes are proportional to the cube of the scale factor between them.

Applying Measurements (Lesson 11-7)

- Population density is the measurement of population per unit of area.

Study Organizer

Use your Foldable to review the chapter. Working with a partner can be helpful. Ask for clarification of concepts as needed.

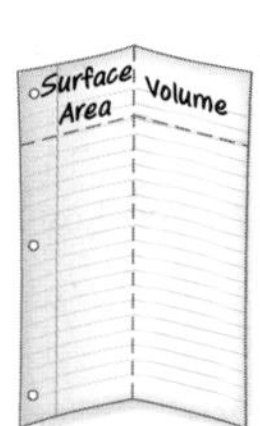

Key Vocabulary

cross section (p. 797)
density (p. 842)
Euclidean geometry (p. 827)
great circle (p. 819)
hemisphere (p. 819)
non-Euclidean geometry (p. 828)
oblique solid (p. 796)
pole (p. 819)
population density (p. 841)
right solid (p. 795)
solid of revolution (p. 798)
spherical geometry (p. 827)

Vocabulary Check

State whether each sentence is *true* or *false*. If *false*, replace the underlined term to make a true sentence.

1. Euclidean geometry deals with a system of points, great circles (lines), and spheres (planes).
2. Similar solids have exactly the same shape but not necessarily the same size.
3. A cross section is the intersection of a solid and a plane.
4. A great circle separates a sphere into two congruent hemispheres.
5. The perpendicular distance from the base of a geometric figure to the opposite vertex, parallel side, or parallel surface is the altitude.
6. Population density is the population per unit of area.
7. A(n) oblique solid has a base that is not perpendicular to the edges.
8. A composite solid is a three-dimensional figure that is composed of simpler figures.

Concept Check

9. Explain how the volumes of two similar solids are related.
10. Define non-Euclidean geometry in terms of Euclidean geometry.
11. Explain how to find the volume of a composite solid.

Study Guide and Review *Continued*

Lesson-by-Lesson Review 

11-1 Cross Sections and Solids of Revolution

Describe each cross section.

12.

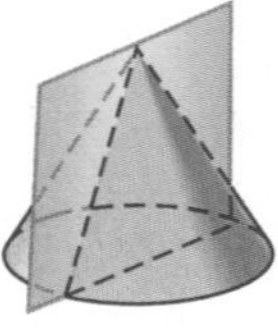

13.

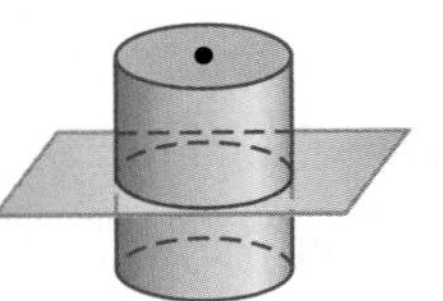

14. **CAKE** The cake shown is cut in half. Describe the cross section of the cake.

Example 1

Describe the cross sections formed when a plane parallel to the base and a plane perpendicular to the base intersect the solid.

When a plane perpendicular to the base intersects the solid, the cross section is a rectangle.
When a plane parallel to the base intersects the solid, the cross section is a circle.

11-2 Volumes of Prisms and Cylinders

15. The volume of a cylinder is 770 cm^3. It has a height of 5 centimeters. Find its radius.

16. Find the volume of the triangular prism.

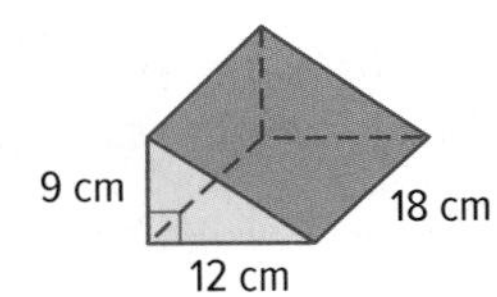

17. **TRAILERS** A semitruck trailer is basically a rectangular prism. A typical height for the inside of these trailers is 108 inches. If the trailer is 8 feet wide and 20 feet long, what is the volume of the trailer?

Example 2

Find the volume of the cylinder.

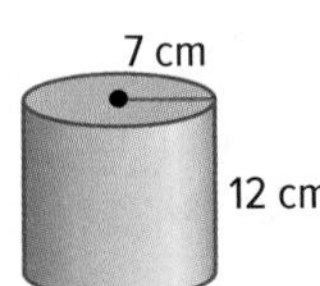

$V = \pi r^2 h$	Volume of a cylinder
$= \pi(7)^2(12)$	$r = 7$ and $h = 12$
≈ 1847.5	Use a calculator.

The volume is approximately 1847.5 cubic centimeters.

11-3 Volumes of Pyramids and Cones

18. Find the volume of a cone that has a radius of 1 centimeter and a height of 3.4 centimeters.

19. Find the volume of the regular pyramid.

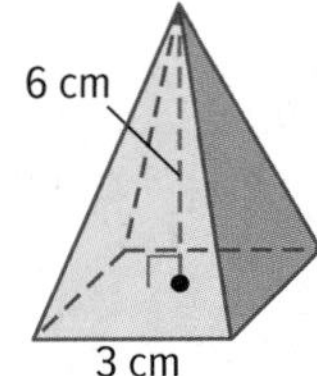

20. **ARCHITECTURE** The Great Pyramid measures 756 feet on each side of the base and the height is 481 feet. Find the volume of the pyramid.

Example 3

Find the volume of the pyramid.

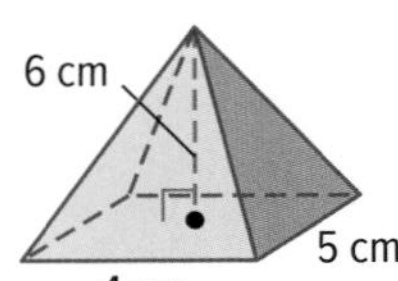

$V = \frac{1}{3}Bh$	Volume of a pyramid
$= \frac{1}{3}(4 \cdot 5)(6)$	$B = 4 \cdot 5$ and $h = 6$
$= 40$	Simplify.

The volume is 40 cubic centimeters.

11-4 Spheres

Find the surface area of each figure.

21.

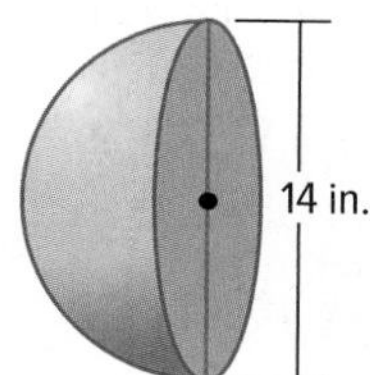

22.

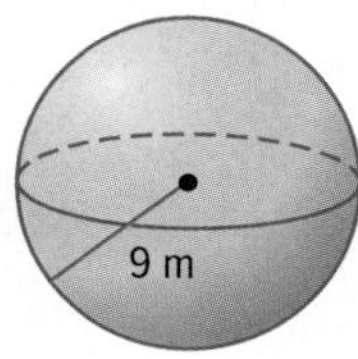

Find the volume of each sphere or hemisphere. Round to the nearest tenth.

23. hemisphere: circumference of great circle $= 24\pi$ m

24. sphere: area of great circle $= 55\pi$ in^2

25. **CONSTRUCTION** Cement is poured into a hemisphere that is 6 feet across. What is the volume of cement used?

Example 4

Find the surface area and volume of the sphere. Round to the nearest tenth.

14 cm

$S = 4\pi r^2$	Surface area of a sphere
$= 4\pi(14)^2$	Substitute.
≈ 2463	Use a calculator.

The surface area is about 2463 square centimeters.

$V = \frac{4}{3}\pi r^3$	Volume of a sphere
$= \frac{4}{3}\pi(14)^3$	Replace r with 14.
$\approx 11{,}494$ cm^3	Use a calculator.

The volume is about 11,494 cubic centimeters.

11-5 Spherical Geometry

Name each of the following on sphere A.

26. two lines containing point C

27. a segment containing point H

28. a triangle containing point B

29. two lines containing point L

30. a segment containing point J

31. a triangle containing point K

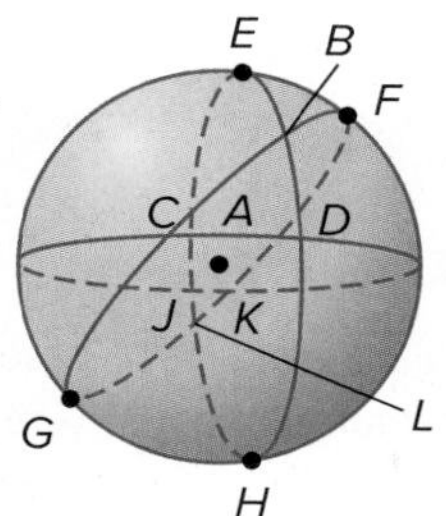

32. **MARBLES** Determine whether figure y on the sphere shown is a line in spherical geometry.

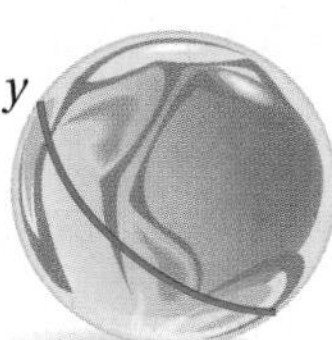

Example 5

Name each of the following on sphere A.

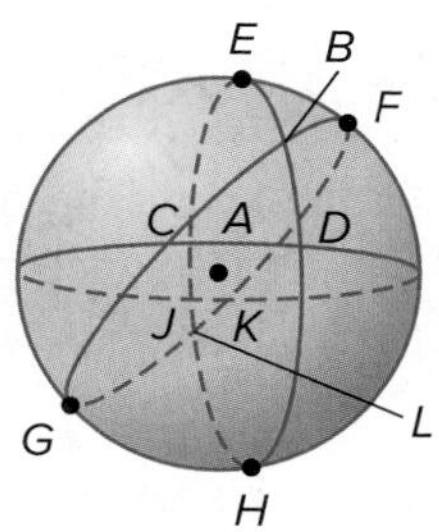

a. two lines containing point D
$\overleftrightarrow{EH}$, $\overleftrightarrow{CK}$

b. a segment containing point E
$\overline{DJ}$

Study Guide and Review *Continued*

11-6 Volume and Nonrigid Transformations

Find x for each pair of similar solids.

33.

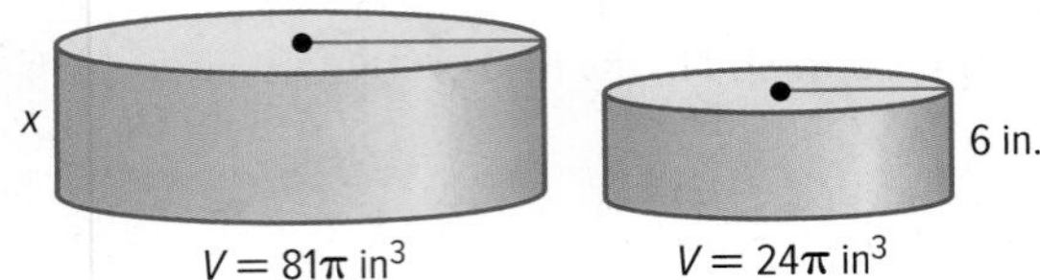

34.

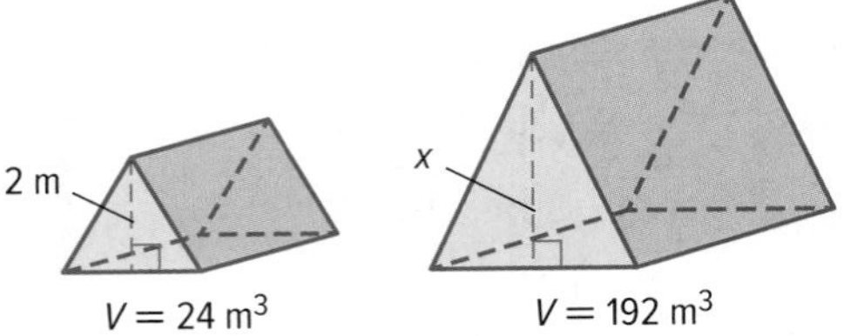

35. **DIMENSIONAL ANALYSIS** Two rectangular prisms are similar. The height of the first prism is 8 inches, and the height of the other prism is 2 feet. If the volume of the first prism is 64 cubic inches, what is the volume of the other prism?

Example 6

The rectangular prisms below are similar. Find the height h of the smaller prism.

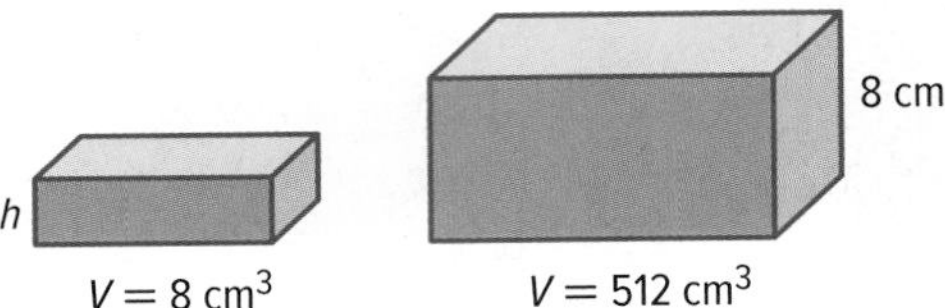

Write a ratio comparing volumes. Then find the scale factor and use it to find h.

$\frac{\text{volume of small prism}}{\text{volume of large prism}} = \frac{8}{512}$ Write a ratio comparing volumes.

$= \frac{1}{64}$ Simplify.

$= \frac{1^3}{4^3}$ The scale factor is 1:4.

Ratio of heights → $\frac{h}{8} = \frac{1}{4}$ ← Scale factor

$h = 2$ Solve for h.

11-7 Applying Measurements

36. **DEER** The table below shows the fall deer population and land area for four different counties. Order the counties from greatest to least population density of deer.

County	Deer Population	Land Area (mi²)
A	9512	232
B	18,720	720
C	55,366	893
D	21,546	1,197

37. Singapore has a population of about 5,535,000 people and a land area of 719 square kilometers. Find the population density.

Example 7

Alaska and New Jersey have the smallest and greatest population density, respectively, of any U.S. state. Determine the population density of each state.

U.S. State	Population	Land Area (mi²)
Alaska	738,432	570,641
New Jersey	8,958,013	7,354

Calculate population density with the formula population density $= \frac{\text{population}}{\text{land area}}$.

The population density of Alaska is $\frac{738{,}432}{570{,}641}$ or about 1.3 people per square mile.

The population density of New Jersey is $\frac{8{,}958{,}013}{7{,}354}$ or about 1218.1 people per square mile.

CHAPTER 11

Practice Test

Identify the shape of the cross section.

1. the intersection of a square pyramid and a plane perpendicular to the base and through the vertex

2. the intersection of a cube and a plane parallel to the base

Identify the solid generated by rotating each two-dimensional shape around the given axis.

3.

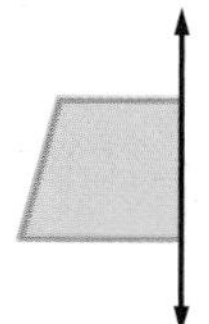

4.

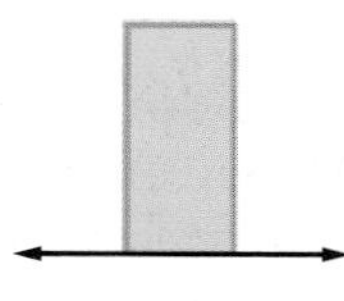

5. Describe the cross section.

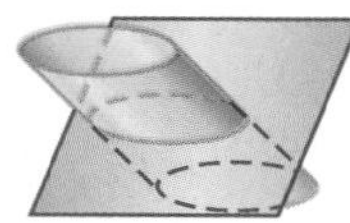

Find the volume of each prism or cylinder. Round to the nearest tenth if necessary.

6.

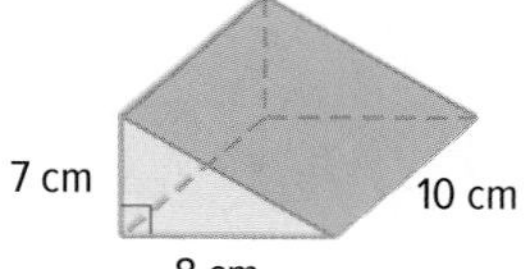

7.

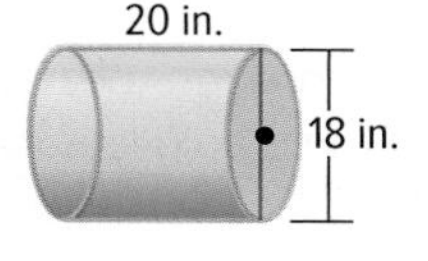

Find the volume of each cone or pyramid. Round to the nearest tenth if necessary.

8.

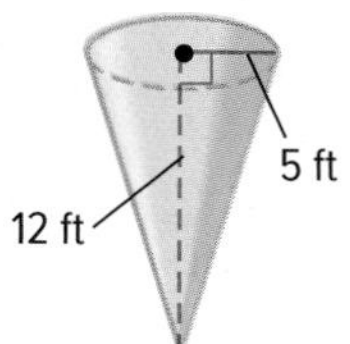

9.

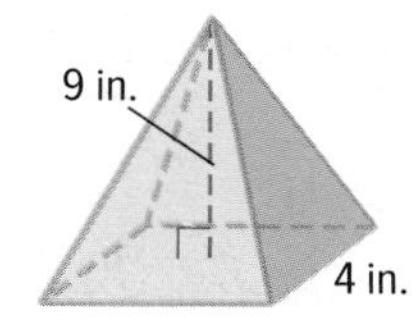

10. **CANDLES** A circular pillar candle is 2.8 inches wide and 6 inches tall. What is the volume of the candle? Round to the nearest tenth if necessary.

11. **TEA** A tea bag is shaped like a regular square pyramid. Each edge of the base is 4 centimeters, and the slant height is 5 centimeters. What is the volume of the tea bag in square centimeters? Round to the nearest tenth if necessary.

12. **SOFTBALL** A regulation softball has a circumference of 12 inches. What is the volume of the softball?

13. **EARTH** Earth's radius is approximately 6400 kilometers. What are the surface area and volume of the Earth? Round to the nearest whole number.

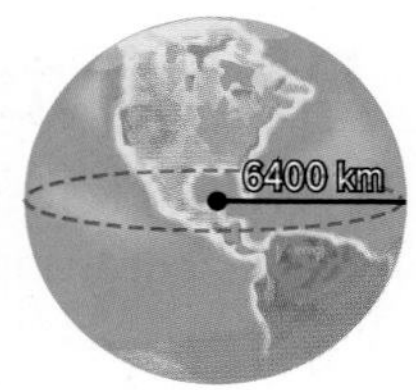

Name each of the following on sphere A.

14. two lines containing point S

15. a segment containing point L

16. a triangle

17. two lines containing point D

18. a segment containing point P

19. Are these two cubes *similar*, *congruent*, or *neither*? Explain your reasoning.

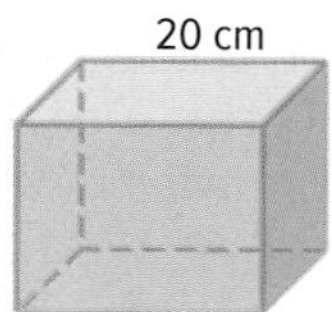

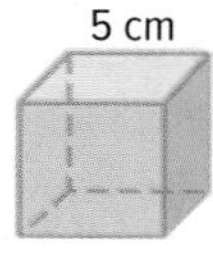

20. Two similar cylinders have heights of 75 centimeters and 25 centimeters. What is the ratio of the volume of the large cylinder to the volume of the small cylinder?

21. **BAKING** Two spherical pieces of cookie dough have radii of 3 centimeters and 5 centimeters, respectively. The pieces are combined to form one large spherical piece of dough. What is the approximate radius of the new sphere of dough? Round to the nearest tenth.

22. **ALGEBRA** A rectangular prism has a base with side lengths x and $x + 3$ and height $2x$. Find the volume of the prism.

23. **TRANSPORTATION** The traffic cone is 19 inches tall and has a radius of 5 inches. Find the volume.

24. **CATTLE** In Lewis County, New York, there were 55,509 cattle and calves on 181,741 acres of farmland in 2012. Find the population density of cattle and calves.

Performance Task

Provide a clear solution to each part of the task. Be sure to show all of your work, include all relevant drawings, and justify your answers.

APPLY MATH Wax & Suds, Inc. makes a variety of soaps and candles.

Part A

Wax & Suds sells a variety of sizes of candles in metal cylindrical tins. The smallest tin is 3 inches tall and 2 inches wide. The medium tin is 4 inches tall and 3 inches wide. The largest tin is 5 inches tall and 4 inches wide. When pouring the candles, they always leave one half inch of space at the top. Write your answers in terms of π.

1. Determine the volume of wax that will be poured into each size tin.

Part B

Wax & Suds make a line of speciality candles by pouring candle wax into a variety of molds. They have a cone-shaped mold that is 4 inches wide at the base and has a volume of 8π cubic inches. They also make a 3-inch wide square pyramid with a volume of 15 cubic inches. They have spherical molds in several different sizes.

2. When they pour the cone and pyramid candles, they insert the wick through the height of the figure and leave 1 inch of wick above the top of the wax. Can the employees cut wicks of the same length for use in each candle? Explain.
3. The smallest sphere they sell has a volume of 4.5π cubic inches. If Wax & Suds adds 3 inches to the diameter of the spherical candle, determine how much greater the volume will be. Write your answers in terms of π.
4. Sometimes, they add a decorative glaze to their candles. Determine the surface area of a spherical candle with a diameter of 4 inches. Write your answer in terms of π.

Part C

Wax & Suds makes bars of soap that are shaped like rectangular prisms in two different sizes. Their regular size is 3 inches long, 2 inches wide, and 1 inch tall. A travel size has dimensions that are exactly half of the larger size.

5. How much greater would the volume of soap be if they added 1 inch to the length of the regular size bar?
6. What is the difference in volume between the travel size bar and the regular bar?

Part D

Wax & Suds uses liquid fragrance to scent their candles. They use 2 ounces of scent per 1 pound of soap. One pound of soap makes 4 regular bars of soap.

7. Determine how many cubic inches of soap are used per ounce of scent.

Test-Taking Strategy

Example

Read the question. Then fill in the correct answer on the answer document provided by your teacher or on a sheet of paper.

Luther is building a model rocket for a science fair project. He attaches a nose cone to a cylindrical body to form the rocket's fuselage. The rocket has a diameter of 4 inches and a total height (including the nose cone) of 29 inches. The nose cone is 7 inches tall. What is the surface area of the rocket?

A 333.0 in^2

B 334.7 in^2

C 422.7 in^2

D 694.7 in^2

7 in.

2 in.

7 in.

2 ft. 5 in.

4 in.

Read the problem statements carefully.

Step 1 **What are you being asked to solve? What information is given?** You are asked to find the surface area of the rocket, which is a composite figure. You are given the total height of the rocket including the nose cone, the height of the nose cone, and the diameter of the rocket.

Step 2 **How can you make your drawing as clear as possible? What can you add to your drawing as you work?**
Label the drawing with all of the information given in the problem. As you make calculations, add the information to the drawing. When the drawing is complete, find the surface area of the cylinder and the surface area of the cone. The surface area of the rocket is the sum of these.

Step 3 **What is the correct answer?**
The correct answer is B.

Test-Taking Tip

Making a Drawing

Making a drawing can be a very helpful way for you to visualize how to solve a problem. Sketch the object and label the measurements that you know as a step in problem solving. The information that is combined in your sketch can then be used to perform calculations.

Apply the Strategy

Read the problem. Identify what you need to know. Then use the information in the problem to solve.

From a single point in her yard, Marti measures and marks distances of 18 feet and 30 feet with stakes for two sides of her garden. How far apart should the two stakes be if the garden will be rectangular shaped?

A 18 ft

B 30 ft

C 35 ft

D 48 ft

Answer the questions below.

a. What are you being asked to solve? What information is given?

b. How can you make your drawing as clear as possible? What can you add to your drawing as you work?

c. What is the correct answer?

CHAPTER 11

Preparing for Assessment

Cumulative Review

Read each question. Then fill in the correct answer on the answer document provided by your teacher or on a sheet of paper.

1. A square pyramid has a base length of 20 meters and a slant height of 14 meters. What is the volume of the pyramid, in cubic meters?

2. The bases of the prism are right isosceles triangles.

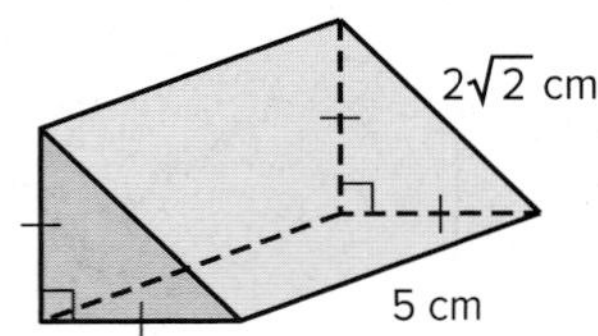

What is the volume of the prism in cubic centimeters?

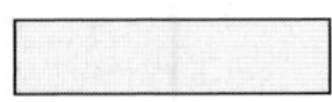

3. The figure shows the dimensions of a tent in the shape of a hemisphere.

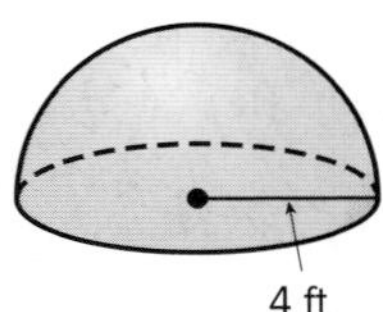

Which is the best estimate of the tent's volume?

- A 134 ft^3
- B 151 ft^3
- C 201 ft^3
- D 268 ft^3

4. The city of Chicago has a total land area of 227.13 square miles. As of 2010, the number of residents living within the city limits was 2,695,598. What is the population density of Chicago rounded to the nearest whole number?

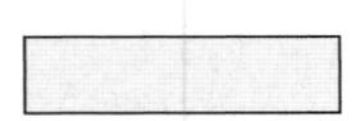

5. Kaden said that in spherical geometry, any two points determine a unique line. Which figure is a counterexample to Kaden's statement?

- A (P, Q)
- C

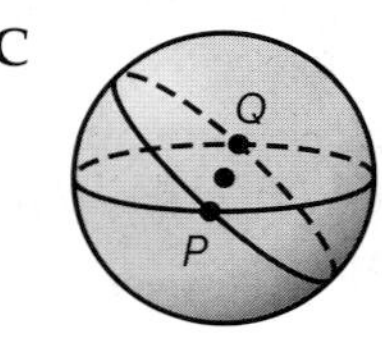

- B (P, Q)
- D

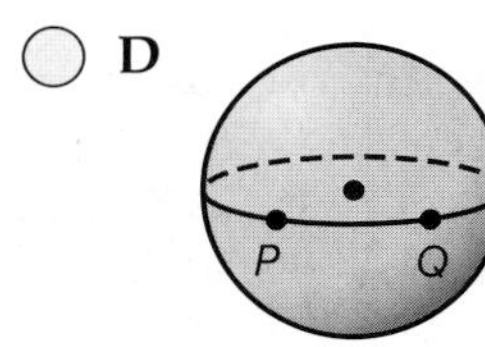

6. The lateral face of a right circular cone makes an angle of 33° with the base of the cone. The cone's radius is 9 inches. Which of the following is closest to the volume of the cone?

- A 1487 in^3
- B 1176 in^3
- C 496 in^3
- D 416 in^3

7. A café sells frozen yogurt in the shape of a square pyramid. LaTasha uses a knife to make one straight cut through her serving of frozen yogurt. Which of the following is not a possible shape of the resulting cross section? Select all that apply.

- A circle
- B square
- C trapezoid
- D triangle
- E rhombus

Test-Taking Tip

Question 6 Make a sketch of the cone and use a trigonometric ratio to find the height of the cone. Then use the volume formula.

Go Online! for Standardized Test Practice

8. What is the area of the curved surface of a hemisphere with diameter 10 inches? Write your answer in terms of pi.

9. Two similar cylinders have surface areas of 320 square feet and 500 square feet. The volume of the smaller cylinder is 640 cubic feet. What is the volume of the larger cylinder?

- A 800 ft^3
- C 1000 ft^3
- B 820 ft^3
- D 1250 ft^3

10. The figure shows the side panel of a skateboard ramp.

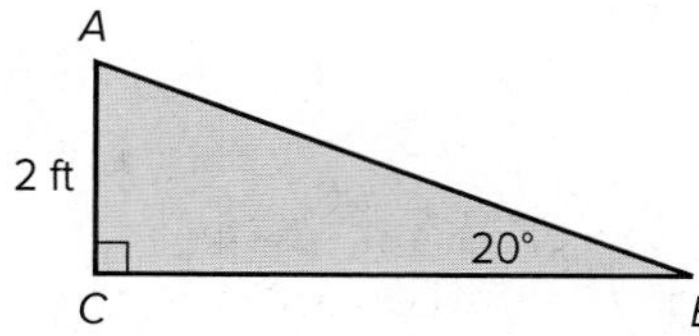

Which of the following is closest to the area of the panel?

- A 0.7 ft^2
- C 5.5 ft^2
- B 2.1 ft^2
- D 5.8 ft^2

11. The figure shows the dimensions of a cylindrical vase. Anthony wants to pour water into the vase so that the volume of the water is 100 cubic inches.

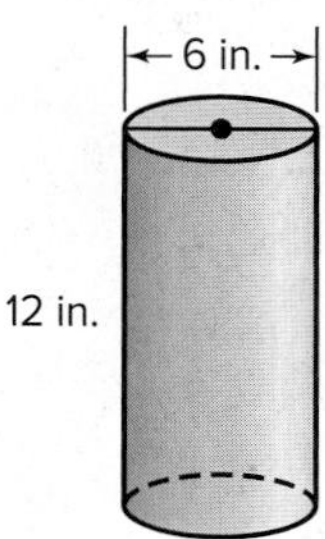

To the nearest tenth of an inch, what should be the height of the water in the vase?

- A 0.9 in.
- C 5.3 in.
- B 3.5 in.
- D 10.6 in.

12.

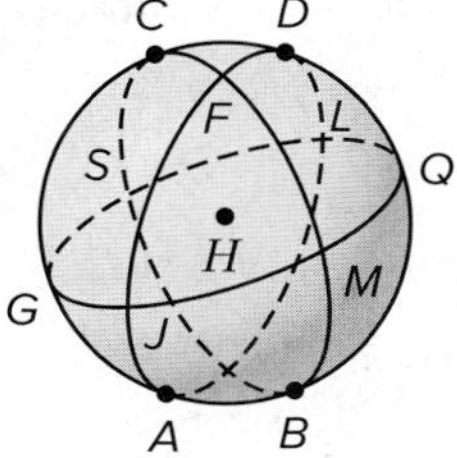

Identify two lines containing point Q, a segment containing point L, and a triangle in the sphere above.

13. Explain how a plane could intersect a square pyramid to produce the following cross sections. Write *not possible* if the cross section would not be possible.

a. hexagon

b. square

c. trapezoid

d. triangle

14. **MULTI-STEP** Marilda has a block of modeling clay that is 8 inches by 5 inches by 3 inches. She is experimenting with different shapes that can be made from the clay. Give your answers to the nearest tenth.

a. What is the volume of her clay?

b. What size cube could she make from the clay?

c. If she rolled the clay into a ball, what would its diameter be?

d. If she made two balls of the same size, what would their diameter be?

e. If she rolled a cylinder with a diameter of 3 inches, how long would it be?

f. If she rolled a cylinder with a diameter of 6 inches, how long would it be?

g. Suppose Marilda wants to make a square pyramid with a 6 × 6 in. base. How tall would it be?

h. If Marilda made a cone of the same height instead of a pyramid, would the diameter of the base be less than, equal to, or greater than 6 inches?

Need Extra Help?

If you missed Question...	1	2	3	4	5	6	7	8	9	10	11	12	13	14
Go to Lesson...	11-3	11-3	11-4	11-7	11-5	11-3	11-1	11-4	11-2	8-4	11-2	11-5	11-1	11-2, 11-3, 11-4

CHAPTER 12

Probability

THEN

You learned about experiments, outcomes, and events. You also found probabilities of simple events.

NOW

In this chapter, you will:

- Represent sample spaces.
- Use permutations and combinations with probability.
- Find probabilities by using length and area.

 WHY

Games There are many games in which you can increase your chances of winning by being able to draw logical conclusions about which outcomes are most likely.

Use the Mathematical Practices to complete the activity.

1. Sense-Making What are the possible outcomes of a spin or roll?

2. Reasoning If you know the likelihood of all possible outcomes, how can you use that information?

3. Applying Math If you roll a number cube, how often would you predict that you will roll an even number?

4. Use Tools Use the Number Cube tool to roll a number cube 10 times.

5. Discuss Did your prediction match your results? Why or why not?

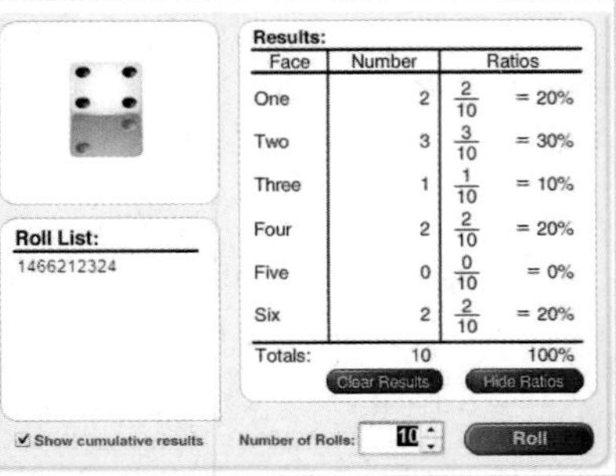

Andor Bujdoso/Alamy

Go Online to Guide Your Learning

Explore & Explain

Coin Toss

Use the **Coin Toss** tool to conduct experiments in which you explore probabilities and ratios.

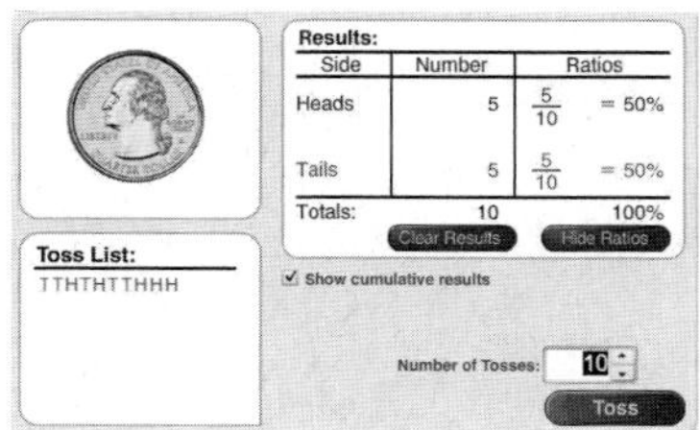

Spinner

Use the **Spinner** tool to explore probability, as discussed in Lesson 12-2.

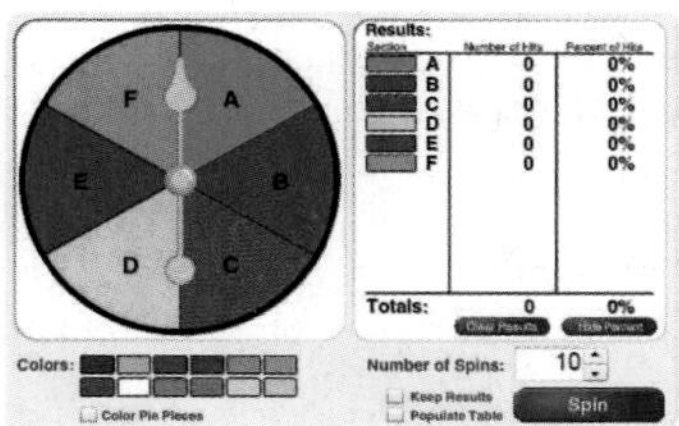

eBook

Interactive Student Guide

Before starting the chapter, answer the **Chapter Focus** preview questions. Check your answers as you complete each lesson. At the end of the chapter, try the **Performance Task**.

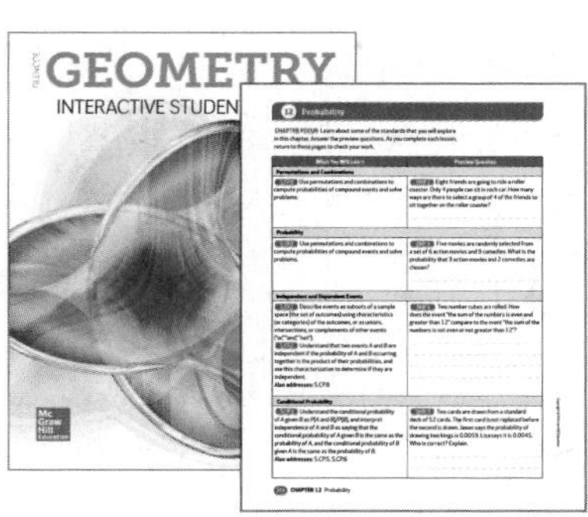

Organize

Foldables

Get organized! Create a **Probability and Measurement Foldable** before you start the chapter to organize your notes about probability.

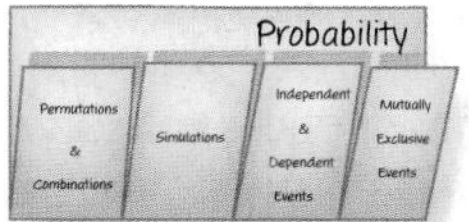

Collaborate

Chapter Project

In the **Fair Games** project, you will use what you have learned about probability and expected values to complete a project that addresses environmental literacy.

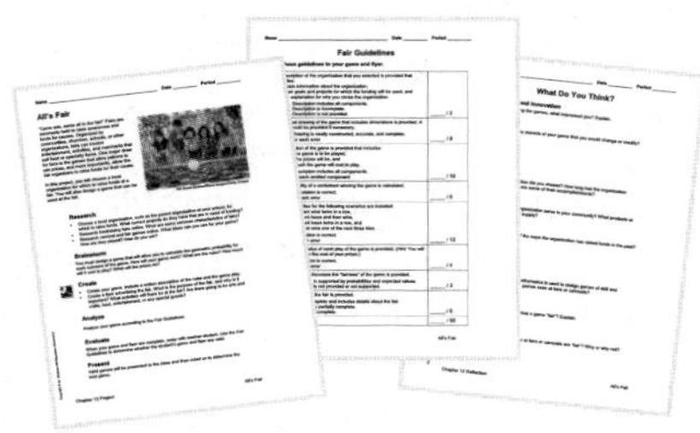

Focus

LEARNSMART®

Need help studying? Complete the **Applications of Probability** domain in LearnSmart to review for the chapter test.

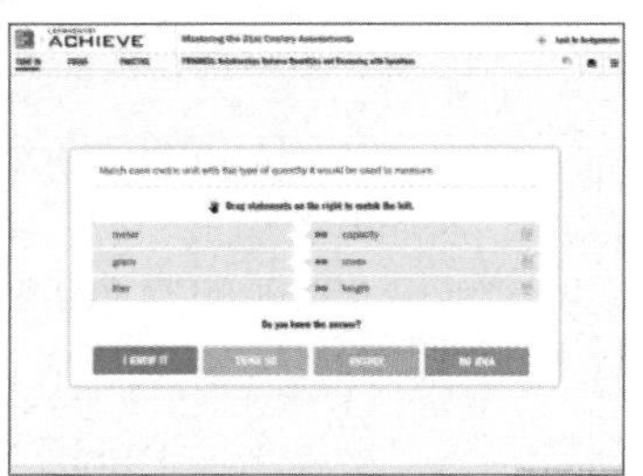

ALEKS®

You can use the **Polygons and Circles** topic in ALEKS to explore what you know about extending area and what you are ready to learn.*

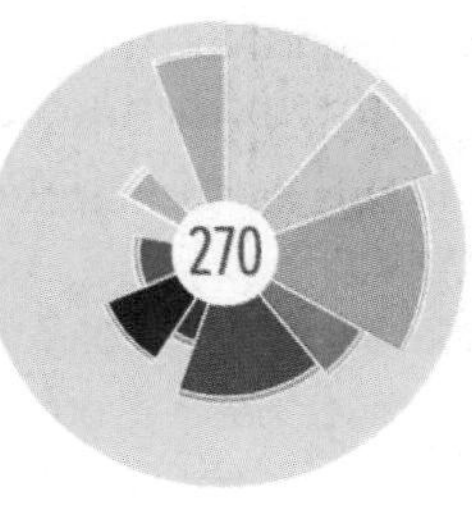

* Ask your teacher if this is part of your program.

Get Ready for the Chapter

Go Online! for Vocabulary Review Games and key vocabulary in 13 languages.

Connecting Concepts

Concept Check

Review the concepts used in this chapter by answering each question below.

1. Given $\frac{6}{9} \cdot \frac{1}{2}$, what would be the first step to simplify?
2. Suppose a number cube is rolled. What is the formula to determine the probability of rolling less than a five?
3. If a number cube is rolled, what is the probability that a value greater than 1 will be rolled?
4. If a number cube is rolled, what is the probability that an odd value will be rolled?
5. If a number cube is rolled, what is the probability that a 1 or a 6 will be rolled?
6. The table shows the results of an experiment in which a spinner numbered 1–4 was spun. What is the formula to determine the experimental probability that the spinner will land on 4?

<table>
<tr><th>Outcome</th><th>Tally</th><th>Frequency</th></tr>
<tr><td>1</td><td>|||</td><td>3</td></tr>
<tr><td>2</td><td>卌 ||</td><td>7</td></tr>
<tr><td>3</td><td>卌 |</td><td>6</td></tr>
<tr><td>4</td><td>||||</td><td>4</td></tr>
</table>

7. Based on the values in the table, what is the experimental probability that the spinner will land on a 4?
8. Based on the values in the table, what is the experimental probability that the spinner will land on an even number?

Performance Task Preview

You can use the concepts and skills in the chapter to plan schedules for a school. Understanding probabilities will help you finish the Performance Task at the end of the chapter.

In this Performance Task you will:

- make sense of problems
- construct viable arguments
- model with mathematics
- attend to precision
- look for and express regularity in repeated reasoning

New Vocabulary

English		Español
sample space	p. 859	espacio muestral
tree diagram	p. 859	diagrama de árbol
two-stage experiment	p. 860	experimento de dos pasos
multi-stage experiment	p. 860	experimentos multietápicos
Fundamental Counting Principle	p. 861	principio fundamental de contar
intersection	p. 866	intersección
complement	p. 867	complemento
permutation	p. 872	permutación
factorial	p. 872	factorial
circular permutation	p. 875	permutación circular
combination	p. 876	combinación
geometric probability	p. 881	probabilidad geométrica
compound event	p. 889	evento compuesto
independent events	p. 889	eventos independientes
dependent events	p. 889	eventos dependientes
mutually exclusive	p. 897	mutuamente exclusivos
conditional probability	p. 903	probabilidad condicional
two-way frequency table	p. 909	tabla de doble enrada o de frecuencias
marginal frequencies	p. 909	frecuencias marginales
joint frequencies	p. 909	frecuencias conjuntas
relative frequency	p. 910	frecuencia relativa

Review Vocabulary

event evento one or more outcomes of an experiment

experiment experimento a situation involving chance such as flipping a coin or rolling a number cube

LESSON 1

Representing Sample Spaces

Then	Now	Why?
● You calculated experimental probability.	● 1 Use lists, tables, and tree diagrams to represent sample spaces. 2 Use the Fundamental Counting Principle to count outcomes.	● Many sports games start with a coin toss to determine who will start with the possession of the ball. The coin can come up heads or tails.

New Vocabulary

sample space
tree diagram
two-stage experiment
multistage experiment
Fundamental Counting Principle

Mathematical Practices

1 Make sense of problems and persevere in solving them.

2 Reason abstractly and quantitatively.

1 Represent a Sample Space

You have learned the following about experiments, outcomes, and events.

Definition	Example
An *experiment* is a situation involving chance that leads to results called *outcomes*.	In the situation above, the experiment is tossing the coin.
An *outcome* is the result of a single performance or *trial* of an experiment.	The possible outcomes are landing on heads or tails.
An *event* is one or more outcomes of an experiment.	One event of this experiment is the coin landing on tails.

The **sample space** of an experiment is the set of all possible outcomes. You can represent a sample space by using an organized list, a table, or a **tree diagram**.

Example 1 Represent a Sample Space

A coin is tossed twice. Represent the sample space for this experiment by making an organized list, a table, and a tree diagram.

For each coin toss, there are two possible outcomes, heads H or tails T.

Organized List

Pair each possible outcome from the first toss with the possible outcomes from the second toss.

H, H   T, T
H, T   T, H

Table

List the outcomes of the first toss in the left column and those of the second toss in the top row.

Outcomes	Heads	Tails
Heads	H, H	H, T
Tails	T, H	T, T

Tree Diagram

	Outcomes			
First Toss	H		T	
Second Toss	H	T	H	T
Sample Space	H, H	H, T	T, H	T, T

Guided Practice

1. A coin is tossed and then a number cube is rolled. Represent the sample space for this experiment by making an organized list, a table, and a tree diagram.

The experiment in Example 1 is an example of a **two-stage experiment**, which is an experiment with two stages or events. Experiments with more than two stages are called **multistage experiments**.

> **Watch Out!**
> **MP Structure** The words *and/or* in the third question for Example 2 suggest an additional stage in the ordering process. By making separate stages for choosing with or without tomato and with or without pickles, you allow for the possibility of choosing *both* tomato and pickles.

Real-World Example 2 Multistage Tree Diagrams

HAMBURGERS To take a hamburger order, Keandra asks each customer the questions from the script shown. Draw a tree diagram to represent the sample space for hamburger orders.

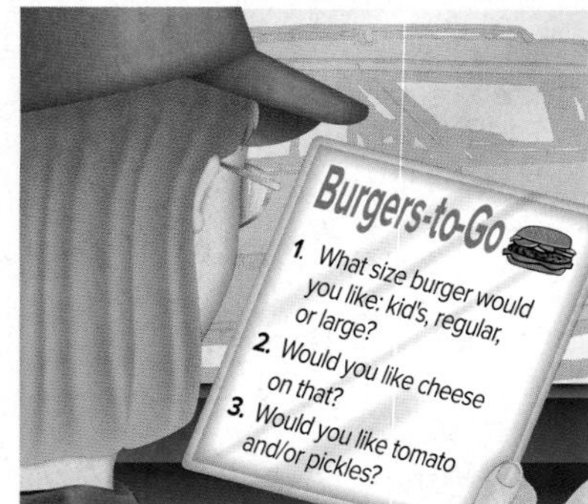

The sample space is the result of four stages.

- burger size (K, R, or L)
- cheese (C or NC)
- tomato (T or NT)
- pickles (P or NP)

Draw a tree diagram with four stages.

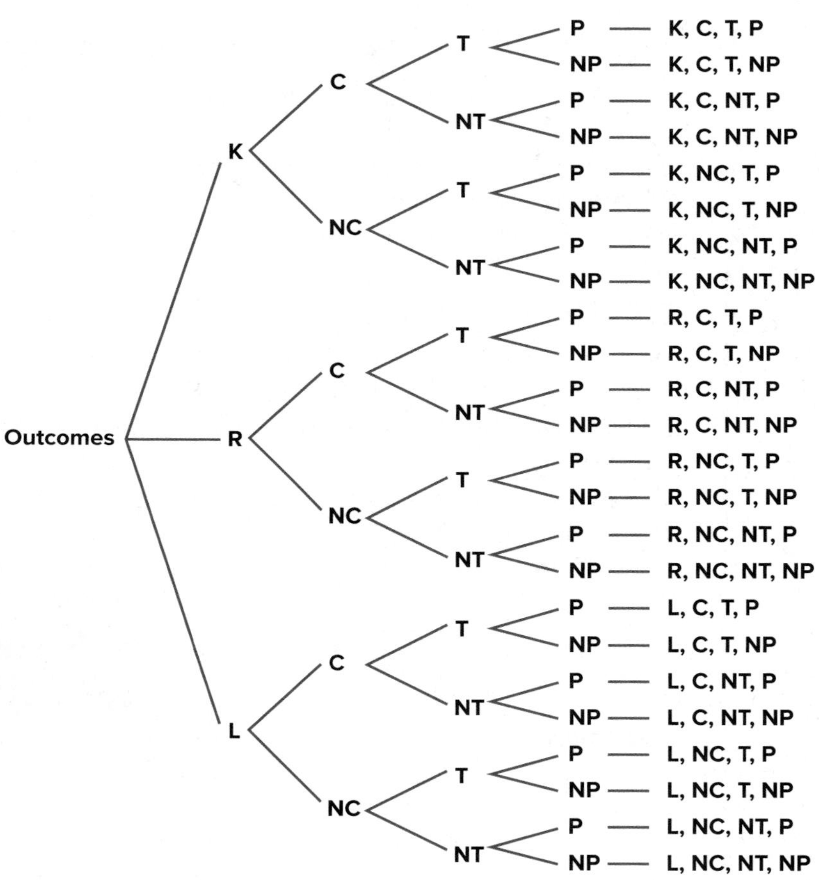

> **Reading Math**
> **Tree Diagram Notation** Choose notation for outcomes in your tree diagrams that will eliminate confusion. In Example 2, *C* stands for *cheese*, while *NC* stands for *no cheese*. Likewise, *NT* and *NP* stand for *no tomato* and *no pickles*, respectively.

Guided Practice

2. **MUSIC** Yoki can choose a portable hard drive with 4 or 8 terabytes in black, teal, sage, or red. She can also get a case and/or enhanced data transfer rates to go with it. Make a tree diagram to represent the sample space for this situation.

2 Fundamental Counting Principle

For some two-stage or multistage experiments, listing the entire sample space may not be practical or necessary. To find the *number* of possible outcomes, you can use the **Fundamental Counting Principle**.

Study Tip

Multiplication Rule The Fundamental Counting Principle is sometimes called the *Multiplication Rule for Counting* or the *Counting Principle.*

Key Concept Fundamental Counting Principle

Words The number of possible outcomes in a sample space can be found by multiplying the number of possible outcomes from each stage or event.

Symbols In a k-stage experiment, let

n_1 = the number of possible outcomes for the first stage.

n_2 = the number of possible outcomes for the second stage after the first stage has occurred.

⋮

n_k = the number of possible outcomes for the kth stage after the first $k - 1$ stages have occurred.

Then the total possible outcomes of this k-stage experiment is

$$n_1 \cdot n_2 \cdot n_3 \cdot \ldots \cdot n_k$$

Real-World Link

Most high school students order a traditional ring style, which includes the name of the school, a stone, and the graduation year.

Real-World Example 3 Use the Fundamental Counting Principle

CLASS RINGS Haley has selected a size and overall style for her class ring. Now she must choose from the ring options shown. How many different rings could Haley create in her chosen style and size?

Ring Options	Number of Choices
metals	10
finishes	2
stone colors	12
stone cuts	5
side 1 activity logos	20
side 2 activity logos	20
band styles	2

Use the Fundamental Counting Principle.

metals		finishes		stone colors		stone cuts		side 1 logos		side 2 logos		band styles		possible outcomes
10	×	2	×	12	×	5	×	20	×	20	×	2	=	960,000

So, Haley could create 960,000 different rings.

Guided Practice

3. Find the number of possible outcomes for each situation.

A. The answer sheet shown is completed.

B. A dot cube is rolled four times.

C. **SHOES** A pair of women's shoes comes in whole sizes 5 through 11 in red, navy, brown, or black. They can be leather or suede and are available in three different widths.

Answer Sheet

1. Ⓐ Ⓑ Ⓒ Ⓓ
2. Ⓐ Ⓑ Ⓒ Ⓓ
3. Ⓐ Ⓑ Ⓒ Ⓓ
4. Ⓐ Ⓑ Ⓒ Ⓓ
5. Ⓐ Ⓑ Ⓒ Ⓓ
6. Ⓐ Ⓑ Ⓒ Ⓓ
7. Ⓣ Ⓕ
8. Ⓣ Ⓕ
9. Ⓣ Ⓕ
10. Ⓣ Ⓕ

Check Your Understanding

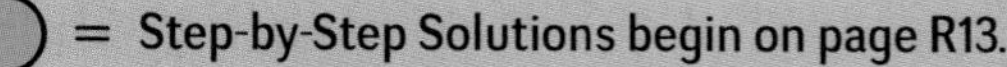

Example 1 **Represent the sample space for each experiment by making an organized list, a table, and a tree diagram.**

1. For each at bat, a player can either get on base or make an out. Suppose a player bats twice.

2. Quinton sold the most tickets in his school for the annual Autumn Festival. As a reward, he gets to choose twice from a grab bag with tickets that say "free juice" or "free notebook."

Example 2

3. **TUXEDOS** Patrick is renting a prom tuxedo from the catalog shown. Draw a tree diagram to represent the sample space for this situation.

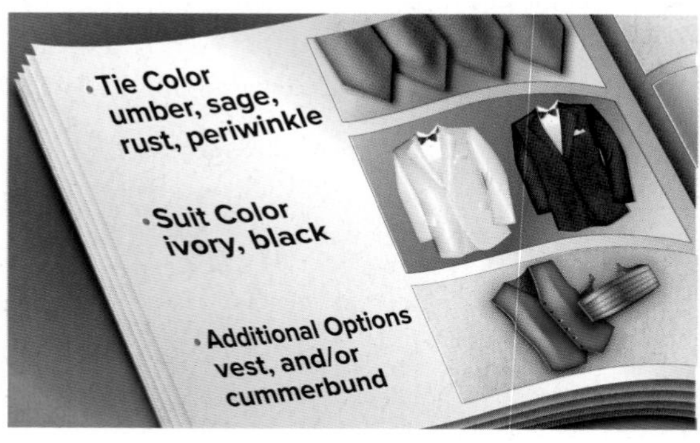

Example 3 **Find the number of possible outcomes for each situation.**

4. Desirée is preparing for homecoming and must decide what to wear. Assume one of each is chosen.

Options	Number of Choices
dress	15
shoes	5
purse	3
earrings	4
necklace	2

5. Marcos is creating a new menu for his restaurant. Assume one of each item is ordered.

Menu Titles	Number of Choices
appetizer	8
soup	4
salad	6
entree	12
dessert	9

Practice and Problem Solving

Extra Practice is on page R12.

Example 1 **STRUCTURE Represent the sample space for each experiment by making an organized list, a table, and a tree diagram.**

6. Gina is a junior and has a choice for the next two years of either playing volleyball or basketball during the winter quarter.

7. Two different history classes in New York City are taking a trip to either the Smithsonian or the Museum of Natural History.

8. Simeon has an opportunity to travel abroad as a foreign exchange student during each of his last two years of college. He can choose between Ecuador or Italy.

9. A new club is formed, and a meeting time must be chosen. The possible meeting times are Monday or Thursday at 5:00 or 6:00 P.M.

10. An exam with multiple versions has exercises with triangles. In the first exercise, there is an obtuse triangle or an acute triangle. In the second exercise, there is an isosceles triangle or a scalene triangle.

11. **PAINTING** In an art class, students are working on two projects where they can use one of two different types of paints for each project. Represent the sample space for this experiment by making an organized list, a table, and a tree diagram.

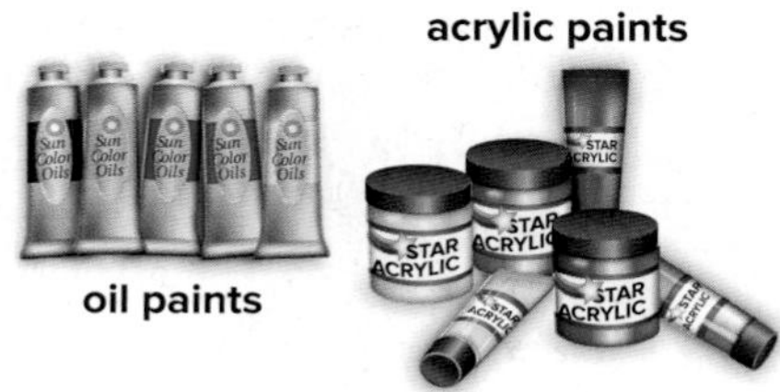

Example 2 **Draw a tree diagram to represent the sample space for each situation.**

12. **BURRITOS** At a burrito stand, customers have the choice of beans, pork, or chicken with rice or no rice, and cheese and/or salsa.

13. **TRANSPORTATION** Blake is buying a vehicle and has a choice of sedan, truck, or van with leather or fabric interior, and a GPS and/or sunroof.

14. **TREATS** Ping and her friends go to a frozen yogurt parlor which has a sign like the one at the right. Draw a tree diagram for all possible combinations of cones with peanuts and/or sprinkles.

Example 3 **MP MODELING In Exercises 15–18, find the number of possible outcomes for each situation.**

15. In the Junior Student Council elections, there are 3 people running for secretary, 4 people running for treasurer, 5 people running for vice president, and 2 people running for class president.

16. When signing up for classes during his first semester of college at Texas A&M, Frederico has 4 class spots to fill with a choice of 4 literature classes, 2 math classes, 6 history classes, and 3 film classes.

17. Niecy is choosing one each of 6 colleges, 5 majors, 2 minors, and 4 clubs.

18. Evita works at a restaurant where she has to wear a white blouse, black pants or skirt, and black shoes. She has 5 blouses, 4 pants, 3 skirts, and 6 pairs of black shoes.

19. **ART** For an art class assignment, Mr. Green gives students their choice of two quadrilaterals to use as a base. One must have sides of equal length, and the other must have at least one set of parallel sides. Represent the sample space by making an organized list, a table, and a tree diagram.

20. **BREAKFAST** A hotel restaurant serves omelets with a choice of vegetables, ham, or sausage that come with a side of hash browns, grits, or toast.

    a. How many different outcomes of omelet and one side are there if a vegetable omelet comes with just one vegetable?

    b. Find the number of possible outcomes for a vegetable omelet if you can get any or all vegetables on any omelet.

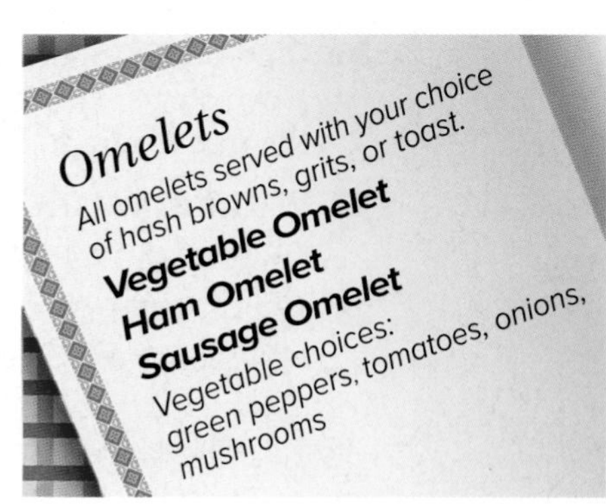

21. **COMPOSITE FIGURES** Carlito is calculating the area of the composite figure at the right. List six different ways he can do this.

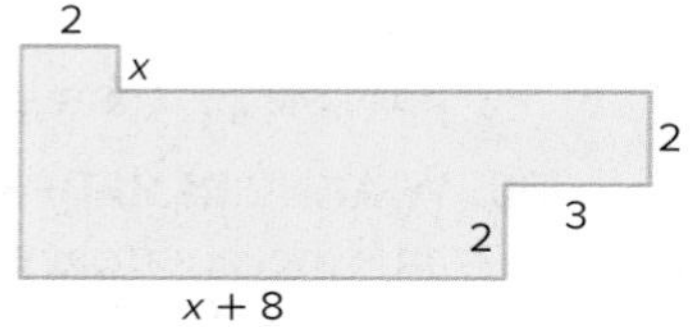

22. **TRANSPORTATION** Miranda got a new bicycle lock that has a four-number combination. Each number in the combination is from 0 to 9.
   a. How many combinations are possible if there are no restrictions on the number of times Miranda can use each number?
   b. How many combinations are possible if Miranda can use each number only once? Explain.

23. **GAMES** Cody and Monette are playing a board game in which you roll two dot cubes per turn.
   a. In one turn, how many outcomes result in a sum of 8?
   b. How many outcomes in one turn result in an odd sum?

24. **MULTIPLE REPRESENTATIONS** In this problem, you will investigate a sequence of events. In the first stage of a two-stage experiment, you spin Spinner 1 below. If the result is red, you flip a coin. If the result is yellow, you roll a dot cube. If the result is green, you roll a number cube. If the result is blue, you spin Spinner 2.

Spinner 1

Spinner 2

   a. **Geometric** Draw a tree diagram to represent the sample space for the experiment.
   b. **Logical** Draw a Venn diagram to represent the possible outcomes of the experiment.
   c. **Analytical** How many possible outcomes are there?
   d. **Verbal** Could you use the Fundamental Counting Principle to determine the number of outcomes? Explain.

H.O.T. Problems Use Higher-Order Thinking Skills

25. MP **CONSTRUCT ARGUMENTS** A box contains n different objects. If you remove three objects from the box, one at a time, without putting the previous object back, how many possible outcomes exist? Explain your reasoning.

26. **CHALLENGE** Sometimes a tree diagram for an experiment is not symmetrical. Describe a two-stage experiment where the tree diagram is asymmetrical. Include a sketch of the tree diagram. Explain.

27. **WRITING IN MATH** Explain why it is not possible to represent the sample space for a multistage experiment by using a table.

28. MP **CONSTRUCT ARGUMENTS** Determine if the following statement is *sometimes*, *always*, or *never* true. Explain your reasoning.

When an outcome falls outside the sample space, it is a failure.

29. **CHALLENGE** A multistage experiment has n possible outcomes at each stage. If the experiment is performed with k stages, write an equation for the total number of possible outcomes P. Explain.

30. **WRITING IN MATH** Explain when it is necessary to show all of the possible outcomes of an experiment by using a tree diagram and when using the Fundamental Counting Principle is sufficient.

31. Nathaniel performs an experiment that involves tossing a coin two times. First he plots the point $Q(3, 1)$. Then he tosses a coin. If the coin lands on heads, he translates point Q along $\langle 1, 1\rangle$. If the coin lands on tails, he translates point Q along $\langle -1, -1\rangle$. Then he tosses the coin again and repeats the process on the image of point Q. Nathaniel notes the final image of the point. Which of the following is not in the sample space for this experiment? MP 6

- ○ **A** (5, 3)
- ○ **B** (4, 2)
- ○ **C** (3, 1)
- ○ **D** (1, −1)

32. Brad's password must be five digits long, use the numbers 0–9, and the digits must not repeat. What is the maximum number of different passwords that Brad can have? MP 2

33. Becky performs an experiment that involves tossing a coin three times. First she plots the point $R(-4, 2)$. She tosses a coin and transforms point R according to the rules in the table. Then she tosses the coin again and transforms the image of point R. Then she tosses the coin a third time and transforms the most recent image. Becky notes the final image of the point. Which of the following is not a possible outcome for this experiment? MP 6

Result of Toss	Transformation
Heads	$(x, y) \rightarrow (-y, x)$
Tails	$(x, y) \rightarrow (x + 2, y)$

- ○ **A** (2, 2)
- ○ **B** (−2, 0)
- ○ **C** (4, 0)
- ○ **D** (−2, 2)
- ○ **E** (0, −2)

34. Amani plots the point $P(2, -1)$. Then she tosses a coin. If the coin lands on heads, she reflects point P in the x-axis. If the coin lands on tails, she reflects point P in the y-axis. Then she tosses the coin again and repeats the process on the image of point P. How many different final images are possible? MP 6

- ○ **A** 1
- ○ **B** 2
- ○ **C** 3
- ○ **D** 4

35. There are 3 trails leading to Camp A from your starting position. There are 3 trails from Camp A to Camp B. How many different routes are there from the starting position to Camp B? Draw a tree diagram to illustrate your answer. MP 1

36. Bag A contains 10 marbles of which 2 are red and 8 are black. Bag B contains 12 marbles of which 4 are red and 8 are black. A coin is tossed and if it comes up heads, a marble is drawn from bag A and its color is noted. If it comes up tails, a marble is drawn from bag B and its color is noted. Find the sample space for this experiment. MP 1

37. A bag contains four apples and six bananas. A fruit is taken from the bag and eaten. Then another fruit is taken and eaten. This process continues. MP 1

a. Find the sample space for the following experiments.

i. one fruit is eaten.

ii. two fruits are eaten.

iii. three fruits are eaten.

b. For $n \leq 4$, write an expression for the number of elements in the sample space when n fruits are eaten.

LESSON 2

Probability and Counting

Then

- You represented sample spaces using tables, lists, and tree diagrams.

Now

1. Describe events as subsets of sample spaces by using intersections and unions.
2. Find probabilities of complements.

Why?

- Most music players offer a shuffle feature. When this feature is selected, the probability that the next song you hear is a rap song or a song that is longer than 4 minutes is related to the union of two sets.

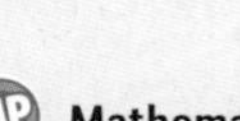

New Vocabulary

union
intersection
complement

Mathematical Practices

1 Make sense of problems and persevere in solving them.
2 Reason abstractly and quantitatively.
4 Model with mathematics.

1 Use Unions and Intersections

An event is a subset of a sample space, consisting of an outcome or a set of outcomes. You can use unions and intersections to describe events.

Key Concept Union and Intersection of Events

The **union** of two events is the set of all outcomes that are in either event. The union of events A and B is represented by $A \cup B$.

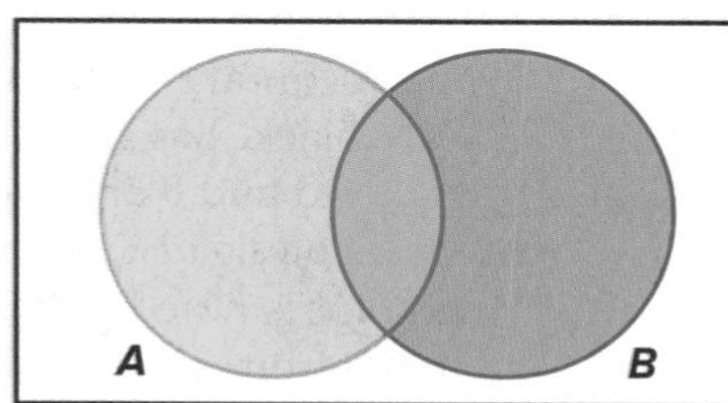

The **intersection** of two events is the set of all outcomes that are common to both events. The intersection of events A and B is represented by $A \cap B$.

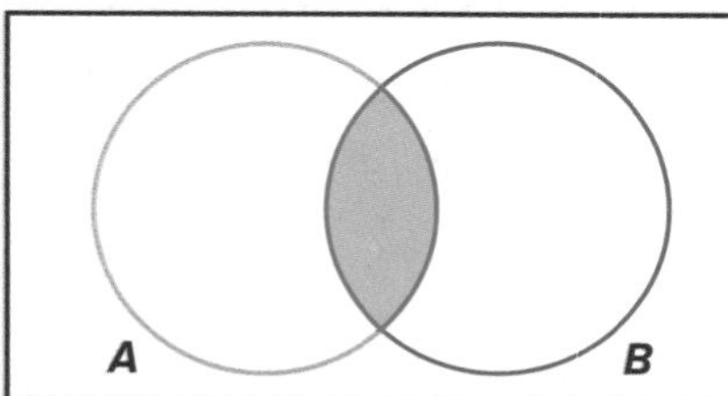

In an experiment with equally likely outcomes, the probability of an event E is the ratio of the number of outcomes in the event to the number of outcomes in the sample space S. You can write this as $P(E) = \frac{n(E)}{n(S)}$, where $P(E)$ is the probability of the event, $n(E)$ is the number of outcomes in the event, and $n(S)$ is the number of outcomes in the sample space.

Example 1 Find the Union of Events

A spinner is divided into 8 equal sections that are numbered 1 through 8 as shown. Let A be the event that the spinner lands on an even number. Let B be the event that it lands on a number between 5 and 8, inclusive.

a. Find $A \cup B$.

$A = \{2, 4, 6, 8\}$

$B = \{5, 6, 7, 8\}$

Make a Venn diagram. List the outcomes that are in Event A or in Event B.

$A \cup B = \{2, 4, 5, 6, 7, 8\}$

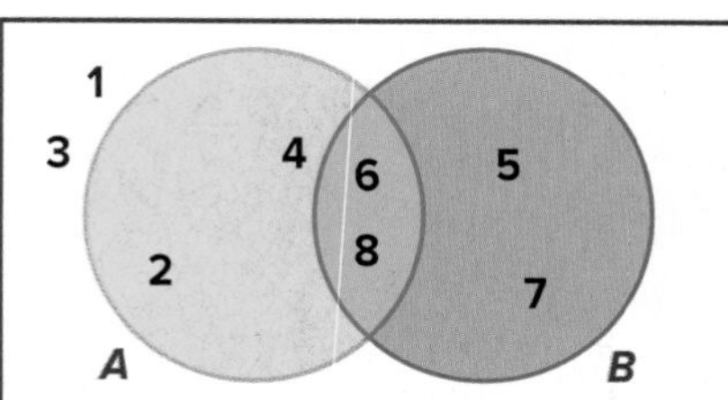

Noel Hendrickson/Blend Images/Getty Images

b. Find the probability that event A or event B will occur.

The probability that event A or event B will occur is $P(A \cup B)$.

$$P(A \cup B) = \frac{n(A \cup B)}{n(S)}$$

$$= \frac{6}{8} \quad \text{There are 6 outcomes in } A \cup B \text{ and 8 outcomes in the sample space.}$$

$$= \frac{3}{4}$$

Study Tip

Unions and Intersections The word *or* is a clue that a union of sets may be needed. The word *and* is a clue that an intersection of sets may be needed.

Guided Practice

Let A be the event that the spinner lands on a number less than 4. Let B be the event that it lands on an odd number.

1A. Find $A \cup B$.

1B. Find the probability that event A or event B will occur.

The probability of the intersection of two events is the probability that both events occur.

Example 2 Find the Intersection of Events

The Venn diagram shows members of two school sports teams who serve on the student council. One of the students will be chosen at random to attend a meeting. Let A be the event that a student plays basketball and let B be the event that a student plays tennis.

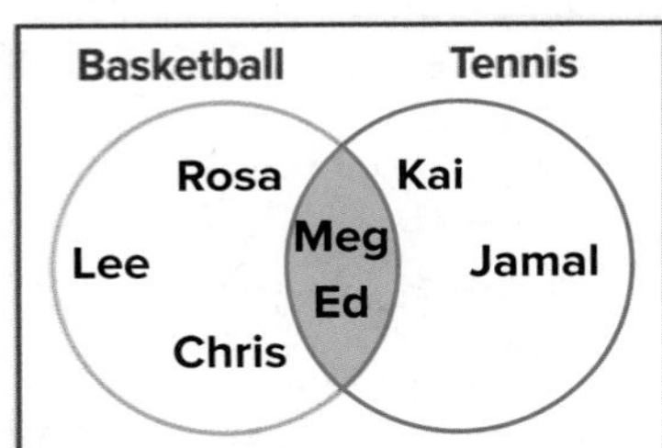

a. Find $A \cap B$.

$A \cap B$ consists of students who play both basketball and tennis.

$A \cap B = \{\text{Meg, Ed}\}$

b. What is the probability that the student who is chosen plays both basketball and tennis?

$$P(A \cap B) = \frac{n(A \cap B)}{n(S)}$$

$$= \frac{2}{7} \quad \text{There are 2 outcomes in } A \cap B \text{ and 7 outcomes in the sample space.}$$

Guided Practice

A standard number cube is rolled. Let A be the event that a number greater than 3 is rolled and let B be the event that an even number is rolled.

2A. Find $A \cap B$.

2B. What is the probability of rolling a number that is greater than 3 and even?

2 Use Complements

The **complement** of an event A consists of all the outcomes in the sample space that are not included as outcomes of event A.

When a standard number cube is rolled, the probability of rolling a 4 is $\frac{1}{6}$. What is the probability of *not* rolling a 4? There are five outcomes for this event: 1, 2, 3, 5, and 6. So, $P(\text{not } 4) = \frac{5}{6}$. Notice that this probability is also $1 - \frac{1}{6}$ or $1 - P(4)$.

Key Concept Probability of the Complement of an Event

Words The probability that an event will not occur is equal to 1 minus the probability that the event will occur.

Symbols For an event A, $P(\text{not } A) = 1 - P(A)$.

Reading Math

Complement Everyday use—something that fills up, completes, or makes perfect. Math meaning—all the outcomes in the sample space that are not included in the outcomes of an event. The complement of event A can be written as A^C.

Example 3 Use Complementary Events

RAFFLE Francesca bought 20 raffle tickets, hoping to win a $100 gift card to her favorite clothing store. If a total of 300 raffle tickets were sold, what is the probability that Francesca will not win the gift card?

Let event A represent selecting one of Francesca's tickets. Then find the probability of the complement of A.

$P(\text{not } A) = 1 - P(A)$ Probability of a complement

$= 1 - \frac{20}{300}$ Substitution

$= \frac{280}{300}$ or $\frac{14}{15}$ Subtract and simplify.

The probability that one of Francesca's tickets will not be selected is $\frac{14}{15}$ or about 93%.

Guided Practice

3. If the chance of rain is 70%, what is the probability that it will not rain?

Real-World Example 4 Find Probabilities of Events

MOVIES Outside a movie theater, 15 people were asked whether they had seen a comedy film or an action film in the past week. The Venn diagram shows the results of the survey. The values show the number of people who gave each response. One person who took the survey will be chosen at random to win a free movie ticket.

Comedy | Action
Comedy only: 5; Both: 2; Action only: 2; Neither: 6

Find the probability that the winner will be someone who saw a comedy film or an action film in the past week.

Add the values in the relevant sections of the Venn diagram to find the number of people who saw a comedy film or an action film: $5 + 2 + 2 = 9$.

$P(\text{comedy} \cup \text{action}) = \frac{n(\text{comedy} \cup \text{action})}{n(S)}$ Probability of a union

$= \frac{9}{15}$ Substitution

$= \frac{3}{5}$ or 60% Simplify.

Guided Practice

4. Find the probability that the winner will be someone who did not see an action film in the past week.

Check Your Understanding

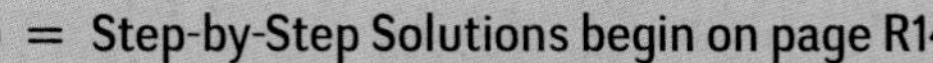
= Step-by-Step Solutions begin on page R14.

Examples 1–2 **Consider the experiment of rolling a standard number cube. Find the following probability of rolling each of the following.**

1. a number less than 3 or an even number
2. an even number or a number divisible by 4
3. a multiple of 3 and an even number
4. an odd number that is a prime
5. a 6 or a 4
6. a perfect square and a multiple of 2

Example 3

7. Consider the experiment of rolling a pair of number cubes. Let *A* be the event of rolling a sum of 8 or less.
   - **a.** What outcomes are in the complement of event *A*?
   - **b.** Find the probability of the complement of event *A*.
   - **c.** Find the probability of event *A*.

Example 4 **FITNESS A local gym requires new members to fill out a questionnaire about whether they want to lose weight (*L*) or improve fitness (*F*). The Venn diagram shows the results. Find the probability of the following events when a new gym member is chosen at random.**

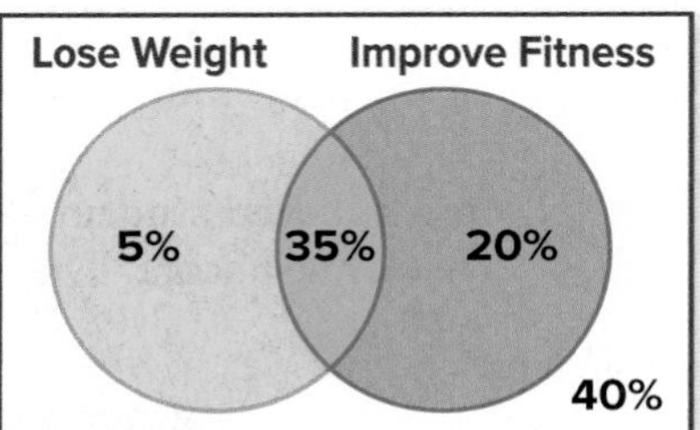

8. 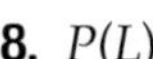 $P(L)$
9. $P(F)$
10. $P(L \text{ or } F)$
11. $P(L \text{ and } F)$
12. $P(\text{not } L)$
13. 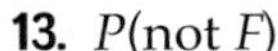 $P(\text{not } F)$

Practice and Problem Solving

Extra Practice is on page R12.

Examples 1–2 **Consider the experiment of picking one of the seven days of the week at random. Find the probability of picking each of the following.**

14. Friday or a day of the weekend
15. a day that starts with the letter *T* and a day that ends with the letter *Y*
16. a day that contains 8 or more letters and a day that contains the letter *D*

Example 3

17. **AFTER-SCHOOL PROGRAM** At Crossdale High School, 49 incoming freshmen were asked whether they prefer music or art. Students were then randomly selected to be placed in an after-school program. The two-way frequency table shows the results of the survey.

	Girl	Boy	Totals
Music	12	17	29
Art	9	11	20
Totals	21	28	49

   - **a.** What is the probability of selecting a girl who prefers music?
   - **b.** What is the probability of not selecting a girl who prefers music?

Example 4 **MARBLES A jar contains 105 marbles. There are 36 red marbles, 44 white marbles, 20 green marbles, and 5 orange marbles. You reach into the jar and choose one marble without looking.**

18. Find the probability of choosing an orange or red marble.
19. Find the probability of choosing a white or green marble.
20. Find the probability of not choosing a green marble.
21. Find the probability that the marble you choose is not red and not orange.
22. Find the probability that the marble you choose is not red or not orange.

Consider the experiment of selecting one card at random from a standard deck of 52 playing cards. Find the probability of selecting each of the following.

23. a card that is not a spade

24. a card that is not a 7

25. a card that is a diamond or a 3

26. a card that is neither a king nor a queen

27. **MP SENSE-MAKING** A sample of 370 magazines was analyzed to determine the types of electronics that were advertised in the magazines. The results are shown in the Venn diagram.

a. What does the 10 represent in the lower right-hand corner of the Venn diagram?

b. If you choose a magazine at random, what is the probability that it will contain an advertisement for a smartphone and a DVR?

Type of Electronics

Digital Media Player
DVR
80
50
30
40
20
30
110
10
Smartphone

One hundred students were asked how many hours they studied per week and their average test score. The table shows the results.

	Less than 70	70–90	Greater than 90	Totals
Studied 8 h or more per week	12	15	37	64
Studied less than 8 h per week	12	16	8	36
Totals	24	31	45	100

28. What is the probability that a student has an average test score greater than 90 and did not study 8 hours or more per week?

29. What is the probability that a student has an average test score between 70 and 90 and studied less than 8 hours per week?

30. What is the probability that a student does not have an average test score greater than 90?

H.O.T. Problems Use Higher-Order Thinking Skills

31. **OPEN-ENDED** Describe two events A and B such that $P(A \text{ and } B) = \frac{1}{2}$.

32. **MP REASONING** There are n outcomes in a sample space and m outcomes in event E, where $m \leq n$. Write an expression for $P(\text{not } E)$.

33. **ERROR ANALYSIS** The Venn diagram shows the number of outcomes in events A and B. Latricia and Amelie were given this information and were asked to find $P(\text{not } A)$. Is either of them correct? Explain your reasoning.

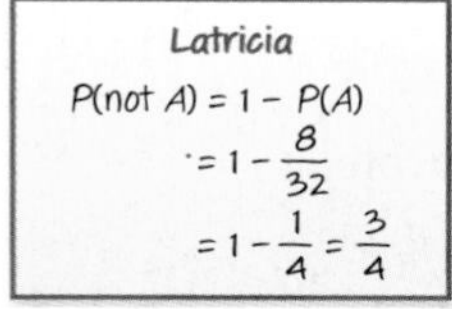

Amelie

$P(\text{not } A) = 1 - P(A)$

$= 1 - \frac{1}{13}$

$= \frac{12}{13}$

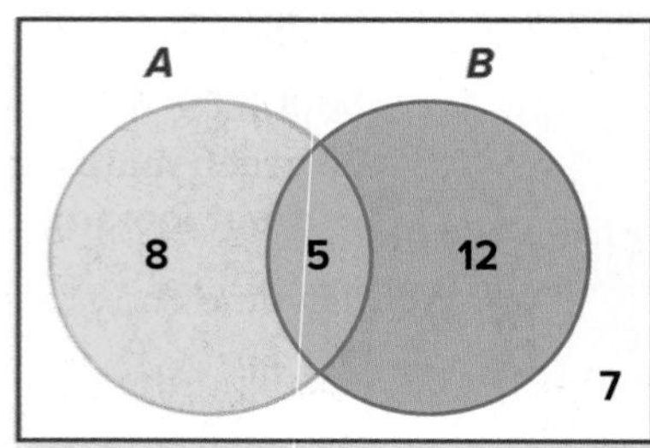

34. **WRITING IN MATH** Explain when you might find the probability of an event by first finding the probability of its complement.

35. A spinner has 10 equal sections that are labeled 1 through 10. Let A be the event that the spinner lands on a prime number and let B be the event that the spinner lands on an even number. What is $A \cup B$? MP 6

- ○ **A** {2}
- ○ **B** {2, 3, 5, 7}
- ○ **C** {2, 3, 4, 5, 6, 7, 8, 10}
- ○ **D** {2, 4, 6, 8}

36. A jar contains 25 ping pong balls that are numbered 1 through 25. Wei chooses one of the ping pong balls without looking. Let A be the event that she chooses an odd number and let B be the event that she chooses a multiple of 5. Which of the following outcomes are in the event $A \cap B$? MP 1

- ☐ **A** 5
- ☐ **B** 7
- ☐ **C** 10
- ☐ **D** 15
- ☐ **E** 19
- ☐ **F** 20

37. There are 32 students in Mr. Ibarra's class. Nine of the students are in the chess club only, 5 of the students are in the drama club only, and 6 of the students are in both clubs. Mr. Ibarra chooses one student at random. What is the probability that he does not choose a student in the chess club? MP 1, 4

- ○ **A** $\frac{9}{32}$
- ○ **B** $\frac{15}{32}$
- ○ **C** $\frac{17}{32}$
- ○ **D** $\frac{21}{32}$

38. There are 8 bottles of juice in a cooler. If 5 of them are apple juice, what is the probability of choosing a bottle at random and not choosing apple juice? Write your answer as a decimal. MP 6

P(not apple juice) = ☐

39. What is the sample space for choosing a prime number less than 14 at random? MP 6

a. {1, 2, 3, 5, 7, 11, 13, 14}

b. {2, 3, 5, 7, 11, 13, 14}

c. {2, 3, 5, 7, 11, 13}

d. {1, 3, 5, 7, 9, 11, 13}

40. Stacie surveyed some of her friends to find out whether they had been bowling in the past week (event A) or if they had been to the skate park in the past week (event B). The Venn diagram shows the results. What is $P(A \cap B)$? MP 1, 4

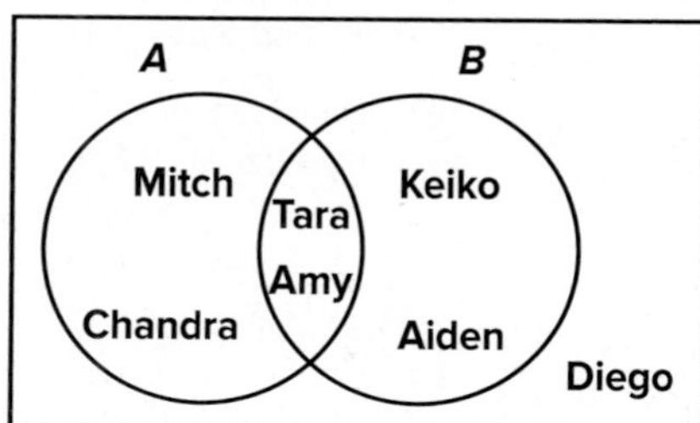

- ○ **A** $\frac{1}{7}$
- ○ **B** $\frac{2}{7}$
- ○ **C** $\frac{4}{7}$
- ○ **D** $\frac{6}{7}$

41. Omar has a bag of 26 tiles, each containing a different letter of the alphabet. He reaches into the bag and chooses one tile without looking. What is the best estimate of the probability that he chooses a vowel or the letter Y? MP 1

- ○ **A** 4%
- ○ **B** 19%
- ○ **C** 23%
- ○ **D** 81%

42. **MULTI-STEP** Julia has a set of 15 cards that are numbered 1 through 15. She shuffles the cards and chooses one card at random. Let A be the event that she chooses a multiple of 3 and let B be the event that she chooses an even number. MP 1

a. Make a Venn diagram to represent the sample space and the two events. List the outcomes in the appropriate sections of the diagram.

b. What is $A \cup B$?

c. Find $P(A \cup B)$.

LESSON 3

Probability with Permutations and Combinations

Then

- You used the Fundamental Counting Principle.

Now

1. Use permutations with probability.
2. Use combinations with probability.

Why?

- Lina, Troy, Davian, and Mary are being positioned for a photograph. There are 4 choices for who can stand on the far left, leaving 3 choices for who can stand in the second position. For the third position, just 2 choices remain, and for the last position just 1 is possible.

New Vocabulary
permutation
factorial
circular permutation
combination

Mathematical Practices

1 Make sense of problems and persevere in solving them.

4 Model with mathematics.

1 Probability Using Permutations

A **permutation** is an arrangement of objects in which order is important. One permutation of the four friends above is Troy, Davian, Mary, and then Lina. Using the Fundamental Counting Principle, there are $4 \cdot 3 \cdot 2 \cdot 1$ or 24 possible ordered arrangements of the friends.

The expression $4 \cdot 3 \cdot 2 \cdot 1$ used to calculate the number of permutations of these four friends can be written as 4!, which is read *4 factorial.*

Key Concept Factorial

Words The **factorial** of a positive integer n, written $n!$, is the product of the positive integers less than or equal to n.

Symbols $n! = n \cdot (n-1) \cdot (n-2) \cdot \ldots \cdot 2 \cdot 1$, where $0! = 1$

Example 1 Probability and Permutations of n Objects

SPORTS Chanise and Renee are members of the lacrosse team. If the 20 girls on the team are each assigned a jersey number from 1 to 20 at random, what is the probability that Chanise's jersey number will be 1 and Renee's will be 2?

Step 1 Find the number of possible outcomes in the sample space. This is the number of permutations of the 20 girls' names, or 20!.

Step 2 Find the number of favorable outcomes. This is the number of permutations of the other girls' names given that Chanise's jersey number is 1 and Renee's is 2: $(20-2)!$ or 18!.

Step 3 Calculate the probability.

$P(\text{Chanise 1, Renee 2}) = \frac{18!}{20!}$ ← number of favorable outcomes / ← number of possible outcomes

$= \frac{\overset{1}{\cancel{18!}}}{20 \cdot 19 \cdot \underset{1}{\cancel{18!}}}$ Expand 20! and divide out common factors.

$= \frac{1}{380}$ Simplify.

Guided Practice

1. **PHOTOGRAPHY** In the opening paragraph, what is the probability that Troy is chosen to stand on the far left and Davian on the far right for the photograph?

Digital Vision/Alamy

In the opening paragraph, suppose 6 friends were available, but the photographer wanted only 4 people in the picture. Using the Fundamental Counting Principle, the number of permutations of 4 friends taken from a group of 6 friends is $6 \cdot 5 \cdot 4 \cdot 3$ or 360.

Another way of describing this situation is the number of permutations of 6 friends taken 4 at a time, denoted ${}_6P_4$. This number can also be computed using factorials.

$${}_6P_4 = 6 \cdot 5 \cdot 4 \cdot 3 = \frac{6 \cdot 5 \cdot 4 \cdot 3 \cdot 2 \cdot 1}{2 \cdot 1} = \frac{6!}{2!} = \frac{6!}{(6-4)!}$$

This suggests the following formula.

Reading Math

MP Modeling The phrase *distinct objects* means that the objects are distinguishable as being different in some way.

Key Concept Permutations

Symbols The number of permutations of n distinct objects taken r at a time is denoted by ${}_nP_r$ and given by ${}_nP_r = \frac{n!}{(n-r)!}$.

Example The number of permutations of 5 objects taken 2 at a time is ${}_5P_2 = \frac{5!}{(5-2)!} = \frac{5 \cdot 4 \cdot \cancel{3!}}{\cancel{3!}}$ or 20.

Study Tip

Randomness When outcomes are decided at random, they are equally likely to occur and their probabilities can be calculated using permutations and combinations.

Example 2 Probability and ${}_nP_r$

A class is divided into teams each made up of 15 students. Each team is directed to select team members to be officers. If Sam, Valencia, and Deshane are on a team, and the positions are decided at random, what is the probability that they are selected as president, vice president, and secretary, respectively?

Step 1 Because choosing officers is a way of ranking team members, order in this situation is important. The number of possible outcomes in the sample space is the number of permutations of 15 people taken 3 at a time, ${}_{15}P_3$.

$${}_{15}P_3 = \frac{15!}{(15-3)!} = \frac{15 \cdot 14 \cdot 13 \cdot \cancel{12!}}{\cancel{12!}} \text{ or } 2730$$

Step 2 The number of favorable outcomes is the number of permutations of the 3 students in their specific positions. This is 1!, or 1.

Step 3 So the probability of Sam, Valencia, and Deshane being selected as the three officers is $\frac{1}{2730}$.

Guided Practice

2. A student identification card consists of 4 digits selected from 10 possible digits from 0 to 9. Digits cannot be repeated.

A. How many possible identification numbers are there?

B. Find the probability that a randomly generated card has the exact number 4213.

In a game, you must try to create a word using randomly selected letter tiles. Suppose you select the tiles shown. If you consider the letters O and O to be distinct, then there are 5! or 120 permutations of these letters.

Four of these possible arrangements are listed below.

POOLS POOLS SPOOL SPOOL

Notice that unless the Os are colored, several of these arrangements would look the same. Because there are 2 Os that can be arranged in 2! or 2 ways, the number of permutations of the letters O, P, O, L, and S can be written as $\frac{5!}{2!}$.

Key Concept Permutations with Repetition

The number of distinguishable permutations of n objects in which one object is repeated r_1 times, another is repeated r_2 times, and so on, is

$$\frac{n!}{r_1! \cdot r_2! \cdot \ldots \cdot r_k!}.$$

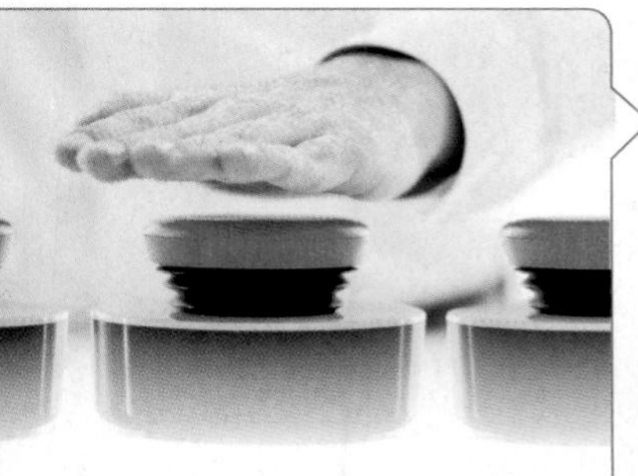

Real-World Link

Created in 1956, *The Price is Right* is the longest-running game show in the United States.

Source: IMDB

Example 3 Probability and Permutations with Repetition

GAME SHOW On a game show, you are given the following letters and asked to unscramble them to name a U.S. river. If you selected a permutation of these letters at random, what is the probability that they would spell the correct answer of MISSISSIPPI?

Step 1 There is a total of 11 letters. Of these letters, I occurs 4 times, S occurs 4 times, and P occurs 2 times. So, the number of distinguishable permutations of these letters is

$$\frac{11!}{4! \cdot 4! \cdot 2!} = \frac{39{,}916{,}800}{1152} \text{ or } 34{,}650.$$ Use a calculator.

Step 2 There is only 1 favorable arrangement—MISSISSIPPI.

Step 3 The probability that a permutation of these letters selected at random spells Mississippi is $\frac{1}{34{,}650}$.

Guided Practice

3. **TELEPHONE NUMBERS** What is the probability that a 7-digit telephone number with the digits 5, 1, 6, 5, 2, 1, and 5 is the number 556-5211?

So far, you have been studying objects that are arranged in *linear* order. Notice that when the spices below are arranged in a line, shifting each spice one position to the right produces a different permutation—curry is now first instead of salt. There are 5! distinct permutations of these spices.

TommL/iStock/Getty Images

Study Tip

MP Sense-Making If the circular object looks the same when it is turned over, such as a plain key ring, then the number of permutations must be divided by 2.

In a **circular permutation**, objects are arranged in a circle or loop. Consider the arrangements of these spices when placed on a turntable. Notice that rotating the turntable clockwise one position does *not* produce a different permutation—the order of the spices relative to each other remains unchanged.

Because 5 rotations of the turntable will produce the same permutation, the number of distinct permutations on the turntable is $\frac{1}{5}$ of the total number of arrangements when the spices are placed in a line.

$$\frac{1}{5} \cdot 5! = \frac{5 \cdot 4!}{5} \text{ or } 4!, \text{ which is } (5 - 1)!$$

Key Concept Circular Permutations

The number of distinguishable permutations of n objects arranged in a circle with no fixed reference point is

$$\frac{n!}{n} \text{ or } (n - 1)!.$$

If the n objects are arranged relative to a fixed reference point, then the arrangements are treated as linear, making the number of permutations $n!$.

Real-World Career

Statisticians Statisticians collect statistical data for various subject areas, including sports and games. They use computer software to analyze, interpret, and summarize the data. Most statisticians have a master's degree.

Stevecoleimages/E+/Getty Images

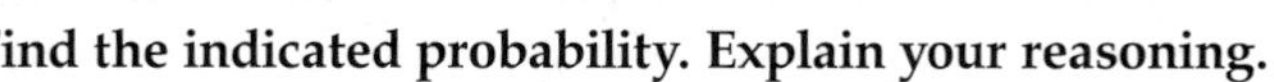

Example 4 Probability and Circular Permutations

Find the indicated probability. Explain your reasoning.

a. JEWELRY If the 6 charms on the bracelet shown are arranged at random, what is the probability that the arrangement shown is produced?

Because there is no fixed reference point, this is a circular permutation. So, there are $(6 - 1)!$ or $5!$ distinguishable permutations of the charms. Thus, the probability that the exact arrangement shown is produced is $\frac{1}{5!}$ or $\frac{1}{120}$.

b. DINING You are seating a party of 4 people at a round table. One of the chairs around this table is next to a window. If the diners are seated at random, what is the probability that the person paying the bill is seated next to the window?

Because the people are seated around a table with a fixed reference point, this is a linear permutation. So there are $4!$ or 24 ways in which the people can be seated around the table. The number of favorable outcomes is the number of permutations of the other 3 diners given that the person paying the bill sits next to the window, $3!$ or 6.

So, the probability that the person paying the bill is seated next to the window is $\frac{6}{24}$ or $\frac{1}{4}$.

Guided Practice

4. FOOTBALL A team's 11 football players huddle together before a play.

A. What is the probability that the fullback stands to the right of the quarterback if the team huddles together at random? Explain your reasoning.

B. If a referee stands directly behind the huddle, what is the probability that the referee stands directly behind the halfback? Explain your reasoning.

2 Probability Using Combinations

A **combination** is an arrangement of objects in which order is *not* important. Suppose you need to pack 3 of your 8 different pairs of socks for a trip. The order in which the socks are chosen does not matter, so the 3! or 6 groups of socks shown below would *not* be considered different. So, you would use combinations to determine the number of possible different sock choices.

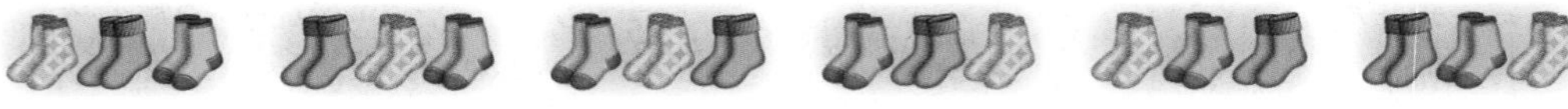

A combination of n objects taken r at a time, or ${}_nC_r$, is calculated by dividing the number of permutations ${}_nP_r$ by the number of arrangements containing the same elements, $r!$.

Study Tip

MP Reasoning Use permutations when the order of an arrangement of objects is important and combinations when order is not important.

Key Concept Combinations

Symbols The number of combinations of n distinct objects taken r at a time is denoted by ${}_nC_r$ and is given by ${}_nC_r = \frac{n!}{(n-r)!\,r!}$.

Example The number of combinations of 8 objects taken 3 at a time is

$${}_8C_3 = \frac{8!}{(8-3)!\,3!} = \frac{8!}{5!3!} = \frac{8 \cdot 7 \cdot \cancel{6} \cdot \cancel{5!}}{\cancel{5!} \cdot \cancel{6}} \text{ or } 56.$$

Go Online!

Many students find problems involving permutations and combinations challenging. Watch the Personal Tutor describe how to solve these problems with a partner. Then try describing how to solve a problem for your partner. Have your partner ask questions to help your understanding.

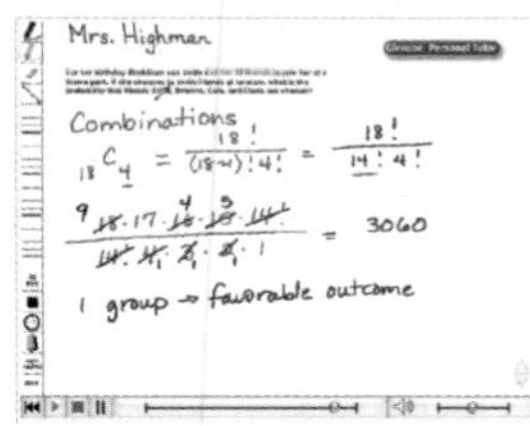

Example 5 Probability and ${}_nC_r$

INVITATIONS For her birthday, Monica can invite 6 of her 20 friends to join her at a theme park. If she chooses to invite friends at random, what is the probability that friends Tessa, Guido, Brendan, Faith, Charlotte, and Rhianna are chosen?

Step 1 Because the order in which the friends are chosen does not matter, the number of possible outcomes in the sample space is the number of combinations of 20 people taken 6 at a time, ${}_{20}C_6$.

$${}_{20}C_6 = \frac{20!}{(20-6)!\,6!} = \frac{\cancel{20} \cdot 19 \cdot \cancel{18} \cdot 17 \cdot \overset{8}{\cancel{16}} \cdot 15 \cdot \cancel{14!}}{\cancel{14!} \cdot \cancel{6} \cdot \cancel{5} \cdot \cancel{4} \cdot \cancel{3} \cdot \cancel{2}} \text{ or } 38{,}760$$

Step 2 There is only 1 favorable outcome—that the six students listed above are chosen. The order in which they are chosen is not important.

Step 3 So the probability of these six friends being chosen is $\frac{1}{38{,}760}$.

Guided Practice

5. GEOMETRY If three points are randomly chosen from those named on the rectangle shown, what is the probability that they all lie on the same line segment?

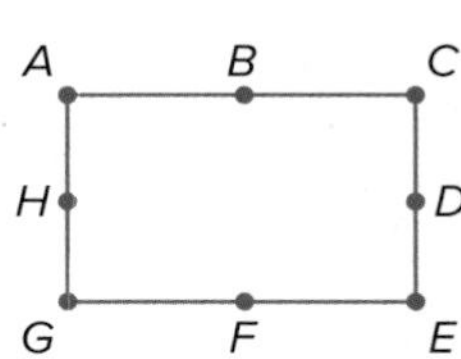

Check Your Understanding

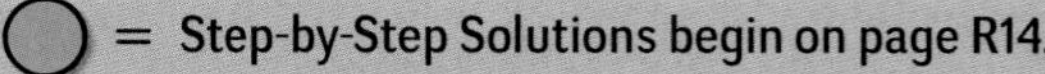

Example 1

1. **GEOMETRY** Five students are asked to randomly select and name a polygon from the group shown below. What is the probability that the first two students choose the triangle and quadrilateral, in that order?

   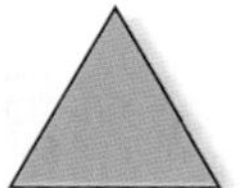 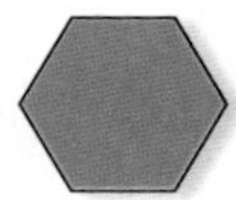

Example 2

2. **PLAYS** A high school performs a production of *A Raisin in the Sun* with each freshman English class of 18 students. If the three members of the crew are decided at random, what is the probability that Chase is selected for lighting, Jaden is selected for props, and Emelina for spotlighting?

Example 3

3. **DRIVING** What is the probability that a license plate using the letters C, F, and F and numbers 3, 3, 3, and 1 will be CFF3133?

Example 4

4. **CHEMISTRY** In chemistry lab, you need to test six samples that are randomly arranged on a circular tray.
   a. What is the probability that the arrangement shown at the right is produced?
   b. What is the probability that test tube 2 will be in the top middle position?

Example 5

5. Five hundred boys, including Josh and Sokka, entered a drawing for two football game tickets. What is the probability that the tickets were won by Josh and Sokka?

Practice and Problem Solving

Extra Practice is on page R12.

Example 1

6. **CONCERTS** Nia and Chad are going to a concert with their high school's key club. If they choose a seat on the row below at random, what is the probability that Chad will be in seat C11 and Nia will be in C12?

7. **FAIRS** Alfonso and Colin each bought one raffle ticket at the state fair. If 50 tickets were randomly sold, what is the probability that Alfonso got ticket 14 and Colin got ticket 23?

Example 2

8. **MODELING** The table shows the finalists for a floor exercises competition. The order in which they will perform will be chosen randomly.
   a. What is the probability that Cecilia, Annie, and Kimi are the first three gymnasts to perform, in any order?
   b. What is the probability that Cecilia is first, Annie is second, and Kimi is third?

Floor Exercises Finalists
Eliza Hernandez
Kimi Kanazawa
Cecilia Long
Annie Montgomery
Shenice Malone
Caroline Smith
Jessica Watson

9. **JOBS** A store randomly assigns their employees work identification numbers to track productivity. Each number consists of 5 digits ranging from 1–9. If the digits cannot repeat, find the probability that a randomly generated number is 25,938.

10. **GROUPS** Two people are chosen randomly from a group of ten. What is the probability that Jimmy was selected first and George second?

Example 3

11. **MAGNETS** Santiago bought some letter magnets that he can arrange to form words on his fridge. If he randomly selected a permutation of the letters shown below, what is the probability that they would form the word BASKETBALL?

12. **ZIP CODES** What is the probability that a zip code randomly generated from among the digits 3, 7, 3, 9, 5, 7, 2, and 3 is the number 39,372?

Example 4

13. **GROUPS** Keith is randomly arranging desks into circles for group activities. If there are 7 desks in his circle, what is the probability that Keith will be in the desk closest to the door?

14. **AMUSEMENT PARKS** Sylvie is at an amusement park with her friends. They go on a ride that has bucket seats in a circle. If there are 8 seats, what is the probability that Sylvie will be in the seat farthest from the entrance to the ride?

Example 5

15. **PHOTOGRAPHY** If you are randomly placing 24 photos in a photo album and you can place four photos on the first page, what is the probability that you choose the photos at the right?

16. **ROAD TRIPS** Rita is going on a road trip across the United States. She needs to choose from 15 cities where she will stay for one night. If she randomly pulls 3 city brochures from a pile of 15, what is the probability that she chooses Austin, Cheyenne, and Savannah?

17. MP **STRUCTURE** Use the figure below. Assume that the balls are aligned at random.

a. What is the probability that in a row of 8 pool balls, the solid 2 and striped 11 would be first and second from the left?

b. What is the probability that if the 8 pool balls were mixed up at random, they would end up in the order shown?

c. What is the probability that in a row of seven balls, with three 8-balls, three 9-balls, and one 6-ball, the three 8-balls would be to the left of the 6-ball and the three 9-balls would be on the right?

d. If the eight original balls were randomly rearranged and formed a circle, what is the probability that the 6-ball is next to the 7-ball?

18. How many lines are determined by 10 randomly selected points, no 3 of which are collinear? Explain your calculation.

19. Suppose 7 points on a circle are chosen at random, as shown at the right.

a. Using the letters A through E, how many ways can the points on the circle be named?

b. If one point on the circle is fixed, how many arrangements are possible?

20. **RIDES** A carousel has 7 horses and one bench seat that will hold two people. One of the horses does not move up or down.

a. How many ways can the seats on the carousel be randomly filled by 9 people?

b. If the carousel is filled randomly, what is the probability that you and your friend will end up in the bench seat?

c. If 6 of the 9 people randomly filling the carousel are under the age of 8, what is the probability that a person under the age of 8 will end up on the one horse that does not move up or down?

21. **LICENSES** A camera positioned above a traffic light photographs cars that fail to stop at a red light. In one unclear photograph, an officer could see that the first letter of the license plate was a Q, the second letter was an M or an N, and the third letter was a B, P, or D. The first number was a 0, but the last two numbers were blurry. How many possible license plates fit this description?

22. **MULTI-STEP** Corie is interested in playing a state lottery game in which she needs to choose a certain three-digit number to win. The different game options are shown in the table. Which option should she choose? Explain your solution process.

Option	Requirements	Examples
Straight	Numbers must match in exact order.	431
3-Way Box	Two digits must be identical and order does not matter.	433, 343, 334
6-Way Box	Digits must be unique, and order does not matter.	431, 413, 134, 143, 314, 341
Straight-Box	Can win straight or boxed, and digits must be unique.	431, 413, 134, 143, 314, 341

H.O.T. Problems Use Higher-Order Thinking Skills

23. **MP MODELING** Fifteen boys and fifteen girls entered a drawing for four free movie tickets. What is the probability that all four tickets were won by girls?

24. **MP REASONING** A student claimed that permutations and combinations were related by $r! \cdot {}_nC_r = {}_nP_r$. Use algebra to show that this is true. Then explain why ${}_nC_r$ and ${}_nP_r$ differ by the factor $r!$.

25. **OPEN-ENDED** Describe a situation in which the probability is given by $\frac{1}{{}_7C_3}$.

26. **MP CONSTRUCT ARGUMENTS** Is the following statement *sometimes, always,* or *never* true? Explain.

$${}_nP_r = {}_nC_r$$

27. **PROOF** Prove that $nC_{n-r} = nC_r$.

28. **WRITING IN MATH** Compare and contrast permutations and combinations.

29. Steven, Yasmin, Tyler, Vanessa, and Zack have been nominated to be club officers (president and vice president). Assuming the officers are chosen at random from among the five nominees, what is the probability that Steven is chosen to be the president and Yasmin is chosen to be the vice president? MP 8

- A $\frac{2}{5}$
- B $\frac{1}{5}$
- C $\frac{1}{10}$
- D $\frac{1}{20}$

30. Manuel must choose two books from a list of seven to read for book reports in his literature class. How many different pairs of books could he choose? MP 6

- A 10
- B 21
- C 42
- D 5040

31. Every student at Shellie's school is assigned a 4-digit PIN to access the school's computers. The first digit of the code is 0 or 1. The remaining digits are chosen from the numbers 2 through 9, and no number may be used more than once in a code. What is the probability that Shellie is assigned the PIN 1234? MP 6

- A $\frac{1}{56}$
- B $\frac{1}{112}$
- C $\frac{1}{336}$
- D $\frac{1}{672}$
- E $\frac{1}{5040}$

32. The coach is going to select a team of 5 from among 10 players. Find the probability that John, a particular player, is on the team. MP 1

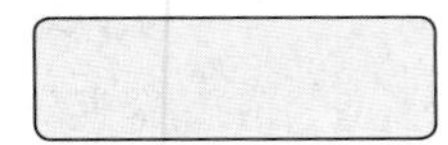

33. Marsala inserts three \$5-bills into 3 of 10 envelopes. The other 7 she leaves empty. She has each of her 10 soccer teammates choose one of the envelopes. How many different ways can the money be won by her teammates? MP 6

- A 120
- B 35
- C 720
- D 3,628,800

34. Midori has tiles with the letters *J, K, L, M, N, P,* and *Q*. She chooses three of the tiles at random. What is the probability that the tiles name three collinear points on the coordinate plane shown here? MP 6

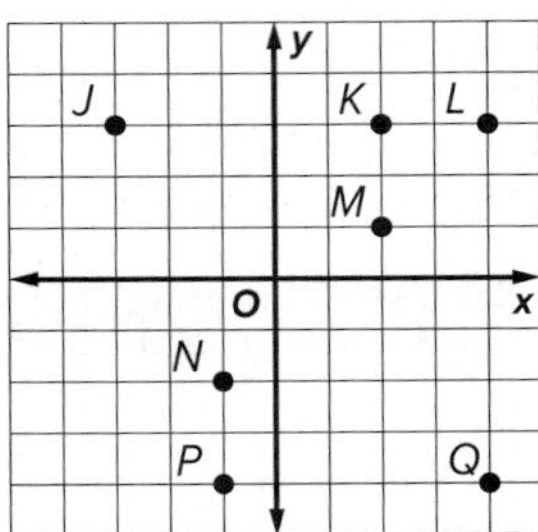

- A $\frac{1}{105}$
- B $\frac{1}{35}$
- C $\frac{2}{35}$
- D $\frac{3}{7}$

35. **MULTI-STEP** Assuming that any arrangement of letters forms a 'word', what is the probability that a 'word' formed from the letters of the word SQUARE: MP 1

a. Starts with E?

b. Starts with E and ends with A? 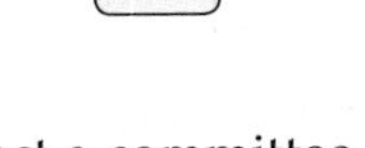

36. 15 student council members select a committee of 3 for a project, then one of the three to be the liaison for the project. MP 1

a. In how many ways is this possible?

b. What is the probability that Carmen is in the committee?

c. What is the probability that Carmen is the liaison for the project?

LESSON 4

Geometric Probability

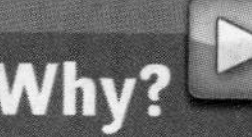

Then

- You found probabilities of simple events.

Now

1. Find probabilities by using length.
2. Find probabilities by using area.

Why?

- The object of the popular carnival game shown is to collect points by rolling a ball up an incline and into one of several circular target areas. The point value of each area is assigned based on the probability of a person landing a ball in that area.

New Vocabulary

geometric probability

Mathematical Practices

1 Make sense of problems and persevere in solving them.

2 Reason abstractly and quantitatively.

1 Probability with Length

The probability of winning the carnival game depends on the area of the target. Probability that involves a geometric measure such as length or area is called **geometric probability**.

Key Concept Length Probability Ratio

Words If a line segment (1) contains another segment (2) and a point on segment (1) is chosen at random, then the probability that the point is on segment (2) is

$$\frac{\text{length of segment (2)}}{\text{length of segment (1)}}.$$

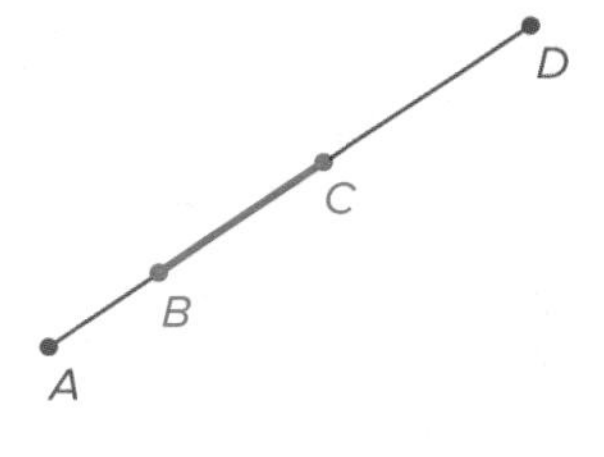

Example If a point E on $\overline{AD}$ is chosen at random, then $P(E \text{ is on } \overline{BC}) = \frac{BC}{AD}$.

Example 1 Use Lengths to Find Geometric Probability

Point X is chosen at random on $\overline{JM}$. Find the probability that X is on $\overline{KL}$.

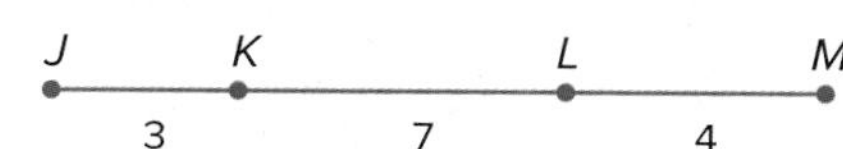

$P(X \text{ is on } \overline{KL}) = \frac{KL}{JM}$ Length probability ratio

$= \frac{7}{14}$ $KL = 7$ and $JM = 3 + 7 + 4$ or 14

$= \frac{1}{2}$, 0.5, or 50% Simplify.

Guided Practice

Point X is chosen at random on $\overline{JM}$. Find the probability of each event.

1A. $P(X \text{ is on } \overline{LM})$

1B. $P(X \text{ is on } \overline{KM})$

Geometric probability can be used in many real-world situations that involve an infinite number of outcomes.

Real-World Example 2 Model Real-World Probabilities

TRANSPORTATION Use the information at the left. Assuming that you arrive at Addison on the Red Line at a random time, what is the probability that you will have to wait 3 or more minutes for a train?

We can use a number line to model this situation. Because the trains arrive every 15 minutes, the next train will arrive in 15 minutes or less. On the number line below, the event of waiting 5 or more minutes is modeled by $\overline{BD}$.

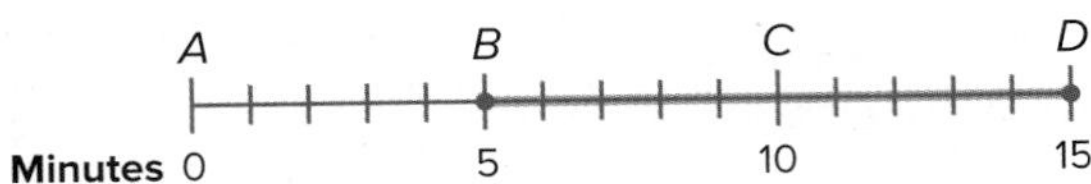

Find the probability of this event.

$$P(\text{waiting 3 or more minutes}) = \frac{BD}{AD} \quad \text{Length probability ratio}$$

$$= \frac{10}{15} \text{ or } \frac{2}{3} \quad BD = 10 \text{ and } AD = 15$$

So, the probability of waiting 5 or more minutes for the next train is $\frac{2}{3}$ or about 67%.

Real-World Link

A Chicago Transit Authority train arrives or departs a station like Addison on the Red Line every 15 minutes.

Source: Chicago Transit Authority

Guided Practice

2. **TEA** Iced tea at a cafeteria-style restaurant is made in 8-gallon containers. Once the level gets below 2 gallons, the flavor of the tea becomes weak.

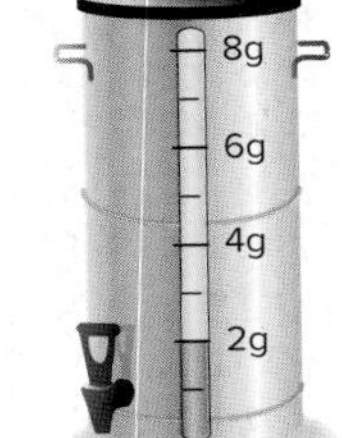

A. What is the probability that when someone tries to pour a glass of tea from the container, it is below 2 gallons?

B. What is the probability that the amount of tea in the container at any time is between 2 and 3 gallons?

2 Probability with Area

Geometric probability can also involve area. The ratio for calculating geometric probability involving area is shown below.

Key Concept Area Probability Ratio

Words If a region A contains a region B and a point E in region A is chosen at random, then the probability that point E is in region B is $\frac{\textbf{area of region } B}{\textbf{area of region } A}$.

Example If a point E is chosen at random in rectangle A, then $P(\text{point } E \text{ is in circle } B) = \frac{\textbf{area of region } B}{\textbf{area of region } A}$.

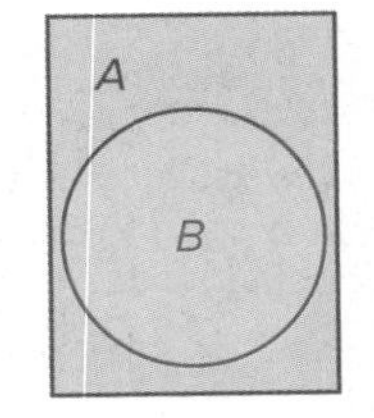

When determining geometric probabilities with targets, we assume

- that the object lands within the target area, and
- it is equally likely that the object will land anywhere in the region.

Go Online!

You can use a spreadsheet to solve probability problems. Learn how with the Spreadsheet Activity Worksheet in ConnectED.

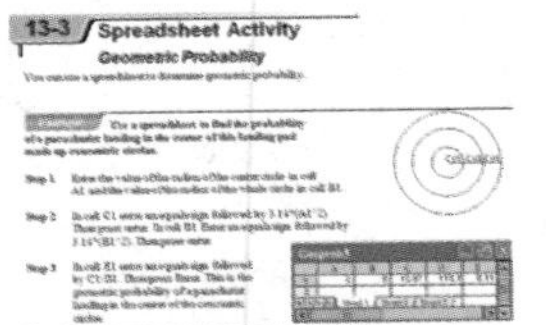

Real-World Link

Champion accuracy skydivers routinely land less than two inches away from the center of a target.

Source: *SkyDiving News*

Real-World Example 3 Use Area to Find Geometric Probability

SKYDIVING Suppose a skydiver must land on a target of three concentric circles. If the diameter of the center circle is 2 yards and the circles are spaced 1 yard apart, what is the probability that the skydiver will land in the red circle?

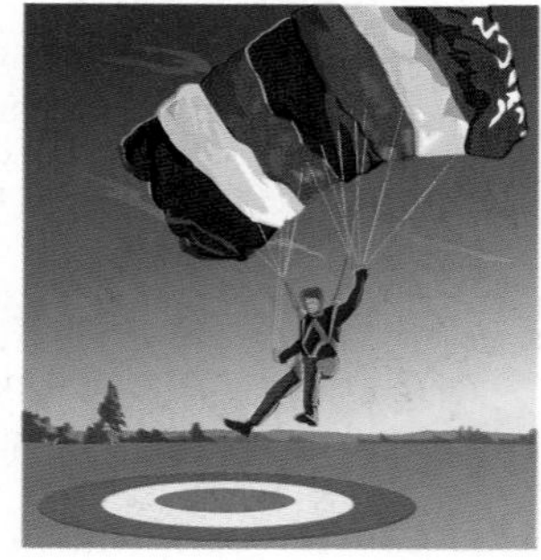

You need to find the ratio of the area of the red circle to the area of the entire target. The radius of the red circle is 1 yard, while the radius of the entire target is $1 + 1 + 1$ or 3 yards.

$P(\text{skydiver lands in red circle}) = \dfrac{\text{area of red circle}}{\text{area of target}}$ Area probability ratio

$= \dfrac{\pi(1)^2}{\pi(3)^2}$ $A = \pi r^2$

$= \dfrac{\pi}{9\pi}$ or $\dfrac{1}{9}$ Simplify.

The probability that the skydiver will land in the red circle is $\frac{1}{9}$ or about 11%.

Guided Practice

3. SKYDIVING Find each probability using the example above.

A. P(skydiver lands in the blue region)

B. P(skydiver lands in white region)

You can also use an angle measure to find geometric probability. The ratio of the area of a sector of a circle to the area of the entire circle is the same as the ratio of the sector's central angle to 360. You will prove this in Exercise 27.

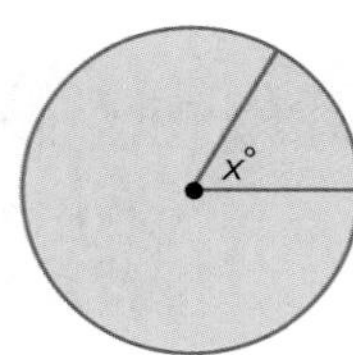

Example 4 Use Angle Measures to Find Geometric Probability

Use the spinner to find each probability.

a. *P*(pointer landing on yellow)

The angle measure of the yellow region is 45.

$P(\text{pointer landing on yellow}) = \dfrac{45}{360}$ or 12.5%

b. *P*(pointer landing on purple)

The angle measure of the purple region is 105.

$P(\text{pointer landing on purple}) = \dfrac{105}{360}$ or about 29%

c. *P*(pointer landing on neither red nor blue)

The combined angle measures of the red and blue region are $50 + 70$ or 120.

$P(\text{pointer landing on neither red nor blue}) = \dfrac{360 - 120}{360}$ or about 67%

Study Tip

MP Sense-Making In Example 4b, the area of the purple sector is a little less than $\frac{1}{3}$ or 33% of the spinner. Therefore, an answer of 29% is reasonable.

Guided Practice

4A. P(pointer landing on blue) **4B.** P(pointer not landing on green)

Check Your Understanding

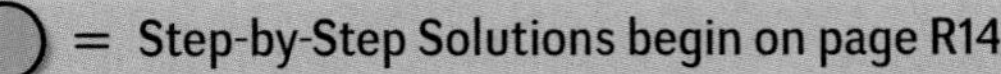
= Step-by-Step Solutions begin on page R14.

Example 1 **Point X is chosen at random on $\overline{AD}$. Find the probability of each event.**

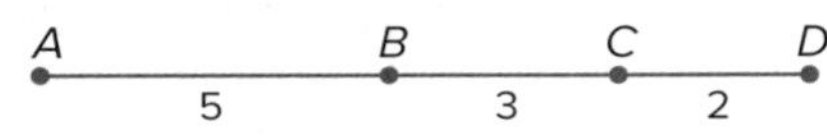

1. $P(X \text{ is on } \overline{BD})$
2. $P(X \text{ is on } \overline{BC})$

Example 2

3. **CARDS** In a game of cards, 43 cards are used, including one joker. Four players are each dealt 10 cards and the rest are put in a pile. If Greg doesn't have the joker, what is the probability that either his partner or the pile have the joker?

Examples 3–4

4. **ARCHERY** An archer aims at a target with 10 concentric circles whose diameters decrease by 12.2 centimeters as they get closer to the center. Find the probability that the archer will hit the center. Assume every point on the target is equally likely to be hit.

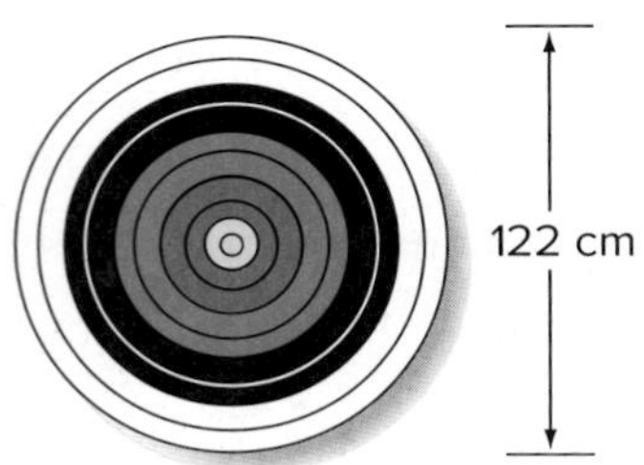

5. **NAVIGATION** A camper lost in the woods points his compass in a random direction. Find the probability that the camper is heading in the N to NE direction.

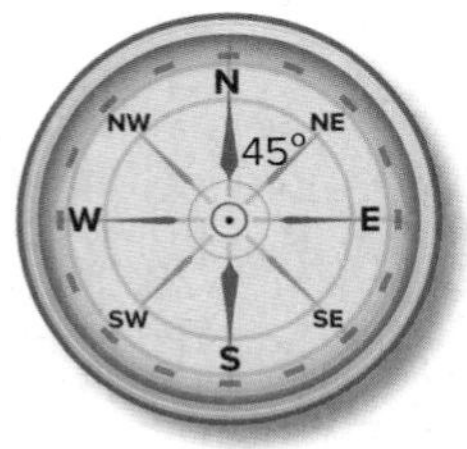

Practice and Problem Solving

Extra Practice is on page R12.

Example 1 **CHALLENGE Point X is chosen at random on $\overline{FK}$. Find the probability of each event.**

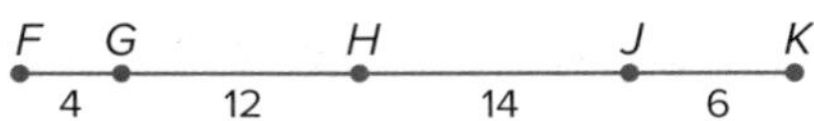

6. $P(X \text{ is on } \overline{FH})$
7. $P(X \text{ is on } \overline{GJ})$
8. $P(X \text{ is on } \overline{HK})$
9. $P(X \text{ is on } \overline{FG})$

10. **BIRDS** Four birds are sitting on a telephone wire. What is the probability that a fifth bird landing at a randomly selected point between birds 1 and 4 will sit at some point between birds 3 and 4?

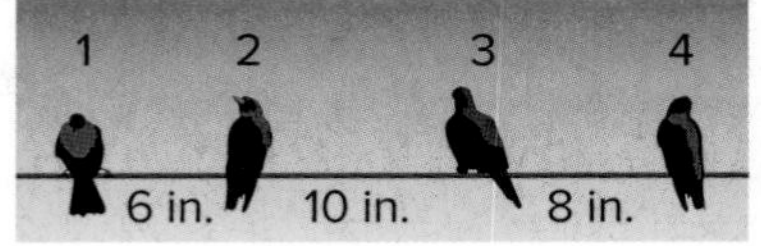

Example 2

11. **TELEVISION** Julio is watching television and sees an ad for a video game that he knows his friend wants for her birthday. If the ad replays at a random time in each 3-hour interval, what is the probability that he will see the ad again during his favorite 30-minute sitcom the next day?

Example 3 **Find the probability that a point chosen at random lies in the shaded region. Assume that figures that seem to be regular and congruent are regular and congruent.**

12. 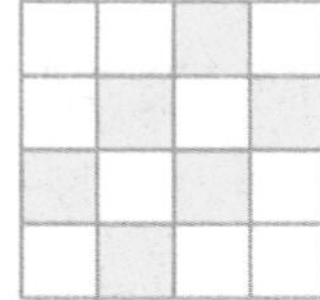

13. 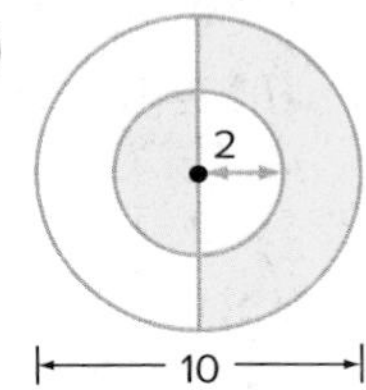

14. 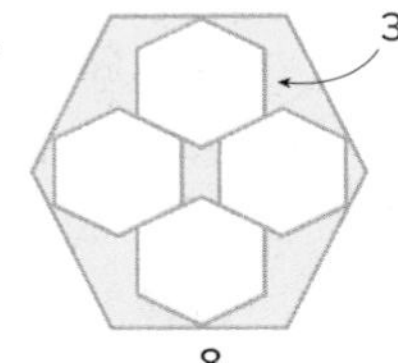

Example 4 S.MD.7

Use the spinner to find each probability. If the spinner lands on a line, it is spun again.

15. P(pointer landing on yellow)

16. P(pointer landing on blue)

17. P(pointer not landing on green)

18. P(pointer landing on red)

19. P(pointer landing on neither red nor yellow)

44° 30° 84° 110° 92°

Describe an event with a 33% probability for each model.

20. 

21. 10 20 30 40

22. 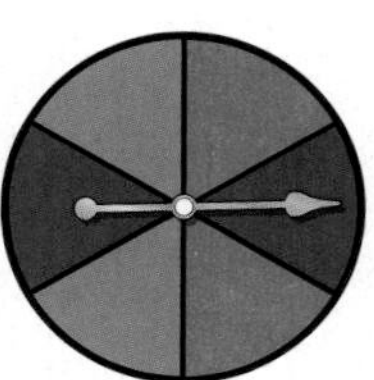

Find the probability that a point chosen at random lies in the shaded region.

23. 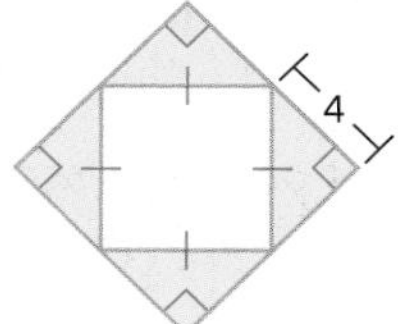

24. 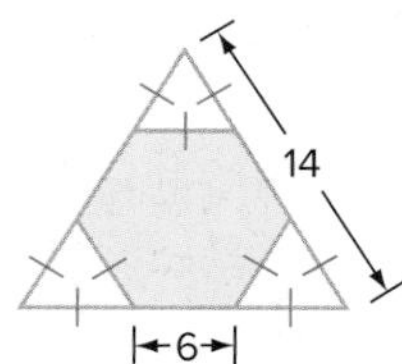

25. 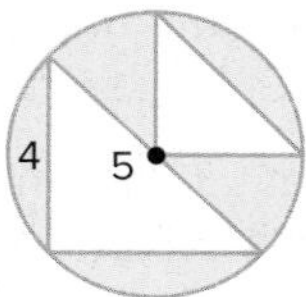

26. **FARMING** The layout for a farm is shown with each square representing a plot. Estimate the area of each field to answer each question.

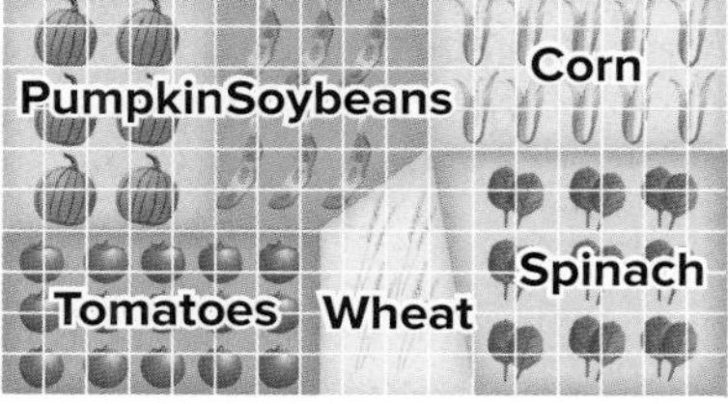

a. What is the approximate combined area of the spinach and corn fields?

b. Find the probability that a randomly chosen plot is used to grow soybeans.

27. **ALGEBRA** Prove that the probability that a randomly chosen point in the circle will lie in the shaded region is equal to $\frac{x}{360}$.

$x°$

28. **COORDINATE GEOMETRY** If a point is chosen at random in the coordinate grid shown at the right, find each probability. Round to the nearest hundredth.

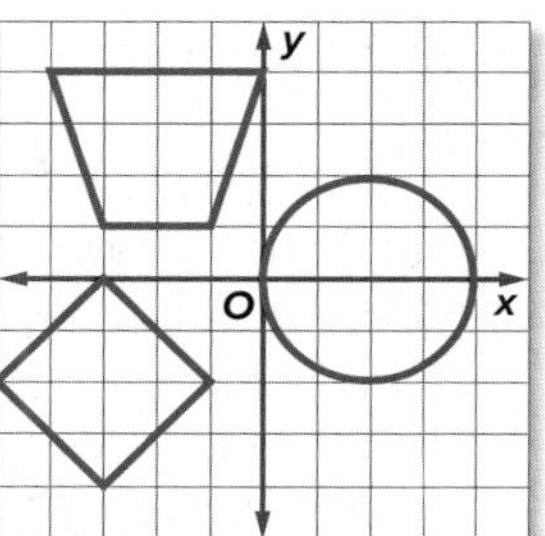

a. P(point inside the circle)

b. P(point inside the trapezoid)

c. P(point inside the trapezoid, square, or circle)

MP STRUCTURE Find the probability that a point chosen at random lies in a shaded region.

29. 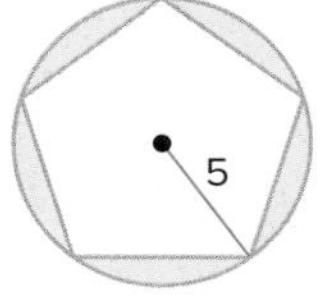

30. 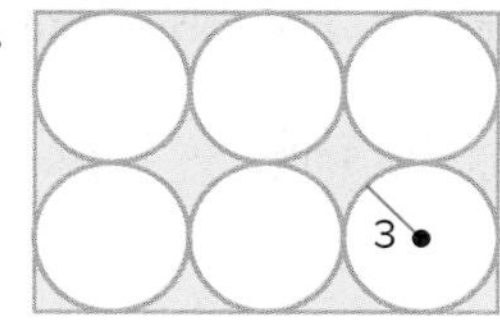

31. 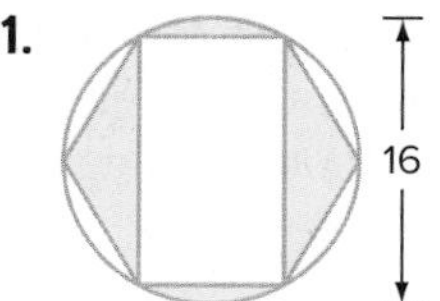

32. **COORDINATE GEOMETRY** Consider a system of inequalities, $1 \leq x \leq 6$, $y \leq x$, and $y \geq 1$. If a point (x, y) in the system is chosen at random, what is the probability that $(x-1)^2 + (y-1)^2 \geq 16$?

33. **VOLUME** The polar bear exhibit at a local zoo has a pool with the side profile shown. If the pool is 20 feet wide, what is the probability that a bear that is equally likely to swim anywhere in the pool will be in the incline region?

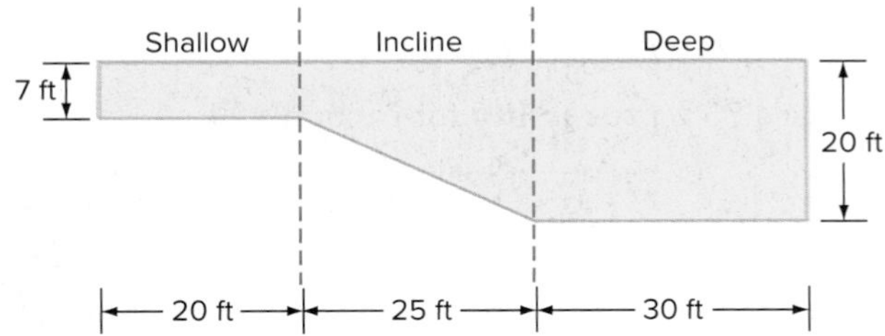

34. **DECISION MAKING** Meleah's flight was delayed and she is running late to make it to a national science competition. She is planning on renting a car at the airport and prefers car rental company A over car rental company B. The courtesy van for car rental company A arrives every 7 minutes, while the courtesy van for car rental company B arrives every 12 minutes.

 a. What is the probability that Meleah will have to wait 5 minutes or less to see each van? Explain your reasoning. (*Hint:* Use an area model.)

 b. What is the probability that Meleah will have to wait 5 minutes or less to see one of the vans? Explain your reasoning.

 c. Meleah can wait no more than 5 minutes without risking being late for the competition. If the van from company B should arrive first, should she wait for the van from company A or take the van from company B? Explain your reasoning.

H.O.T. Problems Use Higher-Order Thinking Skills

35. MP **MODELING** Find the probability that a point chosen at random would lie in the shaded area of the figure. Round to the nearest tenth of a percent.

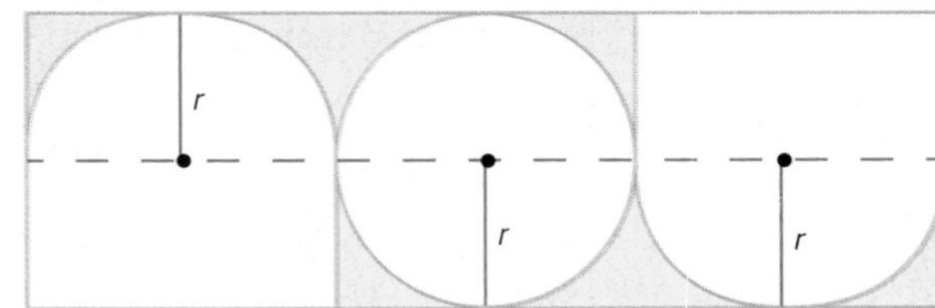

36. MP **CONSTRUCT ARGUMENTS** An isosceles triangle has a perimeter of 32 centimeters. If the lengths of the sides of the triangle are integers, what is the probability that the area of the triangle is exactly 48 square centimeters? Explain.

37. **WRITING IN MATH** Can athletic events be considered random events? Explain.

38. **OPEN-ENDED** Represent a probability of 20% using three different geometric figures.

39. **WRITING IN MATH** Explain why the probability of a randomly chosen point falling in the shaded region of either of the squares shown is the same.

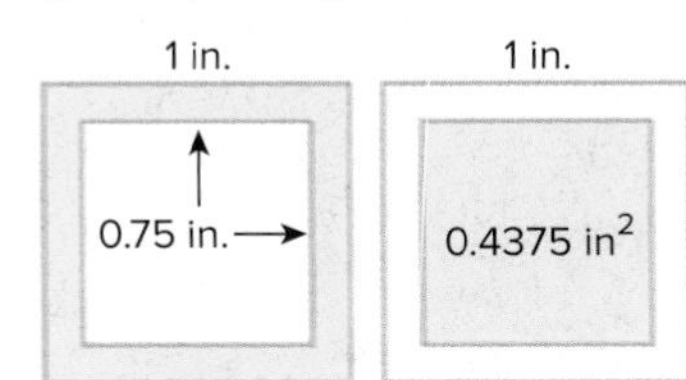

Preparing for Assessment

40. Visitors to a school fair get a chance to use a random number generator to choose coordinates for a point on a square board. They win a prize if the point lies on or in a triangle in the middle of the board, as shown. What is the probability that a visitor wins a prize? Express the probability as a decimal rounded to the nearest hundredth. MP 2

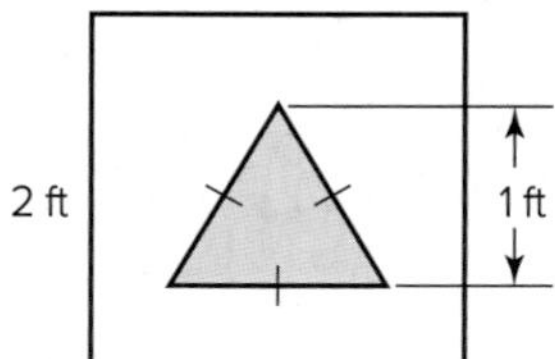

41. Concentric circles have radii of 4 centimeters and 8 centimeters. What is the probability that a grain of rice dropped onto the circles at random lands outside the circle with the 4-centimeter radius and inside the circle with the radius of 8 centimeters? MP 6

- A 4%
- B 25%
- C 50%
- D 75%

42. An archery target consists of a square inscribed in a circle with a radius of 15 inches, as shown. An arrow lands at a random point on the target.

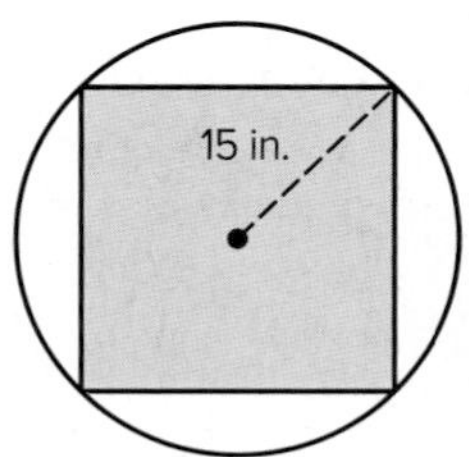

Which expression represents the probability that the arrow lands in the square? MP 2

- A $\frac{15}{\pi}$
- B $\frac{4}{\pi}$
- C $\frac{2}{\pi}$
- D $\frac{1}{\pi}$
- E $\frac{1}{15\pi}$

43. A box contains 7 blue marbles, 6 red marbles, 2 white marbles, and 3 black marbles. If one marble is chosen at random, what is the probability that it will be red? Round to the nearest hundredths. MP 1

44. Ashton is designing a dartboard. He wants to paint a circular target on a square piece of wood with the dimensions shown. He assumes darts will hit the piece of wood randomly. What radius should he use for the circle so that the probability of a dart landing in the circle is 60%? Express the answer in feet and round to the nearest hundredth. MP 2

45. A big square has a side length of 24 units, and has a small square within it with area 16 square units. Find the probability of a dart hitting the small square. MP 2

46. **MULTI-STEP** A point is chosen at random on $\overline{AD}$.

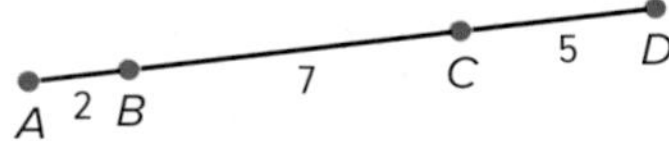

Find the probability: MP 1

a. that the point is on $\overline{BC}$

b. that the point is not on $\overline{AB}$

47. An archery target is made up of three concentric circles with radii 5, 10 and 20 cm respectively. MP 2

a. Find the probability that the arrow lands in the innermost circle.

b. Find the probability that the arrow lands in the outer ring.

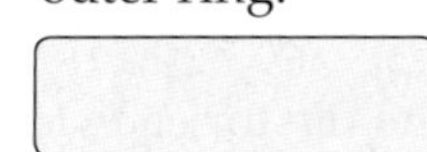

c. The radius of the outer circle needs to be changed to make the probability of the arrow landing in the outer ring $\frac{5}{9}$. What is the new radius of the outer circle?

CHAPTER 12

Mid-Chapter Quiz

Lessons 12-1 through 12-4

1. **LUNCH** A deli has a lunch special, which consists of a sandwich, soup, dessert, and a drink for $4.99. The choices are in the table below. (Lesson 12-1)

Sandwich	Soup	Dessert	Drink
chicken salad	tomato	cookie	tea
ham	chicken noodle	pie	coffee
tuna	vegetable		cola
roast beef			diet cola
			milk

   a. How many different lunches can be created from the items shown in the table?

   b. If a soup and two desserts were added, how many different lunches could be created?

2. **FLAGS** How many different signals can be made with 5 flags from 8 flags of different colors? (Lesson 12-1)

3. **CLOTHING** Marcy has six colors of shirts: red, blue, yellow, green, pink, and orange. She has each color in short-sleeved and long-sleeved styles. Represent the sample space for Marcy's shirt choices by making an organized list, a table, and a tree diagram. (Lesson 12-1)

4. In a class of 26 students, 19 have black hair and 17 have brown eyes. Four students have neither brown hair nor brown eyes. A student is chosen at random from the class. Find the probability that the student has: (Lesson 12-2)

   a. brown eyes

   b. has brown eyes and black hair

   c. has brown eyes, but not black hair

5. A dot cube is rolled and a coin is tossed. Find the probability that: (Lesson 12-2)

   a. the outcome on the dot cube is prime and the coin comes up heads

   b. the outcome on the dot cube is prime or the coin comes up heads

6. **SPELLING** A bag contains one tile for each letter of the word TRAINS. If you selected a permutation of these letters at random, what is the probability that they would spell TRAINS? (Lesson 12-3)

7. **COINS** Ten coins are tossed simultaneously. In how many of the outcomes will the third coin turn up a head? (Lesson 12-3)

8. A 320 meter long tightrope is suspended between two poles. Assume that the line has an equal chance of breaking anywhere along its length. (Lesson 12-4)

   a. Determine the probability that a break will occur in the first 50 meters of the tightrope.

   b. Determine the probability that the break will occur within 20 meters of a pole.

Point A is chosen at random on $\overline{BE}$. Find the probability of each event. (Lesson 12-4)

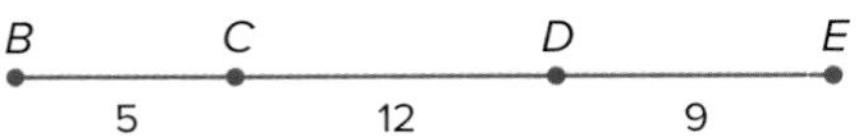

9. $P(A \text{ is on } \overline{CD})$

10. $P(A \text{ is on } \overline{BD})$

11. $P(A \text{ is on } \overline{CE})$

12. $P(A \text{ is on } \overline{DE})$

Use the spinner to find each probability. If the spinner lands on a line, it is spun again. (Lesson 12-4)

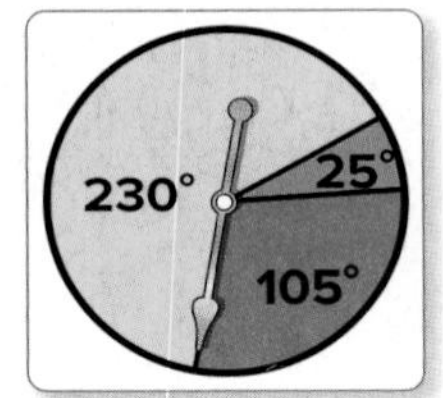

13. P(pointer landing on yellow)

14. P(pointer landing on blue)

15. P(pointer landing on red)

16. **GAMES** At a carnival, the object of a game is to throw a dart at the board and hit region III. (Lesson 12-4)

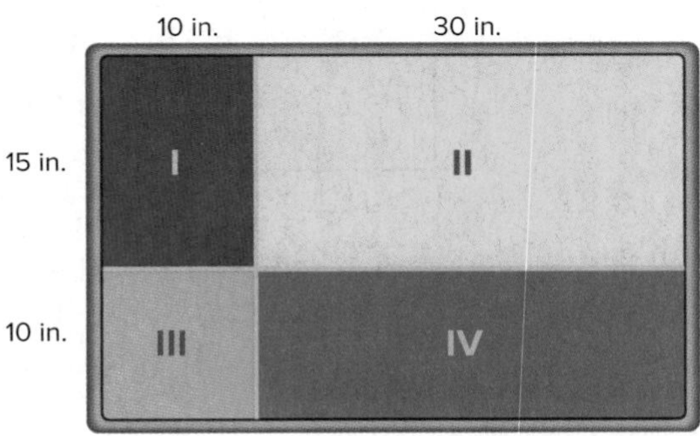

   a. What is the probability that it hits region I?

   b. What is the probability that it hits region II?

   c. What is the probability that it hits region III?

   d. What is the probability that it hits region IV?

LESSON 5

Probability and the Multiplication Rule

Then

- You found simple probabilities.

Now

1. Apply the multiplication rule to situations involving independent events.
2. Apply the multiplication rule to situations involving dependent events.

Why?

- The 18 students in Mrs. Turner's chemistry class are drawing names to determine who will give his or her presentation first. James is hoping to be chosen first and his friend Arturo wants to be second.

New Vocabulary

compound event
independent events
dependent events

Mathematical Practices

1 Make sense of problems and persevere in solving them.
4 Model with mathematics.

1 Probability of Independent Events

A **compound event** or *composite event* consists of two or more simple events. In the example above, James and Arturo being chosen to give their presentations is a compound event. It consists of the event that James is chosen and the event that Arturo is chosen.

Compound events can be independent or dependent.

- Events A and B are **independent events** if the occurrence of A does not affect the probability that B occurs.
- Events A and B are **dependent events** if the occurrence of A in some way changes the probability that B occurs.

Consider choosing objects one at a time from a group of objects. If you replace the object each time, choosing additional objects are independent events. If you do not replace the object each time, choosing additional objects are dependent events.

Example 1 Identify Independent and Dependent Events

Determine whether the events are *independent* or *dependent*. Explain your reasoning.

a. A coin lands heads up, and then a second coin lands tails up.

The outcome of the first coin toss in no way affects the outcome of the second coin toss. Therefore, these two events are *independent.*

b. In the class presentation example above, James is chosen first and then Arturo is chosen second.

After James is chosen, his name cannot be selected again. This affects the probability that Arturo is chosen, because the sample space is reduced by one name. Therefore, these two events are *dependent.*

c. Both Wednesday's lottery numbers and Saturday's lottery numbers are 1-2-3-4-5.

The outcome of Wednesday's drawing has no effect on Saturday's drawing. Therefore, these two events are *independent*.

Guided Practice

1A. A red card is selected from a standard deck of cards and put back. Then an ace is selected.

1B. Andrea selects a red shirt from her closet to wear on Monday and then a blue shirt to wear on Tuesday.

Steve Debenport/E+/Getty Images

Independent events can also be defined in terms of probability. If the probability of two events occurring together is equal to the product of the probabilities of the individual events, then the events are independent.

Example 2 Use Probability to Identify Independent Events

On each turn of a board game, a player rolls a dot cube and spins a spinner. On Dion's first turn, he rolls a 3 and the spinner lands on red. Are these events independent? Explain using probability.

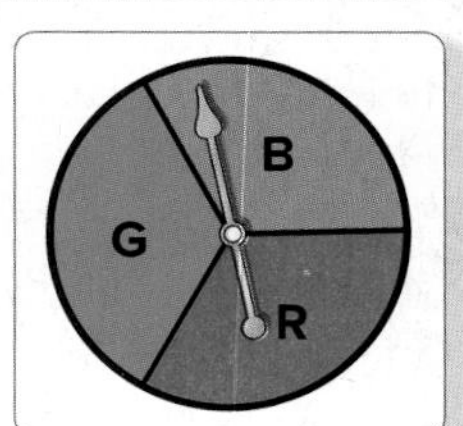

Make an organized list to show the sample space of all possible outcomes for the dot cube and the spinner. Then find the probability that Dion rolls a 3 and the spinner lands on red.

1G	2G	3G	4G	5G	6G
1B	2B	3B	4B	5B	6B
1R	2R	**3R**	4R	5R	6R

$P(3R) = \frac{1}{18}$ There are 18 equally likely outcomes and 1 successful outcome.

Next, find the individual probabilities.

$P(3) = \frac{1}{6}$ $P(R) = \frac{1}{3}$

The product of the individual probabilities is $\frac{1}{6} \times \frac{1}{3} = \frac{1}{8}$, which is equal to the probability of the two events occurring together. Therefore, the two events—rolling a 3 and the spinner landing on red—are independent.

Guided Practice

2. A bag contains a white marble, a blue marble, a yellow marble, and a green marble. Naomi chooses 2 marbles from the bag at the same time. One is blue, and one is yellow. Are these events independent? Explain using probability.

The previous example illustrates the first of two Multiplication Rules for Probability.

Key Concept Probability of Independent Events

Words If two events A and B are independent, then the probability of both events occurring is the product of the probability of A and the probability of B.

Symbols $P(A \text{ and } B) = P(A) \cdot P(B)$

Model

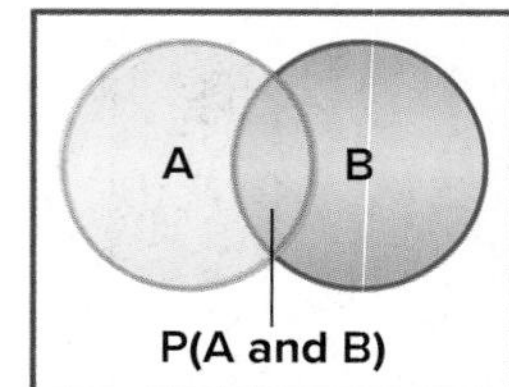

This rule can be extended to any number of events.

Reading Math

and The word *and* is a key word indicating to multiply probabilities.

Real-World Example 3 Probability of Independent Events

Study Tip

Use an Area Model You can also use the area model shown below to calculate the probability that both slips are blue. The blue region represents the probability of drawing two successive blue slips. The area of this region is $\frac{9}{64}$ of the entire model.

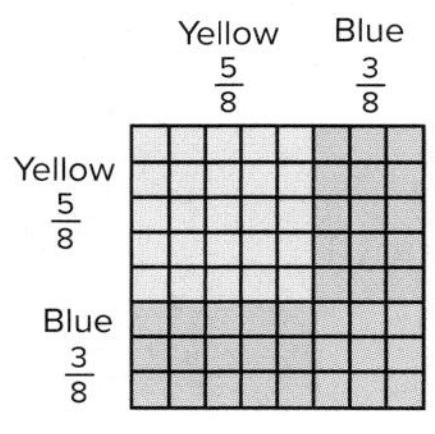

TRANSPORTATION Marisol and her friends are going to a concert. They put the slips of paper shown into a bag. If a person draws a yellow slip, he or she will ride in the van to the concert. A blue slip means he or she will ride in the car.

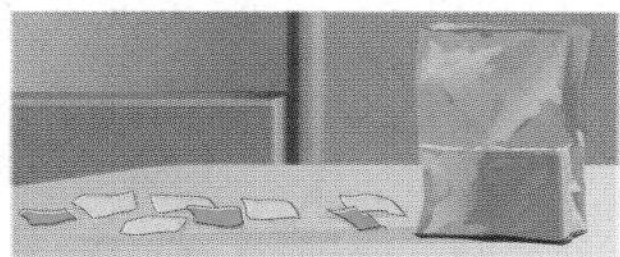

Suppose Marisol draws the first slip, puts it back and draws again. What is the probability that on each draw her slip is blue?

These events are independent since Marisol replaced the first slip. Let B represent a blue slip and Y a yellow slip.

Draw 1: $P(B) = \frac{3}{8}$ ← number of blue slips / ← total numbers of slips

Draw 2: $P(B) = \frac{3}{8}$ Because the first slip is replaced, $P(B)$ is the same for the second draw.

	Draw 1		Draw 2	
$P(B \text{ and } B) =$	$P(B)$	$\cdot$	$P(B)$	Probability of independent events
$=$	$\frac{3}{8}$	$\cdot$	$\frac{3}{8}$	$P(B) = \frac{3}{8}$
$=$	$\frac{9}{64}$			Simplify.

So, the probability that Marisol draws two blue slips is $\frac{9}{64}$ or about 14%.

Guided Practice

Find each probability.

3A. A coin is tossed and a number cube is rolled. What is the probability that the coin lands heads up and the number rolled is a 6?

3B. Suppose you toss a coin four times. What is the probability of getting four tails?

2 Probability of Dependent Events

The second of the Multiplication Rules of Probability addresses the probability of two dependent events.

Key Concept Probability of Dependent Events

Words	If two events A and B are dependent, then the probability of both events occurring is the product of the probability of A and the probability of B after A occurs.
Symbols	$P(A \text{ and } B) = P(A) \cdot P(B \text{ following } A)$

This rule can be extended to any number of events.

Example 4 Probability of Dependent Events

CARDS Cynthia randomly draws three cards from a standard deck one at a time without replacing them. Find the probability that the cards are drawn in the given order.

***P*(diamond, spade, diamond)**

The events are dependent because Cynthia does not replace the cards that she draws.

First card: $P(\text{diamond}) = \frac{13}{52}$ or $\frac{1}{4}$ ← number of diamonds / total number of cards

Second card: $P(\text{spade}) = \frac{13}{51}$ ← number of spades / number of cards remaining

Third card: $P(\text{diamond}) = \frac{12}{50}$ or $\frac{6}{25}$ ← number of diamonds remaining / number of cards remaining

$P(\text{diamond, spade, diamond}) = P(\text{diamond}) \cdot P(\text{spade}) \cdot P(\text{diamond})$

$= \frac{1}{4} \cdot \frac{13}{51} \cdot \frac{6}{25}$ or $\frac{13}{850}$ Substitution

The probability is $\frac{13}{850}$ or about 1.5%.

Guided Practice

Find each probability.

4A. *P*(two, five, not a five)

4B. *P*(heart, not a heart, heart)

Real-World Example 5 Use Probability to Analyze Decisions

CARNIVAL GAMES In a carnival game, 3 cards are marked with stars and 3 cards are marked with circles. All 6 cards are placed face down, and the player turns over 2 cards, one at a time. If the first card is a star and the second card is a circle, the player wins a prize.

Alec decides to play the game only if he has at least a 25% chance of winning. Should he play the game? Justify your answer using probability.

To win, a player must turn over a star followed by a circle. These events are dependent because the first card is removed. Let *S* represent a star and *C* represent a circle.

Card 1 Card 2

$P(S \text{ and } C) = \frac{3}{6} \cdot \frac{3}{5}$ After the first card is turned over, 5 cards remain, and 3 are marked with circles.

$= \frac{9}{30}$ or $\frac{3}{10}$ Simplify.

The probability that Alec will win is $\frac{3}{10}$ or 30%. This probability is at least 25%, so Alec should play the game.

Problem-Solving Tip

Sense-Making Acting out the situation can help you understand the problem. Make a set of cards to model the game described in the problem.

Guided Practice

5. Suppose the rules of the game change so that a player wins a prize only if *both* cards are stars. Should Alec play the game now? Justify your answer using probability.

Check Your Understanding

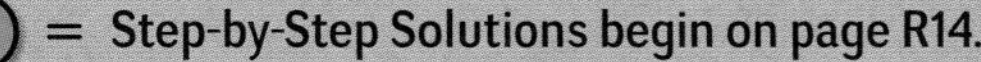

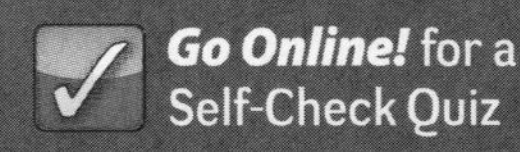

Example 1 **Determine whether the events are *independent* or *dependent*. Explain.**

1. A number cube is rolled twice and an odd number is rolled each time.

2. A bag contains several marbles. Alita selects a black marble, does not replace it, and then selects a yellow marble.

Tell whether the events are independent. Explain using probability.

Example 2

3. José, Pippa, and Raymond are scheduled to give speeches. The order of the speeches is assigned at random. José is assigned to go first, and Raymond is assigned to go second.

4. Sara spins a spinner twice. The spinner has 5 equally-sized sections numbered 1 to 5. The spinner lands on an odd number first and an even number second.

Example 3

5. **CARDS** A card is randomly chosen from a deck of 52 cards, replaced, and a second card is chosen. What is the probability of choosing both of the cards shown at the right?

Example 4

6. **TRANSPORTATION** Isaiah is getting on the bus after work. It costs $0.50 to ride the bus to his house. If he has 3 quarters, 5 dimes, and 2 nickels in his pocket, find the probability that he will randomly pull out two quarters in a row. Assume that the events are equally likely to occur.

Example 5

7. **GAMES** A board game has 12 red cards and 20 blue cards. Three of the blue cards are wild cards and 6 of the red cards are wild cards. Maurice may choose two cards at random, one at a time, to add to his hand. For the greatest possible probability of choosing two wild cards, should Maurice choose 2 red cards, 2 blue cards, or 1 red card and 1 blue card? Justify your answer using probability.

Practice and Problem Solving

Extra Practice is on page R12.

Examples 1–4 MP **REASONING** **Describe whether the events are *independent* or *dependent*. Then find the probability.**

8. In a game, you roll an even number on a number cube and then spin a spinner numbered 1 through 5 and get an odd number.

9. An ace is drawn, without replacement, from a deck of 52 cards. Then, a second ace is drawn.

10. In a bag of 3 green and 4 blue marbles, a blue marble is drawn and not replaced. Then, a second blue marble is drawn.

11. You roll two dot cubes and get a 5 each time.

Describe whether the events are independent. Explain using probability.

12. A computer program randomly assigns a letter from A to E and a number from 1 to 4 as a user's temporary password. The program assigns C2 as Rebecca's temporary password.

13. A box contains a blueberry muffin, a pumpkin muffin, and a cinnamon muffin. Without looking, Andy takes 2 muffins from the box at the same time. One of Andy's muffins is pumpkin, and the other is cinnamon.

14. **GAMES** In a game, the spinner at the right is spun and a coin is tossed. What is the probability of getting an even number on the spinner and the coin landing on tails?

15. **GIFTS** Tisha's class is having a gift exchange. Tisha will draw first and her friend Brandi second. If there are 18 students participating, what is the probability that Brandi and Tisha draw each other's names?

16. **VACATION** A work survey found that 8 out of every 10 employees went on vacation last summer. If 3 employees' names are randomly chosen, with replacement, what is the probability that all 3 employees went on vacation last summer?

17. **CAMPAIGNS** The table shows the number of each color of Student Council campaign buttons Clemente has to give away. If given away at random, what is the probability that the first and second buttons given away are both red?

Button Color	Amount
blue	20
white	15
red	25
black	10

Example 5 **For Exercises 18 and 19, justify your answer using probability.**

18. **CARNIVAL GAMES** A carnival game involves a spinner with 12 equally sized sections. Two of the sections are labeled "Penguin." The player spins the spinner twice. If the spinner lands on a "Penguin" section both times, the player wins a stuffed penguin. Steph decides to play the game only if she has at least a 10% chance of winning a stuffed penguin. Should she play the game?

19. **FESTIVALS** An outdoor music festival is planned for Friday and Saturday. The probability of rain on Friday is 50%. If it rains on Friday, the probability of rain on Saturday is 70%, and if it does not rain on Friday, the probability of rain on Saturday is 40%. The organizer will postpone the festival if the probability of rain on both days is at least 50%. Should the organizer postpone the festival?

20. **CLASSES** The probability that a student takes both geometry and French at Satomi's school is 1.6%. The probability that a student takes French is 8%. If the two events (taking geometry and taking French) are independent, what is the probability that a student takes geometry?

CANDY A box of chocolates contains 10 milk chocolates, 8 dark chocolates, and 12 white chocolates. Sophie randomly chooses a chocolate, eats it, and randomly chooses another chocolate. Find each probability.

21. P(milk and dark)

22. P(dark and white)

23. P(white and dark)

24. P(milk and milk)

25 **SOCKS** Damon has 14 white socks, 6 black socks, and 4 blue socks in his drawer. He chooses two socks at random. What is the probability that both socks are white?

26. **PROJECTS** Angela, Emery, Rico, and Taylor are working on a class project. Each student is assigned one of the following project roles at random, with no roles repeated: leader, presenter, recorder, and timekeeper.

a. Is the assignment of project roles a set of independent events or dependent events? Explain.

b. Find the probability that Angela is the presenter, Emery is the timekeeper, Rico is the leader, and Taylor is the recorder by using the multiplication rule for independent events or for dependent events.

c. Find the probability in part **b** by using combinations or permutations. Compare your result to your answer for part **b**.

27 **TENNIS** In tennis, the serving player has two chances to land his or her serve in the other player's service box without stepping on or over the service line. A fault occurs when the server fails to do this, and a double fault occurs when the server fails on both attempts. The probability that Kelly's first serve is good is 40%, and the probability that her second serve is good is 70%.

a. What is the probability that Kelly will double fault?

b. What is the probability that Kelly will fault on her first serve and her second serve will be good?

28. **VACATION** A travel website surveyed families to determine their vacation destinations. The results indicated that $P(B) = 0.6$ and $P(M) = 0.3$.

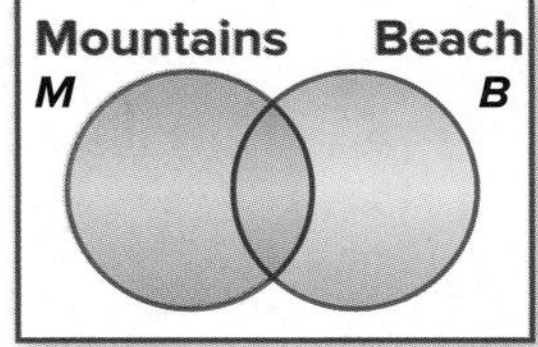

a. What is the probability that a randomly selected family vacations both in the mountains and at the beach?

b. What is the probability that a randomly selected family does not vacation at either destination?

29. **CANDY** A bag contains 10 red jelly beans, 6 green jelly beans, 7 yellow jelly beans, and 5 orange jelly beans. What is the probability of randomly choosing a red jelly bean, replacing it, randomly choosing another red jelly bean, replacing it, and then randomly choosing an orange jelly bean?

30. MP **REASONING** Consider whether the multiplication rule for dependent events can be used to find the probability of independent events.

a. Choose three different pairs of independent events. Find the probability of each pair of events using both multiplication rules. What do you notice?

b. Use mathematical reasoning to explain your observation, and make a conjecture about whether the multiplication rule for dependent events can be used to find the probability of independent events.

31. MP **REASONING** Consider whether the rule for finding the probability of two dependent events A and B, $P(A \text{ and } B) = P(A) \cdot P(B \text{ following } A)$, can also be written as $P(A \text{ and } B) = P(B) \cdot P(A \text{ following } B)$.

a. Suppose a card is randomly selected from a standard deck, and then a second card is selected without replacing the first card. Use both rules given above to find the probability that the first card is a 3 and the second card is an 8. How do the results from the two rules compare?

b. Choose two other pairs of dependent events that do not involve cards. Find the probability of each pair of events using both rules. What do you notice?

c. Make a conjecture about whether the rules are equivalent.

H.O.T. Problems Use Higher-Order Thinking Skills

32. MP **SENSE-MAKING** In some cases, if one bulb in a string of holiday lights fails to work, the whole string will not light. If each bulb in a set has a 99.5% chance of working, what is the maximum number of lights that can be strung together with at least a 90% chance that the whole string will light?

33. MP **CONSTRUCT ARGUMENTS** There are n different objects in a bag. The probability of drawing object A and then object B without replacement is about 2.4%. What is the value of n? Explain.

34. MP **REASONING** If $P(A \text{ following } B)$ is the same as $P(A)$, and $P(B \text{ following } A)$ is the same as $P(B)$, what can be said about the relationship between events A and B?

35. **OPEN-ENDED** Describe a pair of independent events and a pair of dependent events. Explain your reasoning.

36. **WRITING IN MATH** Explain how you could determine whether smoking and having a parent who smokes are independent events.

37. Francisco reaches into the bag shown here and chooses one of the marbles without looking. He puts the marble aside, and then Nevaeh reaches into the bag and chooses a marble without looking. What is the probability that they both choose a black marble? MP 1, 2

- A $\frac{1}{12}$
- B $\frac{1}{9}$
- C $\frac{1}{3}$
- D $\frac{7}{12}$

38. Xavier has a deck of 10 cards that are numbered 1 through 10. He chooses a card at random, notes the number on the card, and places it back in the deck. Then he shuffles the deck and chooses another card. What is the probability that both of the cards Xavier chooses are multiples of 3? MP 1, 2

- A $\frac{1}{100}$
- B $\frac{1}{15}$
- C $\frac{9}{100}$
- D $\frac{3}{10}$
- E $\frac{3}{5}$

39. A box contains 4 black tiles and x white tiles. Michelle chooses a tile without looking, puts it aside, and then chooses a second tile. Which expression represents the probability that both of the tiles that Michelle chooses are white? MP 1, 2, 4

- A $\frac{4}{x+4} \cdot \frac{4}{x+4}$
- B $\frac{4}{x+4} \cdot \frac{3}{x+3}$
- C $\frac{x}{x+4} \cdot \frac{x}{x+4}$
- D $\frac{x}{x+4} \cdot \frac{x-1}{x+3}$

40. LeBron rolls a number cube three times. The cube's faces are numbered 1 through 6. What is the probability as a decimal that he rolls an odd number on all three rolls? MP 1, 2

41. Lila has a standard deck of cards. She selects 4 cards at random, one at a time without replacing them. Is the probability that she selects 4 aces greater than or less than 1 out of 1 million? MP 1, 2

42. At a carnival game, a wading pool contains 50 rubber ducks, and 20 ducks are marked on the bottom with a star. A player selects a duck, does not replace it, and then selects another duck. If both ducks are marked with a star, the player wins a prize. Bella decides to play only if her probability of winning is at least 20%. Should she play the game? Use probability to justify your answer. MP 1, 3

43. **MULTI-STEP** Calvin rolls two number cubes. MP 1, 3, 4

a. Make a table or an organized list to represent the sample space.

b. Use your answer to part **a** to find the probability that both numbers rolled are greater than 4.

c. Use probability to show that these two events—rolling a number greater than 4 and rolling another number greater than 4—are independent.

44. A box contains 6 white balls and 4 red balls. We randomly (and without replacement) draw two balls from the box. What is the probability that the second ball selected is red? MP 2

45. Three cards are dealt successively at random and without replacement from a standard deck of 52 playing cards. What is the probability of receiving, in order, a king, a queen, and a jack? MP 2

LESSON 6

Probability and the Addition Rule

Then

- You found probabilities of independent and dependent events.

Now

1. Apply the addition rule to situations involving mutually exclusive events.
2. Apply the addition rule to situations involving events that are not mutually exclusive.

Why?

- At Wayside High School, freshmen, sophomores, juniors, and seniors can all run for Student Council president. Dominic wants either a junior or a senior candidate to win the election. Travis wants either a sophomore or a female to win, but says, "If the winner is sophomore Katina Smith, I'll be thrilled!"

New Vocabulary
mutually exclusive events

Mathematical Practices
1 Make sense of problems and persevere in solving them.
3 Construct viable arguments and critique the reasoning of others.

1 Mutually Exclusive Events

In Lesson 12-5, you examined probabilities involving the intersection of two or more events. In this lesson, you will examine probabilities involving the union of two or more events.

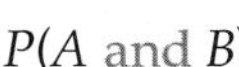

$P(A \text{ and } B)$  Indicates an intersection of two sample spaces.

$P(A \text{ or } B)$  Indicates a union of two sample spaces.

To find the probability that one event occurs *or* another event occurs, you must know how the two events are related. If the two events cannot happen at the same time, they are said to be **mutually exclusive**. That is, the two events have no outcomes in common.

Real-World Example 1 Identify Mutually Exclusive Events

ELECTIONS Refer to the application above. Determine whether the events are mutually exclusive. Explain your reasoning.

a. a junior winning the election or a senior winning the election

These events are mutually exclusive. There are no common outcomes—a student cannot be both a junior and a senior.

b. a sophomore winning the election or a female winning the election

These events are not mutually exclusive. A female student who is a sophomore is an outcome that both events have in common.

c. drawing an ace or a club from a standard deck of cards

Because the ace of clubs represents both events, they are not mutually exclusive.

Guided Practice

Determine whether the events are mutually exclusive. Explain your reasoning.

1A. selecting a number at random from the integers from 1 to 100 and getting a number divisible by 5 or a number divisible by 10

1B. drawing a card from a standard deck and getting a 5 or a heart

1C. getting a sum of 6 or 7 when two number cubes are rolled

Image Source/Photodisc/Getty Images

One way of finding the probability of two mutually exclusive events occurring is to examine their sample space.

When a number cube is rolled, what is the probability of getting a 3 or a 4? From the Venn diagram, you can see that there are two outcomes that satisfy this condition, 3 and 4. So,

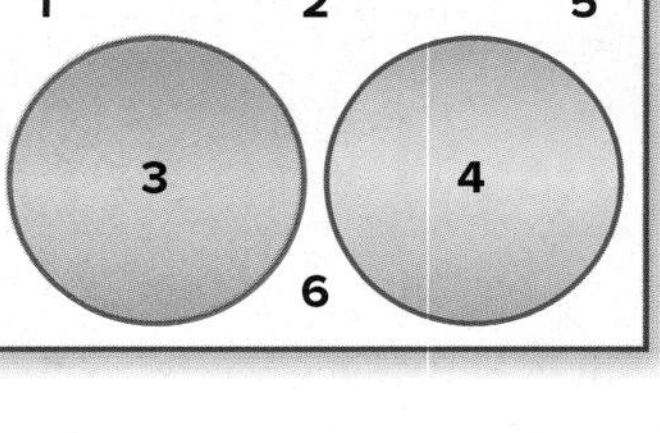

$$P(3 \text{ or } 4) = \frac{2}{6} = \frac{1}{3}.$$

Notice that this same probability can be found by adding the probabilities of each simple event.

$$P(3) = \frac{1}{6} \qquad P(4) = \frac{1}{6} \qquad P(3 \text{ or } 4) = \frac{1}{6} + \frac{1}{6} = \frac{2}{6} \text{ or } \frac{1}{3}$$

This example illustrates the first of two Addition Rules for Probability.

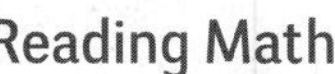

Reading Math

or The word *or* is a key word indicating that at least one of the events occurs. $P(A \text{ or } B)$ is read as *the probability that A occurs or that B occurs.*

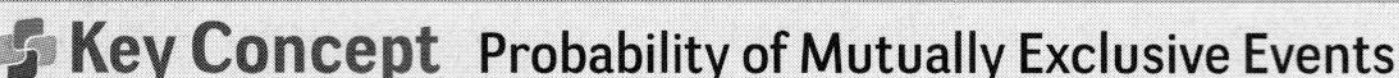

Key Concept Probability of Mutually Exclusive Events

Words If two events A and B are mutually exclusive, then the probability that A or B occurs is the sum of the probabilities of each individual event.

Example If two events A or B are mutually exclusive, then
$P(A \text{ or } B) = P(A) + P(B)$.

Math History Link

Leonhard Euler (1707–1783) Euler introduced graph theory in 1736 in a paper titled *Seven Bridges of Konigsburg*, a famous solved mathematics problem inspired by an actual place and situation. Also, Euler's formula relating the number of edges, vertices, and faces of a convex polyhedron is the origin of graph theory.

Real-World Example 2 Mutually Exclusive Events

MUSIC Ramiro makes a playlist that consists of songs from three different albums by his favorite artist. If he lets his digital media player select the songs from this list at random, what is the probability that the first song played is from Album 1 or Album 2?

Ramiro's Playlist

Album	Number of Songs
1	10
2	12
3	13

These are mutually exclusive events, since the songs selected cannot be from both Album 1 and Album 2.

Let event A1 represent selecting a song from Album 1.
Let event A2 represent selecting a song from Album 2.
There are a total of 10 + 12 + 13 or 35 songs.

$$P(A1 \text{ or } A2) = P(A1) + P(A2) \qquad \text{Probability of mutually exclusive events}$$
$$= \frac{10}{35} + \frac{12}{35} \qquad P(A1) = \frac{10}{35} \text{ and } P(A2) = \frac{12}{35}$$
$$= \frac{22}{35} \qquad \text{Add.}$$

So, the probability that the first song played is from Album 1 or Album 2 is $\frac{22}{35}$ or about 63%.

Guided Practice

2A. Two dice are rolled. What is the probability that doubles are rolled or that the sum is 9?

2B. **CARNIVAL GAMES** If you win the ring toss game at a certain carnival, you receive a stuffed animal. If the stuffed animal is selected at random from among 15 puppies, 16 kittens, 14 frogs, 25 snakes, and 10 unicorns, what is the probability that a winner receives a puppy, a kitten, or a unicorn?

2 Events That Are Not Mutually Exclusive

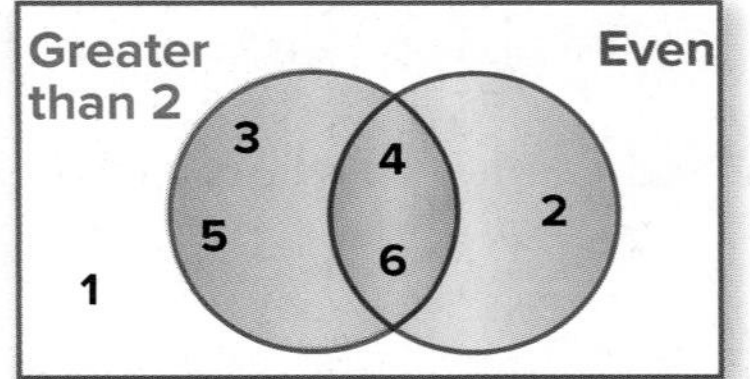

When a number cube is rolled, what is the probability of getting a number greater than 2 or an even number? From the Venn diagram, you can see that there are 5 numbers that are either greater than 2 or are an even number: 2, 3, 4, 5, and 6. So,

$$P(\text{greater than 2 or even}) = \frac{5}{6}.$$

Because it is possible to roll a number that is greater than 2 *and* an even number, these events are not mutually exclusive. Consider the probabilities of each individual event.

$$P(\text{greater than 2}) = \frac{4}{6} \qquad P(\text{even}) = \frac{3}{6}$$

If these probabilities were added, the probability of two outcomes, 4 and 6, would be counted twice—once for being numbers greater than 2 and once for being even numbers. You must subtract the probability of these common outcomes.

$$P(\text{greater than 2 or even}) = P(\text{greater than 2}) + P(\text{even}) - P(\text{greater than 2 and even})$$

$$= \frac{4}{6} + \frac{3}{6} - \frac{2}{6} \text{ or } \frac{5}{6}$$

This leads to the second of the Addition Rules for Probability.

Key Concept Probability of Events That Are Not Mutually Exclusive

Words	If two events *A* and *B* are not mutually exclusive, then the probability that *A* or *B* occurs is the sum of their individual probabilities minus the probability that both *A* and *B* occur.
Symbols	If two events A and B are not mutually exclusive, then $P(A \text{ or } B) = P(A) + P(B) - P(A \text{ and } B)$.

Real-World Example 3 Events That Are Not Mutually Exclusive

ART The table shows the number and type of paintings Namiko has created. If she randomly selects a painting to submit to an art contest, what is the probability that she selects a portrait or an oil painting?

Namiko's Paintings			
Media	Still Life	Portrait	Landscape
watercolor	4	5	3
oil	1	3	2
acrylic	3	2	1
pastel	1	0	5

Because some of Namiko's paintings are both portraits and oil paintings, these events are not mutually exclusive. Use the rule for two events that are not mutually exclusive. The total number of paintings from which to choose is 30.

$$P(\text{oil or portrait}) = P(\text{oil}) + P(\text{portrait}) - P(\text{oil and portrait})$$

$$= \frac{1+3+2}{30} + \frac{5+3+2+0}{30} - \frac{3}{30} \quad \text{Substitution}$$

$$= \frac{6}{30} + \frac{10}{30} - \frac{3}{30} \text{ or } \frac{13}{30} \quad \text{Simplify.}$$

The probability that Namiko selects a portrait or an oil painting is $\frac{13}{30}$ or about 43%.

Guided Practice

3. What is the probability of drawing a king or a diamond from a standard deck of 52 cards?

Real-World Link

Juried art shows are shows in which artists are called to submit pieces and a panel of judges decides which art will be shown. They originated in the early 1800s to exhibit the work of current artists and educate the public.

Source: Humanities Web

REALISTIC REFLECTIONS

Check Your Understanding

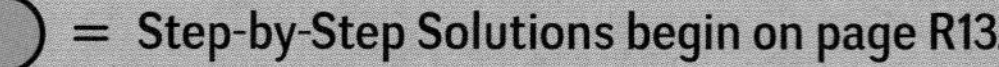

Example 1 **Determine whether the events are mutually exclusive. Explain your reasoning.**

1. drawing a card from a standard deck and getting a jack or a club
2. adopting a cat or a dog

Example 2

3. **JOBS** Adelaide is the employee of the month at her job. Her reward is to select at random from 4 gift cards, 6 coffee mugs, 7 DVDs, 10 picture frames, and 3 gift baskets. What is the probability that Adelaide receives a gift card, coffee mug, or picture frame?
4. **SPORTS CARDS** Dario owns 145 baseball cards, 102 football cards, and 48 basketball cards. He selects a card at random to give to his brother. What is the probability that he selects a baseball or a football card?

Example 3

5. **CLUBS** According to the table, what is the probability that a student in a club is a junior or on the debate team?

Club	Soph.	Junior	Senior
Key	12	14	8
Debate	2	6	3
Math	7	4	5
French	11	15	13

6. **KITTENS** Ruby's cat had 8 kittens. The litter included 2 gray females, 3 mixed-color females, 1 gray male, and 2 mixed-color males. Ruby wants to keep one kitten. What is the probability that she randomly chooses a kitten that is female or gray?

Practice and Problem Solving

Extra Practice is on page R12.

Examples 1–3 **Determine whether the events are mutually exclusive. Then find the probability. Round to the nearest tenth of a percent, if necessary.**

7. rolling a pair of dice and getting doubles or a sum of 8
8. drawing a card from a standard deck and getting a jack or a six
9. selecting a number at random from integers 1 to 20 and getting an even number or a number divisible by 3
10. tossing a coin and getting heads or tails
11. drawing an ace or a heart from a standard deck of 52 cards
12. rolling a pair of number cubes and getting a sum of either 6 or 10
13. **SPORTS** The table includes all of the programs offered at a sports complex and the number of participants aged 14–16. What is the probability that a player is 14 or plays basketball?

Graceland Sports Complex			
Age	Soccer	Baseball	Basketball
14	28	36	42
15	30	26	33
16	35	41	29

14. MP **MODELING** An exchange student is moving back to Italy, and her homeroom class wants to get her a going-away present. The teacher takes a survey of the class of 32 students and finds that 10 people choose a card, 12 choose a T-shirt, 6 choose a video, and 4 choose a bracelet. If the teacher randomly selects the present, what is the probability that the exchange student will get a card or a bracelet?
15. Talia is playing a board game where rolling two dice determines the number of spaces she moves. In Talia's current position, she needs to roll at least a sum of 9 to win. What is the probability that Talia will win on her next turn?
16. A bag contains six red coins labeled 1 through 6 and six green coins labeled 5 through 10. What is the probability of picking a coin labeled with a 5?

CARDS **Suppose you pull a card from a standard 52-card deck. Find the probability of each event.**

17. The card is a 2 or a queen.

18. The card is a diamond or a heart.

19. The card is a 7 or a club.

20. The card is a spade or an ace.

21. The card is a 5 or a prime number.

22. The card is red or an ace.

NACHO CHIPS **A restaurant serves red, blue, and yellow tortilla chips. The bowl of chips Gabriel receives has 10 red chips, 8 blue chips, and 12 yellow chips. Gabriel chooses a chip at random. Find each probability.**

23. P(red or blue)

24. P(blue or yellow)

25. P(yellow or not blue)

26. P(red or not yellow)

27. **EDUCATION** Max surveyed 200 students at his school to determine how many nights per week they do homework. His results are shown in the table.

a. What is the probability that a randomly chosen student does homework at least 3 nights per week?

b. What is the probability that a randomly chosen student does homework no more than 3 nights per week?

Number of Nights	Number of Students
0	10
1	30
2	50
3	90
4	10
5 or more	10

28. **TILES** Kirsten and José are playing a game. Kirsten places tiles numbered 1 to 50 in a bag. José selects a tile at random. If he selects a prime number or a number greater than 40, then he wins the game. What is the probability that José will win on his first turn?

H.O.T. Problems — Use Higher-Order Thinking Skills

29. **ERROR ANALYSIS** George and Aliyah are determining the probability of randomly choosing a blue or red marble from a bag of 8 blue marbles, 6 red marbles, 8 yellow marbles, and 4 white marbles. Is either of them correct? Explain.

George

$$P(\text{blue or red}) = P(\text{blue}) \cdot P(\text{red}) = \frac{8}{26} \cdot \frac{6}{26} = \frac{48}{676}$$

about 7%

Aliyah

$$P(\text{blue or red}) = P(\text{blue}) + P(\text{red}) = \frac{8}{26} + \frac{6}{26} = \frac{14}{26}$$

about 54%

REASONING **Determine whether the following are mutually exclusive. Explain.**

30. choosing a quadrilateral that is a square and a quadrilateral that is a rectangle

31. choosing a triangle that is equilateral and a triangle that is equiangular

32. choosing a complex number and choosing a natural number

33. **OPEN-ENDED** Describe a pair of events that are mutually exclusive and a pair of events that are not mutually exclusive.

34. **WRITING IN MATH** Explain why the sum of the probabilities of two mutually exclusive events is not always 1.

35. Cindy's bowling records indicate that for any frame, the probability that she will bowl a strike is 30%, a spare 45%, and neither 25%. What is the probability that she will bowl either a spare or a strike for any given frame? MP 4

36. Visitors to a school carnival throw a dart at a rectangular target in order to win prizes. The prizes are determined by the row and column in which the dart lands, as shown in the diagram. Tamiko throws a dart that lands at random on the target. Which is closest to the probability that she wins 3 tickets or a T-shirt? MP 1, 2

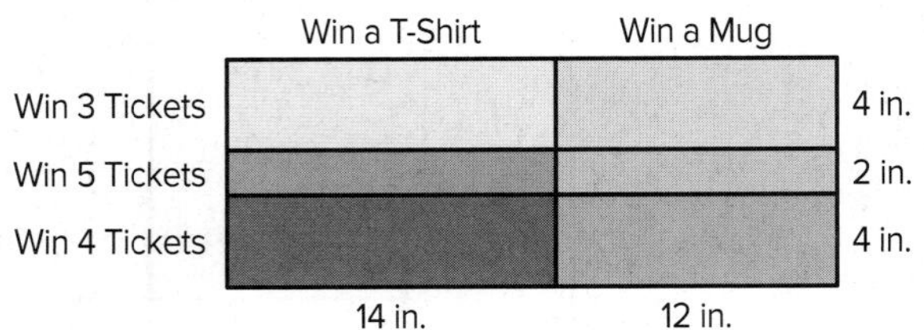

○ **A** 22%

○ **B** 40%

○ **C** 54%

○ **D** 72%

37. The spinner shown here is divided into 8 equal sectors. Elliott spins the spinner one time. What is the probability that the pointer lands on an odd number or a blue sector? MP 1, 2

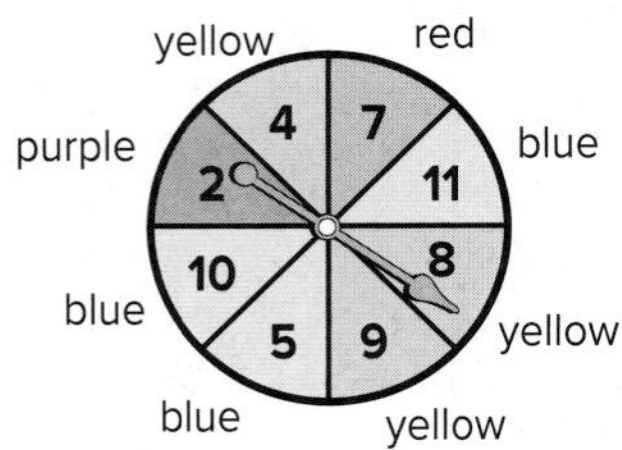

○ **A** $\frac{3}{16}$

○ **B** $\frac{5}{24}$

○ **C** $\frac{1}{4}$

○ **D** $\frac{5}{8}$

○ **E** $\frac{7}{8}$

38. Chelsea has a piece of fabric with the dimensions shown below. She spreads out the fabric on a table and then accidentally lets a drop of ink fall onto the fabric.

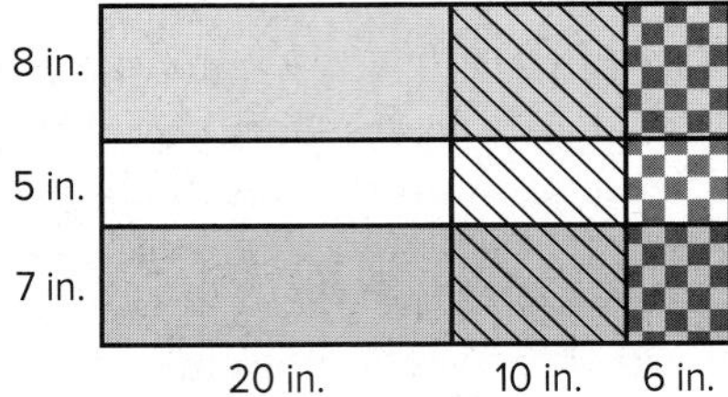

Assuming the ink lands at a random point on the fabric, which is closest to the probability that it lands in the white row or checkerboard column? MP 1, 2

○ **A** 42%

○ **B** 38%

○ **C** 25%

○ **D** 4%

39. A single number cube is rolled. Find each probability. MP 6

a. $P(3 \text{ or } 5) =$

b. $P(\text{at least } 4) =$

40. **MULTI-STEP** The extracurricular activities in which members of the senior class at Valley View High School participate are shown in the Venn diagram. MP 1, 2, 6

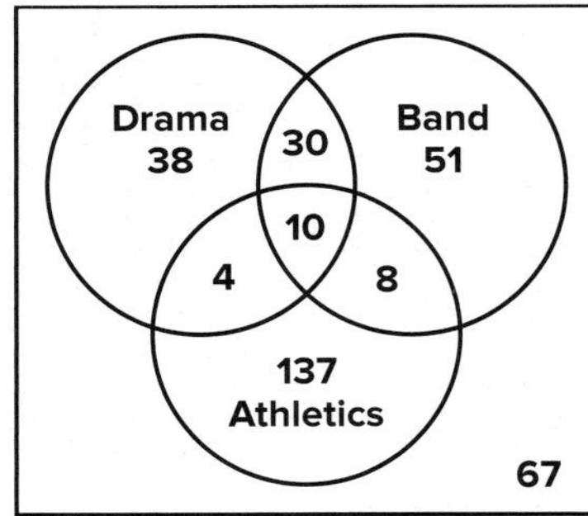

a. How many students are in the senior class?

b. How many seniors participate in athletics?

c. If a senior is randomly chosen, what is the probability that he or she participates in athletics or drama?

d. If a senior is randomly chosen, what is the probability that he or she participates in only drama and band?

LESSON 7

Conditional Probability

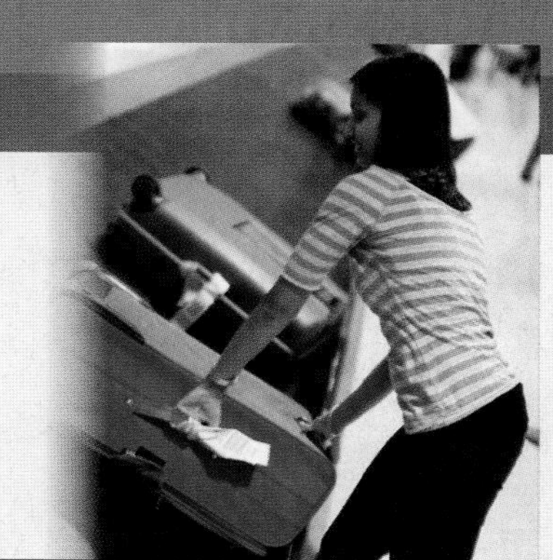

Then

- You applied the addition rule to events that were and were not mutually exclusive.

Now

1. Find the probability of events given the occurrence of other events.
2. Explain conditional probability and independence of everyday events.

Why?

- An airline reports that 5% of its passengers experience flight delays or lost luggage. Evita's flight from Austin to Honolulu has been delayed. Evita wonders whether the airline is now more likely to lose her luggage, given the delay.

New Vocabulary

conditional probability

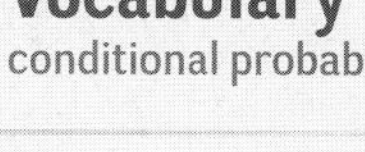

Mathematical Practices

2 Reason abstractly and quantitatively.

4 Model with mathematics.

1 Conditional Probability

Conditional probability is the probability of an event given that another event has already occurred. The notation $P(B|A)$ is read *the probability that event B occurs given that event A has already occurred or the probability of B given A.*

Conditional probability can be used when additional information is known about an event. Suppose a dot cube is rolled and it is known that the number rolled is odd. What is the probability that the number rolled is a 5 given that the number rolled is odd?

There are only three odd numbers that can be rolled, so our sample space is reduced from {1, 2, 3, 4, 5, 6} to {1, 3, 5}. So, the probability that the number rolled is a 5 is $P(5|\text{odd}) = \frac{1}{3}$.

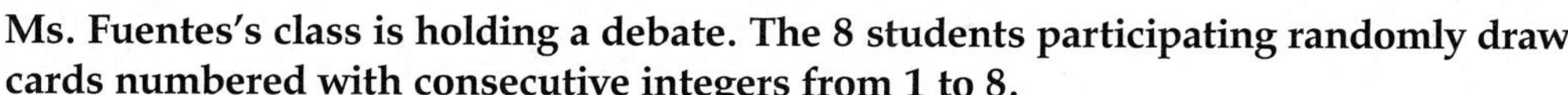

Example 1 Conditional Probability

Ms. Fuentes's class is holding a debate. The 8 students participating randomly draw cards numbered with consecutive integers from 1 to 8.

- **Students who draw odd numbers will be on the Proposition Team.**
- **Students who draw even numbers will be on the Opposition Team.**

If Jonathan is on the Opposition Team, what is the probability that he drew the number 2?

Read the Item

Because Jonathan is on the Opposition Team, he must have drawn an even number. So you need to find the probability that the number drawn was 2 given that the number drawn was even. This is a conditional probability problem.

Solve the Item

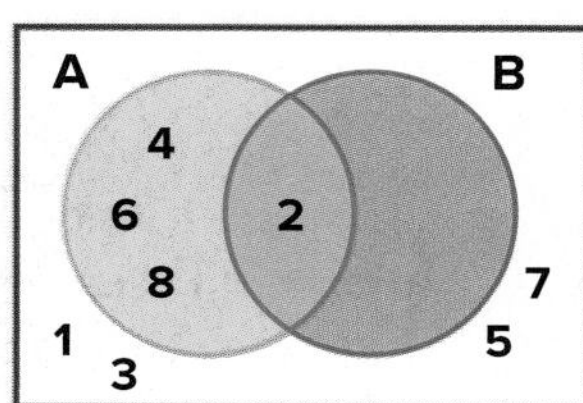

Let A be the event that an even number is drawn.
Let B be the event that the number 2 is drawn.

Draw a Venn diagram to represent this situation. There are only four even numbers in the sample space, and only one out of these numbers is a 2. Therefore, the $P(B|A) = \frac{1}{4}$.

Guided Practice

1. When two dice are rolled, what is the probability that one die is a 4, given that the sum of the two dice is 9?

   A $\frac{1}{6}$ B $\frac{1}{4}$ C $\frac{1}{3}$ D $\frac{1}{2}$

Go Online!

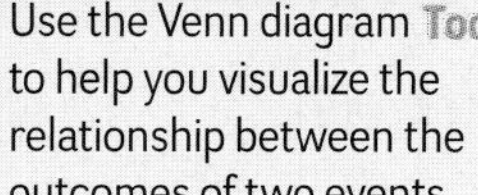

Use the Venn diagram Tool to help you visualize the relationship between the outcomes of two events.

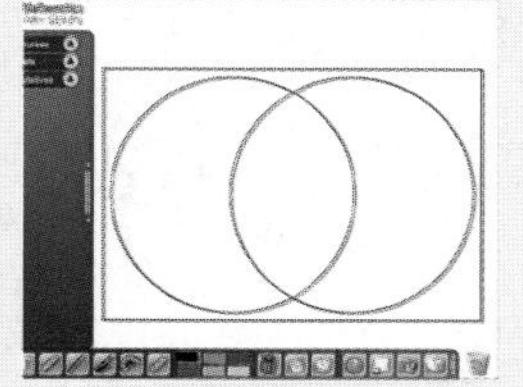

Because conditional probability reduces the sample space, the Venn diagram in Example 1 can be simplified as shown, with the intersection of the two events representing those outcomes in A and B. This suggests the following formula.

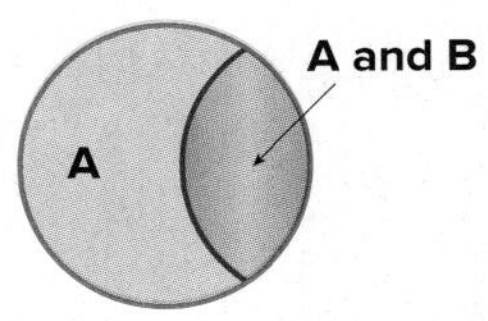

$$P(B|A) = \frac{P(A \text{ and } B)}{P(A)}$$

Key Concept Conditional Probability

The conditional probability of B given A is $P(B|A) = \frac{P(A \text{ and } B)}{P(A)}$, where $P(A) \neq 0$.

Example 2 Using the Conditional Probability Formula

At a grocery store, 32% of the bananas are ripe, 8% of the bananas are bruised, and 3% of the bananas are ripe and bruised. A banana is selected at random. What is the probability that the banana is ripe, given that it is bruised?

$$P(\text{ripe}|\text{bruised}) = \frac{P(\text{ripe and bruised})}{P(\text{bruised})}$$ Conditional probability formula

$$= \frac{0.03}{0.08} \text{ or } \frac{3}{8} \text{ or } 37.5\%$$

Guided Practice

2. What is the probability that a randomly selected banana is bruised, given that it is ripe?

2 Conditional Probability and Independence

Conditional probability can be used to find the probability of dependent events. If A and B are dependent events, then $P(A \text{ and } B) = P(A) \cdot P(B|A)$.

Recall that if events A and B are independent, then the occurrence of one does not affect the probability of the other. In other words, the probability of B given that A has occurred is the same as the probability of B, or $P(B|A) = P(B)$. Similarly, $P(A|B) = P(A)$.

Go Online!

The probability rules in this Concept Summary are important to remember. Log into your **eStudent Edition** to bookmark this page.

Concept Summary Probability Rules

Types of Events	Words	Probability Rule
Independent events	The outcome of a first event *does not affect* the outcome of the second event.	If two events A and B are independent, then $P(A \text{ and } B) = P(A) \cdot P(B)$.
Dependent events	The outcome of a first event *does affect* the outcome of the other event.	If two events A and B are dependent, then $P(A \text{ and } B) = P(A) \cdot P(B|A)$.
Conditional	Additional information is known about the probability of an event.	The conditional probability of A given B is $P(A|B) = \frac{P(A \text{ and } B)}{P(B)}$.
Mutually exclusive events	Events *do not share* common outcomes.	If two events A or B are mutually exclusive, then $P(A \text{ or } B) = P(A) + P(B)$.
Not mutually exclusive events	Events *do share* common outcomes.	If two events A and B are not mutually exclusive, then $P(A \text{ or } B) = P(A) + P(B) - P(A \text{ and } B)$.
Complementary events	The outcomes of one event consist of all the outcomes in the sample space that are not outcomes of the other event.	For an event A, $P(\text{not } A) = 1 - P(A)$.

Real-World Link

About 86% of American motorists and their right-front passengers use a seat belt.

Source: National Highway Traffic Safety Administration

Study Tip

Key Probability Words When determining what type of probability you are dealing with in a situation, look for key words and correctly interpret their meaning.

and → independent or dependent events

or → mutually exclusive or not mutually exclusive

not → complementary events

and then → conditional

at least n → *n* or more

at most n → *n* or less

Digital Vision/Alamy

Real-World Example 3 Identify and Use Probability Rules 

SEAT BELTS Refer to the information at the left. Suppose two people are chosen at random from a group of 100 American motorists and passengers. If this group mirrors the population, what is the probability that at least one of them does not wear a seat belt?

Understand You know that 86% of Americans *do use* a seat belt. The phrase *at least one* means *one or more*. So, you need to find the probability that either

- the first person chosen does not use a seat belt *or*
- the second person chosen does not use a seat belt *or*
- both people chosen do not use a seat belt.

Plan The complement of the event described above is the event that both people chosen *do use* a seat belt. Find the probability of this event, and then find the probability of its complement.

Let event A represent choosing a person who does use a seat belt.

Let event B represent choosing a person who does use a seat belt after the first person has already been chosen.

These are two dependent events, because the outcome of the first event affects the probability of the outcome of the second event.

Solve

$P(A \text{ and } B) = P(A) \cdot P(B|A)$ — Probability of dependent events

$= \frac{86}{100} \cdot \frac{85}{99}$ — $P(A) = \frac{86}{100}, P(B|A) = \frac{85}{99}$

$= \frac{7310}{9900}$ or $\frac{731}{990}$ — Multiply.

$P[\text{not } (A \text{ and } B)] = 1 - P(A \text{ and } B)$ — Probability of a complement

$= 1 - \frac{731}{990}$ — Substitution

$= \frac{259}{990}$ — Subtract.

So, the probability that at least one of the passengers does not use a seat belt is $\frac{259}{990}$ or about 26%.

Check The probability that one person chosen out of 100 does not wear his or her seat belt is (100 − 86)% or 14%. When you choose two people out of 100, the probability that at least one of them does not wear their seatbelt should be greater than 14%. Because 26% > 14%, the answer is reasonable.

In this example, it is more efficient to calculate the probability of the complement than to calculate the probability of each case in the set of desired outcomes.

Guided Practice

3. **CELL PHONES** According to an online poll, 35% of American motorists routinely use their cell phones while driving. Three people are chosen at random from a group of 100 motorists. What is the probability that

A. at least two of them use their cell phone while driving?

B. no more than one use their cell phone while driving?

Check Your Understanding

◯ = Step-by-Step Solutions begin on page R14.

Example 1

1. **GAMES** Every Saturday, 10 friends play dodgeball at a local park. To pick teams, they randomly draw cards with consecutive integers from 1 to 10. Odd numbers are on Team A, and even numbers are Team B. What is the probability that a player on Team B has drawn the number 10?

Example 2

2. In a large city, 70% of the residents are adults, 84% of the residents have traveled out of state, and 9% are adults who have not traveled out of state. What is the probability that a randomly selected resident has not traveled out of state, given that the resident is an adult?

3. **WEATHER** Tomorrow's weather forecast calls for a 25% chance of rain, an 80% chance that the temperature will exceed 80°F, and a 15% chance of both. What is the probability of rain, given that the temperature exceeds 80°F?

Example 3

4. **PROM** In Armando's senior class of 100 students, 91 went to the senior prom. If two people are chosen at random from the entire class, what is the probability that at least one of them did not go to prom?

5. A red marble is selected at random from a bag of 2 blue and 9 red marbles and not replaced. What is the probability that a second marble selected will be red?

Practice and Problem Solving

Extra Practice is on page R12.

Example 1

6. A number cube is rolled and the result is a number greater than 2. What is the probability that the result is a 6?

7. A spinner has 12 equally sized sections numbered 1 through 12. Find the probability that the spinner lands on 11, given that the spinner lands on an odd number.

8. An eight-sided number cube is rolled and the result is an odd number. Find the probability that the result is 5.

9. The perimeter of a quadrilateral is 12 units and the length of each side is an odd integer. What is the probability that the quadrilateral is a rhombus?

10. There are 13 players available to play basketball. The players randomly draw cards numbered with consecutive integers from 1 to 13. Players who draw odd numbers are on Team A. Players who draw even numbers are on Team B.

    a. If Todd is on Team A, what is the probability that he drew the number 5?

    b. If Tyrone is on Team B, what is the probability that he drew the number 6?

Example 2

11. **TECHNOLOGY** At Bell High School, 43% of the students own a smartphone, and 28% own a smartphone and a digital media player. What is the probability that a student owns a digital media player, given that he or she owns a smartphone?

12. **CLASSES** The probability that a student takes geometry and French at Satomi's school is 0.064. The probability that a student takes French is 0.45. What is the probability that a student takes geometry if the student takes French?

13. **PROOF** Use the formula for the probability of two dependent events $P(A \text{ and } B)$ to derive the conditional probability formula for $P(B|A)$.

14. **TENNIS** A double fault in tennis is when the serving player fails to land their serve "in" without stepping on or over the service line in two chances. Kelly's first serve percentage is 40%, while her second serve percentage is 70%.

    a. Draw a probability tree that shows each outcome.

    b. What is the probability that Kelly will double fault?

    c. Design a simulation using a random number generator that can be used to estimate the probability that Kelly double faults on her next serve.

15. Of the T-shirts available at a store, 54% are striped, 32% have a logo, and 13% are striped and have a logo.

   a. What is the probability that a shirt with a logo is striped?
   b. What is the probability that a striped shirt has a logo?

16. In a large town, 73% of the adults are homeowners, 46% have a college degree, and 38% are homeowners with college degrees.

   a. What is the probability that an adult with a college degree owns a home?
   b. What is the probability that a homeowner has a college degree?

17. **SPORTS** At a basketball game, 80% of the fans cheered for the home team. In the same crowd, 20% of the crowd were waving banners and cheering for the home team. What is the probability that a fan who cheered for the home team waved a banner?

Example 3

18. **JOBS** Of young workers aged 18 to 25, 71% are paid by the hour. If two people are randomly chosen from a group of 100 young workers, what is the probability that exactly one is paid by the hour?

19. **RECYCLING** Suppose 31% of Americans recycle. If two Americans are chosen randomly from a group of 50, what is the probability that at most one of them recycles?

20. **MUSIC** A school carried out a survey of 265 students to see which types of music students prefer to hear at school dances. The results are shown in the Venn diagram. Find each probability.

   a. P(country or R&B)
   b. P(rock and country or R&B or rock)
   c. P(R&B but not rock)
   d. P(all three)

21. **VACATION** A random survey was conducted to determine where families vacationed. The results indicated that $P(B) = 0.6$, $P(B \cap M) = 0.2$, and the probability that a family did not vacation at either destination is 0.1.

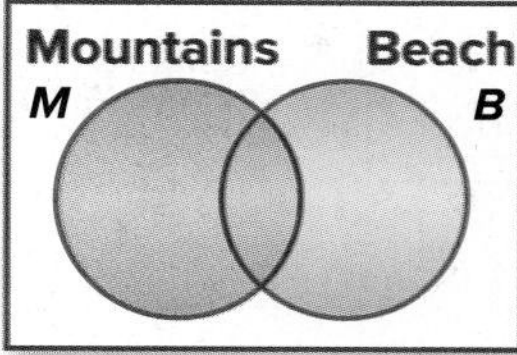

   a. What is the probability that a family vacations only in the mountains?
   b. What is the probability that a family visiting the beach will also visit the mountains?

H.O.T. Problems Use Higher-Order Thinking Skills

22. **PROOF** Use the formula for the probability of two dependent events $P(A \text{ and } B)$ to derive the conditional probability formula for $P(B \mid A)$.

23. MP **PROBLEM SOLVING** You roll 3 number cubes. What is the probability that the outcome on at least two of the cubes will be a number less than or equal to 4? Explain your reasoning.

24. **WRITING IN MATH** Suppose that A and B are mutually exclusive events. What is $P(A \mid B)$ and $P(B \mid A)$? Explain your reasoning.

25. **CONSTRUCT ARGUMENTS** There are n different objects in a bag. The probability of drawing object A and then object **B** without replacement is about 1.4%. What is the value of n? Explain.

26. **REASONING** If $P(A|B)$ is the same as $P(A)$, and $P(B|A)$ is the same as $P(B)$, what can be said about the relationship between events A and B?

27. A deck contains 10 cards that are numbered 1 through 10. Alex chooses a card at random. What is the probability that the number on Alex's card is 5, given that it is a multiple of 5? MP 2

- ○ A $\frac{1}{10}$
- ○ B $\frac{1}{5}$
- ○ C $\frac{1}{2}$
- ○ D $\frac{1}{1}$

28. Leo chooses two marbles at random from the bag shown here. Given that one of Leo's marbles is black, what is the probability that both marbles are black? MP 2

- ○ A $\frac{1}{36}$
- ○ B $\frac{4}{81}$
- ○ C $\frac{1}{15}$
- ○ D $\frac{5}{12}$

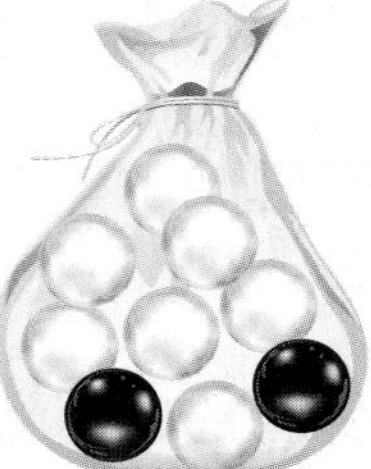

29. LeBron rolls a number greater than or equal to 4 on a dot cube. What is the probability that LeBron has rolled a 5? MP 2

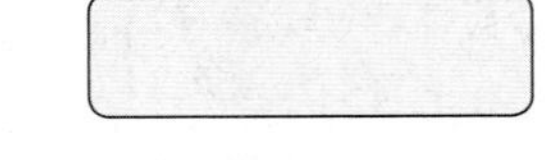

30. At a high school, the probability that a student plays both basketball and football is 0.008. The probability that a student plays football is 0.28. What is the probability that a football player also plays basketball? MP 2

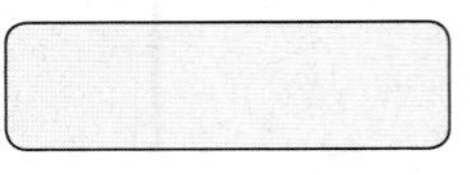

31. What is the probability that the sum of two dice will be greater than 8, given that the first die is 6? MP 2

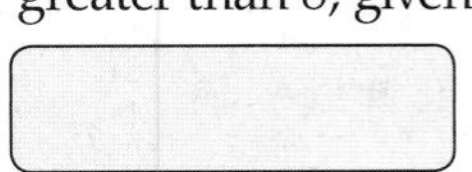

32. What is the probability of drawing two aces from a standard deck of cards, given that the first card is an ace? MP 2

Mr. Marcos organizes a classroom game for 16 students. Each student is randomly assigned a different number from 1 to 16.

- **Students assigned a multiple of 4 or 9 are on Team A.**
- **Students assigned a multiple of 5 or 7 are on Team B.**
- **Students assigned any other number are on Team C.**

Find each probability.

33. P(Maggie was assigned 13 | Maggie is on Team C) MP 2

- ○ A $\frac{3}{128}$
- ○ B $\frac{1}{16}$
- ○ C $\frac{1}{6}$
- ○ D $\frac{3}{8}$

34. P(Nicole was assigned 7 | Nicole is on Team B) 

- ○ A $\frac{1}{16}$
- ○ B $\frac{1}{5}$
- ○ C $\frac{5}{16}$
- ○ D $\frac{3}{8}$

35. **MULTI-STEP** A certain disease affects 2% of the population. The test for this disease is accurate 90% of the time. The accuracy of a patient's test result is independent of whether the patient has the disease. MP 1, 2, 4

a. What is the probability that a patient has the disease and receives a positive (accurate) test result?

b. What is the probability that a patient does not have the disease and receives a positive (inaccurate) test result?

c. What is the probability that a patient receives a positive test result? Explain how you found your answer.

d. Jordan receives a positive test result. What is the probability that Jordan has the disease? Show how you found your answer.

LESSON 8

Two-Way Frequency Tables

Then

- You found conditional probabilities.

Now

- **1** Decide whether events are independent by using two-way frequency tables.
- **2** Approximate conditional probabilities by using two-way frequency tables.

Why?

- Michael surveys students at his school to find out if there is an association between a student's grade level and his or her decision to participate in drama club. Michael can organize and analyze the results in a two-way frequency table.

New Vocabulary

two-way frequency table
marginal frequency
joint frequency
relative frequency

Mathematical Practices

2 Reason abstractly and quantitatively.
4 Model with mathematics.

1 Decide Whether Events are Independent

A **two-way frequency table** or contingency table is used to show the frequencies of data from a survey or experiment in which the data are classified according to two variables. The rows of the table indicate one variable and the columns of the table indicate the other.

Real-World Example 1 Make a Two-Way Frequency Table

FESTIVALS **Michael asks a random sample of 160 upperclassmen at his high school whether or not they plan to participate in drama club. He finds that 44 seniors and 25 juniors plan to attend the festival, while 32 seniors and 59 juniors do not plan to attend. Make a two-way frequency table to organize the data.**

Step 1 Identify the variables.

The students surveyed can be classified according to the variables *class* and *participation*. Because the survey included only upperclassmen, the variable *class* has two categories: senior or junior. The variable *participation* also has two categories: participating or not participating.

Step 2 Create a two-way frequency table.

Let the rows of the table represent *class* and let the columns represent *participation*.

	Participating	Not Participating	Totals
Senior	44	32	76
Junior	25	59	84
Totals	69	91	160

Fill in the cells of the table with the given data.

Step 3 Add a *Totals* row and a *Totals* column to the table and fill in these cells with the correct sums.

Guided Practice

1. **PETS** Karima asks a random sample of 50 visitors to a pet store if they prefer cats or dogs. She finds that 18 women and 17 men prefer dogs, while 6 women and 7 men prefer cats. Make a two-way table to organize the data.

The frequencies reported in the *Totals* row and *Totals* column are called **marginal frequencies**, with the bottom rightmost cell reporting the total number of observations. The frequencies reported in the interior of the table are called **joint frequencies**. These show the frequencies of all possible combinations of the categories for the first variable with the categories for the second variable.

Example 2 Use Marginal and Joint Frequencies

Miguel looked at a random sample of flights for two airlines and collected data on whether the flights arrived on time or late. The data are shown in the two-way frequency table. Answer each question and tell whether you are using a marginal frequency or joint frequency.

	On Time	Late	Totals
AirWorld	13	9	22
RedJet	7	11	18
Totals	20	20	40

a. How many RedJet flights arrived on time?

Look at the intersection of the *On Time* column and the *RedJet* row.

	On Time	Late	Totals
AirWorld	13	9	22
RedJet	7	11	18
Totals	20	20	40

The cell shows that 7 RedJet flights arrived on time.

This is a joint frequency.

b. How many flights arrived late?

This includes data from both airlines, so use the total in the *Late* column.

	On Time	Late	Totals
AirWorld	13	9	22
RedJet	7	11	18
Totals	20	20	40

The cell in the *Totals* row shows that 20 flights arrived late.

This is a marginal frequency.

Guided Practice

Answer each question and tell whether you are using a marginal frequency or joint frequency.

2A. How many AirWorld flights are in the sample?

2B. How many AirWorld flights arrived late?

A **relative frequency** is the ratio of the number of observations in a category to the total number of observations.

Example 3 Determine Relative Frequencies

Convert the table from Example 1 to a table of relative frequencies.

Step 1 Divide the frequency reported in each cell by the total number of respondents, 160.

	Attending the Festival	Not Attending the Festival	Totals
Senior	$\frac{44}{160}$	$\frac{32}{160}$	$\frac{76}{160}$
Junior	$\frac{25}{160}$	$\frac{59}{160}$	$\frac{84}{160}$
Totals	$\frac{69}{160}$	$\frac{91}{160}$	$\frac{160}{160}$

Step 2 Write each fraction as a percent rounded to the nearest tenth.

	Attending the Festival	Not Attending the Festival	Totals
Senior	27.5%	20%	47.5%
Junior	15.6%	36.9%	52.5%
Totals	43.1%	56.9%	100%

Guided Practice

3. Convert the table from Example 2 to a table of relative frequencies.

Study Tip

MP Sense-Making After you convert to relative frequencies, you should still be able to add across rows and down columns to get the percents in the *Totals* cells. This is a good way to check that the percents are correct.

Variable A is considered to be independent of variable B if $P(A \text{ and } B) = P(A) \cdot P(B)$. In a two-way frequency table, you can test for independence of two variables by comparing the joint relative frequencies with the products of the corresponding marginal relative frequencies.

Example 4 Decide Whether Events Are Independent

Use the relative frequency table from Example 3 to determine whether attendance at the festival is independent of class. Explain.

Make a new table. Calculate the expected joint frequencies assuming the two variables are independent. Multiply the marginal relative frequencies to find the joint relative frequencies.

Seniors attending:
$(47.5\%)(43.1\%) \approx$ **20.5%**

Seniors not attending:
$(47.5\%)(56.9\%) \approx$ **27%**

Juniors attending:
$(52.5\%)(43.1\%) \approx$ **22.6%**

Juniors not attending: $(52.5\%)(56.9\%) \approx$ **29.9%**

	Attending the Festival	Not Attending the Festival	Totals
Senior	20.5%	27%	47.5%
Junior	22.6%	29.9%	52.5%
Totals	43.1%	56.9%	100%

Compare these joint relative frequencies to actual joint relative frequencies in the two-way table in Example 3. Because the expected and actual joint relative frequencies are not the same, attendance at the festival is *not* independent of class.

Guided Practice

4. Use the relative frequency table you made in Guided Practice 3 to determine whether on-time arrivals are independent of airline. Explain.

2 Approximate Conditional Probabilities

Recall that the conditional probability of event A given that event B has occurred is given by the formula $P(A \mid B) = \frac{P(A \cap B)}{P(B)}$. In a two-way relative frequency table, $P(A \cap B)$ is a joint relative frequency and $P(B)$ is a marginal relative frequency.

Example 5 Find a Conditional Probability

Use the relative frequency table from Example 3 to find the probability that a surveyed student plans to attend the festival given that he or she is a junior.

Use values from the *Junior* row of the table.

P(attending festival | junior)

$$= \frac{P(\text{attending and junior})}{P\ (\text{junior})}$$

$$\approx \frac{0.156}{0.525} \text{ or } 29.7\%$$

	Attending the Festival	Not Attending the Festival	Totals
Senior	27.5%	20%	47.5%
Junior	15.6%	36.9%	52.5%
Totals	43.1%	56.9%	100%

Study Tip

MP Structure In general, the word *and* corresponds to an intersection or a joint relative frequency. *P*(attending and junior) is found at the intersection of the *Attending* column and the *Junior* row.

Guided Practice

5. Find the probability that a surveyed student is a senior given that he or she is not planning to attend the festival.

Check Your Understanding

= Step-by-Step Solutions begin on page R13.

Go Online! for a Self-Check Quiz

Example 1 **Make a two-way frequency table to organize the given data.**

1. **MEALS** Students in elementary school and high school were surveyed about eating habits. Thirty-eight students in elementary school ate breakfast and 12 skipped breakfast. Twenty-two students in high school ate breakfast and 28 skipped breakfast.

2. **CLASS TRIPS** A random sample of high school seniors was given a survey to determine the location of their class trip. The two choices were an amusement park or an aquarium. The amusement park was the choice of 28 girls and 34 boys. The aquarium was the choice of 9 girls and 8 boys.

Example 2 **Mayumi surveyed a random sample of people who commute to work on a bus or on a train and she asked them whether or not they own a car. The data are shown in the two-way frequency table. Answer each question and tell whether you are using a marginal frequency or joint frequency.**

	Car	No Car	Totals
Bus	8	62	70
Train	16	34	50
Totals	24	96	120

3. How many commuters ride a train and own a car?

4. How many people who commute on a train were surveyed?

5. How many car owners were surveyed?

6. How many car owners ride a bus?

Example 3

7. Convert the two-way frequency table of data about commuters to a table of relative frequencies.

Example 4

8. Determine whether commuting to work on a bus or train is independent of owning a car. Explain.

Example 5 **Use the two-way frequency table of data about commuters to find each conditional probability.**

9. the probability that a commuter rides a bus given that he or she owns a car

10. the probability that a commuter does not own a car given that he or she rides a train

11. the probability that a car owner rides a train

12. the probability that a train rider owns a car

Practice and Problem Solving

Extra Practice is on page R12.

Example 1 **Make a two-way frequency table to organize the given data.**

13. **SLEEP HABITS** A group of doctors and nurses were surveyed about the number of hours of sleep they get each day. Forty doctors got at least 7 hours of sleep each day and 35 doctors got less than 7 hours of sleep each day. Among the nurses, 24 got at least 7 hours of sleep each day and 63 got less than 7 hours of sleep each day.

14. **HOBBIES** A random sample of students in the film club and chess club were asked about whether they prefer to take photos or paint. Twenty-two students in the film club chose taking photos and 9 chose painting. A total of 18 members of the chess club were surveyed and the results split evenly between taking photos and painting.

15. **COOKING** Alison surveyed 100 people at a cooking school and asked them whether they prefer to cook on a gas range or an electric range. Fifty-five men participated in the survey and 21 of them chose a gas range. Of the women who participated, 19 chose an electric range.

Example 2 **Russell surveyed members of his school's soccer team and softball team to find out whether they prefer to drink water or a sports drink while training. The data are shown in the two-way frequency table. Answer each question and tell whether you are using a marginal frequency or joint frequency.**

	Water	Sports Drink	Totals
Soccer	36	24	60
Softball	54	36	90
Totals	90	60	150

16. How many soccer players prefer a sports drink?

17. How many members of the softball team were surveyed?

18. How many team members who prefer water are soccer players?

19. How many softball players do not prefer a sports drink?

Example 3 20. Convert the two-way frequency table of data about sports teams and beverages to a able of relative frequencies.

Example 4 21. Determine whether a team member's beverage preference is independent of his or her team. Explain.

Example 5 **Use the two-way frequency table of data about sports teams and beverages to find each conditional probability.**

22. the probability that a team member prefers water given that he or she is on the soccer team

23. the probability that a team member who prefers a sports drink is on the softball team

24. the probability that a member of the soccer team prefers a sports drink

25. the probability that a team member who prefers water is not on the soccer team

26. **FLOWERS** A florist surveyed a random sample of customers by asking 20 men and 20 women whether they would rather receive one dozen roses or one dozen tulips. Of the women surveyed, 15 chose one dozen tulips.

   a. What is the probability that a customer chose tulips given that she is a woman?
   b. What is the joint relative frequency of women who chose roses?
   c. Can you determine the conditional probability of a customer choosing one dozen roses given that he is a man? If so, explain how to find the probability. If not, explain why not and describe any additional information you would need.

MP **REASONING** **The two-ways frequency tables show data from a school survey about preferred school colors. Copy and complete each table.**

27.

	Blue	Green	Totals
Boys	40		75
Girls		18	50
Totals		53	

28.

	Blue	Green	Totals
Seniors			46
Juniors		43	
Totals	70		125

29.

	Red	Yellow	Totals
Teachers	23		
Students	18		51
Totals			81

30.

	Blue	Purple	Totals
Parents		44	85
Students	17		
Totals		59	

31. **MP REASONING** In a school with 400 students, 150 students are currently taking advanced placement (AP) classes, and 40 students are seniors taking advanced placement classes. There are 100 seniors in all.

   a. Use the given information to create a two-way frequency table. Include the categories *Seniors* and *Not Seniors*, and *AP Classes* and *No AP Classes*.
   b. Convert the table to a table of relative frequencies.
   c. What is the probability that a randomly chosen student at the school is not a senior and not taking advanced placement classes?
   d. Is a randomly chosen student at the school more likely to be taking advanced placement classes or not taking advanced placement classes? How does your table of relative frequencies show this?

32. **ANIMAL SHELTERS** An employee at an animal shelter collected data on a random sample of cats and dogs as they were brought to the shelter. Copy and complete the table assuming that having fleas is independent of being a cat or a dog.

	Fleas	No Fleas	Totals
Cat			80
Dog			
Totals	72		240

33. **MP SENSE-MAKING** Darius surveyed a random sample of 250 employees at the JQP Corporation to find out what the employees do for lunch. Of the 122 men that he surveyed, 43 bring their lunch, 56 eat at the company cafeteria, and the rest go out to get lunch. Of the women that he surveyed, 46 bring their lunch and 30 go out to get lunch.

   a. Make a two-way frequency table to organize the data.
   b. What is the probability that a randomly-chosen employee is a man, given that the employee eats lunch at the company cafeteria?

H.O.T. Problems Use Higher-Order Thinking Skills

34. **OPEN-ENDED** Make a two-way frequency table that shows the results of a survey in which 420 people were surveyed and 30% of the people surveyed were male.

35. **ERROR ANALYSIS** The two-way frequency table shows data about a random sample of apples at a supermarket. Kaci was asked to find the probability that a randomly-chosen apple is red, given that it is organic. Her work is shown below. Explain her error and find the correct answer.

	Organic	Not Organic	Totals
Red	18%	42%	60
Green	12%	28%	40
Totals	30%	70%	100

$$P(\text{red} \backslash \text{organic}) = \frac{0.18}{0.60}$$
$$= 0.3$$
$$= 30\%$$

36. **MP REASONING** A standard two-way frequency table, in which each variable has two categories, contains 9 cells with numerical values. What is the minimum number of these values that you need to know in order to fill in the rest of the table?

37. **MP CHALLENGE** In a two-way relative frequency table in which each variable has two categories, all of the joint relative frequencies are equal. What is the value of the joint relative frequencies?

38. **WRITING IN MATH** Describe the steps for using a two-way frequency table to determine whether two variables are independent of each other.

39. Out of 280 men and women who were surveyed, 95 preferred to watch football rather than soccer, and 40 of those surveyed were women who preferred football. Fifty of those surveyed were men who preferred soccer. What is the relative frequency of men who prefer football? MP 2, 4

- ○ **A** 19.6%
- ○ **B** 33.9%
- ○ **C** 52.4%
- ○ **D** 57.9%

40. The two-way frequency table shows the results of a survey in which college math majors and history majors were asked whether they prefer to work on a laptop computer or a tablet. What is the probability that a student prefers to work on a tablet given that he or she is a math major? MP 4, 6

	Math Major	History Major	Totals
Laptop	69	21	90
Tablet	34	61	95
Totals	103	82	185

- ○ **A** 18.4%
- ○ **B** 33.0%
- ○ **C** 35.8%
- ○ **D** 67.0%

41. Jesse surveys a random sample of 91 adults and children to find out if they prefer a day at the beach or a day in the mountains. He surveys 45 adults and finds that 12 of them prefer a day in the mountains. Twenty-five children prefer a day at the beach. Jesse makes a two-way frequency table of this data. Which of the following values should appear in his table as marginal frequencies? MP 2, 4

- ☐ **A** 12
- ☐ **B** 21
- ☐ **C** 25
- ☐ **D** 33
- ☐ **E** 46
- ☐ **F** 58

42. Kayley surveyed 80 friends to find out whether or not they have a pet dog and whether or not they have a pet cat. She found that 18 friends have a pet dog, 8 friends have a pet cat, and 6 friends have both. What is the probability p that a friend does not have a pet dog, given that he or she has a pet cat? MP 2, 4

$P =$ ☐

43. The manager of a theater asked members of a ballet audience and members of a jazz concert audience whether they would like to learn about upcoming events by a text message or by an email. The variables in the table represent relative frequencies for these data. Which expression can the manager use to find the probability that someone is a member of the ballet audience given that he or she prefers a text message? MP 4, 7

	Text	Email	Totals
Ballet	a	b	c
Jazz	d	e	f
Totals	g	h	j

- ○ **A** $\frac{a}{j}$
- ○ **B** $\frac{a}{c}$
- ○ **C** $\frac{a}{g}$
- ○ **D** $\frac{g}{j}$

44. MULTI-STEP Malik surveyed vegetarians and nonvegetarians about their peanut butter preferences. He recorded some of the data in the two-way frequency table shown. MP 2, 4

	Crunchy	Smooth	Totals
Vegetarian	—	—	32
Nonvegetarian	—	36	—
Totals	20	—	80

a. Copy and complete the table.

b. Convert the table to a table of relative frequencies.

c. Determine whether someone's peanut butter preference is independent of whether or not he or she is a vegetarian. Explain.

Study Guide and Review

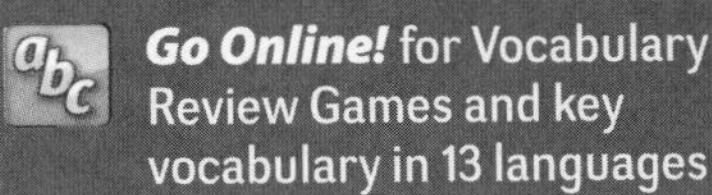

Study Guide

Key Concepts

Representing Sample Spaces (Lesson 12-1 and 12-2)

- The sample space of an experiment is the set of all possible outcomes. It can be determined by using an organized list, a table, or a tree diagram.
- The intersection of two sample spaces includes events contained in both sample spaces.
- The union of two sample spaces includes events contained in either sample space.

Probability with Permutations and Combinations (Lesson 12-3)

- A permutation of n objects taken r at a time is given by $_nP_r = \frac{n!}{(n-r)!}$. Order is important.
- A combination of n objects taken r at a time is given by $_nC_r = \frac{n!}{(n-r)!r!}$. Order is not important.

Geometric Probability (Lesson 12-4)

- If a region A contains a region B and a point E in region A is chosen at random, then the probability that point E is in region B is $\frac{\text{area of region } B}{\text{area of region } A}$.

Probabilities of Compound Events (Lessons 12-5 and 12-6)

- If event A does not affect the outcome of event B, then the events are independent and $P(A \text{ and } B) = P(A) \cdot P(B)$.
- If A and B are dependent, $P(A \text{ and } B) = P(A) \cdot P(B|A)$.
- If two events A and B cannot happen at the same time, they are mutually exclusive and $P(A \text{ or } B) = P(A) + P(B)$.
- If two events A and B are not mutually exclusive, then $P(A \text{ or } B) = P(A) + P(B) - P(A \text{ and } B)$.

Conditional Probability (Lesson 12-7)

- The conditional probability of A given B is $P(A|B) = \frac{P(A \text{ and } B)}{P(B)}$

Two-Way Frequency Tables (Lesson 12-8)

- Compare expected and actual joint relative frequencies to determine if events are independent.

FOLDABLES® Study Organizer

Use your Foldable to review the chapter. Working with a partner can be helpful. Ask for clarification of concepts as needed.

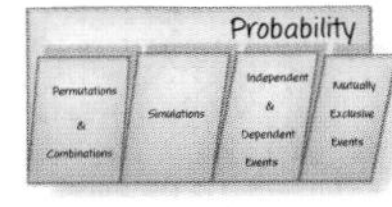

Key Vocabulary

circular permutation (p. 875)
combination (p. 876)
compound events (p. 889)
conditional probability (p. 891)
dependent events (p. 889)
factorial (p. 872)
Fundamental Counting Principle (p. 861)
geometric probability (p. 881)
independent events (p. 889)
intersection (p. 889)
joint frequencies (p. 909)
marginal frequencies (p. 909)
mutually exclusive events (p. 897)
permutation (p. 872)
probability tree (p. 891)
relative frequencies (p. 909)
sample space (p. 859)
tree diagram (p. 859)
two-way frequency table (p. 909)
union (p. 889)

Vocabulary Check

State whether each sentence is *true* or *false*. If *false*, replace the underlined term to make a true sentence.

1. A tree diagram uses line segments to display possible outcomes.
2. Tossing a coin and then tossing another coin is an example of dependent events.
3. Geometric probability involves a geometric measure such as length or area.
4. $6! = 6 \cdot 5 \cdot 4 \cdot 3 \cdot 2 \cdot 1$ is an example of a factorial.
5. The set of all possible outcomes is the sample space.
6. Combining a coin toss and a roll of a die makes a simple event.
7. Drawing two socks out of a drawer without replacing them are examples of mutually exclusive events.

Concept Check

8. Explain when permutations would be more appropriate to use than combinations.
9. Explain the difference between the intersection and the union of two sample spaces.

Lesson-by-Lesson Review 

12-1 Representing Sample Spaces

10. POPCORN A movie theater sells small (S), medium (M), and large (L) size popcorn with the choice of no butter (NB), butter (B), and extra butter (EB). Represent the sample space for popcorn orders by making an organized list, a table, and a tree diagram.

11. SHOES A pair of men's shoes comes in whole sizes 5 through 13 in navy, brown, or black. How many different pairs could be selected?

Example 1

Three coins are tossed. Represent the sample space for this experiment by making an organized list.

Pair each possible outcome from the first toss with the possible outcomes from the second toss and third toss.

HHH, HHT, HTH, HTT, THH, THT, TTH, TTT

12-2 Probability and Counting

12. PETS The Venn diagram shows the results of a pet store survey to determine the pets customers owned.

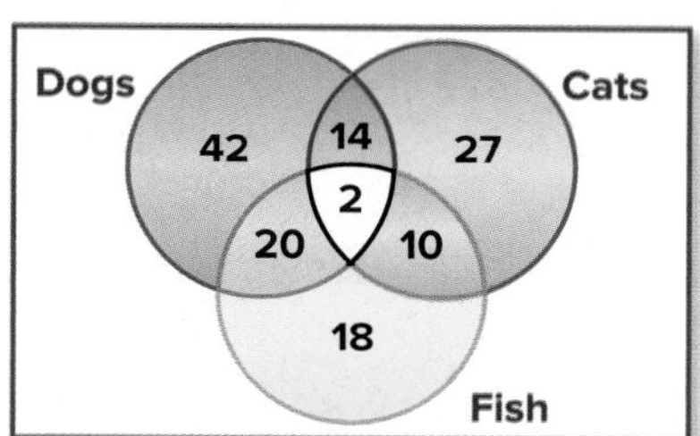

a. How many customers had only fish?

b. How many had only cats and dogs?

c. How many had dogs as well as fish?

Example 2

The Venn diagram shows the number of students who plan to join the soccer or baseball teams.

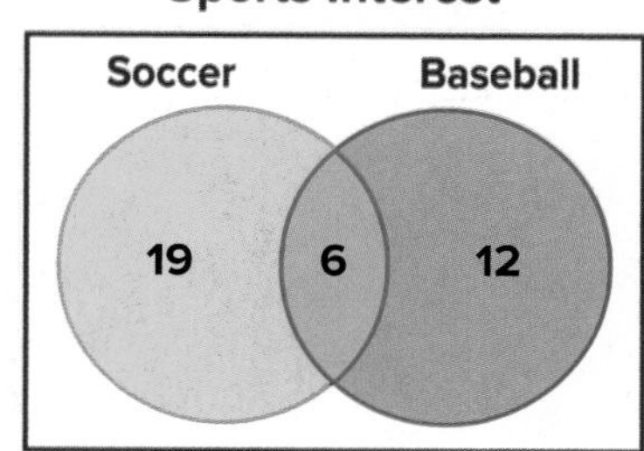

a. How many students plan on joining the soccer or baseball teams?

The students who plan on joining the soccer or baseball teams are represented by the union of the two sets. There are 19 + 6 + 12 = 37 students who plan on joining either team.

b. How many students plan on joining both the soccer and baseball teams?

The students who plan on joining both the soccer and baseball teams are represented by the intersection of the two sets. There are 6 students who plan on joining both teams.

12-3 Probability with Permutations and Combinations

13. DINING Three boys and three girls go out to eat together. The restaurant only has round tables. Fred does not want any girl next to him and Gena does not want any boy next to her. How many arrangements are possible?

14. DANCE The dance committee consisted of 10 students. The committee will select three officers at random. What is the probability that Alice, David, and Carlene are selected?

15. COMPETITION From 32 students, 4 are to be randomly chosen for an academic challenge team. In how many ways can this be done?

Example 3

For a party, Lucita needs to seat four people at a round table. How many combinations are possible?

Because there is no fixed reference point, this is a circular permutation.

$P_n = (n - 1)!$ Formula for circular permutation

$P_4 = (4 - 1)!$ $n = 4$

$= 3!$ or 6 Simplify.

So, there are 6 ways for Lucita to seat four people at a round table.

CHAPTER 12
Study Guide and Review *Continued*

12-4 Geometric Probability

16. **GAMES** Measurements for a beanbag game are shown. What is the probability of each event?
    a. P(hole)
    b. P(no hole)

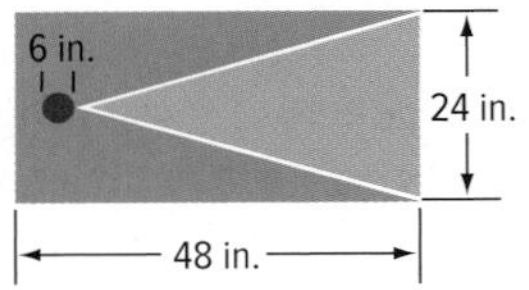

17. **POOL** Morgan, Phil, Callie, and Tyreese are sitting on the side of a pool in that order. Morgan is 2 feet from Phil. Phil is 4 feet from Callie. Callie is 3 feet from Tyreese. Oscar joins them.
    a. Find the probability that Oscar sits between Morgan and Phil.
    b. Find the probability that Oscar sits between Phil and Tyreese.

Example 4

A carnival game is shown.

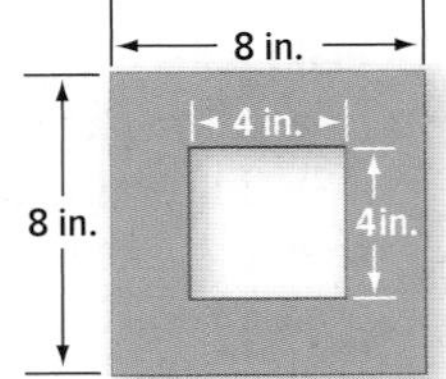

a. If Khianna threw 10 beanbags at the board, what is the probability that the beanbag went in the hole?

Area of hole $= 4 \cdot 4 = 16$

Area of board $= (8 \cdot 8) - 16 = 64 - 16$ or 48

$P(\text{hole}) = \frac{16}{64}$ or about 25%

b. What is the probability that the beanbag did not go in the hole?

$P(\text{no hole}) = \frac{48}{64}$ or about 75%

12-5 Probability and the Multiplication Rule

18. **MARBLES** A box contains 3 white marbles and 4 black marbles. What is the probability of drawing 2 black marbles and 1 white marble in a row without replacing any marbles?

19. **CARDS** Two cards are randomly chosen from a standard deck of cards with replacement. What is the probability of successfully drawing, in order, a three and then a queen?

20. **PIZZA** A nationwide survey found that 72% of people in the United States like pizza. If 3 people are randomly selected, what is the probability that all three like pizza?

Example 5

A bag contains 3 red, 2 white, and 6 blue marbles. What is the probability of drawing, in order, 1 red and 1 blue marble without replacement?

Because the marbles are not being replaced, the events are dependent events. Apply the multiplication rule for dependent events.

$$P(\text{red, red, blue}) = P(\text{red}) \cdot P(\text{blue}|\text{red})$$
$$= \frac{3}{11} \cdot \frac{6}{10}$$
$$= \frac{9}{55} \text{ or about } 16.4\%$$

12-6 Probability and the Addition Rule

21. **ROLLING CUBES** Two dot cubes are rolled. What is the probability that the sum of the numbers is 7 or 11?

22. **CARDS** A card is drawn from a deck of cards. Find the probability of drawing a 10 or a diamond.

23. **RAFFLE** A bag contains 40 raffle tickets numbered 1 through 40.

a. What is the probability that a ticket chosen is an even number or less than 5?

b. What is the probability that a ticket chosen is greater than 30 or less than 10?

Example 6

Two number cubes are rolled. What is the probability that the sum is 6 or doubles are rolled?

These are not mutually exclusive events because the sum of doubles can equal 6. Apply the addition rule for events that are not mutually exclusive.

$$P(\text{sum is 6 or doubles}) = P(\text{sum is 6}) + P(\text{doubles}) - P(\text{sum is 6 and doubles})$$
$$= \frac{5}{36} + \frac{6}{36} - \frac{1}{36}$$
$$= \frac{5}{18} \text{ or about 27.8\%}$$

12-7 Conditional Probability

24. **CARDS** A card is drawn from a deck of cards. If the card is a king, find the probability that it is a red card.

25. **MARBLES** A blue marble is selected at random from a bag of 4 blue and 9 green marbles and not replaced. What is the probability that a second marble selected will be green?

26. **TRANSPORTATION** The probability that a city bus arrives late is 0.24. The probability that the bus arrives late and it is raining is 0.02. What is the probability that it is raining given that the bus arrives late?

27. What is the probability that the sum of two dice will be at least 9, given that the first die is 5?

28. Eileen took two tests. The probability of her passing both tests is 0.6. The probability of her passing the first test is 0.8. What is the probability of her passing the second test given that she has passed the first test?

Example 7

A card is drawn from a deck of cards.

a. **If the card is a diamond card, find the probability that it is an ace.**

There are 13 cards in the sample space, and only one out of these cards is the ace. Therefore, $P(\text{ace}|\text{diamond}) = \frac{1}{13}$.

b. **If the card is an ace, find the probability that it is a diamond card.**

There are 4 cards in the sample space, and only one out of these cards is a diamond. Therefore, $P(\text{diamond}|\text{ace}) = \frac{1}{4}$.

Study Guide and Review *Continued*

12-8 Two-Way Frequency Tables

29. VOTING A political science group conducts a survey of 18–24-year-olds and 25–34-year-olds asking whether or not they are planning on voting in the next election. The relative frequency table below shows the results.

Age Grup	Planning to Vote	Not Planning to Vote	Totals
18–24	21.6%	25.2%	46.8%
25–34	12.6%	40.6%	53.2%
Totals	34.2%	65.8%	100%

a. Find the probability that a survey respondent is planning to vote given that he or she is in the 18–24 age group.

b. Find the probability that a survey respondent is in the 18–24 age group given that he or she is planning to vote.

c. Determine whether planning to vote is independent of age group.

30. The two-way table shows the gender and hair color of the students in a class.

	Brown Hair	Blonde Hair	Red Hair
Girl	3	6	3
Boy	5	4	1

a. Find the total number of students in the class.

b. A student is chosen at random from the class. Find the probability that the student

(i) is a girl

(ii) has blonde hair

(iii) is a girl or has red hair

Example 8

An art teacher conducts a survey of the junior and senior classes to see whether students are interested in a photography seminar. The two-way relative frequency table below shows the results. Determine whether interest in the photography seminar is independent of class.

Class	Interested	Not Interested	Totals
Junior	41.2%	15.6%	56.8%
Senior	24.9%	18.3%	43.2%
Totals	66.1%	33.9%	100%

56.8% of respondents were juniors and 66.1% of respondents were interested in the photography seminar, so one would expect $56.8\% \cdot 66.1\%$ or about 37.5% of respondents to be interested juniors.

Because the expected and actual joint relative frequencies are not the same, interest in the photography seminar is not independent of class.

CHAPTER 12

Practice Test

Point X is chosen at random on $\overline{AE}$. Find the probability of each event.

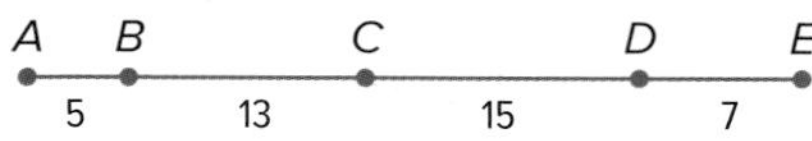

1. $P(X \text{ is on } \overline{AC})$
2. $P(X \text{ is on } \overline{CD})$
3. **BASEBALL** A baseball team fields 9 players. How many possible batting orders are there for the 9 players?
4. **TRAVEL** A traveling salesperson needs to visit four cities in her territory. How many distinct itineraries are there for visiting each city once?

Represent the sample space for each experiment by making an organized list, a table, and a tree diagram.

5. A box has 1 red ball, 1 green ball, and 1 blue ball. Two balls are drawn from the box one after the other, without replacement.
6. Shinsuke wants to adopt a pet and goes to his local humane society to find a dog or cat. While he is there, he decides to adopt two pets.
7. **ENGINEERING** An engineer is analyzing three factors that affect the quality of semiconductors: temperature, humidity, and material selection. There are 6 possible temperature settings, 4 possible humidity settings, and 6 choices of materials. How many combinations of settings are there?
8. A number cube is rolled. If the number rolled is less than 4, find the probability that it is a 1.
9. **PAINTBALL** Cordell is shooting a paintball gun at the target. What is the probability that he will shoot the shaded region?

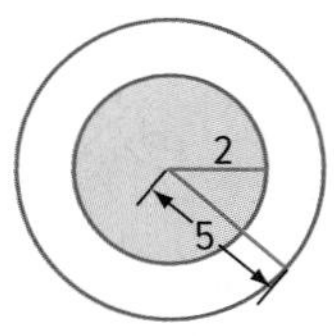

10. What is the probability that a phone number using the numbers 7, 7, 7, 2, 2, 2, and 6 will be 622-2777?
11. **TICKETS** Fifteen people entered the drawing at the right. What is the probability that Jodi, Dan, and Pilar all won the tickets?

Determine whether the events are *independent* or *dependent*. Then find the probability.

12. A deck of cards has 5 yellow, 5 pink, and 5 orange cards. Two cards are chosen from the deck with replacement. Find P(the first card is pink and the second card is pink).
13. There are 6 green, 2 red, 2 brown, 4 navy, and 2 purple marbles in a hat. Sadie picks 2 marbles from the hat without replacement. What is the probability that the first marble is brown and the second marble is not purple?
14. **SPORTS** Refer to the Venn diagram that represents the sports students chose to play at South High School last year.

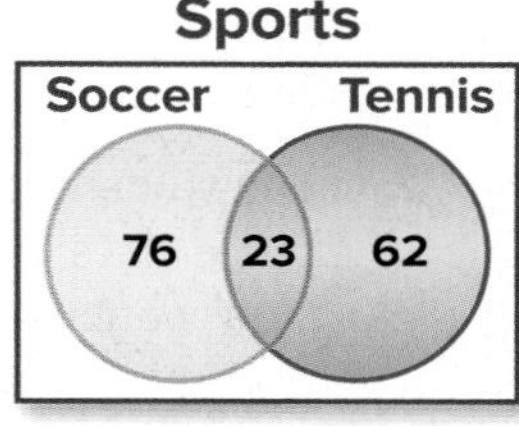

   a. Describe the sports that the students in the nonintersecting portion of the tennis region chose.

   b. How many students played soccer and tennis?

Determine whether the events are *mutually exclusive* or *not mutually exclusive*. Then determine the probability.

15. A card is drawn from a deck of cards. Find the probability of drawing an ace or a red card.
16. Two number cubes are rolled. Find the probability of getting a sum of 10 or 12.
17. If the chance of snow is 0.05, what is the probability that it will not snow?
18. **HOUSING** A real estate agent conducts a survey of men and women on whether they own a home or rent. Of the 50 men in the survey, 12 responded that they own a home. Of the 75 women in the survey, 18 responded that they own a home.

   a. Create a two-way frequency table and relative frequency table.

   b. Find the probability that a survey respondent is a woman given that the respondent owns a home.

   c. Find the probability that a survey respondent owns a home given that the respondent is a woman.

   d. Determine whether owning a home is independent of gender. Explain your reasoning.

CHAPTER 12

Preparing for Assessment

Performance Task

Provide a clear solution to each part of the task. Be sure to show all of your work, include all relevant drawings, and justify your answers.

School Events A high school is planning its calendar for the upcoming school year.

Part A

The school is creating individual teacher and student schedules. There are seven periods in one school day.

1. **Sense-Making** If a teacher teaches a different class all seven periods, determine the number of different schedule combinations that are possible for one teacher.
2. If one student is taking seven different classes, including one math class and one science class, determine the number of schedule combinations that are possible if the school wants to make sure that the math and science classes are not back-to-back.

Part B

The school is also planning their cafeteria options. They plan to offer customizable sandwiches daily. Students can choose between whole wheat and white bread. Students can also choose between turkey, roast beef, and hummus. If the student would like cheese, he or she can choose between Swiss and cheddar.

3. **Tools** Draw a tree diagram to represent the sample space for sandwich orders.

Part C

The school is creating a schedule for which teachers have bus duty each morning. They are planning to split the number of weeks evenly among the teachers who do not have a first-period class. One person proposed they use a spinner to determine which teacher is on duty each of the weeks. Another person proposed they take turns drawing pieces of paper with the week number on them.

4. Explain, in terms of independent and dependent events, which method the school should use.

Part D

The school is thinking about offering a new elective on musical history, but the only place it would fit in the schedule is at the same time as band class. The school wants to see how many students would be interested in this new class, so they take a survey of the students who are eligible to take the elective and record the results in the two-way table to the right.

5. Determine the probability that a randomly selected student is both interested in the class and does not take band. Round your answer to the nearest tenth, if necessary.
6. If the school waits until next semester, they can offer the new course during a period different than band. Explain why you think the school should or should not wait.

Test-Taking Strategy

Example

Read the problem. Identify what you need to know. Then use the information in the problem to solve.

Of the students who speak a foreign language at Marie's school, 18 speak Spanish, 14 speak French, and 16 speak German. There are 8 students who only speak Spanish, 7 who speak only German, 3 who speak Spanish and French, 2 who speak French and German, and 4 who speak all three languages. If a student is selected at random, what is the probability that he or she speaks Spanish or German, but not French?

A $\frac{7}{12}$ B $\frac{9}{16}$ C $\frac{2}{5}$ D $\frac{5}{18}$

Step 1 **Does the problem contain data? What would be the best display to use to organize the data? Why?**
Yes. A Venn diagram would be best because there are some students who speak more than one language.

Step 2 **How can you make your display as organized as possible? What can you add to your display as you work?**
I can label the display with all of the information given in the problem. After I see what is missing, I'll make any calculations I can to fill in the missing information.

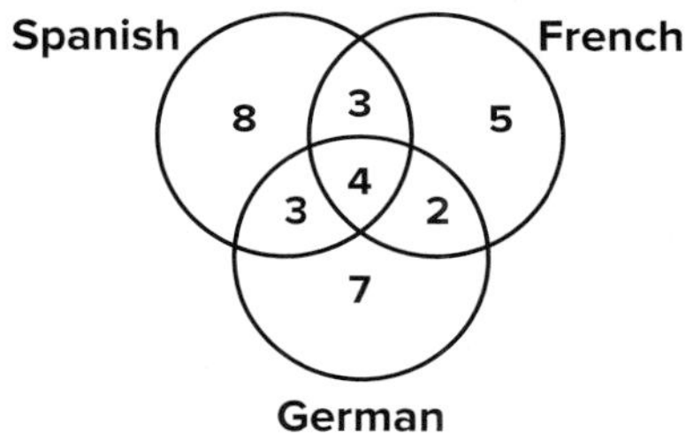

Step 3 **What is the correct answer?**
The correct answer is B.

Test-Taking Tip

Organizing Data
Sometimes you may be given a set of data that you need to analyze in order to solve items on a standardized test. Use this section to practice organizing data and to help you solve problems.

Apply the Strategy

Read the problem. Identify what you need to know. Then use the information in the problem to solve.

There are 10 sophomore, 8 junior, and 9 senior members on the student council. Each member is assigned to help plan one school activity during the year. There are 4 sophomores working on the field day and 6 working on the pep rally. Of the juniors, 2 are working on the field day and 5 are working on the school dance. There are 2 seniors working on the pep rally. If each activity has a total of 9 students helping to plan it, what is the probability that a randomly selected student council member is a junior or is working on the field day?

A $\frac{1}{5}$ B $\frac{4}{18}$ C $\frac{5}{9}$ D $\frac{2}{3}$

Answer the questions below.

a. Does the problem contain data? What would be the best display to use to organize the data?

b. How can you make your display as organized as possible? What can you add to your display as you work?

c. What is the correct answer?

CHAPTER 12

Preparing for Assessment

Cumulative Review

Read each question. Then fill in the correct answer on the answer document provided by your teacher or on a sheet of paper.

1. Kathleen goes on a business trip and takes with her 4 shirts, 3 pairs of slacks, 2 blazers, and 2 pairs of shoes. Assuming she wears one of each at a time, how many different combinations are possible?

2. A target for a penny toss consists of a square board that is 20 inches on each side with a painted square, as shown.

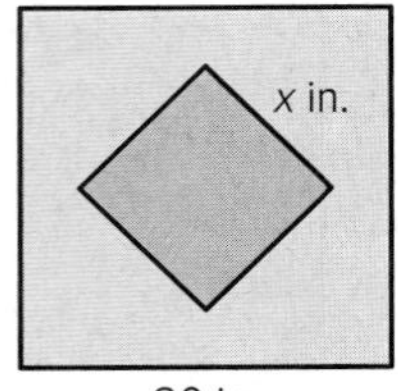

Assume pennies land on the board at random. What should be the value of x so that the probability of a penny landing in the painted square is 50%? Round to the nearest tenth.

3. Mark each of the following events as *independent* or *dependent*.

   a. A child rolls a number cube three times. What is the probability she will roll at 6 all three times?

   b. A woman has a male baby. What is the probability her second child will be female?

   c. A man does a load of laundry containing 14 socks, 2 of which are black. He pulls out two socks, one after the other. What is the probability they will both be black?

   d. A woman accidentally throws 4 used batteries in with her 12 new batteries. She pulls out a battery, plugs it into her remote, and realizes it was one of the used ones. She throws it away and pulls out another battery. What is the probability this one will also be a used one?

4. Point D is located on segment AC below, such that the probability of a random point being on BD is $\frac{1}{3}$ and the probability of the point being on AD is $\frac{3}{4}$. Plot point D on the number line.

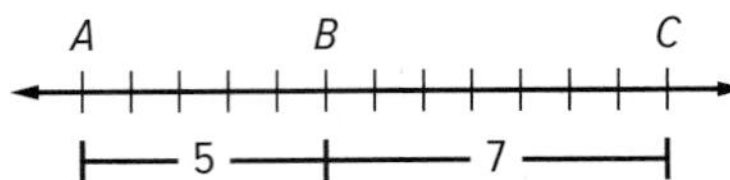

5. Megan has a deck of 10 cards that are numbered 0 through 9. She shuffles the deck and chooses 3 cards at random without replacement. What is the probability that she chooses the cards numbered 7, 8, and 9?

   ○ A $\frac{1}{720}$

   ○ B $\frac{1}{120}$

   ○ C $\frac{1}{40}$

   ○ D $\frac{3}{10}$

Test-Taking Tip

Question 5 The order in which the cards are chosen does not matter, so you should use combinations rather than permutations to solve the problem.

6. A bag has b blue marbles and r red marbles. Enrique chooses a marble from the bag without looking, notes the color, and then replaces it. Then he chooses another marble from the bag. Which expression represents the probability that he chooses a red marble both times?

   ○ A $\frac{2r}{b+r}$

   ○ B $\frac{r}{b+r} \cdot \frac{r-1}{b+r-1}$

   ○ C $\frac{b}{b+r} \cdot \frac{b}{b+r}$

   ○ D $\frac{r}{b+r} \cdot \frac{r}{b+r}$

7. A die is rolled and a coin is tossed. What is the probability that the die shows a 5 and the coin comes up Heads?

8. Determine whether each of the events below is *mutually* exclusive or not *mutually exclusive*. Explain your reasoning. Then find the probability of the event occurring. Round your answer to the nearest tenth.

a. drawing a card from a standard deck and getting an ace or a spade

b. drawing a card from a standard deck and getting a king or a queen

9. Serena throws a dart at the rectangular target, and it hits the target at random. Serena gets 20 points if the dart lands in the blue region and 35 points if the dart lands in the red region.

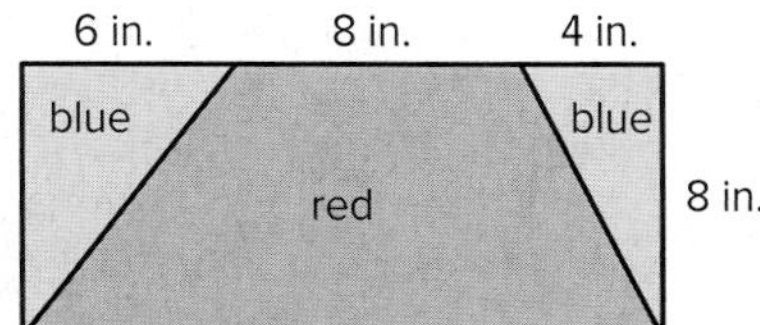

Which is the best estimate of the expected value from the throw?

- A 24.2
- B 27.5
- C 28.6
- D 30.8

10. A skydiver plans to land at a random point within a rectangular target. Part of the target is covered in grass and part is covered in gravel. Also, a county line passes through the target, as shown.

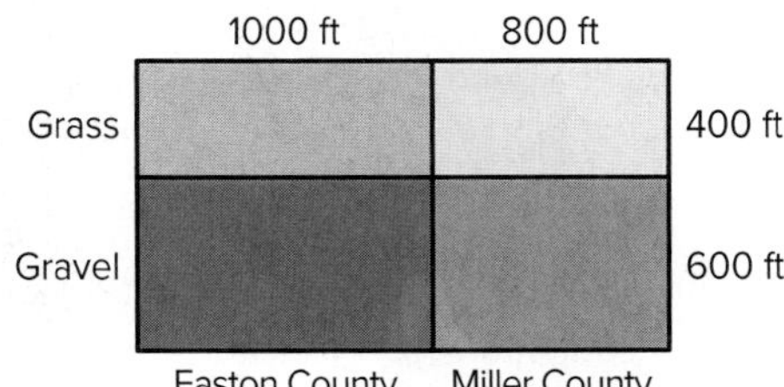

Which is closest to the probability that the skydiver lands on grass or in Miller County?

- A 18%
- B 44%
- C 67%
- D 84%

11. A news reporter is covering a state election. She asks 42 women whether they are supporting candidate A or B. Exactly three sevenths of the women say they support candidate A and the rest support candidate B. The reporter asks 35 men who they are supporting. Three fifths are supporting A and the rest are supporting B.

Complete the two-way table below with the data given in the problem. Then answer the question that follows.

	Supports Candidate A	Supports Candidate B	Totals
Women			
Men			
Totals			

What is the probability that a surveyed respondent is male, given that the respondent supports candidate A? Round your answer to the nearest tenth.

12. Felicity asked 100 students how they came to school one day. Each student walked or came by bicycle or came by car. 49 of the 100 students are girls. 10 of the girls came by car. 16 boys walked. 21 of the 41 students who came by bicycle are boys.

a. Find the total number of students who walked to school.

b. What is the probability that a student walked to school given that the student is a girl?

Need Extra Help?

If you missed Question...	1	2	3	4	5	6	7	8	9	10	11	12
Go to Lesson...	12-1	12-4	12-5	12-4	12-3	12-4	12-4	12-6	12-4	12-4	12-8	12-8

Student Handbook

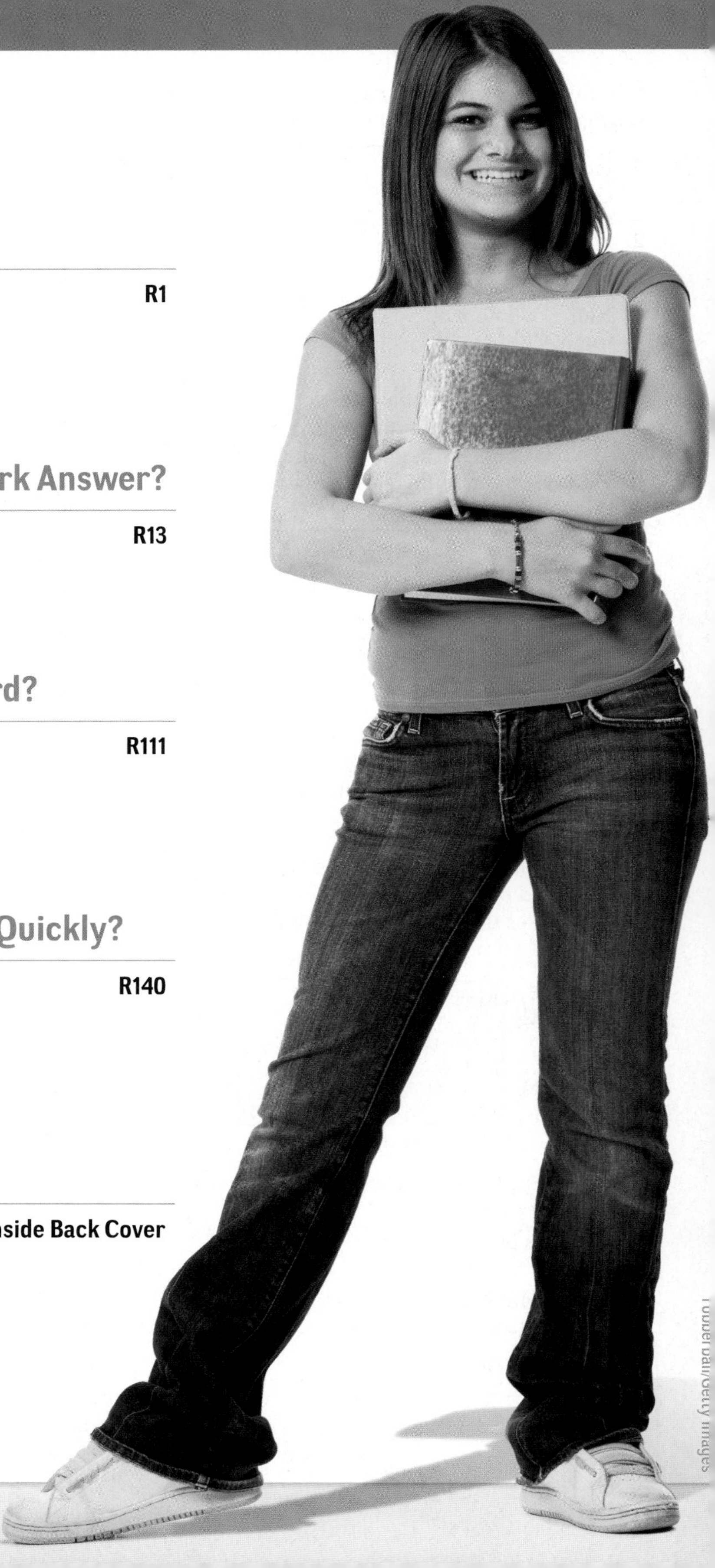

This **Student Handbook** can help you answer these questions.

What if I Need More Practice?

Extra Practice **R1**

The **Extra Practice** section provides additional problems for each lesson so you have ample opportunity to practice new skills.

What if I Need to Check a Homework Answer?

Selected Answers and Solutions **R13**

The answers to odd-numbered problems are included in **Selected Answers and Solutions**.

What if I Forget a Vocabulary Word?

Glossary/Glosario **R111**

The **English-Spanish Glossary** provides definitions and page numbers of important or difficult words used throughout the textbook.

What if I Need to Find Something Quickly?

Index **R140**

The **Index** alphabetically lists the subjects covered throughout the entire textbook and the pages on which each subject can be found.

What if I Forget a Formula?

Formulas and Measures, Symbols and Properties **Inside Back Cover**

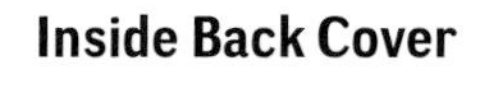

Inside the back cover of your math book is a list of **Formulas and Symbols** that are used in the book.

Extra Practice

CHAPTER 1 Tools of Geometry

Refer to the figure. (Lesson 1-1)

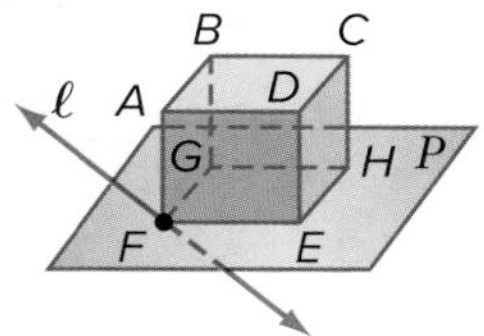

1. How many planes are shown in this figure?

2. Name the intersection of plane *ADE* with plane *P*.

3. Name two noncoplanar lines that do not intersect.

4. Find x and AB if B is between A and C, $AB = 3x$, $BC = 14$, and $AC = 41$. (Lesson 1-2)

5. **PLAYGROUND** The pivot or midpoint of a seesaw on a playground is 13.5 feet from the swing set. If the swing set is 8 feet from the edge of the seesaw when the seesaw is level, how long is the seesaw? (Lesson 1-3)

6. Find P on $\overline{MN}$ if P is $\frac{1}{3}$ the distance from $M(-3, -4)$ to $N(1.5, 2.5)$. (Lesson 1-3)

Copy the diagram shown, and extend each ray. Classify each angle as *right*, *acute*, or *obtuse*. Then use a protractor to measure the angle to the nearest degree. (Lesson 1-4)

7. $\angle AFC$

8. $\angle DFB$

9. $\angle CFD$

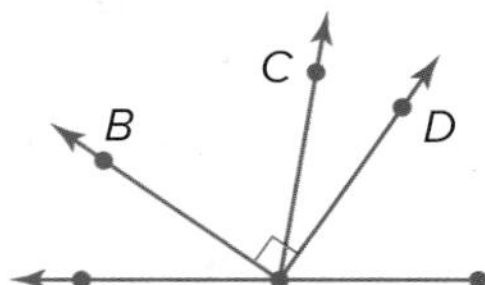

Determine whether each statement can be assumed from the figure. Explain. (Lesson 1-5)

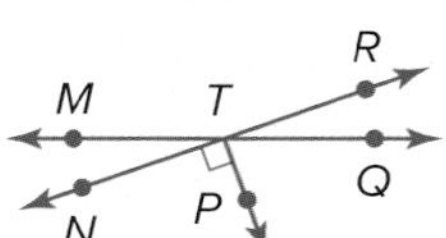

10. $\angle RTQ$ and $\angle MTN$ are vertical angles.

11. $\overrightarrow{PT}$ is perpendicular to $\overrightarrow{TM}$.

Find the perimeter or circumference and area of each figure. Round to the nearest tenth, if necessary. (Lesson 1-6)

12.

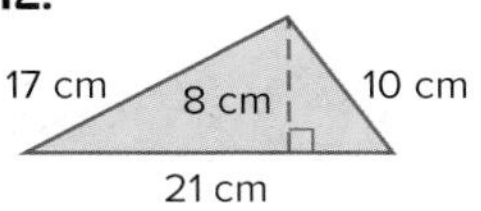

13.

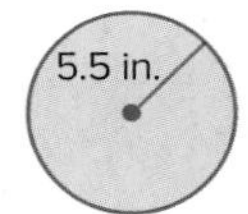

14. **PLANNING** Tim is responsible for renting tables for an awards banquet. He needs 20 tables, and he can choose between circular tables with a diameter of 5.5 feet and square tables with a side length of 5 feet. Which option should Tim choose so that the tables cover the *smallest* area? (Lesson 1-6)

15. Identify the type of congruence transformation shown as a *reflection*, *translation*, or *rotation*. (Lesson 1-7)

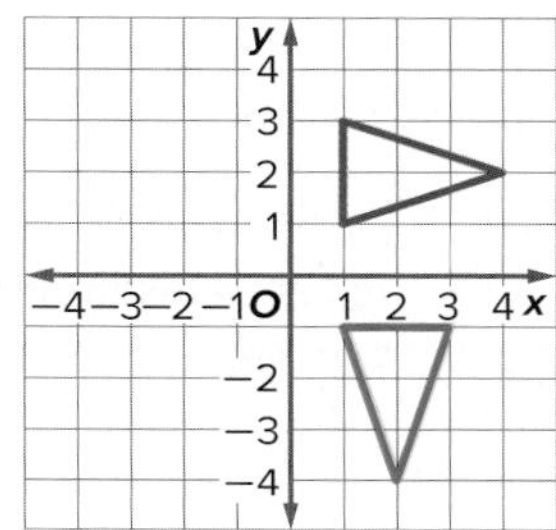

16. **FISH** Sarah has a fish tank with the dimensions shown. (Lesson 1-8)

    a. What is the surface area of the fish tank?

    b. If Sarah fills the tank to a depth of 17 inches, what will be the volume of the water in the tank?

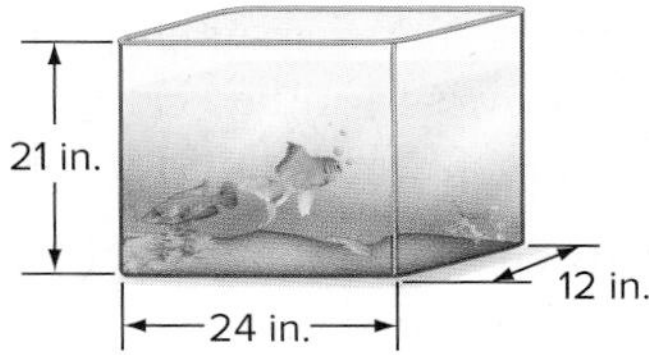

17. Identify the figure that can be formed from the given net. (Lesson 1-9)

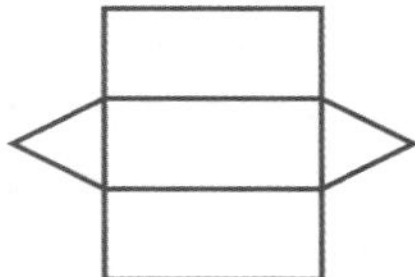

18. Determine the number of significant figures in the measurement 0.00205. (Lesson 1-10)

19. Find the absolute error of the measurement 10 meters. Explain its meaning. (Lesson 1-10)

20. Find the relative error of the measurement $3\frac{1}{8}$ inches. (Lesson 1-10)

1. Determine whether the statement below is *true* or *false*. If false, provide a counterexample. (Lesson 2-1)

   If $\overline{AB}$ is perpendicular to $\overleftrightarrow{CD}$, then exactly two right angles are formed.

Use the following statements to write a compound statement for each conjunction or disjunction. Then find its truth value. Explain your reasoning. (Lesson 2-2)
p: A prism has two bases.
q: A pyramid has two bases.
r: A sphere has no bases.

2. $p \wedge r$
3. $q \vee \sim r$

Determine the truth value of each conditional statement. If *true*, explain your reasoning. If *false*, give a counterexample. (Lesson 2-2)

4. If a number is divisible by 6, then it is divisible by 3.
5. If $x^4 = 16$, then $x = 2$.
6. **GRADES** Draw a valid conclusion from the statement below, if possible. Then state whether your conclusion was drawn using the Law of Detachment or the Law of Syllogism. If no valid conclusion can be drawn, write *no valid conclusion* and explain your reasoning. (Lesson 2-3)

   If Amy scores above a 90 on her chemistry test, she will make an A. If she makes an A, her parents will take her to the theater. Amy scored a 93 on her chemistry test.

Determine whether each conclusion is based on *inductive* or *deductive* reasoning. (Lesson 2-3)

7. Andrew's mom makes spaghetti for dinner on Mondays. Today is Monday. He concludes that his mom will make spaghetti for dinner.
8. If Beth turns in her report early, she receives extra credit. Beth turned in her report early. She concludes that she will receive extra credit.

Determine whether each statement is *always*, *sometimes*, or *never* true. Explain your reasoning. (Lesson 2-4)

9. If planes A and B intersect, then their intersection is a line.
10. If point A lies in plane P, then $\overleftrightarrow{AB}$ lies in plane P.
11. Lines p and q intersect in points M and N.
12. **PROOF** Prove the following. (Lesson 2-5)
    Given: $\overline{AC} \cong \overline{BD}$
    $\overline{EC} \cong \overline{ED}$
    Prove: $\overline{AE} \cong \overline{BE}$

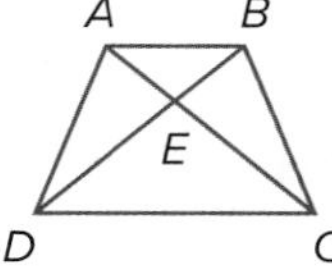

Find the measure of each numbered angle, and name the theorems that justify your work. (Lesson 2-6)

13. $\angle 2$ and $\angle 3$ are supplementary; $m\angle 2 = 149$

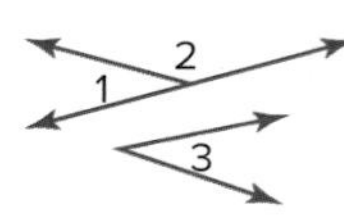

14. 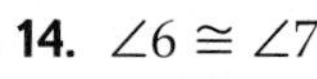 $\angle 6 \cong \angle 7$

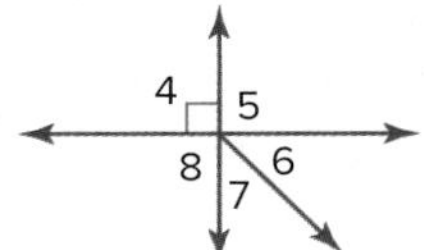

15. **PROOF** Given that $\angle BEC$ is a right angle, prove that $\angle AEB$ and $\angle CED$ are complementary. (Lesson 2-6)

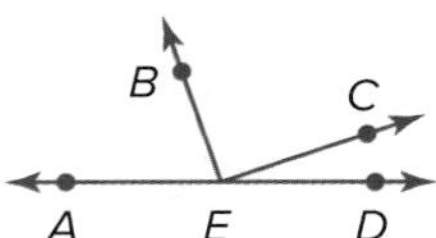

Find the value of the variables in each figure. Explain your reasoning. (Lesson 2-7)

16.

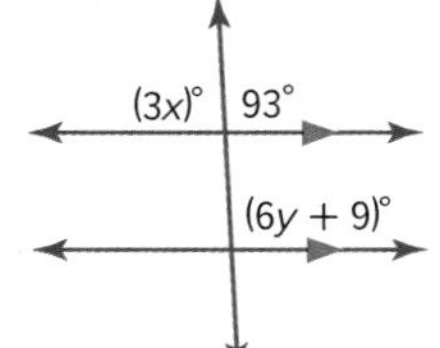

17.

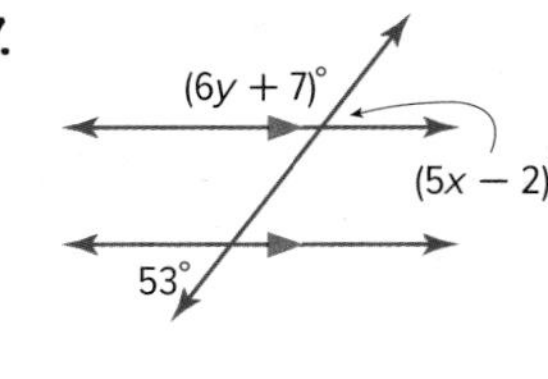

Write an equation in point-slope form of the line having the given slope that contains the given point. Then graph the line. (Lesson 2-8)

18. $m = -\frac{4}{5}$, $(10, -14)$
19. $m = 5$, $(-1, -3)$

Write an equation of the line through each pair of points in slope-intercept form. (Lesson 2-8)

20. $(4, 0)$ and $(-3, -7)$
21. $(-1, 8)$ and $(5, -4)$
22. **PROOF** Write a two-column proof. (Lesson 2-9)
    Given: $\angle 1$ and $\angle 8$ are supplementary.
    Prove: $\ell \parallel m$

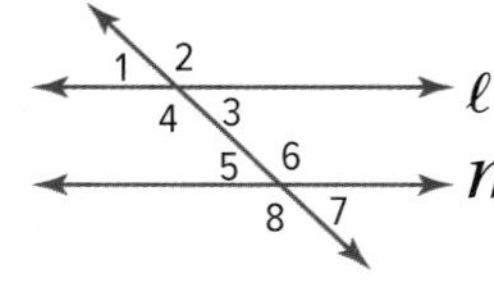

Find the distance from P to ℓ. (Lesson 2-10)

23. Line ℓ contains points $(0, 3)$ and $(-4, -9)$. Point P has coordinates $(-6, -5)$.
24. Line ℓ contains points $(-5, 6)$ and $(1, -6)$. Point P has coordinates $(6, 4)$.

Graph each figure and its image under the given reflection. (Lesson 3-1)

1. $\triangle ABC$ with vertices $A(-1, -4)$, $B(-5, 3)$, and $C(0, 5)$ in the line $y = x$

2. quadrilateral $WXYZ$ with vertices $W(-3, -2)$, $X(-4, 1)$, $Y(1, 4)$, and $Z(2, -2)$ in the y-axis

3. Given rectangles with vertices $A(6, 3)$, $B(5, 3)$, $C(5, -1)$, and $D(6, -1)$ and $M(6, -3)$, $N(5, -3)$, $O(5, 1)$, and $P(6, 1)$, describe the transformation that maps $ABCD$ to $MNOP$ using coordinate notation. (Lesson 3-1)

4. **CAMPING** Jin and Tom plan to hike to the rock bridge one day, return to camp, and then hike to the falls on the next day. Where along the trail should they place their camp in order to minimize the distance they must hike? (Lesson 3-1)

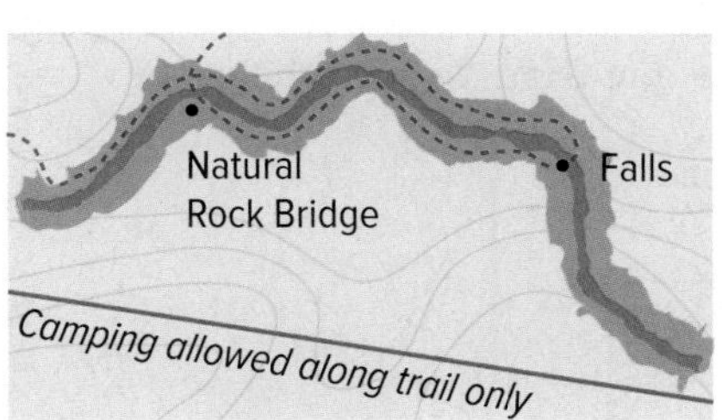

5. **MAPS** Caleb's house and several places he visits are shown on the grid. (Lesson 3-2)

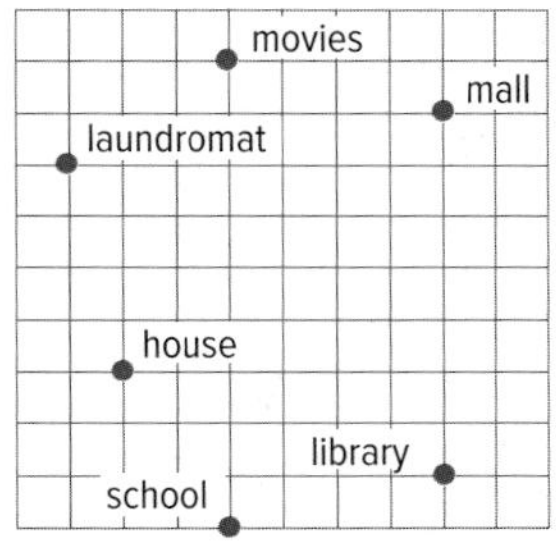

   a. If Caleb leaves his house and goes one block west and four blocks north, what is his new location?

   b. Write a translation vector that will take Caleb from the library to the movies.

6. Given triangles with vertices $D(4, 3)$, $E(1, 3)$, and $F(6, -1)$ and $P(-2, 2)$, $Q(-5, 2)$, and $R(0, -2)$, describe the transformation that maps $\triangle DEF$ to $\triangle PQR$ using coordinate notation. (Lesson 3-2)

Graph each figure and its image after the specified rotation about the origin. (Lesson 3-3)

7. $\triangle PQR$ with vertices $P(-1, -2)$, $Q(-5, -4)$, and $R(-3, -6)$; $90°$

8. parallelogram $WXYZ$ with vertices $W(-3, 3)$, $X(-2, 7)$, $Y(4, 5)$ and $Z(3, 1)$; $180°$

9. Given triangles with vertices $G(6, 3)$, $H(2, 4)$, and $J(3, -1)$ and $W(-3, 6)$, $X(-4, 2)$, and $Y(1, 3)$, describe the transformation that maps $\triangle GHJ$ to $\triangle WXY$ using coordinate notation. (Lesson 3-3)

10. **CLOCKS** Anna looks at a clock at 11:05. When she looks at the clock for a second time during the same hour, the minute hand has rotated 270°. At what time does Anna look at the clock for the second time? (Lesson 3-3)

Graph each figure and its image after the indicated glide reflection. (Lesson 3-4)

11. $\triangle DEF$: $D(-5, 1)$, $E(-3, 5)$, $F(0, 3)$
    translation: along $\langle -4, 3 \rangle$; reflection: in x-axis

12. $\triangle MNP$: $M(2, 5)$, $N(6, 2)$, $P(8, 6)$
    translation: along $\langle 2, 4 \rangle$; reflection: in $y = x$

13. Given rectangles with vertices $D(4, 4)$, $E(2, 4)$, $F(2, 7)$, and $G(4, 7)$ and $W(-6, 5)$, $X(-4, 5)$, $Y(-4, 8)$, and $Z(6, 8)$ describe the transformation that maps $DEFG$ to $WXYZ$ using coordinate notation. (Lesson 3–3)

Graph each figure and its image after the indicated composition of transformations. (Lesson 3-4)

14. $\overline{XY}$: $X(7, 9)$ and $Y(2, 1)$
    counterclockwise rotation of 90°; translation along $\langle -5, -2 \rangle$

15. $\overline{AB}$: $A(-4, -6)$ and $B(-2, 5)$
    reflection in x-axis; counterclockwise rotation of 270°

ALPHABET Determine whether each letter below has *line* symmetry, *rotational* symmetry, *both*, or *neither*. Draw all lines of symmetry and state their number. Then determine the center of rotational symmetry and state the order and magnitude of rotational symmetry. (Lesson 3-5)

16. B

17. X

Extra Practice

1. **BICYCLING** In the bicycle shown, $m\angle C = 60$ and $m\angle D = 105$. Find $m\angle B$. (Lesson 4-1)

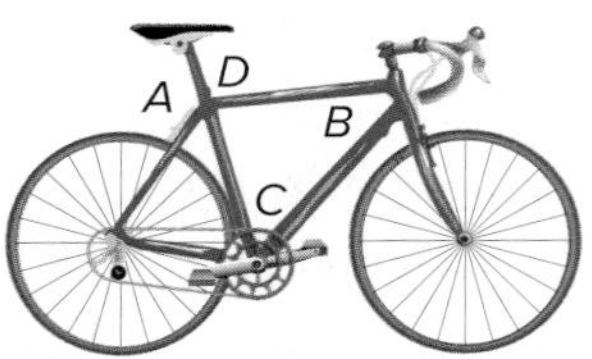

Find each measure. (Lesson 4-1)

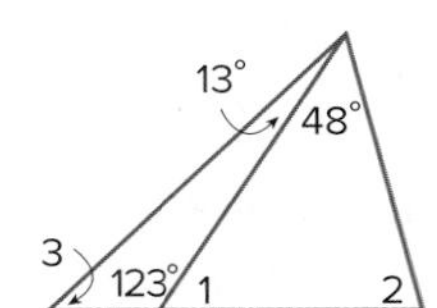

2. $m\angle 1$
3. $m\angle 2$
4. $m\angle 3$
5. Show that the polygons are congruent by using rigid motions and by identifying all congruent corresponding parts. Then write a congruence statement. (Lesson 4-2)

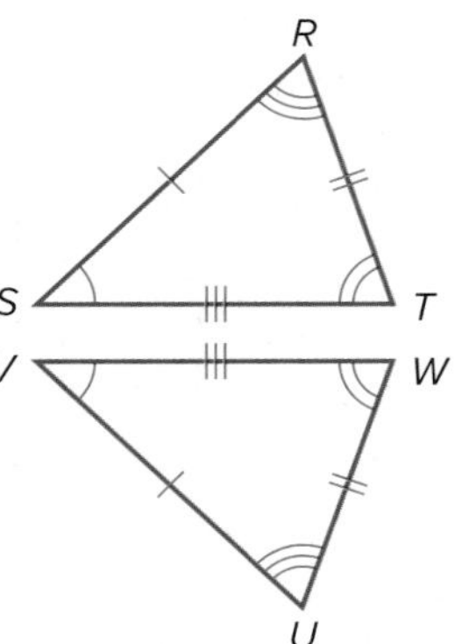

In the figure, $\triangle ABC \cong \triangle DEF$. Find each value. (Lesson 4-2)

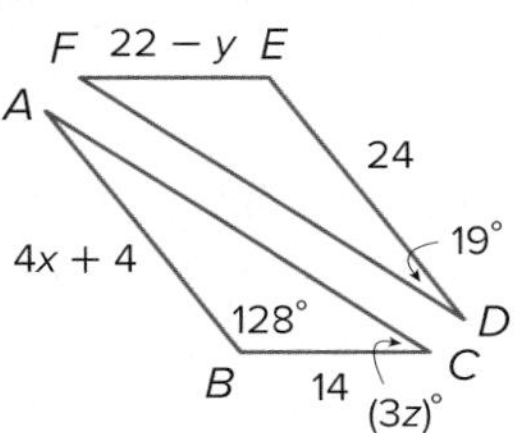

6. x
7. y
8. z

Determine whether $\triangle PQR \cong \triangle XYZ$. Explain. (Lesson 4-3)

9. $P(-4, 2)$, $Q(2, 2)$, $R(2, 8)$; $X(-1, -3)$, $Y(5, -3)$, $Z(5, 4)$
10. $P(-2, 4)$, $Q(-7, 3)$, $R(0, 9)$; $X(3, 6)$, $Y(2, 1)$, $Z(8, 8)$
11. **PROOF** Write a two-column proof. (Lesson 4-3)

Given: $\overline{AB} \cong \overline{DE}$,

$\overline{AC} \cong \overline{DF}$,

$\overline{AB} \parallel \overline{DE}$

Prove: $\triangle ABC \cong \triangle DEF$

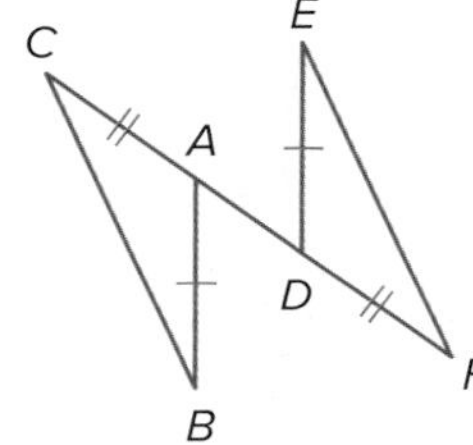

12. **PROOF** Write a paragraph proof. (Lesson 4-4)

Given: $m\angle PRQ = 90$

$m\angle RPT = 90$

$\angle Q \cong \angle T$

Prove: $\triangle QRP \cong \triangle TPR$

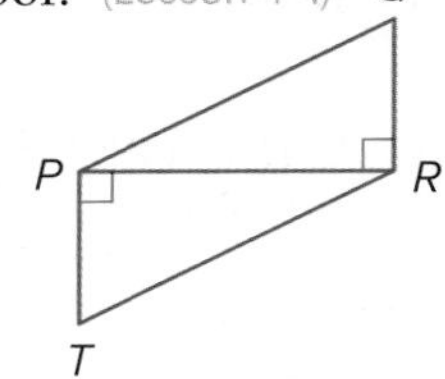

13. **NATURE** What is the approximate width of the creek shown below? Explain your reasoning. (Lesson 4-4)

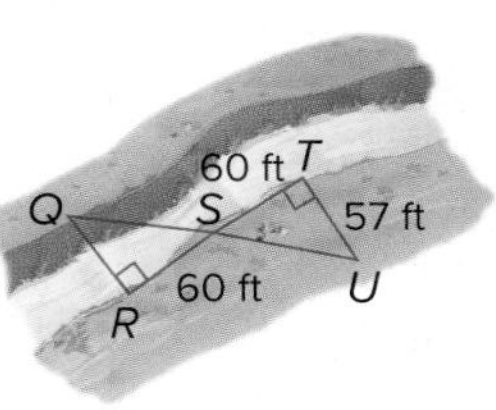

Determine whether each pair of triangles is congruent. If yes, include the theorem or postulate that applies, and describe the series of rigid motions that map one triangle onto the other. (Lesson 4-5)

14.

15.

16.

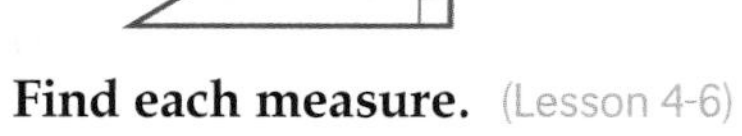

Find each measure. (Lesson 4-6)

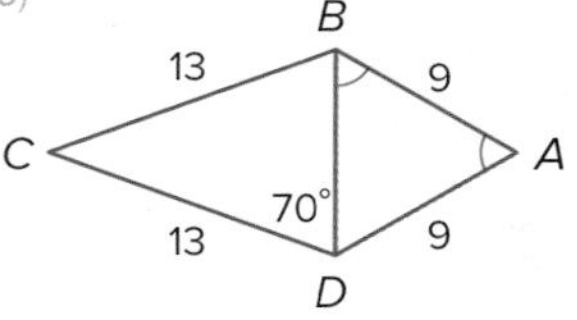

17. $m\angle BDA$
18. BD
19. $m\angle BCD$
20. **CAMPUS** The gym at Alex's school is located 70 feet south and 90 feet west of the main building. The vocational building is located 120 feet south and 200 feet east of the main building. Show that the triangle formed by these three buildings is scalene. Explain your reasoning. (Lesson 4-7)

1. **OFFICE DESIGN** The copy machine C, filing cabinet F, and supply shelves S are positioned in an office as shown. Copy the diagram, and find the location for the center of a work table so that it is the same distance from all three points. (Lesson 5-1)

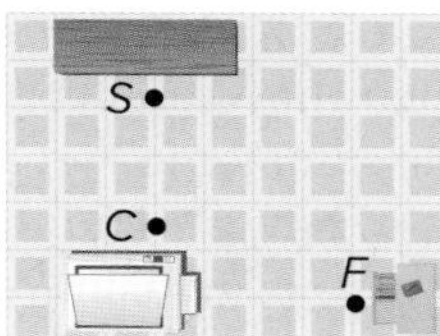

Point T is the incenter of $\triangle MNP$. Find each measure below. Round to the nearest tenth, if necessary. (Lesson 5-1)

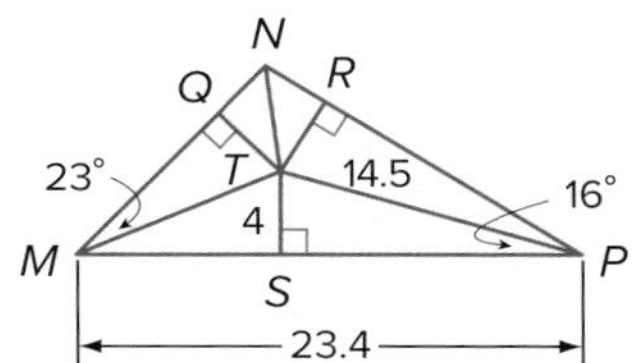

2. QT
3. MT
4. $m\angle PNT$

In $\triangle ABC$, $CH = 70\frac{2}{3}$, $AG = 85$, and $DH = 20\frac{1}{3}$. Find each length. (Lesson 5-2)

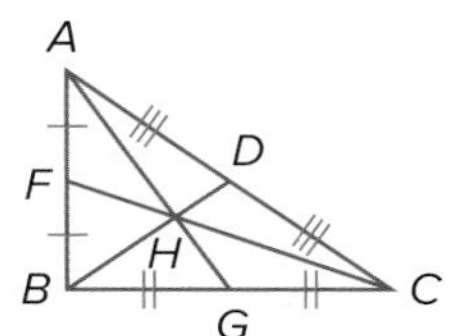

5. FH
6. BD
7. AH

Find the indicated point of concurrency for each triangle with the given vertices. (Lesson 5-2)

8. centroid; $A(-4, -5)$, $B(-1, 2)$, $C(2, -4)$
9. orthocenter; $M(-9, -2)$, $N(1, 8)$, $P(9, -8)$

List the angles and sides of each triangle in order from smallest to largest. (Lesson 5-3)

10. 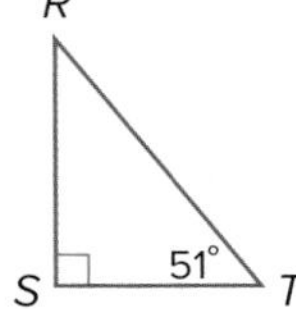

11. 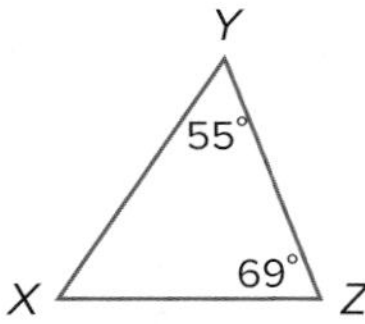

12. **MAPS** Three cities lie in a triangle as shown. Which two cities are the farthest apart? (Lesson 5-3)

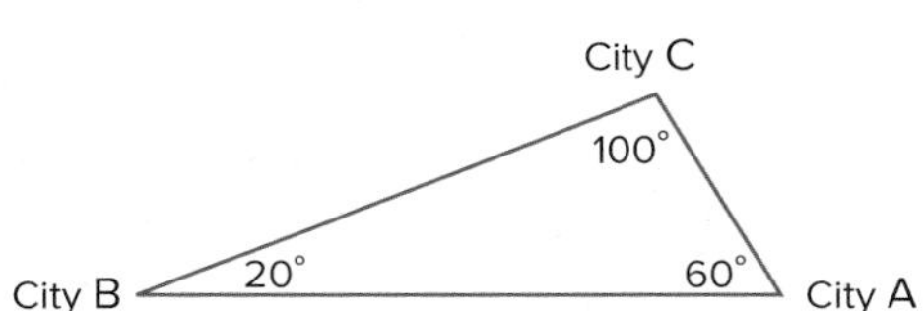

Write an indirect proof of each statement. (Lesson 5-4)

13. If $-7 + 6x < 11$, then $x < 3$.
14. If $5 - 2x < -13$, then $x > 9$.
15. **JOBS** Nate worked for 6 hours today with two breaks. Use indirect reasoning to show that he worked for longer than two hours without a break at some point during his shift. (Lesson 5-4)
16. Write an indirect proof of the statement below. (Lesson 5-4)

    A triangle can have at most one obtuse angle.

17. Is it possible to form a triangle with side lengths 3 centimeters, 8 centimeters, and 11 centimeters? If not, explain why not. (Lesson 5-5)

Find the range for the measure of the third side of a triangle given the measures of two sides. (Lesson 5-5)

18. 9 ft, 16 ft
19. 1.7 m, 2.9 m

20. Determine the possible values of x. (Lesson 5-5)

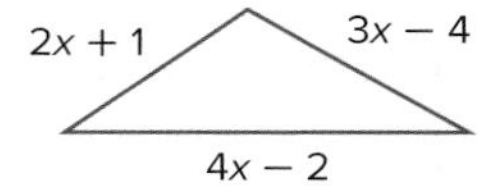

21. **SNOWSHOEING** Two groups of friends leave the same cabin to go snowshoeing. Group A goes 2.5 miles due north and then turns 79° east of north and travels an additional 2.5 miles. Group B goes 2.5 miles due south and then turns 86° east of south and travels an additional 2.5 miles. Who is now closer to the cabin? Use a diagram to explain your reasoning. (Lesson 5-6)

Find the range of values containing x. (Lesson 5-6)

22. 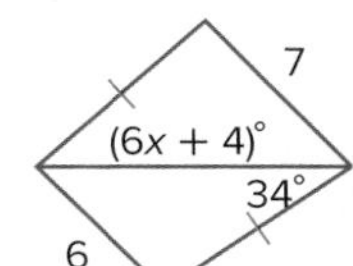

23. 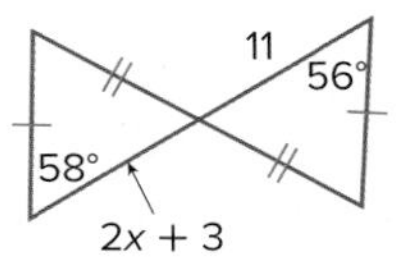

24. **PROOF** Write a two-column proof. (Lesson 5-6)

    Given: $\overline{AB} \cong \overline{CD}$,
    $m\angle CDB > m\angle BAC$
    $m\angle BDA > m\angle CAD$

    Prove: $AC > DB$

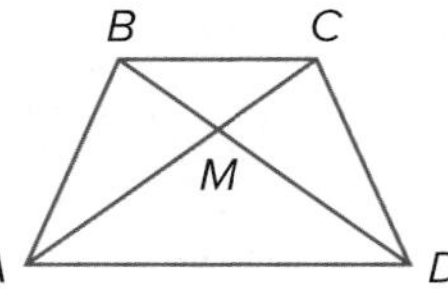

Extra Practice

Find the value of x in each diagram. (Lesson 6-1)

1. $(9x-3)°$ $(13x+2)°$ $(7x-6)°$ $(4x+4)°$

2.

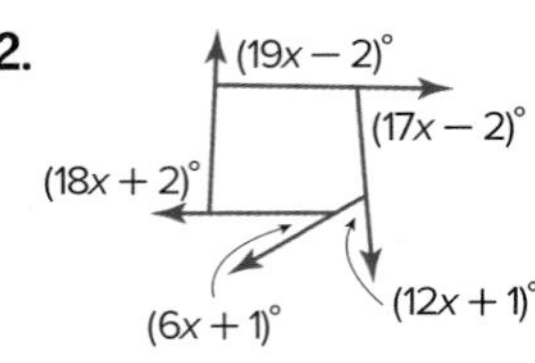

Use parallelogram $ABCD$ to find each measure. (Lesson 6-2)

3. $m\angle A$

4. AD

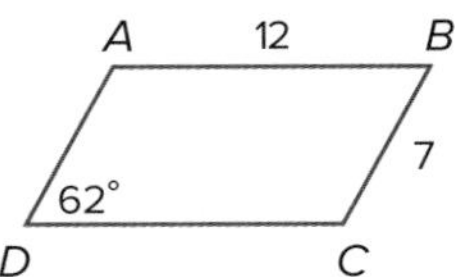

5. AUTO REPAIR A scissor jack is used to lift part of a car to make repairs. $ABCD$ is a parallelogram. As the jack is raised, $m\angle A$ and $m\angle C$ increase. Explain what must happen to $m\angle B$ and $m\angle D$. (Lesson 6-2)

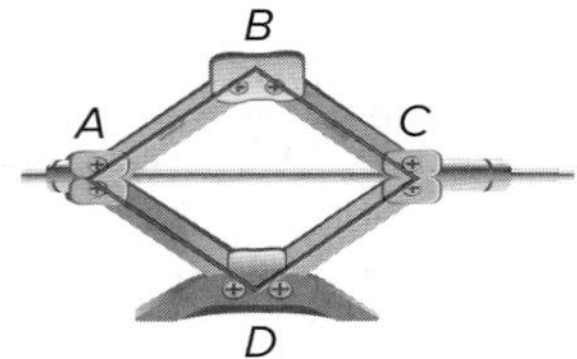

Find the value of each variable in each parallelogram. (Lesson 6-2)

6.

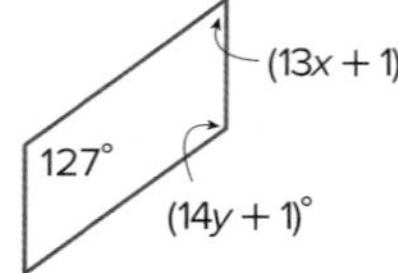

7.

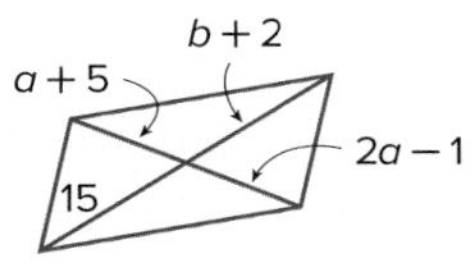

Determine whether each quadrilateral is a parallelogram. Justify your answer. (Lesson 6-3)

8.

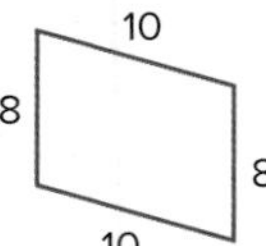

9.

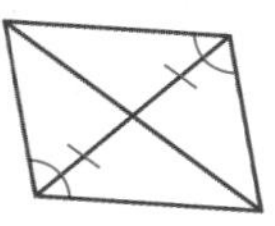

10. DESIGN Valerie is using the design shown for the front of a sweater. If all of the blocks of color are congruent parallelograms, write a two-column proof to show that $ACEG$ is a parallelogram. (Lesson 6-3)

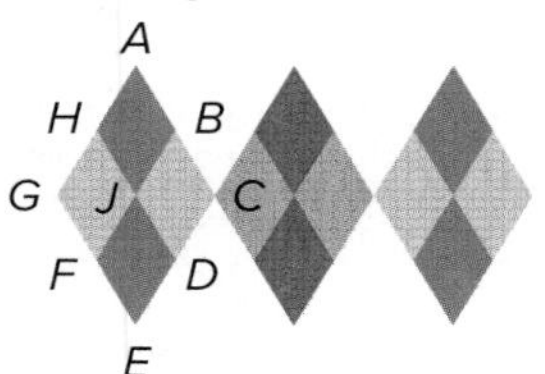

11. Graph the quadrilateral with vertices $A(0, 3)$, $B(10, 4)$, $C(6, -1)$, and $D(-4, -2)$. Determine whether $ABCD$ is a parallelogram. Justify your answer. (Lesson 6-3)

12. SNOW Four boys are throwing snowballs, such that their positions are the vertices of a rectangle as shown. If Carl is 100 feet from Ben, and Abe and Ben throw snowballs along the diagonals at the same time and the snowballs collide in mid air, how far had the snowballs traveled at the time of collision? (Lesson 6-4)

Quadrilateral $MNPQ$ is a rectangle. (Lesson 6-4)

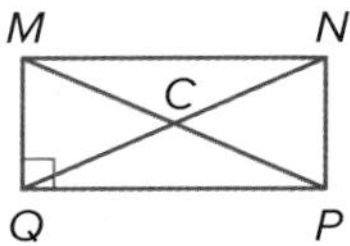

13. If $m\angle MNQ = 3x + 3$ and $m\angle QNP = 10x - 4$, find $m\angle MQN$.

14. If $MC = 4.5x - 2.5$ and $NC = 4x + 1.5$, find QN.

15. PROOF Write a proof to show that the four smaller triangles formed by the diagonals of a rectangle are isosceles. (Lesson 6-4)

Quadrilateral $ABCD$ is a rhombus. Find each value or measure. (Lesson 6-5)

16. $m\angle DBC$

17. DC

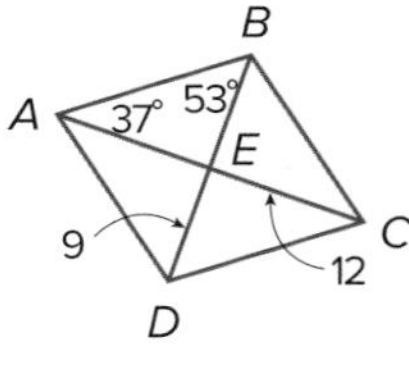

18. Determine whether ▱ $WXYZ$ with verticies $W(-2, 0)$, $X(1, 1)$, $Y(2, -2)$, $Z(-1, -3)$ is a *rhombus*, a *rectangle*, or a *square*. List all that apply. Explain. (Lesson 6-5)

Find each measure. (Lesson 6-6)

19. $m\angle A$

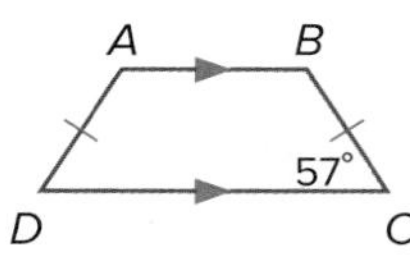

20. MR

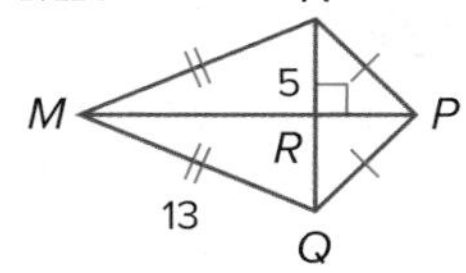

1. **NATURE** The diameter of a snowflake is 2 millimeters. If the diameter appears to be 3 centimeters when viewed under a microscope, what magnification setting (scale factor) was used? (Lesson 7-1)

Graph the image of each polygon with the given vertices after a dilation centered at the origin with the given scale factor. (Lesson 7-1)

2. $A(-5, -4)$, $B(-2, -3)$, $C(-1, -6)$, $D(-4, -8)$; $k = \frac{1}{2}$
3. $X(2, 4)$, $Y(4, 0)$, $Z(5, 5)$; $k = 1.5$

Determine whether each pair of figures is similar. If so, write the similarity statement and scale factor. If not, explain your reasoning. (Lesson 7-2)

4. 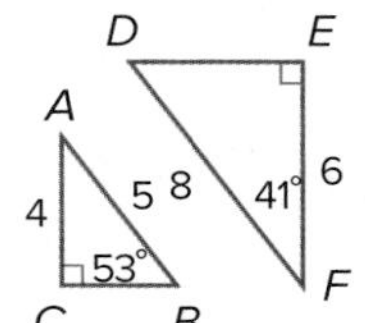

5. 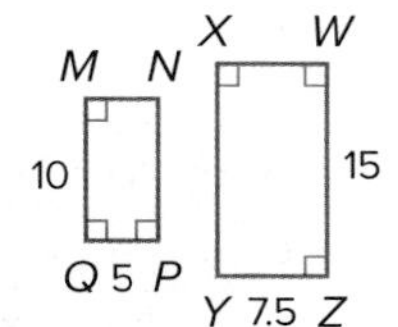

Each pair of polygons is similar. Find the value of x. (Lesson 7-3)

6. 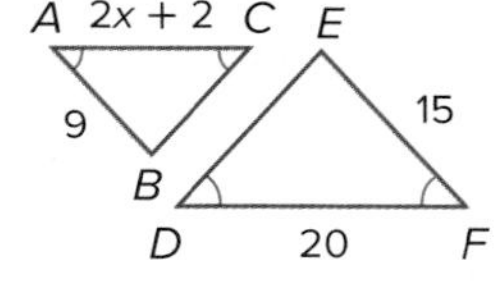

7. 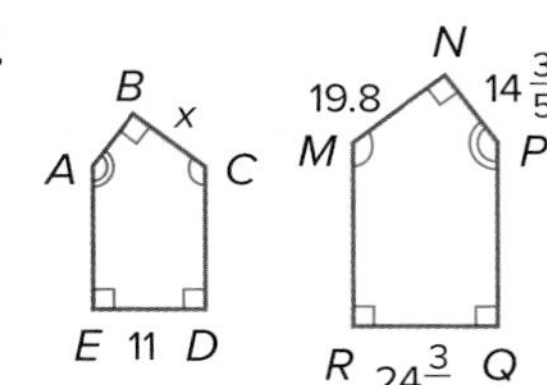

8. **FOOSBALL** Jason wants to determine if his foosball table is similar to his school's soccer field. Both the field and the table are rectangular. The dimensions of the table are 30 inches by $55\frac{1}{2}$ inches, and the dimensions of the field are 60 yards by 110 yards. Are the table and the field similar? Explain. (Lesson 7-2)

9. **HEIGHT** When Rachel stands next to her cousin, Rachel's shadow is 2 feet long and her cousin's shadow is 1 foot long. If Rachel is 5 feet 6 inches tall, how tall is her cousin? (Lesson 7-3)

Determine whether the triangles are similar. If so, write a similarity statement. Explain your reasoning. (Lessons 7-3 and 7-4)

10. 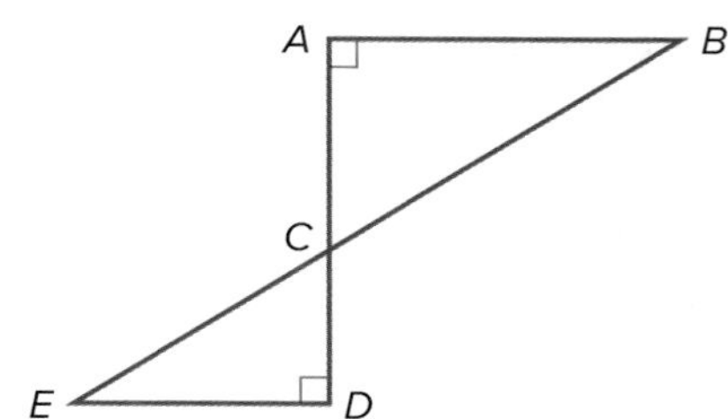

11. 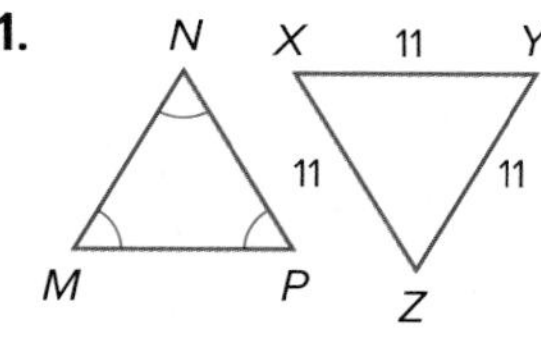

12. 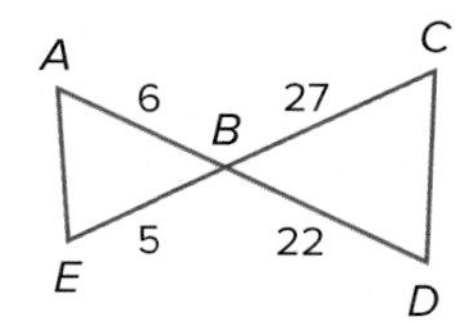

13. 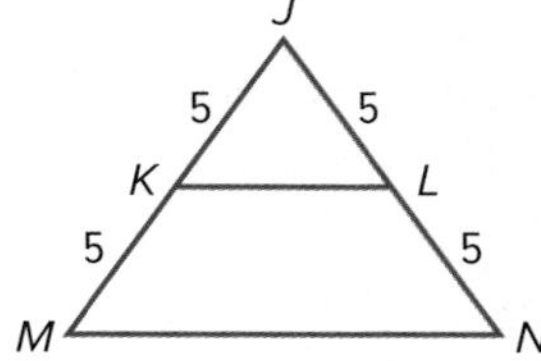

Determine whether the polygons with the given vertices are similar. Explain. (Lesson 7-4)

14. $A(0, -2)$, $B(-1, 3)$, $C(3, -2)$; $X(0, -6)$, $Y(-3, 9)$, $Z(9, -6)$
15. $D(-6, 6)$, $E(6, 2)$, $F(-2, -4)$; $M(-3, 3)$, $N(3, 1)$, $P(-1, -2)$

Refer to the figure shown. (Lesson 7-5)

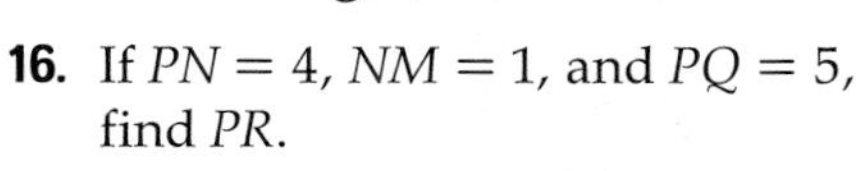

16. If $PN = 4$, $NM = 1$, and $PQ = 5$, find PR.
17. If $PR = 13$, $PQ = 9$, and $NM = 3$, find PN.

$\overline{DE}$, $\overline{EF}$, and $\overline{FD}$ are midsegments of $\triangle ABC$. Find the value of x. (Lesson 7-5)

18. 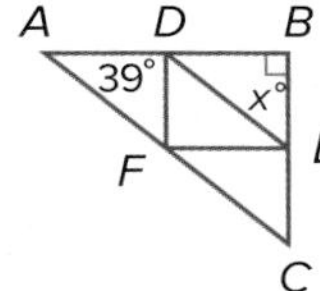

19. 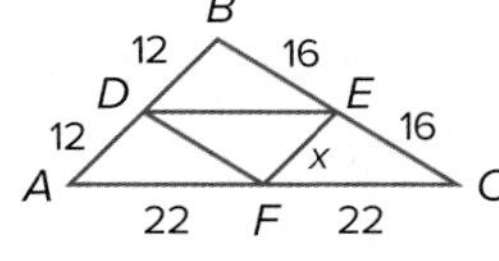

Find x. (Lesson 7-6)

20. 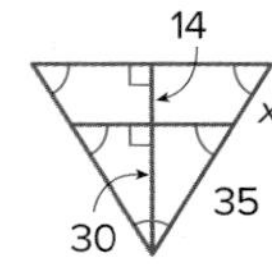

21. 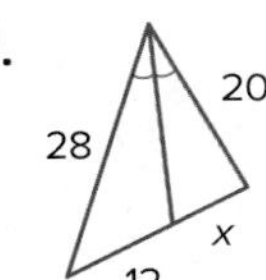

Extra Practice

Find the geometric mean between each pair of numbers. (Lesson 8-1)

1. 7 and 12
2. 8 and 36

Find x, y, and z. (Lesson 8-1)

3.
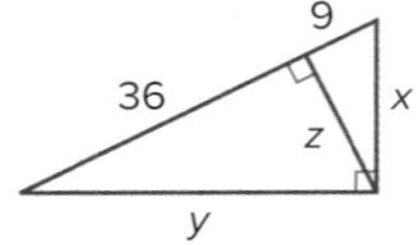

4.
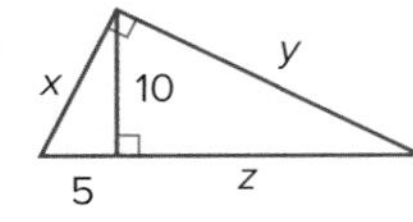

Find x. (Lesson 8-2)

5.
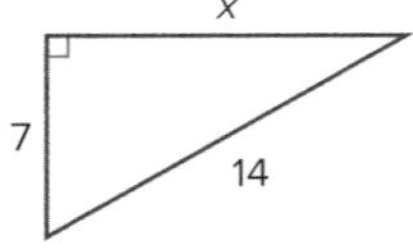

6.
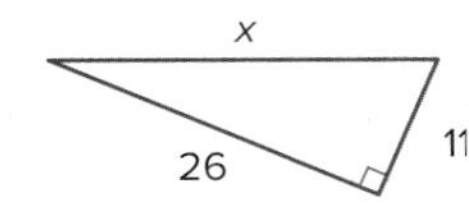

Determine whether each set of numbers can be the measures of the sides of a triangle. If so, classify the triangle as *acute*, *obtuse*, or *right*. Justify your answer. (Lesson 8-2)

7. 24, 32, 41
8. 17.5, 60, 62.5

Find x. (Lesson 8-3)

9.
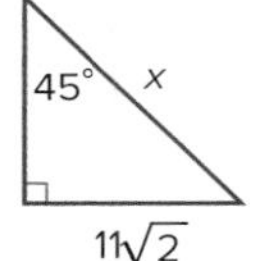

10.
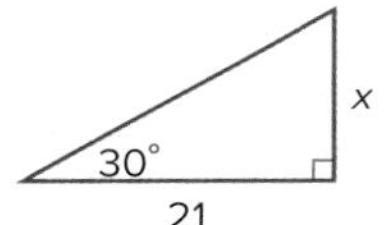

11.
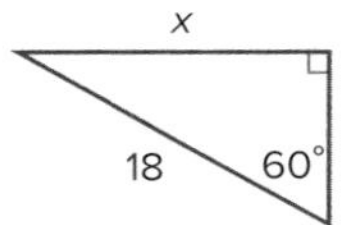

12.
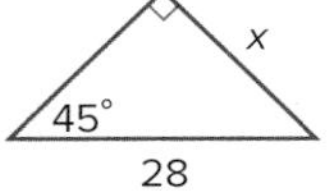

Find x. Round to the nearest tenth, if necessary. (Lesson 8-4)

13.
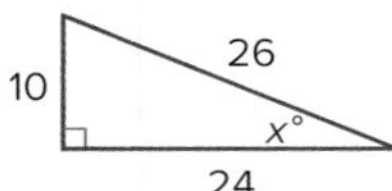

14.
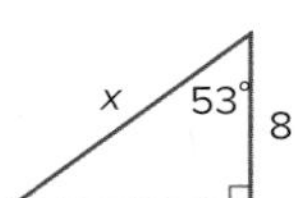

15. **SKATEBOARDING** Lindsey is building a skateboard ramp. She wants the ramp to be 1 foot tall at the end and she wants it to make a 15° angle with the ground. What length of board should she buy for the ramp itself? Round to the nearest foot. (Lesson 8-4)

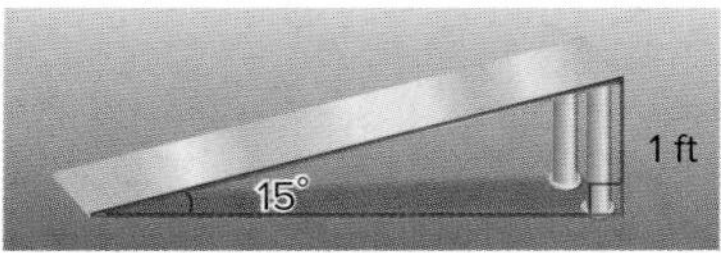

16. **BUILDINGS** Kara is standing about 50 feet from the base of her apartment building, looking up at it with an angle of elevation of 75°. What is the approximate height of Kara's building? (Lesson 8-5)

17. **ROLLER COASTERS** Evan is looking down the hill of a roller coaster from a height of 75 feet with an angle of depression of about 70°. What is the approximate horizontal distance from the top of the hill to the bottom of the hill? (Lesson 8-5)

18. **MOVIES** Kim is sitting in the row behind her friend Somi at the movies. Kim is looking at the screen with an angle of elevation of about 27° and Somi's angle of elevation is about 29°. If there are 3 feet between each row of seats, about how tall is the movie screen? (Lesson 8-5)

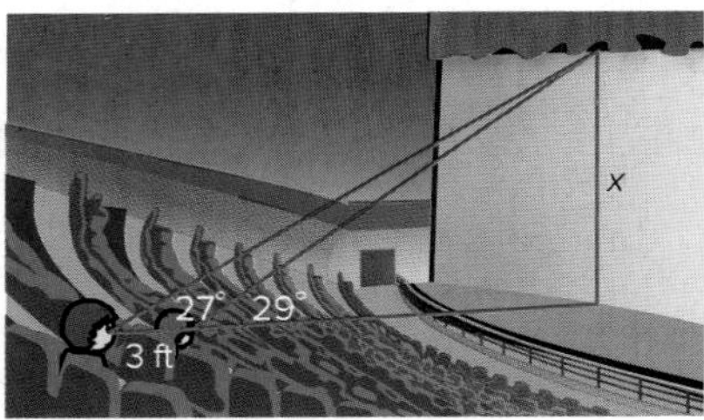

Solve each triangle. Round side lengths to the nearest tenth and angle measures to the nearest degree. (Lesson 8-6)

19.
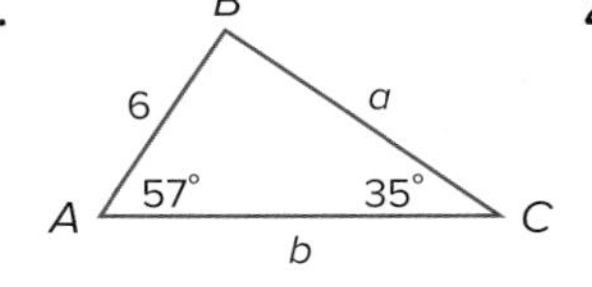

20.
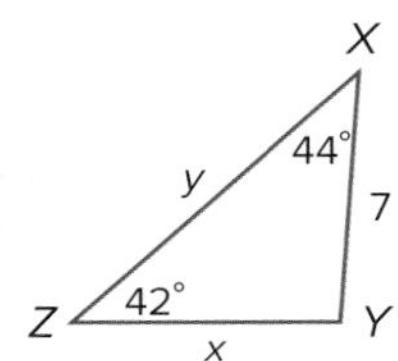

Solve each triangle. Round side lengths to the nearest tenth and angle measures to the nearest degree. (Lesson 8-7)

21.
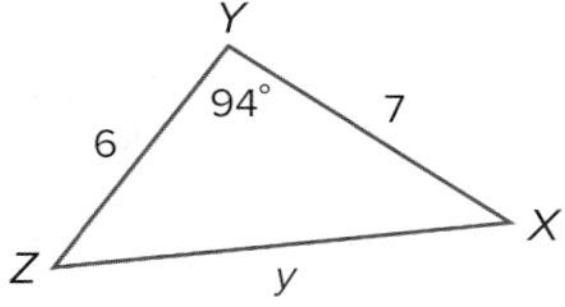

22.
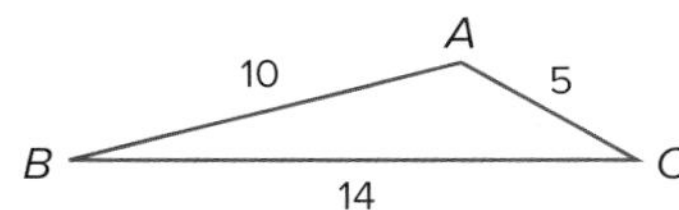

The diameter of the smaller circle centered at A is 3 inches, and the diameter of the larger circle centered at A is 9 inches. The diameter of $\odot D$ is 11 inches. Find each measure. (Lesson 9-1)

1. BC
2. CD

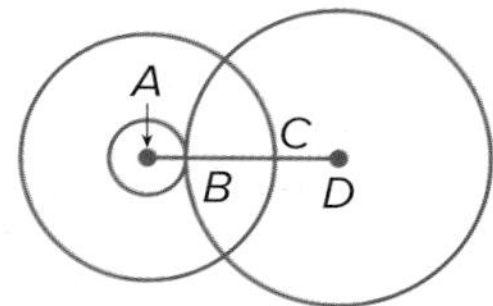

3. **DECORATIONS** To decorate for homecoming, Brittany estimates that she will need to purchase enough streamers to go around the school's circular fountain twice. If the diameter of the fountain is 88 inches, about how many feet of streamers should she buy? (Lesson 9-1)

Use $\odot C$ to find the length of each arc. Round to the nearest hundredth. (Lesson 9-2)

4. $\widehat{XY}$, if the radius is 5 feet
5. $\widehat{YZ}$, if the diameter is 8 meters

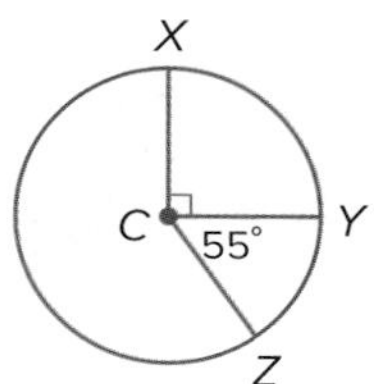

6. **TRANSPORTATION** The graph shows the results of a survey in which students at a high school were asked how they get to school. (Lesson 9-2)

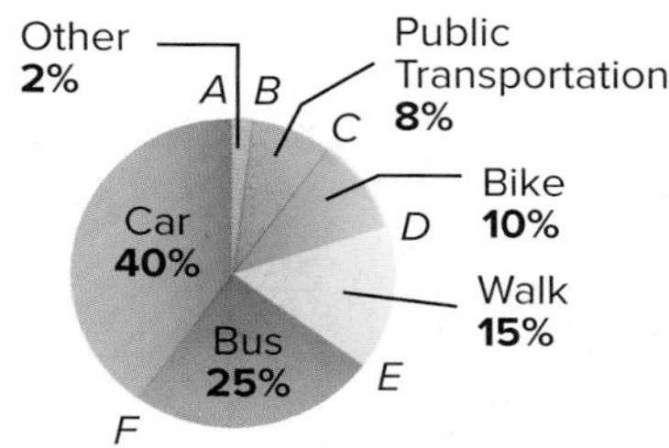

a. Find $m\widehat{CD}$.

b. Find $m\widehat{BC}$.

Find the value of x. (Lesson 9-3)

7.

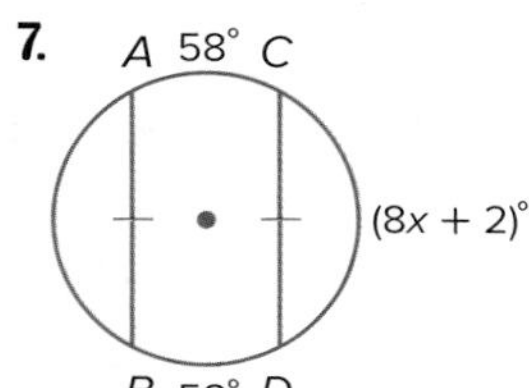

8.

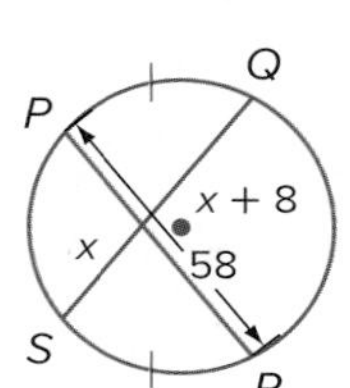

In $\odot M$, $MZ = 12$ and $WY = 20$. Find each measure. Round to the nearest hundredth. (Lesson 9-3)

9. CM
10. XC

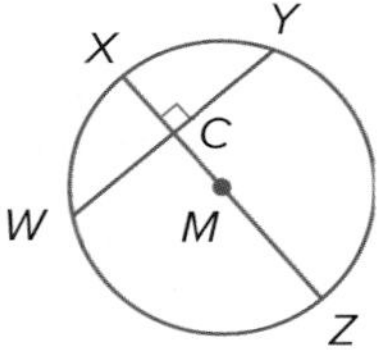

Find each measure. (Lesson 9-4)

11. $m\angle N$

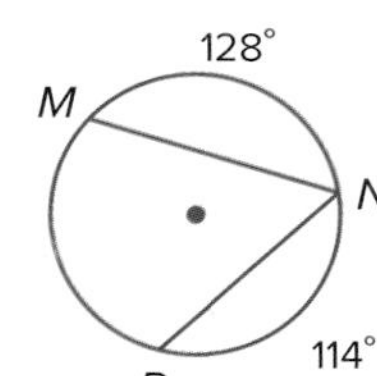

12. $m\angle B$

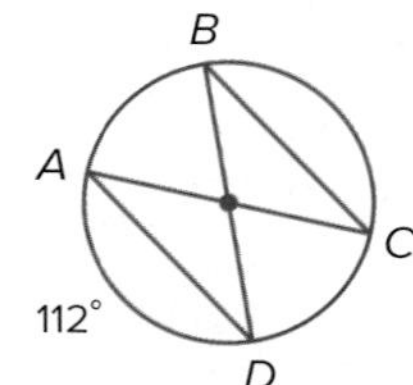

13. Find x. Assume that segments that appear to be tangent are tangent. (Lesson 9-5)

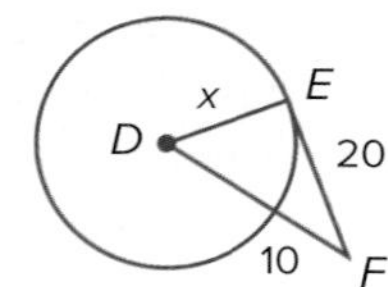

14. Quadrilateral $ABCD$ is circumscribed about $\odot H$. Find m. (Lesson 9-5)

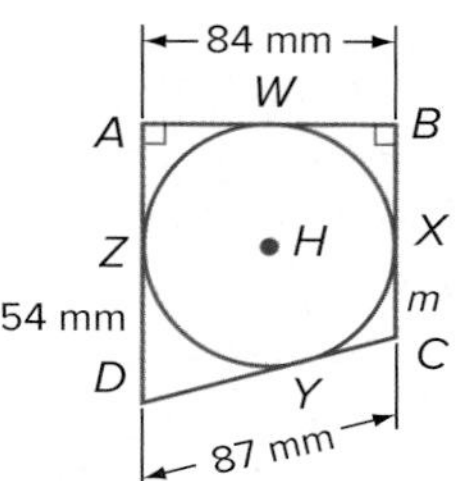

Find each measure. Assume that segments that appear to be tangent are tangent. (Lesson 9-6)

15. $m\widehat{XYZ}$

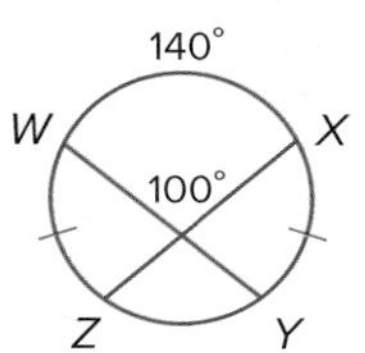

16. $\widehat{MP}$

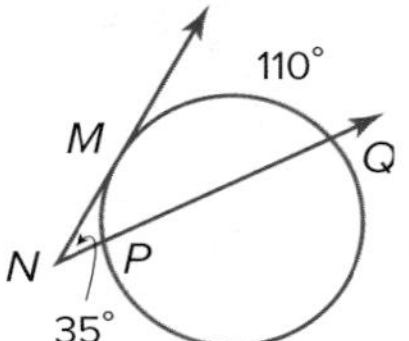

17. **CELL PHONES** A cell phone tower covers a circular area with a radius of 15 miles. (Lesson 9-7)

a. If the tower is located at the origin, write an equation for this circular area of coverage.

b. Will a person 11 miles west and 12 miles south of the tower have coverage? Explain.

18. Write an equation of a circle that contains points $A(-1, 5)$, $B(-5, 9)$, and $C(-9, 5)$. Then graph the equation. (Lesson 9-7)

19. Write an equation of the parabola with focus (0, 2) and directrix $y = -2$. (Lesson 9-8)

20. Write an equation of the parabola with focus $(4, -3)$ and vertex $(1, -3)$. (Lesson 9-8)

Extra Practice

Find the perimeter and area of each figure. Round to the nearest tenth if necessary. (Lesson 10-1)

1. 12 cm, 45°, 7 cm

2. 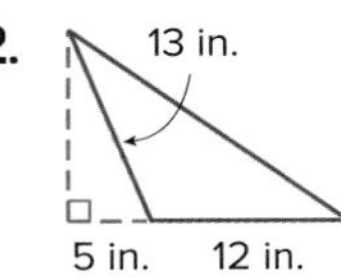

3. The height of a parallelogram is three times its base. If the area of the parallelogram is 108 square meters, find its base and height. (Lesson 10-1)

4. The height of a triangle is three feet less than its base. If the area of the triangle is 275 square feet, find its base and height. (Lesson 10-1)

Find the area of each trapezoid, rhombus, or kite. (Lesson 10-2)

5. 6. 7.

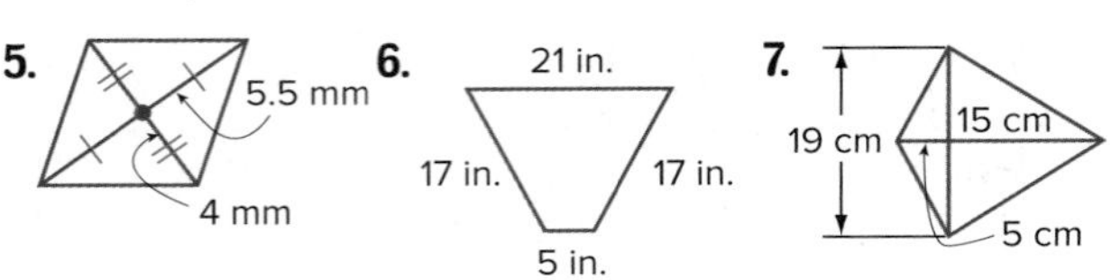

8. **MODELS** Joni is designing a mural for the side of a building. The wall is 15 feet high and 50 feet long. If she covers the wall with a kite as shown, what is the area of the kite? (Lesson 10-2)

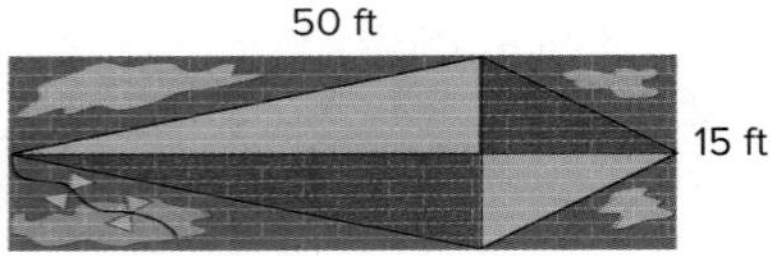

9. A trapezoid has a height of 12 inches, a base length of 9 inches, and an area of 150 square inches. What is the length of the other base? (Lesson 10-2)

Find the indicated measure. Round to the nearest tenth. (Lesson 10-3)

10. The area of a circle is 201 square meters. Find the radius.

11. Find the diameter of a circle with an area of 79 square feet.

Find the area of each shaded sector. Round to the nearest tenth if necessary. (Lesson 10-3)

12. 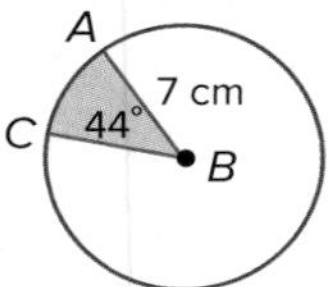

13. 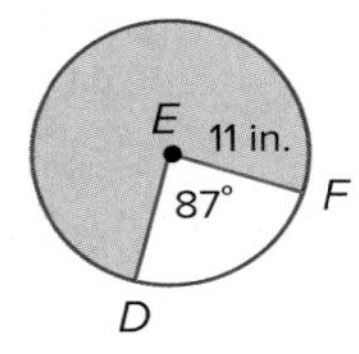

14. **GRAPHS** Len created a circle graph using the survey results shown in the table. (Lesson 10-3)

Preferred Type of Exercise	
treadmill	62%
stationary bike	8%
swimming	7%
aerobics	12%
other	11%

a. What is the angle measure of the sector representing swimming?

b. If the graph has a 3-inch diameter, what is the area of the sector representing treadmill?

Find the area of each regular polygon. Round to the nearest tenth if necessary. (Lesson 10-4)

15. 14 mm

16. 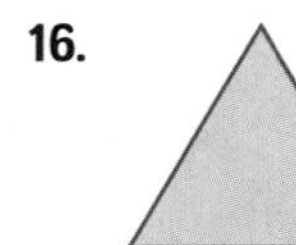

17. 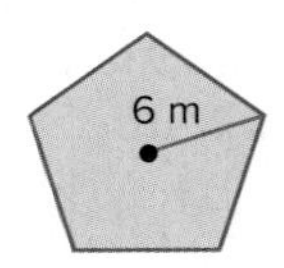

Find the area of each figure. Round to the nearest tenth if necessary. (Lesson 10-4)

18. 9 cm, 21 cm

19. 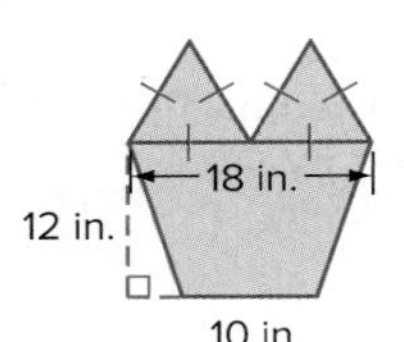

For each pair of similar figures, find the area of the green figure. (Lesson 10-5)

20. 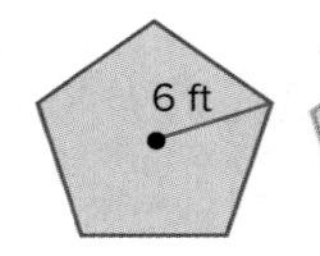

4 ft

$A = 78$ ft^2

21. 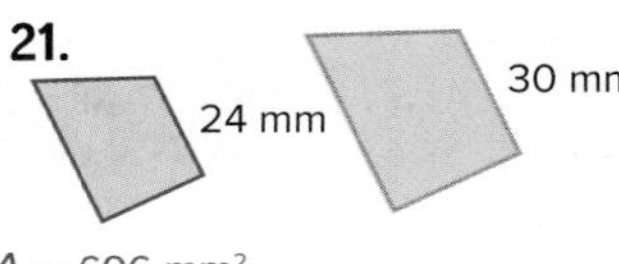

$A = 696$ mm^2

For each pair of similar figures, use the given areas to find the scale factor from the blue to the green figure. Then find *x*. (Lesson 10-5)

22. 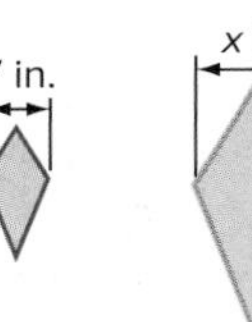

$A = 174$ in^2 $A = 696$ in^2

23. 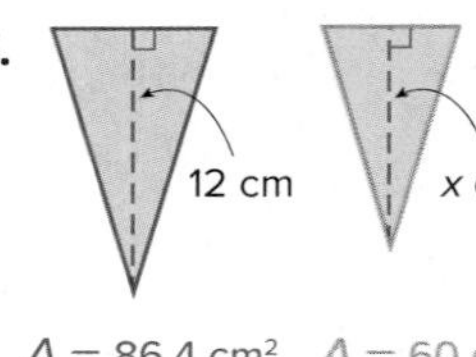

$A = 86.4$ cm^2 $A = 60$ cm^2

24. **PACKAGING** A shoebox in the shape of a rectangular prism is 15 inches long, 8 inches wide, and 5 inches tall. What is the surface area of the shoebox? (Lesson 10-6)

Determine the shape of each cross section of the solids. (Lesson 11-1)

1. plane parallel to a base

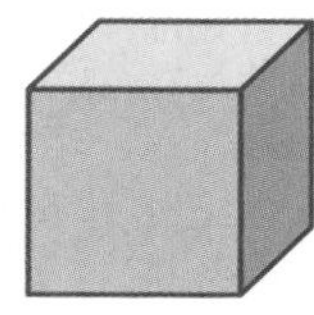

2. plane through the vertex and perpendicular to the base

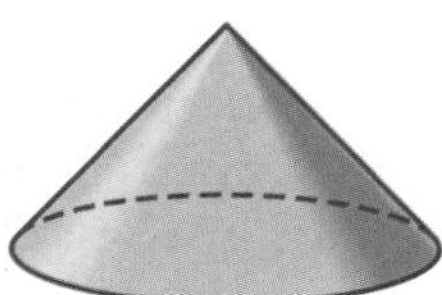

Describe the three-dimensional solid generated by rotating each two-dimensional shape around the given axis. (Lesson 11-1)

3. rectangle

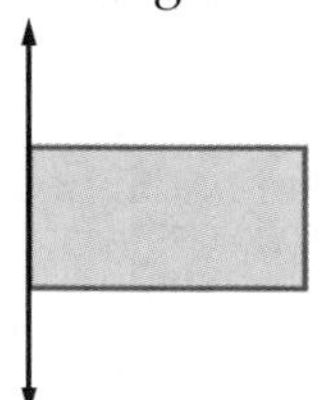

4. circle

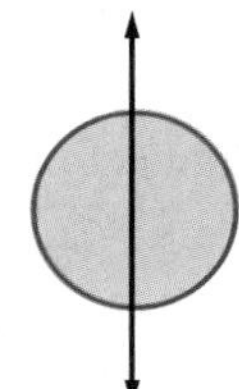

Find the volume of each solid. Round to the nearest tenth if necessary. (Lesson 11-2)

5.

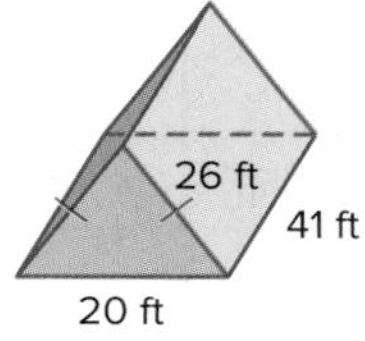

6.

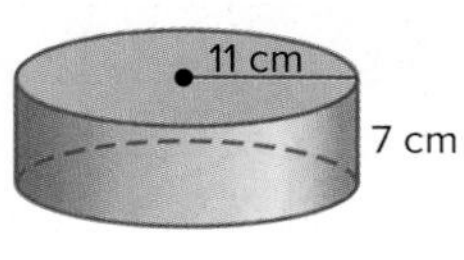

7. **ADVERTISING** A company advertises that their juice boxes contain 20% more juice than their competitor's. If the base dimensions of the boxes are the same, how much taller are the larger boxes? (Lesson 11-2)

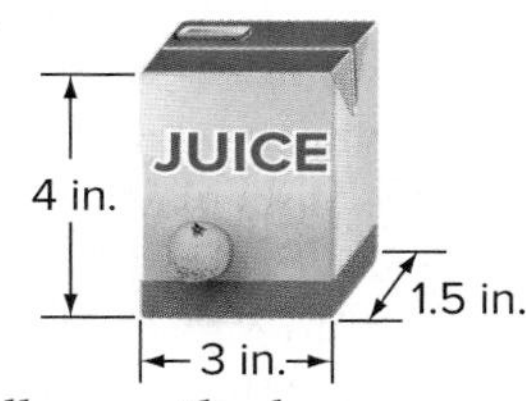

Find the volume of each solid. Round to the nearest tenth if necessary. (Lesson 11-3)

8.

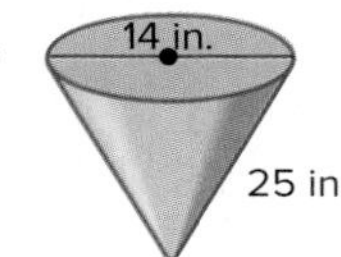

9.

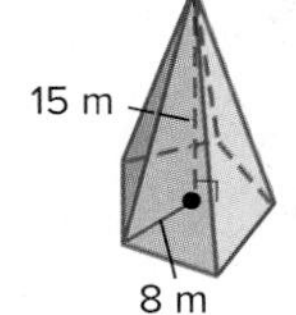

Find the surface area and volume of each sphere or hemisphere. Round to the nearest tenth. (Lesson 11-4)

10.

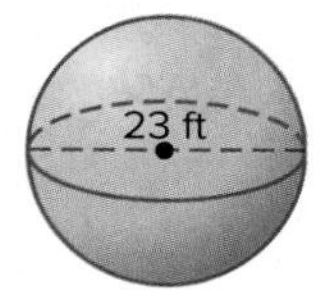

11.

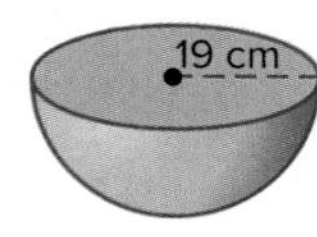

12. **SPORTS** The diameter of a tennis ball is 2.7 inches, and the diameter of a baseball is 2.9 inches. How many times as great is the volume of the baseball as the volume of the tennis ball? (Lesson 11-4)

Name each of the following on sphere *A*. (Lesson 11-5)

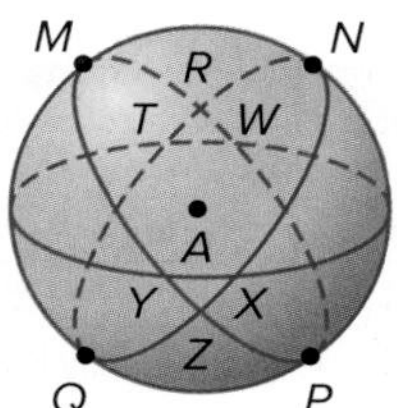

13. a triangle

14. two segments on the same great circle

Determine whether each pair of solids is *similar*, *congruent*, or *neither*. If the solids are similar, state the scale factor. (Lesson 11-6)

15.

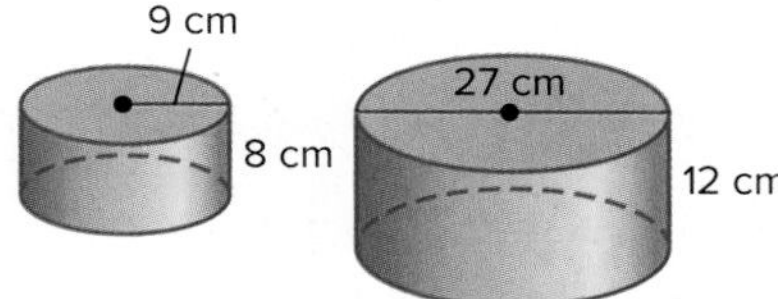

16.

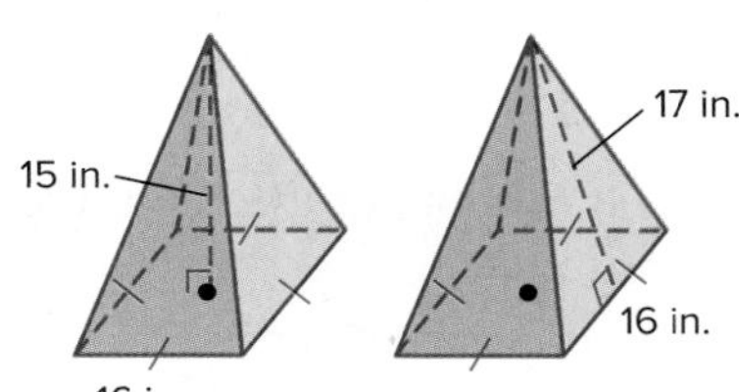

17. A cube has edges that are each 11 centimeters long. The mass of the cube is 564 grams. Find the density of the cube to the nearest hundredth. (Lesson 11-7)

Extra Practice

1. **FITNESS** Laura wants to go to a fitness class tomorrow. She can choose a 5:00 or a 7:30 class and spin or water aerobics. Represent the sample space for the situation by making an organized list, a table, and a tree diagram. (Lesson 12-1)

2. **SCHOOL UNIFORMS** Susan's school dress code allows her to wear a polo shirt or an oxford shirt and a skirt or a pair of pants. She also has a sweater that she can wear if she chooses. Draw a tree diagram to represent the sample space for Susan's uniform. (Lesson 12-1)

3. **CONSTRUCTION** Bert's family is building a house in a new neighborhood, and they must choose one option listed below for each feature. What is the number of possible outcomes for the situation? (Lesson 12-1)

Feature	Options
floor plan	elevation 1, elevation 2
counters	formica, granite
cabinets	French antique glazed, oak, cherry
basement	unfinished, partially finished, finished
garage	none, one car, two car

4. A spinner is divided into 12 equal sections that are numbered 1 through 12. Let A be the event that the spinner lands on a number greater than 8. Let B be the event that it lands on an odd number. Find $P(A \text{ or } B)$. (Lesson 12-2)

5. **NUMBERS** Charlie's phone number is 555-3703. If he places each of the digits in a bowl and randomly selects one number at a time without replacement, what is the probability that he will choose his phone number? (Lesson 12-3)

6. **RAFFLES** Participants in a raffle received tickets 1101 through 1125. If four winners are chosen, what is the probability that the winning tickets are 1103, 1111, 1118, and 1122? (Lesson 12-3)

7. Point X is chosen at random on $\overline{AE}$. Find the probability that X is on $\overline{CE}$. (Lesson 12-4)

Find the probability that a point chosen at random lies in the shaded region. (Lesson 12-4)

8. 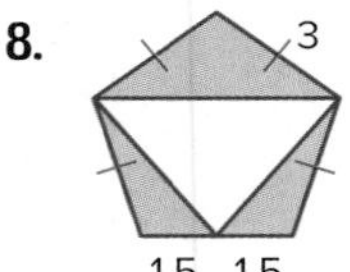

9. 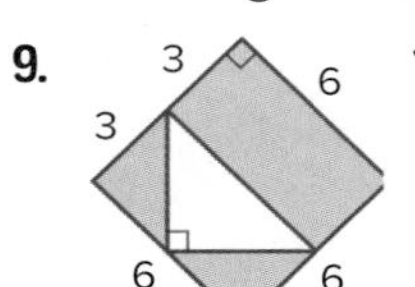

10. 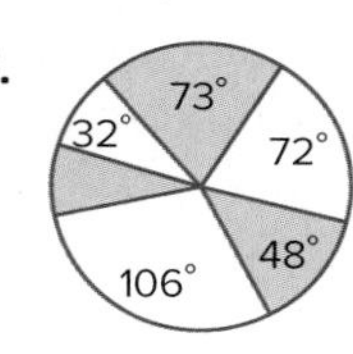

11. A die is rolled twice. What is the probability that the first number rolled is a 3 and the second number rolled is a 5? (Lesson 12-5)

12. Three cards are randomly chosen from a deck of 52 cards without replacement. What is the probability that they will all be red? (Lesson 12-5)

13. A spinner numbered 1 through 6 is spun. Find the probability that the number spun is a 3 given that it was less than 4. (Lesson 12-5)

BOOKS **The table shows the number and type of books that Sarah owns. Find each probability.** (Lesson 12-6)

Medium	Classic	Mystery	Biography
print	29	8	32
audio	3	6	10
electronic	8	3	43

14. A randomly chosen title is a print or audio book.

15. A randomly chosen title is not a biography.

16. **DOGS** The table shows the ages and genders of the dogs at an animal shelter. What is the probability that a randomly chosen dog is a female or over 5 years old? (Lesson 12-6)

Age	Male	Female
under 1 year	6	5
1–5 years	8	7
6–10 years	4	6
over 10 years	3	5

17. At Marco's school, 41% of the students take advanced placement (AP) classes, 12% of the students play an instrument, and 9% of the students take AP classes and play an instrument. A student is selected at random. What is the probability that the student takes AP classes given that he or she plays an instrument? (Lesson 12-7)

YOGURT **The owner of a frozen yogurt shop asks a random sample of 150 customers whether they would prefer peach or banana yogurt as a new flavor. She finds that 33 men and 28 women prefer peach, while 42 men and 47 women prefer banana.** (Lesson 12-8)

18. Make a two-way frequency table of the data.

19. Convert the table to relative frequencies.

20. Find the probability that a surveyed customer prefers banana given that the customer is a man.

Selected Answers and Solutions

CHAPTER 0

Preparing for Geometry

Lesson 0-1

1. cm **3.** kg **5.** mL **7.** 10 **9.** 10,000 **11.** 0.18 **13.** 2.5 **15.** 24 **17.** 0.370 **19.** 4 **21.** 5 **23.** 16 **25.** 208 **27.** 9050

Lesson 0-2

1. 20 **3.** 9.9 **5.** 16 **7.** 12 **9.** 5.4 **11.** 22.47 **13.** 1.125 **15.** 5.4 **17.** 15 **19.** 367.9 g **21.** 735.8 g

Lesson 0-3

1. $\frac{1}{3}$ or 33% **3.** $\frac{2}{3}$ or 67% **5.** $\frac{1}{3}$ or 33% **7.** $\frac{13}{28}$ or about 46% **9.** $\frac{11}{14}$ or about 79% **11.** $\frac{9}{70}$ or about 13% **13.** $\frac{9}{28}$ or about 32% **15.** $\frac{1}{28}$ or about 3.6% **17.** $\frac{13}{28}$ or about 46% **19.** $\frac{13}{14}$ or about 93% **21.** $\frac{1}{10}$ or 10%; $\frac{1}{8}$ or 12.5% **23.** Sample answer: Assign each friend a different colored marble: red, blue, or green. Place all the marbles in a bag and, without looking, select a marble from the bag. Whoever's marble is chosen gets to go first.

Lesson 0-4

1. 3 **3.** -2 **5.** -1 **7.** -26 **9.** 26 **11.** 15

Lesson 0-5

1. -8 **3.** 15 **5.** -72 **7.** $-\frac{15}{2}$ **9.** $\frac{7}{2}$ **11.** -15 **13.** -7 **15.** -7 **17.** -1 **19.** 60 **21.** -4 **23.** 4 **25.** 15 **27.** 21 **29.** -2 **31.** $-\frac{29}{2}$ **33.** -6 **35.** 1

Lesson 0-6

1. $\{x|x < 13\}$ **3.** $\{y|y < 5\}$ **5.** $\{t|t > -42\}$ **7.** $\{d|d \leq 4\}$ **9.** $\{k|k \geq -3\}$ **11.** $\{z|z < -2\}$ **13.** $\{m|m < 29\}$ **15.** $\{b|b \geq -16\}$ **17.** $\{z|z > -2\}$ **19.** $\{b|b \leq 10\}$ **21.** $\{q|q \geq 2\}$ **23.** $\left\{w|w \geq -\frac{7}{3}\right\}$

Lesson 0-7

1. $(-2, 3)$ **3.** $(2, 2)$ **5.** $(-3, 1)$ **7.** $(4, 1)$ **9.** $(-1, -1)$ **11.** $(3, 0)$ **13.** $(2, -4)$ **15.** $(-4, 2)$ **17.** none **19.** IV **21.** I **23.** III

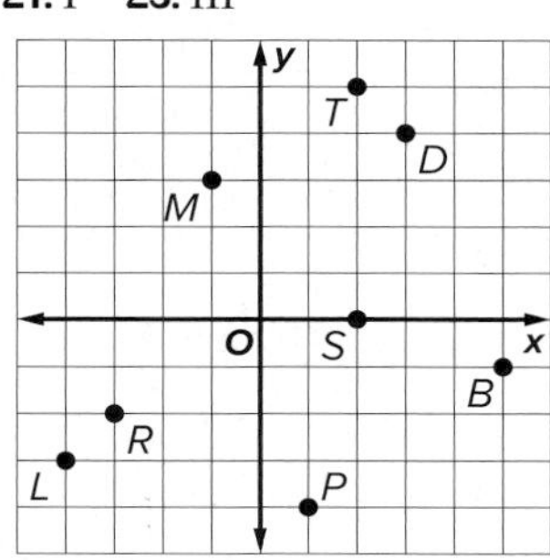

25.

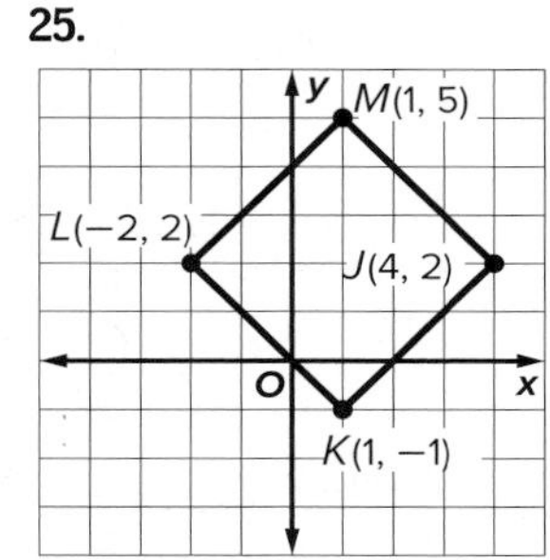

27.

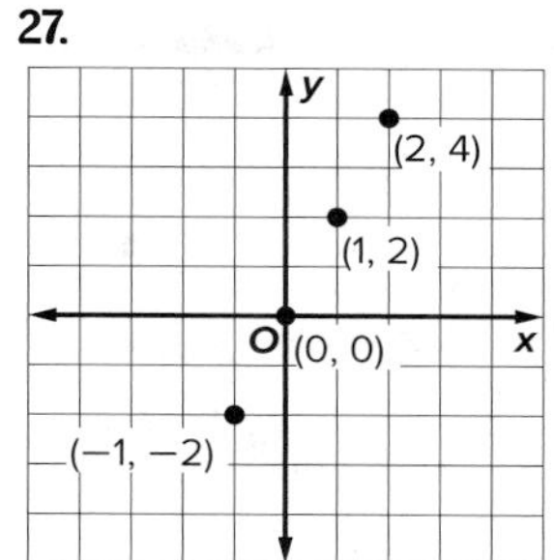

29.

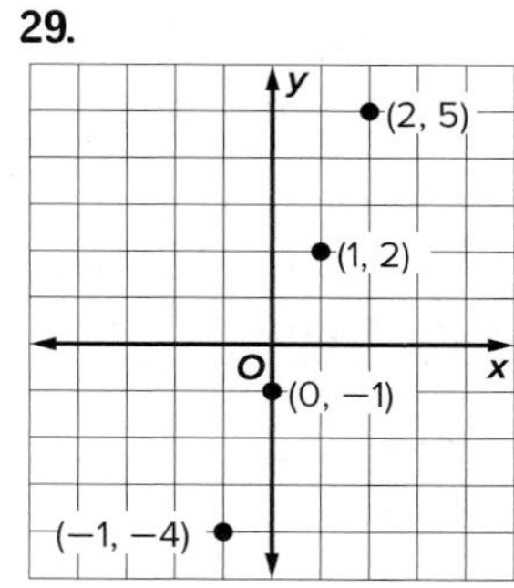

Lesson 0-8

1. $(2, 0)$ **3.** no solution **5.** $(2, -5)$ **7.** $\left(-\frac{4}{3}, 3\right)$ **9.** $(4, 1)$ **11.** elimination, no solution **13.** elimination or substitution, $(3, 0)$ **15.** elimination or substitution, $(-6, 4)$

Lesson 0-9

1. $4\sqrt{2}$ **3.** $10\sqrt{5}$ **5.** 6 **7.** $7x|y^3|\sqrt{2x}$ **9.** $\frac{9}{7}$ **11.** $\frac{3\sqrt{14}}{4}$ **13.** $\frac{p\sqrt{30p}}{9}$ **15.** $\frac{20 + 8\sqrt{3}}{13}$ **17.** $\frac{\sqrt{3}}{4}$ **19.** $\frac{6\sqrt{5} + 3\sqrt{10}}{2}$

CHAPTER 1

Tools of Geometry

Chapter 1 Concept Check

1. You could start at the origin of the coordinate plane. **3.** e5; one block left and two blocks up **5.** 12

Lesson 1-1

1. Sample answer: m **3.** B **5.** plane
7. Sample answer:

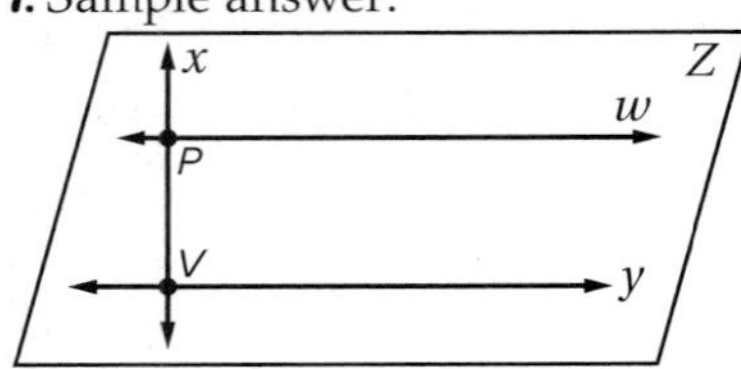

9. Sample answer: A, H, and B **11.** Yes; points B, D, and F lie in plane BDF.

13. Sample answer: n and q **15.** R

17 Sample answer: Points A, B, and C are contained in plane R. Because point P is not contained in plane R, it is not coplanar with points A, B, and C.

19. points A and P **21.** Yes; line n intersects line q when the lines are extended. **23.** plane; intersecting lines **25.** two planes intersecting in a line **27.** point **29.** line **31.** intersecting planes

33. Sample answer: **35.** Sample answer:

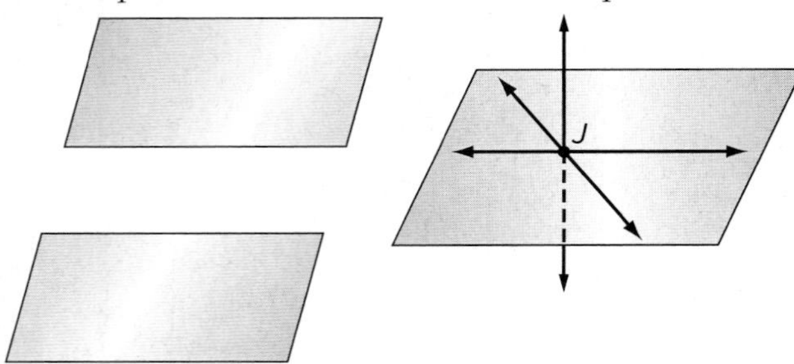

37. Sample answer: **39.** Sample answer:

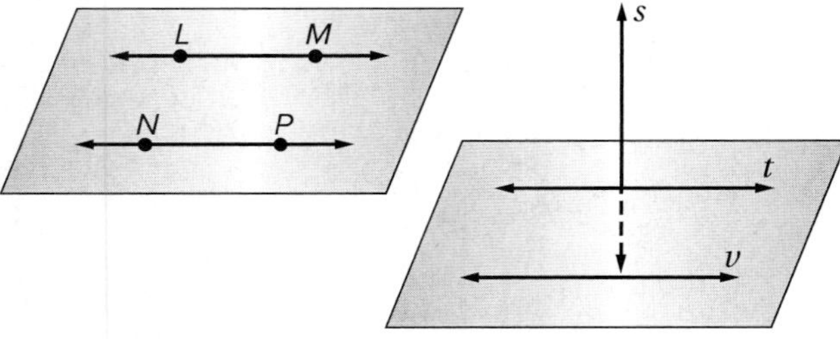

41. edges **43.** Sample answer: M and N

45 The planes appear to be parallel. Because they do not have any lines in common, they do not intersect.

47. No; V does not lie in the same plane.

49 **a.** The intersection between the signs and the pole is represented by a line. **b.** The two planes intersect in a point.

51a. Sample answer:

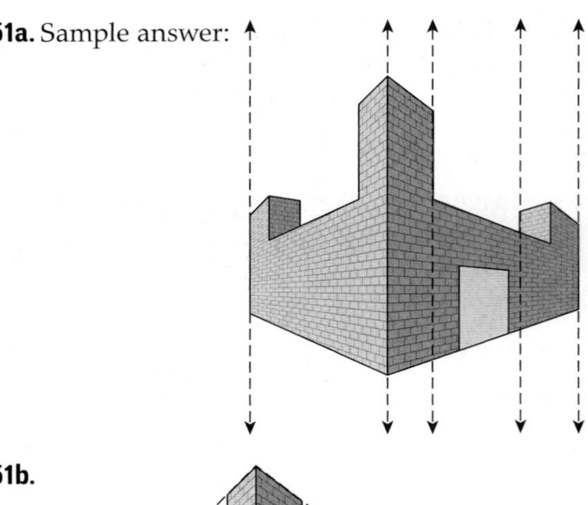

51b.

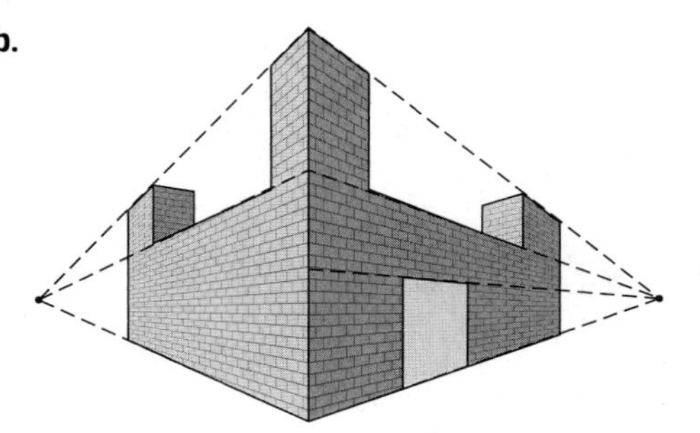

51c. Sample answer: They get closer together. **53.** Sample answer: The airplanes are in different horizontal planes.

55 **a.** There are four ways to choose three points: FGH, FGK, GHK, and FHK. Only one way, FGH, has three points collinear. So, the probability is $\frac{1}{4}$.

b. There is exactly one plane through any three noncollinear points and infinitely many planes through three collinear points. Therefore, the probability that the three points chosen are coplanar is 1.

57. Sample answer:

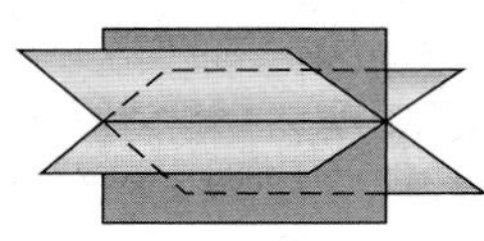

59. 4 **61.** Sample answer: A table is a finite plane. It is not possible to have a real-life object that is an infinite plane because all real-life objects have boundaries.

63. 6 **65.** B **67.** C

Lesson 1-2

1. 3.8 in.

3

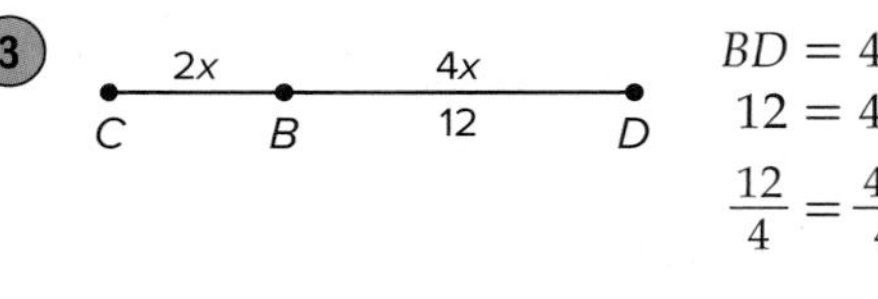

$BD = 4x$	Given
$12 = 4x$	Substitution
$\frac{12}{4} = \frac{4x}{4}$	Divide each side by 4.
$3 = x$	Simplify.

$BC = 2x$	Given
$= 2(3)$	$x = 3$
$= 6$	Simplify.

5. 8 **7.** $\sqrt{148}$ or about 12.2 units **9.** $\sqrt{58}$ or about 7.6 units **11.** 1.1 cm **13.** 1.5 in. **15.** 4.2 cm **17.** 2009:

20 cans, 2010: 35 cans, 2011: 30 cans; Subtract the number of cans the girls brought in from the total number of cans brought in by the girls and the boys. **19.** $b = 12.5$; $YZ = 100$ **21.** $d = 2$; $YZ = 16$ **23.** $a = 6$; $YZ = 38$ **25.** 3 **27.** 2 **29.** 5

31. $\sqrt{45}$ or about 6.7 units
33. 5 units **35.** $\sqrt{200}$ or about 14.1 units **37.** $\sqrt{20}$ or about 4.5 units **39.** $\sqrt{37}$ or about 6.1 units **41.** $\sqrt{29}$ or about 5.4 units

43 Yes; $KJ = 4$ in. and $HL = 4$ in. Because the segments have the same measure, they are congruent.

45. no **47.** yes **49.** $\overline{AB} \cong \overline{BC} \cong \overline{CD} \cong \overline{DE} \cong \overline{DG} \cong \overline{BG} \cong \overline{CG}$, $\overline{AC} \cong \overline{EC}$, $\overline{AH} \cong \overline{HG} \cong \overline{GF} \cong \overline{FE}$, $\overline{AG} \cong \overline{HF} \cong \overline{GE}$, $\overline{BH} \cong \overline{DF}$

51 The lengths, $BD = CE = PQ$, $YZ = JK$, $PQ = RS$, and $GK = KL$. Therefore, $\overline{BD} \cong \overline{CE}$; $\overline{BD} \cong \overline{PQ}$; $\overline{YZ} \cong \overline{JK}$; $\overline{PQ} \cong \overline{RS}$; $\overline{GK} \cong \overline{KL}$. Note that you can find congruent segments other than those given, too.

53a. A: Kim, B: Jen or Mandy, C: Mandy or Jen, D: Randi, E: Makayla **53b.** I know that Makayla can throw the shortest distance of all five team members and that Randi can throw the farthest. So I chose to place these players at points *E* and *D*, respectively. Jen and Mandy can each throw 25 feet (half the width of the court), so I placed them at either point *B* or point *C*. Kim will be at point *A*, and the team will be able to complete the play provided Kim can throw at least 33.3 feet. **53c.** Sample answer: I assumed that the player at point *B* would be 25 feet from point *C*. **55.** If point *B* is between points *A* and *C*, and you know *AB* and *BC*, add *AB* and *BC* to find *AC*. If you know *AB* and *AC*, subtract *AB* from *AC* to find *BC*.
57. Always; if point *M* is between points *C* and *D*, then $CM + MD = CD$. Because measures cannot be negative, *CD*, which represents the whole, must always be greater than either of the lengths of its parts, *CM* or *MD*.
59. Sample answer: The Pythagorean Theorem relates the lengths of the legs of a right triangle to the length of the hypotenuse using the formula $c^2 = a^2 + b^2$. If you take the square root of the formula, you get $c = \sqrt{a^2 + b^2}$. Think of the hypotenuse of the triangle as the distance between the two points, the *a* value as the horizontal distance $x_2 - x_1$, and the *b* value as the vertical distance $y_2 - y_1$. If you substitute, the Pythagorean Theorem becomes the Distance Formula, $c = \sqrt{(x_2 - x_1)^2 + (y_2 - y_1)^2}$. **61.** B
63. 17.1 **65a.** Draw any line segment with two points. **65b.** Draw one point anywhere not on the line segment and create a line with this point. Measure the segment using the compass. Place the endpoint of the compass on the endpoint drawn and draw an arc on the line. Label the point and this is a line segment.
65c. Draw one point anywhere not on the line segment and create a line with this point. Measure the segment using the compass. Estimate one half of the distance of the segment. Place the endpoint of the compass on the endpoint drawn and draw an arc on the line. Label the point and this is a line segment. **65d.** Draw one point anywhere not on the line segment and create a line with this point. Measure the segment using the compass. Draw three segments with the same length using the process of drawing one segment.

Lesson 1-3

1. 4.3 cm **3.** −2 **5.** (4, −5.5)

7 Let *G* be (x_1, y_1) and *J* be (x_2, y_2) in the Midpoint Formula.

$$F\left(\frac{x_1 + 6}{2}, \frac{y_1 + (-2)}{2}\right) = F(1, 3.5) \quad (x_2, y_2) = (6, -2)$$

Write two equations to find the coordinates of *G*.

$\frac{x_1 + 6}{2} = 1$ Midpoint Formula $\frac{y_1 + (-2)}{2} = 3.5$

$x_1 + 6 = 2$ Multiply each side by 2. $y_1 + (-2) = 7$

$x_1 = -4$ Simplify. $y_1 = 9$

The coordinates of *G* are $(-4, 9)$.

9. 1 **11.** (−1, −3) **13.** (3, 2) **15.** 6 **17.** −4.5 **19.** 3

21 $M\left(\frac{x_1 + x_2}{2}, \frac{y_1 + y_2}{2}\right)$ Midpoint Formula

$= M\left(\frac{22 + 15}{2}, \frac{4 + 7}{2}\right)$ $(x_1, y_1) = (22, 4)$ and $(x_2, y_2) = (15, 7)$

$= M\left(\frac{37}{2}, \frac{11}{2}\right)$ Simplify.

$= M(18.5, 5.5)$ Simplify.

The coordinates of the midpoint are (18.5, 5.5).

23. (−6.5, −3) **25.** (−4.2, −10.4) **27.** $\left(-1, -\frac{1}{2}\right)$

29. *A*(1, 6) **31.** *C*(16, −4) **33.** *C*(−12, 13.25)

35. 58 **37.** 4.5 **39.** −1 **41.** 2.6 **43.** $\left(-3, 1\frac{1}{3}\right)$

45. $\left(1\frac{2}{3}, 3\frac{1}{3}\right)$

47 The *y*-coordinate of points on the *x*-axis is 0. So, the points would be of the form (*x*, 0). Use the Distance Formula to find an expression for the distance between the points (*x*, 0) and (1, 8) and equate it to 10.

$\sqrt{(x_2 - x_1)^2 + (y_2 - y_1)^2} = d$	Distance Formula
$\sqrt{(1 - x)^2 + (8 - 0)^2} = 10$	Substitution
$\left(\sqrt{(1 - x)^2 + (8 - 0)^2}\right)^2 = 10^2$	Square each side.
$(1 - x)^2 + (8 - 0)^2 = 100$	Simplify.
$1 - 2x + x^2 + 64 = 100$	Square terms.
$x^2 + 2x + 65 = 100$	Simplify
$x^2 + 2x + 65 - 100 = 100 - 100$	Subtract 100 from each side.
$x^2 + 2x - 35 = 0$	Simplify.
$(x - 7)(x + 5) = 0$	Factor.
$x = 7$ or $x = -5$	Solve for *x*.

There are two possible values for x, -5 and 7. So the two points are $(-5, 0)$ and $(7, 0)$.

49. $\left(-1\frac{1}{2}, -1\right)$ **51.** No; I placed my compass point at *A* and set the width to be greater than half of the segment. I drew arcs above and below the segment. Without changing the compass setting, I placed the point at *D* and drew arcs that intersected the first set. The line segment drawn through the arcs does not intersect point *B*.

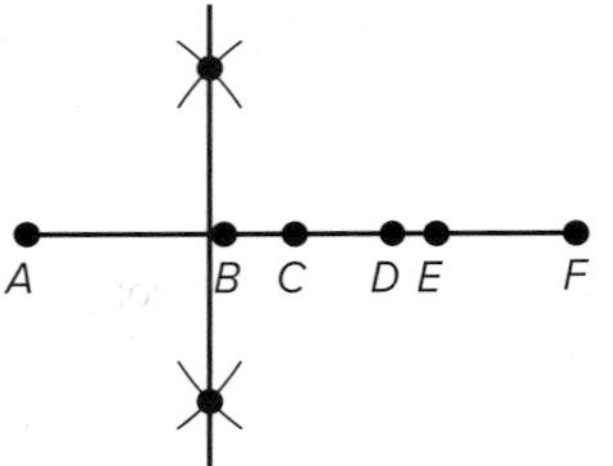

53 **a.** Sample answer: A —— B

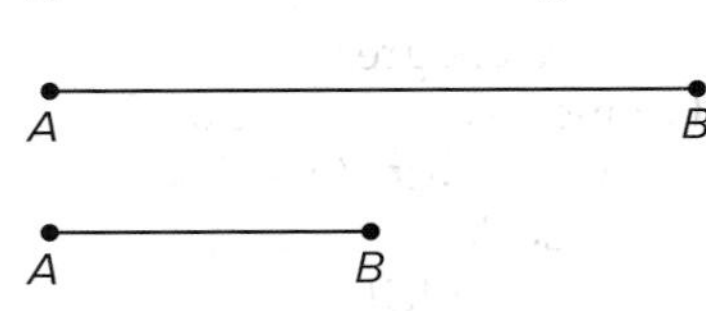

b. Sample answer: A D C B

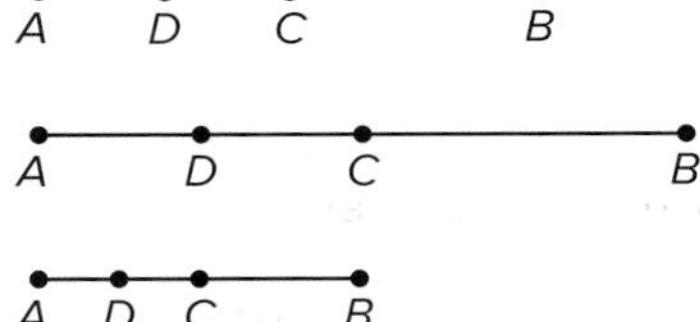

c. Sample answer:

line	*AB* (cm)	*AC* (cm)	*AD* (cm)
1	4	2	1
2	6	3	1.5
3	3	1.5	0.75

d.

$AC = \frac{1}{2}AB$	Definition of midpoint
$AC = \frac{1}{2}x$	$AB = x$
$AD = \frac{1}{2}AC$	Definition of midpoint
$AD = \frac{1}{2}\left(\frac{1}{2}x\right)$	Substitution
$AD = \frac{1}{4}x$	Simplify.

e. Look for a pattern.

Number of Midpoints	Length of Smallest Segment
1	$\frac{1}{2}x$
2	$\frac{1}{2} \cdot \frac{1}{2}x = \frac{1}{2(2)}x$
3	$\frac{1}{2} \cdot \frac{1}{2(2)}x = \frac{1}{2^3}x$
4	$\frac{1}{2} \cdot \frac{1}{2(3)}x = \frac{1}{2^4}x$
n	$\frac{1}{2^n}x$

Sample answer: If n midpoints are found, then the smallest segment will have a measure of $\frac{1}{2^n}x$.

55. Sample answer: sometimes; when the point (x_1, y_1) has coordinates $(0, 0)$.

57. Sample answer:

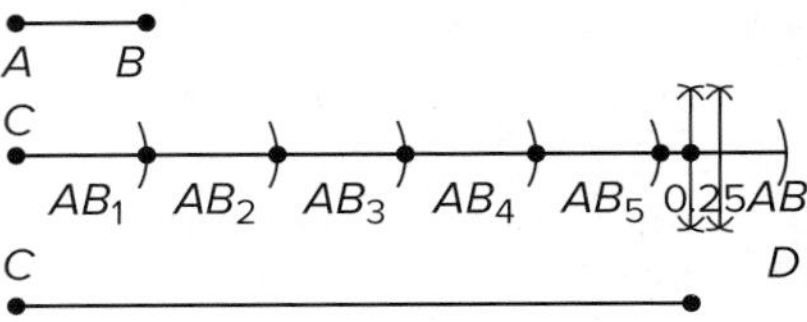

Draw $\overline{AB}$. Next, draw a construction line and place point *C* on it. From point *C*, strike 6 arcs in succession of length *AB*. On the sixth $\overline{AB}$ length, perform a segment bisector two times to create a $\frac{1}{4}AB$ length. Label the endpoint *D*. **59.** B **61.** A **63.** (1, 16) **65a.** $\overline{JK}$, 5.4 units **65b.** $\overline{JL}$, 7.1 units **65c.** 19.5 units

Lesson 1-4

1. *U* **3.** $\angle XYU$, $\angle UYX$ **5.** acute; 40 **7.** right; 90 **9.** 156 **11a.** 45; When joined together, the angles form a right angle, which measures 90. If the two angles that form this right angle are congruent, then the measure of each angle is 90 ÷ 2 or 45. The angle of the cut is an acute angle. **11b.** The joint is the angle bisector of the frame angle. **13.** *P* **15.** *M* **17.** $\overrightarrow{NV}$, $\overrightarrow{NM}$ **19.** $\overrightarrow{RP}$, $\overrightarrow{RQ}$ **21.** $\angle TPQ$ **23.** $\angle TPN$, $\angle NPT$, $\angle TPM$, $\angle MPT$ **25.** $\angle 4$ **27.** *S*, *Q*

29 Sample answer: $\angle MPR$ and $\angle PRQ$ share points *P* and *R*.

31. 90, right **33.** 45, acute **35.** 135, obtuse

37

$m\angle ABE = m\angle EBF$	Definition of ≅ ∡
$2n + 7 = 4n - 13$	Substitution
$7 = 2n - 13$	Subtract 2*n* from each side.
$20 = 2n$	Add 13 to each side.
$10 = n$	Divide each side by 2.

$m\angle ABE = 2n + 7$	Given
$= 2(10) + 7$	$n = 10$
$= 20 + 7$ or 27	Simplify.

39. 16 **41.** 47 **43a.** about 50 **43b.** about 140 **43c.** about 20 **43d.** 0 **45.** acute

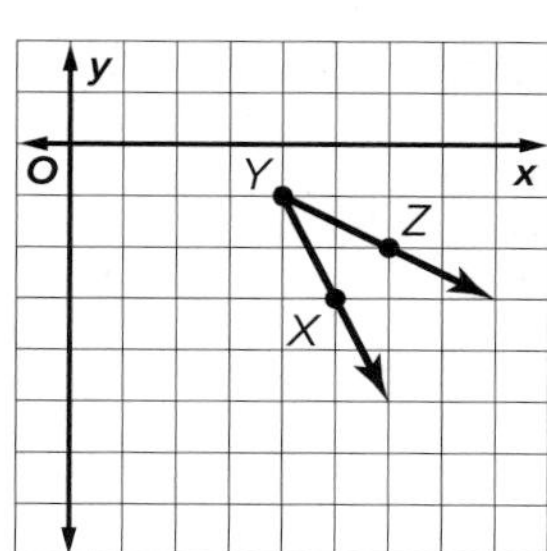

47 **a.** $m\angle 1 \approx 110$; Because $110 > 90$, the angle is obtuse. **b.** $m\angle 2 \approx 85$; since $85 < 90$, the angle is acute. **c.** about 15; If the original path of the light is extended, the measure of the angle the original path makes with the refracted path represents the number of degrees the path of the light changed. The sum of the measure of this angle and the measure of $\angle 3$ is 180. The measure of $\angle 3$ is $360 - (110 + 85)$ or 165, so the measure of the angle the original path makes with the refracted path is $180 - 165$ or 15.

49. The two angles formed are acute angles. Sample answer: With my compass at point A, I drew an arc in the interior of the angle. With the same compass setting, I drew an arc from point C that intersected the arc from point A. From the vertex, I drew $\overrightarrow{BD}$. I used the same compass setting to draw the intersecting arcs, so $\overrightarrow{BD}$ divides $\angle ABC$ so that the measurement of $\angle ABD$ and $\angle DBC$ are equal. Therefore, $\overrightarrow{BD}$ bisects $\angle ABC$.

51. sometimes; sample answer: For example, if you add an angle measure of 4 and an angle measure of 6, you will have an angle measure of 10, which is still acute. But if you add angles with measure of 50 and 60, you will have an obtuse angle with a measure of 110.

53. Sample answer: To measure an acute angle, you can fold the corner of the paper so that the edges meet. This would bisect the angle, allowing you to determine whether the angle was between 0° and 45° or between 45° and 90°. If the paper is folded two more times in the same manner and cut off this corner of the paper, the fold lines would form the increments of a homemade protractor that starts at 0° on one side and progress in $90 \div 8$ or 11.25° increments, ending at the adjacent side, which would indicate 90°. You can estimate halfway between each fold line, which would give you an accuracy of $11.25° \div 2$ or about 6°. The actual measure of the angle shown is 52°. An estimate between 46° and 58° would be acceptable. **55.** D **57.** C **59.** 21

Lesson 1-5

1. $\angle ZVY$, $\angle WVU$ **3a.** vertical **3b.** 15

5 If x ⊥ y, then $m\angle 2 = 90$ and $m\angle 3 = 90$.

$m\angle 2 = 3a - 27$	Given
$90 = 3a - 27$	Substitution
$117 = 3a$	Add 27 to each side.
$39 = a$	Divide each side by 3.
$m\angle 3 = 2b + 14$	Given
$90 = 2b + 14$	Substitution
$76 = 2b$	Subtract 14 from each side.
$38 = b$	Divide each side by 3.

7. Yes: they share a common side and vertex, so they are adjacent. Because $m\angle EDB + m\angle BDA + m\angle ADC = 90$, $\angle EDB$ and $\angle BDA$ cannot be complementary or supplementary. **9.** Sample answer: $\angle BFC$, $\angle DFE$

11. $\angle FDG$, $\angle GDE$ **13.** Sample answer: $\angle CBF$, $\angle ABF$ **15.** $\angle GDE$ **17.** $\angle CAE$ **19.** 65

21

$2x + 25 = 3x - 10$	Vertical ∠s are ≅ and have equal measures.
$25 = x - 10$	Subtract $2x$ from each side.
$35 = x$	Add 10 to each side.
$3x - 10 + y = 180$	Def. of supplementary ∠s
$3(35) - 10 + y = 180$	Substitution
$105 - 10 + y = 180$	Multiply.
$95 + y = 180$	Simplify.
$y = 85$	Subtract 95 from each side.

23. $x = 48$; $y = 21$ **25.** $m\angle F = 63$; $m\angle E = 117$ **27.** 40

29 If $\angle KNM$ is a right angle, then $m\angle KNM = 90$.

$m\angle KNL + m\angle LNM = m\angle KNM$	Sum of parts = whole
$6x - 4 + 4x + 24 = 90$	Substitution
$10x + 20 = 90$	Combine like terms.
$10x = 70$	Subtract 20 from each side.
$x = 7$	Divide each side by 10.

31. 92 **33.** 53; 37 **35.** $a = 8$; $b = 54$ **37.** Yes; the angles form a linear pair. **39.** No; the measures of each angle are unknown. **41.** No; the angles are not adjacent. **43.** Sample answer: $\angle 1$ and $\angle 3$

45 $\angle 1$ and $\angle 3$ are vertical angles, so they are congruent; $m\angle 3 = m\angle 1 = 110$. $\angle 1$ and $\angle 4$ are a linear pair, so they are supplementary.

$m\angle 4 + m\angle 1 = 180$	Def. of supplementary ∠s
$m\angle 4 + 110 = 180$	Substitution
$m\angle 4 = 70$	Subtract 110 from each side.

47. Sample answer: Yes; if the wing is not rotated at all, then all of the angles are right angles, which are neither acute nor obtuse. **49.** Yes; angles that are right or obtuse do not have complements because their measures are greater than or equal to 90. **51a.** Line a is perpendicular to plane P. **51b.** Line m is in plane P. **51c.** Any plane containing line a is perpendicular to plane P. **53a.** C **53b.** bisect $\angle LQR$ **53c.** bisect $\angle MQR$ **55.** D **57.** 100°

Lesson 1-6

1. pentagon; concave; irregular **3.** octagon; regular **5.** hexagon; irregular **7.** ≈ 40.2 cm; ≈ 128.7 cm² **9.** C **11.** triangle; convex; regular

13 The polygon has 8 sides, so it is an octagon. All of the lines containing the sides of the polygon will pass through the interior of the octagon, so it is concave. Because the polygon is not convex, it is irregular.

15. hendecagon; concave; irregular **17.** 7.8 m; ≈ 3.1 m² **19.** 26 in.; 42.3 in²

21

$c^2 = a^2 + b^2$	Pythagorean Theorem
$c^2 = 6.5^2 + 4.5^2$	$a = 6.5$, $b = 4.5$
$c^2 = 62.5$ or ≈ 7.9	Simplify.
$P = a + b + c$	$A = \frac{1}{2}bh$
$\approx 6.5 + 4.5 + 7.9$	$= \frac{1}{2}(4.5)(6.5)$
≈ 18.9 cm	≈ 14.6 cm²

23. Yes; she can trim the square tablecloth, but not the circular tablecloth.

25. triangle; $P = 5 + \sqrt{32} + \sqrt{17} \approx 14.78$ units; $A = 10$ units^2

27. quadrilateral or square; $P = 20$ units; $A = 25$ units^2

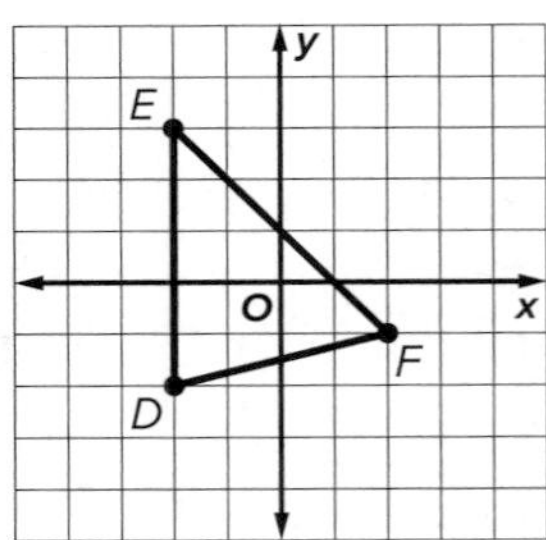

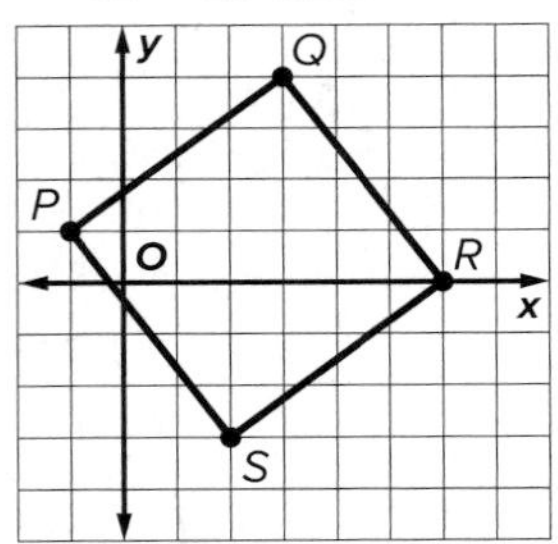

29a. 14 ft **29b.** 12 ft^2 **29c.** The perimeter increases by twice the length of the original rectangle; the area doubles. The perimeter of a rectangle with dimensions 3 ft and 8 ft is 22 ft, which is 8 ft more than the perimeter of the original figure because 8 + 14 ft = 22 ft. The area of a rectangle with dimensions 3 ft and 8 ft is 24 ft^2, which is twice the area of the original figure because 2 • 12 ft^2 = 24 ft^2. **29d.** The perimeter doubles; the area quadruples. The perimeter of a rectangle with dimensions 6 ft and 8 ft is 28 ft, which is twice the perimeter of the original figure because 2 • 14 ft = 28 ft. The area of a rectangle with dimensions 6 ft and 8 ft is 48 ft^2, which is four times the area of the original figure because 4 • 12 ft^2 = 48 ft^2. **31.** 60 yd, 6 yd

33 $C = \pi d$ Circumference $C = \pi d$
$= \pi(8)$ $= \pi(10)$
≈ 25.1 Simplify. ≈ 31.4

minimum circumference: 25.1 in.;
maximum circumference: 31.4 in.

$A = \pi r^2$ Area of a circle
$= \pi(4)^2$ $r = 4$
≈ 50.3 Simplify.

$A = \pi r^2$ Area of a circle
$= \pi(5)^2$ $r = 5$
≈ 78.5 Simplify.

minimum area: 50.3 in^2; maximum area: 78.5 in^2

35. 21.2 m **37.** $2\pi\sqrt{32}$ or about 35.5 units **39.** $12\sqrt{6}$ or about 29.4 in. **41.** 108 in.; 729 in^2

43a–b. Sample answer:

Object	d (cm)	C (cm)	$\frac{C}{d}$
1	3	9.4	3.13
2	9	28.3	3.14
3	4.2	13.2	3.14
4	12	37.7	3.14
5	4.5	14.1	3.13
6	2	6.3	3.15
7	8	25.1	3.14
8	0.7	2.2	3.14
9	1.5	4.7	3.13
10	2.8	8.8	3.14

43c. Sample answer:

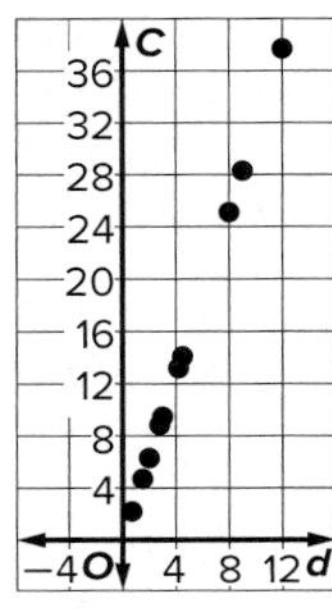

43d. Sample answer: $C = 3.14d$; the equation represents a formula for approximating the circumference of a circle. The slope represents an approximation for pi. **45.** 290.93 units^2

47. Sample answer: The pentagon is convex, because no points of the lines drawn on the edges are in the interior. The pentagon is regular because all of the angles and sides were constructed with the same measurement, making them congruent to each other. **49.** Sample answer: If a convex polygon is equiangular but not also equilateral, then it is not a regular polygon. Likewise, if a polygon is equiangular and equilateral but not convex, then it is not a regular polygon. **51.** E **53.** A **55a.** length: $\frac{40}{3}$; width: $\frac{20}{3}$ **55b.** $\frac{800}{9}$ **57.** D

Lesson 1-7

1. translation **3a.** $X'(-9, -4)$, $Y'(-9, 3)$, and $Z'(-1, 3)$ **b.** $X'(-5, 2)$, $Y'(-5, 9)$, and $Z'(3, 9)$ **c.** $X'(4, 9)$, $Y'(-3, 9)$, and $Z'(-3, 1)$ **5** The green figure is the image of the blue figure reflected in the x-axis.
7. rotation **9.** rotation

11 Reflecting in the y-axis changes the sign of each x-coordinate.
$M(-7, -1) \rightarrow M'(7, -1)$, $P(-7, -7) \rightarrow P'(7, -7)$, $R(-1, -4) \rightarrow R'(1, -4)$

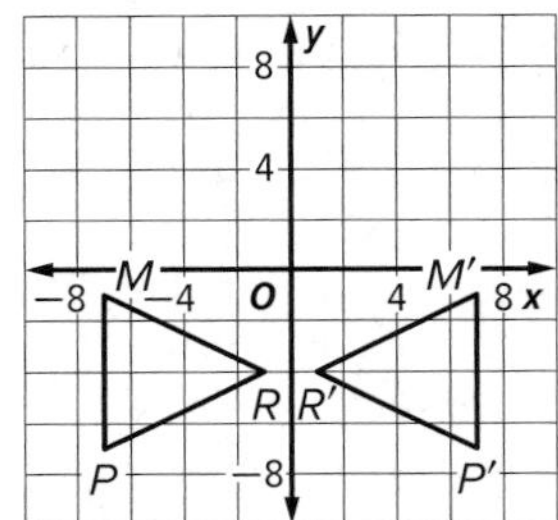

13.

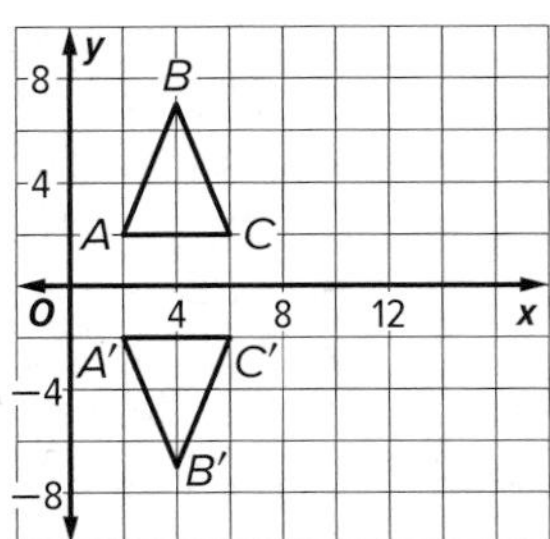

$A(2, 2) \rightarrow A'(2, -2)$, $B(4, 7) \rightarrow B'(4, -7)$, $C(6, 2) \rightarrow C'(6, -2)$

15. translation

17. rotation; The knob is the center of rotation.

19 **a.** For a reflection in the y-axis, $(x, y) \rightarrow (-x, y)$
$A(-4, 4) \rightarrow A'(4, 4)$
$B(-2, 8) \rightarrow B'(2, 8)$
$C(2, 6) \rightarrow C'(-2, 6)$
b. The distance of each vertex from the y-axis is the absolute value of the vertex's x-coordinate. So the

distance of A from the y-axis is $|-4| = 4$ units, the distance of B from the y-axis is $|-2| = 2$ units, and the distance of C from the y-axis is $|2| = 2$ units.
19c. The distances of the images of the vertices are the same as the distances of the vertices of the preimage.
21a. translation along a vector <a, 0>, reflection across the y-axis **21b.** Sample answer: The triangles must be either isosceles or equilateral. When triangles are isosceles or equilateral, they have a line of symmetry, so reflections result in the same figure. **23.** Sample answer: A person looking in a mirror sees a reflection of himself or herself. **25.** Sample answer: A faucet handle rotates when you turn the water on.
27. C **29.** B **31.** B

Lesson 1-8

1. not a polyhedron; cylinder

3 $T = PH + 2B$ Surface area of a prism
$= (14)(3) + 2(12)$ $P = 14$ cm, $h = 3$ cm, $B = 12$ cm^2
$= 66$ cm^2 Simplify.

$V = BH$ Volume of a prism
$= (12)(3)$ $B = 12$ cm^2, $h = 3$ cm
$= 36$ cm^3 Simplify.

5a. ≈ 16.0 in^3 **5b.** ≈ 29.1 in^2 **7.** pyramid; a polyhedron **9.** rectangular prism; a polyhedron **11.** cylinder; not a polyhedron **13.** not a polyhedron; cone **15.** not a polyhedron; sphere **17.** a polyhedron; pentagonal pyramid; base: $JHGFD$; faces: $JHGFD$, $\triangle JEH$, $\triangle HEG$, $\triangle GEF$, $\triangle FED$, $\triangle EDJ$; edges: $\overline{HG}$, $\overline{GF}$, $\overline{FD}$, $\overline{DJ}$, $\overline{JH}$, $\overline{EJ}$, $\overline{EH}$, $\overline{EG}$, $\overline{EF}$, $\overline{ED}$; vertices: J, H, G, F, D, E **19.** 121.5 m^2; 91.1 m^3

21 $T = PH + 2B$ $V = BH$
$= (24)(5) + 2(24)$ $= (24)(5)$
$= 168$ cm^2 $= 120$ cm^3

23. 150π or about 471.2 mm^2; 250π or about 785.4 mm^3

25 **a.** $V = \pi r^2 h$ Volume of a cylinder
$= \pi(7)^2(8.5)$ $r = 7$ in., $h = 8.5$ in.
≈ 1308.5 in^3 Simplify.

b. $S = 2\pi rh + 2\pi r^2$ Surface area of a cylinder
$= 2\pi(7)(8.5) + 2\pi(7)^2$ $r = 7$ in., $h = 8.5$ in.
≈ 681.7 in^2 Simplify.

27. 3 in. **29.** 1200 in^2; 1776 in^3 **31a.** 96 in^2 **31b.** 113.1 in^2 **31c.** prism: 2 cans; cylinder: 3 cans **31d.** 2.18 in.; if the height is 10 in., then the surface area of the rectangular cake is 152 in^2. To find the radius of a cylindrical cake with the same height, solve the equation $152 = \pi r^2 + 20\pi r$. The solutions are $r = -22.18$ or $r = 2.18$. Using a radius of 2.18 in. gives surface area of about 152 in^2.

33 $1 \text{ ft}^3 = (12 \text{ in.})^3 = 1728 \text{ in}^3$
$4320 \text{ in}^3 \cdot \frac{1 \text{ ft}^3}{1728 \text{ in}^3} = 2.5 \text{ ft}^3$

35. The volume of the original prism is 4752 cm^3. The volume of the new prism is 38,016 cm^3. The volume increased by a factor of 8 when each dimension was doubled. **37.** Neither; sample answer: the surface area is twice the sum of the areas of the top, front, and left side of the prism or $2(5 \cdot 3 + 5 \cdot 4 + 3 \cdot 4)$, which is 94 in^2. **39a.** cone **39b.** cylinder **41.** 27 mm^3 **43a.** B **43b.** $375\pi - 125 \approx 1053.097$ cm^3 **43c.** Yes, the radius of the bowl is $288\pi \approx 904.77$ whereas the water left over is $375\pi - 125 \approx 1053.097$ cm^3 **45.** C **47.** 4; 8

Lesson 1-9

1.

3 cylinder

5.

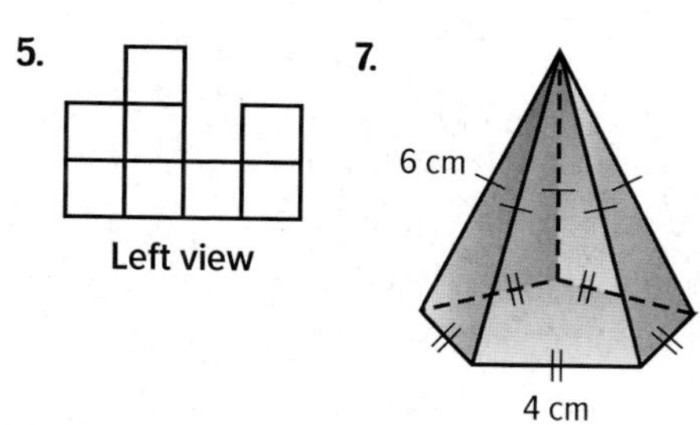

9.

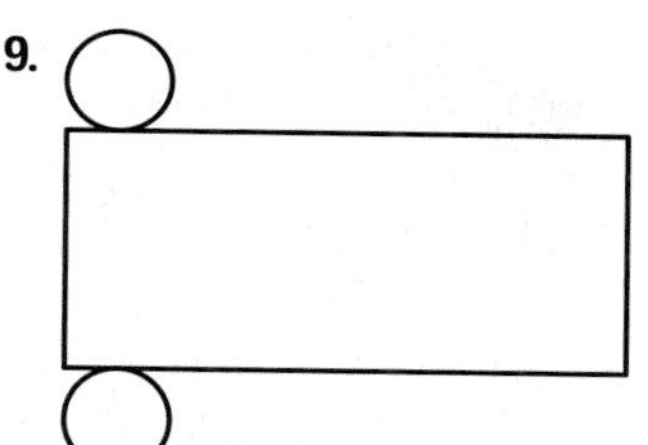

11.

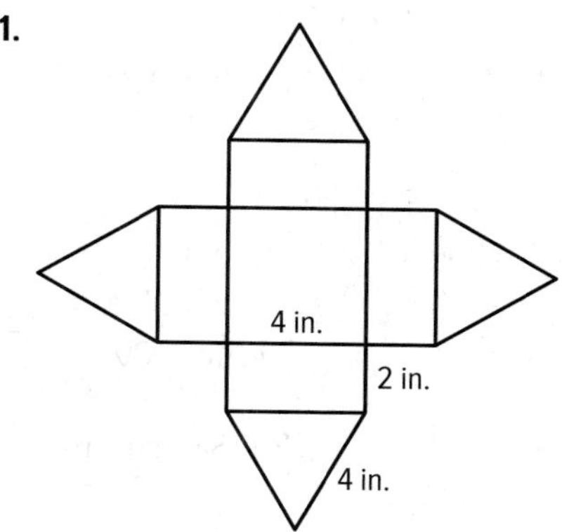

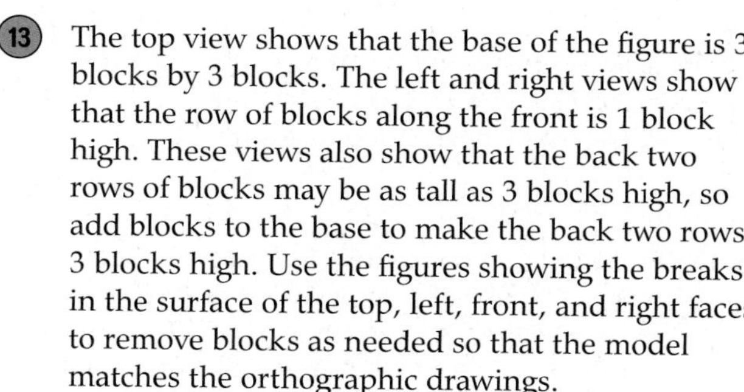

13 The top view shows that the base of the figure is 3 blocks by 3 blocks. The left and right views show that the row of blocks along the front is 1 block high. These views also show that the back two rows of blocks may be as tall as 3 blocks high, so add blocks to the base to make the back two rows 3 blocks high. Use the figures showing the breaks in the surface of the top, left, front, and right faces to remove blocks as needed so that the model matches the orthographic drawings.

Selected Answers and Solutions

15. A **17a.** circle **17b.** rectangle **17c.** The diameter of the circles are the diameter of the can. The length of the rectangle is the distance around the can, or the circumference of each circle. The width of the rectangle is the height of the can.

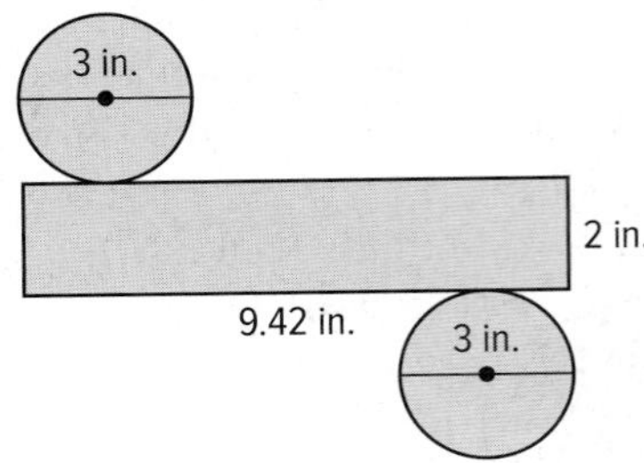

19a. icosahedron **19b.** octahedron
19c. dodecahedron
21. D **23.** B **25.** A, D

Lesson 1-10

1. $\frac{1}{16}$ ft; The exact measurement could be between $2\frac{1}{16}$ ft and $2\frac{3}{16}$ ft. **3.** $\frac{1}{2}$ mi; The exact measurement could be between $8\frac{1}{2}$ mi and $9\frac{1}{2}$ mi. **5.** 5 **7.** 50% **9.** $\frac{1}{32}$ in.; The exact measurement could be between $50\frac{7}{32}$ in. and $50\frac{9}{32}$ in. **11.** 0.0005 cm; The exact measurement could be between 2.7585 cm and 2.7595 cm. **13.** 0.0005 m; The exact measurement could be between 8.0005 m and 8.0015 cm. **15.** 2

17 Use the rules for determining the number of significant digits. This is a whole number. In whole numbers, zeros are significant if they fall between nonzero digits. In this measurement, the zeros fall between nonzero digits, so they are significant. The nonzero digits are always significant, so this measurement has 6 significant digits.

19. 4 **21.** about 1.1% and about 1.6% **23.** about 2.5% **25.** about 8.3% **27.** same precision; 25 mi **29.** $18\frac{1}{4}$ in.; 125 yd

31. 1 ft; Suppose a mountain is 4000 ft tall. If this height is measured to the nearest foot, the relative error would be $\frac{0.5 \text{ ft}}{4000 \text{ ft}}$ or about 0.01%. If measured to the nearest inch, the relative error would be $\frac{0.5 \text{ in.}}{4000 \text{ ft}} = \frac{0.5 \text{ in.}}{48{,}000 \text{ in.}}$ or about 0.001%. If measured to the nearest $\frac{1}{16}$ in., the relative error would be $\frac{0.03125 \text{ in.}}{48{,}000 \text{ in.}}$ or about 0.0007%. While measuring to the nearest $\frac{1}{16}$ in. is a more accurate measure, this level of accuracy is not necessary; about a 1% level of accuracy is sufficient, so measuring to the nearest foot is sufficient.

33. $3\frac{1}{2}$ in. **35.** 9 cm

37 To find the greatest possible length of a side, increase 440 cubits by 0.05%. Convert 0.05% to a decimal by moving the decimal point two units to the left: 0.05% = 0.0005. To increase 440 by 0.05%, multiply by 1 + 0.0005 or 1.0005. 440(1.0005) = 440.22 cubits. To find the least possible length of a side, decrease 440 cubits by 0.05%. To decrease 440 by 0.05%, multiply by 1 − 0.0005 or 0.9995. 440(0.9995) = 439.78 cubits.

39a. Eduardo should use a ruler or tape measure marked in sixteenths of an inch. The absolute error will be $\frac{1}{2}\left(\frac{1}{16}\right) = \frac{1}{32}$. **39b.** $29\frac{31}{32}$ in.; $30\frac{1}{32}$ in.
41. 6.61 km **43.** 40 cm **45.** Manuel is correct. The beetle appears to be $1\frac{11}{16}$ in. long. Because the ruler is marked in $\frac{1}{16}$-inch increments, the absolute error is $\frac{1}{32}$ in. So the actual length of the beetle is between $1\frac{21}{32}$ in. and $1\frac{23}{32}$ in., which is between $1\frac{5}{8}$ in. and $1\frac{3}{4}$ in.
47. Sample answer: The precision of the measurement depends on the situation. **49.** A, C, D, F **51.** C **53.** D
55. A, C, D

Chapter 1 Study Guide and Review

1. plane **3.** four **5.** point P **7.** point W
9. line **11.** $x = 6$, $XP = 27$ **13.** $\sqrt{244} \approx 15.6$
15. $\sqrt{136} \approx 11.7$ **17.** 1.5 mi **19.** (16, −6.5) **21.** (−27, 16)
23. (4.8, 3) **25.** G **27.** $\overrightarrow{CA}$ and $\overrightarrow{CH}$
29. Sample answer: $\angle A$ and $\angle B$ are right, $\angle E$ and $\angle C$ are obtuse, and $\angle D$ is acute. **31.** Sample answer: $\angle QWP$ and $\angle XWV$ **33.** 66° **35.** dodecagon, concave, irregular **37.** Option 1 = 12,000 ft^2, Option 2 = 12,100 ft^2, Option 3 ≈ 15,393.8 ft^2. Option 3 provides the greatest area. **39.** reflection
41. rotation **43.** square pyramid; base: $\square ABCD$; faces: $\square ABCD$, $\triangle XAB$, $\triangle XBC$, $\triangle XCD$, $\triangle XDA$; edges, $\overline{AB}$, $\overline{BC}$, $\overline{CD}$, $\overline{DA}$, $\overline{XA}$, $\overline{XB}$, $\overline{XC}$, $\overline{XD}$; vertices: A, B, C, D, X
45. 603.2 cm^2, 1131.0 cm^3 **47.** 75.4 ft^2; 37.7 ft^3
49. 18 ft^3 **51.** a pyramid

53.

front top right

55. 3 **57.** 4

59. 0.2% **61.** 0.79%

CHAPTER 2

Logical Arguments and Line Relationships

Chapter 2 Concept Check

1. Substitute 6 for x everywhere x appears in the expression. **3.** Multiply $(11x - 7)$ by 3. **5.** They are obtuse angles. **7.** They form a linear pair.

Lesson 2-1

1. Each cost is \$2.25 more than the previous cost; \$11.25.
3. The shading moves to the next point clockwise.

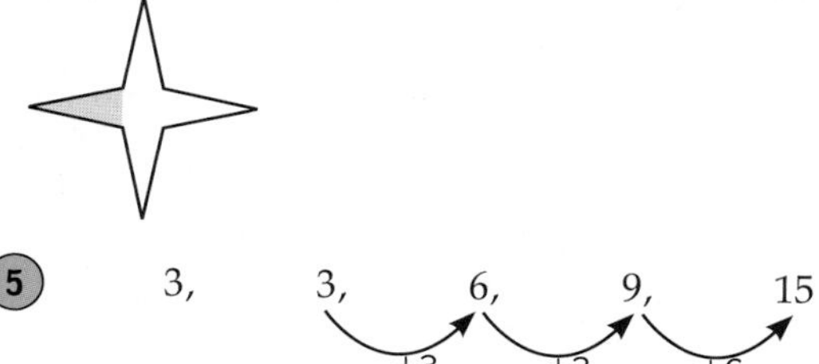

5 3, 3, 6, 9, 15 (+3, +3, +6)

Beginning with the third element, each element in the pattern is the sum of the previous two elements. So, the next element will be 15 + 9 or 24.
7. The product of two even numbers is an even number.
9. The set of points in a plane equidistant from point A is a circle.

11a. Smartphone Users by Year

Users (millions): 0, 50, 75, 100, 125, 150, 175, 200
Year: '10, '11, '12, '13, '14, '15

11b. Sample answer: About 272,000,000 Americans will use smartphones in 2018.

13 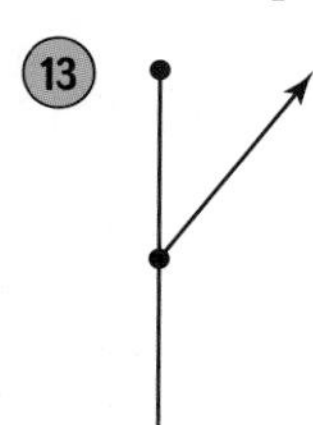

If a ray intersects a segment at its midpoint and forms adjacent angles that are not right angles, then the ray is not perpendicular to the segment. **15.** Each element in the pattern is three more than the previous element; 18.

17. Each element has an additional two as part of the number; 22,222. **19.** Each element is one half the previous element; $\frac{1}{16}$. **21.** Each percentage is 7% less than the previous percentage; 79%. **23.** Each meeting is two months after the previous meeting; July.

25. 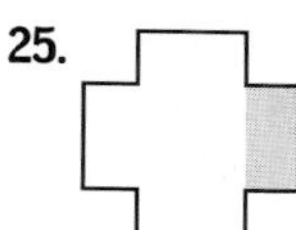
The shading moves to the next area of the figure counterclockwise.

27. 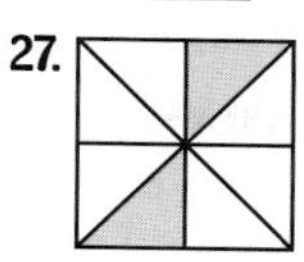
The shading of the lower triangle in the upper right quadrant of the first figure moves clockwise through each set of triangles from one figure to the next.

29. Sample answer: It is drier in the west and hotter in the south than other parts of the country, so less water would be readily available.

31 First, list examples: $1 \cdot 3 = 3$, $3 \cdot 5 = 15$, $7 \cdot 9 = 63$, $11 \cdot 11 = 121$. All the products are odd numbers. So, a conjecture about the product of two odd numbers is that the product is an odd number.
33. They are equal. **35.** The points equidistant from A and B form the perpendicular bisector of $\overline{AB}$. **37.** The area of the rectangle is two times the area of the square.

39a. 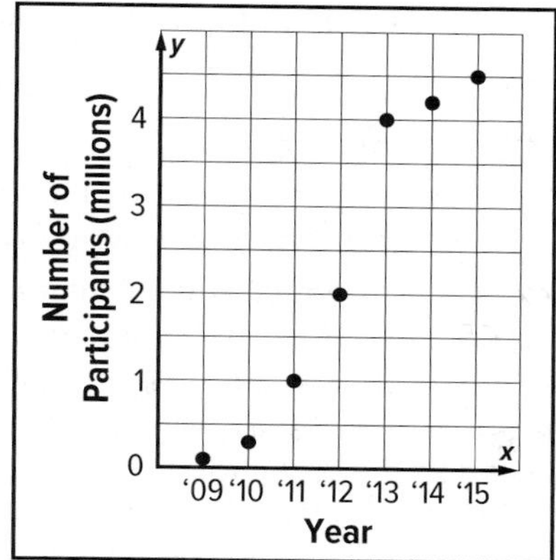

39b. The graph shows that the number of participants rose sharply from 2009 to 2013 and then continued to rise but at a slower rate. A reasonable conjecture is that the number of participants will continue to rise at a modest rate in the future.
41. False; sample answer: Suppose $x = 2$, then $-x = -2$.

43. False; sample answer: 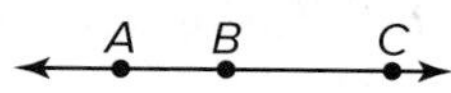

45. False; sample answer: The length could be 4 m, and the width could be 5 m. **47a.** 1, 4, 9, 16

47b. Sample answer: Start by adding 3 to 1 to get the second number, 4. Continue adding the next odd number to the previous number to get the next number in the sequence. **47c.** Sample answer: Each figure is the previous figure with an additional row and column of points added, which is 2 (position number) − 1. One is subtracted because 2 (position number) counts the corner point twice. 2 (position number) − 1 is always an odd number.

47d. 25, 36 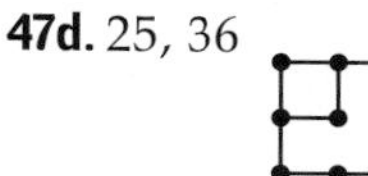 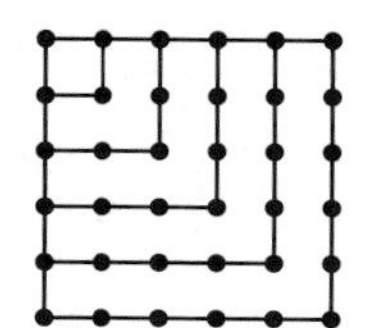

49a. 1, 5, 12, 22 **49b.** Sample answer: Start by adding 4 to 1 to get the second number, 5. Increase the amount added to the previous number by 3 each time to get the next number in sequence. So, add $4 + 3$ or 7 to 5 to get 12, and add $4 + 3 + 3$ or 10 to 12 to get 22. **49c.** Sample answer: The second figure is the previous figure with 4 points added to make a pentagon. The third figure is the previous figure with 7 more points added, which is 3 more than the last number of points added. The fourth figure is the previous figure with 10 points added, which is 3 more than the last number of points added.
49d. 35, 51

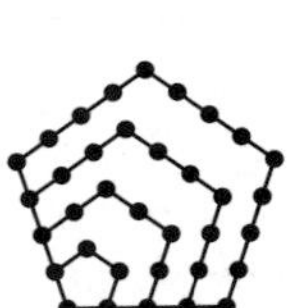

 **a.** For each even number from 10 to 20, write the number as the sum of two primes. Sample answer: $10 = 5 + 5$, $12 = 5 + 7$, $14 = 7 + 7$, $16 = 5 + 11$, $18 = 7 + 11$, $20 = 7 + 13$ **b.** The number 3 can be written as $0 + 3$ and as $1 + 2$. Because neither 0 nor 1 is a prime number, 3 cannot be written as the sum of two primes. So, the conjecture is false.

53. In the sequence of perimeters, each measure is twice the previous measure. Therefore, doubling the side length of a regular hexagon appears to also double its perimeter. In the sequence of areas, each measure is four times the previous measure. Therefore, doubling the side length of a regular hexagon appears to quadruple its area.

55. Jack; 2 is an even prime number.

57. False; sample answer: If the two points do not create a straight angle with the third point, then the conjecture is false. **59.** B **61.** D

63a.

x	x^2	$(x-1)(x+1)$
1	1	0
2	4	3
3	9	8
4	16	15
5	25	24

63b. The square of a whole number is 1 greater than the product of the whole number before it and the whole number after it. **63c.** $80 \times 80 = 6400$; $6400 - 1 = 6399$. **63d.** Sample answer: I think the rule will work for all real numbers. For example:

$x = 1.5$; $x^2 = 2.25$; $(x - 1)(x + 1) = 0.5 \times 2.5 = 1.25 = 2.25 - 1$

$x = -0.2$; $x^2 = 0.04$; $(x - 1)(x + 1) = -1.2 \times 0.8 = -0.96 = 0.04 - 1$

Lesson 2-2

1. A week has seven days, and there are 60 minutes in an hour. p and r is true, because p is true and r is true.

3 $q \vee r$: There are 20 hours in a day, or there are 60 minutes in an hour. A disjunction is true if at least one of the statements is true. So, $q \vee r$ is true because r is true. It does not matter that q is false.

5. A week has seven days, or there are 60 minutes in an hour. $p \vee r$ is true, because p is true and r is true.

7 **hypothesis:** You are sixteen years old.
conclusion: You are eligible to drive.
statement in if-then form: If you are sixteen years old, then you are eligible to drive.

9. If the angle is acute, then its measure is between 0 and 90. **11.** false; $x = -4$

13. True; Friday is the day after Thursday. **15.** True; if the hypothesis is false, then the conditional statement is automatically true. **17.** True; the number 15 is one more than 14.

19. Converse: If a number is divisible by 2, then it is divisible by 4; false. Sample answer: 6 is divisible by 2 but is not divisible by 4. Inverse: If a number is not divisible by 4, then it is not divisible by 2; false. Sample answer: 6 is not divisible by 4 but is divisible by 2. Contrapositive: If a number is not divisible by 2, then it is not divisible by 4; true. **21.** The sum of the measures of two angles is 90° if and only if the angles are complementary; true. **23.** An angle has a measure between 0° and 90° if and only if the angle is an acute angle; true. **25.** A number is an integer if and only if the number is a rational number; false. **27.** A polygon is a dodecagon if and only if it has 12 sides; true.

29. Points C, D, and B are collinear, or $\overrightarrow{DB}$ is the angle bisector of $\angle ADC$; true; q is false, but p is true.

31. $\overline{AD} \cong \overline{DC}$, and points C, D, and B are collinear; false; r is true, but q is false. **33.** $\overrightarrow{DB}$ is not the angle bisector of $\angle ADC$, and $\overline{AD} \ncong \overline{DC}$; false; both $\sim p$ and $\sim r$ are false. **35.** Springfield is the capital of Illinois, and Illinois borders the Atlantic Ocean. $p \wedge q$ is false, because q is false. **37.** Illinois shares a border with Kentucky, or Illinois borders the Atlantic Ocean. $r \vee q$ is true, because r is true. **39.** Illinois is not to the west of Missouri, or Springfield is not the capital of Illinois. $\sim s \vee \sim p$ is true, because $\sim s$ is true. **41.** If you were at the party, then you received a gift. **43.** If a figure is a circle, then the area is πr^2. **45.** If an angle is right, then the angle measures 90 degrees. **47** To show that a conditional is false, you only need to find one counterexample. 9 is an odd number, but is not divisible by 5. The hypothesis of the conditional is true, but the conclusion is false. So, this counterexample shows that the conditional statement is false. **49.** False; ; the angle drawn is

an acute angle whose measure is not 45. **51.** True; when this hypothesis is true, the conclusion is also true, because an angle and its complement's sum is 90. So, the conditional statement is true.
53. True; the hypothesis is false, because red and blue paint make purple paint. A conditional with a false hypothesis is always true, so this conditional statement is true. **55.** False; the animal could be a falcon.

57. False; these lines intersect but do not form right angles. **59.** Converse: If you live in Illinois, then you live in Chicago. False: You can live in Springfield. Inverse: If you do not live in Chicago, then you do not live in Illinois. False: You can live in Springfield. Contrapositive: If you do not live in Illinois, then you do not live in Chicago; true. **61.** Converse: If two angles are congruent, then they have the same measure; true. Inverse: If two angles do not have the same measure, then the angles are not congruent; true. Contrapositive: If two angles are not congruent, then they do not have the same measure; true.
63. Converse: If segments have the same length, then they are congruent; true. Inverse: If segments are not congruent, then they do not have the same length; true. Contrapositive: If segments do not have the same length, then they are not congruent; true. **65.** Points lie in the same plane if and only if they are coplanar; true. **67.** A point is a midpoint of a segment if and only if it bisects the segment; true. **69.** Lines are perpendicular if and only if they meet at right angles; true. **71.** If an animal is a zebra, then it has stripes; true.

73 The inverse is formed by negating both the hypothesis and the conclusion of the conditional. Inverse: If an animal does not have stripes, then it is not a zebra; true.

75. Conditional: If a triangle is equilateral, then all sides have the same length; true. Converse: If all the sides of a triangle have the same length, then the triangle is equilateral; true. Biconditional: Triangles are equilateral if and only if all the sides have the same length. **77.** Conditional: If a number is an integer, then it is a rational number; true. Converse: If a number is a rational number, then it is an integer. False; $\frac{1}{2}$ is a rational number, and it is not an integer. **79.** If the museum is the Andy Warhol Museum, then most of the collection is Andy Warhol's work. **83.** No. The conditional is not true because x can also be -6. So, a true biconditional cannot be formed. **81a.** Sample answer: If a football team makes a touchdown, they get 6 points; if a football team makes a two-point conversion, they get 2 points; if a football team makes a safety, they get 2 points.
81b. Sample answer: If a football team gets 6 points, they made a touchdown; true. If a football team gets 2 points, they made a two-point conversion; false. They could have made a safety. If a football team gets 2 points, they made a safety; false. They could have made a two-point conversion. **83.** No. The conditional is not true because x can also be -6. So, a true biconditional cannot be formed. **85.** In $p \to q$, the symbol $\to$ means that p implies q (if p, then q). In $p \leftrightarrow q$, the symbol $\leftrightarrow$ means that p implies q and q implies p (p if and only if q). **87.** Every segment has a midpoint. **89.** There exists at least one square that is not a rectangle. **91.** For every even number x, $2x - 2 \neq x$. **93.** Sample answer: Kiri; when the hypothesis of a conditional is false, the conditional is always true.
95. True; because the conclusion is false, the converse of the statement must be true. The converse and inverse are logically equivalent, so the inverse is also true. **97.** The inverse of a conditional statement is $\sim p \to \sim q$, so p is *I did not receive a detention* and q is *I arrived at school on time*. Conditional A ($p \to q$): If I did not receive a detention, then I arrived at school on time. Converse ($q \to p$): If I arrived at school on time, then I did not receive a detention. Contrapositive ($\sim q \to \sim p$): If I did not arrive at school on time, then I received a detention. **99.** If a biconditional is true, then the conditional and the converse are both true. If a biconditional is false, then the conditional or the converse is false.
101a. I: If two lines are perpendicular, then the lines intersect; if two lines intersect, then the lines are perpendicular. II: If a figure is a triangle, then it has an acute angle. If a figure has an acute angle, then it is a triangle. III: If two angles are supplementary, then the sum of the measures of the angles is 180°; if the sum of the measures of two angles is 180°, then the angles are supplementary; true. IV: If an angle is a right angle, then it has a measurement of 90°; if an angle has a measurement of 90°, then it is a right angle; true. **101b.** C, D **101c.** C, D
103. D **105a.** A triangle has two congruent sides or a triangle has no congruent sides or a triangle is equilateral; true. **105b.** A triangle has two congruent sides and a triangle has no congruent sides; false.

Lesson 2-3

1 Olivia is basing her conclusion on facts provided to her by her high school, not on a pattern of observations, so she is using deductive reasoning.

3. valid; Law of Detachment **5.** Invalid: Bayview could be inside or outside the Public circle.

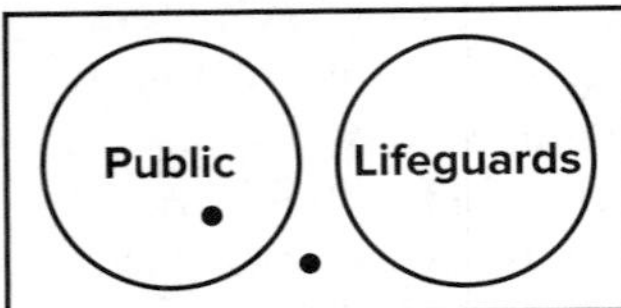

7. Invalid; the conclusion is of the form $r \rightarrow p$, so it does not follow the Law of Syllogism. **9.** Invalid; the conclusion is of the form $q \rightarrow r$, so it does not follow the Law of Syllogism. **11.** No valid conclusion; $\angle 1$ and $\angle 2$ do not have to be vertical in order to be congruent.
13. inductive reasoning **15.** deductive reasoning
17. inductive reasoning

19 The given statement *Figure ABCD has four right angles* satisfies the conclusion of the true conditional. However, having a true conditional and a true conclusion does not make the hypothesis true. The figure could be a rectangle. So, the conclusion is invalid.

21. Invalid; your battery could be dead because it was old. **23.** valid; Law of Detachment
25. Invalid; Monday is outside the circle for the days when the temperature drops below –15°F, and it could be inside or outside the circle for the days when school is cancelled. So, the conclusion is invalid.

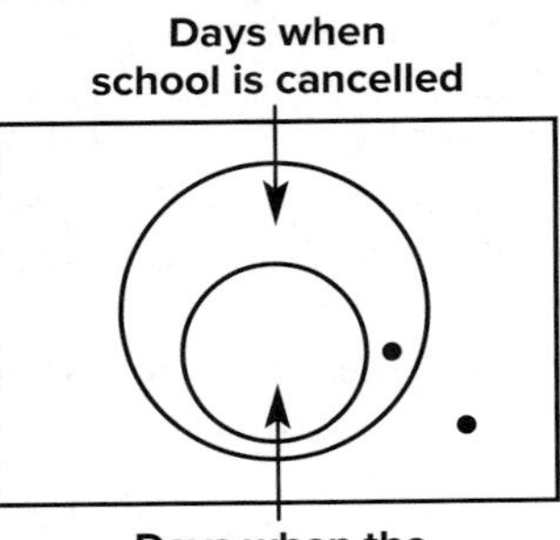

27. Invalid; Sabrina could be inside the Nurses circle or inside the intersection of the circles, so the conclusion is invalid.

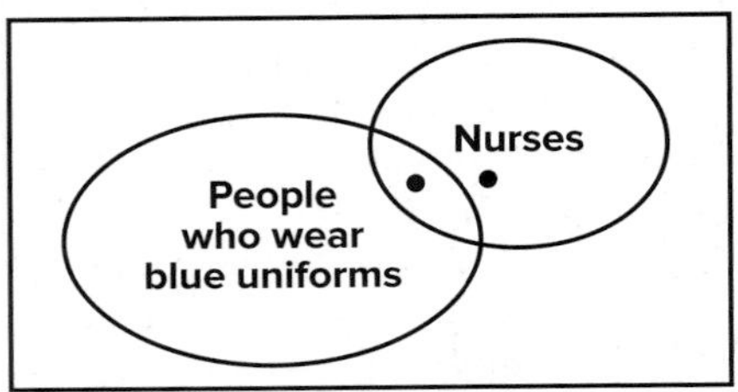

29 The given statement *Ms. Rodriguez has just purchased a vehicle that has four-wheel drive* satisfies the conclusion of the true conditional. However, having a true conditional and a true conclusion does not make the hypothesis true. Ms. Rodriguez's car might be in the Four-wheel-drive section of the diagram that is not a sport-utility vehicle. So, the conclusion is invalid.

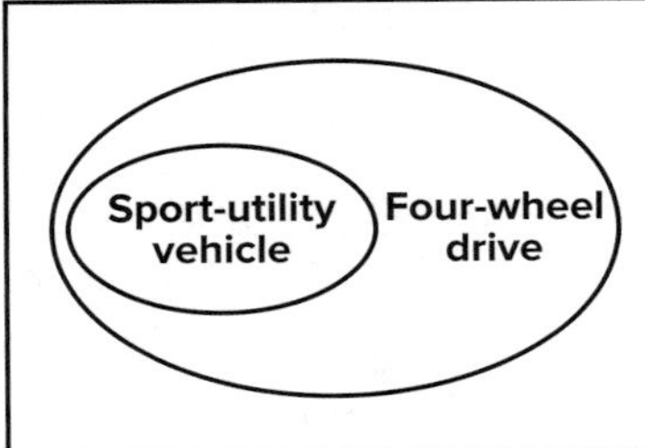

31. no valid conclusion **33.** no valid conclusion
35. If two lines in a plane are not parallel, then they intersect in a point. **37.** Figure *ABCD* has all sides congruent; Law of Detachment.

39 You can reword the first given statement: If you are a ballet dancer, then you like classical music. Statement (1): If you are a ballet dancer, then you like classical music. Statement (2): If you like classical music, then you enjoy the opera. Because the conclusion of Statement (1) is the hypothesis of Statement (2), you can apply the Law of Syllogism. A valid conclusion: If you are a ballet dancer, then you enjoy the opera.

41. No valid conclusion; knowing a conclusion is true does not imply the hypothesis will be true.
43a.

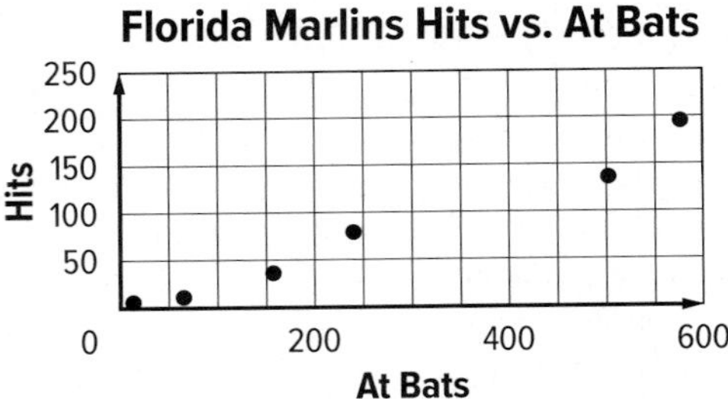

43b. Sample answer: about 91; inductive; a pattern was used to reach the conclusion. **43c.** Sample answer: The player with 240 at bats got more hits; deductive; the facts provided in the table were used to reach the conclusion.
45. Law of Detachment: $[(p \rightarrow q) \wedge p] \rightarrow q$; Law of Syllogism: $[(p \rightarrow q) \wedge (q \rightarrow r)] \rightarrow (p \rightarrow r)$ **47.** Jonah's statement can be restated as, "Jonah is in group B, and Janeka is in group B." For this compound statement to be true, both parts of the statement must be true. If Jonah were in group A, he would not be able to say that he is in group B, because students in group A must always tell the truth. Therefore, the statement that Jonah is in group B is true. For the compound statement to be false, the statement that Janeka is in group B must be false. Therefore, Jonah is in group B, and Janeka is in group A. **49.** A
51a. the Law of Syllogism **51b.** B **51c.** A, D **51d.** B, F

Lesson 2-4

1. The left side and front side have a common edge, line *r*. Planes *P* and *Q* only intersect at line *r*. Postulate 2.7 states that if two planes intersect, then their intersection is a line. **3.** The front bottom edge of the figure is line *n*, which contains points *D*, *C*, and *E*. Postulate 2.3 states that a line contains at least two points. **5.** Points *D* and *E*, which are on line *n*, lie in plane *Q*. Postulate 2.5 states that if two points lie in a plane, then the entire line containing those points lies in that plane. **7.** Never; noncollinear points do not lie on the same line by definition.
9. Postulate 2.2 states that through any three noncollinear points, there is exactly one plane.
11. Postulate 2.1 states that through any two points, there is exactly one line.

13 The first statement is the given information. In the second statement, 12 is added to each side of the equation. The reason this is possible is the Addition Property of Equality. In the third statement, $5x$ is substituted for $5x - 12 + 12$, and 20 is substituted for $8 + 12$. The reason this is possible is the Substitution Property of Equality. To complete the proof, divide each side of $5x = 20$ by 5 to get $x = 4$.

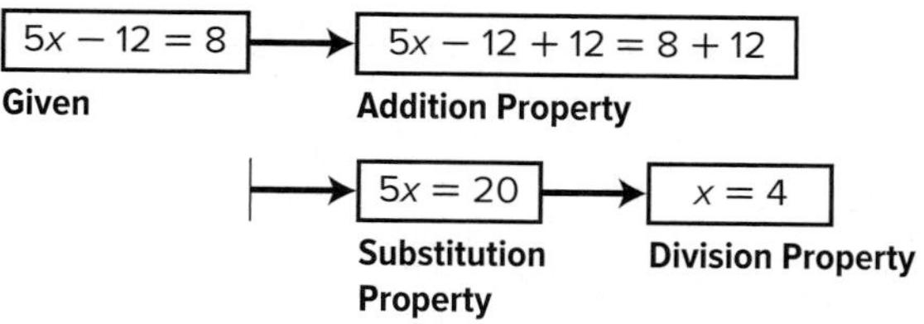

15. Because Y is the midpoint of $\overline{XZ}$, $\overline{XY} \cong \overline{YZ}$ by the Midpoint Theorem. By the Symmetric Property, $\overline{XY} \cong \overline{WY}$ can be written as $\overline{WY} \cong \overline{XY}$. Therefore, $\overline{WY} \cong \overline{YZ}$ by the Transitive Property. **17.** The edges of the sides of the bottom layer of the cake intersect. Plane P and Q of this cake intersect only once in line m. Postulate 2.7; if two planes intersect, then their intersection is a line. **19.** The top edge of the bottom layer of the cake is a straight line n. Points C, D, and K lie along this edge, so they lie along line n. Postulate 2.3; a line contains at least two points. **21** The bottom right part of the cake is a side. The side contains points K, E, F, and G and forms a plane. Postulate 2.2, which states that through any three noncollinear points, there is exactly one plane, shows that this is true. **23.** The top edges of the bottom layer form intersecting lines. Lines h and g of this cake intersect only once at point J. Postulate 2.6; if two lines intersect, then their intersection is exactly one point. **25.** Never; Postulate 2.1 states that through any two points, there is exactly one line. **27.** Always; Postulate 2.5 states that if two points lie in a plane, then the entire line containing those points lies in that plane. **29.** Sometimes; the points must be noncollinear.
31.

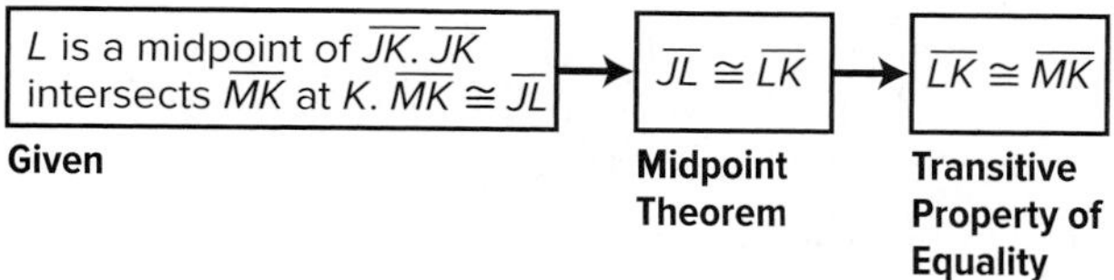

33. It is given that there were 11 bikes and skateboards, so $b + s = 11$ and $s = 11 - b$ by the Subtraction Property. It is given that there were 36 wheels, so $2b + 4s = 36$ and by the Substitution Property, $2b + 4(11 - b) = 36$. By the Distributive Property, $2b + 44 - 4b = 36$. So, $44 - 2b = 36$ by the Substitution Property. By the Subtraction Property, $-2b = -8$, and so $b = 4$ by the Division Property. By substitution, $s = 11 - 4$, or 7, by the Substitution Property. **35** Points E, F, and G lie along the same line. Postulate 2.3 states that a line contains at least two points. **37.** Postulate 2.5; if two points lie in a plane, then the entire line containing those points lies in that plane.

39. Postulate 2.2; through any three noncollinear points, there is exactly one plane. **41.** Postulate 2.6; if two lines intersect, then their intersection is exactly one point. **43.** Sample answer: Because Den is designing a watershed roof, the pitch of the roof should be a minimum of 4 inches per foot. The pitch of the roof in Den's design is 2 inches per foot, which is less than 4 inches per foot. Therefore, the pitch of the roof in Den's design is not steep enough. **45a.** All of the exits are the same distance from the center **45b.** All points of a circle are equidistant from the center. **45c.** radius **47a.** Plane Q is perpendicular to plane P. **47b.** Line a is perpendicular to plane P. **49.** Blake and Katrina used different algebraic methods, but both proofs are correct. **51.** A, C, D, F **53.** C **55.** A

Lesson 2-5

1. Given: $\overline{LK} \cong \overline{NM}$, $\overline{KJ} \cong \overline{MJ}$
Prove: $\overline{LJ} \cong \overline{NJ}$
Proof:
Statements (Reasons)
a. $\overline{LK} \cong \overline{NM}$, $\overline{KJ} \cong \overline{MJ}$ (Given)
b. $LK = NM$, $KJ = MJ$ (Def. of ≅ segs.)
c. $LK + KJ = NM + MJ$ (Add. Prop.)
d. $LJ = LK + KJ$; $NJ = NM + MJ$ (Seg. Add. Post.)
e. $LJ = NJ$ (Subst.)
f. $\overline{LJ} \cong \overline{NJ}$ (Def. of ≅ segs.)

3 Use the definition of congruent segments and the Substitution Property of Equality.
Given: $\overline{AR} \cong \overline{CR}$; $\overline{DR} \cong \overline{BR}$
Prove: $AR + DR = CR + BR$
Proof:
Statements (Reasons)
1. $\overline{AR} \cong \overline{CR}$, $\overline{DR} \cong \overline{BR}$ (Given)
2. $AR = CR$, $DR = BR$ (Definition of ≅ segments)
3. $AR + DR = CR + DR$ (Addition Property)
4. $AR + DR = CR + BR$ (Substitution Property)

5. Given: $\overline{AB} \cong \overline{CD}$, $AB + CD = EF$
Prove: $2AB = EF$
Proof:
Statements (Reasons)
1. $\overline{AB} \cong \overline{CD}$, $AB + CD = EF$ (Given)
2. $AB = CD$ (Def. of ≅ segs.)
3. $AB + AB = EF$ (Subst.)
4. $2AB = EF$ (Subst.)

7 Use the Reflexive Property of Equality and the definition of congruent segments.
Given: $\overline{AB}$
Prove: $\overline{AB} \cong \overline{AB}$
Proof:
Statements (Reasons)
1. $\overline{AB}$ (Given)
2. $AB = AB$ (Reflexive Property)
3. $\overline{AB} \cong \overline{AB}$ (Definition of ≅ segments)

9. Given: $\overline{SC} \cong \overline{HR}$ and $\overline{HR} \cong \overline{AB}$
Prove: $\overline{SC} \cong \overline{AB}$
Proof:
<u>**Statements (Reasons)**</u>
1. $\overline{SC} \cong \overline{HR}$ and $\overline{HR} \cong \overline{AB}$ (Given)
2. $SC = HR$ and $HR = AB$ (Def. of $\cong$ segs.)
3. $SC = AB$ (Trans. Prop.)
4. $\overline{SC} \cong \overline{AB}$ (Def. of $\cong$ segs.)

11. Given: E is the midpoint of $\overline{DF}$ and $\overline{CD} \cong \overline{FG}$.
Prove: $\overline{CE} \cong \overline{EG}$
Proof:
<u>**Statements (Reasons)**</u>
1. E is the midpoint of $\overline{DF}$ and $\overline{CD} \cong \overline{FG}$. (Given)
2. $DE = EF$ (Def. of midpoint)
3. $CD = FG$ (Def. of $\cong$ segs.)
4. $CD + DE = EF + FG$ (Add. Prop.)
5. $CE = CD + DE$ and $EG = EF + FG$ (Seg. Add. Post.)
6. $CE = EG$ (Subst.)
7. $\overline{CE} \cong \overline{EG}$ (Def. of $\cong$ segs.)

13a. Given: $\overline{AC} \cong \overline{GI}$, $\overline{FE} \cong \overline{LK}$, $AC + CF + FE = GI + IL + LK$
Prove: $\overline{CF} \cong \overline{IL}$
Proof:
<u>**Statements (Reasons)**</u>
1. $\overline{AC} \cong \overline{GI}$, $\overline{FE} \cong \overline{LK}$, $AC + CF + FE = GI + IL + LK$ (Given)
2. $AC + CF + FE = AC + IL + LK$ (Subst.)
3. $AC - AC + CF + FE = AC - AC + IL + LK$ (Subt. Prop.)
4. $CF + FE = IL + LK$ (Subst.)
5. $CF + FE = IL + FE$ (Subs.)
6. $CF + FE - FE = IL + FE - FE$ (Subt. Prop.)
7. $CF = IL$ (Subst.)
8. $\overline{CF} \cong \overline{IL}$ (Def. of $\cong$ segs.)

13b. Sample answer: I measured $\overline{CF}$ and $\overline{IL}$, and both were 1.5 inches long, so the two segments are congruent.

15 **a.** Use the definition of midpoint and the Segment Addition Postulate.
Given: $\overline{SH} \cong \overline{TF}$; P is the midpoint of $\overline{SH}$ and $\overline{TF}$.
Prove: $\overline{SP} \cong \overline{TP}$
Proof:
<u>**Statements (Reasons)**</u>
1. $\overline{SH} \cong \overline{TF}$, P is the midpoint of $\overline{SH}$, P is the midpoint of $\overline{TF}$. (Given)
2. $SH = TF$ (Definition of $\cong$ segments)
3. $SP = PH$, $TP = PF$ (Definition of midpoint)
4. $SH = SP + PH$, $TF = TP + PF$ (Segment Addition Postulate)
5. $SP + PH = TP + PF$ (Substitution)
6. $SP + SP = TP + TP$ (Substitution)
7. $2SP = 2TP$ (Substitution)
8. $SP = TP$ (Division Property)
9. $\overline{SP} \cong \overline{TP}$ (Definition of $\cong$ segments)

b. $SP = \frac{1}{2}SH$ — Definition of midpoint
$= \frac{1}{2}(127.3)$ or 63.54 — Substitution

Because $\overline{TF} \cong \overline{SH}$, then $FP = SP = 63.54$.

Because $\triangle SPF$ is a right triangle, use the Pythagorean Theorem to find SF, the distance from first base to second base.

$c^2 = a^2 + b^2$	Pythagorean Theorem
$SF^2 = SP^2 + FP^2$	Substitution
$SF^2 = 63.54^2 + 63.54^2$	Substitution
$SF^2 \approx 8074.6632$	Simplify.
$SF \approx 90$	Take the positive square root of each side.

The distance from first base to second base is about 90 feet.

17. Neither; because $\overline{AB} \cong \overline{CD}$ and $\overline{CD} \cong \overline{BF}$, then $\overline{AB} \cong \overline{BF}$ by the Transitive Property of Congruence.
19. No; congruence refers to segments. Segments cannot be added, only the measures of segments.
21.

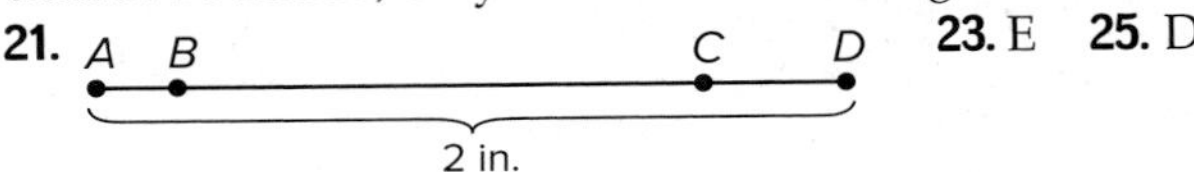

23. E **25.** D

27a. C, D, E **27b.** B **27c.** C

Lesson 2-6

1 $m\angle 1 = 90$ because $\angle 1$ is a right angle.

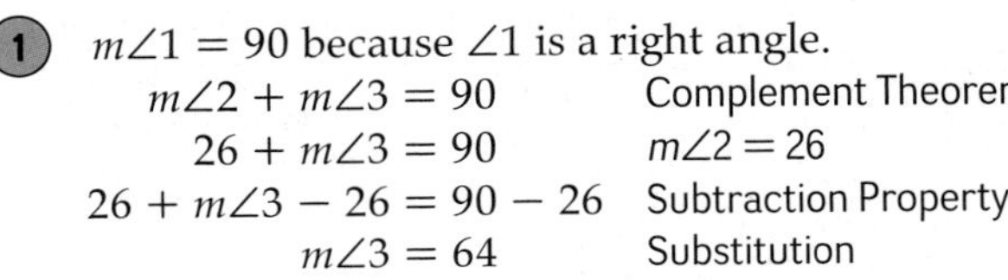

$m\angle 2 + m\angle 3 = 90$	Complement Theorem
$26 + m\angle 3 = 90$	$m\angle 2 = 26$
$26 + m\angle 3 - 26 = 90 - 26$	Subtraction Property
$m\angle 3 = 64$	Substitution

3. $m\angle 4 = 114$, $m\angle 5 = 66$; Suppl. Thm.

5. Given: $\angle 2 \cong \angle 6$
Prove: $\angle 4 \cong \angle 8$
Proof:
<u>**Statements (Reasons)**</u>
1. $\angle 2 \cong \angle 6$ (Given)
2. $m\angle 2 + m\angle 4 = 180$, $m\angle 6 + m\angle 8 = 180$ (Suppl. Thm.)
3. $m\angle 2 + m\angle 8 = 180$ (Subst.)
4. $m\angle 2 - m\angle 2 + m\angle 4 = 180 - m\angle 2$, $m\angle 2 - m\angle 2 + m\angle 8 = 180 - m\angle 2$ (Subt. Prop.)
5. $m\angle 4 = 180 - m\angle 2$, $m\angle 8 = 180 - m\angle 2$ (Subt. Prop.)
6. $m\angle 4 = m\angle 8$ (Subst.)
7. $\angle 4 \cong \angle 8$ (Def. $\cong$ ∡)

7. Given: $\angle 4 \cong \angle 7$
Prove: $\angle 5 \cong \angle 6$
Proof:
<u>**Statements (Reasons)**</u>
1. $\angle 4 \cong \angle 7$ (Given)
2. $\angle 4 \cong \angle 5$ and $\angle 6 \cong \angle 7$ (Vert. ∡ Thm.)
3. $\angle 7 \cong \angle 5$ (Subst.)
4. $\angle 5 \cong \angle 6$ (Subst.)

9. $m\angle 3 = 62$, $m\angle 1 = m\angle 4 = 45$ (Comp. and Suppl. Thm.) **11.** $m\angle 9 = 156$, $m\angle 10 = 24$ (Suppl. Thm.)

13

$m\angle 6 + m\angle 7 = 180$	Supplement Thm.
$2x - 21 + 3x - 34 = 180$	Substitution
$5x - 55 = 180$	Substitution
$5x - 55 + 55 = 180 + 55$	Addition Property
$5x = 235$	Substitution
$\frac{5x}{5} = \frac{235}{5}$	Division Property
$x = 47$	Substitution

$m\angle 6 = 2x - 21$	Given
$m\angle 6 = 2(47) - 21$ or 73	Substitution

$m\angle 7 = 3x - 34$	Given
$m\angle 7 = 3(47) - 34$ or 107	Substitution

$\angle 8 \cong \angle 6$	Vertical Angles Theorem
$m\angle 8 = m\angle 6$	Definition of $\cong$ $\angle$s
$= 73$	Substitution

15. Given: $\angle 5 \cong \angle 6$

Prove: $\angle 4$ and $\angle 6$ are supplementary.

Proof:

Statements (Reasons)

1. $\angle 5 \cong \angle 6$ (Given)
2. $m\angle 5 = m\angle 6$ (Def. of $\cong$ $\angle$s)
3. $\angle 4$ and $\angle 5$ are supplementary. (Def. of linear pairs)
4. $m\angle 4 + m\angle 5 = 180$ (Def. of supp. $\angle$s)
5. $m\angle 4 + m\angle 6 = 180$ (Subst.)
6. $\angle 4$ and $\angle 6$ are supplementary. (Def. of supp. $\angle$s)

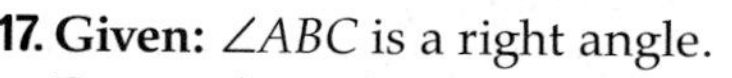

17. Given: $\angle ABC$ is a right angle.

Prove: $\angle 1$ and $\angle 2$ are complementary angles.

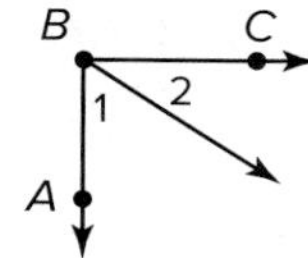

Proof:

Statements (Reasons)

1. $\angle ABC$ is a right angle. (Given)
2. $m\angle ABC = 90$ (Def. of rt. $\angle$s)
3. $m\angle ABC = m\angle 1 + m\angle 2$ ($\angle$ Add. Post.)
4. $90 = m\angle 1 + m\angle 2$ (Subst.)
5. $\angle 1$ and $\angle 2$ are complementary angles. (Def. of comp. $\angle$s)

19. Given: $\angle 1 \cong \angle 2$, $\angle 2 \cong \angle 3$

Prove: $\angle 1 \cong \angle 3$

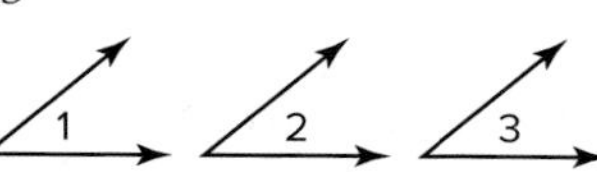

Proof:

Statements (Reasons)

1. $\angle 1 \cong \angle 2$, $\angle 2 \cong \angle 3$ (Given)
2. $m\angle 1 = m\angle 2$, $m\angle 2 = m\angle 3$ (Def. of $\cong$ $\angle$s)
3. $m\angle 1 = m\angle 3$ (Trans. Prop.)
4. $\angle 1 \cong \angle 3$ (Def. of $\cong$ $\angle$s)

21. Given: $\angle 1 \cong \angle 4$

Prove: $\angle 2 \cong \angle 3$

Proof:

Statements (Reasons)

1. $\angle 1 \cong \angle 4$ (Given)
2. $\angle 1 \cong \angle 2$, $\angle 3 \cong \angle 4$ (Vert. $\angle$s are $\cong$.)
3. $\angle 1 \cong \angle 3$ (Trans. Prop.)
4. $\angle 2 \cong \angle 3$ (Subst.)

23. Given: $\angle 1$ and $\angle 2$ are rt. $\angle$s.

Prove: $\angle 1 \cong \angle 2$

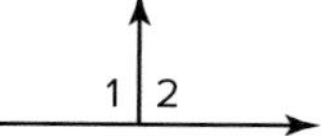

Proof:

Statements (Reasons)

1. $\angle 1$ and $\angle 2$ are rt. $\angle$s. (Given)
2. $m\angle 1 = 90$, $m\angle 2 = 90$ (Def. of rt. $\angle$s)
3. $m\angle 1 = m\angle 2$ (Subst.)
4. $\angle 1 \cong \angle 2$ (Def. of $\cong$ $\angle$s)

25. Given: $\angle 1 \cong \angle 2$, $\angle 1$ and $\angle 2$ are supplementary.

Prove: $\angle 1$ and $\angle 2$ are rt. $\angle$s.

Proof:

Statements (Reasons)

1. $\angle 1 \cong \angle 2$, $\angle 1$ and $\angle 2$ are supplementary. (Given)
2. $m\angle 1 + m\angle 2 = 180$ (Def. of supp. $\angle$s)
3. $m\angle 1 = m\angle 2$ (Def. of $\cong$ $\angle$s)
4. $m\angle 1 + m\angle 1 = 180$ (Subst.)
5. $2(m\angle 1) = 180$ (Subst.)
6. $m\angle 1 = 90$ (Div. Prop.)
7. $m\angle 2 = 90$ (Subst. (steps 3, 6))
8. $\angle 1$ and $\angle 2$ are rt. $\angle$s. (Def. of rt. $\angle$s)

27. Because the path of the pendulum forms a right angle, $\angle ABC$ is a right angle, or measures 90°. $\overrightarrow{BR}$ divides $\angle ABC$ into $\angle ABR$ and $\angle CBR$. By the Angle Addition Postulate, $m\angle ABR + m\angle CBR = m\angle ABC$, and, using substitution, $m\angle ABR + m\angle CBR = 90$. Substituting again, $m\angle 1 + m\angle 2 = 90$. We are given that $m\angle 1$ is 45°, so, substituting, $45 + m\angle 2 = 90$. Using the Subtraction Property, $45 - 45 + m\angle 2 = 90 - 45$, or $m\angle 2 = 45$. Because $m\angle 1$ and $m\angle 2$ are equal, $\overrightarrow{BR}$ is the bisector of $\angle ABC$ by the definition of angle bisector.

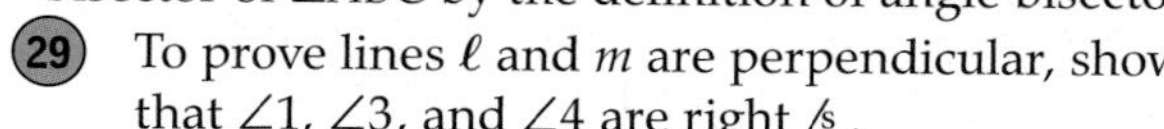

29 To prove lines ℓ and m are perpendicular, show that $\angle 1$, $\angle 3$, and $\angle 4$ are right $\angle$s.

Given: $\angle 2$ is a right angle.

Prove: $\ell \perp m$

Proof:

Statements (Reasons)

1. $\angle 2$ is a right angle. (Given)
2. $m\angle 2 = 90$ (Definition of a rt. $\angle$)
3. $\angle 2 \cong \angle 3$ (Vert. $\angle$s are $\cong$.)
4. $m\angle 3 = 90$ (Substitution)
5. $m\angle 1 + m\angle 2 = 180$ (Supplement Theorem)
6. $m\angle 1 + 90 = 180$ (Substitution)
7. $m\angle 1 + 90 - 90 = 180 - 90$ (Subtraction Property)
8. $m\angle 1 = 90$ (Substitution)
9. $\angle 1 \cong \angle 4$ (Vertical $\angle$s are $\cong$.)
10. $\angle 4 \cong \angle 1$ (Symmetric Property)
11. $m\angle 4 = m\angle 1$ (Definition of $\cong$ $\angle$s)
12. $m\angle 4 = 90$ (Substitution)
13. $\ell \perp m$ ($\perp$ lines intersect to form four rt. $\angle$s.)

31. Given: $\overrightarrow{XZ}$ bisects $\angle WXY$, and $m\angle WXZ = 45$.

Prove: $\angle WXY$ is a right angle.

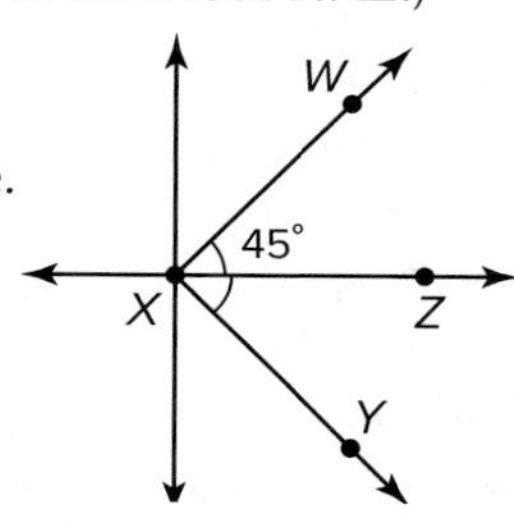

Proof:

Statements (Reasons)

1. $\overrightarrow{XZ}$ bisects $\angle WXY$ and $m\angle WXZ = 45$. (Given)

Selected Answers and Solutions

2. $\angle WXZ \cong \angle ZXY$ (Def. of $\angle$ bisector)
3. $m\angle WXZ = m\angle ZXY$ (Def. of $\cong$ ∠s)
4. $m\angle ZXY = 45$ (Subst.)
5. $m\angle WXY = m\angle WXZ + m\angle ZXY$ ($\angle$ Add. Post.)
6. $m\angle WXY = 45 + 45$ (Subst.)
7. $m\angle WXY = 90$ (Subst.)
8. $\angle WXY$ is a right angle. (Def. of rt. ∠s)

33. Each of these theorems uses the words "or to congruent angles" indicating that this case of the theorem must also be proven true. The other proofs only addressed the "to the same angle" case of the theorem.

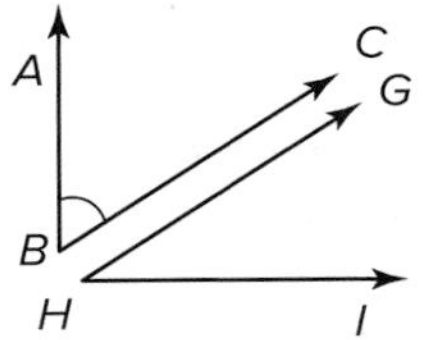

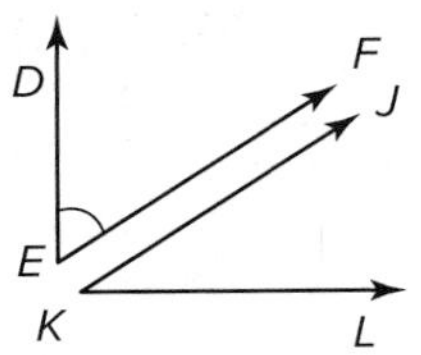

Given: $\angle ABC \cong \angle DEF$, $\angle GHI$ is complementary to $\angle ABC$, $\angle JKL$ is complementary to $\angle DEF$.
Prove: $\angle GHI \cong \angle JKL$
Proof:
Statements (Reasons)
1. $\angle ABC \cong \angle DEF$, $\angle GHI$ is complementary to $\angle ABC$, $\angle JKL$ is complementary to $\angle DEF$. (Given)
2. $m\angle ABC + m\angle GHI = 90$, $\angle DEF + \angle JKL = 90$ (Def. of compl. ∠s)
3. $m\angle ABC + m\angle JKL = 90$ (Subst.)
4. $90 = m\angle ABC + m\angle JKL$ (Symm. Prop.)
5. $m\angle ABC + m\angle GHI = m\angle ABC + m\angle JKL$ (Trans. Prop.)
6. $m\angle ABC - m\angle ABC + m\angle GHI = m\angle ABC - m\angle ABC + m\angle JKL$ (Subt.)
7. $m\angle GHI = m\angle JKL$ (Subst.)
8. $\angle GHI \cong \angle JKL$ (Def. of $\cong$ ∠s)

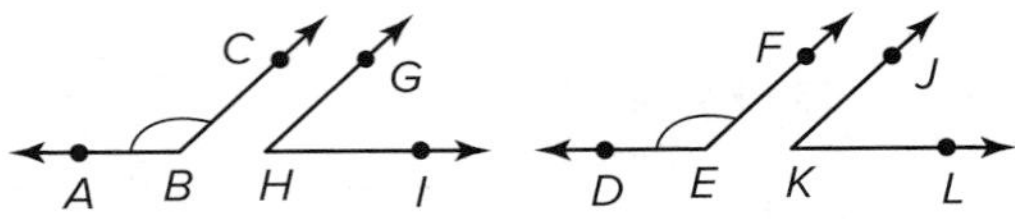

Given: $\angle ABC \cong \angle DEF$, $\angle GHI$ is supplementary to $\angle ABC$, $\angle JKL$ is supplementary to $\angle DEF$.
Prove: $\angle GHI \cong \angle JKL$
Proof:
Statements (Reasons)
1. $\angle ABC \cong \angle DEF$, $\angle GHI$ is supplementary to $\angle ABC$, $\angle JKL$ is supplementary to $\angle DEF$. (Given)
2. $m\angle ABC + m\angle GHI = 180$, $m\angle DEF + m\angle JKL = 180$ (Def. of suppl. ∠s)
3. $m\angle ABC + m\angle JKL = 180$ (Subst.)
4. $180 = m\angle ABC + m\angle JKL$ (Symm. Property)
5. $m\angle ABC + m\angle GHI = m\angle ABC + m\angle JKL$ (Trans. Prop.)
6. $m\angle ABC - m\angle ABC + m\angle GHI = m\angle ABC - m\angle ABC + m\angle JKL$ (Subt.)
7. $m\angle GHI = m\angle JKL$ (Subst.)
8. $\angle GHI \cong \angle JKL$ (Def. of $\cong$ ∠s)

35. Sample answer: Since protractors have the scale for both acute and obtuse angles along the top, the supplement is the measure of the given angle on the other scale. **37a.** $4x - 5 + x = 180$ **37b.** 37°, 143° **39.** B **41.** B

Lesson 2-7

1 $\angle 1$ and $\angle 8$ are nonadjacent exterior angles that lie on opposite sides of the transversal. So, they are alternate exterior angles. **3.** alternate interior angles **5.** 94; Corresponding Angle Postulate **7.** 86; Corresponding Angle Postulate and Supplement Angle Thm. **9.** 79; Vertical Angle Thm. and Cons. Int. ∠s Thm. **11.** $m\angle 2 = 93$, $m\angle 3 = 87$, $m\angle 4 = 87$ **13.** $x = 114$ by the Alt. Ext. ∠s Thm. **15.** corresponding **17.** alternate interior **19.** alternate exterior **21.** consecutive interior **23.** alternate exterior **25.** 62; Corr ∠s Post. **27.** 118; Def Supp ∠s **29.** 38; Corr ∠s Post. **31.** 142; Supplementary Angles Thm. **33.** 38; Alt Ext ∠s Thm

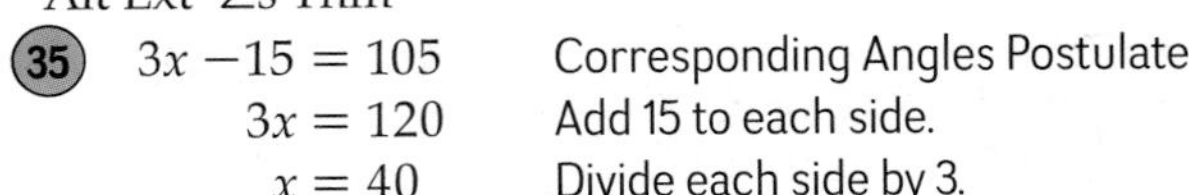

35

$3x - 15 = 105$	Corresponding Angles Postulate
$3x = 120$	Add 15 to each side.
$x = 40$	Divide each side by 3.
$(3x - 15) + (y + 25) = 180$	Supplement Theorem
$105 + y + 25 = 180$	Substitution
$y + 130 = 180$	Simplify.
$y = 50$	Subtract 130 from each side.

So, $x = 40$ by the Corresponding Angles Postulate; $y = 50$ by the Supplement Theorem.
37. $x = 42$, $y = 14$
39. $x = 60$, $y = 10$ **41.** congruent; Alternate Interior Angles **43.** congruent; vertical angles are congruent.

45. Given: $\ell \parallel m$
Prove: $\angle 1 \cong \angle 8$
$\angle 2 \cong \angle 7$
Proof:
Statements (Reasons)
1. $\ell \parallel m$ (Given)
2. $\angle 1 \cong \angle 5$, $\angle 2 \cong \angle 6$ (Corr. ∠s Post.)
3. $\angle 5 \cong \angle 8$, $\angle 6 \cong \angle 7$ (Vertical ∠s Thm.)
4. $\angle 1 \cong \angle 8$, $\angle 2 \cong \angle 7$ (Trans. Prop.)

47.
Given: $m \parallel n$, $t \perp m$
Prove: $t \perp n$
Proof:

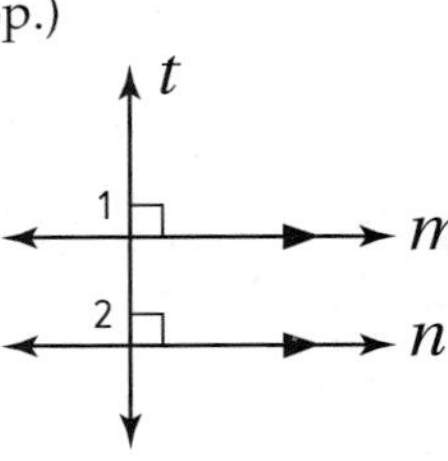

Statements (Reasons)
1. $m \parallel n$, $t \perp m$ (Given)
2. $\angle 1$ is a right angle. (Def. of $\perp$)
3. $m\angle 1 = 90$ (Def. of rt. ∠s)
4. $\angle 1 \cong \angle 2$ (Corr. $\angle$ Post.)
5. $m\angle 1 = m\angle 2$ (Def. of $\cong$ ∠s)
6. $m\angle 2 = 90$ (Subs.)
7. $\angle 2$ is a right angle. (Def. of rt. $\angle$)
8. $t \perp n$ (Def. of $\perp$ lines)

49 Draw a line parallel to the two given lines and label angles 1, 2, 3, and 4. So, $x = m\angle 2 + m\angle 3$. $m\angle 1 = 105$ because vertical angles are congruent and their measures are equal. $\angle 1$ and $\angle 2$ are supplementary by the Consecutive Interior Angles Theorem.

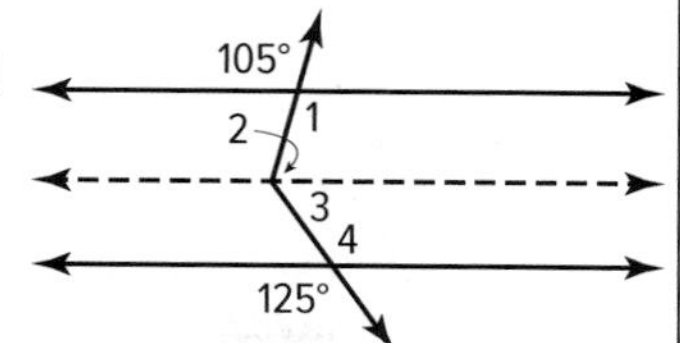

$m\angle 1 + m\angle 2 = 180$	Definition of supplementary angles
$105 + m\angle 2 = 180$	Substitution
$m\angle 2 = 75$	Subtract 105 from each side.

$m\angle 4 = 125$ because vertical angles are congruent and their measures are equal. $\angle 3$ and $\angle 4$ are supplementary by the Consecutive Interior Angles Theorem.

$m\angle 3 + m\angle 4 = 180$	Definition of supplementary angles
$m\angle 3 + 125 = 180$	Substitution
$m\angle 3 = 55$	Subtract 125 from each side.
$m\angle 2 + m\angle 3 = x$	Angle Addition Postulate
$75 + 55 = x$	Substitution
$130 = x$	Simplify.

51a. Sample answer for m and n:

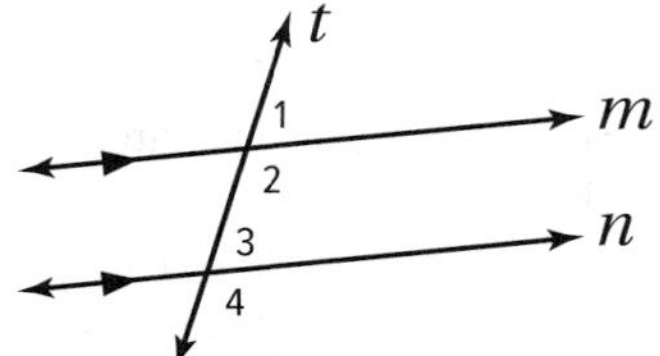

51b. Sample answer:

$m\angle 1$	$m\angle 2$	$m\angle 3$	$m\angle 4$
60	120	60	120
45	135	45	135
70	110	70	110
90	90	90	90
25	155	25	155

51c. Sample answer: Angles on the exterior of a pair of parallel lines located on the same side of the transversal are supplementary. **51d.** Inductive; a pattern was used to make a conjecture. **51e. Given:** parallel lines m and n cut by transversal t

Prove: $\angle 1$ and $\angle 4$ are supplementary.

Proof:

Statements (Reasons)

1. Lines m and n are parallel and cut by transversal t. (Given)

2. $m\angle 1 + m\angle 2 = 180$ (Suppl. Thm.)

3. $\angle 2 \cong \angle 4$ (Corr. $\angle$s are $\cong$.)

4. $m\angle 2 = m\angle 4$ (Def. of congruence.)

5. $m\angle 1 + m\angle 4 = 180$ (Subs.)

6. $\angle 1$ and $\angle 4$ are supplementary. (Def. of supplementary $\angle$s.)

53. In both theorems, a pair of angles is formed when two parallel lines are cut by a transversal. However, in the Alternate Interior Angles Theorem, each pair of alternate interior angles that is formed are congruent, whereas in the Consecutive Interior Angles Theorem, each pair of angles formed is supplementary. **55.** $x = 171$ or $x = 155$, $y = 3$ or $y = 5$ **57.** C **59.** C **61.** E, F

Lesson 2-8

1. -1 **3.** $\frac{6}{5}$ **5.** $y + 3 = \frac{1}{4}(x + 2)$

7

$\overrightarrow{WX} = \frac{y_2 - y_1}{x_2 - x_1}$	Slope formula
$= \frac{5 - 4}{4 - 2}$	$(x_1, y_1) = (2, 4)$, $(x_2, y_2) = (4, 5)$
$= \frac{1}{2}$	Simplify.
$\overrightarrow{YZ} = \frac{y_2 - y_1}{x_2 - x_1}$	Slope formula
$= \frac{-7 - 1}{8 - 4}$	$(x_1, y_1) = (4, 1)$, $(x_2, y_2) = (8, -7)$
$= -\frac{8}{4}$ or -2	Simplify.
$\frac{1}{2}(-2) = -1$	Product of slopes

Because the product of the slopes is -1, $\overleftrightarrow{WX}$ is perpendicular to $\overleftrightarrow{YZ}$.;

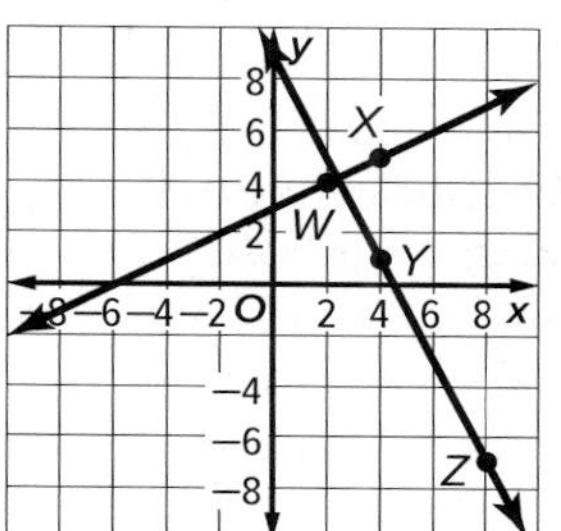

9. parallel;

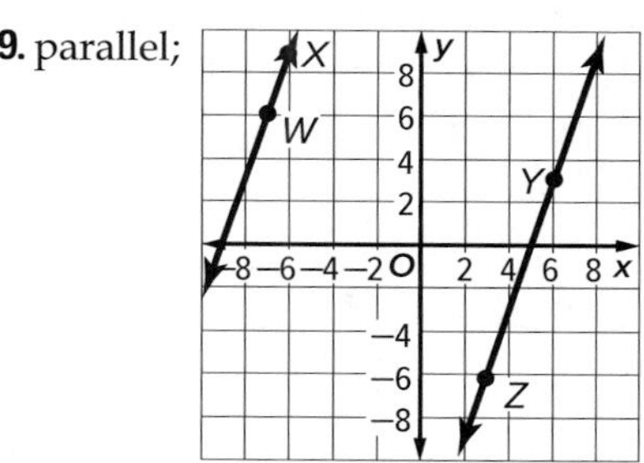

11. $y = \frac{1}{2}x + \frac{1}{2}$ **13** Substitute $(-3, 1)$ for (x_1, y_1) and $(4, -2)$ for (x_2, y_2).

$m = \frac{y_2 - y_1}{x_2 - x_1} = \frac{-2 - 1}{4 - (-3)} = -\frac{3}{7}$

15. 0 **17.** 1 **19.** 0 **21.** undefined **23.** $-\frac{1}{6}$

25. $y - 8 = 4(x + 4)$

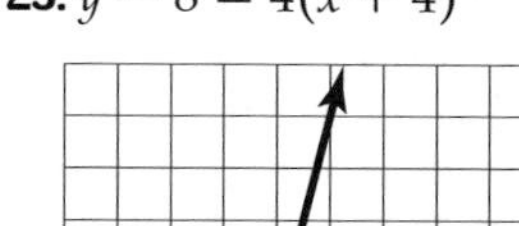

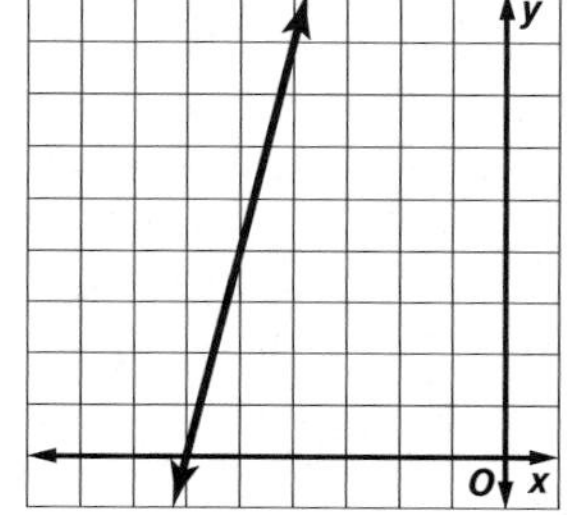

27 $y - y_1 = m(x - x_1)$ Point-slope form
$y - 11 = 2(x - 3)$ $m = 2, (x_1, y_1) = (3, 11)$
$y = 2x + 5$ Simplify.

Graph the given point (3, 11). Use the slope 2 or $\frac{2}{1}$ to find another point 2 units up and 1 unit to the right.

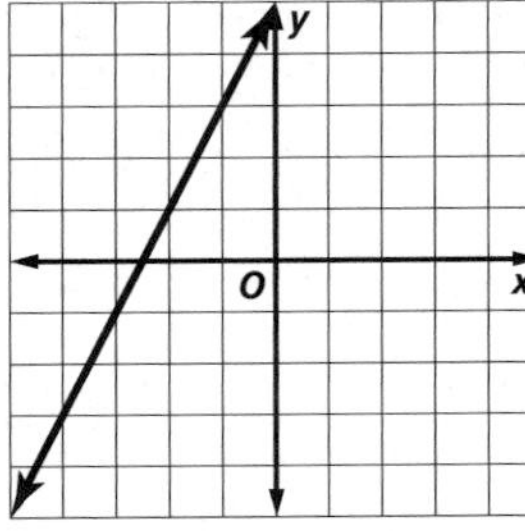

29. $y + 12 = -2.4(x - 14)$ **31.** parallel **33.** perpendicular **35.** neither **37.** p **39.** n, p, or r **41.** perpendicular **43.** neither

45 The line containing (3, 2) and (−7, 2) is a horizontal line because the y-coordinates of the points are the same. The line perpendicular to the line containing these points must be vertical. Because this line passes through (−8, 12), all of the x-coordinates of points on the line are −8, so the equation of the line is $x = -8$. **47a.** $C(t) = 4t + 5.5$ **47b.** $D(t) = 4t + 6.25$ **47c.** The graphs are parallel lines. The lines are parallel because they have the same slope, 4. **49.** rectangle; in slope-intercept form, the equations are $y = \frac{1}{2}x + 3$, $y = -2x + 13$, $y = \frac{1}{2}x - 2$, and $y = -2x - 2$. The slopes of the lines are $\frac{1}{2}$, −2, $\frac{1}{2}$, and −2. This means there are two pairs of parallel lines that intersect at right angles, so the shape enclosed by the lines is a rectangle. **51.** 14 **53.** Sample answer: $y = 2x - 1$, $y = -\frac{1}{2}x - \frac{17}{2}$ **55.** Sample answer: When given the slope and y-intercept, the slope-intercept form is easier to use. When given two points, the point-slope form is easier to use. When given the slope and a point, the point-slope form is easier to use. **57.** C **59a.** $\frac{3}{4}$ **59b.** $-\frac{4}{3}$ **61a.** $y = -3x + 1$ **61b.** $y = \frac{1}{3}x + 1$ **61c.** (0, 1); this point is the intersection of the lines $y = -3x + 1$ and $y = \frac{1}{3}x + 1$. **61d.** $y = -3x - 4$

Lesson 2-9

1. $j \parallel k$; Converse of Corresponding Angles Postulate

3 Angle 3 and ∠10 are alternate exterior angles of lines ℓ and m. Because ∠3 ≅ ∠10, ℓ ∥ m by the Alternate Exterior Angles Converse Theorem.

5. 20 **7.** Yes; sample answer: Because the alternate exterior angles are congruent, the backrest and footrest are parallel. **9.** $u \parallel v$; Alternate Exterior ∠s Converse **11.** $r \parallel s$; Consecutive Interior ∠s Converse **13.** $u \parallel v$; Alternate Interior ∠s Converse **15.** $r \parallel s$; Corresponding ∠s Converse **17.** 22; Conv. Corr. ∠s Post.

19 The angles are consecutive interior angles. For lines m and n to be parallel, consecutive interior angles must be supplementary, according to the Consecutive Interior Angles Converse Theorem.

$(7x - 2) + (10 - 3x) = 180$ Definition of supp. ∠s
$4x + 8 = 180$ Simplify.
$4x = 172$ Subtract 8 from each side.
$x = 43$ Divide each side by 4.

21. 36; Alt. Ext. ∠s Conv. **23a.** ∠1 and ∠2 are supplementary. **23b.** Def. of linear pair **23c.** ∠2 and ∠3 are supplementary; Suppl. Thm. **23d.** ≅ Suppl. Thm. **23e.** Converse of Corr. ∠s Post.

25. Given: ∠1 ≅ ∠3, $\overline{AC} \parallel \overline{BD}$
Prove: $\overline{AB} \parallel \overline{CD}$
Proof:
Statements (Reasons)
1. ∠1 ≅ ∠3, $\overline{AC} \parallel \overline{BD}$ (Given)
2. ∠2 ≅ ∠3 (Corr. ∠s Post.)
3. ∠1 ≅ ∠2 (Trans. Prop.)
4. $\overline{AB} \parallel \overline{CD}$ (If alt. int. ∠s are ≅, then lines are ∥.)

27. Given: ∠ABC ≅ ∠ADC, m∠A + m∠ABC = 180
Prove: $\overline{AB} \parallel \overline{CD}$
Proof:
Statements (Reasons)
1. ∠ABC ≅ ∠ADC, m∠A + m∠ABC = 180 (Given)
2. m∠ABC = m∠ADC (Def. of ≅ ∠s)
3. m∠A + m∠ADC = 180 (Substitution)
4. ∠A and ∠ADC are supplementary. (Def. of supplementary ∠s)
5. $\overline{AB} \parallel \overline{CD}$ (If consec. int. ∠s are supplementary, then lines are ∥.)

29 The Converse of the Perpendicular Transversal Theorem states that two coplanar lines perpendicular to the same line are parallel. Because the slots, or the bottom of each rectangular opening, are perpendicular to each of the sides, the slots are parallel. Because any pair of slots is perpendicular to the sides, they are also parallel.

31. Given: ∠1 ≅ ∠2
Prove: ℓ ∥ m
Proof:
Statements (Reasons)
1. ∠1 ≅ ∠2 (Given)
2. ∠2 ≅ ∠3 (Vertical ∠s are ≅)
3. ∠1 ≅ ∠3 (Transitive Prop.)
4. ℓ ∥ m (If corr ∠s are ≅, then lines are ∥.)

33. $r \parallel s$; Sample answer: The corresponding angles are congruent. Because the measures of the angles are equal, the lines are parallel. **35.** $r \parallel s$; Sample answer: The alternate exterior angles are congruent. Because the measures of the angles are equal, the lines are parallel. **37.** Daniela; ∠1 and ∠2 are alternate interior angles for $\overline{WX}$ and $\overline{YZ}$, so if alternate interior angles are congruent, then the lines are parallel.

39. Given: $a \parallel b$ and $b \parallel c$
Prove: $a \parallel c$
Statements (Reasons)
1. $a \parallel b$ and $b \parallel c$ (Given)
2. ∠1 ≅ ∠3 (Alt. Int. ∠s Thm.)
3. ∠3 ≅ ∠2 (Vert. ∠s are ≅)

4. $\angle 2 \cong \angle 4$ (Alt. Int. ∠s Thm.)
5. $\angle 1 \cong \angle 4$ (Trans. Prop.)
6. $a \parallel c$ (Alt. Int. ∠s Conv. Thm.)

41a. We know that $m\angle 1 + m\angle 2 = 180$. Because $\angle 2$ and $\angle 3$ are linear pairs, $m\angle 2 + m\angle 3 = 180$. By substitution, $m\angle 1 + m\angle 2 = m\angle 2 + m\angle 3$. By subtracting $m\angle 2$ from both sides we get $m\angle 1 = m\angle 3$. $\angle 1 \cong \angle 3$, by the definition of congruent angles. Therefore, $a \parallel c$ since the corresponding angles are congruent. **41b.** We know that $a \parallel c$ and $m\angle 1 + m\angle 3 = 180$. Because $\angle 1$ and $\angle 3$ are corresponding angles, they are congruent and their measures are equal. By substitution, $m\angle 3 + m\angle 3 = 180$ or $2m\angle 3 = 180$. By dividing both sides by 2, we get $m\angle 3 = 90$. Therefore, $t \perp c$ because they form a right angle. **43.** Yes; sample answer: A pair of angles can be both supplementary and congruent if the measure of both angles is 90, because the sum of the angle measures would be 180. **45a.** 24 **45b.** B **47.** B **49.** D

Lesson 2-10

1.

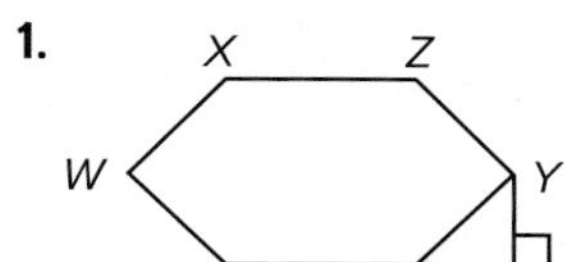

3. The formation should be that of two parallel lines that are also parallel to the 50-yard line; the band members have formed two lines that are equidistant from the 50-yard line, so by Theorem 3.9, the two lines formed are parallel. **5.** $\sqrt{10}$ units

7 Let ℓ represent $y = -2x + 4$ and let m represent $y = -2x + 14$. The slope of the lines is -2. Write an equation for line p. The slope of p is the opposite reciprocal of -2, or $\frac{1}{2}$. Use the y-intercept of line ℓ, $(0, 4)$, as one of the endpoints of the perpendicular segment.

$(y - y_1) = m(x - x_1)$ Point-slope form

$(y - 4) = \frac{1}{2}(x - 0)$ $(x_1, y_1) = (0, 4), m = \frac{1}{2}$

$y - 4 = \frac{1}{2}x$ Simplify.

$y = \frac{1}{2}x + 4$ Add 4 to each side.

Use a system of equations to find the point of intersection of lines m and p.

Equation for m: $y = -2x + 14$

Equation for p: $y = \frac{1}{2}x + 4$

$y = -2x + 14$ Equation for m

$\frac{1}{2}x + 4 = -2x + 14$ Use Equation for p to substitute $\frac{1}{2}x + 4$ for y.

$\frac{5}{2}x + 4 = 14$ Add $2x$ to each side.

$\frac{5}{2}x = 10$ Subtract 4 from each side.

$x = 4$ Multiply each side by $\frac{2}{5}$.

$y = \frac{1}{2}x + 4$ Equation for p

$y = \frac{1}{2}(4) + 4$ Substitute 4 for x.

$y = 6$ Simplify.

The point of intersection is $(4, 6)$. Use the Distance Formula to find the distance between $(0, 4)$ and $(4, 6)$.

$d = \sqrt{(x_2 - x_1)^2 + (y_2 - y_1)^2}$ Distance Formula

$= \sqrt{(4 - 0)^2 + (6 - 4)^2}$ $x_1 = 0, y_1 = 4, x_2 = 4, y_2 = 6$

$= \sqrt{20}$ or $2\sqrt{5}$ Simplify.

The distance between the lines is $2\sqrt{5}$ units.

9.

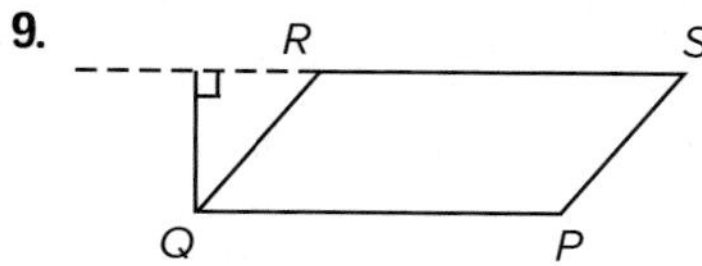

11.

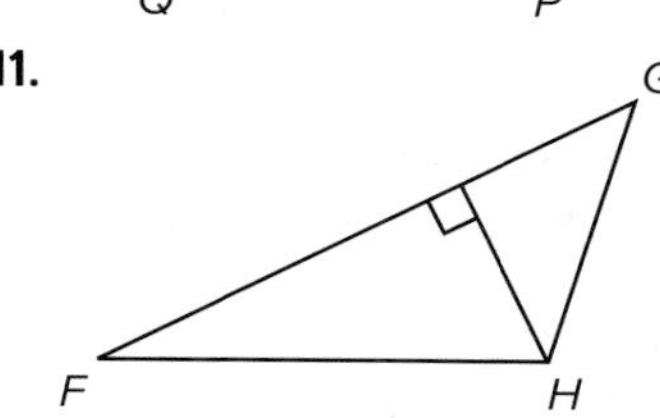

13. No; a driveway perpendicular to the road would be the shortest. The angle the driveway makes with the road is less than 90°, so it is not the shortest possible driveway.

15 Find the slope and y-intercept of line ℓ and write the equation for the line.

$m = \frac{y_2 - y_1}{x_2 - x_1} = \frac{4 - (-3)}{7 - 0}$ or 1

Because ℓ contains $(0, -3)$, the y-intercept is -3. So, the equation for line ℓ is $y = 1x + (-3)$ or $y = x - 3$. The slope of a line perpendicular to ℓ, line w, is -1. Write the equation of line w through $(4, 3)$ with slope -1.

$y = mx + b$ Slope-intercept form
$3 = -1(4) + b$ $m = -1, (x, y) = (4, 3)$
$3 = -4 + b$ Simplify.
$7 = b$ Add 4 to each side.

So, the equation for line w is $y = -x + 7$. Solve the system of equations to determine the point of intersection.

line ℓ: $y = x - 3$
line w: $(+)\ y = -x + 7$
$2y = 4$ Add the two equations.
$y = 2$ Divide each side by 2.

Solve for x.

$y = x - 3$ Equation for line ℓ
$2 = x - 3$ $y = 2$
$5 = x$ Add 3 to each side.

The point of intersection is $(5, 2)$. Let this be point Q. Use the Distance Formula to determine the distance between $P(4, 3)$ and $Q(5, 2)$.

$d = \sqrt{(x_2 - x_1)^2 + (y_2 - y_1)^2}$ Distance Formula

$= \sqrt{(5-4)^2 + (2-3)^2}$ $\quad x_1 = 4, y_1 = 3,$ $x_2 = 5, y_2 = 2$

$= \sqrt{2}$ $\quad$ Simplify.

The distance between the lines is $\sqrt{2}$ units.

17. 6 units **19.** $\sqrt{10}$ units **21.** 6 units **23.** $\sqrt{26}$ units **25.** 21 units **27.** $4\sqrt{17}$ units **29.** $\sqrt{14.76}$ units **31.** 5 units **33.** 6 units

35 He can conclude that the right and left sides of the bulletin board are not parallel, because the perpendicular distance between one line and any point on the other line must be equal anywhere on the lines for the two lines to be parallel. In this case, the length of the top of the bulletin board is not equal to the length of the bottom of the bulletin board.

39b. Place point C any place on line m. The area of the triangle is $\frac{1}{2}$ the height of the triangle times the length of the base of the triangle. The numbers stay constant regardless of the location of C on line m. **39c.** 16.5 in^2

41 To find out if the lines will intersect, determine if they are parallel. If they are parallel, then the perpendicular distance between the two lines at any point will be equal. Use a ruler to measure the distance between points A and C and points B and D. $AC = 1.2$ centimeters and $BD = 1.35$ centimeters. The lines are not parallel, so they will intersect. Shenequa is correct.

43. $a = \pm 1$; $y = \frac{1}{2}x + 6$ and $y = \frac{1}{2}x + \frac{7}{2}$ or $y = -\frac{1}{2}x + 6$ and $y = -\frac{1}{2}x + \frac{7}{2}$

45a. Sample answer:

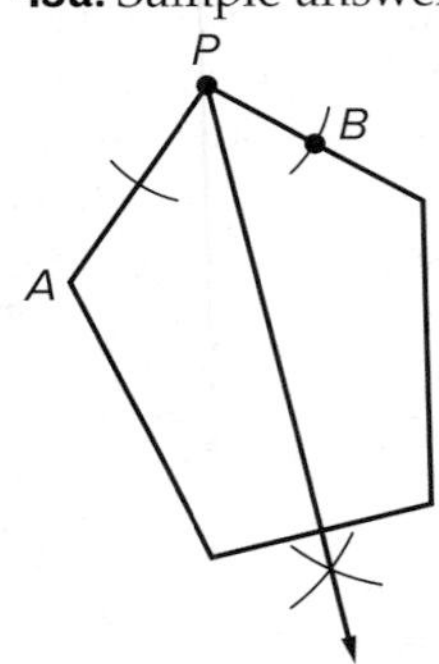

45b. Sample answer: Using a protractor, the measurement of the constructed angle is equal to 90. So, the line constructed from vertex P is perpendicular to the nonadjacent side chosen. **45c.** Sample answer: The same compass setting was used to construct points A and B. Then the same compass setting was used to construct the perpendicular line to the side chosen. Because the compass setting was equidistant in both steps, a perpendicular line was constructed.

47. Because parallel lines are equidistant everywhere, find the distance between a pair of points, which form a line segment perpendicular to the given lines. The Distance Formula is used to determine the distance between the pair of intersection points. Repeat this process for a second pair of points on the lines. Use the Distance Formula to find the distance between these points. This distance should be the same as the distance between the first pair of points. **49.** C **51.** B **53.** A

Chapter 2 Study Guide and Review

1. false; hypothesis **3.** true **5.** false; negation **7.** false; point-slope **9.** Inductive reasoning is a conjecture reached based on observations of previous patterns, while deductive reasoning uses the law of mathematics to reach logical conclusions from given statements. **11.** False; two nonadjacent supplementary angles **13.** Sample answer: Dogs or other pets may threaten or chase wildlife that might not be present in his local park. **15.** A plane contains at least three noncollinear points and the sum of the measures of two complementary angles is not 180; true. **17.** False; if the integer is 0, then 0^2 is not a positive integer. **19.** Converse: If two angles have the same degree measure, then they are congruent. True. Inverse: If two angles are not congruent, then they do not have the same degree measure. True. Contrapositive: If two angles do not have the same degree measure, then they are not congruent. True. **21.** Invalid; the Law of Syllogism does not apply since the conclusion of the first statement is not the hypothesis of the second statement. **23.** Never; if two planes intersect, then they form a line. **25.** Always; if a plane contains a line, then every point of that line lies in the plane. **27.** 15 handshakes;

29. Statements (Reasons)
1. $AB = DC$ (Given)
2. $BC = BC$ (Refl. Prop.)
3. $AB + BC = DC + BC$ (Add. Prop.)
4. $AB + BC = AC$, $DC + BC = DB$ (Seg. Add. Post.)
5. $AC = DB$ (Subs.)

31. 90 **33.** 53 **35.** corresponding **37.** alternate exterior **39.** skew lines **41.** 57; $\angle 5 \cong \angle 13$ by Corr. $\angle$s Post. and $\angle 13$ and $\angle 14$ form a linear pair. **43.** 123; $\angle 11 \cong \angle 5$ by Alt. Int. $\angle$s Thm. and $\angle 5 \cong \angle 1$ by Alt. Ext. $\angle$s Thm. **45.** 57; $\angle 1 \cong \angle 3$ by Corr. $\angle$s Post. and $\angle 3$ and $\angle 6$ form a linear pair. **47.** perpendicular **49.** parallel

51.

53. $y + 9 = 2(x - 4)$ **55.** $y = 5x - 3$ **57.** $y = -\frac{2}{3}x + 10$ **59.** $C = 20h + 50$ **61.** none **63.** $v \parallel z$; Alternate Exterior Angles Converse Thm. **65.** 135

67.

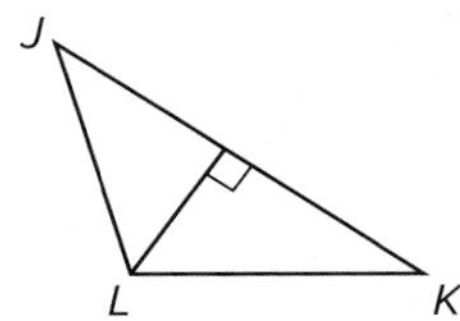

Rigid Transformations and Symmetry

Chapter 3 Concept Check

1. rotation **3.** translation **5.** $A'(1, -3)$, $B'(1, -1)$, $C'(4, -1)$, $D'(4, -3)$

Lesson 3-1

1.

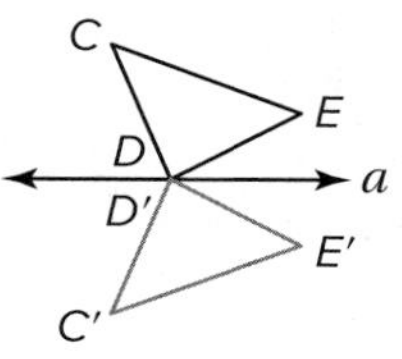

3.

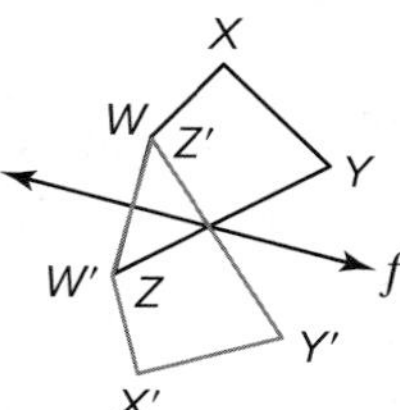

5.

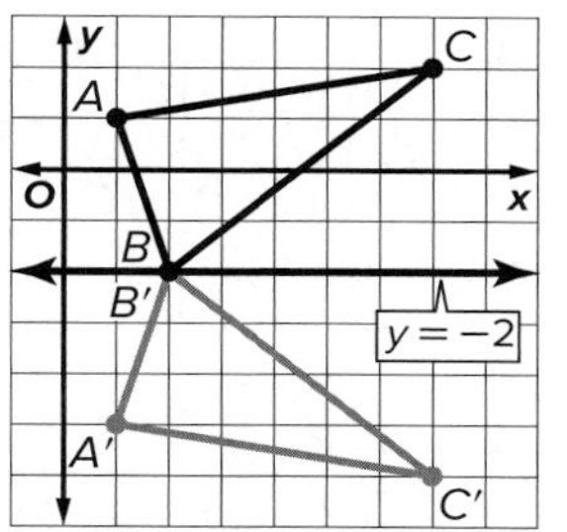

7.

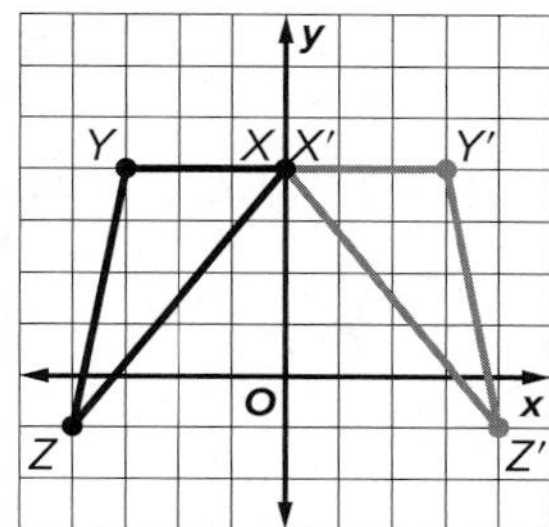

9. $(x, y) \rightarrow (y, x)$; reflection in the line $y = x$

11.

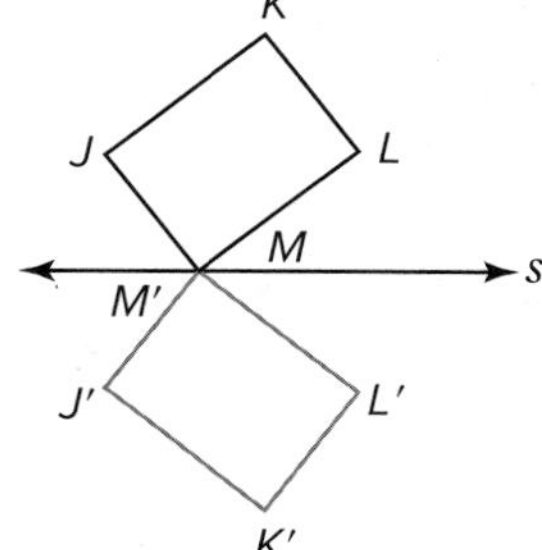

13 Use a protractor to draw perpendiculars from A, B, C, and D to line ℓ. Extend the perpendiculars to the opposite side of line ℓ. Measure the distance along the perpendicular from A to line ℓ. Use this distance to locate A' on the perpendicular on the opposite side of line ℓ. Repeat the process to locate B', C', and D'. Use a straightedge to connect A', B', C', and D'.

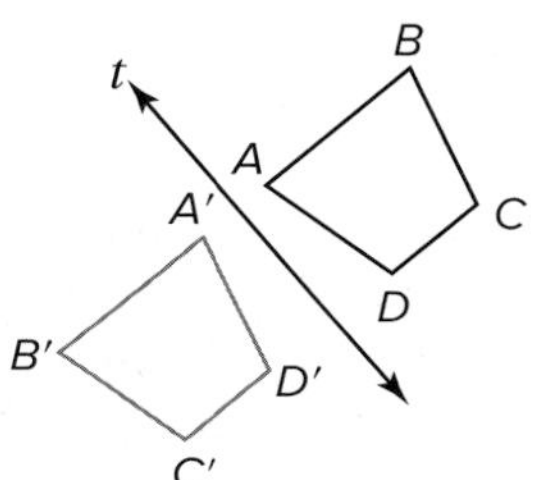

15.

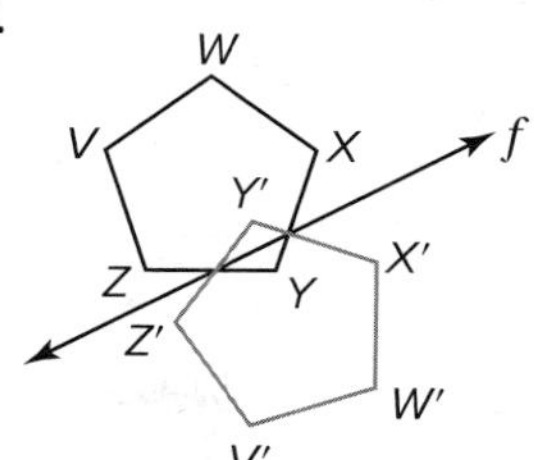

17.

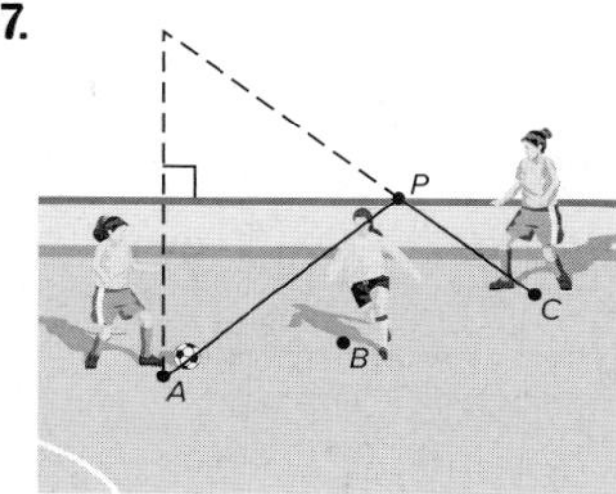

19.

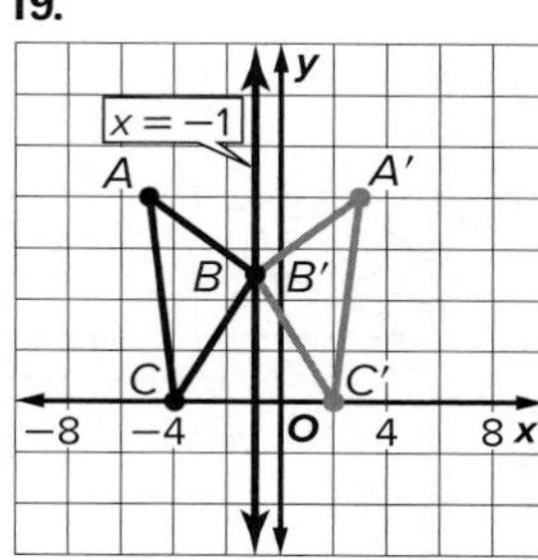

21.

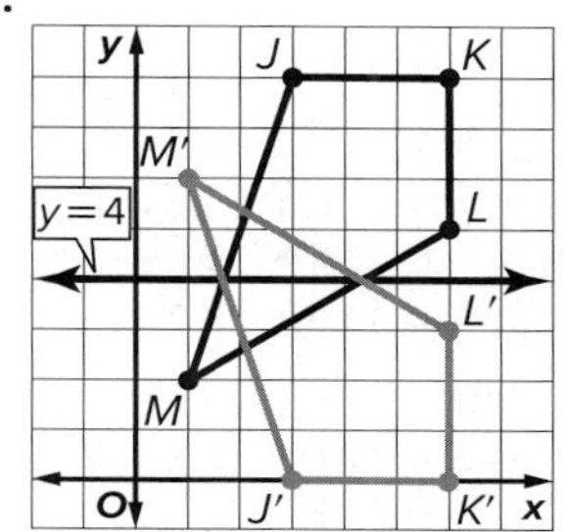

23.

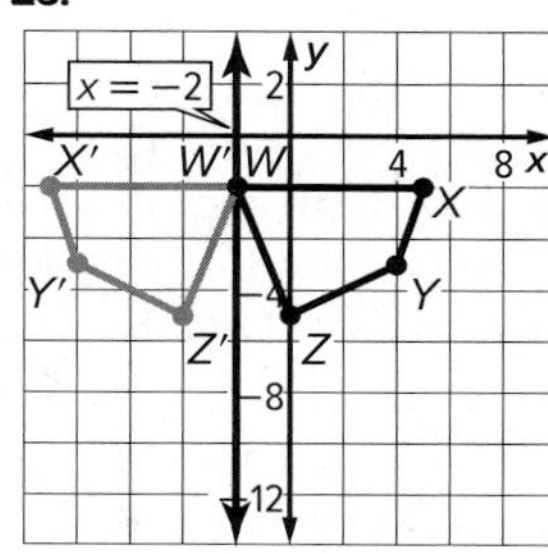

25 Plot the vertices of square $JKLM$. Find the image of each vertex after a reflection in the y-axis using the rule $(x, y) \rightarrow (-x, y)$.

$J(-4, 6)$	$\rightarrow$	$J'(4, 6)$
$K(0, 6)$	$\rightarrow$	$K'(0, 6)$
$L(0, 2)$	$\rightarrow$	$L'(0, 2)$
$M(-4, 2)$	$\rightarrow$	$M'(4, 2)$

Plot the points J', K', L', and M' and connect them to draw the image of square $JKLM$.

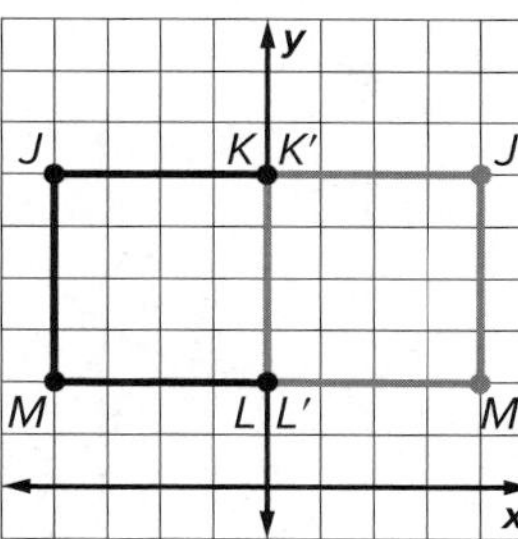

27.

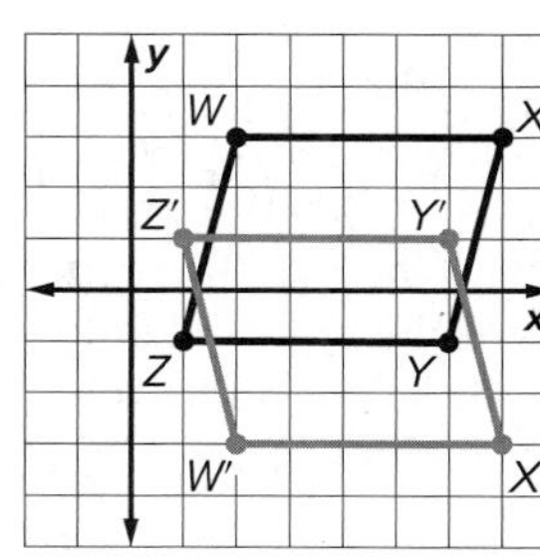

29. $(x, y) \rightarrow (y, x)$; reflection in the line $y = x$

Selected Answers and Solutions

31.

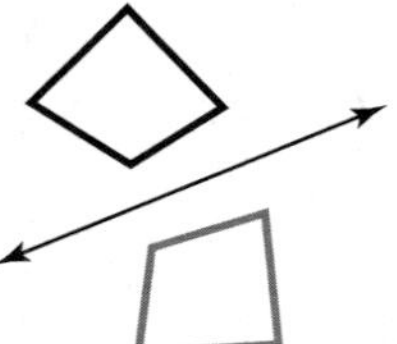

33.

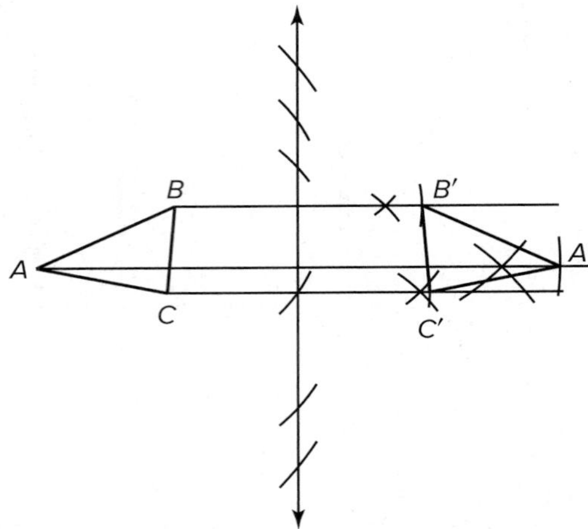

35 **a.** $\overleftrightarrow{BE}$; The figures on either side of this line are mirror images of each other. **b.** Sample answer: A and C, D and F, $\overline{AB}$ and $\overline{CD}$, $\overline{AD}$ and $\overline{CF}$.

37.

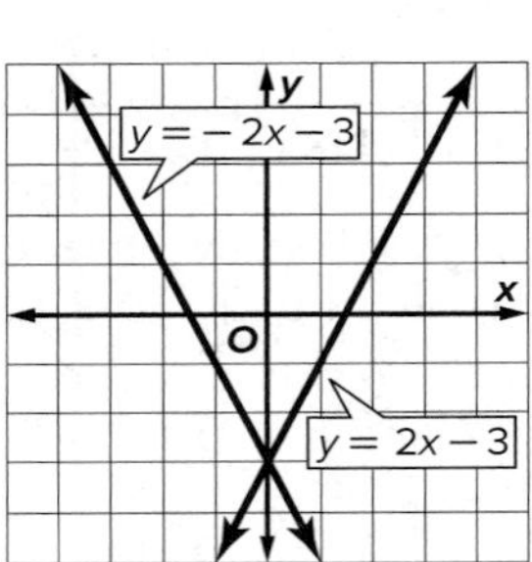

39.

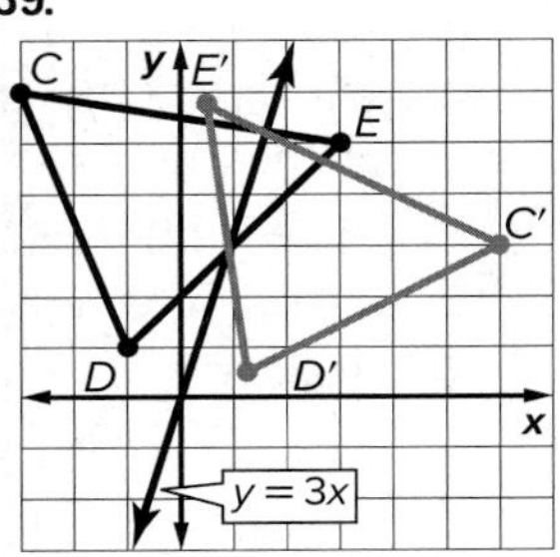

41 The graph is reflected across the x-axis, which changes the coefficient of the x^2-term to its opposite.

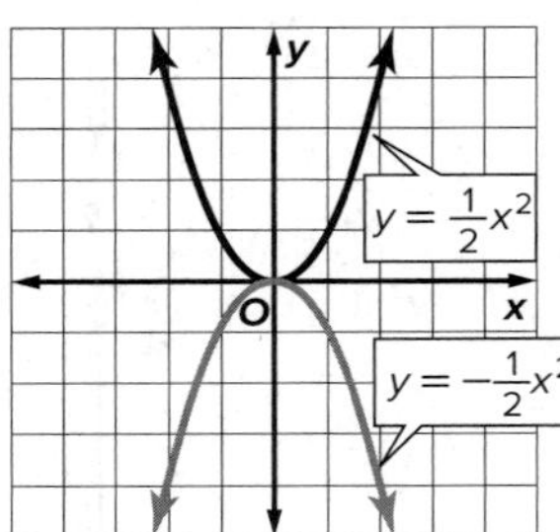

43.

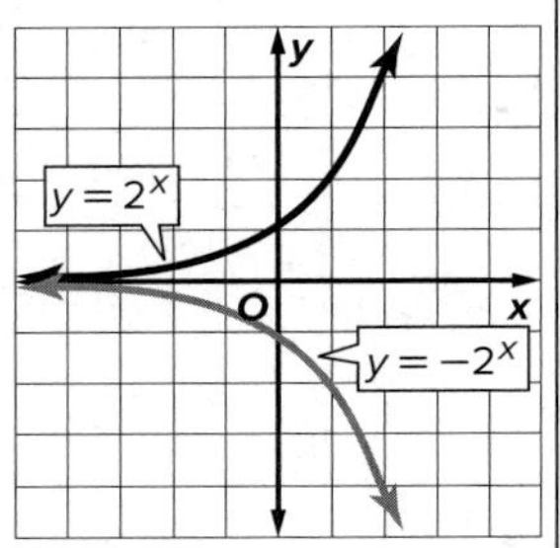

45. Jamil; sample answer: When you reflect a point in the x-axis, the reflected point is in the same place horizontally, but not vertically. When (2, 3) is reflected in the x-axis, the coordinates of the reflected point are (2, −3) because it is in the same location horizontally, but on the other side of the x-axis vertically. **47.** (b, a)

49. The slope of the line connecting the two points is $\frac{3}{5}$. The Midpoint Formula can be used to find the midpoint between the two points, which is $\left(\frac{3}{2}, \frac{3}{2}\right)$. Using point-slope form, the equation of the line is $y = -\frac{5}{3}x = 4$. (The slope of the bisector is $-\frac{5}{3}$ because it is the negative reciprocal of the slope, $\frac{3}{5}$.) **51.** Construct P, Q, R collinear with Q between P and R. Draw line ℓ, then construct perpendicular lines from P, Q, and R to line ℓ. Show equidistance or similarity of slope. **53.** B **55.** −2 **57.** B
59a. Quadrant II **59b.** $(x, y) \rightarrow (-x, y)$
59c. Sample answer: reflection in the line $y = -1$

Lesson 3-2

1. **3.**

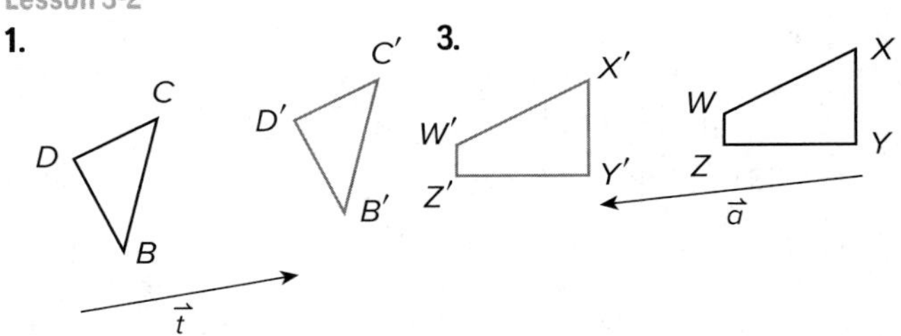

5.

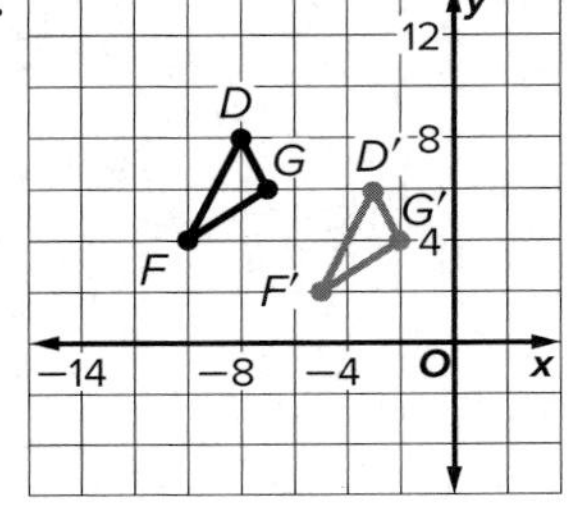

7. $(x, y) \rightarrow (x + 3, y - 5)$

9.

11.

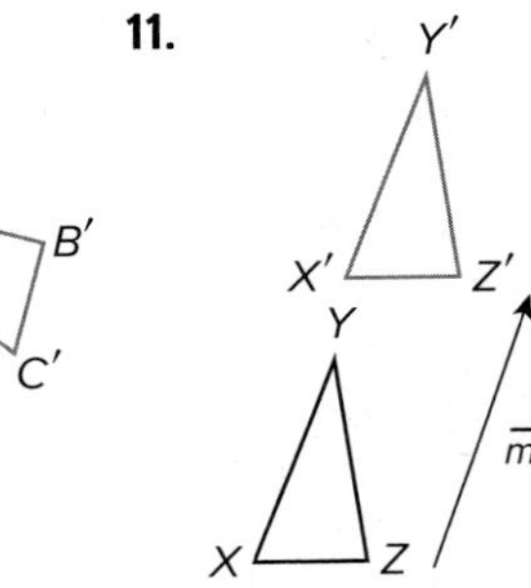

13.

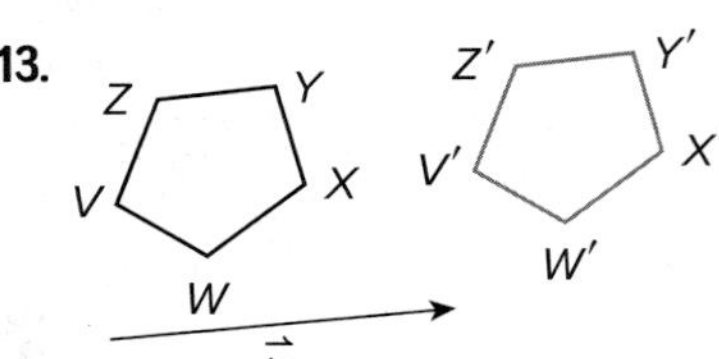

15 Plot the vertices of $\triangle MNP$ at $M(4, -5)$, $N(5, -8)$, and $P(8, -6)$; to find the coordinates of the vertices of the image of $\triangle MNP$, subtract 2 from the x-coordinate of each vertex and add 5 to the y-coordinate of each vertex.

$M(4, -5)$	$\rightarrow$	$M'(2, 0)$
$N(5, -8)$	$\rightarrow$	$N'(3, -3)$
$P(8, -6)$	$\rightarrow$	$P'(6, -1)$

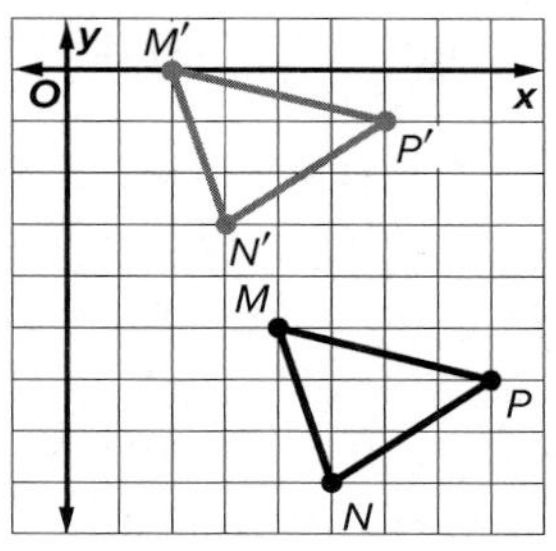

17.

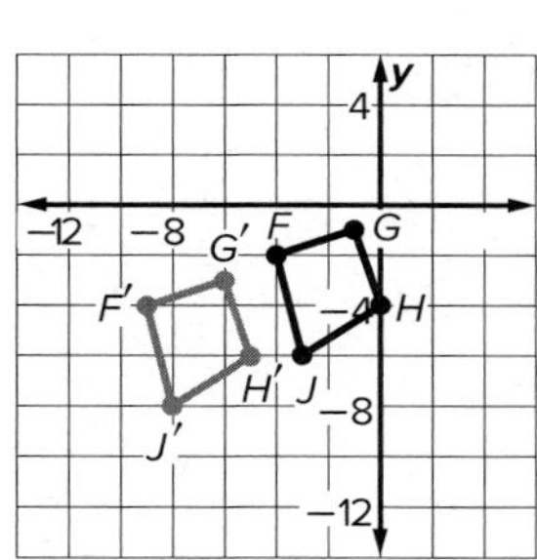

19.

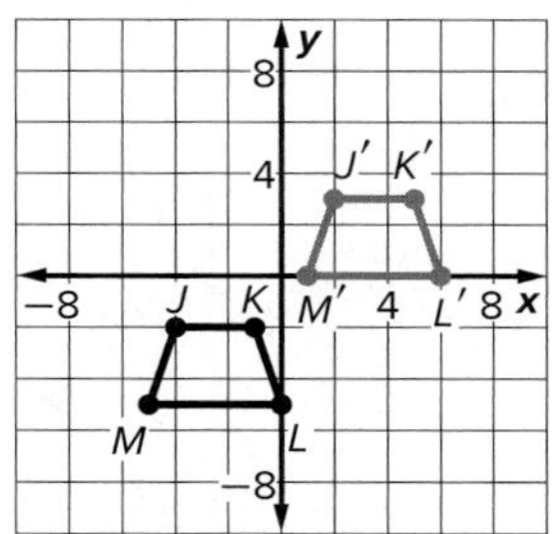

21 ⟨−12, 17⟩; Moving left on the coordinate plane is represented by a negative number (−12), and going down field closer to the goal line represents moving in a positive direction (+17).

23. ⟨3, −5⟩

25. They move to the right 13 seats and back one row; ⟨13, −1⟩.

27 The vector ⟨−2, 0⟩ represents a translation 2 units to the left. To transform the equation of a function so that its graph is translated 2 units to the left, replace x with $x + 2$. The equation of the translated image is $y = -(x + 2)^3$.

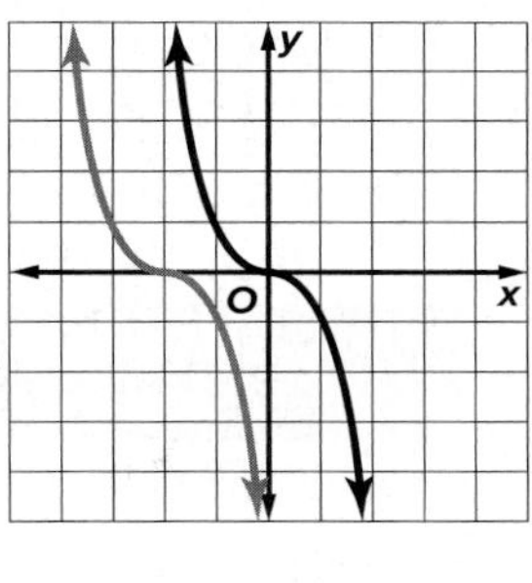

29a.

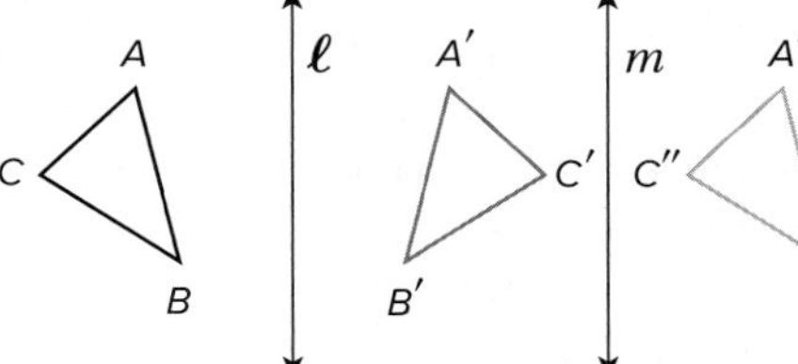

b.

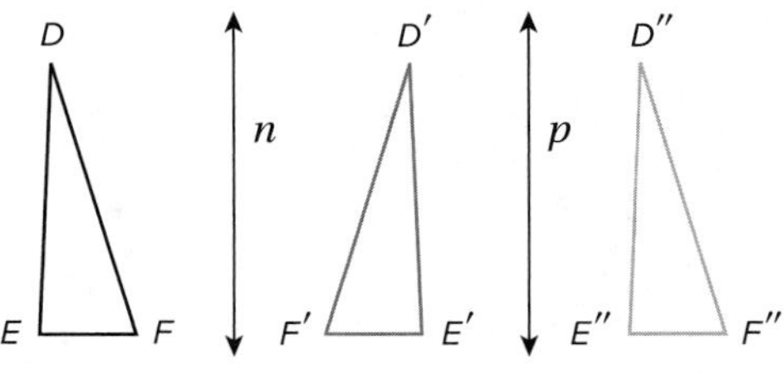

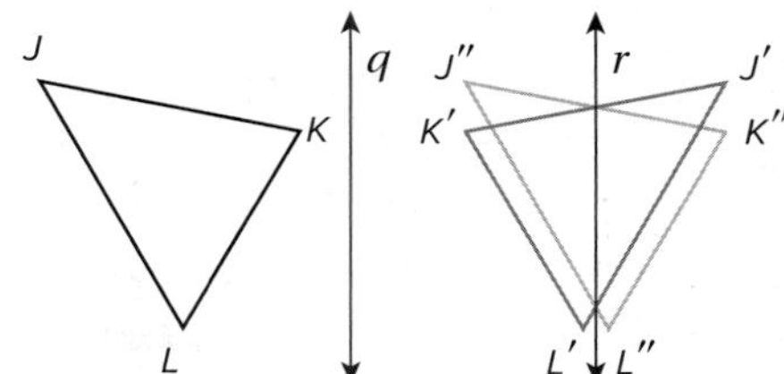

c.

Distance Between Corresponding Points (cm)		Distance Between Parallel Lines (cm)	
A and *A″*, *B* and *B″*, *C* and *C″*	4.4	ℓ and *m*	2.2
D and *D″*, *E* and *E″*, *F* and *F″*	5.6	*n* and *p*	2.8
J and *J″*, *K* and *K″*, *L* and *L″*	2.8	*q* and *r*	1.4

d. Sample answer: This can be described by a horizontal translation that is twice the distance between the two vertical lines.

31. $\langle x + a + c, y + b + d\rangle$

33. Sample answer: Both vector notation and coordinate notation describe the distance a figure is translated in the horizontal and vertical directions. Vector notation does not give a rule in terms of initial location, but coordinate notation does. For example, the translation a units to the right and b units up from the point (x, y) would be written $\langle a, b\rangle$ in vector notation and $(x, y) \rightarrow (x + a, y + b)$ in coordinate notation. **35a.** (0, 4), (1, 2), (−3, 1) **35b.** D **37.** C **39.** D **41.** D

Lesson 3-3

1.

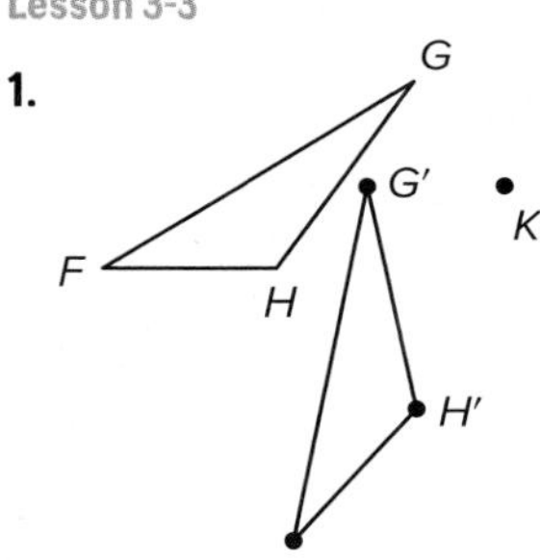

3 To find the vertices of the image of $\triangle DFG$, use the rule for a 180° rotation about the origin, $(x, y) \rightarrow (-x, -y)$

$D(-2, 6)$	→	$D'(2, -6)$
$F(2, 8)$	→	$F'(-2, -8)$
$G(2, 3)$	→	$G'(-2, -3)$

3.

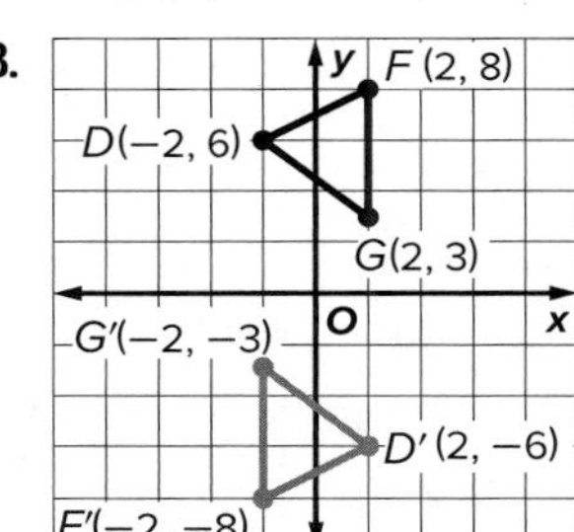

Selected Answers and Solutions

5.

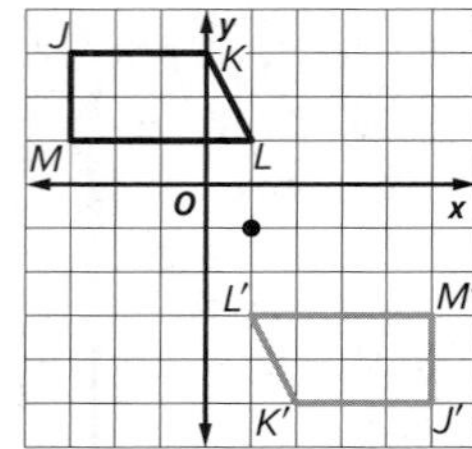

7.

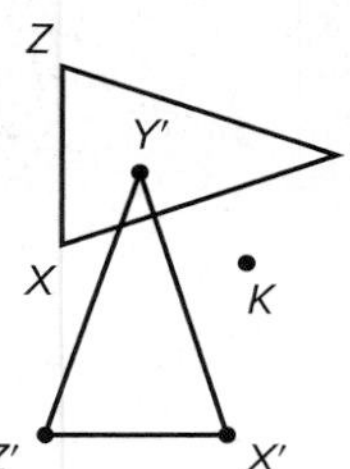

9.

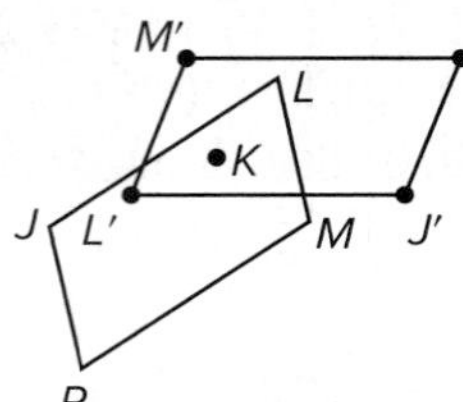

11.

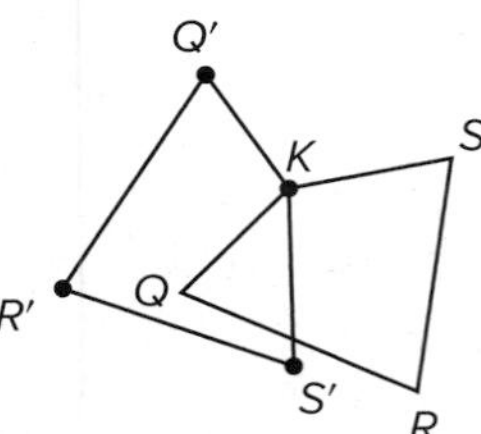

13. 120°; 360° ÷ 6 petals = 60° per petal. Two petal turns is 2 • 60° or 120°.

15. 154.2°; 360° ÷ 7 petals = 51.4° per petal. Three petal turns is 3 • 51.4° or 154.2°.

17.

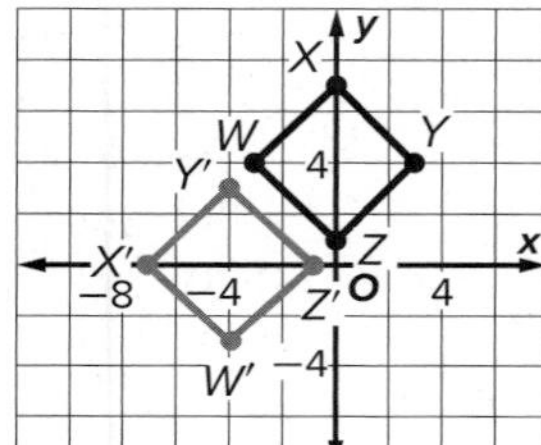

19.

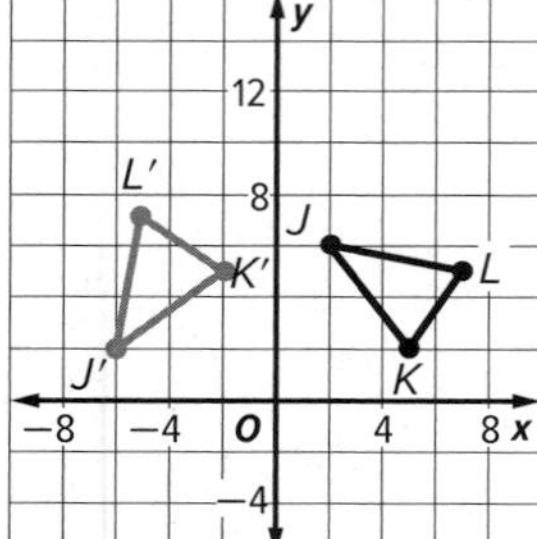

21.

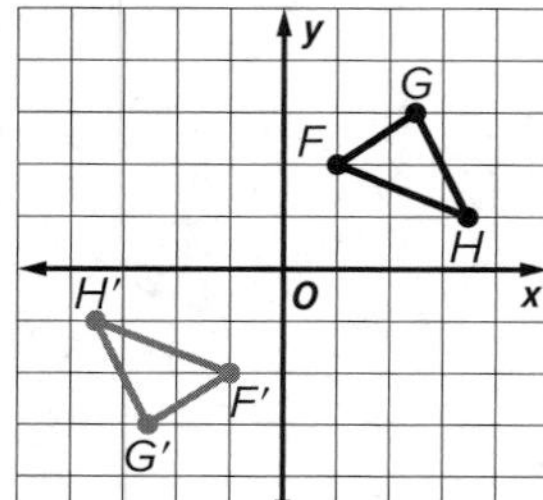

23a. 10° **23b.** about 1.7 seconds

25.

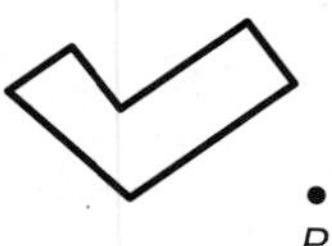

27. $y = -x + 2$; parallel **29.** $y = -x - 2$; collinear

31. x-intercept: $y = 2x + 4$; y-intercept: $y = 2x + 4$

33 After 31 seconds, each car makes $0.5(31) = 17.5$ revolutions. So Jane will be at the point at the top of the circular car. After 31 seconds, the ride makes $0.25(31) = 7.75$ rotations, so the ride will be $\frac{3}{4}$ of a full rotation from its starting position. Jane's final position after 31 seconds is $(2, -4)$.

35 **a.** Sample answer:

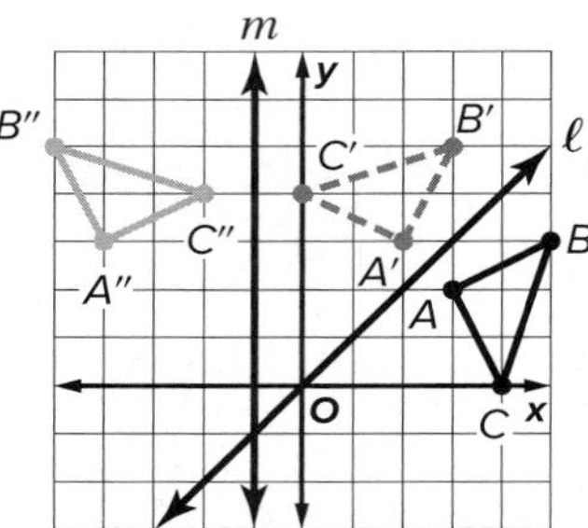

b. Sample answers:

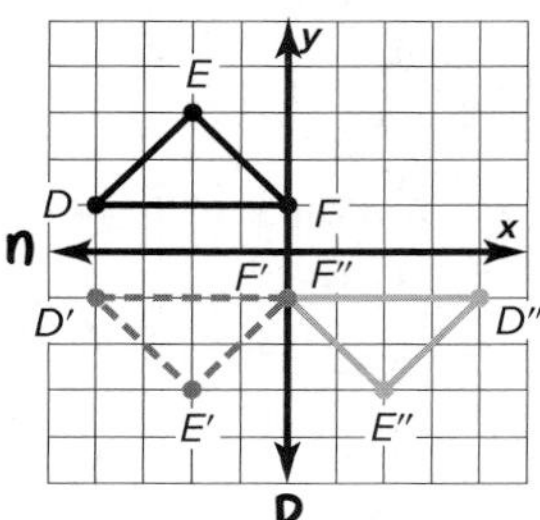

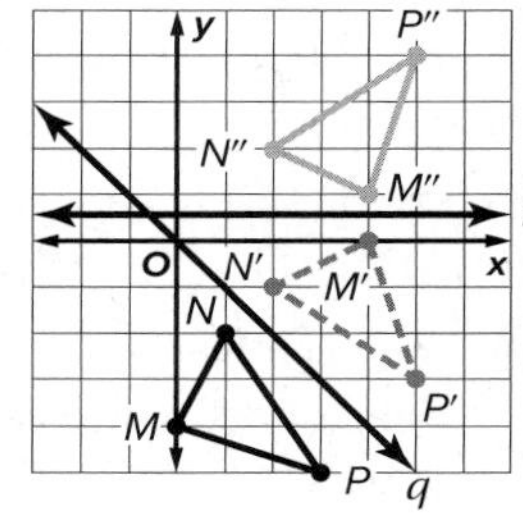

35c.

Angle of Rotation Between Figures		Angle Between Intersecting Lines	
$\triangle ABC$ and $\triangle A''B''C''$	90°	ℓ and m	45°
$\triangle DEF$ and $\triangle D''E''F''$	180°	n and p	90°
$\triangle MNP$ and $\triangle M''N'P'$	90°	q and r	45°

35d. The angle measures on the left side of the table are all twice the corresponding angle measures on the right side of the table. This leads to the following conjecture: The measure of the angle of rotation about the point where the lines intersect is twice the measure of the angle between the two intersecting lines.

37. Sample answer: $(-1, 2)$; Because $\triangle CC'P$ is isosceles and the vertex angle of the triangle is formed by the angle of rotation, both $m\angle PCC'$ and $m\angle PC'C$ are 40° because the base angles of isosceles triangles are congruent. When you construct a 40° angle with a vertex at C and a 40° angle with a vertex at C', the intersection of the rays forming the two angles intersect at the point of rotation, or $(-1, 2)$.

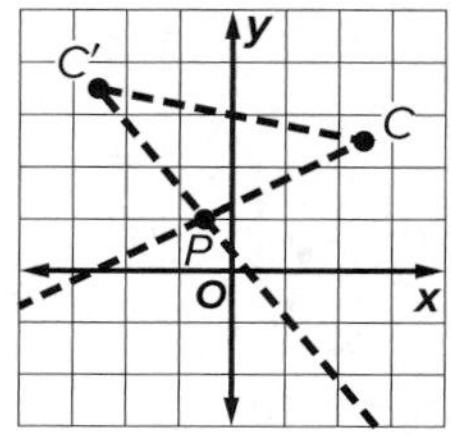

39. No; sample answer: When a figure is reflected about the x-axis, the x-coordinates of the transformed figure remain the same, and the y-coordinates are negated. When a figure is rotated 180° about the origin, both the x- and y-coordinates are negated. Therefore, the transformations are not equivalent.

41. B **43.** A **45.** A, C, D, E

Lesson 3-4

1.

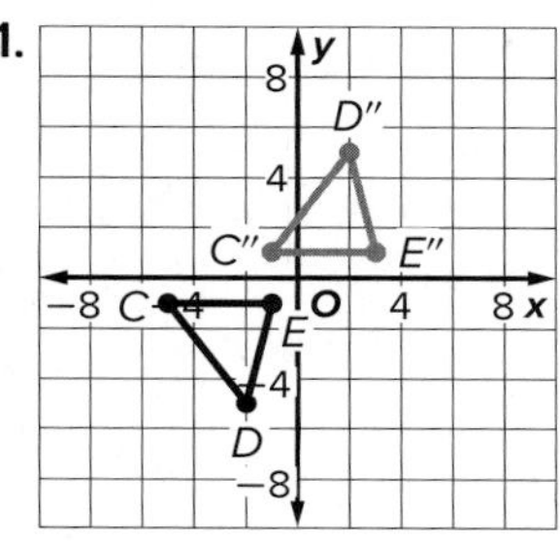

3.

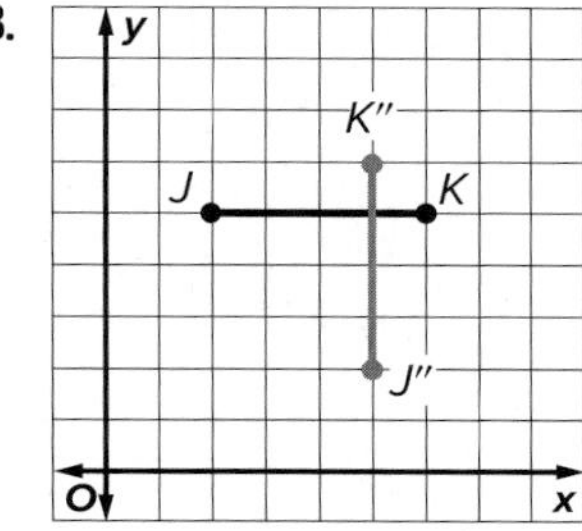

5. rotation clockwise 100° about the point where lines m and p intersect.

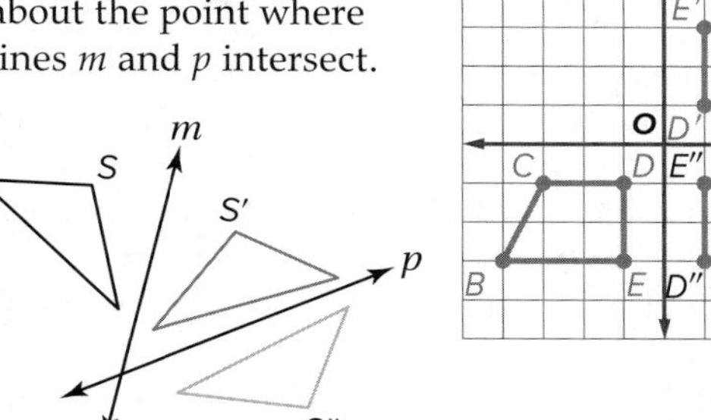
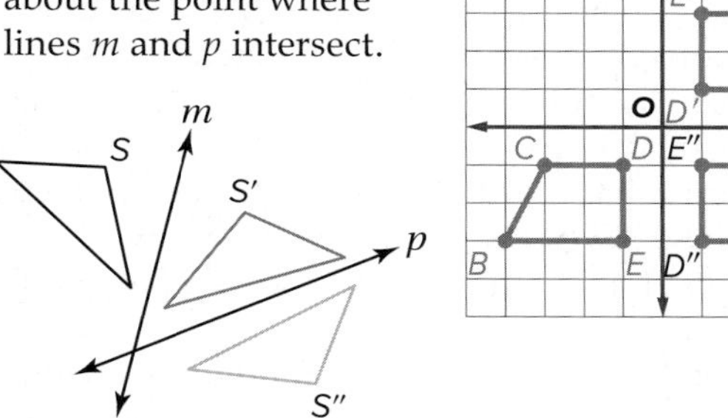

7.

9 translation along ⟨2, 0⟩ | reflection in x-axis

(x, y)	→	$(x + 2, y)$	(x, y)	→	$(x, -y)$
$R(1, -4)$	→	$R'(3, -4)$	$R'(3, -4)$	→	$R''(3, 4)$
$S(6, -4)$	→	$S'(8, -4)$	$S'(8, -4)$	→	$S''(8, 4)$
$T(5, -1)$	→	$T'(7, -1)$	$T'(7, -1)$	→	$T''(7, 1)$

Graph $\triangle RST$ and its image $\triangle R''S''T''$.

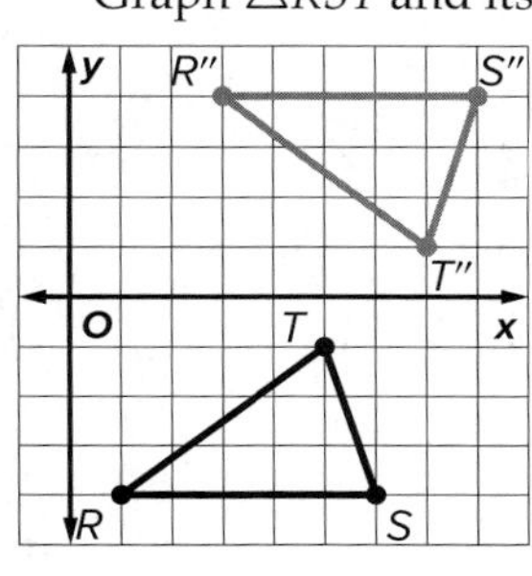

11.

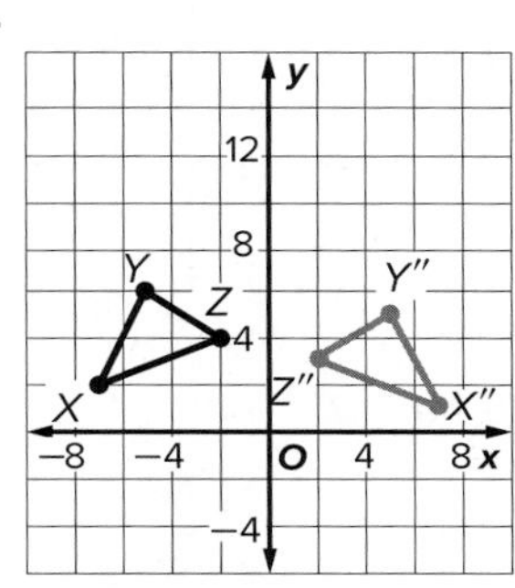

13 Reflection in the x-axis:

$(x, y) \rightarrow (x, -y)$;
$W(-4, 6) \rightarrow W'(-4, -6)$;
$X(-4, 1) \rightarrow X'(-4, -1)$;
Draw a graph.

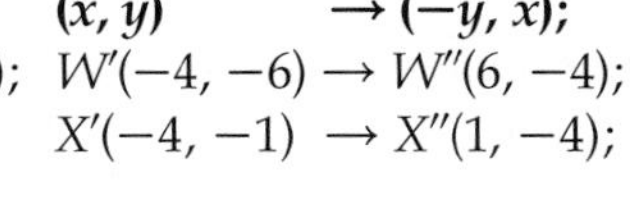

Rotation of 90° about the origin:

$(x, y) \rightarrow (-y, x)$;
$W'(-4, -6) \rightarrow W''(6, -4)$;
$X'(-4, -1) \rightarrow X''(1, -4)$;

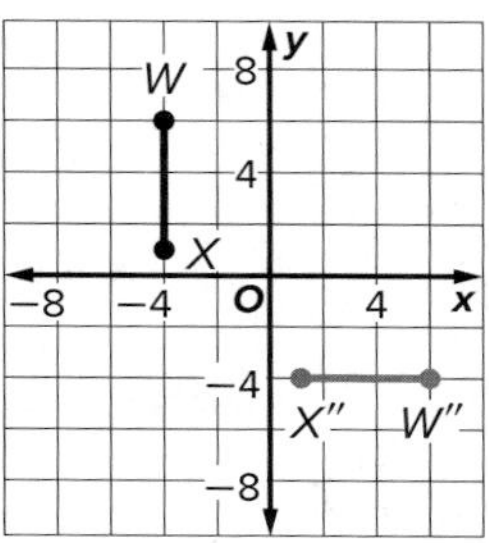

15.

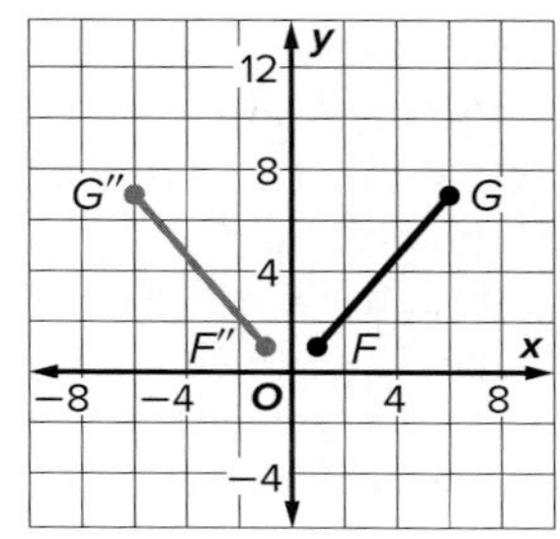

17.

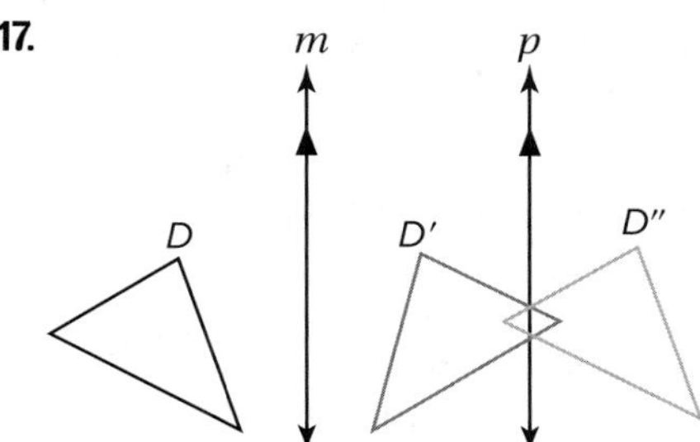

horizontal translation 4 cm to the right

19.

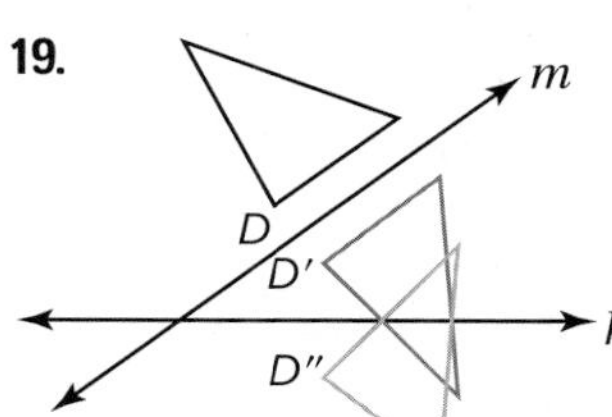

70° rotation about the point where lines m and p intersect

21.

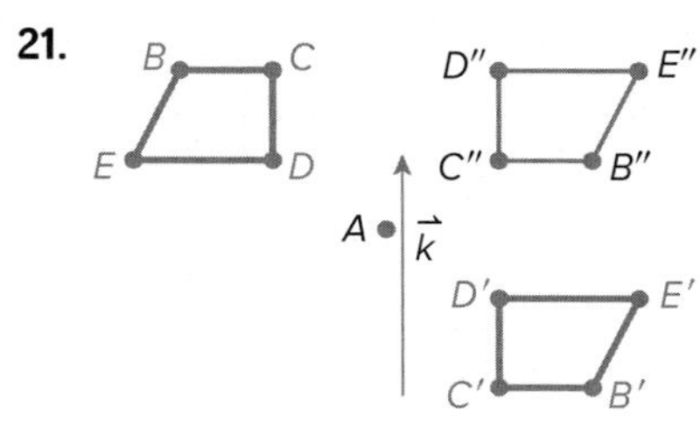

23. translation
25. reflection, then translation

27 rotation 90° about origin | reflection in x-axis

(x, y)	→	$(-y, x)$	(x, y)	→	$(x, -y)$
$(-2, -5)$	→	$(5, -2)$	$(5, -2)$	→	$(5, 2)$
$(0, 1)$	→	$(-1, 0)$	$(-1, 0)$	→	$(-1, 0)$

Graph line y' through points at $(5, -2)$ and $(-1, 0)$.
Graph line y'' through points at $(5, 2)$ and $(-1, 0)$.

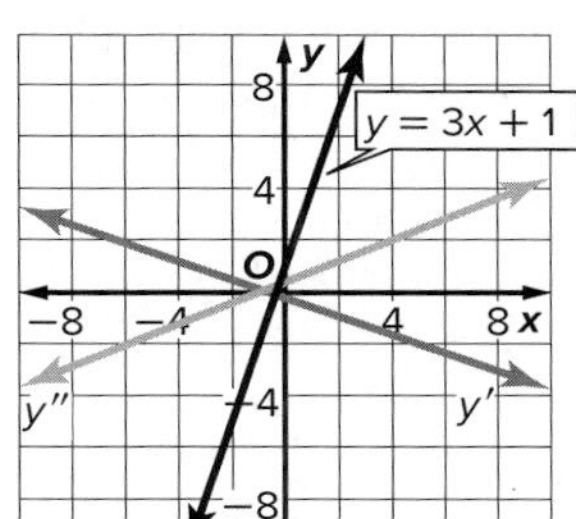

29. $A(-3, 1)$, $B(-2, 3)$, $C(-1, 0)$
31. double reflection
33. rotation 180° about the origin and reflection in the x-axis

35. Given: Lines ℓ and m intersect at point P. A is any point not on ℓ or m.
Prove: a. If you reflect point A in line m, and then reflect its image A' in line ℓ, A'' is the image of A after a rotation about point P.

b. $m\angle APA'' = 2(m\angle SPR)$

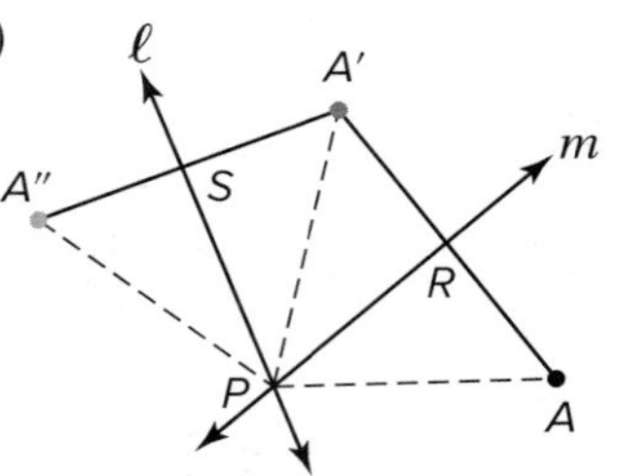

Proof: We are given that line ℓ and line m intersect at point P and that A is not on line ℓ or line m. Reflect A over line m to A' and reflect A' over line ℓ to A''. By the definition of reflection, line m is the perpendicular bisector of $\overline{AA'}$ at R, and line ℓ is the perpendicular bisector of $\overline{A'A''}$ at S. $\overline{AR} \cong \overline{A'R}$ and $\overline{A'S} \cong \overline{A''S}$ by the definition of a perpendicular bisector. Through any two points there is exactly one line, so we can draw auxiliary segments $\overline{AP}$, $\overline{A'P}$, and $\overline{A''P}$. $\angle ARP$, $\angle A'RP$, $\angle A'SP$ and $\angle A''SP$ are right angles by the definition of perpendicular bisectors. $\overline{RP} \cong \overline{RP}$ and $\overline{SP} \cong \overline{SP}$ by the Reflexive Property. $\triangle ARP \cong \triangle A'RP$ and $\triangle A'SP \cong \triangle A''SP$ by the SAS Congruence Postulate. Using CPCTC, $\overline{AP} \cong \overline{A'P}$ and $\overline{A'P} \cong \overline{A''P}$, and $\overline{AP} \cong \overline{A''P}$ by the Transitive Property. By the definition of a rotation, A'' is the image of A after a rotation about point P. Also using CPCTC, $\angle APR \cong \angle A'PR$ and $\angle A'PS \cong \angle A''PS$. By the definition of congruence, $m\angle APR = m\angle A'PR$ and $m\angle A'PS = m\angle A''PS$. $m\angle APR + m\angle A'PR + m\angle A'PS + m\angle A''PS = m\angle APA''$ and $m\angle A'PS + m\angle A'PR = m\angle SPR$ by the Angle Addition Postulate. $m\angle A'PR + m\angle A'PR + m\angle A'PS + m\angle A'PS = m\angle APA''$ by Substitution, which simplifies to $2(m\angle A'PR + m\angle A'PS) = m\angle APA''$. By Substitution, $2(m\angle SPR) = m\angle APA''$.

37. Sample answer: No; there are no invariant points in a glide reflection because all of the points are translated along a vector. Perhaps for compositions of transformations, there may be invariant points when a figure is rotated and reflected, rotated twice, or reflected twice. **39.** Yes; sample answer: If a segment with endpoints (a, b) and (c, d) is to be reflected about the x-axis, the coordinates of the endpoints of the reflected image are $(a, -b)$ and $(c, -d)$. If the segment is then reflected about the line $y = x$, the coordinates of the endpoints of the final image are $(-b, a)$ and $(-d, c)$. If the original image is first reflected about $y = x$, the coordinates of the endpoints of the reflected image are (b, a) and (d, c). If the segment is then reflected about the x-axis, the coordinates of the endpoints of the final image are $(b, -a)$ and $(d, -c)$.

41 Sample answer: When two rotations are performed on a single image, the order of the rotations does not affect the final image when the two rotations are centered at the same point. For example, if $\triangle ABC$ is rotated 45° clockwise about the origin and then rotated 60° clockwise about the origin, $\triangle A''B''C''$ is the same as if the figure were first rotated 60° clockwise about the origin and then rotated 45° clockwise about the origin. If $\triangle ABC$ is rotated 45° clockwise about the origin and then rotated 60° clockwise about $P(2, 3)$, $\triangle A''B''C''$ is different than if the figure were first rotated 60° clockwise about $P(2, 3)$ and then rotated 45° clockwise about the origin. So, the order of the rotations sometimes affects the location of the final image.

43. C **45a.** (1, 3), (1, −1), (3, 0) **45b.** (1, −3), (1, 1), (3, 0) **45c.** A

Lesson 3-5

1. yes; 4 **3.** yes; 1 **5.** yes; 2; 180°

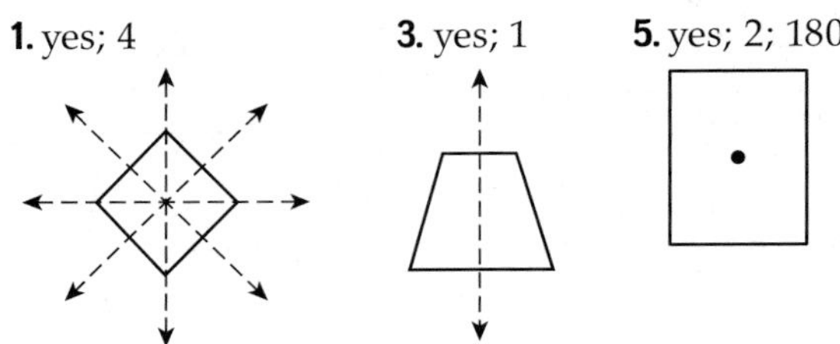

7 There are four lines of symmetry: a vertical line, a horizontal line, and two lines through opposite vertices. The vertical line of symmetry is the y-axis or the line $x = 0$.
The horizontal line of symmetry is the line $y = -1$.
The line through the vertices (−2, −3) and (2, 1) has slope 1 and y-intercept −1, so its equation is $y = x - 1$.
The line through the vertices (−2, 1) and (2, −3) has slope −1 and y-intercept −1, so its equation is $y = -x - 1$.
The square has rotational symmetry. The center of rotation is the intersection of the lines of symmetry that pass through opposite vertices. This point is (0, −1). So the rotations of 90°, 180°, and 270° around the point (0, −1) map the square onto itself.

9. no **11.** yes; 6 **13.** yes; 1

15. no **17.** yes; 1

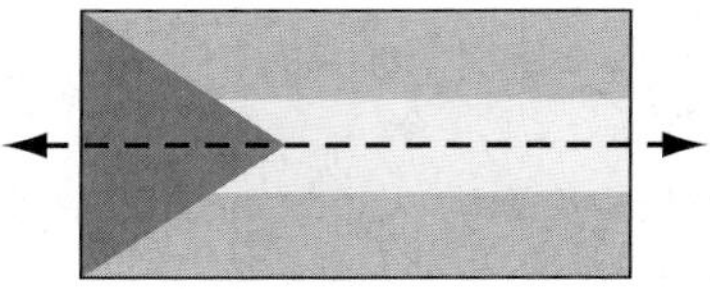

19. yes; 3; 120°

21 No, there is no rotation between 0 and 360 degrees that would result in the moon being mapped onto itself.

23. yes; 8; 45°

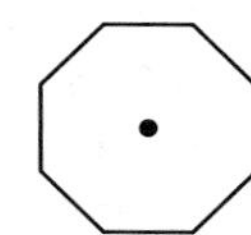

25. yes; 8; 45° **27.** no symmetry

29. line symmetry; the reflection in the line $y = 1.5$ maps the trapezoid onto itself.

31a. Sample answer: There is a horizontal line of symmetry between the tower and its reflection. There is a vertical line of symmetry through the center of the photo.

31b. Yes; there is 180° rotational symmetry; the center of symmetry is the intersection of the horizontal and vertical lines of symmetry.

33 Plot the vertices. The figure is a square. There are four lines of symmetry (the x- and y-axes and the lines $y = x$ and $y = -x$). The figure also has rotational symmetry, since a 90° rotation maps the figure onto itself.

35. line **37.** line; $x = 0$

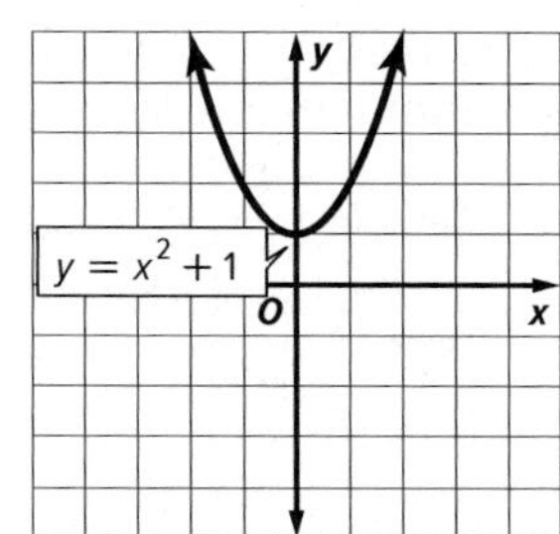

39a. $y = 0.5x + 1$, $y = -2x + 3.5$

39b. The equations of the lines of symmetry do not change; although the rectangle does not map onto itself under this rotation, the lines of symmetry are mapped to each other.

41. neither; Figure A has both rotational and line symmetry. **43.** circle; Every line through the center of a circle is a line of symmetry, and there are infinitely many such lines.

45. Sample answer: In both rotational and line symmetry a figure is mapped onto itself. However, in line symmetry the figure is mapped onto itself by a reflection, and in rotational symmetry a figure is mapped onto itself by a rotation. A figure can have line symmetry and rotational symmetry.

47. B, E **49.** B **51a.** (−3, 3), (7, 3), (7, 1) **51b.** 2
51c. Sample answer: (0, 0), (0, 4), (4, 4), (4, 0)

Chapter 3 Study Guide and Review

1. composition of transformations **3.** dilation
5. line of reflection **7.** translation **9.** rotational

11.

13.

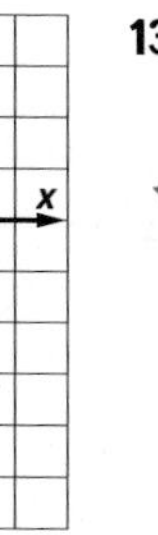

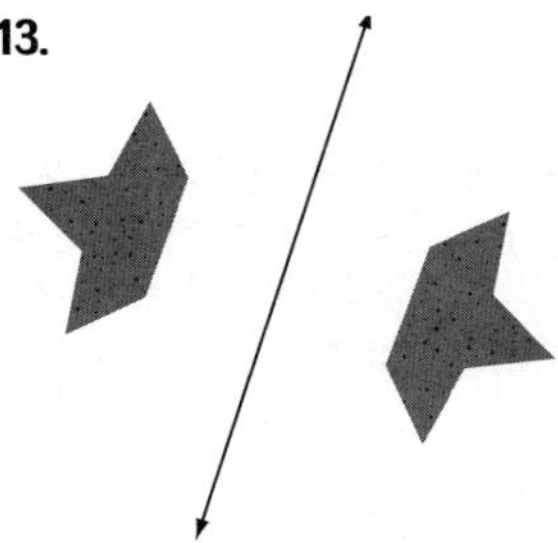

15.

17.

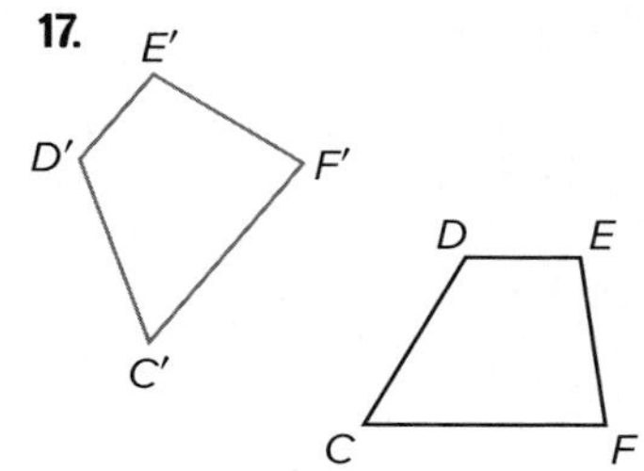

19.

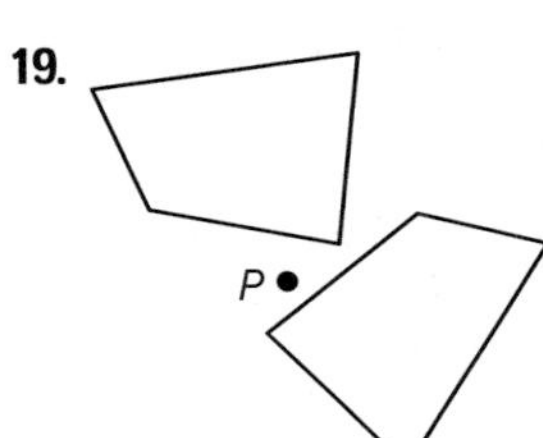

21.

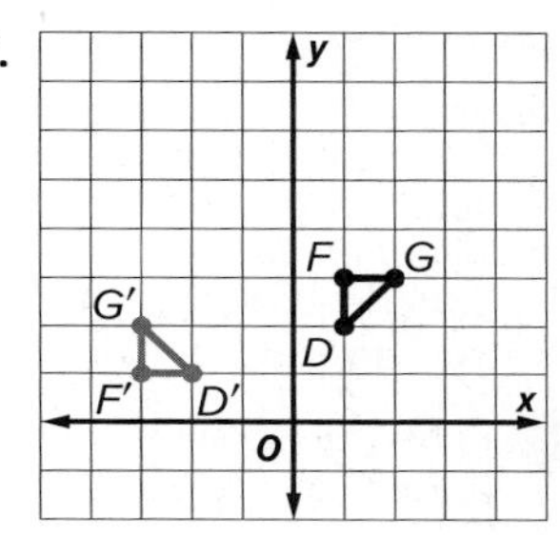

23. rotation 180° clockwise about the origin

25.

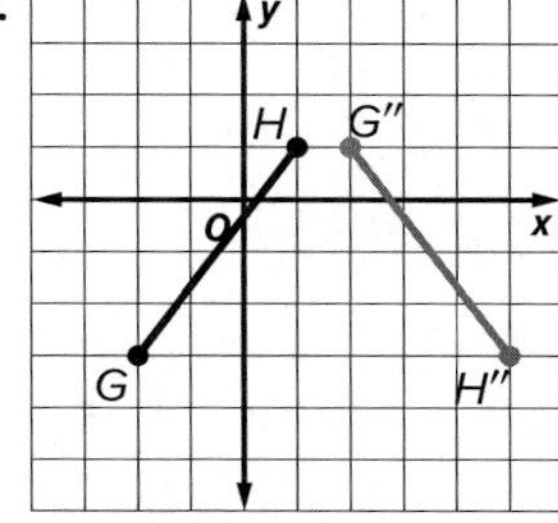

27. translate left **29.** yes; 2 **31.** no

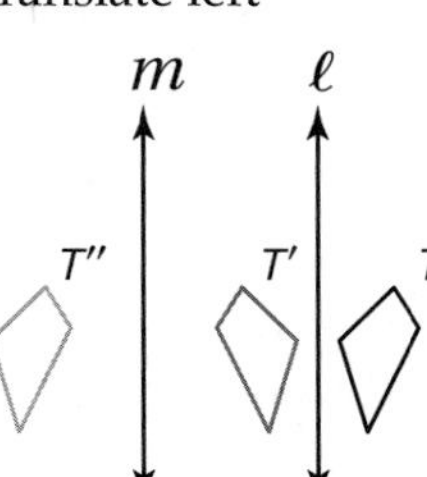

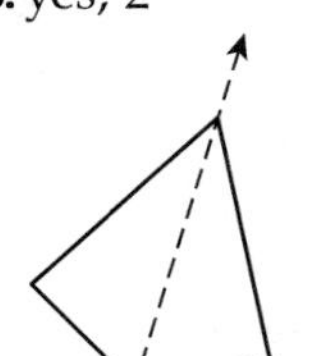

CHAPTER 4

Triangles and Congruence

Chapter 4 Concept Check

1. $\angle B$, $\angle A$, $\angle C$
3. Sample answer: 45°, 45°, 90°
5. Yes; If the transversal is perpendicular to the parallel lines, all angles are right angles.
7. Pythagorean Theorem or Distance Formula

Lesson 4-1

1. 58 **3.** 80 **5.** 40 **7.** 85 **9.** 61 **11.** 151 **13.** 52

15

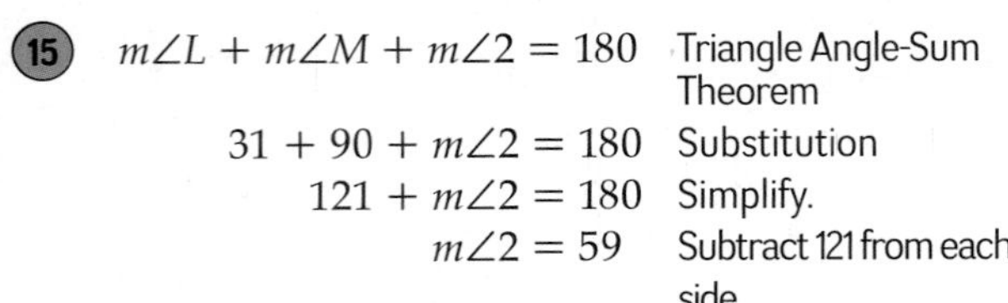

$m\angle L + m\angle M + m\angle 2 = 180$	Triangle Angle-Sum Theorem
$31 + 90 + m\angle 2 = 180$	Substitution
$121 + m\angle 2 = 180$	Simplify.
$m\angle 2 = 59$	Subtract 121 from each side.

$\angle 1$ and $\angle 2$ are congruent vertical angles. So, $m\angle 1 = 59$.

$m\angle 1 + m\angle 3 + m\angle P = 180$	Triangle Angle-Sum Theorem
$59 + m\angle 3 + 22 = 180$	Substitution
$81 + m\angle 3 = 180$	Simplify.
$m\angle 3 = 99$	Subtract 118 from each side.

$m\angle 1 = 59$, $m\angle 2 = 59$, $m\angle 3 = 99$

17. 79 **19.** 21

21

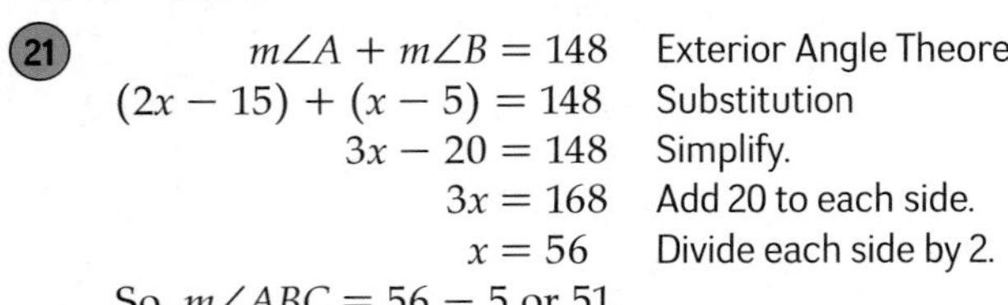

$m\angle A + m\angle B = 148$	Exterior Angle Theorem
$(2x - 15) + (x - 5) = 148$	Substitution
$3x - 20 = 148$	Simplify.
$3x = 168$	Add 20 to each side.
$x = 56$	Divide each side by 2.

So, $m\angle ABC = 56 - 5$ or 51.

23. 78 **25.** 39 **27.** 55 **29.** 35 **31.** $x = 30$; 30, 60

33 In $\triangle ABC$, $\angle B$ and $\angle C$ are congruent, so $m\angle B = m\angle C$.

$m\angle A = 3(m\angle B)$	$m\angle A$ is to be 3 times $m\angle B$.
$m\angle A + m\angle B + m\angle C = 180$	Triangle Angle-Sum Theorem
$3(m\angle B) + m\angle B + m\angle B = 180$	Substitution
$5(m\angle B) = 180$	Simplify.
$m\angle B = 36$	Divide each side by 5.

$m\angle C = m\angle B = 36$ $m\angle A = 3m\angle B = 3(36)$ or 108

35. Given: $\triangle MNO$; $\angle M$ is a right angle.
Prove: There can be at most one right angle in a triangle.
Proof: In $\triangle MNO$, M is a right angle. $m\angle M + m\angle N + m\angle O = 180$. $m\angle M = 90$, so $m\angle N + m\angle O = 90$. If N were a right angle, then $m\angle O = 0$. Because that is impossible, there cannot be two right angles in a triangle.
Given: $\triangle PQR$; $\angle P$ is obtuse.
Prove: There can be at most one obtuse angle in a triangle.
Proof: In $\triangle PQR$, $\angle P$ is obtuse. So $m\angle P > 90$. $m\angle P + m\angle Q + m\angle R = 180$. It must be that $m\angle Q + m\angle R < 90$. So, $\angle Q$ and $\angle R$ must be acute.

37. $m\angle 1 = 65$, $m\angle 2 = 20$, $m\angle 3 = 95$, $m\angle 4 = 40$, $m\angle 5 = 110$, $m\angle 6 = 45$, $m\angle 7 = 70$, $m\angle 8 = 65$ **39.** 67°, 23° **41.** $z < 23$; Sample answer: Because the sum of the measures of the angles of a triangle is 180 and $m\angle X = 157$, $157 + m\angle Y + m\angle Z = 180$, so $m\angle Y + m\angle Z = 23$. If $m\angle Y$ were 0, then $m\angle Z$ would equal 23. But because an angle must have a measure greater than 0, $m\angle Z$ must be less than 23, so $z < 23$.

43 Use the Triangle Angle-Sum Theorem and the Addition Property to prove the statement is true.
Given: $RSTUV$ is a pentagon.
Prove: $m\angle S + m\angle STU + m\angle TUV + m\angle V + m\angle VRS = 540$
Proof:
Statements (Reasons)

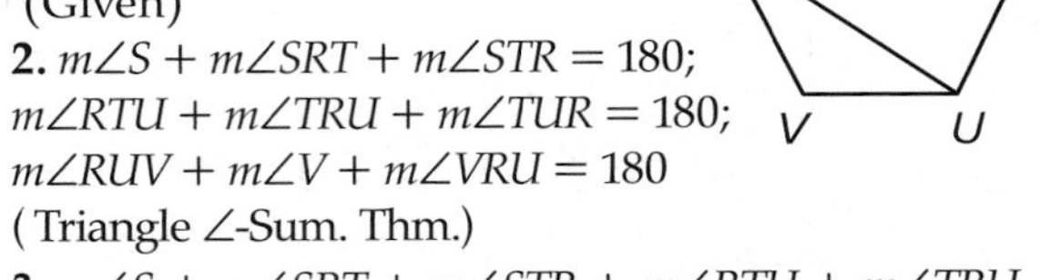

1. $RSTUV$ is a pentagon. (Given)
2. $m\angle S + m\angle SRT + m\angle STR = 180$; $m\angle RTU + m\angle TRU + m\angle TUR = 180$; $m\angle RUV + m\angle V + m\angle VRU = 180$ (Triangle $\angle$-Sum. Thm.)
3. $m\angle S + m\angle SRT + m\angle STR + m\angle RTU + m\angle TRU + m\angle TUR + m\angle RUV + m\angle V + m\angle VRU = 540$ (Addition Property)
4. $m\angle VRS = m\angle SRT + m\angle TRU + m\angle VRU$; $m\angle TUV = m\angle TUR + m\angle RUV$; $m\angle STU = m\angle STR + m\angle RTU$ ($\angle$ Addition)
5. $m\angle S + m\angle STU + m\angle TUV + m\angle V + m\angle VRS = 540$ (Substitution)

45a. Sample answer:

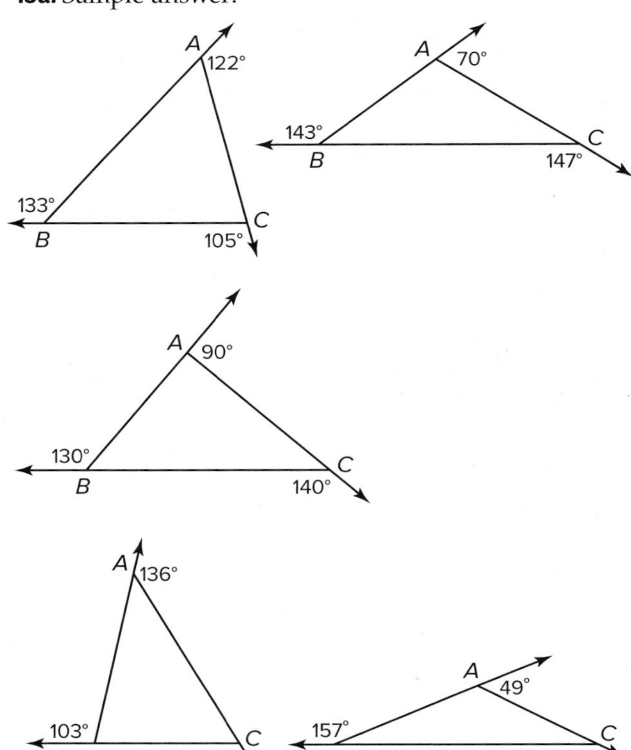

45b. Sample answer:

$m\angle 1$	$m\angle 2$	$m\angle 3$	Sum of Angle Measures
122	105	133	360
70	147	143	360
90	140	130	360
136	121	103	360
49	154	157	360

45c. Sample answer: The sum of the measures of the exterior angles of a triangle is 360.

45d. $m\angle 1 + m\angle 2 + m\angle 3 = 360$

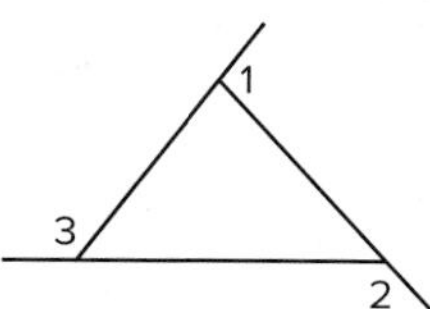

45e. The Exterior Angle Theorem tells us that $m\angle 3 = m\angle BAC + m\angle BCA$, $m\angle 2 = m\angle BAC + m\angle CBA$, $m\angle 1 = m\angle CBA + m\angle BCA$. Through substitution, $m\angle 1 + m\angle 2 + m\angle 3 = m\angle CBA + m\angle BCA + m\angle BAC + m\angle CBA + m\angle BAC + m\angle BCA$. This can be simplified to $m\angle 1 + m\angle 2 + m\angle 3 = 2m\angle CBA + 2m\angle BCA + 2m\angle BAC$. The Distributive Property can be applied and gives $m\angle 1 + m\angle 2 + m\angle 3 = 2(m\angle CBA + m\angle BCA + m\angle BAC)$. The Triangle Angle-Sum Theorem tells us that $m\angle CBA + m\angle BCA + m\angle BAC = 180$. Through substitution we have $m\angle 1 + m\angle 2 + m\angle 3 = 2(180) = 360$.

47. We want to prove that $m\angle BAC + m\angle ABC + m\angle ACB = 180$. By the Angle Addition Postulate and the definition of a straight angle, $m\angle ACB + m\angle ACE + m\angle ECD = 180$. Because $\overrightarrow{AB} \parallel \overrightarrow{EC}$, $\angle BAC \cong \angle ACE$ by the Alternate Interior Angles Theorem. Similarly, $\angle ABC \cong \angle ECD$ by the Corresponding Angles Postulate. Thus, by substitution, $m\angle BAC + m\angle ABC + m\angle ACB = 180$. **49.** $y = 13$, $z = 14$ **51.** Sample answer: Because an exterior angle is acute, the adjacent angle must be obtuse. Because another exterior angle is right, the adjacent angle must be right. A triangle cannot contain both a right and an obtuse angle because it would be more than 180 degrees. Therefore, a triangle cannot have an obtuse, acute, and a right exterior angle. **53a.** 66 **53b.** 118 **53c.** 52 **55.** 70 **57.** Given; $m\angle D + m\angle E = m\angle F = 180°$; Division Property of Equality

Lesson 4-2

1. $\angle Y \cong \angle S$, $\angle X \cong \angle R$, $\angle XZY \cong \angle RZS$, $\overline{YX} \cong \overline{SR}$, $\overline{YZ} \cong \overline{SZ}$, $\overline{XZ} \cong \overline{RZ}$; $\triangle YXZ \cong \triangle SRZ$ **3.** $\frac{1}{2}$ in.; Sample answer: The nut is congruent to the opening for the $\frac{1}{2}$-in. socket.

$\angle M \cong \angle R$	CPCTC
$m\angle M = m\angle R$	Definition of congruence
$y + 10 = 2y - 40$	Substitution
$10 = y - 40$	Subtract y from each side.
$50 = y$	Add 40 to each side.

7. 16; $\angle N$ corresponds to $\angle X$. By the Third Angles Theorem, $m\angle N = 64$, so $4x = 64$. **9.** $\angle X \cong \angle A$, $\angle Y \cong \angle B$, $\angle Z \cong \angle C$, $\overline{XY} \cong \overline{AB}$, $\overline{XZ} \cong \overline{AC}$, $\overline{YZ} \cong \overline{BC}$; $\triangle XYZ \cong \triangle ABC$ **11.** $\angle R \cong \angle J$, $\angle T \cong \angle K$, $\angle S \cong \angle L$, $\overline{RT} \cong \overline{JK}$, $\overline{TS} \cong \overline{KL}$, $\overline{RS} \cong \overline{JL}$; $\triangle RTS \cong \triangle JKL$ **13.** 20

$\overline{ED} \cong \overline{UT}$	CPCTC
$ED = UT$	Definition of congruence
$3z + 10 = z + 16$	Substitution
$2z + 10 = 16$	Subtract z from each side.
$2z = 6$	Subtract 10 from each side.
$z = 3$	Divide each side by 3.

17a. $\triangle DEF \cong \triangle PQR$

17b. $\overline{DE} \cong \overline{PQ}$, $\overline{EF} \cong \overline{QR}$, $\overline{DF} \cong \overline{PR}$

17c. $\angle D \cong \angle P$, $\angle E \cong \angle Q$, $\angle F \cong \angle R$

$148 + 18 + a = 180$	Triangle Angle-Sum Theorem
$166 + a = 180$	Simplify.
$a = 14$	Subtract 166 from each side.

If two angles of one triangle are congruent to two angles of another triangle, then the third angles of the triangles are congruent. So, $3x + y = 14$ and $5x - y = 18$. Solve the system of equations.

$3x + y = 14$	
$(+)\ 5x - y = 18$	
$8x = 32$	Add the equations.
$x = 4$	Divide each side by 8.
$3x + y = 14$	Original equation
$3(4) + y = 14$	$x = 4$
$12 + y = 14$	Simplify.
$y = 2$	Subtract 12 from each side.

21. Given: $\angle A \cong \angle D$
$\angle B \cong \angle E$

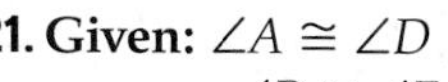

Prove: $\angle C \cong \angle F$

Proof:

Statements (Reasons)

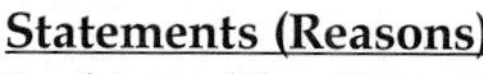

1. $\angle A \cong \angle D$, $\angle B \cong \angle E$ (Given)
2. $m\angle A = m\angle D$, $m\angle B = m\angle E$ (Def. of $\cong$ ∡s)
3. $m\angle A + m\angle B + m\angle C = 180$, $m\angle D + m\angle E + m\angle F = 180$ ($\angle$ Sum Theorem)
4. $m\angle A + m\angle B + m\angle C = m\angle D + m\angle E + m\angle F$ (Trans. Prop.)
5. $m\angle D + m\angle E + m\angle C = m\angle D + m\angle E + m\angle F$ (Subst.)
6. $m\angle C = m\angle F$ (Subt. Prop.)
7. $\angle C \cong \angle F$ (Def. of $\cong$ ∡s)

23. Given: $\overline{BD}$ bisects $\angle B$.
$\overline{BD} \perp \overline{AC}$

Prove: $\angle A \cong \angle C$

Proof:

Statements (Reasons)

1. $\overline{BD}$ bisects $\angle B$, $\overline{BD} \perp \overline{AC}$. (Given)
2. $\angle ABD \cong \angle DBC$ (Def. of angle bisector)
3. $\angle ADB$ and $\angle BDC$ are right angles. ($\perp$ lines form rt. ∡s.)

4. $\angle ADB \cong \angle BDC$ (All rt. ∠s are ≅.)
5. $\angle A \cong \angle C$ (Third ∠s Thm.)

25. Sample answer: Both of the punched flowers are congruent to the flower on the stamp, because it was used to create the images. According to the Transitive Property of Polygon Congruence, the two stamped images are congruent to each other because they are both congruent to the flowers on the punch.

27. Given: $\triangle DEF$
Prove: $\triangle DEF \cong \triangle DEF$

E
D F

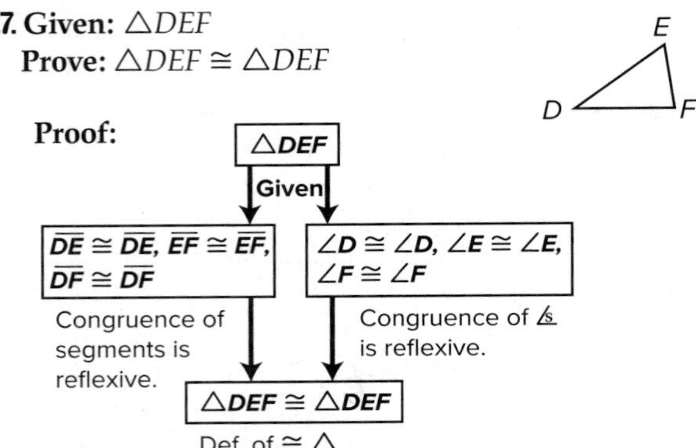

29. M 10y° N 49° L   S 70° T (4x + 9)° R $x = 13$; $y = 7$

31 **a.** All the longer sides of the triangles are congruent and all the shorter sides are congruent. Sample answer: $\overline{AB} \cong \overline{CB}$, $\overline{AB} \cong \overline{DE}$, $\overline{AB} \cong \overline{FE}$, $\overline{CB} \cong \overline{DE}$, $\overline{CB} \cong \overline{FE}$, $\overline{DE} \cong \overline{FE}$, $\overline{AC} \cong \overline{DF}$ **b.** If the area is a square, then each of the four sides measures $\sqrt{100}$ or 10 feet. So, the perimeter of the square is 4(10) or 40 ft. The pennant string will need to be 40 ft long. **c.** Each pennant and the distance to the next pennant is 6 in. or 0.5 ft. So, the number of pennants is $40 \div 0.5$ or 80.

33a. The sides of the first triangle are congruent to the corresponding sides of the second triangle.

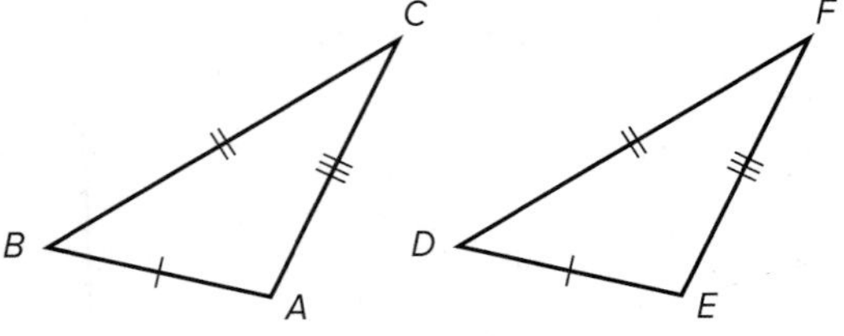

Two of the sides and the included angle are congruent.

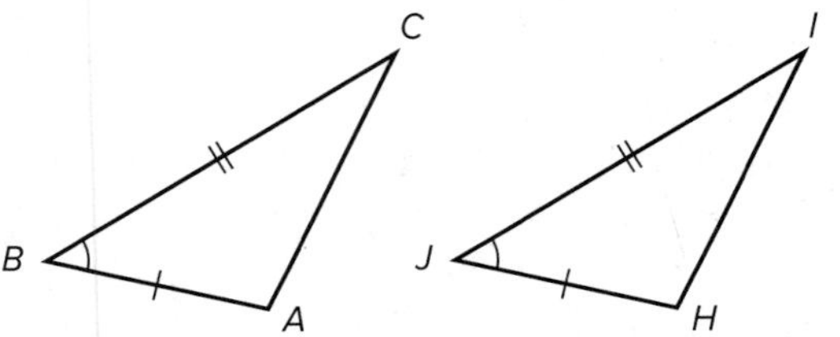

Two angles and the included side are congruent.

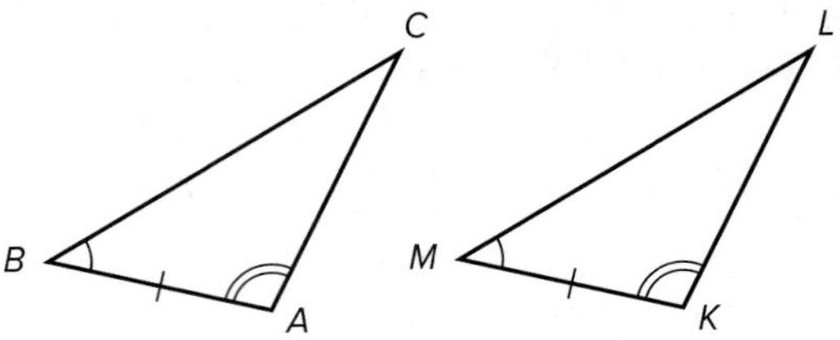

Two angles and a nonincluded side are congruent.

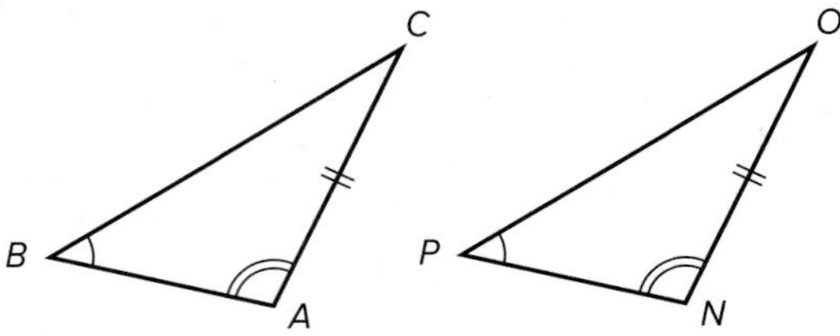

Two sides and a nonincluded angle are congruent.

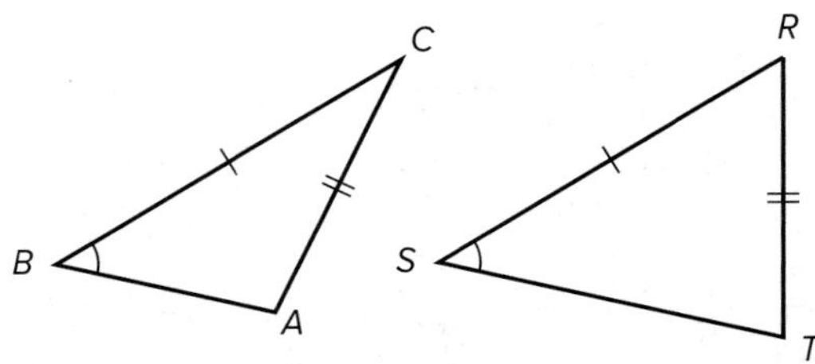

The angles of one are congruent to the corresponding angles of the second triangle.

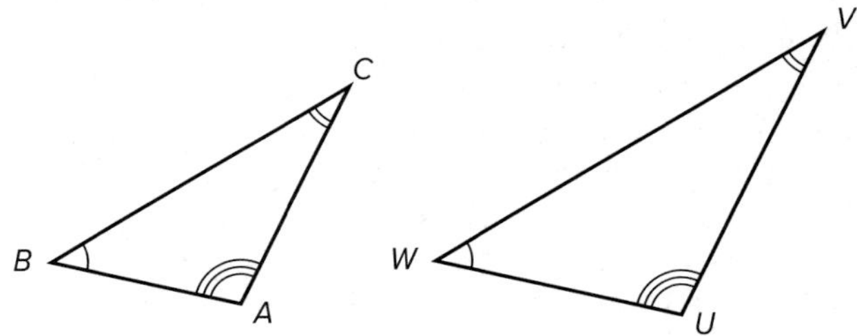

33b. Congruent triangles resulted from these criteria: corresponding sides of each triangle are congruent, two sides and included angle are congruent, two angles and the included side are congruent, and two angles and nonincluded side are congruent. These criteria did not result in congruent triangles: two sides and a nonincluded angle are congruent and corresponding angles are congruent. **33c.** Two triangles are congruent if all three sides are congruent, two sides and an included angle, two angles and the included side, and two angles and a nonincluded side. **35.** diameter, radius, or circumerence; Sample answer: Two circles are the same size if they have the same diameter, radius, or circumference, so she can determine if the hoops are congruent if she measures any of them. **37.** Both; Sample answer: $\angle A$ corresponds with $\angle Y$, $\angle B$ corresponds with $\angle X$, and $\angle C$ corresponds with $\angle Z$.

$\triangle CAB$ is the same triangle as $\triangle ABC$ and $\triangle ZXY$ is the same triangle as $\triangle XYZ$. **39.** $x = 16$, $y = 8$

41. False; $\angle A \cong \angle X$, $\angle B \cong \angle Y$, $\angle C \cong \angle Z$, but corresponding sides are not congruent.

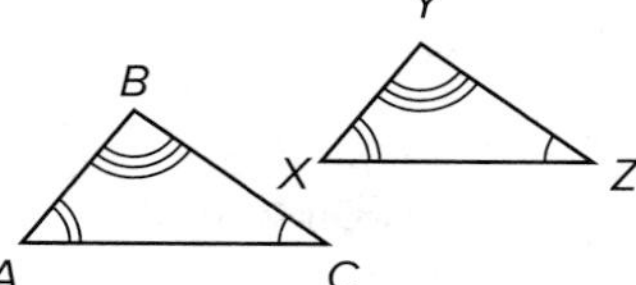

43. Sometimes; equilateral triangles will be congruent if one pair of corresponding sides are congruent. **45.** C **47.** $x = 11$; $y = 3$ **49a.** reflection across the x-axis and translation 5 units right **49b.** after reflection: $F'(-3, 5)$, $G'(-4, 1)$, $H'(-1, 3)$; after translation: $F''(2, 5)$, $G''(1, 1)$, $H''(4, 3)$ **49c.** 79; $m\angle H = 180 - 59 - 42 = 79$; because $H \cong \angle L$, $m\angle L = 79$.

Lesson 4-3

1a. two

1b. Given: $ABCD$ is a square

Prove: $\triangle ABC \cong \triangle CDA$

Proof:

Statements (Reasons)

1. $ABCD$ is a square (Given)
2. $\overline{AB} \cong \overline{CD}$, $\overline{BC} \cong \overline{DA}$ (Def. of a square)
3. $\overline{AC} \cong \overline{CA}$ (Reflex. Prop. $\cong$)
4. $\triangle ABC \cong \triangle CDA$ (SSS)

1c. Sample answer: $\overleftrightarrow{AB} \parallel \overleftrightarrow{CD}$; $\overleftrightarrow{AC}$ is a transversal to $\overleftrightarrow{AB}$ and $\overleftrightarrow{CD}$, so $\angle CAB$ and $\angle ACD$ are alternate interior angles. Because $\triangle ABC \cong \triangle CDA$, $\angle CAB$ and $\angle ACD$ are congruent corresponding angles. Therefore, the lines are parallel.

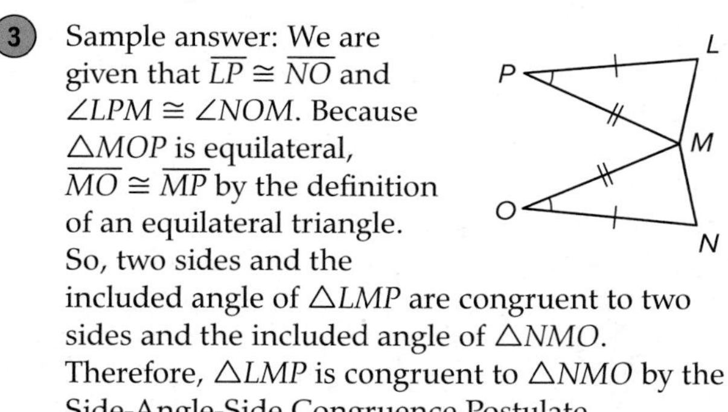

3 Sample answer: We are given that $\overline{LP} \cong \overline{NO}$ and $\angle LPM \cong \angle NOM$. Because $\triangle MOP$ is equilateral, $\overline{MO} \cong \overline{MP}$ by the definition of an equilateral triangle. So, two sides and the included angle of $\triangle LMP$ are congruent to two sides and the included angle of $\triangle NMO$. Therefore, $\triangle LMP$ is congruent to $\triangle NMO$ by the Side-Angle-Side Congruence Postulate.

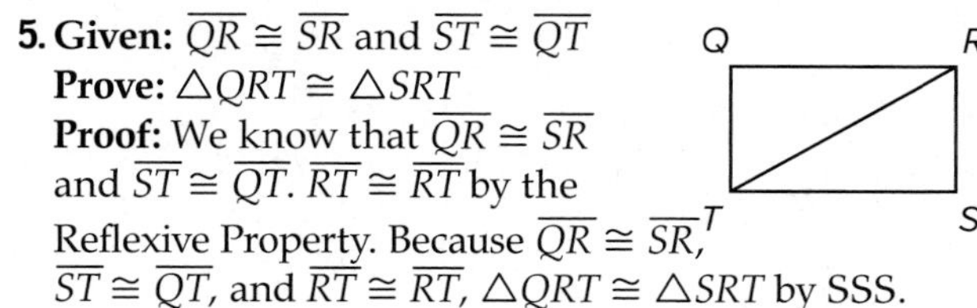

5. Given: $\overline{QR} \cong \overline{SR}$ and $\overline{ST} \cong \overline{QT}$

Prove: $\triangle QRT \cong \triangle SRT$

Proof: We know that $\overline{QR} \cong \overline{SR}$ and $\overline{ST} \cong \overline{QT}$. $\overline{RT} \cong \overline{RT}$ by the Reflexive Property. Because $\overline{QR} \cong \overline{SR}$, $\overline{ST} \cong \overline{QT}$, and $\overline{RT} \cong \overline{RT}$, $\triangle QRT \cong \triangle SRT$ by SSS.

7. Given: $\overline{AB} \cong \overline{ED}$, $\angle ABC$ and $\angle EDC$ are right angles, and C is the midpoint of $\overline{BD}$.

Prove: $\triangle ABC \cong \triangle EDC$

Proof:

Statements (Reasons)

1. $\overline{AB} \cong \overline{ED}$, $\angle ABC$ and $\angle EDC$ are right angles, and C is the midpoint of $\overline{BD}$. (Given)
2. $\angle ABC \cong \angle EDC$ (All rt. $\angle$s $\cong$)
3. $\overline{BC} \cong \overline{DC}$ (Midpoint Thm.)
4. $\triangle ABC \cong \triangle EDC$ (SAS)

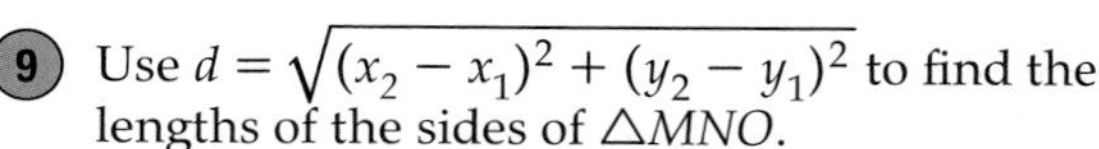

9 Use $d = \sqrt{(x_2 - x_1)^2 + (y_2 - y_1)^2}$ to find the lengths of the sides of $\triangle MNO$.

$MN = \sqrt{(-1 - 0)^2 + [-4 - (-1)]^2}$ $(x_1, y_1) = (0, -1)$, $(x_2, y_2) = (-1, -4)$

$= \sqrt{1 + 9}$ or $\sqrt{10}$ Simplify.

$NO = \sqrt{[-4 - (-1)]^2 + [-3 - (-4)]^2}$ $(x_1, y_1) = (-1, -4)$, $(x_2, y_2) = (-4, -3)$

$= \sqrt{9 + 1}$ or $\sqrt{10}$ Simplify.

$MO = \sqrt{(-4 - 0)^2 + [-3 - (-1)]^2}$ $(x_1, y_1) = (0, -1)$, $(x_2, y_2) = (-4, -3)$

$= \sqrt{16 + 4}$ or $\sqrt{20}$ Simplify.

Find the lengths of the sides of $\triangle QRS$.

$QR = \sqrt{(4 - 3)^2 + [-4 - (-3)]^2}$ $(x_1, y_1) = (3, -3)$, $(x_2, y_2) = (4, -4)$

$= \sqrt{1 + 1}$ or $\sqrt{2}$ Simplify.

$RS = \sqrt{(3 - 4)^2 + [3 - (-4)]^2}$ $(x_1, y_1) = (4, -4)$, $(x_2, y_2) = (3, 3)$

$= \sqrt{1 + 49}$ or $\sqrt{50}$ Simplify.

$QS = \sqrt{(3 - 3)^2 + [3 - (-3)]^2}$ $(x_1, y_1) = (3, -3)$, $(x_2, y_2) = (3, 3)$

$= \sqrt{0 + 36}$ or 6 Simplify.

$MN = \sqrt{10}$, $NO = \sqrt{10}$, $MO = \sqrt{20}$, $QR = \sqrt{2}$, $RS = \sqrt{50}$, and $QS = 6$. The corresponding sides are not congruent, so the triangles are not congruent.

11. $MN = \sqrt{10}$, $NO = \sqrt{10}$, $MO = \sqrt{20}$, $QR = \sqrt{10}$, $RS = \sqrt{10}$, and $QS = \sqrt{20}$. Each pair of corresponding sides has the same measure, so they are congruent. $\triangle MNO \cong \triangle QRS$ by SSS.

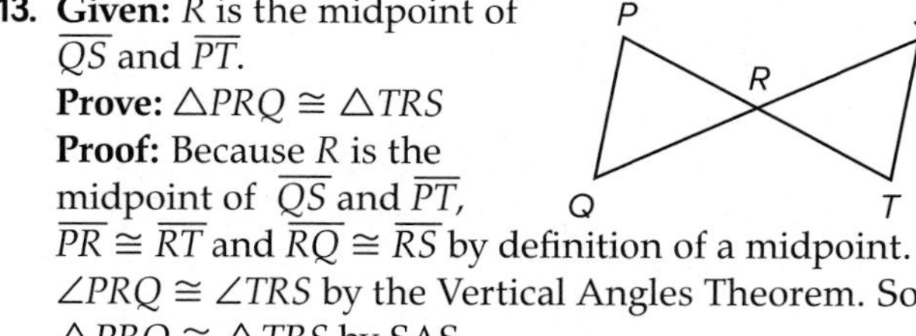

13. Given: R is the midpoint of $\overline{QS}$ and $\overline{PT}$.

Prove: $\triangle PRQ \cong \triangle TRS$

Proof: Because R is the midpoint of $\overline{QS}$ and $\overline{PT}$, $\overline{PR} \cong \overline{RT}$ and $\overline{RQ} \cong \overline{RS}$ by definition of a midpoint. $\angle PRQ \cong \angle TRS$ by the Vertical Angles Theorem. So, $\triangle PRQ \cong \triangle TRS$ by SAS.

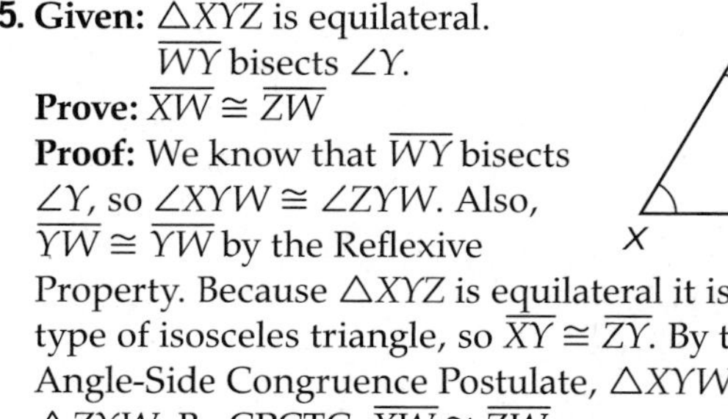

15. Given: $\triangle XYZ$ is equilateral. $\overline{WY}$ bisects $\angle Y$.

Prove: $\overline{XW} \cong \overline{ZW}$

Proof: We know that $\overline{WY}$ bisects $\angle Y$, so $\angle XYW \cong \angle ZYW$. Also, $\overline{YW} \cong \overline{YW}$ by the Reflexive Property. Because $\triangle XYZ$ is equilateral it is a special type of isosceles triangle, so $\overline{XY} \cong \overline{ZY}$. By the Side-Angle-Side Congruence Postulate, $\triangle XYW \cong \triangle ZYW$. By CPCTC, $\overline{XW} \cong \overline{ZW}$.

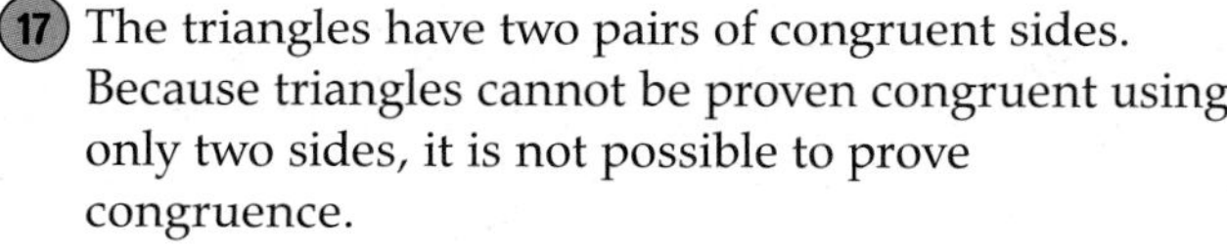

17 The triangles have two pairs of congruent sides. Because triangles cannot be proven congruent using only two sides, it is not possible to prove congruence.

19. SAS; rotation

21. Given: $\overline{MJ} \cong \overline{ML}$;
K is the midpoint of $\overline{JL}$.
Prove: $\triangle MJK \cong \triangle MLK$

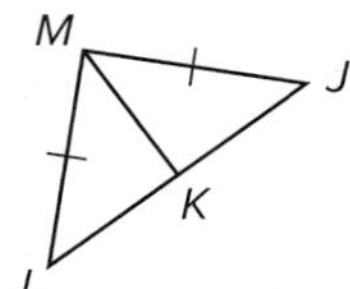

Proof:

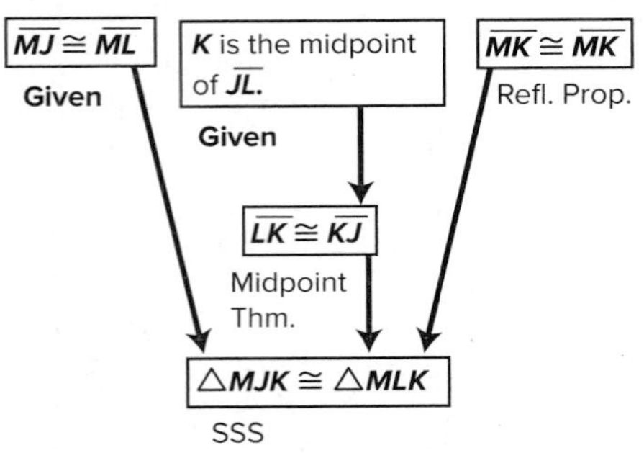

23a. Given: $\overline{TS} \cong \overline{SF} \cong \overline{FR} \cong \overline{RT}$; $\angle TSF$, $\angle SFR$, $\angle FRT$, and $\angle RTS$ are right angles.
Prove: $\overline{RS} \cong \overline{TF}$
Proof:
<u>**Statements (Reasons)**</u>
1. $\overline{TS} \cong \overline{SF} \cong \overline{FR} \cong \overline{RT}$ (Given)
2. $\angle TSF$, $\angle SFR$, $\angle FRT$, and $\angle RTS$ are right angles. (Given)
3. $\angle STR \cong \angle TRF$ (All rt. ∠s are ≅.)
4. $\triangle STR \cong \triangle TRF$ (SAS)
5. $\overline{RS} \cong \overline{TF}$ (CPCTC)

23b. Given: $\overline{TS} \cong \overline{SF} \cong \overline{FR} \cong \overline{RT}$; $\angle TSF$, $\angle SFR$, $\angle FRT$, and $\angle RTS$ are right angles.
Prove: $\angle SRT \cong \angle SRF$
Proof:
<u>**Statements (Reasons)**</u>
1. $\overline{TS} \cong \overline{SF} \cong \overline{FR} \cong \overline{RT}$ (Given)
2. $\angle TSF$, $\angle SFR$, $\angle FRT$, and $\angle RTS$ are right angles. (Given)
3. $\angle STR \cong \angle SFR$ (All rt. ∠s are ≅.)
4. $\triangle STR \cong \triangle SFR$ (SAS)
5. $\angle SRT \cong \angle SRF$ (CPCTC)

25. Given: $\triangle EAB \cong \triangle DCB$
Prove: $\triangle EAD \cong \triangle DCE$
Proof:
<u>**Statements (Reasons)**</u>
1. $\triangle EAB \cong \triangle DCB$ (Given)
2. $\overline{EA} \cong \overline{DC}$ (CPCTC)
3. $\overline{ED} \cong \overline{DE}$ (Reflex. Prop.)
4. $\overline{AB} \cong \overline{CB}$ (CPCTC)
5. $\overline{DB} \cong \overline{EB}$ (CPCTC)
6. $AB = CB$, $DB = EB$ (Def. ≅ segments)
7. $AB + DB = CB + EB$ (Add. Prop. =)
8. $AD = AB + DB$, $CE = CB + EB$ (Seg. addition)
9. $AD = CE$ (Subst. Prop. =)
10. $\overline{AD} \cong \overline{CE}$ (Def. ≅ segments)
11. $\triangle EAD \cong \triangle DCE$ (SSS)

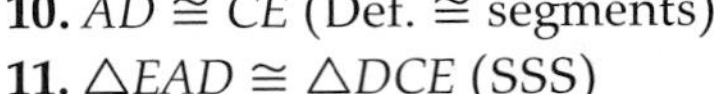

27. Sample answer: Step 1, Copy the triangles onto a sheet of paper. Step 2, copy and label $\triangle ABC$ onto a sheet of tracing paper. Step 3, translate the paper until $\overline{AB}$, $\overline{AC}$, and $\overline{BC}$ lie exactly on top of $\overline{XY}$, $\overline{XZ}$, and $\overline{YZ}$. The activity establishes a rigid motion that maps $\overline{AB}$ onto $\overline{XY}$, $\overline{AC}$ onto $\overline{XZ}$, and $\overline{BC}$ onto $\overline{YZ}$ ensuring that $\overline{AB} \cong \overline{XY}$, $\overline{AC} \cong \overline{XZ}$, and $\overline{BC} \cong \overline{YZ}$. From these statements we know that *A* is mapped onto *X*, *B* is mapped onto *Y*, and *C* is mapped onto *Z*. Because angle measures are preserved in a rigid motion, we know that $\angle A \cong \angle X$, $\angle B \cong \angle Y$, and $\angle C \cong \angle Z$. Therefore, $\triangle ABC$ is mapped exactly onto $\triangle XYZ$, so $\triangle ABC \cong \triangle XYZ$.

29a. Sample answer: Method 1: You could use the Distance Formula to find the length of each of the sides, and then use the Side-Side-Side Congruence Postulate to prove the triangles congruent. Method 2: You could find the slopes of $\overline{ZX}$ and $\overline{WY}$ to prove that they are perpendicular and that $\angle WYZ$ and $\angle WYX$ are both right angles. You can use the Distance Formula to prove that $\overline{XY}$ is congruent to $\overline{ZY}$. Because the triangles share the leg $\overline{WY}$, you can use the Side-Angle-Side Congruence Postulate; sample answer: I think that method 2 is more efficient, because you only have two steps instead of three. **29b.** Sample answer: Yes; the slope of $\overline{WY}$ is −1 and the slope of $\overline{ZX}$ is 1, and −1 and 1 are opposite reciprocals, so $\overline{WY}$ is perpendicular to $\overline{ZX}$. Because they are perpendicular, $\angle WYZ$ and $\angle WYX$ are both 90°. Using the Distance Formula, the length of $\overline{ZY}$ is $\sqrt{(4-1)^2 + (5-2)^2}$ or $3\sqrt{2}$, and the length of $\overline{XY}$ is $\sqrt{(7-4)^2 + (8-5)^2}$ or $3\sqrt{2}$. Because $\overline{WY}$ is congruent to $\overline{WY}$, $\triangle WYZ$ is congruent to $\triangle WYX$ by the Side-Angle-Side Congruence Postulate. **31.** Shada; for SAS the angle must be the included angle and here it is not included. **33.** Case 1: You know the hypotenuses are congruent and two corresponding legs are congruent. Then the Pythagorean Theorem says that the other legs are congruent so the triangles are congruent by SSS. Case 2: You know the legs are congruent and the right angles are congruent, so the triangles are congruent by SAS. **35a.** $\overline{AB} \cong \overline{DC}$ **35b.** $\angle ACB \cong \angle DBC$ **35c.** reflection **35d.** SAS **35e.** We are given that $\overline{AC} \cong \overline{DB}$ and $\angle ACB \cong \angle DBC$. We know $\overline{BC} \cong \overline{CB}$ by the Reflexive Prop. So $\triangle ABC \cong \triangle DCB$ by SAS. **37.** C

Lesson 4-4

1. Given: $\overline{CB}$ bisects $\angle ABD$ and $\angle ACD$.
Prove: $\triangle ABC \cong \triangle DCB$

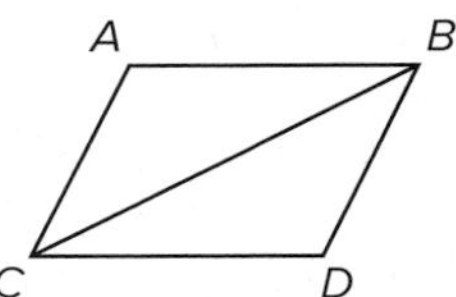

Proof:
<u>**Statements (Reasons)**</u>
1. $\overline{CB}$ bisects $\angle ABD$ and $\angle ACD$. (Given)
2. $\angle ABC \cong \angle DBC$ (Def. of ∠ bisector)

3. $\overline{BC} \cong \overline{BC}$ (Refl. prop.)
4. $\angle ACB \cong \angle DCB$ (Def. of $\angle$ bisector)
5. $\triangle ABC \cong \triangle DCB$ (ASA)

3. Given: $\angle K \cong \angle M$, $\overline{JK} \cong \overline{JM}$, $\overline{JL}$ bisects $\angle KLM$.
Prove: $\triangle JKL \cong \triangle JML$
Proof: We are given $\angle K \cong \angle M$, $\overline{JK} \cong \overline{JM}$, and $\overline{JL}$ bisects $\angle KLM$. Since $\overline{JL}$ bisects $\angle KLM$, we know $\angle KLJ \cong \angle MLJ$. So, $\triangle JKL \cong \triangle JML$ by the AAS Congruence Theorem.

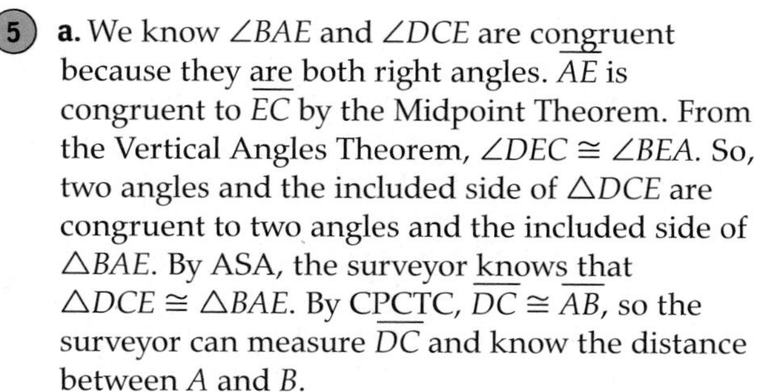

5 **a.** We know $\angle BAE$ and $\angle DCE$ are congruent because they are both right angles. $\overline{AE}$ is congruent to $\overline{EC}$ by the Midpoint Theorem. From the Vertical Angles Theorem, $\angle DEC \cong \angle BEA$. So, two angles and the included side of $\triangle DCE$ are congruent to two angles and the included side of $\triangle BAE$. By ASA, the surveyor knows that $\triangle DCE \cong \triangle BAE$. By CPCTC, $\overline{DC} \cong \overline{AB}$, so the surveyor can measure $\overline{DC}$ and know the distance between A and B.

b.

$\triangle DCE \cong \triangle BAE$	SAS
$\overline{DC} \cong \overline{AB}$	CPCTC
$DC = AB$	Definition of congruence
$550 = AB$	Substitution

So, by the definition of congruence, $AB = 550$ m.

7. Given: $\angle W \cong \angle Y$, $\overline{WZ} \cong \overline{YZ}$, $\overline{XZ}$ bisects $\angle WZY$.
Prove: $\triangle XWZ \cong \triangle XYZ$
Proof: It is given that $\angle W \cong \angle Y$, $\overline{WZ} \cong \overline{YZ}$, and $\overline{XZ}$ bisects $\angle WZY$. By the definition of angle bisector, $\angle WZX \cong \angle YZX$. The Angle-Side-Angle Congruence Postulate tells us that $\triangle XWZ \cong \triangle XYZ$.

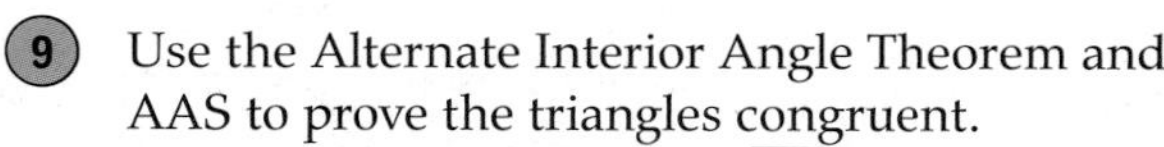

9 Use the Alternate Interior Angle Theorem and AAS to prove the triangles congruent.
Given: V is the midpoint of $\overline{YW}$; $\overline{UY} \parallel \overline{XW}$.
Prove: $\triangle UVY \cong \triangle XVW$
Proof:
Statements (Reasons)
1. V is the midpoint of $\overline{YW}$; $\overline{UY} \parallel \overline{XW}$. (Given)
2. $\overline{YV} \cong \overline{VW}$ (Midpoint Theorem)
3. $\angle VWX \cong \angle VYU$ (Alt. Int. $\angle$s Thm.)
4. $\angle VUY \cong \angle VXW$ (Alt. Int. $\angle$s Thm.)
5. $\triangle UVY \cong \triangle XVW$ (AAS)

11. Given: $\angle A$ and $\angle C$ are right angles. $\angle ABE \cong \angle CBD$, $\overline{AE} \cong \overline{CD}$
Prove: $\overline{BE} \cong \overline{BD}$

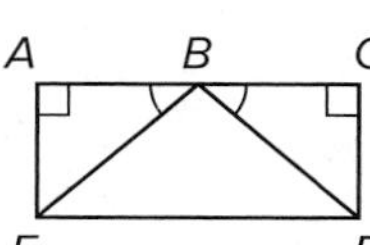

Proof:

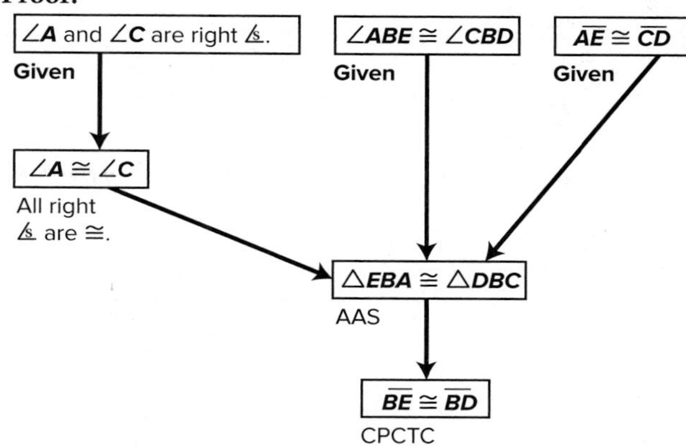

13a. $\angle HJK \cong \angle GFK$ because all right angles are congruent. We are given that $\overline{JK} \cong \overline{KF}$. $\angle HKJ$ and $\angle FKG$ are vertical angles, so $\angle HKJ \cong \angle FKG$ by the Vertical Angles Theorem. By ASA, $\triangle HJK \cong \triangle GFK$, so $\overline{FG} \cong \overline{HJ}$ by CPCTC. **13b.** No; $HJ = 1350$ m, so $FG = 1350$ m. If the regatta is to be 1500 m, the lake is not long enough, because $1350 < 1500$.

15 If the triangles are congruent, $HJ = QJ$. Solve for y.

$HJ = QJ$	CPCTC
$9 = 2y - 1$	$HJ = 9$, $QJ = 2y - 1$
$y = 5$	Simplify.

17. Given: $\overline{AE} \perp \overline{DE}$, $\overline{EA} \perp \overline{AB}$, C is the midpoint of $\overline{AE}$.
Prove: $\overline{CD} \cong \overline{CB}$

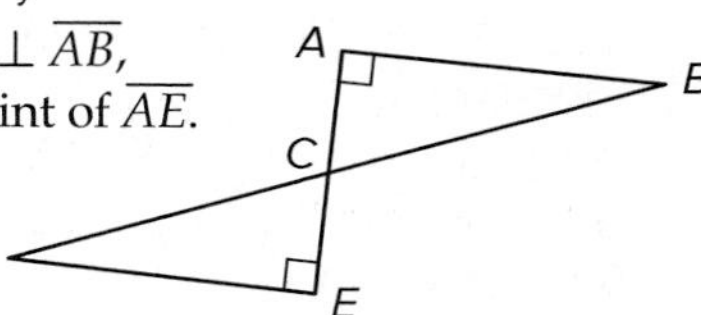

Proof: We are given that $\overline{AE}$ is perpendicular to $\overline{DE}$, $\overline{EA}$ is perpendicular to $\overline{AB}$, and C is the midpoint of $\overline{AE}$. Because $\overline{AE}$ is perpendicular to $\overline{DE}$, $m\angle CED = 90$. Because $\overline{EA}$ is perpendicular to $\overline{AB}$, $m\angle BAC = 90$. $\angle CED \cong \angle BAC$ because all right angles are congruent. $\overline{AC} \cong \overline{CE}$ from the Midpt. Thm. $\angle ECD \cong \angle ACB$ because they are vertical angles. Angle-Side-Angle gives us that $\triangle CED \cong \triangle CAB$. $\overline{CD} \cong \overline{CB}$ because corresponding parts of congruent triangles are congruent.

19. Given: $\angle K \cong \angle M$, $\overline{KP} \perp \overline{PR}$, $\overline{MR} \perp \overline{PR}$
Prove: $\angle KPL \cong \angle MRL$
Proof:
Statements (Reasons)
1. $\angle K \cong \angle M$, $\overline{KP} \perp \overline{PR}$, $\overline{MR} \perp \overline{PR}$ (Given)
2. $\angle KPR$ and $\angle MRP$ are both right angles. (Def. of $\perp$)
3. $\angle KPR \cong \angle MRP$ (All rt. $\angle$s are congruent.)
4. $\overline{PR} \cong \overline{PR}$ (Refl. Prop.)
5. $\triangle KPR \cong \triangle MRP$ (AAS)
6. $\overline{KP} \cong \overline{MR}$ (CPCTC)
7. $\angle KLP \cong \angle MLR$ (Vertical angles are $\cong$.)
8. $\triangle KLP \cong \triangle MLR$ (AAS)
9. $\angle KPL \cong \angle MRL$ (CPCTC)

21 Because $m\angle ACB = m\angle ADB = 44$, and $m\angle CBA = m\angle DBA = 68$, then $\angle ACB \cong \angle ADB$ and $\angle CBA \cong \angle DBA$ by the definition of congruence. $\overline{AB} \cong \overline{AB}$ by the Reflexive Property, so $\triangle ACB \cong \triangle ADB$ by AAS. Then $\overline{AC} \cong \overline{AD}$ by CPCTC. Because $AC = AD$ by the definition of congruence, the two seat stays are the same length.

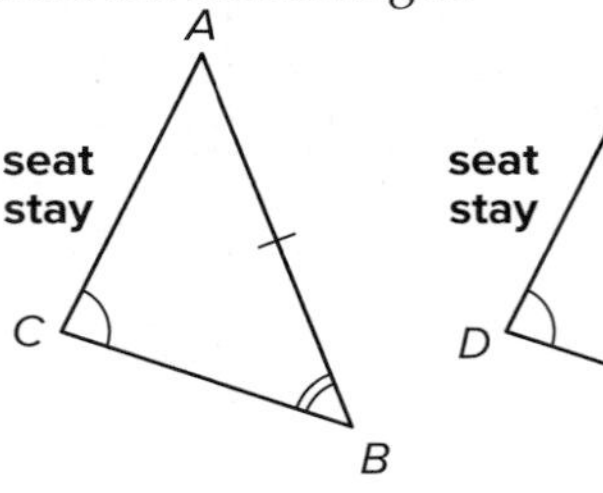

23a. If two angles of a triangle are congruent, then the opposite sides are congruent.
23b. If two sides of a triangle are congruent, then the opposite angles are congruent.
25. Sample answer: Step 1, copy the triangles onto a sheet of paper. Step 2, copy and label $\triangle ABC$ onto a sheet of tracing paper. Step 3, translate the paper until $\angle A$, $\overline{AB}$, and $\angle B$ lie exactly on top of $\angle X$, $\overline{XY}$, and $\angle Y$. The activity establishes a rigid motion that maps $\angle A$ onto $\angle X$, $\overline{AB}$ onto $\overline{XY}$, and $\angle B$ onto $\angle Y$ ensuring that $\angle A \cong \angle X$, $\overline{AB} \cong \overline{XY}$, and $\angle B \cong \angle Y$. From these statements we know that A is mapped onto X, B is mapped onto Y, and C is mapped onto Z. Because distances between points are preserved in a rigid motion, we know that $\overline{BC} \cong \overline{YZ}$. Because angle measures are preserved in a rigid motion, we know that $\angle A \cong \angle X$ and $\angle C \cong \angle Z$.Therefore, $\triangle ABC$ is mapped exactly onto $\triangle XYZ$, so $\triangle ABC \cong \triangle XYZ$.
27a. Sample answer: Proof: We are given that $\angle J \cong \angle M$ and $\angle K \cong \angle N$. By the Distance Formula, $JK = \sqrt{[-2-(-5)]^2 + (3-5)^2} = \sqrt{13}$ and $MN = \sqrt{(3-5)^2 + (-3-0)^2} = \sqrt{13}$. So, $\overline{JK} \cong \overline{MN}$ by the definition of congruent segments. Therefore, $\angle JKL \cong \angle MNP$ by ASA.
27b. Sample answer: Rotate JKL about the origin 90° clockwise, and then translate it 5 units down.
27c. Sample answer: $J'K'L'$: $J'(5, 5)$, $K'(3, 2)$, $L'(1, 3)$; $J''K''L''$: $J''(5, 0)$, $K''(3, -3)$, $L''(1, -2)$

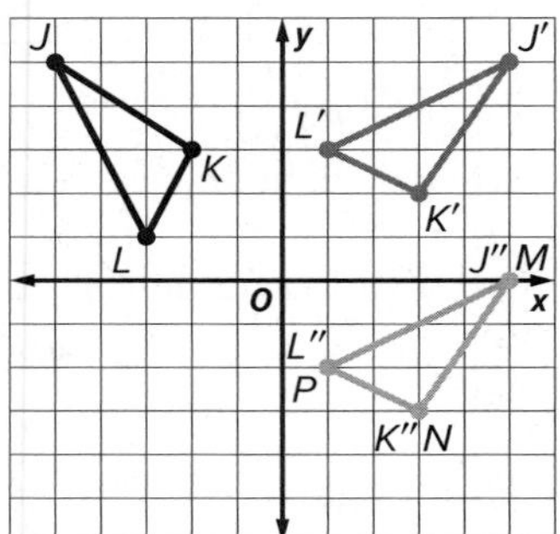

29. Proof:
__Statements (Reasons)__
1. $\overline{BC}$ is perpendicular to $\overline{AD}$. (Given)
2. $\angle ACB$ and $\angle DCB$ are right angles. (Def. of ⊥ segs.)
3. $\angle ACB \cong \angle DCB$ (All rt. angles are ≅.)
4. $\angle A \cong \angle D$ (Given)
5. $\overline{BC} \cong \overline{BC}$ (Refl. Prop.)
6. $\angle ACB \cong \angle DCB$ (AAS)

31. 81 **33.** $\overline{AB} \cong \overline{XY}$

Lesson 4-5

1. Yes; LA; horizontal reflection
3 __Statements (Reasons)__
1. $\overline{FH} \perp \overline{EG}$ (Given)
2. $\angle FHG$ and $\angle FHE$ are right angles. (⊥ lines form rt ∠s.)
3. $\triangle EFH$ and $\triangle EFH$ are rt. △s (Def. of rt △s)
4. $\overline{FH}$ bisects $\overline{EG}$. (Given)
5. $\overline{EG} \cong \overline{GH}$ (Definition of segment bisector)
6. $\triangle EFH \cong \triangle GFH$ (LL)

5. Yes; HA; reflection in the hypotenuse
7. Proof:
__Statements (Reasons)__
1. $\overline{AB} \perp \overline{BC}$, $\overline{DC} \perp \overline{BC}$ (Given)
2. $\angle ABC$ is a right angle, $\angle DCB$ is a right angle. (⊥ lines form rt. ∠s)
3. $\triangle ABC$ is a right triangle, $\triangle DCB$ is a right triangle. (Def. of rt. △)
4. $\overline{AC} \cong \overline{BD}$ (Given)
5. $\overline{BC} \cong \overline{BC}$ (Refl. Prop. of Congruence)
6. $\triangle ABC \cong \triangle DCB$ (HL)
7. $\overline{AB} \cong \overline{DC}$ (CPCTC)

9. Proof:
__Statements (Reasons)__
1. $\overline{AB} \perp \overline{BC}$, $\overline{DC} \perp \overline{BC}$ (Given)
2. $\angle ABC$ is a right angle, $\angle DCB$ is a right angle. (⊥ lines form rt. ∠s)
3. $\triangle ABC$ is a right triangle, $\triangle DCB$ is a right triangle. (Def. of rt. △)
4. $\overline{AB} \cong \overline{DC}$ (Given)
5. $\overline{BC} \cong \overline{BC}$ (Refl. Prop. of Congruence)
6. $\triangle ABC \cong \triangle DCB$ (LL)
7. $\angle A \cong \angle D$ (CPCTC)

11. $\overline{AC} \cong \overline{MP}$ and $\overline{CB} \cong \overline{PN}$
13. Sample answer: $\overline{AB} \cong \overline{MN}$ and $\overline{CB} \cong \overline{PN}$
15. Proof:
__Statements (Reasons)__
1. $\triangle TUV$ and $\triangle XYZ$ are rt. △s, $\overline{TU} \cong \overline{XY}$, $\overline{TV} \cong \overline{XZ}$ (Given)
2. $TU = XY$, $TV = XZ$ (Def. of ≅)
3. $(TU)^2 + (UV)^2 = (VT)^2$, $(XY)^2 + (YZ)^2 = (ZX)^2$ (Pythagorean Thm.)
4. $(TU)^2 + (UV)^2 = (XY)^2 + (YZ)^2$ (Subs. Prop.)
5. $(TU)^2 + (UV)^2 = (TU)^2 + (YZ)^2$ (Subs. Prop.)
6. $(UV)^2 = (YZ)^2$ (Subt. Prop.)
7. $UV = YZ$ (A property of square roots)
8. $\overline{UV} = \overline{YZ}$ (Def. of ≅ segments)
9. $\triangle TUV \cong \triangle XYZ$ (SSS)

17. Proof:

Case 1:

Statements (Reasons)

1. $\triangle LMN$ and $\triangle ABC$ are rt. $\triangle$s (Given)
2. $\overline{LN} \cong \overline{AC}$ (Given)
3. $\angle L \cong \angle A$ (Given)
4. $\angle N$ and $\angle C$ are rt. $\angle$s (Def. of rt. $\triangle$s)
5. $\angle N \cong \angle C$ (all rt. $\angle$s are $\cong$)
6. $\triangle LMN \cong \triangle ABC$ (ASA)

Case 2:

Statements (Reasons)

1. $\triangle LMN$ and $\triangle ABC$ are rt. $\triangle$s (Given)
2. $\overline{MN} \cong \overline{BC}$ (Given)
3. $\angle L \cong \angle A$ (Given)
4. $\angle N$ and $\angle C$ are rt. $\angle$s (Def. of rt $\triangle$s)
5. $\angle N \cong \angle C$ (all rt. $\angle$s are $\cong$)
6. $\triangle LMN \cong \triangle ABC$ (AAS)

19. Yes; the wall of the house is perpendicular to the ground, so the triangles formed by the house, ground, and ladders are right triangles. The hypotenuses are congruent because the ladders are the same length. The corresponding legs along the ground are congruent because the ladders are placed the same distance from the house. So the triangles are congruent by HL. The corresponding legs along the wall are congruent by CPCTC, so the ladders reach to the same height on the house.

21. It is given that $\overline{AD}$ divides $\triangle ABC$ into two right triangles with acute angles of 30° and 60°, so $\triangle ACD \cong \triangle ABD$. $\overline{CD} \cong \overline{BD}$ by CPCTC, and $CD = BD$ by the definition of congruent segments. $CD + BD = CB$ and $CD = x$, so by substitution, $CB = 2x$. Because $\triangle ABC$ is equilateral, $\overline{CB} \cong \overline{AC}$, and $CB = AC$ by the definition of congruent segments. AC is the hypotenuse of $\triangle ADC$, so $(CB)^2 + (AD)^2 = (AC)^2$ by the Pythagorean Theorem. That gives $x^2 + (AD)^2 = (2x)^2$ by substitution, which simplifies to $AD = \sqrt{3x} = x\sqrt{3}$.

23. Either $\angle A \cong \angle D$, $\angle B \cong \angle E$, $\overline{AC} \cong \overline{DF}$, or $\overline{BC} \cong \overline{EF}$.

25. B **27a.** not enough information **27b.** HA **27c.** HL **27d.** LA **27e.** LL **29.** Not enough information is given. The congruency marks are not on corresponding parts, so they triangles may be congruent, but they may not be. **31.** Yes; $x = 4$, so the corresponding legs have the same length and the triangles are congruent by HL.

Lesson 4-6

1. $\angle BAC$ and $\angle BCA$ **3.** 12 **5.** 12

7. Given: $\triangle ABC$ is isosceles; $\overline{EB}$ bisects $\angle ABC$.

Prove: $\triangle ABE \cong \triangle CBE$

Proof:

Statements (Reasons)

1. $\triangle ABC$ is isosceles; $\overline{EB}$ bisects $\angle ABC$. (Given)
2. $\overline{AB} \cong \overline{BC}$ (Def. of isosceles)
3. $\angle ABE \cong \angle CBE$ (Def. of $\angle$ bisector)
4. $\overline{BE} \cong \overline{BE}$ (Refl. Prop.)
5. $\triangle ABE \cong \triangle CBE$ (SAS)

9. $\angle ABE$ is opposite $\overline{AE}$ and $\angle AEB$ is opposite $\overline{AB}$. Because $\overline{AE} \cong \overline{AB}$, $\angle ABE \cong \angle AEB$.

11. $\angle ACD$ and $\angle ADC$ **13.** $\overline{BF}$ and $\overline{BC}$ **15.** 60 **17.** 4

19. The triangle is equiangular, so it is also equilateral. All the sides are congruent.

$2x + 11 = 6x - 9$	Definition of congruence
$11 = 4x - 9$	Subtract $2x$ from each side.
$20 = 4x$	Add 9 to each side.
$5 = x$	Divide each side by 4.

21. $x = 11$, $y = 11$

23. Given: $\triangle HJM$ is an isosceles triangle and $\triangle HKL$ is an equilateral triangle. $\angle JKH$, $\angle HKL$ and $\angle HLK$, $\angle MLH$ are supplementary.

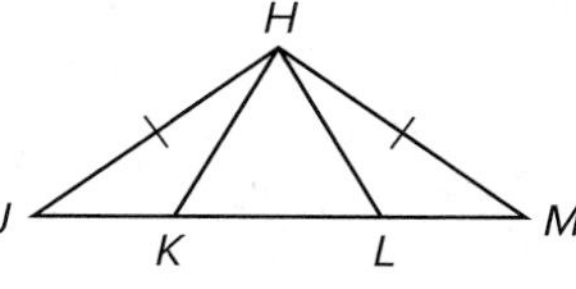

Prove: $\angle JHK \cong \angle MHL$

Proof: We are given that $\triangle HJM$ is an isosceles triangle and $\triangle HKL$ is an equilateral triangle, $\angle JKH$ and $\angle HKL$ are supplementary and $\angle HLK$ and $\angle MLH$ are supplementary. From the Isosceles Triangle Theorem, we know that $\angle HJK \cong \angle HML$. Because $\triangle HKL$ is an equilateral triangle, we know $\angle HLK \cong \angle LKH \cong \angle KHL$ and $\overline{HL} \cong \overline{KL} \cong \overline{HK}$. $\angle JKH$, $\angle HKL$ and $\angle HLK$, $\angle MLH$ are supplementary, and $\angle HKL \cong \angle HLK$, we know $\angle JKH \cong \angle MLH$ by the Congruent Supplements Theorem. By AAS, $\triangle JHK \cong \triangle MLH$. By CPCTC, $\angle JHK \cong \angle MHL$.

25a. 65; because $\triangle ABC$ is isosceles, $\angle ABC \cong \angle ACB$, so $180 - 50 = 130$ and $\frac{130}{2} = 65$.

25b. Given: $\overline{BE} \cong \overline{CD}$

Prove: $\triangle AED$ is isosceles.

Proof:

Statements (Reasons)

1. $\overline{AB} \cong \overline{AC}$, $\overline{BE} \cong \overline{CD}$ (Given)
2. $AB = AC$, $BE = CD$ (Def. of congruence)
3. $AB + BE = AE$, $AC + CD = AD$ (Seg. Add. Post.)
4. $AB + BE = AC + CD$ (Add. Prop. of Eq.)
5. $AE = AD$ (Subst.)
6. $\overline{AE} \cong \overline{AD}$ (Def. of congruence)
7. $\triangle AED$ is isosceles. (Def. of isosceles)

25c. Given: $\overline{BC} \parallel \overline{ED}$ and $\overline{ED} \cong \overline{AD}$

Prove: $\triangle ADE$ is equilateral.

Proof:

Statements (Reasons)

1. $\overline{AB} \cong \overline{AC}$, $\overline{BC} \parallel \overline{ED}$ and $\overline{ED} \cong \overline{AD}$ (Given)
2. $\angle ABC \cong \angle ACB$ (Isos. $\triangle$ Thm.)
3. $m\angle ABC = m\angle ACB$ (Def. of $\cong$ $\angle$s)
4. $\angle ABC \cong \angle AED$, $\angle ACB \cong \angle ADE$ (Corr. $\angle$s Thm.)
5. $m\angle ABC = m\angle AED$, $m\angle ACB = m\angle ADE$ (Def. of $\cong$ $\angle$s)
6. $m\angle AED = m\angle ACB$ (Subst.)
7. $m\angle AED = m\angle ADE$ (Subst.)

8. $\angle AED \cong \angle ADE$ (Def. of $\cong$ ∡)
9. $\overline{AD} \cong \overline{AE}$ (Conv. of Isos. △ Thm.)
10. $\triangle ADE$ is equilateral. (Def. of equilateral △)

25d. One pair of congruent corresponding sides and one pair of congruent corresponding angles; because you know that the triangle is isosceles, if one leg is congruent to a leg of $\triangle ABC$, then you know that both pairs of legs are congruent. Because the base angles of an isosceles triangle are congruent, if you know that $\angle K \cong \angle B$ you know that $\angle K \cong \angle L$, $\angle B \cong \angle C$, and $\angle C \cong \angle L$. Therefore, with one pair of congruent corresponding sides and one pair of congruent corresponding angles, the triangles can be proved congruent using either ASA or SAS.

27.

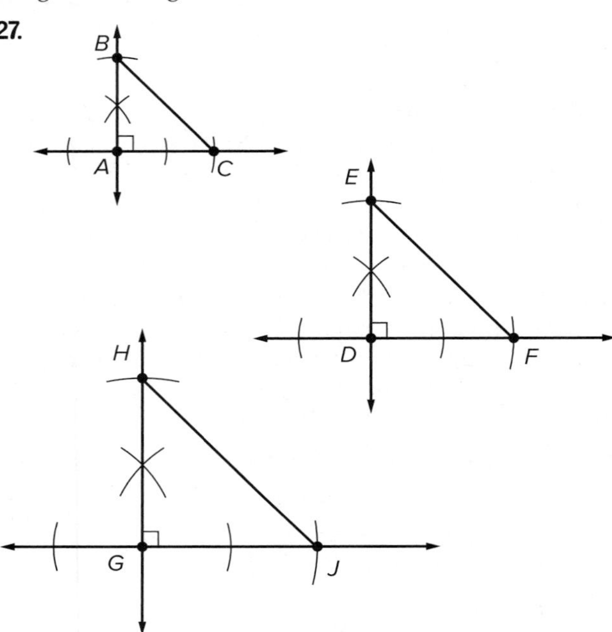

Sample answer: I constructed a pair of perpendicular segments and then used the same compass setting to mark points equidistant from their intersection. I measured both legs for each triangle. Because $AB = AC = 1.3$ cm, $DE = DF = 1.9$ cm, and $GH = GJ = 2.3$ cm, the triangles are isosceles. I used a protractor to confirm that $\angle A$, $\angle D$, and $\angle G$ are all right angles.

29 Because $\overline{AD} \cong \overline{CD}$, base angles CAD and ACD are congruent by the Isosceles Triangle Theorem. So, $m\angle CAD = m\angle ACD$.

$m\angle CAD + m\angle ACD + m\angle D = 180$	Triangle Angle-Sum Theorem
$m\angle CAD + m\angle CAD + 92 = 180$	Substitution
$2m\angle CAD + 92 = 180$	Simplify.
$2m\angle CAD = 88$	Subtract 92 from each side.
$m\angle CAD = 44$	Simplify.

31. 136

33. Given: Each triangle is isosceles, $\overline{BG} \cong \overline{HC}$, $\overline{HD} \cong \overline{JF}$, $\angle G \cong \angle H$, and $\angle H \cong \angle J$.
Prove: The distance from B to F is three times the distance from D to F.

Proof:
Statements (Reasons)
1. Each triangle is isosceles, $\overline{BG} \cong \overline{HC}$, $\overline{HD} \cong \overline{JF}$, $\angle G \cong \angle H$, and $\angle H \cong \angle J$. (Given)
2. $\angle G \cong \angle J$ (Trans. Prop.)
3. $\overline{BG} \cong \overline{CG}$, $\overline{HC} \cong \overline{HD}$, $\overline{JD} \cong \overline{JF}$ (Def. of Isosceles)
4. $\overline{BG} \cong \overline{JD}$ (Trans. Prop.)
5. $\overline{HC} \cong \overline{JD}$ (Trans. Prop.)
6. $\overline{CG} \cong \overline{JF}$ (Trans. Prop.)
7. $\triangle BCG \cong \triangle CDH \cong \triangle DFJ$ (SAS)
8. $\overline{BC} \cong \overline{CD} \cong \overline{DF}$ (CPCTC)
9. $BC = CD = DF$ (Def. of congruence)
10. $BC + CD + DF = BF$ (Seg. Add. Post.)
11. $DF + DF + DF = BF$ (Subst.)
12. $3DF = BF$ (Addition)

35. Case I
Given: $\triangle ABC$ is an equilateral triangle.
Prove: $\triangle ABC$ is an equiangular triangle.

B
A C

Proof:
Statements (Reasons)
1. $\triangle ABC$ is an equilateral triangle. (Given)
2. $\overline{AB} \cong \overline{AC} \cong \overline{BC}$ (Def. of equilateral △)
3. $\angle A \cong \angle B \cong \angle C$ (Isosceles △ Th.)
4. $\triangle ABC$ is an equiangular triangle. (Def. of equiangular)

Case II
Given: $\triangle ABC$ is an equiangular triangle.
Prove: $\triangle ABC$ is an equilateral triangle.

B
A C

Proof:
Statements (Reasons)
1. $\triangle ABC$ is an equiangular triangle. (Given)
2. $\angle A \cong \angle B \cong \angle C$ (Def. of equiangular △)
3. $\overline{AB} \cong \overline{AC} \cong \overline{BC}$ (If 2 ∡ of a △ are $\cong$ then the sides opp. those ∡ are $\cong$.)
4. $\triangle ABC$ is an equilateral triangle. (Def. of equilateral)

37. Given: $\triangle ABC$, $\angle A \cong \angle C$
Prove: $\overline{AB} \cong \overline{CB}$

B
A C
D

Proof:
Statements (Reasons)
1. Let $\overrightarrow{BD}$ bisect $\angle ABC$. (Protractor Post.)
2. $\angle ABD \cong \angle CBD$ (Def. of $\angle$ bisector)
3. $\angle A \cong \angle C$ (Given)
4. $\overline{BD} \cong \overline{BD}$ (Refl. Prop.)
5. $\triangle ABD \cong \triangle CBD$ (AAS)
6. $\overline{AB} \cong \overline{CB}$ (CPCTC)

39. 14

41

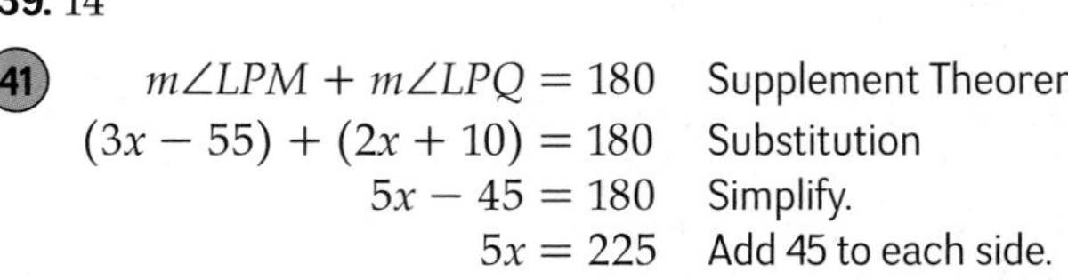

$m\angle LPM + m\angle LPQ = 180$	Supplement Theorem
$(3x - 55) + (2x + 10) = 180$	Substitution
$5x - 45 = 180$	Simplify.
$5x = 225$	Add 45 to each side.
$x = 45$	Divide each side by 45.

$m\angle LPM = 3x - 55$ Given
$= 3(45) - 55$ Substitution
$= 135 - 55$ or 80 Simplify.

Because $\overline{LM} \cong \overline{LP}$, base angles LMP and LPM are congruent by the Isosceles Triangle Theorem. So, $m\angle LMP = m\angle LPM = 80$.

43. 80

45. Given: $\triangle WJZ$ is equilateral, and $\angle ZWP \cong \angle WJM \cong \angle JZL$.

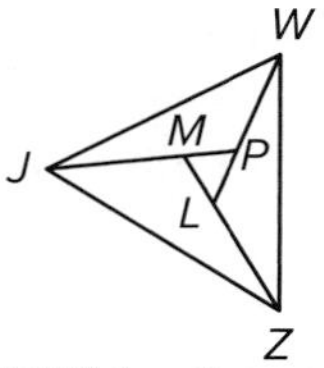

Prove: $\overline{WP} \cong \overline{ZL} \cong \overline{JM}$

Proof: We know that $\triangle WJZ$ is equilateral, because an equilateral $\triangle$ is equiangular, $\angle ZWJ \cong \angle WJZ \cong \angle JZW$. So, $m\angle ZWJ = m\angle WJZ = m\angle JZW$, by the definition of congruence. Because $\angle ZWP \cong \angle WJM \cong \angle JZL$, $m\angle ZWP = m\angle WJM = m\angle JZL$, by the definition of congruence. By the Angle Addition Postulate, $m\angle ZWJ = m\angle ZWP + m\angle PWJ$, $m\angle WJZ = m\angle WJM + m\angle MJZ$, $m\angle JZW = m\angle JZL + m\angle LZW$. By substitution, $m\angle ZWP + m\angle PWJ = m\angle WJM + m\angle MJZ = m\angle JZL + m\angle LZW$. Again by substitution, $m\angle ZWP + m\angle PWJ = m\angle ZWP + m\angle PJZ = m\angle ZWP + m\angle LZW$. By the Subtraction Property, $m\angle PWJ = m\angle PJZ = m\angle LZW$. By the definition of congruence, $\angle PWJ \cong \angle PJZ \cong \angle LZW$. So, by ASA, $\triangle WZL \cong \triangle ZJM \cong \triangle JWP$. By CPCTC, $\overline{WP} \cong \overline{ZL} \cong \overline{JM}$.

47. Never; the measure of the vertex angle will be $180 - 2$(measure of the base angle) so if the base angles are integers, then 2(measure of the base angle) will be even and $180 - 2$(measure of the base angle) will be even. **49.** It is not possible because a triangle cannot have more than one obtuse angle.

51. Sample answer: If a triangle is already classified, you can use the previously proven properties of that type of triangle in the proof. Doing this can save you steps when writing the proof. **53.** D **55a.** 36 **55b.** 54 **55c.** 54 **57.** 16 **59.** C

Lesson 4-7

1.

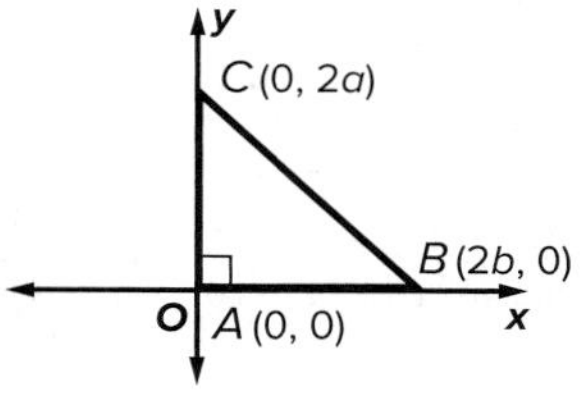

3. $T(2a, 0)$

5. $DC = \sqrt{[-a - (-a)]^2 + (b - 0)^2}$ or b

$GH = \sqrt{(a - a)^2 + (b - 0)^2}$ or b

Because $DC = GH$, $\overline{DC} \cong \overline{GH}$.

$DF = \sqrt{(0 - a)^2 + \left(\frac{b}{2} - b\right)^2}$ or $\sqrt{a^2 + \frac{b^2}{4}}$

$GF = \sqrt{(a - 0)^2 + \left(b - \frac{b}{2}\right)^2}$ or $\sqrt{a^2 + \frac{b^2}{4}}$

$CF = \sqrt{(0 - a)^2 + \left(\frac{b}{2} - 0\right)^2}$ or $\sqrt{a^2 + \frac{b^2}{4}}$

$HF = \sqrt{(a - 0)^2 + \left(0 - \frac{b}{2}\right)^2}$ or $\sqrt{a^2 + \frac{b^2}{4}}$

Because $DF = GF = CF = HF$, $\overline{DF} \cong \overline{GF} \cong \overline{CF} \cong \overline{HF}$. $\triangle FGH \cong \triangle FDC$ by SSS.

7.

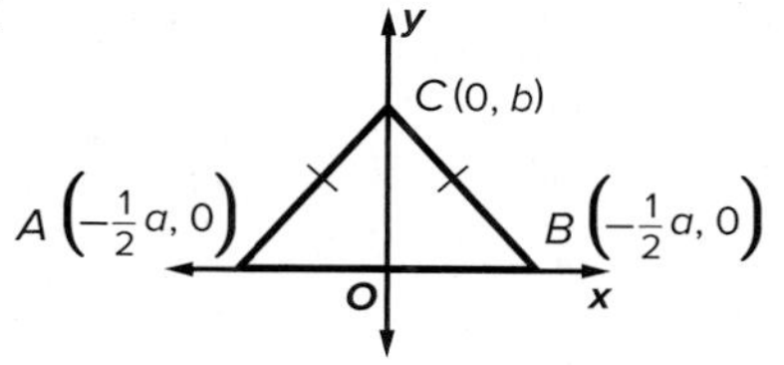

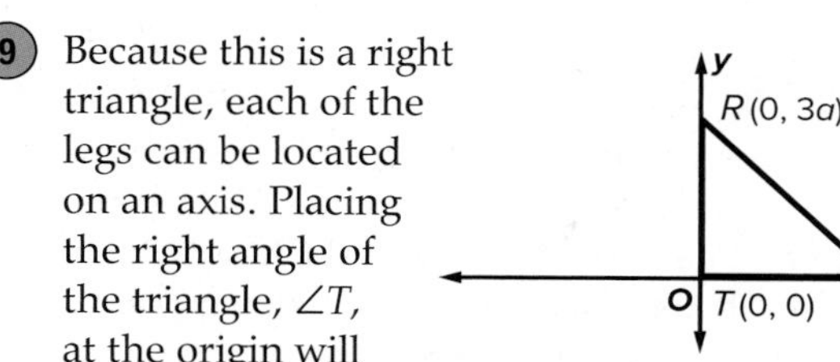

9 Because this is a right triangle, each of the legs can be located on an axis. Placing the right angle of the triangle, $\angle T$, at the origin will allow the two legs to be along the x- and y-axes. Position the triangle in the first quadrant. Because R is on the y-axis, its x-coordinate is 0. Its y-coordinate is $3a$ because the leg is $3a$ units long. Because S is on the x-axis, its y-coordinate is 0. Its x-coordinate is $3a$ because the leg is $3a$ units long.

11.

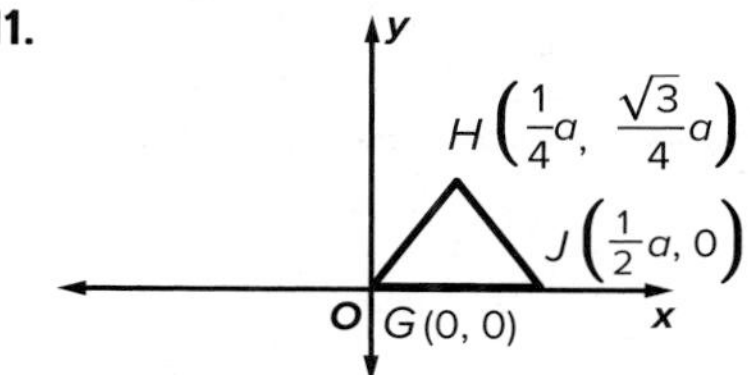

13. $C(a, a)$, $Y(a, 0)$

15 Vertex N is positioned at the origin. So, its coordinates are $(0, 0)$. Vertex L is on the x-axis, so its y-coordinate is 0. The coordinates of vertex L are $(3a, 0)$. $\triangle NJL$ is isosceles, so the x-coordinate of J is located halfway between 0 and $3a$, or $1.5a$. The coordinates of vertex J are $(1.5a, b)$. So, the vertices are $N(0, 0)$, $J(1.5a, b)$, $L(3a, 0)$.

17. $H(2b, 2b\sqrt{3})$, $N(0, 0)$, $D(4b, 0)$

19. Given: Isosceles $\triangle ABC$ with $\overline{AC} \cong \overline{BC}$; R and S are midpoints of legs $\overline{AC}$ and $\overline{BC}$.

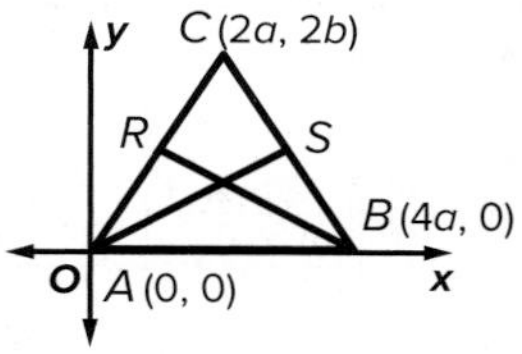

Prove: $\overline{AS} \cong \overline{BR}$

Proof:

The coordinates of S are $\left(\frac{2a + 4a}{2}, \frac{2b + 0}{2}\right)$ or $(3a, b)$.

The coordinates of R are $\left(\frac{2a + 0}{2}, \frac{2b + 0}{2}\right)$ or (a, b).

$AS = \sqrt{(3a - 0)^2 + (b - 0)^2}$ or $\sqrt{9a^2 + b^2}$

$BR = \sqrt{(4a - a)^2 + (0 - b)^2}$ or $\sqrt{9a^2 + b^2}$

Because $AS = BR$, $\overline{AS} \cong \overline{BR}$.

21. Given: Right $\triangle ABC$ with right $\angle BAC$; P is the midpoint of $\overline{BC}$.
Prove: $AP = \frac{1}{2}BC$

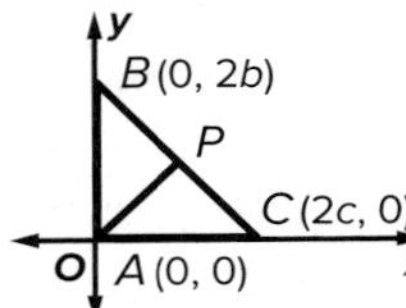

Proof:
Midpoint P is $\left(\frac{0+2c}{2}, \frac{2b+0}{2}\right)$ or (c, b).

$AP = \sqrt{(c-0)^2 + (b-0)^2}$ or $\sqrt{c^2 + b^2}$

$BC = \sqrt{(2c-0)^2 + (0-2b)^2} = \sqrt{4c^2 + 4b^2}$ or $2\sqrt{c^2 + b^2}$

$\frac{1}{2}BC = \sqrt{c^2 + b^2}$

So, $AP = \frac{1}{2}BC$.

23. The distance between Raleigh and Durham is about 0.32 units, between Raleigh and Chapel Hill is about 0.41 units, and between Durham and Chapel Hill is about 0.15 units. Because none of these distances are the same, the Research Triangle is scalene.

25. slope of $\overline{XY} = 1$, slope of $\overline{YZ} = -1$, slope of $\overline{ZX} = 0$; because $1\,(-1) = -1$, $\overline{XY} \perp \overline{YZ}$. Therefore, $\triangle XYZ$ is a right triangle. **27a.** Club B should set up camp at (12, 9). **27b.** I assumed that the ranger's station was located at the origin and that Club B set up camp east of Club A and the ranger's station. To determine where Club B should set up camp we need to determine the slopes of the lines connecting their camp to Club A's camp and their camp to the ranger's station.

Slope between Club A's tent and Club B's tent = $\frac{25-9}{0-x} = -\frac{16}{x}$.

Slope between Club B's tent and ranger's station = $\frac{9-0}{x-0} = \frac{9}{x}$.

For the legs of the triangle to form a right angle the slopes must multiply to give -1.

$-1 = -\frac{16}{x} \cdot \frac{9}{x}$

$-1 = -\frac{144}{x^2}$

$x^2 = 144$

$x = \pm 12$

Because I assumed that Club B camped east of Club A, their coordinates are (12, 9).

29 **a.**

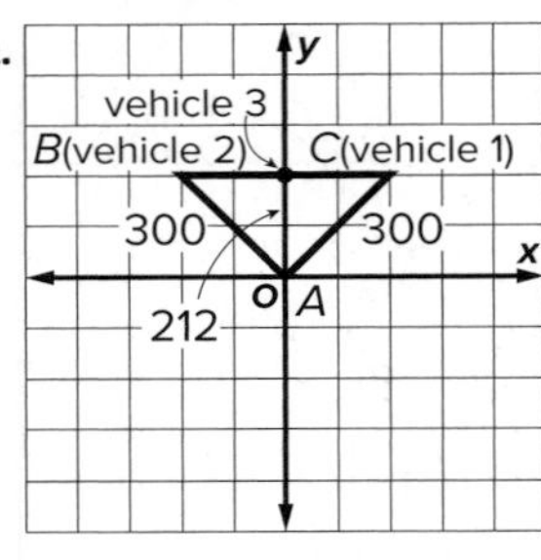

The equation of the line along which the first vehicle lies is $y = x$. The slope is 1 because the vehicle travels the same number of units north as it does east of the origin and the y-intercept is 0. The equation of the line along which the second vehicle lies is $y = -x$. The slope is -1 because the vehicle travels the same number of units north as it does west of the origin and the y-intercept is 0.

b. The paths taken by both the first and second vehicles are 300 yards long. Therefore, the paths are congruent. If two sides of a triangle are congruent, then the triangle is isosceles. You can also write a coordinate proof to prove the triangle formed is isosceles.
Given: $\triangle ABC$
Prove: $\triangle ABC$ is an isosceles right triangle.
Proof: By the Distance Formula,

$AB = \sqrt{(-a-0)^2 + (a-0)^2}$ or $\sqrt{2a^2}$ and

$AC = \sqrt{(a-0)^2 + (a-0)^2}$ or $\sqrt{2a^2}$. So, $AB = AC$

and $\overline{AB} \cong \overline{AC}$. The triangle is isosceles. By the Slope Formula, the slope of $\overline{AB}$ is $\frac{a}{-a}$ or -1 and the slope of $\overline{AC}$ is $\frac{a}{a}$ or 1. Because the slopes are negative reciprocals, the sides of the triangle are perpendicular and therefore form a right angle. So, $\triangle ABC$ is an isosceles right triangle.

c. The paths taken by the first two vehicles form the hypotenuse of isosceles right triangles.

$a^2 + b^2 = c^2$	Pythagorean Theorem
$a^2 + a^2 = 300^2$	$b = a$
$2a^2 = 90{,}000$	Simplify.
$a^2 = 45{,}000$	Divide each side by 2.
$a = 150\sqrt{2}$	Take the positive square root of each side.

First vehicle: $(a, a) = (150\sqrt{2}, 150\sqrt{2})$;
second vehicle: $(-a, a) = (-150\sqrt{2}, 150\sqrt{2})$
The third vehicle travels due north and therefore remains on the y-axis; third vehicle: (0, 212)

d. The y-coordinates of the first two vehicles are $150\sqrt{2} \approx 212.13$, while the y-coordinate of the third vehicle is 212. Because all three vehicles have approximately the same y-coordinate, they are approximately collinear. The midpoint between the first and second vehicles is

$\left(\frac{150\sqrt{2} + (-150\sqrt{2})}{2}, \frac{212+212}{2}\right)$

or approximately (0, 212.13). This is the approximate location of the third vehicle.

31. Sample answer: $(a, 0)$ **33.** Sample answer: $(4a, 0)$

35. Given: $\triangle ABC$ with coordinates $A(0, 0)$, $B(a, b)$, and $C(c, d)$ and $\triangle DEF$ with coordinates $D(0+n, 0+m)$, $E(a+n, b+m)$, and $F(c+n, d+m)$
Prove: $\triangle DEF \cong \triangle ABC$
Proof:

$AB = \sqrt{(a-0)^2 + (b-0)^2}$ or $\sqrt{a^2 + b^2}$

$DE = \sqrt{[a+n-(0+n)]^2 + [b+m-(0+m)]^2}$ or $\sqrt{a^2 + b^2}$

Because $AB = DE$, $\overline{AB} \cong \overline{DE}$.

$BC = \sqrt{(c - a)^2 + (d - b)^2}$ or

$\sqrt{c^2 - 2ac + a^2 + d^2 - 2bd + b^2}$

$EF = \sqrt{[c + n - (a + n)]^2 + [d + m - (b + m)]^2}$ or

$\sqrt{c^2 - 2ac + a^2 + d^2 - 2bd + b^2}$

Because $BC = EF$, $\overline{BC} \cong \overline{EF}$.

$CA = \sqrt{(c - 0)^2 + (d - 0)^2}$ or $\sqrt{c^2 + d^2}$

$FD = \sqrt{[0 + n - (c + n)]^2 + [0 + m - (d + m)]^2}$ or $\sqrt{c^2 + d^2}$

Because $CA = FD$, $\overline{CA} \cong \overline{FD}$.

Therefore, $\triangle DEF \cong \triangle ABC$ by the SSS Postulate.

37a. Using the origin as a vertex of the triangle makes calculations easier because the coordinates are (0, 0). **37b.** Placing at least one side of the triangle on the *x*- or *y*-axis makes it easier to calculate the length of the side because one of the coordinates will be 0. **37c.** Keeping a triangle within the first quadrant makes all of the coordinates positive, and makes the calculations easier. **39.** B **41.** B

Chapter 4 Study Guide & Review

1. false; base **3.** true **5.** false; coordinate proof **7.** false, HA Theorem **9.** Yes, the SAS Postulate requires the same information as the LL Theorem. **11.** The HA Theorem involves the right angle as well as one acute angle, and the hypotenuse is a side, so it uses an angle, an angle, and a side just like the AAS Theorem. **13.** 110° **15.** 104 **17.** rotation and translation; $\angle X \cong \angle J$, $\angle Y \cong \angle K$, $\angle Z \cong \angle L$, $\overline{XY} \cong \overline{JK}$, $\overline{YZ} \cong \overline{KL}$, $\overline{XZ} \cong \overline{JL}$; $\triangle XYZ \cong \triangle JKL$ **19.** Yes, by SSS. $AB = XY$, $BC = YZ = \sqrt{26}$, $ZX = CA = \sqrt{29}$

21. SAS; reflection

23. Given: $\overline{AB} \parallel \overline{DC}$, $\overline{AB} \cong \overline{DC}$

Prove: $\triangle ABE \cong \triangle CDE$

Proof:

Statements (Reasons)

1. $\overline{AB} \parallel \overline{DC}$ (Given)
2. $\angle A \cong \angle DCE$ (Alt. Int. ∠s Thm.)
3. $\overline{AB} \parallel \overline{DC}$ (Given)
4. $\angle ABE \cong \angle D$ (Alt. Int. ∠s Thm.)
5. $\triangle ABE \cong \triangle CDE$ (ASA)

25. AAS, ASA, HL, HA; rotation **27.** 77.5

29.

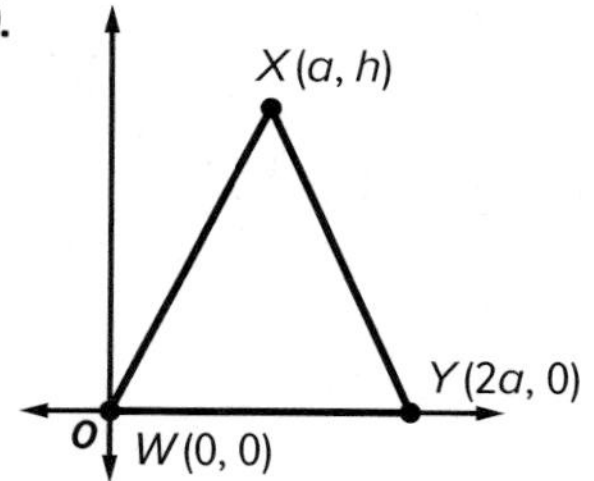

Selected Answers and Solutions

CHAPTER 5

Relationships in Triangles

Chapter 5 Concept Check

1. Sample answer: $m\angle B = m\angle C = m\angle A = 60$
3. isosceles or equilateral **5.** The sides are congruent.
7. $JK = KL$

Lesson 5-1

1. 12 **3.** 15 **5.** 8 **7.** 12

9 $\overrightarrow{MP}$ is the perpendicular bisector of $\overline{LN}$.

$LP = NP$	Perpendicular Bisector Theorem
$2x - 4 = x + 5$	Substitution
$x - 4 = 5$	Subtract x from each side.
$x = 9$	Add 4 to each side.

$NP = 9 + 5$ or 14

11. 6 **13.** 4 **15.** The fair planners could use a map and make each of the locations a vertex of the triangle. Then they could draw the three perpendicular bisectors of the triangle. The point where the perpendicular bisectors meet is the circumcenter, which is equidistant from all three locations. This is where the portable restrooms should be placed.

17. $\overline{CD}$, $\overline{BD}$ **19.** $\overline{BH}$ **21.** 11

23 Because $\overline{QM} \perp \overrightarrow{NM}$, $\overline{QP} \perp \overrightarrow{NP}$, and $QM = QP$, Q is equidistant from the sides of $\angle PNM$. By the Converse of the Angle Bisector Theorem, $\overrightarrow{NQ}$ bisects $\angle PNM$.

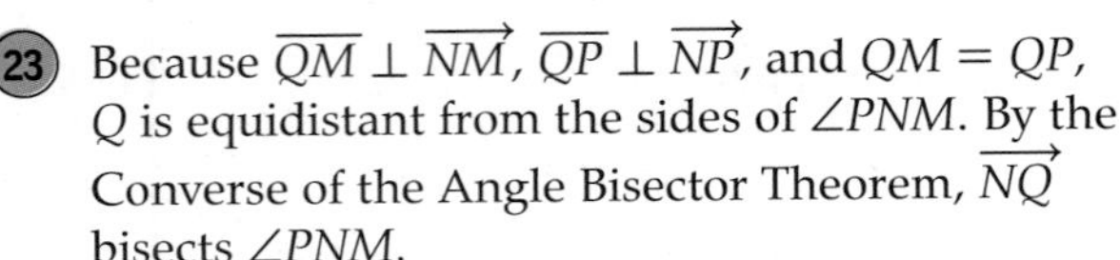

$\angle PNQ \cong \angle QNM$	Definition of angle bisector
$m\angle PNQ = m\angle QNM$	Definition of congruent angles
$4x - 8 = 3x + 5$	Substitution
$x - 8 = 5$	Subtract $3x$ from each side.
$x = 13$	Add 8 to each side.

$m\angle PNM = m\angle PNQ + m\angle QNM$	Angle Addition Postulate
$= (4x - 8) + (3x + 5)$	Substitution
$= 7x - 3$	Simplify.
$= 7(13) - 3$ or 88	$x = 13$

25. 42 **27.** 7.1 **29.** 33

31 Sketch the table and draw the three angle bisectors of the triangle.

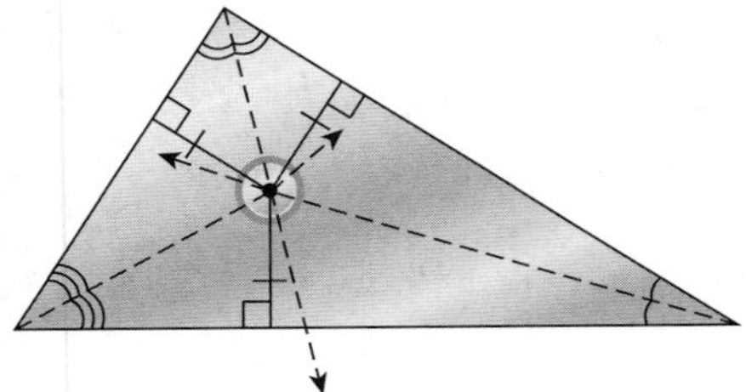

Find the point of concurrency of the angle bisectors of the triangle, the incenter. This point is equidistant from each side of the triangle. So, the centerpiece should be placed at the incenter.

33. No; we need to know whether the perpendicular segments are congruent to each other. **35.** No; we need to know whether the hypotenuses of the triangles are congruent.

37. Given: $\overline{CA} \cong \overline{CB}$, $\overline{AD} \cong \overline{BD}$
Prove: C and D are on the perpendicular bisector of $\overline{AB}$.
Proof:

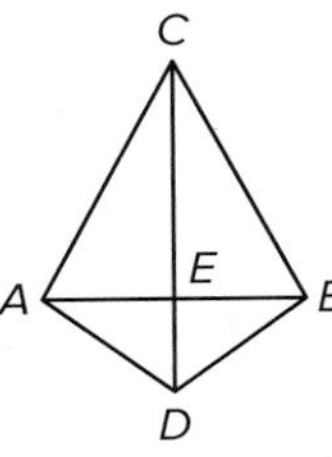

Statements (Reasons)
1. $\overline{CA} \cong \overline{CB}$, $\overline{AD} \cong \overline{BD}$ (Given)
2. $\overline{CD} \cong \overline{CD}$ (Congruence of segments is reflexive.)
3. $\triangle ACD \cong \triangle BCD$ (SSS)
4. $\angle ACD \cong \angle BCD$ (CPCTC)
5. $\overline{CE} \cong \overline{CE}$ (Congruence of segments is reflexive.)
6. $\triangle CEA \cong \triangle CEB$ (SAS)
7. $\overline{AE} \cong \overline{BE}$ (CPCTC)
8. E is the midpoint of $\overline{AB}$. (Def. of midpoint)
9. $\angle CEA \cong \angle CEB$ (CPCTC)
10. $\angle CEA$ and $\angle CEB$ form a linear pair. (Def. of linear pair)
11. $\angle CEA$ and $\angle CEB$ are supplementary. (Supplement Theorem)
12. $m\angle CEA + m\angle CEB = 180$ (Def. of supplementary)
13. $m\angle CEA + m\angle CEA = 180$ (Substitution Prop.)
14. $2m\angle CEA = 180$ (Substitution Prop.)
15. $m\angle CEA = 90$ (Division Prop.)
16. $\angle CEA$ and $\angle CEB$ are rt. $\angle$s. (Def. of rt. $\angle$)
17. $\overline{CD} \perp \overline{AB}$ (Def. of $\perp$)
18. $\overline{CD}$ is the perpendicular bisector of $\overline{AB}$. (Def. of $\perp$ bisector)
19. C and D are on the perpendicular bisector of $\overline{AB}$. (Def. of point on a line)

39. Given: $\overline{CD}$ is the $\perp$ bisector of $\overline{AB}$. E is a point on $\overline{CD}$.
Prove: $EA = EB$

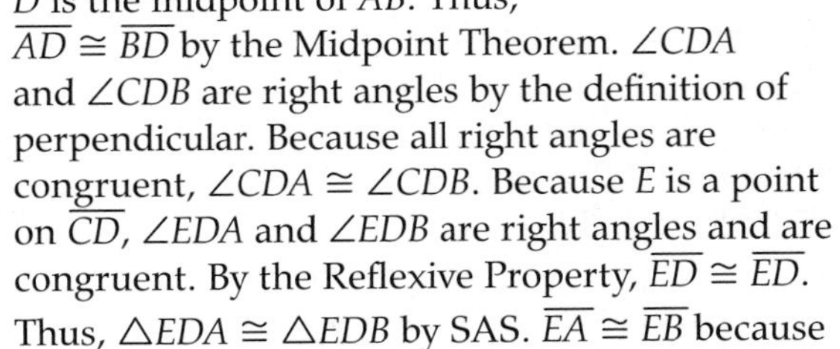

Proof: $\overline{CD}$ is the $\perp$ bisector of $\overline{AB}$. By definition of $\perp$ bisector, D is the midpoint of $\overline{AB}$. Thus, $\overline{AD} \cong \overline{BD}$ by the Midpoint Theorem. $\angle CDA$ and $\angle CDB$ are right angles by the definition of perpendicular. Because all right angles are congruent, $\angle CDA \cong \angle CDB$. Because E is a point on $\overline{CD}$, $\angle EDA$ and $\angle EDB$ are right angles and are congruent. By the Reflexive Property, $\overline{ED} \cong \overline{ED}$. Thus, $\triangle EDA \cong \triangle EDB$ by SAS. $\overline{EA} \cong \overline{EB}$ because CPCTC, and by definition of congruence, $EA = EB$.

41. $y = -\frac{7}{2}x + \frac{15}{4}$; The perpendicular bisector bisects the segment at the midpoint of the segment. The midpoint is $\left(\frac{1}{2}, 2\right)$. The slope of the given segment is $\frac{2}{7}$, so the slope of the perpendicular bisector is $-\frac{7}{2}$.

43. Given: $\overline{PX}$ bisects $\angle QPR$. $\overline{XY} \perp \overline{PQ}$ and $\overline{XZ} \perp \overline{PR}$

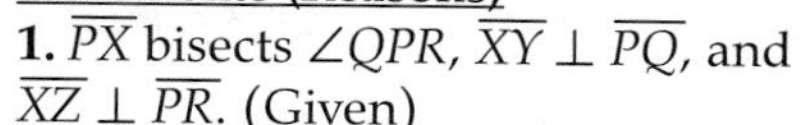

Prove: $\overline{XY} \cong \overline{XZ}$

Proof:

Statements (Reasons)

1. $\overline{PX}$ bisects $\angle QPR$, $\overline{XY} \perp \overline{PQ}$, and $\overline{XZ} \perp \overline{PR}$. (Given)
2. $\angle YPX \cong \angle ZPX$ (Definition of angle bisector)
3. $\angle PYX$ and $\angle PZX$ are right angles. (Definition of perpendicular)
4. $\angle PYX \cong \angle PZX$ (Right angles are congruent.)
5. $\overline{PX} \cong \overline{PX}$ (Reflexive Property)
6. $\triangle PYX \cong \triangle PZX$ (AAS)
7. $\overline{XY} \cong \overline{XZ}$ (CPCTC)

45 The circumcenter is the point where the perpendicular bisectors of a triangle intersect. You can find the circumcenter by locating the point of intersection of two of the perpendicular bisectors. The equation of the perpendicular bisector of $\overline{AB}$ is $y = 3$. The equation of the perpendicular bisector of $\overline{AC}$ is $x = 5$. These lines intersect at (5, 3). The circumcenter is located at (5, 3).

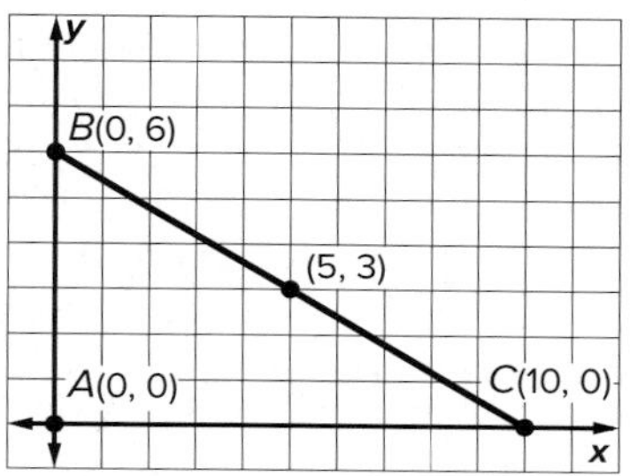

47. a plane perpendicular to the plane in which $\overline{CD}$ lies and bisecting $\overline{CD}$

49. Sample answer:

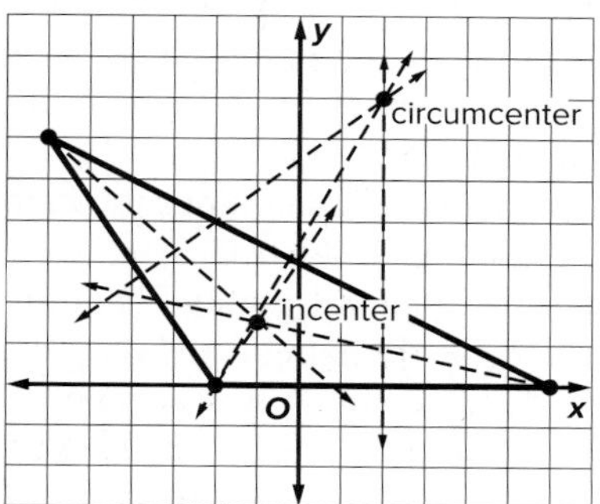

51. always

Given: $\triangle ABC$ is isosceles with legs $\overline{AB}$ and $\overline{BC}$; $\overline{BD}$ is the $\perp$ bisector of $\overline{AC}$.

Prove: $\overline{BD}$ is the angle bisector of $\angle ABC$.

Proof:

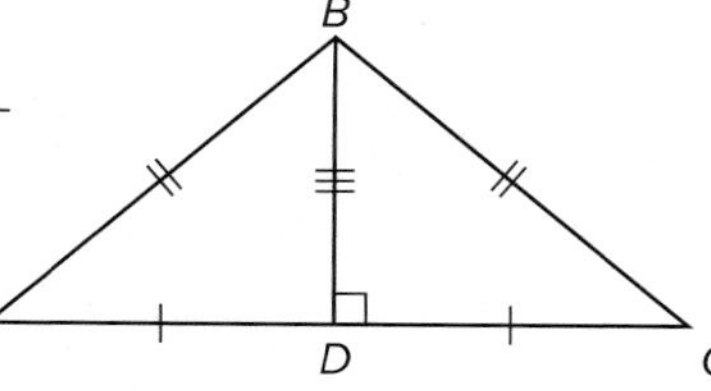

Statements (Reasons)

1. $\triangle ABC$ is isosceles with legs $\overline{AB}$ and $\overline{BC}$. (Given)
2. $\overline{AB} \cong \overline{BC}$ (Def. of isosceles $\triangle$)
3. $\overline{BD}$ is the $\perp$ bisector of $\overline{AC}$. (Given)
4. D is the midpoint of $\overline{AC}$. (Def. of segment bisector)
5. $\overline{AD} \cong \overline{DC}$ (Def. of midpoint)
6. $\overline{BD} \cong \overline{BD}$ (Reflexive Property)
7. $\triangle ABD \cong \triangle CBD$ (SSS)
8. $\angle ABD \cong \angle CBD$ (CPCTC)
9. $\overline{BD}$ is the angle bisector of $\angle ABC$. (Def. of $\angle$ bisector)

53. Given: Plane Z is an angle bisector of $\angle KJH$. $\overline{KJ} \cong \overline{HJ}$

Prove: $\overline{MH} \cong \overline{MK}$

Proof:

Statements (Reasons)

1. Plane Z is an angle bisector of $\angle KJH$; $\overline{KJ} \cong \overline{HJ}$ (Given)
2. $\angle KJM \cong \angle HJM$ (Definition of angle bisector)
3. $\overline{JM} \cong \overline{JM}$ (Reflexive Property)
4. $\triangle KJM \cong \triangle HJM$ (SAS)
5. $\overline{MH} \cong \overline{MK}$ (CPCTC)

55. C **57.** D **59a.** 0.9 **59b.** $1.8 - 0.9 = 0.9$, so $MQ = NQ$. This means $\overline{MQ} \cong \overline{NQ}$. $\angle MQP \cong \angle NQP$ because they are both right angles. $\overline{PQ}$ is congruent to itself. This gives side-angle-side congruence via SAS Theorem. **59c.** C

Lesson 5-2

1

$PC = \frac{2}{3}FC$	Centroid Theorem
$PC = \frac{2}{3}(PF + PC)$	Segment Addition and Substitution
$PC = \frac{2}{3}(6 + PC)$	$PF = 6$
$PC = 4 + \frac{2}{3}PC$	Distributive Property
$\frac{1}{3}PC = 4$	Subtract $\frac{2}{3}PC$ from each side.
$PC = 12$	Multiply each side by 3.

3. (5, 6) **5.** 4.5 **7.** 13.5 **9.** 6 **11.** (3, 6)

13 The centroid is the point of balance for a triangle. Use the Midpoint Theorem to find the midpoint M of the side with endpoints at (0, 8) and (6, 4). The centroid is two-thirds the distance from the opposite vertex to that midpoint.

$$M\left(\frac{0+6}{2}, \frac{8+4}{2}\right) = M(3, 6)$$

The distance from $M(3, 6)$ to the point at (3, 0) is $6 - 0$ or 6 units. If P is the centroid of the triangle, then $P = \frac{2}{3}(6)$ or 4 units up from the point at (3, 0). The coordinates of P are $(3, 0 + 4)$ or (3, 4).

15. $(-4, -4)$ **17.** median **19.** median **21.** 3 **23.** $\frac{1}{2}$

25

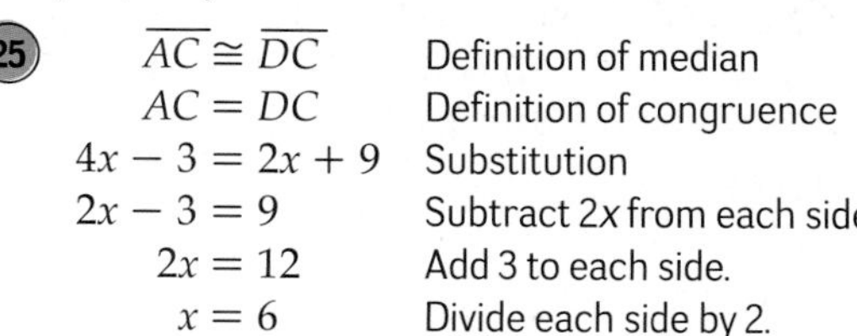

$\overline{AC} \cong \overline{DC}$	Definition of median
$AC = DC$	Definition of congruence
$4x - 3 = 2x + 9$	Substitution
$2x - 3 = 9$	Subtract $2x$ from each side.
$2x = 12$	Add 3 to each side.
$x = 6$	Divide each side by 2.

$m\angle ECA = 15x + 2$	Given
$= 15(6) + 2$	$x = 6$
$= 90 + 2$ or 92	Simplify.

$\overline{EC}$ is not an altitude of $\triangle AED$ because $m\angle ECA = 92$. If $\overline{EC}$ were an altitude, then $m\angle ECA$ must be 90.

Selected Answers and Solutions

27. altitude **29.** median

31. Given: $\triangle XYZ$ is isosceles. $\overline{WY}$ bisects $\angle Y$.

Prove: $\overline{WY}$ is a median.

Proof: Because $\triangle XYZ$ is isosceles, $\overline{XY} \cong \overline{ZY}$. By the definition of angle bisector, $\angle XYW \cong \angle ZYW$. $\overline{YW} \cong \overline{YW}$ by the Reflexive Property. So, by SAS, $\triangle XYW \cong \triangle ZYW$. By CPCTC, $\overline{XW} \cong \overline{ZW}$. By the definition of a midpoint, W is the midpoint of $\overline{XZ}$. By the definition of a median, $\overline{WY}$ is a median.

33a.

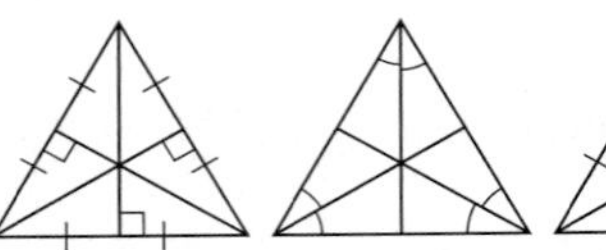

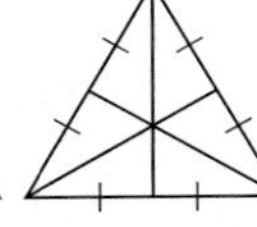

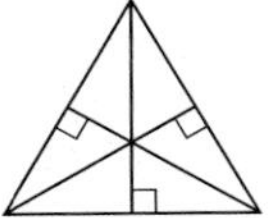

33b. Sample answer: The four points of concurrency of an equilateral triangle are all the same point. **33c.**

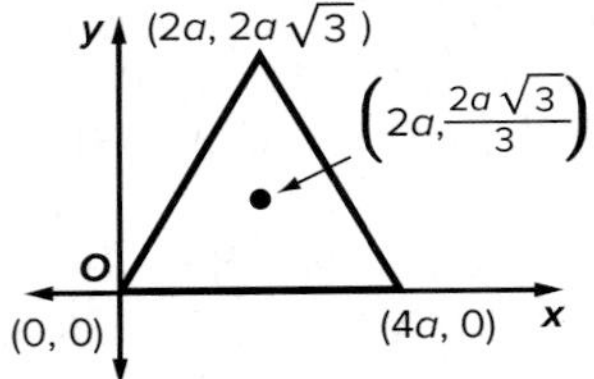

35. 7

37. Sample answer: Kareem is correct. According to the Centroid Theorem, $AP = \frac{2}{3}AD$. The segment lengths are transposed.

39. $\left(1, \frac{5}{3}\right)$; Sample answer: I found the midpoint of $\overline{AC}$ and used it to find the equation for the line that contains point B and the midpoint of $\overline{AC}$, $y = \frac{10}{3}x - \frac{5}{3}$. I also found the midpoint of $\overline{BC}$ and the equation for the line between point A and the midpoint of $\overline{BC}$, $y = -\frac{1}{3}x + 2$. I solved the system of two equations for x and y to get the coordinates of the centroid, $\left(1, \frac{5}{3}\right)$.

41. $2\sqrt{13}$ **43.** Sample answer: Each median divides the triangle into two smaller triangles of equal area, so the triangle can be balanced along any one of those lines. To balance the triangle on one point, you need to find the point where these three balance lines intersect. The balancing point for a rectangle is the intersection of the segments connecting the midpoints of the opposite sides, because each segment connecting these midpoints of a pair of opposite sides divides the rectangle into two parts with equal area. **45.** No; the point at which the medians intersect is always in the interior of the triangle. **47.** C **49.** B **51a.** $(0, \sqrt{3})$ **51b.** $(0, \sqrt{3})$ **51c.** $(0, \sqrt{3})$ **51d.** $(0, \sqrt{3})$ **51e.** The coordinates are the same, because the triangle is an equilateral triangle.

Lesson 5-3

1. $\angle 1$, $\angle 2$ **3.** $\angle 4$, $\angle 6$, $\angle 9$ **5.** $\angle A$, $\angle C$, $\angle B$; $\overline{BC}$, $\overline{AB}$, $\overline{AC}$ **7.** $\overline{BC}$; Sample answer: Because the angle across from segment $\overline{BC}$ is larger than the angle across from $\overline{AC}$, $\overline{BC}$ is longer. **9.** $\angle 1$, $\angle 2$ **11.** $\angle 1$, $\angle 3$, $\angle 6$, $\angle 7$ **13.** $\angle 5$, $\angle 9$

15 The sides from shortest to longest are $\overline{RT}$, $\overline{RS}$, $\overline{ST}$. The angles opposite these sides are $\angle S$, $\angle T$, and $\angle R$, respectively. So the angles from smallest to largest are $\angle S$, $\angle T$, and $\angle R$.

17. $\angle L$, $\angle P$, $\angle M$; $\overline{PM}$, $\overline{ML}$, $\overline{PL}$ **19.** $\angle C$, $\angle D$, $\angle E$; $\overline{DE}$, $\overline{CE}$, $\overline{CD}$

21 To use Theorem 5.9, first show that the measure of the angle opposite $\overline{YZ}$ is greater than the measure of the angle opposite $\overline{XZ}$. If $m\angle X = 90$, then $m\angle Y + m\angle Z = 90$, so $m\angle Y < 90$ by the definition of inequality. So $m\angle X > m\angle Y$. According to Theorem 5.9, if $m\angle X > m\angle Y$, then the length of the side opposite $\angle X$ must be greater than the length of the side opposite $\angle Y$. Because $\overline{YZ}$ is opposite $\angle X$, and $\overline{XZ}$ is opposite $\angle Y$, $YZ > XZ$. So YZ, the length of the top surface of the ramp, must be greater than the length of the ramp.

23. $\angle P$, $\angle Q$, $\angle M$; $\overline{MQ}$, $\overline{PM}$, $\overline{PQ}$ **25.** $\angle 2$ **27.** $\angle 3$ **29.** $\angle 8$ **31.** $m\angle BCF > m\angle CFB$ **33.** $m\angle DBF < m\angle BFD$ **35.** $RP > MP$ **37.** $RM > RQ$

39 Use the Distance Formula $d = \sqrt{(x_2 - x_1)^2 + (y_2 - y_1)^2}$ to find the lengths of the sides.

$AB = \sqrt{[-2 - (-4)]^2 + (1 - 6)^2}$	$x_1 = -4, x_2 = -2,$ $y_1 = 6, y_2 = 1$
$= \sqrt{29}$	Simplify.
≈ 5.4	Use a calculator.
$BC = \sqrt{[5 - (-2)]^2 + (6 - 1)^2}$	$x_1 = -2, x_2 = 5,$ $y_1 = 1, y_2 = 6$
$= \sqrt{74}$	Simplify.
≈ 8.6	Use a calculator.
$AC = \sqrt{[5 - (-4)]^2 + (6 - 6)^2}$	$x_1 = -4, x_2 = 5,$ $y_1 = 6, y_2 = 6$
$= \sqrt{81}$ or 9	Simplify.

Because $AB < BC < AC$, $\angle C < \angle A < \angle B$. The angles in order from smallest to largest are $\angle C$, $\angle A$, $\angle B$.

41. AB, BC, AC, CD, BD; In $\triangle ABC$, $AB < BC < AC$ and in $\triangle BCD$, $BC < CD < BD$. By the figure $AC < CD$, so $BC < AC < CD$. **43.** Sample answer: $\angle R$ is an exterior angle to $\triangle PQR$, so by the Exterior Angle Inequality, $m\angle R$ must be greater than $m\angle Q$. The markings indicate that $\angle R \cong \angle Q$, indicating that $m\angle R = m\angle Q$. This is a contradiction of the Exterior Angle Inequality Theorem, so the markings are incorrect. **45.** Sample answer: 10; $m\angle C > m\angle B$, so if $AB > AC$, Theorem 5.10 is satisfied. Because $10 > 6$, $AB > AC$. **47.** $m\angle 1$, $m\angle 2 = m\angle 5$, $m\angle 4$, $m\angle 6$, $m\angle 3$; Sample answer: The side opposite $\angle 5$ is the smallest side in that triangle and $m\angle 2 = m\angle 5$, so we know that $m\angle 4$ and $m\angle 6$ are both greater than $m\angle 2$ and $m\angle 5$. The side opposite $\angle 6$ is greater than the side opposite $\angle 4$. Because the side opposite $\angle 2$ is greater than the side opposite $\angle 1$, we know that $m\angle 1 < m\angle 2$ and $m\angle 5$. Because $m\angle 2 = m\angle 5$, $m\angle 1 + m\angle 3 = m\angle 4 + m\angle 6$. Because $m\angle 1 < m\angle 4$, then $m\angle 3 > m\angle 6$. **49.** B **51.** B **53.** D **55a.** $\angle 2$, $\angle 3$, $\angle 4$, $\angle 5$ **55b.** $\angle 7$ **55c.** According to the Exterior Angle Theorem, $m\angle 8 = m\angle 2 + m\angle 3 + m\angle 4 + m\angle 5$. If $\angle 2 \cong \angle 3 \cong \angle 4 \cong m\angle 5$, then the measure of each angle is $116 \div 4 = 29°$. The sum of the angle measures of a triangle equals 180°, so $m\angle 6 = 180 - 29 - 29 = 122°$.

1. $\overline{AB} \not\cong \overline{CD}$

3 The conclusion of the conditional statement is $x < 6$. If $x < 6$ is false, then x must be greater than or equal to 6. The negation of the conclusion is $x \geq 6$.

5. **Given:** $2x + 3 < 7$

Prove: $x < 2$

Indirect Proof: Step 1

Assume that $x > 2$ or $x = 2$ is true.

Step 2

x	2	3	4	5	6
2x+3	7	9	11	13	15

When $x > 2$, $2x + 3 > 7$ and when $x = 2$, $2x + 3 = 7$.

Step 3 In both cases, the assumption leads to the contradiction of the given information that $2x + 3 < 7$. Therefore, the assumption that $x \geq 2$ must be false, so the original conclusion that $x < 2$ must be true.

7. Use $a = \text{average}$ or $\dfrac{\text{number of points scored}}{\text{number of games played}}$.

Proof:

Indirect Proof: Step 1 Assume that Christina's average points per game was greater than or equal to 3, $a \geq 3$.

Step 2 CASE 1 | CASE 2

CASE 1: $a = 3$; $3 \stackrel{?}{=} \dfrac{13}{6}$; $3 \neq 2.2$

CASE 2: $a > 3$; $\dfrac{13}{6} \stackrel{?}{>} 3$; $2.2 \not> 3$

Step 3 The conclusions are false, so the assumption must be false. Therefore, Christina's average points per game was less than 3.

9. **Given:** $\triangle ABC$ is a right triangle; $\angle C$ is a right angle.

Prove: $AB > BC$ and $AB > AC$

Indirect Proof: Step 1 Assume that the hypotenuse of a right triangle is not the longest side. That is, $AB < BC$ and $AB < AC$.

Step 2 If $AB < BC$, then $m\angle C < m\angle A$. Because $m\angle C = 90$, $m\angle A > 90$. So, $m\angle C + m\angle A > 180$. By the same reasoning, $m\angle C + m\angle B > 180$.

Step 3 Both relationships contradict the fact that the sum of the measures of the angles of a triangle equals 180. Therefore, the hypotenuse must be the longest side of a right triangle.

11. $x \leq 8$ 13. The lines are not parallel. 15. The triangle is equiangular.

17 To write an indirect proof, first identify the conclusion. Find the negation of the conclusion and assume that it is true. Make a table of values to show that the negation of the conclusion is false. Because the assumption leads to a contradiction, you can conclude that the original conclusion must be true.

Given: $2x - 7 > -11$

Prove: $x > -2$

Indirect Proof: Step 1 The negation of $x > -2$ is $x \leq -2$. So, assume that $x \leq -2$ is true.

Step 2 Make a table with several possibilities for x assuming $x < -2$ or $x = -2$.

x	−6	−5	−4	−3	−2
2x−7	−19	−17	−15	−13	−11

When $x < -2$, $2x - 7 < -11$ and when $x = -2$, $2x - 7 = -11$.

Step 3 In both cases, the assumption leads to the contradiction of the given information that $2x - 7 > -11$. Therefore, the assumption that $x \leq -2$ must be false, so the original conclusion that $x > -2$ must be true.

19. **Given:** $-3x + 4 < 7$

Prove: $x > -1$

Indirect Proof: Step 1 Assume that $x \leq -1$ is true.

Step 2 When $x < -1$, $-3x + 4 > 7$ and when $x = -1$, $-3x + 4 = 7$.

x	−5	−4	−3	−2	−1
−3x+4	19	16	13	10	7

Step 3 In both cases, the assumption leads to the contradiction of the given information that $-3x + 4 < 7$. Therefore, the assumption that $x \leq -1$ must be false, so the original conclusion that $x > -1$ must be true.

21. Let the cost of one game be x and the other be y.

Given: $x + y > 80$

Prove: $x > 40$ or $y > 40$

Indirect Proof: Step 1 Assume that $x \leq 40$ and $y \leq 40$.

Step 2 If $x \leq 40$ and $y \leq 40$, then $x + y \leq 40 + 40$ or $x + y \leq 80$. This is a contradiction because we know that $x + y > 80$.

Step 3 Because the assumption that $x \leq 40$ and $y \leq 40$ leads to a contradiction of a known fact, the assumption must be false. Therefore, the conclusion that $x > 40$ or $y > 40$ must be true. Thus, at least one of the games had to cost more than $40.

23. **Given:** xy is an odd integer.

Prove: x and y are odd integers.

Indirect Proof: Step 1 Assume that x and y are not both odd integers. That is, assume that either x or y is an even integer.

Step 2 You only need to show that the assumption that x is an even integer leads to a contradiction, because the argument for y is an even integer follows the same reasoning. So, assume that x is an even integer and y is an odd integer. This means that $x = 2k$ for some integer k and $y = 2m + 1$ for some integer m.

$xy = (2k)(2m + 1)$ Subst. of assumption

$= 4km + 2k$ Dist. Prop.

$= 2(km + k)$ Dist. Prop.

Because k and m are integers, $km + k$ is also an integer. Let p represent the integer $km + k$. So xy can be represented by $2p$, where p is an integer.

This means that xy is an even integer, but this contradicts the given that xy is an odd integer.
Step 3 Because the assumption that x is an even integer and y is an odd integer leads to a contradiction of the given, the original conclusion that x and y are both odd integers must be true.

25. Given: x is an odd number.
Prove: x is not divisible by 4.
Indirect Proof: Step 1 Assume x is divisible by 4. In other words, 4 is a factor of x.
Step 2 Let $x = 4n$, for some integer n.
$x = 2(2n)$
So, 2 is a factor of x which means x is an even number, but this contradicts the given information.
Step 3 Because the assumption that x is divisible by 4 leads to a contradiction of the given, the original conclusion x is not divisible by 4 must be true.

27. Given: $XZ > YZ$
Prove: $\angle X \not\cong \angle Y$

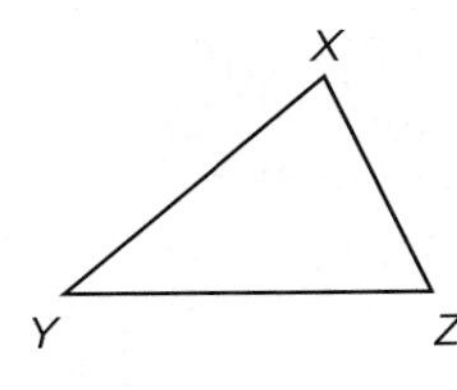

Indirect Proof: Step 1
Assume that $\angle X \cong \angle Y$.
Step 2 $\overline{XZ} \cong \overline{YZ}$ by the converse of the isosceles $\triangle$ theorem.
Step 3 This contradicts the given information that $XZ > YZ$. Therefore, the assumption $\angle X \cong \angle Y$ must be false, so the original conclusion $\angle X \not\cong \angle Y$ must be true.

29. Given: $\triangle ABC$ is isosceles.
Prove: Neither of the base angles is a right angle.
Indirect Proof: Step 1
Assume that $\angle B$ is a right angle.
Step 2 By the Isosceles $\triangle$ Theorem, $\angle C$ is also a right angle.
Step 3 This contradicts the fact that a triangle can have no more than one right angle. Therefore, the assumption that $\angle B$ is a right angle must be false, so the original conclusion neither of the base angles is a right angle must be true.

31. Given: $m\angle A > m\angle ABC$
Prove: $BC > AC$

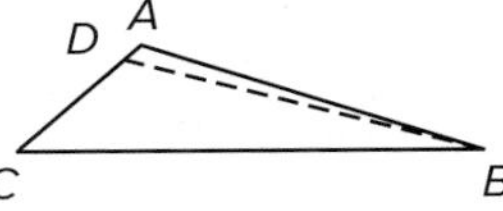

Proof:
Assume $BC \not> AC$. By the Comparison Property, $BC = AC$ or $BC < AC$.
Case 1: If $BC = AC$, then $\angle ABC \cong \angle A$ by the Isosceles Triangle Theorem. (If two sides of a triangle are congruent, then the angles opposite those sides are congruent.) But, $\angle ABC \cong \angle A$ contradicts the given statement that $m\angle A > m\angle ABC$. So, $BC \neq AC$.
Case 2: If $BC < AC$, then there must be a point D between A and C so that $\overline{DC} \cong \overline{BC}$. Draw the auxiliary segment $\overline{BD}$. Because $DC = BC$, by the Isosceles Triangle Theorem $\angle BDC \cong \angle DBC$. Now $\angle BDC$ is an exterior angle of $\triangle BAD$ and by the Exterior Angles Inequality Theorem (the measure of an exterior angle of a triangle is greater than the measure of either corresponding remote interior angle) $m\angle BDC > m\angle A$. By the Angle Addition Postulate, $m\angle ABC = m\angle ABD + m\angle DBC$. Then by the definition of inequality, $m\angle ABC > m\angle DBC$. By Substitution and the Transitive Property of Inequality, $m\angle ABC > m\angle A$. But this contradicts the given statement that $m\angle A > m\angle ABC$. In both cases, a contradiction was found, and hence our assumption must have been false. Therefore, $BC > AC$.

33. We know that the other team scored 3 points and that Katsu thinks that they made a three-point shot. We also know that a player can score 3 points by making a two-point basket and a foul shot.
Step 1 Assume that a player for the other team made a two-point basket and a foul shot.
Step 2 The other team's score before Katsu left was 26, so their score after a two-point basket and a foul shot would be 26 + 3 or 29.
Step 3 The score is correct when we assume that the other team made a two-point basket and a foul shot, so Katsu's assumption may not be correct. The other team could have made a three-point basket or a two-point basket and a foul shot.

35 **a.** To write an indirect proof, first identify the conclusion. Find the negation of the conclusion and assume that it is true. Then use the data to prove that the assumption is false.
Step 1 50% is half, and the statement says more than half of the teens polled said that they recycle, so assume that less than 50% recycle.
Step 2 The data shows that 51% of teens said that they recycle, and 51% > 50%, so the number of teens that recycle is not less than half.
Step 3 This contradicts the data given. Therefore, the assumption is false, and the conclusion that more than half of the teens polled said they recycle must be true.
b. According to the data, 23% of 400 teenagers polled said that they participate in Earth Day. Verify that 23% of 400 is 92.

$400 \cdot 23\% \stackrel{?}{=} 92$
$400 \cdot 0.23 \stackrel{?}{=} 92$
$92 = 92$

 **Given:** $\overline{AB} \perp$ line p

Prove: $\overline{AB}$ is the shortest segment from A to line p.

Indirect Proof: Step 1 Assume $\overline{AB}$ is not the shortest segment from A to p.

Step 2 Because $\overline{AB}$ is not the shortest segment from A to p, there is a point C such that $\overline{AC}$ is the shortest distance. $\triangle ABC$ is a right triangle with hypotenuse $\overline{AC}$, which is the longest side of $\triangle ABC$ because it is across from the largest angle in $\triangle ABC$ by the Angle-Side Relationships in Triangles Theorem.

Step 3 This contradicts the fact that $\overline{AC}$ is the shortest side. Therefore, the assumption is false, and the conclusion that $\overline{AB}$ is the shortest side, must be true.

39a. $n^3 + 3$ **39b.** Sample answer:

n	$n^3 + 3$
2	11
3	30
10	1003
11	1334
24	13,827
25	15,628
100	1,000,003
101	1,030,304
526	145,531,579
527	146,363,186

39c. Sample answer: When $n^3 + 3$ is even, n is odd.

39d. Indirect Proof: Step 1 Assume that n is even. Let $n = 2k$, where k is some integer.

Step 2

$n^3 + 3 = (2k)^3 + 3$	Substitute assumption
$= 8k^3 + 3$	Simplify.
$= (8k^3 + 2) + 1$	Replace 3 with 2 + 1 and group the first two terms.
$= 2(4k^3 + 1) + 1$	Distributive Property

Because k is an integer, $4k^3 + 1$ is also an integer. Therefore, $n^3 + 3$ is odd.

Step 3 This contradicts the given information that $n^3 + 3$ is even. Therefore, the assumption is false, so the conclusion that n is odd must be true.

41. Sample answer: $\triangle ABC$ is scalene.

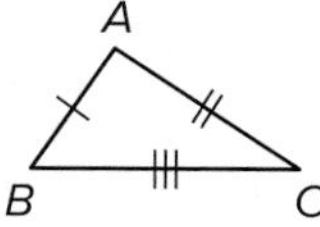

Given: $\triangle ABC$; $AB \neq BC$; $BC \neq AC$; $AB \neq AC$

Prove: $\triangle ABC$ is scalene.

Indirect Proof:

Step 1 Assume that $\triangle ABC$ is not scalene.

Case 1: $\triangle ABC$ is isosceles.

Step 2 If $\triangle ABC$ is isosceles, then $AB = BC$, $BC = AC$, or $AB = AC$.

Step 3 This contradicts the given information, so $\triangle ABC$ is not isosceles.

Case 2: $\triangle ABC$ is equilateral.

In order for a triangle to be equilateral, it must also be isosceles, and Case 1 proved that $\triangle ABC$ is not isosceles. Thus, $\triangle ABC$ is not equilateral. Therefore, $\triangle ABC$ is scalene. **43.** Neither; sample answer: Because the hypothesis is true when the conclusion is false, the statement is false. **45.** C **47.** B **49a.** Assume that $\angle 1$ is not an acute angle. **49b.** Triangle Angle-Sum Theorem **49c.** Sample answer: Assume $\angle 2$ is a right angle. Then $m\angle 2 = 90$. By the Triangle Angle-Sum Theorem, $m\angle 1 + m\angle 2 + 110 = 180$, so $m\angle 1 + 90 + 110 = 180$ or $m\angle 1 + 200 = 180$, and $m\angle 1 < 0$, which contradicts the fact that the measure of an angle is greater than 0, so $\angle 2$ cannot be a right angle.

Lesson 5-5

 Yes; check each inequality.

$5 + 7 \overset{?}{>} 10$ $5 + 10 \overset{?}{>} 7$ $7 + 10 \overset{?}{>} 5$

$12 > 10$ ✓ $15 > 7$ ✓ $17 > 5$ ✓

Because the sum of each pair of side lengths is greater than the third side length, sides with lengths 5 cm, 7 cm, and 10 cm will form a triangle.

3. yes; $6 + 14 > 10$, $6 + 10 > 14$, and $10 + 14 > 6$

5. Given: $\overline{XW} \cong \overline{YW}$

Prove: $YZ + ZW > XW$

Proof:

Statements (Reasons)

1. $\overline{XW} \cong \overline{YW}$ (Given)
2. $XW = YW$ (Def. of $\cong$ segs.)
3. $YZ + ZW > YW$ ($\triangle$ Inequal. Thm.)
4. $YZ + ZW > XW$ (Subst.)

7. yes **9.** no; $2.1 + 4.2 \not> 7.9$ **11.** yes **13.** 6 m $< n <$ 16 m **15.** 5.4 in. $< n <$ 13 in. **17.** $5\frac{1}{3}$ yd $< n <$ 10 yd

19. Given: $\overline{JL} \cong \overline{LM}$

Prove: $KJ + KL > LM$

Proof:

Statements (Reasons)

1. $\overline{JL} \cong \overline{LM}$ (Given)
2. $JL = LM$ (Def. of $\cong$ segments)
3. $KJ + KL > JL$ ($\triangle$ Inequal. Thm.)
4. $KJ + KL > LM$ (Subst.)

21

$$XY + YZ > XZ$$
$$(4x - 1) + (2x + 7) > x + 13$$
$$6x + 6 > x + 13$$
$$5x > 7$$
$$x > \frac{7}{5}$$

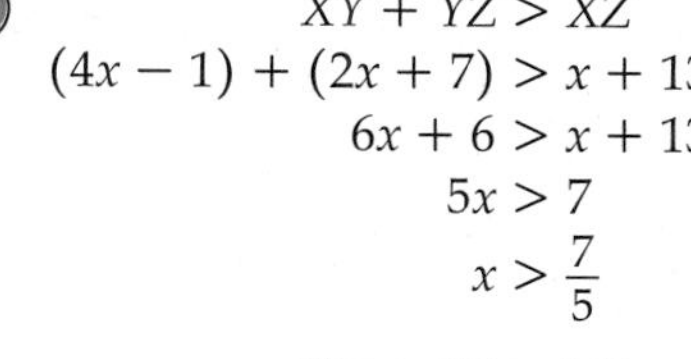

$$XY + XZ > YZ$$
$$(4x - 1) + (x + 13) > 2x + 7$$
$$5x + 12 > 2x + 7$$
$$3x > -5$$
$$x > -\frac{5}{3}$$

$$YZ + XZ > XY$$
$$(2x + 7) + (x + 13) > 4x - 1$$
$$3x + 20 > 4x - 1$$
$$21 > x$$

Because x must be greater than $\frac{7}{5}$, greater than $-\frac{5}{3}$, and less than 21, possible values of x are $\frac{7}{5} < x < 21$.

23. Given: $\triangle ABC$

Prove: $AC + BC > AB$

Proof:

Statements (Reasons)

1. Construct $\overline{CD}$ so that C is between B and D and $\overline{CD} \cong \overline{AC}$. (Ruler Post.)
2. $CD = AC$ (Def. of $\cong$ segments)
3. $\angle CAD \cong \angle ADC$ (Isos. $\triangle$ Thm.)
4. $m\angle CAD = m\angle ADC$ (Def. of $\cong$ $\angle$s)
5. $m\angle BAC + m\angle CAD = m\angle BAD$ ($\angle$ Add. Post.)
6. $m\angle BAC + m\angle ADC = m\angle BAD$ (Subst.)
7. $m\angle ADC < m\angle BAD$ (Def. of inequality)
8. $AB < BD$ (Thoerem 5.10)
9. $BD = BC + CD$ (Seg. Add. Post.)
10. $AB < BC + CD$ (Subst. (Steps 8, 9))
11. $AB < BC + AC$ (Subst. (Steps 2, 10))

25. $2 < x < 10$ **27.** $1 < x < 11$ **29.** $x > 0$ **31.** Yes; sample answer: The measurements on the drawing do not form a triangle. According to the Triangle Inequality Theorem, the sum of the lengths of any two sides of a triangle is greater than the length of the third side. The lengths in the drawing are 1 ft, $3\frac{7}{8}$ ft, and $6\frac{3}{4}$ ft. Because $1 + 3\frac{7}{8} \not> 6\frac{3}{4}$, the triangle is impossible. They should recalculate their measurements before they cut the wood.

33 A triangle 3 feet by 4 feet by x feet is formed. The length of the third side x must be less than the sum of the lengths of the other two sides. So, $x < 3 + 4$ or $x < 7$. Because the awning drapes 6 inches or 0.5 feet over the front, the total length should be less than $7 + 0.5$ or 7.5 feet. She should buy no more than 7.5 feet.

35. Yes; $\sqrt{99} \approx 9.9$ because $\sqrt{100} = 10$, $\sqrt{48} \approx 6.9$ because $\sqrt{49} = 7$, and $\sqrt{65} \approx 8.1$ because $\sqrt{64} = 8$. $6.9 + 8.1 > 9.9$, so it is possible.

37. no; $\sqrt{122} \approx 11.1$ because $\sqrt{121} = 11$, $\sqrt{5} \approx 2.1$ because $\sqrt{4} = 2$, and $\sqrt{26} \approx 5.1$ because $\sqrt{25} = 5$. So, $2.1 + 5.1 \not> 11.1$.

39 Use the Distance Formula $d = \sqrt{(x_2 - x_1)^2 + (y_2 - y_1)^2}$ to find the lengths of the sides.

$FG = \sqrt{[3 - (-4)]^2 + (-3 - 3)^2}$ $\quad x_1 = -4, x_2 = 3, y_1 = 3, y_2 = -3$

$= \sqrt{85}$ $\quad$ Simplify.

≈ 9.2 $\quad$ Use a calculator.

$GH = \sqrt{(4 - 3)^2 + [6 - (-3)]^2}$ $\quad x_1 = 3, x_2 = 4, y_1 = -3, y_2 = 6$

$= \sqrt{82}$ $\quad$ Simplify.

≈ 9.1 $\quad$ Use a calculator.

$FH = \sqrt{[4 - (-4)]^2 + (6 - 3)^2}$ $\quad x_1 = -4, x_2 = 4, y_1 = 3, y_2 = 6$

$= \sqrt{73}$ $\quad$ Simplify.

≈ 8.5 $\quad$ Use a calculator.

$FG + GH \overset{?}{>} FH$	$FG + FH \overset{?}{>} GH$	$GH + FH \overset{?}{>} FG$
$9.2 + 9.1 \overset{?}{>} 8.5$	$9.2 + 8.5 \overset{?}{>} 9.1$	$9.1 + 8.5 \overset{?}{>} 9.2$
$18.3 > 8.5$	$17.7 > 9.1$	$17.6 > 9.2$

Because $FG + GH > FH$, $FG + FH > GH$, and $GH + FH > FG$, the coordinates are the vertices of a triangle.

41. yes; $QR + QS > RS$, $QR + RS > QS$, and $QS + RS > QR$ **43.** The perimeter is greater than 36 and less than 64. Sample answer: From the diagram we know that $\overline{AC} \cong \overline{EC}$ and $\overline{DC} \cong \overline{BC}$, and $\angle ACB \cong \angle ECD$ because vertical angles are congruent, so $\triangle ACB \cong \triangle ECD$. Using the Triangle Inequality Theorem, the minimum value of AB and ED is 2 and the maximum value is 16. Therefore, the minimum value of the perimeter is greater than $2(2 + 7 + 9)$ or 36, and the maximum value of the perimeter is less than $2(16 + 7 + 9)$ or 64.

45. Sample answers: whether or not the side lengths actually form a triangle, what the smallest and largest angles are, whether the triangle is equilateral, isosceles, or scalene

47.

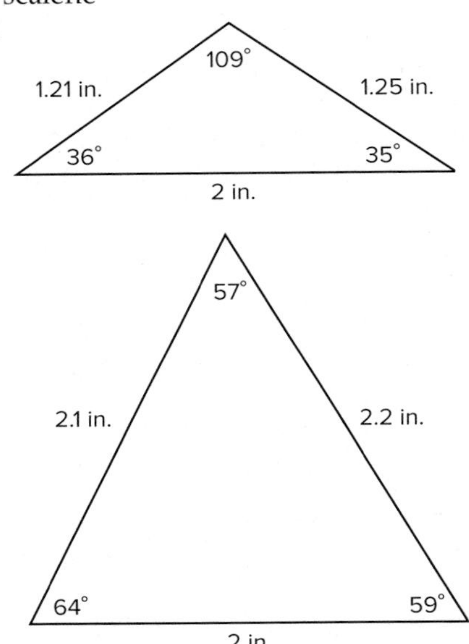

49. D **51.** A **53.** B, C

Lesson 5-6

1. $m\angle ACB > m\angle GDE$

3 $\overline{QR} \cong \overline{SR}$, $\overline{TR} \cong \overline{TR}$, and $m\angle QRT < m\angle SRT$. By the Hinge Theorem, $QT < ST$.

5a. $\overline{AB} \cong \overline{DE}$, $\overline{AC} \cong \overline{DF}$ **5b.** $\angle D$; Sample answer: Because $EF > BC$, according to the converse of the Hinge Theorem, $m\angle D > m\angle A$

7. $\frac{5}{3} < x < B$

9. Given: $\overline{AD} \cong \overline{CB}$, $DC < AB$

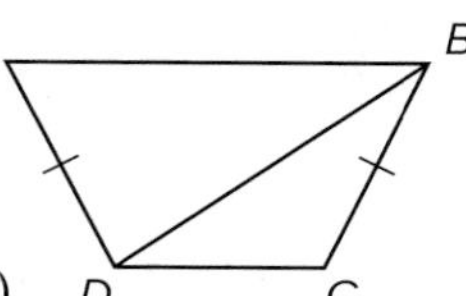

Prove: $m\angle CBD < m\angle ADB$

Statements (Reasons)

1. $\overline{AD} \cong \overline{CB}$ (Given)
2. $\overline{DB} \cong \overline{DB}$ (Reflexive Property)
3. $DC < AB$ (Given)
4. $m\angle CBD < m\angle ADB$ (SSS Inequality)

11. $m\angle MLP < m\angle TSR$

13 $\overline{TU} \cong \overline{VU}$, $\overline{WU} \cong \overline{WU}$, and $WT < WV$. By the Converse of the Hinge Theorem, $m\angle TUW < m\angle VUW$.

15. $JK > HJ$ **17.** $2 < x < 6$

19 Two sides of one triangle are congruent to two sides of the other triangle and the third side of the first triangle is less than the third side of the second triangle. So, by the Converse of the Hinge Theorem, the included angle of the first triangle is less than the included angle of the second triangle.

$0 < x + 20 < 41$ Converse of Hinge Theorem
$-20 < x < 21$ Subtract 20 from each.

The range of values containing x is $-20 < x < 21$.

21. $\overline{RS}$; sample answer: The height of the crane is the same and the length of the crane arm is fixed, so according to the Hinge Theorem, the side opposite the smaller angle is shorter. Because $29° < 52°$, $RS < MN$.

23. Given: $\overline{LK} \cong \overline{JK}$, $\overline{RL} \cong \overline{RJ}$, K is the midpoint of $\overline{QS}$, $m\angle SKL > m\angle QKJ$

Prove: $RS > QR$

Statements (Reasons)

1. $\overline{LK} \cong \overline{JK}$, $\overline{RL} \cong \overline{RJ}$, K is the midpoint of $\overline{QS}$, $m\angle SKL > m\angle QKJ$ (Given)
2. $SK = QK$ (Def. of midpoint)
3. $SL > QJ$ (Hinge Thm.)
4. $RL = RJ$ (Def. of $\cong$ segs.)
5. $SL + RL > RL + RJ$ (Add. Prop.)
6. $SL + RL > QJ + RJ$ (Subst.)
7. $RS = SL + RL$, $QR = QJ + RJ$ (Seg. Add. Post.)
8. $RS > QR$ (Subst.)

25. Given: $\overline{XU} \cong \overline{VW}$, $VW > XW$, $\overline{XU} \parallel \overline{VW}$

Prove: $m\angle XZU > m\angle UZV$

Statements (Reasons)

1. $\overline{XU} \cong \overline{VW}$, $\overline{XU} \parallel \overline{VW}$ (Given)
2. $\angle UXV \cong \angle XVW$, $\angle XUW \cong \angle UWV$ (Alt. Int. ∠s Thm.)
3. $\triangle XZU \cong \triangle VZW$ (ASA)
4. $\overline{XZ} \cong \overline{VZ}$ (CPCTC)
5. $\overline{WZ} \cong \overline{WZ}$ (Refl. Prop.)
6. $VW > XW$ (Given)
7. $m\angle VZW > m\angle XZW$ (Converse of Hinge Thm.)
8. $\angle VZW \cong \angle XZU$, $\angle XZW \cong \angle VZU$ (Vert. ∠s are $\cong$.)
9. $m\angle VZW = m\angle XZU$, $m\angle XZW = m\angle VZU$ (Def. of $\cong$ ∠s)
10. $m\angle XZU > m\angle UZV$ (Subst.)

27 **a.** Sample answer: Use a ruler to measure the distance from her shoulder to her fist for each position. The distance is 1.6 cm for Position 1 and 2 cm for Position 2. Therefore, the distance from her shoulder to her fist is greater in Position 2.
b. Sample answer: In each position, a triangle formed. The distance from her shoulder to her elbow and from her elbow to her wrist is the same in both triangles. Using the measurements in part **a** and the Converse of the Hinge Theorem, you know that the measure of the angle opposite the larger side is larger, so the angle formed by Anica's elbow is greater in Position 2.

29. Given: $\overline{PR} \cong \overline{PQ}$, $SQ > SR$

Prove: $m\angle 1 < m\angle 2$

Statements (Reasons)

1. $\overline{PR} \cong \overline{PQ}$ (Given)
2. $\angle PRQ \cong \angle PQR$ (Isos. △ Thm.)
3. $m\angle PRQ = m\angle 1 + m\angle 4$, $m\angle PQR = m\angle 2 + m\angle 3$ (Angle Add. Post.)
4. $m\angle PRQ = m\angle PQR$ (Def. of $\cong$ ∠s)
5. $m\angle 1 + m\angle 4 = m\angle 2 + m\angle 3$ (Subst.)
6. $SQ > SR$ (Given)
7. $m\angle 4 > m\angle 3$ (Angle Side Relationship Thm.)
8. $m\angle 4 = m\angle 3 + x$ (Def. of inequality)
9. $m\angle 1 + m\angle 4 - m\angle 4 = m\angle 2 + m\angle 3 - (m\angle 3 + x)$ (Subt. Prop.)
10. $m\angle 1 = m\angle 2 - x$ (Subst.)
11. $m\angle 1 + x = m\angle 2$ (Add. Prop.)
12. $m\angle 1 < m\angle 2$ (Def. of inequality)

31. $CB < AB$ **33.** $m\angle BGC < m\angle FBA$

35 $\overline{WZ} \cong \overline{YZ}$, $\overline{ZU} \cong \overline{ZU}$, and $m\angle WZU > m\angle YZU$. By the Hinge Theorem, $WU > YU$.

37a.

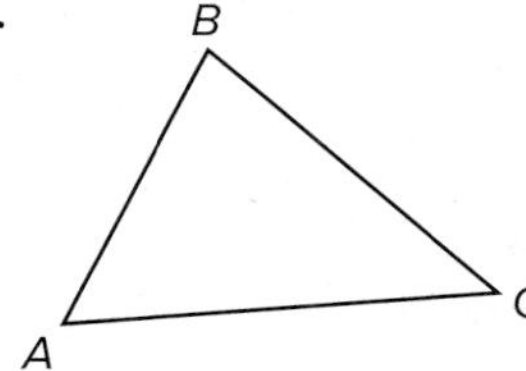

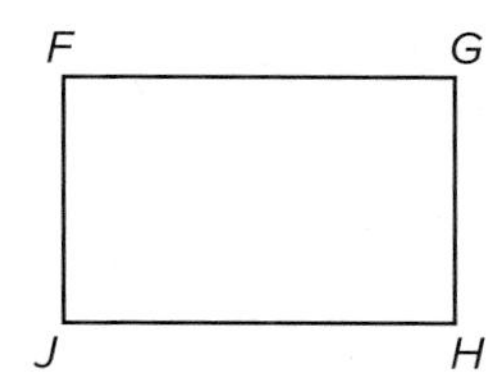

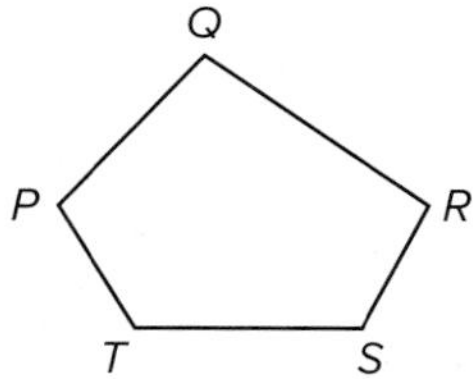

37b. ∠ measures: 59, 76, 45; 90, 90, 90, 90; 105, 100, 96, 116, 123; Sum of ∠s: 180, 360, 540
37c. Sample answer: The sum of the angles of the polygon is equal to 180 times two less than the number of sides of the polygon.
37d. Inductive; sample answer: Because I used a pattern to determine the relationship, the reasoning I used was inductive.
37e. $(n - 2)180$

39. 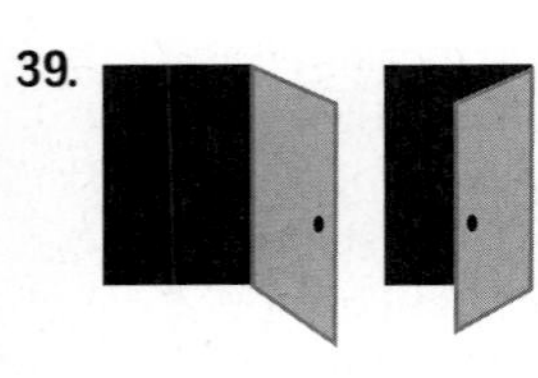 A door; as the door opens, the door opening increases as the angle made by the hinge increases. As the door closes, the door opening decreases as the angle made by the hinge decreases. This is similar to the side opposite the angle in a triangle, because as the side opposite an angle increases the measure of the angle also increases. As the side decreases, the angle also decreases. **41.** Never; from the Converse of the Hinge Theorem, $\angle ADB < \angle BDC$. $\angle ADB$ and $\angle BDC$ form a linear pair. So, $m\angle ADB + m\angle BDC = 180$. Because $m\angle BDC > m\angle ADB$, $m\angle BDC$ must be greater than 90 and $m\angle ADB$ must be less than 90. So, by the definition of obtuse and acute angles, $m\angle BDC$ is always obtuse and $m\angle ADB$ is always acute. **43.** A **45.** B **47.** 30

Chapter 5 Study Guide and Review

1. false; orthocenter **3.** true **5.** true **7.** false; the vertex opposite that side **9.** Assume that the conclusion is false and show that this assumption leads to a contradiction. **11.** 5 **13.** 34 **15.** (2, 3) **17.** $\angle S$, $\angle R$, $\angle T$; $\overline{RT}$, $\overline{TS}$, $\overline{SR}$ **19.** The shorter path is for Sarah to get Irene and then go to Anna's house. **21.** $\triangle FGH$ is not congruent to $\triangle MNO$. **23.** $y \geq 4$

25. Let the cost of one item be x, and the cost of the other item be y.

Given: $x + y > 10$

Prove: $x > 5$ or $y > 5$

Indirect proof:

Step 1 Assume that $x \leq 5$ and $y \leq 5$.

Step 2 If $x \leq 5$ and $y \leq 5$, then $x + y \leq 5 + 5$, or $x + y \leq 10$. This is a contradiction because we know that $x + y > 10$.

Step 3 Because the assumption that $x \leq 5$ and $y \leq 5$ leads to a contradiction of a known fact, the assumption must be false. Therefore, the conclusion that $x > 5$ or $y > 5$ must be true. Thus, at least one item had to be over \$5.

27. no; $3 + 4 < 8$ **29.** Let x be the length of the third side. $6.5 \text{ cm} < x < 14.5 \text{ cm}$ **31.** $m\angle ABC > m\angle DEF$ **33.** Rose

CHAPTER 6

Triangles and Congruence

Chapter 6 Concept Check

1. Exterior Angles Theorem **3.** Triangle Angle Sum Theorem **5.** The lengths are equal. **7.** Substitute the value of x into the expression for the base; substitute x into one of the expressions for the congruent sides.

Lesson 6-1

1. 1440 **3.** $m\angle X = 36$, $m\angle Y = 72$, $m\angle Z = 144$, $m\angle W = 108$

5 $(n-2) \cdot 180 = (16-2) \cdot 180$ $n = 6$
$= 14 \cdot 180$ or 2520 Simplify.

The sum of the interior angle measures is 2520. So, the measure of one interior angle is 2520 ÷ 16 or 157.5.

7. 36 **9.** 68 **11.** 45 **13.** 3240 **15.** 5400

17 $(n-2) \cdot 180 = (4-2) \cdot 180$ $n = 4$
$= 2 \cdot 180$ or 360 Simplify.

The sum of the interior angle measures is 360.

$360 = m\angle J + m\angle K + m\angle L + m\angle M$	Sum of interior angle measures
$360 = (3x-6) + (x+10) + x + (2x-8)$	Substitution
$360 = 7x - 4$	Combine like terms.
$364 = 7x$	Add 4 to each side.
$52 = x$	Simplify.

$m\angle J = 3x - 6 = 3(52) - 6$ or 150 $m\angle K = x + 10 = 52 + 10$ or 62

$m\angle L = x = 52$ $m\angle M = 2x - 8 = 2(52) - 8$ or 96

19. $m\angle U = 60$, $m\angle V = 193$, $m\angle W = 76$, $m\angle Y = 68$, $m\angle Z = 143$ **21.** 150 **23.** 144 **25a.** 720 **25b.** Yes, 120; sample answer: Because the hexagon is regular, the measures of the angles are equal. That means each angle is 720 ÷ 6 or 120. **27.** 4 **29.** 15

31
$$21 + 42 + 29 + (x + 14) + x + (x - 10) + (x - 20) = 360$$
$$4x + 76 = 360$$
$$4x = 284$$
$$x = 71$$

33. 37 **35.** 72 **37.** 24 **39.** 51.4, 128.6 **41.** 25.7, 154.3 **43.** Consider the sum of the measures of the exterior angles N for an n-gon.

N = sum of measures of linear pairs − sum of measures of interior angles
$= 180n - 180(n-2)$
$= 180n - 180n + 360$
$= 360$

So, the sum of the exterior angle measures is 360 for any convex polygon. **45.** 105, 110, 120, 130, 135, 140, 160, 170, 180, 190

47 **a.** 60 ft ÷ 8 = 7.5 ft   Perimeter ÷ number of sides

b. $(n-2) \cdot 180 = (8-2) \cdot 180$ $n = 8$
$= 6 \cdot 180$ or 1080 Simplify.

Sample answer: The sum of the interior angle measures is 1080. So, the measure of each angle of a regular octagon is 1080 ÷ 8 or 135. So if each side of the board makes up half of the angle, each one measures 135 ÷ 2 or 67.5.

49. Liam; by the Exterior Angles Sum Theorem, the sum of the measures of any convex polygon is 360.

51. Always; by the Exterior Angles Sum Theorem, $m\angle QPR = 60$ and $m\angle QRP = 60$. Because the sum of the interior angle measures of a triangle is 180, the measure of $\angle PQR = 180 - m\angle QPR - m\angle QRP = 180 - 60 - 60 = 60$. So, $\triangle PQR$ is an equilateral triangle.

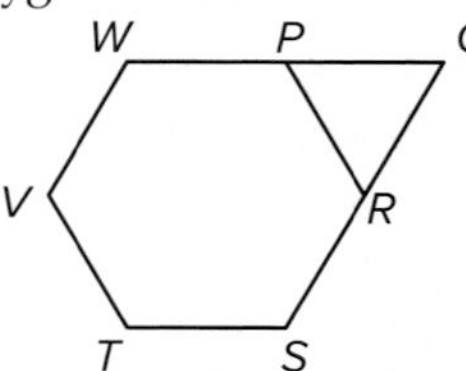

53. The Interior Angles Sum Theorem is derived from the pattern between the number of sides in a polygon and the number of triangles. The formula is the product of the sum of the measures of the angles in a triangle, 180, and the number of triangles in the polygon. **55.** A **57a.** 3 **57b.** 68 **57c.** $A = 215°$, $B = 53°$, $C = 146°$, $D = 68°$, $E = 58°$

Lesson 6-2

1a. 148 **1b.** 125 **1c.** 4 **3.** 15 **5.** $w = 5$, $b = 4$

7. Given: $\square ABCD$, $\angle A$ is a right angle.

Prove: $\angle B$, $\angle C$, and $\angle D$ are right angles. (Theorem 6.6)

Proof: By definition of a parallelogram, $\overline{AB} \parallel \overline{CD}$. Because $\angle A$ is a right angle, $\overline{AC} \perp \overline{AB}$. By the Perpendicular Transversal Theorem, $\overline{AC} \perp \overline{CD}$. $\angle C$ is a right angle, because perpendicular lines form a right angle. $\angle B \cong \angle C$ and $\angle A \cong \angle D$ because opposite angles in a parallelogram are congruent. $\angle C$ and $\angle D$ are right angles, because all right angles are congruent.

9

$m\angle R + m\angle Q = 180$	Consecutive angles are supplementary.
$m\angle R + 128 = 180$	Substitution
$m\angle R = 52$	Subtract 128 from each side.

11. 5

13

a. $\overline{JH} \cong \overline{FG}$	Opposite sides are congruent.
$JH = FG$	Definition of congruence
$= 1$ in.	Substitution
b. $\overline{GH} \cong \overline{FJ}$	Opposite sides are congruent.
$GH = FJ$	Definition of congruence
$= \frac{3}{4}$ in.	Substitution
c. $\angle JFG \cong \angle JHG$	Opposite angles are congruent.
$m\angle JFG = m\angle JHG$	Definition of congruence
$= 62$	Substitution
d. $m\angle FJH + m\angle JHG = 180$	Consecutive angles are supplementary.
$m\angle FJH + 62 = 180$	Substitution
$m\angle FJH = 118$	Subtract 62 from each side.

15. $a = 7$, $b = 11$ **17.** $x = 5$, $y = 17$ **19.** $x = 58$, $y = 63.5$ **21.** (2.5, 2.5)

23. Given: *WXTV* and *ZYVT* are parallelograms.
Prove: $\overline{WX} \cong \overline{ZY}$

Proof:
Statements (Reasons)
1. *WXTV* and *ZYVT* are parallelograms. (Given)
2. $\overline{WX} \cong \overline{VT}$, $\overline{VT} \cong \overline{YZ}$ (Opp. sides of a ▱ are ≅.)
3. $\overline{WX} \cong \overline{ZY}$ (Trans. Prop.)

25. Given: $\triangle ACD \cong \triangle CAB$
Prove: $\overline{DP} \cong \overline{PB}$

Proof:
Statements (Reasons)
1. $\triangle ACD \cong \triangle CAB$ (Given)
2. $\angle ACD \cong \angle CAB$ (CPCTC)

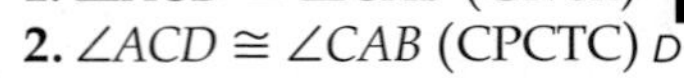

3. $\angle DPC \cong \angle BPA$ (Vert. ∠s are ≅.)
4. $\overline{AB} \cong \overline{CD}$ (CPCTC)
5. $\triangle ABP \cong \triangle CDP$ (AAS)
6. $\overline{DP} \cong \overline{PB}$ (CPCTC)

27. Given: ▱*WXYZ*
Prove: $\triangle WXZ \cong \triangle YZX$ (Theorem 6.8)
Proof:
Statements (Reasons)
1. ▱*WXYZ* (Given)
2. $\overline{WX} \cong \overline{ZY}$, $\overline{WZ} \cong \overline{XY}$ (Opp. sides of a ▱ are ≅.)
3. $\angle ZWX \cong \angle XYZ$ (Opp. ∠s of a ▱ are ≅.)
4. $\triangle WXZ \cong \triangle YZX$ (SAS)

29. Given: *ACDE* is a parallellogram.
Prove: $\overline{EC}$ bisects $\overline{AD}$. (Theorem 6.7)
Proof: It is given that *ACDE* is a parallelogram. Because opposite sides of a parallelogram are congruent, $\overline{EA} \cong \overline{DC}$. By definition of a parallelogram, $\overline{EA} \parallel \overline{DC}$. $\angle AEB \cong \angle DCB$ and $\angle EAB \cong \angle CDB$ because alternate interior angles are congruent. $\triangle EBA \cong \triangle CBD$ by ASA. $\overline{EB} \cong \overline{BC}$ and $\overline{AB} \cong \overline{BD}$ by CPCTC. By the definition of segment bisector, $\overline{EC}$ bisects $\overline{AD}$ and $\overline{AD}$ bisects $\overline{EC}$.

31. 3

33 $\angle AFB$ and $\angle BFC$ form a linear pair.

$\angle AFB + \angle BFC = 180$	Supplement Theorem
$\angle AFB + 49 = 180$	Substitution
$\angle AFB = 131$	Subtract 49 from each side.

35. 29 **37.** $(-1, -1)$; Sample answer: Opposite sides of a parallelogram are parallel. Because the slope of $\overline{BC} = \frac{-6}{2}$, the slope of $\overline{AD}$ must also be $\frac{-6}{2}$. To locate vertex *D*, start from vertex *A* and move down 6 and right 2.

39 First, use properties involving opposite angles and opposite sides of a parallelogram to help prove that $\triangle YUZ$ and $\triangle VXW$ are right angles. Then use the Hypotenuse-Angle Congruence Theorem to prove the triangles are congruent.
Given: ▱*YWVZ*, $\overline{VX} \perp \overline{WY}$, $\overline{YU} \perp \overline{VZ}$
Prove: $\triangle YUZ \cong \triangle VXW$
Proof:
Statements (Reasons)
1. ▱*YWVZ*, $\overline{VX} \perp \overline{WY}$, $\overline{YU} \perp \overline{VZ}$ (Given)
2. $\angle Z \cong \angle W$ (Opp. ∠s of a ▱ are ≅.)
3. $\overline{WV} \cong \overline{ZY}$ (Opp. sides of a ▱ are ≅.)
4. $\angle VXW$ and $\angle YUZ$ are rt. ∠s. (⊥ lines form four rt. ∠s.)
5. $\triangle VXW$ and $\triangle YUZ$ are rt. △s. (Def. of rt. △s)
6. $\triangle YUZ \cong \triangle VXW$ (HA)

41. 7

43.

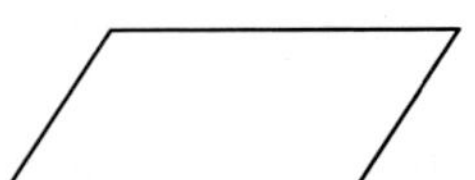

45. Sample answer: In a parallelogram, the opposite sides and angles are congruent. Two consecutive angles in a parallelogram are supplementary. If one angle of a parallelogram is right, then all the angles are right. The diagonals of a parallelogram bisect each other.

47. C **49a.** 11° **49b.** 125° **51a.** Theorem 6.7 **51b.** 34

51c. Proof:
Statements (Reasons)
1. *PQRS* is a parallelogram. (Given)
2. $\overline{PT} \cong \overline{TR}$; $\overline{ST} \cong \overline{TQ}$ (Diag. of a ▱ bisect each other.)
3. $\angle PTS \cong \angle QTR$ (Vert. ∠s are ≅.)
4. $\triangle PST \cong \triangle QRT$ (SAS)

53. D

Lesson 6-3

1. Yes; each pair of opposite angles are congruent.
3. $AP = CP$, $BP = DP$; sample answer: If the diagonals of a quadrilateral bisect each other, then the quadrilateral is a parallelogram, so if $AP = CP$ and $BP = DP$, then the string forms a parallelogram.

5 If both pairs of opposite sides are congruent, then the quadrilateral is a parallelogram.

$2x + 3 = x + 7$	Congruent sides have equal measures.
$x + 3 = 7$	Subtract *x* from each side.
$x = 4$	Subtract 3 from each side.

$3y - 5 = y + 11$	Congruent sides have equal measures.
$2y - 5 = 11$	Subtract *y* from each side.
$2y = 16$	Add 5 to each side.
$y = 8$	Divide each side by 2.

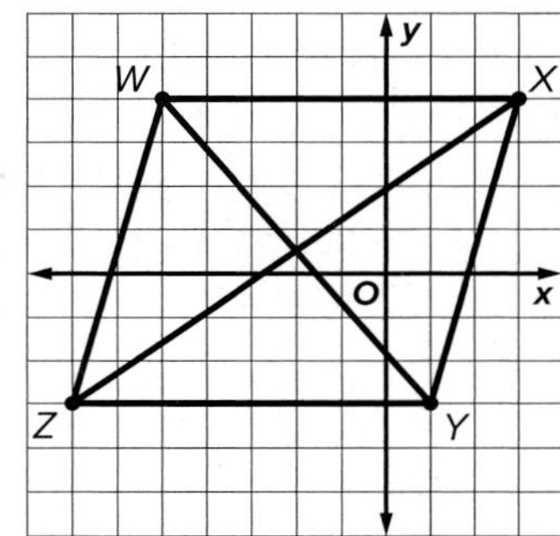

If the diagonals bisect each other, then it is a parallelogram.

Selected Answers and Solutions

midpoint of $\overline{WY}$: $\left(\frac{-5+1}{2}, \frac{4+(-3)}{2}\right) = \left(-2, \frac{1}{2}\right)$

midpoint of $\overline{XZ}$: $\left(\frac{3+(-7)}{2}, \frac{4+(-3)}{2}\right) = \left(-2, \frac{1}{2}\right)$

The midpoint of $\overline{WY}$ and $\overline{XZ}$ is $\left(-2, \frac{1}{2}\right)$. Because the diagonals bisect each other, $WXYZ$ is a parallelogram.

9. Yes; both pairs of opposite sides are congruent. **11.** No; none of the tests for parallelograms are fulfilled. **13.** Yes; the diagonals bisect each other.

15.

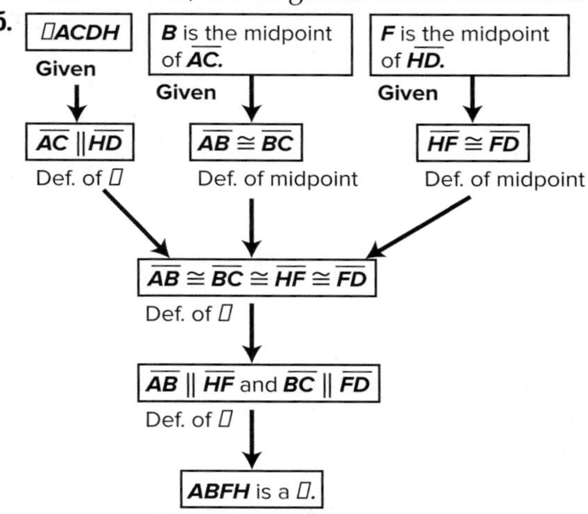

17. Given: $ABEF$ is a parallelogram; $BCDE$ is a parallelogram.
Prove: $ACDF$ is a parallelogram.

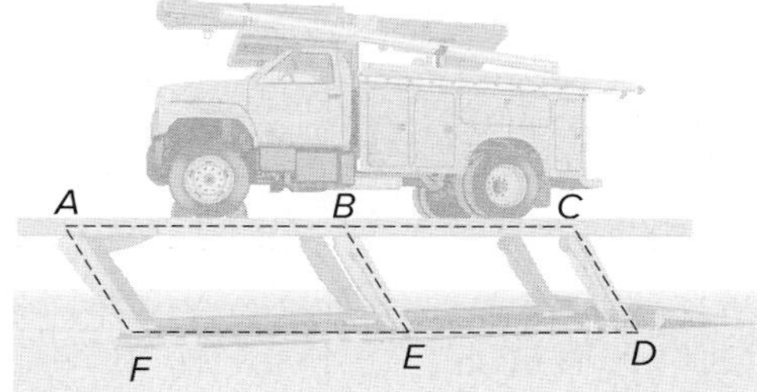

Proof:

Statements (Reasons)

1. $ABEF$ is a parallelogram; $BCDE$ is a parallelogram. (Given)

2. $\overline{AF} \cong \overline{BE}$, $\overline{BE} \cong \overline{CD}$, $\overline{AF} \parallel \overline{BE}$, $\overline{BE} \parallel \overline{CD}$ (Def. of ▱)

3. $\overline{AF} \cong \overline{CD}$, $\overline{AF} \parallel \overline{CD}$ (Trans. Prop.)

4. $ACDF$ is a parallelogram. (If one pair of opp. sides is ≅ and ∥, then the quad. is a ▱.)

19. $x = 8, y = 9$ **21.** $x = 11, y = 7$ **23.** $x = 4, y = 3$

25. No; both pairs of opposite sides must be congruent. The distance between K and L is $\sqrt{53}$. The distance between L and M is $\sqrt{37}$. The distance between M and J is $\sqrt{50}$. The distance between J and K is $\sqrt{26}$. Because both pairs of opposite sides are not congruent, $JKLM$ is not a parallelogram.

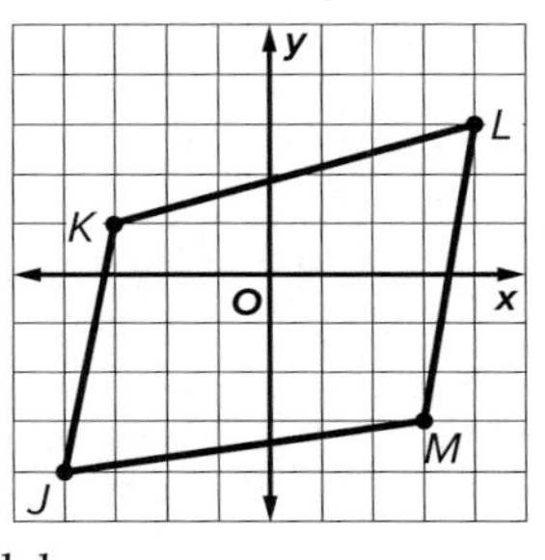

27. Yes; a pair of opposite sides must be parallel and congruent. Slope of $\overline{QR} = \frac{7}{2} =$ slope of $\overline{ST}$, so $\overline{QR} \parallel \overline{ST}$. $QR = ST = \sqrt{53}$, so $\overline{QR} \cong \overline{ST}$. So, $QRST$ is a parallelogram.

29. Given: $ABCD$ is a parallelogram. $\angle A$ is a right angle.
Prove: $\angle B$, $\angle C$, and $\angle D$ are right angles.

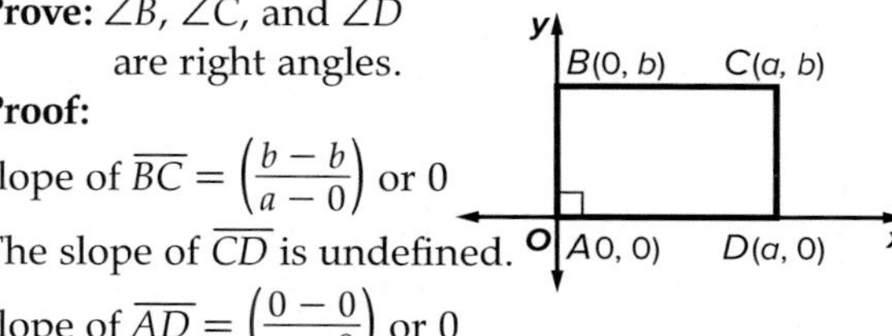

Proof:

slope of $\overline{BC} = \left(\frac{b-b}{a-0}\right)$ or 0

The slope of $\overline{CD}$ is undefined.

slope of $\overline{AD} = \left(\frac{0-0}{a-0}\right)$ or 0

The slope of $\overline{AB}$ is undefined.

Therefore, $\overline{BC} \perp \overline{CD}$, $\overline{CD} \perp \overline{AD}$, and $\overline{AB} \perp \overline{BC}$. So, $\angle B$, $\angle C$, and $\angle D$ are right angles.

31 **a.** Use the Segment Addition Postulate to rewrite AC and CF each as the sum of two measures. Then use substitution, the Subtraction Property, and the Transitive Property to show that $\overline{BC} \cong \overline{DE}$. By proving that $BCDE$ is a parallelogram, you can prove that $\overline{BE} \parallel \overline{CD}$.

Given: $\overline{AC} \cong \overline{CF}$, $\overline{AB} \cong \overline{CD} \cong \overline{BE}$, and $\overline{DF} \cong \overline{DE}$

Prove: $\overline{BE} \parallel \overline{CD}$

Proof: We are given that $\overline{AC} \cong \overline{CF}$, $\overline{AB} \cong \overline{CD} \cong \overline{BE}$, and $\overline{DF} \cong \overline{DE}$. $AC = CF$ by the definition of congruence. $AC = AB + BC$ and $CF = CD + DF$ by the Segment Addition Postulate and $AB + BC = CD + DF$ by substitution. Using substitution again, $AB + BC = AB + DF$, and $BC = DF$ by the Subtraction Property. $\overline{BC} \cong \overline{DF}$ by the definition of congruence, and $\overline{BC} \cong \overline{DE}$ by the Transitive Property. If both pairs of opposite sides of a quadrilateral are congruent, then the quadrilateral is a parallelogram, so $BCDE$ is a parallelogram. By the definition of a parallelogram, $\overline{BE} \parallel \overline{CD}$.

b. Let x = width of the copy. If $AB = 12$, then $CD = BE = 12$. So, $CF = 12 + 8$ or 20.

$\frac{CF}{BE} = \frac{\text{width of copy}}{\text{width of original}}$

$\frac{20}{12} = \frac{x}{5.5}$ Substitution

$20(5.5) = 12x$ Cross multiply.

$110 = 12x$ Simplify.

$9.2 \approx x$ Divide each side by 12.

The width of the original object is about 9.2 in.

Selected Answers and Solutions

33. Given: $\overline{AB} \cong \overline{DC}$, $\overline{AB} \parallel \overline{DC}$
Prove: $ABCD$ is a parallelogram.
Proof:
Statements (Reasons)
1. $\overline{AB} \cong \overline{DC}$, $\overline{AB} \parallel \overline{DC}$ (Given)
2. Draw $\overline{AC}$. (Two points determine a line.)
3. $\angle 1 \cong \angle 2$ (If two lines are $\parallel$, then alt. int. $\measuredangle$ are $\cong$.)
4. $\overline{AC} \cong \overline{AC}$ (Refl. Prop.)
5. $\triangle ABC \cong \triangle CDA$ (SAS)
6. $\overline{AD} \cong \overline{BC}$ (CPCTC)
7. $ABCD$ is a parallelogram. (If both pairs of opp. sides are $\cong$, then the quad. is $\square$.)

35 Opposite sides of a parallelogram are parallel and congruent. Because the slope of $\overline{AB}$ is 0, the slope of $\overline{DC}$ must be 0. The y-coordinate of vertex D must be c. The length of $\overline{AB}$ is $(a + b) - 0$ or $a + b$. So the length of $\overline{DC}$ must be $a + b$. If the x-coordinate of D is $-b$ and the x-coordinate of C is a, then the length of $\overline{DC}$ is $a - (-b)$ or $a + b$. So, the coordinates are $C(a, c)$ and $D(-b, c)$.

37 Sample answer: Because the two vertical rails are both perpendicular to the ground, he knows that they are parallel to each other. If he measures the distance between the two rails at the top of the steps and at the bottom of the steps, and they are equal, then one pair of sides of the quadrilateral formed by the handrails is both parallel and congruent, so the quadrilateral is a parallelogram. Because the quadrilateral is a parallelogram, the two handrails are parallel by definition.

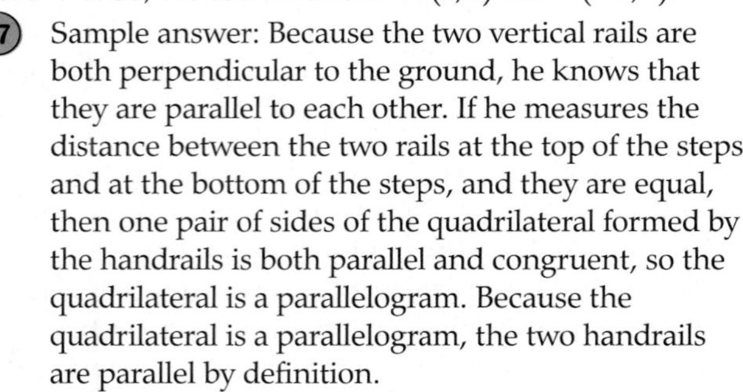

39a. Sample answer:

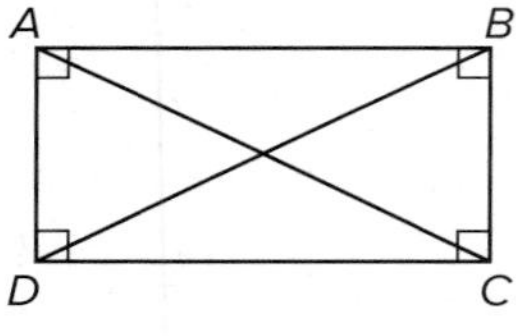

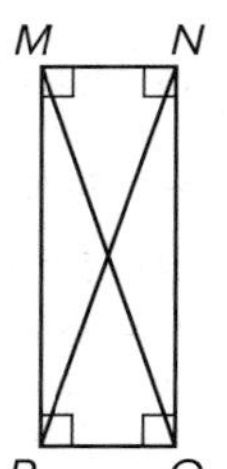

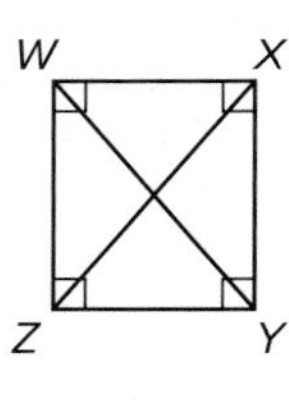

39b. Sample answer:

Rectangle	Side	Length
ABCD	$\overline{AC}$	3.3 cm
	$\overline{BD}$	3.3 cm
MNOP	$\overline{MO}$	2.8 cm
	$\overline{NP}$	2.8 cm
WXYZ	$\overline{WY}$	2.0 cm
	$\overline{XZ}$	2.0 cm

39c. Sample answer: The diagonals of a rectangle are congruent.
41. Sample answer: The theorems are converses of each other. The hypothesis of Theorem 6-3 is "a figure is a parallelogram," and the hypothesis of 6-9 is "both pairs of opposite sides of a quadrilateral are congruent." The conclusion of Theorem 6-3 is "opposite sides are congruent," and the conclusion of 6-9 is "the quadrilateral is a parallelogram."

43.

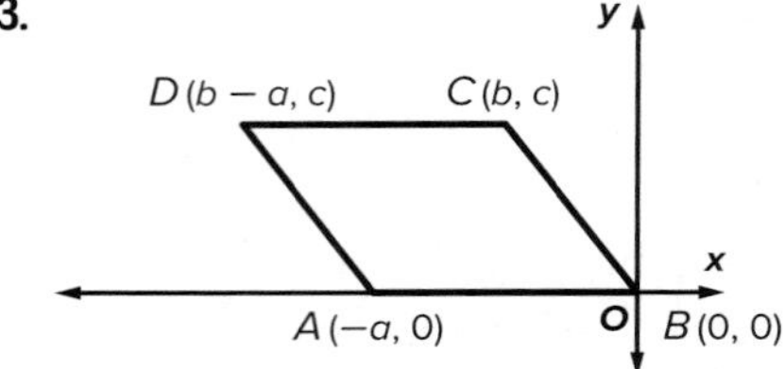

45. Sample answer: You can show that both pairs of opposite sides are congruent or parallel, both pairs of opposite angles are congruent, diagonals bisect each other, or one pair of opposite sides is both congruent and parallel. **47.** C
49a. 56° **49b.** 68° **49c.** 56° **49d.** 56° **49e.** Sample answer: Yes. Since $\angle MNO \cong \angle OPQ$, by the alternate interior angle theorem, the line segments are parallel. **49f.** Sample answer: Yes, since both sets of opposite sides are parallel, $NPQR$ is a parallelogram.
51. B, C, E

Lesson 6-4

1. 7 ft **3.** 33.5 **5.** 11
7. Given: $ABDE$ is a rectangle; $\overline{BC} \cong \overline{DC}$.
Prove: $\overline{AC} \cong \overline{EC}$
Proof:

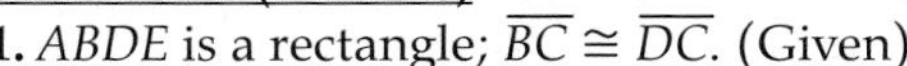

Statements (Reasons)
1. $ABDE$ is a rectangle; $\overline{BC} \cong \overline{DC}$. (Given)
2. $ABDE$ is a parallelogram. (Def. of rectangle)
3. $\overline{AB} \cong \overline{DE}$ (Opp. sides of a $\square$ are $\cong$.)
4. $\angle B$ and $\angle D$ are right angles. (Def. of rectangle)
5. $\angle B \cong \angle D$ (All rt. $\measuredangle$ are $\cong$.)
6. $\triangle ABC \cong \triangle EDC$ (SAS)
7. $\overline{AC} \cong \overline{EC}$ (CPCTC)

9. Yes; $AB = 5 = CD$ and $BC = 8 = AD$. So, $ABCD$ is a parallelogram. $BD = \sqrt{89} = AC$, so the diagonals are congruent. Thus, $ABCD$ is a rectangle.

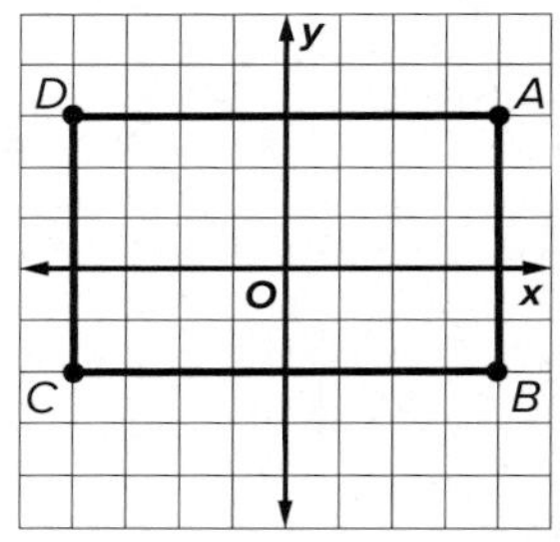

11 Because all angles in a rectangle are right angles, $\angle ABD$ is a right triangle.

$AD^2 + AB^2 = BD^2$	Pythagorean Theorem
$2^2 + 6^2 = BD^2$	Substitution
$40 = BD^2$	Simplify.
$6.3 \approx BD$	Take the positive square root of each side.

$BD \approx 6.3$ ft
13. 25 **15.** 43 **17.** 38 **19.** 46
21. Given: $QTVW$ is a rectangle; $\overline{QR} \cong \overline{ST}$.
Prove: $\triangle SWQ \cong \triangle RVT$
Proof:

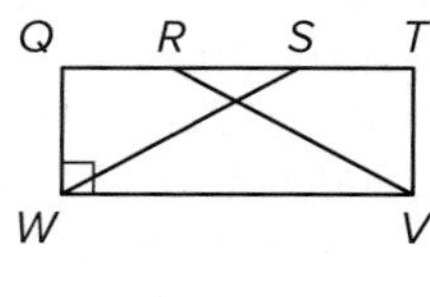

Statements (Reasons)
1. $QTVW$ is a rectangle; $\overline{QR} \cong \overline{ST}$. (Given)
2. $QTVW$ is a parallelogram. (Def. of rectangle)

3. $\overline{WQ} \cong \overline{VT}$ (Opp sides of a ▱ are ≅.)
4. $\angle Q$ and $\angle T$ are right angles. (Def. of rectangle)
5. $\angle Q \cong \angle T$ (All rt. ∠s are ≅.)
6. $QR = ST$ (Def. of ≅ segs.)
7. $\overline{RS} \cong \overline{RS}$ (Refl. Prop.)
8. $RS = RS$ (Def. of ≅ segs.)
9. $QR + RS = RS + ST$ (Add. prop.)
10. $QS = QR + RS$, $RT = RS + ST$ (Seg. Add. Post.)
11. $QS = RT$ (Subst.)
12. $\overline{QS} \cong \overline{RT}$ (Def. of ≅ segs.)
13. $\triangle SWQ \cong \triangle RVT$ (SAS)

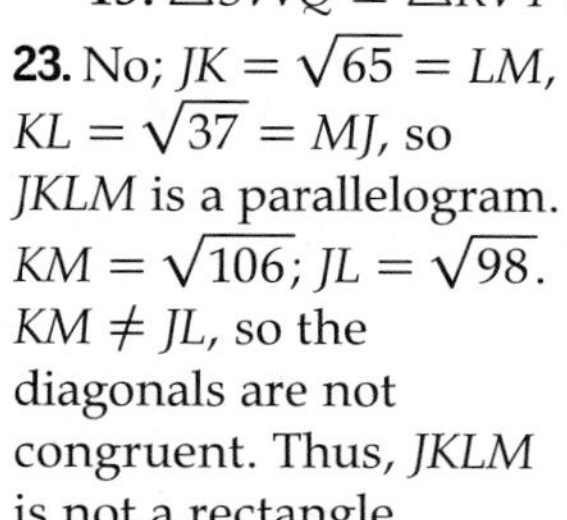

23. No; $JK = \sqrt{65} = LM$, $KL = \sqrt{37} = MJ$, so $JKLM$ is a parallelogram. $KM = \sqrt{106}$; $JL = \sqrt{98}$. $KM \neq JL$, so the diagonals are not congruent. Thus, $JKLM$ is not a rectangle.

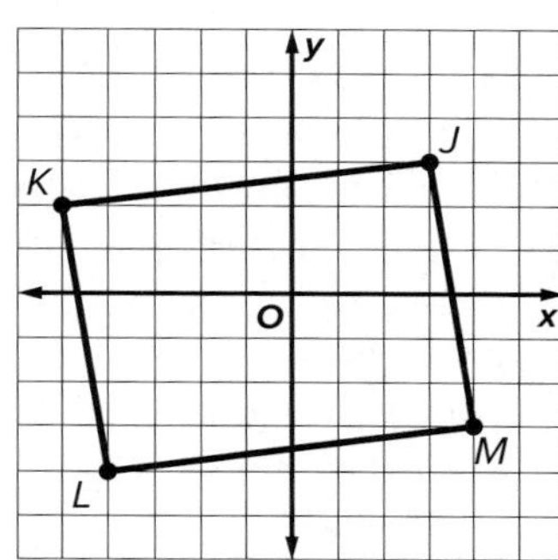

25. No; slope of $\overline{GH} = \frac{1}{8}$ = slope of $\overline{JK}$ and slope of $\overline{HJ} = -6$ = slope of $\overline{KG}$. So, $GHJK$ is a parallelogram. The product of the slopes of consecutive sides $\neq -1$, so the consecutive sides are not perpendicular. Thus, $GHJK$ is not a rectangle. **27.** 40

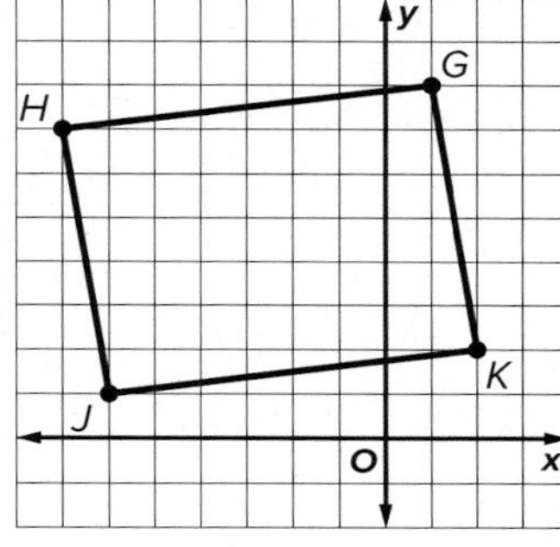

29 A rectangle has four right angles, so $m\angle CDB = 90$. Because a rectangle is a parallelogram, opposite sides are parallel. Alternate interior angles of parallel lines are congruent. So, $\angle 3 \cong \angle 2$ and $m\angle 3 = m\angle 2 = 40$. $m\angle 4 = 90 - 40$ or 50. Because diagonals of a rectangle are congruent and bisect each other, the triangle with angles 4, 5, and 6 is isosceles with $m\angle 6 = m\angle 4$.

$m\angle 4 + m\angle 5 + m\angle 6 = 180$	Triangle Angle-Sum Theorem
$m\angle 4 + m\angle 5 + m\angle 4 = 180$	$m\angle 6 = m\angle 4$
$50 + m\angle 5 + 50 = 180$	Substitution
$100 + m\angle 5 = 180$	Simplify.
$m\angle 5 = 80$	Subtract 100 from each side.

31. 100

33. Given: $WXYZ$ is a rectangle with diagonals $\overline{WY}$ and $\overline{XZ}$.
Prove: $\overline{WY} \cong \overline{XZ}$
Proof:

Statements (Reasons)
1. $WXYZ$ is a rectangle with diagonals $\overline{WY}$ and $\overline{XZ}$. (Given)
2. $\overline{WX} \cong \overline{ZY}$ (Opp. sides of a ▱ are ≅.)
3. $\overline{WZ} \cong \overline{WZ}$ (Refl. Prop.)
4. $\angle XWZ$ and $\angle YZW$ are right angles. (Def. of ▭)
5. $\angle XWZ \cong \angle YZW$ (All right ∠s are ≅.)
6. $\triangle XWZ \cong \triangle YZW$ (SAS)
7. $\overline{WY} \cong \overline{XZ}$ (CPCTC)

35. $ABCD$ is a parallelogram, and $\angle B$ is a right angle. Because $ABCD$ is a parallelogram and has one right angle, then it has four right angles. So by the definition of a rectangle, $ABCD$ is a rectangle.

37. Sample answer: Because $\overline{RP} \perp \overline{PQ}$ and $\overline{SQ} \perp \overline{PQ}$, $m\angle P = m\angle Q = 90$. Lines that are perpendicular to the same line are parallel, so $\overline{RP} \parallel \overline{SQ}$. The same compass setting was used to locate points R and S, so $\overline{RP} \cong \overline{SQ}$. If one pair of opposite sides of a quadrilateral is both parallel and congruent, then the quadrilateral is a parallelogram. A parallelogram with right angles is a rectangle. Thus, $PRSQ$ is a rectangle. **39.** 5 **41a.** 55 ft **41b.** Sample answer: I used the Pythagorean Theorem to find the diagonal length of her existing garden. $56^2 + 40^2 = d^2$, so the length of the diagonal of her existing garden is approximately 68.8 feet. This means that Samantha needs to buy a 75-foot hose for that garden. The width of the other garden is 43 feet and the length of its diagonal can be no more than 75 − 5 or 70 feet in order for Samantha to still be able to buy the 75-foot hose and reach the water faucet that is 5 feet away. Using the Pythagorean Theorem again, $43^2 + \ell^2 < 70^2$. So the maximum length of the other garden, to the nearest foot, is 55 feet.

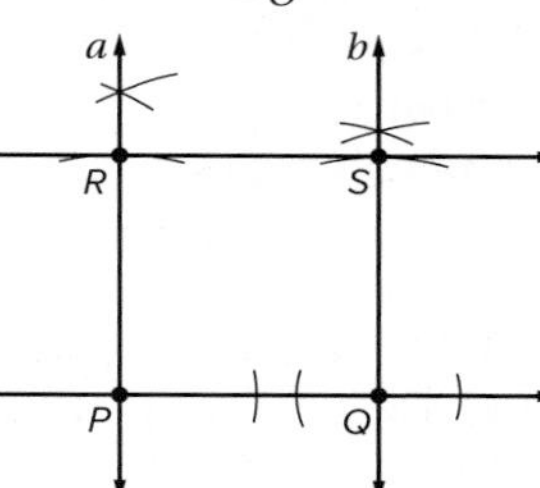

43 Draw ▱ $ABCD$ on the coordinate plane.
Find the slopes of $\overline{AD}$ and $\overline{AB}$ and show that the lines are perpendicular.

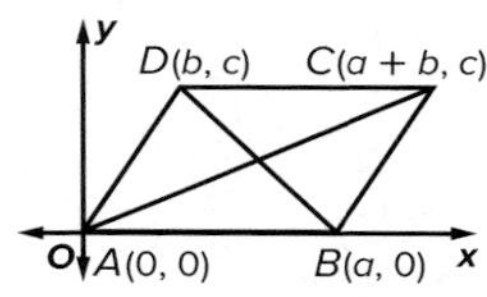

Given: ▱$ABCD$ and $\overline{AC} \cong \overline{BD}$
Prove: ▱$ABCD$ is a rectangle.
Proof:
$AC = \sqrt{(a + b - 0)^2 + (c - 0)^2}$
$BD = \sqrt{(b - a)^2 + (c - 0)^2}$
But $AC = BD$ and
$\sqrt{(a + b - 0)^2 + (c - 0)^2} = \sqrt{(b - a)^2 + (c - 0)^2}$.

$$(a + b - 0)^2 + (c - 0)^2 = (b - a)^2 + (c - 0)^2$$
$$(a + b)^2 + c^2 = (b - a)^2 + c^2$$
$$a^2 + 2ab + b^2 + c^2 = b^2 - 2ab + a^2 + c^2$$
$$2ab = -2ab$$
$$4ab = 0$$
$$a = 0 \text{ or } b = 0$$

Because A and B are different points, $a \neq 0$. Then $b = 0$. The slope of $\overline{AD}$ is undefined and the slope of $\overline{AB} = 0$. Thus, $\overline{AD} \perp \overline{AB}$. $\angle DAB$ is a right angle and $ABCD$ is a rectangle.

Selected Answers and Solutions

45. $x = 6, y = -10$ **47.** 6 **49.** Sample answer: All rectangles are parallelograms because, by definition, both pairs of opposite sides are parallel. Parallelograms with right angles are rectangles, so some parallelograms are rectangles, but others with non-right angles are not. **51.** B **53.** D **55.** B **57.** E **59.** A

Lesson 6-5

1. 32

3. Given: $ABCD$ is a rhombus with diagonal $\overline{DB}$.
Prove: $\overline{AP} \cong \overline{CP}$
Proof:
Statements (Reasons)
1. $ABCD$ is a rhombus with diagonal $\overline{DB}$. (Given)
2. $\angle ABP \cong \angle CBP$ (Diag. of rhombus bisects $\angle$)
3. $\overline{PB} \cong \overline{PB}$ (Refl. Prop.)
4. $\overline{AB} \cong \overline{CB}$ (Def. of rhombus)
5. $\triangle APB \cong \triangle CPB$ (SAS)
6. $\overline{AP} \cong \overline{CP}$ (CPCTC)

5. Rectangle, rhombus, square; consecutive sides are perpendicular, all sides are congruent. **7.** 14 **9.** 28

11
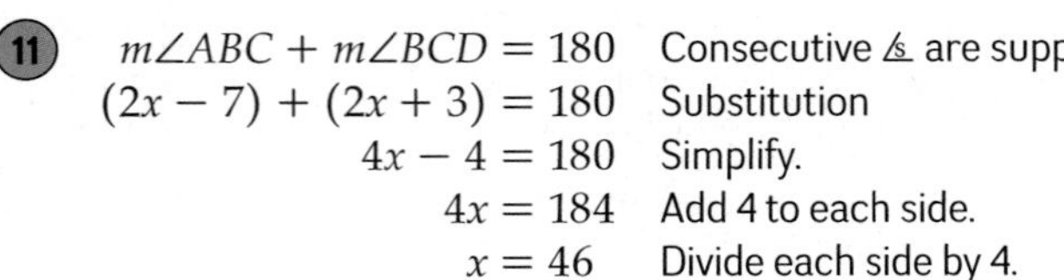

$m\angle ABC + m\angle BCD = 180$	Consecutive ∠s are supp.
$(2x - 7) + (2x + 3) = 180$	Substitution
$4x - 4 = 180$	Simplify.
$4x = 184$	Add 4 to each side.
$x = 46$	Divide each side by 4.

$m\angle BCD = 2x + 3$	Given
$= 2(46) + 3$ or 95	Substitution

$\angle DAB \cong \angle BCD$	Opposite angles are congruent.
$m\angle DAB = m\angle BCD$	Definition of congruence
$= 95$	Substitution

13. Given: $\overline{WZ} \parallel \overline{XY}$, $\overline{WX} \parallel \overline{ZY}$, $\overline{WZ} \cong \overline{ZY}$
Prove: $WXYZ$ is a rhombus.
Proof:
Statements (Reasons)
1. $\overline{WZ} \parallel \overline{XY}$, $\overline{WX} \parallel \overline{ZY}$, $\overline{WZ} \cong \overline{ZY}$ (Given)
2. $WXYZ$ is a ▱. (Both pairs of opp. sides are ∥.)
3. $WXYZ$ is a rhombus. (If one pair of consecutive sides of a ▱ are ≅, the ▱ is a rhombus.)

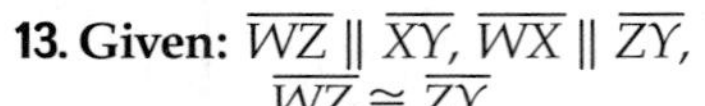

15. Given: $ABCD$ is a parallelogram. $\triangle ABC \cong \triangle ADC$
Prove: $ABCD$ is a rhombus.
Proof:
Statements (Reasons)
1. $\triangle ABC \cong \triangle ADC$ (Given)
2. $\angle BAC \cong \angle DAC$ (CPCTC)
3. $\overline{AC}$ bisects $\angle BAD$. (Def. of $\angle$ bisector)
4. $\angle ACB \cong \angle ACD$ (CPCTC)
5. $\overline{AC}$ bisects $\angle BCD$. (Def. of $\angle$ bisector)
6. $ABCD$ is a rhombus (If one diagonal of a ▱ bisects a pair of opp. ∠s, then the ▱ is a rhombus.)

17. Rhombus; Sample answer: The measure of the angle formed between the two streets is 29, and vertical angles are congruent, so the measure of one angle of the quadrilateral is 29. Because the crosswalks are the same length, the sides of the quadrilateral are congruent. Therefore, they form a rhombus. **19.** Rhombus; the diagonals are perpendicular. **21.** None; the diagonals are not congruent or perpendicular.

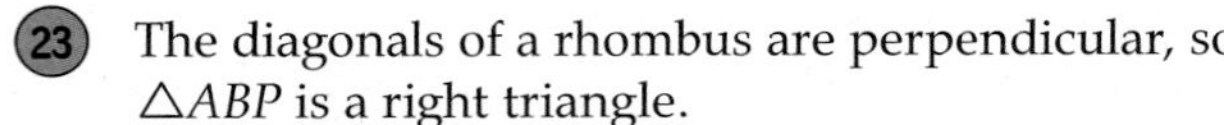

23 The diagonals of a rhombus are perpendicular, so $\triangle ABP$ is a right triangle.

$AP^2 + PB^2 = AB^2$	Pythagorean Theorem
$AP^2 + 12^2 = 15^2$	Substitution
$AP^2 + 144 = 225$	Simplify.
$AP^2 = 81$	Subtract 144 from each side.
$AP = 9$	Take the square root of each side.

25. 24 **27.** 6 **29.** 90 **31.** square **33.** rectangle

35. Given: $ABCD$ is a parallelogram; $\overline{AC} \perp \overline{BD}$.
Prove: $ABCD$ is a rhombus.

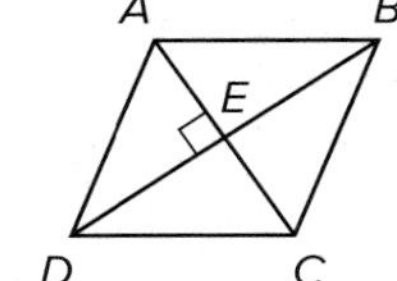

Proof: We are given that $ABCD$ is a parallelogram. The diagonals of a parallelogram bisect each other, so $\overline{AE} \cong \overline{EC}$. $\overline{BE} \cong \overline{BE}$ because congruence of segments is reflexive. We are also given that $\overline{AC} \perp \overline{BD}$. Thus, $\angle AEB$ and $\angle BEC$ are right angles by the definition of perpendicular lines. Then $\angle AEB \cong \angle BEC$ because all right angles are congruent. Therefore, $\triangle AEB \cong \triangle CEB$ by SAS. $\overline{AB} \cong \overline{CB}$ by CPCTC. Opposite sides of parallelograms are congruent, so $\overline{AB} \cong \overline{CD}$ and $\overline{BC} \cong \overline{AD}$. Then because congruence of segments is transitive, $\overline{AB} \cong \overline{CD} \cong \overline{BC} \cong \overline{AD}$. All four sides of $ABCD$ are congruent, so $ABCD$ is a rhombus by definition.

37. Given: $ABCD$ is a parallelogram; $\overline{AB} \cong \overline{BC}$.
Prove: $ABCD$ is a rhombus.
Proof: Opposite sides of a parallelogram are congruent, so $\overline{BC} \cong \overline{AD}$ and $\overline{AB} \cong \overline{CD}$. We are given that $\overline{AB} \cong \overline{BC}$. So, by the Transitive Property, $\overline{BC} \cong \overline{CD}$. So, $\overline{BC} \cong \overline{CD} \cong \overline{AB} \cong \overline{AD}$. Thus, $ABCD$ is a rhombus by definition.

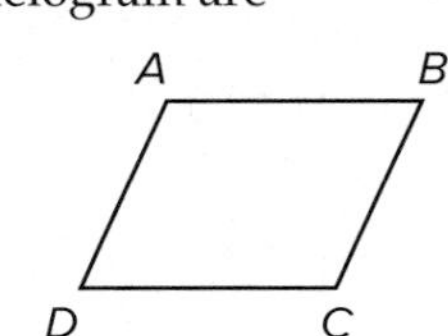

39. Sample answer: The diagonals bisect each other, so the quadrilateral is a parallelogram. Because the diagonals of the parallelogram are perpendicular to each other, the parallelogram is a rhombus.

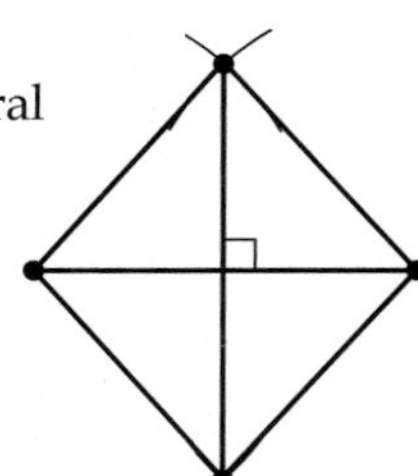

41. Given: $ABCD$ is a square.
Prove: $\overline{AC} \perp \overline{DB}$
Proof:

slope of $\overline{DB} = \frac{0 - a}{a - 0}$ or -1

slope of $\overline{AC} = \frac{0 - a}{0 - a}$ or 1

The slope of $\overline{AC}$ is the negative reciprocal of the slope of $\overline{DB}$, so they are perpendicular.

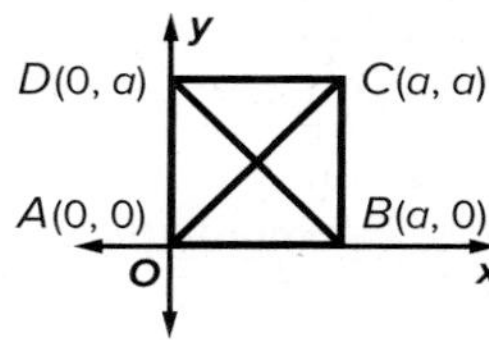

43 Use the properties of exterior angles to classify the quadrilaterals. Sample answer: Because the octagons are regular, each side is congruent, and the quadrilaterals share common sides with the octagons, so the quadrilaterals are either rhombuses or squares. The vertices of the quadrilaterals are formed by the exterior angles of the sides of the octagons adjacent to the vertices. The sum of the measures of the exterior angles of a polygon is always 360, and because a regular octagon has 8 congruent exterior angles, each one measures $360 \div 8$ or 45. As shown in the diagram, each angle of the quadrilaterals in the pattern measures $45 + 45$ or 90. Therefore, the quadrilateral is a square.

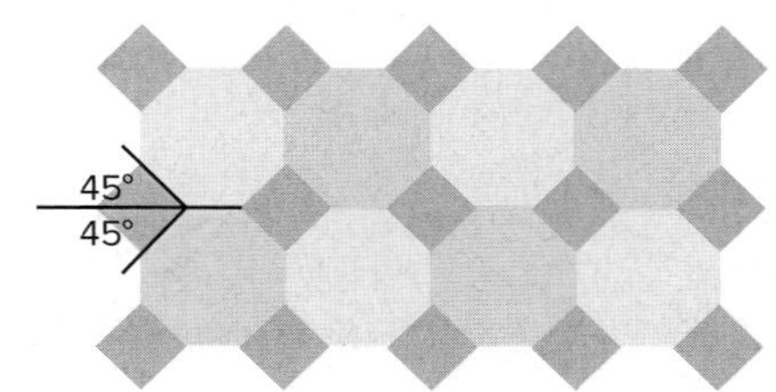

45a.

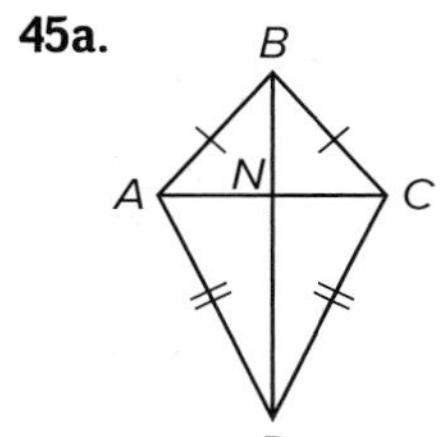

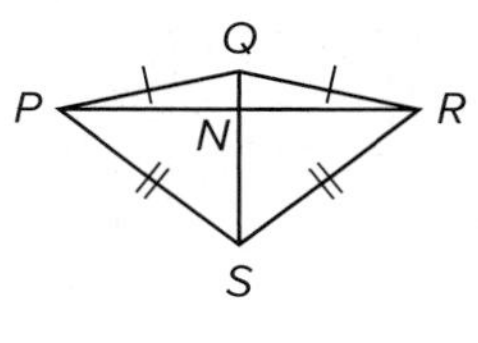

45b.

Figure	Distance from *N* to each vertex along shorter diagonal		Distance from *N* to each vertex along longer diagonal	
ABCD	0.8 cm	0.8 cm	0.9 cm	1.5 cm
PQRS	1.2 cm	1.2 cm	0.3 cm	0.9 cm
WXYZ	0.2 cm	0.2 cm	1.1 cm	0.4 cm

45c. Sample answer: The shorter diagonal of a kite is bisected by the longer diagonal. **47.** True; sample answer: A rectangle is a quadrilateral with four right angles, and a square is both a rectangle and a rhombus, so a square is always a rectangle.
Converse: If a quadrilateral is a rectangle, then it is a square. False; sample answer: A rectangle is a quadrilateral with four right angles. It is not necessarily a rhombus, so it is not necessarily a square.
Inverse: If a quadrilateral is not a square, then it is not a rectangle. False; sample answer: A quadrilateral that has four right angles and two pairs of congruent sides is not a square, but it is a rectangle.
Contrapositive: If a quadrilateral is not a rectangle, then it is not a square. True; sample answer: If a quadrilateral is not a rectangle, it is also not a square by definition.

49. Sample answer: (0, 0), (6, 0), (0, 6), (6, 6) ; the diagonals are perpendicular, and any four points on the lines equidistant from the intersection of the diagonals will be the vertices of a square. **51.** B **53.** C **55.** B **57.** B **59.** 60

Lesson 6-6

1. 101 **3.** $\overline{BC} \parallel \overline{AD}$, $\overline{AB} \nparallel \overline{CD}$; *ABCD* is a trapezoid. **5.** 1.2 **7.** 70 **9.** 70

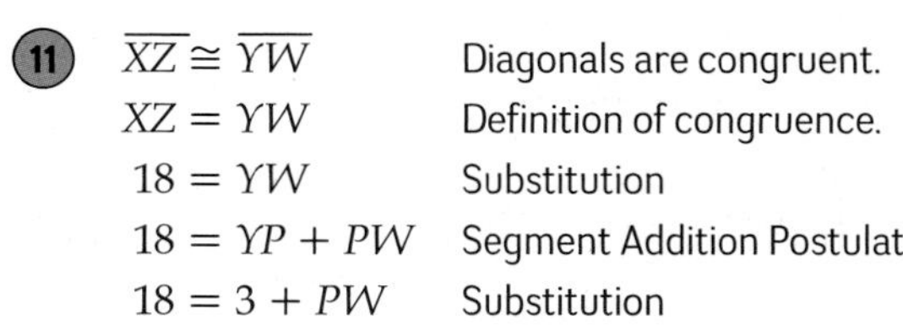

11

$\overline{XZ} \cong \overline{YW}$	Diagonals are congruent.
$XZ = YW$	Definition of congruence.
$18 = YW$	Substitution
$18 = YP + PW$	Segment Addition Postulate
$18 = 3 + PW$	Substitution
$15 = PW$	Subtract 3 from each side.

13. $\overline{JK} \parallel \overline{LM}$, $\overline{KL} \nparallel \overline{JM}$; *JKLM* is a trapezoid, but it's not isosceles because $KL = \sqrt{26}$ and $JM = 5$. **15.** $\overline{XY} \parallel \overline{WZ}$, $\overline{WX} \nparallel \overline{YZ}$; *WXYZ* is a trapezoid, but it's not isosceles because $XZ = \sqrt{74}$ and $WY = \sqrt{68}$. **17.** 10 **19.** 8 **21.** 17

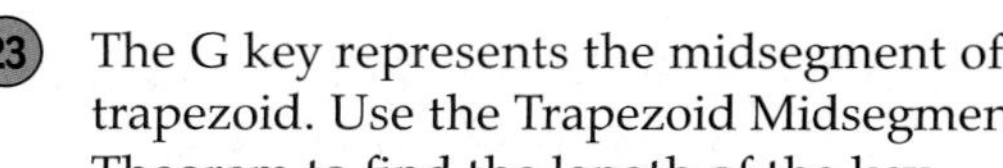

23 The G key represents the midsegment of the trapezoid. Use the Trapezoid Midsegment Theorem to find the length of the key.

length of G key $= \frac{1}{2}$(length of C key + length of D key)

$= \frac{1}{2}(6 + 1.8)$ Substitution

$= \frac{1}{2}(7.8)$ or 3.9 Simplify.

The length of the G key is 3.9 in.

25. $\sqrt{20}$ **27.** 75

29. Given: *ABCD* is a trapezoid; $\angle D \cong \angle C$.
Prove: Trapezoid *ABCD* is isosceles.

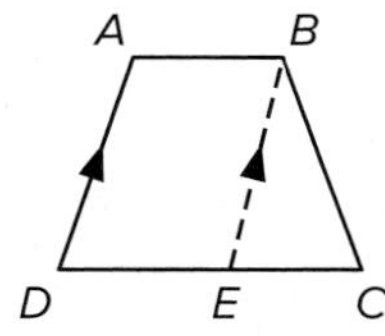

Proof: By the Parallel Postulate, we can draw the auxillary line $\overline{EB} \parallel \overline{AD}$. $\angle D \cong \angle BEC$, by the Corr. ∠s Thm. We are given that $\angle D \cong \angle C$, so by the Trans. Prop, $\angle BEC \cong \angle C$. So, $\triangle EBC$ is isosceles and $\overline{EB} \cong \overline{BC}$. From the definition of a trapezoid, $\overline{AB} \parallel \overline{DE}$. Because both pairs of opposite sides are parallel, *ABED* is a parallelogram. So, $\overline{AD} \cong \overline{EB}$. By the Transitive Property, $\overline{BC} \cong \overline{AD}$. Thus, *ABCD* is an isosceles trapezoid.

31. Given: *ABCD* is a kite with $\overline{AB} \cong \overline{BC}$ and $\overline{AD} \cong \overline{DC}$.
Prove: $\overline{BD} \perp \overline{AC}$

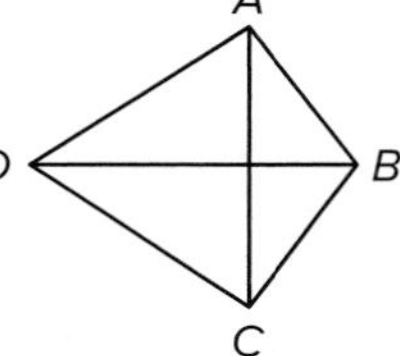

Proof: We know that $\overline{AB} \cong \overline{BC}$ and $\overline{AD} \cong \overline{DC}$. So, *B* and *D* are both equidistant from *A* and *C*. If a point is equidistant from the endpoints of a segment, then it is on the perpendicular bisector of the segment. The line that contains B and D is the perpendicular bisector of $\overline{AC}$, because only one line exists through two points. Thus, $\overline{BD} \perp \overline{AC}$.

33. Given: $ABCD$ is a trapezoid with median $\overline{EF}$.
Prove: $\overline{EF} \parallel \overline{AB}$ and $\overline{EF} \parallel \overline{DC}$ and
$EF = \frac{1}{2}(AB + DC)$

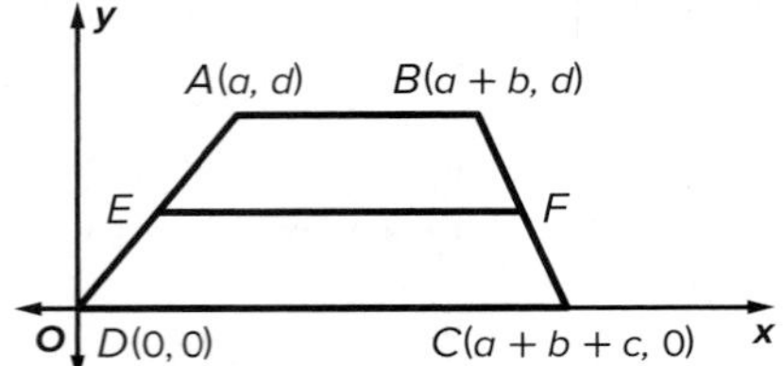

Proof:
By the definition of the median of a trapezoid, E is the midpoint of $\overline{AD}$ and F is the midpoint of $\overline{BC}$.

Midpoint E is $\left(\frac{a+0}{2}, \frac{d+0}{2}\right)$ or $\left(\frac{a}{2}, \frac{d}{2}\right)$.

Midpoint F is $\left(\frac{a+b+a+b+c}{2}, \frac{d+0}{2}\right)$ or $\left(\frac{2a+2b+c}{2}, \frac{d}{2}\right)$.

The slope of $\overline{AB} = 0$, the slope of $\overline{EF} = 0$, and the slope of $\overline{DC} = 0$. Thus, $\overline{EF} \parallel \overline{AB}$ and $\overline{EF} \parallel \overline{DC}$.

$AB = \sqrt{[(a+b)-a]^2 + (d-d)^2} = \sqrt{b^2}$ or b

$DC = \sqrt{[(a+b+c)-0]^2 + (0-0)^2}$
$= \sqrt{(a+b+c)^2}$ or $a+b+c$

$EF = \sqrt{\left(\frac{2a+2b+c-a}{2}\right)^2 + \left(\frac{d}{2} - \frac{d}{2}\right)^2}$
$= \sqrt{\left(\frac{a+2b+c}{2}\right)^2}$ or $\frac{a+2b+c}{2}$

$\frac{1}{2}(AB + DC) = \frac{1}{2}[b + (a+b+c)]$
$= \frac{1}{2}(a + 2b + c)$
$= \frac{a+2b+c}{2}$
$= EF$

Thus, $\frac{1}{2}(AB + DC) = EF$.

35. 15 **37.** 28 ft **39.** 70 **41.** 2 **43.** 20 **45.** 10 in. **47.** 105

49 $\angle ZWX \cong \angle ZYX$ because one pair of opposite sides of a kite is congruent and $\angle ZWX \cong \angle ZYX$. So, $m\angle ZYX = m\angle ZWX = 10x$.

$m\angle ZWX + m\angle WXY + m\angle ZYX + m\angle WZY = 360$	Polygon Interior Angles Sum Theorem
$10x + 120 + 10x + 4x = 360$	Substitution
$24x + 120 = 360$	Simplify.
$24x = 240$	Subtract 120 from each side.
$x = 10$	Divide each side by 24.

So, $m\angle ZYX = 10x$ or 100.

51. Given: $ABCD$ is an isosceles trapezoid.
Prove: $\angle DAC \cong \angle CBD$
Proof:
Statements (Reasons)
1. $ABCD$ is an isosceles trapezoid. (Given)
2. $\overline{AD} \cong \overline{BC}$ (Def. of isos. trap.)
3. $\overline{DC} \cong \overline{DC}$ (Refl. Prop.)
4. $\overline{AC} \cong \overline{BD}$ (Diags. of isos. trap. are $\cong$.)
5. $\triangle ADC \cong \triangle BCD$ (SSS)
6. $\angle DAC \cong \angle CBD$ (CPCTC)

53. Sometimes; opp ∠s are supplementary in an isosceles trapezoid. **55.** Always; by def. a square is a quadrilateral with 4 rt. ∠s and 4 $\cong$ sides. Because by def., a rhombus is a quadrilateral with 4 $\cong$ sides, a square is always a rhombus. **57.** Sometimes; only if the parallelogram has 4 rt. ∠s and/or congruent diagonals is it a rectangle.

59.

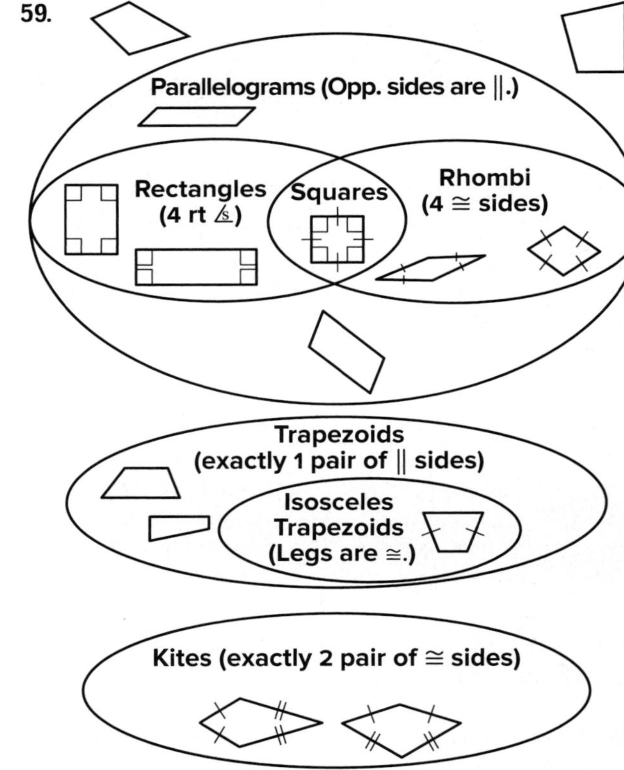

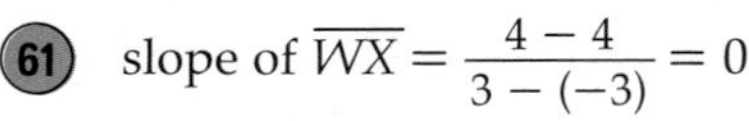

61 slope of $\overline{WX} = \frac{4-4}{3-(-3)} = 0$

slope of $\overline{XY} = \frac{3-4}{5-3} = -\frac{1}{2}$

slope of $\overline{YZ} = \frac{1-3}{-5-5} = \frac{1}{5}$

slope of $\overline{WZ} = \frac{1-4}{-5-(-3)} = \frac{3}{2}$

Because the figure has no parallel sides, it is just a quadrilateral.

63. Given: isosceles trapezoid $ABCD$ with $\overline{AD} \cong \overline{BC}$

Prove: $\overline{BD} \cong \overline{AC}$

Proof:

$DB = \sqrt{(a-b)^2 + (0-c)^2}$ or $\sqrt{(a-b)^2 + c^2}$

$AC = \sqrt{[(a-b)-0]^2 + (c-0)^2}$ or $\sqrt{(a-b)^2 + c^2}$

$BD = AC$ and $\overline{BD} \cong \overline{AC}$

65. Belinda; $m\angle D = m\angle B$. So $m\angle A + m\angle B + m\angle C + m\angle D = 360$ or $m\angle A + 100 + 45 + 100 = 360$. So, $m\angle A = 115$. **67.** Never; a square has all 4 sides ≅, while a kite does not have any opposite sides congruent. **69.** A quadrilateral must have exactly one pair of sides parallel to be a trapezoid. If the legs are congruent, then the trapezoid is an isosceles trapezoid. If a quadrilateral has exactly two pairs of consecutive congruent sides with the opposite sides not congruent, the quadrilateral is a kite. A trapezoid and a kite both have four sides. In a trapezoid and isosceles trapezoid, both have exactly one pair of parallel sides. **71.** B **73.** C **75a.** trapezoid **75b.** 5 **75c.** 35 **75d.** 13 **75e.** A, B, E

Chapter 6 Study Guide and Review

1. false, both pairs of base angles **3.** false, diagonal **5.** true **7.** false, is always **9.** Show that the parallelogram is both a rectangle and a rhombus. **11.** 1440 **13.** 720 **15.** 26 **17.** 18 **19.** 115° **21.** $x = 37$, $y = 6$ **23.** yes, Theorem 6.12

25. Given: ▱$ABCD$, $\overline{AE} \cong \overline{CF}$

Prove: Quadrilateral $EBFD$ is a parallelogram

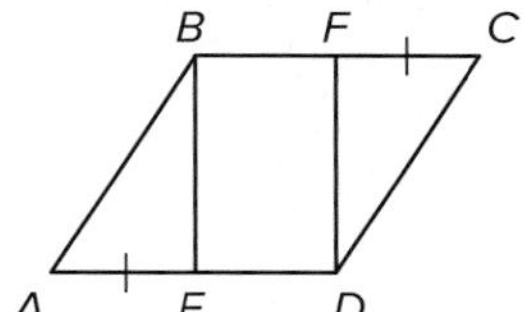

Statements (Reasons)

1. $ABCD$ is a parallelogram, $\overline{AE} \cong \overline{CF}$ (Given)
2. $AE = CF$ (Def. of ≅ segs)
3. $\overline{BC} \cong \overline{AD}$ (Opp. sides of a ▱ are ≅)
4. $BC = AD$ (Def. of ≅ segs)
5. $BC = BF + CF$, $AD = AE + ED$ (Seg. Add. Post.)
6. $BF + CF = AE + ED$ (Subst.)
7. $BF + AE = AE + ED$ (Subst.)
8. $BF = ED$ (Subt. Prop.)
9. $\overline{BF} \cong \overline{ED}$ (Def. of ≅ segs)
10. $\overline{BF} \parallel \overline{ED}$ (Def. of ▱)
11. Quadrilateral $EBFD$ is a parallelogram. (If one pair of opposite sides is parallel and congruent, then it is a parallelogram.)

27. $x = 5$, $y = 12$ **29.** 33 **31.** 64 **33.** 6 **35.** 55 **37.** 35 **39.** Rectangle, rhombus, square; all sides are ≅, consecutive sides are ⊥. **41.** 19.2 **43a.** Sample answer: The legs of the trapezoids are part of the diagonals of the square. The diagonals of a square bisect opposite angles, so each base angle of a trapezoid measures 45°. One pair of sides is parallel and the base angles are congruent.
43b. $16 + 8\sqrt{2} \approx 27.3$ in.

CHAPTER 7

Similarity

Chapter 7 Concept Check

1. cross-multiplication **3.** $m\angle SQR$ is twice $m\angle TQR$. **5.** substitution **7.** The value of x is known and value of $m\angle TQR$ in terms of x is known. Substitute and solve.

Lesson 7-1

1.

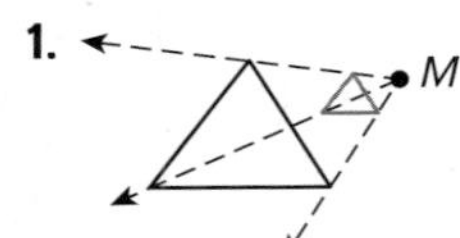

3 The figure increases in size from B to B', so it is an enlargement.

$$\frac{\text{image length}}{\text{preimage length}} = \frac{QB'}{QB} = \frac{8}{6} \text{ or } \frac{4}{3}$$

$$QB + BB' = QB'$$
$$6 + x = 8$$
$$x = 2$$

5.

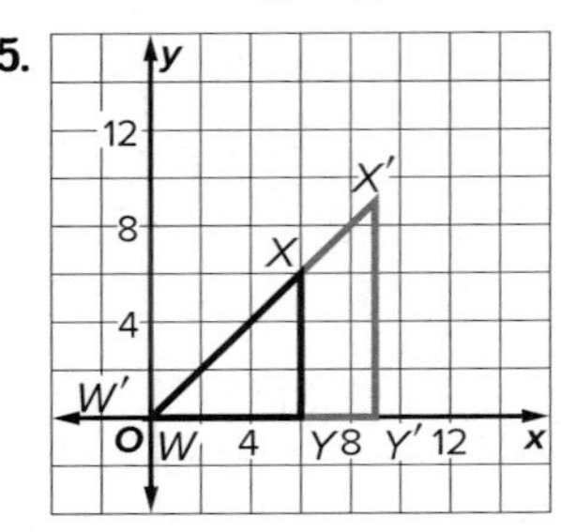

7.

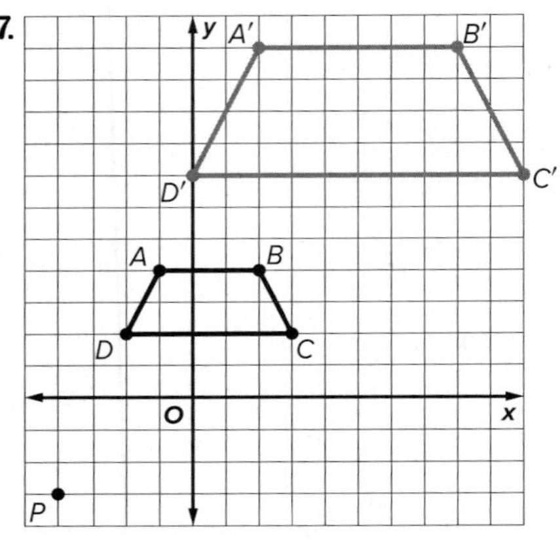

9.

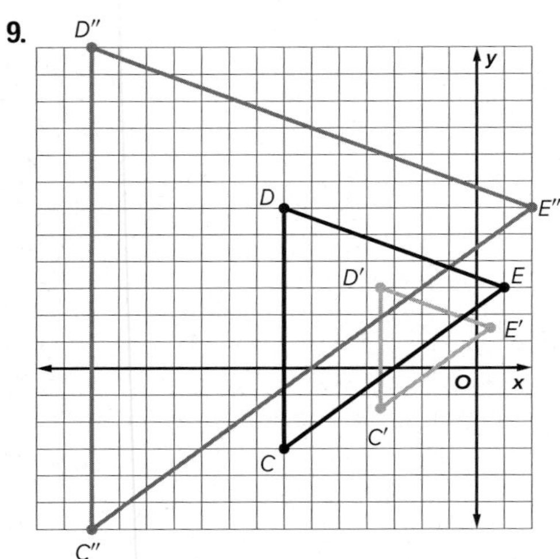

11.

13.

15. enlargement; 2; 4.5 **17.** reduction; $\frac{3}{4}$; 3.5

19. 15×; The insect's image length in millimeters is 3.75 • 10 or 37.5 mm. The scale factor of the dilation is $\frac{37.5}{2.5}$ or 15.

21 Multiply the x- and y-coordinates of each vertex by the scale factor, 0.5.

(x, y)	→	$(0.5x, 0.5y)$
$J(-8, 0)$	→	$J'(-4, 0)$
$K(-4, 4)$	→	$K'(-2, 2)$
$L(-2, 0)$	→	$L'(-1, 0)$

Graph JKL and its image $J'K'L'$.

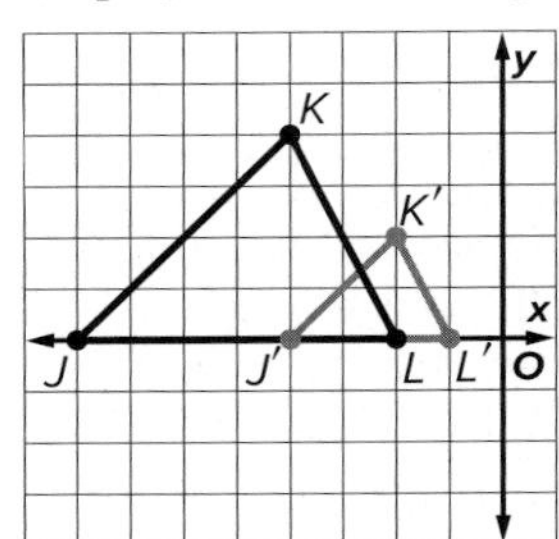

23.

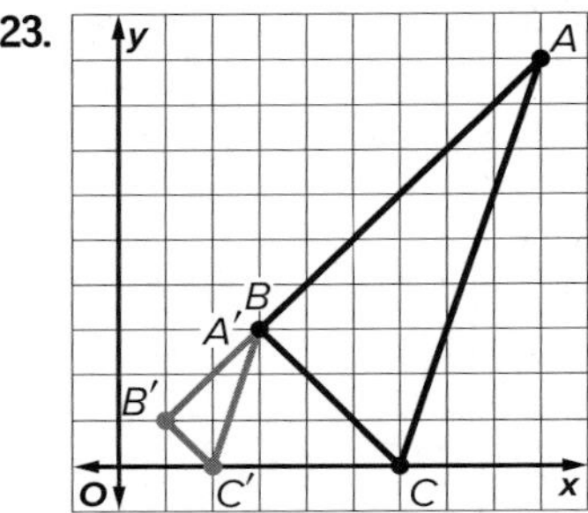

25.

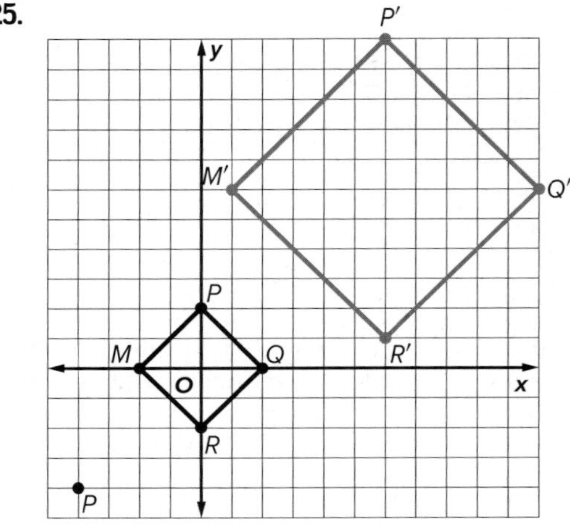

27.

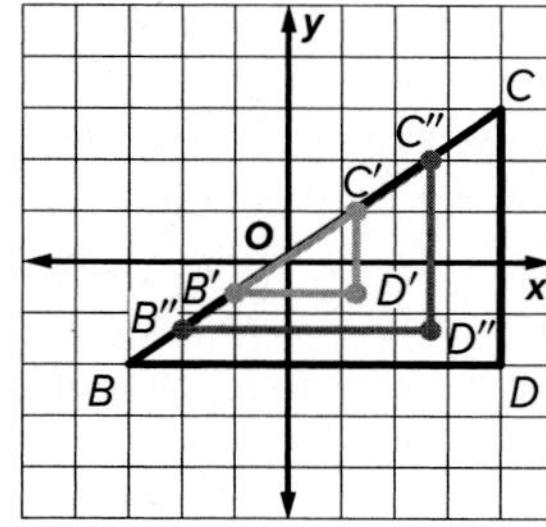

29.

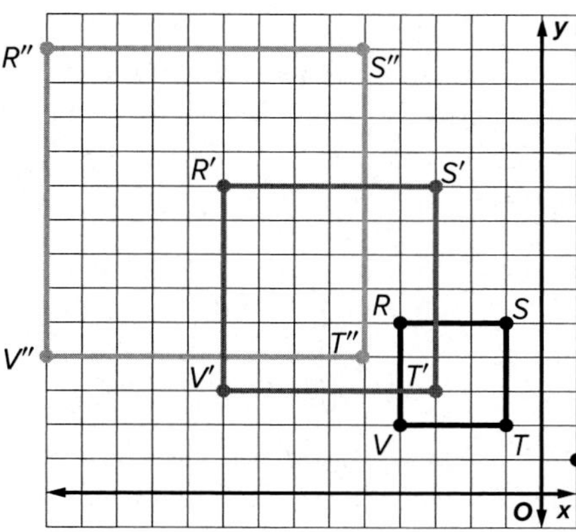

31.

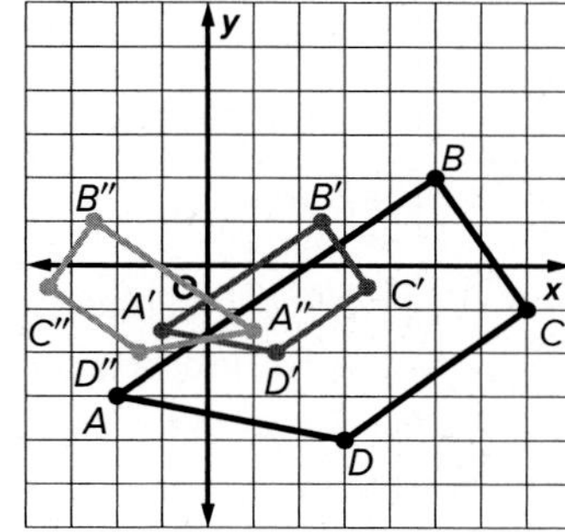

33 **a.** $S = Ph + 2B$ Surface area of a prism

$= (16)(4) + 2(12)$ $P = 16$ cm, $h = 4$ cm, $B = 12$ cm^2

$= 88$ cm^2 Simplify.

$V = Bh$ Volume of a prism

$= (12)(4)$ $B = 12$ cm^2, $h = 4$ cm

$= 48$ cm^3 Simplify.

b. Multiply the dimensions by the scale factor 2: length $= 6 \cdot 2$ or 12 cm, width $= 2 \cdot 2$ or 4 cm, height $= 4 \cdot 2$ or 8 cm.

$S = Ph + 2B$ Surface area of a prism

$= (32)(8) + 2(48)$ $P = 32$ cm, $h = 8$ cm, $B = 48$ cm^2

$= 352$ cm^2 Simplify.

$V = Bh$ Volume of a prism

$= (48)(8)$ $B = 48$ cm^2, $h = 8$ cm

$= 384$ cm^3 Simplify.

c. Multiply the dimensions by the scale factor $\frac{1}{2}$: length $= 6 \cdot \frac{1}{2}$ or 3 cm, width $= 2 \cdot \frac{1}{2}$ or 1 cm, height $= 4 \cdot \frac{1}{2}$ or 2 cm.

$S = Ph + 2B$ Surface area of a prism

$= (8)(2) + 2(3)$ $P = 8$ cm, $h = 2$ cm, $B = 3$ cm^2

$= 22$ cm^2 Simplify.

$V = Bh$ Volume of a prism

$= (3)(2)$ $B = 3$ cm^2, $h = 2$ cm

$= 6$ cm^3 Simplify.

d. surface area of preimage: 88 cm^2

surface area of image with scale factor 2: 352 cm^2 or $(88 \cdot 4)$ cm^2

surface area of image with scale factor $\frac{1}{2}$: 22 cm^2 or $\left(88 \cdot \frac{1}{4}\right)$ cm^2

The surface area is 4 times greater after dilation with scale factor 2, $\frac{1}{4}$ as great after dilation with scale factor $\frac{1}{2}$.

volume of preimage: 48 cm^3

volume of image with scale factor 2: 384 cm^3 or $(48 \cdot 8)$ cm^3

volume of image with scale factor $\frac{1}{2}$: 6 cm^3 or $\left(48 \cdot \frac{1}{8}\right)$ cm^2

The volume is 8 times greater after dilation with scale factor 2; $\frac{1}{8}$ as great after dilation with scale factor $\frac{1}{2}$.

e. The surface area of the preimage would be multiplied by r^2. The volume of the preimage would be multiplied by r^3.

35a.

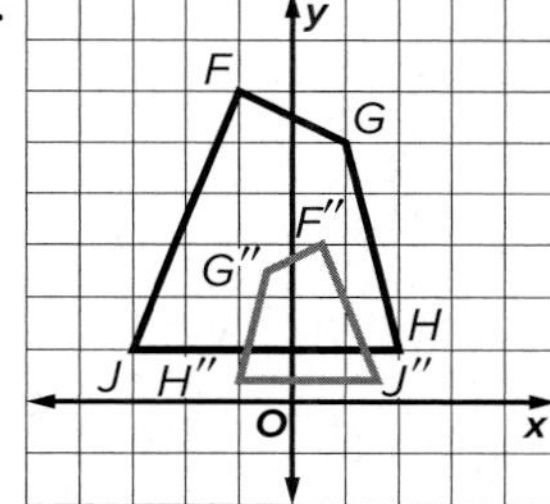

35b.

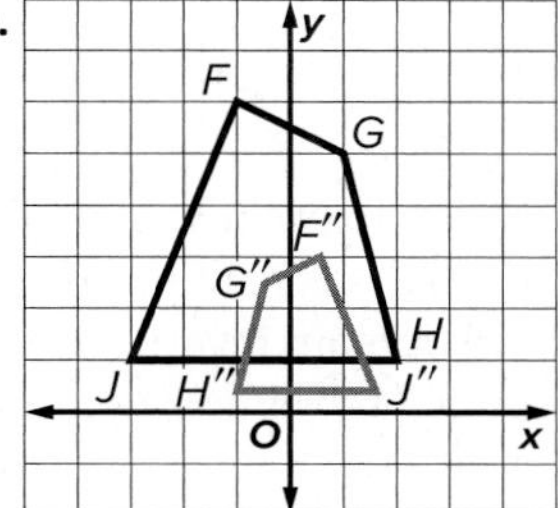

35c. no **35d.** Sometimes; sample answer: For the order of a composition of a dilation centered at the origin and a reflection to be unimportant, the line of reflection must contain the origin, or must be of the form $y = mx$.

37 **a.** $k = \dfrac{\text{diameter of image}}{\text{diameter of preimage}}$

$= \dfrac{2 \text{ mm}}{1.5 \text{ mm}}$

$$= \frac{2}{1\frac{1}{2}}$$
$$= 2 \cdot \frac{2}{3}$$
$$= \frac{4}{3} \text{ or } 1\frac{1}{3}$$

b. $A = \pi r^2$ — Area of a circle
$= \pi(0.75)^2$ — $r = 1.5 \div 2$ or 0.75
$\approx 1.77 \text{ mm}^2$ — Use a calculator.

$A = \pi r^2$ — Area of a circle
$= \pi(1)^2$ — $r = 2 \div 2$ or 1
$\approx 3.14 \text{ mm}^2$ — Use a calculator.

39. $\frac{11}{5}$

P

41 The submitted image is distorted because it was not reduced proportionally. Divide the width of the submitted photo by the width of the original photo to determine that the scale factor is 0.6. Multiply the original height of the photo by the scale factor to find the height of the image that should be submitted is 2.7 in. Becca should submit an image sized to 1.8 in. by 2.7 in.

43. ≈ 23.1

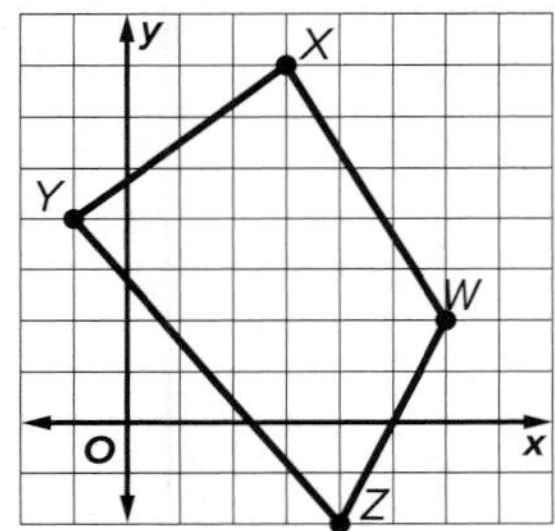

45. $y = 4x - 3$ **47a.** Always; sample answer: Because a dilation of 1 maps an image onto itself, all four vertices will remain invariant under the dilation. **47b.** Always; sample answer: Because the rotation is centered at B, point B will always remain invariant under the rotation. **47c.** Sometimes: sample answer: If one of the vertices is on the x-axis, then that point will remain invariant under reflection. If two vertices are on the x-axis, then the two vertices located on the x-axis will remain invariant under reflection. **47d.** Never; when a figure is translated, all points move an equal distance. Therefore, no points can remain invariant under translation.
47e. Sometimes; sample answer: If one of the vertices of the triangle is located at the origin, then that vertex would remain invariant under the dilation. If none of the points on $\triangle XYZ$ are located at the origin, then no points will remain invariant under the dilation.

49. Sample answer: Translations, reflections, and rotations produce congruent figures because the sides and angles of the preimage are congruent to the corresponding sides and angles of the image. Dilations produce similar figures, because the angles of the preimage and the image are congruent and the sides of the preimage are proportional to the corresponding sides of the image. A dilation with a scale factor of 1 produces an equal figure because the image is mapped onto its corresponding parts in the preimage.
51. B **53.** A

Lesson 7-2

1. Yes; map $DEFG$ to $JKLM$ using a dilation centered at the origin with scale factor 3 followed by a translation along (6, 0). **3.** $\angle A \cong \angle Z$, $\angle B \cong \angle Y$, $\angle C \cong \angle X$; $\frac{AC}{ZX} = \frac{BC}{YX} = \frac{AB}{ZY}$ **5.** no; $\frac{NQ}{WZ} \neq \frac{QR}{WX}$ **7.** 6 **9.** 22 ft
11. Yes; map $PQRS$ to $WXYZ$ using a dilation centered at the origin with scale factor 0.5 followed by a translation along (13, 2). **13.** $\angle J \cong \angle P$, $\angle F \cong \angle S$, $\angle M \cong \angle T$, $\angle H \cong \angle Q$; $\frac{PQ}{JH}, \frac{TS}{MF}, \frac{SQ}{FH} = \frac{TP}{MJ}$; **15** Yes; $\triangle LTK \cong \triangle MTK$ because $\triangle LTK \cong \triangle MTK$; scale factor: 1. **17.** Yes; sample answer: The ratio of the longer dimensions of the screens is approximately 1.1 and the ratio of the shorter dimensions of the screens is approximately 1.1.

19
$\frac{SB}{JH} = \frac{BP}{HT}$ — Similarity proportion
$\frac{2}{3} = \frac{x+3}{2x+2}$ — $SB = 2$, $JH = 3$, $BP = x + 3$, $HT = 2x + 2$
$2(2x + 2) = 3(x + 3)$ — Cross Products Property
$4x + 4 = 3x + 9$ — Distributive Property
$x + 4 = 9$ — Subtract $3x$ from each side.
$x = 5$ — Subtract 4 from each side.

21. 3 **23.** 10.8 **25.** 18.9 **27.** 40 m

29. **Given:** $\triangle ABC \sim \triangle DEF$ and $\frac{AB}{DE} = \frac{m}{n}$
Prove: $\frac{\text{perimeter of } \triangle ABC}{\text{perimeter of } \triangle DEF} = \frac{m}{n}$

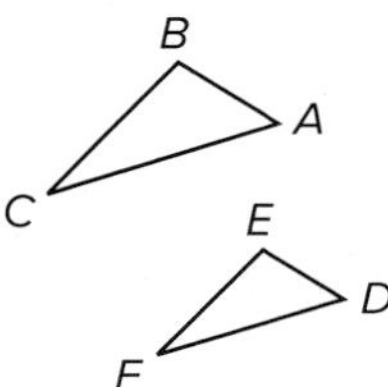

Proof: Because $\triangle ABC \sim \triangle DEF$, $\frac{AB}{DE} = \frac{BC}{EF} = \frac{AC}{DF}$. So $\frac{AB}{DE} = \frac{BC}{EF} = \frac{AC}{DF} = \frac{m}{n}$. Cross products yield $AB = DE\left(\frac{m}{n}\right)$, $BC = EF\left(\frac{m}{n}\right)$, and $AC = DF\left(\frac{m}{n}\right)$.

Using substitution, the perimeter of $\triangle ABC =$ $DE\left(\frac{m}{n}\right) + EF\left(\frac{m}{n}\right) + DF\left(\frac{m}{n}\right)$, or $\frac{m}{n}(DE + EF + DF)$.

The ratio of the two perimeters $= \frac{\frac{m}{n}(DE + EF + DF)}{DE + EF + DF}$ or $\frac{m}{n}$.

31 **a.** The ratio of the areas is the square of the ratio of the sides. $\left(\frac{4}{1}\right)^2 = \frac{16}{1}$ so the ratio of the areas is 16:1. **b.** Because both rectangles had all sides tripled, the ratio is 3(4):3(1) = 12:1 which is equal to 4:1. The ratio of the sides doesn't change. **c.** The ratio of the areas is the square of the ratio of the sides. $\left(\frac{4}{1}\right)^2 = \frac{16}{1}$, so the ratio of the areas is 16:1. **d.** Because both rectangles corresponding sides are doubled, the ratio is the same, 4:1.

33a. Sample answer:

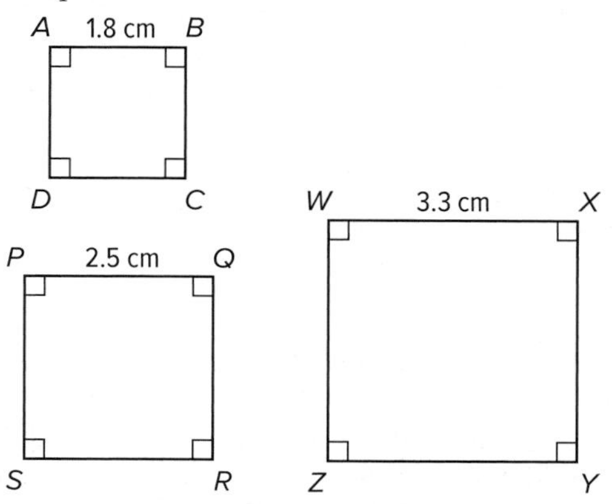

33b.

ABCD and PQRS		PQRS and WXYZ		WXYZ and ABCD	
AB:PQ	0.72	PQ:WX	0.76	WX:AB	1.8
BC:QR	0.72	QR:XY	0.76	XY:BC	1.8
CD:RS	0.72	RS:YZ	0.76	YZ:CD	1.8
AD:SP	0.72	SP:ZW	0.76	ZW:DA	1.8

ABCD is similar to *PQRS*; *PQRS* is similar to *WXYZ*; *WXYZ* is similar to *ABCD*.

33c. Sample answer: All squares are similar.

35. Sample answer:

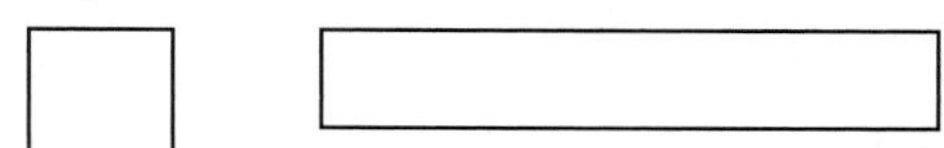

37. Sample answer: The figures could be described as congruent if they are the same size and shape, similar if their corresponding angles are congruent and their corresponding sides are proportional, and equal if they are the same exact figure. **39.** D **41.** 39 **43a.** Map *DEFG* to *HJKL* using a dilation centered at the origin with scale factor 0.5 followed by a translation along <6, 5>. **43b.** $DEFG \cong HJKL$ **43c.** $\frac{DE}{HJ} = \frac{EF}{JK} = \frac{FG}{KL} = \frac{GD}{LH}$ **43d.** 3.1 cm; because the scale factor of the dilation is 0.5, the perimeter of *HJKL* is 0.5 times the perimeter of *DEFG*.

Lesson 7-3

1. Yes; $\triangle YXZ \sim \triangle VWZ$ by AA Similarity. **3.** No; the angles are not congruent. **5.** C **7.** $\triangle QVS \sim \triangle RTS$; 20 **9.** Yes; $\triangle ACE \sim \triangle BCD$ by AA Similarity. **11** Yes; $\angle WVX$ and $\angle VTU$ are both right angles, so they are congruent. $\angle WXV$ and $\angle VUT$ are marked congruent. So, the triangles are similar by AA Similarity. **13.** No; not enough information is given to determine that the triangles are similar. **15.** Reflect or rotate one triangle about the shared vertex at *B*, then dilate one triangle to show the triangles are similar.

17 Because $\overline{RS} \parallel \overline{PT}$, $\angle QRS \cong \angle QPT$ and $\angle QSR \cong \angle QTP$ because they are corresponding angles. By AA Similarity, $\triangle QRS \sim \triangle QPT$.

$\frac{RS}{PT} = \frac{QS}{QT}$	Definition of similar polygons
$\frac{12}{16} = \frac{x}{20}$	$RS = 12, PT = 16, QS = x, QT = 20$
$12 \cdot 20 = 16 \cdot x$	Cross Products Property
$240 = 16x$	Simplify.
$15 = x$	Divide each side by 16.

Because $QS + ST = 20$ and $QS = 15$, $ST = 5$.

19. $\triangle HJK \sim \triangle NQP$; 15, 10 **21.** $\triangle GHJ \sim \triangle GDH$; 14, 20 **23.** 12.8 ft

25. **Reflexive Property of Similarity**

Given: $\triangle ABC$

Prove: $\triangle ABC \sim \triangle ABC$

Proof: Statements (Reasons)

1. $\triangle ABC$ (Given)
2. $\angle A \cong \angle A$, $\angle B \cong \angle B$ (Refl. Prop.)
3. $\triangle ABC \sim \triangle ABC$ (AA Similarity)

Symmetric Property of Similarity

Given: $\triangle ABC \sim \triangle DEF$

Prove: $\triangle DEF \sim \triangle ABC$

Proof: Statements (Reasons)

1. $\triangle ABC \sim \triangle DEF$ (Given)
2. $\angle A \cong \angle D$, $\angle B \cong \angle E$ (Def. of ~ polygons)
3. $\angle D \cong \angle A$, $\angle E \cong \angle B$ (Symm. Prop.)
4. $\triangle DEF \sim \triangle ABC$ (AA Similarity)

Transitive Property of Similarity

Given: $\triangle ABC \sim \triangle DEF$ and $\triangle DEF \sim \triangle GHI$

Prove: $\triangle ABC \sim \triangle GHI$

Proof: Statements (Reasons)

1. $\triangle ABC \sim \triangle DEF$, $\triangle DEF \sim \triangle GHI$ (Given)
2. $\angle A \cong \angle D$, $\angle B \cong \angle E$, $\angle D \cong \angle G$, $\angle E \cong \angle H$ (Def. of ~ polygons)
3. $\angle A \cong \angle G$, $\angle B \cong \angle H$ (Trans. Prop.)
4. $\triangle ABC \sim \triangle GHI$ (AA Similarity)

27. Given: $\triangle XYZ$ and $\triangle ABC$ are right triangles; $\angle Z \cong \angle C$

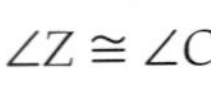

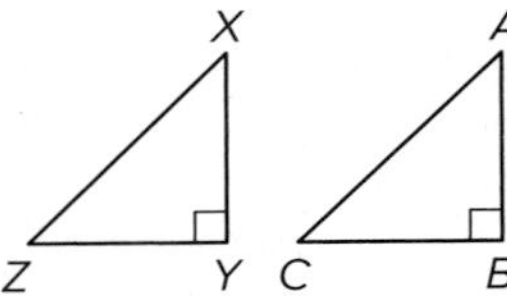

Prove: $\triangle YXZ \sim \triangle BAC$

Proof:

Statements (Reasons)

1. $\triangle XYZ$ and $\triangle ABC$ are right triangles. (Given)
2. $\angle XYZ$ and $\angle ABC$ are right angles. (Def. of rt. $\triangle$)
3. $\angle XYZ \cong \angle ABC$ (All rt. ∠s are ≅.)
4. $\angle Z \cong \angle C$ (Given)
5. $\triangle YXZ \sim \triangle BAC$ (AA Similarity)

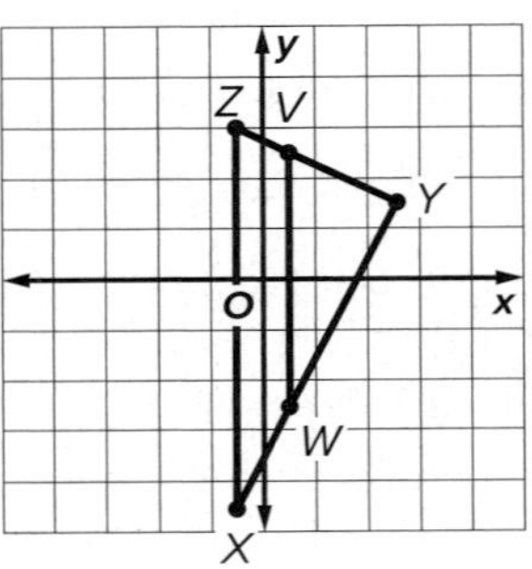

$XY = \sqrt{12^2 + 6^2} = \sqrt{180}$ or $6\sqrt{5}$;
$YZ = \sqrt{3^2 + (-6)^2} = \sqrt{45}$ or $3\sqrt{5}$;
$ZX = 6 - (-9) = 15$; $VW = 5 - (-5) = 10$;
$WY = \sqrt{8^2 + 4^2} = \sqrt{80}$ or $4\sqrt{5}$;
$YV = \sqrt{2^2 + (-4)^2} = \sqrt{20} = 2\sqrt{5}$.
$\frac{XY}{WY} = \frac{6\sqrt{5}}{4\sqrt{5}}$ or $\frac{3}{2}$, $\frac{YZ}{YV} = \frac{3\sqrt{5}}{2\sqrt{5}}$ or $\frac{3}{2}$,
$\frac{ZX}{VW} = \frac{15}{10}$ or $\frac{3}{2}$. Because $\frac{XY}{WY} = \frac{YZ}{YV} = \frac{ZX}{VW} = \frac{3}{2}$,
$\triangle XYZ \sim \triangle WYV$ by SSS Similarity.

31. about 61 in.

33a. The scale factor from $\triangle ABC$ to $\triangle JKL$ is $\frac{1}{2}$. So $JK = \frac{1}{2}AB$, $KL = \frac{1}{2}BC$, and $JL = \frac{1}{2}AC$. Thus the perimeter of $\triangle JKL$ is $\frac{1}{2}AB + \frac{1}{2}BC + \frac{1}{2}AC = \frac{1}{2}$ (perimeter of $\triangle ABC$) $= \frac{1}{2}$ (40 in.) or 20 in. The ratio of the areas is $\frac{1}{2}$, which is the scale factor.
b. The scale factor from $\triangle ABC$ to $\triangle JKL$ is $\frac{1}{3}$. So $JK = \frac{1}{3}AB$, $KL = \frac{1}{3}BC$, and $JL = \frac{1}{3}AC$. Thus the perimeter of $\triangle JKL$ is $\frac{1}{3}AB + \frac{1}{3}BC + \frac{1}{3}AC = \frac{1}{3}$ (perimeter of $\triangle ABC$) $= \frac{1}{3}$ (21 in.) or 7 in. The ratio of the areas is $\frac{1}{3}$, which is the scale factor.

35a. Sample answer:

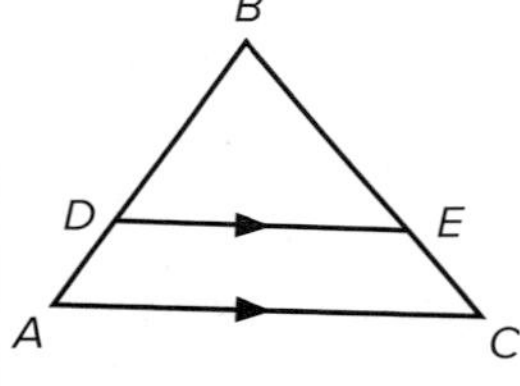

35b. Sample answer:

Lengths		Ratios	
AD	0.9 cm	$\frac{AD}{DB}$	$\frac{1}{2}$
DB	1.8 cm		
CE	1.1 cm	$\frac{CE}{EB}$	$\frac{1}{2}$
EB	2.2 cm		

35c. Sample answer: The segments created by a line ∥ to one side of a $\triangle$ and intersecting the other two sides are proportional. **37.** Yes; Because it is an altitude, $\overline{YW} \perp \overline{XZ}$. So $\angle XWY$ and $\angle ZWY$ are right angles. It is given that $\angle XYZ$ is a right angle. Thus $\angle XWY \cong \angle XYZ$ and $\angle XYZ \cong \angle ZWY$ because all right angles are congruent. $\angle WXY \cong \angle YXZ$ by the Reflexive property. So $\triangle WXY \sim \triangle YXZ$ by AA Similarity. $\angle XZY \cong \angle WZY$ by the Reflexive property. So $\triangle YXZ \sim \triangle WYZ$ by AA Similarity. Therefore $\triangle WXY \sim \triangle WYZ$ by the Transitive property.

39. Sample answer:
$\triangle A'B'C' \sim \triangle ABC$ because the measures of each side have a scale factor of 0.5 and the measures of corresponding angles are equal.

41. 6 **43a.** D **43b.** C

Lesson 7-4

1. Yes; $\triangle LMN \sim \triangle OPN$ by SAS Similarity. **3.** No; corresponding sides are not proportional. **5.** $\triangle LMN$ is not similar to $\triangle PQR$. **7a.** No; need congruent included angles. **7b.** Yes; SSS Similarity. **7c.** Yes; SSS Similarity. **7d.** No; need congruent included angles. **9.** Yes; $\triangle XUZ \sim \triangle WUY$ by SSS Similarity. **11.** Yes; $\triangle CBA \sim \triangle DBF$ by SAS Similarity. **13.** No; not enough information to determine. If $JH = 3$ or $WY = 24$, then $\triangle JHK \sim \triangle XWY$ by SSS Similarity. **15.** Yes; $\triangle JLK \sim \triangle PLM$ by AA Similarity.

17 No; there is not enough information to determine. If sides $\overline{AF}$ and $\overline{DF}$ were known to be proportional, the triangles would be congruent by SAS Similarity. If either $\angle C$ and $\angle B$ or $\angle A$ and $\angle D$ were congruent, the triangles would be congruent by AA Similarity.

19. $\triangle DEF \sim \triangle GHT$ by SSS Similarity because $\frac{DE}{GH} = \frac{EF}{HI} = \frac{DF}{GI} = \frac{1}{3}$ **21.** $\triangle JKL \sim \triangle MNO$ by SSS Similarity because $\frac{JK}{MN} = \frac{KL}{NO} = \frac{JL}{MO} = 4$
23. $\triangle FGH \sim \triangle FJK$ by SAS Similarity because $\angle GFH \cong \angle JFK$ and $\frac{FG}{FJ} = \frac{FH}{FK} = 3$ **25.** $\triangle MNP \not\sim \triangle MRT$

27. Proof:

Statements (Reasons)

1. $\triangle XYZ$ and $\triangle ABC$ are right triangles. (Given)
2. $\angle XYZ$ and $\angle ABC$ are right angles. (Def. of rt. $\triangle$)
3. $\angle XYZ \cong \angle ABC$ (All rt. ∠s are ≅.)
4. $\frac{XY}{AB} = \frac{YZ}{BC}$ (Given)
5. $\triangle YXZ \sim \triangle BAC$ (SAS Similarity)

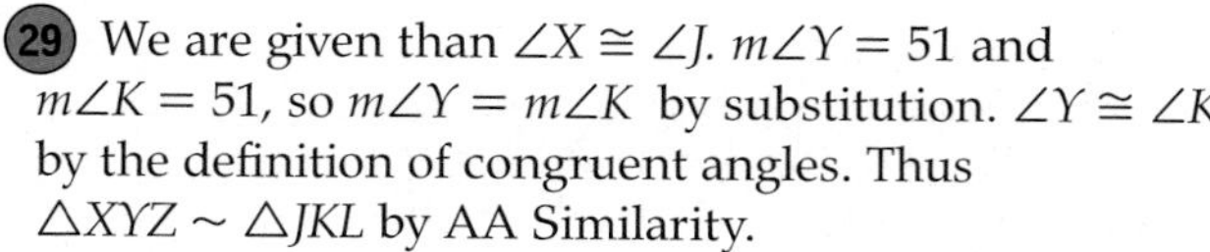

29 We are given than $\angle X \cong \angle J$. $m\angle Y = 51$ and $m\angle K = 51$, so $m\angle Y = m\angle K$ by substitution. $\angle Y \cong \angle K$ by the definition of congruent angles. Thus $\triangle XYZ \sim \triangle JKL$ by AA Similarity.

$$\frac{XY}{JK} = \frac{YZ}{KL}$$
$$\frac{5}{4} = \frac{15}{x}$$
$$5x = 60$$
$$x = 12$$

31. Proof:

<u>Statements (Reasons)</u>

1. $\triangle ABC$ and $\triangle DEF$ are right triangles. (Given)
2. $\angle B$ and $\angle E$ are right angles. (Def. of rt. triangle)
3. $\angle B \cong \angle E$ (All rt. angles are congruent.)
4. $DE = \frac{2}{3}AB$, $EF = \frac{2}{3}BC$ (Given)
5. $\frac{DE}{AB} = \frac{2}{3}, \frac{EF}{BC} = \frac{2}{3}$ (Div. Prop. of =)
6. $\frac{DE}{AB} = \frac{EF}{BC}$ (Substitution)
7. $\triangle ABC \sim \triangle DEF$ (SAS Similarity Theorem)
8. $\frac{DF}{AC} = \frac{DE}{AB}$ (Corr. sides of $\sim \triangle$s are proportional.)

33a.

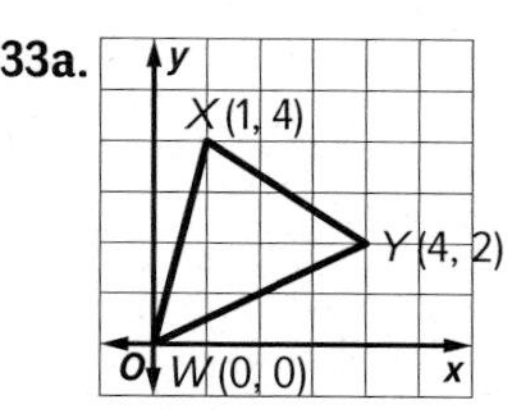

33b. $W'(0, 0)$, $X'(2, 8)$, and $Y'(8, 4)$ **33c.** Use the Distance Formula to find the side lengths of the preimage triangle and of the image triangle. Then find the ratio of the side lengths for each pair of corresponding sides. Use coordinates $W'(0, 0)$, $X'(2, 8)$, and $Y'(8, 4)$ for the image triangle.

$$WX = \sqrt{(1-0)^2 + (4-0)^2} = \sqrt{17}$$
$$W'X' = \sqrt{(2-0)^2 + (8-0)^2} = 2\sqrt{17}$$

So, $\frac{W'X'}{WX} = \frac{2\sqrt{17}}{\sqrt{17}}$ or 2.

$$XY = \sqrt{(4-1)^2 + (2-4)^2} = \sqrt{13}$$
$$X'Y' = \sqrt{(8-2)^2 + (4-8)^2} = 2\sqrt{13}$$

So, $\frac{X'Y'}{XY} = \frac{2\sqrt{13}}{\sqrt{13}}$ or 2.

$$WX = \sqrt{(2-0)^2 + (4-0)^2} = 2\sqrt{5}$$
$$W'X' = \sqrt{(4-0)^2 + (8-0)^2} = 4\sqrt{5}$$

So, $\frac{X'Z'}{XZ} = \frac{4\sqrt{5}}{2\sqrt{5}}$ or 2.

Because all three pairs of corresponding sides have a ratio of 2 to 1, the triangles are similar by the SSS Similarity Theorem.

35a. 10 in^2; The ratio of the areas is the square of the scale factor. **35b.** 7 in^2; The ratio of the areas is the square of the scale factor.

37a.

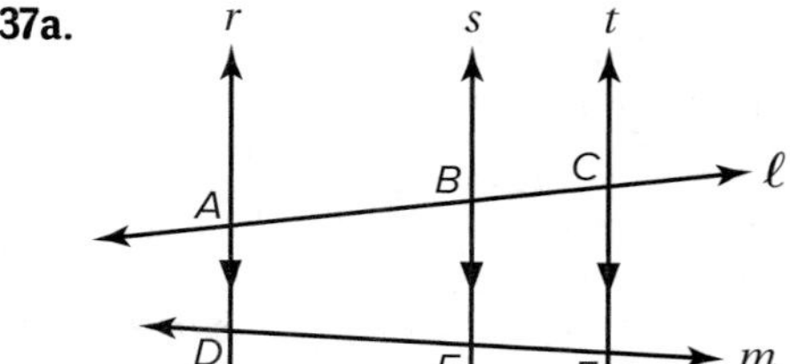

37b. Sample answer:

Lengths		Ratios	
AB	13 mm	$\frac{AB}{BC}$	1.625
BC	8 mm		
DE	12 mm	$\frac{DE}{EF}$	1.714
EF	7 mm		

37c. Sample answer: If three parallel lines intersect two transversals, then they divide the transversals proportionally. **39.** Sample answer: $\triangle EFG$ and $\triangle DBC$ are right triangles. Use the Pythagorean Theorem to find EG, or 6, and DB, or $2\sqrt{2}$. $\angle EFG \cong \angle DBC$ because all right angles are congruent and $\frac{EF}{DB} = \frac{FG}{BC} = \frac{3}{2}$. So $\triangle EFG \sim \triangle DBC$ by the SAS Similarity Theorem.

41. Sample answer:

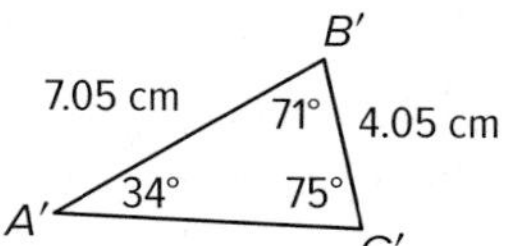

43. Sample answer: Because one pair of corresponding angles are congruent, compare the ratios of the sides for which the angle is the included angle, or $\frac{KM}{ON} = \frac{30}{25}$ or $\frac{6}{5}$; $\frac{LM}{PN} = \frac{36}{30}$ or $\frac{6}{5}$. Because $\frac{KM}{ON} = \frac{LM}{PN}$, the SAS Similarity Theorem applies. To find KL, solve the proportion $\frac{KL}{OP} = \frac{6}{5}$. $\frac{KL}{OP} = \frac{6}{5}$; $\frac{4x}{3x+2} = \frac{6}{5}$; $x = 6$; Thus $KL = 24$, and $OP = 20$. **45a.** C **45b.** D **45c.** B, D, E **47.** D

Lesson 7-5

1. 10 **3.** Yes; $\frac{AD}{DC} = \frac{BE}{EC} = \frac{2}{3}$, so $\overline{DE} \parallel \overline{AB}$. **5.** 11 **7.** 2360.3 ft **9.** $x = 20$; $y = 2$

11 If $AB = 12$ and $AC = 16$, then $BC = 4$.

$\frac{AB}{BC} = \frac{AE}{ED}$	Triangle Proportionality Theorem
$\frac{12}{4} = \frac{AE}{5}$	Substitute.
$12 \cdot 5 = 4 \cdot AE$	Cross Products Property
$60 = 4AE$	Multiply.
$15 = AE$	Divide each side by 4.

13. 10 **15.** yes; $\frac{ZV}{VX} = \frac{WY}{YX} = \frac{11}{5}$ **17.** no; $\frac{ZV}{VX} \neq \frac{WY}{YX}$

19

$m\angle PHM + m\angle PHJ + m\angle JHL = 180$	Definition of a straight angle
$44 + m\angle PHJ + 76 = 180$	Substitution
$120 + m\angle PHJ = 180$	Simplify.

$m\angle PHJ = 60$ Subtract 120 from each side.

By the Triangle Midsegment Theorem, $\overline{PH} \parallel \overline{KL}$.

$\angle PHJ \cong \angle JHL$	Alternate Interior Angles Theorem
$m\angle PHJ = m\angle JHL$	Definition of congruence
$60 = x$	Substitution

21. 1.35 **23.** 1.2 in. **25.** $x = 18$; $y = 3$ **27.** $x = 48$; $y = 72$

29. Given: $\overline{AD} \parallel \overline{BE} \parallel \overline{CF}$, $\overline{AB} \cong \overline{BC}$
Prove: $\overline{DE} \cong \overline{EF}$
Proof:
From Corollary 7.1, $\frac{AB}{BC} = \frac{DE}{EF}$.
Because $\overline{AB} \cong \overline{BC}$, $AB = BC$ by definition of congruence.
Therefore, $\frac{AB}{BC} = 1$.
By substitution, $1 = \frac{DE}{EF}$. Thus, $DE = EF$. By definition of congruence, $\overline{DE} \cong \overline{EF}$.

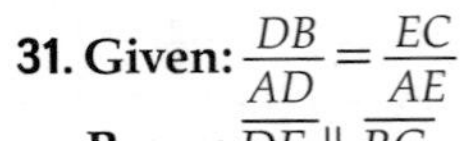

31. Given: $\frac{DB}{AD} = \frac{EC}{AE}$
Prove: $\overline{DE} \parallel \overline{BC}$

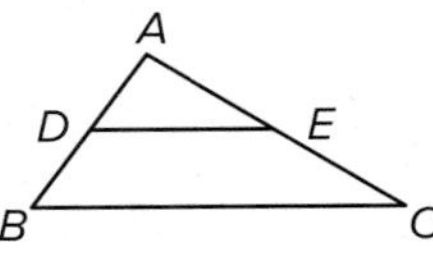

Proof:
Statements (Reasons)
1. $\frac{DB}{AD} = \frac{EC}{AE}$ (Given)
2. $\frac{AD}{AD} + \frac{DB}{AD} = \frac{AE}{AE} + \frac{EC}{AE}$ (Add. Prop.)
3. $\frac{AD + DB}{AD} = \frac{AE + EC}{AE}$ (Subst.)
4. $AB = AD + DB$, $AC = AE + EC$ (Seg. Add. Post.)
5. $\frac{AB}{AD} = \frac{AC}{AE}$ (Subst.)
6. $\angle A \cong \angle A$ (Refl. Prop.)
7. $\triangle ADE \sim \triangle ABC$ (SAS Similarity)
8. $\angle ADE \cong \angle ABC$ (Def. of $\sim$ polygons)
9. $\overline{DE} \parallel \overline{BC}$ (If corr. $\angle$s are $\cong$, then the lines are $\parallel$.)

33. 9

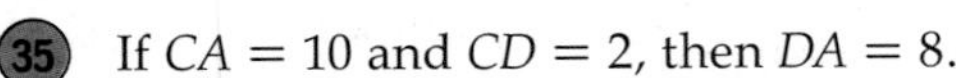

35 If $CA = 10$ and $CD = 2$, then $DA = 8$.

$\frac{CE}{EB} = \frac{CD}{DA}$	Triangle Proportionality Theorem
$\frac{t-2}{t+1} = \frac{2}{8}$	Substitute.
$(t-2)(8) = (t+1)(2)$	Cross Products Property
$8t - 16 = 2t + 2$	Distributive Property
$6t - 16 = 2$	Subtract $2t$ from each side.
$6t = 18$	Add 16 to each side.
$t = 3$	Divide each side by 6.

If $t = 3$, then $CE = 3 - 2$ or 1.

37. 8, 7.5 **39.** Because $\overline{FD} \parallel \overline{AC}$, $\angle EFD \cong \angle EAC$. By the Reflexive Property of Congruence, $\angle AEC \cong \angle AEC$. Therefore, $\triangle EFD \sim \triangle EAC$ by the AA Similarity Postulate. By the Triangle Proportionality Theorem, ED is proportional to EC. Because the segments stay parallel as the gauge is repositioned, ED remains proportional to EC. **41.** 6

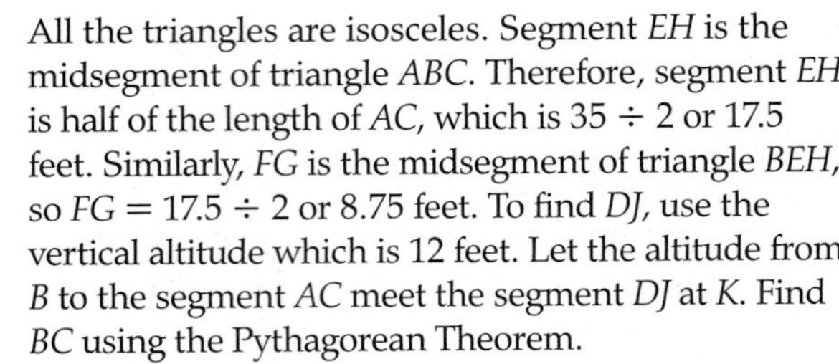

43 All the triangles are isosceles. Segment EH is the midsegment of triangle ABC. Therefore, segment EH is half of the length of AC, which is $35 \div 2$ or 17.5 feet. Similarly, FG is the midsegment of triangle BEH, so $FG = 17.5 \div 2$ or 8.75 feet. To find DJ, use the vertical altitude which is 12 feet. Let the altitude from B to the segment AC meet the segment DJ at K. Find BC using the Pythagorean Theorem.

$$BC^2 = BK^2 + KC^2$$
$$BC^2 = 12^2 + 17.5^2$$
$$BC = \sqrt{12^2 + 17.5^2}$$
$$BC \approx 21.22 \text{ ft}$$

Because the width of each piece of siding is the same, $BJ = \frac{3}{4}BC$, which is about $\frac{3}{4}(21.22)$ or 15.92 ft. Now, use the Triangle Proportionality Theorem.

$$\frac{AC}{BC} = \frac{DJ}{BJ}$$
$$\frac{35}{21.22} = \frac{DJ}{15.92}$$
$$21.22(DJ) = (15.92)(35)$$
$$21.22(DJ) = 557.2$$
$$DJ \approx 26.25 \text{ ft}$$

45. Sample answer:

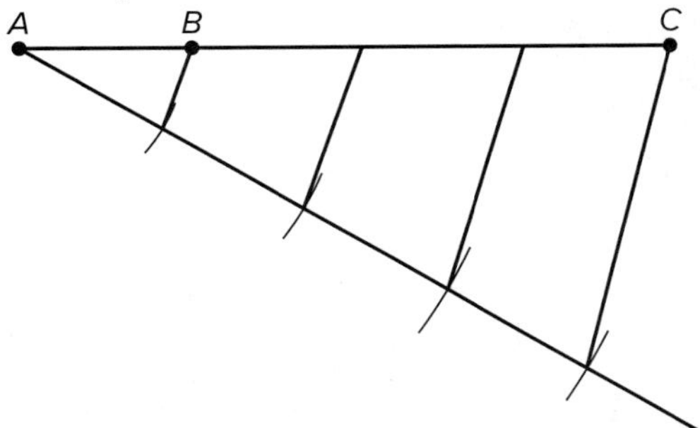

47a. Sample answer:

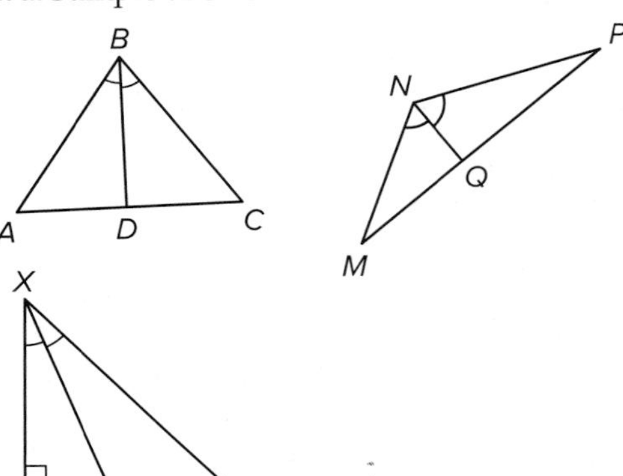

47b.

Triangle	Length		Ratio	
ABC	*AD*	1.1 cm	$\frac{AD}{CD}$	1.0
	CD	1.1 cm		
	AB	2.0 cm	$\frac{AB}{CB}$	1.0
	CB	2.0 cm		
MNP	*MQ*	1.4 cm	$\frac{MQ}{PQ}$	0.8
	PQ	1.7 cm		
	MN	1.6 cm	$\frac{MN}{PN}$	0.8
	PN	2.0 cm		
WXY	*WZ*	0.8 cm	$\frac{WZ}{YZ}$	0.7
	YZ	1.2 cm		
	WX	2.0 cm	$\frac{WX}{YX}$	0.7
	YX	2.9 cm		

47c. Sample answer: The proportion of the segments created by the angle bisector of a triangle is equal to the proportion of their respective consecutive sides.

49. Always; sample answer: $\overline{FH}$ is a midsegment. Let $BC = x$, then $FH = \frac{1}{2}x$. *FHCB* is a trapezoid, so $DE = \frac{1}{2}(BC + FH) = \frac{1}{2}\left(x + \frac{1}{2}x\right) = \frac{1}{2}x + \frac{1}{4}x = \frac{3}{4}x$. Therefore, $DE = \frac{3}{4}BC$.

51.

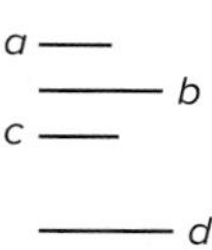

a —
— b
c —
— d

By Corollary 7.1, $\frac{a}{b} = \frac{c}{d}$.

53. B **55.** C **57.** 50 **59.** $\frac{20}{3}$

Lesson 7-6

1 The triangles are similar by AA Similarity.

$\frac{x}{10} = \frac{12}{15}$	$\sim\triangle$s have corr. medians proportional to the corr. sides.
$x \cdot 15 = 10 \cdot 12$	Cross Products Property
$15x = 120$	Simplify.
$x = 8$	Divide each side by 15.

3. 35.7 in. **5.** 20 **7.** 8.5 **9.** 18

11

$\frac{15}{27} = \frac{28 - b}{b}$	Triangle Angle Bisector Theorem
$15 \cdot b = 27(28 - b)$	Cross Products Property
$15b = 756 - 27b$	Multiply.
$42b = 756$	Add $27b$ to each side.
$b = 18$	Divide each side by 42.

13. 15

15

$\frac{AB}{JK} = \frac{AD}{JM}$	$\sim\triangle$s have corr. altitudes proportional to the corr. sides.
$\frac{9}{21} = \frac{4x - 8}{5x + 3}$	Substitute.
$9(5x + 3) = 21(4x - 8)$	Cross Products Property
$45x + 27 = 84x - 168$	Distributive Property
$27 = 39x - 168$	Subtract $45x$ from each side.
$195 = 39x$	Add 168 to each side.
$5 = x$	Divide each side by 39.

17. 4

19. Given: $\triangle ABC \sim \triangle RST$
$\overline{AD}$ is a median of $\triangle ABC$.
$\overline{RU}$ is a median of $\triangle RST$.

Prove: 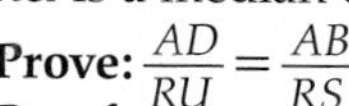 $\frac{AD}{RU} = \frac{AB}{RS}$

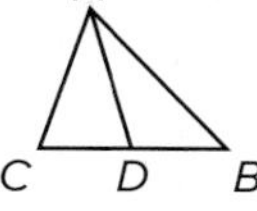

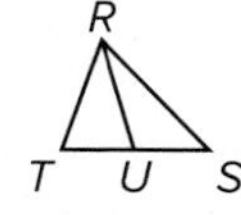

Proof:

Statements (Reasons)

1. $\triangle ABC \sim \triangle RST$; $\overline{AD}$ is a median of $\triangle ABC$; $\overline{RU}$ is a median of $\triangle RST$. (Given)
2. $CD = DB$; $TU = US$ (Def. of median)
3. $\frac{AB}{RS} = \frac{CB}{TS}$ (Def. of $\sim\triangle$s)
4. $CB = CD + DB$; $TS = TU + US$ (Seg. Add. Post.)
5. $\frac{AB}{RS} = \frac{CD + DB}{TU + US}$ (Subst.)
6. $\frac{AB}{RS} = \frac{DB + DB}{US + US}$ or $\frac{2(DB)}{2(US)}$ (Subst.)
7. $\frac{AB}{RS} = \frac{DB}{US}$ (Subst.)
8. $\angle B \cong \angle S$ (Def. of $\sim\triangle$s)
9. $\triangle ABD \sim \triangle RSU$ (SAS Similarity)
10. $\frac{AD}{RU} = \frac{AB}{RS}$ (Def. of $\sim\triangle$s)

21. 3 **23.** 70

25. Given: $\overline{CD}$ bisects $\angle ACB$.
By construction, $\overline{AE} \parallel \overline{CD}$.

Prove: $\frac{AD}{DB} = \frac{AC}{BC}$

E, C, 3, 1, 2, A, D, B

Proof:

Statements (Reasons)

1. $\overline{CD}$ bisects $\angle ACB$; By construction, $\overline{AE} \parallel \overline{CD}$. (Given)
2. $\frac{AD}{DB} = \frac{EC}{BC}$ ($\triangle$ Prop. Thm.)
3. $\angle 1 \cong \angle 2$ (Def. of $\angle$ Bisector)
4. $\angle 3 \cong \angle 1$ (Alt. Int. $\angle$s Thm.)
5. $\angle 2 \cong \angle E$ (Corr. $\angle$s Post.)
6. $\angle 3 \cong \angle E$ (Trans. Prop.)
7. $\overline{EC} \cong \overline{AC}$ (Converse of Isos. $\triangle$ Thm.)
8. $EC = AC$ (Def. of $\cong$ segs.)
9. $\frac{AD}{DB} = \frac{AC}{BC}$ (Subst.)

27. Given: $\triangle QTS \sim \triangle XWZ$,
$\overline{TR}$, $\overline{WY}$ are $\angle$ bisectors.

Prove: $\frac{TR}{WY} = \frac{QT}{XW}$

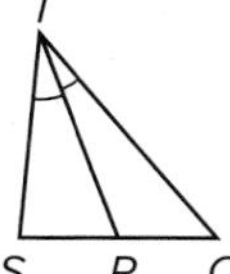

Proof:

Statements (Reasons)

1. $\triangle QTS \sim \triangle XWZ$, $\overline{TR}$ and $\overline{WY}$ are angle bisectors. (Given)
2. $\angle QTS \cong \angle XWZ$, $\angle Q \cong \angle X$ (Def of $\sim\triangle$s)
3. $\angle STR \cong \angle QTR$, $\angle ZWY \cong \angle XWY$ (Def. $\angle$ bisector)
4. $m\angle STQ = m\angle STR + m\angle QTR$, $m\angle ZWX = m\angle ZWY + m\angle XWY$ ($\angle$ Add. Post.)

5. $m\angle STQ = 2m\angle QTR$, $m\angle ZWX = 2m\angle XWY$ (Subst.)
6. $2m\angle QTR = 2m\angle XWY$ (Subst.)
7. $m\angle QTR = m\angle XWY$ (Div. Prop.)
8. $\angle QTR \cong \angle XWY$ (Def. of $\cong$ Angles)
9. $\triangle QTR \sim \triangle XWY$ (AA Similarity)
10. $\frac{TR}{WY} = \frac{QT}{XW}$ (Def. of $\sim$ $\triangle$s)

29 Because the segment from Trevor to Ricardo is an angle bisector, the segments from Ricardo to Craig and from Ricardo to Eli are proportional to the segments from Trevor to Craig and from Trevor to Eli. Because Craig is closer to Trevor than Eli is, Craig is also closer to Ricardo than Eli is. So, Craig will reach Ricardo first.

31. Chun; by the Angle Bisector Theorem, the correct proportion is $\frac{5}{8} = \frac{15}{x}$. **33.** $PS = 18.4$, $RS = 24$ **35.** Both theorems have a segment that bisects an angle and have proportionate ratios. The Triangle Angle Bisector Theorem pertains to one triangle, while Theorem 7.9 pertains to similar triangles. Unlike the Triangle Angle Bisector Theorem, which separates the opposite side into segments that have the same ratio as the other two sides, Theorem 7.9 relates the angle bisector to the measures of the sides. **37.** D **39.** 36

Chapter 7 Study Guide and Review

1. f; midsegment. **3.** g; dilation **5.** i; reduction **7.** j; Transitive Property of Similarity **9.** The line segment must be parallel to the third side and its length is one half the length of the third side. **11.** reduction; 0.45; 8.25 **13.** Yes, the rectangles are similar because all of the corresponding angles are congruent and the corresponding sides are proportional in a 3:2 ratio. **15.** The triangles are not similar. **17.** 34.2 ft **19.** Yes, $\triangle IJK \sim \triangle HFG$ by the SSS $\sim$ Thm. **21.** 22.5 **23.** 6 **25.** 633 mi

CHAPTER 8

Right Triangles and Trigonometry

Chapter 8 Concept Check

1. Multiply by $\frac{\sqrt{3}}{\sqrt{3}}$. **3.** $a^2 + b^2 = c^2$ **5.** Take the positive square root of each side. **7.** 68.8 in. **9.** Plot points A and B and then connect the plotted points.

Lesson 8-1

1. 10 **3.** $10\sqrt{6}$ or 24.5 **5.** $x = 6$; $y = 3\sqrt{5} \approx 6.7$; $z = 6\sqrt{5} \approx 13.4$ **7.** 18 ft 11 in.

9

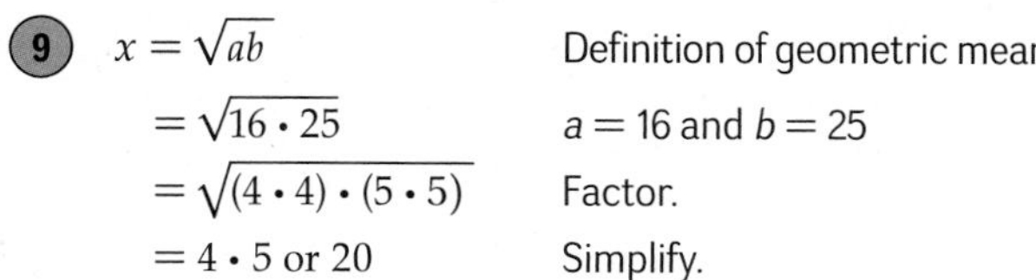

$x = \sqrt{ab}$	Definition of geometric mean
$= \sqrt{16 \cdot 25}$	$a = 16$ and $b = 25$
$= \sqrt{(4 \cdot 4) \cdot (5 \cdot 5)}$	Factor.
$= 4 \cdot 5$ or 20	Simplify.

11. $12\sqrt{6} \approx 29.4$ **13.** $3\sqrt{3} \approx 5.2$ **15.** $\triangle WXY \sim \triangle XZY \sim \triangle WZX$ **17.** $\triangle HGF \sim \triangle HIG \sim \triangle GIF$

19

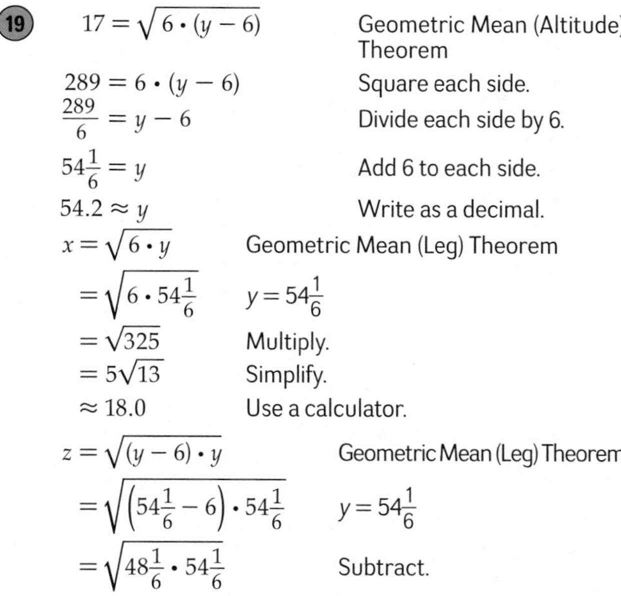

$17 = \sqrt{6 \cdot (y - 6)}$	Geometric Mean (Altitude) Theorem
$289 = 6 \cdot (y - 6)$	Square each side.
$\frac{289}{6} = y - 6$	Divide each side by 6.
$54\frac{1}{6} = y$	Add 6 to each side.
$54.2 \approx y$	Write as a decimal.
$x = \sqrt{6 \cdot y}$	Geometric Mean (Leg) Theorem
$= \sqrt{6 \cdot 54\frac{1}{6}}$	$y = 54\frac{1}{6}$
$= \sqrt{325}$	Multiply.
$= 5\sqrt{13}$	Simplify.
≈ 18.0	Use a calculator.
$z = \sqrt{(y - 6) \cdot y}$	Geometric Mean (Leg) Theorem
$= \sqrt{\left(54\frac{1}{6} - 6\right) \cdot 54\frac{1}{6}}$	$y = 54\frac{1}{6}$
$= \sqrt{48\frac{1}{6} \cdot 54\frac{1}{6}}$	Subtract.
≈ 51.1	Use a calculator.

21. $x \approx 4.7$; $y \approx 1.8$; $z \approx 13.1$ **23.** $x = 24\sqrt{2} \approx 33.9$; $y = 8\sqrt{2} \approx 11.3$; $z = 32$ **25.** 161.8 ft **27.** $\frac{\sqrt{30}}{7}$ or 0.8 **29.** $x = \frac{3\sqrt{3}}{2} \approx 2.6$; $y = \frac{3}{2}$; $z = 3$ **31.** 11 **33.** 3.5 ft **35.** 5 **37.** 4

39. Given: $\angle PQR$ is a right angle. $\overline{QS}$ is an altitude of $\triangle PQR$.

Prove: $\triangle PSQ \sim \triangle PQR$
$\triangle PQR \sim \triangle QSR$
$\triangle PSQ \sim \triangle QSR$

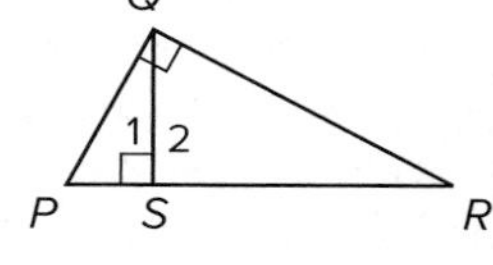

Proof:

Statements (Reasons)

1. $\angle PQR$ is a right angle. $\overline{QS}$ is an altitude of $\triangle PQR$. (Given)
2. $\overline{QS} \perp \overline{RP}$ (Definition of altitude)
3. $\angle 1$ and $\angle 2$ are right angles. (Definition of perpendicular lines)
4. $\angle 1 \cong \angle PQR$, $\angle 2 \cong \angle PQR$ (All right ∠s are ≅.)
5. $\angle P \cong \angle P$, $\angle R \cong \angle R$ (Congruence of angles is reflexive.)
6. $\triangle PSQ \sim \triangle PQR$, $\triangle PQR \sim \triangle QSR$ (AA Similarity Statements 4 and 5)
7. $\triangle PSQ \sim \triangle QSR$ (Similarity of triangles is transitive.)

41. Given: $\angle ADC$ is a right angle. $\overline{DB}$ is an altitude of $\triangle ADC$.

Prove:

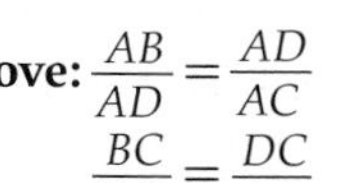

$\frac{AB}{AD} = \frac{AD}{AC}$
$\frac{BC}{DC} = \frac{DC}{AC}$

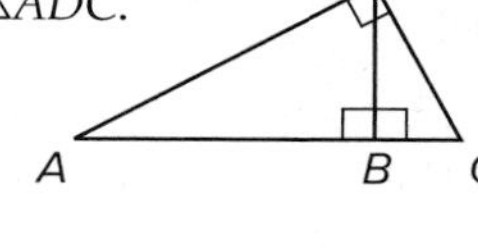

Proof:

Statements (Reasons)

1. $\angle ADC$ is a right angle. $\overline{DB}$ is an altitude of $\triangle ADC$. (Given)
2. $\triangle ADC$ is a right triangle. (Definition of right triangle)
3. $\triangle ABD \sim \triangle ADC$, $\triangle DBC \sim \triangle ADC$ (If the altitude is drawn from the vertex of the rt. ∠ to the hypotenuse of a rt. △, then the 2 △s formed are similar to the given △ and to each other.)
4. $\frac{AB}{AD} = \frac{AD}{AC}$, $\frac{BC}{DC} = \frac{DC}{AC}$ (Def. of similar triangles)

43

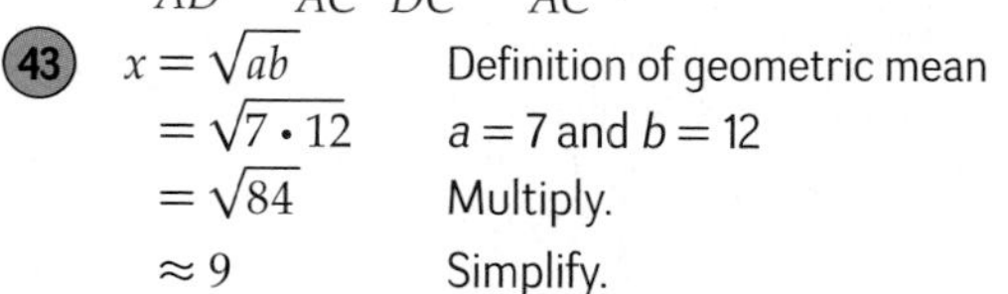

$x = \sqrt{ab}$	Definition of geometric mean
$= \sqrt{7 \cdot 12}$	$a = 7$ and $b = 12$
$= \sqrt{84}$	Multiply.
≈ 9	Simplify.

The average rate of return is about 9%.

45 Sample answer: The geometric mean of two consecutive integers is $\sqrt{x(x+1)}$ and the average of two consecutive integers is $\frac{x + (x + 1)}{2}$.

$$\sqrt{x(x+1)} \stackrel{?}{=} \frac{x + (x + 1)}{2}$$
$$\sqrt{x^2 + x} \stackrel{?}{=} \frac{2x + 1}{2}$$
$$\sqrt{x^2 + x} \stackrel{?}{=} x + \frac{1}{2}$$
$$x^2 + x \stackrel{?}{=} \left(x + \frac{1}{2}\right)^2$$
$$x^2 + x \stackrel{?}{=} x^2 + x + \frac{1}{4}$$
$$0 \neq \frac{1}{4}$$

If you set the two expressions equal to each other, the equation has no solution. So, the statement is never true.

Selected Answers and Solutions

47. Sometimes; sample answer: When the product of the two integers is a perfect square, the geometric mean will be a positive integer. **49.** Neither; sample answer: On the similar triangles created by the altitude, the leg that is x units long on the smaller triangle corresponds with the leg that is 8 units long on the larger triangle, so the correct proportion is $\frac{4}{x} = \frac{x}{8}$ and x is about 5.7. **51.** Sample answer: 9 and 4, 8 and 8; In order for two whole numbers to result in a whole-number geometric mean, their product must be a perfect square. **53.** Sample answer: Both the arithmetic and the geometric mean calculate a value between two given numbers. The arithmetic mean of two numbers a and b is $\frac{a+b}{2}$, and the geometric mean of two numbers a and b is $\sqrt{ab}$. The two means will be equal when $a = b$.

Justification:

$$\frac{a+b}{2} = \sqrt{ab}$$
$$\left(\frac{a+b}{2}\right)^2 = ab$$
$$\frac{(a+b)^2}{4} = ab$$
$$(a+b)^2 = 4ab$$
$$a^2 + 2ab + b^2 = 4ab$$
$$a^2 - 2ab + b^2 = 0$$
$$(a-b)^2 = 0$$
$$a - b = 0$$
$$a = b$$

55. C **57.** D **59a.** 2:3 **59b.** 28, 63

Lesson 8-2

1. 12

3 The side opposite the right angle is the hypotenuse, so $c = 16$.

$a^2 + b^2 = c^2$	Pythagorean Theorem
$4^2 + x^2 = 16^2$	$a = 4$ and $b = x$
$16 + x^2 = 256$	Simplify.
$x^2 = 240$	Subtract 16 from each side.
$x = \sqrt{240}$	Take the positive square root of each side.
$x = 4\sqrt{15}$	Simplify.
$x \approx 15.5$	Use a calculator.

5. D **7.** yes; obtuse
$26^2 \stackrel{?}{=} 16^2 + 18^2$
$676 > 256 + 324$

9. 20 **11.** $\sqrt{21} \approx 4.6$

13. $\frac{\sqrt{10}}{5} \approx 0.6$

15 16 and 30 are both multiples of 2: $16 = 2 \cdot 8$ and $30 = 2 \cdot 15$. Because 8, 15, 17 is a Pythagorean triple, the missing hypotenuse is $2 \cdot 17$ or 34.

17. 70 **19.** about 3 ft

21. yes; obtuse
$21^2 \stackrel{?}{=} 7^2 + 15^2$
$441 > 49 + 225$

23. yes; right
$20.5^2 \stackrel{?}{=} 4.5^2 + 20^2$
$420.25 = 20.25 + 400$

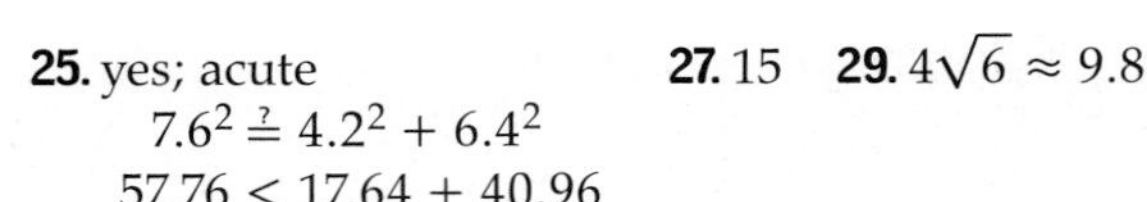

25. yes; acute
$7.6^2 \stackrel{?}{=} 4.2^2 + 6.4^2$
$57.76 < 17.64 + 40.96$

27. 15 **29.** $4\sqrt{6} \approx 9.8$

31. acute; $XY = \sqrt{29}$, $YZ = \sqrt{20}$, $XZ = \sqrt{13}$; $(\sqrt{29})^2 < (\sqrt{20})^2 + (\sqrt{13})^2$ **33.** right; $XY = 6$, $YZ = 10$, $XZ = 8$; $6^2 + 8^2 = 10^2$

35. Given: $\triangle ABC$ with sides of measure a, b, and c, where $c^2 = a^2 + b^2$

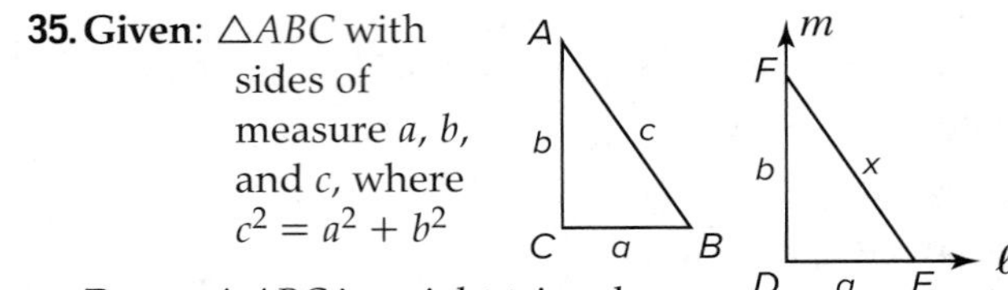

Prove: $\triangle ABC$ is a right triangle.

Proof:

Draw $\overline{DE}$ on line ℓ with measure equal to a. At D, draw line $m \perp \overline{DE}$. Locate point F on m so that $DF = b$. Draw $\overline{FE}$ and call its measure x. Because $\triangle FED$ is a right triangle, $a^2 + b^2 = x^2$. But $a^2 + b^2 = c^2$, so $x^2 = c^2$ or $x = c$. Thus, $\triangle ABC \cong \triangle FED$ by SSS. This means $\angle C \cong \angle D$. Therefore, $\angle C$ must be a right angle, making $\triangle ABC$ a right triangle.

37. Given: In $\triangle ABC$, $c^2 > a^2 + b^2$ where c is the length of the longest side.

C, a, b, B, c, A; P, a, x, R, b, Q

Prove: $\triangle ABC$ is an obtuse triangle.

Proof:

Statements (Reasons)

1. In $\triangle ABC$, $c^2 > a^2 + b^2$ where c is the length of the longest side. In $\triangle PQR$, $\angle R$ is a right angle. (Given)
2. $a^2 + b^2 = x^2$ (Pythagorean Theorem)
3. $c^2 > x^2$ (Substitution Property)
4. $c > x$ (A property of square roots)
5. $m\angle R = 90$ (Definition of a right angle)
6. $m\angle C > m\angle R$ (Converse of the Hinge Theorem)
7. $m\angle C > 90$ (Substitution Property of Equality)
8. $\angle C$ is an obtuse angle. (Definition of an obtuse angle)
9. $\triangle ABC$ is an obtuse triangle. (Definition of an obtuse triangle)

39. $P = 36$ units; $A = 60$ square units^2 **41.** 15

43 **Scale** **Width to Length**

$\frac{16}{9} = \frac{41 \text{ in.}}{x \text{ in.}}$	Write a proportion.
$16 \cdot x = 9 \cdot 41$	Cross Product Property
$16x = 369$	Simplify.
$x = \frac{369}{16}$	Divide each side by 16.

The length of the television is about 23 inches.

$a^2 + b^2 = c^2$	Pythagorean Theorem
$\left(\frac{369}{16}\right)^2 + 41^2 = c^2$	$a = \frac{369}{16}$ and $b = 41$
$\sqrt{\left(\frac{369}{16}\right)^2 + 41^2} = c$	Take the positive square root of each side.
$47.0 \approx c$	Use a calculator.

The screen size is about 47 inches.

(45) The side opposite the right angle is the hypotenuse, so $c = x$.

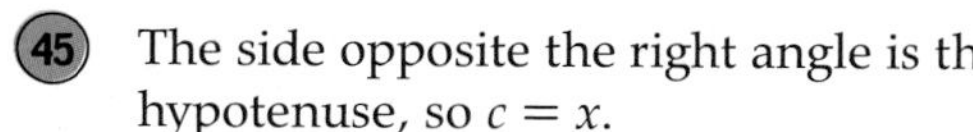

$a^2 + b^2 = c^2$	Pythagorean Theorem
$8^2 + (x - 4)^2 = x^2$	$a = 8$ and $b = x - 4$
$64 + x^2 - 8x + 16 = x^2$	Find 8^2 and $(x - 4)^2$.
$-8x + 80 = 0$	Simplify.
$80 = 8x$	Add $8x$ to each side.
$10 = x$	Divide each side by 8.

47. $\frac{1}{2}$ **49.** 5.4 **51.** Right; sample answer: If you double or halve the side lengths, all three sides of the new triangles are proportional to the sides of the original triangle. Using the Side-Side-Side Similarity Theorem, you know that both of the new triangles are similar to the original triangle, so they are both right.

3 5 4

53. C **55.** 85 **57.** B

Lesson 8-3

1. $5\sqrt{2}$ **3.** 22 **5.** $x = 14$; $y = 7\sqrt{3}$ **7.** Yes; sample answer: The height of the triangle is about $3\frac{1}{2}$ in., so because the height of the plaque is less than the diameter of the opening, it will fit. **9.** $\frac{15\sqrt{2}}{2}$ or $7.5\sqrt{2}$

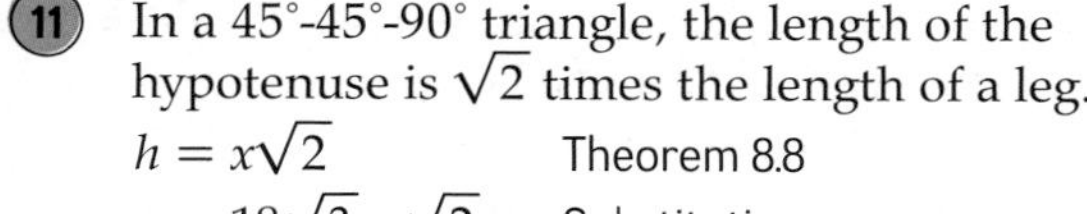

(11) In a 45°-45°-90° triangle, the length of the hypotenuse is $\sqrt{2}$ times the length of a leg.

$h = x\sqrt{2}$	Theorem 8.8
$= 18\sqrt{3} \cdot \sqrt{2}$	Substitution
$= 18\sqrt{6}$	$\sqrt{3} \cdot \sqrt{2} = \sqrt{6}$

13. $20\sqrt{2}$ **15.** $\frac{11\sqrt{2}}{2}$ **17.** $8\sqrt{2}$ or 11.3 cm **19.** $x = 10$; $y = 20$ **21.** $x = \frac{17\sqrt{3}}{2}$; $y = \frac{17}{2}$ **23.** $x = \frac{14\sqrt{3}}{3}$; $y = \frac{28\sqrt{3}}{3}$ **25.** $16\sqrt{3}$ or 27.7 ft **27.** 22.6 ft

(29) In a 45°-45°-90° triangle, the length of the hypotenuse is 2 times the length of a leg.

$h = x\sqrt{2}$	Theorem 8.8
$6 = x\sqrt{2}$	Substitution
$\frac{6}{\sqrt{2}} = x$	Divide each side by $\sqrt{2}$.
$\frac{6}{\sqrt{2}} \cdot \frac{\sqrt{2}}{\sqrt{2}} = x$	Rationalize the denominator.
$\frac{6 \cdot \sqrt{2}}{\sqrt{2} \cdot \sqrt{2}} = x$	Multiply.
$\frac{6\sqrt{2}}{2} = x$	$\sqrt{2} \cdot \sqrt{2} = 2$
$3\sqrt{2} = x$	Simplify.
$h = x\sqrt{2}$	Theorem 8.8
$y = 6\sqrt{2}$	Substitution

31. $x = 5$; $y = 10$ **33.** $x = 45$; $y = 12\sqrt{2}$

(35) In a 30°-60°-90° triangle, the length of the hypotenuse is 2 times the length of the shorter leg.

$h = 2s$	Theorem 8.9
$= 2(25)$ or 50	Substitution

The zip line's length is 50 feet.

37. $x = 9\sqrt{2}$; $y = 6\sqrt{3}$; $z = 12\sqrt{3}$ **39.** 7.5 ft; 10.6 ft; 13.0 ft **41.** (6, 9) **43.** (4, −2)

(45) **a.**

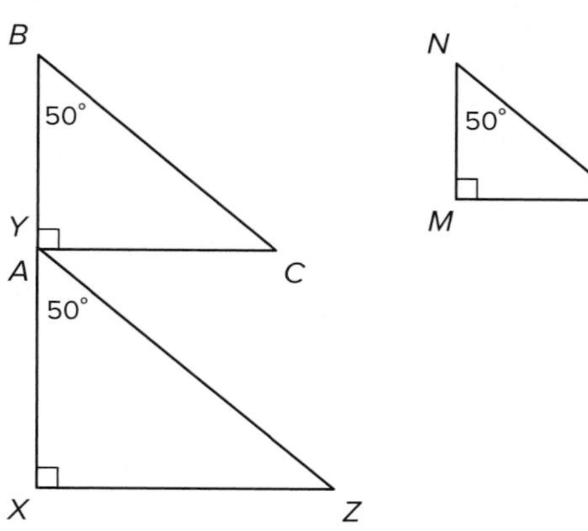

b. Measure the sides of the triangles to the nearest tenth of a centimeter. Find the ratios to the nearest tenth of a centimeter. Sample answer:

Triangle	Length				Ratio	
ABC	AC	2.4 cm	BC	3.2 cm	$\frac{BC}{AC}$	1.3
MNP	MP	1.7 cm	NP	2.2 cm	$\frac{NP}{MP}$	1.3
XYZ	XZ	3.0 cm	YZ	3.9 cm	$\frac{YZ}{XZ}$	1.3

c. Sample answer: In a right triangle with a 50° angle, the ratio of the leg opposite the 50° angle to the hypotenuse will always be the same, 1.3.

47. Sample answer: Let ℓ represent the length. $\ell^2 + w^2 = (2w)^2$; $\ell^2 = 3w^2$; $\ell = w\sqrt{3}$.

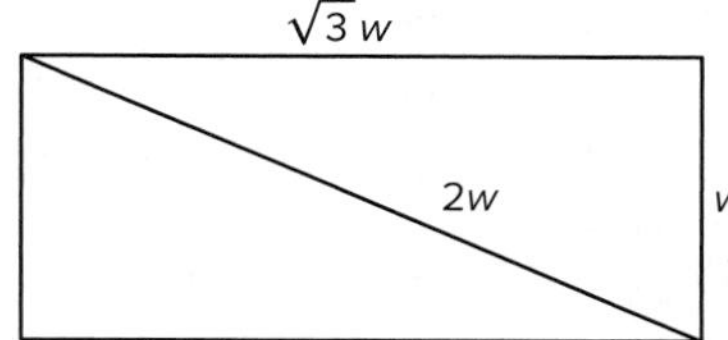

49. 37.9 **51.** C **53.** $20\sqrt{2}$ units **55.** C **57a.** $6\sqrt{2}$ **57b.** $4\sqrt{3}$ **59.** 12

Lesson 8-4

1. $\frac{16}{20} = 0.80$ **3.** $\frac{12}{20} = 0.60$ **5.** $\frac{16}{20} = 0.80$ **7.** $\frac{\sqrt{3}}{2} \approx 0.87$ **9.** 27.44 **11.** about 1.2 ft **13.** 44.4 **15.** $RS \approx 6.7$; $m\angle R \approx 42$; $m\angle T \approx 48$

(17) $\sin J = \frac{\text{opp}}{\text{hyp}} = \frac{56}{65} \approx 0.86$ $\cos J = \frac{\text{adj}}{\text{hyp}} = \frac{33}{65} \approx 0.51$

$\tan J = \frac{\text{opp}}{\text{adj}}$ $= \frac{56}{33}$ ≈ 1.70

$\sin L = \frac{\text{opp}}{\text{hyp}}$ $= \frac{33}{65}$ ≈ 0.51

$\cos L = \frac{\text{adj}}{\text{hyp}}$ $= \frac{56}{65}$ ≈ 0.86

$\tan L = \frac{\text{opp}}{\text{adj}}$ $= \frac{33}{56}$ ≈ 0.59

19. $\frac{84}{85} = 0.99; \frac{13}{85} = 0.15; \frac{84}{13} = 6.46; \frac{13}{85} = 0.15; \frac{84}{85} = 0.99; \frac{13}{84} = 0.15$ **21.** $\frac{\sqrt{3}}{2} = 0.87; \frac{2\sqrt{2}}{4\sqrt{2}} = 0.50; \frac{2\sqrt{6}}{2\sqrt{2}} = \sqrt{3} = 1.73; \frac{2\sqrt{2}}{4\sqrt{2}} = 0.50; \frac{\sqrt{3}}{2} = 0.87; \frac{\sqrt{3}}{3} = 0.58$

23. $\frac{\sqrt{3}}{2} \approx 0.87$ **25.** $\frac{1}{2}$ or 0.5 **27.** $\frac{1}{2}$ or 0.5 **29.** 28.7 **31.** 57.2 **33.** 17.4

35 Let $m\angle A = 55$ and let x be the height of the roller coaster.

$\sin A = \frac{\text{opp}}{\text{hyp}}$	Definition of sine ratio
$\sin 55 = \frac{x}{98}$	Substitution
$98 \cdot \sin 55 = x$	Multiply each side by 98.
$80 \approx x$	Use a calculator.

The height of the roller coaster is about 80 feet.

37. 61.4 **39.** 28.5 **41.** 21.8 **43.** $WX = 15.1$; $XZ = 9.8$; $m\angle W = 33$ **45.** $ST = 30.6$; $m\angle R = 58$; $m\angle T = 32$

47 $JL = \sqrt{[-2-(-2)]^2 + [4-(-3)]^2} = 7$

$KJ = \sqrt{[-2-(-7)]^2 + [-3-(-3)]^2} = 5$

$\tan K = \frac{\text{opp}}{\text{adj}}$	Definition of tangent ratio
$= \frac{7}{5}$	Substitution
$m\angle K = \tan^{-1}\left(\frac{7}{5}\right) \approx 54.5$	Use a calculator.

49. 51.3 **51.** 13.83 in.; 7.51 in^2 **53.** 8.45 ft; 3.06 ft^2 **55.** 0.92

57 The triangle is isosceles, so two sides measure 32 and the two smaller triangles each have a side that measures x. Let $m\angle A = 54$.

$\cos A = \frac{\text{adj}}{\text{hyp}}$	Definition of cosine ratio
$\cos 54 = \frac{x}{32}$	Substitution
$32 \cdot \cos 54 = x$	Multiply each side by 32.
$18.8 \approx x$	Simplify.
$\sin A = \frac{\text{opp}}{\text{hyp}}$	Definition of sine ratio
$\sin 54 = \frac{y}{32}$	Substitution
$32 \cdot \sin 54 = y$	Multiply each side by 32.
$25.9 \approx y$	Simplify.

59. $x = 9.2$; $y = 11.7$

61a.

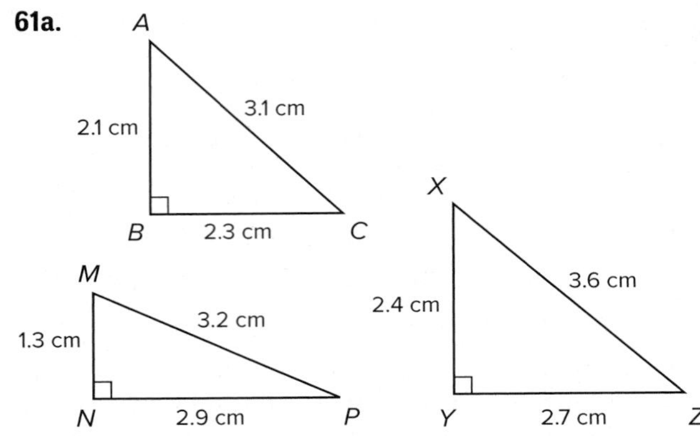

61b. Sample answer:

Triangle	Trigonometric Ratios				Sum of Ratios Squared	
ABC	cos A	0.677	sin A	0.742	$(\cos A)^2 + (\sin A)^2$	1
	cos C	0.742	sin C	0.677	$(\cos C)^2 + (\sin C)^2$	1
MNP	cos M	0.406	sin M	0.906	$(\cos M)^2 + (\sin M)^2$	1
	cos P	0.906	sin P	0.406	$(\cos P)^2 + (\sin P)^2$	1
XYZ	cos X	0.667	sin X	0.75	$(\cos X)^2 + (\sin X)^2$	1
	cos Z	0.75	sin Z	0.667	$(\cos Z)^2 + (\sin Z)^2$	1

61c. Sample answer: The sum of the cosine squared and the sine squared of an acute angle of a right triangle is 1. **61d.** $(\sin X)^2 + (\cos X)^2 = 1$ **61e.** Sample answer:

$(\sin A)^2 + (\cos A)^2 \stackrel{?}{=} 1$	Conjecture
$\left(\frac{y}{r}\right)^2 + \left(\frac{x}{r}\right)^2 \stackrel{?}{=} 1$	$\sin A = \frac{y}{r}$, $\cos A = \frac{x}{r}$
$\frac{y^2}{r^2} + \frac{x^2}{r^2} \stackrel{?}{=} 1$	Simplify.
$\frac{y^2 + x^2}{r^2} \stackrel{?}{=} 1$	Combine fractions with like denominators.
$\frac{r^2}{r^2} \stackrel{?}{=} 1$	Pythagorean Theorem
$1 = 1$	Simplify.

63. Sample answer: Yes; because the values of sine and cosine are both calculated by dividing one of the legs of a right triangle by the hypotenuse, and the hypotenuse is always the longest side of a right triangle, the values will always be less than 1. You will always be dividing the smaller number by the larger number. **65.** Sample answer: To find the measure of an acute angle of a right triangle, you can find the ratio of the leg opposite the angle to the hypotenuse and use a calculator to find the inverse sine of the ratio, you can find the ratio of the leg adjacent to the angle to the hypotenuse and use a calculator to find the inverse cosine of the ratio, or you can find the ratio of the leg opposite the angle to the leg adjacent to the angle and use a calculator to find the inverse tangent of the ratio. **67a.** 47 **67b.** 1 **69.** A **71.** C

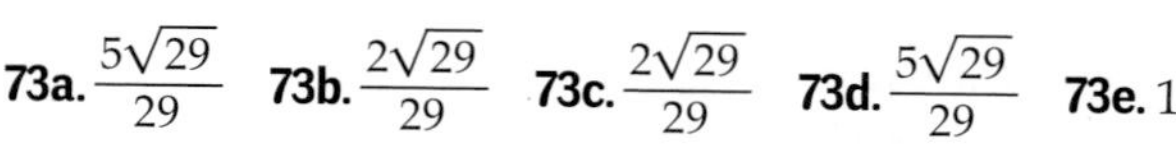

73a. $\frac{5\sqrt{29}}{29}$ **73b.** $\frac{2\sqrt{29}}{29}$ **73c.** $\frac{2\sqrt{29}}{29}$ **73d.** $\frac{5\sqrt{29}}{29}$ **73e.** 1

Lesson 8-5

1. 27.5 ft **3.** 14.2 ft

5 Make a sketch.

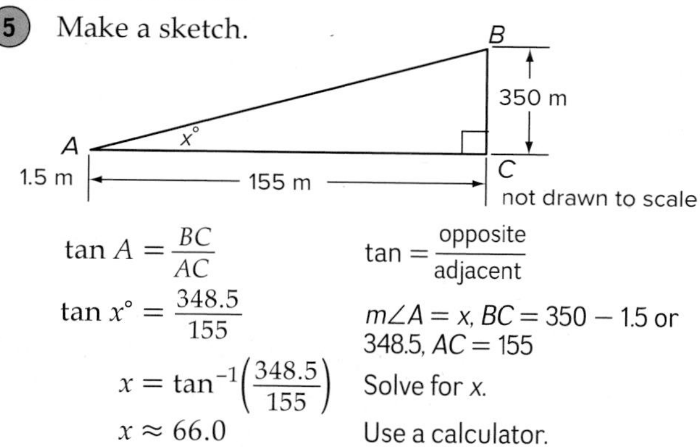

$\tan A = \frac{BC}{AC}$ $\tan = \frac{\text{opposite}}{\text{adjacent}}$

$\tan x° = \frac{348.5}{155}$ $m\angle A = x$, $BC = 350 - 1.5$ or 348.5, $AC = 155$

$x = \tan^{-1}\left(\frac{348.5}{155}\right)$ Solve for x.

$x \approx 66.0$ Use a calculator.

The angle of elevation is about 66°.

7. 14.8°

9 Make a sketch.

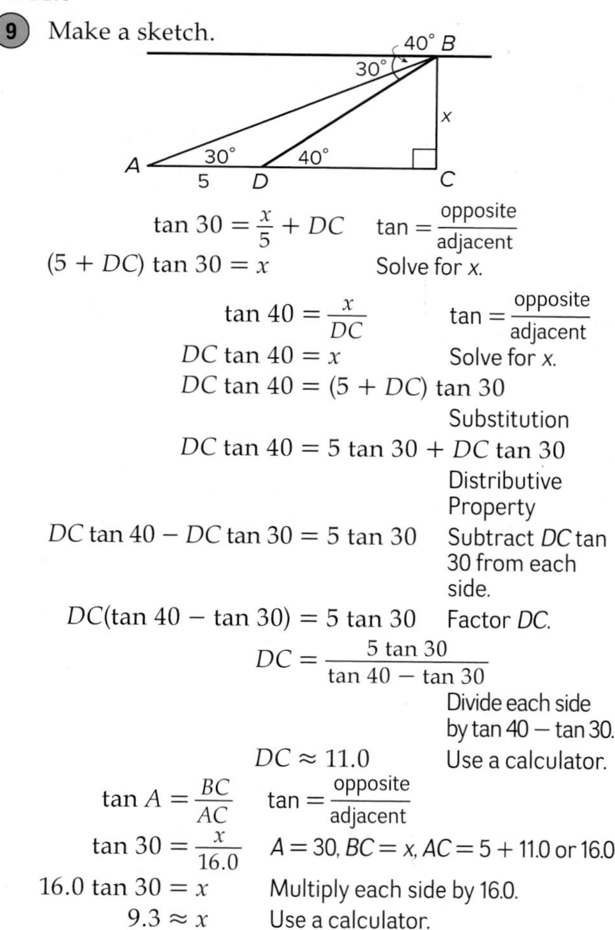

$\tan 30 = \frac{x}{5} + DC$ $\tan = \frac{\text{opposite}}{\text{adjacent}}$

$(5 + DC) \tan 30 = x$ Solve for x.

$\tan 40 = \frac{x}{DC}$ $\tan = \frac{\text{opposite}}{\text{adjacent}}$

$DC \tan 40 = x$ Solve for x.

$DC \tan 40 = (5 + DC) \tan 30$ Substitution

$DC \tan 40 = 5 \tan 30 + DC \tan 30$ Distributive Property

$DC \tan 40 - DC \tan 30 = 5 \tan 30$ Subtract $DC \tan 30$ from each side.

$DC(\tan 40 - \tan 30) = 5 \tan 30$ Factor DC.

$DC = \frac{5 \tan 30}{\tan 40 - \tan 30}$ Divide each side by $\tan 40 - \tan 30$.

$DC \approx 11.0$ Use a calculator.

$\tan A = \frac{BC}{AC}$ $\tan = \frac{\text{opposite}}{\text{adjacent}}$

$\tan 30 = \frac{x}{16.0}$ $A = 30$, $BC = x$, $AC = 5 + 11.0$ or 16.0

$16.0 \tan 30 = x$ Multiply each side by 16.0.

$9.3 \approx x$ Use a calculator.

The platform is about 9.3 feet high.

11. about 1309 ft **13.** 16.6° **15a.** about 2.17 ft

15b. Sample answer: The overhang is used to keep sunlight out of the windows in the summer, but allow the sunlight in during the winter. He would want the overhang to be long enough to block all of the sunlight on the longest day when the Sun is at the greatest elevation. For El Paso, this angle is 81.2°. The overhang begins 2 feet above the windows, so it will need to cover 14 feet. The length of the overhang is $L = \frac{14}{\tan 81.2}$ or about 2.17 feet. If the overhang is any longer, then less sunlight will get into the house in the winter.

17 Make a sketch.

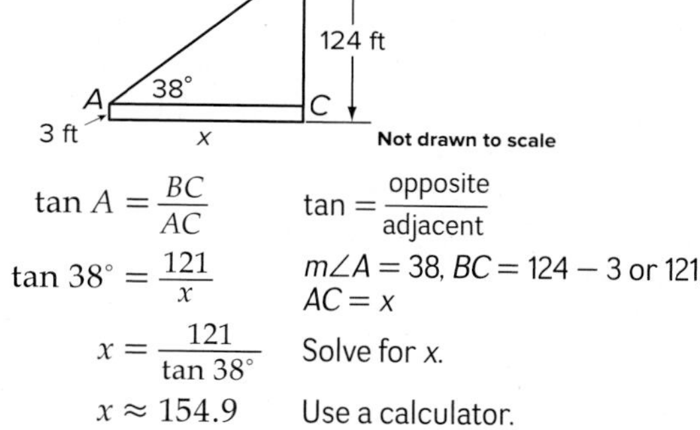

$\tan A = \frac{BC}{AC}$ $\tan = \frac{\text{opposite}}{\text{adjacent}}$

$\tan 38° = \frac{121}{x}$ $m\angle A = 38$, $BC = 124 - 3$ or 121, $AC = x$

$x = \frac{121}{\tan 38°}$ Solve for x.

$x \approx 154.9$ Use a calculator.

You should place the tripod about 154.9 feet from the monument.

19a. ≈75.1° **19b.** ≈110.1 m

21 Make two sketches.

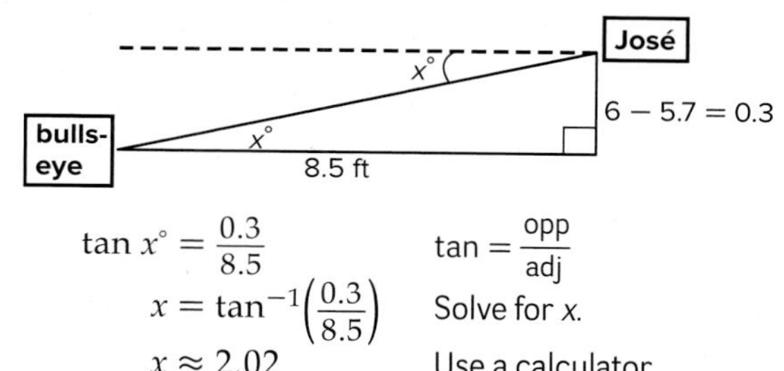

$\tan x° = \frac{0.3}{8.5}$ $\tan = \frac{\text{opp}}{\text{adj}}$

$x = \tan^{-1}\left(\frac{0.3}{8.5}\right)$ Solve for x.

$x \approx 2.02$ Use a calculator.

José throws at an angle of depression of 2.02°.

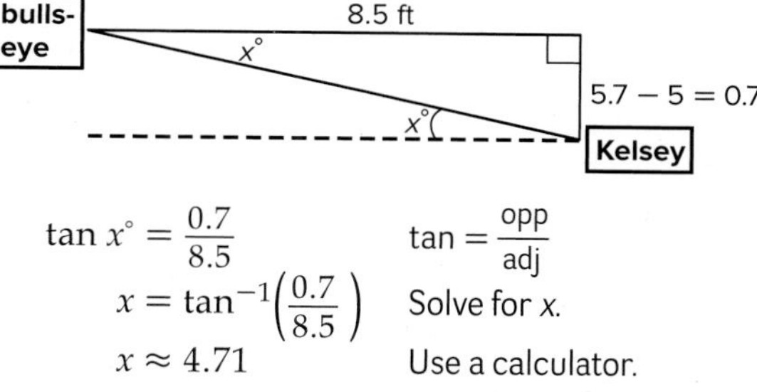

$\tan x° = \frac{0.7}{8.5}$ $\tan = \frac{\text{opp}}{\text{adj}}$

$x = \tan^{-1}\left(\frac{0.7}{8.5}\right)$ Solve for x.

$x \approx 4.71$ Use a calculator.

Kelsey throws at an angle of elevation of 4.71°.

23. Rodrigo; sample answer: Because your horizontal line of sight is parallel to the other person's horizontal line of sight, the angles of elevation and depression are congruent according to the Alternate Interior Angles Theorem.

25. True; sample answer: As a person moves closer to an object, the horizontal distance decreases, but the height of the object is constant. The tangent ratio will increase, and therefore the measure of the angle also increases.

27. Sample answer: If you sight something with a 45° angle of elevation, you don't have to use trigonometry to determine the height of the object. Because the legs of a 45°-45°-90° are congruent, the height of the object will be the same as your horizontal distance from the object.

29. $\frac{70\sqrt{3}}{3}$ ft **31.** D **33.** 66.7 **35.** 56 ft

Lesson 8-6

1. 27.9 mm^2

3 Area $= \frac{1}{2}bc \sin A$ — Area Formula

$= \frac{1}{2}(11)(6) \sin 40°$ — Substitution

$\approx 21.2 \text{ cm}^2$ — Simplify.

5. $E = 107°, d \approx 7.9, f \approx 7.0$ **7.** $F = 60°, f \approx 12.3, h \approx 9.1$
9. no solution **11.** one; $B = 90°, C = 60°, c \approx 5.2$
13. 10.6 km^2 **15.** 36.8 m^2 **17.** 5.9 ft^2 **19.** 65.2 m^2
21. $C = 30°, b \approx 11.1, c \approx 5.8$
23. $L = 74°, m \approx 4.9, n \approx 3.1$

25 $m\angle K = 180 - (53 + 20)$ or $107°$

$\frac{\sin H}{h} = \frac{\sin J}{j}$ — Law of Sines

$\frac{\sin 53°}{31} = \frac{\sin 20°}{j}$ — Substitution

$j = \frac{31 \sin 20°}{\sin 53°}$ — Solve for j.

$j \approx 13.3$ — Use a calculator.

$\frac{\sin H}{h} = \frac{\sin K}{k}$ — Law of Sines

$\frac{\sin 53°}{31} = \frac{\sin 107°}{k}$ — Substitution

$k = \frac{31 \sin 107°}{\sin 53°}$ — Solve for k.

$k \approx 37.1$ — Use a calculator.

27. $B = 63°, b \approx 2.9, c \approx 3.0$ **29.** one; $B \approx 25°$, $C \approx 55°, c \approx 5.8$ **31.** one; $B \approx 32°, C \approx 110°, c \approx 32.1$ **33.** two; $B \approx 53°, C \approx 85°, c \approx 7.4$; $B \approx 127°, C \approx 11°$, $c \approx 1.4$ **35.** no solution **37.** about 28°

39

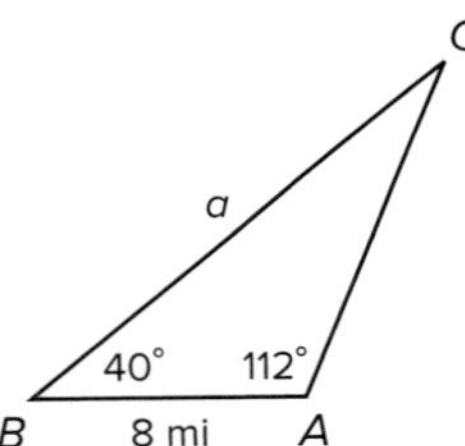

$m\angle C = 180 - (40 + 112)$ or $28°$

$\frac{\sin A}{a} = \frac{\sin C}{c}$ — Law of Sines

$\frac{\sin 112°}{a} = \frac{\sin 28°}{8}$ — Substitution

$a = \frac{8 \sin 112°}{\sin 28°}$ — Solve for a.

$a \approx 15.8$ — Use a calculator.

Sirens B and C are about 15.8 miles apart.

41a. Sample answer:

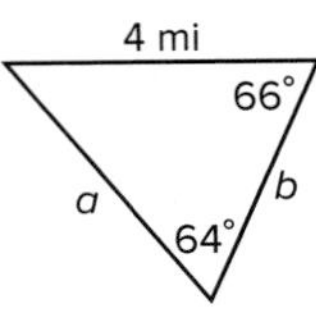

41b. Sample answer: $\frac{\sin 66°}{a} = \frac{\sin 64°}{4}$; $\frac{\sin 50°}{b} = \frac{\sin 64°}{4}$

41c. about 11.5 mi **43.** Cameron; R is acute and $r > t$, so there is one solution.

45. Sample answer:

$\sin A = \frac{\textit{opposite}}{\textit{hypotenuse}}$ — Definition of sine

$\sin A = \frac{h}{c}$ — $h =$ opposite side, $c =$ hypotenuse

$c \sin A = h$ — Multiply both sides by c.

area $= \frac{1}{2} \cdot$ base $\cdot$ height — Area of a triangle

area $= \frac{1}{2}bh$ — $b =$ base, $h =$ height

area $= \frac{1}{2}bc \sin A$ — Substitution

47. Sample answer: In the triangle, $B = 115°$. Using the Law of Sines, $\frac{\sin 50°}{a} = \frac{\sin 115°}{b}$. This equation cannot be solved because there are two unknowns. To solve a triangle using the Law of Sines, two sides and an angle must be given or two angles and a side opposite one of the angles must be given.

49. 6 cm^2 **51.** C, F

53a.

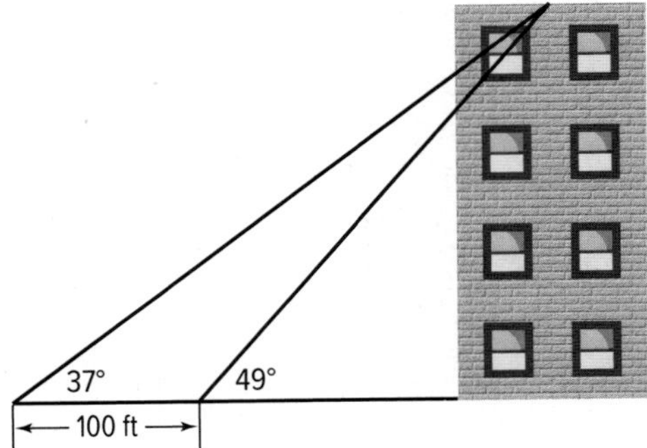

53b. 289.9 ft **53c.** 218.5 ft

Lesson 8-7

1. $A \approx 36°, C \approx 52°, b \approx 5.1$ **3.** $A \approx 18°, B \approx 29°$, $C \approx 133°$ **5.** Sines; $B \approx 40°, C \approx 33°, c \approx 6.9$

7 Because the lengths of two sides and the measure of the included angle are known, first use the Law of Cosines to find the missing side length.

$r^2 = s^2 + t^2 - 2st \cos R$ — Law of Cosines

$r^2 = 16^2 + 9^2 - 2(16)(9) \cos 35°$ — $s = 16, t = 9, R = 35°$

$r^2 \approx 101.1$ — Use a calculator.

$r \approx 10.1$ — Take the positive square root of each side.

$\frac{\sin R}{r} = \frac{\sin T}{t}$ — Law of Sines

$\frac{\sin 35°}{10.1} = \frac{\sin T}{9}$ — Substitution

$\frac{9 \sin 35°}{10.1} = \sin T$ — Multiply each side by 9.

$31° \approx T$ — Use the $\sin^{-1}$ function.

$m\angle S = 180 - (35° + 31°)$ or $114°$

9. $A \approx 70°, B \approx 40°, c \approx 3.0$ **11.** $A \approx 31°, B \approx 108°$, $C \approx 41°$ **13.** $a \approx 6.9, B \approx 41°, C \approx 23°$ **15.** $F \approx 65°$, $G \approx 94°, H \approx 21°$ **17.** Sines; $C \approx 45°, A \approx 85°, a \approx 18.2$ **19.** Cosines; $A \approx 27°, B \approx 115°, C \approx 38°$

21. Sines; $A \approx 17°$, $B \approx 79°$, $b \approx 6.9$

23 $d^2 = 338^2 + 520^2 - 2(338)(520)\cos 70°$
$d^2 \approx 264{,}417.1$
$d \approx 514.2$ m

25. 81°, 36°, 63° **27.** about 13,148 yd^2

29a. Sample answer:

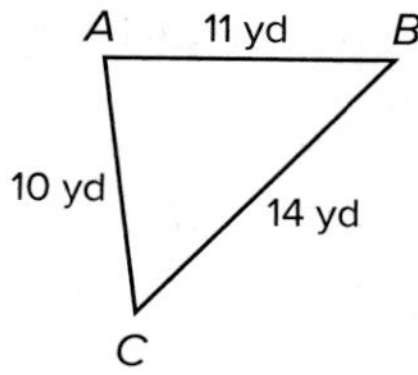

29b. Sample answer: Use the Law of Cosines to find the measure of $\angle A$. Then use the formula Area $= \frac{1}{2}bc \sin A$. **29c.** 54.6 yd^2

31

$\frac{\sin A}{a} = \frac{\sin B}{b}$ Law of Sines

$\frac{\sin 104°}{12.4} = \frac{\sin B}{8.1}$ Substitution

$\frac{8.1 \sin 104°}{12.4} = \sin B$ Multiply each side by 8.1.

$39° \approx B$ Use the $\sin^{-1}$ function.

$m\angle C \approx 180 - (39° + 104°)$ or $37°$

$\frac{\sin A}{a} = \frac{\sin C}{c}$ Law of Sines

$\frac{\sin 104°}{12.4} = \frac{\sin 37°}{c}$ Substitution

$c = \frac{12.4 \sin 37°}{\sin 104°}$ Solve for c.

$c \approx 7.7$ Use a calculator.

33. $F \approx 42°$, $G \approx 72°$, $H \approx 66°$ **35.** The longest side is 14.5 centimeters. Use the Law of Cosines to find the measure of the angle opposite the longest side; 102°. **37.** When two angles and a side are given or when two sides and an angle opposite one of the sides are given, you can use the Law of Sines to solve a triangle. When two sides and an included angle are given or when three sides are given, you can use the Law of Cosines to solve a triangle. **39.** A **41.** 43 **43a.** 95 degrees **43b.** 9.0 **43c.** 15.6 **43d.** 36.6 **45.** 2

Chapter 8 Study Guide and Review

1. false, geometric **3.** false, sum **5.** false, Law of Cosines **7.** The angle of elevation is equal to the angle of depression. **9.** 6 **11.** $\frac{8}{3}$ **13.** 50 ft **15.** $9\sqrt{3} \approx 15.6$ **17.** yes; acute
$16^2 \stackrel{?}{=} 13^2 + 15^2$
$256 < 169 + 225$

19. 18.4 m **21.** $x = 4\sqrt{2}$, $y = 45°$ **23.** $\frac{5}{13}$, 0.38 **25.** $\frac{12}{13}$, 0.92 **27.** $\frac{5}{12}$, 0.42 **29.** 32.2 **31.** 63.4° and 26.6° **33.** 86.6 feet **35.** two solutions; first solution; $C = 30°$, $B = 125°$, $b = 29.1$; second solution; $C = 150°$, $B = 5°$, $b = 3.1$ **37.** 98.9 ft **39.** Sines; $B \approx 52°$, $C \approx 48°$, $c \approx 11.3$ **41.** Sines; $B \approx 75°$, $C \approx 63°$, $c \approx 12.0$ or $B \approx 105°$, $C \approx 33°$, $c \approx 7.3$ **43.** about 750.5 ft **45.** 226 ft

Selected Answers and Solutions

CHAPTER 9

Circles

Chapter 9 Concept Check

1. Change 26% to a decimal. **3.** isosceles right triangle

5. Pythagorean Theorem $x^2 + x^2 = 20^2$ **7.** completing the square

Lesson 9-1

1. $\odot N$ **3.** 8 cm **5.** 14 in. **7.** 22 ft; 138.23 ft
9. $4\pi\sqrt{13}$ cm **11.** $\overline{SU}$ **13.** 8.1 cm

15 $d = 2r$ Diameter formula
$= 2(14)$ or 28 in. Substitute and simplify.

17. 3.7 cm **19.** 14.6 **21.** 30.6 **23.** 13 in.; 81.68 in.
25. 39.47 ft; 19.74 ft **27.** 830.23 m; 415.12 m

29
$a^2 + b^2 = c^2$ Pythagorean Theorem
$(6\sqrt{2})^2 + (6\sqrt{2})^2 = c^2$ Substitution
$144 = c^2$ Simplify.
$12 = c$ Take the positive square root of each side.

The diameter is 12π feet.
$C = \pi d$ Circumference formula
$= \pi(12)$ Substitution
$= 12\pi$ ft Simplify.

31. 10π in. **33.** 14π yd **35a.** 31.42 ft **35b.** 4 ft
37. 22.80 ft; 71.63 ft **39.** $0.25x$; $0.79x$ **41.** neither

43 The radius is 3 units, or $3 \cdot 25 = 75$ feet.
$C = 2\pi r$ Circumference formula
$= 2\pi(75)$ Substitution
$= 150\pi$ Simplify.
≈ 471.2 ft Use a calculator.

45a. Sample answer:

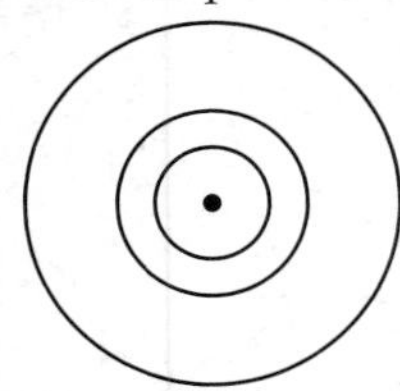

45b.

Circle Radius (cm)	Circumference (cm)
0.5	3.14
1	6.28
2	12.57

45c. They all have the same shape–circular.
45d. The ratio of their circumferences is also 2.
45e. $(C_B) = \frac{b}{a}(C_A)$ **45f.** 4 in.

47 **a.** $C = 2\pi r$ Circumference formula
$= 2\pi(30)$ Substitution
$= 60\pi$ Simplify.
$C = 2\pi r$ Circumference formula
$= 2\pi(5)$ Substitution
$= 10\pi$ Simplify.
$60\pi - 10\pi = 50\pi \approx 157.1$ mi

b. If $r = 5$, $C = 10\pi$; if $r = 10$, $C = 20\pi$; if $r = 15$, $C = 30\pi$, and so on. So, as r increases by 5, C increases by 10π or by about 31.4 miles.

49a. $8r$ and $6r$; Twice the radius of the circle, $2r$ is the side length of the square, so the perimeter of the square is $4(2r)$ or $8r$. The regular hexagon is made up of six equilateral triangles with side length r, so the perimeter of the hexagon is $6(r)$ or $6r$. **49b.** less; greater; $6r < C < 8r$ **49c.** $3d < C < 4d$; The circumference of the circle is between 3 and 4 times its diameter. **49d.** These limits will approach a value of πd, implying that $C = \pi d$.
51. Always; a radius is a segment drawn between the center of the circle and a point on the circle. A segment drawn from the center to a point inside the circle will always have a length less than the radius of the circle.
53. $\frac{8\pi}{\sqrt{3}}$ or $\frac{8\pi\sqrt{3}}{3}$ in. **55.** 25.13 **57.** 50.27 **59a.** circle M
59b. $\overline{MJ}$ or $\overline{ML}$ **59c.** 6π

Lesson 9-2

1. 170 **3.** major arc; 270 **5.** semicircle; 180 **7.** 147
9. 123 **11.** 13.74 cm **13.** 3.14 cm

15 $65 + 70 + x = 360$ Sum of central angles
$135 + x = 360$ Simplify.
$x = 225$ Subtract 135 from each side.

17. 40 **19.** major arc; 125 **21.** major arc; 305
23. semicircle; 180 **25.** major arc; 270 **27a.** 90; 100.8 **27b.** minor; minor **27c.** No; no categories share the same percentage of the circle. **29.** 60
31. 300 **33.** 180 **35.** 220 **37.** 120

39 $\ell = \frac{x}{360} \cdot 2\pi r$ Arc length equation
$= \frac{112}{360} \cdot 2\pi(4.5)$ Substitution.
≈ 8.80 cm Use a calculator.

41. 17.02 in. **43.** 12.04 m **45.** The length of the arc would double. **47.** 40.83 in. **49.** 9.50 ft **51.** 142

53 **a.** $m\widehat{AB} = m\angle ACB$ $\widehat{AB}$ is a minor arc.
$= 180 - (22 + 22)$ Angle Addition Postulate
$= 180 - 44$ or 136 Simplify.
b. $\ell = \frac{x}{360} \cdot 2\pi r$ Arc length equation
$= \frac{136}{360} \cdot 2\pi(62)$ Substitution.
≈ 147.17 ft Use a calculator.

55 **a.**

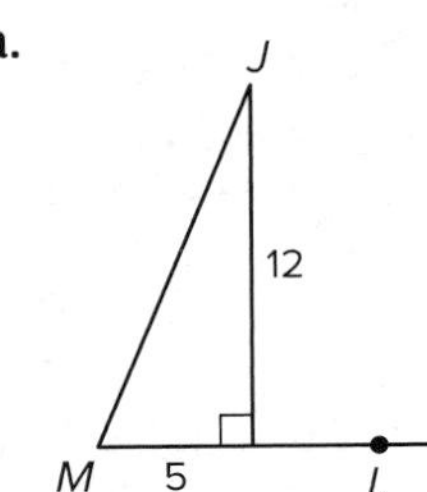

$\tan \angle JML = \frac{12}{5}$
$m\angle JML = \tan^{-1}\left(\frac{12}{5}\right)$
$\approx 67.4°$
$m\widehat{JL} = m\angle JML \approx 67.4°$

b.

K
5
M
12
L

$\tan \angle KML = \frac{5}{12}$
$m\angle KML = \tan^{-1}\left(\frac{5}{12}\right)$
$\approx 22.6°$
$m\widehat{KL} = m\angle KML \approx 22.6°$

c. $m\angle JMK = m\angle JML - m\angle KML$

$\approx 67.4 - 22.6$

$\approx 44.8^\circ$

$m\widehat{JK} = m\angle JMK \approx 44.8^\circ$

d. $r = \sqrt{(x_2 - x_1)^2 + (y_2 - y_1)^2}$ Distance Formula

$= \sqrt{(5 - 0)^2 + (12 - 0)^2}$ $(x_1, y_1) = (0, 0)$ and $(x_2, y_2) = (5, 12)$

$= 13$ Simplify.

$\ell = \frac{x}{360} \cdot 2\pi r$ Arc length equation

$= \frac{67.4}{360} \cdot 2\pi(13)$ Substitution

≈ 15.29 units Use a calculator.

e. $\ell = \frac{x}{360} \cdot 2\pi r$ Arc length equation

$= \frac{44.8}{360} \cdot 2\pi(13)$ Substitution

≈ 10.16 units Use a calculator.

57. Selena; the circles are not congruent because they do not have congruent radii. So, the arcs are not congruent. **59.** Never; obtuse angles intersect arcs that measure between 90° and 180° **61.** $m\widehat{LM} = 150$, $m\widehat{MN} = 90$, $m\widehat{NL} = 120$ **63.** 175 **65.** C **67.** 51 **69.** C

Lesson 9-3

1 $\widehat{ST}$ is a minor arc, so $m\widehat{ST} = 93$. $\overline{RS}$ and $\overline{ST}$ are congruent chords, so the corresponding arcs $\widehat{RS}$ and $\widehat{ST}$ are congruent.

$\widehat{RS} \cong \widehat{ST}$ Corresponding arcs are congruent.

$m\widehat{RS} = m\widehat{ST}$ Definition of congruent arcs

$x = 93$ Substitution

3. 3 **5.** 3.32 **7.** 21 **9.** 127 **11.** 7

13 $\overline{KL}$ and $\overline{AJ}$ are congruent chords in congruent circles, so the corresponding arcs $\widehat{KL}$ and $\widehat{AJ}$ are congruent.

$\widehat{KL} \cong \widehat{AJ}$ Corresponding arcs are congruent.

$m\widehat{KL} = m\widehat{AJ}$ Definition of congruent arcs

$5x = 3x + 54$ Substitution

$2x = 54$ Subtract $3x$ from each side.

$x = 27$ Divide each side by 2.

15. 122.5° **17.** 5.34 **19.** 6.71

21 $DE + EC = DC$ Segment Addition Postulate

$15 + EC = 88$ Substitution

$EC = 73$ Subtract 15 from each side.

$EC^2 + EB^2 = CB^2$ Pythagorean Theorem

$73^2 + EB^2 = 88^2$ Substitution

$EB^2 = 2415$ Subtract 73^2 from each side.

$EB \approx 49.14$ The positive square root of each side

$EB = \frac{1}{2}AB$ $\overline{DC} \perp \overline{AB}$, so $\overline{DC}$ bisects $\overline{AB}$.

$2EB = AB$ Multiply each side by 2.

$2(49.14) \approx AB$ Substitution

$98.3 \approx AB$ Simplify.

23. 4

25. Proof:

Because all radii are congruent, $\overline{QP} \cong \overline{PR} \cong \overline{SP} \cong \overline{PT}$. You are given that $\overline{QR} \cong \overline{ST}$, so $\triangle PQR \cong \triangle PST$ by SSS. Thus, $\angle QPR \cong \angle SPT$ by CPCTC. Because the central angles have the same measure, their intercepted arcs have the same measure and are therefore congruent. Thus, $\widehat{QR} \cong \widehat{ST}$.

27. Each arc is 90°, and each chord is 2.12 ft.

29. Given: $\odot L$, $\overline{LX} \perp \overline{FG}$, $\overline{LY} \perp \overline{JH}$, $\overline{LX} \cong \overline{LY}$

Prove: $\overline{FG} \cong \overline{JH}$

Proof:

Statements (Reasons)

1. $\overline{LG} \cong \overline{LH}$ (All radii of a $\odot$ are $\cong$.)
2. $\overline{LX} \perp \overline{FG}$, $\overline{LY} \perp \overline{JH}$, $\overline{LX} \cong \overline{LY}$ (Given)
3. $\angle LXG$ and $\angle LYH$ are right $\angle$s. (Def. of $\perp$ lines)
4. $\triangle XGL \cong \triangle YHL$ (HL)
5. $\overline{XG} \cong \overline{YH}$ (CPCTC)
6. $XG = YH$ (Def. of $\cong$ segments)
7. $2(XG) = 2(YH)$ (Multiplication Property)
8. $\overline{LX}$ bisects $\overline{FG}$; $\overline{LY}$ bisects $\overline{JH}$. (A radius $\perp$ to a chord bisects the chord.)
9. $FG = 2(XG)$, $JH = 2(YH)$ (Def. of seg. bisector)
10. $FG = JH$ (Substitution)
11. $\overline{FG} \cong \overline{JH}$ (Def. of $\cong$ segments)

31 Because $\overline{AB} \perp \overline{CE}$ and $\overline{DF} \perp \overline{CE}$, $\overline{CE}$ bisects $\overline{AB}$ and $\overline{DF}$.

Because $\overline{AB} \cong \overline{DF}$, $AB = DF$.

$CB = \frac{1}{2}AB$ Definition of bisector

$CB = \frac{1}{2}DF$ Substitution

$CB = DE$ Definition of bisector

$9x = 2x + 14$ Substitution

$7x = 14$ Subtract $2x$ from each side.

$x = 2$ Divide each side by 7.

33. 5 **35.** About 17.3; P and Q are equidistant from the endpoints of $\overline{AB}$, so they lie on the perpendicular bisector of $\overline{AB}$; so, $\overline{PQ}$ is the perpendicular bisector of $\overline{AB}$. Hence, both segments of $\overline{AB}$ are 5. Because $\overline{PS}$ is perpendicular to chord $\overline{AB}$, $\angle PSA$ is a right angle. So, $\triangle PSA$ is a right triangle.

By the Pythagorean Theorem, $PS = \sqrt{(PA)^2 - (AS)^2}$.

By substitution, $PS = \sqrt{11^2 - 5^2}$ or $\sqrt{96}$.

Similarly, $\triangle ASQ$ is a right triangle with $SQ = \sqrt{(AQ)^2 - (AS)^2} = \sqrt{9^2 - 5^2}$ or $\sqrt{56}$.

Because $PQ = PS + SQ$, $PQ = \sqrt{96} + \sqrt{56}$ or about 17.3.

37a. Given: $\overline{CD}$ is the perpendicular bisector of chord $\overline{AB}$ in $\odot X$.

Prove: $\overline{CD}$ contains point X.

Proof:

Suppose X is not on $\overline{CD}$. Draw $\overline{XE}$ and radii $\overline{XA}$ and $\overline{XB}$. Because $\overline{CD}$ is the perpendicular bisector of $\overline{AB}$, E is the midpoint of $\overline{AB}$ and

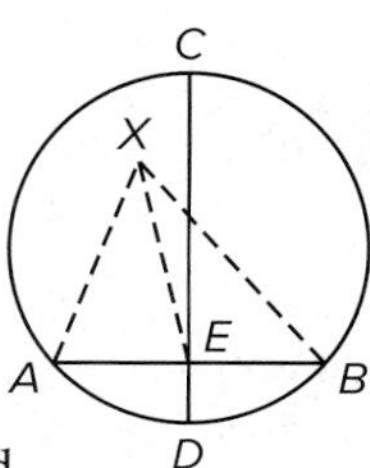

$\overline{AE} \cong \overline{EB}$. Also, $\overline{XA} \cong \overline{XB}$, because all radii of a $\odot$ are $\cong$. $\overline{XE} \cong \overline{XE}$ by the Reflexive Property. So, $\triangle AXE \cong \triangle BXE$ by SSS. By CPCTC, $\angle XEA \cong \angle XEB$. Because they also form a linear pair $\angle XEA$ and $\angle XEB$ are right angles. So, $\overline{XE} \perp \overline{AB}$. By definition $\overline{XE}$ is the perpendicular bisector of $\overline{AB}$. But $\overline{CD}$ is also the perpendicular bisector of $\overline{AB}$. This contradicts the uniqueness of a perpendicular bisector of a segment. Thus, the assumption is false, and center X must be on $\overline{CD}$.

37b. Given: In $\odot X$, X is on $\overline{CD}$ and $\overline{FG}$ bisects $\overline{CD}$ at O.
Prove: Point O is point X.
Proof:
Because point X is on $\overline{CD}$ and C and D are on $\odot X$, $\overline{CD}$ is a diameter of $\odot X$. Because $\overline{FG}$ bisects $\overline{CD}$ at O, O is the midpoint of $\overline{CD}$. Because the midpoint of a diameter is the center of a circle, O is the center of the circle. Therefore, point O is point X.

39. No; sample answer: In a circle with a radius of 12, an arc with a measure of 60 determines a chord of length 12. (the triangle related to a central angle of 60 is equilateral.) If the measure of the arc is tripled to 180, then the chord determined by the arc is a diameter and has a lenth of 2(12) or 24, which is not three times as long as the original chord. **41.** D **43a.** $\triangle JLM$ or $\triangle JLK$ **43b.** $\overline{LM}$ is $\frac{1}{2}$ of 24 feet, or 12 feet; the chord is bisected because it is perpendicular to the diameter. $\overline{JM}$ is the radius, which is $\frac{1}{2}$ the diameter of 36 feet, or 18 feet. **43c.** $JL^2 + 12^2 = 18^2$ **43d.** $6\sqrt{5}$ ft
45. B

Lesson 9-4

1. 30 **3.** 66 **5.** 54
7. Given: $\overline{RT}$ bisects $\overline{SU}$.
Prove: $\triangle RVS \cong \triangle UVT$
Proof:
Statements (Reasons)
1. $\overline{RT}$ bisects $\overline{SU}$. (Given)
2. $\overline{SV} \cong \overline{VU}$ (Def. of segment bisector)
3. $\angle SRT$ intercepts $\widehat{ST}$. $\angle SUT$ intercepts $\widehat{ST}$. (Def. of intercepted arc)
4. $\angle SRT \cong \angle SUT$ (Inscribed ≰ of same arc are $\cong$.)
5. $\angle RVS \cong \angle UVT$ (Vertical ≰ are $\cong$.)
6. $\triangle RVS \cong \triangle UVT$ (AAS)

9. 25 **11.** 162

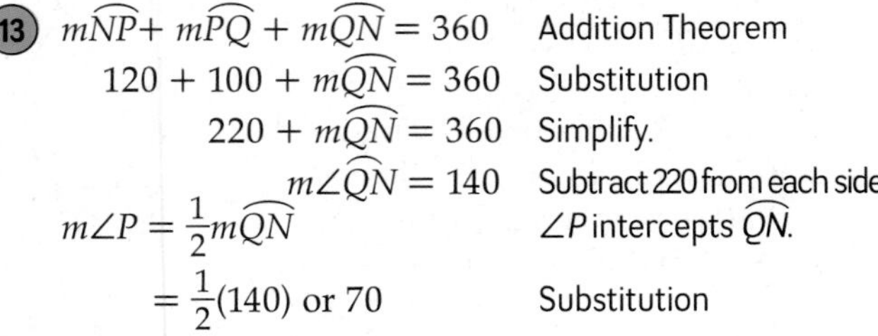
13 $m\widehat{NP} + m\widehat{PQ} + m\widehat{QN} = 360$ Addition Theorem
$120 + 100 + m\widehat{QN} = 360$ Substitution
$220 + m\widehat{QN} = 360$ Simplify.
$m\angle\widehat{QN} = 140$ Subtract 220 from each side.
$m\angle P = \frac{1}{2}m\widehat{QN}$ $\angle P$ intercepts $\widehat{QN}$.
$= \frac{1}{2}(140)$ or 70 Substitution

15. 140 **17.** 32 **19.** 20
21. Given: $m\angle T = \frac{1}{2}m\angle S$
Prove: $m\widehat{TUR} = 2m\widehat{URS}$
Proof:
$m\angle T = \frac{1}{2}m\angle S$ means that $m\angle S = 2m\angle T$. Because $m\angle S = \frac{1}{2}m\widehat{TUR}$ and $m\angle T = \frac{1}{2}m\widehat{URS}$, the equation becomes $\frac{1}{2}m\widehat{TUR} = 2\left(\frac{1}{2}m\widehat{URS}\right)$. Multiplying each side of the equation by 2 results in $m\widehat{TUR} = 2m\widehat{URS}$.

23. 30 **25.** 12.75 **27.** 135 **29.** 106
31. Given: Quadrilateral $ABCD$ is inscribed in $\odot O$.
Prove: $\angle A$ and $\angle C$ are supplementary. $\angle B$ and $\angle D$ are supplementary.
Proof: By arc addition and the definitions of arc measure and the sum of central angles, $m\widehat{DCB} + m\widehat{DAB} = 360$. Because by Theorem 9.6, $m\angle C = \frac{1}{2}m\widehat{DAB}$ and

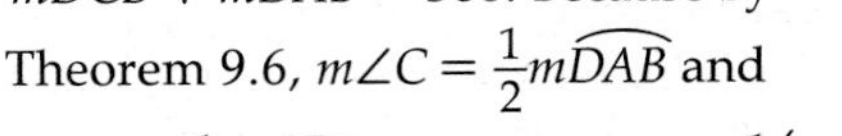

$m\angle A = \frac{1}{2}m\widehat{DCB}$, $m\angle C + m\angle A = \frac{1}{2}\left(m\widehat{DCB} + m\widehat{DAB}\right)$, but $m\widehat{DCB} + m\widehat{DAB} = 360$, so $m\angle C + m\angle A = \frac{1}{2}(360)$ or 180. This makes $\angle C$ and $\angle A$ supplementary. Because the sum of the measures of the interior angles of a quadrilateral is 360, $m\angle A + m\angle C + m\angle B + m\angle D = 360$. But $m\angle A + m\angle C = 180$, so $m\angle B + m\angle D = 180$, making them supplementary also.

33 Because all the sides of the sign are congruent, all the corresponding arcs are congruent.
$8m\widehat{QR} = 360$, so $m\widehat{QR} = \frac{360}{8}$ or 45.
$m\angle RLQ = \frac{1}{2}m\widehat{QR}$
$= \frac{1}{2}(45)$ or 22.5

35. 135
37. Proof:
Statements (Reasons)
1. $m\angle ABC = m\angle ABD + m\angle DBC$ ($\angle$ Addition Postulate)
2. $m\angle ABD = \frac{1}{2}m\widehat{AD}$
$m\angle DBC = \frac{1}{2}m\widehat{DC}$ (The measure of an inscribed $\angle$ whose side is a diameter is half the measure of the intercepted arc (Case 1).)
3. $m\angle ABC = \frac{1}{2}m\widehat{AD} + \frac{1}{2}m\widehat{DC}$ (Substitution)
4. $m\angle ABC = \frac{1}{2}\left(m\widehat{AD} + m\widehat{DC}\right)$ (Factor)
5. $m\widehat{AD} + m\widehat{DC} = m\widehat{AC}$ (Arc Addition Postulate)
6. $m\angle ABC = \frac{1}{2}m\widehat{AC}$ (Substitution)

39 Use the Inscribed Angle Theorem to find the measures of $\angle FAE$ and $\angle CBD$. Then use the definition of congruent arcs and the Multiplication Property of Equality to help prove that the angles are congruent.

Given: $\angle FAE$ and $\angle CBD$ are inscribed; $\widehat{EF} \cong \widehat{DC}$
Prove: $\angle FAE = \angle CBD$
Proof:
Statements (Reasons)
1. $\angle FAE$ and $\angle CBD$ are inscribed; $\widehat{EF} \cong \widehat{DC}$ (Given)
2. $m\angle FAE = \frac{1}{2}m\widehat{EF}$; $m\angle CBD = \frac{1}{2}m\widehat{DC}$ (Measure of an inscribed $\angle$ = half measure of intercepted arc.)
3. $m\widehat{EF} = m\widehat{DC}$ (Def. of $\cong$ arcs)
4. $\frac{1}{2}m\widehat{EF} = \frac{1}{2}m\widehat{DC}$ (Mult. Prop.)
5. $m\angle FAE = m\angle CBD$ (Substitution)
6. $\angle FAE \cong \angle CBD$ (Def. of $\cong$ $\angle$s)

41a.

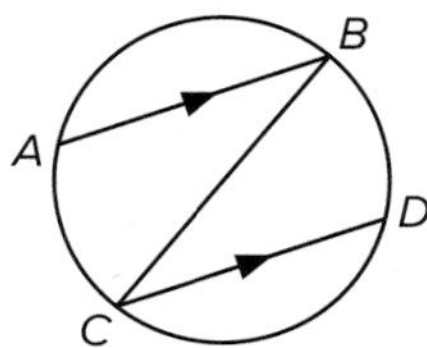

41b. Sample answer: $m\angle A = 30$, $m\angle D = 30$; $m\widehat{AC} = 60$, $m\widehat{BD} = 60$; The arcs are congruent because they have equal measures. **41c.** Sample answer: In a circle, two parallel chords cut congruent arcs. **41d.** 70; 70 **43.** Always; rectangles have right angles at each vertex, therefore each pair of opposite angles will be supplementary and inscribed in a circle. **45.** Sometimes; a rhombus can be inscribed in a circle as long as it is a square. Because the opposite angles of rhombi that are not squares are not supplementary, they can not be inscribed in a circle. **47.** $\frac{\pi}{2}$ **51.** 108 **53.** D

Lesson 9-5

1. no common tangent

3 $FG^2 + GE^2 \stackrel{?}{=} FE^2$
$36^2 + 15^2 \stackrel{?}{=} (24 + 15)^2$
$1521 = 1521$

$\triangle EFG$ is a right triangle with right angle EGF. So $\overline{FG}$ is perpendicular to radius $\overline{EG}$ at point G. Therefore, by Theorem 10.10, $\overline{FG}$ is tangent to $\odot E$.

5. 16 **7.** $x = 4$; $y = 15$

9.

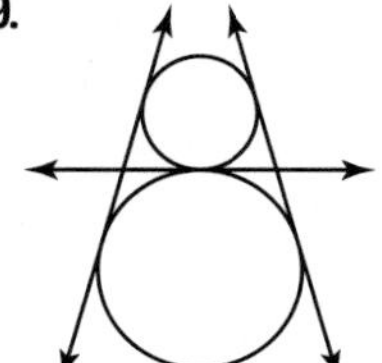

11.

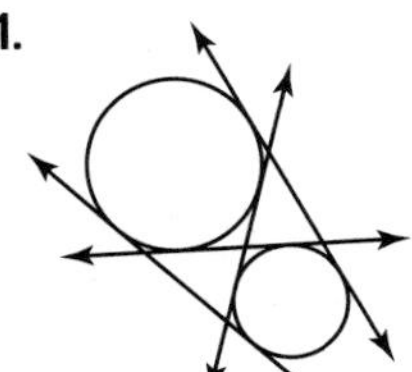

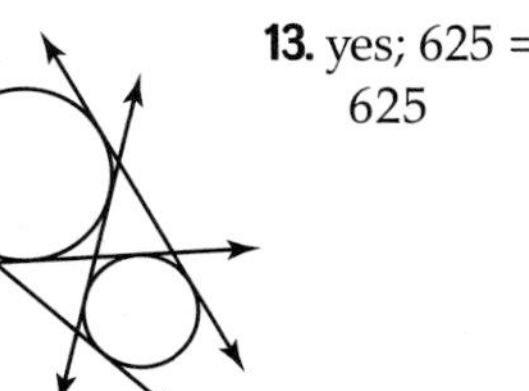

13. yes; 625 = 625

15 $XY^2 + YZ^2 \stackrel{?}{=} XZ^2$
$8^2 + 5^2 \stackrel{?}{=} (3 + 5)^2$
$89 \neq 64$

Because $\triangle XYZ$ is not a right triangle, $\overline{XY}$ is not perpendicular to radius $\overline{YZ}$. So, $\overline{XY}$ is not tangent to $\odot Z$.

17 $\overrightarrow{QP}$ is tangent to $\odot N$ at P. So, $\overrightarrow{QP} \perp \overline{PN}$ and $\triangle PQN$ is a right triangle.

$QP^2 + PN^2 = QN^2$	Pythagorean Theorem
$24^2 + 10^2 = x^2$	$QP = 24$, $PN = 10$, and $QN = x$
$576 + 100 = x^2$	Multiply.
$676 = x^2$	Simplify.
$26 = x$	Take the positive square root of each side.

19. 9 **21.** 4 **23a.** 37.95 in. **23b.** 37.95 in. **25.** 8; 52 cm **27.** 8.06

29. Given: Quadrilateral $ABCD$ is circumscribed about $\odot P$.
Prove: $AB + CD = AD + BC$
Statements (Reasons)
1. Quadrilateral $ABCD$ is circumscribed about $\odot P$. (Given)
2. Sides $\overline{AB}$, $\overline{BC}$, $\overline{CD}$, and $\overline{DA}$ are tangent to $\odot P$ at points H, G, F, and E, respectively. (Def. of circumscribed)
3. $\overline{EA} \cong \overline{AH}$; $\overline{HB} \cong \overline{BG}$; $\overline{GC} \cong \overline{CF}$; $\overline{FD} \cong \overline{DE}$ (Two segments tangent to a circle from the same exterior point are $\cong$.)
4. $AB = AH + HB$, $BC = BG + GC$, $CD = CF + FD$, $DA = DE + EA$ (Segment Addition)
5. $AB + CD = AH + HB + CF + FD$; $DA + BC = DE + EA + BG + GC$ (Substitution)
6. $AB + CD = AH + BG + GC + FD$; $DA + BC = FD + AH + BG + GC$ (Substitution)
7. $AB + CD = FD + AH + BG + GC$ (Commutative Prop. of Add.)
8. $AB + CD = DA + BC$ (Substitution)

31

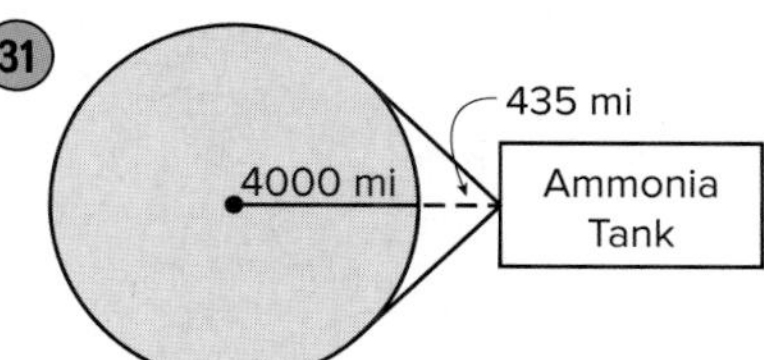

$4000^2 + x^2 = (4000 + 435)^2$	Pythagorean Theorem
$x^2 = 3{,}669{,}225$	Subtract 4000^2 from each side.
$x \approx 1916$ mi	Take the positive square root of each side.

33. Proof: Assume that ℓ is not tangent to $\odot S$. Because ℓ intersects $\odot S$ at T, it must intersect the circle in another place. Call this point Q. Then $ST = SQ$. $\triangle STQ$ is isosceles, so $\angle T \cong \angle Q$. Because $\overline{ST} \perp \ell$, $\angle T$ and $\angle Q$ are right angles. This contradicts that a triangle can only have one right angle. Therefore, ℓ is tangent to $\odot S$.

35. Sample answer: Using the Pythagorean Theorem, $2^2 + x^2 = 10^2$, so $x \approx 9.8$. Because $PQST$ is a rectangle, $PQ = x = 9.8$.

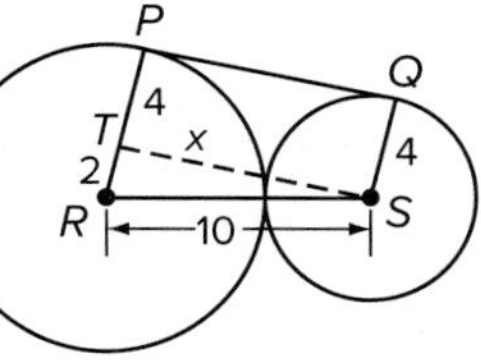

Selected Answers and Solutions

37. Sample answer:

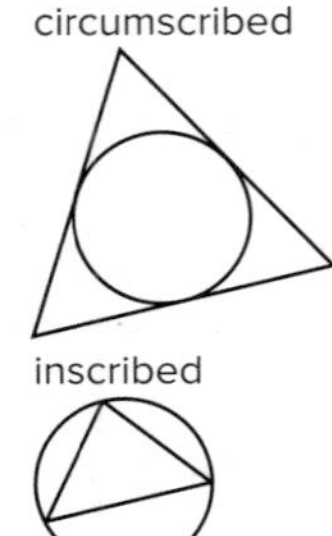

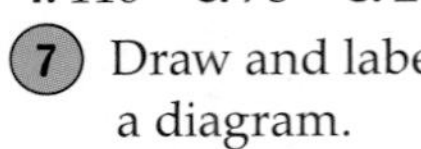

39. No; sample answer: From a point outside the circle, two tangents can be drawn. From a point on the circle, one tangent can be drawn. From a point inside the circle, no tangents can be drawn because a line would intersect the circle in two points.

41. 8.5 in. **43.** C

Lesson 9-6

1. 110 **3.** 73 **5.** 248

7 Draw and label a diagram.

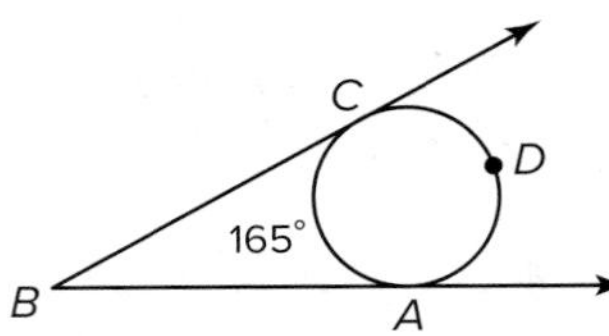

$m\angle B = \frac{1}{2}(m\widehat{CDA} - m\widehat{CA})$ Theorem 10.14

$= \frac{1}{2}[(360 - 165) - 165]$ Substitution

$= \frac{1}{2}(195 - 165)$ or 15 Simplify.

9. 71.5

11

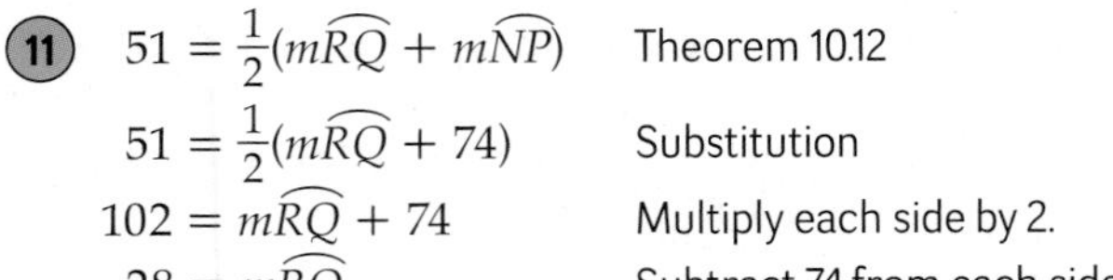

$51 = \frac{1}{2}(m\widehat{RQ} + m\widehat{NP})$ Theorem 10.12

$51 = \frac{1}{2}(m\widehat{RQ} + 74)$ Substitution

$102 = m\widehat{RQ} + 74$ Multiply each side by 2.

$28 = m\widehat{RQ}$ Subtract 74 from each side.

13. 144 **15.** 125 **17a.** 100 **17b.** 20 **19.** 74 **21.** 185 **23.** 22 **25.** 168

27

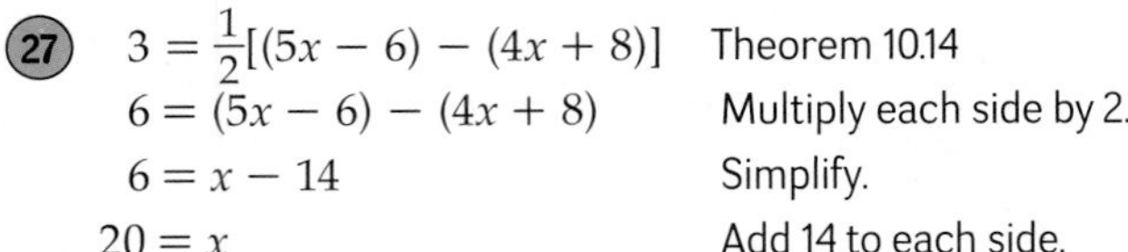

$3 = \frac{1}{2}[(5x - 6) - (4x + 8)]$ Theorem 10.14

$6 = (5x - 6) - (4x + 8)$ Multiply each side by 2.

$6 = x - 14$ Simplify.

$20 = x$ Add 14 to each side.

29a. 145 **29b.** 30

31. __Statements (Reasons)__

1. $\overrightarrow{FM}$ is a tangent to the circle and $\overrightarrow{FL}$ is a secant to the circle. (Given)
2. $m\angle FLH = \frac{1}{2}m\widehat{HG}$, $m\angle LHM = \frac{1}{2}m\widehat{LH}$ (The meas. of an inscribed $\angle = \frac{1}{2}$ the measure of its intercepted arc.)
3. $m\angle LHM = m\angle FLH + m\angle F$ (Exterior ⦻ Th.)
4. $\frac{1}{2}m\widehat{LH} = \frac{1}{2}m\widehat{HG} + m\angle F$ (Substitution)
5. $\frac{1}{2}m\widehat{LH} - \frac{1}{2}m\widehat{HG} = m\angle F$ (Subtraction Prop.)
6. $\frac{1}{2}(m\widehat{LH} - m\widehat{HG}) = m\angle F$ (Distributive Prop.)

33a. Proof: By Theorem 9.10, $\overline{OA} \perp \overline{AB}$. So, $\angle FAE$ is a right $\angle$ with measure 90, and $\widehat{FCA}$ is a semicircle with measure of 180. Because $\angle CAE$ is acute, C is in the interior of $\angle FAE$. By the Angle and Arc Addition Postulates, $m\angle FAE = m\angle FAC + m\angle CAE$ and $m\widehat{FCA} = m\widehat{FC} + m\widehat{CA}$. By substitution, $90 = m\angle FAC + m\angle CAE$ and $180 = m\widehat{FC} + m\widehat{CA}$. So, $90 = \frac{1}{2}m\widehat{FC} + \frac{1}{2}m\widehat{CA}$ by Division Prop., and $m\angle FAC + m\angle CAE = \frac{1}{2}m\widehat{FC} + \frac{1}{2}m\widehat{CA}$ by substitution. $m\angle FAC = \frac{1}{2}m\widehat{FC}$ because $\angle FAC$ is inscribed, so substitution yields $\frac{1}{2}m\widehat{FC} + m\angle CAE = \frac{1}{2}m\widehat{FC} + \frac{1}{2}m\widehat{CA}$. By Subt. Prop., $m\angle CAE = \frac{1}{2}m\widehat{CA}$.

33b. Given: $\angle CAB$ is obtuse.

Prove: $m\angle CAB = \frac{1}{2}m\,\widehat{CDA}$

Proof: Using the Angle and Arc Addition Postulates, $m\angle CAB = m\angle CAF + m\angle FAB$ and $m\widehat{CDA} = m\widehat{CF} + m\widehat{FDA}$. Because $\overline{OA} \perp \overline{AB}$ and $\overline{FA}$ is a diameter, $\angle FAB$ is a right angle with a measure of 90 and $\widehat{FDA}$ is a semicircle with a measure of 180. By substitution, $m\angle CAB = m\angle CAF + 90$ and $m\widehat{CDA} = m\widehat{CF} + 180$. Because $\angle CAF$ is inscribed, $m\angle CAF = \frac{1}{2}m\widehat{CF}$ and by substitution, $m\angle CAB = \frac{1}{2}m\widehat{CF} + 90$. Using the Division and Subtraction Properties on the Arc Addition equation yields $\frac{1}{2}m\widehat{CDA} - \frac{1}{2}m\widehat{CF} = 90$. By substituting for 90, $m\angle CAB = \frac{1}{2}m\widehat{CF} + \frac{1}{2}m\widehat{CDA} - \frac{1}{2}m\widehat{CF}$. By subtraction, $m\angle CAB = \frac{1}{2}m\widehat{CDA}$.

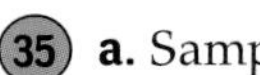

35 **a.** Sample answer:

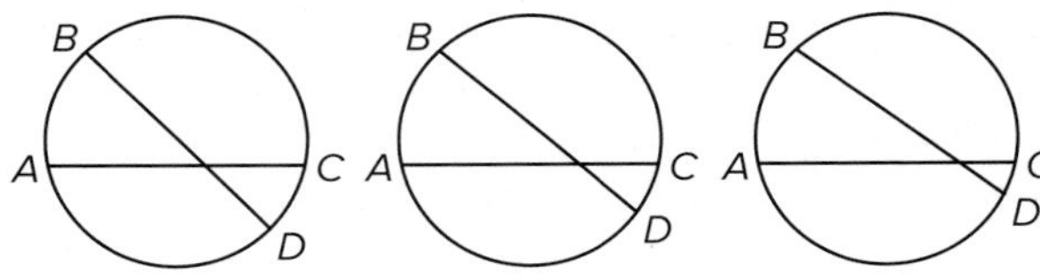

b. Sample answer:

	Circle 1	Circle 2	Circle 3
$\widehat{CD}$	25	15	5
$\widehat{AB}$	50	50	50
x	37.5	32.5	27.5

c. As the measure of $\widehat{CD}$ gets closer to 0, the measure of x approaches half of $m\widehat{AB}$; $\angle AEB$ becomes an inscribed angle.

d. Theorem 9.12 states that if two chords intersect in the interior of a circle, then the measure of an angle formed is one half the sum of the measure of the arcs intercepted by the angle and its vertical angle. Use this theorem to write an equation relating x, $m\widehat{AB}$, and $m\widehat{CD}$. Then let $m\widehat{CD} = 0$ and simplify. The result is Theorem 9.6, the Inscribed Angle Theorem.

$x = \frac{1}{2}(m\widehat{AB} + m\widehat{CD})$

$x = \frac{1}{2}(m\widehat{AB} + 0)$

$x = \frac{1}{2}m\widehat{AB}$

37. 15 **39a.** $m\angle G \leq 90$; $m\angle G < 90$ for all values except when $\overrightarrow{JG} \perp \overrightarrow{GH}$ at G, then $m\angle G = 90$. **39b.** $m\widehat{KH} = 56$; $m\widehat{HJ} = 124$; Because a diameter is involved, the intercepted arcs measure $(180 - x)$ and x degrees. Hence, solving $\frac{180 - x - x}{2} = 34$ leads to the answer. **41.** Sample answer: Using Theorem 9.14, $60° = \frac{1}{2}[(360° - x) - x]$ or $120°$; repeat for $50°$ to get $130°$. The third arc can be found by adding $50°$ and $60°$ and subtracting from $360°$ to get $110°$. **43.** 35 **45.** A

Lesson 9-7

1. $(x - 9)^2 + y^2 = 25$

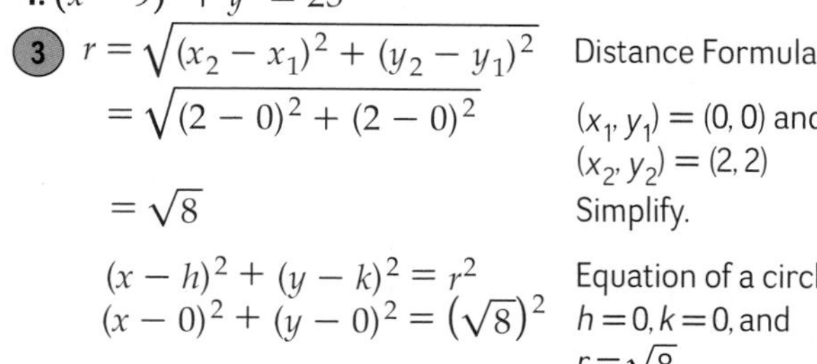

3

$r = \sqrt{(x_2 - x_1)^2 + (y_2 - y_1)^2}$	Distance Formula
$= \sqrt{(2 - 0)^2 + (2 - 0)^2}$	$(x_1, y_1) = (0, 0)$ and $(x_2, y_2) = (2, 2)$
$= \sqrt{8}$	Simplify.
$(x - h)^2 + (y - k)^2 = r^2$	Equation of a circle
$(x - 0)^2 + (y - 0)^2 = (\sqrt{8})^2$	$h = 0, k = 0$, and $r = \sqrt{8}$
$x^2 + y^2 = 8$	Simplify.

5. $(x - 2)^2 + (y - 1)^2 = 4$

7. $(3, -2)$; 4

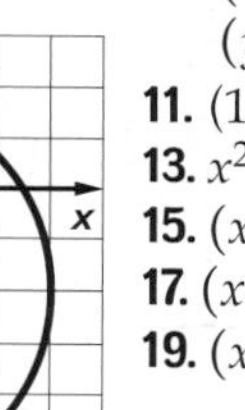

9. $(2, -1)$; $(x - 2)^2 + (y + 1)^2 = 40$

11. $(1, 2)$, $(-1, 0)$

13. $x^2 + y^2 = 16$

15. $(x + 2)^2 + y^2 = 64$

17. $(x + 3)^2 + (y - 6)^2 = 9$

19. $(x + 5)^2 + (y + 1)^2 = 9$

21 The third ring has a radius of 15 + 15 + 15 or 45 miles.

$(x - h)^2 + (y - k)^2 = r^2$	Equation of a circle
$(x - 0)^2 + (y - 0)^2 = 45^2$	$h = 0, k = 0$, and $r = 45$
$x^2 + y^2 = 2025$	Simplify.

23. $(0, 0)$; 6

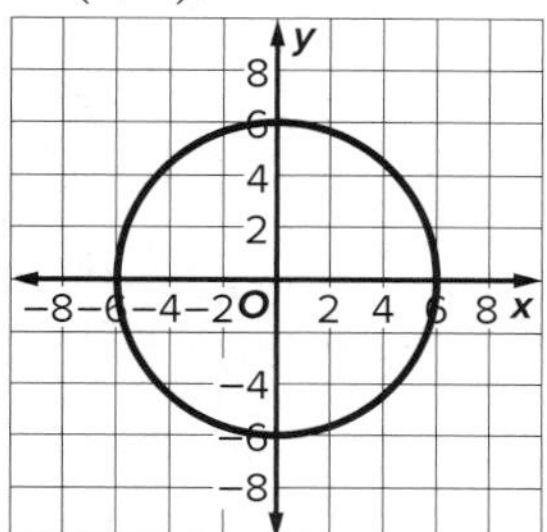

25 Write equation in standard form.

$x^2 + y^2 + 8x - 4y = -4$	Original equation
$x^2 + 8x + y^2 - 4y = -4$	Isolate and group like terms.
$x^2 + 8x + 16 + y^2 - 4y + 4 = -4 + 20$	Complete the squares.
$(x + 4)^2 + (y - 2)^2 = 16$	Factor and simplify.
$[x - (-4)]^2 + (y - 2)^2 = 4^2$	Write + 4 as − (−4) and 16 as 4^2.

So $h = -4$, $k = 2$, and $r = 4$. The center is at $(-4, 2)$ and the radius is 4.

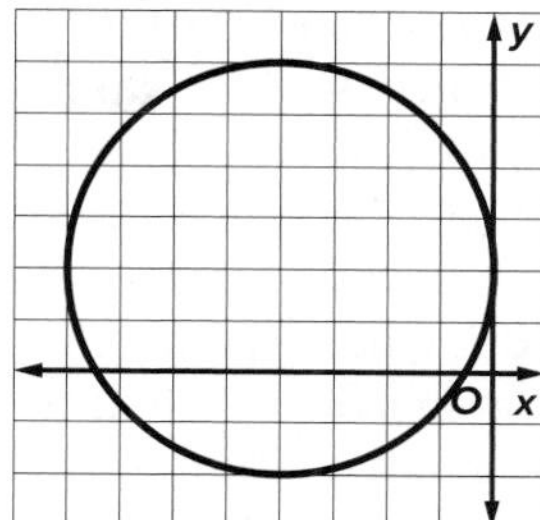

27. $(x - 3)^2 + (y - 3)^2 = 13$

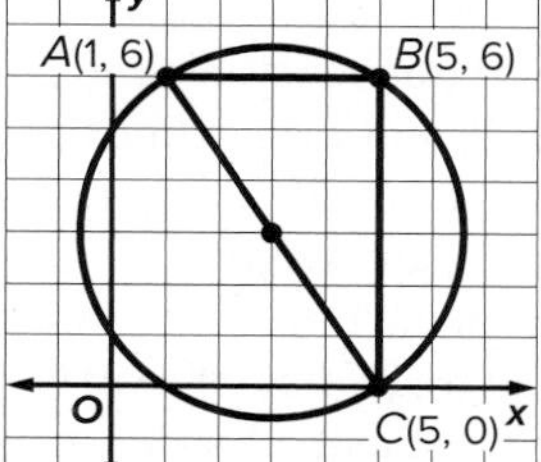

29. $(-2, -1)$, $(2, 1)$

31. $(-2, -4)$, $(2, 0)$

33. $\left(\frac{\sqrt{2}}{2}, \frac{3\sqrt{2}}{2}\right)$, $\left(-\frac{\sqrt{2}}{2}, -\frac{3\sqrt{2}}{2}\right)$

35. $(x - 3)^2 + y^2 = 25$

37a. $x^2 + y^2 = 810{,}000$ **37b.** 3000 ft

39a. No, her friend's house is outside the free delivery area.

39b. I used a coordinate grid with Consuela's house at (0, 0), and the pizza restaurant at (−4, 5). By the Pythagorean Theorem, Consuela's house is $\sqrt{41}$, or approximately 6.4 miles away from the pizza place. The equation of the circle for free delivery is $(x + 4)^2 + (y - 5)^2 = 41$. Consuela's friend's house is at (−1, −1). Substitute this ordered pair into the equation of the circle for free delivery.

$(x + 4)^2 + (y - 5)^2 = 41$	Equation of circle for free delivery
$(-1 + 4)^2 + (-1 - 5)^2 \stackrel{?}{=} 41$	Substitute (−1, −1) for (x, y).
$3^2 + (-6)^2 \stackrel{?}{=} 41$	Simplify.
$9 + 36 \stackrel{?}{=} 41$	Square each term.
$45 > 41$	Compare.

This means that Consuela's friend's house is farther away from the pizza place than Consuela's house. Because Consuela's house is at the edge of the free-delivery area, her friend's house is outside of the free-delivery area.

41 The radius of a circle centered at the origin and containing the point (0, −3) is 3 units. Therefore, the equation of the circle is $(x - 0)^2 + (y - 0)^2 = 3^2$ or $x^2 + y^2 = 9$. The point $(1, 2\sqrt{2})$ lies on the circle, because evaluating $x^2 + y^2 = 9$ for $x = 1$ and $y = 2\sqrt{2}$ results in a true equation.

$$1^2 + (2\sqrt{2})^2 = 9$$
$$1 + 8 = 9$$
$$9 = 9 \checkmark$$

43. $x^2 + y^2 = 16$ **43a.** outside **43b.** on **43c.** inside **43d.** on **45.** $(x + 5)^2 + (y - 2)^2 = 36$
47. $(x - 8)^2 + (y - 2)^2 = 16$; the first circle has its center at $(5, -7)$. If the circle is shifted 3 units right and 9 units up, the new center is at $(8, 2)$, so the new equation becomes $(x - 8)^2 + (y - 2)^2 = 16$. **49a.** 4
49b–c. Method 1: Draw a circle of radius 200 miles centered on each station. Method 2: Use the Pythagorean Theorem to identify pairs of stations that are more than 200 miles apart. Using Method 2, plot the points representing the stations on a graph. Stations that are more than 4 units apart on the graph will be more than 200 miles apart and will thus be able to use the same frequency. Assign station A to the first frequency. Station B is within 4 units of station A, so it must be assigned the second frequency. Station C is within 4 units of stations A and B, so it must be assigned a third frequency. Station D is also within 4 units of stations A, B, and C, so it must be assigned a fourth frequency. Station E is $\sqrt{29}$, or about 5.4 units away from station A, so it can share the first frequency. Station F is $\sqrt{29}$, or about 5.4 units away from station B, so it can share the second frequency. Station G is $\sqrt{32}$, or about 5.7 units away from station C, so it can share the third frequency. Therefore, the least number of frequencies that can be assigned is 4.
51. $(-6.4, 4.8)$ **53.** A **55.** C

Lesson 9-8

1. $x^2 = 24y$ **3.** $y^2 = -8x$ **5.** $\left(x, \frac{9}{16}\right)^2 = 8(y - 1)$
7. Sample answer: $x^2 = 7.2y$

9.

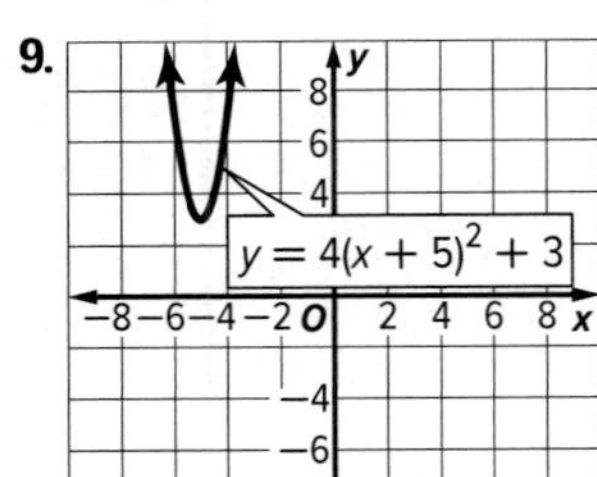

11

$$x = -\frac{1}{4}(y - 4)^2 - 2$$
$$x + 2 = -\frac{1}{4}(y - 4)^2$$
$$-4(x + 2) = (y - 4)^2$$
$$(y - 4)^2 = 4(-1)(-1)(x - (-2))$$

Thus, $h = -2$, $k = 4$, and $p = -1$.
focus, $(h + p, k)$: $(-2 + (-1), 4)$ or $(-3, 4)$
directrix, $x = h - p$: $x = -2 - (-1)$ or -1
axis of symmetry, $y = k$: $y = 4$
vertex, (h, k): $(-2, 4)$
direction of opening: left, because $p < 0$

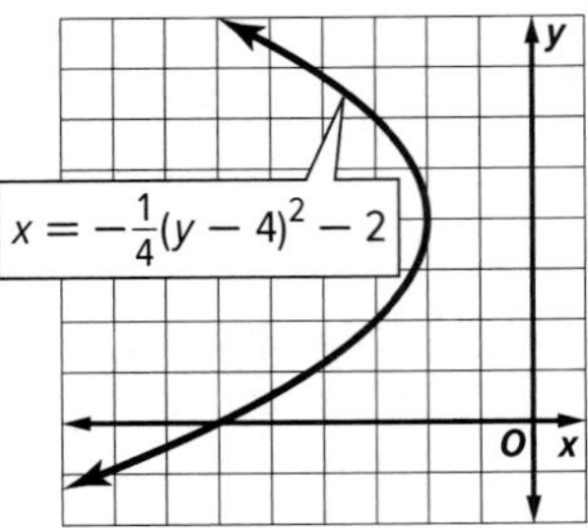

13. $x^2 = -7y$ **15.** $x^2 = -40y$ **17.** $(x - 2)^2 = -8(y - 1)$
19. $(y + 1)^2 = -8(x - 6)$ **21.** $y^2 = -12x$
23. $(x - 1)2 = -8(y + 1)$ **25.** $(y - 1)2 = 4(x + 3)$
27. $y^2 = -20(x + 7)$

29.

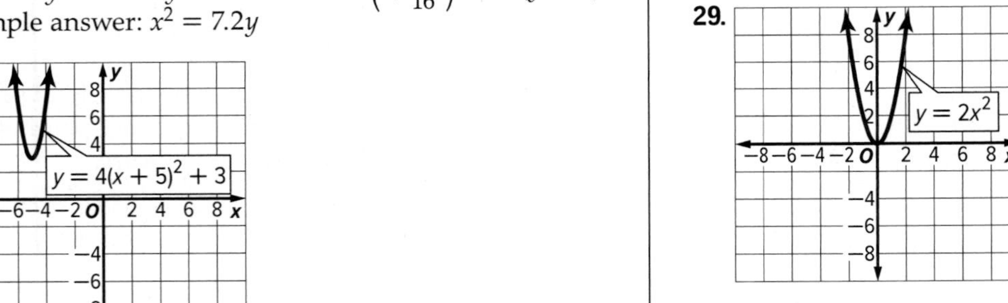

31.

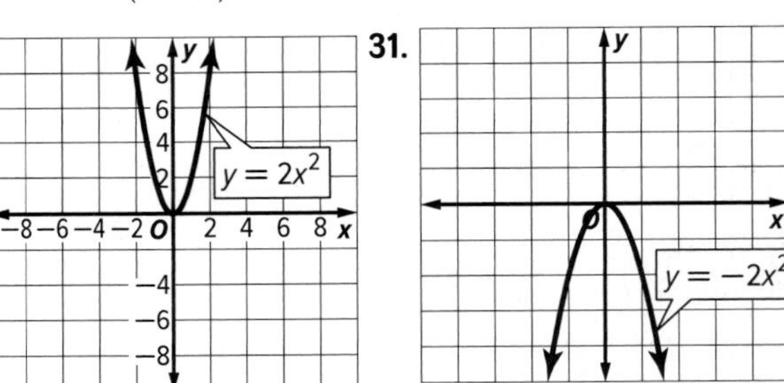

33. **35.**

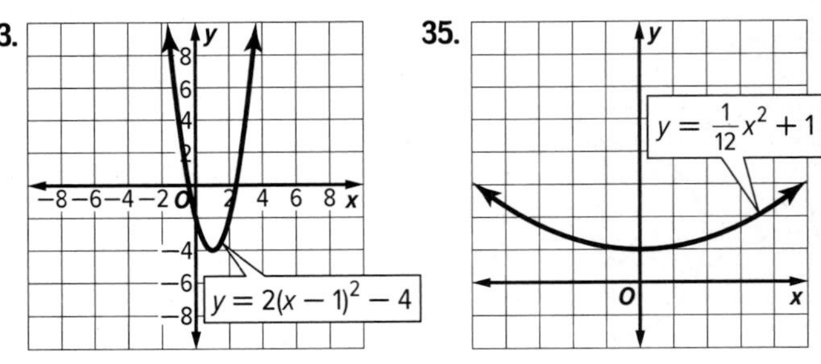

37. **39.**

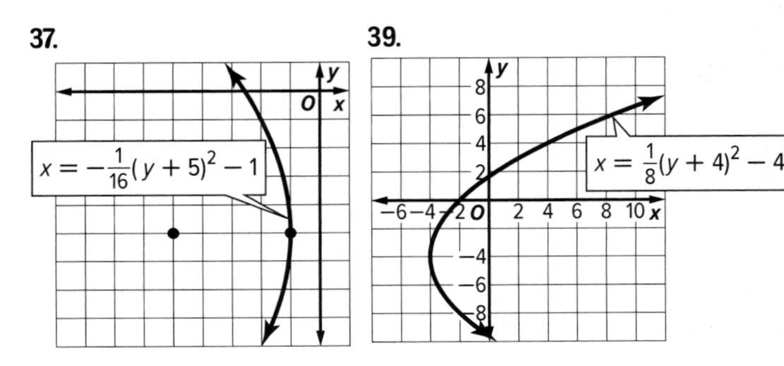

Selected Answers and Solutions

41. $F(-2, 5)$, $y = 1$

43. $F(3, 1)$, $x = -9$

45. $y^2 = 6x$

47. $x^2 = \frac{22}{3}y$

49. $y^2 = 16x$

51. Conjecture: The equation of the new parabola is $(y - 2)^2 = 16(x - 3)$. The graph verifies that $(y - 2)^2 = 16(x - 3)$ is a translation of $y^2 - 16x = 0$ by 3 units to the right and 2 units up.

53. $x^2 = 28y$ **55.** $6\left(x - \frac{3}{4}\right) = (y + 2)^2$

57. vertex: $(4, -2)$, focus: $\left(4\frac{1}{48}, -2\right)$; directrix: $x = 3\frac{47}{48}$

59. vertex: $(4, -6)$, focus: $\left(4\frac{1}{4}, -6\right)$; directrix: $x = 3\frac{3}{4}$

61. vertex: $(0, 4)$, focus: $\left(0, 4\frac{1}{4}\right)$; directrix: $y = 3\frac{3}{4}$

63a. Sample answer: $(0, 4)$, $(2, 0)$ **63b.** $x = 0$ or $x = 2$ **63c.** $y = 4$ or $y = 0$ **63d.** $(0, 4)$, $(2, 0)$; They are the same as the intersection points.

65a.

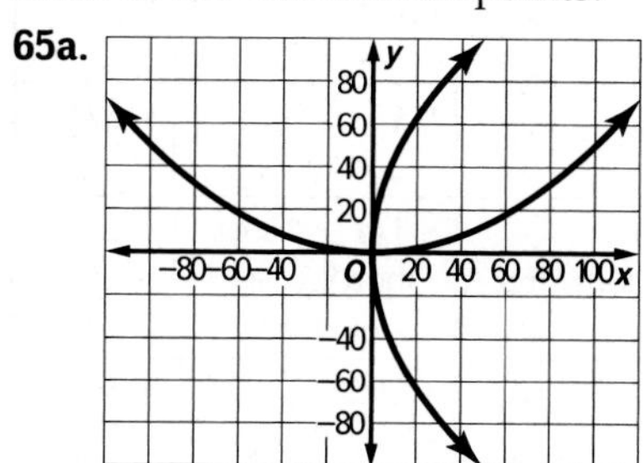

65b. $y^2 = 200x$; $x^2 = 200y$ **65c.** No. The depth of the antenna is the same for both sketches. It is equal to the distance between the focus and the vertex, 50 inches.

67. Sample answer: If you rewrite the equation of the parabola as $y^2 = \frac{1}{4}x$, it is of the form $y^2 = 4px$. Therefore, $4p = \frac{1}{4}$, and $p = \frac{1}{16}$. The value of p represents the distance from the focus to the vertex and from the vertex to the directrix. So, the distance from the focus to the directrix must be $2p$, or $\frac{1}{8}$ in this case.

69. Sample answer: Make a sketch of the parabola by graphing the vertex and focus. Since they are on the same horizontal line and the focus is to the right of the vertex, the parabola opens to the right and will be of the form $(y - k)^2 = 4p(x - h)$ with vertex (h, k). So, $h = 3$ and $k = -2$. The distance between the focus and the vertex is the value of p, or 3. Substitute the values into $(y - k)^2 = 4p(x - h)$ to get the equation, $(y + 2)^2 = 4(3)(x - 3)$, or $(y + 2)^2 = 12(x - 3)$.

71a. Top half of a parabola that has vertex $(0, 0)$ and is open to the right

71b. Domain: $x \geq 0$; Range: $y \geq 0$

71c. Sample answer: If you square both sides of $y = \sqrt{x}$, you get $y^2 = x$. This is related to $y^2 = 4px$ because the coefficient of x in $y^2 = 4px$ is $4p$. So they are part of the same family of parabolas $y^2 = x$. If $p = \frac{1}{4}$, then the equations are identical.

71d. Sample answer: Make a list of values for p, including values less than 1, and then use the calculator to graph the family of functions of the form $y = \sqrt{4px}$ and $y = -\sqrt{4px}$ to see the complete parabolas.
73. Sample answer: As p increases, the focus gets further away from the vertex of the graph of $y^2 = 4px$, and the distance between the focus and directrix increases. As p decreases, the focus gets closer to the vertex of the graph of $y^2 = 4px$, and the distance between the focus and directrix decreases. This is verified by graphing.
75. Sample answer: Graph the three points to see that they are located on a parabola with vertex $(0, 0)$ that is opening downward. Because the vertex is $(0, 0)$, substitute the coordinates into $x^2 = 4py$, and solve for p. If p is the same for both points, then the points are solutions to the same parabola. The parabola containing $(-8, 4)$ has $p = -4$. The parabola containing $\left(-2\sqrt{2}, -\frac{\sqrt{2}}{8}\right)$ has $p = -\frac{16}{\sqrt{2}}$. Because the value of p is different for each point, the points are not on the same parabola. **77a.** $(x + 4)^2 = 12(y - 2)$ **77b.** A **77c.** A, D, E **77d.** A **77e.** The vertex is translated 4 units left and 2 units up. **79a.** $(-5, 1)$ **79b.** $(-7, 1)$ **79c.** $x = -3$ **81.** $x^2 = 34y$ or $y^2 = 34x$

Chapter 9 Study Guide and Review

1. false; chord **3.** true **5.** true **7.** false; congruent **9.** A tangent line intersects the circle at exactly one point, while a secant line intersects the circle at exactly two points. **11.** $\overline{DM}$ or $\overline{DP}$ **13.** 13.69 cm; 6.84 cm **15.** 34.54 ft; 17.27 ft **17.** 163 **19a.** 100.8 **19b.** 18 **19c.** minor arc **21.** 131 **23.** 50.4 **25.** 56

27.

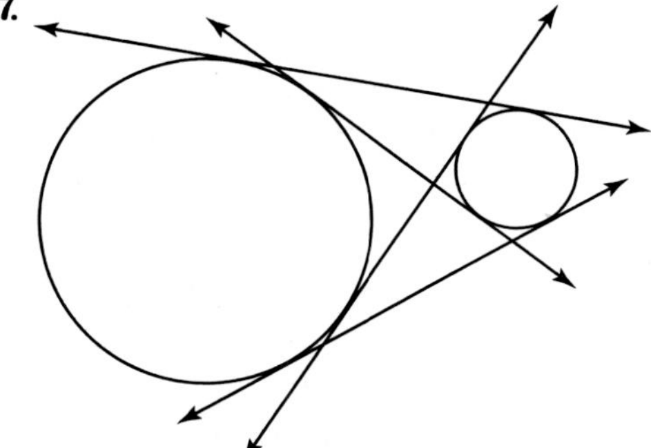

29. 97

31. 214

33. $(x - 1)^2 + (y - 2)^2 = 49$

35. $(1, -5)$; $r = 2$

37. $(3, 1)$; $r = 3$

39. $(-8, -2)$; $r = 7$

41. $(3, 3)$, $(-2, -2)$

43. $x^2 + y^2 = 1156$

45. $y^2 = 16x$

47. $(y + 3)^2 = 12(x - 1)$

Selected Answers and Solutions

CHAPTER 10

Extending Area

Chapter 10 Concept Check

1. $A = \ell \times w$ **3.** substitution **5.** hypotenuse
7. 45°-45°-90° triangle

Lesson 10-1

1. 56 in., 180 in^2 **3.** 64 cm, 207.8 cm^2

5. 43.5 in., 20 in^2

7. 32.5 in., 33.8 in^2 **9.** $15 + \sqrt{65}$ or about 23.1 units; 20 units^2

11 perimeter = 21 + 17 + 21 + 17 or 76 ft
Use the Pythagorean Theorem to find the height.

$8^2 + h^2 = 17^2$ Pythagorean Theorem
$64 + h^2 = 289$ Simplify.
$h^2 = 225$ Subtract 64 from each side.
$h = 15$ Take the positive square root of each side.
$A = bh$ Area of a parallelogram
$= 21(15)$ or 315 ft^2 $b = 21$ and $h = 15$

13. 69.9 m, 129.9 m^2 **15.** 174.4 m, 1520 m^2 **17.** 727.5 ft^2

19. 338.4 cm^2 **21.** 480 m^2

23

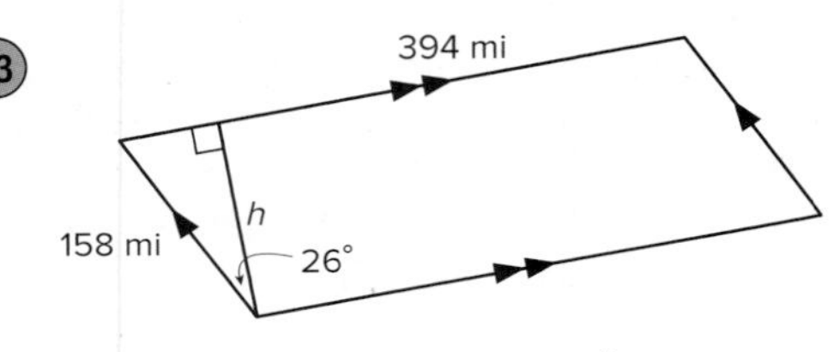

$\cos 26° = \frac{h}{158}$ $\cos = \frac{\text{adjacent}}{\text{hypotenuse}}$
$158 \cos 26° = h$ Multiply each side by 158.
$142 \approx h$ Use a calculator.

$A = bh$ Area of a parallelogram
$\approx 394(142)$ or 55,948 mi^2 $b = 394$ and $h = 142$

25. $6 + 5\sqrt{2} + \sqrt{26}$ or about 18.2 units; 15 units^2

27 Graph the parallelogram, then measure the length of the base and the height and calculate the perimeter and area.

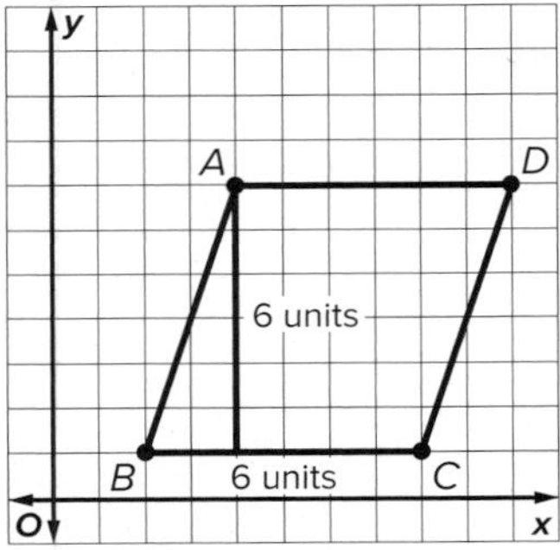

$P = AB + BC + CD + AD$
$= 2\sqrt{10} + 6 + 2\sqrt{10} + 6$
$= 12 + 4\sqrt{10}$ or about 24.6 units

$A = bh$ Area of a parallelogram
$\approx 6(6)$ or 36 units^2 $b = 6$ and $h = 6$

29a. yellow: 1 gal, 1 qt, and 3 8-oz bottles; blue: 2 qt; red: 3 8-oz bottles; purple: 1 8-oz bottle **29b.** Sample answer: I created two tables to organize the information provided and my calculations. The first table shows the relationship between each size paint container and the amount of area each will cover.

Size	8 oz	1 qt (32 oz in a qt)	1 gal (4 qt in a gal)
Cost ($)	3.75	14	30
Area covered (ft^2)	21.875	87.5	350

The second table shows the relationship between the paint color, the area to be covered, and the possible purchase options and cost.

Color	Area (ft^2)	Area to Paint (ft^2)	Possible Purchase	Cost ($)
Red	$A_R = \frac{1}{2}(5)(6) = 15$	(15)(3) = 45	• 3 8-oz bottles	• 11.25
Purple	$A_P = (1)(4) = 4$	(4)(3) = 12	• 1 8-oz bottle	• 3.75
Blue	$A_B = (12)(5) - 4 = 56$	(56)(3) = 168	• 1 qt + 4 8-oz bottles • 2 quarts	• 29 • 28
Yellow	$A_Y = (12)(20) - (56 + 4 + 15) = 165$	(165)(3) = 495	• 2 gal • 1 gal + 2 qt • 1 gal + 1 qt + 3 8-oz bottles	• 60 • 58 • 55.25

Madison should buy 3 8-oz bottles of red paint, 1 8-oz bottle of purple paint, 2 qts of blue paint, and 1 gal, 1 qt, and 3 8-oz bottles of yellow paint.
31. 9.19 in.; 4.79 in^2 **33.** b = 14 ft; h = 7 ft

35a. 10.9 units^2

35b.
$$\sqrt{s(s-a)(s-b)(s-c)} \stackrel{?}{=} \frac{1}{2}bh$$
$$\sqrt{15(15-5)(15-12)(15-13)} \stackrel{?}{=} \frac{1}{2}(5)(12)$$
$$\sqrt{15(10)(3)(2)} \stackrel{?}{=} 30$$
$$\sqrt{900} \stackrel{?}{=} 30$$
$$30 = 30$$

37. 15 units^2; Sample answer: I inscribed the triangle in a 6-by-6 square. I found the area of the square and subtracted the areas of the three right triangles inside the square that were positioned around the given triangle. The area of the given triangle is the difference, or 15 units^2. **39.** Sample answer: The area will not change as K moves along line p. Because lines m and p are parallel, the perpendicular distance between them is constant. That means that no matter where K is on line p, the perpendicular distance to line p, or the height of the triangle, is always the same. Because points J and L are not moving, the distance between them, or the length of the base, is constant. Because the height of the triangle and the base of the triangle are both constant, the area will always be the same. **41.** Sample answer: To find the area of the parallelogram, you can measure the height $\overline{PT}$ and then measure one of the bases $\overline{PQ}$ or $\overline{SR}$ and multiply the height by the base to get the area. You can also measure the height $\overline{SW}$ and measure one of the bases $\overline{QR}$ or $\overline{PS}$ and then multiply the height by the base to get the area. It doesn't matter which side you choose to use as the base, as long as you use the height that is perpendicular to that base to calculate the area. **43.** A **45.** D **47.** 18 **49a.** $6h^2$ **49b.** $12h$ **49c.** 216 **49d.** 72

Lesson 10-2

1. 132 ft^2 **3.** 178.5 m^2 **5.** 8 cm **7.** 6.3 ft **9.** 678.5 ft^2 **11.** 136 in^2 **13.** 137.5 ft^2

15 $A = \frac{1}{2}d_1d_2$ Area of a kite

$= \frac{1}{2}(4.8)(10.2)$ $d_1 = 4.8$ and $d_2 = 10.2$

$= 24.48$ Simplify.

The area is about 24.5 square microns.

17. 784 ft^2

19 Let x represent the length of one diagonal. Then the length of the other diagonal is $3x$.

$A = \frac{1}{2}d_1d_2$ Area of a rhombus

$168 = \frac{1}{2}(x)(3x)$ $A = 168$, $d_1 = x$, and $d_2 = 3x$

$168 = \frac{3}{2}x^2$ Simplify.

$112 = x^2$ Multiply each side by $\frac{2}{3}$.

$\sqrt{112} = x$ Take the positive square root of each side.

So the lengths of the diagonals are $\sqrt{112}$ or about 10.6 centimeters and $3(\sqrt{112})$ or about 31.7 centimeters.

21. 4 m **23.** The area of $\triangle HJF = \frac{1}{2}d_1\left(\frac{1}{2}d_2\right)$ and the area of $\triangle HGF = \frac{1}{2}d_1\left(\frac{1}{2}d_2\right)$. Therefore, the area of $\triangle HJF = \frac{1}{4}d_1d_2$, and the area of $\triangle HGF = \frac{1}{4}d_1d_2$. The area of kite $FGHJ$ is equal to the area of $\triangle HJF$ + the area of $\triangle HGF$ or $\frac{1}{4}d_1d_2 + \frac{1}{4}d_1d_2$. After simplification, the area of kite $FGHJ$ is equal to $\frac{1}{2}d_1d_2$. **25a.** 24 in^2 each of yellow, red, orange, green, and blue; 20 in^2 of purple **25b.** Yes; her kite has an area of 140 in^2, which is less than 200 in^2. **27.** 18 sq. units **29.** The area of a trapezoid is $\frac{1}{2}h(b_1 + b_2)$. So, $A = \frac{1}{2}(x + y)(x + y)$ or $\frac{1}{2}(x^2 + xy + y^2)$. The area of $\triangle 1 = \frac{1}{2}(y)(x)$, $\triangle 2 = \frac{1}{2}(z)(z)$, and $\triangle 3 = \frac{1}{2}(x)(y)$. The area of $\triangle 1 + \triangle 2 + \triangle 3 = \frac{1}{2}xy + \frac{1}{2}z^2 + \frac{1}{2}xy$. Set the area of the trapezoid equal to the combined areas of the triangles to get $\frac{1}{2}(x^2 + 2xy + y^2) = \frac{1}{2}xy + \frac{1}{2}z^2 + \frac{1}{2}xy$. Multiply by 2 on each side: $x^2 + 2xy + y^2 = 2xy + z^2$. When simplified, $x^2 + y^2 = z^2$.

31 The length of the base of the triangle is $\frac{12 - 8}{2}$ or 2. Use trigonometry to find the height of the triangle (and trapezoid).

$\tan 30 = \frac{\text{opposite}}{\text{adjacent}}$

$\frac{\sqrt{3}}{3} = \frac{2}{h}$

$\sqrt{3}h = 6$

$h = \frac{6}{\sqrt{3}}$

$h = 2\sqrt{3}$

Use the Pythagorean Theorem to find the hypotenuse of the triangle.

$a^2 + b^2 = c^2$

$2^2 + (2\sqrt{3})^2 = c^2$

$4 + 12 = c^2$

$16 = c^2$

$4 = c$

Find the perimeter and area of the trapezoid.

perimeter $= 12 + 8 + 4 + 4$

$= 28$ in.

$= \frac{28}{12}$ ft

≈ 2.3 ft

area $= \frac{1}{2}(b_1 + b_2)h$

$= \frac{1}{2}(8 + 12)(2\sqrt{3})$

$= 20\sqrt{3\text{in}^2}$

$= 20\sqrt{3\text{in}^2} \cdot \frac{1 \text{ ft}}{12 \text{ in.}} \cdot \frac{1 \text{ ft}}{12 \text{ in.}}$

$= \frac{20\sqrt{3}}{144}$ ft^2

≈ 0.2 ft^2

33a.

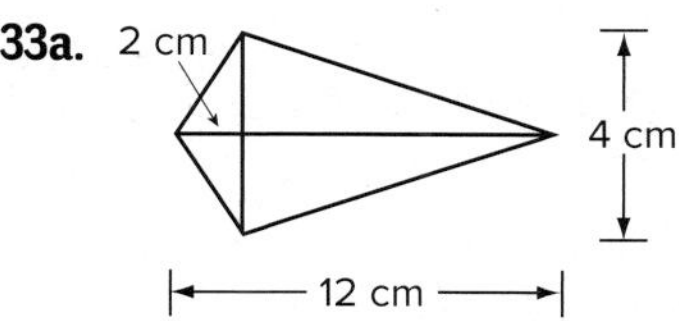

33b.

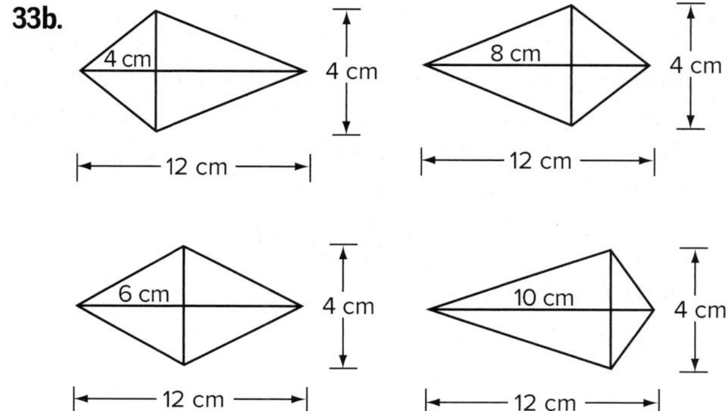

33c.

x	P
2 cm	26.1 cm
4 cm	25.4 cm
6 cm	25.3 cm
8 cm	25.4 cm
10 cm	26.1 cm

33d.

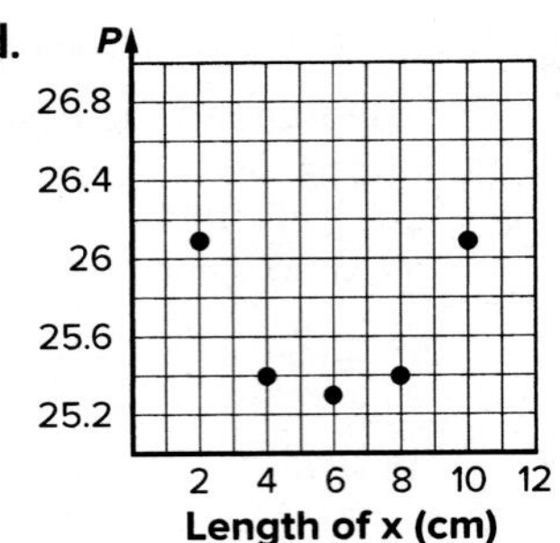

33e. Sample answer: Based on the graph, the perimeter will be minimized when $x = 6$. This value is significant because when $x = 6$, the figure is a rhombus. **35.** 7.2 **37.** Sometimes; sample answer: If the areas are equal, it means that the products of the diagonals are equal. The only time that the perimeters will be equal is when the diagonals are also equal, or when the two rhombi are congruent. **39.** D **41.** 12 m and 48 m **43.** C

Lesson 10-3

1. 1385.4 yd^2

3

$A = \pi r^2$	Area of a circle
$74 = \pi r^2$	$A = 74$
$23.55 \approx r^2$	Divide each side by π.
$4.85 \approx r$	Take the positive square root of each side.

So, the diameter is $2 \cdot 4.85$ or about 9.7 millimeters.

5. 4.5 in^2 **7a.** 10.6 in^2 **7b.** \$48 **9.** 78.5 yd^2 **11.** 14.2 in^2 **13.** 78.5 ft^2 **15.** 10.9 mm **17.** 8.1 ft

19

$A = \frac{x}{360} \cdot \pi r^2$	Area of a sector
$= \frac{72}{360} \cdot \pi(8)^2$	$x = 72$ and $r = 8$
$\approx 40.2\ cm^2$	Use a calculator.

21. 322 m^2 **23.** 284 in^2 **25a.** 1.7 cm^2 **25b.** about 319.4 mg **27.** 13 **29.** 9.8

31 **a.**

$C = \pi d$	Circumference of a circle
$36 = \pi d$	$C = 36$
$11.5\ ft \approx d$	Divide each side by π.

b. age = diameter · growth factor
= 11.5 · 130 or 1495 yrs

33. 53.5 m^2 **35.** 10.7 cm^2 **37.** 7.9 in^2 **39.** 30 mm^2

41 The area equals the area of the large semicircle with a radius of 6 in. plus the area of a small semicircle minus 2 times the area of a small semicircle. The radius of each small semicircle is 2 in.

$$A = \frac{1}{2}\pi(6)^2 + \frac{1}{2}\pi(2)^2 - 2\left[\frac{1}{2}\pi(2)^2\right]$$
$$= 18\pi + 2\pi - 4\pi$$
$$= 16\pi$$
$$\approx 50.3\ in^2$$

43a. $A = \frac{x\pi r^2}{360} - r^2\left[\sin\left(\frac{x}{2}\right)\cos\left(\frac{x}{2}\right)\right]$

43b.

x	A
10	0.1
20	0.5
30	1.7
40	4.0
45	5.6
50	7.7
60	13.0
70	20.3
80	29.6
90	41.1

43c.

Area and Central Angles

Area of Segment: 0, 5, 10, 15, 20, 25, 30, 35, 40, 45, 50

Central Angle Measure: 10 20 30 40 50 60 70 80 90

43d. Sample answer: From the graph, it looks like the area would be about 15.5 when x is 63°. Using the formula, the area is 15.0 when x is 63°. The values are very close because I used the formula to create the graph. **45.** 449.0 cm^2 **47.** Sample answer: You can find the shaded area of the circle by subtracting x from 360° and using the resulting measure in the formula for the area of a sector. You could also find the shaded area by finding the area of the entire circle, finding the area of the unshaded sector using the formula for the area of a sector, and subtracting the area of the unshaded sector from the area of the entire circle. The method in which you find the ratio of the area of a sector to the area of the whole circle is more efficient. It requires fewer steps, is faster, and there is a lower probability for error. **49.** Sample answer: If the radius of the circle doubles, the area will not double. If the radius of the circle doubles, the area will be four times as great. Because the radius is squared, if you multiply the radius by 2, you multiply the area by 2^2, or 4. If the arc length of a sector is doubled, the area of the sector is doubled. Because the arc length is not raised to a power, if the arc length is doubled, the area would also be twice as large. **51.** B **53.** A **55.** C **57a.** 8 in. **57b.** 33.5 in^2 **57c.** 27.7 in^2 **57d.** 5.8 in^2

Lesson 10-4

1. center: point P, radius: $\overline{PC}$, apothem: $\overline{PR}$, central angle: $\angle BPC$, ≈51.4 **3.** 162 in^2 **5.** 120 ft^2

7 **a.** The blue area equals the area of the center circle with a radius of 3 ft plus 2 times the quantity of the area a rectangle 19 ft by 12 ft minus the area of a semicircle with a radius of 6 ft.

Area
= Area of circle + 2 · Area of rectangle − Area of semicircle
$= \pi r^2 + 2 \cdot \left(\ell w - \frac{1}{2}\pi r^2\right)$
$= \pi(3)^2 + 2\left[(19(12) - \frac{1}{2}\pi(6)^2\right]$
$= 9\pi + 2(228 - 18\pi)$
$= 9\pi + 456 - 36\pi$

$= 456 - 27\pi$

$\approx 371 \text{ ft}^2$

b. The red area equals the area of the center circle with a radius of 6 ft minus the center circle with a radius of 3 ft plus 2 times the area of a circle with a radius of 6 ft.

Area

= Area of large circle − Area of small circle + 2 • Area of circle

$= \pi r^2 - \pi r^2 + 2 \cdot \pi r^2$

$= \pi(6)^2 - \pi(3)^2 + 2\pi(6)^2$

$= 36\pi - 9\pi + 72\pi$

$= 99\pi$

$\approx 311 \text{ ft}^2$

9. center: point R, radius: $\overline{RO}$, apothem: $\overline{RT}$, central angle: $\angle ORN$, 45 **11.** 59.4 cm^2 **13.** 584.2 in^2

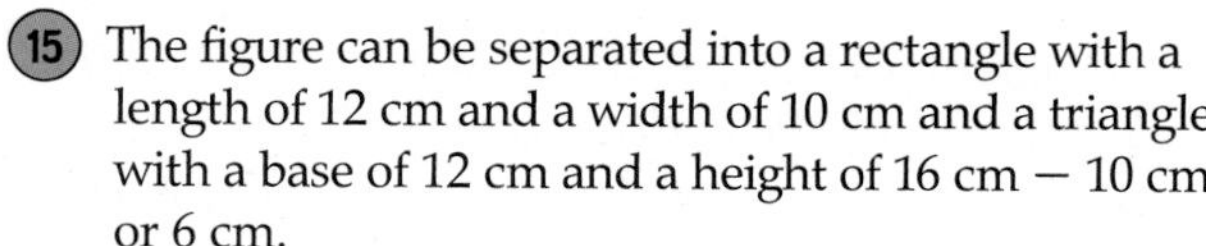

15 The figure can be separated into a rectangle with a length of 12 cm and a width of 10 cm and a triangle with a base of 12 cm and a height of 16 cm − 10 cm or 6 cm.

Area of figure = Area of rectangle + Area of triangle

$= \ell w + \frac{1}{2}bh$

$= 12(10) + \frac{1}{2}(12)(6)$

$= 120 + 36 \text{ or } 156 \text{ cm}^2$

17. 55.6 in^2 **19a.** 29.7 in., 52.3 in^2 **19b.** 16 **21.** $\approx$354 ft^2 **23.** 1.9 in^2 **25a.** 50.9 ft^2 **25b.** 4 boxes **27.** 58.1 mm; 232.4 mm^2

29 To find the area of the shaded region, find the area of the rectangle 8 units by 4 units minus the area of the semicircle with a radius of 2 units minus the area of the trapezoid with bases 4 units and 2 units and height 2 units.

Area of figure

= Area of rectangle − Area of semicircle − Area of trapezoid

$= \ell w - \frac{1}{2}\pi r^2 - \frac{1}{2}h(b_1 + b_2)$

$= 8(4) - \frac{1}{2}\pi(2)^2 - \frac{1}{2}(2)(4 + 2)$

$= 32 - 2\pi - 6$

$= 26 - 2\pi$

$\approx 19.7 \text{ units}^2$

31. 24 units^2 **33.** 0.43 in^2; 0.56 in^2; 0.62 in^2; 0.65 in^2; Sample answer: When the perimeter of a regular polygon is constant, as the number of sides increases, the area of the polygon increases.

35. Chloe; sample answer: The measure of each angle of a regular hexagon is 120°, so the segments from the center to each vertex form 60° angles. The triangles formed by the segments from the center to each vertex are equilateral, so each side of the hexagon is 11 in. The perimeter of the hexagon is 66 in. Using trigonometry, the length of the apothem is about 9.5 in. Putting the values into the formula for the area of a regular polygon and simplifying, the area is about 313.5 in^2.

37. Sample answer:

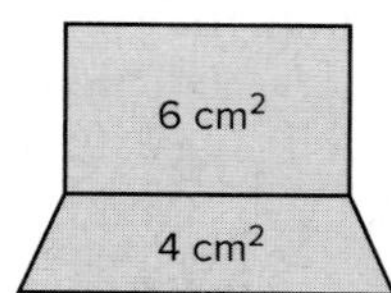

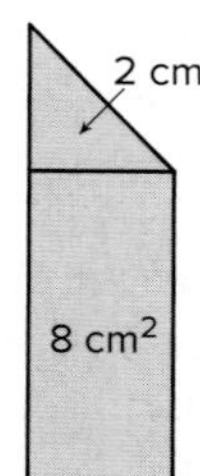

39. Sample answer: You can decompose the figure into shapes for which you know the area formulas. Then, you can sum all of the areas to find the total area of the figure. **41.** A **43.** 84.3 square inches **45.** B **47.** 30.9 square inches

Lesson 10-5

1. 9 yd^2 **3.** $\frac{5}{3}$; 35 **5.** 5.28 in^2

7 The scale factor between the parallelograms is $\frac{7.5}{15}$ or $\frac{1}{2}$, so the ratio of their areas is $\left(\frac{1}{2}\right)^2$ or $\frac{1}{4}$.

$\frac{\text{area of small figure}}{\text{area of large figure}} = \frac{1}{4}$ Write a proportion.

$\frac{60}{\text{area of large figure}} = \frac{1}{4}$ Substitution

$60 \cdot 4 = \text{area of large figure} \cdot 1$ Cross multiply.

$240 = \text{area of large figure}$ Simplify.

So the area of the large parallelogram is 240 ft^2.

9. 672 cm^2 **11.** $\frac{4}{5}$; 17.5 **13.** $\frac{3}{2}$; 36 **15a.** If the area is doubled, the radius changes from 24 in. to 33.9 in. **15b.** If the area is tripled, the radius changes from 24 in. to 41.6 in. **15c.** If the area changes by a factor of x, then the radius changes from 24 in. to $24\sqrt{x}$ in. **17a.** 4 in. **17b.** Larger; sample answer: The area of a circular pie pan with an 8 in. diameter is about 50 in^2. The area of the larger pan is 52.6 in^2, and the area of the smaller pan is 41.6 in^2. The area of the larger pan is closer to the area of the circle, so Kaitlyn should choose the larger pan to make the recipe.

19 Area of $\triangle JKL = \frac{1}{2}bh$

$= \frac{1}{2}(5)(6)$ or 15 square units

The scale factor between the triangles is $\frac{5}{3}$, so the ratio of their areas is $\left(\frac{5}{3}\right)^2$ or $\frac{25}{9}$.

$\frac{\text{area of } \triangle JKL}{\text{area of } \triangle J'K'L'} = \frac{25}{9}$ Write a proportion.

$\frac{15}{\triangle J'K'L'} = \frac{25}{9}$ Area of $\triangle JKL = 15$

$15 \cdot 9 = \text{area of } \triangle J'K'L' \cdot 25$ Cross multiply.

$5.4 = \text{area of } \triangle J'K'L'$ Divide each side by 25.

So the area of $\triangle J'K'L'$ is 5.4 units^2.

21. area of $ABCD = 18$; area of $A'B'C'D' \approx 56.2$

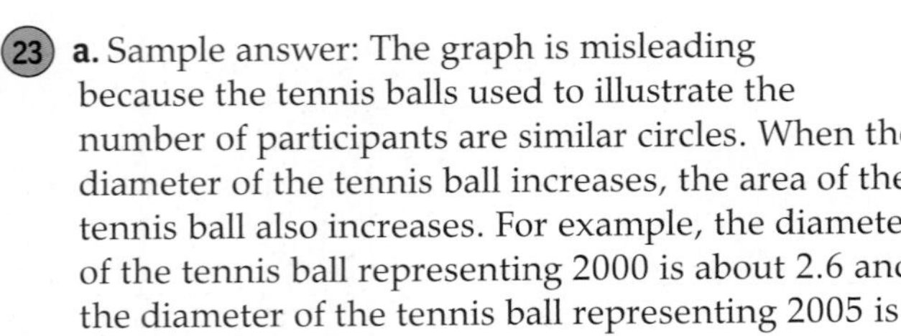

23 **a.** Sample answer: The graph is misleading because the tennis balls used to illustrate the number of participants are similar circles. When the diameter of the tennis ball increases, the area of the tennis ball also increases. For example, the diameter of the tennis ball representing 2000 is about 2.6 and the diameter of the tennis ball representing 2005 is about 3. So, the rate of increase in the diameters is $\frac{3 - 2.6}{2005 - 2000}$ or about 8%. The area of the circle representing 2000 is $\pi(1.3)^2$ and the area of the circle representing 2005 is $\pi(1.5)^2$. So, the rate of increase in the areas is $\frac{2.25\pi - 1.69\pi}{2005 - 2000}$ or about 35%. The area of the tennis ball increases at a greater rate than the diameter of the tennis ball, so it looks like the number of participants in high school tennis is increasing more than it actually is.

b. Sample answer: If you use a figure with a constant width to represent the participation in each year and only change the height, the graph would not be misleading. For example, use rectangles of equal width and height that varies.

25. Neither; sample answer: In order to find the area of the enlarged circle, you can multiply the radius by the scale factor and substitute it into the area formula, or you can multiply the area formula by the scale factor squared. The formula for the area of the enlargment is $A = \pi(kr)^2$ or $A = k^2\pi r^2$.
27. $P_{\text{enlarged}} = Q\sqrt{R}$ **29.** Sample answer: If you know the area of the original polygon and the scale factor of the enlargement, you can find the area of the enlarged polygon by multiplying the original area by the scale factor squared. **31.** A **33.** scale factor: $\frac{4}{3}$; $x = 12$
35a. $\frac{14}{11}$ **35b.** $\frac{\text{area of } KLMN}{\text{area of } FGHJ} = \left(\frac{14}{11}\right)^2$ **35c.** area of $KLMN = \frac{196}{121} \cdot 200 \approx 324$; the area of $KLMN$ is about 324 mm².

Lesson 10-6

1. 640 cm² **3.** ≈ 571.9 cm² **5.** 336 ft² **7.** ≈ 571.8 in²
9. 840 ft²

11 $S = L + 2B$
Find the missing edge length x of the base.
$3^2 + 4^2 = x^2$ by the Pythagorean Theorem.
$25 = x^2$, so $x = 5$ ft
B is the area of a base. The area of a right triangle is $\frac{1}{2}bh$. For the base, $b = 3$ and $h = 4$.
$B = \frac{1}{2}bh = \frac{1}{2}(3)(4) = 6$
L is the total area of the three lateral faces.
$L = 2 \cdot 3 + 2 \cdot 4 + 2 \cdot 5 = 6 + 8 + 10 = 24$
$S = L + 2B = 24 + 2(6) = 36$ ft²
13. ≈32.8 cm² **15.** ≈236.6 ft² **17.** ≈43.8 cm²

19. ≈255 mm² **21.** ≈256.4 in² **23.** ≈339.3 mm²
25. ≈311.2 ft² **27.** ≈427.6 in² **29.** ≈8.4 km²

31 The lateral area of a regular pyramid is $L = \frac{1}{2}P\ell$. The base of the pyramid is a square, so $P = 4(165) = 660$ yd.
To find the slant height ℓ, find the length of the hypotenuse of a right triangle with legs of length 20 yd and $\frac{1}{2}(165) = 82.5$ yd.
$\ell^2 = 20^2 = 82.5^2$
$\ell^2 = 7206.25$
$\ell \approx 84.8896$
$L = \frac{1}{2}(660)(84.8896) \approx 28{,}013.6$ yd²

33. about 380.1 ft² **35a.** about 12 ft
35b. Sample answer: First, find the sum of the surface areas of each individual section. The rectangular section of the front and back is $2 \times 20 \times g$ or $40g$ ft². The sides cover $2 \times 40 \times g$ or $80g$ ft². The triangular tops of the front and back of the greenhouse cover $2(0.5)(4)(20)$ or 80 ft². The slant of the roof is $\sqrt{116} \approx 10.77$. Thus, the roof covers $2(40)(10.77)$ or 861 ft². The total surface area is $861 + 80 + 120g$ ft². Hector can use up to $60{,}000 \div 25$ or 2400 ft². Therefore, g is approximately 12.1. Rounding down, we get a height of 12 ft. **35c.** Sample answer: Hector used the entire available plot. There was no glass used for the base. The entrance was made of glass. The top of the roof ran along the 40-ft length of the greenhouse.

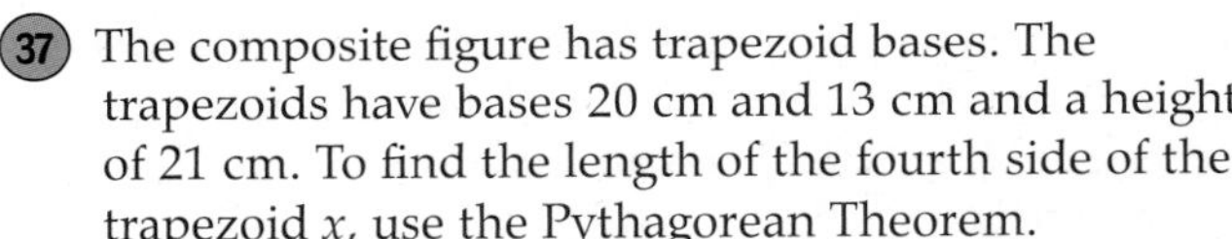

37 The composite figure has trapezoid bases. The trapezoids have bases 20 cm and 13 cm and a height of 21 cm. To find the length of the fourth side of the trapezoid x, use the Pythagorean Theorem.

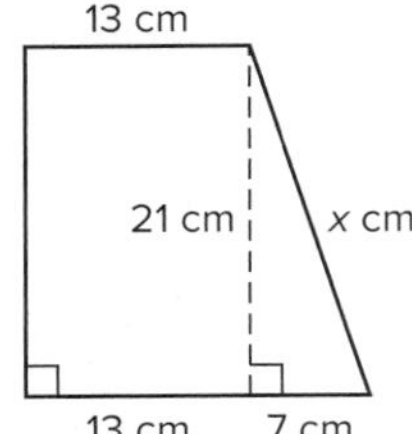

$x^2 = 21^2 + 7^2$	Pythagorean Theorem
$x^2 = 490$	Simplify.
$x \approx 22.136$	Take the square root of each side.

$S = Ph + 2B$	Surface area of a prism
$\approx (21 + 13 + 22.136 + 20)(28) +$	
$2\left[\frac{1}{2}(21)(20 + 13)\right]$	Substitution
$\approx 2131.8 + 693$ or 2824.8 cm²	Simplify.

39. 4524.8 ft^2 **41.** about 299.1 cm^2 **43.** They are not equal. The slant height of the cone is $\frac{2\sqrt{\pi}}{\pi}$ or about 1.13 times greater than the slant height of the square pyramid. 45. Always; if the heights and radii are the same, the surface area of the cylinder will be greater because it has two circular bases and additional lateral area.
47. $\frac{\sqrt{3}}{2}\ell^2 + 3\ell h$; the area of the equilateral triangle of side ℓ is $\frac{\sqrt{3}}{4}\ell^2$ and the perimeter of the triangle is 3ℓ. So, the total surface area is $\frac{\sqrt{3}}{2}\ell^2 + 3\ell h$.
49. D **51.** C **53.** B

Chapter 10 Study Guide and Review

1. false; height **3.** false; radius **5.** true **7.** true **9.** If two polygons are similar, then their areas are proportional to the square of the scale factor between them. **11.** $P = 50$ cm; $A = 60$ cm^2
13. $P = 13.2$ mm; $A = 6$ mm^2 **15.** 132 ft^2 **17.** 96 cm^2
19. 336 cm^2 **21.** 1.5 m^2 **23.** 59 in^2 **25.** 166.3 ft^2
27. 65.0 m^2 **29.** 7.66 cm^2 **31.** $\frac{1}{2}$; 8
33. area of $\triangle RST = 18$ square units; area of $\triangle R'S'T' =$ 4.5 square units **35.** 75 mi^2 **37.** Sample answer: 160 ft^2; 202 ft^2 **39.** 113.1 cm^2; 169.6 cm^2 **41.** 354.4 cm^2; 432.9 cm^2

Selected Answers and Solutions

CHAPTER 11

Extending Volume

Chapter 11 Concept Check

1a. Take one-half the product of the lengths of the diagonals. **1b.** 176 in^2 **3.** 64π **5.** Evaluate the exponents.

Lesson 11-1

1. rectangle **3.** rectangle

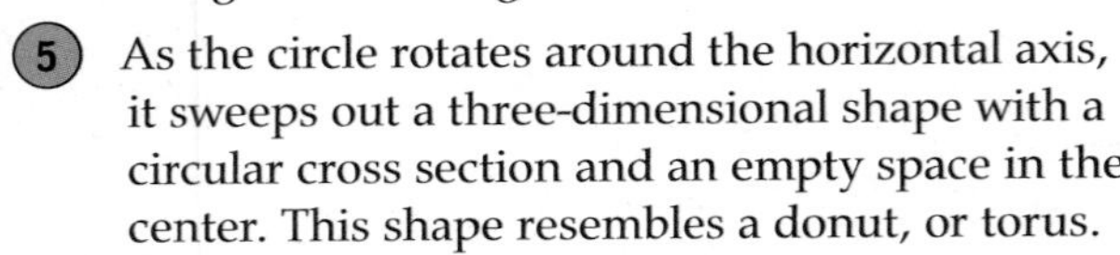

5 As the circle rotates around the horizontal axis, it sweeps out a three-dimensional shape with a circular cross section and an empty space in the center. This shape resembles a donut, or torus.

7. ellipse **9.** circle

11 As the rectangle rotates around the vertical axis, it sweeps out a three-dimensional shape with a rectangular cross section and an empty space in the center. This shape resembles a tube, or open cylinder.

13a. slice perpendicular to the base **13b.** slice parallel to the base **13c.** slice at an angle **15.** triangle **17.** A sample sketch is shown.

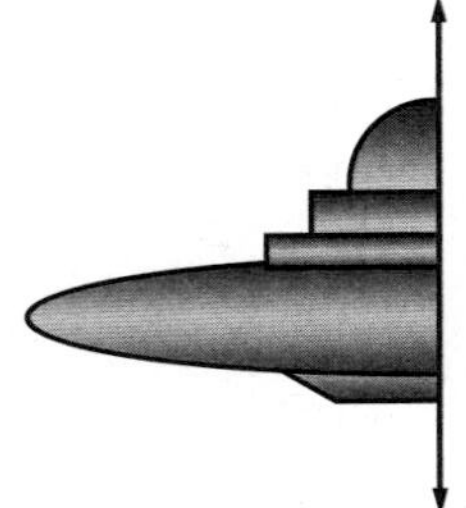

19. Sample answer: The curve would have the same shape as the edge of the vase. The curve would look like the letter *S* stretched vertically. **21a.** rectangle **21b.** Cut off a corner of the clay. **23.** queen, bishop, rook, pawn **25.** a right trapezoid **27.** rectangle **29.** circle **31.** D **33.** A **35.** Rotate a circle around an axis. The outer edge of the circle should be 6 inches from the axis. **37.** Sample answer: A plane intersects a square pyramid at an angle through opposite faces. A plane intersects a triangular pyramid at an angle through opposite faces.

Lesson 11-2

1. 108 cm^3 **3.** 26.95 m^3 **5.** 206.4 ft^3 **7.** 1025.4 cm^3 **9.** D

11
$V = Bh$ — Volume of a prism
$= 38.5(14)$ — $B = \frac{1}{2}(11)(7)$ or 38.5, $h = 14$
$= 539$ m^3 — Simplify.

13. 58.14 ft^3 **15.** 1534.25 in^3

17
$V = \pi r^2 h$ — Volume of a cylinder
$= \pi(6)^2(3.6)$ — Replace r with $12 \div 2$ or 6 and h with 3.6.
≈ 407.2 cm^3 — Use a calculator.

19. 2686.1 mm^3 **21.** 521.5 cm^3 **23.** 31 in^3 **25.** 15 in.; 15.4 in. **27.** 120 m^3

29 Find the volume of the original cylinder.

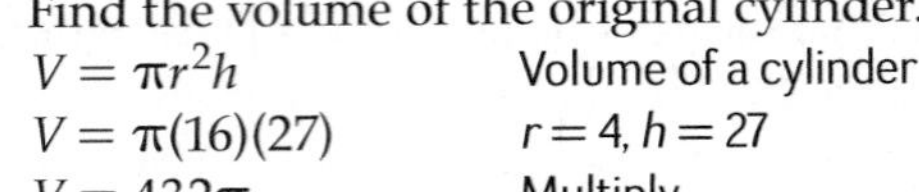

$V = \pi r^2 h$ — Volume of a cylinder
$V = \pi(16)(27)$ — $r = 4, h = 27$
$V = 432\pi$ — Multiply.

The new cylinder is 30% larger than the original. Find 30% of the volume and add to the original.
new volume $= 0.30(432\pi) + 432\pi$ or 561.6π
Because the new can has the same radius as the original, use the volume formula to find the height.

$V = \pi r^2 h$ — Volume of a cylinder
$561.6\pi = 16\pi h$ — Substitution
$35.1 = h$ — Divide.

The height of the larger can is 35.1 cm.

31. 678.6 in^3 **33.** 3934.9 cm^3 **35a.** 0.0019 lb/in^3 **35b.** The plant should grow well in this soil because the bulk density of 0.0019 lb/in^3 is close to the desired bulk density of 0.0018 lb/in^3. **35c.** 8.3 lb **37.** 3,190,680.0 cm^3

39

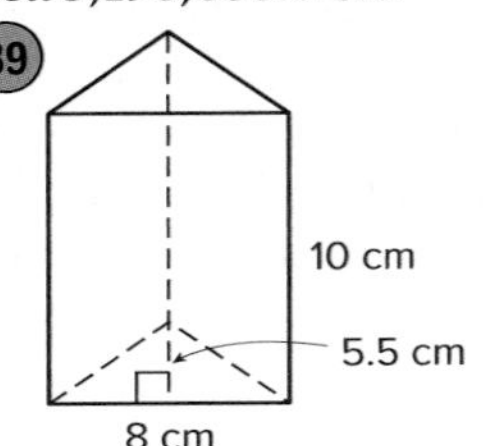

Each triangular prism has a base area of $\frac{1}{2}(8)(5.5)$ or 22 cm^2 and a height of 10 cm. The volume of each triangular prism is $22 \cdot 10$ or 220 cm^3. So, the volume of five triangular prisms is $220 \cdot 5$ or 1100 cm^3.

41a.

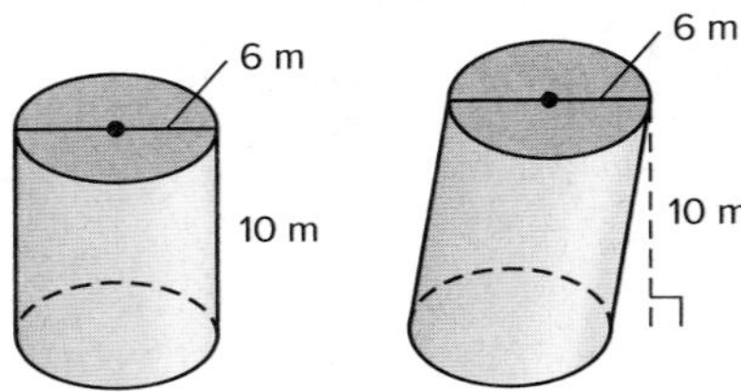

41b. Greater than; a square with a side length of 6 m has an area of 36 m^2. A circle with a diameter of 6 m has an area of 9π or 28.3 m^2. Because the heights are the same, the volume of the square prism is greater. **41c.** Multiplying the radius by x; because the volume is represented by $\pi r^2 h$, multiplying the height by x makes the volume x times greater. Multiplying the radius by x makes the volume x^2 times greater, assuming $x > 1$. **43a.** base 3 in. by 5 in., height 4π in. **43b.** base 5 in. per side, height $\frac{12}{5}\pi$ in. **43c.** base with legs measuring 3 in. and 4 in., height 10π in. **45.** Sample answer:

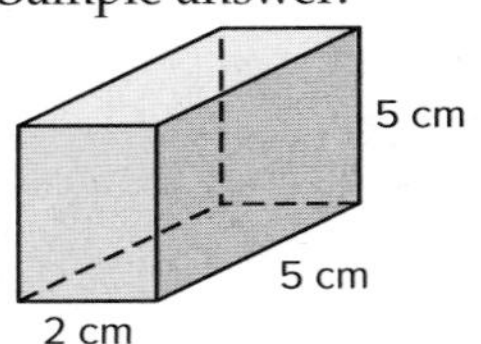

47. Sample answer: Both formulas involve multiplying the area of the base by the height. The base of a prism is polygon, so the expression representing the area varies, depending on the type of polygon it is. The base of a cylinder is a circle, so its area is πr^2. **49.** 9 **51.** 89.4 **53.** 64 in^3

Lesson 11-3

1. 75 in^3 **3.** 62.4 m^3 **5.** 51.3 in^3 **7.** 28.1 mm^3
9. about 16,755 ft^3

11 $V = \frac{1}{3}Bh$ Volume of a pyramid

$= \frac{1}{3}(36.9)(8.6)$ $B = \frac{1}{2} \cdot 9 \cdot 8.2$ or 36.9, $h = 8.6$

≈ 105.8 mm^3 Simplify.

13. 233.8 cm^3 **15.** 35.6 cm^3 **17.** 235.6 in^3
19. 1473.1 cm^3 **21.** 1072.3 in^3

23 $V = \frac{1}{3}\pi r^2 h$ Volume of a cone

$= \frac{1}{3}\pi(4)^2(14)$ Replace r with $\frac{8}{2}$ or 4 and h with 14.

≈ 234.6 cm^3 Use a calculator.

25. 32.2 ft^3 **27.** 3190.6 m^3 **29.** about 13,333 BTUs
31a. The volume is doubled. **31b.** The volume is multiplied by 2^2 or 4. **31c.** The volume is multiplied by 2^3 or 8.

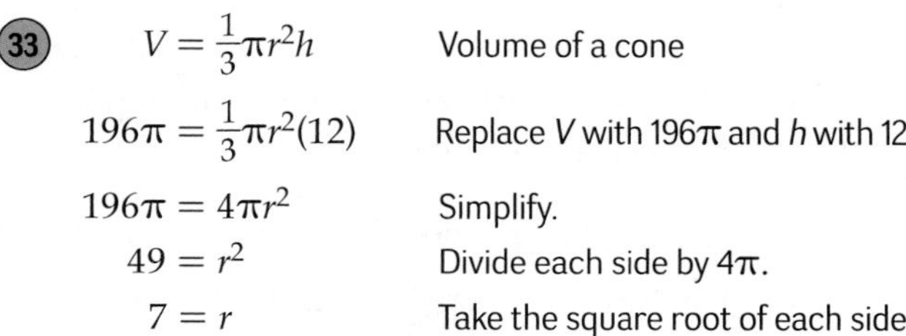

33 $V = \frac{1}{3}\pi r^2 h$ Volume of a cone

$196\pi = \frac{1}{3}\pi r^2(12)$ Replace V with 196π and h with 12.

$196\pi = 4\pi r^2$ Simplify.

$49 = r^2$ Divide each side by 4π.

$7 = r$ Take the square root of each side.

The radius of the cone is 7 inches, so the diameter is $7 \cdot 2$ or 14 inches.

35a. Sample answer:

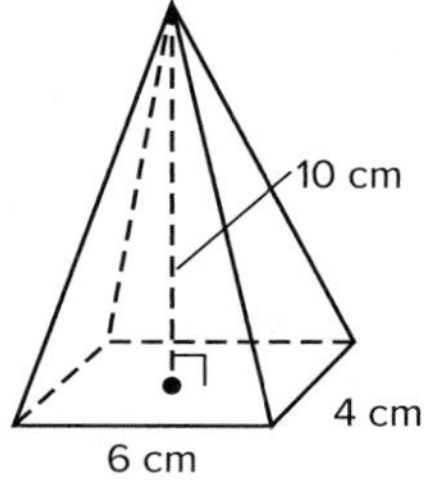

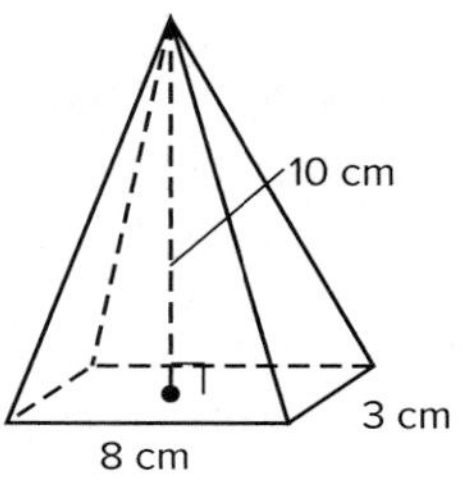

35b. The volumes are the same. The volume of a pyramid equals one third times the base area times the height. So, if the base areas of two pyramids are equal and their heights are equal, then their volumes are equal. **35c.** If the base area is multiplied by 5, the volume is multiplied by 5. If the height is multiplied by 5, the volume is multiplied by 5. If both the base area and the height are multiplied by 5, the volume is multiplied by $5 \cdot 5$ or 25. **37.** Cornelio; Alexandra incorrectly used the slant height. **39.** Sample answer: a square pyramid with a base area of 16 and a height of 12, a prism with a square base of area 16 and height of 4; if a pyramid and prism have the same base, then in order to have the same volume, the height of the pyramid must be 3 times as great as the height of the prism. **41.** C **43.** B **45.** 891π cm^3

Lesson 11-4

1. 1017.9 m^2 **3.** 452.4 yd^2 **5.** 4188.8 ft^3 **7.** 3619.1 m^3
9. 277.0 in^2 **11.** 113.1 cm^2 **13.** 680.9 in^2 **15.** 128 ft^2
17. 530.1 mm^2

19 $V = \frac{4}{3}\pi r^3$ Volume of a sphere

$= \frac{4}{3}\pi(1)^3$ $r = \frac{2}{2}$ or 1

≈ 4.2 cm^3 Use a calculator.

21. 2712.3 cm^3 **23.** 179.8 in^3 **25.** 77.9 m^3
27. 860,289.5 ft^3

29 Surface area $= \frac{1}{2} \cdot$ Area of sphere + Lateral area of cylinder + Area of circle

$= \frac{1}{2}(4\pi r^2) + 2\pi rh + \pi r^2$

$= \frac{1}{2}(4\pi)(4)^2 + 2\pi(4)(5) + \pi(4)^2$

≈ 276.5 in^2

Volume = Volume of hemisphere + Volume of cylinder

$= \frac{1}{2}\left(\frac{4}{3}\pi r^3\right) + \pi r^2 h$

$= \frac{1}{2}\left(\frac{4}{3}\pi \cdot 4^3\right) + \pi(4)^2(5)$

≈ 385.4 in^3

31a. 594.6 cm^2; 1282.8 cm^3 **31b.** 148.7 cm^2; 160.4 cm^3 **33.** $\overline{DC}$ **35.** $\overline{AB}$ **37.** $\odot S$
39a. $\sqrt{r^2 - x^2}$ **39b.** $\pi\left(\sqrt{r^2 - x^2}\right)^2 \cdot y$ or $\pi yr^2 - \pi yx^2$
39c. The volume of the disc from the cylinder is $\pi r^2 y$ or πyr^2. The volume of the disc from the two cones is $\pi x^2 y$ or πyx^2. Subtract the volumes of the discs from the cylinder and cone to get $\pi yr^2 - \pi yx^2$, which is the expression for the volume of the disc from the sphere at height x. **39d.** Cavalieri's Principle **39e.** The volume of the cylinder is $\pi r^2(2r)$ or $2\pi r^3$. The volume of one cone is $\frac{1}{3}\pi r^2(r)$ or $\frac{1}{3}\pi r^3$, so the volume of the double-napped cone is $2 \cdot \frac{1}{3}\pi r^3$ or $\frac{2}{3}\pi r^3$. Therefore, the volume of the hollowed-out cylinder—and thus the sphere—is $2\pi r^3 - \frac{2}{3}\pi r^3$ or $\frac{4}{3}\pi r^3$.

41. Vertical: There is an infinite number of vertical planes that produce reflection symmetry. When any vertical plane intersects the hemisphere through a diameter, both sides of the hemisphere are mirror images. Horizontal: There are no horizontal planes that produce reflection symmetry. When any horizontal plane intersects the hemisphere, the part on top will always be slightly smaller than the bottom. Rotation: There is an infinite number of angles of rotation. When the axis of rotation passes through the center of the sphere perpendicular to its base, the hemisphere can be mapped onto itself by a rotation of any angle between 0° and 360° in the axis.

(43) The surface area is divided by 3^2 or 9. The volume is divided by 3^3 or 27. **45.** 587.7 in^3

47.

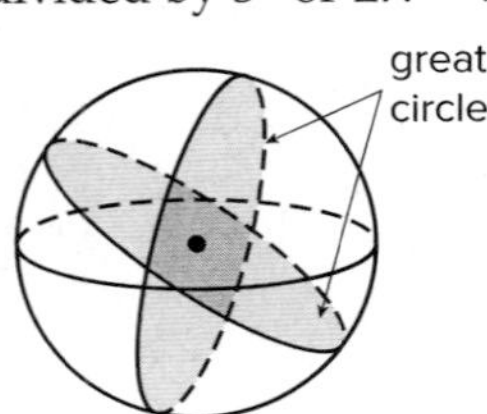

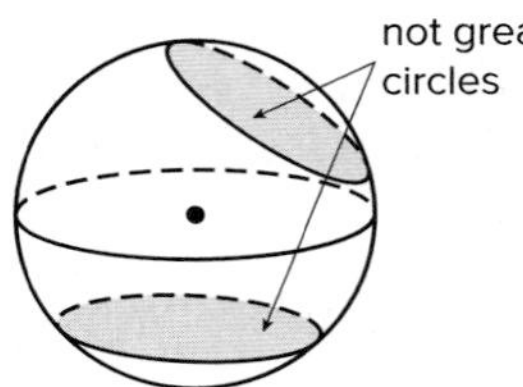

49. B **51.** B **53a.** $\frac{32\pi}{3}$ cm^3 **53b.** $\frac{256\pi}{3}$ cm^3 **53c.** 16π cm^2 **53d.** 64π cm^2 **53e.** $\frac{1}{8}$ **53f.** $\frac{1}{4}$ **55a.** 14.0 in. **55b.** 7.0 in. **55c.** 616.2 in^2 **55d.** 1438.5 in^3

Lesson 11-5

1. $\overleftrightarrow{DH}$, $\overleftrightarrow{FJ}$ **3.** $\triangle JKQ$, $\triangle LMP$

(5) Figure x does not go through the poles of the sphere. So, figure X is not a great circle and not a line in spherical geometry.

7. The points on any great circle or arc of a great circle can be put into one-to-one correspondence with real numbers. **9.** Sample answers: $\overleftrightarrow{WZ}$ and $\overleftrightarrow{XY}$, $\overline{RY}$ or $\overline{TZ}$, $\triangle RST$ or $\triangle MPL$ **11a.** $\overleftrightarrow{AD}$ and $\overleftrightarrow{FC}$ **11b.** Sample answers: $\overline{BG}$ and $\overline{AH}$ **11c.** Sample answers: $\triangle BCD$ and $\triangle ABF$ **11d.** $\overline{QD}$ and $\overline{BL}$ **11e.** $\overleftrightarrow{MJ}$ **11f.** $\overleftrightarrow{MB}$ and $\overleftrightarrow{KF}$ **13.** no

(15) Every great circle (line) is finite and returns to its original starting point. Thus, there exists no great circle that goes on infinitely in two directions.

17. Yes; if three points are collinear, any one of the three points is between the other two. **19.** 14.0 in.; because 100 degrees is $\frac{5}{18}$ of 360 degrees, $\frac{5}{18}$ × circumference of the great circle ≈ 14.0.

21a. about 912 mi; The cities are 13.2° apart on the same great circle, so $\frac{13.2}{360} \times 2\pi \times 3963$ gives the distance between them. **21b.** Yes; sample answer: Because the cities lie on a great circle, the distance between the cities can be expressed as the major arc or the minor arc. The sum of the two values is the circumference of Earth. **21c.** No; sample answer: Because lines of latitude do not go through opposite poles of the sphere, they are not great circles. Therefore, the distance cannot be calculated in the same way. **21d.** Sample answer: infinite locations; If Phoenix were a point on the sphere, then there are infinite points that are equidistant from that point.

(23) **a.** No; if $\overline{CD}$ were perpendicular to $\overline{DA}$, then $\overline{DA}$ would be parallel to $\overline{CB}$. This is not possible, because there are no parallel lines in spherical geometry. **b.** $DA < CB$ because $\overline{CB}$ appears to lie on a great circle. **c.** No; because there are no parallel lines in spherical geometry, the sides of a figure cannot be parallel. So, a rectangle, as defined in Euclidean geometry, cannot exist in non-Euclidean geometry.

25. Sample answer: In plane geometry, the sum of the measures of the angles of a triangle is 180. In spherical geometry, the sum of the measures of the angles of a triangle is greater than 180. In hyperbolic geometry, the sum of the measures of the angles of a triangle is less than 180. **27.** Sometimes; sample answer: Because small circles cannot go through opposite poles, it is possible for them to be parallel, such as lines of latitude. It is also possible for them to intersect when two small circles can be drawn through three points, where they have one point in common and two points that occur on one small circle and not the other.

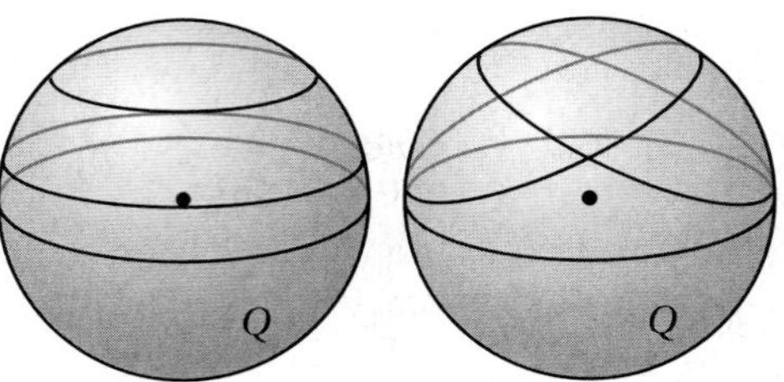

29. False; sample answer: Spherical geometry is non-Euclidean, so it cannot be a subset of Euclidean geometry. **31.** B **33.** 3 **35.** D

Lesson 11-6

1. similar; 4:3 **3.** 414.7 in^2 **5.** 220,893.2 cm^3 **7.** neither **9.** similar; 6:5

(11) $\frac{\text{height of large cylinder}}{\text{height of small cylinder}} = \frac{35}{25}$ or $\frac{7}{5}$

The scale factor is $\frac{7}{5}$. If the scale factor is $\frac{a}{b}$, then the ratio of volumes is $\frac{a^3}{b^3}$. So, $\frac{a^3}{b^3} = \frac{7^3}{5^3}$ or $\frac{343}{125}$. Use this ratio to find the volume of the shorter cylinder; $\frac{343}{125} = \frac{x}{125\pi}$. Thus, the volume of the shorter cylinder is 343π cubic meters.

13. 0.5 ft^3 **15a.** 10:13 **15b.** 419.6 cm^3

17. scale factor $= \frac{12 \text{ ft}}{0.75 \text{ in.}}$ Write a ratio comparing the lengths.

$= \frac{144 \text{ in.}}{0.75 \text{ in.}}$ 12 ft = 12 • 12 or 144 in.

$= \frac{192}{1}$ Simplify.

The scale factor is 192:1.

19. 4.1 in. **21.** 2439.6 cm^3 **23.** about 5.08 to 1

(25) $\frac{\text{area of smaller tent}}{\text{area of larger tent}} = \frac{9}{12.25}$ Write a ratio comparing the floor areas.

$= \frac{3^2}{3.5^2}$ Write as $\frac{a^2}{b^2}$.

The scale factor is 3:3.5.

ratio of diameters → $\frac{6}{d} = \frac{3}{3.5}$ ← scale factor

$6 \cdot 3.5 = d \cdot 3$ Find the cross products.

$7 = d$ Solve for d.

So, the diameter of the larger tent is 7 feet.

$V = \frac{1}{2}\left(\frac{4}{3}\pi r^3\right)$ Volume of a hemisphere

$= \frac{1}{2}\left(\frac{4}{3}\pi \cdot 3.5^3\right)$ Radius $= \frac{7}{2}$ or 3.5

≈ 89.8 Use a calculator.

The volume of the larger tent is about 89.8 ft^3.

27. Laura; Laura compared corresponding parts of the similar figures, while Paloma incorrectly compared the diameter of X to the radius of Y. **29.** Because the scale factor is 15 : 9 or 5 : 3, the ratio of the surface areas is 25 : 9 and the ratio of the volumes is 125 : 27. So, the surface area of the larger prism is $\frac{25}{9}$ or about 2.8 times the surface area of the smaller prism. The volume of the larger prism is $\frac{125}{27}$ or about 4.6 times the volume the smaller prism. **31.** 14 cm **33.** B **35.** C **37.** C **39a.** 12 ft **39b.** 15 ft **39c.** 12.8 ft^2 **39d.** 500 ft^3

Lesson 11-7

1. 0.13 rabbits/ft^2

3 To find the density, divide the population by the area. $17 \div 2.4 = 7.08$. No, they will not approve the club. The population density is about 7.1 pairs/mi^2, which is below the average density of 8.3 pairs/mi^2.

5. 10.4 lb/ft^3 **7.** 19,205,750 **9.** 15,846.5 persons/km^2 **11.** 30.1 persons/km^2 **13.** 3.2 g/cm^3 **15.** Block B has density 40 lb/ft^3 compared to 34.2 lb/ft^3 for Block A. **17.** rubber **19.** cardboard

21 To find the area, find the smallest rectangle/square that surrounds the pentagon, and then subtract the areas of the triangular areas that are not shaded. The area of the square is 64 km^2. There are four triangles in the corners with areas 3 km^2, 3 km^2, 5 km^2, and 6 km^2. The area of the shaded pentagon is 64 km^2 − 3 km^2 − 3 km^2 − 5 km^2 − 6 km^2 = 47 km^2. To find population density, divide the population by the area: $55{,}323 \div 47 \approx 1177.1$ persons/km^2.

23. No; she found the ratio of area to population rather than the ratio of population to area; 389.7 persons/km^2 **25.** 4 **27.** Sample answer: Steel has a greater density than plastic; this means that if a piece of steel and a piece of plastic have the same volume (i.e. the same size), the piece of steel will have a greater mass than the piece of plastic. **29.** B **31.** A, E, F **33.** D

Chapter 11 Study Guide and Review

1. false; spherical geometry **3.** true **5.** true **7.** true **9.** If two solids are similar, then their volumes are proportional to the cube of the scale factor between them. **11.** The volume of the composite solid is the sum of the volumes of each of the simpler solids that make up the composite solid. **13.** circle **5.** 7 cm **17.** 1440 ft^3 **19.** 18 cm^3 **21.** 461.8 in^2 **23.** 3619.1 m^3 **25.** 56.5 cm^3 **27.** $\overline{DL}$ **29.** $\overleftrightarrow{HE}$, $\overleftrightarrow{GF}$ **31.** $\triangle JKL$ **33.** 9 in. **35.** 1728 in^3 **37.** approximately 7698 people per km^2

CHAPTER 12

Probability

Chapter 12 Concept Check

1. Multiply the numerators and the denominators.

3. $\frac{5}{6}$ or 83% **5.** $\frac{1}{3}$ or 33% **7.** $\frac{1}{5}$ or 20%

Lesson 12-1

1. S, S O, O
S, O O, S

Outcomes	Safe	Out
Safe	S, S	S, O
Out	O, S	O, O

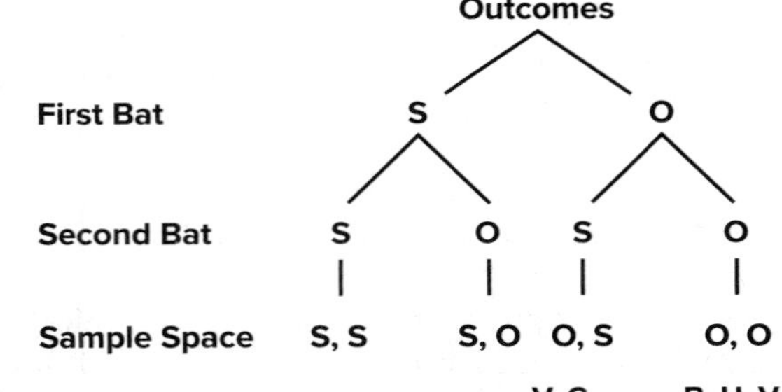

3.

Outcomes
Black
U
V, C — B, U, V, C
V, NC — B, U, V, NC
NV, C — B, U, NV, C
S
V, C — B, S, V, C
V, NC — B, S, V, NC
NV, C — B, S, NV, C
R
V, C — B, R, V, C
V, NC — B, R, V, NC
NV, C — B, R, NV, C
P
V, C — B, P, V, C
V, NC — B, P, V, NC
NV, C — B, P, NV, C
Ivory
U
V, C — I, U, V, C
V, NC — I, U, V, NC
NV, C — I, U, NV, C
S
V, C — I, S, V, C
V, NC — I, S, V, NC
NV, C — I, S, NV, C
R
V, C — I, R, V, C
V, NC — I, R, V, NC
NV, C — I, R, NV, C
P
V, C — I, P, V, C
V, NC — i, P, V, NC
NV, C — I, P, NV, C

5 Possible Outcomes
= Appetizers × Soups × Salads × Entrees × Desserts
= 8 × 4 × 6 × 12 × 9 or 20,736

7. S, S N, N
S, N N, S

Outcomes	Smithsonian	Natural
Smithsonian	S, S	S, N
Natural	N, S	N, N

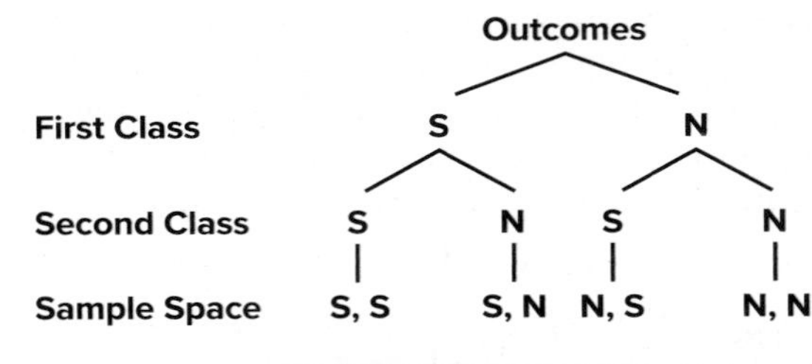

9. M, 5 T, 5
M, 6 T, 6

Outcomes	5	6
Monday	M, 5	M, 6
Thursday	T, 5	T, 6

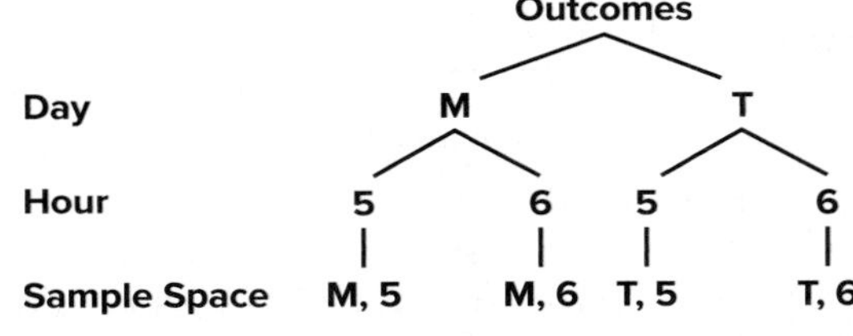

11. O, O A, A
O, A A, O

Outcomes	Oil	Acrylic
Oil	O, O	O, A
Acrylic	A, O	A, A

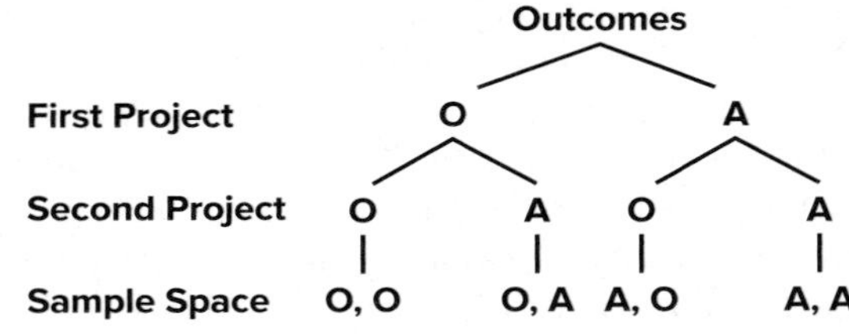

13. S = sedan, T = truck, V = van, L = leather, F = fabric, G = GPS, NG = no GPS, R = sunroof, NR = no sunroof

Sample Space

Outcomes
S
L
G
R — S, L, G, R
NR — S, L, G, NR
NG
R — S, L, NG, R
NR — S, L, NG, NR
F
G
R — S, F, G, R
NR — S, F, G, NR
NG
R — S, F, NG, R
NR — S, F, NG, NR
T
L
G
R — T, L, G, R
NR — T, L, G, NR
NG
R — T, L, NG, R
NR — T, L, NG, NR
F
G
R — T, F, G, R
NR — T, F, G, NR
NG
R — T, F, NG, R
NR — T, F, NG, NR
V
L
G
R — V, L, G, R
NR — V, L, G, NR
NG
R — V, L, NG, R
NR — V, L, NG, NR
F
G
R — V, F, G, R
NR — V, F, G, NR
NG
R — V, F, NG, R
NR — V, F, NG, NR

15 Possible Outcomes = Secretary × Treasurer × Vice President × President
$= 3 \times 4 \times 5 \times 2$ or 120

17. 240 **19.** H = rhombus, P = parallelogram, R = rectangle, S = square, T = trapezoid; H, P; H, R; H, S; H, T; H, H; S, P; S, R; S, S; S, T; S, H

Outcomes	Rhombus	Square
Parallelogram	H, P	S, P
Rectangle	H, R	S, R
Square	H, S	S, S
Trapezoid	H, T	S, T
Rhombus	H, H	S, H

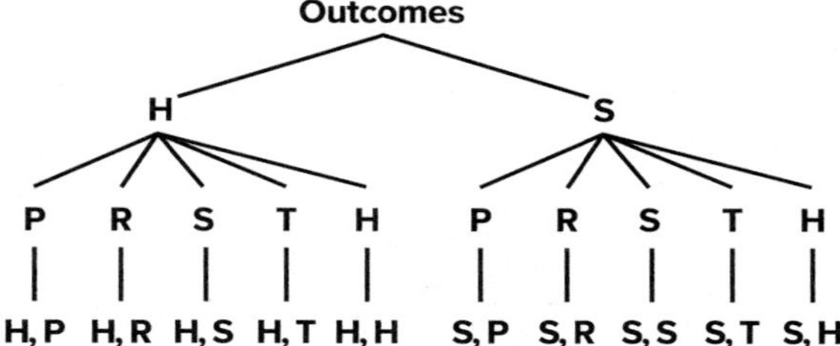

21. Sample answer: 6 different ways:
$4(x + 6) + 2(3) + 2(x + 4)$;
$2(x + 11) + 2(x + 8) + 2(x)$;
$2(x + 4) + 2(x + 9) + 2(x + 6)$;
$2(x) + 2(3) + 4(x + 8)$;
$2(x) + 2(x + 8) + 2(3) + 2(x + 8)$;
$2(x) + 2(3) + 2(4) + 2(x + 6) + 2(x + 6)$

23 **a.** The rolls that result in a sum of 8 are 2 and 6, 3 and 5, 4 and 4, 5 and 3, 6 and 2. So, there are 5 outcomes. **b.** The rolls that result in an odd sum are shown.

1, 2	1, 4	1, 6
2, 1	2, 3	2, 5
3, 2	3, 4	3, 6
4, 1	4, 3	4, 5
5, 2	5, 4	5, 6
6, 1	6, 3	6, 5

So, there are 18 outcomes.

25. $n^3 - 3n^2 + 2n$; Sample answer: There are n objects in the box when you remove the first object, so after you remove one object, there are $n - 1$ possible outcomes. After you remove the second object, there are $n - 2$ possible outcomes. The number of possible outcomes is the product of the number of outcomes of each experiment or $n(n - 1)(n - 2)$.

27. Sample answer: You can list the possible outcomes for one stage of an experiment in the columns and the possible outcomes for the other stage of the experiment in the rows. Since a table is two-dimensional, it would be impossible to list the possible outcomes for three or more stages of an experiment. Therefore, tables can only be used to represent the sample space for a two-stage experiment.

29. $P = n^k$; Sample answer: The total number of possible outcomes is the product of the number of outcomes for each of the stages 1 through k. Since there are k stages, you are multiplying n by itself k times, which is n^k. **31.** B **33.** D **35.** 9

Lesson 12-2

1. $\frac{2}{3}$ **3.** $\frac{1}{6}$ **5.** $\frac{1}{3}$ **7a.** 9, 10, 11, 12 **7b.** $\frac{5}{18}$ **7c.** $\frac{13}{18}$ **9.** 55% **11.** 35% **13.** 45% **15.** $\frac{2}{7}$

17 **a.** The intersection of the column for Girl and the row for Music shows 12. The total number of students is 49. The probability of selecting a girl who prefers music is $\frac{12}{49}$. **b.** The number of girls who do not prefer music is the total number of students minus the number of girls who prefer music: $49 - 12 = 37$. The probability of selecting a girl who does not prefer music is $\frac{37}{49}$. **19.** $\frac{64}{105}$ **21.** $\frac{64}{105}$ **23.** $\frac{3}{4}$ **25.** $\frac{4}{13}$

27 **a.** The number 10 is outside of the regions representing advertisements for specific kinds of electronics. So, 10 of the magazines had no advertisements for any of these types of electronics. **b.** The circle for smartphones and the circle for DVRs overlap in a region that contains the numbers 30 and 40. So there were advertisements for both a smartphone and a DVR in $30 + 40 = 70$ magazines. The total number of magazines is 370. The probability is $\frac{70}{370} = \frac{7}{37}$. **29.** 16% **31.** Sample answer: Rolling a standard number cube with A = rolling a number greater than 1 and B = rolling a number less than 5. **33.** No; Latricia did not include the 5 outcomes common to events A and B when she found $P(A)$; Amelie did not calculate $P(A)$ correctly; $P(A)$ should be $\frac{13}{32}$, so $P(\text{not } A) = \frac{19}{32}$. **35.** C **37.** C **39.** C **41.** C

Lesson 12-3

1. $\frac{1}{20}$ **3.** $\frac{1}{420}$ **5.** $\frac{1}{124{,}750}$

7 The number of possible outcomes is 50!. The number of favorable outcomes is $(50 - 2)!$ or 48!.
$P(\text{Alfonso 14, Colin 23})$

$= \frac{48!}{50!}$ — $\frac{\text{Number of favorable outcomes}}{\text{Number of possible outcomes}}$

$= \frac{\cancel{48!}^{1}}{50 \cdot 49 \cdot \cancel{48!}_{1}}$ — Expand 48! and divide out common factors.

$= \frac{1}{2450}$ — Simplify.

9. $\frac{1}{15{,}120}$

11 There is a total of 10 letters. Of these letters, B occurs 2 times, A occurs 2 times, and L occurs 2 times. So, the number of distinguishable permutations of these letters is

$\frac{10!}{2! \cdot 2! \cdot 2!} = \frac{3,628,800}{8}$ or 453,600 Use a calculator.

There is only 1 favorable arrangement–BASKETBALL. So, the probability that a permutation of these letters selected at random spells basketball is $\frac{1}{453,600}$.

13. $\frac{1}{7}$ **15.** $\frac{1}{10,626}$ **17a.** $\frac{1}{56}$ **17b.** $\frac{1}{40,320}$ **17c.** $\frac{1}{140}$ **17d.** $\frac{2}{7}$ **19a.** 720 **19b.** 5040

21 Find the number of ways to choose the second letter times the number of ways to choose the third letter times the number of ways to choose the last two numbers.

possible license plates $= {}_2C_1 \cdot {}_3C_1 \cdot {}_{10}C_1 \cdot {}_{10}C_1$

$= 2 \cdot 3 \cdot 10 \cdot 10$ or 600

23. $\frac{13}{261}$ **25.** Sample answer: A bag contains seven marbles that are red, orange, yellow, green, blue, purple, and black. The probability that the orange, blue, and black marbles will be chosen if three marbles are drawn at random can be calculated using a combination.

27. $C(n, n-r) \stackrel{?}{=} C(n, r)$

$\frac{n!}{[n-(n-r)]!(n-r)!} \stackrel{?}{=} \frac{n!}{(n-r)!r!}$

$\frac{n!}{r!(n-r)!} \stackrel{?}{=} \frac{n!}{(n-r)!r!}$

$\frac{n!}{(n-r)!r!} = \frac{n!}{(n-r)!r!}$ ✓

29. D **31.** D **33.** A **35a.** $\frac{1}{6}$ **35b.** $\frac{1}{30}$

Lesson 12-4

1. $\frac{1}{2}$, 0.5, or 50% **3.** $\frac{13}{33}$, 0.39, or about 39% **5.** $\frac{1}{8}$, 0.125, or 12.5% **7.** $\frac{13}{18}$, 0.72, or 72% **9.** $\frac{1}{9}$, 0.11, or 11% **11.** $\frac{1}{6}$, 0.17, or about 17%

13 You need to find the ratio of the area of the shaded region to the area of the entire region. The area of shaded region equals the area of the large semicircle minus the area of the small semicircle plus the area of the small semicircle. So, the area of the shaded region equals the area of the large semicircle. Since the area of the large semicircle equals half the total area, P(landing in shaded region) $= \frac{1}{2}$, 0.5, or 50%.

15 P(pointer landing on yellow) $= \frac{44}{360}$ or about 12.2%

17. 69.4% **19.** 62.2% **21.** Sample answer: a point between 10 and 20 **23.** $\frac{1}{2}$, 0.5, or 50% **25.** 53.5%

27. Sample answer: The probability that a randomly chosen point will lie in the shaded region is ratio of the area of the sector to the area of the circle.

$P(\text{point lies in sector}) = \frac{\text{area of sector}}{\text{area of circle}}$

$\frac{x}{360} \stackrel{?}{=} \frac{\frac{x}{360} \cdot \pi r^2}{\pi r^2}$

$\frac{x}{360} = \frac{x}{360}$ ✓

29. 0.24 or 24% **31.** 0.33 or 33%

33 volume of shallow region $= Bh = (7 \cdot 20) \cdot 20$ or 2800 ft^3

volume of incline region $= Bh = \frac{1}{2}(25)(7 + 20) \cdot 20$ or 6750 ft^3

volume of deep region $= Bh = (20 \cdot 30) \cdot 20$ or 12,000 ft^3

P(bear swims in the incline region)

$= \frac{\text{volume of incline region}}{\text{volume of pool}}$

$= \frac{6750}{2800 + 6750 + 12,000}$

≈ 0.31 or 31%

35. 14.3% **37.** No; sample answer: Athletic events should not be considered random because there are other factors involved, such as pressure and ability, that have an impact on the success of the event. **39.** Sample answer: The probability of a randomly chosen point lying in the shaded region of the square on the left is found by subtracting the area of the unshaded square from the area of the larger square and finding the ratio of the difference of the areas to the area of the larger square. The probability is $\frac{1^2 - 0.75^2}{1^2}$ or 43.75%. The probability of a randomly chosen point lying in the shaded region of the square on the right is the ratio of the area of the shaded square to the area of the larger square, which is $\frac{0.4375}{1}$ or 43.75%. Therefore, the probability of a randomly chosen point lying in the shaded area of either square is the same. **41.** D **43.** 0.33 **45.** $\frac{1}{36}$ **47a.** $\frac{1}{16}$ **47b.** $\frac{3}{4}$ **47c.** 15 cm

Lesson 12-5

1. The outcome of the first roll does not affect the probabilities of the outcomes for the second roll. Therefore, these events are independent. **3.** The events are not independent. The sample space has 6 equally likely outcomes: {JPR, JRP, PJR, PRJ, RJP, RPJ}. So P(J 1st and R 2nd) $= \frac{1}{6}$, P(J 1st) $= \frac{1}{3}$, and P(R 2nd) $= \frac{1}{3}$, but $\frac{1}{3} \cdot \frac{1}{3} \neq \frac{1}{6}$. **5.** $\frac{1}{2704}$ or 3.7×10^{-4}

7. Maurice should select two blue cards. Let R represent selecting a wild card from the red deck and B represent selecting a wild card from the blue deck. Then $P(R \text{ and } R) = \frac{3}{12} \cdot \frac{2}{11} = \frac{1}{22} \approx 45\%$;
$P(B \text{ and } B) = \frac{6}{20} \cdot \frac{5}{9} = \frac{3}{38} \approx 7.9\%$; and
$P(B \text{ and } R) = \frac{3}{12} \cdot \frac{6}{20} = \frac{3}{40} \approx 7.5\%$.

9 Since the card is not replaced, the events are dependent. $P(\text{Ace}) = \frac{4}{12}$. After that ace is removed, there are only 51 cards left to choose from, 3 of which are aces, so the $P(\text{another Ace}) = \frac{3}{51}$. So, $P(\text{Ace and Ace}) = \frac{4}{52} \cdot \frac{3}{51} = \frac{1}{221}$, or about 0.005, which is 0.5%.

11. independent; $\frac{1}{36}$ or about 3%. **13.** The events are not independent. The sample space has 3 equally likely outcomes: {BP, BC, PC}. So $P(\text{PC}) = \frac{1}{3}$, $P(\text{P}) = \frac{1}{3}$, and $P(\text{C}) = \frac{1}{3}$, but $\frac{1}{3} \cdot \frac{1}{3} \neq \frac{1}{3}$. **15.** $\frac{1}{306}$ or about 0.3% **17.** $\frac{20}{161}$ or about 12% **19.** No; $P(F \text{ and } S) = P(F) \cdot P(S \text{ after } F) = \frac{5}{10} \cdot \frac{7}{10} = \frac{35}{100}$ or 35%; this is less than 50%. **21.** $\frac{8}{87}$ or about 9% **23.** $\frac{16}{145}$ or about 11%

25 $P(\text{1st white}) = \frac{14}{24}$; $P(\text{2nd white}) = \frac{13}{23}$; P(white, then white) $= \frac{14}{24} \cdot \frac{13}{23} = \frac{182}{552} = \frac{91}{276}$ or about 33%

27 **a.** $P(\text{1st good}) = 40\%$ or 0.4, so $P(\text{1st fault}) = 0.6$. $P(\text{2nd good}) = 70\%$ or 0.7, so $P(\text{2nd fault}) = 0.3$. $P(\text{double fault}) = P(\text{1st fault, then 2nd fault}) = 0.6 \cdot 0.3 = 0.18$ or 18% **27b.** $P(\text{1st fault, then 2nd good}) = 0.6 \cdot 0.7 = 0.42$ or 42% **29.** $\frac{125}{5488}$ or about 2%

31a. Both rules show that $P(3 \text{ and } 8) = \frac{4}{663}$ or about 0.6%. **31b.** Sample answer: A jar contains 5 pennies and 6 dimes. One coin is chosen at random, and then a second coin is chosen without replacing the first. Find the probability that the first coin is a penny and the second coin is a dime. $P(P \text{ and } D) = P(P) \cdot P(D \text{ following } P) = \frac{5}{11} \cdot \frac{6}{10} = \frac{30}{110} = \frac{3}{11}$. $P(P \text{ and } D) = P(D) \cdot P(P \text{ following } D) = \frac{6}{11} \cdot \frac{5}{10} = \frac{30}{110} = \frac{3}{11}$.

31c. $P(A \text{ and } B) = P(A) \cdot P(B \text{ following } A)$ and $P(A \text{ and } B) = P(B) \cdot P(A \text{ following } B)$ are equivalent. **33.** 7; Sample answer: The probability of drawing object A is $\frac{1}{n}$, and the probability of drawing object B when object A is not replaced is $\frac{1}{n-1}$. Since we know that the probability is 2.4%, $\frac{1}{n} \cdot \frac{1}{n-1} = \frac{2.4}{100}$ or 0.024.
Solve this equation to determine that n is 7.
35. Sample answer: The results of two coin flips represent a pair of independent events. Regardless of the outcome of the first flip, the probability of getting heads or tails on the second flip does not change. Drawing two colored marbles out of a bag without replacing the first marble represents a pair of dependent events. Based on the color of the first marble, the probability that the second marble will be a specific color will change. **37.** A **39.** D **41.** greater than **43a.** Sample answer: {(1, 1), (1, 2), (1, 3), (1, 4), (1, 5), (1, 6), (2, 1), (2, 2), (2, 3), (2, 4), (2, 5), (2, 6), (3, 1), (3, 2), (3, 3), (3, 4), (3, 5), (3, 6), (4, 1), (4, 2), (4, 3), (4, 4), (4, 5), (4, 6), (5, 1), (5, 2), (5, 3), (5, 4), (5, 5), (5, 6), (6, 1), (6, 2), (6, 3), (6, 4), (6, 5), (6, 6)} **43b.** $\frac{1}{9}$ or about 11% **43c.** $P(> 4 \text{ and } > 4) = \frac{1}{9}$; $P(> 4) = \frac{1}{3}$; $P(> 4) = \frac{1}{3}$; $\frac{1}{3} \cdot \frac{1}{3} = \frac{1}{9}$ **45.** $\frac{1}{16{,}575}$

Lesson 12-6

1. not mutually exclusive; A jack of clubs is both a jack and a club. **3.** $\frac{2}{3}$ or about 67% **5.** $\frac{11}{35}$ or about 44%

7 Since rolling two fours is both getting doubles and getting a sum of 8, the events are not mutually exclusive.

$P(\text{doubles or a sum of 8})$

$= P(\text{doubles}) + P(\text{a sum of 8}) - P(\text{doubles and a sum of 8})$

$= \frac{6}{36} + \frac{5}{36} - \frac{1}{36}$

$= \frac{10}{36}$ or about 27.8%

9. Not mutually exclusive; $\frac{13}{20}$ or 65% **11.** not mutually exclusive $\frac{4}{13}$ or 30.8% **13.** 56% **15.** $\frac{5}{18}$ or about 28%

17 There are four 2s in a deck and 4 queens in a deck. There are 52 total cards. A card cannot be both a 2 and a queen, so the events are mutually exclusive. Add the probabilities of the individual events. $\frac{4}{52} + \frac{4}{52} = \frac{8}{52} = \frac{2}{13}$ or about 15% **19.** $\frac{4}{13}$ or about 31% **21.** $\frac{4}{13}$ or about 31% **23.** $\frac{3}{5}$ or 60% **25.** $\frac{11}{15}$ or about 73%
27a. $\frac{11}{20}$ or 55% **27b.** $\frac{9}{10}$ or 90% **29.** Aliyah; to find the probability of blue or red, the individual probabilities should be added because the events are mutually exclusive. **31.** Not mutually exclusive; sample answer: If a triangle is equilateral, it is also equiangular. The two can never be mutually exclusive. **33.** Sample answer: If you pull a card from a deck, it can be either a 3 or a 5. The two events are mutually exclusive. If you pull a card from a deck, it can be a 3 and it can be red. The two events are not mutually exclusive.
35. $\frac{3}{4}$ or 75% **37.** D **39a.** 1/3 or about 33%
39b. $\frac{1}{2}$ or 50%

Lesson 12-7

1. 0.20 **3.** $\frac{3}{16}$ or 18.75% **5.** $\frac{4}{5}$ or 80% **7.** $\frac{1}{6}$ or 17%
9. $\frac{1}{5}$ or 20%

11 $P(\text{own digital media player} \mid \text{own smartphone})$

$$= \frac{P(\text{own digital media player and smartphone})}{P(\text{own smartphone})}$$

$$= \frac{0.28}{0.43}$$

$$\approx 0.65$$

13. $P(A \text{ and } B) = P(A) \cdot P(B \mid A)$ Formula for $P(A \text{ and } B)$.

$\frac{P(A \text{ and } B)}{P(A)} = P(B \mid A)$ Divide each side by $P(A)$.

15a. $\frac{13}{32}$ or about 41% **15b.** $\frac{13}{54}$ or about 24%

17 $P(\text{banner}|\text{home team}) = \frac{P(\text{banner and home team})}{P(\text{home team})}$

$$= \frac{0.2}{0.8} = \frac{1}{4} \text{ or } 25\%$$

19. About 90.4% **21a.** 0.3 **21b.** 0.33 **23.** $\frac{20}{27}$ or about 74%; Sample answer: The probability that a number cube shows a number less than or equal to 4 is $\frac{4}{6}$. So the probability that the first two number cubes show a number less than or equal to 4 and the last cube does not is $\frac{4}{6} \cdot \frac{4}{6} \cdot \frac{2}{6}$. Because the number greater than 4 can occur on any of the 3 cubes, the probability of a number less than or equal to 4 on exactly two cubes is $3\left(\frac{4}{6} \cdot \frac{4}{6} \cdot \frac{2}{6}\right)$ or $\frac{4}{9}$. The probability of showing a number less than or equal to 4 on all three cubes is $\frac{4}{6} \cdot \frac{4}{6} \cdot \frac{4}{6}$ or $\frac{8}{27}$. Thus the probability of showing a number less than or equal to 4 on two or more cubes is $\frac{4}{9} + \frac{8}{27} = \frac{20}{27}$ or about 74%. **25.** 7; Sample answer: The probability of drawing object A is $\frac{1}{n}$, and the probability of drawing object B when A is not replaced is $\frac{1}{n-1}$. Since we know that the probability is about 1.4%, $\frac{1}{n} \cdot \frac{1}{n-1} \approx \frac{1.4}{100}$. Solve this equation to determine that n is 9.

27. C **29.** $\frac{1}{3}$ **31.** $\frac{2}{3}$ **33.** C **35a.** 0.018 **35b.** 0.098

35c. Sample answer: A patient who receives a positive test result either has the disease or does not have the disease. These are mutually exclusive events, so add the probabilities. From part **a**, $P(\text{positive test and disease}) = 0.018$. From part **b**, $P(\text{positive test and no disease}) = 0.098$. So $P(\text{positive test}) = 0.018 + 0.098 = 0.116$.

35d. $P(\text{disease} \mid \text{positive test})$

$$= \frac{P(\text{disease and positive test})}{P(\text{positive test})}$$

$$= \frac{0.018}{0.116} \approx 0.155.$$

Lesson 12-8

1.

	Breakfast	No Breakfast	Totals
Elementary	38	12	50
High School	22	28	50
Totals	60	40	100

3. 16; joint frequency **5.** 24; marginal frequency

7.

	Car	No Car	Totals
Bus	6.7%	51.7%	58.3%
Train	13.3%	28.3%	41.7%
Totals	20%	80%	100%

9. 33.3% **11.** 66.7%

13.

	At Least 7 Hours	Less Than 7 Hours	Totals
Doctors	40	35	75
Nurses	24	63	87
Totals	64	98	162

15 Make rows: Men, Women, and Totals. Make columns: Gas, Electric, Totals. In the bottom right cell, enter 100 for the total number of people surveyed. Enter 55 in the total column and row for Men. Enter 21 in the column for Gas and row for Men. Enter 19 in the column for Electric and row for Women. The number of men who chose a gas range is 55 − 21 = 34. The total number of women is 100 − 55 = 45. The number of women who chose a gas range is 45 − 19 = 26. Add the men and women who chose a gas range: 21 + 26 = 47. Add the men and women who chose an electric range: 34 + 19 = 53.

	Gas	Electric	Totals
Men	21	34	55
Women	26	19	45
Totals	47	53	100

17. 90; marginal frequency **19.** 54; joint frequency **21.** Yes; the expected and actual joint relative frequencies are the same. **23.** 60% **25.** 60%

	Blue	Green	Totals
Boys	40	35	75
Girls	32	18	50
Totals	72	53	125

Boys who chose green: 75 − 40 = 35
Girls who chose blue: 50 − 18 = 32
Total Blue: 40 + 32 = 72
Total Green: 35 + 18 = 53
Total: 72 + 53 = 75 + 50 = 125

29.

	Red	Yellow	Totals
Teachers	23	7	30
Students	18	33	51
Totals	41	40	81

31a.

	AP Classes	No AP Classes	Totals
Senior	40	60	100
Not Senior	110	190	300
Totals	150	250	400

Total: 400
Total AP Classes: 150
Seniors taking AP Classes: 40
Total seniors: 100
Seniors not taking AP: 100 − 40 = 60
Not seniors taking AP: 150 − 40 = 110
Total not seniors: 400 − 100 = 300
Not seniors not taking AP: 300 − 110 = 190
Total not taking AP: 60 + 190 = 250

31b.

	AP Classes	No AP Classes	Totals
Senior	10%	15%	25%
Not Senior	27.5%	47.5%	75%
Totals	37.5%	62.5%	100%

Divide each value from the table in part a by the total number of students, 400. Write the quotient as a percent.
31c. The intersection of the row for students who are not seniors and the column for students not taking AP classes is 47.5%.
31d. There are 250 students not taking AP Classes and 150 students taking AP classes, so it is more likely that a student is not taking advanced-placement classes. The marginal relative frequency for *No AP Classes* is 62.5%, which is greater than the marginal relative frequency for *AP Classes.*

33a.

	Bring Lunch	Cafeteria	Go Out	Totals
Men	43	56	23	122
Women	46	52	30	128
Totals	89	108	53	250

33b. 51.9% **35.** Kaci found the probability that a randomly-chosen apple is organic, given that it is red. The correct answer is $\frac{0.18}{0.30} = 60\%$. **37.** 25%
39. A **41.** D, E, F **43.** C

1. true **3.** true **5.** true **7.** false; dependent events
9. The intersection includes events contained in both sample spaces, while the union includes events in either sample space. **11.** 27 **13.** 4 **15.** 35,960
17a. $\frac{2}{9}$ **17b.** $\frac{7}{9}$ **19.** $\frac{1}{169}$ **21.** $\frac{2}{9}$ **23a.** $\frac{11}{20}$ **23b.** $\frac{19}{40}$
25. $\frac{3}{4}$ or 75% **27.** $\frac{1}{2}$ **29a.** 46.2% **29b.** 63.2%
29c. Not independent; 46.8% of respondents are in the 18-24 age group and 34.2% of respondents were planning on voting, so one would expect 46.8% • 34.2% or about 16% of respondents in the 18–24 age group to plan on voting. The expected joint relative frequency of 16% and the actual joint relative frequency of 21.6% are not the same, so planning to vote is not independent of age group.

Glossary/Glosario

Multilingual eGlossary

Go to connectED.mcgraw-hill.com for a glossary of terms in these additional languages:

Arabic	Chinese	Hmong	Spanish	Vietnamese
Bengali	English	Korean	Tagalog	
Brazilian Portuguese	Haitian Creole	Russian	Urdu	

English / Español

A

absolute error (p. 91) The absolute error of a measurement is equal to one half the unit of measure.

error absoluto El error absoluto de una medida es igual a un medio de la unidad de medida.

accuracy (p. 92) The closeness of a measurement to its true value.

exactitud La cercanía de una medida a su valor verdadero.

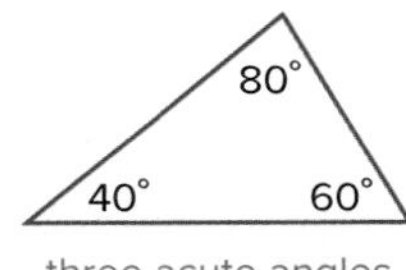

three acute angles

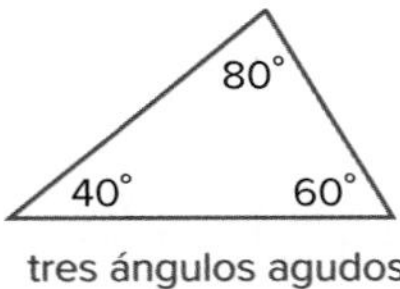

tres ángulos agudos

acute angle (p. 38) An angle with a degree measure less than 90.

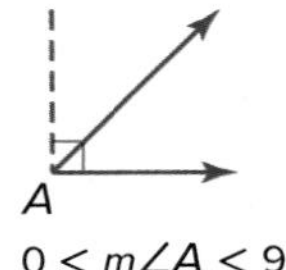

$0 < m\angle A < 90$

ángulo agudo Ángulo cuya medida en grados es menos de 90.

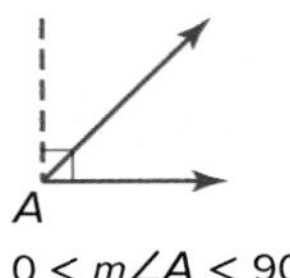

$0 < m\angle A < 90$

adjacent angles (p. 46) Two angles that lie in the same plane and have a common vertex and a common side, but no common interior points.

ángulos adyacentes Dos ángulos que yacen sobre el mismo plano, tienen el mismo vértice y un lado en común, pero ningún punto interior en común.

adjacent arcs (p. 654) Arcs in a circle that have exactly one point in common.

arcos adyacentes Arcos en un círculo que tienen un solo punto en común.

algebraic proof (p. 143) A proof that is made up of a series of algebraic statements. The properties of equality provide justification for many statements in algebraic proofs.

demostración algebraica Demostración que se realiza con una serie de enunciados algebraicos. Las propiedades de la igualdad proveen justificación para muchas enunciados en demostraciones algebraicas.

alternate exterior angles (p. 169) In the figure, transversal t intersects lines ℓ and m. $\angle 5$ and $\angle 3$, and $\angle 6$ and $\angle 4$ are alternate exterior angles.

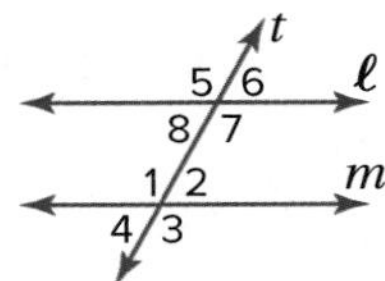

ángulos alternos externos En la figura, la transversal t interseca las rectas ℓ y m. $\angle 5$ y $\angle 3$, y $\angle 6$ y $\angle 4$ son ángulos alternos externos.

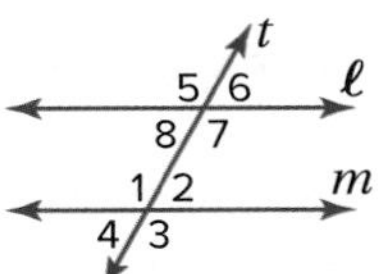

alternate interior angles (p. 169) In the figure for alternate exterior angles, transversal t intersects lines ℓ and m. $\angle 1$ and $\angle 7$, and $\angle 2$ and $\angle 8$ are alternate interior angles.

ángulos alternos internos En la figura anterior, la transversal t interseca las rectas ℓ y m. $\angle 1$ y $\angle 7$, y $\angle 2$ y $\angle 8$ son ángulos alternos internos.

altitude **1.** (p. 367) In a triangle, a segment from a vertex of the triangle to the line containing the opposite side and perpendicular to that side. **2.** (p. 770) In a prism or cylinder, a segment perpendicular to the bases with an endpoint in each plane. **3.** (pp. 773, 775) In a pyramid or cone, the segment that has the vertex as one endpoint and is perpendicular to the base.

altura **1.** En un triángulo, segmento trazado desde uno de los vértices del triángulo hasta el lado opuesto y que es perpendicular a dicho lado. **2.** En un prisma o un cilindro, segmento perpendicular a las bases con un extremo en cada plano. **3.** En una pirámide o un cono, segmento que tiene un extremo en el vértice y que es perpendicular a la base.

ambiguous case (p. 618) A situation in which more than one solution for a triangle exists.

caso ambiguo Una situación en la que existe más de una solución para un triángulo.

ambiguous case of the Law of Sines (p. 816) Given the measures of two sides and a nonincluded angle, there exist two possible triangles.

caso ambiguo de la ley de los senos Dadas las medidas de dos lados y de un ángulo no incluido, existen dos triángulos posibles.

angle (p. 36) The intersection of two noncollinear rays at a common endpoint. The rays are called *sides* and the common endpoint is called the *vertex*.

ángulo La intersección de dos rayos no colineales en un extremo común. Las rayos se llaman *lados* y el punto común se llama *vértice*.

angle bisector (p. 39) A ray that divides an angle into two congruent angles.

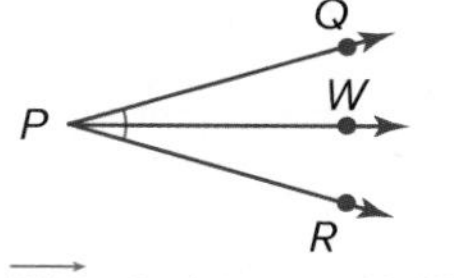

$\overrightarrow{PW}$ is the bisector of $\angle P$.

bisectriz de un ángulo Rayo que divide un ángulo en dos ángulos congruentes.

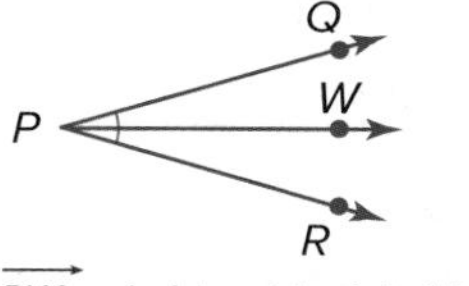

$\overrightarrow{PW}$ es la bisectriz del $\angle P$.

angle of depression (p. 608) The angle between the line of sight and the horizontal when an observer looks downward.

ángulo de depresión Ángulo formado por la horizontal y la línea de visión de un observador que mira hacia abajo.

angle of elevation (p. 608) The angle between the line of sight and the horizontal when an observer looks upward.

ángulo de elevación Ángulo formado por la horizontal y la línea de visión de un observador que mira hacia arriba.

angle of rotation (p. 240) The angle through which a preimage is rotated to form the image.

ángulo de rotación Ángulo a través del cual se rota una preimagen para formar la imagen.

apothem (pp. 751, 752) A segment that is drawn from the center of a regular polygon perpendicular to a side of the polygon.

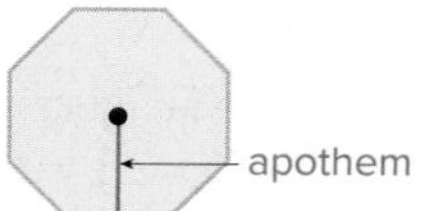

apotema Segmento trazado desde el centro de un polígono regular hasta uno de sus lados y que es perpendicular a dicho lado.

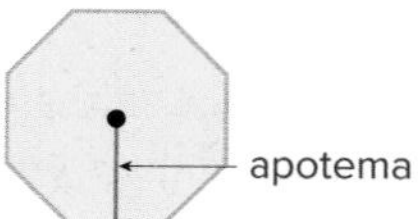

arc (p. 652) A part of a circle that is defined by two endpoints.

arco Parte d e un círculo definida por dos extremos.

arc length (p. 654) The distance between the endpoints along an arc, measured in linear units.

longitud del arco La distancia entre los extremos de un arco, medida en unidades lineales.

area (p. 58) The number of square units needed to cover a surface.

área Número de unidades cuadradas para cubrir una superficie.

auxiliary line (p. 282) An extra line or segment drawn in a figure to help complete a proof.

línea auxiliar Recta o segmento de recta adicional que es traza en una figura para ayudar a completar una demostración.

axiom (p. 141) A statement that is accepted as true.

axioma Enunciado que se acepta como verdadero.

axis **1.** (p. 771) In a cylinder, the segment with endpoints that are the centers of the bases. **2.** (p. 856) In a cone, the segment with endpoints that are the vertex and the center of the base.

eje **1.** En un cilindro, el segmento cuyos extremos son el centro de las bases. **2.** En un cono, el segmento cuyos extremos son el vértice y el centro de la base.

B

base angle of an isosceles triangle (p. 325) See *isosceles triangle* and *isosceles trapezoid.*

ángulo de la base de un triángulo isósceles Ver *triángulo isósceles* y *trapecio isósceles.*

base angle of a trapezoid (p. 469) Angles of a trapezoid that are formed by the base and one of the legs.

ángulo de la base de un trapecio Los ángulos de un trapecio que están formados por la base y uno de los catetos.

base edges (p. 770) The intersection of the lateral faces and bases in a solid figure.

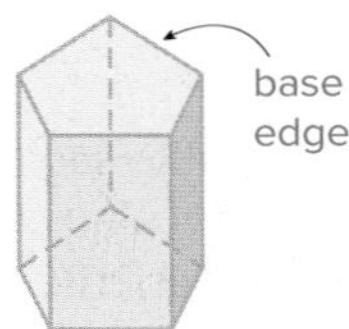

aristas de las bases Intersección de las base con las caras laterales en una figura sólida.

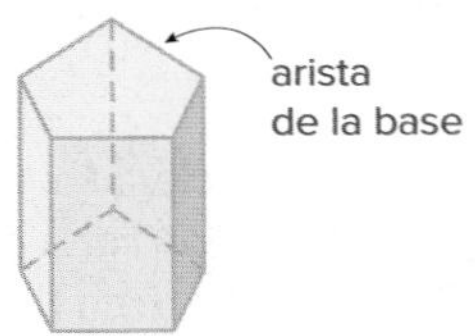

base of a parallelogram (p. 725) Any side of a parallelogram.

base de un paralelogramo Cualquier lado de un paralelogramo.

base of a polyhedron (p. 76) The two parallel congruent faces of a polyhedron.

base de poliedro Las dos caras paralelas y congruentes de un poliedro.

base of a trapezoid (p. 469) The two parallel sides of a trapezoid.

base de un trapecio Los dos lados paralelos de un trapecio.

base of a triangle (p. 727) Any side of a triangle.

base de un triángulo Cualquier lado de un triángulo.

between (p. 14) For any two points A and B on a line, there is another point C between A and B if and only if A, B, and C are collinear and $AC + CB = AB$.

entre Para cualquier par de puntos A y B de una recta, existe un punto C ubicado entre A y B si y sólo si A, B y C son colineales y $AC + CB = AB$.

betweenness of points (p. 14) See *between.*

intermediación de puntos Ver *entre.*

biconditional (p. 124) The conjunction of a conditional statement and its converse.

bicondicional Conjunción entre un enunciado condicional y su recíproco.

C

center of a circle (p. 643) The central point where radii form a locus of points called a circle.

centro de un círculo Punto central desde el cual los radios forman un lugar geométrico de puntos llamado círculo.

center of dilation (p. 492) The center point from which dilations are performed.

centro de la homotecia Punto fijo en torno al cual se realizan las homotecias.

center of a regular polygon (p. 752) The center of the circle that circumscribes the polygon.

centro de un polígono regular El centro del círculo que circunscribe el polígono.

center of rotation (p. 240) A fixed point around which shapes move in a circular motion to a new position.

centro de rotación Punto fijo alrededor del cual gira una figura hasta alcanzar una posición dada.

center of symmetry (p. 260) The point in the center of a figure about which the figure can be mapped onto itself by a rotation.

centro de la simetría Vea *el punto de simetría.*

central angle (p. 652) An angle that intersects a circle in two points and has its vertex at the center of the circle.

ángulo central Ángulo que interseca un círculo en dos puntos y cuyo vértice está en el centro del círculo.

central angle of a regular polygon (p. 752) An angle that has its vertex at the center of a polygon and with sides that pass through consecutive vertices of the polygon.

ángulo central de un polígono regular Ángulo cuyo vértice esta en el centro del polígono y cuyos lados pasan por vértices consecutivas del polígono.

centroid (p. 365) The point of concurrency of the medians of a triangle.

baricentro Punto de intersección de las medianas de un triángulo.

chord **1.** (p. 643) For a given circle, a segment with endpoints that are on the circle. **2.** (p. 880) For a given sphere, a segment with endpoints that are on the sphere.

cuerda **1.** Para cualquier círculo, segmento cuyos extremos están en el círculo. **2.** Para cualquier esfera, segmento cuyos extremos están en la esfera.

circle (p. 643) The locus of all points in a plane equidistant from a given point called the *center* of the circle.

P is the center of the circle.

círculo Lugar geométrico formado por todos los puntos en un plano, equidistantes de un punto dado llamado *centro* del círculo.

P es el centro del círculo.

circular permutation (p. 875) A permutation of objects that are arranged in a circle or loop.

permutación circular Permutación de objetos que se arreglan en un círculo o un bucle.

circumcenter (p. 355) The point of concurrency of the perpendicular bisectors of a triangle.

circuncentro Punto de intersección de las mediatrices de un triángulo.

circumference (pp. 58, 645) The distance around a circle.

circunferencia Distancia alrededor de un círculo.

circumscribed (p. 646) A circle is circumscribed about a polygon if the circle contains all the vertices of the polygon.

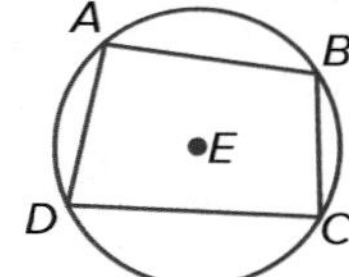

⊙*E* is circumscribed about quadrilateral *ABCD*.

circunscrito Un polígono está circunscrito a un círculo si todos sus vértices están contenidos en el círculo.

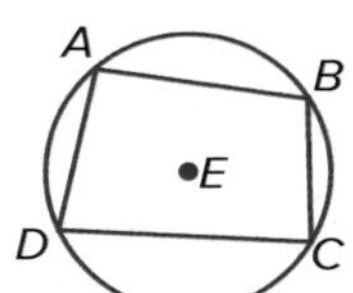

⊙*E* está circunscrito al cuadrilátero *ABCD*.

Glossary/Glosario

collinear (p. 5) Points that lie on the same line.

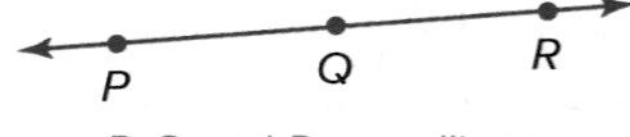

P, *Q*, and *R* are collinear.

colineal Puntos que yacen sobre la misma recta.

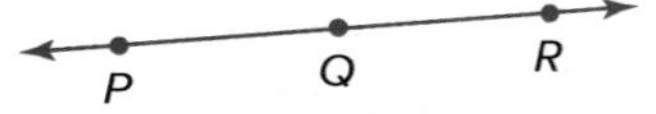

P, *Q* y *R* son colineales.

combination (p. 876) An arrangement or listing in which order is not important.

combinación Arreglo o lista en que el orden no es importante.

common tangent (p. 678) A line or segment that is tangent to two circles in the same plane.

tangente común Recta o segmento de recta tangente a dos círculos en el mismo plano.

complement (p. 867) The complement of an event *A* consists of all the outcomes in the sample space that are not included as outcomes of event *A*.

complemento El complemento de un evento *A* consiste en todos los resultados en el espacio muestral que no se incluyen como resultados del evento *A*.

complementary angles (p. 47) Two angles with measures that have a sum of 90.

ángulos complementarios Dos ángulos cuyas medidas suman 90.

component form (p. 233) A vector expressed as an ordered pair, (change in *x*, change in *y*).

componente Vector expresado en forma de par ordenado, (cambio en *x*, cambio en *y*).

composite figure (p. 754) A figure that can be separated into regions that are basic figures.

figura compuesta Figura que se puede separar en regiones formas de figuras básicas.

composite solid (p. 778) A three-dimensional figure that is composed of simpler figures.

solido compuesto Figura tridimensional formada por figuras más simples.

composition of transformations (p. 249) The resulting transformation when a transformation is applied to a figure and then another transformation is applied to its image.

composición de transformaciones Transformación que resulta cuando se aplica una transformación a una figura y luego se le aplica otra transformación a su imagen.

compound event (p. 889) An event that consists of two or more simple events.

evento compuesto Evento que consiste de dos o más eventos simples.

compound statement (p. 119) A statement formed by joining two or more statements.

enunciado compuesto Enunciado formado por la unión de dos o más enunciados.

concave polygon (p. 56) A polygon for which there is a line containing a side of the polygon that also contains a point in the interior of the polygon.

polígono cóncavo Polígono para el cual existe una recta que contiene un lado del polígono y un punto en el interior del polígono.

concentric circles (p. 644) Coplanar circles with the same center.

círculos concéntricos Círculos coplanarios con el mismo centro.

conclusion (p. 121) In a conditional statement, the statement that immediately follows the word *then*.

conclusión Parte de un enunciado condicional que está escrito justo después de la palabra *entonces*.

concurrent lines (p. 355) Three or more lines that intersect at a common point.

rectas concurrentes Tres o más rectas que se intersecan en un punto común.

conditional probability (p. 903) The probability of an event under the condition that some preceding event has occurred.

probabilidad condicional La probabilidad de un acontecimiento bajo condición que ha ocurrido un cierto acontecimiento precedente.

conditional statement (p. 121) A statement that can be written in *if-then* form.

enunciado condicional Enunciado escrito en la forma *si-entonces*.

cone (p. 76) A solid with a circular base, a vertex not contained in the same plane as the base, and a lateral surface area composed of all points in the segments connecting the vertex to the edge of the base.

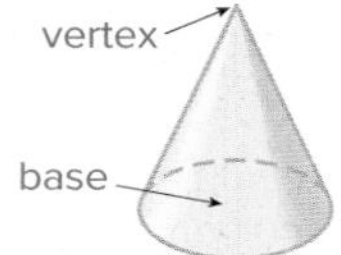

cono Sólido de base circular cuyo vértice no yace en el mismo plano que la base y cuya área de superficie lateral está formada por todos los puntos en los segmentos que conectan el vértice con el bonde de la base.

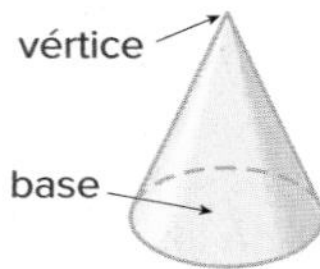

congruence transformations (p. 67) A mapping for which a geometric figure and its image are congruent.

transformaciones de congruencia Aplicación en la cual una figura geométrica y su imagen son congruentes.

congruent (p. 15) Having the same measure.

congruente Que tienen la misma medida.

congruent arcs (p. 653) Arcs in the same circle or in congruent circles that have the same measure.

arcos congruentes Arcos que tienen la misma medida y que pertenecen al mismo círculo o a círculos congruentes.

congruent polygons (p. 291) Polygons in which all matching parts are congruent.

polígonos congruentes Polígonos cuyas partes correspondientes son todas congruentes.

congruent segments (p. 15) Two segments with the same measure.

segmentos congruentes Dos segmentos que tienen la misma medida.

congruent solids (p. 835) Two solids with the same shape, size, and scale factor of 1:1.

sólidos congruentes Dos sólidos con la misma forma, tamaño y factor de escala de 1:1.

conjecture (p. 111) An educated guess based on known information.

conjetura Juicio basado en información conocida.

conjunction (p. 119) A compound statement formed by joining two or more statements with the word *and*.

conjunción Enunciado compuesto que se obtiene al unir dos o más enunciados con la palabra *y*.

consecutive interior angles (p. 169) In the figure, transversal t intersects lines ℓ and m. There are two pairs of consecutive interior angles: $\angle 8$ and $\angle 1$, and $\angle 7$ and $\angle 2$.

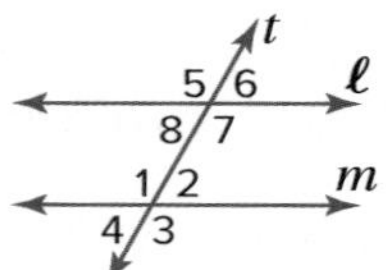

ángulos internos consecutivos En la figura, la transversal t interseca las rectas ℓ y m. La figura presenta dos pares de ángulos internos consecutivos; $\angle 8$ y $\angle 1$; y $\angle 7$ y $\angle 2$.

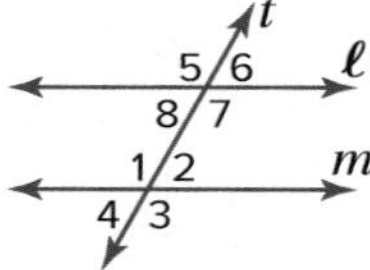

constructions (p. 16) A method of creating geometric figures without the benefit of measuring tools. Generally, only a pencil, straightedge, and compass are used.

construcción Método para dibujar figuras geométricas sin el uso de instrumentos de medición. En general, sólo requiere de un lápiz, una regla y un compás.

contrapositive (p. 122) The statement formed by negating both the hypothesis and conclusion of the converse of a conditional statement.

antítesis Enunciado formado por la negación tanto de la hipótesis como de la conclusión del recíproco de un enunciado condicional.

converse (p. 122) The statement formed by exchanging the hypothesis and conclusion of a conditional statement.

recíproco Enunciado que se obtiene al intercambiar la hipótesis y la conclusión de un enunciado condicional dado.

convex polygon (p. 56) A polygon for which there is no line that contains both a side of the polygon and a point in the interior of the polygon.

coordinate proofs (p. 334) Proofs that use figures in the coordinate plane and algebra to prove geometric concepts.

coplanar (p. 5) Points that lie in the same plane.

corner view (p. 84) The view from a corner of a three-dimensional figure, also called the *isometric view*.

corollary (p. 285) A statement that can be easily proved using a theorem is called a corollary of that theorem.

corresponding angles (p. 169) In the figure, transversal t intersects lines ℓ and m. There are four pairs of corresponding angles: $\angle 5$ and $\angle 1$, $\angle 8$ and $\angle 4$, $\angle 6$ and $\angle 2$, and $\angle 7$ and $\angle 3$.

corresponding parts (p. 291) Matching parts of congruent polygons.

cosecant (p. 606) The reciprocal of the sine of an angle in a right triangle.

cosine (p. 596) For an acute angle of a right triangle, the ratio of the measure of the leg adjacent to the acute angle to the measure of the hypotenuse.

cotangent (p. 606) For an acute angle of a right triangle, the ratio of the adjacent to the opposite side of a right triangle.

counterexample (p. 114) An example used to show that a given statement is not always true.

cross section (p. 797) The intersection of a solid and a plane.

cylinder (p. 76) A figure with bases that are formed by congruent circles in parallel planes.

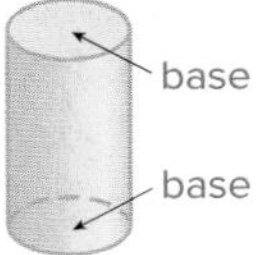

polígono convexo Polígono para el cual no existe recta alguna que contenga un lado del polígono y un punto en el interior del polígono.

demostraciones en coordinadas Demostraciones que usan figuras en el plano de coordinados y álgebra para demostrar conceptos geométricos.

coplanar Puntos que yacen en el mismo plano.

vista de esquina Vista desde una de las esquinas de una figura tridimensional. También se conoce como *vista en perspectiva*.

corolario Un enunciado que se puede demostrar fácilmente usando un teorema se conoce como corolario de dicho teorema.

ángulos correspondientes En la figura, la transversal t interseca las rectas ℓ y m. La figura muestra cuatro pares de ángulos correspondientes: $\angle 5$ y $\angle 1$, $\angle 8$ y $\angle 4$, $\angle 6$ y $\angle 2$; y $\angle 7$ y $\angle 3$.

partes correspondientes Partes que coinciden de polígonos congruentes.

cosecante Recíproco del seno de un ángulo en un triángulo rectangulo.

coseno Para cualquier ángulo agudo de un triángulo rectángulo, razón de la medida del cateto adyacente al ángulo agudo a la medida de la hipotenusa.

cotangente Razón de la medida del cateto adyacente a la medida de cateto opuesto de un triángulo rectángulo.

contraejemplo Ejemplo que se usa para demostrar que un enunciado dado no siempre es verdadero.

sección transversal Intersección de un sólido con un plano.

cilindro Figura cuyas bases son círculos congruentes ubicados en planos paralelos.

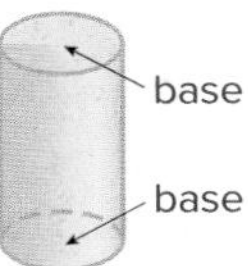

D

deductive argument (p. 143) A proof formed by a group of algebraic steps used to solve a problem.

argumento deductivo Demostración que consta de un conjunto de pasos algebraicos que se usan para resolver un problema.

deductive reasoning (p. 131) A system of reasoning that uses facts, rules, definitions, or properties to reach logical conclusions.

razonamiento deductivo Sistema de razonamiento que emplea hechos, reglas, definiciones o propiedades para obtener conclusiones lógicas.

degree (p. 37) A unit of measure used in measuring angles and arcs. An arc of a circle with a measure of 1° is $\frac{1}{360}$ of the entire circle.

grado Unidad de medida que se usa para medir ángulos y arcos. El arco de un círculo que mide 1° equivale a $\frac{1}{360}$ del círculo completo.

density (p. 842) The mass of an object per unit of volume.

densidad La masa de un objeto por unidad de volumen.

dependent events (p. 889) Two or more events in which the outcome of one event affects the outcome of the other events.

eventos dependientes Dos o más eventos en que el resultado de un evento afecta el resultado de los otros eventos.

diagonal (p. 423) In a polygon, a segment that connects nonconsecutive vertices of the polygon.

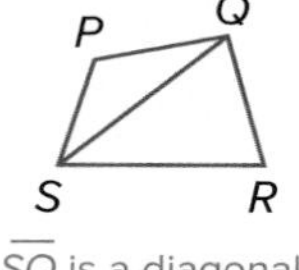

$\overline{SQ}$ is a diagonal.

diagonal Recta que conecta vértices no consecutivos de un polígono.

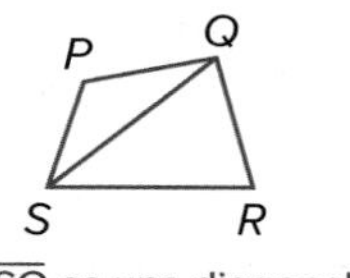

$\overline{SQ}$ es una diagonal.

diameter **1.** (p. 643) In a circle, a chord that passes through the center of the circle. **2.** (p. 818) In a sphere, a segment that contains the center of the sphere, and has endpoints that are on the sphere.

diámetro **1.** En un círculo cuerda que pasa por el centro. **2.** En una estera segmento que incluye el centro de la esfera y cuyos extremos están ubicados en la esfera.

dilation (p. 834) A transformation that enlarges or reduces the original figure proportionally. A dilation with center C and positive scale factor k, $k \neq 1$, is a function that maps a point P in a figure to its image such that

- if point P and C coincide, then the image and preimage are the same point, or
- if point P is not the center of dilation, then P' lies on $\overrightarrow{CP}$ and $CP' = k(CP)$.

If $k < 0$, P' is the point on the ray opposite $\overrightarrow{CP}$ such that $CP' = |k|(CP)$.

homotecia Transformación que amplía o disminuye proporcionalmente el tamaño de una figura. Una homotecia con centro C y factor de escala positivo k, $k \neq 1$, es una función que aplica un punto P a su imagen, de modo que si el punto P coincide con el punto C, entonces la imagen y la preimagen son el mismo punto, o si el punto P no es el centro de la homotecia, entonces P' yace sobre $\overrightarrow{CP}$ y $CP' = k(CP)$. Si $k < 0$, P' es el punto sobre el rayo opuesto a $\overrightarrow{CP}$, tal que $CP' = |k|(CP)$.

direct isometry (p. 69) An isometry in which the image of a figure is found by moving the figure intact within the plane.

isometría directa Isometría en la cual se obtiene la imagen de una figura, al mover la figura intacta dentro del plano.

directrix (p. 703) The fixed line in a parabola that is equidistant from the locus of all points in a plane.

directriz Línea fija en una parábola que está equidistante del lugar geométrico de todos los puntos en un plano.

disjunction (p. 120) A compound statement formed by joining two or more statements with the word *or*.

disyunción Enunciado compuesto que se forma al unir dos o más enunciados con la palabra *o*.

distance between two points (p. 16) The length of the segment between two points.

distancia entre dos puntos Longitud del segmento entre dos puntos.

edge (p. 76) A line that connects two nodes in a network.

edge of a polyhedron (p. 76) A line segment where the faces of a polyhedron intersect.

elimination (p. P18) The use of addition or subtraction in combination with multiplication or division to eliminate one variable and solve a system of equations.

equiangular polygon (p. 57) A polygon with all congruent angles.

equiangular triangle (p. 326) A triangle with all angles congruent.

equidistant (p. 197) The distance between two lines measured along a perpendicular line is always the same.

equilateral polygon (p. 57) A polygon with all congruent sides.

equilateral triangle (p. 326) A triangle with all sides congruent.

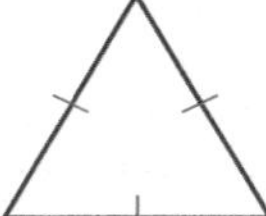

Euclidean geometry (p. 827) A geometrical system in which a plane is a flat surface made up of points that extend infinitely in all directions.

event (p. P8) A specific outcome or type of outcome.

experiment (p. P8) A situation involving chance such as flipping a coin or rolling a die.

experimental probability (p. P9) The ratio of the number of positive outcomes to the total number of events or trials in a probability experiment.

exterior (p. 36) A point is in the exterior of an angle if it is neither on the angle nor in the interior of the angle.

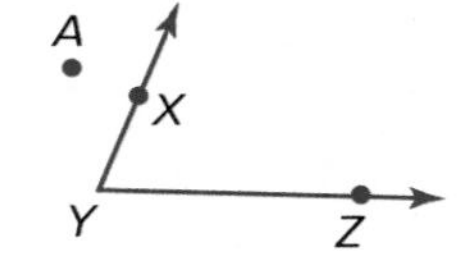

A is in the exterior of $\angle XYZ$.

arista Recta que conecta dos nodos en una red.

arista de un poliedro Segmento de recta donde se intersecan las caras de un poliedro.

eliminación El uso de la suma o la resta en combinación con la multiplicación o la división para eliminar una variable y resolver un sistema de ecuaciones.

polígono equiangular Polígono cuyos ángulos son todos congruentes.

triángulo equiangular Triángulo cuyos ángulos son todos congruentes.

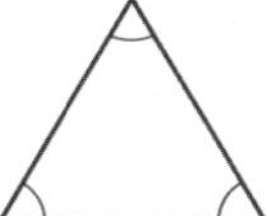

equidistante La distancia entre dos rectas que siempre permanece constante cuando se mide a lo largo de una perpendicular.

polígono equilátero Polígono cuyos lados son todos congruentes.

triángulo equilátero Triángulo cuyos lados son todos congruentes.

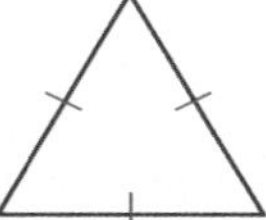

geometría euclidiana Sistema en el cual un plano es una superficie plana formada por puntos que se extienden infinitamente en todas las direcciones.

suceso Un resultado o tipo de resultado específico.

experimento Una situación que involucra la probabilidad, como lanzar una moneda o un dado.

probabilidad experimental La razón de la cantidad de resultados positivos a la cantidad total de sucesos o pruebas en un experimento de probabilidad.

exterior Un punto yace en el exterior de un ángulo si no se ubica ni en el ángulo ni en el interior del ángulo.

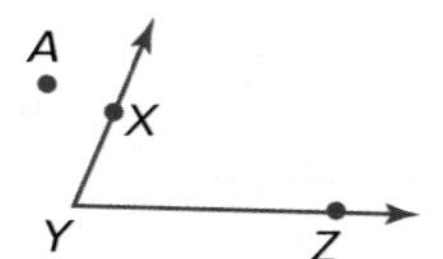

A está en el exterior del $\angle XYZ$.

exterior angle (p. 169) An angle formed by one side of a triangle and the extension of another side.

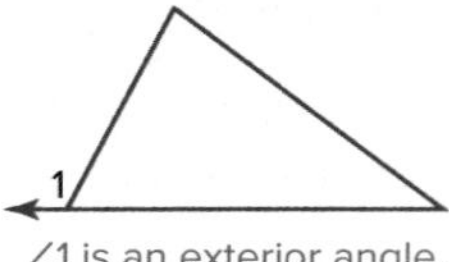

∠1 is an exterior angle.

ángulo externo Ángulo formado por un lado de un triángulo y la prolongación de otro de sus lados.

∠1 es un ángulo externo.

exterior angle (p. 284) An angle that lies in the region that is not between two transversals that intersect the same line.

ángulo externo Un ángulo que está en la región que no está entre dos transversals que cruzan la misma línea.

F

face of a polyhedron (p. 76) A flat surface of a polyhedron.

cara de un poliedro Superficie plana de un poliedro.

factorial (p. 872) The product of the integers less than or equal to a positive integer *n*, written as *n*!

factorial Producto de los enteros menores o iguales a un número positivo *n*, escrito como *n*!

finite plane (p. 10) A plane that has boundaries or does not extend indefinitely.

plano finito Plano que tiene límites o que no se extiende indefinidamente.

flow proof (pp. 144, 284) A proof that organizes statements in logical order, starting with the given statements. Each statement is written in a box with the reason verifying the statement written below the box. Arrows are used to indicate the order of the statements.

demostración de flujo Demostración que organiza los enunciados en orden lógico, comenzando con los enunciados dados. Cada enunciado se escribe en una casilla y debajo de cada casilla se escribe el argumento que verifica dicho enunciado. El orden de los enunciados se indica con flechas.

focus (p. 703) The fixed point in a parabola that is equidistant from the locus of all points in a plane.

foco Punto fijo en una parábola que está equidistante del lugar geométrico de todos los puntos en un plano.

fractal (p. 552) A figure generated by repeating a special sequence of steps infinitely often. Fractals often exhibit self-similarity.

fractal Figura que se obtiene mediante la repetición infinita de una sucesión particular de pasos. Los fractales a menudo exhiben autosemejanza.

Fundamental Counting Principle (p. 861) A method used to determine the number of possible outcomes in a sample space by multiplying the number of possible outcomes from each stage or event.

principio fundamental de contar Método para determinar el número de resultados posibles en un espacio muestral multiplicando el número de resultados posibles de cada etapa o evento.

G

geometric mean (p. 565) For any positive numbers *a* and *b*, the positive number *x* such that $\frac{a}{x} = \frac{x}{b}$.

media geométrica Para todo número positivo *a* y *b*, existe un número positivo *x* tal que $\frac{a}{x} = \frac{x}{b}$.

geometric probability (p. 881) Using the principles of length and area to find the probability of an event.

probabilidad geométrica Uso de los principios de longitud y área para calcular la probabilidad de un evento.

glide reflection (p. 249) The composition of a translation followed by a reflection in a line parallel to the translation vector.

reflexión del deslizamiento Composición de una traslación seguida por una reflexión en una recta paralela al vector de la traslación.

great circle (p. 819) A circle formed when a plane intersects a sphere with its center at the center of the sphere.

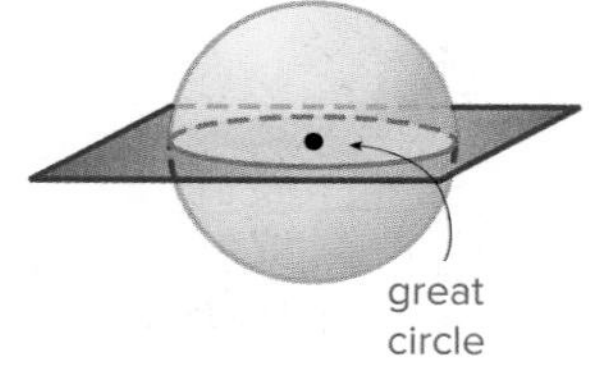

círculo mayor Círculo que se forma cuando un plano interseca una esfera y cuyo centro es el mismo que el centro de la esfera.

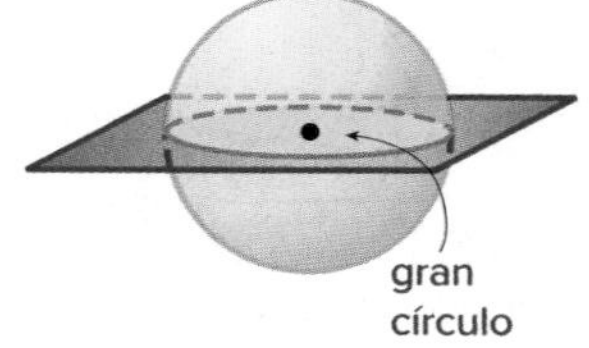

H

height of a parallelogram (p. 725) The length of an altitude of a parallelogram.

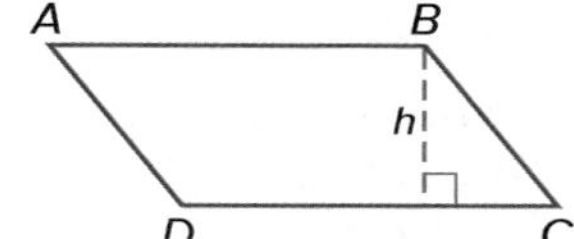

h is the height of parallelogram *ABCD*.

altura de un paralelogramo Longitud del segmento perpendicular que va desde la base hasta el vértice opuesto a ella.

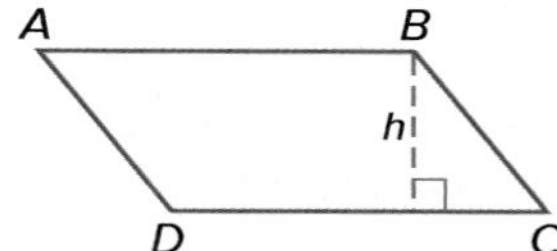

h es la altura del paralelogramo *ABCD*.

height of a solid figure (p. 770) The height is the length of the altitude.

altura de un cuerpo geométrico La altura es la longitud de la altitud.

height of a trapezoid (p. 735) The perpendicular distance between the bases of a trapezoid.

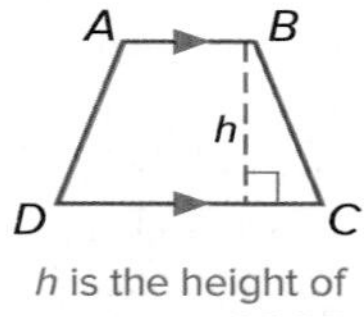

h is the height of trapezoid *ABCD*.

altura de un trapecio Distancia perpendicular entre las bases de un trapecio.

h es la altura del trapecio *ABCD*.

height of a triangle (p. 727) The length of an altitude drawn to a given base of a triangle.

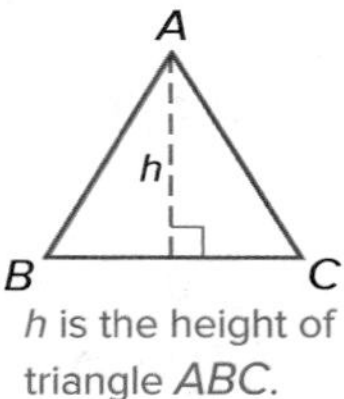

h is the height of triangle *ABC*.

altura de un triángulo Longitud de una altura trazada a una base dada de un triángulo.

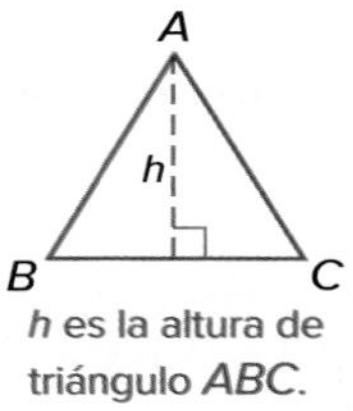

h es la altura de triángulo *ABC*.

hemisphere (p. 819) One of the two congruent parts into which a great circle separates a sphere.

hemisferio Una de las dos partes congruentes en las cuales un círculo mayor divide una esfera.

hypothesis (p. 121) In a conditional statement, the statement that immediately follows the word *if*.

hipótesis Enunciado escrito inmediatamente después de la palabra *si* en un enunciado condicional.

I

if-then statement (p. 121) A compound statement of the form "if *p*, then *q*," where *p* and *q* are statements.

enunciado si-entonces Enunciado compuesto de la forma "si *p*, entonces *q*," donde *p* y *q* son enunciados.

image (p. 67) A figure that results from the transformation of a geometric figure.

imagen Figura que resulta de la transformación de una figura geométrica.

incenter (p. 358) The point of concurrency of the angle bisectors of a triangle.

incentro Punto de intersección de las bisectrices interiores de un triángulo.

included angle (p. 302) In a triangle, the angle formed by two sides is the included angle for those two sides.

ángulo incluido En un triángulo, el ángulo formado por dos lados es el ángulo incluido de esos dos lados.

included side (p. 311) The side of a polygon that is a side of each of two angles.

lado incluido Lado de un polígono común a dos de sus ángulos.

independent events (p. 889) Two or more events in which the outcome of one event does not affect the outcome of the other events.

eventos independientes El resultado de un evento no afecta el resultado del otro evento.

indirect isometry (p. 69) An isometry that cannot be performed by maintaining the orientation of the points, as in a direct isometry.

isometría indirecta Tipo de isometría que no se puede obtener manteniendo la orientación de los puntos, como ocurre con la isometría directa.

indirect proof (p. 385) In an indirect proof, one assumes that the statement to be proved is false. One then uses logical reasoning to deduce that a statement contradicts a postulate, theorem, or one of the assumptions. Once a contradiction is obtained, one concludes that the statement assumed false must in fact be true.

demostración indirecta En una demostración indirecta, se supone que el enunciado a demostrar es falso. Después, se deduce lógicamente que existe un enunciado que contradice un postulado, un teorema o una de las conjeturas. Una vez hallada una contradicción, se concluye que el enunciado que se suponía falso debe ser, en realidad, verdadero.

indirect reasoning (p. 385) Reasoning that assumes that the conclusion is false and then shows that this assumption leads to a contradiction of the hypothesis like a postulate, theorem, or corollary. Then, since the assumption has been proved false, the conclusion must be true.

razonamiento indirecto Razonamiento en que primero se supone que la conclusión es falsa y luego se demuestra que esta conjetura lleva a una contradicción de la hipótesis como un postulado, un teorema o un corolario. Finalmente, como se ha demostrado que la conjetura es falsa, la conclusión debe ser verdadera.

inductive reasoning (p. 111) Reasoning that uses a number of specific examples to arrive at a plausible generalization or prediction. Conclusions arrived at by inductive reasoning lack the logical certainty of those arrived at by deductive reasoning.

razonamiento inductivo Razonamiento que usa varios ejemplos específicos para lograr una generalización o una predicción pausible. Las conclusiones obtenidas por razonamiento inductivo carecen de la certeza lógica de aquellas obtenidas por razonamiento deductivo.

inscribed (p. 646) A polygon is inscribed in a circle if each of its vertices lie on the circle.

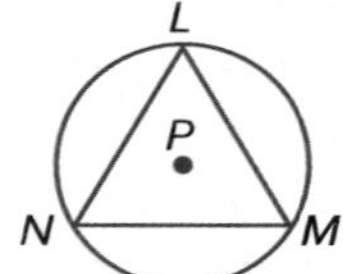

$\triangle LMN$ is inscribed in $\odot P$.

inscrito Un polígono está inscrito en un círculo si todos sus vértices yacen en el círculo.

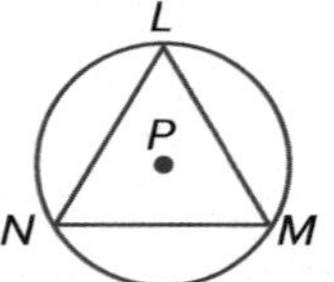

$\triangle LMN$ está inscrito en $\odot P$.

inscribed angle (p. 669) An angle that has a vertex on a circle and sides that contain chords of the circle.

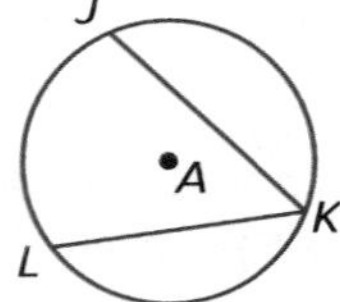

In $\odot A$, $\angle JKL$ is an inscribed angle.

ángulo inscrito Ángulo cuyo vértice esté en un círculo y cuyos lados contienen cuerdas del círculo.

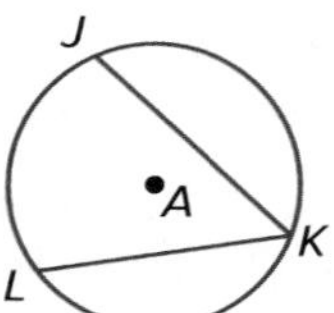

En $\odot A$, $\angle JKL$ es un ángulo inscrito.

intercepted arc (p. 669) An angle intercepts an arc if and only if each of the following conditions are met.

1. The endpoints of the arc lie on the angle.
2. All points of the arc except the endpoints are in the interior of the circle.
3. Each side of the angle has an endpoint on the arc.

arco intersecado Un ángulo interseca un arco si y sólo si se cumple cada una de las siguientes condiciones.

1. Los extremos del arco yacen en el ángulo.
2. Todos los puntos del arco, excepto los extremos, yacen en el interior del círculo.
3. Cada lado del ángulo tiene un extremo del arco.

interior (p. 36) A point is in the interior of an angle if it does not lie on the angle itself and it lies on a segment with endpoints that are on the sides of the angle.

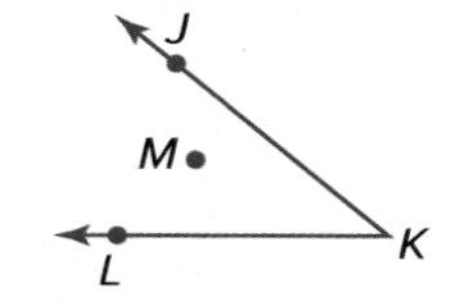

M is in the interior of $\angle JKL$.

interior angles (p. 169) Angles that lie between two transversals that intersect the same line.

intersection 1. (p. 6) A set of points common to two or more geometric figures. 2. (p. 866) For the intersection of event A and event B, the set of all outcomes that are common to both events; represented by $A \cap B$.

inverse (p. 122) The statement formed by negating both the hypothesis and conclusion of a conditional statement.

inverse cosine (p. 599) The inverse function of cosine, or $\cos^{-1}$. If the cosine of an acute $\angle A$ is equal to *x*, then $\cos^{-1} x$ is equal to the measure of $\angle A$.

inverse sine (p. 599) The inverse function of sine, or $\sin^{-1}$. If the sine of an acute $\angle A$ is equal to *x*, then $\sin^{-1} x$ is equal to the measure of $\angle A$.

inverse tangent (p. 599) The inverse function of tangent, or $\tan^{-1}$. If the tangent of an acute $\angle A$ is equal to *x*, then $\tan^{-1} x$ is equal to the measure of $\angle A$.

irrational number (p. 17) A number that cannot be expressed as a terminating or repeating decimal.

irregular figure (p. 57) A polygon with sides and angles that are not all congruent.

isometry (p. 67) A mapping for which the original figure and its image are congruent.

isosceles trapezoid (p. 469) A trapezoid in which the legs are congruent, both pairs of base angles are congruent, and the diagonals are congruent.

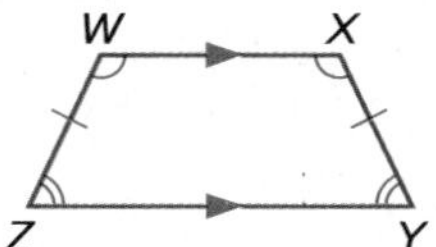

interior Un punto se encuenta en el interior de un ángulo si no yace en el ángulo como tal y si está en un segmento cuyos extremos están en los lados del ángulo.

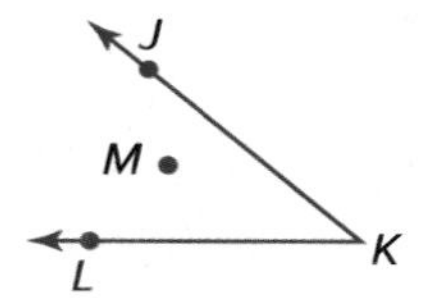

M está en el interior del $\angle JKL$.

ángulos interiores Ángulos que yacen entre dos transversales que intersecan la misma recta.

intersección 1. Conjunto de puntos comunes a dos o más figuras geométricas. 2. Para la intersección del suceso A y el suceso B, el conjunto de todos los resultados que son comunes a ambos sucesos; se representa con $A \cap B$.

inverso Enunciado que se obtiene al negar tanto la hipótesis como la conclusión de un enunciado condicional.

inverso del coseno Función inversa del coseno, o $\cos^{-1}$. Si el coseno de un $\angle A$ agudo es igual a *x*, entonces $\cos^{-1} x$ es igual a la medida del $\angle A$.

inverso del seno Función inversa del seno, o $\sin^{-1}$. Si el seno de un $\angle A$ agudo es igual a *x*, entonces $\sin^{-1} x$ es igual a la medida del *A*.

inverse del tangente Función inversa de la tangente, o $\tan^{-1}$. Si la tangente de un $\angle A$ agudo es igual a *x*, entonces $\tan^{-1} x$ es igual a la medida del $\angle A$.

número irracional Número que no se puede expresar como un decimal terminal o periódico.

figura irregular Polígono cuyos lados y ángulos no son todo congruentes.

isometría Aplicación en la cual la figura original y su imagen son congruentes.

trapecio isósceles Trapecio cuyos catetos son congruentes, ambos pares de ángulos de las bases son congruentes y las diagonales son congruentes.

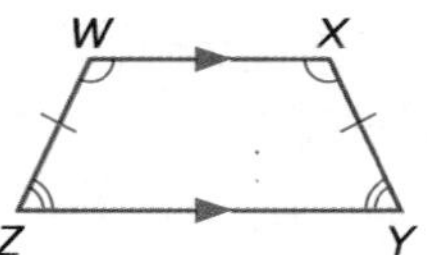

isosceles triangle (p. 325) A triangle with at least two sides congruent. The congruent sides are called *legs*. The angles opposite the legs are *base angles*. The angle formed by the two legs is the *vertex angle*. The side opposite the vertex angle is the *base*.

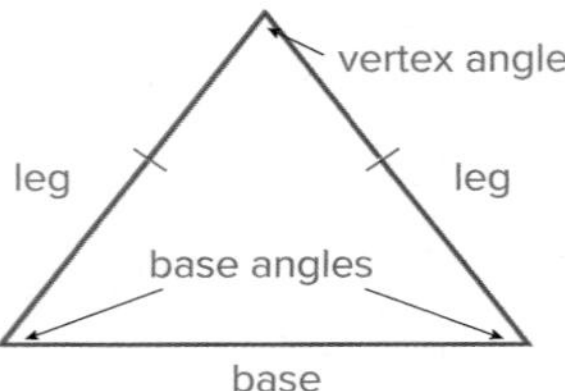

triángulo isósceles Triángulo que tiene por lo menos dos lados congruentes. Los lados congruentes se llaman *catetos*. Los ángulos opuestos a los catetos son los *ángulos de la base*. El ángulo formado por los dos catetos es el *ángulo del vértice*. El lado opuesto al ángulo del vértice es la *base*.

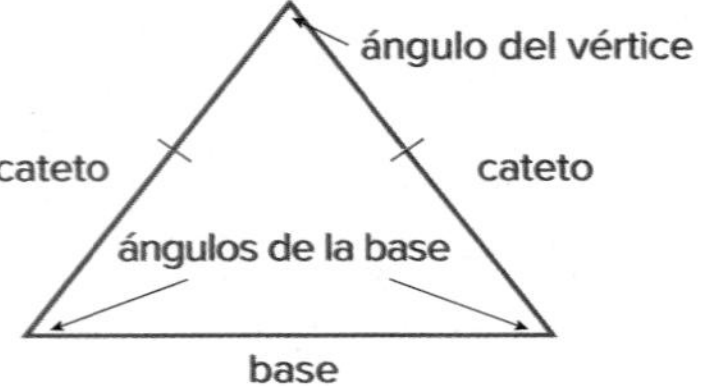

iteration (p. 552) A process of repeating the same procedure over and over again.

iteración Proceso de repetir el mismo procedimiento una y otra vez.

J

joint frequencies (p. 909) In a two-way frequency table, the frequencies reported in the cells in the interior of the table.

frecuencias conjuntas En una tabla de double entrada o de frecuencias, las frecuencias reportadas en las celdas en el interior de la tabla.

K

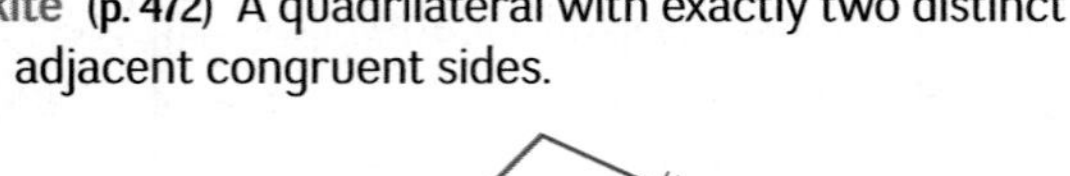

kite (p. 472) A quadrilateral with exactly two distinct pairs of adjacent congruent sides.

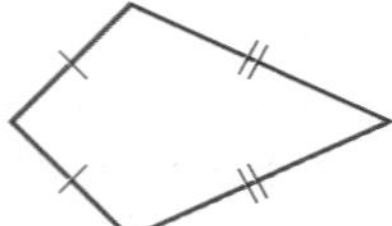

cometa Cuadrilátero que tiene exactamente dos pares differentes de lados congruentes y adyacentes.

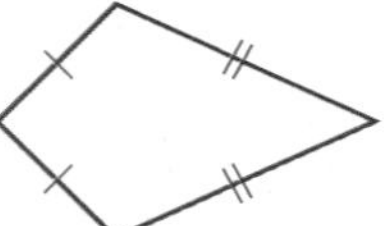

L

lateral area (p. 770) For prisms, pyramids, cylinders, and cones, the area of the faces of the figure not including the bases.

área lateral En prismas, pirámides, cilindros y conos, es el área de la caras de la figura sin incluir el área de las bases.

lateral edges (p. 770) In a prism, the intersection of two adjacent lateral faces. In a pyramid, lateral edges are the edges of the lateral faces that join the vertex to vertices of the base.

aristas laterales En un prisma, la intersección de dos caras laterales adyacentes. En una pirámide, las aristas de las caras laterales que unen el vértice de la pirámide con los vértices de la base.

lateral faces (p. 770) In a prism, the faces that are not bases. In a pyramid, faces that intersect at the vertex.

caras laterales En un prisma, las caras que no forman las bases. En una pirámide, las caras que se intersecan en el vértice.

latitude (p. 833) A measure of distance north or south of the equator.

latitud Medida de la distancia al norte o al sur del ecuador.

Law of Cosines (p. 624) Let $\triangle ABC$ be any triangle with a, b, and c representing the measures of sides opposite the angles with measures A, B, and C, respectively. Then the following equations are true.

$a^2 = b^2 + c^2 - 2bc \cos A$

$b^2 = a^2 + c^2 - 2ac \cos B$

$c^2 = a^2 + b^2 - 2ab \cos C$

ley de los cosenos Sea $\triangle ABC$ cualquier triángulo donde a, b y c son las medidas de los lados opuestos a los ángulos que miden A, B, y C, respectivamente. Entonces las siguientes ecuaciones son verdaderas.

$a^2 = b^2 + c^2 - 2bc \cos A$

$b^2 = a^2 + c^2 - 2ac \cos B$

$c^2 = a^2 + b^2 - 2ab \cos C$

Law of Detachment (p. 131) If $p \rightarrow q$ is a true conditional and p is true, then q is also true.

ley de indiferencia Si $p \rightarrow q$ es un enunciado condicional verdadero y p es verdadero, entonces q también es verdadero.

Law of Sines (p. 617) Let $\triangle ABC$ be any triangle with a, b, and c representing the measures of sides opposite the angles with measures A, B, and C, respectively.

Then, $\frac{\sin A}{a} = \frac{\sin B}{b} = \frac{\sin C}{c}$.

ley de los senos Sea $\triangle ABC$ cualquier triángulo donde a, b y c representan las medidas de los lados opuestos a los ángulos que miden A, B, y C, respectivamente.

Entonces, $\frac{\text{scn } A}{a} = \frac{\text{scn } B}{b} = \frac{\text{scn } C}{c}$.

Law of Syllogism (p. 133) If $p \rightarrow q$ and $q \rightarrow r$ are true conditionals, then $p \rightarrow r$ is also true.

ley del silogismo Si $p \rightarrow q$ y $q \rightarrow r$ son enunciados condicionales verdaderos, entonces $p \rightarrow r$ también es verdadero.

legs of an isosceles triangle (p. 325) The two congruent sides of an isosceles triangle.

catetos de un triángulo isósceles Las dos lados congruentes de un triángulo isósceles.

legs of a trapezoid (p. 469) The nonparallel sides of a trapezoid.

catetos de un trapecio Los lados no paralelos de un trapecio.

line (p. 5) A basic undefined term of geometry. A line is made up of points and has no thickness or width. In a figure, a line is shown with an arrowhead at each end. Lines are usually named by lowercase script letters or by writing capital letters for two points on the line, with a double arrow over the pair of letters.

recta Término geometrico basico no definido. Una recta está formada por puntos y carece de grosor o ancho. En una figura, una recta se representa con una flecha en cada extremo. Generalmente se designan con letras minúsculas o con las dos letras mayúsculas de dos puntos sobre la recta y una flecha doble sobre el par de letras.

line of reflection (p. 221) A line in which each point on the preimage and its corresponding point on the image are the same distance from this line.

línea de reflexión Una línea en la cual cada punto en el preimage y el su corresponder senalañ en la imagen es la misma distancia de esta línea.

line of symmetry (p. 259) A line that can be drawn through a plane figure so that the figure on one side is the reflection image of the figure on the opposite side.

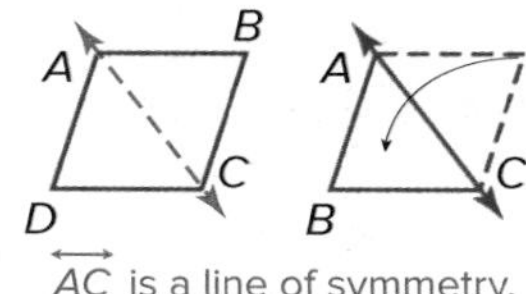

$\overleftrightarrow{AC}$ is a line of symmetry.

eje de simetría Recta que se traza a través de una figura plana, de modo que un lado de la figura es la imagen reflejada del lado opuesto.

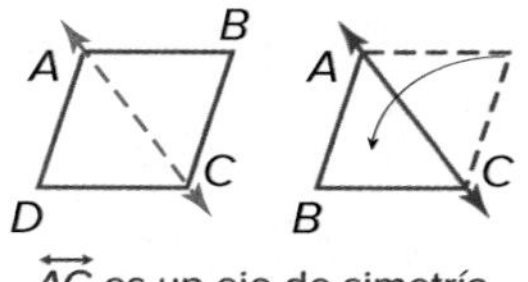

$\overleftrightarrow{AC}$ es un eje de simetría.

line segment (p. 14) A measurable part of a line that consists of two points, called endpoints, and all of the points between them.

segmento de recta Sección medible de una recta que consta de dos puntos, llamados extremos, y todos los puntos entre ellos.

line symmetry (p. 259) If a figure can be mapped onto itself by a reflection in a line, the figure has reflectional symmetry or line symmetry.

línea de simetría Si doblamos una figura a lo largo de una recta y las dos partes coinciden, entonces la figura tiene simetría reflexiva o simetría axial.

linear pair (p. 46) A pair of adjacent angles whose noncommon sides are opposite rays.

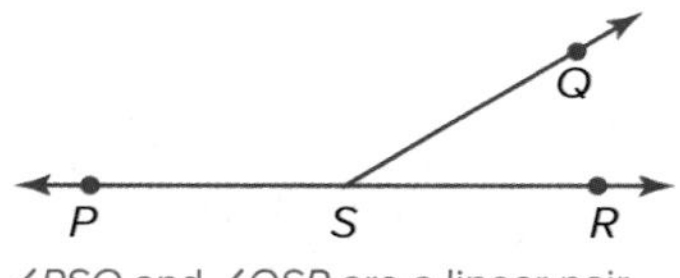

$\angle PSQ$ and $\angle QSR$ are a linear pair.

par lineal Par de ángulos adyacentes cuyos lados no comunes forman rayos opuestos.

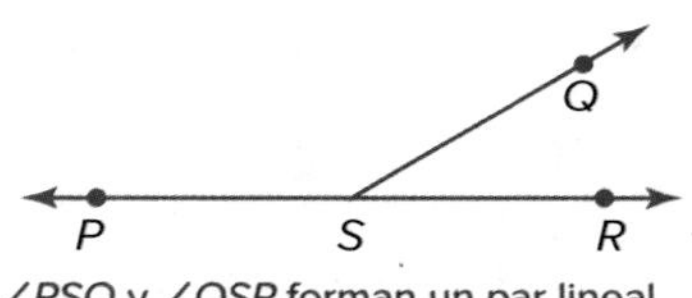

$\angle PSQ$ y $\angle QSR$ forman un par lineal.

locus (p. 11) The set of points that satisfy a given condition.

lugar geométrico Conjunto de puntos que satisfacen una condición dada.

logically equivalent (p. 123) Statements that have the same truth values.

lógicamente equivalentes Enunciados que poseen los mismos valores verdaderos.

longitude (p. 833) A measure of distance east or west of the Prime Meridian.

longitud Medida de la distancia del este o al oeste del Primer Meridiano.

M

magnitude of symmetry (p. 260) The smallest angle through which a figure can be rotated so that it maps onto itself.

magnitud de la simetria El angulo mas pequeno con el cual una figura puede serrotada de modo que traz sobre si mismo.

major arc (p. 653) An arc with a measure greater than 180. $\widehat{ACB}$ is a major arc.

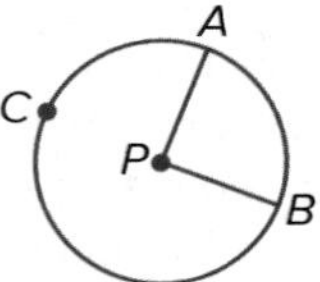

arco mayor Arco que mide más de 180. $\widehat{ACB}$ es un arco mayor.

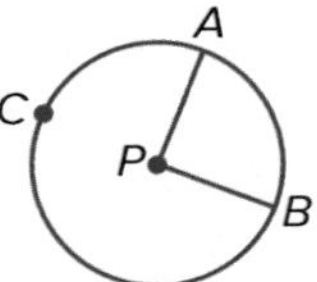

marginal frequencies (p. 909) In a two-way frequency table, the accumulated frequencies reported in the Totals row and Totals column.

frecuencias marginales En una tabla de double entrada o de frecuencias, las frecuencias acumuladas que se reportan en la hilera de los totales y en la columna de los totales.

matrix logic (p. 383) A rectangular array in which learned clues are recorded in order to solve a logic or reasoning problem.

lógica matricial Arreglo rectangular en que las claves aprendidas se escriben en orden para resolver un problema de lógica o razonamiento.

median (p. 365) In a triangle, a line segment with endpoints that are a vertex of a triangle and the midpoint of the side opposite the vertex.

mediana En un triángulo, Segmento de recta de cuyos extremos son un vértice del triángulo y el punto medio del lado opuesto a dicho vértice.

meridians (p. 833) Imaginary vertical lines drawn around the Earth through the North and South Poles.

meridianos Líneas verticales imaginarias dibujadas alrededor de la Tierra que von del polo norte al polo sur.

midpoint (p. 26) The point on a segment exactly halfway between the endpoints of the segment.

punto medio Punto en un segmento que yace exactamente en la mitad, entre los extremos del segmento.

midsegment of a trapezoid (p. 471) A segment that connects the midpoints of the legs of a trapezoid.

segmento medio de un trapecio Segmento que conecta los puntos medios de los catetos de un trapecio.

midsegment of a triangle (p. 535) A segment with endpoints that are the midpoints of two sides of a triangle.

segmento medio de un triángulo Segmento cuyas extremos son los puntos medianos de dos lados de un triángulo.

minor arc (p. 653) An arc with a measure less than 180. $\widehat{AB}$ is a minor arc.

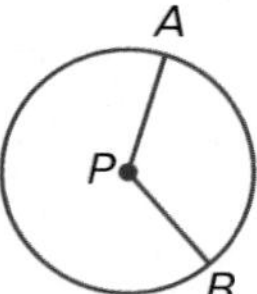

arco menor Arco que mide menos de 180. $\widehat{AB}$ es un arco menor.

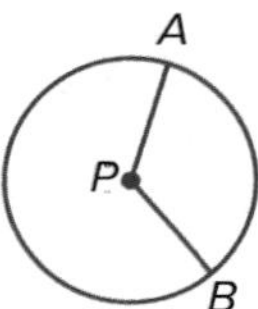

multistage experiments (p. 860) Experiments with more than two stages.

experimentos multietápicos Experimentos con más de dos etapas.

mutually exclusive (p. 897) Two events that have no outcomes in common.

mutuamente exclusivos Eventos que no tienen resultados en común.

N

negation (p. 119) If a statement is represented by *p*, then *not p* is the negation of the statement.

negación Si *p* representa un enunciado, entonces *no p* es la negación del enunciado.

net (p. 85) A two-dimensional figure that when folded forms the surfaces of a three-dimensional object.

red Figura bidimensional que al ser plegada forma las superficies de un objeto tridimensional.

***n*-gon** (p. 57) A polygon with *n* sides.

enágono Polígono con *n* lados.

non-Euclidean geometry (p. 828) The study of geometrical systems that are not in accordance with the Parallel Postulate of Euclidean geometry.

geometría no euclidiana El estudio de sistemas geométricos que no satisfacen el postulado de las paralelas de la geometría euclidiana.

O

oblique cone (p. 775) A cone that is not a right cone.

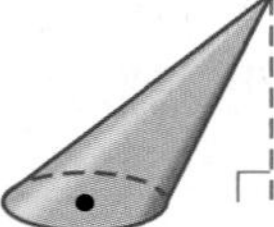

cono oblicuo Cono que no es un cono recto.

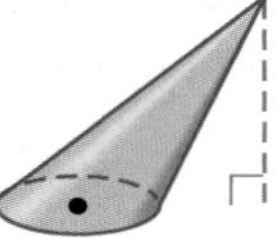

oblique cylinder (p. 772) A cylinder that is not a right cylinder.

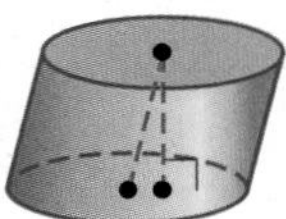

cilindro oblicuo Cilindro que no es un cilindro recto.

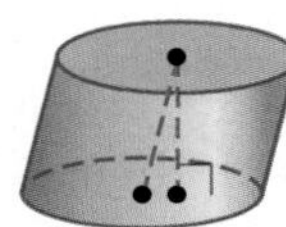

oblique prism (p. 770) A prism in which the lateral edges are not perpendicular to the bases.

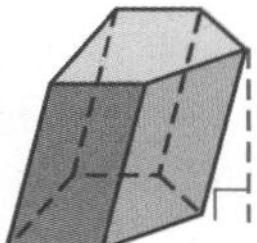

prisma oblicuo Prisma cuyas aristas laterales no son perpendiculares a las bases.

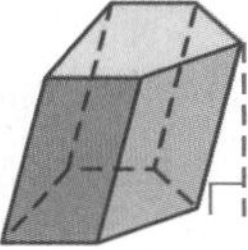

oblique solid (p. 796) A solid with base(s) that are not perpendicular to the edges connecting the two bases or vertex.

sólido oblicuo Sólido con base o bases que no son perpendiculares a las aristas, las cuales conectan las dos bases o vértice.

obtuse angle (p. 38) An angle with degree measure greater than 90 and less than 180.

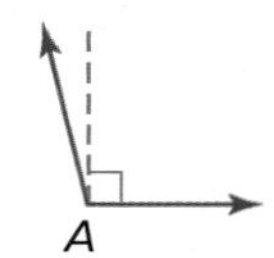

$90 < m\angle A < 180$

ángulo obtuso Ángulo que mide más de 90 y menos de 180.

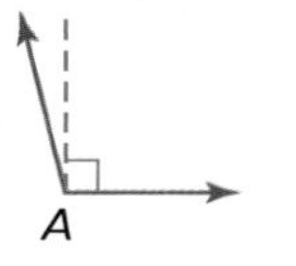

$90 < m\angle A < 180$

Glossary/Glosario

opposite rays (p. 36) Two rays $\overrightarrow{BA}$ and $\overrightarrow{BC}$ such that B is between A and C.

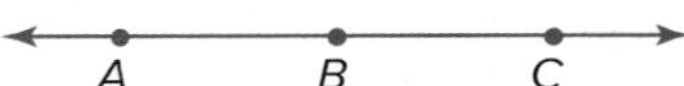

rayos opuestos Dos rayos $\overrightarrow{BA}$ y $\overrightarrow{BC}$ donde B esta entre A y C.

order of symmetry (p. 260) The number of times a figure can map onto itself as it rotates from 0° to 360°.

orden de la simetría Número de veces que una figura se puede aplicar sobre sí misma mientras gira de 0° a 360°.

ordered pairs (p. P15) A pair of numbers used to locate a point in the coordinate plane or the solution of an equation in two variables. An ordered pair is written in the form (x-coordinate, y-coordinate).

pares ordenados Un par de números que se usa para ubicar un punto en el plano de coordenadas o la solución de una ecuación con dos variables. Un par ordenado se escribe en la forma (coordenada x, coordenada y).

ordered triple (p. 584) Three numbers given in a specific order used to locate points in space.

triple ordenado Tres números dados en un orden específico que sirven para ubicar puntos en el espacio.

origin (p. P15) The point (0, 0) in a coordinate plane where the x-axis and the y-axis intersect.

origen El punto (0, 0) en un plano de coordinadas, donde se intersecan el eje x y el eje y.

orthocenter (p. 367) The point of concurrency of the altitudes of a triangle.

ortocentro Punto de intersección de las alturas de un triángulo.

orthographic drawing (p. 84) The two-dimensional top view, left view, front view, and right view of a three-dimensional object.

proyección ortogonal Vista bidimensional superior, del lado izquierda, frontal y del lado derecho de un objeto tridimensionl.

outcome (p. P8) One possible result of a probability event. Example: 4 is an outcome when a number cube is rolled.

resultado Una consecuencia posible de un suceso de probabilidad. Ejemplo: 4 es un resultado cuando se lanza un cubo numérico.

P

parabola (p. 703) The graph of a quadratic function. The set of all points in a plane that are the same distance from a given point, called the focus, and a given line, called the directrix.

parábola La grafica de una funcion cuadratica. Conjunto de todos los puntos de un plano que estan a la misma distancia de un punto dado, llamado foco, y de una recta dada, llamada directriz.

paragraph proof (p. 145) An informal proof written in the form of a paragraph that explains why a conjecture for a given situation is true.

demostración de párrafo Demostración informal escrita en párrafo que explica por qué una conjetura para una situación dada es verdadera.

parallel lines (p. 170) Coplanar lines that do not intersect.

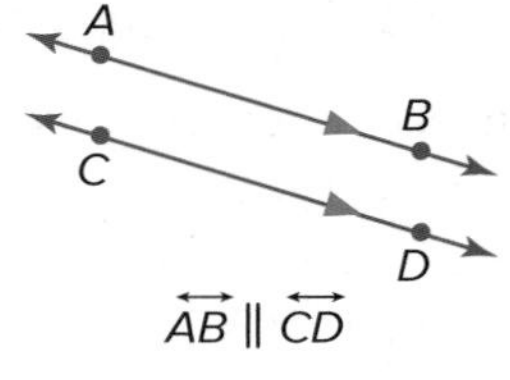

rectas paralelas Rectas coplanares que no se intersecan.

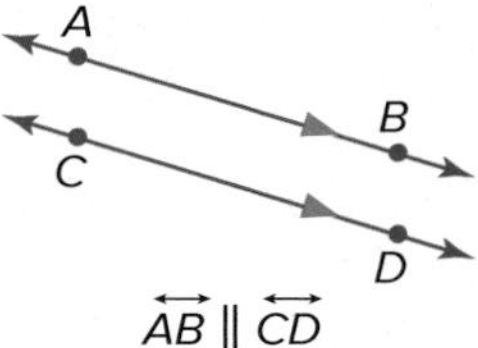

parallel planes (p. 169) Planes that do not intersect.

planos paralelos Planos que no se intersecan.

parallelogram (p. 433) A quadrilateral with parallel opposite sides. Any side of a parallelogram may be called a *base*.

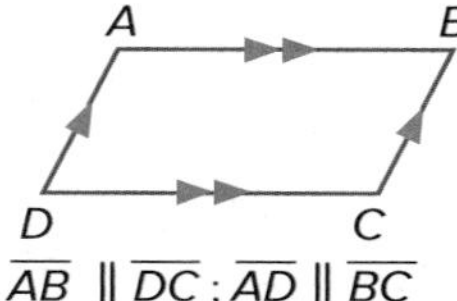

$\overline{AB} \parallel \overline{DC}$; $\overline{AD} \parallel \overline{BC}$

paralelogramo Cuadrilátero cuyos lados opuestos son paralelos y cuya *base* puede ser cualquier de sus lados.

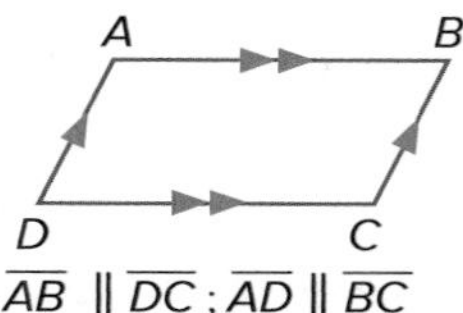

$\overline{AB} \parallel \overline{DC}$; $\overline{AD} \parallel \overline{BC}$

parallels (p. 833) Imaginary horizontal lines parallel to the equator.

paralelos Rectas horizontales imaginarias paralelas al ecuador.

perimeter (p. 58) The sum of the lengths of the sides of a polygon.

perímetro Suma de la longitud de los lados de un polígono.

permutation (p. 872) An arrangement of objects in which order is important.

permutación Disposicion de objetos en la cual el orden es importante.

perpendicular bisector (p. 354) In a triangle, a line, segment, or ray that passes through the midpoint of a side and is perpendicular to that side.

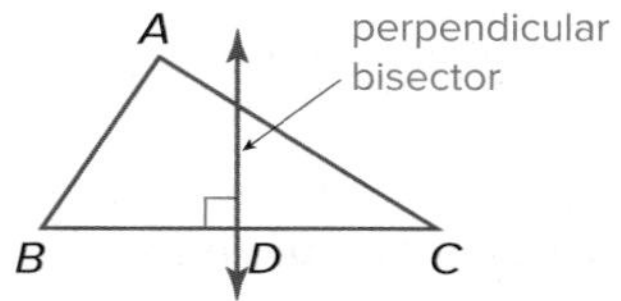

D is the midpoint of $\overline{BC}$.

mediatriz Recta, segmento de recta o rayo perpendicular que corta un lado del triángulo en su punto medio.

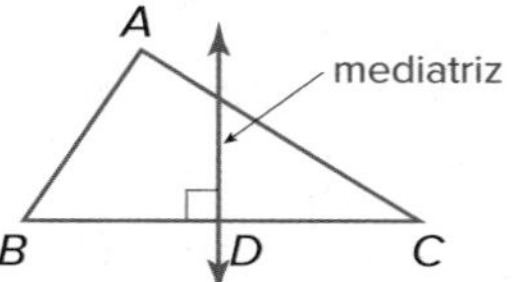

D es el punto medio de $\overline{BC}$.

perpendicular lines (p. 48) Lines that form right angles.

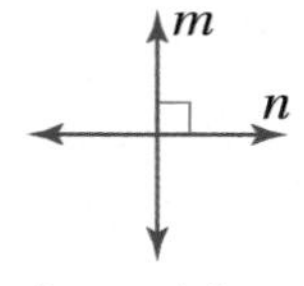

line $m \perp$ line n

rectas perpendiculares Rectas que forman ángulos rectos.

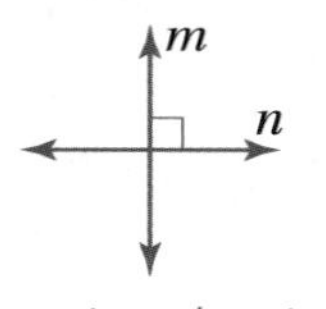

recta $m \perp$ recta n

pi (π) (p. 645) An irrational number represented by the ratio of the circumference of a circle to the diameter of the circle.

pi (π) Número irracional representado por la razón de la circunferencia de un círculo al diámetro del mismo.

plane (p. 5) A basic undefined term of geometry. A plane is a flat surface made up of points that has no depth and extends indefinitely in all directions. In a figure, a plane is often represented by a shaded, slanted four-sided figure. Planes are usually named by a capital script letter or by three noncollinear points on the plane.

plano Término geométrico básico no definido. Superficie plana sin espesor formada por puntos y que se extiende hasta el infinito en todas direcciones. En una figura, los planos a menudo se representan con una figura inclinada y sombreada y se designan con una letra mayúscula o con tres puntos no colineales del plano.

Platonic solids (p. 77) The five regular polyhedra: tetrahedron, hexahedron, octahedron, dodecahedron, or icosahedron.

sólidos platónicos Los cinco poliedros regulares siguientes: tetraedro, hexaedro, octaedro, dodecaedro e icosaedro.

point (p. 5) A basic undefined term of geometry. A point is a location. In a figure, points are represented by a dot. Points are named by capital letters.

punto Término geométrico básico no definido. Un punto representa un lugar o ubicación. En una figura, se representa con una marca puntual y se designan con letras mayúsculas.

point of concurrency (p. 355) The point of intersection of concurrent lines.

punto de concurrencia Punto de intersección de rectas concurrentes.

point of symmetry (p. 260) A figure that can be mapped onto itself by a rotation of 180°.

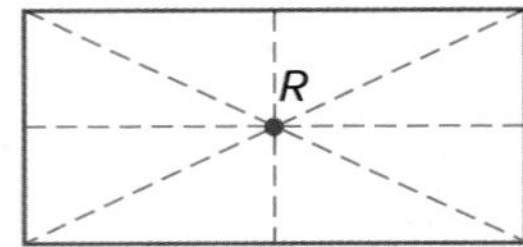

R is a point of symmetry.

punto de simetría Una figura que se puede traz sobre sí mismo por una rotación de 180°.

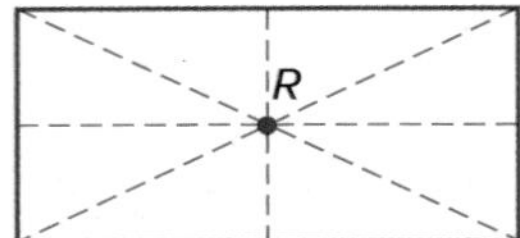

R es un punto de simetría.

point of tangency (p. 678) For a line that intersects a circle in only one point, the point at which they intersect.

punto de tangencia Punto de intersección de una recta en un círculo en un solo punto.

point-slope form (p. 179) An equation of the form $y - y_1 = m(x - x_1)$, where (x_1, y_1) are the coordinates of any point on the line and m is the slope of the line.

forma punto-pendiente Ecuación de la forma $y - y_1 = m(x - x_1)$, donde (x_1, y_1) representan las coordenadas de un punto cualquiera sobre la recta y m representa la pendiente de la recta.

poles (p. 819) The endpoints of the diameter of a great circle.

postes Las extremos del diámetro de un círculo mayor.

polygon (p. 56) A closed figure formed by a finite number of coplanar segments called *sides* such that the following conditions are met:

1. The sides that have a common endpoint are noncollinear.
2. Each side intersects exactly two other sides, but only at their endpoints, called the *vertices*.

polígono Figura cerrada formada por un número finito de segmentos coplanares llamados *lados*, tal que satisface las siguientes condiciones:

1. Los lados que tienen un extremo común son no colineales.
2. Cada lado interseca exactamente dos lados mas, pero sólo en sus extremos, llamados *vértices*.

polyhedrons (p. 76) Closed three-dimensional figures made up of flat polygonal regions. The flat regions formed by the polygons and their interiors are called *faces*. Pairs of faces intersect in segments called *edges*. Points where three or more edges intersect are called *vertices*.

poliedros Figuras tridimensionales cerrada formadas por regiones poligonales planas. Las regiones planas definidas por un polígono y sus interiores se llaman *caras*. Cada intersección entre dos caras se llama *arista*. Los puntos donde se intersecan tres o más aristas se llaman *vértices*.

population density (p. 841) A measurement of population per unit of area.

densidad demográfica Medida de la población por unidad de área.

postulate (p. 141) A statement that describes a fundamental relationship between the basic terms of geometry. Postulates are accepted as true without proof.

postulado Enunciado que describe una relación fundamental entre los términos geométricos básicos. Los postulados se aceptan como verdaderos sin necesidad de demostración.

precision (p. 91) The preciseness of a measurement depends on the unit of measure. The smaller the unit, the more precise the measurement.

precisión La precisión de una medida depende de la unidad de medida. Cuanto más pequeña es la unidad, más precisa es la medida.

preimage (p. 67) The graph of an object before a transformation.

preimagen Gráfica de una figura antes de una transformación.

principle of superposition (p. 291) Two figures are congruent if and only if there is a rigid motion or a series of rigid motions that maps one figure exactly onto the other.

principio de superposición Dos figuras son congruentes si y sólo si existe un movimiento rígido o una serie de movimientos rígidos que aplican una de las figuras exactamente sobre la otra.

prism (p. 76) A solid with the following characteristics:

1. Two faces, called *bases*, are formed by congruent polygons that lie in parallel planes.

prisma Sólido con las siguientes características:

1. Dos caras llamadas *bases*, formadas por polígonos congruentes que yacen en planos paralelos.

2. The faces that are not bases, called *lateral faces*, are formed by parallelograms.

3. The intersections of two adjacent lateral faces are called *lateral edges* and are parallel segments.

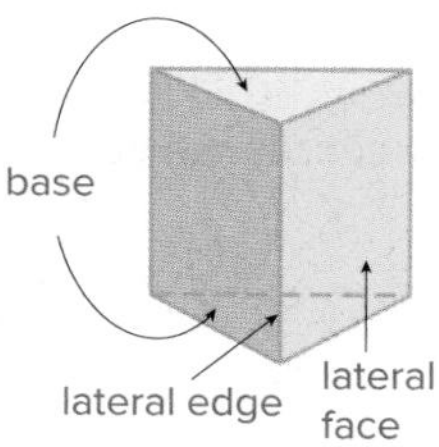

triangular prism

probability (p. P8) The ratio of the number of favorable outcomes for an event to the number of possible outcomes of the event. P(a) = Number of favorable outcomes/total number of possible outcomes

proof (p. 143) A logical argument in which each statement you make is supported by a statement that is accepted as true.

proof by contradiction (p. 385) An indirect proof in which one assumes that the statement to be proved is false. One then uses logical reasoning to deduce a statement that contradicts a postulate, theorem, or one of the assumptions. Once a contradiction is obtained, one concludes that the statement assumed false must in fact be true.

proportion An equation of the form $\frac{a}{b} = \frac{c}{d}$ that states that two ratios are equal.

pyramid (p. 76) A solid with the following characteristics:

1. All of the faces, except one face, intersect at a point called the *vertex*.

2. The face that does not contain the vertex is called the *base* and is a polygonal region.

3. The faces meeting at the vertex are called *lateral faces* and are triangular regions.

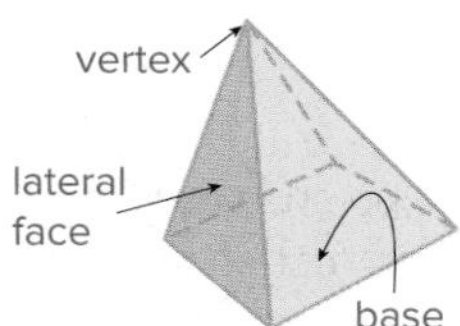

rectangular pyramid

2. Las caras que no son las bases, llamadas *caras laterales*, son paralelogramos.

3. Las intersecciones de dos caras laterales adyacentes se llaman *aristas laterales* y son segmentos paralelos.

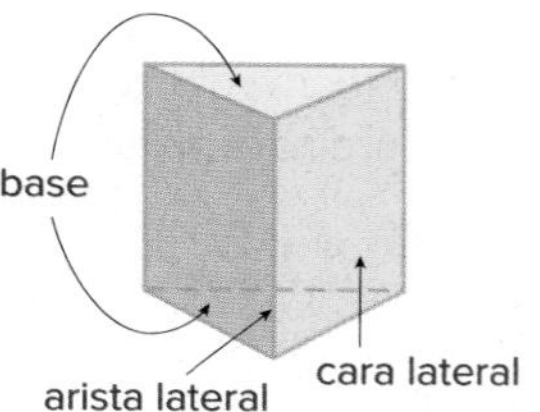

prisma triangular

probabilidad La razón de la cantidad de resultados favorables de un suceso a la cantidad de resultados posibles de ese suceso. P(a) = Cantidad total de resultados favorables/cantidad total de resultados posibles

demostración Argumento lógico en el cual cada enunciado que se hace está respaldado por un enunciado que se acepta como verdadero.

demostración por contradicción Demostración indirecta en la cual se supone que el enunciado a demostrarse es falso. Luego, se usa el razonamiento lógico para inferir un enunciado que contradiga el postulado, teorema o una de las conjeturas. Una vez que se obtiene una contradicción, se concluye que el enunciado que se supuso falso es, en realidad, verdadero.

proporción Ecuación de la forma $\frac{a}{b} = \frac{c}{d}$ que establece que dos razones son iguales.

pirámide Sólido con las siguientes características:

1. Todas las caras, excepto una, se intersecan en un punto llamado *vértice*.

2. La cara sin el vértice se llama *base* y es una región poligonal.

3. Las caras que se encuentran en los vértices se llaman *caras laterales* y son regiones triangulares.

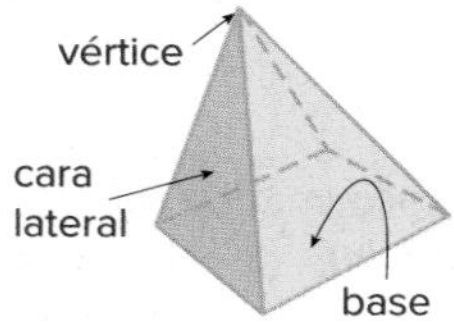

pirámide rectangular

Pythagorean triple (p. 576) A group of three whole numbers that satisfies the equation $a^2 + b^2 = c^2$, where c is the greatest number.

triplete pitágorico Grupo de tres números enteros que satisfacen la ecuación $a^2 + b^2 = c^2$, donde c es el número mayor.

Q

quadrant (p. P15) One of four regions into which the x- and y-axes separate the coordinate plane.

cuadrante Una de las cuatro regiones en las que los ejes x e y dividen el plano de coordenadas.

R

radian measure (p. 655) The radian measure, θ, of a central angle is the ratio of the arc length to the radius of the circle: $\theta = \frac{\ell}{r}$ radians.

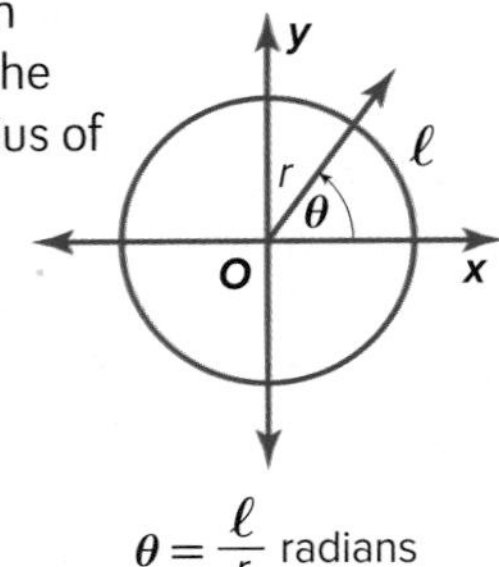

$\theta = \frac{\ell}{r}$ radians

medida del radián La medida en radianes de un ángulo central es igual a la proporción entre la longitud del arco y el radio del círculo: $\theta = \frac{\ell}{r}$ rad.

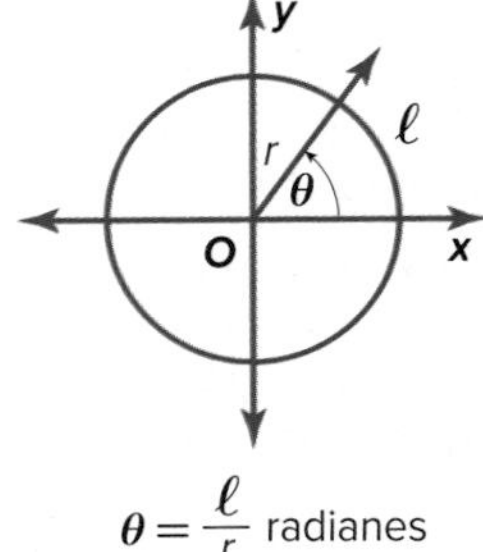

$\theta = \frac{\ell}{r}$ radianes

radius **1.** (p. 643) In a circle, any segment with endpoints that are the center of the circle and a point on the circle. **2.** (p. 880) In a sphere, any segment with endpoints that are the center and a point on the sphere.

radio **1.** En un círculo, cualquier segmento cuyos extremos son en el centro y un punto del círculo. **2.** En una esfera, cualquier segmento cuyos extremos son el centro y un punto de la esfera.

radius of a regular polygon (p. 752) The radius of a circle circumscribed about a polygon.

radio de un polígono regular Radio de un círculo circunscrito alrededor de un polígono.

rate of change (p. 178) Describes how a quantity is changing over time.

tasa de cambio Describe cómo cambia una cantidad a través del tiempo.

ratio A comparison of two quantities using division.

razón Comparación de dos cantidades mediante división.

ray (p. 36) $\overrightarrow{PQ}$ is a ray if it is the set of points consisting of $\overline{PQ}$ and all points S for which Q is between P and S.

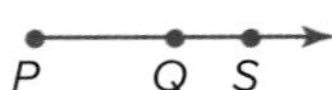

rayo $\overrightarrow{PQ}$ es un rayo si se el conjunto de puntos formado por $\overline{PQ}$ y todos los puntos S para los cuales Q se ubica entre P y S.

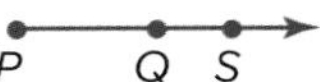

rectangle (p. 453) A quadrilateral with four right angles.

rectángulo Cuadrilátero con cuatro ángulos rectos.

reduction (p. 493) An image that is smaller than the original figure.

reducción Imagen más pequeña que la figura original.

reflection (p. 67) A transformation representing the flip of a figure over a point, line, or plane. A reflection in a line is a function that maps a point to its image such that

- if the point is on the line, then the image and preimage are the same point, or
- if the point is not on the line, then the line is the perpendicular bisector of the segment joining the two points.

reflexión Transformación en la cual una figura se "voltea" a través de un punto, una recta o un plano. Una reflexión en una recta es una función que aplica un punto a su imagen, de modo que si el punto yace sobre la recta, entonces la imagen y la preimagen son el mismo punto, o si el punto no yace sobre la recta, la recta es la mediatriz del segmento que une los dos puntos.

regular polygon (p. 57) A convex polygon in which all of the sides are congruent and all of the angles are congruent.

polígono regular Polígono convexo cuyos los lados y ángulos son congruentes.

regular polyhedron (p. 77) A polyhedron in which all of the faces are regular congruent polygons.

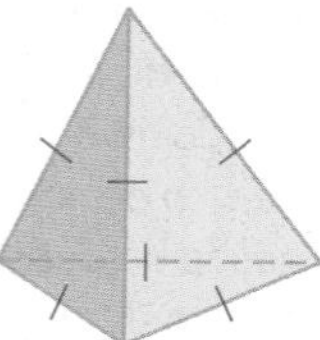

regular pyramid (p. 773) A pyramid with a base that is a regular polygon.

related conditionals (p. 122) Statements that are based on a given conditional statement.

relative error (p. 92) The ratio of the absolute error to the expected measure.

relative frequency (p. 910) In a frequency table, the ratio of the number of observations in a category to the total number of observations.

remote interior angles (p. 284) The angles of a triangle that are not adjacent to a given exterior angle.

rhombus (p. 460) A quadrilateral with all four sides congruent.

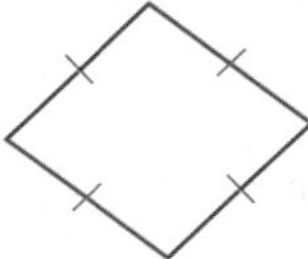

right angle (p. 38) An angle with a degree measure of 90.

right cone (p. 775) A cone with an axis that is also an altitude.

right cylinder (p. 772) A cylinder with an axis that is also an altitude.

right prism (p. 770) A prism with lateral edges that are also altitudes.

right solid (p. 795) A solid with base(s) that are perpendicular to the edges connecting them or connecting the base and the vertex of the solid.

right triangle (p. 319) A triangle with a right angle. The side opposite the right angle is called the *hypotenuse*. The other two sides are called *legs*.

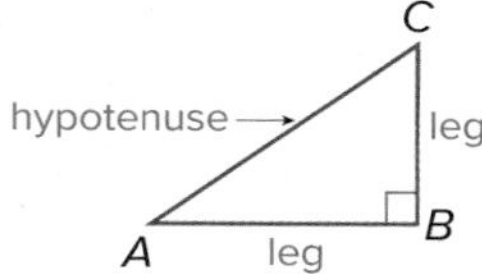

rigid transformation (pp. 15, 67) A transformation in which the position of the image may differ from that of the preimage, but the two figures remain congruent.

poliedro regular Poliedro cuyas caras son polígonos regulares congruentes.

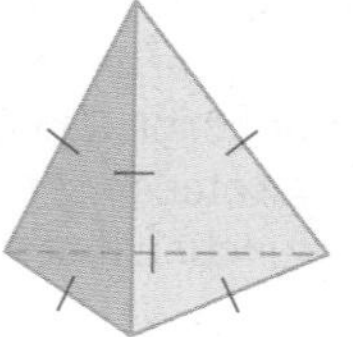

pirámide regular Pirámide cuya base es un polígono regular.

condicionales relacionados Enunciados que se basan en un enunciado condicional dado.

error relativo La razón del error absoluto a la medida esperada.

frecuencia relativa En una tabla de frecuencias, la razón del número de observaciones en una categoría al número total de observaciones.

ángulos internos no adyacentes Ángulos de un triángulo que no son adyacentes a un ángulo exterior dado.

rombo Cuadrilátero con cuatro lados congruentes.

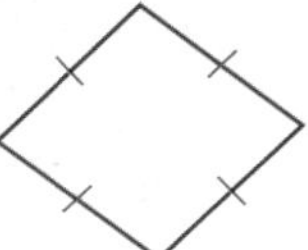

ángulo recto Ángulo que mide 90.

cono recto Cono cuyo eje es también su altura.

cilindro recto Cilindro cuyo eje es también su altura.

prisma recto Prisma cuyas aristas laterales también son su altura.

sólido recto Sólido con base o bases perpendiculares a las aristas, conectándolas entre sí o conectando la base y el vértice del sólido.

triángulo rectángulo Triángulo con un ángulo recto. El lado opuesto al ángulo recto se conoce como *hipotenusa*. Los otros dos lados se llaman *catetos*.

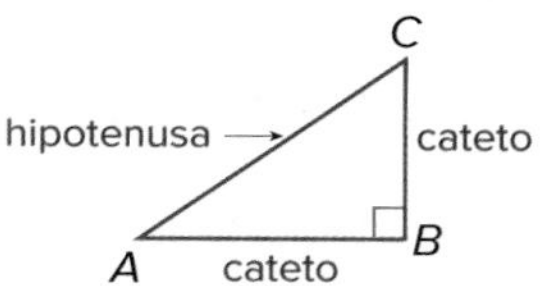

transformación rígida Es la transformación en que la posición de una imagen puede diferir con la posición de la imagen previa, pero las dos figuras siguen siendo congruentes.

rotation (pp. 67, 239) A transformation that turns every point of a preimage through a specified angle and direction about a fixed point, called the *center of rotation*. A rotation about a fixed point through an angle of $x°$ is a function that maps a point to its image such that

- if the point is the center of rotation, then the image and preimage are the same point, or
- if the point is not the center of rotation, then the image and preimage are the same distance from the center of rotation and the measure of the angle of rotation formed by the preimage, center of rotation, and image points is x.

rotación Transformación en la cual se hace girar cada punto de la preimagen a través de un ángulo y una dirección determinadas alrededor de un punto llamado *centro de rotación*. La rotación de $x°$ es una función que aplica un punto a su imagen, de modo que si el punto es el centro de rotación, entonces la imagen y la preimagen están a la misma distancia del centro de rotación y la medida del ángulo formado por los puntos de la preimagen, centro de rotación e imagen es x.

rotational symmetry (p. 260) If a figure can be rotated less than 360° about a point so that the image and the preimage are indistinguishable, then the figure has rotational symmetry.

simetría rotacional Si una imagen se puede girar menos de 360° alrededor de un punto, de modo que la imagen y la preimagen sean idénticas, entonces la figura tiene simetría rotacional.

S

sample space (p. 859) The set of all possible outcomes of an experiment.

espacio muestral El conjunto de todos los resultados posibles de un experimento.

scale factor (pp. 504, 834) The ratio of corresponding measurements of two similar figures.

factor de escala La razón de las medidas correspondientes de dos figuras semejantes.

scale factor of dilation (p. 504) The ratio of a length on an image to a corresponding length on the preimage.

factor de escala de homotecia Razon de una longitud en la imagen a una longitud correspondiente en la preimagen.

secant (pp. 606, 687) Any line that intersects a circle in exactly two points.

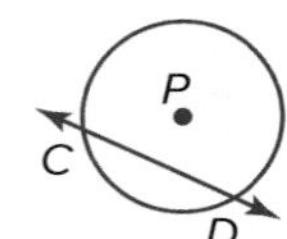

$\overleftrightarrow{CD}$ is a secant of $\odot P$.

secante Cualquier recta que interseca un círculo exactamente en dos puntos.

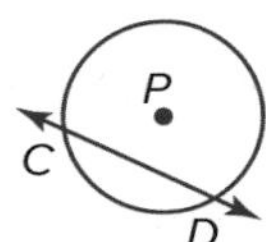

$\overleftrightarrow{CD}$ es una secante de $\odot P$.

sector of a circle (p. 744) A region of a circle bounded by a central angle and its intercepted arc.

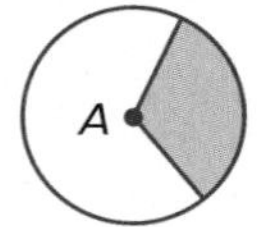

The shaded region is a sector of $\odot A$.

sector circular Región de un círculo limitada por un ángulo central y su arco de intersección.

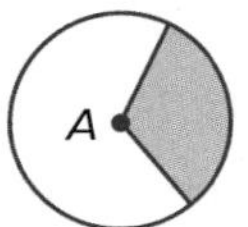

La región sombreada es un sector de $\odot A$.

segment (p. 14) See *line segment*.

segmento Ver *segmento de recta*.

segment bisector (p. 27) A segment, line, or plane that intersects a segment at its midpoint.

bisector del segmento Segmento, recta o plano que interseca un segmento en su punto medio.

segment of a circle (p. 748) The region of a circle bounded by an arc and a chord.

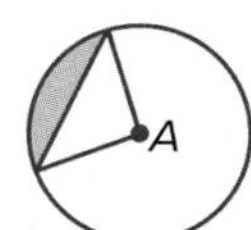

The shaded region is a segment of ⊙A.

segmento de un círculo Región de un círculo limitada por un arco y una cuerda.

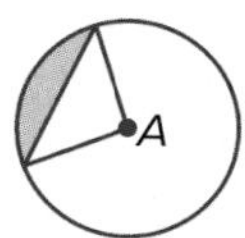

La región sombreada es un segmento de ⊙A.

self-similar (p. 552) If any parts of a fractal image are replicas of the entire image, the image is self-similar.

autosemejante Si cualquier parte de una imagen fractal es una réplica de la imagen completa, entonces la imagen es autosemejante.

semicircle (p. 653) An arc that measures 180.

semicírculo Arco que mide 180.

sides of an angle (p. 36) The rays of an angle.

lados de un ángulo Los rayos de un ángulo.

Sierpinski Triangle (p. 552) A self-similar fractal described by Waclaw Sierpinski. The figure was named for him.

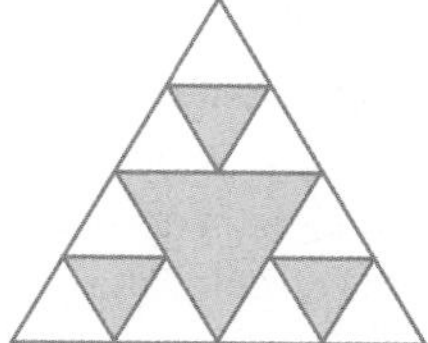

triángulo de Sierpinski Fractal autosemejante descrito por el matemático Waclaw Sierpinski. La figura se nombró en su honor.

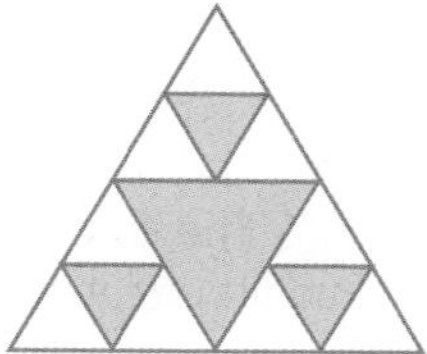

significant digits (p. 92) All of the digits of a measurement that are known to be accurate, plus one estimated digit.

dígitos significativos Todos los dígitos de una medida que se sabe que son exactos más un dígito estimado.

similar polygons (p. 502) Polygons that have the same shape, but not necessarily the same size.

polígonos semejantes Polígonos que tienen la misma forma, pero no necesariamente el mismo tamaño.

similar solids (p. 834) Solids that have exactly the same shape, but not necessarily the same size.

sólidos semejantes Sólidos que tienen exactamente la misma forma, pero no necesariamente el mismo tamaño.

similarity ratio (p. 504) The scale factor between two similar polygons.

razón de semejanza Factor de escala entre dos polígonos semejantes.

similarity transformation (p. 502) When a figure and its transformation image are similar.

transformación de semejanza cuando una figura y su imagen transformada son semejantes.

sine (p. 596) For an acute angle of a right triangle, the ratio of the measure of the leg opposite the acute angle to the measure of the hypotenuse.

seno Para un ángulo agudo de un triángulo rectángulo, razón entre la medida del cateto opuesto al ángulo agudo a la medida de la hipotenusa.

skew lines (p. 170) Lines that do not intersect and are not coplanar.

rectas alabeadas Rectas que no se intersecan y que no son coplanares.

slant height (p. 773) The height of the lateral side of a pyramid or cone.

altura oblicua Altura de la cara lateral de una pirámide o un cono.

slope (p. 178) For a (nonvertical) line containing two points (x_1, y_1) and (x_2, y_2), the number m given by the formula $m = \frac{y_2 - y_1}{x_2 - x_1}$ where $x_2 \neq x_1$.

pendiente Para una recta (no vertical) que contiene dos puntos (x_1, y_1) y (x_2, y_2), tel número m viene dado por la fórmula $m = \frac{y_2 - y_1}{x_2 - x_1}$ donde $x_2 \neq x_1$.

slope-intercept form (p. 179) A linear equation of the form $y = mx + b$. The graph of such an equation has slope m and y-intercept b.

solid of revolution (p. 798) A three-dimensional figure obtained by rotating a plane figure about a line.

solving a triangle (p. 617) Finding the measures of all of the angles and sides of a triangle.

space (p. 7) A boundless three-dimensional set of all points.

sphere (p. 76) In space, the set of all points that are a given distance from a given point, called the *center*.

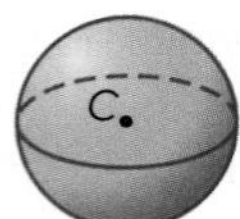

C is the center of the sphere.

spherical geometry (p. 827) The branch of geometry that deals with a system of points, great circles (lines), and spheres (planes).

square (p. 461) A quadrilateral with four right angles and four congruent sides.

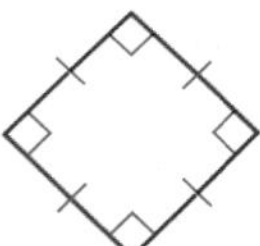

statement (p. 119) Any sentence that is either true or false, but not both.

substitution (p. P17) The process of solving one equation for a variable and substituting the resulting expression for that variable in another equation to solve a system of equations.

supplementary angles (p. 47) Two angles with measures that have a sum of 180.

surface area (p. 78) The sum of the areas of all faces and side surfaces of a three-dimensional figure.

symmetry (p. 259) A figure has symmetry if there exists a rigid motion—reflection, translation, rotation, or glide reflection—that maps the figure onto itself.

system of equations (p. P17) A set of two or more equations with the same variables.

forma pendiente-intersección Ecuación lineal de la forma $y = mx + b$ donde, la pendiente es m y la intersección y es b.

sólido de revolución Figura tridimensional que se obtiene al rotar una figura plana alrededor de una recta.

resolver un triángulo Calcular las medidas de todos los ángulos y todos los lados de un triángulo.

espacio Conjunto tridimensional no acotado de todos los puntos.

esfera En el espacio, conjunto de todos los puntos a cierta distancia de un punto dado llamado *centro*.

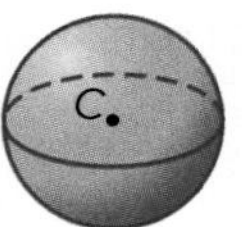

C es el centro de la esfera.

geometría esférica Rama de la geometría que estudia los sistemas de puntos, los círculos mayores (rectas) y las esferas (planos).

cuadrado Cuadrilátero con cuatro ángulos rectos y cuatro lados congruentes.

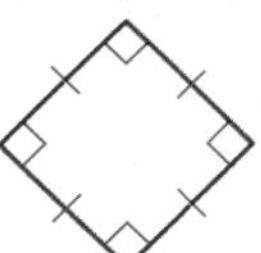

enunciado Cualquier suposición que puede ser falsa o verdadera, pero no ambas.

sustitución El proceso de resolver una ecuación para hallar una variable y sustituir esa variable por la expresión resultante en otra ecuación para resolver un sistema de ecuaciones.

ángulos suplementarios Dos ángulos cuya suma es igual a 180.

área de superficie Suma de las áreas de todas las caras y superficies laterales de una figura tridimensional.

simetría Una figura tiene simetría si existe un movimiento rígido (reflexión, translación, rotación, o reflexión con deslizamiento) que aplica la figura sobre sí misma.

sistema de ecuaciones Un conjunto de dos o más ecuaciones que tienen las mismas variables.

tangent **1.** (p. 596) For an acute angle of a right triangle, the ratio of the measure of the leg opposite the acute angle to the measure of the leg adjacent to the acute angle. **2.** (p. 678) A line in the plane of a circle that intersects the circle in exactly one point. The point of intersection is called the *point of tangency.* **3.** (p. 818) A line that intersects a sphere in exactly one point.

tangente **1.** Para un ángulo agudo de un triángulo rectángulo, razón de la medida del cateto opuesto al ángulo agudo a la medida del cateto adyacente al ángulo agudo. **2.** Recta en el plano de un círculo que interseca el círculo en exactamente un punto. El punto de intersección se conoce como *punto de tangencia.* **3.** Recta que interseca una esfera en exatamente un punto.

theorem (p. 144) A statement or conjecture that can be proven true by undefined terms, definitions, and postulates.

teorema Enunciado o conjetura que se puede demostrar como verdadera mediante términos geométricos básicos, definiciones y postulados.

theoretical probability (p. P9) The ratio of the number of favorable outcomes to the total number of possible outcomes.

probabilidad teórica La razón de la cantidad de resultados favorables a la cantidad total de resultados posibles.

transformation (p. 67) In a plane, a mapping for which each point has exactly one image point and each image point has exactly one preimage point.

transformación En un plano, aplicación para la cual cada punto del plano tiene un único punto de la imagen y cada punto de la imagen tiene un único punto de la preimagen.

translation (p. 67) A transformation that moves a figure the same distance in the same direction. A translation is a function that maps each point to its image along a vector such that each segment joining a point and its image has the same length as the vector, and this segment is also parallel to the vector.

traslación Transformación que mueve una figura la misma distancia en la misma dirección. Una traslación es una función que aplica cada punto a su imagen a lo largo de un vector, de modo que cada segmento que une un punto a su imagen tiene la misma longitud que el vector y este segmento es también paralelo al vector.

translation vector (p. 232) The vector in which a translation maps each point to its image.

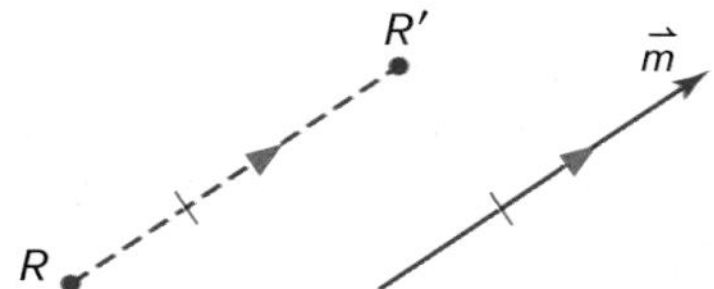

Point R', is a translation of point R along translation vector m.

vector de traslación Vector en el cual una traslación aplica cada punto a su imagen.

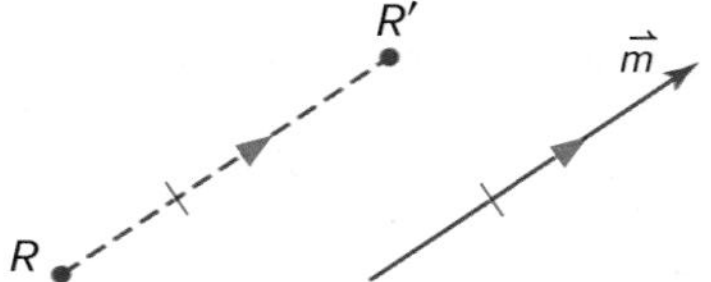

El punta R', es la traslación del punto R a lo largo del vector m de traslación.

transversal (p. 169) A line that intersects two or more lines in a plane at different points.

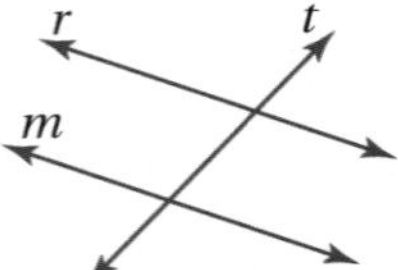

Line *t* is a transversal.

transversal Recta que interseca dos o más rectas en el diferentes puntos del mismo plano.

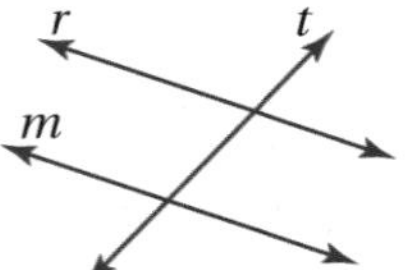

La recta *t* es una transversal.

trapezoid (p. 469) A quadrilateral with exactly one pair of parallel sides. The parallel sides of a trapezoid are called *bases*. The nonparallel sides are called *legs*. The pairs of angles with their vertices at the endpoints of the same base are called *base angles*.

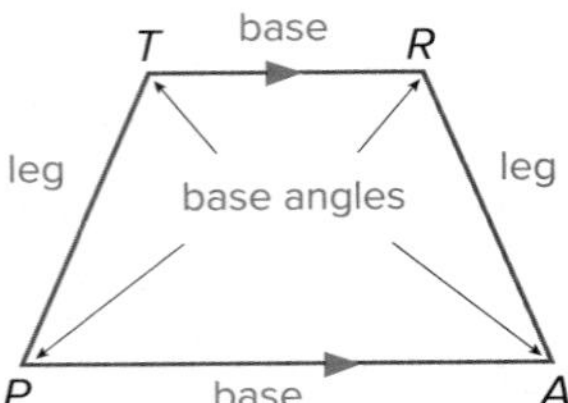

trapecio Cuadrilátero con sólo un par de lados paralelos. Los lados paralelos del trapecio se llaman *bases*. Los lados no paralelos se llaman *catetos*. Los pares de ángulos cuyos vértices coinciden en los extremos de la misma base son los *ángulos de la base*.

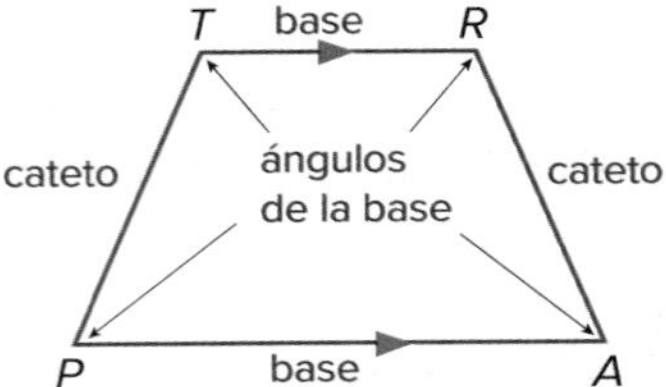

tree diagram (p. 859) An organized table of line segments (branches) that shows possible experiment outcomes.

diagrama del árbol Tabla organizada de segmentos de recta (ramas) que muestra los resultados posibles de un experimento.

trial (p. P8) A single performance of an experiment such as rolling a die one time.

prueba Una sola realización de un experimento, como lanzar un dado una vez.

trigonometric ratio (p. 596) A ratio of the lengths of sides of a right triangle.

razón trigonométrica Razón de las longitudes de los lados de un triángulo rectángulo.

trigonometry (p. 596) The study of the properties of triangles and trigonometric functions and their applications.

trigonometría Estudio de las propiedades de los triángulos, de las funciones trigonométricas y sus aplicaciones.

truth value (p. 119) The truth or falsity of a statement.

valor de verdad Condición de un enunciado de ser verdadero o falso.

two-column proof (p. 152) A formal proof that contains statements and reasons organized in two columns. Each step is called a *statement*, and the properties that justify each step are called *reasons*.

demostración de dos columnas Demonstración formal que contiene enunciados y razones organizadas en dos columnas. Cada paso se llama *enunciado* y las propiedades que lo justifican son las *razones*.

two-stage experiment (p. 860) An experiment with two stages or events.

experimento de dos pasos Experimento que consta de dos pasos o eventos.

two-way frequency table (p. 909) A table that is used to show the frequencies or relative frequencies of data from a survey or experiment classified according to two variables, with the rows indicating one variable and the columns indicating the other.

tabla de doble entrada o de frecuencias Tabla que se usa para mostrar las frecuencias o frecuencias relativas de los datos de una encuesta o experimento clasificado de acuerdo con dos variables, en la cual las hileras indican una variable y las columnas indican la otra variable.

U

undefined terms (p. 5) Words, usually readily understood, that are not formally explained by means of more basic words and concepts. The basic undefined terms of geometry are *point*, *line*, and *plane*.

término geométrico básico no definido Palabras que por lo general se entienden fácilmente y que no se explican formalmente mediante palabras o conceptos más básicos. Los términos geométricos básicos no definidos son el *punto*, la *recta* y el *plano*.

union (p. 866) For the union of event A and event B, the set of all outcomes in either event; represented by $A \cup B$.

unión Para la unión del suceso A y el suceso B, el conjunto de todos los resultados de cualquiera de los dos sucesos; se representa con $A \cup B$.

V

vector (p. 600) A directed segment representing a quantity that has both magnitude (length) and direction.

vector Segmento dirigido que representa una cantidad, la cual posee tanto magnitud (longitud) como dirección.

vertex angle of an isosceles triangle (p. 285) See *isosceles triangle.*

ángulo del vértice un triángulo isosceles Ver *triángulo isósceles.*

vertex of an angle (p. 36) The common endpoint of an angle.

vértice de un ángulo Extremo común de un ángulo.

vertex of a polygon (p. 56) The vertex of each angle of a polygon.

vertice de un polígono Vértice de cada ángulo de un polígono.

vertex of a polyhedron (p. 76) The intersection of three edges of a polyhedron.

vértice de un poliedro Intersección de las aristas de un poliedro.

vertical angles (p. 46) Two nonadjacent angles formed by two intersecting lines.

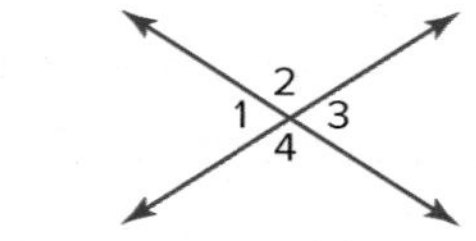

∠1 and ∠3 are vertical angles.
∠2 and ∠4 are vertical angles.

ángulos opuestos por el vértice Dos ángulos no adyacentes formados por dos rectas que se intersecan.

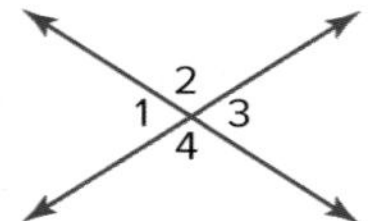

∠1 y ∠3 son ángulos opuestos por el vértice.
∠2 y ∠4 son ángulos opuestos por el vértice.

volume (p. 78) A measure of the amount of space enclosed by a three-dimensional figure.

volumen La medida de la cantidad de espacio contiene una figura tridimensional.

X

***x*-coordinate** (p. P15) The first number in an ordered pair.

coordenada *x* El primer número de un par ordenado.

Y

***y*-coordinate** (p. P15) The second number in an ordered pair.

coordenada *y* El segundo número de un par ordenado.

Index

B

C

D

Index

E

F

G

H

I

Index

M

N

O

P

S

U

Z

Symbols

$\neq$	is not equal to	$\parallel$	is parallel to	$\lvert\overrightarrow{AB}\rvert$	magnitude of the vector from *A* to *B*
$\approx$	is approximately equal to	$\nparallel$	is not parallel to	A'	the image of preimage *A*
$\cong$	is congruent to	$\perp$	is perpendicular to	$\rightarrow$	is mapped onto
$\sim$	is similar to	$\triangle$	triangle	$\odot A$	circle with center *A*
$\angle$, $\measuredangle$	angle, angles	$>, \geq$	is greater than, is greater than or equal to	π	pi
$m\angle A$	degree measure of $\angle A$	$<, \leq$	is less than, is less than or equal to	$\widehat{AB}$	minor arc with endpoints *A* and *B*
$^\circ$	degree	$\square$	parallelogram	$\widehat{ABC}$	major arc with endpoints *A* and *C*
$\overleftrightarrow{AB}$	line containing points *A* and *B*	*n*-gon	polygon with *n* sides	$m\widehat{AB}$	degree measure of arc *AB*
$\overline{AB}$	segment with endpoints *A* and *B*	$a:b$	ratio of *a* to *b*	$f(x)$	*f* of *x*, the value of *f* at *x*
$\overrightarrow{AB}$	ray with endpoint *A* containing *B*	(x, y)	ordered pair	!	factorial
AB	measure of $\overline{AB}$, distance between points *A* and *B*	(x, y, z)	ordered triple	$_nP_r$	permutation of *n* objects taken *r* at a time
$\sim p$	negation of *p*, not *p*	$\sin x$	sine of *x*	$_nC_r$	combination of *n* objects taken *r* at a time
$p \wedge q$	conjunction of *p* and *q*	$\cos x$	cosine of *x*	$P(A)$	probability of *A*
$p \vee q$	disjunction of *p* and *q*	$\tan x$	tangent of *x*	$P(A \vert B)$	the probability of *A* given that *B* has already occurred
$p \longrightarrow q$	conditional statement, if *p* then *q*	$\vec{a}$	vector *a*		
$p \longleftrightarrow q$	biconditional statement, *p* if and only if *q*	$\overrightarrow{AB}$	vector from *A* to *B*		

Measures

Metric	**Customary**
Length	
1 kilometer (km) = 1000 meters (m) 1 meter = 100 centimeters (cm) 1 centimeter = 10 millimeters (mm)	1 mile (mi) = 1760 yards (yd) 1 mile = 5280 feet (ft) 1 yard = 3 feet 1 yard = 36 inches (in.) 1 foot = 12 inches
Volume and Capacity	
1 liter (L) = 1000 milliliters (mL) 1 kiloliter (kL) = 1000 liters	1 gallon (gal) = 4 quarts (qt) 1 gallon = 128 fluid ounces (fl oz) 1 quart = 2 pints (pt) 1 pint = 2 cups (c) 1 cup = 8 fluid ounces
Weight and Mass	
1 kilogram (kg) = 1000 grams (g) 1 gram = 1000 milligrams (mg) 1 metric ton (t) = 1000 kilograms	1 ton (T) = 2000 pounds (lb) 1 pound = 16 ounces (oz)

Formulas

Coordinate Geometry

Slope	$m = \frac{y_2 - y_1}{x_2 - x_1}$
Distance on a number line	$d = \lvert a - b \rvert$
Distance on a coordinate plane	$d = \sqrt{(x_2 - x_1)^2 + (y_2 - y_1)^2}$
Distance in space	$d = \sqrt{(x_2 - x_1)^2 + (y_2 - y_1)^2 + (z_2 - z_1)^2}$
Arc length	$\ell = \frac{x}{360} \cdot 2\pi r$
Midpoint on a number line	$M = \frac{a + b}{2}$
Midpoint on a coordinate plane	$M = \left(\frac{x_1 + x_2}{2}, \frac{y_1 + y_2}{2}\right)$
Midpoint in space	$M = \left(\frac{x_1 + x_2}{2}, \frac{y_1 + y_2}{2}, \frac{z_1 + z_2}{2}\right)$

Perimeter and Circumference

square	$P = 4s$	**rectangle**	$P = 2\ell + 2w$	**circle**	$C = 2\pi r$ or $C = \pi d$

Area

square	$A = s^2$	**triangle**	$A = \frac{1}{2}bh$
rectangle	$A = \ell w$ or $A = bh$	**regular polygon**	$A = \frac{1}{2}Pa$
parallelogram	$A = bh$	**circle**	$A = \pi r^2$
trapezoid	$A = \frac{1}{2}h(b_1 + b_2)$	**sector of a circle**	$A = \frac{x}{360} \cdot \pi r^2$
rhombus	$A = \frac{1}{2}d_1d_2$ or $A = bh$		

Lateral Surface Area

prism	$L = Ph$	**regular pyramid**	$L = \frac{1}{2}P\ell$
cylinder	$L = 2\pi rh$	**cone**	$L = \pi r\ell$

Total Surface Area

prism	$S = Ph + 2B$	**cone**	$S = \pi r\ell + \pi r^2$
cylinder	$S = 2\pi rh + 2\pi r^2$	**sphere**	$S = 4\pi r^2$
regular pyramid	$S = \frac{1}{2}P\ell + B$		

Volume

cube	$V = s^3$	**pyramid**	$V = \frac{1}{3}Bh$
rectangular prism	$V = \ell wh$	**cone**	$V = \frac{1}{3}\pi r^2 h$
prism	$V = Bh$	**sphere**	$V = \frac{4}{3}\pi r^3$
cylinder	$V = \pi r^2 h$		

Equations for Figures on a Coordinate Plane

slope-intercept form of a line	$y = mx + b$	**circle**	$(x - h)^2 + (y - k)^2 = r^2$
point-slope form of a line	$y - y_1 = m(x - x_1)$		

Trigonometry

Law of Sines	$\frac{\sin A}{a} = \frac{\sin B}{b} = \frac{\sin C}{c}$	**Law of Cosines**	$a^2 = b^2 + c^2 - 2bc \cos A$ $b^2 = a^2 + c^2 - 2ac \cos B$ $c^2 = a^2 + b^2 - 2ab \cos C$
Pythagorean Theorem	$a^2 + b^2 = c^2$		